EXPONENTIAL AND LOGARITHMIC FUNCTIONS

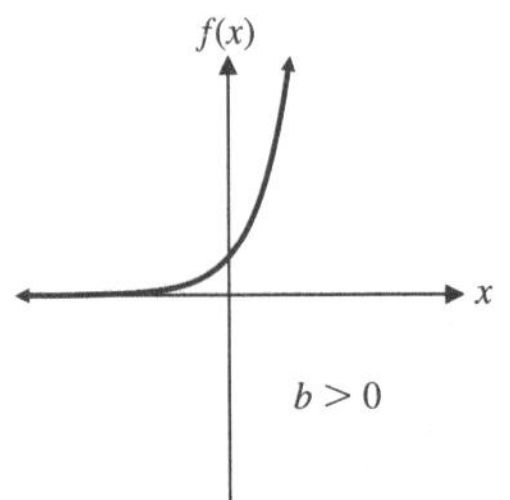

Exponential function
$f(x) = b^x$

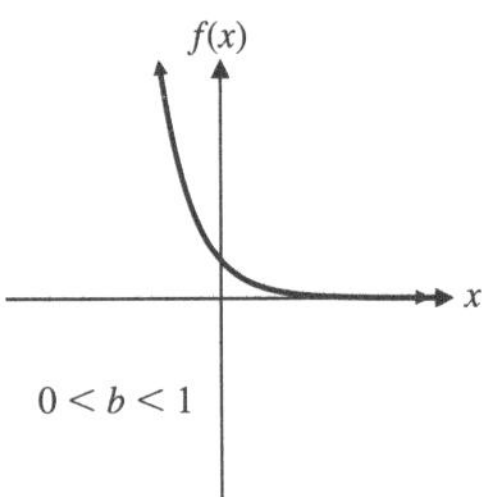

Exponential function
$f(x) = b^x$

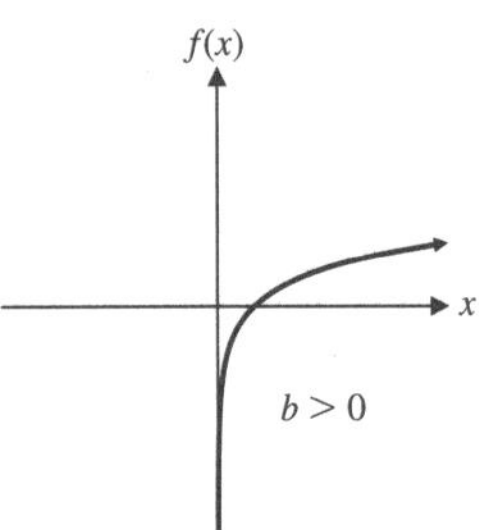

Logarithmic function
$f(x) = \log_b x$

REPRESENTATIVE POLYNOMIAL FUNCTIONS (DEGREE > 2)

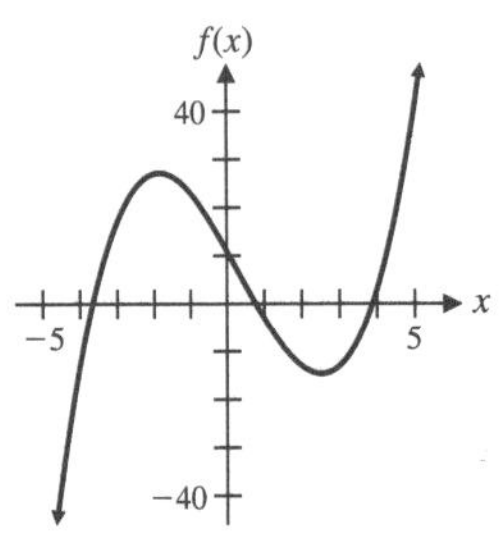

Third-degree polynomial
$f(x) = x^3 - x^2 - 14x + 11$

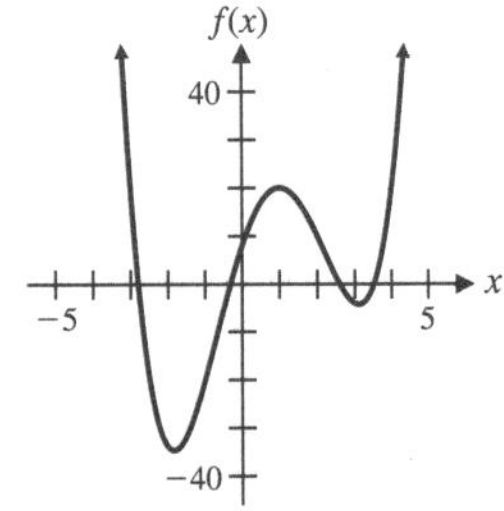

Fourth-degree polynomial
$f(x) = x^4 - 3x^3 - 9x^2 + 23x + 8$

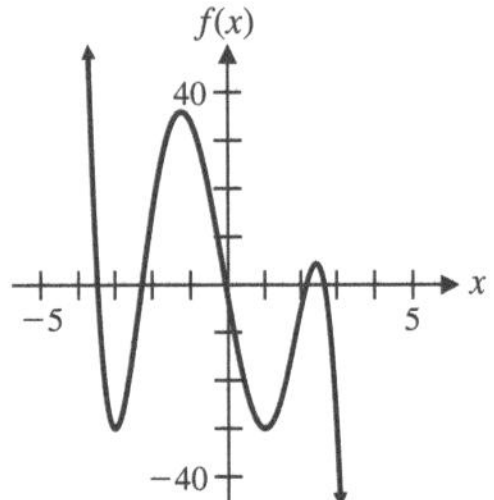

Fifth-degree polynomial
$f(x) = -x^5 - x^4 + 14x^3 + 6x^2 - 45x - 3$

REPRESENTATIVE RATIONAL FUNCTIONS

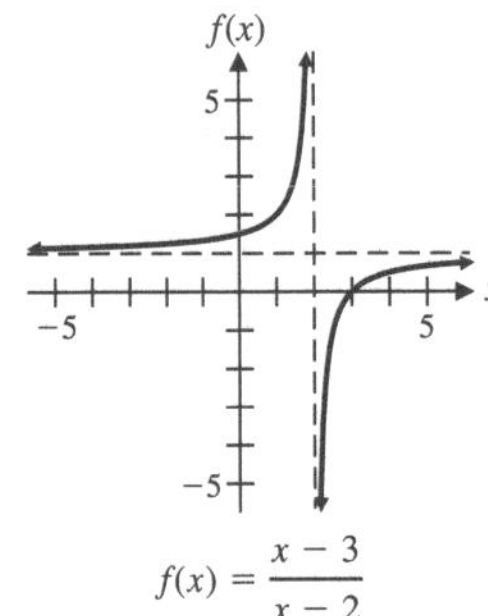

$f(x) = \dfrac{x - 3}{x - 2}$

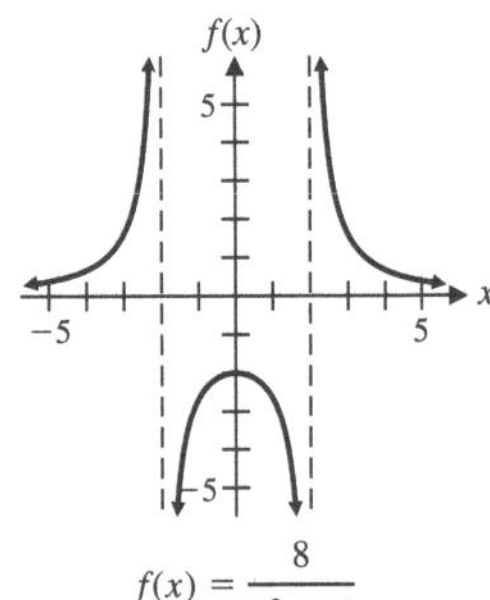

$f(x) = \dfrac{8}{x^2 - 4}$

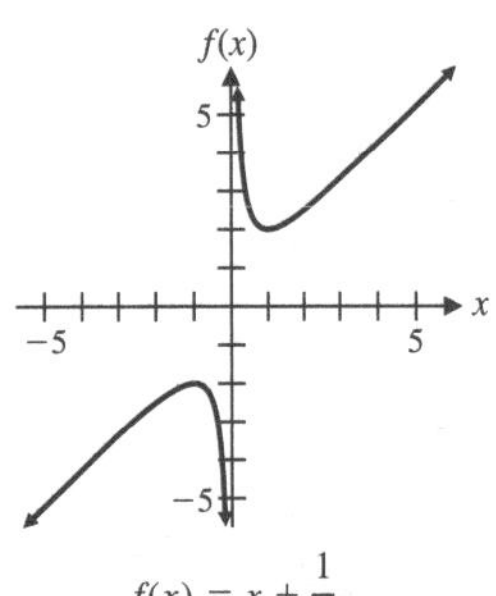

$f(x) = x + \dfrac{1}{x}$

GRAPH TRANSFORMATIONS

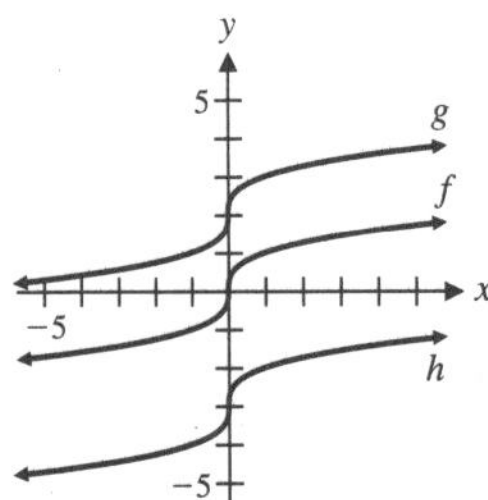

Vertical translation
$g(x) = f(x) + 2$
$h(x) = f(x) - 3$

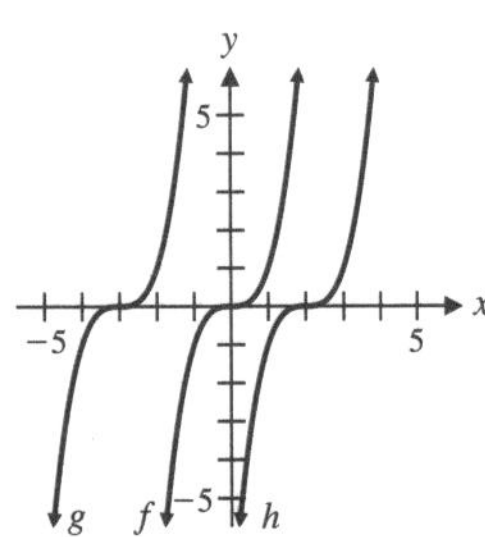

Horizontal translation
$g(x) = f(x + 3)$
$h(x) = f(x - 2)$

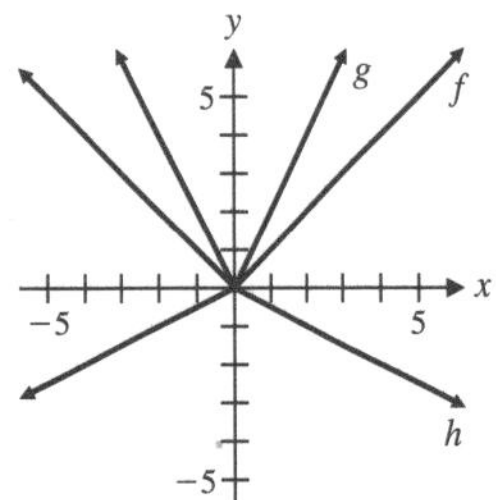

Expansion, contraction, and reflection
$g(x) = 2f(x)$
$h(x) = -0.5f(x)$

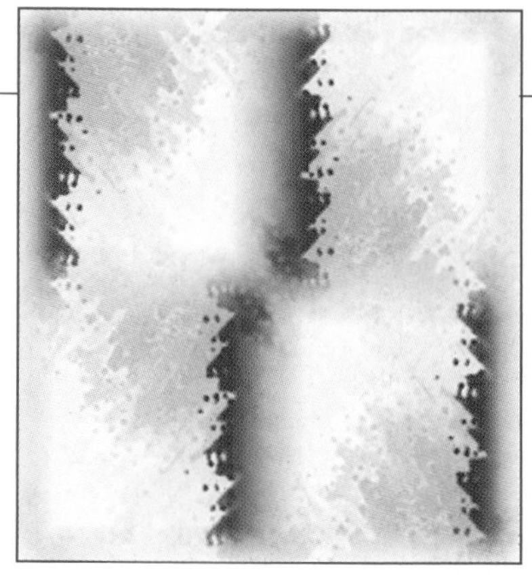

College Mathematics

For Business, Economics, Life Sciences, and Social Sciences

Eighth Edition

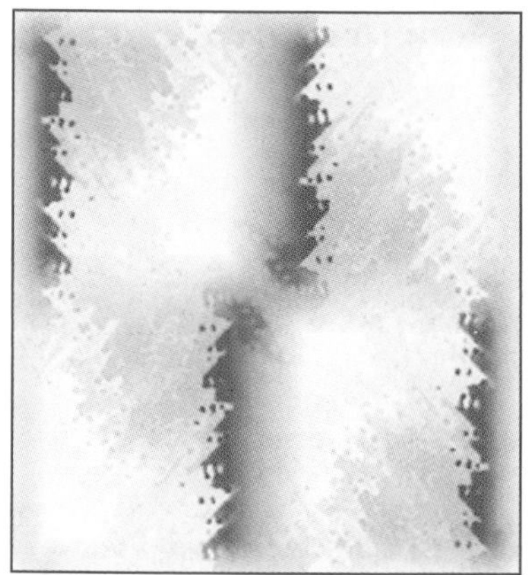

Raymond A. Barnett
Merritt College

Michael R. Ziegler
Marquette University

Karl E. Byleen
Marquette University

Prentice Hall
Upper Saddle River, New Jersey 07458

LIBRARY OF CONGRESS CATALOGING-IN-PUBLICATION DATA
Barnett, Raymond A.
College mathematics for business, economics, life sciences, and social sciences / Raymond A. Barnett, Michael R. Ziegler, Karl E. Byleen.—8th ed.
p. cm.
Includes indexes.
ISBN 0-13-079768-5 (hardcover)
1. Mathematics. 2. Social sciences—Mathematics. 3. Biomathematics. I. Ziegler, Michael R. II. Byleen, Karl. III. Title.
QA37.2.B364 1998 98-24476
510—dc21 CIP

Executive Acquisitions Editor: Sally Simpson
Sponsoring Editor: Gina M. Huck
Marketing Manager: Patrice Lumumba Jones
Editor-in-Chief: Jerome Grant
Editorial Director: Tim Bozik
Production Editor: Phyllis Niklas
Senior Managing Editor: Linda Mihatov Behrens
Executive Managing Editor: Kathleen Schiaparelli
Assistant Vice President of Production and Manufacturing: David W. Riccardi
Manufacturing Buyer: Alan Fischer
Manufacturing Manager: Trudy Pisciotti
Editorial Assistant: Joanne Wendelken
Supplements Editor: Gina M. Huck
Art Director: Maureen Eide
Associate Creative Director: Amy Rosen
Director of Creative Services: Paula Maylahn
Assistant to Art Director: John Christiana
Art Manager: Gus Vibal
Art Editor: Grace Hazeldine
Interior/Cover Designer: Jill Little
Cover Photo: W. Yanagida/IMA USA/Photonica Division
Art Studio: Scientific Illustrators

Simon & Schuster/A Viacom Company
Upper Saddle River, NJ 07458

Printed in the United States of America
10 9 8 7 6 5 4 3 2 1

ISBN 0-13-079768-5

Prentice-Hall International (UK) Limited, *London*
Prentice-Hall of Australia Pty. Limited, *Sydney*
Prentice-Hall Canada, Inc., *Toronto*
Prentice-Hall Hispanoamericano, S. A., *Mexico*
Prentice-Hall of India Private Limited, *New Delhi*
Prentice-Hall of Japan, Inc., *Tokyo*
Simon & Schuster Asia Pte. Ltd., *Singapore*
Editora Prentice-Hall do Brasil, Ltda., *Rio de Janeiro*

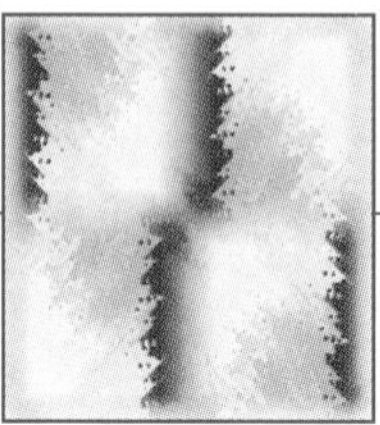

CONTENTS

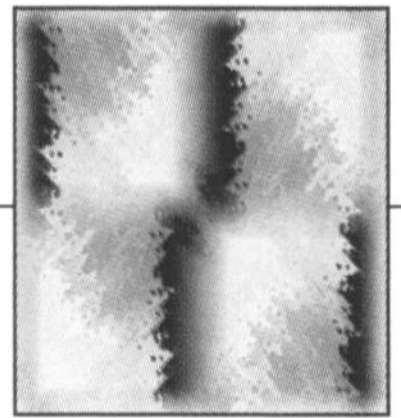

Chapter Dependencies

PART ONE A LIBRARY OF ELEMENTARY FUNCTIONS*

1 A Beginning Library of Elementary Functions → 2 Additional Elementary Functions

PART TWO FINITE MATHEMATICS

3 Mathematics of Finance

4 Systems of Linear Equations; Matrices → 5 Linear Inequalities and Linear Programming

4 Systems of Linear Equations; Matrices → 7 Markov Chains

6 Probability → 7 Markov Chains

PART THREE CALCULUS

8 The Derivative → 9 Graphing and Optimization → 10 Additional Derivative Topics → 11 Integration

11 Integration → 12 Additional Integration Topics

11 Integration → 13 Multivariable Calculus

APPENDIXES

A Self-Test Basic Algebra Review

B Special Topics

*Selected topics from Part One may be referred to as needed in Parts Two or Three or reviewed systematically before starting Part Two.

PREFACE

The eighth edition of *College Mathematics for Business, Economics, Life Sciences, and Social Sciences* is designed for a two-term (or condensed one-term) course in finite mathematics and calculus and for students who have had $1\frac{1}{2}$–2 years of high school algebra or the equivalent. The choice and independence of topics make the text readily adaptable to a variety of courses (see the Chapter Dependency Chart on page vii). It is one of eight books in the authors' college mathematics series.

Improvements in this edition evolved out of the generous response from a large number of users of the last and previous editions as well as survey results from instructors, mathematics departments, course outlines, and college catalogs. Fundamental to a book's growth and effectiveness is classroom use and feedback. Now in its eighth edition, *College Mathematics for Business, Economics, Life Sciences, and Social Sciences* has had the benefit of having a substantial amount of both.

■ EMPHASIS AND STYLE

The text is **written for student comprehension.** Great care has been taken to write a book that is mathematically correct and accessible to students. Emphasis is on computational skills, ideas, and problem solving rather than mathematical theory. Most derivations and proofs are omitted except where their inclusion adds significant insight into a particular concept. General concepts and results are usually presented only after particular cases have been discussed.

■ EXAMPLES AND MATCHED PROBLEMS

Over 400 completely worked examples are used to introduce concepts and to demonstrate problem-solving techniques. Many examples have multiple parts, significantly increasing the total number of worked examples. Each example is followed by a similar **matched problem for the student to work** while reading the material. This actively involves the student in the learning process. The answers to these matched problems are included at the end of each section for easy reference.

■ EXPLORATION AND DISCUSSION

Every section contains **Explore–Discuss** problems interspersed at appropriate places to encourage the student to think about a relationship or process before a result is stated, or to investigate additional consequences of a development in the text. **Verbalization** of mathematical concepts, results, and processes is encouraged in these Explore–Discuss problems, as well as in some matched problems, and in some problems in almost every exercise set. The Explore–Discuss material also can be used as in-class or out-of-class **group activities.** In addition, at the end of every chapter, we have included two special **chapter group activities** that involve several of the concepts discussed in the chapter. Problems in the exercise sets that require verbalization are indicated by color problem numbers.

Exercise Sets

The book contains over 5,600 problems. Many problems have multiple parts, significantly increasing the total number of problems. Each exercise set is designed so that an average or below-average student will experience success and a very capable student will be challenged. Exercise sets are mostly divided into A (routine, easy mechanics), B (more difficult mechanics), and C (difficult mechanics and some theory) levels.

Applications

A major objective of this book is to give the student substantial experience in **modeling and solving real-world problems.** Enough applications are included to convince even the most skeptical student that mathematics is really useful (see the Applications Index inside the back cover). Worked examples involving applications are identified by . **Almost every exercise set contains application problems,** usually divided into business and economics, life science, and social science groupings. An instructor with students from all three disciplines can let them choose applications from their own field of interest; if most students are from one of the three areas, then special emphasis can be placed there. Most of the applications are simplified versions of actual real-world problems taken from professional journals and books. No specialized experience is required to solve any of the applications.

Technology

The generic term **graphing utility** is used to refer to any of the various graphing calculators or computer software packages that might be available to a student using this book. (See the description of the software accompanying this book later in this Preface.) Although **access to a graphing utility is not assumed,** it is likely that many students will want to make use of one of these devices. To assist these students, **optional graphing utility activities** are included in appropriate places in the book. These include brief discussions in the text, examples or portions of examples solved on a graphing utility, problems for the student to solve, and a **group activity that involves the use of technology** at the end of each chapter. Beginning with the group activity at the end of Chapter 1, and continuing throughout the text, **linear regression** on a graphing utility is used at appropriate points to illustrate **mathematical modeling with real data.** All the optional graphing utility material is clearly identified by either or C and can be omitted without loss of continuity, if desired.

Graphs

All graphs are computer-generated to ensure mathematical accuracy. Graphing utility screens displayed in the text are actual output from a graphing calculator.

■ Additional Pedagogical Features

Annotation of examples and developments, in small color type, is found throughout the text to help students through critical stages (see Sections 1-1 and 4-2). **Think boxes** (dashed boxes) are used to enclose steps that are usually performed mentally (see Sections 1-1 and 4-1). **Boxes** are used to highlight important definitions, results, and step-by-step processes (see Sections 1-1 and 1-4). **Caution** statements appear throughout the text where student errors often occur (see Sections 4-3 and 4-5). **Functional use of color** improves the clarity of many illustrations, graphs, and developments, and guides students through certain critical steps (see Sections 1-1 and 4-2). **Boldface type** is used to introduce new terms and highlight important comments. **Chapter review** sections include a review of all important terms and symbols and a comprehensive review exercise. **Answers to most review exercises,** keyed to appropriate sections, are included in the back of the book. Answers to all other odd-numbered problems are also in the back of the book. Answers to application problems in linear programming include both the mathematical model and the numeric answer.

■ Content

The text begins with the development of a library of elementary functions in Chapters 1 and 2, including their properties and uses. We encourage students to investigate mathematical ideas and processes **graphically** and **numerically,** as well as **algebraically.** This development lays a firm foundation for studying mathematics both in this book and in future endeavors. Depending on the syllabus for the course and the background of the students, some or all of this material can be covered at the beginning of a course, or selected portions can be referred to as needed later in the course.

The material in Part Two (Finite Mathematics) can be thought of as four units: **mathematics of finance** (Chapter 3); linear algebra, including **matrices, linear systems, and linear programming** (Chapters 4 and 5); **probability** (Chapter 6); and applications of linear algebra and probability to **Markov chains** (Chapter 7). The first three units are independent of each other, while the last chapter is dependent on some of the earlier chapters (see the Chapter Dependency Chart preceding this Preface).

Chapter 3 presents a thorough treatment of simple and compound interest and present and future value of ordinary annuities. Appendix B contains a section on arithmetic and geometric sequences that can be covered in conjunction with this chapter, if desired.

Chapter 4 covers linear systems and matrices with an **emphasis on using row operations and Gauss–Jordan elimination** to solve systems and to find matrix inverses. This chapter also contains numerous applications of **mathematical modeling** utilizing systems and matrices. To assist students in formulating solutions, **all the answers in the back of the book to application problems** in Exercises 4-3, 4-5, and the chapter Review Exercise **contain both the mathematical model and its solution.** The row operations discussed in Sections 4-2 and 4-3 are required for the simplex method in Chapter 5. Matrix multiplication, matrix inverses, and systems of equations are required for Markov chains in Chapter 7.

Chapter 5 provides **broad and flexible coverage of linear programming.** The first two sections cover two-variable graphing techniques. Instructors who wish to emphasize techniques can cover the basic simplex method in Sections 5-3 and 5-4 and then discuss any or all of the following: the dual method (Section 5-5), the big *M* method (Section 5-6), or the two-phase simplex method (Group Activity 1). Those who want to emphasize modeling can discuss the formation of the mathematical model for any of the application examples in Sections 5-4, 5-5, and 5-6, and either omit the solution or use software to find the solution (see the description of the software that accompanies this text later in this Preface). To facilitate this approach, **all the answers in the back of the book to application problems** in Exercises 5-4, 5-5, and 5-6, and the chapter Review Exercise **contain both the mathematical model and its solution.**

Chapter 6 covers **counting techniques and basic probability,** including Bayes' formula and random variables. Appendix A contains a review of basic set theory and notation to support the use of sets in probability.

Chapter 7 ties together concepts developed in earlier chapters and applies them to **Markov chains.** This provides an excellent unifying conclusion to the finite mathematics portion of the text.

The material in Part Three (Calculus) consists of **differential calculus** (Chapters 8–10), **integral calculus** (Chapters 11–12), and **multivariable calculus** (Chapter 13). In general, Chapters 8–11 must be covered in sequence; however, certain sections can be omitted or given brief treatments, as pointed out in the discussion that follows (see the Chapter Dependency Chart on page vii).

Chapter 8 introduces the **derivative,** covers the **limit properties** essential to understanding the definition of the derivative, develops the **rules of differentiation** (including the chain rule for power forms), and introduces **applications** of derivatives in business and economics. The interplay between graphical, numerical, and algebraic concepts is emphasized here and throughout the text.

Chapter 9 focuses on **graphing** and **optimization.** The first three sections cover continuity and first-derivative and second-derivative graph properties, while emphasizing **polynomial graphing. Rational function** graphing is covered in Section 9-4. In a course that does not include graphing rational functions, this section can be omitted or given a brief treatment. Optimization is covered in Section 9-5, including examples and problems involving end-point solutions.

The first three sections of Chapter 10 extend the derivative concepts discussed in Chapters 8 and 9 to **exponential and logarithmic functions** (including the general form of the chain rule). This material is required for all the remaining chapters. **Implicit differentiation** is introduced in Section 10-4 and applied to **related rate problems** in Section 10-5. These topics are not referred to elsewhere in the text and can be omitted.

Chapter 11 introduces **integration.** The first two sections cover **antidifferentiation** techniques essential to the remainder of the text. Section 11-3 discusses some applications involving **differential equations** that can be omitted. Sections 11-4 and 11-5 discuss the **definite integral** in terms of **Riemann sums,** including **approximations** with various types of sums and some **simple error estimation.** As before, the interplay between the graphical, numeric, and algebraic properties is emphasized. These two sections also are required for the remaining chapters in the text.

Chapter 12 covers **additional integration topics** and is organized to provide maximum flexibility for the instructor. The first section extends the **area** concepts introduced in Chapter 11 to the area between two curves and related applications. Section 12-2 covers three more **applications** of integration, and Sections 12-3 and 12-4 deal with additional **techniques of integration.** Any or all of the topics in Chapter 12 can be omitted.

The first five sections of Chapter 13 deal with **differential multivariable calculus** and can be covered any time after Section 10-3 has been completed. Section 13-6 requires the **integration** concepts discussed in Chapter 11.

Appendix A contains a **self-test** and a **concise review of basic algebra** that also may be covered as part of the course or referred to as needed. As mentioned above, Appendix B contains additional topics that can be covered in conjunction with certain sections in the text, if desired.

Supplements for the Student

1. A **Student Solutions Manual** by Garret J. Etgen is available through your book store. The manual includes detailed solutions to all odd-numbered problems and all review exercises.
2. **Computer software and documentation** for IBM-compatible computers are packaged with the *Student Solutions Manual. Explorations in Finite Mathematics* and *Visual Calculus* by David Schneider each contain over twenty routines that provide additional insight into the topics discussed in the text. Although these software packages have much of the computing power of standard mathematical software packages, they are primarily teaching tools that focus on understanding mathematical concepts, rather than on computing. All the routines in these software packages are menu-driven and very easy to use. Included in *Explorations in Finite Mathematics* are routines for Gaussian elimination, matrix inversion, solution of linear programming problems by both the geometric method and the simplex method, Markov chains, probability and statistics, and mathematics of finance. The matrix routines use and display rational numbers, and matrices may be saved and printed. The *Visual Calculus* routines incorporate graphics whenever possible to illustrate topics such as secant lines; tangent lines; velocity; optimization; the relationship between the graphs of f, f', and f''; and the various approaches to approximating definite integrals. The software will run on DOS or Windows platforms.
3. A **Graphing Calculator Manual** by Carolyn L. Meitler contains examples illustrating the use of a graphics calculator to solve problems similar to those discussed in the text. The manual follows the chapter organization of the text, making it easy to find examples in the manual illustrating appropriate calculator solution methods for problems in the text. The manual includes keystrokes for the TI-82, TI-83, TI-85, and TI-86 calculators. However, the examples and techniques can be used with any graphing utility.
4. The **PH Companion Website,** designed to complement and expand upon the text, offers a variety of teaching and learning tools, including links to related websites, practice work for students, and the ability for instructors to monitor and evaluate students' work on the website. For more information, contact your local Prentice Hall representative.
www.prenhall.com/barnett

Supplements for the Instructor

For a summary of all available supplementary materials and detailed information regarding examination copy requests and orders, see page xvii.

1. **PH Custom Test, a menu-driven random test system** for either Windows or Macintosh is available to instructors. The test system has been greatly expanded and now offers **on-line testing.** Carefully constructed algorithms

use random-number generators to produce different, yet equivalent, versions of each of these problems. In addition, the system incorporates a unique **editing function** that allows the instructor to create additional problems, or alter any of the existing problems in the test, using a full set of mathematical notation. The test system offers **free-response, multiple-choice, and mixed exams.** An almost unlimited number of quizzes, review exercises, chapter tests, midterms, and final examinations, each different from the other, can be generated quickly and easily. At the same time, the system will produce answer keys, student worksheets, and a gradebook for the instructor, if desired.

2. A **Test Item File,** prepared by Laurel Technical Services, provides a hard copy of the test items available in PH Custom Test.
3. An **Instructor's Resource Manual** provides over 100 transparency masters and all the answers not included in the text. This manual is available to instructors without charge.
4. A **Student Solutions Manual** by Garret J. Etgen (see Supplements for the Student) is available to instructors.
5. **Computer software and documentation** for *Explorations in Finite Mathematics* and *Visual Calculus* by David Schneider are available to instructors. The software and documentation are packaged with the *Student Solutions Manual* (see Supplements for the Student). In addition to providing students with the opportunity to use the computer as an effective tool in the learning process, instructors will find the software very useful for activities such as preparing examples for class, constructing test questions, and classroom demonstrations.
6. A **Graphing Calculator Manual** by Carolyn L. Meitler (see Supplements for the Student) is available to instructors. The manual contains all the necessary information for a student with no previous experience with a graphing calculator, eliminating the need for the instructor to prepare materials related to calculator usage. In particular, separate appendixes for the TI-82, TI-83, TI-85, and TI-86 graphing calculators contain detailed instructions, including calculator-specific keystrokes, for performing the various operations required to effectively use each of these calculators to solve problems in the text. Furthermore, the methods illustrated for these calculators are easily adapted to other graphing utilities. The manual is very effective both for a class where all students purchase the same calculator and in a setting where students are using a variety of different calculators—an important consideration as more and more students arrive at college having already purchased a graphing calculator.
7. The **PH Companion Website,** designed to complement and expand upon the text, offers a variety of interactive teaching and learning tools, including links to related websites, practice work for students, and the ability for instructors to monitor and evaluate students' work on the website. For more information, contact your local Prentice Hall representative. *www.prenhall.com/barnett*

■ Error Check

Because of the careful checking and proofing by a number of mathematics instructors (acting independently), the authors and publisher believe this book to be substantially error-free. For any errors remaining, the authors would be grateful if they were sent to: Michael R. Ziegler, 509 W. Dean Court, Fox Point, WI 53217; or, by e-mail, to: michaelziegler@execpc.com

■ Acknowledgments

In addition to the authors, many others are involved in the successful publication of a book.

We wish to thank the following reviewers of the seventh edition:

Celeste Carter, Richland College
Lou D'Alotto, St. Johns University
John Dickerson, Valencia Community College—West
Joel Haack, University of Northern Iowa
Paul Hutchins, Florissant Community College
Inessa Levi, University of Louisville
Wayne Miller, Lee College
Michael Montano, Riverside Community College
Shala Peterman, University of Missouri
Dix Petty, University of Missouri
John Ryan, Texas A & M University
Larry Small, Los Angeles Pierce College

We also wish to thank our colleagues who have provided input on previous editions:

Chris Boldt, Bob Bradshaw, Bruce Chaffee, Robert Chaney, Dianne Clark, Charles E. Cleaver, Barbara Cohen, Richard L. Conlon, Catherine Cron, Madhu Deshpande, Kenneth A. Dodaro, Michael W. Ecker, Jerry R. Ehman, Lucina Gallagher, Martha M. Harvey, Sue Henderson, Lloyd R. Hicks, Louis F. Hoelzle, Paul Hutchins, K. Wayne James, Robert H. Johnston, Robert Krystock, James T. Loats, Frank Lopez, Roy H. Luke, Mel Mitchell, Ronald Persky, Kenneth A. Peters, Jr., Tom Plavchak, Bob Prielipp, Stephen Rodi, Arthur Rosenthal, Sheldon Rothman, Elaine Russell, Daniel E. Scanlon, George R. Schriro, Arnold L. Schroeder, Hari Shanker, Joan Smith, Steven Terry, Delores A. Williams, Caroline Woods, Charles W. Zimmerman, and Pat Zrolka.

We also express our thanks to:

Carolyn Meitler, Stephen Merrill, Robert Mullins, and Caroline Woods for providing a careful and thorough check of all the mathematical calculations in the book, the *Student Solutions Manual,* and the *Instructor's Resource Manual* (a tedious but extremely important job).

Garret Etgen, Carolyn Meitler, and David Schneider for developing the supplemental manuals that are so important to the success of a text.

Jeanne Wallace for accurately and efficiently producing most of the manuals that supplement the text.

George Morris and his staff at Scientific Illustrators for their effective illustrations and accurate graphs.

Phyllis Niklas for guiding the book smoothly through all publication details.

All the people at Prentice Hall who contributed their efforts to the production of this book, especially Sally Simpson, our executive acquisitions editor, and Gina Huck, our sponsoring editor.

Producing this new edition with the help of all these extremely competent people has been a most satisfying experience.

R. A. Barnett
M. R. Ziegler
K. E. Byleen

Ordering Information

When requesting examination copies or placing orders for this text or any of the related supplementary material listed below, please refer to the corresponding ISBN numbers.

Title	ISBN Number
College Mathematics for Business, Economics, Life Sciences, and Social Sciences, Eighth Edition	0-13-079768-5
Computer-generated random test system for College Mathematics, Eighth Edition:	
Test Item File	0-13-082770-3
PH Custom Test Windows	0-13-961202-5
PH Custom Test MAC	0-13-961210-6
Instructor's Resource Manual to accompany College Mathematics, Eighth Edition	0-13-096120-5
Student Solutions Manual to accompany College Mathematics, Eighth Edition, and *Explorations in Finite Mathematics* and *Visual Calculus* (3.5 inch disk and documentation)	0-13-961236-X
Graphing Calculator Manual to accompany College Mathematics, Eighth Edition	0-13-961228-9

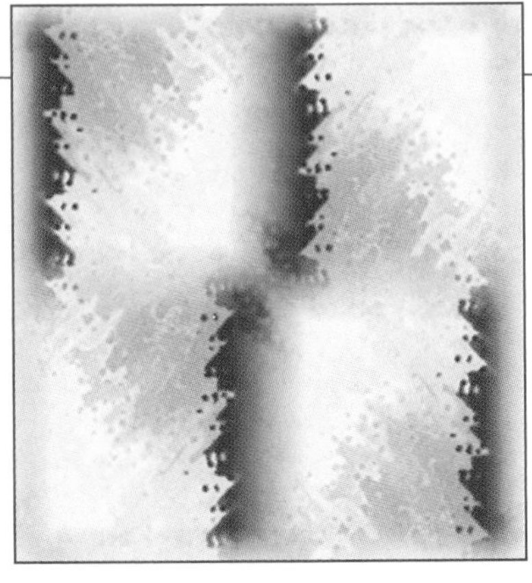

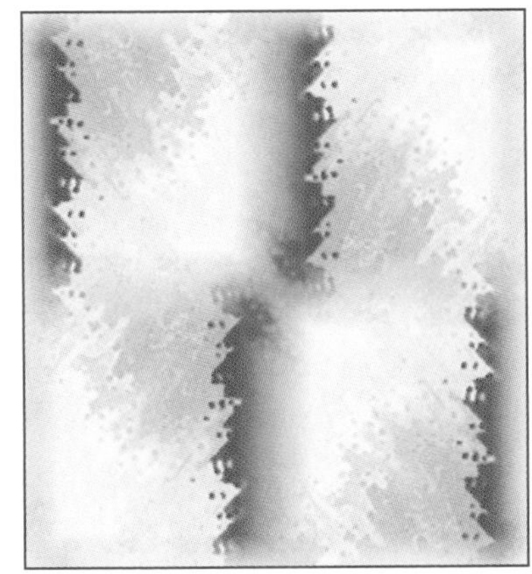

Part One

A Library of Elementary Functions

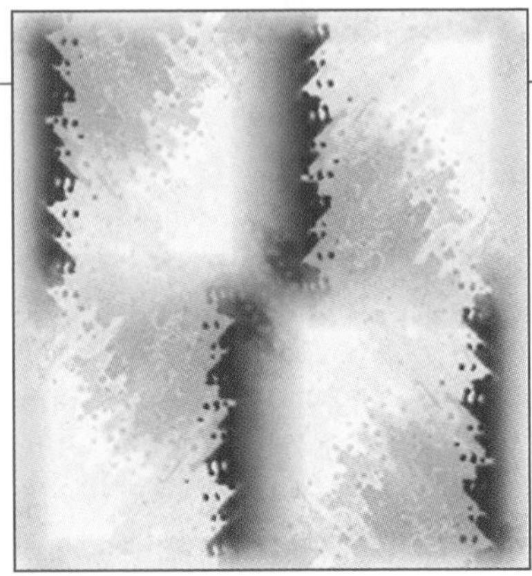

CHAPTER 1

A Beginning Library of Elementary Functions

INTRODUCTION

The function concept is one of the most important ideas in mathematics. The study of mathematics beyond the elementary level requires a firm understanding of a basic list of elementary functions, their properties, and their graphs. See the inside front cover of this book for a list of the functions that form our library of elementary functions. Most functions in the list will be introduced to you by the end of Chapter 2 and should become a part of your mathematical toolbox for use in this and most future courses or activities that involve mathematics. A few more elementary functions may be added to these in other courses, but the functions listed inside the front cover are more than sufficient for all the applications in this text.

SECTION 1-1

Functions

- Cartesian Coordinate System
- Graphing: Point-by-Point
- Definition of a Function
- Functions Specified by Equations
- Function Notation
- Applications

After a brief review of the Cartesian (rectangular) coordinate system in the plane and point-by-point graphing, we discuss the concept of function, one of the most important ideas in mathematics.

CARTESIAN COORDINATE SYSTEM

Recall that to form a **Cartesian** or **rectangular coordinate system,** we select two real number lines, one horizontal and one vertical, and let them cross through their origins as indicated in Figure 1. Up and to the right are the usual choices for the positive directions. These two number lines are called the **horizontal axis** and the **vertical axis,** or, together, the **coordinate axes.** The horizontal axis is usually referred to as the ***x* axis** and the vertical axis as the ***y* axis,** and each is labeled accordingly. Other labels may be used in certain situations. The coordinate axes divide the plane into four parts called **quadrants,** which are numbered counterclockwise from I to IV (see Fig. 1).

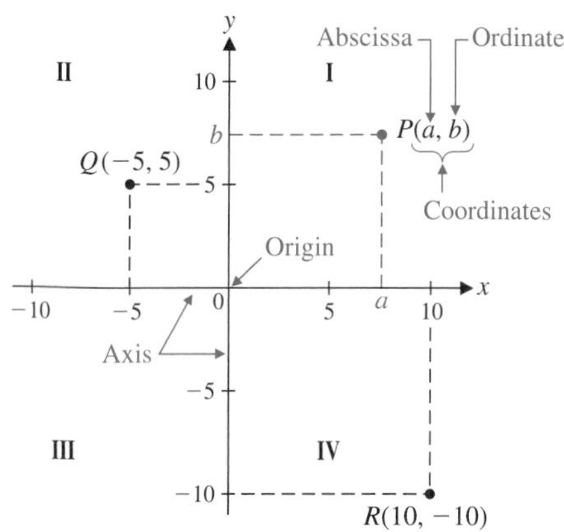

FIGURE 1
The Cartesian (rectangular) coordinate system

Now we want to assign *coordinates* to each point in the plane. Given an arbitrary point P in the plane, pass horizontal and vertical lines through the point (Fig. 1). The vertical line will intersect the horizontal axis at a point with coordinate a, and the horizontal line will intersect the vertical axis at a point with coordinate b. These two numbers written as the **ordered pair** (a, b) form the **coordinates** of the point P. The first coordinate, a, is called the **abscissa** of P; the second coordinate, b, is called the **ordinate** of P. The abscissa of Q in Figure 1 is -5, and the ordinate of Q is 5. The coordinates of a point can also be referenced in terms of the axis labels. The ***x* coordinate** of R in Figure 1 is 10, and the ***y* coordinate** of R is -10. The point with coordinates $(0, 0)$ is called the **origin.**

The procedure we have just described assigns to each point P in the plane a unique pair of real numbers (a, b). Conversely, if we are given an ordered pair of real numbers (a, b), then, reversing this procedure, we can determine a unique point P in the plane. Thus:

There is a one-to-one correspondence between the points in a plane and the elements in the set of all ordered pairs of real numbers.

This is often referred to as the **fundamental theorem of analytic geometry.**

■ Graphing: Point-by-Point

The fundamental theorem of analytic geometry allows us to look at algebraic forms geometrically and to look at geometric forms algebraically. We begin by considering an algebraic form, an equation in two variables:

$$y = 9 - x^2 \tag{1}$$

A **solution** to equation (1) is an ordered pair of real numbers (a, b) such that

$$b = 9 - a^2$$

The **solution set** for equation (1) is the set of all these ordered pairs.

To find a solution to equation (1), we replace x with a number and calculate the value of y. For example, if $x = 2$, then $y = 9 - 2^2 = 5$, and the ordered pair (2, 5) is a solution of equation (1). Similarly, if $x = -3$, then $y = 9 - (-3)^2 = 0$, and $(-3, 0)$ is a solution. Since any real number substituted for x in equation (1) will produce a solution, the solution set must have an infinite number of elements. We use a rectangular coordinate system to provide a geometric representation of this set.

The **graph of an equation** is the graph of all the ordered pairs in its solution set. To **sketch the graph of an equation,** we plot enough points from its solution set in a rectangular coordinate system so that the total graph is apparent and then connect these points with a smooth curve. This process is called **point-by-point plotting.**

Example 1 ➠ **Point-by-Point Plotting** Sketch a graph of $y = 9 - x^2$.

Solution Make up a table of solutions—that is, ordered pairs of real numbers that satisfy the given equation. For easy mental calculation, choose integer values for x.

x	−4	−3	−2	−1	0	1	2	3	4
y	−7	0	5	8	9	8	5	0	−7

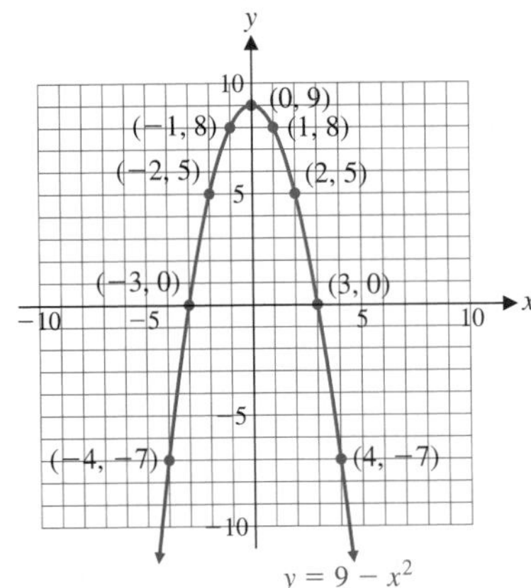

FIGURE 2

After plotting these solutions, if there are any portions of the graph that are unclear, plot additional points until the shape of the graph is apparent. Then join all the plotted points with a smooth curve as shown in Figure 2. Arrowheads are used to indicate that the graph continues beyond the portion shown here with no significant changes in shape. ■

*Matched Problem 1** ➠ Sketch a graph of $y = x^2 - 4$ using point-by-point plotting. ■

---

*Answers to matched problems are found near the end of each section, before the exercise set.

Explore–Discuss 1

To graph the equation $y = -x^3 + 3x$, we use point-by-point plotting to obtain:

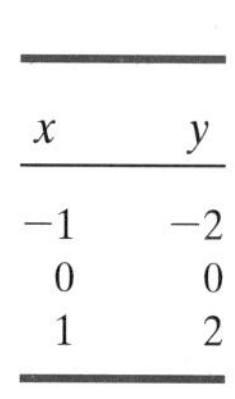

x	y
-1	-2
0	0
1	2

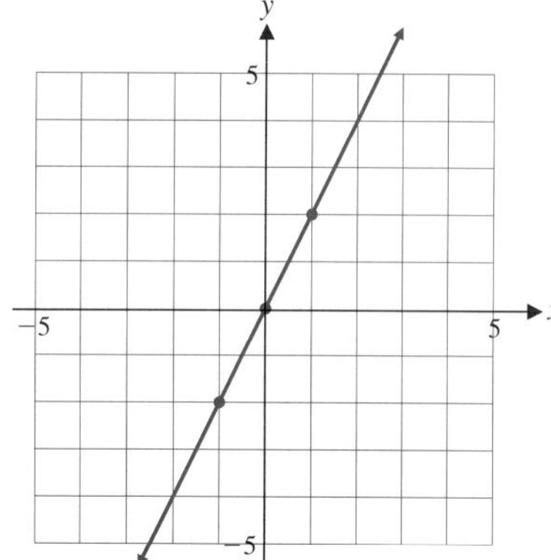

(A) Do you think this is the correct graph of the equation? If so, why? If not, why?
(B) Add points on the graph for $x = -2, -1.5, -0.5, 0.5, 1.5$, and 2.
(C) Now, what do you think the graph looks like? Sketch your version of the graph, adding more points as necessary.
(D) Graph this equation on a graphing utility and compare it with your graph from part (C).

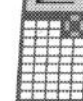

The icons in the margin are used throughout this text to identify optional graphing utility activities that are intended to give you additional insight into the concepts under discussion. You may have to consult the manual for your graphing utility or the supplement that accompanies this text (see Preface) for the details necessary to carry out these activities. For example, to graph the equation in Explore–Discuss 1 on most graphing utilities, you first have to enter the equation (Fig. 3A) and the window variables (Fig. 3B).

(A)

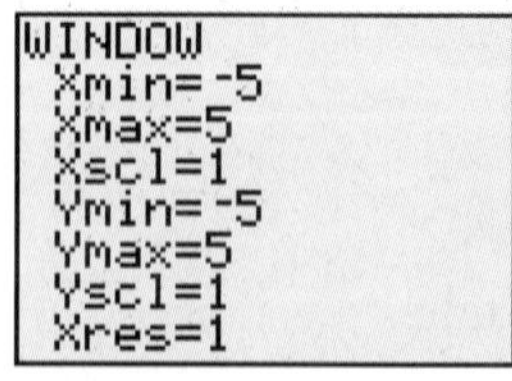

(B)

FIGURE 3

As Explore–Discuss 1 illustrates, the shape of a graph may not be "apparent" from your first choice of points on the graph. One of the objectives of this chapter is to provide you with a library of basic equations that will aid you in sketching graphs. For example, the curve in Figure 2 is called a *parabola*. Notice that if we fold the paper along the y axis, the right side will match the left side. We say that the graph is *symmetric with respect to the y axis* and call the y axis the *axis of the parabola*. Later in this chapter, we will see that all parabolas have graphs with similar properties. Identifying a given equation as one whose graph will be a parabola simplifies graphing the equation by hand.

■ Definition of a Function

Central to the concept of function is correspondence. You have already had experiences with correspondences in daily living. For example:

To each person there corresponds an annual income.

To each item in a supermarket there corresponds a price.

To each student there corresponds a grade-point average.
To each day there corresponds a maximum temperature.
For the manufacture of x items there corresponds a cost.
For the sale of x items there corresponds a revenue.
To each square there corresponds an area.
To each number there corresponds its cube.

One of the most important aspects of any science is the establishment of correspondences among various types of phenomena. Once a correspondence is known, predictions can be made. A cost analyst would like to predict costs for various levels of output in a manufacturing process; a medical researcher would like to know the correspondence between heart disease and obesity; a psychologist would like to predict the level of performance after a subject has repeated a task a given number of times; and so on.

What do all the above examples have in common? Each describes the matching of elements from one set with the elements in a second set. Consider the tables of the cube, square, and square root given in Tables 1–3.

Table 1

DOMAIN *Number*	RANGE *Cube*
-2 →	-8
-1 →	-1
0 →	0
1 →	1
2 →	8

Table 2

DOMAIN *Number*	RANGE *Square*
-2 →	4
-1 →	1
0 →	0
1 →	1
2 →	4

Table 3

DOMAIN *Number*	RANGE *Square root*
0 →	0
1 →	1
1 →	-1
4 →	2
4 →	-2
9 →	3
9 →	-3

Tables 1 and 2 specify functions, but Table 3 does not. Why not? The definition of the term *function* will explain.

Definition of a Function

A **function** is a rule (process or method) that produces a correspondence between two sets of elements such that to each element in the first set there corresponds one and only one element in the second set.

The first set is called the **domain,** and the set of corresponding elements in the second set is called the **range.**

Tables 1 and 2 specify functions, since to each domain value there corresponds exactly one range value (for example, the cube of -2 is -8 and no other number). On the other hand, Table 3 does not specify a function, since to at least one domain value there corresponds more than one range value (for example, to the domain value 9 there corresponds -3 and 3, both square roots of 9).

Explore–Discuss 2

Consider the set of students enrolled in a college and the set of faculty members of that college. Suppose we define a correspondence between the two sets by saying that a student corresponds to a faculty member if the student is currently enrolled in a course taught by that faculty member. Is this correspondence a function? Discuss.

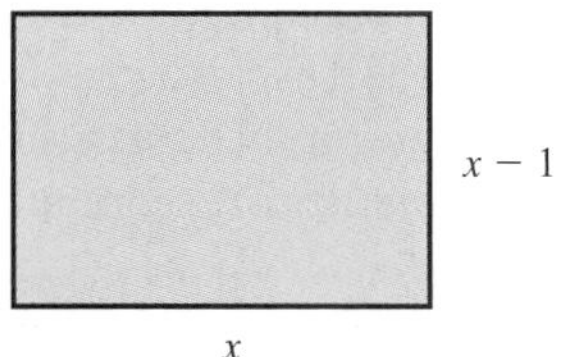

FIGURE 4

Functions Specified by Equations

Most of the domains and ranges included in this text will be (infinite) sets of real numbers, and the rules associating range values with domain values will be equations in two variables. Consider, for example, the equation for the area of a rectangle with width 1 inch less than its length (Fig. 4). If x is the length, then the area y is given by

$$y = x(x - 1) \qquad x \geq 1$$

For each **input** x (length), we obtain an **output** y (area). For example:

If $x = 5$, then $y = 5(5 - 1) = 5 \cdot 4 = 20$.
If $x = 1$, then $y = 1(1 - 1) = 1 \cdot 0 = 0$.
If $x = \sqrt{5}$, then $y = \sqrt{5}(\sqrt{5} - 1) = 5 - \sqrt{5} \approx 2.76$.

The input values are domain values, and the output values are range values. The equation (a rule) assigns each domain value x a range value y. The variable x is called an *independent variable* (since values can be "independently" assigned to x from the domain), and y is called a *dependent variable* (since the value of y "depends" on the value assigned to x). In general, any variable used as a placeholder for domain values is called an **independent variable;** any variable that is used as a placeholder for range values is called a **dependent variable.**

When does an equation specify a function?

Functions Defined by Equations

If in an equation in two variables, we get exactly one output (value for the dependent variable) for each input (value for the independent variable), then the equation defines a function.

If we get more than one output for a given input, the equation does not define a function.

Example 2 **Functions and Equations** Determine which of the following equations specify functions with independent variable x.

(A) $4y - 3x = 8$, x a real number (B) $y^2 - x^2 = 9$, x a real number

Solution (A) Solving for the dependent variable y, we have

$$\begin{aligned} 4y - 3x &= 8 \\ 4y &= 8 + 3x \\ y &= 2 + \tfrac{3}{4}x \end{aligned} \qquad (2)$$

Since each input value x corresponds to exactly one output value $\left(y = 2 + \frac{3}{4}x\right)$, we see that equation (2) specifies a function.

(B) Solving for the dependent variable y, we have

$$y^2 - x^2 = 9 \tag{3}$$
$$y^2 = 9 + x^2$$
$$y = \pm\sqrt{9 + x^2}$$

Since $9 + x^2$ is always a positive real number for any real number x and since each positive real number has two square roots,* to each input value x there corresponds two output values $(y = -\sqrt{9 + x^2}$ and $y = \sqrt{9 + x^2})$. For example, if $x = 4$, then equation (3) is satisfied for $y = 5$ and for $y = -5$. Thus, equation (3) does not specify a function.

Matched Problem 2 Determine which of the following equations specify functions with independent variable x.

(A) $y^2 - x^4 = 9$, x a real number (B) $3y - 2x = 3$, x a real number

Since the graph of an equation is the graph of all the ordered pairs that satisfy the equation, it is very easy to determine whether an equation specifies a function by examining its graph. The graphs of the two equations we considered in Example 2 are shown in Figure 5.

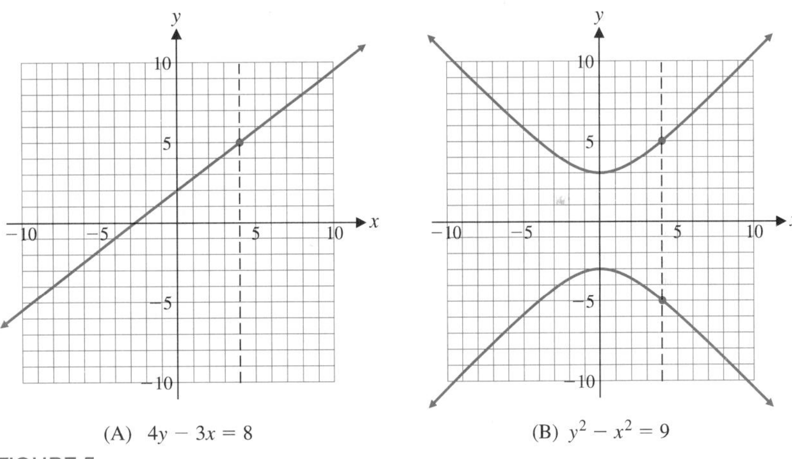

(A) $4y - 3x = 8$ (B) $y^2 - x^2 = 9$

FIGURE 5

In Figure 5A notice that any vertical line will intersect the graph of the equation $4y - 3x = 8$ in exactly one point. This shows that to each x value there corresponds exactly one y value and confirms our conclusion that this equation specifies a function. On the other hand, Figure 5B shows that there exist vertical lines that intersect the graph of $y^2 - x^2 = 9$ in two points. This indicates that there exist x values to which there correspond two different y values and verifies our conclusion that this equation does not specify a function. These observations are generalized in Theorem 1.

*Recall that each positive real number N has two square roots: $\sqrt{N}$, the principal square root, and $-\sqrt{N}$, the negative of the principal square root (see Appendix A-7).

Theorem 1 ■■ VERTICAL-LINE TEST FOR A FUNCTION

An equation defines a function if each vertical line in the coordinate system passes through at most one point on the graph of the equation.

If any vertical line passes through two or more points on the graph of an equation, then the equation does not define a function. ■■

Explore–Discuss 3

The definition of a function specifies that to each element in the domain there corresponds one and only one element in the range.

(A) Give an example of a function such that to each element of the range there correspond exactly two elements of the domain.
(B) Give an example of a function such that to each element of the range there corresponds exactly one element of the domain.

In Example 2, the domains were explicitly stated along with the given equations. In many cases, this will not be done. Unless stated to the contrary, we shall adhere to the following convention regarding domains and ranges for functions specified by equations:

Agreement on Domains and Ranges

If a function is specified by an equation and the domain is not indicated, then we assume that the domain is the set of all real number replacements of the independent variable (inputs) that produce real values for the dependent variable (outputs). The range is the set of all outputs corresponding to input values.

In many applied problems the domain is determined by practical considerations within the problem (see Example 7).

Example 3 **Finding a Domain** Find the domain of the function specified by the equation $y = \sqrt{4 - x}$, assuming x is the independent variable.

SOLUTION For y to be real, $4 - x$ must be greater than or equal to 0; that is,

$$\begin{aligned} 4 - x &\geqslant 0 \\ -x &\geqslant -4 \\ x &\leqslant 4 \end{aligned}$$

Sense of inequality reverses when both sides are divided by -1.

Thus,

Domain: $x \leqslant 4$ (inequality notation) or $(-\infty, 4]$ (interval notation)

Matched Problem 3 Find the domain of the function specified by the equation $y = \sqrt{x - 2}$, assuming x is the independent variable.

■ Function Notation

We have just seen that a function involves two sets, a domain and a range, and a rule of correspondence that enables us to assign to each element in the domain exactly one element in the range. We use different letters to denote names for numbers; in essentially the same way, we will now use different letters to denote names for functions. For example, f and g may be used to name the functions specified by the equations $y = 2x + 1$ and $y = x^2 + 2x - 3$:

$$\begin{aligned} f&:\quad y = 2x + 1 \\ g&:\quad y = x^2 + 2x - 3 \end{aligned} \tag{4}$$

If x represents an element in the domain of a function f, then we frequently use the symbol

$$\boldsymbol{f(x)}$$

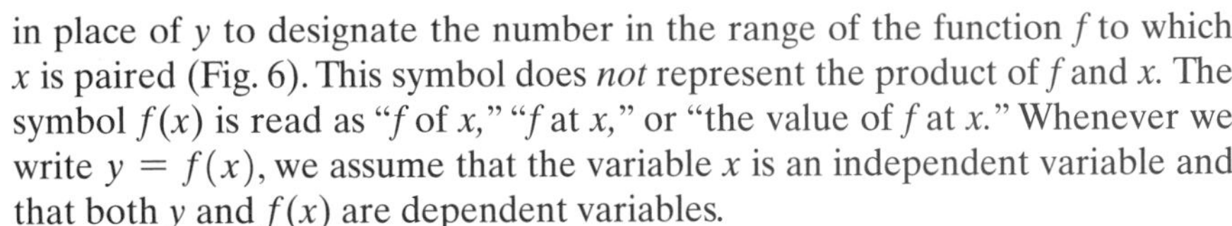

in place of y to designate the number in the range of the function f to which x is paired (Fig. 6). This symbol does *not* represent the product of f and x. The symbol $f(x)$ is read as "f of x," "f at x," or "the value of f at x." Whenever we write $y = f(x)$, we assume that the variable x is an independent variable and that both y and $f(x)$ are dependent variables.

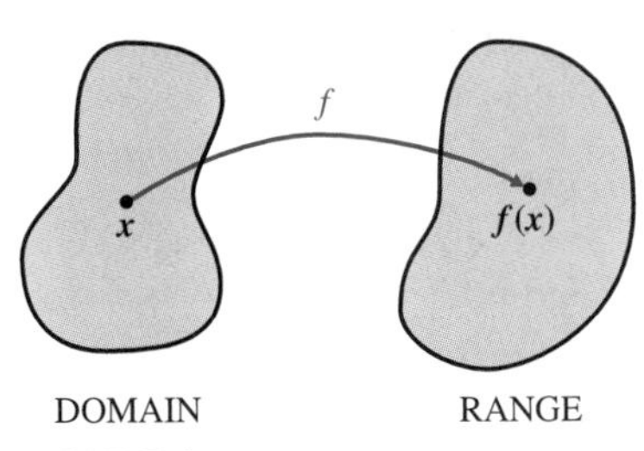

FIGURE 6

Using function notation, we can now write functions f and g in (4) in the form

$$f(x) = 2x + 1 \qquad \text{and} \qquad g(x) = x^2 + 2x - 3$$

Let us find $f(3)$ and $g(-5)$. To find $f(3)$, we replace x with 3 wherever x occurs in $f(x) = 2x + 1$ and evaluate the right side:

$$\begin{aligned} f(x) &= 2x + 1 \\ f(\mathbf{3}) &= 2 \cdot \mathbf{3} + 1 \\ &= 6 + 1 = 7 \qquad \text{For input 3, the output is 7.} \end{aligned}$$

Thus,

$$f(3) = 7 \qquad \text{The function } f \text{ assigns the range value 7 to the domain value 3.}$$

To find $g(-5)$, we replace x by -5 wherever x occurs in $g(x) = x^2 + 2x - 3$ and evaluate the right side:

$$\begin{aligned} g(x) &= x^2 + 2x - 3 \\ g(\mathbf{-5}) &= (\mathbf{-5})^2 + 2(\mathbf{-5}) - 3 \\ &= 25 - 10 - 3 = 12 \qquad \text{For input } -5\text{, the output is 12.} \end{aligned}$$

Thus,

$$g(-5) = 12 \qquad \text{The function } g \text{ assigns the range value 12 to the domain value } -5.$$

It is very important to understand and remember the definition of $f(x)$:

The Symbol $f(x)$

For any element x in the domain of the function f, the symbol $\boldsymbol{f(x)}$ represents the element in the range of f corresponding to x in the domain of f. If x is an input value, then $f(x)$ is the corresponding output value. If x is an element that is not in the domain of f, then f is *not defined at* x and $f(x)$ *does not exist.*

Example 4 **Function Evaluation** If

$$f(x) = \frac{12}{x - 2} \qquad g(x) = 1 - x^2 \qquad h(x) = \sqrt{x - 1}$$

then:

(A) $f(\mathbf{6}) = \boxed{\frac{12}{\mathbf{6} - 2}}^{*} = \frac{12}{4} = 3$

(B) $g(\mathbf{-2}) = \boxed{1 - (\mathbf{-2})^2} = 1 - 4 = -3$

(C) $h(\mathbf{-2}) = \boxed{\sqrt{\mathbf{-2} - 1}} = \sqrt{-3}$

But $\sqrt{-3}$ is not a real number. Since we have agreed to restrict the domain of a function to values of x that produce real values for the function, -2 is not in the domain of h and $h(-2)$ does not exist.

(D) $f(\mathbf{0}) + g(\mathbf{1}) - h(\mathbf{10}) = \boxed{\frac{12}{\mathbf{0} - 2} + (1 - \mathbf{1}^2) - \sqrt{\mathbf{10} - 1}}$

$$= \frac{12}{-2} + 0 - \sqrt{9}$$

$$= -6 - 3 = -9$$

Matched Problem 4 Use the functions in Example 4 to find:

(A) $f(-2)$ (B) $g(-1)$ (C) $h(-8)$ (D) $\frac{f(3)}{h(5)}$

Example 5 **Finding Domains** Find the domains of functions *f*, *g*, and *h*:

$$f(x) = \frac{12}{x - 2} \qquad g(x) = 1 - x^2 \qquad h(x) = \sqrt{x - 1}$$

SOLUTION *Domain of f.* $12/(x - 2)$ represents a real number for all replacements of x by real numbers except for $x = 2$ (division by 0 is not defined). Thus, $f(2)$ does not exist, and the domain of f is the set of all real numbers except 2. We often indicate this by writing

$$f(x) = \frac{12}{x - 2} \qquad x \neq 2$$

Domain of g. The domain is R, the set of all real numbers, since $1 - x^2$ represents a real number for all replacements of x by real numbers.

Domain of h. The domain is the set of all real numbers x such that $\sqrt{x - 1}$ is a real number—that is, such that

$$x - 1 \geqslant 0$$

$$x \geqslant 1 \quad \text{or} \quad [1, \infty)$$

*Dashed boxes are used throughout the book to represent steps that are usually performed mentally.

Matched Problem 5 Find the domains of functions F, G, and H:

$$F(x) = x^2 - 3x + 1 \qquad G(x) = \frac{5}{x + 3} \qquad H(x) = \sqrt{2 - x}$$

In addition to evaluating functions at specific numbers, it is important to be able to evaluate functions at expressions that involve one or more variables. For example, the **difference quotient**

$$\frac{f(x + h) - f(x)}{h} \qquad x \text{ and } x + h \text{ in the domain of } f,\ h \neq 0$$

is studied extensively in calculus.

Explore–Discuss 4

Let x and h be real numbers.

(A) If $f(x) = 4x + 3$, which of the following is true?
   (1) $f(x + h) = 4x + 3 + h$
   (2) $f(x + h) = 4x + 4h + 3$
   (3) $f(x + h) = 4x + 4h + 6$

(B) If $g(x) = x^2$, which of the following is true?
   (1) $g(x + h) = x^2 + h$
   (2) $g(x + h) = x^2 + h^2$
   (3) $g(x + h) = x^2 + 2hx + h^2$

(C) If $M(x) = x^2 + 4x + 3$, describe the operations that must be performed to evaluate $M(x + h)$.

Example 6 **Using Function Notation** For $f(x) = x^2 - 2x + 7$, find:

(A) $f(a)$ (B) $f(a + h)$ (C) $\dfrac{f(a + h) - f(a)}{h}$

SOLUTION (A) $f(a) = a^2 - 2a + 7$

(B) $f(a + h) = (a + h)^2 - 2(a + h) + 7$

$$= a^2 + 2ah + h^2 - 2a - 2h + 7$$

(C)

$$\frac{f(a + h) - f(a)}{h} = \frac{(a^2 + 2ah + h^2 - 2a - 2h + 7) - (a^2 - 2a + 7)}{h}$$

$$= \frac{2ah + h^2 - 2h}{h} = \frac{h(2a + h - 2)}{h} = 2a + h - 2$$

Matched Problem 6 Repeat Example 6 for $f(x) = x^2 - 4x + 9$.

Applications

We now turn to the important concepts of **break-even** and **profit–loss** analysis, which we will return to a number of times in this text. Any manufacturing company has **costs,** C, and **revenues,** R. The company will have a **loss** if $R < C$, will **break even** if $R = C$, and will have a **profit** if $R > C$. Costs include **fixed costs** such as plant overhead, product design, setup, and promotion; and **variable costs,** which are dependent on the number of items produced at a certain cost per item. In addition, **price–demand** functions, usually established by financial departments using historical data or sampling techniques, play an important part in profit–loss analysis. We will let x, the number of units manufactured and sold, represent the independent variable. Cost functions, revenue functions, profit functions, and price–demand functions are often stated in the following forms, where a, b, m, and n are constants determined from the context of a particular problem:

Cost Function

$$\begin{aligned} C &= (\text{Fixed costs}) + (\text{Variable costs}) \\ &= a + bx \end{aligned}$$

Price–Demand Function

$$p = m - nx \quad \text{x is the number of items that can be sold at \$$p$ per item.}$$

Revenue Function

$$\begin{aligned} R &= (\text{Number of items sold}) \times (\text{Price per item}) \\ &= xp = x(m - nx) \end{aligned}$$

Profit Function

$$\begin{aligned} P &= R - C \\ &= x(m - nx) - (a + bx) \end{aligned}$$

Example 7 and Matched Problem 7 explore the relationships among the algebraic definition of a function, the numerical values of the function, and the graphical representation of the function. The interplay among algebraic, numeric, and graphic viewpoints is an important aspect of our treatment of functions and their use. In Example 7, we also see how a function can be used to describe data from the real world, a process that is often referred to as *mathematical modeling.* The material in this example will be returned to in subsequent sections so that we can analyze it in greater detail and from different points of view.

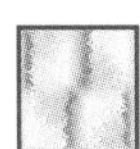

Example 7 **Price–Demand and Revenue Modeling** A manufacturer of a popular automatic camera wholesales the camera to retail outlets throughout the United States. Using statistical methods, the financial department in the company produced the price–demand data in Table 4 where p is the wholesale price per camera at which x million cameras are sold. Notice that as the price goes down, the number sold goes up.

Table 4

PRICE–DEMAND

x (MILLION)	p (\$)
2	87
5	68
8	53
12	37

Table 5
REVENUE

x (MILLION)	$R(x)$ (MILLION \$)
1	90
3	
6	
9	
12	
15	

Using special analytical techniques (regression analysis), an analyst arrived at the following price–demand function that models the Table 4 data:

$$p(x) = 94.8 - 5x \qquad 1 \leqslant x \leqslant 15 \tag{5}$$

(A) Plot the data in Table 4. Then sketch a graph of the price–demand function in the same coordinate system.
(B) What is the company's revenue function for this camera, and what is the domain of this function?
(C) Complete Table 5, computing revenues to the nearest million dollars.
(D) Plot the data in Table 5. Then sketch a graph of the revenue function using these points.
C (E) Plot the revenue function on a graphing utility.

SOLUTION (A)

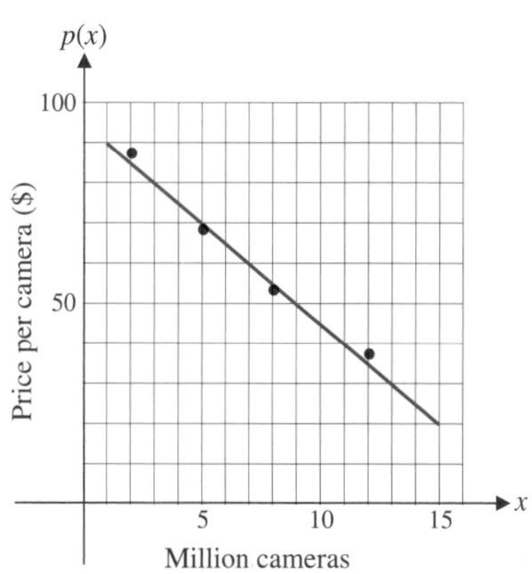

FIGURE 7
Price–demand

In Figure 7, notice that the model approximates the actual data in Table 4, and it is assumed that it gives realistic and useful results for all other values of x between 1 million and 15 million.

(B) $R(x) = xp(x) = x(94.8 - 5x)$ million dollars
Domain: $1 \leqslant x \leqslant 15$
[Same domain as the price–demand function, equation (5).]

(C) Table 5
REVENUE

x (MILLION)	$R(x)$ (MILLION \$)
1	90
3	239
6	389
9	448
12	418
15	297

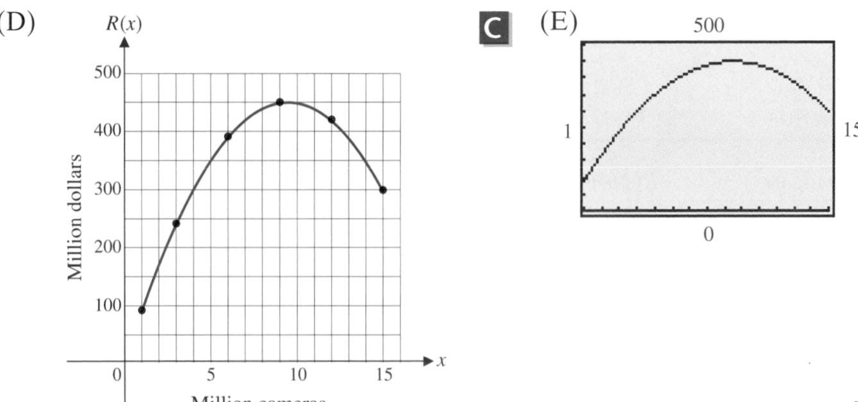

Matched Problem 7 The financial department in Example 7, using statistical techniques, produced the data in Table 6, where $C(x)$ is the cost in millions of dollars for manufacturing and selling x million cameras.

Table 6

Cost Data

x (MILLION)	$C(x)$ (MILLION \$)
1	175
5	260
8	305
12	395

Using special analytical techniques (regression analysis), an analyst produced the following cost function to model the data:

$$C(x) = 156 + 19.7x \qquad 1 \leqslant x \leqslant 15 \tag{6}$$

(A) Plot the data in Table 6. Then sketch a graph of equation (6) in the same coordinate system.

(B) What is the company's profit function for this camera, and what is its domain?

(C) Complete Table 7, computing profits to the nearest million dollars.

Table 7

Profit

x (MILLION)	$P(x)$ (MILLION \$)
1	−86
3	
6	
9	
12	
15	

(D) Plot the points from part (C). Then sketch a graph of the profit function through these points.

(E) Plot the profit function on a graphing utility.

Answers to Matched Problems

1.

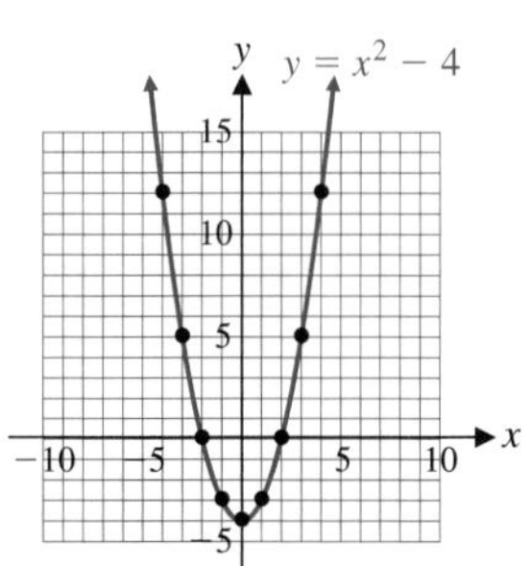

2. (A) Does not specify a function
(B) Specifies a function

3. $x \geq 2$ (inequality notation) or $[2, \infty)$ (interval notation)

4. (A) -3 (B) 0 (C) Does not exist (D) 6

5. Domain of F: R; Domain of G: All real numbers except -3;
Domain of H: $x \leq 2$ (inequality notation) or $(-\infty, 2]$ (interval notation)

6. (A) $a^2 - 4a + 9$ (B) $a^2 + 2ah + h^2 - 4a - 4h + 9$
(C) $2a + h - 4$

7. (A)

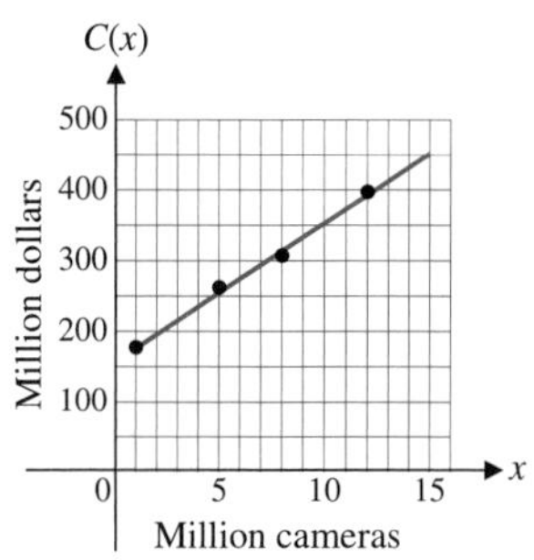

(B) $P(x) = R(x) - C(x) = x(94.8 - 5x) - (156 + 19.7x)$;
Domain: $1 \leq x \leq 15$

(C) Table 7

PROFIT

x (MILLION)	$P(x)$ (MILLION \$)
1	−86
3	24
6	115
9	115
12	25
15	−155

(D)

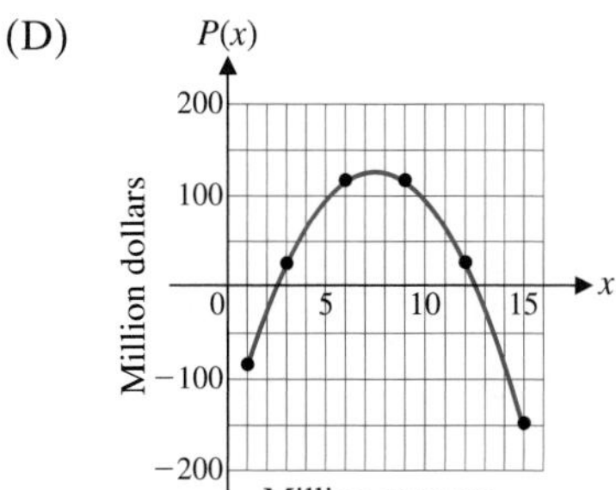

C (E)

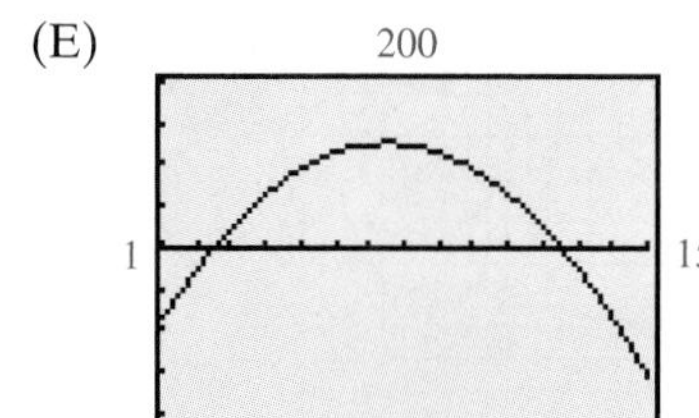

Exercise 1-1

A *Indicate whether each table specifies a function.*

1.

Domain	Range
3 →	0
5 →	1
7 →	2

2.

Domain	Range
−1 →	5
−2 →	7
−3 →	9

3.

Domain	Range
3 →	5
3 →	6
4 →	7
5 →	8

4.

Domain	Range
8 →	0
9 →	1
9 →	2
10 →	3

5.

Domain	Range
3 →	5
6 →	5
9 →	6
12 →	6

6.

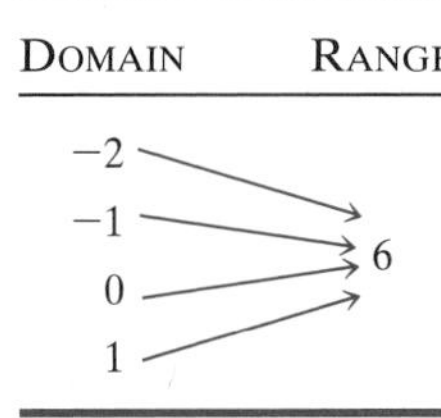

Domain	Range
−2 →	6
−1 →	6
0 →	6
1 →	6

Indicate whether each graph in Problems 7–12 specifies a function

7.

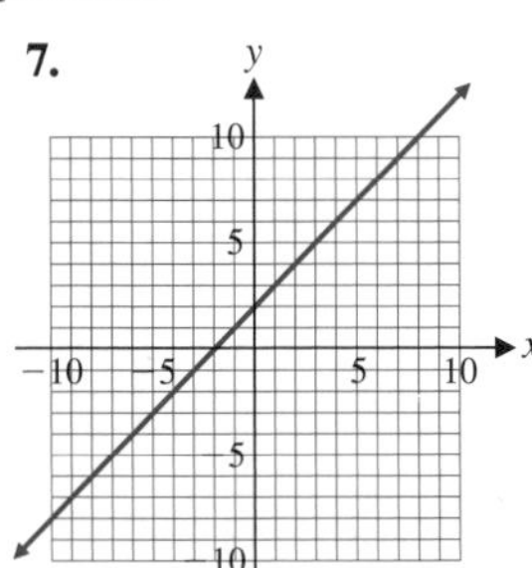

8.

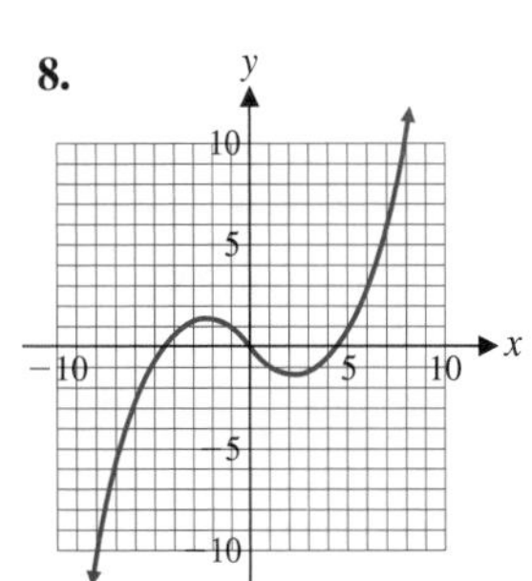

9.

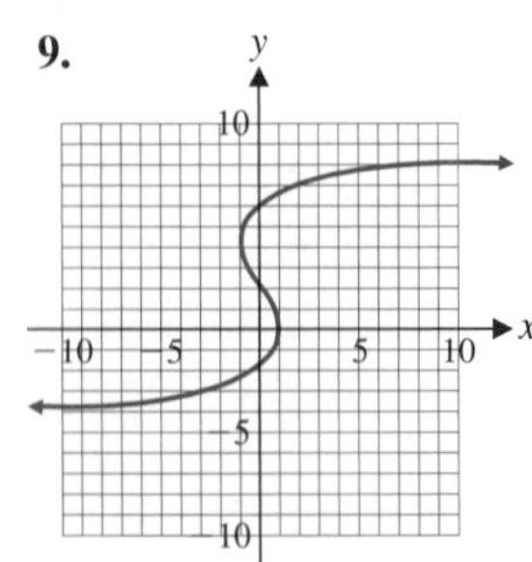

10.

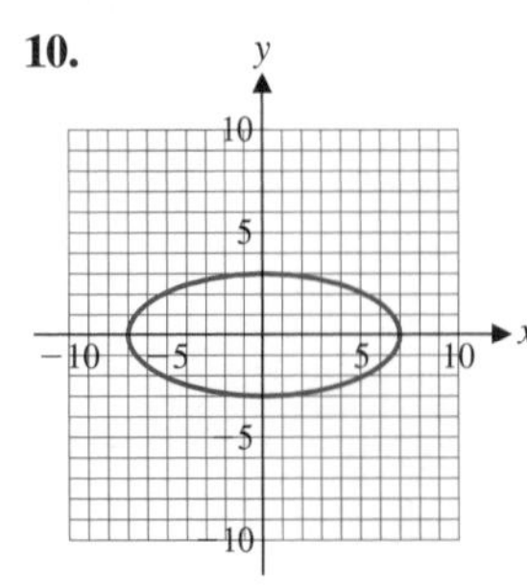

11.

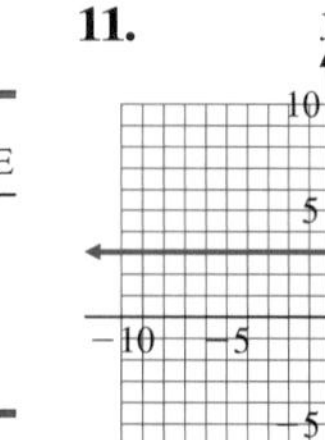

12.

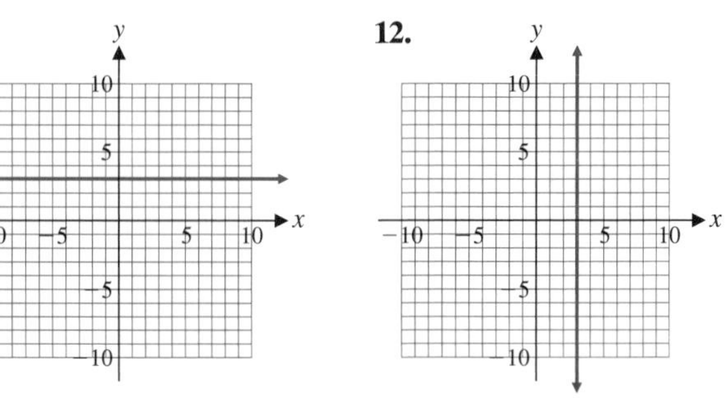

If $f(x) = 3x - 2$ and $g(x) = x - x^2$, find each of the expressions in Problems 13–30.

13. $f(2)$ **14.** $f(1)$ **15.** $f(-1)$

16. $f(-2)$ **17.** $g(3)$ **18.** $g(1)$

19. $f(0)$ **20.** $f(\frac{1}{3})$ **21.** $g(-3)$

22. $g(-2)$ **23.** $f(1) + g(2)$ **24.** $g(1) + f(2)$

25. $g(2) - f(2)$ **26.** $f(3) - g(3)$ **27.** $g(3) \cdot f(0)$

28. $g(0) \cdot f(-2)$ **29.** $\dfrac{g(-2)}{f(-2)}$ **30.** $\dfrac{g(-3)}{f(2)}$

In Problems 31–38, use the following graph of a function f to determine x or y to the nearest integer, as indicated. Some problems may have more than one answer.

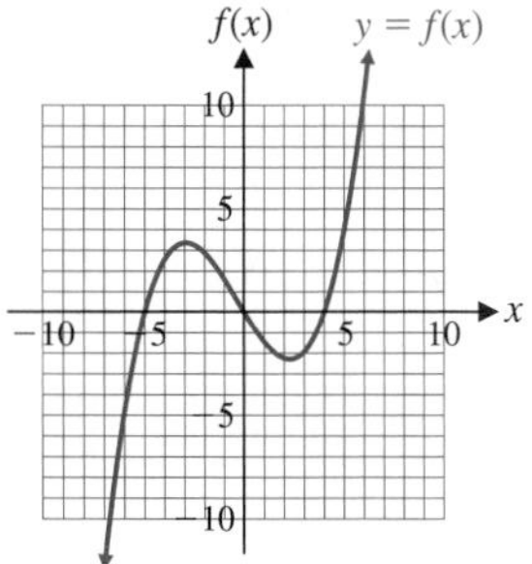

31. $y = f(-5)$ **32.** $y = f(4)$ **33.** $y = f(5)$

34. $y = f(-2)$ **35.** $0 = f(x)$ **36.** $3 = f(x), x < 0$

37. $-4 = f(x)$ **38.** $4 = f(x)$

B *In Problems 39–48, find the domain of each function.*

39. $F(x) = 2x^3 - x^2 + 3$ **40.** $H(x) = 7 - 2x^2 - x^4$

41. $f(x) = \frac{x - 2}{x + 4}$ **42.** $g(x) = \frac{x + 1}{x - 2}$

43. $F(x) = \frac{x + 2}{x^2 + 3x - 4}$ **44.** $G(x) = \frac{x - 7}{x^2 + x - 6}$

45. $g(x) = \sqrt{7 - x}$ **46.** $f(x) = \sqrt{5 + x}$

47. $G(x) = \frac{1}{\sqrt{7 - x}}$ **48.** $F(x) = \frac{1}{\sqrt{5 + x}}$

49. Two people are discussing the function

$$f(x) = \frac{x^2 - 4}{x^2 - 9}$$

and one says to the other, "$f(2)$ exists but $f(3)$ does not." Explain what they are talking about.

50. Referring to the function in Problem 49, do $f(-2)$ and $f(-3)$ exist? Explain.

The verbal statement "function f multiplies the square of the domain element by 3 and then subtracts 7 from the result" and the algebraic statement "$f(x) = 3x^2 - 7$" define the same function. In Problems 51–54, translate each verbal definition of a function into an algebraic definition.

51. Function g subtracts 5 from twice the cube of the domain element.

52. Function f multiplies the domain element by -3 and adds 4 to the result.

53. Function G multiplies the square root of the domain element by 2 and subtracts the square of the domain element from the result.

54. Function F multiplies the cube of the domain element by -8 and adds 3 times the square root of 3 to the result.

In Problems 55–58, translate each algebraic definition of the function into a verbal definition.

55. $f(x) = 2x - 3$ **56.** $g(x) = -2x + 7$

57. $F(x) = 3x^3 - 2\sqrt{x}$ **58.** $G(x) = 4\sqrt{x} - x^2$

Determine which of the equations in Problems 59–68 specify functions with independent variable x. For those that do, find the domain. For those that do not, find a value of x to which there corresponds more than one value of y.

59. $4x - 5y = 20$ **60.** $3y - 7x = 15$

61. $x^2 - y = 1$ **62.** $x - y^2 = 1$

63. $x + y^2 = 10$ **64.** $x^2 + y = 10$

65. $xy - 4y = 1$ **66.** $xy + y - x = 5$

67. $x^2 + y^2 = 25$ **68.** $x^2 - y^2 = 16$

69. If $F(t) = 4t + 7$, find: $\frac{F(3 + h) - F(3)}{h}$

70. If $G(r) = 3 - 5r$, find: $\frac{G(2 + h) - G(2)}{h}$

71. If $Q(x) = x^2 - 5x + 1$, find: $\frac{Q(2 + h) - Q(2)}{h}$

72. If $P(x) = 2x^2 - 3x - 7$, find: $\frac{P(3 + h) - P(3)}{h}$

C *In Problems 73–80, find and simplify:*

$$\frac{f(a + h) - f(a)}{h}$$

73. $f(x) = 4x - 3$ **74.** $f(x) = -3x + 9$

75. $f(x) = 4x^2 - 7x + 6$ **76.** $f(x) = 3x^2 + 5x - 8$

77. $f(x) = x^3$ **78.** $f(x) = x^3 - x$

79. $f(x) = \sqrt{x}$ **80.** $f(x) = \frac{1}{x}$

Problems 81–84 refer to the area A and perimeter P of a rectangle with length l and width w (see the figure).

$A = lw$

$P = 2l + 2w$

w

l

81. The area of a rectangle is 25 square inches. Express the perimeter $P(w)$ as a function of the width w, and state the domain of this function.

82. The area of a rectangle is 81 square inches. Express the perimeter $P(l)$ as a function of the length l, and state the domain of this function.

83. The perimeter of a rectangle is 100 meters. Express the area $A(l)$ as a function of the length l, and state the domain of this function.

84. The perimeter of a rectangle is 160 meters. Express the area $A(w)$ as a function of the width w, and state the domain of this function.

APPLICATIONS

Business & Economics

85. *Price–demand.* A company manufactures memory chips for microcomputers. Its marketing research department, using statistical techniques, collected the data shown in Table 8, where p is the wholesale price per chip at which x million chips can be sold. Using special analytical techniques (regression analysis), an analyst produced the following price–demand function to model the data:

$$p(x) = 119 - 6x \qquad 1 \leqslant x \leqslant 15$$

Table 8
PRICE–DEMAND

x (MILLION)	p ($)
1	115
6	80
10	65
15	31

Plot the data points in Table 8, and sketch a graph of the price–demand function in the same coordinate system. What would be the estimated price per chip for a demand of 8 million chips? For a demand of 11 million chips?

86. *Price–demand.* A company manufactures "Notebook" computers. Its marketing research department, using statistical techniques, collected the data shown in Table 9, where p is the wholesale price per computer at which x thousand computers can be sold. Using special analytical techniques (regression analysis), an analyst produced the following price–demand function to model the data:

$$p(x) = 1{,}190 - 36x \qquad 1 \leqslant x \leqslant 25$$

Table 9
PRICE–DEMAND

x (THOUSAND)	p ($)
2	1,110
5	1,030
10	815
14	695
21	435

Plot the data points in Table 9, and sketch a graph of the price–demand function in the same coordinate system. What would be the estimated price per computer for a demand of 9 thousand computers? For a demand of 18 thousand computers?

87. *Revenue.*
(A) Using the price–demand function

$$p(x) = 119 - 6x \qquad 1 \leqslant x \leqslant 15$$

from Problem 85, write the company's revenue function and indicate its domain.
(B) Complete Table 10, computing revenues to the nearest million dollars.

Table 10
REVENUE

x (MILLION)	$R(x)$ (MILLION $)
1	113
3	
6	
9	
12	
15	

(C) Plot the points from part (B) and sketch a graph of the revenue function through these points. Choose millions for the units on the horizontal and vertical axes. (C A graphing utility can be used as an aid.)

88. *Revenue.*
(A) Using the price–demand function

$$p(x) = 1{,}190 - 36x \qquad 1 \leqslant x \leqslant 25$$

from Problem 86, write the company's revenue function and indicate its domain.
(B) Complete Table 11, computing revenues to the nearest thousand dollars.

Table 11
REVENUE

x (THOUSAND)	$R(x)$ (THOUSAND $)
2	2,236
5	
10	
15	
20	
25	

(C) Plot the points from part (B) and sketch a graph of the revenue function through these points. Choose thousands for the units on the horizontal and vertical axes. (C A graphing utility can be used as an aid.)

89. *Profit.* The financial department for the company in Problems 85 and 87 established the following cost function for producing and selling x million memory chips:

$$C(x) = 234 + 23x \text{ million dollars}$$

(A) Write a profit function for producing and selling x million memory chips, and indicate its domain.

(B) Complete Table 12, computing profits to the nearest million dollars.

Table 12

PROFIT

x (MILLION)	$P(x)$ (MILLION \$)
1	−144
3	
6	
9	
12	
15	

(C) Plot the points in part (B) and sketch a graph of the profit function through these points. (C A graphing utility can be used as an aid.)

90. *Profit.* The financial department for the company in Problems 86 and 88 established the following cost function for producing and selling x thousand "Notebook" computers:

$$C(x) = 4{,}320 + 146x \text{ thousand dollars}$$

(A) Write a profit function for producing and selling x thousand "Notebook" computers, and indicate the domain of this function.

(B) Complete Table 13, computing profits to the nearest thousand dollars.

Table 13

PROFIT

x (THOUSAND)	$P(x)$ (THOUSAND \$)
2	−2,376
5	
10	
15	
20	
25	

(C) Plot the points in part (B) and sketch a graph of the profit function through these points. (C A graphing utility can be used as an aid.)

91. *Packaging.* A candy box is to be made out of a piece of cardboard that measures 8 by 12 inches. Equal-sized squares x inches on a side will be cut out of each corner, and then the ends and sides will be folded up to form a rectangular box.

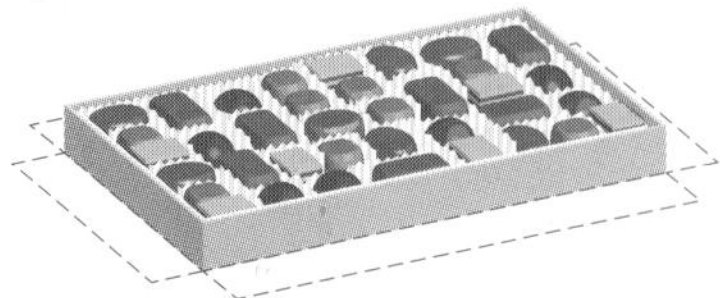

(A) Express the volume of the box $V(x)$ in terms of x.

(B) What is the domain of the function V (determined by the physical restrictions)?

(C) Complete Table 14.

Table 14

VOLUME

x	$V(x)$
1	
2	
3	

(D) Plot the points in part (C) and sketch a graph of the volume function through these points. (C A graphing utility can be used as an aid.)

92. *Packaging.* Refer to Problem 91.

(A) Table 15 shows the volume of the box for some values of x between 1 and 2. Use these values to estimate to one decimal place the value of x between 1 and 2 that would produce a box with a volume of 65 cubic inches.

Table 15

VOLUME

x	$V(x)$
1.1	62.524
1.2	64.512
1.3	65.988
1.4	66.976
1.5	67.5
1.6	67.584
1.7	67.252

(B) Describe how you could refine this table to estimate x to two decimal places.

C (C) Carry out the refinement you described in part (B) and approximate x to two decimal places.

93. *Packaging.* Refer to Problems 91 and 92.

(A) Examine the graph of $V(x)$ from Problem 91D and discuss the possible locations of other values of x that would produce a box with a volume of 65 cubic inches. Construct a table like Table 15 to estimate any such value to one decimal place.

C (B) Refine the table you constructed in part (A) to provide an approximation to two decimal places.

94. *Packaging.* A parcel delivery service will only deliver packages with length plus girth (distance around) not exceeding 108 inches. A rectangular shipping box with square ends x inches on a side is to be used.

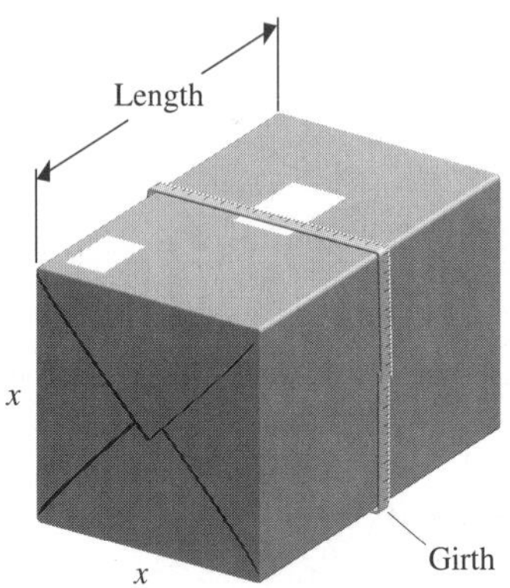

(A) If the full 108 inches is to be used, express the volume of the box $V(x)$ in terms of x.

(B) What is the domain of the function V (determined by the physical restrictions)?

(C) Complete Table 16.

Table 16
VOLUME

x	$V(x)$
5	
10	
15	
20	
25	

(D) Plot the points in part (C) and sketch a graph of the volume function through these points. (C A graphing utility can be used as an aid.)

Life Sciences

95. *Muscle contraction.* In a study of the speed of muscle contraction in frogs under various loads, noted British biophysicist and Nobel prize winner A. W. Hill determined that the weight w (in grams) placed on the muscle and the speed of contraction v (in centimeters per second) are approximately related by an equation of the form

$$(w + a)(v + b) = c$$

where a, b, and c are constants. Suppose that for a certain muscle, $a = 15$, $b = 1$, and $c = 90$. Express v as a function of w. Find the speed of contraction if a weight of 16 grams is placed on the muscle.

Social Sciences

96. *Politics.* The percentage s of seats in the House of Representatives won by Democrats and the percentage v of votes cast for Democrats (when expressed as decimal fractions) are related by the equation

$$5v - 2s = 1.4 \qquad 0 < s < 1, \quad 0.28 < v < 0.68$$

(A) Express v as a function of s, and find the percentage of votes required for the Democrats to win 51% of the seats.

(B) Express s as a function of v, and find the percentage of seats won if Democrats receive 51% of the votes.

SECTION 1-2

Elementary Functions: Graphs and Transformations

- A Beginning Library of Elementary Functions
- Vertical and Horizontal Shifts
- Reflections, Expansions, and Contractions
- Piecewise-Defined Functions

The functions

$$g(x) = x^2 - 4 \qquad h(x) = (x - 4)^2 \qquad k(x) = -4x^2$$

all can be expressed in terms of the function $f(x) = x^2$ as follows:

$$g(x) = f(x) - 4 \qquad h(x) = f(x - 4) \qquad k(x) = -4f(x)$$

In this section we will see that the graphs of functions g, h, and k are closely related to the graph of function f. Insight gained by understanding these relationships will help us analyze and interpret the graphs of many different functions.

A Beginning Library of Elementary Functions

As you progress through this book, and most any other mathematics course beyond this one, you will repeatedly encounter a relatively small list of elementary functions. We will identify these functions, study their basic properties, and include them in a library of elementary functions (see the inside front cover). This library will become an important addition to your mathematical toolbox and can be used in any course or activity where mathematics is applied.

Figure 1 shows six basic functions that you will encounter frequently. You should know the definition, domain, and range of each, and be able to recognize their graphs. For Figure 1B, recall the definition of *absolute value:*

$$|x| = \begin{cases} -x & \text{if } x < 0 \\ x & \text{if } x \geq 0 \end{cases}$$

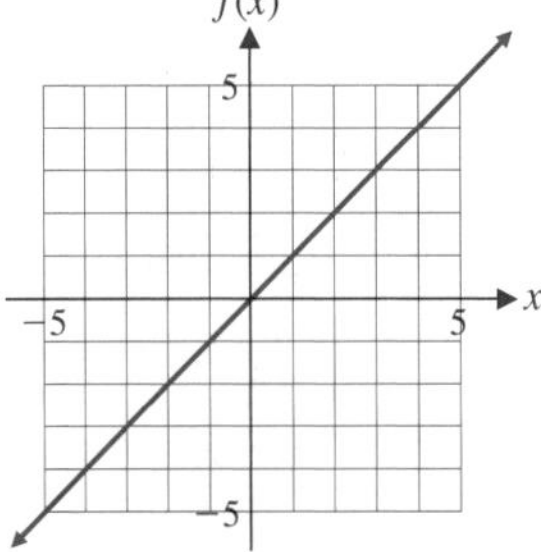

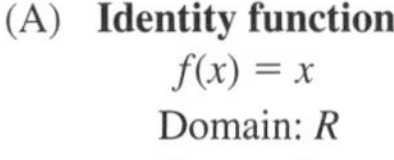
(A) **Identity function**
$f(x) = x$
Domain: R
Range: R

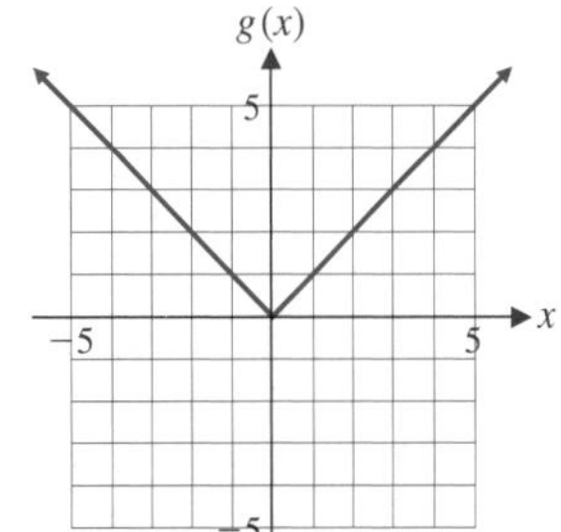

(B) **Absolute value function**
$g(x) = |x|$
Domain: R
Range: $[0, \infty)$

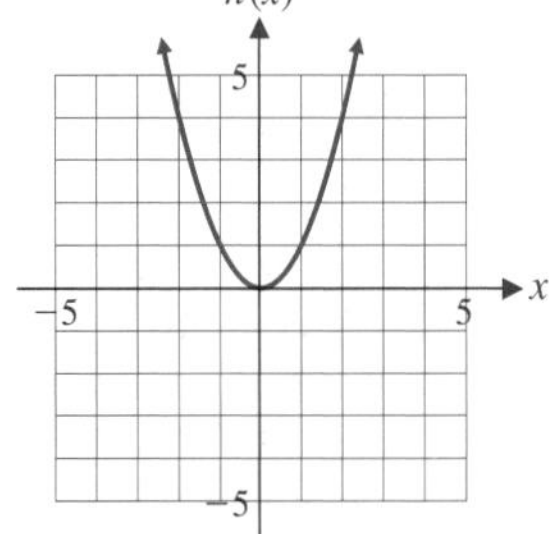

(C) **Square function**
$h(x) = x^2$
Domain: R
Range: $[0, \infty)$

FIGURE 1
Some basic functions and their graphs.
Note: Letters used to designate these functions may vary from context to context; R is the set of all real numbers.

(continued)

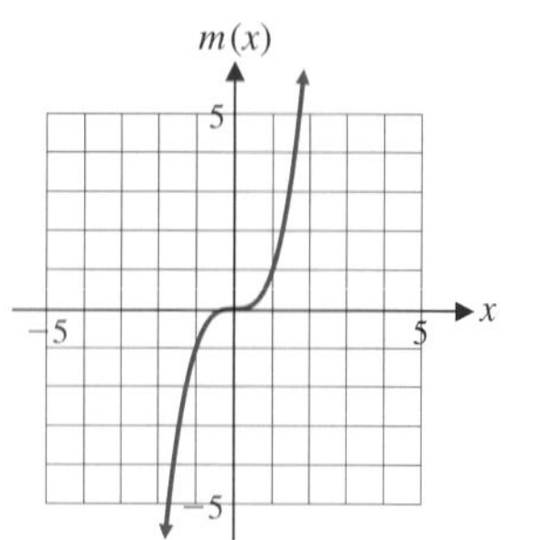

(D) **Cube function**
$m(x) = x^3$
Domain: R
Range: R

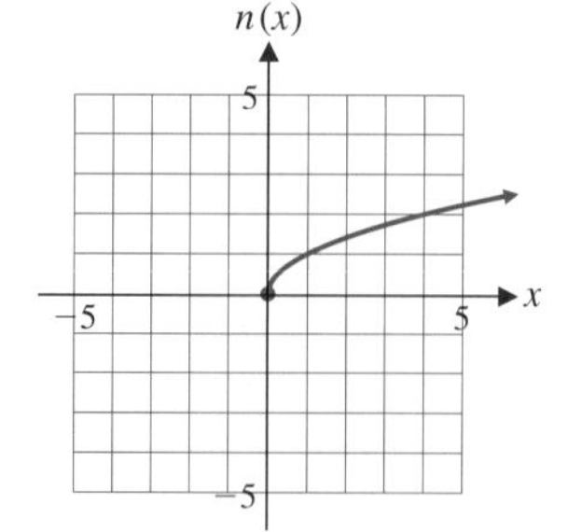

(E) **Square root function**
$n(x) = \sqrt{x}$
Domain: $[0, \infty)$
Range: $[0, \infty)$

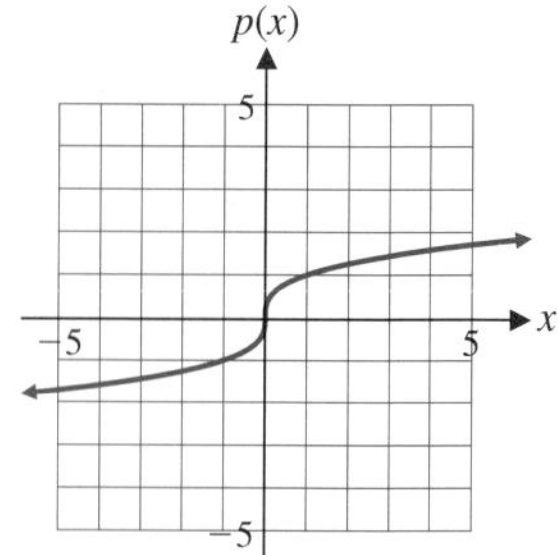

(F) **Cube root function**
$p(x) = \sqrt[3]{x}$
Domain: R
Range: R

FIGURE 1 (*continued*)

■ Vertical and Horizontal Shifts

If a new function is formed by performing an operation on a given function, then the graph of the new function is called a **transformation** of the graph of the original function. For example, graphs of both $y = f(x) + k$ and $y = f(x + h)$ are transformations of the graph of $y = f(x)$.

Explore–Discuss 1

Let $f(x) = x^2$.

(A) Graph $y = f(x) + k$ for $k = -4, 0$, and 2 simultaneously in the same coordinate system. Describe the relationship between the graph of $y = f(x)$ and the graph of $y = f(x) + k$ for k any real number.

(B) Graph $y = f(x + h)$ for $h = -4, 0$, and 2 simultaneously in the same coordinate system. Describe the relationship between the graph of $y = f(x)$ and the graph of $y = f(x + h)$ for h any real number.

Example 1 ➠ **Vertical and Horizontal Shifts**

(A) How are the graphs of $y = |x| + 4$ and $y = |x| - 5$ related to the graph of $y = |x|$? Confirm your answer by graphing all three functions simultaneously in the same coordinate system.

(B) How are the graphs of $y = |x + 4|$ and $y = |x - 5|$ related to the graph of $y = |x|$? Confirm your answer by graphing all three functions simultaneously in the same coordinate system.

Solution (A) The graph of $y = |x| + 4$ is the same as the graph of $y = |x|$ shifted upward 4 units, and the graph of $y = |x| - 5$ is the same as the graph of $y = |x|$ shifted downward 5 units. Figure 2 confirms these conclusions. [It appears that the graph of $y = f(x) + k$ is the graph of $y = f(x)$ shifted up if k is positive and down if k is negative.]

(B) The graph of $y = |x + 4|$ is the same as the graph of $y = |x|$ shifted to the left 4 units, and the graph of $y = |x - 5|$ is the same as the graph of $y = |x|$ shifted to the right 5 units. Figure 3 confirms these conclusions. [It appears that the graph of $y = f(x + h)$ is the graph of $y = f(x)$ shifted right if h is negative and left if h is positive—the opposite of what you might expect.]

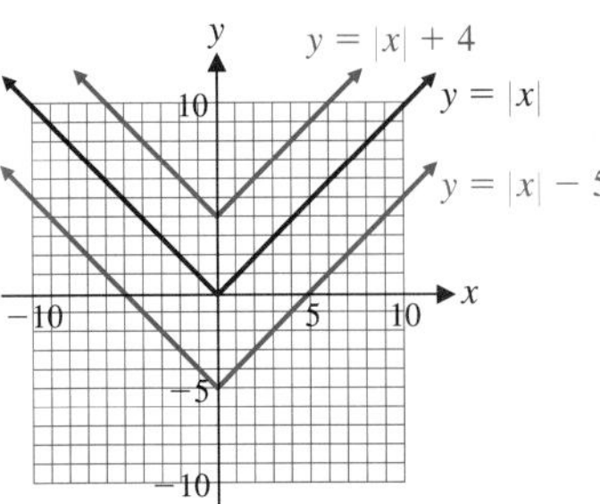

FIGURE 2
Vertical shifts

FIGURE 3
Horizontal shifts

Matched Problem 1 ➡ (A) How are the graphs of $y = \sqrt{x} + 5$ and $y = \sqrt{x} - 4$ related to the graph of $y = \sqrt{x}$? Confirm your answer by graphing all three functions simultaneously in the same coordinate system.

(B) How are the graphs of $y = \sqrt{x + 5}$ and $y = \sqrt{x - 4}$ related to the graph of $y = \sqrt{x}$? Confirm your answer by graphing all three functions simultaneously in the same coordinate system.

Comparing the graphs of $y = f(x) + k$ with the graph of $y = f(x)$, we see that the graph of $y = f(x) + k$ can be obtained from the graph of $y = f(x)$ by **vertically translating** (shifting) the graph of the latter upward k units if k is positive and downward $|k|$ units if k is negative. Comparing the graphs of $y = f(x + h)$ with the graph of $y = f(x)$, we see that the graph of $y = f(x + h)$ can be obtained from the graph of $y = f(x)$ by **horizontally translating** (shifting) the graph of the latter h units to the left if h is positive and $|h|$ units to the right if h is negative.

Example 2 ➡ **Vertical and Horizontal Translations (Shifts)** The graphs in Figure 4 are either horizontal or vertical shifts of the graph of $f(x) = x^2$. Write appropriate equations for functions H, G, M, and N in terms of f.

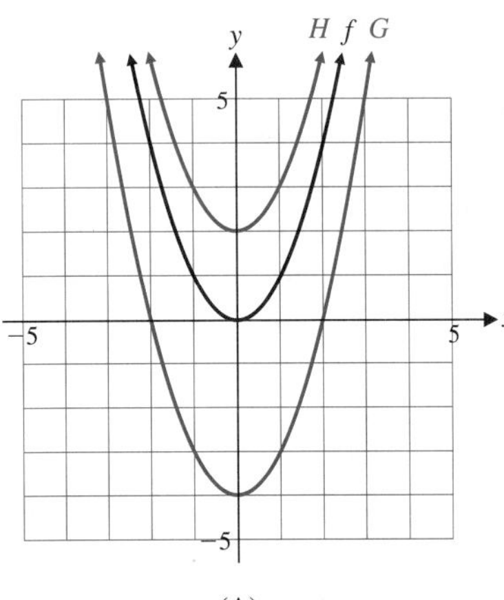

(A)

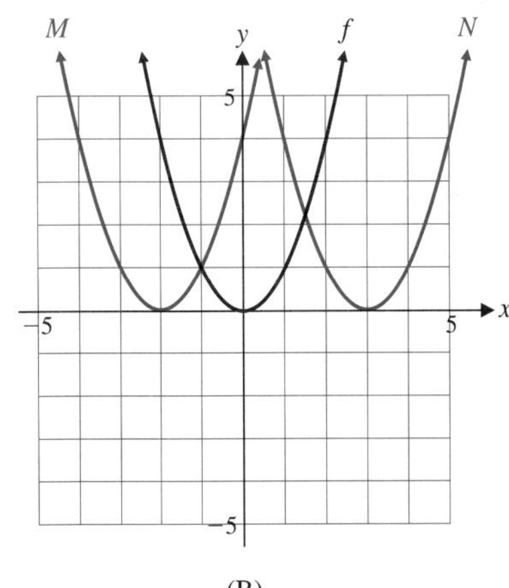

(B)

FIGURE 4
Vertical and horizontal shifts

Solution Functions H and G are vertical shifts given by

$$H(x) = x^2 + 2 \qquad G(x) = x^2 - 4$$

Functions M and N are horizontal shifts given by

$$M(x) = (x + 2)^2 \qquad N(x) = (x - 3)^2$$

Matched Problem 2 The graphs in Figure 5 are either horizontal or vertical shifts of the graph of $f(x) = \sqrt[3]{x}$. Write appropriate equations for functions H, G, M, and N in terms of f.

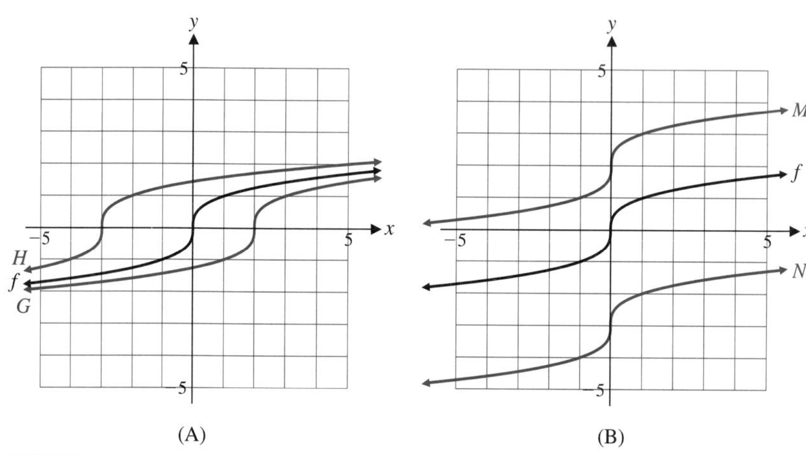

FIGURE 5
Vertical and horizontal shifts

Reflections, Expansions, and Contractions

We now investigate how the graph of $y = Af(x)$ is related to the graph of $y = f(x)$ for different real numbers A.

Explore–Discuss 2

(A) Graph $y = Ax^2$ for $A = 1, 4$, and $\frac{1}{4}$ simultaneously in the same coordinate system.
(B) Graph $y = Ax^2$ for $A = -1, -4$, and $-\frac{1}{4}$ simultaneously in the same coordinate system.
(C) Describe the relationship between the graph of $h(x) = x^2$ and the graph of $G(x) = Ax^2$ for A any real number.

Comparing $y = Af(x)$ to $y = f(x)$, we see that the graph of $y = Af(x)$ can be obtained from the graph of $y = f(x)$ by multiplying each ordinate value of the latter by A. The result is a **vertical expansion** of the graph of $y = f(x)$ if $A > 1$, a **vertical contraction** of the graph of $y = f(x)$ if $0 < A < 1$, and a **reflection in the *x* axis** if $A = -1$. If A is a negative number other than -1, then the result is a combination of a reflection in the x axis and either a vertical expansion or a vertical contraction.

Example 3 **Reflections, Expansions, and Contractions**

(A) How are the graphs of $y = 2|x|$ and $y = 0.5|x|$ related to the graph of $y = |x|$? Confirm your answer by graphing all three functions simultaneously in the same coordinate system.

(B) How is the graph of $y = -2|x|$ related to the graph of $y = |x|$? Confirm your answer by graphing both functions simultaneously in the same coordinate system.

SOLUTION (A) The graph of $y = 2|x|$ is a vertical expansion of the graph of $y = |x|$ by a factor of 2, and the graph of $y = 0.5|x|$ is a vertical contraction of the graph of $y = |x|$ by a factor of 0.5. Figure 6 confirms this conclusion.

(B) The graph of $y = -2|x|$ is a reflection in the x axis and a vertical expansion of the graph of $y = |x|$. Figure 7 confirms this conclusion.

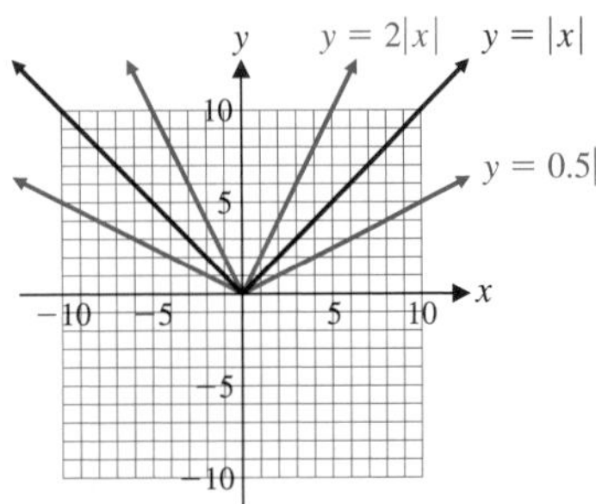

FIGURE 6
Vertical expansion and contraction

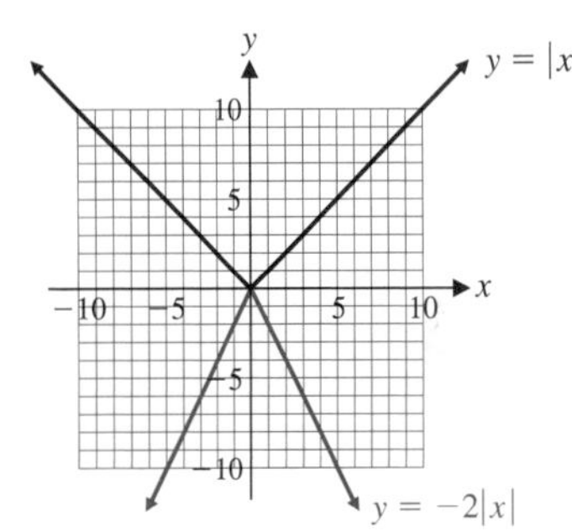

FIGURE 7
Reflection and vertical expansion

Matched Problem 3

(A) How are the graphs of $y = 2x$ and $y = 0.5x$ related to the graph of $y = x$? Confirm your answer by graphing all three functions simultaneously in the same coordinate system.

(B) How is the graph of $y = -0.5x$ related to the graph of $y = x$? Confirm your answer by graphing both functions in the same coordinate system.

The various transformations considered above are summarized in the following box for easy reference:

Graph Transformations (Summary)

Vertical Translation:

$$y = f(x) + k \quad \begin{cases} k > 0 & \text{Shift graph of } y = f(x) \text{ up } k \text{ units.} \\ k < 0 & \text{Shift graph of } y = f(x) \text{ down } |k| \text{ units.} \end{cases}$$

Horizontal Translation:

$$y = f(x + h) \quad \begin{cases} h > 0 & \text{Shift graph of } y = f(x) \text{ left } h \text{ units.} \\ h < 0 & \text{Shift graph of } y = f(x) \text{ right } |h| \text{ units.} \end{cases}$$

(continued)

Graph Transformations (Summary) (*continued*)

Reflection:

$y = -f(x)$ Reflect the graph of $y = f(x)$ in the x axis.

Vertical Expansion and Contraction:

$$y = Af(x) \quad \begin{cases} A > 1 & \text{Vertically expand graph of } y = f(x) \text{ by multiplying each ordinate value by } A. \\ 0 < A < 1 & \text{Vertically contract graph of } y = f(x) \text{ by multiplying each ordinate value by } A. \end{cases}$$

Explore–Discuss 3

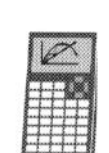

Use a graphing utility, if available, to explore the graph of $y = A(x + h)^2 + k$ for various values of the constants A, h, and k. Discuss how the graph of $y = A(x + h)^2 + k$ is related to the graph of $y = x^2$.

Example 4 **Combining Graph Transformations** Discuss how the graph of $y = -|x - 3| + 1$ is related to the graph of $y = |x|$. Confirm your answer by graphing both functions simultaneously in the same coordinate system.

Solution The graph of $y = -|x - 3| + 1$ is a reflection in the x axis, a horizontal translation of 3 units to the right, and a vertical translation of 1 unit upward of the graph of $y = |x|$. Figure 8 confirms this description.

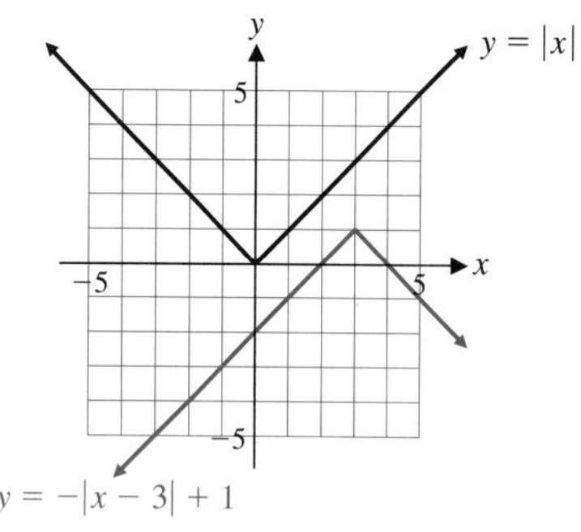

FIGURE 8
Combined transformations

Matched Problem 4 The graph of $y = G(x)$ in Figure 9 involves a reflection and a translation of the graph of $y = x^3$. Describe how the graph of function G is related to the graph of $y = x^3$ and find an equation of the function G.

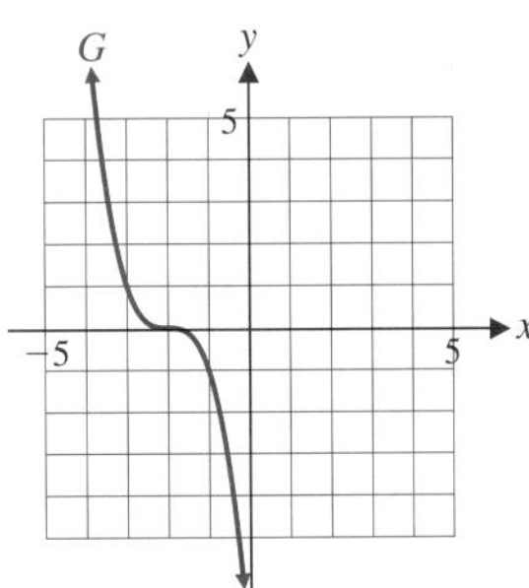

FIGURE 9
Combined transformations

■ Piecewise-Defined Functions

Earlier we noted that the absolute value of a real number x is defined as

$$|x| = \begin{cases} -x & \text{if } x < 0 \\ x & \text{if } x \geq 0 \end{cases}$$

Notice that this function is defined by different rules for different parts of its domain. Functions whose definitions involve more than one rule are called **piecewise-defined functions.** As the next example illustrates, piecewise-defined functions occur naturally in many applications.

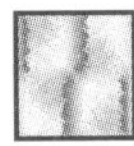

Example 5 **Discount Pricing** Nordic Office Products uses volume discounting to promote sales. Table 1 shows the price per box for $\frac{1}{4}$ inch staples. The discounted prices for volume purchases apply to all boxes purchased.

Table 1

Price per Box

Number of boxes	1	10	20
Price ($)	2.99	2.28	1.96

(A) Write a piecewise definition for the cost $C(x)$ of ordering x boxes of staples.
(B) Graph C.
(C) Explain why a customer would never want to order 8 or 9 boxes of staples.

Solution (A) The purchase cost of these staples is the price per box times the number of boxes purchased. Since the price per box depends on the volume purchased, the cost function cannot be given by a single rule and must be written as a piecewise-defined function:

$$C(x) = \begin{cases} 2.99x & \text{if } 1 \leq x < 10 \\ 2.28x & \text{if } 10 \leq x < 20 \\ 1.96x & \text{if } 20 \leq x \end{cases}$$

(B) Note that this cost function is only defined for nonnegative integer values of x (an order for a partial box of staples is not practical). However, to facilitate graphing of this cost function, it is convenient to assume that $C(x)$ is defined for all $x \geq 1$. To graph C, first note that each rule in the definition of C represents a transformation of the identity function $f(x) = x$. Graphing each transformation over the indicated interval produces the graph of C shown in Figure 10. We use this graph to represent the cost function, with the understanding that noninteger values of x may require special interpretation.

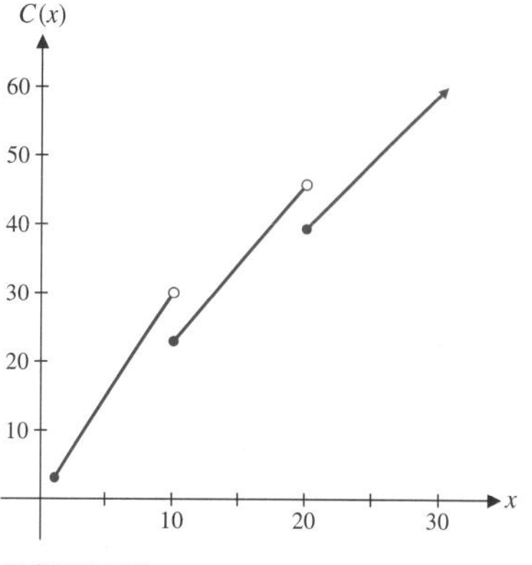

Note: A solid dot on the graph indicates that the point is part of the graph; an open dot indicates that the point is not part of the graph.

FIGURE 10
Cost of purchasing x boxes of staples

Table 2

x	$C(x)$ ($)
8	23.92
9	26.91
10	22.80

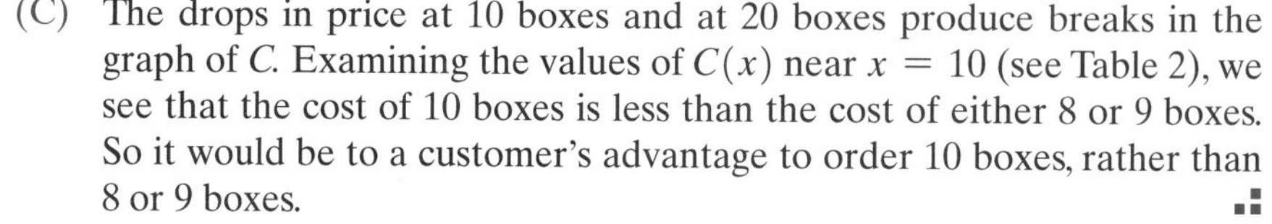

(C) The drops in price at 10 boxes and at 20 boxes produce breaks in the graph of C. Examining the values of $C(x)$ near $x = 10$ (see Table 2), we see that the cost of 10 boxes is less than the cost of either 8 or 9 boxes. So it would be to a customer's advantage to order 10 boxes, rather than 8 or 9 boxes.

Matched Problem 5

Table 3 shows the price per pack for packs of 3 × 5 inch index cards at Nordic Office Products.

Table 3
Price per Pack

Number of Packs	1	5	10
Price ($)	2.19	1.99	1.78

(A) Write a piecewise definition for the cost $C(x)$ of ordering x packs of index cards.
(B) Graph C.
(C) Explain why a customer would never want to order 9 packs of index cards.

Functions like the cost function C in Figure 10 are said to be **discontinuous** at the points where breaks occur. Thus, C is discontinuous at $x = 10$ and at $x = 20$. (The study of calculus includes a more formal presentation of this topic.)

Explore–Discuss 4

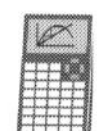

How can we graph the piecewise-defined function C in Example 5 on a graphing utility (see Fig. 10)? With your graphing utility set in the connected mode, enter y_1, y_2, and y_3, as shown in Figure 11A and graph for $0 \leqslant x \leqslant 30, 0 \leqslant y \leqslant 70$. Then modify the definitions of y_1, y_2, and y_3, as shown in Figure 11B, and graph $y_4 = y_1 + y_2 + y_3$. Now try graphing y_4 in the dot mode. Which graph gives the most useful representation of Figure 10? Which method would be most useful for identifying points of discontinuity?

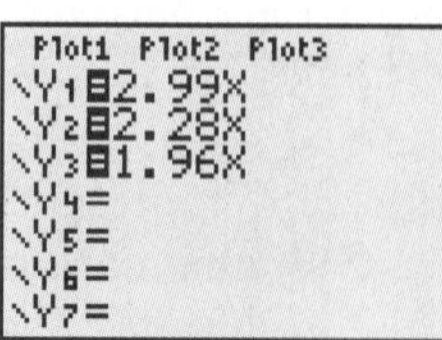

(A)

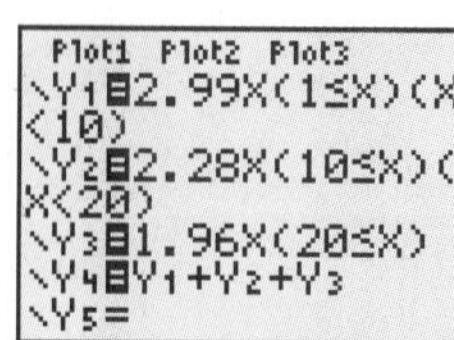

(B)

FIGURE 11

Answers to Matched Problems

1. (A) The graph of $y = \sqrt{x} + 5$ is the same as the graph of $y = \sqrt{x}$ shifted upward 5 units, and the graph of $y = \sqrt{x} - 4$ is the same as the graph of $y = \sqrt{x}$ shifted downward 4 units. The figure confirms these conclusions.

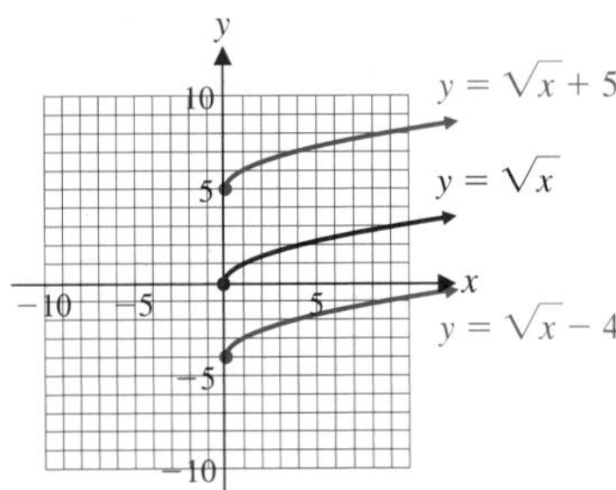

(B) The graph of $y = \sqrt{x + 5}$ is the same as the graph of $y = \sqrt{x}$ shifted to the left 5 units, and the graph of $y = \sqrt{x - 4}$ is the same as the graph of $y = \sqrt{x}$ shifted to the right 4 units. The figure confirms these conclusions.

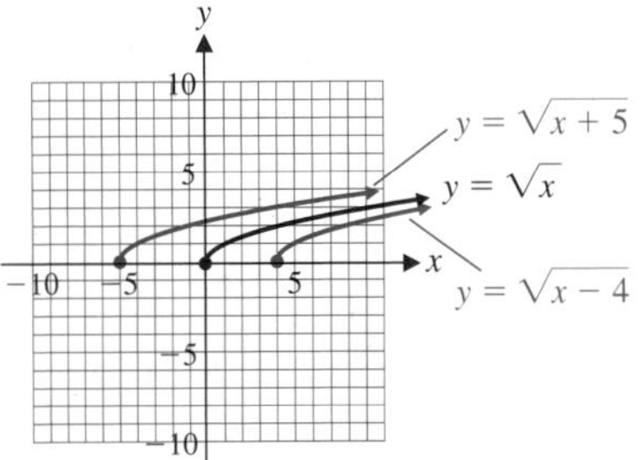

2. $H(x) = \sqrt[3]{x + 3}$, $G(x) = \sqrt[3]{x - 2}$, $M(x) = \sqrt[3]{x} + 2$, $N(x) = \sqrt[3]{x} - 3$

3. (A) The graph of $y = 2x$ is a vertical expansion of the graph of $y = x$, and the graph of $y = 0.5x$ is a vertical contraction of the graph of $y = x$. The figure confirms these conclusions.

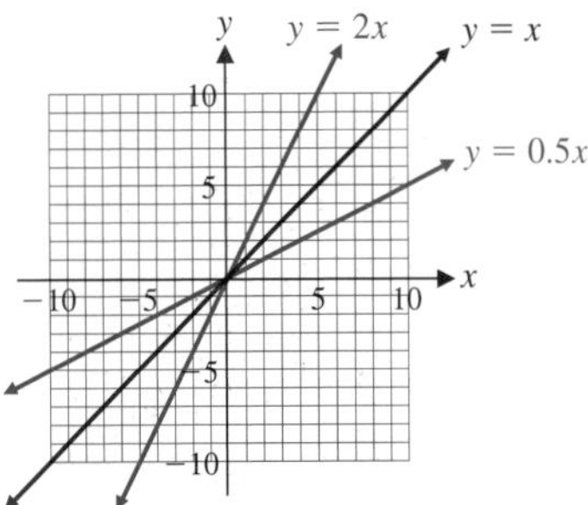

(B) The graph of $y = -0.5x$ is a vertical contraction and a reflection in the x axis of the graph of $y = x$. The figure confirms this conclusion.

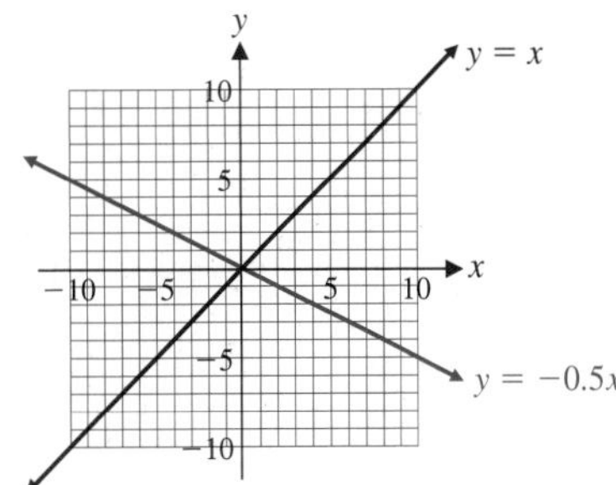

4. The graph of function G is a reflection in the x axis and a horizontal translation of 2 units to the left of the graph of $y = x^3$. An equation for G is $G(x) = -(x + 2)^3$.

5. (A) $C(x) = \begin{cases} 2.19x & \text{if } 1 \leq x < 5 \\ 1.99x & \text{if } 5 \leq x < 10 \\ 1.78x & \text{if } 10 \leq x \end{cases}$

(B)

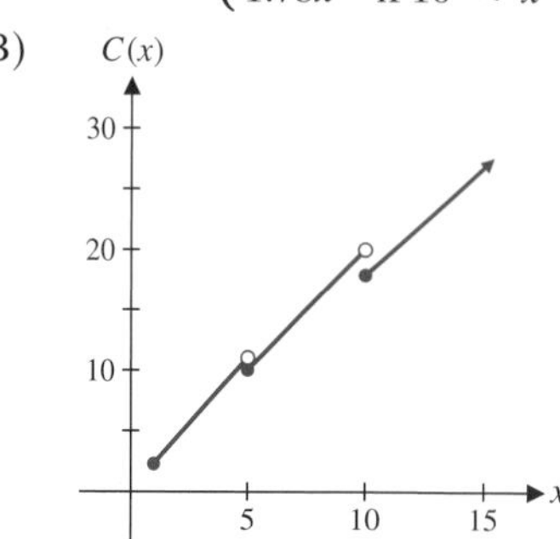

(C) The cost of 9 packs is $17.91 and the cost of 10 packs is $17.80. So it is always to the customer's advantage to order 10 packs instead of 9.

EXERCISE 1-2

A *Without looking back in the text, indicate the domain and range of each of the following functions. (Making rough sketches on scratch paper may help.)*

1. $f(x) = 0.4x$
2. $g(x) = 3x$
3. $h(x) = -x^2$
4. $m(x) = -|x|$
5. $g(x) = -2\sqrt{x}$
6. $f(x) = -0.5\sqrt[3]{x}$
7. $F(x) = -0.1x^3$
8. $G(x) = 5x^3$

Graph each of the functions in Problems 9–20 using the graphs of functions f and g below:

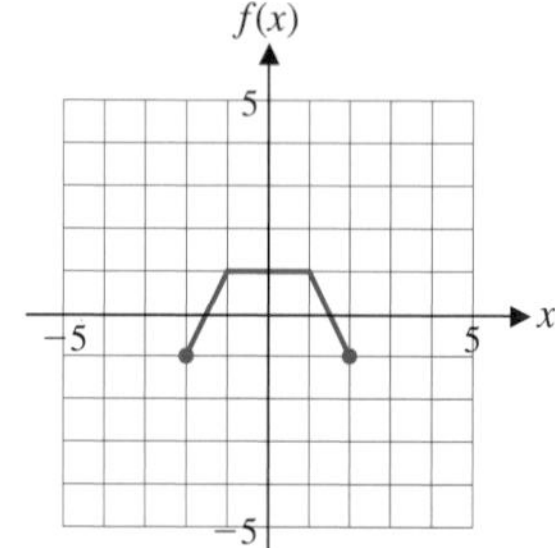

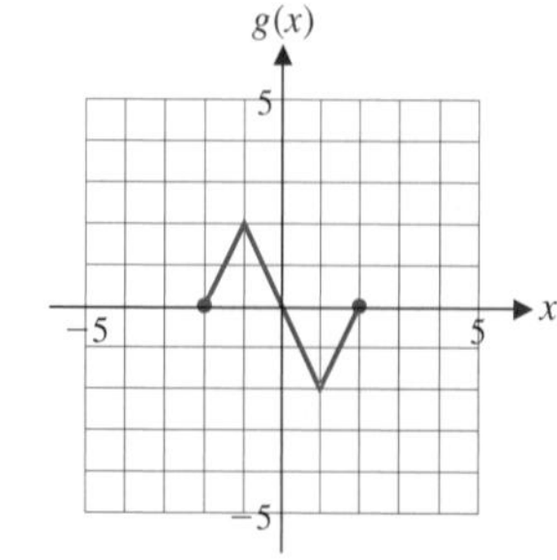

9. $y = f(x) + 2$
10. $y = g(x) - 1$
11. $y = f(x + 2)$
12. $y = g(x - 1)$
13. $y = g(x - 3)$
14. $y = f(x + 3)$
15. $y = g(x) - 3$
16. $y = f(x) + 3$
17. $y = -f(x)$
18. $y = -g(x)$
19. $y = 0.5g(x)$
20. $y = 2f(x)$

B *In Problems 21–28, indicate how the graph of each function is related to the graph of one of the six basic functions in Figure 1 (at the beginning of this section). Sketch a graph of each function.*

21. $g(x) = -|x + 3|$
22. $h(x) = -|x - 5|$
23. $f(x) = (x - 4)^2 - 3$
24. $m(x) = (x + 3)^2 + 4$
25. $f(x) = 7 - \sqrt{x}$
26. $g(x) = -6 + \sqrt[3]{x}$
27. $h(x) = -3|x|$
28. $m(x) = -0.4x^2$

Check your descriptions and graphs in Problems 21–28 by graphing each function on a graphing utility.

Each graph in Problems 29–36 is the result of applying a sequence of transformations to the graph of one of the six basic functions in Figure 1 (at the beginning of this section). Identify the basic function and describe the transformation verbally. Write an equation for the given graph.

29.

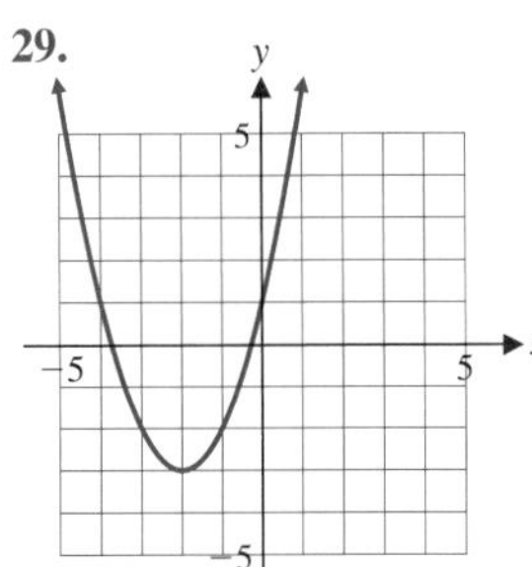

30.

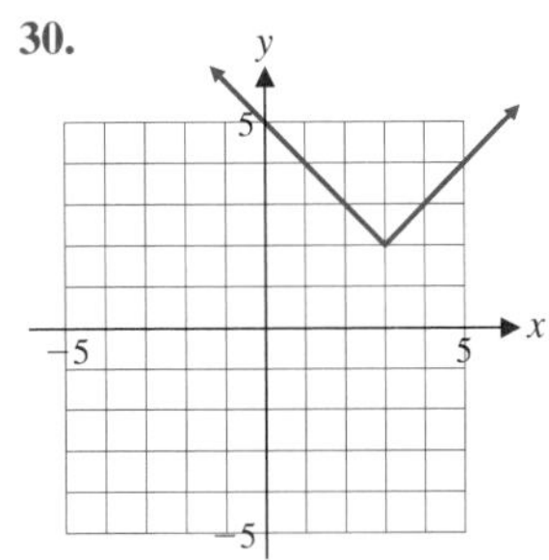

31.

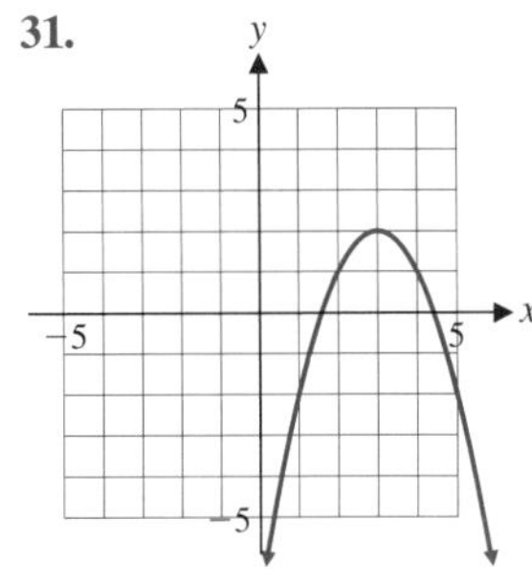

32.

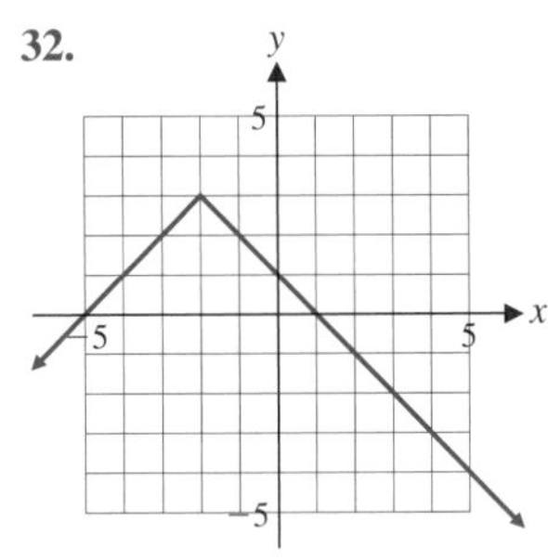

33.

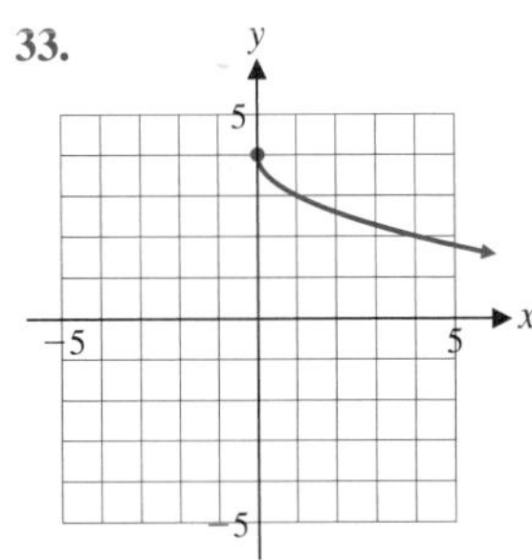

34.

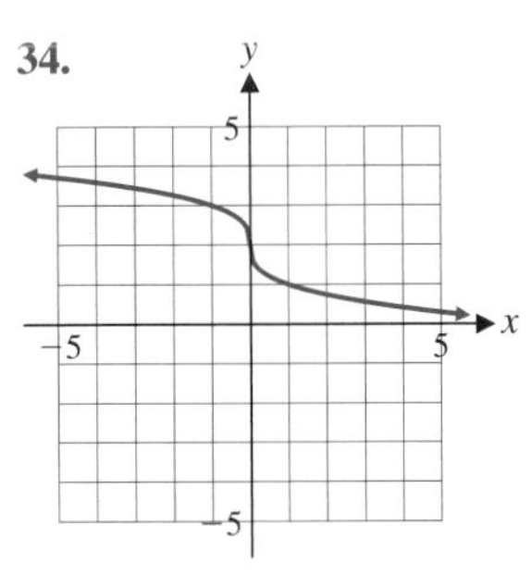

35.

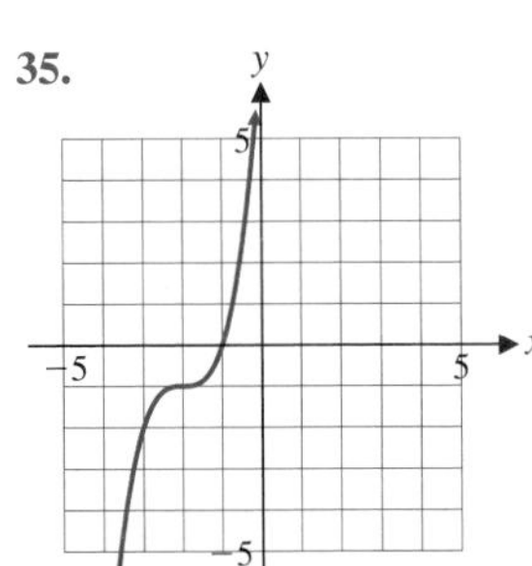

36.

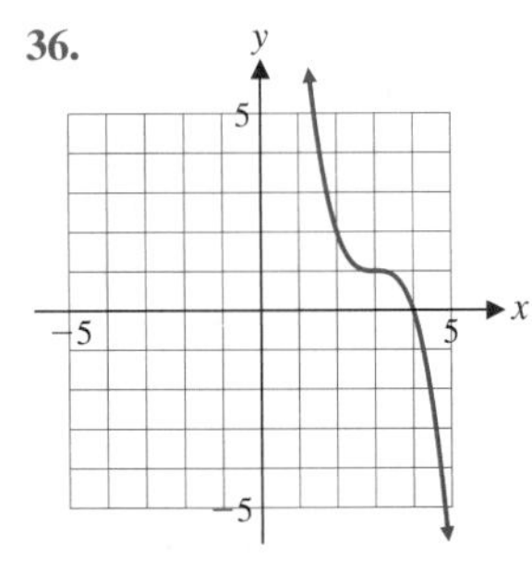

Check your equations in Problems 29–36 by graphing each on a graphing utility.

In Problems 37–42, the graph of the function g is formed by applying the indicated sequence of transformations to the given function f. Find an equation for the function g and graph g using $-5 \leqslant x \leqslant 5$ and $-5 \leqslant y \leqslant 5$.

37. The graph of $f(x) = \sqrt{x}$ is shifted 2 units to the right and 3 units down.

38. The graph of $f(x) = \sqrt[3]{x}$ is shifted 3 units to the left and 2 units up.

39. The graph of $f(x) = |x|$ is reflected in the x axis and shifted to the left 3 units.

40. The graph of $f(x) = |x|$ is reflected in the x axis and shifted to the right 1 unit.

41. The graph of $f(x) = x^3$ is reflected in the x axis and shifted 2 units to the right and down 1 unit.

42. The graph of $f(x) = x^2$ is reflected in the x axis and shifted to the left 2 units and up 4 units.

Graph each function in Problems 43–48 and find any points of discontinuity.

43. $f(x) = \begin{cases} x + 1 & \text{if } x < 0 \\ x - 1 & \text{if } x \geqslant 0 \end{cases}$

44. $g(x) = \begin{cases} x^2 - 1 & \text{if } x < 0 \\ 2 - x^2 & \text{if } x \geqslant 0 \end{cases}$

45. $h(x) = \begin{cases} -x & \text{if } x \leqslant 0 \\ \sqrt{x} & \text{if } x > 0 \end{cases}$

46. $k(x) = \begin{cases} \sqrt[3]{x} & \text{if } x \leqslant 0 \\ x^3 & \text{if } x > 0 \end{cases}$

47. $p(x) = \begin{cases} -2x & \text{if } x \leqslant -1 \\ x^2 & \text{if } -1 < x < 1 \\ \sqrt{x - 1} & \text{if } 1 \leqslant x \end{cases}$

48. $q(x) = \begin{cases} \sqrt[3]{x + 2} & \text{if } x < -2 \\ \frac{1}{2}x & \text{if } -2 \leqslant x \leqslant 2 \\ \sqrt[3]{x - 2} & \text{if } 2 < x \end{cases}$

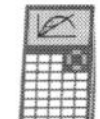

Check Problems 43–48 by graphing on a graphing utility.

C *Each of the graphs in Problems 49–54 involves a reflection in the x axis and/or a vertical expansion or contraction of one of the basic functions in Figure 1 (at the beginning of this section). Identify the basic function, and describe the transformation verbally. Write an equation for the given graph.*

49.

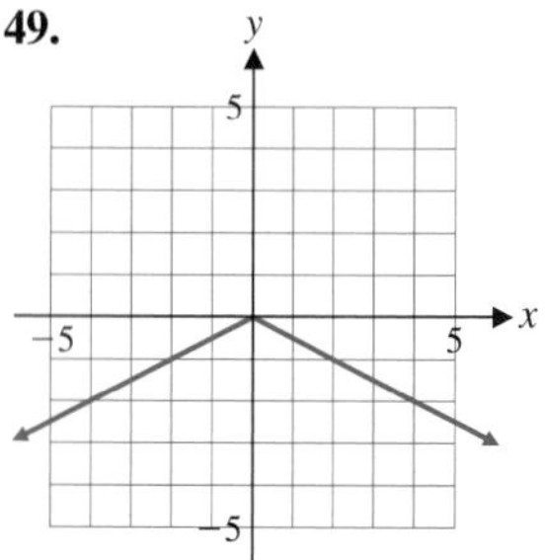

50.

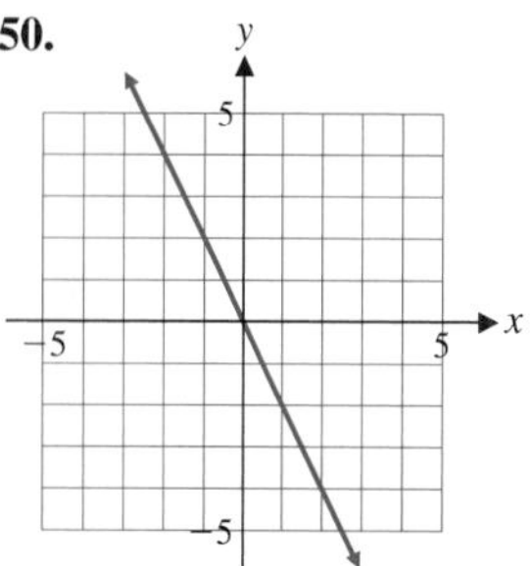

51.

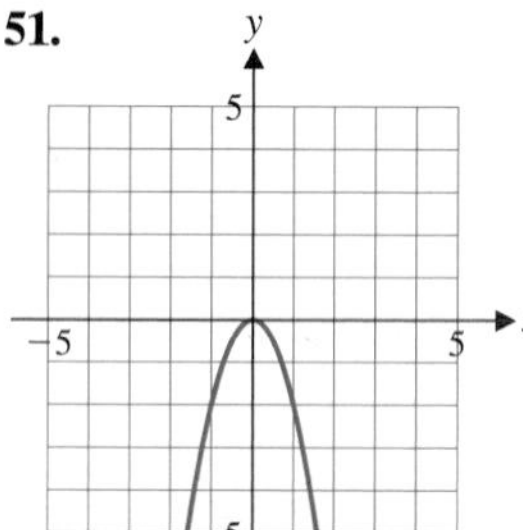

52.

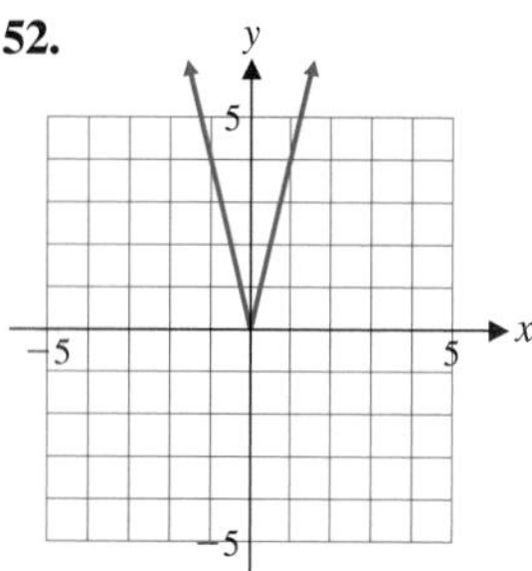

53.

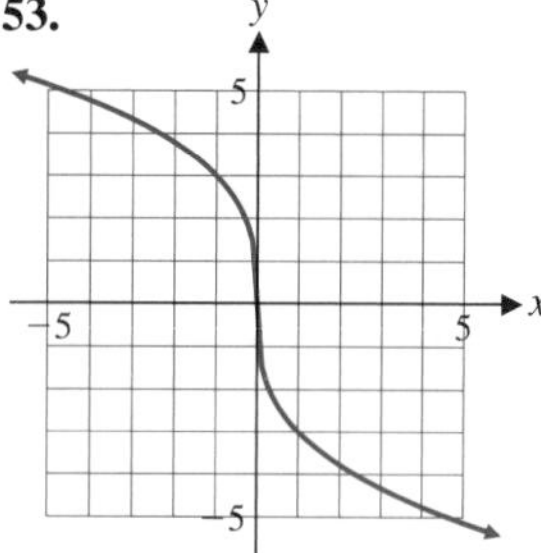

54.

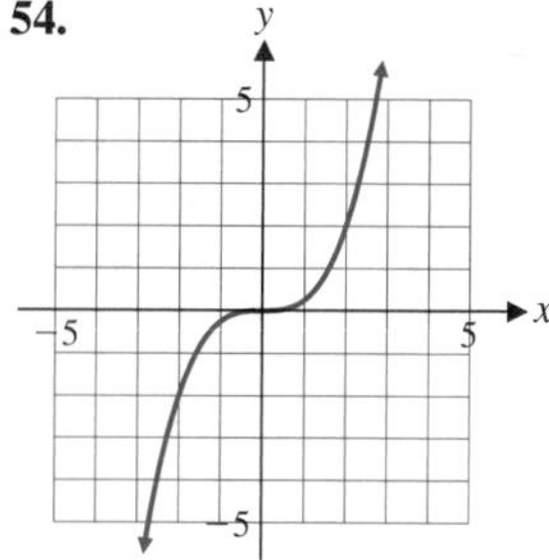

Check your equations in Problems 49–54 by graphing each on a graphing utility.

Changing the order in a sequence of transformations may change the final result. Investigate each pair of transformations in Problems 55–60 to determine if reversing their order can produce a different result. Support your conclusions with specific examples and/or mathematical arguments.

55. Vertical shift; horizontal shift

56. Vertical shift; reflection in y axis

57. Vertical shift; reflection in x axis

58. Vertical shift; expansion

59. Horizontal shift; reflection in y axis

60. Horizontal shift; contraction

APPLICATIONS

Business & Economics

61. *Price–demand.* A retail chain sells CD players. The retail price $p(x)$ (in dollars) and the weekly demand x for a particular model are related by

$$p(x) = 115 - 4\sqrt{x} \qquad 9 \le x \le 289$$

(A) Describe how the graph of function p can be obtained from the graph of one of the basic functions in Figure 1 (at the beginning of this section).
(B) Sketch a graph of function p using part (A) as an aid.

62. *Price–supply.* The manufacturers of the CD players in Problem 61 are willing to supply x players at a price of $p(x)$ as given by the equation

$$p(x) = 4\sqrt{x} \qquad 9 \le x \le 289$$

(A) Describe how the graph of function p can be obtained from the graph of one of the basic functions in Figure 1 (at the beginning of this section).
(B) Sketch a graph of function p using part (A) as an aid.

63. *Hospital costs.* Using statistical methods, the financial department of a hospital arrived at the cost equation

$$C(x) = 0.00048(x - 500)^3 + 60{,}000 \qquad 100 \le x \le 1{,}000$$

where $C(x)$ is the cost in dollars for handling x cases per month.

(A) Describe how the graph of function C can be obtained from the graph of one of the basic functions in Figure 1 (at the beginning of this section).
C (B) Sketch a graph of function C using part (A) and a graphing utility as aids.

64. *Price–demand.* A company manufactures and sells in-line skates. Their financial department has established the price–demand function

$$p(x) = 190 - 0.013(x - 10)^2 \qquad 10 \le x \le 100$$

where $p(x)$ is the price at which x thousand pairs of skates can be sold.

(A) Describe how the graph of function p can be obtained from the graph of one of the basic functions in Figure 1 (at the beginning of this section).
C (B) Sketch a graph of function p using part (A) and a graphing utility as aids.

65. *Discount pricing.* The prices for boxes of double-pocket portfolios at Nordic Office Products are given in Table 4.
   (A) Write a piecewise definition for the cost $C(x)$ of ordering x boxes of portfolios.
   (B) Graph C.
   (C) Would a customer ever want to purchase 3 boxes of portfolios? 9 boxes of portfolios? Explain.

Table 4

PRICE PER BOX

NUMBER OF BOXES	1	4	10
PRICE ($)	10.06	8.52	7.31

66. *Discount pricing.* The prices for purchasing rolls of transparent tape at Nordic Office Products are given in Table 5.
   (A) Write a piecewise definition for the cost $C(x)$ of ordering x rolls of tape.
   (B) Graph C.
   (C) Would a customer ever want to purchase 11 rolls of tape? 22 or 23 rolls of tape? Explain.

Table 5

PRICE PER ROLL

NUMBER OF ROLLS	1	12	24
PRICE ($)	2.39	2.19	1.99

67. *Discount pricing.* A car rental company charges $0.50 per mile for the first 50 miles driven, $0.35 per mile for the next 50 miles, and $0.20 per mile for any additional miles.
   (A) Write a piecewise definition for the total charges $C(x)$ for driving x miles.
   (B) Graph C.
   (C) How does this pricing method differ from the discount pricing used by Nordic Office Products in Problems 65 and 66? What effect does this have on the graph of C?

68. *Discount pricing.* A telephone company charges $0.06 per call for the first 60 calls during a 1 month billing period, $0.04 for the next 60 calls, and $0.02 for any additional calls.
   (A) Write a piecewise definition for the total charges $C(x)$ for making x calls.
   (B) Graph C.
   (C) How does this pricing method differ from the discount pricing used by Nordic Office Products in Problems 65 and 66? What effect does this have on the graph of C?

Life Sciences

69. *Physiology.* A good approximation of the normal weight of a person 60 inches or taller but not taller than 80 inches is given by $w(x) = 5.5x - 220$, where x is height in inches and $w(x)$ is weight in pounds.

   (A) Describe how the graph of function w can be obtained from the graph of one of the basic functions in Figure 1 (at the beginning of this section).
   (B) Sketch a graph of function w using part (A) as an aid.

70. *Physiology.* The average weight of a particular species of snake is given by $w(x) = 463x^3$, $0.2 \leq x \leq 0.8$, where x is length in meters and $w(x)$ is weight in grams.
   (A) Describe how the graph of function w can be obtained from the graph of one of the basic functions in Figure 1 (at the beginning of this section).
   (B) Sketch a graph of function w using part (A) as an aid.

Social Sciences

71. *Safety research.* Under ideal conditions, if a person driving a vehicle slams on the brakes and skids to a stop, the speed of the vehicle $v(x)$ (in miles per hour) is given approximately by $v(x) = C\sqrt{x}$, where x is the length of skid marks (in feet) and C is a constant that depends on the road conditions and the weight of the vehicle. For a particular vehicle, $v(x) = 7.08\sqrt{x}$ and $4 \leq x \leq 144$.
   (A) Describe how the graph of function v can be obtained from the graph of one of the basic functions in Figure 1 (at the beginning of this section).
   (B) Sketch a graph of function v using part (A) as an aid.

72. *Learning.* A production analyst has found that on the average it takes a new person $T(x)$ minutes to perform a particular assembly operation after x performances of the operation, where $T(x) = 10 - \sqrt[3]{x}$, $0 \leq x \leq 155$.
   (A) Describe how the graph of function T can be obtained from the graph of one of the basic functions in Figure 1 (at the beginning of this section).
   (B) Sketch a graph of function T using part (A) as an aid.

SECTION 1-3

Linear Functions and Straight Lines

- INTERCEPTS
- LINEAR FUNCTIONS, EQUATIONS, AND INEQUALITIES
- GRAPHS OF $Ax + By = C$
- SLOPE OF A LINE
- EQUATIONS OF LINES—SPECIAL FORMS
- APPLICATIONS

In this section we will add another important class of functions to our basic list of elementary functions. These functions are called *linear functions* and include the identity function $f(x) = x$ as a special case. We will investigate the relationship between linear functions and the solutions to linear equations and inequalities. (A detailed treatment of algebraic solutions to linear equations and inequalities can be found in Appendix A-8. Finally, we will review the concept of slope and some of the standard equations of straight lines. These new tools will be applied to a variety of significant applications, including cost and price–demand functions.

INTERCEPTS

Figure 1 illustrates the graphs of three functions *f*, *g*, and *h*.

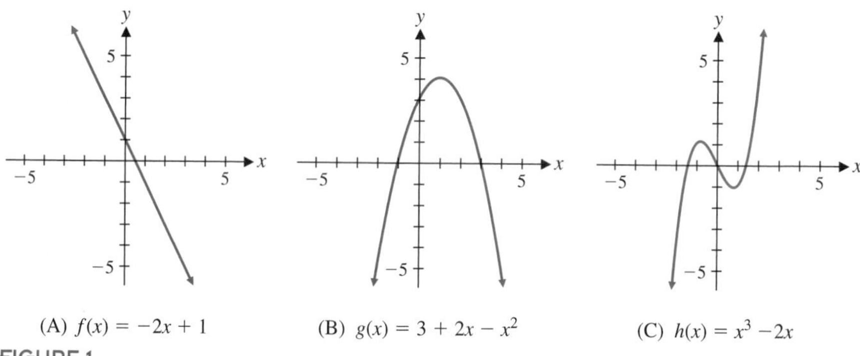

FIGURE 1
Graphs of several functions

If the graph of a function f crosses the x axis at a point with x coordinate a, then a is called an **x intercept** of f. If the graph of f crosses the y axis at a point with y coordinate b, then b is called the **y intercept.** It is common practice to refer to both the numbers a and b and the points $(a, 0)$ and $(0, b)$ as the x and y intercepts. If the y intercept exists, then 0 must be in the domain of f and the y intercept is simply $f(0)$. Thus, the graph of a function can have at most one y intercept. The x intercepts are all real solutions or roots of $f(x) = 0$, which may vary from none to an unlimited number. In Figure 1, function f has one y intercept and one x intercept; function g has one y intercept and two x intercepts; and function h has one y intercept and three x intercepts.

LINEAR FUNCTIONS, EQUATIONS, AND INEQUALITIES

In Figure 1, the graph of $f(x) = -2x + 1$ is a straight line, and because of this, we choose to call this type of function a *linear function.* In general:

> **Linear and Constant Functions**
>
> A function f is a **linear function** if
>
> $$f(x) = mx + b \qquad m \neq 0$$
>
> where m and b are real numbers. The **domain** is the set of all real numbers, and the **range** is the set of all real numbers. If $m = 0$, then f is called a **constant function,**
>
> $$f(x) = b$$
>
> which has the set of all real numbers as its **domain** and the constant b as its **range.**

Since $mx + b, m \neq 0$, is a first-degree polynomial, linear functions are also called **first-degree functions.** Figure 2 shows the graphs of two linear functions f and g, and a constant function h.

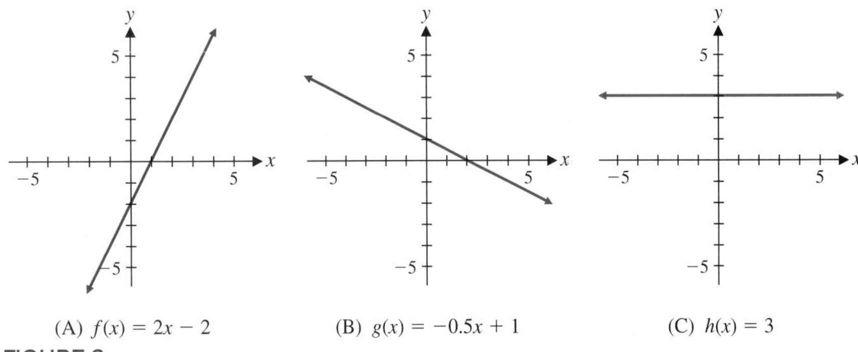

FIGURE 2
Two linear functions and a constant function

In Section 1-2 we saw that the graph of the identity function $f(x) = x$ is a straight line. The graph of a linear function $g(x) = mx + b, m \neq 0$, is the same as the graph of $f(x) = x$ reflected in the x axis if m is negative, vertically expanded or contracted by a factor of $|m|$, and shifted up or down $|b|$ units. In general, it can be shown that:

The graph of a linear function is a straight line that is neither horizontal nor vertical.

The graph of a constant function is a horizontal straight line.

What about vertical lines? Recall from Section 1-1 that the graph of a function cannot contain two points with the same x coordinate and different

y coordinates. Since *all* points on a vertical line have the same x coordinate, the graph of a function can never be a vertical line. Later in this section we will discuss equations of vertical lines, but these equations never define functions.

Explore–Discuss 1

(A) Is it possible for a linear function to have two x intercepts? No x intercepts? If either of your answers is yes, give an example.
(B) Is it possible for a linear function to have two y intercepts? No y intercept? If either of your answers is yes, give an example.
(C) Discuss the possible number of x and y intercepts for a constant function.

Example 1 **Intercepts, Equations, and Inequalities**

(A) Graph $f(x) = \frac{3}{2}x - 4$ in a rectangular coordinate system.
(B) Find the x and y intercepts algebraically to two decimal places.
C (C) Graph $f(x) = \frac{3}{2}x - 4$ in a standard viewing window.
C (D) Find the x and y intercepts to two decimal places using trace and zoom or an appropriate built-in routine in your graphing utility.
(E) Solve $\frac{3}{2}x - 4 \leq 0$ graphically to two decimal places using parts (A) and (B) or (C) and (D).

Solution

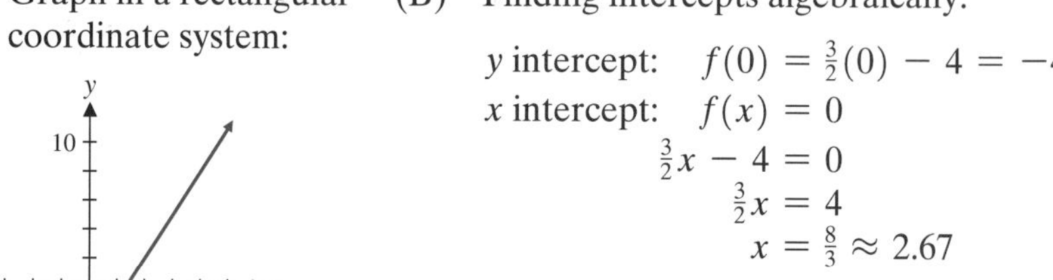

(A) Graph in a rectangular coordinate system:

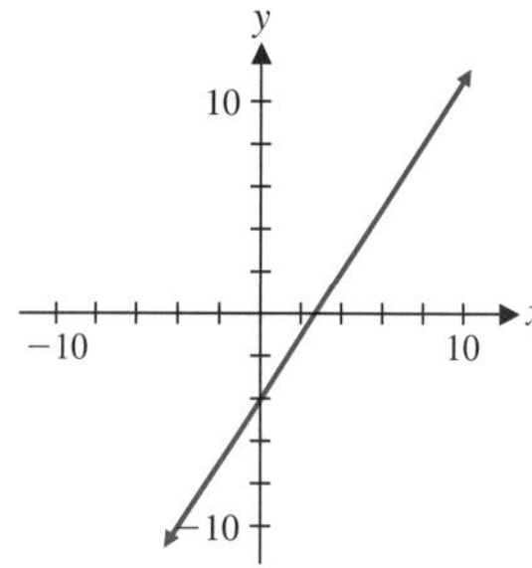

(B) Finding intercepts algebraically:

y intercept: $f(0) = \frac{3}{2}(0) - 4 = -4$
x intercept: $f(x) = 0$

$$\frac{3}{2}x - 4 = 0$$
$$\frac{3}{2}x = 4$$
$$x = \frac{8}{3} \approx 2.67$$

C (C) Graph in a graphing utility:

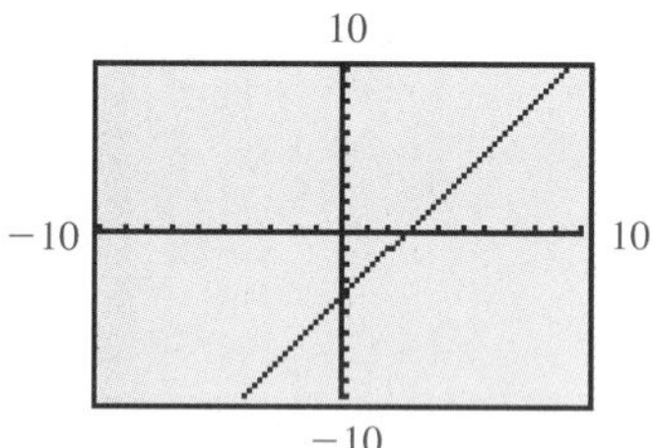

C (D) Finding intercepts graphically in a graphing utility:

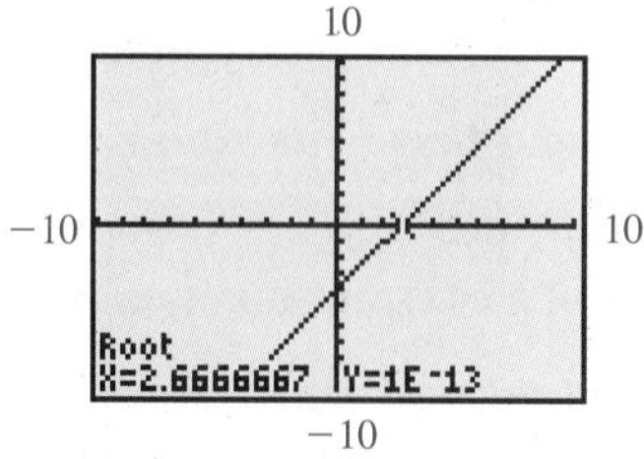

x intercept: 2.67

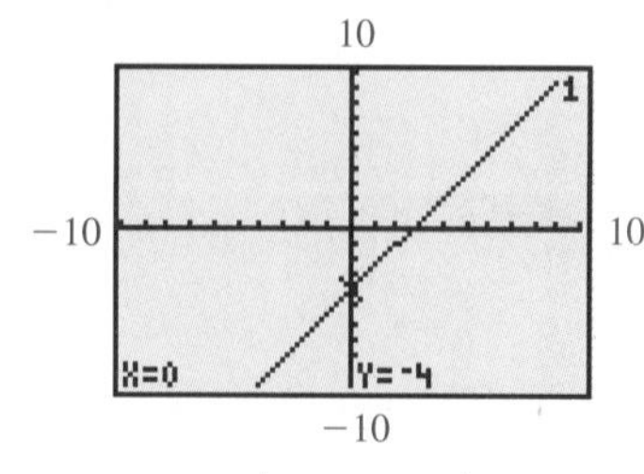

y intercept: -4

(E) Solving $\frac{3}{2}x - 4 \leqslant 0$ graphically using parts (A) and (B) or (C) and (D): The linear inequality $\frac{3}{2}x - 4 \leqslant 0$ holds for those values of x for which the graph of $f(x) = \frac{3}{2}x - 4$ in the figure in part (A) or (C) is at or below the x axis. This happens for x less than or equal to the x intercept found in parts (B) or (D). Thus, the solution set for the linear inequality is $x \leqslant 2.67$ or $(-\infty, 2.67]$.

Matched Problem 1

(A) Graph $f(x) = -\frac{4}{3}x + 5$ in a rectangular coordinate system.
(B) Find the x and y intercepts algebraically to two decimal places.

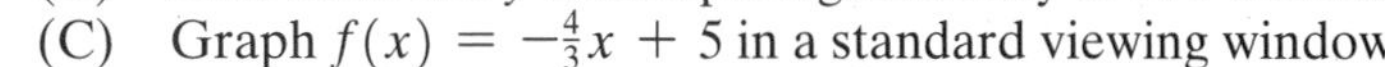

C (C) Graph $f(x) = -\frac{4}{3}x + 5$ in a standard viewing window.

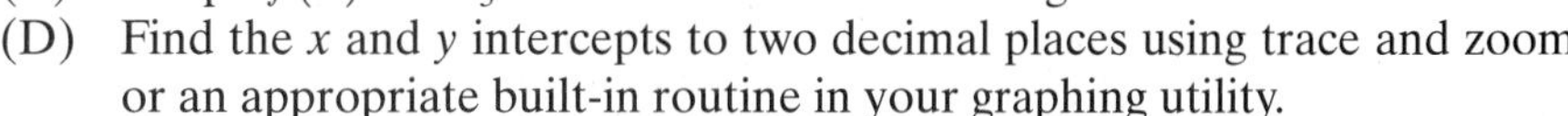

C (D) Find the x and y intercepts to two decimal places using trace and zoom or an appropriate built-in routine in your graphing utility.
(E) Solve $-\frac{4}{3}x + 5 \geqslant 0$ graphically to two decimal places using parts (A) and (B) or (C) and (D).

Graphs of $Ax + By = C$

We now investigate graphs of linear, or first-degree, equations in two variables:

$$Ax + By = C \tag{1}$$

where A and B are not both 0. Depending on the values of A and B, this equation defines a linear function, a constant function, or no function at all. If $A \neq 0$ and $B \neq 0$, then equation (1) can be written as

$$y = -\frac{A}{B}x + \frac{C}{B} \quad \text{Linear function (slanted line)} \tag{2}$$

which is in the form $f(x) = mx + b$, $m \neq 0$, and hence is a linear function. If $A = 0$ and $B \neq 0$, then equation (1) can be written as

$$\begin{aligned} 0x + By &= C \\ y &= \frac{C}{B} \quad \text{Constant function (horizontal line)} \end{aligned} \tag{3}$$

which is in the form $g(x) = b$, and hence is a constant function. If $A \neq 0$ and $B = 0$, then equation (1) can be written as

$$\begin{aligned} Ax + 0y &= C \\ x &= \frac{C}{A} \quad \text{Not a function (vertical line)} \end{aligned} \tag{4}$$

We can see that the graph of (4) is a vertical line, since the equation is satisfied for any value of y as long as x is the constant C/A. Hence, this form does not define a function.

The following theorem is a generalization of the above discussion:

Theorem 1 GRAPH OF A LINEAR EQUATION IN TWO VARIABLES

In a Cartesian plane, the graph of any equation of the form

$$\textbf{Standard Form} \quad Ax + By = C \tag{5}$$

where A, B, and C are real constants (A and B not both 0), is a straight line. Every straight line in a Cartesian plane coordinate system is the graph of an equation of this type.

(continued)

Theorem 1 (*continued*) Vertical and horizontal lines have particularly simple equations, which are special cases of equation (5):

Horizontal line with y intercept $C/B = b$: $y = b$
Vertical line with x intercept $C/A = a$: $x = a$ ■■

Explore–Discuss 2

Graph the following three special cases of $Ax + By = C$ in the same coordinate system:

(A) $3x + 2y = 6$ (B) $0x - 3y = 12$ (C) $2x + 0y = 10$

Which cases define functions? Explain why, or why not.

C Graph each case in the same viewing window using a graphing utility. (Check your manual on how to graph vertical lines.)

Sketching the graphs of equations of either form

$$Ax + By = C \quad \text{or} \quad y = mx + b$$

is very easy, since the graph of each equation is a straight line. All that is necessary is to plot any two points from the solution set and use a straightedge to draw a line through these two points. The x and y intercepts are usually the easiest to find.

Example 2 **Sketching Graphs of Lines**

(A) Graph $x = -4$ and $y = 6$ simultaneously in the same rectangular coordinate system. C Also, graph in a graphing utility.
(B) Write the equations of the vertical and horizontal lines that pass through the point $(7, -5)$.
(C) Graph the equation $2x - 3y = 12$ by hand. C Also, graph in a graphing utility.

SOLUTION (A) Graphing $x = -4$ and $y = 6$:

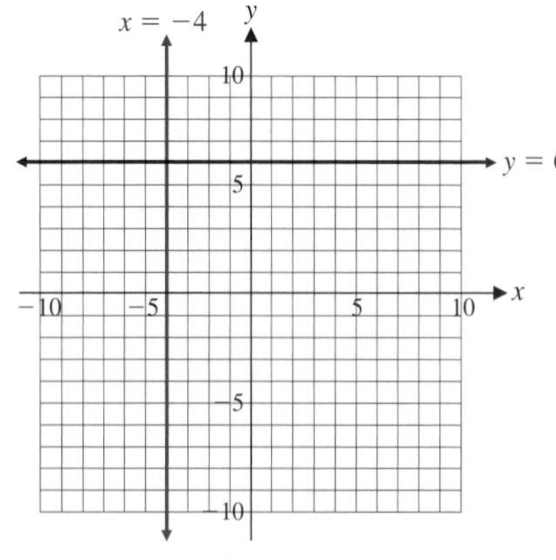

By hand

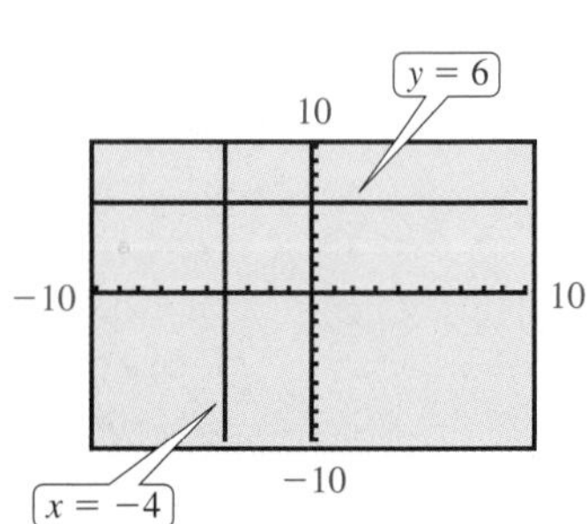

C In a graphing utility

(B) Horizontal line through $(7, -5)$: $y = -5$
Vertical line through $(7, -5)$: $x = 7$

(C) Graphing $2x - 3y = 12$: For the hand-drawn graph, find the intercepts by first letting $x = 0$ and solving for y and then letting $y = 0$ and solving for x. Then draw a line through the intercepts. C To graph in a graphing utility, solve the equation for y in terms of x and enter the result.

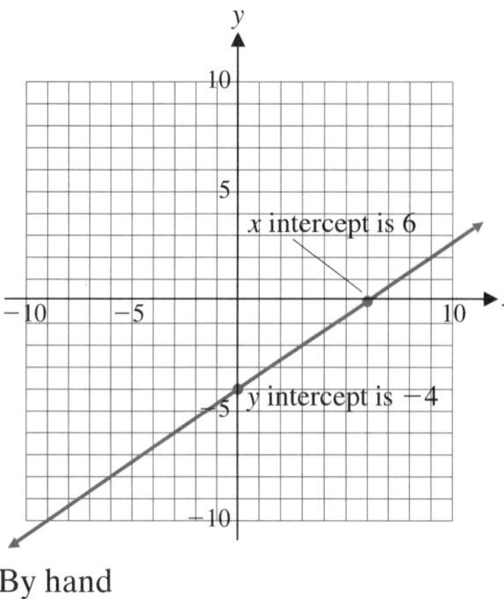

By hand

$y_1 = (2/3)x - 4$

10

−10 10

−10

C In a graphing utility

Matched Problem 2

(A) Graph $x = 5$ and $y = -3$ simultaneously in the same rectangular coordinate system. C Also, graph in a graphing utility.

(B) Write the equations of the vertical and horizontal lines that pass through the point $(-8, 2)$.

(C) Graph the equation $3x + 4y = 12$ by hand. C Also, graph in a graphing utility.

Slope of a Line

If we take two points $P_1(x_1, y_1)$ and $P_2(x_2, y_2)$, on a line, then the ratio of the change in y to the change in x as the point moves from point P_1 to point P_2 is called the **slope** of the line. In a sense, slope provides a measure of the "steepness" of a line relative to the x axis. The change in x is often called the **run** and the change in y the **rise**.

Slope of a Line

If a line passes through two distinct points $P_1(x_1, y_1)$ and $P_2(x_2, y_2)$, then its slope is given by the formula

$$m = \frac{y_2 - y_1}{x_2 - x_1} \qquad x_1 \neq x_2$$

$$= \frac{\text{Vertical change (Rise)}}{\text{Horizontal change (Run)}}$$

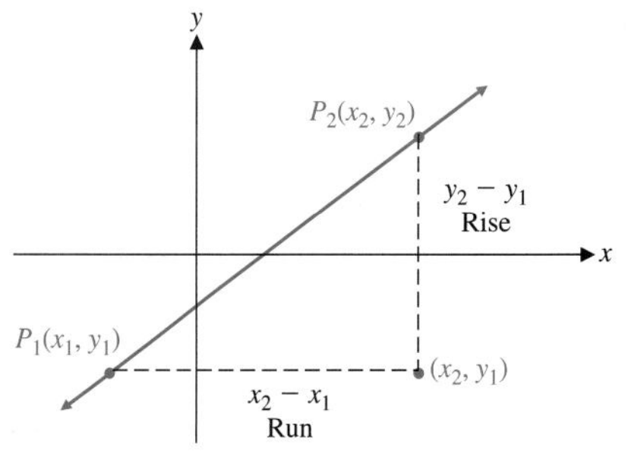

For a horizontal line, y does not change; hence, its slope is 0. For a vertical line, x does not change; hence, $x_1 = x_2$ and its slope is not defined. In general, the slope of a line may be positive, negative, 0, or not defined. Each case is illustrated geometrically in Table 1.

Table 1
GEOMETRIC INTERPRETATION OF SLOPE

LINE	SLOPE	EXAMPLE
Rising as x moves from left to right	Positive	
Falling as x moves from left to right	Negative	
Horizontal	0	
Vertical	Not defined	

In using the formula to find the slope of the line through two points, it does not matter which point is labeled P_1 or P_2, since changing the labeling will change the sign in both the numerator and denominator of the slope formula, resulting in equivalent expressions. In addition, it is important to note that the definition of slope does not depend on the two points chosen on the line as long as they are distinct. This follows from the fact that the ratios of corresponding sides of similar triangles are equal.

Example 3 **Finding Slopes** Sketch a line through each pair of points, and find the slope of each line.

(A) $(-3, -2), (3, 4)$ (B) $(-1, 3), (2, -3)$
(C) $(-2, -3), (3, -3)$ (D) $(-2, 4), (-2, -2)$

SOLUTION (A)

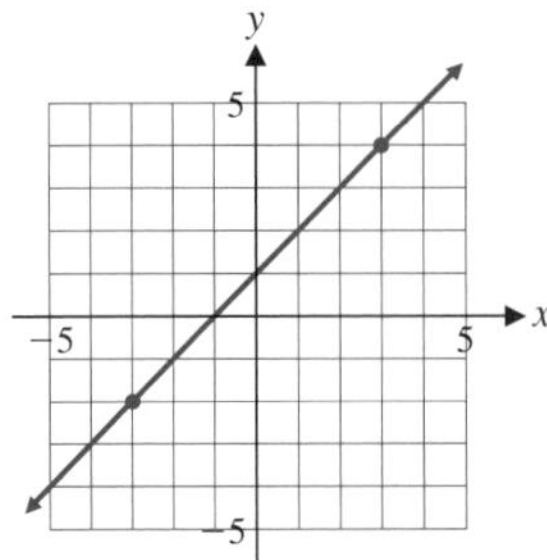

$$m = \frac{4 - (-2)}{3 - (-3)} = \frac{6}{6} = 1$$

(B)

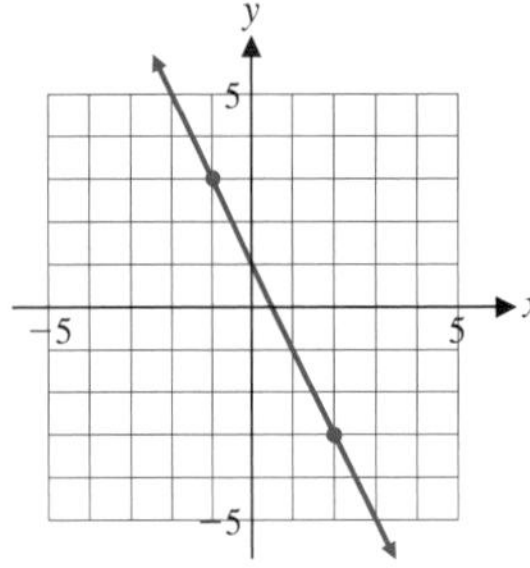

$$m = \frac{-3 - 3}{2 - (-1)} = \frac{-6}{3} = -2$$

(C)

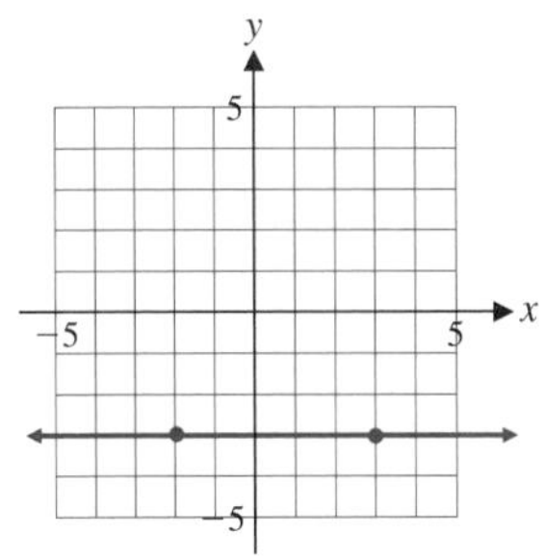

$$m = \frac{-3 - (-3)}{3 - (-2)} = \frac{0}{5} = 0$$

(D)

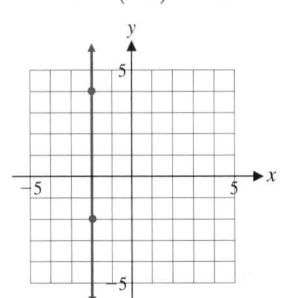

$$m = \frac{-2 - 4}{-2 - (-2)} = \frac{-6}{0}$$

Slope is not defined

Matched Problem 3 Find the slope of the line through each pair of points.

(A) $(-2, 4), (3, 4)$ (B) $(-2, 4), (0, -4)$
(C) $(-1, 5), (-1, -2)$ (D) $(-1, -2), (2, 1)$

■ EQUATIONS OF LINES—SPECIAL FORMS

Let us start by investigating why $y = mx + b$ is called the *slope–intercept form* for a line.

Explore–Discuss 3

(A) Graph $y = x + b$ for $b = -5, -3, 0, 3,$ and 5 simultaneously in the same coordinate system. Verbally describe the geometric significance of b.

(B) Graph $y = mx - 1$ for $m = -2, -1, 0, 1,$ and 2 simultaneously in the same coordinate system. Verbally describe the geometric significance of m.

C (C) Using a graphing utility, explore the graph of $y = mx + b$ for different values of m and b.

As you can see from the above exploration, constants m and b in $y = mx + b$ have special geometric significance, which we now explicitly state.

If we let $x = 0$, then $y = b$, and we observe that the graph of $y = mx + b$ crosses the y axis at $(0, b)$. The constant b is the *y intercept.* For example, the y intercept of the graph of $y = -4x - 1$ is -1.

To determine the geometric significance of m, we proceed as follows: If $y = mx + b$, then by setting $x = 0$ and $x = 1$, we conclude that $(0, b)$ and $(1, m + b)$ lie on its graph (a line). Hence, the slope of this graph (line) is given by:

$$\text{Slope} = \frac{y_2 - y_1}{x_2 - x_1} = \frac{(m + b) - b}{1 - 0} = m$$

Thus, m is the slope of the line given by $y = mx + b$.

Slope–Intercept Form

The equation

$$y = mx + b \qquad m = \text{Slope}, b = y \text{ intercept} \tag{6}$$

is called the **slope–intercept form** of an equation of a line.

Example 4 **Using the Slope–Intercept Form**

(A) Find the slope and y intercept, and graph $y = -\frac{2}{3}x - 3$.
(B) Write the equation of the line with slope $\frac{2}{3}$ and y intercept -2.

SOLUTION (A) Slope $= m = -\frac{2}{3}$
y intercept $= b = -3$

(B) $m = \frac{2}{3}$ and $b = -2$;
thus, $y = \frac{2}{3}x - 2$

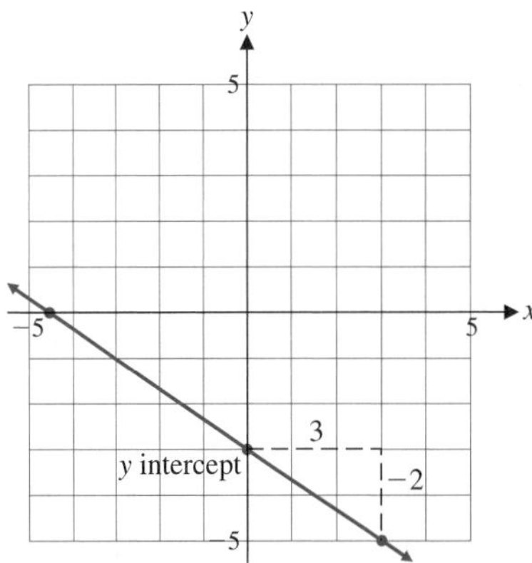

Matched Problem 4 Write the equation of the line with slope $\frac{1}{2}$ and y intercept -1. Graph.

Suppose a line has slope m and passes through a fixed point (x_1, y_1). If the point (x, y) is any other point on the line (Fig. 3), then

$$\frac{y - y_1}{x - x_1} = m$$

That is,

$$y - y_1 = m(x - x_1) \tag{7}$$

We now observe that (x_1, y_1) also satisfies equation (7) and conclude that equation (7) is an equation of a line with slope m that passes through (x_1, y_1).

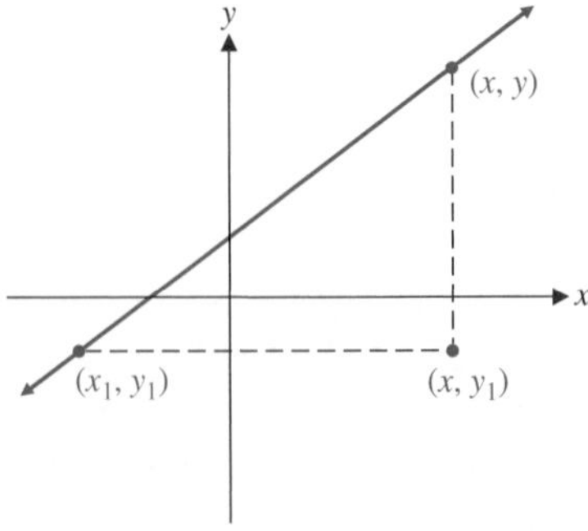

FIGURE 3

Point–Slope Form

An equation of a line with slope m that passes through (x_1, y_1) is

$$y - y_1 = m(x - x_1) \tag{7}$$

which is called the **point–slope form** of an equation of a line.

The point–slope form is extremely useful, since it enables us to find an equation for a line if we know its slope and the coordinates of a point on the line or if we know the coordinates of two points on the line.

Example 5 **Using the Point–Slope Form**

(A) Find an equation for the line that has slope $\frac{1}{2}$ and passes through $(-4, 3)$. Write the final answer in the form $Ax + By = C$.

(B) Find an equation for the line that passes through the two points $(-3, 2)$ and $(-4, 5)$. Write the resulting equation in the form $y = mx + b$.

Solution (A) Use $y - y_1 = m(x - x_1)$. Let $m = \frac{1}{2}$ and $(x_1, y_1) = (-4, 3)$. Then

$$\begin{aligned} y - 3 &= \tfrac{1}{2}[x - (-4)] \\ y - 3 &= \tfrac{1}{2}(x + 4) \qquad \text{Multiply by 2.} \\ 2y - 6 &= x + 4 \\ -x + 2y &= 10 \quad \text{or} \quad x - 2y = -10 \end{aligned}$$

(B) First, find the slope of the line by using the slope formula:

$$m = \frac{y_2 - y_1}{x_2 - x_1} = \frac{5 - 2}{-4 - (-3)} = \frac{3}{-1} = -3$$

Now use $y - y_1 = m(x - x_1)$ with $m = -3$ and $(x_1, y_1) = (-3, 2)$:

$$\begin{aligned} y - 2 &= -3[x - (-3)] \\ y - 2 &= -3(x + 3) \\ y - 2 &= -3x - 9 \\ y &= -3x - 7 \end{aligned}$$

Matched Problem 5

(A) Find an equation for the line that has slope $\frac{2}{3}$ and passes through $(6, -2)$. Write the resulting equation in the form $Ax + By = C$, $A > 0$.

(B) Find an equation for the line that passes through $(2, -3)$ and $(4, 3)$. Write the resulting equation in the form $y = mx + b$.

The various forms of the equation of a line that we have discussed are summarized in Table 2 for convenient reference.

Table 2

Equations of a Line

Standard form	$Ax + By = C$	A and B not both 0
Slope–intercept form	$y = mx + b$	Slope: m; y intercept: b
Point–slope form	$y - y_1 = m(x - x_1)$	Slope: m; Point: (x_1, y_1)
Horizontal line	$y = b$	Slope: 0
Vertical line	$x = a$	Slope: Undefined

APPLICATIONS

We will now see how equations of lines occur in certain applications.

Example 6 **Cost Equation** The management of a company that manufactures roller skates has fixed costs (costs at 0 output) of \$300 per day and total costs of \$4,300 per day at an output of 100 pairs of skates per day. Assume that cost C is linearly related to output x.

(A) Find the slope of the line joining the points associated with outputs of 0 and 100; that is, the line passing through (0, 300) and (100, 4,300).

(B) Find an equation of the line relating output to cost. Write the final answer in the form $C = mx + b$.

(C) Graph the cost equation from part (B) for $0 \leq x \leq 200$.

SOLUTION (A) $m = \frac{y_2 - y_1}{x_2 - x_1} = \frac{4{,}300 - 300}{100 - 0} = \frac{4{,}000}{100} = 40$

(B) We must find an equation of the line that passes through (0, 300) with slope 40. We use the slope–intercept form:

$$C = mx + b$$
$$C = 40x + 300$$

(C)

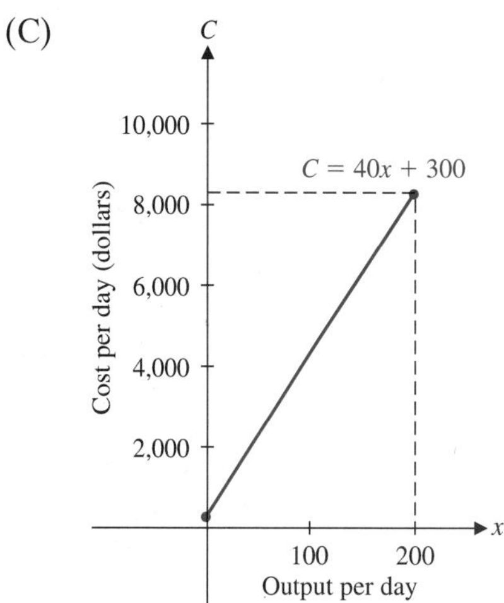

In Example 6, the **fixed cost** of \$300 per day covers plant cost, insurance, and so on. This cost is incurred whether or not there is any production. The **variable cost** is $40x$, which depends on the day's output. Note that the slope 40 is the cost of producing one pair of skates; that is, the cost per unit output.

Matched Problem 6 Answer parts (A) and (B) in Example 6 for fixed costs of \$250 per day and total costs of \$3,450 per day at an output of 80 pairs of skates per day.

Example 7 **Price–Demand** A company manufactures and sells a specialty watch. The financial research department, using statistical and analytical methods, determined that at a price of \$88 each, the demand would be 2 thousand watches, and at \$38 each, 12 thousand watches. Assuming a linear relationship between

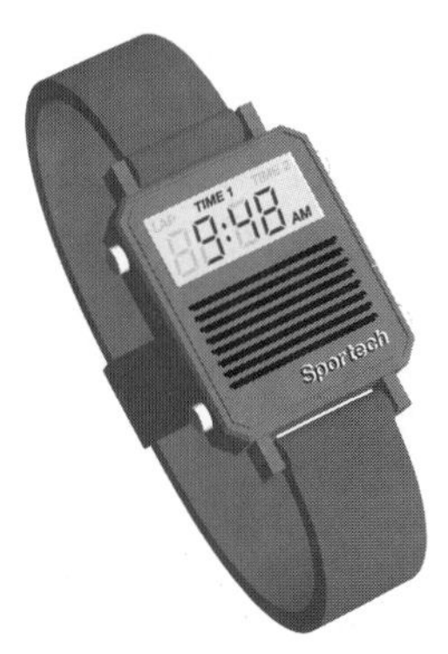

price and demand, find a linear function that models the price–demand relationship in the form $p(x) = mx + b$. What would be the price at a demand of 8 thousand watches? 15 thousand watches?

SOLUTION Find the equation of the line that passes through (2, 88) and (12, 38). We first find the slope of the line:

$$m = \frac{38 - 88}{12 - 2} = \frac{-50}{10} = -5$$

Use the point–slope form to find the equation of the line:

$$\begin{aligned} y - y_1 &= m(x - x_1) \\ y - 88 &= -5(x - 2) \\ y - 88 &= -5x + 10 \\ y &= -5x + 98 \end{aligned}$$

or

$$\begin{aligned} p(x) &= -5x + 98 && \text{Price–demand equation} \\ p(8) &= -5(8) + 98 = \$58 \\ p(15) &= -5(15) + 98 = \$23 \end{aligned}$$

Thus, the price is \$58 when the demand is 8,000 and \$23 when the demand is 15,000.

Matched Problem 7 The company in Example 7 also manufactures and sells a watch designed for sailboat racing. The financial analyst found that the company could sell 3 hundred of these watches at a wholesale price of \$140 each, and 11 hundred watches at a wholesale price of \$92 each. Assuming a linear relationship between price and demand, find a linear function that models the price–demand relationship in the form $p(x) = mx + b$. What would be the price at a demand of 7 hundred watches? 12 hundred watches?

Answers to Matched Problems

1. (A)

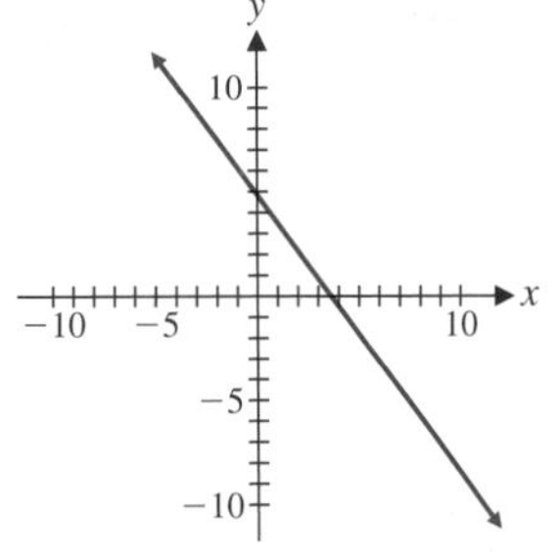

(B) x intercept: 3.75; y intercept: 5

C (C)

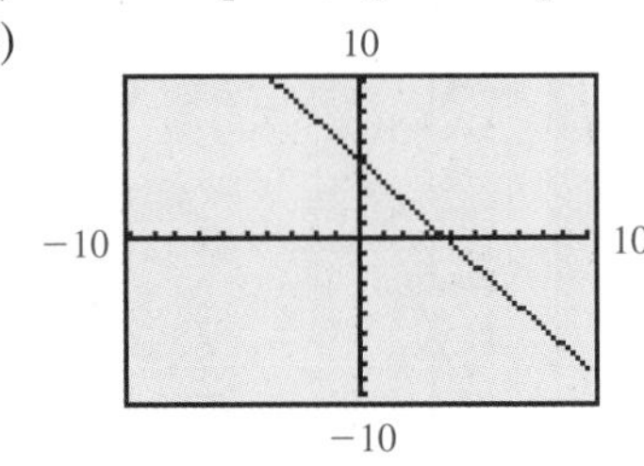

C (D) x intercept: 3.75; y intercept: 5

(E) $x \leqslant 3.75$ or $(-\infty, 3.75]$

2. (A)

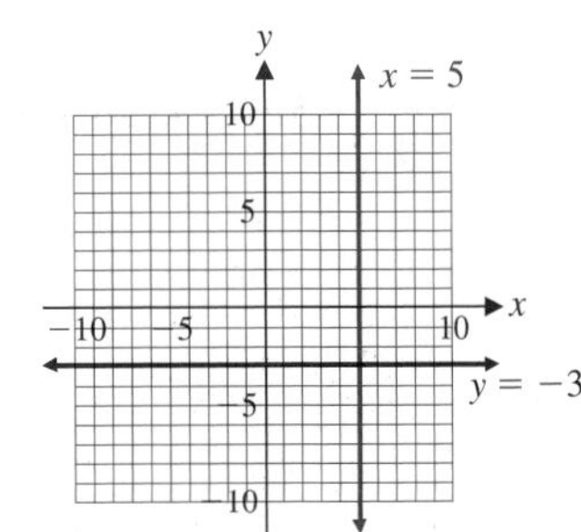

By hand

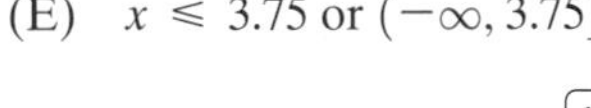

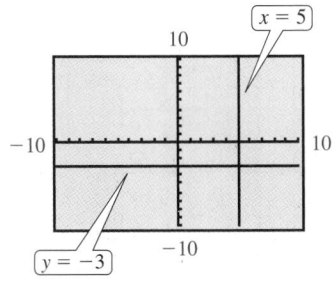

C In graphing utility

(B) Horizontal line: $y = 2$; Vertical line: $x = -8$

(C)

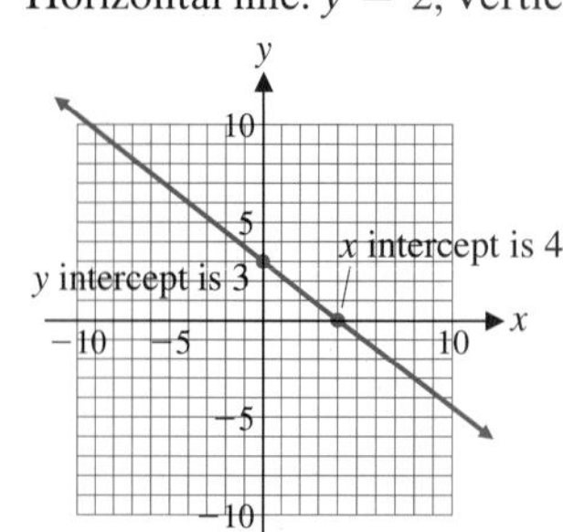

By hand

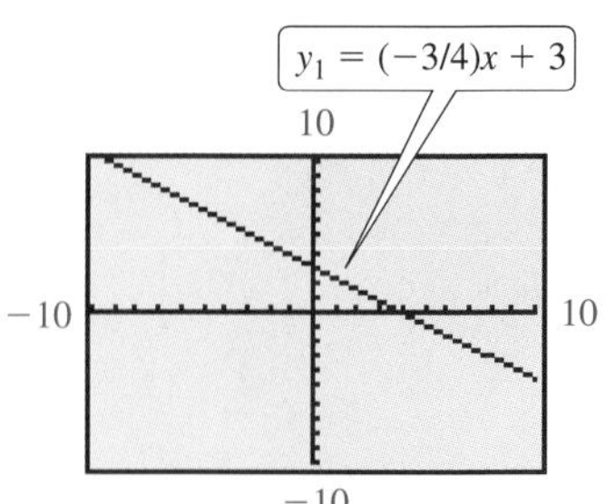

In graphing utility

3. (A) 0 (B) −4 (C) Not defined (D) 1

4. $y = \frac{1}{2}x - 1$

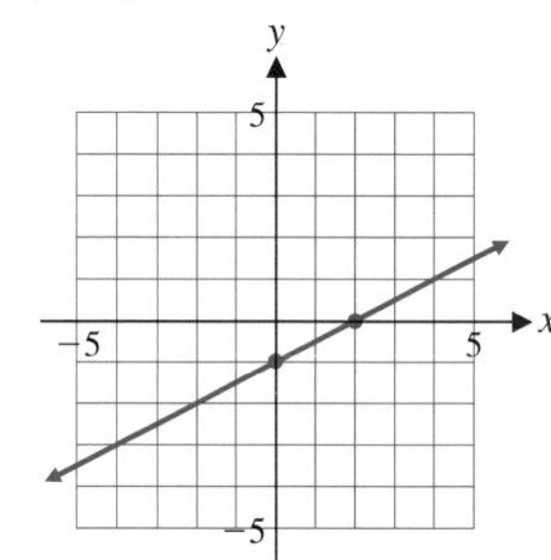

5. (A) $2x - 3y = 18$ (B) $y = 3x - 9$

6. (A) $m = 40$ (B) $C = 40x + 250$

7. $p(x) = -6x + 158$; $p(7) = \$116$; $p(12) = \$86$

EXERCISE 1-3

A *Problems 1–4 refer to graphs (A)–(D).*

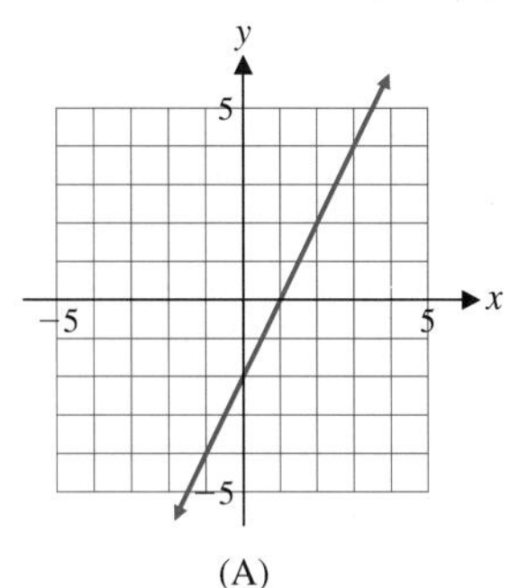

(A)

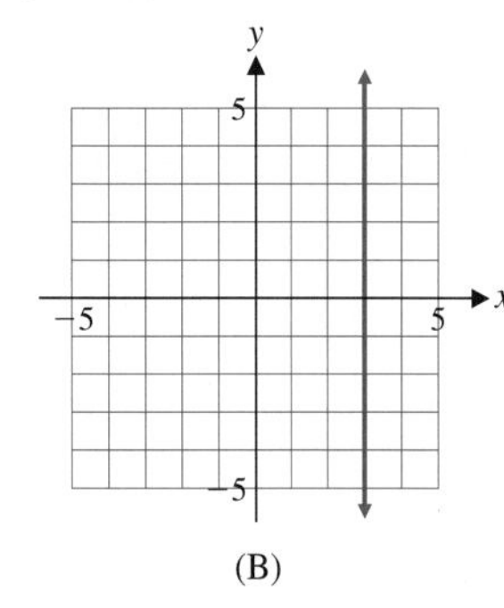

(B)

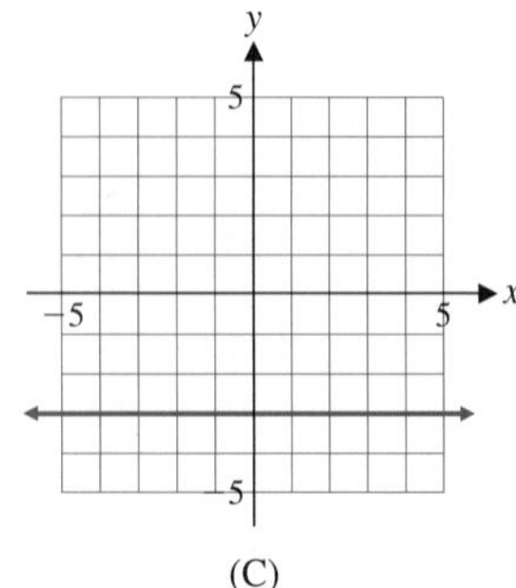

(C)

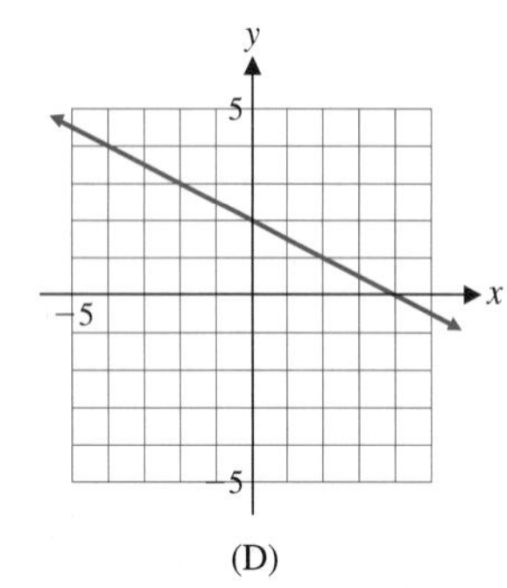

(D)

1. Identify the graph(s) of linear functions with a negative slope.
2. Identify the graph(s) of linear functions with a positive slope.
3. Identify the graph(s) of any constant functions. What is the slope of the graph?
4. Identify any graphs that are not the graphs of functions. What can you say about their slopes?

Sketch a graph of each equation in a rectangular coordinate system.

5. $y = 2x - 3$
6. $y = \frac{x}{2} + 1$
7. $2x + 3y = 12$
8. $8x - 3y = 24$

Find the slope and y intercept of the graph of each equation.

9. $y = 2x - 3$
10. $y = \frac{x}{2} + 1$
11. $y = -\frac{2}{3}x + 2$
12. $y = \frac{3}{4}x - 2$

Write an equation of the line with the indicated slope and y intercept.

13. Slope $= -2$
y intercept $= 4$

14. Slope $= -\frac{2}{3}$
y intercept $= -2$

15. Slope $= -\frac{3}{5}$
y intercept $= 3$

16. Slope $= 1$
y intercept $= -2$

B *Sketch a graph of each equation or pair of equations in a rectangular coordinate system.*

17. $y = -\frac{2}{3}x - 2$

18. $y = -\frac{3}{2}x + 1$

19. $3x - 2y = 10$

20. $5x - 6y = 15$

21. $x = 3; y = -2$

22. $x = -3; y = 2$

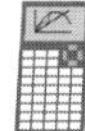

Check your graphs for Problems 17–22 by graphing each in a graphing utility.

In Problems 23–26, find the slope of the graph of each equation. (First write the equation in the form $y = mx + b$.)

23. $3x + y = 5$

24. $2x - y = -3$

25. $2x + 3y = 12$

26. $3x - 2y = 10$

27. (A) Graph $f(x) = 1.2x - 4.2$ in a rectangular coordinate system.

(B) Find the x and y intercepts algebraically to one decimal place.

(C) Graph $f(x) = 1.2x - 4.2$ in a graphing utility.

(D) Find the x and y intercepts to one decimal place using trace and zoom or an appropriate built-in routine in your graphing utility.

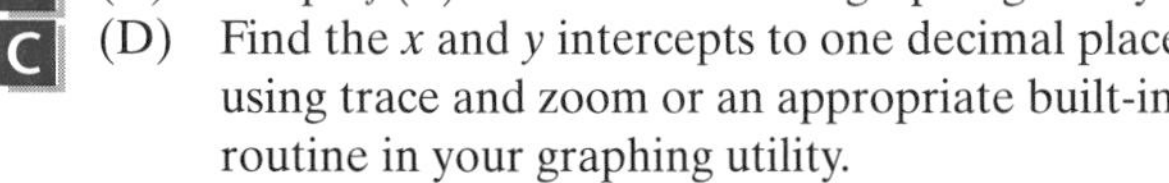

(E) Using the results of parts (A) and (B) or (C) and (D), find the solution set for the linear inequality

$$1.2x - 4.2 > 0$$

28. (A) Graph $f(x) = -0.8x + 5.2$ in a rectangular coordinate system.

(B) Find the x and y intercepts algebraically to one decimal place.

(C) Graph $f(x) = -0.8x + 5.2$ in a graphing utility.

(D) Find the x and y intercepts to one decimal place using trace and zoom or an appropriate built-in routine in your graphing utility.

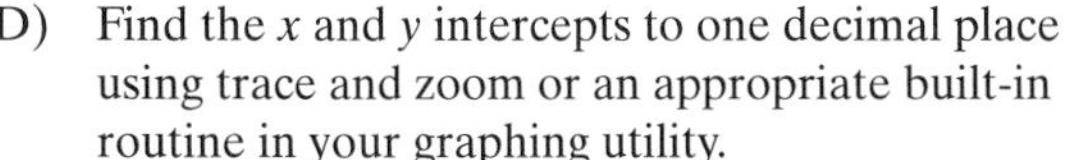

(E) Using the results of parts (A) and (B) or (C) and (D), find the solution set for the linear inequality

$$-0.8x + 5.2 < 0$$

Write the equations of the vertical and horizontal lines through each point.

29. $(3, -5)$

30. $(-2, 7)$

31. $(-1, -3)$

32. $(96, -4)$

Write the equation of the line through each indicated point with the indicated slope. Write the final answer in the form $y = mx + b$.

33. $m = -3; (4, -1)$

34. $m = -2; (-3, 2)$

35. $m = \frac{2}{3}; (-6, -5)$

36. $m = \frac{1}{2}; (-4, 3)$

37. $m = 0; (3, -5)$

38. $m = 0; (-4, 7)$

Find the slope of the line that passes through the given points.

39. $(1, 3)$ and $(7, 5)$

40. $(2, 1)$ and $(10, 5)$

41. $(-5, -2)$ and $(5, -4)$

42. $(3, 7)$ and $(-6, 4)$

43. $(2, 7)$ and $(2, -3)$

44. $(-2, 3)$ and $(-2, -1)$

45. $(2, 3)$ and $(-5, 3)$

46. $(-3, -3)$ and $(0, -3)$

Write an equation of the line through each indicated pair of points. Write the final answer in the form $Ax + By = C$.

47. $(1, 3)$ and $(7, 5)$

48. $(2, 1)$ and $(10, 5)$

49. $(-5, -2)$ and $(5, -4)$

50. $(3, 7)$ and $(-6, 4)$

51. $(2, 7)$ and $(2, -3)$

52. $(-2, 3)$ and $(-2, -1)$

53. $(2, 3)$ and $(-5, 3)$

54. $(-3, -3)$ and $(0, -3)$

In Problems 55–60, graph the equations obtained from Problems 49–54 and indicate which define a linear function, a constant function, or no function at all.

55. Problem 49

56. Problem 50

57. Problem 51

58. Problem 52

59. Problem 53

60. Problem 54

61. Discuss the relationship among the graphs of the lines with equation $y = mx + 2$, where m is any real number.

62. Discuss the relationship among the graphs of the lines with equation $y = -0.5x + b$, where b is any real number.

C

63. (A) Graph the following equations in the same coordinate system:

$$3x + 2y = 6 \qquad 3x + 2y = 3$$
$$3x + 2y = -6 \qquad 3x + 2y = -3$$

(B) From your observations in part (A), describe the family of lines obtained by varying C in $Ax + By = C$ while holding A and B fixed.

64. (A) Graph the following two equations in the same coordinate system:

$$3x + 4y = 12 \qquad 4x - 3y = 12$$

(B) Graph the following two equations in the same coordinate system:

$$2x + 3y = 12 \qquad 3x - 2y = 12$$

(C) From your observations in parts (A) and (B), describe the apparent relationship of the graphs of

$$Ax + By = C \quad \text{and} \quad Bx - Ay = C$$

65. Describe the relationship between the graphs of $f(x) = mx + b$ and $g(x) = |mx + b|$, $m \neq 0$, and illustrate with examples. Is $g(x)$ always, sometimes, or never a linear function?

66. Describe the relationship between the graphs of $f(x) = mx + b$ and $g(x) = m|x| + b$, $m \neq 0$, and illustrate with examples. Is $g(x)$ always, sometimes, or never a linear function?

APPLICATIONS

Business & Economics

67. *Simple interest.* If \$$P$ (the principal) is invested at an interest rate of r, then the amount A that is due after t years is given by

$$A = Prt + P$$

If \$100 is invested at 6% ($r = 0.06$), then $A = 6t + 100$, $t \geq 0$.

(A) What will \$100 amount to after 5 years? After 20 years?

(B) Sketch a graph of $A = 6t + 100$ for $0 \leq t \leq 20$.

(C) Find the slope of the graph and interpret verbally.

68. *Simple interest.* Use the simple interest formula from Problem 67. If \$1,000 is invested at 7.5% ($r = 0.075$), then $A = 75t + 1{,}000$, $t \geq 0$.

(A) What will \$1,000 amount to after 5 years? After 20 years?

(B) Sketch a graph of $A = 75t + 1{,}000$ for $0 \leq t \leq 1{,}000$.

(C) Find the slope of the graph and interpret verbally.

69. *Cost function.* The management of a company that manufactures surfboards has fixed costs (at 0 output) of \$200 per day and total costs of \$3,800 per day at a daily output of 20 boards.

(A) Assuming the total cost per day, $C(x)$, is linearly related to the total output per day, x, write an equation for the cost function.

(B) What are the total costs for an output of 12 boards per day?

(C) Graph the cost function for $0 \leq x \leq 20$.

70. *Cost function.* Repeat Problem 69 if the company has fixed costs of \$300 per day and total costs per day at an output of 20 boards of \$5,100.

71. *Price–demand function.* A manufacturing company is interested in introducing a new power mower. Its market research department gave the management the price–demand forecast listed in Table 3.

Table 3

PRICE–DEMAND

DEMAND x	WHOLESALE PRICE (\$) $p(x)$
0	200
2,400	160
4,800	120
7,800	70

(A) Plot these points, letting $p(x)$ represent the price at which x number of mowers can be sold (demand). Label the horizontal axis x.

(B) Note that the points in part (A) lie along a straight line. Find an equation for the price–demand function.

(C) What would be the price for a demand of 3,000 units?

(D) Write a brief verbal interpretation of the slope of the line found in part (B).

72. *Depreciation.* Office equipment was purchased for \$20,000 and is assumed to have a scrap value of \$2,000 after 10 years. If its value is depreciated linearly (for tax purposes) from \$20,000 to \$2,000:

(A) Find the linear equation that relates value (V) in dollars to time (t) in years.

(B) What would be the value of the equipment after 6 years?

(C) Graph the equation for $0 \leq t \leq 10$.

(D) Write a brief verbal interpretation of the slope of the line found in part (A).

Merck & Co., Inc. is the world's largest pharmaceutical company. Problems 73 and 74 refer to the data in Table 4 taken from the company's 1993 annual report.

Table 4

SELECTED FINANCIAL DATA FOR MERCK & CO., INC.

(BILLION $)	1988	1989	1990	1991	1992
SALES	5.9	6.5	7.7	8.6	9.7
NET INCOME	1.2	1.5	1.8	2.1	2.4

73. *Sales analysis.* A mathematical model for Merck's sales is given by

$$f(x) = 5.74 + 0.97x$$

where $x = 0$ corresponds to 1988.

(A) Complete the following table. Round values of $f(x)$ to one decimal place.

x	0	1	2	3	4
SALES	5.9	6.5	7.7	8.6	9.7
$f(x)$					

(B) Sketch the graph of f and the sales data on the same axes.

(C) Use the modeling equation to estimate the sales in 1993. In 2000.

(D) Write a brief verbal description of the company's sales from 1988 to 1992.

74. *Income analysis.* A mathematical model for Merck's income is given by

$$f(x) = 1.2 + 0.3x$$

where $x = 0$ corresponds to 1988.

(A) Complete the following table. Round values of $f(x)$ to one decimal place.

x	0	1	2	3	4
NET INCOME	1.2	1.5	1.8	2.1	2.4
$f(x)$					

(B) Sketch the graph of f and the income data on the same axes.

(C) Use the modeling equation to estimate the income in 1993. In 2000.

(D) Write a brief verbal description of the company's income from 1988 to 1992.

Life Sciences

75. *Nutrition.* In a nutrition experiment, a biologist wants to prepare a special diet for the experimental animals. Two food mixes, A and B, are available. If mix A contains 20% protein and mix B contains 10% protein, what combination of each mix will provide exactly 20 grams of protein? Let x be the amount of A used and let y be the amount of B used. Then write a linear equation relating x, y, and 20. Graph this equation for $x \geq 0$ and $y \geq 0$.

76. *Ecology.* As one descends into the ocean, pressure increases linearly. The pressure is 15 pounds per square inch on the surface and 30 pounds per square inch 33 feet below the surface.

(A) If p is the pressure in pounds and d is the depth below the surface in feet, write an equation that expresses p in terms of d. [*Hint:* Find an equation of the line that passes through $(0, 15)$ and $(33, 30)$.]

(B) What is the pressure at 12,540 feet (the average depth of the ocean)?

(C) Graph the equation for $0 \leq d \leq 12{,}540$.

(D) Write a brief verbal interpretation of the slope of the line found in part (A).

Social Sciences

77. *Psychology.* In an experiment on motivation, J. S. Brown trained a group of rats to run down a narrow passage in a cage to obtain food in a goal box. Using a harness, he then connected the rats to an overhead wire that was attached to a spring scale. A rat was placed at different distances d (in centimeters) from the goal box, and the pull p (in grams) of the rat toward the food was measured. Brown found that the relationship between these two variables was very close to being linear and could be approximated by the equation

$$p = -\tfrac{1}{5}d + 70 \qquad 30 \leq d \leq 175$$

(See J. S. Brown, *Journal of Comparative and Physiological Psychology,* 1948, 41:450–465.)

(A) What was the pull when $d = 30$? When $d = 175$?

(B) Graph the equation.

(C) What is the slope of the line?

SECTION 1-4

Quadratic Functions

- QUADRATIC FUNCTIONS, EQUATIONS, AND INEQUALITIES
- PROPERTIES OF QUADRATIC FUNCTIONS AND THEIR GRAPHS
- APPLICATIONS

If the degree of a linear function is increased by one, we obtain a *second-degree function,* usually called a *quadratic function,* another basic function that we will need in our library of elementary functions. We will investigate relationships between quadratic functions and the solutions to quadratic equations and inequalities. (A detailed treatment of algebraic solutions to quadratic equations can be found in Appendix A-9.) Other important properties of quadratic functions will also be investigated, including maximum and minimum properties. We will then be in a position to solve important practical problems such as finding production levels that will produce maximum revenue or maximum profit.

QUADRATIC FUNCTIONS, EQUATIONS, AND INEQUALITIES

A quadratic function is defined as follows:

Quadratic Function

A function f is a **quadratic function** if

$$f(x) = ax^2 + bx + c \qquad a \neq 0$$

where a, b, and c are real numbers. The domain of a quadratic function is the set of all real numbers.

The graphs of three quadratic functions are shown in Figure 1.

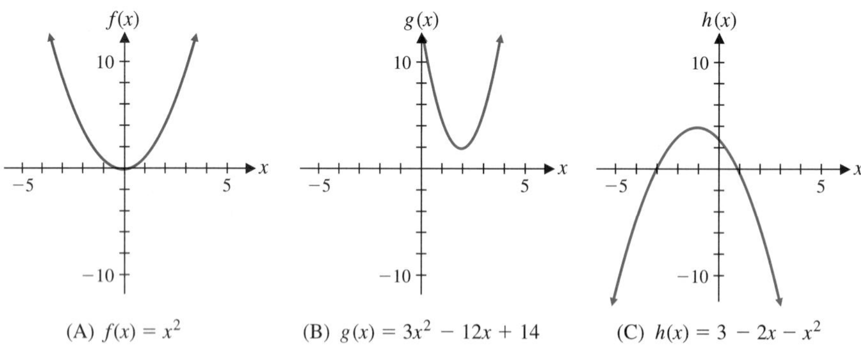

(A) $f(x) = x^2$ (B) $g(x) = 3x^2 - 12x + 14$ (C) $h(x) = 3 - 2x - x^2$

FIGURE 1
Graphs of quadratic functions

The graph of a quadratic function is called a **parabola.** We will discuss this in more detail later in this section.

Example 1 **Intercepts, Equations, and Inequalities**

(A) Sketch a graph of $f(x) = -x^2 + 5x + 3$ in a rectangular coordinate system.
(B) Find x and y intercepts algebraically to two decimal places.
C (C) Graph $f(x) = -x^2 + 5x + 3$ in a standard viewing window.
C (D) Find the x and y intercepts to two decimal places using trace and zoom or an appropriate built-in routine in your graphing utility.
(E) Solve the quadratic inequality $-x^2 + 5x + 3 \geq 0$ graphically to two decimal places using the results of parts (A) and (B) or (C) and (D).
C (F) Solve the equation $-x^2 + 5x + 3 = 4$ graphically to two decimal places using trace and zoom or an appropriate built-in routine in your graphing utility.

SOLUTION (A) Hand-sketching a graph of f:

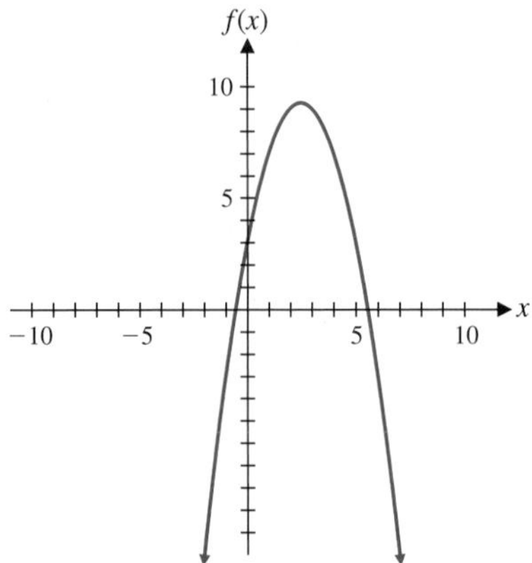

(B) Finding intercepts algebraically:

y intercept: $f(0) = -(0)^2 + 5(0) + 3 = 3$
x intercepts: $f(x) = 0$

$$-x^2 + 5x + 3 = 0 \qquad \text{Quadratic equation}$$

$$x = \frac{-b \pm \sqrt{b^2 - 4ac}}{2a} \qquad \text{Quadratic formula (see Appendix A-9)}$$

$$x = \frac{-(5) \pm \sqrt{5^2 - 4(-1)(3)}}{2(-1)}$$

$$= \frac{-5 \pm \sqrt{37}}{-2} = -0.54 \quad \text{or} \quad 5.54$$

C (C) Graphing in a graphing utility:

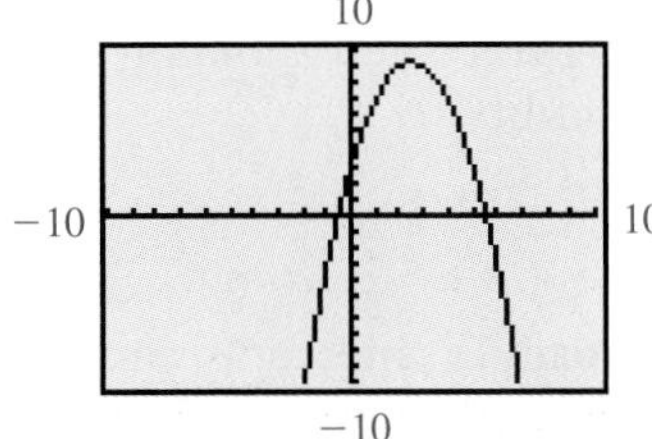

(D) Finding intercepts graphically using a graphing utility:

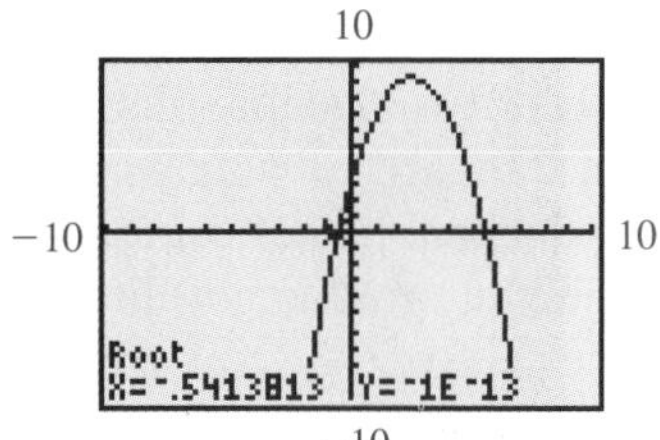

x intercept: −0.54

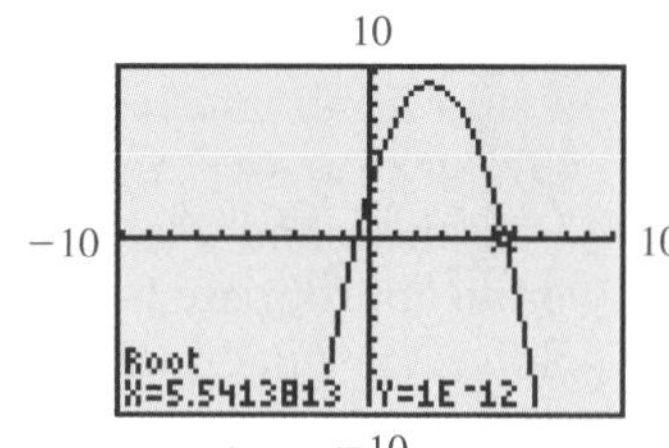

x intercept: 5.54

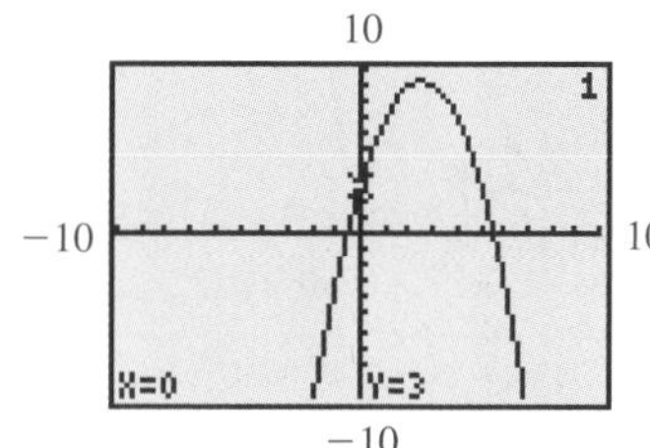

y intercept: 3

(E) Solving $-x^2 + 5x + 3 \geq 0$ graphically: The quadratic inequality

$$-x^2 + 5x + 3 \geq 0$$

holds for those values of x for which the graph of $f(x) = -x^2 + 5x + 3$ in the figures in parts (A) and (C) is at or above the x axis. This happens for x between the two x intercepts [found in parts (B) or (D)], including the two x intercepts. Thus, the solution set for the quadratic inequality is $-0.54 \leq x \leq 5.54$ or $[-0.54, 5.54]$.

(F) Solving the equation $-x^2 + 5x + 3 = 4$ using a graphing utility:

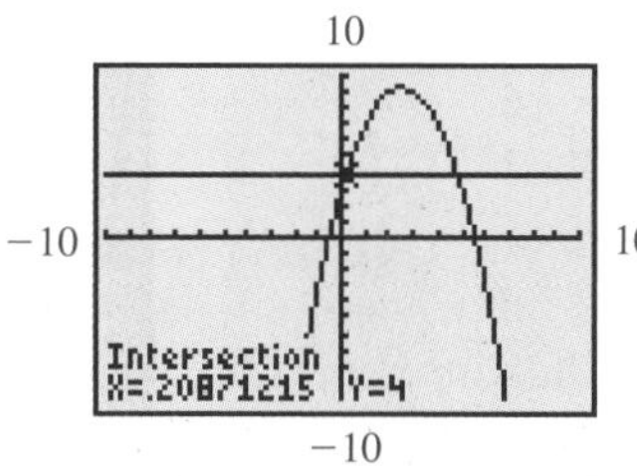

$-x^2 + 5x + 3 = 4$ at $x = 0.21$

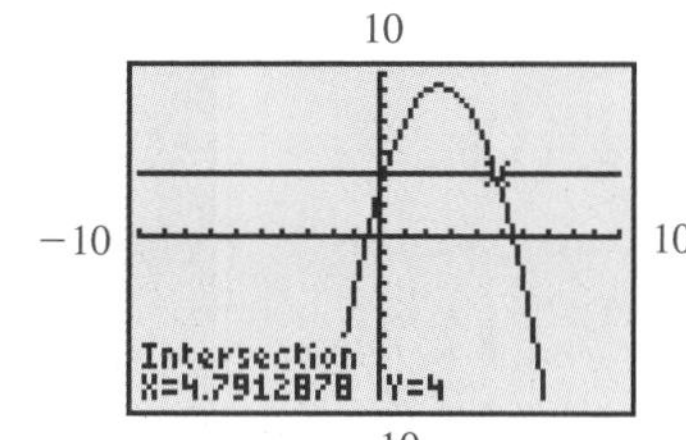

$-x^2 + 5x + 3 = 4$ at $x = 4.79$

Matched Problem 1

(A) Sketch a graph of $g(x) = 2x^2 - 5x - 5$ in a rectangular coordinate system.
(B) Find x and y intercepts algebraically to two decimal places.
(C) Graph $g(x) = 2x^2 - 5x - 5$ in a standard viewing window.
(D) Find the x and y intercepts to two decimal places using trace and zoom or an appropriate built-in routine in your graphing utility.
(E) Solve $2x^2 - 5x - 5 \geq 0$ graphically to two decimal places using the results of parts (A) and (B) or (C) and (D).
(F) Solve the equation $2x^2 - 5x - 5 = -3$ graphically to two decimal places using trace and zoom or an appropriate built-in routine in your graphing utility.

Explore–Discuss 1

How many x intercepts can the graph of a quadratic function have? How many y intercepts? Explain your reasoning.

■ Properties of Quadratic Functions and Their Graphs

Many useful properties of the quadratic function can be uncovered by transforming

$$f(x) = ax^2 + bx + c \qquad a \neq 0 \tag{1}$$

into the form

$$f(x) = a(x - h)^2 + k \tag{2}$$

The process of *completing the square* (see Appendix A-9) is central to the transformation. We illustrate the process through a specific example and then generalize the results.

Consider the quadratic function given by

$$f(x) = -2x^2 + 16x - 24 \tag{3}$$

We start by transforming equation (3) into the form (2) by completing the square:

$$\begin{aligned} f(x) &= -2x^2 + 16x - 24 && \text{Factor the coefficient of } x^2 \text{ out of the first two terms.} \\ &= -2(x^2 - 8x) - 24 \\ &= -2(x^2 - 8x + \textbf{?}) - 24 && \text{Add 16 to complete the square inside the parentheses. Because of the } -2 \text{ outside the parentheses, we have actually added } -32, \text{ so we must add 32 to the outside.} \\ &= -2(x^2 - 8x + \mathbf{16}) - 24 + \mathbf{32} \\ &= -2(x - 4)^2 + 8 && \text{The transformation is complete and can be checked by multiplying out.} \end{aligned}$$

Thus,

$$f(x) = -2(x - 4)^2 + 8 \tag{4}$$

If $x = 4$, then $-2(x - 4)^2 = 0$ and $f(4) = 8$. For any other value of x, the negative number $-2(x - 4)^2$ is added to 8, making it smaller. (Think about this.) Therefore,

$$f(4) = 8$$

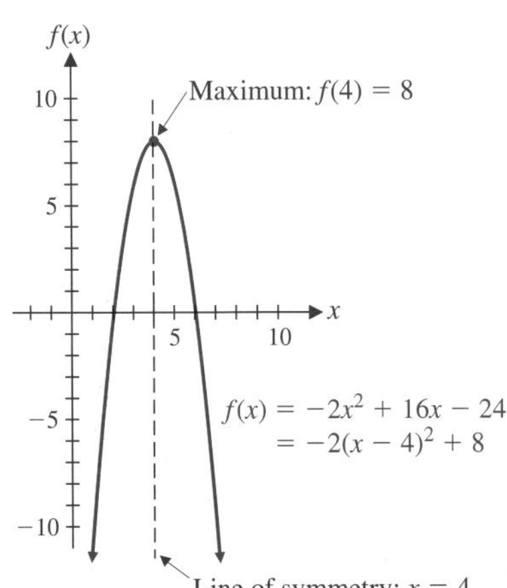

FIGURE 2
Graph of a quadratic function

is the *maximum value* of $f(x)$ for all x—a very important result! Furthermore, if we choose any two x values that are the same distance from 4, we will obtain the same function value. For example, $x = 3$ and $x = 5$ are each one unit from $x = 4$ and their function values are

$$\begin{aligned} f(3) &= -2(3 - 4)^2 + 8 = 6 \\ f(5) &= -2(5 - 4)^2 + 8 = 6 \end{aligned}$$

Thus, the vertical line $x = 4$ is a line of symmetry. That is, if the graph of equation (3) is drawn on a piece of paper and the paper is folded along the line $x = 4$, then the two sides of the parabola will match exactly. All these results are illustrated by graphing equations (3) and (4) and the line $x = 4$ simultaneously in the same coordinate system (Fig. 2).

From the preceding discussion, we see that as x moves from left to right, $f(x)$ is increasing on $(-\infty, 4]$, and decreasing on $[4, \infty)$, and that $f(x)$ can assume no value greater than 8. Thus,

$$\text{Range of } f\text{:} \quad y \leq 8 \quad \text{or} \quad (-\infty, 8]$$

In general, the graph of a quadratic function is a parabola with line of symmetry parallel to the vertical axis. The lowest or highest point on the parabola, whichever exists, is called the **vertex.** The maximum or minimum value of a quadratic function always occurs at the vertex of the parabola. The line of symmetry through the vertex is called the **axis** of the parabola. In the above example, $x = 4$ is the axis of the parabola and $(4, 8)$ is its vertex.

Applying the graph transformation properties discussed in Section 1-2 to the transformed equation,

$$\begin{aligned} f(x) &= -2x^2 + 16x - 24 \\ &= -2(x - 4)^2 + 8 \end{aligned}$$

we see that the graph of $f(x) = -2x^2 + 16x - 24$ is the graph of $g(x) = x^2$ vertically expanded by a factor of 2, reflected in the x axis, and shifted to the right 4 units and up 8 units, as shown in Figure 3.

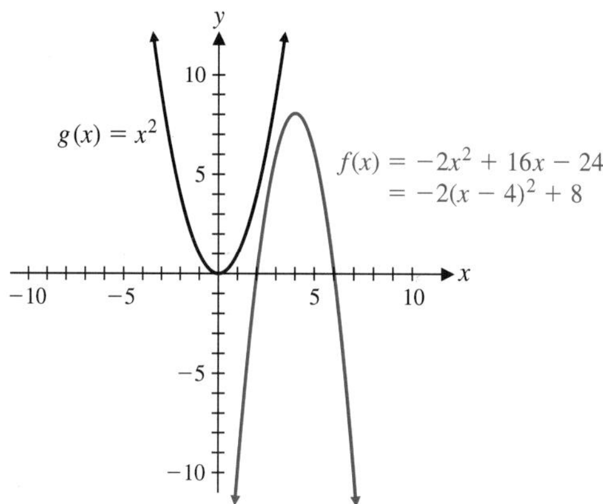

FIGURE 3
Graph of f is the graph of g transformed

Note the important results we have obtained by transforming equation (3) into equation (4):

- The vertex of the parabola
- The axis of the parabola
- The maximum value of $f(x)$
- The range of the function f
- The relationship between the graph of $g(x) = x^2$ and the graph of $f(x) = -2x^2 + 16x - 24$

Now, let us explore the effects of changing the constants a, h, and k on the graph of $f(x) = a(x - h)^2 + k$.

Explore–Discuss 2

(A) Let $a = 1$ and $h = 5$. Graph $f(x) = a(x - h)^2 + k$ for $k = -4, 0$, and 3 simultaneously in the same coordinate system. Explain the effect of changing k on the graph of f.

(B) Let $a = 1$ and $k = 2$. Graph $f(x) = a(x - h)^2 + k$ for $h = -4$, 0, and 5 simultaneously in the same coordinate system. Explain the effect of changing h on the graph of f.

(C) Let $h = 5$ and $k = -2$. Graph $f(x) = a(x - h)^2 + k$ for $a = 0.25$, 1, and 3 simultaneously in the same coordinate system. Graph function f for $a = 1, -1$, and -0.25 simultaneously in the same coordinate system. Explain the effect of changing a on the graph of f.

C (D) Discuss parts (A)–(C) using a graphing utility and a standard viewing window.

The preceding discussion is generalized for all quadratic functions in the following box:

Properties of a Quadratic Function and Its Graph

Given a quadratic function

$$f(x) = ax^2 + bx + c \qquad a \neq 0$$

and the form obtained by completing the square,

$$f(x) = a(x - h)^2 + k$$

we summarize general properties as follows:

1. The graph of f is a parabola:

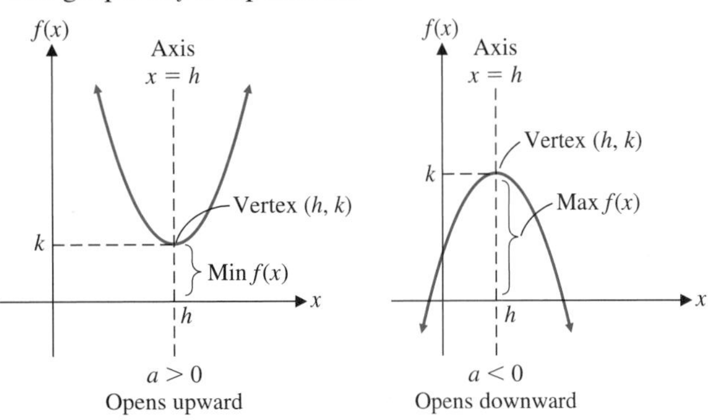

2. Vertex: (h, k) (Parabola increases on one side of the vertex and decreases on the other)
3. Axis (of symmetry): $x = h$ (Parallel to y axis)
4. $f(h) = k$ is the minimum if $a > 0$ and the maximum if $a < 0$
5. Domain: All real numbers
   Range: $(-\infty, k]$ if $a < 0$ or $[k, \infty)$ if $a > 0$
6. The graph of f is the graph of $g(x) = ax^2$ translated horizontally h units and vertically k units.

Example 2 **Analyzing a Quadratic Function** Given the quadratic function

$$f(x) = 0.5x^2 - 6x + 21$$

(A) Find the vertex and the maximum or minimum (to the nearest integer) algebraically by completing the square. State the range of f.

(B) Referring to the completed square form in part (A), describe how the graph of function f can be obtained from the graph of $g(x) = x^2$ using transformations discussed in Section 1-2.

(C) Using parts (A) and/or (B), sketch a graph of function f in a rectangular coordinate system.

C (D) Graph function f using a suitable viewing window.

C (E) Find the vertex and the maximum or minimum (to the nearest integer) graphically using trace and zoom or an appropriate built-in routine. State the range of f.

SOLUTION (A) Finding the vertex, minimum, and range algebraically: Complete the square:

$$\begin{aligned} f(x) &= 0.5x^2 - 6x + 21 \\ &= 0.5(x^2 - 12x + ?) + 21 \\ &= 0.5(x^2 - 12x + 36) + 21 - 18 \\ &= 0.5(x - 6)^2 + 3 \end{aligned}$$

From the last form, we see that $h = 6$ and $k = 3$. Thus, vertex: $(6, 3)$; minimum: $f(6) = 3$; range: $y \geq 3$ or $[3, \infty)$.

(B) The graph of $f(x) = 0.5(x - 6)^2 + 3$ is the same as the graph of $g(x) = x^2$ vertically contracted by a factor of 0.5, and shifted to the right 6 units and up 3 units.

(C) Graph in a rectangular coordinate system:

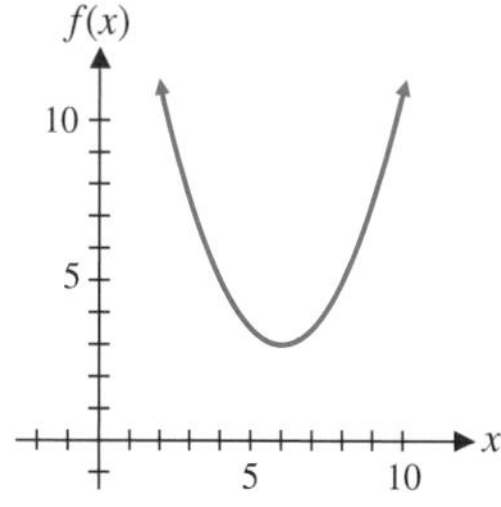

C (D) Graph in a graphing utility:

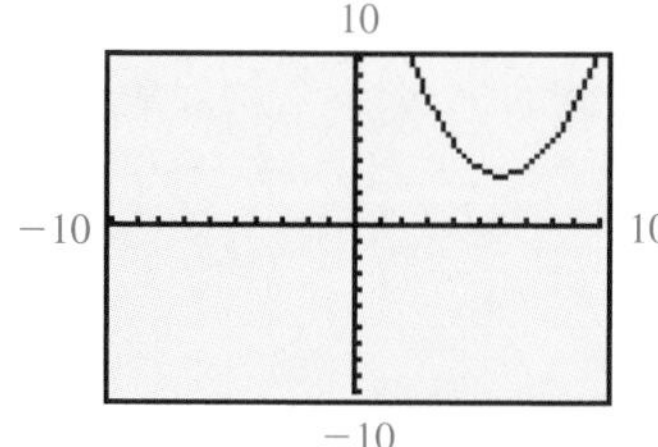

C (E) Finding the vertex, minimum, and range graphically using a graphing utility:

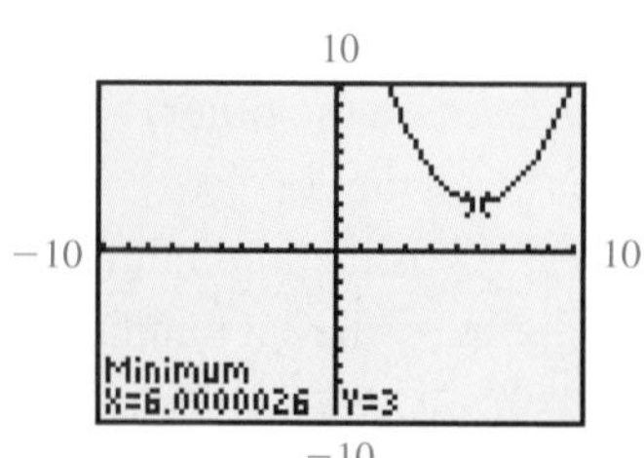

Vertex: $(6, 3)$; minimum: $f(6) = 3$; range: $y \geq 3$ or $[3, \infty)$.

Matched Problem 2 ➡ Given the quadratic function: $f(x) = -0.25x^2 - 2x + 2$

(A) Find the vertex and the maximum or minimum (to the nearest integer) algebraically by completing the square. State the range of f.

(B) Referring to the completed square form in part (A), describe how the graph of function f can be obtained from the graph of $g(x) = x^2$ using transformations discussed in Section 1-2.

(C) Using parts (A) and/or (B), sketch a graph of function f in a rectangular coordinate system.

C (D) Graph function f using a suitable viewing window.

C (E) Find the vertex and the maximum or minimum (to the nearest integer) graphically using trace and zoom or an appropriate built-in routine. State the range of f.

Applications

Example 3 ➡ **Maximum Revenue** This is a continuation of Example 7 in Section 1-1. Recall that the financial department in the company that produces an automatic camera arrived at the following price–demand function and the corresponding revenue function:

$$p(x) = 94.8 - 5x \qquad \text{Price–demand function}$$
$$R(x) = xp(x) = x(94.8 - 5x) \qquad \text{Revenue function}$$

where $p(x)$ is the wholesale price per camera at which x million cameras can be sold and $R(x)$ is the corresponding revenue (in million dollars). Both functions have domain $1 \leq x \leq 15$.

(A) Find the output to the nearest thousand cameras that will produce the maximum revenue. What is the maximum revenue to the nearest thousand dollars? Solve the problem algebraically by completing the square.

(B) What is the wholesale price per camera (to the nearest dollar) that produces the maximum revenue?

C (C) Graph the revenue function using an appropriate viewing window.

C (D) Find the output to the nearest thousand cameras that will produce the maximum revenue. What is the maximum revenue to the nearest thousand dollars? Solve the problem graphically using trace and zoom or an appropriate built-in routine.

Solution (A) Algebraic solution:

$$\begin{aligned} R(x) &= x(94.8 - 5x) \\ &= -5x^2 + 94.8x \\ &= -5(x^2 - 18.96x + ?) \\ &= -5(x^2 - 18.96x + 89.8704) + 449.352 \\ &= -5(x - 9.48)^2 + 449.352 \end{aligned}$$

The maximum revenue of 449.352 million dollars ($449,352,000) occurs when $x = 9.480$ million cameras (9,480,000 cameras).

(B) Finding the wholesale price per camera: Use the price–demand function for an output of 9.480 million cameras:

$$\begin{aligned} p(x) &= 94.8 - 5x \\ p(9.480) &= 94.8 - 5(9.480) \\ &= \$47 \end{aligned}$$

(C) Graph in a graphing utility:

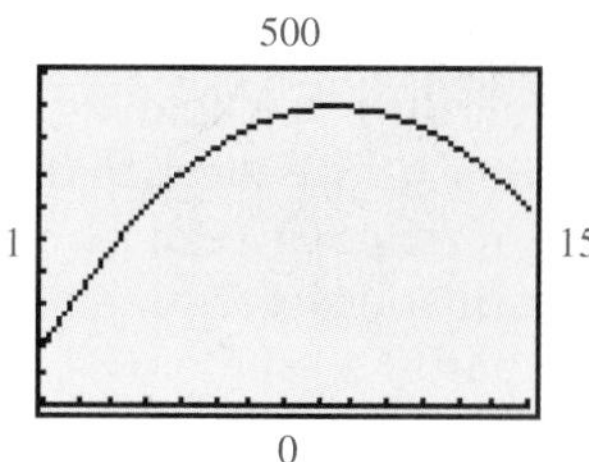

(D) Graphical solution using a graphing utility:

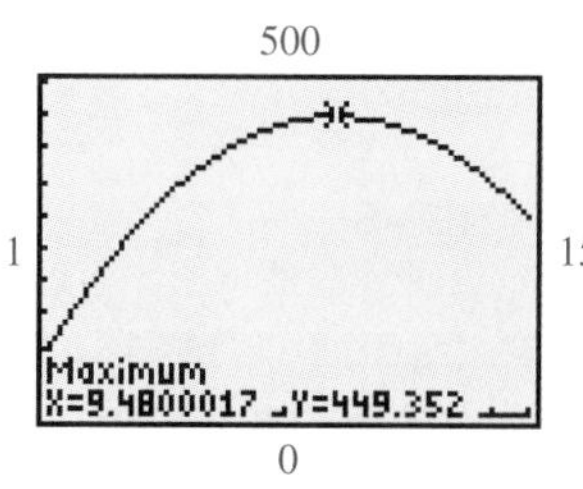

An output of 9.480 million cameras (9,480,000 cameras) will produce a maximum revenue of 449.352 million dollars ($449,352,000).

Matched Problem 3 The financial department in Example 3, using statistical and analytical techniques (see Matched Problem 7 in Section 1-1), arrived at the cost function

$$C(x) = 156 + 19.7x \quad \text{Cost function}$$

where $C(x)$ is the cost (in million dollars) for manufacturing and selling x million cameras.

(A) Using the revenue function from Example 3 and the cost function above, write an equation for the profit function.
(B) Find the output to the nearest thousand cameras that will produce the maximum profit. What is the maximum profit to the nearest thousand dollars? Solve the problem algebraically by completing the square.
(C) What is the wholesale price per camera (to the nearest dollar) that produces the maximum profit?

(D) Graph the profit function using an appropriate viewing window.
(E) Find the output to the nearest thousand cameras that will produce the maximum profit. What is the maximum profit to the nearest thousand dollars? Solve the problem graphically using trace and zoom or an appropriate built-in routine.

Example 4 **Break-Even Analysis** Use the revenue function from Example 3 and the cost function from Matched Problem 3:

$$R(x) = x(94.8 - 5x) \quad \text{Revenue function}$$
$$C(x) = 156 + 19.7x \quad \text{Cost function}$$

Both have domain $1 \leq x \leq 15$.

(A) Sketch the graphs of both functions in the same coordinate system.
(B) **Break-even points** are the production levels at which $R(x) = C(x)$. Find the break-even points algebraically to the nearest thousand cameras.

C (C) Plot both functions simultaneously in the same viewing window.

C (D) Find the break-even points graphically to the nearest thousand cameras using trace and zoom or an appropriate built-in routine.

(E) Recall that a loss occurs if $R(x) < C(x)$ and a profit occurs if $R(x) > C(x)$. For what outputs (to the nearest thousand cameras) will a loss occur? A profit?

SOLUTION (A) Sketch of functions:

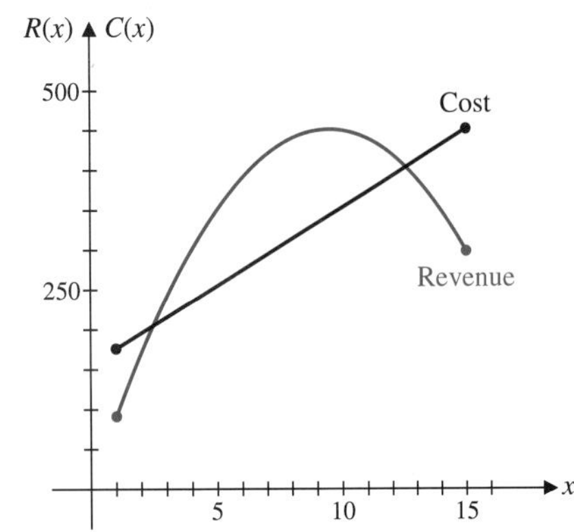

(B) Find x such that $R(x) = C(x)$:

$$x(94.8 - 5x) = 156 + 19.7x$$
$$-5x^2 + 75.1x - 156 = 0$$
$$x = \frac{-75.1 \pm \sqrt{75.1^2 - 4(-5)(-156)}}{2(-5)} \qquad \text{Quadratic formula}$$
$$= \frac{-75.1 \pm \sqrt{2{,}520.01}}{-10}$$
$$x = 2.490 \quad \text{or} \quad 12.530$$

The company breaks even at $x = 2.490$ and 12.530 million cameras.

C (C) Graph in a graphing utility:

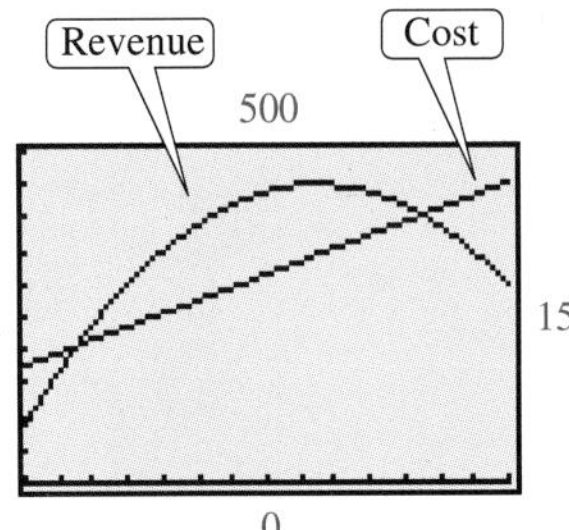

C (D) Graphical solution:

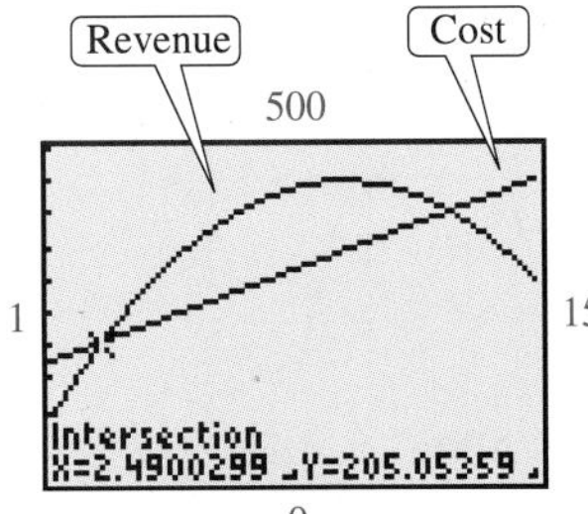

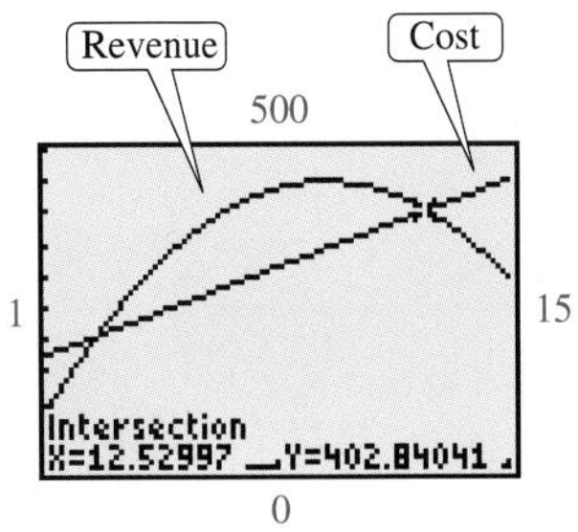

The company breaks even at $x = 2.490$ and 12.530 million cameras.

(E) Use the results from parts (A) and (B) or (C) and (D):

Loss: $1 \leqslant x < 2.490$ or $12.530 < x \leqslant 15$
Profit: $2.490 < x < 12.530$

Matched Problem 4 Use the profit equation from Matched Problem 3:

$$P(x) = R(x) - C(x)$$
$$= -5x^2 + 75.1x - 156 \quad \text{Profit function}$$

Domain: $1 \leqslant x \leqslant 15$

(A) Sketch a graph of the profit function in a rectangular coordinate system.
(B) Break-even points occur when $P(x) = 0$. Find the break-even points algebraically to the nearest thousand cameras.
(C) Plot the profit function in an appropriate viewing window.
(D) Find the break-even points graphically to the nearest thousand cameras using trace and zoom or an appropriate built-in routine.
(E) A loss occurs if $P(x) < 0$, and a profit occurs if $P(x) > 0$. For what outputs (to the nearest thousand cameras) will a loss occur? A profit?

Answers to Matched Problems

1. (A)

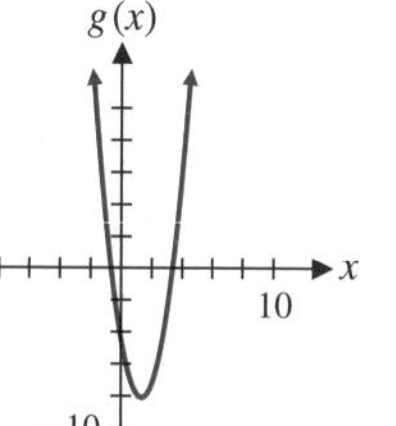

(B) x intercepts: $-0.77, 3.27$; y intercept: -5

(C)

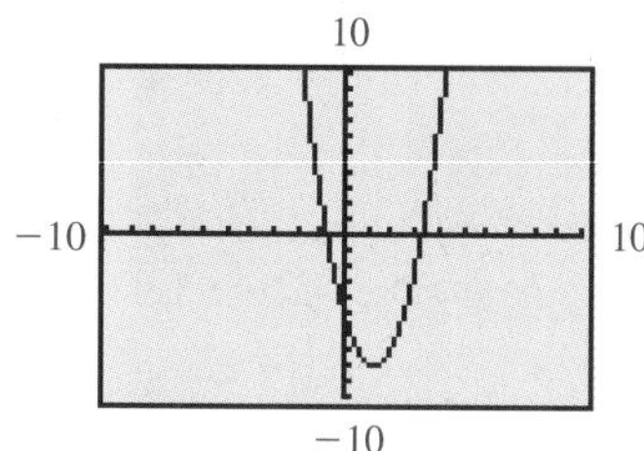

(D) x intercepts: $-0.77, 3.27$; y intercept: -5
(E) $x \leqslant -0.77$ or $x \geqslant 3.27$; or $(-\infty, -0.77]$ or $[3.27, \infty)$
(F) $x = -0.35, 2.85$

2. (A) $f(x) = -0.25(x + 4)^2 + 6$; Vertex: $(-4, 6)$; Maximum: $f(-4) = 6$; Range: $y \leqslant 6$ or $(-\infty, 6]$
(B) The graph of $f(x) = -0.25(x + 4)^2 + 6$ is the same as the graph of $g(x) = x^2$ vertically contracted by a factor of 0.25, reflected in the x axis, and shifted 4 units to the left and 6 units up.
(C)

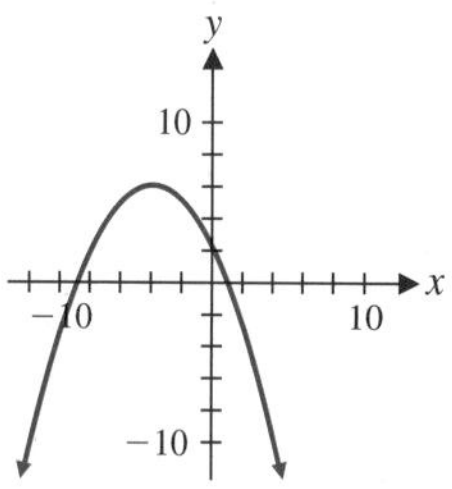

(D)

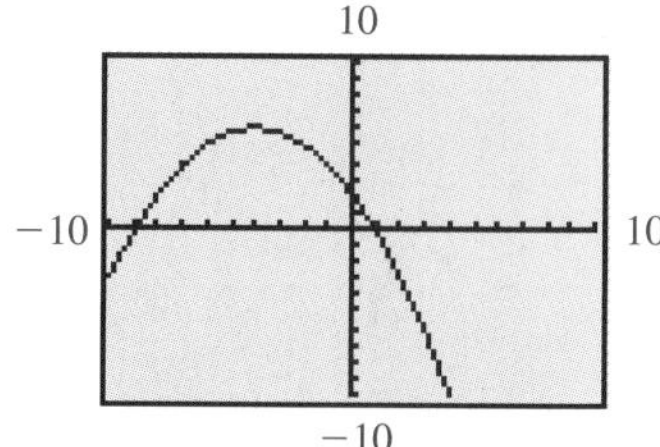

(E) Vertex: $(-4, 6)$; Maximum: $f(-4) = 6$; Range: $y \leqslant 6$ or $(-\infty, 6]$

3. (A) $P(x) = R(x) - C(x) = -5x^2 + 75.1x - 156$
(B) $P(x) = R(x) - C(x) = -5(x - 7.51)^2 + 126.0005$; output of 7.510 million cameras will produce a maximum profit of 126.001 million dollars.
(C) $p(7.510) = \$57$

(D) 300, 1, 15, −300

(E) 300, 1, 15, −300

Maximum
X=7.5099984 Y=126.0005

An output of 7.51 million cameras will produce a maximum profit of 126.001 million dollars. (Notice that maximum profit does not occur at the same output where maximum revenue occurs.)

4. (A)

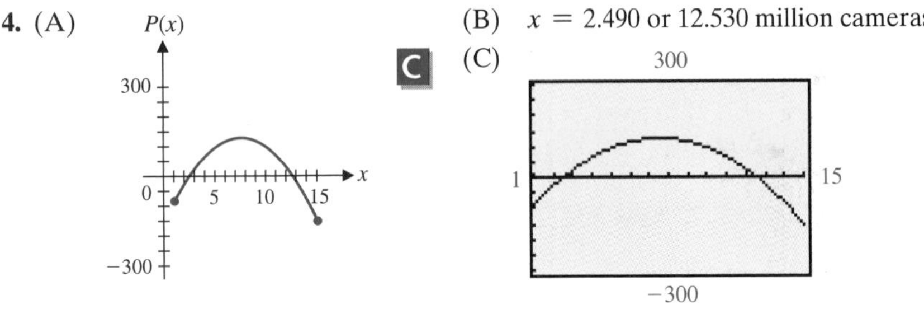

(B) $x = 2.490$ or 12.530 million cameras

(C)

(D) $x = 2.490$ or 12.530 million cameras

(E) Loss: $1 \leq x < 2.490$ or $12.530 < x \leq 15$; Profit: $2.490 < x < 12.530$

EXERCISE 1-4

A

1. Indicate which equations define a quadratic function with x the independent variable.

(A) $y = 3 - x^2$ (B) $y^2 + x^2 = 4$

(C) $y = (2 - 3x)^2$ (D) $2x - y = 3$

(E) $y = 2x(3 - x)$ (F) $y = -2x^2 + 5x - 1$

2. Indicate which equations define a quadratic function with x the independent variable.

(A) $x^2 - y^2 = 9$ (B) $y = x(3x - 5)$

(C) $y = 1 - \sqrt{x}$ (D) $y = 4 - 3x - 2x^2$

(E) $y = 2(x - 3)^2$ (F) $y = 8.3 - 0.3x^2$

3. Match each equation with a graph of one of the functions f, g, m, or n in the figure.

(A) $y = -(x + 2)^2 + 1$ (B) $y = (x - 2)^2 - 1$

(C) $y = (x + 2)^2 - 1$ (D) $y = -(x - 2)^2 + 1$

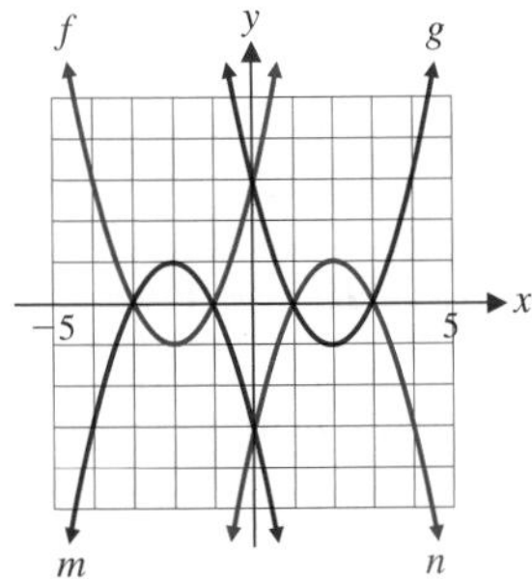

Figure for 3

4. Match each equation with a graph of one of the functions f, g, m, or n in the figure.

(A) $y = (x - 3)^2 - 4$ (B) $y = -(x + 3)^2 + 4$

(C) $y = -(x - 3)^2 + 4$ (D) $y = (x + 3)^2 - 4$

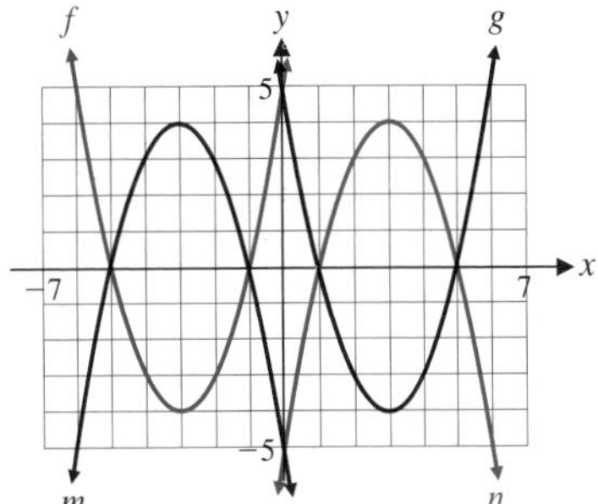

Figure for 4

For the functions indicated in Problems 5–8, find each of the following to the nearest integer by referring to the graphs for Problems 3 and Problem 4.

(A) Intercepts *(B) Vertex*

(C) Maximum or minimum *(D) Range*

(E) Increasing interval *(F) Decreasing interval*

5. Function n in the figure for Problem 3

6. Function m in the figure for Problem 4

7. Function f in the figure for Problem 3

8. Function g in the figure for Problem 4

In Problems 9–12, find each of the following algebraically (to the nearest integer) without referring to any graphs.
(A) Intercepts *(B) Vertex*
(C) Maximum or minimum *(D) Range*

9. $f(x) = -(x-2)^2 + 1$ **10.** $g(x) = -(x+3)^2 + 4$

11. $M(x) = (x+2)^2 - 1$ **12.** $N(x) = (x-3)^2 - 4$

 **B** *In Problems 13–16, write an equation for each graph in the form $y = a(x-h)^2 + k$, where a is either 1 or −1 and h and k are integers.*

13.

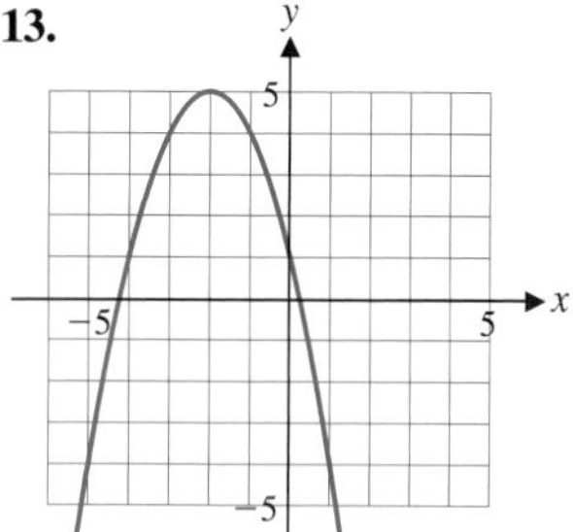

14.

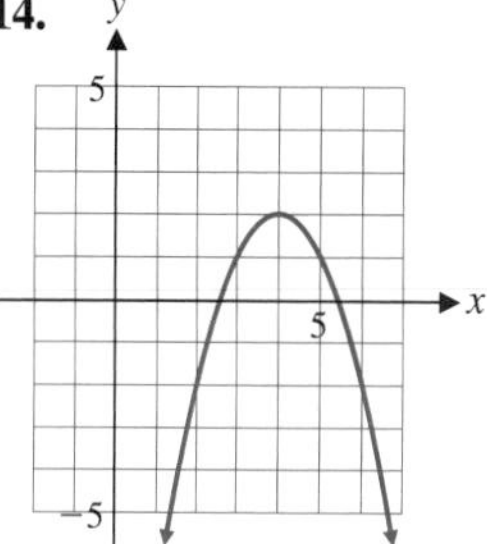

15.

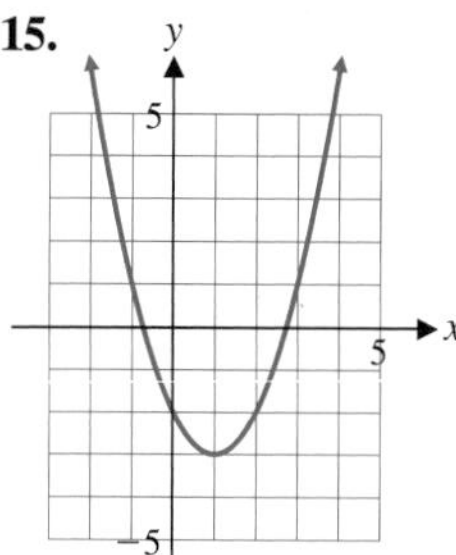

16.

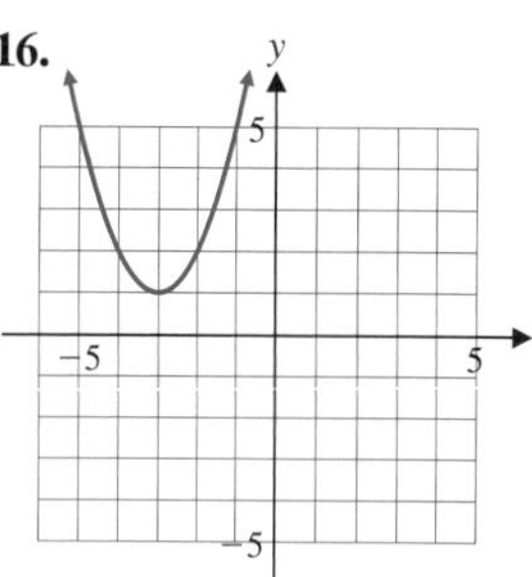

In Problems 17–22, first write each function in the form $f(x) = a(x-h)^2 + k$; then find each of the following algebraically (to one decimal place) without using any graphs:
(A) Intercepts *(B) Vertex*
(C) Maximum or minimum *(D) Range*

17. $f(x) = x^2 - 8x + 13$ **18.** $g(x) = x^2 + 10x + 20$

19. $M(x) = 1 - 6x - x^2$ **20.** $N(x) = -10 + 8x - x^2$

21. $G(x) = 0.5x^2 - 4x + 10$

22. $H(x) = -0.5x^2 - 2x - 3$

 *Solve Problems 17–22 using a graphing utility without completing the square. Solve graphically using trace and zoom or an appropriate built-in routine.*

23. C Let $f(x) = 0.3x^2 - x - 8$. Solve each equation graphically to two decimal places.
(A) $f(x) = 4$ (B) $f(x) = -1$ (C) $f(x) = -9$

24. C Let $g(x) = -0.6x^2 + 3x + 4$. Solve each equation graphically to two decimal places.
(A) $g(x) = -2$ (B) $g(x) = 5$ (C) $g(x) = 8$

25. Explain under what graphical conditions a quadratic function has exactly one real zero.

26. Explain under what graphical conditions a quadratic function has no real zeros.

C *In Problems 27–30, first write each function in the form $f(x) = a(x-h)^2 + k$; then find each of the following algebraically (to two decimal places) without using any graphs:*
(A) Intercepts *(B) Vertex*
(C) Maximum or minimum *(D) Range*

27. $g(x) = 0.25x^2 - 1.5x - 7$

28. $m(x) = 0.20x^2 - 1.6x - 1$

29. $f(x) = -0.12x^2 + 0.96x + 1.2$

30. $n(x) = -0.15x^2 - 0.90x + 3.3$

 *Solve Problems 31–36 graphically to two decimal places using a graphing utility. Do not change the form of any equation or inequality.*

31. $2 - 5x - x^2 = 0$ **32.** $7 + 3x - 2x^2 = 0$

33. $1.9x^2 - 1.5x - 5.6 < 0$ **34.** $3.4 + 2.9x - 1.1x^2 \geq 0$

35. $2.8 + 3.1x - 0.9x^2 \leq 0$ **36.** $1.8x^2 - 3.1x - 4.9 > 0$

37. Given that f is a quadratic function with minimum $f(x) = f(2) = 4$, find the axis, vertex, range, and x intercepts.

38. Given that f is a quadratic function with maximum $f(x) = f(-3) = -5$, find the axis, vertex, range, and x intercepts.

In Problems 39–42:
(A) Graph f and g in the same coordinate system.
(B) Solve $f(x) = g(x)$ algebraically to two decimal places.
(C) Solve $f(x) > g(x)$ using parts (A) and (B).
(D) Solve $f(x) < g(x)$ using parts (A) and (B).

39. $f(x) = -0.4x(x-10)$
$g(x) = 0.3x + 5$
$0 \leq x \leq 10$

40. $f(x) = -0.7x(x-7)$
$g(x) = 0.5x + 3.5$
$0 \leq x \leq 7$

41. $f(x) = -0.9x^2 + 7.2x$
$g(x) = 1.2x + 5.5$
$0 \leq x \leq 8$

42. $f(x) = -0.7x^2 + 6.3x$
$g(x) = 1.1x + 4.8$
$0 \leq x \leq 9$

43. Give a simple example of a quadratic function that has no real zeros. Explain how its graph is related to the x axis.

44. Give a simple example of a quadratic function that has exactly one real zero. Explain how its graph is related to the x axis.

Applications

Business & Economics

45. *Tire mileage.* An automobile tire manufacturer collected the data in the table relating tire pressure x (in pounds per square inch) and mileage (in thousands of miles):

x	28	30	32	34	36
MILEAGE	45	52	55	51	47

A mathematical model for the data is given by

$$f(x) = -0.518x^2 + 33.3x - 481$$

(A) Complete the following table. Round values of $f(x)$ to one decimal place.

x	28	30	32	34	36
MILEAGE	45	52	55	51	47
$f(x)$					

(B) Sketch the graph of f and the mileage data in the same coordinate system.

C (C) Use values of the modeling function rounded to two decimal places to estimate the mileage for a tire pressure of 31 pounds per square inch. For 35 pounds per square inch.

(D) Write a brief description of the relationship between tire pressure and mileage.

46. *Automobile production.* The table lists General Motors' total U.S. vehicle production in millions of units from 1989 to 1993:

YEAR	1989	1990	1991	1992	1993
PRODUCTION	4.7	4.1	3.5	3.7	5.0

A mathematical model for GM's production data is given by

$$f(x) = 0.33x^2 - 1.3x + 4.8$$

where $x = 0$ corresponds to 1989.

(A) Complete the following table. Round values of $f(x)$ to one decimal place.

x	0	1	2	3	4
PRODUCTION	4.7	4.1	3.5	3.7	5.0
$f(x)$					

(B) Sketch the graph of f and the production data in the same coordinate system.

C (C) Use values of the modeling function f rounded to two decimal places to estimate the production in 1994. In 1995.

(D) Write a brief verbal description of GM's production from 1989 to 1993.

47. *Revenue.* Refer to Problems 85 and 87, Exercise 1-1. We found that the marketing research department for the company that manufactures and sells memory chips for microcomputers established the following price–demand and revenue functions:

$p(x) = 119 - 6x$ — Price–demand function

$R(x) = xp(x) = x(119 - 6x)$ — Revenue function

where $p(x)$ is the wholesale price in dollars at which x million chips can be sold, and $R(x)$ is in millions of dollars. Both functions have domain $1 \leq x \leq 15$.

(A) Sketch a graph of the revenue function in a rectangular coordinate system.

(B) Find the output (to the nearest thousand chips) that will produce the maximum revenue. What is the maximum revenue to the nearest thousand dollars? Solve the problem algebraically by completing the square.

C (C) Graph the revenue function using an appropriate viewing window.

C (D) Find the output (to the nearest thousand chips) that will produce the maximum revenue. What is the maximum revenue to the nearest thousand dollars? Solve the problem graphically using trace and zoom or an appropriate built-in routine.

(E) What is the wholesale price per chip (to the nearest dollar) that produces the maximum revenue?

48. *Revenue.* Refer to Problems 86 and 88, Exercise 1-1. We found that the marketing research department for the company that manufactures and sells "Notebook" computers established the following price–demand and revenue functions:

$$p(x) = 1{,}190 - 36x \qquad \text{Price–demand function}$$
$$R(x) = xp(x) \qquad \text{Revenue function}$$
$$= x(1{,}190 - 36x)$$

where $p(x)$ is the wholesale price in dollars at which x thousand computers can be sold, and $R(x)$ is in thousands of dollars. Both functions have domain $1 \le x \le 25$.

(A) Sketch a graph of the revenue function in a rectangular coordinate system.
(B) Find the output (to the nearest hundred computers) that will produce the maximum revenue. What is the maximum revenue to the nearest thousand dollars? Solve the problem algebraically by completing the square.
C (C) Graph the revenue function using an appropriate viewing window.
C (D) Find the output (to the nearest hundred computers) that will produce the maximum revenue. What is the maximum revenue to the nearest thousand dollars? Solve the problem graphically using trace and zoom or an appropriate built-in routine.
(E) What is the wholesale price per computer (to the nearest dollar) that produces the maximum revenue?

49. *Break-even analysis.* Use the revenue function from Problem 47 in this exercise and the cost function from Problem 89, Exercise 1-1:

$$R(x) = x(119 - 6x) \qquad \text{Revenue function}$$
$$C(x) = 234 + 23x \qquad \text{Cost function}$$

where x is in millions of chips, and $R(x)$ and $C(x)$ are in millions of dollars. Both functions have domain $1 \le x \le 15$.

(A) Sketch a graph of both functions in the same rectangular coordinate system.
(B) Find the break-even points algebraically to the nearest thousand chips.
C (C) Graph both functions simultaneously in the same viewing window.
C (D) Find the break-even points graphically to the nearest thousand chips using trace and zoom or an appropriate built-in routine.
(E) For what outputs will a loss occur? A profit?

50. *Break-even analysis.* Use the revenue function from Problem 48 in this exercise and the cost function from Problem 90, Exercise 1-1:

$$R(x) = x(1{,}190 - 36x) \qquad \text{Revenue function}$$
$$C(x) = 4{,}320 + 146x \qquad \text{Cost function}$$

where x is thousands of computers, and $R(x)$ and $C(x)$ are in thousands of dollars. Both functions have domain $1 \le x \le 25$.

(A) Sketch a graph of both functions in the same rectangular coordinate system.
(B) Find the break-even points algebraically to the nearest hundred computers.
C (C) Graph both functions simultaneously in the same viewing window.
C (D) Find the break-even points graphically to the nearest hundred computers using trace and zoom or an appropriate built-in routine.
(E) For what outputs will a loss occur? Will a profit occur?

51. *Profit–loss analysis.* Use the revenue function from Problem 47 in this exercise and the cost function from Problem 89, Exercise 1-1:

$$R(x) = x(119 - 6x) \qquad \text{Revenue function}$$
$$C(x) = 234 + 23x \qquad \text{Cost function}$$

where x is in millions of chips, and $R(x)$ and $C(x)$ are in millions of dollars. Both functions have domain $1 \le x \le 15$.

(A) Form a profit function P, and graph R, C, and P in the same rectangular coordinate system.
C (B) Graph R, C, and P simultaneously in an appropriate viewing window.
(C) Discuss the relationship between the intersection points of the graphs of R and C and the x intercepts of P.
(D) Find the x intercepts of P algebraically to the nearest thousand chips. Find the break-even points to the nearest thousand chips.
C (E) Solve part (D) graphically.
(F) Refer to the graph drawn in part (A) or (B). Does the maximum profit appear to occur at the same output level as the maximum revenue? Are the maximum profit and the maximum revenue equal? Explain.
(G) Verify your conclusion in part (F) by completing the square for the profit function to find the output (to the nearest thousand chips) that produces the maximum profit. Find the maximum profit (to the nearest thousand dollars), and compare with Problem 47D.
C (H) Solve part (G) graphically.

52. *Profit–loss analysis.* Use the revenue function from Problem 48 in this exercise and the cost function from Problem 90, Exercise 1-1:

$$R(x) = x(1{,}190 - 36x) \qquad \text{Revenue function}$$
$$C(x) = 4{,}320 + 146x \qquad \text{Cost function}$$

where x is thousands of computers, and $R(x)$ and $C(x)$ are in thousands of dollars. Both functions have domain $1 \le x \le 25$.

(A) Form a profit function P, and graph R, C, and P in the same rectangular coordinate system.

C (B) Graph R, C, and P simultaneously in an appropriate viewing window.

(C) Discuss the relationship between the intersection points of the graphs of R and C and the x intercepts of P.

(D) Find the x intercepts of P algebraically to the nearest hundred computers. Find the break-even points to the nearest hundred computers.

C (E) Solve part (D) graphically.

(F) Refer to the graph drawn in part (A) or (B). Does the maximum profit appear to occur at the same output level as the maximum revenue? Are the maximum profit and the maximum revenue equal? Explain.

(G) Verify your conclusion in part (F) by completing the square for the profit function to find the output (to the nearest hundred computers) that produces the maximum profit. Find the maximum profit (to the nearest thousand dollars), and compare with Problem 48D.

C (H) Solve part (G) graphically.

Life Sciences

53. *Medicine.* The French physician Poiseuille was the first to discover that blood flows faster near the center of an artery than near the edge. Experimental evidence has shown that the rate of flow v (in centimeters per second) at a point x centimeters from the center of an artery (see the figure) is given by

$$v = f(x) = 1{,}000(0.04 - x^2) \qquad 0 \leq x \leq 0.2$$

(A) Find the distance from the center that the rate of flow is 20 centimeters per second. Solve algebraically to two decimal places.

C (B) Use a graphing utility to solve part (A).

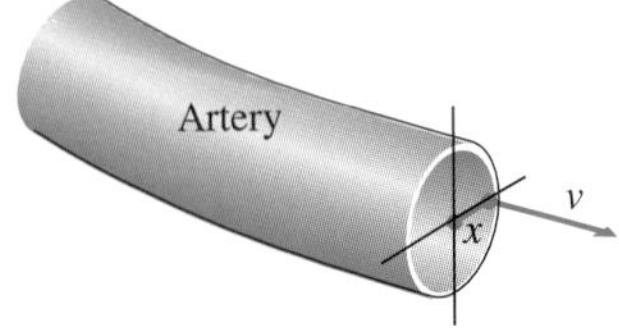

Figure for 53 and 54

54. *Medicine.* Refer to Problem 53.

(A) Find the distance from the center that the rate of flow is 30 centimeters per second. Solve algebraically to two decimal places.

C (B) Use a graphing utility to solve part (A).

Important Terms and Symbols

1-1 *Functions.* Cartesian or rectangular coordinate system; coordinate axes; quadrants; ordered pairs; coordinates; abscissa; ordinate; fundamental theorem of analytic geometry; solution and solution set for an equation in two variables; graph of an equation in two variables; point-by-point plotting; function; domain; range; functions specified by equations; input; output; independent variable; dependent variable; vertical-line test; function notation; difference quotient; cost function; price–demand function; revenue function; profit function

$f(x)$; $C = a + bx$; $p = m - nx$;
$R = xp$; $P = R - C$

1-2 *Elementary Functions: Graphs and Transformations.* Six basic functions: identity, absolute value, square, cube, square root, cube root; transformations; horizontal shift or translation; vertical shift or translation; reflection in the x axis; vertical expansion; vertical contraction; piecewise-defined function; discontinuity

$f(x) = x$; $g(x) = |x|$; $h(x) = x^2$;
$m(x) = x^3$; $n(x) = \sqrt{x}$; $p(x) = \sqrt[3]{x}$

See the inside front cover for the graphs of these functions.

1-3 *Linear Functions and Straight Lines.* x intercepts; y intercept; linear function; constant function; first-degree function; graph of a linear function; graph of a constant function; graphs of linear equations in two variables; standard form for the equation of a line; equations of vertical lines and horizontal lines; slope; slope–intercept form; point–slope form

$f(x) = mx + b, m \neq 0$; $f(x) = b$; $x = a$;
$Ax + By = C$; $y = mx + b$; $y - y_1 = m(x - x_1)$;
$m = \dfrac{y_2 - y_1}{x_2 - x_1}, x_1 \neq x_2$

1-4 *Quadratic Functions.* Quadratic function; parabola; finding intercepts; vertex; axis; maximum or minimum; range; break-even points

$f(x) = ax^2 + bx + c, a \neq 0$;
$f(x) = a(x - h)^2 + k, a \neq 0$

Review Exercise

Work through all the problems in this chapter review and check your answers in the back of the book. Answers to all review problems are there along with section numbers in italics to indicate where each type of problem is discussed. Where weaknesses show up, review appropriate sections in the text.

A

1. Use point-by-point plotting to sketch a graph of $y = 5 - x^2$. Use integer values for x from -3 to 3.

2. Indicate whether each graph specifies a function:

(A)

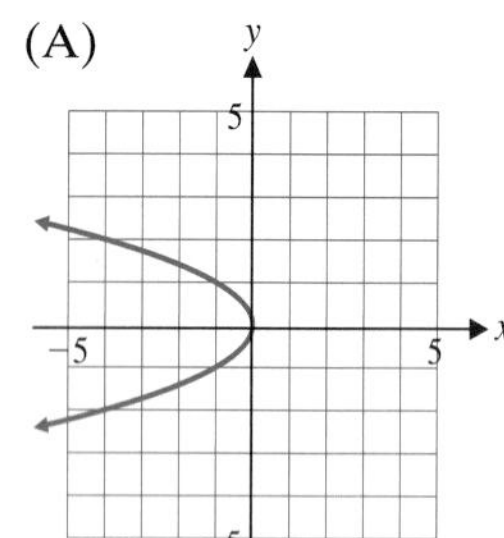

(B)

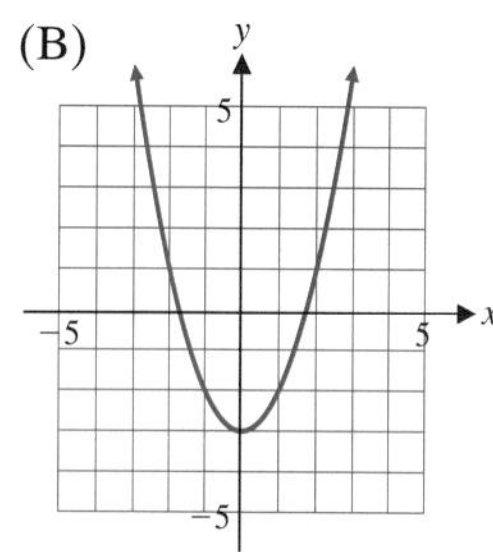

(C)

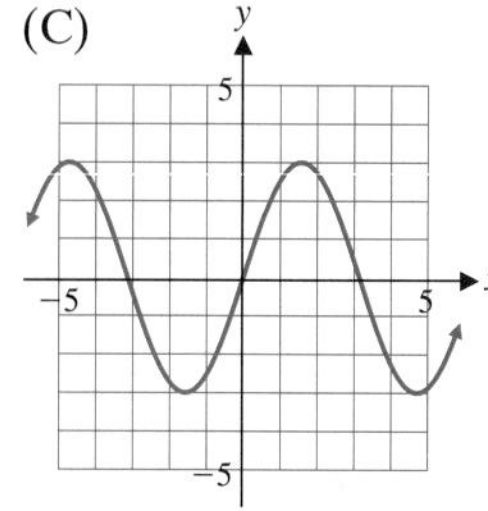

(D)

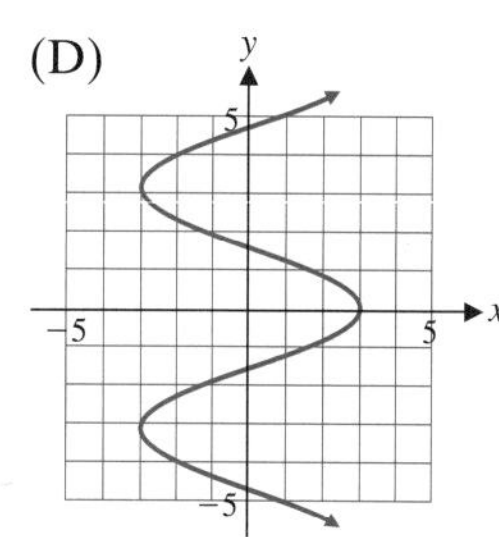

3. For $f(x) = 2x - 1$ and $g(x) = x^2 - 2x$, find:

(A) $f(-2) + g(-1)$ (B) $f(0) \cdot g(4)$

(C) $\dfrac{g(2)}{f(3)}$ (D) $\dfrac{f(3)}{g(2)}$

4. Use the graph of function f in the figure to determine (to the nearest integer) x or y as indicated.

(A) $y = f(0)$ (B) $4 = f(x)$

(C) $y = f(3)$ (D) $3 = f(x)$

(E) $y = f(-6)$ (F) $-1 = f(x)$

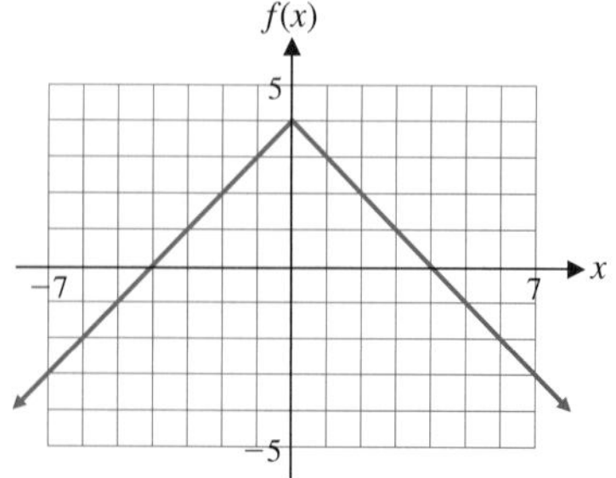

Figure for 4

5. Sketch a graph of each of the functions in parts (A)–(D) using the graph of function f in the figure below.

(A) $y = -f(x)$ (B) $y = f(x) + 4$

(C) $y = f(x - 2)$ (D) $y = -f(x + 3) - 3$

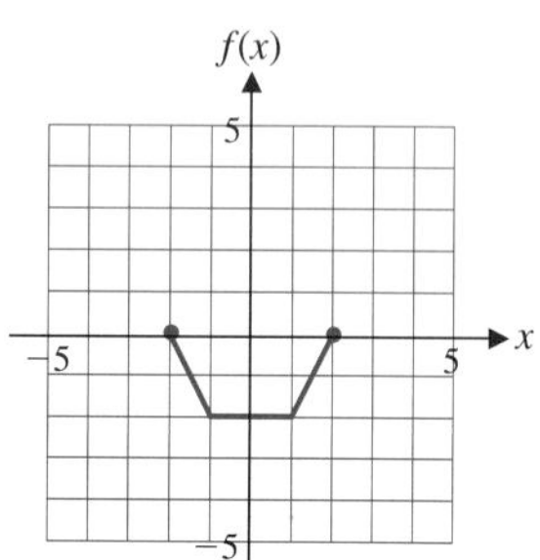

Figure for 5

6. Refer to the figure below.

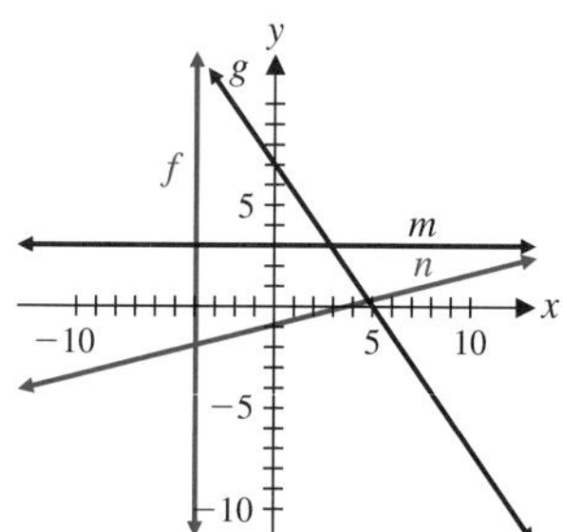

Figure for 6

(A) Identify the graphs of any linear functions with positive slopes.

(B) Identify the graphs of any linear functions with negative slopes.

(C) Identify the graphs of any constant functions. What are their slopes?

(D) Identify any graphs that are not graphs of functions. What can you say about their slopes?

7. Write an equation in the form $y = mx + b$ for a line with slope $-\frac{2}{3}$ and y intercept 6.

8. Write the equations of the vertical line and the horizontal line that pass through $(-6, 5)$.

9. Sketch a graph of $2x - 3y = 18$. What are the intercepts and slope of the line?

10. Indicate which equations define a quadratic function with x the independent variable.

(A) $y^2 = 4 - x^2$ (B) $x^2 - 3 = y$

(C) $y = -2(x - 3)^2 + 1$ (D) $y = x(x + 7)$

(E) $y = 3x + 9$ (F) $y = 0.2x^2 - 5.7$

11. Match each equation with a graph of one of the functions f, g, m, or n in the figure.
(A) $y = (x - 2)^2 - 4$ (B) $y = -(x + 2)^2 + 4$
(C) $y = -(x - 2)^2 + 4$ (D) $y = (x + 2)^2 - 4$

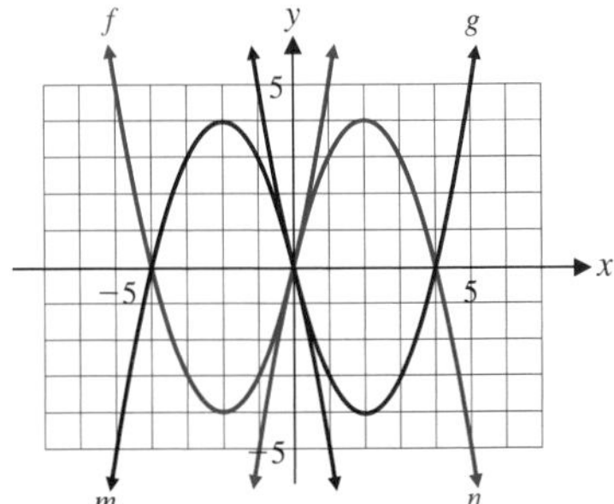

Figure for 11

12. Referring to the graph of function f in the figure for Problem 11 and using known properties of quadratic functions, find each of the following to the nearest integer:
(A) Intercepts (B) Vertex
(C) Maximum or minimum (D) Range
(E) Increasing interval (F) Decreasing interval

B

13. Indicate which of the following equations define a linear function or a constant function:
(A) $2x - 3y = 5$ (B) $x = -2$
(C) $y = 4 - 3x$ (D) $y = -5$
(E) $x = 3y + 5$ (F) $\frac{x}{2} - \frac{y}{3} = 1$

14. Find the domain of each function:
(A) $f(x) = \frac{2x - 5}{x^2 - x - 6}$ (B) $g(x) = \frac{3x}{\sqrt{5 - x}}$

15. The function g is defined by $g(x) = 2x - 3\sqrt{x}$. Translate into a verbal definition.

16. Describe the graphs of $x = -3$ and $y = 2$. Graph both simultaneously in the same rectangular coordinate system.

17. Let $f(x) = 0.4x(x + 4)(2 - x)$.
C (A) Sketch a graph of f on graph paper by first plotting points using odd integer values of x from -3 to 3. Then complete the graph using a graphing utility.
(B) Discuss the number of solutions of each of the following equations:

$$f(x) = 3 \qquad f(x) = 2 \qquad f(x) = 1$$

(C) Approximate the solutions of each equation in part (B) to two decimal places.

18. Find $\frac{f(2 + h) - f(2)}{h}$ for $f(x) = 3 - 2x$.

19. Find $\frac{f(a + h) - f(a)}{h}$ for $f(x) = x^2 - 3x + 1$.

20. Explain how the graph of $m(x) = -|x - 4|$ is related to the graph of $y = |x|$.

21. Explain how the graph of $g(x) = 0.3x^3 + 3$ is related to the graph of $y = x^3$.

22. The following graph is the result of applying a sequence of transformations to the graph of $y = x^2$. Describe the transformations verbally and write an equation for the given graph.

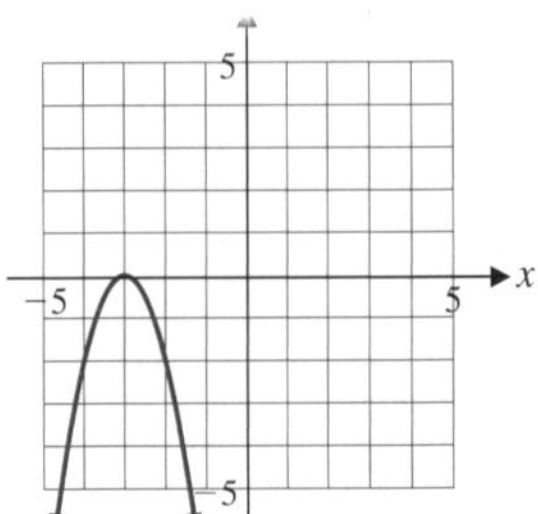

Figure for 22

23. The graph of a function f is formed by vertically expanding the graph of $y = \sqrt{x}$ by a factor of 2, and shifting it to the left 3 units and down 1 unit. Find an equation for function f and graph it for $-5 \leq x \leq 5$ and $-5 \leq y \leq 5$.

24. Let

$$f(x) = \begin{cases} -x - 2 & \text{for } x < 0 \\ 0.2x^2 & \text{for } x \geq 0 \end{cases}$$

(A) Sketch the graph of f.
(B) Find any points of discontinuity.
C (C) Check by graphing on a graphing utility.

25. Write the equation of a line through each indicated point with the indicated slope. Write the final answer in the form $y = mx + b$.
(A) $m = -\frac{2}{3}$; $(-3, 2)$ (B) $m = 0$; $(3, 3)$

26. Write the equation of the line through the two indicated points. Write the final answer in the form $Ax + By = C$.
(A) $(-3, 5), (1, -1)$ (B) $(-1, 5), (4, 5)$
(C) $(-2, 7), (-2, -2)$

27. Write an equation for the graph shown in the form $y = a(x - h)^2 + k$, where a is either -1 or $+1$ and h and k are integers.

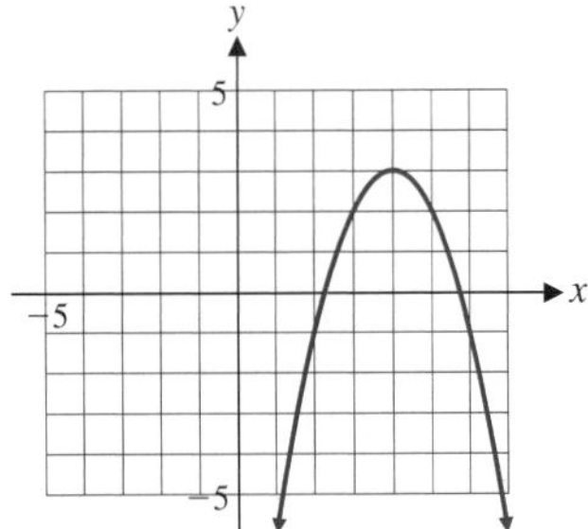

Figure for 27

28. Given $f(x) = -0.4x^2 + 3.2x + 1.2$, find the following algebraically (to one decimal place) without referring to a graph:
(A) Intercepts (B) Vertex
(C) Maximum or minimum (D) Range

29. Graph $f(x) = -0.4x^2 + 3.2x + 1.2$ in a graphing utility and find the following (to one decimal place) using trace and zoom or an appropriate graphing utility:
(A) Intercepts (B) Vertex
(C) Maximum or minimum (D) Range

C

30. The following graph is the result of applying a sequence of transformations to the graph of $y = \sqrt[3]{x}$. Describe the transformations verbally, and write an equation for the graph.

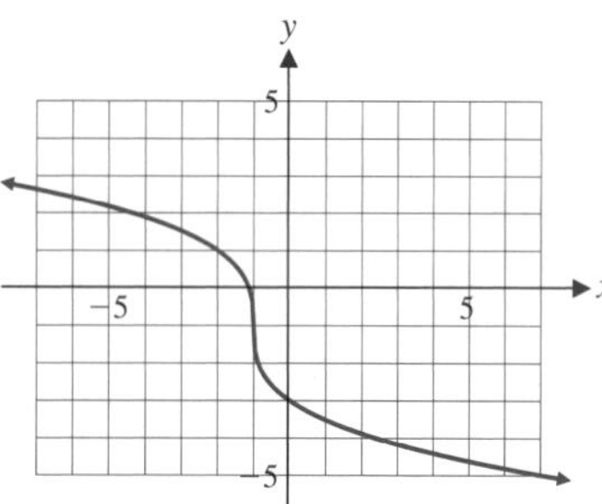

Figure for 30

31. Graph

$$y = mx + b \quad \text{and} \quad y = -\frac{1}{m}x + b$$

simultaneously in the same coordinate system for b fixed and several different values of m, $m \neq 0$. Describe the apparent relationship between the graphs of the two equations.

32. Find and simplify $\dfrac{f(x+h) - f(x)}{h}$ for each of the following functions:

(A) $f(x) = \sqrt{x}$ (B) $f(x) = \dfrac{1}{x}$

33. Given $G(x) = 0.3x^2 + 1.2x - 6.9$, find the following algebraically (to one decimal place) without the use of a graph:
(A) Intercepts (B) Vertex
(C) Maximum or minimum (D) Range
(E) Increasing and decreasing intervals

34. Graph $G(x) = 0.3x^2 + 1.2x - 6.9$ in a standard viewing window. Then find each of the following (to one decimal place) using trace or zoom or an appropriate built-in routine:
(A) Intercepts (B) Vertex
(C) Maximum or minimum (D) Range
(E) Increasing and decreasing intervals

Applications

Business & Economics

35. *Linear depreciation.* A computer system was purchased by a small business for \$12,000 and, for tax purposes, is assumed to have a salvage value of \$2,000 after 8 years. If its value is depreciated linearly from \$12,000 to \$2,000:
(A) Find the linear equation that relates the value V in dollars to the time t in years. Then graph the equation in a rectangular coordinate system.
(B) What would be the value of the system after 5 years?

36. *Compound interest.* If \$1,000 is invested at $100r\%$ compounded annually, at the end of 3 years it will grow to $A = 1{,}000(1 + r)^3$.
(A) Graph the equation in a graphing utility for $0 \leq r \leq 0.25$; that is, for money invested at between 0% and 25% compounded annually.
(B) At what rate of interest would \$1,000 have to be invested to amount to \$1,500 in 3 years? Solve for r graphically (to four decimal places) using trace and zoom or an appropriate built-in routine.

37. *Markup.* A sporting goods store sells a tennis racket that cost \$130 for \$208 and court shoes that cost \$50 for \$80.

(A) If the markup policy of the store for items that cost over \$10 is assumed to be linear and is reflected in the pricing of these two items, write an equation that relates retail price R to cost C.
(B) What would be the retail price of a pair of in-line skates that cost \$120?
(C) What would be the cost of a pair of cross-country skis that had a retail price of \$176?
(D) What is the slope of the graph of the equation found in part (A)? Interpret the slope relative to the problem.

38. *Demand.* Egg consumption has been decreasing for some time, presumably because of increasing awareness of the high cholesterol content in egg yolks. The table lists the annual per capita consumption of eggs in the United States.

YEAR	1970	1975	1980	1985	1990
NUMBER OF EGGS	309	276	271	255	233

Source: U.S. Department of Agriculture.

A mathematical model for the data is given by

$$f(x) = 303.4 - 3.46x$$

where $x = 0$ corresponds to 1970.

(A) Complete the following table. Round values of $f(x)$ to the nearest integer.

x	0	5	10	15	20
CONSUMPTION	309	276	271	255	233
$f(x)$					

(B) Graph $y = f(x)$ and the data in the table in the same coordinate system.

(C) Use the modeling function f to estimate the per capita egg consumption in 1995. In 2000.

(D) Based on the information in the table, write a brief verbal description of egg consumption from 1970 to 1990.

39. *Pricing.* An office supply store sells ballpoint pens for \$0.49 each. For an order of 3 dozen or more pens, the price per pen for all pens ordered is reduced to \$0.44, and for an order of 6 dozen or more, the price per pen for all pens ordered is reduced to \$0.39.

(A) If $C(x)$ is the total cost in dollars for an order of x pens, write a piecewise definition for C.

(B) Graph $y = C(x)$ for $0 \leq x \leq 108$ and identify any points of discontinuity.

40. *Break-even analysis.* A video production company is planning to produce an instructional videotape. The producer estimates that it will cost \$84,000 to shoot the video and \$15 per unit to copy and distribute the tape. The wholesale price of the tape is \$50 per unit.

(A) Write cost and revenue equations, and graph both simultaneously in a rectangular coordinate system.

(B) Algebraically determine when $R = C$. Then, with the aid of part (A), determine when $R < C$ and $R > C$.

C (C) Using a graphing utility, determine when $R = C$, $R < C$, and $R > C$.

41. *Break-even analysis.* The research department in a company that manufacturers AM/FM clock radios established the following price–demand, cost, and revenue functions:

$$\begin{aligned} p(x) &= 50 - 1.25x && \text{Price–demand function} \\ C(x) &= 160 + 10x && \text{Cost function} \\ R(x) &= xp(x) \\ &= x(50 - 1.25x) && \text{Revenue function} \end{aligned}$$

where x is in thousands of units, and $C(x)$ and $R(x)$ are in thousands of dollars. All three functions have domain $1 \leq x \leq 40$.

(A) Graph the cost function and the revenue function simultaneously in the same coordinate system.

(B) Algebraically determine when $R = C$. Then, with the aid of part (A), determine when $R < C$ and $R > C$ to the nearest unit.

(C) Algebraically determine the maximum revenue (to the nearest thousand dollars) and the output (to the nearest unit) that produces the maximum revenue. What is the wholesale price of the radio (to the nearest dollar) at this output?

42. C *Profit–loss analysis.* Use the cost and revenue functions from Problem 41.

(A) Write a profit function and graph it in a graphing utility.

(B) Graphically determine when $P = 0$, $P < 0$, and $P > 0$ to the nearest unit.

(C) Graphically determine the maximum profit (to the nearest thousand dollars) and the output (to the nearest unit) that produces the maximum profit. What is the wholesale price of the radio (to the nearest dollar) at this output? [Compare with Problem 41C.]

43. *Construction.* A construction company has 840 feet of chain-link fence that is used to enclose storage areas for equipment and materials at construction sites. The supervisor wants to set up two identical rectangular storage areas sharing a common fence (see the figure):

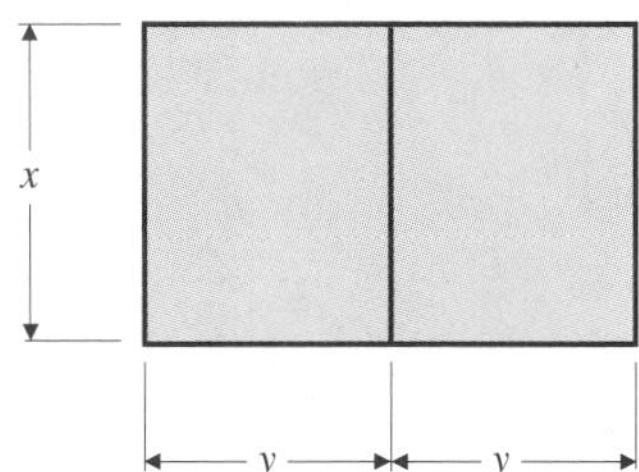

Figure for 43

Assuming all fencing is used:

(A) Express the total area $A(x)$ enclosed by both pens as a function of x.

(B) From physical considerations, what is the domain of the function A?

(C) Graph function A in a rectangular coordinate system.

(D) Use the graph to discuss the number and approximate locations of values of x that would produce storage areas with a combined area of 25,000 square feet.

C (E) Approximate graphically (to the nearest foot) the values of x that would produce storage areas with a combined area of 25,000 square feet.

(F) Algebraically determine the dimensions of the storage areas that have the maximum total combined area. What is the maximum area?

Life Sciences

44. *Air pollution.* On an average summer day in a large city, the pollution index at 8:00 AM is 20 parts per million, and it increases linearly by 15 parts per million each hour until 3:00 PM. Let $P(x)$ be the amount of pollutants in the air x hours after 8:00 AM.

(A) Express $P(x)$ as a linear function of x.

(B) What is the air pollution index at 1:00 PM?

(C) Graph the function P for $0 \leq x \leq 7$.

(D) What is the slope of the graph? (The slope is the amount of increase in pollution for each additional hour of time.)

Social Sciences

45. *Psychology—sensory perception.* One of the oldest studies in psychology concerns the following question: Given a certain level of stimulation (light, sound, weight lifting, electric shock, and so on), how much should the stimulation be increased for a person to notice the difference? In the middle of the nineteenth century, E. H. Weber (a German physiologist) formulated a law that still carries his name: If Δs is the change in stimulus that will just be noticeable at a stimulus level s, then the ratio of Δs to s is a constant:

$$\frac{\Delta s}{s} = k$$

Hence, the amount of change that will be noticed is a linear function of the stimulus level, and we note that the greater the stimulus, the more it takes to notice a difference. In an experiment on weight lifting, the constant k for a given individual was found to be $\frac{1}{30}$.

(A) Find Δs (the difference that is just noticeable) at the 30 pound level. At the 90 pound level.

(B) Graph $\Delta s = s/30$ for $0 \leq s \leq 120$.

(C) What is the slope of the graph?

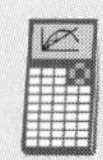

Group Activity 1 Introduction to Regression Analysis

In real-world applications collected data may be assembled in table form and then examined to find a function to model the data. A very powerful mathematical tool called *regression analysis* is frequently used for this purpose. Several of the modeling functions stated in examples and exercises in Chapter 1 were constructed using *regression techniques* and a graphing utility.

Regression analysis is the process of fitting a function to a set of data points. This process is also referred to as **curve fitting.** In this group activity, we will restrict our attention to **linear regression**—that is, to fitting data with linear functions. Other types of regression analysis will be discussed in the next chapter.

At this time, we will not be interested in discussing the underlying mathematical methods used to construct a particular regression equation. Instead, we will concentrate on the mechanics of using a graphing utility to apply regression techniques to data sets.

Table 1
PRICE–DEMAND

x (MILLION)	p (\$)
2	87
5	68
8	53
12	37

In Example 7, Section 1-1, we were given the data in Table 1 and the corresponding modeling function

$$p(x) = 94.8 - 5x \tag{1}$$

where p is the wholesale price of a camera and x is the demand in millions.

Now we want to see how the function p was determined using linear regression on a graphing utility. On most graphing utilities, this process can be broken down into three steps:

Step 1. Enter the data set.

Step 2. Compute the desired regression equation.

Step 3. Graph the data set and the regression equation in the same viewing window.

The details for carrying out each step vary greatly from one graphing utility to another. Consult your manual. If you are using a Texas Instruments graphing calculator or a spreadsheet on a computer, you may also want to consult the supplements for this text (see Preface).

Figure 1 shows the results of applying this process to the data in Table 1 on a graphing calculator, and Figure 2 shows the same process on a spreadsheet.

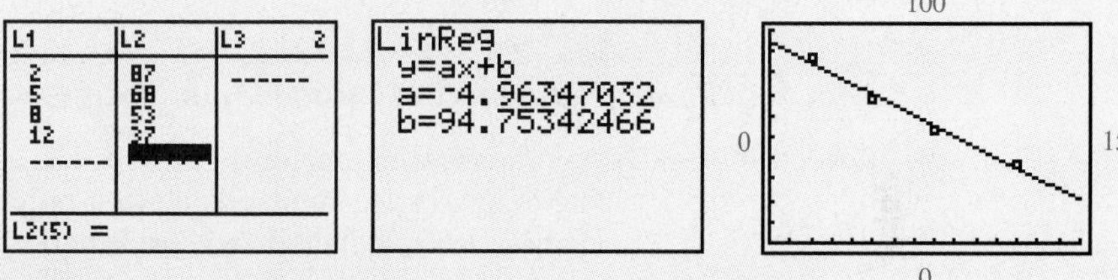

FIGURE 1
Linear regression on a graphing calculator

Price-Demand	
x (millions)	p ($)
2	87
5	68
8	53
12	37
Slope	y intercept
-4.9634703	94.7534247

FIGURE 2
Linear regression on a spreadsheet

Notice how well the line graphed in Figures 1 and 2 appears to fit this set of data. At this time, we will not discuss any mathematical techniques for determining how well a curve fits a given set of data. Instead, we will simply assume that, in some sense, the linear regression line for a given data set is the "best fit" for that data.

Most graphing utilities compute coefficients of regression equations to many more decimal places than we will need for our purposes. Normally, when a regression equation is written, the coefficients are rounded. For example, if we round each coefficient in Figures 1 or 2 to one decimal place, we will obtain the modeling function f we used in Example 7, Section 1-1 [see equation (1)].

(A) If you have not already done so, carry out steps 1–3 for the data in Table 1. Write a brief summary of the details for performing these steps on your graphing utility and keep it for future reference. (Some of the optional exercises in subsequent chapters will require you to compute regression equations.)

(B) Each of the following examples and exercises from Chapter 1 contains a data set and an equation that models the data. In each case, use a graphing utility to compute a linear regression equation for the data. Plot the data and graph the equation in the same viewing rectangle. Discuss the difference between the graphing utility's regression equation and the modeling function stated in the exercise.

- Section 1-1, Matched Problem 7
- Exercise 1-1, Problems 85 and 86
- Exercise 1-3, Problems 73 and 74

Group Activity 2 Mathematical Modeling in Business

A company manufactures and sells mountain bikes. The management would like to have price–demand and cost functions for break-even and profit–loss analysis. Price–demand and cost functions are often established by collecting appropriate data at different levels of output, and then finding a model in the form of a basic elementary function (from our library of elementary functions) that "closely fits" the collected data. The financial department, using statistical techniques, arrived at the price–demand and cost data in Tables 1 and 2, where p is the wholesale price of a bike for a demand of x hundred bikes, $0 \le x \le 220$, and C is the cost (in hundreds of dollars) of producing and selling x bikes, $0 \le x \le 220$.

Table 1
PRICE–DEMAND

x (HUNDREDS)	p ($)
0	525
64	370
125	270
185	130

Table 2
COST

x (HUNDREDS)	C (HUNDRED $)
0	8,470
30	13,510
120	19,140
180	22,580
220	28,490

(A) *Building a Mathematical Model for Price–Demand.* Plot the data in Table 1 and observe that the relationship between p and x appears to be linear.

1. If you have completed Group Activity 1, use your graphing utility to find the linear regression line for the data in Table 1. Graph the line and the data in the same viewing window.
2. The linear regression line found in part 1 is a mathematical model for price–demand and is given by

$$p(x = -2.09x + 519 \quad \text{Price–demand equation}$$

Graph the data points from Table 1 and the price–demand equation in the same rectangular coordinate system.

3. The linear regression line defines a linear function. Interpret the slope of the line. Discuss its domain and range. Using the mathematical model, determine the price for a demand of 10,000 bikes. For a demand of 20,000 bikes.

(B) *Building a Mathematical Model for Cost.* Plot the data in Table 2 in a rectangular coordinate system. Which type of function appears to best fit the data?

C 1. If you have completed Group Activity 1, use your graphing utility to find the linear regression line for the data in Table 2. Graph the line and the data in the same viewing window.

2. The linear regression line found in part 1 is a mathematical model for cost and is given by

$$C(x) = 81x + 9{,}498 \quad \text{Cost equation}$$

Graph the data points from Table 2 and the cost equation in the same rectangular coordinate system.

3. Interpret the slope of the cost equation function. Discuss its domain and range. Using the mathematical model, determine the cost for an output and sales of 10,000 bikes. For an output and sales of 20,000 bikes.

(C) *Break-Even and Profit–Loss Analysis.* Refer to the price–demand equation in part (A) and the cost equation in part (B). Write an equation for the revenue function and state its domain. Write an equation for the profit function and state its domain.

1. Graph the revenue function and the cost function simultaneously in the same rectangular coordinate system. Algebraically determine at what outputs (to the nearest unit) the break-even points occur. Determine the outputs where costs exceed revenues and where revenues exceed costs.

C 2. Graph the revenue function and the cost function simultaneously in the same viewing window. Graphically determine at what outputs (to the nearest unit) break-even occurs, costs exceed revenues, and revenues exceed costs.

3. Graph the profit function in a rectangular coordinate system. Algebraically determine at what outputs (to the nearest unit) the break-even points occur. Determine where profits occur and where losses occur. At what output and price will a maximum profit occur? Does the maximum revenue and maximum profit occur for the same output? Discuss.

C 4. Graph the profit function in a graphing utility. Graphically determine at what outputs (to the nearest unit) break-even occurs, losses occur, and profits occur. At what output and price will a maximum profit occur? Does the maximum revenue and maximum profit occur for the same output? Discuss.

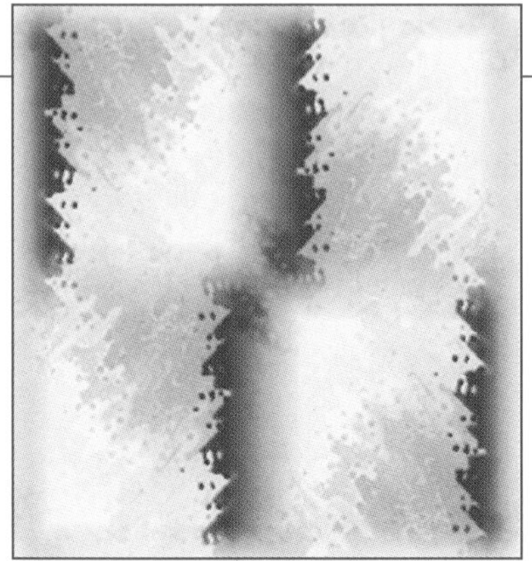

CHAPTER 2

ADDITIONAL ELEMENTARY FUNCTIONS

INTRODUCTION

In this chapter we add the following four general classes of functions to the beginning library of elementary functions started in Chapter 1: polynomial, rational, exponential, and logarithmic. This expanded library of elementary functions should take care of most of your function needs in this and many other courses. The linear and quadratic functions studied in Chapter 1 are special cases of the more general class of functions called *polynomials.*

SECTION 2-1 Polynomial and Rational Functions

- POLYNOMIAL FUNCTIONS
- POLYNOMIAL ROOT APPROXIMATION
- REGRESSION POLYNOMIALS
- RATIONAL FUNCTIONS
- APPLICATION

POLYNOMIAL FUNCTIONS

In Chapter 1 you were introduced to the basic functions

$$f(x) = b \qquad \text{Constant function}$$
$$f(x) = ax + b \qquad a \neq 0 \qquad \text{Linear function}$$
$$f(x) = ax^2 + bx + c \qquad a \neq 0 \qquad \text{Quadratic function}$$

as well as some special cases of

$$f(x) = ax^3 + bx^2 + cx + d \qquad a \neq 0 \qquad \text{Cubic function}$$

Most of the earlier applications we considered, including cost, revenue, profit, loss, and packaging applications, made use of these functions. Notice the evolving pattern going from the constant function to the cubic function—the terms in each equation are of the form ax^n, where n is a nonnegative integer and a is a real number. All these functions are special cases of the general class of functions called *polynomial functions:*

Polynomial Function

A **polynomial function** is a function of the form

$$f(x) = a_n x^n + a_{n-1}x^{n-1} + \cdots + a_1 x + a_0$$

for n a nonnegative integer, called the **degree** of the polynomial. The coefficients $a_0, a_1, \ldots, a_n$ are real numbers with $a_n \neq 0$. The **domain** of a polynomial function is the set of all real numbers.

The shape of the graph of a polynomial function is connected to the degree of the polynomial. The shapes of odd-degree polynomial functions have something in common, and the shapes of even-degree polynomial functions have something in common. Figure 1 shows graphs of representative

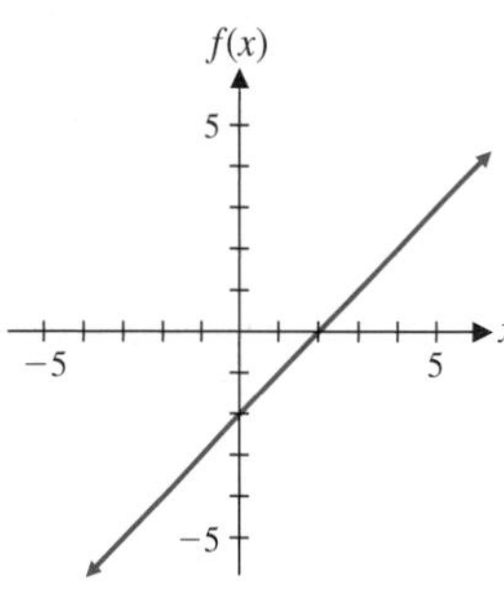

(A) $f(x) = x - 2$

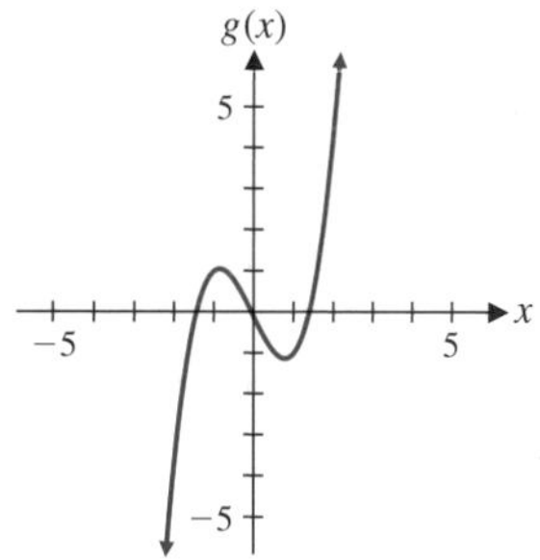

(B) $g(x) = x^3 - 2x$

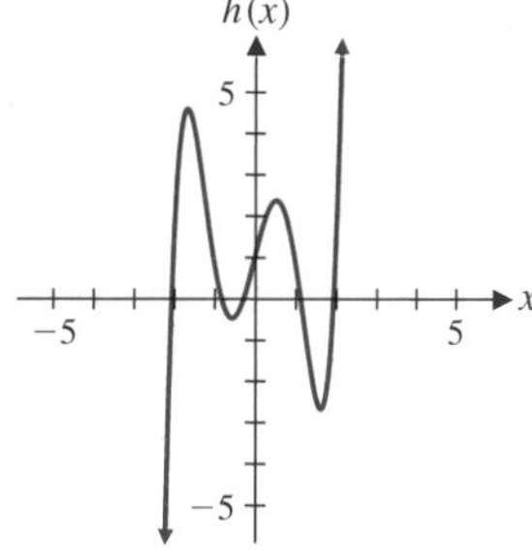

(C) $h(x) = x^5 - 5x^3 + 4x + 1$

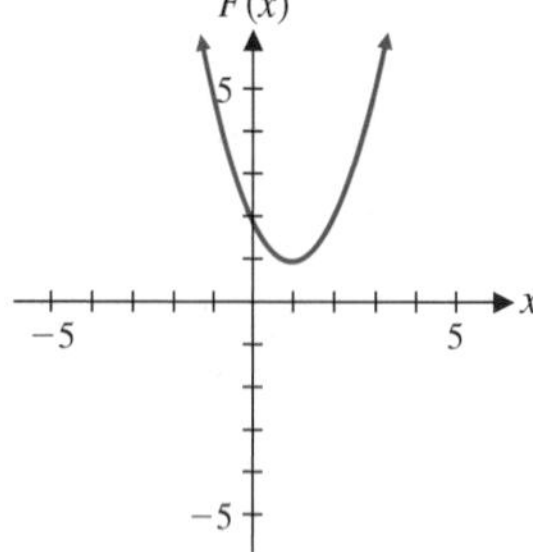

(D) $F(x) = x^2 - 2x + 2$

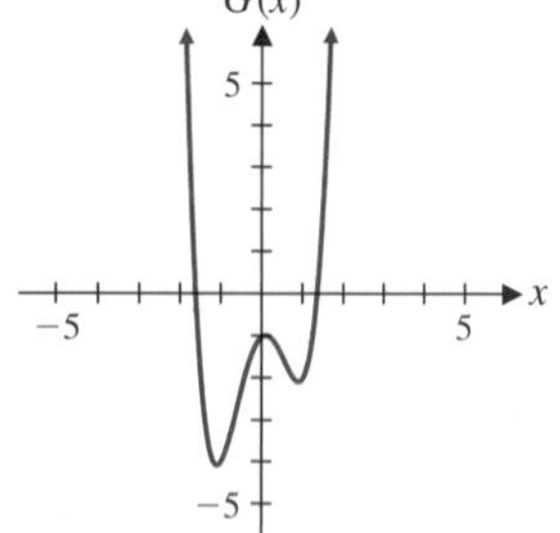

(E) $G(x) = 2x^4 - 4x^2 + x - 1$

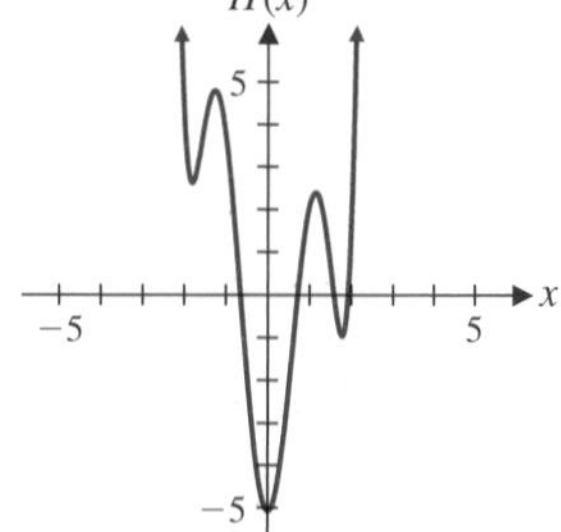

(F) $H(x) = x^6 - 7x^4 + 14x^2 - x - 5$

FIGURE 1
Graphs of polynomial functions

polynomial functions from degrees 1 to 6, and suggests some general properties of graphs of polynomial functions.

Notice that the odd-degree polynomial graphs start negative, end positive, and cross the x axis at least once. The even-degree polynomial graphs start positive, end positive, and may not cross the x axis at all. In all cases in Figure 1, the coefficient of the highest-degree term was chosen positive. If any leading coefficient had been chosen negative, then we would have a similar graph but flipped over.

The graph of a polynomial function is **continuous,** with no holes or breaks. That is, the graph can be drawn without removing a pen from the paper. Also, the graph of a polynomial has no sharp corners. Figure 2 shows the graphs of two functions, one that is not continuous, and the other that is continuous, but with a sharp corner. Neither function is a polynomial.

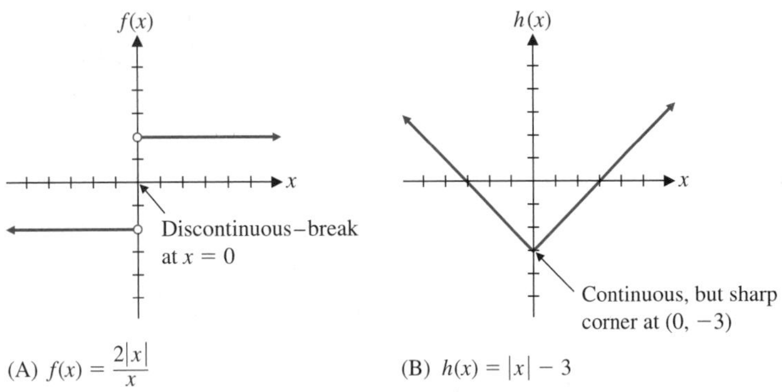

(A) $f(x) = \frac{2|x|}{x}$ (B) $h(x) = |x| - 3$

FIGURE 2
Discontinuous and sharp-corner functions

Figure 1 gives examples of polynomial functions with graphs containing the maximum number of *turning points* possible for a polynomial of that degree. A **turning point** on a continuous graph is a point that separates an increasing portion from a decreasing portion or vice versa. In general, it can be shown that:

The graph of a polynomial function of positive degree n can have at most $(n - 1)$ turning points and can cross the x axis at most n times.

Explore–Discuss 1

(A) What is the least number of turning points an odd-degree polynomial function can have? An even-degree polynomial function?
(B) What is the maximum number of x intercepts the graph of a polynomial function of degree n can have?
(C) What is the maximum number of real solutions an nth-degree polynomial equation can have?
(D) What is the least number of x intercepts the graph of a polynomial function of odd degree can have? Of even degree?
(E) What is the least number of real solutions a polynomial function of odd degree can have? Of even degree?

We now compare the graphs of two polynomial functions relative to points close to the origin and then "zoom out" to compare points distant from the origin. Compare the graphs in Figure 3.

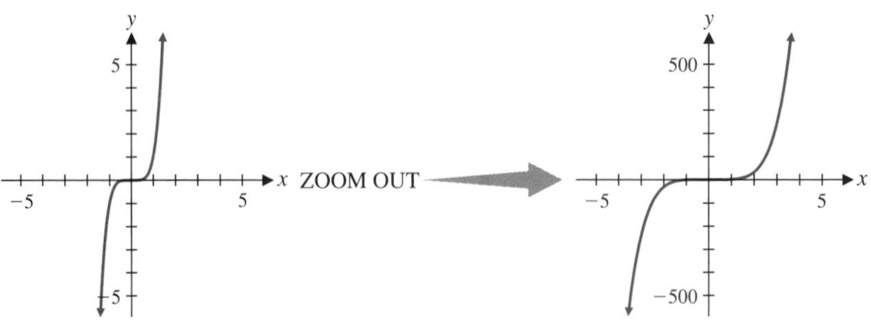

(A) $y = x^5$ (B) $y = x^5$

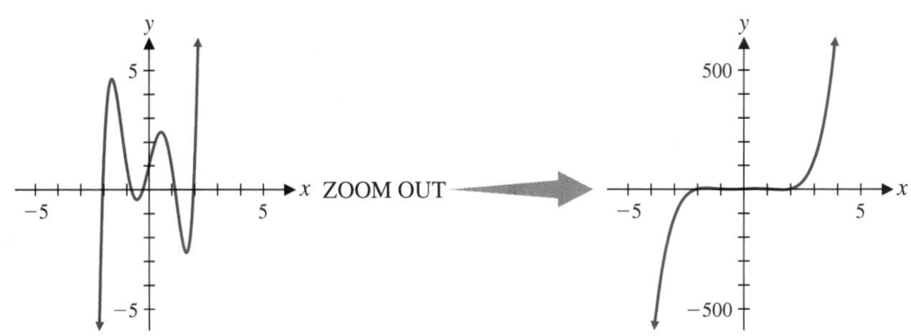

(C) $y = x^5 - 5x^3 + 4x + 1$ (D) $y = x^5 - 5x^3 + 4x + 1$

FIGURE 3
Close and distant comparisons

Figure 3 clearly shows that the highest-degree term in the polynomial dominates all other terms combined in the polynomial. As we "zoom out," the graph of $y = x^5 - 5x^3 + 4x + 1$ looks more and more like the graph of $y = x^5$. This is a general property of polynomial functions.

Explore–Discuss 2

Compare the graphs of $y = x^6$ and $y = x^6 - 7x^4 + 14x^2 - x - 5$ in the following two viewing windows:

(A) $-5 \leq x \leq 5, -5 \leq y \leq 5$
(B) $-5 \leq x \leq 5, -500 \leq y \leq 500$

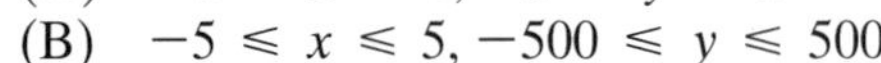

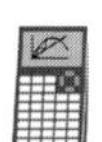

Polynomial Root Approximation

An x intercept of a function f is also called a **zero*** of the function and a **root** of the equation $f(x) = 0$. Approximating the zeros of a function with a graphing utility is a simple matter, provided that we can find a window that shows where the graph of the function touches or crosses the x axis. But how can we be sure that a particular window shows all the zeros? For polynomial

*Only real numbers can be x intercepts. Functions may have roots or zeros that are not real numbers, but these will not appear as x intercepts. Since we have agreed to discuss only functions whose domains are sets of real numbers, we will not consider any complex zeros or roots that ar not real numbers.

functions, Theorem 1 can be used to locate a window that contains all the zeros of the polynomial. Theorem 1 is due to A. L. Cauchy (1789–1857), a French mathematician who has many theorems and concepts associated with his name, and illustrates that classical mathematical results are often useful tools in modern applications.

Theorem 1 ■■ LOCATING ZEROS OF A POLYNOMIAL

If r is a zero of the polynomial

$$P(x) = x^n + a_{n-1}x^{n-1} + a_{n-2}x^{n-2} + \cdots + a_1x + a_0$$

then*

$$|r| < 1 + \max\{|a_{n-1}|, |a_{n-2}|, \ldots, |a_1|, |a_0|\}$$ ■■

In a polynomial, the coefficient of the term containing the highest power of x is often referred to as the **leading coefficient.** Notice that Theorem 1 requires that this leading coefficient must be a 1.

Example 1 **Approximating the Zeros of a Polynomial** Approximate (to two decimal places) the real zeros of

$$P(x) = 2x^4 - 5x^3 - 4x^2 + 3x + 6$$

SOLUTION Since the leading coefficient of $P(x)$ is 2, Theorem 1 cannot be applied to $P(x)$. But it can be applied to

$$\begin{aligned} Q(x) = \tfrac{1}{2}P(x) &= \tfrac{1}{2}(2x^4 - 5x^3 - 4x^2 + 3x + 6) \\ &= x^4 - \tfrac{5}{2}x^3 - 2x^2 + \tfrac{3}{2}x + 3 \end{aligned}$$

Since multiplying a function by a positive constant expands or contracts the graph (see Section 1-2), but does not change the x intercepts, $P(x)$ and $Q(x)$ have the same zeros. Thus, Theorem 1 implies that any zero, r, of $P(x)$ must satisfy

$$|r| < 1 + \max\{|-\tfrac{5}{2}|, |-2|, |\tfrac{3}{2}|, |3|\} = 1 + 3 = 4$$

and all zeros of $P(x)$ must be between -4 and 4. Graphing $P(x)$ (Fig. 4A), we see two real zeros of $P(x)$ and we can be certain that there are no zeros outside this window. Using a built-in approximation routine, the real zeros of $P(x)$ (to two decimal places) are 1.15 (Fig. 4B) and 2.89 (Fig. 4C). Note also that $P(x)$ has three turning points in this window, the maximum allowable for a fourth-degree polynomial. Thus, there can't be any turning points outside this window.

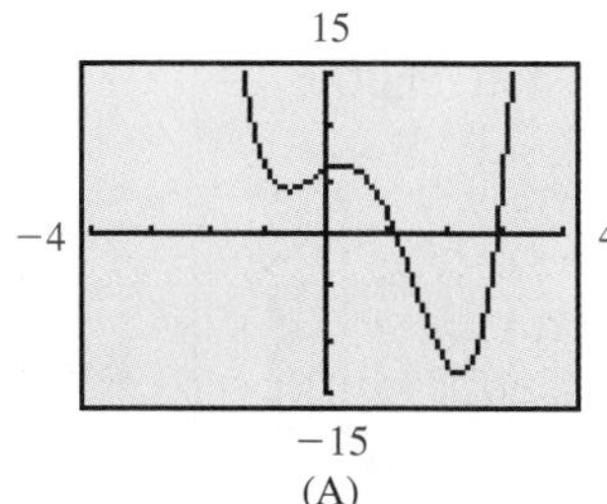

(A)

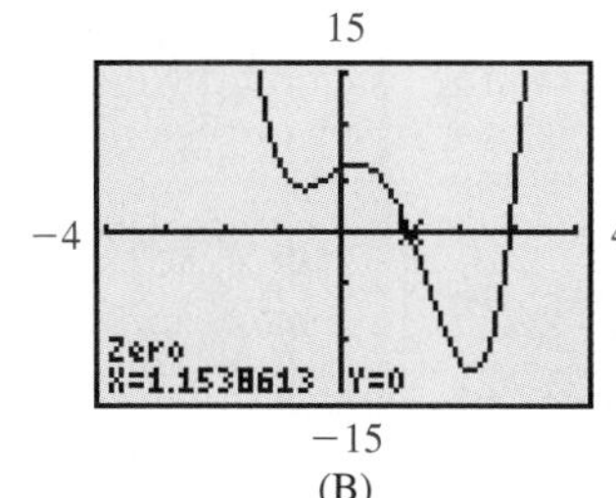

(B)

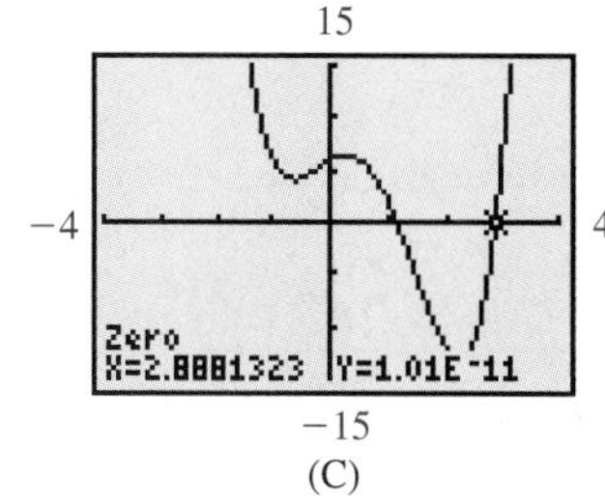

(C)

FIGURE 4 ■

*Note: $\max\{|a_{n-1}|, |a_{n-2}|, \ldots, |a_1|, |a_0|\}$ is the largest number in the list $|a_{n-1}|, |a_{n-2}|, \ldots, |a_0|$

Matched Problem 1 ➡ Approximate (to two decimal places) the real zeros of

$$P(x) = 3x^3 + 12x^2 + 9x + 4$$

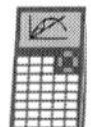

■ Regression Polynomials

In Group Activity 1 at the end of Chapter 1, we saw that regression techniques can be used to fit a straight line to a set of data. Linear functions are not the only ones that can be applied in this manner. Most graphing utilities have the ability to fit a variety of curves to a given set of data. We will discuss polynomial regression models in this section and other types of regression models in later sections.

Example 2 ➡ **Estimating the Weight of a Fish** Using the length of a fish to estimate its weight is of interest to both scientists and sport anglers. The data in Table 1 give the average weights of lake trout for certain lengths. Use the data and regression techniques to find a polynomial model that can be used to estimate the weight of a lake trout for any length. Estimate (to the nearest ounce) the weights of lake trout of lengths 39, 40, 41, 42, and 43 inches, respectively.

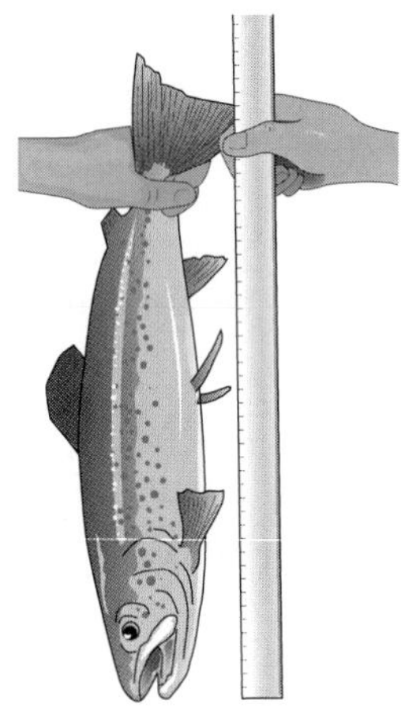

Table 1

Lake Trout

Length (in.)	Weight (oz)	Length (in.)	Weight (oz)
x	y	x	y
10	5	30	152
14	12	34	226
18	26	38	326
22	56	44	536
26	96		

Source: www.thefishernet.com

Solution The graph of the data in Table 1 (Fig. 5A) indicates that a linear regression model would not be appropriate in this case. And, in fact, we would not expect a linear relationship between length and weight. Instead, it is more likely that the weight would be related to the cube of the length. We use a cubic regression polynomial to model the data (Fig. 5B). (Consult your manual for the details of calculating regression polynomials on your graphing utility.) Figure 5C adds the graph of the polynomial model to the graph of the data. The graph in Figure 5C shows that this cubic polynomial does provide a good fit for the data. (We will have more to say about the choice of functions and the accuracy of the fit provided by regression analysis later in the text.) Figure 5D shows the estimated weights for the requested lengths.

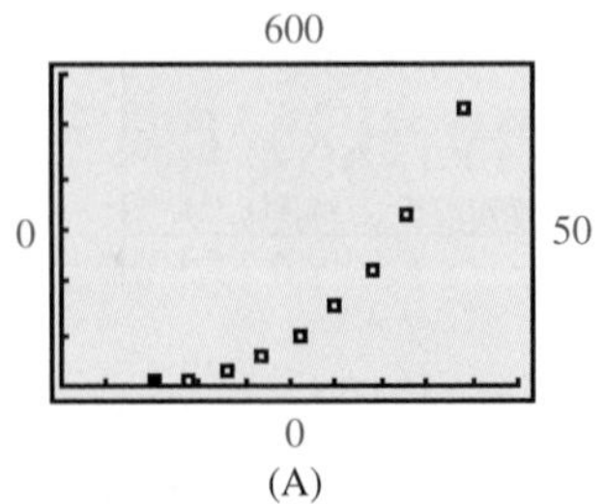

(A)

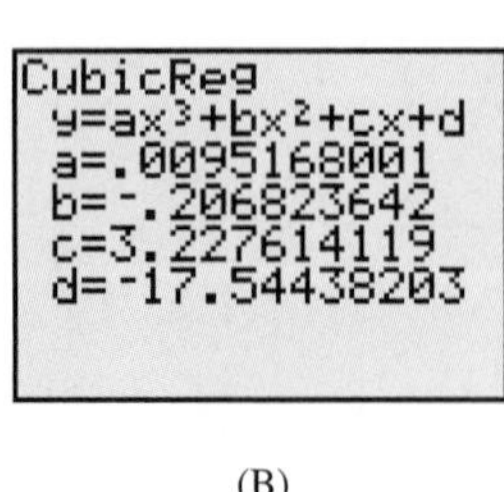

(B)

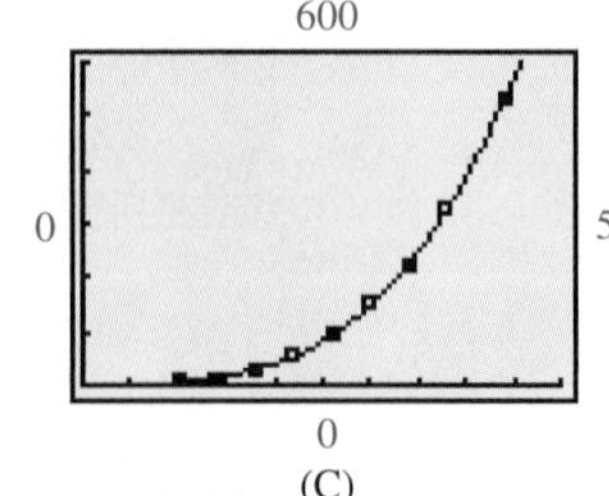

(C)

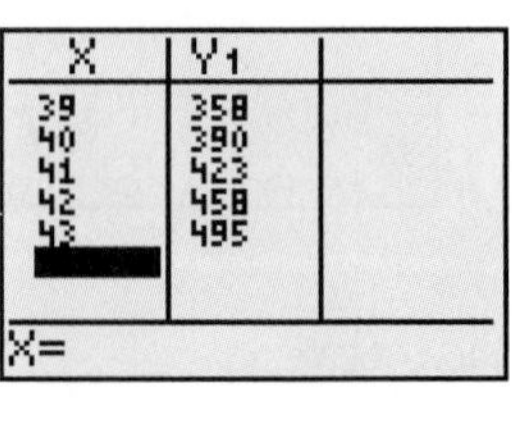

(D)

FIGURE 5

Matched Problem 2 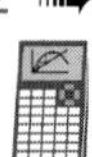 The data in Table 2 give the average weights of pike for certain lengths. Use a cubic regression polynomial to model the data. Estimate (to the nearest ounce) the weights of pike of lengths 39, 40, 41, 42, and 43 inches, respectively.

Table 2

PIKE

LENGTH (IN.)	WEIGHT (OZ)	LENGTH (IN.)	WEIGHT (OZ)
x	y	x	y
10	5	30	108
14	12	34	154
18	26	38	210
22	44	44	326
26	72	52	522

Source: www.thefishernet.com

RATIONAL FUNCTIONS

Just as rational numbers are defined in terms of quotients of integers, *rational functions* are defined in terms of quotients of polynomials. The following equations define rational functions:

$$f(x) = \frac{1}{x} \qquad g(x) = \frac{x-2}{x^2-x-6} \qquad h(x) = \frac{x^3-8}{x}$$

$$p(x) = 3x^2 - 5x \qquad q(x) = 7 \qquad r(x) = 0$$

Rational Function

A **rational function** is any function of the form

$$f(x) = \frac{n(x)}{d(x)} \qquad d(x) \neq 0$$

where $n(x)$ and $d(x)$ are polynomials. The **domain** is the set of all real numbers such that $d(x) \neq 0$. We assume $n(x)/d(x)$ is reduced to lowest terms.

Example 3 **Domain and Intercepts** Find the domain and intercepts for the rational function

$$f(x) = \frac{x-2}{x+1}$$

SOLUTION *Domain:* The denominator is 0 at $x = -1$. Therefore, the domain is the set of all real numbers except -1. The graph of f cannot cross the vertical line $x = -1$.

x intercepts: Find x such that $f(x) = 0$. This happens only if $x - 2 = 0$, that is, at $x = 2$. Thus, 2 is the only x intercept.

y-intercept: The y intercept is

$$f(0) = \frac{0-2}{0+1} = -2$$

Matched Problem 3 Find the domain and intercepts for the rational function: $g(x) = \dfrac{2x}{x - 2}$

In the next example we investigate the graph of $f(x) = (x - 2)/(x + 1)$ near the *point of discontinuity,* $x = -1$, and the behavior of the graph as x increases or decreases without bound. Using this information, we can complete a sketch of the graph of function f with little additional trouble. The investigation uncovers some characteristic features of graphs of rational functions.

Example 4 **Graph of a Rational Function** Given function f from Example 3:

$$f(x) = \frac{x - 2}{x + 1}$$

(A) Investigate the graph of f near the point of discontinuity, $x = -1$.
(B) Investigate the graph of f as x increases or decreases without bound.
(C) Sketch a graph of function f.

SOLUTION (A) *Let x approach −1 from the left:*

x	−2	−1.1	−1.01	−1.001	−1.0001	−1.00001
$f(x)$	4	31	301	3,001	30,001	300,001

We see that $f(x)$ increases without bound as x approaches -1 from the left. Symbolically,

$$f(x) \to \infty \quad \text{as} \quad x \to -1^-$$

Let x approach −1 from the right:

x	0	−0.9	−0.99	−0.999	−0.9999	−0.99999
$f(x)$	−2	−29	−299	−2,999	−29,999	−299,999

We see that $f(x)$ decreases without bound as x approaches -1 from the right. Symbolically,

$$f(x) \to -\infty \quad \text{as} \quad x \to -1^+$$

The vertical line $x = -1$ is called a *vertical asymptote.* The graph of f gets closer to this line as x gets closer to -1. Sketching the vertical asymptote provides an aid to drawing the graph of f near the asymptote (Fig. 6).

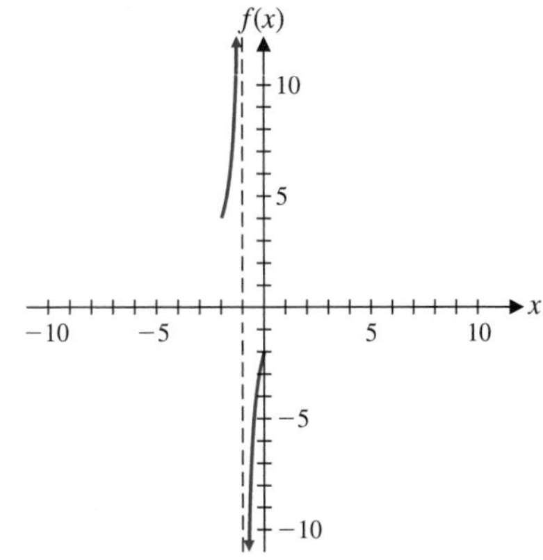

FIGURE 6
Graph near vertical asymptote $x = -1$

(B) Divide each term in the numerator and denominator of $f(x)$ by x, the highest power of x to occur in the numerator and denominator:

$$f(x) = \frac{x - 2}{x + 1} = \frac{1 - \dfrac{2}{x}}{1 + \dfrac{1}{x}}$$

As x increases or decreases without bound, $2/x$ and $1/x$ approach 0 and $f(x)$ gets closer and closer to 1. The horizontal line $y = 1$ is called a *horizontal asymptote.* The graph of $y = f(x)$ gets closer to this line as

x decreases or increases without bound. But how does the graph of $y = f(x)$ approach the horizontal line $y = 1$? Does it approach from above? From below? Or from both? To answer these questions we investigate table values as follows:

Let x approach ∞:

x	10	100	1,000	10,000	100,000
$f(x)$	0.72727	0.97030	0.99700	0.99970	0.99997

The graph of $y = f(x)$ approaches the line $y = 1$ from below as x increases without bound.

Let x approach $-\infty$:

x	−10	−100	−1,000	−10,000	−100,000
$f(x)$	1.33333	1.03030	1.00300	1.00030	1.00003

The graph of $y = f(x)$ approaches the line $y = 1$ from above as x decreases without bound.

Sketching the horizontal asymptote first provides an aid to drawing the graph of f as x moves away from the origin (Fig. 7).

(C) It is now easy to complete the sketch of the graph of f using the intercepts from Example 3 and filling in any points of uncertainty (Fig. 8).

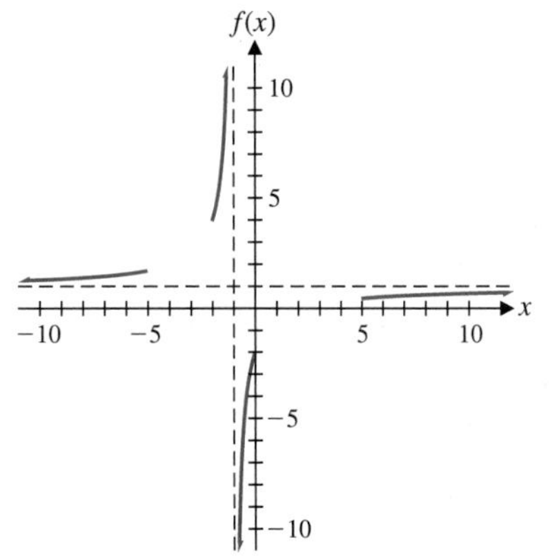

FIGURE 7
Graph of $y = f(x)$ near horizontal asymptote $y = 1$

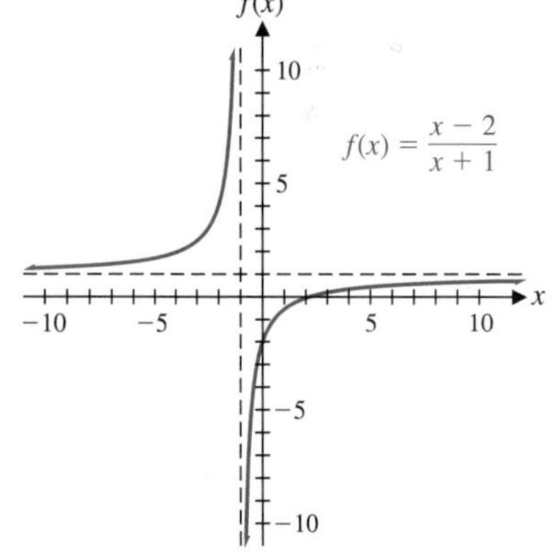

FIGURE 8
Sketch of the rational function f

Matched Problem 4 Given function g from Matched Problem 3: $g(x) = \frac{2x}{x - 2}$

(A) Investigate the graph of g near the point of discontinuity, $x = 2$.
(B) Investigate the graph of g as x increases or decreases without bound.
(C) Sketch a graph of function g.

Graphing rational functions is considerably aided by locating vertical and horizontal asymptotes first, if they exist. The following general procedures are suggested by Example 4:

Vertical and Horizontal Asymptotes and Rational Functions

Given the rational function

$$f(x) = \frac{n(x)}{d(x)}$$

where $n(x)$ and $d(x)$ are polynomials without common factors:

1. If a is a real number such that $d(a) = 0$, then the line $x = a$ is a **vertical asymptote** of the graph of $y = f(x)$.
2. **Horizontal asymptotes,** if any exist, can be found by dividing each term of the numerator $n(x)$ and denominator $d(x)$ by the highest power of x that appears in the numerator and denominator, and then proceeding as in Example 4.

Example 5 **Graphing Rational Functions** Given the rational function:

$$f(x) = \frac{3x}{x^2 - 4}$$

(A) Find intercepts and equations for any vertical and horizontal asymptotes.

(B) Using the information from part (A) and additional points as necessary, sketch a graph of f for $-7 \leq x \leq 7$ and $-7 \leq y \leq 7$.

SOLUTION (A) *x intercepts:* $f(x) = 0$ only if $3x = 0$, or $x = 0$. Thus, the only x intercept is 0.

y intercept:

$$f(0) = \frac{3 \cdot 0}{0^2 - 4} = \frac{0}{-4} = 0$$

Thus, the y intercept is 0.

Vertical asymptotes:

$$f(x) = \frac{3x}{x^2 - 4} = \frac{3x}{(x - 2)(x + 2)}$$

The denominator is 0 at $x = -2$ and $x = 2$; hence, $x = -2$ and $x = 2$ are vertical asymptotes.

Horizontal asymptotes: Divide each term in the numerator and denominator by x^2, the highest power of x in the numerator and denominator:

$$f(x) = \frac{3x}{x^2 - 4} = \frac{\frac{3x}{x^2}}{\frac{x^2}{x^2} - \frac{4}{x^2}} = \frac{\frac{3}{x}}{1 - \frac{4}{x^2}}$$

As x increases or decreases without bound, the numerator tends to 0 and the denominator tends to 1; thus, $f(x)$ tends to 0. The line $y = 0$ is a horizontal asymptote.

(B) Use the information from part (A) and plot additional points as necessary to complete the graph, as shown in Figure 9.

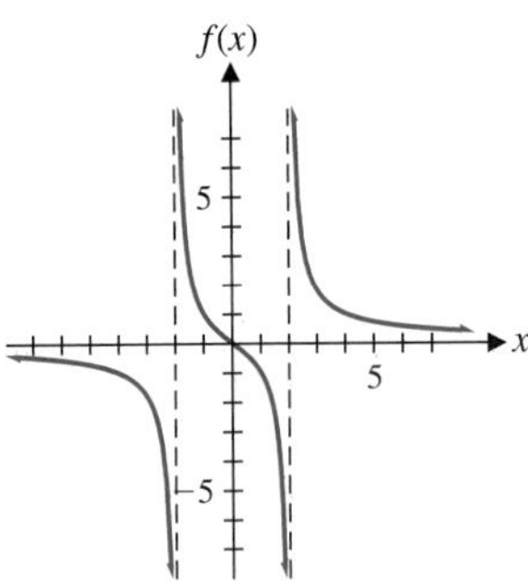

FIGURE 9

Matched Problem 5 Given the rational function: $g(x) = \dfrac{3x + 3}{x^2 - 9}$

(A) Find all intercepts and equations for any vertical and horizontal asymptotes.

(B) Using the information from part (A) and additional points as necessary, sketch a graph of g for $-10 \leq x \leq 10$ and $-10 \leq y \leq 10$.

APPLICATION

Rational functions occur naturally in many types of applications.

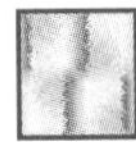

Example 6 **Employee Training** A company that manufactures computers has established that, on the average, a new employee can assemble $N(t)$ components per day after t days of on-the-job training, as given by

$$N(t) = \frac{50t}{t + 4} \qquad t \geq 0$$

Sketch a graph of N, $0 \leq t \leq 100$, including any vertical or horizontal asymptotes. What does $N(t)$ approach as t increases without bound?

SOLUTION *Vertical asymptotes:* None for $t \geq 0$

Horizontal asymptote:

$$N(t) = \frac{50t}{t + 4} = \frac{50}{1 + \dfrac{4}{t}}$$

$N(t)$ approaches 50 as t increases without bound. Thus, $y = 50$ is a horizontal asymptote.

Sketch of graph: The graph is shown in the margin.

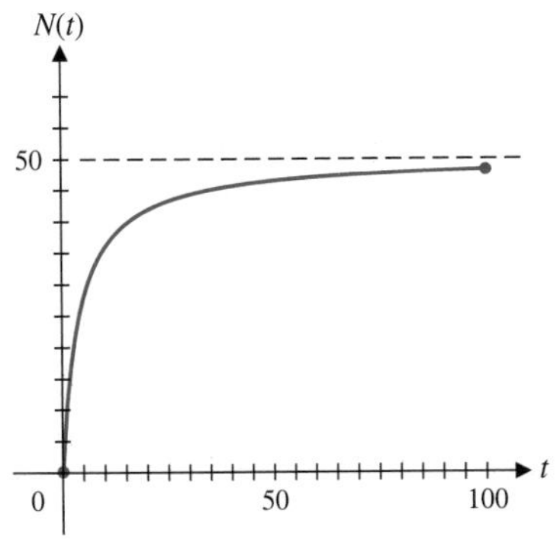

$N(t)$ approaches 50 as t increases without bound. It appears that 50 components per day would be the upper limit that an employee would be expected to assemble.

Matched Problem 6 Repeat Example 6 for: $N(t) = \dfrac{25t + 5}{t + 5} \qquad t \geq 0$

Answers to Matched Problems

1. -3.19

2.

X	Y_1
39	229
40	246
41	264
42	283
43	303

X=

3. Domain: All real numbers except 2; x intercept: 0; y intercept: 0

4. (A) Vertical asymptote: $x = 2$ (B) Horizontal asymptote: $y = 2$

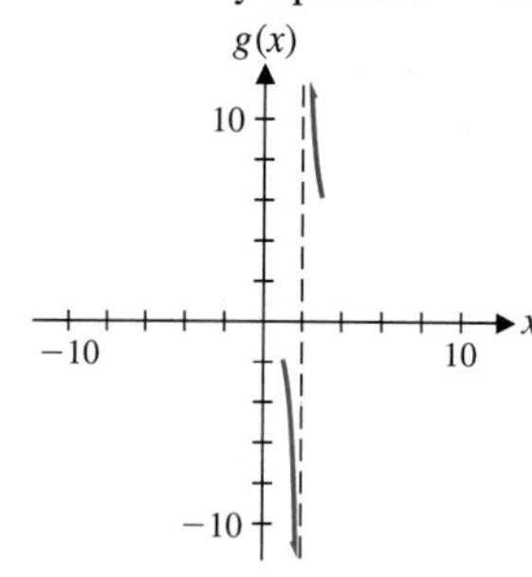

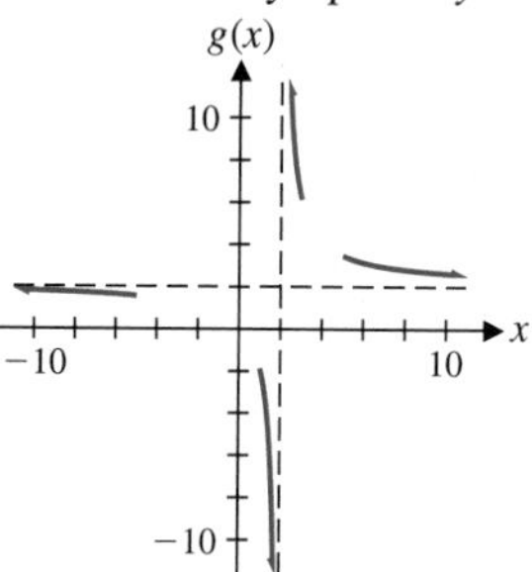

(C)

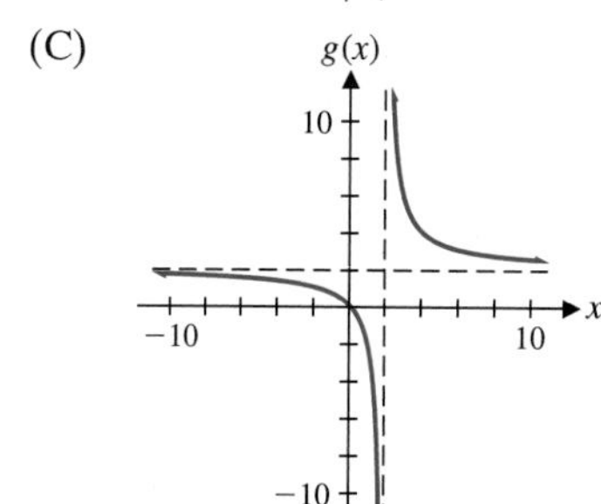

5. (A) x intercept: -1; y intercept; $-\frac{1}{3}$;
Vertical asymptotes: $x = -3$ and $x = 3$;
Horizontal asymptote: $y = 0$

(B)

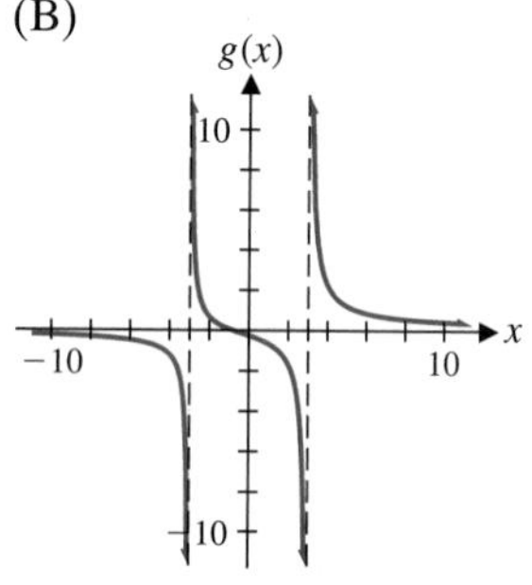

6. No vertical asymptotes for $t \geq 0$; $y = 25$ is a horizontal asymptote. $N(t)$ approaches 25 as t increases without bound. It appears that 25 components per day would be the upper limit that an employee would be expected to assemble.

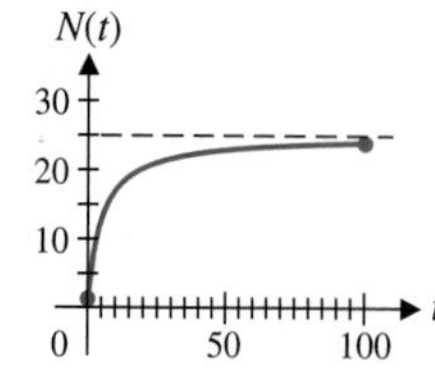

EXERCISE 2-1

A *For each polynomial function in Problems 1–6, find the following:*

(A) Degree of the polynomial
(B) Maximum number of turning points of the graph
(C) Maximum number of x intercepts of the graph
(D) Minimum number of x intercepts of the graph
(E) Maximum number of y intercepts of the graph
(F) Minimum number of y intercepts of the graph

1. $f(x) = ax^2 + bx + c, a \neq 0$

2. $f(x) = ax + b, a \neq 0$

3. $f(x) = ax^5 + bx^4 + cx^3 + dx^2 + ex + f, a \neq 0$

4. $f(x) = ax^4 + bx^3 + cx^2 + dx + e, a \neq 0$

5. $f(x) = ax^6 + bx^5 + cx^4 + dx^3 + ex^2 + fx + g,$ $a \neq 0$

6. $f(x) = ax^3 + bx^2 + cx + d, a \neq 0$

Each graph in Problems 7–14 is the graph of a polynomial function. Answer the following questions for each graph:

(A) How many turning points are on the graph?
(B) What is the minimum degree of a polynomial function that could have the graph?
(C) Is the leading coefficient of the polynomial negative or positive?

7.

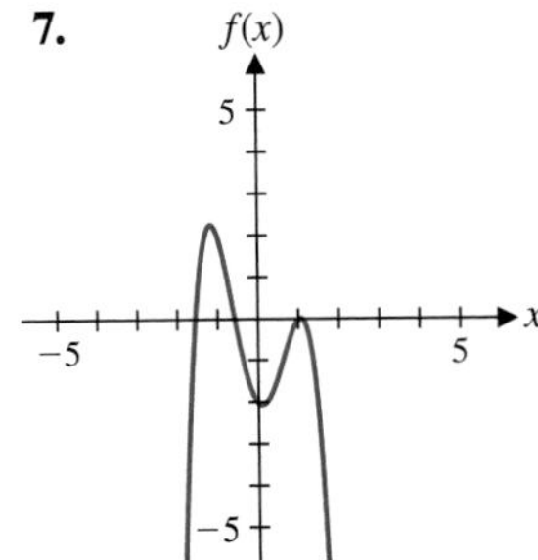

8.

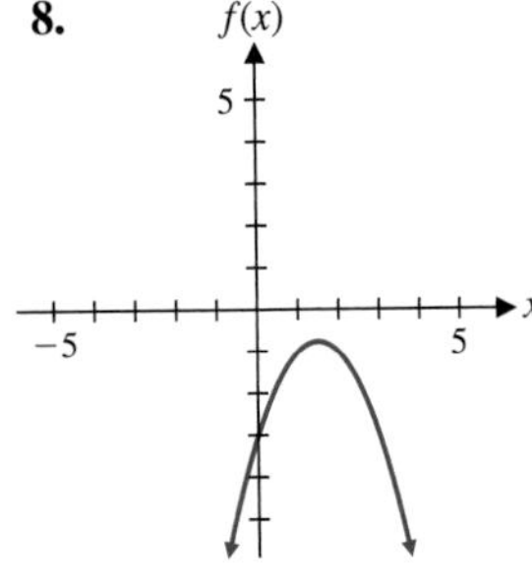

9.

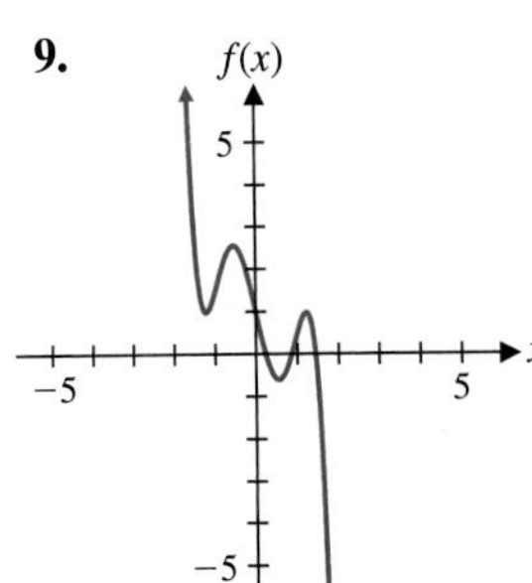

10.

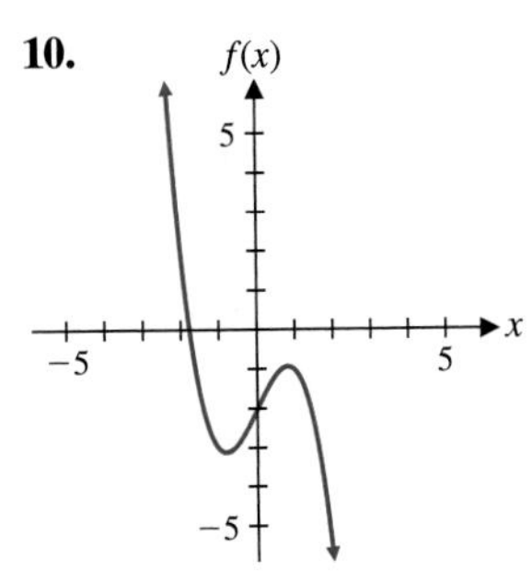

11.

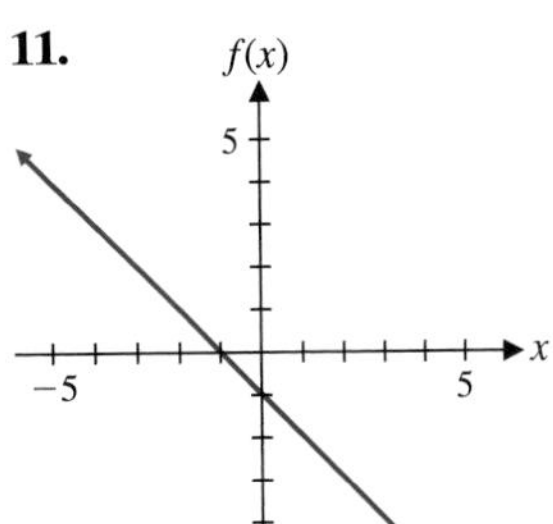

12.

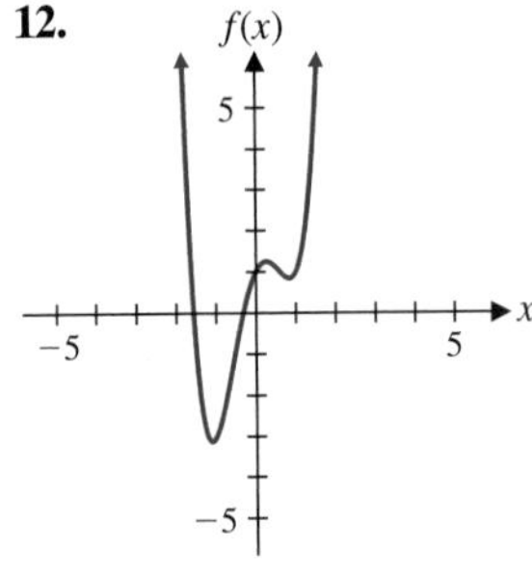

13.

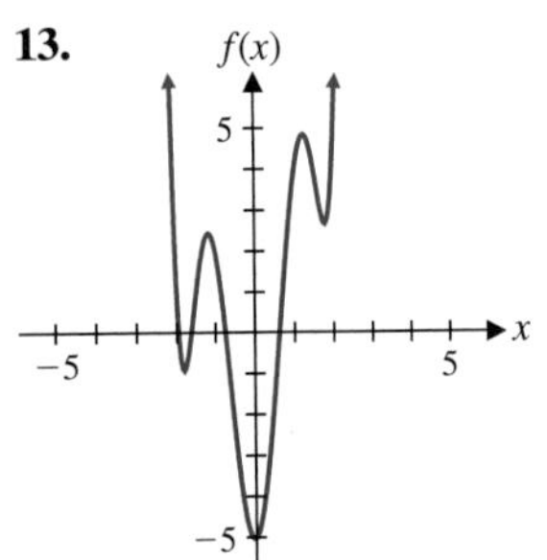

14.

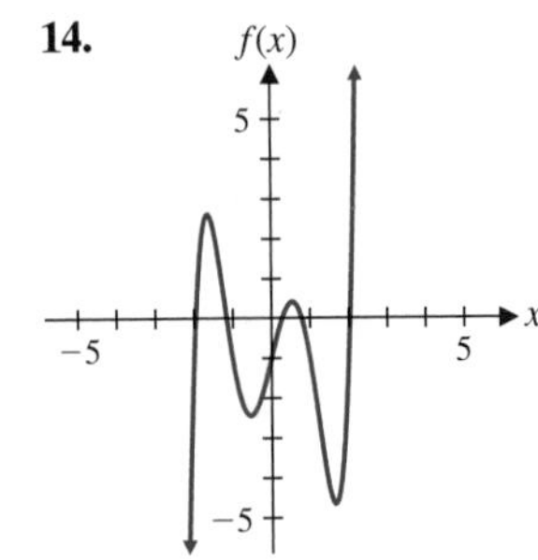

B *For each rational function in Problems 15–20:*

(A) Find the intercepts for the graph.
(B) Determine the domain.
(C) Find any vertical or horizontal asymptotes for the graph.
(D) Sketch any asymptotes as dashed lines. Then sketch a graph of $y = f(x)$ for $-10 \leq x \leq 10$ and $-10 \leq y \leq 10$.
(E) Graph $y = f(x)$ in a standard viewing window using a

C *graphing utility.*

15. $f(x) = \dfrac{x + 2}{x - 2}$

16. $f(x) = \dfrac{x - 3}{x + 3}$

17. $f(x) = \dfrac{3x}{x + 2}$

18. $f(x) = \dfrac{2x}{x - 3}$

19. $f(x) = \dfrac{4 - 2x}{x - 4}$

20. $f(x) = \dfrac{3 - 3x}{x - 2}$

21. How does the graph of $f(x) = 2x^4 - 5x^2 + x + 2$ compare to the graph of $y = 2x^4$ as we "zoom out" (see Fig. 3)?

22. How does the graph of $f(x) = x^3 - 2x + 2$ compare to the graph of $y = x^3$ as we "zoom out"?

23. How does the graph of $f(x) = -x^5 + 4x^3 - 4x + 1$ compare to the graph of $y = -x^5$ as we "zoom out"?

24. How does the graph of $f(x) = -x^5 + 5x^3 + 4x - 1$ compare to the graph of $y = -x^5$ as we "zoom out"?

25. Compare the graph of $y = 2x^4$ to the graph of $y = 2x^4 - 5x^2 + x + 2$ in the following two viewing windows:
(A) $-5 \leq x \leq 5, -5 \leq y \leq 5$
(B) $-5 \leq x \leq 5, -500 \leq y \leq 500$

26. Compare the graph of $y = x^3$ to the graph of $y = x^3 - 2x + 2$ in the following two viewing windows:
(A) $-5 \leq x \leq 5, -5 \leq y \leq 5$
(B) $-5 \leq x \leq 5, -500 \leq y \leq 500$

27. Compare the graph of $y = -x^5$ to the graph of $y = -x^5 + 4x^3 - 4x + 1$ in the following two viewing windows:
(A) $-5 \leq x \leq 5, -5 \leq y \leq 5$
(B) $-5 \leq x \leq 5, -500 \leq y \leq 500$

28. Compare the graph of $y = -x^5$ to the graph of $y = -x^5 + 5x^3 - 5x + 2$ in the following two viewing windows:
(A) $-5 \leq x \leq 5, -5 \leq y \leq 5$
(B) $-5 \leq x \leq 5, -500 \leq y \leq 500$

In Problems 29–34, approximate the real zeros of P(x) to two decimal places.

29. $P(x) = 2x^3 - x^2 - 8x - 7$

30. $P(x) = 3x^3 + 12x^2 + 8x + 5$

31. $P(x) = x^4 + 2x^3 - 3x^2 - 2x + 3$

32. $P(x) = x^4 - 3x^3 - x^2 + 2x + 2$

33. $P(x) = x^5 - 12x^4 + 8x^3 + 11$

34. $P(x) = x^5 + 14x^4 - 12x^2 - 13$

35. Graph the line $y = 0.5x + 3$. Choose any two distinct points on this line and find the linear regression model for the data set consisting of the two points you chose. Experiment with other lines of your choice. Discuss the relationship between a linear regression model for two points and the line that goes through the two points.

36. Graph the parabola $y = x^2 - 5x$. Choose any three distinct points on this parabola and find the quadratic regression model for the data set consisting of the three points you chose. Experiment with other parabolas of your choice. Discuss the relationship between a quadratic regression model for three noncollinear points and the parabola that goes through the three points.

C *For each rational function in Problems 37–42:*
(A) Find any intercepts for the graph.
(B) Find any vertical and horizontal asymptotes for the graph.
(C) Sketch any asymptotes as dashed lines. Then sketch a graph of f for $-10 \leq x \leq 10$ and $-10 \leq y \leq 10$.
(D) Graph the function in a standard viewing window using a graphing utility.

37. $f(x) = \dfrac{2x^2}{x^2 - x - 6}$

38. $f(x) = \dfrac{3x^2}{x^2 + x - 6}$

39. $f(x) = \dfrac{6 - 2x^2}{x^2 - 9}$

40. $f(x) = \dfrac{3 - 3x^2}{x^2 - 4}$

41. $f(x) = \dfrac{-4x}{x^2 + x - 6}$

42. $f(x) = \dfrac{5x}{x^2 + x - 12}$

43. Write an equation for the lowest-degree polynomial function with the graph and intercepts shown in the figure.

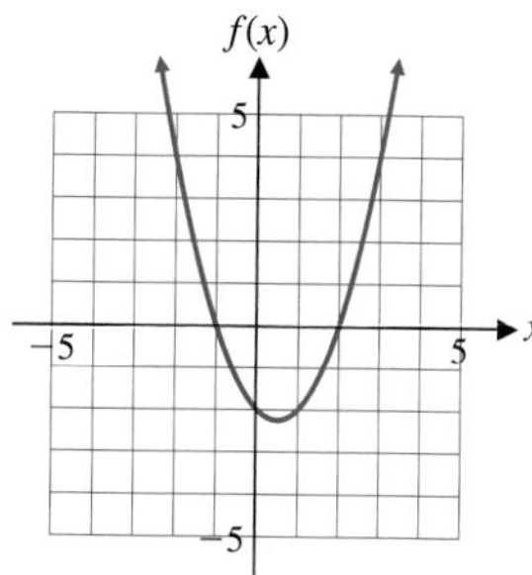

Figure for 43

44. Write an equation for the lowest-degree polynomial function with the graph and intercepts shown in the figure.

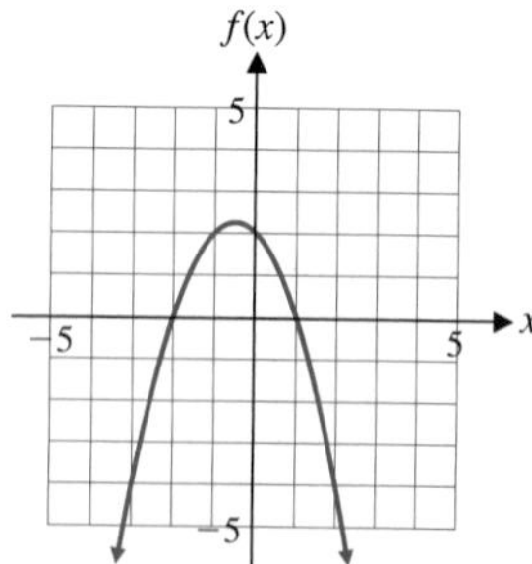

Figure for 44

45. Write an equation for the lowest-degree polynomial function with the graph and intercepts shown in the figure.

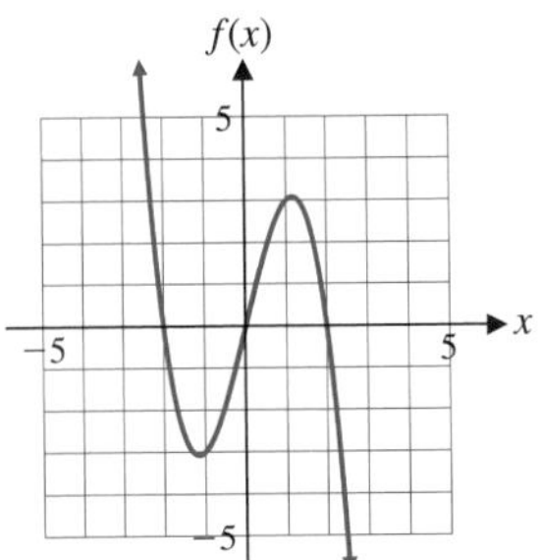

46. Write an equation for the lowest-degree polynomial function with the graph and intercepts shown in the figure.

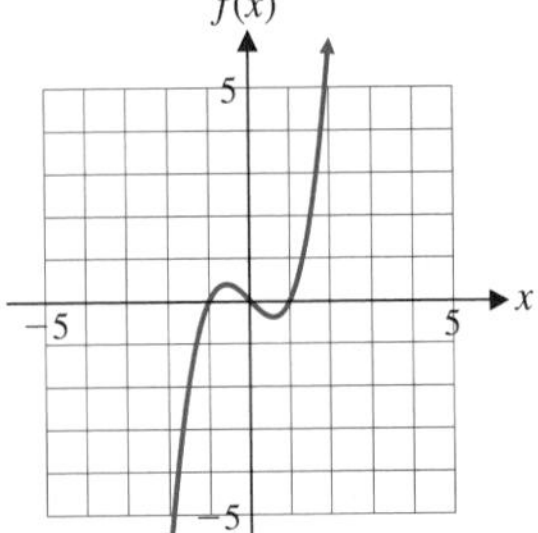

APPLICATIONS

Business & Economics

47. *Average cost.* A company manufacturing snowboards has fixed costs of \$200 per day and total costs of \$3,800 per day at a daily output of 20 boards.

(A) Assuming the total cost per day, $C(x)$, is linearly related to the total output per day, x, write an equation for the cost function.

(B) The average cost per board for an output of x boards is given by $\overline{C}(x) = C(x)/x$. Find the average cost function.

(C) Sketch a graph of the average cost function, including any asymptotes, for $1 \leq x \leq 30$.

(D) What does the average cost per board tend to as production increases?

48. *Average cost.* A company manufacturing surfboards has fixed costs of \$300 per day and total costs of \$5,100 per day at a daily output of 20 boards.

(A) Assuming the total cost per day, $C(x)$, is linearly related to the total output per day, x, write an equation for the cost function.

(B) The average cost per board for an output of x boards is given by $\overline{C}(x) = C(x)/x$. Find the average cost function.

(C) Sketch a graph of the average cost function, including any asymptotes, for $1 \leq x \leq 30$.

(D) What does the average cost per board tend to as production increases?

49. *Replacement time.* An office copier has an initial price of \$2,500. A service contract costs \$200 for the first year and increases \$50 per year thereafter. It can be shown that the total cost of the copier after n years is given by

$$C(n) = 2{,}500 + 175n + 25n^2$$

The average cost per year for n years is $\overline{C}(n) = C(n)/n$.

(A) Find the rational function $\overline{C}$.

(B) Sketch a graph of $\overline{C}$ for $2 \leq n \leq 20$.

(C) When is the average cost per year at a minimum, and what is the minimum average annual cost? [*Hint:* Refer to the sketch in part (B) and evaluate $\overline{C}(n)$ at appropriate integer values until a minimum value is found.] The time when the average cost is minimum is frequently referred to as the **replacement time** for the piece of equipment.

C (D) Graph the average cost function $\overline{C}$ in a graphing utility and use trace and zoom or an appropriate built-in routine to find when the average annual cost is at a minimum.

50. *Minimum average cost.* Financial analysts in a company that manufactures audio CD players arrived at the following daily cost equation for manufacturing x CD players per day:

$$C(x) = x^2 + 2x + 2{,}000$$

The average cost per unit at a production level of x players per day is $\overline{C}(x) = C(x)/x$.

(A) Find the rational function $\overline{C}$.

(B) Sketch a graph of $\overline{C}$ for $5 \leq x \leq 150$.

(C) For what daily production level (to the nearest integer) is the average cost per unit at a minimum, and what is the minimum average cost per player (to the nearest cent)? [*Hint:* Refer to the sketch in part (B) and evaluate $\overline{C}(x)$ at appropriate integer values until a minimum value is found.]

C (D) Graph the average cost function $\overline{C}$ in a graphing utility and use trace and zoom or an appropriate built-in routine to find the daily production level (to the nearest integer) at which the average cost per player is at a minimum. What is the minimum average cost to the nearest cent?

51. *Minimum average cost.* A consulting firm, using statistical methods, provided a veterinary clinic with the cost equation

C

$$C(x) = 0.00048(x - 500)^3 + 60{,}000 \qquad 100 \le x \le 1{,}000$$

where $C(x)$ is the cost in dollars for handling x cases per month. The average cost per case is given by $\overline{C}(x) = C(x)/x$.

(A) Write the equation for the average cost function $\overline{C}$.

(B) Graph $\overline{C}$ on a graphing utility.

(C) Use trace and zoom or an appropriate built-in routine to find the monthly caseload for the minimum average cost per case. What is the minimum average cost per case?

52. *Minimum average cost.* The financial department of a hospital, using statistical methods, arrived at the cost equation

C

$$C(x) = 20x^3 - 360x^2 + 2{,}300x - 1{,}000 \qquad 1 \le x \le 12$$

where $C(x)$ is the cost in thousands of dollars for handling x thousand cases per month. The average cost per case is given by $\overline{C}(x) = C(x)/x$.

(A) Write the equation for the average cost function $\overline{C}$.

(B) Graph $\overline{C}$ on a graphing utility.

(C) Use trace and zoom or an appropriate built-in routine to find the monthly caseload for the minimum average cost per case. What is the minimum average cost per case to the nearest dollar?

*Problems 53 and 54 involve the following concepts: In a free competitive market, the price of a product is determined by the relationship between supply and demand. If $p = D(x)$ and $p = S(x)$ are the price–demand and price–supply equations, respectively, for a product and if $(\overline{x}, \overline{p})$ is the point of intersection of these equations, then $(\overline{x}, \overline{p})$ is the **equilibrium point,** $\overline{p}$ is the **equilibrium price,** and $\overline{x}$ is the **equilibrium quantity.** The equilibrium price is the price at which the supply and the demand are the same.*

53. *Equilibrium point.* A particular CD is sold through a chain of stores in a city. A marketing company has established price–demand and price–supply tables for selected prices for this CD (Tables 3 and 4), where x is the daily number of CD's people are willing to buy and the store is willing to sell at a price of p dollars per CD.

C

Table 3

PRICE–DEMAND

x	$p = D(x)$ ($)
25	19.50
100	14.25
175	10.00
250	8.25

Table 4

PRICE–SUPPLY

x	$p = S(x)$ ($)
25	2.10
100	3.80
175	8.50
250	15.70

(A) Use a linear regression equation to model the data in Table 3 and a quadratic regression equation to model the data in Table 4.

(B) Find the point of intersection of the two equations from part (A). Write the equilibrium price to the nearest cent and the equilibrium quantity to the nearest unit.

54. *Equilibrium point.* Repeat Problem 53 with the price–demand and price–supply tables below, except use a quadratic regression model for the data in Table 5 and a linear regression model for the data in Table 6.

C

Table 5

PRICE–DEMAND

x	$p = D(x)$ ($)
0	24
40	23
65	20
115	11

Table 6

PRICE–SUPPLY

x	$p = S(x)$ ($)
0	5
40	10
65	12
115	16

Life Sciences

55. *Health care.* Table 7 shows the total national expenditures (in billion dollars) and the per capita expenditures (in dollars) for selected years since 1960.

C

(A) Let x represent the number of years since 1960 and find a cubic regression polynomial for the total national expenditures.

(B) Use the polynomial model from part (A) to estimate the total national expenditures (to the nearest tenth of a billion) for 1995.

Table 7

NATIONAL HEALTH EXPENDITURES

DATE	TOTAL EXPENDITURES (BILLION $)	PER CAPITA EXPENDITURES ($)
1960	27.1	143
1965	41.6	204
1970	74.4	346
1975	132.9	592
1980	247.2	1,002
1985	428.2	1,666
1990	697.5	2,588

Source: U.S. Bureau of the Census

56. *Health care.* Refer to Table 7.

C

(A) Let x represent the number of years since 1960 and find a cubic regression polynomial for the per capita expenditures.

(B) Use the polynomial model from part (A) to estimate the per capita expenditures (to the nearest dollar) for 1995.

57. *Physiology.* In a study on the speed of muscle contraction in frogs under various loads, researchers W. O. Fems and J. Marsh found that the speed of contraction decreases with increasing loads. In particular, they found that the relationship between speed of contraction v (in centimeters per second) and load x (in grams) is given approximately by

$$v(x) = \frac{26 + 0.06x}{x} \qquad x \geqslant 5$$

(A) What does $v(x)$ approach as x increases?
(B) Sketch a graph of function v.

Social Sciences

58. *Learning theory.* In 1917, L. L. Thurstone, a pioneer in quantitative learning theory, proposed the rational function

$$f(x) = \frac{a(x + c)}{(x + c) + b}$$

to model the number of successful acts per unit time that a person could accomplish after x practice sessions. Suppose that for a particular person enrolled in a typing class,

$$f(x) = \frac{55(x + 1)}{(x + 8)} \qquad x \geqslant 0$$

where $f(x)$ is the number of words per minute the person is able to type after x weeks of lessons.

(A) What does $f(x)$ approach as x increases?
(B) Sketch a graph of function f, including any vertical or horizontal asymptotes.

59. C *Marriage.* Table 8 shows the marriage and divorce rates per 1,000 population for selected years since 1950.
(A) Let x represent the number of years since 1950 and find a cubic regression polynomial for the marriage rate.
(B) Use the polynomial model from part (A) to estimate the marriage rate (to one decimal place) for 1995.

Table 8

Marriages and Divorces (per 1,000 Population)

Date	Marriages	Divorces
1950	11.1	2.6
1955	9.3	2.3
1960	8.5	2.2
1965	9.3	2.5
1970	10.6	3.5
1975	10.0	4.8
1980	10.6	5.2
1985	10.1	5.0
1990	9.8	4.7

Source: U.S. Bureau of the Census

60. C *Divorce.* Refer to Table 8.
(A) Let x represent the number of years since 1950 and find a cubic regression polynomial for the divorce rate.
(B) Use the polynomial model from part (A) to estimate the divorce rate (to one decimal place) for 1995.

Section 2-2 Exponential Functions

- Exponential Functions
- Base e Exponential Functions
- Growth and Decay Applications
- Compound Interest
- Continuous Compound Interest

This section introduces the important class of functions called *exponential functions.* These functions are used extensively in modeling and solving a wide variety of real-world problems, including growth of money at compound interest; growth of populations of people, animals, and bacteria; radioactive decay; and learning associated with the mastery of such devices as a new computer or an assembly process in a manufacturing plant.

Exponential Functions

We start by noting that

$$f(x) = 2^x \qquad \text{and} \qquad g(x) = x^2$$

are not the same function. Whether a variable appears as an exponent with a constant base or as a base with a constant exponent, makes a big difference. The function g is a quadratic function, which we have already discussed. The function f is a new type of function called an *exponential function.* In general:

Exponential Function

The equation

$$f(x) = b^x \qquad b > 0, b \neq 1$$

defines an **exponential function** for each different constant b, called the **base.** The **domain** of f is the set of all real numbers, and the **range** of f is the set of all positive real numbers.

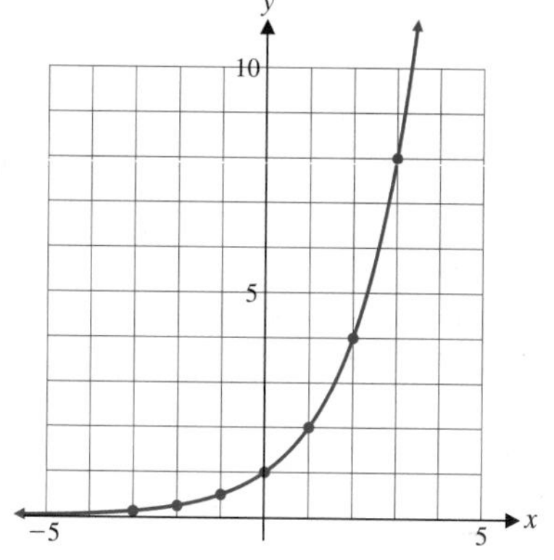

FIGURE 1
$y = 2^x$

We require the base b to be positive to avoid imaginary numbers such as $(-2)^{1/2} = \sqrt{-2} = i\sqrt{2}$. We exclude $b = 1$ as a base, since $f(x) = 1^x = 1$ is a constant function, which we have already considered.

Many students, if asked to hand sketch graphs of equations such as $y = 2^x$ or $y = 2^{-x}$, would not hesitate at all. [*Note:* $2^{-x} = 1/2^x = (1/2)^x$.] They would likely make up tables by assigning integers to x, plot the resulting points, and then join these points with a smooth curve as in Figure 1. The only catch is that we have not defined 2^x for all real numbers. From Appendix A-7, we know what $2^5, 2^{-3}, 2^{2/3}, 2^{-3/5}, 2^{1.4}$, and $2^{-3.14}$ mean (that is, 2^p, where p is a rational number), but what does

$$2^{\sqrt{2}}$$

mean? The question is not easy to answer at this time. In fact, a precise definition of $2^{\sqrt{2}}$ must wait for more advanced courses, where it is shown that

$$2^x$$

names a positive real number for x any real number, and that the graph of $y = 2^x$ is indeed as indicated in Figure 1.

It is useful to compare the graphs of $y = 2^x$ and $y = 2^{-x}$ by plotting both on the same set of coordinate axes, as shown in Figure 2A. The graph of

$$f(x) = b^x \qquad b > 1 \text{ (Fig. 2B)}$$

looks very much like the graph of $y = 2^x$, and the graph of

$$f(x) = b^x \qquad 0 < b < 1 \text{ (Fig. 2B)}$$

looks very much like the graph of $y = 2^{-x}$. Note that in both cases the x axis is a horizontal asymptote for the graphs.

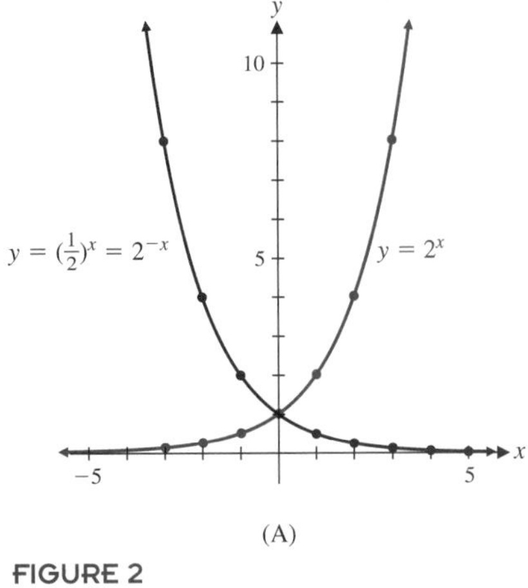

(A)

y
$y = b^x$
$0 < b < 1$
$y = b^x$
$b > 1$
x
Domain = $(-\infty, \infty)$
Range = $(0, \infty)$

(B)

FIGURE 2
Exponential functions

The graphs in Figure 2 suggest the following important general properties of exponential functions, which we state without proof:

> **Basic Properties of the Graph of $f(x) = b^x, b > 0, b \neq 1$**
>
> 1. All graphs will pass through the point (0, 1). *$b^0 = 1$ for any permissible base b.*
> 2. All graphs are continuous curves, with no holes or jumps.
> 3. The x axis is a horizontal asymptote.
> 4. If $b > 1$, then b^x increases as x increases.
> 5. If $0 < b < 1$, then b^x decreases as x increases.

The use of a calculator with the key $\boxed{y^x}$, or its equivalent, makes the graphing of exponential functions almost routine. Example 1 illustrates the process.

Example 1 **Graphing Exponential Functions** Sketch a graph of $y = (\frac{1}{2})4^x$, $-2 \leq x \leq 2$.

SOLUTION Use a calculator to create the table of values shown. Plot these points, and then join them with a smooth curve as in Figure 3.

x	y
−2	0.031
−1	0.125
0	0.50
1	2.00
2	8.00

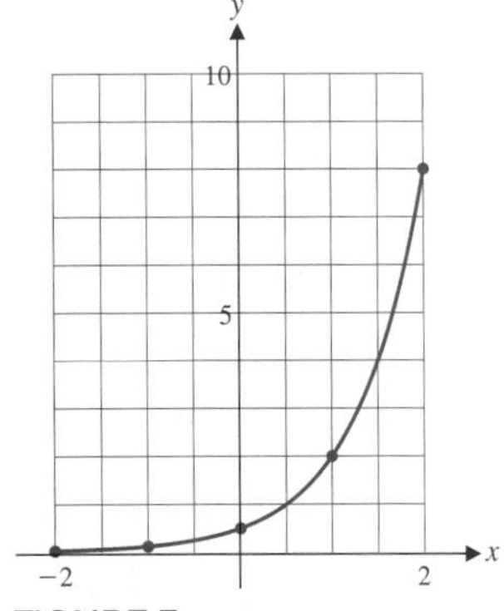

FIGURE 3
Graph of $y = (\frac{1}{2})4^x$

Matched Problem 1 ➡ Sketch a graph of $y = (\frac{1}{2})4^{-x}, -2 \leq x \leq 2$.

Explore–Discuss 1

Graph the functions $f(x) = 2^x$ and $g(x) = 3^x$ on the same set of coordinate axes. At which values of x do the graphs intersect? For which values of x is the graph of f above the graph of g? Below the graph of g? Are the graphs close together as x increases without bound? Are the graphs close together as x decreases without bound? Discuss.

Exponential functions, whose domains include irrational numbers, obey the familiar laws of exponents discussed in Appendix A-7 for rational exponents. We summarize these exponent laws here and add two other important and useful properties.

Exponential Function Properties

For a and b positive, $a \neq 1, b \neq 1$, and x and y real:

1. Exponent laws:

$$a^x a^y = a^{x+y} \qquad \frac{a^x}{a^y} = a^{x-y} \qquad \frac{4^{2y}}{4^{5y}} = \boxed{4^{2y-5y}} = 4^{-3y}$$

$$(a^x)^y = a^{xy} \qquad (ab)^x = a^x b^x \qquad \left(\frac{a}{b}\right)^x = \frac{a^x}{b^x}$$

2. $a^x = a^y$ if and only if $x = y$ — If $7^{5t+1} = 7^{3t-3}$, then $5t + 1 = 3t - 3$, and $t = -2$.

3. For $x \neq 0$,

   $a^x = b^x$ if and only if $a = b$ — If $a^5 = 2^5$, then $a = 2$.

■ Base e Exponential Functions

Of all the possible bases b we can use for the exponential function $y = b^x$, which ones are the most useful? If you look at the keys on a calculator, you will likely see $\boxed{10^x}$ and $\boxed{e^x}$. It is clear why base 10 would be important, because our number system is a base 10 system. But what is e, and why is it included as a base? It turns out that base e is used more frequently than all other bases combined. The reason for this is that certain formulas and the results of certain processes found in calculus and more advanced mathematics take on their simplest form if this base is used. This is why you will see e used extensively in expressions and formulas that model real-world phenomena. In fact, its use is so prevalent that you will often hear people refer to $y = e^x$ as *the* exponential function.

The base e is an irrational number, and like π, it cannot be represented exactly by any finite decimal fraction. However, e can be approximated as closely as we like by evaluating the expression

$$\left(1 + \frac{1}{x}\right)^x \tag{1}$$

Table 1

x	$\left(1 + \frac{1}{x}\right)^x$
1	2
10	2.593 74...
100	2.704 81...
1,000	2.716 92...
10,000	2.718 14...
100,000	2.718 27...
1,000,000	2.718 28...
⋮	⋮

for sufficiently large x. What happens to the value of expression (1) as x increases without bound? Think about this for a moment before proceeding. Maybe you guessed that the value approaches 1, because

$$1 + \frac{1}{x}$$

approaches 1, and 1 raised to any power is 1. Let us see if this reasoning is correct by actually calculating the value of the expression for larger and larger values of x. Table 1 summarizes the results.

Interestingly, the value of expression (1) is never close to 1, but seems to be approaching a number close to 2.7183. In fact, as x increases without bound, the value of expression (1) approaches an irrational number that we call ***e***. The irrational number e to twelve decimal places is

$$e = \mathbf{2.718\ 281\ 828\ 459}$$

Compare this value of e with the value of e^1 from a calculator. Exactly who discovered the constant e is still being debated. It is named after the great Swiss mathematician Leonhard Euler (1707–1783).

Exponential Function with Base *e*

Exponential functions with base e and base $1/e$, respectively, are defined by

$y = e^x$ and $y = e^{-x}$
Domain: $(-\infty, \infty)$
Range: $(0, \infty)$

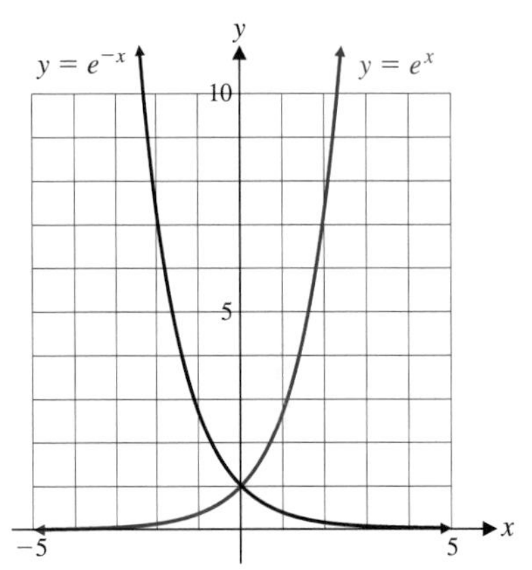

Explore–Discuss 2

Graph the functions $f(x) = e^x$, $g(x) = 2^x$, and $h(x) = 3^x$ on the same set of coordinate axes. At which values of x do the graphs intersect? For positive values of x, which of the three graphs lies above the other two? Below the other two? How does your answer change for negative values of x?

Growth and Decay Applications

Most exponential growth and decay problems are modeled using base e exponential functions. We present two applications here and many more in Exercise 2-2.

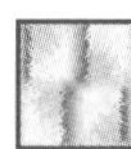

Example 2 **Exponential Growth** Cholera, an intestinal disease, is caused by a cholera bacterium that multiplies exponentially by cell division as given approximately by

$$N = N_0 e^{1.386t}$$

where N is the number of bacteria present after t hours and N_0 is the number of bacteria present at the start ($t = 0$). If we start with 25 bacteria, how many bacteria (to the nearest unit) will be present:

(A) In 0.6 hour? (B) In 3.5 hours?

SOLUTION Substituting $N_0 = 25$ into the above equation, we obtain

$$N = 25e^{1.386t}$$ The graph is shown in Figure 4.

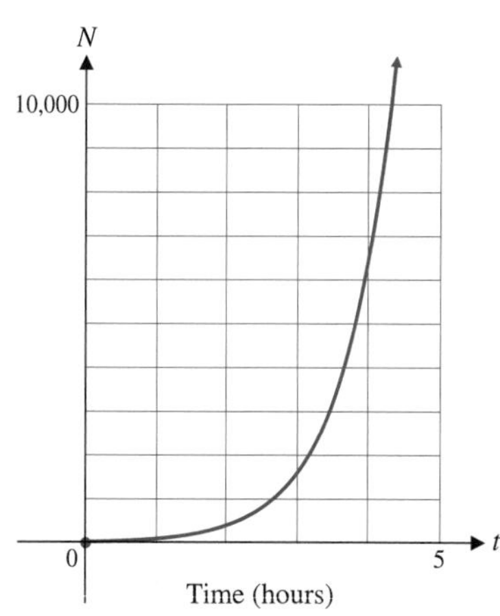

FIGURE 4

(A) Solve for N when $t = 0.6$:

$$N = 25e^{1.386(0.6)}$$ Use a calculator.
$$= 57 \text{ bacteria}$$

(B) Solve for N when $t = 3.5$:

$$N = 25e^{1.386(3.5)}$$ Use a calculator.
$$= 3{,}197 \text{ bacteria}$$

Matched Problem 2 Refer to the exponential growth model for cholera in Example 2. If we start with 55 bacteria, how many bacteria (to the nearest unit) will be present:

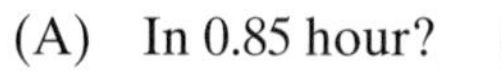

(A) In 0.85 hour? (B) In 7.25 hours?

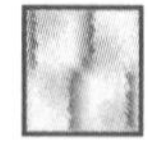

Example 3 **Exponential Decay** Cosmic-ray bombardment of the atmosphere produces neutrons, which in turn react with nitrogen to produce radioactive carbon-14 (^{14}C). Radioactive ^{14}C enters all living tissues through carbon dioxide, which is first absorbed by plants. As long as a plant or animal is alive, ^{14}C is maintained in the living organism at a constant level. Once the organism dies, however, ^{14}C decays according to the equation

$$A = A_0 e^{-0.000124t}$$

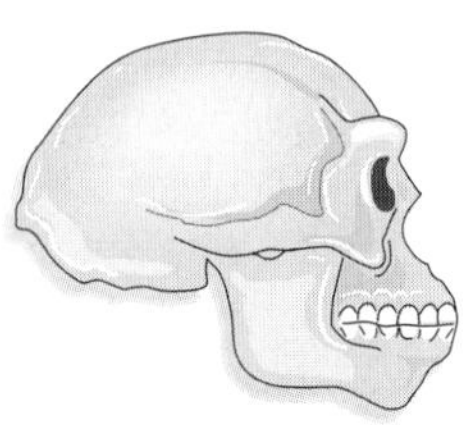

where A is the amount present after t years and A_0 is the amount present at time $t = 0$. If 500 milligrams of ^{14}C are present in a sample from a skull at the time of death, how many milligrams will be present in the sample in:

(A) 15,000 years? (B) 45,000 years?

Compute answers to two decimal places.

Solution Substituting $A_0 = 500$ in the decay equation, we have

$$A = 500e^{-0.000124t}$$ See the graph in Figure 5.

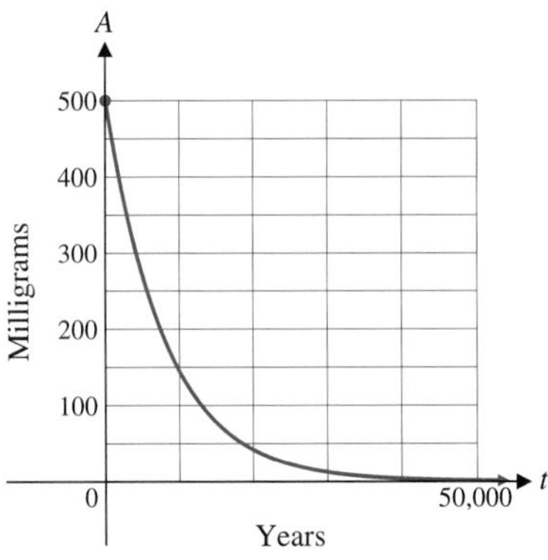

FIGURE 5

(A) Solve for A when $t = 15{,}000$:

$$A = 500e^{-0.000124(15{,}000)}$$ Use a calculator.
$$= 77.84 \text{ milligrams}$$

(B) Solve for A when $t = 45{,}000$:

$$A = 500e^{-0.000124(45{,}000)}$$ Use a calculator.
$$= 1.89 \text{ milligrams}$$

Matched Problem 3 Refer to the exponential decay model in Example 3. How many milligrams of ^{14}C would have to be present at the beginning in order to have 25 milligrams present after 18,000 years? Compute the answer to the nearest milligram.

Explore–Discuss 3

(A) On the same set of coordinate axes, graph the three decay equations $A = A_0e^{-0.35t}$, $t \geq 0$, for $A_0 = 10, 20$, and 30.
(B) Identify any asymptotes for the three graphs in part (A).
(C) Discuss the long-term behavior for the equations in part (A).

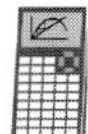

Example 4 **Depreciation** Table 2 gives the market value of a minivan (in dollars) x years after its purchase. Find an exponential regression model of the form $y = ab^x$ for this data set. Estimate the purchase price of the van. Estimate the value of the van 10 years after its purchase. Round answers to the nearest dollar.

Table 2

x	1	2	3	4	5	6
VALUE ($)	12,575	9,455	8,115	6,845	5,225	4,485

Solution Enter the data into a graphing utility (Fig. 6A) and find the exponential regression equation (Fig. 6B). The estimated purchase price is $y_1(0) = \$14{,}910$. The data set and the regression equation are graphed in Figure 6C. Using the trace feature, we see that the estimated value after 10 years is \$1,959.

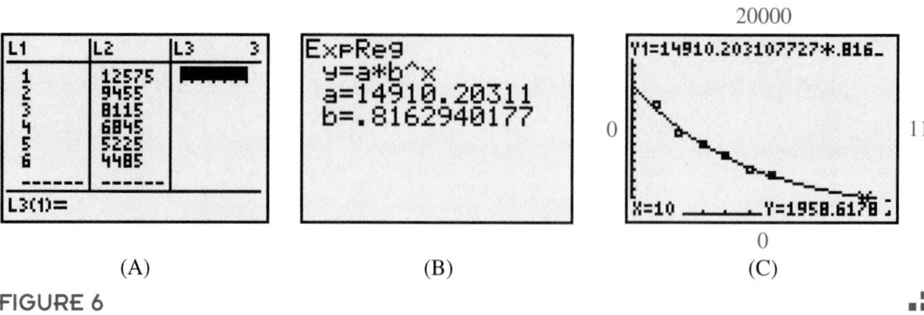

(A) (B) (C)

FIGURE 6

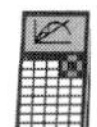

Matched Problem 4 Table 3 gives the market value of a luxury sedan (in dollars) x years after its purchase. Find an exponential regression model of the form $y = ab^x$ for this data set. Estimate the purchase price of the sedan. Estimate the value of the sedan 10 years after its purchase. Round answers to the nearest dollar.

Table 3

x	1	2	3	4	5	6
Value (\$)	23,125	19,050	15,625	11,875	9,450	7,125

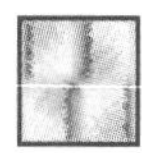

Compound Interest

We now turn to the growth of money at compound interest. The fee paid to use another's money is called **interest.** It is usually computed as a percent (called **interest rate**) of the principal over a given period of time. If, at the end of a payment period, the interest due is reinvested at the same rate, then the interest earned as well as the principal will earn interest during the next payment period. Interest paid on interest reinvested is called **compound interest,** and may be calculated using the following compound interest formula:

Compound Interest

If a **principal P (present value)** is invested at an annual **rate r** (expressed as a decimal) compounded m times a year, then the **amount A (future value)** in the account at the end of t years is given by

$$A = P\left(1 + \frac{r}{m}\right)^{mt}$$

[*Note:* P could be replaced by A_0, but convention dictates otherwise.]

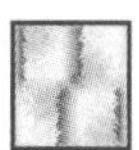

Example 5 **Compound Growth** If \$1,000 is invested in an account paying 10% compounded monthly, how much will be in the account at the end of 10 years? Compute the answer to the nearest cent.

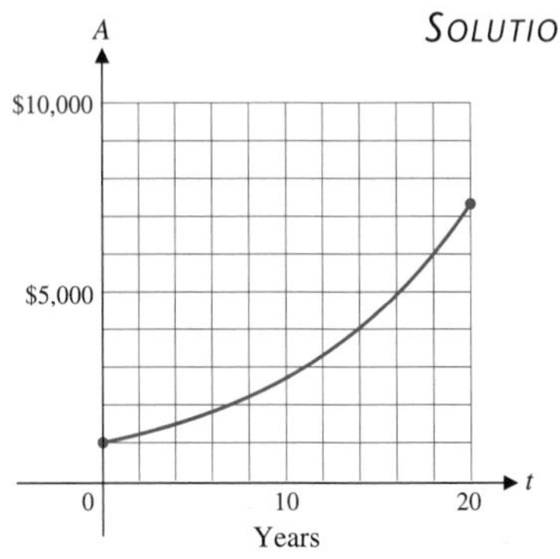

FIGURE 7

Solution We use the compound interest formula as follows:

$$A = P\left(1 + \frac{r}{m}\right)^{mt}$$

$$= 1{,}000\left(1 + \frac{0.10}{12}\right)^{(12)(10)} \qquad \text{Use a calculator.}$$

$$= \$2{,}707.04$$

The graph of

$$A = 1{,}000\left(1 + \frac{0.10}{12}\right)^{12t}$$

for $0 \leq t \leq 20$ is shown in Figure 7.

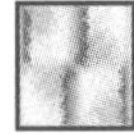

Matched Problem 5 If you deposit $5,000 in an account paying 9% compounded daily, how much will you have in the account in 5 years? Compute the answer to the nearest cent.

Explore–Discuss 4

Suppose that $1,000 is deposited in a savings account at an annual rate of 5%. Guess the amount in the account at the end of 1 year if interest is compounded (1) quarterly, (2) monthly, (3) daily, (4) hourly. Use the compound interest formula to compute the amounts at the end of 1 year to the nearest cent. Discuss the accuracy of your initial guesses.

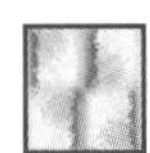

Continuous Compound Interest

Returning to the compound interest formula,

$$A = P\left(1 + \frac{r}{m}\right)^{mt}$$

suppose the principal P, the annual rate r, and the time t are held fixed, and the number of compounding periods per year m is increased without bound. Will the amount A increase without bound, or will it tend to some limiting value?

Starting with $P = \$100$, $r = 0.08$, and $t = 2$ years, we construct Table 4 for several values of m with the aid of a calculator. Notice that the largest gain appears in going from annual to semiannual compounding. Then, the gains slow down as m increases. It appears that A gets closer and closer to $117.35 as m gets larger and larger.

Table 4

Compounding Frequency	m	$A = 100\left(1 + \frac{0.08}{m}\right)^{2m}$
Annually	1	$116.6400
Semiannually	2	116.9859
Quarterly	4	117.1659
Weekly	52	117.3367
Daily	365	117.3490
Hourly	8,760	117.3510

It can be shown that

$$P\left(1 + \frac{r}{m}\right)^{mt}$$

gets closer and closer to Pe^{rt} as the number of compounding periods m gets larger and larger. The latter is referred to as the **continuous compound interest formula,** a formula that is widely used in business, banking, and economics.

Continuous Compound Interest Formula

If a principal P is invested at an annual rate r (expressed as a decimal) compounded continuously, then the amount A in the account at the end of t years is given by

$$A = Pe^{rt}$$

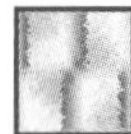

Example 6 **Compounding Daily and Continuously** What amount will an account have after 2 years if \$5,000 is invested at an annual rate of 8%:

(A) Compounded daily? (B) Compounded continuously?

Compute answers to the nearest cent.

Solution (A) Use the compound interest formula

$$A = P\left(1 + \frac{r}{m}\right)^{mt}$$

with $P = 5{,}000$, $r = 0.08$, $m = 365$, and $t = 2$:

$$A = 5{,}000\left(1 + \frac{0.08}{365}\right)^{(365)(2)} \quad \text{Use a calculator.}$$
$$= \$5{,}867.45$$

(B) Use the continuous compound interest formula

$$A = Pe^{rt}$$

with $P = 5{,}000$, $r = 0.08$, and $t = 2$:

$$A = 5{,}000e^{(0.08)(2)} \quad \text{Use a calculator.}$$
$$= \$5{,}867.55$$

Matched Problem 6 What amount will an account have after 1.5 years if \$8,000 is invested at an annual rate of 9%:

(A) Compounded weekly? (B) Compounded continuously?

Compute answers to the nearest cent.

The formulas for simple interest, compound interest, and continuous compound interest are summarized in the box at the top of the next page for convenient reference.

Interest Formulas

Simple interest	$A = P(1 + rt)$
Compound interest	$A = P\left(1 + \frac{r}{m}\right)^{mt}$
Continuous compound interest	$A = Pe^{rt}$

Answers to Matched Problems

1.

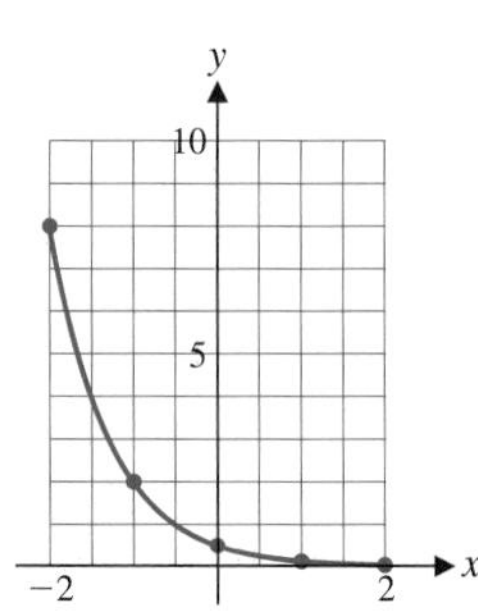

2. (A) 179 bacteria (B) 1,271,659 bacteria

3. 233 mg **4.** Purchase price: $30,363; value after 10 yr: $2,864

5. $7,841.13 **6.** (A) $9,155.23 (B) $9,156.29

Exercise 2-2

A *Graph each function in Problems 1–12 over the indicated interval.*

1. $y = 5^x; [-2, 2]$ **2.** $y = 3^x; [-3, 3]$

3. $y = (\frac{1}{5})^x = 5^{-x}; [-2, 2]$ **4.** $y = (\frac{1}{3})^x = 3^{-x}; [-3, 3]$

5. $f(x) = -5^x; [-2, 2]$ **6.** $g(x) = -3^{-x}; [-3, 3]$

7. $y = -e^{-x}; [-3, 3]$ **8.** $y = -e^x; [-3, 3]$

9. $y = 100e^{0.1x}; [-5, 5]$ **10.** $y = 10e^{0.2x}; [-10, 10]$

11. $g(t) = 10e^{-0.2t}; [-5, 5]$ **12.** $f(t) = 100e^{-0.1t}; [-5, 5]$

Simplify each expression in Problems 13–18.

13. $(4^{3x})^{2y}$ **14.** $10^{3x-1}10^{4-x}$ **15.** $\frac{e^{x-3}}{e^{x-4}}$

16. $\frac{e^x}{e^{1-x}}$ **17.** $(2e^{1.2t})^3$ **18.** $(3e^{-1.4x})^2$

B *In Problems 19–26, describe the transformations that can be used to obtain the graph of g from the graph of f (see Section 1-2).*

19. $g(x) = -2^x; f(x) = 2^x$ **20.** $g(x) = 2^{x-2}; f(x) = 2^x$

21. $g(x) = 3^{x+1}; f(x) = 3^x$ **22.** $g(x) = -3^x; f(x) = 3^x$

23. $g(x) = e^x + 1; f(x) = e^x$

24. $g(x) = e^x - 2; f(x) = e^x$

25. $g(x) = 2e^{-(x+2)}; f(x) = e^{-x}$

26. $g(x) = 0.5e^{-(x-1)}; f(x) = e^{-x}$

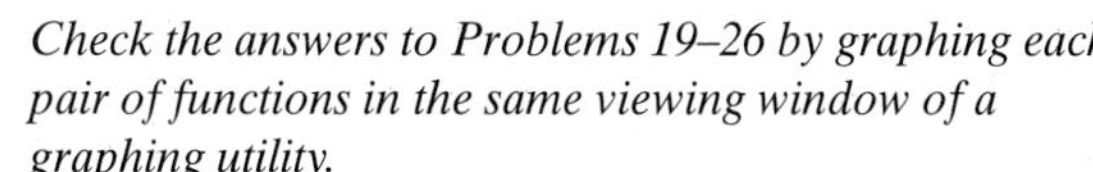

Check the answers to Problems 19–26 by graphing each pair of functions in the same viewing window of a graphing utility.

In Problems 27–36, graph each function over the indicated interval.

27. $f(t) = 2^{t/10}$; $[-30, 30]$

28. $G(t) = 3^{t/100}$; $[-200, 200]$

29. $y = -3 + e^{1+x}$; $[-4, 2]$

30. $y = 2 + e^{x-2}$; $[-1, 5]$

31. $y = e^{|x|}$; $[-3, 3]$

32. $y = e^{-|x|}$; $[-3, 3]$

33. $C(x) = \dfrac{e^x + e^{-x}}{2}$; $[-5, 5]$

34. $M(x) = e^{x/2} + e^{-x/2}$; $[-5, 5]$

35. $y = e^{-x^2}$; $[-3, 3]$

36. $y = 2^{-x^2}$; $[-3, 3]$

37. Find all real numbers a such that $a^2 = a^{-2}$. Explain why this does not violate the second exponential function property in the box on page 96.

38. Find real numbers a and b such that $a \neq b$ but $a^4 = b^4$. Explain why this does not violate the third exponential function property in the box on page 96.

39. A graphing utility was used to graph the exponential functions $f(x) = 2^x$ and $g(x) = e^x$ in the same viewing window, as shown in the figure. Which curve belongs to which function? Explain.

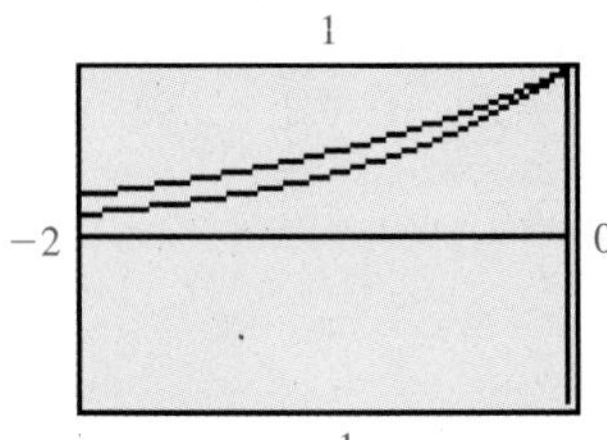

Figure for 39

40. A graphing utility was used to graph the exponential functions $f(x) = 2^x$ and $g(x) = e^x$ in the same viewing window, as shown in the figure. Which curve belongs to which function? Explain.

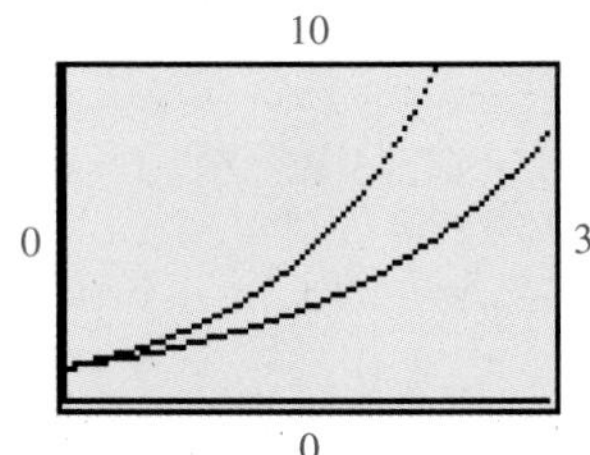

Figure for 40

41. A graphing utility was used to graph the exponential functions $f(x) = 2^{-x}$ and $g(x) = e^{-x}$ in the same viewing window, as shown in the figure. Which curve belongs to which function? Explain.

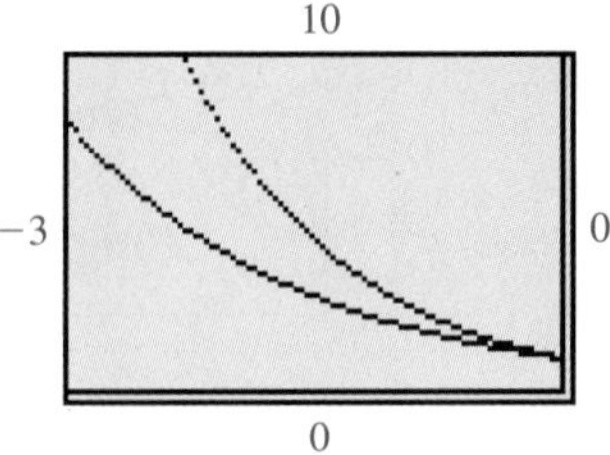

Figure for 41

42. A graphing utility was used to graph the exponential functions $f(x) = 2^{-x}$ and $g(x) = e^{-x}$ in the same viewing window, as shown in the figure. Which curve belongs to which function? Explain.

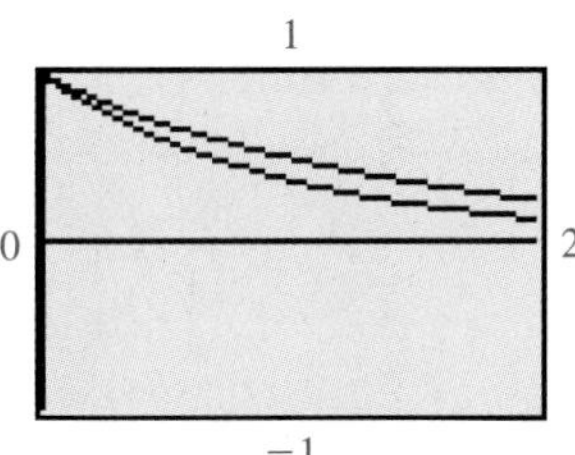

Figure for 42

Solve each equation in Problems 43–48 for x.

43. $10^{2-3x} = 10^{5x-6}$

44. $5^{3x} = 5^{4x-2}$

45. $4^{5x-x^2} = 4^{-6}$

46. $7^{x^2} = 7^{2x+3}$

47. $5^3 = (x + 2)^3$

48. $(1 - x)^5 = (2x - 1)^5$

C *Solve each equation in Problems 49–52 for x. (Remember, $e^x \neq 0$ and $e^{-x} \neq 0$.)*

49. $(x - 3)e^x = 0$

50. $2xe^{-x} = 0$

51. $3xe^{-x} + x^2e^{-x} = 0$

52. $x^2e^x - 5xe^x = 0$

Graph each function in Problems 53–56 over the indicated interval.

53. $h(x) = x(2^x)$; $[-5, 0]$

54. $m(x) = x(3^{-x})$; $[0, 3]$

55. $N = \dfrac{100}{1 + e^{-t}}$; $[0, 5]$

56. $N = \dfrac{200}{1 + 3e^{-t}}$; $[0, 5]$

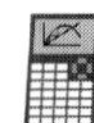

In Problems 57–60, approximate the real zeros of each function to two decimal places.

57. $f(x) = 4^x - 7$

58. $f(x) = 5 - 3^{-x}$

59. $f(x) = 2 + 3x + 10^x$

60. $f(x) = 7 - 2x^2 + 2^{-x}$

APPLICATIONS

Business & Economics

61. *Finance.* Suppose $2,500 is invested at 7% compounded quarterly. How much money will be in the account in:
(A) $\frac{3}{4}$ year? (B) 15 years?

Compute answers to the nearest cent.

62. *Finance.* Suppose $4,000 is invested at 11% compounded weekly. How much money will be in the account in:
(A) $\frac{1}{2}$ year? (B) 10 years?

Compute answers to the nearest cent.

63. *Money growth.* If you invest $7,500 in an account paying 8.35% compounded continuously, how much money will be in the account at the end of:
(A) 5.5 years? (B) 12 years?

64. *Money growth.* If you invest $5,250 in an account paying 11.38% compounded continuously, how much money will be in the account at the end of:
(A) 6.25 years? (B) 17 years?

65. *Finance.* A person wishes to have $15,000 cash for a new car 5 years from now. How much should be placed in an account now, if the account pays 9.75% compounded weekly? Compute the answer to the nearest dollar.

66. *Finance.* A couple just had a baby. How much should they invest now at 8.25% compounded daily in order to have $40,000 for the child's education 17 years from now? Compute the answer to the nearest dollar.

67. *Money growth. Barron's* (a national business and financial weekly) published the following "Top Savings Deposit Yields" for 1 year certificate of deposit accounts:

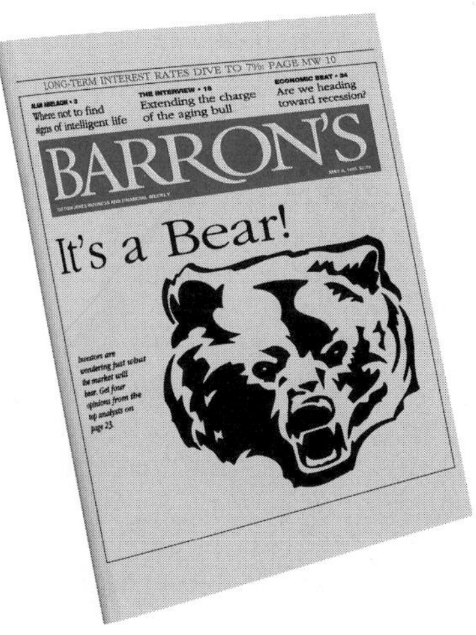

(A) Alamo Savings, 8.25% compounded quarterly
(B) Lamar Savings, 8.05% compounded continuously

Compute the value of $10,000 invested in each account at the end of 1 year.

68. *Money growth.* Refer to Problem 67. In another issue of *Barron's,* $2\frac{1}{2}$ year certificate of deposit accounts included the following:
(A) Gill Saving, 8.30% compounded continuously
(B) Richardson Savings and Loan, 8.40% compounded quarterly
(C) USA Savings, 8.25% compounded daily

Compute the value of $1,000 invested in each account at the end of $2\frac{1}{2}$ years.

69. *Present value.* A promissory note will pay $50,000 at maturity $5\frac{1}{2}$ years from now. How much should you be willing to pay for the note now if money is worth 10% compounded continuously?

70. *Present value.* A promissory note will pay $30,000 at maturity 10 years from now. How much should you be willing to pay for the note now if money is worth 9% compounded continuously?

71. *Advertising.* A company is trying to introduce a new product to as many people as possible through television advertising in a large metropolitan area with 2 million possible viewers. A model for the number of people N (in millions) who are aware of the product after t days of advertising was found to be

$$N = 2(1 - e^{-0.037t})$$

Graph this function for $0 \leq t \leq 50$. What value does N tend to as t increases without bound?

72. *Learning curve.* People assigned to assemble circuit boards for a computer manufacturing company undergo on-the-job training. From past experience it was found that the learning curve for the average employee is given by

$$N = 40(1 - e^{-0.12t})$$

where N is the number of boards assembled per day after t days of training. Graph this function for $0 \leq t \leq 30$. What is the maximum number of boards an average employee can be expected to produce in 1 day?

73. *Sports salaries.* Table 5 gives the average salary (in thousands of dollars) for players in the National Football League (NFL) and the National Basketball Association (NBA) in selected years since 1980.

(A) Let x represent the number of years since 1980 and find an exponential regression model ($y = ab^x$) for the NFL's average salary data. Estimate the average salary (to the nearest thousand dollars) for 1995 and 2010.

(B) The average salary in the NFL for 1995 was \$637,000. How does this compare with the estimated value from the model in part (A)? How does this 1995 salary information affect the estimated salary in 2010? Explain.

Table 5

AVERAGE SALARIES (THOUSAND \$)

YEAR	1980	1984	1985	1986	1987	1988	1989	1990
NFL	79	178	194	198	203	239	295	352
NBA	170	275	325	375	440	510	603	817

Source: U.S. Bureau of the Census

74. *Sports salaries.* Refer to Table 5.

(A) Let x represent the number of years since 1980 and find an exponential regression model ($y = ab^x$) for the NBA's average salary data. Estimate the average salary (to the nearest thousand dollars) for 1994 and 2010.

(B) The average salary in the NBA for 1994 was \$1,700,000. How does this compare with the estimated value from the model in part (A)? How does this 1994 salary information affect the estimated salary in 2010? Explain.

Life Sciences

75. *Marine biology.* Marine life is dependent upon the microscopic plant life that exists in the *photic zone,* a zone that goes to a depth where about 1% of the surface light still remains. In some waters with a great deal of sediment, the photic zone may go down only 15–20 feet. In some murky harbors, the intensity of light d feet below the surface is given approximately by

$$I = I_0e^{-0.23d}$$

What percentage of the surface light will reach a depth of:
(A) 10 feet? (B) 20 feet?

76. *Marine biology.* Refer to Problem 75. Light intensity I relative to depth d (in feet) for one of the clearest bodies of water in the world, the Sargasso Sea in the West Indies, can be approximated by

$$I = I_0e^{-0.00942d}$$

where I_0 is the intensity of light at the surface. What percent of the surface light will reach a depth of:
(A) 50 feet? (B) 100 feet?

77. *AIDS epidemic.* The U.S. Department of Health and Human Services reports that prior to 1993 there were about 40,000 cases of AIDS among intravenous drug users in the United States. It was estimated that the disease was spreading in this group at an annual rate of 21% compounded continuously. Let 1992 be year 0 and assume this rate does not change.

(A) Write an equation that models the growth of AIDS in this group, starting in 1992.

(B) Based on the model, how many cases (to the nearest thousand) should we expect by the end of the year 2000? The year 2005?

(C) Sketch a graph of the equation found in part (A). Cover the years from 1992 through 2005.

78. *AIDS epidemic.* The U.S. Department of Health and Human Services reports that prior to 1993 there were about 245,000 cases of AIDS among the general population in the United States. It was estimated that the disease was spreading at an annual rate of 18% compounded continuously. Let 1992 be year 0 and assume this rate does not change.

(A) Write an equation that models the growth of AIDS, starting in 1992.

(B) Based on the model, how many cases (to the nearest hundred thousand) should we expect by the end of the year 2000? The year 2005?

(C) Sketch a graph of the equation found in part (A). Cover the years from 1992 through 2005.

Social Sciences

79. *World population growth.* It took from the dawn of humanity to 1830 for the population to grow to the first billion people, just 100 more years (by 1930) for the second billion, and a billion more were added in only 60 more years (by 1990). In 1995, the estimated world population was 5.7 billion. In 1994, the World Bank estimated the world population would be growing at an annual rate of 1.14% compounded continuously until 2030.

(A) Write an equation that models the world population growth, letting 1995 be year 0.

(B) Based on the model, what is the expected world population (to the nearest hundred million) in 2010? In 2030?

(C) Sketch a graph of the equation found in part (A). Cover the years from 1995 through 2030.

80. *Population growth in Ethiopia.* In 1995, the estimated population in Ethiopia was 88 million people. In 1994, the World Bank estimated the population would grow at an annual rate of 1.67% compounded continuously until 2030.

(A) Write an equation that models the population growth in Ethiopia, letting 1995 be year 0.

(B) Based on the model, what is the expected population in Ethiopia (to the nearest million) in 2010? In 2030?

(C) Sketch a graph of the equation found in part (A). Cover the years from 1995 through 2030.

81. C *Internet growth.* An Internet host is a computer that can be reached directly by other computers on the Internet. The number of Internet hosts almost doubled annually from 1990 to 1994 (Table 6).

(A) Let x represent the number of years since 1990. Find an exponential regression model ($y = ab^x$) for this data set and estimate the number of hosts in 1995 and 1996 (to the nearest thousand).

Table 6

INTERNET HOSTS (THOUSANDS)

YEAR	1990	1991	1992	1993	1994
HOSTS	313	617	1,136	2,056	3,864

Source: Network Wizards, www.nw.com

(B) Discuss the implications of this model if the number of Internet hosts continues to grow at this rate.

82. C *National debt growth.* The national debt has been growing rapidly for decades. Table 7 lists the debt (in billions of dollars) for selected years since 1970. Let x represent the number of years since 1970. Find an exponential regression model for this data set and estimate the national debt in 2000 and 2010 (to the nearest billion).

Table 7

NATIONAL DEBT (BILLION $)

YEAR	1970	1975	1980	1985	1990	1995
DEBT	389	577	930	1,946	3,233	4,974

Source: Bureau of the Public Debt, www.savingsbonds.gov

SECTION 2-3

Logarithmic Functions

- INVERSE FUNCTIONS
- LOGARITHMIC FUNCTIONS
- PROPERTIES OF LOGARITHMIC FUNCTIONS
- CALCULATOR EVALUATION OF LOGARITHMS
- APPLICATION

Find the exponential function keys [10ˣ] and [eˣ] on your calculator. Close to these keys you will find [LOG] and [LN] keys. The latter represent *logarithmic functions,* and each is closely related to the exponential function it is near. In fact, the exponential function and the corresponding logarithmic function are said to be *inverses* of each other. In this section we will develop the concept of inverse functions and use it to define a logarithmic function as the inverse of an exponential function. We will then investigate basic properties of logarithmic functions, use a calculator to evaluate them for particular values of x, and apply them to real-world problems.

Logarithmic functions are used in modeling and solving many types of problems. For example, the decibel scale is a logarithmic scale used to measure sound intensity, and the Richter scale is a logarithmic scale used to measure the strength of the force of an earthquake. An important business application has to do with finding the time it takes money to double if it is invested at a certain rate compounded a given number of times a year or compounded continuously. This requires the solution of an exponential equation, and logarithms play a central role in the process.

Inverse Functions

Look at the graphs of $f(x) = \frac{x}{2}$ and $g(x) = \frac{|x|}{2}$ in Figure 1:

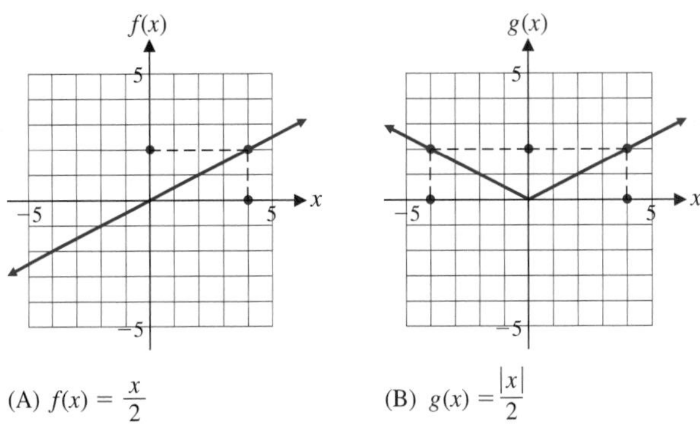

(A) $f(x) = \frac{x}{2}$ (B) $g(x) = \frac{|x|}{2}$

FIGURE 1

Because both f and g are functions, each domain value corresponds to exactly one range value. For which function does each range value correspond to exactly one domain value? This is the case only for function f. Note that for the range value 2, the corresponding domain value is 4. For function g the range value 2 corresponds to both -4 and 4. Function f is said to be *one-to-one.* In general:

One-to-One Functions

A function f is said to be **one-to-one** if each range value corresponds to exactly one domain value.

It can be shown that any continuous function that is either increasing or decreasing for all domain values is one-to-one. If a continuous function increases for some domain values and decreases for others, it cannot be one-to-one. Figure 1 shows an example of each case.

Explore–Discuss 1

Graph $f(x) = 2^x$ and $g(x) = x^2$. For a range value of 4, what are the corresponding domain values for each function? Which of the two functions is one-to-one? Explain why.

Starting with a one-to-one function f we can obtain a new function called the *inverse* of f as follows:

> **Inverse of a Function**
>
> If f is a one-to-one function, then the **inverse** of f is the function formed by interchanging the independent and dependent variables for f. Thus, if (a, b) is a point on the graph of f, then (b, a) is a point on the graph of the inverse of f.
>
> [*Note:* If f is not one-to-one, then f **does not have an inverse.**]

A number of important functions in any library of elementary functions are the inverses of other basic functions in the library. In this course, we are interested in the inverses of exponential functions, called *logarithmic functions.*

Logarithmic Functions

If we start with the exponential function f defined by

$$y = 2^x \tag{1}$$

and interchange the variables, we obtain the inverse of f:

$$x = 2^y \tag{2}$$

We call the inverse the **logarithmic function with base 2,** and write

$$y = \log_2 x \qquad \text{if and only if} \qquad x = 2^y$$

We can graph $y = \log_2 x$ by graphing $x = 2^y$, since they are equivalent. Any ordered pair of numbers on the graph of the exponential function will be on the graph of the logarithmic function if we interchange the order of the components. For example, $(3, 8)$ satisfies equation (1) and $(8, 3)$ satisfies equation (2). The graphs of $y = 2^x$ and $y = \log_2 x$ are shown in Figure 2. Note that if we fold the paper along the dashed line $y = x$ in Figure 2, the two graphs match exactly. The line $y = x$ is a line of symmetry for the two graphs.

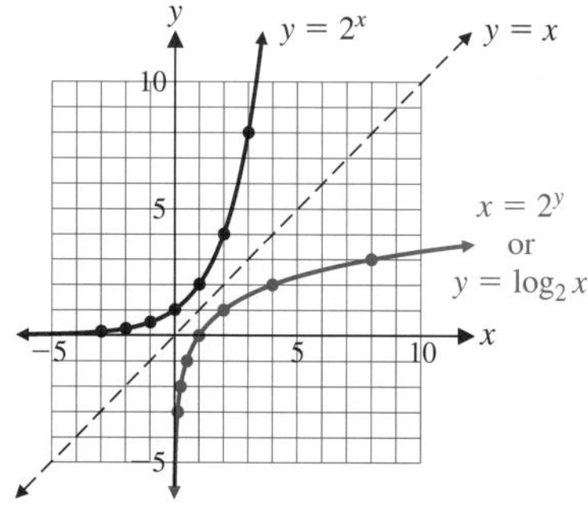

FIGURE 2

EXPONENTIAL FUNCTION x	$y = 2^x$	LOGARITHMIC FUNCTION $x = 2^y$	y
-3	$\frac{1}{8}$	$\frac{1}{8}$	-3
-2	$\frac{1}{4}$	$\frac{1}{4}$	-2
-1	$\frac{1}{2}$	$\frac{1}{2}$	-1
0	1	1	0
1	2	2	1
2	4	4	2
3	8	8	3

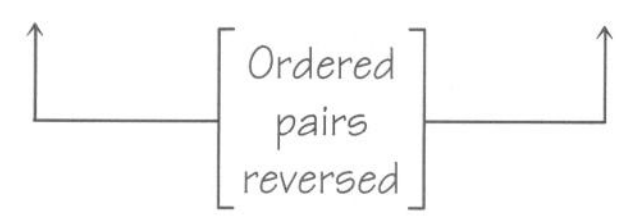

In general, since the graphs of all exponential functions of the form $f(x) = b^x$, $b \neq 1$, $b > 0$, are either increasing or decreasing (see Section 2-2), exponential functions have inverses.

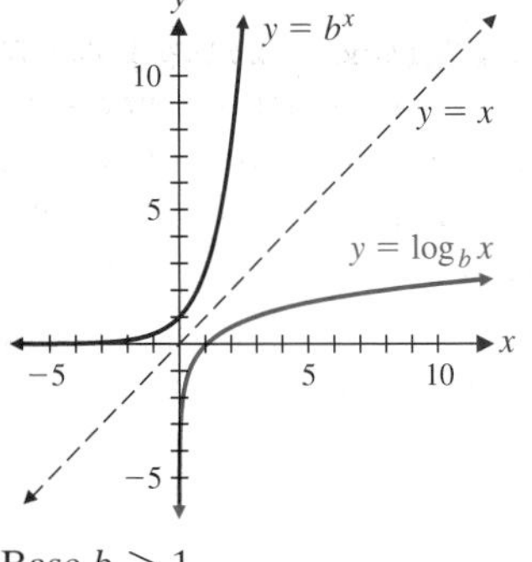

Base $b > 1$

Logarithmic Functions

The inverse of an exponential function is called a **logarithmic function.** For $b > 0$ and $b \neq 1$,

Logarithmic form		Exponential form
$y = \log_b x$	is equivalent to	$x = b^y$

The log to the base *b* of *x* is the exponent to which b must be raised to obtain x. [*Remember:* A logarithm is an exponent.] The **domain** of the logarithmic function is the set of all positive real numbers, which is also the range of the corresponding exponential function; and the **range** of the logarithmic function is the set of all real numbers, which is also the domain of the corresponding exponential function. Typical graphs of an exponential function and its inverse, a logarithmic function, are shown in the figure in the margin.

The following examples involve converting logarithmic forms to equivalent exponential forms and vice versa.

Example 1 **Logarithmic–Exponential Conversions** Change each logarithmic form to an equivalent exponential form:

(A) $\log_5 25 = 2$ (B) $\log_9 3 = \frac{1}{2}$ (C) $\log_2(\frac{1}{4}) = -2$

SOLUTION

(A) $\log_5 25 = 2$ is equivalent to $25 = 5^2$
(B) $\log_9 3 = \frac{1}{2}$ is equivalent to $3 = 9^{1/2}$
(C) $\log_2(\frac{1}{4}) = -2$ is equivalent to $\frac{1}{4} = 2^{-2}$

Matched Problem 1 Change each logarithmic form to an equivalent exponential form:

(A) $\log_3 9 = 2$ (B) $\log_4 2 = \frac{1}{2}$ (C) $\log_3(\frac{1}{9}) = -2$

Example 2 **Exponential–Logarithmic Conversions** Change each exponential form to an equivalent logarithmic form:

(A) $64 = 4^3$ (B) $6 = \sqrt{36}$ (C) $\frac{1}{8} = 2^{-3}$

SOLUTION

(A) $64 = 4^3$ is equivalent to $\log_4 64 = 3$
(B) $6 = \sqrt{36}$ is equivalent to $\log_{36} 6 = \frac{1}{2}$
(C) $\frac{1}{8} = 2^{-3}$ is equivalent to $\log_2(\frac{1}{8}) = -3$

Matched Problem 2 Change each exponential form to an equivalent logarithmic form.

(A) $49 = 7^2$ (B) $3 = \sqrt{9}$ (C) $\frac{1}{3} = 3^{-1}$

To gain a little deeper understanding of logarithmic functions and their relationship to the exponential functions, we consider a few problems where we want to find x, b, or y in $y = \log_b x$, given the other two values. All values are chosen so that the problems can be solved exactly without a calculator.

Example 3 **Solutions of the Equation $y = \log_b x$** Find y, b, or x, as indicated.

(A) Find y: $y = \log_4 16$ (B) Find x: $\log_2 x = -3$
(C) Find y: $y = \log_8 4$ (D) Find b: $\log_b 100 = 2$

Solution (A) $y = \log_4 16$ is equivalent to $16 = 4^y$. Thus,

$$y = 2$$

(B) $\log_2 x = -3$ is equivalent to $x = 2^{-3}$. Thus,

$$x = \frac{1}{2^3} = \frac{1}{8}$$

(C) $y = \log_8 4$ is equivalent to

$$4 = 8^y \qquad \text{or} \qquad 2^2 = 2^{3y}$$

Thus,

$$3y = 2$$
$$y = \tfrac{2}{3}$$

(D) $\log_b 100 = 2$ is equivalent to $100 = b^2$. Thus,

$b = 10$ Recall that b cannot be negative.

Matched Problem 3 Find y, b, or x, as indicated.

(A) Find y: $y = \log_9 27$ (B) Find x: $\log_3 x = -1$
(C) Find b: $\log_b 1{,}000 = 3$

Properties of Logarithmic Functions

Logarithmic functions have many powerful and useful properties. We list eight basic properties in Theorem 1.

Theorem 1 PROPERTIES OF LOGARITHMIC FUNCTIONS

If b, M, and N are positive real numbers, $b \neq 1$, and p and x are real numbers, then:

1. $\log_b 1 = 0$
2. $\log_b b = 1$
3. $\log_b b^x = x$
4. $b^{\log_b x} = x, \quad x > 0$
5. $\log_b MN = \log_b M + \log_b N$
6. $\log_b \dfrac{M}{N} = \log_b M - \log_b N$
7. $\log_b M^p = p \log_b M$
8. $\log_b M = \log_b N$ if and only if $M = N$

The first four properties in Theorem 1 follow directly from the definition of a logarithmic function. Here we will sketch a proof of property 5. The other properties are established in a similar way. Let

$$u = \log_b M \qquad \text{and} \qquad v = \log_b N$$

Or, in equivalent exponential form,

$$M = b^u \qquad \text{and} \qquad N = b^v$$

Now, see if you can provide reasons for each of the following steps:

$$\log_b MN = \log_b b^u b^v = \log_b b^{u+v} = u + v = \log_b M + \log_b N$$

Example 4 **Using Logarithmic Properties**

(A) $\log_b \frac{wx}{yz} = \log_b wx - \log_b yz$

$= \log_b w + \log_b x - (\log_b y + \log_b z)$

$= \log_b w + \log_b x - \log_b y - \log_b z$

(B) $\log_b (wx)^{3/5} = \frac{3}{5} \log_b wx = \frac{3}{5}(\log_b w + \log_b x)$

Matched Problem 4 Write in simpler logarithmic forms, as in Example 4.

(A) $\log_b \frac{R}{ST}$ (B) $\log_b \left(\frac{R}{S}\right)^{2/3}$

The following examples and problems, though somewhat artificial, will give you additional practice in using basic logarithmic properties.

Example 5 **Solving Logarithmic Equations** Find x so that:

$$\tfrac{3}{2} \log_b 4 - \tfrac{2}{3} \log_b 8 + \log_b 2 = \log_b x$$

SOLUTION

$$\begin{aligned} \tfrac{3}{2} \log_b 4 - \tfrac{2}{3} \log_b 8 + \log_b 2 &= \log_b x \\ \log_b 4^{3/2} - \log_b 8^{2/3} + \log_b 2 &= \log_b x && \text{Property 7} \\ \log_b 8 - \log_b 4 + \log_b 2 &= \log_b x \\ \log_b \frac{8 \cdot 2}{4} &= \log_b x && \text{Properties 5 and 6} \\ \log_b 4 &= \log_b x \\ x &= 4 && \text{Property 8} \end{aligned}$$

Matched Problem 5 Find x so that: $3 \log_b 2 + \frac{1}{2} \log_b 25 - \log_b 20 = \log_b x$

Example 6 **Solving Logarithmic Equations** Solve: $\log_{10} x + \log_{10}(x + 1) = \log_{10} 6$

SOLUTION

$$\begin{aligned} \log_{10} x + \log_{10}(x + 1) &= \log_{10} 6 \\ \log_{10}[x(x + 1)] &= \log_{10} 6 && \text{Property 5} \\ x(x + 1) &= 6 && \text{Property 8} \\ x^2 + x - 6 &= 0 && \text{Solve by factoring.} \\ (x + 3)(x - 2) &= 0 \\ x &= -3, 2 \end{aligned}$$

We must exclude $x = -3$, since the domain of the function $\log_{10}(x + 1)$ is $x > -1$ or $(-1, \infty)$; hence, $x = 2$ is the only solution.

Matched Problem 6 Solve: $\log_3 x + \log_3(x - 3) = \log_3 10$

Explore–Discuss 2

Discuss the relationship between each of the following pairs of expressions. If the two expressions are equivalent, explain why. If they are not, give an example.

(A) $\log_b M - \log_b N$; $\dfrac{\log_b M}{\log_b N}$

(B) $\log_b M - \log_b N$; $\log_b \dfrac{M}{N}$

(C) $\log_b M + \log_b N$; $\log_b MN$

(D) $\log_b M + \log_b N$; $\log_b(M + N)$

Calculator Evaluation of Logarithms

Of all possible logarithmic bases, the base e and the base 10 are used almost exclusively. Before we can use logarithms in certain practical problems, we need to be able to approximate the logarithm of any positive number either to base 10 or to base e. And conversely, if we are given the logarithm of a number to base 10 or base e, we need to be able to approximate the number. Historically, tables were used for this purpose, but now calculators make computations faster and far more accurate.

Common logarithms (also called **Briggsian logarithms**) are logarithms with base 10. **Natural logarithms** (also called **Napierian logarithms**) are logarithms with base e. Most calculators have a key labeled "log" (or "LOG") and a key labeled "ln" (or "LN"). The former represents a common (base 10) logarithm and the latter a natural (base e) logarithm. In fact, "log" and "ln" are both used extensively in mathematical literature, and whenever you see either used in this book without a base indicated they will be interpreted as follows:

Logarithmic Notation

Common logarithm: $\log x = \log_{10} x$
Natural logarithm: $\ln x = \log_e x$

Finding the common or natural logarithm using a calculator is very easy. On some calculators, you simply enter a number from the domain of the function and press [LOG] or [LN]. On other calculators, you press either [LOG] or [LN], enter a number from the domain, and then press [ENTER]. Check the user's manual for your calculator.

Example 7 **Calculator Evaluation of Logarithms** Use a calculator to evaluate each to six decimal places:

(A) log 3,184 (B) ln 0.000 349 (C) log(−3.24)

Solution (A) log 3,184 = 3.502 973 (B) ln 0.000 349 = −7.960 439

(C) log(−3.24) = Error* −3.24 is not in the domain of the log function.

*Some calculators use a more advanced definition of logarithms involving complex numbers and will display an ordered pair of real numbers as the value of log(−3.24). You should interpret such a result as an indication that the number entered is not in the domain of the logarithm function as we have defined it.

Matched Problem 7 ➡ Use a calculator to evaluate each to six decimal places:

(A) $\log 0.013\ 529$ (B) $\ln 28.693\ 28$ (C) $\ln(-0.438)$

We now turn to the second problem mentioned above: Given the logarithm of a number, find the number. We make direct use of the logarithmic–exponential relationships, which follow from the definition of logarithmic function given at the beginning of this section.

Logarithmic–Exponential Relationships

$\log x = y$	is equivalent to	$x = 10^y$
$\ln x = y$	is equivalent to	$x = e^y$

Example 8 ➡ **Solving $\log_b x = y$ for x** Find x to four decimal places, given the indicated logarithm:

(A) $\log x = -2.315$ (B) $\ln x = 2.386$

SOLUTION

(A) $\log x = -2.315$ *Change to equivalent exponential form.*
$x = 10^{-2.315}$ *Evaluate with a calculator.*
$x = 0.0048$

(B) $\ln x = 2.386$ *Change to equivalent exponential form.*
$x = e^{2.386}$ *Evaluate with a calculator.*
$x = 10.8699$

Matched Problem 8 ➡ Find x to four decimal places, given the indicated logarithm:

(A) $\ln x = -5.062$ (B) $\log x = 2.0821$

Example 9 ➡ **Solving Exponential Equations** Solve for x to four decimal places:

(A) $10^x = 2$ (B) $e^x = 3$ (C) $3^x = 4$

SOLUTION

(A) $10^x = 2$ *Take common logarithms of both sides.*
$\log 10^x = \log 2$ *Property 3*
$x = \log 2$ *Use a calculator.*
$x = 0.3010$

(B) $e^x = 3$ *Take natural logarithms of both sides.*
$\ln e^x = \ln 3$ *Property 3*
$x = \ln 3$ *Use a calculator.*
$x = 1.0986$

(C) $3^x = 4$ *Take either natural or common logarithms of both sides. (We choose common logarithms.)*
$\log 3^x = \log 4$ *Property 7*
$x \log 3 = \log 4$ *Solve for x.*
$x = \dfrac{\log 4}{\log 3}$ *Use a calculator.*
$x = 1.2619$

Exponential equations can also be solved graphically by graphing both sides of an equation and finding the points of intersection. Figure 3 illustrates this approach for the equations in Example 9.

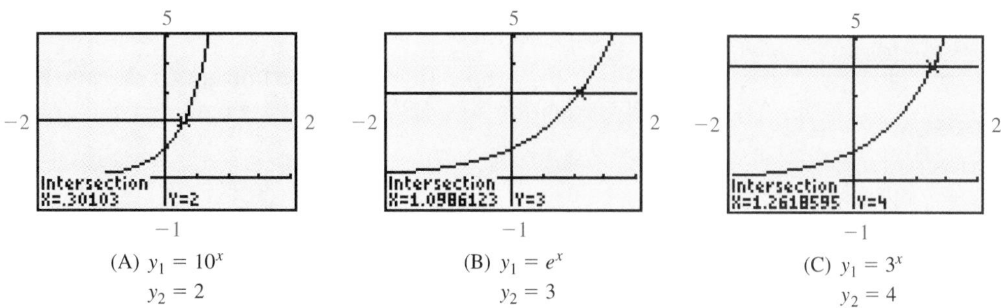

FIGURE 3
Graphical solution of exponential equations

Matched Problem 9 ➡ Solve for x algebraically and graphically (to four decimal places):

(A) $10^x = 7$ (B) $e^x = 6$ (C) $4^x = 5$

Explore–Discuss 3

Discuss how you could find $y = \log_5 38.25$ using either natural or common logarithms on a calculator. [*Hint:* Start by rewriting the equation in exponential form.]

■ Application

A convenient and easily understood way of comparing different investments is to use their **doubling times**—the length of time it takes the value of an investment to double. Logarithm properties, as you will see in Example 10, provide us with just the right tool for solving some doubling-time problems.

Example 10 ➡ **Doubling Time for an Investment** How long (to the next whole year) will it take money to double if it is invested at 20% compounded annually?

Solution We use the compound interest formula discussed in Section 2-2:

$$A = P\left(1 + \frac{r}{m}\right)^{mt} \quad \text{Compound interest}$$

The problem is to find t, given $r = 0.20$, $m = 1$, and $A = 2p$; that is,

$$\begin{aligned} 2P &= P(1 + 0.2)^t \\ 2 &= 1.2^t \\ 1.2^t &= 2 && \text{Solve for } t \text{ by taking the natural or common} \\ \ln 1.2^t &= \ln 2 && \text{logarithm of both sides (we choose the natural logarithm).} \\ t \ln 1.2 &= \ln 2 && \text{Property 7} \\ t &= \frac{\ln 2}{\ln 1.2} && \text{Use a calculator.} \\ &= 3.8 \text{ years} && [\textit{Note:} \ (\ln 2)/(\ln 1.2) \neq \ln 2 - \ln 1.2] \\ &\approx 4 \text{ years} && \text{To the next whole year} \end{aligned}$$

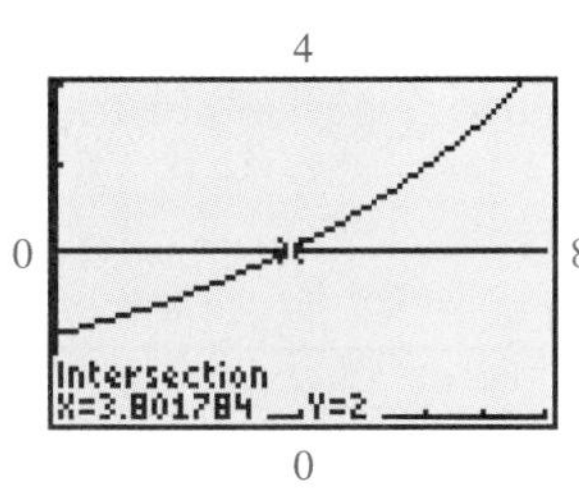

FIGURE 4
$y_1 = 1.2^x$, $y_2 = 2$

When interest is paid at the end of 3 years, the money will not be doubled; when paid at the end of 4 years, the money will be slightly more than doubled.

Example 10 can also be solved graphically by graphing both sides of the equation $2 = 1.2^t$, and finding the intersection point (Fig. 4).

Matched Problem 10 How long (to the next whole year) will it take money to triple if it is invested at 13% compounded annually?

It is interesting and instructive to graph the doubling times for various rates compounded annually. We proceed as follows:

$$
\begin{aligned}
A &= P(1 + r)^t \\
2P &= P(1 + r)^t \\
2 &= (1 + r)^t \\
(1 + r)^t &= 2 \\
\ln(1 + r)^t &= \ln 2 \\
t \ln(1 + r) &= \ln 2 \\
t &= \frac{\ln 2}{\ln(1 + r)}
\end{aligned}
$$

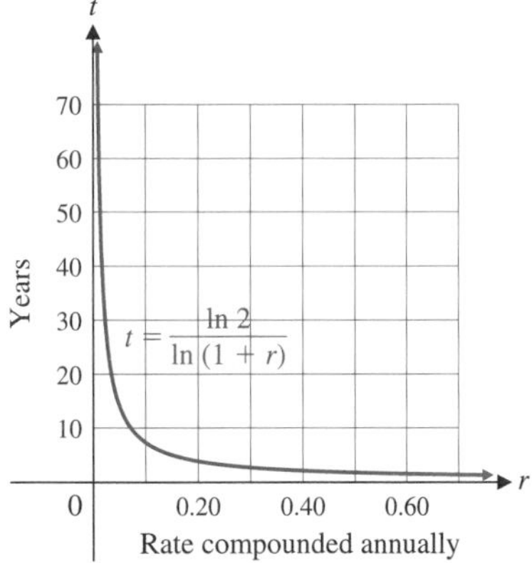

FIGURE 5
Doubling time (in years) at various rates of interest compounded annually.

Figure 5 shows the graph of this equation (doubling time in years) for interest rates compounded annually from 1% to 70% (expressed as decimals). Note the dramatic change in doubling time as rates change from 1% to 20% (from 0.01 to 0.20).

Answers to Matched Problems

1. (A) $9 = 3^2$ (B) $2 = 4^{1/2}$ (C) $\frac{1}{9} = 3^{-2}$
2. (A) $\log_7 49 = 2$ (B) $\log_9 3 = \frac{1}{2}$ (C) $\log_3(\frac{1}{3}) = -1$
3. (A) $y = \frac{3}{2}$ (B) $x = \frac{1}{3}$ (C) $b = 10$
4. (A) $\log_b R - \log_b S - \log_b T$ (B) $\frac{2}{3}(\log_b R - \log_b S)$ **5.** $x = 2$
6. $x = 5$ **7.** (A) $-1.868\ 734$ (B) $3.356\ 663$ (C) Not defined
8. (A) 0.0063 (B) 120.8092
9. (A) 0.8451 (B) 1.7918 (C) 1.1610 **10.** 6 yr

EXERCISE 2-3

A *Rewrite in equivalent exponential form:*

1. $\log_3 27 = 3$ **2.** $\log_2 32 = 5$ **3.** $\log_{10} 1 = 0$
4. $\log_e 1 = 0$ **5.** $\log_4 8 = \frac{3}{2}$ **6.** $\log_9 27 = \frac{3}{2}$

Rewrite in equivalent logarithmic form:

7. $49 = 7^2$ **8.** $36 = 6^2$ **9.** $8 = 4^{3/2}$
10. $9 = 27^{2/3}$ **11.** $A = b^u$ **12.** $M = b^x$

Evaluate each of the following without a calculator:

13. $\log_{10} 1$ **14.** $\log_e 1$ **15.** $\log_e e$
16. $\log_{10} 10$ **17.** $\log_{0.2} 0.2$ **18.** $\log_{13} 13$
19. $\log_{10} 10^3$ **20.** $\log_{10} 10^{-5}$ **21.** $\log_2 2^{-3}$
22. $\log_3 3^5$ **23.** $\log_{10} 1{,}000$ **24.** $\log_6 36$

Write in terms of simpler logarithmic forms, as in Example 4.

25. $\log_b \frac{P}{Q}$ **26.** $\log_b FG$ **27.** $\log_b L^5$

28. $\log_b w^{15}$ **29.** $\log_b \frac{p}{qrs}$ **30.** $\log_b PQR$

B *Find x, y, or b without a calculator.*

31. $\log_3 x = 2$ **32.** $\log_2 x = 2$

33. $\log_7 49 = y$ **34.** $\log_3 27 = y$

35. $\log_b 10^{-4} = -4$ **36.** $\log_b e^{-2} = -2$

37. $\log_4 x = \frac{1}{2}$ **38.** $\log_{25} x = \frac{1}{2}$

39. $\log_{1/3} 9 = y$ **40.** $\log_{49}(\frac{1}{7}) = y$

41. $\log_b 1{,}000 = \frac{3}{2}$ **42.** $\log_b 4 = \frac{2}{3}$

Write in terms of simpler logarithmic forms, going as far as you can with logarithmic properties (see Example 4).

43. $\log_b \frac{x^5}{y^3}$ **44.** $\log_b(x^2y^3)$

45. $\log_b \sqrt[3]{N}$ **46.** $\log_b \sqrt[5]{Q}$

47. $\log_b(x^2 \sqrt[3]{y})$ **48.** $\log_b \sqrt[3]{\frac{x^2}{y}}$

49. $\log_b(50 \cdot 2^{-0.2t})$ **50.** $\log_b(100 \cdot 1.06^t)$

51. $\log_b[P(1 + r)^t]$ **52.** $\log_e Ae^{-0.3t}$

53. $\log_e 100e^{-0.01t}$ **54.** $\log_{10}(67 \cdot 10^{-0.12x})$

Find x.

55. $\log_b x = \frac{2}{3}\log_b 8 + \frac{1}{2}\log_b 9 - \log_b 6$

56. $\log_b x = \frac{2}{3}\log_b 27 + 2\log_b 2 - \log_b 3$

57. $\log_b x = \frac{3}{2}\log_b 4 - \frac{2}{3}\log_b 8 + 2\log_b 2$

58. $\log_b x = 3\log_b 2 + \frac{1}{2}\log_b 25 - \log_b 20$

59. $\log_b x + \log_b(x - 4) = \log_b 21$

60. $\log_b(x + 2) + \log_b x = \log_b 24$

61. $\log_{10}(x - 1) - \log_{10}(x + 1) = 1$

62. $\log_{10}(x + 6) - \log_{10}(x - 3) = 1$

Graph Problems 63 and 64 by converting to exponential form first.

63. $y = \log_2(x - 2)$ **64.** $y = \log_3(x + 2)$

65. Explain how the graph of the equation in Problem 63 can be obtained from the graph of $y = \log_2 x$ using a simple transformation (see Section 1-2).

66. Explain how the graph of the equation in Problem 64 can be obtained from the graph of $y = \log_3 x$ using a simple transformation (see Section 1-2).

67. What are the domain and range of the function defined by $y = 1 + \ln(x + 1)$?

68. What are the domain and range of the function defined by $y = \log(x - 1) - 1$?

Evaluate to five decimal places using a calculator.

69. (A) log 3,527.2 (B) log 0.006 913 2 (C) ln 277.63 (D) ln 0.040 883

70. (A) log 72.604 (B) log 0.033 041 (C) ln 40,257 (D) ln 0.005 926 3

Find x to four decimal places.

71. (A) $\log x = 1.1285$ (B) $\log x = -2.0497$ (C) $\ln x = 2.7763$ (D) $\ln x = -1.8879$

72. (A) $\log x = 2.0832$ (B) $\log x = -1.1577$ (C) $\ln x = 3.1336$ (D) $\ln x = -4.3281$

Solve each equation to four decimal places.

73. $10^x = 12$ **74.** $10^x = 153$

75. $e^x = 4.304$ **76.** $e^x = 0.3059$

77. $1.03^x = 2.475$ **78.** $1.075^x = 1.837$

79. $1.005^{12t} = 3$ **80.** $1.02^{4t} = 2$

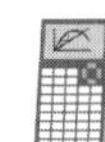

Check Problems 73–80 by solving graphically.

Graph Problems 81–88 using a calculator and point-by-point plotting. Indicate increasing and decreasing intervals.

81. $y = \ln x$ **82.** $y = -\ln x$

83. $y = |\ln x|$ **84.** $y = \ln|x|$

85. $y = 2\ln(x + 2)$ **86.** $y = 2\ln x + 2$

87. $y = 4\ln x - 3$ **88.** $y = 4\ln(x - 3)$

Check Problems 81–88 by graphing on a graphing utility.

C

89. Explain why the calculator display shown does not contradict the logarithmic property

$$\log \frac{M}{N} = \log M - \log N$$

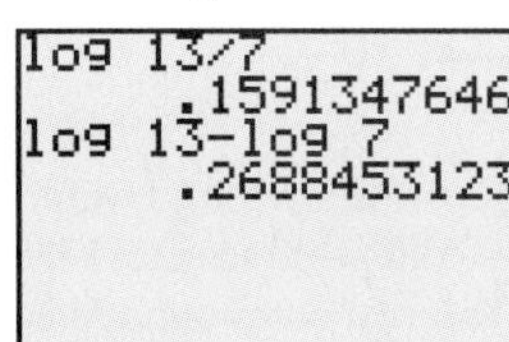

Figure for 89

90. Explain why the calculator display shown does not contradict the logarithmic property

$$\log MN = \log M + \log N$$

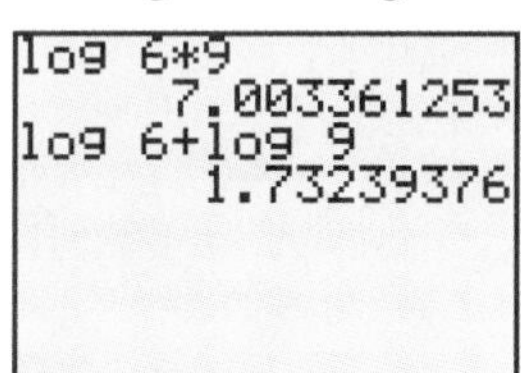

Figure for 90

91. Explain why the logarithm of 1 for any permissible base is 0.

92. Explain why 1 is not a suitable logarithmic base.

93. Write $\log_{10} y - \log_{10} c = 0.8x$ in an exponential form that is free of logarithms.

94. Write $\log_e x - \log_e 25 = 0.2t$ in an exponential form that is free of logarithms.

95. Let $p(x) = \ln x$, $q(x) = \sqrt{x}$, and $r(x) = x$. Use a graphing utility to draw graphs of all three functions in the same viewing window for $1 \leq x \leq 16$. Discuss what it means for one function to be larger than another on an interval, and then order the three functions from largest to smallest for $1 < x \leq 16$.

96. Let $p(x) = \log x$, $q(x) = \sqrt[3]{x}$, and $r(x) = x$. Use a graphing utility to draw graphs of all three functions in the same viewing window for $1 \leq x \leq 16$. Discuss what it means for one function to be smaller than another on an interval, and then order the three functions from smallest to largest for $1 < x \leq 16$.

Applications

Business & Economics

97. *Doubling time.* How long (to the next whole year) will it take money to double if it is invested at 6% interest compounded annually?

98. *Doubling time.* How long (to the next whole year) will it take money to double if it is invested at 3% interest compounded annually?

99. *Investing.* How many years (to two decimal places) will it take \$1,000 to grow to \$1,800 if it is invested at 6% compounded quarterly? Compounded continuously?

100. *Investing.* How many years (to two decimal places) will it take \$5,000 to grow to \$7,500 if it is invested at 8% compounded semiannually? Compounded continuously?

101. *Investment.* A newly married couple wishes to have \$20,000 in 8 years for the down payment on a house. At what rate of interest compounded continuously (to three decimal places) must \$10,000 be invested now to accomplish this goal?

102. *Investment.* The parents of a newborn child want to have \$45,000 for the child's college education 17 years from now. At what rate of interest compounded continuously (to three decimal places) must a grandparent's gift of \$10,000 be invested now to achieve this goal?

103. *Supply and demand.* A cordless screwdriver is sold through a national chain of discount stores. A marketing company established price–demand and price–supply tables (Tables 1 and 2), where x is the number of screwdrivers people are willing to buy and the store is willing to sell each month at a price of p dollars per screwdriver.

Table 1

PRICE–DEMAND

x	$p = D(x)$ (\$)
1,000	91
2,000	73
3,000	64
4,000	56
5,000	53

Table 2

PRICE–SUPPLY

x	$p = S(x)$ (\$)
1,000	9
2,000	26
3,000	34
4,000	38
5,000	41

(A) Find a logarithmic regression model ($y = a + b \ln x$) for the data in Table 1. Estimate the demand (to the nearest unit) at a price level of \$50.

(B) Find a logarithmic regression model ($y = a + b \ln x$) for the data in Table 2. Estimate the supply (to the nearest unit) at a price level of \$50.

(C) Does a price level of \$50 represent a stable condition, or is the price likely to increase or decrease? Explain.

104. *Equilibrium point.* Use the models constructed in Problem 103 to find the equilibrium point. Write the equilibrium price to the nearest cent and the equilibrium quantity to the nearest unit.

Life Sciences

105. *Sound intensity—decibels.* Because of the extraordinary range of sensitivity of the human ear (a range of over 1,000 million millions to 1), it is helpful to use a logarithmic scale, rather than an absolute scale, to measure sound intensity over this range. The unit of measure is called the *decibel,* after the inventor of the telephone, Alexander Graham Bell. If we let N be the number of decibels, I the power of the sound in question (in watts per square centimeter), and I_0 the power of sound just below the threshold of hearing (approximately 10^{-16} watt per square centimeter), then

$$I = I_0 10^{N/10}$$

Show that this formula can be written in the form

$$N = 10 \log \frac{I}{I_0}$$

106. *Sound intensity—decibels.* Use the formula in Problem 105 (with $I_0 = 10^{-16}$ watt/cm^2) to find the decibel ratings of the following sounds:

(A) Whisper: 10^{-13} watt/cm^2

(B) Normal conversation: 3.16×10^{-10} watt/cm^2

(C) Heavy traffic: 10^{-8} watt/cm^2

(D) Jet plane with afterburner: 10^{-1} watt/cm^2

107. C *Agriculture.* Table 3 shows the yield (in bushels per acre) and the total production (in millions of bushels) for corn in the United States for selected years since 1950. Let x represent years since 1900.

Table 3

United States Corn Production

Year	x	Yield (bushels per acre)	Total production (million bushels)
1950	50	37.6	2,782
1960	60	55.6	3,479
1970	70	81.4	4,802
1980	80	97.7	6,867
1990	90	115.6	7,802

Source: U.S. Department of Agriculture

(A) Find a logarithmic regression model ($y = a + b \ln x$) for the yield. Estimate (to one decimal place) the yield in 1996 and in 2010.

(B) The actual yield in 1996 was 127.1 bushels per acre. How does this compare with the estimated yield in part (A)? What effect will this additional 1996 information have on the estimate for 2010? Explain.

108. C *Agriculture.* Refer to Table 3.

(A) Find a logarithmic regression model ($y = a + b \ln x$) for the total production. Estimate (to the nearest million) the production in 1996 and in 2010.

(B) The actual production in 1996 was 7,949 billion bushels. How does this compare with the estimated production in part (A)? What effect will this 1996 production information have on the estimate for 2010? Explain.

Social Sciences

109. *World population.* If the world population is now 5.8 billion people and if it continues to grow at an annual rate of 1.14% compounded continuously, how long (to the nearest year) will it take before there is only 1 square yard of land per person? (The Earth contains approximately 1.68×10^{14} square yards of land.)

110. *Archaeology—carbon-14 dating.* The radioactive carbon-14 (^{14}C) in an organism at the time of its death decays according to the equation

$$A = A_0 e^{-0.000124t}$$

where t is time in years and A_0 is the amount of ^{14}C present at time $t = 0$. (See Example 3 in Section 2-2.) Estimate the age of a skull uncovered in an archaeological site if 10% of the original amount of ^{14}C is still present. [*Hint:* Find t such that $A = 0.1A_0$.]

IMPORTANT TERMS AND SYMBOLS

2-1 *Polynomial and Rational Functions.* Polynomial function; degree; continuity; turning point; root; zero; leading coefficient; rational function; points of discontinuity; vertical and horizontal asymptotes

$$f(x) = a_n x^n + a_{n-1}x^{n-1} + \cdots + a_1 x + a_0,\ a_n \neq 0;$$
$$f(x) = \frac{n(x)}{d(x)},\ d(x) \neq 0$$

2-2 *Exponential Functions.* Exponential function; base; basic graphs; horizontal asymptote; basic properties; irrational number *e;* exponential function with base *e;* exponential growth; exponential decay; compound interest; principal (present value); amount (future value); continuous compound interest

$$f(x) = b^x, b > 0, b \neq 1;\quad y = e^x;\quad N = N_0 e^{kt};$$
$$A = A_0 e^{-kt};\quad A = P\left(1 + \frac{r}{m}\right)^{mt};\quad A = Pe^{rt}$$

2-3 *Logarithmic Functions.* Inverse functions; one-to-one functions; logarithmic function; base; equivalent exponential form; properties; common logarithm; natural logarithm; calculator evaluation; solving logarithmic and exponential equations; doubling time

$y = \log_b x$ is equivalent to $x = b^y$;
$\log_b x, b > 0, b \neq 1;\quad \log x;\quad \ln x$

REVIEW EXERCISE

Work through all the problems in this chapter review and check your answers in the back of the book. Answers to all review problems are there along with section numbers in italics to indicate where each type of problem is discussed. Where weaknesses show up, review appropriate sections in the text.

A

1. Write in logarithmic form using base *e:* $u = e^v$

2. Write in logarithmic form using base 10: $x = 10^y$

3. Write in exponential form using base *e:* $\ln M = N$

4. Write in exponential form using base 10: $\log u = v$

Simplify:

5. $\dfrac{5^{x+4}}{5^{4-x}}$ **6.** $\left(\dfrac{e^u}{e^{-u}}\right)^u$

Solve for x exactly without using a calculator.

7. $\log_3 x = 2$ **8.** $\log_x 36 = 2$

9. $\log_2 16 = x$

Solve for x to three decimal places.

10. $10^x = 143.7$ **11.** $e^x = 503{,}000$

12. $\log x = 3.105$ **13.** $\ln x = -1.147$

For each polynomial function in Problems 14 and 15, find the following:

(A) The degree of the polynomial
(B) The maximum number of turning points of the graph
(C) The maximum number of x intercepts of the graph
(D) The minimum number of x intercepts of the graph
(E) The maximum number of y intercepts of the graph
(F) The minimum number of y intercepts of the graph

14. $p(x) = ax^3 + bx^2 + cx + d, a \neq 0$

15. $p(x) = ax^4 + bx^3 + cx^2 + dx + e, a \neq 0$

Each graph in Problems 16 and 17 is the graph of a polynomial function. Answer the following questions for each graph:

(A) How many turning points are on the graph?
(B) What is the minimum degree of a polynomial function that could have the graph?
(C) Is the leading coefficient of the polynomial negative or positive?

16. **17.**

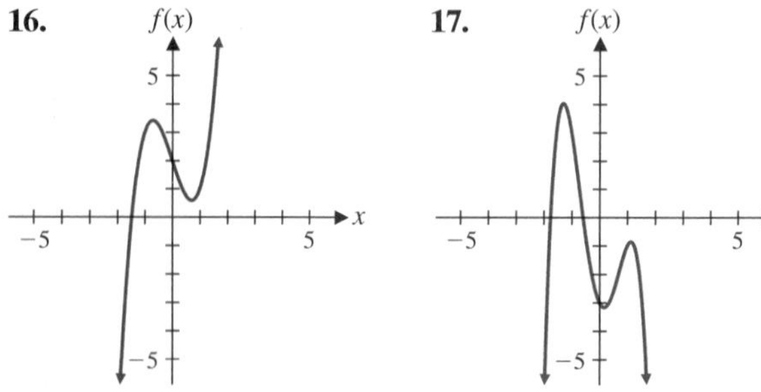

B *For each rational function in Problems 18 and 19:*
(A) Find the intercepts for the graph.
(B) Determine the domain.
(C) Find any vertical or horizonal asymptotes for the graph.
(D) Sketch any asymptotes as dashed lines. Then sketch a graph of f for $-10 \le x \le 10$ and $-10 \le y \le 10$.
(E) Graph $y = f(x)$ in a standard viewing window using a **C** *graphing utility.*

18. $f(x) = \dfrac{x + 4}{x - 2}$ **19.** $f(x) = \dfrac{3x - 4}{2 + x}$

Solve for x exactly without using a calculator.

20. $\log(x + 5) = \log(2x - 3)$
21. $2\ln(x - 1) = \ln(x^2 - 5)$
22. $9^{x-1} = 3^{1+x}$ **23.** $e^{2x} = e^{x^2-3}$
24. $2x^2e^x = 3xe^x$ **25.** $\log_{1/3} 9 = x$
26. $\log_x 8 = -3$ **27.** $\log_9 x = \frac{3}{2}$

Solve Problems 28–37 for x to four decimal places.

28. $x = 3(e^{1.49})$ **29.** $x = 230(10^{-0.161})$
30. $\log x = -2.0144$ **31.** $\ln x = 0.3618$
32. $35 = 7(3^x)$ **33.** $0.01 = e^{-0.05x}$
34. $8{,}000 = 4{,}000(1.08^x)$ **35.** $5^{2x-3} = 7.08$
36. $x = \log_2 7$ **37.** $x = \log_{0.2} 5.321$

38. How does the graph of $f(x) = x^4 - 4x^2 + 1$ compare to the graph of $y = x^4$ as we "zoom out"?

39. **C** Compare the graphs of $y = x^4$ and $y = x^4 - 4x^2 + 1$ in the following two viewing windows:
(A) $-5 \le x \le 5, -5 \le y \le 5$
(B) $-5 \le x \le 5, -500 \le y \le 500$

40. **C** Let $p(x) = 2x^4 - 11x^3 - 15x^2 - 14x - 16$. Approximate the real zeros of $p(x)$ to two decimal places.

41. **C** Let $f(x) = e^x - 1$ and $g(x) = \ln(x + 2)$. Find all points of intersection for the graphs of f and g. Round answers to two decimal places.

Simplify.

42. $e^x(e^{-x} + 1) - (e^x + 1)(e^{-x} - 1)$
43. $(e^x - e^{-x})^2 - (e^x + e^{-x})(e^x - e^{-x})$

Graph Problems 44–46 over the indicated interval. Indicate increasing and decreasing intervals.

44. $y = 2^{x-1}; [-2, 4]$ **45.** $f(t) = 10e^{-0.08t}; t \ge 0$
46. $y = \ln(x + 1); (-1, 10]$

C

47. **C** Noting that $\pi = 3.141\ 592\ 654\ldots$ and $\sqrt{2} = 1.414\ 213\ 562\ldots$, explain why the calculator results shown here are obvious. Discuss similar connections between the natural logarithmic function and the exponential function with base e.

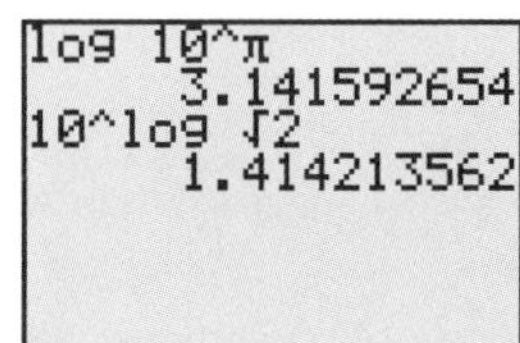

Figure for 47

Solve Problems 48–51 exactly without using a calculator.

48. $\log x - \log 3 = \log 4 - \log(x + 4)$
49. $\ln(2x - 2) - \ln(x - 1) = \ln x$
50. $\ln(x + 3) - \ln x = 2\ln 2$
51. $\log 3x^2 = 2 + \log 9x$

52. Write $\ln y = -5t + \ln c$ in an exponential form free of logarithms. Then solve for y in terms of the remaining variables.

53. Explain why 1 cannot be used as a logarithmic base.

Applications

Business & Economics

The two formulas below will be of use in some of the problems that follow:

$$A = P\left(1 + \frac{r}{m}\right)^{mt} \quad \text{Compound interest}$$

$$A = Pe^{rt} \quad \text{Continuous compound interest}$$

54. *Money growth.* If \$5,000 is invested at 12% compounded weekly, how much (to the nearest cent) will be in the account 6 years from now?

55. *Money growth.* If \$5,000 is invested at 12% compounded continuously, how much (to the nearest cent) will be in the account 6 years from now?

56. *Finance.* Find the tripling time (to the next whole year) for money invested at 15% compounded annually.

57. *Finance.* Find the doubling time (to two decimal places) for money invested at 10% compounded continuously.

58. *Minimum average cost.* The financial department of a company that manufactures in-line skates has fixed costs of \$300 per day and total costs of \$4,300 per day at an output of 100 pairs of skates per day. Assume the cost $C(x)$ is linearly related to output x.

(A) Find an expression for the cost function $C(x)$ and the average cost function $\overline{C}(x) = C(x)/x$.
(B) Sketch a graph of the average cost function for $5 \leq x \leq 200$.
(C) Identify any asymptotes.
(D) What does the average cost approach as production increases?

59. *Minimum average cost.* The cost $C(x)$ in thousands of dollars for operating a hospital for a year is given by

$$C(x) = 20x^3 - 360x^2 + 2{,}300x - 1{,}000$$

where x is the number of cases per year (in thousands). The average cost function $\overline{C}$ is given by $\overline{C}(x) = C(x)/x$.
(A) Write an equation for the average cost function.
(B) Graph the average cost function for $1 \leq x \leq 12$.
(C) Use trace and zoom or a built-in routine to find the number of cases per year the hospital should handle to have the minimum average cost. What is the minimum average cost?

60. *Equilibrium point.* A company is planning to introduce a ten-piece set of nonstick cookware. A marketing company established price–demand and price–supply tables for selected prices (Tables 1 and 2), where x is the number of cookware sets people are willing to buy and the company is willing to sell each month at a price of p dollars per set.

Table 1
PRICE–DEMAND

x	$p = D(x)$ (\$)
985	330
2,145	225
2,950	170
4,225	105
5,100	50

Table 2
PRICE–SUPPLY

x	$p = S(x)$ (\$)
985	30
2,145	75
2,950	110
4,225	155
5,100	190

(A) Find a quadratic regression model for the data in Table 1. Estimate the demand at a price level of \$180.
(B) Find a linear regression model for the data in Table 2. Estimate the supply at a price level of \$180.
(C) Does a price level of \$180 represent a stable condition, or is the price likely to increase or decrease? Explain.
(D) Use the models in parts (A) and (B) to find the equilibrium point. Write the equilibrium price to the nearest cent and the equilibrium quantity to the nearest unit.

Life Sciences

61. *Medicine.* One leukemic cell injected into a healthy mouse will divide into 2 cells in about $\frac{1}{2}$ day. At the end of the day these 2 cells will divide into 4. This doubling continues until 1 billion cells are formed; then the animal dies with leukemic cells in every part of the body.
(A) Write an equation that will give the number N of leukemic cells at the end of t days.
(B) When, to the nearest day, will the mouse die?

62. *Marine biology.* The intensity of light entering water is reduced according to the exponential equation

$$I = I_0 e^{-kd}$$

where I is the intensity d feet below the surface, I_0 is the intensity at the surface, and k is the coefficient of extinction. Measurements in the Sargasso Sea in the West Indies have indicated that half of the surface light reaches a depth of 73.6 feet. Find k (to five decimal places), and find the depth (to the nearest foot) at which 1% of the surface light remains.

63. *Agriculture.* The total United States corn consumption (in millions of bushels) is shown in Table 3 for selected years since 1975. Let x represent years since 1900.

Table 3
CORN CONSUMPTION

YEAR	x	TOTAL CONSUMPTION (MILLION BUSHELS)
1975	75	522
1980	80	659
1985	85	1,152
1990	90	1,373
1995	95	1,690

Source: U.S. Department of Agriculture

(A) Find a logarithmic regression model ($y = a + b \ln x$) for the data. Estimate (to the nearest million bushels) the total consumption in 1996 and in 2010.

(B) The actual consumption in 1996 was 1,583 million bushels. How does this compare with the estimated consumption in part (A)? What effect will this additional 1996 information have on the estimate for 2010? Explain.

Social Sciences

64. *Population growth.* Many countries have a population growth rate of 3% (or more) per year. At this rate, how many years (to the nearest tenth of a year) will it take a population to double? Use the annual compounding growth model $P = P_0(1 + r)^t$.

65. *Population growth.* Repeat Problem 64 using the continuous compounding growth model $P = P_0e^{rt}$.

66.  *Medicare.* The annual expenditures for Medicare (in billions of dollars) by the United States government for selected years since 1980 are shown in Table 4. Let x represent years since 1980.

Table 4

MEDICARE EXPENDITURES

YEAR	BILLION $
1980	37
1985	72
1990	111
1995	181

Source: U.S. Bureau of the Census

(A) Find an exponential regression model ($y = ab^x$) for the data. Estimate (to the nearest billion) the total expenditures in 1996 and in 2010.

(B) When will the total expenditures reach 500 billion dollars?

Group Activity 1 Comparing the Growth of Exponential and Polynomial Functions, and Logarithmic and Root Functions

(A) An exponential function such as $f(x) = 2^x$ increases extremely rapidly for large values of x, more rapidly than any polynomial function. Show that the graphs of $f(x) = 2^x$ and $g(x) = x^2$ intersect three times. The intersection points divide the x axis into four regions. Describe which function is greater than the other relative to each region.

(B) A logarithmic function such as $r(x) = \ln x$ increases extremely slowly for large values of x, more slowly than a function like $s(x) = \sqrt[3]{x}$. Sketch graphs of both functions in the same coordinate system for $x > 0$, and determine how many times the two graphs intersect. Describe which function is greater than the other relative to the regions determined by the intersection points.

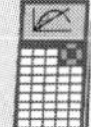

Group Activity 2 Comparing Regression Models

We have used polynomial, exponential, and logarithmic regression models to fit curves to data sets. And there are other equations that can be used for curve fitting. (The TI-83 graphing calculator has twelve different equations on its STAT-CALC menu.) How can we determine which equation provides the best fit for a given set of data? There are two principal ways to select models. The first is to use information about the type of data to help make a choice. For example, we expect the weight of a fish to be related to the cube of its length. And we expect most populations to grow exponentially, at least over the short term. The second method for choosing between equations involves developing a measure of how close an equation fits a given data set. This is best introduced through an example. Consider the data set in Figure 1, where L1 represents the x coordinates and L2 represents the y coordinates. The graph of this data set is shown in Figure 2. Suppose we arbitrarily choose the equation $y_1 = 0.6x + 2$ to model the data (Fig. 3).

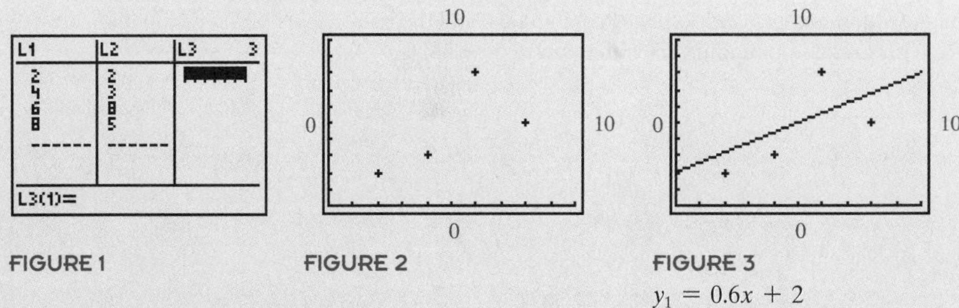

FIGURE 1 FIGURE 2 FIGURE 3 $y_1 = 0.6x + 2$

To measure how well the graph of y_1 fits the data, we examine the difference between the y coordinates in the data set and the corresponding y coordinates on the graph of y_1 (L3 in Figs. 4 and 5). Each of these differences is called a **residual.** The most commonly accepted measure of the fit provided by a given model is the **sum of the squares of the residuals (SSR).** Computing this quantity is a simple matter on a graphing calculator (Fig. 6) or a spreadsheet (Fig. 7).

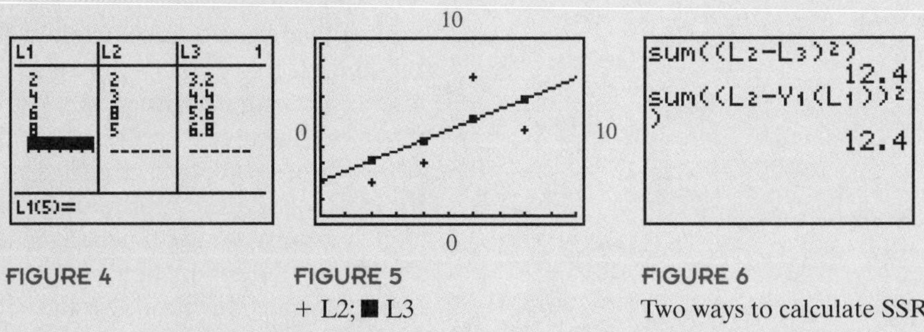

FIGURE 4 FIGURE 5 + L2; ■ L3 FIGURE 6 Two ways to calculate SSR

	A	B	C	D	E
1	Data Set				
2	x	y	y1=0.6x + 2	Residual	Residual^2
3	2	2	3.2	-1.2	1.44
4	4	3	4.4	-1.4	1.96
5	6	8	5.6	2.4	5.76
6	8	5	6.8	-1.8	3.24
7				SSR	12.4

FIGURE 7

(A) Find the linear regression model for the data in Figure 1, compute the SSR for this equation, and compare it with the one we computed for y_1.

It turns out that among all possible linear polynomials, **the linear regression model minimizes the sum of the squares of the residuals.** For this reason, the linear regression model is often called the **least squares line.** A similar statement can be made for polynomials of any fixed degree. That is, the quadratic regression model minimizes the SSR over all quadratic polynomials, the cubic regression model minimizes the SSR over all cubic polynomials, and so on. The same statement cannot be made for exponential or logarithmic regression models. Nevertheless, the SSR can still be used to compare exponential, logarithmic, and polynomial models.

(B) Find the exponential and logarithmic regression models for the data in Figure 1, compute their SSR's, and compare with the linear model.

(C) National annual advertising expenditures for selected years since 1950 are shown in Table 1, where x is years since 1950 and y is total expenditures in billions of dollars. Which regression model would fit the data best: a quadratic model, a cubic model, or an exponential model? Use the SSR's to support your choice.

Table 1

Annual Advertising Expenditures, 1950–1995

x (YEARS)	0	5	10	15	20	25	30	35	40	45
y (BILLION $)	5.7	9.2	12.0	15.3	19.6	27.9	53.6	94.8	128.6	160.9

Source: U.S. Bureau of the Census

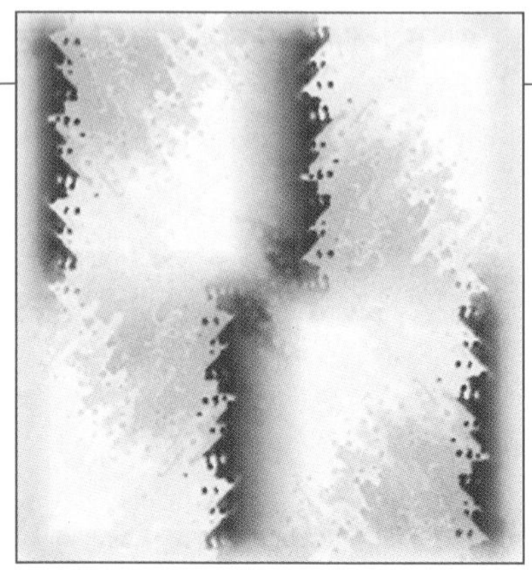

PART TWO

FINITE MATHEMATICS

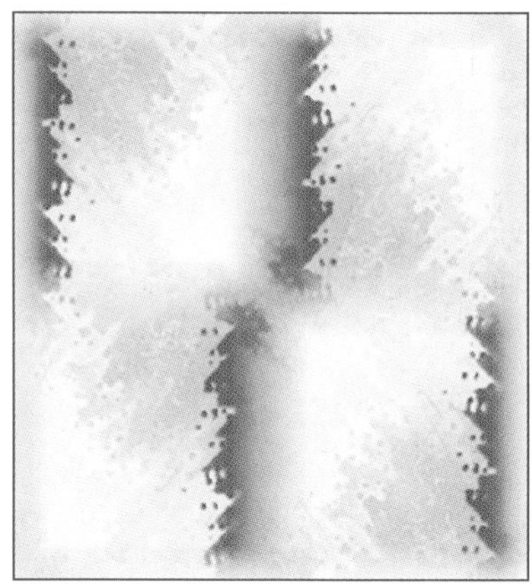

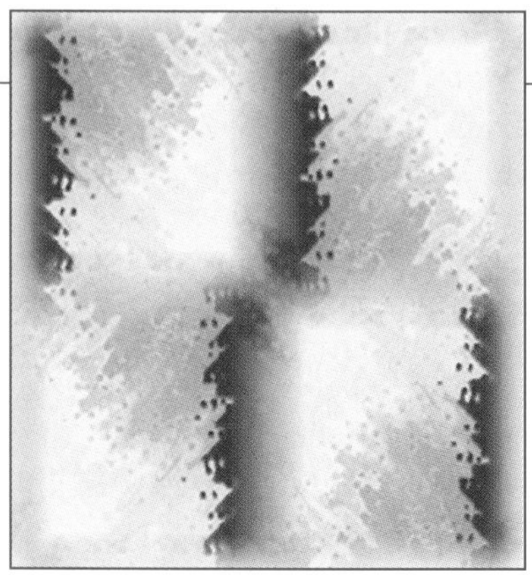

CHAPTER 3

MATHEMATICS OF FINANCE

INTRODUCTION

This chapter is independent of the others; you can study it at any time. In particular, we do not assume that you have studied Chapter 2, where a few of the topics in this chapter were briefly discussed as applications of exponential and logarithmic functions.

The low cost and convenience of the calculators that are currently available make them excellent tools for solving problems on compound interest, annuities, amortization, and so on. Any calculator with logarithmic and exponentiation keys is sufficient for solving the problems in this chapter. A graphing utility offers the additional advantage of enabling us to visualize the rate at which an investment grows, or the rate at which the principal on a loan is amortized.

If time permits, you may wish to cover arithmetic and geometric sequences, discussed in Appendix B-2, before beginning this chapter. Though not necessary, these topics will provide additional insight into some of the topics covered.

To avoid repeating the statement many times, we now point out:

Throughout the chapter, interest rates are to be converted to decimal form before they are used in a formula.

SECTION 3-1 Simple Interest

Simple interest is generally used only on short-term notes—often of duration less than 1 year. The concept of simple interest, however, forms the basis of much of the rest of the material developed in this chapter, for which time periods may be much longer than a year.

If you deposit a sum of money P in a savings account or if you borrow a sum of money P from a lending agent, then P is referred to as the **principal.** When money is borrowed—whether it is a savings institution borrowing from you when you deposit money in your account or you borrowing from a lending agent—a fee is charged for the money borrowed. This fee is rent paid for the use of another's money, just as rent is paid for the use of another's house. The fee is called **interest.** It is usually computed as a percentage (called the **interest rate**)* of the principal over a given period of time. The interest rate, unless otherwise stated, is an annual rate. **Simple interest** is given by the following formula:

Simple Interest

$$I = Prt \tag{1}$$

where

P = Principal
r = Annual simple interest rate (written as a decimal)
t = Time in years

For example, the interest on a loan of \$100 at 12% for 9 months would be

$$\begin{aligned} I &= Prt \\ &= (100)(0.12)(0.75) \\ &= \$9 \end{aligned}$$

Convert 12% to a decimal (0.12) and 9 months to years ($\frac{9}{12} = 0.75$).

At the end of 9 months, the borrower would repay the principal (\$100) plus the interest (\$9), or a total of \$109.

In general, if a principal P is borrowed at a rate r, then after t years the borrower will owe the lender an amount A that will include the principal P (the **face value** of the note) plus the interest I (the rent paid for the use of the money). Since P is the amount that is borrowed now and A is the amount that must be paid back in the future, P is often referred to as the **present value** and A as the **future value.** The formula relating A and P is as follows:

Amount—Simple Interest

$$\begin{aligned} A &= P + Prt \\ &= P(1 + rt) \end{aligned} \tag{2}$$

where

P = Principal, or present value
r = Annual simple interest rate (written as a decimal)
t = Time in years
A = Amount, or future value

*If r is the interest rate written as a decimal, then $100r\%$ is the rate using %. For example, if $r = 0.12$, then using the percent symbol, %, we have $100r\% = 100(0.12)\% = 12\%$. The expressions 0.12 and 12% are equivalent.

Given any three of the four variables A, P, r, and t in (2), we can solve for the fourth. The following examples illustrate several types of common problems that can be solved by using formula (2).

Example 1 **Total Amount Due on a Loan** Find the total amount due on a loan of \$800 at 18% simple interest at the end of 4 months.

SOLUTION To find the amount A (future value) due in 4 months, we use formula (2) with $P = 800$, $r = 0.18$, and $t = \frac{4}{12} = \frac{1}{3}$ year. Thus,

$$\begin{aligned} A &= P(1 + rt) \\ &= 800[1 + 0.18(\tfrac{1}{3})] \\ &= 800(1.06) \\ &= \$848 \end{aligned}$$

Matched Problem 1 Find the total amount due on a loan of \$500 at 12% simple interest at the end of 30 months.

Explore–Discuss 1

(A) Your dear sister has loaned you \$1,000 with the understanding that the principal plus 4% simple interest are to be repaid when you are able. How much would you owe her if you repaid the loan after 1 year? After 2 years? After 5 years? After 10 years?

(B) How is the interest after 10 years related to the interest after 1 year? After 2 years? After 5 years?

(C) Explain why your answers are consistent with the fact that for simple interest the graph of future value as a function of time is a straight line (see Fig. 1).

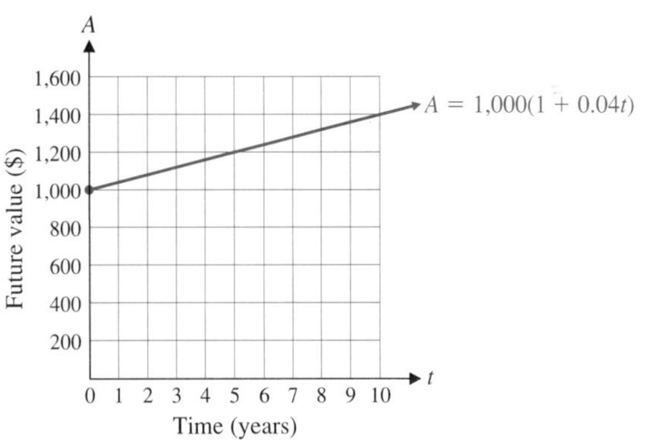

FIGURE 1

Example 2 **Present Value of an Investment** If you want to earn an annual rate of 10% on your investments, how much (to the nearest cent) should you pay for a note that will be worth \$5,000 in 9 months?

SOLUTION We again use formula (2), but now we are interested in finding the principal P (present value), given $A = \$5{,}000$, $r = 0.1$, and $t = \frac{9}{12} = 0.75$ year. Thus,

$$\begin{aligned} A &= P(1 + rt) \\ 5{,}000 &= P[1 + 0.1(0.75)] \\ 5{,}000 &= (1.075)P \\ P &= \$4{,}651.16 \end{aligned}$$

Replace A, r, and t with the given values, and solve for P.

Matched Problem 2 ➡ Repeat Example 2 with a time period of 6 months.

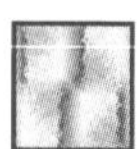

Example 3 ➡ **Interest Rate Earned on a Note** If you must pay \$960 for a note that will be worth \$1,000 in 6 months, what annual simple interest rate will you earn? (Express the answer as a percentage, correct to two decimal places.)

SOLUTION Again we use formula (2), but this time we are interested in finding r, given $P = \$960, A = \$1{,}000$, and $t = \frac{6}{12} = 0.5$ year. Thus,

$$
\begin{aligned}
A &= P(1 + rt) \\
1{,}000 &= 960[1 + r(0.5)] \\
1{,}000 &= 960 + 960r(0.5) \\
40 &= 480r \\
r &= \tfrac{40}{480} \approx 0.0833 \quad \text{or} \quad 8.33\%
\end{aligned}
$$

Replace P, A, and t with the given values, and solve for r.

Matched Problem 3 ➡ Repeat Example 3 assuming you have paid \$952 for the note.

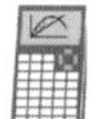

REMARK

It certainly isn't necessary to use a graphing utility to solve Example 3. However, in later sections we will encounter more complicated formulas where using a graphing utility may be desirable and, in some cases, absolutely necessary. There are a number of different ways to solve equations like $1{,}000 = 960(1 + 0.5r)$, depending on the type of graphing utility you are using. Figure 2A shows the standard graphical approach of graphing both sides of the equation and finding the intersection point. Figure 2B illustrates the output from using an equation solver on a graphing calculator, and Figure 2C shows similar output from a spreadsheet. Since the details for using an equation solver vary greatly from one graphing utility to another, you should consult the manual for your graphing utility or one of the manuals that accompany this book (see Preface) for specific instructions. Use some of the problems in this section to practice solution techniques on your graphing utility before encountering more complicated problems later in this book.

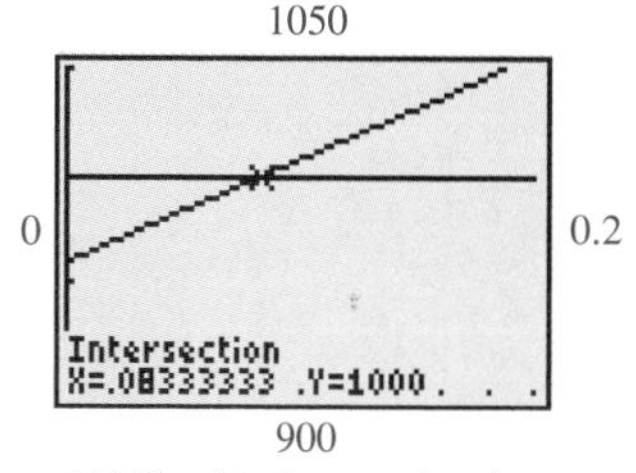

(A) Graphical approximation

A-P(1+RT)=0
A=1000
P=960
R=.08333333333…
T=.5
bound={-1E99, 1…

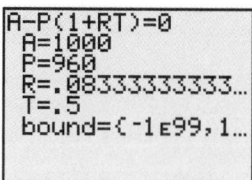

(B) Equation solver

	A	B
1	Amount:	\$1,000.00
2	Principal:	\$960.00
3	Rate:	0.08333333
4	Time:	0.5

(C) Spreadsheet

FIGURE 2
Various methods for solving equations on a graphing utility

Example 4 ➡ **Interest Rate Earned on an Investment** Suppose after buying a new car you decide to sell your old car to a friend. You accept a 270 day note for \$3,500 at 10% simple interest as payment. (Both principal and interest will be paid at the end of 270 days.) Sixty days later you find that you need the money and sell the note to a third party for \$3,550. What annual interest rate will the third party receive for the investment? (Express the answer as a percentage, correct to three decimal places.)

Solution **Step 1.** Find the amount that will be paid at the end of 270 days to the holder of the note. Some financial institutions use a 365 day year and others a 360 day year. In all the problems in this section involving days, we assume a 360 day year.*

$$\begin{aligned} A &= P(1 + rt) \\ &= \$3{,}500[1 + (0.1)(\tfrac{270}{360})] \\ &= \$3{,}762.50 \end{aligned}$$

Step 2. For the third party we are to find the annual rate of interest r required to make \$3,550 grow to \$3,762.50 in 210 days (270 − 60); that is, we are to find r (which is to be converted to $100r\%$), given $A =$ \$3,762.50, $P =$ \$3,550, and $t = \frac{210}{360}$.

$$A = P + Prt \quad \text{Solve for } r.$$

$$r = \frac{A - P}{Pt}$$

$$r = \frac{3{,}762.50 - 3{,}550}{(3{,}550)(\frac{210}{360})} = 0.102\ 62 \quad \text{or} \quad 10.262\%$$

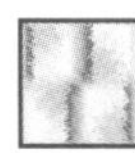

Matched Problem 4 Repeat Example 4 assuming that 90 days after it was initially signed, the note was sold to a third party for \$3,500.

Explore–Discuss 2

(A) Starting with formula (2), derive each of the following formulas:

$$P = \frac{A}{1 + rt} \qquad r = \frac{A - P}{Pt} \qquad t = \frac{A - P}{Pr}$$

(B) Explain why it is unnecessary to memorize the above formulas for P, r, and t if you know formula (2).

Answers to Matched Problems **1.** \$650 **2.** \$4,761.90 **3.** 10.08% **4.** 15.0%

Exercise 3-1

A *In Problems 1–4, make the indicated conversions assuming a 360 day year.*

1. 9.5% = ? (decimal); 60 days = ? year

2. 8.75% = ? (decimal); 3 quarters = ? year

3. 0.18 = ? (percentage); 5 months = ? year

4. 0.0525 = ? (percentage); 240 days = ? year

Using formula (1) for simple interest, find each of the indicated quantities in Problems 5–8.

5. $P =$ \$500; $r = 8\%$; $t =$ 6 months; $I = ?$

6. $P =$ \$900; $r = 10\%$; $t =$ 9 months; $I = ?$

7. $I =$ \$80; $P =$ \$500; $t =$ 2 years; $r = ?$

8. $I =$ \$40; $P =$ \$400; $t =$ 4 years; $r = ?$

B *Use formula (2) in an appropriate form to find the indicated quantities in Problems 9–12.*

9. $P =$ \$100; $r = 8\%$; $t =$ 18 months; $A = ?$

10. $P =$ \$6,000; $r = 6\%$; $t =$ 8 months; $A = ?$

11. $A =$ \$1,000; $r = 10\%$; $t =$ 15 months; $P = ?$

12. $A =$ \$8,000; $r = 12\%$; $t =$ 7 months; $P = ?$

Check Problems 11 and 12 by solving on a graphing utility.

*In other sections we will use a 365 day year. The choice will always be clearly stated.

C *In Problems 13–16, solve each formula for the indicated variable.*

13. $I = Prt$; for r

14. $I = Prt$; for P

15. $A = P + Prt$; for P

16. $A = P + Prt$; for r

17. Discuss the similarities and differences in the graphs of future value A as a function of time t if $1,000 is invested at simple interest at rates of 4%, 8%, and 12%, respectively (see the figure).

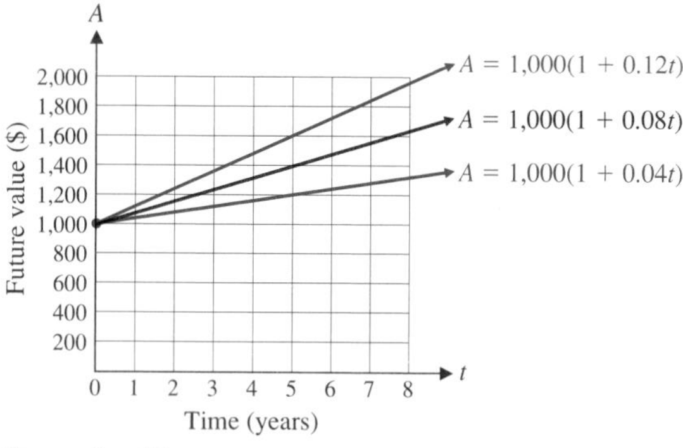

Figure for 17

18. Discuss the similarities and differences in the graphs of future value A as a function of time t for loans of $400, $800, and $1,200, respectively, each at 7.5% simple interest (see the figure).

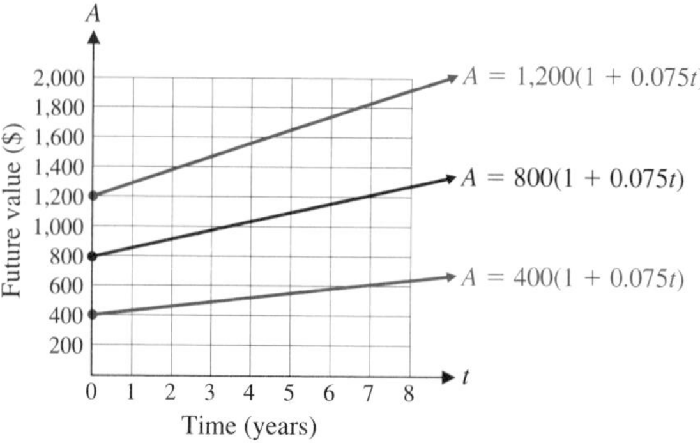

Figure for 18

APPLICATIONS*

Business & Economics

In all problems involving days, a 360 day year is assumed. When annual rates are requested as an answer, express the rate as a percentage, correct to three decimal places.

19. If $3,000 is loaned for 4 months at a 14% annual rate, how much interest is earned?

20. If $5,000 is loaned for 10 months at a 10% annual rate, how much interest is earned?

21. How much interest will you have to pay for a credit card balance of $554 that is 1 month overdue, if a 20% annual rate is charged?

22. A department store charges an 18% annual rate for overdue accounts. How much interest will be owed on an $835 account that is 2 months overdue?

23. A loan of $7,250 was repaid at the end of 8 months. What size repayment check (principal and interest) was written, if a 9% annual rate of interest was charged?

24. A loan of $10,000 was repaid at the end of 14 months. What amount (principal and interest) was repaid, if a 12% annual rate of interest was charged?

25. A loan of $4,000 was repaid at the end of 8 months with a check for $4,270. What annual rate of interest was charged?

26. A check for $3,262.50 was used to retire a 15 month $3,000 loan. What annual rate of interest was charged?

27. If you paid $30 to a loan company for the use of $1,000 for 60 days, what annual rate of interest did they charge?

28. If you paid $120 to a loan company for the use of $2,000 for 90 days, what annual rate of interest did they charge?

29. A radio commercial for a loan company states: "You only pay 50¢ a day for each $500 borrowed." If you borrow $1,500 for 120 days, what amount will you repay, and what annual interest rate is the company actually charging?

30. George finds a company that charges 70¢ per day for each $1,000 borrowed. If he borrows $3,000 for 60 days, what amount will he repay, and what annual interest rate will he be paying the company?

31. You are interested in buying a 13 week T-bill (treasury bill) from the U.S. Treasury Department. If you buy a T-bill with a maturity value of $10,000 for $9,776.94, what annual interest rate will you earn?

*The authors wish to thank Professor Roy Luke of Pierce College for his many useful suggestions of applications in this chapter.

32. If you buy a 26 week T-bill with a maturity value of \$10,000 for \$9,562.56 from the U.S. Treasury Department, what annual interest rate will you earn?

33. If an investor wants to earn an annual interest rate of 12.63% on a 13 week T-bill with a maturity value of \$10,000, how much should the investor pay for the T-bill?

34. If an investor wants to earn an annual interest rate of 10.58% on a 26 week T-bill with a maturity value of \$10,000, how much should the investor pay for the T-bill?

35. For services rendered, an attorney accepts a 90 day note for \$5,500 at 12% simple interest from a client. (Both interest and principal will be repaid at the end of 90 days.) Wishing to be able to use her money sooner, the attorney sells the note to a third party for \$5,540 after 30 days. What annual interest rate will the third party receive for the investment?

36. To complete the sale of a house, the seller accepts a 180 day note for \$10,000 at 10% simple interest. (Both interest and principal will be repaid at the end of 180 days.) Wishing to be able to use the money sooner for the purchase of another house, the seller sells the note to a third party for \$10,100 after 60 days. What annual interest rate will the third party receive for the investment?

The following buying and selling commission schedule is from a well-known discount brokerage house:

DOLLAR RANGE PER TRANSACTION	COMMISSION
\$0–2,500	\$19 + 1.6% of principal amount
\$2,501–6,000	\$44 + 0.6% of principal amount
\$6,001–22,000	\$62 + 0.3% of principal amount
\$22,001–50,000	\$84 + 0.2% of principal amount
\$50,001–500,000	\$134 + 0.1% of principal amount
\$500,000 +	\$234 + 0.08% of principal amount

Example: The commission on 500 shares at \$15 per share is \$84.50.

Taking into consideration the buying and selling commissions in the schedule, find the annual rate of interest earned by each investment in Problems 37–40.

37. An investor purchases 500 shares at \$14.20 a share, holds the stock for 39 weeks, and then sells the stock for \$16.84 a share.

38. An investor purchases 450 shares at \$64.84 a share, holds the stock for 26 weeks, and then sells the stock for \$72.08 a share.

39. An investor purchases 2,000 shares at \$23.75 a share, holds the stock for 300 days, and then sells the stock for \$26.15 a share.

40. An investor purchases 75 shares at \$31.50 a share, holds the stock for 150 days, and then sells the stock for \$35.40 a share.

SECTION 3-2 Compound Interest

- Compound Interest
- Effective Rate
- Growth and Time

COMPOUND INTEREST

If at the end of a payment period the interest due is reinvested at the same rate, then the interest as well as the original principal will earn interest during the next payment period. Interest paid on interest reinvested is called **compound interest.**

For example, suppose you deposit \$1,000 in a bank that pays 8% compounded quarterly. How much will the bank owe you at the end of a year?

Compounding quarterly means that earned interest is paid to your account at the end of each 3 month period and that interest as well as the principal earns interest for the next quarter. Using the simple interest formula (2) from the previous section, we compute the amount in the account at the end of the first quarter after interest has been paid:

$$\begin{aligned} A &= P(1 + rt) \\ &= 1{,}000[1 + 0.08(\tfrac{1}{4})] \\ &= 1{,}000(1.02) = \$1{,}020 \end{aligned}$$

Now, \$1,020 is your new principal for the second quarter. At the end of the second quarter, after interest is paid, the account will have

$$\begin{aligned} A &= \$1{,}020[1 + 0.08(\tfrac{1}{4})] \\ &= \$1{,}020(1.02) = \$1{,}040.40 \end{aligned}$$

Similarly, at the end of the third quarter, you will have

$$\begin{aligned} A &= \$1{,}040.40[1 + 0.08(\tfrac{1}{4})] \\ &= \$1{,}040.40(1.02) = \$1{,}061.21 \end{aligned}$$

Finally, at the end of the fourth quarter, the account will have

$$\begin{aligned} A &= \$1{,}061.21([1 + 0.08(\tfrac{1}{4})] \\ &= \$1{,}061.21(1.02) = \$1{,}082.43 \end{aligned}$$

How does this compound amount compare with simple interest? The amount with simple interest would be

$$\begin{aligned} A &= P(1 + rt) \\ &= \$1{,}000[1 + 0.08(1)] \\ &= \$1{,}000(1.08) = \$1{,}080 \end{aligned}$$

We see that compounding quarterly yields \$2.43 more than simple interest would provide.

Let us look over the above calculations for compound interest to see if we can uncover a pattern that might lead to a general formula for computing compound interest for arbitrary cases:

$$\begin{aligned} A &= 1{,}000(1.02) && \text{End of first quarter} \\ A &= [1{,}000(1.02)](1.02) = 1{,}000(1.02)^2 && \text{End of second quarter} \\ A &= [1{,}000(1.02)^2](1.02) = 1{,}000(1.02)^3 && \text{End of third quarter} \\ A &= [1{,}000(1.02)^3](1.02) = 1{,}000(1.02)^4 && \text{End of fourth quarter} \end{aligned}$$

It appears that at the end of n quarters, we would have

$$A = 1{,}000(1.02)^n \qquad \text{End of } n\text{th quarter}$$

or

$$\begin{aligned} A &= 1{,}000[1 + 0.08(\tfrac{1}{4})]^n \\ &= 1{,}000[1 + \tfrac{0.08}{4}]^n \end{aligned}$$

where $\frac{0.08}{4} = 0.02$ is the interest rate per quarter. Since interest rates are generally quoted as *annual nominal rates*, the **rate per compounding period** is found by dividing the annual nominal rate by the number of compounding periods per year.

In general, if P is the principal earning interest compounded m times a year at an annual rate of r, then (by repeated use of the simple interest formula, using $i = r/m$, the rate per period) the amount A at the end of each period is

$$\begin{aligned} A &= P(1 + i) && \text{End of the first period} \\ A &= [P(1 + i)](1 + i) = P(1 + i)^2 && \text{End of second period} \\ A &= [P(1 + i)^2](1 + i) = P(1 + i)^3 && \text{End of third period} \\ &\vdots \\ A &= [P(1 + i)^{n-1}](1 + i) = P(1 + i)^n && \text{End of } n\text{th period} \end{aligned}$$

We summarize this important result in the following box:

Amount—Compound Interest

$$A = P(1 + i)^n \tag{1}$$

where $i = r/m$ and

r = Annual nominal rate*
m = Number of compounding periods per year
i = Rate per compounding period
n = Total number of compounding periods
P = Principal (present value)
A = Amount (future value) at the end of n periods

*This is often shortened to "annual rate" or just "rate."

Several examples will illustrate different uses of formula (1). If any three of the four variables in (1) are given, we can solve for the fourth using some algebra and a calculator. In particular, if A, P, and i are given, we can solve for the exponent n using properties of the logarithm and a calculator.

Example 1 **Comparing Interest for Various Compounding Periods** If $1,000 is invested at 8% compounded

(A) annually (B) semiannually (C) quarterly (D) monthly

what is the amount after 5 years? Write answers to the nearest cent.

Solution (A) Compounding annually means that there is one interest payment period per year. Thus, $n = 5$ and $i = r = 0.08$.

$$\begin{aligned} A &= P(1 + i)^n \\ &= 1{,}000(1 + 0.08)^5 && \text{Use a calculator.} \\ &= 1{,}000(1.469\ 328) \\ &= \$1{,}469.33 && \text{Interest earned} = A - P = \$469.33 \end{aligned}$$

(B) Compounding semiannually means that there are two interest payment periods per year. Thus, the number of payment periods in 5 years is $n = 2(5) = 10$, and the interest rate per period is

$$i = \frac{r}{m} = \frac{0.08}{2} = 0.04$$

So,

$$\begin{aligned} A &= P(1+i)^n \\ &= 1{,}000(1+0.04)^{10} && \text{Use a calculator.} \\ &= 1{,}000(1.480\ 244) \\ &= \$1{,}480.24 && \text{Interest earned} = A - P = \$480.24 \end{aligned}$$

(C) Compounding quarterly means that there are four interest payments per year. Thus, $n = 4(5) = 20$ and $i = \frac{0.08}{4} = 0.02$. So,

$$\begin{aligned} A &= P(1+i)^n \\ &= 1{,}000(1+0.02)^{20} && \text{Use a calculator.} \\ &= 1{,}000(1.485\ 947) \\ &= \$1{,}485.95 && \text{Interest earned} = A - P = \$485.95 \end{aligned}$$

(D) Compounding monthly means that there are twelve interest payments per year. Thus, $n = 12(5) = 60$ and $i = \frac{0.08}{12} = 0.006\ 66\overline{6}$.* So,

$$\begin{aligned} A &= P(1+i)^n \\ &= 1{,}000\left(1 + \frac{0.08}{12}\right)^{60} && \text{Use a calculator.} \\ &= 1{,}000(1.489\ 846) \\ &= \$1{,}489.85 && \text{Interest earned} = A - P = \$489.85 \end{aligned}$$

Matched Problem 1 Repeat Example 1 with an annual interest rate of 6% over an 8 year period.

Notice the rather significant increase in interest earned in going from annual compounding to monthly compounding. One might wonder what happens if we compound daily, or every minute, or every second, and so on. Figure 1 shows the effect of increasing the number of compounding periods per year when $1,000 is invested at 8% compound interest for 5 years.

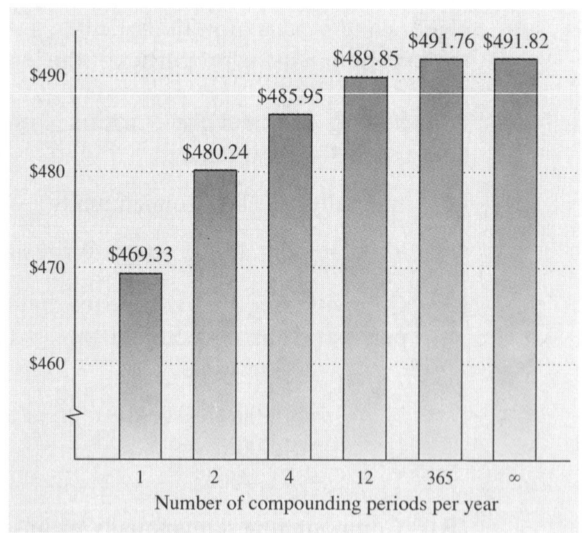

FIGURE 1
Interest on $1,000 for 5 years at 8% with various compounding periods

*Recall that the bar over the 6 indicates a repeating decimal expansion. Rounding i to a small number of decimal places, such as 0.007 or 0.0067, can result in round-off errors. To avoid this, use as many decimal places for i as your calculator is capable of displaying.

Note from Figure 1 that the difference in interest earned between annual and semiannual compounding is \$480.24 − \$469.33 = \$10.91, but the difference between semiannual and quarterly compounding is only \$5.71. Furthermore, the difference between quarterly and monthly compounding is only \$3.90, even though the number of periods is tripled, and the difference between monthly and daily compounding is only \$1.91, even though the number of compounding periods has increased dramatically. These facts suggest that the interest earned approaches a limit. The limit is reached at compounding *continuously,* which yields \$491.82—only 6¢ more than compounding daily. (If interest is compounded continuously, then $A = Pe^{rt}$, as discussed in Section 2-2.) Compare the results in Figure 1 with simple interest earned over the same time period:

$$I = Prt = 1{,}000(0.08)5 = \$400$$

Example 2 ➡ **Visualizing Investments with a Graphing Utility**

(A) Use a graphing utility to graph the growth of investments of \$1,000 at 8% simple interest and of \$1,000 at 8% compounded monthly, both over a period of 5 years.

(B) Use a graphing utility to graph the growth of both investments in part (A) during the fifth year.

SOLUTION (A) To graph the first investment, the formula for simple interest, $A = P(1 + rt)$, is entered as $y_1 = 1000*(1 + .08*x)$, where the variable x represents time in years and y_1 represents the amount after x years. Similarly, to graph the second investment, the formula for compound interest, $A = P(1 + i)^n$, is entered as $y_2 = 1000*(1 + .08/12)\text{^}(12*x)$. Note that n, the number of periods, is entered as $12*x$ (12 months per year times x years), making the variable x have the same meaning in both equations. This allows us to compare y_1 and y_2 at various values of x using the trace feature of the graphing utility, for example. Figure 2 shows the equations that represent each investment, the window variables, and the resulting graphs. (An alternative approach would be to let the variable x represent months rather than years in both equations, and to use $y_1 = 1000*(1 + .08*(x/12))$, $y_2 = 1000*(1 + .08/12)\text{^}x$, Xmin=0, Xmax=60.)

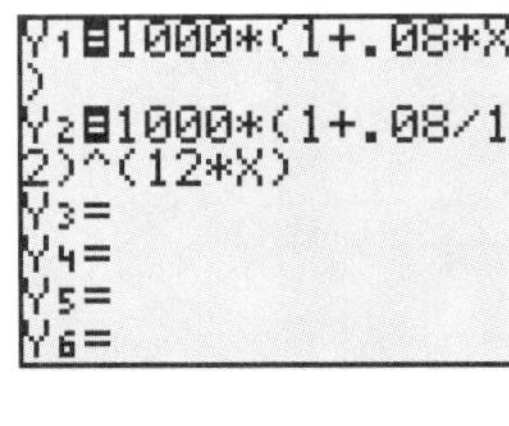

(A)

WINDOW FORMAT
Xmin=0
Xmax=5
Xscl=1
Ymin=1000
Ymax=1500
Yscl=100

(B)

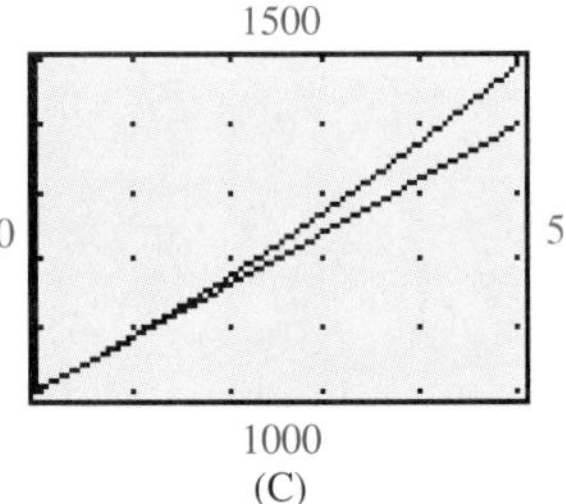

(C)

FIGURE 2

(B) Only the window variables need to be modified. From Figure 2 we see that the investment at simple interest is worth something more than \$1,300 when $x = 4$. By setting Xmin = 4, Xmax = 5, Ymin = 1300, Ymax = 1500, our viewing window will show the growth of both investments during the fifth year. We set Xscl = $\frac{1}{12}$ to obtain a scale for the x axis in months, and set Yscl = 20 to obtain a scale for the y axis in increments of \$20. Figure 3 shows the equations that represent each investment, the window variables, and the resulting graphs.

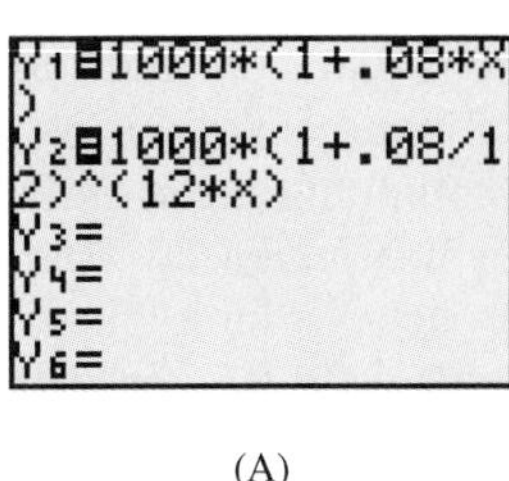

(A)

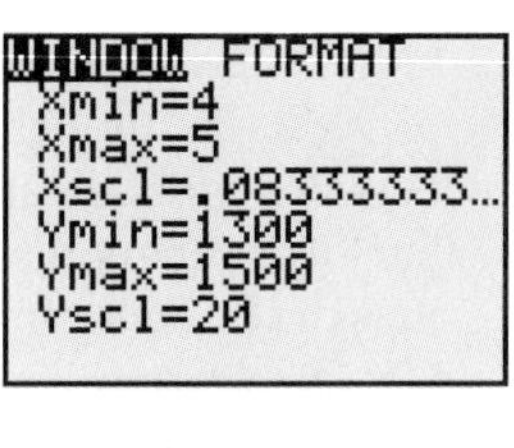

(B)

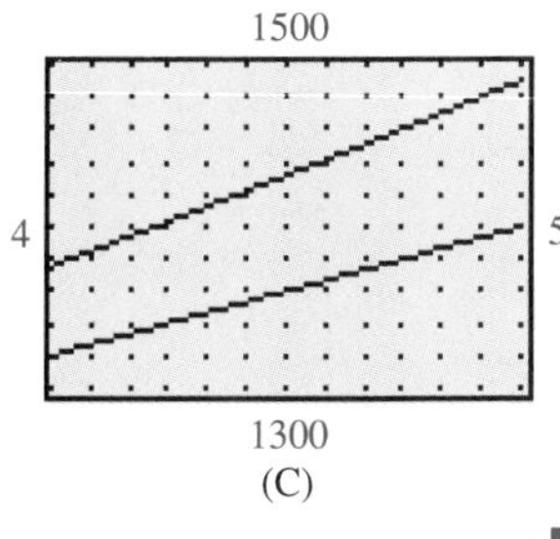

(C)

FIGURE 3

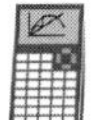

Matched Problem 2 Use a graphing utility to graph the growth of an investment of \$5,000 at 10% compounded semiannually, and of a second investment of \$5,000 at 10% compounded monthly, both over a period of 30 years.

REMARK

Refer to Figure 3C. Since compound interest is paid at the end of each compounding period, the graph of y_2 in Figure 3C is an approximation of the growth of the investment. A more accurate graph can be produced by considering only the integer part of $12x$ and placing the graphing utility in the dot mode, as shown in Figure 4. However, for graphical comparisons, it is common practice to use the approximate graph in Figure 3C instead of the exact graph in Figure 4B.

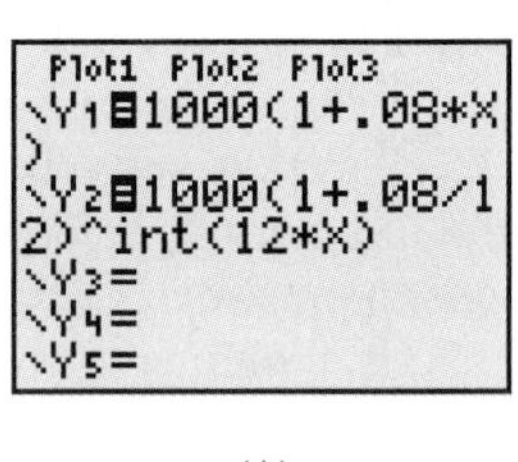

(A)

1500

4 5

1300

(B)

FIGURE 4

Explore–Discuss 1

(A) Which would be the better way to invest \$1,000: at 9% simple interest for 10 years, or at 7% compounded monthly for 10 years?

(B) Explain why the graph of future value as a function of time is a straight line for simple interest, but for compound interest the graph curves upward (see Fig. 5).

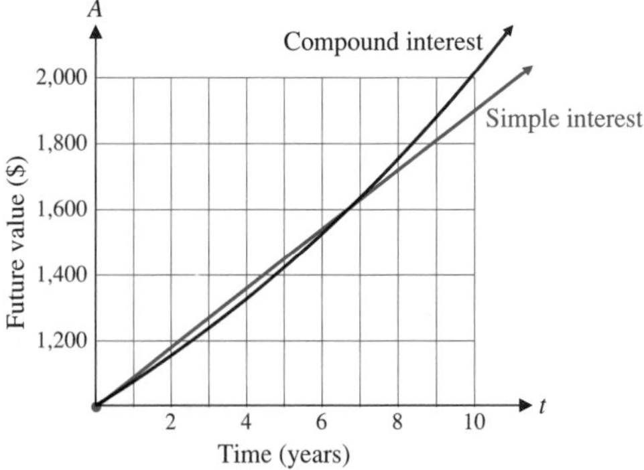

FIGURE 5

Another use of the compound interest formula is in determining how much you should invest now to have a given amount at a future date.

Example 3 **Finding Present Value** How much should you invest now at 10% compounded quarterly to have $8,000 toward the purchase of a car in 5 years?

SOLUTION We are given a future value $A = \$8{,}000$ for a compound interest investment, and we need to find the present value (principal P) given $i = \frac{0.10}{4} = 0.025$ and $n = 4(5) = 20$.

$$\begin{aligned} A &= P(1 + i)^n \\ 8{,}000 &= P(1 + 0.025)^{20} \\ P &= \frac{8{,}000}{(1 + 0.025)^{20}} \qquad \text{Use a calculator.} \\ &= \frac{8{,}000}{1.638\ 616} = \$4{,}882.17 \end{aligned}$$

Thus, your initial investment of $4,882.17 will grow to $8,000 in 5 years.

Matched Problem 3 How much should new parents invest now at 8% compounded semiannually to have $80,000 toward their child's college education in 17 years?

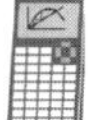

A graphing utility is a useful tool for studying compound interest. In Figure 6, we use a spreadsheet to illustrate the growth of the investment in Example 3 both numerically and graphically. Similar results can be obtained from most graphing calculators. The concepts and formulas discussed in this chapter are used extensively in spreadsheets to produce tables and graphs for a wide variety of business applications.

	A	B	C
1	Period	Interest	Amount
2	0		$4,882.17
3	1	$122.05	$5,004.22
4	2	$125.11	$5,129.33
5	3	$128.23	$5,257.56
6	4	$131.44	$5,389.00
7	5	$134.73	$5,523.73
8	6	$138.09	$5,661.82
9	7	$141.55	$5,803.37
10	8	$145.08	$5,948.45
11	9	$148.71	$6,097.16
12	10	$152.43	$6,249.59
13	11	$156.24	$6,405.83
14	12	$160.15	$6,565.98
15	13	$164.15	$6,730.13
16	14	$168.25	$6,898.38
17	15	$172.46	$7,070.84
18	16	$176.77	$7,247.61
19	17	$181.19	$7,428.80
20	18	$185.72	$7,614.52
21	19	$190.36	$7,804.88
22	20	$195.12	$8,000.00

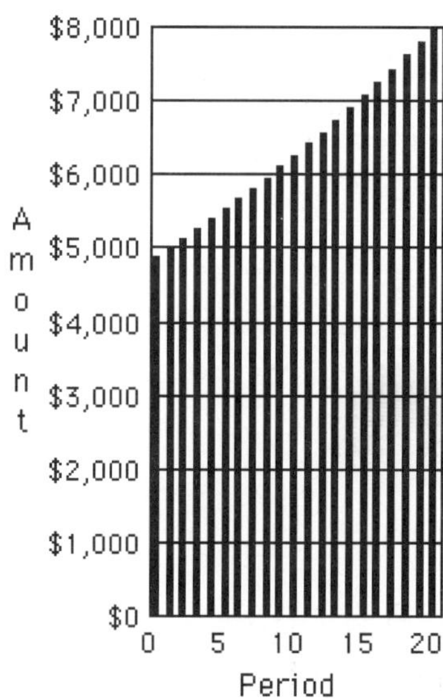

FIGURE 6

Growth of $4,882.17 at 10% compounded quarterly for 5 years

Explore–Discuss 2

(A) To become a millionaire, you intend to deposit an amount P at rate r compounded quarterly on your 25th birthday, and withdraw 1 million dollars on your 75th birthday. How much would you have to deposit if $r = 4\%$? If $r = 8\%$? If $r = 12\%$?

(B) Suppose you deposit \$2,500 on your 25th birthday. What rate r compounded quarterly must your deposit earn in order to grow to 1 million dollars by your 75th birthday?

Effective Rate

Suppose you read in the newspaper that one investment pays 15% compounded monthly and another pays 15.2% compounded semiannually. Which has the better return? A good way to compare investments is to determine their **effective rates**—the simple interest rates that would produce the same returns in *1 year* if the same principal had been invested at simple interest without compounding. (Effective rates are also called **annual yields** or **true interest rates.**)

If principal P is invested at an annual (nominal) rate r compounded m times a year, then in 1 year,

$$A = P\left(1 + \frac{r}{m}\right)^m$$

What simple interest rate will produce the same amount A in 1 year? We call this simple interest rate the effective rate, and denote it by r_e. To find r_e we proceed as follows:

$$\begin{pmatrix}\text{Amount at}\\ \text{simple interest}\\ \text{after 1 year}\end{pmatrix} = \begin{pmatrix}\text{Amount at}\\ \text{compound interest}\\ \text{after 1 year}\end{pmatrix}$$

$$P(1 + r_e) = P\left(1 + \frac{r}{m}\right)^m \qquad \text{Divide both sides by } P.$$

$$1 + r_e = \left(1 + \frac{r}{m}\right)^m \qquad \text{Isolate } r_e \text{ on the left side.}$$

$$r_e = \left(1 + \frac{r}{m}\right)^m - 1$$

Effective Rate

If principal P is invested at the annual (nominal) rate r compounded m times a year, then the effective rate r_e is given by

$$r_e = \left(1 + \frac{r}{m}\right)^m - 1$$

If interest is compounded continuously at the annual rate r, then the effective rate is given in terms of the exponential function with base e (introduced in Section 2-2) by $r_e = e^r - 1$.

Example 4 **Computing Effective Rate** A savings and loan pays 8% compounded quarterly. What is the effective rate? (Express the answer as a percentage, correct to three decimal places.)

SOLUTION

$$\begin{aligned} r_e &= \left(1 + \frac{r}{m}\right)^m - 1 \\ &= \left(1 + \frac{0.08}{4}\right)^4 - 1 \\ &= (1.02)^4 - 1 \qquad \text{Use a calculator.} \\ &= 1.082\ 432 - 1 \\ &= 0.082\ 432 \quad \text{or} \quad 8.243\% \end{aligned}$$

This shows that money invested at 8.243% simple interest earns the same amount of interest in *1 year* as money invested at 8% compounded quarterly. Thus, the effective rate of 8% compounded quarterly is 8.243%.

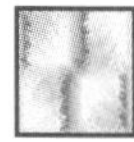

Matched Problem 4 What is the effective rate of money invested at 6% compounded quarterly?

Example 5 **Computing the Annual Nominal Rate Given the Effective Rate** A savings and loan wants to offer a CD (certificate of deposit) with a monthly compounding rate that has an effective rate of 7.5%. What annual nominal rate compounded monthly should they use? **C** Check with a graphing utility.

SOLUTION

$$\begin{aligned} r_e &= \left(1 + \frac{r}{m}\right)^m - 1 \\ 0.075 &= \left(1 + \frac{r}{12}\right)^{12} - 1 \\ 1.075 &= \left(1 + \frac{r}{12}\right)^{12} \\ \sqrt[12]{1.075} &= 1 + \frac{r}{12} \\ \sqrt[12]{1.075} - 1 &= \frac{r}{12} \\ r &= 12(\sqrt[12]{1.075} - 1) \qquad \text{Use a calculator.} \\ &= 0.072\ 539 \quad \text{or} \quad 7.254\% \end{aligned}$$

Thus, an annual nominal rate of 7.254% compounded monthly is equivalent to an effective rate of 7.5%.

C CHECK We use an equation solver on a graphing calculator to check this result (Fig. 7).

```
re=(1+r/m)^m-1
 re=.075
 r=.072539028291
 m=12
 bound={-1E99,1E99}

GRAPH RANGE ZOOM TRACE SOLVE
```

FIGURE 7

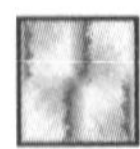

Matched Problem 5 What is the annual nominal rate compounded quarterly for a bond that has an effective rate of 8.8%? **C** Check with a graphing utility.

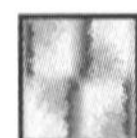

Example 6 **Comparing Two Investments** An investor has an opportunity to purchase two different notes: Note *A* pays 15% compounded monthly, and note *B* pays 15.2% compounded semiannually. Which is the better investment, assuming all else is equal?

SOLUTION Nominal rates with different compounding periods cannot be compared directly. We must first find the effective rate of each nominal rate and then compare the effective rates to determine which investment will yield the larger return.

$$\begin{aligned}
\textit{Effective Rate for Note A:}\quad r_e &= \left(1 + \frac{r}{m}\right)^m - 1 \\
&= \left(1 + \frac{0.15}{12}\right)^{12} - 1 \\
&= (1.0125)^{12} - 1 \qquad \text{Use a calculator.} \\
&= 1.160\ 755 - 1 \\
&= 0.160\ 755 \quad \text{or} \quad 16.076\% \\
\textit{Effective Rate for Note B:}\quad r_e &= \left(1 + \frac{r}{m}\right)^m - 1 \\
&= \left(1 + \frac{0.152}{2}\right)^{2} - 1 \\
&= (1.076)^2 - 1 \\
&= 1.157\ 776 - 1 \\
&= 0.157\ 776 \quad \text{or} \quad 15.778\%
\end{aligned}$$

Since the effective rate for note *A* is greater than the effective rate for note *B*, note *A* is the preferred investment.

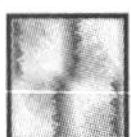

Matched Problem 6 Repeat Example 6 if note *A* pays 9% compounded monthly and note *B* pays 9.2% compounded semiannually.

GROWTH AND TIME

Investments are also compared by computing their **growth time**—the time it takes a given principal to grow to a particular value (the shorter the time, the greater the return on the investment). Example 7 illustrates two methods for making this calculation.

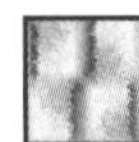

Example 7 **Computing Growth Time** How long will it take \$10,000 to grow to \$12,000 if it is invested at 9% compounded monthly?

SOLUTION **Method 1.** Use logarithms and a calculator:

$$\begin{aligned}
A &= P(1 + i)^n \\
12{,}000 &= 10{,}000\left(1 + \frac{0.09}{12}\right)^n \\
1.2 &= 1.0075^n
\end{aligned}$$

Now, solve for n by taking logarithms of both sides:

$$
\begin{aligned}
\ln 1.2 &= \ln 1.0075^n && \text{Logarithms to any base can be used; we choose the natural} \\
\ln 1.2 &= n \ln 1.0075 && \text{logarithm (base } e\text{) and use the property } \log_b M^p = p \log_b M. \\
n &= \frac{\ln 1.2}{\ln 1.0075} \\
&= 24.40 \approx 25 \text{ months} \quad \text{or} \quad 2 \text{ years and } 1 \text{ month}
\end{aligned}
$$

[*Note:* 24.40 is rounded up to 25 to guarantee reaching \$12,000, since interest is paid at the end of each month.]

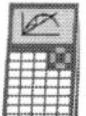

Method 2. Use a graphing utility: To solve this problem using graphical approximation techniques, we graph both sides of the equation $12{,}000 = 10{,}000(1.0075)^n$ and find that the graphs intersect at $x = n = 24.40$ months (Fig. 8A). Thus, the growth time is 25 months. We also come to the same conclusion by using an equation solver (Fig. 8B).

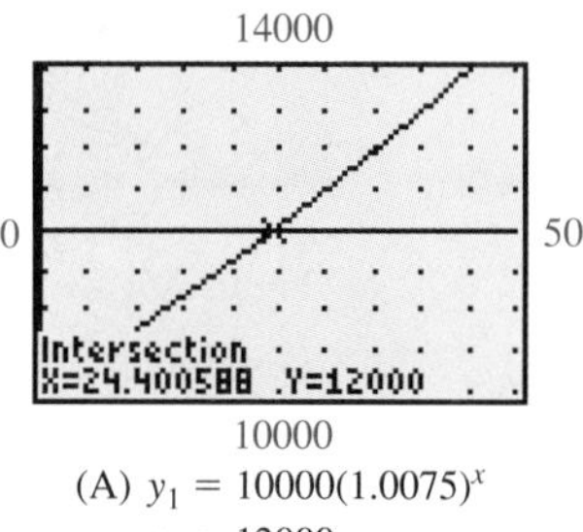

(A) $y_1 = 10000(1.0075)^x$
$y_2 = 12000$

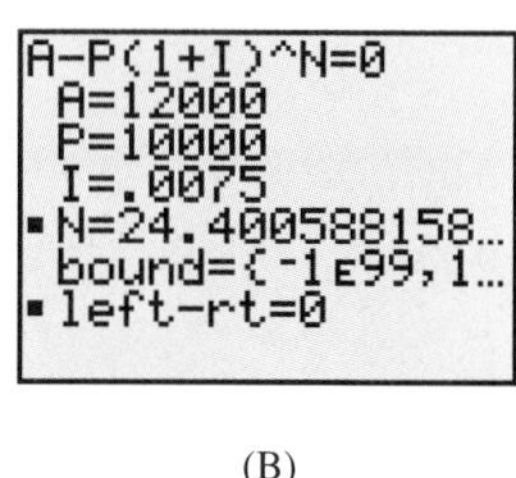

(B)

FIGURE 8

Matched Problem 7 ➡ How long will it take \$10,000 to grow to \$25,000 if it is invested at 18% compounded quarterly?

CAUTION

Each compound interest problem involves two interest rates. Referring to Example 7, $r = 0.09$ or 9% is the annual nominal compounding rate, and $i = r/12 = 0.0075$ or 0.75% is the interest rate per month. Do not confuse these two rates by using r in place of i in the compound interest formula. If interest is compounded annually, then $i = r/1 = r$. In all other cases, r and i are not the same.

Answers to Matched Problems

1. (A) \$1,593.85 (B) \$1,604.71 (C) \$1,610.32 (D) \$1,614.14

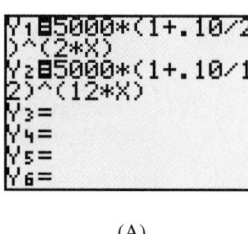

2.

Y1=5000*(1+.10/2)^(2*X)
Y2=5000*(1+.10/12)^(12*X)
Y3=
Y4=
Y5=
Y6=

(A)

100000
0
30
0

(B)

3. \$21,084.17 **4.** 6.136% **5.** 8.524%

6. Note B is better (effective rate of note A is 9.381% and of note B is 9.412%).

7. $20.82 \approx 21$ quarters, or 5 years and 3 months

EXERCISE 3-2

Find all dollar amounts to the nearest cent. When an interest rate is requested as an answer, express the rate as a percentage correct to two decimal places, unless directed otherwise. In all problems involving days, use a 365 day year.

A *In Problems 1–8, use compound interest formula (1) to find each of the indicated values.*

1. $P = \$100;\ i = 0.01;\ n = 12;\ A = ?$
2. $P = \$1{,}000;\ i = 0.015;\ n = 20;\ A = ?$
3. $P = \$800;\ i = 0.06;\ n = 25;\ A = ?$
4. $P = \$10{,}000;\ i = 0.08;\ n = 30;\ A = ?$
5. $A = \$10{,}000;\ i = 0.03;\ n = 48;\ P = ?$
6. $A = \$1{,}000;\ i = 0.015;\ n = 60;\ P = ?$
7. $A = \$18{,}000;\ i = 0.01;\ n = 90;\ P = ?$
8. $A = \$50{,}000;\ i = 0.005;\ n = 70;\ P = ?$

Given the annual rate and the compounding period in Problems 9–12, find i, the interest rate per compounding period.

9. 9% compounded monthly
10. 15% compounded annually
11. 7% compounded quarterly
12. 11% compounded semiannually

Given the rate per compounding period in Problems 13–16, find r, the annual rate.

13. 0.8% per month
14. 5% per year
15. 4.5% per half-year
16. 2.3% per quarter

B

17. If \$100 is invested at 6% compounded
(A) annually (B) quarterly (C) monthly
what is the amount after 4 years? How much interest is earned?

18. If \$2,000 is invested at 7% compounded
(A) annually (B) quarterly (C) monthly
what is the amount after 5 years? How much interest is earned?

19. If \$5,000 is invested at 18% compounded monthly, what is the amount after
(A) 2 years? (B) 4 years?

20. If \$20,000 is invested at 6% compounded monthly, what is the amount after
(A) 5 years? (B) 8 years?

21. Discuss the similarities and the differences in the graphs of future value A as a function of time t if \$1,000 is invested for 8 years and interest is compounded monthly at annual rates of 4%, 8%, and 12%, respectively (see the figure).

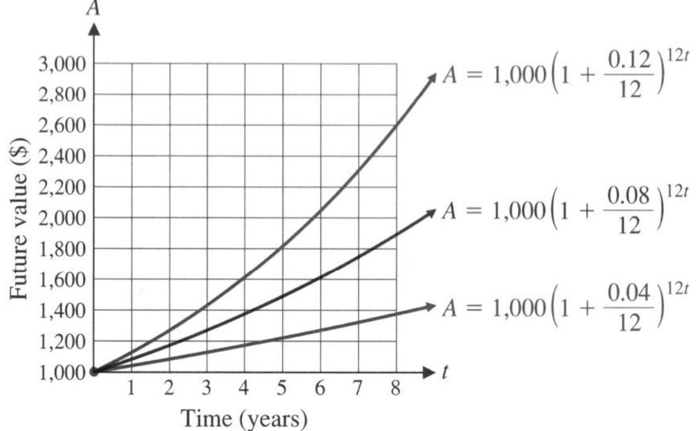

Figure for 21

22. Discuss the similarities and differences in the graphs of future value A as a function of time t for loans of \$4,000, \$8,000, and \$12,000, respectively, each at 7.5% compounded monthly for 8 years (see the figure).

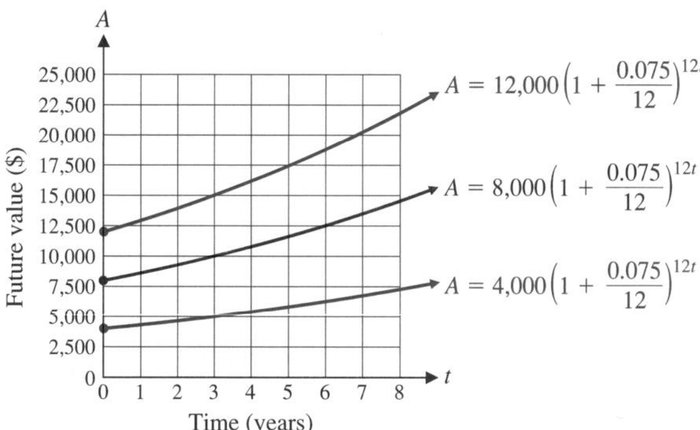

Figure for 22

23. If \$1,000 is invested in an account that earns 9.75% compounded annually for 6 years, find the interest earned during each year and the amount in the account at the end of each year. Organize your results in a table.

24. If \$2,000 is invested in an account that earns 8.25% compounded annually for 5 years, find the interest earned during each year and the amount in the account at the end of each year. Organize your results in a table.

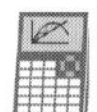

Check Problems 23 and 24 by constructing a table on a graphing utility.

25. If an investment company pays 8% compounded semiannually, how much should you deposit now to have \$10,000
(A) 5 years from now? (B) 10 years from now?

26. If an investment company pays 10% compounded quarterly, how much should you deposit now to have \$6,000
(A) 3 years from now? (B) 6 years from now?

27. What is the effective rate of interest for money invested at
(A) 10% compounded quarterly?
(B) 12% compounded monthly?

28. What is the effective rate of interest for money invested at
(A) 6% compounded monthly?
(B) 14% compounded semiannually?

29. How long will it take \$4,000 to grow to \$9,000 if it is invested at 15% compounded monthly?

30. How long will it take \$5,000 to grow to \$7,000 if it is invested at 8% compounded quarterly?

C *In Problems 31 and 32, use the compound interest formula (1) to find n to the nearest larger integer value.*

31. $A = 2P; i = 0.06; n = ?$

32. $A = 2P; i = 0.05; n = ?$

33. How long will it take money to double if it is invested at
(A) 10% compounded quarterly?
(B) 12% compounded quarterly?

34. How long will it take money to double if it is invested at
(A) 14% compounded semiannually?
(B) 10% compounded semiannually?

Applications

Business & Economics

35. A newborn child receives a \$5,000 gift toward a college education from her grandparents. How much will the \$5,000 be worth in 17 years if it is invested at 9% compounded quarterly?

36. A person with \$8,000 is trying to decide whether to purchase a car now, or to invest the money at 12% compounded semiannually and then buy a more expensive car. How much will be available for the purchase of a car at the end of 3 years?

37. What will a \$110,000 house cost 10 years from now if the inflation rate over that period averages 6% compounded annually?

38. If the inflation rate averages 8% per year compounded annually for the next 5 years, what will a car costing \$10,000 now cost 5 years from now?

39. Rental costs for office space have been going up at 7% per year compounded annually for the past 5 years. If office space rent is now \$20 per square foot per month, what were the rental rates 5 years ago?

40. In a suburb of a city, housing costs have been increasing at 8% per year compounded annually for the past 8 years. A house with a \$160,000 value now would have had what value 8 years ago?

41. If the population in a particular third-world country is growing at 4% compounded annually, how long will it take the population to double? (Round up to the next higher year if not exact.)

42. If the world population is now about 5 billion people and is growing at 2% compounded annually, how long will it take the population to grow to 8 billion people? (Round up to the next higher year if not exact.)

43. Which is the better investment and why: 9% compounded monthly or 9.3% compounded annually?

44. Which is the better investment and why: 8% compounded quarterly or 8.3% compounded annually?

45. (A) If an investment of \$100 were made in the year the Declaration of Independence was signed, and if it earned 3% compounded quarterly, how much would it be worth in 1998?
(B) Discuss the effect of compounding interest monthly, daily, and continuously (rather than quarterly) on the \$100 investment.
C (C) Use a graphing utility to graph the growth of the investment of part (A).

46. (A) Starting with formula (1), derive each of the following formulas:

$$P = \frac{A}{(1 + i)^n} \qquad i = \left(\frac{A}{P}\right)^{1/n} - 1 \qquad n = \frac{\ln A - \ln P}{\ln(1 + i)}$$

(B) Explain why it is unnecessary to memorize the above formulas for *P*, *i*, and *n* if you know formula (1).

47. You have saved \$7,000 toward the purchase of a car costing \$9,000. How long will the \$7,000 have to be invested at 9% compounded monthly to grow to \$9,000? (Round up to the next higher month if not exact.)

48. A newly married couple has \$15,000 toward the purchase of a house. For the type of house they are interested in buying, they estimate that a \$20,000 down payment will be necessary. How long will the money have to be invested at 10% compounded quarterly to grow to \$20,000? (Round up to the next higher quarter if not exact.)

49. An Individual Retirement Account (IRA) has \$20,000 in it, and the owner decides not to add any more money to the account other than interest earned at 8% compounded daily. How much will be in the account 35 years from now when the owner reaches retirement age?

50. If \$1 had been placed in a bank account at the birth of Christ and forgotten until now, how much would be in the account at the end of 2010 if the money earned 2% interest compounded annually? 2% simple interest? (Now you can see the power of compounding and see why inactive accounts are closed after a relatively short period of time.)

51. How long will it take money to double if it is invested at 14% compounded daily? 15% compounded annually? (Compute answers in years to three decimal places.)

52. How long will it take money to triple if it is invested at 10% compounded daily? 11% compounded annually? (Compute answers in years to three decimal places.)

53. In a conversation with a friend, you mention that you have two real estate investments, one that has doubled in value in the past 9 years and another that has doubled in value in the past 12 years. Your friend replies immediately that the first investment has been growing at approximately 8% compounded annually and the second at 6% compounded annually. How did your friend make these estimates? The **rule of 72** states that the annual compound rate of growth r of an investment that doubles in n years can be approximated by $r = 72/n$. Construct a table comparing the exact rate of growth and the approximate rate provided by the rule of 72 for doubling times of $n = 6, 7, \dots, 12$ years. Round both rates to one decimal place.

54. Refer to Problem 53. Show that the exact annual compound rate of growth of an investment that doubles in n years is given by $r = 100(2^{1/n} - 1)$. **C** Graph this equation and the rule of 72 on a graphing utility for $5 \le n \le 20$.

Solve Problems 55–58 using graphical approximation techniques on a graphing utility.

55. How long does it take for a \$2,400 investment at 13% compounded quarterly to be worth more than a \$3,000 investment at 9% compounded quarterly?

56. How long does it take for a \$4,800 investment at 10% compounded monthly to be worth more than a \$5,000 investment at 7% compounded monthly?

57. One investment pays 10% simple interest and another pays 7% compounded annually. Which investment would you choose? Why?

58. One investment pays 9% simple interest and another pays 6% compounded monthly. Which investment would you choose? Why?

59. What is the annual nominal rate compounded daily for a bond that has an effective rate of 7.4%?

60. What is the annual nominal rate compounded monthly for a CD that has an effective rate of 8.2%?

Problems 61–64 refer to zero coupon bonds. A ***zero coupon bond*** *is a bond that is sold now at a discount and will pay its face value at some time in the future when it matures—no interest payments are made.*

61. Parents wishing to have enough money for their child's college education 17 years from now decide to buy a \$30,000 face value zero coupon bond. If money is worth 10% compounded annually, what should they pay for the bond?

62. How much should a \$20,000 face value zero coupon bond, maturing in 10 years, be sold for now if its rate of return is to be 8% compounded annually?

63. If the parents in Problem 61 pay \$6,844.79 for the \$30,000 face value zero coupon bond, what annual compound rate of return will they be receiving?

64. If you pay \$5,893.24 for a \$12,000 face value zero coupon bond that matures in 7 years, what is your annual compound rate of return?

65. *Barron's* (a national business and financial weekly) published the following "Top Savings Deposit Yields" for money market deposit accounts:
(A) Virginia Beach Federal S&L
8.28% compounded monthly
(B) Franklin Savings
8.25% compounded daily
(C) Guaranty Federal Savings
8.25% compounded monthly
What is the effective yield for each?

66. *Barron's* also published the following "Top Savings Deposit Yields" for 1 year CD accounts:
(A) Spindletop Savings
9.00% compounded daily
(B) Alamo Savings of Texas
9.10% compounded quarterly
(C) Atlas Savings and Loan
9.00% compounded quarterly
What is the effective yield for each?

67. If you just sold a stock for $32,456.32 (net) that cost you $24,766.81 (net) 2 years ago, what annual compound rate of return did you make on your investment?

68. If you just sold a stock for $27,339.79 (net) that cost you $14,664.76 (net) 4 years ago, what annual compound rate of return did you make on your investment?

69. What annual nominal rate compounded monthly has the same effective rate as 8% compounded quarterly?

70. What annual nominal rate compounded daily has the same effective rate as 9.6% compounded monthly?

SECTION 3-3 Future Value of an Annuity; Sinking Funds

- Future Value of an Annuity
- Sinking Funds
- C Approximating Interest Rates

FUTURE VALUE OF AN ANNUITY

An **annuity** is any sequence of equal periodic payments. If payments are made at the end of each time interval, then the annuity is called an **ordinary annuity.** We will consider only ordinary annuities in this book. The amount, or **future value,** of an annuity is the sum of all payments plus all interest earned.

Suppose you decide to deposit $100 every 6 months into an account that pays 6% compounded semiannually. If you make six deposits, one at the end of each interest payment period, over 3 years, how much money will be in the account after the last deposit is made? To solve this problem, let us look at it in terms of a time line. Using the compound amount formula $A = P(1 + i)^n$, we can find the value of each deposit after it has earned compound interest up through the sixth deposit, as shown in Figure 1.

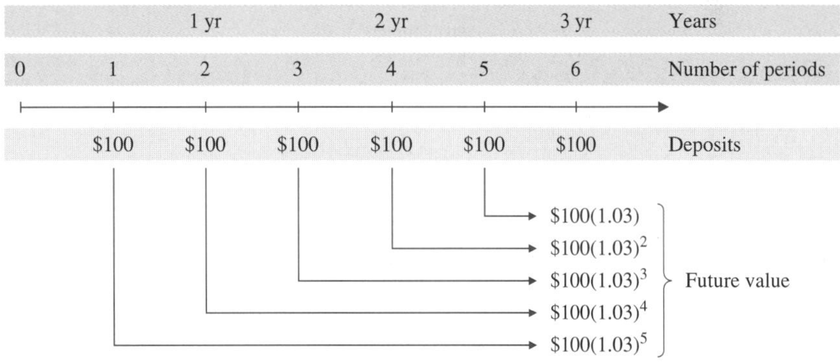

FIGURE 1

We could, of course, evaluate each of the future values in Figure 1 using a calculator and then add the results to find the amount in the account at the time of the sixth deposit—a tedious project at best. Instead, we take another approach that leads directly to a formula that will produce the same result in a few steps (even when the number of deposits is very large). We start by writing the total amount in the account after the sixth deposit in the form

$$S = 100 + 100(1.03) + 100(1.03)^2 + 100(1.03)^3 + 100(1.03)^4 + 100(1.03)^5 \quad (1)$$

We would like a simple way to sum these terms. Let us multiply each side of (1) by 1.03 to obtain

$$1.03S = 100(1.03) + 100(1.03)^2 + 100(1.03)^3 + 100(1.03)^4 + 100(1.03)^5 + 100(1.03)^6 \quad (2)$$

Subtracting (1) from (2), left side from left side and right side from right side, we obtain

$$1.03S - S = 100(1.03)^6 - 100$$ Notice how many terms drop out.

$$0.03S = 100[(1.03)^6 - 1]$$

$$S = 100\frac{(1 + 0.03)^6 - 1}{0.03} \quad (3)$$ We write S in this form to observe a general pattern.

In general, if R is the periodic deposit, i the rate per period, and n the number of periods, then the future value is given by

$$S = R + R(1 + i) + R(1 + i)^2 + \cdots + R(1 + i)^{n-1}$$ Note how this compares to (1).

and proceeding as in the above example, we obtain the general formula for the future value of an ordinary annuity:

$$S = R\frac{(1 + i)^n - 1}{i} \quad (4)$$ Note how this compares to (3).

Explore–Discuss 1

Verify formula (4) by applying to the equation

$$S = R + R(1 + i) + R(1 + i)^2 + \cdots + R(1 + i)^{n-1}$$

the technique we used above to sum equation (1).

Returning to the example above, we use a calculator to complete the problem:

$$S = 100\frac{(1.03)^6 - 1}{0.03}$$

$$= \$646.84$$

For improved accuracy, keep all values in the calculator until the end; then round to the required number of decimal places.

An alternative to the above computation of S is provided by the use of tables of values of certain functions of the mathematics of finance. One such function is the fractional factor of (4), denoted by the symbol $s_{\overline{n}|i}$ (read "*s* angle *n* at *i*"):

$$s_{\overline{n}|i} = \frac{(1 + i)^n - 1}{i}$$

Tables found in books on finance and mathematical handbooks list values of $s_{\overline{n}|i}$ for various values of n and i. To complete the computation of S, R is multiplied by $s_{\overline{n}|i}$ as indicated by (4). (A main advantage of using a calculator rather than tables is that a calculator can handle many more situations than a table, no matter how large the table.)

It is common to use *FV* (future value) for S and *PMT* (payment) for R in formula (4). Making these changes, we have the formula in the box at the top of the next page.

Future Value of an Ordinary Annuity

$$FV = PMT\frac{(1+i)^n - 1}{i} = PMT\,s_{\overline{n}|i} \qquad (5)$$

where

PMT = Periodic payment
i = Rate per period
n = Number of payments (periods)
FV = Future value (amount)

[*Note*: Payments are made at the end of each period.]

Example 1 **Future Value of an Ordinary Annuity** What is the value of an annuity at the end of 20 years if $2,000 is deposited each year into an account earning 8.5% compounded annually? How much of this value is interest?

SOLUTION To find the value of the annuity, use formula (5) with $PMT = \$2{,}000$, $i = r = 0.085$, and $n = 20$.

$$FV = PMT\frac{(1+i)^n - 1}{i}$$

$$= 2{,}000\frac{(1.085)^{20} - 1}{0.085} = \$96{,}754.03 \qquad \text{Use a calculator.}$$

To find the amount of interest earned, subtract the total amount deposited in the annuity (20 payments of $2,000) from the total value of the annuity after the 20th payment.

$$\text{Deposits} = 20(2{,}000) = \$40{,}000$$
$$\text{Interest} = \text{Value} - \text{Deposits} = 96{,}754.03 - 40{,}000 = \$56{,}754.03$$

Figure 2, which was generated using a spreadsheet, illustrates the growth of this account over 20 years.

	A	B	C	D
1	Period	Payment	Interest	Balance
2	1	$2,000.00	$0.00	$2,000.00
3	2	$2,000.00	$170.00	$4,170.00
4	3	$2,000.00	$354.45	$6,524.45
5	4	$2,000.00	$554.58	$9,079.03
6	5	$2,000.00	$771.72	$11,850.75
7	6	$2,000.00	$1,007.31	$14,858.06
8	7	$2,000.00	$1,262.94	$18,120.99
9	8	$2,000.00	$1,540.28	$21,661.28
10	9	$2,000.00	$1,841.21	$25,502.49
11	10	$2,000.00	$2,167.71	$29,670.20
12	11	$2,000.00	$2,521.97	$34,192.17
13	12	$2,000.00	$2,906.33	$39,098.50
14	13	$2,000.00	$3,323.37	$44,421.87
15	14	$2,000.00	$3,775.86	$50,197.73
16	15	$2,000.00	$4,266.81	$56,464.54
17	16	$2,000.00	$4,799.49	$63,264.02
18	17	$2,000.00	$5,377.44	$70,641.47
19	18	$2,000.00	$6,004.52	$78,645.99
20	19	$2,000.00	$6,684.91	$87,330.90
21	20	$2,000.00	$7,423.13	$96,754.03
22	Totals	$40,000.00	$56,754.03	

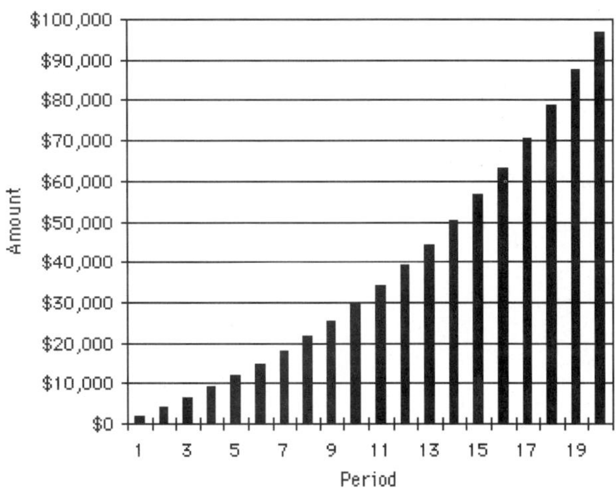

FIGURE 2
Ordinary annuity at 8.5% compounded annually for 20 years

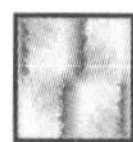

Matched Problem 1 What is the value of an annuity at the end of 10 years if \$1,000 is deposited every 6 months into an account earning 8% compounded semiannually? How much of this value is interest?

The table in Figure 2 is called a **balance sheet.** Let's take a closer look at the construction of this table. The first line is a special case because the payment is made at the end of the period and no interest is earned. Each subsequent line of the table is computed as follows:

Payment	+ Interest	+ Old balance	= New balance	
2,000	+ 0.085(2,000)	+ 2,000	= 4,170	Period 2
2,000	+ 0.085(4,170)	+ 4,170	= 6,524.45	Period 3

And so on. The amounts at the bottom of each column in the balance sheet agree with the results we obtained by using formula (5), as you would expect. Although balance sheets are appropriate for certain situations, we will concentrate on applications of formula (5). There are many important problems that can be solved only by using this formula.

Explore–Discuss 2

(A) Discuss the similarities and differences in the graphs of future value FV as a function of time t for ordinary annuities in which \$100 is deposited each month for 8 years and interest is compounded monthly at annual rates of 4%, 8%, and 12%, respectively (Fig. 3).

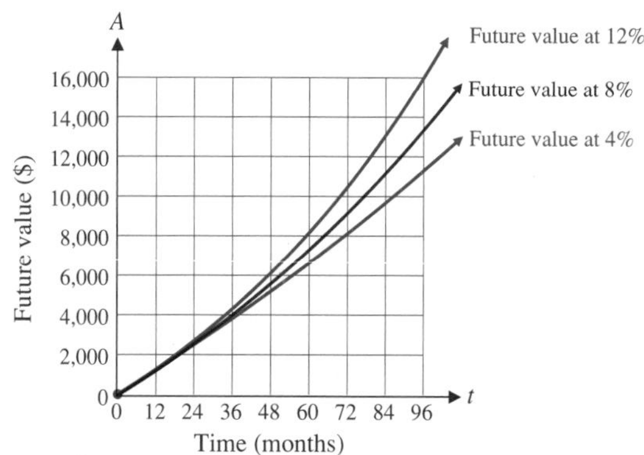

FIGURE 3

(B) Discuss the connections between the graph of the equation $y = 100t$, where t is time in months, and the graphs of part (A).

■ Sinking Funds

The formula for the future value of an ordinary annuity has another important application. Suppose the parents of a newborn child decide that on each of the child's birthdays up to the 17th year, they will deposit \$$PMT$ in an account that pays 6% compounded annually. The money is to be used for college expenses. What should the annual deposit \$$PMT$ be in order for the amount in the account to be \$80,000 after the 17th deposit?

We are given *FV, i,* and *n* in formula (5), and our problem is to find *PMT.* Thus,

$$FV = PMT\frac{(1+i)^n - 1}{i}$$

$$80{,}000 = PMT\frac{(1.06)^{17} - 1}{0.06} \qquad \text{Solve for } PMT.$$

$$PMT = 80{,}000\frac{0.06}{(1.06)^{17} - 1} \qquad \text{Use a calculator.}$$

$$= \$2{,}835.58 \text{ per year}$$

An annuity of 17 annual deposits of $2,835.58 at 6% compounded annually will amount to $80,000 in 17 years.

This is one of many examples of a similar type that are referred to as *sinking fund problems.* In general, any account that is established for accumulating funds to meet future obligations or debts is called a **sinking fund.** If the payments are to be made in the form of an ordinary annuity, then we have only to solve formula (5) for the **sinking fund payment** *PMT:*

$$PMT = FV\frac{i}{(1+i)^n - 1} \tag{6}$$

It is important to understand that formula (6), which is convenient to use, is simply a variation of formula (5). You can always find the sinking fund payment by first substituting the appropriate values into formula (5) and then solving for *PMT,* as we did in the college fund example discussed above. Or you can substitute directly into formula (6), as we do in the next example. Use whichever method is easier for you.

Example 2 **Computing the Payment for a Sinking Fund** A company estimates that it will have to replace a piece of equipment at a cost of $10,000 in 5 years. To have this money available in 5 years, a sinking fund is established by making fixed monthly payments into an account paying 6% compounded monthly. How much should each payment be?

SOLUTION To find *PMT,* we can use either formula (5) or (6). We choose formula (6) with $FV = \$10{,}000$, $i = \frac{0.06}{12} = 0.005$, and $n = 5(12) = 60$:

$$PMT = FV\frac{i}{(1+i)^n - 1}$$

$$= 10{,}000\frac{0.005}{(1.005)^{60} - 1} \qquad \text{Use a calculator.}$$

$$= \$143.33 \text{ per month}$$

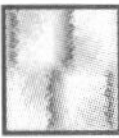

Matched Problem 2 A bond issue is approved for building a marina in a city. The city is required to make regular payments every 6 months into a sinking fund paying 6% compounded semiannually. At the end of 10 years, the bond obligation will be retired at a cost of $5,000,000. What should each payment be?

Explore–Discuss 3

Suppose you intend to establish an ordinary annuity earning 7.5% compounded monthly that will be worth 1 million dollars on your 70th birthday. How much would your monthly payments be if you started making payments at age 20? At age 35? At age 50?

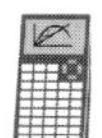

■ Approximating Interest Rates

Algebra can be used to solve the future value formula (5) for *PMT* or *n*, but not for *i*. However, graphical techniques or equation solvers can be used to approximate *i* to as many decimal places as desired.

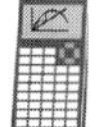

Example 3 ➡ **Approximating an Interest Rate** An individual makes monthly deposits of $100 into an ordinary annuity. After 30 years, the annuity is worth $160,000. What annual rate compounded monthly has this annuity earned during this 30 year period? Express the answer as a percentage, correct to two decimal places.

Solution Substituting $FV = \$160{,}000$, $PMT = \$100$, and $n = 30(12) = 360$ in (5) produces the following equation:

$$160{,}000 = 100\frac{(1 + i)^{360} - 1}{i}$$

We can approximate the solution to this equation by using graphical techniques (Figs. 4A and 4B) or an equation solver (Fig. 4C). From Figure 4B or 4C, we see that $i = 0.006\ 956\ 7$ and $12(i) = 0.083\ 480\ 4$. Thus, the annual rate (to two decimal places) is $r = 8.35\%$.

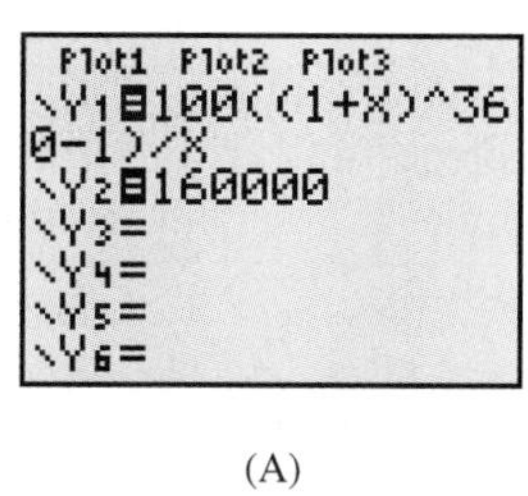

(A)

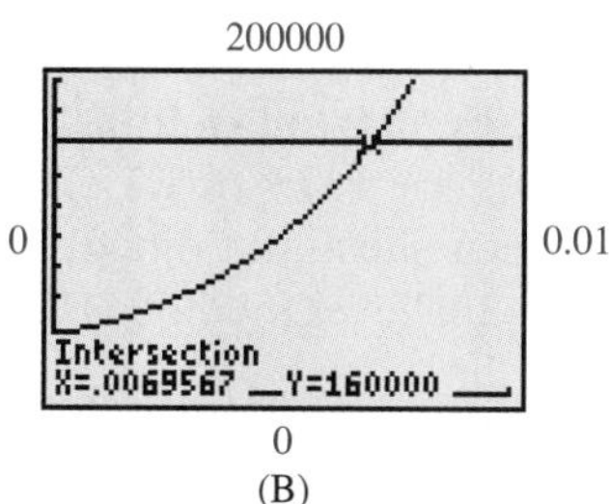

(B)

(C)

FIGURE 4

Matched Problem 3 ➡ An individual makes annual deposits of $1,000 into an ordinary annuity. After 20 years, the annuity is worth $55,000. What annual compound rate has this annuity earned during this 20 year period? Express the answer as a percentage, correct to two decimal places.

Answers to Matched Problems

1. Value: $29,778.08; Interest: $9,778.08 **2.** $186,078.54 every 6 mo for 10 yr
3. 9.64%

Exercise 3-3

In Problems 1–12, use future value formula (5) to find each of the indicated values.

A

1. $n = 20; i = 0.03; PMT = \$500; FV = ?$

2. $n = 25; i = 0.04; PMT = \$100; FV = ?$

3. $n = 40; i = 0.02; PMT = \$1{,}000; FV = ?$

4. $n = 30; i = 0.01; PMT = \$50; FV = ?$

B

5. $FV = \$3{,}000; n = 20; i = 0.02; PMT = ?$

6. $FV = \$8{,}000; n = 30; i = 0.03; PMT = ?$

7. $FV = \$5{,}000; n = 15; i = 0.01; PMT = ?$

8. $FV = \$2{,}500; n = 10; i = 0.08; PMT = ?$

C

9. $FV = \$4{,}000; i = 0.02; PMT = 200; n = ?$

10. $FV = \$8{,}000; i = 0.04; PMT = 500; n = ?$

11. $FV = \$7{,}600; PMT = \$500; n = 10; i = ?$
C (Round answer to two decimal places.)

12. $FV = \$4{,}100; PMT = \$100; n = 20; i = ?$
C (Round answer to two decimal places.)

APPLICATIONS

Business & Economics

13. What is the value of an ordinary annuity at the end of 10 years if \$500 per quarter is deposited into an account earning 8% compounded quarterly? How much of this value is interest?

14. What is the value of an ordinary annuity at the end of 20 years if \$1,000 per year is deposited into an account earning 7% compounded annually? How much of this value is interest?

15. In order to accumulate enough money for a down payment on a house, a couple deposits \$300 per month into an account paying 6% compounded monthly. If payments are made at the end of each period, how much money will be in the account in 5 years?

16. A self-employed person has a Keogh retirement plan. (This type of plan is free of taxes until money is withdrawn.) If deposits of \$7,500 are made each year into an account paying 8% compounded annually, how much will be in the account after 20 years?

17. In 5 years a couple would like to have \$25,000 for a down payment on a house. What fixed amount should be deposited each month into an account paying 9% compounded monthly?

18. A person wishes to have \$200,000 in an account for retirement 15 years from now. How much should be deposited quarterly in an account paying 8% compounded quarterly?

19. A company estimates it will need \$100,000 in 8 years to replace a computer. If it establishes a sinking fund by making fixed monthly payments into an account paying 12% compounded monthly, how much should each payment be?

20. Parents have set up a sinking fund in order to have \$120,000 in 15 years for their children's college education. How much should be paid semiannually into an account paying 10% compounded semiannually?

21. If \$1,000 is deposited at the end of each year for 5 years into an ordinary annuity earning 8.32% compounded annually, construct a balance sheet showing the interest earned during each year and the balance at the end of each year.

22. If \$2,000 is deposited at the end of each quarter for 2 years into an ordinary annuity earning 7.9% compounded quarterly, construct a balance sheet showing the interest earned during each quarter and the balance at the end of each quarter.

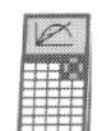

Check Problems 21 and 22 by constructing a balance sheet on a graphing utility.

23. Beginning in January, a person plans to deposit \$100 at the end of each month into an account earning 9% compounded monthly. Each year taxes must be paid on the interest earned during that year. Find the interest earned during each year for the first 3 years.

24. If \$500 is deposited each quarter into an account paying 12% compounded quarterly for 3 years, find the interest earned during each of the 3 years.

25. Why does it make sense to open an Individual Retirement Account (IRA) early in one's life? (For people in certain income brackets, money deposited into an IRA and earnings from an IRA are tax-deferred until withdrawal.) Compare the following:
    (A) Jane deposits \$2,000 a year into an IRA earning 9% compounded annually. She makes her first deposit on her 24th birthday and her last deposit on her 31st birthday (8 deposits in all). Making no additional deposits, she leaves the accumulated amount from the 8 deposits in the account, earning interest at 9% compounded annually, until her 65th birthday. How much (to the nearest dollar) will be in her account on her 65th birthday?
    (B) John procrastinates and does not make his first \$2,000 deposit into an IRA until he is 32, but then he continues to deposit \$2,000 on every birthday until he is 65 (34 deposits in all). If his account also earns 9% compounded annually, how much (to the nearest dollar) will he have in his account when he makes his last deposit on his 65th birthday?
    (*Surprise*—Jane will have more money than John!)

26. Starting on his 24th birthday, and continuing on every birthday up to and including his 65th, a person deposits \$2,000 a year into an IRA. How much (to the nearest dollar) will be in the account on the 65th birthday, if the account earns:
    (A) 6% compounded annually?
    (B) 8% compounded annually?
    (C) 10% compounded annually?
    (D) 12% compounded annually?

27. You wish to have $10,000 in 4 years to buy a car. How much should you deposit each month into an account paying 8% compounded monthly? How much interest will the account earn in the 4 years?

28. A company establishes a sinking fund to upgrade a plant in 5 years at an estimated cost of $1,500,000. How much should be invested each quarter into an account paying 9.15% compounded quarterly? How much interest will the account earn in the 5 years?

29. You can afford monthly deposits of only $150 into an account that pays 8.5% compounded monthly. How long will it be until you have $7,000 to buy a boat? (Round to the next higher month if not exact.)

30. A company establishes a sinking fund for upgrading office equipment with monthly payments of $1,000 into an account paying 10% compounded monthly. How long will it be before the account has $100,000? (Round up to the next higher month if not exact.)

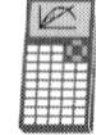

In Problems 31–34, use graphical approximation techniques or an equation solver to approximate the desired interest rate. Express each answer as a percentage, correct to two decimal places.

31. An individual makes annual payments of $1,000 into an ordinary annuity. At the end of 5 years, the amount in the annuity is $6,300. What annual nominal compounding rate has this annuity earned?

32. An individual invests $2,000 annually in an IRA. At the end of 6 years, the amount in the fund is $15,000. What annual nominal compounding rate has this fund earned?

33. At the end of each month, an employee deposits $50 into a Christmas club fund. At the end of the year, the fund contains $620. What annual nominal rate compounded monthly has this fund earned?

34. At the end of each month, an employee deposits $80 into a credit union account. At the end of 2 years, the account contains $2,100. What annual nominal rate compounded monthly has this account earned?

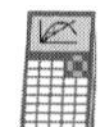

In Problems 35 and 36, use graphical approximation techniques to answer the questions.

35. When would an ordinary annuity consisting of quarterly payments of $500 at 6% compounded quarterly be worth more than a principal of $5,000 invested at 4% simple interest?

36. When would an ordinary annuity consisting of monthly payments of $200 at 5% compounded monthly be worth more than a principal of $10,000 invested at 7.5% compounded monthly?

SECTION 3-4 Present Value of an Annuity; Amortization

- Present Value of an Annuity
- Amortization
- Amortization Schedules
- A General Problem-Solving Strategy

PRESENT VALUE OF AN ANNUITY

How much should you deposit in an account paying 6% compounded semiannually in order to be able to withdraw $1,000 every 6 months for the next 3 years? (After the last payment is made, no money is to be left in the account.)

Actually, we are interested in finding the **present value** of each $1,000 that is paid out during the 3 years. We can do this by solving for P in the compound interest formula:

$$A = P(1 + i)^n$$

$$P = \frac{A}{(1 + i)^n} = A(1 + i)^{-n}$$

The rate per period is $i = \frac{0.06}{2} = 0.03$. The present value P of the first payment is $1{,}000(1.03)^{-1}$, the present value of the second payment is $1{,}000(1.03)^{-2}$, and so on. Figure 1 shows this in terms of a time line.

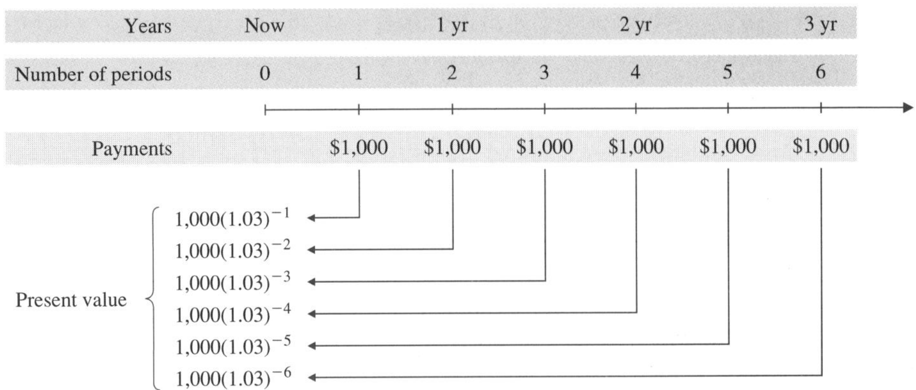

FIGURE 1

We could evaluate each of the present values in Figure 1 using a calculator and add the results to find the total present values of all the payments (which will be the amount that is needed now to buy the annuity). Since this is generally a tedious process, particularly when the number of payments is large, we will use the same device we used in the last section to produce a formula that will accomplish the same result in a couple of steps. We start by writing the sum of the present values in the form

$$P = 1{,}000(1.03)^{-1} + 1{,}000(1.03)^{-2} + \cdots + 1{,}000(1.03)^{-6} \quad (1)$$

Multiplying both sides of (1) by 1.03, we obtain

$$1.03P = 1{,}000 + 1{,}000(1.03)^{-1} + \cdots + 1{,}000(1.03)^{-5} \quad (2)$$

Now subtract (1) from (2):

$$1.03P - P = 1{,}000 - 1{,}000(1.03)^{-6} \qquad \text{Notice how many terms drop out.}$$

$$0.03P = 1{,}000[1 - (1 + 0.03)^{-6}]$$

$$P = 1{,}000\frac{1 - (1 + 0.03)^{-6}}{0.03} \qquad \text{We write } P \text{ in this form to observe a general pattern.} \quad (3)$$

In general, if R is the periodic payment, i the rate per period, and n the number of periods, then the present value of all payments is given by

$$P = R(1 + i)^{-1} + R(1 + i)^{-2} + \cdots + R(1 + i)^{-n} \qquad \text{Note how this compares to (1).}$$

Proceeding as in the above example, we obtain the general formula for the present value of an ordinary annuity:

$$P = R\frac{1 - (1 + i)^{-n}}{i} \qquad \text{Note how this compares to (3).} \quad (4)$$

Explore–Discuss 1

Verify formula (4) by applying the technique we used above to sum equation (1) to

$$P = R(1 + i)^{-1} + R(1 + i)^{-2} + \cdots + R(1 + i)^{-n}$$

Returning to the example above, we use a calculator to complete the problem:

$$P = 1{,}000\frac{1 - (1.03)^{-6}}{0.03}$$
$$= \$5{,}417.19$$

The fractional factor of (4) may be denoted by the symbol $a_{\overline{n}|i}$, read as "a angle n at i":

$$a_{\overline{n}|i} = \frac{1 - (1 + i)^{-n}}{i}$$

An alternative approach to the above computation of P is to use a table that gives values of $a_{\overline{n}|i}$ for various values of n and i. The computation of P is then completed by multiplying R by $a_{\overline{n}|i}$ as indicated by (4). (As we noted earlier, a main advantage of using a calculator is that it can handle many more situations than any table.)

It is common to use PV (present value) for P and PMT (payment) for R in formula (4). Making these changes, we have the following:

Present Value of an Ordinary Annuity

$$PV = PMT\frac{1 - (1 + i)^{-n}}{i} = PMT a_{\overline{n}|i} \qquad (5)$$

where

PMT = Periodic payment
i = Rate per period
n = Number of periods
PV = Present value of all payments

[*Note:* Payments are made at the end of each period.]

Example 1 ➡ **Present Value of an Annuity** What is the present value of an annuity that pays \$200 per month for 5 years if money is worth 6% compounded monthly?

SOLUTION To solve this problem, use formula (5) with $PMT = \$200$, $i = \frac{0.06}{12} = 0.005$, and $n = 12(5) = 60$:

$$PV = PMT\frac{1 - (1 + i)^{-n}}{i}$$
$$= 200\frac{1 - (1.005)^{-60}}{0.005} \qquad \text{Use a calculator.}$$
$$= \$10{,}345.11$$

Matched Problem 1 ➡ How much should you deposit in an account paying 8% compounded quarterly in order to receive quarterly payments of \$1,000 for the next 4 years?

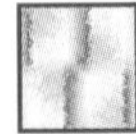

Example 2 **Retirement Planning** A person wants to make annual deposits for 25 years into an ordinary annuity that earns 9.4% compounded annually in order to then make 20 equal annual withdrawals of \$25,000, reducing the balance in the account to zero. How much must be deposited annually to accumulate sufficient funds to provide for these payments? How much total interest is earned during this entire 45 year process?

SOLUTION This problem involves both future and present values. Figure 2 illustrates the flow of money into and out of the annuity.

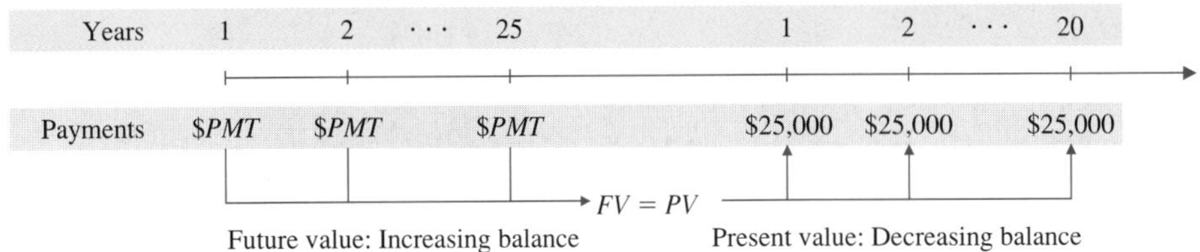

FIGURE 2

Since we are given the required withdrawals, we begin by finding the present value necessary to provide for these withdrawals. Using formula (5) with $PMT = \$25{,}000$, $i = 0.094$, and $n = 20$, we have

$$\begin{aligned} PV &= PMT\frac{1-(1+i)^{-n}}{i} \\ &= 25{,}000\frac{1-(1.094)^{-20}}{0.094} \quad \text{Use a calculator.} \\ &= \$221{,}854.69 \end{aligned}$$

Now we find the deposits that will produce a future value of \$221,854.69 in 25 years. Using formula (6) from Section 3-3 with $FV = \$221{,}854.69$, $i = 0.094$, and $n = 25$, we have

$$\begin{aligned} PMT &= FV\frac{i}{(1+i)^n - 1} \\ &= 221{,}854.69\frac{0.094}{(1.094)^{25}-1} \quad \text{Use a calculator.} \\ &= \$2{,}467.96 \end{aligned}$$

Thus, depositing \$2,467.96 annually for 25 years will provide for 20 annual withdrawals of \$25,000.

The interest earned during the entire 45 year process is

$$\begin{aligned} \text{Interest} &= (\text{Total withdrawals}) - (\text{Total deposits}) \\ &= 20(25{,}000) - 25(2{,}467.96) \\ &= \$438{,}301 \end{aligned}$$

Matched Problem 2 Refer to Example 2. If \$4,000 is deposited annually for the first 25 years, how much can be withdrawn annually for the next 20 years?

■ AMORTIZATION

The present value formula for an ordinary annuity (5) has another important use. Suppose you borrow \$5,000 from a bank to buy a car and agree to repay the loan in 36 equal monthly payments, including all interest due. If the bank charges 1% per month on the unpaid balance (12% per year compounded monthly), how much should each payment be to retire the total debt including interest in 36 months?

Actually, the bank has bought an annuity from you. The question is: If the bank pays you \$5,000 (present value) for an annuity paying them \$$PMT$ per month for 36 months at 12% interest compounded monthly, what are the monthly payments (PMT)? (Note that the value of the annuity at the end of 36 months is zero.) To find PMT, we have only to use formula (5) with $PV = \$5{,}000$, $i = 0.01$, and $n = 36$:

$$PV = PMT\frac{1 - (1 + i)^{-n}}{i}$$

$$5{,}000 = PMT\frac{1 - (1.01)^{-36}}{0.01} \qquad \text{Solve for } PMT \text{ and use a calculator.}$$

$$= \$166.07 \text{ per month}$$

At \$166.07 per month, the car will be yours after 36 months. That is, you have *amortized* the debt in 36 equal monthly payments. (*Mort* means "death"; you have "killed" the loan in 36 months.) In general, **amortizing a debt** means that the debt is retired in a given length of time by equal periodic payments that include compound interest. We are usually interested in computing the equal periodic payments. Solving the present value formula (5) for PMT in terms of the other variables, we obtain the following **amortization formula:**

$$PMT = PV\frac{i}{1 - (1 + i)^{-n}} \qquad (6)$$

Formula (6) is simply a variation of formula (5), and either formula can be used to find the periodic payment PMT.

Example 3 **Monthly Payment and Total Interest on an Amortized Debt** Assume that you buy a television set for \$800 and agree to pay for it in 18 equal monthly payments at $1\frac{1}{2}\%$ interest per month on the unpaid balance.

(A) How much are your payments? (B) How much interest will you pay?

SOLUTION (A) Use formula (6) with $PV = \$800$, $i = 0.015$, and $n = 18$:

$$PMT = PV\frac{i}{1 - (1 + i)^{-n}}$$

$$= 800\frac{0.015}{1 - (1.015)^{-18}} \qquad \text{Use a calculator.}$$

$$= \$51.04 \text{ per month}$$

(B) Total interest paid = (Amount of all payments) − (Initial loan)
= 18(\$51.04) − \$800
= \$118.72

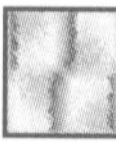

Matched Problem 3 If you sell your car to someone for \$2,400 and agree to finance it at 1% per month on the unpaid balance, how much should you receive each month to amortize the loan in 24 months? How much interest will you receive?

Explore–Discuss 2

To purchase a home, a family plans to sign a mortgage of \$70,000 at 8% on the unpaid balance. Discuss the advantages and disadvantages of a 20 year mortgage as opposed to a 30 year mortgage. Include a comparison of monthly payments and total interest paid.

■ AMORTIZATION SCHEDULES

What happens if you are amortizing a debt with equal periodic payments and at some point decide to pay off the remainder of the debt in one lump-sum payment? This occurs each time a home with an outstanding mortgage is sold. In order to understand what happens in this situation, we must take a closer look at the amortization process. We begin with an example that is simple enough to allow us to examine the effect each payment has on the debt.

Example 4 **Constructing an Amortization Schedule** If you borrow \$500 that you agree to repay in 6 equal monthly payments at 1% interest per month on the unpaid balance, how much of each monthly payment is used for interest and how much is used to reduce the unpaid balance?

SOLUTION First, we compute the required monthly payment using formula (5) or (6). We choose formula (6) with $PV = \$500, i = 0.01$, and $n = 6$:

$$
\begin{aligned}
PMT &= PV\frac{i}{1-(1+i)^{-n}} \\
&= 500\frac{0.01}{1-(1.01)^{-6}} \qquad \text{Use a calculator.} \\
&= \$86.27 \text{ per month}
\end{aligned}
$$

At the end of the first month, the interest due is

$$\$500(0.01) = \$5.00$$

The amortization payment is divided into two parts, payment of the interest due and reduction of the unpaid balance (repayment of principal):

Monthly payment		Interest due		Unpaid balance: reduction
\$86.27	=	\$5.00	+	\$81.27

The unpaid balance for the next month is

Previous unpaid balance		Unpaid balance reduction		New unpaid balance
\$500.00	−	\$81.27	=	\$418.73

At the end of the second month, the interest due on the unpaid balance of \$418.73 is

$$\$418.73(0.01) = \$4.19$$

Thus, at the end of the second month, the monthly payment of \$86.27 covers interest and unpaid balance reduction as follows:

$$\$86.27 = \$4.19 + \$82.08$$

and the unpaid balance for the third month is

$$\$418.73 - \$82.08 = \$336.65$$

This process continues until all payments have been made and the unpaid balance is reduced to zero. The calculations for each month are listed in Table 1, which is referred to as an **amortization schedule.**

Table 1

AMORTIZATION SCHEDULE

PAYMENT NUMBER	PAYMENT	INTEREST	UNPAID BALANCE REDUCTION	UNPAID BALANCE
0				$500.00
1	$ 86.27	$ 5.00	$ 81.27	418.73
2	86.27	4.19	82.08	336.65
3	8(27	3.37	82.90	253.75
4	86.27	2.54	83.73	170.02
5	86.27	1.70	84.57	85.45
6	86.30	0.85	85.45	0.00
Totals	$517.65	$17.65	$500.00	

Notice that the last payment had to be increased by $0.03 in order to reduce the unpaid balance to zero. This small discrepancy is due to round-off errors that occur in the computations. In almost all cases, the last payment must be adjusted slightly in order to obtain a final unpaid balance of exactly zero. ■

Matched Problem 4 ➡ Construct the amortization schedule for a $1,000 debt that is to be amortized in 6 equal monthly payments at 1.25% interest per month on the unpaid balance. ■

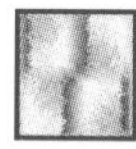

Example 5 ➡ **Equity in a Home** A family purchased a home 10 years ago for $80,000. The home was financed by paying 20% down and signing a 30 year mortgage at 9% on the unpaid balance. The net market value of the house (amount received after subtracting all costs involved in selling the house) is now $120,000, and the family wishes to sell the house. How much equity (to the nearest dollar) does the family have in the house now after making 120 monthly payments? [**Equity** = (Current net market value) − (Unpaid loan balance).]*

SOLUTION How can we find the unpaid loan balance after 10 years or 120 monthly payments? One way to proceed would be to construct an amortization schedule, but this would require a table with 120 lines. Fortunately, there is an easier way. The unpaid balance after 120 payments is the amount of the loan that can be paid off with the remaining 240 monthly payments (20 remaining years on the loan). Since the lending institution views a loan as an annuity

*We use the word *equity* in keeping with common usage. If a family wants to sell a house and buy another more expensive house, then the price of a new house that they can afford to buy will often depend on their equity in the first house, where equity is defined by the equation given here. In refinancing a house or taking out an "equity loan," the new mortgage (or second mortgage) often will be based on the equity in the house. Other, more technical definitions of equity do not concern us here.

that they bought from the family, **the unpaid balance of a loan with n remaining payments is the present value of that annuity and can be computed by using formula (5).** Since formula (5) requires knowledge of the monthly payment, we compute PMT first using formula (6).

Step 1. Find the monthly payment:

$$PMT = PV \frac{i}{1 - (1 + i)^{-n}}$$

$PV = (0.80)(\$80{,}000) = \$64{,}000$
$i = \frac{0.09}{12} = 0.0075$
$n = 12(30) = 360$

$$PMT = 64{,}000 \frac{0.0075}{1 - (1.0075)^{-360}}$$ Use a calculator.

$$= \$514.96 \text{ per month}$$

Step 2. Find the present value of a \$514.96 per month 20 year annuity:

$$PV = PMT \frac{1 - (1 + i)^{-n}}{i}$$

$PMT = \$514.96$
$n = 12(20) = 240$
$i = \frac{0.09}{12} = 0.0075$

$$PV = 514.96 \frac{1 - (1.0075)^{-240}}{0.0075}$$ Use a calculator.

$$= \$57{,}235$$ Unpaid loan balance

Step 3. Find the equity:

$$\begin{aligned} \text{Equity} &= (\text{Current net market value}) - (\text{Unpaid loan balance}) \\ &= \$120{,}000 - \$57{,}235 \\ &= \$62{,}765 \end{aligned}$$

Thus, if the family sells the house for \$120,000 net, they will have \$62,765 after paying off the unpaid loan balance of \$57,235.

Matched Problem 5 A couple purchased a home 20 years ago for \$65,000. The home was financed by paying 20% down and signing a 30 year mortgage at 8% on the unpaid balance. The net market value of the house is now \$130,000, and the couple wishes to sell the house. How much equity (to the nearest dollar) does the couple have in the house now after making 240 monthly payments?

The unpaid loan balance in Example 5 may seem a surprisingly large amount to owe after having made payments for 10 years, but long-term amortizations start out with very small reductions in the unpaid balance. For example, the interest due at the end of the very first period of the loan in Example 5 was

$$\$64{,}000(0.0075) = \$480.00$$

The first monthly payment was divided as follows:

Monthly payment		Interest due		Unpaid balance reduction
\$514.96	$-$	\$480.00	$=$	\$34.96

Thus, only \$34.96 was applied to the unpaid balance.

Explore–Discuss 3

(A) A family has an \$85,000, 30 year mortgage at 9.6% compounded monthly. Show that the monthly payments are \$720.94.

(B) Explain why the equation

$$y = 720.94\frac{1 - (1.008)^{-12(30-x)}}{0.008}$$

gives the unpaid balance of the loan after x years.

(C) Find the unpaid balance after 5 years, after 10 years, and after 15 years.

(D) When is the unpaid balance exactly half of the original \$85,000?

C (E) Solve part (D) using graphical approximation techniques on a graphing utility (see Fig. 3).

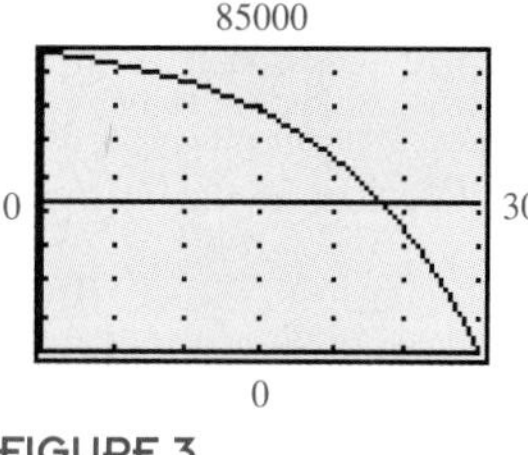

FIGURE 3

■ General Problem-Solving Strategy

After working the problems in Exercise 3-4, it is very important that you work the problems in the Review Exercise. This will give you valuable experience in distinguishing among the various types of problems we have considered in this chapter. It is impossible to completely categorize all the problems you will encounter, but you may find the following guidelines helpful in determining which of the four basic formulas is involved in a particular problem. Be aware that some problems may involve more than one of these formulas and others may not involve any of them.

A Strategy for Solving Mathematics of Finance Problems

Step 1. Determine whether the problem involves a single payment or a sequence of equal periodic payments. Simple and compound interest problems involve a single present value and a single future value. Ordinary annuities may be concerned with a present value or a future value but always involve a sequence of equal periodic payments.

Step 2. If a single payment is involved, determine whether simple or compound interest is used. Simple interest is usually used for durations of a year or less and compound interest for longer periods.

Step 3. If a sequence of periodic payments is involved, determine whether the payments are being made into an account that is increasing in value—a future value problem—or the payments are being made out of an account that is decreasing in value—a present value problem. Remember that amortization problems always involve the present value of an ordinary annuity.

Steps 1–3 will help you choose the correct formula for a problem, as indicated in Figure 4. Then you must determine the values of the quantities in the formula that are given in the problem and those that must be computed, and solve the problem.

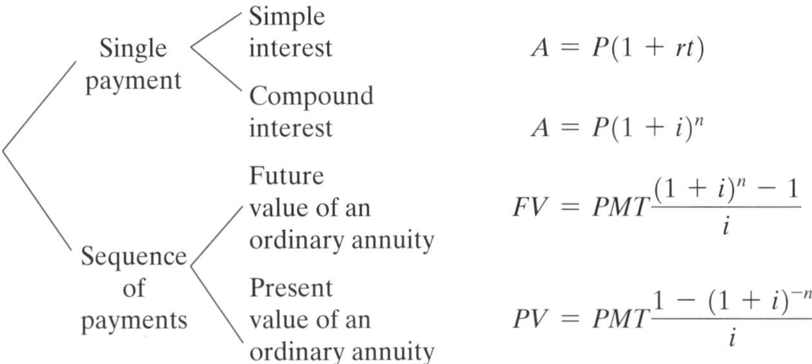

FIGURE 4

Selecting the correct formula for a problem

Answers to Matched Problems

1. \$13,577.71 **2.** \$40,519.27 **3.** PMT = \$112.98/mo; Total interest = \$311.52

4.

PAYMENT NUMBER	PAYMENT	INTEREST	UNPAID BALANCE REDUCTION	UNPAID BALANCE
0				\$1,000.00
1	\$ 174.03	\$12.50	\$ 161.53	838.47
2	174.03	10.48	163.55	674.92
3	174.03	8.44	165.59	509.33
4	174.03	6.37	167.66	341.67
5	174.03	4.27	169.76	171.91
6	174.06	2.15	171.91	0.00
Totals	\$1,044.21	\$44.21	\$1,000.00	

5. \$98,551

EXERCISE 3-4

Use formula (5) or (6) to solve each problem.

A

1. $n = 30; i = 0.04; PMT = \$200; PV = ?$

2. $n = 40; i = 0.01; PMT = \$400; PV = ?$

3. $n = 25; i = 0.025; PMT = \$250; PV = ?$

4. $n = 60; i = 0.0075; PMT = \$500; PV = ?$

B

5. $PV = \$6{,}000; n = 36; i = 0.01; PMT = ?$

6. $PV = \$1{,}200; n = 40; i = 0.025; PMT = ?$

7. $PV = \$40{,}000; n = 96; i = 0.0075; PMT = ?$

8. $PV = \$14{,}000; n = 72; i = 0.005; PMT = ?$

C

9. $PV = \$5{,}000; i = 0.01; PMT = \$200; n = ?$

10. $PV = \$20{,}000; i = 0.0175; PMT = \$500; n = ?$

11. $PV = \$9{,}000; PMT = \$600; n = 20; i = ?$
C (Round answer to three decimal places.)

12. $PV = \$12{,}000; PMT = \$400; n = 40; i = ?$
C (Round answer to three decimal places.)

APPLICATIONS

Business & Economics

13. A relative wills you an annuity paying $4,000 per quarter for the next 10 years. If money is worth 8% compounded quarterly, what is the present value of this annuity?

14. How much should you deposit in an account paying 12% compounded monthly in order to receive $1,000 per month for the next 2 years?

15. Parents of a college student wish to set up an annuity that will pay $350 per month to the student for 4 years. How much should they deposit now at 9% interest compounded monthly to establish this annuity? How much will the student receive in the 4 years?

16. A person pays $120 per month for 48 months for a car, making no down payment. If the loan costs 1.5% interest per month on the unpaid balance, what was the original cost of the car? How much total interest will be paid?

17. (A) If you buy a stereo set for $600 and agree to pay for it in 18 equal installments at 1% interest per month on the unpaid balance, how much are your monthly payments? How much interest will you pay?
    (B) Repeat part (A) for 1.5% interest per month on the unpaid balance.

18. (A) A company buys a large copy machine for $12,000 and finances it at 12% interest compounded monthly. If the loan is to be amortized in 6 years in equal monthly payments, how much is each payment? How much interest will be paid?
    (B) Repeat part (A) with 18% interest compounded monthly.

19. A sailboat costs $16,000. You pay 25% down and amortize the rest with equal monthly payments over a 6 year period. If you must pay 1.5% interest per month on the unpaid balance (18% compounded monthly), what is your monthly payment? How much interest will you pay over the 6 years?

20. A law firm buys a computerized word-processing system costing $10,000. If the firm pays 20% down and amortizes the rest with equal monthly payments over 5 years at 9% compounded monthly, what will be the monthly payment? How much interest will the firm pay?

21. Construct the amortization schedule for a $5,000 debt that is to be amortized in 8 equal quarterly payments at 4.5% interest per quarter on the unpaid balance.

22. Construct the amortization schedule for a $10,000 debt that is to be amortized in 6 equal quarterly payments at 3.5% interest per quarter on the unpaid balance.

23. A woman borrows $6,000 at 12% compounded monthly, which is to be amortized over 3 years in equal monthly payments. For tax purposes, she needs to know the amount of interest paid during each year of the loan. Find the interest paid during the first year, the second year, and the third year of the loan. [*Hint:* Find the unpaid balance after 12 payments and after 24 payments.]

24. A man establishes an annuity for retirement by depositing $50,000 into an account that pays 9% compounded monthly. Equal monthly withdrawals will be made each month for 5 years, at which time the account will have a zero balance. Each year taxes must be paid on the interest earned by the account during that year. How much interest was earned during the first year? [*Hint:* The amount in the account at the end of the first year is the present value of a 4 year annuity.]

25. Some friends tell you that they paid $25,000 down on a new house and are to pay $525 per month for 30 years. If interest is 9.8% compounded monthly, what was the selling price of the house? How much interest will they pay in 30 years?

26. A family is thinking about buying a new house costing $120,000. They must pay 20% down, and the rest is to be amortized over 30 years in equal monthly payments. If money costs 9.6% compounded monthly, what will their monthly payment be? How much total interest will be paid over the 30 years?

27. A student receives a federally backed student loan of $6,000 at 3.5% interest compounded monthly. After finishing college in 2 years, the student must amortize the loan in the next 4 years by making equal monthly payments. What will the payments be and what total interest will the student pay? [*Hint:* This is a two-part problem. First find the amount of the debt at the end of the first 2 years; then amortize this amount over the next 4 years.]

28. A person establishes a sinking fund for retirement by contributing $7,500 per year at the end of each year for 20 years. For the next 20 years, equal yearly payments are withdrawn, at the end of which time the account will have a zero balance. If money is worth 9% compounded annually, what yearly payments will the person receive for the last 20 years?

29. A family has a $75,000, 30 year mortgage at 13.2% compounded monthly. Find the monthly payment. Also find the unpaid balance after:
    (A) 10 years (B) 20 years (C) 25 years

30. A family has a \$50,000, 20 year mortgage at 10.8% compounded monthly. Find the monthly payment. Also find the unpaid balance after:
(A) 5 years (B) 10 years (C) 15 years

31. A family has a \$30,000, 20 year mortgage at 15% compounded monthly.
(A) Find the monthly payment and the total interest paid.
(B) Suppose the family decides to add an extra \$100 to its mortgage payment each month starting with the very first payment. How long will it take the family to pay off the mortgage? How much interest will the family save?

32. At the time they retire, a couple has \$200,000 in an account that pays 8.4% compounded monthly.
(A) If they decide to withdraw equal monthly payments for 10 years, at the end of which time the account will have a zero balance, how much should they withdraw each month?
(B) If they decide to withdraw \$3,000 a month until the balance in the account is zero, how many withdrawals can they make?

33. An ordinary annuity that earns 7.5% compounded monthly has a current balance of \$500,000. The owner of the account is about to retire and has to decide how much to withdraw from the account each month. Find the number of withdrawals under each of the following options:
(A) \$5,000 monthly (B) \$4,000 monthly
(C) \$3,000 monthly

34. Refer to Problem 33. If the account owner decides to withdraw \$3,000 monthly, how much is in the account after 10 years? After 20 years? After 30 years?

35. An ordinary annuity pays 7.44% compounded monthly.
(A) An individual deposits \$100 monthly for 30 years and then makes equal monthly withdrawals for the next 15 years, reducing the balance to zero. What are the monthly withdrawals? How much interest is earned during the entire 45 year process?
(B) If the individual wants to make withdrawals of \$2,000 per month for the last 15 years, how much must be deposited monthly for the first 30 years?

36. An ordinary annuity pays 6.48% compounded monthly.
(A) An individual wants to make equal monthly deposits into the account for 15 years in order to then make equal monthly withdrawals of \$1,500 for the next 20 years, reducing the balance to zero. How much should be deposited each month for the first 15 years? What is the total interest earned during this 35 year process?
(B) If the individual makes monthly deposits of \$1,000 for the first 15 years, how much can be withdrawn monthly for the next 20 years?

37. A couple wishes to borrow money using the equity in their home for collateral. A loan company will loan them up to 70% of their equity. They purchased their home 12 years ago for \$79,000. The home was financed by paying 20% down and signing a 30 year mortgage at 12% on the unpaid balance. Equal monthly payments were made to amortize the loan over the 30 year period. The net market value of the house is now \$100,000. After making their 144th payment, they applied to the loan company for the maximum loan. How much (to the nearest dollar) will they receive?

38. A person purchased a house 10 years ago for \$100,000. The house was financed by paying 20% down and signing a 30 year mortgage at 9.6% on the unpaid balance. Equal monthly payments were made to amortize the loan over a 30 year period. The owner now (after the 120th payment) wishes to refinance the house because of a need for additional cash. If the loan company agrees to a new 30 year mortgage of 80% of the new appraised value of the house, which is \$136,000, how much cash (to the nearest dollar) will the owner receive after repaying the balance of the original mortgage?

39. Discuss the similarities and differences in the graphs of unpaid balance as a function of time for 30 year mortgages of \$50,000, \$75,000, and \$100,000, respectively, each at 9% compounded monthly (see the figure). Include computations of the monthly payment and total interest paid in each case.

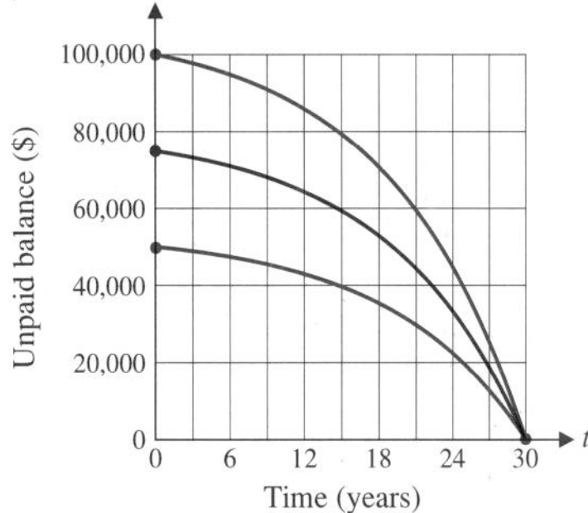

Figure for 39

40. Discuss the similarities and differences in the graphs of unpaid balance as a function of time for 30 year mortgages of \$60,000 at rates of 7%, 10%, and 13%, respectively (see the figure). Include computations of the monthly payment and total interest paid in each case.

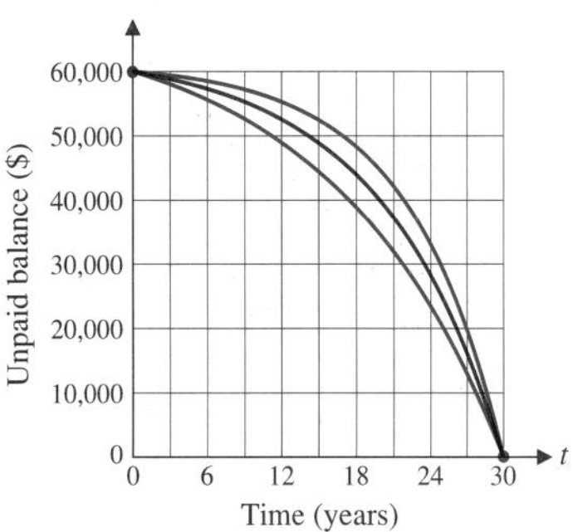

Figure for 40

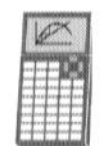

In Problems 41–44, use graphical approximation techniques or an equation solver to approximate the desired interest rate. Express each answer as a percentage, correct to two decimal places.

41. A discount electronics store offers to let you pay for a \$1,000 stereo in 12 equal \$90 installments. The store claims that since you repay \$1,080 in 1 year, the \$80 finance charge represents an 8% annual rate. This would be true if you repaid the loan in a single payment at the end of the year. But since you start repayment after 1 month, this is an amortized loan, and 8% is not the correct rate. What is the annual nominal compounding rate for this loan? [Did you expect the rate to be this high? Loans of this type are called **add-on interest loans** and were very common before Congress enacted the **Truth in Lending Act.** Now credit agreements must fully disclose all credit costs and must express interest rates in terms of the rates used in the amortization process.]

42. A \$2,000 computer can be financed by paying \$100 per month for 2 years. What is the annual nominal compounding rate for this loan?

43. The owner of a small business has received two offers of purchase. The first prospective buyer offers to pay the owner \$100,000 in cash now. The second offers to pay the owner \$10,000 now and monthly payments of \$1,200 for 10 years. In effect, the second buyer is asking the owner for a \$90,000 loan. If the owner accepts the second offer, what annual nominal compounding rate will the owner receive for financing this purchase?

44. At the time they retire, a couple has \$200,000 invested in an annuity. They can take the entire amount in a single payment, or they can receive monthly payments of \$2,000 for 15 years. If they elect to receive the monthly payments, what annual nominal compounding rate will they earn on the money invested in the annuity?

Important Terms and Symbols

3-1 *Simple Interest.* Principal; interest; interest rate; simple interest; face value; present value; future value

$$I = Prt; \quad A = P(1 + rt)$$

3-2 *Compound Interest.* Compound interest; rate per compounding period; principal (present value); amount (future value); annual nominal rate; effective rate (or annual yield); growth time; rule of 72; zero coupon bond

$$A = P(1 + i)^n; \quad i = \frac{r}{m}; \quad r_e = \left(1 + \frac{r}{m}\right)^m - 1$$

3-3 *Future Value of an Annuity; Sinking Funds.* Annuity; ordinary annuity; future value; balance sheet; sinking fund; sinking fund payment

$$\text{Future value:} \quad FV = PMT\frac{(1 + i)^n - 1}{i}$$

$$\text{Sinking fund:} \quad PMT = FV\frac{i}{(1 + i)^n - 1}$$

3-4 *Present Value of an Annuity; Amortization.* Present value; amortizing a debt; amortization schedule; equity; problem-solving strategy

$$\text{Present value:} \quad PV = PMT\frac{1 - (1 + i)^{-n}}{i}$$

$$\text{Amortization:} \quad PMT = PV\frac{i}{1 - (1 + i)^{-n}}$$

Review Exercise

Work through all the problems in this chapter review and check your answers in the back of the book. Answers to all review problems are there along with section numbers in italics to indicate where each type of problem is discussed. Where weaknesses show up, review appropriate sections in the text.

A *In Problems 1–4, find the indicated quantity, given* $A = P(1 + rt)$.

1. $A = ?; P = \$100; r = 9\%; t = 6$ months

2. $A = \$808; P = ?; r = 12\%; t = 1$ month

3. $A = \$212; P = \$200; r = 8\%; t = ?$

4. $A = \$4{,}120; P = \$4{,}000; r = ?; t = 6$ months

B *In Problems 5 and 6, find the indicated quantity, given* $A = P(1 + i)^n$.

5. $A = ?; P = \$1{,}200; i = 0.005; n = 30$

6. $A = \$5{,}000; P = ?; i = 0.0075; n = 60$

In Problems 7 and 8, find the indicated quantity, given

$$FV = PMT\frac{(1 + i)^n - 1}{i}$$

7. $FV = ?; PMT = \$1{,}000; i = 0.005; n = 60$

8. $FV = \$8{,}000; PMT = ?; i = 0.015; n = 48$

In Problems 9 and 10, find the indicated quantity, given

$$PV = PMT\frac{1 - (1 + i)^{-n}}{i}$$

9. $PV = ?; PMT = \$2{,}500; i = 0.02; n = 16$

10. $PV = \$8{,}000; PMT = ?; i = 0.0075; n = 60$

C

11. Solve the equation $2{,}500 = 1{,}000(1.06)^n$ for n to the nearest integer using:
    (A) Logarithms
    C (B) Graphical approximation techniques or an equation solver on a graphing utility

12. Solve the equation

$$5{,}000 = 100\frac{(1.01)^n - 1}{0.01}$$

for n to the nearest integer using:
    (A) Logarithms
    C (B) Graphical approximation techniques or an equation solver on a graphing utility

APPLICATIONS

Business & Economics

Find all dollar amounts correct to the nearest cent. When an interest rate is requested as an answer, express the rate as a percentage, correct to two decimal places.

13. If you borrow \$3,000 at 14% simple interest for 10 months, how much will you owe in 10 months? How much interest will you pay?

14. Grandparents deposited \$6,000 into a grandchild's account toward a college education. How much money (to the nearest dollar) will be in the account 17 years from now if the account earns 9% compounded monthly?

15. How much should you deposit initially in an account paying 10% compounded semiannually in order to have \$25,000 in 10 years?

16. A savings account pays 5.4% compounded annually. Construct a balance sheet showing the interest earned during each year and the balance at the end of each year for 4 years if:
    (A) A single deposit of \$400 is made at the beginning of the first year.
    (B) Four deposits of \$100 are made at the end of each year.

17. C One investment pays 13% simple interest and another 9% compounded annually. Which investment would you choose? Why?

18. A \$10,000 retirement account is left to earn interest at 7% compounded daily. How much money will be in the account 40 years from now when the owner reaches 65? (Use a 365 day year and round answer to the nearest dollar.)

19. Which is the better investment and why: 9% compounded quarterly or 9.25% compounded annually?

20. What is the value of an ordinary annuity at the end of 8 years if \$200 per month is deposited into an account earning 9% compounded monthly? How much of this value is interest?

21. A credit card company charges a 22% annual rate for overdue accounts. How much interest will be owed on a \$635 account 1 month overdue?

22. What will an \$8,000 car cost (to the nearest dollar) 5 years from now if the inflation rate over that period averages 5% compounded annually?

23. What would the \$8,000 car in Problem 22 have cost (to the nearest dollar) 5 years ago if the inflation rate over that period had averaged 5% compounded annually?

24. A loan of \$2,500 was repaid at the end of 10 months with a check for \$2,812.50. What annual rate of interest was charged?

25. A car salesperson tells you that you can buy the car you are looking at for \$3,000 down and \$200 a month for 48 months. If interest is 14% compounded monthly, what is the selling price of the car and how much interest will you pay during the 48 months?

26. You have $2,500 toward the purchase of a boat that will cost $3,000. How long will it take the $2,500 to grow to $3,000 if it is invested at 9% compounded quarterly? (Round up to the next higher quarter if not exact.)

27. How long will it take money to double if it is invested at 12% compounded monthly? 18% compounded monthly? (Round up to the next higher month if not exact.)

28. Starting on his 21st birthday, and continuing on every birthday up to and including his 65th, John deposits $2,000 a year into an IRA. How much (to the nearest dollar) will be in the account on John's 65th birthday, if the account earns:

(A) 7% compounded annually?

(B) 11% compounded annually?

29. If you just sold a stock for $17,388.17 (net) that cost you $12,903.28 (net) 3 years ago, what annual compound rate of return did you make on your investment?

30. If you paid $100 to a loan company for the use of $1,500 for 120 days, what annual rate of interest did they charge? (Use a 360 day year.)

31. A scholarship committee wishes to establish a sinking fund that will pay $1,500 per quarter to a student for 2 years. How much should they deposit now into an account that pays 8% compounded quarterly in order to make the scholarship payments starting 3 months from now?

32. An individual wants to establish an annuity for retirement purposes. He wants to make quarterly deposits for 20 years so that he can then make quarterly withdrawals of $5,000 for 10 years. The annuity earns 12% interest compounded quarterly.

(A) How much will have to be in the account at the time he retires?

(B) How much should be deposited each quarter for 20 years in order to accumulate the required amount?

(C) What is the total amount of interest earned during the 30 year period?

33. A state-of-the-art compact disk stereo system costs $3,000. You pay one-third down and amortize the rest with equal monthly payments over a 2 year period. If you are charged 1.5% interest per month on the unpaid balance, what is your monthly payment? How much interest will you pay over the 2 years?

34. A company decides to establish a sinking fund to replace a piece of equipment in 6 years at an estimated cost of $50,000. To accomplish this, they decide to make fixed monthly payments into an account that pays 9% compounded monthly. How much should each payment be?

35. How long will it take money to double if it is invested at 10% compounded daily? 10% compounded annually? (Express answers in years to two decimal places and assume a 365 day year.)

36. A student receives a student loan for $8,000 at 5.5% interest compounded monthly to help her finish the last 1.5 years of college. Starting 1 year after finishing college, the student must amortize the loan in the next 5 years by making equal monthly payments. What will the payments be and what total interest will the student pay?

37. A company makes a payment of $1,200 each month into a sinking fund that earns 6% compounded monthly. Use graphical approximation techniques on a graphing utility to determine when the fund will be worth $100,000.

38. A couple has a $50,000, 20 year mortgage at 9% compounded monthly. Use graphical approximation techniques on a graphing utility to determine when the unpaid balance will drop below $10,000.

39. A loan company advertises in the paper that you will pay only 8¢ a day for each $100 borrowed. What annual rate of interest are they charging? (Use a 360 day year.)

40. Construct the amortization schedule for a $1,000 debt that is to be amortized in 4 equal quarterly payments at 2.5% interest per quarter on the unpaid balance.

41. You can afford monthly deposits of only $200 into an account that pays 7.98% compounded monthly. How long will it be until you will have $2,500 to purchase a used car? (Round to the next higher month if not exact.)

42. A company establishes a sinking fund for plant retooling in 6 years at an estimated cost of $850,000. How much should be invested semiannually into an account paying 8.76% compounded semiannually? How much interest will the account earn in the 6 years?

43. Security Savings & Loan pays 9.38% compounded monthly and West Lake Savings & Loan pays 9.35% compounded daily. Which is the better investment? For each investment, express the effective rate as a percentage, correct to three decimal places. Use a 365 day year.

44. If you buy a 13 week T-bill with a maturity value of $5,000 for $4,899.08 from the U.S. Treasury Department, what annual interest rate will you earn?

45. In order to save enough money for the down payment on a condominium, a young couple deposits $200 each month into an account that pays 9% interest compounded monthly. If they need $10,000 for a down payment, how many deposits will they have to make?

46. A business borrows $80,000 at 15% interest compounded monthly for 8 years.
(A) What is the monthly payment?
(B) What is the unpaid balance at the end of the first year?
(C) How much interest was paid during the first year?

47. You unexpectedly inherit $10,000 just after you have made the 72nd monthly payment on a 30 year mortgage of $60,000 at 10.25% compounded monthly. Discuss the relative merits of using the inheritance to reduce the principal of the loan, or to buy a certificate of deposit paying 8.75% compounded monthly.

48. Your parents are considering a $75,000, 30 year mortgage to purchase a new home. The bank at which they have done business for many years offers a rate of 12% compounded monthly. A competitor is offering 11.25% compounded monthly. Would it be worthwhile for your parents to switch banks?

49. How much should a $5,000 face value zero coupon bond, maturing in 5 years, be sold for now, if its rate of return is to be 9.5% compounded annually?

50. If you pay $4,476.20 for a $10,000 face value zero coupon bond that matures in 10 years, what is your annual compound rate of return?

51. If an investor wants to earn an annual interest rate of 10.76% on a 26 week T-bill with a maturity value of $5,000, how much should the investor pay for the T-bill?

52. Two years ago you borrowed $10,000 at 12% interest compounded monthly, which was to be amortized over 5 years. Now you have acquired some additional funds and decide that you want to pay off this loan. What is the unpaid balance after making equal monthly payments for 2 years?

53. A savings and loan company pays 9% compounded monthly. What is the effective rate?

54. (A) A man deposits $2,000 in an IRA on his 21st birthday and on each subsequent birthday up to, and including, his 29th (9 deposits in all). The account earns 8% compounded annually. If he then leaves the money in the account without making any more deposits, how much will he have on his 65th birthday, assuming the account continues to earn the same rate of interest?
(B) How much would be in the account (to the nearest dollar) on his 65th birthday if he had started the deposits on his 30th birthday and continued making deposits on each birthday until (and including) his 65th birthday?

55. In a new housing development, the houses are selling for $100,000 and require a 20% down payment. The buyer is given a choice of 30 year or 15 year financing, both at 10.75% compounded monthly.
(A) What is the monthly payment for the 30 year choice? For the 15 year choice?
(B) What is the unpaid balance after 10 years for the 30 year choice? For the 15 year choice?

56. A loan company will loan up to 60% of the equity in a home. A family purchased their home 8 years ago for $83,000. The home was financed by paying 20% down and signing a 30 year mortgage at 11.25% for the balance. Equal monthly payments were made to amortize the loan over the 30 year period. The market value of the house is now $95,000. After making their 96th payment, the family applied to the loan company for the maximum loan. How much (to the nearest dollar) will they receive?

57. C A $600 stereo is financed for 6 months by making monthly payments of $110. What is the annual nominal compounding rate for this loan?

58. C An individual deposits $2,000 each year for 25 years into an IRA. When she retires immediately after making the 25th deposit, the IRA is worth $220,000.
(A) Find the interest rate earned by the IRA over the 25 year period leading up to retirement.
(B) Assume that the IRA continues to earn the interest rate found in part (A). How long can the retiree withdraw $30,000 per year? How long can she withdraw $24,000 per year?

Group Activity 1 Reducing Interest Payments on a Home Mortgage

In appreciation for her parents' financial support of her college education, a recent graduate intends to save her parents some money on the financing of their home. The parents have just made the 180th payment on a 30 year loan of $80,000 at 11%.

(A) How much would be saved in interest payments if the graduate were able to find a lender willing to refinance the unpaid balance over 15 years at 9%? At 8%?

C (B) At what interest rate would the refinancing reduce the total interest payments by $10,000?

(C) Instead of refinancing, the graduate decides to contribute $100 each month to reduce the principal until the loan is paid off. How much will the graduate save her parents in interest payments? How long would she contribute? How much would she contribute?

(D) Refer to part (C). How much would the graduate need to contribute each month in order to save her parents a total of $10,000 in interest payments?

Ask a parent, relative, or friend about the terms of their mortgage. Then do the calculations necessary to develop specific strategies for reducing the interest payments on the loan by refinancing and/or early payment of the principal.

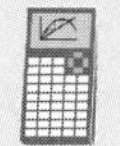

Group Activity 2 Yield to Maturity and Internal Rate of Return

Table 1

YEAR	BONUS
1	$3,000
2	$7,000
3	$4,000
4	$8,000

A salesman receives annual bonuses for 4 years (see Table 1), which he immediately invests in a mutual fund. The interest earned by this fund varies from year to year.

One year after making the last investment, the salesman closes out the fund and receives a payment of $30,000. He would like to find a compound interest rate that represents the interest his investments earned over the entire 4 year period. This rate is called the **yield to maturity.** Since he deposited different amounts each year, none of the annuity formulas discussed earlier can be used in this situation. However, the compound interest formula discussed in Section 3-2 can be used. If r is the yield to maturity, then the compound interest formula expresses the value of each investment at the end of the 4th year in terms of r (Fig. 1).

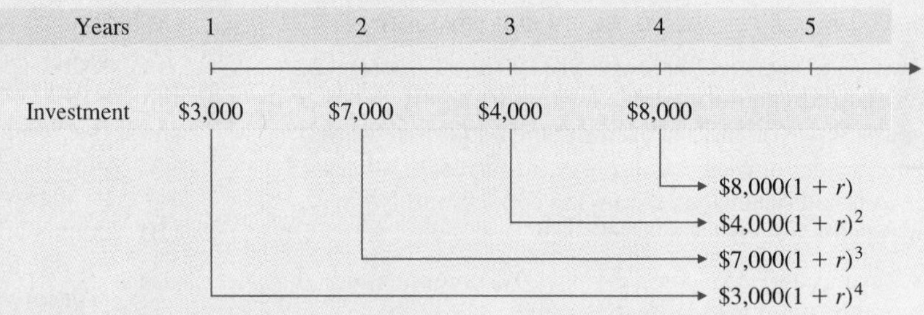

FIGURE 1

The sum of all these values must equal $30,000, the amount in the fund at the end of the 4th year. Thus, the yield to maturity must satisfy the following equation:

$$8{,}000(1 + r) + 4{,}000(1 + r)^2 + 7{,}000(1 + r)^3 + 3{,}000(1 + r)^4 = 30{,}000$$

Using graphical approximation techniques to solve this equation (Fig. 2), the yield to maturity (to two decimal places) is $r = 14.39\%$.

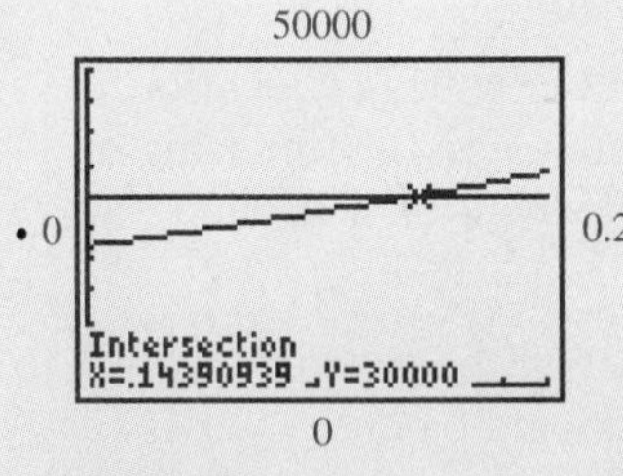

FIGURE 2

(A) In the preceding discussion, we used graphical approximation techniques to find the yield to maturity. If you have access to a graphing utility with an equation solver, use it to find the yield to maturity.

(B) Suppose your grandparents give you \$1,000 in stock certificates on your 16th birthday, \$1,500 on your 17th birthday, and \$2,000 on your 18th birthday. On your 21st birthday, you sell all the stock for \$7,000. What is the yield to maturity for this sequence of investments?

(C) A single investment of \$10,000 in a small business returns payments of \$3,000 at the end of the first year, \$4,000 at the end of the second year, and a final payment of \$5,000 at the end of the third year. If this investment is to return earnings of $100r\%$, then the sum of the present values of the three annual returns at $100r\%$ must equal the original investment. Show that this leads to the equation

$$10{,}000 = \frac{3{,}000}{1 + r} + \frac{4{,}000}{(1 + r)^2} + \frac{5{,}000}{(1 + r)^3}$$

The solution to this equation is called the **internal rate of return.** Use graphical approximation techniques or an equation solver to find the solution.

(D) An individual pays \$160,000 for a two-family housing unit that generates annual profits of \$15,000. After 4 years, the property is sold for \$190,000. Find the internal rate of return for this investment.

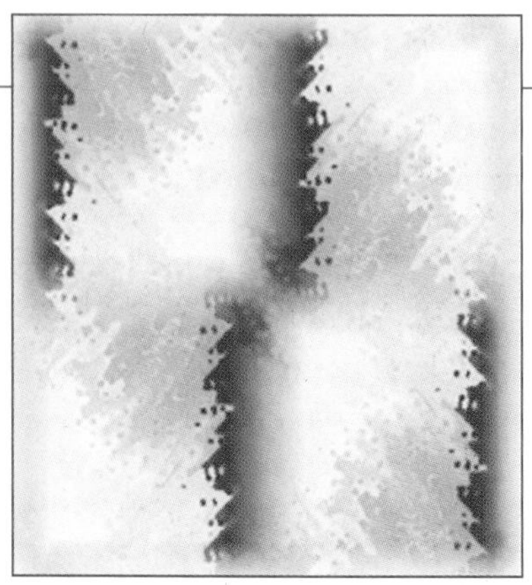

CHAPTER 4

SYSTEMS OF LINEAR EQUATIONS; MATRICES

INTRODUCTION

In this chapter we first review how systems of linear equations involving two variables are solved using techniques learned in elementary algebra. (Basic properties of linear equations and lines are reviewed in Section 1-3 and Appendix A-8.) Because these techniques are not suitable for linear systems involving larger numbers of equations and variables, we then turn to a different method of solution involving the concept of an *augmented matrix,* which arises quite naturally when dealing with larger linear systems. We then study *matrices* and *matrix operations* in their own right as a new mathematical form. With these new operations added to our mathematical toolbox, we return to systems of equations from a fresh point of view. Finally, we discuss Wassily Leontief's Nobel prize-winning application of matrices to an important economics problem.

Many of the matrix methods introduced in this chapter can be performed on a variety of graphing utilities, including graphing calculators, spreadsheets, and software packages such as MatLab or *Explorations in Finite Mathematics* (see Preface). We will see that certain problems can be solved using either graphical approximation methods or matrix methods, whereas others can be solved by only one of these methods. To avoid confusion between the two methods, we will always clearly state which is to be used. Of course, all graphing utility activities continue to be optional and will be so marked, as we have done throughout the rest of the book.

SECTION 4-1

Review: Systems of Linear Equations in Two Variables

- SYSTEMS IN TWO VARIABLES
- GRAPHING
- SUBSTITUTION
- ELIMINATION BY ADDITION
- APPLICATIONS

SYSTEMS IN TWO VARIABLES

To establish basic concepts, consider the following simple example: If 2 adult tickets and 1 child ticket cost \$8, and if 1 adult ticket and 3 child tickets cost \$9, what is the price of each?

Let: $x =$ Price of adult ticket
$y =$ Price of child ticket

Then:
$$2x + y = 8$$
$$x + 3y = 9$$

We now have a system of two linear equations in two variables. It is easy to find ordered pairs (x, y) that satisfy one or the other of these equations. For example, the ordered pair $(4, 0)$ satisfies the first equation, but not the second, and the ordered pair $(6, 1)$ satisfies the second, but not the first. To solve this system, we must find all ordered pairs of real numbers that satisfy both equations at the same time. In general, we have the following definition:

Systems of Two Equations in Two Variables

Given the **linear system**

$$ax + by = h$$
$$cx + dy = k$$

where $a, b, c, d, h,$ and k are real constants, a pair of numbers $x = x_0$ and $y = y_0$ [also written as an ordered pair (x_0, y_0)] is a **solution** of this system if each equation is satisfied by the pair. The set of all such ordered pairs is called the **solution set** for the system. To **solve** a system is to find its solution set.

We will consider three methods of solving such systems: *graphing, substitution,* and *elimination by addition.* Each method has certain advantages, depending on the situation.

GRAPHING

Recall that the graph of a line is a graph of all the ordered pairs that satisfy the equation of the line. To solve the ticket problem by graphing, we graph both equations in the same coordinate system. The coordinates of any points that the graphs have in common must be solutions to the system, since they must satisfy both equations.

Example 1 **Solving a System by Graphing** Solve the ticket problem by graphing:

$$\begin{aligned} 2x + y &= 8 \\ x + 3y &= 9 \end{aligned}$$

SOLUTION Graph both equations in the same coordinate system (both graphs are straight lines). Then estimate the coordinates of any common points on the two lines.

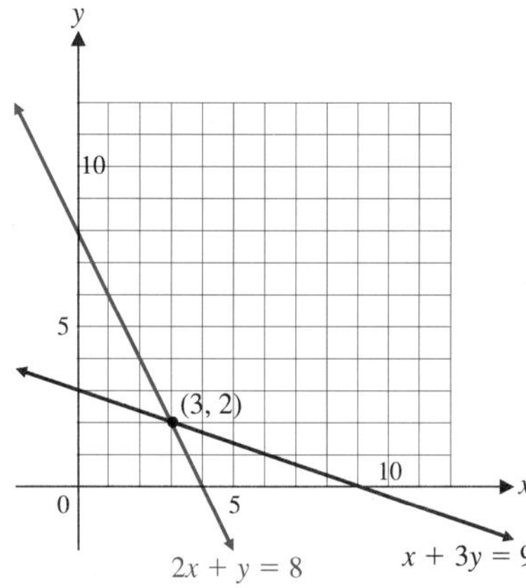

$x = \$3$ Adult ticket
$y = \$2$ Child ticket

CHECK

$$\begin{aligned} 2x + y &= 8 \\ 2(3) + 2 &\stackrel{?}{=} 8 \\ 8 &\stackrel{\checkmark}{=} 8 \end{aligned} \qquad \begin{aligned} x + 3y &= 9 \\ 3 + 3(2) &\stackrel{?}{=} 9 \\ 9 &\stackrel{\checkmark}{=} 9 \end{aligned}$$

(3, 2) must satisfy each original equation for a complete check.

Matched Problem 1 Solve by graphing and check:

$$\begin{aligned} 2x - y &= -3 \\ x + 2y &= -4 \end{aligned}$$

It is clear that Example 1 has exactly one solution, since the lines have exactly one point of intersection. In general, lines in a rectangular coordinate system are related to each other in one of the three ways illustrated in the next example.

Example 2 **Solving Systems by Graphing** Solve each of the following systems by graphing:

(A) $\begin{aligned} x - 2y &= 2 \\ x + y &= 5 \end{aligned}$ (B) $\begin{aligned} x + 2y &= -4 \\ 2x + 4y &= 8 \end{aligned}$ (C) $\begin{aligned} 2x + 4y &= 8 \\ x + 2y &= 4 \end{aligned}$

SOLUTION

(A)

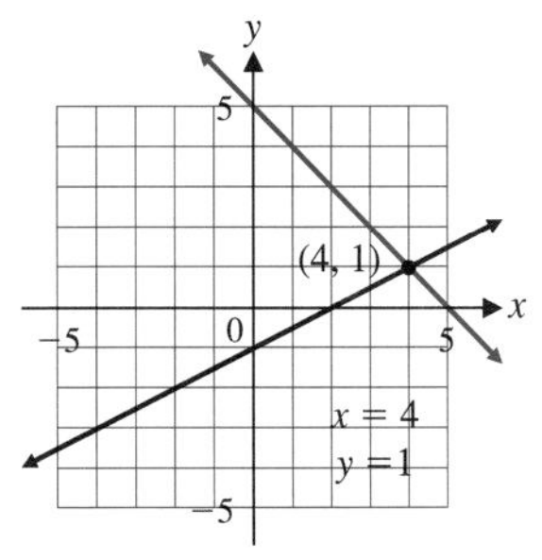

Intersection at one point only—exactly one solution

(B)

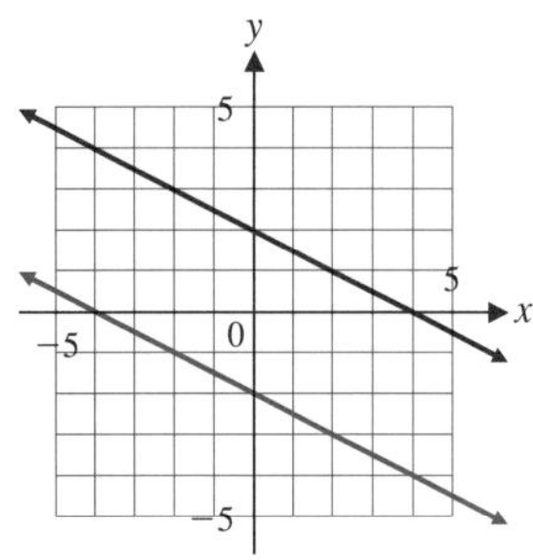

Lines are parallel (each has slope $-\frac{1}{2}$)—no solutions

(C)

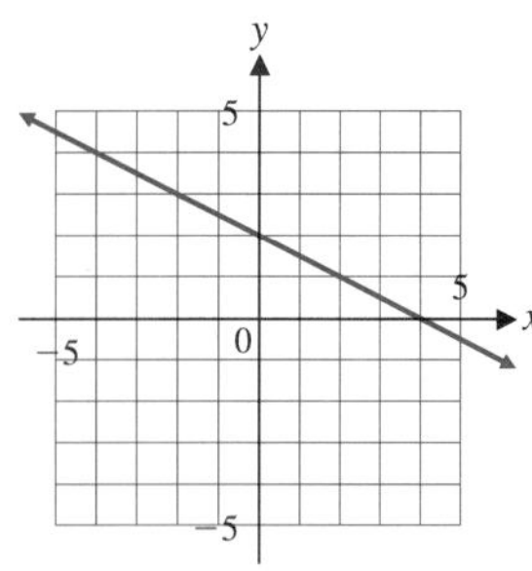

Lines coincide—infinite number of solutions

Matched Problem 2 Solve each of the following systems by graphing:

(A) $x + y = 4$, $2x - y = 2$ (B) $6x - 3y = 9$, $2x - y = 3$ (C) $2x - y = 4$, $6x - 3y = -18$

We now define some terms that we can use to describe the different types of solutions to systems of equations that we will encounter.

Systems of Linear Equations: Basic Terms

A system of linear equations is **consistent** if it has one or more solutions and **inconsistent** if no solutions exist. Furthermore, a consistent system is said to be **independent** if it has exactly one solution (often referred to as the **unique solution**) and **dependent** if it has more than one solution.

Referring to the three systems in Example 2, the system in part (A) is a consistent and independent system with the unique solution $x = 4$, $y = 1$. The system in part (B) is inconsistent. And the system in part (C) is a consistent and dependent system with an infinite number of solutions (all the points on the two coinciding lines).

Explore–Discuss 1

Can a consistent and dependent system have exactly two solutions? Exactly three solutions? Explain.

By geometrically interpreting a system of two linear equations in two variables, we gain useful information about what to expect in the way of solutions to the system. In general, any two lines in a coordinate plane must intersect in exactly one point, be parallel, or coincide (have identical graphs).

Thus, the systems in Example 2 illustrate the only three possible types of solutions for systems of two linear equations in two variables. These ideas are summarized in Theorem 1.

Theorem 1 POSSIBLE SOLUTIONS TO A LINEAR SYSTEM

The linear system

$$\begin{aligned} ax + by &= h \\ cx + dy &= k \end{aligned}$$

must have:

(A) Exactly one solution **Consistent and independent**

Or:

(B) No solution **Inconsistent**

Or:

(C) Infinitely many solutions **Consistent and dependent**

There are no other possibilities.

In the past, one drawback of the graphical solution method was the inaccuracy of hand-drawn graphs. The recent advent of graphing utilities has changed that. Graphical solutions performed on a graphing utility provide both a useful geometrical interpretation and an accurate approximation of the solution to a system of linear equations in two variables. Example 3 will demonstrate such a solution.

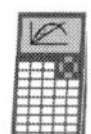

Example 3 **Solving a System Using a Graphing Utility** Solve to two decimal places using graphical approximation techniques on a graphing utility:

$$\begin{aligned} 5x + 2y &= 15 \\ 2x - 3y &= 16 \end{aligned}$$

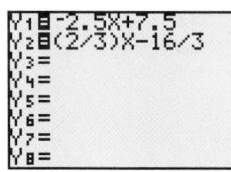

(A) Equation definitions

(B) Intersection point

FIGURE 1

SOLUTION First, solve each equation for y:

$$\begin{aligned} 5x + 2y &= 15 \\ 2y &= -5x + 15 \\ y &= -2.5x + 7.5 \end{aligned} \qquad \begin{aligned} 2x - 3y &= 16 \\ -3y &= -2x + 16 \\ y &= \tfrac{2}{3}x - \tfrac{16}{3} \end{aligned}$$

Next, enter each equation in the graphing utility (Fig. 1A), graph in an appropriate viewing window, and approximate the intersection point (Fig. 1B).

Rounding the values in Figure 1B to two decimal places, we see that the solution is $x = 4.05$ and $y = -2.63$, or $(4.05, -2.63)$.

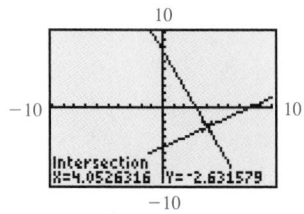

CHECK

$$\begin{aligned} 5x + 2y &= 15 \\ 5(4.05) + 2(-2.63) &\stackrel{?}{=} 15 \\ 14.99 &\approx 15 \end{aligned} \qquad \begin{aligned} 2x - 3y &= 16 \\ 2(4.05) - 3(-2.63) &\stackrel{?}{=} 16 \\ 15.99 &\approx 16 \end{aligned}$$

The checks are not exact because the values of x and y are approximations.

Matched Problem 3 Solve to two decimal places using graphical approximation techniques on a graphing utility:

$$\begin{aligned} 2x - 5y &= -25 \\ 4x + 3y &= 5 \end{aligned}$$

Graphic methods help us visualize a system and its solutions, frequently reveal relationships that might otherwise be hidden, and, with the assistance of a graphing utility, provide very accurate approximations to solutions.

SUBSTITUTION

Now we will review an algebraic method that is easy to use and provides exact solutions to a system of two equations in two variables, provided solutions exist. In this method, first we choose one of two equations in a system and solve for one variable in terms of the other. (We make a choice that avoids fractions, if possible.) Then we **substitute** the result into the other equation and solve the resulting linear equation in one variable. Finally, we substitute this result back into the results of the first step to find the second variable. An example should make the process clear.

Example 4 **Solving a System by Substitution** Solve by substitution:

$$\begin{aligned} 5x + y &= 4 \\ 2x - 3y &= 5 \end{aligned}$$

SOLUTION Solve either equation for one variable in terms of the other; then substitute into the remaining equation. In this problem we can avoid fractions by choosing the first equation and solving for y in terms of x:

$$\begin{aligned} 5x + y &= 4 && \text{Solve the first equation for } y \text{ in terms of } x. \\ y &= 4 - 5x && \text{Substitute into second equation.} \\ 2x - 3y &= 5 && \text{Second equation} \\ 2x - 3(4 - 5x) &= 5 && \text{Solve for } x. \\ 2x - 12 + 15x &= 5 \\ 17x &= 17 \\ x &= 1 \end{aligned}$$

Now, replace x with 1 in $y = 4 - 5x$ to find y:

$$\begin{aligned} y &= 4 - 5x \\ y &= 4 - 5(1) \\ y &= -1 \end{aligned}$$

Thus, the solution is $x = 1$, $y = -1$ or $(1, -1)$.

CHECK

$$\begin{aligned} 5x + y &= 4 \\ 5(1) + (-1) &\stackrel{?}{=} 4 \\ 4 &\stackrel{\checkmark}{=} 4 \end{aligned} \qquad \begin{aligned} 2x - 3y &= 5 \\ 2(1) - 3(-1) &\stackrel{?}{=} 5 \\ 5 &\stackrel{\checkmark}{=} 5 \end{aligned}$$

Matched Problem 4 Solve by substitution:

$$\begin{aligned} 3x + 2y &= -2 \\ 2x - y &= -6 \end{aligned}$$

Explore–Discuss 2

Return to Example 2 and solve each system by substitution. Based on your results, describe how you can recognize a dependent system or an inconsistent system when using substitution.

Elimination by Addition

The methods of graphing and substitution both work well for systems involving two variables. However, neither is easily extended to larger systems. Now we turn to **elimination by addition.** This is probably the most important method of solution. It readily generalizes to larger systems and forms the basis for computer-based solution methods.

You are already familiar with operations that can be performed on a single equation without changing its solution set (see Appendix A-8). Similarly, elimination by addition involves performing appropriate operations on a system of equations to produce new and simpler *equivalent* systems with the same solution set. In general, we say that two systems of equations are **equivalent** if they have the same solution set. Theorem 2 lists three useful operations that produce equivalent systems.

Theorem 2 OPERATIONS THAT PRODUCE EQUIVALENT SYSTEMS

A system of linear equations is transformed into an equivalent system if:

(A) Two equations are interchanged.
(B) An equation is multiplied by a nonzero constant.
(C) A constant multiple of one equation is added to another equation.

Any one of the three operations in Theorem 2 can be used to produce an equivalent system, but the operations in parts (B) and (C) will be of most use to us now. Part (A) becomes useful when we apply the theorem to larger systems. The use of Theorem 2 is best illustrated by examples.

Example 5 **Solving a System Using Elimination by Addition** Solve the following system using elimination by addition:

$$\begin{aligned} 3x - 2y &= 8 \\ 2x + 5y &= -1 \end{aligned}$$

Solution We use Theorem 2 to eliminate one of the variables, thus obtaining a system with an obvious solution:

$$\begin{aligned} 3x - 2y &= 8 \\ 2x + 5y &= -1 \end{aligned}$$

Multiply the top equation by 5 and the bottom equation by 2 (Theorem 2B).

$$\begin{aligned} 5(3x - 2y) &= 5(8) \\ 2(2x + 5y) &= 2(-1) \end{aligned}$$

$$\begin{aligned} 15x - 10y &= 40 \\ 4x + 10y &= -2 \\ \hline 19x \quad &= 38 \end{aligned}$$

Add the top equation to the bottom equation (Theorem 2C), eliminating the y terms.

Divide both sides by 19, which is the same as multiplying the equation by $\frac{1}{19}$ (Theorem 2B).

$$x = 2$$

This equation paired with either of the two original equations produces a system equivalent to the original system.

Knowing that $x = 2$, we substitute this number back into either of the two original equations (we choose the second) to solve for y:

$$\begin{aligned} 2(2) + 5y &= -1 \\ 5y &= -5 \\ y &= -1 \end{aligned}$$

Thus, the solution is $x = 2$, $y = -1$ or $(2, -1)$.

CHECK

$$\begin{aligned} 3x - \quad 2y &= 8 \\ 3(2) - 2(-1) &\stackrel{?}{=} 8 \\ 8 &\stackrel{\checkmark}{=} 8 \end{aligned} \qquad \begin{aligned} 2x + \quad 5y &= -1 \\ 2(2) + 5(-1) &\stackrel{?}{=} -1 \\ -1 &\stackrel{\checkmark}{=} -1 \end{aligned}$$

Matched Problem 5 Solve the following system using elimination by addition:

$$\begin{aligned} 5x - 2y &= 12 \\ 2x + 3y &= 1 \end{aligned}$$

Let us see what happens in the elimination process when a system has either no solution or infinitely many solutions. Consider the following system:

$$\begin{aligned} 2x + 6y &= -3 \\ x + 3y &= 2 \end{aligned}$$

Multiplying the second equation by -2 and adding, we obtain

$$\begin{aligned} 2x + 6y &= -3 \\ -2x - 6y &= -4 \\ \hline 0 &= -7 \end{aligned}$$

Not possible

We have obtained a contradiction. The assumption that the original system has solutions must be false (otherwise, we have proved that $0 = -7$!). Thus, the system has no solutions, and its solution set is the empty set. The graphs of the equations are parallel and the system is inconsistent.

Now consider the system

$$\begin{aligned} x - \tfrac{1}{2}y &= 4 \\ -2x + y &= -8 \end{aligned}$$

If we multiply the top equation by 2 and add the result to the bottom equation, we obtain

$$\begin{array}{r} 2x - y = 8 \\ -2x + y = -8 \\ \hline 0 = 0 \end{array}$$

Obtaining $0 = 0$ by addition implies that the equations are equivalent; that is, their graphs coincide and the system is dependent. If we let $x = k$, where k is any real number, and solve either equation for y, we obtain $y = 2k - 8$. Thus, $(k, 2k - 8)$ is a solution for any real number k. The variable k is called a **parameter,** and replacing k with a real number produces a **particular solution** to the system. For example, some particular solutions to this system are

$k = -1$	$k = 2$	$k = 5$	$k = 9.4$
$(-1, -10)$	$(2, -4)$	$(5, 2)$	$(9.4, 10.8)$

Applications

Many real-world problems are readily solved by constructing a mathematical model consisting of two linear equations in two variables and applying the solution methods we have discussed. We shall examine two applications in detail.

Example 6 **Diet** A woman wants to use milk and orange juice to increase the amount of calcium and vitamin A in her daily diet. An ounce of milk contains 37 milligrams of calcium and 57 micrograms* of vitamin A. An ounce of orange juice contains 5 milligrams of calcium and 65 micrograms of vitamin A. How many ounces of milk and orange juice should the woman drink each day to provide exactly 500 milligrams of calcium and 1,200 micrograms of vitamin A?

Solution First, we define the relevant variables:

x = Number of ounces of milk
y = Number of ounces of orange juice

	Milk	Orange Juice	Total Needed
Calcium	37	5	500
Vitamin A	57	65	1,200

Next, we summarize the given information in the table in the margin. It is convenient to organize the table so that the quantities represented by the variables correspond to columns in the table (rather than to rows), as shown.

Now we use the information in the table to form equations involving x and y:

$$\begin{array}{ccccc} \begin{pmatrix}\text{Calcium in } x \text{ oz}\\ \text{of milk}\end{pmatrix} & + & \begin{pmatrix}\text{Calcium in } y \text{ oz}\\ \text{of orange juice}\end{pmatrix} & = & \begin{pmatrix}\text{Total calcium}\\ \text{needed (mg)}\end{pmatrix} \\ 37x & + & 5y & = & 500 \\ \begin{pmatrix}\text{Vitamin A in } x \text{ oz}\\ \text{of milk}\end{pmatrix} & + & \begin{pmatrix}\text{Vitamin A in } y \text{ oz}\\ \text{of orange juice}\end{pmatrix} & = & \begin{pmatrix}\text{Total vitamin A}\\ \text{needed } (\mu\text{g})\end{pmatrix} \\ 57x & + & 65y & = & 1{,}200 \end{array}$$

*A microgram (μg) is one-millionth (10^{-6}) of a gram.

Thus, we have the following model to solve:

$$\begin{aligned} 37x + 5y &= 500 \\ 57x + 65y &= 1{,}200 \end{aligned}$$

We multiply the first equation by -13 and use elimination by addition:

$$\begin{array}{rcr} -481x - 65y &=& -6{,}500 \\ 57x + 65y &=& 1{,}200 \\ \hline -424x \quad &=& -5{,}300 \\ x &=& \mathbf{12.5} \end{array} \qquad \begin{aligned} 37(\mathbf{12.5}) + 5y &= 500 \\ 5y &= 37.5 \\ y &= \mathbf{7.5} \end{aligned}$$

Drinking 12.5 ounces of milk and 7.5 ounces of orange juice each day will provide the required amounts of calcium and vitamin A.

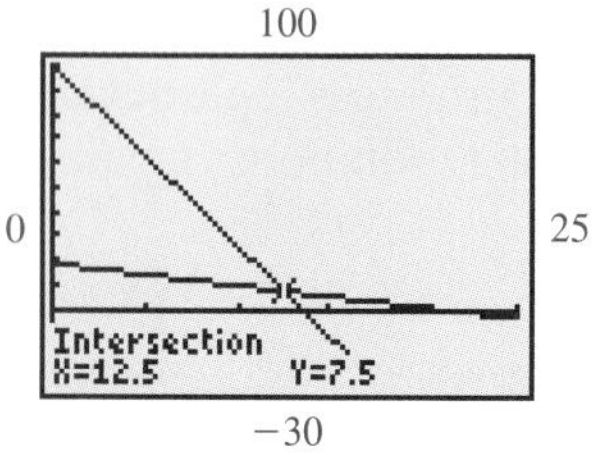

CHECK

$$\begin{aligned} 37x + 5y &= 500 \\ 37(12.5) + 5(7.5) &\stackrel{?}{=} 500 \\ 500 &\stackrel{\checkmark}{=} 500 \end{aligned} \qquad \begin{aligned} 57x + 65y &= 1{,}200 \\ 57(12.5) + 65(7.5) &\stackrel{?}{=} 1{,}200 \\ 1{,}200 &\stackrel{\checkmark}{=} 1{,}200 \end{aligned}$$

100

0 25

Intersection X=12.5 Y=7.5

−30

FIGURE 2

$y_1 = (500 - 37x)/5$
$y_2 = (1{,}200 - 57x)/65$

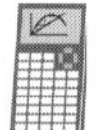

Figure 2 illustrates a solution to Example 6 using graphical approximation techniques.

Matched Problem 6

A man wants to use cottage cheese and yogurt to increase the amount of protein and calcium in his daily diet. An ounce of cottage cheese contains 3 grams of protein and 15 milligrams of calcium. An ounce of yogurt contains 1 gram of protein and 41 milligrams of calcium. How many ounces of cottage cheese and yogurt should he eat each day to provide exactly 62 grams of protein and 760 milligrams of calcium?

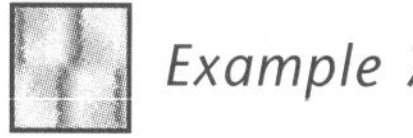

Example 7

Supply and Demand The quantity of a product that people are willing to buy during some period of time depends on its price. Generally, the higher the price, the less the demand; the lower the price, the greater the demand. Similarly, the quantity of a product that a supplier is willing to sell during some period of time also depends on the price. Generally, a supplier will be willing to supply more of a product at higher prices and less of a product at lower prices. The simplest supply and demand model is a linear model where the graphs of a demand equation and a supply equation are straight lines.

Suppose we are interested in analyzing the sale of cherries each day in a particular city. Using special analytical techniques (regression analysis) and collected data, an analyst arrives at the following price–demand and price–supply models:

$$\begin{aligned} p &= -0.2q + 4 && \text{Price–demand equation (consumer)} \\ p &= 0.07q + 0.76 && \text{Price–supply equation (supplier)} \end{aligned}$$

where q represents the quantity of cherries in thousands of pounds and p represents the price in dollars. For example, we see that consumers will purchase 10 thousand pounds ($q = 10$) when the price is $p = -0.2(10) + 4 = \$2$ per pound. On the other hand, suppliers will be willing to supply 17.714 thousand pounds of cherries at \$2 per pound (solve $2 = 0.07q + 0.76$). Thus, at

\$2 per pound the suppliers are willing to supply more cherries than consumers are willing to purchase. The supply exceeds the demand at that price and the price will come down. At what price will cherries stabilize for the day? That is, at what price will supply equal demand? This price, if it exists, is called the **equilibrium price,** and the quantity sold at that price is called the **equilibrium quantity.** The point where the two curves for the price–demand equation and the price–supply equation intersect is called the **equilibrium point.** How do we find these quantities? We solve the linear system

$$\begin{aligned} p &= -0.2q + 4 && \text{Demand equation} \\ p &= 0.07q + 0.76 && \text{Supply equation} \end{aligned}$$

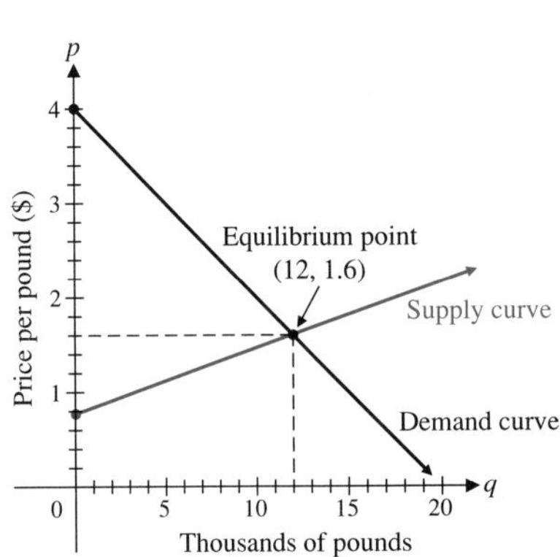

FIGURE 3

by using substitution (substituting $p = -0.2q + 4$ into the second equation):

$$\begin{aligned} -0.2q + 4 &= 0.07q + 0.76 \\ -0.27q &= -3.24 \\ q &= \mathbf{12\ thousand\ pounds} && \text{Equilibrium quantity} \end{aligned}$$

Now substitute $q = 12$ back into either of the original equations in the system and solve for p (we choose the first equation):

$$\begin{aligned} p &= -0.2(12) + 4 \\ p &= \mathbf{\$1.60\ per\ pound} && \text{Equilibrium price} \end{aligned}$$

These results are interpreted geometrically in Figure 3.

If the price is above the equilibrium price of \$1.60 per pound, the supply will exceed the demand and the price will come down. If the price is below the equilibrium price of \$1.60 per pound, the demand will exceed the supply and the price will rise. Thus, the price will reach equilibrium at \$1.60. At this price, suppliers will supply 12 thousand pounds of cherries and consumers will purchase 12 thousand pounds.

Matched Problem 7 Repeat Example 7 (including drawing the graph) given:

$$\begin{aligned} p &= -0.1q + 3 && \text{Demand equation} \\ p &= 0.08q + 0.66 && \text{Supply equation} \end{aligned}$$

Answers to Matched Problems

1. $x = -2$, $y = -1$

Check:

$$\begin{aligned} 2x - y &= -3 \\ 2(-2) - (-1) &\stackrel{?}{=} -3 \\ -3 &\stackrel{\checkmark}{=} -3 \end{aligned}$$

$$\begin{aligned} x + 2y &= -4 \\ (-2) + 2(-1) &\stackrel{?}{=} -4 \\ -4 &\stackrel{\checkmark}{=} -4 \end{aligned}$$

$(-2, -1)$; $2x - y = -3$; $x + 2y = -4$

2. (A) $x = 2$, $y = 2$ (B) Infinitely many solutions (C) No solution

3. $x = -1.92$, $y = 4.23$ **4.** $x = -2$, $y = 2$ **5.** $x = 2$, $y = -1$

6. 16.5 oz of cottage cheese, 12.5 oz of yogurt
7. Equilibrium quantity = 13 thousand pounds; Equilibrium price = \$1.70 per pound

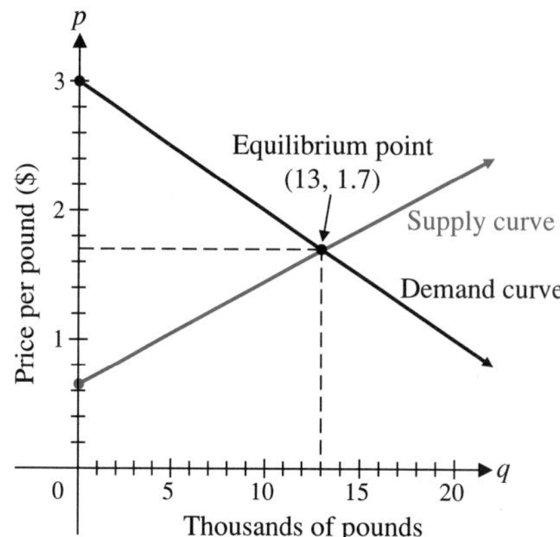

EXERCISE 4-1

A *Match each system in Problems 1–4 with one of the following graphs, and use the graph to solve the system.*

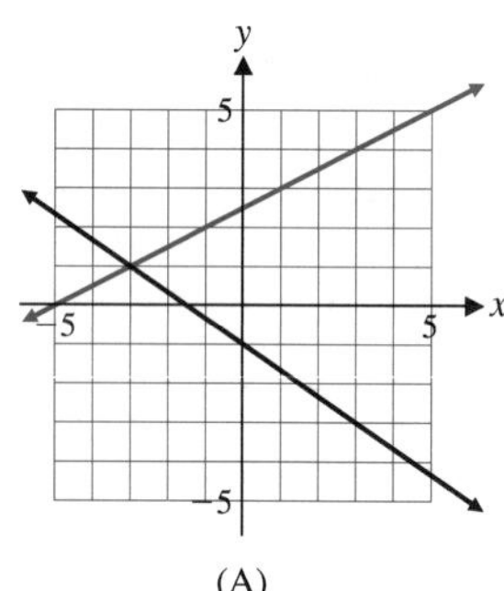
(A)

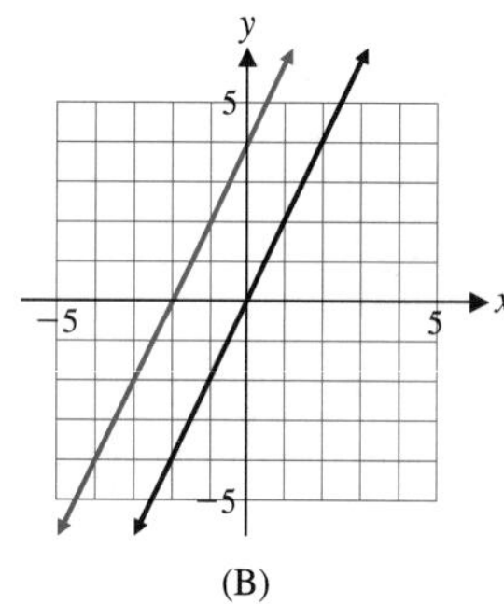
(B)

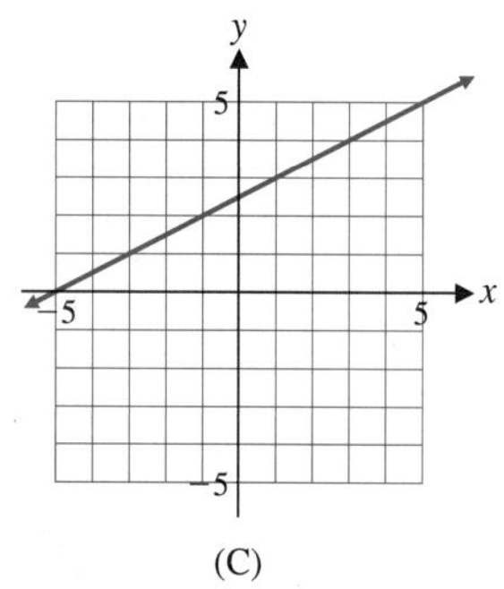
(C)

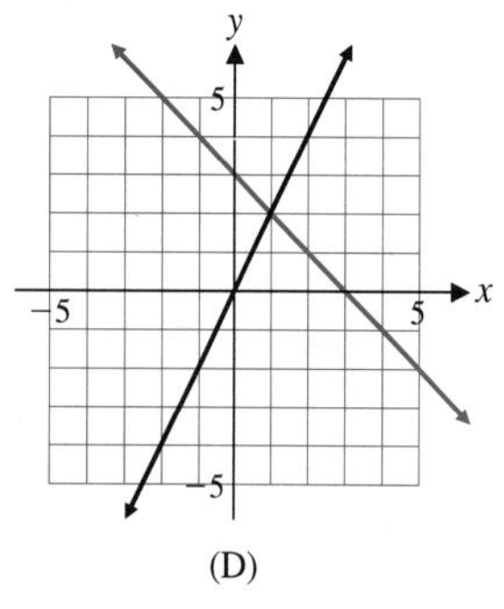
(D)

1. $\begin{aligned} -4x + 2y &= 8 \\ 2x - y &= 0 \end{aligned}$

2. $\begin{aligned} x + y &= 3 \\ 2x - y &= 0 \end{aligned}$

3. $\begin{aligned} -x + 2y &= 5 \\ 2x + 3y &= -3 \end{aligned}$

4. $\begin{aligned} 2x - 4y &= -10 \\ -x + 2y &= 5 \end{aligned}$

Solve Problems 5–8 by graphing.

5. $\begin{aligned} 3x - y &= 2 \\ x + 2y &= 10 \end{aligned}$

6. $\begin{aligned} 3x - 2y &= 12 \\ 7x + 2y &= 8 \end{aligned}$

7. $\begin{aligned} m + 2n &= 4 \\ 2m + 4n &= -8 \end{aligned}$

8. $\begin{aligned} 3u + 5v &= 15 \\ 6u + 10v &= -30 \end{aligned}$

Solve Problems 9–12 using substitution.

9. $\begin{aligned} y &= 2x - 3 \\ x + 2y &= 14 \end{aligned}$

10. $\begin{aligned} y &= x - 4 \\ x + 3y &= 12 \end{aligned}$

11. $\begin{aligned} 2x + y &= 6 \\ x - y &= -3 \end{aligned}$

12. $\begin{aligned} 3x - y &= 7 \\ 2x + 3y &= 1 \end{aligned}$

Solve Problems 13–16 using elimination by addition.

13. $\begin{aligned} 3u - 2v &= 12 \\ 7u + 2v &= 8 \end{aligned}$

14. $\begin{aligned} 2x - 3y &= -8 \\ 5x + 3y &= 1 \end{aligned}$

15. $\begin{aligned} 2m - n &= 10 \\ m - 2n &= -4 \end{aligned}$

16. $\begin{aligned} 2x + 3y &= 1 \\ 3x - y &= 7 \end{aligned}$

B *Solve Problems 17–30 using substitution or elimination by addition.*

17. $\begin{aligned} 9x - 3y &= 24 \\ 11x + 2y &= 1 \end{aligned}$

18. $\begin{aligned} 4x + 3y &= 26 \\ 3x - 11y &= -7 \end{aligned}$

19. $\begin{aligned} 2x - 3y &= -2 \\ -4x + 6y &= 7 \end{aligned}$

20. $\begin{aligned} 3x - 6y &= -9 \\ -2x + 4y &= 12 \end{aligned}$

21. $\begin{aligned} 3x + 8y &= 4 \\ 15x + 10y &= -10 \end{aligned}$

22. $\begin{aligned} 7m + 12n &= -1 \\ 5m - 3n &= 7 \end{aligned}$

23. $\begin{aligned} -6x + 10y &= -30 \\ 3x - 5y &= 15 \end{aligned}$

24. $\begin{aligned} 2x + 4y &= -8 \\ x + 2y &= 4 \end{aligned}$

25. $\begin{aligned} x + y &= 1 \\ 0.3x - 0.4y &= 0 \end{aligned}$

26. $\begin{aligned} x + y &= 1 \\ 0.5x - 0.4y &= 0 \end{aligned}$

27. $\begin{aligned} 0.2x - 0.5y &= 0.07 \\ 0.8x - 0.3y &= 0.79 \end{aligned}$

28. $\begin{aligned} 0.3u - 0.6v &= 0.18 \\ 0.5u + 0.2v &= 0.54 \end{aligned}$

29. $\begin{aligned} \tfrac{2}{5}x + \tfrac{3}{2}y &= 2 \\ \tfrac{7}{3}x - \tfrac{5}{4}y &= -5 \end{aligned}$

30. $\begin{aligned} \tfrac{7}{2}x - \tfrac{5}{6}y &= 10 \\ \tfrac{2}{5}x + \tfrac{4}{3}y &= 6 \end{aligned}$

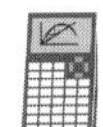

In Problems 31–34, use a graphing calculator to approximate the solution of each system to two decimal places.

31. $\begin{aligned} 3x - 2y &= 5 \\ 4x + 3y &= 13 \end{aligned}$

32. $\begin{aligned} 3x - 7y &= -20 \\ 2x + 5y &= 8 \end{aligned}$

33. $\begin{aligned} -2.4x + 3.5y &= 0.1 \\ -1.7x + 2.6y &= -0.2 \end{aligned}$

34. $\begin{aligned} 4.2x + 5.4y &= -12.9 \\ 6.4x + 3.7y &= -4.5 \end{aligned}$

C *In Problems 35–40, graph all three equations in the same coordinate system, and then find the coordinates of any points where two or more lines intersect.*

35. $\begin{aligned} x - 2y &= -6 \\ 2x + y &= 8 \\ x + 2y &= -2 \end{aligned}$

36. $\begin{aligned} x + y &= 3 \\ x + 3y &= 15 \\ 3x - y &= 5 \end{aligned}$

37. $\begin{aligned} x + y &= 1 \\ x - 2y &= -8 \\ 3x + y &= -3 \end{aligned}$

38. $\begin{aligned} x - y &= 6 \\ x - 2y &= 8 \\ x + 4y &= -4 \end{aligned}$

39. $\begin{aligned} 4x - 3y &= -24 \\ 2x + 3y &= 12 \\ 8x - 6y &= 24 \end{aligned}$

40. $\begin{aligned} 2x + 3y &= 18 \\ 2x - 6y &= -6 \\ 4x + 6y &= -24 \end{aligned}$

41. The coefficients of the three systems given below are very similar. One might guess that the solution sets to the three systems would also be nearly identical. Develop evidence for or against this guess by considering graphs of the systems and solutions obtained using substitution or elimination by addition.

(A) $\begin{aligned} 5x + 4y &= 4 \\ 11x + 9y &= 4 \end{aligned}$ (B) $\begin{aligned} 5x + 4y &= 4 \\ 11x + 8y &= 4 \end{aligned}$

(C) $\begin{aligned} 5x + 4y &= 4 \\ 10x + 8y &= 4 \end{aligned}$

42. Repeat Problem 41 for the following systems:

(A) $\begin{aligned} 6x - 5y &= 10 \\ -13x + 11y &= -20 \end{aligned}$ (B) $\begin{aligned} 6x - 5y &= 10 \\ -13x + 10y &= -20 \end{aligned}$

(C) $\begin{aligned} 6x - 5y &= 10 \\ -12x + 10y &= -20 \end{aligned}$

Applications

Business & Economics

43. *Supply and demand.* Suppose the supply and demand equations for printed T-shirts in a resort town for a particular week are

$$\begin{aligned} p &= 0.7q + 3 && \text{Supply equation} \\ p &= -1.7q + 15 && \text{Demand equation} \end{aligned}$$

where p is the price in dollars and q is the quantity in hundreds.

(A) Find the supply and the demand (to the nearest unit) if T-shirts are priced at \$4 each. Discuss the stability of the T-shirt market at this price level.

(B) Find the supply and the demand (to the nearest unit) if T-shirts are priced at \$9 each. Discuss the stability of the T-shirt market at this price level.

(C) Find the equilibrium price and quantity.

(D) Graph the two equations in the same coordinate system and identify the equilibrium point, supply curve, and demand curve.

44. *Supply and demand.* Suppose the supply and demand for printed baseball caps in a resort town for a particular week are

$$\begin{aligned} p &= 0.4q + 3.2 && \text{Supply equation} \\ p &= -1.9q + 17 && \text{Demand equation} \end{aligned}$$

where p is the price in dollars and q is the quantity in hundreds.

(A) Find the supply and the demand (to the nearest unit) if baseball caps are priced at \$4 each. Discuss the stability of the baseball cap market at this price level.

(B) Find the supply and the demand (to the nearest unit) if baseball caps are priced at \$9 each. Discuss the stability of the baseball cap market at this price level.

(C) Find the equilibrium price and quantity.

(D) Graph the two equations in the same coordinate system and identify the equilibrium point, supply curve, and demand curve.

45. *Supply and demand.* At \$0.60 per bushel, the daily supply for wheat is 450 bushels, and the daily demand is 570 bushels. When the price is raised to \$0.75 per bushel, the daily supply increases to 600 bushels, and the daily demand decreases to 495 bushels. Assume that the supply and demand equations are linear.

(A) Find the supply equation. [*Hint:* Write the supply equation in the form $p = aq + b$ and solve for a and b.]
(B) Find the demand equation.
(C) Find the equilibrium price and quantity.
(D) Graph the two equations in the same coordinate system and identify the equilibrium point, supply curve, and demand curve.

46. *Supply and demand.* At \$1.40 per bushel, the daily supply for oats is 850 bushels, and the daily demand is 580 bushels. When the price falls to \$1.20 per bushel, the daily supply decreases to 350 bushels, and the daily demand increases to 980 bushels. Assume that the supply and demand equations are linear.

(A) Find the supply equation.
(B) Find the demand equation.
(C) Find the equilibrium price and quantity.
(D) Graph the two equations in the same coordinate system and identify the equilibrium point, supply curve, and demand curve.

47. *Break-even analysis.* A small company manufactures portable home computers. The plant has fixed costs (leases, insurance, and so on) of \$48,000 per month and variable costs (labor, materials, and so on) of \$1,400 per unit produced. The computers are sold for \$1,800 each. Thus, the cost and revenue equations are

$$y = 48{,}000 + 1{,}400x \quad \text{Cost equation}$$
$$y = 1{,}800x \quad \text{Revenue equation}$$

where x is the total number of computers produced and sold each month, and the monthly costs and revenue are in dollars.

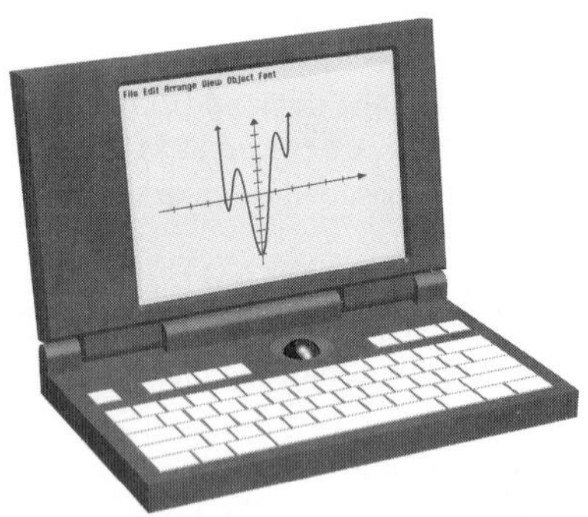

(A) How many units must be manufactured and sold each month for the company to break even?
(B) Graph both equations in the same coordinate system and show the break-even point. Interpret the regions between the lines to the left and to the right of the break-even point.

48. *Break-even analysis.* Repeat Problem 47 with the cost and revenue equations

$$y = 65{,}000 + 1{,}100x \quad \text{Cost equation}$$
$$y = 1{,}600x \quad \text{Revenue equation}$$

49. *Break-even analysis.* A mail-order company markets videotapes that sell for \$19.95, including shipping and handling. The monthly fixed costs (advertising, rent, and so on) are \$24,000, and the variable costs (materials, shipping, and so on) are \$7.45 per tape.

(A) How many tapes must be sold each month for the company to break even?
(B) Graph the cost and revenue equations in the same coordinate system and show the break-even point. Interpret the regions between the lines to the left and to the right of the break-even point.

50. *Break-even analysis.* Repeat Problem 49 if the monthly fixed costs increase to \$27,200, the variable costs increase to \$9.15, and the company raises the selling price of the tapes to \$21.95.

51. *Delivery charges.* United Express, a nationwide package delivery service, charges a base price for overnight delivery of packages weighing 1 pound or less and a surcharge for each additional pound (or fraction thereof). A customer is billed \$27.75 for shipping a 5 pound package and \$64.50 for shipping a 20 pound package. Find the base price and the surcharge for each additional pound.

52. *Delivery charges.* Refer to Problem 51. Federated Shipping, a competing overnight delivery service, informs the customer in Problem 51 that they would ship the 5 pound package for \$29.95 and the 20 pound package for \$59.20.

(A) If Federated Shipping computes its cost in the same manner as United Express, find the base price and the surcharge for Federated Shipping.
(B) Devise a simple rule that the customer can use to choose the cheaper of the two services for each package shipped. Justify your answer.

53. *Resource allocation.* A coffee manufacturer uses Colombian and Brazilian coffee beans to produce two blends, robust and mild. A pound of the robust blend requires 12 ounces of Colombian beans and 4 ounces of Brazilian beans. A pound of the mild blend requires 6 ounces of Colombian beans and 10 ounces of Brazilian beans. Coffee is shipped in 132 pound burlap bags. The company has 50 bags of Colombian beans and 40 bags of Brazilian beans on hand. How many pounds of each blend should they produce in order to use all the available beans?

54. *Resource allocation.* Refer to Problem 53.

(A) If the company decides to discontinue production of the robust blend and produce only the mild blend, how many pounds of the mild blend can they produce and how many beans of each type will they use? Are there any beans that are not used?

(B) Repeat part (A) if the company decides to discontinue production of the mild blend and produce only the robust blend.

Life Sciences

55. *Nutrition.* Animals in an experiment are to be kept under a strict diet. Each animal is to receive, among other things, 20 grams of protein and 6 grams of fat. The laboratory technician is able to purchase two food mixes of the following compositions: Mix A has 10% protein and 6% fat; mix B has 20% protein and 2% fat. How many grams of each mix should be used to obtain the right diet for a single animal?

56. *Nutrition—plants.* A fruit grower can use two types of fertilizer in an orange grove, brand A and brand B. Each bag of brand A contains 8 pounds of nitrogen and 4 pounds of phosphoric acid. Each bag of brand B contains 7 pounds of nitrogen and 6 pounds of phosphoric acid. Tests indicate that the grove needs 720 pounds of nitrogen and 500 pounds of phosphoric acid. How many bags of each brand should be used to provide the required amounts of nitrogen and phosphoric acid?

Social Sciences

57. *Psychology—approach and avoidance.* People often approach certain situations with "mixed emotions." For example, public speaking often brings forth the positive response of recognition and the negative response of failure. Which dominates? J. S. Brown, in an experiment on approach and avoidance, trained rats by feeding them from a goal box. Then the rats received mild electric shocks from the same goal box. This established an approach–avoidance conflict relative to the goal box. Using appropriate apparatus, Brown arrived at the following relationships:

$$p = -\tfrac{1}{5}d + 70 \quad \text{Approach equation}$$
$$p = -\tfrac{4}{3}d + 230 \quad \text{Avoidance equation}$$

where $30 \leq d \leq 172.5$. The approach equation gives the pull (in grams) toward the food goal box when the rat is placed d centimeters away from it. The avoidance equation gives the pull (in grams) away from the shock goal box when the rat is placed d centimeters from it.

(A) Graph the approach equation and the avoidance equation in the same coordinate system.

(B) Find the value of d for the point of intersection of these two equations.

(C) What do you think the rat would do when placed the distance d from the box found in part (B)?

(For additional discussion of this phenomenon, see J. S. Brown, "Gradients of Approach and Avoidance Responses and Their Relation to Motivation," *Journal of Comparative and Physiological Psychology,* 1948, 41:450–465.)

SECTION 4-2

Systems of Linear Equations and Augmented Matrices

- MATRICES
- SOLVING LINEAR SYSTEMS USING AUGMENTED MATRICES
- SUMMARY

Most linear systems of any consequence involve large numbers of equations and variables. It is impractical to try to solve such systems by hand. In the past, these complex systems could be solved only on large computers. Now a wide array of graphing utilities can be used to solve linear systems.

These range from graphing calculators (such as the Texas Instruments TI-83), to software packages [such as *Explorations in Finite Mathematics* (see Preface) or MatLab], to spreadsheets (such as Excel). All these graphing utilities have one thing in common: **The user is expected to be familiar with the techniques used to solve large linear systems.** In the remainder of this chapter we will develop several *matrix methods* for solving systems, with the understanding that these methods are usually used in conjunction with a graphing utility. It is important to keep in mind that we are not presenting these techniques as more efficient methods for solving linear systems manually—there are none. So we will not stress computational shortcuts for hand calculations. Instead, we will emphasize formulation of mathematical models and interpretation of the results—two activities that graphing utilities cannot perform for you.

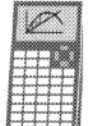

As we mentioned earlier, when referring to the optional problems that should be solved with a graphing utility, we will continue to clearly state whether matrix methods or graphical approximation methods are to be used.

■ MATRICES

In solving systems of equations using elimination by addition, the coefficients of the variables and the constant terms played a central role. The process can be made more efficient for generalization and computer work by the introduction of a mathematical form called a *matrix*. A **matrix** is a rectangular array of numbers written within brackets. Two examples are

$$A = \begin{bmatrix} 1 & -4 & 5 \\ 7 & 0 & -2 \end{bmatrix} \qquad B = \begin{bmatrix} -4 & 5 & 12 \\ 0 & 1 & 8 \\ -3 & 10 & 9 \\ -6 & 0 & -1 \end{bmatrix} \tag{1}$$

Each number in a matrix is called an **element** of the matrix. Matrix A has 6 elements arranged in 2 rows and 3 columns. Matrix B has 12 elements arranged in 4 rows and 3 columns. If a matrix has m rows and n columns, it is called an $\boldsymbol{m \times n}$ **matrix** (read "m by n matrix"). The expression $m \times n$ is called the **size** of the matrix, and the numbers m and n are called the **dimensions** of the matrix. It is important to note that the number of rows is always given first. Referring to (1) above, A is a 2×3 matrix and B is a 4×3 matrix. A matrix with n rows and n columns is called a **square matrix of order** $\boldsymbol{n}$**.** A matrix with only 1 column is called a **column matrix,** and a matrix with only 1 row is called a **row matrix.** These definitions are illustrated by the following:

$$\overset{3 \times 3}{\begin{bmatrix} 0.5 & 0.2 & 1.0 \\ 0.0 & 0.3 & 0.5 \\ 0.7 & 0.0 & 0.2 \end{bmatrix}} \qquad \overset{4 \times 1}{\begin{bmatrix} 3 \\ -2 \\ 1 \\ 0 \end{bmatrix}} \qquad \overset{1 \times 4}{\begin{bmatrix} 2 & \frac{1}{2} & 0 & -\frac{2}{3} \end{bmatrix}}$$

Square matrix of order 3 — Column matrix — Row matrix

The **position** of an element in a matrix is given by the row and column containing the element. This is usually denoted using **double subscript notation a_{ij}**, where i is the row and j is the column containing the element a_{ij}, as illustrated below:

$$A = \begin{bmatrix} 1 & -4 & 5 \\ 7 & 0 & -2 \end{bmatrix} \qquad \begin{matrix} a_{11} = 1, & a_{12} = -4, & a_{13} = 5 \\ a_{21} = 7, & a_{22} = 0, & a_{23} = -2 \end{matrix}$$

Note that a_{12} is read "a sub one two" (*not* "a sub twelve"). The elements $a_{11} = 1$ and $a_{22} = 0$ make up the *principal diagonal* of A. In general, the **principal diagonal** of a matrix A consists of the elements $a_{11}, a_{22}, a_{33}, \ldots$.

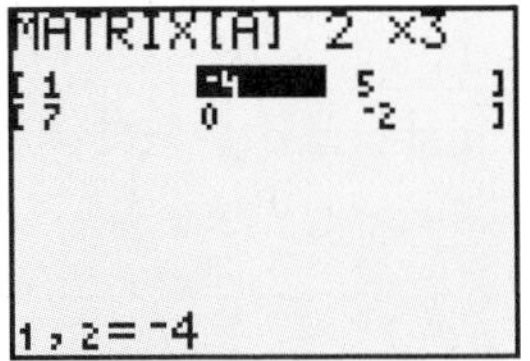

FIGURE 1
Matrix notation on a graphing calculator

REMARK

Most graphing utilities are capable of storing and manipulating matrices. Figure 1 shows matrix A displayed in the editing screen of a graphing calculator. The size of the matrix is given at the top of the screen. The position and value of the currently selected element is given at the bottom. Notice that a comma is used in the notation for the position. This is common practice on many graphing utilities, but not in mathematical literature. In a spreadsheet, matrices are referred to by their location in the spreadsheet (upper left corner to lower right corner), using either row and column numbers (Fig. 2A) or row numbers and column letters (Fig. 2B).

	1	2	3
1	1	-4	5
2	7	0	-2

(A) Location of matrix A: R1C1:R2C3

	A	B	C	D	E	F
1						
2						
3						
4						
5				1	-4	5
6				7	0	-2

(B) Location of matrix A: D5:F6

FIGURE 2
Matrix notation in a spreadsheet

The coefficients and constant terms in a system of linear equations can be used to form several matrices of interest. Related to the system

$$\begin{aligned} 2x - 3y &= 5 \\ x + 2y &= -3 \end{aligned} \tag{2}$$

are the following matrices:

$$\underset{\text{Coefficient matrix}}{\begin{bmatrix} 2 & -3 \\ 1 & 2 \end{bmatrix}} \qquad \underset{\text{Constant matrix}}{\begin{bmatrix} 5 \\ -3 \end{bmatrix}} \qquad \underset{\text{Augmented coefficient matrix}}{\left[\begin{array}{cc|c} 2 & -3 & 5 \\ 1 & 2 & -3 \end{array}\right]}$$

The augmented coefficient matrix contains all the essential parts of the system—both the coefficients and the constants. The vertical bar is included only as a visual aid to help us separate the coefficients from the constant terms. (Matrices entered and displayed on a graphing calculator or computer will not display this line.) Later in this chapter we will make use of the coefficient and constant matrices. For now, we will find the augmented coefficient matrix sufficient for our needs.

For ease of generalization to the larger systems in the following sections, we are now going to change the notation for the variables in (2) to a subscript form. (We would soon run out of letters, but we will not run out of subscripts.) That is, in place of x and y, we use x_1 and x_2, respectively, and (2) is rewritten as

$$\begin{aligned} 2x_1 - 3x_2 &= 5 \\ x_1 + 2x_2 &= -3 \end{aligned}$$

In general, associated with each linear system of the form

$$\begin{aligned} a_{11}x_1 + a_{12}x_2 &= k_1 \\ a_{21}x_1 + a_{22}x_2 &= k_2 \end{aligned} \tag{3}$$

where x_1 and x_2 are variables, is the **augmented matrix** of the system:

Column 1 (C_1)
Column 2 (C_2)
Column 3 (C_3)

$$\left[\begin{array}{cc|c} a_{11} & a_{12} & k_1 \\ a_{21} & a_{22} & k_2 \end{array}\right] \begin{array}{l} \leftarrow \text{Row 1 } (R_1) \\ \leftarrow \text{Row 2 } (R_2) \end{array}$$

This matrix contains the essential parts of system (3). Our objective is to learn how to manipulate augmented matrices in order to solve system (3), if a solution exists. The manipulative process is a direct outgrowth of the elimination process discussed in Section 4-1.

Recall that two linear systems are said to be **equivalent** if they have exactly the same solution set. How did we transform linear systems into equivalent linear systems? We used the operations listed below (Theorem 2, Section 4-1):

Operations That Produce Equivalent Systems

A system of linear equations is transformed into an equivalent system if:

(A) Two equations are interchanged.
(B) An equation is multiplied by a nonzero constant.
(C) A constant multiple of one equation is added to another equation.

Paralleling the earlier discussion, we say that two augmented matrices are **row-equivalent,** denoted by the symbol ~ placed between the two matrices, if they are augmented matrices of equivalent systems of equations. (Think about this.) How do we transform augmented matrices into row-equivalent matrices? We use Theorem 1, which is a direct consequence of the operations listed above.

Theorem 1 ■■ OPERATIONS THAT PRODUCE ROW-EQUIVALENT MATRICES

An augmented matrix is transformed into a row-equivalent matrix by performing any of the following **row operations:**

(A) Two rows are interchanged ($R_i \leftrightarrow R_j$).
(B) A row is multiplied by a nonzero constant ($kR_i \rightarrow R_i$).
(C) A constant multiple of one row is added to another row ($kR_j + R_i \rightarrow R_i$).

[*Note:* The arrow $\rightarrow$ means "replaces."] ■■

■ Solving Linear Systems Using Augmented Matrices

The use of Theorem 1 in solving systems in the form of (3) is best illustrated by examples.

Example 1 **Solving a System Using Augmented Matrix Methods** Solve using augmented matrix methods:

$$\begin{aligned} 3x_1 + 4x_2 &= 1 \\ x_1 - 2x_2 &= 7 \end{aligned} \qquad (4)$$

Solution We start by writing the augmented matrix corresponding to (4):

$$\left[\begin{array}{rr|r} 3 & 4 & 1 \\ 1 & -2 & 7 \end{array}\right] \qquad (5)$$

Our objective is to use row operations from Theorem 1 to try to transform (5) into the form

$$\left[\begin{array}{rr|r} 1 & 0 & m \\ 0 & 1 & n \end{array}\right] \qquad (6)$$

where m and n are real numbers. The solution to system (4) will then be obvious, since matrix (6) will be the augmented matrix of the following system (a row in an augmented matrix always corresponds to an equation in a linear system):

$$\begin{aligned} x_1 &= m \qquad & x_1 + 0x_2 &= m \\ x_2 &= n \qquad & 0x_1 + x_2 &= n \end{aligned}$$

We now proceed to use row operations to transform (5) into form (6).

Step 1. To get a 1 in the upper left corner, we interchange R_1 and R_2 (Theorem 1A):

$$\left[\begin{array}{rr|r} 3 & 4 & 1 \\ 1 & -2 & 7 \end{array}\right] \begin{array}{c} R_1 \leftrightarrow R_2 \\ \sim \end{array} \left[\begin{array}{rr|r} 1 & -2 & 7 \\ 3 & 4 & 1 \end{array}\right]$$

Step 2. To get a 0 in the lower left corner, we multiply R_1 by (-3) and add to R_2 (Theorem 1C)—this changes R_2 but not R_1. Some people find it useful to write $(-3R_1)$ outside the matrix to help reduce errors in arithmetic, as shown:

$$\left[\begin{array}{rr|r} 1 & -2 & 7 \\ 3 & 4 & 1 \end{array}\right] \begin{array}{c} \sim \\ (-3)R_1 + R_2 \rightarrow R_2 \end{array} \left[\begin{array}{rr|r} 1 & -2 & 7 \\ 0 & 10 & -20 \end{array}\right]$$
$$\begin{array}{ccc} -3 & 6 & -21 \end{array}$$

Step 3. To get a 1 in the second row, second column, we multiply R_2 by $\frac{1}{10}$ (Theorem 1B):

$$\left[\begin{array}{rr|r} 1 & -2 & 7 \\ 0 & 10 & -20 \end{array}\right] \begin{array}{c} \sim \\ \frac{1}{10}R_2 \rightarrow R_2 \end{array} \left[\begin{array}{rr|r} 1 & -2 & 7 \\ 0 & 1 & -2 \end{array}\right]$$

Step 4. To get a 0 in the first row, second column, we multiply R_2 by 2 and add the result to R_1 (Theorem 1C)—this changes R_1 but not R_2:

$$\begin{array}{ccc} 0 & 2 & -4 \end{array}$$
$$\left[\begin{array}{rr|r} 1 & -2 & 7 \\ 0 & 1 & -2 \end{array}\right] \begin{array}{c} 2R_2 + R_1 \rightarrow R_1 \\ \sim \end{array} \left[\begin{array}{rr|r} 1 & 0 & 3 \\ 0 & 1 & -2 \end{array}\right]$$

We have accomplished our objective! The last matrix is the augmented matrix for the system

$$\begin{aligned} x_1 &= 3 \qquad x_1 + 0x_2 = 3 \\ x_2 &= -2 \qquad 0x_1 + x_2 = -2 \end{aligned} \tag{7}$$

Since system (7) is equivalent to system (4), our starting system, we have solved (4); that is, $x_1 = 3$ and $x_2 = -2$.

CHECK

$$\begin{aligned} 3x_1 + 4x_2 &= 1 \\ 3(3) + 4(-2) &\stackrel{?}{=} 1 \\ 1 &\stackrel{\checkmark}{=} 1 \end{aligned} \qquad \begin{aligned} x_1 - 2x_2 &= 7 \\ 3 - 2(-2) &\stackrel{?}{=} 7 \\ 7 &\stackrel{\checkmark}{=} 7 \end{aligned}$$

The above process may be written more compactly as follows:

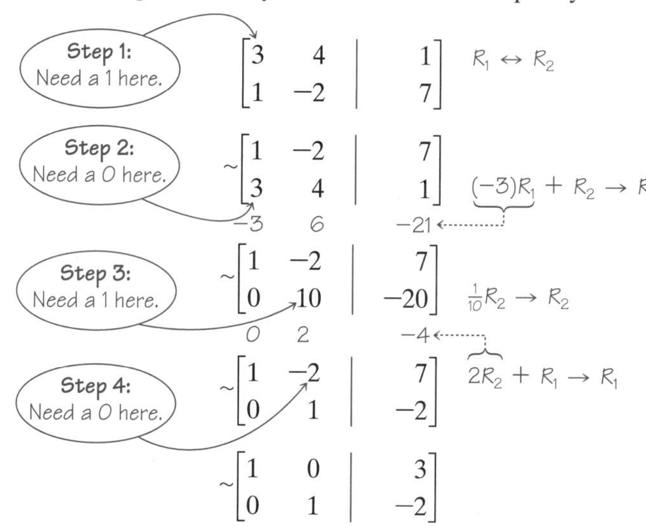

Therefore, $x_1 = 3$ and $x_2 = -2$.

Matched Problem 1 ➡ Solve using augmented matrix methods:

$$\begin{aligned} 2x_1 - x_2 &= -7 \\ x_1 + 2x_2 &= 4 \end{aligned}$$

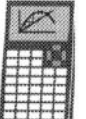

Many graphing utilities can perform row operations. Figure 3 shows the results of performing the row operations used in the solution of Example 1. Consult your manual for the details of performing row operations on your graphing utility.

```
[A]
      [[3 4  1]
       [1 -2 7]]
rowSwap([A],1,2)
→[A]
      [[1 -2 7]
       [3 4  1]]
```

(A) $R_1 \leftrightarrow R_2$

```
[A]
      [[1 -2 7]
       [3 4  1]]
*row+(-3,[A],1,2
)→[A]
    [[1 -2 7  ]
     [0 10 -20]]
```

(B) $(-3)R_1 + R_2 \rightarrow R_2$

```
[A]
    [[1 -2 7  ]
     [0 10 -20]]
*row(.1,[A],2)→[
A]
    [[1 -2 7 ]
     [0 1  -2]]
```

(C) $\frac{1}{10}R_2 \rightarrow R_2$

```
[A]
     [[1 -2 7 ]
      [0 1  -2]]
*row+(2,[A],2,1)
     [[1 0 3 ]
      [0 1 -2]]
```

(D) $2R_2 + R_1 \rightarrow R_1$

FIGURE 3
Row operations on a graphing utility

Explore–Discuss 1

The summary following Example 1 shows five augmented coefficient matrices. Write the linear system that each matrix represents, solve each system graphically, and discuss the relationships among these solutions.

Example 2 **Solving a System Using Augmented Matrix Methods** Solve using augmented matrix methods:

$$\begin{aligned} 2x_1 - 3x_2 &= 6 \\ 3x_1 + 4x_2 &= \tfrac{1}{2} \end{aligned}$$

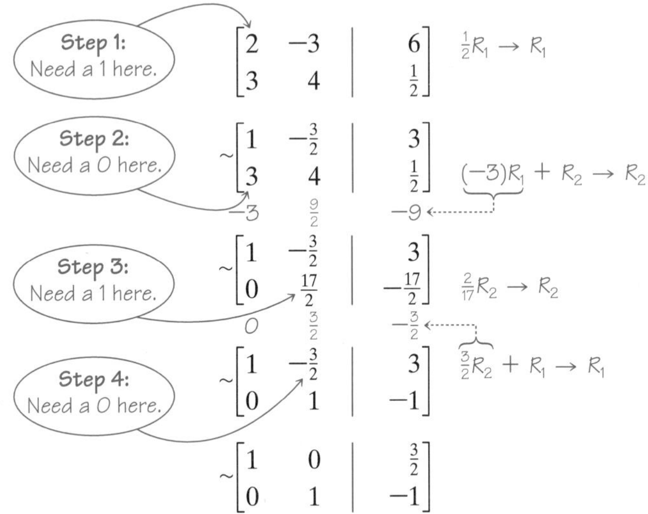

$$\left[\begin{array}{rr|r} 2 & -3 & 6 \\ 3 & 4 & \frac{1}{2} \end{array}\right] \quad \tfrac{1}{2}R_1 \to R_1$$

$$\sim \left[\begin{array}{rr|r} 1 & -\frac{3}{2} & 3 \\ 3 & 4 & \frac{1}{2} \end{array}\right] \quad (-3)R_1 + R_2 \to R_2$$

$$\begin{array}{rr|r} -3 & \frac{9}{2} & -9 \end{array}$$

$$\sim \left[\begin{array}{rr|r} 1 & -\frac{3}{2} & 3 \\ 0 & \frac{17}{2} & -\frac{17}{2} \end{array}\right] \quad \tfrac{2}{17}R_2 \to R_2$$

$$\begin{array}{rr|r} 0 & \frac{3}{2} & -\frac{3}{2} \end{array}$$

$$\sim \left[\begin{array}{rr|r} 1 & -\frac{3}{2} & 3 \\ 0 & 1 & -1 \end{array}\right] \quad \tfrac{3}{2}R_2 + R_1 \to R_1$$

$$\sim \left[\begin{array}{rr|r} 1 & 0 & \frac{3}{2} \\ 0 & 1 & -1 \end{array}\right]$$

Thus, $x_1 = \frac{3}{2}$ and $x_2 = -1$. The check is left to the reader.

Matched Problem 2 Solve using augmented matrix methods:

$$\begin{aligned} 5x_1 - 2x_2 &= 11 \\ 2x_1 + 3x_2 &= \tfrac{5}{2} \end{aligned}$$

Example 3 **Solving a System Using Augmented Matrix Methods** Solve using augmented matrix methods:

$$\begin{aligned} 2x_1 - x_2 &= 4 \\ -6x_1 + 3x_2 &= -12 \end{aligned} \qquad (8)$$

SOLUTION

$$\left[\begin{array}{rr|r} 2 & -1 & 4 \\ -6 & 3 & -12 \end{array}\right] \quad \begin{array}{l} \tfrac{1}{2}R_1 \to R_1 \text{ (to get a 1 in the upper left corner)} \\ \tfrac{1}{3}R_2 \to R_2 \text{ (this simplifies } R_2\text{)} \end{array}$$

$$\sim \left[\begin{array}{rr|r} 1 & -\frac{1}{2} & 2 \\ -2 & 1 & -4 \end{array}\right] \quad 2R_1 + R_2 \to R_2 \text{ (to get a 0 in the lower left corner)}$$

$$\begin{array}{rr|r} 2 & -1 & 4 \end{array}$$

$$\sim \left[\begin{array}{rr|r} 1 & -\frac{1}{2} & 2 \\ 0 & 0 & 0 \end{array}\right]$$

The last matrix corresponds to the system

$$\begin{aligned} x_1 - \tfrac{1}{2}x_2 &= 2 \qquad x_1 - \tfrac{1}{2}x_2 = 2 \\ 0 &= 0 \qquad 0x_1 + 0x_2 = 0 \end{aligned} \tag{9}$$

This system is equivalent to the original system. Geometrically, the graphs of the two original equations coincide and there are infinitely many solutions. In general, if we end up with a row of zeros in an augmented matrix for a two-equation–two-variable system, the system is dependent and there are infinitely many solutions.

We represent the infinitely many solutions using the same method that was used in Section 4-1; that is, by introducing a parameter. We start by solving $x_1 - \frac{1}{2}x_2 = 2$, the first equation in (9), for either variable in terms of the other. We choose to solve for x_1 in terms of x_2 because it is easier:

$$x_1 = \tfrac{1}{2}x_2 + 2 \tag{10}$$

Now we introduce a parameter t (we can use other letters, such as k, s, p, q, and so on, to represent a parameter just as well). If we let $x_2 = t$, then for t any real number,

$$\begin{aligned} x_1 &= \tfrac{1}{2}t + 2 \\ x_2 &= t \end{aligned} \tag{11}$$

represents a solution of system (8). Using ordered pair notation, we may also write:

For any real number t,

$$\left(\tfrac{1}{2}t + 2, t\right) \tag{12}$$

is a solution of system (8). And more formally, we may write:

$$\text{Solution set} = \left\{\left(\tfrac{1}{2}t + 2, t\right) \middle| t \in R\right\} \tag{13}$$

We will generally use the less formal forms (11) and (12) to represent the solution set for problems of this type.

CHECK The following is a check that (11) provides a solution for system (8) for any real number t:

$$\begin{aligned} 2x_1 - x_2 &= 4 \\ 2\left(\tfrac{1}{2}t + 2\right) - t &\stackrel{?}{=} 4 \\ t + 4 - t &\stackrel{?}{=} 4 \\ 4 &\stackrel{\checkmark}{=} 4 \end{aligned} \qquad \begin{aligned} -6x_1 + 3x_2 &= -12 \\ -6\left(\tfrac{1}{2}t + 2\right) + 3t &\stackrel{?}{=} -12 \\ -3t - 12 + 3t &\stackrel{?}{=} -12 \\ -12 &\stackrel{\checkmark}{=} -12 \end{aligned}$$

Matched Problem 3 ➡ Solve using augmented matrix methods:

$$\begin{aligned} -2x_1 + 6x_2 &= 6 \\ 3x_1 - 9x_2 &= -9 \end{aligned}$$

Explore–Discuss 2

The solution of Example 3 involved three augmented coefficient matrices. Write the linear system that each matrix represents, solve each system graphically, and discuss the relationships among these solutions.

Example 4 **Solving a System Using Augmented Matrix Methods** Solve using augmented matrix methods:

$$\begin{aligned} 2x_1 + 6x_2 &= -3 \\ x_1 + 3x_2 &= 2 \end{aligned}$$

SOLUTION

$$\left[\begin{array}{cc|c} 2 & 6 & -3 \\ 1 & 3 & 2 \end{array}\right] \quad R_1 \leftrightarrow R_2$$

$$\sim \left[\begin{array}{cc|c} 1 & 3 & 2 \\ 2 & 6 & -3 \end{array}\right] \quad (-2)R_1 + R_2 \rightarrow R_2$$
$$\begin{array}{ccc} -2 & -6 & -4 \end{array}$$

$$\sim \left[\begin{array}{cc|c} 1 & 3 & 2 \\ 0 & 0 & -7 \end{array}\right] \quad R_2 \text{ implies the contradiction } 0 = -7.$$

This is the augmented matrix of the system

$$\begin{aligned} x_1 + 3x_2 &= 2 & x_1 + 3x_2 &= 2 \\ 0 &= -7 & 0x_1 + 0x_2 &= -7 \end{aligned}$$

The second equation is not satisfied by any ordered pair of real numbers. Hence, as we saw in Section 4-1, the original system is inconsistent and has no solution—otherwise, we have once again proved that $0 = -7$! Thus, if in a row of an augmented matrix we obtain all zeros to the left of the vertical bar and a nonzero number to the right, then the system is inconsistent and there are no solutions.

Matched Problem 4 Solve using augmented matrix methods:

$$\begin{aligned} 2x_1 - x_2 &= 3 \\ 4x_1 - 2x_2 &= -1 \end{aligned}$$

SUMMARY

Summary

Form 1: A Unique Solution (Consistent and Independent)	**Form 2: Infinitely Many Solutions (Consistent and Dependent)**	**Form 3: No Solution (Inconsistent)**
$\left[\begin{array}{cc\|c} 1 & 0 & m \\ 0 & 1 & n \end{array}\right]$	$\left[\begin{array}{cc\|c} 1 & m & n \\ 0 & 0 & 0 \end{array}\right]$	$\left[\begin{array}{cc\|c} 1 & m & n \\ 0 & 0 & p \end{array}\right]$

m, n, p real numbers; $p \neq 0$

The process of solving systems of equations described in this section is referred to as **Gauss–Jordan elimination.** We will formalize this method in the next section so that it will apply to systems of any size, including systems where the number of equations and the number of variables are not the same.

Answers to Matched Problems **1.** $x_1 = -2, x_2 = 3$ **2.** $x_1 = 2, x_2 = -\frac{1}{2}$
3. The system is dependent. For t any real number, a solution is $x_1 = 3t - 3, x_2 = t$.
4. Inconsistent—no solution

EXERCISE 4-2

A *Problems 1–10 refer to the following matrices:*

$$A = \begin{bmatrix} 2 & -4 & 0 \\ 6 & 1 & -5 \end{bmatrix} \qquad B = \begin{bmatrix} -1 & 9 & 0 \\ -4 & 8 & 7 \\ 2 & 4 & 0 \end{bmatrix}$$

$$C = \begin{bmatrix} 2 & -3 & 0 \end{bmatrix} \qquad D = \begin{bmatrix} -5 \\ 8 \end{bmatrix}$$

1. What is the size of A? Of C?

2. What is the size of B? Of D?

3. Identify all row matrices.

4. Identify all column matrices.

5. Identify all square matrices.

6. For matrix B, find b_{21} and b_{13}.

7. For matrix A, find a_{12} and a_{23}.

8. For matrices C and D, find c_{13} and d_{21}.

9. Find the elements on the principal diagonal of matrix B.

10. Find the elements on the principal diagonal of matrix A.

Problems 11 and 12 refer to the matrices shown below.

$$E = \begin{bmatrix} 1 & -2 & 3 & 9 \\ -5 & 0 & 7 & -8 \end{bmatrix} \qquad F = \begin{bmatrix} 4 & -6 \\ 2 & 3 \\ -5 & 7 \end{bmatrix}$$

11. (A) What is the size of E?
(B) How many additional columns would F have to have to be a square matrix?
(C) Find e_{23} and f_{12}.

12. (A) What is the size of F?
(B) How many additional rows would E have to have to be a square matrix?
(C) Find e_{14} and f_{31}.

Perform the row operations indicated in Problems 13–24 on the following matrix:

$$\left[\begin{array}{cc|c} 1 & -3 & 2 \\ 4 & -6 & -8 \end{array}\right]$$

13. $R_1 \leftrightarrow R_2$ **14.** $\frac{1}{2}R_2 \rightarrow R_2$

15. $-4R_1 \rightarrow R_1$ **16.** $-2R_1 \rightarrow R_1$

17. $2R_2 \rightarrow R_2$ **18.** $-1R_2 \rightarrow R_2$

19. $(-4)R_1 + R_2 \rightarrow R_2$ **20.** $(-\frac{1}{2})R_2 + R_1 \rightarrow R_1$

21. $(-2)R_1 + R_2 \rightarrow R_2$ **22.** $(-3)R_1 + R_2 \rightarrow R_2$

23. $(-1)R_1 + R_2 \rightarrow R_2$ **24.** $R_1 + R_2 \rightarrow R_2$

Each of the matrices in Problems 25–30 is the result of performing a single row operation on the matrix A shown below. Identify the row operation.

$$A = \left[\begin{array}{cc|c} -1 & 2 & -3 \\ 6 & -3 & 12 \end{array}\right]$$

25. $\left[\begin{array}{cc|c} -1 & 2 & -3 \\ 2 & -1 & 4 \end{array}\right]$ **26.** $\left[\begin{array}{cc|c} -2 & 4 & -6 \\ 6 & -3 & 12 \end{array}\right]$

27. $\left[\begin{array}{cc|c} -1 & 2 & -3 \\ 0 & 9 & -6 \end{array}\right]$ **28.** $\left[\begin{array}{cc|c} 3 & 0 & 5 \\ 6 & -3 & 12 \end{array}\right]$

29. $\left[\begin{array}{cc|c} 1 & 1 & 1 \\ 6 & -3 & 12 \end{array}\right]$ **30.** $\left[\begin{array}{cc|c} -1 & 2 & -3 \\ 2 & 5 & 0 \end{array}\right]$

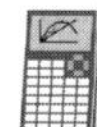
Check Problems 25–30 by performing the row operation you identified on a graphing utility.

Solve Problems 31 and 32 using augmented matrix methods. Write the linear system represented by each augmented matrix in your solution, and solve each of these systems graphically. Discuss the relationships among the solutions of these systems.

31. $x_1 + x_2 = 5$
$x_1 - x_2 = 1$

32. $x_1 - x_2 = 2$
$x_1 + x_2 = 6$

B *Solve Problems 33–52 using augmented matrix methods.*

33. $x_1 - 2x_2 = 1$
$2x_1 - x_2 = 5$

34. $x_1 + 3x_2 = 1$
$3x_1 - 2x_2 = 14$

35. $x_1 - 4x_2 = -2$
$-2x_1 + x_2 = -3$

36. $x_1 - 3x_2 = -5$
$-3x_1 - x_2 = 5$

37. $3x_1 - x_2 = 2$
$x_1 + 2x_2 = 10$

38. $2x_1 + x_2 = 0$
$x_1 - 2x_2 = -5$

39. $x_1 + 2x_2 = 4$
$2x_1 + 4x_2 = -8$

40. $2x_1 - 3x_2 = -2$
$-4x_1 + 6x_2 = 7$

41. $\begin{aligned} 2x_1 + x_2 &= 6 \\ x_1 - x_2 &= -3 \end{aligned}$

42. $\begin{aligned} 3x_1 - x_2 &= -5 \\ x_1 + 3x_2 &= 5 \end{aligned}$

43. $\begin{aligned} 3x_1 - 6x_2 &= -9 \\ -2x_1 + 4x_2 &= 6 \end{aligned}$

44. $\begin{aligned} 2x_1 - 4x_2 &= -2 \\ -3x_1 + 6x_2 &= 3 \end{aligned}$

45. $\begin{aligned} 4x_1 - 2x_2 &= 2 \\ -6x_1 + 3x_2 &= -3 \end{aligned}$

46. $\begin{aligned} -6x_1 + 2x_2 &= 4 \\ 3x_1 - x_2 &= -2 \end{aligned}$

47. $\begin{aligned} 2x_1 + x_2 &= 1 \\ 4x_1 - x_2 &= -7 \end{aligned}$

48. $\begin{aligned} 2x_1 - x_2 &= -8 \\ 2x_1 + x_2 &= 8 \end{aligned}$

49. $\begin{aligned} 4x_1 - 6x_2 &= 8 \\ -6x_1 + 9x_2 &= -10 \end{aligned}$

50. $\begin{aligned} 2x_1 - 4x_2 &= -4 \\ -3x_1 + 6x_2 &= 4 \end{aligned}$

51. $\begin{aligned} -4x_1 + 6x_2 &= -8 \\ 6x_1 - 9x_2 &= 12 \end{aligned}$

52. $\begin{aligned} -2x_1 + 4x_2 &= 4 \\ 3x_1 - 6x_2 &= -6 \end{aligned}$

C *Solve Problems 53–58 using augmented matrix methods.*

53. $\begin{aligned} 3x_1 - x_2 &= 7 \\ 2x_1 + 3x_2 &= 1 \end{aligned}$

54. $\begin{aligned} 2x_1 - 3x_2 &= -8 \\ 5x_1 + 3x_2 &= 1 \end{aligned}$

55. $\begin{aligned} 3x_1 + 2x_2 &= 4 \\ 2x_1 - x_2 &= 5 \end{aligned}$

56. $\begin{aligned} 4x_1 + 3x_2 &= 26 \\ 3x_1 - 11x_2 &= -7 \end{aligned}$

57. $\begin{aligned} 0.2x_1 - 0.5x_2 &= 0.07 \\ 0.8x_1 - 0.3x_2 &= 0.79 \end{aligned}$

58. $\begin{aligned} 0.3x_1 - 0.6x_2 &= 0.18 \\ 0.5x_1 - 0.2x_2 &= 0.54 \end{aligned}$

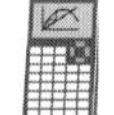

Solve Problems 59–62 using augmented matrix methods. Use a graphing utility to perform the row operations.

59. $\begin{aligned} 0.8x_1 + 2.88x_2 &= 4 \\ 1.25x_1 + 4.34x_2 &= 5 \end{aligned}$

60. $\begin{aligned} 2.7x_1 - 15.12x_2 &= 27 \\ 3.25x_1 - 18.52x_2 &= 33 \end{aligned}$

61. $\begin{aligned} 4.8x_1 - 40.32x_2 &= 295.2 \\ -3.75x_1 + 28.7x_2 &= -211.2 \end{aligned}$

62. $\begin{aligned} 5.7x_1 - 8.55x_2 &= -35.91 \\ 4.5x_1 + 5.73x_2 &= 76.17 \end{aligned}$

SECTION 4-3

Gauss–Jordan Elimination

- REDUCED MATRICES
- SOLVING SYSTEMS BY GAUSS–JORDAN ELIMINATION
- APPLICATION

Now that you have had some experience with row operations on simple augmented matrices, we will consider systems involving more than two variables. In addition, we will not require that a system have the same number of equations as variables. It turns out that the results for two-variable–two-equation linear systems stated in Theorem 1, Section 4-1, actually hold for linear systems of any size.

Possible Solutions to a Linear System

It can be shown that any linear system must have exactly one solution, no solution, or an infinite number of solutions, regardless of the number of equations or number of variables in the system. The terms *unique solution, consistent, inconsistent, dependent,* and *independent* are used to describe these solutions, just as in the two-variable case.

REDUCED MATRICES

In the last section we used row operations to transform the augmented coefficient matrix for a system of two equations in two variables,

$$\left[\begin{array}{cc|c} a_{11} & a_{12} & k_1 \\ a_{21} & a_{22} & k_2 \end{array}\right] \qquad \begin{aligned} a_{11}x_1 + a_{12}x_2 &= k_1 \\ a_{21}x_1 + a_{22}x_2 &= k_2 \end{aligned}$$

into one of the following simplified forms:

$$\begin{array}{ccc} \text{Form 1} & \text{Form 2} & \text{Form 3} \\ \left[\begin{array}{cc|c} 1 & 0 & m \\ 0 & 1 & n \end{array}\right] & \left[\begin{array}{cc|c} 1 & m & n \\ 0 & 0 & 0 \end{array}\right] & \left[\begin{array}{cc|c} 1 & m & n \\ 0 & 0 & p \end{array}\right] \end{array} \tag{1}$$

where m, n, and p are real numbers, $p \neq 0$. Each of these reduced forms represents a system that has a different type of solution set, and no two of these forms are row-equivalent. Thus, we consider each of these to be a different simplified form. Now we want to consider larger systems with more variables and more equations.

Explore–Discuss 1

Forms 1, 2, and 3 in (1) represent systems that have a unique solution, an infinite number of solutions, and no solution, respectively. Discuss the number of solutions for the systems of three equations in three variables represented by the following augmented coefficient matrices:

(A) $\left[\begin{array}{ccc|c} 1 & 2 & 3 & 5 \\ 0 & 0 & 0 & 6 \\ 0 & 0 & 0 & 0 \end{array}\right]$ (B) $\left[\begin{array}{ccc|c} 1 & 2 & 3 & 5 \\ 0 & 0 & 0 & 0 \\ 0 & 0 & 0 & 0 \end{array}\right]$

(C) $\left[\begin{array}{ccc|c} 1 & 0 & 0 & 5 \\ 0 & 1 & 0 & 6 \\ 0 & 0 & 1 & 7 \end{array}\right]$

Since there is no upper limit on the number of variables or the number of equations in a linear system, it is not feasible to explicitly list all possible "simplified forms" for larger systems, as we did for systems of two equations in two variables. Instead, we state a general definition of a simplified form called a *reduced matrix* that can be applied to all matrices and systems, regardless of size.

Reduced Matrix

A matrix is in **reduced form** if:

1. Each row consisting entirely of zeros is below any row having at least one nonzero element.
2. The leftmost nonzero element in each row is 1.
3. All other elements in the column containing the leftmost 1 of a given row are zeros.
4. The leftmost 1 in any row is to the right of the leftmost 1 in the row above.

The following matrices are in reduced form. Check each one carefully to convince yourself that the conditions in the definition are met.

$$\left[\begin{array}{cc|c} 1 & 0 & 2 \\ 0 & 1 & -3 \end{array}\right] \quad \left[\begin{array}{ccc|c} 1 & 0 & 0 & 2 \\ 0 & 1 & 0 & -1 \\ 0 & 0 & 1 & 3 \end{array}\right] \quad \left[\begin{array}{cc|c} 1 & 0 & 3 \\ 0 & 1 & -1 \\ 0 & 0 & 0 \end{array}\right]$$

$$\left[\begin{array}{cccc|c} 1 & 4 & 0 & 0 & -3 \\ 0 & 0 & 1 & 0 & 2 \\ 0 & 0 & 0 & 1 & 6 \end{array}\right] \quad \left[\begin{array}{ccc|c} 1 & 0 & 4 & 0 \\ 0 & 1 & 3 & 0 \\ 0 & 0 & 0 & 1 \end{array}\right]$$

Example 1 **Reduced Forms** The matrices below are not in reduced form. Indicate which condition in the definition is violated for each matrix. State the row operation(s) required to transform the matrix into reduced form and find the reduced form.

(A) $\left[\begin{array}{cc|c} 0 & 1 & -2 \\ 1 & 0 & 3 \end{array}\right]$ (B) $\left[\begin{array}{ccc|c} 1 & 2 & -2 & 3 \\ 0 & 0 & 1 & -1 \end{array}\right]$

(C) $\left[\begin{array}{cc|c} 1 & 0 & -3 \\ 0 & 0 & 0 \\ 0 & 1 & -2 \end{array}\right]$ (D) $\left[\begin{array}{ccc|c} 1 & 0 & 0 & -1 \\ 0 & 2 & 0 & 3 \\ 0 & 0 & 1 & -5 \end{array}\right]$

SOLUTION (A) Condition 4 is violated: The leftmost 1 in Row 2 is not to the right of the leftmost 1 in Row 1. Perform the row operation $R_1 \leftrightarrow R_2$ to obtain

$$\left[\begin{array}{cc|c} 1 & 0 & 3 \\ 0 & 1 & -2 \end{array}\right]$$

(B) Condition 3 is violated: The column containing the leftmost 1 in Row 2 has a nonzero element above the 1. Perform the row operation $2R_2 + R_1 \rightarrow R_1$ to obtain

$$\left[\begin{array}{ccc|c} 1 & 2 & 0 & 1 \\ 0 & 0 & 1 & -1 \end{array}\right]$$

(C) Condition 1 is violated: The second row contains all zeros and it is not below any row having at least one nonzero element. Perform the row operation $R_2 \leftrightarrow R_3$ to obtain

$$\left[\begin{array}{cc|c} 1 & 0 & -3 \\ 0 & 1 & -2 \\ 0 & 0 & 0 \end{array}\right]$$

(D) Condition 2 is violated: The leftmost nonzero element in Row 2 is not a 1. Perform the row operation $\frac{1}{2}R_2 \rightarrow R_2$ to obtain

$$\left[\begin{array}{ccc|c} 1 & 0 & 0 & -1 \\ 0 & 1 & 0 & \frac{3}{2} \\ 0 & 0 & 1 & -5 \end{array}\right]$$

Matched Problem 1 ➡ The matrices below are not in reduced form. Indicate which condition in the definition is violated for each matrix. State the row operation(s) required to transform the matrix into reduced form and find the reduced form.

(A) $\left[\begin{array}{cc|c} 1 & 0 & 2 \\ 0 & 3 & -6 \end{array}\right]$ (B) $\left[\begin{array}{ccc|c} 1 & 5 & 4 & 3 \\ 0 & 1 & 2 & -1 \\ 0 & 0 & 0 & 0 \end{array}\right]$

(C) $\left[\begin{array}{ccc|c} 0 & 1 & 0 & -3 \\ 1 & 0 & 0 & 0 \\ 0 & 0 & 1 & 2 \end{array}\right]$ (D) $\left[\begin{array}{ccc|c} 1 & 2 & 0 & 3 \\ 0 & 0 & 0 & 0 \\ 0 & 0 & 1 & 4 \end{array}\right]$

Solving Systems by Gauss–Jordan Elimination

We are now ready to outline the Gauss–Jordan method for solving systems of linear equations. The method systematically transforms an augmented matrix into a reduced form. The system corresponding to a reduced augmented coefficient matrix is called a **reduced system.** As we shall see, reduced systems are easy to solve.

The Gauss–Jordan elimination method is named after the German mathematician Carl Friedrich Gauss (1777–1885) and the German geodesist Wilhelm Jordan (1842–1899). Gauss, one of the greatest mathematicians of all time, used a method of solving systems of equations in his astronomical work that was later generalized by Jordan to solve problems in large-scale surveying.

Example 2 ➡ **Solving a System Using Gauss–Jordan Elimination** Solve by Gauss–Jordan elimination:

$$\begin{aligned} 2x_1 - 2x_2 + x_3 &= 3 \\ 3x_1 + x_2 - x_3 &= 7 \\ x_1 - 3x_2 + 2x_3 &= 0 \end{aligned}$$

Solution Write the augmented matrix and follow the steps indicated at the right.

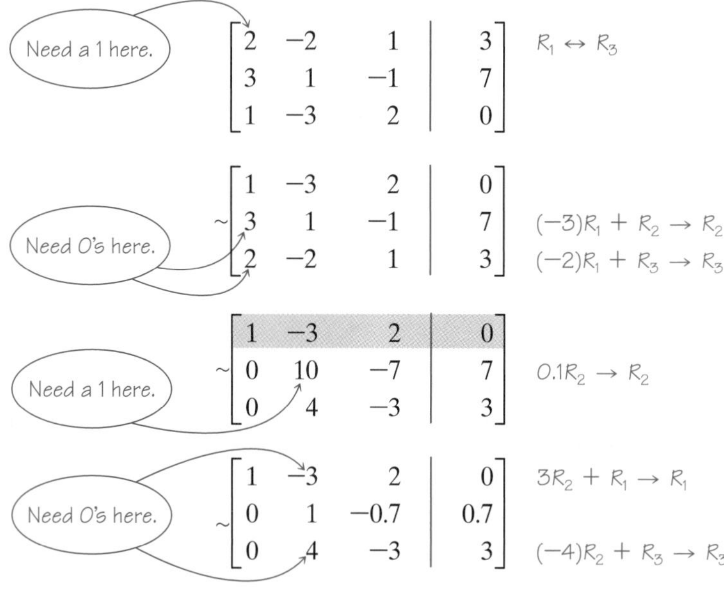

$$\left[\begin{array}{ccc|c} 2 & -2 & 1 & 3 \\ 3 & 1 & -1 & 7 \\ 1 & -3 & 2 & 0 \end{array}\right] \quad R_1 \leftrightarrow R_3$$

Step 1. Choose the leftmost nonzero column and get a 1 at the top.

$$\sim \left[\begin{array}{ccc|c} 1 & -3 & 2 & 0 \\ 3 & 1 & -1 & 7 \\ 2 & -2 & 1 & 3 \end{array}\right] \quad \begin{array}{l} (-3)R_1 + R_2 \rightarrow R_2 \\ (-2)R_1 + R_3 \rightarrow R_3 \end{array}$$

Step 2. Use multiples of the row containing the 1 from step 1 to get zeros in all remaining places in the column containing this 1.

$$\sim \left[\begin{array}{ccc|c} 1 & -3 & 2 & 0 \\ 0 & 10 & -7 & 7 \\ 0 & 4 & -3 & 3 \end{array}\right] \quad 0.1R_2 \rightarrow R_2$$

Step 3. Repeat step 1 with the *submatrix* formed by (mentally) deleting the top row.

$$\sim \left[\begin{array}{ccc|c} 1 & -3 & 2 & 0 \\ 0 & 1 & -0.7 & 0.7 \\ 0 & 4 & -3 & 3 \end{array}\right] \quad \begin{array}{l} 3R_2 + R_1 \rightarrow R_1 \\ (-4)R_2 + R_3 \rightarrow R_3 \end{array}$$

Step 4. Repeat step 2 with the *entire matrix.*

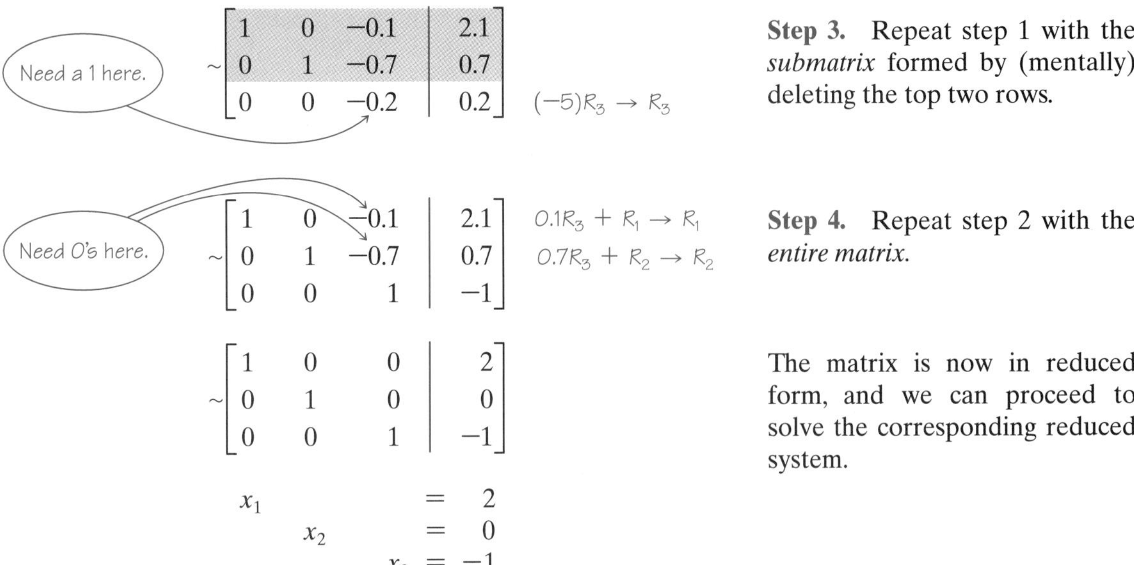

$$\sim\left[\begin{array}{rrr|r} 1 & 0 & -0.1 & 2.1 \\ 0 & 1 & -0.7 & 0.7 \\ 0 & 0 & -0.2 & 0.2 \end{array}\right] \quad (-5)R_3 \to R_3$$

Step 3. Repeat step 1 with the *submatrix* formed by (mentally) deleting the top two rows.

$$\sim\left[\begin{array}{rrr|r} 1 & 0 & -0.1 & 2.1 \\ 0 & 1 & -0.7 & 0.7 \\ 0 & 0 & 1 & -1 \end{array}\right] \quad \begin{array}{l} 0.1R_3 + R_1 \to R_1 \\ 0.7R_3 + R_2 \to R_2 \end{array}$$

Step 4. Repeat step 2 with the *entire matrix.*

$$\sim\left[\begin{array}{rrr|r} 1 & 0 & 0 & 2 \\ 0 & 1 & 0 & 0 \\ 0 & 0 & 1 & -1 \end{array}\right]$$

The matrix is now in reduced form, and we can proceed to solve the corresponding reduced system.

$$\begin{aligned} x_1 \qquad\qquad &= 2 \\ x_2 \qquad &= 0 \\ x_3 &= -1 \end{aligned}$$

The solution to this system is $x_1 = 2$, $x_2 = 0$, $x_3 = -1$. You should check this solution in the original system. ■

Gauss–Jordan Elimination

Step 1. Choose the leftmost nonzero column and use appropriate row operations to get a 1 at the top.

Step 2. Use multiples of the row containing the 1 from step 1 to get zeros in all remaining places in the column containing this 1.

Step 3. Repeat step 1 with the **submatrix** formed by (mentally) deleting the row used in step 2 and all rows above this row.

Step 4. Repeat step 2 with the **entire matrix,** including the mentally deleted rows. Continue this process until it is impossible to go further.

[*Note:* If at any point in this process we obtain a row with all zeros to the left of the vertical line and a nonzero number to the right, we can stop, since we will have a contradiction: $0 = n$, $n \neq 0$. We can then conclude that the system has no solution.]

Remarks

1. Even though each matrix has a unique reduced form, the sequence of steps (algorithm) presented here for transforming a matrix into a reduced form is not unique. That is, other sequences of steps (using row operations) can produce a reduced matrix. (For example, it is possible to use row operations in such a way that computations involving fractions are minimized.) But we emphasize again that we are not interested in the most efficient hand methods for transforming small matrices into

reduced forms. Our main interest is in giving you a little experience with a method that is suitable for solving large-scale systems on a graphing utility.

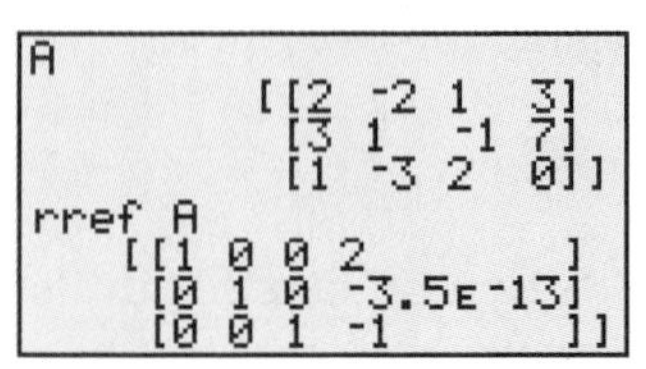

FIGURE 1
Gauss–Jordan elimination on a graphing calculator.

2. Most graphing utilities have the ability to find reduced forms, either directly or with some programming. Figure 1 illustrates the solution of Example 2 on a graphing calculator that has a built-in routine for finding reduced forms. Notice that in Row 2 and Column 4 of the reduced form the graphing calculator has displayed the very small number $-3.5\text{E} - 13$, instead of the exact value 0. This is a common occurrence on a graphing calculator and causes no problems. Just replace any very small numbers displayed in scientific notation with 0.

Matched Problem 2 Solve by Gauss–Jordan elimination:

$$\begin{aligned} 3x_1 + x_2 - 2x_3 &= 2 \\ x_1 - 2x_2 + x_3 &= 3 \\ 2x_1 - x_2 - 3x_3 &= 3 \end{aligned}$$

Example 3 **Solving a System Using Gauss–Jordan Elimination** Solve by Gauss–Jordan elimination:

$$\begin{aligned} 2x_1 - 4x_2 + x_3 &= -4 \\ 4x_1 - 8x_2 + 7x_3 &= 2 \\ -2x_1 + 4x_2 - 3x_3 &= 5 \end{aligned}$$

SOLUTION

$$\left[\begin{array}{rrr|r} 2 & -4 & 1 & -4 \\ 4 & -8 & 7 & 2 \\ -2 & 4 & -3 & 5 \end{array}\right] \quad 0.5R_1 \to R_1$$

$$\sim \left[\begin{array}{rrr|r} 1 & -2 & 0.5 & -2 \\ 4 & -8 & 7 & 2 \\ -2 & 4 & -3 & 5 \end{array}\right] \quad \begin{array}{l} (-4)R_1 + R_2 \to R_2 \\ 2R_1 + R_3 \to R_3 \end{array}$$

$$\sim \left[\begin{array}{rrr|r} 1 & -2 & 0.5 & -2 \\ 0 & 0 & 5 & 10 \\ 0 & 0 & -2 & 1 \end{array}\right] \quad 0.2R_2 \to R_2$$

Note that Column 3 is the leftmost nonzero column in this submatrix

$$\sim \left[\begin{array}{rrr|r} 1 & -2 & 0.5 & -2 \\ 0 & 0 & 1 & 2 \\ 0 & 0 & -2 & 1 \end{array}\right] \quad \begin{array}{l} (-0.5)R_2 + R_1 \to R_1 \\ 2R_2 + R_3 \to R_3 \end{array}$$

$$\sim \left[\begin{array}{rrr|r} 1 & -2 & 0 & -3 \\ 0 & 0 & 1 & 2 \\ 0 & 0 & 0 & 5 \end{array}\right]$$

We stop the Gauss–Jordan elimination, even though the matrix is not in reduced form, since the last row produces a contradiction.

Matched Problem 3 Solve by Gauss–Jordan elimination:

$$\begin{aligned} 2x_1 - 4x_2 - x_3 &= -8 \\ 4x_1 - 8x_2 + 3x_3 &= 4 \\ -2x_1 + 4x_2 + x_3 &= 11 \end{aligned}$$

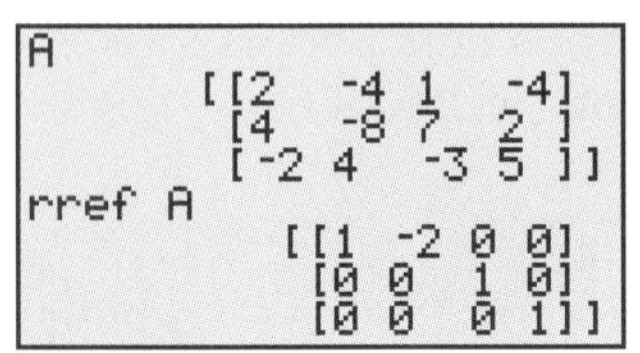

FIGURE 2
Recognizing contradictions on a graphing calculator

CAUTION

Figure 2 shows the solution to Example 3 on a graphing calculator with a built-in reduced form routine. Notice that the graphing calculator does not stop when a contradiction first occurs, but continues on to find the reduced form. Nevertheless, the last row in the reduced form still produces a contradiction. Do not confuse this type of reduced form with one that represents a consistent system (see Fig. 1).

Example 4 **Solving a System Using Gauss–Jordan Elimination** Solve by Gauss–Jordan elimination:

$$\begin{aligned} 3x_1 + 6x_2 - 9x_3 &= 15 \\ 2x_1 + 4x_2 - 6x_3 &= 10 \\ -2x_1 - 3x_2 + 4x_3 &= -6 \end{aligned}$$

SOLUTION

$$\left[\begin{array}{rrr|r} 3 & 6 & -9 & 15 \\ 2 & 4 & -6 & 10 \\ -2 & -3 & 4 & -6 \end{array}\right] \quad \tfrac{1}{3}R_1 \to R_1$$

$$\sim \left[\begin{array}{rrr|r} 1 & 2 & -3 & 5 \\ 2 & 4 & -6 & 10 \\ -2 & -3 & 4 & -6 \end{array}\right] \quad \begin{array}{l} (-2)R_1 + R_2 \to R_2 \\ 2R_1 + R_3 \to R_3 \end{array}$$

$$\sim \left[\begin{array}{rrr|r} 1 & 2 & -3 & 5 \\ 0 & 0 & 0 & 0 \\ 0 & 1 & -2 & 4 \end{array}\right] \quad R_2 \leftrightarrow R_3$$

Note that we must interchange Rows 2 and 3 to obtain a nonzero entry at the top of the second column of this submatrix.

$$\sim \left[\begin{array}{rrr|r} 1 & 2 & -3 & 5 \\ 0 & 1 & -2 & 4 \\ 0 & 0 & 0 & 0 \end{array}\right] \quad (-2)R_2 + R_1 \to R_1$$

$$\sim \left[\begin{array}{rrr|r} 1 & 0 & 1 & -3 \\ 0 & 1 & -2 & 4 \\ 0 & 0 & 0 & 0 \end{array}\right]$$

The matrix is now in reduced form. Write the corresponding reduced system and solve.

$$\begin{aligned} x_1 \qquad + x_3 &= -3 \\ x_2 - 2x_3 &= 4 \end{aligned}$$

We discard the equation corresponding to the third (all zero) row in the reduced form, since it is satisfied by all values of x_1, x_2, and x_3.

Note that the leftmost variable in each equation appears in one and only one equation. We solve for the leftmost variables x_1 and x_2 in terms of the remaining variable, x_3:

$$\begin{aligned} x_1 &= -x_3 - 3 \\ x_2 &= 2x_3 + 4 \end{aligned}$$

If we let $x_3 = t$, then for any real number t,

$$\begin{aligned} x_1 &= -t - 3 \\ x_2 &= 2t + 4 \\ x_3 &= t \end{aligned}$$

You should check that $(-t - 3, 2t + 4, t)$ is a solution of the original system for any real number t. Some particular solutions are

$t = 0$	$t = -2$	$t = 3.5$
$(-3, 4, 0)$	$(-1, 0, -2)$	$(-6.5, 11, 3.5)$

In general:

If the number of leftmost 1's in a reduced augmented coefficient matrix is less than the number of variables in the system and there are no contradictions, then the system is dependent and has infinitely many solutions.

Describing the solution set to this type of system is not difficult. In a reduced system, the *leftmost variables* correspond to the leftmost 1's in the corresponding reduced augmented matrix. The definition of reduced form for an augmented matrix ensures that each leftmost variable in the corresponding reduced system appears in one and only one equation of the system. Solving for each leftmost variable in terms of the remaining variables and writing a general solution to the system is usually easy. (Example 5 illustrates a slightly more involved case.)

Matched Problem 4 Solve by Gauss–Jordan elimination:

$$\begin{aligned} 2x_1 - 2x_2 - 4x_3 &= -2 \\ 3x_1 - 3x_2 - 6x_3 &= -3 \\ -2x_1 + 3x_2 + x_3 &= 7 \end{aligned}$$

Explore–Discuss 2

Explain why the definition of reduced form ensures that each leftmost variable in a reduced system appears in one and only one equation and no equation contains more than one leftmost variable. Discuss methods for determining whether a consistent system is independent or dependent by examining the reduced form.

Example 5 **Solving a System Using Gauss–Jordan Elimination** Solve by Gauss–Jordan elimination:

$$\begin{aligned} x_1 + 2x_2 + 4x_3 + x_4 - x_5 &= 1 \\ 2x_1 + 4x_2 + 8x_3 + 3x_4 - 4x_5 &= 2 \\ x_1 + 3x_2 + 7x_3 + 3x_5 &= -2 \end{aligned}$$

SOLUTION

$$\left[\begin{array}{ccccc|c} 1 & 2 & 4 & 1 & -1 & 1 \\ 2 & 4 & 8 & 3 & -4 & 2 \\ 1 & 3 & 7 & 0 & 3 & -2 \end{array}\right] \quad \begin{array}{l} \\ (-2)R_1 + R_2 \to R_2 \\ (-1)R_1 + R_3 \to R_3 \end{array}$$

$$\sim \left[\begin{array}{ccccc|c} 1 & 2 & 4 & 1 & -1 & 1 \\ 0 & 0 & 0 & 1 & -2 & 0 \\ 0 & 1 & 3 & -1 & 4 & -3 \end{array}\right] \quad R_2 \leftrightarrow R_3$$

$$\sim \left[\begin{array}{ccccc|c} 1 & 2 & 4 & 1 & -1 & 1 \\ 0 & 1 & 3 & -1 & 4 & -3 \\ 0 & 0 & 0 & 1 & -2 & 0 \end{array}\right] \quad \begin{array}{l} (-2)R_2 + R_1 \to R_1 \\ \\ \\ \end{array}$$

$$\sim\left[\begin{array}{ccccc|c} 1 & 0 & -2 & 3 & -9 & 7 \\ 0 & 1 & 3 & -1 & 4 & -3 \\ 0 & 0 & 0 & 1 & -2 & 0 \end{array}\right] \qquad \begin{array}{l} (-3)R_3 + R_1 \to R_1 \\ R_3 + R_2 \to R_2 \end{array}$$

$$\sim\left[\begin{array}{ccccc|c} 1 & 0 & -2 & 0 & -3 & 7 \\ 0 & 1 & 3 & 0 & 2 & -3 \\ 0 & 0 & 0 & 1 & -2 & 0 \end{array}\right] \qquad \text{Matrix is in reduced form.}$$

$$\begin{aligned} x_1 \quad - 2x_3 \quad - 3x_5 &= 7 \\ x_2 + 3x_3 \quad + 2x_5 &= -3 \\ x_4 - 2x_5 &= 0 \end{aligned}$$

Solve for the leftmost variables x_1, x_2, and x_4 in terms of the remaining variables x_3 and x_5:

$$\begin{aligned} x_1 &= 2x_3 + 3x_5 + 7 \\ x_2 &= -3x_3 - 2x_5 - 3 \\ x_4 &= 2x_5 \end{aligned}$$

If we let $x_3 = s$ and $x_5 = t$, then for any real numbers s and t,

$$\begin{aligned} x_1 &= 2s + 3t + 7 \\ x_2 &= -3s - 2t - 3 \\ x_3 &= s \\ x_4 &= 2t \\ x_5 &= t \end{aligned}$$

is a solution. The check is left for you to perform.

Matched Problem 5 Solve by Gauss–Jordan elimination:

$$\begin{aligned} x_1 - x_2 + 2x_3 \qquad - 2x_5 &= 3 \\ -2x_1 + 2x_2 - 4x_3 - x_4 + x_5 &= -5 \\ 3x_1 - 3x_2 + 7x_3 + x_4 - 4x_5 &= 6 \end{aligned}$$

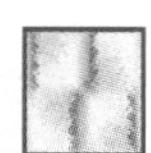

APPLICATION

Dependent systems of linear equations provide an excellent opportunity to discuss mathematical modeling in a little more detail. The process of using mathematics to solve real-world problems can be broken down into three steps (Fig. 3):

Step 1. *Construct* a mathematical model whose solution will provide information about the real-world problem.

Step 2. *Solve* the mathematical model.

Step 3. *Interpret* the solution to the mathematical model in terms of the original real-world problem.

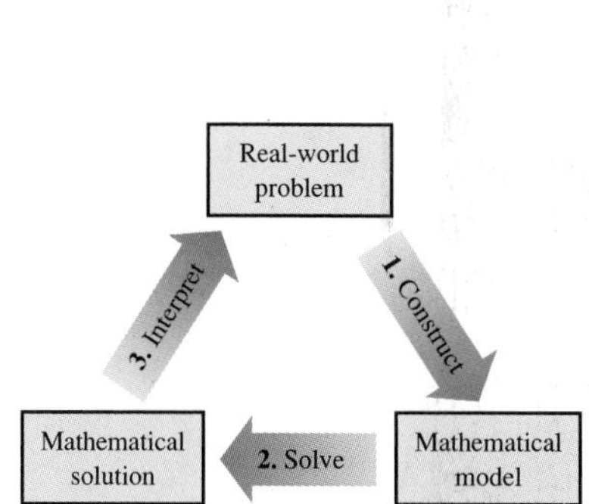

FIGURE 3

In more complex problems, this cycle may have to be repeated several times to obtain the required information about the real-world problem.

 *Example 6* **Production Scheduling** A casting company produces three different bronze sculptures. The casting department has available a maximum of 140 labor-hours per week, and the finishing department has a maximum of 180 labor-hours available per week. Sculpture A requires 30 hours for casting and 10 hours for finishing; sculpture B requires 10 hours for casting and 10 hours for finishing; and sculpture C requires 10 hours for casting and 30 hours for finishing. If the plant is to operate at maximum capacity, how many of each sculpture should be produced each week?

SOLUTION First, we summarize the relevant manufacturing data in a table:

	Labor-hours per sculpture A	B	C	Maximum labor-hours available per week
Casting department	30	10	10	140
Finishing department	10	10	30	180

Let: x_1 = Number of sculpture A produced per week
x_2 = Number of sculpture B produced per week
x_3 = Number of sculpture C produced per week

Then an appropriate mathematical model for this problem is

$$\begin{aligned} 30x_1 + 10x_2 + 10x_3 &= 140 \quad \text{Casting department} \\ 10x_1 + 10x_2 + 30x_3 &= 180 \quad \text{Finishing department} \end{aligned} \tag{2}$$

Now we can form the augmented matrix of the system and solve by using Gauss–Jordan elimination:

$$\left[\begin{array}{ccc|c} 30 & 10 & 10 & 140 \\ 10 & 10 & 30 & 180 \end{array}\right] \quad \begin{array}{l} \frac{1}{10}R_1 \to R_1 \quad \text{Simplify each row.} \\ \frac{1}{10}R_2 \to R_2 \end{array}$$

$$\sim \left[\begin{array}{ccc|c} 3 & 1 & 1 & 14 \\ 1 & 1 & 3 & 18 \end{array}\right] \quad R_1 \leftrightarrow R_2$$

$$\sim \left[\begin{array}{ccc|c} 1 & 1 & 3 & 18 \\ 3 & 1 & 1 & 14 \end{array}\right] \quad (-3)R_1 + R_2 \to R_2$$

$$\sim \left[\begin{array}{ccc|c} 1 & 1 & 3 & 18 \\ 0 & -2 & -8 & -40 \end{array}\right] \quad (-\tfrac{1}{2})R_2 \to R_2$$

$$\sim \left[\begin{array}{ccc|c} 1 & 1 & 3 & 18 \\ 0 & 1 & 4 & 20 \end{array}\right] \quad (-1)R_2 + R_1 \to R_1$$

$$\sim \left[\begin{array}{ccc|c} 1 & 0 & -1 & -2 \\ 0 & 1 & 4 & 20 \end{array}\right] \quad \text{Matrix is in reduced form.}$$

$$\begin{aligned} x_1 \quad - x_3 &= -2 \\ x_2 + 4x_3 &= 20 \end{aligned} \quad \text{or} \quad \begin{aligned} x_1 &= x_3 - 2 \\ x_2 &= -4x_3 + 20 \end{aligned}$$

Let $x_3 = t$. Then for t any real number,

$$\begin{aligned} x_1 &= t - 2 \\ x_2 &= -4t + 20 \\ x_3 &= t \end{aligned} \tag{3}$$

is a solution to the mathematical model in (2).

Now we must interpret this solution set in terms of the original problem. Since the variables x_1, x_2, and x_3 represent numbers of sculptures, they must be nonnegative numbers. And if we assume that we cannot produce a fractional number of sculptures, then each must be a nonnegative whole number. Since $t = x_3$, it follows that t must also be a nonnegative whole number. The first and second equations in (3) place further restrictions on the values that t can assume:

$$\begin{array}{lll} x_1 = t - 2 \geqslant 0 & \text{implies that} & t \geqslant 2 \\ x_2 = 20 - 4t \geqslant 0 & \text{implies that} & t \leqslant 5 \end{array}$$

Thus, the only possible values of t that will produce meaningful solutions to the original problem are 2, 3, 4, and 5. That is, the only possible production schedules that utilize the full capacity of the plant are $x_1 = t - 2$ type A, $x_2 = 20 - 4t$ type B, and $x_3 = t$ type C sculptures, where $t = 2, 3, 4$, or 5. A table is a convenient way to display these solutions:

	SCULPTURE A	SCULPTURE B	SCULPTURE C
t	x_1	x_2	x_3
2	0	12	2
3	1	8	3
4	2	4	4
5	3	0	5

Matched Problem 6 Repeat Example 6 given a casting capacity of 120 labor-hours per week and a finishing capacity of 200 labor-hours per week.

Answers to Matched Problems

1. (A) Condition 2 is violated: The 3 in Row 2 and Column 2 should be a 1. Perform the operation $\frac{1}{3}R_2 \to R_2$ to obtain

$$\left[\begin{array}{cc|c} 1 & 0 & 2 \\ 0 & 1 & -2 \end{array}\right]$$

(B) Condition 3 is violated: The 5 in Row 1 and Column 2 should be a 0. Perform the operation $(-5)R_2 + R_1 \to R_1$ to obtain

$$\left[\begin{array}{ccc|c} 1 & 0 & -6 & 8 \\ 0 & 1 & 2 & -1 \\ 0 & 0 & 0 & 0 \end{array}\right]$$

(C) Condition 4 is violated. The leftmost 1 in the second row is not to the right of the leftmost 1 in the first row. Perform the operation $R_1 \leftrightarrow R_2$ to obtain

$$\left[\begin{array}{ccc|c} 1 & 0 & 0 & 0 \\ 0 & 1 & 0 & -3 \\ 0 & 0 & 1 & 2 \end{array}\right]$$

(D) Condition 1 is violated: The all-zero second row should be at the bottom. Perform the operation $R_2 \leftrightarrow R_3$ to obtain

$$\left[\begin{array}{ccc|c} 1 & 2 & 0 & 3 \\ 0 & 0 & 1 & 4 \\ 0 & 0 & 0 & 0 \end{array}\right]$$

2. $x_1 = 1, x_2 = -1, x_3 = 0$ **3.** Inconsistent; no solution
4. $x_1 = 5t + 4, x_2 = 3t + 5, x_3 = t$, t any real number
5. $x_1 = s + 7, x_2 = s, x_3 = t - 2, x_4 = -3t - 1, x_5 = t$, s and t any real numbers
6. $x_1 = t - 4$ type A, $x_2 = 24 - 4t$ type B, and $x_3 = t$ type C sculptures, where $t = 4, 5$, or 6

EXERCISE 4-3

A *In Problems 1–10, if a matrix is in reduced form, say so. If not, explain why and indicate the row operation(s) necessary to transform the matrix into reduced form.*

1. $\left[\begin{array}{cc|c} 1 & 0 & 2 \\ 0 & 1 & -1 \end{array}\right]$

2. $\left[\begin{array}{cc|c} 0 & 1 & 2 \\ 1 & 0 & -1 \end{array}\right]$

3. $\left[\begin{array}{ccc|c} 1 & 0 & 2 & 3 \\ 0 & 0 & 0 & 0 \\ 0 & 1 & -1 & 4 \end{array}\right]$

4. $\left[\begin{array}{ccc|c} 1 & 0 & 0 & -2 \\ 0 & 1 & 0 & 0 \\ 0 & 0 & 1 & 1 \end{array}\right]$

5. $\left[\begin{array}{ccc|c} 0 & 1 & 0 & 2 \\ 0 & 0 & 3 & -1 \\ 0 & 0 & 0 & 0 \end{array}\right]$

6. $\left[\begin{array}{ccc|c} 1 & 2 & -3 & 1 \\ 0 & 0 & 1 & 4 \\ 0 & 0 & 0 & 0 \end{array}\right]$

7. $\left[\begin{array}{ccc|c} 1 & 1 & 0 & 1 \\ 0 & 0 & 1 & 1 \\ 0 & 0 & 0 & 0 \end{array}\right]$

8. $\left[\begin{array}{ccc|c} 1 & 0 & -1 & 3 \\ 0 & 2 & 1 & 1 \\ 0 & 0 & 0 & 0 \end{array}\right]$

9. $\left[\begin{array}{cccc|c} 1 & 0 & -2 & 0 & 1 \\ 0 & 0 & 1 & 1 & 0 \end{array}\right]$

10. $\left[\begin{array}{cccc|c} 1 & -2 & 0 & 0 & 1 \\ 0 & 0 & 1 & 1 & 0 \end{array}\right]$

Write the linear system corresponding to each reduced augmented matrix in Problems 11–18 and solve.

11. $\left[\begin{array}{ccc|c} 1 & 0 & 0 & -2 \\ 0 & 1 & 0 & 3 \\ 0 & 0 & 1 & 0 \end{array}\right]$

12. $\left[\begin{array}{cccc|c} 1 & 0 & 0 & 0 & -2 \\ 0 & 1 & 0 & 0 & 0 \\ 0 & 0 & 1 & 0 & 1 \\ 0 & 0 & 0 & 1 & 3 \end{array}\right]$

13. $\left[\begin{array}{ccc|c} 1 & 0 & -2 & 3 \\ 0 & 1 & 1 & -5 \\ 0 & 0 & 0 & 0 \end{array}\right]$

14. $\left[\begin{array}{ccc|c} 1 & -2 & 0 & -3 \\ 0 & 0 & 1 & 5 \\ 0 & 0 & 0 & 0 \end{array}\right]$

15. $\left[\begin{array}{cc|c} 1 & 0 & 0 \\ 0 & 1 & 0 \\ 0 & 0 & 1 \end{array}\right]$

16. $\left[\begin{array}{cc|c} 1 & 0 & 5 \\ 0 & 1 & -3 \\ 0 & 0 & 0 \end{array}\right]$

17. $\left[\begin{array}{cccc|c} 1 & -2 & 0 & -3 & -5 \\ 0 & 0 & 1 & 3 & 2 \end{array}\right]$

18. $\left[\begin{array}{cccc|c} 1 & 0 & -2 & 3 & 4 \\ 0 & 1 & -1 & 2 & -1 \end{array}\right]$

B *Use row operations to change each matrix in Problems 19–24 to reduced form.*

19. $\left[\begin{array}{cc|c} 1 & 2 & -1 \\ 0 & 1 & 3 \end{array}\right]$

20. $\left[\begin{array}{cc|c} 1 & 3 & 1 \\ 0 & 2 & -4 \end{array}\right]$

21. $\left[\begin{array}{ccc|c} 1 & 0 & -3 & 1 \\ 0 & 1 & 2 & 0 \\ 0 & 0 & 3 & -6 \end{array}\right]$

22. $\left[\begin{array}{ccc|c} 1 & 0 & 4 & 0 \\ 0 & 1 & -3 & -1 \\ 0 & 0 & -2 & 2 \end{array}\right]$

23. $\left[\begin{array}{ccc|c} 1 & 2 & -2 & -1 \\ 0 & 3 & -6 & 1 \\ 0 & -1 & 2 & -\frac{1}{3} \end{array}\right]$

24. $\left[\begin{array}{ccc|c} 0 & -2 & 8 & 1 \\ 2 & -2 & 6 & -4 \\ 0 & -1 & 4 & \frac{1}{2} \end{array}\right]$

Solve Problems 25–44 using Gauss–Jordan elimination.

25. $\begin{aligned} 2x_1 + 4x_2 - 10x_3 &= -2 \\ 3x_1 + 9x_2 - 21x_3 &= 0 \\ x_1 + 5x_2 - 12x_3 &= 1 \end{aligned}$

26. $\begin{aligned} 3x_1 + 5x_2 - x_3 &= -7 \\ x_1 + x_2 + x_3 &= -1 \\ 2x_1 + 11x_3 &= 7 \end{aligned}$

27. $\begin{aligned} 3x_1 + 8x_2 - x_3 &= -18 \\ 2x_1 + x_2 + 5x_3 &= 8 \\ 2x_1 + 4x_2 + 2x_3 &= -4 \end{aligned}$

28. $\begin{aligned} 2x_1 + 6x_2 + 15x_3 &= -12 \\ 4x_1 + 7x_2 + 13x_3 &= -10 \\ 3x_1 + 6x_2 + 12x_3 &= -9 \end{aligned}$

29. $\begin{aligned} 2x_1 - x_2 - 3x_3 &= 8 \\ x_1 - 2x_2 &= 7 \end{aligned}$

30. $\begin{aligned} 2x_1 + 4x_2 - 6x_3 &= 10 \\ 3x_1 + 3x_2 - 3x_3 &= 6 \end{aligned}$

31. $\begin{aligned} 2x_1 - x_2 &= 0 \\ 3x_1 + 2x_2 &= 7 \\ x_1 - x_2 &= -1 \end{aligned}$

32. $\begin{aligned} 2x_1 - x_2 &= 0 \\ 3x_1 + 2x_2 &= 7 \\ x_1 - x_2 &= -2 \end{aligned}$

33. $\begin{aligned} 3x_1 - 4x_2 - x_3 &= 1 \\ 2x_1 - 3x_2 + x_3 &= 1 \\ x_1 - 2x_2 + 3x_3 &= 2 \end{aligned}$

34. $\begin{aligned} 3x_1 + 7x_2 - x_3 &= 11 \\ x_1 + 2x_2 - x_3 &= 3 \\ 2x_1 + 4x_2 - 2x_3 &= 10 \end{aligned}$

35. $\begin{aligned} 3x_1 - 2x_2 + x_3 &= -7 \\ 2x_1 + x_2 - 4x_3 &= 0 \\ x_1 + x_2 - 3x_3 &= 1 \end{aligned}$

36. $\begin{aligned} 2x_1 + 3x_2 + 5x_3 &= 21 \\ x_1 - x_2 - 5x_3 &= -2 \\ 2x_1 + x_2 - x_3 &= 11 \end{aligned}$

37. $\begin{aligned} 2x_1 + 4x_2 - 2x_3 &= 2 \\ -3x_1 - 6x_2 + 3x_3 &= -3 \end{aligned}$

38. $\begin{aligned} 3x_1 - 9x_2 + 12x_3 &= 6 \\ -2x_1 + 6x_2 - 8x_3 &= -4 \end{aligned}$

39. $\begin{aligned} 4x_1 - x_2 + 2x_3 &= 3 \\ -4x_1 + x_2 - 3x_3 &= -10 \\ 8x_1 - 2x_2 + 9x_3 &= -1 \end{aligned}$

40. $\begin{aligned} 4x_1 - 2x_2 + 2x_3 &= 5 \\ -6x_1 + 3x_2 - 3x_3 &= -2 \\ 10x_1 - 5x_2 + 9x_3 &= 4 \end{aligned}$

41. $\begin{aligned} 2x_1 - 5x_2 - 3x_3 &= 7 \\ -4x_1 + 10x_2 + 2x_3 &= 6 \\ 6x_1 - 15x_2 - x_3 &= -19 \end{aligned}$

42. $\begin{aligned} -4x_1 + 8x_2 + 10x_3 &= -6 \\ 6x_1 - 12x_2 - 15x_3 &= 9 \\ -8x_1 + 14x_2 + 19x_3 &= -8 \end{aligned}$

43. $\begin{aligned} 5x_1 - 3x_2 + 2x_3 &= 13 \\ 2x_1 - x_2 - 3x_3 &= 1 \\ 4x_1 - 2x_2 + 4x_3 &= 12 \end{aligned}$

44. $\begin{aligned} 4x_1 - 2x_2 + 3x_3 &= 3 \\ 3x_1 - x_2 - 2x_3 &= -10 \\ 2x_1 + 4x_2 - x_3 &= -1 \end{aligned}$

45. Consider a consistent system of three linear equations in three variables. Discuss the nature of the system and its solution set if the reduced form of the augmented coefficient matrix has:

(A) One leftmost 1 (B) Two leftmost 1's
(C) Three leftmost 1's (D) Four leftmost 1's

46. Consider a system of three linear equations in three variables. Give examples of two reduced forms that are not row-equivalent if the system is:

(A) Consistent and dependent
(B) Inconsistent

In Problems 47–50, discuss the relationship between the number of solutions of the system and the constant k.

47. $\begin{aligned} x_1 - x_2 &= 4 \\ 3x_1 + kx_2 &= 7 \end{aligned}$

48. $\begin{aligned} x_1 + 2x_2 &= 4 \\ -2x_1 + kx_2 &= -8 \end{aligned}$

49. $\begin{aligned} x_1 + kx_2 &= 3 \\ 2x_1 + 6x_2 &= 6 \end{aligned}$

50. $\begin{aligned} x_1 + kx_2 &= 3 \\ 2x_1 + 4x_2 &= 8 \end{aligned}$

C *Solve Problems 51–56 using Gauss–Jordan elimination.*

51. $\begin{aligned} x_1 + 2x_2 - 4x_3 - x_4 &= 7 \\ 2x_1 + 5x_2 - 9x_3 - 4x_4 &= 16 \\ x_1 + 5x_2 - 7x_3 - 7x_4 &= 13 \end{aligned}$

52. $\begin{aligned} 2x_1 + 4x_2 + 5x_3 + 4x_4 &= 8 \\ x_1 + 2x_2 + 2x_3 + x_4 &= 3 \end{aligned}$

53. $\begin{aligned} x_1 - x_2 + 3x_3 - 2x_4 &= 1 \\ -2x_1 + 4x_2 - 3x_3 + x_4 &= 0.5 \\ 3x_1 - x_2 + 10x_3 - 4x_4 &= 2.9 \\ 4x_1 - 3x_2 + 8x_3 - 2x_4 &= 0.6 \end{aligned}$

54. $\begin{aligned} x_1 + x_2 + 4x_3 + x_4 &= 1.3 \\ -x_1 + x_2 - x_3 &= 1.1 \\ 2x_1 + x_3 + 3x_4 &= -4.4 \\ 2x_1 + 5x_2 + 11x_3 + 3x_4 &= 5.6 \end{aligned}$

55. $\begin{aligned} x_1 - 2x_2 + x_3 + x_4 + 2x_5 &= 2 \\ -2x_1 + 4x_2 + 2x_3 + 2x_4 - 2x_5 &= 0 \\ 3x_1 - 6x_2 + x_3 + x_4 + 5x_5 &= 4 \\ -x_1 + 2x_2 + 3x_3 + x_4 + x_5 &= 3 \end{aligned}$

56. $\begin{aligned} x_1 - 3x_2 + x_3 + x_4 + 2x_5 &= 2 \\ -x_1 + 5x_2 + 2x_3 + 2x_4 - 2x_5 &= 0 \\ 2x_1 - 6x_2 + 2x_3 + 2x_4 + 4x_5 &= 4 \\ -x_1 + 3x_2 - x_3 - x_5 &= -3 \end{aligned}$

Applications

Construct a mathematical model for each of the following problems. (The answers in the back of the book include both the mathematical model and the interpretation of its solution.) Use Gauss–Jordan elimination to solve the model and then interpret the solution.

Business & Economics

57. *Production scheduling.* A small manufacturing plant makes three types of inflatable boats: one-person, two-person, and four-person models. Each boat requires the services of three departments, as listed in the table. The cutting, assembly, and packaging departments have available a maximum of 380, 330, and 120 labor-hours per week, respectively.

Department	One-Person Boat	Two-Person Boat	Four-Person Boat
Cutting	0.5 hr	1.0 hr	1.5 hr
Assembly	0.6 hr	0.9 hr	1.2 hr
Packaging	0.2 hr	0.3 hr	0.5 hr

(A) How many boats of each type must be produced each week for the plant to operate at full capacity?
(B) How is the production schedule in part (A) affected if the packaging department is no longer used?
(C) How is the production schedule in part (A) affected if the four-person boat is no longer produced?

58. *Production scheduling.* Repeat Problem 57 assuming the cutting, assembly, and packaging departments have available a maximum of 350, 330, and 115 labor-hours per week, respectively.

59. *Purchasing.* A chemical manufacturer wants to purchase a fleet of 24 railroad tank cars with a combined carrying capacity of 250,000 gallons. Tank cars with three different carrying capacities are available: 6,000 gallons, 8,000 gallons, and 18,000 gallons. How many of each type of tank car should be purchased?

60. *Purchasing.* A commuter airline wants to purchase a fleet of 30 airplanes with a combined carrying capacity of 960 passengers. The three available types of planes carry 18, 24, and 42 passengers, respectively. How many of each type of plane should be purchased?

61. *Income tax.* A corporation has a taxable income of $7,650,000. At this income level, the federal income tax rate is 50%, the state tax rate is 20%, and the local tax rate is 10%. If each tax rate is applied to the total taxable income, the resulting tax liability for the corporation would be 80% of taxable income. However, it is customary to deduct taxes paid to one agency before computing taxes for the other agencies. Assume that the federal taxes are based on the income that remains after the state and local taxes are deducted, and that state and local taxes are computed in a similar manner. What is the tax liability of the corporation (as a percentage of taxable income) if these deductions are taken into consideration?

62. *Income tax.* Repeat Problem 61 if local taxes are not allowed as a deduction for federal and state taxes.

63. C *Taxable income.* As a result of several mergers and acquisitions, stock in four companies has been distributed among the companies. Each row of the following table gives the percentage of stock in the four companies that a particular company owns and the annual net income of each company (in millions of dollars):

	Percentage of Stock Owned in Company				Annual Net Income
Company	*A*	*B*	*C*	*D*	*Million $*
A	71	8	3	7	3.2
B	12	81	11	13	2.6
C	11	9	72	8	3.8
D	6	2	14	72	4.4

Thus, company *A* holds 71% of its own stock, 8% of the stock in company *B*, 3% of the stock in company *C*, and so on. For the purpose of assessing a state tax on corporate income, the taxable income of each company is defined to be its share of its own annual net income plus its share of the taxable income of each of the other companies, as determined by the percentages in the table. What is the taxable income of each company (to the nearest thousand dollars)?

64. C *Taxable income.* Repeat Problem 63 if tax law is changed so that the taxable income of a company is defined to be all of its own annual net income plus its share of the taxable income of each of the other companies.

Life Sciences

65. *Nutrition.* A dietitian in a hospital is to arrange a special diet composed of three basic foods. The diet is to include exactly 340 units of calcium, 180 units of iron, and 220 units of vitamin A. The number of units per ounce of each special ingredient for each of the foods is indicated in the table.

	UNITS PER OUNCE		
	Food A	*Food B*	*Food C*
CALCIUM	30	10	20
IRON	10	10	20
VITAMIN A	10	30	20

(A) How many ounces of each food must be used to meet the diet requirements?
(B) How is the diet in part (A) affected if food C is not used?
(C) How is the diet in part (A) affected if the vitamin A requirement is dropped?

66. *Nutrition.* Repeat Problem 65 if the diet is to include exactly 400 units of calcium, 160 units of iron, and 240 units of vitamin A.

67. *Nutrition—plants.* A farmer can buy four types of plant food. Each barrel of mix A contains 30 pounds of phosphoric acid, 50 pounds of nitrogen, and 30 pounds of potash; each barrel of mix B contains 30 pounds of phosphoric acid, 75 pounds of nitrogen, and 20 pounds of potash; each barrel of mix C contains 30 pounds of phosphoric acid, 25 pounds of nitrogen, and 20 pounds of potash; and each barrel of mix D contains 60 pounds of phosphoric acid, 25 pounds of nitrogen, and 50 pounds of potash. Soil tests indicate that a particular field needs 900 pounds of phosphoric acid, 750 pounds of nitrogen, and 700 pounds of potash. How many barrels of each type of food should the farmer mix together to supply the necessary nutrients for the field?

68. *Nutrition—animals.* In a laboratory experiment, rats are to be fed 5 packets of food containing a total of 80 units of vitamin E. There are four different brands of food packets that can be used. A packet of brand A contains 5 units of vitamin E, a packet of brand B contains 10 units of vitamin E, a packet of brand C contains 15 units of vitamin E, and a packet of brand D contains 20 units of vitamin E. How many packets of each brand should be mixed and fed to the rats?

Social Sciences

69. *Sociology.* Two sociologists have grant money to study school busing in a particular city. They wish to conduct an opinion survey using 600 telephone contacts and 400 house contacts. Survey company A has personnel to do 30 telephone and 10 house contacts per hour; survey company B can handle 20 telephone and 20 house contacts per hour. How many hours should be scheduled for each firm to produce exactly the number of contacts needed?

70. *Sociology.* Repeat Problem 69 if 650 telephone contacts and 350 house contacts are needed.

71. *Traffic flow.* The rush-hour traffic flow for a network of four one-way streets in a city is shown in the figure. The numbers next to each street indicate the number of vehicles per hour that enter and leave the network on that street. The variables x_1, x_2, x_3, and x_4 represent the flow of traffic between the four intersections in the network.

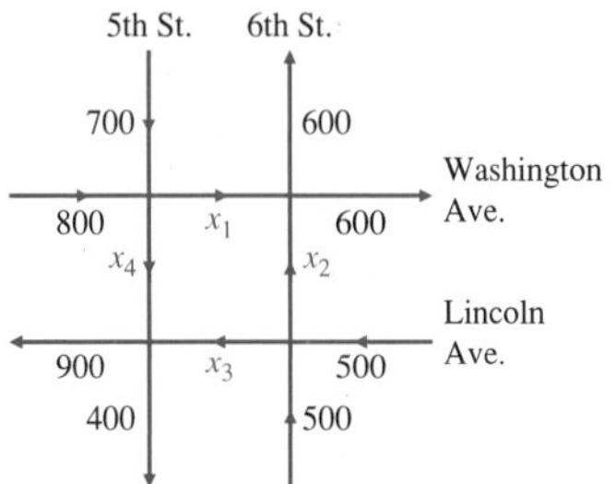

Figure for 71

(A) For a smooth traffic flow, the number of vehicles entering each intersection should always equal the number leaving. For example, since 1,500 vehicles enter the intersection of 5th Street and Washington Avenue each hour and $x_1 + x_4$ vehicles leave this intersection, we see that $x_1 + x_4 = 1{,}500$. Find the equations determined by the traffic flow at each of the other three intersections.
(B) Find the solution to the system in part (A).
(C) What is the maximum number of vehicles that can travel from Washington Avenue to Lincoln Avenue on 5th Street? What is the minimum number?
(D) If traffic lights are adjusted so that 1,000 vehicles per hour travel from Washington Avenue to Lincoln Avenue on 5th Street, determine the flow around the rest of the network.

72. *Traffic flow.* Refer to Problem 71. Closing Washington Avenue east of 6th Street for construction changes the traffic flow for the network as indicated in the figure. Repeat parts (A)–(D) of Problem 71 for this traffic flow.

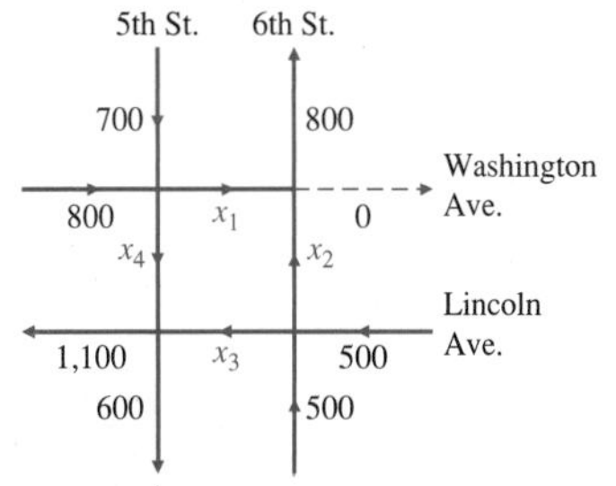

Figure for 72

SECTION 4-4

Matrices: Basic Operations

- ADDITION AND SUBTRACTION
- PRODUCT OF A NUMBER k AND A MATRIX M
- MATRIX PRODUCT

In the last two sections we introduced the important new idea of matrices. In this and the following sections, we will develop this concept further. Matrices are both a very ancient and a very current mathematical concept. References to matrices and systems of equations can be found in Chinese manuscripts dating back to around 200 B.C. More recently, the advent of computers has made matrices a very useful tool for a wide variety of applications. Most graphing calculators and computers are capable of performing calculations with matrices.

As we will see, matrix addition and multiplication are similar to real number addition and multiplication in many respects, but there are some important differences. A brief review of Appendix A-2, where the basic properties of real number operations are discussed, will help you understand the similarities and the differences.

ADDITION AND SUBTRACTION

Before we can discuss arithmetic operations for matrices, we have to define equality for matrices. Two matrices are **equal** if they have the same size and their corresponding elements are equal. For example,

$$\underset{2\times 3}{\begin{bmatrix} a & b & c \\ d & e & f \end{bmatrix}} = \underset{2\times 3}{\begin{bmatrix} u & v & w \\ x & y & z \end{bmatrix}} \quad \text{if and only if} \quad \begin{matrix} a = u & b = v & c = w \\ d = x & e = y & f = z \end{matrix}$$

The **sum of two matrices of the same size** is the matrix with elements that are the sum of the corresponding elements of the two given matrices.

Addition is not defined for matrices of different sizes.

Example 1 **Matrix Addition**

(A) $\begin{bmatrix} a & b \\ c & d \end{bmatrix} + \begin{bmatrix} w & x \\ y & z \end{bmatrix} = \begin{bmatrix} (a + w) & (b + x) \\ (c + y) & (d + z) \end{bmatrix}$

(B) $\begin{bmatrix} 2 & -3 & 0 \\ 1 & 2 & -5 \end{bmatrix} + \begin{bmatrix} 3 & 1 & 2 \\ -3 & 2 & 5 \end{bmatrix} = \begin{bmatrix} 5 & -2 & 2 \\ -2 & 4 & 0 \end{bmatrix}$

(C) $\begin{bmatrix} 5 & 0 & -2 \\ 1 & -3 & 8 \end{bmatrix} + \begin{bmatrix} -1 & 7 \\ 0 & 6 \\ -2 & 8 \end{bmatrix}$ Not defined

Matched Problem 1 Add: $\begin{bmatrix} 3 & 2 \\ -1 & -1 \\ 0 & 3 \end{bmatrix} + \begin{bmatrix} -2 & 3 \\ 1 & -1 \\ 2 & -2 \end{bmatrix}$

```
[A]
      [[2 -3 0 ]
       [1 2  -5]]
[B]
      [[3  1 2]
       [-3 2 5]]
[A]+[B]
      [[5  -2 2]
       [-2 4  0]]
```

FIGURE 1
Addition on a graphing calculator

Graphing utilities can be used to solve problems involving matrix operations. Figure 1 illustrates the solution to Example 1B on a graphing calculator.

Because we add two matrices by adding their corresponding elements, it follows from the properties of real numbers that matrices of the same size are commutative and associative relative to addition. That is, if A, B, and C are matrices of the same size, then

Commutative: $A + B = B + A$
Associative: $(A + B) + C = A + (B + C)$

A matrix with elements that are all zeros is called a **zero matrix.** For example,

$$\begin{bmatrix} 0 & 0 & 0 \end{bmatrix} \qquad \begin{bmatrix} 0 & 0 \\ 0 & 0 \end{bmatrix} \qquad \begin{bmatrix} 0 \\ 0 \\ 0 \\ 0 \end{bmatrix} \qquad \begin{bmatrix} 0 & 0 & 0 & 0 \\ 0 & 0 & 0 & 0 \\ 0 & 0 & 0 & 0 \end{bmatrix}$$

are zero matrices of different sizes. [*Note:* The simpler notation "0" is often used to denote the zero matrix of an arbitrary size.] The **negative of a matrix** $\boldsymbol{M}$**,** denoted by $\boldsymbol{-M}$**,** is a matrix with elements that are the negatives of the elements in M. Thus, if

$$M = \begin{bmatrix} a & b \\ c & d \end{bmatrix} \quad \text{then} \quad -M = \begin{bmatrix} -a & -b \\ -c & -d \end{bmatrix}$$

Note that $M + (-M) = 0$ (a zero matrix).

If A and B are matrices of the same size, then we define **subtraction** as follows:

$$\mathbf{A - B = A + (-B)}$$

Thus, to subtract matrix B from matrix A, we simply add the negative of B to A.

Example 2 ➡ **Matrix Subtraction**

$$\begin{bmatrix} 3 & -2 \\ 5 & 0 \end{bmatrix} - \begin{bmatrix} -2 & 2 \\ 3 & 4 \end{bmatrix} = \begin{bmatrix} 3 & -2 \\ 5 & 0 \end{bmatrix} + \begin{bmatrix} 2 & -2 \\ -3 & -4 \end{bmatrix} = \begin{bmatrix} 5 & -4 \\ 2 & -4 \end{bmatrix}$$

Matched Problem 2 ➡ Subtract: $\begin{bmatrix} 2 & -3 & 5 \end{bmatrix} - \begin{bmatrix} 3 & -2 & 1 \end{bmatrix}$

■ Product of a Number k and a Matrix M

The **product of a number** $\boldsymbol{k}$ **and a matrix** $\boldsymbol{M}$**,** denoted by $\boldsymbol{kM}$**,** is a matrix formed by multiplying each element of M by k.

Example 3 ➡ **Multiplication of a Matrix by a Number**

$$-2\begin{bmatrix} 3 & -1 & 0 \\ -2 & 1 & 3 \\ 0 & -1 & -2 \end{bmatrix} = \begin{bmatrix} -6 & 2 & 0 \\ 4 & -2 & -6 \\ 0 & 2 & 4 \end{bmatrix}$$

Matched Problem 3 ➡ Find: $10\begin{bmatrix} 1.3 \\ 0.2 \\ 3.5 \end{bmatrix}$

Explore–Discuss 1

Multiplication of two numbers can be interpreted as repeated addition if one of the numbers is a positive integer. That is,

$$2a = a + a \qquad 3a = a + a + a \qquad 4a = a + a + a + a$$

and so on. Discuss this interpretation for the product of an integer k and matrix M. Use specific examples to illustrate your remarks.

The next example illustrates the use of matrix operations in an applied setting.

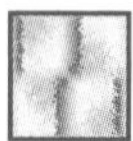

Example 4 **Sales Commissions** Ms. Smith and Mr. Jones are salespeople in a new-car agency that sells only two models. August was the last month for this year's models, and next year's models were introduced in September. Gross dollar sales for each month are given in the following matrices:

$$\begin{array}{l} \\ \text{Ms. Smith} \\ \text{Mr. Jones} \end{array} \overset{\text{August sales}}{\overset{\text{Compact}\quad\text{Luxury}}{\begin{bmatrix} \$54{,}000 & \$88{,}000 \\ \$126{,}000 & 0 \end{bmatrix}}} = A \qquad \overset{\text{September sales}}{\overset{\text{Compact}\quad\text{Luxury}}{\begin{bmatrix} \$228{,}000 & \$368{,}000 \\ \$304{,}000 & \$322{,}000 \end{bmatrix}}} = B$$

(For example, Ms. Smith had \$54,000 in compact sales in August, and Mr. Jones had \$322,000 in luxury car sales in September.)

(A) What were the combined dollar sales in August and September for each salesperson and each model?
(B) What was the increase in dollar sales from August to September?
(C) If both salespeople receive 5% commissions on gross dollar sales, compute the commission for each person for each model sold in September.

SOLUTION (A) $A + B = \overset{\text{Compact}\quad\text{Luxury}}{\begin{bmatrix} \$282{,}000 & \$456{,}000 \\ \$430{,}000 & \$322{,}000 \end{bmatrix}} \begin{array}{l} \text{Ms. Smith} \\ \text{Mr. Jones} \end{array}$

(B) $B - A = \overset{\text{Compact}\quad\text{Luxury}}{\begin{bmatrix} \$174{,}000 & \$280{,}000 \\ \$178{,}000 & \$322{,}000 \end{bmatrix}} \begin{array}{l} \text{Ms. Smith} \\ \text{Mr. Jones} \end{array}$

(C) $0.05B = \begin{bmatrix} (0.05)(\$228{,}000) & (0.05)(\$368{,}000) \\ (0.05)(\$304{,}000) & (0.05)(\$322{,}000) \end{bmatrix}$

$= \begin{bmatrix} \$11{,}400 & \$18{,}400 \\ \$15{,}200 & \$16{,}100 \end{bmatrix} \begin{array}{l} \text{Ms. Smith} \\ \text{Mr. Jones} \end{array}$

Matched Problem 4 Repeat Example 4 with

$$A = \begin{bmatrix} \$45{,}000 & \$77{,}000 \\ \$106{,}000 & \$22{,}000 \end{bmatrix} \quad \text{and} \quad B = \begin{bmatrix} \$190{,}000 & \$345{,}000 \\ \$266{,}000 & \$276{,}000 \end{bmatrix}$$

Figure 2 illustrates a solution for Example 4 on a spreadsheet.

	1	2	3	4	5	6	7
1		August Sales		September Sales		September Commissions	
2		Compact	Luxury	Compact	Luxury	Compact	Luxury
3	Smith	$54,000	$88,000	$228,000	$368,000	$11,400	$18,400
4	Jones	$126,000	$0	$304,000	$322,000	$15,200	$16,100
5		Combined Sales		Sales Increase			
6	Smith	$282,000	$456,000	$174,000	$280,000		
7	Jones	$430,000	$322,000	$178,000	$322,000		

FIGURE 2

Matrix Product

Now we are going to introduce a matrix multiplication that will seem rather strange at first. Despite this apparent strangeness, this operation is well-founded in the general theory of matrices and, as we will see, is extremely useful in many practical problems.

Historically, matrix multiplication was introduced by the English mathematician Arthur Cayley (1821–1895) in studies of systems of linear equations and linear transformations. In Section 4-6, you will see that matrix multiplication is central to the process of expressing systems of linear equations as matrix equations and to the process of solving matrix equations. Matrix equations and their solutions provide us with an alternate method of solving linear systems with the same number of variables as equations.

We start by defining the product of two special matrices, a row matrix and a column matrix.

Product of a Row Matrix and a Column Matrix

The **product** of a $1 \times n$ row matrix and an $n \times 1$ column matrix is a 1×1 matrix given by

$$\underset{1 \times n}{[a_1\ a_2\ \dots\ a_n]} \underset{n \times 1}{\begin{bmatrix} b_1 \\ b_2 \\ \vdots \\ b_n \end{bmatrix}} = [a_1b_1 + a_2b_2 + \cdots + a_nb_n]$$

Note that the number of elements in the row matrix and in the column matrix must be the same for the product to be defined.

Example 5 **Product of a Row Matrix and a Column Matrix**

$$\begin{aligned} [2\quad -3\quad 0]\begin{bmatrix} -5 \\ 2 \\ -2 \end{bmatrix} &= [(2)(-5) + (-3)(2) + (0)(-2)] \\ &= [-10 - 6 + 0] = [-16] \end{aligned}$$

Matched Problem 5 $\begin{bmatrix} -1 & 0 & 3 & 2 \end{bmatrix} \begin{bmatrix} 2 \\ 3 \\ 4 \\ -1 \end{bmatrix} = ?$

Refer to Example 5. The distinction between the real number -16 and the 1×1 matrix $[-16]$ is a technical one, and it is common to see 1×1 matrices written as real numbers without brackets. In the work that follows, we will frequently refer to 1×1 matrices as real numbers and omit the brackets whenever it is convenient to do so.

Example 6 **Labor Costs** A factory produces a slalom water ski that requires 3 labor-hours in the assembly department and 1 labor-hour in the finishing department. Assembly personnel receive \$9 per hour and finishing personnel receive \$6 per hour. Total labor cost per ski is given by the product:

$$\begin{bmatrix} 3 & 1 \end{bmatrix} \begin{bmatrix} 9 \\ 6 \end{bmatrix} = [(3)(9) + (1)(6)] = [27 + 6] = [33] \quad \text{or} \quad \$33 \text{ per ski}$$

Matched Problem 6 If the factory in Example 6 also produces a trick water ski that requires 5 labor-hours in the assembly department and 1.5 labor-hours in the finishing department, write a product between appropriate row and column matrices that will give the total labor cost for this ski. Compute the cost.

We now use the product of a $1 \times n$ row matrix and an $n \times 1$ column matrix to extend the definition of matrix product to more general matrices.

Matrix Product

If A is an $m \times p$ matrix and B is a $p \times n$ matrix, then the **matrix product** of A and B, denoted AB, is an $m \times n$ matrix whose element in the ith row and jth column is the real number obtained from the product of the ith row of A and the jth column of B. If the number of columns in A does not equal the number of rows in B, then the matrix product AB is **not defined.**

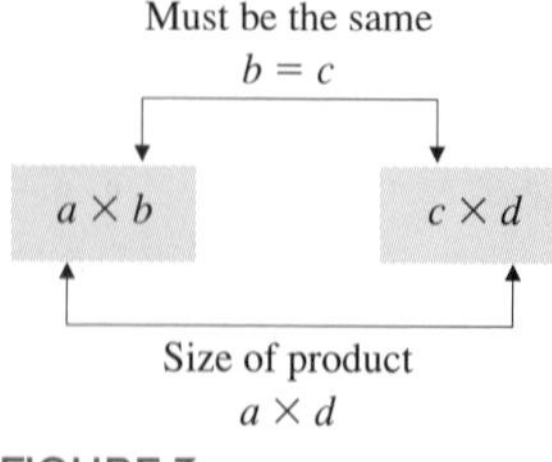

FIGURE 3

It is important to check sizes before starting the multiplication process. If A is an $a \times b$ matrix and B is a $c \times d$ matrix, then if $b = c$, the product AB will exist and will be an $a \times d$ matrix (see Fig. 3.). If $b \neq c$, then the product AB does not exist. The definition is not as complicated as it might first seem. An example should help clarify the process.

For

$$A = \begin{bmatrix} 2 & 3 & -1 \\ -2 & 1 & 2 \end{bmatrix} \quad \text{and} \quad B = \begin{bmatrix} 1 & 3 \\ 2 & 0 \\ -1 & 2 \end{bmatrix}$$

A is 2×3 and B is 3×2, so AB is 2×2. To find the first row of AB, we take the product of the first row of A with every column of B and write each result as a real number, not as a 1×1 matrix. The second row of AB is computed in the same manner. The four products of row and column matrices used to produce the four elements in AB are shown in the following dashed box. These products are usually calculated mentally or with the aid of a calculator, and need not be written out. The shaded portions highlight the steps involved in computing the element in the first row and second column of AB.

$$\underset{2\times 3}{\begin{bmatrix} 2 & 3 & -1 \\ -2 & 1 & 2 \end{bmatrix}} \underset{3\times 2}{\begin{bmatrix} 1 & 3 \\ 2 & 0 \\ -1 & 2 \end{bmatrix}} = \begin{bmatrix} [2 \;\; 3 \;\; -1]\begin{bmatrix} 1 \\ 2 \\ -1 \end{bmatrix} & [2 \;\; 3 \;\; -1]\begin{bmatrix} 3 \\ 0 \\ 2 \end{bmatrix} \\ [-2 \;\; 1 \;\; 2]\begin{bmatrix} 1 \\ 2 \\ -1 \end{bmatrix} & [-2 \;\; 1 \;\; 2]\begin{bmatrix} 3 \\ 0 \\ 2 \end{bmatrix} \end{bmatrix} = \underset{2\times 2}{\begin{bmatrix} 9 & 4 \\ -2 & -2 \end{bmatrix}}$$

Example 7 **Matrix Multiplication**

(A) $$\underset{3\times 2}{\begin{bmatrix} 2 & 1 \\ 1 & 0 \\ -1 & 2 \end{bmatrix}} \underset{2\times 4}{\begin{bmatrix} 1 & -1 & 0 & 1 \\ 2 & 1 & 2 & 0 \end{bmatrix}} = \underset{3\times 4}{\begin{bmatrix} 4 & -1 & 2 & 2 \\ 1 & -1 & 0 & 1 \\ 3 & 3 & 4 & -1 \end{bmatrix}}$$

(B) $$\underset{2\times 4}{\begin{bmatrix} 1 & -1 & 0 & 1 \\ 2 & 1 & 2 & 0 \end{bmatrix}} \underset{3\times 2}{\begin{bmatrix} 2 & 1 \\ 1 & 0 \\ -1 & 2 \end{bmatrix}}$$

Not defined.

(C) $$\begin{bmatrix} 2 & 6 \\ -1 & -3 \end{bmatrix}\begin{bmatrix} 1 & 2 \\ 3 & 6 \end{bmatrix} = \begin{bmatrix} 20 & 40 \\ -10 & -20 \end{bmatrix}$$

(D) $$\begin{bmatrix} 1 & 2 \\ 3 & 6 \end{bmatrix}\begin{bmatrix} 2 & 6 \\ -1 & -3 \end{bmatrix} = \begin{bmatrix} 0 & 0 \\ 0 & 0 \end{bmatrix}$$

(E) $$[2 \;\; -3 \;\; 0]\begin{bmatrix} -5 \\ 2 \\ -2 \end{bmatrix} = [-16]$$

(F) $$\begin{bmatrix} -5 \\ 2 \\ -2 \end{bmatrix}[2 \;\; -3 \;\; 0] = \begin{bmatrix} -10 & 15 & 0 \\ 4 & -6 & 0 \\ -4 & 6 & 0 \end{bmatrix}$$

Matched Problem 7 Find each product, if it is defined:

(A) $$\begin{bmatrix} -1 & 0 & 3 & -2 \\ 1 & 2 & 2 & 0 \end{bmatrix}\begin{bmatrix} -1 & 1 \\ 2 & 3 \\ 1 & 0 \end{bmatrix}$$

(B) $$\begin{bmatrix} -1 & 1 \\ 2 & 3 \\ 1 & 0 \end{bmatrix}\begin{bmatrix} -1 & 0 & 3 & -2 \\ 1 & 2 & 2 & 0 \end{bmatrix}$$

(C) $\begin{bmatrix} 1 & 2 \\ -1 & -2 \end{bmatrix}\begin{bmatrix} -2 & 4 \\ 1 & -2 \end{bmatrix}$ (D) $\begin{bmatrix} -2 & 4 \\ 1 & -2 \end{bmatrix}\begin{bmatrix} 1 & 2 \\ -1 & -2 \end{bmatrix}$

(E) $[3 \quad -2 \quad 1]\begin{bmatrix} 4 \\ 2 \\ 3 \end{bmatrix}$ (F) $\begin{bmatrix} 4 \\ 2 \\ 3 \end{bmatrix}[3 \quad -2 \quad 1]$

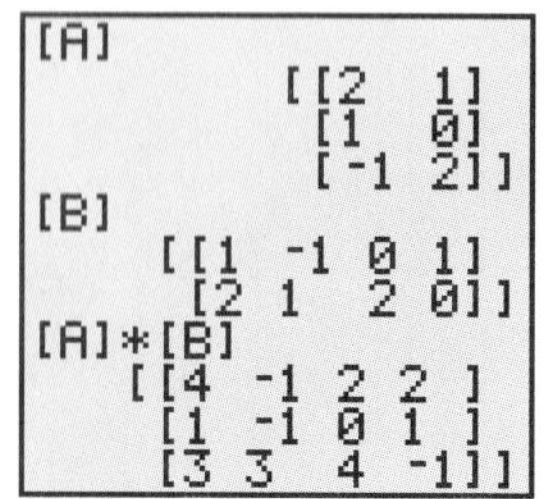

FIGURE 4
Multiplication on a graphing calculator

Figure 4 illustrates a graphing utility solution to Example 7A. What would you expect to happen if you tried to solve Example 7B on a graphing utility?

In the arithmetic of real numbers it does not matter in which order we multiply; for example, $5 \times 7 = 7 \times 5$. In matrix multiplication, however, it does make a difference. That is, AB does not always equal BA, even if both multiplications are defined and both products are the same size (see Examples 7C and 7D). Thus:

Matrix multiplication is not commutative.

Also, AB may be zero with neither A nor B equal to zero (see Example 7D). Thus:

The zero property does not hold for matrix multiplication.

(See Appendix A-2 for a discussion of the zero property for real numbers.)

Explore–Discuss 2

In addition to the commutative and zero properties, there are other significant differences between real number multiplication and matrix multiplication.

(A) In real number multiplication, the only real number whose square is 0 is the real number 0 ($0^2 = 0$). Find at least one 2×2 matrix A with all elements nonzero such that $A^2 = 0$,* where 0 is the 2×2 zero matrix.

(B) In real number multiplication the only nonzero real number that is equal to its square is the real number 1 ($1^2 = 1$). Find at least one 2×2 matrix A with all elements nonzero such that $A^2 = A$.

We will continue our discussion of properties of matrix multiplication later in this chapter. Now we consider an application of matrix multiplication.

Example 8 **Labor Costs** Let us combine the time requirements for slalom and trick water skis discussed in Example 6 and Matched Problem 6 into one matrix:

	Labor-hours per ski	
	Assembly department	Finishing department
Trick ski	5 hr	1.5 hr
Slalom ski	3 hr	1 hr

$$\begin{bmatrix} 5 \text{ hr} & 1.5 \text{ hr} \\ 3 \text{ hr} & 1 \text{ hr} \end{bmatrix} = L$$

*Following standard algebraic notation, we write $A^2 = AA$, $A^3 = AAA$, and so on.

Now suppose the company has two manufacturing plants, one in California and the other in Wisconsin, and that their hourly rates for each department are given in the following matrix:

$$\begin{array}{r} \\ \text{Assembly department} \\ \text{Finishing department} \end{array} \begin{array}{c} \text{Hourly wages} \\ \begin{array}{cc} \text{California} & \text{Wisconsin} \end{array} \\ \begin{bmatrix} \$12 & \$9 \\ \$8 & \$6 \end{bmatrix} \end{array} = H$$

Since H and L are both 2×2 matrices, we can take the product of H and L in either order and the result will be a 2×2 matrix:

$$HL = \begin{bmatrix} 12 & 9 \\ 8 & 6 \end{bmatrix} \begin{bmatrix} 5 & 1.5 \\ 3 & 1 \end{bmatrix} = \begin{bmatrix} 87 & 27 \\ 58 & 18 \end{bmatrix}$$

$$LH = \begin{bmatrix} 5 & 1.5 \\ 3 & 1 \end{bmatrix} \begin{bmatrix} 12 & 9 \\ 8 & 6 \end{bmatrix} = \begin{bmatrix} 72 & 54 \\ 44 & 33 \end{bmatrix}$$

How can we interpret the elements in these products? Let's begin with the product HL. The element 87 in the first row and first column of HL is the product of the first row matrix of H and the first column matrix of L:

$$\overset{\text{CA}\quad\text{WI}}{\begin{bmatrix} 12 & 9 \end{bmatrix}} \begin{bmatrix} 5 \\ 3 \end{bmatrix} \begin{array}{l} \text{Trick} \\ \text{Slalom} \end{array} = 12(5) + 9(3) = 60 + 27 = 87$$

Notice that \$60 is the labor cost for assembling a trick ski at the California plant and \$27 is the labor cost for assembling a slalom ski at the Wisconsin plant. Although both numbers represent labor costs, it makes no sense to add them together. They do not pertain to the same type of ski or to the same plant. Thus, even though the product HL happens to be defined mathematically, it has no useful interpretation in this problem.

Now let's consider the product LH. The element 72 in the first row and first column of LH is given by the following product:

$$\overset{\text{Assembly}\quad\text{Finishing}}{\begin{bmatrix} 5 & 1.5 \end{bmatrix}} \begin{bmatrix} 12 \\ 8 \end{bmatrix} \begin{array}{l} \text{Assembly} \\ \text{Finishing} \end{array} = 5(12) + 1.5(8) = 60 + 12 = 72$$

This time, \$60 is the labor cost for assembling a trick ski at the California plant and \$12 is the labor cost for finishing a trick ski at the California plant. Thus, the sum is the total labor cost for producing a trick ski at the California plant. The other elements in LH also represent total labor costs, as indicated by the row and column labels shown below:

$$LH = \begin{array}{c} \text{Labor costs per ski} \\ \begin{array}{cc} \text{CA} & \text{WI} \end{array} \\ \begin{bmatrix} \$72 & \$54 \\ \$44 & \$33 \end{bmatrix} \end{array} \begin{array}{l} \\ \\ \text{Trick} \\ \text{Slalom} \end{array}$$

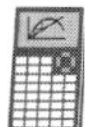

Figure 5 shows a solution to Example 8 on a spreadsheet.

	A	B	C	D	E	F
1		Labor-hours per ski			Hourly wages	
2		Assembly	Finishing		California	Wisconsin
3	Trick ski	5	1.5	Assembly	\$12	\$9
4	Slalom ski	3	1	Finishing	\$8	\$6
5		Labor costs per ski				
6		California	Wisconsin			
7	Trick ski	\$72	\$54			
8	Slalom ski	\$44	\$33			

FIGURE 5

Matrix multiplication in a spreadsheet:
The command MMULT(B3: C4, E3: F4) produces the matrix in B7:C8

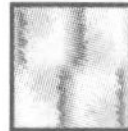

Matched Problem 8 Refer to Example 8. The company wants to know how many hours to schedule in each department in order to produce 2,000 trick skis and 1,000 slalom skis. These production requirements can be represented by either of the following matrices:

$$P = \overset{\begin{matrix}\text{Trick} & \text{Slalom}\\ \text{skis} & \text{skis}\end{matrix}}{\begin{bmatrix} 2{,}000 & 1{,}000 \end{bmatrix}} \qquad Q = \begin{bmatrix} 2{,}000 \\ 1{,}000 \end{bmatrix} \begin{matrix}\text{Trick skis}\\ \text{Slalom skis}\end{matrix}$$

Using the labor-hour matrix L from Example 8, find PL or LQ, whichever has a meaningful interpretation for this problem, and label the rows and columns accordingly.

CAUTION

Example 8 and Matched Problem 8 illustrate an important point about matrix multiplication. Even if you are using a graphing utility to perform the calculations in a matrix product, it is still necessary for you to know the definition of matrix multiplication so that you can interpret the results correctly.

Answers to Matched Problems

1. $\begin{bmatrix} 1 & 5 \\ 0 & -2 \\ 2 & 1 \end{bmatrix}$ **2.** $[-1 \quad -1 \quad 4]$ **3.** $\begin{bmatrix} 13 \\ 2 \\ 35 \end{bmatrix}$

4. (A) $\begin{bmatrix} \$235{,}000 & \$422{,}000 \\ \$372{,}000 & \$298{,}000 \end{bmatrix}$ (B) $\begin{bmatrix} \$145{,}000 & \$268{,}000 \\ \$160{,}000 & \$254{,}000 \end{bmatrix}$

(C) $\begin{bmatrix} \$9{,}500 & \$17{,}250 \\ \$13{,}300 & \$13{,}800 \end{bmatrix}$

5. $[8]$ **6.** $[5 \quad 1.5]\begin{bmatrix} 9 \\ 6 \end{bmatrix} = [54]$, or \$54

7. (A) Not defined (B) $\begin{bmatrix} 2 & 2 & -1 & 2 \\ 1 & 6 & 12 & -4 \\ -1 & 0 & 3 & -2 \end{bmatrix}$ (C) $\begin{bmatrix} 0 & 0 \\ 0 & 0 \end{bmatrix}$

(D) $\begin{bmatrix} -6 & -12 \\ 3 & 6 \end{bmatrix}$ (E) $[11]$ (F) $\begin{bmatrix} 12 & -8 & 4 \\ 6 & -4 & 2 \\ 9 & -6 & 3 \end{bmatrix}$

8. $PL = \overset{\begin{matrix}\text{Assembly} & \text{Finishing}\end{matrix}}{[13{,}000 \quad 4{,}000]}$ Labor-hours

EXERCISE 4-4

A *Perform the indicated operations in Problems 1–14, if possible.*

1. $\begin{bmatrix} 2 & -1 \\ 3 & 0 \end{bmatrix} + \begin{bmatrix} -3 & 1 \\ 2 & -3 \end{bmatrix}$

2. $\begin{bmatrix} -3 & 5 \\ 2 & 0 \\ 1 & 4 \end{bmatrix} + \begin{bmatrix} 2 & 1 \\ -6 & 3 \\ 0 & -5 \end{bmatrix}$

3. $\begin{bmatrix} 4 & -1 & 0 \\ 2 & 1 & 3 \end{bmatrix} + \begin{bmatrix} 2 & 1 \\ -6 & 3 \\ 0 & -5 \end{bmatrix}$

4. $\begin{bmatrix} -3 & 5 \\ 2 & 0 \\ 1 & 4 \end{bmatrix} + \begin{bmatrix} -2 & 1 & 3 \\ 5 & 6 & -8 \end{bmatrix}$

5. $\begin{bmatrix} 4 & -5 \\ 1 & 0 \\ 1 & -3 \end{bmatrix} - \begin{bmatrix} -1 & 2 \\ 6 & -2 \\ 1 & -7 \end{bmatrix}$

6. $\begin{bmatrix} 6 & 2 & -3 \\ 0 & -4 & 5 \end{bmatrix} - \begin{bmatrix} 4 & -1 & 2 \\ -5 & 1 & -2 \end{bmatrix}$

7. $5\begin{bmatrix} 1 & -2 & 0 & 4 \\ -3 & 2 & -1 & 6 \end{bmatrix}$

8. $10\begin{bmatrix} 2 & -1 & 3 \\ 0 & -4 & 5 \end{bmatrix}$

9. $\begin{bmatrix} 3 & 4 \\ -1 & -2 \end{bmatrix}\begin{bmatrix} -1 \\ 2 \end{bmatrix}$

10. $\begin{bmatrix} -1 & 1 \\ 2 & -3 \end{bmatrix}\begin{bmatrix} 4 \\ -2 \end{bmatrix}$

11. $\begin{bmatrix} 2 & -3 \\ 1 & 2 \end{bmatrix}\begin{bmatrix} 1 & -1 \\ 0 & -2 \end{bmatrix}$

12. $\begin{bmatrix} -3 & 2 \\ 4 & -1 \end{bmatrix}\begin{bmatrix} -2 & 5 \\ -1 & 3 \end{bmatrix}$

13. $\begin{bmatrix} 1 & -1 \\ 0 & -2 \end{bmatrix}\begin{bmatrix} 2 & -3 \\ 1 & 2 \end{bmatrix}$

14. $\begin{bmatrix} -2 & 5 \\ -1 & 3 \end{bmatrix}\begin{bmatrix} -3 & 2 \\ 4 & -1 \end{bmatrix}$

B *Find the products in Problems 15–22.*

15. $\begin{bmatrix} 5 & -2 \end{bmatrix}\begin{bmatrix} -3 \\ -4 \end{bmatrix}$

16. $\begin{bmatrix} -4 & 3 \end{bmatrix}\begin{bmatrix} -2 \\ 1 \end{bmatrix}$

17. $\begin{bmatrix} -3 \\ -4 \end{bmatrix}\begin{bmatrix} 5 & -2 \end{bmatrix}$

18. $\begin{bmatrix} -2 \\ 1 \end{bmatrix}\begin{bmatrix} -4 & 3 \end{bmatrix}$

19. $\begin{bmatrix} 3 & -2 & -4 \end{bmatrix}\begin{bmatrix} 1 \\ 2 \\ -3 \end{bmatrix}$

20. $\begin{bmatrix} 1 & -2 & 2 \end{bmatrix}\begin{bmatrix} 2 \\ -1 \\ 1 \end{bmatrix}$

21. $\begin{bmatrix} 1 \\ 2 \\ -3 \end{bmatrix}\begin{bmatrix} 3 & -2 & -4 \end{bmatrix}$

22. $\begin{bmatrix} 2 \\ -1 \\ 1 \end{bmatrix}\begin{bmatrix} 1 & -2 & 2 \end{bmatrix}$

Problems 23–40 refer to the following matrices:

$$A = \begin{bmatrix} 2 & -1 & 3 \\ 0 & 4 & -2 \end{bmatrix} \qquad B = \begin{bmatrix} -3 & 1 \\ 2 & 5 \end{bmatrix}$$

$$C = \begin{bmatrix} -1 & 0 & 2 \\ 4 & -3 & 1 \\ -2 & 3 & 5 \end{bmatrix} \qquad D = \begin{bmatrix} 3 & -2 \\ 0 & -1 \\ 1 & 2 \end{bmatrix}$$

Perform the indicated operations, if possible.

23. AC **24.** CA **25.** AB

26. BA **27.** B^2 **28.** C^2

29. $B + AD$ **30.** $C + DA$ **31.** $(0.1)DB$

32. $(0.2)CD$ **33.** $(3)BA + (4)AC$

34. $(2)DB + (5)CD$ **35.** $(-2)BA + (6)CD$

36. $(-1)AC + (3)DB$ **37.** ACD **38.** CDA

39. DBA **40.** BAD

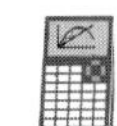

In Problems 41 and 42, use a graphing utility to calculate $B, B^2, B^3, \ldots$ and $AB, AB^2, AB^3, \ldots$. Describe any patterns you observe in each sequence of matrices.

41. $A = \begin{bmatrix} 0.3 & 0.7 \end{bmatrix}$ and $B = \begin{bmatrix} 0.4 & 0.6 \\ 0.2 & 0.8 \end{bmatrix}$

42. $A = \begin{bmatrix} 0.4 & 0.6 \end{bmatrix}$ and $B = \begin{bmatrix} 0.9 & 0.1 \\ 0.3 & 0.7 \end{bmatrix}$

43. Find a, b, c, and d so that

$$\begin{bmatrix} a & b \\ c & d \end{bmatrix} + \begin{bmatrix} 2 & -3 \\ 0 & 1 \end{bmatrix} = \begin{bmatrix} 1 & -2 \\ 3 & -4 \end{bmatrix}$$

44. Find w, x, y, and z so that

$$\begin{bmatrix} 4 & -2 \\ -3 & 0 \end{bmatrix} + \begin{bmatrix} w & x \\ y & z \end{bmatrix} = \begin{bmatrix} 2 & -3 \\ 0 & 5 \end{bmatrix}$$

45. Find x and y so that

$$\begin{bmatrix} 2x & 4 \\ -3 & 5x \end{bmatrix} + \begin{bmatrix} 3y & -2 \\ -2 & -y \end{bmatrix} = \begin{bmatrix} -5 & 2 \\ -5 & 13 \end{bmatrix}$$

46. Find x and y so that

$$\begin{bmatrix} 5 & 3x \\ 2x & -4 \end{bmatrix} + \begin{bmatrix} 1 & -4y \\ 7y & 4 \end{bmatrix} = \begin{bmatrix} 6 & -7 \\ 5 & 0 \end{bmatrix}$$

C

47. Find x and y so that

$$\begin{bmatrix} x & -1 \\ 1 & 0 \end{bmatrix}\begin{bmatrix} 2 & 1 \\ 4 & 1 \end{bmatrix} = \begin{bmatrix} y & y \\ 2 & 1 \end{bmatrix}$$

48. Find x and y so that

$$\begin{bmatrix} 1 & 3 \\ -2 & -2 \end{bmatrix}\begin{bmatrix} x & 1 \\ 3 & 2 \end{bmatrix} = \begin{bmatrix} y & 7 \\ y & -6 \end{bmatrix}$$

49. Find a, b, c, and d so that

$$\begin{bmatrix} 1 & -2 \\ 2 & -3 \end{bmatrix}\begin{bmatrix} a & b \\ c & d \end{bmatrix} = \begin{bmatrix} 1 & 0 \\ 3 & 2 \end{bmatrix}$$

50. Find a, b, c, and d so that

$$\begin{bmatrix} 1 & 3 \\ 1 & 4 \end{bmatrix}\begin{bmatrix} a & b \\ c & d \end{bmatrix} = \begin{bmatrix} 6 & -5 \\ 7 & -7 \end{bmatrix}$$

51. A square matrix is a **diagonal matrix** if all elements not on the principal diagonal are zero. Thus, a 2×2 diagonal matrix has the form

$$A = \begin{bmatrix} a & 0 \\ 0 & d \end{bmatrix}$$

where a and d are any real numbers. Discuss the validity of each of the following statements. If the statement is always true, explain why. If not, give examples.

(A) If A and B are 2×2 diagonal matrices, then $A + B$ is a 2×2 diagonal matrix.

(B) If A and B are 2×2 diagonal matrices, then $A + B = B + A$.

(C) If A and B are 2×2 diagonal matrices, then AB is a 2×2 diagonal matrix.

(D) If A and B are 2×2 diagonal matrices, then $AB = BA$.

52. A square matrix is an **upper triangular matrix** if all elements below the principal diagonal are zero. Thus, a 2×2 upper triangular matrix has the form

$$A = \begin{bmatrix} a & b \\ 0 & d \end{bmatrix}$$

where a, b, and d are any real numbers. Discuss the validity of each of the following statements. If the statement is always true, explain why. If not, give examples.

(A) If A and B are 2×2 upper triangular matrices, then $A + B$ is a 2×2 upper triangular matrix.

(B) If A and B are 2×2 upper triangular matrices, then $A + B = B + A$.

(C) If A and B are 2×2 upper triangular matrices, then AB is a 2×2 upper triangular matrix.

(D) If A and B are 2×2 upper triangular matrices, then $AB = BA$.

Applications

Business & Economics

53. *Cost analysis.* A company with two different plants manufactures guitars and banjos. Its production costs for each instrument are given in the following matrices:

$$\begin{array}{r} \\ \text{Materials} \\ \text{Labor} \end{array}\begin{array}{c} \text{Plant } X \\ \text{Guitar Banjo} \\ \begin{bmatrix} \$30 & \$25 \\ \$60 & \$80 \end{bmatrix} = A \end{array} \qquad \begin{array}{c} \text{Plant } Y \\ \text{Guitar Banjo} \\ \begin{bmatrix} \$36 & \$27 \\ \$54 & \$74 \end{bmatrix} = B \end{array}$$

Find $\frac{1}{2}(A + B)$, the average cost of production for the two plants.

54. *Cost analysis.* If both labor and materials at plant X in Problem 53 are increased by 20%, find $\frac{1}{2}(1.2A + B)$, the new average cost of production for the two plants.

55. *Markup.* An import car dealer sells three models of a car. The retail prices and the current dealer invoice prices (costs) for the basic models and indicated options are given in the following two matrices (where "Air" means air-conditioning):

Retail price

$$\begin{array}{r} \\ \text{Model A} \\ \text{Model B} \\ \text{Model C} \end{array}\begin{array}{c} \begin{array}{cccc} \text{Basic car} & \text{Air} & \text{AM/FM radio} & \text{Cruise control} \end{array} \\ \begin{bmatrix} \$10{,}900 & \$683 & \$253 & \$195 \\ \$13{,}000 & \$738 & \$382 & \$206 \\ \$16{,}300 & \$867 & \$537 & \$225 \end{bmatrix} = M \end{array}$$

Dealer invoice price

$$\begin{array}{r} \\ \text{Model A} \\ \text{Model B} \\ \text{Model C} \end{array}\begin{array}{c} \begin{array}{cccc} \text{Basic car} & \text{Air} & \text{AM/FM radio} & \text{Cruise control} \end{array} \\ \begin{bmatrix} \$9{,}400 & \$582 & \$195 & \$160 \\ \$11{,}500 & \$621 & \$295 & \$171 \\ \$14{,}100 & \$737 & \$420 & \$184 \end{bmatrix} = N \end{array}$$

We define the markup matrix to be $M - N$ (**markup** is the difference between the retail price and the dealer invoice price). Suppose the value of the dollar has had a sharp decline and the dealer invoice price is to have an across-the-board 15% increase next year. In order to stay competitive with domestic cars, the dealer increases the retail prices only 10%. Calculate a markup matrix for next year's models and the indicated options. (Compute results to the nearest dollar.)

56. *Markup.* Referring to Problem 55, what is the markup matrix resulting from a 20% increase in dealer invoice prices and an increase in retail prices of 15%? (Compute results to the nearest dollar.)

57. *Labor costs.* A company with manufacturing plants located in different parts of the country has labor-hour and wage requirements for the manufacture of three types of inflatable boats as given in the following two matrices:

Labor-hours per boat

M	Cutting department	Assembly department	Packaging department	
	0.6 hr	0.6 hr	0.2 hr	One-person boat
	1.0 hr	0.9 hr	0.3 hr	Two-person boat
	1.5 hr	1.2 hr	0.4 hr	Four-person boat

Hourly wages

N	Plant I	Plant II	
	$8	$9	Cutting department
	$10	$12	Assembly department
	$5	$6	Packaging department

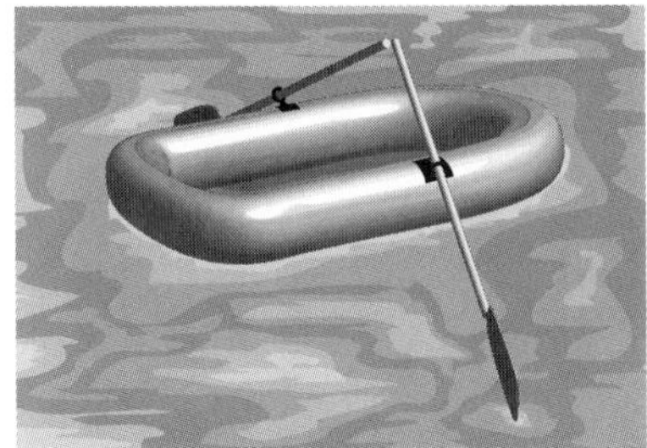

(A) Find the labor costs for a one-person boat manufactured at plant I.

(B) Find the labor costs for a four-person boat manufactured at plant II.

(C) Discuss possible interpretations of the elements in the matrix products MN and NM.

(D) If either of the products MN or NM has a meaningful interpretation, find the product and label its rows and columns.

58. *Inventory value.* A personal computer retail company sells five different computer models through three stores located in a large metropolitan area. The inventory of each model on hand in each store is summarized in matrix M. Wholesale (W) and retail (R) values of each model computer are summarized in matrix N.

Model

M	A	B	C	D	E	
	4	2	3	7	1	Store 1
	2	3	5	0	6	Store 2
	10	4	3	4	3	Store 3

N	W	R	Model
	$700	$840	A
	$1,400	$1,800	B
	$1,800	$2,400	C
	$2,700	$3,300	D
	$3,500	$4,900	E

(A) What is the retail value of the inventory at store 2?

(B) What is the wholesale value of the inventory at store 3?

(C) Discuss possible interpretations of the elements in the matrix products MN and NM.

(D) If either of the products MN or NM has a meaningful interpretation, find the product and label its rows and columns.

(E) Discuss methods of matrix multiplication that can be used to find the total inventory of each model on hand at all three stores. State the matrices that can be used and perform the necessary operations.

(F) Discuss methods of matrix multiplication that can be used to find the total inventory of all five models at each store. State the matrices that can be used and perform the necessary operations.

59. *Air freight.* A nationwide air freight service has connecting flights between five cities as illustrated in the diagram. To represent this schedule in matrix form, we construct a 5 × 5 **incidence matrix** A, where the rows represent the origin of each flight and the columns represent the destinations. We place a 1 in the ith row and jth column of this matrix if there is a connecting flight from the ith city to the jth city; otherwise, insert a 0. We also place 0's on the principal diagonal, because a connecting flight with the same origin and destination makes no sense. With the schedule represented in the mathematical form of a matrix, we can perform operations on this matrix to obtain information about the schedule.

$$\begin{array}{c|ccccc} & 1 & 2 & 3 & 4 & 5 \\ \hline 1 & 0 & 1 & 0 & 1 & 0 \\ 2 & 0 & 0 & 1 & 0 & 0 \\ 3 & 1 & 0 & 0 & 0 & 1 \\ 4 & 0 & 0 & 1 & 0 & 0 \\ 5 & 0 & 0 & 0 & 1 & 0 \end{array} = A$$

(A) Find A^2. What does the 1 in Row 2 and Column 1 of A^2 indicate about the schedule? What does the 2 in Row 1 and Column 3 indicate about the schedule? In general, how would you interpret each element not on the principal diagonal of A^2? [*Hint*: Examine the diagram for possible connections between the ith city and the jth city.]

(B) Find A^3. What does the 1 in Row 4 and Column 2 of A^3 indicate about the schedule? What does the

2 in Row 1 and Column 5 indicate about the schedule? In general, how would you interpret each element not on the principal diagonal of A^3?

(C) Compute $A, A + A^2, A + A^2 + A^3, \ldots,$ until you obtain a matrix with no 0 elements (except possibly on the principal diagonal), and interpret.

60. *Air freight.* Find the incidence matrix A for the flight schedule illustrated in the diagram. Compute $A, A + A^2, A + A^2 + A^3, \ldots,$ until you obtain a matrix with no 0 elements (except possibly on the principal diagonal), and interpret.

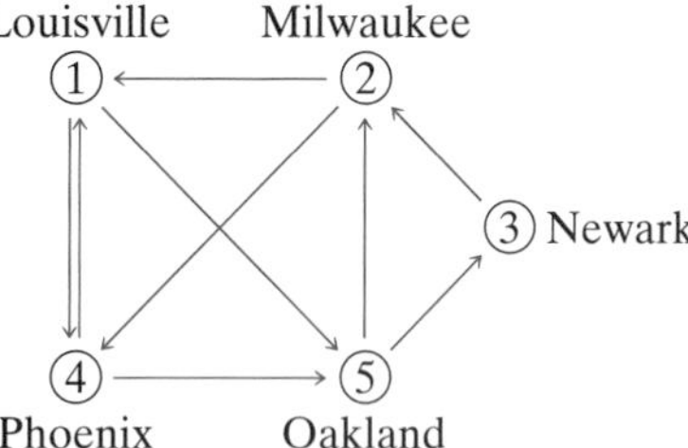

Life Sciences

61. *Nutrition.* A nutritionist for a cereal company blends two cereals in three different mixes. The amounts of protein, carbohydrate, and fat (in grams per ounce) in each cereal are given by matrix M. The amounts of each cereal used in the three mixes are given by matrix N.

$$M = \begin{bmatrix} 4 \text{ g/oz} & 2 \text{ g/oz} \\ 20 \text{ g/oz} & 16 \text{ g/oz} \\ 3 \text{ g/oz} & 1 \text{ g/oz} \end{bmatrix} \begin{matrix} \text{Protein} \\ \text{Carbohydrate} \\ \text{Fat} \end{matrix}$$

(columns: Cereal A, Cereal B)

$$N = \begin{bmatrix} 15 \text{ oz} & 10 \text{ oz} & 5 \text{ oz} \\ 5 \text{ oz} & 10 \text{ oz} & 15 \text{ oz} \end{bmatrix} \begin{matrix} \text{Cereal A} \\ \text{Cereal B} \end{matrix}$$

(columns: Mix X, Mix Y, Mix Z)

(A) Find the amount of protein in mix X.

(B) Find the amount of fat in mix Z.

(C) Discuss possible interpretations of the elements in the matrix products MN and NM.

(D) If either of the products MN or NM has a meaningful interpretation, find the product and label its rows and columns.

62. *Heredity.* Gregor Mendel (1822–1884), an Austrian monk and botanist, made discoveries that revolutionized the science of genetics. In one experiment, he crossed dihybrid yellow round peas (yellow and round are dominant characteristics; the peas also contained genes for the recessive characteristics green and wrinkled) and obtained 560 peas of the types indicated in the matrix:

$$\begin{matrix} \text{Yellow} \\ \text{Green} \end{matrix} \begin{bmatrix} 319 & 101 \\ 108 & 32 \end{bmatrix} = M$$

(columns: Round, Wrinkled)

Suppose he carried out a second experiment of the same type and obtained 640 peas of the types indicated in this matrix:

$$\begin{matrix} \text{Yellow} \\ \text{Green} \end{matrix} \begin{bmatrix} 370 & 124 \\ 110 & 36 \end{bmatrix} = N$$

(columns: Round, Wrinkled)

If the results of the two experiments are combined, write the resulting matrix $M + N$. Compute the percentage of the total number of peas (1,200) in each category of the combined results.

Social Sciences

63. *Politics.* In a local California election, a group hired a public relations firm to promote its candidate in three ways: telephone calls, house calls, and letters. The cost per contact is given in matrix M, and the number of contacts of each type made in two adjacent cities is given in matrix N.

$$M = \begin{bmatrix} \$0.40 \\ \$1.00 \\ \$0.35 \end{bmatrix} \begin{matrix} \text{Telephone call} \\ \text{House call} \\ \text{Letter} \end{matrix}$$

(column: Cost per contact)

$$N = \begin{bmatrix} 1{,}000 & 500 & 5{,}000 \\ 2{,}000 & 800 & 8{,}000 \end{bmatrix} \begin{matrix} \text{Berkeley} \\ \text{Oakland} \end{matrix}$$

(columns: Telephone call, House call, Letter)

(A) Find the total amount spent in Berkeley.

(B) Find the total amount spent in Oakland.

(C) Discuss possible interpretations of the elements in the matrix products MN and NM.

(D) If either of the products MN or NM has a meaningful interpretation, find the product and label its rows and columns.

(E) Discuss methods of matrix multiplication that can be used to find the total number of telephone calls, house calls, and letters. State the matrices that can be used and perform the necessary operations.

(F) Discuss methods of matrix multiplication that can be used to find the total number of contacts in Berkeley and in Oakland. State the matrices that can be used and perform the necessary operations.

64. *Averaging tests.* A teacher has given four tests to a class of five students and stored the results in the following matrix:

$$\begin{array}{l} \\ \\ \text{Ann} \\ \text{Bob} \\ \text{Carol} \\ \text{Dan} \\ \text{Eric} \end{array} \begin{array}{c} \text{Tests} \\ \begin{array}{cccc} 1 & 2 & 3 & 4 \end{array} \\ \begin{bmatrix} 78 & 84 & 81 & 86 \\ 91 & 65 & 84 & 92 \\ 95 & 90 & 92 & 91 \\ 75 & 82 & 87 & 91 \\ 83 & 88 & 81 & 76 \end{bmatrix} \end{array} = M$$

Discuss methods of matrix multiplication that the teacher can use to obtain the information indicated below. In each case, state the matrices to be used and then perform the necessary operations.

(A) The average on all four tests for each student, assuming that all four tests are given equal weight

(B) The average on all four tests for each student, assuming that the first three tests are given equal weight and the fourth is given twice this weight

(C) The class average on each of the four tests

65. *Dominance relation.* In order to rank players for an upcoming tennis tournament, a club decides to have each player play one set with every other player. The results are given in the table.

PLAYER	DEFEATED
1. Aaron	Charles, Dan, Elvis
2. Bart	Aaron, Dan, Elvis
3. Charles	Bart, Dan
4. Dan	Frank
5. Elvis	Charles, Dan, Frank
6. Frank	Aaron, Bart, Charles

(A) Express the outcomes as an incidence matrix A by placing a 1 in the ith row and jth column of A if player i defeated player j and a 0 otherwise (see Problem 59).

(B) Compute the matrix $B = A + A^2$.

(C) Discuss matrix multiplication methods that can be used to find the sum of the rows in B. State the matrices that can be used and perform the necessary operations.

(D) Rank the players from strongest to weakest. Explain the reasoning behind your ranking.

66. *Dominance relation.* Each member of a chess team plays one match with every other player. The results are given in the table.

PLAYER	DEFEATED
1. Anne	Diane
2. Bridget	Anne, Carol, Diane
3. Carol	Anne
4. Diane	Carol, Erlene
5. Erlene	Anne, Bridget, Carol

(A) Express the outcomes as an incidence matrix A by placing a 1 in the ith row and jth column of A if player i defeated player j and a 0 otherwise (see Problem 59).

(B) Compute the matrix $B = A + A^2$.

(C) Discuss matrix multiplication methods that can be used to find the sum of the rows in B. State the matrices that can be used and perform the necessary operations.

(D) Rank the players from strongest to weakest. Explain the reasoning behind your ranking.

SECTION 4-5 Inverse of a Square Matrix

- IDENTITY MATRIX FOR MULTIPLICATION
- INVERSE OF A SQUARE MATRIX
- APPLICATION: CRYPTOGRAPHY

IDENTITY MATRIX FOR MULTIPLICATION

We know that

$$1a = a1 = a \qquad \text{for all real numbers } a$$

The number 1 is called the **identity** for real number multiplication. Does the set of all matrices of a given size have an identity element for multiplication? That is, if M is an arbitrary $m \times n$ matrix, does M have an identity element

I such that $IM = MI = M$? The answer, in general, is no. However, the set of all **square matrices of order n** (matrices with n rows and n columns) does have an identity, and it is given as follows:

> The **identity element for multiplication** for the set of all square matrices of order n is the square matrix of order n, denoted by I, with 1's along the principal diagonal (from the upper left corner to the lower right) and 0's elsewhere.

```
identity 2
        [[1 0]
         [0 1]]
identity 3
      [[1 0 0]
       [0 1 0]
       [0 0 1]]
```

FIGURE 1
Identity matrices

For example,

$$\begin{bmatrix} 1 & 0 \\ 0 & 1 \end{bmatrix} \quad \text{and} \quad \begin{bmatrix} 1 & 0 & 0 \\ 0 & 1 & 0 \\ 0 & 0 & 1 \end{bmatrix}$$

are the identity matrices for all square matrices of order 2 and 3, respectively.

Most graphing utilities have a built-in command for generating the identity matrix of a given order (see Fig. 1)

Example 1 ➡ **Identity Matrix Multiplication**

(A) $\begin{bmatrix} 1 & 0 & 0 \\ 0 & 1 & 0 \\ 0 & 0 & 1 \end{bmatrix}\begin{bmatrix} a & b & c \\ d & e & f \\ g & h & i \end{bmatrix} = \begin{bmatrix} a & b & c \\ d & e & f \\ g & h & i \end{bmatrix}$

(B) $\begin{bmatrix} a & b & c \\ d & e & f \\ g & h & i \end{bmatrix}\begin{bmatrix} 1 & 0 & 0 \\ 0 & 1 & 0 \\ 0 & 0 & 1 \end{bmatrix} = \begin{bmatrix} a & b & c \\ d & e & f \\ g & h & i \end{bmatrix}$

(C) $\begin{bmatrix} 1 & 0 \\ 0 & 1 \end{bmatrix}\begin{bmatrix} a & b & c \\ d & e & f \end{bmatrix} = \begin{bmatrix} a & b & c \\ d & e & f \end{bmatrix}$

(D) $\begin{bmatrix} a & b & c \\ d & e & \mathrm{f} \end{bmatrix}\begin{bmatrix} 1 & 0 & 0 \\ 0 & 1 & 0 \\ 0 & 0 & 1 \end{bmatrix} = \begin{bmatrix} a & b & c \\ d & e & f \end{bmatrix}$

Matched Problem 1 ➡ Multiply:

(A) $\begin{bmatrix} 1 & 0 \\ 0 & 1 \end{bmatrix}\begin{bmatrix} 2 & -3 \\ 5 & 7 \end{bmatrix}$ and $\begin{bmatrix} 2 & -3 \\ 5 & 7 \end{bmatrix}\begin{bmatrix} 1 & 0 \\ 0 & 1 \end{bmatrix}$

(B) $\begin{bmatrix} 1 & 0 & 0 \\ 0 & 1 & 0 \\ 0 & 0 & 1 \end{bmatrix}\begin{bmatrix} 4 & 2 \\ 3 & -5 \\ 6 & 8 \end{bmatrix}$ and $\begin{bmatrix} 4 & 2 \\ 3 & -5 \\ 6 & 8 \end{bmatrix}\begin{bmatrix} 1 & 0 \\ 0 & 1 \end{bmatrix}$

In general, we can show that if M is a square matrix of order n and I is the identity matrix of order n, then

$$IM = MI = M$$

If M is an $m \times n$ matrix that is not square ($m \neq n$), then it is still possible to multiply M on the left and on the right by an identity matrix, but not with the same size identity matrix (see Examples 1C and 1D). In order to avoid the complications involved with associating two different identity matrices with each nonsquare matrix, we restrict our attention in this section to square matrices.

Explore–Discuss 1

The only real number solutions to the equation $x^2 = 1$ are $x = 1$ and $x = -1$.

(A) Show that $A = \begin{bmatrix} 0 & 1 \\ 1 & 0 \end{bmatrix}$ satisfies $A^2 = I$, where I is the 2×2 identity.

(B) Show that $B = \begin{bmatrix} 0 & -1 \\ -1 & 0 \end{bmatrix}$ satisfies $B^2 = I$.

(C) Find a 2×2 matrix with all elements nonzero whose square is the 2×2 identity matrix.

Inverse of a Square Matrix

In the set of real numbers, we know that for each real number a (except 0) there exists a real number a^{-1} such that

$$a^{-1}a = 1$$

The number a^{-1} is called the **inverse** of the number a relative to multiplication, or the **multiplicative inverse** of a. For example, 2^{-1} is the multiplicative inverse of 2, since $2^{-1} \cdot 2 = 1$. We use this idea to define the inverse of a square matrix.

Inverse of a Square Matrix

Let M be a square matrix of order n and I be the identity matrix of order n. If there exists a matrix M^{-1} (read "M inverse") such that

$$M^{-1}M = MM^{-1} = I$$

then M^{-1} is called the **multiplicative inverse of M** or, more simply, the **inverse of M.**

The multiplicative inverse of a nonzero real number a also can be written as $1/a$. This notation is not used for matrix inverses.

Let us use the above definition to find M^{-1} for

$$M = \begin{bmatrix} 2 & 3 \\ 1 & 2 \end{bmatrix}$$

We are looking for

$$M^{-1} = \begin{bmatrix} a & c \\ b & d \end{bmatrix}$$

such that

$$MM^{-1} = M^{-1}M = I$$

Thus, we write

$$\overset{M}{\begin{bmatrix} 2 & 3 \\ 1 & 2 \end{bmatrix}} \overset{M^{-1}}{\begin{bmatrix} a & c \\ b & d \end{bmatrix}} = \overset{I}{\begin{bmatrix} 1 & 0 \\ 0 & 1 \end{bmatrix}}$$

and try to find a, b, c, and d so that the product of M and M^{-1} is the identity matrix I. Multiplying M and M^{-1} on the left side, we obtain

$$\begin{bmatrix} (2a + 3b) & (2c + 3d) \\ (a + 2b) & (c + 2d) \end{bmatrix} = \begin{bmatrix} 1 & 0 \\ 0 & 1 \end{bmatrix}$$

which is true only if

$$\begin{aligned} 2a + 3b &= 1 \\ a + 2b &= 0 \end{aligned} \qquad \begin{aligned} 2c + 3d &= 0 \\ c + 2d &= 1 \end{aligned}$$

Solving these two systems, we find that $a = 2$, $b = -1$, $c = -3$, and $d = 2$. Thus,

$$M^{-1} = \begin{bmatrix} 2 & -3 \\ -1 & 2 \end{bmatrix}$$

as is easily checked:

$$\overset{M}{\begin{bmatrix} 2 & 3 \\ 1 & 2 \end{bmatrix}} \overset{M^{-1}}{\begin{bmatrix} 2 & -3 \\ -1 & 2 \end{bmatrix}} = \overset{I}{\begin{bmatrix} 1 & 0 \\ 0 & 1 \end{bmatrix}} = \overset{M^{-1}}{\begin{bmatrix} 2 & -3 \\ -1 & 2 \end{bmatrix}} \overset{M}{\begin{bmatrix} 2 & 3 \\ 1 & 2 \end{bmatrix}}$$

Unlike nonzero real numbers, inverses do not always exist for square matrices. For example, if

$$N = \begin{bmatrix} 2 & 1 \\ 4 & 2 \end{bmatrix}$$

then, proceeding as above, we are led to the systems

$$\begin{aligned} 2a + b &= 1 \\ 4a + 2b &= 0 \end{aligned} \qquad \begin{aligned} 2c + d &= 0 \\ 4c + 2d &= 1 \end{aligned}$$

These are both inconsistent and have no solution. Hence, N^{-1} does not exist.

Being able to find inverses, when they exist, leads to direct and simple solutions to many practical problems. In the next section, we shall show how inverses can be used to solve systems of linear equations.

The method outlined above for finding M^{-1}, if it exists, gets very involved for matrices of order larger than 2. Now that we know what we are looking for, we can introduce the idea of the augmented matrix (considered in Sections 4-2 and 4-3) to make the process more efficient.

Example 2 ➡ **Finding the Inverse of a Matrix** Find the inverse, if it exists, of the matrix

$$M = \begin{bmatrix} 1 & -1 & 1 \\ 0 & 2 & -1 \\ 2 & 3 & 0 \end{bmatrix}$$

SOLUTION We start as before and write

$$\overset{M}{\begin{bmatrix} 1 & -1 & 1 \\ 0 & 2 & -1 \\ 2 & 3 & 0 \end{bmatrix}} \overset{M^{-1}}{\begin{bmatrix} a & d & g \\ b & e & h \\ c & f & i \end{bmatrix}} = \overset{I}{\begin{bmatrix} 1 & 0 & 0 \\ 0 & 1 & 0 \\ 0 & 0 & 1 \end{bmatrix}}$$

which is true only if

$$\begin{aligned} a - b + c &= 1 \\ 2b - c &= 0 \\ 2a + 3b \quad &= 0 \end{aligned} \qquad \begin{aligned} d - e + f &= 0 \\ 2e - f &= 1 \\ 2d + 3e \quad &= 0 \end{aligned} \qquad \begin{aligned} g - h + i &= 0 \\ 2h - i &= 0 \\ 2g + 3h \quad &= 1 \end{aligned}$$

Now we write augmented matrices for each of the three systems:

$$\overset{\text{First}}{\left[\begin{array}{rrr|r} 1 & -1 & 1 & 1 \\ 0 & 2 & -1 & 0 \\ 2 & 3 & 0 & 0 \end{array}\right]} \qquad \overset{\text{Second}}{\left[\begin{array}{rrr|r} 1 & -1 & 1 & 0 \\ 0 & 2 & -1 & 1 \\ 2 & 3 & 0 & 0 \end{array}\right]} \qquad \overset{\text{Third}}{\left[\begin{array}{rrr|r} 1 & -1 & 1 & 0 \\ 0 & 2 & -1 & 0 \\ 2 & 3 & 0 & 1 \end{array}\right]}$$

Since each matrix to the left of the vertical bar is the same, exactly the same row operations can be used on each augmented matrix to transform it into a reduced form. We can speed up the process substantially by combining all three augmented matrices into the single augmented matrix form below:

$$\left[\begin{array}{rrr|rrr} 1 & -1 & 1 & 1 & 0 & 0 \\ 0 & 2 & -1 & 0 & 1 & 0 \\ 2 & 3 & 0 & 0 & 0 & 1 \end{array}\right] = [M|I] \tag{1}$$

We now try to perform row operations on matrix (1) until we obtain a row-equivalent matrix that looks like matrix (2):

$$\left[\begin{array}{ccc|ccc} \multicolumn{3}{c|}{I} & \multicolumn{3}{c}{B} \\ 1 & 0 & 0 & a & d & g \\ 0 & 1 & 0 & b & e & h \\ 0 & 0 & 1 & c & f & i \end{array}\right] = [I|B] \tag{2}$$

If this can be done, then the new matrix B to the right of the vertical bar will be M^{-1}! Now let us try to transform (1) into a form like (2). We will follow the same sequence of steps as we did in the solution of linear systems by Gauss–Jordan elimination (see Section 4-3).

$$\begin{array}{c} \\ \\ \\ \\ \sim \\ \\ \\ \sim \\ \\ \\ \sim \\ \\ \\ \sim \\ \\ \\ \sim \end{array}
\overset{\qquad M \qquad\qquad\qquad I}{\left[\begin{array}{rrr|rrr} 1 & -1 & 1 & 1 & 0 & 0 \\ 0 & 2 & -1 & 0 & 1 & 0 \\ 2 & 3 & 0 & 0 & 0 & 1 \end{array}\right]} \quad (-2)R_1 + R_3 \to R_3$$

$$\sim \left[\begin{array}{rrr|rrr} 1 & -1 & 1 & 1 & 0 & 0 \\ 0 & 2 & -1 & 0 & 1 & 0 \\ 0 & 5 & -2 & -2 & 0 & 1 \end{array}\right] \quad \tfrac{1}{2}R_2 \to R_2$$

$$\sim \left[\begin{array}{rrr|rrr} 1 & -1 & 1 & 1 & 0 & 0 \\ 0 & 1 & -\frac{1}{2} & 0 & \frac{1}{2} & 0 \\ 0 & 5 & -2 & -2 & 0 & 1 \end{array}\right] \quad \begin{array}{l} R_2 + R_1 \to R_1 \\ \\ (-5)R_2 + R_3 \to R_3 \end{array}$$

$$\sim \left[\begin{array}{rrr|rrr} 1 & 0 & \frac{1}{2} & 1 & \frac{1}{2} & 0 \\ 0 & 1 & -\frac{1}{2} & 0 & \frac{1}{2} & 0 \\ 0 & 0 & \frac{1}{2} & -2 & -\frac{5}{2} & 1 \end{array}\right] \quad 2R_3 \to R_3$$

$$\sim \left[\begin{array}{rrr|rrr} 1 & 0 & \frac{1}{2} & 1 & \frac{1}{2} & 0 \\ 0 & 1 & -\frac{1}{2} & 0 & \frac{1}{2} & 0 \\ 0 & 0 & 1 & -4 & -5 & 2 \end{array}\right] \quad \begin{array}{l} (-\frac{1}{2})R_3 + R_1 \to R_1 \\ \frac{1}{2}R_3 + R_2 \to R_2 \end{array}$$

$$\sim \left[\begin{array}{rrr|rrr} 1 & 0 & 0 & 3 & 3 & -1 \\ 0 & 1 & 0 & -2 & -2 & 1 \\ 0 & 0 & 1 & -4 & -5 & 2 \end{array}\right] = [I|B]$$

Converting back to systems of equations equivalent to our three original systems (we will not have to do this step in practice), we have

$$\begin{array}{lll} a = 3 & d = 3 & g = -1 \\ b = -2 & e = -2 & h = 1 \\ c = -4 & f = -5 & i = 2 \end{array}$$

And these are just the elements of M^{-1} that we are looking for! Hence,

$$M^{-1} = \begin{bmatrix} 3 & 3 & -1 \\ -2 & -2 & 1 \\ -4 & -5 & 2 \end{bmatrix}$$

Note that this is the matrix to the right of the vertical line in the last augmented matrix. That is, $M^{-1} = B$.

Since the definition of matrix inverse requires that

$$M^{-1}M = I \qquad \text{and} \qquad MM^{-1} = I \tag{3}$$

it appears that we must compute both $M^{-1}M$ and MM^{-1} to check our work. However, it can be shown that if one of the equations in (3) is satisfied, then the other is also satisfied. Thus, for checking purposes, it is sufficient to compute either $M^{-1}M$ or MM^{-1}; we do not need to do both.

CHECK $$M^{-1}M = \begin{bmatrix} 3 & 3 & -1 \\ -2 & -2 & 1 \\ -4 & -5 & 2 \end{bmatrix} \begin{bmatrix} 1 & -1 & 1 \\ 0 & 2 & -1 \\ 2 & 3 & 0 \end{bmatrix} = \begin{bmatrix} 1 & 0 & 0 \\ 0 & 1 & 0 \\ 0 & 0 & 1 \end{bmatrix} = I$$

Matched Problem 2 Let: $M = \begin{bmatrix} 3 & -1 & 1 \\ -1 & 1 & 0 \\ 1 & 0 & 1 \end{bmatrix}$

(A) Form the augmented matrix $[M|I]$.
(B) Use row operations to transform $[M|I]$ into $[I|B]$.
(C) Verify by multiplication that $B = M^{-1}$ (that is, show that $BM = I$).

The procedure shown in Example 2 can be used to find the inverse of any square matrix, if the inverse exists, and will also indicate when the inverse does not exist. These ideas are summarized in Theorem 1.

Theorem 1 INVERSE OF A SQUARE MATRIX M

If $[M|I]$ is transformed by row operations into $[I|B]$, then the resulting matrix B is M^{-1}. However, if we obtain 0's in one or more rows to the left of the vertical line, then M^{-1} does not exist.

Explore–Discuss 2

(A) Suppose that the square matrix M has a row of all zeros. Explain why M has no inverse.
(B) Suppose that the square matrix M has a column of all zeros. Explain why M has no inverse.

Example 3 **Finding a Matrix Inverse** Find M^{-1}, given: $M = \begin{bmatrix} 4 & -1 \\ -6 & 2 \end{bmatrix}$

SOLUTION

$$\left[\begin{array}{cc|cc} 4 & -1 & 1 & 0 \\ -6 & 2 & 0 & 1 \end{array}\right] \quad \tfrac{1}{4}R_1 \to R_1$$

$$\sim \left[\begin{array}{cc|cc} 1 & -\frac{1}{4} & \frac{1}{4} & 0 \\ -6 & 2 & 0 & 1 \end{array}\right] \quad 6R_1 + R_2 \to R_2$$

$$\sim \left[\begin{array}{cc|cc} 1 & -\frac{1}{4} & \frac{1}{4} & 0 \\ 0 & \frac{1}{2} & \frac{3}{2} & 1 \end{array}\right] \quad 2R_2 \to R_2$$

$$\sim \left[\begin{array}{cc|cc} 1 & -\frac{1}{4} & \frac{1}{4} & 0 \\ 0 & 1 & 3 & 2 \end{array}\right] \quad \tfrac{1}{4}R_2 + R_1 \to R_1$$

$$\sim \left[\begin{array}{cc|cc} 1 & 0 & 1 & \frac{1}{2} \\ 0 & 1 & 3 & 2 \end{array}\right]$$

Thus,

$$M^{-1} = \begin{bmatrix} 1 & \frac{1}{2} \\ 3 & 2 \end{bmatrix}$$

Check by showing $M^{-1}M = I$.

Matched Problem 3 Find M^{-1}, given: $M = \begin{bmatrix} 2 & -6 \\ 1 & -2 \end{bmatrix}$

Most graphing utilities can compute matrix inverses, as illustrated in Figure 2 for the solution to Example 3.

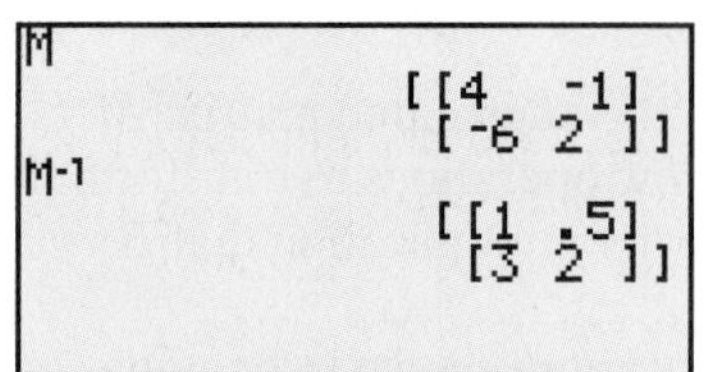

(A) The command M^{-1} produces the inverse on this graphing calculator

	A	B	C	D	E	F	G
1		M			M Inverse		
2		4	-1		1	0.5	
3		-6	2		3	2	
4							

(B) The command MINVERSE (B2:C3) produces the inverse in this spreadsheet

FIGURE 2
Finding a matrix inverse on a graphing utility

Explore–Discuss 3

The inverse of

$$A = \begin{bmatrix} a & b \\ c & d \end{bmatrix}$$

is

$$A^{-1} = \begin{bmatrix} \dfrac{d}{ad - bc} & \dfrac{-b}{ad - bc} \\ \dfrac{-c}{ad - bc} & \dfrac{a}{ad - bc} \end{bmatrix} = \frac{1}{D}\begin{bmatrix} d & -b \\ -c & a \end{bmatrix} \qquad D = ad - bc$$

provided $D \neq 0$.

(A) Use matrix multiplication to verify this formula. What can you conclude about A^{-1} if $D = 0$?

(B) Use this formula to find the inverse of matrix A in Example 3.

Example 4 **Finding a Matrix Inverse** Find M^{-1}, given: $M = \begin{bmatrix} 2 & -4 \\ -3 & 6 \end{bmatrix}$

SOLUTION

$$\left[\begin{array}{rr|rr} 2 & -4 & 1 & 0 \\ -3 & 6 & 0 & 1 \end{array}\right] \quad \tfrac{1}{2}R_1 \to R_1$$

$$\sim \left[\begin{array}{rr|rr} 1 & -2 & \frac{1}{2} & 0 \\ -3 & 6 & 0 & 1 \end{array}\right] \quad 3R_1 + R_2 \to R_2$$

$$\sim \left[\begin{array}{rr|rr} 1 & -2 & \frac{1}{2} & 0 \\ 0 & 0 & \frac{3}{2} & 1 \end{array}\right]$$

We have all 0's in the second row to the left of the vertical bar; therefore, the inverse does not exist.

Matched Problem 4 Find N^{-1}, given: $N = \begin{bmatrix} 3 & 1 \\ 6 & 2 \end{bmatrix}$

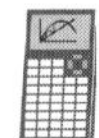

CAUTION

Matrices that do not have inverses are called **singular matrices.** Graphing calculators and computers usually recognize singular matrices and display an appropriate error message. However, in some cases, roundoff error may cause a graphing calculator or computer to display an incorrect inverse when the matrix is singular or to display an error message when the inverse does, in fact, exist. Figure 3 illustrates the results of solving Example 4 and Matched Problem 4 on a particular graphing calculator. Notice that the calculator gives the correct answer to Example 4, but displays an inverse for the singular matrix N in Matched Problem 4. The extremely large numbers in N^{-1} suggest that an error has occurred, and a quick check on the graphing calculator shows that NN^{-1} is not the identity. Whenever the results of finding an inverse on a graphing calculator or computer are in doubt, use Theorem 1 to find the inverse.

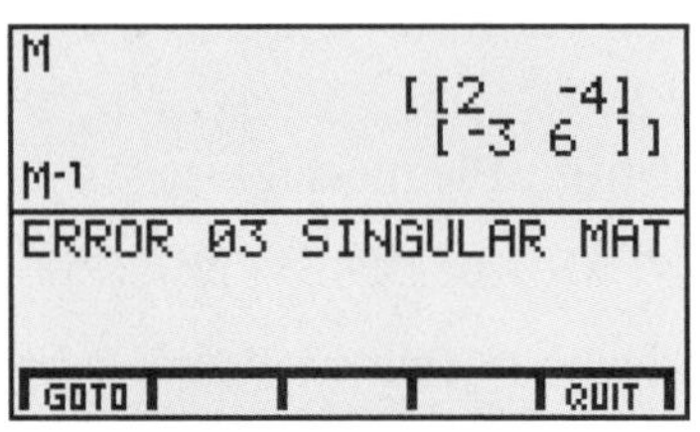

(A) Solution to Example 4

(B) Solution to Matched Problem 4

FIGURE 3

■ APPLICATION: CRYPTOGRAPHY

Matrix inverses can provide a simple and effective procedure for encoding and decoding messages. To begin, assign the numbers 1–26 to the letters in the alphabet, as shown below. Also assign the number 27 to a blank to provide for space between words. (A more sophisticated code could include both capital and lowercase letters and punctuation symbols.)

A	B	C	D	E	F	G	H	I	J	K	L	M	N	O	P	Q	R	S	T	U	V	W	X	Y	Z	Blank
1	2	3	4	5	6	7	8	9	10	11	12	13	14	15	16	17	18	19	20	21	22	23	24	25	26	27

Thus, the message "SECRET CODE" corresponds to the sequence

19 5 3 18 5 20 27 3 15 4 5

Any matrix whose elements are positive integers and whose inverse exists can be used as an **encoding matrix.** For example, to use the 2×2 matrix

$$A = \begin{bmatrix} 4 & 3 \\ 1 & 1 \end{bmatrix}$$

to encode the above message, first we divide the numbers in the sequence into groups of 2 and use these groups as the columns of a matrix B with 2 rows:

$$B = \begin{bmatrix} 19 & 3 & 5 & 27 & 15 & 5 \\ 5 & 18 & 20 & 3 & 4 & 27 \end{bmatrix}$$

Proceed down the columns, not across the rows.

(Notice that we added an extra blank at the end of the message to make the columns come out even.) Then we multiply this matrix on the left by A:

$$AB = \begin{bmatrix} 4 & 3 \\ 1 & 1 \end{bmatrix} \begin{bmatrix} 19 & 3 & 5 & 27 & 15 & 5 \\ 5 & 18 & 20 & 3 & 4 & 27 \end{bmatrix}$$
$$= \begin{bmatrix} 91 & 66 & 80 & 117 & 72 & 101 \\ 24 & 21 & 25 & 30 & 19 & 32 \end{bmatrix}$$

The coded message is

91 24 66 21 80 25 117 30 72 19 101 32

This message can be decoded simply by putting it back into matrix form and multiplying on the left by the **decoding matrix** A^{-1}. Since A^{-1} is easily determined if A is known, the encoding matrix A is the only key needed to decode messages encoded in this manner. Although simple in concept, codes of this type can be very difficult to crack.

Example 5 **Cryptography** The message

46 84 85 55 101 100 31 59 64 57 102 99 29 57 38 65 111 122

was encoded with the matrix A shown below. Decode this message.

$$A = \begin{bmatrix} 1 & 1 & 1 \\ 2 & 1 & 2 \\ 2 & 3 & 1 \end{bmatrix}$$

SOLUTION Since the encoding matrix A is 3×3, we begin by entering the coded message in the columns of a matrix C with three rows:

$$C = \begin{bmatrix} 46 & 55 & 31 & 57 & 29 & 65 \\ 84 & 101 & 59 & 102 & 57 & 111 \\ 85 & 100 & 64 & 99 & 38 & 122 \end{bmatrix}$$

If B is the matrix containing the uncoded message, then B and C are related by $C = AB$. To recover B, we find A^{-1} (details omitted) and multiply both sides of the equation $C = AB$ by A^{-1}:

$$B = A^{-1}C$$
$$= \begin{bmatrix} -5 & 2 & 1 \\ 2 & -1 & 0 \\ 4 & -1 & -1 \end{bmatrix} \begin{bmatrix} 46 & 55 & 31 & 57 & 29 & 65 \\ 84 & 101 & 59 & 102 & 57 & 111 \\ 85 & 100 & 64 & 99 & 38 & 122 \end{bmatrix}$$
$$= \begin{bmatrix} 23 & 27 & 27 & 18 & 7 & 19 \\ 8 & 9 & 3 & 12 & 1 & 19 \\ 15 & 19 & 1 & 27 & 21 & 27 \end{bmatrix}$$

Writing the numbers in the columns of this matrix in sequence and using the correspondence between numbers and letters noted earlier produces the decoded message:

23 8 15 27	9 19 27	3 1 18 12 27	7 1 21 19 19 27
W H O	I S	C A R L	G A U S S

(The answer to this question can be found earlier in this chapter.)

 *Matched Problem 5* The message below was also encoded with the matrix A in Example 5. Decode this message:

46 84 85 55 101 100 59 95 132 25 42 53 52 91 90 43 71 83 19 37 25

Answers to Matched Problems

1. (A) $\begin{bmatrix} 2 & -3 \\ 5 & 7 \end{bmatrix}$ (B) $\begin{bmatrix} 4 & 2 \\ 3 & -5 \\ 6 & 8 \end{bmatrix}$

2. (A) $\left[\begin{array}{rrr|rrr} 3 & -1 & 1 & 1 & 0 & 0 \\ -1 & 1 & 0 & 0 & 1 & 0 \\ 1 & 0 & 1 & 0 & 0 & 1 \end{array}\right]$

(B) $\left[\begin{array}{rrr|rrr} 1 & 0 & 0 & 1 & 1 & -1 \\ 0 & 1 & 0 & 1 & 2 & -1 \\ 0 & 0 & 1 & -1 & -1 & 2 \end{array}\right]$

(C) $\begin{bmatrix} 1 & 1 & -1 \\ 1 & 2 & -1 \\ -1 & -1 & 2 \end{bmatrix}\begin{bmatrix} 3 & -1 & 1 \\ -1 & 1 & 0 \\ 1 & 0 & 1 \end{bmatrix} = \begin{bmatrix} 1 & 0 & 0 \\ 0 & 1 & 0 \\ 0 & 0 & 1 \end{bmatrix}$

3. $\begin{bmatrix} -1 & 3 \\ -\frac{1}{2} & 1 \end{bmatrix}$ **4.** Does not exist **5.** WHO IS WILHELM JORDAN

EXERCISE 4-5

A *Perform the indicated operations in Problems 1–8.*

1. $\begin{bmatrix} 1 & 0 \\ 0 & 1 \end{bmatrix}\begin{bmatrix} 2 & -3 \\ 4 & 5 \end{bmatrix}$

2. $\begin{bmatrix} 1 & 0 \\ 0 & 1 \end{bmatrix}\begin{bmatrix} -1 & 6 \\ 0 & 2 \end{bmatrix}$

3. $\begin{bmatrix} 2 & -3 \\ 4 & 5 \end{bmatrix}\begin{bmatrix} 1 & 0 \\ 0 & 1 \end{bmatrix}$

4. $\begin{bmatrix} -1 & 6 \\ 0 & 2 \end{bmatrix}\begin{bmatrix} 1 & 0 \\ 0 & 1 \end{bmatrix}$

5. $\begin{bmatrix} 1 & 0 & 0 \\ 0 & 1 & 0 \\ 0 & 0 & 1 \end{bmatrix}\begin{bmatrix} -2 & 1 & 3 \\ 2 & 4 & -2 \\ 5 & 1 & 0 \end{bmatrix}$

6. $\begin{bmatrix} 1 & 0 & 0 \\ 0 & 1 & 0 \\ 0 & 0 & 1 \end{bmatrix}\begin{bmatrix} 3 & -4 & 0 \\ 1 & 2 & -5 \\ 6 & -3 & -1 \end{bmatrix}$

7. $\begin{bmatrix} -2 & 1 & 3 \\ 2 & 4 & -2 \\ 5 & 1 & 0 \end{bmatrix}\begin{bmatrix} 1 & 0 & 0 \\ 0 & 1 & 0 \\ 0 & 0 & 1 \end{bmatrix}$

8. $\begin{bmatrix} 3 & -4 & 0 \\ 1 & 2 & -5 \\ 6 & -3 & -1 \end{bmatrix}\begin{bmatrix} 1 & 0 & 0 \\ 0 & 1 & 0 \\ 0 & 0 & 1 \end{bmatrix}$

In Problems 9–18, examine the product of the two matrices to determine if each is the inverse of the other.

9. $\begin{bmatrix} 3 & -4 \\ -2 & 3 \end{bmatrix}; \begin{bmatrix} 3 & 4 \\ 2 & 3 \end{bmatrix}$

10. $\begin{bmatrix} -2 & -1 \\ -4 & 2 \end{bmatrix}; \begin{bmatrix} 1 & -1 \\ 2 & -2 \end{bmatrix}$

11. $\begin{bmatrix} 2 & 2 \\ -1 & -1 \end{bmatrix}; \begin{bmatrix} 1 & 1 \\ -1 & -1 \end{bmatrix}$

12. $\begin{bmatrix} 5 & -7 \\ -2 & 3 \end{bmatrix}; \begin{bmatrix} 3 & 7 \\ 2 & 5 \end{bmatrix}$

13. $\begin{bmatrix} -5 & 2 \\ -8 & 3 \end{bmatrix}; \begin{bmatrix} 3 & -2 \\ 8 & -5 \end{bmatrix}$

14. $\begin{bmatrix} 7 & 4 \\ -5 & -3 \end{bmatrix}; \begin{bmatrix} 3 & 4 \\ -5 & -7 \end{bmatrix}$

15. $\begin{bmatrix} 1 & 2 & 0 \\ 0 & 1 & 0 \\ -1 & -1 & 1 \end{bmatrix}; \begin{bmatrix} 1 & -2 & 0 \\ 0 & 1 & 0 \\ 1 & -1 & 0 \end{bmatrix}$

16. $\begin{bmatrix} 1 & 0 & 1 \\ -3 & 1 & -2 \\ 0 & 0 & 1 \end{bmatrix}; \begin{bmatrix} 1 & 0 & -1 \\ 3 & 1 & -1 \\ 0 & 0 & 1 \end{bmatrix}$

17. $\begin{bmatrix} 1 & -1 & 1 \\ 0 & 2 & -1 \\ 2 & 3 & 0 \end{bmatrix}; \begin{bmatrix} 3 & 3 & -1 \\ -2 & -2 & 1 \\ -4 & -5 & 2 \end{bmatrix}$

18. $\begin{bmatrix} 1 & 0 & -1 \\ 3 & 1 & -1 \\ 0 & 0 & 0 \end{bmatrix}; \begin{bmatrix} 1 & 0 & -1 \\ -3 & 1 & -2 \\ 0 & 0 & 1 \end{bmatrix}$

B *Given M in Problems 19–28, find M^{-1} and show that $M^{-1}M = I$.*

19. $\begin{bmatrix} -1 & 0 \\ -3 & 1 \end{bmatrix}$

20. $\begin{bmatrix} 1 & -5 \\ 0 & -1 \end{bmatrix}$

21. $\begin{bmatrix} 1 & 2 \\ 1 & 3 \end{bmatrix}$

22. $\begin{bmatrix} 2 & 1 \\ 5 & 3 \end{bmatrix}$

23. $\begin{bmatrix} 1 & 3 \\ 2 & 7 \end{bmatrix}$

24. $\begin{bmatrix} 2 & 1 \\ 1 & 1 \end{bmatrix}$

25. $\begin{bmatrix} 1 & -3 & 0 \\ 0 & 1 & 1 \\ 2 & -1 & 4 \end{bmatrix}$

26. $\begin{bmatrix} 2 & 3 & 0 \\ 1 & 2 & 3 \\ 0 & -1 & -5 \end{bmatrix}$

27. $\begin{bmatrix} 1 & 1 & 0 \\ 2 & 3 & -1 \\ 1 & 0 & 2 \end{bmatrix}$

28. $\begin{bmatrix} 1 & 0 & -1 \\ 2 & -1 & 0 \\ 1 & 1 & -2 \end{bmatrix}$

Find the inverse of each matrix in Problems 29–34, if it exists.

29. $\begin{bmatrix} 4 & 3 \\ -3 & -2 \end{bmatrix}$

30. $\begin{bmatrix} -4 & 3 \\ -5 & 4 \end{bmatrix}$

31. $\begin{bmatrix} 2 & 6 \\ 3 & 9 \end{bmatrix}$

32. $\begin{bmatrix} 2 & -4 \\ -3 & 6 \end{bmatrix}$

33. $\begin{bmatrix} 2 & 1 \\ 4 & 3 \end{bmatrix}$

34. $\begin{bmatrix} -5 & 3 \\ 2 & -2 \end{bmatrix}$

C *Find the inverse of each matrix in Problems 35–42, if it exists.*

35. $\begin{bmatrix} -5 & -2 & -2 \\ 2 & 1 & 0 \\ 1 & 0 & 1 \end{bmatrix}$

36. $\begin{bmatrix} 2 & -2 & 4 \\ 1 & 1 & 1 \\ 1 & 0 & 1 \end{bmatrix}$

37. $\begin{bmatrix} 2 & 1 & 1 \\ 1 & 1 & 0 \\ -1 & -1 & 0 \end{bmatrix}$

38. $\begin{bmatrix} 1 & -1 & 0 \\ 2 & -1 & 1 \\ 0 & 1 & 1 \end{bmatrix}$

39. $\begin{bmatrix} -1 & -2 & 2 \\ 4 & 3 & 0 \\ 4 & 0 & 4 \end{bmatrix}$

40. $\begin{bmatrix} 4 & 2 & 2 \\ 4 & 2 & 0 \\ 5 & 0 & 5 \end{bmatrix}$

41. $\begin{bmatrix} 2 & -1 & -2 \\ -4 & 2 & 8 \\ 6 & -2 & -1 \end{bmatrix}$

42. $\begin{bmatrix} -1 & -1 & 4 \\ 3 & 3 & -22 \\ -2 & -1 & 19 \end{bmatrix}$

43. Show that $(A^{-1})^{-1} = A$ for: $A = \begin{bmatrix} 4 & 3 \\ 3 & 2 \end{bmatrix}$

44. Show that $(AB)^{-1} = B^{-1}A^{-1}$ for:

$$A = \begin{bmatrix} 4 & 3 \\ 3 & 2 \end{bmatrix} \quad \text{and} \quad B = \begin{bmatrix} 2 & 5 \\ 3 & 7 \end{bmatrix}$$

45. Discuss the existence of M^{-1} for 2×2 diagonal matrices of the form

$$M = \begin{bmatrix} a & 0 \\ 0 & d \end{bmatrix}$$

Generalize your conclusions to $n \times n$ diagonal matrices.

46. Discuss the existence of M^{-1} for 2×2 upper triangular matrices of the form

$$M = \begin{bmatrix} a & b \\ 0 & d \end{bmatrix}$$

Generalize your conclusions to $n \times n$ upper triangular matrices.

In Problems 47–49, find A^{-1} and A^2.

47. $A = \begin{bmatrix} 3 & 2 \\ -4 & -3 \end{bmatrix}$

48. $A = \begin{bmatrix} -2 & -1 \\ 3 & 2 \end{bmatrix}$

49. $A = \begin{bmatrix} 4 & 3 \\ -5 & -4 \end{bmatrix}$

50. Based on your observations in Problems 47–49, if $A = A^{-1}$ for a square matrix A, what is A^2? Give a mathematical argument to support your conclusion.

In Problems 51–54, use a graphing utility to find the inverse of each matrix, if it exists.

51. $\begin{bmatrix} 6 & 2 & 0 & 4 \\ 5 & 3 & 2 & 1 \\ 0 & -1 & 1 & -2 \\ 2 & -3 & 1 & 0 \end{bmatrix}$

52. $\begin{bmatrix} -2 & 4 & 0 & -1 \\ 2 & -1 & 2 & 5 \\ 0 & 2 & -1 & 7 \\ 2 & -3 & 0 & 5 \end{bmatrix}$

53. $\begin{bmatrix} 3 & 2 & 3 & 4 & 4 \\ 5 & 4 & 3 & 2 & 1 \\ -1 & -1 & 2 & -2 & 3 \\ 3 & -3 & 1 & 0 & 1 \\ 1 & 1 & 2 & 0 & 2 \end{bmatrix}$

54. $\begin{bmatrix} 1 & 2 & 3 & 4 & 5 \\ 2 & 6 & 4 & 5 & 6 \\ -1 & -2 & -1 & 2 & 3 \\ 1 & 6 & 1 & 6 & 4 \\ 1 & -4 & 3 & -7 & -4 \end{bmatrix}$

APPLICATIONS

Social Sciences

Problems 55–58 refer to the encoding matrix:

$$A = \begin{bmatrix} 1 & 2 \\ 1 & 3 \end{bmatrix}$$

55. *Cryptography.* Encode the message "THE SUN ALSO RISES" using matrix A.

56. *Cryptography.* Encode the message "THE GRAPES OF WRATH" using matrix A.

57. *Cryptography.* The following message was encoded with matrix A. Decode this message:

37 52 24 29 73 96 49 69 62 89 36 44 59 86 41 50 22 26

58. *Cryptography.* The following message was encoded with matrix A. Decode this message:

9 13 40 49 29 34 29 30 22 26 33 36 43 57 29 34 54 74

Problems 59–62 require the use of a graphing calculator or a computer. To use the 5×5 encoding matrix B given below, form a matrix with 5 rows and as many columns as necessary to accommodate each message.

$$B = \begin{bmatrix} 1 & 0 & 1 & 0 & 1 \\ 0 & 1 & 1 & 0 & 3 \\ 2 & 1 & 1 & 1 & 1 \\ 0 & 0 & 1 & 0 & 2 \\ 1 & 1 & 1 & 2 & 1 \end{bmatrix}$$

59. *Cryptography.* Encode the message "THE BEST YEARS OF OUR LIVES" using matrix B.

60. *Cryptography.* Encode the message "THE BRIDGE ON THE RIVER KWAI" using matrix B.

61. *Cryptography.* The following message was encoded with matrix B. Decode this message:

32 34 87 19 94 24 21 67 11 69 54 71 112 43 112 56 92 109 55 109 48 66 98 41 89 62 135 124 81 143

62. *Cryptography.* The following message was encoded with matrix B. Decode this message:

28 22 83 11 90 57 108 102 69 105 57 78 91 57 89 66 111 112 79 125

SECTION 4-6

Matrix Equations and Systems of Linear Equations

- MATRIX EQUATIONS
- MATRIX EQUATIONS AND SYSTEMS OF LINEAR EQUATIONS
- APPLICATION

The identity matrix and inverse matrix discussed in the last section can be put to immediate use in the solution of certain simple matrix equations. Being able to solve a matrix equation gives us another important method of solving systems of equations, provided the system is independent and has the same number of variables as equations. If the system is dependent or if it has either fewer or more variables than equations, we must return to the Gauss–Jordan method of elimination.

MATRIX EQUATIONS

Before we discuss the solution of matrix equations, you will probably find it helpful to briefly review the basic properties of real numbers discussed in Appendix A-2 and the discussion of linear equations in Appendix A-8.

Explore–Discuss 1

Let a, b, and c be real numbers, with $a \neq 0$. Solve each equation for x.

(A) $ax = b$ (B) $ax + b = c$

Solving simple matrix equations follows very much the same procedures used in solving real number equations. We have, however, less freedom with matrix equations, because matrix multiplication is not commutative. In solving matrix equations, we will be guided by the properties of matrices summarized in Theorem 1.

Theorem 1 BASIC PROPERTIES OF MATRICES

Assuming all products and sums are defined for the indicated matrices A, B, C, I, and 0, then:

Addition Properties

Associative: $(A + B) + C = A + (B + C)$
Commutative: $A + B = B + A$
Additive Identity: $A + 0 = 0 + A = A$
Additive Inverse: $A + (-A) = (-A) + A = 0$

Multiplication Properties

Associative Property: $A(BC) = (AB)C$
Multiplicative Identity: $AI = IA = A$
Multiplicative Inverse: If A is a square matrix and A^{-1} exists, then $AA^{-1} = A^{-1}A = I$.

Combined Properties

Left Distributive: $A(B + C) = AB + AC$
Right Distributive: $(B + C)A = BA + CA$

Equality

Addition: If $A = B$, then $A + C = B + C$.
Left Multiplication: If $A = B$, then $CA = CB$.
Right Mulplication: If $A = B$, then $AC = BC$.

The use of these properties in the solution of matrix equations is best illustrated by an example.

Example 1 **Solving a Matrix Equation** Given an $n \times n$ matrix A and $n \times 1$ column matrices B and X, solve $AX = B$ for X. Assume all necessary inverses exist.

SOLUTION We are interested in finding a column matrix X that satisfies the matrix equation $AX = B$. To solve this equation, we multiply both sides on the left by A^{-1}, assuming it exists, to isolate X on the left side.

$$
\begin{aligned}
AX &= B && \text{Use the left multiplication property.}\\
A^{-1}(AX) &= A^{-1}B && \text{Use the associative property.}\\
(A^{-1}A)X &= A^{-1}B && A^{-1}A = I\\
IX &= A^{-1}B && IX = X\\
X &= A^{-1}B
\end{aligned}
$$

Matched Problem 1 Given an $n \times n$ matrix A and $n \times 1$ column matrices B, C, and X, solve $AX + C = B$ for X. Assume all necessary inverses exist.

CAUTION

Do not mix the left multiplication property and the right multiplication property. If $AX = B$, then

$$A^{-1}(AX) \neq BA^{-1}$$

MATRIX EQUATIONS AND SYSTEMS OF LINEAR EQUATIONS

We now show how independent systems of linear equations with the same number of variables as equations can be solved. First, convert the system into a matrix equation of the form $AX = B$, and then use $X = A^{-1}B$ as obtained in Example 1.

Example 2 **Using Inverses to Solve Systems of Equations** Use matrix inverse methods to solve the system:

$$\begin{aligned} x_1 - x_2 + x_3 &= 1 \\ 2x_2 - x_3 &= 1 \\ 2x_1 + 3x_2 &= 1 \end{aligned} \tag{1}$$

SOLUTION The inverse of the coefficient matrix

$$A = \begin{bmatrix} 1 & -1 & 1 \\ 0 & 2 & -1 \\ 2 & 3 & 0 \end{bmatrix}$$

provides an efficient method for solving this system. To see how, we convert system (1) into a matrix equation:

$$\overset{A}{\begin{bmatrix} 1 & -1 & 1 \\ 0 & 2 & -1 \\ 2 & 3 & 0 \end{bmatrix}} \overset{X}{\begin{bmatrix} x_1 \\ x_2 \\ x_3 \end{bmatrix}} = \overset{B}{\begin{bmatrix} 1 \\ 1 \\ 1 \end{bmatrix}}$$

Check that matrix equation (2) is equivalent to system (1) by finding the product of the left side and then equating corresponding elements on the left with those on the right. Now you see another important reason for defining matrix multiplication as we did.

We are interested in finding a column matrix X that satisfies the matrix equation $AX = B$. In Example 1 we found that if A^{-1} exists, then

$$X = A^{-1}B$$

The inverse of A was found in Example 2, Section 4-5, to be

$$A^{-1} = \begin{bmatrix} 3 & 3 & -1 \\ -2 & -2 & 1 \\ -4 & -5 & 2 \end{bmatrix}$$

Thus,

$$\overset{X}{\begin{bmatrix} x_1 \\ x_2 \\ x_3 \end{bmatrix}} = \overset{A^{-1}}{\begin{bmatrix} 3 & 3 & -1 \\ -2 & -2 & 1 \\ -4 & -5 & 2 \end{bmatrix}} \overset{B}{\begin{bmatrix} 1 \\ 1 \\ 1 \end{bmatrix}} = \begin{bmatrix} 5 \\ -3 \\ -7 \end{bmatrix}$$

and we can conclude that $x_1 = 5, x_2 = -3$, and $x_3 = -7$. Check this result in system (1).

Matched Problem 2 Use matrix inverse methods to solve the system:

$$\begin{aligned} 3x_1 - x_2 + x_3 &= 1 \\ -x_1 + x_2 \qquad &= 3 \\ x_1 \qquad + x_3 &= 2 \end{aligned}$$

[*Note:* The inverse of the coefficient matrix was found in Matched Problem 2, Section 4-5.]

At first glance, using matrix inverse methods seems to require the same amount of effort as using Gauss–Jordan elimination. In either case, row operations must be applied to an augmented matrix involving the coefficients of the system. The advantage of the inverse matrix method becomes readily apparent when solving a number of systems with a common coefficient matrix and different constant terms.

Example 3 **Using Inverses to Solve Systems of Equations** Use matrix inverse methods to solve each of the following systems:

(A)
$$\begin{aligned} x_1 - x_2 + x_3 &= 3 \\ 2x_2 - x_3 &= 1 \\ 2x_1 + 3x_2 \qquad &= 4 \end{aligned}$$

(B)
$$\begin{aligned} x_1 - x_2 + x_3 &= -5 \\ 2x_2 - x_3 &= 2 \\ 2x_1 + 3x_2 \qquad &= -3 \end{aligned}$$

SOLUTION Notice that both systems have the same coefficient matrix A as system (1) in Example 2. Only the constant terms have changed. Thus, we can use A^{-1} to solve these systems just as we did in Example 2.

(A)
$$\overset{X}{\begin{bmatrix} x_1 \\ x_2 \\ x_3 \end{bmatrix}} = \overset{A^{-1}}{\begin{bmatrix} 3 & 3 & -1 \\ -2 & -2 & 1 \\ -4 & -5 & 2 \end{bmatrix}} \overset{B}{\begin{bmatrix} 3 \\ 1 \\ 4 \end{bmatrix}} = \begin{bmatrix} 8 \\ -4 \\ -9 \end{bmatrix}$$

Thus, $x_1 = 8, x_2 = -4$, and $x_3 = -9$.

(B)
$$\overset{X}{\begin{bmatrix} x_1 \\ x_2 \\ x_3 \end{bmatrix}} = \overset{A^{-1}}{\begin{bmatrix} 3 & 3 & -1 \\ -2 & -2 & 1 \\ -4 & -5 & 2 \end{bmatrix}} \overset{B}{\begin{bmatrix} -5 \\ 2 \\ -3 \end{bmatrix}} = \begin{bmatrix} -6 \\ 3 \\ 4 \end{bmatrix}$$

Thus, $x_1 = -6, x_2 = 3$, and $x_3 = 4$.

Matched Problem 3 Use matrix inverse methods to solve each of the following systems (see Matched Problem 2):

(A)
$$\begin{aligned} 3x_1 - x_2 + x_3 &= 3 \\ -x_1 + x_2 \quad\quad &= -3 \\ x_1 \quad\quad + x_3 &= 2 \end{aligned}$$

(B)
$$\begin{aligned} 3x_1 - x_2 + x_3 &= -5 \\ -x_1 + x_2 \quad\quad &= 1 \\ x_1 \quad\quad + x_3 &= -4 \end{aligned}$$

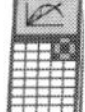

As Examples 2 and 3 illustrate, inverse methods are very convenient for hand calculations because once the inverse is found, it can be used to solve any new system formed by changing only the constant terms. Since most graphing utilities can compute the inverse of a matrix, this method also adapts readily to graphing utility solutions (see Fig. 1). However, if your graphing utility also has a built-in procedure for finding the reduced form of an augmented coefficient matrix, then it is just as convenient to use Gauss–Jordan elimination. Furthermore, Gauss–Jordan elimination can be used in all cases and, as noted below, matrix inverse methods cannot always be used.

	A	B	C	D	E	F	G	H	I	J
1			A			B	X		B	X
2		1	-1	1		3	8		-5	-6
3		0	2	-1		1	-4		2	3
4		2	3	0		4	-9		-3	4

FIGURE 1
Using inverse methods on a spreadsheet:
The values in G2:G4 are produced by the command
MMULT(MINVERSE(B2:D4),F2:F4)

Using Inverse Methods to Solve Systems of Equations

If the number of equations in a system equals the number of variables and the coefficient matrix has an inverse, then the system will always have a unique solution that can be found by using the inverse of the coefficient matrix to solve the corresponding matrix equation.

Matrix equation	Solution
$AX = B$	$X = A^{-1}B$

REMARK

What happens if the coefficient matrix does not have an inverse? In this case, it can be shown that the system does not have a unique solution and is either dependent or inconsistent. Gauss–Jordan elimination must be used to determine which is the case. Also, as we mentioned earlier, Gauss–Jordan elimination always must be used if the number of variables is not the same as the number of equations.

APPLICATION

The following application will illustrate the usefulness of the inverse matrix method for solving systems of equations.

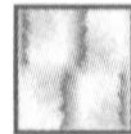

Example 4 **Investment Analysis** An investment advisor currently has two types of investments available for clients: a conservative investment A that pays 10% per year and an investment B of higher risk that pays 20% per year. Clients may divide their investments between the two to achieve any total return desired between 10% and 20%. However, the higher the desired return, the higher the risk. How should each client listed in the table invest to achieve the indicated return?

	CLIENT			
	1	2	3	k
TOTAL INVESTMENT	\$20,000	\$50,000	\$10,000	k_1
ANNUAL RETURN DESIRED	\$ 2,400 (12%)	\$ 7,500 (15%)	\$ 1,300 (13%)	k_2

SOLUTION We will solve the problem for an arbitrary client k by finding an inverse matrix. Then we will apply the result to the three specific clients.

Let: x_1 = Amount invested in A
x_2 = Amount invested in B

Then we have the following mathematical model:

$$\begin{aligned} x_1 + \quad x_2 &= k_1 \quad \text{Total Invested} \\ 0.1x_1 + 0.2x_2 &= k_2 \quad \text{Total annual return desired} \end{aligned}$$

Write as a matrix equation:

$$\overset{A}{\begin{bmatrix} 1 & 1 \\ 0.1 & 0.2 \end{bmatrix}} \overset{X}{\begin{bmatrix} x_1 \\ x_2 \end{bmatrix}} = \overset{B}{\begin{bmatrix} k_1 \\ k_2 \end{bmatrix}}$$

If A^{-1} exists, then

$$X = A^{-1}B$$

We now find A^{-1} by starting with the augmented matrix $[A|I]$ and proceeding as discussed in Section 4-5:

$$\left[\begin{array}{cc|cc} 1 & 1 & 1 & 0 \\ 0.1 & 0.2 & 0 & 1 \end{array}\right] \quad 10R_2 \to R_2$$

$$\sim \left[\begin{array}{cc|cc} 1 & 1 & 1 & 0 \\ 1 & 2 & 0 & 10 \end{array}\right] \quad (-1)R_1 + R_2 \to R_2$$

$$\sim \left[\begin{array}{cc|cc} 1 & 1 & 1 & 0 \\ 0 & 1 & -1 & 10 \end{array}\right] \quad (-1)R_2 + R_1 \to R_1$$

$$\sim \left[\begin{array}{cc|cc} 1 & 0 & 2 & -10 \\ 0 & 1 & -1 & 10 \end{array}\right]$$

Thus,

$$A^{-1} = \begin{bmatrix} 2 & -10 \\ -1 & 10 \end{bmatrix} \qquad \text{Check: } \overset{A^{-1}}{\begin{bmatrix} 2 & -10 \\ -1 & 10 \end{bmatrix}} \overset{A}{\begin{bmatrix} 1 & 1 \\ 0.1 & 0.2 \end{bmatrix}} = \overset{I}{\begin{bmatrix} 1 & 0 \\ 0 & 1 \end{bmatrix}}$$

and

$$\overset{X}{\begin{bmatrix} x_1 \\ x_2 \end{bmatrix}} = \overset{A^{-1}}{\begin{bmatrix} 2 & -10 \\ -1 & 10 \end{bmatrix}} \overset{B}{\begin{bmatrix} k_1 \\ k_2 \end{bmatrix}}$$

To solve each client's investment problem, we replace k_1 and k_2 with appropriate values from the table and multiply by A^{-1}:

Client 1

$$\begin{bmatrix} x_1 \\ x_2 \end{bmatrix} = \begin{bmatrix} 2 & -10 \\ -1 & 10 \end{bmatrix} \begin{bmatrix} 20{,}000 \\ 2{,}400 \end{bmatrix} = \begin{bmatrix} 16{,}000 \\ 4{,}000 \end{bmatrix}$$

Solution: $x_1 = \$16{,}000$ in investment A, $x_2 = \$4{,}000$ in investment B

Client 2

$$\begin{bmatrix} x_1 \\ x_2 \end{bmatrix} = \begin{bmatrix} 2 & -10 \\ -1 & 10 \end{bmatrix} \begin{bmatrix} 50{,}000 \\ 7{,}500 \end{bmatrix} = \begin{bmatrix} 25{,}000 \\ 25{,}000 \end{bmatrix}$$

Solution: $x_1 = \$25{,}000$ in investment A, $x_2 = \$25{,}000$ in investment B

Client 3

$$\begin{bmatrix} x_1 \\ x_2 \end{bmatrix} = \begin{bmatrix} 2 & -10 \\ -1 & 10 \end{bmatrix} \begin{bmatrix} 10{,}000 \\ 1{,}300 \end{bmatrix} = \begin{bmatrix} 7{,}000 \\ 3{,}000 \end{bmatrix}$$

Solution: $x_1 = \$7{,}000$ in investment A, $x_2 = \$3{,}000$ in investment B

Matched Problem 4 Repeat Example 4 with investment A paying 8% and investment B paying 24%.

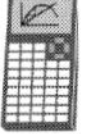

Figure 2 illustrates a solution to Example 4 on a spreadsheet.

	A	B	C	D	E	F	G
1			Clients				
2		1	2	3		A	
3	Total Investment	$20,000	$50,000	$10,000		1	1
4	Annual Return	$2,400	$7,500	$1,300		0.1	0.2
5	Amount Invested in A	$16,000	$25,000	$7,000			
6	Amount Invested in B	$4,000	$25,000	$3,000			

FIGURE 2

Explore–Discuss 2

Refer to the mathematical model in Example 4:

$$\overset{A}{\begin{bmatrix} 1 & 1 \\ 0.1 & 0.2 \end{bmatrix}} \overset{X}{\begin{bmatrix} x_1 \\ x_2 \end{bmatrix}} = \overset{B}{\begin{bmatrix} k_1 \\ k_2 \end{bmatrix}} \quad (3)$$

(A) Does equation (3) always have a solution for any constant matrix B?

(B) Do all these solutions make sense for the original problem? If not, give examples.

(C) If the total investment is $k_1 = \$10{,}000$, describe all possible annual returns k_2.

Answers to Matched Problems

1.

$$\begin{aligned} AX + C &= B \\ (AX + C) - C &= B - C \\ AX + (C - C) &= B - C \\ AX + 0 &= B - C \\ AX &= B - C \\ A^{-1}(AX) &= A^{-1}(B - C) \\ (A^{-1}A)X &= A^{-1}(B - C) \\ IX &= A^{-1}(B - C) \\ X &= A^{-1}(B - C) \end{aligned}$$

2. $x_1 = 2, x_2 = 5, x_3 = 0$

3. (A) $x_1 = -2, x_2 = -5, x_3 = 4$ (B) $x_1 = 0, x_2 = 1, x_3 = -4$

4. $A^{-1} = \begin{bmatrix} 1.5 & -6.25 \\ -0.5 & 6.25 \end{bmatrix}$; Client 1: \$15,000 in A and \$5,000 in B; Client 2: \$28,125 in A and \$21,875 in B; Client 3: \$6,875 in A and \$3,125 in B

EXERCISE 4-6

A *Write Problems 1–4 as systems of linear equations without matrices.*

1. $\begin{bmatrix} 3 & 1 \\ 2 & -1 \end{bmatrix} \begin{bmatrix} x_1 \\ x_2 \end{bmatrix} = \begin{bmatrix} 5 \\ -4 \end{bmatrix}$

2. $\begin{bmatrix} -2 & 1 \\ -3 & 4 \end{bmatrix} \begin{bmatrix} x_1 \\ x_2 \end{bmatrix} = \begin{bmatrix} -5 \\ 7 \end{bmatrix}$

3. $\begin{bmatrix} -3 & 1 & 0 \\ 2 & 0 & 1 \\ -1 & 3 & -2 \end{bmatrix} \begin{bmatrix} x_1 \\ x_2 \\ x_3 \end{bmatrix} = \begin{bmatrix} 3 \\ -4 \\ 2 \end{bmatrix}$

4. $\begin{bmatrix} 2 & -1 & 0 \\ -2 & 3 & -1 \\ 4 & 0 & 3 \end{bmatrix} \begin{bmatrix} x_1 \\ x_2 \\ x_3 \end{bmatrix} = \begin{bmatrix} 6 \\ -4 \\ 7 \end{bmatrix}$

Write each system in Problems 5–8 as a matrix equation of the form $AX = B$.

5. $\begin{aligned} 3x_1 - 4x_2 &= 1 \\ 2x_1 + x_2 &= 5 \end{aligned}$

6. $\begin{aligned} 2x_1 + x_2 &= 8 \\ -5x_1 + 3x_2 &= -4 \end{aligned}$

7. $\begin{aligned} x_1 - 3x_2 + 2x_3 &= -3 \\ -2x_1 + 3x_2 &= 1 \\ x_1 + x_2 + 4x_3 &= -2 \end{aligned}$

8. $\begin{aligned} 3x_1 + 2x_3 &= 9 \\ -x_1 + 4x_2 + x_3 &= -7 \\ -2x_1 + 3x_2 &= 6 \end{aligned}$

Find x_1 and x_2 in Problems 9–12.

9. $\begin{bmatrix} x_1 \\ x_2 \end{bmatrix} = \begin{bmatrix} 3 & -2 \\ 1 & 4 \end{bmatrix}\begin{bmatrix} -2 \\ 1 \end{bmatrix}$

10. $\begin{bmatrix} x_1 \\ x_2 \end{bmatrix} = \begin{bmatrix} -2 & 1 \\ -1 & 2 \end{bmatrix}\begin{bmatrix} 3 \\ -2 \end{bmatrix}$

11. $\begin{bmatrix} x_1 \\ x_2 \end{bmatrix} = \begin{bmatrix} -2 & 3 \\ 2 & -1 \end{bmatrix}\begin{bmatrix} 3 \\ 2 \end{bmatrix}$

12. $\begin{bmatrix} x_1 \\ x_2 \end{bmatrix} = \begin{bmatrix} 3 & -1 \\ 0 & 2 \end{bmatrix}\begin{bmatrix} -2 \\ 1 \end{bmatrix}$

In Problems 13–16, find x_1 and x_2.

13. $\begin{bmatrix} 1 & -1 \\ 1 & -2 \end{bmatrix}\begin{bmatrix} x_1 \\ x_2 \end{bmatrix} = \begin{bmatrix} 5 \\ 7 \end{bmatrix}$

14. $\begin{bmatrix} 1 & 3 \\ 1 & 4 \end{bmatrix}\begin{bmatrix} x_1 \\ x_2 \end{bmatrix} = \begin{bmatrix} 9 \\ 6 \end{bmatrix}$

15. $\begin{bmatrix} 1 & 1 \\ 2 & -3 \end{bmatrix}\begin{bmatrix} x_1 \\ x_2 \end{bmatrix} = \begin{bmatrix} 15 \\ 10 \end{bmatrix}$

16. $\begin{bmatrix} 1 & 1 \\ 3 & -2 \end{bmatrix}\begin{bmatrix} x_1 \\ x_2 \end{bmatrix} = \begin{bmatrix} 10 \\ 20 \end{bmatrix}$

B *In Problems 17–24, write each system as a matrix equation and solve by using inverses. [Note: The inverses were found in Problems 21–28 in Exercise 4-5.]*

17. $x_1 + 2x_2 = k_1$
$x_1 + 3x_2 = k_2$
(A) $k_1 = 1, k_2 = 3$
(B) $k_1 = 3, k_2 = 5$
(C) $k_1 = -2, k_2 = 1$

18. $2x_1 + x_2 = k_1$
$5x_1 + 3x_2 = k_2$
(A) $k_1 = 2, k_2 = 13$
(B) $k_1 = 2, k_2 = 4$
(C) $k_1 = 1, k_2 = -3$

19. $x_1 + 3x_2 = k_1$
$2x_1 + 7x_2 = k_2$
(A) $k_1 = 2, k_2 = -1$
(B) $k_1 = 1, k_2 = 0$
(C) $k_1 = 3, k_2 = -1$

20. $2x_1 + x_2 = k_1$
$x_1 + x_2 = k_2$
(A) $k_1 = -1, k_2 = -2$
(B) $k_1 = 2, k_2 = 3$
(C) $k_1 = 2, k_2 = 0$

21. $x_1 - 3x_2 = k_1$
$x_2 + x_3 = k_2$
$2x_1 - x_2 + 4x_3 = k_3$
(A) $k_1 = 1, k_2 = 0, k_3 = 2$
(B) $k_1 = -1, k_2 = 1, k_3 = 0$
(C) $k_1 = 2, k_2 = -2, k_3 = 1$

22. $2x_1 + 3x_2 = k_1$
$x_1 + 2x_2 + 3x_3 = k_2$
$- x_2 - 5x_3 = k_3$
(A) $k_1 = 0, k_2 = 2, k_3 = 1$
(B) $k_1 = -2, k_2 = 0, k_3 = 1$
(C) $k_1 = 3, k_2 = 1, k_3 = 0$

23. $x_1 + x_2 = k_1$
$2x_1 + 3x_2 - x_3 = k_2$
$x_1 + 2x_3 = k_3$
(A) $k_1 = 2, k_2 = 0, k_3 = 4$
(B) $k_1 = 0, k_2 = 4, k_3 = -2$
(C) $k_1 = 4, k_2 = 2, k_3 = 0$

24. $x_1 - x_3 = k_1$
$2x_1 - x_2 = k_2$
$x_1 + x_2 - 2x_3 = k_3$
(A) $k_1 = 4, k_2 = 8, k_3 = 0$
(B) $k_1 = 4, k_2 = 0, k_3 = -4$
(C) $k_1 = 0, k_2 = 8, k_3 = -8$

In Problems 25–30, explain why the system cannot be solved by matrix inverse methods. Discuss methods that could be used and then solve the system.

25. $-2x_1 + 4x_2 = -5$
$6x_1 - 12x_2 = 15$

26. $-2x_1 + 4x_2 = 5$
$6x_1 - 12x_2 = 15$

27. $x_1 - 3x_2 - 2x_3 = -1$
$-2x_1 + 6x_2 + 4x_3 = 3$

28. $x_1 - 3x_2 - 2x_3 = -1$
$-2x_1 + 7x_2 + 3x_3 = 3$

29. $x_1 - 2x_2 + 3x_3 = 1$
$2x_1 - 3x_2 - 2x_3 = 3$
$x_1 - x_2 - 5x_3 = 2$

30. $x_1 - 2x_2 + 3x_3 = 1$
$2x_1 - 3x_2 - 2x_3 = 3$
$x_1 - x_2 - 5x_3 = 4$

C *For $n \times n$ matrices A and B, and $n \times 1$ column matrices C, D, and X, solve each matrix equation in Problems 31–36 for X. Assume all necessary inverses exist.*

31. $AX - BX = C$

32. $AX + BX = C$

33. $AX + X = C$

34. $AX - X = C$

35. $AX - C = D - BX$

36. $AX + C = BX + D$

37. Use matrix inverse methods to solve the following system for the indicated values of k_1 and k_2.

$$x_1 + 2.001x_2 = k_1$$
$$x_1 + 2x_2 = k_2$$

(A) $k_1 = 1, k_2 = 1$
(B) $k_1 = 1, k_2 = 0$
(C) $k_1 = 0, k_2 = 1$

Discuss the effect of small changes in the constant terms on the solution set of this system.

38. Repeat Problem 37 for the following system:

$$x_1 - 3.001x_2 = k_1$$
$$x_1 - 3x_2 = k_2$$

In Problems 39–42, write each system as a matrix equation and solve by using the inverse coefficient matrix. Use a graphing utility to perform the necessary calculations.

39. $$\begin{aligned} x_1 + 8x_2 + 7x_3 &= 135 \\ 6x_1 + 6x_2 + 8x_3 &= 155 \\ 3x_1 + 4x_2 + 6x_3 &= 75 \end{aligned}$$

40. $$\begin{aligned} 5x_1 + 3x_2 - 2x_3 &= 112 \\ 7x_1 + 5x_2 &= 70 \\ 3x_1 + x_2 - 9x_3 &= 96 \end{aligned}$$

41. $$\begin{aligned} 6x_1 + 9x_2 + 7x_3 + 5x_4 &= 250 \\ 6x_1 + 4x_2 + 7x_3 + 3x_4 &= 195 \\ 4x_1 + 5x_2 + 3x_3 + 2x_4 &= 145 \\ 4x_1 + 3x_2 + 8x_3 + 2x_4 &= 125 \end{aligned}$$

42. $$\begin{aligned} 3x_1 + 3x_2 + 6x_3 + 5x_4 &= 10 \\ 4x_1 + 5x_2 + 8x_3 + 2x_4 &= 15 \\ 3x_1 + 6x_2 + 7x_3 + 4x_4 &= 30 \\ 4x_1 + x_2 + 6x_3 + 3x_4 &= 25 \end{aligned}$$

APPLICATIONS

Construct a mathematical model for each of the following problems. (The answers in the back of the book include both the mathematical model and the interpretation of its solution.) Use matrix inverse methods to solve the model and then interpret the solution.

Business & Economics

43. *Resource allocation.* A concert hall has 10,000 seats and two categories of ticket prices, \$4 and \$8. Assume all seats in each category can be sold.

	CONCERT		
	1	*2*	*3*
TICKETS SOLD	10,000	10,000	10,000
RETURN REQUIRED	\$56,000	\$60,000	\$68,000

(A) How many tickets of each category should be sold to bring in each of the returns indicated in the table?
(B) Is it possible to bring in a return of \$90,000? Of \$30,000? Explain.
(C) Describe all the possible returns.

44. *Parking receipts.* Parking fees at a municipal zoo are \$5.00 for local residents and \$7.50 for all others. At the end of each day, the total number of vehicles parked that day and the gross receipts for the day are recorded, but the number of vehicles in each category is not. The following table contains the relevant information for a recent 4 day period:

	DAY			
	1	*2*	*3*	*4*
VEHICLES PARKED	1,200	1,550	1,740	1,400
GROSS RECEIPTS	\$7,125	\$9,825	\$11,100	\$8,650

(A) How many vehicles in each category used the zoo's parking facilities each day?
(B) If 1,200 vehicles are parked in one day, is it possible to take in gross receipts of \$5,000? Of \$10,000? Explain.
(C) Describe all possible gross receipts on a day when 1,200 vehicles are parked.

45. *Production scheduling.* A supplier for the automobile industry manufactures car and truck frames at two different plants. The production rates (in frames per hour) for each plant are given in the table:

PLANT	CAR FRAMES	TRUCK FRAMES
A	10	5
B	8	8

How many hours should each plant be scheduled to operate to exactly fill each of the orders in the following table?

	ORDERS		
	1	*2*	*3*
CAR FRAMES	3,000	2,800	2,600
TRUCK FRAMES	1,600	2,000	2,200

46. *Production scheduling.* Labor and material costs for manufacturing two guitar models are given in the table:

GUITAR MODEL	LABOR COST	MATERIAL COST
A	\$30	\$20
B	\$40	\$30

(A) If a total of $3,000 a week is allowed for labor and material, how many of each model should be produced each week to use exactly each of the allocations of the $3,000 indicated in the following table?

	Weekly Allocation		
	1	*2*	*3*
Labor	$1,800	$1,750	$1,720
Material	$1,200	$1,250	$1,280

(B) Is it possible to use an allocation of $1,600 for labor and $1,400 for material? Of $2,000 for labor and $1,000 for material? Explain.

47. *Incentive plan.* A small company provides an incentive plan for its top executives. Each executive receives as a bonus a percentage of the portion of the annual profit that remains after the bonuses for the other executives have been deducted (see the table). If the company has an annual profit of $2 million, find the bonus for each executive. Round each bonus to the nearest hundred dollars.

Officer	Bonus
President	3%
Executive vice-president	2.5%
Associate vice-president	2%
Assistant vice-president	1.5%

48. *Incentive plan.* Repeat Problem 47 if the company decides to include a 1% bonus for the sales manager in the incentive plan.

Life Sciences

49. *Diets.* A biologist has available two commercial food mixes containing the percentages of protein and fat given in the table:

Mix	Protein (%)	Fat (%)
A	20	2
B	10	6

(A) How many ounces of each mix should be used to prepare each of the diets listed in the following table?

	Diet		
	1	*2*	*3*
Protein	20 oz	10 oz	10 oz
Fat	6 oz	4 oz	6 oz

(B) Is it possible to prepare a diet consisting of 20 ounces of protein and 14 ounces of fat? Of 20 ounces of protein and 1 ounce of fat? Explain.

Social Sciences

50. *Education—resource allocation.* A state university system is planning to hire new faculty at the rank of lecturer or instructor for several of its 2 year community colleges. The number of sections taught and the annual salary (in thousands of dollars) for each rank are given in the table:

	Rank	
	Lecturer	*Instructor*
Sections taught	3	4
Annual salary (thousand $)	20	25

The number of sections that must be taught by the new faculty and the amount budgeted for salaries (in thousands of dollars) at each of the colleges are given in the following table. How many faculty of each rank should be hired at each college to exactly meet the demand for sections and completely exhaust the salary budget?

	Community College		
	1	*2*	*3*
Demand for sections	30	33	35
Salary budget (thousand $)	200	210	220

SECTION 4-7 Leontief Input–Output Analysis

- Two-Industry Model
- Three-Industry Model

A very important application of matrices and their inverses is found in the branch of applied mathematics called **input–output analysis.** Wassily Leontief, the primary force behind these new developments, was awarded the Nobel prize in economics in 1973 because of the significant impact his work had on economic planning for industrialized countries. Among other things, he conducted a comprehensive study of how 500 sectors of the American economy interacted with each other. Of course, large-scale computers played an important role in this analysis.

Our investigation will be more modest. In fact, we will start with an economy comprised of only two industries. From these humble beginnings, ideas and definitions will evolve that can be readily generalized for more realistic economies. Input–output analysis attempts to establish equilibrium conditions under which industries in an economy have just enough output to satisfy each other's demands in addition to final (outside) demands. Given the internal demands within the industries for each other's output, the problem is to determine output levels that will meet various levels of final (outside) demands.

Two-Industry Model

To make the problem concrete, let us start with a hypothetical economy with only two industries—electric company E and water company W. Output for both companies is measured in dollars. The electric company uses both electricity and water (input) in the production of electricity (output), and the water company uses both electricity and water (input) in the production of water (output). Suppose that the production of each dollar's worth of electricity requires \$0.30 worth of electricity and \$0.10 worth of water, and the production of each dollar's worth of water requires \$0.20 worth of electricity and \$0.40 worth of water. If the final demand from the outside sector of the economy (the demand from all other users of electricity and water) is

$$d_1 = \$12 \text{ million for electricity}$$
$$d_2 = \$8 \text{ million for water}$$

how much electricity and water should be produced to meet this final demand?

To begin, suppose the electric company produces \$12 million worth of electricity and the water company produces \$8 million worth of water (the final demand). Then the production processes of the companies would require

Electricity required to produce electricity		Electricity required to produce water	
$0.3(12)$	$+$	$0.2(8)$	$= \$5.2$ million of electricity

and

Water required to produce electricity		Water required to produce water	
$0.1(12)$	$+$	$0.4(8)$	$= \$4.4$ million of water

leaving only \$6.8 million of electricity and \$3.6 million of water to satisfy the final demand of the outside sector. Thus, in order to meet the internal demands of both companies and to end up with enough electricity for the final outside demand, both companies must produce more than just the amount demanded by the outside sector. In fact, they must produce exactly enough to meet their own internal demands plus that demanded by the outside sector.

Explore–Discuss 1

Suppose the electric company doubles its production to \$24 million of electricity and the water company doubles its production to \$16 million of water. How much electricity and water is consumed in the production process? How much is left for the final demand of the outside sector? Try other values to see if you can determine by trial and error the production levels that will result in a final demand of \$12 million for electricity and \$8 million for water.

We now state the main problem of input–output analysis.

Basic Input–Output Problem

Given the internal demands for each industry's output, determine output levels for the various industries that will meet a given final (outside) level of demand as well as the internal demand.

If

x_1 = Total output from electric company
x_2 = Total output from water company

then reasoning as above, the internal demands are

$0.3x_1 + 0.2x_2$ Internal demand for electricity
$0.1x_1 + 0.4x_2$ Internal demand for water

Combining the internal demand with the final demand produces the following system of equations:

Total output		Internal demand		Final demand
x_1	$=$	$0.3x_1 + 0.2x_2$	$+$	d_1
x_2	$=$	$0.1x_1 + 0.4x_2$	$+$	d_2

$$\begin{aligned} x_1 &= 0.3x_1 + 0.2x_2 + d_1 \\ x_2 &= 0.1x_1 + 0.4x_2 + d_2 \end{aligned} \qquad (1)$$

or, in matrix form,

$$\begin{bmatrix} x_1 \\ x_2 \end{bmatrix} = \begin{bmatrix} 0.3 & 0.2 \\ 0.1 & 0.4 \end{bmatrix} \begin{bmatrix} x_1 \\ x_2 \end{bmatrix} + \begin{bmatrix} d_1 \\ d_2 \end{bmatrix}$$

or

$$X = MX + D \tag{2}$$

where

$$D = \begin{bmatrix} d_1 \\ d_2 \end{bmatrix} \quad \text{Final demand matrix}$$

$$X = \begin{bmatrix} x_1 \\ x_2 \end{bmatrix} \quad \text{Output matrix}$$

$$M = \begin{matrix} & \begin{matrix} E & W \end{matrix} \\ \begin{matrix} E \\ W \end{matrix} & \begin{bmatrix} 0.3 & 0.2 \\ 0.1 & 0.4 \end{bmatrix} \end{matrix} \quad \text{Technology matrix}$$

The **technology matrix** is the heart of input–output analysis. The elements in the technology matrix are determined as follows (read from left to right and then up):

$$\text{Input} \quad \begin{matrix} & \text{Output} \\ & \begin{matrix} E & & W \end{matrix} \\ \begin{matrix} E \to \\ \\ W \end{matrix} & \begin{bmatrix} \begin{pmatrix} \text{Input from } E \\ \text{to produce \$1} \\ \text{of electricity} \end{pmatrix} & \begin{pmatrix} \text{Input from } E \\ \text{to produce \$1} \\ \text{of water} \end{pmatrix} \\ \begin{pmatrix} \text{Input from } W \\ \text{to produce \$1} \\ \text{of electricity} \end{pmatrix} & \begin{pmatrix} \text{Input from } W \\ \text{to produce \$1} \\ \text{of water} \end{pmatrix} \end{bmatrix} = M \end{matrix}$$

Now our problem is to solve equation (2) for X. We proceed as in Section 4-6:

$$\begin{aligned} X &= MX + D \\ X - MX &= D \\ IX - MX &= D \qquad I = \begin{bmatrix} 1 & 0 \\ 0 & 1 \end{bmatrix} \\ (I - M)X &= D \\ X &= (I - M)^{-1}D \qquad \text{Assuming } I - M \text{ has an inverse} \end{aligned} \tag{3}$$

We find

$$I - M = \begin{bmatrix} 0.7 & -0.2 \\ -0.1 & 0.6 \end{bmatrix} \quad \text{and} \quad (I - M)^{-1} = \begin{bmatrix} 1.5 & 0.5 \\ 0.25 & 1.75 \end{bmatrix}$$

Then we have

$$\begin{aligned} \begin{bmatrix} x_1 \\ x_2 \end{bmatrix} &= \begin{bmatrix} 1.5 & 0.5 \\ 0.25 & 1.75 \end{bmatrix} \begin{bmatrix} d_1 \\ d_2 \end{bmatrix} \\ &= \begin{bmatrix} 1.5 & 0.5 \\ 0.25 & 1.75 \end{bmatrix} \begin{bmatrix} 12 \\ 8 \end{bmatrix} = \begin{bmatrix} 22 \\ 17 \end{bmatrix} \end{aligned} \tag{4}$$

Therefore, the electric company must have an output of \$22 million and the water company must have an output of \$17 million so that each company can meet both internal and final demands.

CHECK We use equation (2) to check our work:

$$X = MX + D$$

$$\begin{bmatrix} 22 \\ 17 \end{bmatrix} \stackrel{?}{=} \begin{bmatrix} 0.3 & 0.2 \\ 0.1 & 0.4 \end{bmatrix} \begin{bmatrix} 22 \\ 17 \end{bmatrix} + \begin{bmatrix} 12 \\ 8 \end{bmatrix}$$

$$\begin{bmatrix} 22 \\ 17 \end{bmatrix} \stackrel{?}{=} \begin{bmatrix} 10 \\ 9 \end{bmatrix} + \begin{bmatrix} 12 \\ 8 \end{bmatrix}$$

$$\begin{bmatrix} 22 \\ 17 \end{bmatrix} \stackrel{\checkmark}{=} \begin{bmatrix} 22 \\ 17 \end{bmatrix}$$

Actually, (4) solves the original problem for arbitrary final demands d_1 and d_2. This is very useful, since (4) gives a quick solution not only for the final demands stated in the original problem, but also for various other projected final demands. If we had solved system (1) by Gauss–Jordan elimination, then we would have to start over for each new set of final demands.

Suppose in the original problem that the projected final demands 5 years from now are $d_1 = 24$ and $d_2 = 16$. To determine each company's output for this projection, we simply substitute these values into (4) and multiply:

$$\begin{bmatrix} x_1 \\ x_2 \end{bmatrix} = \begin{bmatrix} 1.5 & 0.5 \\ 0.25 & 1.75 \end{bmatrix} \begin{bmatrix} 24 \\ 16 \end{bmatrix} = \begin{bmatrix} 44 \\ 34 \end{bmatrix}$$

We summarize these results for convenient reference.

Solution to a Two-Industry Input–Output Problem

Given two industries, C_1 and C_2, with

$$\underset{\text{Technology matrix}}{M = \begin{matrix} C_1 \\ C_2 \end{matrix}\overset{\begin{matrix} C_1 & C_2 \end{matrix}}{\begin{bmatrix} a_{11} & a_{12} \\ a_{21} & a_{22} \end{bmatrix}}} \qquad \underset{\text{Output matrix}}{X = \begin{bmatrix} x_1 \\ x_2 \end{bmatrix}} \qquad \underset{\text{Final demand matrix}}{D = \begin{bmatrix} d_1 \\ d_2 \end{bmatrix}}$$

where a_{ij} is the input required from C_i to produce a dollar's worth of output for C_j, the solution to the input–output matrix equation

$$\underset{\text{Total output}}{X} = \underset{\text{Internal demand}}{MX} + \underset{\text{Final demand}}{D} \tag{2}$$

is

$$X = (I - M)^{-1}D \tag{3}$$

assuming $I - M$ has an inverse.

To solve an input–output problem on a graphing utility, simply store matrices M, D, and I in memory; then use equation (3) to find X and equation (2) to check your results. Figure 1 illustrates this process on a graphing calculator.

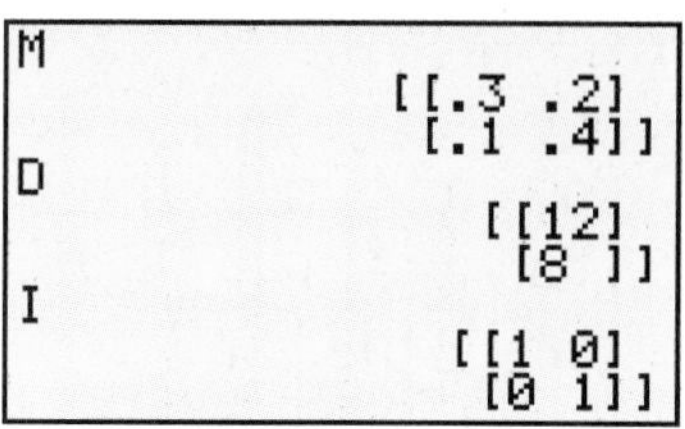

(A) Store M, D, and I in the graphing calculator's memory

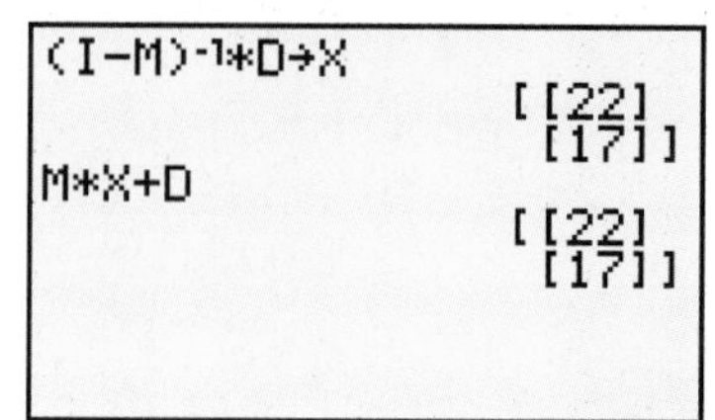

(B) Compute X and check in equation (2)

FIGURE 1

Three-Industry Model

Equations (2) and (3) in the solution to a two-industry input–output problem are the same for a three-industry economy, a four-industry economy, or an economy with n industries (where n is any natural number). The steps we took going from (2) to (3) hold for arbitrary matrices as long as the matrices have the correct sizes and $(I - M)^{-1}$ exists.

Explore–Discuss 2

If equations (2) and (3) are valid for an economy with n industries, discuss the size of all the matrices in each equation.

The next example illustrates the application of equations (2) and (3) to a three-industry economy.

Example 1 **Input–Output Analysis** An economy is based on three sectors, agriculture (A), energy (E), and manufacturing (M). Production of a dollar's worth of agriculture requires an input of \$0.20 from the agriculture sector and \$0.40 from the energy sector. Production of a dollar's worth of energy requires an input of \$0.20 from the energy sector and \$0.40 from the manufacturing sector. Production of a dollar's worth of manufacturing requires an input of \$0.10 from the agriculture sector, \$0.10 from the energy sector, and \$0.30 from the manufacturing sector. Find the output from each sector that is needed to satisfy a final demand of \$20 billion for agriculture, \$10 billion for energy, and \$30 billion for manufacturing.

Solution Since this is a three-industry problem, the technology matrix will be a 3×3 matrix, and the output and final demand matrices will be 3×1 column matrices. Using the information given in the problem, we can write

Technology matrix (columns A, E, M; rows A, E, M) — Final demand matrix — Output matrix

$$M = \begin{matrix} A \\ E \\ M \end{matrix}\begin{bmatrix} 0.2 & 0 & 0.1 \\ 0.4 & 0.2 & 0.1 \\ 0 & 0.4 & 0.3 \end{bmatrix} \qquad D = \begin{bmatrix} 20 \\ 10 \\ 30 \end{bmatrix} \qquad X = \begin{bmatrix} x_1 \\ x_2 \\ x_3 \end{bmatrix}$$

where M, X, and D satisfy the input–output equation $X = MX + D$. Since the solution to this equation is $X = (I - M)^{-1}D$, we must first find $I - M$ and then $(I - M)^{-1}$. Omitting the details of the calculations, we have

$$I - M = \begin{bmatrix} 0.8 & 0 & -0.1 \\ -0.4 & 0.8 & -0.1 \\ 0 & -0.4 & 0.7 \end{bmatrix}$$

and

$$(I - M)^{-1} = \begin{bmatrix} 1.3 & 0.1 & 0.2 \\ 0.7 & 1.4 & 0.3 \\ 0.4 & 0.8 & 1.6 \end{bmatrix}$$

Thus, the output matrix X is given by

$$\overset{X}{\begin{bmatrix} x_1 \\ x_2 \\ x_3 \end{bmatrix}} = \overset{(I-M)^{-1}}{\begin{bmatrix} 1.3 & 0.1 & 0.2 \\ 0.7 & 1.4 & 0.3 \\ 0.4 & 0.8 & 1.6 \end{bmatrix}} \overset{D}{\begin{bmatrix} 20 \\ 10 \\ 30 \end{bmatrix}} = \begin{bmatrix} 33 \\ 37 \\ 64 \end{bmatrix}$$

An output of \$33 billion for agriculture, \$37 billion for energy, and \$64 billion for manufacturing will meet the given final demands. You should check this result in equation (2).

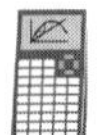

Figure 2 illustrates a spreadsheet solution for Example 1.

	A	B	C	D	E	F	G	H	I	J	K	L	M
1		Technology Matrix M								Final		Output	
2		A	E	M			I - M			Demand			
3	A	0.2	0	0.1		0.8	0	-0.1		20		33	
4	E	0.4	0.2	0.1		-0.4	0.8	-0.1		10		37	
5	M	0	0.4	0.3		0	-0.4	0.7		30		64	

FIGURE 2
The command MMULT(MINVERSE(F3:H5), J3:J5) produces the output in L3:L5

Matched Problem 1 An economy is based on three sectors, coal, oil, and transportation. Production of a dollar's worth of coal requires an input of \$0.20 from the coal sector and \$0.40 from the transportation sector. Production of a dollar's worth of oil requires an input of \$0.10 from the oil sector and \$0.20 from the transportation sector. Production of a dollar's worth of transportation requires an input of \$0.40 from the coal sector, \$0.20 from the oil sector, and \$0.20 from the transportation sector.

(A) Find the technology matrix M.
(B) Find $(I - M)^{-1}$.
(C) Find the output from each sector that is needed to satisfy a final demand of \$30 billion for coal, \$10 billion for oil, and \$20 billion for transportation.

Answers to Matched Problem

1. (A) $\begin{bmatrix} 0.2 & 0 & 0.4 \\ 0 & 0.1 & 0.2 \\ 0.4 & 0.2 & 0.2 \end{bmatrix}$ (B) $\begin{bmatrix} 1.7 & 0.2 & 0.9 \\ 0.2 & 1.2 & 0.4 \\ 0.9 & 0.4 & 1.8 \end{bmatrix}$

(C) \$71 billion for coal, \$26 billion for oil, and \$67 billion for transportation

EXERCISE 4-7

A *Problems 1–6 pertain to the following input–output model: Assume an economy is based on two industrial sectors, agriculture (A) and energy (E). The technology matrix M and final demand matrices (in billions of dollars) are*

$$\begin{matrix} & A & E \\ A & 0.4 & 0.2 \\ E & 0.2 & 0.1 \end{matrix} = M$$

$$D_1 = \begin{bmatrix} 6 \\ 4 \end{bmatrix} \qquad D_2 = \begin{bmatrix} 8 \\ 5 \end{bmatrix} \qquad D_3 = \begin{bmatrix} 12 \\ 9 \end{bmatrix}$$

1. How much input from A and E are required to produce a dollar's worth of output for A?

2. How much input from A and E are required to produce a dollar's worth of output for E?

3. Find $I - M$ and $(I - M)^{-1}$.

4. Find the output for each sector that is needed to satisfy the final demand D_1.

5. Repeat Problem 4 for D_2.

6. Repeat Problem 4 for D_3.

B *Problems 7–12 pertain to the following input–output model: Assume an economy is based on three industrial sectors, agriculture (A), building (B), and energy (E). The technology matrix M and final demand matrices (in billions of dollars) are*

$$\begin{matrix} & A & B & E \\ A & 0.3 & 0.2 & 0.2 \\ B & 0.1 & 0.1 & 0.1 \\ E & 0.2 & 0.1 & 0.1 \end{matrix} = M$$

$$D_1 = \begin{bmatrix} 5 \\ 10 \\ 15 \end{bmatrix} \qquad D_2 = \begin{bmatrix} 20 \\ 15 \\ 10 \end{bmatrix}$$

7. How much input from A, B, and E are required to produce a dollar's worth of output for B?

8. How much of each of B's output dollars is required as input for each of the three sectors?

9. Show that

$$I - M = \begin{bmatrix} 0.7 & -0.2 & -0.2 \\ -0.1 & 0.9 & -0.1 \\ -0.2 & -0.1 & 0.9 \end{bmatrix}$$

10. Given

$$(I - M)^{-1} = \begin{bmatrix} 1.6 & 0.4 & 0.4 \\ 0.22 & 1.18 & 0.18 \\ 0.38 & 0.22 & 1.22 \end{bmatrix}$$

show that $(I - M)^{-1}(I - M) = I$.

11. Use $(I - M)^{-1}$ in Problem 10 to find the output for each sector that is needed to satisfy the final demand D_1.

12. Repeat Problem 11 for D_2.

In Problems 13–16, find $(I - M)^{-1}$ and X.

13. $M = \begin{bmatrix} 0.2 & 0.2 \\ 0.3 & 0.3 \end{bmatrix}$; $D = \begin{bmatrix} 10 \\ 25 \end{bmatrix}$

14. $M = \begin{bmatrix} 0.4 & 0.1 \\ 0.2 & 0.3 \end{bmatrix}$; $D = \begin{bmatrix} 15 \\ 20 \end{bmatrix}$

C

15. $M = \begin{bmatrix} 0.3 & 0.1 & 0.3 \\ 0.2 & 0.1 & 0.2 \\ 0.1 & 0.1 & 0.1 \end{bmatrix}$; $D = \begin{bmatrix} 20 \\ 5 \\ 10 \end{bmatrix}$

16. $M = \begin{bmatrix} 0.3 & 0.2 & 0.3 \\ 0.1 & 0.1 & 0.1 \\ 0.1 & 0.2 & 0.1 \end{bmatrix}$; $D = \begin{bmatrix} 10 \\ 25 \\ 15 \end{bmatrix}$

17. The technology matrix for an economy based on agriculture (A) and manufacturing (M) is

$$M = \begin{matrix} & A & M \\ A & 0.3 & 0.25 \\ M & 0.1 & 0.25 \end{matrix}$$

(A) Find the output for each sector that is needed to satisfy a final demand of \$40 million for agriculture and \$40 million for manufacturing.

(B) Discuss the effect on the final demand if the agriculture output in part (A) is increased by \$20 million and manufacturing output remains unchanged.

18. The technology matrix for an economy based on energy (E) and transportation (T) is

$$M = \begin{array}{c} \\ E \\ T \end{array} \begin{array}{c} \begin{array}{cc} E & T \end{array} \\ \begin{bmatrix} 0.25 & 0.25 \\ 0.4 & 0.2 \end{bmatrix} \end{array}$$

(A) Find the output for each sector that is needed to satisfy a final demand of \$50 million for energy and \$50 million for transportation.

(B) Discuss the effect on the final demand if the transportation output in part (A) is increased by \$40 million and the energy output remains unchanged.

19. The technology matrix for an economy based on energy (E) and mining (M) is

$$M = \begin{array}{c} \\ E \\ M \end{array} \begin{array}{c} \begin{array}{cc} E & M \end{array} \\ \begin{bmatrix} 0.2 & 0.3 \\ 0.4 & 0.3 \end{bmatrix} \end{array}$$

The management of these two sectors would like to set the total output level so that the final demand is always 40% of the total output. Discuss methods that could be used to accomplish this objective.

20. The technology matrix for an economy based on automobiles (A) and construction (C) is

$$M = \begin{array}{c} \\ A \\ C \end{array} \begin{array}{c} \begin{array}{cc} A & C \end{array} \\ \begin{bmatrix} 0.1 & 0.4 \\ 0.1 & 0.1 \end{bmatrix} \end{array}$$

The management of these two sectors would like to set the total output level so that the final demand is always 70% of the total output. Discuss methods that could be used to accomplish this objective.

21. All the technology matrices in the text have elements between 0 and 1. Why is this the case? Would you ever expect to find an element in a technology matrix that is negative? That is equal to 0? That is equal to 1? That is greater than 1?

22. The sum of the elements in a column of any of the technology matrices in the text is less than 1. Why is this the case? Would you ever expect to find a column with a sum equal to 1? Greater than 1? How would you describe an economic system where the sum of the elements in every column of the technology matrix is 1?

APPLICATIONS

Business & Economics

23. An economy is based on two industrial sectors, coal and steel. Production of a dollar's worth of coal requires an input of \$0.10 from the coal sector and \$0.20 from the steel sector. Production of a dollar's worth of steel requires an input of \$0.20 from the coal sector and \$0.40 from the steel sector. Find the output for each sector that is needed to satisfy a final demand of \$20 billion for coal and \$10 billion for steel.

24. An economy is based on two sectors, transportation and manufacturing. Production of a dollar's worth of transportation requires \$0.10 of input from each sector and production of a dollar's worth of manufacturing requires an input of \$0.40 from each sector. Find the output for each sector that is needed to satisfy a final demand of \$5 billion for transportation and \$20 billion for manufacturing.

25. The economy of a small island nation is based on two sectors, agriculture and tourism. Production of a dollar's worth of agriculture requires an input of \$0.20 from agriculture and \$0.15 from tourism. Production of a dollar's worth of tourism requires an input of \$0.40 from agriculture and \$0.30 from tourism. Find the output from each sector that is needed to satisfy a final demand of \$60 million for agriculture and \$80 million for tourism.

26. The economy of a country is based on two sectors, agriculture and oil. Production of a dollar's worth of agriculture requires an input of \$0.40 from agriculture and \$0.35 from oil. Production of a dollar's worth of oil requires an input of \$0.20 from agriculture and \$0.05 from oil. Find the output from each sector that is needed to satisfy a final demand of \$40 million for agriculture and \$250 million for oil.

27. An economy is based on three sectors, agriculture, manufacturing, and energy. Production of a dollar's worth of agriculture requires inputs of \$0.20 from agriculture, \$0.20 from manufacturing, and \$0.20 from energy. Production of a dollar's worth of manufacturing requires inputs of \$0.40 from agriculture, \$0.10 from manufacturing, and \$0.10 from energy. Production of a dollar's worth of energy requires inputs of \$0.30 from agriculture, \$0.10 from manufacturing, and \$0.10 from energy. Find the output for each sector that is needed to satisfy a final demand of \$10 billion for agriculture, \$15 billion for manufacturing, and \$20 billion for energy.

28. A large energy company produces electricity, natural gas, and oil. The production of a dollar's worth of electricity requires inputs of \$0.30 from electricity, \$0.10 from natural gas, and \$0.20 from oil. Production of a dollar's worth of natural gas requires inputs of \$0.30 from electricity, \$0.10 from natural gas, and \$0.20 from oil. Production of a dollar's worth of oil requires inputs of \$0.10 from each sector. Find the output for each sector that is needed to satisfy a final demand of \$25 billion for electricity, \$15 billion for natural gas, and \$20 billion for oil.

Use a graphing utility to solve Problems 29 and 30.

29. An economy is based on four sectors, agriculture (A), energy (E), labor (L), and manufacturing (M). The table gives the input requirements for a dollar's worth of output for each sector, along with the projected final demand (in billions of dollars) for a 3 year period. Find the output for each sector that is needed to satisfy each of these final demands. Round answers to the nearest billion dollars.

		Output				Final Demand		
		A	*E*	*L*	*M*	*1*	*2*	*3*
Input	*A*	0.05	0.17	0.23	0.09	23	32	55
	E	0.07	0.12	0.15	0.19	41	48	62
	L	0.25	0.08	0.03	0.32	18	21	25
	M	0.11	0.19	0.28	0.16	31	33	35

30. Repeat Problem 29 with the following table:

		Output				Final Demand		
		A	*E*	*L*	*M*	*1*	*2*	*3*
Input	*A*	0.07	0.09	0.27	0.12	18	22	37
	E	0.14	0.07	0.21	0.24	26	31	42
	L	0.17	0.06	0.02	0.21	12	19	28
	M	0.15	0.13	0.31	0.19	41	45	49

Important Terms and Symbols

4-1 *Review: Systems of Linear Equations in Two Variables.* Linear system; solution of a system; solution set; graphing method; consistent; inconsistent; independent; unique solution; dependent; substitution method; elimination by addition; equivalent systems; parameter; particular solution; equilibrium price; equilibrium quantity; equilibrium point

4-2 *Systems of Linear Equations and Augmented Matrices.* Matrix; element; size; $m \times n$ matrix; dimensions; square matrix of order n; column matrix; row matrix; position of an element; double subscript notation, a_{ij}; principal diagonal; coefficient matrix; constant matrix; augmented matrix; equivalent systems; row-equivalent matrices; row operations; Gauss–Jordan elimination

$$R_i \leftrightarrow R_j; \quad kR_i \to R_i; \quad kR_j + R_i \to R_i$$

4-3 *Gauss–Jordan Elimination.* Reduced form; reduced system; Gauss–Jordan elimination; submatrix

4-4 *Matrices: Basic Operations.* Equal matrices; sum of two matrices; zero matrix; negative of a matrix; subtraction of matrices; product of a number and a matrix; product of a row matrix and a column matrix; product of two matrices; diagonal matrix; upper triangular matrix; markup; incidence matrix

4-5 *Inverse of a Square Matrix.* Identity element for multiplication; multiplicative inverse; singular matrix; encoding matrix; decoding matrix

4-6 *Matrix Equations and Systems of Linear Equations.* Matrix equation; basic properties of matrices

4-7 *Leontief Input–Output Analysis.* Input–output analysis; technology matrix; final demand matrix; output matrix

Review Exercise

Work through all the problems in this chapter review and check your answers in the back of the book. Answers to all review problems are there along with section numbers in italics to indicate where each type of problem is discussed. Where weaknesses show up, review appropriate sections in the text.

A

1. Solve the following system by graphing:

$$\begin{aligned} 2x - y &= 4 \\ x - 2y &= -4 \end{aligned}$$

2. Solve the system in Problem 1 by substitution.

3. If a matrix is in reduced form, say so. If not, explain why and state the row operation(s) necessary to transform the matrix into reduced form.

(A) $\left[\begin{array}{cc|c} 0 & 1 & 2 \\ 1 & 0 & 3 \end{array}\right]$ (B) $\left[\begin{array}{cc|c} 1 & 0 & 2 \\ 0 & 3 & 3 \end{array}\right]$

(C) $\left[\begin{array}{ccc|c} 1 & 0 & 1 & 2 \\ 0 & 1 & 1 & 3 \end{array}\right]$ (D) $\left[\begin{array}{ccc|c} 1 & 1 & 0 & 2 \\ 0 & 1 & 1 & 3 \end{array}\right]$

4. Given matrices A and B:

$$A = \begin{bmatrix} 5 & 3 & -1 & 0 & 2 \\ -4 & 8 & 1 & 3 & 0 \end{bmatrix} \qquad B = \begin{bmatrix} -3 & 2 \\ 0 & 4 \\ -1 & 7 \end{bmatrix}$$

(A) What is the size of A? Of B?
(B) Find a_{24}, a_{15}, b_{31}, and b_{22}.
(C) Is AB defined? Is BA defined?

5. Find x_1 and x_2:

$$\begin{bmatrix} 1 & -2 \\ 1 & -3 \end{bmatrix} \begin{bmatrix} x_1 \\ x_2 \end{bmatrix} = \begin{bmatrix} 4 \\ 2 \end{bmatrix}$$

In Problems 6–14, perform the operations that are defined, given the following matrices:

$$A = \begin{bmatrix} 1 & 2 \\ 3 & 1 \end{bmatrix} \qquad B = \begin{bmatrix} 2 & 1 \\ 1 & 1 \end{bmatrix}$$

$$C = \begin{bmatrix} 2 & 3 \end{bmatrix} \qquad D = \begin{bmatrix} 1 \\ 2 \end{bmatrix}$$

6. $A + B$ **7.** $B + D$ **8.** $A - 2B$
9. AB **10.** AC **11.** AD
12. DC **13.** CD **14.** $C + D$

15. Find the inverse of the matrix A given below by appropriate row operations on $[A|I]$. Show that $A^{-1}A = I$.

$$A = \begin{bmatrix} 4 & 3 \\ 3 & 2 \end{bmatrix}$$

16. Solve the following system using elimination by addition:

$$\begin{aligned} 4x_1 + 3x_2 &= 3 \\ 3x_1 + 2x_2 &= 5 \end{aligned}$$

17. Solve the system in Problem 16 by performing appropriate row operations on the augmented matrix of the system.

18. Solve the system in Problem 16 by writing the system as a matrix equation and using the inverse of the coefficient matrix (see Problem 15). Also, solve the system if the constants 3 and 5 are replaced by 7 and 10, respectively. By 4 and 2, respectively.

B *In Problems 19–24, perform the operations that are defined, given the following matrices:*

$$A = \begin{bmatrix} 2 & -2 \\ 1 & 0 \\ 3 & 2 \end{bmatrix} \qquad B = \begin{bmatrix} -1 \\ 2 \\ 3 \end{bmatrix} \qquad C = \begin{bmatrix} 2 & 1 & 3 \end{bmatrix}$$

$$D = \begin{bmatrix} 3 & -2 & 1 \\ -1 & 1 & 2 \end{bmatrix} \qquad E = \begin{bmatrix} 3 & -4 \\ -1 & 0 \end{bmatrix}$$

19. $A + D$ **20.** $E + DA$ **21.** $DA - 3E$
22. BC **23.** CB **24.** $AD - BC$

25. Find the inverse of the matrix A given below by appropriate row operations on $[A|I]$. Show that $A^{-1}A = I$.

$$A = \begin{bmatrix} 1 & 2 & 3 \\ 2 & 3 & 4 \\ 1 & 2 & 1 \end{bmatrix}$$

26. Solve by Gauss–Jordan elimination:

(A)
$$\begin{aligned} x_1 + 2x_2 + 3x_3 &= 1 \\ 2x_1 + 3x_2 + 4x_3 &= 3 \\ x_1 + 2x_2 + x_3 &= 3 \end{aligned}$$

(B)
$$\begin{aligned} x_1 + 2x_2 - x_3 &= 2 \\ 2x_1 + 3x_2 + x_3 &= -3 \\ 3x_1 + 5x_2 &= -1 \end{aligned}$$

27. Solve the system in Problem 26A by writing the system as a matrix equation and using the inverse of the coefficient matrix (see Problem 25). Also, solve the system if the constants 1, 3, and 3 are replaced by 0, 0, and -2, respectively. By -3, -4, and 1, respectively.

28. Discuss the relationship between the number of solutions of the following system and the constant k.

$$\begin{aligned} 2x_1 - 6x_2 &= 4 \\ -x_1 + kx_2 &= -2 \end{aligned}$$

29. Given the technology matrix M and the final demand matrix D (in billions of dollars), find $(I - M)^{-1}$ and the output matrix X:

$$M = \begin{bmatrix} 0.2 & 0.15 \\ 0.4 & 0.3 \end{bmatrix} \qquad D = \begin{bmatrix} 30 \\ 20 \end{bmatrix}$$

30. C Use zoom and trace or other graphical approximation techniques on a graphing utility to find the solution of the following system to two decimal places:

$$\begin{aligned} x - 5y &= -5 \\ 2x + 3y &= 12 \end{aligned}$$

C

31. Find the inverse of the matrix A given below. Show that $A^{-1}A = I$.

$$A = \begin{bmatrix} 4 & 5 & 6 \\ 4 & 5 & -4 \\ 1 & 1 & 1 \end{bmatrix}$$

32. Solve the system

$$\begin{aligned} 0.04x_1 + 0.05x_2 + 0.06x_3 &= 360 \\ 0.04x_1 + 0.05x_2 - 0.04x_3 &= 120 \\ x_1 + x_2 + x_3 &= 7{,}000 \end{aligned}$$

by writing it as a matrix equation and using the inverse of the coefficient matrix. (Before starting, multiply the first two equations by 100 to eliminate decimals. Also, see Problem 31.)

33. Solve Problem 32 by Gauss–Jordan elimination.

34. Given the technology matrix M and the final demand matrix D (in billions of dollars), find $(I - M)^{-1}$ and the output matrix X:

$$M = \begin{bmatrix} 0.2 & 0 & 0.4 \\ 0.1 & 0.3 & 0.1 \\ 0 & 0.4 & 0.2 \end{bmatrix} \qquad D = \begin{bmatrix} 40 \\ 20 \\ 30 \end{bmatrix}$$

35. Discuss the number of solutions for a system of n equations in n variables if the coefficient matrix:
(A) Has an inverse (B) Does not have an inverse

36. Discuss the number of solutions for the system corresponding to the reduced form shown below if:
(A) $m \neq 0$ (B) $m = 0$ and $n \neq 0$
(C) $m = 0$ and $n = 0$

$$\left[\begin{array}{ccc|c} 1 & 0 & -2 & 5 \\ 0 & 1 & 3 & 3 \\ 0 & 0 & m & n \end{array}\right]$$

37. One solution to the input–output equation $X = MX + D$ is given by $X = (I - M)^{-1}D$. Discuss the validity of each step in the following solutions of this equation. (Assume all necessary inverses exist.) Are both solutions correct?

(A)
$$\begin{aligned} X &= MX + D \\ X - MX &= D \\ X(I - M) &= D \\ X &= D(I - M)^{-1} \end{aligned}$$

(B)
$$\begin{aligned} X &= MX + D \\ -D &= MX - X \\ -D &= (M - I)X \\ X &= (M - I)^{-1}(-D) \end{aligned}$$

Applications

Business & Economics

38. *Break-even analysis.* A cookware manufacturer is preparing to market a new pasta machine. The company's fixed costs for research, development, tooling, etc., are $243,000 and the variable costs are $22.45 per machine. The company sells the pasta machine for $59.95.
(A) Find the cost and revenue equations.
(B) Find the break-even point.
(C) Graph both equations in the same coordinate system and show the break-even point. Use the graph to determine the production levels that will result in a profit and in a loss.

39. *Resource allocation.* A mining company has two mines with ore composition as given in the table. How many tons of each ore should be used to obtain exactly 4.5 tons of nickel and 10 tons of copper? Solve using Gauss–Jordan elimination.

Ore	Nickel (%)	Copper (%)
A	1	2
B	2	5

40. *Resource allocation.*
(A) Set up Problem 39 as a matrix equation and solve using the inverse of the coefficient matrix.
(B) Solve Problem 39 as in part (A) if 2.3 tons of nickel and 5 tons of copper are needed.

41. *Purchasing.* A soft-drink distributor has budgeted \$300,000 for the purchase of 12 new delivery trucks. If a model A truck costs \$18,000, a model B truck costs \$22,000, and a model C truck costs \$30,000, how many trucks of each model should the distributor purchase to use exactly all the budgeted funds?

42. *Material costs.* A manufacturer wishes to make two different bronze alloys in a metal foundry. The quantities of copper, tin, and zinc needed are indicated in matrix M. The costs for these materials (in dollars per pound) from two suppliers are summarized in matrix N. The company must choose one supplier or the other.

$$M = \begin{array}{c} \begin{array}{ccc} \text{Copper} & \text{Tin} & \text{Zinc} \end{array} \\ \begin{bmatrix} 4{,}800 \text{ lb} & 600 \text{ lb} & 300 \text{ lb} \\ 6{,}000 \text{ lb} & 1{,}400 \text{ lb} & 700 \text{ lb} \end{bmatrix} \begin{array}{l} \text{Alloy 1} \\ \text{Alloy 2} \end{array} \end{array}$$

$$N = \begin{array}{c} \begin{array}{cc} \text{Supplier A} & \text{Supplier B} \end{array} \\ \begin{bmatrix} \$0.75 & \$0.70 \\ \$6.50 & \$6.70 \\ \$0.40 & \$0.50 \end{bmatrix} \begin{array}{l} \text{Copper} \\ \text{Tin} \\ \text{Zinc} \end{array} \end{array}$$

(A) Discuss possible interpretations of the elements in the matrix products MN and NM.
(B) If either of the products MN or NM has a meaningful interpretation, find the product and label its rows and columns.
(C) Discuss methods of matrix multiplication that can be used to determine the supplier that will provide the necessary materials at the lowest cost.

43. *Labor costs.* A company with manufacturing plants in California and Texas has labor-hour and wage requirements for the manufacture of two inexpensive calculators as given in matrices M and N below:

Labor-hours per calculator

$$M = \begin{array}{c} \begin{array}{ccc} \text{Fabricating department} & \text{Assembly department} & \text{Packaging department} \end{array} \\ \begin{bmatrix} 0.15 \text{ hr} & 0.10 \text{ hr} & 0.05 \text{ hr} \\ 0.25 \text{ hr} & 0.20 \text{ hr} & 0.05 \text{ hr} \end{bmatrix} \begin{array}{l} \text{Model A} \\ \text{Model B} \end{array} \end{array}$$

Hourly wages

$$N = \begin{array}{c} \begin{array}{cc} \text{California plant} & \text{Texas plant} \end{array} \\ \begin{bmatrix} \$15 & \$12 \\ \$12 & \$10 \\ \$\ 4 & \$\ 4 \end{bmatrix} \begin{array}{l} \text{Fabricating department} \\ \text{Assembly department} \\ \text{Packaging department} \end{array} \end{array}$$

(A) Find the labor cost for producing one model B calculator at the California plant.
(B) Discuss possible interpretations of the elements in the matrix products MN and NM.
(C) If either of the products MN or NM has a meaningful interpretation, find the product and label its rows and columns.

44. *Investment analysis.* A person has \$5,000 to invest, part at 5% and the rest at 10%. How much should be invested at each rate to yield \$400 per year? Solve using augmented matrix methods.

45. *Investment analysis.* Solve Problem 44 by using a matrix equation and the inverse of the coefficient matrix.

46. *Investment analysis.* In Problem 44, is it possible to have an annual yield of \$200? Of \$600? Describe all possible annual yields.

47. *Resource allocation.* An outdoor amphitheater has 25,000 seats. Ticket prices are \$8, \$12, and \$20, and the number of tickets priced at \$8 must equal the number priced at \$20. How many tickets of each type should be sold (assuming all seats can be sold) to bring in each of the returns indicated in the table? Solve using the inverse of the coefficient matrix.

	CONCERT		
	1	2	3
TICKETS SOLD	25,000	25,000	25,000
RETURN REQUIRED	\$320,000	\$330,000	\$340,000

48. *Resource allocation.* Discuss the effect on the solutions to Problem 47 if it is no longer required to have an equal number of \$8 tickets and \$20 tickets.

49. *Input–output analysis.* An economy is based on two industrial sectors, agriculture and fabrication. Production of a dollar's worth of agriculture requires an input of \$0.30 from the agriculture sector and \$0.20 from the fabricating sector. Production of a dollar's worth of fabrication requires \$0.10 from the agriculture sector and \$0.40 from the fabrication sector.
(A) Find the output for each sector that is needed to satisfy a final demand of \$50 billion for agriculture and \$20 billion for fabrication.
(B) Find the output for each sector that is needed to satisfy a final demand of \$80 billion for agriculture and \$60 billion for fabrication.

Social Sciences

50. *Cryptography.* The following message was encoded with the matrix B shown below. Decode the message:

25 8 26 24 25 33 21 41 48 41 30 50 21 32 41 52 52 79

$$B = \begin{bmatrix} 1 & 1 & 0 \\ 1 & 0 & 1 \\ 1 & 1 & 1 \end{bmatrix}$$

51. *Traffic flow.* The rush-hour traffic flow (in vehicles per hour) for a network of four one-way streets is shown in the figure.

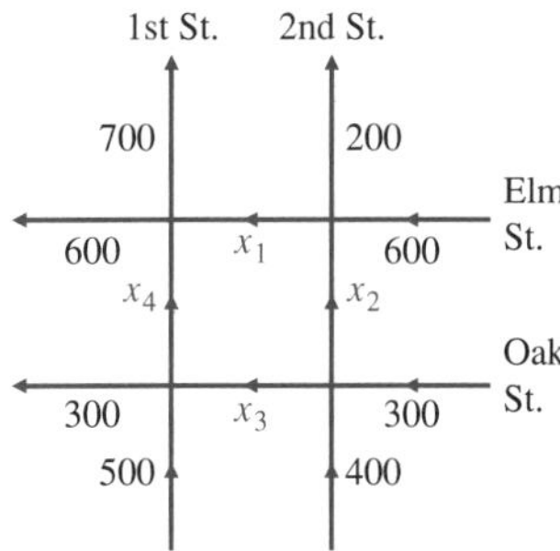

Figure for 51

(A) Write the system of equations determined by the flow of traffic through the four intersections.
(B) Find the solution of the system in part (A).
(C) What is the maximum number of vehicles per hour that can travel from Oak Street to Elm Street on 1st Street? What is the minimum number?
(D) If traffic lights are adjusted so that 500 vehicles per hour travel from Oak Street to Elm Street on 1st Street, determine the flow around the rest of the network.

52. *Dominance relation.* In a tournament between four football teams, each team plays one game with every other team with the following results: Team *A* defeats team *B*, team *B* defeats teams *C* and *D*, team *C* defeats teams *A* and *D*, and team *D* defeats team *A*.
(A) Represent the results of the tournament as an incidence matrix *M*.
(B) Compute $M + M^2$ and use this matrix to rank the four teams from high to low. Explain the reasoning behind your ranking.

Group Activity 1 *Using Matrices to Find Cost, Revenue, and Profit*

A toy distributor purchases model train components from various suppliers and packages these components in three different ready-to-run train sets, the Limited, the Empire, and the Comet. The components used in each set are listed in Table 1. For convenience, the total labor time required to prepare a set for shipping is included as a component.

Table 1
PRODUCT COMPONENTS

	TRAIN SET		
COMPONENT	*Limited*	*Empire*	*Comet*
Locomotive	1	1	2
Car	5	6	8
Track piece	20	24	32
Track switch	1	2	4
Power pack	1	1	1
Labor (minutes)	15	18	24

The current costs of the components are given in Table 2, and the distributor's selling prices for the sets are given in Table 3.

Table 2
COMPONENT COSTS

COMPONENT	COST PER UNIT ($)
Locomotive	12.52
Car	1.43
Track piece	0.25
Track switch	2.29
Power pack	12.54
Labor (per minute)	0.15

Table 3
SELLING PRICES

SET	PRICE ($)
Limited	54.60
Empire	62.28
Comet	81.15

Table 4

CUSTOMER ORDER

SET	QUANTITY
Limited	48
Empire	24
Comet	12

The distributor has just received the order shown in Table 4 from a retail toy store.

The distributor wants to store the information in each table in a matrix and use matrix operations to find the following information:

1. The inventory (parts and labor) required to fill the order
2. The cost (parts and labor) of filling the order
3. The revenue (sales) received from the customer
4. The profit realized on the order

(A) Use a single letter to designate the matrix representing each table and write matrix expressions in terms of these letters that will provide the required information. Discuss the size of the matrix you must use to represent each table so that all the pertinent matrix operations are defined.

(B) Evaluate the matrix expressions in part (A).

Shortly after filling the order in Table 4, a supplier informs the distributor that the cars and locomotives used in these train sets are no longer available. The distributor currently has 30 locomotives and 134 cars in stock.

(C) How many train sets of each type can the distributor produce using all the available locomotives and cars? Assume that the distributor has unlimited quantities of the other components used in these sets.

(D) How much profit will the distributor make if all these sets are sold? If there is more than one way to use all the available locomotives and cars, which one will produce the largest profit?

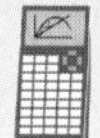

Group Activity 2 Direct and Indirect Operating Costs

An electronics firm has three production departments that produce three different finished products: copiers, printers, and fax machines. The company also has four other departments, payroll, advertising, research, and maintenance, that provide services for the production departments. Some of the service departments also provide services for the other service departments and themselves. The monthly direct cost of operating each department is given in Table 1, along with the percentage of service time each department receives from each service department (including themselves). The total cost of operating each department consists of its direct costs plus the cost of services received, which are determined by the percentages in Table 1.

Table 1

		PERCENTAGE OF SERVICE TIME RECEIVED FROM:			
DEPARTMENT	DIRECT COSTS (MONTHLY, $)	*Payroll*	*Advertising*	*Research*	*Maintenance*
Payroll	100,000	15	0	0	10
Advertising	150,000	15	0	0	10
Research	120,000	15	0	25	20
Maintenance	60,000	10	0	0	0
Copiers	270,000	15	20	40	20
Printers	140,000	15	20	20	20
Fax machines	190,000	15	60	15	20

Let x_1, x_2, x_3, and x_4 denote the total cost of operating the payroll, advertising, research, and maintenance departments, respectively, and y_1, y_2, and y_3 denote the total cost of operating the copier, printer, and fax machine departments, respectively.

(A) The total costs for the payroll department, x_1, must satisfy

$$x_1 = 100{,}000 + 0.15x_1 + 0.1x_4$$

Write similar equations for the other three service departments.

(B) Express the system in part (A) as a matrix equation of the form $X = D + AX$.

(C) The total costs for the copier department, y_1, must satisfy

$$y_1 = 270{,}000 + 0.15x_1 + 0.2x_2 + 0.4x_3 + 0.2x_4$$

Write similar equations for the other two production departments.

(D) Express the system in part (C) as a matrix equation of the form $Y = C + BX$.

(E) Use matrix inverse methods to solve the equation in part (B) and use this solution to solve the equation in part (D).

(F) Compare the sum of the direct costs of all seven departments as listed in Table 1 with the sum of the total costs of the three production departments found in part (E), and interpret.

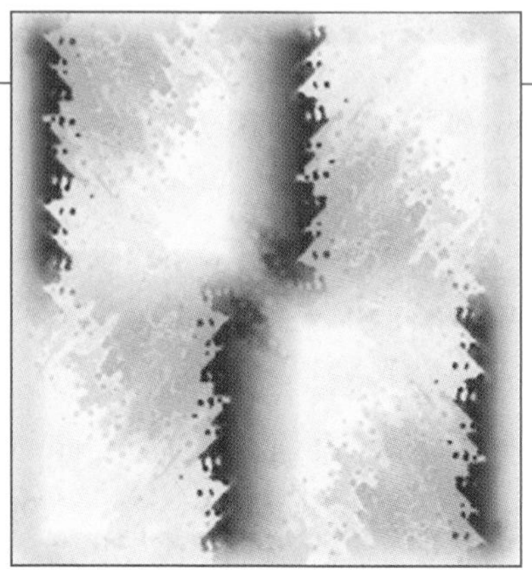

Chapter 5

Linear Inequalities and Linear Programming

INTRODUCTION

In this chapter we will discuss linear inequalities in two and more variables. In addition, we will introduce a relatively new and powerful mathematical tool called *linear programming,* which will be used to solve a variety of interesting practical problems. The row operations on matrices introduced in Chapter 4 will be particularly useful in Sections 5-4, 5-5, and 5-6.

Section 5-1 Systems of Linear Inequalities in Two Variables

- Graphing Linear Inequalities in Two Variables
- Solving Systems of Linear Inequalities Graphically
- Applications

Many applications of mathematics involve systems of inequalities rather than systems of equations. A graph is often the most convenient way to represent the solutions of a system of linear inequalities in two variables. In this section we discuss techniques for graphing both a single linear inequality in two variables and a system of linear inequalities in two variables.

Graphing Linear Inequalities in Two Variables

We know how to graph first-degree equations such as

$$y = 2x - 3 \qquad \text{and} \qquad 2x - 3y = 5$$

but how do we graph first-degree inequalities such as the following?

$$y \leq 2x - 3 \qquad \text{and} \qquad 2x - 3y > 5$$

We will find that graphing these inequalities is almost as easy as graphing the equations, but first we must discuss some important subsets of a plane in a rectangular coordinate system.

A line divides the plane into two halves called **half-planes.** A vertical line divides it into **left** and **right half-planes;** a nonvertical line divides it into **upper** and **lower half-planes** (Fig. 1).

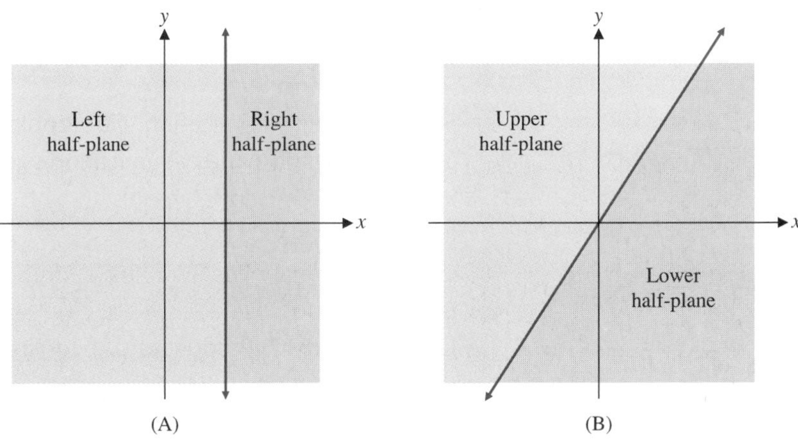

FIGURE 1

Explore–Discuss 1

Consider the following linear equation and related linear inequalities:

$$(1)\ 3x - 4y = 24 \qquad (2)\ 3x - 4y < 24 \qquad (3)\ 3x - 4y > 24$$

(A) Graph the line with equation (1).

(B) Find the point on this line with x coordinate 4 and draw a vertical line through this point. Discuss the relationship between the y coordinates of the points on this line and statements (1), (2), and (3).

(C) Repeat part (B) for $x = -4$. For $x = 12$.

(D) Based on your observations in parts (B) and (C), write a verbal description of all the points in the plane that satisfy equation (1), those that satisfy inequality (2), and those that satisfy inequality (3).

To investigate the half-planes determined by a linear equation such as $y - x = -2$, we rewrite the equation as $y = x - 2$. For any given value of x, there is exactly one value for y such that (x, y) lies on the line. For example, for $x = 4$, we have $y = 4 - 2 = 2$. For the same x and smaller values of y, the point (x, y) will lie below the line, since $y < x - 2$. Thus, the lower half-plane corresponds to the solution of the inequality $y < x - 2$. Similarly, the upper half-plane corresponds to $y > x - 2$, as shown in Figure 2.

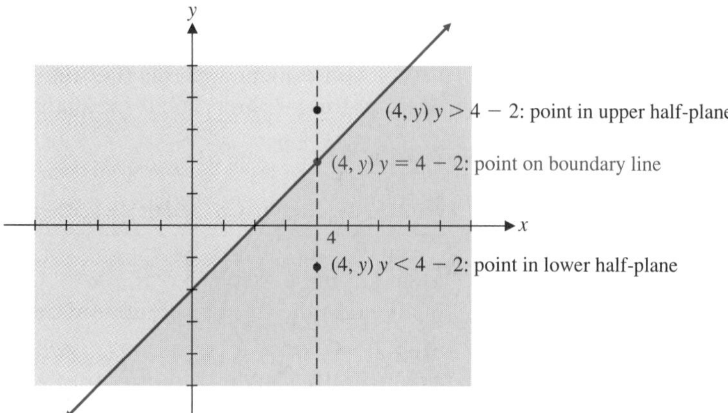

FIGURE 2

The four inequalities formed from $y = x - 2$ by replacing the $=$ sign by $>, \geq, <$, and $\leq$, respectively, are

$$y > x - 2 \qquad y \geq x - 2 \qquad y < x - 2 \qquad y \leq x - 2$$

The graph of each is a half-plane, excluding the boundary line for $<$ and $>$, and including the boundary line for $\leq$ and $\geq$. In Figure 3, the half-planes are indicated with small arrows on the graph of $y = x - 2$ and then graphed as shaded regions. Excluded boundary lines are shown as dashed lines, and included boundary lines are shown as solid lines.

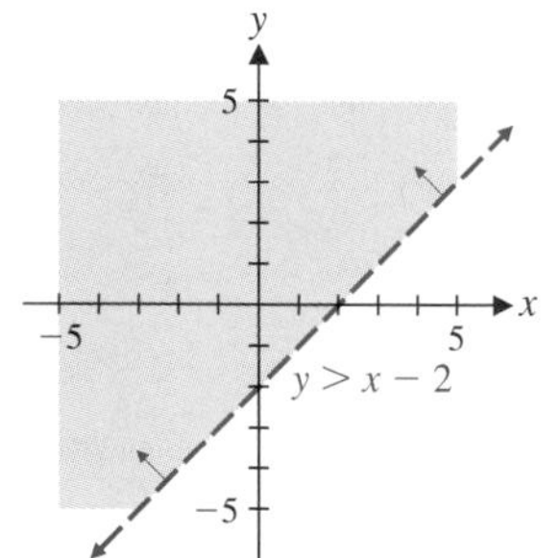

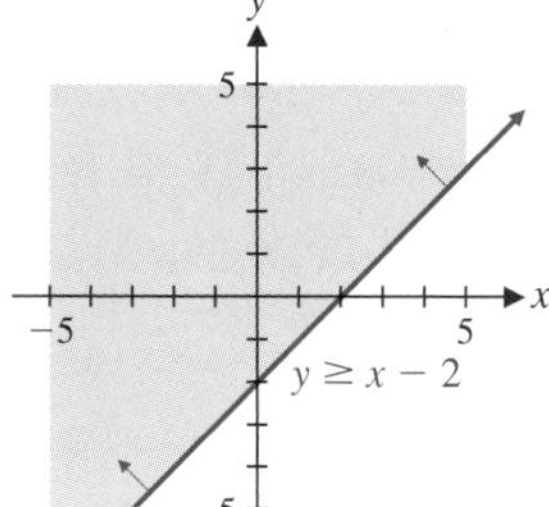

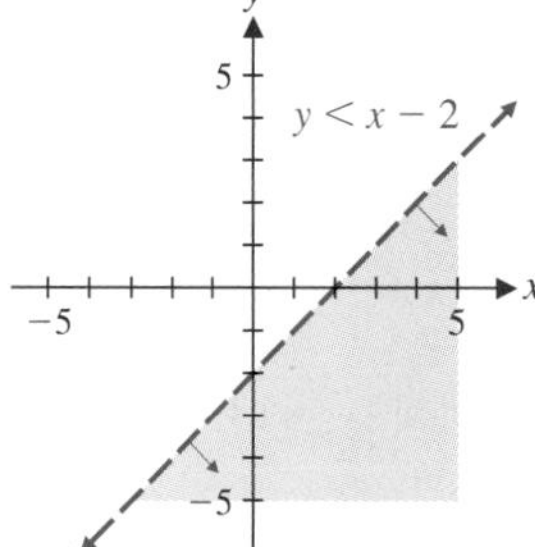

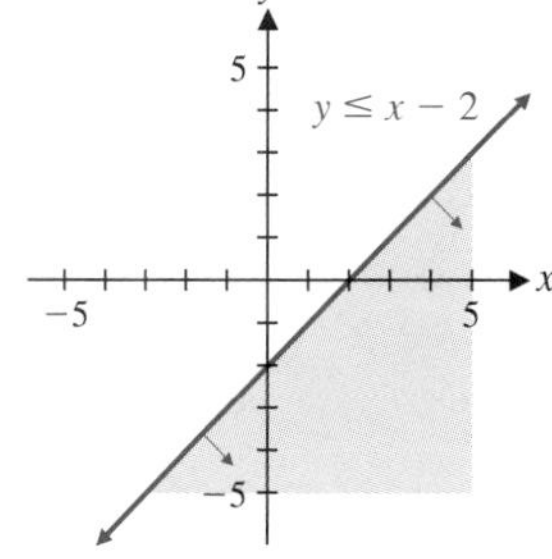

FIGURE 3

The preceding discussion suggests the following theorem, which is stated without proof:

Theorem 1 ■■ GRAPHS OF LINEAR INEQUALITIES

The graph of the linear inequality

$$Ax + By < C \qquad \text{or} \qquad Ax + By > C$$

with $B \neq 0$, is either the upper half-plane or the lower half-plane (but not both) determined by the line $Ax + By = C$.

(continued)

If $B = 0$ and $A \neq 0$, the graph of

$$Ax < C \qquad \text{or} \qquad Ax > C$$

is either the left half-plane or the right half-plane (but not both) determined by the line $Ax = C$. ■■

As a consequence of this theorem, we state a simple and fast mechanical procedure for graphing linear inequalities.

Procedure for Graphing Linear Inequalities

Step 1. First graph $Ax + By = C$ as a dashed line if equality is not included in the original statement or as a solid line if equality is included.

Step 2. Choose a test point anywhere in the plane not on the line [the origin (0, 0) usually requires the least computation] and substitute the coordinates into the inequality.

Step 3. The graph of the original inequality includes the half-plane containing the test point if the inequality is satisfied by that point or the half-plane not containing the test point if the inequality is not satisfied by that point.

Example 1 **Graphing a Linear Inequality** Graph $2x - 3y \leq 6$.

C Check on a graphing utility.

SOLUTION **Step 1.** Graph $2x - 3y = 6$ as a solid line, since equality is included in the original statement:

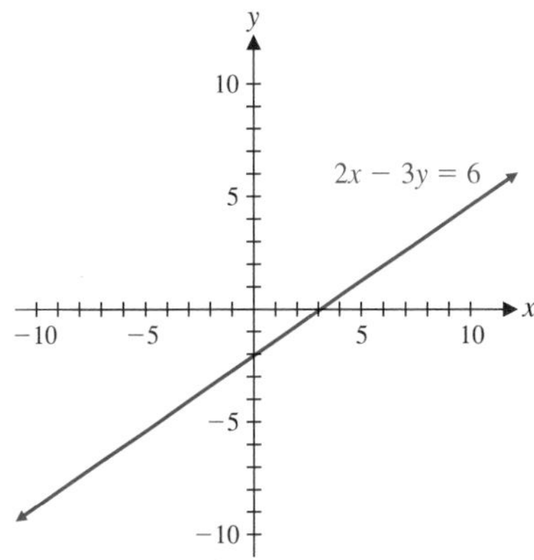

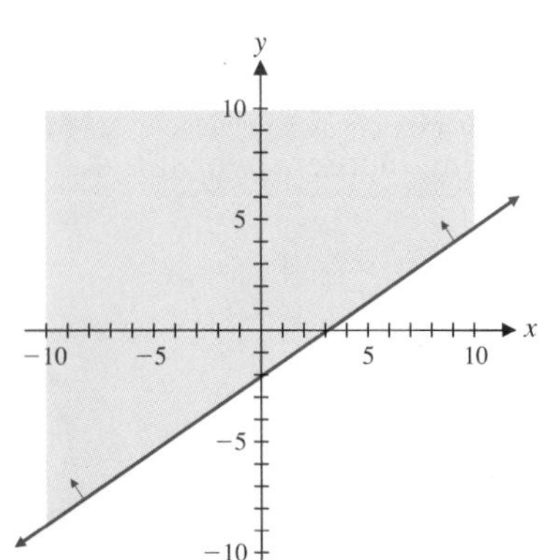

FIGURE 4

Step 2. Pick a convenient test point above or below the line. The origin (0, 0) requires the least computation, so substituting (0, 0) into the inequality, we get

$$2x - 3y \leq 6$$
$$2(0) - 3(0) = 0 \leq 6$$

This is a true statement; therefore, the point (0, 0) is in the solution set.

Step 3. The line $2x - 3y = 6$ and the half-plane containing the origin form the graph of $2x - 3y \leq 6$, as shown in Figure 4.

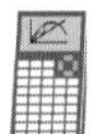

CHECK Figure 5 shows a check of this solution on a graphing utility. In Figure 5A, the small triangle to the left of y_1 indicates that the option to shade above the graph was selected. Consult the manual to see how to shade graphs on your graphing utility.

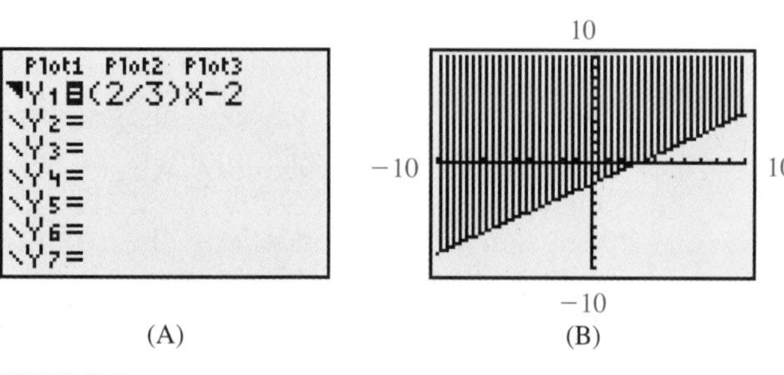

(A) (B)

FIGURE 5

Matched Problem 1 Graph $6x - 3y > 18$.

Check on a graphing utility.

Example 2 **Graphing Inequalities** Graph:

(A) $y > -3$ (B) $2x \leq 5$ (C) $x \leq 3y$

SOLUTION (A)

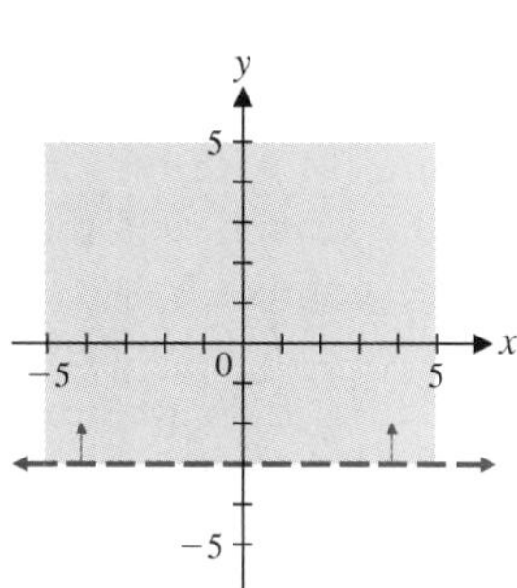

(B)

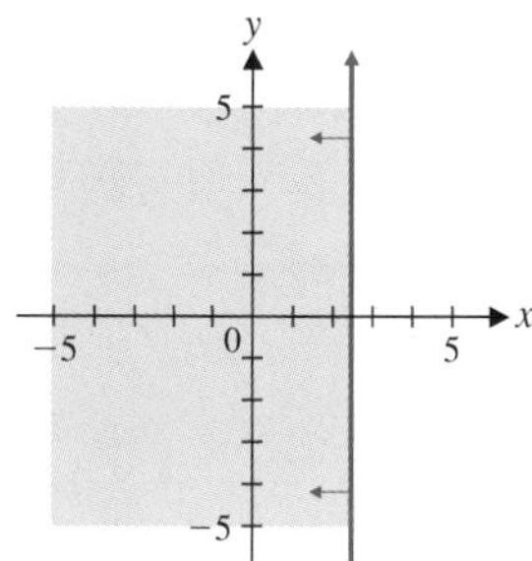

(C)

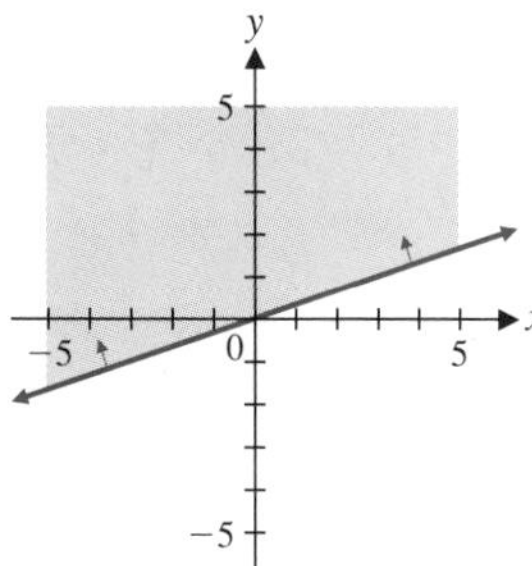

Matched Problem 2 Graph:

(A) $y < 4$ (B) $4x \geq -9$ (C) $3x \geq 2y$

Solving Systems of Linear Inequalities Graphically

We now consider systems of linear inequalities such as

$$\begin{aligned} x + y &\geq 6 \\ 2x - y &\geq 0 \end{aligned} \quad \text{and} \quad \begin{aligned} 2x + y &\leq 22 \\ x + y &\leq 13 \\ 2x + 5y &\leq 50 \\ x &\geq 0 \\ y &\geq 0 \end{aligned}$$

We wish to **solve** such systems **graphically;** that is, to find the graph of all ordered pairs of real numbers (x, y) that simultaneously satisfy all the inequalities in the system. The graph is called the **solution region** for the system. (In many applications, the solution region is also called the **feasible region.**) To find the solution region, we graph each inequality in the system and then take the intersection of all the graphs. To simplify the discussion that follows, **we will consider only systems of linear inequalities where equality is included in each statement in the system.**

Example 3 **Solving a System of Linear Inequalities Graphically** Solve the following system of linear inequalities graphically:

$$\begin{aligned} x + y &\geq 6 \\ 2x - y &\geq 0 \end{aligned}$$

Solution Graph the line $x + y = 6$ and shade the region that satisfies the linear inequality $x + y \geq 6$. This region is shaded with gray lines in Figure 6A. Next, graph the line $2x - y = 0$ and shade the region that satisfies the inequality $2x - y \geq 0$. This region is shaded with blue lines in Figure 6A. The solution region for the system of inequalities is the intersection of these two regions. This is the region shaded in both gray and blue in Figure 6A and redrawn in Figure 6B with only the solution region shaded. The coordinates of any point in the shaded region of Figure 6B specify a solution to the system. For example, the points (2, 4), (6, 3), and (7.43, 8.56) are three of infinitely many solutions, as can be easily checked. The intersection point (2, 4) is obtained by solving the equations $x + y = 6$ and $2x - y = 0$ simultaneously using any of the techniques discussed in Chapter 4.

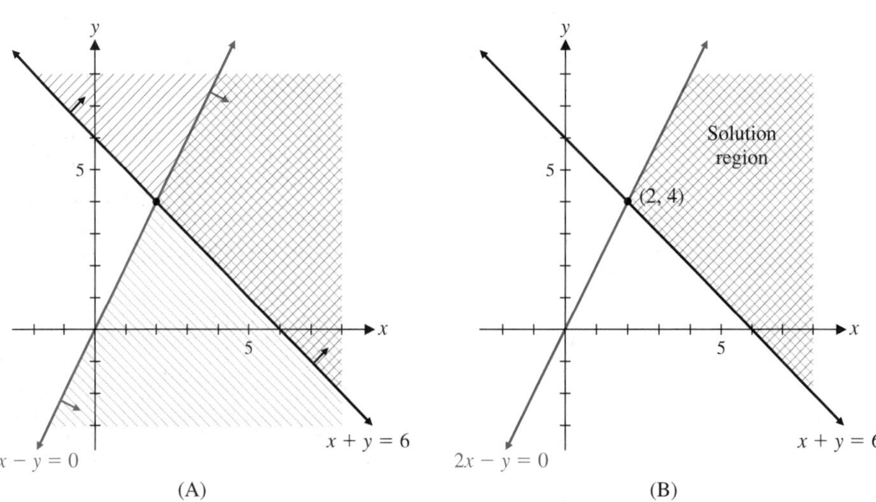

FIGURE 6

Matched Problem 3 ➡ Solve the following system of linear inequalities graphically:

$$\begin{aligned} 3x + y &\leq 21 \\ x - 2y &\leq 0 \end{aligned}$$

Explore–Discuss 2

Refer to Example 3. Graph each boundary line and shade the region containing the points that do *not* satisfy each inequality. That is, shade the region of the plane that corresponds to the inequality $x + y < 6$ and then shade the region that corresponds to the inequality $2x - y < 0$. What portion of the plane is left unshaded? Compare this method with the one used in the solution to Example 3.

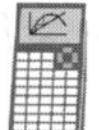

The method of solving inequalities investigated in Explore–Discuss 2 works very well on a graphing utility that allows the user to shade above and below a graph. Referring to Example 3, the unshaded region in Figure 7B corresponds to the solution region in Figure 6B.

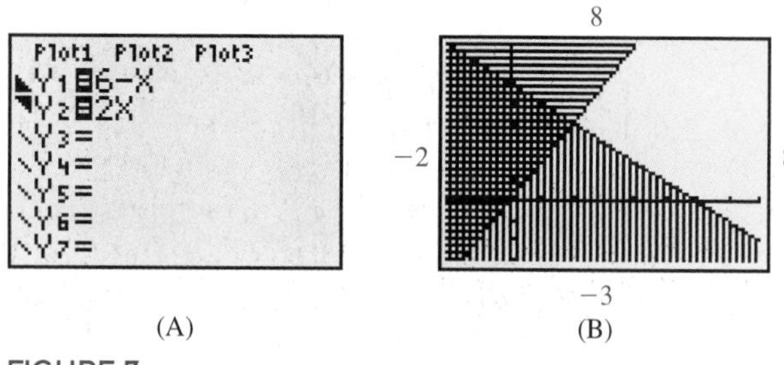

(A) (B)

FIGURE 7

The points of intersection of the lines that form the boundary of a solution region will play a fundamental role in the solution of linear programming problems, which are discussed in the next section.

Corner Point

A **corner point** of a solution region is a point in the solution region that is the intersection of two boundary lines.

For example, the point (2, 4) is the only corner point of the solution region in Example 3 (Fig. 6).

Example 4 ➡ **Solving a System of Linear Inequalities Graphically** Solve the following system of linear inequalities graphically, and find the corner points:

$$\begin{aligned} 2x + y &\leq 22 \\ x + y &\leq 13 \\ 2x + 5y &\leq 50 \\ x &\geq 0 \\ y &\geq 0 \end{aligned}$$

SOLUTION The inequalities $x \geqslant 0$ and $y \geqslant 0$ indicate that the solution region will lie in the first quadrant.* Thus, we can restrict our attention to that portion of the plane. First, we graph the lines

$$\begin{aligned} 2x + y &= 22 \\ x + y &= 13 \\ 2x + 5y &= 50 \end{aligned}$$

Find the x and y intercepts of each line; then sketch the line through these points.

Next, choosing (0, 0) as a test point, we see that the graph of each of the first three inequalities in the system consists of its corresponding line and the half-plane lying below the line, as indicated by the small arrows in Figure 8. Thus, the solution region of the system consists of the points in the first quadrant that simultaneously lie on or below all three of these lines (see the shaded region in Figure 8).

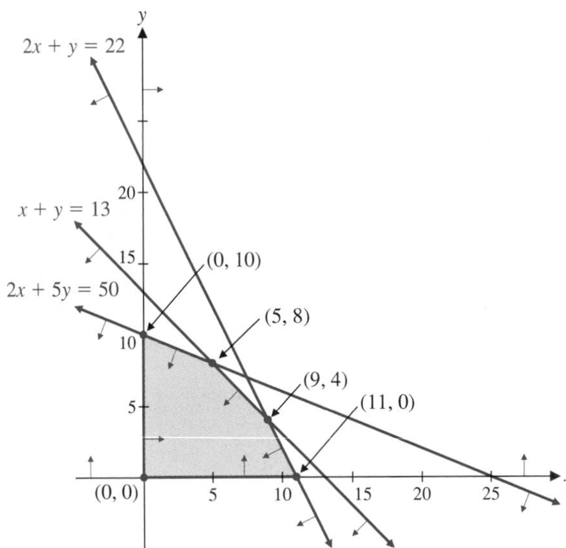

FIGURE 8

The corner points (0, 0), (0, 10), and (11, 0) can be determined from the graph. The other two corner points are determined as follows:

Solve the system	Solve the system
$2x + 5y = 50$	$2x + y = 22$
$x + y = 13$	$x + y = 13$
to obtain (5, 8).	to obtain (9, 4).

Note that the lines $2x + 5y = 50$ and $2x + y = 22$ also intersect, but the intersection point is not part of the solution region, and hence, is not a corner point.

*The inequalities $x \geqslant 0$ and $y \geqslant 0$ occur frequently in applications involving systems of inequalities, since x and y often represent quantities that cannot be negative (number of units produced, number of hours worked, etc.).

Matched Problem 4 Solve the following system of linear inequalities graphically, and find the corner points:

$$\begin{aligned} 5x + y &\geq 20 \\ x + y &\geq 12 \\ x + 3y &\geq 18 \\ x &\geq 0 \\ y &\geq 0 \end{aligned}$$

If we compare the solution regions of Example 3 and Example 4, we see that there is a fundamental difference between these two regions. We can draw a circle around the solution region in Example 4; however, it is impossible to include all the points in the solution region in Example 3 in any circle, no matter how large we draw it. This leads to the following definition:

> **Bounded and Unbounded Solution Regions**
>
> A solution region of a system of linear inequalities is **bounded** if it can be enclosed within a circle. If it cannot be enclosed within a circle, then it is **unbounded.**

Thus, the solution region for Example 4 is bounded, and the solution region for Example 3 is unbounded. This definition will be important in the next section.

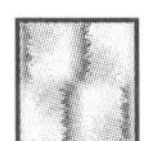

Applications

Example 5 **Medicine** A patient in a hospital is required to have at least 84 units of drug *A* and 120 units of drug *B* each day (assume that an overdosage of either drug is harmless). Each gram of substance *M* contains 10 units of drug *A* and 8 units of drug *B*, and each gram of substance *N* contains 2 units of drug *A* and 4 units of drug *B*. How many grams of substances *M* and *N* can be mixed to meet the minimum daily requirements?

Solution To clarify relationships, we summarize the information in the following table:

	Amount of drug per gram		Minimum daily
	Substance M	*Substance N*	requirement
Drug *A*	10 units	2 units	84 units
Drug *B*	8 units	4 units	120 units

Let: x = Number of grams of substance *M* used
y = Number of grams of substance *N* used

Then: $10x$ = Number of units of drug *A* in x grams of substance *M*
$2y$ = Number of units of drug *A* in y grams of substance *N*
$8x$ = Number of units of drug *B* in x grams of substance *M*
$4y$ = Number of units of drug *B* in y grams of substance *N*

The following conditions must be satisfied to meet daily requirements:

$$\begin{pmatrix}\text{Number of units of}\\ \text{drug } A\\ \text{in } x \text{ grams of substance } M\end{pmatrix} + \begin{pmatrix}\text{Number of units of}\\ \text{drug } A\\ \text{in } y \text{ grams of substance } N\end{pmatrix} \geq 84$$

$$\begin{pmatrix}\text{Number of units of}\\ \text{drug } B\\ \text{in } x \text{ grams of substance } M\end{pmatrix} + \begin{pmatrix}\text{Number of units of}\\ \text{drug } B\\ \text{in } y \text{ grams of substance } N\end{pmatrix} \geq 120$$

$$(\text{Number of grams of substance } M \text{ used}) \geq 0$$
$$(\text{Number of grams of substance } N \text{ used}) \geq 0$$

Converting these verbal statements into symbolic statements by using the variables x and y introduced above, we obtain the following model consisting of a system of linear inequalities:

$$\begin{aligned} 10x + 2y &\geq 84 && \text{Drug A restriction} \\ 8x + 4y &\geq 120 && \text{Drug B restriction} \\ x &\geq 0 && \text{Cannot use a negative amount of M} \\ y &\geq 0 && \text{Cannot use a negative amount of N} \end{aligned}$$

Graphing this system of linear inequalities, we obtain the set of feasible solutions, or the feasible region (solution region), as shown in Figure 9. Thus, any point in the shaded area (including the straight-line boundaries) will meet the daily requirements; any point outside the shaded area will not. For example, 4 units of drug M and 23 units of drug N will meet the daily requirements, but 4 units of drug M and 21 units of drug N will not. (Note that the feasible region is unbounded.)

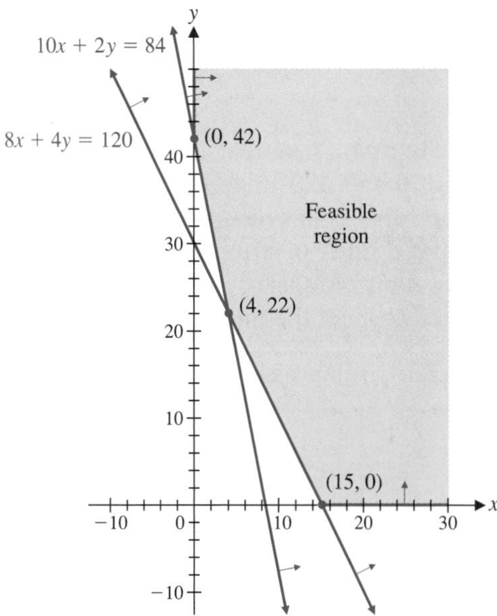

FIGURE 9

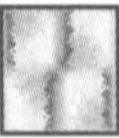

Matched Problem 5 **Resource Allocation** A manufacturing plant makes two types of inflatable boats, a two-person boat and a four-person boat. Each two-person boat requires 0.9 labor-hour in the cutting department and 0.8 labor-hour in the assembly department. Each four-person boat requires 1.8 labor-hours in the cutting department and 1.2 labor-hours in the assembly department. The maximum labor-hours available each month in the cutting and assembly departments are 864 and 672, respectively.

(A) Summarize this information in a table.
(B) If x two-person boats and y four-person boats are manufactured each month, write a system of linear inequalities that reflect the conditions indicated. Find the set of feasible solutions graphically.

REMARK

Refer to Example 5 and Matched Problem 5. In Example 5, it is certainly reasonable to consider noninteger values for x and y. The situation is not as clear in Matched Problem 5. How do we interpret a production schedule of 214.5 two-person boats and 347.75 four-person boats? It is not possible to manufacture a fraction of a boat. But it is possible to *average* 214.5 two-person and 347.75 four-person boats per month. In general, we will assume that all points in the feasible region represent acceptable solutions, even though noninteger solutions might require special interpretation.

Answers to Matched Problems

1. Graph $6x - 3y = 18$ as a dashed line (since equality is not included). Choosing the origin $(0, 0)$ as a test point, we see that $6(0) - 3(0) > 18$ is a false statement; thus, the lower half-plane determined by $6x - 3y = 18$ is the graph of $6x - 3y > 18$.

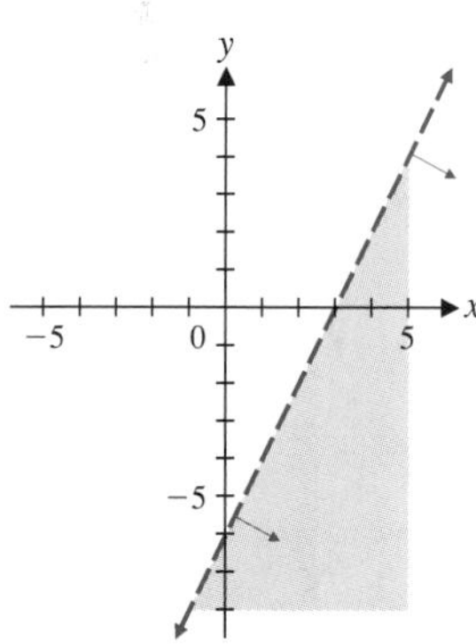

2. (A)

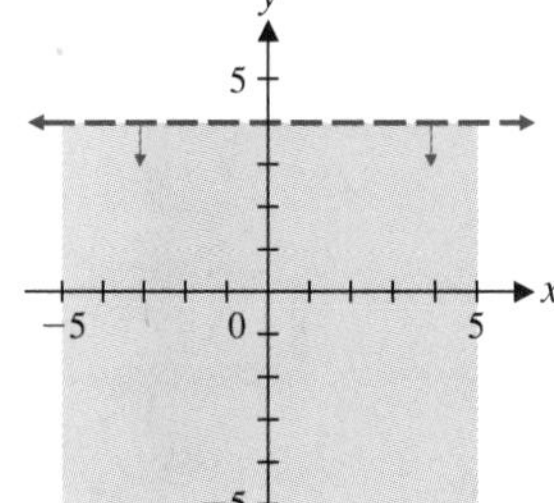

(B)

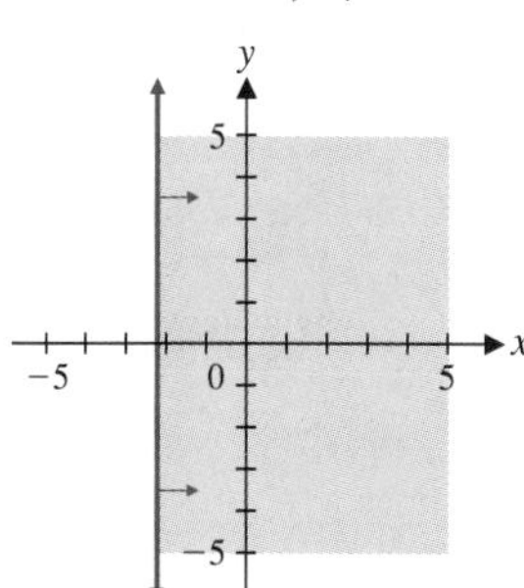

(C)

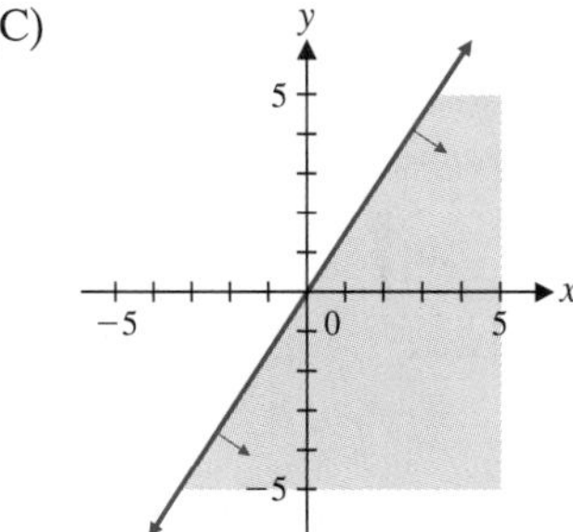

3.

y
20
Solution region 10
(6, 3)
−20
10 20
x
−10
$x - 2y = 0$
$3x + y = 21$

4.

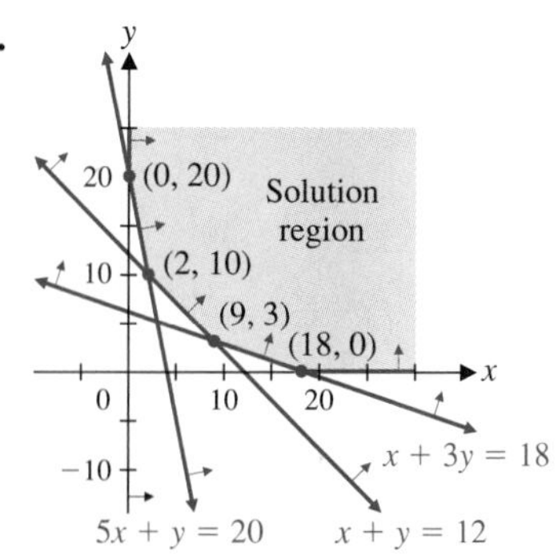

5. (A)

	LABOR-HOURS REQUIRED *Two-person boat*	*Four-person boat*	MAXIMUM LABOR-HOURS AVAILABLE PER MONTH
CUTTING DEPARTMENT	0.9	1.8	864
ASSEMBLY DEPARTMENT	0.8	1.2	672

(B) $0.9x + 1.8y \leq 864$
$0.8x + 1.2y \leq 672$
$x \geq 0$
$y \geq 0$

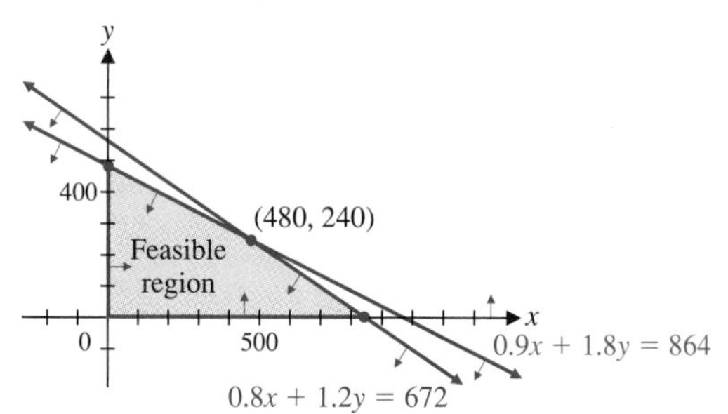

EXERCISE 5-1

A *Graph each inequality in Problems 1–10.*

1. $y \leq x - 1$ **2.** $y > x + 1$

3. $3x - 2y > 6$ **4.** $2x - 5y \leq 10$

5. $x \geq -4$ **6.** $y < 5$

7. $6x + 4y \geq 24$ **8.** $4x + 8y \geq 32$

9. $5x \leq -2y$ **10.** $6x \geq 4y$

In Problems 11–14, match the solution region of each system of linear inequalities with one of the four regions shown in the figure.

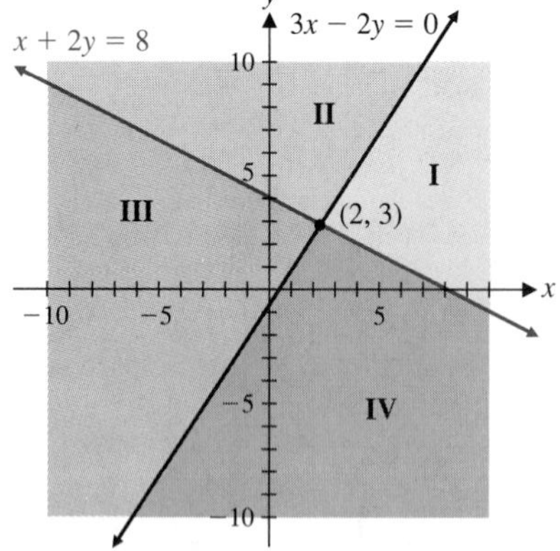

Figure for 11–14

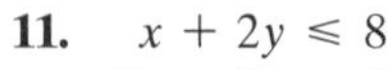

11. $x + 2y \leq 8$
$3x - 2y \geq 0$

12. $x + 2y \geq 8$
$3x - 2y \leq 0$

13. $x + 2y \geq 8$
$3x - 2y \geq 0$

14. $x + 2y \leq 8$
$3x - 2y \leq 0$

In Problems 15–18, solve each system of linear inequalities graphically.

15. $3x + y \geq 6$, $x \leq 4$

16. $3x + 4y \leq 12$, $y \geq -3$

17. $x - 2y \leq 12$, $2x + y \geq 4$

18. $2x + 5y \leq 20$, $x - 5y \geq -5$

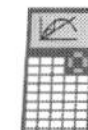

Problems 19–22 require a graphing utility that gives the user the option of shading above or below a graph.

(A) Graph the boundary lines in a standard viewing window and shade the region that contains the points that satisfy each inequality.

(B) Repeat part (A), but this time shade the region that contains the points that do not satisfy each inequality. (See Explore–Discuss 2 and Fig. 7.)

Explain how you can recognize the solution region in each graph.

19. $x + y \leq 5$, $2x - y \leq 1$

20. $x - 2y \leq 1$, $x + 3y \geq 12$

21. $2x + y \geq 4$, $3x - y \leq 7$

22. $3x + y \geq -2$, $x - 2y \geq -6$

B *In Problems 23–26, match the solution region of each system of linear inequalities with one of the four regions shown in the figure. Identify the corner points of each solution region.*

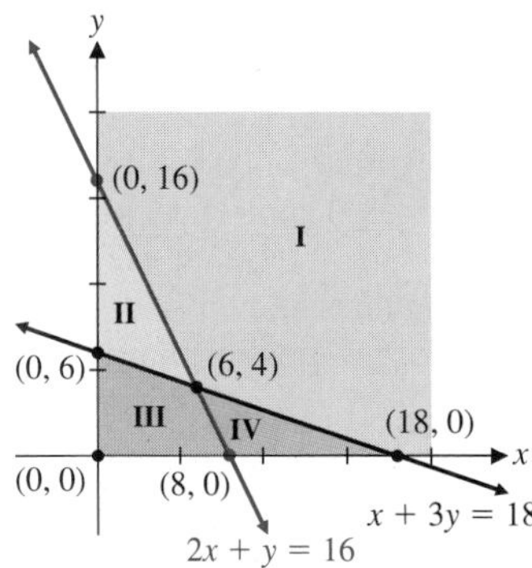

Figure for 23–26

23. $x + 3y \leq 18$, $2x + y \geq 16$, $x \geq 0$, $y \geq 0$

24. $x + 3y \leq 18$, $2x + y \leq 16$, $x \geq 0$, $y \geq 0$

25. $x + 3y \geq 18$, $2x + y \geq 16$, $x \geq 0$, $y \geq 0$

26. $x + 3y \geq 18$, $2x + y \leq 16$, $x \geq 0$, $y \geq 0$

Solve the systems in Problems 27–36 graphically, and indicate whether each solution region is bounded or unbounded. Find the coordinates of each corner point.

27. $2x + 3y \leq 12$, $x \geq 0$, $y \geq 0$

28. $3x + 4y \leq 24$, $x \geq 0$, $y \geq 0$

29. $2x + y \leq 10$, $x + 2y \leq 8$, $x \geq 0$, $y \geq 0$

30. $6x + 3y \leq 24$, $3x + 6y \leq 30$, $x \geq 0$, $y \geq 0$

31. $2x + y \geq 10$, $x + 2y \geq 8$, $x \geq 0$, $y \geq 0$

32. $4x + 3y \geq 24$, $3x + 4y \geq 8$, $x \geq 0$, $y \geq 0$

33. $2x + y \leq 10$, $x + y \leq 7$, $x + 2y \leq 12$, $x \geq 0$, $y \geq 0$

34. $3x + y \leq 21$, $x + y \leq 9$, $x + 3y \leq 21$, $x \geq 0$, $y \geq 0$

35. $2x + y \geq 16$, $x + y \geq 12$, $x + 2y \geq 14$, $x \geq 0$, $y \geq 0$

36. $3x + y \geq 24$, $x + y \geq 16$, $x + 3y \geq 30$, $x \geq 0$, $y \geq 0$

Check Problems 27–36 on a graphing utility.

C *Solve the systems in Problems 37–46 graphically, and indicate whether each solution region is bounded or unbounded. Find the coordinates of each corner point.*

37. $x + 4y \leq 32$, $3x + y \leq 30$, $4x + 5y \geq 51$

38. $x + y \leq 11$, $x + 5y \geq 15$, $2x + y \geq 12$

39. $4x + 3y \leq 48$, $2x + y \geq 24$, $x \leq 9$

40. $2x + 3y \geq 24$, $x + 3y \leq 15$, $y \geq 4$

41. $x - y \leq 0$, $2x - y \leq 4$, $0 \leq x \leq 8$

42. $2x + 3y \geq 12$, $-x + 3y \leq 3$, $0 \leq y \leq 5$

43. $-x + 3y \geq 1$, $5x - y \geq 9$, $x + y \leq 9$, $x \leq 5$

44. $x + y \leq 10$, $5x + 3y \geq 15$, $-2x + 3y \leq 15$, $2x - 5y \leq 6$

45. C $16x + 13y \leq 119$, $12x + 16y \geq 101$, $-4x + 3y \leq 11$

46. C $8x + 4y \leq 41$, $-15x + 5y \leq 19$, $2x + 6y \geq 37$

Problems 47 and 48 introduce an algebraic process for finding the corner points of a solution region without drawing a graph. We will have a great deal more to say about this process later in this chapter.

47. Consider the following system of inequalities and corresponding boundary lines:

$$\begin{aligned} 3x + 4y &\le 36 \\ 3x + 2y &\le 30 \\ x &\ge 0 \\ y &\ge 0 \end{aligned} \qquad \begin{aligned} 3x + 4y &= 36 \\ 3x + 2y &= 30 \\ x &= 0 \\ y &= 0 \end{aligned}$$

(A) Use algebraic methods to find the intersection points (if any exist) for each possible pair of boundary lines. (There are six different possible pairs.)
(B) Test each intersection point in all four inequalities to determine which are corner points.

48. Repeat Problem 47 for

$$\begin{aligned} 2x + y &\le 16 \\ 2x + 3y &\le 36 \\ x &\ge 0 \\ y &\ge 0 \end{aligned} \qquad \begin{aligned} 2x + y &= 16 \\ 2x + 3y &= 36 \\ x &= 0 \\ y &= 0 \end{aligned}$$

Applications

Business & Economics

49. *Manufacturing—resource allocation.* A manufacturing company makes two types of water skis, a trick ski and a slalom ski. The trick ski requires 6 labor-hours for fabricating and 1 labor-hour for finishing. The slalom ski requires 4 labor-hours for fabricating and 1 labor-hour for finishing. The maximum labor-hours available per day for fabricating and finishing are 108 and 24, respectively. If x is the number of trick skis and y is the number of slalom skis produced per day, write a system of linear inequalities that indicates appropriate restraints on x and y. Find the set of feasible solutions graphically for the number of each type of ski that can be produced.

50. *Manufacturing—resource allocation.* A furniture manufacturing company manufactures dining room tables and chairs. A table requires 8 labor-hours for assembling and 2 labor-hours for finishing. A chair requires 2 labor-hours for assembling and 1 labor-hour for finishing. The maximum labor-hours available per day for assembly and finishing are 400 and 120, respectively. If x is the number of tables and y is the number of chairs produced per day, write a system of linear inequalities that indicates appropriate restraints on x and y. Find the set of feasible solutions graphically for the number of tables and chairs that can be produced.

51. *Manufacturing—resource allocation.* Refer to Problem 49. The company makes a profit of \$50 on each trick ski and a profit of \$60 on each slalom ski.
(A) If the company makes 10 trick skis and 10 slalom skis per day, the daily profit will be \$1,100. Are there other production schedules that will result in a daily profit of \$1,100? How are these schedules related to the graph of the line $50x + 60y = 1{,}100$?
(B) Find a production schedule that will produce a daily profit greater than \$1,100 and repeat part (A) for this schedule.
(C) Discuss methods for using lines like those in parts (A) and (B) to find the largest possible daily profit.

52. *Manufacturing—resource allocation.* Refer to Problem 50. The company makes a profit of \$50 on each table and a profit of \$15 on each chair.
(A) If the company makes 20 tables and 20 chairs per day, the daily profit will be \$1,300. Are there other production schedules that will result in a daily profit of \$1,300? How are these schedules related to the graph of the line $50x + 15y = 1{,}300$?
(B) Find a production schedule that will produce a daily profit greater than \$1,300 and repeat part (A) for this schedule.
(C) Discuss methods for using lines like those in parts (A) and (B) to find the largest possible daily profit.

Life Sciences

53. *Nutrition—plants.* A farmer can buy two types of plant food, mix A and mix B. Each cubic yard of mix A contains 20 pounds of phosphoric acid, 30 pounds of nitrogen, and 5 pounds of potash. Each cubic yard of mix B contains 10 pounds of phosphoric acid, 30 pounds of nitrogen, and 10 pounds of potash. The minimum monthly requirements are 460 pounds of phosphoric acid, 960 pounds of nitrogen, and 220 pounds of potash. If x is the number of cubic yards of mix A used and y is the number of cubic yards of mix B used, write a system of linear inequalities that indicates appropriate restraints on x and y. Find the set of feasible solutions graphically for the amounts of mix A and mix B that can be used.

54. *Nutrition—people.* A dietitian in a hospital is to arrange a special diet using two foods. Each ounce of food M contains 30 units of calcium, 10 units of iron, and 10 units of vitamin A. Each ounce of food N contains 10 units of calcium, 10 units of iron, and 30 units of vitamin A. The minimum requirements in the diet are 360 units of calcium, 160 units of iron, and 240 units of vitamin A. If x is the number of ounces of food M used and y is the number of ounces of food N used, write a system of linear inequalities that reflects the conditions indicated. Find the set of feasible solutions graphically for the amount of each kind of food that can be used.

Social Sciences

55. *Psychology.* In an experiment on conditioning, a psychologist uses two types of Skinner (conditioning) boxes with mice and rats. Each mouse spends 10 minutes per day in box A and 20 minutes per day in box B. Each rat spends 20 minutes per day in box A and 10 minutes per day in box B. The total maximum time available per day is 800 minutes for box A and 640 minutes for box B. We are interested in the various numbers of mice and rats that can be used in the experiment under the conditions stated. If we let x be the number of mice used and y the number of rats used, write a system of linear inequalities that indicates appropriate restrictions on x and y. Find the set of feasible solutions graphically.

SECTION 5-2 Linear Programming in Two Dimensions—A Geometric Approach

- A Linear Programming Problem
- Linear Programming—A General Description
- Geometric Solution of Linear Programming Problems
- Applications

Several problems discussed in the last section are related to a more general type of problem called a *linear programming problem.* Linear programming is a mathematical process that has been developed to help management in decision-making, and it has become one of the most widely used and best-known tools of management science. We will introduce this topic by considering an example in detail, using an intuitive geometric approach. Insight gained from this approach will prove invaluable when we later consider an algebraic approach that is less intuitive but necessary in solving most real-world problems.

NOTATION CHANGE

For ease of generalization to the larger problems in later sections, we will now change variable notation from letters such as x and y to subscript forms such as x_1 and x_2.

A Linear Programming Problem

We begin our discussion with a concrete example. The geometric method of solution will suggest two important theorems and a simple general geometric procedure for solving linear programming problems in two variables.

Example 1 **Production Scheduling** A manufacturer of lightweight mountain tents makes a standard model and an expedition model for national distribution. Each standard tent requires 1 labor-hour from the cutting department and 3 labor-hours from the assembly department. Each expedition tent requires 2 labor-hours from the cutting department and 4 labor-hours from the assembly department. The maximum labor-hours available per day in the cutting department and the assembly department are 32 and 84, respectively. If the company makes a profit of \$50 on each standard tent and \$80 on each expedition tent, how many tents of each type should be manufactured each day to maximize the total daily profit (assuming all tents can be sold)?

Solution This is an example of a linear programming problem. To see relationships more clearly, we summarize the manufacturing requirements, objectives, and restrictions in Table 1.

Table 1

	Labor-hours per tent		Maximum labor-hours available per day
	Standard model	*Expedition model*	
Cutting department	1	2	32
Assembly department	3	4	84
Profit per tent	\$50	\$80	

We now proceed to formulate a mathematical model for the problem and then to solve it by using geometric methods.

Objective Function The *objective* of management is to *decide* how many of each tent model should be produced each day in order to maximize profit.

Let x_1 = Number of standard tents produced per day
x_2 = Number of expedition tents produced per day } Decision variables

The following equation gives the total profit for x_1 standard tents and x_2 expedition tents manufactured each day, assuming all tents manufactured are sold:

$$P = 50x_1 + 80x_2 \quad \text{Objective function}$$

Mathematically, the manufacturer needs to decide on values for the **decision variables** (x_1, x_2) that achieve its objective—that is, maximize the **objective function** (profit) $P = 50x_1 + 80x_2$. (Note that P is a function of two independent variables.) It appears that the profit can be made as large as we like by manufacturing more and more tents—or can it?

CONSTRAINTS Any manufacturing company, no matter how large or small, has manufacturing limits imposed by available resources, plant capacity, demand, and so forth. These limits are referred to as **problem constraints.**

CUTTING DEPARTMENT CONSTRAINT

$$\begin{pmatrix}\text{Daily cutting}\\ \text{time for } x_1\\ \text{standard tents}\end{pmatrix} + \begin{pmatrix}\text{Daily cutting}\\ \text{time for } x_2\\ \text{expedition tents}\end{pmatrix} \leq \begin{pmatrix}\text{Maximum labor-}\\ \text{hours available}\\ \text{per day}\end{pmatrix}$$

$$1x_1 \quad + \quad 2x_2 \quad \leq \quad 32$$

ASSEMBLY DEPARTMENT CONSTRAINT

$$\begin{pmatrix}\text{Daily assembly}\\ \text{time for } x_1\\ \text{standard tents}\end{pmatrix} + \begin{pmatrix}\text{Daily assembly}\\ \text{time for } x_2\\ \text{expedition tents}\end{pmatrix} \leq \begin{pmatrix}\text{Maximum labor-}\\ \text{hours available}\\ \text{per day}\end{pmatrix}$$

$$3x_1 \quad + \quad 4x_2 \quad \leq \quad 84$$

NONNEGATIVE CONSTRAINTS It is not possible to manufacture a negative number of tents; thus, we have the **nonnegative constraints**

$$x_1 \geq 0$$
$$x_2 \geq 0$$

which we usually write in the form

$$x_1, x_2 \geq 0$$

MATHEMATICAL MODEL We now have a **mathematical model** for the problem under consideration:

$$\begin{aligned} \text{Maximize} \quad P &= 50x_1 + 80x_2 && \text{Objective function} \\ \text{Subject to} \quad x_1 + 2x_2 &\leq 32 && \text{Problem constraints} \\ 3x_1 + 4x_2 &\leq 84 && \\ x_1, x_2 &\geq 0 && \text{Nonnegative constraints} \end{aligned}$$

GRAPHIC SOLUTION **Solving** the set of linear inequality constraints **graphically** (see the last section), we obtain the feasible region for production schedules (Figure 1).

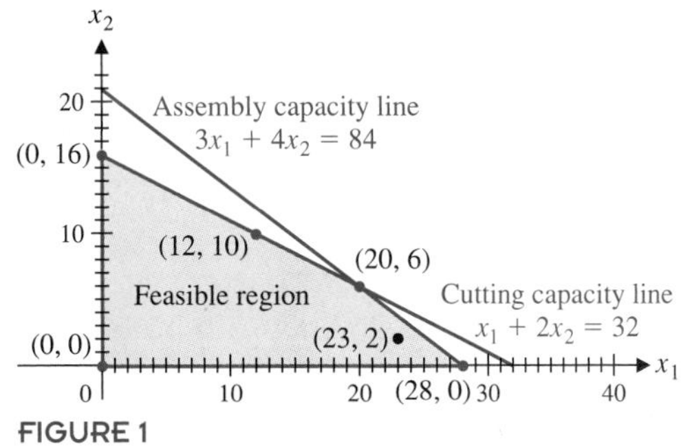

FIGURE 1

By choosing a production schedule (x_1, x_2) from the feasible region, a profit can be determined using the objective function

$$P = 50x_1 + 80x_2$$

For example, if $x_1 = 12$ and $x_2 = 10$, then the profit for the day would be

$$\begin{aligned} P &= 50(12) + 80(10) \\ &= \$1{,}400 \end{aligned}$$

Or if $x_1 = 23$ and $x_2 = 2$, then the profit for the day would be

$$\begin{aligned} P &= 50(23) + 80(2) \\ &= \$1,310 \end{aligned}$$

But the question is, out of all possible production schedules (x_1, x_2) from the feasible region, which schedule(s) produces the *maximum* profit? Thus, we have a **maximization problem.** Since point-by-point checking is impossible (there are infinitely many points to check), we must find another way.

By assigning P in $P = 50x_1 + 80x_2$ a particular value and plotting the resulting equation in the coordinate system shown in Figure 1, we obtain a **constant-profit line (isoprofit line).** Every point in the feasible region on this line represents a production schedule that will produce the same profit. By doing this for a number of values for P, we obtain a family of constant-profit lines (Fig. 2) that are parallel to each other, since they all have the same slope. To see this, we write $P = 50x_1 + 80x_2$ in the slope–intercept form

$$x_2 = -\frac{5}{8}x_1 + \frac{P}{80}$$

and note that for any profit P, the constant-profit line has slope $-\frac{5}{8}$. We also observe that as the profit P increases, the x_2 intercept $(P/80)$ increases, and the line moves away from the origin.

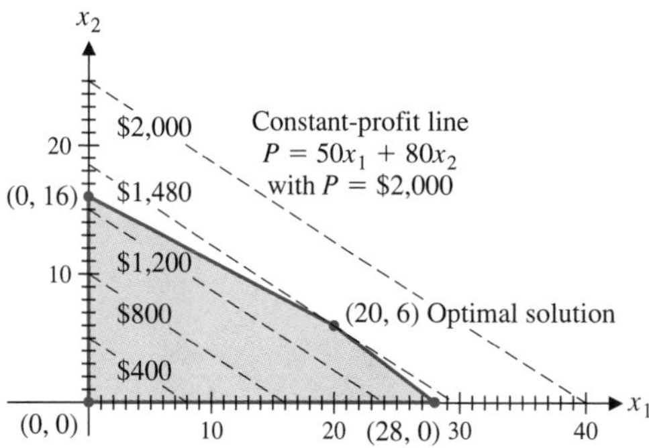

FIGURE 2
Constant-profit lines

Thus, the maximum profit occurs at a point where a constant-profit line is the farthest from the origin but still in contact with the feasible region. In this example, this occurs at (20, 6), as is seen in Figure 2. Thus, if the manufacturer makes 20 standard tents and 6 expedition tents per day, the profit will be maximized at

$$\begin{aligned} P &= 50(20) + 80(6) \\ &= \$1,480 \end{aligned}$$

The point (20, 6) is called an **optimal solution** to the problem, because it maximizes the objective (profit) function and is in the feasible region. In general, it appears that a maximum profit occurs at one of the corner points. We also note that the minimum profit $(P = 0)$ occurs at the corner point $(0, 0)$. ▪

Matched Problem 1 A manufacturing plant makes two types of inflatable boats, a two-person boat and a four-person boat. Each two-person boat requires 0.9 labor-hour from the cutting department and 0.8 labor-hour from the assembly department. Each four-person boat requires 1.8 labor-hours from the cutting department and 1.2 labor-hours from the assembly department. The maximum labor-hours available per month in the cutting department and the assembly department are 864 and 672, respectively. The company makes a profit of \$25 on each two-person boat and \$40 on each four-person boat.

(A) Summarize the relevant material in a table similar to Table 1 in Example 1.
(B) Identify the decision variables.
(C) Write the objective function P.
(D) Write the problem constraints and the nonnegative constraints.
(E) Graph the feasible region. Include graphs of the objective function for $P = \$5{,}000, P = \$10{,}000, P = \$15{,}000$, and $P = \$21{,}600$.
(F) From the graph and constant-profit lines, determine how many boats should be manufactured each month to maximize the profit. What is the maximum profit?

Explore–Discuss 1

Refer to the feasible region S shown in the figure.

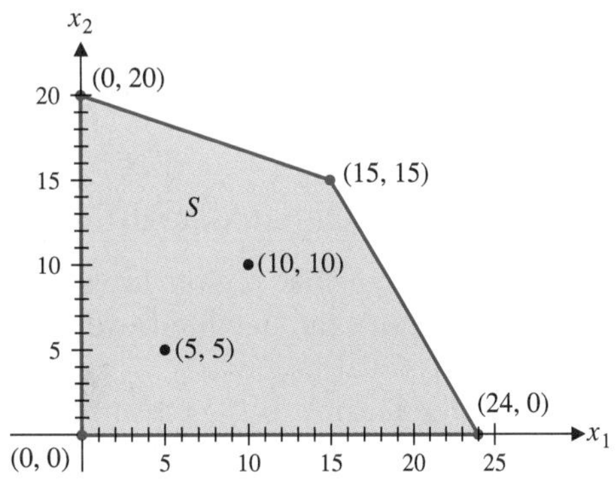

(A) Let $P = x_1 + x_2$. Graph the isoprofit lines through the points $(5, 5)$ and $(10, 10)$. Place a straightedge along the line with the smaller profit and slide it in the direction of increasing profit, without changing its slope. What is the maximum value of P? Where does this maximum value occur?
(B) Repeat part (A) for $P = x_1 + 10x_2$.
(C) Repeat part (A) for $P = 10x_1 + x_2$.

Linear Programming—A General Description

In Example 1 and Matched Problem 1, the optimal solution occurs at a corner point of the feasible region. Is this always the case? The answer is a qualified yes, as will be seen in Theorem 1. First, we give a few general definitions.

A **linear programming problem** is one that is concerned with finding the **optimal value** (maximum or minimum value) of a linear **objective function** of the form

$$z = c_1x_1 + c_2x_2 + \cdots + c_nx_n$$

where the **decision variables** $x_1, x_2, \ldots, x_n$ are subject to **problem constraints** in the form of linear inequalities and equations. In addition, the decision variables must satisfy the **nonnegative constraints** $x_i \geqslant 0, i = 1, 2, \ldots, n$. The set of points satisfying both the problem constraints and the nonnegative constraints is called the **feasible region** for the problem. Any point in the feasible region that produces the optimal value of the objective function over the feasible region is called an **optimal solution.**

Theorem 1 ■■ FUNDAMENTAL THEOREM OF LINEAR PROGRAMMING—VERSION 1

If the optimal value of the objective function in a linear programming problem exists, then that value must occur at one (or more) of the corner points of the feasible region. ■■

Theorem 1 provides a simple procedure for solving a linear programming problem, *provided the problem has an optimal solution—not all do.* In order to use Theorem 1, we must know that the problem under consideration has an optimal solution. Theorem 2 provides some conditions that will ensure that a linear programming problem has an optimal solution.

Theorem 2 ■■ EXISTENCE OF OPTIMAL SOLUTIONS

(A) If the feasible region for a linear programming problem is bounded, then both the maximum value and the minimum value of the objective function always exist.

(B) If the feasible region is unbounded and the coefficients of the objective function are positive, then the minimum value of the objective function exists, but the maximum value does not.

(C) If the feasible region is empty (that is, there are no points that satisfy all the constraints), then both the maximum value and the minimum value of the objective function do not exist. ■■

Theorem 2 does not cover all possibilities. For example, what happens if the feasible region is unbounded and one (or both) of the coefficients of the objective function are negative? Problems of this type must be solved by carefully examining the graph of the objective function for various feasible solutions, as we did in Example 1.

■ GEOMETRIC SOLUTION OF LINEAR PROGRAMMING PROBLEMS

The discussion above leads to the following procedure for the geometric solution of linear programming problems with two decision variables:

Geometric Solution of a Linear Programming Problem with Two Decision Variables

Step 1. For an applied problem, summarize relevant material in table form (see Table 1 in Example 1).

Step 2. Form a mathematical model for the problem:

(A) Introduce decision variables, and write a linear objective function.
(B) Write problem constraints using linear inequalities and/or equations.
(C) Write nonnegative constraints.

Step 3. Graph the feasible region. Then, if an optimal solution exists according to Theorem 2, find the coordinates of each corner point.

Step 4. Make a table listing the value of the objective function at each corner point.

Step 5. Determine the optimal solution(s) from the table in step 4.

Step 6. For an applied problem, interpret the optimal solution(s) in terms of the original problem.

Before we consider additional applications, let us use this procedure to solve some linear programming problems where the model has already been determined.

Example 2 **Solving a Linear Programming Problem**

(A) Minimize and maximize

$$z = 3x_1 + x_2$$

Subject to

$$\begin{aligned} 2x_1 + x_2 &\leqslant 20 \\ 10x_1 + x_2 &\geqslant 36 \\ 2x_1 + 5x_2 &\geqslant 36 \\ x_1, x_2 &\geqslant 0 \end{aligned}$$

(B) Minimize and maximize

$$z = 10x_1 + 20x_2$$

Subject to

$$\begin{aligned} 6x_1 + 2x_2 &\geqslant 36 \\ 2x_1 + 4x_2 &\geqslant 32 \\ x_2 &\leqslant 20 \\ x_1, x_2 &\geqslant 0 \end{aligned}$$

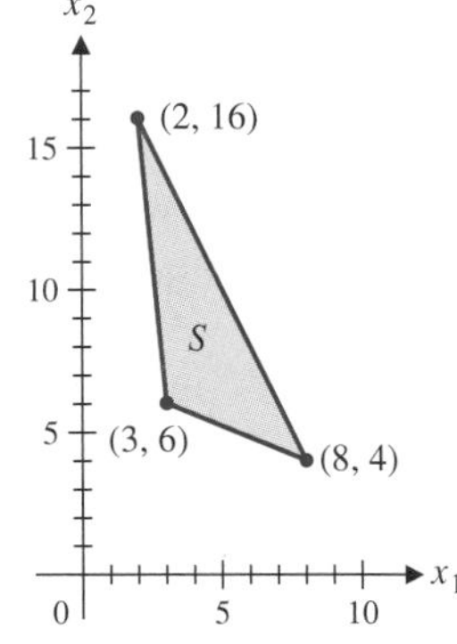

CORNER POINT (x_1, x_2)	$z = 3x_1 + x_2$
(3, 6)	15
(2, 16)	22
(8, 4)	28

SOLUTION (A) We begin with step 3, since step 1 does not apply and step 2 has already been done for us.

Step 3. Graph the feasible region S. Then, after checking Theorem 2 to determine that an optimal solution exists, find the coordinates of each corner point. Since S is bounded, z will have both a maximum and a minimum on S (Theorem 2A) and these will both occur at corner points (Theorem 1).

Step 4. Evaluate the objective function at each corner point, as shown in the table in the margin.

Step 5. Determine the optimal solutions from step 4. Examining the values in the table, we see that the minimum value of z is 15 at (3, 6) and the maximum value of z is 28 at (8, 4).

(B) Again, we can begin with step 3.

Step 3. Graph the feasible region S. Then, after checking Theorem 2 to determine that an optimal solution exists, find the coordinates of each corner point. Since S is unbounded and the coefficients of the objective function are positive, z has a minimum value on S but no maximum value (Theorem 2B).

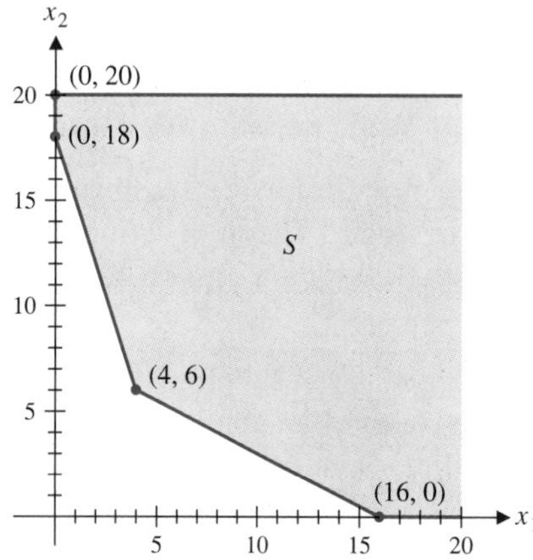

CORNER POINT (x_1, x_2)	$z = 10x_1 + 20x_2$
(0, 20)	400
(0, 18)	360
(4, 6)	160
(16, 0)	160

Step 4. Evaluate the objective function at each corner point, as shown in the table in the margin.

Step 5. Determine the optimal solution from step 4. The minimum value of z is 160 at $(4, 6)$ and at $(16, 0)$.

The solution to Example 2 is a **multiple optimal solution.**

In general, if two corner points are both optimal solutions to a linear programming problem, then any point on the line segment joining them is also an optimal solution.

This is the only time that optimal solutions also occur at noncorner points.

Matched Problem 2 (A) Maximize and minimize $z = 4x_1 + 2x_2$ subject to the constraints given in Example 2A.

(B) Maximize and minimize $z = 20x_1 + 5x_2$ subject to the constraints given in Example 2B.

Explore–Discuss 2

In Example 2B we saw that there was no optimal solution for the problem of maximizing the objective function P over the feasible region S. We want to add an additional constraint to modify the feasible region so that an optimal solution for the maximization problem does exist. Which of the following constraints will accomplish this objective?

(A) $x_1 \leq 20$ (B) $x_2 \geq 4$ (C) $x_1 \leq x_2$ (D) $x_2 \leq x_1$

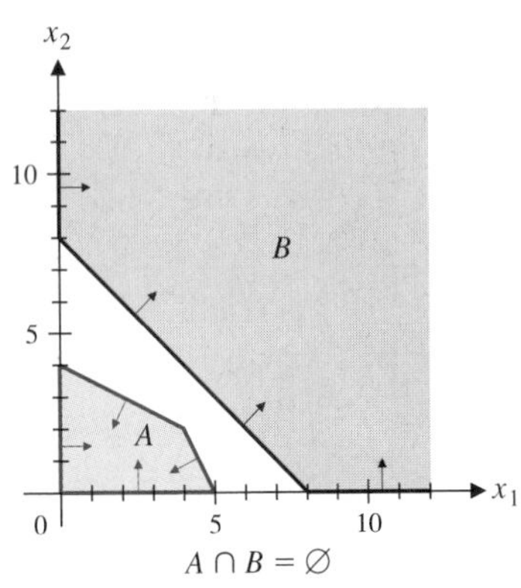

FIGURE 3

For an illustration of Theorem 2C, consider the following:

$$\begin{aligned} \text{Maximize} \quad & P = 2x_1 + 3x_2 \\ \text{Subject to} \quad & x_1 + x_2 \geq 8 \\ & x_1 + 2x_2 \leq 8 \\ & 2x_1 + x_2 \leq 10 \\ & x_1, x_2 \geq 0 \end{aligned}$$

The intersection of the graphs of the constraint inequalities is the empty set (Fig. 3); hence, the *feasible region is empty* (see Appendix A-1). If this happens, then the problem should be reexamined to see if it has been formulated properly. If it has, then the management may have to reconsider items such as labor-hours, overtime, budget, and supplies allocated to the project in order to obtain a nonempty feasible region and a solution to the original problem.

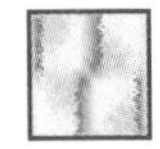

■ APPLICATIONS

Example 3 **Medicine** We now convert Example 5 from the preceding section into a linear programming problem. A patient in a hospital is required to have at least 84 units of drug D_1 and 120 units of drug D_2 each day (assume that an overdosage of either drug is harmless). Each gram of substance M contains 10 units of drug D_1 and 8 units of drug D_2, and each gram of substance N contains 2 units of drug D_1 and 4 units of drug D_2. Now, suppose both M and N contain an undesirable drug D_3, 3 units per gram in M and 1 unit per gram in N. How many grams of each of substances M and N should be mixed to meet the minimum daily requirements and at the same time minimize the intake of drug D_3? How many units of the undesirable drug D_3 will be in this mixture?

SOLUTION **Step 1.** Summarize relevant material in table form:

	AMOUNT OF DRUG PER GRAM		MINIMUM DAILY
	Substance M	*Substance N*	REQUIREMENT
DRUG D_1	10 units	2 units	84 units
DRUG D_2	8 units	4 units	120 units
DRUG D_3	3 units	1 unit	

Step 2. Form a mathematical model for the problem.

Let: x_1 = Number of grams of substance M used
x_2 = Number of grams of substance N used } Decision variables

We form the linear objective function

$$C = 3x_1 + x_2$$

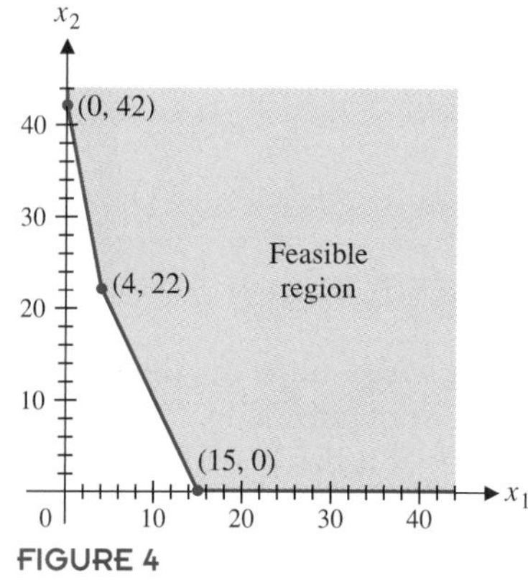

FIGURE 4

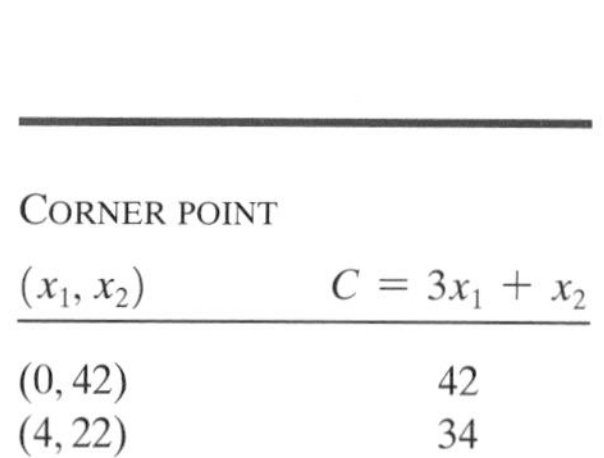

CORNER POINT (x_1, x_2)	$C = 3x_1 + x_2$
(0, 42)	42
(4, 22)	34
(15, 0)	45

which gives the amount of the undesirable drug D_3 in x_1 grams of M and x_2 grams of N. Proceeding as before, we formulate the following mathematical model for the problem:

$$\begin{aligned} \text{Minimize} \quad & C = 3x_1 + x_2 && \textit{Objective function} \\ \text{Subject to} \quad & 10x_1 + 2x_2 \geq 84 && \textit{Drug } D_1 \textit{ constraint} \\ & 8x_1 + 4x_2 \geq 120 && \textit{Drug } D_2 \textit{ constraint} \\ & x_1, x_2 \geq 0 && \textit{Nonnegative constraints} \end{aligned}$$

Step 3. Graph the feasible region. Then, after checking Theorem 2 to determine that an optimal solution exists, find the coordinates of each corner point. Solving the system of constraint inequalities graphically, we obtain the feasible region shown in Figure 4. Since the feasible region is unbounded and the coefficients of the objective function are positive, this minimization problem has a solution.

Step 4. Evaluate the objective function at each corner point, as shown in the table.

Step 5. Determine the optimal solution from step 4. The optimal solution is $C = 34$ at the corner point $(4, 22)$.

Step 6. Interpret the optimal solution in terms of the original problem. If we use 4 grams of substance M and 22 grams of substance N, we will supply the minimum daily requirements for drugs D_1 and D_2 and minimize the intake of the undesirable drug D_3 at 34 units. (Any other combination of M and N from the feasible region will result in a larger amount of the undesirable drug D_3.)

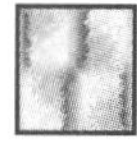

Matched Problem 3

Agriculture A chicken farmer can buy a special food mix A at 20¢ per pound and a special food mix B at 40¢ per pound. Each pound of mix A contains 3,000 units of nutrient N_1 and 1,000 units of nutrient N_2; each pound of mix B contains 4,000 units of nutrient N_1 and 4,000 units of nutrient N_2. If the minimum daily requirements for the chickens collectively are 36,000 units of nutrient N_1 and 20,000 units of nutrient N_2, how many pounds of each food mix should be used each day to minimize daily food costs while meeting (or exceeding) the minimum daily nutrient requirements? What is the minimum daily cost?

REMARK

Refer to Example 3. If we change the minimum requirement for drug D_2 from 120 to 125, the optimal solution changes to 3.6 grams of substance M and 24.1 grams of substance N, correct to one decimal place.

Now refer to Example 1. If we change the maximum labor-hours available per day in the assembly department from 84 to 79, the solution changes to 15 standard tents and 8.5 expedition tents.

We can measure 3.6 grams of substance M and 24.1 grams of substance N, but how can we make 8.5 tents? Should we make 8 tents? Or 9 tents? If the solutions to a problem must be integers and the optimal solution found graphically involves decimals, rounding the decimal value to the nearest integer does not always produce the *optimal integer solution* (see Problem 36 in Exercise 5-2). Finding optimal integer solutions to a linear programming problem is called *integer programming* and requires special techniques that are beyond the scope of this book. As mentioned earlier, if we encounter a solution like 8.5 tents per day, we will interpret this as an *average* value over many days of production.

Answers to Matched Problems

1. (A)

	Labor-hours required: Two-person boat	Labor-hours required: Four-person boat	Maximum labor-hours available per month
Cutting department	0.9	1.8	864
Assembly department	0.8	1.2	672
Profit per boat	\$25	\$40	

(B) x_1 = Number of two-person boats produced each month
x_2 = Number of four-person boats produced each month

(C) $P = 25x_1 + 40x_2$ (D) $0.9x_1 + 1.8x_2 \leq 864$
$0.8x_1 + 1.2x_2 \leq 672$
$x_1, x_2 \geq 0$

(E)

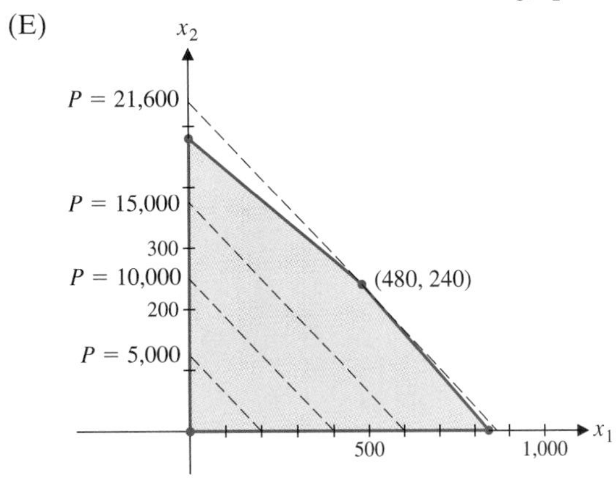

(F) 480 two-person boats, 240 four-person boats; Max P = \$21,600 per month

2. (A) Min $z = 24$ at $(3, 6)$; Max $z = 40$ at $(2, 16)$ and $(8, 4)$ (multiple optimal solution)
(B) Min $z = 90$ at $(0, 18)$; no maximum value

3. 8 lb of mix A, 3 lb of mix B; Min C = \$2.80 per day

Exercise 5-2

A *Find the maximum value of each objective function in Problems 1–4 over the feasible region S shown in the figure.*

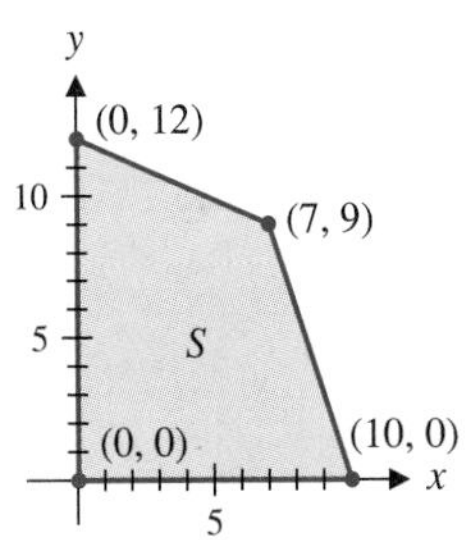

1. $z = x + y$ **2.** $z = 4x + y$

3. $z = 3x + 7y$ **4.** $z = 9x + 3y$

Find the minimum value of each objective function in Problems 5–8 over the feasible region T shown in the figure.

y
(0, 12)
10
(0, 8)
5
T
(4, 3)
(12, 0)
0 5 10 x

5. $z = 7x + 4y$ **6.** $z = 7x + 9y$

7. $z = 3x + 8y$ **8.** $z = 5x + 4y$

B *Solve the linear programming problems stated in Problems 9–26.*

9. Maximize $P = 5x_1 + 5x_2$
Subject to
$$\begin{aligned} 2x_1 + x_2 &\leqslant 10 \\ x_1 + 2x_2 &\leqslant 8 \\ x_1, x_2 &\geqslant 0 \end{aligned}$$

10. Maximize $P = 3x_1 + 2x_2$
Subject to
$$\begin{aligned} 6x_1 + 3x_2 &\leqslant 24 \\ 3x_1 + 6x_2 &\leqslant 30 \\ x_1, x_2 &\geqslant 0 \end{aligned}$$

11. Minimize and maximize
$$z = 2x_1 + 3x_2$$
Subject to
$$\begin{aligned} 2x_1 + x_2 &\geqslant 10 \\ x_1 + 2x_2 &\geqslant 8 \\ x_1, x_2 &\geqslant 0 \end{aligned}$$

12. Minimize and maximize
$$z = 8x_1 + 7x_2$$
Subject to
$$\begin{aligned} 4x_1 + 3x_2 &\geqslant 24 \\ 3x_1 + 4x_2 &\geqslant 8 \\ x_1, x_2 &\geqslant 0 \end{aligned}$$

13. Maximize $P = 30x_1 + 40x_2$
Subject to
$$\begin{aligned} 2x_1 + x_2 &\leqslant 10 \\ x_1 + x_2 &\leqslant 7 \\ x_1 + 2x_2 &\leqslant 12 \\ x_1, x_2 &\geqslant 0 \end{aligned}$$

14. Maximize $P = 20x_1 + 10x_2$
Subject to
$$\begin{aligned} 3x_1 + x_2 &\leqslant 21 \\ x_1 + x_2 &\leqslant 9 \\ x_1 + 3x_2 &\leqslant 21 \\ x_1, x_2 &\geqslant 0 \end{aligned}$$

15. Minimize and maximize
$$z = 10x_1 + 30x_2$$
Subject to
$$\begin{aligned} 2x_1 + x_2 &\geqslant 16 \\ x_1 + x_2 &\geqslant 12 \\ x_1 + 2x_2 &\geqslant 14 \\ x_1, x_2 &\geqslant 0 \end{aligned}$$

16. Minimize and maximize
$$z = 400x_1 + 100x_2$$
Subject to
$$\begin{aligned} 3x_1 + x_2 &\geqslant 24 \\ x_1 + x_2 &\geqslant 16 \\ x_1 + 3x_2 &\geqslant 30 \\ x_1, x_2 &\geqslant 0 \end{aligned}$$

17. Minimize and maximize
$$P = 30x_1 + 10x_2$$
Subject to
$$\begin{aligned} 2x_1 + 2x_2 &\geqslant 4 \\ 6x_1 + 4x_2 &\leqslant 36 \\ 2x_1 + x_2 &\leqslant 10 \\ x_1, x_2 &\geqslant 0 \end{aligned}$$

18. Minimize and maximize
$$P = 2x_1 + x_2$$
Subject to
$$\begin{aligned} x_1 + x_2 &\geqslant 2 \\ 6x_1 + 4x_2 &\leqslant 36 \\ 4x_1 + 2x_2 &\leqslant 20 \\ x_1, x_2 &\geqslant 0 \end{aligned}$$

19. Minimize and maximize
$$P = 3x_1 + 5x_2$$
Subject to
$$\begin{aligned} x_1 + 2x_2 &\leqslant 6 \\ x_1 + x_2 &\leqslant 4 \\ 2x_1 + 3x_2 &\geqslant 12 \\ x_1, x_2 &\geqslant 0 \end{aligned}$$

20. Minimize and maximize
$$P = -x_1 + 3x_2$$
Subject to
$$\begin{aligned} 2x_1 - x_2 &\geqslant 4 \\ -x_1 + 2x_2 &\leqslant 4 \\ x_2 &\leqslant 6 \\ x_1, x_2 &\geqslant 0 \end{aligned}$$

21. Minimize and maximize
$$P = 20x_1 + 10x_2$$
Subject to
$$\begin{aligned} 2x_1 + 3x_2 &\geqslant 30 \\ 2x_1 + x_2 &\leqslant 26 \\ -2x_1 + 5x_2 &\leqslant 34 \\ x_1, x_2 &\geqslant 0 \end{aligned}$$

22. Minimize and maximize
$$P = 12x_1 + 14x_2$$
Subject to
$$\begin{aligned} -2x_1 + x_2 &\geqslant 6 \\ x_1 + x_2 &\leqslant 15 \\ 3x_1 - x_2 &\geqslant 0 \\ x_1, x_2 &\geqslant 0 \end{aligned}$$

23. Maximize $P = 20x_1 + 30x_2$
Subject to
$$\begin{aligned} 0.6x_1 + 1.2x_2 &\leqslant 960 \\ 0.03x_1 + 0.04x_2 &\leqslant 36 \\ 0.3x_1 + 0.2x_2 &\leqslant 270 \\ x_1, x_2 &\geqslant 0 \end{aligned}$$

24. Minimize $C = 30x_1 + 10x_2$
Subject to
$$\begin{aligned} 1.8x_1 + 0.9x_2 &\geq 270 \\ 0.3x_1 + 0.2x_2 &\geq 54 \\ 0.01x_1 + 0.03x_2 &\geq 3.9 \\ x_1, x_2 &\geq 0 \end{aligned}$$

25. Maximize $P = 525x_1 + 478x_2$
C Subject to
$$\begin{aligned} 275x_1 + 322x_2 &\leq 3{,}381 \\ 350x_1 + 340x_2 &\leq 3{,}762 \\ 425x_1 + 306x_2 &\leq 4{,}114 \\ x_1, x_2 &\geq 0 \end{aligned}$$

26. Maximize $P = 300x_1 + 460x_2$
C Subject to
$$\begin{aligned} 245x_1 + 452x_2 &\leq 4{,}181 \\ 290x_1 + 379x_2 &\leq 3{,}888 \\ 390x_1 + 299x_2 &\leq 4{,}407 \\ x_1, x_2 &\geq 0 \end{aligned}$$

In Problems 27 and 28, explain why Theorem 2 cannot be used to conclude that a maximum or minimum value exists. Graph the feasible regions and use graphs of the objective function $z = x_1 - x_2$ for various values of z to discuss the existence of a maximum value and a minimum value.

27. Minimize and maximize
$$z = x_1 - x_2$$
Subject to
$$\begin{aligned} x_1 - 2x_2 &\leq 0 \\ 2x_1 - x_2 &\leq 6 \\ x_1, x_2 &\geq 0 \end{aligned}$$

28. Minimize and maximize
$$z = x_1 - x_2$$
Subject to
$$\begin{aligned} x_1 - 2x_2 &\geq -6 \\ 2x_1 - x_2 &\geq 0 \\ x_1, x_2 &\geq 0 \end{aligned}$$

C

29. The corner points for the bounded feasible region determined by the system of linear inequalities
$$\begin{aligned} x_1 + 2x_2 &\leq 10 \\ 3x_1 + x_2 &\leq 15 \\ x_1, x_2 &\geq 0 \end{aligned}$$
are $O = (0, 0)$, $A = (0, 5)$, $B = (4, 3)$, and $C = (5, 0)$. If $P = ax_1 + bx_2$ and $a, b > 0$, determine conditions on a and b that will ensure that the maximum value of P occurs:
(A) Only at A
(B) Only at B
(C) Only at C
(D) At both A and B
(E) At both B and C

30. The corner points for the feasible region determined by the system of linear inequalities
$$\begin{aligned} x_1 + 4x_2 &\geq 30 \\ 3x_1 + x_2 &\geq 24 \\ x_1, x_2 &\geq 0 \end{aligned}$$
are $A = (0, 24)$, $B = (6, 6)$, and $D = (30, 0)$. If $C = ax_1 + bx_2$ and $a, b > 0$, determine conditions on a and b that will ensure that the minimum value of C occurs:
(A) Only at A
(B) Only at B
(C) Only at D
(D) At both A and B
(E) At both B and D

APPLICATIONS

Business & Economics

31. *Manufacturing—resource allocation.* A manufacturing company makes two types of water skis, a trick ski and a slalom ski. The relevant manufacturing data are given in the table.

DEPARTMENT	LABOR-HOURS PER SKI *Trick ski*	*Slalom ski*	MAXIMUM LABOR-HOURS AVAILABLE PER DAY
FABRICATING	6	4	108
FINISHING	1	1	24

(A) If the profit on a trick ski is \$40 and the profit on a slalom ski is \$30, how many of each type of ski should be manufactured each day to realize a maximum profit? What is the maximum profit?
(B) Discuss the effect on the production schedule and the maximum profit if the profit on a slalom ski decreases to \$25.
(C) Discuss the effect on the production schedule and the maximum profit if the profit on a slalom ski increases to \$45.

32. *Manufacturing—resource allocation.* A furniture manufacturing company manufactures dining room tables and chairs. The relevant manufacturing data are given in the table.

DEPARTMENT	LABOR-HOURS PER UNIT *Table*	LABOR-HOURS PER UNIT *Chair*	MAXIMUM LABOR-HOURS AVAILABLE PER DAY
ASSEMBLY	8	2	400
FINISHING	2	1	120
PROFIT PER UNIT	$90	$25	

(A) How many tables and chairs should be manufactured each day to realize a maximum profit? What is the maximum profit?

(B) Discuss the effect on the production schedule and the maximum profit if the marketing department of the company decides that the number of chairs produced should be at least four times the number of tables produced.

33. *Manufacturing—production scheduling.* A furniture company has two plants that produce the lumber used in manufacturing tables and chairs. In 1 day of operation, plant *A* can produce the lumber required to manufacture 20 tables and 60 chairs, and plant *B* can produce the lumber required to manufacture 25 tables and 50 chairs. The company needs enough lumber to manufacture at least 200 tables and 500 chairs.

(A) If it costs $1,000 to operate plant *A* for 1 day and $900 to operate plant *B* for 1 day, how many days should each plant be operated in order to produce a sufficient amount of lumber at a minimum cost? What is the minimum cost?

(B) Discuss the effect on the operating schedule and the minimum cost if the daily cost of operating plant *A* is reduced to $600 and all other data in part (A) remain the same.

(C) Discuss the effect on the operating schedule and the minimum cost if the daily cost of operating plant *B* is reduced to $800 and all other data in part (A) remain the same.

34. *Manufacturing—resource allocation.* An electronics firm manufactures two types of personal computers, a standard model and a portable model. The production of a standard computer requires a capital expenditure of $400 and 40 hours of labor. The production of a portable computer requires a capital expenditure of $250 and 30 hours of labor. The firm has $20,000 capital and 2,160 labor-hours available for production of standard and portable computers.

(A) What is the maximum number of computers the company is capable of producing?

(B) If each standard computer contributes a profit of $320 and each portable model contributes a profit of $220, how much profit will the company make by producing the maximum number of computers determined in part (A)? Is this the maximum profit? If not, what is the maximum profit?

35. *Transportation.* The officers of a high school senior class are planning to rent buses and vans for a class trip. Each bus can transport 40 students, requires 3 chaperones, and costs $1,200 to rent. Each van can transport 8 students, requires 1 chaperone, and costs $100 to rent. Since there are 400 students in the senior class that may be eligible to go on the trip, the officers must plan to accommodate at least 400 students. Since only 36 parents have volunteered to serve as chaperones, the officers must plan to use at most 36 chaperones. How many vehicles of each type should the officers rent in order to minimize the transportation costs? What are the minimal transportation costs?

36. *Transportation.* Refer to Problem 35. If each van can transport 7 people and there are 35 available chaperones, show that the optimal solution found graphically involves decimals. Find all feasible solutions with integer coordinates and identify the one that minimizes the transportation costs. Can this optimal integer solution be obtained by rounding the optimal decimal solution? Explain.

37. *Investment.* An investor has $60,000 to invest in a CD and a mutual fund. The CD yields 5% and the mutual fund yields on the average 9%. The mutual fund requires a minimum investment of $10,000 and the investor requires that twice as much should be invested in CD's as in the mutual fund. How much should be invested in CD's and how much in the mutual fund to maximize the return? What is the maximum return?

38. *Investment.* An investor has $24,000 to invest in bonds of AAA and B qualities. The AAA bonds yield on the average 6% and the B bonds yield 10%. The investor requires that at least three times as much money should be invested in AAA bonds as in B bonds. How much should be invested in each type of bond to maximize the return? What is the maximum return?

39. *Pollution control.* Because of new federal regulations on pollution, a chemical plant introduced a new, more expensive process to supplement or replace an older process used in the production of a particular chemical. The older process emitted 20 grams of sulfur dioxide and 40 grams of particulate matter into the atmosphere for each gallon of chemical produced. The new process emits 5 grams of sulfur dioxide and 20 grams of particulate matter for each gallon produced. The company makes a profit of 60¢ per gallon and 20¢ per gallon on the old and new processes, respectively.

(A) If the government allows the plant to emit no more than 16,000 grams of sulfur dioxide and 30,000 grams of particulate matter daily, how many gallons of the chemical should be produced by each process to maximize daily profit? What is the maximum daily profit?

(B) Discuss the effect on the production schedule and the maximum profit if the government decides to restrict emissions of sulfur dioxide to 11,500 grams daily and all other data remain unchanged.

(C) Discuss the effect on the production schedule and the maximum profit if the government decides to restrict emissions of sulfur dioxide to 7,200 grams daily and all other data remain unchanged.

40. *Capital expansion.* A fast-food chain plans to expand by opening several new restaurants. The chain operates two types of restaurants, drive-through and full-service. A drive-through restaurant costs $100,000 to construct, requires 5 employees, and has an expected annual revenue of $200,000. A full-service restaurant costs $150,000 to construct, requires 15 employees, and has an expected annual revenue of $500,000. The chain has $2,400,000 in capital available for expansion. Labor contracts require that they hire no more than 210 employees, and licensing restrictions require that they open no more than 20 new restaurants. How many restaurants of each type should the chain open in order to maximize the expected revenue? What is the maximum expected revenue? How much of their capital will they use and how many employees will they hire?

Life Sciences

41. *Nutrition—plants.* A fruit grower can use two types of fertilizer in his orange grove, brand *A* and brand *B*. The amounts (in pounds) of nitrogen, phosphoric acid, and chlorine in a bag of each brand are given in the table. Tests indicate that the grove needs at least 1,000 pounds of phosphoric acid and at most 400 pounds of chlorine.

	POUNDS PER BAG	
	Brand A	*Brand B*
NITROGEN	8	3
PHOSPHORIC ACID	4	4
CHLORINE	2	1

(A) If the grower wants to maximize the amount of nitrogen added to the grove, how many bags of each mix should be used? How much nitrogen will be added?

(B) If the grower wants to minimize the amount of nitrogen added to the grove, how many bags of each mix should be used? How much nitrogen will be added?

42. *Nutrition—people.* A dietitian in a hospital is to arrange a special diet composed of two foods, *M* and *N*. Each ounce of food *M* contains 30 units of calcium, 10 units of iron, 10 units of vitamin A, and 8 units of cholesterol. Each ounce of food *N* contains 10 units of calcium, 10 units of iron, 30 units of vitamin A, and 4 units of cholesterol. If the minimum daily requirements are 360 units of calcium, 160 units of iron, and 240 units of vitamin A, how many ounces of each food should be used to meet the minimum requirements and at the same time minimize the cholesterol intake? What is the minimum cholesterol intake?

43. *Nutrition—plants.* A farmer can buy two types of plant food, mix *A* and mix *B*. Each cubic yard of mix *A* contains 20 pounds of phosphoric acid, 30 pounds of nitrogen, and 5 pounds of potash. Each cubic yard of mix *B* contains 10 pounds of phosphoric acid, 30 pounds of nitrogen, and 10 pounds of potash. The minimum monthly requirements are 460 pounds of phosphoric acid, 960 pounds of nitrogen, and 220 pounds of potash. If mix *A* costs $30 per cubic yard and mix *B* costs $35 per cubic yard, how many cubic yards of each mix should the farmer blend to meet the minimum monthly requirements at a minimal cost? What is this cost?

44. *Nutrition—animals.* A laboratory technician in a medical research center is asked to formulate a diet from two commercially packaged foods, food *A* and food *B*, for a group of animals. Each ounce of food *A* contains 8 units of fat, 16 units of carbohydrate, and 2 units of protein. Each ounce of food *B* contains 4 units of fat, 32 units of carbohydrate, and 8 units of protein. The minimum daily requirements are 176 units of fat, 1,024 units of carbohydrate, and 384 units of protein. If food *A* costs 5¢ per ounce and food *B* costs 5¢ per ounce, how many ounces of each food should be used to meet the minimum daily requirements at the least cost? What is the cost for this amount of food?

Social Sciences

45. *Psychology.* In an experiment on conditioning, a psychologist uses two types of Skinner boxes with mice and rats. The amount of time (in minutes) each mouse and each rat spends in each box per day is given in the table. What is the maximum number of mice and rats that can be used in this experiment? How many mice and how many rats produce this maximum?

	TIME		MAXIMUM TIME AVAILABLE PER DAY
	Mice	*Rats*	
SKINNER BOX *A*	10 min	20 min	800 min
SKINNER BOX *B*	20 min	10 min	640 min

46. *Sociology.* A city council voted to conduct a study on inner-city community problems. A nearby university was contacted to provide sociologists and research assistants. Allocation of time and costs per week are given in the table. How many sociologists and how many research assistants should be hired to minimize the cost and meet the weekly labor-hour requirements? What is the minimum weekly cost?

	Labor-Hours		Minimum Labor-Hours Needed per Week
	Sociologist	*Research assistant*	
Fieldwork	10	30	180
Research Center	30	10	140
Costs per Week	\$500	\$300	

Section 5-3

A Geometric Introduction to the Simplex Method

- Standard Maximization Problems in Standard Form
- Slack Variables
- Basic and Nonbasic Variables; Basic Solutions and Basic Feasible Solutions
- Basic Feasible Solutions and the Simplex Method

The geometric method of solving linear programming problems provides us with an overview of the subject and some useful terminology. But, practically speaking, the method is useful only for problems involving two decision variables and relatively few problem constraints. What happens when we need more decision variables and more problem constraints? We use an algebraic method called the *simplex method,* which was developed by George B. Dantzig in 1947 while on assignment to the U.S. Department of the Air Force. Ideally suited to computer use, the method is used routinely on applied problems involving hundreds and even thousands of variables and problem constraints.

The algebraic procedures utilized in the simplex method require the problem constraints to be written as equations rather than inequalities. This new form of the linear programming problem also prompts the use of some new terminology. We introduce this new form and associated terminology through a simple example and an appropriate geometric interpretation. From this example we can illustrate what the simplex method does geometrically before we immerse ourselves in the algebraic details of the process.

Standard Maximization Problems in Standard Form

We now return to the tent production problem in Example 1 from the last section. Recall the mathematical model for the problem:

$$
\begin{array}{lrl}
\text{Maximize} & P = 50x_1 + 80x_2 & \text{Objective function} \\
\text{Subject to} & x_1 + 2x_2 \leq 32 & \text{Cutting department constraint} \\
 & 3x_1 + 4x_2 \leq 84 & \text{Assembly department constraint} \\
 & x_1, x_2 \geq 0 & \text{Nonnegative constraints}
\end{array}
\qquad (1)
$$

where the decision variables x_1 and x_2 are the number of standard and expedition tents, respectively, produced each day.

Notice that the problem constraints involve $\leq$ inequalities with positive constants to the right of the inequality. Maximization problems that satisfy this condition are called *standard maximization problems.* In this and the next section we will restrict our attention to standard maximization problems.

A Standard Maximization Problem in Standard Form

A linear programming problem is said to be a **standard maximization problem in standard form** if its mathematical model is of the following form:

Maximize the objective function

$$P = c_1x_1 + c_2x_2 + \cdots + c_nx_n$$

Subject to problem constraints of the form

$$a_1x_1 + a_2x_2 + \cdots + a_nx_n \leq b \qquad b \geq 0$$

With nonnegative constraints

$$x_1, x_2, \ldots, x_n \geq 0$$

[*Note:* Mathematical model (1) above is a standard maximization problem in standard form. Also note that the coefficients of the objective function can be any real numbers.]

Explore–Discuss 1

Find an example of a standard maximization problem in standard form involving two variables and one problem constraint such that:

(A) The feasible region is bounded.
(B) The feasible region is unbounded.

Is it possible for a standard maximization problem to have no solution? Explain.

Slack Variables

To adapt a linear programming problem to the matrix methods used in the simplex process (as discussed in the next section), we convert the problem constraint inequalities into a system of linear equations by using a simple device called a *slack variable.* In particular, to convert the system of problem constraint inequalities from (1),

$$\begin{aligned} x_1 + 2x_2 &\leq 32 \quad \text{Cutting department constraint} \\ 3x_1 + 4x_2 &\leq 84 \quad \text{Assembly department constraint} \end{aligned} \tag{2}$$

into a system of equations, we add variables s_1 and s_2 to the left sides of (2) to obtain

$$\begin{aligned} x_1 + 2x_2 + s_1 \qquad &= 32 \\ 3x_1 + 4x_2 \qquad + s_2 &= 84 \end{aligned} \tag{3}$$

The variables s_1 and s_2 are called **slack variables** because each makes up the difference (takes up the slack) between the left and right sides of the inequalities in (2). For example, if we produced 20 standard tents ($x_1 = 20$) and 5 expedition tents ($x_2 = 5$), then the number of labor-hours used in the cutting department would be $20 + 2(5) = 30$, leaving a slack of 2 unused labor-hours out of the 32 available. Thus, s_1 would have the value of 2.

Notice that if the decision variables x_1 and x_2 satisfy the system of constraint inequalities (2), then the slack variables s_1 and s_2 are nonnegative. We will have more to say about this later in this discussion.

Basic and Nonbasic Variables; Basic Solutions and Basic Feasible Solutions

Observe that system (3) has infinitely many solutions—just solve for s_1 and s_2 in terms of x_1 and x_2, and then assign x_1 and x_2 arbitrary values. Certain solutions of system (3), called *basic solutions,* are related to the intersection points of the (extended) boundary lines of the feasible region in Figure 1.

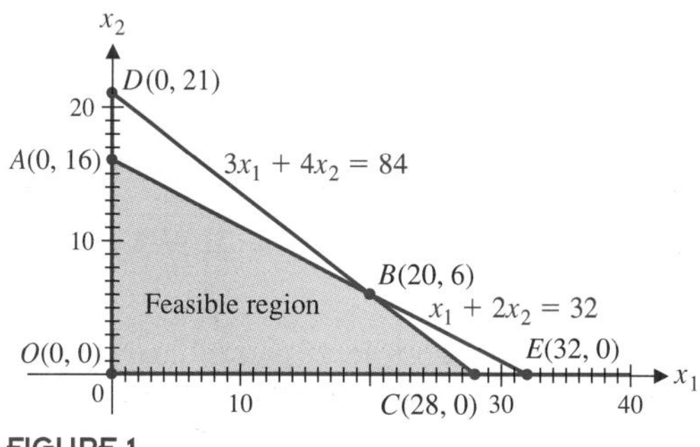

FIGURE 1

How are basic solutions to system (3) determined? System (3) involves four variables and two equations. We divide the four variables into two groups, called *basic variables* and *nonbasic variables,* as follows: Basic variables are selected arbitrarily with the restriction that there be as many basic variables as there are equations. The remaining variables are nonbasic variables.

Since system (3) has two equations, we can select any two of the four variables as basic variables. The remaining two variables are then nonbasic variables. A solution found by setting the two nonbasic variables equal to 0 and solving for the two basic variables is a basic solution. [Note that setting two variables equal to 0 in system (3) results in a system of two equations with two variables, which has (from Chapter 4) exactly one solution, infinitely many solutions, or no solution.]

Example 1 **Basic Solutions and the Feasible Region**

(A) Find two basic solutions for system (3) by first selecting s_1 and s_2 as basic variables, and then by selecting x_2 and s_1 as basic variables.

(B) Associate each basic solution found in part (A) with an intersection point of the (extended) boundary lines of the feasible region in Figure 1, and indicate which boundary lines produce each intersection point.

(C) Indicate which of the intersection points found in part (B) are in the feasible region.

SOLUTION (A) With s_1 and s_2 selected as basic variables, x_1 and x_2 are nonbasic variables. A basic solution is found by setting the nonbasic variables equal to 0 and solving for the basic variables. If $x_1 = 0$ and $x_2 = 0$, then system (3) becomes

$$\begin{aligned} s_1 &= 32 \\ s_2 &= 84 \end{aligned} \qquad \begin{array}{ccccccc} 0 & & 0 & & & & \\ x_1 & + & 2x_2 & + & s_1 & & = 32 \\ 3x_1 & + & 4x_2 & & & + s_2 & = 84 \end{array}$$

and the basic solution is

$$x_1 = 0, \quad x_2 = 0, \quad s_1 = 32, \quad s_2 = 84 \tag{4}$$

If we select x_2 and s_1 as basic variables, then x_1 and s_2 are nonbasic variables. Setting the nonbasic variables equal to 0, system (3) becomes

$$\begin{aligned} 2x_2 + s_1 &= 32 \\ 4x_2 &= 84 \end{aligned} \qquad \begin{array}{ccccccc} 0 & & & & & 0 & \\ x_1 & + 2x_2 & + s_1 & & & & = 32 \\ 3x_1 & + 4x_2 & & & + & s_2 & = 84 \end{array}$$

Solving, we see that $x_2 = 21$ and $s_1 = -10$, and the basic solution is

$$x_1 = 0, \quad x_2 = 21, \quad s_1 = -10, \quad s_2 = 0 \tag{5}$$

(B) Basic solution (4)—since $x_1 = 0$ and $x_2 = 0$—corresponds to the origin $O(0, 0)$ in Figure 1, which is the intersection of the boundary lines $x_1 = 0$ and $x_2 = 0$. Basic solution (5)—since $x_1 = 0$ and $x_2 = 21$—corresponds to the intersection point $D(0, 21)$, which is the intersection of the boundary lines $x_1 = 0$ and $3x_1 + 4x_2 = 84$.

(C) The intersection point $O(0, 0)$ is in the feasible region; hence, it is natural to call the corresponding basic solution a *basic feasible solution.* The intersection point $D(0, 21)$ is not in the feasible region; hence, the corresponding basic solution is not feasible.

Matched Problem 1 (A) Find two basic solutions for system (3) by first selecting x_1 and s_1 as basic variables, and then by selecting x_1 and s_2 as basic variables.

(B) Associate each basic solution found in part (A) with an intersection point of the (extended) boundary lines of the feasible region in Figure 1, and indicate which boundary lines produce each intersection point.

(C) Indicate which of the intersection points found in part (B) are in the feasible region.

Proceeding systematically as in Example 1 and Matched Problem 1, we can obtain all basic solutions to system (3). The results are summarized in Table 1, which also includes geometric interpretations of the basic solutions relative to Figure 1. Figure 2 summarizes these interpretations. A careful study of Table 1 and Figure 2 is very worthwhile. (Note that to be sure we have listed all possible basic solutions in Table 1, it is convenient to organize the table in terms of the zero values of the nonbasic variables.)

Table 1

BASIC SOLUTIONS

BASIC SOLUTIONS				INTERSECTION	INTERSECTING	
x_1	x_2	s_1	s_2	POINT	BOUNDARY LINES	FEASIBLE
0	0	32	84	$O(0, 0)$	$x_1 = 0$ $x_2 = 0$	Yes
0	16	0	20	$A(0, 16)$	$x_1 = 0$ $x_1 + 2x_2 = 32$	Yes
0	21	−10	0	$D(0, 21)$	$x_1 = 0$ $3x_1 + 4x_2 = 84$	No
32	0	0	−12	$E(32, 0)$	$x_2 = 0$ $x_1 + 2x_2 = 32$	No
28	0	4	0	$C(28, 0)$	$x_2 = 0$ $3x_1 + 4x_2 = 84$	Yes
20	6	0	0	$B(20, 6)$	$x_1 + 2x_2 = 32$ $3x_1 + 4x_2 = 84$	Yes

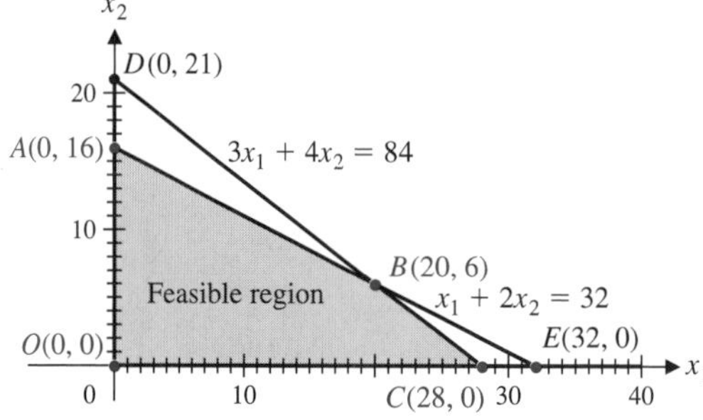

FIGURE 2
Basic feasible solutions: O, A, B, C
Basic solutions that are not feasible: D, E

OBSERVATIONS FROM TABLE 1 AND FIGURE 2 IMPORTANT TO THE DEVELOPMENT OF THE SIMPLEX METHOD

1. In Table 1, observe that a basic solution that is not feasible includes at least one negative value and that a basic feasible solution does not include any negative values. That is, we can determine the feasibility of a basic solution simply by examining the signs of all the variables in the solution.
2. In Table 1 and Figure 2, observe that basic feasible solutions are associated with the corner points of the feasible region, which include the optimal solution to the original linear programming problem.

Before proceeding further, let us formalize the definitions alluded to in the discussion above so that they apply to standard maximization problems in general, without any reference to geometric forms.

Basic Variables and Nonbasic Variables; Basic Solutions and Basic Feasible Solutions

Given a system of linear equations associated with a linear programming problem (such a system will always have more variables than equations):

> The variables are divided into two (mutually exclusive) groups, as follows: **Basic variables** are selected arbitrarily with the one restriction that there be as many basic variables as there are equations. The remaining variables are called **nonbasic variables.**
>
> A solution found by setting the nonbasic variables equal to 0 and solving for the basic variables is called a **basic solution.** If a basic solution has no negative values, it is a **basic feasible solution.**

Example 2 **Basic Variables and Basic Solutions** Suppose there is a system of three problem constraint equations with eight (slack and decision) variables associated with a standard maximization problem.

(A) How many basic variables and how many nonbasic variables are associated with the system?
(B) Setting the nonbasic variables equal to 0 will result in a system of how many linear equations with how many variables?
(C) If a basic solution has all nonnegative elements, is it feasible or not feasible?

SOLUTION
(A) Since there are three equations in the system, there should be three basic variables and five nonbasic variables.
(B) Three equations with three variables
(C) Feasible

Matched Problem 2 Suppose there is a system of five problem constraint equations with eleven (slack and decision) variables associated with a standard maximization problem.

(A) How many basic variables and how many nonbasic variables are associated with the system?
(B) Setting the nonbasic variables equal to 0 will result in a system of how many linear equations with how many variables?
(C) If a basic solution has one or more negative elements, is it feasible or not feasible?

Basic Feasible Solutions and the Simplex Method

The following important theorem, which is equivalent to the fundamental theorem (Theorem 1 in the preceding section), is stated without proof:

Theorem 1 FUNDAMENTAL THEOREM OF LINEAR PROGRAMMING—VERSION 2

If the optimal value of the objective function in a linear programming problem exists, then that value must occur at one (or more) of the basic feasible solutions.

Now you can understand why the concepts of basic and nonbasic variables and basic solutions and basic feasible solutions are so important—these concepts are central to the process of finding optimal solutions to linear programming problems.

Explore–Discuss 2

If we know that a standard maximization problem has an optimal solution, how could we use a table of basic solutions like Table 1 to find the optimal solution?

We have taken the first step toward finding a general method of solving linear programming problems involving any number of variables and problem constraints. That is, **we have found a method of identifying all the corner points (basic feasible solutions) of a feasible region without drawing its graph.** This is a critical step if we want to consider problems with more than two decision variables. Unfortunately, the number of corner points increases dramatically as the number of variables and constraints increases. In real-world problems, it is not practical to find all the corner points in order to find the optimal solution. Thus, the next step is to find a method of locating the optimal solution without finding every corner point. The procedure for doing this is the simplex method mentioned at the beginning of this section.

The simplex method, using a special matrix and row operations, automatically moves from one basic feasible solution to another—that is, from one corner point of the feasible region to another—each time getting closer to an optimal solution (if one exists), until an optimal solution is reached. Then the process stops. A remarkable property of the simplex method is that in large linear programming problems it usually arrives at an optimal solution (if one exists) by testing relatively few of the large number of basic feasible solutions (corner points) available.

With this background, we are now ready to discuss the algebraic details of the simplex method in the next section.

Answers to Matched Problems

1. (A) Basic solution corresponding to basic variables x_1 and s_1: $x_1 = 28, x_2 = 0, s_1 = 4, s_2 = 0$. Basic solution corresponding to basic variables x_1 and s_2: $x_1 = 32, x_2 = 0, s_1 = 0, s_2 = -12$.

(B) The first basic solution corresponds to $C(28, 0)$, which is the intersection of the boundary lines $x_2 = 0$ and $3x_1 + 4x_2 = 84$. The second basic solution corresponds to $E(32, 0)$, which is the intersection of the boundary lines $x_2 = 0$ and $x_1 + 2x_2 = 32$.

(C) $C(28, 0)$ is in the feasible region (hence, the corresponding basic solution is a basic feasible solution); $E(32, 0)$ is not in the feasible region (hence, the corresponding basic solution is not feasible).

2. (A) Five basic variables and six nonbasic variables

(B) Five equations and five variables (C) Not feasible

EXERCISE 5-3

A

1. Discuss the relationship between a standard maximization problem with two problem constraints and three decision variables, and the associated system of problem constraint equations. In particular, find each of the following quantities and explain how each was determined:
(A) The number of slack variables that must be introduced to form the system of problem constraint equations
(B) The number of basic variables and the number of nonbasic variables associated with the system
(C) The number of linear equations and the number of variables in the system formed by setting the nonbasic variables equal to 0

2. Repeat Problem 1 for a standard maximization problem with three problem constraints and four decision variables.

3. Discuss the relationship between a standard maximization problem and the associated system of problem constraint equations, if the system of problem constraint equations has nine variables, including five slack variables. In particular, find each of the following quantities and explain how each was determined:
(A) The number of constraint equations in the system
(B) The number of decision variables in the system
(C) The number of basic variables and the number of nonbasic variables associated with the system
(D) The number of linear equations and the number of variables in the system formed by setting the nonbasic variables equal to 0

4. Repeat Problem 3 if the system of problem constraint equations has ten variables, including four slack variables.

5. Listed in the table below are all the basic solutions for the system

$$\begin{aligned} 2x_1 + 3x_2 + s_1 \quad &= 24 \\ 4x_1 + 3x_2 \quad + s_2 &= 36 \end{aligned}$$

For each basic solution, identify the nonbasic variables and the basic variables. Then classify each basic solution as feasible or not feasible.

	x_1	x_2	s_1	s_2
(A)	0	0	24	36
(B)	0	8	0	12
(C)	0	12	−12	0
(D)	12	0	0	−12
(E)	9	0	6	0
(F)	6	4	0	0

6. Repeat Problem 5 for the system

$$\begin{aligned} 2x_1 + x_2 + s_1 \quad &= 30 \\ x_1 + 5x_2 \quad + s_2 &= 60 \end{aligned}$$

whose basic solutions are given in the following table:

	x_1	x_2	s_1	s_2
(A)	0	0	30	60
(B)	0	30	0	−90
(C)	0	12	18	0
(D)	15	0	0	45
(E)	60	0	−90	0
(F)	10	10	0	0

7. Listed in the table below are all the possible choices of nonbasic variables for the system

$$\begin{aligned} 2x_1 + x_2 + s_1 \quad &= 50 \\ x_1 + 2x_2 \quad + s_2 &= 40 \end{aligned}$$

In each case, find the values of the basic variables and determine whether the basic solution is feasible.

	x_1	x_2	s_1	s_2
(A)	0	0	?	?
(B)	0	?	0	?
(C)	0	?	?	0
(D)	?	0	0	?
(E)	?	0	?	0
(F)	?	?	0	0

8. Repeat Problem 7 for the system

$$\begin{aligned} x_1 + 2x_2 + s_1 \quad &= 12 \\ 3x_1 + 2x_2 \quad + s_2 &= 24 \end{aligned}$$

B *Graph the systems of inequalities in Problems 9–12. Introduce slack variables to convert each system of inequalities to a system of equations, and find all the basic solutions of the system. Construct a table (similar to Table 1) listing each basic solution, the corresponding point on the graph, and whether the basic solution is feasible. (You do not need to list the intersecting lines.)*

9. $$\begin{aligned} x_1 + x_2 &\le 16 \\ 2x_1 + x_2 &\le 20 \\ x_1, x_2 &\ge 0 \end{aligned}$$

10. $$\begin{aligned} 5x_1 + x_2 &\le 35 \\ 4x_1 + x_2 &\le 32 \\ x_1, x_2 &\ge 0 \end{aligned}$$

11. $$\begin{aligned} 2x_1 + x_2 &\le 22 \\ x_1 + x_2 &\le 12 \\ x_1 + 2x_2 &\le 20 \\ x_1, x_2 &\ge 0 \end{aligned}$$

12. $$\begin{aligned} 4x_1 + x_2 &\le 28 \\ 2x_1 + x_2 &\le 16 \\ x_1 + x_2 &\le 13 \\ x_1, x_2 &\ge 0 \end{aligned}$$

SECTION 5-4

The Simplex Method: Maximization with Problem Constraints of the Form ≤

- INITIAL SYSTEM
- THE SIMPLEX TABLEAU
- THE PIVOT OPERATION
- INTERPRETING THE SIMPLEX PROCESS GEOMETRICALLY
- THE SIMPLEX METHOD SUMMARIZED
- APPLICATION

We are now ready to develop the simplex method for a standard maximization problem. Specific details in the presentation of the method generally vary from from one book to another, even though the underlying process is the same. The presentation developed here emphasizes concept development and understanding.

As pointed out in the last section, the simplex method is most useful when used with computers. Consequently, it is not intended that you become expert in manually solving linear programming problems using the simplex method. But it is important that you become proficient in constructing the models for linear programming problems so that they can be solved using a computer, and it is also important that you develop skill in interpreting the results. One way to gain this proficiency and interpretive skill is to set up and manually solve a number of fairly simple linear programming problems using the simplex method. This is the main goal here and in Sections 5-5 and 5-6. To assist you in learning to develop the models, the answer sections for Exercises 5-4, 5-5, and 5-6 contain both the model and its solution. The software that accompanies *Explorations in Finite Mathematics* (see Preface) can be used to solve the linear programming problems in this chapter, as can spreadsheets, such as Excel.

INITIAL SYSTEM

We will introduce the concepts and procedures involved in the simplex method through an example—the tent production example we have discussed earlier. We restate the problem here in standard form for convenient reference:

$$\begin{array}{llr} \text{Maximize} & P = 50x_1 + 80x_2 & \text{Objective function} \\ \text{Subject to} & \left.\begin{array}{r} x_1 + 2x_2 \le 32 \\ 3x_1 + 4x_2 \le 84 \end{array}\right\} & \text{Problem constraints} \\ & x_1, x_2 \ge 0 & \text{Nonnegative constraints} \end{array} \qquad (1)$$

Introducing slack variables s_1 and s_2, we convert the problem constraint inequalities in (1) into the following system of problem constraint equations:

$$\begin{aligned} x_1 + 2x_2 + s_1 \quad\;\; &= 32 \\ 3x_1 + 4x_2 \quad\;\; + s_2 &= 84 \\ x_1, x_2, s_1, s_2 &\ge 0 \end{aligned} \qquad (2)$$

Since a basic solution of (2) is not feasible if it contains any negative values, we have also included the nonnegative constraints for both the decision variables x_1 and x_2 and the slack variables s_1 and s_2. From our discussion in the last section, we know that out of the infinitely many solutions to system (2), an optimal solution is among the basic feasible solutions, which correspond to the corner points of the feasible region.

As part of the simplex method we add the objective function equation $P = 50x_1 + 80x_2$ in the form $-50x_1 - 80x_2 + P = 0$ to system (2) to create what is called the **initial system:**

$$\begin{aligned} x_1 + 2x_2 + s_1 \qquad\qquad &= 32 \\ 3x_1 + 4x_2 \qquad + s_2 \qquad &= 84 \\ -50x_1 - 80x_2 \qquad\qquad + P &= 0 \\ x_1, x_2, s_1, s_2 &\geq 0 \end{aligned} \tag{3}$$

When we add the objective function equation to system (2), we must slightly modify the earlier definitions of basic solution and basic feasible solution so that they apply to the initial system (3).

Basic Solutions and Basic Feasible Solutions for Initial Systems

1. The objective function variable P is always selected as a basic variable and is never selected as a nonbasic variable.
2. Note that a basic solution of system (3) is also a basic solution of system (2) after P is deleted.
3. If a basic solution of system (3) is a basic feasible solution of system (2) after deleting P, then the basic solution of (3) is called a **basic feasible solution** of (3).
4. A basic feasible solution of system (3) can contain a negative number, but only if it is the value of P, the objective function variable.

These changes lead to a small change in the second version of the fundamental theorem (see Theorem 1 in Section 5-3).

Theorem 1 ■■ FUNDAMENTAL THEOREM OF LINEAR PROGRAMMING—VERSION 3

If the optimal value of the objective function in a linear programming problem exists, then that value must occur at one (or more) of the basic feasible solutions of the initial system. ■■

With these adjustments understood, we start the simplex process with a basic feasible solution of the initial system (3), which we will refer to as an **initial basic feasible solution.** An initial basic feasible solution that is easy to find is the one associated with the origin.

Since system (3) has three equations and five variables, it has three basic variables and two nonbasic variables. Looking at the system, we see that x_1 and x_2 appear in all equations and s_1, s_2, and P each appears only once and each in a different equation. A basic solution can be found by inspection by selecting s_1, s_2, and P as the basic variables (remember, P is always selected as a basic variable) and x_1 and x_2 as the nonbasic variables to be set equal to 0. Setting x_1 and x_2 equal to 0 and solving for the basic variables, we obtain the basic solution:

$$x_1 = 0, \quad x_2 = 0, \quad s_1 = 32, \quad s_2 = 84, \quad P = 0$$

This basic solution is feasible since none of the variables (excluding P) are negative. Thus, this is the initial basic feasible solution we seek.

Now you can see why we wanted to add the objective function equation to system (2): A basic feasible solution of (3) not only includes a basic feasible solution of (2), but, in addition, it includes the value of P for that basic feasible solution of (2).

The initial basic feasible solution we just found is associated with the origin. Of course, if we do not produce any tents, we do not expect a profit, so $P = \$0$. Starting with this easily obtained initial basic feasible solution, the simplex process moves through each iteration (each repetition) to another basic feasible solution, each time improving the profit, and the process continues until the maximum profit is reached. Then the process stops.

■ The Simplex Tableau

To facilitate the search for the optimal solution, we now turn to matrix methods discussed in Chapter 4. Our first step is to write the augmented matrix for the initial system (3). This matrix is called the **initial simplex tableau,** and it is simply a tabulation of the coefficients in system (3).

$$\begin{array}{c} \\ s_1 \\ s_2 \\ P \end{array} \begin{array}{c} \begin{array}{ccccc|c} x_1 & x_2 & s_1 & s_2 & P & \end{array} \\ \left[\begin{array}{ccccc|c} 1 & 2 & 1 & 0 & 0 & 32 \\ 3 & 4 & 0 & 1 & 0 & 84 \\ \hdashline -50 & -80 & 0 & 0 & 1 & 0 \end{array}\right] \end{array} \quad \text{Initial simplex tableau} \qquad (4)$$

In tableau (4), the row below the dashed line always corresponds to the objective function. Each of the basic variables we selected above, s_1, s_2, and P, is also placed on the left of the tableau so that the intersection element in its row and column is not 0. For example, we place the basic variable s_1 on the left so that the intersection element of the s_1 row and the s_1 column is 1 and not 0. The basic variable s_2 is similarly placed. The objective function variable P is always placed at the bottom. The reason for writing the basic variables on the left in this way is that this placement makes it possible to read certain basic feasible solutions directly from the tableau. If $x_1 = 0$ and $x_2 = 0$, then the basic variables on the left of tableau (4) are lined up with their corresponding values, 32, 84, and 0, to the right of the vertical line.

Looking at tableau (4) relative to the choice of s_1, s_2, and P as basic variables, we see that each basic variable is above a column that has all

0 elements except for a single 1 and that no two such columns contain 1's in the same row. These observations lead to a formalization of the process of selecting basic and nonbasic variables that is an important part of the simplex method:

Selecting Basic and Nonbasic Variables for the Simplex Method

Given a simplex tableau:

Step 1. Determine the number of basic variables and the number of nonbasic variables. These numbers do not change during the simplex process.

Step 2. *Selecting Basic Variables.* A variable can be selected as a basic variable only if it corresponds to a column in the tableau that has exactly one nonzero element (usually 1) and the nonzero element in the column is not in the same row as the nonzero element in the column of another basic variable. (This procedure always selects P as a basic variable, since the P column never changes during the simplex process.)

Step 3. *Selecting Nonbasic Variables.* After the basic variables are selected in step 2, the remaining variables are selected as the nonbasic variables. (The tableau columns under the nonbasic variables usually contain more than one nonzero element.)

The earlier selection of s_1, s_2, and P as basic variables and x_1 and x_2 as nonbasic variables conforms to this prescribed convention of selecting basic and nonbasic variables for the simplex process.

The Pivot Operation

The simplex method will now switch one of the nonbasic variables, x_1 or x_2, for one of the basic variables, s_1 or s_2 (but not P), as a step toward improving the profit. For a nonbasic variable to be classified as a basic variable we need to perform appropriate row operations on the tableau so that the newly selected basic variable will end up with exactly one nonzero element in its column. In this process, the old basic variable will usually gain additional nonzero elements in its column as it becomes nonbasic.

Which nonbasic variable should we select to become basic? It makes sense to select the nonbasic variable that will increase the profit the most per unit change in that variable. Looking at the objective function

$$P = 50x_1 + 80x_2$$

we see that if x_1 stays a nonbasic variable (set equal to 0) and if x_2 becomes a new basic variable, then

$$P = 50(0) + 80x_2 = 80x_2$$

and for each unit increase in x_2, P will increase \$80. If x_2 stays a nonbasic variable and x_1 becomes a new basic variable, then (reasoning in the same way) for each unit increase in x_1, P will increase only \$50. So, we select the nonbasic variable x_2 to enter the set of basic variables, and call it the **entering variable.** (The basic variable leaving the set of basic variables to become a nonbasic variable is called the **exiting variable.** Exiting variables will be discussed shortly.)

We call the column corresponding to the entering variable the **pivot column.** Looking at the bottom row in tableau (4)—the objective function row below the dashed line—we see that the pivot column is associated with the column to the left of the P column that has the most negative bottom element. In general, the most negative element in the bottom row to the left of the P column *indicates* the variable above it that will produce the greatest increase in P for a unit increase in that variable. For this reason, we call the elements in the bottom row of the tableau to the left of the P column **indicators.**

We illustrate the indicators, the pivot column, the entering variable, and the initial basic feasible solution below:

Entering variable ↓

$$\begin{array}{c} \\ s_1 \\ s_2 \\ P \end{array}\begin{array}{c} \begin{array}{cccccc} x_1 & x_2 & s_1 & s_2 & P & \end{array} \\ \left[\begin{array}{rrrrr|r} 1 & 2 & 1 & 0 & 0 & 32 \\ 3 & 4 & 0 & 1 & 0 & 84 \\ \hline -50 & -80 & 0 & 0 & 1 & 0 \end{array}\right] \end{array} \qquad (5)$$

Initial simplex tableau

Indicators are shown in color.

↑ Pivot column

$x_1 = 0, \quad x_2 = 0, \quad s_1 = 32, \quad s_2 = 84, \quad P = 0$ Initial basic feasible solution

Now that we have chosen the nonbasic variable x_2 as the entering variable (the nonbasic variable to become basic), which of the two basic variables, s_1 or s_2, should we choose as the exiting variable (the basic variable to become nonbasic)? We saw above that for $x_1 = 0$, each unit increase in the entering variable x_2 results in an increase of \$80 for P. Can we increase x_2 without limit? No! A limit is imposed by the nonnegative requirements for s_1 and s_2. (Remember that if any of the basic variables except P become negative, we no longer have a feasible solution.) So we rephrase the question and ask: How much can x_2 be increased when $x_1 = 0$ without causing s_1 or s_2 to become negative? To see how much x_2 can be increased, we refer to tableau (5) or system (3) and write the two problem constraint equations with $x_1 = 0$:

$$\begin{aligned} 2x_2 + s_1 &= 32 \\ 4x_2 + s_2 &= 84 \end{aligned}$$

Solving for s_1 and s_2, we have

$$\begin{aligned} s_1 &= 32 - 2x_2 \\ s_2 &= 84 - 4x_2 \end{aligned}$$

For s_1 and s_2 to be nonnegative, x_2 must be chosen so that both $32 - 2x_2$ and $84 - 4x_2$ are nonnegative. That is, so that

$$\begin{aligned} 32 - 2x_2 &\geq 0 \\ -2x_2 &\geq -32 \\ x_2 &\leq \tfrac{32}{2} = 16 \end{aligned} \qquad \text{and} \qquad \begin{aligned} 84 - 4x_2 &\geq 0 \\ -4x_2 &\geq -84 \\ x_2 &\leq \tfrac{84}{4} = 21 \end{aligned}$$

For both inequalities to be satisfied, x_2 must be less than or equal to the smaller of the values, which is 16. Thus, x_2 can increase to 16 without either s_1 or s_2 becoming negative. Now, observe how each value (16 and 21) can be obtained directly from tableau (6) below:

Entering variable ↓ (column x_2)

$$\begin{array}{c} \\ s_1 \\ s_2 \\ P \end{array} \begin{array}{c} \begin{array}{cccccc} x_1 & x_2 & s_1 & s_2 & P & \end{array} \\ \left[\begin{array}{rrrrr|r} 1 & 2 & 1 & 0 & 0 & 32 \\ 3 & 4 & 0 & 1 & 0 & 84 \\ \hdashline -50 & -80 & 0 & 0 & 1 & 0 \end{array}\right] \end{array} \begin{array}{l} \\ \frac{32}{2} = 16 \text{ (smallest)} \\ \frac{84}{4} = 21 \\ \\ \end{array} \qquad (6)$$

↑ Pivot column

From (6) we can determine the amount the entering variable can increase by choosing the smallest of the quotients obtained by dividing each element in the last column above the dashed line by the corresponding *positive* element in the pivot column. The row with the smallest quotient is called the **pivot row,** and the variable to the left of the pivot row is the exiting variable. In this case, s_1 will be the exiting variable, and the roles of x_2 and s_1 will be interchanged. The element at the intersection of the pivot column and the pivot row is called the **pivot element,** and we circle this element for ease of recognition. Since a negative or 0 element in the pivot column places no restriction on the amount an entering variable can increase, it is not necessary to compute the quotient for negative or 0 values in the pivot column.

A negative or 0 element is never selected for the pivot element.

Tableau (7) illustrates this process, which is summarized in the next box.

Entering variable ↓ (column x_2); Pivot element → circled 2

$$\begin{array}{r} \\ \text{Exiting variable} \rightarrow s_1 \\ s_2 \\ P \end{array} \begin{array}{c} \begin{array}{cccccc} x_1 & x_2 & s_1 & s_2 & P & \end{array} \\ \left[\begin{array}{rrrrr|r} 1 & \textcircled{2} & 1 & 0 & 0 & 32 \\ 3 & 4 & 0 & 1 & 0 & 84 \\ \hdashline -50 & -80 & 0 & 0 & 1 & 0 \end{array}\right] \end{array} \begin{array}{l} \\ \leftarrow \text{Pivot row} \\ \\ \\ \end{array} \qquad (7)$$

↑ Pivot column

Selecting the Pivot Element

Step 1. Locate the most negative indicator in the bottom row of the tableau to the left of the P column (the negative number with the largest absolute value). The column containing this element is the *pivot column*. If there is a tie for the most negative, choose either.

Step 2. Divide each *positive* element in the pivot column above the dashed line into the corresponding element in the last column. The *pivot row* is the row corresponding to the smallest quotient obtained. If there is a tie for the smallest quotient, choose either. If the pivot column above the dashed line has no positive elements, then there is no solution, and we stop.

Step 3. The *pivot* (or *pivot element*) is the element in the intersection of the pivot column and pivot row. [*Note:* The pivot element is always positive and is never in the bottom row.]

[*Remember:* The entering variable is at the top of the pivot column, and the exiting variable is at the left of the pivot row.]

In order for x_2 to be classified as a basic variable, we perform row operations on tableau (7) so that the pivot element is transformed into 1 and all other elements in the column into 0's. This procedure for transforming a nonbasic variable into a basic variable is called a *pivot operation,* or *pivoting,* and is summarized in the box.

Performing a Pivot Operation

A **pivot operation,** or **pivoting,** consists of performing row operations as follows:

Step 1. Multiply the pivot row by the reciprocal of the pivot element to transform the pivot element into a 1. (If the pivot element is already a 1, omit this step.)

Step 2. Add multiples of the pivot row to other rows in the tableau to transform all other nonzero elements in the pivot column into 0's.

[*Note:* Rows are not to be interchanged while performing a pivot operation. The only way the (positive) pivot element can be transformed into 1 (if it is not a 1 already) is for the pivot row to be multiplied by the reciprocal of the pivot element.]

Performing a pivot operation has the following effects:

1. The (entering) nonbasic variable becomes a basic variable.
2. The (exiting) basic variable becomes a nonbasic variable.
3. The value of the objective function is increased, or, in some cases, remains the same.

We now carry out the pivot operation on tableau (7). (To facilitate the process, we do not repeat the variables after the first tableau, and we use "Enter" and "Exit" for "Entering variable" and "Exiting variable," respectively.)

$$
\begin{array}{c}
\\
\\
\text{Exit} \rightarrow s_1 \\
s_2 \\
P
\end{array}
\begin{array}{c}
\qquad\qquad \text{Enter} \\
\qquad\qquad \downarrow \\
\begin{array}{ccccc|c} x_1 & x_2 & s_1 & s_2 & P & \end{array} \\
\left[\begin{array}{ccccc|c}
1 & \textcircled{2} & 1 & 0 & 0 & 32 \\
3 & 4 & 0 & 1 & 0 & 84 \\
\hline
-50 & -80 & 0 & 0 & 1 & 0
\end{array}\right]
\end{array}
\quad \tfrac{1}{2}R_1 \rightarrow R_1
$$

$$
\sim \left[\begin{array}{ccccc|c}
\frac{1}{2} & \textcircled{1} & \frac{1}{2} & 0 & 0 & 16 \\
3 & 4 & 0 & 1 & 0 & 84 \\
\hline
-50 & -80 & 0 & 0 & 1 & 0
\end{array}\right]
\quad
\begin{array}{l}
(-4)R_1 + R_2 \rightarrow R_2 \\
80R_1 + R_3 \rightarrow R_3
\end{array}
$$

$$
\sim \left[\begin{array}{ccccc|c}
\frac{1}{2} & 1 & \frac{1}{2} & 0 & 0 & 16 \\
1 & 0 & -2 & 1 & 0 & 20 \\
\hline
-10 & 0 & 40 & 0 & 1 & 1{,}280
\end{array}\right]
$$

We have completed the pivot operation, and now we must insert appropriate variables for this new tableau. Since x_2 replaced s_1, the basic variables are now x_2, s_2, and P, as indicated by the labels on the left side of the new tableau. Note that this selection of basic variables agrees with the procedure outlined on page 305 for selecting basic variables. We write the new basic feasible solution by setting the nonbasic variables x_1 and s_1 equal to 0 and solving for the basic variables by inspection. (Remember, the values of the basic variables listed on the left are the corresponding numbers to the right of the vertical line. To see this, substitute $x_1 = 0$ and $s_1 = 0$ in the corresponding system shown next to the simplex tableau.)

$$
\begin{array}{c}
\\
x_2 \\
s_2 \\
P
\end{array}
\begin{array}{c}
\begin{array}{ccccc|c} x_1 & x_2 & s_1 & s_2 & P & \end{array} \\
\left[\begin{array}{ccccc|c}
\frac{1}{2} & 1 & \frac{1}{2} & 0 & 0 & 16 \\
1 & 0 & -2 & 1 & 0 & 20 \\
\hline
-10 & 0 & 40 & 0 & 1 & 1{,}280
\end{array}\right]
\end{array}
\qquad
\begin{array}{rcl}
\frac{1}{2}x_1 + x_2 + \frac{1}{2}s_1 & & = 16 \\
x_1 \qquad - 2s_1 + s_2 & & = 20 \\
-10x_1 + \qquad 40s_1 & + P & = 1{,}280
\end{array}
$$

$$x_1 = 0, \quad x_2 = 16, \quad s_1 = 0, \quad s_2 = 20, \quad P = \$1{,}280$$

A profit of \$1,280 is a marked improvement over the \$0 profit produced by the initial basic feasible solution. But we can improve P still further, since a negative indicator still remains in the bottom row. To see why, we write out the objective function:

$$-10x_1 + 40s_1 + P = 1{,}280$$

or

$$P = 10x_1 - 40s_1 + 1{,}280$$

If s_1 stays a nonbasic variable (set equal to 0) and x_1 becomes a new basic variable, then

$$P = 10x_1 - 40(0) + 1{,}280 = 10x_1 + 1{,}280$$

and for each unit increase in x_1, P will increase \$10.

We now go through another iteration of the simplex process (that is, we repeat the above sequence of steps) using another pivot element. The pivot element and the entering and exiting variables are shown in the following tableau:

$$\begin{array}{r} \\ x_2 \\ \text{Exit} \rightarrow s_2 \\ P \end{array}
\begin{array}{c} \text{Enter} \\ \downarrow \\ \begin{array}{ccccc|c} x_1 & x_2 & s_1 & s_2 & P & \end{array} \\ \left[\begin{array}{ccccc|c} \frac{1}{2} & 1 & \frac{1}{2} & 0 & 0 & 16 \\ \textcircled{1} & 0 & -2 & 1 & 0 & 20 \\ \hline -10 & 0 & 40 & 0 & 1 & 1{,}280 \end{array}\right] \end{array}
\begin{array}{l} \\ \\ \frac{16}{1/2} = 32 \\ \frac{20}{1} = 20 \\ \\ \end{array}$$

We now pivot on (the circled) 1. That is, we perform a pivot operation using this 1 as the pivot element. Since the pivot element is 1, we do not need to perform the first step in the pivot operation, so we proceed to the second step to get 0's above and below the pivot element 1. As before, to facilitate the process, we omit writing the variables, except for the first tableau.

$$\begin{array}{r} \\ x_2 \\ \text{Exit} \rightarrow s_2 \\ P \end{array}
\begin{array}{c} \text{Enter} \\ \downarrow \\ \begin{array}{ccccc|c} x_1 & x_2 & s_1 & s_2 & P & \end{array} \\ \left[\begin{array}{ccccc|c} \frac{1}{2} & 1 & \frac{1}{2} & 0 & 0 & 16 \\ \textcircled{1} & 0 & -2 & 1 & 0 & 20 \\ \hline -10 & 0 & 40 & 0 & 1 & 1{,}280 \end{array}\right] \end{array}
\begin{array}{l} \\ \\ (-\frac{1}{2})R_2 + R_1 \rightarrow R_1 \\ \\ 10R_2 + R_3 \rightarrow R_3 \end{array}$$

$$\sim \left[\begin{array}{ccccc|c} 0 & 1 & \frac{3}{2} & -\frac{1}{2} & 0 & 6 \\ 1 & 0 & -2 & 1 & 0 & 20 \\ \hline 0 & 0 & 20 & 10 & 1 & 1{,}480 \end{array}\right]$$

Since there are no more negative indicators in the bottom row, we are through. Let us insert the appropriate variables for this last tableau and write the corresponding basic feasible solution. The basic variables are now x_1, x_2, and P, so to get the corresponding basic feasible solution, we set the nonbasic variables s_1 and s_2 equal to 0 and solve for the basic variables by inspection.

$$\begin{array}{r} \\ x_2 \\ x_1 \\ P \end{array}
\begin{array}{c} \begin{array}{ccccc|c} x_1 & x_2 & s_1 & s_2 & P & \end{array} \\ \left[\begin{array}{ccccc|c} 0 & 1 & \frac{3}{2} & -\frac{1}{2} & 0 & 6 \\ 1 & 0 & -2 & 1 & 0 & 20 \\ \hline 0 & 0 & 20 & 10 & 1 & 1{,}480 \end{array}\right] \end{array}$$

$$x_1 = 20, \quad x_2 = 6, \quad s_1 = 0, \quad s_2 = 0, \quad P = 1{,}480$$

To see why this is the maximum, we rewrite the objective function from the bottom row:

$$\begin{aligned} 20s_1 + 10s_2 + P &= 1{,}480 \\ P &= 1{,}480 - 20s_1 - 10s_2 \end{aligned}$$

Since s_1 and s_2 cannot be negative, any increase of either from 0 will make the profit smaller.

Finally, returning to our original problem, we conclude that a production schedule of 20 standard tents and 6 expedition tents will produce a maximum profit of $1,480 per day, which is the same as the geometric solution obtained in Section 5-2. The fact that the slack variables are both 0 means that for this production schedule, the plant will operate at full capacity—there is no slack in either the cutting department or the assembly department.

INTERPRETING THE SIMPLEX PROCESS GEOMETRICALLY

We can now interpret the simplex process just completed geometrically in terms of the feasible region graphed in the preceding section. Table 1 lists the three basic feasible solutions we just found using the simplex method (in the order they were found). The table also includes the corresponding corner points of the feasible region illustrated in Figure 1.

Table 1

BASIC FEASIBLE SOLUTIONS (OBTAINED ABOVE)

x_1	x_2	s_1	s_2	P (\$)	CORNER POINT
0	0	32	84	0	$O(0, 0)$
0	16	0	20	1,280	$A(0, 16)$
20	6	0	0	1,480	$B(20, 6)$

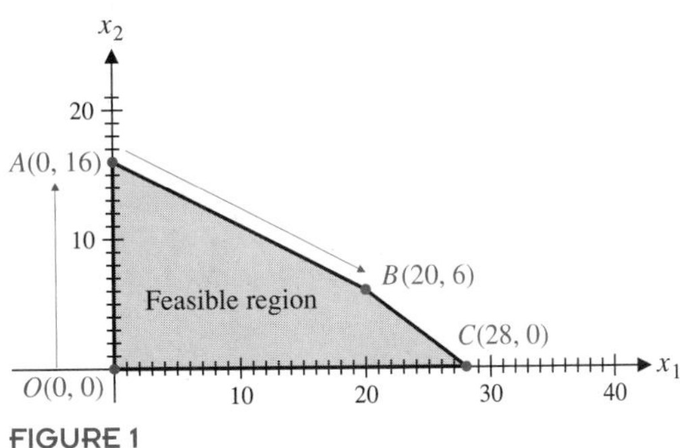

FIGURE 1

Looking at Table 1 and Figure 1, we see that the simplex process started at the origin, moved to the adjacent corner point $A(0, 16)$, and then to the optimal solution $B(20, 6)$ at the next adjacent corner point. This is typical of the simplex process.

■ The Simplex Method Summarized

Before commencing with further examples, we summarize the important parts of the simplex method schematically in Figure 2.

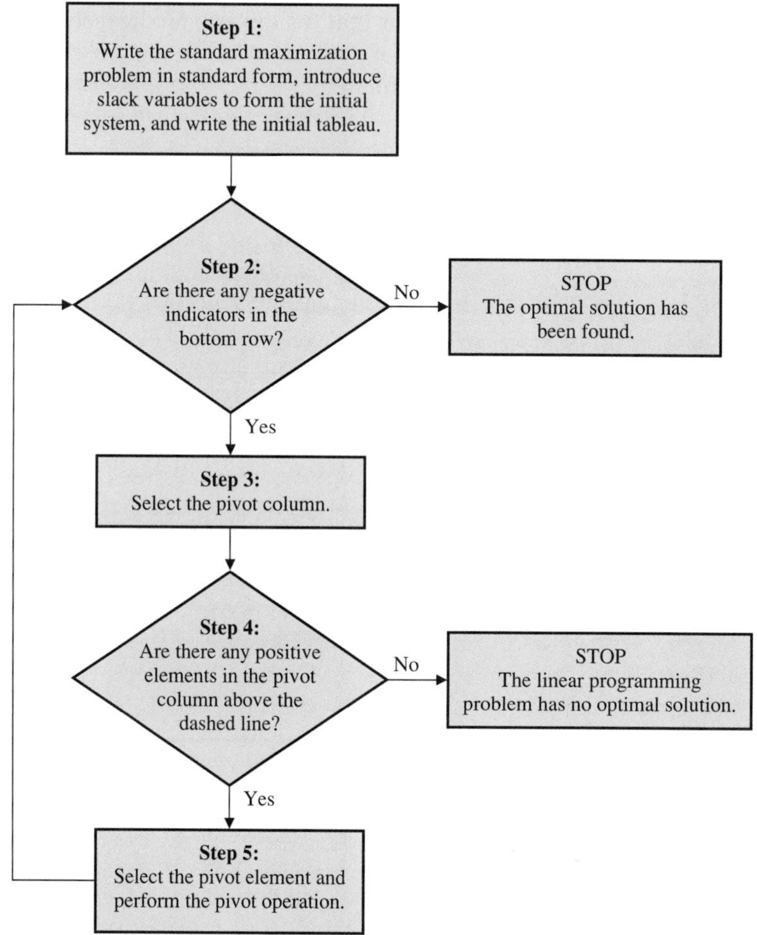

FIGURE 2
Simplex algorithm for standard maximization problems
(Problem constraints are of the $\leq$ form with nonnegative constants on the right. The coefficients of the objective function can be any real numbers.)

Example 1 **Using the Simplex Method** Solve the following linear programming problem using the simplex method:

$$\begin{aligned} \text{Maximize} \quad & P = 10x_1 + 5x_2 \\ \text{Subject to} \quad & 4x_1 + x_2 \leq 28 \\ & 2x_1 + 3x_2 \leq 24 \\ & x_1, x_2 \geq 0 \end{aligned}$$

SOLUTION Introduce slack variables s_1 and s_2, and write the initial system:

$$
\begin{aligned}
4x_1 + x_2 + s_1 \qquad\qquad &= 28 \\
2x_1 + 3x_2 \qquad + s_2 \qquad &= 24 \\
-10x_1 - 5x_2 \qquad\qquad + P &= 0 \\
x_1, x_2, s_1, s_2 &\geqslant 0
\end{aligned}
$$

Write the simplex tableau, and identify the first pivot element and the entering and exiting variables:

$$
\begin{array}{r}
\\ \\ \text{Exit} \rightarrow s_1 \\ s_2 \\ P
\end{array}
\begin{array}{c}
\text{Enter} \\ \downarrow \\ \\ \\ \\
\end{array}
\begin{array}{c}
\begin{array}{cccccc}
x_1 & x_2 & s_1 & s_2 & P & \\
\end{array} \\
\left[\begin{array}{ccccc|c}
\textcircled{4} & 1 & 1 & 0 & 0 & 28 \\
2 & 3 & 0 & 1 & 0 & 24 \\
\hdashline
-10 & -5 & 0 & 0 & 1 & 0
\end{array}\right]
\end{array}
\begin{array}{l}
\frac{28}{4} = 7 \\
\frac{24}{2} = 12 \\
\\
\end{array}
$$

Perform the pivot operation:

$$
\begin{array}{r}
\\ \\ \text{Exit} \rightarrow s_1 \\ s_2 \\ P
\end{array}
\begin{array}{c}
\text{Enter} \\ \downarrow \\ \\ \\ \\
\end{array}
\begin{array}{c}
\begin{array}{cccccc}
x_1 & x_2 & s_1 & s_2 & P & \\
\end{array} \\
\left[\begin{array}{ccccc|c}
\textcircled{4} & 1 & 1 & 0 & 0 & 28 \\
2 & 3 & 0 & 1 & 0 & 24 \\
\hdashline
-10 & -5 & 0 & 0 & 1 & 0
\end{array}\right]
\end{array}
\begin{array}{l}
\frac{1}{4}R_1 \rightarrow R_1 \\
\\
\\
\end{array}
$$

$$
\sim \left[\begin{array}{ccccc|c}
\textcircled{1} & 0.25 & 0.25 & 0 & 0 & 7 \\
2 & 3 & 0 & 1 & 0 & 24 \\
\hdashline
-10 & -5 & 0 & 0 & 1 & 0
\end{array}\right]
\begin{array}{l}
\\
(-2)R_1 + R_2 \rightarrow R_2 \\
10R_1 + R_3 \rightarrow R_3
\end{array}
$$

$$
\begin{array}{r}
x_1 \\ \sim s_2 \\ P
\end{array}
\left[\begin{array}{ccccc|c}
1 & 0.25 & 0.25 & 0 & 0 & 7 \\
0 & 2.5 & -0.5 & 1 & 0 & 10 \\
\hdashline
0 & -2.5 & 2.5 & 0 & 1 & 70
\end{array}\right]
$$

Since there is still a negative indicator in the last row, we repeat the process by finding a new pivot element:

$$
\begin{array}{r}
\\ \\ x_1 \\ \text{Exit} \rightarrow s_2 \\ P
\end{array}
\begin{array}{c}
\begin{array}{cccccc}
 & \text{Enter} & & & & \\
 & \downarrow & & & & \\
x_1 & x_2 & s_1 & s_2 & P & \\
\end{array} \\
\left[\begin{array}{ccccc|c}
1 & 0.25 & 0.25 & 0 & 0 & 7 \\
0 & \textcircled{2.5} & -0.5 & 1 & 0 & 10 \\
\hdashline
0 & -2.5 & 2.5 & 0 & 1 & 70
\end{array}\right]
\end{array}
\begin{array}{l}
\frac{7}{0.25} = 28 \\
\frac{10}{2.5} = 4 \\
\\
\end{array}
$$

Performing the pivot operation, we obtain

Enter ↓ (column x_2); Exit → s_2

$$\begin{array}{r} \\ x_1 \\ \text{Exit} \to s_2 \\ P \end{array}\begin{array}{c} \begin{array}{ccccc|c} x_1 & x_2 & s_1 & s_2 & P & \end{array} \\ \left[\begin{array}{ccccc|c} 1 & 0.25 & 0.25 & 0 & 0 & 7 \\ 0 & \textcircled{2.5} & -0.5 & 1 & 0 & 10 \\ \hline 0 & -2.5 & 2.5 & 0 & 1 & 70 \end{array}\right] \end{array} \quad \tfrac{1}{2.5}R_2 \to R_2$$

$$\sim \left[\begin{array}{ccccc|c} 1 & 0.25 & 0.25 & 0 & 0 & 7 \\ 0 & \textcircled{1} & -0.2 & 0.4 & 0 & 4 \\ \hline 0 & -2.5 & 2.5 & 0 & 1 & 70 \end{array}\right] \quad \begin{array}{l} (-0.25)R_2 + R_1 \to R_1 \\ \\ 2.5R_2 + R_3 \to R_3 \end{array}$$

$$\begin{array}{r} x_1 \\ \sim x_2 \\ P \end{array}\left[\begin{array}{ccccc|c} 1 & 0 & 0.3 & -0.1 & 0 & 6 \\ 0 & 1 & -0.2 & 0.4 & 0 & 4 \\ \hline 0 & 0 & 2 & 1 & 1 & 80 \end{array}\right]$$

Since all the indicators in the last row are nonnegative, we stop and read the optimal solution:

$$\text{Max } P = 80 \quad \text{at} \quad x_1 = 6, \quad x_2 = 4, \quad s_1 = 0, \quad s_2 = 0$$

(To see why this makes sense, write the objective function corresponding to the last row to see what happens to P when you try to increase s_1 or s_2.)

Matched Problem 1 Solve the following linear programming problem using the simplex method:

$$\begin{aligned} \text{Maximize} \quad & P = 2x_1 + x_2 \\ \text{Subject to} \quad & 5x_1 + x_2 \leqslant 9 \\ & x_1 + x_2 \leqslant 5 \\ & x_1, x_2 \geqslant 0 \end{aligned}$$

Explore–Discuss 1

Graph the feasible region for the linear programming problem in Example 1 and trace the path to the optimal solution determined by the simplex method.

Example 2 **Using the Simplex Method** Solve using the simplex method:

$$\begin{aligned} \text{Maximize} \quad & P = 6x_1 + 3x_2 \\ \text{Subject to} \quad & -2x_1 + 3x_2 \leqslant 9 \\ & -x_1 + 3x_2 \leqslant 12 \\ & x_1, x_2 \geqslant 0 \end{aligned}$$

SOLUTION Write the initial system using the slack variables s_1 and s_2:

$$\begin{aligned} -2x_1 + 3x_2 + s_1 \qquad\qquad &= 9 \\ -x_1 + 3x_2 \qquad + s_2 \qquad &= 12 \\ -6x_1 - 3x_2 \qquad\qquad + P &= 0 \end{aligned}$$

Write the simplex tableau and identify the first pivot element:

$$\begin{matrix} & x_1 & x_2 & s_1 & s_2 & P & \\ s_1 & -2 & 3 & 1 & 0 & 0 & 9 \\ s_2 & -1 & 3 & 0 & 1 & 0 & 12 \\ P & -6 & -3 & 0 & 0 & 1 & 0 \end{matrix}$$

↑
Pivot column

Since both elements in the pivot column above the dashed line are negative, we are unable to select a pivot row. We stop and conclude that there is no optimal solution.

Matched Problem 2 Solve using t e simplex method:

$$\begin{aligned} \text{Maximize} \quad & P = 2x_1 + 3x_2 \\ \text{Subject to} \quad & -3x_1 + 4x_2 \leq 12 \\ & x_2 \leq 2 \\ & x_1, x_2 \geq 0 \end{aligned}$$

Explore–Discuss 2

In Example 2 we encountered a tableau with a pivot column and no pivot row, indicating a problem with no optimal solution.

(A) What happens if you violate the rule for selecting the pivot row, select the negative element -1 for the pivot, and then continue with the simplex method?

(B) There was another negative indicator in the tableau in Example 2. What happens if you violate the rule for selecting the pivot column, select the second column for a pivot column, and then continue with the simplex method?

(C) Graph the feasible region in Example 2, graph the lines corresponding to $P = 36$ and $P = 66$, and explain why this problem has no optimal solution.

Refer to Examples 1 and 2. In Example 1 we concluded that we had found the optimal solution because we could not select a pivot column. In Example 2 we concluded that the problem had no optimal solution because we selected a pivot column and then could not select a pivot row. Notice that we do not try to continue with the simplex method by selecting a negative pivot element or using a different column for the pivot column. Remember:

If it is not possible to select a pivot column, the simplex method stops and we conclude that the optimal solution has been found. If the pivot column has been selected and it is not possible to select a pivot row, the simplex method stops and we conclude that there is no optimal solution.

■ **APPLICATION**

Example 3 ➡ **Agriculture** A farmer owns a 100 acre farm and plans to plant at most three crops. The seed for crops A, B, and C costs \$40, \$20, and \$30 per acre, respectively. A maximum of \$3,200 can be spent on seed. Crops A, B, and C require 1, 2, and 1 workdays per acre, respectively, and there are a maximum of 160 workdays available. If the farmer can make a profit of \$100 per acre on crop A, \$300 per acre on crop B, and \$200 per acre on crop C, how many acres of each crop should be planted to maximize profit?

SOLUTION Let:
x_1 = Number of acres of crop A
x_2 = Number of acres of crop B
x_3 = Number of acres of crop C
P = Total profit

Then we have the following linear programming problem:

$$\begin{array}{llr} \text{Maximize} & P = 100x_1 + 300x_2 + 200x_3 & \text{Objective function} \\ \text{Subject to} & \left.\begin{array}{r} x_1 + x_2 + x_3 \le 100 \\ 40x_1 + 20x_2 + 30x_3 \le 3{,}200 \\ x_1 + 2x_2 + x_3 \le 160 \end{array}\right\} & \text{Problem constraints} \\ & x_1, x_2, x_3 \ge 0 & \text{Nonnegative constraints} \end{array}$$

Next, we introduce slack variables and form the initial system:

$$\begin{aligned} x_1 + x_2 + x_3 + s_1 \qquad\qquad &= 100 \\ 40x_1 + 20x_2 + 30x_3 \quad + s_2 \qquad &= 3{,}200 \\ x_1 + 2x_2 + x_3 \qquad + s_3 \quad &= 160 \\ -100x_1 - 300x_2 - 200x_3 \qquad\qquad + P &= 0 \\ x_1, x_2, x_3, s_1, s_2, s_3 &\ge 0 \end{aligned}$$

Notice that the initial system has $7 - 4 = 3$ nonbasic variables and 4 basic variables. Now we form the simplex tableau and solve by the simplex method:

Enter ↓ (x_2)

		x_1	x_2	x_3	s_1	s_2	s_3	P		
	s_1	1	1	1	1	0	0	0	100	
	s_2	40	20	30	0	1	0	0	3,200	
Exit →	s_3	1	(2)	1	0	0	1	0	160	$0.5R_3 \to R_3$
	P	−100	−300	−200	0	0	0	1	0	

~

x_1	x_2	x_3	s_1	s_2	s_3	P		
1	1	1	1	0	0	0	100	$(-1)R_3 + R_1 \to R_1$
40	20	30	0	1	0	0	3,200	$(-20)R_3 + R_2 \to R_2$
0.5	(1)	0.5	0	0	0.5	0	80	
−100	−300	−200	0	0	0	1	0	$300R_3 + R_4 \to R_4$

$$
\begin{array}{r}
\\
\text{Exit} \rightarrow s_1 \\ s_2 \\ x_2 \\ P
\end{array}
\begin{array}{c}
\begin{array}{ccccccc} x_1 & x_2 & \overset{\text{Enter}}{\overset{\downarrow}{x_3}} & s_1 & s_2 & s_3 & P \end{array} \\
\left[\begin{array}{ccccccc|c}
0.5 & 0 & \textcircled{0.5} & 1 & 0 & -0.5 & 0 & 20 \\
30 & 0 & 20 & 0 & 1 & -10 & 0 & 1{,}600 \\
0.5 & 1 & 0.5 & 0 & 0 & 0.5 & 0 & 80 \\
\hline
50 & 0 & -50 & 0 & 0 & 150 & 1 & 24{,}000
\end{array}\right]
\end{array}
\quad 2R_1 \rightarrow R_1
$$

$$
\sim
\left[\begin{array}{ccccccc|c}
1 & 0 & \textcircled{1} & 2 & 0 & -1 & 0 & 40 \\
30 & 0 & 20 & 0 & 1 & -10 & 0 & 1{,}600 \\
0.5 & 1 & 0.5 & 0 & 0 & 0.5 & 0 & 80 \\
\hline
50 & 0 & -50 & 0 & 0 & 150 & 1 & 24{,}000
\end{array}\right]
\quad
\begin{array}{l}
\\ (-20)R_1 + R_2 \rightarrow R_2 \\ (-0.5)R_1 + R_3 \rightarrow R_3 \\ 50R_1 + R_4 \rightarrow R_4
\end{array}
$$

$$
\sim
\begin{array}{r} x_3 \\ s_2 \\ x_2 \\ P \end{array}
\left[\begin{array}{ccccccc|c}
1 & 0 & 1 & 2 & 0 & -1 & 0 & 40 \\
10 & 0 & 0 & -40 & 1 & 10 & 0 & 800 \\
0 & 1 & 0 & -1 & 0 & 1 & 0 & 60 \\
\hline
100 & 0 & 0 & 100 & 0 & 100 & 1 & 26{,}000
\end{array}\right]
$$

All indicators in the bottom row are nonnegative, and we can now read the optimal solution:

$$x_1 = 0, \quad x_2 = 60, \quad x_3 = 40, \quad s_1 = 0, \quad s_2 = 800, \quad s_3 = 0, \quad P = \$26{,}000$$

Thus, if the farmer plants 60 acres in crop *B*, 40 acres in crop *C*, and no crop *A*, the maximum profit of \$26,000 will be realized. The fact that $s_2 = 800$ tells us (look at the second row in the equations at the start) that this maximum profit is reached by using only \$2,400 of the \$3,200 available for seed; that is, we have a slack of \$800 that can be used for some other purpose. ∎

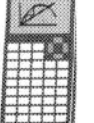

Figure 3 illustrates a solution to Example 3 in Excel, a popular spreadsheet for personal computers. The software in *Explorations in Finite Mathematics* (see Preface) can also be used to solve this problem.

	A	B	C	D	E	F
1	Resources	Crop A	Crop B	Crop C	Available	Used
2	Acres	1	1	1	100	100
3	Seed($)	40	20	30	3,200	2,400
4	Workdays	1	2	1	160	160
5	Profit Per Acre	100	300	200	26,000	<-Total
6	Acres to plant	0	60	40		Profit

FIGURE 3

Matched Problem 3 ➡ Repeat Example 3 modified as follows:

	INVESTMENT PER ACRE *Crop A*	*Crop B*	*Crop C*	MAXIMUM AVAILABLE
SEED COST	\$24	\$40	\$30	\$3,600
WORKDAYS	1	2	2	160
PROFIT	\$140	\$200	\$160	

∎

REMARKS

1. It can be shown that the feasible region for the linear programming problem in Example 3 has eight corner points, yet the simplex method found the solution in only two steps. Now you begin to see the power of the simplex method. In larger problems, the difference between the total number of corner points and the number of steps required by the simplex method is even more dramatic. A feasible region may have hundreds or even thousands of corner points, yet the simplex method will often find the optimal solution in 10 or 15 steps.
2. Refer to the second problem constraint in the model for Example 3:

$$40x_1 + 20x_2 + 30x_3 \leq 3{,}200$$

Multiplying both sides of this inequality by $\frac{1}{10}$ before introducing a slack variable simplifies subsequent calculations. However, performing this operation has a side effect—it changes the units of the slack variable from dollars to tens of dollars. To see why this happens, compare the equations

$$40x_1 + 20x_2 + 30x_3 + s_2 = 3{,}200 \qquad s_2 \text{ represents dollars}$$

and

$$4x_1 + 2x_2 + 3x_3 + s_2' = 320 \qquad s_2' \text{ represents tens of dollars}$$

In general, if you multiply a problem constraint by a number, remember to take this into account when you interpret the value of the slack variable for that constraint.

3. It is important to realize that in order to keep this introduction as simple as possible, we have purposely avoided certain degenerate cases that lead to difficulties. Discussion and resolution of these problems is left to a more advanced treatment of the subject.

Answers to Matched Problems

1. Max $P = 6$ when $x_1 = 1$ and $x_2 = 4$ 2. No optimal solution
3. 40 acres of crop A, 60 acres of crop B, no crop C; Max $P = \$17{,}600$ (since $s_2 = 240$, \$240 out of the \$3,600 will not be spent)

EXERCISE 5-4

A *For the simplex tableaux in Problems 1–4:*
(A) Identify the basic and nonbasic variables.
(B) Find the corresponding basic feasible solution.
(C) Determine whether the optimal solution has been found, an additional pivot is required, or the problem has no optimal solution.

1.
$$\begin{array}{ccccc|c} x_1 & x_2 & s_1 & s_2 & P & \\ \hline 2 & 1 & 0 & 3 & 0 & 12 \\ 3 & 0 & 1 & -2 & 0 & 15 \\ \hline -4 & 0 & 0 & 4 & 1 & 20 \end{array}$$

2.
$$\begin{array}{ccccc|c} x_1 & x_2 & s_1 & s_2 & P & \\ \hline 1 & 4 & -2 & 0 & 0 & 10 \\ 0 & 2 & 3 & 1 & 0 & 25 \\ \hline 0 & 5 & 6 & 0 & 1 & 35 \end{array}$$

3.
$$\begin{array}{ccccccc|c} x_1 & x_2 & x_3 & s_1 & s_2 & s_3 & P & \\ \hline -2 & 0 & 1 & 3 & 1 & 0 & 0 & 5 \\ 0 & 1 & 0 & -2 & 0 & 0 & 0 & 15 \\ -1 & 0 & 0 & 4 & 1 & 1 & 0 & 12 \\ \hline -4 & 0 & 0 & 2 & 4 & 0 & 1 & 45 \end{array}$$

4.
$$\begin{array}{ccccccc|c} x_1 & x_2 & x_3 & s_1 & s_2 & s_3 & P & \\ \hline 0 & 2 & -1 & 1 & 4 & 0 & 0 & 5 \\ 0 & 1 & 2 & 0 & -2 & 1 & 0 & 2 \\ 1 & 3 & 0 & 0 & 5 & 0 & 0 & 11 \\ \hline 0 & -5 & 4 & 0 & -3 & 0 & 1 & 27 \end{array}$$

In Problems 5–8, find the pivot element, identify the entering and exiting variables, and perform one pivot operation.

5.
$$\begin{array}{c} \\ \\ \end{array}\begin{array}{ccccc} x_1 & x_2 & s_1 & s_2 & P \end{array}$$
$$\left[\begin{array}{ccccc|c} 1 & 4 & 1 & 0 & 0 & 4 \\ 3 & 5 & 0 & 1 & 0 & 24 \\ \hline -8 & -5 & 0 & 0 & 1 & 0 \end{array}\right]$$

6.
$$\begin{array}{ccccc} x_1 & x_2 & s_1 & s_2 & P \end{array}$$
$$\left[\begin{array}{ccccc|c} 1 & 6 & 1 & 0 & 0 & 36 \\ 3 & 1 & 0 & 1 & 0 & 5 \\ \hline -1 & -2 & 0 & 0 & 1 & 0 \end{array}\right]$$

7.
$$\begin{array}{cccccc} x_1 & x_2 & s_1 & s_2 & s_3 & P \end{array}$$
$$\left[\begin{array}{cccccc|c} 2 & 1 & 1 & 0 & 0 & 0 & 4 \\ 3 & 0 & 1 & 1 & 0 & 0 & 8 \\ 0 & 0 & 2 & 0 & 1 & 0 & 2 \\ \hline -4 & 0 & -3 & 0 & 0 & 1 & 5 \end{array}\right]$$

8.
$$\begin{array}{cccccc} x_1 & x_2 & s_1 & s_2 & s_3 & P \end{array}$$
$$\left[\begin{array}{cccccc|c} 0 & 0 & 2 & 1 & 1 & 0 & 2 \\ 1 & 0 & -4 & 0 & 1 & 0 & 3 \\ 0 & 1 & 5 & 0 & 2 & 0 & 11 \\ \hline 0 & 0 & -6 & 0 & -5 & 1 & 18 \end{array}\right]$$

In Problems 9–12:

(A) Using slack variables, write the initial system for each linear programming problem.

(B) Write the simplex tableau, circle the first pivot, and identify the entering and exiting variables.

(C) Use the simplex method to solve the problem.

9. Maximize $P = 15x_1 + 10x_2$
Subject to
$$\begin{aligned} 2x_1 + x_2 &\leq 10 \\ x_1 + 3x_2 &\leq 10 \\ x_1, x_2 &\geq 0 \end{aligned}$$

10. Maximize $P = 3x_1 + 2x_2$
Subject to
$$\begin{aligned} 5x_1 + 2x_2 &\leq 20 \\ 3x_1 + 2x_2 &\leq 16 \\ x_1, x_2 &\geq 0 \end{aligned}$$

11. Repeat Problem 9 with the objective function changed to $P = 30x_1 + x_2$.

12. Repeat Problem 10 with the objective function changed to $P = x_1 + 3x_2$.

B *Solve the linear programming problems in Problems 13–28 using the simplex method.*

13. Maximize $P = 30x_1 + 40x_2$
Subject to
$$\begin{aligned} 2x_1 + x_2 &\leq 10 \\ x_1 + x_2 &\leq 7 \\ x_1 + 2x_2 &\leq 12 \\ x_1, x_2 &\geq 0 \end{aligned}$$

14. Maximize $P = 15x_1 + 20x_2$
Subject to
$$\begin{aligned} 2x_1 + x_2 &\leq 9 \\ x_1 + x_2 &\leq 6 \\ x_1 + 2x_2 &\leq 10 \\ x_1, x_2 &\geq 0 \end{aligned}$$

15. Maximize $P = 2x_1 + 3x_2$
Subject to
$$\begin{aligned} -2x_1 + x_2 &\leq 2 \\ -x_1 + x_2 &\leq 5 \\ x_2 &\leq 6 \\ x_1, x_2 &\geq 0 \end{aligned}$$

16. Repeat Problem 15 with $P = -x_1 + 3x_2$.

17. Maximize $P = -x_1 + 2x_2$
Subject to
$$\begin{aligned} -x_1 + x_2 &\leq 2 \\ -x_1 + 3x_2 &\leq 12 \\ x_1 - 4x_2 &\leq 4 \\ x_1, x_2 &\geq 0 \end{aligned}$$

18. Repeat Problem 17 with $P = x_1 + 2x_2$.

19. Maximize $P = 5x_1 + 2x_2 - x_3$
Subject to
$$\begin{aligned} x_1 + x_2 - x_3 &\leq 10 \\ 2x_1 + 4x_2 + 3x_3 &\leq 30 \\ x_1, x_2, x_3 &\geq 0 \end{aligned}$$

20. Maximize $P = 4x_1 - 3x_2 + 2x_3$
Subject to
$$\begin{aligned} x_1 + 2x_2 - x_3 &\leq 5 \\ 3x_1 + 2x_2 + 2x_3 &\leq 22 \\ x_1, x_2, x_3 &\geq 0 \end{aligned}$$

21. Maximize $P = 2x_1 + 3x_2 + 4x_3$
Subject to
$$\begin{aligned} x_1 + x_3 &\leq 4 \\ x_2 + x_3 &\leq 3 \\ x_1, x_2, x_3 &\geq 0 \end{aligned}$$

22. Maximize $P = x_1 + x_2 + 2x_3$
Subject to
$$\begin{aligned} x_1 - 2x_2 + x_3 &\leq 9 \\ 2x_1 + x_2 + 2x_3 &\leq 28 \\ x_1, x_2, x_3 &\geq 0 \end{aligned}$$

23. Maximize $P = 4x_1 + 3x_2 + 2x_3$
Subject to
$$\begin{aligned} 3x_1 + 2x_2 + 5x_3 &\leq 23 \\ 2x_1 + x_2 + x_3 &\leq 8 \\ x_1 + x_2 + 2x_3 &\leq 7 \\ x_1, x_2, x_3 &\geq 0 \end{aligned}$$

24. Maximize $P = 4x_1 + 2x_2 + 3x_3$
Subject to
$$\begin{aligned} x_1 + x_2 + x_3 &\leq 11 \\ 2x_1 + 3x_2 + x_3 &\leq 20 \\ x_1 + 3x_2 + 2x_3 &\leq 20 \\ x_1, x_2, x_3 &\geq 0 \end{aligned}$$

C

25. Maximize $P = 20x_1 + 30x_2$
Subject to
$$\begin{aligned} 0.6x_1 + 1.2x_2 &\leq 960 \\ 0.03x_1 + 0.04x_2 &\leq 36 \\ 0.3x_1 + 0.2x_2 &\leq 270 \\ x_1, x_2 &\geq 0 \end{aligned}$$

26. Repeat Problem 25 with $P = 20x_1 + 20x_2$.

27. Maximize $P = x_1 + 2x_2 + 3x_3$
Subject to
$$\begin{aligned} 2x_1 + 2x_2 + 8x_3 &\leq 600 \\ x_1 + 3x_2 + 2x_3 &\leq 600 \\ 3x_1 + 2x_2 + x_3 &\leq 400 \\ x_1, x_2, x_3 &\geq 0 \end{aligned}$$

28. Maximize $P = 10x_1 + 50x_2 + 10x_3$
Subject to
$$\begin{aligned} 3x_1 + 3x_2 + 3x_3 &\leq 66 \\ 6x_1 - 2x_2 + 4x_3 &\leq 48 \\ 3x_1 + 6x_2 + 9x_3 &\leq 108 \\ x_1, x_2, x_3 &\geq 0 \end{aligned}$$

In Problems 29 and 30, first solve the linear programming problem by the simplex method, keeping track of the basic feasible solutions at each step. Then graph the feasible region and illustrate the path to the optimal solution determined by the simplex method.

29. Maximize $P = 2x_1 + 5x_2$
Subject to
$$\begin{aligned} x_1 + 2x_2 &\leq 40 \\ x_1 + 3x_2 &\leq 48 \\ x_1 + 4x_2 &\leq 60 \\ x_2 &\leq 14 \\ x_1, x_2 &\geq 0 \end{aligned}$$

30. Maximize $P = 5x_1 + 3x_2$
Subject to
$$\begin{aligned} 5x_1 + 4x_2 &\leq 100 \\ 2x_1 + x_2 &\leq 28 \\ 4x_1 + x_2 &\leq 42 \\ x_1 &\leq 10 \\ x_1, x_2 &\geq 0 \end{aligned}$$

In Problems 31–34, there is a tie for the choice of the first pivot column. Use the simplex method to solve each problem two different ways: first by choosing Column 1 as the first pivot column, and then by choosing Column 2 as the first pivot column. Discuss the relationship between these two solutions.

31. Maximize $P = x_1 + x_2$
Subject to
$$\begin{aligned} 2x_1 + x_2 &\leq 16 \\ x_1 &\leq 6 \\ x_2 &\leq 10 \\ x_1, x_2 &\geq 0 \end{aligned}$$

32. Maximize $P = x_1 + x_2$
Subject to
$$\begin{aligned} x_1 + 2x_2 &\leq 10 \\ x_1 &\leq 6 \\ x_2 &\leq 4 \\ x_1, x_2 &\geq 0 \end{aligned}$$

33. Maximize $P = 3x_1 + 3x_2 + 2x_3$
Subject to
$$\begin{aligned} x_1 + x_2 + 2x_3 &\leq 20 \\ 2x_1 + x_2 + 4x_3 &\leq 32 \\ x_1, x_2, x_3 &\geq 0 \end{aligned}$$

34. Maximize $P = 2x_1 + 2x_2 + x_3$
Subject to
$$\begin{aligned} x_1 + x_2 + 3x_3 &\leq 10 \\ 2x_1 + 4x_2 + 5x_3 &\leq 24 \\ x_1, x_2, x_3 &\geq 0 \end{aligned}$$

APPLICATIONS

In Problems 35–48, construct a mathematical model in the form of a linear programming problem. (The answers in the back of the book for these application problems include the model.) Then solve the problem using the simplex method. Include an interpretation of any nonzero slack variables in the optimal solution.

Business & Economics

35. *Manufacturing—resource allocation.* A small company manufactures three different electronic components for computers. Component *A* requires 2 hours of fabrication and 1 hour of assembly; component *B* requires 3 hours of fabrication and 1 hour of assembly; and component *C* requires 2 hours of fabrication and 2 hours of assembly. The company has up to 1,000 labor-hours of fabrication time and 800 labor-hours of assembly time available per week. The profit on each component, *A*, *B*, and *C*, is \$7, \$8, and \$10, respectively. How many components of each type should the company manufacture each week in order to maximize its profit (assuming all components that it manufactures can be sold)? What is the maximum profit?

36. *Manufacturing—resource allocation.* Solve Problem 35 with the additional restriction that the combined total number of components produced each week cannot exceed 420. Discuss the effect of this restriction on the solution to Problem 35.

37. *Investment.* An investor has at most \$100,000 to invest in government bonds, mutual funds, and money market funds. The average yields for government bonds, mutual funds, and money market funds are 8%, 13%, and 15%, respectively. The investor's policy requires that the total

amount invested in mutual and money market funds not exceed the amount invested in government bonds. How much should be invested in each type of investment in order to maximize the return? What is the maximum return?

38. *Investment.* Repeat Problem 37 under the additional assumption that no more than $30,000 can be invested in money market funds.

39. *Advertising.* A department store chain has up to $20,000 to spend on television advertising for a sale. All ads will be placed with one television station, where a 30 second ad costs $1,000 on daytime TV and is viewed by 14,000 potential customers, $2,000 on prime-time TV and is viewed by 24,000 potential customers, and $1,500 on late-night TV and is viewed by 18,000 potential customers. The television station will not accept a total of more than 15 ads in all three time periods. How many ads should be placed in each time period in order to maximize the number of potential customers who will see the ads? How many potential customers will see the ads? (Ignore repeated viewings of the ad by the same potential customer.)

40. *Advertising.* Repeat Problem 39 if the department store increases its budget to $24,000 and requires that at least half of the ads be placed in prime-time shows.

41. *Construction—resource allocation.* A contractor is planning to build a new housing development consisting of colonial, split-level, and ranch-style houses. A colonial house requires $\frac{1}{2}$ acre of land, $60,000 capital, and 4,000 labor-hours to construct, and returns a profit of $20,000. A split-level house requires $\frac{1}{2}$ acre of land, $60,000 capital, and 3,000 labor-hours to construct, and returns a profit of $18,000. A ranch house requires 1 acre of land, $80,000 capital, and 4,000 labor-hours to construct, and returns a profit of $24,000. The contractor has available 30 acres of land, $3,200,000 capital, and 180,000 labor-hours.
   (A) How many houses of each type should be constructed to maximize the contractor's profit? What is the maximum profit?
   (B) A decrease in demand for colonial houses causes the profit on a colonial house to drop from $20,000 to $17,000. Discuss the effect of this change on the number of houses built and on the maximum profit.
   (C) An increase in demand for colonial houses causes the profit on a colonial house to rise from $20,000 to $25,000. Discuss the effect of this change on the number of houses built and on the maximum profit.

42. *Manufacturing—resource allocation.* A company manufactures three-speed, five-speed, and ten-speed bicycles. Each bicycle passes through three departments, fabrication, painting & plating, and final assembly. The relevant manufacturing data are given in the table.

	Labor-Hours per Bicycle			Maximum Labor-Hours Available per Day
	Three-speed	*Five-speed*	*Ten-speed*	
Fabrication	3	4	5	120
Painting & Plating	5	3	5	130
Final Assembly	4	3	5	120
Profit per Bicycle ($)	80	70	100	

   (A) How many bicycles of each type should the company manufacture per day in order to maximize its profit? What is the maximum profit?
   (B) Discuss the effect on the solution to part (A) if the profit on a ten-speed bicycle increases to $110 and all other data in part (A) remain the same.
   (C) Discuss the effect on the solution to part (A) if the profit on a five-speed bicycle increases to $110 and all other data in part (A) remain the same.

43. *Packaging—product mix.* A candy company makes three types of candy, solid-center, fruit-filled, and cream-filled, and packages these candies in three different assortments. A box of assortment I contains 4 solid-center, 4 fruit-filled, and 12 cream-filled candies, and sells for $9.40. A box of assortment II contains 12 solid-center, 4 fruit-filled, and 4 cream-filled candies, and sells for $7.60. A box of assortment III contains 8 solid-center, 8 fruit-filled, and 8 cream-filled candies, and sells for $11.00. The manufacturing costs per piece of candy are $0.20 for solid-center, $0.25 for fruit-filled, and $0.30 for cream-filled. The company can manufacture 4,800 solid-center, 4,000 fruit-filled, and 5,600 cream-filled candies weekly.
   (A) How many boxes of each type should the company produce each week in order to maximize their profit? What is the maximum profit?
   (B) Discuss the effect on the solution to part (A) if the number of fruit-filled candies manufactured weekly is increased to 5,000 and all other data in part (A) remain the same.
   (C) Discuss the effect on the solution to part (A) if the number of fruit-filled candies and the number of solid-center candies manufactured weekly are each increased to 6,000 and all other data in part (A) remain the same.

44. *Scheduling—resource allocation.* A small accounting firm prepares tax returns for three types of customers: individual, commercial, and industrial. The tax preparation process begins with a 1 hour interview with the customer. The data collected during this interview are entered into a time-sharing computer system, which produces the customer's tax return. It takes 1 hour to enter the data for an individual customer, 2 hours for a commercial customer, and $1\frac{1}{2}$ hours for an industrial customer. It takes 10 minutes of computer time to process an individual return, 25 minutes to process a commercial return, and 20 minutes to process an industrial return. The firm has one employee who conducts the initial interview and two who enter the data into the computer. The interviewer can work a maximum of 50 hours a week, and each of the data-entry employees can work a maximum of 40 hours a week. The computer is available for a maximum of 1,025 minutes a week. The firm makes a profit of $50 on each individual customer, $65 on each commercial customer, and $60 on each industrial customer.
    (A) How many customers of each type should the firm schedule each week in order to maximize its profit? What is the maximum profit?
    (B) Discuss the effect on the solution to part (A) if the maximum number of hours the interviewer works is decreased to 35 hours per week and all other data in part (A) remain the same.
    (C) Discuss the effect on the solution to part (A) if the maximum number of minutes of available computer time is decreased to 450 minutes per week and all other data in part (A) remain the same.

Life Sciences

45. *Nutrition—animals.* The natural diet of a certain animal consists of three foods, *A, B,* and *C.* The number of units of calcium, iron, and protein in 1 gram of each food and the average daily intake are given in the table. A scientist wants to investigate the effect of increasing the protein in the animal's diet while not allowing the units of calcium and iron to exceed their average daily intakes. How many grams of each food should be used to maximize the amount of protein in the diet? What is the maximum amount of protein?

	Units per Gram			Average Daily Intake
	Food A	*Food B*	*Food C*	(Units)
Calcium	1	3	2	30
Iron	2	1	2	24
Protein	3	4	5	60

46. *Nutrition—animals.* Repeat Problem 45 if the scientist wants to maximize the daily calcium intake while not allowing the intake of iron or protein to exceed the average daily intake.

Social Sciences

47. *Opinion survey.* A political scientist has received a grant to fund a research project involving voting trends. The budget of the grant includes $3,200 for conducting door-to-door interviews the day before an election. Undergraduate students, graduate students, and faculty members will be hired to conduct the interviews. Each undergraduate student will conduct 18 interviews and be paid $100. Each graduate student will conduct 25 interviews and be paid $150. Each faculty member will conduct 30 interviews and be paid $200. Due to limited transportation facilities, no more than 20 interviewers can be hired. How many undergraduate students, graduate students, and faculty members should be hired in order to maximize the number of interviews that will be conducted? What is the maximum number of interviews?

48. *Opinion survey.* Repeat Problem 47 if one of the requirements of the grant is that at least 50% of the interviewers be undergraduate students.

Section 5-5 The Dual; Minimization with Problem Constraints of the Form $\geqslant$

- Formation of the Dual Problem
- Solution of Minimization Problems
- Application: A Transportation Problem
- Summary of Problem Types and Solution Methods

In the last section we restricted ourselves to standard maximization problems (problem constraints of the form $\leqslant$, with nonnegative constants on the right and objective function coefficients any real numbers). Now we will consider minimization problems with ≥ problem constraints. These two types of problems turn out to be very closely related.

■ Formation of the Dual Problem

Associated with each minimization problem with ≥ constraints is a maximization problem called the **dual problem.** To illustrate the procedure for forming the dual problem, consider the following minimization problem:

$$\begin{aligned} \text{Minimize} \quad & C = 16x_1 + 45x_2 \\ \text{Subject to} \quad & 2x_1 + 5x_2 \geq 50 \\ & x_1 + 3x_2 \geq 27 \\ & x_1, x_2 \geq 0 \end{aligned} \tag{1}$$

The first step in forming the dual problem is to construct a matrix by using the problem constraints and the objective function written in the following form:

$$\begin{aligned} 2x_1 + 5x_2 &\geq 50 \\ x_1 + 3x_2 &\geq 27 \\ 16x_1 + 45x_2 &= C \end{aligned} \qquad A = \left[\begin{array}{cc|c} 2 & 5 & 50 \\ 1 & 3 & 27 \\ \hline 16 & 45 & 1 \end{array}\right]$$

Caution

Do not confuse matrix A with the simplex tableau. We use a solid horizontal line in matrix A to help distinguish the dual matrix from the simplex tableau. No slack variables are involved in matrix A, and the coefficient of C is in the same column as the constants from the problem constraints.

Now we will form a second matrix called the *transpose of A*. In general, the **transpose** of a given matrix A is the matrix A^T formed by interchanging the rows and corresponding columns of A (first row with first column, second row with second column, and so on).

$$A = \left[\begin{array}{cc|c} 2 & 5 & 50 \\ 1 & 3 & 27 \\ \hline 16 & 45 & 1 \end{array}\right] \qquad \begin{array}{l} R_1 \text{ in } A = C_1 \text{ in } A^T \\ R_2 \text{ in } A = C_2 \text{ in } A^T \\ R_3 \text{ in } A = C_3 \text{ in } A^T \end{array}$$

$$A^T = \left[\begin{array}{cc|c} 2 & 1 & 16 \\ 5 & 3 & 45 \\ \hline 50 & 27 & 1 \end{array}\right] \qquad A^T \text{ is the transpose of } A.$$

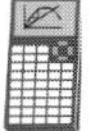

If you are using a graphing utility for matrix operations, consult your manual for information on finding the transpose of a matrix (see Fig. 1).

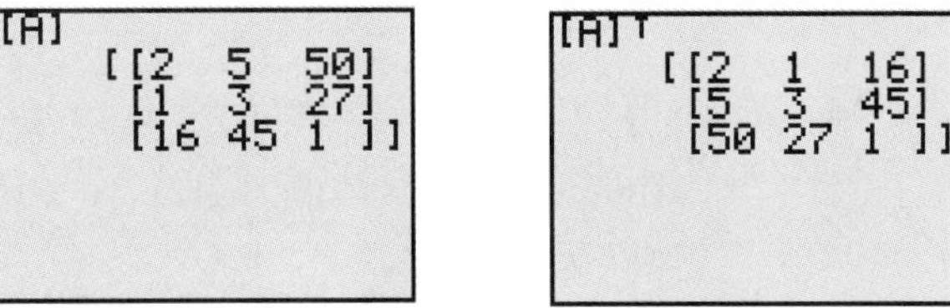

FIGURE 1
Finding A^T on a graphing utility

Now, we use the rows of A^{T} to define a new linear programming problem. This new problem will always be a maximization problem with $\leqslant$ problem constraints. To avoid confusion, we shall use different variables in this new problem:

$$\begin{aligned} 2y_1 + y_2 &\leqslant 16 \\ 5y_1 + 3y_2 &\leqslant 45 \\ 50y_1 + 27y_2 &= P \end{aligned} \qquad A^{\mathrm{T}} = \begin{array}{c} \begin{matrix} y_1 & y_2 & \end{matrix} \\ \left[\begin{array}{cc|c} 2 & 1 & 16 \\ 5 & 3 & 45 \\ \hline 50 & 27 & 1 \end{array}\right] \end{array}$$

The dual of the minimization problem (1) is the following maximization problem:

$$\begin{aligned} \text{Maximize} \quad & P = 50y_1 + 27y_2 \\ \text{Subject to} \quad & 2y_1 + y_2 \leqslant 16 \\ & 5y_1 + 3y_2 \leqslant 45 \\ & y_1, y_2 \geqslant 0 \end{aligned} \tag{2}$$

Explore–Discuss 1

Excluding the nonnegative constraints, the components of a linear programming problem can be divided into three categories: the coefficients of the objective function, the coefficients of the problem constraints, and the constants on the right side of the problem constraints. Write a verbal description of the relationship between the components of the original minimization problem (1) and the dual maximization problem (2).

The procedure for forming the dual problem is summarized in the box below:

Formation of the Dual Problem

Given a minimization problem with $\geqslant$ problem constraints:

Step 1. Use the coefficients and constants in the problem constraints and the objective function to form a matrix A with the coefficients of the objective function in the last row.

Step 2. Interchange the rows and columns of matrix A to form the matrix A^{T}, the transpose of A.

Step 3. Use the rows of A^{T} to form a maximization problem with $\leqslant$ problem constraints.

Example 1 **Forming the Dual Problem** Form the dual problem:

$$\begin{aligned} \text{Minimize} \quad & C = 40x_1 + 12x_2 + 40x_3 \\ \text{Subject to} \quad & 2x_1 + x_2 + 5x_3 \geqslant 20 \\ & 4x_1 + x_2 + x_3 \geqslant 30 \\ & x_1, x_2, x_3 \geqslant 0 \end{aligned}$$

SOLUTION **Step 1.** Form the matrix A:

$$A = \left[\begin{array}{ccc|c} 2 & 1 & 5 & 20 \\ 4 & 1 & 1 & 30 \\ \hline 40 & 12 & 40 & 1 \end{array}\right]$$

Step 2. Form the matrix A^{T}, the transpose of A:

$$A^{\mathrm{T}} = \left[\begin{array}{cc|c} 2 & 4 & 40 \\ 1 & 1 & 12 \\ 5 & 1 & 40 \\ \hline 20 & 30 & 1 \end{array}\right]$$

Step 3. State the dual problem:

$$\begin{aligned} \text{Maximize} \quad & P = 20y_1 + 30y_2 \\ \text{Subject to} \quad & 2y_1 + 4y_2 \leq 40 \\ & y_1 + y_2 \leq 12 \\ & 5y_1 + y_2 \leq 40 \\ & y_1, y_2 \geq 0 \end{aligned}$$

Matched Problem 1 ➡ Form the dual problem:

$$\begin{aligned} \text{Minimize} \quad & C = 16x_1 + 9x_2 + 21x_3 \\ \text{Subject to} \quad & x_1 + x_2 + 3x_3 \geq 12 \\ & 2x_1 + x_2 + x_3 \geq 16 \\ & x_1, x_2, x_3 \geq 0 \end{aligned}$$

■ SOLUTION OF MINIMIZATION PROBLEMS

The following theorem establishes the relationship between the solution of a minimization problem and the solution of its dual:

Theorem 1 ■■ THE FUNDAMENTAL PRINCIPLE OF DUALITY

A minimization problem has a solution if and only if its dual problem has a solution. If a solution exists, then the optimal value of the minimization problem is the same as the optimal value of the dual problem. ■■

The proof of Theorem 1 is beyond the scope of this text. However, we can illustrate Theorem 1 by solving minimization problem (1) and its dual maximization problem (2) geometrically. (Note that Theorem 2 in Section 5-2 guarantees that both problems have solutions.)

Original Problem (1)

Minimize $C = 16x_1 + 45x_2$
Subject to $2x_1 + 5x_2 \geq 50$
$x_1 + 3x_2 \geq 27$
$x_1, x_2 \geq 0$

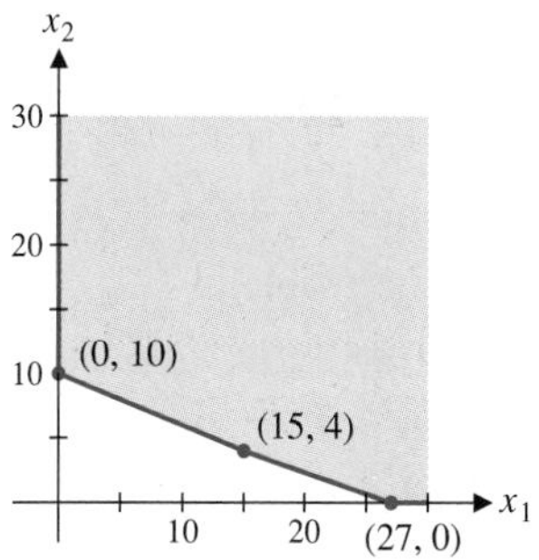

Corner Point (x_1, x_2)	$C = 16x_1 + 45x_2$
(0, 10)	450
(15, 4)	420
(27, 0)	432

Min $C = 420$ at (15, 4)

Dual Problem (2)

Maximize $P = 50y_1 + 27y_2$
Subject to $2y_1 + y_2 \leq 16$
$5y_1 + 3y_2 \leq 45$
$y_1, y_2 \geq 0$

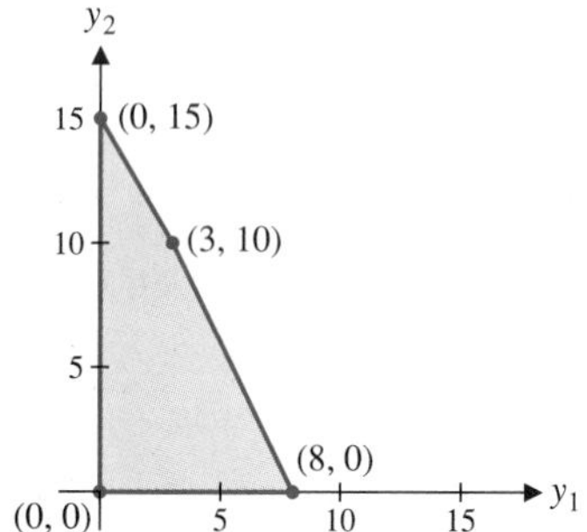

Corner Point (y_1, y_2)	$P = 50y_1 + 27y_2$
(0, 0)	0
(0, 15)	405
(3, 10)	420
(8, 0)	400

Max $P = 420$ at (3, 10)

Thus, the minimum value of C in problem (1) is the same as the maximum value of P in problem (2). Notice that the optimal solutions that produce this optimal value are different: (15, 4) is the optimal solution for problem (1), and (3, 10) is the optimal solution for problem (2). Theorem 1 only guarantees that the optimal values of a minimization problem and its dual are equal, not that the optimal solutions are the same. In general, it is not possible to determine an optimal solution for a minimization problem by examining the feasible set for the dual problem. However, it is possible to apply the simplex method to the dual problem and find both the optimal value and an optimal solution to the original minimization problem. To see how this is done, we will now solve problem (2) by the simplex method.

For reasons that will become clear later, we will use the variables x_1 and x_2 from the original problem as the slack variables in the dual problem:

$$\begin{aligned} 2y_1 + y_2 + x_1 \qquad &= 16 \\ 5y_1 + 3y_2 \qquad + x_2 &= 45 \\ -50y_1 - 27y_2 \qquad + P &= 0 \end{aligned}$$

Initial system for the dual problem

$$
\begin{array}{c} \\ x_1 \\ x_2 \\ P \end{array}
\begin{array}{c}
\begin{array}{ccccc|c} y_1 & y_2 & x_1 & x_2 & P & \\ \end{array} \\
\left[\begin{array}{ccccc|c}
\textcircled{2} & 1 & 1 & 0 & 0 & 16 \\
5 & 3 & 0 & 1 & 0 & 45 \\ \hline
-50 & -27 & 0 & 0 & 1 & 0
\end{array}\right]
\end{array}
\quad 0.5R_1 \to R_1
$$

$$
\sim \left[\begin{array}{ccccc|c}
\textcircled{1} & 0.5 & 0.5 & 0 & 0 & 8 \\
5 & 3 & 0 & 1 & 0 & 45 \\ \hline
-50 & -27 & 0 & 0 & 1 & 0
\end{array}\right]
\quad \begin{array}{l} (-5)R_1 + R_2 \to R_2 \\ 50R_1 + R_3 \to R_3 \end{array}
$$

$$
\begin{array}{c} y_1 \\ \sim x_2 \\ P \end{array}
\left[\begin{array}{ccccc|c}
1 & 0.5 & 0.5 & 0 & 0 & 8 \\
0 & \textcircled{0.5} & -2.5 & 1 & 0 & 5 \\ \hline
0 & -2 & 25 & 0 & 1 & 400
\end{array}\right]
\quad 2R_2 \to R_2
$$

$$
\sim \left[\begin{array}{ccccc|c}
1 & 0.5 & 0.5 & 0 & 0 & 8 \\
0 & \textcircled{1} & -5 & 2 & 0 & 10 \\ \hline
0 & -2 & 25 & 0 & 1 & 400
\end{array}\right]
\quad \begin{array}{l} (-0.5)R_2 + R_1 \to R_1 \\ \\ 2R_2 + R_3 \to R_3 \end{array}
$$

$$
\begin{array}{c} y_1 \\ \sim y_2 \\ P \end{array}
\left[\begin{array}{ccccc|c}
1 & 0 & 3 & -1 & 0 & 3 \\
0 & 1 & -5 & 2 & 0 & 10 \\ \hline
0 & 0 & 15 & 4 & 1 & 420
\end{array}\right]
$$

Since all indicators in the bottom row are nonnegative, the solution to the dual problem is

$$y_1 = 3, \quad y_2 = 10, \quad x_1 = 0, \quad x_2 = 0, \quad P = 420$$

which agrees with our earlier geometric solution. Furthermore, examining the bottom row of the final simplex tableau, we see the same optimal solution to the minimization problem that we obtained directly by the geometric method:

$$\text{Min } C = 420 \qquad \text{at} \qquad x_1 = 15, \quad x_2 = 4$$

This is no accident.

An optimal solution to a minimization problem always can be obtained from the bottom row of the final simplex tableau for the dual problem.

Now we can see that using x_1 and x_2 as slack variables in the dual problem makes it easy to identify the solution of the original problem.

Explore–Discuss 2

The simplex method can be used to solve any standard maximization problem. Which of the following minimization problems have dual problems that are standard maximization problems? (Do not solve the problems.)

(A) Minimize $C = 2x_1 + 3x_2$
Subject to
$$\begin{aligned} 2x_1 - 5x_2 &\geq 4 \\ x_1 - 3x_2 &\geq -6 \\ x_1, x_2 &\geq 0 \end{aligned}$$

(B) Minimize $C = 2x_1 - 3x_2$
Subject to
$$\begin{aligned} -2x_1 + 5x_2 &\geq 4 \\ -x_1 + 3x_2 &\geq 6 \\ x_1, x_2 &\geq 0 \end{aligned}$$

In general, what conditions must a minimization problem satisfy so that its dual problem is a standard maximization problem?

The procedure for solving a minimization problem by applying the simplex method to its dual problem is summarized in the following box:

Solution of a Minimization Problem

Given a minimization problem with nonnegative coefficients in the objective function:

Step 1. Write all problem constraints as $\geq$ inequalities. (This may introduce negative numbers on the right side of some problem constraints.)

Step 2. Form the dual problem.

Step 3. Write the initial system of the dual problem, using the variables from the minimization problem as slack variables.

Step 4. Use the simplex method to solve the dual problem.

Step 5. Read the solution of the minimization problem from the bottom row of the final simplex tableau in step 4. [*Note:* If the dual problem has no optimal solution, then the minimization problem has no optimal solution.]

Example 2 **Solving a Minimization Problem** Solve the following minimization problem by maximizing the dual:

$$\begin{aligned} \text{Minimize} \quad & C = 40x_1 + 12x_2 + 40x_3 \\ \text{Subject to} \quad & 2x_1 + x_2 + 5x_3 \geq 20 \\ & 4x_1 + x_2 + x_3 \geq 30 \\ & x_1, x_2, x_3 \geq 0 \end{aligned}$$

Solution From Example 1 the dual is

$$\begin{aligned} \text{Maximize} \quad & P = 20y_1 + 30y_2 \\ \text{Subject to} \quad & 2y_1 + 4y_2 \leq 40 \\ & y_1 + y_2 \leq 12 \\ & 5y_1 + y_2 \leq 40 \\ & y_1, y_2 \geq 0 \end{aligned}$$

Using x_1, x_2, and x_3 for slack variables, we obtain the initial system for the dual:

$$\begin{aligned} 2y_1 + 4y_2 + x_1 \qquad\qquad\quad &= 40 \\ y_1 + y_2 \qquad + x_2 \qquad\quad &= 12 \\ 5y_1 + y_2 \qquad\qquad + x_3 \quad &= 40 \\ -20y_1 - 30y_2 \qquad\qquad\qquad + P &= 0 \end{aligned}$$

Now we form the simplex tableau and solve the dual problem:

$$
\begin{array}{c}
\\
x_1 \\ x_2 \\ x_3 \\ P
\end{array}
\begin{array}{c}
\begin{array}{cccccc|c} y_1 & y_2 & x_1 & x_2 & x_3 & P & \end{array} \\
\left[\begin{array}{cccccc|c}
2 & \textcircled{4} & 1 & 0 & 0 & 0 & 40 \\
1 & 1 & 0 & 1 & 0 & 0 & 12 \\
5 & 1 & 0 & 0 & 1 & 0 & 40 \\
\hline
-20 & -30 & 0 & 0 & 0 & 1 & 0
\end{array}\right]
\end{array}
\quad \tfrac{1}{4}R_1 \to R_1
$$

$$
\sim \left[\begin{array}{cccccc|c}
\frac{1}{2} & \textcircled{1} & \frac{1}{4} & 0 & 0 & 0 & 10 \\
1 & 1 & 0 & 1 & 0 & 0 & 12 \\
5 & 1 & 0 & 0 & 1 & 0 & 40 \\
\hline
-20 & -30 & 0 & 0 & 0 & 1 & 0
\end{array}\right]
\quad
\begin{array}{l}
\\
(-1)R_1 + R_2 \to R_2 \\
(-1)R_1 + R_3 \to R_3 \\
30R_1 + R_4 \to R_4
\end{array}
$$

$$
\sim
\begin{array}{c}
y_2 \\ x_2 \\ x_3 \\ P
\end{array}
\left[\begin{array}{cccccc|c}
\frac{1}{2} & 1 & \frac{1}{4} & 0 & 0 & 0 & 10 \\
\textcircled{\tfrac{1}{2}} & 0 & -\frac{1}{4} & 1 & 0 & 0 & 2 \\
\frac{9}{2} & 0 & -\frac{1}{4} & 0 & 1 & 0 & 30 \\
\hline
-5 & 0 & \frac{15}{2} & 0 & 0 & 1 & 300
\end{array}\right]
\quad
\begin{array}{l}
\\
2R_2 \to R_2 \\
\\
\\
\end{array}
$$

$$
\sim \left[\begin{array}{cccccc|c}
\frac{1}{2} & 1 & \frac{1}{4} & 0 & 0 & 0 & 10 \\
\textcircled{1} & 0 & -\frac{1}{2} & 2 & 0 & 0 & 4 \\
\frac{9}{2} & 0 & -\frac{1}{4} & 0 & 1 & 0 & 30 \\
\hline
-5 & 0 & \frac{15}{2} & 0 & 0 & 1 & 300
\end{array}\right]
\quad
\begin{array}{l}
(-\frac{1}{2})R_2 + R_1 \to R_1 \\
\\
(-\frac{9}{2})R_2 + R_3 \to R_3 \\
5R_2 + R_4 \to R_4
\end{array}
$$

$$
\sim
\begin{array}{c}
y_2 \\ y_1 \\ x_3 \\ P
\end{array}
\left[\begin{array}{cccccc|c}
0 & 1 & \frac{1}{2} & -1 & 0 & 0 & 8 \\
1 & 0 & -\frac{1}{2} & 2 & 0 & 0 & 4 \\
0 & 0 & 2 & -9 & 1 & 0 & 12 \\
\hline
0 & 0 & 5 & 10 & 0 & 1 & 320
\end{array}\right]
$$

From the bottom row of this tableau, we see that

$$\text{Min } C = 320 \quad \text{at} \quad x_1 = 5, \quad x_2 = 10, \quad x_3 = 0$$

Matched Problem 2 Solve the following minimization problem by maximizing the dual (see Matched Problem 1):

$$
\begin{array}{ll}
\text{Minimize} & C = 16x_1 + 9x_2 + 21x_3 \\
\text{Subject to} & x_1 + x_2 + 3x_3 \geqslant 12 \\
& 2x_1 + x_2 + x_3 \geqslant 16 \\
& x_1, x_2, x_3 \geqslant 0
\end{array}
$$

CAUTION

In the preceding section, we noted that multiplying a problem constraint by a number (usually in order to simplify calculations) changes the units of the slack variable. This requires some special interpretation of the value of the slack variable in the optimal solution, but causes no serious problems. However, when using the dual method, multiplying a problem constraint in the dual problem by a number can have some very serious consequences—the bottom row of the final simplex tableau may no longer give the correct solution to the minimization problem. To see this, refer to the first problem constraint of the dual problem in Example 2:

$$2y_1 + 4y_2 \leqslant 40$$

If we multiply this constraint by $\frac{1}{2}$ and then solve, the final tableau is (verify this):

$$\begin{array}{c} \begin{matrix} y_1 & y_2 & x_1 & x_2 & x_3 & P \end{matrix} \\ \left[\begin{array}{cccccc|c} 0 & 1 & 1 & -1 & 0 & 0 & 8 \\ 1 & 0 & -1 & 2 & 0 & 0 & 4 \\ 0 & 0 & 4 & -9 & 1 & 0 & 12 \\ \hdashline 0 & 0 & 10 & 10 & 0 & 1 & 320 \end{array}\right] \end{array}$$

The bottom row of this tableau indicates that the optimal solution to the minimization problem is $C = 320$ at $x_1 = 10$ and $x_2 = 10$. This is not the correct answer ($x_1 = 5$ is the correct answer). Thus, **you should never multiply a problem constraint in a maximization problem by a number if that maximization problem is being used to solve a minimization problem.** You may still simplify problem constraints in a minimization problem before forming the dual problem.

Example 3 **Solving a Minimization Problem** Solve the following minimization problem by maximizing the dual:

$$\begin{aligned} \text{Minimize} \quad & C = 5x_1 + 10x_2 \\ \text{Subject to} \quad & x_1 - x_2 \geqslant 1 \\ & -x_1 + x_2 \geqslant 2 \\ & x_1, x_2 \geqslant 0 \end{aligned}$$

SOLUTION

$$A = \left[\begin{array}{cc|c} 1 & -1 & 1 \\ -1 & 1 & 2 \\ \hline 5 & 10 & 1 \end{array}\right] \qquad A^{T} = \left[\begin{array}{cc|c} 1 & -1 & 5 \\ -1 & 1 & 10 \\ \hline 1 & 2 & 1 \end{array}\right]$$

The dual problem is

$$\begin{aligned} \text{Maximize} \quad & P = y_1 + 2y_2 \\ \text{Subject to} \quad & y_1 - y_2 \leqslant 5 \\ & -y_1 + y_2 \leqslant 10 \\ & y_1, y_2 \geqslant 0 \end{aligned}$$

Introduce slack variables x_1 and x_2, and form the initial system for the dual:

$$\begin{aligned} y_1 - y_2 + x_1 \qquad\qquad &= 5 \\ -y_1 + y_2 \qquad + x_2 \qquad &= 10 \\ -y_1 - 2y_2 \qquad\qquad + P &= 0 \end{aligned}$$

Form the simplex tableau and solve:

$$\begin{array}{c} \\ x_1 \\ x_2 \\ P \end{array}\begin{array}{c} \begin{array}{ccccc} y_1 & y_2 & x_1 & x_2 & P \end{array} \\ \left[\begin{array}{rrrrr|r} 1 & -1 & 1 & 0 & 0 & 5 \\ -1 & \textcircled{1} & 0 & 1 & 0 & 10 \\ \hline -1 & -2 & 0 & 0 & 1 & 0 \end{array}\right] \end{array} \begin{array}{l} R_2 + R_1 \to R_1 \\ \\ 2R_2 + R_3 \to R_3 \end{array}$$

$$\sim \left[\begin{array}{rrrrr|r} 0 & 0 & 1 & 1 & 0 & 15 \\ -1 & 1 & 0 & 1 & 0 & 10 \\ \hline -3 & 0 & 0 & 2 & 1 & 20 \end{array}\right] \quad \text{No positive elements above dashed line in pivot column}$$

↑ Pivot column

The -3 in the bottom row indicates that Column 1 is the pivot column. Since no positive elements appear in the pivot column above the dashed line, we are unable to select a pivot row. We stop the pivot operation and conclude that this maximization problem has no optimal solution (see the flowchart in Fig. 2, Section 5-4). Theorem 1 now implies that the original minimization problem has no solution. The graph of the inequalities in the minimization problem (Fig. 2) shows that the feasible region is empty; thus, it is not surprising that an optimal solution does not exist.

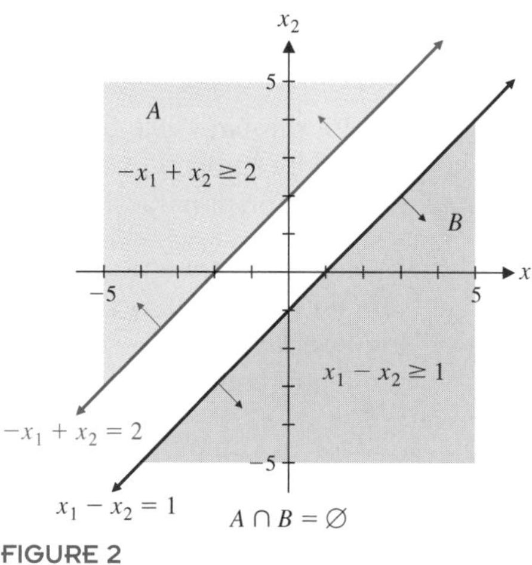

FIGURE 2

Matched Problem 3 ➡ Solve the following minimization problem by maximizing the dual:

$$\begin{aligned} \text{Minimize} \quad & C = 2x_1 + 3x_2 \\ \text{Subject to} \quad & x_1 - 2x_2 \geq 2 \\ & -x_1 + x_2 \geq 1 \\ & x_1, x_2 \geq 0 \end{aligned}$$

■ APPLICATION: A TRANSPORTATION PROBLEM

One of the first applications of linear programming was to the problem of minimizing the cost of transporting materials. Problems of this type are referred to as **transportation problems.**

Example 4 **A Transportation Problem** A computer manufacturing company has two assembly plants, plant A and plant B, and two distribution outlets, outlet I and outlet II. Plant A can assemble at most 700 computers a month, and plant B can assemble at most 900 computers a month. Outlet I must have at least 500 computers a month, and outlet II must have at least 1,000 computers a month. Transportation costs for shipping one computer from each plant to each outlet are as follows: \$6 from plant A to outlet I; \$5 from plant A to outlet II; \$4 from plant B to outlet I; \$8 from plant B to outlet II. Find a shipping schedule that will minimize the total cost of shipping the computers from the assembly plants to the distribution outlets. What is this minimum cost?

SOLUTION First, we summarize the relevant data in a table:

	Distribution Outlet *I*	Distribution Outlet *II*	Assembly Capacity
Plant *A*	\$6	\$5	700
Plant *B*	\$4	\$8	900
Minimum Required	500	1,000	

In order to find a shipping schedule, we must determine the number of computers that should be shipped from each plant to each outlet. This will require the use of four decision variables:

x_1 = Number of computers shipped from plant A to outlet I
x_2 = Number of computers shipped from plant A to outlet II
x_3 = Number of computers shipped from plant B to outlet I
x_4 = Number of computers shipped from plant B to outlet II

The total number of computers shipped from plant A is $x_1 + x_2$. Since this cannot exceed the assembly capacity at A, we have

$$x_1 + x_2 \leqslant 700 \quad \text{Number shipped from plant } A$$

Similarly, the total number shipped from plant B must satisfy

$$x_3 + x_4 \leqslant 900 \quad \text{Number shipped from plant } B$$

The total number shipped to each outlet must satisfy

$$x_1 + x_3 \geqslant 500 \quad \text{Number shipped to outlet I}$$

and

$$x_2 + x_4 \geqslant 1{,}000 \quad \text{Number shipped to outlet II}$$

Using the shipping charges in the table, the total shipping charges are

$$C = 6x_1 + 5x_2 + 4x_3 + 8x_4$$

Thus, we must solve the following linear programming problem:

$$\begin{aligned} &\text{Minimize} && C = 6x_1 + 5x_2 + 4x_3 + 8x_4 \\ &\text{Subject to} && x_1 + x_2 \leq 700 && \text{Available from A} \\ &&& x_3 + x_4 \leq 900 && \text{Available from B} \\ &&& x_1 + x_3 \geq 500 && \text{Required at I} \\ &&& x_2 + x_4 \geq 1{,}000 && \text{Required at II} \\ &&& x_1, x_2, x_3, x_4 \geq 0 \end{aligned}$$

Before we can solve this problem, we must multiply the first two constraints by -1 so that all the problem constraints are of the $\geq$ type. This will introduce negative constants into the minimization problem but not into the dual. Since the coefficients of C are nonnegative, the constants in the dual problem will be nonnegative and the dual will be a standard maximization problem. The problem can now be stated as

$$\begin{aligned} &\text{Minimize} && C = 6x_1 + 5x_2 + 4x_3 + 8x_4 \\ &\text{Subject to} && -x_1 - x_2 \geq -700 \\ &&& -x_3 - x_4 \geq -900 \\ &&& x_1 + x_3 \geq 500 \\ &&& x_2 + x_4 \geq 1{,}000 \\ &&& x_1, x_2, x_3, x_4 \geq 0 \end{aligned}$$

$$A = \left[\begin{array}{cccc|c} -1 & -1 & 0 & 0 & -700 \\ 0 & 0 & -1 & -1 & -900 \\ 1 & 0 & 1 & 0 & 500 \\ 0 & 1 & 0 & 1 & 1{,}000 \\ \hline 6 & 5 & 4 & 8 & 1 \end{array}\right]$$

$$A^{\mathrm{T}} = \left[\begin{array}{cccc|c} -1 & 0 & 1 & 0 & 6 \\ -1 & 0 & 0 & 1 & 5 \\ 0 & -1 & 1 & 0 & 4 \\ 0 & -1 & 0 & 1 & 8 \\ \hline -700 & -900 & 500 & 1{,}000 & 1 \end{array}\right]$$

The dual problem is

$$\begin{aligned} &\text{Maximize} && P = -700y_1 - 900y_2 + 500y_3 + 1{,}000y_4 \\ &\text{Subject to} && -y_1 + y_3 \leq 6 \\ &&& -y_1 + y_4 \leq 5 \\ &&& -y_2 + y_3 \leq 4 \\ &&& -y_2 + y_4 \leq 8 \\ &&& y_1, y_2, y_3, y_4 \geq 0 \end{aligned}$$

Introduce slack variables x_1, x_2, x_3, and x_4, and form the initial system for the dual:

$$\begin{aligned} -y_1 + y_3 + x_1 &= 6 \\ -y_1 + y_4 + x_2 &= 5 \\ -y_2 + y_3 + x_3 &= 4 \\ -y_2 + y_4 + x_4 &= 8 \\ 700y_1 + 900y_2 - 500y_3 - 1{,}000y_4 + P &= 0 \end{aligned}$$

Form the simplex tableau and solve:

$$
\begin{array}{c}
\\ x_1 \\ x_2 \\ x_3 \\ x_4 \\ P
\end{array}
\begin{array}{c}
\begin{array}{ccccccccc} y_1 & y_2 & y_3 & y_4 & x_1 & x_2 & x_3 & x_4 & P \end{array} \\
\left[\begin{array}{rrrrrrrrr|r}
-1 & 0 & 1 & 0 & 1 & 0 & 0 & 0 & 0 & 6 \\
-1 & 0 & 0 & \textcircled{1} & 0 & 1 & 0 & 0 & 0 & 5 \\
0 & -1 & 1 & 0 & 0 & 0 & 1 & 0 & 0 & 4 \\
0 & -1 & 0 & 1 & 0 & 0 & 0 & 1 & 0 & 8 \\
\hdashline
700 & 900 & -500 & -1{,}000 & 0 & 0 & 0 & 0 & 1 & 0
\end{array}\right]
\end{array}
\begin{array}{l}
\\ \\ \\ \\ (-1)R_2 + R_4 \to R_4 \\ 1{,}000R_2 + R_5 \to R_5
\end{array}
$$

$$
\begin{array}{c}
x_1 \\ y_4 \\ \sim x_3 \\ x_4 \\ P
\end{array}
\left[\begin{array}{rrrrrrrrr|r}
-1 & 0 & 1 & 0 & 1 & 0 & 0 & 0 & 0 & 6 \\
-1 & 0 & 0 & 1 & 0 & 1 & 0 & 0 & 0 & 5 \\
0 & -1 & \textcircled{1} & 0 & 0 & 0 & 1 & 0 & 0 & 4 \\
1 & -1 & 0 & 0 & 0 & -1 & 0 & 1 & 0 & 3 \\
\hdashline
-300 & 900 & -500 & 0 & 0 & 1{,}000 & 0 & 0 & 1 & 5{,}000
\end{array}\right]
\begin{array}{l}
(-1)R_3 + R_1 \to R_1 \\ \\ \\ \\ 500R_3 + R_5 \to R_5
\end{array}
$$

$$
\begin{array}{c}
x_1 \\ y_4 \\ \sim y_3 \\ x_4 \\ P
\end{array}
\left[\begin{array}{rrrrrrrrr|r}
-1 & 1 & 0 & 0 & 1 & 0 & -1 & 0 & 0 & 2 \\
-1 & 0 & 0 & 1 & 0 & 1 & 0 & 0 & 0 & 5 \\
0 & -1 & 1 & 0 & 0 & 0 & 1 & 0 & 0 & 4 \\
\textcircled{1} & -1 & 0 & 0 & 0 & -1 & 0 & 1 & 0 & 3 \\
\hdashline
-300 & 400 & 0 & 0 & 0 & 0 & 0 & 0 & 1 & 7{,}000
\end{array}\right]
\begin{array}{l}
R_4 + R_1 \to R_1 \\ R_4 + R_2 \to R_2 \\ \\ \\ 300R_4 + R_5 \to R_5
\end{array}
$$

$$
\begin{array}{c}
x_1 \\ y_4 \\ \sim y_3 \\ y_1 \\ P
\end{array}
\left[\begin{array}{rrrrrrrrr|r}
0 & 0 & 0 & 0 & 1 & -1 & -1 & 1 & 0 & 6 \\
0 & -1 & 0 & 1 & 0 & 0 & 0 & 1 & 0 & 5 \\
0 & -1 & 1 & 0 & 0 & 0 & 1 & 0 & 0 & 4 \\
1 & -1 & 0 & 0 & 0 & -1 & 0 & 1 & 0 & 3 \\
\hdashline
0 & 100 & 0 & 0 & 0 & 700 & 500 & 300 & 1 & 7{,}900
\end{array}\right]
$$

From the bottom row of this tableau, we have

$$\text{Min } C = 7{,}900 \quad \text{at} \quad x_1 = 0, \quad x_2 = 700, \quad x_3 = 500, \quad x_4 = 300$$

The shipping schedule that minimizes the shipping charges is 700 from plant A to outlet II, 500 from plant B to outlet I, and 300 from plant B to outlet II. The total shipping cost is \$7,900. ▪

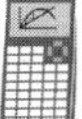

Figure 3 shows a solution to Example 4 in Excel, a popular spreadsheet for personal computers. Notice that Excel permits the user to organize the original data and the solution in a format that is clear and easy to read. This is one of the main advantages of using spreadsheets to solve linear programming problems. The software in *Explorations in Finite Mathematics* (see Preface) can also be used to solve this problem.

	A	B	C	D	E	F	G	H	I
1		DATA					SHIPPING SCHEDULE		
2		DISTRIBUTION					DISTRIBUTION		
3		OUTLET		ASSEMBLY			OUTLET		
4		I	II	CAPACITY			I	II	TOTAL
5	PLANT A	\$6	\$5	700		PLANT A	0	700	700
6	PLANT B	\$4	\$8	900		PLANT B	500	300	800
7	MINIMUM					TOTAL	500	1,000	
8	REQUIRED	500	1,000					TOTAL COST	\$7,900

FIGURE 3

Matched Problem 4 ➡ Repeat Example 4 if the shipping charge from plant A to outlet I is increased to \$7 and the shipping charge from plant B to outlet II is decreased to \$3.

Summary of Problem Types and Solution Methods

In this and the preceding sections, we have solved both maximization and minimization problems, but with certain restrictions on the problem constraints, constants on the right, and/or objective function coefficients. Table 1 summarizes the types of problems and methods of solution we have considered so far.

Table 1

Summary of Problem Types and Simplex Solution Methods

Problem Type	Problem Constraints	Right-Side Constants	Coefficients of Objective Function	Method of Solution
1. Maximization	$\leqslant$	Nonnegative	Any real numbers	Simplex method with slack variables
2. Minimization	$\geqslant$	Any real numbers	Nonnegative	Form dual and solve by simplex method with slack variables

The next section develops a generalized version of the simplex method that can handle both maximization and minimization problems with any combination of $\leqslant$, $\geqslant$, and $=$ problem constraints.

Answers to Matched Problems

1. Maximize $P = 12y_1 + 16y_2$
Subject to
$$\begin{aligned} y_1 + 2y_2 &\leqslant 16 \\ y_1 + y_2 &\leqslant 9 \\ 3y_1 + y_2 &\leqslant 21 \\ y_1, y_2 &\geqslant 0 \end{aligned}$$

2. Min $C = 136$ at $x_1 = 4, x_2 = 8, x_3 = 0$

3. Dual problem:
Maximize $P = 2y_1 + y_2$
Subject to
$$\begin{aligned} y_1 - y_2 &\leqslant 2 \\ -2y_1 + y_2 &\leqslant 3 \\ y_1, y_2 &\geqslant 0 \end{aligned}$$
No optimal solution

4. 600 from plant A to outlet II, 500 from plant B to outlet I, 400 from plant B to outlet II; total shipping cost is \$6,200

Exercise 5-5

A *In Problems 1–8, find the transpose of each matrix.*

1. $\begin{bmatrix} -5 & 0 & 3 & -1 & 8 \end{bmatrix}$

2. $\begin{bmatrix} 1 & 0 & -7 & 3 & -2 \end{bmatrix}$

3. $\begin{bmatrix} 1 \\ -2 \\ 0 \\ 4 \end{bmatrix}$

4. $\begin{bmatrix} 9 \\ 5 \\ -4 \\ 0 \end{bmatrix}$

5. $\begin{bmatrix} 2 & 1 & -6 & 0 & -1 \\ 5 & 2 & 0 & 1 & 3 \end{bmatrix}$

6. $\begin{bmatrix} 7 & 3 & -1 & 3 \\ -6 & 1 & 0 & -9 \end{bmatrix}$

7. $\begin{bmatrix} 1 & 2 & -1 \\ 0 & 2 & -7 \\ 8 & 0 & 1 \\ 4 & -1 & 3 \end{bmatrix}$

8. $\begin{bmatrix} 1 & -1 & 3 & 2 \\ 1 & -4 & 0 & 2 \\ 4 & -5 & 6 & 1 \\ -3 & 8 & 0 & -1 \\ 2 & 7 & -3 & 1 \end{bmatrix}$

In Problems 9 and 10:
(A) Form the dual problem.
(B) Write the initial system for the dual problem.
(C) Write the initial simplex tableau for the dual problem and label the columns of the tableau.

9. Minimize $C = 8x_1 + 9x_2$
Subject to
$$\begin{aligned} x_1 + 3x_2 &\geq 4 \\ 2x_1 + x_2 &\geq 5 \\ x_1, x_2 &\geq 0 \end{aligned}$$

10. Minimize $C = 12x_1 + 5x_2$
Subject to
$$\begin{aligned} 2x_1 + x_2 &\geq 7 \\ 3x_1 + x_2 &\geq 9 \\ x_1, x_2 &\geq 0 \end{aligned}$$

In Problems 11 and 12, a minimization problem, the corresponding dual problem, and the final simplex tableau in the solution of the dual problem are given.
(A) Find the optimal solution of the dual problem.
(B) Find the optimal solution of the minimization problem.

11. Minimize $C = 21x_1 + 50x_2$
Subject to
$$\begin{aligned} 2x_1 + 5x_2 &\geq 12 \\ 3x_1 + 7x_2 &\geq 17 \\ x_1, x_2 &\geq 0 \end{aligned}$$

Maximize $P = 12y_1 + 17y_2$
Subject to
$$\begin{aligned} 2y_1 + 3y_2 &\leq 21 \\ 5y_1 + 7y_2 &\leq 50 \\ y_1, y_2 &\geq 0 \end{aligned}$$

$$\begin{array}{c} \begin{matrix} y_1 & y_2 & x_1 & x_2 & P & \end{matrix} \\ \left[\begin{array}{ccccc|c} 0 & 1 & 5 & -2 & 0 & 5 \\ 1 & 0 & -7 & 3 & 0 & 3 \\ \hline 0 & 0 & 1 & 2 & 1 & 121 \end{array}\right] \end{array}$$

12. Minimize $C = 16x_1 + 25x_2$
Subject to
$$\begin{aligned} 3x_1 + 5x_2 &\geq 30 \\ 2x_1 + 3x_2 &\geq 19 \\ x_1, x_2 &\geq 0 \end{aligned}$$

Maximize $P = 30y_1 + 19y_2$
Subject to
$$\begin{aligned} 3y_1 + 2y_2 &\leq 16 \\ 5y_1 + 3y_2 &\leq 25 \\ y_1, y_2 &\geq 0 \end{aligned}$$

$$\begin{array}{c} \begin{matrix} y_1 & y_2 & x_1 & x_2 & P & \end{matrix} \\ \left[\begin{array}{ccccc|c} 0 & 1 & 5 & -3 & 0 & 5 \\ 1 & 0 & -3 & 2 & 0 & 2 \\ \hline 0 & 0 & 5 & 3 & 1 & 155 \end{array}\right] \end{array}$$

In Problems 13–20:
(A) Form the dual problem.
(B) Find the solution to the original problem by applying the simplex method to the dual problem.

13. Minimize $C = 9x_1 + 2x_2$
Subject to
$$\begin{aligned} 4x_1 + x_2 &\geq 13 \\ 3x_1 + x_2 &\geq 12 \\ x_1, x_2 &\geq 0 \end{aligned}$$

14. Minimize $C = x_1 + 4x_2$
Subject to
$$\begin{aligned} x_1 + 2x_2 &\geq 5 \\ x_1 + 3x_2 &\geq 6 \\ x_1, x_2 &\geq 0 \end{aligned}$$

15. Minimize $C = 7x_1 + 12x_2$
Subject to
$$\begin{aligned} 2x_1 + 3x_2 &\geq 15 \\ x_1 + 2x_2 &\geq 8 \\ x_1, x_2 &\geq 0 \end{aligned}$$

16. Minimize $C = 3x_1 + 5x_2$
Subject to
$$\begin{aligned} 2x_1 + 3x_2 &\geq 7 \\ x_1 + 2x_2 &\geq 4 \\ x_1, x_2 &\geq 0 \end{aligned}$$

17. Minimize $C = 11x_1 + 4x_2$
Subject to
$$\begin{aligned} 2x_1 + x_2 &\geq 8 \\ -2x_1 + 3x_2 &\geq 4 \\ x_1, x_2 &\geq 0 \end{aligned}$$

18. Minimize $C = 40x_1 + 10x_2$
Subject to
$$\begin{aligned} 2x_1 + x_2 &\geq 12 \\ 3x_1 - x_2 &\geq 3 \\ x_1, x_2 &\geq 0 \end{aligned}$$

19. Minimize $C = 7x_1 + 9x_2$
Subject to
$$\begin{aligned} -3x_1 + x_2 &\geq 6 \\ x_1 - 2x_2 &\geq 4 \\ x_1, x_2 &\geq 0 \end{aligned}$$

20. Minimize $C = 10x_1 + 15x_2$
Subject to
$$\begin{aligned} -4x_1 + x_2 &\geq 12 \\ 12x_1 - 3x_2 &\geq 10 \\ x_1, x_2 &\geq 0 \end{aligned}$$

B *Solve the linear programming problems in Problems 21–32 by applying the simplex method to the dual problem.*

21. Minimize $C = 3x_1 + 9x_2$
Subject to
$$\begin{aligned} 2x_1 + x_2 &\geq 8 \\ x_1 + 2x_2 &\geq 8 \\ x_1, x_2 &\geq 0 \end{aligned}$$

22. Minimize $C = 2x_1 + x_2$
Subject to
$$\begin{aligned} x_1 + x_2 &\geq 8 \\ x_1 + 2x_2 &\geq 4 \\ x_1, x_2 &\geq 0 \end{aligned}$$

23. Minimize $C = 7x_1 + 5x_2$
Subject to
$$\begin{aligned} x_1 + x_2 &\geq 4 \\ x_1 - 2x_2 &\geq -8 \\ -2x_1 + x_2 &\geq -8 \\ x_1, x_2 &\geq 0 \end{aligned}$$

24. Minimize $C = 10x_1 + 4x_2$
Subject to
$$\begin{aligned} 2x_1 + x_2 &\geq 6 \\ x_1 - 4x_2 &\geq -24 \\ -8x_1 + 5x_2 &\geq -24 \\ x_1, x_2 &\geq 0 \end{aligned}$$

25. Minimize $C = 10x_1 + 30x_2$
Subject to
$$\begin{aligned} 2x_1 + x_2 &\geq 16 \\ x_1 + x_2 &\geq 12 \\ x_1 + 2x_2 &\geq 14 \\ x_1, x_2 &\geq 0 \end{aligned}$$

26. Minimize $C = 40x_1 + 10x_2$
Subject to
$$\begin{aligned} 3x_1 + x_2 &\geq 24 \\ x_1 + x_2 &\geq 16 \\ x_1 + 4x_2 &\geq 30 \\ x_1, x_2 &\geq 0 \end{aligned}$$

27. Minimize $C = 5x_1 + 7x_2$
Subject to
$$\begin{aligned} x_1 &\geq 4 \\ x_1 + x_2 &\geq 8 \\ x_1 + 2x_2 &\geq 10 \\ x_1, x_2 &\geq 0 \end{aligned}$$

28. Minimize $C = 4x_1 + 5x_2$
Subject to
$$\begin{aligned} 2x_1 + x_2 &\geq 12 \\ x_1 + x_2 &\geq 9 \\ + x_2 &\geq 4 \\ x_1, x_2 &\geq 0 \end{aligned}$$

29. Minimize $C = 10x_1 + 7x_2 + 12x_3$
Subject to
$$\begin{aligned} x_1 + x_2 + 2x_3 &\geq 7 \\ 2x_1 + x_2 + x_3 &\geq 4 \\ x_1, x_2, x_3 &\geq 0 \end{aligned}$$

30. Minimize $C = 14x_1 + 8x_2 + 20x_3$
Subject to
$$\begin{aligned} x_1 + x_2 + 3x_3 &\geq 6 \\ 2x_1 + x_2 + x_3 &\geq 9 \\ x_1, x_2, x_3 &\geq 0 \end{aligned}$$

31. Minimize $C = 5x_1 + 2x_2 + 2x_3$
Subject to
$$\begin{aligned} x_1 - 4x_2 + x_3 &\geq 6 \\ -x_1 + x_2 - 2x_3 &\geq 4 \\ x_1, x_2, x_3 &\geq 0 \end{aligned}$$

32. Minimize $C = 6x_1 + 8x_2 + 3x_3$
Subject to
$$\begin{aligned} -3x_1 - 2x_2 + x_3 &\geq 4 \\ x_1 + x_2 - x_3 &\geq 2 \\ x_1, x_2, x_3 &\geq 0 \end{aligned}$$

33. A minimization problem has 4 variables and 2 problem constraints. How many variables and problem constraints are in the dual problem?

34. A minimization problem has 3 variables and 5 problem constraints. How many variables and problem constraints are in the dual problem?

35. If you want to solve a minimization problem by applying the geometric method to the dual problem, how many variables and problem constraints must be in the original problem?

36. If you want to solve a minimization problem by applying the geometric method to the original problem, how many variables and problem constraints must be in the original problem?

In Problems 37–40, determine whether a minimization problem with the indicated condition can be solved by applying the simplex method to the dual problem. If your answer is yes, describe any necessary modifications that must be made before forming the dual problem. If your answer is no, explain why.

37. A coefficient of the objective function is negative.

38. A coefficient of a problem constraint is negative.

39. A problem constraint is of the $\leq$ form.

40. A problem constraint has a negative constant on the right side.

C *Solve the linear programming problems in Problems 41–44 by applying the simplex method to the dual problem.*

41. Minimize $C = 16x_1 + 8x_2 + 4x_3$
Subject to
$$\begin{aligned} 3x_1 + 2x_2 + 2x_3 &\geq 16 \\ 4x_1 + 3x_2 + x_3 &\geq 14 \\ 5x_1 + 3x_2 + x_3 &\geq 12 \\ x_1, x_2, x_3 &\geq 0 \end{aligned}$$

42. Minimize $C = 6x_1 + 8x_2 + 12x_3$
Subject to
$$\begin{aligned} x_1 + 3x_2 + 3x_3 &\geq 6 \\ x_1 + 5x_2 + 5x_3 &\geq 4 \\ 2x_1 + 2x_2 + 3x_3 &\geq 8 \\ x_1, x_2, x_3 &\geq 0 \end{aligned}$$

43. Minimize $C = 5x_1 + 4x_2 + 5x_3 + 6x_4$
Subject to
$$\begin{aligned} x_1 + x_2 &\leq 12 \\ x_3 + x_4 &\leq 25 \\ x_1 + x_3 &\geq 20 \\ x_2 + x_4 &\geq 15 \\ x_1, x_2, x_3, x_4 &\geq 0 \end{aligned}$$

44. Repeat Problem 43 with $C = 4x_1 + 7x_2 + 5x_3 + 6x_4$.

APPLICATIONS

In Problems 45–52, construct a mathematical model in the form of a linear programming problem. (The answers in the back of the book for these application problems include the model.) Then solve the problem by applying the simplex method to the dual problem.

Business & Economics

45. *Manufacturing—production scheduling.* A food processing company produces regular and deluxe ice cream at three plants. Per hour of operation, the plant in Cedarburg produces 20 gallons of regular ice cream and 10 gallons of deluxe ice cream, the Grafton plant produces 10 gallons of regular and 20 gallons of deluxe, and the West Bend plant produces 20 gallons of regular and 20 gallons of deluxe. It costs $70 per hour to operate the Cedarburg plant, $75 per hour to operate the Grafton plant, and $90 per hour to operate the West Bend plant.

(A) The company needs at least 300 gallons of regular ice cream and at least 200 gallons of deluxe ice cream each day. How many hours per day should each plant be scheduled to operate in order to produce the required amounts of ice cream and minimize the cost of production? What is the minimum production cost?

(B) Discuss the effect on the production schedule and the minimum production cost if the demand for deluxe ice cream increases to 300 gallons per day and all other data in part (A) remain the same.

(C) Repeat part (B) if the demand for deluxe ice cream increases to 400 gallons per day.

46. *Mining—production scheduling.* A mining company operates two mines, each of which produces three grades of ore. The West Summit mine can produce 2 tons of low-grade ore, 3 tons of medium-grade ore, and 1 ton of high-grade ore per hour of operation. The North Ridge mine can produce 2 tons of low-grade ore, 1 ton of medium-grade ore, and 2 tons of high-grade ore per hour of operation. To satisfy existing orders, the company needs at least 100 tons of low-grade ore, 60 tons of medium-grade ore, and 80 tons of high-grade ore. The cost of operating each mine varies, depending on the conditions encountered while extracting the ore.

(A) If it costs $400 per hour to operate the West Summit mine and $600 per hour to operate the North Ridge mine, how many hours should each mine be operated to supply the required amounts of ore and minimize the cost of production? What is the minimum production cost?

(B) Discuss the effect on the production schedule and the minimum production cost if it costs $300 per hour to operate the West Summit mine, $700 per hour to operate the North Ridge mine, and all other data in part (A) remain the same.

(C) Repeat part (B) if it costs $800 per hour to operate the West Summit mine and $200 per hour to operate the North Ridge mine.

47. *Purchasing.* Acme Micros markets computers with single-sided and double-sided disk drives. The disk drives are supplied by two other companies, Associated Electronics and Digital Drives. Associated Electronics charges $250 for a single-sided disk drive and $350 for a double-sided disk drive. Digital Drives charges $290 for a single-sided disk drive and $320 for a double-sided disk drive. Each month, Associated Electronics can supply at most 1,000 disk drives in any combination of single-sided and double-sided drives. The combined monthly total supplied by Digital Drives cannot exceed 2,000 disk drives. Acme Micros needs at least 1,200 single-sided drives and at least 1,600 double-sided drives each month. How many disk drives of each type should Acme Micros order from each supplier in order to meet its monthly demand and minimize the purchase cost? What is the minimum purchase cost?

48. *Transportation.* A feed company stores grain in elevators located in Ames, Iowa, and Bedford, Indiana. Each month the grain is shipped to processing plants in Columbia, Missouri, and Danville, Illinois. The monthly supply (in tons) of grain at each elevator, the monthly demand (in tons) at each processing plant, and the cost per ton for transporting the grain are given in the table. Determine a shipping schedule that will minimize the cost of transporting the grain. What is the minimum cost?

	SHIPPING COST ($ PER TON)		SUPPLY
	Columbia	*Danville*	(TONS)
AMES	22	38	700
BEDFORD	46	24	500
DEMAND (TONS)	400	600	

Life Sciences

49. *Nutrition—people.* A dietitian in a hospital is to arrange a special diet using three foods, L, M, and N. Each ounce of food L contains 20 units of calcium, 10 units of iron, 10 units of vitamin A, and 20 units of cholesterol. Each ounce of food M contains 10 units of calcium, 10 units of iron, 15 units of vitamin A, and 24 units of cholesterol. Each ounce of food N contains 10 units of calcium, 10 units of iron, 10 units of vitamin A, and 18 units of cholesterol. If the minimum daily requirements are 300 units of calcium, 200 units of iron, and 240 units of vitamin A, how many ounces of each food should be used to meet the minimum requirements and at the same time minimize the cholesterol intake? What is the minimum cholesterol intake?

50. *Nutrition—plants.* A farmer can buy three types of plant food, mix A, mix B, and mix C. Each cubic yard of mix A contains 20 pounds of phosphoric acid, 10 pounds of nitrogen, and 10 pounds of potash. Each cubic yard of mix B contains 10 pounds of phosphoric acid, 10 pounds of nitrogen, and 15 pounds of potash. Each cubic yard of mix C contains 20 pounds of phosphoric acid, 20 pounds of nitrogen, and 5 pounds of potash. The minimum monthly requirements are 480 pounds of phosphoric acid, 320 pounds of nitrogen, and 225 pounds of potash. If mix A costs \$30 per cubic yard, mix B costs \$36 per cubic yard, and mix C costs \$39 per cubic yard, how many cubic yards of each mix should the farmer blend to meet the minimum monthly requirements at a minimal cost? What is the minimum cost?

Social Sciences

51. *Education—resource allocation.* A metropolitan school district has two high schools that are overcrowded and two that are underenrolled. In order to balance the enrollment, the school board has decided to bus students from the overcrowded schools to the underenrolled schools. North Division High School has 300 more students than it should have, and South Division High School has 500 more students than it should have. Central High School can accommodate 400 additional students, and Washington High School can accommodate 500 additional students. The weekly cost of busing a student from North Division to Central is \$5, from North Division to Washington is \$2, from South Division to Central is \$3, and from South Division to Washington is \$4. Determine the number of students that should be bused from each of the overcrowded schools to each of the underenrolled schools in order to balance the enrollment and minimize the cost of busing the students. What is the minimum cost?

52. *Education—resource allocation.* Repeat Problem 51 if the weekly cost of busing a student from North Division to Washington is \$7 and all the other information remains the same.

SECTION 5-6 Maximization and Minimization with Mixed Problem Constraints

- An Introduction to the Big M Method
- The Big M Method
- Minimization by the Big M Method
- Summary of Methods of Solution
- Larger Problems—A Refinery Application

In the preceding two sections, we have seen how to solve both maximization and minimization problems, but with rather severe restrictions on problem constraints, right-side constants, and/or objective function coefficients (see the summary in Table 1 of the last section). In this section we will present a generalized version of the simplex method that will solve both maximization and minimization problems with any combination of $\leq$, $\geq$, and $=$ problem constraints. The only requirement is that each problem constraint have a nonnegative constant on the right side. (This restriction is easily accommodated, as you will see.)

An Introduction to the Big M Method

We introduce the *big M method* through a simple maximization problem with mixed problem constraints. The key parts of the method will then be summarized and applied to more complex problems.

Consider the following problem:

$$\begin{aligned} \text{Maximize} \quad & P = 2x_1 + x_2 \\ \text{Subject to} \quad & x_1 + x_2 \leqslant 10 \\ & -x_1 + x_2 \geqslant 2 \\ & x_1, x_2 \geqslant 0 \end{aligned} \tag{1}$$

To form an equation out of the first inequality, we introduce a slack variable s_1, as before, and write

$$x_1 + x_2 + s_1 = 10$$

How can we form an equation out of the second inequality? We introduce a second variable s_2 and subtract it from the left side so that we can write

$$-x_1 + x_2 - s_2 = 2$$

The variable s_2 is called a **surplus variable,** because it is the amount (surplus) by which the left side of the inequality exceeds the right side.

We now express the linear programming problem (1) as a system of equations:

$$\begin{aligned} x_1 + x_2 + s_1 \qquad\qquad & = 10 \\ -x_1 + x_2 \qquad - s_2 \qquad & = 2 \\ -2x_1 - x_2 \qquad\qquad + P & = 0 \\ x_1, x_2, s_1, s_2 & \geqslant 0 \end{aligned} \tag{2}$$

It can be shown that a basic solution of (2) is not feasible if any of the variables (excluding P) are negative. Thus, **a surplus variable is required to satisfy the nonnegative constraint.**

The basic solution found by setting the nonbasic variables x_1 and x_2 equal to 0 is

$$x_1 = 0, \quad x_2 = 0, \quad s_1 = 10, \quad s_2 = -2, \quad P = 0$$

But this basic solution is not feasible, since the surplus variable s_2 is negative (which is a violation of the nonnegative requirements of all variables except P). The simplex method works only when the basic solution for a tableau is feasible, so we cannot solve this problem simply by writing the tableau for (2) and starting pivot operations.

Explore–Discuss 1

To see that the simplex method will not work when the basic solution is not feasible, write the tableau for (2) and perform a pivot operation. Is the new basic solution feasible? Is it possible to select another pivot element?

In order to use the simplex method on problems with mixed constraints, we turn to an ingenious device called an *artificial variable.* This variable has no physical meaning in the original problem (which explains the use of the word "artificial") and is introduced solely for the purpose of obtaining a basic feasible solution so that we can apply the simplex method. An **artificial variable** is a variable introduced into each equation that has a surplus variable. As before, to ensure that we consider only feasible basic solutions, **an artificial variable is required to satisfy the nonnegative constraint.** (As we shall see later, artificial variables are also used to augment equality problem constraints when they are present.)

Returning to the problem at hand, we introduce an artificial variable a_1 into the equation involving the surplus variable s_2:

$$-x_1 + x_2 - s_2 + a_1 = 2$$

To prevent an artificial variable from becoming part of an optimal solution to the original problem, a very large "penalty" is introduced into the objective function. This penalty is created by choosing a positive constant M so large that the artificial variable is forced to be 0 in any final optimal solution of the original problem. (Since the constant M can be made as large as we wish in computer solutions, M is often selected as the largest number the computer can hold!) We then add the term $-Ma_1$ to the objective function:

$$P = 2x_1 + x_2 - Ma_1$$

We now have a new problem, which we call the **modified problem:**

$$\begin{aligned} &\text{Maximize} \quad P = 2x_1 + x_2 - Ma_1 \\ &\text{Subject to} \quad \begin{aligned} x_1 + x_2 + s_1 \qquad\qquad &= 10 \\ -x_1 + x_2 \qquad - s_2 + a_1 &= 2 \\ x_1, x_2, s_1, s_2, a_1 &\geq 0 \end{aligned} \end{aligned} \tag{3}$$

The initial system for the modified problem (3) is

$$\begin{aligned} x_1 + x_2 + s_1 \qquad\qquad\qquad &= 10 \\ -x_1 + x_2 \qquad - s_2 + \quad a_1 \qquad &= 2 \\ -2x_1 - x_2 \qquad\qquad + Ma_1 + P &= 0 \\ x_1, x_2, s_1, s_2, a_1 &\geq 0 \end{aligned} \tag{4}$$

We next write the augmented coefficient matrix for (4), which we call the **preliminary simplex tableau** for the modified problem. (The reason we call it the "preliminary" simplex tableau instead of the "initial" simplex tableau will be made clear shortly.)

$$\begin{array}{c} \begin{matrix} x_1 & x_2 & s_1 & s_2 & a_1 & P \end{matrix} \\ \left[\begin{array}{rrrrrr|r} 1 & 1 & 1 & 0 & 0 & 0 & 10 \\ -1 & 1 & 0 & -1 & 1 & 0 & 2 \\ \hline -2 & -1 & 0 & 0 & M & 1 & 0 \end{array}\right] \end{array} \tag{5}$$

To start the simplex process, including any necessary pivot operations, the preliminary simplex tableau should either meet the two requirements given in the following box or be transformed by row operations into a tableau that meets these two requirements.

Initial Simplex Tableau Requirements

For a system tableau to be considered an **initial simplex tableau,** it must satisfy the following two requirements:

1. The requisite number of basic variables must be selectable by the process described in Section 5-4. That is, a variable can be selected as a basic variable only if it corresponds to a column in the tableau that has exactly one nonzero element and the nonzero element in the column is not in the same row as the nonzero element in the column of another basic variable. The remaining variables are then selected as nonbasic variables to be set equal to 0 in determining a basic solution.
2. The basic solution found by setting the nonbasic variables equal to 0 is feasible.

Tableau (5) satisfies the first initial simplex tableau requirement, since s_1, s_2, and P can be selected as basic variables according to the criterion stated. (Not all preliminary simplex tableaux satisfy the first requirement; see Example 2.) However, tableau (5) does not satisfy the second initial simplex tableau requirement, since the basic solution is not feasible ($s_2 = -2$). To use the simplex method, we must first use row operations to transform (5) into an equivalent matrix that satisfies both initial simplex tableau requirements. **Note that this transformation is not a pivot operation.**

To get an idea of how to proceed, notice in tableau (5) that -1 in the s_2 column is in the same row as 1 in the a_1 column. This is not an accident! The artificial variable a_1 was introduced so that this would happen. If we eliminate M from the bottom of the a_1 column, then the nonbasic variable a_1 will become a basic variable and the troublesome basic variable s_2 will become a nonbasic variable. Thus, we proceed to eliminate M from the a_1 column using row operations:

$$\begin{array}{c} \\ \\ \\ \\ \end{array}\begin{matrix} x_1 & x_2 & s_1 & s_2 & a_1 & P & \\ \end{matrix}$$

$$\left[\begin{array}{cccccc|c} 1 & 1 & 1 & 0 & 0 & 0 & 10 \\ -1 & 1 & 0 & -1 & 1 & 0 & 2 \\ \hline -2 & -1 & 0 & 0 & M & 1 & 0 \end{array}\right] \quad (-M)R_2 + R_3 \to R_3$$

$$\sim \left[\begin{array}{cccccc|c} 1 & 1 & 1 & 0 & 0 & 0 & 10 \\ -1 & 1 & 0 & -1 & 1 & 0 & 2 \\ \hline M-2 & -M-1 & 0 & M & 0 & 1 & -2M \end{array}\right]$$

From this last matrix we see that the basic variables are s_1, a_1, and P. The basic solution found by setting the nonbasic variables x_1, x_2, and s_2 equal to 0 is

$$x_1 = 0, \quad x_2 = 0, \quad s_1 = 10, \quad s_2 = 0, \quad a_1 = 2, \quad P = -2M$$

The basic solution is feasible (remember, P can be negative), and both requirements for an initial simplex tableau are met. We can now commence with the simplex process using pivot operations.

The pivot column is determined by the most negative indicator in the bottom row of the tableau. Since M is a positive number, $-M - 1$ is certainly a negative indicator. What about the indicator $M - 2$? Remember that M is a very large positive number. We will assume that M is so large that any expression of the form $M - k$ is positive. Thus, the only negative indicator in the bottom row is $-M - 1$.

$$
\begin{array}{r}
\\
\\
\text{Pivot row} \rightarrow \\
\\
\end{array}
\begin{array}{c}
\begin{matrix} x_1 & x_2 & s_1 & s_2 & a_1 & P \end{matrix} \\
\left[\begin{array}{cccccc|c}
1 & 1 & 1 & 0 & 0 & 0 & 10 \\
-1 & \textcircled{1} & 0 & -1 & 1 & 0 & 2 \\
\hdashline
M - 2 & -M - 1 & 0 & M & 0 & 1 & -2M
\end{array}\right]
\end{array}
\begin{array}{l}
\\
\frac{10}{1} = 10 \\
\frac{2}{1} = 2 \\
\\
\end{array}
$$

$\uparrow$ Pivot column

Having identified the pivot element, we now begin pivoting:

$$
\begin{array}{c}
\begin{matrix} & x_1 & x_2 & s_1 & s_2 & a_1 & P \end{matrix} \\
\begin{matrix} s_1 \\ a_1 \\ P \end{matrix}
\left[\begin{array}{cccccc|c}
1 & 1 & 1 & 0 & 0 & 0 & 10 \\
-1 & \textcircled{1} & 0 & -1 & 1 & 0 & 2 \\
\hdashline
M - 2 & -M - 1 & 0 & M & 0 & 1 & -2M
\end{array}\right]
\end{array}
\begin{array}{l}
(-1)R_2 + R_1 \rightarrow R_1 \\
\\
(M + 1)R_2 + R_3 \rightarrow R_3
\end{array}
$$

$$
\begin{matrix} s_1 \\ \sim x_2 \\ P \end{matrix}
\left[\begin{array}{cccccc|c}
\textcircled{2} & 0 & 1 & 1 & -1 & 0 & 8 \\
-1 & 1 & 0 & -1 & 1 & 0 & 2 \\
\hdashline
-3 & 0 & 0 & -1 & M + 1 & 1 & 2
\end{array}\right]
\begin{array}{l}
\frac{1}{2}R_1 \rightarrow R_1 \\
\\
\\
\end{array}
$$

$$
\sim
\left[\begin{array}{cccccc|c}
\textcircled{1} & 0 & \frac{1}{2} & \frac{1}{2} & -\frac{1}{2} & 0 & 4 \\
-1 & 1 & 0 & -1 & 1 & 0 & 2 \\
\hdashline
-3 & 0 & 0 & -1 & M + 1 & 1 & 2
\end{array}\right]
\begin{array}{l}
\\
R_1 + R_2 \rightarrow R_2 \\
3R_1 + R_3 \rightarrow R_3
\end{array}
$$

$$
\begin{matrix} x_1 \\ \sim x_2 \\ P \end{matrix}
\left[\begin{array}{cccccc|c}
1 & 0 & \frac{1}{2} & \frac{1}{2} & -\frac{1}{2} & 0 & 4 \\
0 & 1 & \frac{1}{2} & -\frac{1}{2} & \frac{1}{2} & 0 & 6 \\
\hdashline
0 & 0 & \frac{3}{2} & \frac{1}{2} & M - \frac{1}{2} & 1 & 14
\end{array}\right]
$$

Since all the indicators in the last row are nonnegative $\left(M - \frac{1}{2}\right.$ is nonnegative because M is a very large positive number$\left.\right)$, we can stop and write the optimal solution:

$$\text{Max } P = 14 \quad \text{at} \quad x_1 = 4, \quad x_2 = 6, \quad s_1 = 0, \quad s_2 = 0, \quad a_1 = 0$$

This is an optimal solution to the modified problem (3). How is it related to the original problem (2)? Since $a_1 = 0$ in this solution,

$$x_1 = 4, \quad x_2 = 6, \quad s_1 = 0, \quad s_2 = 0, \quad P = 14 \tag{6}$$

is certainly a feasible solution for (2). [You can verify this by direct substitution into (2).] Surprisingly, it turns out that (6) is an optimal solution to the original problem. To see that this is true, suppose we were able to find feasible values of x_1, x_2, s_1, and s_2 that satisfy the original system (2) and produce a value of $P > 14$. Then by using these same values in (3) along with $a_1 = 0$, we have found a feasible solution of (3) with $P > 14$. This contradicts the fact that $P = 14$ is the maximum value of P for the modified problem. Thus, (6) is an optimal solution for the original problem.

As this example illustrates, if $a_1 = 0$ is an optimal solution for the modified problem, then deleting a_1 produces an optimal solution for the original problem. What happens if $a_1 \neq 0$ in the optimal solution for the modified problem? In this case, it can be shown that the original problem has no optimal solution because its feasible set is empty.

Explore–Discuss 2

Graph the feasible region for problem (1), plot the (x_1, x_2) coordinates of the basic solution for each simplex tableau in the solution of this problem, and illustrate the path to the optimal solution.

In larger problems, each $\geq$ problem constraint will require the introduction of a surplus variable and an artificial variable. If one of the problem constraints is an equation rather than an inequality, there is no need to introduce a slack or surplus variable. However, each $=$ problem constraint will require the introduction of another artificial variable to prevent the initial basic solution from violating the equality constraint—the decision variables are often 0 in the initial basic solution (see Example 2). Finally, each artificial variable also must be included in the objective function for the modified problem. The same constant M can be used for each artificial variable. Because of the role that the constant M plays in this approach, this method is often called the **big M method.**

THE BIG M METHOD

We now summarize the key steps of the big M method and use them to solve several problems.

The Big M Method—Introducing Slack, Surplus, and Artificial Variables to Form the Modified Problem

Step 1. If any problem constraints have negative constants on the right side, multiply both sides by -1 to obtain a constraint with a nonnegative constant. (If the constraint is an inequality, this will reverse the direction of the inequality.)

Step 2. Introduce a slack variable in each $\leq$ constraint.

Step 3. Introduce a surplus variable and an artificial variable in each $\geq$ constraint.

Step 4. Introduce an artificial variable in each $=$ constraint.

Step 5. For each artificial variable a_i, add $-Ma_i$ to the objective function. Use the same constant M for all artificial variables.

Example 1 **Finding the Modified Problem** Find the modified problem for the following linear programming problem. (Do not attempt to solve the problem.)

$$\begin{aligned}
\text{Maximize} \quad & P = 2x_1 + 5x_2 + 3x_3 \\
\text{Subject to} \quad & x_1 + 2x_2 - x_3 \leq 7 \\
& -x_1 + x_2 - 2x_3 \leq -5 \\
& x_1 + 4x_2 + 3x_3 \geq 1 \\
& 2x_1 - x_2 + 4x_3 = 6 \\
& x_1, x_2, x_3 \geq 0
\end{aligned}$$

SOLUTION First, we multiply the second constraint by -1 to change -5 to 5:

$$(-1)(-x_1 + x_2 - 2x_3) \geq (-1)(-5)$$
$$x_1 - x_2 + 2x_3 \geq 5$$

Next, we introduce the slack, surplus, and artificial variables according to the rules stated in the box:

$$\begin{aligned}
x_1 + 2x_2 - x_3 + s_1 &= 7 \\
x_1 - x_2 + 2x_3 - s_2 + a_1 &= 5 \\
x_1 + 4x_2 + 3x_3 - s_3 + a_2 &= 1 \\
2x_1 - x_2 + 4x_3 + a_3 &= 6
\end{aligned}$$

Finally, we add $-Ma_1$, $-Ma_2$, and $-Ma_3$ to the objective function:

$$P = 2x_1 + 5x_2 + 3x_3 - Ma_1 - Ma_2 - Ma_3$$

The modified problem is

$$\begin{aligned}
\text{Maximize}\quad & P = 2x_1 + 5x_2 + 3x_3 - Ma_1 - Ma_2 - Ma_3 \\
\text{Subject to}\quad & x_1 + 2x_2 - x_3 + s_1 = 7 \\
& x_1 - x_2 + 2x_3 - s_2 + a_1 = 5 \\
& x_1 + 4x_2 + 3x_3 - s_3 + a_2 = 1 \\
& 2x_1 - x_2 + 4x_3 + a_3 = 6 \\
& x_1, x_2, x_3, s_1, s_2, s_3, a_1, a_2, a_3 \geq 0
\end{aligned}$$

Matched Problem 1 Repeat Example 1 for:

$$\begin{aligned}
\text{Maximize}\quad & P = 3x_1 - 2x_2 + x_3 \\
\text{Subject to}\quad & x_1 - 2x_2 + x_3 \geq 5 \\
& -x_1 - 3x_2 + 4x_3 \leq -10 \\
& 2x_1 + 4x_2 + 5x_3 \leq 20 \\
& 3x_1 - x_2 - x_3 = -15 \\
& x_1, x_2, x_3 \geq 0
\end{aligned}$$

We now list the key steps for solving a problem using the big M method. The various steps and remarks are based on a number of important theorems, which we assume without proof. In particular, step 2 is based on the fact that (except for some degenerate cases not considered here) if the modified linear programming problem has an optimal solution, then the preliminary simplex tableau will be transformed into an initial simplex tableau by eliminating the M's from the columns corresponding to the artificial variables in the preliminary simplex tableau. Having obtained an initial simplex tableau, we can then commence with pivot operations.

The Big M Method—Solving the Problem

Step 1. Form the preliminary simplex tableau for the modified problem.

Step 2. Use row operations to eliminate the M's in the bottom row of the preliminary simplex tableau in the columns corresponding to the artificial variables. The resulting tableau is the initial simplex tableau.

Step 3. Solve the modified problem by applying the simplex method to the initial simplex tableau found in step 2.

(continued)

The Big M Method—Solving the Problem (*continued*)

Step 4. Relate the optimal solution of the modified problem to the original problem.

(A) If the modified problem has no optimal solution, then the original problem has no optimal solution.
(B) If all artificial variables are 0 in the optimal solution to the modified problem, then delete the artificial variables to find an optimal solution to the original problem.
(C) If any artificial variables are nonzero in the optimal solution to the modified problem, then the original problem has no optimal solution.

Example 2 **Using the Big M Method** Solve the following linear programming problem using the big M method:

$$\begin{array}{ll} \text{Maximize} & P = x_1 - x_2 + 3x_3 \\ \text{Subject to} & x_1 + x_2 \leq 20 \\ & x_1 + x_3 = 5 \\ & x_2 + x_3 \geq 10 \\ & x_1, x_2, x_3 \geq 0 \end{array}$$

SOLUTION State the modified problem:

$$\begin{array}{ll} \text{Maximize} & P = x_1 - x_2 + 3x_3 - Ma_1 - Ma_2 \\ \text{Subject to} & x_1 + x_2 + s_1 = 20 \\ & x_1 + x_3 + a_1 = 5 \\ & x_2 + x_3 - s_2 + a_2 = 10 \\ & x_1, x_2, x_3, s_1, s_2, a_1, a_2 \geq 0 \end{array}$$

Write the preliminary simplex tableau for the modified problem, and find the initial simplex tableau by eliminating the M's from the artificial variable columns:

$$\begin{array}{c} \\ \\ \\ \\ \end{array}
\begin{array}{cccccccc|c} x_1 & x_2 & x_3 & s_1 & a_1 & s_2 & a_2 & P & \\ 1 & 1 & 0 & 1 & 0 & 0 & 0 & 0 & 20 \\ 1 & 0 & 1 & 0 & 1 & 0 & 0 & 0 & 5 \\ 0 & 1 & 1 & 0 & 0 & -1 & 1 & 0 & 10 \\ \hline -1 & 1 & -3 & 0 & M & 0 & M & 1 & 0 \end{array}$$

Eliminate M from the a_1 column. $(-M)R_2 + R_4 \to R_4$

$$\sim \left[\begin{array}{cccccccc|c} 1 & 1 & 0 & 1 & 0 & 0 & 0 & 0 & 20 \\ 1 & 0 & 1 & 0 & 1 & 0 & 0 & 0 & 5 \\ 0 & 1 & 1 & 0 & 0 & -1 & 1 & 0 & 10 \\ \hline -M-1 & 1 & -M-3 & 0 & 0 & 0 & M & 1 & -5M \end{array}\right]$$

Eliminate M from the a_2 column. $(-M)R_3 + R_4 \to R_4$

$$\sim \left[\begin{array}{cccccccc|c} 1 & 1 & 0 & 1 & 0 & 0 & 0 & 0 & 20 \\ 1 & 0 & 1 & 0 & 1 & 0 & 0 & 0 & 5 \\ 0 & 1 & 1 & 0 & 0 & -1 & 1 & 0 & 10 \\ \hline -M-1 & -M+1 & -2M-3 & 0 & 0 & M & 0 & 1 & -15M \end{array}\right]$$

From this last matrix we see that the basic variables are s_1, a_1, a_2, and P. The basic solution found by setting the nonbasic variables x_1, x_2, x_3, and s_2 equal to 0 is

$$x_1 = 0, \quad x_2 = 0, \quad x_3 = 0, \quad s_1 = 20, \quad a_1 = 5, \quad s_2 = 0, \quad a_2 = 10, \quad P = -15M$$

The basic solution is feasible, and both requirements for an initial simplex tableau are met. We can now commence with pivot operations to find the optimal solution.

	x_1	x_2	x_3	s_1	a_1	s_2	a_2	P		
s_1	1	1	0	1	0	0	0	0	20	
a_1	1	0	(1)	0	1	0	0	0	5	
a_2	0	1	1	0	0	-1	1	0	10	$(-1)R_2 + R_3 \to R_3$
P	$-M - 1$	$-M + 1$	$-2M - 3$	0	0	M	0	1	$-15M$	$(2M + 3)R_2 + R_4 \to R_4$

$\sim$

	x_1	x_2	x_3	s_1	a_1	s_2	a_2	P		
s_1	1	1	0	1	0	0	0	0	20	$(-1)R_3 + R_1 \to R_1$
x_3	1	0	1	0	1	0	0	0	5	
a_2	-1	(1)	0	0	-1	-1	1	0	5	
P	$M + 2$	$-M + 1$	0	0	$2M + 3$	M	0	1	$-5M + 15$	$(M - 1)R_3 + R_4 \to R_4$

$\sim$

	x_1	x_2	x_3	s_1	a_1	s_2	a_2	P	
s_1	2	0	0	1	1	1	-1	0	15
x_3	1	0	1	0	1	0	0	0	5
x_2	-1	1	0	0	-1	-1	1	0	5
P	3	0	0	0	$M + 4$	1	$M - 1$	1	10

Since the bottom row has no negative indicators, we can stop and write the optimal solution to the modified problem:

$$x_1 = 0, \quad x_2 = 5, \quad x_3 = 5, \quad s_1 = 15, \quad a_1 = 0, \quad s_2 = 0, \quad a_2 = 0, \quad P = 10$$

Since $a_1 = 0$ and $a_2 = 0$, the solution to the original problem is

$$\text{Max } P = 10 \quad \text{at} \quad x_1 = 0, \quad x_2 = 5, \quad x_3 = 5$$

Matched Problem 2 ➡ Solve the following linear programming problem using the big M method:

$$\begin{aligned} \text{Maximize} \quad & P = x_1 + 4x_2 + 2x_3 \\ \text{Subject to} \quad & x_2 + x_3 \leq 4 \\ & x_1 - x_3 = 6 \\ & x_1 - x_2 - x_3 \geq 1 \\ & x_1, x_2, x_3 \geq 0 \end{aligned}$$

Example 3 ➡ **Using the Big M Method** Solve the following linear programming problem using the big M method:

$$\begin{aligned} \text{Maximize} \quad & P = 3x_1 + 5x_2 \\ \text{Subject to} \quad & 2x_1 + x_2 \leq 4 \\ & x_1 + 2x_2 \geq 10 \\ & x_1, x_2 \geq 0 \end{aligned}$$

Solution Introducing slack, surplus, and artificial variables, we obtain the modified problem:

$$\begin{aligned} 2x_1 + x_2 + s_1 &&= 4 \\ x_1 + 2x_2 - s_2 + a_1 &&= 10 \\ -3x_1 - 5x_2 + Ma_1 + P &&= 0 \end{aligned}$$ Modified problem

Preliminary simplex tableau

$$\begin{array}{c} \\ \\ \\ \end{array}\begin{matrix} x_1 & x_2 & s_1 & s_2 & a_1 & P & \end{matrix}$$

$$\left[\begin{array}{cccccc|c} 2 & 1 & 1 & 0 & 0 & 0 & 4 \\ 1 & 2 & 0 & -1 & 1 & 0 & 10 \\ \hline -3 & -5 & 0 & 0 & M & 1 & 0 \end{array}\right]$$

Eliminate M in the a_1 column.

$(-M)R_2 + R_3 \to R_3$

Initial simplex tableau

$$\begin{array}{c} s_1 \\ \sim a_1 \\ P \end{array}\left[\begin{array}{cccccc|c} 2 & \textcircled{1} & 1 & 0 & 0 & 0 & 4 \\ 1 & 2 & 0 & -1 & 1 & 0 & 10 \\ \hline -M-3 & -2M-5 & 0 & M & 0 & 1 & -10M \end{array}\right]$$

Begin pivot operations.

$(-2)R_1 + R_2 \to R_2$

$(2M + 5)R_1 + R_3 \to R_3$

$$\begin{array}{c} x_2 \\ \sim a_1 \\ P \end{array}\left[\begin{array}{cccccc|c} 2 & 1 & 1 & 0 & 0 & 0 & 4 \\ -3 & 0 & -2 & -1 & 1 & 0 & 2 \\ \hline 3M+7 & 0 & 2M+5 & M & 0 & 1 & -2M+20 \end{array}\right]$$

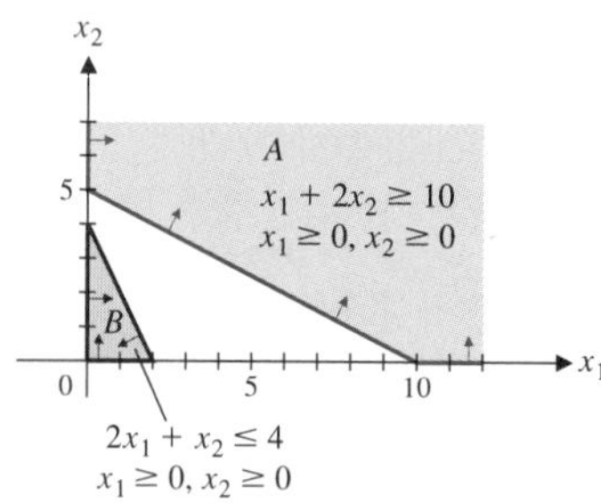

FIGURE 1

The optimal solution of the modified problem is

$$x_1 = 0, \quad x_2 = 4, \quad s_1 = 0, \quad s_2 = 0, \quad a_1 = 2, \quad P = -2M + 20$$

Since $a_1 \neq 0$, the original problem has no optimal solution. Figure 1 shows that the feasible region for the original problem is empty.

Matched Problem 3 Solve the following linear programming problem using the big M method:

$$\begin{aligned} \text{Maximize} \quad & P = 3x_1 + 2x_2 \\ \text{Subject to} \quad & x_1 + 5x_2 \leqslant 5 \\ & 2x_1 + x_2 \geqslant 12 \\ & x_1, x_2 \geqslant 0 \end{aligned}$$

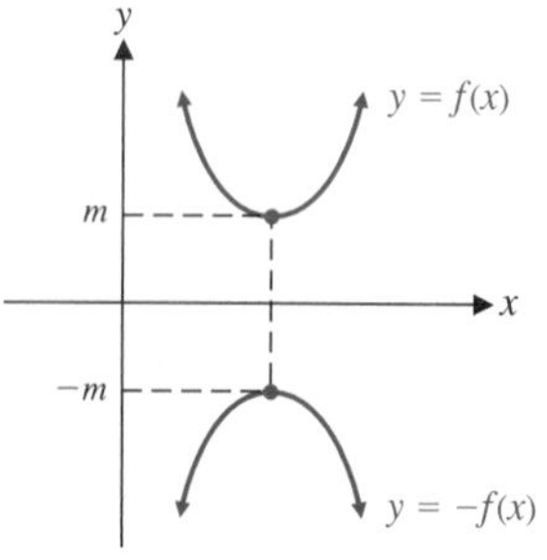

FIGURE 2

Minimization by the Big M Method

In addition to solving any maximization problem, the big M method can be used to solve minimization problems. To minimize an objective function, we have only to maximize its negative. Figure 2 illustrates the fact that the minimum value of a function f occurs at the same point as the maximum value of the function $-f$. Furthermore, if m is the minimum value of f, then $-m$ is the maximum value of $-f$, and conversely. Thus, we can find the minimum value of a function f by finding the maximum value of $-f$ and then changing the sign of the maximum value.

Example 4 **Production Scheduling—A Minimization Problem** A small jewelry manufacturing company employs a person who is a highly skilled gem cutter, and it wishes to use this person at least 6 hours per day for this purpose. On the other hand, the polishing facilities can be used in any amounts up to 10 hours per day. The company specializes in three kinds of semiprecious gemstones, *J*, *K*, and *L*. Relevant cutting, polishing, and cost requirements are listed in the table. How many gemstones of each type should be processed each day to minimize the cost of the finished stones? What is the minimum cost?

	J	*K*	*L*
Cutting	1 hr	1 hr	1 hr
Polishing	2 hr	1 hr	2 hr
Cost per stone	$30	$30	$10

Solution If we let x_1, x_2, and x_3 represent the number of type *J*, *K*, and *L* stones finished per day, respectively, then we have the following linear programming problem to solve, where *C* is the cost of the stones:

$$\begin{aligned} &\text{Minimize} & C &= 30x_1 + 30x_2 + 10x_3 && \text{Objective function} \\ &\text{Subject to} & x_1 + x_2 + x_3 &\geq 6 && \text{Problem constraints} \\ && 2x_1 + x_2 + 2x_3 &\leq 10 && \\ && x_1, x_2, x_3 &\geq 0 && \text{Nonnegative constraints} \end{aligned}$$

We convert this to a maximization problem by letting

$$P = -C = -30x_1 - 30x_2 - 10x_3$$

Thus, we get

$$\begin{aligned} &\text{Maximize} & P &= -30x_1 - 30x_2 - 10x_3 \\ &\text{Subject to} & x_1 + x_2 + x_3 &\geq 6 \\ && 2x_1 + x_2 + 2x_3 &\leq 10 \\ && x_1, x_2, x_3 &\geq 0 \end{aligned}$$

and Min $C = -$Max P. To solve, we first state the modified problem:

$$\begin{aligned} x_1 + x_2 + x_3 - s_1 + a_1 \qquad\qquad &= 6 \\ 2x_1 + x_2 + 2x_3 \qquad\qquad + s_2 \qquad &= 10 \\ 30x_1 + 30x_2 + 10x_3 \qquad + Ma_1 \qquad + P &= 0 \\ x_1, x_2, x_3, s_1, s_2, a_1 &\geq 0 \end{aligned}$$

$$\begin{array}{c} \\ \\ \\ \end{array}\left[\begin{array}{ccccccc|c} x_1 & x_2 & x_3 & s_1 & a_1 & s_2 & P & \\ 1 & 1 & 1 & -1 & 1 & 0 & 0 & 6 \\ 2 & 1 & 2 & 0 & 0 & 1 & 0 & 10 \\ \hline 30 & 30 & 10 & 0 & M & 0 & 1 & 0 \end{array}\right]$$

Eliminate *M* in the a_1 column.

$(-M)R_1 + R_3 \to R_3$

Begin pivot operations. Assume *M* is so large that $-M + 30$ and $-M + 10$ are negative.

$$\begin{array}{c} \\ a_1 \\ \sim s_2 \\ P \end{array}\left[\begin{array}{ccccccc|c} x_1 & x_2 & x_3 & s_1 & a_1 & s_2 & P & \\ 1 & 1 & 1 & -1 & 1 & 0 & 0 & 6 \\ 2 & 1 & \textcircled{2} & 0 & 0 & 1 & 0 & 10 \\ \hline -M+30 & -M+30 & -M+10 & M & 0 & 0 & 1 & -6M \end{array}\right] \quad 0.5R_2 \to R_2$$

$$\sim\left[\begin{array}{ccccccc|c} 1 & 1 & 1 & -1 & 1 & 0 & 0 & 6 \\ 1 & 0.5 & \textcircled{1} & 0 & 0 & 0.5 & 0 & 5 \\ \hline -M+30 & -M+30 & -M+10 & M & 0 & 0 & 1 & -6M \end{array}\right] \quad \begin{array}{l} (-1)R_2 + R_1 \to R_1 \\ \\ (M-10)R_2 + R_3 \to R_3 \end{array}$$

	x_1	x_2	x_3	s_1	a_1	s_2	P		
a_1	0	(0.5)	0	−1	1	−0.5	0	1	$2R_1 \to R_1$
x_3	1	0.5	1	0	0	0.5	0	5	
P	20	$-0.5M + 25$	0	M	0	$0.5M - 5$	1	$-M - 50$	

$\sim$

	x_1	x_2	x_3	s_1	a_1	s_2	P		
	0	(1)	0	−2	2	−1	0	2	
	1	0.5	1	0	0	0.5	0	5	$(-0.5)R_1 + R_2 \to R_2$
	20	$-0.5M + 25$	0	M	0	$0.5M - 5$	1	$-M - 50$	$(0.5M - 25)R_1 + R_3 \to R_3$

$\sim$

	x_1	x_2	x_3	s_1	a_1	s_2	P	
x_2	0	1	0	−2	2	−1	0	2
x_3	1	0	1	1	−1	1	0	4
P	20	0	0	50	$M - 50$	20	1	−100

The bottom row has no negative indicators, so the optimal solution for the modified problem is

$$x_1 = 0, \quad x_2 = 2, \quad x_3 = 4, \quad s_1 = 0, \quad a_1 = 0, \quad s_2 = 0, \quad P = -100$$

Since $a_1 = 0$, deleting a_1 produces the optimal solution to the original maximization problem and also to the minimization problem. Thus,

$$\text{Min } C = -\text{Max } P = -(-100) = 100 \qquad \text{at} \qquad x_1 = 0, \quad x_2 = 2, \quad x_3 = 4$$

That is, a minimum cost of \$100 for gemstones will be realized if no type *J*, 2 type *K*, and 4 type *L* stones are processed each day.

Matched Problem 4 ➡ Repeat Example 4 if the gem cutter is to be used for at least 8 hours a day and all other data remain the same.

■ Summary of Methods of Solution

The big *M* method can be used to solve any minimization problem, including those that can be solved by the dual method. (Note that Example 4 could have been solved, by the dual method.) Both methods of solving minimization problems are important. You will be instructed to solve most minimization problems in Exercise 5-6 by the big *M* method in order to gain more experience with this method. If the method of solution is not specified, then the dual method is usually easier.

Table 1 should help you select the proper method of solution for any linear programming problem.

Table 1

Summary of Problem Types and Simplex Solution Methods

Problem Type	Problem Constraints	Right-Side Constants	Coefficients of Objective Function	Method of Solution
1. Maximization	$\leq$	Nonnegative	Any real numbers	Simplex method with slack variables
2. Minimization	$\geq$	Any real numbers	Nonnegative	Form dual and solve by the preceding method
3. Maximization	Mixed ($\leq, \geq, =$)	Nonnegative	Any real numbers	Form modified problem with slack, surplus, and artificial variables, and solve by the big *M* method
4. Minimization	Mixed ($\leq, \geq, =$)	Nonnegative	Any real numbers	Maximize negative of objective function by the preceding method

■ Larger Problems—A Refinery Application

Up to this point, all the problems we have considered could be solved by hand. However, the real value of the simplex method lies in its ability to solve problems with a large number of variables and constraints, where a computer is generally used to perform the actual pivot operations. As a final application, we will consider a problem that would require the use of a computer to complete the solution.

Example 5 **Petroleum Blending** A refinery produces two grades of gasoline, regular and premium, by blending together two components, A and B. Component A has an octane rating of 90 and costs $28 a barrel. Component B has an octane rating of 110 and costs $32 a barrel. The octane rating for regular gasoline must be at least 95, and the octane rating for premium must be at least 105. Regular gasoline sells for $34 a barrel and premium sells for $40 a barrel. Currently, the company has 30,000 barrels of component A and 20,000 barrels of component B. It also has orders for 20,000 barrels of regular and 10,000 barrels of premium that must be filled. Assuming that all the gasoline produced can be sold, determine the maximum possible profit.

Solution First we organize the information given in the problem in Table 2.

Table 2

Component	Octane Rating	Cost ($)	Available Supply
A	90	28	30,000 barrels
B	110	32	20,000 barrels
Grade	**Minimum Octane Rating**	**Selling Price ($)**	**Existing Orders**
Regular	95	34	20,000 barrels
Premium	105	40	10,000 barrels

Let: x_1 = Number of barrels of component A used in regular gasoline
x_2 = Number of barrels of component A used in premium gasoline
x_3 = Number of barrels of component B used in regular gasoline
x_4 = Number of barrels of component B used in premium gasoline

The total amount of component A used is $x_1 + x_2$. This cannot exceed the available supply. Thus, one constraint is

$$x_1 + x_2 \leq 30{,}000$$

The corresponding inequality for component B is

$$x_3 + x_4 \leq 20{,}000$$

The amounts of regular and premium gasoline produced must be sufficient to meet the existing orders:

$$x_1 + x_3 \geq 20{,}000 \quad \text{Regular}$$
$$x_2 + x_4 \geq 10{,}000 \quad \text{Premium}$$

Now consider the octane ratings. The octane rating of a blend is simply the proportional average of the octane ratings of the components. Thus, the octane rating for regular gasoline is

$$90\frac{x_1}{x_1 + x_3} + 110\frac{x_3}{x_1 + x_3}$$

where $x_1/(x_1 + x_3)$ is the percentage of component A used in regular gasoline and $x_3/(x_1 + x_3)$ is the percentage of component B. The final octane rating of regular gasoline must be at least 95; thus,

$$\begin{aligned} 90\frac{x_1}{x_1 + x_3} + 110\frac{x_3}{x_1 + x_3} &\geq 95 && \text{Multiply by } x_1 + x_3. \\ 90x_1 + 110x_3 &\geq 95(x_1 + x_3) && \text{Collect like terms on the left side.} \\ -5x_1 + 15x_3 &\geq 0 && \text{Octane rating for regular} \end{aligned}$$

The corresponding inequality for premium gasoline is

$$\begin{aligned} 90\frac{x_2}{x_2 + x_4} + 110\frac{x_4}{x_2 + x_4} &\geq 105 \\ 90x_2 + 110x_4 &\geq 105(x_2 + x_4) \\ -15x_2 + 5x_4 &\geq 0 && \text{Octane rating for premium} \end{aligned}$$

The cost of the components used is

$$C = 28(x_1 + x_2) + 32(x_3 + x_4)$$

The revenue from selling all the gasoline is

$$R = 34(x_1 + x_3) + 40(x_2 + x_4)$$

and the profit is

$$\begin{aligned} P &= R - C \\ &= 34(x_1 + x_3) + 40(x_2 + x_4) - 28(x_1 + x_2) - 32(x_3 + x_4) \\ &= (34 - 28)x_1 + (40 - 28)x_2 + (34 - 32)x_3 + (40 - 32)x_4 \\ &= 6x_1 + 12x_2 + 2x_3 + 8x_4 \end{aligned}$$

To find the maximum profit, we must solve the following linear programming problem:

$$\begin{aligned} \text{Maximize } \quad & P = 6x_1 + 12x_2 + 2x_3 + 8x_4 && \text{Profit} \\ \text{Subject to } \quad & x_1 + x_2 \leq 30{,}000 && \text{Available A} \\ & x_3 + x_4 \leq 20{,}000 && \text{Available B} \\ & x_1 + x_3 \geq 20{,}000 && \text{Required regular} \\ & x_2 + x_4 \geq 10{,}000 && \text{Required premium} \\ & -5x_1 + 15x_3 \geq 0 && \text{Octane for regular} \\ & -15x_2 + 5x_4 \geq 0 && \text{Octane for premium} \\ & x_1, x_2, x_3, x_4 \geq 0 \end{aligned}$$

The tableau for this problem would have seven rows and sixteen columns. Solving this problem by hand is possible but would require considerable effort. In actual practice, computers are used to solve most linear programming problems. Table 3 displays typical input and output for a computer solution of this problem. [This problem could also be solved with a spreadsheet or with the software in *Explorations in Finite Mathematics* (see Preface).]

Table 3

INPUT TO PROGRAM

```
        * - LINEAR PROGRAMMING - *
            PROBLEM DISPLAY

MAXIMIZE
P = 6X1 + 12X2 + 2X3 + 8X4
SUBJECT TO:
   X1 +     X2                <= 30000
                  X3 +   X4 <= 20000
   X1           +    X3       >= 20000
            X2          +  X4 >= 10000
-5X1          + 15X3          >=     0
        -15X2          + 5X4 >=     0
XI >= 0, I = 1 TO 4
```

OUTPUT FROM PROGRAM

```
        * - LINEAR PROGRAMMING - *
             SOLUTION DISPLAY

DECISION VARIABLES     SLACK VARIABLES
X1 = 26250             S1 = 0
X2 = 3750              S2 = 0
X3 = 8750              SURPLUS VARIABLES
X4 = 11250             S3 = 15000
                       S4 = 5000
                       S5 = 0
                       S6 = 0
MAXIMUM VALUE OF OBJECTIVE FUNCTION
310000
```

According to the output in Table 3, the refinery should blend 26,250 barrels of component A and 8,750 barrels of component B to produce 35,000 barrels of regular. They should blend 3,750 barrels of component A and 11,250 barrels of component B to produce 15,000 barrels of premium. This will result in a maximum profit of \$310,000.

Explore–Discuss 3

Interpret the values of the slack and surplus variables in the computer solution in Table 3.

Matched Problem 5 ➡ Suppose the refinery in Example 5 has 35,000 barrels of component A, which costs \$25 a barrel, and 15,000 barrels of component B, which costs \$35 a barrel. If all the other information is unchanged, formulate a linear programming problem whose solution is the maximum profit. Do not attempt to solve the problem (unless you have access to software that solves linear programming problems).

Answers to Matched Problems

1. Maximize $P = 3x_1 - 2x_2 + x_3 - Ma_1 - Ma_2 - Ma_3$

Subject to

$$\begin{aligned} x_1 - 2x_2 + x_3 - s_1 + a_1 \qquad\qquad &= 5 \\ x_1 + 3x_2 - 4x_3 \qquad - s_2 + a_2 \qquad &= 10 \\ 2x_1 + 4x_2 + 5x_3 \qquad\qquad + s_3 \quad &= 20 \\ -3x_1 + x_2 + x_3 \qquad\qquad\qquad + a_3 &= 15 \\ x_1, x_2, x_3, s_1, a_1, s_2, a_2, s_3, a_3 &\geq 0 \end{aligned}$$

2. Max $P = 22$ at $x_1 = 6$, $x_2 = 4$, $x_3 = 0$ **3.** No optimal solution

4. A minimum cost of \$200 is realized when no type J, 6 type K, and 2 type L stones are processed each day.

5. Maximize $P = 9x_1 + 15x_2 - x_3 + 5x_4$

Subject to

$$\begin{aligned}
x_1 + x_2 &\le 35{,}000\\
x_3 + x_4 &\le 15{,}000\\
x_1 + x_3 &\ge 20{,}000\\
x_2 + x_4 &\ge 10{,}000\\
-5x_1 + 15x_3 &\ge 0\\
-15x_2 + 5x_4 &\ge 0\\
x_1, x_2, x_3, x_4 &\ge 0
\end{aligned}$$

EXERCISE 5-6

A *In Problems 1–8:*

(A) *Introduce slack, surplus, and artificial variables and form the modified problem.*

(B) *Write the preliminary simplex tableau for the modified problem and find the initial simplex tableau.*

(C) *Find the optimal solution of the modified problem by applying the simplex method to the initial simplex tableau.*

(D) *Find the optimal solution of the original problem, if it exists.*

1. Maximize $P = 5x_1 + 2x_2$
Subject to $x_1 + 2x_2 \le 12$
$x_1 + x_2 \ge 4$
$x_1, x_2 \ge 0$

2. Maximize $P = 3x_1 + 7x_2$
Subject to $2x_1 + x_2 \le 16$
$x_1 + x_2 \ge 6$
$x_1, x_2 \ge 0$

3. Maximize $P = 3x_1 + 5x_2$
Subject to $2x_1 + x_2 \le 8$
$x_1 + x_2 = 6$
$x_1, x_2 \ge 0$

4. Maximize $P = 4x_1 + 3x_2$
Subject to $x_1 + 3x_2 \le 24$
$x_1 + x_2 = 12$
$x_1, x_2 \ge 0$

5. Maximize $P = 4x_1 + 3x_2$
Subject to $-x_1 + 2x_2 \le 2$
$x_1 + x_2 \ge 4$
$x_1, x_2 \ge 0$

6. Maximize $P = 3x_1 + 4x_2$
Subject to $x_1 - 2x_2 \le 2$
$x_1 + x_2 \ge 5$
$x_1, x_2 \ge 0$

7. Maximize $P = 5x_1 + 10x_2$
Subject to $x_1 + x_2 \le 3$
$2x_1 + 3x_2 \ge 12$
$x_1, x_2 \ge 0$

8. Maximize $P = 4x_1 + 6x_2$
Subject to $x_1 + x_2 \le 2$
$3x_1 + 5x_2 \ge 15$
$x_1, x_2 \ge 0$

B *Use the big M method to solve Problems 9–22.*

9. Minimize and maximize $P = 2x_1 - x_2$
Subject to $x_1 + x_2 \le 8$
$5x_1 + 3x_2 \ge 30$
$x_1, x_2 \ge 0$

10. Minimize and maximize $P = -4x_1 + 16x_2$
Subject to $3x_1 + x_2 \le 28$
$x_1 + 2x_2 \ge 16$
$x_1, x_2 \ge 0$

11. Maximize $P = 2x_1 + 5x_2$
Subject to $x_1 + 2x_2 \le 18$
$2x_1 + x_2 \le 21$
$x_1 + x_2 \ge 10$
$x_1, x_2 \ge 0$

12. Maximize $P = 6x_1 + 2x_2$
Subject to $x_1 + 2x_2 \le 20$
$2x_1 + x_2 \le 16$
$x_1 + x_2 \ge 9$
$x_1, x_2 \ge 0$

13. Maximize $P = 10x_1 + 12x_2 + 20x_3$
Subject to $3x_1 + x_2 + 2x_3 \ge 12$
$x_1 - x_2 + 2x_3 = 6$
$x_1, x_2, x_3 \ge 0$

14. Maximize $P = 5x_1 + 7x_2 + 9x_3$
Subject to $x_1 - x_2 + x_3 \ge 20$
$2x_1 + x_2 + 5x_3 = 35$
$x_1, x_2, x_3 \ge 0$

15. Minimize $C = -5x_1 - 12x_2 + 16x_3$
Subject to $x_1 + 2x_2 + x_3 \le 10$
$2x_1 + 3x_2 + x_3 \ge 6$
$2x_1 + x_2 - x_3 = 1$
$x_1, x_2, x_3 \ge 0$

16. Minimize $C = -3x_1 + 15x_2 - 4x_3$
Subject to
$$\begin{aligned} 2x_1 + x_2 + 3x_3 &\le 24 \\ x_1 + 2x_2 + x_3 &\ge 6 \\ x_1 - 3x_2 + x_3 &= 2 \\ x_1, x_2, x_3 &\ge 0 \end{aligned}$$

17. Maximize $P = 3x_1 + 5x_2 + 6x_3$
Subject to
$$\begin{aligned} 2x_1 + x_2 + 2x_3 &\le 8 \\ 2x_1 + x_2 - 2x_3 &= 0 \\ x_1, x_2, x_3 &\ge 0 \end{aligned}$$

18. Maximize $P = 3x_1 + 6x_2 + 2x_3$
Subject to
$$\begin{aligned} 2x_1 + 2x_2 + 3x_3 &\le 12 \\ 2x_1 - 2x_2 + x_3 &= 0 \\ x_1, x_2, x_3 &\ge 0 \end{aligned}$$

19. Maximize $P = 2x_1 + 3x_2 + 4x_3$
Subject to
$$\begin{aligned} x_1 + 2x_2 + x_3 &\le 25 \\ 2x_1 + x_2 + 2x_3 &\le 60 \\ x_1 + 2x_2 - x_3 &\ge 10 \\ x_1, x_2, x_3 &\ge 0 \end{aligned}$$

20. Maximize $P = 5x_1 + 2x_2 + 9x_3$
Subject to
$$\begin{aligned} 2x_1 + 4x_2 + x_3 &\le 150 \\ 3x_1 + 3x_2 + x_3 &\le 90 \\ -x_1 + 5x_2 + x_3 &\ge 120 \\ x_1, x_2, x_3 &\ge 0 \end{aligned}$$

21. Maximize $P = x_1 + 2x_2 + 5x_3$
Subject to
$$\begin{aligned} x_1 + 3x_2 + 2x_3 &\le 60 \\ 2x_1 + 5x_2 + 2x_3 &\ge 50 \\ x_1 - 2x_2 + x_3 &\ge 40 \\ x_1, x_2, x_3 &\ge 0 \end{aligned}$$

22. Maximize $P = 2x_1 + 4x_2 + x_3$
Subject to
$$\begin{aligned} 2x_1 + 3x_2 + 5x_3 &\le 280 \\ 2x_1 + 2x_2 + x_3 &\ge 140 \\ 2x_1 + x_2 &\ge 150 \\ x_1, x_2, x_3 &\ge 0 \end{aligned}$$

23. Solve Problems 5 and 7 by the geometric method. Compare the conditions in the big M method that indicate no optimal solution exists with the conditions stated in Theorem 2 in Section 5-2.

24. Repeat Problem 23 with Problems 6 and 8.

C *Problems 25–32 are mixed. Some can be solved by the methods presented in Sections 5-4 and 5-5, while others must be solved by the big M method.*

25. Minimize $C = 10x_1 - 40x_2 - 5x_3$
Subject to
$$\begin{aligned} x_1 + 3x_2 &\le 6 \\ 4x_2 + x_3 &\le 3 \\ x_1, x_2, x_3 &\ge 0 \end{aligned}$$

26. Maximize $P = 7x_1 - 5x_2 + 2x_3$
Subject to
$$\begin{aligned} x_1 - 2x_2 + x_3 &\ge -8 \\ x_1 - x_2 + x_3 &\le 10 \\ x_1, x_2, x_3 &\ge 0 \end{aligned}$$

27. Maximize $P = -5x_1 + 10x_2 + 15x_3$
Subject to
$$\begin{aligned} 2x_1 + 3x_2 + x_3 &\le 24 \\ x_1 - 2x_2 - 2x_3 &\ge 1 \\ x_1, x_2, x_3 &\ge 0 \end{aligned}$$

28. Minimize $C = -5x_1 + 10x_2 + 15x_3$
Subject to
$$\begin{aligned} 2x_1 + 3x_2 + x_3 &\le 24 \\ x_1 - 2x_2 - 2x_3 &\ge 1 \\ x_1, x_2, x_3 &\ge 0 \end{aligned}$$

29. Minimize $C = 10x_1 + 40x_2 + 5x_3$
Subject to
$$\begin{aligned} x_1 + 3x_2 &\ge 6 \\ 4x_2 + x_3 &\ge 3 \\ x_1, x_2, x_3 &\ge 0 \end{aligned}$$

30. Maximize $P = 8x_1 + 2x_2 - 10x_3$
Subject to
$$\begin{aligned} x_1 + x_2 - 3x_3 &\le 6 \\ 4x_1 - x_2 + 2x_3 &\le -7 \\ x_1, x_2, x_3 &\ge 0 \end{aligned}$$

31. Maximize $P = 12x_1 + 9x_2 + 5x_3$
Subject to
$$\begin{aligned} x_1 + 3x_2 + x_3 &\le 40 \\ 2x_1 + x_2 + 3x_3 &\le 60 \\ x_1, x_2, x_3 &\ge 0 \end{aligned}$$

32. Minimize $C = 10x_1 + 12x_2 + 28x_3$
Subject to
$$\begin{aligned} 4x_1 + 2x_2 + 3x_3 &\ge 20 \\ 3x_1 - x_2 - 4x_3 &\le 10 \\ x_1, x_2, x_3 &\ge 0 \end{aligned}$$

APPLICATIONS

In Problems 33–40, construct a mathematical model in the form of a linear programming problem. (The answers in the back of the book for these application problems include the model.) Then solve the problem using the big M method.

Business & Economics

33. *Manufacturing—resource allocation.* An electronics company manufactures two types of add-on memory modules for microcomputers, a 16k module and a 64k module. Each 16k module requires 10 minutes for assembly and 2 minutes for testing. Each 64k module requires 15 minutes for assembly and 4 minutes for testing. The company makes a profit of \$18 on each 16k module and \$30 on each 64k module. The assembly department can work a maximum of 2,200 minutes per day, and the testing department can work a maximum of 500 minutes a day. In order to satisfy current orders, the company must produce at least 50 of the 16k modules per day. How many units of each module should the company manufacture each day in order to maximize the daily profit? What is the maximum profit?

34. *Manufacturing—resource allocation.* Discuss the effect on the solution to Problem 33 if the assembly department can work only 2,100 minutes daily.

35. *Advertising.* A company planning an advertising campaign to attract new customers wants to place a total of at most 10 ads in 3 newspapers. Each ad in the *Sentinel* costs $200 and will be read by 2,000 people. Each ad in the *Journal* costs $200 and will be read by 500 people. Each ad in the *Tribune* costs $100 and will be read by 1,500 people. The company wants at least 16,000 people to read its ads. How many ads should it place in each paper in order to minimize the advertising costs? What is the minimum cost?

36. *Advertising.* Discuss the effect on the solution to Problem 35 if the *Tribune* will not accept more than 4 ads from the company.

Life Sciences

37. *Nutrition—people.* An individual on a high-protein, low-carbohydrate diet requires at least 100 units of protein and at most 24 units of carbohydrates daily. The diet will consist entirely of three special liquid diet foods, *A*, *B*, and *C*. The contents and costs of the diet foods are given in the table. How many bottles of each brand of diet food should be consumed daily in order to meet the protein and carbohydrate requirements at minimal cost? What is the minimum cost?

	Units per Bottle		
	A	*B*	*C*
Protein	10	10	20
Carbohydrates	2	3	4
Cost per Bottle ($)	0.60	0.40	0.90

38. *Nutrition—people.* Discuss the effect on the solution to Problem 37 if the cost of brand *C* liquid diet food increases to $1.50 per bottle.

39. *Nutrition—plants.* A farmer can use three types of plant food, mix *A*, mix *B*, and mix *C*. The amounts (in pounds) of nitrogen, phosphoric acid, and potash in a cubic yard of each mix are given in the table. Tests performed on the soil in a large field indicate that the field needs at least 800 pounds of potash. The tests also indicate that no more than 700 pounds of phosphoric acid should be added to the field. The farmer plans to plant a crop that requires a great deal of nitrogen. How many cubic yards of each mix should be added to the field in order to satisfy the potash and phosphoric acid requirements and maximize the amount of nitrogen added? What is the maximum amount of nitrogen?

	Pounds per Cubic Yard		
	A	*B*	*C*
Nitrogen	12	16	8
Phosphoric Acid	12	8	16
Potash	16	8	16

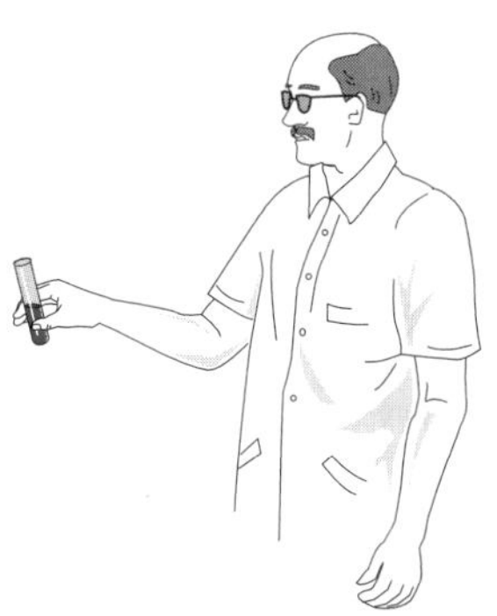

40. *Nutrition—plants.* Discuss the effect on the solution to Problem 39 if the limit on phosphoric acid is increased to 1,000 pounds.

In Problems 41–49, construct a mathematical model in the form of a linear programming problem. Do not solve.

Business & Economics

41. *Manufacturing—production scheduling.* A company manufactures car and truck frames at plants in Milwaukee and Racine. The Milwaukee plant has a daily operating budget of $50,000 and can produce at most 300 frames daily in any combination. It costs $150 to manufacture a car frame and $200 to manufacture a truck frame at the Milwaukee plant. The Racine plant has a daily operating budget of $35,000, can produce a maximum combined total of 200 frames daily, and produces a car frame at a cost of $135 and a truck frame at a cost of $180. Based on past demand, the company wants to limit production to a maximum of 250 car frames and 350 truck frames per day. If the company realizes a profit of $50 on each car frame and $70 on each truck frame, how many frames of each type should be produced at each plant to maximize the daily profit?

42. *Finances—loan distributions.* A savings and loan company has $3 million to lend. The types of loans and annual returns offered by the company are given in the table. State laws require that at least 50% of the money loaned for mortgages must be for first mortgages and

that at least 30% of the total amount loaned must be for either first or second mortgages. Company policy requires that the amount of signature and automobile loans cannot exceed 25% of the total amount loaned and that signature loans cannot exceed 15% of the total amount loaned. How much money should be allocated to each type of loan in order to maximize the company's return?

TYPE OF LOAN	ANNUAL RETURN (%)
Signature	18
First mortgage	12
Second mortgage	14
Automobile	16

43. *Blending—petroleum.* A refinery produces two grades of gasoline, regular and premium, by blending together three components, *A*, *B*, and *C*. Component *A* has an octane rating of 90 and costs $28 a barrel, component *B* has an octane rating of 100 and costs $30 a barrel, and component *C* has an octane rating of 110 and costs $34 a barrel. The octane rating for regular must be at least 95 and the octane rating for premium must be at least 105. Regular gasoline sells for $38 a barrel and premium sells for $46 a barrel. The company has 40,000 barrels of component *A*, 25,000 barrels of component *B*, and 15,000 barrels of component *C*, and must produce at least 30,000 barrels of regular and 25,000 barrels of premium. How should the components be blended in order to maximize profit?

44. *Blending—food processing.* A company produces two brands of trail mix, regular and deluxe, by mixing dried fruits, nuts, and cereal. The recipes for the mixes are given in the table. The company has 1,200 pounds of dried fruits, 750 pounds of nuts, and 1,500 pounds of cereal to be used in producing the mixes. The company makes a profit of $0.40 on each pound of regular mix and $0.60 on each pound of deluxe mix. How many pounds of each ingredient should be used in each mix in order to maximize the company's profit?

TYPE OF MIX	INGREDIENTS
Regular	At least 20% nuts At most 40% cereal
Deluxe	At least 30% nuts At most 25% cereal

45. *Investment strategy.* An investor is planning to divide her investments among high-tech mutual funds, global mutual funds, corporate bonds, municipal bonds, and CD's. Each of these investments has an estimated annual yield and a risk factor (see the table). The risk level for each choice is the product of its risk factor and the percentage of the total funds invested in that choice. The total risk level is the sum of the risk levels for all the investments. The investor wants at least 20% of her investments to be in CD's and does not want the risk level to exceed 1.8. What percentage of her total investments should be invested in each choice to maximize the return?

INVESTMENT	ANNUAL RETURN (%)	RISK FACTOR
High-tech funds	0.11	2.7
Global funds	0.1	1.8
Corporate bonds	0.09	1.2
Muncipal bonds	0.08	0.5
CD's	0.05	0

46. *Investment strategy.* Refer to Problem 45. Suppose the investor decides that she would like to minimize the total risk factor, as long as her return does not fall below 9%. What percentage of her total investments should be invested in each choice to minimize the total risk level?

Life Sciences

47. *Nutrition—people.* A dietitian in a hospital is to arrange a special diet using the foods *L*, *M*, and *N*. The table gives the nutritional contents and the cost of 1 ounce of each food. The daily requirements for the diet are at least 400 units of calcium, at least 200 units of iron, at least 300 units of vitamin A, at most 150 units of cholesterol, and at most 900 calories. How many ounces of each food should be used in order to meet the requirements of the diet at minimal cost?

	UNITS PER BOTTLE		
	L	*M*	*N*
CALCIUM	30	10	30
IRON	10	10	10
VITAMIN A	10	30	20
CHOLESTEROL	8	4	6
CALORIES	60	40	50
COST PER OUNCE ($)	0.40	0.60	0.80

48. *Nutrition—feed mixtures.* A farmer grows three crops, corn, oats, and soybeans, which he mixes together to feed his cows and pigs. At least 40% of the feed mix for the cows must be corn. The feed mix for the pigs must contain at least twice as much soybeans as corn. He has harvested 1,000 bushels of corn, 500 bushels of oats, and 1,000 bushels of soybeans. He needs 1,000 bushels of each feed mix for his livestock. The unused corn, oats, and soybeans can be sold for \$4, \$3.50, and \$3.25 a bushel, respectively (thus, these amounts also represent the cost of the crops used to feed the livestock). How many bushels of each crop should be used in each feed mix in order to produce sufficient food for the livestock at minimal cost?

Social Sciences

49. *Education—resource allocation.* Three towns are forming a consolidated school district with two high schools. Each high school has a maximum capacity of 2,000 students. Town A has 500 high school students, town B has 1,200, and town C has 1,800. The weekly costs of transporting a student from each town to each school are given in the table. In order to keep the enrollment balanced, the school board has decided that each high school must enroll at least 40% of the total student population. Furthermore, no more than 60% of the students in any town should be sent to the same high school. How many students from each town should be enrolled in each school in order to meet these requirements and minimize the cost of transporting the students?

	WEEKLY TRANSPORTATION COST PER STUDENT (\$)	
	School I	*School II*
TOWN A	4	8
TOWN B	6	4
TOWN C	3	9

IMPORTANT TERMS AND SYMBOLS

5-1 *Systems of Linear Inequalities in Two Variables.* Graph of a linear inequality in two variables; left half-plane; right half-plane; upper half-plane; lower half-plane; graphical solution of systems of linear inequalities; solution region; feasible region; corner point; bounded regions; unbounded regions

5-2 *Linear Programming in Two Dimensions—A Geometric Approach.* Linear programming problem; decision variables; objective function; problem constraints; nonnegative constraints; mathematical model; graphical solution; maximization problem; constant-profit line; optimal solution; feasible region; fundamental theorem of linear programming; multiple optimal solution; empty feasible region; unbounded objective function

5-3 *A Geometric Introduction to the Simplex Method.* Standard maximization problem in standard form; slack variables; basic variables; nonbasic variables; basic solution; basic feasible solution; fundamental theorem of linear programming

5-4 *The Simplex Method: Maximization with Problem Constraints of the Form $\leqslant$.* Initial system; basic solutions and basic feasible solutions for initial systems; fundamental theorem of linear programming; initial basic feasible solution; simplex tableau; initial simplex tableau; selecting basic and nonbasic variables for the simplex method; pivot operation; entering variable; exiting variable; pivot column; indicators; pivot row; pivot element; pivoting

5-5 *The Dual; Minimization with Problem Constraints of the Form $\geqslant$.* Dual problem; transpose; A^{T}; solution of a minimization problem; fundamental principal of duality; transportation problem

5-6 *Maximization and Minimization with Mixed Problem Constraints.* Surplus variable; artificial variable; modified problem; preliminary simplex tableau; initial simplex tableau; big M method

SUMMARY OF PROBLEM TYPES AND SIMPLEX SOLUTION METHODS

PROBLEM TYPE	PROBLEM CONSTRAINTS	RIGHT-SIDE CONSTANTS	COEFFICIENTS OF OBJECTIVE FUNCTION	METHOD OF SOLUTION
1. Maximization	$\leqslant$	Nonnegative	Any real numbers	Simplex method with slack variables
2. Minimization	$\geqslant$	Any real numbers	Nonnegative	Form dual and solve by the preceding method
3. Maximization	Mixed ($\leqslant, \geqslant, =$)	Nonnegative	Any real numbers	Form modified problem with slack, surplus, and artificial variables, and solve by the big M method
4. Minimization	Mixed ($\leqslant, \geqslant, =$)	Nonnegative	Any real numbers	Maximize negative of objective function by the preceding method

Review Exercise

Work through all the problems in this chapter review and check your answers in the back of the book. Answers to all review problems are there along with section numbers in italics to indicate where each type of problem is discussed. Where weaknesses show up, review appropriate sections in the text.

A *Solve the systems in Problems 1 and 2 graphically, and indicate whether each solution region is bounded or unbounded. Find the coordinates of each corner point.*

1. $$\begin{aligned} 2x_1 + x_2 &\le 8 \\ 3x_1 + 9x_2 &\le 27 \\ x_1, x_2 &\ge 0 \end{aligned}$$

2. $$\begin{aligned} 3x_1 + x_2 &\ge 9 \\ 2x_1 + 4x_2 &\ge 16 \\ x_1, x_2 &\ge 0 \end{aligned}$$

3. Solve the linear programming problem geometrically:

$$\begin{aligned} \text{Maximize} \quad & P = 6x_1 + 2x_2 \\ \text{Subject to} \quad & 2x_1 + x_2 \le 8 \\ & x_1 + 2x_2 \le 10 \\ & x_1, x_2 \ge 0 \end{aligned}$$

4. Convert the problem constraints in Problem 3 into a system of equations using slack variables.

5. How many basic variables and how many nonbasic variables are associated with the system in Problem 4?

6. Find all basic solutions for the system in Problem 4, and determine which basic solutions are feasible.

7. Write the simplex tableau for Problem 3, and circle the pivot element. Indicate the entering and exiting variables.

8. Solve Problem 3 using the simplex method.

9. For the simplex tableau below, identify the basic and nonbasic variables. Find the pivot element, the entering and exiting variables, and perform one pivot operation.

$$\begin{array}{c} \begin{matrix} x_1 & x_2 & x_3 & s_1 & s_2 & s_3 & P & \end{matrix} \\ \left[\begin{array}{rrrrrrr|r} 2 & 1 & 3 & -1 & 0 & 0 & 0 & 20 \\ 3 & 0 & 4 & 1 & 1 & 0 & 0 & 30 \\ 2 & 0 & 5 & 2 & 0 & 1 & 0 & 10 \\ \hline -8 & 0 & -5 & 3 & 0 & 0 & 1 & 50 \end{array}\right] \end{array}$$

10. Find the basic solution for each tableau. Determine whether the optimal solution has been reached, additional pivoting is required, or the problem has no optimal solution.

(A)
$$\begin{array}{c} \begin{matrix} x_1 & x_2 & s_1 & s_2 & P & \end{matrix} \\ \left[\begin{array}{rrrrr|r} 4 & 1 & 0 & 0 & 0 & 2 \\ 2 & 0 & 1 & 1 & 0 & 5 \\ \hline -2 & 0 & 3 & 0 & 1 & 12 \end{array}\right] \end{array}$$

(B)
$$\begin{array}{c} \begin{matrix} x_1 & x_2 & s_1 & s_2 & P & \end{matrix} \\ \left[\begin{array}{rrrrr|r} -1 & 3 & 0 & 1 & 0 & 7 \\ 0 & 2 & 1 & 0 & 0 & 0 \\ \hline -2 & 1 & 0 & 0 & 1 & 22 \end{array}\right] \end{array}$$

(C)
$$\left[\begin{array}{rrrrr|r} 1 & -2 & 0 & 4 & 0 & 6 \\ 0 & 2 & 1 & 6 & 0 & 15 \\ \hline 0 & 3 & 0 & 2 & 1 & 10 \end{array}\right]$$

11. Solve the linear programming problem geometrically:

$$\begin{aligned} \text{Minimize} \quad & C = 5x_1 + 2x_2 \\ \text{Subject to} \quad & x_1 + 3x_2 \ge 15 \\ & 2x_1 + x_2 \ge 20 \\ & x_1, x_2 \ge 0 \end{aligned}$$

12. Form the dual of Problem 11.

13. Write the initial system for the dual in Problem 12.

14. Write the first simplex tableau for the dual in Problem 12 and label the columns.

15. Use the simplex method to find the optimal solution of the dual in Problem 12.

16. Use the final simplex tableau from Problem 15 to find the optimal solution of Problem 11.

B

17. Solve the linear programming problem geometrically:

$$\begin{aligned} \text{Maximize} \quad & P = 3x_1 + 4x_2 \\ \text{Subject to} \quad & 2x_1 + 4x_2 \le 24 \\ & 3x_1 + 3x_2 \le 21 \\ & 4x_1 + 2x_2 \le 20 \\ & x_1, x_2 \ge 0 \end{aligned}$$

18. Solve Problem 17 using the simplex method.

19. Solve the linear programming problem geometrically:

$$\begin{aligned} \text{Minimize} \quad & C = 3x_1 + 8x_2 \\ \text{Subject to} \quad & x_1 + x_2 \ge 10 \\ & x_1 + 2x_2 \ge 15 \\ & x_2 \ge 3 \\ & x_1, x_2 \ge 0 \end{aligned}$$

20. Form the dual of Problem 19.

21. Solve Problem 19 by applying the simplex method to the dual in Problem 20.

Solve the linear programming Problems 22 and 23.

22. $$\begin{aligned} \text{Maximize} \quad & P = 5x_1 + 3x_2 - 3x_3 \\ \text{Subject to} \quad & x_1 - x_2 - 2x_3 \le 3 \\ & 2x_1 + 2x_2 - 5x_3 \le 10 \\ & x_1, x_2, x_3 \ge 0 \end{aligned}$$

23. $$\begin{aligned} \text{Maximize} \quad & P = 5x_1 + 3x_2 - 3x_3 \\ \text{Subject to} \quad & x_1 - x_2 - 2x_3 \le 3 \\ & x_1 + x_2 \le 5 \\ & x_1, x_2, x_3 \ge 0 \end{aligned}$$

In Problems 24 and 25:

(A) Introduce slack, surplus, and artificial variables and form the modified problem.

(B) Write the preliminary simplex tableau for the modified problem and find the initial simplex tableau.

(C) Find the optimal solution of the modified problem by applying the simplex method to the initial simplex tableau.

(D) Find the optimal solution of the original problem, if it exists.

24. Maximize $P = x_1 + 3x_2$
Subject to
$$x_1 + x_2 \geq 6$$
$$x_1 + 2x_2 \leq 8$$
$$x_1, x_2 \geq 0$$

25. Maximize $P = x_1 + x_2$
Subject to
$$x_1 + x_2 \geq 5$$
$$x_1 + 2x_2 \leq 4$$
$$x_1, x_2 \geq 0$$

26. Find the modified problem for the following linear programming problem. (Do not solve.)

Maximize $P = 2x_1 + 3x_2 + x_3$
Subject to
$$x_1 - 3x_2 + x_3 \leq 7$$
$$-x_1 - x_2 + 2x_3 \leq -2$$
$$3x_1 + 2x_2 - x_3 = 4$$
$$x_1, x_2, x_3 \geq 0$$

Write a brief verbal description of the type of linear programming problem that can be solved by the method indicated in Problems 27–30. Include the type of optimization, the number of variables, the type of constraints, and any restrictions on the coefficients and constants.

27. Geometric method

28. Basic simplex method with slack variables

29. Dual method

30. Big M method

C

31. Solve the following linear programming problem by the simplex method, keeping track of the obvious basic solution at each step. Then graph the feasible region and illustrate the path to the optimal solution determined by the simplex method.

Maximize $P = 2x_1 + 3x_2$
Subject to
$$x_1 + 2x_2 \leq 22$$
$$3x_1 + x_2 \leq 26$$
$$x_1 \leq 8$$
$$x_2 \leq 10$$
$$x_1, x_2 \geq 0$$

32. Solve by the dual method:

Minimize $C = 3x_1 + 2x_2$
Subject to
$$2x_1 + x_2 \leq 20$$
$$2x_1 + x_2 \geq 9$$
$$x_1 + x_2 \geq 6$$
$$x_1, x_2 \geq 0$$

33. Solve Problem 32 by the big M method.

34. Solve by the dual method:

Minimize $C = 15x_1 + 12x_2 + 15x_3 + 18x_4$
Subject to
$$x_1 + x_2 \leq 240$$
$$x_3 + x_4 \leq 500$$
$$x_1 + x_3 \geq 400$$
$$x_2 + x_4 \geq 300$$
$$x_1, x_2, x_3, x_4 \geq 0$$

APPLICATIONS

In Problems 35–39, construct a mathematical model in the form of a linear programming problem. (The answers in the back of the book for these application problems include the model.) Then solve the problem by the indicated method.

Business & Economics

35. *Manufacturing—resource allocation.* South Shore Sail Loft manufactures regular and competition sails. Each regular sail takes 2 hours to cut and 4 hours to sew. Each competition sail takes 3 hours to cut and 10 hours to sew. There are 150 hours available in the cutting department and 380 hours available in the sewing department. Use the geometric method.

(A) If the Loft makes a profit of \$100 on each regular sail and \$200 on each competition sail, how many sails of each type should the company manufacture to maximize their profit? What is the maximum profit?

(B) An increase in the demand for competition sails causes the profit on a competition sail to rise to \$260. Discuss the effect of this change on the number of sails manufactured and on the maximum profit.

(C) A decrease in the demand for competition sails causes the profit on a competition sail to drop to \$140. Discuss the effect of this change on the number of sails manufactured and on the maximum profit.

36. *Investment.* An investor has \$150,000 to invest in oil stock, steel stock, and government bonds. The bonds are guaranteed to yield 5%, but the yield for each stock can vary. To protect against major losses, the investor decides that the amount invested in oil stock should not exceed \$50,000 and that the total amount invested in stock cannot exceed the amount invested in bonds by more than \$25,000. Use the simplex method to answer the following questions.
    (A) If the oil stock yields 12% and the steel stock yields 9%, how much money should be invested in each alternative in order to maximize the return? What is the maximum return?
    (B) Repeat part (A) if the oil stock yields 9% and the steel stock yields 12%.

37. *Transportation—shipping schedule.* A company produces motors for washing machines at factory A and factory B. The motors are then shipped to either plant X or plant Y, where the washing machines are assembled. The maximum number of motors that can be produced at each factory monthly, the minimum number required monthly for each plant to meet anticipated demand, and the shipping charges for one motor are given in the table. Determine a shipping schedule that will minimize the cost of transporting the motors from the factories to the assembly plants. Use the dual method.

	PLANT X	PLANT Y	MAXIMUM PRODUCTION
FACTORY A	\$5	\$8	1,500
FACTORY B	\$9	\$7	1,000
MINIMUM REQUIREMENT	900	1,200	

38. *Blending—food processing.* A company blends long-grain rice and wild rice to produce two brands of rice mixes, brand A, which is marketed under the company's name, and brand B, which is marketed as a generic brand. Brand A must contain at least 10% wild rice, and brand B must contain at least 5% wild rice. Long-grain rice costs the company \$0.70 per pound, and wild rice costs \$3.40 per pound. The company sells brand A for \$1.50 a pound and brand B for \$1.20 a pound. The company has 8,000 pounds of long-grain rice and 500 pounds of wild rice on hand. How should the company use the available rice to maximize their profit on these two brands of rice? What is the maximum profit? Use the simplex method.

Life Sciences

39. *Nutrition—animals.* A special diet for laboratory animals is to contain at least 850 units of vitamins, 800 units of minerals, and 1,150 calories. There are two feed mixes available, mix A and mix B. A gram of mix A contains 2 units of vitamins, 2 units of minerals, and 4 calories. A gram of mix B contains 5 units of vitamins, 4 units of minerals, and 5 calories. Use the geometric method.
    (A) If mix A costs \$0.04 per gram and mix B costs \$0.09 per gram, how many grams of each mix should be used to satisfy the requirements of the diet at minimal cost? What is the minimum cost?
    (B) If the price of mix B decreases to \$0.06 per gram, discuss the effect of this change on the solution in part (A).
    (C) If the price of mix B increases to \$0.12 per gram, discuss the effect of this change on the solution in part (A).

Group Activity 1 *The Two-Phase Method (An Alternative to the Big M Method)*

In the big M method, we added artificial variables to $\geq$ and $=$ constraints in order to obtain a basic feasible solution. Then we added terms of the form $-Ma_i$ to the objective function and assumed that M was an arbitrarily large number, which forced the values of the artificial variables to be zero in any optimal solution. There is another commonly used method, called the *two-phase method,* that treats the artificial variables in a different manner. To illustrate this method, consider the following example:

$$\begin{aligned} \text{Maximize} \quad & P = 2x_1 + x_2 \\ \text{Subject to} \quad & x_1 + x_2 \leq 10 \\ & x_1 \geq 3 \\ & x_2 \geq 4 \\ & x_1, x_2 \geq 0 \end{aligned}$$

(A) Graph the feasible region and solve this problem by the geometric method for comparison with the new method that follows.

If we add slack and surplus variables, we obtain the following system:

$$\begin{aligned} x_1 + x_2 + s_1 \qquad\qquad &= 10 \\ x_1 \qquad - s_2 \qquad &= 3 \\ x_2 \qquad - s_3 &= 4 \\ -2x_1 - x_2 \qquad\qquad + P &= 0 \end{aligned} \tag{1}$$

The basic solution for this system is

$$x_1 = 0, \quad x_2 = 0, \quad s_1 = 10, \quad s_2 = -3, \quad s_3 = -4, \quad P = 0$$

This solution is not feasible, and we cannot start the simplex process. As before, we add artificial variables to obtain a basic feasible solution, but this time we do not modify the objective function:

$$\begin{aligned} x_1 + x_2 + s_1 \qquad\qquad\qquad &= 10 \\ x_1 \qquad - s_2 + a_1 \qquad\qquad &= 3 \\ x_2 \qquad\qquad - s_3 + a_2 &= 4 \\ -2x_1 - x_2 \qquad\qquad\qquad + P &= 0 \end{aligned} \tag{2}$$

The basic solution for this system is

$$x_1 = 0, \quad x_2 = 0, \quad s_1 = 10, \quad s_2 = 0, \quad a_1 = 3, \quad s_3 = 0, \quad a_2 = 4, \quad P = 0$$

This solution is feasible, and we can start the simplex process.

(B) Write the simplex tableau for system (2) and apply the simplex procedure. Compare your result with the geometric solution from part (A).

In general, applying the simplex method to system (2) will produce an optimal solution of (2) with one or more of the artificial variables having nonzero values. In most cases, simply deleting the artificial variables will not produce a basic feasible solution to (1). However, if we can find a basic feasible solution of (2) with the artificial variables all equal to zero, then we can delete the artificial variables and find a basic feasible solution to (1). In the big M method, we accomplished this by adding terms to the objective function. In this new method, we consider another objective function:

$$C = a_1 + a_2$$

If we minimize C and the minimum value turns out to be zero, then, since a_1 and a_2 are nonnegative, they will both be zero. To minimize C, we maximize

$$T = -C = -a_1 - a_2$$

We cannot add T to system (2) in this form. To preserve the necessary form for a simplex tableau, we must express T in terms of the nonbasic variables. Solving the second and third equations in (2) for a_1 and a_2, respectively, and adding, we obtain

$$T = x_1 + x_2 - s_2 - s_3 - 7$$

Transposing all the variables to the left side, we add this equation to (2) to form system (3):

$$\begin{aligned}
x_1 + x_2 + s_1 \qquad\qquad\qquad\qquad\qquad\qquad &= 10 \\
x_1 \qquad - s_2 + a_1 \qquad\qquad\qquad\qquad &= 3 \\
x_2 \qquad\qquad - s_3 + a_2 \qquad\qquad &= 4 \qquad (3) \\
-2x_1 - x_2 \qquad\qquad\qquad\qquad + P \qquad &= 0 \\
-x_1 - x_2 + \qquad s_2 \qquad + s_3 \qquad\qquad + T &= -7
\end{aligned}$$

(C) Write the simplex tableau for system (3) and apply the simplex method, with the following restriction: *Do not choose any elements in the objective function row as pivot elements.* When you find the optimal solution, write the corresponding system, which is shown below.

$$\begin{aligned}
s_1 + s_2 - a_1 + s_3 - a_2 \qquad\qquad &= 3 \\
x_1 \qquad - s_2 + a_1 \qquad\qquad\qquad &= 3 \\
x_2 \qquad\qquad - s_3 + a_2 \qquad &= 4 \qquad (4) \\
- 2s_2 + 2a_1 - s_3 + a_2 + P \qquad &= 10 \\
a_1 \qquad + a_2 \qquad + T &= 0
\end{aligned}$$

The basic feasible solution for (4) is

$$x_1 = 3, \quad x_2 = 4, \quad s_1 = 3, \quad s_2 = 0, \quad a_1 = 0,$$
$$s_3 = 0, \quad a_2 = 0, \quad P = 10, \quad T = 0$$

Note that the maximum value of T is 0 and that it occurs for $a_1 = a_2 = 0$. Thus, deleting the equation for T and the columns of the artificial variables gives us system (5), which is equivalent to system (1) but has a basic solution that is feasible:

$$\begin{aligned}
s_1 + s_2 + s_3 \qquad &= 3 \\
x_1 \qquad - s_2 \qquad\qquad &= 3 \\
x_2 \qquad\qquad - s_3 \qquad &= 4 \qquad (5) \\
- 2s_2 - s_3 + P &= 10
\end{aligned}$$

(D) Write the simplex tableau for (5) and apply the simplex procedure to find the optimal solution. Compare the results with the solution in part (A).

This method is called the *two-phase method* because it involves solving two problems:

Phase 1. Minimize the sum of the artificial variables to obtain a basic feasible solution with all artificial variables equal to zero.

Phase 2. Delete the artificial variables and complete the solution of the original problem.

(E) Write a step-by-step procedure for using the two-phase method to solve linear programming problems. Discuss situations where this method will not produce an optimal solution. (Considering some simple examples might be helpful.) Use your procedure to solve Problems 13, 15, 17, and 19 in Exercise 5-6.

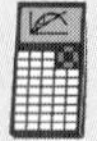

Group Activity 2 *Production Scheduling*

A company manufactures a line of outdoor furniture consisting of small tables, regular chairs, rocking chairs, chaise longues, and large tables. Each piece of furniture passes through three different production departments: fabrication, assembly, and finishing. Table 1 gives the time each piece must spend in each department, the labor-hours available in each department, the profit from each piece, and the existing orders for each product. The company must make a sufficient number of pieces of furniture to fill all the existing orders.

Table 1

PRODUCT	PRODUCTION DEPARTMENT (LABOR-HOURS PER UNIT)			PROFIT PER UNIT ($)	ORDERS
	Fabrication	*Assembly*	*Finishing*		
Small table	1	1	1	8	200
Regular chair	1	2	3	17	160
Rocking chair	2	2	3	24	180
Chaise longue	3	4	2	31	240
Large table	5	4	5	52	100
Available labor-hours	2,500	3,000	3,500		

(A) Construct a mathematical model in the form of a linear programming problem and use a graphing utility to show that the maximum profit of $27,830 is realized when the company produces 200 small tables, 270 regular chairs, 180 rocking chairs, 240 chaise longues, and 190 large tables.

(B) Discuss the effect on the optimal solution in part (A) if the profit on a small table increases to $12.

(C) Discuss the effect on the optimal solution in part (A) if the profit on a large table decreases to $30.

(D) Discuss the effect on the optimal solution in part (A) if the orders for rocking chairs are canceled.

(E) Discuss the effect on the optimal solution in part (A) if the orders for chaise longues are increased to 300.

(F) Discuss the effect on the optimal solution in part (A) if the available hours in the assembly department are increased to 3,300.

(G) Discuss the effect on the optimal solution in part (A) if the available hours in the fabrication department are increased to 2,800.

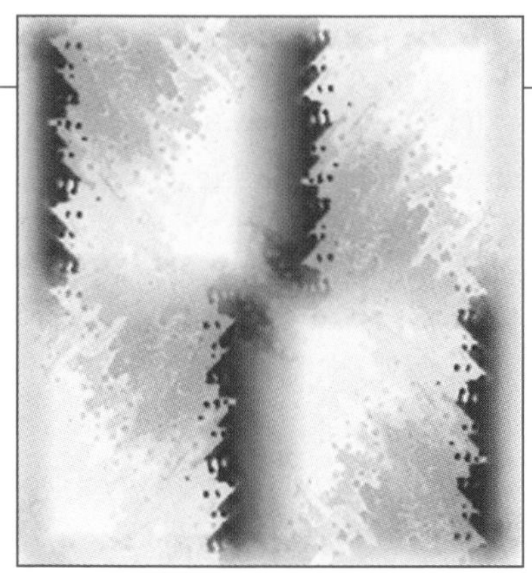

Chapter 6

Probability

INTRODUCTION

Probability, like many branches of mathematics, evolved out of practical considerations. Girolamo Cardano (1501–1576), a gambler and physician, produced some of the best mathematics of his time, including a systematic analysis of gambling problems. In 1654, another gambler, the Chevalier de Méré, plagued with bad luck, approached the well-known French philosopher and mathematician Blaise Pascal (1623–1662) regarding certain dice problems. Pascal became interested in these problems, studied them, and discussed them with Pierre de Fermat (1601–1665), another French mathematician. Thus, out of the gaming rooms of western Europe the study of probability was born.

In spite of this lowly birth, probability has matured into a highly respected and immensely useful branch of mathematics. It is used in practically every field. Probability theory can be thought of as the science of uncertainty. If, for example, a card is drawn from a deck of 52 playing cards, it is uncertain which card will be drawn. But suppose a card is drawn and replaced in the deck and a card is again drawn and replaced, and this action is repeated a large number of times. A particular card, say the ace of spades, will be drawn over the long run with a relative frequency that is approximately predictable. Probability theory is concerned with determining the long-run frequency of the occurrence of a given event.

How do we assign probabilities to events? There are two basic approaches to this problem, one theoretical and the other empirical. An example will illustrate the difference between the two approaches.

Suppose you were asked, "What is the probability of obtaining a 2 on a single throw of a die?" Using a *theoretical approach*, we would reason as follows: Since there are 6 *equally likely* ways the die can turn up (assuming the die is fair) and there is only 1 way a 2 can turn up, then the probability of obtaining a 2 is $\frac{1}{6}$. Here, we have arrived at a probability assignment without even rolling a die once; we have used certain assumptions and a reasoning process.

What does the result have to do with reality? We would expect that in the long run (after rolling a die many times), the 2 would appear approximately $\frac{1}{6}$ of the time.

With the *empirical approach*, we make no assumption about the equally likely ways in which the die can turn up. We simply set up an experiment and roll the die a large number of times. Then we compute the percentage of times the 2 appears and use this number as an estimate of the probability of obtaining a 2 on a single roll of the die. Each approach has advantages and drawbacks; these will be discussed in the following sections.

We will first consider the theoretical approach and develop procedures that will lead to the solution of a large variety of interesting problems. These procedures require counting the number of ways certain events can happen, and this is not always easy. However, powerful mathematical tools can assist us in this counting task. The development of these tools is the subject matter of the next two sections.

SECTION 6-1 Basic Counting Principles

- ADDITION PRINCIPLE
- VENN DIAGRAMS
- MULTIPLICATION PRINCIPLE

ADDITION PRINCIPLE

If the enrollment in a college chemistry class consists of 13 males and 15 females, then it is easy to see that there are a total of 28 students enrolled in the class. This is a simple example of a **counting technique**—a method for determining the number of elements in a set without actually enumerating the elements one by one. Set operations play a fundamental role in many counting techniques (see Appendix A-1 for a review of set operations). For example, if M is the set of male students in the chemistry class and F is the set of female students, then the *union* of sets M and F, denoted $M \cup F$, is the set of all students in the class. Since these sets have no elements in common, the *intersection* of sets M and F, denoted $M \cap F$, is the *empty set* $\varnothing$; we then say that M and F are *disjoint sets.* The total number of students enrolled in the class is the **number of elements** in $M \cup F$, denoted by $n(M \cup F)$ and given by

$$\begin{aligned} n(M \cup F) &= n(M) + n(F) \\ &= 13 + 15 = 28 \end{aligned} \tag{1}$$

Thus, we see that in this example the number of elements in the union of sets M and F is the sum of the number of elements in M and in F. However, this does not work for all pairs of sets. To see why, consider another example. Suppose that the enrollment in a mathematics class consists of 22 math majors and 16 physics majors, and that 7 of these students have majors in both subjects. If M represents the set of math majors and P represents the set of physics majors, then $M \cap P$ represents the set of double majors. It is tempting to proceed as before and conclude that there are $22 + 16 = 38$ students in the class, but this is incorrect. We have counted the double majors twice, once as math majors and again as physics majors. To correct for this double counting, we subtract the number of double majors from this sum. Thus, the total number of students enrolled in this class is given by

$$\begin{aligned} n(M \cup P) &= n(M) + n(P) - n(M \cap P) \\ &= 22 + 16 - 7 = 31 \end{aligned} \tag{2}$$

Equation (1) for the chemistry class and equation (2) for the math class are illustrations of the *addition principle* for counting the elements in the union of two sets.

Addition Principle (for Counting)

For any two sets A and B,

$$n(A \cup B) = n(A) + n(B) - n(A \cap B) \tag{3}$$

If A and B are disjoint, then

$$n(A \cup B) = n(A) + n(B) \tag{4}$$

Example 1 **Employee Benefits** According to a survey of business firms in a certain city, 750 firms offer their employees health insurance, 640 offer dental insurance, and 280 offer health insurance and dental insurance. How many firms offer their employees health insurance or dental insurance?

SOLUTION If H is the set of firms that offer their employees health insurance and D is the set that offer dental insurance, then

$H \cap D$ = Set of firms that offer health insurance **and** dental insurance
$H \cup D$ = Set of firms that offer health insurance **or** dental insurance

Thus,

$$n(H) = 750 \qquad n(D) = 640 \qquad n(H \cap D) = 280$$

and

$$\begin{aligned} n(H \cup D) &= n(H) + n(D) - n(H \cap D) \\ &= 750 + 640 - 280 = 1{,}110 \end{aligned}$$

Thus, 1,110 firms offer their employees health insurance or dental insurance.

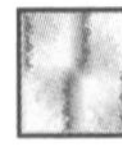

Matched Problem 1 The survey in Example 1 also indicated that 345 firms offer their employees group life insurance, 285 offer long-term disability insurance, and 115 offer group life insurance and long-term disability insurance. How many firms offer their employees group life insurance or long-term disability insurance?

VENN DIAGRAMS

The next example introduces some additional set concepts that provide useful counting techniques.

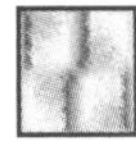

Example 2 **Market Research** A city has two daily newspapers, the *Sentinel* and the *Journal.* The following information was obtained from a survey of 100 residents of the city: 35 people subscribe to the *Sentinel,* 60 subscribe to the *Journal,* and 20 subscribe to both newspapers.

(A) How many people in the survey subscribe to the *Sentinel* but not to the *Journal?*
(B) How many subscribe to the *Journal* but not to the *Sentinel?*
(C) How many do not subscribe to either paper?
(D) Organize this information in a table.

SOLUTION Let U be the group of people surveyed, let S be the set of people who subscribe to the *Sentinel,* and let J be the set of people who subscribe to the *Journal.* Since U contains all the elements under consideration, it is the *universal set* for this problem. The *complement* of S, denoted S', is the set of people in the survey group U who do not subscribe to the *Sentinel.* Similarly, J' is the set of people in the group who do not subscribe to the *Journal.* Using the sets S and J, their complements, and set intersection, we can divide the set U into the four disjoint subsets defined below and illustrated in the *Venn diagram* in Figure 1.

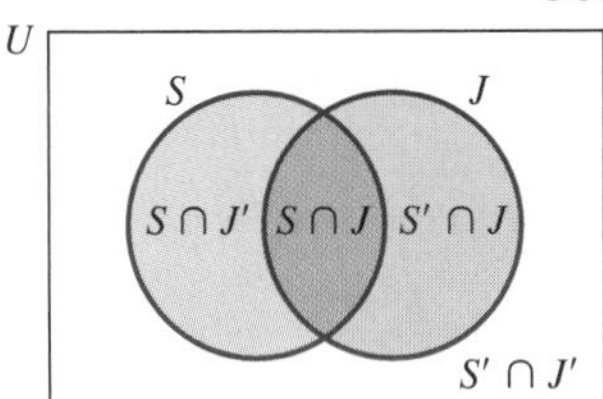

FIGURE 1
Venn diagram for the newspaper survey

$S \cap J$ = Set of people who subscribe to both papers
$S \cap J'$ = Set of people who subscribe to the *Sentinel* but not the *Journal*
$S' \cap J$ = Set of people who subscribe to the *Journal* but not the *Sentinel*
$S' \cap J'$ = Set of people who do not subscribe to either paper

The given survey information can be expressed in terms of set notation as

$$n(U) = 100 \qquad n(S) = 35 \qquad n(J) = 60 \qquad n(S \cap J) = 20$$

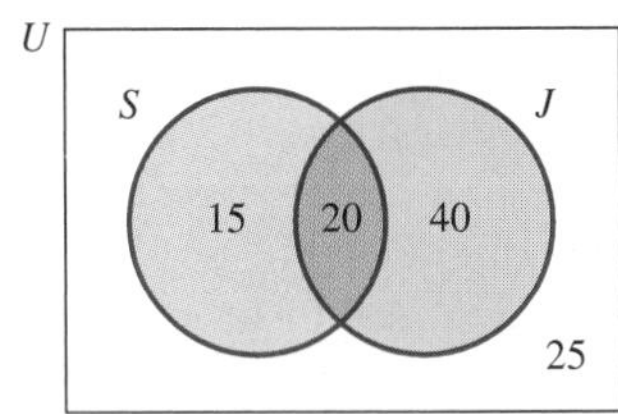

FIGURE 2
Newspaper survey results

We can use this information and a Venn diagram to answer parts (A)–(C). To begin, we place 20 in $S \cap J$ in the diagram (see Fig. 2). As we proceed through parts (A)–(C), we add each answer to the diagram.

(A) Since 35 people subscribe to the *Sentinel* and 20 subscribe to both papers, the number of people who subscribe to the *Sentinel* but not to the *Journal* is

$$n(S \cap J') = 35 - 20 = 15$$

(B) In a similar manner, the number of people who subscribe to the *Journal* but not to the *Sentinel* is

$$n(S' \cap J) = 60 - 20 = 40$$

(C) The total number of newspaper subscribers is $20 + 15 + 40 = 75$. Thus, the number of people who do not subscribe to either paper is

$$n(S' \cap J') = 100 - 75 = 25$$

(D) Venn diagrams are useful tools for determining the number of elements in the various sets in a survey, but often the results must be presented in the form of a table, rather than a diagram. The following table contains the same information as Figure 2, but organized in a different format:

		JOURNAL		
		Subscriber, J	*Nonsubscriber, J'*	*Totals*
SENTINEL	*Subscriber, S*	20	15	35
	Nonsubscriber, S'	40	25	65
	Totals	60	40	100

Matched Problem 2 A small town has two radio stations, an AM station and an FM station. A survey of 100 residents of the town produced the following results: In the last 30 days, 65 people have listened to the AM station, 45 have listened to the FM station, and 30 have listened to both stations.

(A) How many people in the survey have listened to the AM station, but not to the FM station, in this 30 day period?
(B) How many have listened to the FM station, but not to the AM station?
(C) How many have not listened to either station?
(D) Organize this information in a table.

Explore–Discuss 1

Let *A*, *B*, and *C* be three sets. Use a Venn diagram to explain the following equation:

$$n(A \cup B \cup C) = n(A) + n(B) + n(C) - n(A \cap B) - n(A \cap C) - n(B \cap C) + n(A \cap B \cap C)$$

MULTIPLICATION PRINCIPLE

As we have just seen, if the elements of a set are determined by the union operation, then addition and subtraction are used to count the number of elements in the set. Now we want to consider sets whose elements are determined by a sequence of operations. We will see that multiplication is used to count the number of elements in sets formed in this manner. The best way to see how this works is to start with an example.

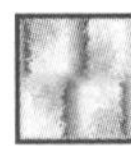

Example 3 **Product Mix** A retail store stocks windbreaker jackets in small, medium, large, and extra large, and all are available in blue or red. What are the combined choices and how many combined choices are there?

Solution To solve the problem we use a tree diagram:

Size Choices (Outcomes)	*Color Choices (Outcomes)*	*Combined Choices (Outcomes)*
Start → S	B	(S, B)
	R	(S, R)
Start → M	B	(M, B)
	R	(M, R)
Start → L	B	(L, B)
	R	(L, R)
Start → XL	B	(XL, B)
	R	(XL, R)

Thus, there are 8 possible combined choices (outcomes). There are 4 ways a size can be chosen and 2 ways a color can be chosen. The first element in the ordered pair represents a size choice, and the second element represents a color choice.

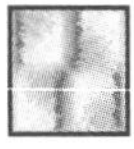

Matched Problem 3 A company offers its employees health plans from three different companies, *R*, *S*, and *T*. Each company offers two levels of coverage, *A* and *B*, with one level requiring additional employee contributions. What are the combined choices, and how many choices are there? Solve using a tree diagram.

Now suppose you asked, "From the 26 letters in the alphabet, how many ways can 3 letters appear in a row on a license plate if no letter is repeated?" To try to count the possibilities using a tree diagram would be extremely tedious, to say the least. The following **multiplication principle** will enable us to solve this problem easily; in addition, it forms the basis for several other counting devices that are developed in the next section.

Multiplication Principle (for Counting)

1. If two operations O_1 and O_2 are performed in order, with N_1 possible outcomes for the first operation and N_2 possible outcomes for the second operation, then there are

   $$N_1 \cdot N_2$$

   possible combined outcomes of the first operation followed by the second.
2. In general, if n operations $O_1, O_2, \ldots, O_n$ are performed in order, with possible number of outcomes $N_1, N_2, \ldots, N_n$, respectively, then there are

   $$N_1 \cdot N_2 \cdot \cdots \cdot N_n$$

   possible combined outcomes of the operations performed in the given order.

In Example 3, we see that there are 4 possible outcomes in choosing a size (the first operation) and 2 possible outcomes in choosing a color (the second operation); hence, by the multiplication principle, there are $4 \cdot 2 = 8$ possible combined outcomes. Use the multiplication principle to solve Matched Problem 3. [*Answer:* $3 \cdot 2 = 6$]

To answer the license plate question: There are 26 ways the first letter can be chosen; after a first letter is chosen, there are 25 ways a second letter can be chosen; and after 2 letters are chosen, there are 24 ways a third letter can be chosen. Hence, using the multiplication principle, there are $26 \cdot 25 \cdot 24 = 15{,}600$ possible ways 3 letters can be chosen from the alphabet without repeats.

Explore–Discuss 2

A state is about to adopt a 6-character format for its license plates, consisting of a block of letters followed by a block of numbers. The format must accommodate up to 20 million vehicles. What size would you recommend for each block?

Example 4 **Computer-Assisted Testing** Many colleges and universities are now using computer-assisted testing procedures. Suppose a screening test is to consist of 5 questions, and a computer stores 5 comparable questions for the first test question, 8 for the second, 6 for the third, 5 for the fourth, and 10 for the fifth. How many different 5-question tests can the computer select? (Two tests are considered different if they differ in one or more questions.)

SOLUTION

O_1: Selecting the first question — N_1: 5 ways
O_2: Selecting the second question — N_2: 8 ways
O_3: Selecting the third question — N_3: 6 ways
O_4: Selecting the fourth question — N_4: 5 ways
O_5: Selecting the fifth question — N_5: 10 ways

Thus, the computer can generate

$$5 \cdot 8 \cdot 6 \cdot 5 \cdot 10 = 12{,}000 \text{ different tests}$$

Matched Problem 4 Each question on a multiple-choice test has 5 choices. If there are 5 such questions on a test, how many different response sheets are possible if only 1 choice is marked for each question?

Example 5 **Code Words** How many 3-letter code words are possible using the first 8 letters of the alphabet if:

(A) No letter can be repeated? (B) Letters can be repeated?
(C) Adjacent letters cannot be alike?

SOLUTION To form 3-letter code words from the 8 letters available, we select a letter for the first position, one for the second position, and one for the third position. Altogether, there are three operations.

(A) No letter can be repeated:

O_1: Selecting the first letter — N_1: 8 ways
O_2: Selecting the second letter — N_2: 7 ways — Since 1 letter has been used
O_3: Selecting the third letter — N_3: 6 ways — Since 2 letters have been used

Thus, there are

$$8 \cdot 7 \cdot 6 = 336 \text{ possible code words}$$ *Possible combined operations*

(B) Letters can be repeated:

O_1: Selecting the first letter	N_1: 8 ways		
O_2: Selecting the second letter	N_2: 8 ways	Repeats allowed	
O_3: Selecting the third letter	N_3: 8 ways	Repeats allowed	

Thus, there are

$$8 \cdot 8 \cdot 8 = 8^3 = 512 \text{ possible code words}$$

(C) Adjacent letters cannot be alike:

O_1: Selecting the first letter	N_1: 8 ways	
O_2: Selecting the second letter	N_2: 7 ways	Cannot be the same as the first
O_3: Selecting the third letter	N_3: 7 ways	Cannot be the same as the second, but can be the same as the first

Thus, there are

$$8 \cdot 7 \cdot 7 = 392 \text{ possible code words}$$

Matched Problem 5 How many 4-letter code words are possible using the first 10 letters of the alphabet under the three different conditions stated in Example 5?

Answers to Matched Problems

1. 515

2. (A) 35 (B) 15 (C) 20

(D)

		FM		
		Listener	*Nonlistener*	*Totals*
AM	*Listener*	30	35	65
	Nonlistener	15	20	35
	Totals	45	55	100

3. There are 6 combined choices:

	Company Choices (Outcomes)	Coverage Choices (Outcomes)	Combined Choices (Outcomes)
Start	R	A	(R, A)
		B	(R, B)
	S	A	(S, A)
		B	(S, B)
	T	A	(T, A)
		B	(T, B)

4. 5^5, or 3,125

5. (A) $10 \cdot 9 \cdot 8 \cdot 7 = 5{,}040$ (B) $10 \cdot 10 \cdot 10 \cdot 10 = 10{,}000$
(C) $10 \cdot 9 \cdot 9 \cdot 9 = 7{,}290$

EXERCISE 6-1

A *Problems 1–12 refer to the following Venn diagram. Find the number of elements in each of the indicated sets.*

U; A; B; 75; 40; 95; 90

1. A 2. B 3. U
4. A' 5. B' 6. $A \cap B$
7. $A \cup B$ 8. $A \cap B'$ 9. $A' \cap B$
10. $A' \cap B'$ 11. $(A \cap B)'$ 12. $(A \cup B)'$

Solve Problems 13–16 two ways: (A) using a tree diagram and (B) using the multiplication principle.

13. How many ways can 2 coins turn up—heads, H, or tails, T—if the combined outcome (H, T) is to be distinguished from the outcome (T, H)?
14. How many 2-letter code words can be formed from the first 3 letters of the alphabet if no letter can be used more than once?
15. A coin is tossed with possible outcomes heads, H, or tails, T. Then a single die is tossed with possible outcomes 1, 2, 3, 4, 5, or 6. How many combined outcomes are there?
16. In how many ways can 3 coins turn up—heads, H, or tails, T—if combined outcomes such as (H, T, H), (H, H, T), and (T, H, H) are to be considered different?

B *In Problems 17–20, use the given information to determine the number of elements in each of the four disjoint subsets in the following Venn diagram.*

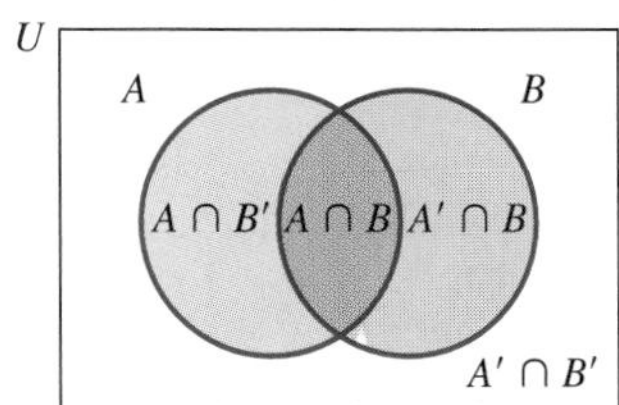

17. $n(A) = 80,\quad n(B) = 50,\quad n(A \cap B) = 20,\quad n(U) = 200$
18. $n(A) = 45,\quad n(B) = 35,\quad n(A \cap B) = 15,\quad n(U) = 100$
19. $n(A) = 25,\quad n(B) = 55,\quad n(A \cup B) = 60,\quad n(U) = 100$
20. $n(A) = 70,\quad n(B) = 90,\quad n(A \cup B) = 120,\quad n(U) = 200$

In Problems 21–24, use the given information to complete the following table.

	A	A'	*Totals*
B	?	?	?
B'	?	?	?
Totals	?	?	?

21. $n(A) = 70,\quad n(B) = 90,\quad n(A \cap B) = 30,\quad n(U) = 200$
22. $n(A) = 55,\quad n(B) = 65,\quad n(A \cap B) = 35,\quad n(U) = 100$
23. $n(A) = 45,\quad n(B) = 55,\quad n(A \cup B) = 80,\quad n(U) = 100$
24. $n(A) = 80,\quad n(B) = 70,\quad n(A \cup B) = 110,\quad n(U) = 200$

In Problems 25 and 26, discuss the validity of each statement. If the statement is always true, explain why. If not, give a counterexample.

25. (A) If A or B is the empty set, then A and B are disjoint.
    (B) If A and B are disjoint, then A or B is the empty set.
26. (A) If A and B are disjoint, then $n(A \cap B) = n(A) + n(B)$.
    (B) If $n(A \cup B) = n(A) + n(B)$, then A and B are disjoint.

27. A particular new car model is available with 5 choices of color, 3 choices of transmission, 4 types of interior, and 2 types of engine. How many different variations of this model car are possible?
28. A delicatessen serves meat sandwiches with the following options: 3 kinds of bread, 5 kinds of meat, and lettuce or sprouts. How many different sandwiches are possible, assuming one item is used out of each category?
29. How many 4-letter code words are possible from the first 6 letters of the alphabet if no letter is repeated? If letters are repeated? If adjacent letters must be different?
30. How many 5-letter code words are possible from the first 7 letters of the alphabet if no letter is repeated? If letters are repeated? If adjacent letters must be different?
31. A combination lock has 5 wheels, each labeled with the 10 digits from 0 to 9. How many 5-digit opening combinations are possible if no digit is repeated? If digits are repeated? If successive digits must be different?

32. A small combination lock on a suitcase has 3 wheels, each labeled with the 10 digits from 0 to 9. How many 3-digit combinations are possible if no digit is repeated? If digits are repeated? If successive digits must be different?

33. How many different license plates are possible if each contains 3 letters (out of the 26 letters of the alphabet) followed by 3 digits (from 0 to 9)? How many of these license plates contain no repeated letters and no repeated digits?

34. How many 5-digit ZIP code numbers are possible? How many of these numbers contain no repeated digits?

35. In Example 3 does it make any difference in which order the selection operations are performed? That is, if we select a jacket color first and then select a size, are there as many combined choices available as selecting a size first and then a color? Justify your answer using tree diagrams and the multiplication principle.

36. Explain how three sets, A, B, and C, can be related to each other in order for the following equation to hold (Venn diagrams may be helpful):

$$n(A \cup B \cup C) = n(A) + n(B) + n(C) - n(A \cap C) - n(B \cap C)$$

C

37. A group of 75 people includes 32 who play tennis, 37 who play golf, and 8 who play both tennis and golf. How many people in the group play neither sport?

38. A class of 30 music students includes 13 who play the piano, 16 who play the guitar, and 5 who play both the piano and the guitar. How many students in the class play neither instrument?

39. A group of 100 people touring Europe includes 42 people who speak French, 55 who speak German, and 17 who speak neither language. How many people in the group speak both French and German?

40. A high school football team with 40 players includes 16 players who played offense last year, 17 who played defense, and 12 who were not last year's team. How many players from last year played both offense and defense?

Applications

Business & Economics

41. *Management selection.* A management selection service classifies its applicants (using tests and interviews) as high-IQ, middle-IQ, or low-IQ and as aggressive or passive. How many combined classifications are possible?
(A) Solve by using a tree diagram.
(B) Solve by using the multiplication principle.

42. *Management selection.* A corporation plans to fill 2 different positions for vice-president, V_1 and V_2, from administrative officers in 2 of its manufacturing plants. Plant A has 6 officers and plant B has 8. How many ways can these 2 positions be filled if the V_1 position is to be filled from plant A and the V_2 position from plant B? How many ways can the 2 positions be filled if the selection is made without regard to plant?

43. *Transportation.* A sales representative who lives in city A wishes to start from home and fly to 3 different cities, B, C, and D. If there are 2 choices of local transportation (drive her own car to and from the airport or use a taxi for both trips), and if all cities are interconnected by airlines, how many travel plans can be constructed to visit each city exactly once and return home?

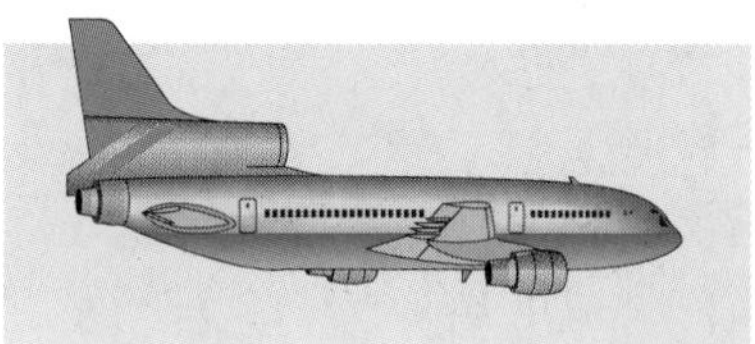

44. *Transportation.* A manufacturing company in city A wishes to truck its product to 4 different cities, B, C, D, and E. If the cities are all interconnected by roads, how many different route plans can be constructed so that a single truck, starting from A, will visit each city exactly once, then return home?

45. *Market research.* A survey of 1,200 people in a certain city indicates that 850 own microwave ovens, 740 own VCR's, and 580 own microwave ovens and VCR's.

(A) How many people in the survey own either a microwave oven or a VCR?

(B) How many own neither a microwave oven nor a VCR?

(C) How many own a microwave oven and do not own a VCR?

46. *Market research.* A survey of 800 small businesses indicates that 250 own photocopiers, 420 own fax machines, and 180 own photocopiers and fax machines.

(A) How many businesses in the survey own either a photocopier or a fax machine?

(B) How many own neither a photocopier nor a fax machine?

(C) How many own a fax machine and do not own a photocopier?

47. *Communications.* A cable television company has 8,000 subscribers in a suburban community. The company offers two premium channels, HBO and Showtime. If 2,450 subscribers receive HBO, 1,940 receive Showtime, and 5,180 do not receive any premium channel, how many subscribers receive both HBO and Showtime?

48. *Communications.* A local telephone company offers its 10,000 customers two special services: call forwarding and call waiting. If 3,770 customers use call forwarding, 3,250 use call waiting, and 4,530 do not use either of these services, how many customers use both call forwarding and call waiting?

49. *Minimum wage.* The table below gives the number of male and female workers (in thousands) earning at or below the minimum wage for several age categories.

	Workers per Age Group (in thousands)			
	16–19	*20–24*	*25+*	*Totals*
Males at minimum wage	343	154	237	734
Males below minimum wage	118	102	159	379
Females at minimum wage	367	186	503	1,056
Females below minimum wage	251	202	540	993
Totals	1,079	644	1,439	3,162

(A) How many males are age 20–24 and below minimum wage?

(B) How many females are age 20 or older and at minimum wage?

(C) How many workers are age 16–19 or males at minimum wage?

(D) How many workers are below minimum wage?

50. *Minimum wage.* Refer to the table in Problem 49.

(A) How many females are age 16–19 and at minimum wage?

(B) How many males are age 16–24 and below minimum wage?

(C) How many workers are age 20–24 or females below minimum wage?

(D) How many workers are at minimum wage?

Life Sciences

51. *Medicine.* A medical researcher classifies subjects according to male or female; smoker or nonsmoker; and underweight, average weight, or overweight. How many combined classifications are possible?

(A) Solve using a tree diagram.

(B) Solve using the multiplication principle.

52. *Family planning.* A couple is planning to have 3 children. How many boy–girl combinations are possible? Distinguish between combined outcomes such as (B, B, G), (B, G, B), and (G, B, B).

(A) Solve by using a tree diagram.

(B) Solve by using the multiplication principle.

Social Sciences

53. *Politics.* A politician running for a third term in office is planning to contact all contributors to her first two campaigns. If 1,475 individuals contributed to the first campaign, 2,350 individuals contributed to the second campaign, and 920 individuals contributed to the first and second campaigns, how many individuals have contributed to the first or second campaign?

54. *Politics.* If 12,457 people voted for a politician in his first election, 15,322 voted for him in his second election, and 9,345 voted for him in the first and second elections, how many people voted for this politician in the first or second election?

SECTION 6-2

Permutations and Combinations

- FACTORIAL
- PERMUTATIONS
- COMBINATIONS
- APPLICATIONS

The multiplication principle discussed in the preceding section can be used to develop two additional devices for counting that are extremely useful in more complicated counting problems. Both of these devices use a function called a *factorial function,* which we introduce first.

FACTORIAL

When using the multiplication principle, we encountered expressions of the form

$$26 \cdot 25 \cdot 24 \qquad 8 \cdot 7 \cdot 6$$

where each natural number factor is decreased by 1 as we move from left to right. The factors in the following product continue to decrease by 1 until a factor of 1 is reached:

$$5 \cdot 4 \cdot 3 \cdot 2 \cdot 1$$

Products of this type are encountered so frequently in counting problems that it is useful to be able to express them in a concise notation. The product of the first n natural numbers is called ***n* factorial** and is denoted by ***n*!**. Also, we define **zero factorial, 0!,** to be 1. Symbolically:

Factorial

For n a natural number,

$$n! = n(n-1)(n-2) \cdot \cdots \cdot 2 \cdot 1 \qquad 4! = 4 \cdot 3 \cdot 2 \cdot 1$$
$$0! = 1$$
$$n! = n \cdot (n-1)!$$

[*Note:* Many calculators have an [n!] key or its equivalent.]

Example 1 **Computing Factorials**

(A) $5! = 5 \cdot 4 \cdot 3 \cdot 2 \cdot 1 = 120$

(B) $\dfrac{7!}{6!} = \dfrac{7 \cdot \cancel{6!}}{\cancel{6!}} = 7$

(C) $\dfrac{8!}{5!} = \dfrac{8 \cdot 7 \cdot 6 \cdot \cancel{5!}}{\cancel{5!}} = 8 \cdot 7 \cdot 6 = 336$

(D) $\dfrac{52!}{5!47!} = \dfrac{52 \cdot 51 \cdot 50 \cdot 49 \cdot 48 \cdot \cancel{47!}}{5 \cdot 4 \cdot 3 \cdot 2 \cdot 1 \cdot \cancel{47!}} = 2{,}598{,}960$

Matched Problem 1 Find:

(A) $6!$ (B) $\frac{10!}{9!}$ (C) $\frac{10!}{7!}$ (D) $\frac{5!}{0!3!}$ (E) $\frac{20!}{3!17!}$

It is interesting and useful to note that $n!$ grows very rapidly. Compare the following:

$$5! = 120 \qquad 10! = 3{,}628{,}800 \qquad 15! = 1{,}307{,}674{,}368{,}000$$

Try 69!, 70!, and 71! on your calculator.

PERMUTATIONS

A particular (horizontal) arrangement of a set of paintings on a wall is called a *permutation* of the set of paintings. In general:

A Permutation of a Set of Objects

A **permutation** of a set of distinct objects is an arrangement of the objects in a specific order without repetition.

Suppose 4 pictures are to be arranged from left to right on one wall of an art gallery. How many permutations (ordered arrangements) are possible? Using the multiplication principle, there are 4 ways of selecting the first picture; after the first picture is selected, there are 3 ways of selecting the second picture; after the first 2 pictures are selected, there are 2 ways of selecting the third picture; and after the first 3 pictures are selected, there is only 1 way to select the fourth. Thus, the number of permutations (ordered arrangements) of the set of 4 pictures is

$$4 \cdot 3 \cdot 2 \cdot 1 = 4! = 24$$

In general, how many permutations of a set of n distinct objects are possible? Reasoning as above, there are n ways in which the first object can be chosen, there are $n - 1$ ways in which the second object can be chosen, and so on. Using the multiplication principle, we have the following:

The Number of Permutations of n Objects

The number of permutations of n distinct objects without repetition, denoted by $P_{n,n}$, is

$$P_{n,n} = n(n-1) \cdot \cdots \cdot 2 \cdot 1 = n! \quad n \text{ factors}$$

Example: The number of permutations of 7 objects is

$$P_{7,7} = 7 \cdot 6 \cdot 5 \cdot 4 \cdot 3 \cdot 2 \cdot 1 = 7! \quad 7 \text{ factors}$$

Now suppose the director of the art gallery decides to use only 2 of the 4 available paintings, and they will be arranged on the wall from left to right. We are now talking about a particular arrangement of 2 paintings out of the 4, which is called a *permutation of 4 objects taken 2 at a time.* In general:

A Permutation of n Objects Taken r at a Time

A permutation of a set of n distinct objects taken r at a time without repetition is an arrangement of r of the n objects in a specific order.

How many ordered arrangements of 2 pictures can be formed from the 4? That is, how many permutations of 4 objects taken 2 at a time are there? There are 4 ways the first picture can be selected; after selecting the first picture, there are 3 ways the second picture can be selected. Thus, the number of permutations of a set of 4 objects taken 2 at a time, which is denoted by $P_{4,2}$, is given by

$$P_{4,2} = 4 \cdot 3$$

In terms of factorials, we have

$$P_{4,2} = 4 \cdot 3 = \frac{4 \cdot 3 \cdot 2!}{2!} = \frac{4!}{2!} \quad \text{Multiplying } 4 \cdot 3 \text{ by 1 in the form } 2!/2!$$

Reasoning in the same way as in the example, we find that the number of permutations of n distinct objects taken r at a time without repetition $(0 \leq r \leq n)$ is given by

$$\begin{aligned} P_{n,r} &= n(n-1)(n-2) \cdot \cdots \cdot (n-r+1) && r \text{ factors} \\ P_{9,6} &= 9(9-1)(9-2) \cdot \cdots \cdot (9-6+1) && 6 \text{ factors} \\ &= 9 \cdot 8 \cdot 7 \cdot 6 \cdot 5 \cdot 4 \end{aligned}$$

Multiplying the right side of the equation for $P_{n,r}$ by 1 in the form $(n-r)!/(n-r)!$, we obtain a factorial form for $P_{n,r}$:

$$P_{n,r} = n(n-1)(n-2) \cdot \cdots \cdot (n-r+1)\frac{(n-r)!}{(n-r)!}$$

But, since

$$n(n-1)(n-2) \cdot \cdots \cdot (n-r+1)(n-r)! = n!$$

The above expression simplifies to

$$P_{n,r} = \frac{n!}{(n-r)!}$$

We summarize these results in the following box:

Number of Permutations of n Objects Taken r at a Time

The number of permutations of n distinct objects taken r at a time without repetition is given by*

$$P_{n,r} = n(n-1)(n-2) \cdot \cdots \cdot (n-r+1) \qquad r \text{ factors}$$

$$P_{5,2} = 5 \cdot 4 \qquad 2 \text{ factors}$$

or

$$P_{n,r} = \frac{n!}{(n-r)!} \qquad 0 \le r \le n \qquad P_{5,2} = \frac{5!}{(5-2)!} = \frac{5!}{3!}$$

[*Note*: $P_{n,n} = \dfrac{n!}{(n-n)!} = \dfrac{n!}{0!} = n!$ permutations of n objects taken n at a time. Remember, by definition, $0! = 1$.]

*In place of the symbol $P_{n,r}$, the symbols P_r^n, ${}_nP_r$, and $P(n, r)$ are often used.

Example 2 **Permutations** Given the set $\{A, B, C\}$, how many permutations are there of this set of 3 objects taken 2 at a time? Answer the question:

(A) Using a tree diagram (B) Using the multiplication principle
(C) Using the two formulas for $P_{n,r}$

SOLUTION (A) Using a tree diagram:

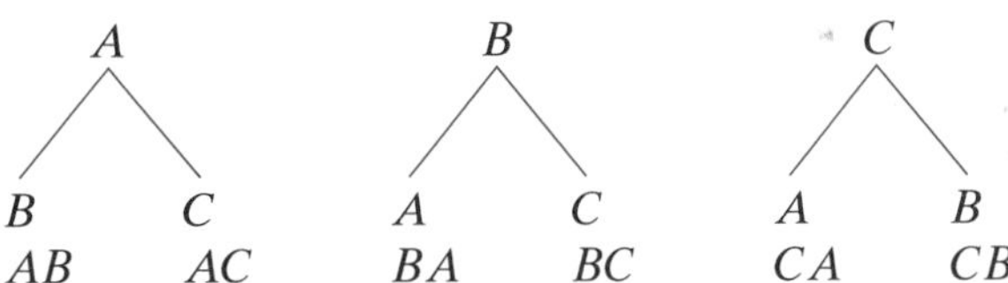

There are 6 permutations of 3 objects taken 2 at a time.

(B) Using the multiplication principle:

O_1: Fill the first position $\qquad N_1$: 3 ways
O_2: Fill the second position $\qquad N_2$: 2 ways

Thus, there are

$$3 \cdot 2 = 6 \text{ permutations of 3 objects taken 2 at a time}$$

(C) Using the two formulas for $P_{n,r}$:

$$\overset{\text{2 factors}}{P_{3,2} = 3 \cdot 2 = 6} \qquad \text{or} \qquad P_{3,2} = \frac{3!}{(3-2)!} = \frac{3 \cdot 2 \cdot 1}{1} = 6$$

Thus, there are 6 permutations of 3 objects taken 2 at a time.

Of course, all three methods produce the same answer.

Matched Problem 2 ➡ Given the set $\{A, B, C, D\}$, how many permutations are there of this set of 4 objects taken 2 at a time? Answer the question:

(A) Using a tree diagram (B) Using the multiplication principle
(C) Using the two formulas for $P_{n,r}$

In Example 2 you probably found the multiplication principle the easiest method to use. But for large values of n and r you will find that the factorial formula is more convenient. In fact, many calculators have functions that compute $n!$ and $P_{n,r}$ directly. See Figure 1 in Example 3 and the user's manual for your calculator.

Example 3 ➡ **Permutations** Find the number of permutations of 13 objects taken 8 at a time. Compute the answer using a calculator.

Solution We use the factorial formula for $P_{n,r}$:

$$P_{13,8} = \frac{13!}{(13-8)!} = \frac{13!}{5!} = 51{,}891{,}840$$

```
13!
      6227020800
13!/5!
        51891840
13 nPr 8
        51891840
```

FIGURE 1

Using a tree diagram to solve this problem would involve a monumental effort. Using the multiplication principle would involve multiplying $13 \cdot 12 \cdot 11 \cdot 10 \cdot 9 \cdot 8 \cdot 7 \cdot 6$ (8 factors), which is not too bad. A calculator can provide instant results. See Figure 1.

Matched Problem 3 ➡ Find the number of permutations of 30 objects taken 4 at a time. Compute the answer exactly using a calculator.

Combinations

Now suppose that an art museum owns 8 paintings by a given artist and another art museum wishes to borrow 3 of these paintings for a special show. In selecting 3 of the 8 paintings for shipment, the order would not matter, and we would simply be selecting a 3 element subset from the set of 8 paintings. That is, we would be selecting what is called *a combination of 8 objects taken 3 at a time.* In general:

A Combination of n Objects Taken r at a Time

A **combination** of a set of n distinct objects taken r at a time without repetition is an r element subset of the set of n objects. The arrangement of the elements in the subset does not matter.

How many ways can the 3 paintings be selected for shipment out of the 8 available? That is, what is the number of combinations of 8 objects taken 3 at a time? To answer this question, and to get a better insight into the general problem, we return to Example 2.

In Example 2 we were given the set $\{A, B, C\}$ and found the number of permutations of 3 objects taken 2 at a time using a tree diagram. From this tree diagram we also can determine the number of combinations of 3 objects taken 2 at a time (the number of 2 element subsets from a 3 element set), and compare it with the number of permutations (see Fig. 2)

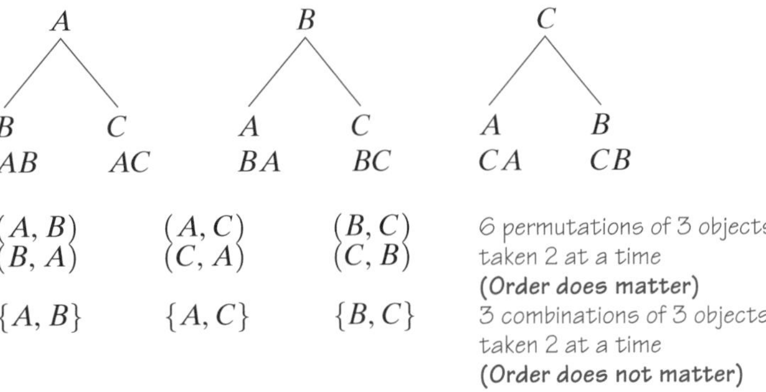

FIGURE 2

There are fewer combinations than permutations, as we would expect. To each subset (combination) there corresponds two ordered pairs (permutations). We denote the number of combinations in Figure 2 by

$$C_{3,2} \qquad \text{or} \qquad \binom{3}{2}$$

Our final goal is to find a factorial formula for $C_{n,r}$, the number of combinations of n objects taken r at a time. But first, we will develop a formula for $C_{3,2}$, and then we will generalize from this experience.

We know the number of permutations of 3 objects taken 2 at a time is given by $P_{3,2}$, and we have a formula for computing this number. Now, suppose we think of $P_{3,2}$ in terms of two operations:

O_1: Selecting a subset of 2 elements $\qquad N_1$: $C_{3,2}$ ways
O_2: Arranging the subset in a given order $\qquad N_2$: 2! ways

The combined operation, O_1 followed by O_2, produces a permutation of 3 objects taken 2 at a time. Thus,

$$P_{3,2} = C_{3,2} \cdot 2! \qquad \text{or} \qquad C_{3,2} = \frac{P_{3,2}}{2!}$$

To find $C_{3,2}$, the number of combinations of 3 objects taken 2 at a time, we substitute

$$P_{3,2} = \frac{3!}{(3-2)!}$$

and solve for $C_{3,2}$:

$$C_{3,2} = \frac{\mathbf{3!}}{\mathbf{2!(3-2)!}} = \frac{3 \cdot 2 \cdot 1}{(2 \cdot 1)(1)} = 3$$

This result agrees with the result we got using a tree diagram. Note that the number of combinations of 3 objects taken 2 at a time is the same as the number of permutations of 3 objects taken 2 at a time divided by the number of permutations of the elements in a 2 element subset. This observation also can be made in Figure 2.

Reasoning the same way as in the example, the number of combinations of n objects taken r at a time $(0 \leq r \leq n)$ is given by

$$C_{n,r} = \frac{P_{n,r}}{r!}$$

$$= \frac{n!}{r!(n-r)!} \qquad \text{Since } P_{n,r} = \frac{n!}{(n-r)!}$$

In summary:

Number of Combinations of *n* Objects Taken *r* at a Time

The number of combinations of n distinct objects taken r at a time without repetition is given by*

$$C_{n,r} = \binom{n}{r} \qquad\qquad C_{52,5} = \binom{52}{5}$$

$$= \frac{P_{n,r}}{r!} \qquad\qquad = \frac{P_{52,5}}{5!}$$

$$= \frac{n!}{r!(n-r)!} \qquad 0 \leq r \leq n \qquad = \frac{52!}{5!(52-5)!}$$

*In place of the symbols $C_{n,r}$ and $\binom{n}{r}$, the symbols C_r^n, ${}_nC_r$, and $C(n, r)$ are often used.

Now we can answer the question posed earlier in the museum example. There are

$$C_{8,3} = \frac{8!}{3!(8-3)!} = \frac{8!}{3!5!} = \frac{8 \cdot 7 \cdot 6 \cdot \cancel{5!}}{3 \cdot 2 \cdot 1 \cdot \cancel{5!}} = 56$$

ways the 3 paintings can be selected for shipment. That is, there are 56 combinations of 8 objects taken 3 at a time.

Example 4 **Permutations and Combinations** From a committee of 10 people:

(A) In how many ways can we choose a chairperson, a vice-chairperson, and a secretary, assuming that one person cannot hold more than one position?

(B) In how many ways can we choose a subcommittee of 3 people?

SOLUTION Note how parts (A) and (B) differ. In part (A), order of choice makes a difference in the selection of the officers. In part (B), the ordering does not matter in choosing a 3 person subcommittee. Thus, in part (A), we are interested in the number of *permutations* of 10 objects taken 3 at a time; and in part (B), we are interested in the number of *combinations* of 10 objects taken 3 at a time. These quantities are computed as follows (and since the numbers are not large, we do not need to use a calculator):

$$\text{(A)}\quad P_{10,3} = \frac{10!}{(10-3)!} = \frac{10!}{7!} = \boxed{\frac{10 \cdot 9 \cdot 8 \cdot \cancel{7!}}{\cancel{7!}}} = 720 \text{ ways}$$

$$\text{(B)}\quad C_{10,3} = \frac{10!}{3!(10-3)!} = \frac{10!}{3!7!} = \boxed{\frac{10 \cdot 9 \cdot 8 \cdot \cancel{7!}}{3 \cdot 2 \cdot 1 \cdot \cancel{7!}}} = 120 \text{ ways}$$

Matched Problem 4 From a committee of 12 people:

(A) In how many ways can we choose a chairperson, a vice-chairperson, a secretary, and a treasurer, assuming that one person cannot hold more than one position?

(B) In how many ways can we choose a subcommittee of 4 people?

If n and r are other than small numbers (as in Example 4), a calculator is a useful aid in evaluating expressions involving factorials. Many calculators have a function that computes $C_{n,r}$ directly. (See Fig. 3 below.)

Example 5 **Combinations** Find the number of combinations of 13 objects taken 8 at a time. Compute the answer exactly, using a calculator.

SOLUTION

$$C_{13,8} = \binom{13}{8} = \frac{13!}{8!(13-8)!} = \frac{13!}{8!5!} = 1{,}287$$

Compare the result in Example 5 with that obtained in Example 3, and note that $C_{13,8}$ is substantially smaller than $P_{13,8}$. See Figure 3.

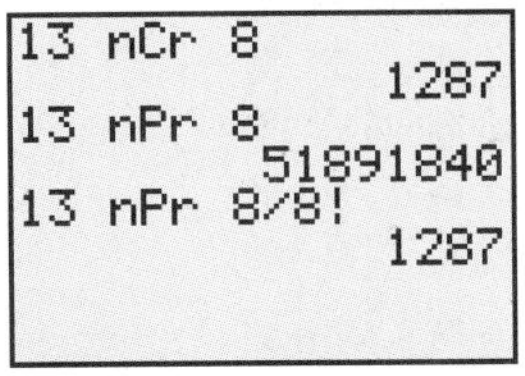

```
13 nCr 8
              1287
13 nPr 8
          51891840
13 nPr 8/8!
              1287
```

FIGURE 3

Matched Problem 5 Find the number of combinations of 30 objects taken 4 at a time. Compute the answer exactly using a calculator.

REMEMBER

In a permutation, order matters.
In a combination, order does not matter.

To determine whether permutations or combinations are involved in a problem, see if rearranging the collection or listing produces a different object. If so, use permutations; if not, use combinations.

Explore–Discuss 1

(A) List alphabetically by the first letter, all 3-letter license plate codes consisting of 3 different letters chosen from M, A, T, H. Discuss how this list relates to $P_{n,r}$.

(B) Reorganize the list from part (A) so that now all codes without M come first, then all codes without A, then all codes without T, and finally all codes without H. Discuss how this list illustrates the formula $P_{n,r} = r!C_{n,r}$.

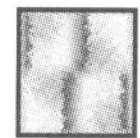

Applications

We now consider some applications of the concepts discussed above. Several applications in this and the following sections involve a standard 52-card deck of playing cards, which is described below:

Standard 52-Card Deck of Playing Cards

A standard deck of 52 cards (see Fig. 4) has four 13-card suits: diamonds, hearts, clubs, and spades. The diamonds and hearts are red, and the clubs and spades are black. Each 13-card suit contains cards numbered from 2 to 10, a jack, a queen, a king, and an ace. The jack, queen, and king are called *face cards.* Depending on the game, the ace may be counted as the lowest and/or the highest card in the suit.

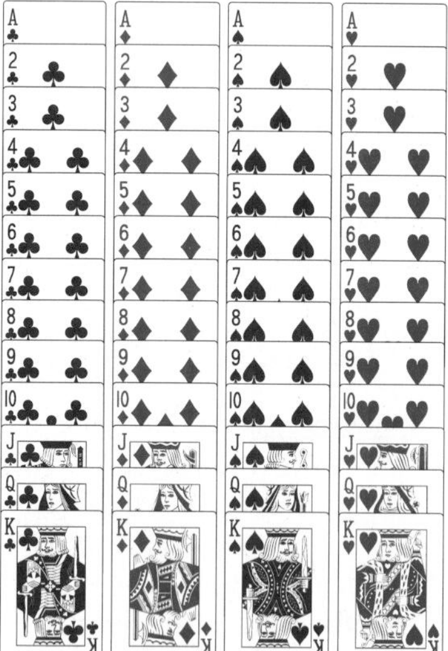

FIGURE 4

Example 6 **Counting Techniques** How many 5-card hands will have 3 aces and 2 kings?

Solution The solution involves both the multiplication principle and combinations. Think of selecting the 5-card hand in terms of the following two operations:

O_1: Choosing 3 aces out of 4 possible (order is not important) N_1: $C_{4,3}$
O_2: Choosing 2 kings out of 4 possible (order is not important) N_2: $C_{4,2}$

Using the multiplication principle, we have

$$\begin{aligned}\text{Number of hands} &= C_{4,3} \cdot C_{4,2} \\ &= \frac{4!}{3!(4-3)!} \cdot \frac{4!}{2!(4-2)!} \\ &= 4 \cdot 6 = 24\end{aligned}$$

Matched Problem 6 How many 5-card hands will have 3 hearts and 2 spades?

Example 7 **Counting Techniques** Serial numbers for a product are to be made using 2 letters followed by 3 numbers. If the letters are to be taken from the first 8 letters of the alphabet with no repeats and the numbers are to be taken from the 10 digits (0–9) with no repeats, how many serial numbers are possible?

Solution The solution involves both the multiplication principle and permutations. Think of selecting a serial number in terms of the following two operations:

O_1: Choosing 2 letters out of 8 available (order is important) N_1: $P_{8,2}$
O_2: Choosing 3 numbers out of 10 available (order is important) N_2: $P_{10,3}$

Using the multiplication principle, we have

$$\begin{aligned}\text{Number of serial numbers} &= P_{8,2} \cdot P_{10,3} \\ &= \frac{8!}{(8-2)!} \cdot \frac{10!}{(10-3)!} \\ &= 56 \cdot 720 = 40{,}320\end{aligned}$$

Matched Problem 7 Repeat Example 7 under the same conditions, except the serial numbers are now to have 3 letters followed by 2 digits (no repeats).

Example 8 **Counting Techniques** A company has 7 senior and 5 junior officers. An ad hoc legislative committee is to be formed. In how many ways can a 4-officer committee be formed so that it is composed of:

(A) Any 4 officers?
(B) 4 senior officers?
(C) 3 senior officers and 1 junior officer?
(D) 2 senior and 2 junior officers?
(E) At least 2 senior officers?

SOLUTION (A) Since there are a total of 12 officers in the company, the number of different 4-member committees is

$$C_{12,4} = \frac{12!}{4!(12-4)!} = \frac{12!}{4!8!} = 495$$

(B) If only senior officers can be on the committee, the number of different committees is

$$C_{7,4} = \frac{7!}{4!(7-4)!} = \frac{7!}{4!3!} = 35$$

(C) The 3 senior officers can be selected in $C_{7,3}$ ways, and the 1 junior officer can be selected in $C_{5,1}$ ways. Applying the multiplication principle, the number of ways that 3 senior officers and 1 junior officer can be selected is

$$C_{7,3} \cdot C_{5,1} = \frac{7!}{3!(7-3)!} \cdot \frac{5!}{1!(5-1)!} = \frac{7!5!}{3!4!1!4!} = 175$$

(D) $$C_{7,2} \cdot C_{5,2} = \frac{7!}{2!(7-2)!} \cdot \frac{5!}{2!(5-2)!} = \frac{7!5!}{2!5!2!3!} = 210$$

(E) The committees with *at least* 2 senior officers can be divided into three disjoint collections:

1. Committees with 4 senior officers and 0 junior officers
2. Committees with 3 senior officers and 1 junior officer
3. Committees with 2 senior officers and 2 junior officers

The number of committees of types 1, 2, and 3 were computed in parts (B), (C), and (D), respectively. The total number of committees of all three types is the sum of these quantities:

$$\underset{\text{Type 1}}{C_{7,4}} + \underset{\text{Type 2}}{C_{7,3} \cdot C_{5,1}} + \underset{\text{Type 3}}{C_{7,2} \cdot C_{5,2}} = 35 + 175 + 210 = 420$$

Matched Problem 8 Given the information in Example 8, answer the following questions:

(A) How many 4-officer committees with 1 senior officer and 3 junior officers can be formed?
(B) How many 4-officer committees with 4 junior officers can be formed?
(C) How many 4-officer committees with at least 2 junior officers can be formed?

Explore–Discuss 2

(A) Compute the numbers $C_{n,0}, C_{n,1}, C_{n,2}, \ldots, C_{n,n}$ for $n = 4, 5, 6$. Discuss the patterns you observe.
(B) Compute the sum

$$C_{n,0} + C_{n,1} + C_{n,2} + \cdots + C_{n,n}$$

and the alternating sum

$$C_{n,0} - C_{n,1} + C_{n,2} - \cdots + (-1)^n C_{n,n}$$

for $n = 4, 5, 6$. Guess the sum and alternating sum for $n = 7$, and verify your guess.

Answers to Matched Problems

1. (A) 720 (B) 10 (C) 720 (D) 20 (E) 1,140

2. (A)

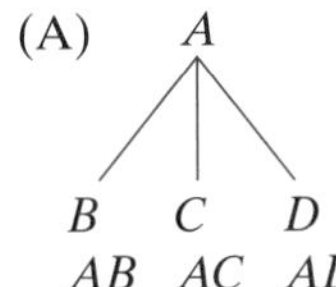

A	B	C	D
B C D	A C D	A B D	A B C
AB AC AD	BA BC BD	CA CB CD	DA DB DC

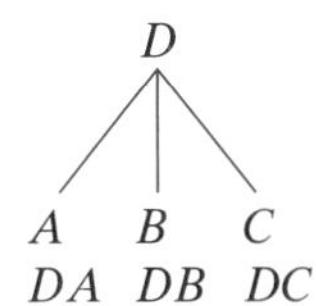

12 permutations of 4 objects taken 2 at a time

(B) O_1: Fill first position N_1: 4 ways
O_2: Fill second position N_2: 3 ways
$4 \cdot 3 = 12$

(C) $P_{4,2} = 4 \cdot 3 = 12$; $P_{4,2} = \dfrac{4!}{(4-2)!} = 12$

3. $P_{30,4} = \dfrac{30!}{(30-4)!} = 657{,}720$

4. (A) $P_{12,4} = \dfrac{12!}{(12-4)!} = 11{,}880$ ways (B) $C_{12,4} = \dfrac{12!}{4!(12-4)!} = 495$ ways

5. $C_{30,4} = \dfrac{30!}{4!(30-4)!} = 27{,}405$ 6. $C_{13,3} \cdot C_{13,2} = 22{,}308$

7. $P_{8,3} \cdot P_{10,2} = 30{,}240$

8. (A) $C_{7,1} \cdot C_{5,3} = 70$ (B) $C_{5,4} = 5$
(C) $C_{7,2} \cdot C_{5,2} + C_{7,1} \cdot C_{5,3} + C_{5,4} = 285$

EXERCISE 6-2

A *Evaluate the expressions in Problems 1–20.*

1. $4!$ **2.** $6!$ **3.** $\dfrac{9!}{8!}$

4. $\dfrac{14!}{13!}$ **5.** $\dfrac{11!}{8!}$ **6.** $\dfrac{14!}{12!}$

7. $\dfrac{5!}{2!3!}$ **8.** $\dfrac{6!}{4!2!}$ **9.** $\dfrac{7!}{4!(7-4)!}$

10. $\dfrac{8!}{3!(8-3)!}$ **11.** $\dfrac{7!}{7!(7-7)!}$ **12.** $\dfrac{8!}{0!(8-0)!}$

13. $P_{5,3}$ **14.** $P_{4,2}$ **15.** $P_{52,4}$

16. $P_{52,2}$ **17.** $C_{5,3}$ **18.** $C_{4,2}$

19. $C_{52,4}$ **20.** $C_{52,2}$

In Problems 21 and 22, would you consider the selection to be a combination or a permutation? Explain your reasoning.

21. (A) The recently elected university president named 3 new officers: a vice-president of finance, a vice-president of academic affairs, and a vice-president of student affairs.
(B) A university president selected 2 of the vice-presidents to attend the dedication ceremony of a new branch campus.

22. (A) A student checked out 4 novels from the library to read over the holidays.
(B) A student did some holiday shopping by buying 4 books, 1 for his father, 1 for his mother, 1 for his younger sister, and 1 for his older brother.

23. In a horse race, how many different finishes among the first 3 places are possible if 10 horses are running? (Exclude ties.)

24. In a long-distance foot race, how many different finishes among the first 5 places are possible if 50 people are running? (Exclude ties.)

25. How many ways can a 3-person subcommittee be selected from a committee of 7 people? How many ways can a president, vice-president, and secretary be chosen from a committee of 7 people?

26. Nine cards are numbered with the digits from 1 to 9. A 3-card hand is dealt, 1 card at a time. How many hands are possible where:
(A) Order is taken into consideration?
(B) Order is not taken into consideration?

B

27. Discuss the relative growth rates of $x!$, 3^x, and x^3.

28. Discuss the relative growth rates of $x!$, 2^x, and x^2.

29. From a standard 52-card deck, how many 5-card hands will have all hearts?

30. From a standard 52-card deck, how many 5-card hands will have all face cards? All face cards, but no kings?

31. From a standard 52-card deck, how many 7-card hands have exactly 5 spades and 2 hearts?

32. From a standard 52-card deck, how many 5-card hands will have 2 clubs and 3 hearts?

33. A catering service offers 8 appetizers, 10 main courses, and 7 desserts. A banquet committee is to select 3 appetizers, 4 main courses, and 2 desserts. How many ways can this be done?

34. Three departments have 12, 15, and 18 members, respectively. If each department is to select a delegate and an alternate to represent the department at a conference, how many ways can this be done?

In Problems 35 and 36, refer to the table in the graphing calculator display below, which shows $y_1 = P_{n,r}$ and $y_2 = C_{n,r}$ for $n = 6$.

X	Y1	Y2
0	1	1
1	6	6
2	30	15
3	120	20
4	360	15
5	720	6
6	720	1

Y2=6 nCr X

35. Discuss and explain the symmetry of the numbers in the y_2 column of the table.

36. Explain how the table illustrates the formula

$$P_{n,r} = r!C_{n,r}$$

C

37. Eight distinct points are selected on the circumference of a circle.
    (A) How many chords can be drawn by joining the points in all possible ways?
    (B) How many triangles can be drawn using these 8 points as vertices?
    (C) How many quadrilaterals can be drawn using these 8 points as vertices?

38. Five distinct points are selected on the circumference of a circle.
    (A) How many chords can be drawn by joining the points in all possible ways?
    (B) How many triangles can be drawn using these 5 points as vertices?

39. How many ways can 2 people be seated in a row of 5 chairs? 3 people? 4 people? 5 people?

40. Each of 2 countries sends 5 delegates to a negotiating conference. A rectangular table is used with 5 chairs on each long side. If each country is assigned a long side of the table (operation 1), how many seating arrangements are possible?

41. A basketball team has 5 distinct positions. Out of 8 players, how many starting teams are possible if:
    (A) The distinct positions are taken into consideration?
    (B) The distinct positions are not taken into consideration?
    (C) The distinct positions are not taken into consideration, but either Mike or Ken (but not both) must start?

42. How many 4-person committees are possible from a group of 9 people if:
    (A) There are no restrictions?
    (B) Both Jim and Mary must be on the committee?
    (C) Either Jim or Mary (but not both) must be on the committee?

43. Find the largest integer k such that your calculator can compute $k!$ without an overflow error.

44. Find the largest integer k such that your calculator can compute $C_{2k,k}$ without an overflow error.

45. **C** Note from the table in the graphing calculator display below that the largest value of $C_{n,r}$ when $n = 20$ is $C_{20,10} = 184{,}756$. Use a similar table to find the largest value of $C_{n,r}$ when $n = 24$.

X	Y1
7	77520
8	125970
9	167960
10	184756
11	167960
12	125970
13	77520

Y1=20 nCr X

46. **C** Note from the table in the graphing calculator display below that the largest value of $C_{n,r}$ when $n = 21$ is $C_{21,10} = C_{21,11} = 352{,}716$. Use a similar table to find the largest value of $C_{n,r}$ when $n = 17$.

X	Y1
7	116280
8	203490
9	293930
10	352716
11	352716
12	293930
13	203490

Y1=21 nCr X

Applications

Business & Economics

47. *Quality control.* A computer store receives a shipment of 24 laser printers, including 5 that are defective. Three of these printers are selected to be displayed in the store.
(A) How many selections can be made?
(B) How many of these selections will contain no defective printers?

48. *Quality control.* An electronics store receives a shipment of 30 graphing calculators, including 6 that are defective. Four of these calculators are selected to be sent to a local high school.
(A) How many selections can be made?
(B) How many of these selections will contain no defective calculators?

49. *Business closings.* A jewelry store chain with 8 stores in Georgia, 12 in Florida, and 10 in Alabama is planning to close 10 of these stores.
(A) How many ways can this be done?
(B) The company decides to close 2 stores in Georgia, 5 in Florida, and 3 in Alabama. How many ways can this be done?

50. *Employee layoffs.* A real estate company with 14 employees in their central office, 8 in their north office, and 6 in their south office is planning to lay off 12 employees.
(A) How many ways can this be done?
(B) The company decides to lay off 5 employees from the central office, 4 from the north office, and 3 from the south office. How many ways can this be done?

51. *Personnel selection.* Suppose 6 female and 5 male applicants have been successfully screened for 5 positions. In how many ways can the following compositions be selected?
(A) 3 females and 2 males
(B) 4 females and 1 male
(C) 5 females
(D) 5 people regardless of sex
(E) At least 4 females

52. *Committee selection.* A 4-person grievance committee is to be selected out of 2 departments, A and B, with 15 and 20 people, respectively. In how many ways can the following committees be selected?
(A) 3 from A and 1 from B
(B) 2 from A and 2 from B
(C) All from A
(D) 4 people regardless of department
(E) At least 3 from department A

Life Sciences

53. *Medicine.* There are 8 standard classifications of blood type. An examination for prospective laboratory technicians consists of having each candidate determine the type for 3 blood samples. How many different examinations can be given if no 2 of the samples provided for the candidate have the same type? If 2 or more samples can have the same type?

54. *Medical research.* Because of the limited funds, 5 research centers are to be chosen out of 8 suitable ones for a study on heart disease. How many choices are possible?

Social Sciences

55. *Politics.* A nominating convention is to select a president and vice-president from among 4 candidates. Campaign buttons, listing a president and a vice-president, are to be designed for each possible outcome before the convention. How many different kinds of buttons should be designed?

56. *Politics.* In how many different ways can 6 candidates for an office be listed on a ballot?

Section 6-3 Sample Spaces, Events, and Probability

- Experiments
- Sample Spaces and Events
- Probability of an Event
- Equally Likely Assumption

This section provides a relatively brief and informal introduction to probability. More detailed and formal treatments can be found in books and courses devoted entirely to the subject. Probability studies involve many subtle ideas, and care must be taken at the beginning to understand the fundamental concepts.

■ Experiments

Our first step in constructing a mathematical model for probability studies is to describe the type of experiments on which probability studies are based. Some experiments do not yield the same results each time they are performed no matter how carefully they are repeated under the same conditions. These experiments are called **random experiments.** Familiar examples of random experiments are flipping coins, rolling dice, observing the frequency of defective items from an assembly line, or observing the frequency of deaths in a certain age group.

Probability theory is a branch of mathematics that has been developed to deal with outcomes of random experiments, both real and conceptual. In the work that follows, we will simply use the word **experiment** to mean a random experiment.

■ Sample Spaces and Events

FIGURE 1

Associated with outcomes of experiments are *sample spaces* and *events.* Our second step in constructing a mathematical model for probability studies is to define these two terms. Set concepts, which are reviewed in Appendix A-1, are useful in this regard.

Consider the experiment, "A wheel with 18 numbers on the perimeter (Fig. 1) is spun and allowed to come to rest so that a pointer points within a numbered sector."

What outcomes might we observe? When the wheel stops, we might be interested in which number is next to the pointer, or whether that number is an odd number, or whether that number is divisible by 5, or whether that number is prime, or whether the pointer is in a shaded or white sector, and so on. The list of possible outcomes appears endless. In general, there is no unique method of analyzing all possible outcomes of an experiment. Therefore, before conducting an experiment, it is important to decide just what outcomes are of interest.

Suppose we limit our interest to the set of numbers on the wheel and to various subsets of these numbers, such as the set of prime numbers on the wheel or the set of odd numbers. Having decided what to observe, we make a list of outcomes of the experiment, called *simple outcomes* or *simple events,* such that in each trial of the experiment (each spin of the wheel), one and only one of the outcomes on the list will occur. For our stated interests, we choose each number on the wheel as a simple event and form the set

$$S = \{1, 2, 3, \ldots, 17, 18\}$$

The set of simple events S for the experiment is called a *sample space* for the experiment.

Now consider the outcome, "When the wheel comes to rest, the number next to the pointer is divisible by 4." This outcome is not a simple outcome (or simple event), since it is not associated with one and only one element in the sample space S. The outcome will occur whenever any one of the simple events 4, 8, 12, or 16 occurs; that is, whenever an element in the subset

$$E = \{4, 8, 12, 16\}$$

occurs. Subset E is called a *compound event* (and the outcome, a *compound outcome*).

In general:

Sample Spaces and Events

If we formulate a set S of outcomes (events) of an experiment in such a way that in each trial of the experiment one and only one of the outcomes (events) in the set will occur, then we call the set S a **sample space** for the experiment. Each element in S is called a **simple outcome,** or **simple event.**

An **event** $\boldsymbol{E}$ is defined to be any subset of S (including the empty set $\varnothing$ and the sample space S). Event E is a **simple event** if it contains only one element and a **compound event** if it contains more than one element. We say that **an event** $\boldsymbol{E}$ **occurs** if any of the simple events in E occurs.

We use the terms *event* and *outcome of an experiment* interchangeably. Technically, an event is the mathematical counterpart of an outcome of an experiment, but we will not insist on a strict adherence to this distinction in our development of probability.

REAL WORLD	MATHEMATICAL MODEL
Experiment (real or conceptual)	Sample space (set S)
Outcome (simple or compound)	Event (subset of S; simple or compound)

Example 1 **Simple and Compound Events** Relative to the number wheel experiment (see Fig. 1) and the sample space

$$S = \{1, 2, 3, \ldots, 17, 18\}$$

what is the event E (subset of the sample space S) that corresponds to each of the following outcomes? Indicate whether the event is a simple event or a compound event.

(A) The outcome is a prime number. (B) The outcome is the square of 4.

SOLUTION (A) The outcome is a prime number if any of the simple events 2, 3, 5, 7, 11, 13, or 17 occurs.* Thus, to say the event "A prime number occurs" is the same as saying the experiment has an outcome in the set

$$E = \{2, 3, 5, 7, 11, 13, 17\}$$

Since event E has more than one element, it is a compound event.

*Technically, we should write $\{2\}, \{3\}, \{5\}, \{7\}, \{11\}, \{13\}$, and $\{17\}$ for the simple events, since there is a logical distinction between an element of a set and a subset consisting of only that element. But we will just keep this in mind and drop the braces for simple events to simplify the notation.

(B) The outcome is the square of 4 if 16 occurs. Thus, to say the event "The square of 4 occurs" is the same as saying the experiment has an outcome in the set

$$E = \{16\}$$

Since E has only one element, it is a simple event.

Matched Problem 1 Repeat Example 1 for:

(A) The outcome is a number divisible by 12.
(B) The outcome is an even number greater than 15.

Example 2 **Sample Spaces** A nickel and a dime are tossed. How shall we identify a sample space for this experiment? There are a number of possibilities, depending on our interest. We shall consider three.

(A) If we are interested in whether each coin falls heads (H) or tails (T), then, using a tree diagram, we can easily determine an appropriate sample space for the experiment:

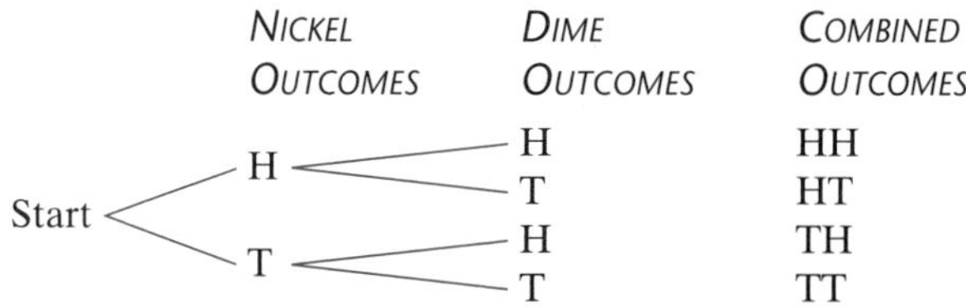

Thus,

$$S_1 = \{\text{HH}, \text{HT}, \text{TH}, \text{TT}\}$$

and there are 4 simple events in the sample space.

(B) If we are interested only in the number of heads that appear on a single toss of the two coins, then we can let

$$S_2 = \{0, 1, 2\}$$

and there are 3 simple events in the sample space.

(C) If we are interested in whether the coins match (M) or do not match (D), then we can let

$$S_3 = \{M, D\}$$

and there are only 2 simple events in the sample space.

In Example 2, which sample space would be appropriate for all three interests? Sample space S_1 contains more information than either S_2 or S_3. If we know which outcome has occurred in S_1, then we know which outcome has occurred in S_2 and S_3. However, the reverse is not true. (Note that the simple events in S_2 and S_3 are compound events in S_1.) In this sense, we say that S_1 is a more **fundamental sample space** than either S_2 or S_3. Thus, we would choose S_1 as an appropriate sample space for all three expressed interests.

Choosing Sample Spaces

There is no one correct sample space for a given experiment. When specifying a sample space for an experiment, we include as much detail as is necessary to answer *all* questions of interest regarding the outcomes of the experiment. When in doubt, choose a sample space with more elements rather than fewer.

Matched Problem 2 An experiment consists of recording the boy–girl composition of 2-child families.

(A) What is an appropriate sample space if we are interested in the sex of each child in the order of their births? Draw a tree diagram.
(B) What is an appropriate sample space if we are interested only in the number of girls in a family?
(C) What is an appropriate sample space if we are interested only in whether the sexes are alike (A) or different (D)?
(D) What is an appropriate sample space for all three interests expressed in parts (A)–(C)?

Example 3 

Sample Spaces and Events Consider an experiment of rolling two dice. A convenient sample space that will enable us to answer many questions about interesting events is shown in Figure 2. Let S be the set of all ordered pairs in the figure. The simple event $(3, 2)$ is to be distinguished from the simple event $(2, 3)$. The former indicates that a 3 turned up on the first die and a 2 on the second, while the latter indicates that a 2 turned up on the first die and a 3 on the second.

SECOND DIE

FIRST DIE	1	2	3	4	5	6
1	(1, 1)	(1, 2)	(1, 3)	(1, 4)	(1, 5)	(1, 6)
2	(2, 1)	(2, 2)	(2, 3)	(2, 4)	(2, 5)	(2, 6)
3	(3, 1)	(3, 2)	(3, 3)	(3, 4)	(3, 5)	(3, 6)
4	(4, 1)	(4, 2)	(4, 3)	(4, 4)	(4, 5)	(4, 6)
5	(5, 1)	(5, 2)	(5, 3)	(5, 4)	(5, 5)	(5, 6)
6	(6, 1)	(6, 2)	(6, 3)	(6, 4)	(6, 5)	(6, 6)

FIGURE 2

What is the event (subset of the sample space S) that corresponds to each of the following outcomes?

(A) A sum of 7 turns up. (B) A sum of 11 turns up.
(C) A sum less than 4 turns up. (D) A sum of 12 turns up.

Solution (A) By "A sum of 7 turns up," we mean that the sum of all dots on both turned-up faces is 7. This outcome corresponds to the event

$$\{(6, 1), (5, 2), (4, 3), (3, 4), (2, 5), (1, 6)\}$$

(B) "A sum of 11 turns up" corresponds to the event

$$\{(6, 5), (5, 6)\}$$

(C) "A sum less than 4 turns up" corresponds to the event

$$\{(1, 1), (2, 1), (1, 2)\}$$

(D) "A sum of 12 turns up" corresponds to the event

$$\{(6, 6)\}$$

Matched Problem 3 Refer to the sample space shown in Figure 2. What is the event that corresponds to each of the following outcomes?

(A) A sum of 5 turns up.
(B) A sum that is a prime number greater than 7 turns up.

As indicated earlier, we often use the terms *event* and *outcome of an experiment* interchangeably. Thus, in Example 3 we might say "the event, 'A sum of 11 turns up' " in place of "the outcome, 'A sum of 11 turns up,' " or even write

$$E = \text{A sum of 11 turns up} = \{(6, 5), (5, 6)\}$$

Probability of an Event

The next step in developing our mathematical model for probability studies is the introduction of a *probability function.* This is a function that assigns to an arbitrary event associated with a sample space a real number between 0 and 1, inclusive. We start by discussing ways in which probabilities are assigned to simple events in the sample space S.

Probabilities for Simple Events

Given a sample space

$$S = \{e_1, e_2, \ldots, e_n\}$$

with n simple events, to each simple event e_i we assign a real number, denoted by $P(e_i)$, called the **probability of the event e_i.** These numbers can be assigned in an arbitrary manner as long as the following two conditions are satisfied:

(continued)

Probabilities for Simple Events (*continued*)

1. The probability of a simple event is a number between 0 and 1, inclusive. That is,

$$0 \leq P(e_i) \leq 1$$

2. The sum of the probabilities of all simple events in the sample space is 1. That is,

$$P(e_1) + P(e_2) + \cdots + P(e_n) = 1$$

Any probability assignment that meets conditions 1 and 2 is said to be an **acceptable probability assignment.**

Our mathematical theory does not explain how acceptable probabilities are assigned to simple events. These assignments are generally based on the expected or actual percentage of times a simple event occurs when an experiment is repeated a large number of times. Assignments based on this principle are called **reasonable.**

Let an experiment be the flipping of a single coin, and let us choose a sample space S to be

$$S = \{\text{H}, \text{T}\}$$

If a coin appears to be fair, we are inclined to assign probabilities to the simple events in S as follows:

$$P(\text{H}) = \tfrac{1}{2} \qquad \text{and} \qquad P(\text{T}) = \tfrac{1}{2}$$

These assignments are based on reasoning that, since there are 2 ways a coin can land, in the long run, a head will turn up half the time and a tail will turn up half the time. These probability assignments are acceptable, since both conditions for acceptable probability assignments stated in the box are satisfied:

1. $0 \leq P(\text{H}) \leq 1, \qquad 0 \leq P(\text{T}) \leq 1$
2. $P(\text{H}) + P(\text{T}) = \tfrac{1}{2} + \tfrac{1}{2} = 1$

If we were to flip a coin 1,000 times, we would expect a head to turn up approximately, but not exactly, 500 times. The random number feature on a graphing utility can be used to simulate 1,000 flips of a coin. Figure 3 shows the results of 3 such simulations: 497 heads the first time, 495 heads the second, and 504 heads the third.

```
randBin(1000,.5,
3)
  {497 495 504}
```

FIGURE 3

If, however, we get only 376 heads in 1,000 flips of a coin, we might suspect the coin is not fair. Then we might assign the simple events in the sample space S the following probabilities, based on our experimental results:

$$P(\text{H}) = .376 \qquad \text{and} \qquad P(\text{T}) = .624$$

This is also an acceptable assignment. However, the probability assignment

$$P(\text{H}) = 1 \qquad \text{and} \qquad P(\text{T}) = 0$$

though acceptable, is not reasonable (unless the coin has 2 heads). And the assignment

$$P(\text{H}) = .6 \qquad \text{and} \qquad P(\text{T}) = .8$$

is not acceptable, since $.6 + .8 = 1.4$, which violates condition 2 in the box. [*Note:* In probability studies, the 0 to the left of the decimal is usually omitted. Thus, we write .6 and .8 instead of 0.6 and 0.8.]

It is important to keep in mind that out of the infinitely many possible acceptable probability assignments to simple events in a sample space, we are generally inclined to choose one assignment over another based on reasoning or experimental results.

Given an acceptable probability assignment for simple events in a sample space S, how do we define the probability of an arbitrary event E associated with S?

Probability of an Event E

Given an acceptable probability assignment for the simple events in a sample space S, we define the **probability of an arbitrary event E,** denoted by $\boldsymbol{P(E)}$, as follows:

(A) If E is the empty set, then $P(E) = 0$.
(B) If E is a simple event, then $P(E)$ has already been assigned.
(C) If E is a compound event, then $P(E)$ is the sum of the probabilities of all the simple events in E.
(D) If E is the sample space S, then $P(E) = P(S) = 1$ [this is a special case of part (C)].

Example 4 **Probabilities of Events** Let us return to Example 2, the tossing of a nickel and a dime, and the sample space

$$S = \{\text{HH}, \text{HT}, \text{TH}, \text{TT}\}$$

Since there are 4 simple outcomes and the coins are assumed to be fair, it would appear that each outcome would occur 25% of the time, in the long run. Let us assign the same probability of $\frac{1}{4}$ to each simple event in S:

SIMPLE EVENT e_i	HH	HT	TH	TT
$P(e_i)$	$\frac{1}{4}$	$\frac{1}{4}$	$\frac{1}{4}$	$\frac{1}{4}$

This is an acceptable assignment according to conditions 1 and 2, and it is a reasonable assignment for ideal (perfectly balanced) coins or coins close to ideal.

(A) What is the probability of getting 1 head (and 1 tail)?
(B) What is the probability of getting at least 1 head?
(C) What is the probability of getting at least 1 head or at least 1 tail?
(D) What is the probability of getting 3 heads?

SOLUTION (A) E_1 = Getting 1 head = {HT, TH}

Since E_1 is a compound event, we use part (C) in the box and find $P(E_1)$ by adding the probabilities of the simple events in E_1:

$$P(E_1) = P(\text{HT}) + P(\text{TH}) = \tfrac{1}{4} + \tfrac{1}{4} = \tfrac{1}{2}$$

(B) E_2 = Getting at least 1 head = {HH, HT, TH}

$$P(E_2) = P(\text{HH}) + P(\text{HT}) + P(\text{TH}) = \tfrac{1}{4} + \tfrac{1}{4} + \tfrac{1}{4} = \tfrac{3}{4}$$

(C) $E_3 = \{\text{HH, HT, TH, TT}\} = S$

$$P(E_3) = P(S) = 1 \quad \tfrac{1}{4} + \tfrac{1}{4} + \tfrac{1}{4} + \tfrac{1}{4} = 1$$

(D) E_4 = Getting 3 heads = $\varnothing$ Empty set

$$P(\varnothing) = 0$$

Steps for Finding the Probability of an Event *E*

Step 1. Set up an appropriate sample space S for the experiment.
Step 2. Assign acceptable probabilities to the simple events in S.
Step 3. To obtain the probability of an arbitrary event E, add the probabilities of the simple events in E.

The function P defined in steps 2 and 3 is a **probability function** whose domain is all possible events (subsets) in the sample space S and whose range is a set of real numbers between 0 and 1, inclusive.

Matched Problem 4 Suppose in Example 4 that, after flipping the nickel and dime 1,000 times, we find that HH turns up 273 times, HT turns up 206 times, TH turns up 312 times, and TT turns up 209 times. On the basis of this evidence, we assign probabilities to the simple events in S as follows:

SIMPLE EVENT e_i	HH	HT	TH	TT
$P(e_i)$	.273	.206	.312	.209

This is an acceptable and reasonable probability assignment for the simple events in S. What are the probabilities of the following events?

(A) E_1 = Getting at least 1 tail
(B) E_2 = Getting 2 tails
(C) E_3 = Getting at least 1 head or at least 1 tail

Example 4 and Matched Problem 4 illustrate two important ways in which acceptable and reasonable probability assignments are made for simple events in a sample space S. Each approach has its advantage in certain situations:

1. *Theoretical.* We use assumptions and a deductive reasoning process to assign probabilities to simple events. No experiments are actually conducted. This is what we did in Example 4.
2. *Empirical.* We assign probabilities to simple events based on the results of actual experiments. This is what we did in Matched Problem 4.

Empirical probability concepts are stated more precisely as follows: If we conduct an experiment n times and event E occurs with **frequency $f(E)$,** then the ratio $f(E)/n$ is called the **relative frequency** of the occurrence of event E in n trials. We define the **empirical probability** of E, denoted by **$P(E)$,** by the number (if it exists) that the relative frequency $f(E)/n$ approaches as n gets larger and larger. For any particular n, the relative frequency $f(E)/n$ is also called the **approximate empirical probability** of event E.

Empirical Probability Approximation

$$P(E) \approx \frac{\text{Frequency of occurrence of } E}{\text{Total number of trials}} = \frac{f(E)}{n}$$

(The larger n is, the better the approximation.)

For most of this section we will emphasize the theoretical approach. In the next section we will return to the empirical approach.

Equally Likely Assumption

In tossing a nickel and a dime (Example 4), we assigned the same probability, $\frac{1}{4}$, to each simple event in the sample space $S = \{\text{HH, HT, TH, TT}\}$. By assigning the same probability to each simple event in S, we are actually making the assumption that each simple event is as likely to occur as any other. We refer to this as an **equally likely assumption.** In general, we have the following:

Probability of a Simple Event Under an Equally Likely Assumption

If, in a sample space

$$S = \{e_1, e_2, \ldots, e_n\}$$

with n elements, we assume each simple event e_i is as likely to occur as any other, then we assign the probability $1/n$ to each. That is,

$$P(e_i) = \frac{1}{n}$$

Under an equally likely assumption, we can develop a very useful formula for finding probabilities of arbitrary events associated with a sample space S. Consider the following example.

If a single die is rolled and we assume each face is as likely to come up as any other, then for the sample space

$$S = \{1, 2, 3, 4, 5, 6\}$$

we assign a probability of $\frac{1}{6}$ to each simple event, since there are 6 simple events. The probability of

$$E = \text{Rolling a prime number} = \{2, 3, 5\}$$

is

$$P(E) = P(2) + P(3) + P(5) = \tfrac{1}{6} + \tfrac{1}{6} + \tfrac{1}{6} = \tfrac{3}{6} = \tfrac{1}{2}$$

(Number of elements in E: 3 — Number of elements in S: 6)

Thus, under the assumption that each simple event is as likely to occur as any other, the computation of the probability of the occurrence of any event E in a sample space S is the number of elements in E divided by the number of elements in S.

Theorem 1 ■■ **PROBABILITY OF AN ARBITRARY EVENT UNDER AN EQUALLY LIKELY ASSUMPTION**

If we assume each simple event in sample space S is as likely to occur as any other, then the probability of an arbitrary event E in S is given by

$$P(E) = \frac{\text{Number of elements in } E}{\text{Number of elements in } S} = \frac{n(E)}{n(S)}$$ ■■

Example 5 **Probabilities and Equally Likely Assumptions** Let us again consider rolling two dice, and assume each simple event in the sample space shown in Figure 2 (page 393) is as likely as any other. Find the probabilities of the following events:

(A) E_1 = A sum of 7 turns up
(B) E_2 = A sum of 11 turns up
(C) E_3 = A sum less than 4 turns up
(D) E_4 = A sum of 12 turns up

SOLUTION Referring to Figure 2 (page 393) and the results found in Example 3, we find:

(A) $P(E_1) = \frac{n(E_1)}{n(S)} = \frac{6}{36} = \frac{1}{6}$ $\quad E_1 = \{(6,1), (5,2), (4,3), (3,4), (2,5), (1,6)\}$

(B) $P(E_2) = \frac{n(E_2)}{n(S)} = \frac{2}{36} = \frac{1}{18}$ $\quad E_2 = \{(6,5), (5,6)\}$

(C) $P(E_3) = \frac{n(E_3)}{n(S)} = \frac{3}{36} = \frac{1}{12}$ $\quad E_3 = \{(1,1), (2,1), (1,2)\}$

(D) $P(E_4) = \frac{n(E_4)}{n(S)} = \frac{1}{36}$ $\quad E_4 = \{(6,6)\}$

Matched Problem 5 ➡ Under the conditions in Example 5, find the probabilities of the following events (each event refers to the sum of the dots facing up on both dice):

(A) E_5 = A sum of 5 turns up
(B) E_6 = A sum that is a prime number greater than 7 turns up

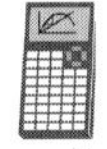

Example 6 ➡ **Simulation and Empirical Probabilities** Use output from the random number feature of a graphing utility to simulate 100 rolls of two dice. Determine the empirical probabilities of the following events, and compare with the theoretical probabilities:

(A) E_1 = A sum of 7 turns up (B) E_2 = A sum of 11 turns up

SOLUTION A graphing utility can be used to select a random integer from 1 to 6. Each of the six integers in the given range is equally likely to be selected. Therefore, by selecting a random integer from 1 to 6 and adding it to a second random integer from 1 to 6, we simulate rolling two dice and recording the sum (see the first command in Figure 4A and your user's manual). The second command in Figure 4A simulates 100 rolls of two dice; the sums are stored in list L_1. From the statistical plot of L_1 in Figure 4B we obtain the empirical probabilities:*

(A) The empirical probability of E_1 is $\frac{15}{100} = .15$; the theoretical probability of E_1 (see Example 5A) is $\frac{6}{36} = .167$.
(B) The empirical probability of E_2 is $\frac{6}{100} = .06$; the theoretical probability of E_2 (see Example 5B) is $\frac{2}{36} = .056$.

(A)

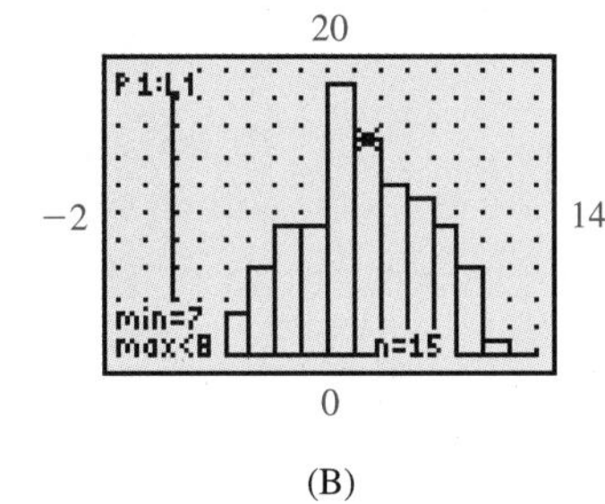

(B)

FIGURE 4

Matched Problem 6 ➡ Use the graphing utility output in Figure 4B to determine the empirical probabilities of the following events, and compare with the theoretical probabilities:

(A) E_3 = A sum less than 4 turns up (B) E_4 = A sum of 12 turns up

Explore–Discuss 1

A shipment box contains 12 graphing calculators, out of which 2 are defective. A calculator is drawn at random from the box and then, without replacement, a second calculator is drawn. Discuss whether the equally likely assumption would be appropriate for the sample space $S = \{GG, GD, DG, DD\}$, where G is a good calculator and D is a defective one.

*If you simulate this experiment on your graphing utility, you should not expect to get the same empirical probabilities.

We now turn to some examples that make use of the counting techniques developed in the preceding sections.

Example 7 **Probability and Equally Likely Assumption** In drawing 5 cards from a 52-card deck without replacement, what is the probability of getting 5 spades?

SOLUTION Let the sample space S be the set of all 5-card hands from a 52-card deck. Since the order in a hand does not matter, $n(S) = C_{52,5}$. Let event E be the set of all 5-card hands from 13 spades. Again, the order does not matter and $n(E) = C_{13,5}$. Thus, assuming each 5-card hand is as likely as any other,

$$P(E) = \frac{n(E)}{n(S)} = \frac{C_{13,5}}{C_{52,5}} = \frac{1,287}{2,598,960} \approx .0005$$

Matched Problem 7 In drawing 7 cards from a 52-card deck without replacement, what is the probability of getting 7 hearts?

Example 8 **Probability and Equally Likely Assumption** The board of regents of a university is made up of 12 men and 16 women. If a committee of 6 is chosen at random, what is the probability that it will contain 3 men and 3 women?

SOLUTION Let S be the set of all 6-person committees out of 28 people. Then

$$n(S) = C_{28,6}$$

Let E be the set of all 6-person committees with 3 men and 3 women. To find $n(E)$, we use the multiplication principle and the following two operations:

O_1: Select 3 men out of the 12 available $\qquad N_1$: $C_{12,3}$
O_2: Select 3 women out of the 16 available $\qquad N_2$: $C_{16,3}$

Thus,

$$n(E) = N_1 \cdot N_2 = C_{12,3} \cdot C_{16,3}$$

and

$$P(E) = \frac{n(E)}{n(S)} = \frac{C_{12,3} \cdot C_{16,3}}{C_{28,6}} \approx .327$$

Matched Problem 8 What is the probability that the committee in Example 8 will have 4 men and 2 women?

It needs to be pointed out that there are many counting problems for which it is not possible to produce a simple formula that will yield the number of possible cases. In situations of this type, we often revert back to tree diagrams and counting branches.

Explore–Discuss 2

Five football teams play each other in the Central Division. At the beginning of the season we are interested in discussing what will be the relative standing of the five teams at the end of the season, excluding ties. Describe an appropriate sample space for this discussion. How many simple events does it contain? If each possible final standing of the five teams is as likely as any other, what would be the probability of the occurrence of a particular final standing? Discuss the reasons that a sports writer would or would not use the equally likely assumption in predicting the final standing of the five teams.

Answers to Matched Problems

1. (A) $E = \{12\}$; simple event (B) $E = \{16, 18\}$; compound event

2. (A) $S_1 = \{BB, BG, GB, GG\}$;

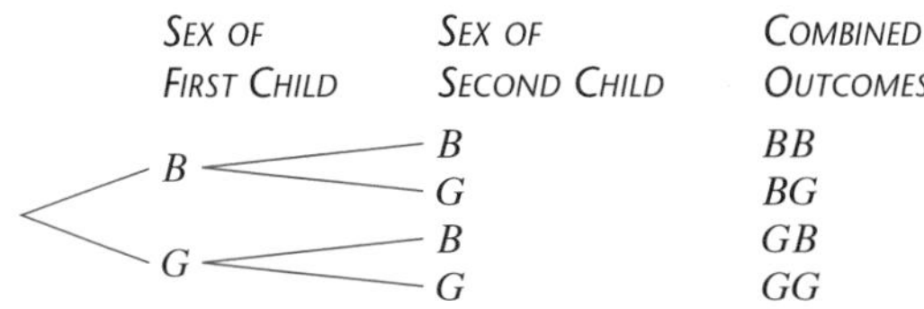

(B) $S_2 = \{0, 1, 2\}$ (C) $S_3 = \{A, D\}$ (D) S_1

3. (A) $\{(4, 1), (3, 2), (2, 3), (1, 4)\}$ (B) $\{(6, 5), (5, 6)\}$

4. (A) .727 (B) .209 (C) 1 **5.** (A) $P(E_5) = \frac{1}{9}$ (B) $P(E_6) = \frac{1}{18}$

6. (A) $\frac{9}{100} = .09$ (empirical); $\frac{3}{36} = .083$ (theoretical)
(B) $\frac{1}{100} = .01$ (empirical); $\frac{1}{36} = .028$ (theoretical)

7. $C_{13,7}/C_{52,7} \approx 1.3 \times 10^{-5}$ **8.** $C_{12,4}C_{16,2}/C_{28,6} \approx .158$

EXERCISE 6-3

A

1. How would you interpret $P(E) = 1$?
2. How would you interpret $P(E) = 0$?
3. In a family with 2 children, excluding multiple births, what is the probability of having 2 children of the opposite sex? Assume a girl is as likely as a boy at each birth.
4. In a family with 2 children, excluding multiple births, what is the probability of having 2 girls? Assume a girl is as likely as a boy at each birth.
5. A store carries four brands of CD players, *J, G, P,* and *S.* From past records, the manager found that the relative frequency of brand choice among customers varied. Which of the following probability assignments for a particular customer choosing a particular brand of CD player would have to be rejected? Why?

   (A) $P(J) = .15, P(G) = -.35, P(P) = .50,$ $P(S) = .70$
   (B) $P(J) = .32, P(G) = .28, P(P) = .24,$ $P(S) = .30$
   (C) $P(J) = .26, P(G) = .14, P(P) = .30,$ $P(S) = .30$

6. Using the probability assignments in Problem 5C, what is the probability that a random customer will not choose brand *S*?
7. Using the probability assignments in Problem 5C, what is the probability that a random customer will choose brand *J* or brand *P*?
8. Using the probability assignments in Problem 5C, what is the probability that a random customer will not choose brand *J* or brand *P*?

B

9. In a family with 3 children, excluding multiple births, what is the probability of having 2 boys and 1 girl, in that order? Assume a boy is as likely as a girl at each birth.
10. In a family with 3 children, excluding multiple births, what is the probability of having 2 boys and 1 girl, in any order? Assume a boy is as likely as a girl at each birth.

11. A small combination lock on a suitcase has 3 wheels, each labeled with the 10 digits from 0 to 9. If an opening combination is a particular sequence of 3 digits with no repeats, what is the probability of a person guessing the right combination?

12. A combination lock has 5 wheels, each labeled with the 10 digits from 0 to 9. If an opening combination is a particular sequence of 5 digits with no repeats, what is the probability of a person guessing the right combination?

Refer to the description of a standard deck of 52 cards and Figure 4 on page 384. An experiment consists of dealing 5 cards from a standard 52-card deck. In Problems 13–16, what is the probability of being dealt:

13. 5 black cards? **14.** 5 hearts?

15. 5 face cards? **16.** 5 nonface cards?

17. Twenty thousand students are enrolled at a state university. A student is selected at random and his or her birthday (month and day, not year) is recorded. Describe an appropriate sample space for this experiment, and assign acceptable probabilities to the simple events. What are your assumptions in making this assignment?

18. In a hotly contested three-way race for the U.S. Senate, polls indicate the two leading candidates are running neck-and-neck while the third candidate is receiving half the support of either of the others. Registered voters are chosen at random and are asked which of the three will get their vote. Describe an appropriate sample space for this random survey experiment and assign acceptable probabilities to the simple events.

19. Suppose 5 thank-you notes are written and 5 envelopes are addressed. Accidentally, the notes are randomly inserted into the envelopes and mailed without checking the addresses. What is the probability that all the notes will be inserted into the correct envelopes?

20. Suppose 6 people check their coats in a checkroom. If all claim checks are lost and the 6 coats are randomly returned, what is the probability that all the people will get their own coats back?

An experiment consists of rolling two fair dice and adding the dots on the two sides facing up. Using the sample space shown in Figure 2 (page 393) and assuming each simple event is as likely as any other, find the probability of the sum of the dots indicated in Problems 21–36.

21. Sum is 2. **22.** Sum is 10.

23. Sum is 6. **24.** Sum is 8.

25. Sum is less than 5. **26.** Sum is greater than 8.

27. Sum is not 7 or 11. **28.** Sum is not 2, 4, or 6.

29. Sum is 1. **30.** Sum is 13.

31. Sum is divisible by 3. **32.** Sum is divisible by 4.

33. Sum is 7 or 11 (a "natural").

34. Sum is 2, 3, or 12 ("craps").

35. Sum is divisible by 2 or 3.

36. Sum is divisible by 2 and 3.

An experiment consists of tossing three fair (not weighted) coins, except one of the three coins has a head on both sides. Compute the probability of obtaining the indicated results in Problems 37–42.

37. 1 head **38.** 2 heads

39. 3 heads **40.** 0 heads

41. More than 1 head **42.** More than 1 tail

43. (A) Is it possible to get 19 heads in 20 flips of a fair coin? Explain.
(B) If you flipped a coin 40 times and got 37 heads, would you suspect that the coin was unfair? Why or why not? If you suspect an unfair coin, what empirical probabilities would you assign to the simple events of the sample space?

44. (A) Is it possible to get 7 double 6's in 10 rolls of a pair of fair dice? Explain.
(B) If you rolled a pair of dice 36 times and got 11 double 6's, would you suspect that the dice were unfair? Why or why not? If you suspect loaded dice, what empirical probability would you assign to the event of rolling a double 6?

C *An experiment consists of rolling two fair (not weighted) dice and adding the dots on the two sides facing up. Each die has the number 1 on two opposite faces, the number 2 on two opposite faces, and the number 3 on two opposite faces. Compute the probability of obtaining the indicated sums in Problems 45–52.*

45. 2 **46.** 3 **47.** 4

48. 5 **49.** 6 **50.** 7

51. An odd sum **52.** An even sum

Refer to the description of a standard deck of 52 cards on page 384. An experiment consists of dealing 5 cards from a standard 52-card deck. In Problems 53–60, what is the probability of being dealt:

53. 5 face cards or aces?

54. 5 numbered cards (2 through 10)?

55. 4 aces?

56. Four of a kind (4 queens, 4 kings, and so on)?

57. A 10, jack, queen, king, and ace, all in the same suit?

58. A 2, 3, 4, 5, and 6, all in the same suit?

59. 2 aces and 3 queens?

60. 2 kings and 3 aces?

In Problems 61–64, several experiments are simulated using the random number feature on a graphing utility. For example, the roll of a fair die can be simulated by selecting a random integer from 1 to 6, and 50 rolls of a fair die by selecting 50 random integers from 1 to 6 (see Fig. A for Problem 61 and your user's manual).

61. From the statistical plot of the outcomes of rolling a fair die 50 times (see Fig. B), we see, for example, that the number 4 was rolled exactly 11 times.

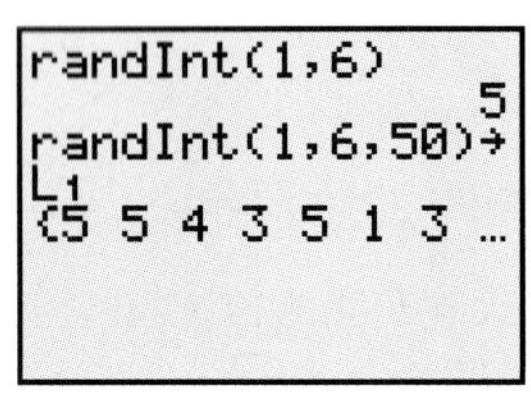

(A)

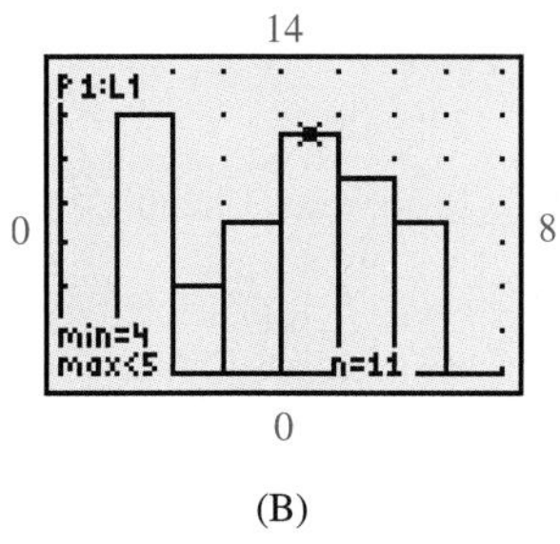

(B)

Figure for 61

(A) What is the empirical probability that the number 6 was rolled?

(B) What is the probability that a 6 is rolled under the equally likely assumption?

(C) Use a graphing utility to simulate 100 rolls of a fair die and determine the empirical probabilities of the six outcomes.

62. Use a graphing utility to simulate 200 tosses of a nickel and dime, representing the outcomes HH, HT, TH, and TT by 1, 2, 3, and 4, respectively.

(A) Find the empirical probabilities of the four outcomes.

(B) What is the probability of each outcome under the equally likely assumption?

63. (A) Explain how a graphing utility can be used to simulate 500 tosses of a coin.

(B) Carry out the simulation and find the empirical probabilities of the two outcomes.

(C) What is the probability of each outcome under the equally likely assumption?

64. From a box containing a dozen balls numbered 1 through 12, one ball is drawn at random.

(A) Explain how a graphing utility can be used to simulate 400 repetitions of this experiment.

(B) Carry out the simulation and find the empirical probability of drawing the 8 ball.

(C) What is the probability of drawing the 8 ball under the equally likely assumption?

Applications

Business & Economics

65. *Consumer testing.* Twelve popular brands of beer are to be used in a blind taste study for consumer recognition.

(A) If 4 distinct brands are chosen at random from the 12 and if a consumer is not allowed to repeat any answers, what is the probability that all 4 brands could be identified by just guessing?

(B) If repeats are allowed in the 4 brands chosen at random from the 12 and if the consumer is allowed to repeat answers, what is the probability of correct identification of all 4 by just guessing?

66. *Consumer testing.* Six popular brands of cola are to be used in a blind taste study for consumer recognition.

(A) If 3 distinct brands are chosen at random from the 6 and if a consumer is not allowed to repeat any answers, what is the probability that all 3 brands could be identified by just guessing?

(B) If repeats are allowed in the 3 brands chosen at random from the 6 and if the consumer is allowed to repeat answers, what is the probability of correct identification of all 3 by just guessing?

67. *Personnel selection.* Suppose 6 female and 5 male applicants have been successfully screened for 5 positions. If the 5 positions are filled at random from the 11 finalists, what is the probability of selecting:

(A) 3 females and 2 males?

(B) 4 females and 1 male?

(C) 5 females?

(D) At least 4 females?

68. *Committee selection.* A 4-person grievance committee is to be composed of employees in 2 departments, A and B, with 15 and 20 employees, respectively. If the 4 people are selected at random from the 35 employees, what is the probability of selecting:

(A) 3 from A and 1 from B?

(B) 2 from A and 2 from B?

(C) All from A?

(D) At least 3 from A?

Life Sciences

69. *Medicine.* A prospective laboratory technician is to be tested on identifying blood types from 8 standard classifications.
    (A) If 3 distinct samples are chosen at random from the 8 types and if the examinee is not allowed to repeat any answers, what is the probability that all 3 could be correctly identified by just guessing?
    (B) If repeats are allowed in the 3 blood types chosen at random from the 8 and if the examinee is allowed to repeat answers, what is the probability of correct identification of all 3 by just guessing?

70. *Medical research.* Because of limited funds, 5 research centers are to be chosen out of 8 suitable ones for a study on heart disease. If the selection is made at random, what is the probability that 5 particular research centers will be chosen?

Social Sciences

71. *Membership selection.* A town council has 11 members, 6 Democrats and 5 Republicans.
    (A) If the president and vice-president are selected at random, what is the probability that they are both Democrats?
    (B) If a 3-person committee is selected at random, what is the probability that Republicans make up the majority?

SECTION 6-4 Union, Intersection, and Complement of Events; Odds

- Union and Intersection
- Complement of an Event
- Odds
- Applications to Empirical Probability

Recall that in Section 6-3 we said that given a sample space

$$S = \{e_1, e_2, \ldots, e_n\}$$

any function P defined on S such that

$$0 \le P(e_i) \le 1 \qquad i = 1, 2, \ldots, n$$

and

$$P(e_1) + P(e_2) + \cdots + P(e_n) = 1$$

is called a *probability function.* In addition, we said that any subset of S is called an *event* E, and we defined the probability of E to be the sum of the probabilities of the simple events in E.

Union and Intersection

Since events are subsets of a sample space, the union and intersection of events are simply the union and intersection of sets as defined in Appendix A-1 and in the following box. In this section we will concentrate on the union of events and consider only simple cases of intersection. The latter will be investigated in more detail in the next section.

Union and Intersection of Events*

If A and B are two events in a sample space S, then the **union** of A and B, denoted by $\mathbf{A \cup B}$, and the **intersection** of A and B, denoted by $\mathbf{A \cap B}$, are defined as follows:

$$A \cup B = \{e \in S \mid e \in A \textbf{ or } e \in B\} \qquad A \cap B = \{e \in S \mid e \in A \textbf{ and } e \in B\}$$

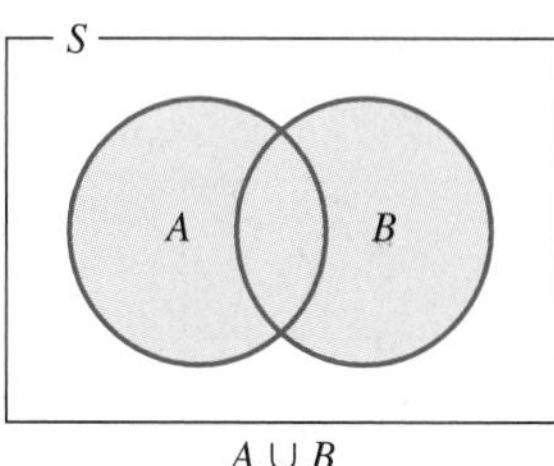

$A \cup B$

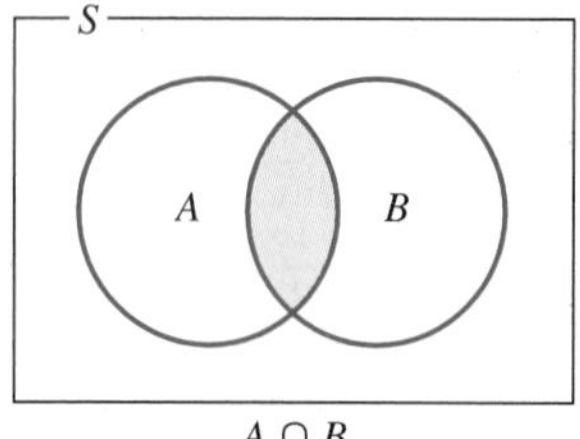

$A \cap B$

Furthermore, we define:

The **event A or B** to be $A \cup B$
The **event A and B** to be $A \cap B$

*See Appendix A-1 for a discussion of set notation.

Example 1 **Probability Involving Union and Intersection** Consider the sample space of equally likely events for the rolling of a single fair die:

$$S = \{1, 2, 3, 4, 5, 6\}$$

(A) What is the probability of rolling a number that is odd **and** exactly divisible by 3?
(B) What is the probability of rolling a number that is odd **or** exactly divisible by 3?

SOLUTION (A) Let A be the event of rolling an odd number, B the event of rolling a number divisible by 3, and F the event of rolling a number that is odd **and** divisible by 3. Then

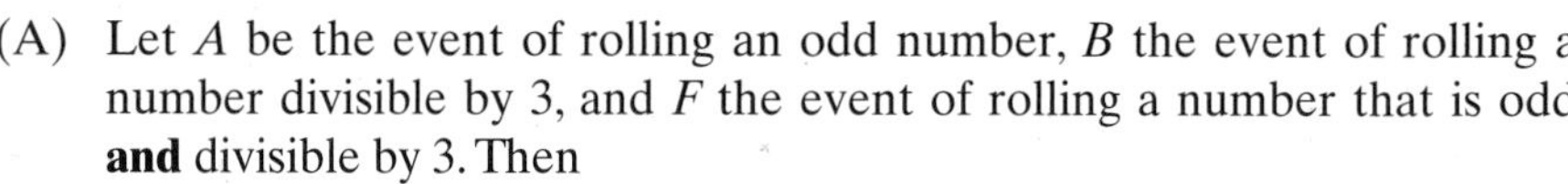

$$A = \{1, 3, 5\} \qquad B = \{3, 6\} \qquad F = A \cap B = \{3\}$$

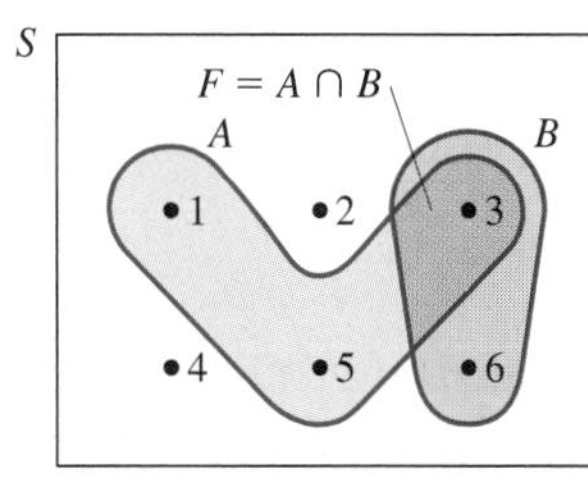

Thus, the probability of rolling a number that is odd **and** exactly divisible by 3 is

$$P(F) = P(A \cap B) = \frac{n(A \cap B)}{n(S)} = \frac{1}{6}$$

(B) Let A and B be the same events as in part (A), and let E be the event of rolling a number that is odd **or** divisible by 3. Then

$$A = \{1, 3, 5\} \qquad B = \{3, 6\} \qquad E = A \cup B = \{1, 3, 5, 6\}$$

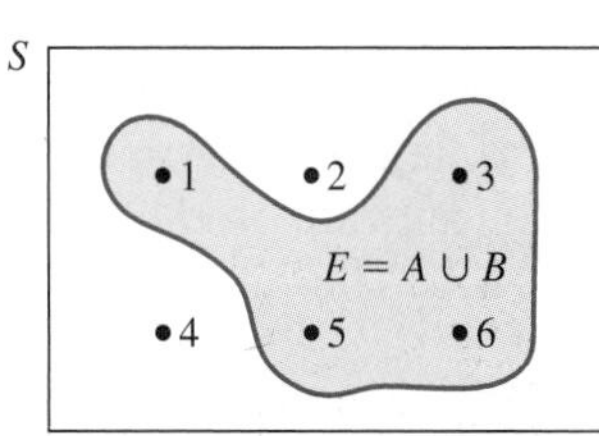

Thus, the probability of rolling a number that is odd **or** exactly divisible by 3 is

$$P(E) = P(A \cup B) = \frac{n(A \cup B)}{n(S)} = \frac{4}{6} = \frac{2}{3}$$

Matched Problem 1 Use the sample space in Example 1 to answer the following:

(A) What is the probability of rolling an odd number **and** a prime number?
(B) What is the probability of rolling an odd number **or** a prime number?

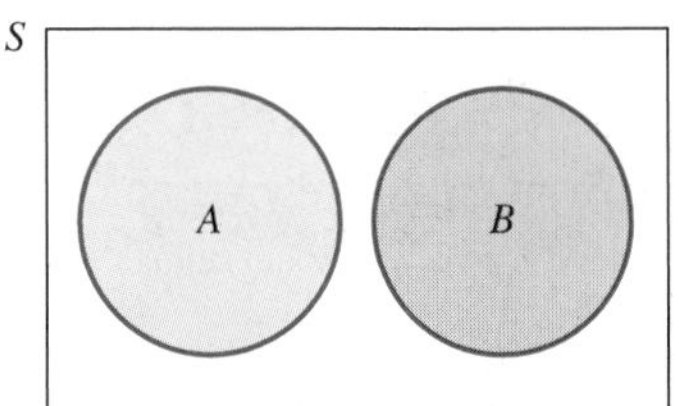

FIGURE 1
Mutually exclusive: $A \cap B = \varnothing$

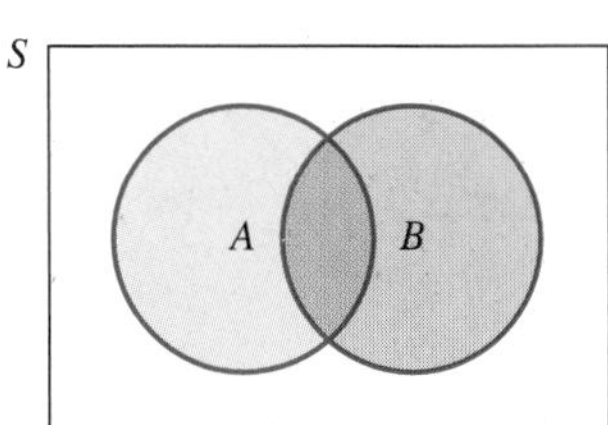

FIGURE 2
Not mutually exclusive: $A \cap B \neq \varnothing$

Suppose

$$E = A \cup B$$

Can we find $P(E)$ in terms of A and B? The answer is almost yes, but we must be careful. It would be nice if

$$P(A \cup B) = P(A) + P(B) \tag{1}$$

This turns out to be true if events A and B are **mutually exclusive (disjoint);** that is, if $A \cap B = \varnothing$ (Fig. 1). In this case, $P(A \cup B)$ is the sum of all the probabilities of simple events in A added to the sum of all the probabilities of simple events in B. But what happens if events A and B are not mutually exclusive; that is, if $A \cap B \neq \varnothing$ (see Fig. 2)? If we simply add the probabilities of the elements in A to the probabilities of the elements in B, we are including some of the probabilities twice, namely those for elements that are in both A and B. To compensate for this double counting, we subtract $P(A \cap B)$ from equation (1) to obtain

$$P(A \cup B) = P(A) + P(B) - P(A \cap B) \tag{2}$$

We notice that (2) holds for both cases, $A \cap B \neq \varnothing$ and $A \cap B = \varnothing$, since (2) reduces to (1) for the latter case $[P(A \cap B) = P(\varnothing) = 0]$. It is better to use (2) if there is any doubt that A and B are mutually exclusive. We summarize this discussion in the box for convenient reference.

Probability of a Union of Two Events

For any events A and B,

$$P(A \cup B) = P(A) + P(B) - P(A \cap B) \tag{2}$$

If A and B are mutually exclusive, then

$$P(A \cup B) = P(A) + P(B) \tag{1}$$

Example 2 **Probability Involving Union and Intersection** Suppose two fair dice are rolled:

(A) What is the probability that a sum of 7 or 11 turns up?
(B) What is the probability that both dice turn up the same or that a sum less than 5 turns up?

SOLUTION (A) If A is the event that a sum of 7 turns up and B is the event that a sum of 11 turns up, then (see Fig. 3) the event that a sum of 7 or 11 turns up is $A \cup B$, where

$$A = \{(1, 6), (2, 5), (3, 4)\ (4, 3), (5, 2), (6, 1)\}$$
$$B = \{(5, 6), (6, 5)\}$$

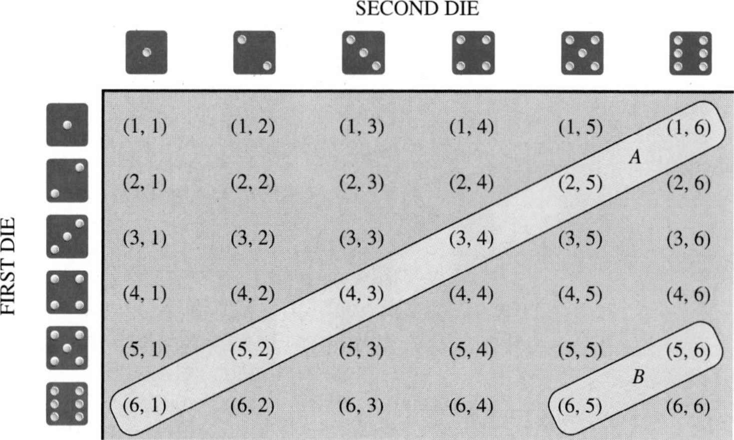

FIGURE 3

Since events A and B are mutually exclusive, we can use equation (1) to calculate $P(A \cup B)$:

$$\begin{aligned} P(A \cup B) &= P(A) + P(B) \\ &= \tfrac{6}{36} + \tfrac{2}{36} \\ &= \tfrac{8}{36} = \tfrac{2}{9} \end{aligned}$$

In this equally likely sample space, $n(A) = 6$, $n(B) = 2$, and $n(S) = 36$.

(B) If A is the event that both dice turn up the same and B is the event that the sum is less than 5, then (see Fig. 4) the event that both dice turn up the same or the sum is less than 5 is $A \cup B$, where

$$A = \{(1, 1), (2, 2), (3, 3), (4, 4), (5, 5), (6, 6)\}$$
$$B = \{(1, 1), (1, 2), (1, 3), (2, 1), (2, 2), (3, 1)\}$$

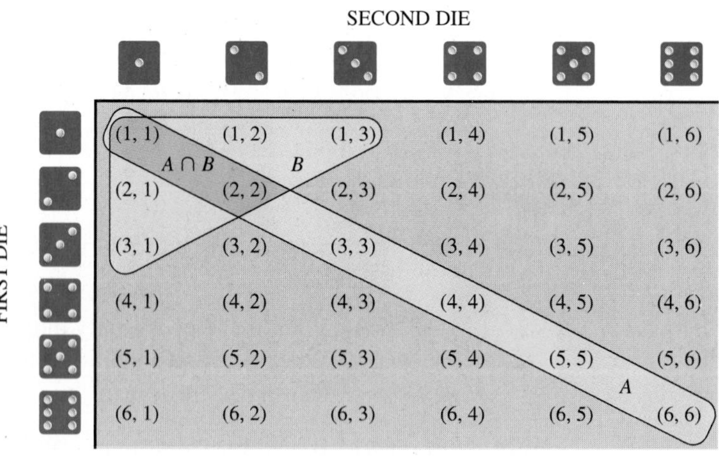

FIGURE 4

Matched Problem 4 A shipment of 40 precision parts, including 8 that are defective, is sent to an assembly plant. The quality control division selects 10 at random for testing and rejects the whole shipment if 1 or more in the sample are found defective. What is the probability that the shipment will be rejected?

Example 5 **Birthday Problem** In a group of n people, what is the probability that at least 2 people have the same birthday (the same month and day, excluding February 29)? (Make a guess for a class of 40 people, and check your guess with the conclusion of this example.)

SOLUTION If we form a list of the birthdays of all the people in the group, then we have a simple event in the sample space

$$S = \text{Set of all lists of } n \text{ birthdays}$$

For any person in the group, we will assume that any birthday is as likely as any other, so that the simple events in S are equally likely. How many simple events are in the set S? Since any person could have any one of 365 birthdays (excluding February 29), the multiplication principle implies that the number of simple events in S is

$$\begin{aligned} n(S) &= \underset{\text{1st person}}{365} \cdot \underset{\text{2nd person}}{365} \cdot \underset{\text{3rd person}}{365} \cdot \cdots \cdot \underset{n\text{th person}}{365} \\ &= 365^n \end{aligned}$$

Now, let E be the event that at least 2 people in the group have the same birthday. Then E' is the event that no 2 people have the same birthday. The multiplication principle also can be used to determine the number of simple events in E':

$$\begin{aligned} n(E') &= \underset{\text{1st person}}{365} \cdot \underset{\text{2nd person}}{364} \cdot \underset{\text{3rd person}}{363} \cdot \cdots \cdot \underset{n\text{th person}}{(366 - n)} \\ &= \frac{[365 \cdot 364 \cdot 363 \cdot \cdots \cdot (366 - n)](365 - n)!}{(365 - n)!} \\ &= \frac{365!}{(365 - n)!} \end{aligned}$$

Multiply numerator and denominator by $(365 - n)!$

Since we have assumed that S is an equally likely sample space,

$$P(E') = \frac{n(E')}{n(S)} = \frac{\dfrac{365!}{(365 - n)!}}{365^n} = \frac{365!}{365^n(365 - n)!}$$

Thus,

$$\begin{aligned} P(E) &= 1 - P(E') \\ &= 1 - \frac{365!}{365^n(365 - n)!} \end{aligned} \qquad (4)$$

Equation (4) is valid for any n satisfying $1 \leq n \leq 365$. [What is $P(E)$ if $n > 365$?] For example, in a group of 6 people,

$$\begin{aligned} P(E) &= 1 - \frac{365!}{(365)^6 359!} \\ &= 1 - \frac{\cancel{365} \cdot 364 \cdot 363 \cdot 362 \cdot 361 \cdot 360 \cdot \cancel{359!}}{\cancel{365} \cdot 365 \cdot 365 \cdot 365 \cdot 365 \cdot 365 \cdot \cancel{359!}} \\ &= .04 \end{aligned}$$

It is interesting to note that as the size of the group increases, $P(E)$ increases more rapidly than you might expect. Figure 6 shows the graph of $P(E)$ for $1 \leq n \leq 39$. Table 1 gives the value of $P(E)$ for selected values of n. Notice that for a group of only 23 people, the probability that 2 or more have the same birthday is greater than $\frac{1}{2}$. [See Problem 51 in Exercise 6-4 for a discussion of the form of equation (4) used to produce the graph in Fig. 6.]

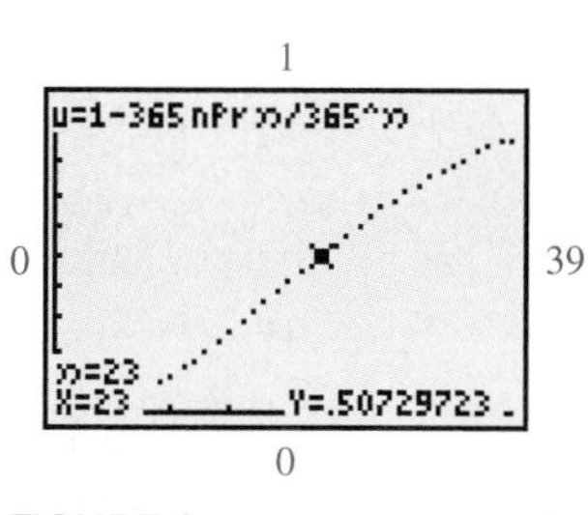

FIGURE 6

Table 1

THE BIRTHDAY PROBLEM

NUMBER OF PEOPLE IN GROUP n	PROBABILITY THAT 2 OR MORE HAVE SAME BIRTHDAY $P(E)$
5	.027
10	.117
15	.253
20	.411
23	**.507**
30	.706
40	.891
50	.970
60	.994
70	.999

Matched Problem 5 ➡ Use equation (4) to evaluate $P(E)$ for $n = 4$.

Explore–Discuss 2

Determine the smallest number n such that in a group of n people the probability that 2 or more have a birthday in the same month is greater than .5. Discuss the assumptions underlying your computation.

■ ODDS

When the probability of an event E is known, it is often customary (particularly in gaming situations) to speak of *odds* for (or against) the event E, rather than the *probability* of the event E. The relationship between these two designations is outlined in the box at the top of the next page.

From Probability to Odds

If $P(E)$ is the probability of the event E, then we define:

(A) **Odds for E** $= \dfrac{P(E)}{1 - P(E)} = \dfrac{P(E)}{P(E')}$ $\quad P(E) \neq 1$

(B) **Odds against E** $= \dfrac{P(E')}{P(E)}$ $\quad P(E) \neq 0$

[*Note:* When possible, odds are expressed as ratios of whole numbers.]

Example 6 **Probability and Odds** If you roll a fair die once, the probability of rolling a 4 is $\frac{1}{6}$, whereas the odds in favor of rolling a 4 are

$$\frac{P(E)}{P(E')} = \frac{\frac{1}{6}}{\frac{5}{6}} = \frac{1}{5} \quad \text{Read as “1 to 5” and also written as “1:5”}$$

and

$$\text{Odds against rolling a 4} = \frac{5}{1}$$

In terms of a fair game, if you bet \$1 on a 4 turning up, you would lose \$1 to the house if any number other than 4 turns up and you would be paid \$5 by the house (and in addition your bet of \$1 returned) if a 4 turns up. (An experienced gambler would say that the house pays 5 to 1 on a 4 turning up on a single roll of a die.)

In general, if the odds for an event E are $a{:}b$, then **a game is fair** if your bet of \$$a$ is lost if event E does not happen, but you win \$$b$ (and in addition your bet of \$$a$ is returned) if event E does happen. (Gamblers would say that the house pays b to a on event E happening.)

Matched Problem 6
(A) What are the odds for rolling a sum of 8 in a single roll of two fair dice?
(B) If you bet \$5 that a sum of 8 will turn up, what should the house pay (plus returning your \$5 bet) if a sum of 8 does turn up for the game to be fair?

Now we will go in the other direction: If we are given the odds for an event, what is the probability of the event? (The verification of the following formula is left to Problem 55 in Exercise 6-4.)

From Odds to Probability

If the odds for event E are a/b, then the probability of E is

$$P(E) = \frac{a}{a + b}$$

Example 7 **Odds and Probability** If in repeated rolls of two fair dice the odds for rolling a 5 before rolling a 7 are 2 to 3, then the probability of rolling a 5 before rolling a 7 is

$$P(E) = \frac{a}{a+b} = \frac{2}{2+3} = \frac{2}{5}$$

Matched Problem 7 If in repeated rolls of two fair dice the odds against rolling a 6 before rolling a 7 are 6 to 5, then what is the probability of rolling a 6 before rolling a 7? (Be careful! Read the problem again.)

Applications to Empirical Probability

In the following discussions, the term *empirical probability* will mean the probability of an event determined by a sample that is used to approximate the probability of the corresponding event in the total population. How does the approximate empirical probability of an event determined from a sample relate to the actual probability of an event relative to the total population? In mathematical statistics an important theorem called the **law of large numbers** (or the **law of averages**) is proved. Informally, it states that the approximate empirical probability can be made as close to the actual probability as we please by making the sample sufficiently large.

Example 8 **Market Research** From a survey involving 1,000 people in a certain city, it was found that 500 people had tried a certain brand of diet cola, 600 had tried a certain brand of regular cola, and 200 had tried both types of cola. If a resident of the city is selected at random, what is the (empirical) probability that:

(A) The resident has tried the diet or the regular cola? What are the (empirical) odds for this event?

(B) The resident has tried one of the colas but not both? What are the (empirical) odds against this event?

Solution Let D be the event that a person has tried the diet cola and R the event that a person has tried the regular cola. The events D and R can be used to partition the residents of the city into four mutually exclusive subsets (a collection of subsets is mutually exclusive if the intersection of any two of them is the empty set):

$D \cap R$ = Set of people who have tried both colas
$D \cap R'$ = Set of people who have tried the diet cola but not the regular cola
$D' \cap R$ = Set of people who have tried the regular cola but not the diet cola
$D' \cap R'$ = Set of people who have not tried either cola

These sets are displayed in the Venn diagram in Figure 7.

The sample population of 1,000 residents is also partitioned into four mutually exclusive sets, with $n(D) = 500$, $n(R) = 600$, and $n(D \cap R) = 200$. By using a Venn diagram (Fig. 8), we can determine the number of sample points in the sets $D \cap R'$, $D' \cap R$, and $D' \cap R'$ (see Example 2 in Section 6-1).

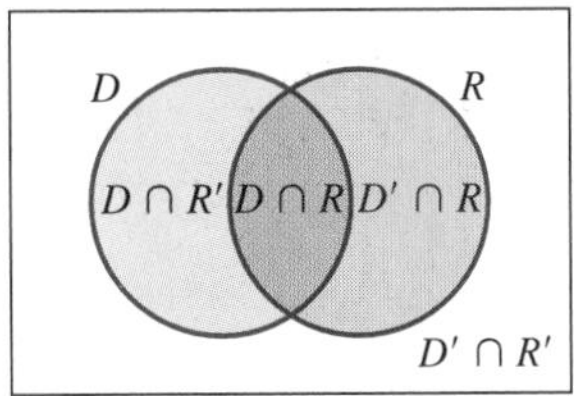

FIGURE 7
Total population

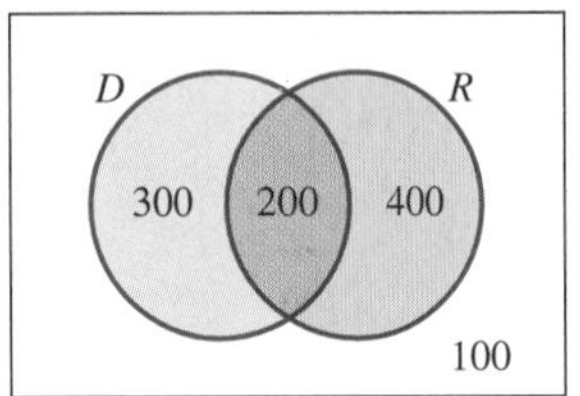

FIGURE 8
Sample population

These frequencies can be conveniently displayed in a table:

		REGULAR R	NO REGULAR R'	*Totals*
DIET	D	200	300	500
NO DIET	D'	400	100	500
	Totals	600	400	1,000

Assuming that each sample point is equally likely, we form a probability table by dividing each entry in this table by 1,000, the total number surveyed. These are empirical probabilities for the sample population, which we can use to approximate the corresponding probabilities for the total population.

		REGULAR R	NO REGULAR R'	*Totals*
DIET	D	.2	.3	.5
NO DIET	D'	.4	.1	.5
	Totals	.6	.4	1.0

Now we are ready to compute the required probabilities.

(A) The event that a person has tried the diet or the regular cola is $E = D \cup R$. We compute $P(E)$ two ways.

Method 1. Directly:

$$\begin{aligned} P(E) &= P(D \cup R) \\ &= P(D) + P(R) - P(D \cap R) \\ &= .5 + .6 - .2 = .9 \end{aligned}$$

Method 2. Using the complement of E:

$$\begin{aligned} P(E) &= 1 - P(E') \\ &= 1 - P(D' \cap R') \quad\quad E' = (D \cup R)' = D' \cap R' \text{ (see Fig. 7)} \\ &= 1 - .1 = .9 \end{aligned}$$

In either case,

$$\text{Odds for } E = \frac{P(E)}{P(E')} = \frac{.9}{.1} = \frac{9}{1} \quad \text{or} \quad 9:1$$

(B) The event that a person has tried one cola but not both is the event that the person has tried diet and not regular cola or has tried regular and not diet cola. In terms of sets, this is event $E = (D \cap R') \cup (D' \cap R)$. Since $D \cap R'$ and $D' \cap R$ are mutually exclusive (look at the Venn diagram in Fig. 7),

$$\begin{aligned} P(E) &= P[(D \cap R') \cup (D' \cap R)] \\ &= P(D \cap R') + P(D' \cap R) \\ &= .3 + .4 = .7 \end{aligned}$$

$$\text{Odds against } E = \frac{P(E')}{P(E)} = \frac{.3}{.7} = \frac{3}{7} \quad \text{or} \quad 3:7$$

Matched Problem 8 ➡ If a resident from the city in Example 8 is selected at random, what is the (empirical) probability that:

(A) The resident has not tried either cola? What are the (empirical) odds for this event?

(B) The resident has tried the diet cola or has not tried the regular cola? What are the (empirical) odds against this event?

Answers to Matched Problems

1. (A) $\frac{1}{3}$ (B) $\frac{2}{3}$ **2.** (A) $\frac{1}{12}$ (B) $\frac{7}{18}$ **3.** $\frac{47}{140} \approx .336$ **4.** .92

5. .016 **6.** (A) 5:31 (B) \$31 **7.** $\frac{5}{11} \approx .455$

8. (A) $P(D' \cap R') = .1$; odds for $D' \cap R' = \frac{1}{9}$ or 1:9

(B) $P(D \cup R') = .6$; odds against $D \cup R' = \frac{2}{3}$ or 2:3

EXERCISE 6-4

A

1. If a manufactured item has the probability .003 of failing within 90 days, what is the probability that the item will not fail in that time period?
2. In a particular cross of two plants the probability that the flowers will be red is .25. What is the probability that they will not be red?

A spinner is numbered from 1 through 10, and each number is as likely to come up as any other. In Problems 3–6, use equation (1) or (2), indicating which is used, to compute the probability that in a single spin the spinner will stop at:

3. A number less than 3 or larger than 7
4. A 2 or a number larger than 6
5. An even number or a number divisible by 3
6. An odd number or a number divisible by 3

Problems 7–18 refer to the Venn diagram for events A and B in an equally likely sample space S shown below. Find each of the indicated probabilities.

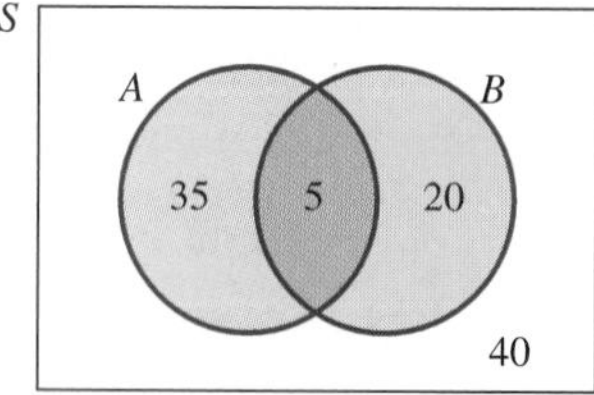

7. $P(A)$	**8.** $P(A')$	**9.** $P(B)$
10. $P(B')$	**11.** $P(A \cap B)$	**12.** $P(A \cap B')$
13. $P(A' \cap B)$	**14.** $P(A' \cap B')$	**15.** $P(A \cup B)$
16. $P(A \cup B')$	**17.** $P(A' \cup B)$	**18.** $P(A' \cup B')$

In Problems 19–22, use the equally likely sample space in Example 2 and equation (1) or (2), indicating which is used, to compute the probability of the following events:

19. A sum of 5 or 6 **20.** A sum of 9 or 10

21. The number on the first die is a 1 or the number on the second die is a 1.

22. The number on the first die is a 1 or the number on the second die is less than 3.

23. Given the following probabilities for an event E, find the odds for and against E:
(A) $\frac{3}{8}$ (B) $\frac{1}{4}$ (C) .4 (D) .55

24. Given the following probabilities for an event E, find the odds for and against E:
(A) $\frac{3}{5}$ (B) $\frac{1}{7}$ (C) .6 (D) .35

25. Compute the probability of event E if the odds in favor of E are:
(A) $\frac{3}{8}$ (B) $\frac{11}{7}$ (C) $\frac{4}{1}$ (D) $\frac{49}{51}$

26. Compute the probability of event E if the odds in favor of E are:
(A) $\frac{5}{9}$ (B) $\frac{4}{3}$ (C) $\frac{3}{7}$ (D) $\frac{23}{77}$

In Problems 27 and 28, discuss the validity of each statement. If the statement is always true, explain why. If not, give a counterexample.

27. (A) The empirical probability of an event is less than or equal to its theoretical probability.
(B) If events E and F are mutually exclusive, then

$$P(E) + P(F) = P(E \cup F) + P(E \cap F)$$

28. (A) If $P(E) = \frac{1}{2}$, then the odds for E equal the odds against E.
(B) If the odds for E are $a{:}b$, then the odds against E' are $b{:}a$.

B *In Problems 29–32, compute the odds in favor of obtaining:*

29. A head in a single toss of a coin

30. A number divisible by 3 in a single roll of a die

31. At least 1 head when a single coin is tossed 3 times

32. 1 head when a single coin is tossed twice

In Problems 33–36, compute the odds against obtaining:

33. A number greater than 4 in a single roll of a die

34. 2 heads when a single coin is tossed twice

35. A 3 or an even number in a single roll of a die

36. An odd number or a number divisible by 3 in a single roll of a die

37. (A) What are the odds for rolling a sum of 5 in a single roll of two fair dice?
(B) If you bet $1 that a sum of 5 will turn up, what should the house pay (plus returning your $1 bet) if a sum of 5 turns up for the game to be fair?

38. (A) What are the odds for rolling a sum of 10 in a single roll of two fair dice?
(B) If you bet $1 that a sum of 10 will turn up, what should the house pay (plus returning your $1 bet) if a sum of 10 turns up for the game to be fair?

A pair of dice are rolled 1,000 times with the following frequencies of outcomes:

SUM	2	3	4	5	6	7	8	9	10	11	12
FREQUENCY	10	30	50	70	110	150	170	140	120	80	70

Use these frequencies to calculate the approximate empirical probabilities and odds for the events in Problems 39 and 40.

39. (A) The sum is less than 4 or greater than 9.
(B) The sum is even or exactly divisible by 5.

40. (A) The sum is a prime number or is exactly divisible by 4.
(B) The sum is an odd number or exactly divisible by 3.

In Problems 41–44, a single card is drawn from a standard 52-card deck. Calculate the probability of and odds for each event.

41. A face card or a club is drawn.

42. A king or a heart is drawn.

43. A black card or an ace is drawn.

44. A heart or a number less than 7 (count an ace as 1) is drawn.

45. What is the probability of getting at least 1 diamond in a 5-card hand dealt from a standard 52-card deck?

46. What is the probability of getting at least 1 black card in a 7-card hand dealt from a standard 52-card deck?

47. What is the probability that a number selected at random from the first 1,000 positive integers is (exactly) divisible by 6 or 8?

48. What is the probability that a number selected at random from the first 600 positive integers is (exactly) divisible by 6 or 9?

49. Explain how the three events A, B, and C from a sample space S are related to each other in order for the following equation to hold:

$$P(A \cup B \cup C) = P(A) + P(B) + P(C) - P(A \cap B)$$

50. Explain how the three events A, B, and C from a sample space S are related to each other in order for the following equation to hold:

$$P(A \cup B \cup C) = P(A) + P(B) + P(C)$$

51. Show that the solution to the birthday problem in Example 5 can be written in the form

$$P(E) = 1 - \frac{P_{365,\,n}}{365^n}$$

For a calculator that has a $\boxed{P_{n,r}}$ key, explain why this form may be better for direct evaluation than the other form used in the solution to Example 5. Try direct evaluation of both forms on a calculator for $n = 25$.

52. Many (but not all) calculators experience an overflow error when computing $P_{365,n}$ for $n > 39$ and when computing 365^n. Explain how you would evaluate $P(E)$ for any $n > 39$ on such a calculator.

C

53. In a group of n people ($n \leq 12$), what is the probability that at least 2 of them have the same birth month? (Assume any birth month is as likely as any other.)

54. In a group of n people ($n \leq 100$), each person is asked to select a number between 1 and 100, write the number on a slip of paper, and place the slip in a hat. What is the probability that at least 2 of the slips in the hat have the same number written on them?

55. If the odds in favor of an event E occurring are a to b, show that

$$P(E) = \frac{a}{a + b}$$

[*Hint:* Solve the equation $P(E)/P(E') = a/b$ for $P(E)$.]

56. If $P(E) = c/d$, show that odds in favor of E occurring are c to $d - c$.

57. C The command in Figure A was used on a graphing utility to simulate 50 repetitions of rolling a pair of dice and recording their sum. A statistical plot of the results is shown in Figure B.

(A) Use Figure B to find the empirical probability of rolling a 7 or 8.

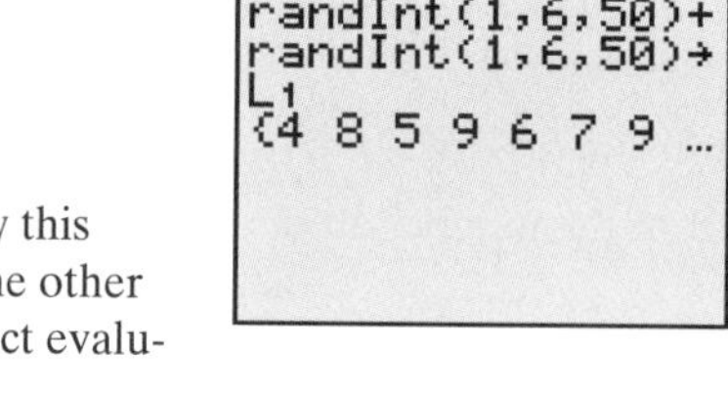

(A)

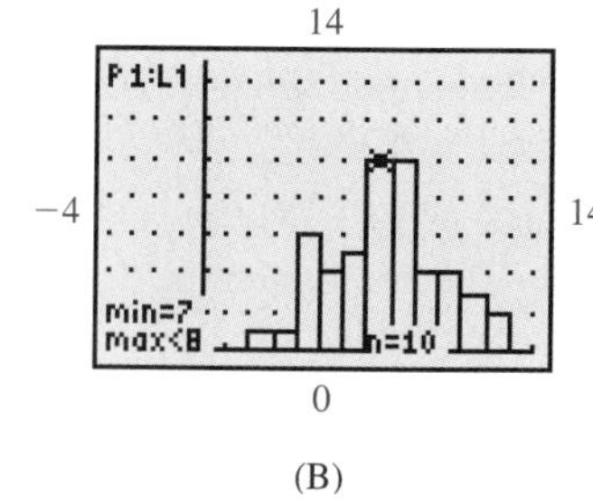

(B)

Figure for 57

(B) What is the theoretical probability of rolling a 7 or 8?

(C) Use a graphing utility to simulate 200 repetitions of rolling a pair of dice and recording their sum, and find the empirical probability of rolling a 7 or 8.

58. C Consider the command in Figure A and the associated statistical plot in Figure B.

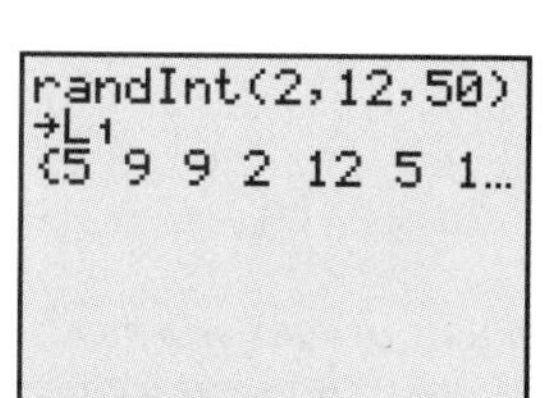

(A)

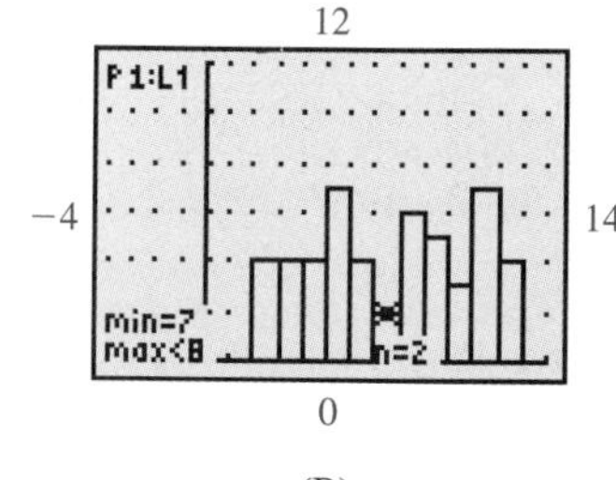

(B)

Figure for 58

(A) Explain why the command does not simulate 50 repetitions of rolling a pair of dice and recording their sum.

(B) Describe an experiment that is simulated by this command.

(C) Simulate 200 repetitions of the experiment you described in part (B), and find the empirical probability of recording a 7 or 8.

(D) What is the theoretical probability of recording a 7 or 8?

Applications

Business & Economics

59. *Market research.* From a survey involving 1,000 students at a large university, a market research company found that 750 students owned stereos, 450 owned cars, and 350 owned cars and stereos. If a student at the university is selected at random, what is the (empirical) probability that:

(A) The student owns either a car or a stereo?

(B) The student owns neither a car nor a stereo?

60. *Market research.* If a student at the university in Problem 59 is selected at random, what is the (empirical) probability that:

(A) The student does not own a car?

(B) The student owns a car but not a stereo?

61. *Insurance.* By examining the past driving records of drivers in a certain city, an insurance company has determined the (empirical) probabilities in the table shown at the top of the next page.

		Miles Driven Per Year			
		Less than 10,000, M_1	*10,000–15,000 inclusive,* M_2	*More than 15,000,* M_3	*Totals*
Accident	A	.05	.1	.15	.3
No Accident	A'	.15	.2	.35	.7
	Totals	.2	.3	.5	1.0

If a driver in this city is selected at random, what is the probability that:

(A) He or she drives less than 10,000 miles per year or has an accident?

(B) He or she drives 10,000 or more miles per year and has no accidents?

62. *Insurance.* Use the (empirical) probabilities in Problem 61 to find the probability that a driver in the city selected at random:

(A) Drives more than 15,000 miles per year or has an accident

(B) Drives 15,000 or fewer miles per year and has an accident

63. *Manufacturing.* Manufacturers of a portable computer provide a 90 day limited warranty covering only the keyboard and the disk drive. Their records indicate that during the warranty period, 6% of their computers are returned because they have defective keyboards, 5% are returned because they have defective disk drives, and 1% are returned because both the keyboard and the disk drive are defective. What is the (empirical) probability that a computer will not be returned during the warranty period?

64. *Product testing.* In order to test a new car, an automobile manufacturer wants to select 4 employees to test drive the car for 1 year. If 12 management and 8 union employees volunteer to be test drivers and the selection is made at random, what is the probability that at least 1 union employee is selected?

65. *Quality control.* A shipment of 60 inexpensive digital watches, including 9 that are defective, is sent to a department store. The receiving department selects 10 at random for testing and rejects the whole shipment if 1 or more in the sample are found defective. What is the probability that the shipment will be rejected?

66. *Quality control.* An automated manufacturing process produces 40 computer circuit boards, including 7 that are defective. The quality control department selects 10 at random (from the 40 produced) for testing and will shut down the plant for trouble shooting if 1 or more in the sample are found defective. What is the probability that the plant will be shut down?

Life Sciences

67. *Medicine.* In order to test a new drug for adverse reactions, the drug was administered to 1,000 test subjects with the following results: 60 subjects reported that their only adverse reaction was a loss of appetite, 90 subjects reported that their only adverse reaction was a loss of sleep, and 800 subjects reported no adverse reactions at all. If this drug is released for general use, what is the (empirical) probability that a person using the drug will suffer both a loss of appetite and a loss of sleep?

68. *Medicine.* Thirty animals are to be used in a medical experiment on diet deficiency: 3 male and 7 female rhesus monkeys, 6 male and 4 female chimpanzees, and 2 male and 8 female dogs. If one animal is selected at random, what is the probability of getting:

(A) A chimpanzee or a dog?

(B) A chimpanzee or a male?

(C) An animal other than a female monkey?

Social Sciences

Problems 69 and 70 refer to the data in the table below, obtained from a random survey of 1,000 residents of a state. The participants were asked their political affiliations and their preferences in an upcoming gubernatorial election. (In the table, D = *Democrat,* R = *Republican, and* U = *Unaffiliated.)*

		D	*R*	*U*	*Totals*
Candidate *A*	*A*	200	100	85	385
Candidate *B*	*B*	250	230	50	530
No Preference	*N*	50	20	15	85
	Total	500	350	150	1,000

69. *Politics.* If a resident of the state is selected at random, what is the (empirical) probability that the resident is:

(A) Not affiliated with a political party or has no preference? What are the odds for this event?

(B) Affiliated with a political party and prefers candidate *A*? What are the odds against this event?

70. *Politics.* If a resident of the state is selected at random, what is the (empirical) probability that the resident is:

(A) A Democrat or prefers candidate *B*? What are the odds for this event?

(B) Not a Democrat and has no preference? What are the odds against this event?

71. *Sociology.* A group of 5 Blacks, 5 Asians, 5 Latinos, and 5 Whites was used in a study on racial influence in small group dynamics. If 3 people are chosen at random, what is the probability that at least 1 is Black?

SECTION 6-5

Conditional Probability, Intersection, and Independence

- CONDITIONAL PROBABILITY
- INTERSECTION OF EVENTS—PRODUCT RULE
- PROBABILITY TREES
- INDEPENDENT EVENTS
- SUMMARY

In the previous section we asked if the probability of the union of two events could be expressed in terms of the probabilities of the individual events, and we found the answer to be a qualified yes (see page 407). Now we ask the same question for the intersection of two events; that is, can the probability of the intersection of two events be represented in terms of the probabilities of the individual events? The answer again is a qualified yes. But before we find out how and under what conditions, we must investigate a related concept called *conditional probability.*

CONDITIONAL PROBABILITY

The probability of an event may change if we are told of the occurrence of another event. For example, if an adult (a person 21 years or older) is selected at random from all adults in the United States, the probability of that person having lung cancer would not be too high. However, if we are told that the person is also a heavy smoker, then we would certainly want to revise the probability upward.

In general, the probability of the occurrence of an event A, given the occurrence of another event B, is called a **conditional probability** and is denoted by $P(A|B)$.

In the above illustration, events A and B would be

$A =$ Adult has lung cancer
$B =$ Adult is a heavy smoker

And $P(A|B)$ would represent the probability of an adult having lung cancer, given that he or she is a heavy smoker.

Our objective now is to try to formulate a precise definition of $P(A|B)$. It is helpful to start with a relatively simple problem, solve it intuitively, and then generalize from this experience.

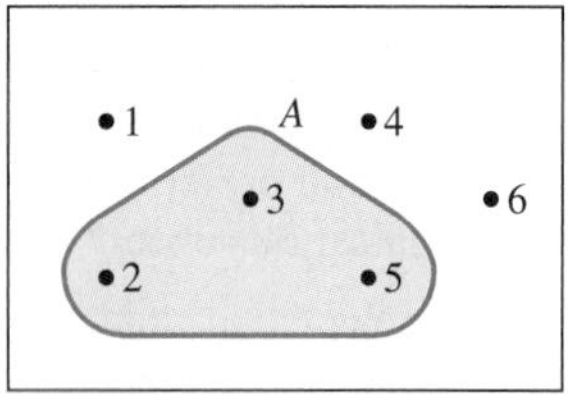

FIGURE 1

What is the probability of rolling a prime number (2, 3, or 5) in a single roll of a fair die? Let

$$S = \{1, 2, 3, 4, 5, 6\}$$

Then the event of rolling a prime number is (see Fig. 1)

$$A = \{2, 3, 5\}$$

Thus, since we assume each simple event in the sample space is equally likely,

$$P(A) = \frac{n(A)}{n(S)} = \frac{3}{6} = \frac{1}{2}$$

Now suppose you are asked, "In a single roll of a fair die, what is the probability that a prime number has turned up if we are given the additional information that an odd number has turned up?" The additional knowledge that another event has occurred, namely,

$$B = \text{Odd number turns up}$$

puts the problem in a new light. We are now interested only in the part of event A (rolling a prime number) that is in event B (rolling an odd number). Event B, since we know it has occurred, becomes the new sample space. The Venn diagrams in Figure 2 illustrate the various relationships. Thus, the probability of A given B is the number of A elements in B divided by the total number of elements in B. Symbolically,

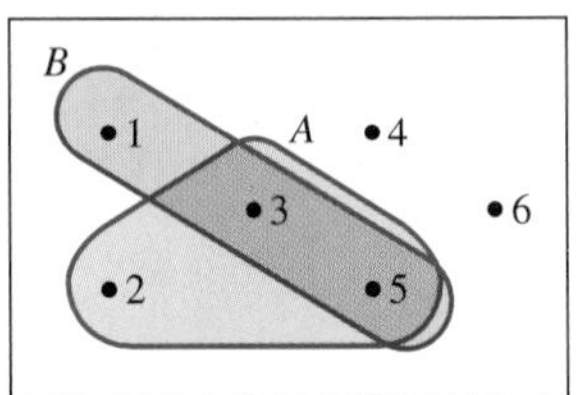

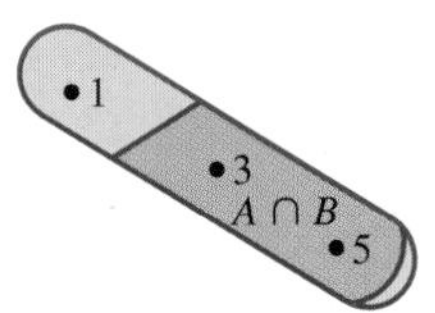

FIGURE 2
B is the new sample space

$$P(A|B) = \frac{n(A \cap B)}{n(B)} = \frac{2}{3}$$

Dividing the numerator and denominator of $n(A \cap B)/n(B)$ by $n(S)$, the number of elements in the original sample space, we can express $P(A|B)$ in terms of $P(A \cap B)$ and $P(B)$:*

$$P(A|B) = \frac{n(A \cap B)}{n(B)} = \frac{\dfrac{n(A \cap B)}{n(S)}}{\dfrac{n(B)}{n(S)}} = \frac{P(A \cap B)}{P(B)}$$

Using the right side to compute $P(A|B)$ for the example above, we obtain the same result (as we should):

$$P(A|B) = \frac{P(A \cap B)}{P(B)} = \frac{\frac{2}{6}}{\frac{3}{6}} = \frac{2}{3}$$

We use the above formula to motivate the following definition of *conditional probability,* which applies to any sample space, including those having simple events that are not equally likely (see Example 1).

Conditional Probability

For events A and B in an arbitrary sample space S, we define the conditional probability of A given B by

$$P(A|B) = \frac{P(A \cap B)}{P(B)} \qquad P(B) \neq 0 \qquad (1)$$

Conditional Probability A pointer is spun once on the circular spinner shown in Figure 3. The probability assigned to the pointer landing on a given integer (from 1 to 6) is the ratio of the area of the corresponding circular sector to the area of the whole circle, as given in the table:

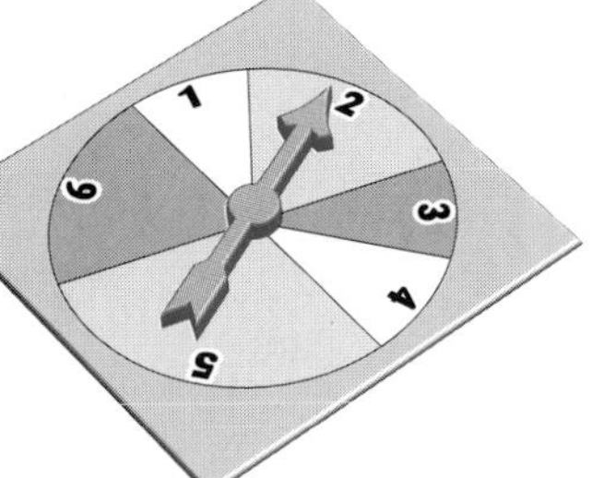

FIGURE 3

e_i	1	2	3	4	5	6
$P(e_i)$	.1	.2	.1	.1	.3	.2

$S = \{1, 2, 3, 4, 5, 6\}$

*Note that $P(A|B)$ is a probability based on the new sample space B, while $P(A \cap B)$ and $P(B)$ are both probabilities based on the original sample space S.

(A) What is the probability of the pointer landing on a prime number?
(B) What is the probability of the pointer landing on a prime number, given that it landed on an odd number?

SOLUTION Let the events E and F be defined as follows:

$$E = \text{Pointer lands on a prime number} = \{2, 3, 5\}$$
$$F = \text{Pointer lands on an odd number} = \{1, 3, 5\}$$

(A)
$$P(E) = P(2) + P(3) + P(5) = .2 + .1 + .3 = .6$$

(B) First note that $E \cap F = \{3, 5\}$.

$$P(E|F) = \frac{P(E \cap F)}{P(F)} = \frac{P(3) + P(5)}{P(1) + P(3) + P(5)} = \frac{.1 + .3}{.1 + .1 + .3} = \frac{.4}{.5} = .8$$

Matched Problem 1 Refer to the spinner and table in Example 1.

(A) What is the probability of the pointer landing on a number greater than 4?
(B) What is the probability of the pointer landing on a number greater than 4, given that it landed on an even number?

Explore–Discuss 1

Compare the spinner problem in Example 1 to the die problem at the beginning of the section. Discuss their similarities and differences.

Example 2 **Safety Research** Suppose past records in a large city produced the following probability data on a driver being in an accident on the last day of a Memorial Day weekend:

		ACCIDENT A	NO ACCIDENT A'	*Totals*
RAIN	R	.025	.335	.360
NO RAIN	R'	.015	.625	.640
	Totals	.040	.960	1.000

$S = \{RA, RA', R'A, R'A'\}$

(A) Find the probability of an accident, rain or no rain.
(B) Find the probability of rain, accident or no accident.
(C) Find the probability of an accident and rain.
(D) Find the probability of an accident, given rain.

SOLUTION (A) Let $A = \{RA, R'A\}$ Event: "accident"

$$P(A) = P(RA) + P(R'A) = .025 + .015 = .040$$

(B) Let $R = \{RA, RA'\}$ Event: "rain"

$$P(R) = P(RA) + P(RA') = .025 + .335 = .360$$

(C) $A \cap R = \{RA\}$ Event: "accident and rain"

$$P(A \cap R) = P(RA) = .025$$

(D) $P(A|R) = \frac{P(A \cap R)}{P(R)} = \frac{.025}{.360} = .069$ Event: "accident, given rain"

Compare this result with that in part (A). Note that $P(A|R) \neq P(A)$, and the probability of an accident, given rain, is higher than the probability of an accident without the knowledge of rain.

Matched Problem 2 Referring to the table in Example 2, determine the following:

(A) Probability of no rain
(B) Probability of an accident and no rain
(C) Probability of an accident, given no rain [Use formula (1) and the results of parts (A) and (B).]

Intersection of Events—Product Rule

We now return to the original problem of this section; that is, representing the probability of an intersection of two events in terms of the probabilities of the individual events. If $P(A) \neq 0$ and $P(B) \neq 0$, then using formula (1) we can write

$$P(A|B) = \frac{P(A \cap B)}{P(B)} \quad \text{and} \quad P(B|A) = \frac{P(B \cap A)}{P(A)}$$

Solving the first equation for $P(A \cap B)$ and the second equation for $P(B \cap A)$, we have

$$P(A \cap B) = P(B)P(A|B) \quad \text{and} \quad P(B \cap A) = P(A)P(B|A)$$

Since $A \cap B = B \cap A$ for any sets A and B, it follows that

$$P(A \cap B) = P(B)P(A|B) = P(A)P(B|A)$$

and we have the **product rule:**

Product Rule

For events A and B with nonzero probabilities in a sample space S,

$$P(A \cap B) = P(A)P(B|A) = P(B)P(A|B) \quad (2)$$

and we can use either $P(A)P(B|A)$ or $P(B)P(A|B)$ to compute $P(A \cap B)$.

Example 3 **Consumer Survey** If 60% of a department store's customers are female and 75% of the female customers have charge accounts at the store, what is the probability that a customer selected at random is a female and has a charge account?

SOLUTION Let: S = All store customers
F = Female customers
C = Customers with a charge account

If 60% of the customers are female, then the probability that a customer selected at random is a female is

$$P(F) = .60$$

Since 75% of the female customers have charge accounts, the probability that a customer has a charge account, given that the customer is a female, is

$$P(C|F) = .75$$

Using equation (2), the probability that a customer is a female and has a charge account is

$$P(F \cap C) = P(F)P(C|F) = (.60)(.75) = .45$$

Matched Problem 3 If 80% of the male customers of the department store in Example 3 have charge accounts, what is the probability that a customer selected at random is a male and has a charge account?

PROBABILITY TREES

We used tree diagrams in Section 6-1 to help us count the number of combined outcomes in a sequence of experiments. In much the same way we will now use probability trees to help us compute the probabilities of combined outcomes in a sequence of experiments. An example will help make the process of forming and using probability trees clear.

Example 4 **Probability Tree** Two balls are drawn in succession, without replacement, from a box containing 3 blue and 2 white balls (Fig. 4). What is the probability of drawing a white ball on the second draw?

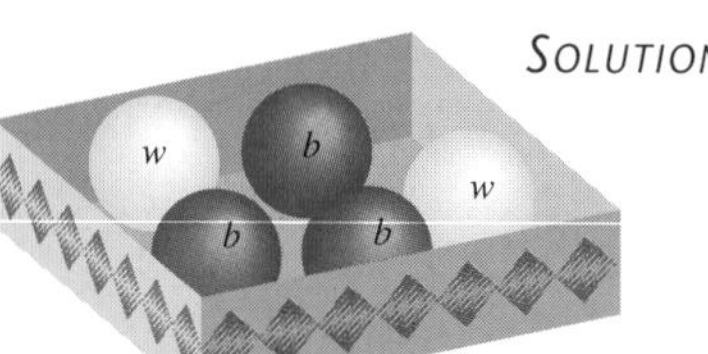

FIGURE 4

SOLUTION We start with a tree diagram (Fig. 5) showing the combined outcomes of the two experiments (first draw and second draw):

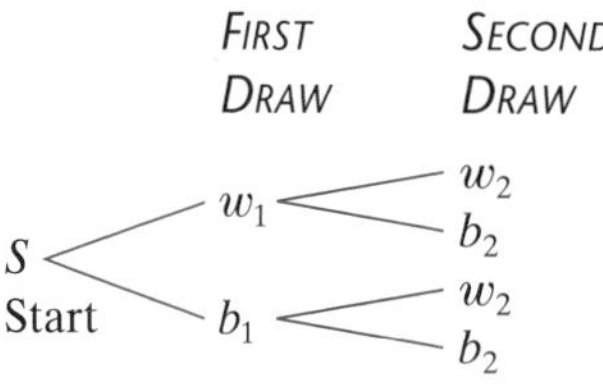

FIGURE 5

We now assign a probability to each branch on the tree (Fig. 6). For example, we assign the probability $\frac{2}{5}$ to the branch Sw_1, since this is the probability of drawing a white ball on the first draw (there are 2 white balls and 3 blue balls in the box). What probability should be assigned to the branch w_1w_2? This is the conditional probability $P(w_2|w_1)$; that is, the probability of drawing a white ball on the second draw given that a white ball was drawn on the first draw and not replaced. Since the box now contains 1 white ball and

3 blue balls, the probability is $\frac{1}{4}$. Continuing in the same way, we assign probabilities to the other branches of the tree and obtain Figure 6.

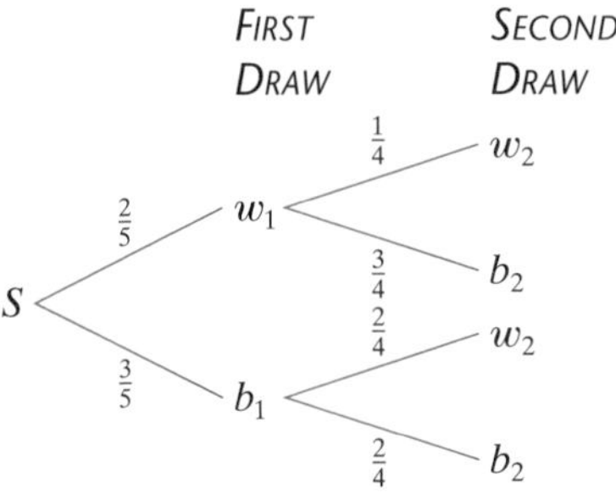

FIGURE 6

What is the probability of the combined outcome $w_1 \cap w_2$; that is, of drawing a white ball on the first draw and a white ball on the second draw?* Using the product rule (2), we have

$$\begin{aligned} P(w_1 \cap w_2) &= P(w_1)P(w_2|w_1) \\ &= \left(\tfrac{2}{5}\right)\left(\tfrac{1}{4}\right) = \tfrac{1}{10} \end{aligned}$$

The combined outcome $w_1 \cap w_2$ corresponds to the unique path Sw_1w_2 in the tree diagram, and we see that the probability of reaching w_2 along this path is just the product of the probabilities assigned to the branches on the path. Reasoning in the same way, we obtain the probability of each remaining combined outcome by multiplying the probabilities assigned to the branches on the path corresponding to the given combined outcomes. These probabilities are often written at the ends of the paths to which they correspond (Fig. 7).

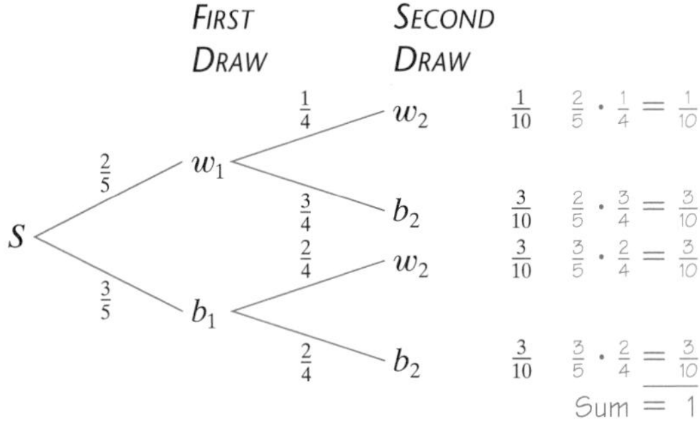

FIGURE 7

Now it is an easy matter to complete the problem. A white ball drawn on the second draw corresponds to either the combined outcome $w_1 \cap w_2$ or $b_1 \cap w_2$ occurring. Thus, since these combined outcomes are mutually exclusive,

$$\begin{aligned} P(w_2) &= P(w_1 \cap w_2) + P(b_1 \cap w_2) \\ &= \tfrac{1}{10} + \tfrac{3}{10} = \tfrac{4}{10} = \tfrac{2}{5} \end{aligned}$$

which is just the sum of the probabilities listed at the ends of the two paths terminating in w_2.

---

*The sample space for the combined outcomes is $S = \{w_1w_2, w_1b_2, b_1w_2, b_1b_2\}$. If we let $w_1 = \{w_1w_2, w_1b_2\}$ and $w_2 = \{w_1w_2, b_1w_2\}$, then $w_1 \cap w_2 = \{w_1w_2\}$.

Matched Problem 4 ➡ Two balls are drawn in succession without replacement from a box containing 4 red and 2 white balls. What is the probability of drawing a red ball on the second draw?

The sequence of two experiments in Example 4 is an example of a *stochastic process.* In general, a **stochastic process** involves a sequence of experiments where the outcome of each experiment is not certain. Our interest is in making predictions about the process as a whole. The analysis in Example 4 generalizes to stochastic processes involving any finite sequence of experiments. We summarize the procedures used in Example 4 for general application:

Constructing Probability Trees

Step 1. Draw a tree diagram corresponding to all combined outcomes of the sequence of experiments.

Step 2. Assign a probability to each tree branch. (This is the probability of the occurrence of the event on the right end of the branch subject to the occurrence of all events on the path leading to the event on the right end of the branch. The probability of the occurrence of a combined outcome that corresponds to a path through the tree is the product of all branch probabilities on the path.*)

Step 3. Use the results in steps 1 and 2 to answer various questions related to the sequence of experiments as a whole.

*If we form a sample space S such that each simple event in S corresponds to one path through the tree, and if the probability assigned to each simple event in S is the product of the branch probabilities on the corresponding path, then it can be shown that this is not only an acceptable assignment (all probabilities for the simple events in S are nonnegative and their sum is 1), but it is the only assignment consistent with the method used to assign branch probabilities within the tree.

Explore–Discuss 2

Refer to the table on rain and accidents in Example 2 and use formula (1), where appropriate, to complete the following probability tree:

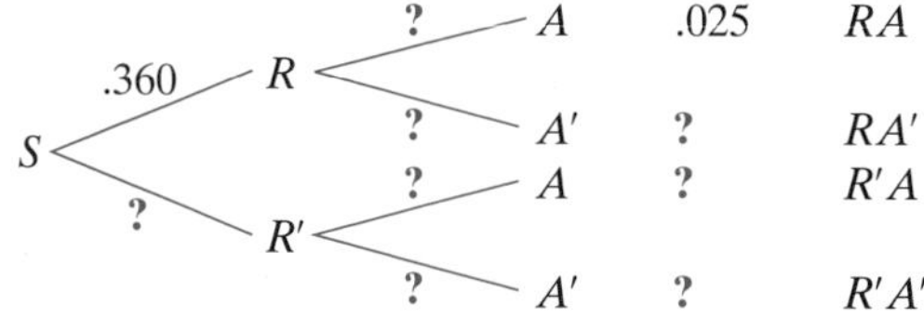

Discuss the difference between $P(R \cap A)$ and $P(A|R)$.

Example 5 ➡ **Product Defects** A large computer company A subcontracts the manufacturing of its circuit boards to two companies, 40% to company B and 60% to company C. Company B in turn subcontracts 70% of the orders it receives from company A to company D and the remaining 30% to company E, both subsidiaries of company B. When the boards are completed by companies D, E, and C, they are shipped to company A to be used in various computer models. It has been found that 1.5%, 1%, and .5% of the boards

from D, E, and C, respectively, prove defective during the 90 day warranty period after a computer is first sold. What is the probability that a given board in a computer will be defective during the 90 day warranty period?

SOLUTION Draw a tree diagram and assign probabilities to each branch (Fig. 8):

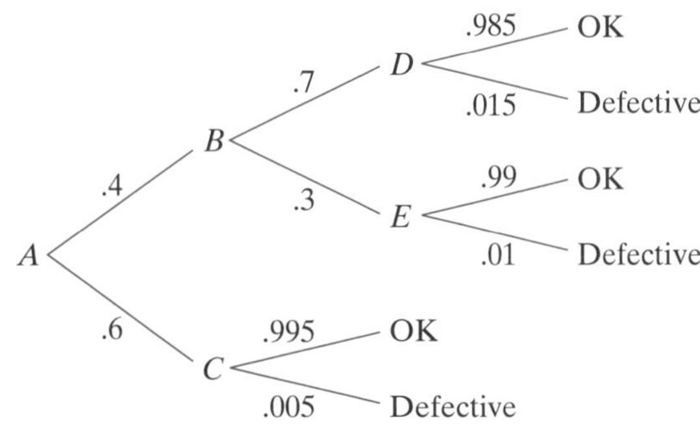

FIGURE 8

There are three paths leading to defective (the board will be defective within the 90 day warranty period). We multiply the branch probabilities on each path and add the three products:

$$P(\text{Defective}) = (.4)(.7)(.015) + (.4)(.3)(.01) + (.6)(.005)$$
$$= .0084$$

Matched Problem 5 In Example 5, what is the probability that a circuit board in a completed computer came from company E or C?

INDEPENDENT EVENTS

We return to Example 4, which involved drawing two balls in succession without replacement from a box of 3 blue and 2 white balls (Fig. 4). What difference does "without replacement" and "with replacement" make? Figure 9 shows probability trees corresponding to each case. Go over the probability assignments for the branches in part (B) to convince yourself of their correctness.

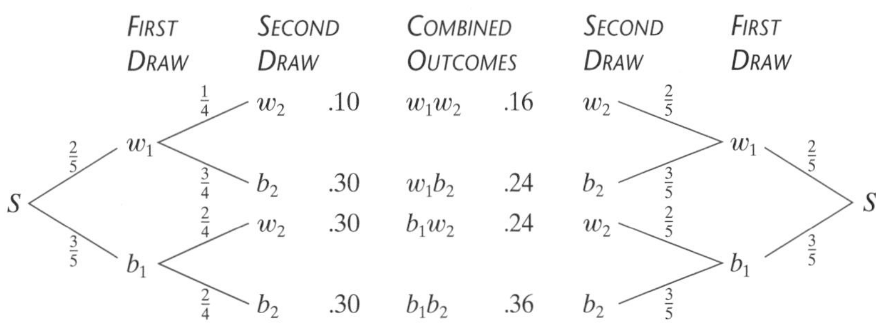

FIGURE 9

$S = \{w_1w_2, w_1b_2, b_1w_2, b_1b_2\}$

Let: A = White ball on second draw $= \{w_1w_2, b_1w_2\}$
B = White ball on first draw $= \{w_1w_2, w_1b_2\}$

We now compute $P(A|B)$ and $P(A)$ for each case in Figure 9.

Case 1. *Without Replacement:*

$$P(A|B) = \frac{P(A \cap B)}{P(B)} = \frac{P\{w_1w_2\}}{P\{w_1w_2, w_1b_2\}} = \frac{.10}{.10 + .30} = .25$$

(This is the assignment to branch w_1w_2 we made by looking in the box and counting.)

$$P(A) = P\{w_1w_2, b_1w_2\} = .10 + .30 = .40$$

Note that $P(A|B) \neq P(A)$, and we conclude that the probability of A is affected by the occurrence of B.

Case 2. *With Replacement:*

$$P(A|B) = \frac{P(A \cap B)}{P(B)} = \frac{P\{w_1w_2\}}{P\{w_1w_2, w_1b_2\}} = \frac{.16}{.16 + .24} = .40$$

(This is the assignment to branch w_1w_2 we made by looking in the box and counting.)

$$P(A) = P\{w_1w_2, b_1w_2\} = .16 + .24 = .40$$

Note that $P(A|B) = P(A)$, and we conclude that the probability of A is not affected by the occurrence of B.

Intuitively, if $P(A|B) = P(A)$, then it appears that event A is "independent" of B. Let us pursue this further. If events A and B are such that

$$P(A|B) = P(A)$$

then replacing the left side by its equivalent from equation (1), we obtain

$$\frac{P(A \cap B)}{P(B)} = P(A)$$

After multiplying both sides by $P(B)$, the last equation becomes

$$P(A \cap B) = P(A)P(B)$$

This result motivates the following definition of *independence:*

Independence

If A and B are any events in a sample space S, we say that **A and B are independent** if and only if

$$P(A \cap B) = P(A)P(B) \qquad (3)$$

Otherwise, A and B are said to be **dependent.**

From the definition of independence one can prove (see Problems 43 and 44 in Exercise 6-5) the following theorem:

Theorem 1 ■■ ON INDEPENDENCE

If A and B are independent events with nonzero probabilities in a sample space S, then

$$P(A|B) = P(A) \quad \text{and} \quad P(B|A) = P(B) \qquad (4)$$

If either equation in (4) holds, then A and B are independent. ■■

In practice, one often has correct intuitive feelings about independence. For example, if you toss a coin twice, the second toss is independent of the first (a coin has no memory); if a card is drawn from a deck twice, with replacement, the second draw is independent of the first (the deck has no memory); if you roll a pair of dice twice, the second roll is independent of the first (dice have no memory); and so on. However, there are pairs of events that are not obviously independent or dependent, so we need equations (3) or (4) to decide these cases. Example 6 considers two events that are obviously independent, and Example 7 considers events where independence or dependence is not obvious.

Example 6 **Testing for Independence** In two tosses of a single fair coin show that the events "A head on the first toss" and "A head on the second toss" are independent.

SOLUTION Consider the sample space of equally likely outcomes for the tossing of a fair coin twice,

$$S = \{\text{HH, HT, TH, TT}\}$$

and the two events,

$$A = \text{A head on the first toss} = \{\text{HH, HT}\}$$
$$B = \text{A head on the second toss} = \{\text{HH, TH}\}$$

Then

$$P(A) = \tfrac{2}{4} = \tfrac{1}{2} \qquad P(B) = \tfrac{2}{4} = \tfrac{1}{2} \qquad P(A \cap B) = \tfrac{1}{4}$$

Thus,

$$P(A \cap B) = \tfrac{1}{4} = \tfrac{1}{2} \cdot \tfrac{1}{2} = P(A)P(B)$$

and the two events are independent. (The theory agrees with our intuition—a coin has no memory.)

Matched Problem 6 In Example 6, compute $P(B|A)$ and compare with $P(B)$.

Example 7 **Testing for Independence** A single card is drawn from a standard 52-card deck. Test the following events for independence (try guessing the answer to each part before looking at the solution):

(A) E = The drawn card is a spade.
F = The drawn card is a face card.

(B) G = The drawn card is a club.
H = The drawn card is a heart.

SOLUTION (A) To test E and F for independence, we compute $P(E \cap F)$ and $P(E)P(F)$. If they are equal, then events E and F are independent; if they are not equal, then events E and F are dependent.

$$P(E \cap F) = \tfrac{3}{52} \qquad P(E)P(F) = \left(\tfrac{13}{52}\right)\left(\tfrac{12}{52}\right) = \tfrac{3}{52}$$

Events E and F are independent. (Did you guess this?)

(B) Proceeding as in part (A), we see that

$$P(G \cap H) = P(\varnothing) = 0 \qquad P(G)P(H) = \left(\tfrac{13}{52}\right)\left(\tfrac{13}{52}\right) = \tfrac{1}{16}$$

Events G and H are dependent. (Did you guess this?)

CAUTION

Students often confuse *mutually exclusive (disjoint) events* with *independent events.* One does not necessarily imply the other. In fact, it is not difficult to show (see Problem 47, Exercise 6-5) that any two mutually exclusive events A and B, with nonzero probabilities, are always dependent!

Matched Problem 7 A single card is drawn from a standard 52-card deck. Test the following events for independence:

(A) E = The drawn card is a red card.
F = The drawn card's number is divisible by 5 (face cards are not assigned values).

(B) G = The drawn card is a king.
H = The drawn card is a queen.

Explore–Discuss 3

In college basketball, would it be reasonable to assume that the following events are independent? Explain why or why not.

A = The Golden Eagles win in the first round of the NCAA tournament.
B = The Golden Eagles win in the second round of the NCAA tournament.

The notion of independence can be extended to more than two events:

Independent Set of Events

A set of events is said to be **independent** if for each finite subset $\{E_1, E_2, \ldots, E_k\}$

$$P(E_1 \cap E_2 \cap \cdots \cap E_k) = P(E_1)P(E_2) \cdot \cdots \cdot P(E_k) \qquad (5)$$

The next example makes direct use of this definition.

Example 8 **Computer Control Systems** A space shuttle has four independent computer control systems. If the probability of failure (during flight) of any one system is .001, what is the probability of failure of all four systems?

SOLUTION Let: E_1 = Failure of system 1 $\quad E_3$ = Failure of system 3
E_2 = Failure of system 2 $\quad E_4$ = Failure of system 4

Then, since events E_1, E_2, E_3, and E_4 are given to be independent,

$$\begin{aligned} P(E_1 \cap E_2 \cap E_3 \cap E_4) &= P(E_1)P(E_2)P(E_3)P(E_4) \\ &= (.001)^4 \\ &= .000\ 000\ 000\ 001 \end{aligned}$$

Matched Problem 8 A single die is rolled 6 times. What is the probability of getting the sequence, 1, 2, 3, 4, 5, 6?

Summary

The key results in this section are summarized in the following box for an overview and ease of reference:

Summary of Key Concepts

Conditional Probability

$$P(A|B) = \frac{P(A \cap B)}{P(B)} \qquad P(B|A) = \frac{P(B \cap A)}{P(A)}$$

[*Note*: $P(A|B)$ is a probability based on the new sample space B, while $P(A \cap B)$ and $P(B)$ are probabilities based on the original sample space S.]

Product Rule

$$P(A \cap B) = P(A)P(B|A) = P(B)P(A|B)$$

Independent Events

- A and B are independent if and only if

$$P(A \cap B) = P(A)P(B)$$

- If A and B are independent events with nonzero probabilities, then

$$P(A|B) = P(A) \quad \text{and} \quad P(B|A) = P(B)$$

- If A and B are events with nonzero probabilities and either $P(A|B) = P(A)$ or $P(B|A) = P(B)$, then A and B are independent.
- If $E_1, E_2, \ldots, E_n$ are independent, then

$$P(E_1 \cap E_2 \cap \cdots \cap E_n) = P(E_1)P(E_2) \cdot \cdots \cdot P(E_n)$$

Answers to Matched Problems

1. (A) .5 (B) .4

2. (A) $P(R') = .640$ (B) $P(A \cap R') = .015$

(C) $P(A|R') = \dfrac{P(A \cap R')}{P(R')} = .023$

3. $P(M \cap C) = P(M)P(C|M) = .32$

4. $\frac{2}{3}$ **5.** .72

6. $P(B|A) = \dfrac{P(A \cap B)}{P(A)} = \dfrac{\frac{1}{4}}{\frac{1}{2}} = \dfrac{1}{2} = P(B)$

7. (A) E and F are independent

(B) G and H are dependent

8. $(\frac{1}{6})^6 \approx .000\ 021\ 4$

EXERCISE 6-5

A *Given the probabilities in the table below for events in a sample space S, solve Problems 1–16 relative to these probabilities.*

	A	*B*	*C*	*Totals*
D	.10	.04	.06	.20
E	.40	.26	.14	.80
Totals	.50	.30	.20	1.00

In Problems 1–8, find each probability directly from the table.

1. $P(A)$ **2.** $P(C)$ **3.** $P(D)$

4. $P(E)$ **5.** $P(A \cap D)$ **6.** $P(A \cap E)$

7. $P(C \cap D)$ **8.** $P(C \cap E)$

In Problems 9–12 compute each probability using equation (1) and appropriate table values:

9. $P(A|D)$ **10.** $P(A|E)$ **11.** $P(C|D)$

12. $P(C|E)$

In Problems 13–16, test each pair of events for independence:

13. A and D **14.** A and E

15. C and D **16.** C and E

17. A fair coin is tossed 8 times.
(A) What is the probability of tossing a head on the 8th toss, given that the preceding 7 tosses were heads?
(B) What is the probability of getting 8 heads or 8 tails?

18. A fair die is rolled 5 times.
(A) What is the probability of getting a 6 on the 5th roll, given that a 6 turned up on the preceding 4 rolls?
(B) What is the probability that the same number turns up every time?

19. A pointer is spun once on the circular spinner shown at the top of the next column. The probability assigned to the pointer landing on a given integer (from 1 to 5) is the ratio of the area of the corresponding circular sector to the area of the whole circle, as given in the table:

e_i	1	2	3	4	5
$P(e_i)$	.3	.1	.2	.3	.1

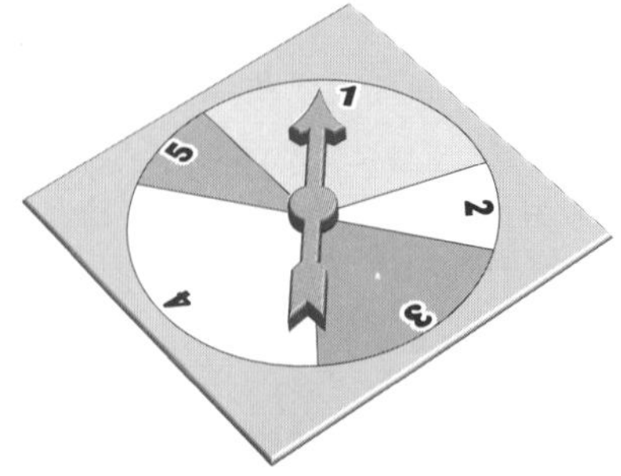

Given the events:

E = Pointer lands on an even number
F = Pointer lands on a number less than 4

(A) Find $P(F|E)$.
(B) Test events E and F for independence.

20. Repeat Problem 19 with the following events:

E = Pointer lands on an odd number
F = Pointer lands on a prime number

Compute the indicated probabilities in Problems 21 and 22 by referring to the following probability tree:

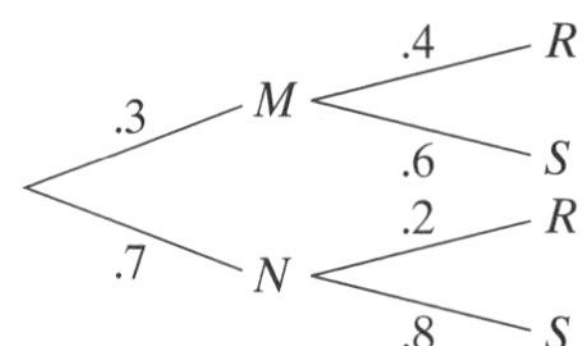

21. (A) $P(M \cap S)$ (B) $P(R)$

22. (A) $P(N \cap R)$ (B) $P(S)$

B

23. A fair coin is tossed twice. Let S = {HH, HT, TH, TT} be the sample space of equally likely simple events. We are interested in the following events:

E_1 = A head on the first toss
E_2 = A tail on the first toss
E_3 = A tail on the second toss
E_4 = A head on the second toss

For each pair of events, discuss whether they are independent and whether they are mutually exclusive.
(A) E_1 and E_4 (B) E_1 and E_2

24. For each pair of events (see Problem 23), discuss whether they are independent and whether they are mutually exclusive.
(A) E_1 and E_3 (B) E_3 and E_4

25. In 2 throws of a fair die, what is the probability that you will get an even number on each throw? An even number on the first or second throw?

26. In 2 throws of a fair die, what is the probability that you will get at least 5 on each throw? At least 5 on the first or second throw?

27. Two cards are drawn in succession from a standard 52-card deck. What is the probability that the first card is a club and the second card is a heart:
(A) If the cards are drawn without replacement?
(B) If the cards are drawn with replacement?

28. Two cards are drawn in succession from a standard 52-card deck. What is the probability that both cards are red:
(A) If the cards are drawn without replacement?
(B) If the cards are drawn with replacement?

29. A card is drawn at random from a standard 52-card deck. Events G and H are:

G = The drawn card is black.
H = The drawn card is divisible by 3 (face cards are not valued).

(A) Find $P(H|G)$.
(B) Test H and G for independence.

30. A card is drawn at random from a standard 52-card deck. Events M and N are:

M = The drawn card is a diamond.
N = The drawn card is even (face cards are not valued).

(A) Find $P(N|M)$.
(B) Test M and N for independence.

31. Let A be the event that all of a family's children are the same sex, and let B be the event that the family has at most 1 boy. Assuming the probability of having a girl is the same as the probability of having a boy (both .5), test events A and B for independence if:
(A) The family has 2 children.
(B) The family has 3 children.

32. An experiment consists of tossing n coins. Let A be the event that at least 2 heads turn up, and let B be the event that all the coins turn up the same. Test A and B for independence if:
(A) 2 coins are tossed. (B) 3 coins are tossed.

Problems 33–36 refer to the following experiment: 2 balls are drawn in succession out of a box containing 2 red and 5 white balls. Let R_i be the event that the ith ball is red, and let W_i be the event that the ith ball is white.

33. Construct a probability tree for this experiment and find the probability of each of the events $R_1 \cap R_2$, $R_1 \cap W_2$, $W_1 \cap R_2$, $W_1 \cap W_2$, given that the first ball drawn was:
(A) Replaced before the second draw
(B) Not replaced before the second draw

34. Find the probability that the second ball was red, given that the first ball was:
(A) Replaced before the second draw
(B) Not replaced before the second draw

35. Find the probability that at least 1 ball was red, given that the first ball was:
(A) Replaced before the second draw
(B) Not replaced before the second draw

36. Find the probability that both balls were the same color, given that the first ball was:
(A) Replaced before the second draw
(B) Not replaced before the second draw

In Problems 37 and 38, discuss the validity of each statement. If the statement is always true, explain why. If not, give a counterexample.

37. (A) If A and B are independent events, then

$$P(A|B) = P(B|A)$$

(B) If $P(A \cap B) = P(A)P(B|A)$, then A and B are independent.

38. (A) If two balls are drawn in succession, with replacement, from a box containing m red and n white balls ($m \geqslant 1$ and $n \geqslant 1$), then

$$P(W_1 \cap R_2) = P(R_1 \cap W_2)$$

(B) If two balls are drawn in succession, without replacement, from a box containing m red and n white balls ($m \geqslant 1$ and $n \geqslant 1$), then

$$P(W_1 \cap R_2) = P(R_1 \cap W_2)$$

C

39. A box contains 2 red, 3 white, and 4 green balls. Two balls are drawn out of the box in succession without replacement. What is the probability that both balls are the same color?

40. For the experiment in Problem 39, what is the probability that no white balls are drawn?

41. An urn contains 2 one-dollar bills, 1 five-dollar bill, and 1 ten-dollar bill. A player draws bills one at a time without replacement from the urn until a ten-dollar bill is drawn. Then the game stops. All bills are kept by the player.
(A) What is the probability of winning $16?
(B) What is the probability of winning all bills in the urn?
(C) What is the probability of the game stopping at the second draw?

42. Ann and Barbara are playing a tennis match. The first player to win 2 sets wins the match. For any given set, the probability that Ann wins that set is $\frac{2}{3}$. Find the probability that:
(A) Ann wins the match. (B) 3 sets are played.
(C) The player who wins the first set goes on to win the match.

43. Show that if A and B are independent events with nonzero probabilities in a sample space S, then

$$P(A|B) = P(A) \quad \text{and} \quad P(B|A) = P(B)$$

44. Show that if A and B are events with nonzero probabilities in a sample space S, and either $P(A|B) = P(A)$ or $P(B|A) = P(B)$, then events A and B are independent.

45. Show that $P(A|A) = 1$ when $P(A) \neq 0$.

46. Show that $P(A|B) + P(A'|B) = 1$.

47. Show that A and B are dependent if A and B are mutually exclusive and $P(A) \neq 0$, $P(B) \neq 0$.

48. Show that $P(A|B) = 1$ if B is a subset of A and $P(B) \neq 0$.

Applications

Business & Economics

49. *Labor relations.* In a study to determine employee voting patterns in a recent strike election, 1,000 employees were selected at random and the following tabulation was made:

		Salary Classification			
		Hourly (H)	*Salary (S)*	*Salary + Bonus (B)*	*Totals*
To Strike	*Yes (Y)*	400	180	20	600
	No (N)	150	120	130	400
	Totals	550	300	150	1,000

(A) Convert this table to a probability table by dividing each entry by 1,000.
(B) What is the probability of an employee voting to strike, given the person is paid hourly?
(C) What is the probability of an employee voting to strike, given the person receives a salary plus bonus?
(D) What is the probability of an employee being on straight salary (S)? Of being on straight salary given he or she voted in favor of striking?
(E) What is the probability of an employee being paid hourly? Of being paid hourly given he or she voted in favor of striking?
(F) What is the probability of an employee being in a salary plus bonus position and voting against striking?
(G) Are events S and Y independent?
(H) Are events H and Y independent?
(I) Are events B and N independent?

50. *Quality control.* An automobile manufacturer produces 37% of its cars at plant A. If 5% of the cars manufactured at plant A have defective emission control devices, what is the probability that one of this manufacturer's cars was manufactured at plant A and has a defective emission control device?

51. *Bonus incentives.* If a salesperson has gross sales of over \$600,000 in a year, he or she is eligible to play the company's bonus game: A black box contains 1 twenty-dollar bill, 2 five-dollar bills, and 1 one-dollar bill. Bills are drawn out of the box one at a time without replacement until a twenty-dollar bill is drawn. Then the game stops. The salesperson's bonus is 1,000 times the value of the bills drawn.
(A) What is the probability of winning a \$26,000 bonus?
(B) What is the probability of winning the maximum bonus, \$31,000, by drawing out all bills in the box?
(C) What is the probability of the game stopping at the third draw?

52. *Personnel selection.* To transfer into a particular technical department, a company requires an employee to pass a screening test. A maximum of 3 attempts are allowed at 6 month intervals between trials. From past records it is found that 40% pass on the first trial; of those that fail the first trial and take the test a second time, 60% pass; and of those that fail on the second trial and take the test a third time, 20% pass. For an employee wishing to transfer:
(A) What is the probability of passing the test on the first or second try?
(B) What is the probability of failing on the first 2 trials and passing on the third?
(C) What is the probability of failing on all 3 attempts?

Life Sciences

53. *Food and Drug Administration.* An ice cream company wishes to use a red dye to enhance the color in its strawberry ice cream. The Food and Drug Administration (FDA) requires the dye to be tested for cancer-producing potential in a laboratory using laboratory rats. The results of one test on 1,000 rats are summarized in the following table:

		DEVELOPED CANCER C	NO CANCER C'	*Totals*
ATE RED DYE	R	60	440	500
DID NOT EAT RED DYE	R'	20	480	500
	Totals	80	920	1,000

(A) Convert the table into a probability table by dividing each entry by 1,000.
(B) Are "developing cancer" and "eating red dye" independent events?
(C) Should the FDA approve or ban the use of the dye? Explain why or why not using $P(C|R)$ and $P(C)$.
(D) Suppose the number of rats that ate red dye and developed cancer was 20, but the number that developed cancer was still 80 and the number that ate red dye was still 500. What should the FDA do, based on these results? Explain why.

54. *Genetics.* In a study to determine frequency and dependency of color-blindness relative to females and males, 1,000 people were chosen at random and the following results were recorded:

		FEMALE F	MALE F'	*Totals*
COLOR-BLIND	C	2	24	26
NORMAL	C'	518	456	974
	Totals	520	480	1,000

(A) Convert this table to a probability table by dividing each entry by 1,000.
(B) What is the probability that a person is a woman, given that the person is color-blind?
(C) What is the probability that a person is color-blind, given that the person is a male?
(D) Are the events color-blindness and male independent?
(E) Are the events color-blindness and female independent?

Social Sciences

55. *Psychology.* In a study to determine the frequency and dependency of IQ ranges relative to males and females, 1,000 people were chosen at random and the following results were recorded:

		IQ Below 90 (A)	IQ 90–120 (B)	IQ Above 120 (C)	*Totals*
FEMALE	F	130	286	104	520
MALE	F'	120	264	96	480
	Totals	250	550	200	1,000

(A) Convert this table to a probability table by dividing each entry by 1,000.
(B) What is the probability of a person having an IQ below 90, given that the person is a female? A male?
(C) What is the probability of a person having an IQ above 120, given that the person is a female? A male?
(D) What is the probability of a person having an IQ below 90?
(E) What is the probability of a person having an IQ between 90 and 120? Of a person having an IQ between 90 and 120, given that the person is a male?
(F) What is the probability of a person being female and having an IQ above 120?
(G) Are any of the events A, B, or C dependent relative to F or F'?

56. *Voting patterns.* A survey of the residents of a precinct in a large city revealed that 55% of the residents were members of the Democratic party and that 60% of the Democratic party members voted in the last election. What is the probability that a person selected at random from the residents of this precinct is a member of the Democratic party and voted in the last election?

SECTION 6-6 Bayes' Formula

In the preceding section we discussed the conditional probability of the occurrence of an event, given the occurrence of an earlier event. Now we are going to reverse the problem and try to find the probability of an earlier event conditioned on the occurrence of a later event. As you will see before the discussion is over, a number of practical problems are of this form. First, let us consider a relatively simple problem that will provide the basis for a generalization.

Example 1 **Probability of an Earlier Event Given a Later Event** One urn has 3 blue and 2 white balls; a second urn has 1 blue and 3 white balls (Fig. 1). A single fair die is rolled and if 1 or 2 comes up, a ball is drawn out of the first urn; otherwise, a ball is drawn out of the second urn. If the drawn ball is blue, what is the probability that it came out of the first urn? Out of the second urn?

Urn U_1

Urn U_2

FIGURE 1

SOLUTION We form a probability tree, letting U_1 represent urn 1, U_2 urn 2, B a blue ball, and W a white ball. Then, on the various outcome branches we assign appropriate probabilities. For example, $P(U_1) = \frac{1}{3}$, $P(B|U_1) = \frac{3}{5}$, and so on:

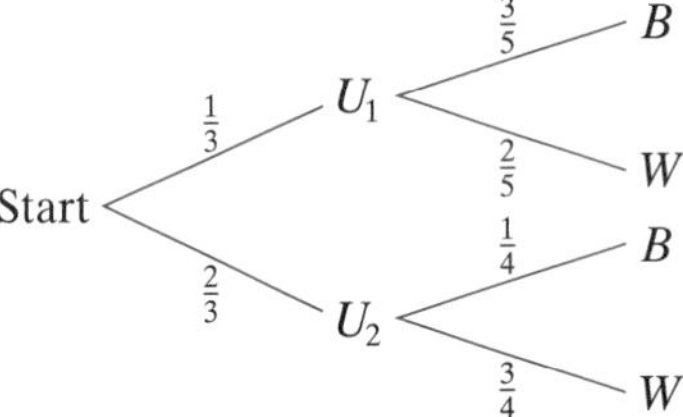

Now we are interested in finding $P(U_1|B)$; that is, the probability that the ball came out of urn 1, given the drawn ball is blue. Using equation (1) from Section 6-5, we can write

$$P(U_1|B) = \frac{P(U_1 \cap B)}{P(B)} \tag{1}$$

If we look at the tree diagram, we can see that B is at the end of two different branches; thus,

$$P(B) = P(U_1 \cap B) + P(U_2 \cap B) \tag{2}$$

After substituting equation (2) into equation (1), we get

$$\begin{aligned} P(U_1|B) &= \frac{P(U_1 \cap B)}{P(U_1 \cap B) + P(U_2 \cap B)} \qquad P(A \cap B) = P(A)P(B|A) \\ &= \frac{P(U_1)P(B|U_1)}{P(U_1)P(B|U_1) + P(U_2)P(B|U_2)} \\ &= \frac{P(B|U_1)P(U_1)}{P(B|U_1)P(U_1) + P(B|U_2)P(U_2)} \end{aligned} \tag{3}$$

Formula (3) is really a lot simpler to use than it looks. You do not need to memorize it; you simply need to understand its form relative to the probability tree above. Referring to the probability tree, we see that

$$\begin{aligned} P(B|U_1)P(U_1) &= \text{Product of branch probabilites leading to } B \text{ through } U_1 \\ &= \left(\tfrac{3}{5}\right)\left(\tfrac{1}{3}\right) \qquad \text{We usually start at } B \text{ and work back through } U_1. \\ P(B|U_2)P(U_2) &= \text{Product of branch probabilities leading to } B \text{ through } U_2 \\ &= \left(\tfrac{1}{4}\right)\left(\tfrac{2}{3}\right) \qquad \text{We usually start at } B \text{ and work back through } U_2. \end{aligned}$$

Equation (3) now can be interpreted in terms of the probability tree as follows:

$$P(U_1|B) = \frac{\text{Product of branch probabilities leading to } B \text{ through } U_1}{\text{Sum of all branch products leading to } B}$$

$$= \frac{(\frac{3}{5})(\frac{1}{3})}{(\frac{3}{5})(\frac{1}{3}) + (\frac{1}{4})(\frac{2}{3})} = \frac{6}{11} \approx .55$$

Similarly,

$$P(U_2|B) = \frac{\text{Product of branch probabilities leading to } B \text{ through } U_2}{\text{Sum of all branch products leading to } B}$$

$$= \frac{(\frac{1}{4})(\frac{2}{3})}{(\frac{3}{5})(\frac{1}{3}) + (\frac{1}{4})(\frac{2}{3})} = \frac{5}{11} \approx .45$$

[*Note*: We also could have obtained $P(U_2|B)$ by subtracting $P(U_1|B)$ from 1. Why?]

Matched Problem 1 Repeat Example 1, but find $P(U_1|W)$ and $P(U_2|W)$.

Explore–Discuss 1

Given the probability tree

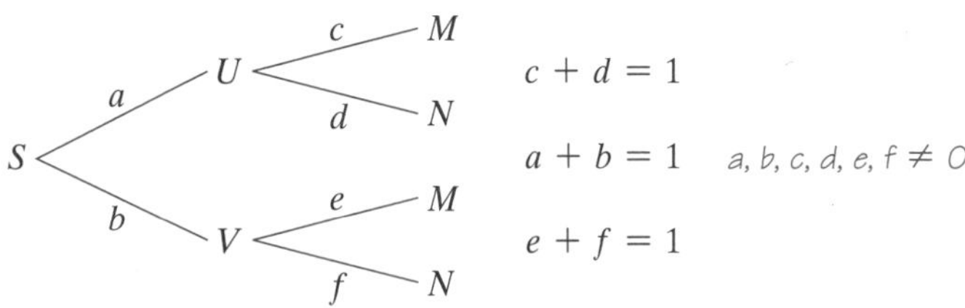

(A) Discuss the difference between $P(M|U)$ and $P(U|M)$, and between $P(N|V)$ and $P(V|N)$, in terms of *a, b, c, d, e,* and *f.*

(B) Show that $ac + ad + be + bf = 1$. What is the significance of this result?

In generalizing the results in Example 1, it is helpful to look at its structure in terms of the Venn diagram shown in Figure 2. We note that U_1 and U_2 are mutually exclusive (disjoint), and their union forms *S*. The following two equations can now be interpreted in terms of this diagram:

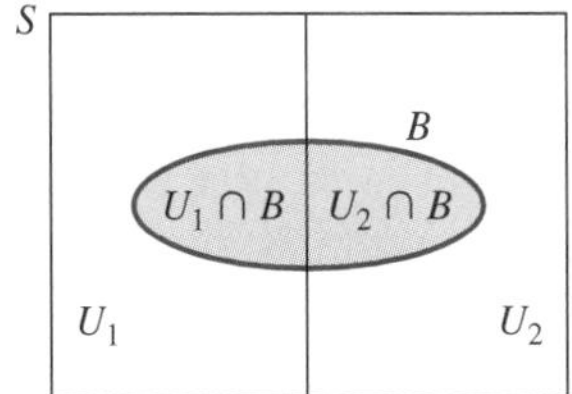

FIGURE 2

$$P(U_1|B) = \frac{P(U_1 \cap B)}{P(B)} = \frac{P(U_1 \cap B)}{P(U_1 \cap B) + P(U_2 \cap B)}$$

$$P(U_2|B) = \frac{P(U_2 \cap B)}{P(B)} = \frac{P(U_2 \cap B)}{P(U_1 \cap B) + P(U_2 \cap B)}$$

Look over the equations and the diagram carefully.

Of course, there is no reason to stop here. Suppose U_1, U_2, and U_3 are three mutually exclusive events whose union is the whole sample space *S*. Then, for an arbitrary event *E* in *S*, with $P(E) \neq 0$, the corresponding Venn diagram looks like Figure 3, and

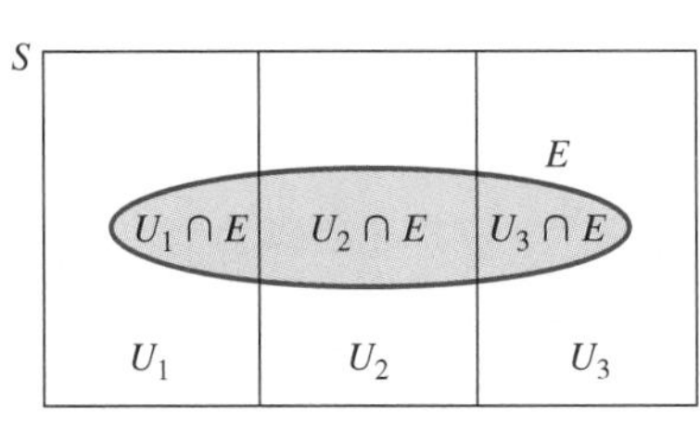

FIGURE 3

$$P(U_1|E) = \frac{P(U_1 \cap E)}{P(E)} = \frac{P(U_1 \cap E)}{P(U_1 \cap E) + P(U_2 \cap E) + P(U_3 \cap E)}$$

Similar results hold for U_2 and U_3.

Reasoning in the same way, we arrive at the following famous theorem, which was first stated by the Presbyterian minister Thomas Bayes (1702–1763):

Theorem 1 BAYES' FORMULA

Let $U_1, U_2, \ldots, U_n$ be n mutually exclusive events whose union is the sample space S. Let E be an arbitrary event in S such that $P(E) \neq 0$. Then

$$\begin{aligned} P(U_1|E) &= \frac{P(U_1 \cap E)}{P(E)} \\ &= \frac{P(U_1 \cap E)}{P(U_1 \cap E) + P(U_2 \cap E) + \cdots + P(U_n \cap E)} \\ &= \frac{P(E|U_1)P(U_1)}{P(E|U_1)P(U_1) + P(E|U_2)P(U_2) + \cdots + P(E|U_n)P(U_n)} \end{aligned}$$

Similar results hold for $U_2, U_3, \ldots, U_n$.

You do not need to memorize Bayes' formula. In practice, it is often easier to draw a probability tree and use the following:

Bayes' Formula and Probability Trees

$$P(U_1|E) = \frac{\text{Product of branch probabilities leading to } E \text{ through } U_1}{\text{Sum of all branch products leading to } E}$$

Similar results hold for $U_2, U_3, \ldots, U_n$.

Example 2 **Tuberculosis Screening** A new, inexpensive skin test is devised for detecting tuberculosis. To evaluate the test before it is put into use, a medical researcher randomly selects 1,000 people. Using precise but more expensive methods already available, it is found that 8% of the 1,000 people tested have tuberculosis. Now each of the 1,000 subjects is given the new skin test and the following results are recorded: The test indicates tuberculosis in 96% of those who have it and in 2% of those who do not. Based on these results, what is the probability of a randomly chosen person having tuberculosis given that the skin test indicates the disease? What is the probability of a person not having tuberculosis given that the skin test indicates the disease? (That is, what is the probability of the skin test giving a *false positive result?*)

SOLUTION Now we will see the power of Bayes' formula in an important application. To start, we form a tree diagram and place appropriate probabilities on each branch:

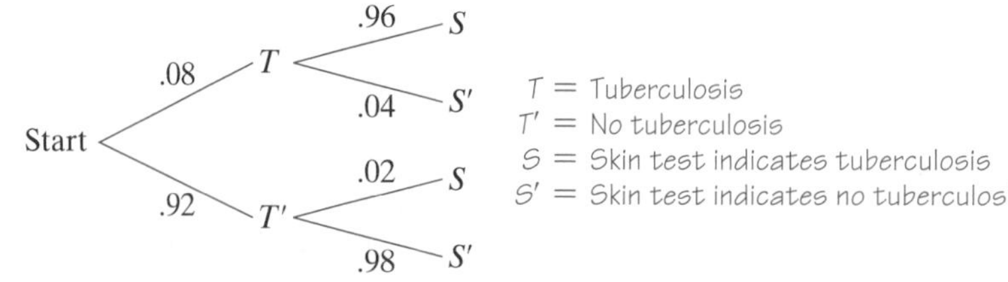

We are interested in finding $P(T|S)$; that is, the probability of a person having tuberculosis given that the skin test indicates the disease. Bayes' formula for this case is

$$P(T|S) = \frac{\text{Product of branch probabilities leading to } S \text{ through } T}{\text{Sum of all branch products leading to } S}$$

Substituting appropriate values from the probability tree, we obtain

$$P(T|S) = \frac{(.08)(.96)}{(.08)(.96) + (.92)(.02)} = .81$$

The probability of a person not having tuberculosis given that the skin test indicates the disease, denoted by $P(T'|S)$, is

$$P(T'|S) = 1 - P(T|S) = 1 - .81 = .19 \qquad P(T|S) + P(T'|S) = 1$$

Other important questions that need to be answered are indicated in Matched Problem 2.

Matched Problem 2 What is the probability that a person has tuberculosis given that the test indicates no tuberculosis is present? (That is, what is the probability of the skin test giving a *false negative result?*) What is the probability that a person does not have tuberculosis given that the test indicates no tuberculosis is present?

Example 3 **Product Defects** A company produces 1,000 refrigerators a week at three plants. Plant A produces 350 refrigerators a week, plant B produces 250 refrigerators a week, and plant C produces 400 refrigerators a week. Production records indicate that 5% of the refrigerators produced at plant A will be defective, 3% of those produced at plant B will be defective, and 7% of those produced at plant C will be defective. All the refrigerators are shipped to a central warehouse. If a refrigerator at the warehouse is found to be defective, what is the probability that it was produced at plant A?

Solution We begin by constructing a tree diagram:

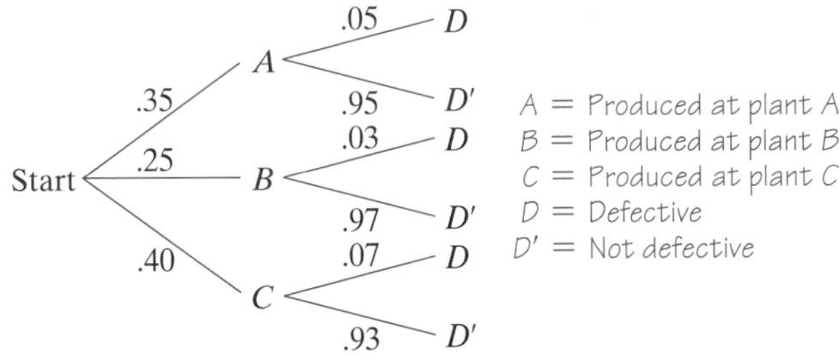

The probability that a defective refrigerator was produced at plant A is $P(A|D)$. Bayes' formula for this case is

$$P(A|D) = \frac{\text{Product of branch probabilities leading to } D \text{ through } A}{\text{Sum of all branch products leading to } D}$$

Using the values from the probability tree, we have

$$P(A|D) = \frac{(.35)(.05)}{(.35)(.05) + (.25)(.03) + (.40)(.07)} \approx .33$$

Matched Problem 3 In Example 3, what is the probability that a defective refrigerator in the warehouse was produced at plant *B*? At plant *C*?

Explore–Discuss 2

Given the probability tree:

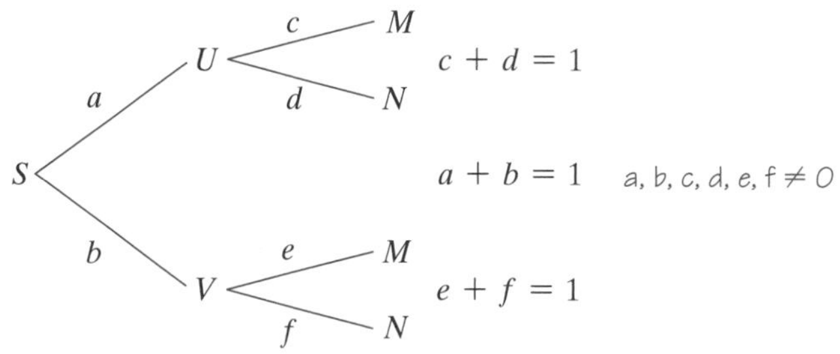

Suppose *U* and *M* are independent events. Discuss the restrictions that their independence forces on the probabilities *a, b, c, d, e,* and *f*.

Answers to Matched Problems

1. $P(U_1|W) = \frac{4}{19} \approx .21$; $P(U_2|W) = \frac{15}{19} \approx .79$
2. $P(T|S') = .004$; $P(T'|S') = .996$
3. $P(B|D) \approx .14$; $P(C|D) \approx .53$

EXERCISE 6-6

A *Find the probabilities in Problems 1–6 by referring to the tree diagram below.*

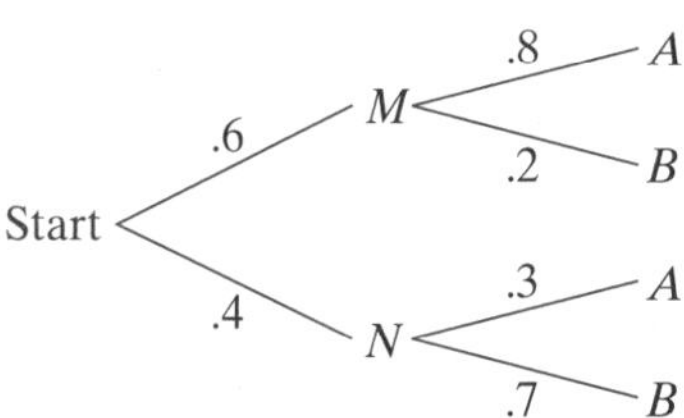

1. $P(M \cap A) = P(M)P(A|M)$
2. $P(N \cap B) = P(N)P(B|N)$
3. $P(A) = P(M \cap A) + P(N \cap A)$
4. $P(B) = P(M \cap B) + P(N \cap B)$
5. $P(M|A) = \dfrac{P(M \cap A)}{P(M \cap A) + P(N \cap A)}$
6. $P(N|B) = \dfrac{P(N \cap B)}{P(N \cap B) + P(M \cap B)}$

Find the probabilities in Problems 7–10 by referring to the following Venn diagram and using Bayes' formula (assume that the simple events in S are equally likely):

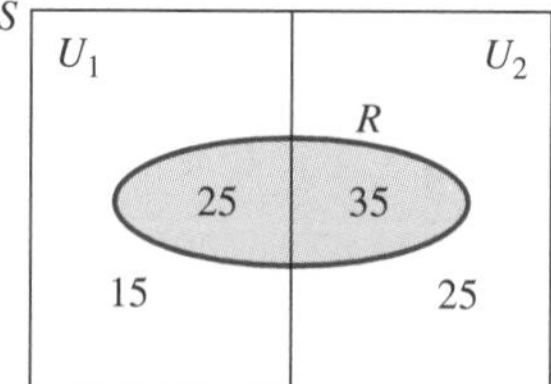

7. $P(U_1|R)$
8. $P(U_2|R)$
9. $P(U_1|R')$
10. $P(U_2|R')$

B *Find the probabilities in Problems 11–16 by referring to the following tree diagram and using Bayes' formula:*

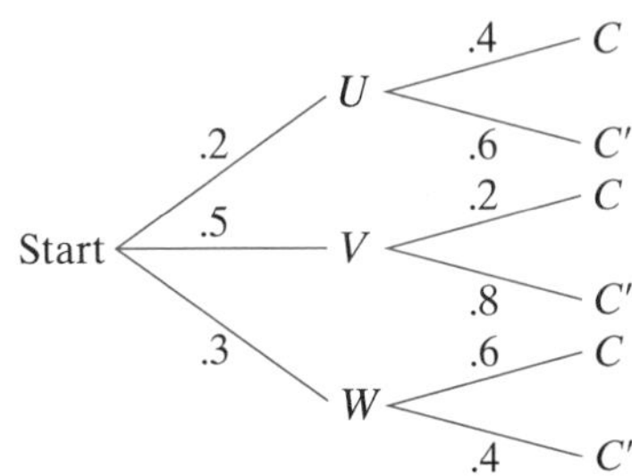

11. $P(U|C)$ **12.** $P(V|C')$ **13.** $P(W|C)$

14. $P(U|C')$ **15.** $P(V|C)$ **16.** $P(W|C')$

Find the probabilities in Problems 17–22 by referring to the following Venn diagram and using Bayes' formula (assume that the simple events in S are equally likely):

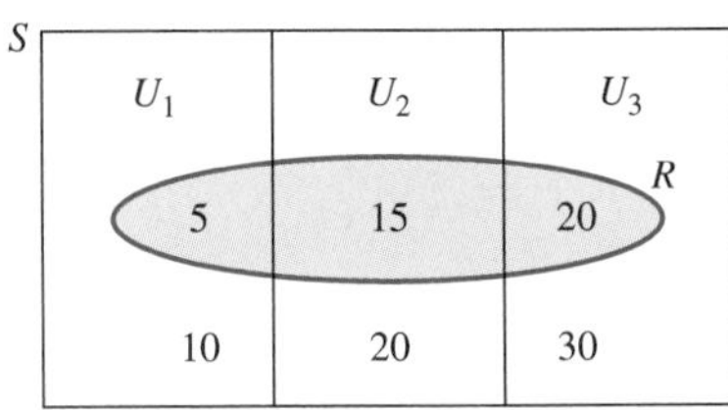

17. $P(U_1|R)$ **18.** $P(U_2|R')$ **19.** $P(U_3|R)$

20. $P(U_1|R')$ **21.** $P(U_2|R)$ **22.** $P(U_3|R')$

In Problems 23 and 24, use the probabilities in the first tree diagram to find the probability of each branch of the second tree diagram.

23.

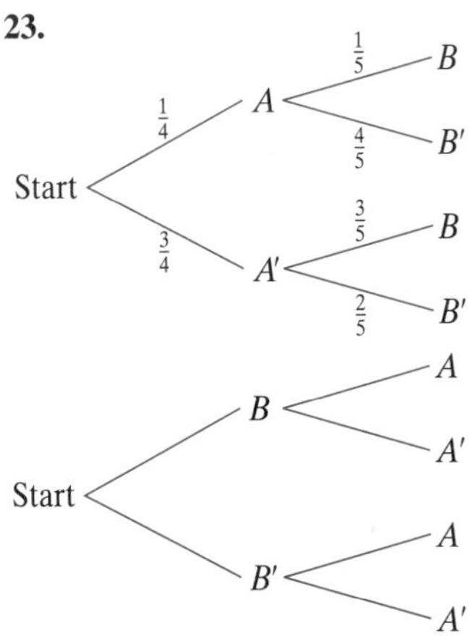

24.

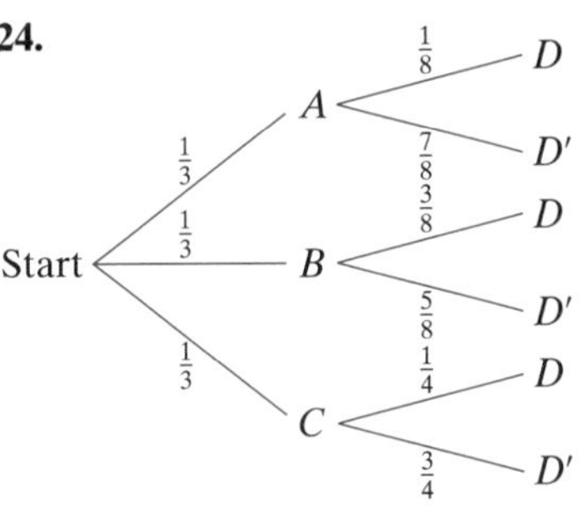

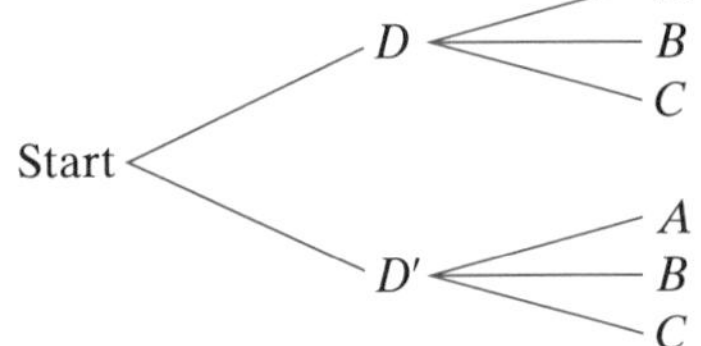

In Problems 25–28, one of two urns is chosen at random with one as likely to be chosen as the other. Then a ball is withdrawn from the chosen urn. Urn 1 contains 1 white and 4 red balls, and urn 2 has 3 white and 2 red balls.

25. If a white ball is drawn, what is the probability that it came from urn 1?

26. If a white ball is drawn, what is the probability that it came from urn 2?

27. If a red ball is drawn, what is the probability that it came from urn 2?

28. If a red ball is drawn, what is the probability that it came from urn 1?

In Problems 29 and 30, an urn contains 4 red and 5 white balls. Two balls are drawn in succession without replacement.

29. If the second ball is white, what is the probability that the first ball was white?

30. If the second ball is red, what is the probability that the first ball was red?

In Problems 31 and 32, urn 1 contains 7 red and 3 white balls. Urn 2 contains 4 red and 5 white balls. A ball is drawn from urn 1 and placed in urn 2. Then a ball is drawn from urn 2.

31. If the ball drawn from urn 2 is red, what is the probability that the ball drawn from urn 1 was red?

32. If the ball drawn from urn 2 is white, what is the probability that the ball drawn from urn 1 was white?

In Problems 33 and 34, refer to the following probability tree:

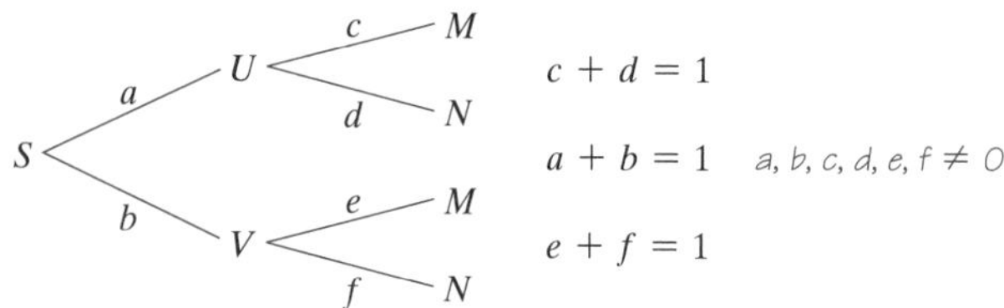

33. Suppose that $c = e$. Discuss the dependence or independence of events U and M.

34. Suppose that $c = d = e = f$. Discuss the dependence or independence of events M and N.

In Problems 35 and 36, two balls are drawn in succession from an urn containing m blue balls and n white balls ($m \geq 2$ and $n \geq 2$). Discuss the validity of each statement. If the statement is always true, explain why. If not, give a counterexample.

35. (A) If the two balls are drawn with replacement, then $P(B_1|B_2) = P(B_2|B_1)$.
    (B) If the two balls are drawn without replacement, then $P(B_1|B_2) = P(B_2|B_1)$.

36. (A) If the two balls are drawn with replacement, then $P(B_1|W_2) = P(W_2|B_1)$.
    (B) If the two balls are drawn without replacement, then $P(B_1|W_2) = P(W_2|B_1)$.

C

37. If 2 cards are drawn in succession from a standard 52-card deck without replacement and the second card is a heart, what is the probability that the first card is a heart?

38. A box contains 10 balls numbered 1 through 10. Two balls are drawn in succession without replacement. If the second ball drawn has the number 4 on it, what is the probability that the first ball had a smaller number on it? An even number on it?

39. Show that $P(U_1|R) + P(U_1'|R) = 1$.

40. If U_1 and U_2 are two mutually exclusive events whose union is the equally likely sample space S and if E is an arbitrary event in S such that $P(E) \neq 0$, show that

$$P(U_1|E) = \frac{n(U_1 \cap E)}{n(U_1 \cap E) + n(U_2 \cap E)}$$

Applications

In the following applications, the word "probability" is often understood to mean "approximate empirical probability."

Business & Economics

41. *Employee screening.* The management of a company finds that 30% of the secretaries hired are unsatisfactory. The personnel director is instructed to devise a test that will improve the situation. One hundred employed secretaries are chosen at random and are given a newly constructed test. Out of these, 90% of the successful secretaries pass the test and 20% of the unsuccessful secretaries pass. Based on these results, if a person applies for a secretarial job, takes the test, and passes it, what is the probability that he or she is a good secretary? If the applicant fails the test, what is the probability that he or she is a good secretary?

42. *Employee rating.* A company has rated 75% of its employees as satisfactory and 25% as unsatisfactory. Personnel records indicate that 80% of the satisfactory workers had previous work experience, while only 40% of the unsatisfactory workers had any previous work experience. If a person with previous work experience is hired, what is the probability that this person will be a satisfactory employee? If a person with no previous work experience is hired, what is the probability that this person will be a satisfactory employee?

43. *Product defects.* A manufacturer obtains clock–radios from three different subcontractors: 20% from *A*, 40% from *B*, and 40% from *C*. The defective rates for these subcontractors are 1%, 3%, and 2%, respectively. If a defective clock–radio is returned by a customer, what is the probability that it came from subcontractor *A*? From *B*? From *C*?

44. *Product defects.* A computer store sells three types of microcomputers, brand *A*, brand *B*, and brand *C*. Of the computers they sell, 60% are brand *A*, 25% are brand *B*, and 15% are brand *C*. They have found that 20% of the brand *A* computers, 15% of the brand *B* computers, and 5% of the brand *C* computers are returned for service during the warranty period. If a computer is returned for service during the warranty period, what is the probability that it is a brand *A* computer? A brand *B* computer? A brand *C* computer?

Life Sciences

45. *Cancer screening.* A new, simple test has been developed to detect a particular type of cancer. The test must be evaluated before it is put into use. A medical researcher selects a random sample of 1,000 adults and finds (by other means) that 2% have this type of cancer. Each of the 1,000 adults is given the test, and it is found that the test indicates cancer in 98% of those who have it and in 1% of those who do not. Based on these results, what is the probability of a randomly chosen person having cancer given that the test indicates cancer? Of a person having cancer given that the test does not indicate cancer?

46. *Pregnancy testing.* In a random sample of 200 women who suspect that they are pregnant, 100 turn out to be pregnant. A new pregnancy test given to these women indicated pregnancy in 92 of the 100 pregnant women and in 12 of the 100 nonpregnant women. If a woman suspects she is pregnant and this test indicates that she is pregnant, what is the probability that she is pregnant? If the test indicates that she is not pregnant, what is the probability that she is not pregnant?

47. *Medical research.* In a random sample of 1,000 people, it is found that 7% have a liver ailment. Of those who have a liver ailment, 40% are heavy drinkers, 50% are moderate drinkers, and 10% are nondrinkers. Of those who do not have a liver ailment, 10% are heavy drinkers, 70% are moderate drinkers, and 20% are nondrinkers. If a person is chosen at random and it is found that he or she is a heavy drinker, what is the probability of that person having a liver ailment? What is the probability for a nondrinker?

48. *Tuberculosis screening.* A test for tuberculosis was given to 1,000 subjects, 8% of whom were known to have tuberculosis. For the subjects who had tuberculosis, the test indicated tuberculosis in 90% of the subjects, was inconclusive for 7%, and indicated no tuberculosis in 3%. For the subjects who did not have tuberculosis, the test indicated tuberculosis in 5% of the subjects, was inconclusive for 10%, and indicated no tuberculosis in the remaining 85%. What is the probability of a randomly selected person having tuberculosis given that the test indicates tuberculosis? Of not having tuberculosis given that the test was inconclusive?

Social Sciences

49. *Police science.* A new lie-detector test has been devised and must be tested before it is put into use. One hundred people are selected at random, and each person draws and keeps a card from a box of 100 cards. Half the cards instruct the person to lie and the others instruct the person to tell the truth. The test indicates lying in 80% of those who lied and in 5% of those who did not. What is the probability that a randomly chosen subject will have lied given that the test indicates lying? That the subject will not have lied given that the test indicates lying?

50. *Politics.* In a given county, records show that of the registered voters, 45% are Democrats, 35% are Republicans, and 20% are independents. In an election, 70% of the Democrats, 40% of the Republicans, and 80% of the independents voted in favor of a parks and recreation bond proposal. If a registered voter chosen at random is found to have voted in favor of the bond, what is the probability that the voter is a Republican? An independent? A Democrat?

SECTION 6-7 Random Variable, Probability Distribution, and Expected Value

- RANDOM VARIABLE; PROBABILITY DISTRIBUTION
- EXPECTED VALUE OF A RANDOM VARIABLE
- DECISION-MAKING AND EXPECTED VALUE

■ RANDOM VARIABLE; PROBABILITY DISTRIBUTION

When performing a random experiment, a sample space S is selected in such a way that all probability problems of interest relative to the experiment can be solved. In many situations we may not be interested in each simple event in the sample space S but in some numerical value associated with the event.

For example, if 3 coins are tossed, we may be interested in the number of heads that turn up rather than in the particular pattern that turns up. Or, in selecting a random sample of students, we may be interested in the proportion that are women rather than which particular students are women. In the same way, a "craps" player is usually interested in the sum of the dots on the showing faces of the dice rather than the pattern of dots on each face.

In each of these examples, we have a rule that assigns to each simple event in S a single real number. Mathematically speaking, we are dealing with a function (see Section 1-1). Historically, this particular type of function has been called a "random variable."

Random Variable

A **random variable** is a function that assigns a numerical value to each simple event in a sample space S.

The term *random variable* is an unfortunate choice, since it is neither random nor a variable—it is a function with a numerical value, and it is defined on a sample space. But the terminology has stuck and is now standard, so we shall have to live with it. Capital letters, such as X, are used to represent random variables.

Table 1

Number of Heads in the Toss of 3 Coins

Sample Space S	Number of Heads $X(e_j)$
e_1: TTT	0
e_2: TTH	1
e_3: THT	1
e_4: HTT	1
e_5: THH	2
e_6: HTH	2
e_7: HHT	2
e_8: HHH	3

Let us return to the experiment of tossing 3 coins. A sample space S of equally likely simple events is indicated in Table 1. Suppose we are interested in the number of heads (0, 1, 2, or 3) appearing on each toss of the 3 coins and the probability of each of these events. We introduce a random variable X (a function) that indicates the number of heads for each simple event in S (see the second column in Table 1). For example, $X(e_1) = 0$, $X(e_2) = 1$, and so on. The random variable X assigns a numerical value to each simple event in the sample space S.

We are interested in the probability of the occurrence of each image or range value of X; that is, in the probability of the occurrence of 0 heads, 1 head, 2 heads, or 3 heads in the single toss of 3 coins. We indicate this probability by

$$p(x) \qquad \text{where} \quad x \in \{0, 1, 2, 3\}$$

The function p is called the **probability function* of the random variable X.**

What is $p(2)$, the probability of getting exactly 2 heads on the single toss of 3 coins? "Exactly 2 heads occur" is the event

$$E = \{\text{THH, HTH, HHT}\}$$

Thus,

$$p(2) = \frac{n(E)}{n(S)} = \frac{3}{8}$$

Proceeding similarly for $p(0)$, $p(1)$, and $p(3)$, we obtain the results in Table 2. This table is called a **probability distribution for the random variable X.** Probability distributions are also represented graphically, as shown in Figure 1. The graph of a probability distribution is often called a **histogram.**

*The probability function p of the random variable X is defined by $p(x) = P(\{e_i \in S \mid X(e_i) = x\})$, which, because of its cumbersome nature, is usually simplified to $p(x) = P(X = x)$ or, simply, $p(x)$. We will use the simplified notation.

Table 2

PROBABILITY DISTRIBUTION

NUMBER OF HEADS x	0	1	2	3
PROBABILITY $p(x)$	$\frac{1}{8}$	$\frac{3}{8}$	$\frac{3}{8}$	$\frac{1}{8}$

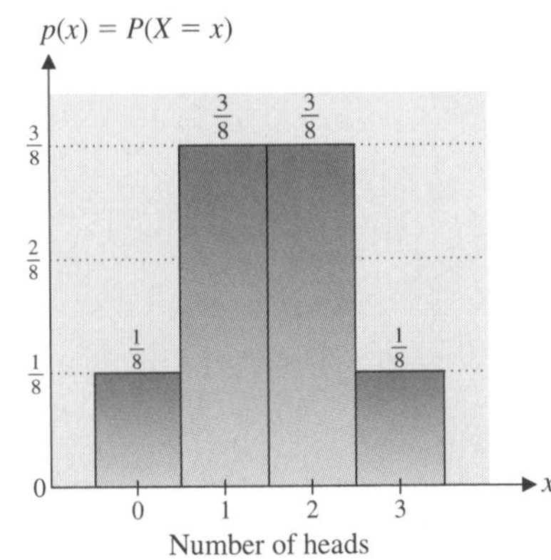

FIGURE 1
Histogram for a probability distribution

Note from Table 2 or Figure 1 that

1. $0 \leq p(x) \leq 1, \quad x \in \{0, 1, 2, 3\}$
2. $p(0) + p(1) + p(2) + p(3) = \frac{1}{8} + \frac{3}{8} + \frac{3}{8} + \frac{1}{8} = 1$

These are general properties that any probability distribution of a random variable X associated with a finite sample space must have.

Probability Distribution of a Random Variable X

A probability function $P(X = x) = p(x)$ is a **probability distribution of the random variable X** if

1. $0 \leq p(x) \leq 1, \quad x \in \{x_1, x_2, \ldots, x_n\}$
2. $p(x_1) + p(x_2) + \cdots + p(x_n) = 1$

where $\{x_1, x_2, \ldots, x_n\}$ are the (range) values of X (see Fig. 2).

Figure 2 illustrates the process of forming a probability distribution of a random variable.

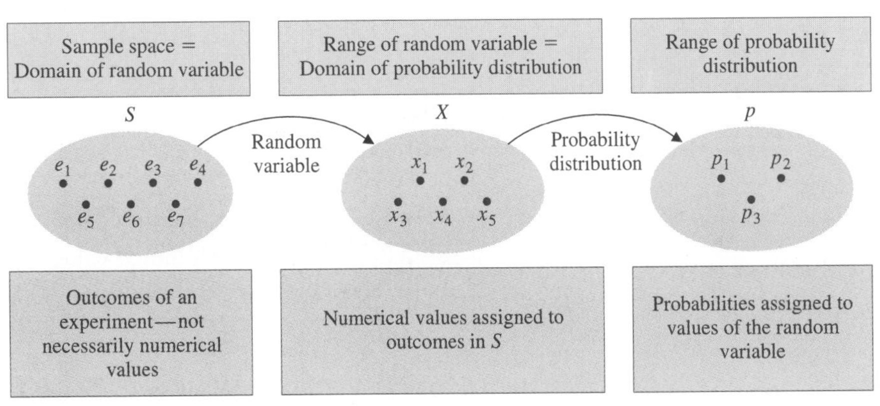

FIGURE 2
Probability distribution of a random variable for a finite sample space

Expected Value of a Random Variable

Suppose the experiment of tossing 3 coins was repeated a large number of times. What would be the average number of heads per toss (the total number of heads in all tosses divided by the total number of tosses)? Consulting the probability distribution in Table 2 or Figure 1, we see that we would expect to toss 0 heads $\frac{1}{8}$ of the time, 1 head $\frac{3}{8}$ of the time, 2 heads $\frac{3}{8}$ of the time, and 3 heads $\frac{1}{8}$ of the time. Thus, in the long run, we would expect the average number of heads per toss of the 3 coins, or the *expected value* $E(X)$, to given by

$$E(X) = 0(\tfrac{1}{8}) + 1(\tfrac{3}{8}) + 2(\tfrac{3}{8}) + 3(\tfrac{1}{8}) = \tfrac{12}{8} = 1.5$$

It is important to note that the expected value is not a value that will necessarily occur in a single experiment (1.5 heads cannot occur in the toss of 3 coins), but it is an average of what occurs over a large number of experiments. Sometimes we will toss more than 1.5 heads and sometimes less, but if the experiment is repeated many times, the average number of heads per experiment should be close to 1.5.

We now make the above discussion more precise through the following definition of expected value:

Expected Value of a Random Variable X

Given the probability distribution for the random variable X,

x_i	x_1	x_2	$\dots$	x_n
p_i	p_1	p_2	$\dots$	p_n

where $p_i = p(x_i)$, we define the **expected value of X,** denoted **$E(X)$,** by the formula

$$E(X) = x_1p_1 + x_2p_2 + \cdots + x_np_n$$

We again emphasize that the expected value is not to be expected to occur in a single experiment; it is a long-run average of repeated experiments—it is the weighted average of the possible outcomes, each weighted by its probability.

Steps for Computing the Expected Value of a Random Variable X

Step 1. Form the probability distribution of the random variable X.

Step 2. Multiply each image value of X, x_i, by its corresponding probability of occurrence p_i; then add the results.

Example 1 ➡ **Expected Value** What is the expected value (long-run average) of the number of dots facing up for the roll of a single die?

Solution If we choose

$$S = \{1, 2, 3, 4, 5, 6\}$$

as our sample space, then each simple event is a numerical outcome reflecting our interest, and each is equally likely. The random variable X in this case is just the identity function (each number is associated with itself). Thus, the probability distribution for X is

x_i	1	2	3	4	5	6
p_i	$\frac{1}{6}$	$\frac{1}{6}$	$\frac{1}{6}$	$\frac{1}{6}$	$\frac{1}{6}$	$\frac{1}{6}$

Hence,

$$E(X) = 1(\tfrac{1}{6}) + 2(\tfrac{1}{6}) + 3(\tfrac{1}{6}) + 4(\tfrac{1}{6}) + 5(\tfrac{1}{6}) + 6(\tfrac{1}{6})$$
$$= \tfrac{21}{6} = 3.5$$

Matched Problem 1 Suppose the die in Example 1 is not fair and we obtain (empirically) the following probability distribution for X:

x_i	1	2	3	4	5	6
p_i	.14	.13	.18	.20	.11	.24

[*Note:* Sum = 1.]

What is the expected value of X?

Explore–Discuss 1

In a class of 10 students, 1 student scored 95 on the first exam, 3 scored 85, 1 scored 82, 2 scored 75, 2 scored 73, and 1 scored 65. The probability distribution of an exam score for a student selected at random from the class is as follows:

x_i	65	73	75	82	85	95
p_i	.1	.2	.2	.1	.3	.1

Discuss the relationship between the expected value of the probability distribution and the class average for the exam.

Example 2 **Expected Value** A carton of 20 calculator batteries contains 2 dead ones. A random sample of 3 is selected from the 20 and tested. Let X be the random variable associated with the number of dead batteries found in a sample.

(A) Find the probability distribution of X.
(B) Find the expected number of dead batteries in a sample.

Solution (A) The number of ways of selecting a sample of 3 from 20 (order is not important) is $C_{20,3}$. This is the number of simple events in the experiment, each as likely as the other. A sample will have either 0, 1, or 2 dead batteries. These are the values of the random variable in which we are interested. The probability distribution is computed as follows:

$$p(0) = \frac{C_{18,3}}{C_{20,3}} \approx .716 \qquad p(1) = \frac{C_{2,1}C_{18,2}}{C_{20,3}} \approx .268 \qquad p(2) = \frac{C_{2,2}C_{18,1}}{C_{20,3}} \approx .016$$

We summarize the above results in a convenient table:

x_i	0	1	2
p_i	.716	.268	.016

[*Note:* $.716 + .268 + .016 = 1.$]

(B) The expected number of dead batteries in a sample is readily computed as follows:

$$E(X) = (0)(.716) + (1)(.268) + (2)(.016) = .3$$

The expected value is not one of the random variable values; rather, it is a number that the average number of dead batteries in a sample would approach as the experiment is repeated without end.

Matched Problem 2 Repeat Example 2 using a random sample of 4.

Example 3 **Expected Value of a Game** A spinner device is numbered from 0 to 5, and each of the 6 numbers is as likely to come up as any other. A player who bets \$1 on any given number wins \$4 (and gets the bet back) if the pointer comes to rest on the chosen number; otherwise, the \$1 bet is lost. What is the expected value of the game (long-run average gain or loss per game)?

Solution The sample space of equally likely events is

$$S = \{0, 1, 2, 3, 4, 5\}$$

Each sample point occurs with a probability of $\frac{1}{6}$. The random variable X assigns \$4 to the winning number and −\$1 to each of the remaining numbers. Thus, the probability distribution for X, called a **payoff table,** is as shown in the margin. The probability of winning \$4 is $\frac{1}{6}$ and of losing \$1 is $\frac{5}{6}$. We can now compute the expected value of the game:

Payoff Table (Probability Distribution for X)

x_i	\$4	−\$1
p_i	$\frac{1}{6}$	$\frac{5}{6}$

$$E(X) = \$4\left(\tfrac{1}{6}\right) + (-\$1)\left(\tfrac{5}{6}\right) = -\$\tfrac{1}{6} \approx -\$0.1667 \approx -17¢ \text{ per game}$$

Thus, in the long run the player will lose an average of about 17¢ per game.

Using the definition of a fair game in Section 6-4, it can be shown that **a game is fair if and only if $E(X) = 0$.** The game in Example 3 is not fair.

Matched Problem 3 Repeat Example 3 with the player winning \$5 instead of \$4 if the chosen number turns up. The loss is still \$1 if any other number turns up. Is this now a fair game?

Explore–Discuss 2

A coin is tossed twice and the number of heads recorded. You pay \$1 to play. You keep your dollar if no heads turn up and lose your dollar if exactly 1 head turns up. What should the house pay if 2 heads turn up in order for the game to be fair? Discuss the process and reasoning used to arrive at your answer.

Example 4 **Expected Value and Insurance** Suppose you are interested in insuring a car stereo system for \$500 against theft. An insurance company charges a premium of \$60 for coverage for 1 year, claiming an empirically determined probability of .1 that the stereo will be stolen some time during the year. What is your expected return from the insurance company if you take out this insurance?

SOLUTION This is actually a game of chance in which your stake is \$60. You have a .1 chance of receiving \$440 from the insurance company (\$500 minus your stake of \$60) and a .9 chance of losing your stake of \$60. What is the expected value of this "game"? We form a payoff table (the probability distribution for X) as shown in the margin. Then we compute the expected value as follows:

PAYOFF TABLE

x_i	\$440	−\$60
p_i	.1	.9

$$E(X) = (\$440)(.1) + (-\$60)(.9) = -\$10$$

This means that if you insure with this company over many years and the circumstances remain the same, you would have an average net loss to the insurance company of \$10 per year.

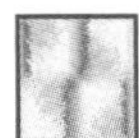

Matched Problem 4 Find the expected value in Example 4 from the insurance company's point of view.

■ DECISION-MAKING AND EXPECTED VALUE

We conclude this section with an example in decision-making.

Example 5 **Decision Analysis** An outdoor concert featuring a very popular musical group is scheduled for a Sunday afternoon in a large open stadium. The promoter, worrying about being rained out, contacts a long-range weather forecaster who predicts the chance of rain on that Sunday to be .24. If it does not rain, the promoter is certain to net \$100,000; if it does rain, the promoter estimates that the net will be only \$10,000. An insurance company agrees to insure the concert for \$100,000 against rain at a premium of \$20,000. Should the promoter buy the insurance?

SOLUTION The promoter has a choice between two courses of action; A_1: Insure and A_2: Do not insure. As an aid in making a decision, the expected value is computed for each course of action. Probability distributions are indicated in the following payoff table (read vertically):

PAYOFF TABLE

	A_1: INSURE	A_2: DO NOT INSURE
p_i	x_i	x_i
.24 (rain)	\$90,000	\$ 10,000
.76 (no rain)	\$80,000	\$100,000

Note that the \$90,000 entry comes from the insurance company's payoff (\$100,000) minus the premium (\$20,000) plus gate receipts (\$10,000). The reasons for the other entries should be obvious. The expected value for each course of action is computed as follows:

A_1: INSURE

$$E(X) = x_1p_1 + x_2p_2 = (\$90{,}000)(.24) + (\$80{,}000)(.76) = \$82{,}400$$

A_2: DO NOT INSURE

$$E(X) = (\$10{,}000)(.24) + (\$100{,}000)(.76) = \$78{,}400$$

It appears that the promoter's best course of action is to buy the insurance at \$20,000. The promoter is using a long-run average to make a decision about a single event—a common practice in making decisions in areas of uncertainty.

Matched Problem 5 In Example 5, what is the insurance company's expected value if it writes the policy?

Answers to Matched Problems

1. $E(X) = 3.73$
2. (A)

x_i	0	1	2
p_i	.632	.337	.032*

*Note: Due to roundoff error, sum = 1.001 ≈ 1.

(B) .4

3. $E(X) =$ \$0; the game is fair
4. $E(X) = (-\$440)(.1) + (\$60)(.9) = \$10$ (This amount, of course, is necessary to cover expenses and profit.)
5. $E(X) = (-\$80{,}000)(.24) + (\$20{,}000)(.76) = -\$4{,}000$ (This means the insurance company had other information regarding the weather than the promoter; otherwise, the company would not have written this policy.)

EXERCISE 6-7

Where possible, construct a probability distribution or payoff table for a suitable random variable X; then complete the problem.

A

1. If the probability distribution for the random variable X is given in the table, what is the expected value of X?

x_i	−3	0	4
p_i	.3	.5	.2

2. If the probability distribution for the random variable X is given in the table, what is the expected value of X?

x_i	−2	−1	0	1	2
p_i	.1	.2	.4	.2	.1

3. In tossing 2 fair coins, what is the expected number of heads?
4. In a family with 2 children, excluding multiple births and assuming a boy is as likely as a girl at each birth, what is the expected number of boys?
5. A fair coin is flipped. If a head turns up, you win \$1. If a tail turns up, you lose \$1. What is the expected value of the game? Is the game fair?
6. Repeat Problem 5, assuming an unfair coin with the probability of a head being .55 and a tail being .45.

B

7. After paying \$4 to play, a single fair die is rolled and you are paid back the number of dollars corresponding to the number of dots facing up. For example, if a 5 turns up, \$5 is returned to you for a net gain, or payoff, of \$1; if a 1 turns up, \$1 is returned for a net gain of −\$3; and so on. What is the expected value of the game? Is the game fair?
8. Repeat Problem 7 with the same game costing \$3.50 for each play.

9. Two coins are flipped. You win \$2 if either 2 heads or 2 tails turn up; you lose \$3 if a head and a tail turn up. What is the expected value of the game?

10. In Problem 9, for the game to be fair, how much *should* you lose if a head and a tail turn up?

11. A friend offers the following game: She wins \$1 from you if on four rolls of a single die, a 6 turns up at least once; otherwise, you win \$1 from her. What is the expected value of the game to you? To her?

12. On three rolls of a single die, you will lose \$10 if a 5 turns up at least once, and you will win \$7 otherwise. What is the expected value of the game?

13. A single die is rolled once. You win \$5 if a 1 or 2 turns up and \$10 if a 3, 4, or 5 turns up. How much should you lose if a 6 turns up for the game to be fair? Describe the steps you took to arrive at your answer.

14. A single die is rolled once. You lose \$12 if a number divisible by 3 turns up. How much should you win if a number not divisible by 3 turns up for the game to be fair? Describe the process and reasoning used to arrive at your answer.

15. A pair of dice is rolled once. Suppose you lose \$10 if a 7 turns up and win \$11 if an 11 or 12 turns up. How much should you win or lose if any other number turns up in order for the game to be fair?

16. A coin is tossed three times. Suppose you lose \$3 if 3 heads appear, lose \$2 if 2 heads appear, and win \$3 if 0 heads appear. How much should you win or lose if 1 head appears in order for the game to be fair?

17. The payoff table for two courses of action, A_1 or A_2, is given below. Which of the two actions will produce the largest expected value? What is it?

	A_1	A_2
p_i	x_i	x_i
.1	−\$200	−\$100
.2	\$100	\$200
.4	\$400	\$300
.3	\$100	\$200

18. The payoff table for three possible courses of action is given below. Which of the three actions will produce the largest expected value? What is it?

	A_1	A_2	A_3
p_i	x_i	x_i	x_i
.2	\$ 500	\$ 400	\$ 300
.4	\$1,200	\$1,100	\$1,000
.3	\$1,200	\$1,800	\$1,700
.1	\$1,200	\$1,800	\$2,400

19. Roulette wheels in Nevada generally have 38 equally spaced slots numbered 00, 0, 1, 2, . . . , 36. A player who bets \$1 on any given number wins \$35 (and gets the bet back) if the ball comes to rest on the chosen number; otherwise, the \$1 bet is lost. What is the expected value of this game?

20. In roulette (see Problem 19) the numbers from 1 to 36 are evenly divided between red and black. A player who bets \$1 on black wins \$1 (and gets the bet back) if the ball comes to rest on black; otherwise (if the ball lands on red, 0 or 00), the \$1 bet is lost. What is the expected value of the game?

21. A game has an expected value to you of \$100. It costs \$100 to play, but if you win you receive \$100,000 (including your \$100 bet), for a net gain of \$99,900. What is the probability of winning? Would you play this game? Discuss the factors that would influence your decision.

22. A game has an expected value to you of −\$0.50. It costs \$2 to play, but if you win you receive \$20 (including your \$2 bet), for a net gain of \$18. What is the probability of winning? Would you play this game? Discuss the factors that would influence your decision.

C

23. Five thousand tickets are sold at \$1 each for a charity raffle. Tickets are to be drawn at random and monetary prizes awarded as follows: 1 prize of \$500; 3 prizes of \$100, 5 prizes of \$20, and 20 prizes of \$5. What is the expected value of this raffle if you buy 1 ticket?

24. Ten thousand raffle tickets are sold at \$2 each for a local library benefit. Prizes are awarded as follows: 2 prizes of \$1,000, 4 prizes of \$500, and 10 prizes of \$100. What is the expected value of this raffle if you purchase 1 ticket?

25. A box of 10 flashbulbs contains 3 defective bulbs. A random sample of 2 is selected and tested. Let X be the random variable associated with the number of defective bulbs in the sample.
(A) Find the probability distribution of X.
(B) Find the expected number of defective bulbs in a sample.

26. A box of 8 flashbulbs contains 3 defective bulbs. A random sample of 2 is selected and tested. Let X be the random variable associated with the number of defective bulbs in a sample.
(A) Find the probability distribution of X.
(B) Find the expected number of defective bulbs in a sample.

27. One thousand raffle tickets are sold at \$1 each. Three tickets will be drawn at random (without replacement) and each will pay \$200. Suppose you buy 5 tickets.
(A) Create a payoff table for 0, 1, 2, and 3 winning tickets among the 5 tickets you purchased. (If you do not have any winning tickets, you lose \$5; if you have 1 winning ticket, you net \$195, since your initial \$5 will not be returned to you; and so on.)
(B) What is the expected value of the raffle to you?

28. Repeat Problem 27 with the purchase of 10 tickets.

29. To simulate roulette on a graphing utility, a random integer between -1 and 36 is selected (-1 represents 00; see Problem 19). The command in Figure A simulates 200 games.

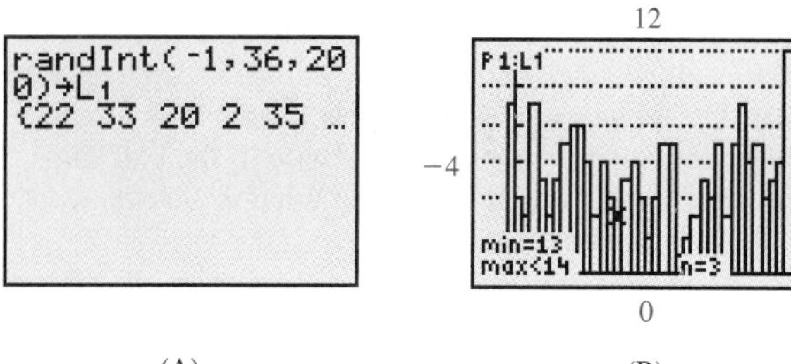

Figure for 29

(A) Use the statistical plot in Figure B to determine the net gain or loss of placing a $1 bet on the number 13 in each of the 200 games.

(B) Compare the results of part (A) with the expected value of the game.

(C) Use a graphing utility to simulate betting $1 on the number 7 in each of 500 games of roulette, and compare the simulated and expected gains or losses.

30. Use a graphing utility to simulate the results of placing a $1 bet on black in each of 400 games of roulette (see Problems 20 and 29), and compare the simulated and expected gains or losses.

Applications

Business & Economics

31. *Insurance.* The annual premium for a $5,000 insurance policy against the theft of a painting is $150. If the (empirical) probability that the painting will be stolen during the year is .01, what is your expected return from the insurance company if you take out this insurance?

32. *Insurance.* Repeat Problem 31 from the point of view of the insurance company.

33. *Decision analysis.* An oil company, after careful testing and analysis, is considering drilling in two different sites. It is estimated that site A will net $30 million if successful (probability .2) and lose $3 million if not (probability .8); site B will net $70 million if successful (probability .1) and lose $4 million if not (probability .9). Which site should the company choose according to the expected return from each site?

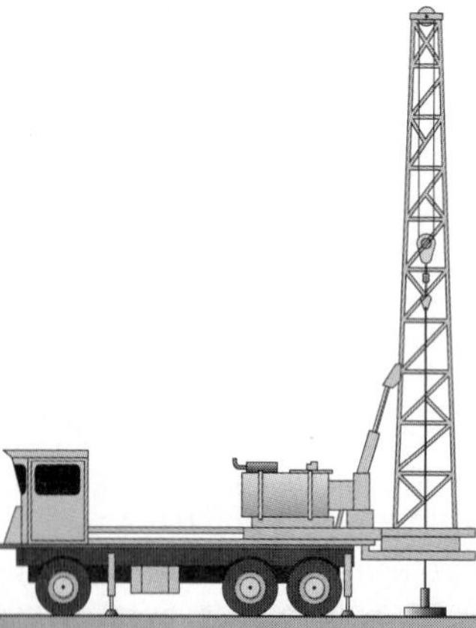

34. *Decision analysis.* Repeat Problem 33, assuming additional analysis caused the estimated probability of success in field B to be changed from .1 to .11.

Life Sciences

35. *Genetics.* Suppose that, at each birth, having a girl is not as likely as having a boy. The probability assignments for the number of boys in a 3-child family are approximated empirically from past records and are given in the table. What is the expected number of boys in a 3-child family?

Number of Boys x_i	p_i
0	.12
1	.36
2	.38
3	.14

36. *Genetics.* A pink-flowering plant is of genotype RW. If two such plants are crossed, we obtain a red plant (RR) with probability .25, a pink plant (RW or WR) with probability .50, and a white plant (WW) with probability .25, as shown in the table. What is the expected number of W genes present in a crossing of this type?

Number of W Genes Present x_i	p_i
0	.25
1	.50
2	.25

Social Sciences

37. *Politics.* A money drive is organized by a campaign committee for a candidate running for public office. Two approaches are considered:

A_1: A general mailing with a follow-up mailing
A_2: Door-to-door solicitation with follow-up telephone calls

From campaign records of previous committees, average donations and their corresponding probabilities are estimated to be:

A_1		A_2	
x_i *(Return per person)*	p_i	x_i *(Return per person)*	p_i
\$10	.3	\$15	.3
5	.2	3	.1
0	.5	0	.6
	1.0		1.0

What are the expected returns? Which course of action should be taken according to the expected returns?

Important Terms and Symbols

6-1 *Basic Counting Principles.* Counting technique; number of elements in a set; addition principle; tree diagram; multiplication principle

$$n(A);\quad n(A \cup B) = n(A) + n(B) - n(A \cap B);$$
$$n(A \cup B) = n(A) + n(B), \text{ if } A \cap B = \varnothing$$

6-2 *Permutations and Combinations.* n factorial; zero factorial; permutation; permutation of n objects; permutation of n objects taken r at a time; combination; combination of n objects taken r at a time

$$n! = n(n-1)(n-2) \cdot \cdots \cdot 2 \cdot 1;\quad 0! = 1;$$
$$P_{n,n} = n!;\quad P_{n,r} = \frac{n!}{(n-r)!};$$
$$C_{n,r} = \binom{n}{r} = \frac{P_{n,r}}{r!} = \frac{n!}{r!(n-r)!},\ 0 \le r \le n$$

6-3 *Sample Spaces, Events, and Probability.* Random experiment; experiment; sample space; event; simple outcome; simple event; compound event; fundamental sample space; probability of an event; acceptable probability assignment; reasonable probability assignment; probability function; frequency; relative frequency; empirical probability; approximate empirical probability; equally likely assumption

$$P(E);\quad P(E) = \frac{f(E)}{n};\quad n(E)$$

6-4 *Union, Intersection, and Complement of Events; Odds.* Union; intersection; event A **or** event B; event A **and** event B; mutually exclusive; disjoint; complement of an event; odds; fair game; law of large numbers

$$A \cup B;\quad A \cap B;$$
$$P(A \cup B) = P(A) + P(B) - P(A \cap B);$$
$$P(A \cup B) = P(A) + P(B), \text{ if } A \cap B = \varnothing;$$
$$A';\quad P(A') = 1 - P(A);\quad \frac{P(E)}{P(E')} \text{ (odds for } E)$$

6-5 *Conditional Probability, Intersection, and Independence.* Conditional probability of A given B; product rule; stochastic process; independent events; dependent events

$$P(A|B) = \frac{P(A \cap B)}{P(B)}, \text{ when } P(B) \ne 0;$$
$$P(A \cap B) = P(A)P(B|A);$$
$P(A \cap B) = P(A)P(B)$, if and only if A and B are independent;

$P(A|B) = P(A)$ and $P(B|A) = P(B)$, if A and B are independent and $P(A) \ne 0$, $P(B) \ne 0$;

$P(E_1 \cap E_2 \cap \cdots \cap E_n) = P(E_1)P(E_2) \cdot \cdots \cdot P(E_n)$, if $E_1, E_2, \ldots, E_n$ are independent

6-6 *Bayes' Formula.*

$$P(U_1|E) = \frac{P(E|U_1)P(U_1)}{P(E|U_1)P(U_1) + P(E|U_2)P(U_2) + \cdots + P(E|U_n)P(U_n)}$$
$$= \frac{\left(\text{Product of branch probabilities leading to } E \text{ through } U_1\right)}{\text{Sum of all branch products leading to } E}$$

6-7 *Random Variable, Probability Distribution, and Expected Value.* Random variable; probability function of a random variable X; probability distribution for a random variable; histogram; expected value of a random variable; payoff table; fair game

$$E(X) = x_1p_1 + x_2p_2 + \cdots + x_np_n$$

Review Exercise

Work through all the problems in this chapter review and check your answers in the back of the book. Answers to all review problems are there along with section numbers in italics to indicate where each type of problem is discussed. Where weaknesses show up, review appropriate sections in the text.

A

1. A single die is rolled and a coin is flipped. How many combined outcomes are possible? Solve:
   (A) By using a tree diagram
   (B) By using the multiplication principle

2. Use the Venn diagram to find the number of elements in each of the following sets:
   (A) A (B) B
   (C) $A \cap B$ (D) $A \cup B$
   (E) U (F) A'
   (G) $(A \cap B)'$ (H) $(A \cup B)'$

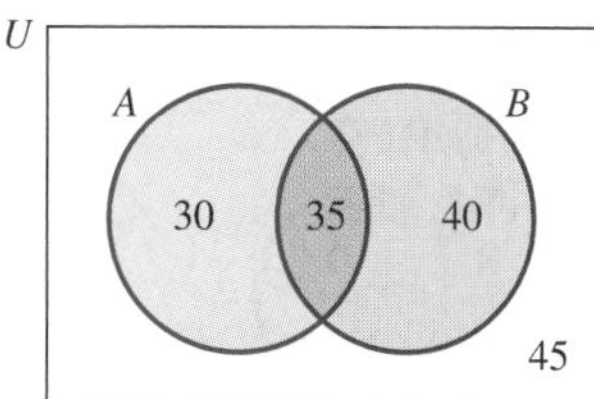

3. Evaluate $C_{6,2}$ and $P_{6,2}$.

4. How many seating arrangements are possible with 6 people and 6 chairs in a row? Solve by using the multiplication principle.

5. Solve Problem 4 using permutations or combinations, whichever is applicable.

6. In a single deal of 5 cards from a standard 52-card deck, what is the probability of being dealt 5 clubs?

7. Betty and Bill are members of a 15-person ski club. If the president and treasurer are selected by lottery, what is the probability that Betty will be president and Bill will be treasurer? (A person cannot hold more than one office.)

8. Each of the first 10 letters of the alphabet is printed on a separate card. What is the probability of drawing 3 cards and getting the code word *dig* by drawing *d* on the first draw, *i* on the second draw, and *g* on the third draw? What is the probability of being dealt a 3-card hand containing the letters *d, i,* and *g* in any order?

9. A drug has side effects for 50 out of 1,000 people in a test. What is the approximate empirical probability that a person using the drug will have side effects?

10. A spinning device has 5 numbers, 1, 2, 3, 4, and 5, each as likely to turn up as the other. A person pays $3 and then receives back the dollar amount corresponding to the number turning up on a single spin. What is the expected value of the game? Is the game fair?

11. If A and B are events in a sample space S and $P(A) = .3$, $P(B) = .4$, and $P(A \cap B) = .1$, find:
    (A) $P(A')$ (B) $P(A \cup B)$

12. A spinner lands on R with probability .3, on G with probability .5, and on B with probability .2. Find the probability and odds for the spinner landing on either R or G.

13. If in repeated rolls of two fair dice the odds for rolling a sum of 8 before rolling a sum of 7 are 5 to 6, then what is the probability of rolling a sum of 8 before rolling a sum of 7?

Answer Problems 14–22 using the table of probabilities shown below.

	X	Y	Z	*Totals*
S	.10	.25	.15	.50
T	.05	.20	.02	.27
R	.05	.15	.03	.23
Totals	.20	.60	.20	1.00

14. Find $P(T)$.
15. Find $P(Z)$.
16. Find $P(T \cap Z)$.
17. Find $P(R \cap Z)$.
18. Find $P(R|Z)$.
19. Find $P(Z|R)$.
20. Find $P(T|Z)$.
21. Are T and Z independent?
22. Are S and X independent?

Answer Problems 23–30 using the following probability tree:

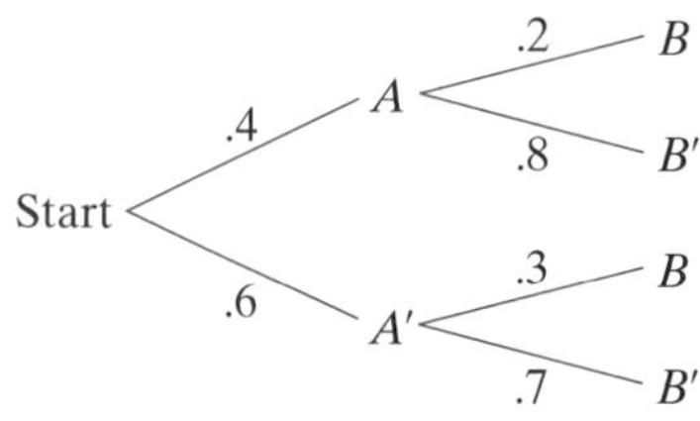

23. $P(A)$
24. $P(B|A)$
25. $P(B|A')$
26. $P(A \cap B)$
27. $P(A' \cap B)$
28. $P(B)$
29. $P(A|B)$
30. $P(A|B')$

31. (A) If 10 out of 32 students in a class were born in June, July, or August, what is the approximate empirical probability of any student being born in June, July, or August?
(B) If one is as likely to be born in any of the 12 months of a year as any other, what is the theoretical probability of being born in either June, July, or August?
(C) Discuss the discrepancy between the answers to parts (A) and (B).

In Problems 32 and 33, discuss the validity of each statement. If the statement is always true, explain why. If not, give a counterexample.

32. (A) If A or B is the empty set, then A and B are mutually exclusive.
(B) If A and B are mutually exclusive events, then

$$P(A \cap B) = P(A)P(B)$$

33. (A) If A and B are independent events, then

$$P(A \cup B) = P(A) + P(B)$$

(B) The events A and A' are independent.

B

34. A player tosses two coins and receives \$5 if 2 heads turn up, loses \$4 if 1 head turns up, and wins \$2 if 0 heads turn up. (Would you play this game?) Compute the expected value of the game. Is the game fair?

35. A spinning device has 3 numbers, 1, 2, and 3, each as likely to turn up as the other. If the device is spun twice, what is the probability that:
(A) The same number turns up both times?
(B) The sum of the numbers turning up is 5?

36. In a single draw from a standard 52-card deck, what are the probability and odds for drawing:
(A) A jack or a queen? (B) A jack or a spade?
(C) A card other than an ace?

37. (A) What are the odds for rolling a sum of 5 on the single roll of two fair dice?
(B) If you bet \$1 that a sum of 5 will turn up, what should the house pay (plus return your \$1 bet) for the game to be fair?

38. Two coins are flipped 1,000 times with the following frequencies:

2 heads	210
1 head	480
0 heads	310

(A) Compute the empirical probability for each outcome.
(B) Compute the theoretical probability for each outcome.
(C) Using the theoretical probabilities computed in part (B), compute the expected frequency of each outcome, assuming fair coins.

39. A man has 5 children. Each of those children has 3 children, who in turn each have 2 children. Discuss the number of descendants that the man has.

40. A fair coin is tossed 10 times. On each of the first 9 tosses the outcome is heads. Discuss the probability of a head on the 10th toss.

41. An experiment consists of rolling a pair of fair dice. Let X be the random variable associated with the sum of the values that turn up.
(A) Find the probability distribution for X.
(B) Find the expected value of X.

42. Two dice are rolled. The sample space is chosen as the set of all ordered pairs of integers taken from $\{1, 2, 3, 4, 5, 6\}$. What is the event A that corresponds to the sum being divisible by 4? What is the event B that corresponds to the sum being divisible by 6? What are $P(A), P(B), P(A \cap B)$, and $P(A \cup B)$?

43. A person tells you that the following approximate empirical probabilities apply to the sample space $\{e_1, e_2, e_3, e_4\}$: $P(e_1) \approx .1, P(e_2) \approx -.2, P(e_3) \approx .6, P(e_4) \approx 2$. There are three reasons why P cannot be a probability function. Name them.

44. Use the following information to complete the frequency table below:

$$n(A) = 50, \quad n(B) = 45,$$
$$n(A \cup B) = 80, \quad n(U) = 100$$

	A	A'	*Totals*
B			
B'			
Totals			

45. A pointer is spun on a circular spinner. The probabilities of the pointer landing on the integers from 1 to 5 are given in the table below.

e_i	1	2	3	4	5
p_i	.2	.1	.3	.3	.1

(A) What is the probability of the pointer landing on an odd number?
(B) What is the probability of the pointer landing on a number less than 4 given that it landed on an odd number?

46. A card is drawn at random from a standard 52-card deck. If E is the event "The drawn card is red" and F is the event "The drawn card is an ace," then:
(A) Find $P(F|E)$.
(B) Test E and F for independence.

47. How many 3-letter code words are possible using the first 8 letters of the alphabet if no letter can be repeated? If letters can be repeated? If adjacent letters cannot be alike?

48. Solve the following problems using $P_{n,r}$ or $C_{n,r}$:
(A) How many 3-digit opening combinations are possible on a combination lock with 6 digits if the digits cannot be repeated?
(B) Five tennis players have made the finals. If each of the 5 players is to play every other player exactly once, how many games must be scheduled?

49. C Use graphical techniques on a graphing utility to find the largest value of $C_{n,r}$ when $n = 25$.

In Problems 50–54, urn U_1 contains 2 white balls and 3 red balls; urn U_2 contains 2 white balls and 1 red ball.

50. Two balls are drawn out of urn U_1 in succession. What is the probability of drawing a white ball followed by a red ball if the first ball is:
(A) Replaced? (B) Not replaced?

51. Which of the two parts in Problem 50 involve dependent events?

52. In Problem 50, what is the expected number of red balls if the first ball is:
(A) Replaced? (B) Not replaced?

53. An urn is selected at random by flipping a fair coin; then a ball is drawn from the urn. Compute:
(A) $P(R|U_1)$ (B) $P(R|U_2)$
(C) $P(R)$ (D) $P(U_1|R)$
(E) $P(U_2|W)$ (F) $P(U_1 \cap R)$

54. In Problem 53, are the events "Selecting urn U_1" and "Drawing a red ball" independent?

55. From a standard deck of 52 cards, what is the probability of obtaining a 5-card hand:
(A) Of all diamonds?
(B) Of 3 diamonds and 2 spades?
Write answers in terms of $C_{n,r}$ or $P_{n,r}$; do not evaluate.

56. A group of 10 people includes one married couple. If 4 people are selected at random, what is the probability that the married couple is selected?

57. If 3 operations O_1, O_2, O_3 are performed in order, with possible number of outcomes N_1, N_2, N_3, respectively, determine the number of branches in the corresponding tree diagram.

58. A 5-card hand is drawn from a standard deck. Discuss how you can tell that the following two events are dependent without any computation.

S = Hand consists entirely of spades
H = Hand consists entirely of hearts

59. C The command in Figure A was used on a graphing utility to simulate 50 repetitions of rolling a pair of dice and recording the minimum of the two numbers. A statistical plot of the results is shown in Figure B.

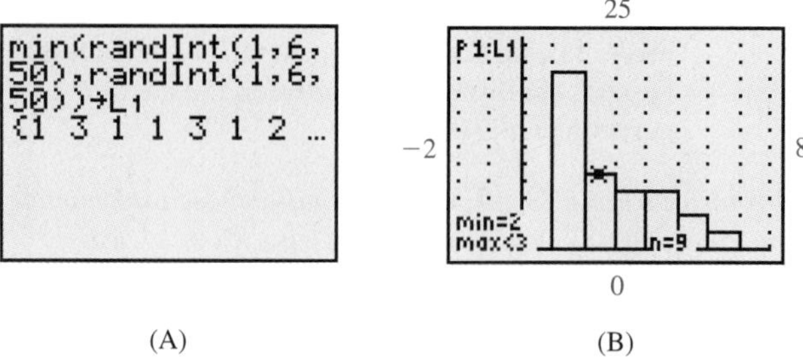

Figure for 59

(A) Use Figure B to find the empirical probability that the minimum is 2.
(B) What is the theoretical probability that the minimum is 2?
(C) Use a graphing utility to simulate 200 rolls of a pair of dice, determine the empirical probability that the minimum is 4, and compare with the theoretical probability.

60. C A card is drawn at random from a standard 52-card deck. Use a graphing utility to simulate 800 such draws, determine the empirical probability that the card is a black jack, and compare with the theoretical probability.

C

61. Three fair coins are tossed 1,000 times with the following frequencies of outcomes:

Number of Heads	0	1	2	3
Frequency	120	360	350	170

(A) What is the approximate empirical probability of obtaining 2 heads?
(B) What is the theoretical probability of obtaining 2 heads?
(C) What is the expected frequency of obtaining 2 heads?

62. You bet a friend \$1 that you will get 1 or more double 6's on 24 rolls of a pair of fair dice. What is your expected value for this game? What is your friend's expected value? Is the game fair?

63. A software development department consists of 6 women and 4 men.
(A) How many ways can they select a chief programmer, a backup programmer, and a programming librarian?

(B) If the positions in part (A) are selected by lottery, what is the probability that women are selected for all 3 positions?
(C) How many ways can they select a team of 3 programmers to work on a particular project?
(D) If the selections in part (C) are made by lottery, what is the probability that a majority of the team members will be women?

64. A group of 150 people includes 52 who play chess, 93 who play checkers, and 28 who play both chess and checkers. How many people in the group play neither game?

65. If 3 people are selected from a group of 7 men and 3 women, what is the probability that at least 1 woman is selected?

Two cards are drawn in succession without replacement from a standard 52-card deck. In Problems 66 and 67, compute the indicated probabilities.

66. The second card is a heart given the first card is a heart.

67. The first card is a heart given the second card is a heart.

68. Two fair (not weighted) dice are each numbered with a 3 on one side, a 2 on two sides, and a 1 on three sides. The dice are rolled and the numbers on the two up faces are added. If X is the random variable associated with the sample space $S = \{2, 3, 4, 5, 6\}$:
(A) Find the probability distribution of X.
(B) Find the expected value of X.

69. If you pay $3.50 to play the game in Problem 68 (the dice are rolled once) and you are returned the dollar amount corresponding to the sum on the faces, what is the expected value of the game? Is the game fair? If it is not fair, how much should you pay to make the game fair?

70. How many different 5-child families are possible where the sex of each child in the order of their birth is taken into consideration [that is, birth sequences such as (B, G, G, B, B) and (G, B, G, B, B) produce different families]? How many families are possible if the order pattern is not taken into account?

71. Suppose 3 white balls and 1 black ball are placed in a box. Balls are drawn in succession without replacement until a black ball is drawn, and then the game is over. You win if the black ball is drawn on the fourth draw.
(A) What are the probability and odds for winning?
(B) If you bet $1, what should the house pay you for winning (plus return your $1 bet) if the game is to be fair?

72. If each of 5 people is asked to identify his or her favorite book from a list of 10 best-sellers, what is the probability that at least 2 of them identify the same book?

Applications

Business & Economics

73. *Transportation.* A distribution center A wishes to distribute its products to five different retail stores, B, C, D, E, and F, in a city. How many different route plans can be constructed so that a single truck, starting from A, will deliver to each store exactly once, and then return to the center?

74. *Market research.* A survey of 1,000 people indicates that 340 have invested in stocks, 480 have invested in bonds, and 210 have invested in stocks and bonds.
(A) How many people in the survey have invested in stocks or bonds?
(B) How many have invested in neither stocks nor bonds?
(C) How many have invested in bonds and not in stocks?

75. *Market research.* From a survey of 100 residents of a city, it was found that 40 read the daily morning paper, 70 read the daily evening paper, and 30 read both papers. What is the (empirical) probability that a resident selected at random:
(A) Reads a daily paper?
(B) Does not read a daily paper?
(C) Reads exactly one daily paper?

76. *Market research.* A market research firm has determined that 40% of the people in a certain area have seen the advertising for a new product and that 85% of those who have seen the advertising have purchased the product. What is the probability that a person in this area has seen the advertising and purchased the product?

77. *Market analysis.* A clothing company selected 1,000 persons at random and surveyed them to determine a relationship between age of purchaser and annual purchases of jeans. The results are given in the table.

Age	Jeans Purchased Annually 0	1	2	Above 2	*Totals*
Under 12	60	70	30	10	170
12–18	30	100	100	60	290
19–25	70	110	120	30	330
Over 25	100	50	40	20	210
Totals	260	330	290	120	1,000

Given the events:

$A =$ Person buys 2 pairs of jeans
$B =$ Person is between 12 and 18 years old
$C =$ Person does not buy more than 2 pairs of jeans
$D =$ Person buys more than 2 pairs of jeans

(A) Find $P(A), P(B), P(A \cap B), P(A|B), P(B|A)$.
(B) Are events A and B independent? Explain.
(C) Find $P(C), P(D), P(C \cap D), P(C|D), P(D|C)$.
(D) Are events C and D mutually exclusive? Independent? Explain.

78. *Decision analysis.* A company sales manager, after careful analysis, presents two sales plans. It is estimated that plan A will net \$10 million if successful (probability .8) and lose \$2 million if not (probability .2); plan B will net \$12 million if successful (probability .7) and lose \$2 million if not (probability .3). What is the expected return for each plan? Which plan should be chosen based on the expected return?

79. *Insurance.* A \$300 bicycle is insured against theft for an annual premium of \$30. If the probability that the bicycle will be stolen during the year is .08 (empirically determined), what is the expected value of the policy?

80. *Quality control.* Twelve precision parts, including 2 that are substandard, are sent to an assembly plant. The plant will select 4 at random and will return the whole shipment if 1 or more of the sample are found to be substandard. What is the probability that the shipment will be returned?

81. *Quality control.* A dozen computer circuit boards, including 2 that are defective, are sent to a computer service center. A random sample of 3 is selected and tested. Let X be the random variable associated with the number of circuit boards in a sample that is defective.
(A) Find the probability distribution of X.
(B) Find the expected number of defective boards in a sample.

Life Sciences

82. *Medicine—cardiogram test.* By testing a large number of individuals, it has been determined that 82% of the population have normal hearts, 11% have some minor heart problems, and 7% have severe heart problems. Ninety-five percent of the persons with normal hearts, 30% of those with minor problems, and 5% of those with severe problems will pass a cardiogram test. What is the probability that a person who passes the cardiogram test has a normal heart?

83. *Genetics.* Six men in 100 and 1 woman in 100 are color-blind. A person is selected at random and is found to be color-blind. What is the probability that this person is a man? (Assume the total population contains the same number of women as men.)

Social Sciences

84. *Voter preference.* In a straw poll, the 30 students in a mathematics class are asked to indicate their preference for president of the student government. Approximate empirical probabilities are assigned on the basis of the poll: candidate A should receive 53% of the vote, candidate B should receive 37%, and candidate C should receive 10%. One week later, candidate B wins the election. Discuss the factors that may account for the discrepancy between the poll and the election results.

Group Activity 1 Car and Crying Towels

The host on a television game show gives you (the contestant) a choice of three closed doors labeled *A*, *B*, and *C*. Behind one of the doors is a car, and behind each of the other two, a crying towel. After you have selected a door, which is not opened, the host opens one of the two doors not selected by you. The host, knowing which door hides the car, always chooses a crying towel door. The host then gives you a chance to stick to your original choice or to switch to the other unopened door. Which of the following two strategies would you choose and why?

Strategy 1. Stick to your original choice.

Strategy 2. Switch to the other unopened door.

(A) *Solve by "Common Sense" Reasoning.* Let each person in your group indicate their feeling about which strategy would be best or if it does not matter. Present the problem to friends and ask which strategy they would prefer and why. Summarize the results of these discussions.

(B) *Solve by Simulation.* To simulate this problem use a spinner with three equally likely outcomes, and suppose the car is hidden behind door *C*. Your original door choice is made by spinning the spinner once and making the choice indicated by the spinner. [*Note:* A single die can be used in place of the spinner by choosing door *A* if 1 or 2 dots turn up, door *B* for 3 or 4 dots, and door *C* for 5 or 6 dots.]

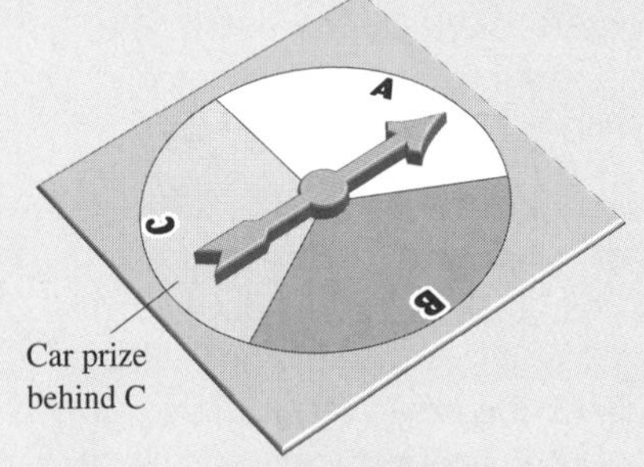

We now outline a simulation of the Stick strategy, and leave it to you to do the same for the Switch strategy.

Stick Strategy Simulation.

1. Suppose the spinner lands on door *A*. The host opens door *B*, knowing the car is behind door *C*, and reveals a crying towel. You stick to your original choice *A* and lose; getting a crying towel when door *A* is opened.
2. Suppose the spinner lands on door *B*. The host opens door *A*, knowing the car is behind *C*, and reveals a crying towel. You stick to *B* and lose; getting a crying towel when door *B* is opened.
3. Suppose the spinner lands on door *C*. The host opens either door *A* or *B*, revealing a crying towel. You stick to door *C* and win, getting a car when door *C* is opened.

Switch Strategy Simulation. Modify the simulation for the Stick strategy to apply to the Switch strategy.

Each student in your group should run at least 100 trials for each strategy simulation and record the wins and losses. Pool the results of all students in your group. Use the relative frequency of wins for each strategy to write approximate empirical probabilities for a win for that strategy. Now, which strategy do you think you would use and why?

(C) *Solve Theoretically Using Probability Trees.* The car is assumed to be behind door *C*.

Again, we outline a probability tree for the Stick strategy on the next page, and leave it to you to do the same for the Switch strategy.

Stick Strategy. Complete the following probability tree and compute the probability of a win for the Stick strategy.

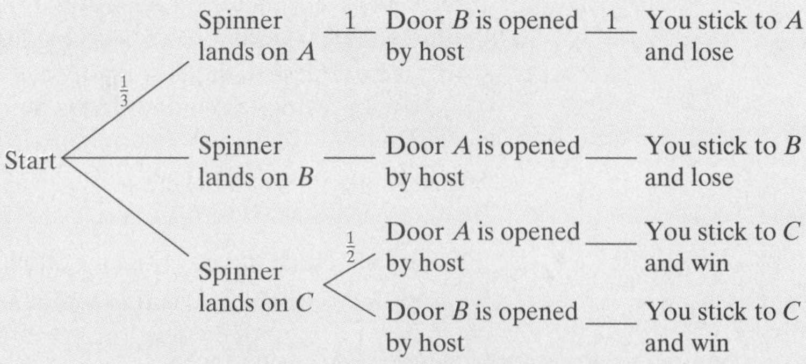

Switch Strategy. Construct a probability tree for the Switch strategy, and compute the probability of a win for this strategy.

Compare the results of parts (A), (B), and (C). Which strategy would you now choose and why? Present the problem to a friend and try to convince your friend that the strategy you have chosen is the best strategy.

Now consider the **three cards problem:** Three cards are in a box. Both sides of one card are white, both sides of a second card are black, and one side of a third card is white and the other side is black. A card is drawn from the box at random and placed on the table. A white side is showing. What is the probability that the other side is black? Describe how you would solve this problem by using a simulation approach and by using a theoretical approach. Use each approach to solve the problem.

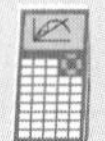

Group Activity 2 *Simulation—A Draft Lottery for Professional Sports*

What is the probability that hurricane Hypatia will hit land north of Cape Hatteras? That the new county jail will be at maximum capacity by year's end? That a supermarket customer will wait more than 7 minutes in one of the checkout lanes?

It would be virtually impossible to answer such questions by assigning probabilities to the simple events of a sample space. The sample space might be too large to be manageable, or it might be extremely difficult to determine probabilities of simple events.

Simulation offers an alternative approach. In several of the exercises in this chapter we have used a graphing utility to simulate flipping a coin, rolling a pair of dice and recording their sum, playing roulette, and so on. We use such simulations and computer simulations of more complex processes to determine the approximate empirical probabilities of events.

Professional sports leagues hold annual draft lotteries. The draft lottery is used to determine the order in which teams will select players in the draft, which is held at a later date. The lottery is weighted in favor of the poorer teams but also involves an element of chance.

Suppose there are 10 teams that did not make the playoffs; call them *A, B, C, D, E, F, G, H, I,* and *J,* arranged from the team with the poorest record

(A) to the team with the best record (J). The draft order for these 10 teams is determined by drawing balls at random from a drum that initially contains 1,000 balls of 10 different colors: the numbers of balls for the 10 teams A–J are 600, 200, 100, 50, 30, 10, 4, 3, 2, and 1, respectively.

If the first ball drawn belongs to team D, then D will select first in the player draft. All other balls belonging to D are removed from the drum before the second ball is drawn. If the second ball drawn belongs to team F, then F will select second in the player draft, and so on. The drawing continues in this fashion until an order, say *DFABCHGJEI*, is established for the player draft.

(A) What is the theoretical probability that team A wins the lottery and selects first in the player draft?
(B) What is the theoretical probability that team A selects second in the player draft?

```
randInt(1,1000)
            812
```

FIGURE 1

We can use a graphing utility to simulate the first draw of a ball from the drum: We use the random number feature to select a random integer from 1 to 1,000 and consider an integer in the range 1–600 as belonging to team A, 601–800 to team B, 801–900 to team C, and so on (see Fig. 1).

Figure 1 indicates that the first ball drawn belongs to team C. Now, instead of removing the balls belonging to C, it is more convenient to imagine that the drum contains all 1,000 balls for the next several draws, with the understanding that if another ball belonging to C is drawn, that draw is ignored. Now, suppose the command randInt(1,1000,5) produces $\{966, 432, 656, 843, 729\}$ as the next set of 5 elements, which belong to E, A, B, C, and B, respectively. Then the order *CEAB* has been established for the first four draft picks (843 and 729 are ignored, since they belong to teams already selected).

Now imagine that the balls belonging to teams A, B, and C have been removed from the drum. Suppose the command randInt(901,1000,5) produces the next set $\{943, 969, 990, 928, 948\}$, whose elements belong to D, E, F, D, and D, respectively. Then the order *CEABDF* has been established for the first six draft picks.

Now imagine that the balls belonging to all 6 of these teams have been removed. If the command randInt(991,1000,5) produces the set $\{993, 995, 1000, 992, 997\}$, whose elements belong to G, H, J, G, and H, respectively, then the order for all 10 teams is determined, *CEABDFGHJI*, and we have completed one simulation of the lottery.

(C) Carry out at least 20 simulations of the draft lottery.
(D) Based on your simulations, what is the empirical probability that team A drafts first? That A drafts second? Compare with the theoretical probabilities.
(E) What is the empirical probability that A drafts fifth or later? That F gets one of the first three draft picks? That J gets one of the first five draft picks?
(F) Discuss the difficulty of computing the theoretical probabilities of the events of part (E).

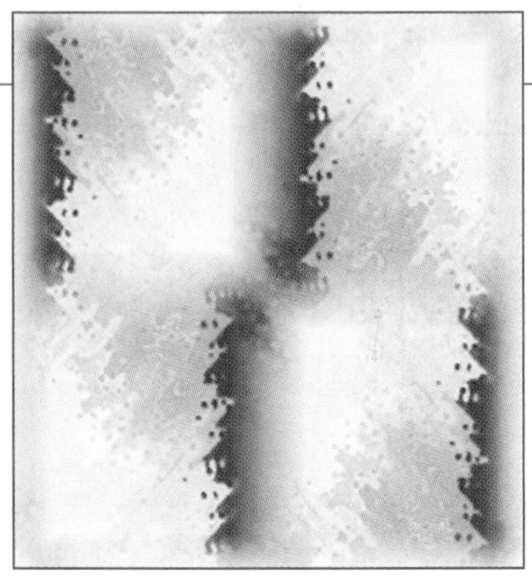

Chapter 7

Markov Chains

INTRODUCTION

In this chapter we consider a mathematical model that combines probability and matrices to analyze a *stochastic process,* which consists of a sequence of trials that satisfy certain conditions. The sequence of trials is called a *Markov chain* after the Russian mathematician Andrei Markov (1856–1922), who is credited with supplying much of the foundation work in stochastic processes. Many of the early applications of Markov chains were in the physical sciences. More recent applications of Markov chains involve a wide variety of topics, including finance, market research, genetics, medicine, demographics, psychology, and political science.

In the first section we introduce the basic properties of Markov chains. In the remaining sections we discuss the long-term behavior of two different types of Markov chains.

Section 7-1 Properties of Markov Chains

- Introduction
- Transition and State Matrices
- Powers of Transition Matrices
- Application

Introduction

In this section we are going to be interested in physical *systems* and their possible *states.* To understand what this means, consider the following examples:

1. A stock listed on the New York Stock Exchange either increases, decreases, or does not change in price each day the exchange is open. The stock can be thought of as a physical system with three possible states: increase, decrease, or no change.

2. A commuter, relative to a rapid transit system, can be thought of as a physical system with two states, a user or a nonuser.
3. A voting precinct casts a simple majority vote during each congressional election for a Republican candidate, a Democratic candidate, or a third-party candidate. The precinct, relative to all congressional elections past, present, and future, constitutes a physical system that can be thought of as being in one (and only one) of three states after each election: Republican, Democratic, or other.

If a system evolves from one state to another in such a way that chance elements are involved in progressing from one state to the next, then the progression of the system through a sequence of states is called a **stochastic process** (*stochos* is the Greek word for "guess"). We will now consider a simple example of a stochastic process in detail, and out of it will arise further definitions and methodology.

A toothpaste company markets a product (brand A) that currently has 10% of the toothpaste market. The company hires a market research firm to estimate the percentage of the market it might acquire in the future if it launches an aggressive sales campaign. The research firm uses test marketing and extensive surveys to predict the effect of the campaign. They find that if a person is using brand A, then the probability is .8 that this person will buy it again when he or she runs out of toothpaste. On the other hand, a person using another brand will switch to brand A with a probability of .6 when he or she runs out of toothpaste. Thus, each toothpaste consumer can be considered to be in one of two possible states:

$$A = \text{Uses brand } A \qquad \text{and} \qquad A' = \text{Uses another brand}$$

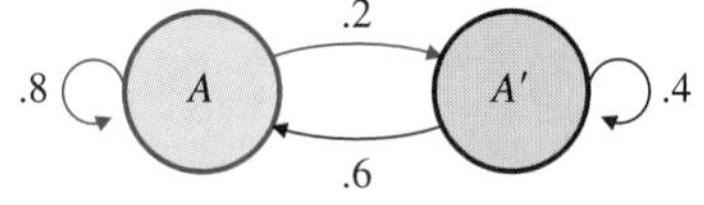

FIGURE 1
Transition diagram

The probabilities determined by the market research firm can be represented graphically in a **transition diagram** as shown in Figure 1.

We can also represent the information from the market research firm numerically in a **transition probability matrix:**

$$\text{Current state} \quad \begin{matrix} & \text{Next state} \\ & \begin{matrix} A & A' \end{matrix} \\ \begin{matrix} A \\ A' \end{matrix} & \begin{bmatrix} .8 & .2 \\ .6 & .4 \end{bmatrix} \end{matrix} = P$$

Explore–Discuss 1

(A) Refer to the transition diagram in Figure 1. What is the probability that a person using brand A will switch to another brand when he or she runs out of toothpaste?

(B) Refer to transition probability matrix P. What is the probability that a person who is not using brand A will not switch to brand A when he or she runs out of toothpaste?

(C) In Figure 1, the sum of the probabilities on the arrows leaving each state is 1. Will this be true for any transition diagram? Explain your answer.

(D) In transition probability matrix P, the sum of the probabilities in each row is 1. Will this be true for any transition probability matrix? Explain your answer.

The toothpaste company's 10% share of the market at the beginning of the sales campaign can be represented as an **initial-state distribution matrix:**

$$S_0 = \begin{matrix} A & A' \\ [.1 & .9] \end{matrix}$$

If a person is chosen at random, then the probability that this person uses brand A (state A) is .1, and the probability that this person does not use brand A (state A') is .9. Thus, S_0 also can be interpreted as an **initial-state probability matrix.**

What are the probabilities of a person being in state A or A' on the first purchase after the start of the sales campaign? Let us look at the probability tree given below. [*Note:* A_0 represents state A at the beginning of the campaign, A'_1 represents state A' on the first purchase after the campaign, and so on.]

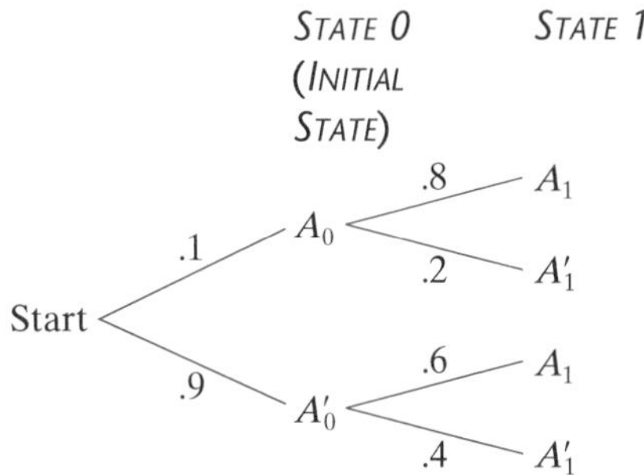

Proceeding as in Chapter 6, we can read the required probabilities directly from the tree:

$$\begin{aligned} P(A_1) &= P(A_0 \cap A_1) + P(A'_0 \cap A_1) \\ &= (.1)(.8) + (.9)(.6) = .62 \end{aligned}$$

$$\begin{aligned} P(A'_1) &= P(A_0 \cap A'_1) + P(A'_0 \cap A'_1) \\ &= (.1)(.2) + (.9)(.4) = .38 \end{aligned}$$

[*Note:* $P(A_1) + P(A'_1) = 1$, as expected.]

Thus, the **first-state matrix** is

$$S_1 = \begin{matrix} A & A' \\ [.62 & .38] \end{matrix}$$

This matrix gives us the probabilities of a randomly chosen person being in state A or A' on the first purchase after the start of the campaign. We see that brand A's market share has increased from 10% to 62%.

Now, if you were asked to find the probabilities of a person being in state A or state A' on the tenth purchase after the start of the campaign, you might start to draw additional branches on the probability tree, but you would soon become discouraged—and for good reason, because the number of branches doubles for each successive purchase. By the tenth purchase, there would be $2^{11} = 2{,}048$ branches! Fortunately, we can convert the summing of branch products to matrix multiplication. In particular, if we multiply the initial-state matrix S_0 by the transition matrix P, we obtain the first-state matrix S_1:

$$\underset{\text{Initial state}}{S_0P = \overset{A \quad A'}{[.1 \quad .9]}} \underset{\text{Transition matrix}}{\begin{bmatrix} .8 & .2 \\ .6 & .4 \end{bmatrix}} = \underbrace{[(.1)(.8) + (.9)(.6) \qquad (.1)(.2) + (.9)(.4)]}_{\text{Compare with the tree computations above}} = \underset{\text{First state}}{\overset{A \quad A'}{[.62 \quad .38]}} = S_1$$

As you might guess, we can get the second-state matrix S_2 (for the second purchase) by multiplying the first-state matrix by the transition matrix:

$$S_1P = \underset{\text{First state}}{\overset{A \quad A'}{[.62 \quad .38]}} \begin{bmatrix} .8 & .2 \\ .6 & .4 \end{bmatrix} = \underset{\text{Second state}}{\overset{A \quad A'}{[.724 \quad .276]}} = S_2$$

The third-state matrix S_3 is computed in a similar manner:

$$S_2P = \underset{\text{Second state}}{\overset{A \quad A'}{[.724 \quad .276]}} \begin{bmatrix} .8 & .2 \\ .6 & .4 \end{bmatrix} = \underset{\text{Third state}}{\overset{A \quad A'}{[.7448 \quad .2552]}} = S_3$$

Examining the values in the first three state matrices, we see that brand A's market share increases after each toothpaste purchase. Will the market share for brand A continue to increase until it approaches 100%, or will it level off at some value less than 100%? These questions will be answered in the next section when we develop techniques for determining the long-run behavior of state matrices.

Transition and State Matrices

The sequence of trials (toothpaste purchases) with the constant transition matrix P described above is a special kind of stochastic process called a *Markov chain.* In general, a **Markov chain,** or **process,** is a sequence of experiments, trials, or observations such that the transition probability matrix from one state to the next is constant. A Markov process has no memory. The various matrices associated with a Markov chain are defined in the next box.

Markov Chains

Given a Markov chain with n states, a ***k*th-state matrix** is a matrix of the form

$$S_k = [s_{k1} \quad s_{k2} \quad \dots \quad s_{kn}]$$

Each entry s_{ki} is the proportion of the population that is in state i after the kth trial, or, equivalently, the probability of a randomly selected element of the population being in state i after the kth trial. The sum of all the entries in the kth state matrix S_k must be 1.

A **transition matrix** is a constant square matrix P of order n such that the entry in the ith row and jth column indicates the probability of the system moving from the ith state to the jth state on the next observation or trial. The sum of the entries in each row must be 1.

REMARKS

1. Since the entries in a kth-state matrix or a transition matrix are probabilities, they must be real numbers between 0 and 1, inclusive.
2. Rearranging the various states and corresponding transition probabilities in a transition matrix will produce a different, but equivalent, transition matrix. For example, both of the following matrices are transition matrices for the toothpaste manufacturer discussed earlier:

$$P = \begin{array}{c} \\ A \\ A' \end{array}\!\!\begin{array}{c} \begin{array}{cc} A & A' \end{array} \\ \begin{bmatrix} .8 & .2 \\ .6 & .4 \end{bmatrix} \end{array} \qquad P' = \begin{array}{c} \\ A' \\ A \end{array}\!\!\begin{array}{c} \begin{array}{cc} A' & A \end{array} \\ \begin{bmatrix} .4 & .6 \\ .2 & .8 \end{bmatrix} \end{array}$$

Such rearrangements will affect the form of the matrices used in the solution of a problem, but will not affect any of the information obtained from these matrices. In Section 7-3 we will encounter situations where it will be helpful to select a transition matrix that has a special form. For now, you can choose any order for the states in a transition matrix.

As we indicated in the preceding discussion, matrix multiplication can be used to compute the various state matrices of a Markov chain, as described in the following box.

Computing State Matrices for a Markov Chain

If S_0 is the initial-state matrix and P is the transition matrix for a Markov chain, then the subsequent state matrices are given by:

$$\begin{aligned} S_1 &= S_0P && \text{First-state matrix} \\ S_2 &= S_1P && \text{Second-state matrix} \\ S_3 &= S_2P && \text{Third-state matrix} \\ &\vdots \\ S_k &= S_{k-1}P && k\text{th-state matrix} \end{aligned}$$

Example 1

Insurance An insurance company found that on the average, over a period of 10 years, 23% of the drivers in a particular community who were involved in an accident one year were also involved in an accident the following year. They also found that only 11% of the drivers who were not involved in an accident one year were involved in an accident the following year. Use these percentages as approximate empirical probabilities for the following:

(A) Draw a transition diagram.
(B) Find the transition matrix P.
(C) If 5% of the drivers in the community are involved in an accident this year, what is the probability that a driver chosen at random from the community will be involved in an accident next year? Year after next?

SOLUTION (A)

(B)

$$\begin{array}{c} \\ \text{This year} \end{array} \begin{array}{c} \\ A \\ A' \end{array} \overset{\begin{array}{c}\text{Next year}\\ A \quad\; A'\end{array}}{\begin{bmatrix} .23 & .77 \\ .11 & .89 \end{bmatrix}} = P \quad \text{Transition matrix}$$

(C) The initial-state matrix S_0 is

$$S_0 = \overset{\begin{array}{cc} A & A' \end{array}}{[.05 \quad .95]} \quad \text{Initial-state matrix}$$

Thus,

$$S_0P = \underset{\substack{\text{This year}\\ \text{(initial state)}}}{\overset{\begin{array}{cc} A & A' \end{array}}{[.05 \quad .95]}} \begin{bmatrix} .23 & .77 \\ .11 & .89 \end{bmatrix} = \underset{\substack{\text{Next year}\\ \text{(first state)}}}{\overset{\begin{array}{cc} A & A' \end{array}}{[.116 \quad .884]}} = S_1$$

$$S_1P = \underset{\substack{\text{Next year}\\ \text{(first state)}}}{\overset{\begin{array}{cc} A & A' \end{array}}{[.116 \quad .884]}} \begin{bmatrix} .23 & .77 \\ .11 & .89 \end{bmatrix} = \underset{\substack{\text{Year after next}\\ \text{(second state)}}}{\overset{\begin{array}{cc} A & A' \end{array}}{[.12392 \quad .87608]}} = S_2$$

The probability of a driver chosen at random from the community having an accident next year is .116 and having an accident year after next is .12392. That is, it is expected that 11.6% of the drivers in the community will have an accident next year and 12.392% the year after.

Matched Problem 1 ➡ An insurance company classifies drivers as low-risk if they are accident-free for 1 year. Past records indicate that 98% of the drivers in the low-risk category (L) one year will remain in that category the next year, and 78% of the drivers that are not in the low-risk category (L') one year will be in the low-risk category the next year.

(A) Draw a transition diagram.
(B) Find the transition matrix P.
(C) If 90% of the drivers in the community are in the low-risk category this year, what is the probability that a driver chosen at random from the community will be in the low-risk category next year? Year after next?

POWERS OF TRANSITION MATRICES

We will now investigate the effective use of the powers of a transition matrix.

Explore–Discuss 2

Given the transition matrix P and initial-state matrix S_0, where

$$P = \begin{matrix} & A & A' \\ A & .9 & .1 \\ A' & .7 & .3 \end{matrix} \quad \text{and} \quad S_0 = \begin{matrix} A & A' \\ [.5 & .5] \end{matrix}$$

(A) Find S_2 and S_4.
(B) Find P^2 and P^4. [Recall that $P^2 = PP$ and $P^4 = P^2P^2$.]
(C) Find S_0P^2 and S_0P^4.
(D) Compare the results of parts (A) and (C). What interpretation of the entries in P^2 and P^4 does this suggest?

The state matrices for a Markov chain are defined **recursively;** that is, each state matrix is defined in terms of the preceding state matrix. For example, to find the fourth-state matrix S_4, it is necessary to compute the preceding three state matrices:

$$S_1 = S_0P \qquad S_2 = S_1P \qquad S_3 = S_2P \qquad S_4 = S_3P$$

Is there any way to compute a given state matrix directly without first computing all the preceding state matrices? If we substitute the equation for S_1 in the equation for S_2, substitute this new equation for S_2 in the equation for S_3, and so on, a definite pattern emerges:

$$\begin{aligned} S_1 &= S_0P \\ S_2 &= S_1P = (S_0P)P = S_0P^2 \\ S_3 &= S_2P = (S_0P^2)P = S_0P^3 \\ S_4 &= S_3P = (S_0P^3)P = S_0P^4 \\ &\vdots \end{aligned}$$

In general, it can be shown that the kth-state matrix is given by $S_k = S_0P^k$. We summarize this important result in Theorem 1.

Theorem 1 ■■ POWERS OF A TRANSITION MATRIX

If P is the transition matrix and S_0 is an initial-state matrix for a Markov chain, then the kth-state matrix is given by

$$S_k = S_0P^k$$

The entry in the ith row and jth column of P^k indicates the probability of the system moving from the ith state to the jth state in k observations or trials. The sum of the entries in each row of P^k is 1. ■■

Example 2 **Using P^k to Compute S_k** Find P^4 and use it to find S_4 for

$$P = \begin{matrix} & A & A' \\ A & .1 & .9 \\ A' & .6 & .4 \end{matrix} \quad \text{and} \quad S_0 = \begin{matrix} A & A' \\ [.2 & .8] \end{matrix}$$

SOLUTION
$$P^2 = PP = \begin{bmatrix} .1 & .9 \\ .6 & .4 \end{bmatrix}\begin{bmatrix} .1 & .9 \\ .6 & .4 \end{bmatrix} = \begin{bmatrix} .55 & .45 \\ .3 & .7 \end{bmatrix}$$
$$P^4 = P^2P^2 = \begin{bmatrix} .55 & .45 \\ .3 & .7 \end{bmatrix}\begin{bmatrix} .55 & .45 \\ .3 & .7 \end{bmatrix} = \begin{bmatrix} .4375 & .5625 \\ .375 & .625 \end{bmatrix}$$
$$S_4 = S_0P^4 = [.2 \quad .8]\begin{bmatrix} .4375 & .5625 \\ .375 & .625 \end{bmatrix} = [.3875 \quad .6125]$$

Matched Problem 2 Find P^4 and use it to find S_4 for

$$P = \begin{matrix} & \begin{matrix} A & A' \end{matrix} \\ \begin{matrix} A \\ A' \end{matrix} & \begin{bmatrix} .8 & .2 \\ .3 & .7 \end{bmatrix} \end{matrix} \quad \text{and} \quad S_0 = \begin{matrix} \begin{matrix} A & A' \end{matrix} \\ [.8 \quad .2] \end{matrix}$$

If a graphing calculator or a computer is available for computing matrix products and powers of a matrix, then finding state matrices for any number of trials becomes a routine calculation.

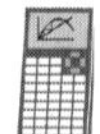

Example 3 **Using a Graphing Utility and P^k to Compute S_k** Use P^8 and a graphing utility to find S_8 for P and S_0 as given in Example 2. Round values in S_8 to six decimal places.

SOLUTION After storing the matrices P and S_0 in the graphing utility's memory, use the equation

$$S_8 = S_0P^8$$

to compute S_8. Figure 2 shows the result on a typical graphing utility. Thus, we see that (to six decimal places)

$$S_8 = [.399219 \quad .600781]$$

FIGURE 2

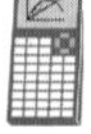

Matched Problem 3 Use P^8 and a graphing utility to find S_8 for P and S_0 as given in Matched Problem 2. Round values in S_8 to six decimal places.

APPLICATION

The next example illustrates the use of Theorem 1 in an applied problem.

Example 4 **University Enrollment** Part-time students admitted to an MBA program in a university are considered to be first-year students until they successfully complete 15 credits. Then they are classified as second-year students and may begin to take more advanced courses and to work on the thesis required for graduation. Past records indicate that at the end of each year 10% of the first-year students (F) drop out of the program (D) and 30% become second-year students (S). Also, 10% of the second-year students drop out of the program and 40% graduate (G) each year. Students that graduate or drop out never return to the program.

(A) Draw a transition diagram.
(B) Find the transition matrix P.
(C) What is the probability that a first-year student graduates within 4 years? Drops out within 4 years?

SOLUTION (A) If 10% of the first-year students drop out and 30% become second-year students, the remaining 60% must continue as first-year students for another year (see the diagram). Similarly, 50% of the second-year students must continue as second-year students for another year. Since students who drop out never return, all the students in state D in one year will continue in that state the next year. We indicate this by placing a 1 on the arrow from D back to D. State G is labeled in the same manner.

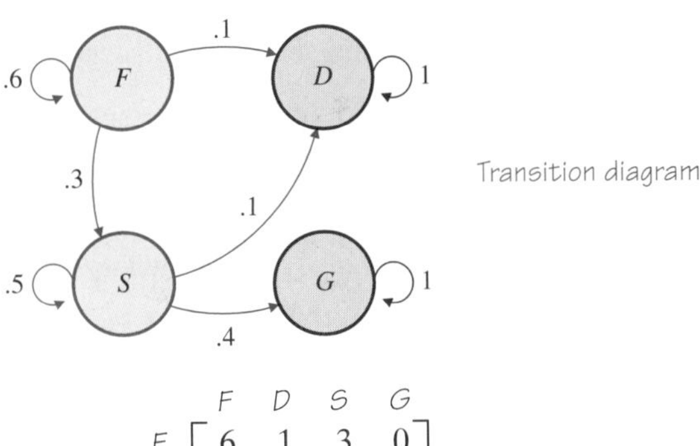

Transition diagram

(B)
$$P = \begin{matrix} & \begin{matrix} F & D & S & G \end{matrix} \\ \begin{matrix} F \\ D \\ S \\ G \end{matrix} & \begin{bmatrix} .6 & .1 & .3 & 0 \\ 0 & 1 & 0 & 0 \\ 0 & .1 & .5 & .4 \\ 0 & 0 & 0 & 1 \end{bmatrix} \end{matrix}$$

Transition matrix

(C) The probability that a first-year student moves from state F to state G within 4 years is the entry in Row 1 and Column 4 of P^4 (Theorem 1). Hand computation of P^4 requires two multiplications:

$$P^2 = \begin{bmatrix} .6 & .1 & .3 & 0 \\ 0 & 1 & 0 & 0 \\ 0 & .1 & .5 & .4 \\ 0 & 0 & 0 & 1 \end{bmatrix} \begin{bmatrix} .6 & .1 & .3 & 0 \\ 0 & 1 & 0 & 0 \\ 0 & .1 & .5 & .4 \\ 0 & 0 & 0 & 1 \end{bmatrix} = \begin{bmatrix} .36 & .19 & .33 & .12 \\ 0 & 1 & 0 & 0 \\ 0 & .15 & .25 & .6 \\ 0 & 0 & 0 & 1 \end{bmatrix}$$

$$P^4 = P^2P^2 = \begin{bmatrix} .36 & .19 & .33 & .12 \\ 0 & 1 & 0 & 0 \\ 0 & .15 & .25 & .6 \\ 0 & 0 & 0 & 1 \end{bmatrix} \begin{bmatrix} .36 & .19 & .33 & .12 \\ 0 & 1 & 0 & 0 \\ 0 & .15 & .25 & .6 \\ 0 & 0 & 0 & 1 \end{bmatrix}$$

$$= \begin{bmatrix} .1296 & .3079 & .2013 & .3612 \\ 0 & 1 & 0 & 0 \\ 0 & .1875 & .0625 & .75 \\ 0 & 0 & 0 & 1 \end{bmatrix}$$

Thus, the probability that a first-year student has graduated within 4 years is .3612. Similarly, the probability that a first-year student has dropped out within 4 years is .3079 (the entry in Row 1 and Column 2 of P^4).

Matched Problem 4 ➡ Refer to Example 4. At the end of each year the faculty examines the progress each second-year student has made in writing the required thesis. Past records indicate that 30% of the second-year students (S) have their theses approved (A) and 10% of the students are dropped from the program for insufficient progress (D), never to return. The remaining students continue to work on their theses.

(A) Draw a transition diagram.
(B) Find the transition matrix P.
(C) What is the probability that a second-year student completes the thesis requirement within 4 years? Is dropped from the program for insufficient progress within 4 years?

Explore–Discuss 3

Refer to Example 4. States D and G are referred to as *absorbing states*, because a student who enters either one of these states never leaves it. Absorbing states are discussed in detail in Section 7-3.

(A) How can absorbing states be recognized from a transition diagram? Draw a transition diagram with two states, one that is absorbing and one that is not, to illustrate.
(B) How can absorbing states be recognized from a transition matrix? Write the transition matrix for the diagram you drew in part (A) to illustrate.

Answers to Matched Problems

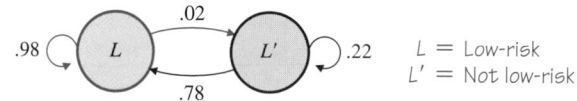

1. (A)

.98 (L) —.02→ (L') .22; (L') —.78→ (L)

L = Low-risk
L′ = Not low-risk

(B)
$$\begin{array}{c} \\ \text{This year} \end{array}\begin{array}{c} \\ L \\ L' \end{array}\overset{\text{Next year}}{\begin{array}{cc} L & L' \\ \end{array}}\begin{bmatrix} .98 & .02 \\ .78 & .22 \end{bmatrix} = P$$

(C) Next year: .96; year after next: .972

2. $P^4 = \begin{bmatrix} .625 & .375 \\ .5625 & .4375 \end{bmatrix}$; $S_4 = [.6125 \quad .3875]$

3. $S_8 = [.600781 \quad .399219]$

4. (A)

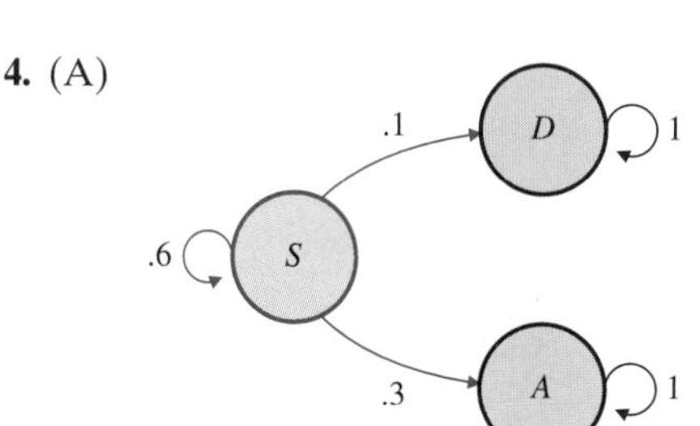

(B)
$$P = \begin{array}{c} \\ S \\ A \\ D \end{array}\overset{\begin{array}{ccc} S & A & D \end{array}}{\begin{bmatrix} .6 & .3 & .1 \\ 0 & 1 & 0 \\ 0 & 0 & 1 \end{bmatrix}}$$

(C) .6528; .2176

Exercise 7-1

A *Problems 1–8 refer to the following transition matrix:*

$$P = \begin{array}{c} \\ A \\ B \end{array}\overset{\begin{array}{cc} A & B \end{array}}{\begin{bmatrix} .8 & .2 \\ .4 & .6 \end{bmatrix}}$$

In Problems 1–4, find S_1 for the indicated initial-state matrix S_0, and interpret with the aid of a tree diagram.

1. $S_0 = [1 \quad 0]$

2. $S_0 = [0 \quad 1]$

3. $S_0 = [.5 \quad .5]$

4. $S_0 = [.3 \quad .7]$

In Problems 5–8, find S_2 for the indicated initial-state matrix S_0, and explain what it represents.

5. $S_0 = [1 \quad 0]$

6. $S_0 = [0 \quad 1]$

7. $S_0 = [.5 \quad .5]$

8. $S_0 = [.3 \quad .7]$

In Problems 9–14, is there a unique way of filling in the missing probabilities in the transition diagram? If so, complete the transition diagram and write the corresponding transition matrix. If not, explain why.

9.

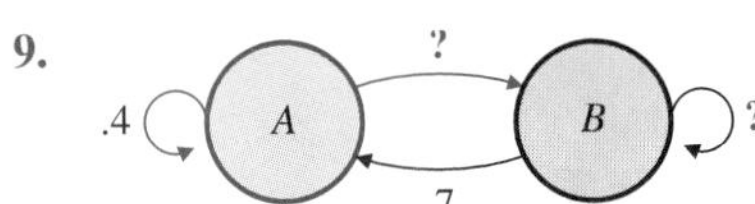

10.

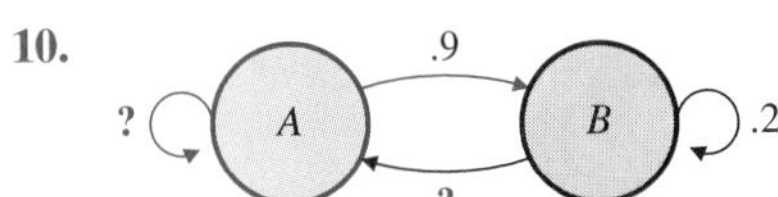

11.

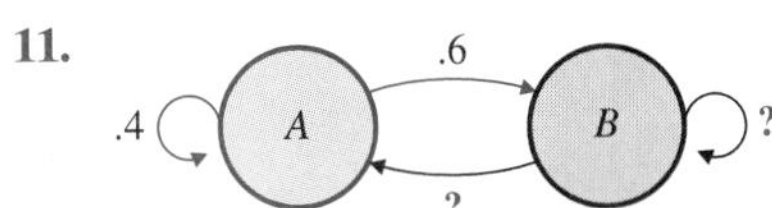

12.

? A ? .3 B .7

13.

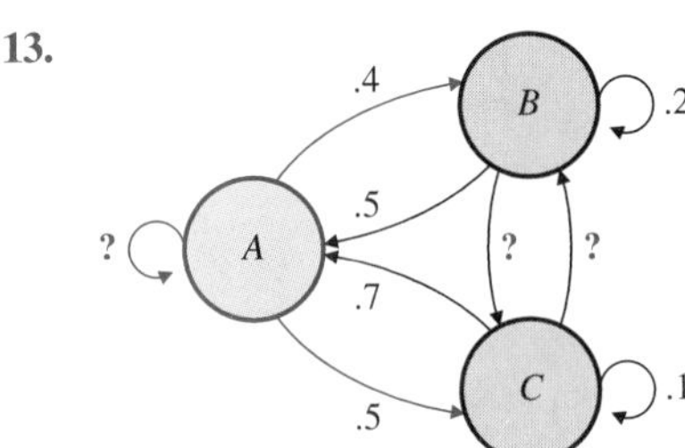

14.

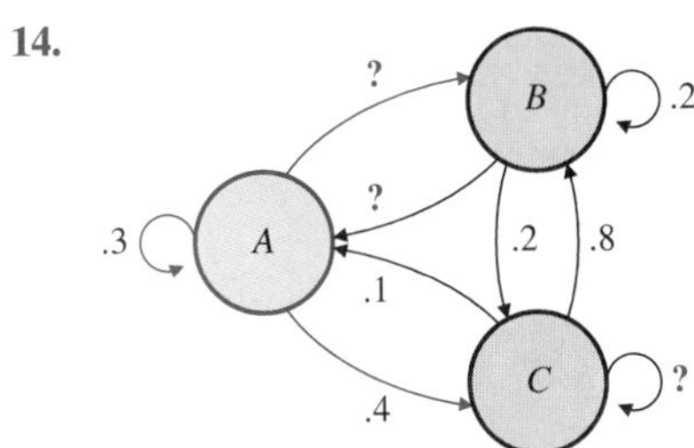

In Problems 15–20, are there unique values of a, b, and c that will make P a transition matrix? If so, complete the transition matrix and draw the corresponding transition diagram. If not, explain why.

15. $P = \begin{array}{c} \\ A \\ B \\ C \end{array}\begin{array}{c} \begin{array}{ccc} A & B & C \end{array} \\ \begin{bmatrix} 0 & .5 & a \\ b & 0 & .4 \\ .2 & c & .1 \end{bmatrix} \end{array}$

16. $P = \begin{array}{c} \\ A \\ B \\ C \end{array}\begin{array}{c} \begin{array}{ccc} A & B & C \end{array} \\ \begin{bmatrix} a & 0 & .9 \\ .2 & .3 & b \\ .6 & c & 0 \end{bmatrix} \end{array}$

17. $P = \begin{array}{c} \\ A \\ B \\ C \end{array}\begin{array}{c} \begin{array}{ccc} A & B & C \end{array} \\ \begin{bmatrix} 0 & a & .3 \\ 0 & b & 0 \\ c & .8 & 0 \end{bmatrix} \end{array}$

18. $P = \begin{array}{c} \\ A \\ B \\ C \end{array}\begin{array}{c} \begin{array}{ccc} A & B & C \end{array} \\ \begin{bmatrix} 0 & 1 & a \\ 0 & 0 & b \\ c & .5 & 0 \end{bmatrix} \end{array}$

19. $P = \begin{array}{c} \\ A \\ B \\ C \end{array}\begin{array}{c} \begin{array}{ccc} A & B & C \end{array} \\ \begin{bmatrix} .2 & .1 & .7 \\ a & .4 & c \\ .5 & b & .4 \end{bmatrix} \end{array}$

20. $P = \begin{array}{c} \\ A \\ B \\ C \end{array}\begin{array}{c} \begin{array}{ccc} A & B & C \end{array} \\ \begin{bmatrix} a & .8 & .1 \\ .3 & b & .4 \\ .6 & .5 & c \end{bmatrix} \end{array}$

B *In Problems 21–24, use the given information to draw the transition diagram and to find the transition matrix.*

21. A Markov process has two states, A and B. The probability of going from state A to state B in one trial is .7, and the probability of going from state B to state A in one trial is .9.

22. A Markov process has two states, A and B. The probability of going from state A to state A in one trial is .6, and the probability of going from state B to state B in one trial is .2.

23. A Markov chain has three states, A, B, and C. The probability of going from state A to state B in one trial is .1, the probability of going from state A to state C in one trial is .3, the probability of going from state B to state A in one trial is .2, the probability of going from state B to state C in one trial is .5, and the probability of going from state C to state C in one trial is 1.

24. A Markov chain has three states, A, B, and C. The probability of going from state A to state B in one trial is 1, the probability of going from state B to state A in one trial is .5, the probability of going from state B to state C in one trial is .5, and the probability of going from state C to state A in one trial is 1.

Problems 25–34 refer to the transition matrix P and the powers of P given below:

$$P = \begin{array}{c} \\ A \\ B \\ C \end{array}\begin{array}{c} \begin{array}{ccc} A & B & C \end{array} \\ \begin{bmatrix} .6 & .3 & .1 \\ .2 & .5 & .3 \\ .1 & .2 & .7 \end{bmatrix} \end{array} \qquad P^2 = \begin{array}{c} \\ A \\ B \\ C \end{array}\begin{array}{c} \begin{array}{ccc} A & B & C \end{array} \\ \begin{bmatrix} .43 & .35 & .22 \\ .25 & .37 & .38 \\ .17 & .27 & .56 \end{bmatrix} \end{array}$$

$$P^3 = \begin{array}{c} \\ A \\ B \\ C \end{array}\begin{array}{c} \begin{array}{ccc} A & B & C \end{array} \\ \begin{bmatrix} .35 & .348 & .302 \\ .262 & .336 & .402 \\ .212 & .298 & .49 \end{bmatrix} \end{array}$$

25. Find the probability of going from state A to state B in two trials.

26. Find the probability of going from state B to state C in two trials.

27. Find the probability of going from state C to state A in three trials.

28. Find the probability of going from state B to state B in three trials.

29. Find S_2 for $S_0 = [1 \quad 0 \quad 0]$, and explain what it represents.

30. Find S_2 for $S_0 = [0 \quad 1 \quad 0]$, and explain what it represents.

31. Find S_3 for $S_0 = [0 \quad 0 \quad 1]$, and explain what it represents.

32. Find S_3 for $S_0 = [1 \quad 0 \quad 0]$, and explain what it represents.

33. Using a graphing utility to compute powers of P, find the smallest positive integer n such that the corresponding entries in P^n and P^{n+1} are equal to two decimal places.

34. Using a graphing utility to compute powers of P, find the smallest positive integer n such that the corresponding entries in P^n and P^{n+1} are equal to three decimal places.

In Problems 35–38, given the transition matrix P and initial-state matrix S_0, find P^4 and use P^4 to find S_4.

35. $P = \begin{matrix} & A & B \\ A & .1 & .9 \\ B & .6 & .4 \end{matrix}$; $S_0 = [.8 \quad .2]$

36. $P = \begin{matrix} & A & B \\ A & .8 & .2 \\ B & .3 & .7 \end{matrix}$; $S_0 = [.4 \quad .6]$

37. $P = \begin{matrix} & A & B & C \\ A & 0 & .4 & .6 \\ B & 0 & 0 & 1 \\ C & 1 & 0 & 0 \end{matrix}$; $S_0 = [.2 \quad .3 \quad .5]$

38. $P = \begin{matrix} & A & B & C \\ A & 0 & 1 & 0 \\ B & .8 & 0 & .2 \\ C & 1 & 0 & 0 \end{matrix}$; $S_0 = [.4 \quad .2 \quad .4]$

39. A Markov process with two states has transition matrix P. If the initial-state matrix is $S_0 = [1 \quad 0]$, discuss the relationship between the entries in the kth-state matrix and the entries in the kth power of P.

40. Repeat Problem 39 if the initial-state matrix is $S_0 = [0 \quad 1]$.

C

41. Given the transition matrix

$$P = \begin{matrix} & A & B & C & D \\ A & .2 & .2 & .3 & .3 \\ B & 0 & 1 & 0 & 0 \\ C & .2 & .2 & .1 & .5 \\ D & 0 & 0 & 0 & 1 \end{matrix}$$

(A) Find P^4.
(B) Find the probability of going from state A to state D in four trials.
(C) Find the probability of going from state C to state B in four trials.
(D) Find the probability of going from state B to state A in four trials.

42. Repeat Problem 41 for the transition matrix

$$P = \begin{matrix} & A & B & C & D \\ A & .5 & .3 & .1 & .1 \\ B & 0 & 1 & 0 & 0 \\ C & 0 & 0 & 1 & 0 \\ D & .1 & .2 & .3 & .4 \end{matrix}$$

*A matrix is called a **probability matrix** if all its entries are real numbers between 0 and 1, inclusive, and the sum of the entries in each row is 1. Thus, transition matrices are square probability matrices and state matrices are probability matrices with one row.*

43. If

$$P = \begin{bmatrix} a & 1-a \\ 1-b & b \end{bmatrix}$$

is a probability matrix, show that P^2 is a probability matrix.

44. If

$$P = \begin{bmatrix} a & 1-a \\ 1-b & b \end{bmatrix} \quad \text{and} \quad S = [c \quad 1-c]$$

are probability matrices, show that SP is a probability matrix.

Use a graphing utility and the formula $S_k = S_0P^k$ (Theorem 1) to compute the required state matrices in Problems 45–48.

45. The transition matrix for a Markov chain is

$$P = \begin{bmatrix} .4 & .6 \\ .2 & .8 \end{bmatrix}$$

(A) If $S_0 = [0 \quad 1]$, find $S_2, S_4, S_8, \ldots$. Can you identify a state matrix S that the matrices S_k seem to be approaching?
(B) Repeat part (A) for $S_0 = [1 \quad 0]$.
(C) Repeat part (A) for $S_0 = [.5 \quad .5]$.
(D) Find SP for any matrix S you identified in parts (A)–(C).
(E) Write a brief verbal description of the long-term behavior of the state matrices of this Markov chain based on your observations in parts (A)–(D).

46. Repeat Problem 45 for $P = \begin{bmatrix} .9 & .1 \\ .4 & .6 \end{bmatrix}$.

47. Refer to Problem 45. Find P^k for $k = 2, 4, 8, \ldots$. Can you identify a matrix Q that the matrices P^k are approaching? If so, how is Q related to the results you discovered in Problem 45?

48. Refer to Problem 46. Find P^k for $k = 2, 4, 8, \ldots$. Can you identify a matrix Q that the matrices P^k are approaching? If so, how is Q related to the results you discovered in Problem 46?

APPLICATIONS

Business & Economics

49. *Scheduling.* An outdoor restaurant in a summer resort closes only on days that it rains. From past records it is found that from May through September, when it rains one day, the probability of rain for the next day is .4; when it does not rain one day, the probability of rain the next day is .06.
(A) Draw a transition diagram.
(B) Write the transition matrix.
(C) If it rains on Thursday, what is the probability that the restaurant will be closed on Saturday? On Sunday?

50. *Scheduling.* Repeat Problem 49 if the probability of rain following a rainy day is .6 and the probability of rain following a nonrainy day is .1.

51. *Advertising.* A vigorous television advertising campaign is conducted during the football season to promote a well-known brand X shaving cream. For each of several weeks, a survey is made, and it is found that each week 80% of those using brand X continue to use it and 20% switch to another brand. It is also found that of those not using brand X, 20% switch to brand X while the other 80% continue using another brand.
(A) Draw a transition diagram.
(B) Write the transition matrix.
(C) If 20% of the people are using brand X at the start of the advertising campaign, what percentage will be using it 1 week later? 2 weeks later?

52. *Car rental.* A car rental agency has rental and return facilities at both Kennedy and LaGuardia airports, two of the principal airports in the New York City area. Assume a car rented at either airport must be returned to one or the other airport. If a car is rented at LaGuardia, the probability that it will be returned there is .8; if a car is rented at Kennedy, the probability that it will be returned there is .7. Assume the company rents all its 100 cars each day and each car is rented (and returned) only once a day. If we start with 50 cars at each airport:
(A) What is the expected distribution the next day?
(B) What is the expected distribution 2 days later?

53. *Homeowners insurance.* The market for homeowners insurance in a particular city is dominated by two companies, National Property and United Family. Currently, National Property insures 50% of the homes in the city, United Family insures 30%, and the remainder are insured by a collection of smaller companies. United Family decides to offer rebates to increase its market share. This has the following effects on insurance purchases for the next several years: each year 25% of National Property's customers switch to United Family and 10% switch to other companies; 10% of United Family's customers switch to National Property and 5% switch to other companies; and 15% of the customers of other companies switch to National Property and 35% switch to United Family.
(A) Draw a transition diagram.
(B) Write the transition matrix.
(C) What percentage of the homes will be insured by National Property next year? The year after next?
(D) What percentage of the homes will be insured by United Family next year? The year after next?

54. *Service contracts.* A small community has two heating services that offer annual service contracts for home heating systems, Alpine Heating and Badger Furnaces. Currently, 25% of the homeowners have service contracts with Alpine, 30% have service contracts with Badger, and the remainder do not have service contracts. Both companies launch aggressive advertising campaigns to attract new customers, with the following effects on service contract purchases for the next several years: each year 35% of the homeowners with no current service contract decide to purchase a contract from Alpine and 40% decide to purchase one from Badger. In addition, 10% of the previous customers at each company decide to switch to the other company, and 5% decide they do not want a service contract.
(A) Draw a transition diagram.
(B) Write the transition matrix.
(C) What percentage of the homes will have service contracts with Alpine next year? The year after next?
(D) What percentage of the homes will have service contracts with Badger next year? The year after next?

55. *Employee training.* A nationwide chain of travel agencies maintains a training program for new travel agents. Initially, all new employees are classified as beginning agents requiring extensive supervision. Every 6 months the performance of each agent is reviewed. Past records indicate that after each semiannual review, 40% of the beginning agents are promoted to intermediate agents requiring only minimal supervision, 10% are terminated for unsatisfactory performance, and the remainder continue as beginning agents. Furthermore, 30% of the intermediate agents are promoted to qualified travel agents requiring no supervision, 10% are terminated for unsatisfactory performance, and the remainder continue as intermediate agents.
(A) Draw a transition diagram.
(B) Write the transition matrix.
(C) What is the probability that a beginning agent is promoted to qualified agent within 1 year? Within 2 years?

56. *Employee training.* All welders in a factory begin as apprentices. Every year the performance of each apprentice is reviewed. Past records indicate that after each review, 10% of the apprentices are promoted to professional welder, 20% are terminated for unsatisfactory performance, and the remainder continue as apprentices.
(A) Draw a transition diagram.
(B) Write the transition matrix.
(C) What is the probability that an apprentice is promoted to professional welder within 2 years? Within 4 years?

Life Sciences

57. *Health insurance.* A midwestern university offers its employees three choices for health care: a clinic-based health maintenance organization (HMO), a preferred provider organization (PPO), and a traditional fee-for-service program (FFS). Each year the university designates an open enrollment period during which employees may change from one health plan to another. Prior to the last open enrollment period, 20% of the employees were enrolled in the HMO, 25% in the PPO, and the remainder in the FFS. During the open enrollment period, 15% of the employees in the HMO switched to the PPO and 5% switched to the FFS; 20% of the employees in the PPO switched to the HMO and 10% to the FFS; and 25% of the employees in the FFS switched to the HMO and 30% switched to the PPO.
(A) Write the transition matrix.
(B) What percentage of employees were enrolled in each plan after the last open enrollment period?
(C) If this trend continues, what percentage of employees will be enrolled in each program after the next open enrollment period?

58. *Dental insurance.* Refer to Problem 57. During the open enrollment period, employees at the university can switch between the two available dental care programs, the low-option plan (LOP) and the high-option plan (HOP). Prior to the last open enrollment period, 40% of the employees were enrolled in the LOP and 60% in the HOP. During the open enrollment program, 30% of the employees in the LOP switched to the HOP and 10% of the employees in the HOP switched to the LOP.
(A) Write the transition matrix.
(B) What percentage of employees were enrolled in each plan after the last open enrollment period?
(C) If this trend continues, what percentage of employees will be enrolled in each program after the next open enrollment period?

Social Sciences

59. *Housing trends.* Home ownership in the United States declined during the 1980's, ending a 40 year period of growth. The 1980 census reported that 45.1% of the households in a particular city were homeowners and the remainder were renters. During the next decade, 12% of the homeowners became renters and the rest continued to be homeowners. Likewise, 5% of the renters became homeowners and the rest continued to rent.
(A) Write the appropriate transition matrix.
(B) According to this transition matrix, what percentage of households were homeowners in 1990?
(C) If the transition matrix remains the same, what percentage of the households will be homeowners in 2000?

60. *Housing trends.* The 1980 census reported that 73.3% of the households in a suburban county were homeowners. During the next decade, 5% of the homeowners became renters and the rest continued to be homeowners. Likewise, 3% of the renters became homeowners and the rest continued to rent.
(A) Write the appropriate transition matrix.
(B) According to this transition matrix, what percentage of households were homeowners in 1990?
(C) If the transition matrix remains the same, what percentage of the households will be homeowners in 2000?

SECTION 7-2 Regular Markov Chains

- STATIONARY MATRICES
- REGULAR MARKOV CHAINS
- APPLICATIONS
-  GRAPHING UTILITY APPROXIMATIONS

Given a Markov chain with transition matrix P and initial-state matrix S_0, the entries in the state matrix S_k are the probabilities of being in the corresponding states after k trials. What happens to these probabilities as the number of trials k increases? In this section we establish conditions on the transition matrix P that enable us to determine the long-run behavior of both the state matrices S_k and the powers of the transition matrix P^k.

■ Stationary Matrices

We begin by considering a concrete example—the toothpaste company discussed in the preceding section. Recall that the transition matrix was given by

$$P = \begin{matrix} & \begin{matrix} A & A' \end{matrix} \\ \begin{matrix} A \\ A' \end{matrix} & \begin{bmatrix} .8 & .2 \\ .6 & .4 \end{bmatrix} \end{matrix} \qquad \begin{aligned} A &= \text{Uses brand } A \text{ toothpaste} \\ A' &= \text{Uses another brand} \end{aligned}$$

Initially, this company had a 10% share of the toothpaste market. If the probabilities in the transition matrix P remain valid over a long period of time, what will happen to the company's market share? Examining the first several state matrices will give us some insight into this situation (matrix multiplication details are omitted):

$$\begin{aligned} S_0 &= [.1 \quad .9] \\ S_1 = S_0P &= [.62 \quad .38] \\ S_2 = S_1P &= [.724 \quad .276] \\ S_3 = S_2P &= [.7448 \quad .2552] \\ S_4 = S_3P &= [.74896 \quad .25104] \\ S_5 = S_4P &= [.749792 \quad .250208] \\ S_6 = S_5P &= [.7499584 \quad .2500416] \end{aligned}$$

It appears that the state matrices are getting closer and closer to $S = [.75 \quad .25]$ as we proceed to higher states. Let us multiply the matrix S (the matrix the other state matrices appear to be approaching) by the transition matrix:

$$SP = [.75 \quad .25]\begin{bmatrix} .8 & .2 \\ .6 & .4 \end{bmatrix} = [.75 \quad .25] = S$$

No change occurs! The matrix $[.75 \quad .25]$ is called a **stationary matrix.** If we reach this state or are very close to it, the system is said to be at steady-state—that is, later states either will not change or will not change very much. In terms of this example, this means that in the long run a person will purchase brand A with a probability of .75; that is, the company can expect to capture 75% of the market, assuming the transition matrix does not change.

The general definition of a stationary matrix is given in the next box.

Stationary Matrix for a Markov Chain

The state matrix $S = [s_1 \quad s_2 \quad \dots \quad s_n]$ is a **stationary matrix** for a Markov chain with transition matrix P if

$$SP = S$$

where $s_i \geq 0, i = 1, \dots, n$, and $s_1 + s_2 + \cdots + s_n = 1$.

Explore–Discuss 1

(A) Suppose the toothpaste company started with only 5% of the market instead of 10%. Write the initial-state matrix, find the next six state matrices, and discuss the behavior of these state matrices as you proceed to higher states.
(B) Repeat part (A) if the company started with 90% of the toothpaste market.

Regular Markov Chains

Does every Markov chain have a unique stationary matrix? And if a Markov chain has a unique stationary matrix, will the successive state matrices always approach this stationary matrix? Unfortunately, the answer to both these questions is no. (See Problems 31–34 in Exercise 7-2.) However, there is one important type of Markov chain for which both questions always can be answered in the affirmative. These are called *regular Markov chains.*

Regular Markov Chains

A transition matrix P is **regular** if some power of P has only positive entries. A Markov chain is a **regular Markov chain** if its transition matrix is regular.

Example 1 **Recognizing Regular Matrices** Which of the following matrices are regular?

(A) $P = \begin{bmatrix} .8 & .2 \\ .6 & .4 \end{bmatrix}$ (B) $P = \begin{bmatrix} 0 & 1 \\ 1 & 0 \end{bmatrix}$ (C) $P = \begin{bmatrix} .5 & .5 & 0 \\ 0 & .5 & .5 \\ 1 & 0 & 0 \end{bmatrix}$

Solution (A) This is the transition matrix for the toothpaste company. Since all the entries in P are positive, we can immediately conclude that P is regular.

(B) P has two 0 entries, so we must examine higher powers of P:

$$P^2 = \begin{bmatrix} 1 & 0 \\ 0 & 1 \end{bmatrix} \quad P^3 = \begin{bmatrix} 0 & 1 \\ 1 & 0 \end{bmatrix} \quad P^4 = \begin{bmatrix} 1 & 0 \\ 0 & 1 \end{bmatrix} \quad P^5 = \begin{bmatrix} 0 & 1 \\ 1 & 0 \end{bmatrix}$$

Since the powers of P oscillate between P and I, the 2×2 identity, all powers of P will contain 0 entries. Hence, P is not regular.

(C) Again, we examine higher powers of P:

$$P^2 = \begin{bmatrix} .25 & .5 & .25 \\ .5 & .25 & .25 \\ .5 & .5 & 0 \end{bmatrix} \quad P^3 = \begin{bmatrix} .375 & .375 & .25 \\ .5 & .375 & .125 \\ .25 & .5 & .25 \end{bmatrix}$$

Since all the entries in P^3 are positive, P is regular.

Matched Problem 1 Which of the following matrices are regular?

(A) $P = \begin{bmatrix} .3 & .7 \\ 1 & 0 \end{bmatrix}$ (B) $P = \begin{bmatrix} 1 & 0 \\ 1 & 0 \end{bmatrix}$ (C) $P = \begin{bmatrix} 0 & 1 & 0 \\ .5 & 0 & .5 \\ .5 & 0 & .5 \end{bmatrix}$

Explore–Discuss 2

Consider the toothpaste company (discussed at the beginning of this section) with regular transition matrix P and stationary matrix S, where

$$P = \begin{bmatrix} .8 & .2 \\ .6 & .4 \end{bmatrix} \quad \text{and} \quad S = [.75 \quad .25]$$

Compare P^2, P^4, and P^8. Discuss any apparent relationships between P^k and S.

The relationships among successive state matrices, powers of the transition matrix, and the stationary matrix for a regular Markov chain are given in Theorem 1. The proof of this theorem is left to more advanced courses.

Theorem 1 ■■ PROPERTIES OF REGULAR MARKOV CHAINS

Let P be the transition matrix for a regular Markov chain.

(A) There is a unique stationary matrix S that can be found by solving the equation

$$SP = S$$

(B) Given any initial-state matrix S_0, the state matrices S_k approach the stationary matrix S.

(C) The matrices P^k approach a **limiting matrix $\overline{P}$,** where each row of $\overline{P}$ is equal to the stationary matrix S. ■■

Example 2 ➡ **Finding the Stationary Matrix** The transition matrix for a Markov chain is

$$P = \begin{bmatrix} .7 & .3 \\ .2 & .8 \end{bmatrix}$$

(A) Find the stationary matrix S.

(B) Discuss the long-run behavior of S_k and P^k.

SOLUTION (A) Since P is regular, the stationary matrix S must exist. To find it, we must solve the equation $SP = S$. Let

$$S = [s_1 \quad s_2]$$

and write

$$[s_1 \quad s_2]\begin{bmatrix} .7 & .3 \\ .2 & .8 \end{bmatrix} = [s_1 \quad s_2]$$

After multiplying the left side, we obtain

$$[(.7s_1 + .2s_2) \quad (.3s_1 + .8s_2)] = [s_1 \quad s_2]$$

which is equivalent to the system

$$\begin{aligned} .7s_1 + .2s_2 &= s_1 \quad \text{or} \quad -.3s_1 + .2s_2 = 0 \\ .3s_1 + .8s_2 &= s_2 \quad \text{or} \quad .3s_1 - .2s_2 = 0 \end{aligned} \tag{1}$$

System (1) is dependent and has an infinite number of solutions. However, we are looking for a solution that is also a state matrix. This gives us another equation that we can add to (1) to obtain a system with a unique solution.

$$\begin{aligned} -.3s_1 + .2s_2 &= 0 \\ .3s_1 - .2s_2 &= 0 \\ s_1 + \quad s_2 &= 1 \end{aligned} \tag{2}$$

System (2) can be solved using matrix methods or elimination to obtain

$$s_1 = .4 \qquad \text{and} \qquad s_2 = .6$$

Thus,

$$S = [.4 \quad .6]$$

is the stationary matrix.

CHECK

$$SP = [.4 \quad .6]\begin{bmatrix} .7 & .3 \\ .2 & .8 \end{bmatrix} = [.4 \quad .6] = S$$

(B) Given any initial-state matrix S_0, Theorem 1 guarantees that the state matrices S_k will approach the stationary matrix S. Furthermore,

$$P^k = \begin{bmatrix} .7 & .3 \\ .2 & .8 \end{bmatrix}^k \quad \text{approaches the limiting matrix} \quad \overline{P} = \begin{bmatrix} .4 & .6 \\ .4 & .6 \end{bmatrix}$$

Matched Problem 2 The transition matrix for a Markov chain is

$$P = \begin{bmatrix} .6 & .4 \\ .1 & .9 \end{bmatrix}$$

Find the stationary matrix S and the limiting matrix $\overline{P}$.

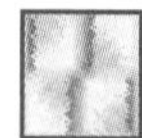

APPLICATIONS

Example 3 **Insurance** Refer to Example 1 in Section 7-1, where we found the following transition matrix for an insurance company:

$$P = \begin{array}{c} \\ A \\ A' \end{array}\!\!\begin{array}{c} \begin{array}{cc} A & A' \end{array} \\ \begin{bmatrix} .23 & .77 \\ .11 & .89 \end{bmatrix} \end{array} \qquad \begin{array}{l} A = \text{Accident} \\ A' = \text{No Accident} \end{array}$$

If these probabilities remain valid over a long period of time, what percentage of drivers can the company expect to have an accident during any given year?

SOLUTION To determine what happens in the long run, we find the stationary matrix by solving the following system:

$$[s_1 \quad s_2]\begin{bmatrix} .23 & .77 \\ .11 & .89 \end{bmatrix} = [s_1 \quad s_2] \qquad \text{and} \qquad s_1 + s_2 = 1$$

which is equivalent to

$$\begin{aligned} .23s_1 + .11s_2 &= s_1 \\ .77s_1 + .89s_2 &= s_2 \\ s_1 + \quad s_2 &= 1 \end{aligned} \qquad \text{or} \qquad \begin{aligned} -.77s_1 + .11s_2 &= 0 \\ .77s_1 - .11s_2 &= 0 \\ s_1 + \quad s_2 &= 1 \end{aligned}$$

Solving this system, we obtain

$$s_1 = .125 \quad \text{and} \quad s_2 = .875$$

The stationary matrix is $[.125 \quad .875]$, which means, in the long run, assuming the transition matrix does not change, about 12.5% of the drivers in the community will have an accident during any given year.

Matched Problem 3 Refer to Matched Problem 1 in Section 7-1, where we found the following transition matrix for an insurance company:

$$P = \begin{array}{c} \\ L \\ L' \end{array}\begin{array}{c} \begin{array}{cc} L & L' \end{array} \\ \begin{bmatrix} .98 & .02 \\ .78 & .22 \end{bmatrix} \end{array} \qquad \begin{array}{l} L = \text{Low-risk} \\ L' = \text{Not low-risk} \end{array}$$

If these probabilities remain valid over a long period of time, what percentage of drivers can the company expect to be in the low-risk category during any given year?

Example 4 **Employee Evaluation** A company rates every employee as below-average, average, or above-average. Past performance indicates that each year 10% of the below-average employees will raise their rating to average and 25% of the average employees will raise their rating to above-average. On the other hand, 15% of the average employees will lower their rating to below-average, and 15% of the above-average employees will lower their rating to average. Company policy prohibits rating changes from below-average to above-average, or conversely, in a single year. Over the long run, what percentage of employees will receive below-average ratings? Average ratings? Above-average ratings?

SOLUTION First, we find the transition matrix:

$$\begin{array}{c} \text{This} \\ \text{year} \end{array} \begin{array}{c} A^- \\ A \\ A^+ \end{array} \overset{\text{Next year}}{\overset{\begin{array}{ccc} A^- & A & A^+ \end{array}}{\begin{bmatrix} .9 & .1 & 0 \\ .15 & .6 & .25 \\ 0 & .15 & .85 \end{bmatrix}}} \qquad \begin{array}{l} A^- = \text{Below-average} \\ A = \text{Average} \\ A^+ = \text{Above-average} \end{array}$$

To determine what happens over the long run, we find the stationary matrix by solving the following system:

$$[s_1 \quad s_2 \quad s_3] \begin{bmatrix} .9 & .1 & 0 \\ .15 & .6 & .25 \\ 0 & .15 & .85 \end{bmatrix} = [s_1 \quad s_2 \quad s_3] \quad \text{and} \quad s_1 + s_2 + s_3 = 1$$

which is equivalent to

$$\begin{aligned} .9s_1 + .15s_2 &= s_1 \\ .1s_1 + .6s_2 + .15s_3 &= s_2 \\ .25s_2 + .85s_3 &= s_3 \\ s_1 + s_2 + s_3 &= 1 \end{aligned} \qquad \text{or} \qquad \begin{aligned} -.1s_1 + .15s_2 &= 0 \\ .1s_1 - .4s_2 + .15s_3 &= 0 \\ .25s_2 - .15s_3 &= 0 \\ s_1 + s_2 + s_3 &= 1 \end{aligned}$$

Using Gauss–Jordan elimination to solve this system of four equations with three variables, we obtain

$$s_1 = .36 \qquad s_2 = .24 \qquad s_3 = .4$$

Thus, in the long run, 36% of the employees will be rated as below-average, 24% as average, and 40% as above-average.

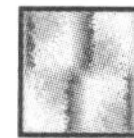

Matched Problem 4 ➡ A mail order company classifies its customers as preferred, standard, or infrequent, depending on the number of orders placed in a year. Past records indicate that each year 5% of the preferred customers are reclassified as standard and 12% as infrequent; 5% of the standard customers are reclassified as preferred and 5% as infrequent; and 9% of the infrequent customers are reclassified as preferred and 10% as standard. Assuming these percentages remain valid, what percentage of customers can the company expect to have in each category in the long run?

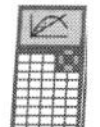

■ Graphing Utility Approximations

If P is the transition matrix for a regular Markov chain, then the powers of P approach the limiting matrix $\overline{P}$, where each row of $\overline{P}$ is equal to the stationary matrix S (Theorem 1C). We can use this result to approximate S by computing P^k for sufficiently large values of k. The next example illustrates this approach on a graphing utility.

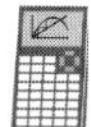

Example 5 ➡ **Approximating the Stationary Matrix** Compute powers of the transition matrix P to approximate $\overline{P}$ and S to four decimal places. Check the approximation in the equation $SP = S$.

$$P = \begin{bmatrix} .5 & .2 & .3 \\ .7 & .1 & .2 \\ .4 & .1 & .5 \end{bmatrix}$$

Solution To approximate $\overline{P}$ to four decimal places we enter P in a graphing utility (Fig. 1A), set the decimal display to four places, and compute powers of P until all three rows of P^k are identical. Examining the output in Figure 1B, we conclude that

$$\overline{P} = \begin{bmatrix} .4943 & .1494 & .3563 \\ .4943 & .1494 & .3563 \\ .4943 & .1494 & .3563 \end{bmatrix} \quad \text{and} \quad S = [.4943 \quad .1494 \quad .3563]$$

Entering S in the graphing utility and computing SP shows that these matrices are correct to four decimal places (see Fig. 1C).

```
P
        [[.5 .2 .3]
         [.7 .1 .2]
         [.4 .1 .5]]
```

(A) P

```
P^9
[[.4943 .1494 .3563]
 [.4943 .1494 .3563]
 [.4943 .1494 .3563]]
```

(B) P^9

```
S
[[.4943 .1494 .3563]]
S*P
[[.4943 .1494 .3563]]
```

(C) Check: $SP = S$

FIGURE 1

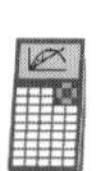

Matched Problem 5 ➡ Repeat Example 5 for: $P = \begin{bmatrix} .3 & .6 & .1 \\ .2 & .3 & .5 \\ .1 & .2 & .7 \end{bmatrix}$

REMARKS

1. We used a relatively small value of k to approximate $\overline{P}$ in Example 5. Many graphing utilities will compute P^k for large values of k almost as rapidly as for small values. However, round-off errors can occur in these calculations. A safe procedure is to start with a relatively small value of k, such as $k = 8$, and then keep doubling k until the rows of P^k are identical to the specified number of decimal places.
2. If any of the entries of P^k are approaching 0, the graphing utility may use scientific notation to display these entries as very small numbers. Figure 2 shows the 100th power of a transition matrix P. The entry in Row 2 and Column 2 of P^{100} is approaching 0, but the graphing utility displays it as 5.1538×10^{-53}. If this occurs, simply change this value to 0 in the corresponding entry in $\overline{P}$. Thus, from the output in Figure 2 we conclude that

FIGURE 2

$$P^k = \begin{bmatrix} 1 & 0 \\ .7 & .3 \end{bmatrix}^k \quad \text{approaches} \quad \overline{P} = \begin{bmatrix} 1 & 0 \\ 1 & 0 \end{bmatrix}$$

Answers to Matched Problems

1. (A) Regular (B) Not regular (C) Regular
2. $S = [.2 \quad .8]$; $\overline{P} = \begin{bmatrix} .2 & .8 \\ .2 & .8 \end{bmatrix}$
3. 97.5%
4. 28% preferred, 43% standard, 29% infrequent
5. $\overline{P} = \begin{bmatrix} .1618 & .2941 & .5441 \\ .1618 & .2941 & .5441 \\ .1618 & .2941 & .5441 \end{bmatrix}$; $S = [.1618 \quad .2941 \quad .5441]$

EXERCISE 7-2

A *In Problems 1–14, determine whether each transition matrix is regular.*

1. $P = \begin{bmatrix} .1 & .9 \\ .7 & .3 \end{bmatrix}$
2. $P = \begin{bmatrix} .8 & .2 \\ .5 & .5 \end{bmatrix}$
3. $P = \begin{bmatrix} 1 & 0 \\ 0 & 1 \end{bmatrix}$
4. $P = \begin{bmatrix} .5 & .5 \\ .5 & .5 \end{bmatrix}$
5. $P = \begin{bmatrix} .2 & .8 \\ 1 & 0 \end{bmatrix}$
6. $P = \begin{bmatrix} .9 & .1 \\ 0 & 1 \end{bmatrix}$
7. $P = \begin{bmatrix} 1 & 0 \\ .6 & .4 \end{bmatrix}$
8. $P = \begin{bmatrix} 0 & 1 \\ .3 & .7 \end{bmatrix}$
9. $P = \begin{bmatrix} .3 & .4 & .3 \\ 0 & 0 & 1 \\ .4 & .2 & .4 \end{bmatrix}$
10. $P = \begin{bmatrix} 0 & 0 & 1 \\ .1 & .1 & .8 \\ .5 & .3 & .2 \end{bmatrix}$
11. $P = \begin{bmatrix} 0 & 1 & 0 \\ .4 & 0 & .6 \\ 0 & 1 & 0 \end{bmatrix}$
12. $P = \begin{bmatrix} .1 & .3 & .6 \\ 0 & 0 & 1 \\ 1 & 0 & 0 \end{bmatrix}$
13. $P = \begin{bmatrix} 0 & 0 & 1 \\ .8 & .1 & .1 \\ 0 & 1 & 0 \end{bmatrix}$
14. $P = \begin{bmatrix} 0 & .5 & .5 \\ 1 & 0 & 0 \\ 1 & 0 & 0 \end{bmatrix}$

B *For each transition matrix P in Problems 15–22, solve the equation $SP = S$ to find the stationary matrix S and the limiting matrix $\overline{P}$.*

15. $P = \begin{bmatrix} .1 & .9 \\ .6 & .4 \end{bmatrix}$
16. $P = \begin{bmatrix} .8 & .2 \\ .3 & .7 \end{bmatrix}$
17. $P = \begin{bmatrix} .5 & .5 \\ .3 & .7 \end{bmatrix}$
18. $P = \begin{bmatrix} .9 & .1 \\ .7 & .3 \end{bmatrix}$
19. $P = \begin{bmatrix} .5 & .1 & .4 \\ .3 & .7 & 0 \\ 0 & .6 & .4 \end{bmatrix}$
20. $P = \begin{bmatrix} .4 & .1 & .5 \\ .2 & .8 & 0 \\ 0 & .5 & .5 \end{bmatrix}$
21. $P = \begin{bmatrix} .8 & .2 & 0 \\ .5 & .1 & .4 \\ 0 & .6 & .4 \end{bmatrix}$
22. $P = \begin{bmatrix} .2 & .8 & 0 \\ .6 & .1 & .3 \\ 0 & .9 & .1 \end{bmatrix}$

In Problems 23 and 24, discuss the validity of each statement. If the statement is always true, explain why. If not, give a counterexample.

23. (A) If two entries of a 2×2 transition matrix P are 0, then P is not regular.
(B) If three entries of a 3×3 transition matrix P are 0, then P is not regular.

24. (A) If P is an $n \times n$ transition matrix for a Markov chain and S is a $1 \times n$ matrix such that $SP = S$, then S is a stationary matrix.
(B) If a transition matrix P for a Markov chain has a stationary matrix S, then P is regular.

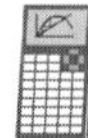

In Problems 25–28, approximate the stationary matrix S for each transition matrix P by computing powers of the transition matrix P. Round matrix entries to four decimal places.

25. $P = \begin{bmatrix} .51 & .49 \\ .27 & .73 \end{bmatrix}$

26. $P = \begin{bmatrix} .68 & .32 \\ .19 & .81 \end{bmatrix}$

27. $P = \begin{bmatrix} .5 & .5 & 0 \\ 0 & .5 & .5 \\ .8 & .1 & .1 \end{bmatrix}$

28. $P = \begin{bmatrix} .2 & .2 & .6 \\ .5 & 0 & .5 \\ .5 & 0 & .5 \end{bmatrix}$

C

29. A red urn contains 2 red marbles and 3 blue marbles, and a blue urn contains 1 red marble and 4 blue marbles. A marble is selected from an urn, the color is noted, and the marble is returned to the urn from which it was drawn. The next marble is drawn from the urn whose color is the same as the marble just drawn. Thus, this is a Markov process with two states: draw from the red urn or draw from the blue urn.
(A) Draw a transition diagram for this process.
(B) Write the transition matrix.
(C) Find the stationary matrix and describe the long-run behavior of this process.

30. Repeat Problem 29 if the red urn contains 5 red and 3 blue marbles, and the blue urn contains 1 red and 3 blue marbles.

31. Given the transition matrix:

$$P = \begin{bmatrix} 0 & 1 \\ 1 & 0 \end{bmatrix}$$

(A) Discuss the behavior of the state matrices $S_1, S_2, S_3, \ldots$ for the initial-state matrix $S_0 = [.2 \quad .8]$.
(B) Repeat part (A) for $S_0 = [.5 \quad .5]$.
(C) Discuss the behavior of P^k, $k = 2, 3, 4, \ldots$.
(D) Which of the conclusions of Theorem 1 are not valid for this matrix? Why is this not a contradiction?

32. Given the transition matrix:

$$P = \begin{bmatrix} 0 & 1 & 0 \\ 0 & 0 & 1 \\ 1 & 0 & 0 \end{bmatrix}$$

(A) Discuss the behavior of the state matrices $S_1, S_2, S_3, \ldots$ for the initial-state matrix $S_0 = [.2 \quad .3 \quad .5]$.
(B) Repeat part (A) for $S_0 = [\frac{1}{3} \quad \frac{1}{3} \quad \frac{1}{3}]$.
(C) Discuss the behavior of P^k, $k = 2, 3, 4, \ldots$.
(D) Which of the conclusions of Theorem 1 are not valid for this matrix? Why is this not a contradiction?

33. The transition matrix for a Markov chain is

$$P = \begin{bmatrix} 1 & 0 & 0 \\ .2 & .2 & .6 \\ 0 & 0 & 1 \end{bmatrix}$$

(A) Show that $R = [1 \quad 0 \quad 0]$ and $S = [0 \quad 0 \quad 1]$ are both stationary matrices for P. Explain why this does not contradict Theorem 1A.
(B) Find another stationary matrix for P. [*Hint:* Consider $T = aR + (1 - a)S$, where $0 < a < 1$.]
(C) How many different stationary matrices does P have?

34. The transition matrix for a Markov chain is

$$P = \begin{bmatrix} .7 & 0 & .3 \\ 0 & 1 & 0 \\ .2 & 0 & .8 \end{bmatrix}$$

(A) Show that $R = [.4 \quad 0 \quad .6]$ and $S = [0 \quad 1 \quad 0]$ are both stationary matrices for P. Explain why this does not contradict Theorem 1A.
(B) Find another stationary matrix for P. [*Hint:* Consider $T = aR + (1 - a)S$, where $0 < a < 1$.]
(C) How many different stationary matrices does P have?

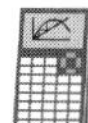

Problems 35 and 36 require the use of a graphing utility.

35. Refer to the transition matrix P in Problem 33. What matrix $\overline{P}$ do the powers of P appear to be approaching? Are the rows of $\overline{P}$ stationary matrices for P?

36. Refer to the transition matrix P in Problem 34. What matrix $\overline{P}$ do the powers of P appear to be approaching? Are the rows of $\overline{P}$ stationary matrices for P?

37. The transition matrix for a Markov chain is

$$P = \begin{bmatrix} .1 & .5 & .4 \\ .3 & .2 & .5 \\ .7 & .1 & .2 \end{bmatrix}$$

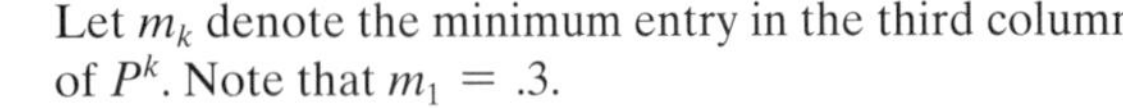

Let M_k denote the maximum entry in the second column of P^k. Note that $M_1 = .5$.

C (A) Find M_2, M_3, M_4, and M_5 to three decimal places.
(B) Explain why $M_k \geqslant M_{k+1}$ for all positive integers k.

38. The transition matrix for a Markov chain is

$$P = \begin{bmatrix} 0 & .2 & .8 \\ .3 & .3 & .4 \\ .6 & .1 & .3 \end{bmatrix}$$

Let m_k denote the minimum entry in the third column of P^k. Note that $m_1 = .3$.

C (A) Find m_2, m_3, m_4, and m_5 to three decimal places.
(B) Explain why $m_k \leqslant m_{k+1}$ for all positive integers k.

APPLICATIONS

Business & Economics

39. *Transportation.* Most railroad cars are owned by individual railroad companies. When a car leaves its home railroad's trackage, it becomes part of a national pool of cars and can be used by other railroads. The rules governing the use of these pooled cars are designed to eventually return the car to the home trackage. A particular railroad found that each month 11% of its box cars on the home trackage left to join the national pool and 29% of its box cars in the national pool were returned to the home trackage. If these percentages remain valid for a long period of time, what percentage of its box cars can this railroad expect to have on its home trackage in the long run?

40. *Transportation.* The railroad in Problem 39 also has a fleet of tank cars. If 14% of the tank cars on the home trackage enter the national pool each month and 26% of the tank cars in the national pool are returned to the home trackage each month, what percentage of its tank cars can the railroad expect to have on its home trackage in the long run?

41. *Labor force.* Table 1 gives the percentage of the female population of the United States who were members of the civilian labor force in the indicated years:

Table 1

YEAR	PERCENT
1970	43.3
1980	51.5
1990	57.5

Source: U.S. Bureau of Labor Statistics

The following transition matrix P is proposed as a model for the data, where L represents females who are in the labor force and L' represents females who are not in the labor force:

Next decade

$$\begin{matrix} & & L & L' \\ \text{Current} & L & .93 & .07 \\ \text{decade} & L' & .2 & .8 \end{matrix} = P$$

(A) Let $S_0 = [.433 \quad .567]$, and find S_1 and S_2. (Compute both matrices exactly and then round entries to three decimal places.)
(B) Construct a new table comparing the results from part (A) with the data in Table 1.
(C) According to this transition matrix, what percentage of the female population will be in the labor force in the long run?

42. *Market research.* Table 2 gives the percentage of households that owned a dog over an 8 year period:

Table 2

YEAR	PERCENT
1983	42.5
1987	38.2
1991	36.5

Source: American Veterinary Medical Association

The following transition matrix P is proposed as a model for the data, where D represents the households that own a dog:

Next 4 year period

$$\begin{matrix} & & D & D' \\ \text{Current} & D & .61 & .39 \\ \text{4 year period} & D' & .21 & .79 \end{matrix} = P$$

(A) Let $S_0 = [.425 \quad .575]$, and find S_1 and S_2. (Compute both matrices exactly and then round entries to three decimal places.)
(B) Construct a new table comparing the results from part (A) with the data in Table 2.
(C) According to this transition matrix, what percentage of the households will own a dog in the long run?

43. *Market share.* Consumers in a certain state can choose between three long-distance telephone services: GTT, NCJ, and Dash. Aggressive marketing by all three companies results in a continual shift of customers among the three services. Each year, GTT loses 5% of its customers to NCJ and 20% to Dash, NCJ loses 15% of its customers to GTT and 10% to Dash, and Dash loses 5% of its customers to GTT and 10% to NCJ. Assuming these percentages remain valid over a long period of time, what is each company's expected market share in the long run?

44. *Market share.* Consumers in a certain area can choose between three package delivery services: APS, GX, and WWP. Each week, APS loses 10% of its customers to GX and 20% to WWP, GX loses 15% of its customers to APS and 10% to WWP, and WWP loses 5% of its customers to APS and 5% to GX. Assuming these percentages remain valid over a long period of time, what is each company's expected market share in the long run?

45. *Insurance.* An auto insurance company classifies its customers in three categories: poor, satisfactory, and preferred. Each year, 40% of those in the poor category are moved to satisfactory, and 20% of those in the satisfactory category are moved to preferred. Also, 20% in the preferred category are moved to the satisfactory category, and 20% in the satisfactory category are moved to the poor category. Customers are never moved from poor to preferred, or conversely, in a single year. Assuming these percentages remain valid over a long period of time, how many customers can the company expect to have in each category in the long run?

46. *Insurance.* Repeat Problem 45 if 40% of the preferred customers are moved to the satisfactory category each year and all other information remains the same.

Problems 47 and 48 require the use of a graphing utility.

47. *Market share.* Acme Soap Company markets one brand of soap, called Standard Acme (SA), and Best Soap Company markets two brands, Standard Best (SB) and Deluxe Best (DB). Currently, Acme has 40% of the market, and the remainder is equally divided between the two Best brands. Acme is considering the introduction of a second brand to get a larger share of the market. A proposed new brand, called brand X, was test-marketed in several large cities, producing the following transition matrix for the consumers' weekly buying habits:

$$P = \begin{array}{c} SB \\ DB \\ SA \\ X \end{array} \overset{\begin{array}{cccc} SB & DB & SA & X \end{array}}{\begin{bmatrix} .4 & .1 & .3 & .2 \\ .3 & .2 & .2 & .3 \\ .1 & .2 & .2 & .5 \\ .3 & .3 & .1 & .3 \end{bmatrix}}$$

Assuming that P represents the consumers' buying habits over a long period of time, use this transition matrix and the initial-state matrix $S_0 = [.3 \quad .3 \quad .4 \quad 0]$ to compute successive state matrices in order to approximate the elements in the stationary matrix correct to two decimal places. If Acme decides to market this new soap, what is the long-run expected total market share for their two soaps?

48. *Market share.* Refer to Problem 47. The chemists at Acme Soap Company have developed a second new soap, called brand Y. Test-marketing this soap against the three established brands produces the following transition matrix:

$$P = \begin{array}{c} SB \\ DB \\ SA \\ Y \end{array} \overset{\begin{array}{cccc} SB & DB & SA & Y \end{array}}{\begin{bmatrix} .3 & .2 & .2 & .3 \\ .2 & .2 & .2 & .4 \\ .2 & .2 & .4 & .2 \\ .1 & .2 & .3 & .4 \end{bmatrix}}$$

Proceed as in Problem 47 to approximate the elements in the stationary matrix correct to two decimal places. If Acme decides to market brand Y, what is the long-run expected total market share for Standard Acme and brand Y? Should Acme market brand X or brand Y?

Life Sciences

49. *Genetics.* A given plant species has red, pink, or white flowers according to the genotypes RR, RW, and WW, respectively. If each of these genotypes is crossed with a pink-flowering plant (genotype RW), then the transition matrix is

$$\text{This generation} \quad \begin{array}{c} \text{Red} \\ \text{Pink} \\ \text{White} \end{array} \overset{\begin{array}{c}\text{Next generation} \\ \begin{array}{ccc} \text{Red} & \text{Pink} & \text{White} \end{array}\end{array}}{\begin{bmatrix} .5 & .5 & 0 \\ .25 & .5 & .25 \\ 0 & .5 & .5 \end{bmatrix}}$$

Assuming the plants of each generation are crossed only with pink plants to produce the next generation, show that regardless of the makeup of the first generation, the genotype composition will eventually stabilize at 25% red, 50% pink, and 25% white. (Find the stationary matrix.)

50. *Gene mutation.* Suppose a gene in a chromosome is of type A or type B. Assume the probability that a gene of type A will mutate to type B in one generation is 10^{-4} and that a gene of type B will mutate to type A is 10^{-6}.
(A) What is the transition matrix?
(B) After many generations, what is the probability that the gene will be of type A? Of type B? (Find the stationary matrix.)

Social Sciences

51. *Rapid transit.* A new rapid transit system has just started operating. In the first month of operation, it is found that 25% of the commuters are using the system, while 75% still travel by automobile. The following transition matrix was determined from records of other rapid transit systems:

$$\begin{array}{ccc} & & \text{Next month} \\ & & \begin{array}{cc}\text{Rapid transit} & \text{Automobile}\end{array} \\ \text{Current month} & \begin{array}{c}\text{Rapid transit} \\ \text{Automobile}\end{array} & \begin{bmatrix} .8 & .2 \\ .3 & .7 \end{bmatrix} \end{array}$$

(A) What is the initial-state matrix?
(B) What percentage of the commuters will be using the new system after 1 month? After 2 months?
(C) Find the percentage of commuters using each type of transportation after it has been in service for a long time.

52. *Politics—filibuster.* The Senate is in the middle of a floor debate and a filibuster is threatened. Senator Hanks, who is still vacillating, has a probability of .1 of changing his mind during the next 5 minutes. If this pattern continues for each 5 minutes that the debate continues, and if a 24 hour filibuster takes place before a vote is taken, what is the probability that Senator Hanks will cast a yes vote? A no vote?
(A) First, complete the following transition matrix:

$$\begin{array}{ccc} & & \text{Next 5 minutes} \\ & & \begin{array}{cc}\text{Yes} & \text{No}\end{array} \\ \text{Current 5 minutes} & \begin{array}{c}\text{Yes} \\ \text{No}\end{array} & \begin{bmatrix} .9 & .1 \\ & \end{bmatrix} \end{array}$$

(B) Find the stationary matrix and answer the two questions.
(C) What is the stationary matrix if the probability of Senator Hanks changing his mind (.1) is replaced with an arbitrary probability p?

The center of population of the 48 contiguous states of the United States is the point where a flat, rigid map of the contiguous states would balance if the location of each person was represented on the map by a weight of equal measure. In 1790 the population center was 23 miles east of Baltimore, Maryland. By 1990, the center had shifted about 800 miles west and 100 miles south to a point in southeast Missouri. To study this shifting population, the U.S. Bureau of the Census divides the states into the four regions shown in the figure. Problems 53 and 54 deal with population shifts among some of these regions.

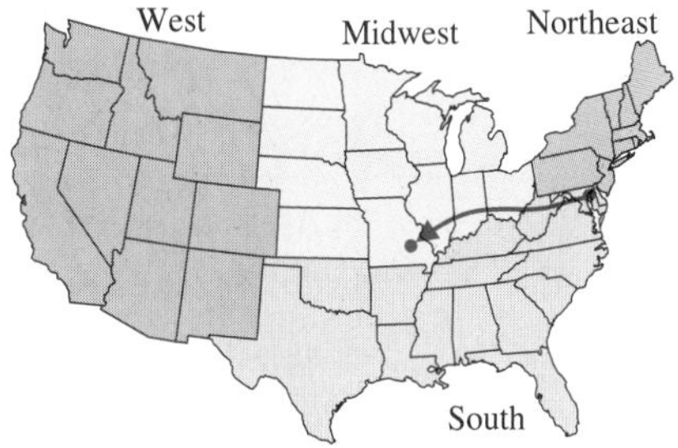

Figure for 53 and 54:
Regions of the United States and the center of population

53. *Population shifts.* Table 3 gives the percentage of the U.S. population living in the south region of the United States in the indicated years:

Table 3

YEAR	PERCENT
1970	30.9
1980	33.3
1990	34.4

Source: U.S. Bureau of the Census

The following transition matrix P is proposed as a model for the data, where S represents the population that lives in the south region:

$$\begin{array}{ccc} & & \text{Next decade} \\ & & \begin{array}{cc}S & S'\end{array} \\ \text{Current decade} & \begin{array}{c}S \\ S'\end{array} & \begin{bmatrix} .61 & .39 \\ .21 & .79 \end{bmatrix} = P \end{array}$$

(A) Let $S_0 = [.309 \quad .691]$, and find S_1 and S_2. (Compute both matrices exactly and then round entries to three decimal places.)

(B) Construct a new table comparing the results from part (A) with the data in Table 3.
(C) According to this transition matrix, what percentage of the population will live in the south region in the long run?

54. *Population shifts.* Table 4 gives the percentage of the U.S. population living in the northeast region of the United States in the indicated years:

Table 4

YEAR	PERCENT
1970	24.1
1980	21.7
1990	20.4

Source: U.S. Bureau of the Census

The following transition matrix P is proposed as a model for the data, where N represents the population that lives in the northeast region:

$$\begin{array}{cc} & \text{Next decade} \\ & \begin{array}{cc} N & N' \end{array} \\ \text{Current decade} \begin{array}{c} N \\ N' \end{array} & \begin{bmatrix} .61 & .39 \\ .09 & .91 \end{bmatrix} = P \end{array}$$

(A) Let $S_0 = [.241 \quad .759]$, and find S_1 and S_2. (Compute both matrices exactly and then round entries to three decimal places.)
(B) Construct a new table comparing the results from part (A) with the data in Table 4.
(C) According to this transition matrix, what percentage of the population will live in the northeast region in the long run?

Section 7-3 Absorbing Markov Chains

- Absorbing States and Absorbing Chains
- Standard Form
- Limiting Matrix
-  Graphing Utility Approximations

In Section 7-2, we saw that the powers of a regular transition matrix always approach a limiting matrix. Not all transition matrices have this property. In this section we discuss another type of Markov chain called an *absorbing Markov chain.* Although regular and absorbing Markov chains have some differences, they have one important similarity: the powers of the transition matrix for an absorbing Markov chain also approach a limiting matrix. After introducing basic concepts, we will develop methods for finding the limiting matrix and discuss the relationship between the states in the Markov chain and the entries in the limiting matrix.

Absorbing States and Absorbing Chains

A state in a Markov chain is called an **absorbing state** if once the state is entered, it is impossible to leave.

Example 1 **Recognizing Absorbing States** Identify any absorbing states for the following transition matrices:

$$\text{(A)}\quad P = \begin{array}{c} \\ A \\ B \\ C \end{array}\begin{array}{c} \begin{array}{ccc} A & B & C \end{array} \\ \begin{bmatrix} 1 & 0 & 0 \\ .5 & .5 & 0 \\ 0 & .5 & .5 \end{bmatrix} \end{array} \qquad \text{(B)}\quad P = \begin{array}{c} \\ A \\ B \\ C \end{array}\begin{array}{c} \begin{array}{ccc} A & B & C \end{array} \\ \begin{bmatrix} 0 & 0 & 1 \\ 0 & 1 & 0 \\ 1 & 0 & 0 \end{bmatrix} \end{array}$$

Solution (A) The probability of going from state A to state A is 1, and the probability of going from state A to either state B or state C is 0. Thus, once state A is entered, it is impossible to leave; hence, A is an absorbing state. Since the probability of going from state B to state A is nonzero, it is possible to leave B, and B is not an absorbing state. Similarly, the probability of going from state C to state B is nonzero, so C is not an absorbing state.

(B) Reasoning as before, the 1 in Row 2 and Column 2 indicates that state B is an absorbing state. The probability of going from state A to state C and the probability of going from state C to state A are both nonzero. Hence, A and C are not absorbing states.

Matched Problem 1 Identify any absorbing states for the following transition matrices:

$$\text{(A)}\quad P = \begin{matrix} & \begin{matrix} A & B & C \end{matrix} \\ \begin{matrix} A \\ B \\ C \end{matrix} & \begin{bmatrix} .5 & 0 & .5 \\ 0 & 1 & 0 \\ 0 & .5 & .5 \end{bmatrix} \end{matrix} \qquad \text{(B)}\quad P = \begin{matrix} & \begin{matrix} A & B & C \end{matrix} \\ \begin{matrix} A \\ B \\ C \end{matrix} & \begin{bmatrix} 0 & 1 & 0 \\ 1 & 0 & 0 \\ 0 & 0 & 1 \end{bmatrix} \end{matrix}$$

The reasoning used in Example 1 to identify absorbing states is generalized in Theorem 1.

Theorem 1 ABSORBING STATES AND TRANSITION MATRICES

A state in a Markov chain is **absorbing** if and only if the row of the transition matrix corresponding to the state has a 1 on the main diagonal and 0's elsewhere.

The presence of an absorbing state in a transition matrix does not guarantee that the powers of the matrix approach a limiting matrix nor that the state matrices in the corresponding Markov chain approach a stationary matrix. For example, if we square the matrix P from Example 1B, we obtain

$$P^2 = \begin{bmatrix} 0 & 0 & 1 \\ 0 & 1 & 0 \\ 1 & 0 & 0 \end{bmatrix} \begin{bmatrix} 0 & 0 & 1 \\ 0 & 1 & 0 \\ 1 & 0 & 0 \end{bmatrix} = \begin{bmatrix} 1 & 0 & 0 \\ 0 & 1 & 0 \\ 0 & 0 & 1 \end{bmatrix} = I$$

Since $P^2 = I$, the 3×3 identity matrix, it follows that

$$P^3 = PP^2 = PI = P \qquad \text{Since } P^2 = I$$
$$P^4 = PP^3 = PP = I \qquad \text{Since } P^3 = P \text{ and } PP = P^2 = I$$

In general, the powers of this transition matrix P oscillate between P and I and do not approach a limiting matrix.

Explore–Discuss 1

(A) For the initial-state matrix $S_0 = [a \quad b \quad c]$, find the first four state matrices, S_1, S_2, S_3, and S_4, in the Markov chain with transition matrix

$$P = \begin{bmatrix} 0 & 0 & 1 \\ 0 & 1 & 0 \\ 1 & 0 & 0 \end{bmatrix}$$

(B) Do the state matrices appear to be approaching a stationary matrix? Discuss.

To ensure that transition matrices for Markov chains with one or more absorbing states have limiting matrices, it is necessary to require the chain to satisfy one additional condition, as stated in the following definition.

> **Absorbing Markov Chains**
>
> A Markov chain is an **absorbing chain** if:
>
> 1. There is at least one absorbing state.
> 2. It is possible to go from each nonabsorbing state to at least one absorbing state in a finite number of steps.

As we saw earlier, absorbing states are easily identified by examining the rows of a transition matrix. It is also possible to use a transition matrix to determine whether a Markov chain is an absorbing chain, but this can be a difficult task, especially if the matrix is large. A transition diagram is often a more appropriate tool for determining whether a Markov chain is absorbing. The next example illustrates this approach for the two matrices discussed in Example 1.

Example 2 **Recognizing Absorbing Markov Chains** Use a transition diagram to determine whether P is the transition matrix for an absorbing Markov chain.

$$\text{(A)}\quad P = \begin{matrix} & \begin{matrix} A & B & C \end{matrix} \\ \begin{matrix} A \\ B \\ C \end{matrix} & \begin{bmatrix} 1 & 0 & 0 \\ .5 & .5 & 0 \\ 0 & .5 & .5 \end{bmatrix} \end{matrix} \qquad \text{(B)}\quad P = \begin{matrix} & \begin{matrix} A & B & C \end{matrix} \\ \begin{matrix} A \\ B \\ C \end{matrix} & \begin{bmatrix} 0 & 0 & 1 \\ 0 & 1 & 0 \\ 1 & 0 & 0 \end{bmatrix} \end{matrix}$$

SOLUTION

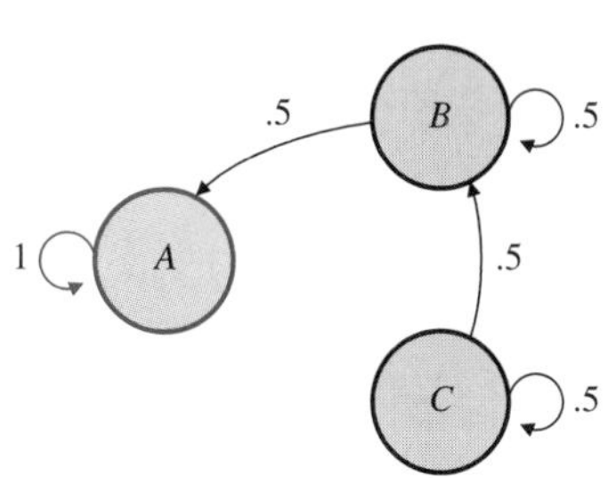

(A) From Example 1A, we know that A is the only absorbing state. The second condition in the definition of an absorbing Markov chain is satisfied if we can show that it is possible to go from the nonabsorbing states B and C to the absorbing state A in a finite number of steps. This is easily determined by drawing a transition diagram. Examining the diagram in the margin, we see that it is possible to go from state B to the absorbing state A in one step and from state C to the absorbing state A in two steps. Thus, P is the transition matrix for an absorbing Markov chain.

(B) Again, we draw the transition diagram for P, as shown. From this diagram it is clear that it is impossible to go from either state A or state C to the absorbing state B. Hence, P is not the transition matrix for an absorbing Markov chain.

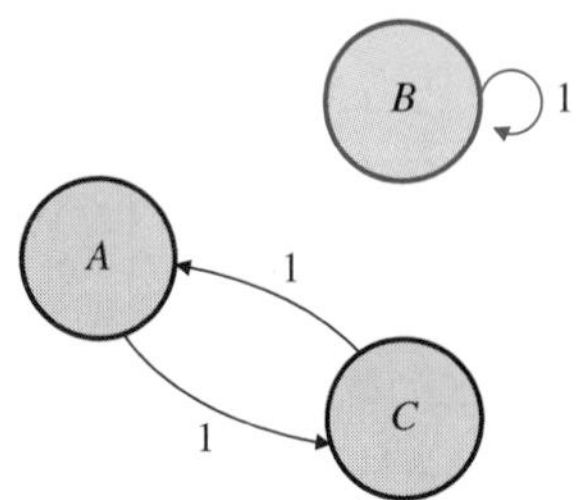

Matched Problem 2 Use a transition diagram to determine whether P is the transition matrix for an absorbing Markov chain.

(A) $P = \begin{array}{c} \\ A \\ B \\ C \end{array}\begin{array}{c} \begin{array}{ccc} A & B & C \end{array} \\ \begin{bmatrix} .5 & 0 & .5 \\ 0 & 1 & 0 \\ 0 & .5 & .5 \end{bmatrix} \end{array}$ (B) $P = \begin{array}{c} \\ A \\ B \\ C \end{array}\begin{array}{c} \begin{array}{ccc} A & B & C \end{array} \\ \begin{bmatrix} 0 & 1 & 0 \\ 1 & 0 & 0 \\ 0 & 0 & 1 \end{bmatrix} \end{array}$

Explore–Discuss 2

Determine whether each statement is true or false. Use examples and verbal arguments to support your conclusions.

(A) A Markov chain with two states, one nonabsorbing and one absorbing, is always an absorbing chain.
(B) A Markov chain with two states, both of which are absorbing, is always an absorbing chain.
(C) A Markov chain with three states, one nonabsorbing and two absorbing, is always an absorbing chain.

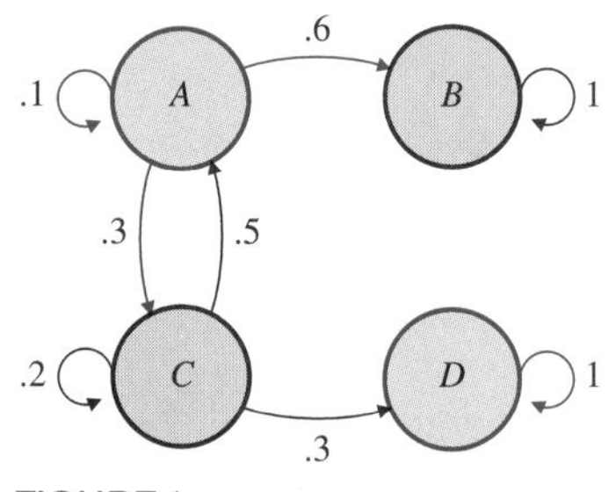

FIGURE 1

Standard Form

The transition matrix for a Markov chain is not unique. Consider the transition diagram in Figure 1. Since there are $4! = 24$ different ways to arrange the four states in this diagram, there are 24 different ways to write a transition matrix. (Some of these matrices may have identical entries, but all are different when the row and column labels are taken into account.) For example, matrices M, N, and P shown below are three different transition matrices for this diagram.

$$M = \begin{array}{c} \\ A \\ B \\ C \\ D \end{array}\begin{array}{c} \begin{array}{cccc} A & B & C & D \end{array} \\ \begin{bmatrix} .1 & .6 & .3 & 0 \\ 0 & 1 & 0 & 0 \\ .5 & 0 & .2 & .3 \\ 0 & 0 & 0 & 1 \end{bmatrix} \end{array} \quad N = \begin{array}{c} \\ D \\ B \\ C \\ A \end{array}\begin{array}{c} \begin{array}{cccc} D & B & C & A \end{array} \\ \begin{bmatrix} 1 & 0 & 0 & 0 \\ 0 & 1 & 0 & 0 \\ .3 & 0 & .2 & .5 \\ 0 & .6 & .3 & .1 \end{bmatrix} \end{array} \quad P = \begin{array}{c} \\ B \\ D \\ A \\ C \end{array}\begin{array}{c} \begin{array}{cccc} B & D & A & C \end{array} \\ \begin{bmatrix} 1 & 0 & 0 & 0 \\ 0 & 1 & 0 & 0 \\ .6 & 0 & .1 & .3 \\ 0 & .3 & .5 & .2 \end{bmatrix} \end{array} \quad (1)$$

In matrices N and P, notice that all the absorbing states precede all the nonabsorbing states. A transition matrix written in this form is said to be a *standard form.* We will find standard forms very useful in determining limiting matrices for absorbing Markov chains. The general definition of standard form is given in the next box.

Standard Forms for Absorbing Markov Chains

A transition matrix for an absorbing Markov chain is a **standard form** if the rows and columns are labeled so that all the absorbing states precede all the nonabsorbing states. (There may be more than one standard form.) Any standard form can always be partitioned into four submatrices:

$$\begin{array}{c} \\ A \\ N \end{array}\begin{array}{c} \begin{array}{cc} A & N \end{array} \\ \left[\begin{array}{c|c} I & 0 \\ \hline R & Q \end{array}\right] \end{array} \qquad \begin{array}{l} A = \text{All absorbing states} \\ N = \text{All nonabsorbing states} \end{array}$$

where I is an identity matrix and 0 is a zero matrix.

Referring to the matrix P in (1), we see that the submatrices in this standard form are

$$I = \begin{bmatrix} 1 & 0 \\ 0 & 1 \end{bmatrix} \qquad 0 = \begin{bmatrix} 0 & 0 \\ 0 & 0 \end{bmatrix} \qquad P = \begin{array}{c} \\ B \\ D \\ A \\ C \end{array} \begin{array}{c} \begin{array}{cccc} B & D & A & C \end{array} \\ \left[\begin{array}{cc|cc} 1 & 0 & 0 & 0 \\ 0 & 1 & 0 & 0 \\ \hline .6 & 0 & .1 & .3 \\ 0 & .3 & .5 & .2 \end{array}\right] \end{array}$$

$$R = \begin{bmatrix} .6 & 0 \\ 0 & .3 \end{bmatrix} \qquad Q = \begin{bmatrix} .1 & .3 \\ .5 & .2 \end{bmatrix}$$

Explore–Discuss 3

We used the diagram in Figure 1 to find the transition matrices M, N, and P in (1). Devise a procedure for transforming matrix M into standard forms N and P by interchanging rows and columns in M without reference to the transition diagram. Discuss the relative merits of using transition diagrams versus using row and column rearrangements to find standard forms.

Limiting Matrix

We are now ready to discuss the long-run behavior of absorbing Markov chains. We begin with an application.

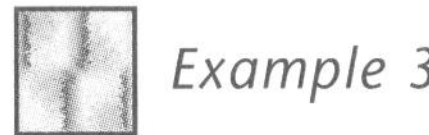

Example 3 **Real Estate Development** Two competing real estate companies are trying to buy all the farms in a particular area for future housing development. Each year, 20% of the farmers decide to sell to company A, 30% decide to sell to company B, and the rest continue to farm their land. Neither company ever sells any of the farms they purchase.

(A) Draw a transition diagram for this Markov process and determine whether the associated Markov chain is absorbing.
(B) Write a transition matrix that is in standard form.
(C) If neither company owns any farms at the beginning of this competitive buying process, estimate the percentage of farms that each company will purchase in the long run.
(D) If company A buys 50% of the farms before company B enters this competitive buying process, estimate the percentage of farms that each company will purchase in the long run.

Solution (A)

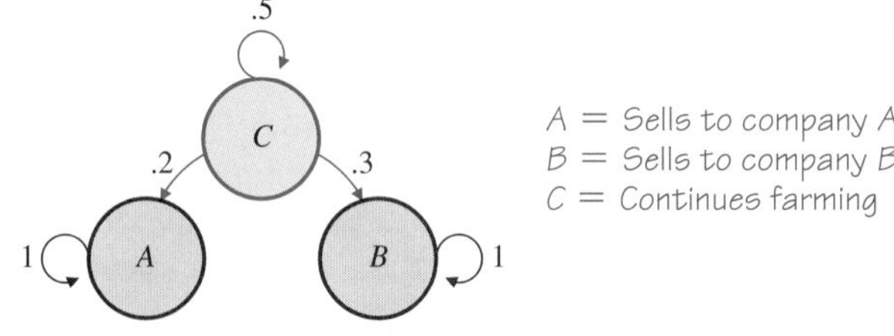

The associated Markov chain is absorbing, since there are two absorbing states, A and B, and it is possible to go from the nonabsorbing state C to either A or B in one step.

(B) We use the transition diagram to write a transition matrix that is in standard form:

$$P = \begin{matrix} & \begin{matrix} A & B & C \end{matrix} \\ \begin{matrix} A \\ B \\ C \end{matrix} & \begin{bmatrix} 1 & 0 & 0 \\ 0 & 1 & 0 \\ .2 & .3 & .5 \end{bmatrix} \end{matrix} \quad \text{Standard form}$$

(C) At the beginning of the competitive buying process all the farmers are in state C (own a farm). Thus, $S_0 = [0 \quad 0 \quad 1]$. The successive state matrices are (multiplication details omitted):

$$\begin{aligned}
S_1 &= S_0P = [.2 \quad .3 \quad .5] \\
S_2 &= S_1P = [.3 \quad .45 \quad .25] \\
S_3 &= S_2P = [.35 \quad .525 \quad .125] \\
S_4 &= S_3P = [.375 \quad .5625 \quad .0625] \\
S_5 &= S_4P = [.3875 \quad .58125 \quad .03125] \\
S_6 &= S_5P = [.39375 \quad .590625 \quad .015625] \\
S_7 &= S_6P = [.396875 \quad .5953125 \quad .0078125] \\
S_8 &= S_7P = [.3984375 \quad .59765625 \quad .00390625] \\
S_9 &= S_8P = [.39921875 \quad .598828125 \quad .001953125]
\end{aligned}$$

It appears that these state matrices are approaching the matrix

$$S = \begin{matrix} A & B & C \\ [.4 & .6 & 0] \end{matrix}$$

This indicates that in the long run, company A will acquire approximately 40% of the farms and company B will acquire the remaining 60%.

(D) This time, at the beginning of the competitive buying process 50% of the farmers are already in state A and the rest are in state C. Thus, $S_0 = [.5 \quad 0 \quad .5]$. The successive state matrices are (multiplication details omitted):

$$\begin{aligned}
S_1 &= S_0P = [.6 \quad .15 \quad .25] \\
S_2 &= S_1P = [.65 \quad .225 \quad .125] \\
S_3 &= S_2P = [.675 \quad .2625 \quad .0625] \\
S_4 &= S_3P = [.6875 \quad .28125 \quad .03125] \\
S_5 &= S_4P = [.69375 \quad .290625 \quad .015625] \\
S_6 &= S_5P = [.696875 \quad .2953125 \quad .0078125] \\
S_7 &= S_6P = [.6984375 \quad .29765625 \quad .00390625] \\
S_8 &= S_7P = [.69921875 \quad .298828125 \quad .001953125]
\end{aligned}$$

These state matrices approach a matrix different from the one in part (A):

$$S' = \begin{matrix} A & B & C \\ [.7 & .3 & 0] \end{matrix}$$

Because of its head start, company A will now acquire approximately 70% of the farms and company B will acquire the remaining 30%.

Matched Problem 3 ➡ Repeat Example 3 if 10% of the farmers sell to company A each year, 40% sell to company B, and the remainder continue farming.

Recall from Theorem 1 in Section 7-2 that the successive state matrices of a regular Markov chain always approach a stationary matrix. Furthermore, this stationary matrix is unique. That is, changing the initial-state matrix does not change the stationary matrix. The successive state matrices for an absorbing Markov chain also approach a stationary matrix, but this matrix is not unique. To confirm this, consider the transition matrix P and the state matrices S and S' from Example 3:

$$P = \begin{matrix} & \begin{matrix} A & B & C \end{matrix} \\ \begin{matrix} A \\ B \\ C \end{matrix} & \begin{bmatrix} 1 & 0 & 0 \\ 0 & 1 & 0 \\ .2 & .3 & .5 \end{bmatrix} \end{matrix} \qquad S = \overset{\begin{matrix} A & B & C \end{matrix}}{[.4 \quad .6 \quad 0]} \qquad S' = \overset{\begin{matrix} A & B & C \end{matrix}}{[.7 \quad .3 \quad 0]}$$

It turns out that S and S' are both stationary matrices, as the following multiplications verify:

$$SP = [.4 \quad .6 \quad 0]\begin{bmatrix} 1 & 0 & 0 \\ 0 & 1 & 0 \\ .2 & .3 & .5 \end{bmatrix} = [.4 \quad .6 \quad 0] = S$$

$$S'P = [.7 \quad .3 \quad 0]\begin{bmatrix} 1 & 0 & 0 \\ 0 & 1 & 0 \\ .2 & .3 & .5 \end{bmatrix} = [.7 \quad .3 \quad 0] = S'$$

In fact, this absorbing Markov chain has an infinite number of stationary matrices (see Problems 41 and 42 in Exercise 7-3).

Thus, changing the initial-state matrix for an absorbing Markov chain can cause the successive state matrices to approach a different stationary matrix.

In Section 7-2 we used the unique stationary matrix for a regular Markov chain to find the limiting matrix $\overline{P}$. Since an absorbing Markov chain can have many different stationary matrices, we cannot expect this approach to work for absorbing chains. However, it turns out that transition matrices for absorbing chains do have limiting matrices, and they are not very difficult to find. Theorem 2 gives us the necessary tools. The proof of this theorem is left for more advanced courses.

Theorem 2 ■■ LIMITING MATRICES FOR ABSORBING MARKOV CHAINS

If a standard form P for an absorbing Markov chain is partitioned as

$$P = \left[\begin{array}{c|c} I & 0 \\ \hline R & Q \end{array}\right]$$

then P^k approaches a limiting matrix $\overline{P}$ as k increases, where

$$\overline{P} = \left[\begin{array}{c|c} I & 0 \\ \hline FR & 0 \end{array}\right]$$

The matrix F is given by $F = (I - Q)^{-1}$ and is called the **fundamental matrix** for P.

The identity matrix used to form the fundamental matrix F must be the same size as the matrix Q. ■■

Example 4 **Finding the Limiting Matrix**

(A) Find the limiting matrix $\overline{P}$ for the standard form P found in Example 3.
(B) Use $\overline{P}$ to find the limit of the successive state matrices for $S_0 = [0 \quad 0 \quad 1]$.
(C) Use $\overline{P}$ to find the limit of the successive state matrices for $S_0 = [.5 \quad 0 \quad .5]$.

SOLUTION (A) From Example 3, we have

$$P = \begin{bmatrix} 1 & 0 & 0 \\ 0 & 1 & 0 \\ .2 & .3 & .5 \end{bmatrix} \quad \begin{bmatrix} I & 0 \\ R & Q \end{bmatrix}$$

where

$$I = \begin{bmatrix} 1 & 0 \\ 0 & 1 \end{bmatrix} \qquad 0 = \begin{bmatrix} 0 \\ 0 \end{bmatrix} \qquad R = [.2 \quad .3] \qquad Q = [.5]$$

If $I = [1]$ is the 1×1 identity matrix, then $I - Q$ is also a 1×1 matrix, and $F = (I - Q)^{-1}$ is simply the multiplicative inverse of the single entry in $I - Q$. Thus,

$$F = ([1] - [.5])^{-1} = [.5]^{-1} = [2]$$
$$FR = [2][.2 \quad .3] = [.4 \quad .6]$$

and the limiting matrix is

$$\overline{P} = \begin{array}{c} \\ A \\ B \\ C \end{array} \begin{array}{c} \begin{array}{ccc} A & B & C \end{array} \\ \begin{bmatrix} 1 & 0 & 0 \\ 0 & 1 & 0 \\ .4 & .6 & 0 \end{bmatrix} \end{array} \quad \begin{bmatrix} I & 0 \\ FR & 0 \end{bmatrix}$$

(B) Since the successive state matrices are given by $S_k = S_0P^k$ (Theorem 1 in Section 7-1) and P^k approaches $\overline{P}$, it follows that S_k approaches

$$S_0\overline{P} = [0 \quad 0 \quad 1]\begin{bmatrix} 1 & 0 & 0 \\ 0 & 1 & 0 \\ .4 & .6 & 0 \end{bmatrix} = [.4 \quad .6 \quad 0]$$

which agrees with the results in part (C) of Example 3.

(C) This time, the successive state matrices approach

$$S_0\overline{P} = [.5 \quad 0 \quad .5]\begin{bmatrix} 1 & 0 & 0 \\ 0 & 1 & 0 \\ .4 & .6 & 0 \end{bmatrix} = [.7 \quad .3 \quad 0]$$

which agrees with the results in part (D) of Example 3.

Matched Problem 4 Repeat Example 4 for the standard form P found in Matched Problem 3.

Recall that the limiting matrix for a regular Markov chain contains the long-run probabilities of going from any state to any other state. This is also true for the limiting matrix of an absorbing Markov chain. Let's compare the transition matrix P and its limiting matrix $\overline{P}$ from Example 4:

$$P = \begin{matrix} & \begin{matrix} A & B & C \end{matrix} \\ \begin{matrix} A \\ B \\ C \end{matrix} & \begin{bmatrix} 1 & 0 & 0 \\ 0 & 1 & 0 \\ .2 & .3 & .5 \end{bmatrix} \end{matrix} \quad \text{approaches} \quad \overline{P} = \begin{matrix} & \begin{matrix} A & B & C \end{matrix} \\ \begin{matrix} A \\ B \\ C \end{matrix} & \begin{bmatrix} 1 & 0 & 0 \\ 0 & 1 & 0 \\ .4 & .6 & 0 \end{bmatrix} \end{matrix}$$

The rows of P and $\overline{P}$ corresponding to the absorbing states A and B are identical. That is, if the probability of going from state A to state A is 1 at the beginning of the chain, this probability will remain 1 for all trials in the chain and for the limiting matrix. The entries in the third row of $\overline{P}$ give the long-run probabilities of going from the nonabsorbing state C to states A, B, or C.

The fundamental matrix F provides some additional information about an absorbing chain. Recall from Example 4 that $F = [2]$. It can be shown that the entries in F determine the average number of trials it takes to go from a given nonabsorbing state to an absorbing state. In the case of Example 4, the single entry 2 in F indicates that it will take an average of 2 years for a farmer to go from state C (owns a farm) to one of the absorbing states (sells the farm). Some will reach an absorbing state in 1 year, some will take more than 2 years, but the average will be 2 years. These observations are summarized in Theorem 3, which we state below without proof.

Theorem 3 ■■ PROPERTIES OF THE LIMITING MATRIX $\overline{P}$

If P is a transition matrix in standard form for an absorbing Markov chain, F is the fundamental matrix, and $\overline{P}$ is the limiting matrix, then:

(A) The entry in Row i and Column j of $\overline{P}$ is the long-run probability of going from state i to state j. For the nonabsorbing states, these probabilities are also the entries in the matrix FR used to form $\overline{P}$.

(B) The sum of the entries in each row of the fundamental matrix F is the average number of trials it will take to go from each nonabsorbing state to some absorbing state.

[Note that the rows of both F and FR correspond to the nonabsorbing states in the order given in the standard form P.] ■■

REMARKS

1. The zero matrix in the lower right corner of the limiting matrix $\overline{P}$ in Theorem 2 indicates that the long-run probability of going from any nonabsorbing state to any other nonabsorbing state is always 0. That is, in the long run, all elements in an absorbing Markov chain end up in one of the absorbing states.
2. If the transition matrix for an absorbing Markov chain is not a standard form, it is still possible to find a limiting matrix (see Problems 37 and 38 in Exercise 7-3). However, it is customary to use a standard form when investigating the limiting behavior of an absorbing chain.

Now that we have developed the necessary tools for analyzing the long-run behavior of an absorbing Markov chain, we apply these tools to an application we considered earlier (see Example 4 in Section 7-1).

Example 5 **University Enrollment** The transition diagram for part-time students enrolled in an MBA program in a university is shown below:

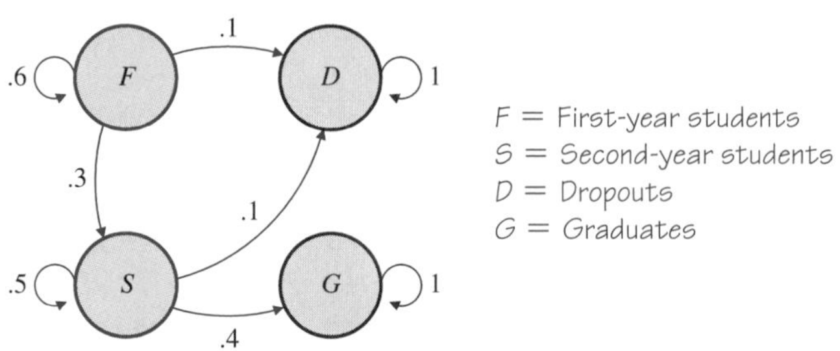

(A) In the long run, what percentage of first-year students will graduate? What percentage of second-year students will not graduate?

(B) What is the average number of years a first-year student will remain in this program? A second-year student?

SOLUTION (A) First, notice that this is an absorbing Markov chain with two absorbing states, state D and state G. A standard form for this absorbing chain is

$$P = \begin{array}{c} \\ D \\ G \\ F \\ S \end{array}\begin{array}{c} \begin{array}{cccc} D & G & F & S \end{array} \\ \left[\begin{array}{cc|cc} 1 & 0 & 0 & 0 \\ 0 & 1 & 0 & 0 \\ \hline .1 & 0 & .6 & .3 \\ .1 & .4 & 0 & .5 \end{array}\right] \end{array} \quad \left[\begin{array}{c|c} I & 0 \\ \hline R & Q \end{array}\right]$$

The submatrices in this partition are

$$I = \begin{bmatrix} 1 & 0 \\ 0 & 1 \end{bmatrix} \qquad 0 = \begin{bmatrix} 0 & 0 \\ 0 & 0 \end{bmatrix} \qquad R = \begin{bmatrix} .1 & 0 \\ .1 & .4 \end{bmatrix} \qquad Q = \begin{bmatrix} .6 & .3 \\ 0 & .5 \end{bmatrix}$$

Thus,

$$F = (I - Q)^{-1} = \left(\begin{bmatrix} 1 & 0 \\ 0 & 1 \end{bmatrix} - \begin{bmatrix} .6 & .3 \\ 0 & .5 \end{bmatrix}\right)^{-1}$$

$$= \begin{bmatrix} .4 & -.3 \\ 0 & .5 \end{bmatrix}^{-1}$$ Use row operations to find this matrix inverse.

$$= \begin{bmatrix} 2.5 & 1.5 \\ 0 & 2 \end{bmatrix}$$

and

$$FR = \begin{bmatrix} 2.5 & 1.5 \\ 0 & 2 \end{bmatrix}\begin{bmatrix} .1 & 0 \\ .1 & .4 \end{bmatrix} = \begin{bmatrix} .4 & .6 \\ .2 & .8 \end{bmatrix}$$

The limiting matrix is

$$\bar{P} = \begin{array}{c} \\ D \\ G \\ F \\ S \end{array}\begin{array}{c} \begin{array}{cccc} D & G & F & S \end{array} \\ \begin{bmatrix} 1 & 0 & 0 & 0 \\ 0 & 1 & 0 & 0 \\ .4 & .6 & 0 & 0 \\ .2 & .8 & 0 & 0 \end{bmatrix} \end{array} \quad \left[\begin{array}{c|c} I & 0 \\ \hline FR & 0 \end{array}\right]$$

From this limiting form, we see that in the long run 60% of the first-year students will graduate and 20% of the second-year students will not graduate.

(B) The sum of the entries in the first row of the fundamental matrix F is $2.5 + 1.5 = 4$. According to Theorem 3, this indicates that a first-year student will spend an average of 4 years in the transient states F and S before reaching one of the absorbing states, D or G. The sum of the entries in the second row of F is $0 + 2 = 2$. Thus, a second-year student spends an average of 2 years in the program before either graduating or dropping out.

Matched Problem 5 Repeat Example 5 for the following transition diagram:

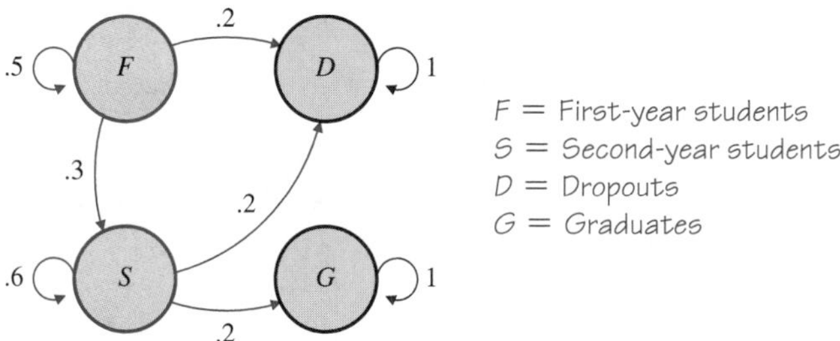

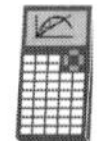

■ Graphing Utility Approximations

Just as was the case for regular Markov chains, the limiting matrix $\overline{P}$ for an absorbing Markov chain with transition matrix P can be approximated by computing P^k on a graphing utility for sufficiently large values of k. For example, computing P^{50} for the standard form P in Example 5 produces the following results:

$$P^{50} = \begin{bmatrix} 1 & 0 & 0 & 0 \\ 0 & 1 & 0 & 0 \\ .1 & 0 & .6 & .3 \\ .1 & .4 & 0 & .5 \end{bmatrix}^{50} = \begin{bmatrix} 1 & 0 & 0 & 0 \\ 0 & 1 & 0 & 0 \\ .4 & .6 & 0 & 0 \\ .2 & .8 & 0 & 0 \end{bmatrix} = \overline{P}$$

where once again we have replaced very small numbers displayed in scientific notation with 0 (see Remark 2 on page 483 in Section 7-2).

Caution

Before you use P^k to approximate $\overline{P}$, be certain to determine that $\overline{P}$ does in fact exist. If you attempt to approximate a limiting matrix when none exists, the results can be misleading. For example, consider the transition matrix

$$P = \begin{bmatrix} 1 & 0 & 0 & 0 & 0 \\ .2 & .2 & 0 & .3 & .3 \\ 0 & 0 & 0 & .5 & .5 \\ 0 & 0 & 1 & 0 & 0 \\ 0 & 0 & 1 & 0 & 0 \end{bmatrix}$$

Computing P^{50} on a graphing utility produces the following matrix:

$$P^{50} = \begin{bmatrix} 1 & 0 & 0 & 0 & 0 \\ .25 & 0 & .625 & .0625 & .0625 \\ 0 & 0 & 1 & 0 & 0 \\ 0 & 0 & 0 & .5 & .5 \\ 0 & 0 & 0 & .5 & .5 \end{bmatrix} \qquad (2)$$

It is tempting to stop at this point and conclude that the matrix in (2) must be a good approximation for $\overline{P}$. But to do so would be incorrect! If P^{50} approximates a limiting matrix $\overline{P}$, then P^{51} should also approximate the same matrix. However, computing P^{51} produces quite a different matrix:

$$P^{51} = \begin{bmatrix} 1 & 0 & 0 & 0 & 0 \\ .25 & 0 & .125 & .3125 & .3125 \\ 0 & 0 & 0 & .5 & .5 \\ 0 & 0 & 1 & 0 & 0 \\ 0 & 0 & 1 & 0 & 0 \end{bmatrix} \qquad (3)$$

Computing additional powers of P shows that the even powers of P approach the matrix in (2) while the odd powers approach the matrix in (3). Thus, the transition matrix P does not have a limiting matrix.

A graphing utility also can be used to perform the matrix calculations necessary to find $\overline{P}$ exactly, as illustrated in Figure 2 for the transition matrix P from Example 5. This approach has the advantage of producing the fundamental matrix F whose row sums provide additional information about the long-run behavior of the chain.

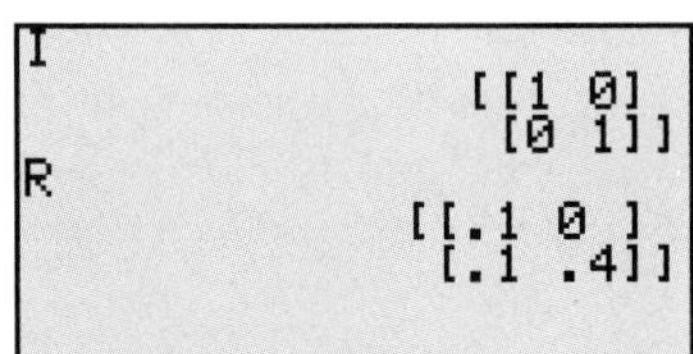

(A) Store I and R in the graphing utility memory

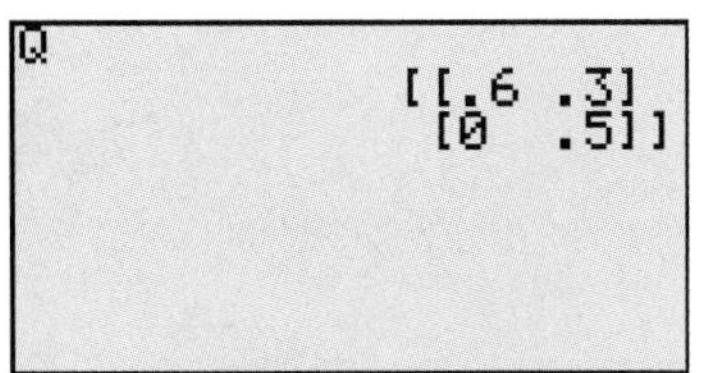

(B) Store Q in the graphing utility memory

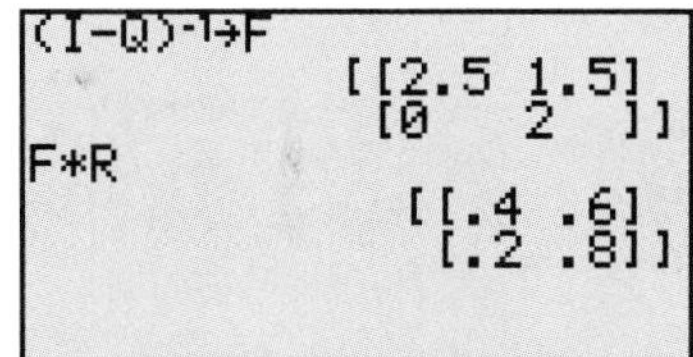

(C) Compute F and FR

FIGURE 2
Matrix calculations

Answers to Matched Problems

1. (A) State B is absorbing. (B) State C is absorbing.
2. (A) Absorbing Markov chain (B) Not an absorbing Markov chain
3. (A)

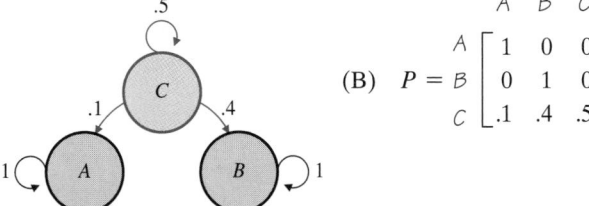

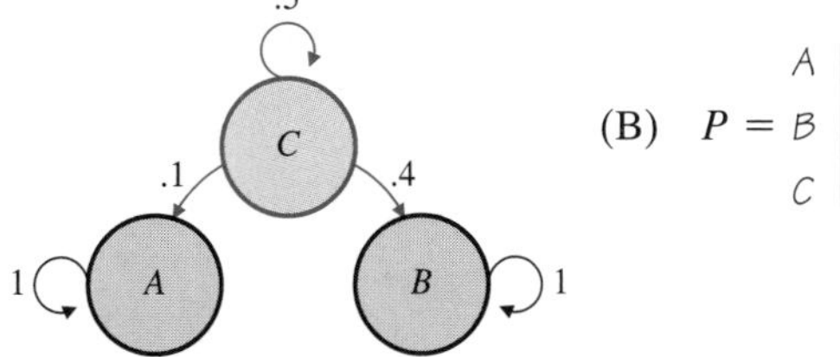

(B) $P = \begin{matrix} & \begin{matrix} A & B & C \end{matrix} \\ \begin{matrix} A \\ B \\ C \end{matrix} & \begin{bmatrix} 1 & 0 & 0 \\ 0 & 1 & 0 \\ .1 & .4 & .5 \end{bmatrix} \end{matrix}$

(C) Company A will purchase 20% of the farms and company B will purchase 80%.

(D) Company A will purchase 60% of the farms and company B will purchase 40%.

4. (A) $\overline{P} = \begin{matrix} & A & B & C \\ A & 1 & 0 & 0 \\ B & 0 & 1 & 0 \\ C & .2 & .8 & 0 \end{matrix}$ (B) [.2 .8 0] (C) [.6 .4 0]

5. (A) 30% of the first-year students will graduate; 50% of the second-year students will not graduate.

(B) A first-year student will spend an average of 3.5 years in the program; a second-year student will spend an average of 2.5 years in the program.

EXERCISE 7-3

A *In Problems 1–6, identify the absorbing states in the indicated transition matrix.*

1. $P = \begin{matrix} & A & B & C \\ A & .6 & .3 & .1 \\ B & 0 & 1 & 0 \\ C & 0 & 0 & 1 \end{matrix}$

2. $P = \begin{matrix} & A & B & C \\ A & 0 & 1 & 0 \\ B & .3 & .2 & .5 \\ C & 0 & 0 & 1 \end{matrix}$

3. $P = \begin{matrix} & A & B & C \\ A & 0 & 0 & 1 \\ B & 1 & 0 & 0 \\ C & 0 & 1 & 0 \end{matrix}$

4. $P = \begin{matrix} & A & B & C \\ A & 1 & 0 & 0 \\ B & .3 & .4 & .3 \\ C & 0 & 0 & 1 \end{matrix}$

5. $P = \begin{matrix} & A & B & C & D \\ A & 1 & 0 & 0 & 0 \\ B & 0 & 0 & 1 & 0 \\ C & .1 & .1 & .5 & .3 \\ D & 0 & 0 & 0 & 1 \end{matrix}$

6. $P = \begin{matrix} & A & B & C & D \\ A & 0 & 1 & 0 & 0 \\ B & 1 & 0 & 0 & 0 \\ C & .1 & .2 & .3 & .4 \\ D & .7 & .1 & .1 & .1 \end{matrix}$

In Problems 7–10, identify the absorbing states for each transition diagram, and determine whether the diagram represents an absorbing Markov chain.

7.

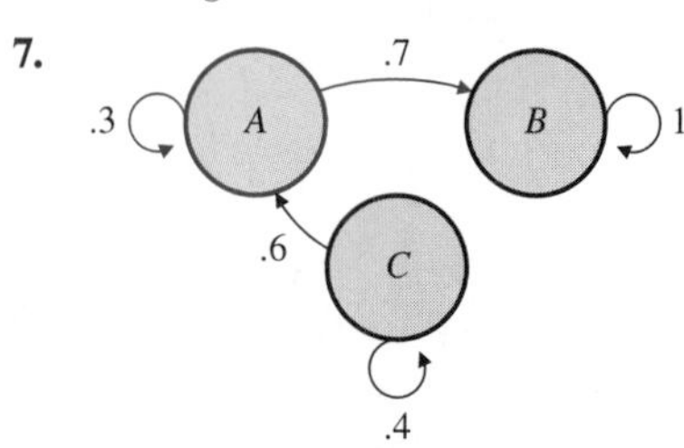

8.

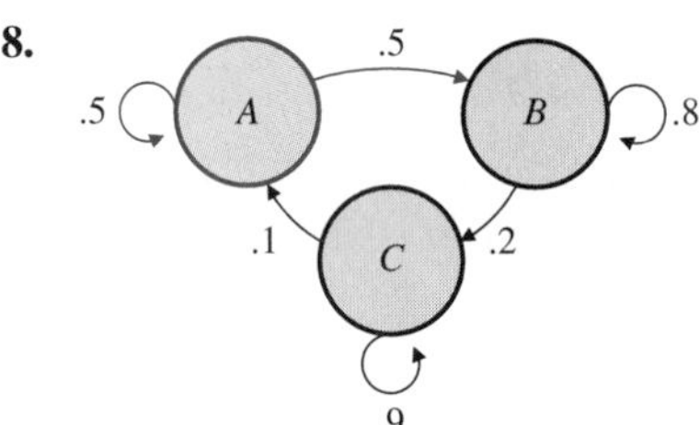

9.

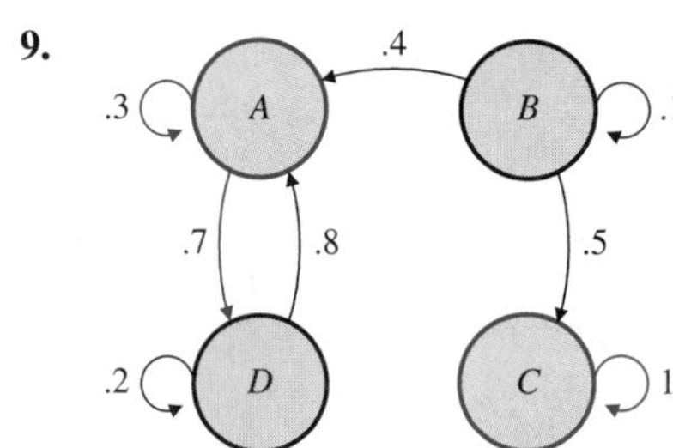

10.

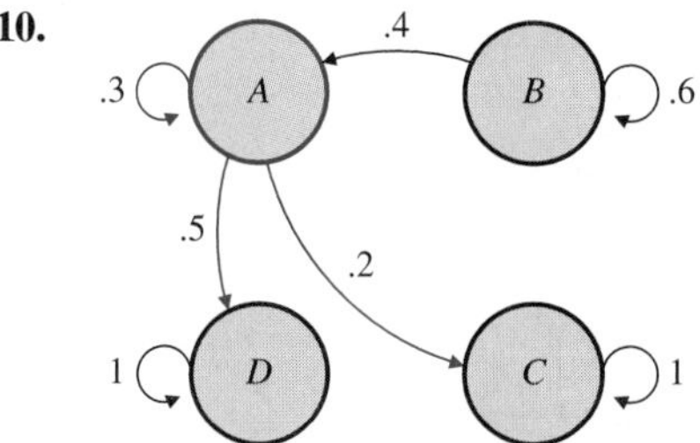

B *In Problems 11–14, find a standard form for the absorbing Markov chain with the indicated transition diagram.*

11.

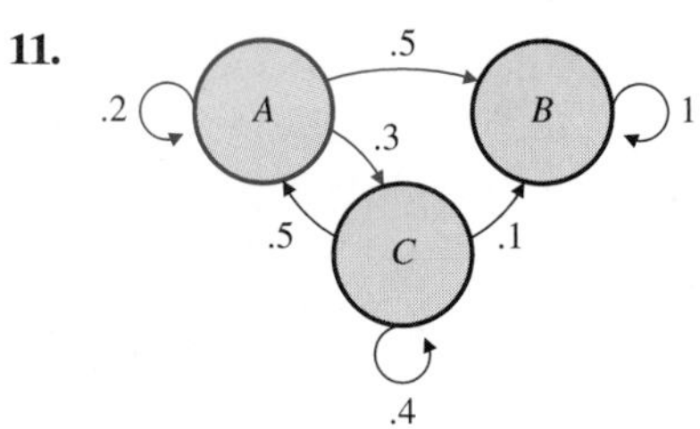

12.

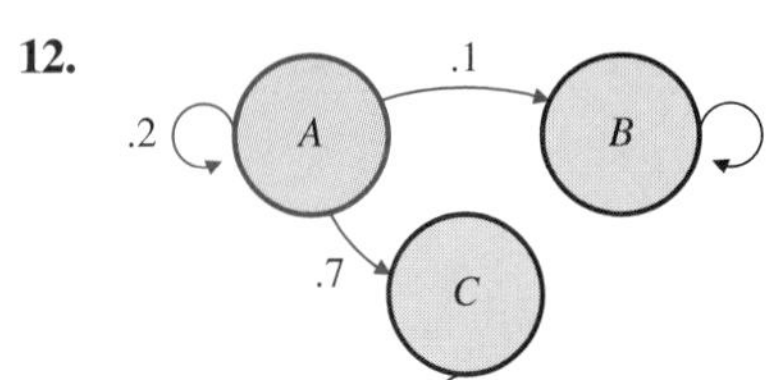

13.

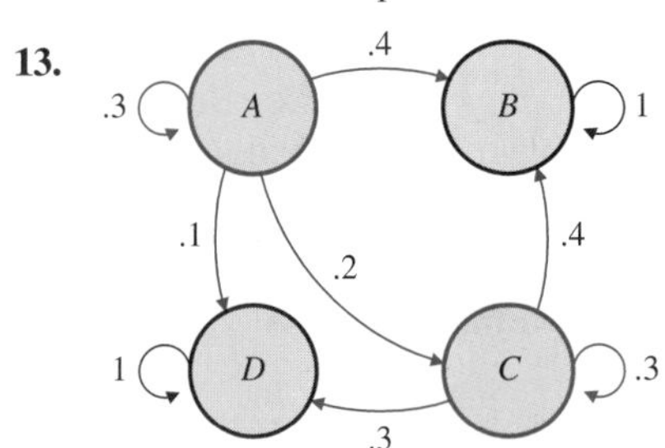

14.

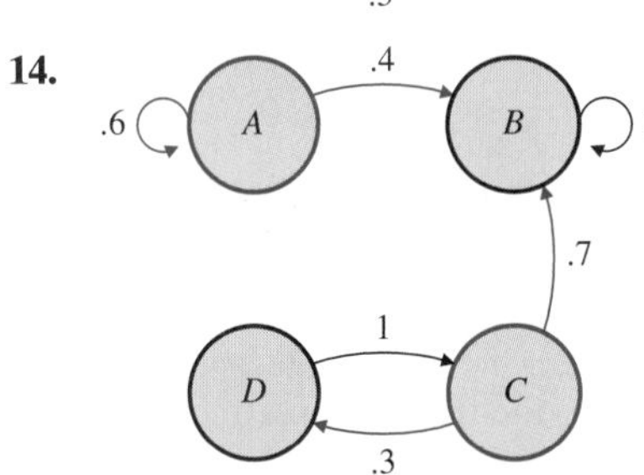

In Problems 15–18, find a standard form for the absorbing Markov chain with the indicated transition matrix.

15. $$P = \begin{matrix} & \begin{matrix} A & B & C \end{matrix} \\ \begin{matrix} A \\ B \\ C \end{matrix} & \begin{bmatrix} .2 & .3 & .5 \\ 1 & 0 & 0 \\ 0 & 0 & 1 \end{bmatrix} \end{matrix}$$

16. $$P = \begin{matrix} & \begin{matrix} A & B & C \end{matrix} \\ \begin{matrix} A \\ B \\ C \end{matrix} & \begin{bmatrix} 0 & 0 & 1 \\ 0 & 1 & 0 \\ .7 & .2 & .1 \end{bmatrix} \end{matrix}$$

17. $$P = \begin{matrix} & \begin{matrix} A & B & C & D \end{matrix} \\ \begin{matrix} A \\ B \\ C \\ D \end{matrix} & \begin{bmatrix} .1 & .2 & .3 & .4 \\ 0 & 1 & 0 & 0 \\ .5 & .2 & .2 & .1 \\ 0 & 0 & 0 & 1 \end{bmatrix} \end{matrix}$$

18. $$P = \begin{matrix} & \begin{matrix} A & B & C & D \end{matrix} \\ \begin{matrix} A \\ B \\ C \\ D \end{matrix} & \begin{bmatrix} 0 & .3 & .3 & .4 \\ 0 & 1 & 0 & 0 \\ 0 & 0 & 1 & 0 \\ .8 & .1 & .1 & 0 \end{bmatrix} \end{matrix}$$

In Problems 19–24, find the limiting matrix for the indicated standard form. Find the long-run probability of going from each nonabsorbing state to each absorbing state and the average number of trials needed to go from each nonabsorbing state to an absorbing state.

19. $$P = \begin{matrix} & \begin{matrix} A & B & C \end{matrix} \\ \begin{matrix} A \\ B \\ C \end{matrix} & \begin{bmatrix} 1 & 0 & 0 \\ 0 & 1 & 0 \\ .1 & .4 & .5 \end{bmatrix} \end{matrix}$$

20. $$P = \begin{matrix} & \begin{matrix} A & B & C \end{matrix} \\ \begin{matrix} A \\ B \\ C \end{matrix} & \begin{bmatrix} 1 & 0 & 0 \\ 0 & 1 & 0 \\ .3 & .2 & .5 \end{bmatrix} \end{matrix}$$

21. $$P = \begin{matrix} & \begin{matrix} A & B & C \end{matrix} \\ \begin{matrix} A \\ B \\ C \end{matrix} & \begin{bmatrix} 1 & 0 & 0 \\ .2 & .6 & .2 \\ .4 & .2 & .4 \end{bmatrix} \end{matrix}$$

22. $$P = \begin{matrix} & \begin{matrix} A & B & C \end{matrix} \\ \begin{matrix} A \\ B \\ C \end{matrix} & \begin{bmatrix} 1 & 0 & 0 \\ .1 & .6 & .3 \\ .2 & .2 & .6 \end{bmatrix} \end{matrix}$$

23. $$P = \begin{matrix} & \begin{matrix} A & B & C & D \end{matrix} \\ \begin{matrix} A \\ B \\ C \\ D \end{matrix} & \begin{bmatrix} 1 & 0 & 0 & 0 \\ 0 & 1 & 0 & 0 \\ .1 & .2 & .6 & .1 \\ .2 & .2 & .3 & .3 \end{bmatrix} \end{matrix}$$

24. $$P = \begin{matrix} & \begin{matrix} A & B & C & D \end{matrix} \\ \begin{matrix} A \\ B \\ C \\ D \end{matrix} & \begin{bmatrix} 1 & 0 & 0 & 0 \\ 0 & 1 & 0 & 0 \\ .1 & .1 & .7 & .1 \\ .3 & .1 & .4 & .2 \end{bmatrix} \end{matrix}$$

Problems 25–30 refer to the matrices in Problems 19–24, as indicated. Use the limiting matrix $\overline{P}$ found for each transition matrix P in Problems 19–24 to determine the long-run behavior of the successive state matrices for the indicated initial-state matrices.

25. For matrix P from Problem 19 with:
(A) $S_0 = [0 \quad 0 \quad 1]$ (B) $S_0 = [.2 \quad .5 \quad .3]$

26. For matrix P from Problem 20 with:
(A) $S_0 = [0 \quad 0 \quad 1]$ (B) $S_0 = [.2 \quad .5 \quad .3]$

27. For matrix P from Problem 21 with:
(A) $S_0 = [0 \quad 0 \quad 1]$ (B) $S_0 = [.2 \quad .5 \quad .3]$

28. For matrix P from Problem 22 with:
(A) $S_0 = [0 \quad 0 \quad 1]$ (B) $S_0 = [.2 \quad .5 \quad .3]$

29. For matrix P from Problem 23 with:
(A) $S_0 = [0 \quad 0 \quad 0 \quad 1]$ (B) $S_0 = [0 \quad 0 \quad 1 \quad 0]$
(C) $S_0 = [0 \quad 0 \quad .4 \quad .6]$ (D) $S_0 = [.1 \quad .2 \quad .3 \quad .4]$

30. For matrix P from Problem 24 with:
(A) $S_0 = [0 \quad 0 \quad 0 \quad 1]$ (B) $S_0 = [0 \quad 0 \quad 1 \quad 0]$
(C) $S_0 = [0 \quad 0 \quad .4 \quad .6]$ (D) $S_0 = [.1 \quad .2 \quad .3 \quad .4]$

In Problems 31 and 32, discuss the validity of each statement. If the statement is always true, explain why. If not, give a counterexample.

31. (A) If every state of a Markov chain is absorbing, then it is an absorbing chain.
(B) If a Markov chain has absorbing states, then it is an absorbing chain.

32. (A) In an absorbing Markov chain, if a nonabsorbing state is exited, it can never be entered again.
(B) An absorbing Markov chain can have an infinite number of stationary matrices.

In Problems 33–36, use a graphing utility to approximate the limiting matrix for the indicated standard form.

33. $P = \begin{array}{c} \\ A \\ B \\ C \\ D \end{array} \begin{array}{c} \begin{array}{cccc} A & B & C & D \end{array} \\ \begin{bmatrix} 1 & 0 & 0 & 0 \\ 0 & 1 & 0 & 0 \\ .5 & .3 & .1 & .1 \\ .6 & .2 & .1 & .1 \end{bmatrix} \end{array}$

34. $P = \begin{array}{c} \\ A \\ B \\ C \\ D \end{array} \begin{array}{c} \begin{array}{cccc} A & B & C & D \end{array} \\ \begin{bmatrix} 1 & 0 & 0 & 0 \\ 0 & 1 & 0 & 0 \\ .1 & .1 & .5 & .3 \\ 0 & .2 & .3 & .5 \end{bmatrix} \end{array}$

35. $P = \begin{array}{c} \\ A \\ B \\ C \\ D \\ E \end{array} \begin{array}{c} \begin{array}{ccccc} A & B & C & D & E \end{array} \\ \begin{bmatrix} 1 & 0 & 0 & 0 & 0 \\ 0 & 1 & 0 & 0 & 0 \\ 0 & .4 & .5 & 0 & .1 \\ 0 & .4 & 0 & .3 & .3 \\ .4 & .4 & 0 & .2 & 0 \end{bmatrix} \end{array}$

36. $P = \begin{array}{c} \\ A \\ B \\ C \\ D \\ E \end{array} \begin{array}{c} \begin{array}{ccccc} A & B & C & D & E \end{array} \\ \begin{bmatrix} 1 & 0 & 0 & 0 & 0 \\ 0 & 1 & 0 & 0 & 0 \\ .5 & 0 & 0 & 0 & .5 \\ 0 & .4 & 0 & .2 & .4 \\ 0 & 0 & .1 & .7 & .2 \end{bmatrix} \end{array}$

C

37. The following matrix P is a nonstandard transition matrix for an absorbing Markov chain:

$$P = \begin{array}{c} \\ A \\ B \\ C \\ D \end{array} \begin{array}{c} \begin{array}{cccc} A & B & C & D \end{array} \\ \begin{bmatrix} .2 & .2 & .6 & 0 \\ 0 & 1 & 0 & 0 \\ .5 & .1 & 0 & .4 \\ 0 & 0 & 0 & 1 \end{bmatrix} \end{array}$$

To find a limiting matrix for P, follow the steps outlined below:

Step 1. Using a transition diagram as an aid, rearrange the columns and rows of P to produce a standard form for this chain.

Step 2. Find the limiting matrix for this standard form.

Step 3. Using a transition diagram as an aid, reverse the process used in step 1 to produce a limiting matrix for the original matrix P.

38. Repeat Problem 37 for

$$P = \begin{array}{c} \\ A \\ B \\ C \\ D \end{array} \begin{array}{c} \begin{array}{cccc} A & B & C & D \end{array} \\ \begin{bmatrix} 1 & 0 & 0 & 0 \\ .3 & .6 & 0 & .1 \\ .2 & .3 & .5 & 0 \\ 0 & 0 & 0 & 1 \end{bmatrix} \end{array}$$

39. **C** Verify the results in Problem 37 by computing P^k on a graphing utility for large values of k.

40. **C** Verify the results in Problem 38 by computing P^k on a graphing utility for large values of k.

41. Show that $S = [x \quad 1 - x \quad 0], 0 \le x \le 1$, is a stationary matrix for the transition matrix

$$P = \begin{array}{c} \\ A \\ B \\ C \end{array} \begin{array}{c} \begin{array}{ccc} A & B & C \end{array} \\ \begin{bmatrix} 1 & 0 & 0 \\ 0 & 1 & 0 \\ .1 & .5 & .4 \end{bmatrix} \end{array}$$

Discuss the generalization of this result to any absorbing Markov chain with two absorbing states and one nonabsorbing state.

42. Show that $S = [x \quad 1 - x \quad 0 \quad 0], 0 \le x \le 1$, is a stationary matrix for the transition matrix

$$P = \begin{array}{c} \\ A \\ B \\ C \\ D \end{array} \begin{array}{c} \begin{array}{cccc} A & B & C & D \end{array} \\ \begin{bmatrix} 1 & 0 & 0 & 0 \\ 0 & 1 & 0 & 0 \\ .1 & .2 & .3 & .4 \\ .6 & .2 & .1 & .1 \end{bmatrix} \end{array}$$

Discuss the generalization of this result to any absorbing Markov chain with two absorbing states and two nonabsorbing states.

43. **C** An absorbing Markov chain has the following matrix P as a standard form:

$$P = \begin{array}{c} \\ A \\ B \\ C \\ D \end{array} \begin{array}{c} \begin{array}{cccc} A & B & C & D \end{array} \\ \begin{bmatrix} 1 & 0 & 0 & 0 \\ .2 & .3 & .1 & .4 \\ 0 & .5 & .3 & .2 \\ 0 & .1 & .6 & .3 \end{bmatrix} \end{array} \quad \left[\begin{array}{c|c} I & 0 \\ \hline R & Q \end{array}\right]$$

Let w_k denote the maximum entry in Q^k. Note that $w_1 = .6$.

(A) Find w_2, w_4, w_8, w_{16}, and w_{32} to three decimal places.

(B) Describe Q^k when k is large.

44. **C** Refer to the matrices P and Q of Problem 43. For k a positive integer, let $T_k = I + Q + Q^2 + \cdots + Q^k$.

(A) Explain why $T_{k+1} = T_k Q + I$.

(B) Using a graphing utility and part (A) to quickly compute the matrices T_k, discover and describe the connection between $(I - Q)^{-1}$ and T_k when k is large.

Applications

Business & Economics

45. *Loans.* A credit union classifies automobile loans into one of four categories: the loan has been paid in full (F), the account is in good standing (G) with all payments up to date, the account is in arrears (A) with one or more missing payments, or the account has been classified as a bad debt (B) and sold to a collection agency. Past records indicate that each month 10% of the accounts in good standing pay the loan in full, 80% remain in good standing, and 10% become in arrears. Furthermore, 10% of the accounts in arrears are paid in full, 40% become accounts in good standing, 40% remain in arrears, and 10% are classified as bad debts.

(A) In the long run, what percentage of the accounts in arrears will pay their loan in full?

(B) In the long run, what percentage of the accounts in good standing will become bad debts?

(C) What is the average number of months an account in arrears will remain in this system before it is either paid in full or classified as a bad debt?

46. *Employee training.* A national chain of automobile muffler and brake repair shops maintains a training program for its mechanics. All new mechanics begin training in muffler repairs. Every 3 months the performance of each mechanic is reviewed. Past records indicate that after each quarterly review, 30% of the muffler repair trainees are rated as qualified to repair mufflers and begin training in brake repairs, 20% are terminated for unsatisfactory performance, and the remainder continue as muffler repair trainees. Also, 30% of the brake repair trainees are rated as fully qualified mechanics requiring no further training, 10% are terminated for unsatisfactory performance, and the remainder continue as brake repair trainees.

(A) In the long run, what percentage of the muffler repair trainees will become fully qualified mechanics?

(B) In the long run, what percentage of the brake repair trainees will be terminated?

(C) What is the average number of quarters a muffler repair trainee will remain in the training program before being either terminated or promoted to fully qualified mechanic?

47. *Marketing.* Three electronics firms are aggressively marketing their graphing calculators to high school and college mathematics departments by offering volume discounts, complimentary display equipment, and assistance with curriculum development. Due to the amount of equipment involved and the necessary curriculum changes, once a department decides to use a particular calculator in their courses, they never switch to another brand or stop using calculators. Each year, 6% of the departments decide to use calculators from company A, 3% decide to use calculators from company B, 11% decide to use calculators from company C, and the remainder decide not to use any calculators in their courses.

(A) In the long run, what is the market share of each company?

(B) On the average, how many years will it take a department to decide to use calculators from one of these companies in their courses?

48. *Pensions.* Once a year employees at a company are given the opportunity to join one of three pension plans, A, B, or C. Once an employee decides to join one of these plans, the employee cannot drop the plan or switch to another plan. Past records indicate that each year 4% of the employees elect to join plan A, 14% elect to join plan B, 7% elect to join plan C, and the remainder do not join any plan.

(A) In the long run, what percentage of the employees will elect to join plan A? Plan B? Plan C?

(B) On the average, how many years will it take an employee to decide to join a plan?

Life Sciences

49. *Medicine.* After bypass surgery, patients are placed in an intensive care unit (ICU) until their condition stabilizes. Then they are transferred to a cardiac care ward (CCW) where they remain until they are released from the hospital. In a particular metropolitan area, a study of hospital records produced the following data: each day 2% of the patients in the ICU died, 52% were transferred to the CCW, and the remainder stayed in the ICU. Furthermore, each day 4% of the patients in the CCW developed complications and were returned to the ICU, 1% died while in the CCW, 22% were released from the hospital, and the remainder stayed in the CCW.

(A) In the long run, what percentage of the patients in the ICU are released from the hospital?

(B) In the long run, what percentage of the patients in the CCW die without ever being released from the hospital?

(C) What is the average number of days a patient in the ICU will stay in the hospital?

50. *Medicine.* The study discussed in Problem 49 also produced the following data for patients who underwent aortic valve replacements: each day 2% of the patients in the ICU died, 60% were transferred to the CCW, and the remainder stayed in the ICU. Furthermore, each day 5% of the patients in the CCW developed complications and were returned to the ICU, 1% died while in the

CCW, 19% were released from the hospital, and the remainder stayed in the CCW.

(A) In the long run, what percentage of the patients in the CCW are released from the hospital?

(B) In the long run, what percentage of the patients in the ICU die without ever being released from the hospital?

(C) What is the average number of days a patient in the CCW will stay in the hospital?

Social Sciences

51. *Psychology.* A rat is placed in room F or room B of the maze shown in the figure. The rat wanders from room to room until it enters one of the rooms containing food, L or R. Assume that the rat chooses an exit from a room at random and that once it enters a room with food it never leaves.

(A) What is the long-run probability that a rat placed in room B ends up in room R?

(B) What is the average number of exits a rat placed in room B will choose until it finds food?

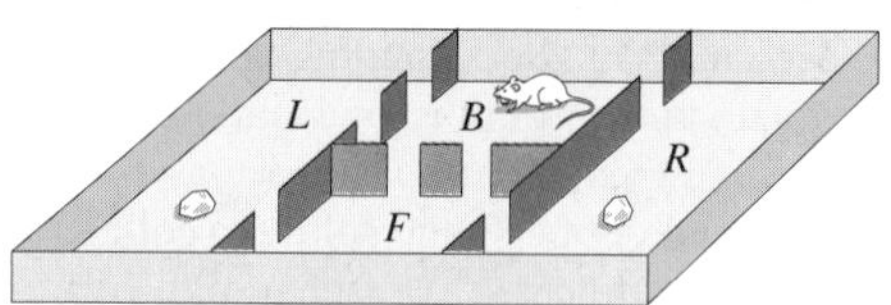

Figure for 51 and 52

52. *Psychology.* Repeat Problem 51 if the exit from room B to room R is blocked.

Important Terms and Symbols

7-1 *Properties of Markov Chains.* Stochastic process; transition diagram; transition probability matrix; initial-state distribution matrix; initial-state probability matrix; first-state matrix; Markov chain or process; kth-state matrix; recursive definition of state matrices; powers of a transition matrix

$$S_k = S_{k-1}P = S_0P^k$$

7-2 *Regular Markov Chains.* Stationary matrix; regular transition matrix; regular Markov chain; limiting matrix

$$SP = S; \quad P^k \text{ approaches } \overline{P}$$

7-3 *Absorbing Markov Chains.* Absorbing state; absorbing Markov chain; standard form; limiting matrix; fundamental matrix

$$P = \left[\begin{array}{c|c} I & 0 \\ \hline R & Q \end{array}\right]; \quad \overline{P} = \left[\begin{array}{c|c} I & 0 \\ \hline FR & 0 \end{array}\right];$$

$$F = (I - Q)^{-1}$$

Review Exercise

Work through all the problems in this chapter review and check your answers in the back of the book. Answers to all review problems are there along with section numbers in italics to indicate where each type of problem is discussed. Where weaknesses show up, review appropriate sections in the text.

A

1. Given the transition matrix P and initial-state matrix S_0 shown below, find S_1 and S_2 and explain what each represents:

$$P = \begin{array}{c} \\ A \\ B \end{array}\begin{array}{c} \begin{array}{cc} A & B \end{array} \\ \begin{bmatrix} .6 & .4 \\ .2 & .8 \end{bmatrix} \end{array} \qquad S_0 = [.3 \quad .7]$$

In Problems 2–6, P is a transition matrix for a Markov chain. Identify any absorbing states and classify the chain as regular, absorbing, or neither.

2. $P = \begin{array}{c} \\ A \\ B \end{array}\begin{array}{c} \begin{array}{cc} A & B \end{array} \\ \begin{bmatrix} 1 & 0 \\ .7 & .3 \end{bmatrix} \end{array}$

3. $P = \begin{array}{c} \\ A \\ B \end{array}\begin{array}{c} \begin{array}{cc} A & B \end{array} \\ \begin{bmatrix} 0 & 1 \\ .7 & .3 \end{bmatrix} \end{array}$

4. $P = \begin{array}{c} \\ A \\ B \end{array}\begin{array}{c} \begin{array}{cc} A & B \end{array} \\ \begin{bmatrix} 0 & 1 \\ 1 & 0 \end{bmatrix} \end{array}$

5. $P = \begin{array}{c} \\ A \\ B \\ C \end{array} \begin{array}{c} \begin{array}{ccc} A & B & C \end{array} \\ \begin{bmatrix} .8 & 0 & .2 \\ 0 & 1 & 0 \\ 0 & 0 & 1 \end{bmatrix} \end{array}$

6. $P = \begin{array}{c} \\ A \\ B \\ C \\ D \end{array} \begin{array}{c} \begin{array}{cccc} A & B & C & D \end{array} \\ \begin{bmatrix} 1 & 0 & 0 & 0 \\ 0 & 1 & 0 & 0 \\ 0 & 0 & .3 & .7 \\ 0 & 0 & .6 & .4 \end{bmatrix} \end{array}$

In Problems 7–10, write a transition matrix for the indicated transition diagram, identify any absorbing states, and classify each Markov chain as regular, absorbing, or neither.

7.

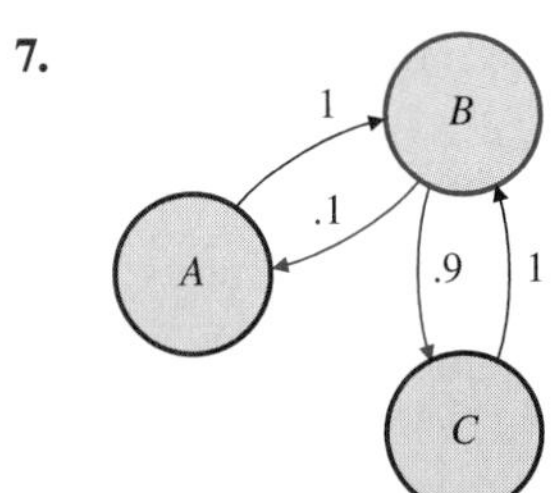

8.

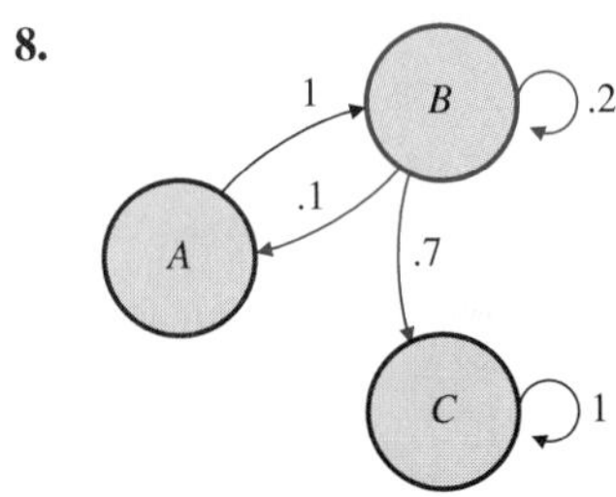

9.

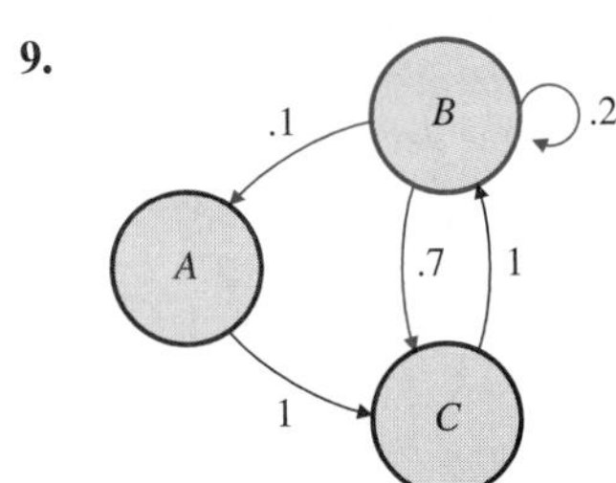

10.

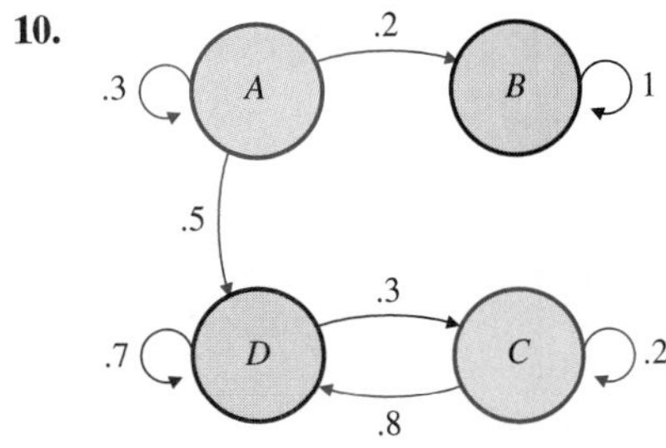

B

11. A Markov chain has three states, A, B, and C. The probability of going from state A to state B in one trial is .2, the probability of going from state A to state C in one trial is .5, the probability of going from state B to state A in one trial is .8, the probability of going from state B to state C in one trial is .2, the probability of going from state C to state A in one trial is .1, and the probability of going from state C to state B in one trial is .3. Draw a transition diagram and write a transition matrix for this chain.

12. Given the transition matrix

$$P = \begin{array}{c} \\ A \\ B \end{array} \begin{array}{c} \begin{array}{cc} A & B \end{array} \\ \begin{bmatrix} .4 & .6 \\ .9 & .1 \end{bmatrix} \end{array}$$

find the probability of:

(A) Going from state A to state B in two trials.

(B) Going from state B to state A in three trials.

In Problems 13 and 14, solve the equation $SP = S$ to find the stationary matrix S and the limiting matrix $\overline{P}$.

13. $P = \begin{array}{c} \\ A \\ B \end{array} \begin{array}{c} \begin{array}{cc} A & B \end{array} \\ \begin{bmatrix} .4 & .6 \\ .2 & .8 \end{bmatrix} \end{array}$

14. $P = \begin{array}{c} \\ A \\ B \\ C \end{array} \begin{array}{c} \begin{array}{ccc} A & B & C \end{array} \\ \begin{bmatrix} .4 & .6 & 0 \\ .5 & .3 & .2 \\ 0 & .8 & .2 \end{bmatrix} \end{array}$

In Problems 15 and 16, find the limiting matrix for the indicated standard form. Find the long-run probability of going from each nonabsorbing state to each absorbing state and the average number of trials needed to go from each nonabsorbing state to an absorbing state.

15. $P = \begin{array}{c} \\ A \\ B \\ C \end{array} \begin{array}{c} \begin{array}{ccc} A & B & C \end{array} \\ \begin{bmatrix} 1 & 0 & 0 \\ 0 & 1 & 0 \\ .3 & .1 & .6 \end{bmatrix} \end{array}$

16. $P = \begin{array}{c} \\ A \\ B \\ C \\ D \end{array} \begin{array}{c} \begin{array}{cccc} A & B & C & D \end{array} \\ \begin{bmatrix} 1 & 0 & 0 & 0 \\ 0 & 1 & 0 & 0 \\ .1 & .5 & .2 & .2 \\ .1 & .1 & .4 & .4 \end{bmatrix} \end{array}$

In Problems 17–20, use a graphing utility to approximate the limiting matrix for the indicated transition matrix.

17. Matrix P from Problem 13
18. Matrix P from Problem 14
19. Matrix P from Problem 15
20. Matrix P from Problem 16

21. Find a standard form for the absorbing Markov chain with transition matrix

$$P = \begin{array}{c} \\ A \\ B \\ C \\ D \end{array}\begin{array}{c} \begin{array}{cccc} A & B & C & D \end{array} \\ \begin{bmatrix} .6 & .1 & .2 & .1 \\ 0 & 1 & 0 & 0 \\ .3 & .2 & .3 & .2 \\ 0 & 0 & 0 & 1 \end{bmatrix} \end{array}$$

In Problems 22 and 23, determine the long-run behavior of the successive state matrices for the indicated transition matrix and initial-state matrices.

22. $P = \begin{array}{c} \\ A \\ B \\ C \end{array}\begin{array}{c} \begin{array}{ccc} A & B & C \end{array} \\ \begin{bmatrix} 0 & 1 & 0 \\ 0 & 0 & 1 \\ .2 & .6 & .2 \end{bmatrix} \end{array}$

(A) $S_0 = [0 \quad 0 \quad 1]$ (B) $S_0 = [.5 \quad .3 \quad .2]$

23. $P = \begin{array}{c} \\ A \\ B \\ C \end{array}\begin{array}{c} \begin{array}{ccc} A & B & C \end{array} \\ \begin{bmatrix} 1 & 0 & 0 \\ 0 & 1 & 0 \\ .2 & .6 & .2 \end{bmatrix} \end{array}$

(A) $S_0 = [0 \quad 0 \quad 1]$ (B) $S_0 = [.5 \quad .3 \quad .2]$

24. Let P be a 2×2 transition matrix for a Markov chain. Can P be regular if two of its entries are 0? Explain.

25. Let P be a 3×3 transition matrix for a Markov chain. Can P be regular if three of its entries are 0? If four of its entries are 0? Explain.

27. Repeat Problem 26 if the blue and green marbles are removed from the red urn.

28. Show that $S = [x \quad y \quad z \quad 0]$, where $x + y + z = 1$, $0 \leq x \leq 1, 0 \leq y \leq 1, 0 \leq z \leq 1$, is a stationary matrix for the transition matrix

$$P = \begin{array}{c} \\ A \\ B \\ C \\ D \end{array}\begin{array}{c} \begin{array}{cccc} A & B & C & D \end{array} \\ \begin{bmatrix} 1 & 0 & 0 & 0 \\ 0 & 1 & 0 & 0 \\ 0 & 0 & 1 & 0 \\ .1 & .3 & .4 & .2 \end{bmatrix} \end{array}$$

Discuss the generalization of this result to any absorbing chain with three absorbing states and one nonabsorbing state.

In Problems 29–35, either give an example of a Markov chain with the indicated properties or explain why no such chain can exist.

29. A regular Markov chain with an absorbing state.

30. An absorbing Markov chain that is regular.

31. A regular Markov chain with two different stationary matrices.

32. An absorbing Markov chain with two different stationary matrices.

33. A Markov chain with no limiting matrix.

34. A regular Markov chain with no limiting matrix.

35. An absorbing Markov chain with no limiting matrix.

C

26. A red urn contains 2 red marbles, 1 blue marble, and 1 green marble. A blue urn contains 1 red marble, 3 blue marbles, and 1 green marble. A green urn contains 6 red marbles, 3 blue marbles, and 1 green marble. A marble is selected from an urn, the color is noted, and the marble is returned to the urn from which it was drawn. The next marble is drawn from the urn whose color is the same as the marble just drawn. Thus, this is a Markov process with three states: draw from the red urn, draw from the blue urn, or draw from the green urn.

(A) Draw a transition diagram for this process.
(B) Write the transition matrix P.
(C) Determine whether this chain is regular, absorbing, or neither.
(D) Find the limiting matrix $\overline{P}$, if it exists, and describe the long-run behavior of this process.

In Problems 36 and 37, use a graphing utility to approximate the entries (to three decimal places) of the limiting matrix, if it exists, of the indicated transition matrix.

36. $P = \begin{array}{c} \\ A \\ B \\ C \\ D \end{array}\begin{array}{c} \begin{array}{cccc} A & B & C & D \end{array} \\ \begin{bmatrix} .2 & .3 & .1 & .4 \\ 0 & 0 & 1 & 0 \\ 0 & .8 & 0 & .2 \\ 0 & 0 & 1 & 0 \end{bmatrix} \end{array}$

37. $P = \begin{array}{c} \\ A \\ B \\ C \\ D \end{array}\begin{array}{c} \begin{array}{cccc} A & B & C & D \end{array} \\ \begin{bmatrix} .1 & 0 & .3 & .6 \\ .2 & .4 & .1 & .3 \\ .3 & .5 & 0 & .2 \\ .9 & .1 & 0 & 0 \end{bmatrix} \end{array}$

APPLICATIONS

Business & Economics

38. *Product switching.* A company's brand (X) has 20% of the market. A market research firm finds that if a person uses brand X, the probability is .7 that he or she will buy it next time. On the other hand, if a person does not use brand X (represented by X'), the probability is .5 that he or she will switch to brand X the next time.

(A) Draw a transition diagram.
(B) Write a transition matrix.
(C) Write the initial-state matrix.
(D) Find the first-state matrix and explain what it represents.
(E) Find the stationary matrix.
(F) What percentage of the market will brand X have in the long run if the transition matrix does not change?

39. *Marketing.* Recent technological advances have led to the development of three new milling machines, brand A, brand B, and brand C. Due to the extensive retooling and startup costs, once a company converts its machine shop to one of these new machines, it never switches to another brand. Each year 6% of the machine shops convert to brand A machines, 8% convert to brand B machines, 11% convert to brand C machines, and the remainder continue to use their old machines.

(A) In the long run, what is the market share of each brand?
(B) What is the average number of years a company waits before converting to one of the new milling machines?

40. *Market research.* Table 1 gives the percentage of households that owned a VCR in the indicated years:

Table 1

YEAR	PERCENT
1984	10.6
1988	58.0
1992	72.5

Source: Television Bureau of Advertising, Inc.

The following transition matrix P is proposed as a model for the data, where V represents the households that own a VCR:

$$\begin{array}{cc} & \text{Four years later} \\ & \begin{array}{cc} V & V' \end{array} \\ \text{Current year} \begin{array}{c} V \\ V' \end{array} & \begin{bmatrix} .853 & .147 \\ .553 & .447 \end{bmatrix} = P \end{array}$$

(A) Let $S_0 = [.106 \quad .894]$, and find S_1 and S_2. (Compute both matrices exactly and then round entries to three decimal places.)
(B) Construct a new table comparing the results from part (A) with the data in Table 1.
(C) According to the transition matrix, what percentage of the households will own a VCR in the long run?

41. *Employee training.* In order to become a Fellow of the Society of Actuaries, an individual must pass a series of ten examinations given by the Society. Passage of the first two preliminary exams is a prerequisite for employment as a trainee in the actuarial department of a large insurance company. Each year 15% of the trainees complete the next three exams in the program and become associates of the Society of Actuaries, 5% leave the company, never to return, and the remainder continue as trainees. Furthermore, each year 17% of the associates complete the remaining five exams and become fellows of the Society of Actuaries, 3% leave the company, never to return, and the remainder continue as associates.

(A) In the long run, what percentage of the trainees will become fellows?
(B) In the long run, what percentage of the associates will leave the company?
(C) What is the average number of years a trainee remains in this program before either becoming a fellow or being discharged?

Life Sciences

42. *Genetics.* A given plant species has red, pink, or white flowers according to the genotypes RR, RW, and WW, respectively. If each of these genotypes is crossed with a red-flowering plant, then the transition matrix is

$$\begin{array}{cc} & \text{Next generation} \\ & \begin{array}{ccc} \text{Red} & \text{Pink} & \text{White} \end{array} \\ \text{This generation} \begin{array}{c} \text{Red} \\ \text{Pink} \\ \text{White} \end{array} & \begin{bmatrix} 1 & 0 & 0 \\ .5 & .5 & 0 \\ 0 & 1 & 0 \end{bmatrix} \end{array}$$

If each generation of the plant is crossed only with red plants to produce the next generation, show that eventually all the flowers produced by the plants will be red. (Find the limiting matrix.)

Social Sciences

43. *Farm population.* Table 2 gives the percentage of the U.S. population living on farms in the indicated years:

Table 2

YEAR	PERCENT
1980	2.8
1985	2.3
1990	1.9

Source: U.S. Bureau of the Census

The following transition matrix P is proposed as a model for the data, where F represents the population that lives on a farm:

$$\begin{array}{c} \\ \text{Current year} \end{array} \begin{array}{c} \\ F \\ F' \end{array} \overset{\text{Five years later}}{\begin{array}{cc} F & F' \\ \end{array}} \begin{bmatrix} .751 & .249 \\ .001 & .999 \end{bmatrix} = P$$

(A) Let $S_0 = [.028 \quad .972]$, and find S_1 and S_2.
(B) Construct a new table comparing the results from part (A) with the data in Table 2.
(C) According to this transition matrix, what percentage of the population will live on a farm in the long run?

Group Activity 1 Social Mobility

The government of a developing country uses annual income to classify its citizens into three classes: lower (L), middle (M), and upper (U). A Markov chain with transition matrix P is used to model the mobility of the population among the three classes:

$$P = \begin{array}{c} L \\ M \\ U \end{array} \overset{\begin{array}{ccc} L & M & U \end{array}}{\begin{bmatrix} .9 & .1 & 0 \\ .15 & .8 & .05 \\ 0 & .1 & .9 \end{bmatrix}}$$

Each year, for example, 10% of the population moves from the lower to the middle class, 5% from the middle to the upper class, and so on.

(A) If 75% of the population is now in the lower class, 20% in the middle class, and 5% in the upper class, predict the proportions in each class after 1, 2, and 3 years.
(B) Explain why P is regular, and find the stationary and limiting matrices.

One of the goals of the government is to increase the size of the middle class. Consequently, the government has directed its ministries of economics, education, and human services to promote policies that will increase the movement from the lower to the middle class—that is, to increase the value of a in the transition matrix P':

$$P' = \begin{array}{c} L \\ M \\ U \end{array} \overset{\begin{array}{ccc} L & M & U \end{array}}{\begin{bmatrix} 1-a & a & 0 \\ .15 & .8 & .05 \\ 0 & .1 & .9 \end{bmatrix}}$$

(C) Find the stationary matrix for P' if $a = .2$. If $a = .3$.
(D) Find the value of a so that in the long run the middle class will be three times the size of the lower class.
(E) Can the lower class be eliminated entirely by increasing a? Explain.

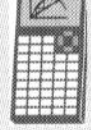

Group Activity 2 Gambler's Ruin

Games of chance between two players that continue until one player goes broke are often referred to as **gambler's ruin problems.** For example, suppose two people who have a total of \$4 between them are playing a simple coin-tossing game. A fair coin is tossed. If the coin comes up heads, player A pays player B \$1. If the coin comes up tails, player B pays player A \$1. The game continues until one player has all the money and the other player is broke, or ruined, hence the name *gambler's ruin.*

This game can be analyzed as an absorbing Markov chain, where the states are the amount of money player A has at each stage of the game. Thus, the possible states are \$0, \$1, \$2, \$3, and \$4. The game ends when player A has \$0 or \$4, so these two states are absorbing. The transition diagram and a standard form for the transition matrix are shown below.

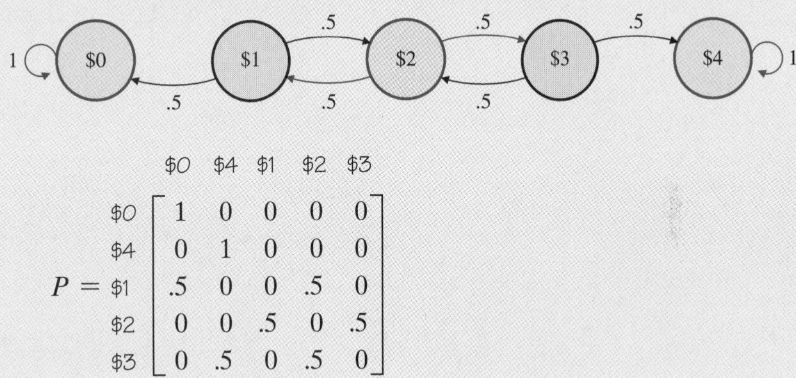

$$P = \begin{array}{c} \\ \$0 \\ \$4 \\ \$1 \\ \$2 \\ \$3 \end{array} \begin{array}{c} \begin{array}{ccccc} \$0 & \$4 & \$1 & \$2 & \$3 \end{array} \\ \begin{bmatrix} 1 & 0 & 0 & 0 & 0 \\ 0 & 1 & 0 & 0 & 0 \\ .5 & 0 & 0 & .5 & 0 \\ 0 & 0 & .5 & 0 & .5 \\ 0 & .5 & 0 & .5 & 0 \end{bmatrix} \end{array}$$

(A) Find the fundamental matrix F and the limiting matrix $\overline{P}$. What is the probability that player A wins all the money if player A starts with \$1? With \$2? With \$3? What is the average number of coin tosses until the game ends if player A starts with \$1? With \$2? With \$3?

(B) Repeat part (A), but assume the players start with a total of \$5 between them. Draw the corresponding transition diagram, and find a standard form for the transition matrix. Be certain to consider all the different amounts of money player A could have at the beginning of the game.

(C) Roulette wheels in Nevada generally have 38 equally spaced slots numbered 00, 0, 1, 2, . . . , 36. The numbers from 1 to 36 are evenly divided between red and black. A player who bets \$1 on black wins \$1 (and gets the bet back) if the ball comes to rest on black; otherwise (if the ball lands on red, 0, or 00), the \$1 bet is lost. A player at a roulette wheel starts with \$3 and plans to continue to make \$1 wagers on black until he either doubles his money or goes broke. Find the probability that the player goes broke. Find the average number of wagers the player will make before either going broke or doubling his money. Solve using Markov chain techniques.

Part Three

Calculus

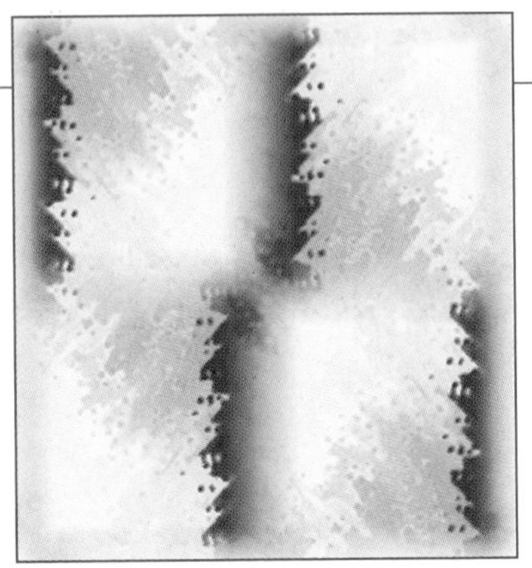

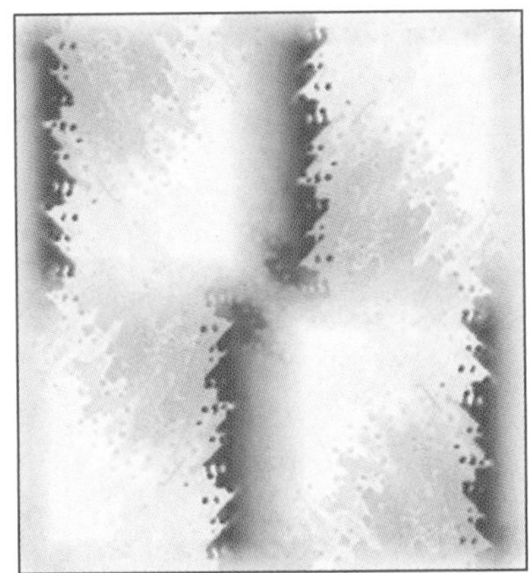

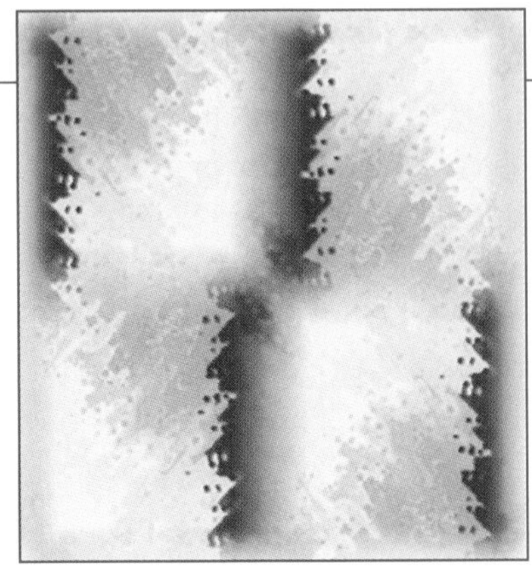

CHAPTER 8

THE DERIVATIVE

INTRODUCTION

How do algebra and calculus differ? The two words *static* and *dynamic* probably come as close as any in expressing the difference between the two disciplines. In algebra, we solve equations for a particular value of a variable—a static notion. In calculus, we are interested in how a change in one variable affects another variable—a dynamic notion.

Parts (A)–(C) of the figure illustrate three basic problems in calculus. It may surprise you to learn that all three problems—as different as they appear—

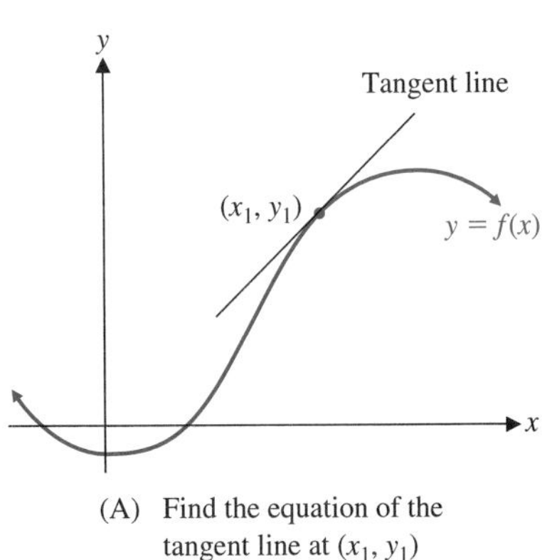

(A) Find the equation of the tangent line at (x_1, y_1) given $y = f(x)$

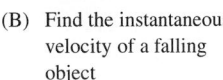

(B) Find the instantaneous velocity of a falling object

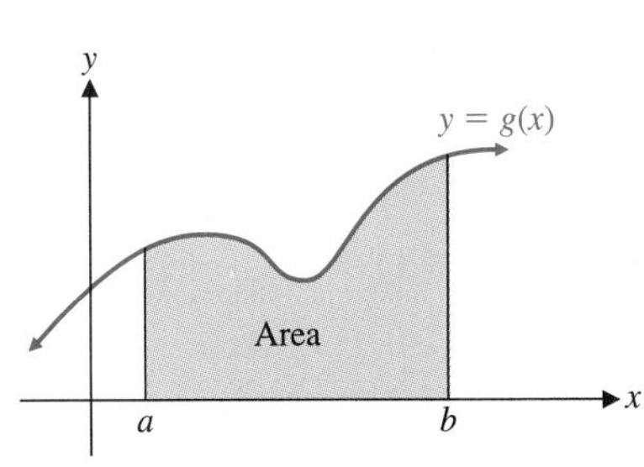

(C) Find the indicated area bounded by $y = g(x)$, $x = a$, $x = b$, and the x axis

are mathematically related. The solutions to these problems and the discovery of their relationship required the creation of a new kind of mathematics. Isaac Newton (1642–1727) of England and Gottfried Wilhelm von Leibniz (1646–1716) of Germany simultaneously and independently developed this new mathematics, called **the calculus**—it was an idea whose time had come.

In addition to solving the problems described in the figure, calculus will enable us to solve many other important problems. Until fairly recently, calculus was used primarily in the physical sciences, but now people in many other disciplines are finding it a useful tool.

SECTION 8-1 Rate of Change and Slope

- RATE OF CHANGE
- SLOPE
- SUMMARY

To provide a feeling and overview for the subject of calculus, a number of key concepts are introduced informally in this section. These concepts will be considered again in subsequent sections.

RATE OF CHANGE

Let us start by considering a simple example.

Example 1 **Revenue Analysis** The revenue (in dollars) from the sale of x plastic planter boxes is given by

$$R(x) = 20x - 0.02x^2 \qquad 0 \leq x \leq 1{,}000$$

which is graphed in Figure 1.

(A) What is the change in revenue if production is changed from 100 planters to 400 planters?
(B) What is the average change in revenue for this change in production?

SOLUTION (A) The change in revenue is given by

$$\begin{aligned} R(400) - R(100) &= 20(400) - 0.02(400)^2 - [20(100) - 0.02(100)^2] \\ &= 4{,}800 - 1{,}800 = \$3{,}000 \end{aligned}$$

Thus, increasing production from 100 planters to 400 planters will increase revenue by \$3,000.

(B) To find the average change in revenue, we divide the change in revenue by the change in production:

$$\frac{R(400) - R(100)}{400 - 100} = \frac{3{,}000}{300} = \$10$$

Thus, the average change in revenue is \$10 per planter when production is increased from 100 to 400 planters.

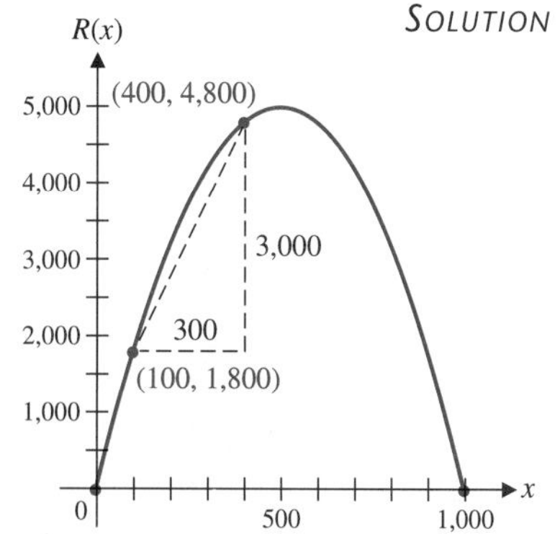

FIGURE 1
$R(x) = 20x - 0.02x^2$

Matched Problem 1 ➠ Refer to the revenue function in Example 1.

(A) What is the change in revenue if production is changed from 600 planters to 800 planters?

(B) What is the average change in revenue for this change in production?

In general, if we are given a function $y = f(x)$ and if x is changed from a to $a + h$, then y will change from $f(a)$ to $f(a + h)$. The *average rate of change is the ratio of the change in y to the change in x.*

Average Rate of Change

For $y = f(x)$, the **average rate of change from $x = a$ to $x = a + h$** is

$$\frac{f(a + h) - f(a)}{(a + h) - a} = \frac{f(a + h) - f(a)}{h} \qquad h \neq 0 \qquad (1)$$

The mathematical expression in (1) is also referred to as the **difference quotient.** The preceding discussion shows that the difference quotient can be interpreted as an average rate of change. The next example illustrates another interpretation of this quotient: velocity of a moving object.

Example 2 ➠ **Average Velocity** A small steel ball dropped from a tower will fall a distance of y feet in x seconds, as given approximately by the formula (from physics)

$$y = f(x) = 16x^2$$

Figure 2 shows the position of the ball on a coordinate line (positive direction down) at the end of 0, 1, 2, and 3 seconds. Find the average velocity from $x = 2$ seconds to $x = 3$ seconds.

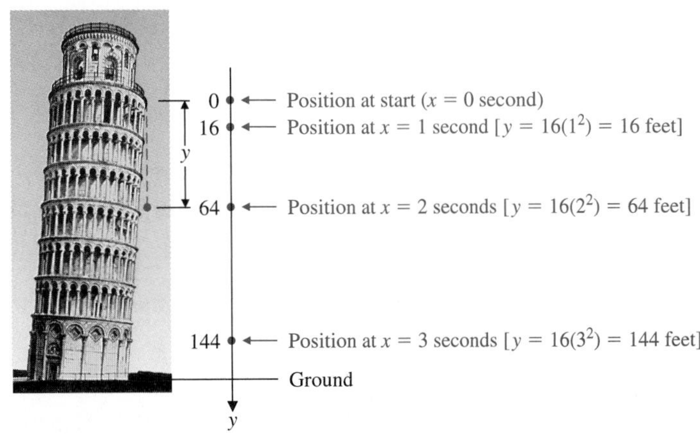

FIGURE 2

Note: Positive y direction is down.

SOLUTION Recall the formula $d = rt$, which can be written in the equivalent form

$$r = \frac{d}{t} = \frac{\text{Distance covered}}{\text{Elapsed time}} = \text{Average velocity (rate)}$$

For example, if a person drives from San Francisco to Los Angeles (a distance of about 420 miles) in 7 hours, then the average velocity is

$$r = \frac{d}{t} = \frac{420}{7} = 60 \text{ miles per hour}$$

Sometimes the person will be traveling faster and sometimes slower, but the average velocity is 60 miles per hour. In our present problem, the average velocity of the steel ball from $x = 2$ seconds to $x = 3$ seconds is

$$\begin{aligned}\text{Average velocity} &= \frac{\text{Distance covered}}{\text{Elapsed time}}\\ &= \frac{f(3) - f(2)}{3 - 2}\\ &= \frac{16(3)^2 - 16(2)^2}{1} = 80 \text{ feet per second}\end{aligned}$$

Thus, we see that if $y = f(x)$ is the position of the falling ball, then the average velocity is simply the average rate of change of $f(x)$ with respect to time x. And we have another interpretation of the difference quotient (1).

Matched Problem 2 For the falling steel ball in Example 2, find the average velocity from $x = 1$ second to $x = 2$ seconds.

In Figure 2, note that the distance the ball falls during each 1 second interval increases with time. Thus, the velocity of the ball is increasing as it falls. Rather than considering average velocity, suppose we are interested in determining the velocity of the ball at the exact instant 2 seconds after it was released. One way to do this is to consider the difference quotient

$$\frac{f(2 + h) - f(2)}{h} \qquad \frac{\text{Distance covered}}{\text{Elapsed time}}$$

which is the average velocity for the change in time from 2 seconds to $2 + h$ seconds if $h > 0$ (or from $2 + h$ seconds to 2 seconds if $h < 0$). We can use these average velocities to approximate the velocity at $x = 2$ seconds. And we would expect that the smaller the choice of h, the better the approximation of this velocity.

Example 3 **Velocity** For the steel ball in Example 2, find the average velocity for $x = 2$ and $h = \pm 0.1, \pm 0.01, \pm 0.001$. What do these average velocities indicate about the velocity at 2 seconds?

SOLUTION The average velocities are shown in Table 1.

Table 1

AVERAGE VELOCITIES

h	-0.1	-0.01	$-0.001 \to 0 \leftarrow 0.001$	0.01	0.1
$\dfrac{f(2+h)-f(2)}{h}$	62.4	63.84	$63.984 \to 64 \leftarrow 64.016$	64.16	65.6

Examining these values, it seems reasonable to conclude that the average velocity from 2 seconds to $2 + h$ seconds is approaching 64 feet per second as h approaches 0. We describe this type of behavior verbally by saying that 64 is *the limit of the average velocity as h approaches 0,* and we express it symbolically by writing

$$\frac{f(2+h)-f(2)}{h} \to 64 \qquad \text{as} \qquad h \to 0$$

or

$$\lim_{h\to 0} \frac{f(2+h)-f(2)}{h} = 64$$

The value of this limit is called the **instantaneous velocity** of the ball at the end of 2 seconds.

Matched Problem 3 For the steel ball in Example 2, find the average velocity for $x = 1$ and $h = \pm 0.1, \pm 0.01, \pm 0.001$. What do these average velocities indicate about the velocity at the end of 1 second?

Explore–Discuss 1

Recall the revenue function discussed in Example 1: $R(x) = 20x - 0.02x^2$. Approximate

$$\lim_{h\to 0} \frac{R(100+h)-R(100)}{h}$$

using $h = \pm 10, \pm 1, \pm 0.1, \pm 0.01$. Discuss possible interpretations of this limit.

The ideas introduced in Example 3 are not confined to just average velocity, but can be applied to the average rate of change of any function.

Instantaneous Rate of Change

For $y = f(x)$, the **instantaneous rate of change at $x = a$** is

$$\lim_{h \to 0} \frac{f(a + h) - f(a)}{h} \tag{2}$$

if the limit exists.

The adjective "instantaneous" is often omitted, with the understanding that the phrase **rate of change** always refers to the instantaneous rate of change and not the average rate of change. In subsequent sections we will develop procedures for finding the exact value of limit (2). For now, we will simply approximate these limits numerically.

Example 4 **Agriculture** Using data from the U.S. Bureau of the Census, an analyst constructed the following function to model the number of farms (in thousands) in the United States:

$$f(t) = 2{,}300 + 4t - 2t^2$$

where t is time in years and $t = 0$ corresponds to 1974.

(A) Find the number of farms in 1984.
(B) Approximate the instantaneous rate of change of the number of farms with respect to time in 1984 by constructing a table similar to that used in Example 3.
(C) Interpret the results in parts (A) and (B).

SOLUTION (A) $f(10) = 2{,}300 + 4(10) - 2(10)^2 = 2{,}140$ or 2.14 million farms
(B) We construct a table of values for $t = 10$ and h small:

h	−0.1	−0.01	−0.001 →	0 ←	0.001	0.01	0.1
$\frac{f(10 + h) - f(10)}{h}$	−35.8	−35.98	−35.998 →	−36 ←	−36.002	−36.02	−36.2

From the table we conclude that the instantaneous rate of change is approximately −36 or −36,000 farms per year.
(C) In 1984 the total number of farms was 2.14 million and was decreasing at the rate of 36,000 farms per year.

Matched Problem 4 Although the total number of farms in the United States has been declining, the number of small farms (under 10 acres) has been growing. The following function gives the number of small farms (in thousands):

$$f(t) = 125 + 10t - 0.4t^2$$

where t is time in years and $t = 0$ corresponds to 1974.

(A) Find the number of small farms in 1984.
(B) Approximate the instantaneous rate of change of the number of small farms with respect to time in 1984 by constructing an appropriate table.
(C) Interpret the results in parts (A) and (B).

One of the major applications of calculus is the calculation and interpretation of instantaneous rates of change. Notice that the brief discussion given in part (C) of Example 4 includes the value of the independent variable (1984), the value of the dependent variable (2.14 million farms), and how this variable is changing (decreasing at the rate of 36,000 farms per year). This information should be included in most interpretations involving rates of change.

SLOPE

So far our interpretations of the difference quotient have been numerical in nature. Now we want to consider a geometric interpretation. A line through two points on the graph of a function is called a **secant line.** If $(a, f(a))$ and $(a + h, f(a + h))$ are two points on the graph of $y = f(x)$, then we can use the slope formula from Section 1-3 to find the slope of the secant line through these points (see Fig. 3).

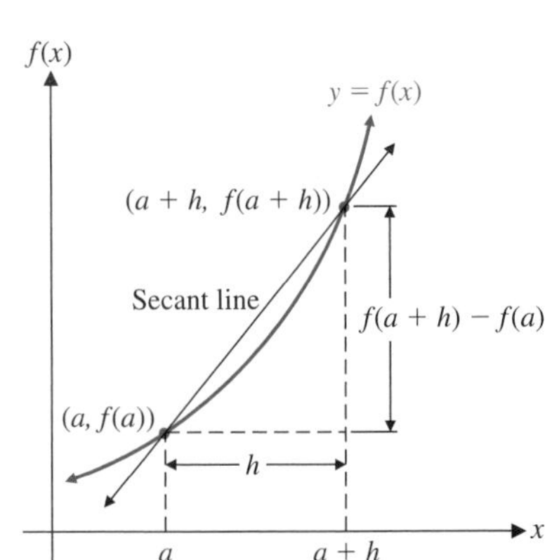

FIGURE 3
Secant line

$$\textbf{Slope of secant line} = \frac{f(a + h) - f(a)}{(a + h) - a} = \frac{f(a + h) - f(a)}{h} \quad \text{Difference quotient}$$

Thus, the difference quotient can be interpreted as both the average rate of change and the slope of the secant line.

Example 5 **Slope of Secant Line** Given $f(x) = 0.5x^2$, find the slopes of the secant lines for $a = 1$ and $h = 2$ and 1, respectively. Graph $y = f(x)$ and the two secant lines.

SOLUTION For $a = 1$ and $h = 2$, the secant line goes through $(1, f(1)) = (1, 0.5)$ and $(3, f(3)) = (3, 4.5)$, and its slope is

$$\frac{f(1 + 2) - f(1)}{2} = \frac{0.5(3)^2 - 0.5(1)^2}{2} = 2$$

For $a = 1$ and $h = 1$, the secant line goes through $(1, f(1)) = (1, 0.5)$ and $(2, f(2)) = (2, 2)$, and its slope is

$$\frac{f(1 + 1) - f(1)}{1} = \frac{0.5(2)^2 - 0.5(1)^2}{1} = 1.5$$

The graphs of $y = f(x)$ and the two secant lines are shown in Figure 4.

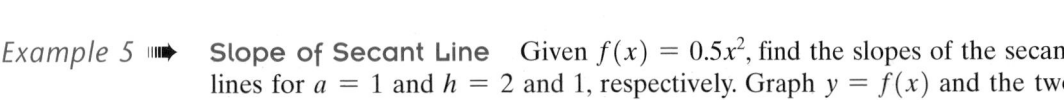

FIGURE 4
Secant lines

Matched Problem 5 Refer to Example 5. Find the slopes of the secant lines for $a = 1$ and $h = -1$ and -0.5, respectively.

In Example 5, suppose we continue to compute the slopes of the secant lines for smaller and smaller values of h (see Table 2):

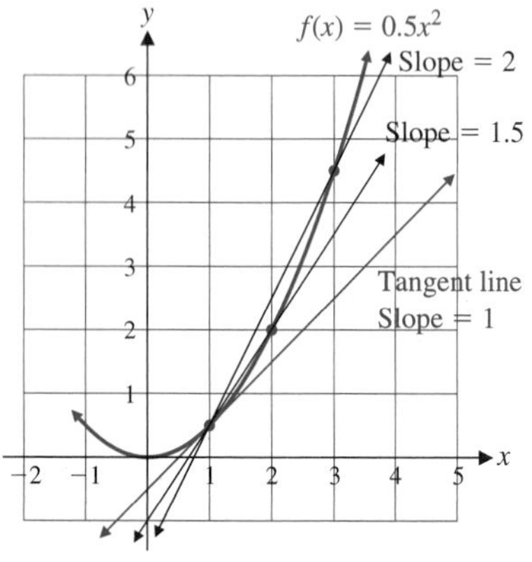

FIGURE 5
Tangent line

Table 2

SLOPES OF SECANT LINES FOR $f(x) = 0.5x^2$ AT $a = 1$

h	-0.1	-0.01	$-0.001 \to 0 \leftarrow 0.001$	0.01	0.1
$\frac{f(1+h) - f(1)}{h}$	0.95	0.995	$0.9995 \to 1 \leftarrow 1.0005$	1.005	1.05

The values in Table 2 suggest that

$$\lim_{h \to 0} \frac{f(1+h) - f(1)}{h} = 1$$

The value of this limit is called the *slope of the graph* of $y = f(x)$ at the point $(1, f(1))$. The line with this slope through the point $(1, f(1))$ is called the *tangent line* (see Fig. 5).

Explore–Discuss 2

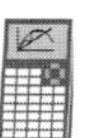

The equation of the line tangent to the graph of $f(x) = 0.5x^2$ at $x = 1$ is $y = x - 0.5$. Graph $y_1 = 0.5x^2$ and $y_2 = x - 0.5$, and zoom in on the point $(1, 0.5)$ repeatedly. Use trace to discuss the relationship between these graphs near the point $(1, 0.5)$.

The ideas introduced in the preceding discussion are summarized below.

Slope of a Graph

Given $y = f(x)$, the **slope of the graph** at the point $(a, f(a))$ is given by

$$\lim_{h \to 0} \frac{f(a+h) - f(a)}{h} \tag{3}$$

provided the limit exists. The slope of the graph is also the **slope of the tangent line** at the point $(a, f(a))$.

From plane geometry, we know that a line tangent to a circle is a line that passes through one and only one point of the circle (see Fig. 6A). Although this definition cannot be extended to graphs of functions in general, the visual relationship between graphs of functions and their tangent lines is similar to the circle case (see Fig. 6B). Limit (3) provides both a mathematically sound definition for the concept of tangent line and a method for approximating the slope of the tangent line.

If a function is defined by an equation, then the slope of the tangent line at a point on the graph can be estimated by computing slopes of secant lines, as we did earlier in Table 2. It is also important to be able to estimate the slope of a tangent line by examining the graph of the function.

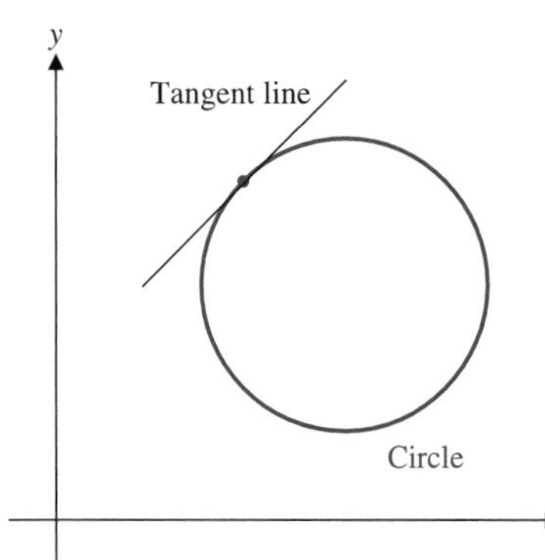

(A) Tangent line for a circle

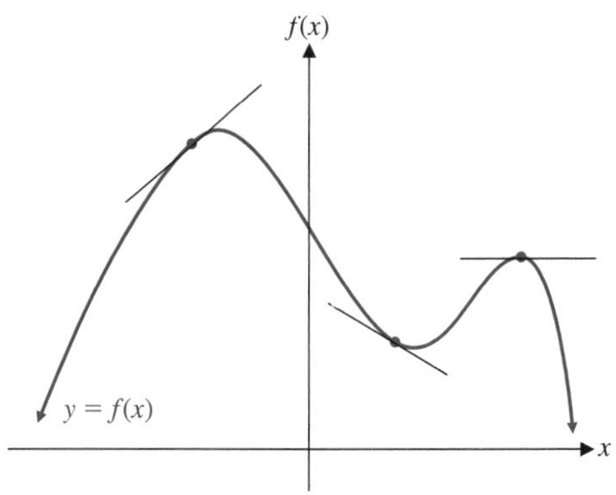

(B) Tangent lines for the graph of a function

FIGURE 6

Example 6 **Estimating the Slope of a Graph** Use the graph of $y = f(x)$ in Figure 7 to estimate the slope of the graph at $x = -3$ and $x = 1$.

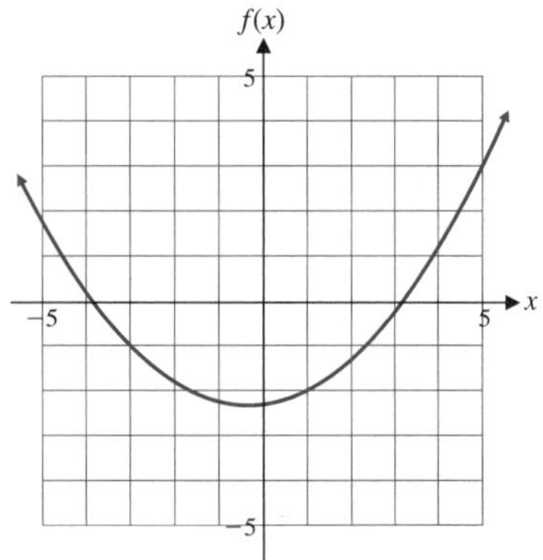

FIGURE 7

Solution Sketch tangent lines at the points $(-3, -1)$ and $(1, -2)$, and estimate the slope of each tangent line by taking the ratio of the rise to the run (see Fig. 8).

$$\text{Slope at } (-3, -1) = \frac{\text{Rise}}{\text{Run}} \approx \frac{-1}{1} = -1$$

$$\text{Slope at } (1, -2) = \frac{\text{Rise}}{\text{Run}} \approx \frac{1}{2}$$

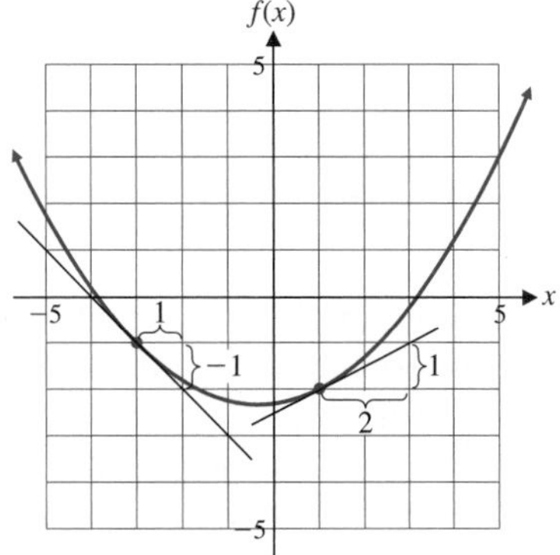

FIGURE 8

Matched Problem 6 ➡ Use the graph of $y = f(x)$ in Figure 9 to estimate the slope of the graph at $x = -2$ and $x = 1$.

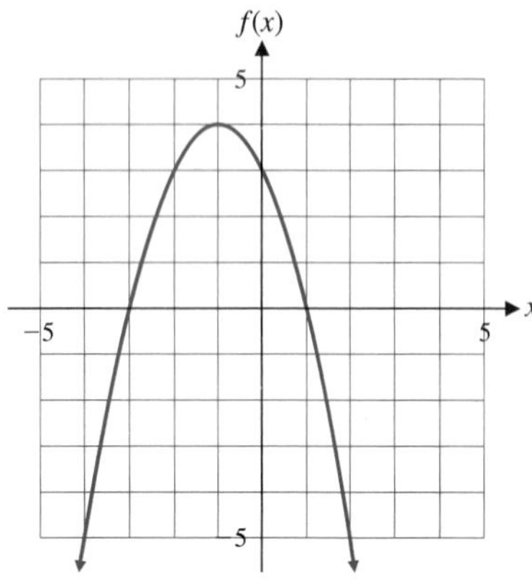

FIGURE 9

Explore–Discuss 3

(A) Sketch the graph of a function that has slope 1 at $(-2, 2)$ and slope -3 at $(3, 1)$.
(B) Sketch the graph of a function that has slope $-\frac{1}{2}$ at $(-3, -1)$ and slope 1 at $(3, 1)$.

■ SUMMARY

We have seen that the difference quotient for a function occurs naturally in a number of different situations (see Table 3).

Table 3

THE DIFFERENCE QUOTIENT AND ITS INTERPRETATIONS

FORMULA OR PROCESS	NUMERIC INTERPRETATION	GEOMETRIC INTERPRETATION
$\frac{f(a+h) - f(a)}{h}$	Average rate of change, or average velocity	Slope of secant line
$\lim_{h \to 0} \frac{f(a+h) - f(a)}{h}$	Instantaneous rate of change, or velocity	Slope of graph and slope of tangent line

We will encounter many more situations involving the interplay between formulas for difference quotients, numeric values, and geometric visualizations. In this introduction our approach has been intuitive and informal. Before we can make the concepts of instantaneous rate of change and slope of a graph more precise, we must discuss the concept of limit in more detail. This is the subject of the next section.

Answers to Matched Problems

1. (A) −$1,600 (B) −$8 per planter 2. 48 ft/sec
3.

h	−0.1	−0.01	−0.001 → 0 ← 0.001	0.01	0.1
$\frac{f(1+h)-f(1)}{h}$	30.4	31.84	31.984 → 32 ← 32.016	32.16	33.6

The instantaneous velocity after 1 sec is approx. 32 ft/sec.

4. (A) 185,000 farms (B) 2,000 farms per year
(C) In 1984 the total number of farms was 185,000 and was increasing at the rate of 2,000 farms per year.
5. For $h = -1$, slope of secant line is 0.5; for $h = -0.5$, slope of secant line is 0.75.
6. Slope at $(-2, 3) \approx 2$; slope at $(1, 0) \approx -4$

EXERCISE 8-1

A *In Problems 1–6, find the indicated quantities for the function $y = f(x) = 3x^2$.*

1. The change in y if x changes from 1 to 4
2. The change in y if x changes from 2 to 5
3. The average rate of change if x changes from 1 to 4
4. The average rate of change if x changes from 2 to 5
5. The slope of the secant line through the points $(1, f(1))$ and $(4, f(4))$ on the graph of $y = f(x)$
6. The slope of the secant line through the points $(2, f(2))$ and $(5, f(5))$ on the graph of $y = f(x)$
7. How are the answers to Problems 3 and 5 related? Is this always the case? Explain.
8. How are the answers to Problems 4 and 6 related? Is this always the case? Explain.

In Problems 9 and 10, complete the following table for $y = f(x) = 3x^2$ and the indicated value of a:

h	−0.1	−0.01	−0.001	→ 0 ←	0.001	0.01	0.1
$\frac{f(a+h)-f(a)}{h}$	?	?	?	→ ? ←	?	?	?

9. $a = 1$
10. $a = 2$

In Problems 11 and 12, approximate (to the nearest integer) the instantaneous rate of change of $y = f(x) = 3x^2$ with respect to x for the indicated value of x.

11. $x = 1$
12. $x = 2$

In Problems 13 and 14, the position of an object moving along the y axis is given by $y = f(x) = 3x^2$, where y is distance in feet and x is time in seconds. Approximate (to the nearest integer) the instantaneous velocity for the indicated value of x.

13. $x = 1$
14. $x = 2$

In Problems 15 and 16, approximate (to the nearest integer) the slope of the graph of $y = f(x) = 3x^2$ at the indicated point.

15. $(1, f(1))$
16. $(2, f(2))$
17. How are the answers to Problems 11, 13, and 15 related? Is this always the case? Explain.
18. How are the answers to Problems 12, 14, and 16 related? Is this always the case? Explain.

B *Problems 19–22 refer to the following situation: An automobile starts from rest and travels down a straight section of road. The distance y (in feet) of the car from the starting position after x seconds is given by $y = f(x) = 10x^2$.*

19. Find the average velocity of the car from $x = 4$ seconds to $x = 6$ seconds.
20. Find the average velocity of the car from $x = 5$ seconds to $x = 7$ seconds.

21. Use a table of average velocities to approximate (to the nearest integer) the instantaneous velocity at $x = 4$ seconds.

22. Use a table of average velocities to approximate (to the nearest integer) the instantaneous velocity at $x = 5$ seconds.

23. Given $y = f(x) = x^2 - 2x - 4$:
(A) Find the slope of the secant line through the points $(2, f(2))$ and $(4, f(4))$.
(B) Find the slope of the secant line through the points $(2, f(2))$ and $(3, f(3))$.
(C) Use a table of secant line slopes to approximate (to the nearest integer) the slope of the graph at $x = 2$.
(D) Graph $y = f(x)$, the secant lines from parts (A) and (B), and the tangent line at $x = 2$.

24. Given $y = f(x) = 5 - x^2$:
(A) Find the slope of the secant line through the points $(1, f(1))$ and $(3, f(3))$.
(B) Find the slope of the secant line through the points $(1, f(1))$ and $(2, f(2))$.
(C) Use a table of secant line slopes to approximate (to the nearest integer) the slope of the graph at $x = 1$.
(D) Graph $y = f(x)$, the secant lines from parts (A) and (B), and the tangent line at $x = 1$.

In Problems 25–28, use the given graph of $y = f(x)$ to approximate (to the nearest integer) the slope of the graph at the points with the indicated x coordinates.

25. $x = -1, x = 3$

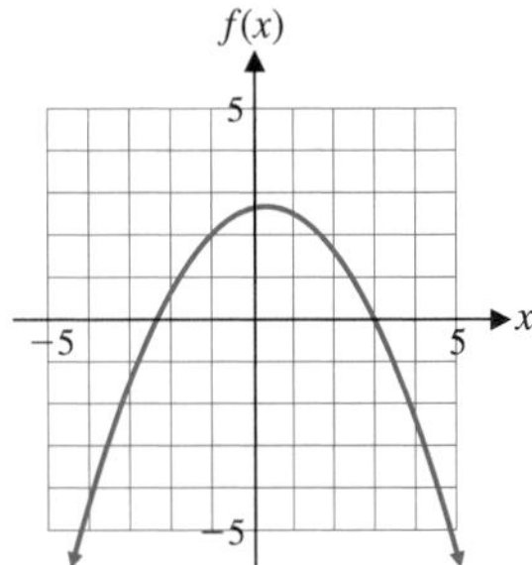

26. $x = -2, x = 2$

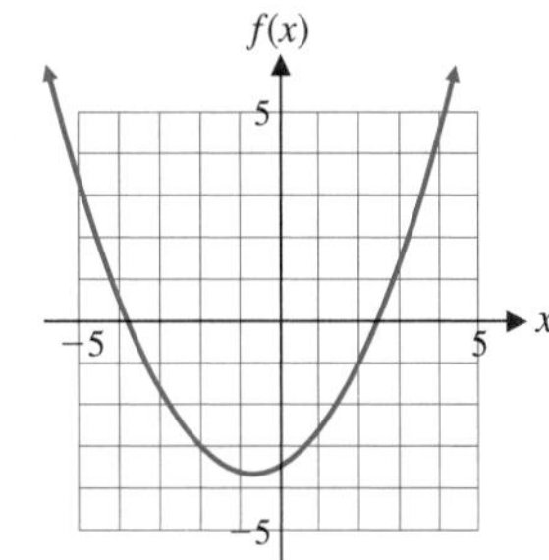

27. $x = -3, x = -1, x = 1, x = 3$

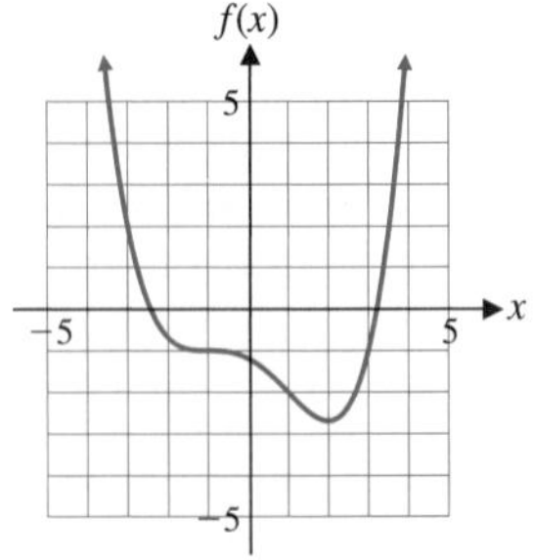

28. $x = -2, x = 0, x = 2, x = 4$

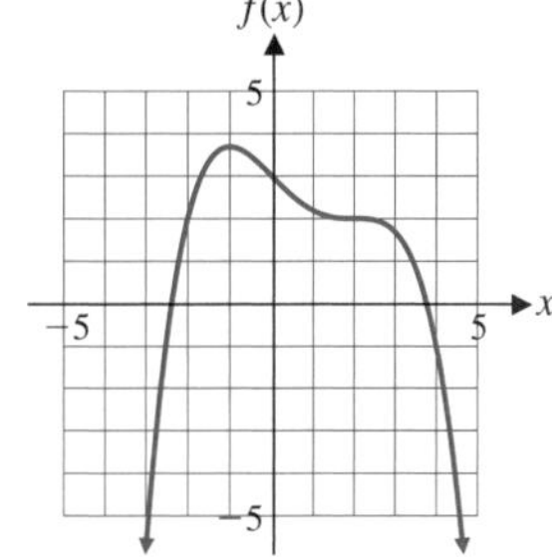

The tables in Problems 29–32 list points on the graph of a function $y = f(x)$ and the slope of the graph at these points. Sketch a possible graph of $y = f(x)$.

29.

Point	Slope
$(-3, 2)$	2
$(-2, 3)$	0
$(0, -1)$	-4

30.

Point	Slope
$(-1, 1)$	-4
$(1, -3)$	0
$(2, -2)$	2

31.

Point	Slope
$(-8, 1)$	3
$(-5, 5)$	0
$(-2, 3)$	-1
$(1, 1)$	0
$(4, 5)$	3

32.

Point	Slope
$(-4, 6)$	-3
$(-1, 2)$	0
$(2, 4)$	1
$(5, 6)$	0
$(8, 2)$	-3

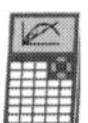

In Problems 33–36, use a graphing utility to approximate (to two decimal places) the slope of the graph of $y = f(x)$ at the point $(a, f(a))$ as follows:

1. *Zoom in on the graph of $y = f(x)$ near $(a, f(a))$ until the graph appears to be a straight line.*
2. *Use trace to find a second point on the graph of $y = f(x)$ near the point $(a, f(a))$.*
3. *Find the slope of the secant line through these two points.*

33. $f(x) = 2x^2; (2, 8)$

34. $f(x) = 2x^2; (-1, 2)$

35. $f(x) = \sqrt{x}; (4, 2)$

36. $f(x) = \sqrt{x}; (1, 1)$

C *In Problems 37 and 38, use the given graph of $y = f(x)$ to approximate (to the nearest integer) the slope of the graph of $y = f(x)$ at $x = -3, -2, -1, 0, 1, 2,$ and 3. Find a function whose values agree with these slopes.*

37.

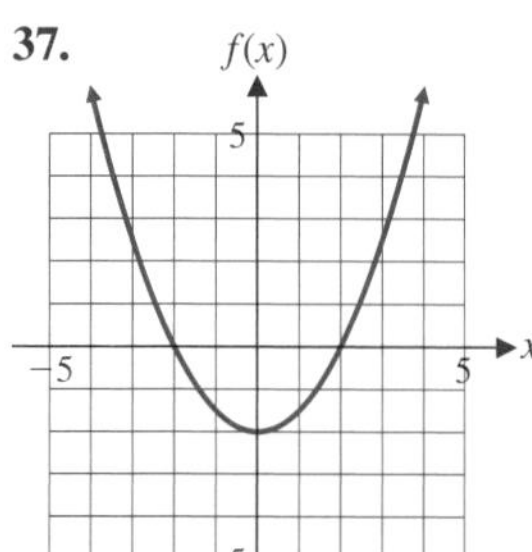

38.

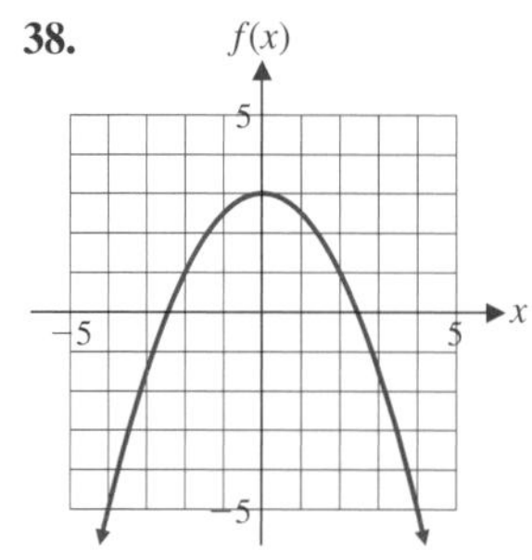

39. Discuss the relationship between the slope of the line $y = mx + b$ and the slope of the graph of the linear function $f(x) = mx + b, m \neq 0$.

40. Discuss the relationship between the slope of the horizontal line $y = b$ and the slope of the graph of the constant function $f(x) = b$.

In Problems 41 and 42, find and simplify the difference quotient for $y = f(x) = 3x^2$ at the indicated value of a. Discuss the relationship between the value of the simplified difference quotient at $h = 0$ and the slope of the graph at $x = a$.

41. $a = 1$

42. $a = 2$

In Problems 43 and 44, construct a table of secant line slopes at $a = 0$ and discuss the slope of the graph at the point $(0, 0)$.

43. $f(x) = |x|$

44. $f(x) = |2x|$

Applications

Business & Economics

45. *Labor.* The average weekly hours and hourly earnings for production workers in the United States from 1970 to 1990 are given in Table 4.

Table 4

YEAR	1970	1975	1980	1985	1990
WEEKLY HOURS	37.1	36.1	35.3	34.9	34.5
HOURLY EARNINGS	\$3.23	\$4.53	\$6.66	\$8.57	\$10.01

(A) Find the average rate of change of weekly hours from 1975 to 1985.

(B) Find the average rate of change of hourly earnings from 1975 to 1990.

46. *Labor.* Refer to Table 4.

(A) Find the average rate of change of weekly hours from 1970 to 1980.

(B) Find the average rate of change of hourly earnings from 1970 to 1985.

47. *Labor.* Refer to Table 4.

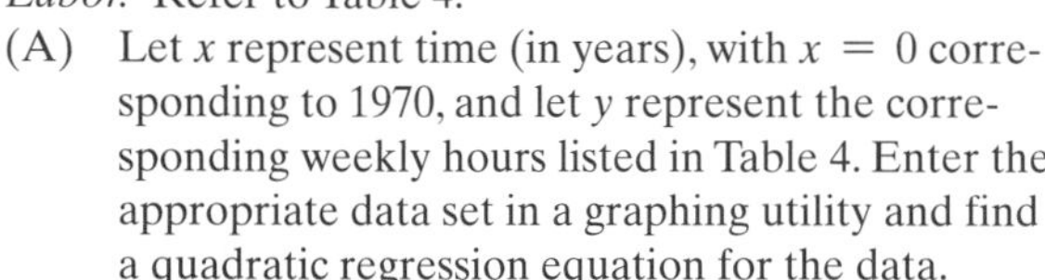

C (A) Let x represent time (in years), with $x = 0$ corresponding to 1970, and let y represent the corresponding weekly hours listed in Table 4. Enter the appropriate data set in a graphing utility and find a quadratic regression equation for the data.

(B) Use the quadratic regression equation to approximate (to two decimal places) the instantaneous rate of change of weekly hours with respect to time in 1980.

48. *Labor.* Refer to Table 4.

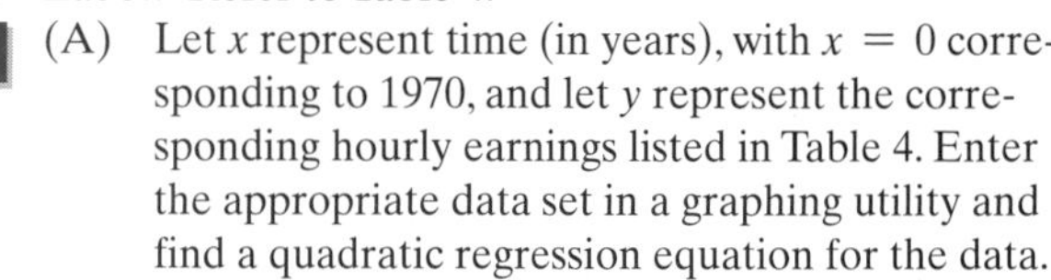

C (A) Let x represent time (in years), with $x = 0$ corresponding to 1970, and let y represent the corresponding hourly earnings listed in Table 4. Enter the appropriate data set in a graphing utility and find a quadratic regression equation for the data.

(B) Use the quadratic regression equation to approximate (to the nearest cent) the instantaneous rate of change of hourly earnings with respect to time in 1980.

49. *Revenue.* The revenue (in dollars) from the sale of x car seats for infants is given by

$$R(x) = 60x - 0.025x^2 \qquad 0 \leq x \leq 2{,}400$$

(A) Find the revenue, and approximate (to the nearest integer) the instantaneous rate of change of revenue at a production level of 1,000 car seats. Write a brief verbal interpretation of these results.

(B) Repeat part (A) for a production level of 1,300 car seats.

50. *Profit.* The profit (in dollars) from the sale of x car seats for infants is given by

$$P(x) = 45x - 0.025x^2 - 5{,}000 \qquad 0 \leq x \leq 2{,}400$$

(A) Find the profit, and approximate (to the nearest integer) the instantaneous rate of change of profit at a production level of 800 car seats. Write a brief verbal interpretation of these results.

(B) Repeat part (A) for a production level of 1,100 car seats.

51. *Mineral production.* The annual U.S. production of sulfur (in thousands of metric tons) is given approximately by

$$f(t) = -150t^2 + 770t + 10{,}400$$

where t is time in years and $t = 0$ corresponds to 1987. Find the annual production in 1990, and approximate (to the nearest integer) the instantaneous rate of change of production in 1990. Write a brief verbal interpretation of these results.

52. *Mineral production.* The annual U.S. production of talc (in thousands of metric tons) is given approximately by

$$f(t) = -30t^2 + 100t + 1{,}170$$

where t is time in years and $t = 0$ corresponds to 1987. Find the annual production in 1991, and approximate (to the nearest integer) the instantaneous rate of change of production in 1991. Write a brief verbal interpretation of these results.

Life Sciences

53. *Health care expenditures.* Table 5 lists U.S. health care expenditures (in billions of dollars) for services and supplies and for research and construction from 1987 to 1991.

Table 5

YEAR	1987	1988	1989	1990	1991
SERVICES AND SUPPLIES	476.9	526.2	583.6	652.4	728.6
RESEARCH AND CONSTRUCTION	17.3	19.8	20.7	22.7	23.1

(A) Find the average rate of change of expenditures for services and supplies from 1988 to 1991.
(B) Find the average rate of change of expenditures for research and construction from 1987 to 1989.

54. *Health care expenditures.* Refer to Table 5.
(A) Find the average rate of change of expenditures for services and supplies from 1989 to 1991.
(B) Find the average rate of change of expenditures for research and construction from 1987 to 1990.

55. *Health care expenditures.* Refer to Table 5.
C (A) Let x represent time (in years), with $x = 0$ corresponding to 1987, and let y represent the corresponding expenditures for services and supplies listed in Table 5. Enter the appropriate data set in a graphing utility and find a quadratic regression equation for the data.
(B) Use the quadratic regression equation to approximate (to two decimal places) the instantaneous rate of change of expenditures for services and supplies with respect to time in 1989.

56. *Health care expenditures.* Refer to Table 5.
 (A) Let x represent time (in years), with $x = 0$ corresponding to 1987, and let y represent the corresponding expenditures for research and construction listed in Table 5. Enter the appropriate data set in a graphing utility and find a quadratic regression equation for the data.
(B) Use the quadratic regression equation to approximate (to two decimal places) the instantaneous rate of change of expenditures for research and construction with respect to time in 1989.

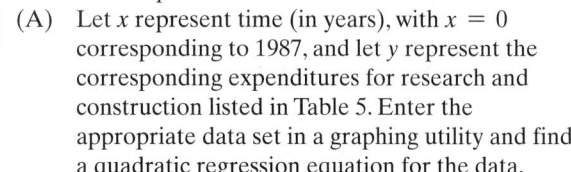

Social Sciences

57. *Infant mortality.* The number of male infant deaths per 100,000 births in the United States is given approximately by

$$f(t) = 0.008t^2 - 0.9t + 29.6$$

where t is time in years and $t = 0$ corresponds to 1960. Find the number of male infant deaths in 1990, and approximate (to the nearest integer) the instantaneous rate of change of the number of male infant deaths in 1990. Write a brief verbal interpretation of these results.

58. *Infant mortality.* The number of female infant deaths per 100,000 births in the United States is given approximately by

$$f(t) = 0.005t^2 - 0.65t + 22.8$$

where t is time in years and $t = 0$ corresponds to 1960. Find the number of female infant deaths in 1990, and approximate (to the nearest integer) the instantaneous rate of change of the number of female infant deaths in 1990. Write a brief verbal interpretation of these results.

SECTION 8-2

Limits

- ■ FUNCTIONS AND GRAPHS—A BRIEF REVIEW
- ■ LIMITS
- ■ LIMIT EVALUATION
- ■ LIMITS OF DIFFERENCE QUOTIENTS

In the preceding section we saw that determining the instantaneous rate of change of a function or the slope of a graph involves a limit of a difference quotient:

$$\lim_{h \to 0} \frac{f(a + h) - f(a)}{h} \tag{1}$$

Up to now we have approximated these limits numerically. Before we can proceed further in the study of calculus, we need to discuss the limit concept in more detail. As we will see, limit evaluation is a basic calculus tool that can be used in many different situations. In this section we will develop a combined numerical, graphical, and algebraic approach that will enable us to evaluate a variety of limit forms, including the very important form (1).

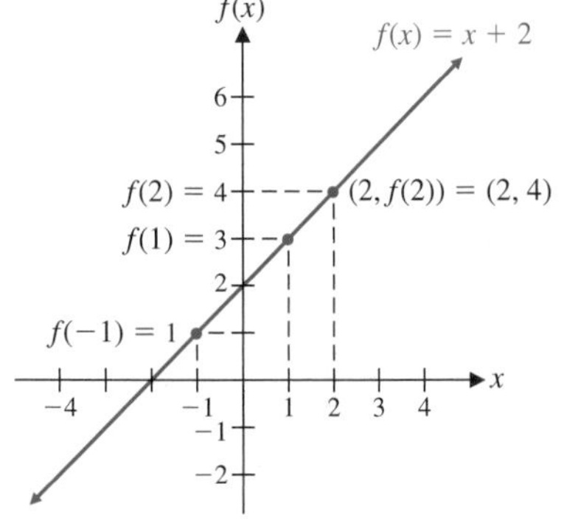

FIGURE 1

■ FUNCTIONS AND GRAPHS—A BRIEF REVIEW

The graph of the function $y = f(x) = x + 2$ is the graph of the set of all ordered pairs $(x, f(x))$. For example, if $x = 2$, then $f(2) = 4$ and $(2, f(2)) = (2, 4)$ is a point on the graph of f. Figure 1 shows $(-1, f(-1))$, $(1, f(1))$, and $(2, f(2))$ plotted on the graph of f. Notice that the domain values -1, 1, and 2 are associated with the x axis, and the range values $f(-1) = 1$, $f(1) = 3$, and $f(2) = 4$ are associated with the y axis.

Given x, it is sometimes useful to be able to read $f(x)$ directly from the graph of f. Example 1 reviews this process.

Example 1 **Finding Values of a Function from Its Graph** Complete the table below using the given graph of the function g.

x	$g(x)$
−2	
1	
3	
4	

Solution To determine $g(x)$, proceed vertically from the x value on the x axis to the graph of g, then horizontally to the corresponding y value, $g(x)$, on the y axis (as indicated by the dashed lines):

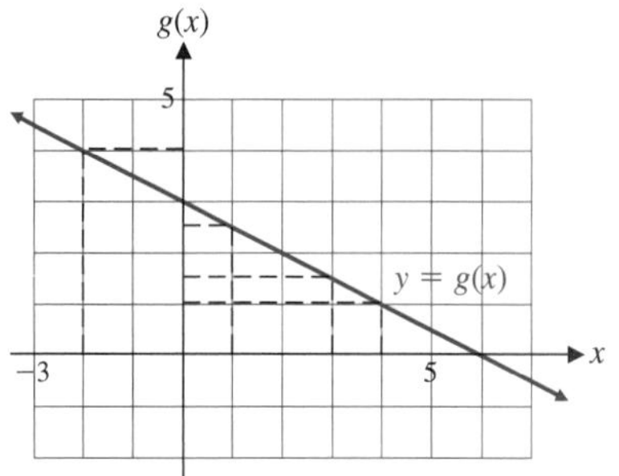

x	$g(x)$
−2	4.0
1	2.5
3	1.5
4	1.0

Matched Problem 1 Complete the table below using the given graph of the function h.

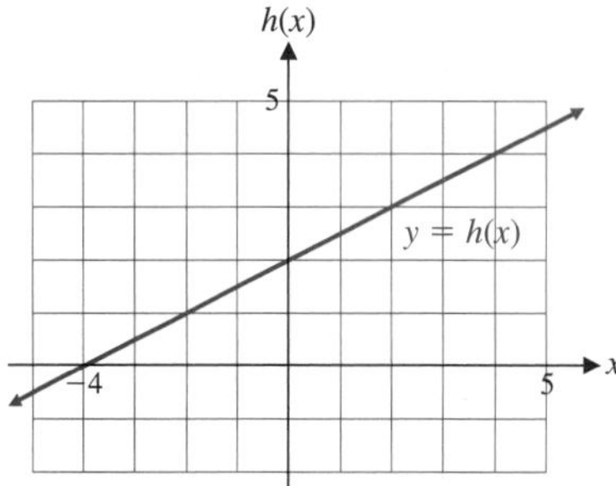

x	$h(x)$
−2	
−1	
0	
1	
2	
3	
4	

Limits

We introduce the important notion of *limit* through two examples, after which the limit concept will be defined.

Example 2 **Analyzing a Limit** Let $f(x) = x + 2$. Discuss the behavior of $f(x)$ numerically, graphically, and algebraically when x is chosen closer and closer to 2, but not equal to 2.

Solution To investigate the behavior of $f(x)$ numerically for x near 2, we construct a table of values (Table 1), as we did in Section 8-1.

Table 1

x	1.5	1.8	1.9	1.99	$1.999 \to 2 \leftarrow 2.001$	2.01	2.1	2.2	2.5
$f(x)$	3.5	3.8	3.9	3.99	$3.999 \to 4 \leftarrow 4.001$	4.01	4.1	4.2	4.5

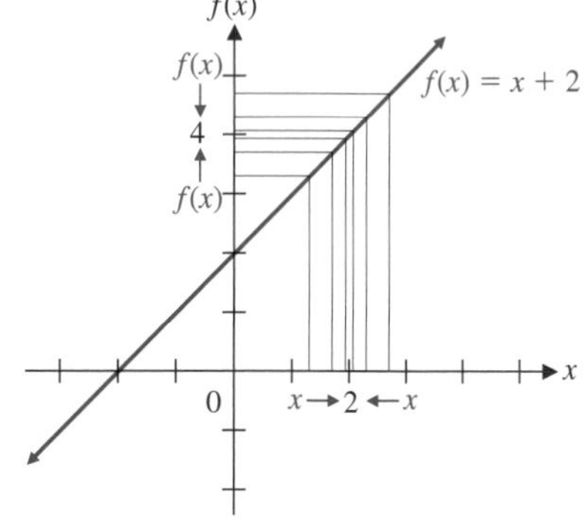

FIGURE 2

To investigate the behavior of $f(x)$ graphically, we draw a graph of f for x near 2 (Fig. 2). Referring to the table and the graph, we see that $f(x)$ approaches 4 as x approaches 2 from either side of 2. To confirm this algebraically, we note that if x approaches 2, then $x + 2$ must approach $2 + 2 = 4$. Collecting all this evidence, we conclude that

$$\lim_{x \to 2}(x + 2) = 4 \qquad \text{or} \qquad x + 2 \to 4 \quad \text{as} \quad x \to 2$$

Also note that $f(2) = 4$. Thus, the value of the function at 2 and the limit of the function at 2 are the same. That is,

$$\lim_{x \to 2}(x + 2) = f(2)$$

Graphically, this means that there is no break or hole in the graph of f at $x = 2$.

Matched Problem 2 ➡ Let $f(x) = x + 1$.

(A) Complete the following table:

x	0.9	0.99	0.999	$\to 1 \leftarrow$	1.001	1.01	1.1
$f(x)$	?	?	?	$\to ? \leftarrow$	?	?	?

(B) Graph $f(x) = x + 1$.
(C) Use the table, the graph, and the algebraic expression $x + 1$ to find

$$\lim_{x \to 1}(x + 1)$$

(D) Use a similar approach to find

$$\lim_{x \to 0}(x + 1) \qquad \text{and} \qquad \lim_{x \to 3}(x + 1)$$

The results found in Example 2 and Matched Problem 2 were fairly obvious. The next example is a little less obvious.

Example 3 ➡ **Analyzing a Limit** Let

$$g(x) = \frac{x^2 - 4}{x - 2} \qquad x \neq 2$$

Even though the function is not defined when $x = 2$ (both the numerator and denominator are 0), we can still ask how $g(x)$ behaves when x is near 2, but not equal to 2. Can you guess what happens to $g(x)$ as x approaches 2? The numerator tending to 0 is a force pushing the fraction toward 0. The denominator tending to 0 is another force pushing the fraction toward larger values. How do these two forces balance out?

SOLUTION Proceeding as before, we construct a table of values of $g(x)$ (Table 2):

Table 2

x	1.5	1.8	1.9	1.99	1.999	$\to 2 \leftarrow$	2.001	2.01	2.1	2.2	2.5
$g(x)$	3.5	3.8	3.9	3.99	3.999	$\to 4 \leftarrow$	4.001	4.01	4.1	4.2	4.5

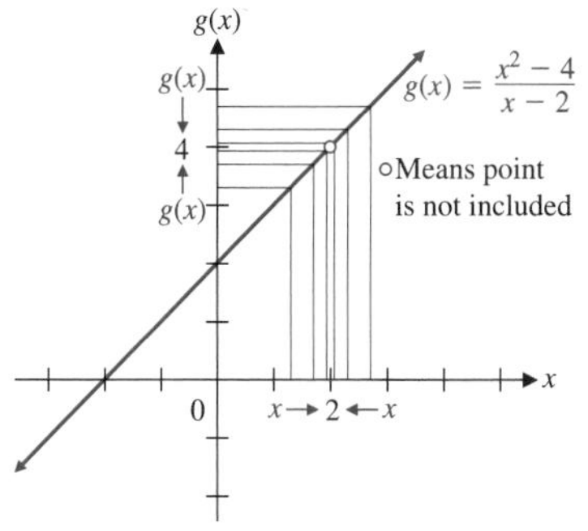

FIGURE 3

Notice that these values agree with the values of $f(x) = x + 2$ in Table 1. This suggests that there is an algebraic relationship between f and g:

$$g(x) = \frac{x^2 - 4}{x - 2} = \frac{(x - 2)(x + 2)}{x - 2} = x + 2 \qquad x \neq 2$$

Thus, we see that $g(x) = f(x)$ for all x, except $x = 2$. This implies that the graph of g is the same as the graph of f (Fig. 2), except that the graph of g has a hole at the point with coordinates (2, 4), as shown in Figure 3.

Since the behavior of $(x^2 - 4)/(x - 2)$ for x near 2, but not equal to 2, is the same as the behavior of $x + 2$ for x near 2, but not equal to 2, we have

$$\lim_{x \to 2} \frac{x^2 - 4}{x - 2} = \lim_{x \to 2}(x + 2) = 4$$

And we see that the limit of the function g at 2 exists even though the function is not defined there (the graph has a hole at $x = 2$).

Matched Problem 3 Proceed as in Example 3 to find: $\lim_{x \to 1} \frac{x^2 - 1}{x - 1}$

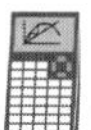

REMARK

If you use a graphing utility to investigate the graph of $g(x) = (x^2 - 4)/(x - 2)$ (from Example 3), you may not see a hole in the graph at $x = 2$ (see Fig. 4). This is due to the difference between the coordinates of a point in the plane (an infinite set) and the coordinates of a pixel on the screen (a finite set). On many graphing utilities, a simple way to make this hole visible is to choose the window parameters so that $x = 2$ is the midpoint of the graphing interval (see Fig. 5). Notice that the graphing utility does not display a y coordinate corresponding to $x = 2$, since $g(2)$ is not defined. See Problems 85 and 86 in Exercise 8-2 for another way to choose window parameters so that holes in graphs will be visible.

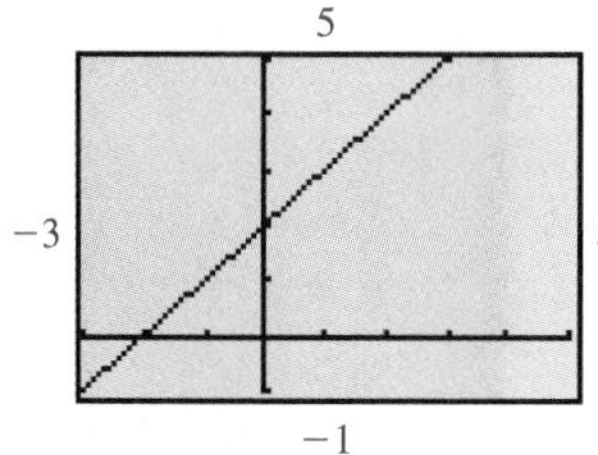

FIGURE 4

$g(x) = \frac{x^2 - 4}{x - 2}$ for $-3 \leq x \leq 5$

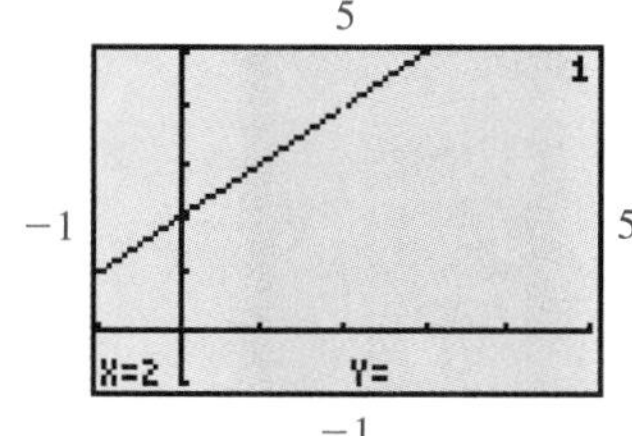

FIGURE 5

$g(x) = \frac{x^2 - 4}{x - 2}$ for $-1 \leq x \leq 5$

We now present an informal definition of the important concept of limit. A precise definition is not needed for our discussion, but one is given in the footnote.*

*To make the informal definition of limit precise, the use of the word *close* must be made more precise. This is done as follows: We write $\lim_{x \to c} f(x) = L$ if for each $e > 0$, there exists a $d > 0$ such that $|f(x) - L| < e$ whenever $0 < |x - c| < d$. This definition is used to establish particular limits and to prove many useful properties of limits that will be helpful to us in finding particular limits. [Even though intuitive notions of limit existed for a long time, it was not until the nineteenth century that a precise definition was given by the German mathematician, Karl Weierstrass (1815–1897).]

> **Limit**
>
> We write
>
> $$\lim_{x \to c} f(x) = L \quad \text{or} \quad f(x) \to L \quad \text{as} \quad x \to c$$
>
> if the functional value $f(x)$ is close to the single real number L whenever x is close to, but not equal to, c (on either side of c).
>
> [*Note:* The existence of a limit at c has nothing to do with the value of the function at c. In fact, c may not even be in the domain of f (see Example 3). However, the function must be defined on both sides of c.]

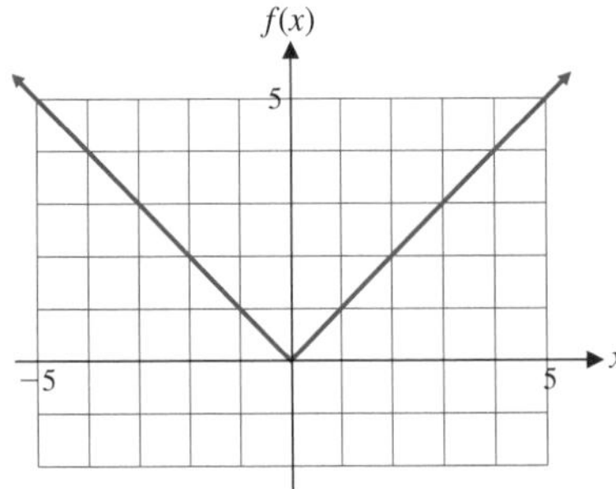

FIGURE 6
$f(x) = |x|$

The next example involves the **absolute value function:**

$$f(x) = |x| = \begin{cases} -x & \text{if } x < 0 \\ x & \text{if } x \geq 0 \end{cases}$$

$f(-2) = |-2| = -(-2) = 2$
$f(3) = |3| = 3$

The graph of f is shown in Figure 6.

Example 4 **Analyzing a Limit** Let $h(x) = |x|/x$. Explore the behavior of $h(x)$ for x near 0, but not equal to 0, using a table and a graph. Find $\lim_{x \to 0} h(x)$, if it exists.

SOLUTION The function h is defined for all real numbers except 0. For example,

$$h(-2) = \frac{|-2|}{-2} = \frac{2}{-2} = -1$$

$$h(0) = \frac{|0|}{0} = \frac{0}{0} \qquad \text{Not defined}$$

$$h(2) = \frac{|2|}{2} = \frac{2}{2} = 1$$

In general, $h(x)$ is -1 for all negative x and 1 for all positive x. Table 3 and Figure 7 illustrate the behavior of $h(x)$ for x near 0.

Table 3

x	-2	-1	-0.1	-0.01	$-0.001 \to$	0	$\leftarrow 0.001$	0.01	0.1	1	2
$h(x)$	-1	-1	-1	-1	-1	$\to -1 \neq 1 \leftarrow$	1	1	1	1	1

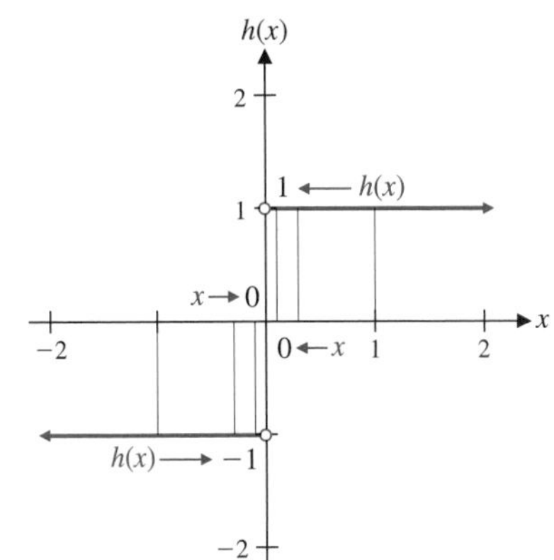

FIGURE 7

When x is near 0 (on either side of 0), is $h(x)$ near one specific number? The answer is "No," because $h(x)$ is -1 for $x < 0$ and 1 for $x > 0$. Consequently, we say that

$$\lim_{x \to 0} \frac{|x|}{x} \quad \text{does not exist}$$

Thus, neither $h(x)$ nor the limit of $h(x)$ exist at $x = 0$. However, the limit from the left and the limit from the right both exist at 0, but they are not equal. (We will discuss this further below.)

Matched Problem 4 Graph

$$h(x) = \frac{x - 2}{|x - 2|}$$

and find $\lim_{x \to 2} h(x)$, if it exists.

In Example 2, we found it helpful to examine the values of the function $f(x)$ as x approached 2 from the left and then from the right. In Example 4, we saw that the values of the function $h(x)$ approached two different numbers, depending on the direction of approach, and it was natural to refer to these values as "the limit from the left" and "the limit from the right." These experiences suggest that the notion of **one-sided limits** will be very useful when discussing basic limit concepts.

We write

$$\lim_{x \to c^-} f(x) = K$$ $x \to c^-$ is read "x approaches c from the left" and means $x \to c$ and $x < c$.

and call K the **limit from the left** (or **left-hand limit**) if $f(x)$ is close to K whenever x is close to c, but to the left of c on the real number line. We write

$$\lim_{x \to c^+} f(x) = L$$ $x \to c^+$ is read "x approaches c from the right" and means $x \to c$ and $x > c$.

and call L the **limit from the right** (or **right-hand limit**) if $f(x)$ is close to L whenever x is close to c, but to the right of c on the real number line.

We now make the following important observation:

On the Existence of a Limit

In order for a limit to exist, the limit from the left and the limit from the right must exist and be equal.

In Example 4,

$$\lim_{x \to 0^-} \frac{|x|}{x} = -1 \quad \text{and} \quad \lim_{x \to 0^+} \frac{|x|}{x} = 1$$

Since the left- and right-hand limits are not te same,

$$\lim_{x \to 0} \frac{|x|}{x} \quad \text{does not exist}$$

Example 5 **Analyzing Limits Graphically** Given the graph of the function f shown in Figure 8, we discuss the behavior of $f(x)$ for x near $-1, 1, 2$, and 3.

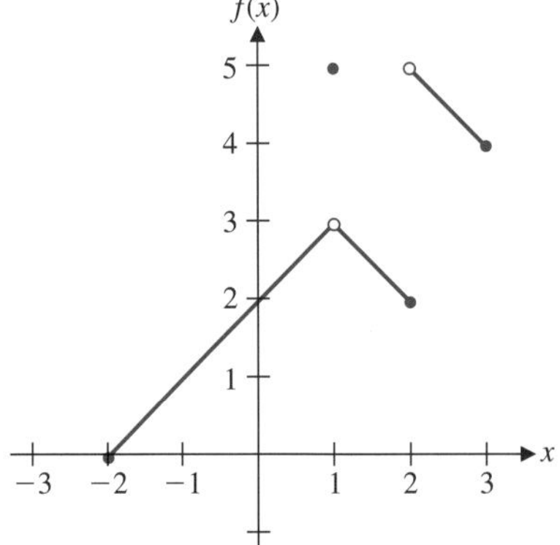

FIGURE 8

(A) Behavior of $f(x)$ for x near -1:

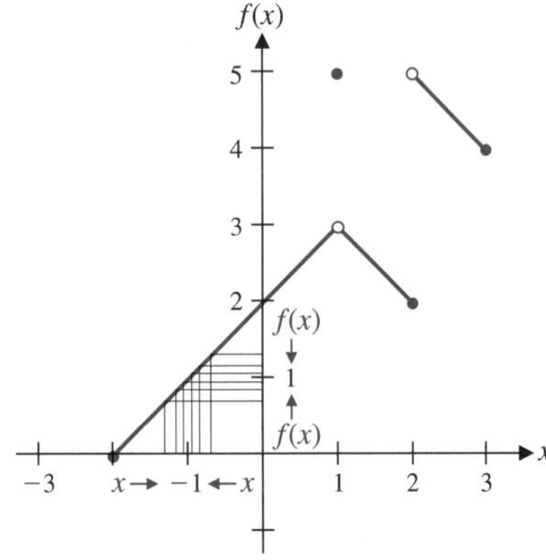

$$\lim_{x \to -1^-} f(x) = 1$$
$$\lim_{x \to -1^+} f(x) = 1$$
$$\lim_{x \to -1} f(x) = 1$$
$$f(-1) = 1$$

(B) Behavior of $f(x)$ for x near 1:

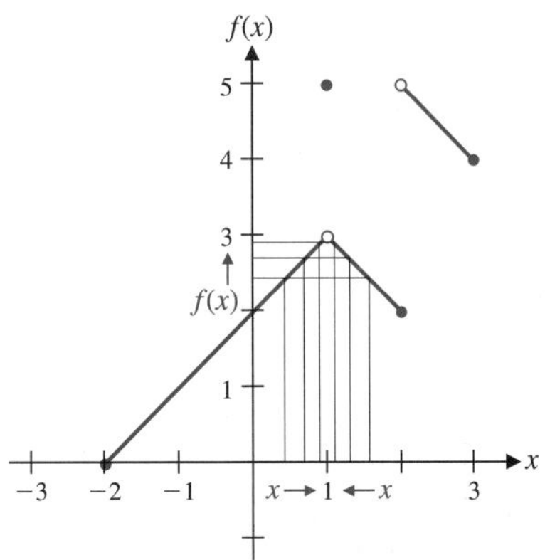

$$\lim_{x \to 1^-} f(x) = 3$$
$$\lim_{x \to 1^+} f(x) = 3$$
$$\lim_{x \to 1} f(x) = 3$$
$$f(1) = 5$$

(C) Behavior of $f(x)$ for x near 2:

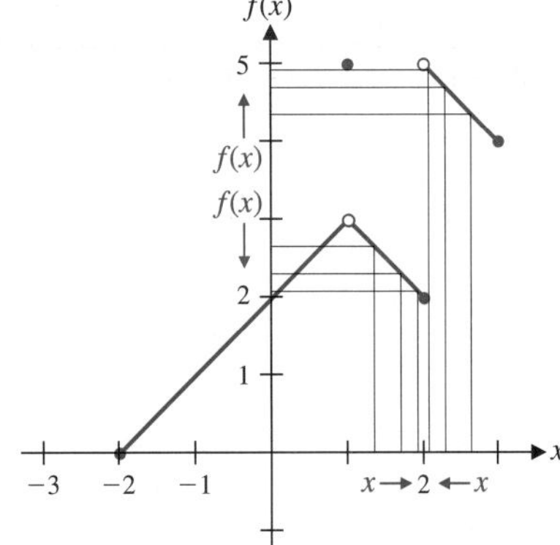

$$\lim_{x \to 2^-} f(x) = 2$$
$$\lim_{x \to 2^+} f(x) = 5$$
$$\lim_{x \to 2} f(x) \quad \text{does not exist}$$
$$f(2) = 2$$

(D) Behavior of $f(x)$ for x near 3:

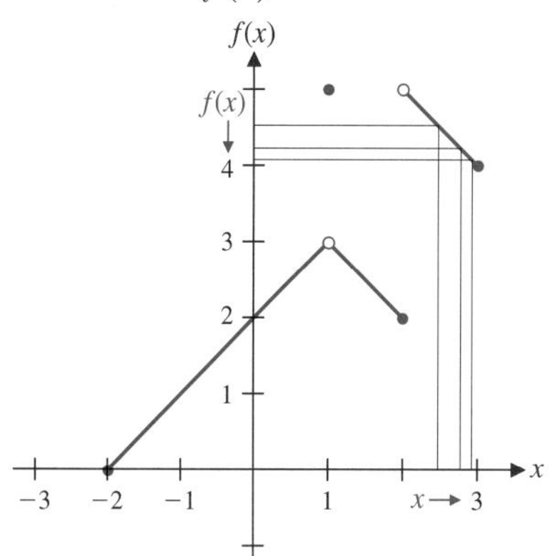

$$\lim_{x \to 3^-} f(x) = 4$$

$\lim_{x \to 3^+} f(x)$ does not exist

f is not defined for x > 3

$\lim_{x \to 3} f(x)$ does not exist

$$f(3) = 4$$

Matched Problem 5 Given the graph of the function f shown in Figure 9, discuss the following, as we did in Example 5:

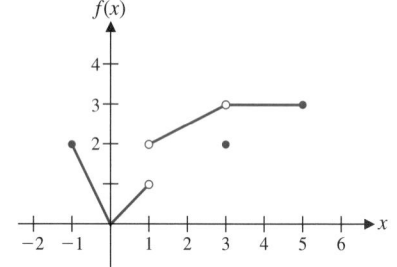

(A) Behavior of $f(x)$ for x near -1
(B) Behavior of $f(x)$ for x near 0
(C) Behavior of $f(x)$ for x near 1
(D) Behavior of $f(x)$ for x near 3

FIGURE 9

LIMIT EVALUATION

Tables and graphs are very useful tools for investigating limits, especially if something unusual happens at the point in question. However, many of the limits encountered in calculus are routine and can be evaluated quickly, using a little algebraic simplification, some intuition, and basic properties of limits. The following list of properties of limits forms the basis for this approach.

Theorem 1 PROPERTIES OF LIMITS

Let f and g be two functions, and assume that

$$\lim_{x \to c} f(x) = L \qquad \lim_{x \to c} g(x) = M$$

where L and M are real numbers (both limits exist). Then:

1. $\lim_{x \to c}[f(x) + g(x)] = \lim_{x \to c} f(x) + \lim_{x \to c} g(x) = L + M$

2. $\lim_{x \to c}[f(x) - g(x)] = \lim_{x \to c} f(x) - \lim_{x \to c} g(x) = L - M$

3. $\lim_{x \to c} kf(x) = k \lim_{x \to c} f(x) = kL$ for any constant k

4. $\lim_{x \to c}[f(x) \cdot g(x)] = \left[\lim_{x \to c} f(x)\right]\left[\lim_{x \to c} g(x)\right] = LM$

5. $\lim_{x \to c} \dfrac{f(x)}{g(x)} = \dfrac{\lim_{x \to c} f(x)}{\lim_{x \to c} g(x)} = \dfrac{L}{M}$ if $M \neq 0$

6. $\lim_{x \to c} \sqrt[n]{f(x)} = \sqrt[n]{\lim_{x \to c} f(x)} = \sqrt[n]{L}$ $L > 0$ for n even

Explore–Discuss 1

The properties listed in Theorem 1 can be paraphrased in brief verbal statements. For example, property 1 simply states that *the limit of a sum is equal to the sum of the limits.* Write brief verbal statements for the remaining properties in Theorem 1.

Example 6 **Using Limit Properties** Find: $\lim_{x\to 3}(x^2 - 4x)$

SOLUTION First, note the following obvious limit:

$$\lim_{x\to 3} x = 3$$

(All this says is that if x approaches 3, then x approaches 3.) Now we make use of this limit and the limit properties:

$$\begin{aligned}\lim_{x\to 3}(x^2 - 4x) &= \lim_{x\to 3} x^2 - \lim_{x\to 3} 4x && \text{Property 2}\\ &= \left(\lim_{x\to 3} x\right)\cdot\left(\lim_{x\to 3} x\right) - 4\lim_{x\to 3} x && \text{Properties 3 and 4}\\ &= 3\cdot 3 - 4\cdot 3 = -3\end{aligned}$$

With a little practice you will soon be able to omit the steps in the dashed boxes and simply write

$$\lim_{x\to 3}(x^2 - 4x) = 3\cdot 3 - 4\cdot 3 = -3$$

Matched Problem 6 Find: $\lim_{x\to -2}(x^2 + 5x)$

What happens if we try to evaluate a limit like the one in Example 6, but with x approaching an unspecified number, such as c? Proceeding as we did in Example 6, we have

$$\lim_{x\to c}(x^2 - 4x) = c\cdot c - 4\cdot c = c^2 - 4c$$

If we let $f(x) = x^2 - 4x$, then we have

$$\lim_{x\to c} f(x) = \lim_{x\to c}(x^2 - 4x) = c^2 - 4c = f(c)$$

That is, this limit can be evaluated simply by evaluating the function f at c. It would certainly simplify the process of evaluating limits if we could identify the functions for which

$$\lim_{x\to c} f(x) = f(c) \tag{2}$$

since we could use this fact to evaluate the limit. It turns out that there are many functions that satisfy (2). We will postpone a detailed discussion of these functions until the next chapter. For now, we note that if

$$f(x) = a_n x^n + a_{n-1}x^{n-1} + \cdots + a_0$$

is a polynomial function, then the properties in Theorem 1 imply that

$$\begin{aligned}\lim_{x\to c} f(x) &= \lim_{x\to c}(a_n x^n + a_{n-1}x^{n-1} + \cdots + a_0)\\ &= a_n c^n + a_{n-1}c^{n-1} + \cdots + a_0 = f(c)\end{aligned}$$

We state this useful result in Theorem 2.

Theorem 2 ■■ LIMIT OF A POLYNOMIAL FUNCTION

If $f(x)$ is a polynomial function and c is any real number, then

$$\lim_{x\to c} f(x) = f(c)$$ ■■

Now let's see how we can use Theorems 1 and 2 together to simply and quickly evaluate limits.

Example 7 **Evaluating Limits** Find each limit:

(A) $\lim_{x\to 2}(x^3 - 5x - 1)$ (B) $\lim_{x\to -1}\sqrt{2x^2 + 3}$ (C) $\lim_{x\to 4}\frac{2x}{3x + 1}$

SOLUTION (A) $\lim_{x\to 2}(x^3 - 5x - 1) = 2^3 - 5\cdot 2 - 1 = -3$ Theorem 2

(B) $\lim_{x\to -1}\sqrt{2x^2 + 3} = \sqrt{\lim_{x\to -1}(2x^2 + 3)}$ Property 6

$= \sqrt{2(-1)^2 + 3}$ Theorem 2

$= \sqrt{5}$

(C) $\lim_{x\to 4}\frac{2x}{3x + 1} = \frac{\lim_{x\to 4} 2x}{\lim_{x\to 4}(3x + 1)}$ Property 5

$= \frac{2\cdot 4}{3\cdot 4 + 1}$ Theorem 2

$= \frac{8}{13}$

Matched Problem 7 Find each limit:

(A) $\lim_{x\to -1}(x^4 - 2x + 3)$ (B) $\lim_{x\to 2}\sqrt{3x^2 - 6}$ (C) $\lim_{x\to -2}\frac{x^2}{x^2 + 1}$

It is important to note that there are restrictions on some of the limit properties. In particular, if

$$\lim_{x\to c} f(x) = 0 \quad \text{and} \quad \lim_{x\to c} g(x) = 0$$

then finding

$$\lim_{x\to c}\frac{f(x)}{g(x)} \tag{3}$$

may present some difficulties, since limit property 5 (the limit of a quotient) does not apply when $\lim_{x\to c} g(x) = 0$. We often have to use algebraic manipulation or other devices to determine the outcome. Recall from Examples 3 and 4 that

$$\lim_{x\to 2}\frac{x^2 - 4}{x - 2} = \lim_{x\to 2}\frac{(x - 2)(x + 2)}{x - 2} = \lim_{x\to 2}(x + 2) = 4$$

and

$$\lim_{x \to 0} \frac{|x|}{x} \quad \text{does not exist}$$

From these two examples, it is clear that knowing only that $\lim_{x \to c} f(x) = 0$ and $\lim_{x \to c} g(x) = 0$ is not enough to determine limit (3). Depending on the choice of functions f and g, the limit (3) may or may not exist. Consequently, if we are given (3) and $\lim_{x \to c} f(x) = 0$ and $\lim_{x \to c} g(x) = 0$, then (3) is said to be **indeterminate,** or, more specifically, a **0/0 indeterminate form.**

CAUTION

The expression 0/0 does not represent a real number and should never be used as the value of a limit. If a limit is a 0/0 indeterminate form, then further investigation is always required to determine whether the limit exists and to find its value, if it does exist.

Explore–Discuss 2

Use algebraic, numerical, and/or graphical techniques to analyze each of the following indeterminate forms:

(A) $\lim_{x \to 1} \frac{x - 1}{x^2 - 1}$ (B) $\lim_{x \to 1} \frac{(x - 1)^2}{x^2 - 1}$ (C) $\lim_{x \to 1} \frac{x^2 - 1}{(x - 1)^2}$

■ LIMITS OF DIFFERENCE QUOTIENTS

Now that we have developed some experience working with limits, we are ready to apply these techniques to the limits of difference quotients.

Example 8 **Limit of a Difference Quotient** Find the following limit for $f(x) = 4x - 5$.

$$\lim_{h \to 0} \frac{f(3 + h) - f(3)}{h}$$

SOLUTION

$$\begin{aligned} \lim_{h \to 0} \frac{f(3 + h) - f(3)}{h} &= \lim_{h \to 0} \frac{[4(3 + h) - 5] - [4(3) - 5]}{h} \\ &= \lim_{h \to 0} \frac{12 + 4h - 5 - 12 + 5}{h} \\ &= \lim_{h \to 0} \frac{4h}{h} = \lim_{h \to 0} 4 = 4 \end{aligned}$$

Since this is a 0/0 indeterminate form and property 5 in Theorem 1 does not apply, we proceed with algebraic simplification.

Matched Problem 8 Find the following limit for $f(x) = 7 - 2x$: $\lim_{h \to 0} \frac{f(4 + h) - f(4)}{h}$

Explore–Discuss 3

The following is an incorrect solution to Example 8 with the invalid statements indicated by $\neq$. Explain why each $\neq$ is used.

$$\begin{aligned}\lim_{h\to 0}\frac{f(3+h)-f(3)}{h} &\neq \lim_{h\to 0}\frac{4(3+h)-5-4(3)-5}{h}\\ &= \lim_{h\to 0}\frac{-10+4h}{h}\\ &\neq \lim_{h\to 0}\frac{-10+4}{1} = -6\end{aligned}$$

Example 9 **Limit of a Difference Quotient** Find the following limit for $f(x) = |x+5|$.

$$\lim_{h\to 0}\frac{f(-5+h)-f(-5)}{h}$$

SOLUTION

$$\begin{aligned}\lim_{h\to 0}\frac{f(-5+h)-f(-5)}{h} &= \lim_{h\to 0}\frac{|(-5+h)+5|-|-5+5|}{h}\\ &= \lim_{h\to 0}\frac{|h|}{h} \quad \text{does not exist}\end{aligned}$$

Since this is a 0/0 indeterminate form and property 5 in Theorem 1 does not apply, we proceed with algebraic simplification. See Example 4.

Matched Problem 9 Find the following limit for $f(x) = |x-1|$: $\lim_{h\to 0}\frac{f(1+h)-f(1)}{h}$

Example 10 **Limit of a Difference Quotient** Find the following limit for $f(x) = \sqrt{x}$.

$$\lim_{h\to 0}\frac{f(2+h)-f(2)}{h}$$

SOLUTION

$$\lim_{h\to 0}\frac{f(2+h)-f(2)}{h} = \lim_{h\to 0}\frac{\sqrt{2+h}-\sqrt{2}}{h}$$

This is a 0/0 indeterminate form, so property 5 in Theorem 1 does not apply. Rationalizing the numerator will be of help.

$$\begin{aligned}&= \lim_{h\to 0}\frac{\sqrt{2+h}-\sqrt{2}}{h}\cdot\frac{\sqrt{2+h}+\sqrt{2}}{\sqrt{2+h}+\sqrt{2}} \qquad (A-B)(A-B) = A^2-B^2\\ &= \lim_{h\to 0}\frac{2+h-2}{h(\sqrt{2+h}+\sqrt{2})}\\ &= \lim_{h\to 0}\frac{1}{\sqrt{2+h}+\sqrt{2}}\\ &= \frac{1}{\sqrt{2}+\sqrt{2}} = \frac{1}{2\sqrt{2}}\end{aligned}$$

Matched Problem 10 Find the following limit for $f(x) = \sqrt{x}$: $\lim_{h\to 0}\frac{f(3+h)-f(3)}{h}$

Example 11 **Slope of a Graph** Find the slope of the graph of $y = f(x) = 2x - x^2$ at the point $(2, 0)$. Graph f and the tangent line at the point $(2, 0)$.

SOLUTION Recall from Section 8-1 that

$$\text{Slope of the graph} = \lim_{h\to 0}\frac{f(2+h)-f(2)}{h}$$

So we have

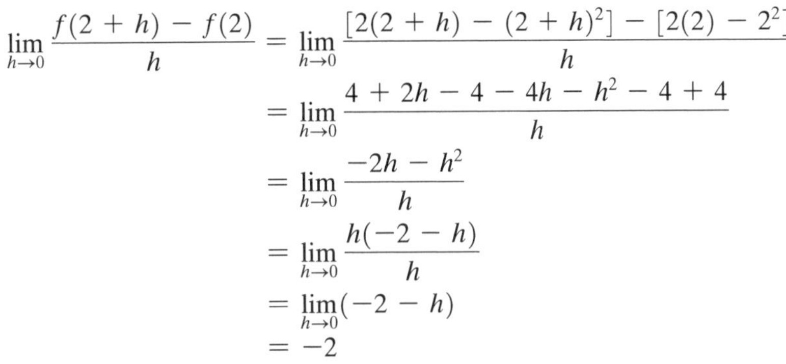

$$
\begin{aligned}
\lim_{h\to 0}\frac{f(2+h)-f(2)}{h} &= \lim_{h\to 0}\frac{[2(2+h)-(2+h)^2]-[2(2)-2^2]}{h}\\
&= \lim_{h\to 0}\frac{4+2h-4-4h-h^2-4+4}{h}\\
&= \lim_{h\to 0}\frac{-2h-h^2}{h}\\
&= \lim_{h\to 0}\frac{h(-2-h)}{h}\\
&= \lim_{h\to 0}(-2-h)\\
&= -2
\end{aligned}
$$

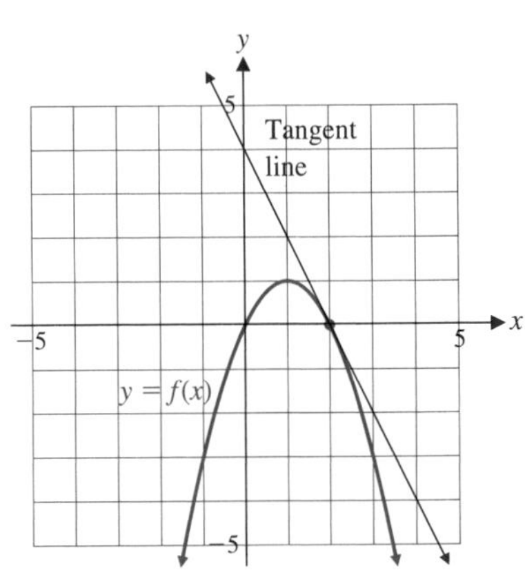

FIGURE 10

Thus, the slope of the graph at the point $(2, 0)$ is -2. The graph of f and the tangent line [the line with slope -2 through the point $(2, 0)$] are shown in Figure 10.

Matched Problem 11 Find the slope of the graph of $y = f(x) = 2x - x^2$ at the point $(-1, -3)$.

Answers to Matched Problems

1.

x	−2	−1	0	1	2	3	4
$h(x)$	1.0	1.5	2.0	2.5	3.0	3.5	4.0

2. (A)

x	0.9	0.99	$0.999 \to 1 \leftarrow 1.001$	1.01	1.1
$f(x)$	1.9	1.99	$1.999 \to 2 \leftarrow 2.001$	2.01	2.1

(B)

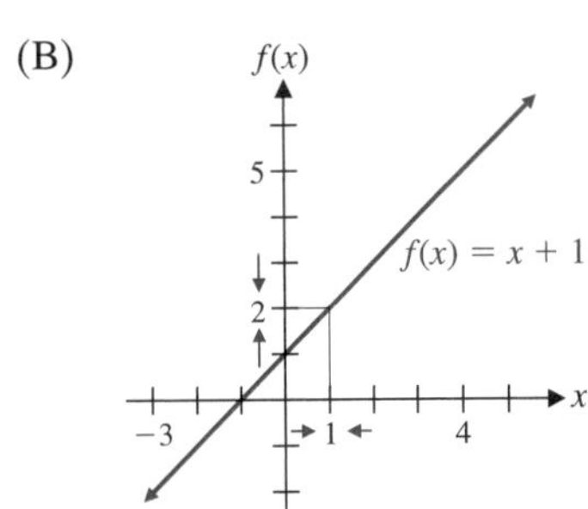

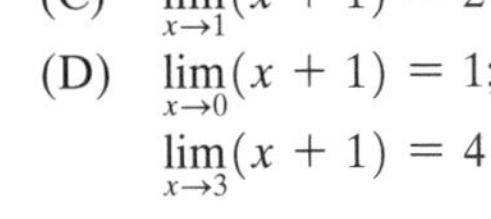

(C) $\lim_{x\to 1}(x+1) = 2$

(D) $\lim_{x\to 0}(x+1) = 1$; $\lim_{x\to 3}(x+1) = 4$

3. 2

4.

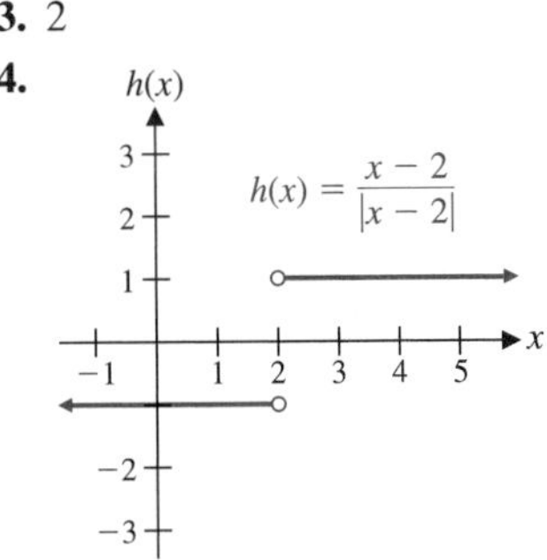

$\lim_{x\to 2}\frac{x-2}{|x-2|}$ does not exist

5. (A) $\lim_{x\to -1^-} f(x)$ does not exist
$\lim_{x\to -1^+} f(x) = 2$
$\lim_{x\to -1} f(x)$ does not exist
$f(-1) = 2$
(B) $\lim_{x\to 0^-} f(x) = 0$
$\lim_{x\to 0^+} f(x) = 0$
$\lim_{x\to 0} f(x) = 0$
$f(0) = 0$
(C) $\lim_{x\to 1^-} f(x) = 1$
$\lim_{x\to 1^+} f(x) = 2$
$\lim_{x\to 1} f(x)$ does not exist
$f(1)$ not defined
(D) $\lim_{x\to 3^-} f(x) = 3$
$\lim_{x\to 3^+} f(x) = 3$
$\lim_{x\to 3} f(x) = 3$
$f(3) = 2$

6. -6 **7.** (A) 6 (B) $\sqrt{6}$ (C) $\frac{4}{5}$ **8.** -2 **9.** Does not exist
10. $1/(2\sqrt{3})$ **11.** 4

Exercise 8-2

A *In Problems 1–14, sketch a possible graph of a function that satisfies the given conditions.*

1. $f(0) = 1$; $\lim_{x\to 0^-} f(x) = 3$; $\lim_{x\to 0^+} f(x) = 1$
2. $f(1) = -2$; $\lim_{x\to 1^-} f(x) = 2$; $\lim_{x\to 1^+} f(x) = -2$
3. $f(-2) = 2$; $\lim_{x\to -2^-} f(x) = 1$; $\lim_{x\to -2^+} f(x) = 1$
4. $f(0) = -1$; $\lim_{x\to 0^-} f(x) = 2$; $\lim_{x\to 0^+} f(x) = 2$

In Problems 5–8, use the graph of the function f shown below to estimate the indicated limits and function values.

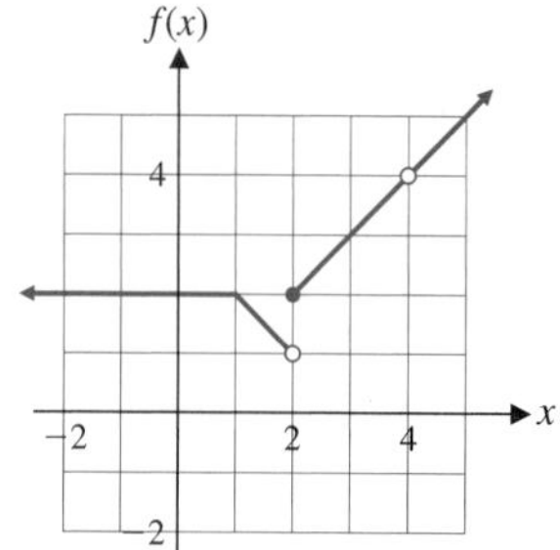

5. (A) $\lim_{x\to 0^-} f(x)$ (B) $\lim_{x\to 0^+} f(x)$
(C) $\lim_{x\to 0} f(x)$ (D) $f(0)$
6. (A) $\lim_{x\to 1^-} f(x)$ (B) $\lim_{x\to 1^+} f(x)$
(C) $\lim_{x\to 1} f(x)$ (D) $f(1)$
7. (A) $\lim_{x\to 2^-} f(x)$ (B) $\lim_{x\to 2^+} f(x)$
(C) $\lim_{x\to 2} f(x)$ (D) $f(2)$
(E) Is it possible to redefine $f(2)$ so that $\lim_{x\to 2} f(x) = f(2)$? Explain.
8. (A) $\lim_{x\to 4^-} f(x)$ (B) $\lim_{x\to 4^+} f(x)$
(C) $\lim_{x\to 4} f(x)$ (D) $f(4)$
(E) Is it possible to define $f(4)$ so that $\lim_{x\to 4} f(x) = f(4)$? Explain.

In Problems 9–12, use the graph of the function g shown below to estimate the indicated limits and function values.

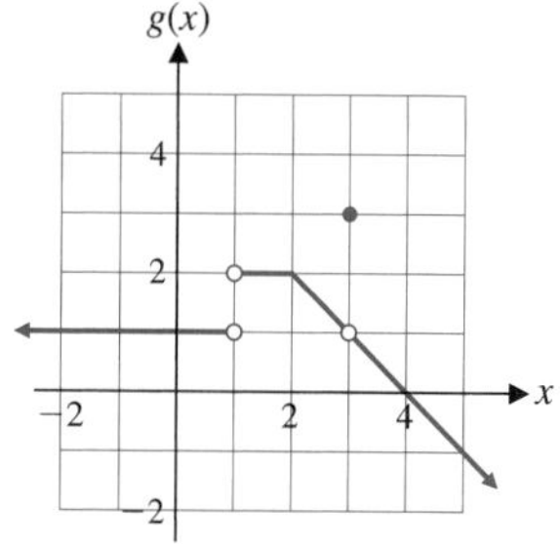

9. (A) $\lim_{x\to 1^-} g(x)$ (B) $\lim_{x\to 1^+} g(x)$
(C) $\lim_{x\to 1} g(x)$ (D) $g(1)$
(E) Is it possible to define $g(1)$ so that $\lim_{x\to 1} g(x) = g(1)$? Explain.
10. (A) $\lim_{x\to 2^-} g(x)$ (B) $\lim_{x\to 2^+} g(x)$
(C) $\lim_{x\to 2} g(x)$ (D) $g(2)$
11. (A) $\lim_{x\to 3^-} g(x)$ (B) $\lim_{x\to 3^+} g(x)$
(C) $\lim_{x\to 3} g(x)$ (D) $g(3)$
(E) Is it possible to redefine $g(3)$ so that $\lim_{x\to 3} g(x) = g(3)$? Explain.
12. (A) $\lim_{x\to 4^-} g(x)$ (B) $\lim_{x\to 4^+} g(x)$
(C) $\lim_{x\to 4} g(x)$ (D) $g(4)$

Given $\lim_{x\to 3} f(x) = 5$ and $\lim_{x\to 3} g(x) = 9$, find the indicated limits in Problems 13–22.

13. $\lim_{x\to 3}[f(x) - g(x)]$
14. $\lim_{x\to 3}[f(x) + g(x)]$
15. $\lim_{x\to 3} 4g(x)$
16. $\lim_{x\to 3}(-2)f(x)$
17. $\lim_{x\to 3} \dfrac{f(x)}{g(x)}$
18. $\lim_{x\to 3}[f(x) \cdot g(x)]$

19. $\lim_{x\to 3} \sqrt{f(x)}$

20. $\lim_{x\to 3} \sqrt{g(x)}$

21. $\lim_{x\to 3} \frac{f(x) + g(x)}{2f(x)}$

22. $\lim_{x\to 3} \frac{g(x) - f(x)}{3g(x)}$

B *In Problems 23–52, use limit properties, algebraic simplification, tables of values, and/or graphs to find each limit, if it exists.*

23. $\lim_{x\to 5}(2x^2 - 3)$

24. $\lim_{x\to 2}(x^2 - 8x + 2)$

25. $\lim_{x\to 2} \frac{5x}{2 + x^2}$

26. $\lim_{x\to 10} \frac{2x + 5}{3x - 5}$

27. $\lim_{x\to 2}(x + 1)^3(2x - 1)^2$

28. $\lim_{x\to 3}(x + 2)^2(2x - 4)$

29. $\lim_{x\to -1} \sqrt{5 - 4x}$

30. $\lim_{x\to 2} \sqrt{5 - 4x}$

31. $\lim_{x\to 6} \sqrt{25 - x^2}$

32. $\lim_{x\to 4} \sqrt{25 - x^2}$

33. $\lim_{x\to -3} \frac{x^2 - 9}{x + 3}$

34. $\lim_{x\to -5} \frac{x^2 - 25}{x + 5}$

35. $\lim_{x\to 1^+} \frac{|x - 1|}{x - 1}$

36. $\lim_{x\to 3^-} \frac{x - 3}{|x - 3|}$

37. $\lim_{x\to 1^-} \frac{|x - 1|}{x - 1}$

38. $\lim_{x\to 3^+} \frac{x - 3}{|x - 3|}$

39. $\lim_{x\to 1} \frac{|x - 1|}{x - 1}$

40. $\lim_{x\to 3} \frac{x - 3}{|x - 3|}$

41. $\lim_{x\to 1} \frac{x - 2}{x^2 - 2x}$

42. $\lim_{x\to 1} \frac{x + 3}{x^2 + 3x}$

43. $\lim_{x\to 2} \frac{x - 2}{x^2 - 2x}$

44. $\lim_{x\to -3} \frac{x + 3}{x^2 + 3x}$

45. $\lim_{x\to 2} \frac{x^2 - x - 6}{x + 2}$

46. $\lim_{x\to 3} \frac{x^2 + x - 6}{x + 3}$

47. $\lim_{x\to -2} \frac{x^2 - x - 6}{x + 2}$

48. $\lim_{x\to -3} \frac{x^2 + x - 6}{x + 3}$

49. $\lim_{x\to 3}\left(\frac{x}{x + 3} + \frac{x - 3}{x^2 - 9}\right)$

50. $\lim_{x\to 2}\left(\frac{1}{x + 2} + \frac{x - 2}{x^2 - 4}\right)$

51. $\lim_{x\to 0}\left(\sqrt{x^2 + 9} - \frac{x^2 + 3x}{x}\right)$

52. $\lim_{x\to 1}\left(\frac{x^2 - 1}{x - 1} + \sqrt{x^2 + 3}\right)$

Compute the following limit for each function in Problems 53–60:

$$\lim_{h\to 0} \frac{f(2 + h) - f(2)}{h}$$

53. $f(x) = 3x + 1$

54. $f(x) = 5x - 1$

55. $f(x) = x^2 + 1$

56. $f(x) = x^2 - 2$

57. $f(x) = \sqrt{x} - 2$

58. $f(x) = 1 + \sqrt{x}$

59. $f(x) = |x - 2| - 3$

60. $f(x) = 2 + |x - 2|$

In Problems 61–64, find the slope of the graph of $y = f(x)$ at the indicated point. Graph f and the tangent line at this point.

61. $f(x) = x^2 - 3; (2, 1)$

62. $f(x) = 5 - x^2; (-1, 4)$

63. $f(x) = \sqrt{x}; (4, 2)$

64. $f(x) = \sqrt{x}; (1, 1)$

In Problems 65 and 66, an automobile starts from rest and travels down a straight section of road. The distance y (in feet) of the car from the starting position after x seconds is given by $y = f(x) = 10x^2$.

65. Find the instantaneous velocity at $x = 4$ seconds.

66. Find the instantaneous velocity at $x = 5$ seconds.

In Problems 67 and 68, use the given graph of f to estimate

$$\lim_{h\to 0} \frac{f(a + h) - f(a)}{h}$$

to the nearest integer for the indicated values of a.

67. $a = -3, a = -1, a = 3$

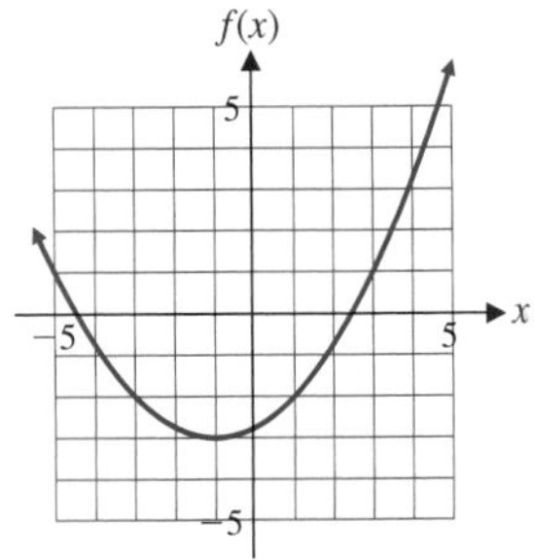

68. $a = -3, a = 1, a = 3$

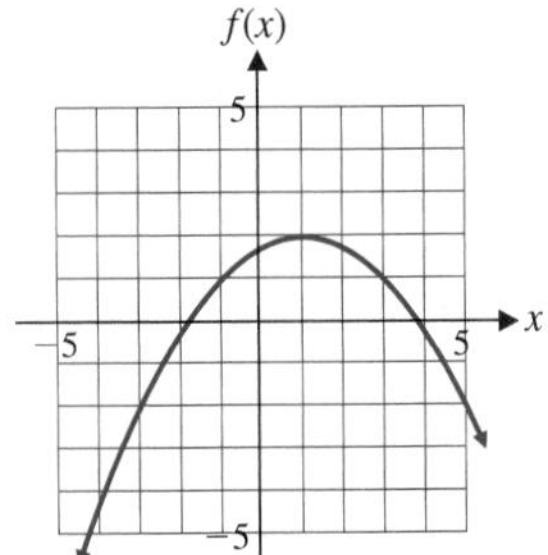

69. Use a table of values to investigate the behavior of the following limits:

(A) $\lim_{x\to 0^-} \frac{1}{x}$ (B) $\lim_{x\to 0^+} \frac{1}{x}$

Write a brief verbal description of each limit.

70. Use a table of values to investigate the behavior of the following limits:

(A) $\lim_{x\to 0^-} \frac{1}{x^2}$ (B) $\lim_{x\to 0^+} \frac{1}{x^2}$

Write a brief verbal description of each limit.

C

71. Let f be defined by

$$f(x) = \begin{cases} 1 + mx & \text{if } x \leqslant 1 \\ 4 - mx & \text{if } x > 1 \end{cases}$$

where m is a constant.

(A) Graph f for $m = 1$, and find

$$\lim_{x\to 1^-} f(x) \quad \text{and} \quad \lim_{x\to 1^+} f(x)$$

(B) Graph f for $m = 2$, and find

$$\lim_{x\to 1^-} f(x) \quad \text{and} \quad \lim_{x\to 1^+} f(x)$$

(C) Find m so that

$$\lim_{x\to 1^-} f(x) = \lim_{x\to 1^+} f(x)$$

and graph f for this value of m.

(D) Write a brief verbal description of each graph. How does the graph in part (C) differ from the graphs in parts (A) and (B)?

72. Let f be defined by

$$f(x) = \begin{cases} -3m + 0.5x & \text{if } x \leqslant 2 \\ 3m - x & \text{if } x > 2 \end{cases}$$

where m is a constant.

(A) Graph f for $m = 0$, and find

$$\lim_{x\to 2^-} f(x) \quad \text{and} \quad \lim_{x\to 2^+} f(x)$$

(B) Graph f for $m = 1$, and find

$$\lim_{x\to 2^-} f(x) \quad \text{and} \quad \lim_{x\to 2^+} f(x)$$

(C) Find m so that

$$\lim_{x\to 2^-} f(x) = \lim_{x\to 2^+} f(x)$$

and graph f for this value of m.

(D) Write a brief verbal description of each graph. How does the graph in part (C) differ from the graphs in parts (A) and (B)?

Find each limit in Problems 73–76, where a is a real constant.

73. $\lim_{h\to 0} \frac{(a + h)^2 - a^2}{h}$

74. $\lim_{h\to 0} \frac{[3(a + h) - 2] - (3a - 2)}{h}$

75. $\lim_{h\to 0} \frac{\sqrt{a + h} - \sqrt{a}}{h}, \quad a > 0$

76. $\lim_{h\to 0} \frac{\frac{1}{a + h} - \frac{1}{a}}{h}, \quad a \neq 0$

77. Let $f(x) = x^2 - 3x + 1$.

(A) Find the slope of the graph of f at $a = 1, a = 2$, and $a = 3$

(B) Find a formula for the slope of the graph of f at any point $(a, f(a))$.

(C) Use the formula from part (B) to find the slope of the graph at $a = 1, a = 2$, and $a = 3$. Write a brief verbal comparison of these two methods for finding the slope of a graph.

78. Let $f(x) = 5x - 2x^2$.

(A) Find the slope of the graph of f at $a = -1, a = 0$, and $a = 1$.

(B) Find a formula for the slope of the graph of f at any point $(a, f(a))$.

(C) Use the formula from part (B) to find the slope of the graph at $a = -1, a = 0$, and $a = 1$. Write a brief verbal comparison of these two methods for finding the slope of a graph.

In Problems 79–84, use a table of values to estimate each limit to three decimal places.

79. $\lim_{x\to 1} \frac{x^{10} - 1}{x - 1}$

80. $\lim_{x\to 1} \frac{x^{15} - 1}{x - 1}$

81. $\lim_{x\to 0} \frac{2^x - 1}{x}$

82. $\lim_{x\to 0} \frac{3^x - 1}{x}$

83. $\lim_{x\to 0}(1 + x)^{1/x}$

84. $\lim_{x\to 0}(1 + 2x)^{1/x}$

Problems 85–90 require the use of a graphing utility. When investigating limits with a graphing utility, it is helpful to choose Xmin and Xmax so that coordinates displayed on the screen have finite decimal expansions. Problems 85 and 86 describe a simple way to accomplish this.

85. Let Xmin = 0, and use trial and error to find an integer n so that if Xmax = n, then the x coordinates of points displayed on the screen are integers.

86. Given any value for Xmin, let Xmax = Xmin + $h \cdot n$, where n is the integer discovered in Problem 85. Discuss the nature of the x coordinates of points displayed on the screen if $h = 0.5$. If $h = 0.1$.

In Problems 87–90, graph each function and use zoom and trace to investigate the left- and right-hand limits at the indicated value(s) of c. Use the ideas discussed in Problems 85 and 86 to choose Xmin and Xmax.

87. $f(x) = \dfrac{x^4 - 10x^2 + 24}{4 - x^2}; c = -2, c = 2$

88. $f(x) = \dfrac{x^4 - 12x^2 + 27}{x^2 - 9}; c = -3, c = 3$

89. $f(x) = \dfrac{x^3 - 9x}{|x^2 - 9|}; c = -3, c = 3$

90. $f(x) = \dfrac{4x - x^3}{|x^2 - 4|}; c = -2, c = 2$

APPLICATIONS

Business & Economics

91. *Revenue.* The revenue (in dollars) from the sale of x variable-speed jigsaws is given by

$$R(x) = 200x - 0.1x^2 \qquad 0 \le x \le 2{,}000$$

(A) Find the revenue and the instantaneous rate of change of revenue at a production level of 900 jigsaws. Write a brief verbal interpretation of these results.

(B) Repeat part (A) for a production level of 1,200 jigsaws.

92. *Profit.* The profit (in dollars) from the sale of x variable-speed jigsaws is given by

$$P(x) = 150x - 0.1x^2 - 5{,}000 \qquad 0 \le x \le 2{,}000$$

(A) Find the profit and the instantaneous rate of change of profit at a production level of 700 jigsaws. Write a brief verbal interpretation of these results.

(B) Repeat part (A) for a production level of 900 jigsaws.

93. *Consumer debt.* Revolving-credit debt (in billions of dollars) in the United States can be described approximately by

$$f(t) = 0.62t^2 - t + 5.1$$

where t is time in years and $t = 0$ corresponds to 1970. Find the debt and the instantaneous rate of change of the debt in 1990. Write a brief verbal interpretation of these results.

94. *Consumer debt.* Credit union debt (in billions of dollars) in the United States can be described approximately by

$$f(t) = 0.5t^2 + 5.6t + 46.6$$

where t is time in years and $t = 0$ corresponds to 1970. Find the debt and the instantaneous rate of change of the debt in 1990. Write a brief verbal interpretation of these results.

Life Sciences

95. *Human physiology.* Many people experience headaches, fatigue, and shortness of breath at high altitudes, due to an inadequate supply of oxygen in the bloodstream. The reduced air pressure at high altitudes results in less oxygen being forced through the membranes of the lungs into the blood. The *alveolar pressure* measures the amount of oxygen that can enter the bloodstream at various altitudes. If y is the ratio of the pressure at an altitude of x thousand feet to the pressure at sea level (expressed as a percentage) for an average size person, then y is given approximately by

$$y = 0.04x^2 - 3.66x + 100$$

Notice that $y = 100\%$ at $x = 0$ (sea level). Find y and find the instantaneous rate of change of y with respect to x at an altitude of 9,000 feet. (This is the altitude at which most people will begin to experience discomfort.) Write a brief verbal interpretation of these results.

96. *Plant physiology.* High altitudes and the corresponding shorter growing seasons have an effect on the size of plants. For a particular species, the relationship between the height y (in inches) of a full-grown plant and the altitude x (in thousands of feet) is given approximately by

$$y = 0.16x^2 - 5.5x + 48$$

Find y and find the instantaneous rate of change of y with respect to x at an altitude of 5,000 feet. Write a brief verbal interpretation of these results.

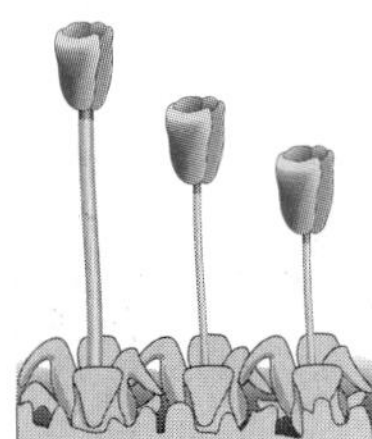

Social Sciences

97. *Education.* For a particular year, the total school-aged population (5–17 years of age) in the United States (in millions) is given approximately by

$$y = -0.03t^2 + 1.5t + 32$$

where t is time in years and $t = 0$ corresponds to 1950.
(A) Find the school-aged population and the instantaneous rate of change of this population in 1970. Write a brief verbal interpretation of these results.
(B) Repeat part (A) for 1990.

98. *Education.* The number of students who graduate from high school each year in the United States (in millions) is given approximately by

$$y = -0.002t^2 + 0.115t + 0.95$$

where t is time in years and $t = 0$ corresponds to 1950.
(A) Find the number of high school graduates and the instantaneous rate of change of this population in 1970. Write a brief verbal interpretation of these results.
(B) Repeat part (A) for 1990.

SECTION 8-3 The Derivative

- THE DERIVATIVE
- NONEXISTENCE OF THE DERIVATIVE
- SUMMARY

In this section we apply the limit techniques developed in Section 8-2 to the difference quotients discussed in Section 8-1. This will lead to the definition of the *derivative,* a fundamental mathematical concept that forms the basis for much of the subject matter of calculus.

THE DERIVATIVE

We begin with an example that will review some of the ideas discussed in the preceding two sections.

Example 1 **Slope of a Tangent Line** Given $f(x) = x^2$:

(A) Find the slope of the tangent line at $a = 1$.
(B) Find the equation of the tangent line at $a = 1$; that is, through $(1, f(1))$.
(C) Sketch the graph of f, the tangent line at $(1, f(1))$, and the secant line passing through $(1, f(1))$ and $(2, f(2))$.

SOLUTION (A) Slope of the tangent line:

Step 1. Write the difference quotient and simplify:

$$\frac{f(1+h) - f(1)}{h} = \frac{(1+h)^2 - 1^2}{h}$$

This is the slope of a secant line passing through $(1, f(1))$ and $(1 + h, f(1 + h))$.

$$= \frac{1 + 2h + h^2 - 1}{h}$$

$$= \frac{2h + h^2}{h} = \frac{h(2+h)}{h} = 2 + h \qquad h \neq 0$$

Step 2. Find the limit of the difference quotient:

$$\text{Slope of tangent line} = \lim_{h\to 0} \frac{f(1+h) - f(1)}{h}$$

$$= \lim_{h\to 0}(2 + h) = 2$$

This is also the slope of the graph of $f(x) = x^2$ at $(1, f(1))$.

(B) The tangent line passes through $(1, f(1)) = (1, 1)$ with slope $m = 2$ [from part (A)]. The point–slope formula from Section 1-3 gives its equation:

$$y - y_1 = m(x - x_1) \quad \text{Point–slope formula}$$
$$y - 1 = 2(x - 1)$$
$$y = 2x - 1 \quad \text{Tangent line equation}$$

(C)

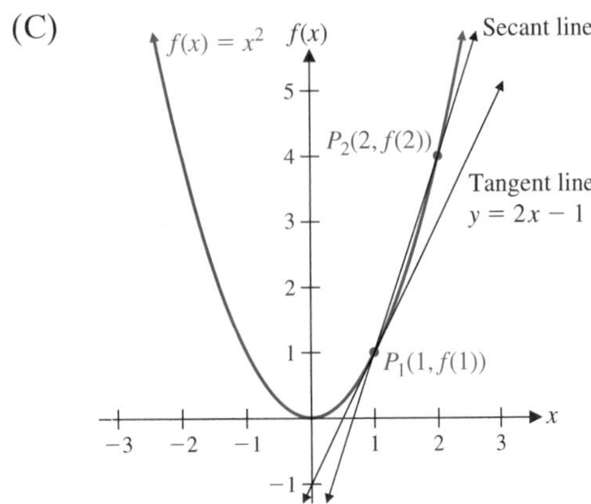

Matched Problem 1 Find the slope of the tangent line for the graph of $f(x) = x^2$ at $a = 2$, and write the equation of the tangent line in the form $y = mx + b$.

Refer to Example 1 and Matched Problem 1. What happens if we want to find the slope of the line tangent to the graph of $y = f(x) = x^2$ at some additional points on the graph of f? Each time we consider a different point, we must evaluate another limit. It would be much more efficient if we could evaluate a single limit that would give us the slope of the tangent line at any point on the graph of f. To do this, we apply the two-step process illustrated in Example 1 at the point $(a, f(a)) = (a, a^2)$, where a is an unspecified, but fixed, real number.

Step 1. Write the difference quotient and simplify:

$$\frac{f(a + h) - f(a)}{h} = \frac{(a + h)^2 - a^2}{h} \quad \text{Slope of the secant line}$$
$$= \frac{a^2 + 2ah + h^2 - a^2}{h}$$
$$= \frac{2ah + h^2}{h} = \frac{h(2a + h)}{h}$$
$$= 2a + h \qquad h \neq 0$$

Step 2. Find the limit of the difference quotient:

$$\text{Slope of tangent line} = \lim_{h \to 0} \frac{f(a + h) - f(a)}{h}$$
$$= \lim_{h \to 0} (2a + h)$$
$$= 2a$$

Thus, we see that if $(a, f(a)) = (a, a^2)$ is any point on the graph of f, then the slope of the tangent line at this point is $2a$. Note that the slope is a function of a, the first coordinate of the point of tangency. Also note that when $a = 1$, the slope is 2, which agrees with the result in Example 1A.

Generalizing the process of finding slopes of tangent lines is not simply a matter of increasing efficiency. Rather, the relationship between a function and the slope of the tangent line at any point on the graph of the function is one of the most fundamental concepts of calculus.

Explore–Discuss 1

(A) Use the graph of the function f in Figure 1 to estimate (to one decimal place) the slopes of the tangent lines at the points whose x coordinates are given in Table 1.

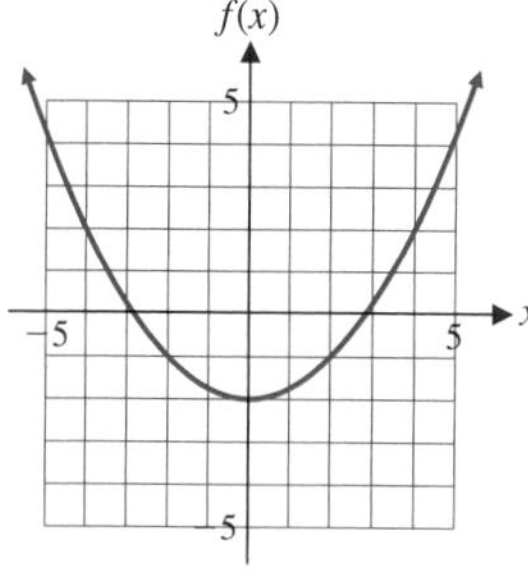

FIGURE 1

Table 1

x	SLOPE OF TANGENT LINE AT $(x, f(x))$
−4	
−3	
−2	
−1	
0	
1	
2	
3	
4	

(B) Find a function whose values agree with the values listed in Table 1.

(C) If $f(x) = 0.25x^2 - 2$ is the function whose graph is shown in Figure 1, find the slope of the tangent line at any x by evaluating

$$\lim_{h \to 0} \frac{f(x + h) - f(x)}{h}$$

Discuss the relationship between the function you found in part (B) and this limit.

We are now ready to define the *derivative of a function.* To follow customary practice, we use x in place of a and think of the difference quotient

$$\frac{f(x + h) - f(x)}{h}$$

as a function of h, with x held fixed as h tends to 0.

The Derivative

For $y = f(x)$, we define the **derivative of f at x,** denoted by **$f'(x)$,** to be

$$f'(x) = \lim_{h \to 0} \frac{f(x + h) - f(x)}{h} \quad \text{if the limit exists}$$

If $f'(x)$ exists for each x in the open interval (a, b), then f is said to be **differentiable** over (a, b).

(Differentiability from the left or from the right is defined using $h \to 0^-$ or $h \to 0^+$, respectively, in place of $h \to 0$ in the above definition.)

The process of finding the derivative of a function is called **differentiation.** That is, the derivative of a function is obtained by **differentiating** the function.

Interpretations of the Derivative

The derivative of a function f is a new function f'. The domain of f' is a subset of the domain of f. The derivative has various applications and interpretations, including:

1. *Slope of the Tangent Line.* For each x in the domain of f', $f'(x)$ is the slope of the line tangent to the graph of f at the point $(x, f(x))$.
2. *Instantaneous Rate of Change.* For each x in the domain of f', $f'(x)$ is the instantaneous rate of change of $y = f(x)$ with respect to x.

For example, if $f(x) = x^2$, then the derivative is

$$f'(x) = \lim_{h \to 0} \frac{f(x + h) - f(x)}{h} = 2x$$

(See page 545 for the two-step process involved in evaluating this limit.) We can interpret this derivative graphically as follows: The slope of the tangent line at any point (x, x^2) is $2x$. On the other hand, if $y = f(x)$ is the position (measured in feet) at x seconds of an object moving along the y axis, then we can also interpret the derivative as a rate of change: The instantaneous velocity of the object at x seconds is $2x$ feet per second.

Example 2 **Finding a Derivative** Find $f'(x)$, the derivative of f at x, for $f(x) = 4x - x^2$.

SOLUTION To find $f'(x)$, we use the two-step process:

Step 1. Form the difference quotient and simplify:

$$\begin{aligned} \frac{f(x + h) - f(x)}{h} &= \frac{[4(x + h) - (x + h)^2] - (4x - x^2)}{h} \\ &= \frac{4x + 4h - x^2 - 2xh - h^2 - 4x + x^2}{h} \\ &= \frac{4h - 2xh - h^2}{h} \\ &= \frac{h(4 - 2x - h)}{h} \\ &= 4 - 2x - h \qquad h \neq 0 \end{aligned}$$

Step 2. Find the limit of the difference quotient:

$$\begin{aligned} f'(x) &= \lim_{h \to 0} \frac{f(x + h) - f(x)}{h} \\ &= \lim_{h \to 0} (4 - 2x - h) = 4 - 2x \end{aligned}$$

Thus, if $f(x) = 4x - x^2$, then $f'(x) = 4 - 2x$. The derivative f' is a new function derived from the function f.

Matched Problem 2 Find $f'(x)$, the derivative of f at x, for $f(x) = 8x - 2x^2$.

Example 3 **Finding Tangent Line Slopes** In Example 2, we started with the function specified by $f(x) = 4x - x^2$ and found the derivative of f at x to be $f'(x) = 4 - 2x$. Thus, the slope of a tangent line to the graph of f at any point $(x, f(x))$ on the graph is

$$m = f'(x) = 4 - 2x$$

(A) Find the slope of the graph of f at $x = 0, x = 2$, and $x = 3$.
(B) Graph $y = f(x) = 4x - x^2$, and use the slopes found in part (A) to make a rough sketch of the tangent lines to the graph at $x = 0, x = 2$, and $x = 3$.

SOLUTION (A) Using $f'(x) = 4 - 2x$, we have

$$f'(0) = 4 - 2(0) = 4 \quad \text{Slope at } x = 0$$
$$f'(2) = 4 - 2(2) = 0 \quad \text{Slope at } x = 2$$
$$f'(3) = 4 - 2(3) = -2 \quad \text{Slope at } x = 3$$

(B)

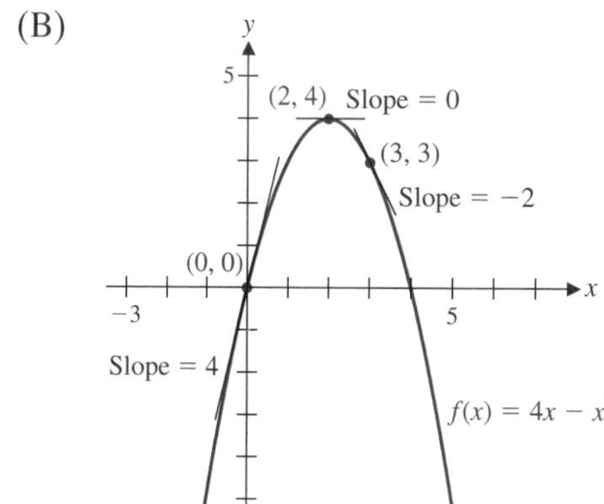

Matched Problem 3 In Matched Problem 2, we started with the function specified by $f(x) = 8x - 2x^2$. Using the derivative found there:

(A) Find the slope of the graph of f at $x = 1, x = 2$, and $x = 4$.
(B) Graph $y = f(x) = 8x - 2x^2$, and use the slopes from part (A) to make a rough sketch of the tangent lines to the graph at $x = 1, x = 2$, and $x = 4$.

Explore–Discuss 2

In Example 2 we found that the derivative of $f(x) = 4x - x^2$ is $f'(x) = 4 - 2x$, and in Example 3 we graphed $f(x)$ and several tangent lines.

(A) Graph f and f' on the same set of axes.
(B) The graph of f' is a straight line. Is it a tangent line for the graph of f? Explain.
(C) Find the x intercept for the graph of f'. What is the slope of the line tangent to the graph of f for this value of x? Write a verbal description of the relationship between the slopes of the tangent lines of a function and the x intercepts of the derivative of the function.

REMARK: DERIVATIVES AND TECHNOLOGY

Refer to the values of $f'(x)$ computed in Example 3A. Some graphing utilities, including most graphing calculators, have a built-in **numerical differentiation** routine that will approximate numerically the value of $f'(x)$ for any given value of x. Figure 2A illustrates this approach on the TI-83 graphing calculator. Other graphing utilities use **symbolic differentiation** to find an algebraic formula for the derivative, and then evaluate this formula at the indicated values of x. Figure 2B illustrates this approach in Maple V, a very powerful mathematical software package.

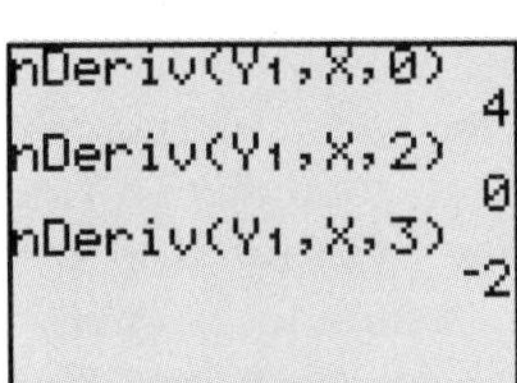

(A) Numerical differentiation:
y1=4x-x^2

```
• f:=4*x-x^2;d:=diff(f,x);
                    2
        f := 4 x - x
         d := 4 - 2 x
• x:=0;d;x:=2;d;x:=3;d;
              x := 0
               4
              x := 2
               0
              x := 3
               -2
```

(B) Symbolic differentiation:
f:=4*x-x^2

FIGURE 2

Differentiation is as basic to calculus as addition is to algebra, and it is important that you learn to perform this operation for yourself, without using technology. However, you certainly can use numerical differentiation on a graphing utility to check your work.

Example 4 **Finding a Derivative** Find $f'(x)$, the derivative of f at x, for $f(x) = \sqrt{x} + 2$.

SOLUTION To find $f'(x)$, we find the following limit using the two-step process.

$$\lim_{h\to 0} \frac{f(x+h) - f(x)}{h}$$

Step 1. Form the difference quotient and simplify:

$$\frac{f(x+h) - f(x)}{h} = \frac{(\sqrt{x+h} + 2) - (\sqrt{x} + 2)}{h} = \frac{\sqrt{x+h} - \sqrt{x}}{h}$$

Since this is a 0/0 indeterminate form, we change the form by rationalizing the numerator:

$$\frac{\sqrt{x+h} - \sqrt{x}}{h} \cdot \frac{\sqrt{x+h} + \sqrt{x}}{\sqrt{x+h} + \sqrt{x}} = \frac{x + h - x}{h(\sqrt{x+h} + \sqrt{x})} = \frac{h}{h(\sqrt{x+h} + \sqrt{x})} = \frac{1}{\sqrt{x+h} + \sqrt{x}} \qquad h \neq 0$$

Step 2. Find the limit of the difference quotient:

$$
\begin{aligned}
f'(x) &= \lim_{h\to 0} \frac{f(x+h) - f(x)}{h} \\
&= \lim_{h\to 0} \frac{1}{\sqrt{x+h} + \sqrt{x}} \\
&= \frac{1}{\sqrt{x} + \sqrt{x}} = \frac{1}{2\sqrt{x}} \qquad x > 0
\end{aligned}
$$

Thus, the derivative of $f(x) = \sqrt{x} + 2$ is $f'(x) = 1/(2\sqrt{x})$, a new function. The domain of f is $[0, \infty)$. Since $f'(0)$ is not defined, the domain of f' is $(0, \infty)$, a subset of the domain of f.

Matched Problem 4 ➠ Find $f'(x)$ for $f(x) = \sqrt{x+4}$.

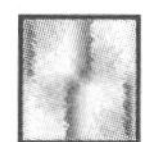

Example 5 ➠ **Sales Analysis** The total sales of a company (in millions of dollars) t months from now are given by $S(t) = \sqrt{t} + 2$. Find $S(25)$ and $S'(25)$, and interpret. Use these results to estimate the total sales after 26 months and after 27 months. **C** Check the values of $S(25)$ and $S'(25)$ on a graphing utility.

Solution The total sales function S has the same form as the function f in Example 4—only the letters used to represent the function and the independent variable have been changed. It follows that S' and f' also have the same form:

$$
\begin{aligned}
S(t) &= \sqrt{t} + 2 & f(x) &= \sqrt{x} + 2 \\
S'(t) &= \frac{1}{2\sqrt{t}} & f'(x) &= \frac{1}{2\sqrt{x}}
\end{aligned}
$$

Evaluating S and S' at $t = 25$, we have

$$
S(25) = \sqrt{25} + 2 = 7 \qquad S'(25) = \frac{1}{2\sqrt{25}} = 0.1
$$

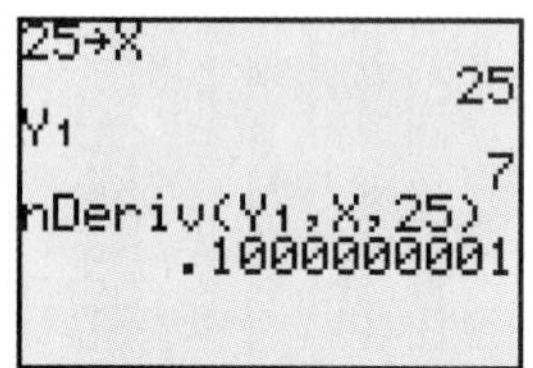

FIGURE 3
$y_1 = \sqrt{x} + 2$

Figure 3 shows a check of these results on a graphing utility. Thus, 25 months from now the total sales are \$7 million and are increasing at the rate of \$0.1 million (\$100,000) per month. If this instantaneous rate of change of sales remained constant, then the sales would grow to \$7.1 million after 26 months, \$7.2 million after 27 months, and so on. Even though $S'(t)$ is not a constant function in this case, these values provide useful estimates of the total sales.

Matched Problem 5 ➠ The total sales of a company (in millions of dollars) t months from now are given by $S(t) = \sqrt{t+4}$. Find $S(12)$ and $S'(12)$, and interpret. Use these results to estimate the total sales after 13 months and after 14 months. (Use the derivative found in Matched Problem 4.) **C** Check with a graphing utility.

Refer to Example 5. It is instructive to compare the estimates of total sales obtained by using the derivative with the corresponding exact values of $S(t)$:

$$
\begin{aligned}
& \text{Exact values} & & \text{Estimated values} \\
S(26) &= \sqrt{26} + 2 = 7.099\ldots & &\approx 7.1 \\
S(27) &= \sqrt{27} + 2 = 7.196\ldots & &\approx 7.2
\end{aligned}
$$

For this function, the estimated values provide very good approximations to the exact values of $S(t)$. For other functions, the approximation might not be as accurate.

Using the instantaneous rate of change of a function at a point to estimate values of the function at nearby points is a simple, but important application of the derivative.

Nonexistence of the Derivative

The existence of a derivative at $x = a$ depends on the existence of a limit at $x = a$; that is, on the existence of

$$f'(a) = \lim_{h\to 0} \frac{f(a+h) - f(a)}{h} \qquad (1)$$

If the limit does not exist at $x = a$, we say that the function f is **nondifferentiable at $x = a$,** or **$f'(a)$ does not exist.**

Explore–Discuss 3

Let $f(x) = |x - 1|$.

(A) Graph f.
(B) Complete the following table:

h	-0.1	-0.01	$-0.001 \to 0 \leftarrow 0.001$	0.01	0.1
$\dfrac{f(1+h) - f(1)}{h}$	?	?	? $\to$? $\leftarrow$?	?	?

(C) Find the following limit, if it exists.

$$\lim_{h\to 0} \frac{f(1+h) - f(1)}{h}$$

(D) Use the results of parts (A)–(C) to discuss the existence of $f'(1)$.

Repeat parts (A)–(D) for $g(x) = \sqrt[3]{x - 1}$.

How can we recognize the points on the graph of f where $f'(a)$ does not exist? It is impossible to describe all the ways that the limit in (1) can fail to exist. However, we can illustrate some common situations where $f'(a)$ does fail to exist (see Fig. 4, at the top of the next page):

1. If the graph of f has a hole or a break at $x = a$, then $f'(a)$ does not exist* (Fig. 4A).
2. If the graph of f has a sharp corner at $x = a$, then $f'(a)$ does not exist and the graph has no tangent line at $x = a$ (Fig. 4B). (In Fig. 4B, the left- and right-hand derivatives exist but are not equal.)

---

*Informally, if the graph of a function has no holes or breaks, then the function is said to be **continuous** (see Section 2-1). We will discuss the formal definition of continuity in Section 9-1. For now, this intuitive formulation will suffice.

3. If the graph of f has a vertical tangent line at $x = a$, then $f'(a)$ does not exist (Fig. 4C and D).

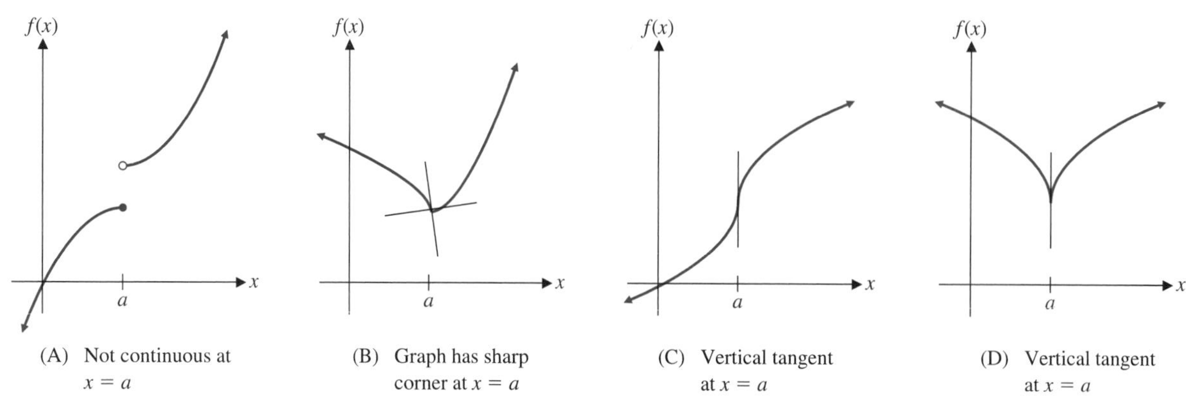

FIGURE 4
The function f is nondifferentiable at $x = a$.

■ Summary

Given a differentiable function f and a fixed number x, we have seen three different ways to find or approximate $f'(x)$:

1. Numerically, by computing

$$\frac{f(x + h) - f(x)}{h}$$

for small values of h.
2. Graphically, by estimating the slope of the tangent line at the point $(x, f(x))$.
3. Algebraically, by using the two-step limiting process to evaluate

$$\lim_{h \to 0} \frac{f(x + h) - f(x)}{h}$$

Each of these approaches has its uses. In fact, in order to effectively use derivative concepts in the many and varied applications we will consider, it is necessary to understand all three approaches. The first two methods produce approximate values of the derivative at a single fixed value of x. These methods are especially useful for functions defined by tables or graphs. However, for functions defined by algebraic expressions, the third method has the distinct advantage of producing an algebraic expression for f'. This expression can be used to find the exact value of $f'(x)$ at any number x. For example, once we know that the derivative of $f(x) = x^2$ is $f'(x) = 2x$, then we know the exact value of $f'(x)$ for any real number x.

At this point in our development, finding the algebraic expression for f' requires using the two-step limiting process for each new function we encounter. In the next three sections we will develop formulas and general properties of derivatives that will enable us to find the derivatives of many functions without having to go through the two-step limiting process each time.

Answers to Matched Problems

1. $y = 4x - 4$ **2.** $f'(x) = 8 - 4x$

3. (A) $f'(1) = 4, f'(2) = 0, f'(4) = -8$

(B)

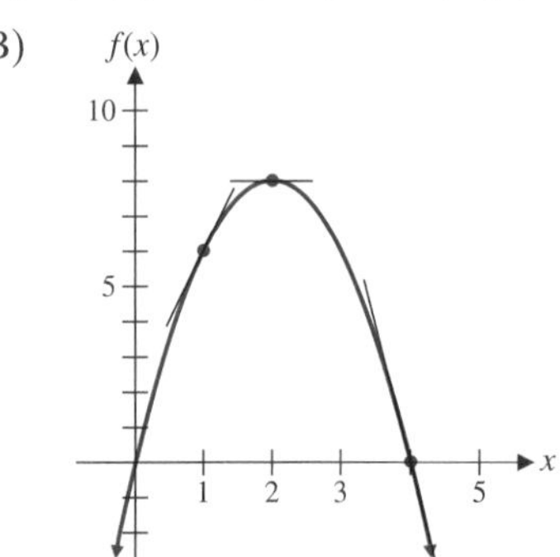

4. $f'(x) = 1/(2\sqrt{x + 4})$

5. $S(12) = 4, S'(12) = 0.125$; 12 months from now the total sales are $4 million and are increasing at the rate of $0.125 million ($125,000) per month. The estimated total sales are $4.125 million after 13 months and $4.25 million after 14 months.

EXERCISE 8-3

A *In Problems 1 and 2, find the indicated quantity for $y = f(x) = 5 - x^2$, and interpret in terms of the graph shown below.*

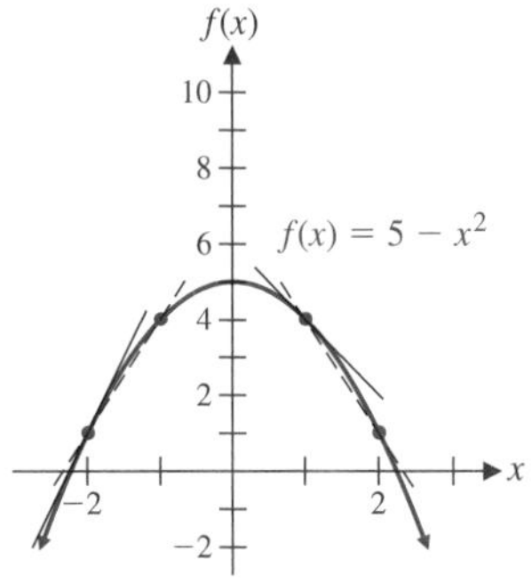

1. (A) $\frac{f(2) - f(1)}{2 - 1}$ (B) $\frac{f(1 + h) - f(1)}{h}$

(C) $\lim_{h \to 0} \frac{f(1 + h) - f(1)}{h}$

2. (A) $\frac{f(-1) - f(-2)}{-1 - (-2)}$ (B) $\frac{f(-2 + h) - f(-2)}{h}$

(C) $\lim_{h \to 0} \frac{f(-2 + h) - f(-2)}{h}$

In Problems 3–6, use the given expression for $f(x + h) - f(x)$ to find $f'(x)$.

3. $f(x + h) - f(x) = 4hx - 3h + 2h^2$

4. $f(x + h) - f(x) = 6hx - 5h + 3h^2$

5. $f(x + h) - f(x) = 3hx^2 - 2hx + 3h^2x - h^2 + h^3$

6. $f(x + h) - f(x)$
$= -6hx^2 + 10hx - 6h^2x + 5h^2 - 2h^3$

In Problems 7–12, find $f'(x)$ using the two-step process:

Step 1. *Simplify:* $\frac{f(x + h) - f(x)}{h}$

Step 2. *Evaluate:* $\lim_{h \to 0} \frac{f(x + h) - f(x)}{h}$

Then find $f'(1)$, $f'(2)$, and $f'(3)$.

7. $f(x) = 4$ **8.** $f(x) = -3$

9. $f(x) = 2x - 3$ **10.** $f(x) = 4x + 3$

11. $f(x) = 2 - x^2$ **12.** $f(x) = 2x^2 + 5$

In Problems 13 and 14, use the graph of $y = f(x)$ to estimate (to the nearest integer) the values of x for which $f'(x) = 0$.

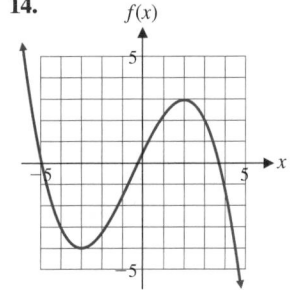

13.

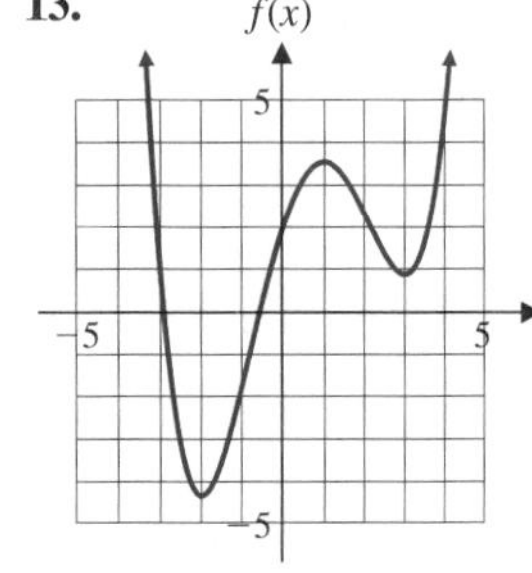

14. f(x) 5 −5 5 x −5

In Problems 15 and 16, use the graph of $y = f(x)$ to estimate $f'(x)$ (to the nearest integer) at $x = 3$ and at $x = -3$.

15.

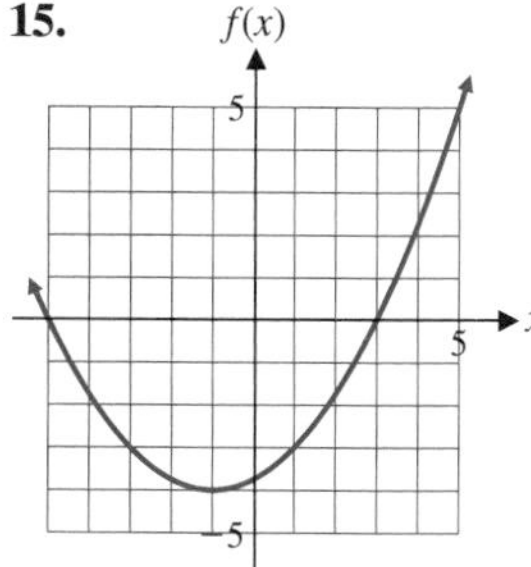

16.

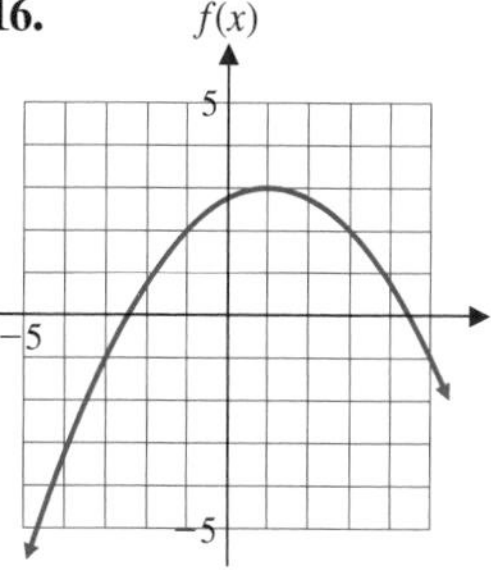

B *Problems 17 and 18 refer to the graph of $y = f(x) = x^2 + x$ shown.*

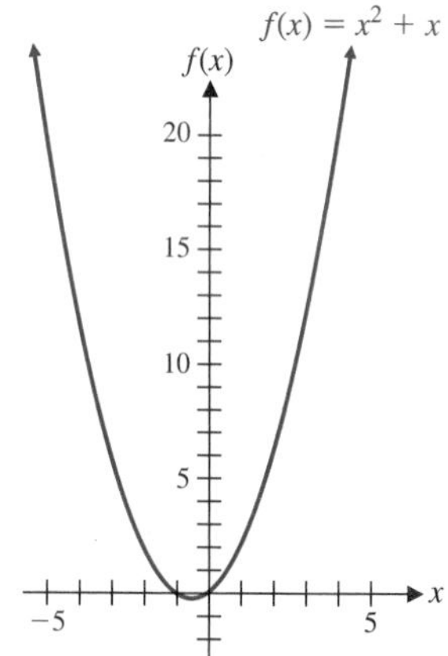

17. (A) Find the slope of the secant line joining $(1, f(1))$ and $(3, f(3))$.
(B) Find the slope of the secant line joining $(1, f(1))$ and $(1 + h, f(1 + h))$.
(C) Find the slope of the tangent line at $(1, f(1))$.
(D) Find the equation of the tangent line at $(1, f(1))$.

18. (A) Find the slope of the secant line joining $(2, f(2))$ and $(4, f(4))$.
(B) Find the slope of the secant line joining $(2, f(2))$ and $(2 + h, f(2 + h))$.
(C) Find the slope of the tangent line at $(2, f(2))$.
(D) Find the equation of the tangent line at $(2, f(2))$.

In Problems 19 and 20, suppose an object moves along the y axis so that its location is $y = f(x) = x^2 + x$ at time x (y is in meters and x is in seconds). Find:

19. (A) The average velocity (the average rate of change of y with respect to x) for x changing from 1 to 3 seconds
(B) The average velocity for x changing from 1 to $1 + h$ seconds
(C) The instantaneous velocity at $x = 1$ second

20. (A) The average velocity (the average rate of change of y with respect to x) for x changing from 2 to 4 seconds
(B) The average velocity for x changing from 2 to $2 + h$ seconds
(C) The instantaneous velocity at $x = 2$ seconds

In Problems 21–26, find $f'(x)$ using the two-step limiting process. Then find $f'(1)$, $f'(2)$, and $f'(3)$.

21. $f(x) = 6x - x^2$

22. $f(x) = 2x - 3x^2$

23. $f(x) = \sqrt{x} - 3$

24. $f(x) = 2 - \sqrt{x}$

25. $f(x) = \dfrac{-1}{x}$

26. $f(x) = \dfrac{1}{x + 1}$

Problems 27–34 refer to the function F in the graph shown. Use the graph to determine whether $F'(x)$ exists at each indicated value of x.

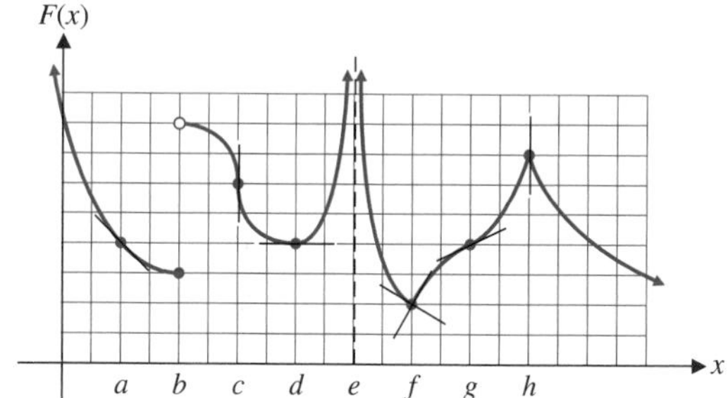

27. $x = a$

28. $x = b$

29. $x = c$

30. $x = d$

31. $x = e$

32. $x = f$

33. $x = g$

34. $x = h$

35. Given $f(x) = x^2 - 4x$:
(A) Find $f'(x)$.
(B) Find the slopes of the tangent lines to the graph of f at $x = 0, 2$, and 4.
(C) Graph f, and sketch in the tangent lines at $x = 0$, 2, and 4.

36. Given $f(x) = x^2 + 2x$:
(A) Find $f'(x)$.
(B) Find the slopes of the tangent line to the graph of f at $x = -2, -1$, and 1.
(C) Graph f, and sketch in the tangent lines at $x = -2, -1$, and 1.

37. If an object moves along a line so that it is at $y = f(x) = 4x^2 - 2x$ at time x (in seconds), find the instantaneous velocity function $v = f'(x)$, and find the velocity at times $x = 1, 3$, and 5 seconds (y is measured in feet).

38. Repeat Problem 37 with $f(x) = 8x^2 - 4x$.

In Problems 39–42, use the information in each table to sketch a possible graph for the function $y = f(x)$.

39.

x	$f(x)$	$f'(x)$
-4	-3	0.5
0	-1	0.5
2	0	0.5

40.

x	$f(x)$	$f'(x)$
-1	1	-2
0	-1	-2
1	-3	-2

41.

x	$f(x)$	$f'(x)$
-3	0	-3
-1	-2	0
3	0	0.75

42.

x	$f(x)$	$f'(x)$
1	-3	-0.75
3	-4	0
4	-3	3

In Problems 43–46, use zoom and trace on a graphing utility to find points on the graph of $y = f(x)$ near $(0, f(0))$. Then use secant line slopes to approximate $f'(0)$ to two decimal places. (If your graphing utility has a numerical differentiation routine that will approximate the value of a derivative, use it to check your answers.)

43. $f(x) = 2^x$

44. $f(x) = 3^x$

45. $f(x) = \sqrt{2 + 2x - x^2}$

46. $f(x) = \sqrt{3 - 2x - x^2}$

47. Let $f(x) = x^2$, $g(x) = x^2 - 3$, and $h(x) = x^2 + 1$.
(A) How are the graphs of these functions related? How would you expect the derivatives of these functions to be related?
(B) Use the two-step process to find the derivative of $m(x) = x^2 + C$, where C is any real number constant.

48. Let $f(x) = 2x$, $g(x) = 2x - 1$, and $h(x) = 2x + 2$.
(A) How are the graphs of these functions related? How would you expect the derivatives of these functions to be related?
(B) Use the two-step process to find the derivative of $m(x) = 2x + C$, where C is any real number constant.

49. (A) Give a geometric explanation of the following statement: If $f(x) = C$ is a constant function, then $f'(x) = 0$.
(B) Use the two-step process to verify the statement in part (A).

50. (A) Give a geometric explanation of the following statement: If $f(x) = mx + b$ is a linear function, then $f'(x) = m$.
(B) Use the two-step process to verify the statement in part (A).

C *In Problems 51–54, sketch the graph of f and determine where f is nondifferentiable.*

51. $f(x) = \begin{cases} 2x & \text{if } x < 1 \\ 2 & \text{if } x \geq 1 \end{cases}$

52. $f(x) = \begin{cases} 2x & \text{if } x < 2 \\ 6 - x & \text{if } x \geq 2 \end{cases}$

53. $f(x) = \begin{cases} x^2 + 1 & \text{if } x < 0 \\ 1 & \text{if } x \geq 0 \end{cases}$

54. $f(x) = \begin{cases} 2 - x^2 & \text{if } x \leq 0 \\ 2 & \text{if } x > 0 \end{cases}$

In Problems 55–58, determine whether f is differentiable at $x = 0$ by considering

$$\lim_{h \to 0} \frac{f(0 + h) - f(0)}{h}$$

55. $f(x) = |x|$

56. $f(x) = 1 - |x|$

57. $f(x) = x^{1/3}$

58. $f(x) = x^{2/3}$

59. Show that $f(x) = 2x - x^2$ is differentiable over the closed interval $[0, 2]$ by showing that each of the following limits exists:

(A) $\displaystyle\lim_{h \to 0} \frac{f(x + h) - f(x)}{h}, \quad 0 < x < 2$

(B) $\displaystyle\lim_{h \to 0^+} \frac{f(0 + h) - f(0)}{h}, \quad x = 0$

(C) $\displaystyle\lim_{h \to 0^-} \frac{f(2 + h) - f(2)}{h}, \quad x = 2$

60. Show that $f(x) = \sqrt{x}$ is differentiable over the open interval $(0, \infty)$ but not over the half-closed interval $[0, \infty)$ by considering

$$\lim_{h \to 0} \frac{f(x + h) - f(x)}{h} \qquad 0 < x < \infty$$

and

$$\lim_{h \to 0^+} \frac{f(0 + h) - f(0)}{h} \qquad x = 0$$

APPLICATIONS

Business & Economics

61. *Sales analysis.* The total sales of a company (in millions of dollars) t months from now are given by

$$S(t) = 2\sqrt{t + 10}$$

(A) Use the two-step process to find $S'(t)$.
(B) Find $S(15)$ and $S'(15)$. Write a brief verbal interpretation of these results.
(C) Use the results in part (B) to estimate the total sales after 16 months and after 17 months.

62. *Sales analysis.* The total sales of a company (in millions of dollars) t months from now are given by

$$S(t) = 2\sqrt{t + 6}$$

(A) Use the two-step process to find $S'(t)$.
(B) Find $S(10)$ and $S'(10)$. Write a brief verbal interpretation of these results.
(C) Use the results in part (B) to estimate the total sales after 11 months and after 12 months.

63. *Compound interest.* If \$100 is invested in an account that earns 6% compounded annually, then the amount in the account after t years is given by

$$A(t) = 100(1.06)^t$$

(A) Find $A(5)$ and use secant line slopes to approximate $A'(5)$ to the nearest cent.
(B) Write a brief verbal interpretation of the results in part (A).

64. *Compound interest.* If \$500 is invested in an account that earns 8% compounded annually, then the amount in the account after t years is given by

$$A(t) = 500(1.08)^t$$

(A) Find $A(7)$ and use secant line slopes to approximate $A'(7)$ to the nearest cent.
(B) Write a brief verbal interpretation of the results in part (A).

65. *Financial analysis.* The price of a stock during a 12 month period is graphed in the figure.

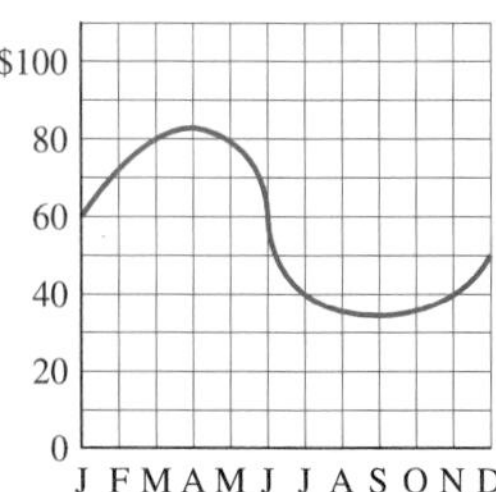

Figure for 65 and 66

(A) Use this graph to estimate (to the nearest \$5) the price of the stock and the rate of change of the price in March. Write a brief verbal interpretation of these results.
(B) When did the stock reach its highest price, and what is the rate of change of the price at this point in time?

66. *Financial analysis.* Refer to the figure.
(A) Use the graph to estimate (to the nearest \$5) the price of the stock and the rate of change of the price in June. Write a brief verbal interpretation of these results.
(B) When did the stock reach its lowest price, and what is the rate of change of the price at this point in time?

67. *Residential energy consumption.* A survey by the Energy Information Administration produced the results in Table 2.

C

Table 2

HOUSEHOLDS WITH TELEVISION SETS

YEAR	BLACK-AND-WHITE (%)	COLOR (%)
1980	51	82
1981	48	82
1982	47	85
1984	43	88
1987	36	93
1990	31	96
1993	20	98

(A) Let x represent time (in years), with $x = 0$ corresponding to 1980, and let y represent the corresponding percentage of households with black-and-white television sets. Enter the appropriate data set in a graphing utility and find a quadratic regression equation for the data.
(B) If $y = B(x)$ denotes the regression equation found in part (A), find $B(15)$ and $B'(15)$, and write a brief verbal interpretation of these results. Express answers as percentages correct to one decimal place.

68. *Residential energy consumption.* Refer to the survey by the Energy Information Administration in Table 2.

C

(A) Let x represent time (in years), with $x = 0$ corresponding to 1980, and let y represent the corresponding percentage of households with color television sets. Enter the appropriate data set in a graphing utility and find a quadratic regression equation for the data.

(B) If $y = C(x)$ denotes the regression equation found in part (A), find $C(15)$ and $C'(15)$, and write a brief verbal interpretation of these results. Express answers as percentages correct to one decimal place.

Life Sciences

69. *Air pollution.* The ozone level (in parts per billion) on a summer day in a metropolitan area is given by

$$P(t) = 80 + 12t - t^2$$

where t is time in hours and $t = 0$ corresponds to 9 AM.

(A) Use the two-step process to find $P'(t)$.

(B) Find $P(3)$ and $P'(3)$. Write a brief verbal interpretation of these results.

70. *Medicine.* The body temperature (in degrees Fahrenheit) of a patient t hours after being given a fever-reducing drug is given by

$$F(t) = 98 + \frac{4}{t + 1}$$

(A) Use the two-step process to find $F'(t)$.

(B) Find $F(3)$ and $F'(3)$. Write a brief verbal interpretation of these results.

Social Sciences

71. *Population.* The number of persons in the United States aged 65 and older can be described by the function

$$P(t) = 26(1.02)^t$$

where P is population in millions and t is time in years since 1980.

(A) Find $P(30)$ and use secant line slopes to approximate $P'(30)$. Round both quantities to one decimal place.

(B) Write a brief verbal interpretation of the results in part (A).

(C) Use the results in part (B) to estimate the 65 or over population in 2011 and in 2012.

72. *Population.* The number of households in the United States can be described by the function

$$H(t) = 80(1.014)^t$$

where H is the number of households in millions and t is time in years since 1980.

(A) Find $H(25)$ and use secant line slopes to approximate $H'(25)$. Round both quantities to one decimal place.

(B) Write a brief verbal interpretation of the results in part (A).

(C) Use the results in part (B) to estimate the number of households in 2006 and in 2007.

SECTION 8-4 Derivatives of Constants, Power Forms, and Sums

- DERIVATIVE OF A CONSTANT FUNCTION
- POWER RULE
- DERIVATIVE OF A CONSTANT TIMES A FUNCTION
- DERIVATIVES OF SUMS AND DIFFERENCES
- APPLICATIONS

In the preceding section, we defined the derivative of f at x as

$$f'(x) = \lim_{h \to 0} \frac{f(x + h) - f(x)}{h}$$

if the limit exists, and we used this definition and a two-step process to find the derivatives of several functions. In this and the next two sections, we will develop some rules based on this definition that will enable us to determine the derivatives of a rather large class of functions without having to go through the two-step process each time.

Before starting on these rules, we list some symbols that are widely used to represent derivatives in the box at the top of the next page.

Derivative Notation

Given $y = f(x)$, then

$$f'(x) \qquad y' \qquad \frac{dy}{dx}$$

all represent the derivative of f at x.

Each of these symbols for the derivative has its particular advantage in certain situations. All of them will become familiar to you after a little experience.

Derivative of a Constant Function

Suppose

$$f(x) = C \qquad C \text{ a constant} \qquad \text{A constant function}$$

Geometrically, the graph of $f(x) = C$ is a horizontal straight line with slope 0 (see Fig. 1); hence, we would expect $f'(x) = 0$. We will show that this is actually the case using the definition of the derivative and the two-step process introduced earlier. We want to find

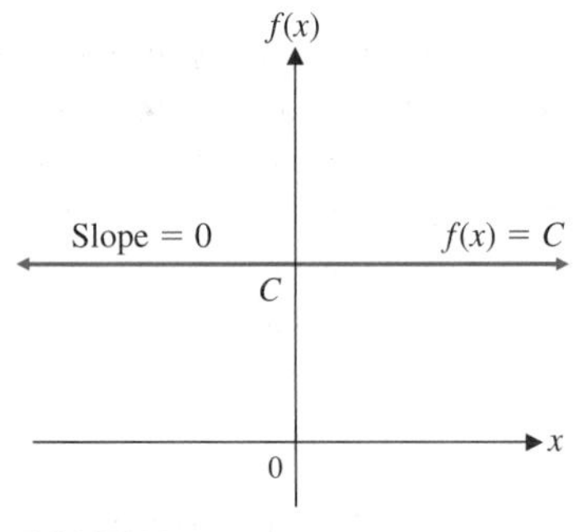

FIGURE 1

$$f'(x) = \lim_{h \to 0} \frac{f(x+h) - f(x)}{h} \qquad \text{Definition of } f'(x)$$

Step 1. $\dfrac{f(x+h) - f(x)}{h} = \dfrac{C - C}{h} = \dfrac{0}{h} = 0 \qquad h \neq 0$

Step 2. $\lim\limits_{h \to 0} 0 = 0$

Thus,

$$f'(x) = 0$$

We conclude that:

The derivative of any constant function is 0.

Derivative of a Constant Function Rule

If $y = f(x) = C$, then

$$f'(x) = 0$$

Also, $y' = 0$ and $dy/dx = 0$.

[*Note:* When we write $C' = 0$ or $\dfrac{d}{dx}C = 0$, we mean $y' = \dfrac{dy}{dx} = 0$ where $y = C$.]

Example 1 **Differentiating Constant Functions**

(A) If $f(x) = 3$, then $f'(x) = 0$. (B) If $y = -1.4$, then $y' = 0$.

(C) If $y = \pi$, then $\dfrac{dy}{dx} = 0$. (D) $\dfrac{d}{dx}23 = 0$

Matched Problem 1 Find:

(A) $f'(x)$ for $f(x) = -24$ (B) y' for $y = 12$

(C) $\dfrac{dy}{dx}$ for $y = -\sqrt{7}$ (D) $\dfrac{d}{dx}(-\pi)$

POWER RULE

A function of the form $f(x) = x^k$, where k is a real number, is called a **power function.** The elementary functions (see the inside front cover) listed below are examples of power functions:

$$f(x) = x \qquad h(x) = x^2 \qquad m(x) = x^3 \qquad (1)$$
$$n(x) = \sqrt{x} \qquad p(x) = \sqrt[3]{x}$$

Explore–Discuss 1

(A) It is clear that the functions f, h, and m in (1) are power functions. Explain why the functions n and p are also power functions.

(B) The domain of a power function depends on the power. Discuss the domain of each of the following power functions:

$$r(x) = x^4 \qquad s(x) = x^{-4} \qquad t(x) = x^{1/4}$$
$$u(x) = x^{-1/4} \qquad v(x) = x^{1/5} \qquad w(x) = x^{-1/5}$$

The definition of the derivative and the two-step process introduced in the preceding section can be used to find the derivatives of many power functions. For example, it can be shown that:

If $f(x) = x^2$, then $f'(x) = 2x$.
If $f(x) = x^3$, then $f'(x) = 3x^2$.
If $f(x) = x^4$, then $f'(x) = 4x^3$.
If $f(x) = x^5$, then $f'(x) = 5x^4$.

Notice the pattern in these derivatives. In each case, the power in f becomes the coefficient in f' and the power in f' is 1 less than the power in f. In general, for any positive integer n:

$$\text{If} \quad f(x) = x^n, \quad \text{then} \quad f'(x) = nx^{n-1}. \qquad (2)$$

In fact, more advanced techniques can be used to show that (2) holds for *any* real number n. We will assume this general result for the remainder of this book.

Power Rule

If $y = f(x) = x^n$, where n is a real number, then

$$f'(x) = nx^{n-1}$$

Also, $y' = nx^{n-1}$ and $dy/dx = nx^{n-1}$.

Explore–Discuss 2

(A) Write a verbal description of the power rule.
(B) If $f(x) = x$, what is $f'(x)$? Discuss how this derivative can be obtained from the power rule.

Example 2 **Differentiating Power Functions**

(A) If $f(x) = x^5$, then $f'(x) = 5x^{5-1} = 5x^4$.
(B) If $y = x^{25}$, then $y' = 25x^{25-1} = 25x^{24}$.
(C) If $y = x^{-3}$, then $\dfrac{dy}{dx} = -3x^{-3-1} = -3x^{-4} = -\dfrac{3}{x^4}$
(D) $\dfrac{d}{dx}x^{5/3} = \dfrac{5}{3}x^{(5/3)-1} = \dfrac{5}{3}x^{2/3}$

Matched Problem 2 Find:

(A) $f'(x)$ for $f(x) = x^6$ (B) y' for $y = x^{30}$
(C) $\dfrac{dy}{dx}$ for $y = x^{-2}$ (D) $\dfrac{d}{dx}x^{3/2}$

In some cases, properties of exponents must be used to rewrite an expression before the power rule is applied.

Example 3 **Differentiating Power Functions**

(A) If $f(x) = 1/x^4$, then we can write $f(x) = x^{-4}$ and

$$f'(x) = -4x^{-4-1} = -4x^{-5} \quad \text{or} \quad \frac{-4}{x^5}$$

(B) If $y = \sqrt{x}$, then we can write $y = x^{1/2}$ and

$$y' = \frac{1}{2}x^{(1/2)-1} = \frac{1}{2}x^{-1/2} \quad \text{or} \quad \frac{1}{2\sqrt{x}}$$

(C) $\dfrac{d}{dx}\dfrac{1}{\sqrt[3]{x}} = \dfrac{d}{dx}x^{-1/3} = -\dfrac{1}{3}x^{(-1/3)-1} = -\dfrac{1}{3}x^{-4/3} \quad \text{or} \quad \dfrac{-1}{3\sqrt[3]{x^4}}$

Matched Problem 3 Find:

(A) $f'(x)$ for $f(x) = \dfrac{1}{x}$ (B) y' for $y = \sqrt[3]{x^2}$ (C) $\dfrac{d}{dx}\dfrac{1}{\sqrt{x}}$

Derivative of a Constant Times a Function

Let $f(x) = ku(x)$, where k is a constant and u is differentiable at x. Then, using the two-step process, we have the following:

Step 1. $\dfrac{f(x+h) - f(x)}{h} = \dfrac{ku(x+h) - ku(x)}{h} = k\left[\dfrac{u(x+h) - u(x)}{h}\right]$

Step 2.

$$\begin{aligned}\lim_{h\to 0}\frac{f(x+h) - f(x)}{h} &= \lim_{h\to 0} k\left[\frac{u(x+h) - u(x)}{h}\right] && \lim_{x\to c} kg(x) = k\lim_{x\to c} g(x)\\ &= k\lim_{h\to 0}\left[\frac{u(x+h) - u(x)}{h}\right] && \text{Definition of } u'(x)\\ &= ku'(x)\end{aligned}$$

Thus:

The derivative of a constant times a differentiable function is the constant times the derivative of the function.

Constant Times a Function Rule

If $y = f(x) = ku(x)$, then

$$f'(x) = ku'(x)$$

Also,

$$y' = ku' \qquad \frac{dy}{dx} = k\frac{du}{dx}$$

Example 4 ➡ **Differentiating a Constant Times a Function**

(A) If $f(x) = 3x^2$, then $f'(x) = 3 \cdot 2x^{2-1} = 6x$.

(B) If $y = \frac{x^3}{6} = \frac{1}{6}x^3$, then $\frac{dy}{dx} = \frac{1}{6} \cdot 3x^{3-1} = \frac{1}{2}x^2$.

(C) If $y = \frac{1}{2x^4} = \frac{1}{2}x^{-4}$, then $y' = \frac{1}{2}(-4x^{-4-1}) = -2x^{-5}$ or $\frac{-2}{x^5}$.

(D) $\frac{d}{dx}\frac{0.4}{\sqrt{x^3}} = \frac{d}{dx}\frac{0.4}{x^{3/2}} = \frac{d}{dx}0.4x^{-3/2} = 0.4\left[-\frac{3}{2}x^{(-3/2)-1}\right]$

$$= -0.6x^{-5/2} \quad \text{or} \quad -\frac{0.6}{\sqrt{x^5}}$$

Matched Problem 4 ➡ Find:

(A) $f'(x)$ for $f(x) = 4x^5$

(B) $\frac{dy}{dx}$ for $y = \frac{x^4}{12}$

(C) y' for $y = \frac{1}{3x^3}$

(D) $\frac{d}{dx}\frac{0.9}{\sqrt[3]{x}}$

■ Derivatives of Sums and Differences

Let $f(x) = u(x) + v(x)$, where $u'(x)$ and $v'(x)$ exist. Then, using the two-step process, we have the following:

Step 1. $$\frac{f(x+h) - f(x)}{h} = \frac{[u(x+h) + v(x+h)] - [u(x) + v(x)]}{h}$$

$$= \frac{u(x+h) + v(x+h) - u(x) - v(x)}{h}$$

$$= \frac{u(x+h) - u(x)}{h} + \frac{v(x+h) - v(x)}{h}$$

Step 2. $$\lim_{h\to 0}\frac{f(x+h)-f(x)}{h} = \lim_{h\to 0}\left[\frac{u(x+h)-u(x)}{h}+\frac{v(x+h)-v(x)}{h}\right]$$

$\lim_{x\to c}[g(x)+h(x)] = \lim_{x\to c} g(x) + \lim_{x\to c} h(x)$

$$= \lim_{h\to 0}\frac{u(x+h)-u(x)}{h} + \lim_{h\to 0}\frac{v(x+h)-v(x)}{h}$$

$$= u'(x) + v'(x)$$

Thus,

The derivative of the sum of two differentiable functions is the sum of the derivatives.

Similarly, we can show that:

The derivative of the difference of two differentiable functions is the difference of the derivatives.

Together, we then have the **sum and difference rule** for differentiation:

Sum and Difference Rule

If $y = f(x) = u(x) \pm v(x)$, then

$$f'(x) = u'(x) \pm v'(x)$$

Also,

$$y' = u' \pm v' \qquad \frac{dy}{dx} = \frac{du}{dx} \pm \frac{dv}{dx}$$

[*Note:* This rule generalizes to the sum and difference of any given number of functions.]

With this and the other rules stated previously, we will be able to compute the derivatives of all polynomials and a variety of other functions.

Example 5 **Differentiating Sums and Differences**

(A) If $f(x) = 3x^2 + 2x$, then

$$f'(x) = (3x^2)' + (2x)' = 3(2x) + 2(1) = 6x + 2$$

(B) If $y = 4 + 2x^3 - 3x^{-1}$, then

$$y' = (4)' + (2x^3)' - (3x^{-1})' = 0 + 2(3x^2) - 3(-1)x^{-2} = 6x^2 + 3x^{-2}$$

(C) If $y = \sqrt[3]{x} - 3x$, then

$$\frac{dy}{dx} = \frac{d}{dx}x^{1/3} - \frac{d}{dx}3x = \frac{1}{3}x^{-2/3} - 3 = \frac{1}{3x^{2/3}} - 3$$

(D) $$\frac{d}{dx}\left(\frac{5}{3x^2} - \frac{2}{x^4} + \frac{x^3}{9}\right) = \frac{d}{dx}\frac{5}{3}x^{-2} - \frac{d}{dx}2x^{-4} + \frac{d}{dx}\frac{1}{9}x^3$$

$$= \frac{5}{3}(-2)x^{-3} - 2(-4)x^{-5} + \frac{1}{9}\cdot 3x^2$$

$$= -\frac{10}{3x^3} + \frac{8}{x^5} + \frac{1}{3}x^2$$

Matched Problem 5 ➡ Find:

(A) $f'(x)$ for $f(x) = 3x^4 - 2x^3 + x^2 - 5x + 7$
(B) y' for $y = 3 - 7x^{-2}$
(C) $\dfrac{dy}{dx}$ for $y = 5x^3 - \sqrt[4]{x}$
(D) $\dfrac{d}{dx}\left(-\dfrac{3}{4x} + \dfrac{4}{x^3} - \dfrac{x^4}{8}\right)$

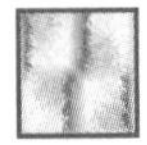

APPLICATIONS

Example 6 ➡ **Instantaneous Velocity** An object moves along the y axis (marked in feet) so that its position at time x (in seconds) is

$$f(x) = x^3 - 6x^2 + 9x$$

(A) Find the instantaneous velocity function v.
(B) Find the velocity at $x = 2$ and $x = 5$ seconds.
(C) Find the time(s) when the velocity is 0.

SOLUTION (A) $v = f'(x) = (x^3)' - (6x^2)' + (9x)' = 3x^2 - 12x + 9$

(B) $f'(2) = 3(2)^2 - 12(2) + 9 = -3$ feet per second
$f'(5) = 3(5)^2 - 12(5) + 9 = 24$ feet per second

(C)
$$\begin{aligned} v = f'(x) = 3x^2 - 12x + 9 &= 0 \\ 3(x^2 - 4x + 3) &= 0 \\ 3(x - 1)(x - 3) &= 0 \\ x &= 1, 3 \end{aligned}$$

Thus, $v = 0$ at $x = 1$ and $x = 3$ seconds.

Matched Problem 6 ➡ Repeat Example 6 for $f(x) = x^3 - 15x^2 + 72x$.

Example 7 ➡ **Tangents** Let $f(x) = x^4 - 6x^2 + 10$.

(A) Find $f'(x)$.
(B) Find the equation of the tangent line at $x = 1$.
(C) Find the values of x where the tangent line is horizontal.

SOLUTION (A)
$$\begin{aligned} f'(x) &= (x^4)' - (6x^2)' + (10)' \\ &= 4x^3 - 12x \end{aligned}$$

(B)
$$\begin{aligned} y - y_1 &= m(x - x_1) && y_1 = f(x_1) = f(1) = (1)^4 - 6(1)^2 + 10 = 5 \\ y - 5 &= -8(x - 1) && m = f'(x_1) = f'(1) = 4(1)^3 - 12(1) = -8 \\ y &= -8x + 13 && \text{Tangent line at } x = 1 \end{aligned}$$

(C) Since a horizontal line has 0 slope, we must solve $f'(x) = 0$ for x:

$$\begin{aligned} f'(x) = 4x^3 - 12x &= 0 \\ 4x(x^2 - 3) &= 0 \\ 4x(x + \sqrt{3})(x - \sqrt{3}) &= 0 \\ x &= 0, -\sqrt{3}, \sqrt{3} \end{aligned}$$

Matched Problem 7 ➡ Repeat Example 7 for $f(x) = x^4 - 8x^3 + 7$.

REMARK

In Example 7, we used algebraic techniques to solve the equation $f'(x) = 0$. A graphing utility also can be used to approximate the solutions to equations of this form. Figure 2 illustrates this process for the equation in Example 7C. Note that the graphing utility gives decimal approximations to $-\sqrt{3}$ and $\sqrt{3}$.

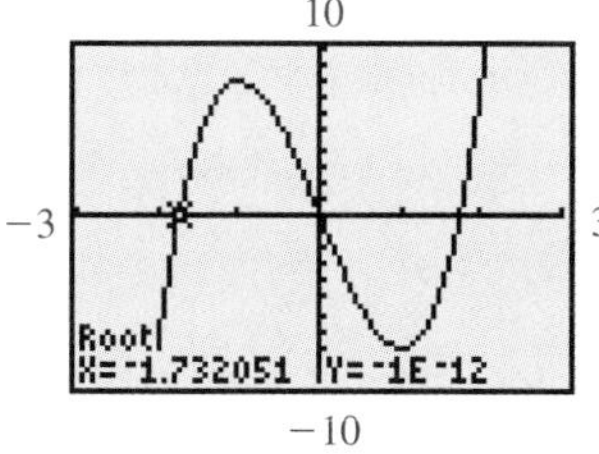

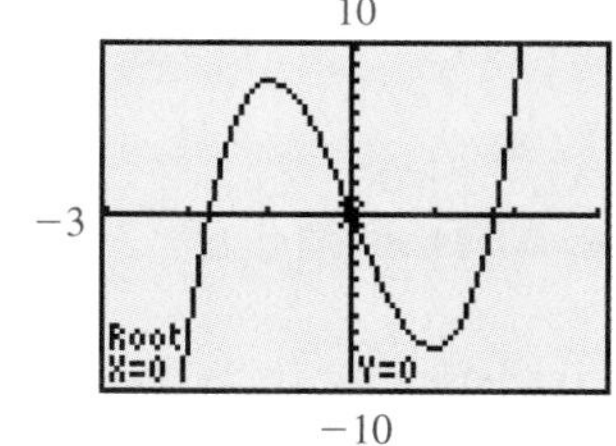

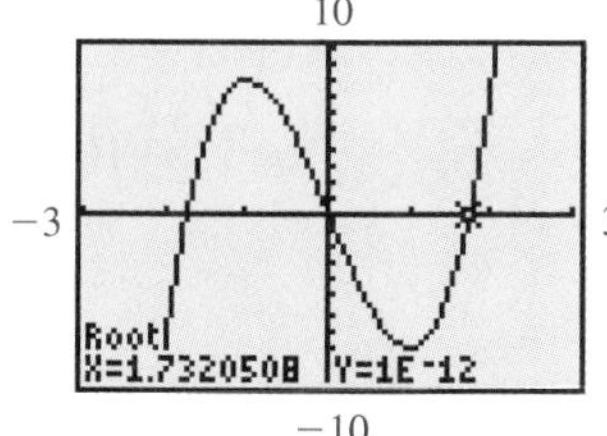

FIGURE 2
$y_1 = 4x^3 - 12x$

In business and economics, the rates at which certain quantities are changing often provide useful insight into various economic systems. A manufacturer, for example, is interested not only in the total cost $C(x)$ at certain production levels, but is also interested in the rate of change of costs at various production levels.

In economics, the word **marginal** refers to a rate of change; that is, to a derivative. Thus, if

$$C(x) = \text{Total cost of producing } x \text{ items}$$

then

$$\begin{aligned} C'(x) &= \textbf{Marginal cost} \\ &= \text{Instantaneous rate of change of total cost } C(x) \text{ with respect to} \\ &\quad\ \text{the number of items produced at a production level of } x \text{ items} \end{aligned}$$

Example 8 **Marginal Cost** Suppose the total cost $C(x)$ (in thousands of dollars) for manufacturing x sailboats per year is given by the function

$$C(x) = 575 + 25x - 0.25x^2 \qquad 0 \leq x \leq 50$$

(see Fig. 3).

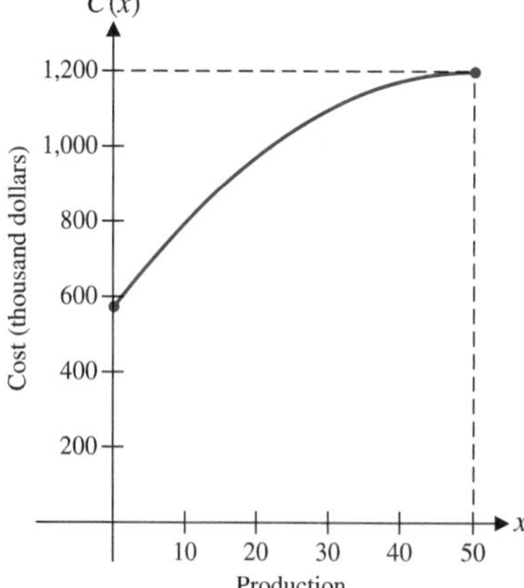

FIGURE 3

(A) Find the marginal cost at a production level of x boats per year.
(B) Find the marginal cost at a production level of 40 boats per year, and interpret the results.

SOLUTION (A) The marginal cost at a production level of x boats is

$$C'(x) = (575)' + (25x)' - (0.25x^2)' = 25 - 0.5x$$

(B) The marginal cost at a production level of 40 boats is

$$C'(40) = 25 - 0.5(40) = 5 \quad \text{or} \quad \$5{,}000 \text{ per boat}$$

At a production level of 40 boats per year, the total cost is increasing at the rate of \$5,000 per boat.

Matched Problem 8 Suppose the total cost $C(x)$ (in thousands of dollars) for manufacturing x sailboats per year is given by the function

$$C(x) = 500 + 24x - 0.2x^2 \qquad 0 \le x \le 50$$

(A) Find the marginal cost at a production level of x boats per year.
(B) Find the marginal cost at a production level of 35 boats per year, and interpret the results.

The interpretation of marginal cost given in Example 8B is the usual way we use the derivative to describe the manner in which a quantity is changing. However, there is a special way to interpret the rate of change of a total cost function that is used extensively in economics. Referring to the results in Example 8B, if the total cost is increasing at the rate of \$5,000 per boat when 40 boats are produced per year, and if production is increased to 41 boats per year, then the total cost will increase by approximately \$5,000. But the increase in total cost is just the cost of producing the next boat. Thus, the cost of producing the 41st boat must be approximately \$5,000, the marginal cost at the production level of 40 boats per year. These ideas are summarized in the following box:

Marginal Cost Function

If $C(x)$ is the total cost of producing x items, then the marginal cost function $C'(x)$ approximates the cost of producing one more item at a production level of x items.

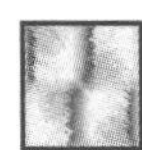

Example 9 **Marginal Cost** Refer to the total cost function given in Example 8:

$$C(x) = 575 + 25x - 0.25x^2 \qquad 0 \le x \le 50$$

(A) Use the marginal cost function to approximate the cost of producing the 31st boat.
(B) Use the total cost function to find the exact cost of producing the 31st boat.

SOLUTION (A) From Example 8, $C'(x) = 25 - 0.5x$. Thus,

$$C'(30) = 25 - 0.5(30) = 10 \quad \text{or} \quad \$10{,}000$$

The cost of producing the 31st boat is approximately \$10,000.

(B) The exact cost of producing the 31st boat is

$$\begin{pmatrix}\text{Total cost of}\\ \text{producing}\\ \text{31 boats}\end{pmatrix} - \begin{pmatrix}\text{Total cost of}\\ \text{producing}\\ \text{30 boats}\end{pmatrix}$$
$$= C(31) \quad - \quad C(30)$$
$$= 1{,}109.75 - 1{,}100 = 9.750 \quad \text{or} \quad \$9{,}750$$

Notice that the marginal cost of \$10,000 per boat is a close approximation to this exact cost.

Matched Problem 9 Refer to the cost function given in Matched Problem 8:

$$C(x) = 500 + 24x - 0.2x^2 \qquad 0 \leqslant x \leqslant 50$$

(A) Use the marginal cost function to approximate the cost of producing the 41st boat.

(B) Use the total cost function to find the exact cost of producing the 41st boat.

REMARK

A derivative can always be interpreted as an instantaneous rate of change, as we did in Example 8B. The interpretation of the marginal cost function as the approximate cost of producing the next item, as in Example 9A, is a special case that applies to total cost functions. This interpretation also applies to total revenue functions and total profit functions, but not to most of the other economic functions we will consider.

Answers to Matched Problems

1. All are 0.
2. (A) $6x^5$ (B) $30x^{29}$ (C) $-2x^{-3} = -2/x^3$ (D) $\frac{3}{2}x^{1/2}$
3. (A) $-x^{-2}$, or $-1/x^2$ (B) $\frac{2}{3}x^{-1/3}$, or $2/(3\sqrt[3]{x})$
   (C) $-\frac{1}{2}x^{-3/2}$, or $-1/(2\sqrt{x^3})$
4. (A) $20x^4$ (B) $x^3/3$ (C) $-x^{-4}$, or $-1/x^4$
   (D) $-0.3x^{-4/3}$, or $-0.3/\sqrt[3]{x^4}$
5. (A) $12x^3 - 6x^2 + 2x - 5$ (B) $14x^{-3}$, or $14/x^3$
   (C) $15x^2 - \frac{1}{4}x^{-3/4}$, or $15x^2 - 1/(4x^{3/4})$ (D) $3/(4x^2) - (12/x^4) - (x^3/2)$
6. (A) $v = 3x^2 - 30x + 72$ (B) $f'(2) = 24$ ft/sec; $f'(5) = -3$ ft/sec
   (C) $x = 4$ and $x = 6$ sec
7. (A) $f'(x) = 4x^3 - 24x^2$ (B) $y = -20x + 20$ (C) $x = 0$ and $x = 6$
8. (A) $C'(x) = 24 - 0.4x$
   (B) $C'(35) = 10$ or \$10,000 per boat; at a production level of 35 boats, total cost is increasing at the rate of \$10,000 per boat.
9. (A) $C'(40) = 8$ or \$8,000; the cost of producing the 41st boat is approximately \$8,000.
   (B) $C(41) - C(40) = 7.8$ or \$7,800

EXERCISE 8-4

A *Find the indicated derivatives in Problems 1–18.*

1. $f'(x)$ for $f(x) = 12$

2. $\frac{dy}{dx}$ for $y = -\sqrt{3}$

3. $\frac{dy}{dx}$ for $y = x^{12}$

4. $\frac{d}{dx}x^5$

5. $f'(x)$ for $f(x) = x$

6. y' for $y = x^7$

7. y' for $y = x^{-7}$

8. $f'(x)$ for $f(x) = x^{-11}$

9. $\frac{dy}{dx}$ for $y = x^{5/2}$

10. $\frac{d}{dx}x^{7/3}$

11. $\frac{d}{dx}\frac{1}{x^5}$

12. $f'(x)$ for $f(x) = \frac{1}{x^9}$

13. $f'(x)$ for $f(x) = 2x^4$

14. $\frac{dy}{dx}$ for $y = -3x$

15. $\frac{d}{dx}(0.2x^6)$

16. y' for $y = 0.6x^4$

17. $\frac{dy}{dx}$ for $y = \frac{x^5}{15}$

18. $f'(x)$ for $f(x) = \frac{x^6}{24}$

Problems 19–24 refer to functions f and g that satisfy $f'(2) = 3$ and $g'(2) = -1$. In each problem, find $h'(2)$ for the indicated function h.

19. $h(x) = 4f(x)$

20. $h(x) = 5g(x)$

21. $h(x) = f(x) + g(x)$

22. $h(x) = g(x) - f(x)$

23. $h(x) = 2f(x) - 3g(x) + 7$

24. $h(x) = -4f(x) + 5g(x) - 9$

B *Find the indicated derivatives in Problems 25–48.*

25. $\frac{d}{dx}(2x^{-5})$

26. y' for $y = -4x^{-1}$

27. $f'(x)$ for $f(x) = \frac{4}{x^4}$

28. $\frac{dy}{dx}$ for $y = \frac{-3}{x^6}$

29. $\frac{d}{dx}\frac{-1}{2x^2}$

30. y' for $y = \frac{1}{6x^3}$

31. $f'(x)$ for $f(x) = -3x^{1/3}$

32. $\frac{dy}{dx}$ for $y = -8x^{1/4}$

33. $\frac{d}{dx}(2.4x^2 - 3.5x + 4)$

34. y' for $y = 3x^2 + 4x - 7$

35. $\frac{dy}{dx}$ for $y = 3x^5 - 2x^3 + 5$

36. $f'(x)$ for $f(x) = 2.6x^3 - 6.7x + 5.2$

37. $\frac{d}{dx}(3x^{-4} + 2x^{-2})$

38. y' for $y = 2x^{-3} - 4x^{-1}$

39. $\frac{dy}{dx}$ for $y = \frac{1}{2x} - \frac{2}{3x^3}$

40. $f'(x)$ for $f(x) = \frac{3}{4x^3} + \frac{1}{2x^5}$

41. $\frac{d}{dx}(3x^{2/3} - 5x^{1/3})$

42. $\frac{d}{dx}(8x^{3/4} + 4x^{-1/4})$

43. $\frac{d}{dx}\left(\frac{3}{x^{3/5}} - \frac{6}{x^{1/2}}\right)$

44. $\frac{d}{dx}\left(\frac{5}{x^{1/5}} - \frac{8}{x^{3/2}}\right)$

45. $\frac{d}{dx}\frac{1}{\sqrt[3]{x}}$

46. y' for $y = \frac{10}{\sqrt[5]{x}}$

47. $\frac{dy}{dx}$ for $y = \frac{1.2}{\sqrt{x}} - 3.2x^{-2} + x$

48. $f'(x)$ for $f(x) = 2.8x^{-3} - \frac{0.6}{\sqrt[3]{x^2}} + 7$

For Problems 49–52, find:

(A) $f'(x)$
(B) The slope of the graph of f at $x = 2$ and $x = 4$
(C) The equations of the tangent lines at $x = 2$ and $x = 4$
(D) The value(s) of x where the tangent line is horizontal

49. $f(x) = 6x - x^2$

50. $f(x) = 2x^2 + 8x$

51. $f(x) = 3x^4 - 6x^2 - 7$

52. $f(x) = x^4 - 32x^2 + 10$

If an object moves along the y axis (marked in feet) so that its position at time x (in seconds) is given by the indicated function in Problems 53–56, find:

(A) The instantaneous velocity function $v = f'(x)$
(B) The velocity when $x = 0$ and $x = 3$ seconds
(C) The time(s) when $v = 0$

53. $f(x) = 176x - 16x^2$

54. $f(x) = 80x - 10x^2$

55. $f(x) = x^3 - 9x^2 + 15x$

56. $f(x) = x^3 - 9x^2 + 24x$

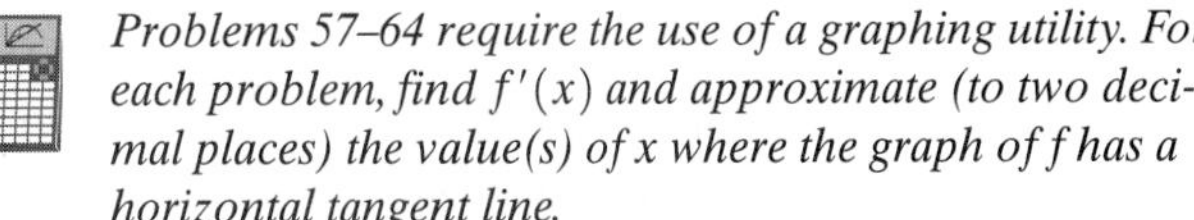

Problems 57–64 require the use of a graphing utility. For each problem, find $f'(x)$ and approximate (to two decimal places) the value(s) of x where the graph of f has a horizontal tangent line.

57. $f(x) = x^2 - 3x - 4\sqrt{x}$

58. $f(x) = x^2 + x - 10\sqrt{x}$

59. $f(x) = 3\sqrt[3]{x^4} - 1.5x^2 - 3x$

60. $f(x) = 3\sqrt[3]{x^4} - 2x^2 + 4x$

61. $f(x) = 0.05x^4 + 0.1x^3 - 1.5x^2 - 1.6x + 3$

62. $f(x) = 0.02x^4 - 0.06x^3 - 0.78x^2 + 0.94x + 2.2$

63. $f(x) = 0.2x^4 - 3.12x^3 + 16.25x^2 - 28.25x + 7.5$

64. $f(x) = 0.25x^4 - 2.6x^3 + 8.1x^2 - 10x + 9$

65. Let $f(x) = ax^2 + bx + c, a \neq 0$. Recall that the graph of $y = f(x)$ is a parabola. Use the derivative $f'(x)$ to derive a formula for the x coordinate of the vertex of this parabola.

66. Now that you know how to find derivatives, explain why it is no longer necessary for you to memorize the formula for the x coordinate of the vertex of a parabola.

67. Give an example of a cubic polynomial function that has:
(A) No horizontal tangents
(B) One horizontal tangent
(C) Two horizontal tangents

68. Can a cubic polynomial function have more than two horizontal tangents? Explain.

C *In Problems 69–72, find each derivative.*

69. $f'(x)$ for $f(x) = \dfrac{10x + 20}{x}$

70. $\dfrac{dy}{dx}$ for $y = \dfrac{x^2 + 25}{x^2}$

71. $\dfrac{d}{dx}\dfrac{x^4 - 3x^3 + 5}{x^2}$

72. y' for $y = \dfrac{2x^5 - 4x^3 + 2x}{x^3}$

In Problems 73 and 74, use the definition of derivative and the two-step process to verify each statement.

73. $\dfrac{d}{dx}x^3 = 3x^2$

74. $\dfrac{d}{dx}x^4 = 4x^3$

75. The domain of the power function $f(x) = x^{1/3}$ is the set of all real numbers. Find the domain of the derivative $f'(x)$. Discuss the nature of the graph of $y = f(x)$ for any x values excluded from the domain of $f'(x)$.

76. The domain of the power function $f(x) = x^{2/3}$ is the set of all real numbers. Find the domain of the derivative $f'(x)$. Discuss the nature of the graph of $y = f(x)$ for any x values excluded from the domain of $f'(x)$.

APPLICATIONS

Business & Economics

77. *Marginal cost.* The total cost (in dollars) of producing x tennis rackets per day is

$$C(x) = 800 + 60x - 0.25x^2 \qquad 0 \le x \le 120$$

(A) Find the marginal cost at a production level of x rackets.
(B) Find the marginal cost at a production level of 60 rackets, and interpret the result.
(C) Find the actual cost of producing the 61st racket, and compare this cost with the result found in part (B).
(D) Find $C'(80)$, and interpret the result.

78. *Marginal cost.* The total cost (in dollars) of producing x portable radios per day is

$$C(x) = 1{,}000 + 100x - 0.5x^2 \qquad 0 \le x \le 100$$

(A) Find the marginal cost at a production level of x radios.
(B) Find the marginal cost at a production level of 80 radios, and interpret the result.
(C) Find the actual cost of producing the 81st radio, and compare this cost with the result found in part (B).
(D) Find $C'(50)$, and interpret the result.

79. **C** *Motor vehicle production.* Annual limousine production levels in the United States for selected years are given in Table 1.

Table 1

YEAR	LIMOUSINE PRODUCTION
1980	2,500
1985	6,500
1990	4,400
1995	3,800
1997	4,000

Source: www.limousinecentral.com

(A) Let x represent time (in years) since 1980, and let y represent the corresponding U.S. production of limousines. Enter the data in a graphing utility and find a cubic regression equation for the data.
(B) If $y = L(x)$ denotes the regression equation found in part (A), find $L(16)$ and $L'(16)$ (to the nearest hundred), and write a brief verbal interpretation of these results.

80. *Motor vehicle operation.* Table 2 lists the total number of registered limousine operators in the United States for selected years.

C

Table 2

Year	Limousine Operators
1985	4,000
1990	7,000
1995	8,900
1997	9,600

Source: www.limousinecentral.com

(A) Let x represent time (in years) since 1980, and let y represent the corresponding number of registered limousine operators in the United States. Enter the data in a graphing utility and find a quadratic regression equation for the data.

(B) If $y = L(x)$ denotes the regression equation found in part (A), find $L(16)$ and $L'(16)$ (to the nearest hundred), and write a brief verbal interpretation of these results.

81. *Marginal cost.* The total cost (in dollars) of producing x microwave ovens per week is shown in the figure. Which is greater, the approximate cost of producing the 101st oven or the approximate cost of producing the 401st oven? Does this graph represent a manufacturing process that is becoming more efficient or less efficient as production levels increase? Explain.

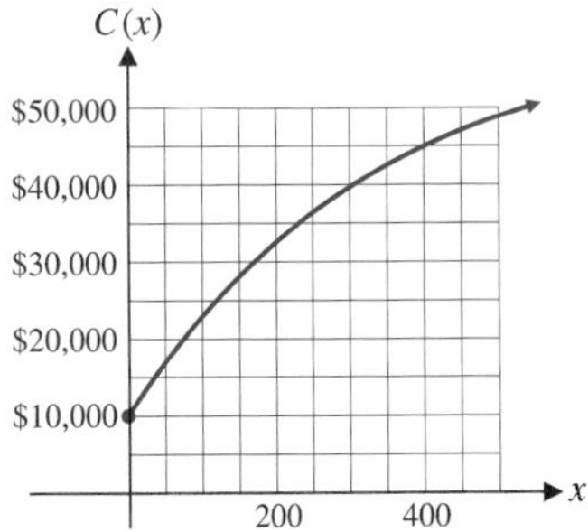

Figure for 81

82. *Marginal cost.* The total cost (in dollars) of producing x electric stoves per week is shown in the figure. Which is greater, the approximate cost of producing the 101st stove or the approximate cost of producing the 401st stove? Does this graph represent a manufacturing process that is becoming more efficient or less efficient as production levels increase? Explain.

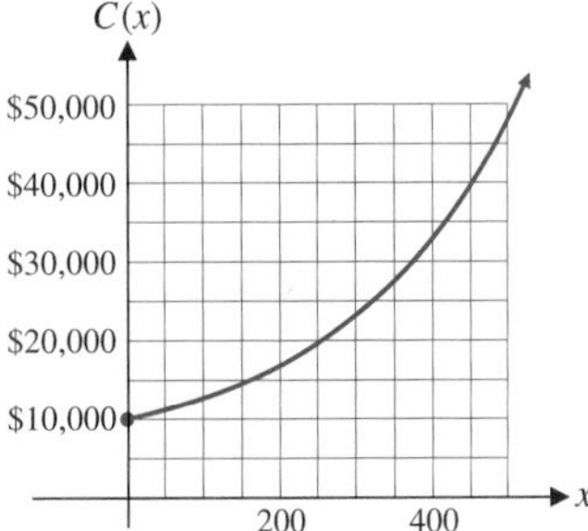

Figure for 82

83. *Advertising.* Using past records, it is estimated that a marine manufacturer will sell $N(x)$ power boats after spending $\$x$ thousand on advertising, as given by

$$N(x) = 1{,}000 - \frac{3{,}780}{x} \qquad 5 \le x \le 30$$

See the figure.

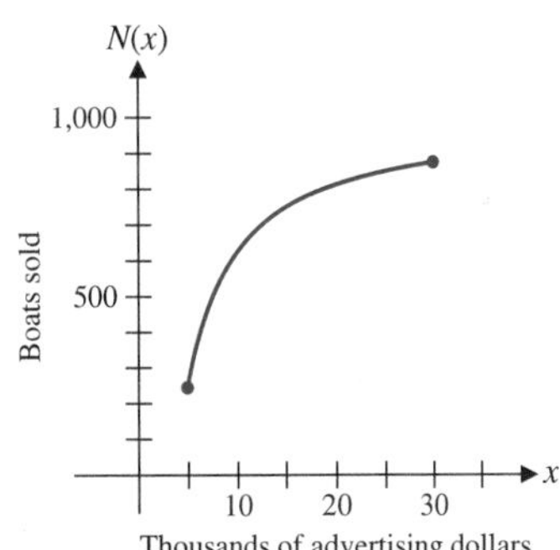

Figure for 83

(A) Find $N'(x)$.

(B) Find $N'(10)$ and $N'(20)$. Write a brief verbal interpretation of these results.

84. *Price–demand equation.* Suppose that in a given gourmet food store, people are willing to buy x pounds of chocolate candy per day at $\$p$ per quarter pound, as given by the price–demand equation

$$x = 10 + \frac{180}{p} \qquad 2 \le p \le 10$$

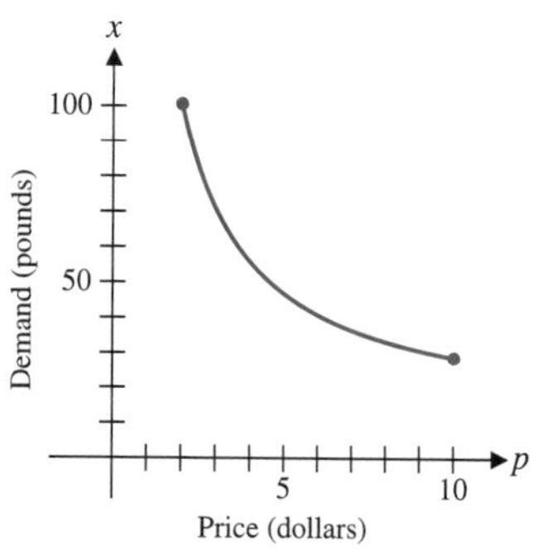

Figure for 84

This function is graphed in the figure. Find the demand and the instantaneous rate of change of demand with respect to price when the price is \$5. Write a brief verbal interpretation of these results.

Life Sciences

85. *Medicine.* A person x inches tall has a pulse rate of y beats per minute, as given approximately by

$$y = 590x^{-1/2} \qquad 30 \leq x \leq 75$$

What is the instantaneous rate of change of pulse rate at the:
(A) 36 inch level? (B) 64 inch level?

86. *Ecology.* A coal-burning electrical generating plant emits sulfur dioxide into the surrounding air. The concentration $C(x)$, in parts per million, is given approximately by

$$C(x) = \frac{0.1}{x^2}$$

where x is the distance from the plant in miles. Find the instantaneous rate of change of concentration at:
(A) $x = 1$ mile (B) $x = 2$ miles

Social Sciences

87. *Learning.* Suppose a person learns y items in x hours, as given by

$$y = 50\sqrt{x} \qquad 0 \leq x \leq 9$$

(see the figure). Find the rate of learning at the end of:
(A) 1 hour (B) 9 hours

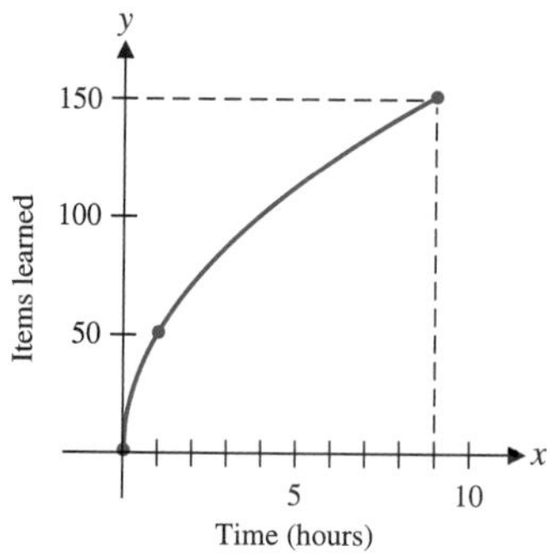

Figure for 87

88. *Learning.* If a person learns y items in x hours, as given by

$$y = 21\sqrt[3]{x^2} \qquad 0 \leq x \leq 8$$

find the rate of learning at the end of:
(A) 1 hour (B) 8 hours

SECTION 8-5 Derivatives of Products and Quotients

- DERIVATIVES OF PRODUCTS
- DERIVATIVES OF QUOTIENTS

The derivative rules discussed in the preceding section added substantially to our ability to compute and apply derivatives to many practical problems. In this and the next section, we will add a few more rules that will increase this ability even further.

■ DERIVATIVES OF PRODUCTS

In Section 8-4 we found that the derivative of a sum is the sum of the derivatives. Is the derivative of a product the product of the derivatives?

Explore–Discuss 1

Let $F(x) = x^2$, $S(x) = x^3$, and $f(x) = F(x)S(x) = x^5$. Which of the following is $f'(x)$?

(A) $F'(x)S'(x)$ (B) $F(x)S'(x)$
(C) $F'(x)S(x)$ (D) $F(x)S'(x) + F'(x)S(x)$

Comparing the various expressions computed in Explore–Discuss 1, we see that the derivative of a product is not the product of the derivatives, but appears to involve a slightly more complicated form.

Using the definition of the derivative and the two-step process, it can be shown that:

The derivative of the product of two functions is the first function times the derivative of the second function plus the second function times the derivative of the first function.

That is:

Product Rule

If

$$y = f(x) = F(x)S(x)$$

and if $F'(x)$ and $S'(x)$ exist, then

$$f'(x) = F(x)S'(x) + S(x)F'(x)$$

Also,

$$y' = FS' + SF' \qquad \frac{dy}{dx} = F\frac{dS}{dx} + S\frac{dF}{dx}$$

Example 1 **Differentiating a Product** Use two different methods to find $f'(x)$ for $f(x) = 2x^2(3x^4 - 2)$.

SOLUTION **Method 1.** Use the product rule:

$$\begin{aligned} f'(x) &= 2x^2(3x^4 - 2)' + (3x^4 - 2)(2x^2)' \\ &= 2x^2(12x^3) + (3x^4 - 2)(4x) \\ &= 24x^5 + 12x^5 - 8x \\ &= 36x^5 - 8x \end{aligned}$$

First times derivative of second plus second times derivative of first

Method 2. Multiply first; then take derivatives:

$$\begin{aligned} f(x) &= 2x^2(3x^4 - 2) = 6x^6 - 4x^2 \\ f'(x) &= 36x^5 - 8x \end{aligned}$$

Matched Problem 1 Use two different methods to find $f'(x)$ for $f(x) = 3x^3(2x^2 - 3x + 1)$.

At this point, all the products we will encounter can be differentiated by either of the methods illustrated in Example 1. In the next and later sections, we will see that there are situations where the product rule must be used. Unless instructed otherwise, you should use the product rule to differentiate all products in this section to gain experience with the use of this important differentiation rule.

Example 2 **Tangent Lines** Let $f(x) = (2x - 9)(x^2 + 6)$.

(A) Find the equation of the line tangent to the graph of $f(x)$ at $x = 3$.
(B) Find the value(s) of x where the tangent line is horizontal.

SOLUTION (A) First, find $f'(x)$:

$$\begin{aligned} f'(x) &= (2x - 9)(x^2 + 6)' + (x^2 + 6)(2x - 9)' \\ &= (2x - 9)(2x) + (x^2 + 6)(2) \end{aligned}$$

Now, find the equation of the tangent line at $x = 3$:

$$\begin{aligned} y - y_1 &= m(x - x_1) && y_1 = f(x_1) = f(3) = -45 \\ y - (-45) &= 12(x - 3) && m = f'(x_1) = f'(3) = 12 \\ y &= 12x - 81 && \text{Tangent line at } x = 3 \end{aligned}$$

(B) The tangent line is horizontal at any value of x such that $f'(x) = 0$, so

$$\begin{aligned} f'(x) = (2x - 9)2x + (x^2 + 6)2 &= 0 \\ 6x^2 - 18x + 12 &= 0 \\ x^2 - 3x + 2 &= 0 \\ (x - 1)(x - 2) &= 0 \\ x &= 1, 2 \end{aligned}$$

The tangent line is horizontal at $x = 1$ and at $x = 2$.

Matched Problem 2 Repeat Example 2 for $f(x) = (2x + 9)(x^2 - 12)$.

As Example 2 illustrates, the way we write $f'(x)$ depends on what we want to do with it. If we are interested only in evaluating $f'(x)$ at specified values of x, the form in part (A) is sufficient. However, if we want to solve $f'(x) = 0$, we must multiply and collect like terms, as we did in part (B).

Derivatives of Quotients

As is the case with a product, the derivative of a quotient of two functions is not the quotient of the derivatives of the two functions.

Explore–Discuss 2

Let $T(x) = x^5$, $B(x) = x^2$, and

$$f(x) = \frac{T(x)}{B(x)} = \frac{x^5}{x^2} = x^3$$

Which of the following is $f'(x)$?

(A) $\dfrac{T'(x)}{B'(x)}$ (B) $\dfrac{T'(x)B(x)}{[B(x)]^2}$ (C) $\dfrac{T(x)B'(x)}{[B(x)]^2}$

(D) $\dfrac{T'(x)B(x)}{[B(x)]^2} - \dfrac{T(x)B'(x)}{[B(x)]^2} = \dfrac{T'(x)B(x) - T(x)B'(x)}{[B(x)]^2}$

The expressions in Explore–Discuss 2 suggest that the derivative of a quotient leads to a more complicated quotient than you might expect.

In general, if $T(x)$ and $B(x)$ are any two differentiable functions and

$$f(x) = \frac{T(x)}{B(x)}$$

then it can be shown that

$$f'(x) = \frac{B(x)T'(x) - T(x)B'(x)}{[B(x)]^2}$$

Thus:

The derivative of the quotient of two functions is the bottom function times the derivative of the top function minus the top function times the derivative of the bottom function, all over the bottom function squared.

Quotient Rule

If

$$y = f(x) = \frac{T(x)}{B(x)}$$

and if $T'(x)$ and $B'(x)$ exist, then

$$f'(x) = \frac{B(x)T'(x) - T(x)B'(x)}{[B(x)]^2}$$

Also,

$$y' = \frac{BT' - TB'}{B^2} \qquad \frac{dy}{dx} = \frac{B\dfrac{dT}{dx} - T\dfrac{dB}{dx}}{B^2}$$

Example 3 **Differentiating Quotients**

(A) If $f(x) = \dfrac{x^2}{2x - 1}$, find $f'(x)$. (B) If $y = \dfrac{x^2 - x}{x^3 + 1}$, find y'.

(C) Find $\dfrac{d}{dx}\dfrac{x^2 - 3}{x^2}$ by using the quotient rule and also by splitting the fraction into two fractions.

SOLUTION (A)

$$\begin{aligned} f'(x) &= \frac{(2x - 1)(x^2)' - x^2(2x - 1)'}{(2x - 1)^2} \\ &= \frac{(2x - 1)(2x) - x^2(2)}{(2x - 1)^2} \\ &= \frac{4x^2 - 2x - 2x^2}{(2x - 1)^2} \\ &= \frac{2x^2 - 2x}{(2x - 1)^2} \end{aligned}$$

The bottom times the derivative of the top minus the top times the derivative of the bottom, all over the square of the bottom

(B)
$$\begin{aligned} y' &= \frac{(x^3+1)(x^2-x)' - (x^2-x)(x^3+1)'}{(x^3+1)^2} \\ &= \frac{(x^3+1)(2x-1) - (x^2-x)(3x^2)}{(x^3+1)^2} \\ &= \frac{2x^4 - x^3 + 2x - 1 - 3x^4 + 3x^3}{(x^3+1)^2} \\ &= \frac{-x^4 + 2x^3 + 2x - 1}{(x^3+1)^2} \end{aligned}$$

(C) **Method 1.** Use the quotient rule:

$$\begin{aligned} \frac{d}{dx}\frac{x^2-3}{x^2} &= \frac{x^2 \frac{d}{dx}(x^2-3) - (x^2-3)\frac{d}{dx}x^2}{(x^2)^2} \\ &= \frac{x^2(2x) - (x^2-3)2x}{x^4} \\ &= \frac{2x^3 - 2x^3 + 6x}{x^4} = \frac{6x}{x^4} = \frac{6}{x^3} \end{aligned}$$

Method 2. Split into two fractions:

$$\frac{x^2-3}{x^2} = \frac{x^2}{x^2} - \frac{3}{x^2} = 1 - 3x^{-2}$$

$$\frac{d}{dx}(1 - 3x^{-2}) = 0 - 3(-2)x^{-3} = \frac{6}{x^3}$$

Comparing methods 1 and 2, we see that it often pays to change an expression algebraically before blindly using a differentiation formula.

Matched Problem 3 Find:

(A) $f'(x)$ for $f(x) = \dfrac{2x}{x^2+3}$ (B) y' for $y = \dfrac{x^3-3x}{x^2-4}$

(C) $\dfrac{d}{dx}\dfrac{2+x^3}{x^3}$ two ways

Explore–Discuss 3

Explain why $\neq$ is used below, and then find the correct derivative.

$$\frac{d}{dx}\frac{x^3}{x^2+3x+4} \neq \frac{3x^2}{2x+3}$$

Example 4 **Sales Analysis** The total sales S (in thousands of games) for a home video game t months after the game is introduced are given by

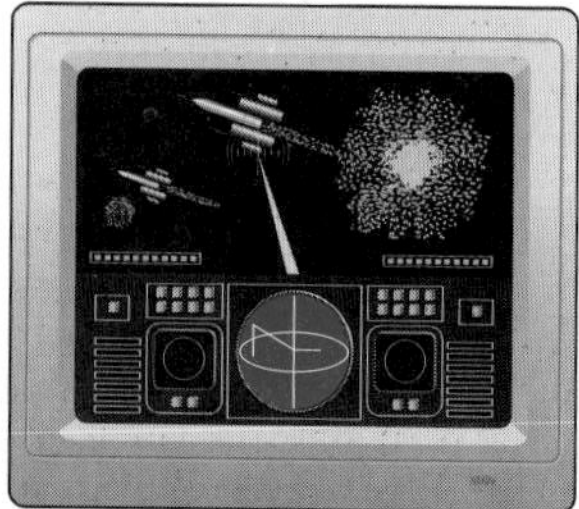

$$S(t) = \frac{125t^2}{t^2+100}$$

(A) Find $S'(t)$.

(B) Find $S(10)$ and $S'(10)$. Write a brief verbal interpretation of these results. **C** Check on a graphing utility.

(C) Use the results from part (B) to estimate the total sales after 11 months.

SOLUTION (A)
$$S'(t) = \frac{(t^2 + 100)(125t^2)' - 125t^2(t^2 + 100)'}{(t^2 + 100)^2}$$
$$= \frac{(t^2 + 100)(250t) - 125t^2(2t)}{(t^2 + 100)^2}$$
$$= \frac{250t^3 + 25{,}000t - 250t^3}{(t^2 + 100)^2}$$
$$= \frac{25{,}000t}{(t^2 + 100)^2}$$

(B) $S(10) = \frac{125(10)^2}{10^2 + 100} = 62.5$ and $S'(10) = \frac{25{,}000(10)}{(10^2 + 100)^2} = 6.25$

```
10→X
                10
Y1
              62.5
nDeriv(Y1,X,10)
              6.25
```

FIGURE 1
$y_1 = \frac{125x^2}{x^2 + 100}$

The total sales after 10 months are 62,500 games, and sales are increasing at the rate of 6,250 games per month. Figure 1 shows a check of these results on a graphing utility.

(C) The total sales will increase by approximately 6,250 games during the next month. Thus, the estimated total sales after 11 months are $62{,}500 + 6{,}250 = 68{,}750$ games.

Matched Problem 4 ➡ Refer to Example 4. Suppose the total sales S (in thousands of games) t months after the game is introduced are given by

$$S(t) = \frac{150t}{t + 3}$$

(A) Find $S'(t)$.

(B) Find $S(12)$ and $S'(12)$. Write a brief verbal interpretation of these results. **C** Check on a graphing utility.

(C) Use the results from part (B) to estimate the total sales after 13 months.

Answers to Matched Problems

1. $30x^4 - 36x^3 + 9x^2$

2. (A) $y = 84x - 297$

(B) $x = -4, x = 1$

3. (A) $\frac{(x^2 + 3)2 - (2x)(2x)}{(x^2 + 3)^2} = \frac{6 - 2x^2}{(x^2 + 3)^2}$

(B) $\frac{(x^2 - 4)(3x^2 - 3) - (x^3 - 3x)(2x)}{(x^2 - 4)^2} = \frac{x^4 - 9x^2 + 12}{(x^2 - 4)^2}$

(C) $-\frac{6}{x^4}$

4. (A) $S'(t) = \frac{450}{(t + 3)^2}$

(B) $S(12) = 120$; $S'(12) = 2$. After 12 months, the total sales are 120,000 games, and sales are increasing at the rate of 2,000 games per month.

(C) 122,000 games

Exercise 8-5

The answers to most of the problems in this exercise set contain both an unsimplified form and a simplified form of the derivative. When checking your work, first check that you applied the rules correctly, and then check that you performed the algebraic simplification correctly. Unless instructed otherwise, when differentiating a product, use the product rule rather than performing the multiplication first.

A *In Problems 1–18, find $f'(x)$ and simplify.*

1. $f(x) = 2x^3(x^2 - 2)$

2. $f(x) = 5x^2(x^3 + 2)$

3. $f(x) = (x - 3)(2x - 1)$

4. $f(x) = (3x + 2)(4x - 5)$

5. $f(x) = \dfrac{x}{x - 3}$

6. $f(x) = \dfrac{3x}{2x + 1}$

7. $f(x) = \dfrac{2x + 3}{x - 2}$

8. $f(x) = \dfrac{3x - 4}{2x + 3}$

9. $f(x) = (x^2 + 1)(2x - 3)$

10. $f(x) = (3x + 5)(x^2 - 3)$

11. $f(x) = (0.4x + 2)(0.5x - 5)$

12. $f(x) = (0.5x - 4)(0.2x + 1)$

13. $f(x) = \dfrac{x^2 + 1}{2x - 3}$

14. $f(x) = \dfrac{3x + 5}{x^2 - 3}$

15. $f(x) = (x^2 + 2)(x^2 - 3)$

16. $f(x) = (x^2 - 4)(x^2 + 5)$

17. $f(x) = \dfrac{x^2 + 2}{x^2 - 3}$

18. $f(x) = \dfrac{x^2 - 4}{x^2 + 5}$

Problems 19–24 refer to functions f and g that satisfy $f(1) = 4$, $f'(1) = -2$, $g(1) = 2$, and $g'(1) = 3$. In each problem, find $h'(1)$ for the indicated function h.

19. $h(x) = f(x)g(x)$

20. $h(x) = \dfrac{f(x)}{g(x)}$

21. $h(x) = \dfrac{g(x)}{f(x)}$

22. $h(x) = f(x)f(x)$

23. $h(x) = \dfrac{1}{f(x)}$

24. $h(x) = \dfrac{1}{g(x)}$

B *In Problems 25–32, find the indicated derivatives and simplify.*

25. $f'(x)$ for $f(x) = (2x + 1)(x^2 - 3x)$

26. y' for $y = (x^3 + 2x^2)(3x - 1)$

27. $\dfrac{dy}{dx}$ for $y = (2.5x - x^2)(4x + 1.4)$

28. $\dfrac{d}{dx}[(3 - 0.4x^3)(0.5x^2 - 2x)]$

29. y' for $y = \dfrac{5x - 3}{x^2 + 2x}$

30. $f'(x)$ for $f(x) = \dfrac{3x^2}{2x - 1}$

31. $\dfrac{d}{dx}\dfrac{x^2 - 3x + 1}{x^2 - 1}$

32. $\dfrac{dy}{dx}$ for $y = \dfrac{x^4 - x^3}{3x - 1}$

In Problems 33–36, find $f'(x)$ and find the equation of the line tangent to the graph of f at $x = 2$.

33. $f(x) = (1 + 3x)(5 - 2x)$

34. $f(x) = (7 - 3x)(1 + 2x)$

35. $f(x) = \dfrac{x - 8}{3x - 4}$

36. $f(x) = \dfrac{2x - 5}{2x - 3}$

In Problems 37–40, find $f'(x)$ and find the value(s) of x where $f'(x) = 0$.

37. $f(x) = (2x - 15)(x^2 + 18)$

38. $f(x) = (2x - 3)(x^2 - 6)$

39. $f(x) = \dfrac{x}{x^2 + 1}$

40. $f(x) = \dfrac{x}{x^2 + 9}$

In Problems 41–44, find $f'(x)$ two ways; by using the product or quotient rule and by simplifying first.

41. $f(x) = x^3(x^4 - 1)$

42. $f(x) = x^4(x^3 - 1)$

43. $f(x) = \dfrac{x^3 + 9}{x^3}$

44. $f(x) = \dfrac{x^4 + 4}{x^4}$

C *In Problems 45–56, find each derivative and simplify.*

45. $f'(x)$ for $f(x) = (2x^4 - 3x^3 + x)(x^2 - x + 5)$

46. $\dfrac{dy}{dx}$ for $y = (x^2 - 3x + 1)(x^3 + 2x^2 - x)$

47. $\dfrac{d}{dx}\dfrac{3x^2 - 2x + 3}{4x^2 + 5x - 1}$

48. y' for $y = \dfrac{x^3 - 3x + 4}{2x^2 + 3x - 2}$

49. $\dfrac{dy}{dx}$ for $y = 9x^{1/3}(x^3 + 5)$

50. $\dfrac{d}{dx}[(4x^{1/2} - 1)(3x^{1/3} + 2)]$

51. $f'(x)$ for $f(x) = \dfrac{6\sqrt[3]{x}}{x^2 - 3}$

52. y' for $y = \dfrac{2\sqrt{x}}{x^2 - 3x + 1}$

53. $\dfrac{d}{dx}\dfrac{x^3 - 2x^2}{\sqrt[3]{x^2}}$

54. $\dfrac{dy}{dx}$ for $y = \dfrac{x^2 - 3x + 1}{\sqrt[4]{x}}$

55. $f'(x)$ for $f(x) = \dfrac{(2x^2 - 1)(x^2 + 3)}{x^2 + 1}$

56. y' for $y = \dfrac{2x - 1}{(x^3 + 2)(x^2 - 3)}$

Problems 57–60 refer to a function of the form $f(x) = [u(x)]^n$, where $u(x)$ is a differentiable function.

57. Use the product rule to show that if $n = 2$, then $f'(x) = 2u(x)u'(x)$.

58. Use the product rule and the result from Problem 57 to show that if $n = 3$, then $f'(x) = 3[u(x)]^2u'(x)$.

59. Based on the results in Problems 57 and 58, write a formula for $f'(x)$ for any n.

60. Use the quotient rule to find $f'(x)$ if $n = -1$. Does this agree with your formula in Problem 59?

In Problems 61–64, approximate (to two decimal places) the value(s) of x where the graph of f has a horizontal tangent line.

61. $f(x) = (x^2 - 5)(x^2 - x)$

62. $f(x) = (x^2 - 3)(x^2 - 2x)$

63. $f(x) = \dfrac{x^3 + 12x + 1}{x^2 + 1}$

64. $f(x) = \dfrac{x^3 + 14x - 1}{x^2 + 1}$

APPLICATIONS

Business & Economics

65. *Sales analysis.* The total sales S (in thousands of CD's) for a compact disk are given by

$$S(t) = \frac{90t^2}{t^2 + 50}$$

where t is the number of months since the release of the CD.

(A) Find $S'(t)$.

(B) Find $S(10)$ and $S'(10)$. Write a brief verbal interpretation of these results.

(C) Use the results from part (B) to estimate the total sales after 11 months.

66. *Sales analysis.* A communications company has installed a cable television system in a city. The total number N (in thousands) of subscribers t months after the installation of the system is given by

$$N(t) = \frac{200t}{t + 5}$$

(A) Find $N'(t)$.

(B) Find $N(15)$ and $N'(15)$. Write a brief verbal interpretation of these results.

(C) Use the results from part (B) to estimate the total number of subscribers after 16 months.

67. *Price–demand equation.* According to classical economic theory, the demand x for a quantity in a free market decreases as the price p increases (see the figure). Suppose that the number x of CD players people are willing to buy per week from a retail chain at a price of \$$p$ is given by

$$x = \frac{4{,}000}{0.1p + 1} \qquad 10 \le p \le 70$$

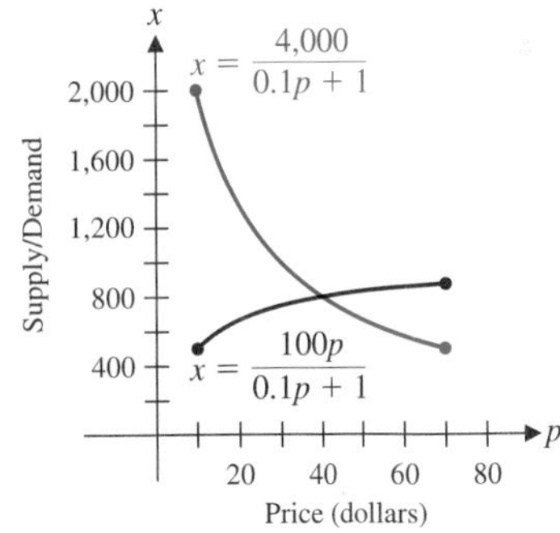

Figure for 67 and 68

(A) Find dx/dp.

(B) Find the demand and the instantaneous rate of change of demand with respect to price when the price is \$40. Write a brief verbal interpretation of these results.

(C) Use the results from part (B) to estimate the demand if the price is increased to \$41.

68. *Price–supply equation.* Also according to classical economic theory, the supply x for a quantity in a free market increases as the price p increases (see the figure). Suppose that the number x of CD players a retail chain is willing to sell per week at a price of \$$p$ is given by

$$x = \frac{100p}{0.1p + 1} \qquad 10 \le p \le 70$$

(A) Find dx/dp.

(B) Find the supply and the instantaneous rate of change of supply with respect to price when the price is \$40. Write a brief verbal interpretation of these results.

(C) Use the results from part (B) to estimate the supply if the price is increased to \$41.

Life Sciences

69. *Medicine.* A drug is injected into the bloodstream of a patient through her right arm. The concentration of the drug (in milligrams per cubic centimeter) in the bloodstream of the left arm t hours after the injection is given by

$$C(t) = \frac{0.14t}{t^2 + 1}$$

(A) Find $C'(t)$.
(B) Find $C'(0.5)$ and $C'(3)$, and interpret the results.

70. *Drug sensitivity.* One hour after x milligrams of a particular drug are given to a person, the change in body temperature $T(x)$, in degrees Fahrenheit, is given approximately by

$$T(x) = x^2\left(1 - \frac{x}{9}\right) \qquad 0 \leq x \leq 7$$

The rate $T'(x)$ at which T changes with respect to the size of the dosage, x, is called the *sensitivity* of the body to the dosage.
(A) Find $T'(x)$ using the product rule.
(B) Find $T'(1)$, $T'(3)$, and $T'(6)$.

Social Sciences

71. *Learning.* In the early days of quantitative learning theory (around 1917), L. L. Thurstone found that a given person successfully accomplished $N(x)$ acts after x practice acts, as given by

$$N(x) = \frac{100x + 200}{x + 32}$$

(A) Find the instantaneous rate of change of learning, $N'(x)$, with respect to the number of practice acts x.
(B) Find $N'(4)$ and $N'(68)$.

SECTION 8-6 Chain Rule: Power Form

- Chain Rule: Power Rule
- Combining Rules of Differentiation

In this section we develop a rule for differentiating powers of functions—a special case of the very important *chain rule,* which we will return to in Chapter 10. Also, for the first time, we will encounter some product forms that cannot be simplified by multiplication and must be differentiated by the power rule.

Chain Rule: Power Rule

We have already made extensive use of the power rule,

$$\frac{d}{dx}x^n = nx^{n-1} \tag{1}$$

Now we want to generalize this rule so that we can differentiate functions of the form $[u(x)]^n$, where $u(x)$ is a differentiable function. Is rule (1) still valid if we replace x with a function $u(x)$?

Explore–Discuss 1

Let $u(x) = 2x^2$ and $f(x) = [u(x)]^3 = 8x^6$. Which of the following is $f'(x)$?

(A) $3[u(x)]^2$ (B) $3[u'(x)]^2$ (C) $3[u(x)]^2u'(x)$

The calculations in Explore–Discuss 1 show that we cannot generalize the power rule simply by replacing x with $u(x)$ in (1).

How can we find a formula for the derivative of $[u(x)]^n$, where $u(x)$ is an arbitrary differentiable function? Let's begin by considering the derivatives of $[u(x)]^2$ and $[u(x)]^3$ to see if a general pattern emerges. Since $[u(x)]^2 = u(x)u(x)$, we use the product rule to write

$$\begin{aligned} \frac{d}{dx}[u(x)]^2 &= \frac{d}{dx}[u(x)u(x)] \\ &= u(x)u'(x) + u(x)u'(x) \\ &= 2u(x)u'(x) \end{aligned} \qquad (2)$$

Since $[u(x)]^3 = [u(x)]^2 u(x)$, we now use the product rule and the result in (2) to write

$$\begin{aligned} \frac{d}{dx}[u(x)]^3 &= \frac{d}{dx}\{[u(x)]^2 u(x)\} \\ &= [u(x)]^2 \frac{d}{dx}u(x) + u(x)\frac{d}{dx}[u(x)]^2 \\ &= [u(x)]^2 u'(x) + u(x)[2u(x)u'(x)] \\ &= 3[u(x)]^2 u'(x) \end{aligned}$$

Use (2) to substitute for $\frac{d}{dx}[u(x)]^2$.

Continuing in this fashion, it can be shown that

$$\frac{d}{dx}[u(x)]^n = n[u(x)]^{n-1}u'(x) \qquad n \text{ a positive integer} \qquad (3)$$

Using more advanced techniques, formula (3) can be established for all real numbers n. Thus, we have the **general power rule:**

General Power Rule

If $u(x)$ is a differentiable function, n is any real number, and

$$y = f(x) = [u(x)]^n$$

then

$$f'(x) = n[u(x)]^{n-1}u'(x)$$

This rule is often written more compactly as

$$y' = nu^{n-1}u' \qquad \text{or} \qquad \frac{d}{dx}u^n = nu^{n-1}\frac{du}{dx} \qquad \text{where } u = u(x)$$

The general power rule is a special case of a very important and useful differentiation rule called the **chain rule.** In essence, the chain rule will enable us to differentiate a composition form $f[g(x)]$ if we know how to differentiate $f(x)$ and $g(x)$. We defer a complete discussion of the chain rule until Chapter 10.

Example 1 **Differentiating Power Forms** Find $f'(x)$:

(A) $f(x) = (3x + 1)^4$ (B) $f(x) = (x^3 + 4)^7$

(C) $f(x) = \dfrac{1}{(x^2 + x + 4)^3}$ (D) $f(x) = \sqrt{3 - x}$

SOLUTION (A) $f(x) = (3x + 1)^4$ Let $u = 3x + 1$, $n = 4$.

$$f'(x) = 4(3x+1)^3(3x+1)' \qquad nu^{n-1}\frac{du}{dx}$$

$$= 4(3x+1)^3 3 \qquad \frac{du}{dx} = 3$$

$$= 12(3x+1)^3$$

(B) $f(x) = (x^3 + 4)^7$ Let $u = (x^3 + 4)$, $n = 7$.

$$f'(x) = 7(x^3+4)^6(x^3+4)' \qquad nu^{n-1}\frac{du}{dx}$$

$$= 7(x^3+4)^6 3x^2 \qquad \frac{du}{dx} = 3x^2$$

$$= 21x^2(x^3+4)^6$$

(C) $f(x) = \dfrac{1}{(x^2 + x + 4)^3} = (x^2 + x + 4)^{-3}$ Let $u = x^2 + x + 4$, $n = -3$.

$$f'(x) = -3(x^2+x+4)^{-4}(x^2+x+4)' \qquad nu^{n-1}\frac{du}{dx}$$

$$= -3(x^2+x+4)^{-4}(2x+1) \qquad \frac{du}{dx} = 2x + 1$$

$$= \frac{-3(2x+1)}{(x^2+x+4)^4}$$

(D) $f(x) = \sqrt{3 - x} = (3 - x)^{1/2}$ Let $u = 3 - x$, $n = \frac{1}{2}$.

$$f'(x) = \frac{1}{2}(3-x)^{-1/2}(3-x)' \qquad nu^{n-1}\frac{du}{dx}$$

$$= \frac{1}{2}(3-x)^{-1/2}(-1) \qquad \frac{du}{dx} = -1$$

$$= -\frac{1}{2(3-x)^{1/2}} \quad \text{or} \quad -\frac{1}{2\sqrt{3-x}}$$

Matched Problem 1 Find $f'(x)$:

(A) $f(x) = (5x + 2)^3$ (B) $f(x) = (x^4 - 5)^5$

(C) $f(x) = \dfrac{1}{(x^2 + 4)^2}$ (D) $f(x) = \sqrt{4 - x}$

Notice that we used two steps to differentiate each function in Example 1. First, we applied the general power rule; then we found du/dx. As you gain experience with the general power rule, you may want to combine these two steps. If you do this, be certain to multiply by du/dx. For example,

$$\frac{d}{dx}(x^5 + 1)^4 = 4(x^5 + 1)^3 5x^4 \qquad \text{Correct}$$

$$\frac{d}{dx}(x^5 + 1)^4 \neq 4(x^5 + 1)^3 \qquad du/dx = 5x^4 \text{ is missing}$$

If we let $u(x) = x$, then $du/dx = 1$, and the general power rule reduces to the (ordinary) power rule discussed in Section 8-4. Compare the following:

$$\frac{d}{dx}x^n = nx^{n-1} \qquad \text{Yes—power rule}$$

$$\frac{d}{dx}u^n = nu^{n-1}\frac{du}{dx} \qquad \text{Yes—general power rule}$$

$$\frac{d}{dx}u^n \neq nu^{n-1} \qquad \text{Unless } u(x) = x + k \text{ so that } du/dx = 1$$

Combining Rules of Differentiation

The following examples illustrate the use of the general power rule in combination with other rules of differentiation.

Example 2 **Tangent Lines** Find the equation of the line tangent to the graph of f at $x = 2$ for $f(x) = x^2\sqrt{2x + 12}$. **C** Check with a graphing utility.

Solution

$$f(x) = x^2\sqrt{2x + 12}$$

$$= x^2(2x + 12)^{1/2} \quad \text{Apply the product rule.}$$

$$f'(x) = x^2 \frac{d}{dx}(2x + 12)^{1/2} + (2x + 12)^{1/2} \frac{d}{dx} x^2$$

$$= x^2\left[\frac{1}{2}(2x + 12)^{-1/2}\right](2) + (2x + 12)^{1/2}(2x)$$

$$= \frac{x^2}{\sqrt{2x + 12}} + 2x\sqrt{2x + 12}$$

Use the general power rule to differentiate $(2x + 12)^{1/2}$ and the ordinary power rule to differentiate x^2.

$$f'(2) = \frac{4}{\sqrt{16}} + 4\sqrt{16} = 1 + 16 = 17$$

$$f(2) = 4\sqrt{16} = 16$$

$$(x_1, y_1) = (2, f(2)) = (2, 16) \quad \text{Point}$$

$$m = f'(2) = 17 \quad \text{Slope}$$

$$y - 16 = 17(x - 2) \quad y - y_1 = m(x - x_1)$$

$$y = 17x - 18 \quad \text{Tangent line}$$

Figure 1A checks the values of $f(2)$ and $f'(2)$, and Figure 1B shows the graphs of $y = f(x)$ and the tangent line at $x = 2$. Note that we graphed the equation $y = 17x - 18$ in order to check our work. We did not use a built-in routine for graphing tangent lines. Such a routine would produce the same graph, but it would not tell us if we have the correct equation for the tangent line.

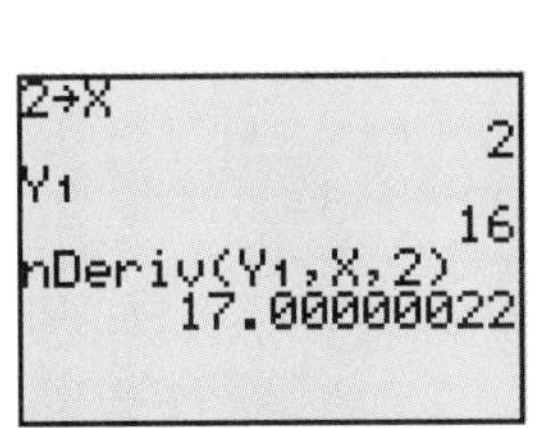

(A)

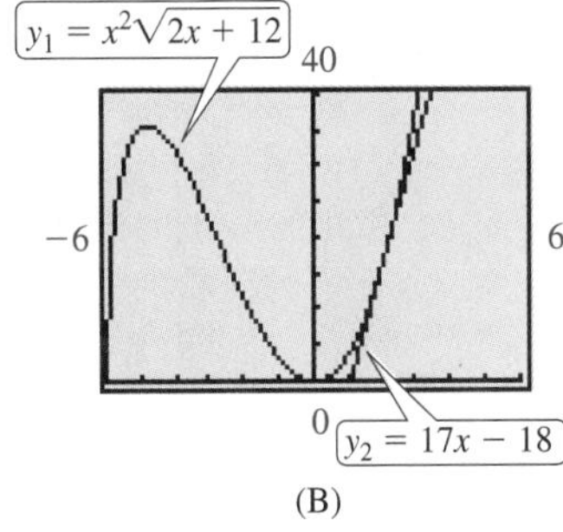

(B)

FIGURE 1

Matched Problem 2 Find $f'(x)$ and the equation of the line tangent to the graph of f at $x = 3$ for $f(x) = x\sqrt{15 - 2x}$. **C** Check with a graphing utility.

Example 3 **Tangent Lines** Find the value(s) of x where the tangent line is horizontal for

$$f(x) = \frac{x^3}{(2 - 3x)^5}$$

SOLUTION Use the quotient rule:

$$f'(x) = \frac{(2-3x)^5 \frac{d}{dx} x^3 - x^3 \frac{d}{dx}(2-3x)^5}{[(2-3x)^5]^2}$$

Use the ordinary power rule to differentiate x^3 and the general power rule to differentiate $(2-3x)^5$.

$$= \frac{(2-3x)^5 3x^2 - x^3 5(2-3x)^4(-3)}{(2-3x)^{10}}$$

$$= \frac{(2-3x)^4 3x^2[(2-3x)+5x]}{(2-3x)^{10}}$$

$$= \frac{3x^2(2+2x)}{(2-3x)^6} = \frac{6x^2(x+1)}{(2-3x)^6}$$

Since a fraction is 0 when the numerator is 0 and the denominator is not, we see that $f'(x) = 0$ at $x = -1$ and $x = 0$. Thus, the graph of f will have horizontal tangent lines at $x = -1$ and $x = 0$.

Matched Problem 3 Find the value(s) of x where the tangent line is horizontal for

$$f(x) = \frac{x^3}{(3x-2)^2}$$

Example 4 **Combining Differentiation Rules** Starting with the function f in Example 3, write f as a product and then differentiate.

SOLUTION

$$f(x) = \frac{x^3}{(2-3x)^5} = x^3(2-3x)^{-5}$$

$$f'(x) = x^3 \frac{d}{dx}(2-3x)^{-5} + (2-3x)^{-5}\frac{d}{dx}x^3$$

$$= x^3(-5)(2-3x)^{-6}(-3) + (2-3x)^{-5}3x^2$$

$$= 15x^3(2-3x)^{-6} + 3x^2(2-3x)^{-5}$$

At this point, we have an unsimplified form for $f'(x)$. This may be satisfactory for some purposes, but not for others. For example, if we need to solve the equation $f'(x) = 0$, we must simplify algebraically:

$$f'(x) = \frac{15x^3}{(2-3x)^6} + \frac{3x^2}{(2-3x)^5} = \frac{15x^3}{(2-3x)^6} + \frac{3x^2(2-3x)}{(2-3x)^6}$$

$$= \frac{15x^3 + 3x^2(2-3x)}{(2-3x)^6} = \frac{3x^2(5x+2-3x)}{(2-3x)^6}$$

$$= \frac{3x^2(2+2x)}{(2-3x)^6} = \frac{6x^2(1+x)}{(2-3x)^6}$$

Matched Problem 4 Refer to the function f in Matched Problem 3. Write f as a product and then differentiate.

As Example 4 illustrates, any quotient can be converted to a product and differentiated by the product rule. However, if the derivative must be simplified, it is usually easier to use the quotient rule. (Compare the algebraic simplifications in Example 4 with those in Example 3.) There is one special case where using negative exponents is the preferred method—a fraction whose numerator is a constant.

Example 5 **Alternate Methods of Differentiation** Find $f'(x)$ two ways for:

$$f(x) = \frac{4}{(x^2 + 9)^3}$$

SOLUTION **Method 1.** Use the quotient rule:

$$\begin{aligned} f'(x) &= \frac{(x^2 + 9)^3 \frac{d}{dx} 4 - 4 \frac{d}{dx}(x^2 + 9)^3}{[(x^2 + 9)^3]^2} \\ &= \frac{(x^2 + 9)^3(0) - 4[3(x^2 + 9)^2(2x)]}{(x^2 + 9)^6} \\ &= \frac{-24x(x^2 + 9)^2}{(x^2 + 9)^6} = \frac{-24x}{(x^2 + 9)^4} \end{aligned}$$

Method 2. Rewrite as a product, and use the general power rule:

$$\begin{aligned} f(x) &= \frac{4}{(x^2 + 9)^3} = 4(x^2 + 9)^{-3} \\ f'(x) &= 4(-3)(x^2 + 9)^{-4}(2x) \\ &= \frac{-24x}{(x^2 + 9)^4} \end{aligned}$$

Which method do you prefer?

Matched Problem 5 Find $f'(x)$ two ways for: $f(x) = \dfrac{5}{(x^3 + 1)^2}$

Answers to Matched Problems

1. (A) $15(5x + 2)^2$ (B) $20x^3(x^4 - 5)^4$ (C) $-4x/(x^2 + 4)^3$
(D) $-1/(2\sqrt{4 - x})$

2. $f'(x) = \sqrt{15 - 2x} - \dfrac{x}{\sqrt{15 - 2x}}$; $y = 2x + 3$ **3.** $x = 0, x = 2$

4. $-6x^3(3x - 2)^{-3} + 3x^2(3x - 2)^{-2} = \dfrac{3x^2(x - 2)}{(3x - 2)^3}$ **5.** $\dfrac{-30x^2}{(x^3 + 1)^3}$

EXERCISE 8-6

The answers to many of the problems in this exercise set contain both an unsimplified form and a simplified form of a derivative. When checking your work, first check that you applied the rules correctly, and then check that you performed the algebraic simplification correctly.

A *In Problems 1–6, replace the* **?** *with an expression that will make the indicated equation valid.*

1. $\dfrac{d}{dx}(3x + 4)^4 = 4(3x + 4)^3$ __?__

2. $\dfrac{d}{dx}(5 - 2x)^6 = 6(5 - 2x)^5$ __?__

3. $\dfrac{d}{dx}(4 - 2x^2)^3 = 3(4 - 2x^2)^2$ __?__

4. $\dfrac{d}{dx}(3x^2 + 7)^5 = 5(3x^2 + 7)^4$ __?__

5. $\dfrac{d}{dx}(1 + 2x + 3x^2)^7 = 7(1 + 2x + 3x^2)^6$ __?__

6. $\dfrac{d}{dx}(4 - 3x - 2x^2)^8 = 8(4 - 3x - 2x^2)^7$ __?__

In Problems 7–20, find $f'(x)$ using the general power rule and simplify.

7. $f(x) = (2x + 5)^3$
8. $f(x) = (3x - 7)^5$
9. $f(x) = (5 - 2x)^4$
10. $f(x) = (9 - 5x)^2$
11. $f(x) = (4 + 0.2x)^5$
12. $f(x) = (6 - 0.5x)^4$
13. $f(x) = (3x^2 + 5)^5$
14. $f(x) = (5x^2 - 3)^6$
15. $f(x) = (x^3 - 2x^2 + 2)^8$
16. $f(x) = (2x^2 + x + 1)^7$
17. $f(x) = (2x - 5)^{1/2}$
18. $f(x) = (4x + 3)^{1/2}$
19. $f(x) = (x^4 + 1)^{-2}$
20. $f(x) = (x^5 + 2)^{-3}$

In Problems 21–24, find $f'(x)$ and the equation of the line tangent to the graph of f at the indicated value of x. Find the value(s) of x where the tangent line is horizontal.

21. $f(x) = (2x - 1)^3$; $x = 1$
22. $f(x) = (3x - 1)^4$; $x = 1$
23. $f(x) = (4x - 3)^{1/2}$; $x = 3$
24. $f(x) = (2x + 8)^{1/2}$; $x = 4$

B *In Problems 25–44, find dy/dx using the general power rule and simplify.*

25. $y = 3(x^2 - 2)^4$
26. $y = 2(x^3 + 6)^5$
27. $y = 2(x^2 + 3x)^{-3}$
28. $y = 3(x^3 + x^2)^{-2}$
29. $y = \sqrt{x^2 + 8}$
30. $y = \sqrt[3]{3x - 7}$
31. $y = \sqrt[3]{3x + 4}$
32. $y = \sqrt{2x - 5}$
33. $y = \sqrt[4]{0.8x + 3.6}$
34. $y = \sqrt[5]{1.6x - 4.5}$
35. $y = (x^2 - 4x + 2)^{1/2}$
36. $y = (2x^2 + 2x - 3)^{1/2}$
37. $y = \dfrac{1}{2x + 4}$
38. $y = \dfrac{1}{3x - 7}$
39. $y = \dfrac{1}{(x^3 + 4)^5}$
40. $y = \dfrac{1}{(x^2 - 3)^6}$
41. $y = \dfrac{1}{4x^2 - 4x + 1}$
42. $y = \dfrac{1}{2x^2 - 3x + 1}$
43. $y = \dfrac{4}{\sqrt{x^2 - 3x}}$
44. $y = \dfrac{3}{\sqrt[3]{x - x^2}}$

In Problems 45–50, find $f'(x)$, and find the equation of the line tangent to the graph of f at the indicated value of x.

45. $f(x) = x(4 - x)^3$; $x = 2$
46. $f(x) = x^2(1 - x)^4$; $x = 2$
47. $f(x) = \dfrac{x}{(2x - 5)^3}$; $x = 3$
48. $f(x) = \dfrac{x^4}{(3x - 8)^2}$; $x = 4$
49. $f(x) = x\sqrt{2x + 2}$; $x = 1$
50. $f(x) = x\sqrt{x - 6}$; $x = 7$

In Problems 51–56, find $f'(x)$, and find the value(s) of x where the tangent line is horizontal.

51. $f(x) = x^2(x - 5)^3$
52. $f(x) = x^3(x - 7)^4$
53. $f(x) = \dfrac{x}{(2x + 5)^2}$
54. $f(x) = \dfrac{x - 1}{(x - 3)^3}$
55. $f(x) = \sqrt{x^2 - 8x + 20}$
56. $f(x) = \sqrt{x^2 + 4x + 5}$

In Problems 57–62, approximate (to two decimal places) the value(s) of x where the graph of f has a horizontal tangent line.

57. $f(x) = (x^2 + 2x - 15)\sqrt{x^2 + 1}$
58. $f(x) = (x^2 - 3x - 20)\sqrt{x^2 + 1}$
59. $f(x) = \dfrac{x^3 - x + 1}{(x^2 + 1)^2}$
60. $f(x) = \dfrac{x^3 - 2x + 2}{(x^2 + 1)^2}$
61. $f(x) = \sqrt{x^4 - 4x^3 + 6x + 10}$
62. $f(x) = \sqrt{x^4 - 5x^2 + x + 9}$

C *In Problems 63–74, find each derivative and simplify.*

63. $\dfrac{d}{dx}[3x(x^2 + 1)^3]$
64. $\dfrac{d}{dx}[2x^2(x^3 - 3)^4]$
65. $\dfrac{d}{dx}\dfrac{(x^3 - 7)^4}{2x^3}$
66. $\dfrac{d}{dx}\dfrac{3x^2}{(x^2 + 5)^3}$
67. $\dfrac{d}{dx}[(2x - 3)^2(2x^2 + 1)^3]$
68. $\dfrac{d}{dx}[(x^2 - 1)^3(x^2 - 2)^2]$
69. $\dfrac{d}{dx}(4x^2\sqrt{x^2 - 1})$
70. $\dfrac{d}{dx}(3x\sqrt{2x^2 + 3})$
71. $\dfrac{d}{dx}\dfrac{2x}{\sqrt{x - 3}}$
72. $\dfrac{d}{dx}\dfrac{x^2}{\sqrt{x^2 + 1}}$
73. $\dfrac{d}{dx}\sqrt{(2x - 1)^3(x^2 + 3)^4}$
74. $\dfrac{d}{dx}\sqrt{\dfrac{4x + 1}{2x^2 + 1}}$

Applications

Business & Economics

75. *Marginal cost.* The total cost (in hundreds of dollars) of producing x calculators per day is

$$C(x) = 10 + \sqrt{2x + 16} \qquad 0 \leq x \leq 50$$

(see the figure).

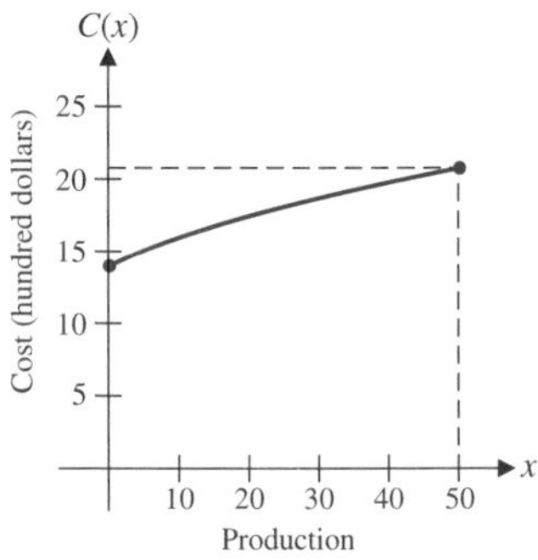

Figure for 75

(A) Find the marginal cost at a production level of x calculators.
(B) Find $C'(24)$ and $C'(42)$, and interpret the results.

76. *Marginal cost.* The total cost (in hundreds of dollars) of producing x cameras per week is

$$C(x) = 6 + \sqrt{4x + 4} \qquad 0 \leq x \leq 30$$

(A) Find the marginal cost at a production level of x cameras.
(B) Find $C'(15)$ and $C'(24)$, and interpret the results.

77. *Price–supply equation.* The number x of stereo speakers a retail chain is willing to sell per week at a price of $\$p$ is given by

$$x = 80\sqrt{p + 25} - 400 \qquad 20 \leq p \leq 100$$

(see the figure).

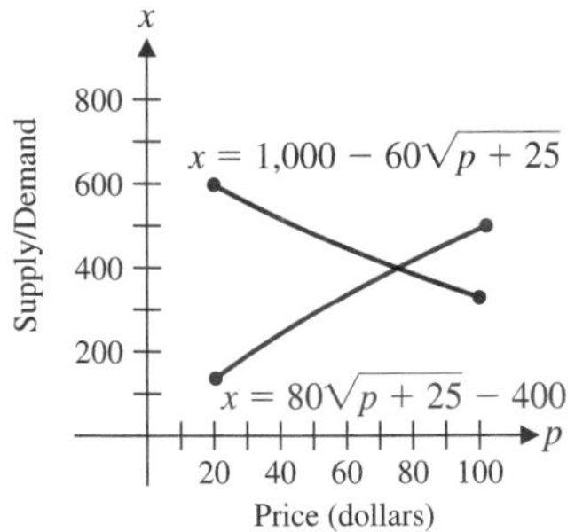

Figure for 77 and 78

(A) Find dx/dp.
(B) Find the supply and the instantaneous rate of change of supply with respect to price when the price is $75. Write a brief verbal interpretation of these results.

78. *Price—demand equation.* The number x of stereo speakers people are willing to buy per week from a retail chain at a price of $\$p$ is given by

$$x = 1{,}000 - 60\sqrt{p + 25} \qquad 20 \leq p \leq 100$$

(see the figure).
(A) Find dx/dp.
(B) Find the demand and the instantaneous rate of change of demand with respect to price when the price is $75. Write a brief verbal interpretation of these results.

79. *Compound interest.* If $1,000 is invested at an annual interest rate r compounded monthly, the amount in the account at the end of 4 years is given by

$$A = 1{,}000\left(1 + \tfrac{1}{12}r\right)^{48}$$

Find the rate of change of the amount A with respect to the interest rate r.

80. *Compound interest.* If $100 is invested at an annual interest rate r compounded semiannually, the amount in the account at the end of 5 years is given by

$$A = 100\left(1 + \tfrac{1}{2}r\right)^{10}$$

Find the rate of change of the amount A with respect to the interest rate r.

Life Sciences

81. *Bacteria growth.* The number y of bacteria in a certain colony after x days is given approximately by

$$y = (3 \times 10^6)\left[1 - \frac{1}{\sqrt[3]{(x^2 - 1)^2}}\right]$$

Find dy/dx.

82. *Pollution.* A small lake in a resort area became contaminated with harmful bacteria because of excessive septic tank seepage. After treating the lake with a bactericide, the Department of Public Health estimated the bacteria concentration (number per cubic centimeter) after t days to be given by

$$C(t) = 500(8 - t)^2 \qquad 0 \leq t \leq 7$$

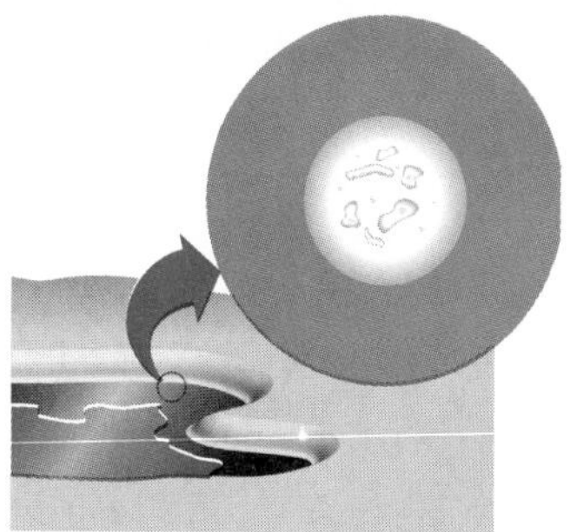

(A) Find $C'(t)$ using the general power rule.
(B) Find $C'(1)$ and $C'(6)$, and interpret the results.

Social Sciences

83. *Learning.* In 1930, L. L. Thurstone developed the following formula to indicate how learning time T depends on the length of a list n:

$$T = f(n) = \frac{c}{k} n\sqrt{n - a}$$

where a, c, and k are empirical constants. Suppose that for a particular person, time T (in minutes) for learning a list of length n is

$$T = f(n) = 2n\sqrt{n - 2}$$

(A) Find dT/dn.
(B) Find $f'(11)$ and $f'(27)$, and interpret the results.

Section 8-7 Marginal Analysis in Business and Economics

- Marginal Cost, Revenue, and Profit
- Application
- Marginal Average Cost, Revenue, and Profit

Marginal Cost, Revenue, and Profit

One important use of calculus in business and economics is in *marginal analysis.* We introduced the concept of *marginal cost* earlier. There is no reason to stop there. Economists also talk about *marginal revenue* and *marginal profit.* Recall that the word "marginal" refers to an instantaneous rate of change—that is, a derivative. Thus, we define the following:

Marginal Cost, Revenue, and Profit

If x is the number of units of a product produced in some time interval, then

Total cost $= C(x)$ Total revenue $= R(x)$
Marginal cost $= C'(x)$ **Marginal revenue** $= R'(x)$

Total profit $= P(x) = R(x) - C(x)$
Marginal profit $= P'(x) = R'(x) - C'(x)$
$=$ (Marginal revenue) $-$ (Marginal cost)

Marginal cost (or revenue or profit) is the instantaneous rate of change of cost (or revenue or profit) relative to production at a given production level.

It is important to remember that whenever we refer to a cost function $C(x)$ it is understood that $C(x)$ represents the *total cost* of producing x items. To find the exact cost of producing a particular item, we use the difference of two successive values of $C(x)$:

$$\text{Total cost of producing } x + 1 \text{ items} = C(x + 1)$$
$$\text{Total cost of producing } x \text{ items} = C(x)$$
$$\text{Exact cost of producing the } (x + 1)\text{st item} = C(x + 1) - C(x)$$

As we noted in Section 8-4, the marginal cost function can be used to approximate this exact cost. To see why, we return to the definition of a derivative:

$$C'(x) = \lim_{h\to 0} \frac{C(x + h) - C(x)}{h} \qquad \text{Marginal cost}$$
$$C'(x) \approx \frac{C(x + h) - C(x)}{h} \qquad h \neq 0$$
$$C'(x) \approx C(x + 1) - C(x) \qquad h = 1$$

Thus, we see that the marginal cost $C'(x)$ approximates $C(x + 1) - C(x)$, the exact cost of producing the $(x + 1)$st item. These observations are summarized below and illustrated in Figure 1.

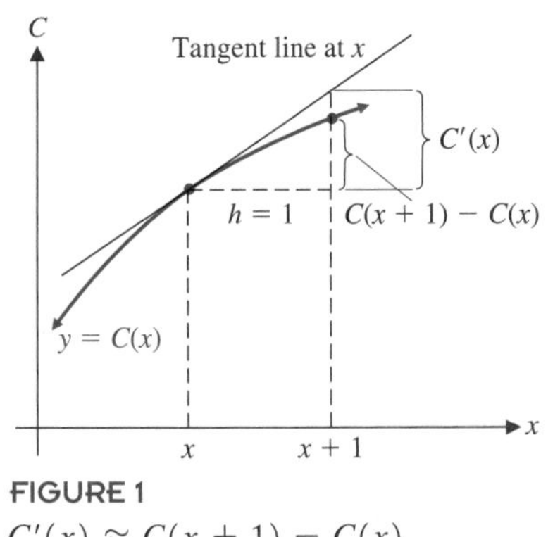

FIGURE 1
$C'(x) \approx C(x + 1) - C(x)$

Marginal Cost and Exact Cost

If $C(x)$ is the total cost of producing x items, then the marginal cost function approximates the exact cost of producing the $(x + 1)$st item:

$$\underset{\text{Marginal cost}}{C'(x)} \approx \underset{\text{Exact cost}}{C(x + 1) - C(x)}$$

Similar interpretations can be made for total revenue functions and total profit functions.

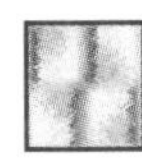

Example 1 **Cost Analysis** A company manufactures fuel tanks for automobiles. The total weekly cost (in dollars) of producing x tanks is given by

$$C(x) = 10{,}000 + 90x - 0.05x^2$$

(A) Find the marginal cost function.
(B) Find $C'(500)$, and discuss the various interpretations of this result.
(C) Find the exact cost of producing the 501st tank, and discuss the relationship between this result and the marginal cost found in part (B).

C Check all calculations on a graphing utility.

SOLUTION (A) $C'(x) = 90 - 0.1x$
(B) $C'(500) = 90 - 0.1(500) = \$40$

The standard interpretation of a derivative certainly applies to marginal cost functions. Thus, at a production level of 500 tanks per week, total production costs are increasing at the rate of \$40 per tank. The special interpretation of a marginal cost function also applies. That is, \$40 is the approximate cost of producing the 501st tank.

```
nDeriv(Y1,X,500)
                40
500→X:Y1
             42500
501→X:Y1
          42539.95
```

FIGURE 2
$y_1 = 10{,}000 + 90x - 0.05x^2$

(C) $C(501) = 10{,}000 + 90(501) - 0.05(501)^2$
$= \$42{,}539.95$ Total cost of producing 501 tanks per week
$C(500) = 10{,}000 + 90(500) - 0.05(500)^2$
$= \$42{,}500.00$ Total cost of producing 500 tanks per week
$C(501) - C(500) = 42{,}539.95 - 42{,}500.00$
$= \$39.95$ Exact cost of producing the 501st tank

Comparing this result with the marginal cost found in part (B), we see that the marginal cost does provide a good approximation to the exact cost of producing the 501st tank.

See Figure 2 for a check of these calculations. ▪

Matched Problem 1 ➡ A company manufactures automatic transmissions for automobiles. The total weekly cost (in dollars) of producing x transmissions is given by

$$C(x) = 50{,}000 + 600x - 0.75x^2$$

(A) Find the marginal cost function.
(B) Find $C'(200)$, and discuss the various interpretations of this result.
(C) Find the exact cost of producing the 201st transmission, and discuss the relationship between this result and the marginal cost found in part (B).

C Check all calculations on a graphing utility. ▪

$C(x)$
\$50,000
\$40,000
\$30,000
\$20,000
\$10,000
Slope = $C'(500)$ = 40
Slope = $C'(200)$ = 70
200 400 600 800 x

FIGURE 3
$C(x) = 10{,}000 + 90x - 0.05x^2$

It is instructive to compute both the approximate cost $C'(x)$ and the exact cost $C(x + 1) - C(x)$ for comparison purposes. However, in actual practice the marginal cost is used much more frequently than the exact cost. One reason for this is that the marginal cost is easily visualized when examining the graph of the total cost function. Figure 3 shows the graph of the cost function discussed in Example 1 with tangent lines added at $x = 200$ and $x = 500$. The graph clearly shows that as production increases, the slope of the tangent line decreases. Thus, the cost of producing the next tank also decreases, a desirable characteristic of a total cost function. We will have much more to say about graphical analysis in the next chapter.

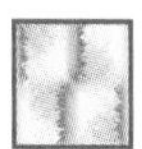

■ Application

We now want to discuss how price, demand, revenue, cost, and profit are tied together in typical applications. Although either price or demand can be used as the independent variable in a price–demand equation, it is common practice to use demand as the independent variable when marginal revenue, cost, and profit are also involved.

Explore–Discuss 1

The market research department of a company used test marketing to determine the demand for a new radio (Table 1).

Table 1

Demand x	Price p
3,000	\$7
6,000	\$4

(A) Assuming that the relationship between price p and demand x is linear, find the price–demand equation and write the result in the form $x = f(p)$. Graph the equation and find the domain of f. Discuss the effect of price increases and decreases on demand.

(B) Solve the equation found in part (A) for p, obtaining an equation of the form $p = g(x)$. Graph this equation and find the domain of g. Discuss the effect of price increases and decreases on demand.

 *Example 2* **Production Strategy** The market research department of a company recommends that the company manufacture and market a new transistor radio. After suitable test marketing, the research department presents the following **price–demand equation:**

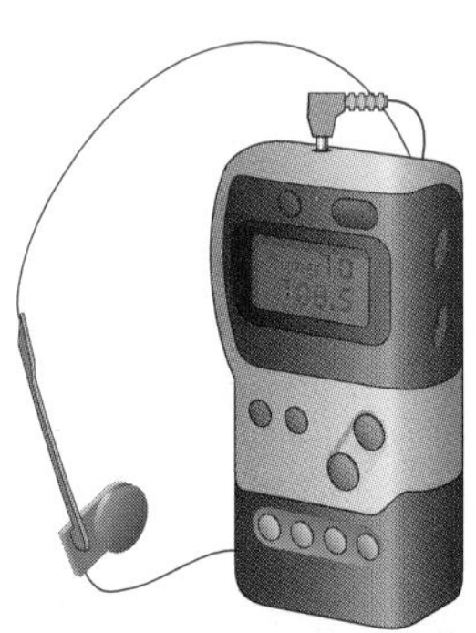

$$x = 10{,}000 - 1{,}000p \quad \text{x is demand at price } p \tag{1}$$

Or, solving (1) for p,

$$p = 10 - 0.001x \tag{2}$$

where x is the number of radios retailers are likely to buy at $\$p$ per radio.

The financial department provides the following **cost function:**

$$C(x) = 7{,}000 + 2x \tag{3}$$

where \$7,000 is the estimate of fixed costs (tooling and overhead) and \$2 is the estimate of variable costs per radio (materials, labor, marketing, transportation, storage, etc.).

(A) Find the domain of the function defined by the price–demand equation in (2).
(B) Find the marginal cost function $C'(x)$ and interpret.
(C) Find the revenue function as a function of x, and find its domain.
(D) Find the marginal revenue at $x = 2{,}000$, 5,000, and 7,000. Interpret these results.
(E) Graph the cost function and the revenue function in the same coordinate system, find the intersection points of these two graphs, and interpret the results.
(F) Find the profit function and its domain, and sketch its graph.
(G) Find the marginal profit at $x = 1{,}000$, 4,000, and 6,000. Interpret these results.

Solution

(A) Since price p and demand x must be nonnegative, we have $x \geqslant 0$ and

$$\begin{aligned} p = 10 - 0.001x &\geqslant 0 \\ 10 &\geqslant 0.001x \\ 10{,}000 &\geqslant x \end{aligned}$$

Thus, the permissible values of x are $0 \leqslant x \leqslant 10{,}000$.

(B) The marginal cost is $C'(x) = 2$. Since this is a constant, it costs an additional \$2 to produce one more radio at any production level.

(C) The **revenue** is the amount of money R received by the company for manufacturing and selling x radios at $\$p$ per radio and is given by

$$R = (\text{Number of radios sold})(\text{Price per radio}) = xp$$

In general, the revenue R can be expressed as a function of p by using equation (1) or as a function of x by using equation (2). As we mentioned earlier, when using marginal functions, we will always use the number of items x as the independent variable. Thus, the **revenue function** is

$$\begin{aligned} R(x) = xp &= x(10 - 0.001x) \quad \text{Using equation (2)} \\ &= 10x - 0.001x^2 \end{aligned} \tag{4}$$

Since equation (2) is defined only for $0 \leqslant x \leqslant 10{,}000$, it follows that the domain of the revenue function is $0 \leqslant x \leqslant 10{,}000$.

(D) The **marginal revenue** is

$$R'(x) = 10 - 0.002x$$

For production levels of $x = 2{,}000, 5{,}000$, and $7{,}000$, we have

$$R'(2{,}000) = 6 \qquad R'(5{,}000) = 0 \qquad R'(7{,}000) = -4$$

This means that at production levels of 2,000, 5,000, and 7,000, the respective approximate changes in revenue per unit change in production are \$6, \$0, and −\$4. That is, at the 2,000 output level, revenue increases as production increases; at the 5,000 output level, revenue does not change with a "small" change in production; and at the 7,000 output level, revenue decreases with an increase in production.

(E) When we graph $R(x)$ and $C(x)$ in the same coordinate system, we obtain Figure 4. The intersection points are called the **break-even points** because revenue equals cost at these production levels—the company neither makes nor loses money, but just breaks even. The break-even points are obtained as follows:

$$\begin{aligned} C(x) &= R(x) \\ 7{,}000 + 2x &= 10x - 0.001x^2 \\ 0.001x^2 - 8x + 7{,}000 &= 0 \\ x^2 - 8{,}000x + 7{,}000{,}000 &= 0 \end{aligned}$$

Solve using the quadratic formula (see Appendix A-9).

$$\begin{aligned} x &= \frac{8{,}000 \pm \sqrt{8{,}000^2 - 4(7{,}000{,}000)}}{2} \\ &= \frac{8{,}000 \pm \sqrt{36{,}000{,}000}}{2} \\ &= \frac{8{,}000 \pm 6{,}000}{2} \\ &= 1{,}000, \quad 7{,}000 \end{aligned}$$

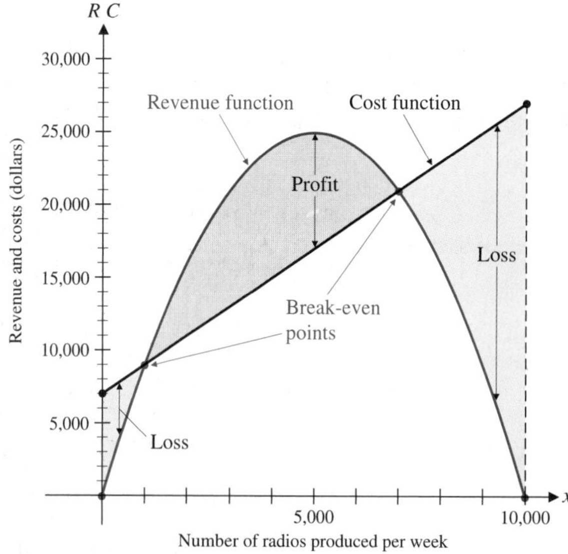

FIGURE 4

$$\begin{aligned} R(1{,}000) &= 10(1{,}000) - 0.001(1{,}000)^2 = 9{,}000 \\ C(1{,}000) &= 7{,}000 + 2(1{,}000) = 9{,}000 \\ R(7{,}000) &= 10(7{,}000) - 0.001(7{,}000)^2 = 21{,}000 \\ C(7{,}000) &= 7{,}000 + 2(7{,}000) = 21{,}000 \end{aligned}$$

Thus, the break-even points are (1,000, 9,000) and (7,000, 21,000), as shown in Figure 4. Further examination of the figure shows that cost is greater than revenue for production levels between 0 and 1,000 and also between 7,000 and 10,000. Consequently, the company incurs a loss at these levels. On the other hand, for production levels between 1,000 and 7,000, revenue is greater than cost and the company makes a profit.

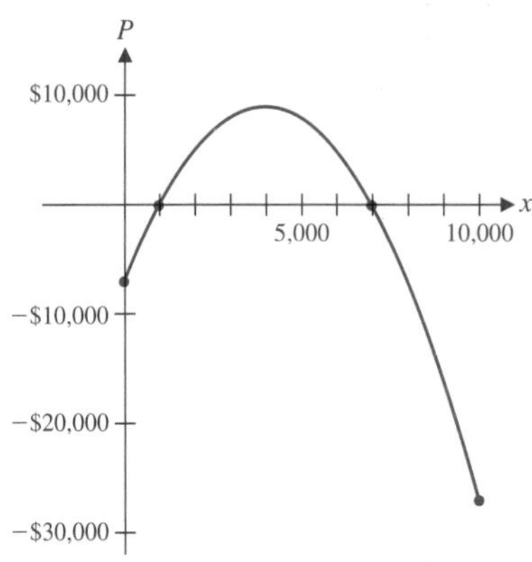

FIGURE 5

(F) The **profit function** is

$$\begin{aligned} P(x) &= R(x) - C(x) \\ &= (10x - 0.001x^2) - (7{,}000 + 2x) \\ &= -0.001x^2 + 8x - 7{,}000 \end{aligned}$$

The domain of the cost function is $x \geq 0$ and the domain of the revenue function is $0 \leq x \leq 10{,}000$. Thus, the domain of the profit function is the set of x values for which both functions are defined; that is, $0 \leq x \leq 10{,}000$. The graph of the profit function is shown in Figure 5. Notice that the x coordinates of the break-even points in Figure 4 are the x intercepts of the profit function. Furthermore, the intervals where cost is greater than revenue and where revenue is greater than cost correspond, respectively, to the intervals where profit is negative and the intervals where profit is positive.

(G) The **marginal profit** is

$$P'(x) = -0.002x + 8$$

For production levels of 1,000, 4,000, and 6,000, we have

$$P'(1{,}000) = 6 \qquad P'(4{,}000) = 0 \qquad P'(6{,}000) = -4$$

This means that at production levels of 1,000, 4,000, and 6,000, the respective approximate changes in profit per unit change in production are \$6, \$0, and −\$4. That is, at the 1,000 output level, profit will be increased if production is increased; at the 4,000 output level, profit does not change for "small" changes in production; and at the 6,000 output level, profits will decrease if production is increased. It seems the best production level to produce a maximum profit is 4,000.

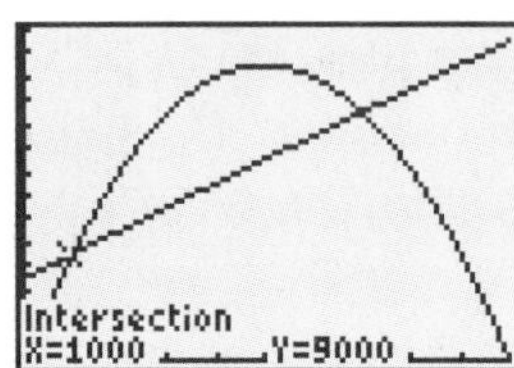

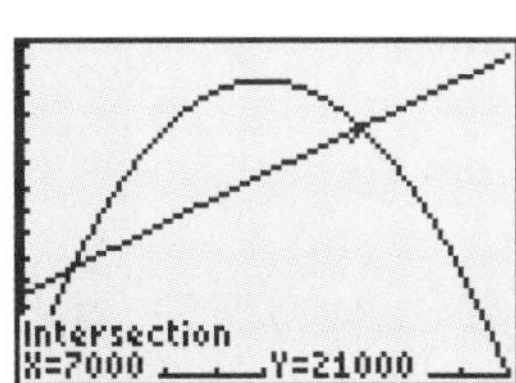

FIGURE 6
$y_1 = 7{,}000 + 2x$
$y_2 = 10x - 0.001x^2$

Break-even points can also be located by using graphical approximation on a graphing utility. The break-even points for Example 2 are shown in Figure 6.

Example 2 warrants careful study, since a number of important ideas in economics and calculus are involved. In the next chapter, we will develop a systematic procedure for finding the production level (and, using the demand equation, the selling price) that will maximize profit.

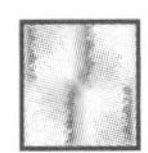

Matched Problem 2 Refer to the revenue and profit equations in Example 2.

(A) Find $R'(3{,}000)$ and $R'(6{,}000)$, and interpret the results.
(B) Find $P'(2{,}000)$ and $P'(7{,}000)$, and interpret the results.

Explore–Discuss 2

Let

$$C(x) = 12{,}000 + 5x \qquad \text{and} \qquad R(x) = 9x - 0.002x^2$$

Explain why $\neq$ is used below. Then find the correct expression for the profit function.

$$P(x) = R(x) - C(x) \neq 9x - 0.002x^2 - 12{,}000 + 5x$$

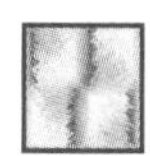

■ Marginal Average Cost, Revenue, and Profit

Sometimes, it is desirable to carry out marginal analysis relative to **average cost (cost per unit), average revenue (revenue per unit), and average profit (profit per unit).** The relevant definitions are summarized in the following box:

Marginal Average Cost, Revenue, and Profit

If x is the number of units of a product produced in some time interval, then

Cost per unit: $\quad$ **Average cost** $= \overline{C}(x) = \dfrac{C(x)}{x}$

Marginal average cost $= \overline{C}'(x) = \dfrac{d}{dx}\overline{C}(x)$

Revenue per unit: $\quad$ **Average revenue** $= \overline{R}(x) = \dfrac{R(x)}{x}$

Marginal average revenue $= \overline{R}'(x) = \dfrac{d}{dx}\overline{R}(x)$

Profit per unit: $\quad$ **Average profit** $= \overline{P}(x) = \dfrac{P(x)}{x}$

Marginal average profit $= \overline{P}'(x) = \dfrac{d}{dx}\overline{P}(x)$

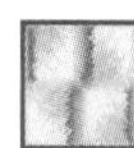

Example 3 **Cost Analysis** A small machine shop manufactures drill bits used in the petroleum industry. The shop manager estimates that the total daily cost (in dollars) of producing x bits is

$$C(x) = 1{,}000 + 25x - 0.1x^2$$

(A) Find $\overline{C}(x)$ and $\overline{C}'(x)$.
(B) Find $\overline{C}(10)$ and $\overline{C}'(10)$, and interpret.
(C) Use the results in part (B) to estimate the average cost per bit at a production level of 11 bits per day.

SOLUTION (A) $\overline{C}(x) = \dfrac{C(x)}{x} = \dfrac{1{,}000 + 25x - 0.1x^2}{x}$

$= \dfrac{1{,}000}{x} + 25 - 0.1x$ *Average cost function*

$\overline{C}'(x) = \dfrac{d}{dx}\overline{C}(x) = -\dfrac{1{,}000}{x^2} - 0.1$ *Marginal average cost function*

(B) $\overline{C}(10) = \dfrac{1{,}000}{10} + 25 - 0.1(10) = \$124$

$\overline{C}'(10) = -\dfrac{1{,}000}{10^2} - 0.1 = -\$10.10$

At a production level of 10 bits per day, the average cost of producing a bit is \$124, and this cost is decreasing at the rate of \$10.10 per bit.

(C) If production is increased by 1 bit, then the average cost per bit will decrease by approximately \$10.10. Thus, the average cost per bit at a production level of 11 bits per day is approximately \$124 − \$10.10 = \$113.90.

Matched Problem 3 Consider the cost function for the production of radios from Example 2:

$C(x) = 7{,}000 + 2x$

(A) Find $\overline{C}(x)$ and $\overline{C}'(x)$.
(B) Find $\overline{C}(100)$ and $\overline{C}'(100)$, and interpret.
(C) Use the results in part (B) to estimate the average cost per radio at a production level of 101 radios.

Explore–Discuss 3

A student produced the following solution to Matched Problem 3:

$C(x) = 7{,}000 + 2x$ *Cost*

$C'(x) = 2$ *Marginal cost*

$\dfrac{C'(x)}{x} = \dfrac{2}{x}$ *"Average" of the marginal cost*

Explain why the last function is not the same as the marginal average cost function.

CAUTION

1. The marginal average cost function must be computed by first finding the average cost function and then finding its derivative. As Explore–Discuss 3 illustrates, reversing the order of these two steps produces a different function that does not have any useful economic interpretations.
2. Recall that the marginal cost function has two interpretations: the usual interpretation of any derivative as an instantaneous rate of change, and the special interpretation as an approximation to the exact cost of the $(x + 1)$st item. This special interpretation does not apply to the marginal

average cost function. Referring to Example 3, it would be incorrect to interpret $\overline{C}'(10) = -\$10.10$ to mean that the average cost of the next bit is approximately $-\$10.10$. In fact, the phrase "average cost of the next bit" does not even make sense. Averaging is a concept applied to a collection of items, not to a single item.

These remarks also apply to revenue and profit functions.

Answers to Matched Problems

1. (A) $C'(x) = 600 - 1.5x$
(B) $C'(200) = 300$. At a production level of 200 transmissions, total costs are increasing at the rate of \$300 per transmission. Also, the approximate cost of producing the 201st transmission is \$300.
(C) $C(201) - C(200) = \$299.25$. The marginal cost from part (B) provides a good approximation to this exact cost.

2. (A) $R'(3{,}000) = 4$. At a production level of 3,000, a unit increase in production will increase revenue by approx. \$4.
$R'(6{,}000) = -2$. At a production level of 6,000, a unit increase in production will decrease revenue by approx. \$2.
(B) $P'(2{,}000) = 4$. At a production level of 2,000, a unit increase in production will increase profit by approx. \$4.
$P'(7{,}000) = -6$. At a production level of 7,000, a unit increase in production will decrease profit by approx. \$6.

3. (A) $\overline{C}(x) = \dfrac{7{,}000}{x} + 2; \overline{C}'(x) = -\dfrac{7{,}000}{x^2}$
(B) $\overline{C}(100) = \$72; \overline{C}'(100) = -\$0.70$. At a production level of 100 radios, the average cost per radio is \$72, and this average cost is decreasing at a rate of \$0.70 per radio.
(C) Approx. \$71.30.

EXERCISE 8-7

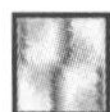

APPLICATIONS

Business & Economics

1. *Cost analysis.* The total cost (in dollars) of producing x food processors is

$$C(x) = 2{,}000 + 50x - 0.5x^2$$

(A) Find the exact cost of producing the 21st food processor.
(B) Use the marginal cost to approximate the cost of producing the 21st food processor.

2. *Cost analysis.* The total cost (in dollars) of producing x electric guitars is

$$C(x) = 1{,}000 + 100x - 0.25x^2$$

(A) Find the exact cost of producing the 51st guitar.
(B) Use the marginal cost to approximate the cost of producing the 51st guitar.

3. *Cost analysis.* The total cost (in dollars) of manufacturing x auto body frames is

$$C(x) = 60{,}000 + 300x$$

(A) Find the average cost per unit if 500 frames are produced.
(B) Find the marginal average cost at a production level of 500 units, and interpret the results.
(C) Use the results from parts (A) and (B) to estimate the average cost per frame if 501 frames are produced.

4. *Cost analysis.* The total cost (in dollars) of printing x dictionaries is

$$C(x) = 20{,}000 + 10x$$

(A) Find the average cost per unit if 1,000 dictionaries are produced.
(B) Find the marginal average cost at a production level of 1,000 units, and interpret the results.
(C) Use the results from parts (A) and (B) to estimate the average cost per dictionary if 1,001 dictionaries are produced.

5. *Revenue analysis.* The total revenue (in dollars) from the sale of x clock radios is

$$R(x) = 100x - 0.025x^2$$

Evaluate the marginal revenue at the given values of x, and interpret the results.
(A) $x = 1{,}600$ (B) $x = 2{,}500$

6. *Revenue analysis.* The total revenue (in dollars) from the sale of x steam irons is

$$R(x) = 50x - 0.05x^2$$

Evaluate the marginal revenue at the given values of x, and interpret the results.
(A) $x = 400$ (B) $x = 650$

7. *Profit analysis.* The total profit (in dollars) from the sale of x skateboards is

$$P(x) = 30x - 0.5x^2 - 250$$

(A) Find the exact profit from the sale of the 26th skateboard.
(B) Use the marginal profit to approximate the profit from the sale of the 26th skateboard.

8. *Profit analysis.* The total profit (in dollars) from the sale of x portable stereos is

$$P(x) = 22x - 0.1x^2 - 400$$

(A) Find the exact profit from the sale of the 41st stereo.
(B) Use the marginal profit to approximate the profit from the sale of the 41st stereo.

9. *Profit analysis.* The total profit (in dollars) from the sale of x video cassettes is

$$P(x) = 5x - 0.005x^2 - 450$$

Evaluate the marginal profit at the given values of x, and interpret the results.
(A) $x = 450$ (B) $x = 750$

10. *Profit analysis.* The total profit (in dollars) from the sale of x cameras is

$$P(x) = 12x - 0.02x^2 - 1{,}000$$

Evaluate the marginal profit at the given values of x, and interpret the results.
(A) $x = 200$ (B) $x = 350$

11. *Profit analysis.* The total profit (in dollars) from the sale of x lawn mowers is

$$P(x) = 30x - 0.03x^2 - 750$$

(A) Find the average profit per mower if 50 mowers are produced.
(B) Find the marginal average profit at a production level of 50 mowers and interpret.
(C) Use the results from parts (A) and (B) to estimate the average profit per mower if 51 mowers are produced.

12. *Profit analysis.* The total profit (in dollars) from the sale of x charcoal grills is

$$P(x) = 20x - 0.02x^2 - 320$$

(A) Find the average profit per grill if 40 grills are produced.
(B) Find the marginal average profit at a production level of 40 grills and interpret.
(C) Use the results from parts (A) and (B) to estimate the average profit per grill if 41 grills are produced.

13. *Revenue, cost, and profit.* The price–demand equation and the cost function for the production of table saws are given, respectively, by

$$p = 200 - \frac{x}{30} \quad \text{and} \quad C(x) = 72{,}000 + 60x$$

where x is the number of saws that can be sold at a price of $\$p$ per saw and $C(x)$ is the total cost (in dollars) of producing x saws.
(A) Find the marginal cost.
(B) Find the revenue function in terms of x.
(C) Find the marginal revenue.
(D) Find $R'(1{,}500)$ and $R'(4{,}500)$, and interpret the results.
(E) Graph the cost function and the revenue function on the same coordinate system for $0 \leq x \leq 6{,}000$. Find the break-even points, and indicate regions of loss and profit.
(F) Find the profit function in terms of x.
(G) Find the marginal profit.
(H) Find $P'(1{,}500)$ and $P'(3{,}000)$, and interpret the results.

14. *Revenue, cost, and profit.* The price–demand equation and the cost function for the production of television sets are given, respectively, by

$$p = 300 - \frac{x}{30} \quad \text{and} \quad C(x) = 150{,}000 + 30x$$

where x is the number of sets that can be sold at a price of \$$p$ per set and $C(x)$ is the total cost (in dollars) of producing x sets.
(A) Find the marginal cost.
(B) Find the revenue function in terms of x.
(C) Find the marginal revenue.
(D) Find $R'(3{,}000)$ and $R'(6{,}000)$, and interpret the results.
(E) Graph the cost function and the revenue function on the same coordinate system for $0 \le x \le 9{,}000$. Find the break-even points, and indicate regions of loss and profit.
(F) Find the profit function in terms of x.
(G) Find the marginal profit.
(H) Find $P'(1{,}500)$ and $P'(4{,}500)$, and interpret the results.

15. *Revenue, cost, and profit.* A company is planning to manufacture and market a new two-slice electric toaster. After conducting extensive market surveys, the research department provides the following estimates: a weekly demand of 200 toasters at a price of \$16 per toaster and a weekly demand of 300 toasters at a price of \$14 per toaster. The financial department estimates that weekly fixed costs will be \$1,400 and variable costs (cost per unit) will be \$4.

(A) Assume that the price–demand equation is linear. Use the research department's estimates to find the price–demand equation.
(B) Find the revenue function in terms of x.
(C) Assume that the cost function is linear. Use the financial department's estimates to find the cost function.
(D) Graph the cost function and the revenue function on the same coordinate system for $0 \le x \le 1{,}000$. Find the break-even points, and indicate regions of loss and profit.
(E) Find the profit function in terms of x.
(F) Evaluate the marginal profit at $x = 250$ and $x = 475$, and interpret the results.

16. *Revenue, cost, and profit.* The company in Problem 15 is also planning to manufacture and market a four-slice toaster. For this toaster, the research department's estimates are a weekly demand of 300 toasters at a price of \$25 per toaster and a weekly demand of 400 toasters at a price of \$20. The financial department's estimates are fixed weekly costs of \$5,000 and variable costs of \$5 per toaster. Assume the price–demand equation and cost function are linear (see Problem 15A and C).
(A) Use the research department's estimates to find the price–demand equation.
(B) Find the revenue function in terms of x.
(C) Use the financial department's estimates to find the cost function in terms of x.
(D) Graph the cost function and the revenue function on the same coordinate system for $0 \le x \le 800$. Find the break-even points, and indicate regions of loss and profit.
(E) Find the profit function in terms of x.
(F) Evaluate the marginal profit at $x = 325$ and $x = 425$, and interpret the results.

17. *Revenue, cost, and profit.* The total cost and the total revenue (in dollars) for the production and sale of x ski jackets are given, respectively, by

$$C(x) = 24x + 21{,}900 \quad \text{and} \quad R(x) = 200x - 0.2x^2 \quad 0 \le x \le 1{,}000$$

(A) Find the value of x where the graph of $R(x)$ has a horizontal tangent line.
(B) Find the profit function $P(x)$.
(C) Find the value of x where the graph of $P(x)$ has a horizontal tangent line.
(D) Graph $C(x)$, $R(x)$, and $P(x)$ on the same coordinate system for $0 \le x \le 1{,}000$. Find the break-even points. Find the x intercepts for the graph of $P(x)$.

18. *Revenue, cost, and profit.* The total cost and the total revenue (in dollars) for the production and sale of x hair dryers are given, respectively, by

$$C(x) = 5x + 2{,}340 \quad \text{and} \quad R(x) = 40x - 0.1x^2 \quad 0 \le x \le 400$$

(A) Find the value of x where the graph of $R(x)$ has a horizontal tangent line.
(B) Find the profit function $P(x)$.
(C) Find the value of x where the graph of $P(x)$ has a horizontal tangent line.
(D) Graph $C(x)$, $R(x)$, and $P(x)$ on the same coordinate system for $0 \le x \le 400$. Find the break-even points. Find the x intercepts for the graph of $P(x)$.

19. C *Break-even analysis.* The price–demand equation and the cost function for the production of garden hoses are given, respectively, by

$$p = 20 - \sqrt{x} \quad \text{and} \quad C(x) = 500 + 2x$$

where x is the number of garden hoses that can be sold at a price of $\$p$ per unit and $C(x)$ is the total cost (in dollars) of producing x garden hoses.

(A) Express the revenue function in terms of x.

(B) Graph the cost function and the revenue function in the same viewing window for $0 \leq x \leq 400$. Use approximation techniques to find the break-even points correct to the nearest unit.

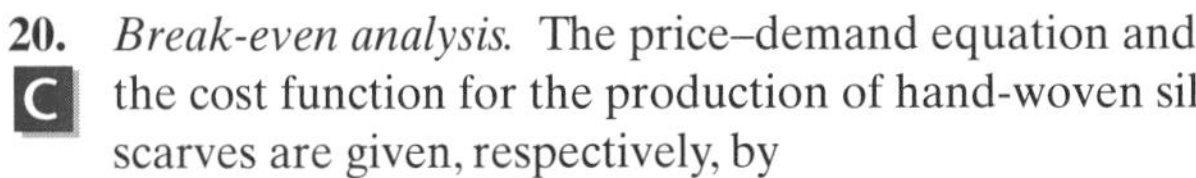

20. *Break-even analysis.* The price–demand equation and the cost function for the production of hand-woven silk scarves are given, respectively, by

$$p = 60 - 2\sqrt{x} \qquad \text{and} \qquad C(x) = 3{,}000 + 5x$$

where x is the number of scarves that can be sold at a price of $\$p$ per unit and $C(x)$ is the total cost (in dollars) of producing x scarves.

(A) Express the revenue function in terms of x.

(B) Graph the cost function and the revenue function in the same viewing window for $0 \leq x \leq 900$. Use approximation techniques to find the break-even points correct to the nearest unit.

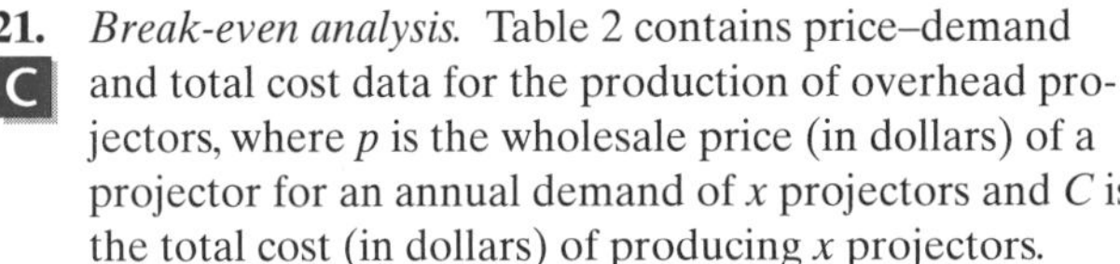

21. *Break-even analysis.* Table 2 contains price–demand and total cost data for the production of overhead projectors, where p is the wholesale price (in dollars) of a projector for an annual demand of x projectors and C is the total cost (in dollars) of producing x projectors.

Table 2

x	p (\$)	C (\$)
3,190	581	1,130,000
4,570	405	1,241,000
5,740	181	1,410,000
7,330	85	1,620,000

(A) Find a quadratic regression equation for the price–demand data using x as the independent variable.

(B) Find a linear regression equation for the cost data using x as the independent variable. Use this equation to estimate the fixed costs and the variable costs per projector. Round answers to the nearest dollar.

(C) Find the break-even points. Round answers to the nearest integer.

(D) Find the price range for which the company will make a profit. Round answers to the nearest dollar.

22. *Break-even analysis.* Table 3 contains price–demand and total cost data for the production of treadmills, where p is the wholesale price (in dollars) of a treadmill for an annual demand of x treadmills and C is the total cost (in dollars) of producing x treadmills.

Table 3

x	p (\$)	C (\$)
2,910	1,435	3,650,000
3,415	1,280	3,870,000
4,645	1,125	4,190,000
5,330	910	4,380,000

(A) Find a linear regression equation for the price–demand data using x as the independent variable.

(B) Find a linear regression equation for the cost data using x as the independent variable. Use this equation to estimate the fixed costs and the variable costs per treadmill. Round answers to the nearest dollar.

(C) Find the break-even points. Round answers to the nearest integer.

(D) Find the price range for which the company will make a profit. Round answers to the nearest dollar.

IMPORTANT TERMS AND SYMBOLS

8-1 *Rate of Change and Slope.* Average rate of change; difference quotient; average velocity; instantaneous velocity; instantaneous rate of change; secant line; slope of a secant line; slope of a graph; slope of a tangent line

$$\frac{f(a+h)-f(a)}{h}; \quad \lim_{h\to 0}\frac{f(a+h)-f(a)}{h}$$

8-2 *Limits.* Limit; absolute value function; one-sided limits; limit from the left; left-hand limit; limit from the right; right-hand limit; properties of limits; limit of a polynomial function; 0/0 indeterminate form

$$\lim_{x\to c} f(x); \quad \lim_{x\to c^-} f(x); \quad \lim_{x\to c^+} f(x)$$

8-3 *The Derivative.* Derivative; differentiation; differentiable; interpretations of the derivative; nondifferentiable; continuous

$$f'(x) = \lim_{h\to 0}\frac{f(x+h)-f(x)}{h}$$

8-4 *Derivatives of Constants, Power Forms, and Sums.* Derivative notation; derivative of a constant function rule; power rule; constant times a function rule; sum and difference rule; marginal cost function

$$f'(x); \quad y'; \quad \frac{dy}{dx}$$

8-5 *Derivatives of Products and Quotients.* Product rule; quotient rule

8-6 *Chain Rule: Power Form.* General power rule; chain rule; combining rules of differentiation

8-7 *Marginal Analysis in Business and Economics.* Marginal cost; marginal revenue; marginal profit; exact cost; price–demand equation; cost function; revenue function; marginal revenue; break-even points; profit function; marginal profit; average cost; marginal average cost; average revenue; marginal average revenue; average profit; marginal average profit

$$C'(x); \quad \overline{C}(x); \quad \overline{C}'(x); \quad R'(x); \quad \overline{R}(x); \quad \overline{R}'(x);$$
$$P'(x); \quad \overline{P}(x); \quad \overline{P}'(x)$$

Summary of Rules of Differentiation

$$\frac{d}{dx}k = 0$$
$$\frac{d}{dx}x^n = nx^{n-1}$$
$$\frac{d}{dx}kf(x) = kf'(x)$$
$$\frac{d}{dx}[u(x) \pm v(x)] = u'(x) \pm v'(x)$$
$$\frac{d}{dx}[F(x)S(x)] = F(x)S'(x) + S(x)F'(x)$$
$$\frac{d}{dx}\frac{T(x)}{B(x)} = \frac{B(x)T'(x) - T(x)B'(x)}{[B(x)]^2}$$
$$\frac{d}{dx}[u(x)]^n = n[u(x)]^{n-1}u'(x)$$

Review Exercise

Work through all the problems in this chapter review and check your answers in the back of the book. Answers to all review problems are there along with section numbers in italics to indicate where each type of problem is discussed. Where weaknesses show up, review appropriate sections in the text.

Many of the problems in this exercise set ask you to find a derivative. Most of the answers to these problems contain both an unsimplified form and a simplified form of the derivative. When checking your work, first check that you applied the rules correctly, and then check that you performed the algebraic simplification correctly.

A

1. Find the indicated quantities for $y = f(x) = 2x^2 + 5$:
   (A) The change in y if x changes from 1 to 3
   (B) The average rate of change of y with respect to x if x changes from 1 to 3
   (C) The slope of the secant line through the points $(1, f(1))$ and $(3, f(3))$ on the graph of $y = f(x)$
   (D) The instantaneous rate of change of y with respect to x at $x = 1$
   (E) The slope of the line tangent to the graph of $y = f(x)$ at $x = 1$
   (F) $f'(1)$

2. Use the graph of f shown below to approximate $f'(-1)$ and $f'(1)$ to the nearest integer.

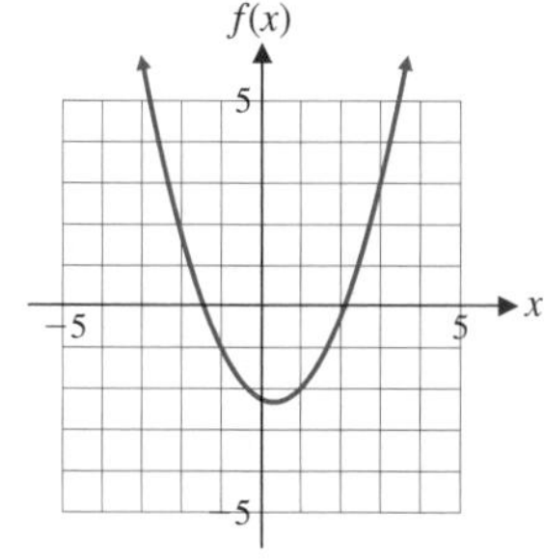

Figure for 2

3. If $\lim_{x\to1} f(x) = 2$ and $\lim_{x\to1} g(x) = 4$, find:
   (A) $\lim_{x\to1}(5f(x) + 3g(x))$ (B) $\lim_{x\to1}[f(x)g(x)]$
   (C) $\lim_{x\to1}\frac{g(x)}{f(x)}$

In Problems 4–6, use the graph of f shown below to estimate the indicated limits and function values.

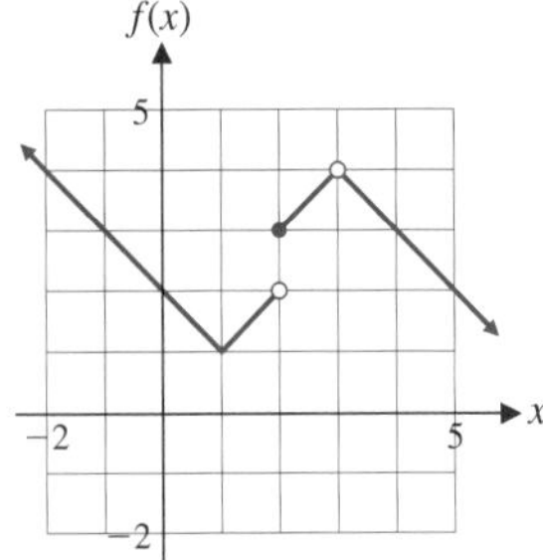

4. (A) $\lim_{x\to 1^-} f(x)$ (B) $\lim_{x\to 1^+} f(x)$
   (C) $\lim_{x\to 1} f(x)$ (D) $f(1)$
5. (A) $\lim_{x\to 2^-} f(x)$ (B) $\lim_{x\to 2^+} f(x)$
   (C) $\lim_{x\to 2} f(x)$ (D) $f(2)$
6. (A) $\lim_{x\to 3^-} f(x)$ (B) $\lim_{x\to 3^+} f(x)$
   (C) $\lim_{x\to 3} f(x)$ (D) $f(3)$

7. Use the following expression and the definition of derivative to find $f'(x)$:
$$f(x+h) - f(x) = 3x^2h + 3xh^2 + h^3 + 2xh$$

8. Given $f(5) = 4$, $f'(5) = -1$, $g(5) = 2$, and $g'(5) = -3$, find $h'(5)$ for each of the following functions:
   (A) $h(x) = 2f(x) + 3g(x)$ (B) $h(x) = f(x)g(x)$
   (C) $h(x) = \dfrac{f(x)}{g(x)}$ (D) $h(x) = [f(x)]^2$

9. Replace the ? in the equation below with an expression that makes the equation valid.
$$\frac{d}{dx}(3x^2 + 4x + 1)^5 = 5(3x^2 + 4x + 1)^4 \underline{\quad ? \quad}$$

In Problems 10–21, find f′(x) and simplify.

10. $f(x) = 3x^4 - 2x^2 + 1$
11. $f(x) = 2x^{1/2} - 3x$
12. $f(x) = 5$
13. $f(x) = \dfrac{1}{2x^2} + \dfrac{x^2}{2}$
14. $f(x) = \dfrac{0.5}{x^4} + 0.25x^4$
15. $f(x) = (2x - 1)(3x + 2)$
16. $f(x) = (x^2 - 1)(x^3 - 3)$
17. $f(x) = (0.2x - 1.5)(0.5x + 0.4)$
18. $f(x) = \dfrac{2x}{x^2 + 2}$
19. $f(x) = \dfrac{1}{3x + 2}$
20. $f(x) = (2x - 3)^3$
21. $f(x) = (x^2 + 2)^{-2}$

B

22. Let $f(x) = 0.5x^2 - 5$.
    (A) Find the slope of the secant line through $(2, f(2))$ and $(4, f(4))$.
    (B) Find the slope of the secant line through $(2, f(2))$ and $(2 + h, f(2 + h))$, $h \neq 0$.
    (C) Find the slope of the tangent line at $x = 2$.

23. Sketch a possible graph of a function $y = f(x)$ passing through the indicated points and having the indicated slope:

Point	Slope
(−3, −3)	−2
(−2, −4)	0
(0, 0)	4
(2, 4)	0
(3, 3)	−2

In Problems 24–32, find the indicated derivative and simplify.

24. $\dfrac{dy}{dx}$ for $y = 3x^4 - 2x^{-3} + 5$
25. y' for $y = (2x^2 - 3x + 2)(x^2 + 2x - 1)$
26. $f'(x)$ for $f(x) = \dfrac{2x - 3}{(x - 1)^2}$
27. y' for $y = 2\sqrt{x} + \dfrac{4}{\sqrt{x}}$
28. $g'(x)$ for $g(x) = 1.8\sqrt[3]{x} + \dfrac{0.9}{\sqrt[3]{x}}$
29. $\dfrac{d}{dx}[(x^2 - 1)(2x + 1)^2]$
30. $\dfrac{d}{dx}\sqrt[3]{x^3 - 5}$
31. $\dfrac{dy}{dx}$ for $y = \dfrac{3x^2 + 4}{x^2}$
32. $\dfrac{d}{dx}\dfrac{(x^2 + 2)^4}{2x - 3}$

33. For $y = f(x) = x^2 + 4$, find:
    (A) The slope of the graph at $x = 1$
    (B) The equation of the tangent line at $x = 1$ in the form $y = mx + b$

34. Repeat Problem 33 for $f(x) = x^3(x + 1)^2$.

In Problems 35–38, find the value(s) of x where the tangent line is horizontal.

35. $f(x) = 10x - x^2$
36. $f(x) = (x + 3)(x^2 - 45)$
37. $f(x) = \dfrac{x}{x^2 + 4}$
38. $f(x) = x^2(2x - 15)^3$

In Problems 39–41, approximate (to two decimal places) the value(s) of x where the graph of f has a horizontal tangent line.

39. $f(x) = x^4 - x^3 - 4x^2 + 5x$

40. $f(x) = \dfrac{x^3 + 20x + 1}{x^2 + 2}$

41. $f(x) = \dfrac{x^2 + x}{(x^2 + 1)^2}$

42. If an object moves along the y axis (scale in feet) so that it is at $y = f(x) = 16x^2 - 4x$ at time x (in seconds), find:
(A) The instantaneous velocity function
(B) The velocity at time $x = 3$ seconds

43. An object moves along the y axis (scale in feet) so that at time x (in seconds) it is at $y = f(x) = 96x - 16x^2$. Find:
(A) The instantaneous velocity function
(B) The time(s) when the velocity is 0

44. Complete the following table for $f(x) = 5^x$, and estimate $f'(0)$ to two decimal places.

h	-0.1	-0.01	-0.001	$\to 0 \leftarrow$	0.001	0.01	0.1
$\dfrac{f(h) - f(0)}{h}$	?	?	?	→?←	?	?	?

45. C Graph $f(x) = 4^{-x}$, zoom in on the graph near $x = 0$, and use secant line slopes to approximate $f'(0)$ to two decimal places.

46. Let $f(x) = x^3$, $g(x) = (x - 4)^3$, and $h(x) = (x + 3)^3$.
(A) How are the graphs of f, g, and h related? Illustrate your conclusion by graphing f, g, and h on the same coordinate axes.
(B) How would you expect the graphs of the derivatives of these functions to be related? Illustrate your conclusion by graphing f', g', and h' on the same coordinate axes.

47. Let $f(x)$ be a differentiable function and let k be a nonzero constant. For each function g, write a brief verbal description of the relationship between the graphs of f and g. Do the same for the graphs of f' and g'.
(A) $g(x) = f(x + k)$ (B) $g(x) = f(x) + k$

In Problems 48–57, find each limit, if it exists.

48. $\lim\limits_{x \to 3} \dfrac{2x - 3}{x + 5}$

49. $\lim\limits_{x \to 3} (2x^2 - x + 1)$

50. $\lim\limits_{x \to 0} \dfrac{2x}{3x^2 - 2x}$

51. $\lim\limits_{x \to 3} \dfrac{x - 3}{x^2 - 9}$

52. $\lim\limits_{x \to 4^-} \dfrac{|x - 4|}{x - 4}$

53. $\lim\limits_{x \to 4^+} \dfrac{|x - 4|}{x - 4}$

54. $\lim\limits_{x \to 4} \dfrac{|x - 4|}{x - 4}$

55. $\lim\limits_{h \to 0} \dfrac{[(2 + h)^2 - 1] - [2^2 - 1]}{h}$

56. $\lim\limits_{h \to 0} \dfrac{f(2 + h) - f(2)}{h}$ for $f(x) = x^2 + 4$

57. $\lim\limits_{h \to 0} \dfrac{f(x + h) - f(x)}{h}$ for $f(x) = \dfrac{1}{x + 2}$

58. C Let

$$f(x) = \frac{x^3 - 4x^2 - 4x + 16}{|x^2 - 4|}$$

Graph f and use zoom and trace to investigate the left- and right-hand limits at the indicated values of c.
(A) $c = -2$ (B) $c = 0$ (C) $c = 2$

In Problems 59 and 60, use the definition of the derivative and the two step-process to find $f'(x)$.

59. $f(x) = x^2 - x$

60. $f(x) = \sqrt{x} - 3$

C *Problems 61–64 refer to the function f in the figure. Determine whether f is differentiable at the indicated value of x.*

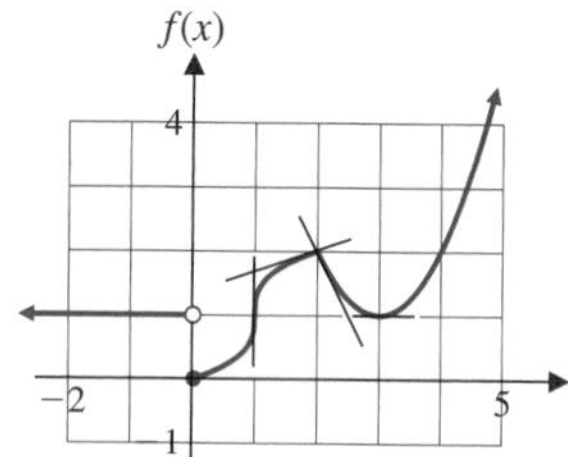

61. $x = 0$ **62.** $x = 1$ **63.** $x = 2$ **64.** $x = 3$

In Problems 65–68, find $f'(x)$ and simplify.

65. $f(x) = (x - 4)^4(x + 3)^3$

66. $f(x) = \dfrac{x^5}{(2x + 1)^4}$

67. $f(x) = \dfrac{\sqrt{x^2 - 1}}{x}$

68. $f(x) = \dfrac{x}{\sqrt{x^2 + 4}}$

69. The domain of the power function $f(x) = x^{1/5}$ is the set of all real numbers. Find the domain of the derivative $f'(x)$. Discuss the nature of the graph of $y = f(x)$ for any x values excluded from the domain of $f'(x)$.

70. Let f be defined by

$$f(x) = \begin{cases} x^2 - m & \text{if } x \le 1 \\ -x^2 + m & \text{if } x > 1 \end{cases}$$

where m is a constant.

(A) Graph f for $m = 0$, and find

$$\lim_{x \to 1^-} f(x) \quad \text{and} \quad \lim_{x \to 1^+} f(x)$$

(B) Graph f for $m = 2$, and find

$$\lim_{x \to 1^-} f(x) \quad \text{and} \quad \lim_{x \to 1^+} f(x)$$

(C) Find m so that

$$\lim_{x \to 1^-} f(x) = \lim_{x \to 1^+} f(x)$$

and graph f for this value of m.

(D) Write a brief verbal description of each graph. How does the graph in part (C) differ from the graphs in parts (A) and (B)?

71. Let $f(x) = 1 - |x - 1|, 0 \leq x \leq 2$ (see the figure).

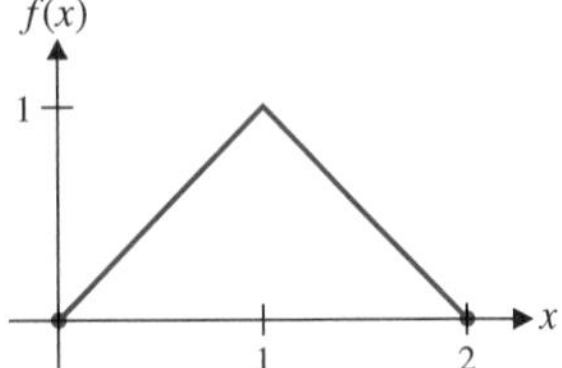

Figure for 71

(A) $\lim_{h \to 0^-} \frac{f(1 + h) - f(1)}{h} = ?$

(B) $\lim_{h \to 0^+} \frac{f(1 + h) - f(1)}{h} = ?$

(C) $\lim_{h \to 0} \frac{f(1 + h) - f(1)}{h} = ?$

(D) Does $f'(1)$ exist?

Applications

Business & Economics

72. *Cost analysis.* The total cost (in dollars) of producing x television sets is

$$C(x) = 10{,}000 + 200x - 0.1x^2$$

(A) Find the exact cost of producing the 101st television set.

(B) Use the marginal cost to approximate the cost of producing the 101st television set.

73. *Cost analysis.* The total cost (in dollars) of producing x bicycles is

$$C(x) = 5{,}000 + 40x + 0.05x^2$$

(A) Find the total cost and the marginal cost at a production level of 100 bicycles, and interpret the results.

(B) Find the average cost and the marginal average cost at a production level of 100 bicycles, and interpret the results.

74. *Cost analysis.* The total cost (in dollars) of producing x laser printers per week is shown in the figure. Which is greater, the approximate cost of producing the 201st printer or the approximate cost of producing the 601st printer? Does this graph represent a manufacturing process that is becoming more efficient or less efficient as production levels increase? Explain.

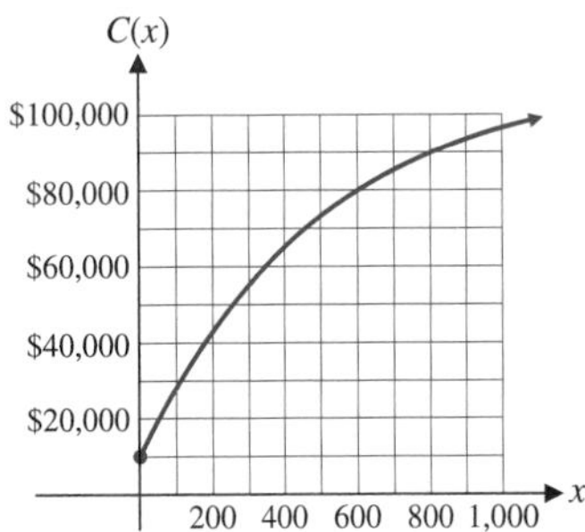

Figure for 74

75. *Cost analysis.* Let

$$p = 25 - 0.01x \quad \text{and} \quad C(x) = 2x + 9{,}000$$
$$0 \leq x \leq 2{,}500$$

be the price–demand equation and the cost function, respectively, for the manufacture of umbrellas.

(A) Find the marginal cost, average cost, and marginal average cost functions.

(B) Express the revenue in terms of x, and find the marginal revenue, average revenue, and marginal average revenue functions.

(C) Find the profit, marginal profit, average profit, and marginal average profit functions.

(D) Find the break-even point(s).

(E) Evaluate the marginal profit at $x = 1{,}000$, 1,150, and 1,400, and interpret the results.

(F) Graph $R = R(x)$ and $C = C(x)$ on the same coordinate system, and locate regions of profit and loss.

76. *Employee training.* A company producing computer components has established that on the average, a new employee can assemble $N(t)$ components per day after t days of on-the-job training, as given by

$$N(t) = \frac{40t}{t + 2}$$

(A) Find the average rate of change of $N(t)$ from 3 days to 6 days.

(B) Find the instantaneous rate of change of $N(t)$ at 3 days.

77. *Sales analysis.* Past sales records for a swimming pool manufacturer indicate that the total number of swimming pools, N (in thousands), sold during a year are given by

$$N(t) = t\sqrt{4 + t}$$

where t is the number of months since the beginning of the year. Find $N(5)$ and $N'(5)$, and interpret.

78. *Compound interest.* If \$5,000 is invested in an account that earns 7% compounded annually, then the amount in the account after t years is given by

$$A(t) = 5{,}000(1.07)^t$$

(A) Find $A(10)$ and use secant line slopes to approximate $A'(10)$ to the nearest dollar.

(B) Write a brief verbal interpretation of the results in part (A).

79. C *Natural gas consumption.* The data in Table 1 give the U.S. consumption of natural gas in trillions of cubic feet.

Table 1

YEAR	NATURAL GAS CONSUMPTION
1960	12.7
1965	16.0
1970	21.9
1975	20.1
1980	19.7
1985	17.2
1990	18.6

(A) Let x represent time (in years), with $x = 0$ corresponding to 1960, and let y represent the corresponding U.S. consumption of natural gas. Enter the data set in a graphing utility and find a cubic regression equation for the data.

(B) If $y = N(x)$ denotes the regression equation found in part (A), find $N(35)$ and $N'(35)$, and write a brief verbal interpretation of these results.

80. C *Break-even analysis.* Table 2 contains price–demand and total cost data from a bakery for the production of kringles (a Danish pastry), where p is the price (in dollars) of a kringle for a daily demand of x kringles and C is the total cost (in dollars) of producing x kringles.

Table 2

x	p (\$)	C (\$)
125	9	740
140	8	785
170	7	850
200	6	900

(A) Find a linear regression equation for the price–demand data using x as the independent variable.

(B) Find a linear regression equation for the cost data using x as the independent variable. Use this equation to estimate the fixed costs and the variable costs per kringle.

(C) Find the break-even points.

(D) Find the price range for which the bakery will make a profit.

In all answers, round dollar amounts to the nearest cent.

Life Sciences

81. *Pollution.* A sewage treatment plant disposes of its effluent through a pipeline that extends 1 mile toward the center of a large lake. The concentration of effluent $C(x)$, in parts per million, x meters from the end of the pipe is given approximately by

$$C(x) = 500(x + 1)^{-2}$$

What is the instantaneous rate of change of concentration at 9 meters? At 99 meters?

82. *Medicine.* The body temperature (in degrees Fahrenheit) of a patient t hours after being given a fever-reducing drug is given by

$$F(t) = 98 + \frac{4}{\sqrt{t + 1}}$$

Find $F(3)$ and $F'(3)$. Write a brief verbal interpretation of these results.

Social Sciences

83. *Learning.* If a person learns N items in t hours, as given by

$$N(t) = 20\sqrt{t}$$

find the rate of learning after:

(A) 1 hour (B) 4 hours

84. *Population.* The number of married couples in the United States can be described by the function

$$M(t) = 40.7(1.01)^t$$

where M is number of married couples (in millions) and t is time in years since 1960.

(A) Find $M(50)$, and use secant line slopes to approximate $M'(50)$. Round both quantities to one decimal place.

(B) Write a brief verbal interpretation of the results in part (A).

(C) Use the results in part (B) to estimate the number of married couples in 2011 and in 2012.

Group Activity 1 *Minimal Average Cost*

If $C(x)$ is the total cost of producing x items, the marginal cost function $C'(x)$ gives the approximate cost of the next item produced, while the average cost function $\overline{C}(x)$ gives the average cost per item for the items already produced. Thus, $C'(x)$ looks forward to the next item, while $\overline{C}(x)$ looks backward at all the items produced thus far. Given this difference in viewpoint, it is somewhat surprising that there is an important relationship between these two functions. As we will see, information gained from comparing the values of $C'(x)$ and $\overline{C}(x)$ can help determine the production level that minimizes average cost.

(A) The total cost (in dollars) of producing x items is given by

$$C(x) = 0.01x^2 + 40x + 3{,}600$$

Find $C'(x)$ and $\overline{C}(x)$, and complete Table 1.

Table 1

x	$\overline{C}(x)$	$C'(x)$
100		
200		
300		
400		
500		
600		
700		
800		
900		
1,000		

(B) Repeat part (A) for

$$C(x) = 0.00016x^3 - 0.12x^2 + 30x + 10{,}000$$

(C) Examine the values in the tables from parts (A) and (B), and write a brief verbal description of the behavior of each function. Does each average cost function appear to have a minimum value? What is the minimal value, and where does it occur? What relationship do you observe between the minimum average cost and the marginal cost at the production level that minimizes average cost?

C (D) If you have access to a graphing utility, confirm your observations in part (C) by examining the graphs of $C'(x)$ and $\overline{C}(x)$.

(E) The following statements can help justify the relationship you observed in part (C). In each case, fill in the blank with "increase" or "decrease" and justify your choice.

1. If $C'(x) < \overline{C}(x)$ (that is, the cost of the next item is less than the average cost of the items already produced), then increasing production by 1 item will ________ the average cost.
2. If $C'(x) > \overline{C}(x)$ (that is, the cost of the next item is more than the average cost of the items already produced), then increasing production by 1 item will ________ the average cost.

(F) Discuss the validity of the following statement for an arbitrary cost function $C(x)$:

If the minimum value of $\overline{C}(x)$ occurs at a production level x, then $C'(x) = \overline{C}(x)$ at that production level.

(G) We used a quadratic function and a cubic function in parts (A) and (B) to illustrate the relationship between $C'(x)$ and $\overline{C}(x)$. But linear cost functions are one of the most important types. To see why we did not choose a linear cost function, try to parallel the above development for the cost function

$$C(x) = 30x + 12{,}000$$

Do any of your findings contradict the statements in parts (E) and (F)?

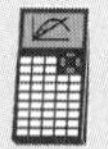

Group Activity 2 Numerical Differentiation on a Graphing Utility

Most graphing utilities have a built-in routine for approximating the derivative of a function, often denoted by nDeriv (check the manual for your graphing utility). For example, nDeriv(x^3,x,a) approximates the derivative of $y = x^3$ at a number a.

(A) Find nDeriv(x^3,x,a) on a graphing utility for $a = 1, 2, 3, 4$, and 5, and compare with the corresponding values of

$$\frac{d}{dx}x^3 = 3x^2$$

(B) Enter $y_1 = x^3$, $y_2 = 3x^2$, and $y_3 =$ nDeriv(y_1,x,x) in the equation editor of a graphing utility. Graph y_2 and y_3 for $-2 \le x \le 2$, $-2 \le y \le 2$, and discuss the relationship between these graphs. Use the trace feature or tables of values to support your conclusions.

Most graphing utilities use the following average of difference quotients with a fixed value of h to approximate a derivative:

$$\frac{1}{2}\left(\frac{f(x+h) - f(x)}{h} + \frac{f(x-h) - f(x)}{-h}\right) = \frac{f(x+h) - f(x-h)}{2h}$$

Thus, for a given number a and a fixed value of h,

$$\text{nDeriv}(f(x),x,a) = \frac{f(a+h) - f(a-h)}{2h} \tag{1}$$

(C) Use (1) to find and simplify nDeriv($f(x)$,x,a) for $f(x) = x^3$. Compare the values of the simplified form with the values of nDeriv you computed in part (A) to see if you can determine the fixed value of h for your graphing utility. Check your manual to see if you can change this value.

(D) Let $f(x) = |x|$. What is nDeriv($f(x)$,x,0)? Is $f(x)$ differentiable at $x = 0$?

(E) Repeat part (D) for $f(x) = 1/x^2$.

(F) When using nDeriv to approximate the derivative of a function $f(x)$, why is it important to know in advance the location of any points where $f'(x)$ does not exist?

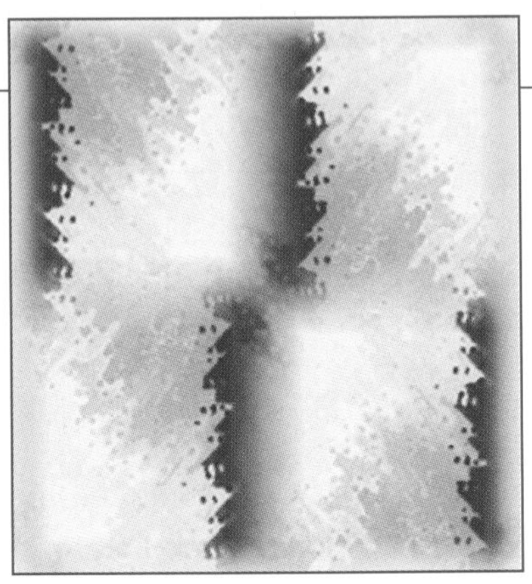

CHAPTER 9

GRAPHING AND OPTIMIZATION

INTRODUCTION

Historically, one of the primary reasons for studying calculus was the development of powerful tools for sketching graphs of functions. In fact, prior to the very recent advent of graphing utilities, hand-sketching was the only way most students could produce graphs. Now things have changed. An accurate graph of a function specified by an equation can be produced quickly and easily on a graphing utility. Does this mean that the old tools are no longer relevant? On the contrary, efficient use of graphing technology still requires a person at the helm, directing the electronic device, and this person still needs a sound foundation in graphing tools and analysis.

In the first four sections of this chapter, we will develop the graphing concepts and tools necessary to analyze, discuss, and sketch by hand the graphs of functions. The functions we consider may be specified by equations, or by information obtained from tables, verbal descriptions, or other graphs. In the last section, we will use these graphing tools to solve optimization problems, one of the most important applications of calculus to real-world problems.

A brief review of Section 2-1 would be helpful before beginning this chapter. In particular, we will make frequent use of the following important property of polynomials:

The graph of a polynomial function of positive degree n can have at most $n - 1$ turning points and at most n real zeros or x intercepts.

SECTION 9-1 Continuity and Graphs

- CONTINUITY
- CONTINUITY PROPERTIES
- INFINITE LIMITS
- SOLVING INEQUALITIES USING CONTINUITY PROPERTIES

In this section we return to the limit concept and use it to study an important property of functions called *continuity.* An understanding of continuity is essential for sketching and analyzing graphs. We will also see that continuity provides a simple and efficient method for solving inequalities, a tool that we will use extensively in later sections.

CONTINUITY

Compare the graphs shown in Figure 1, which were discussed in Examples 2–4 in Section 8-2. Notice that two of the graphs are broken; that is, they cannot be drawn without lifting a pen off the paper. Informally, a function is *continuous over an interval* if its graph over the interval can be drawn without removing a pen from the paper. A function whose graph is broken (disconnected) at $x = c$ is said to be *discontinuous* at $x = c$. Function f (Fig. 1A) is continuous for all x. Function g (Fig. 1B) is discontinuous at $x = 2$, but is continuous over any interval that does not include 2. Function h (Fig. 1C) is discontinuous at $x = 0$, but is continuous over any interval that does not include 0.

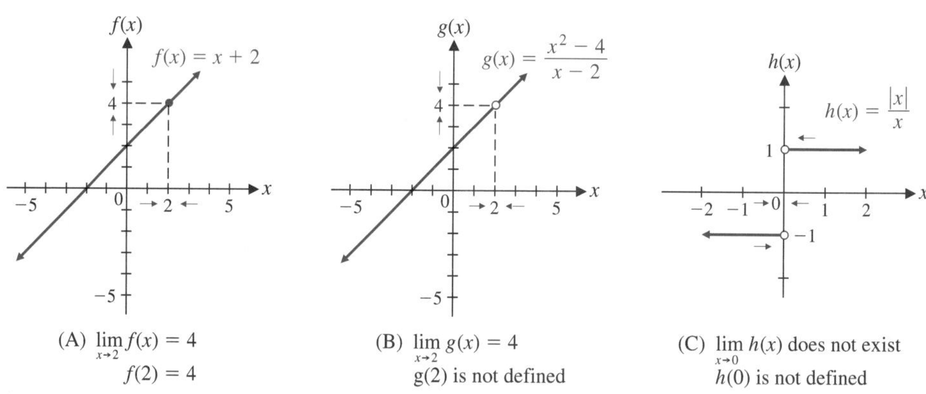

(A) $\lim_{x \to 2} f(x) = 4$, $f(2) = 4$

(B) $\lim_{x \to 2} g(x) = 4$, $g(2)$ is not defined

(C) $\lim_{x \to 0} h(x)$ does not exist, $h(0)$ is not defined

FIGURE 1

Most graphs of natural phenomena are continuous, whereas many graphs in business and economics applications have discontinuities. Figure 2A illustrates temperature variation over a 24 hour period—a continuous phenomenon. Figure 2B illustrates warehouse inventory over a 1 week period—a discontinuous phenomenon.

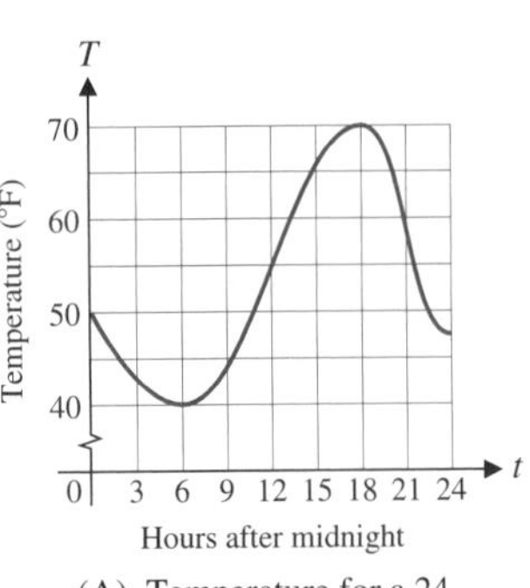

(A) Temperature for a 24 hour period

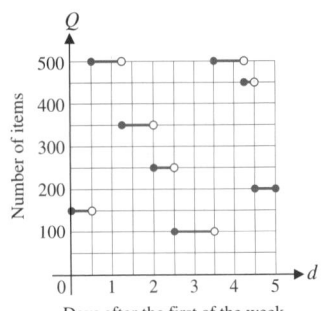

(B) Inventory in a warehouse during 1 week

FIGURE 2

Explore–Discuss 1

(A) Write a brief verbal description of the temperature variation illustrated in Figure 2A, including estimates of the high and low temperatures during this period and the times at which they occurred.
(B) Write a brief verbal description of the changes in inventory illustrated in Figure 2B, including estimates of the changes in inventory and the times at which these changes occurred.

The preceding discussion leads to the following formal definition of continuity:

Continuity

A function f is **continuous at the point $x = c$** if

1. $\lim_{x \to c} f(x)$ exists 2. $f(c)$ exists 3. $\lim_{x \to c} f(x) = f(c)$

A function is **continuous on the open interval* (a, b)** if it is continuous at each point on the interval.

*See Appendix A-8 for a review of interval notation.

If one or more of the three conditions in the definition fails, then the function is **discontinuous** at $x = c$.

Explore–Discuss 2

Sketch a graph of a function that is discontinuous at a point because it fails to satisfy condition 1 in the definition of continuity. Repeat for conditions 2 and 3.

Example 1 **Continuity of a Function Defined by a Graph** Use the definition of continuity to discuss the continuity of the function whose graph is shown in Figure 3.

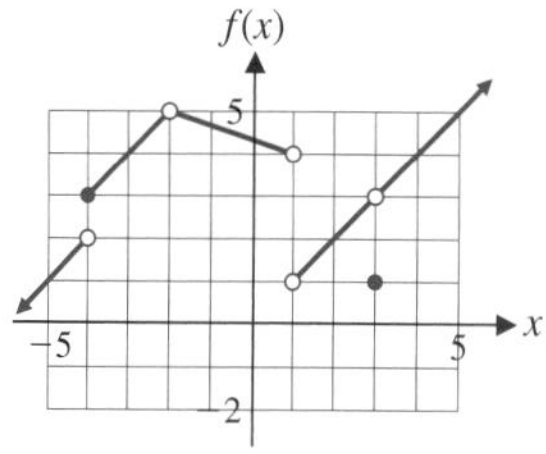

FIGURE 3

Solution We begin by identifying the points of discontinuity. Examining the graph, we see breaks and/or holes in the graph at $x = -4, -2, 1$, and 3. Now we must determine which conditions in the definition of continuity are not satisfied at each of these points. In each case, we find the value of the function and the limit of the function at the point in question.

Discontinuity at $x = -4$:

$$\lim_{x\to -4^-} f(x) = 2$$
$$\lim_{x\to -4^+} f(x) = 3$$
$$\lim_{x\to -4} f(x) \text{ does not exist}$$
$$f(-4) = 3$$

Since the one-sided limits are different, the limit does not exist (Section 8-2).

Thus, f is not continuous at $x = -4$ because condition 1 is not satisfied.

Discontinuity at $x = -2$:

$$\lim_{x\to -2^-} f(x) = 5$$
$$\lim_{x\to -2^+} f(x) = 5$$
$$\lim_{x\to -2} f(x) = 5$$
$$f(-2) \text{ does not exist}$$

The hole at $(-2, 5)$ indicates that 5 is not the value of f at -2. Since there is no solid dot elsewhere on the vertical line $x = -2$, $f(-2)$ is not defined.

Thus, f is not continuous at $x = -2$ because condition 2 is not satisfied.

Discontinuity at $x = 1$:

$$\lim_{x\to 1^-} f(x) = 4$$
$$\lim_{x\to 1^+} f(x) = 1$$
$$\lim_{x\to 1} f(x) \text{ does not exist}$$
$$f(1) \text{ does not exist}$$

This time, f is not continuous at $x = 1$ because both conditions 1 and 2 are not satisfied.

Discontinuity at $x = 3$:

$$\lim_{x\to 3^-} f(x) = 3$$
$$\lim_{x\to 3^+} f(x) = 3$$
$$\lim_{x\to 3} f(x) = 3$$
$$f(3) = 1$$

The solid dot at $(3, 1)$ indicates that $f(3) = 1$.

Conditions 1 and 2 are satisfied, but f is not continuous at $x = 3$ because condition 3 is not satisfied.

Having identified and discussed all points of discontinuity, we can now conclude that f is continuous except at $x = -4, -2, 1$, and 3.

Matched Problem 1 Use the definition of continuity to discuss the continuity of the function whose graph is shown in Figure 4.

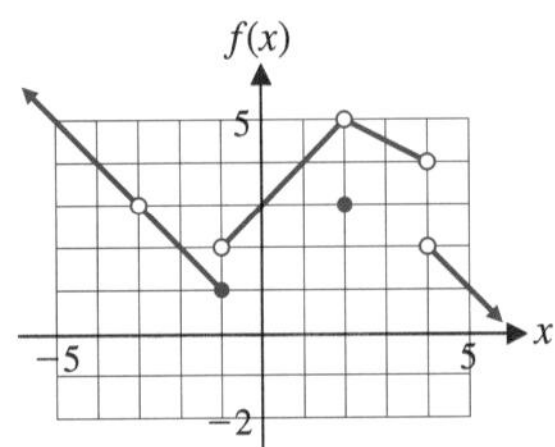

FIGURE 4

For functions defined by equations, it is also important to be able to locate points of discontinuity by examining the equation.

Example 2 **Continuity of Functions Defined by Equations** Using the definition of continuity, discuss the continuity of each function at the indicated point(s).

(A) $f(x) = x + 2$ at $x = 2$ (B) $g(x) = \dfrac{x^2 - 4}{x - 2}$ at $x = 2$

(C) $h(x) = \dfrac{|x|}{x}$ at $x = 0$ and at $x = 1$

SOLUTION (A) f is continuous at $x = 2$, since

$$\lim_{x\to 2} f(x) = 4 = f(2) \quad \text{See Figure 1A.}$$

(B) g is not continuous at $x = 2$, since $g(2) = 0/0$ is not defined. (See Fig. 1B.)

(C) h is not continuous at $x = 0$, since $h(0) = |0|/0$ is not defined; also, $\lim_{x\to 0} h(x)$ does not exist.

h is continuous at $x = 1$, since

$$\lim_{x\to 1} \frac{|x|}{x} = 1 = h(1) \quad \text{See Figure 1C.}$$

Matched Problem 2 Using the definition of continuity, discuss the continuity of each function at the indicated point(s).

(A) $f(x) = x + 1$ at $x = 1$ (B) $g(x) = \dfrac{x^2 - 1}{x - 1}$ at $x = 1$

(C) $h(x) = \dfrac{x - 2}{|x - 2|}$ at $x = 2$ and at $x = 0$

We can also talk about one-sided continuity, just as we talked about one-sided limits. For example, a function is said to be **continuous on the right** at $x = c$ if $\lim_{x\to c^+} f(x) = f(c)$ and **continuous on the left** at $x = c$ if

$\lim_{x \to c^-} f(x) = f(c)$. A function is **continuous on the closed interval [*a*, *b*]** if it is continuous on the open interval (a, b) and is continuous on the right at a and continuous on the left at b.

Figure 5A illustrates a function that is continuous on the closed interval $[-1, 1]$. Figure 5B illustrates a function that is continuous on a half-closed interval $[0, \infty)$.

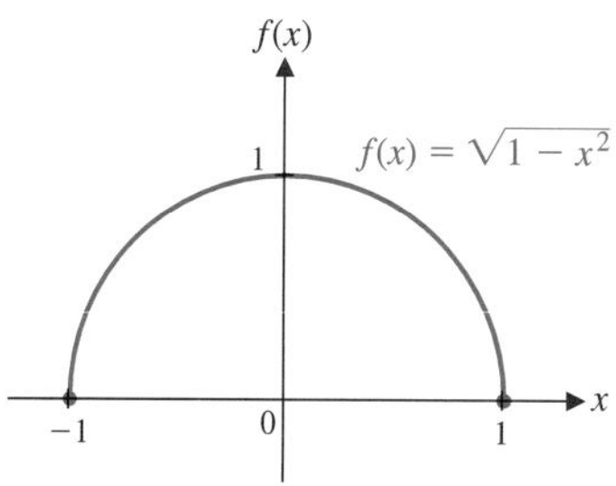

(A) f is continuous on the closed interval $[-1, 1]$

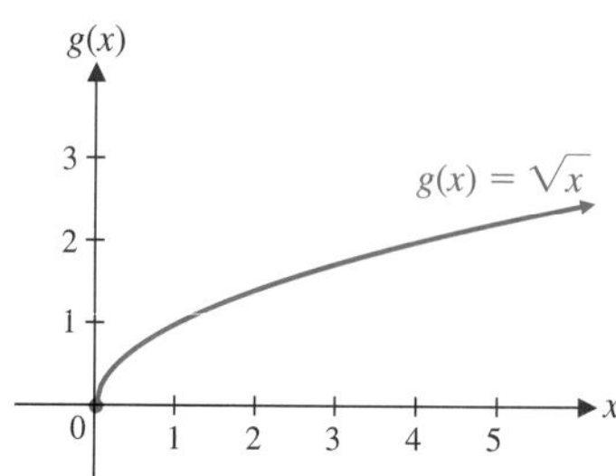

(B) g is continuous on the half-closed interval $[0, \infty)$

FIGURE 5
Continuity on closed and half-closed intervals

CONTINUITY PROPERTIES

Functions have some useful **general continuity properties:**

> **If two functions are continuous on the same interval, then their sum, difference, product, and quotient are continuous on the same interval, except for values of x that make a denominator 0.**

These properties, along with Theorem 1 below, enable us to determine intervals of continuity for some important classes of functions without having to look at their graphs or use the three conditions in the definition.

Theorem 1 CONTINUITY PROPERTIES OF SOME SPECIFIC FUNCTIONS

(A) A constant function $f(x) = k$, where k is a constant, is continuous for all x.

$f(x) = 7$ is continuous for all x.

(B) For n a positive integer, $f(x) = x^n$ is continuous for all x.

$f(x) = x^5$ is continuous for all x.

(C) A polynomial function is continuous for all x.

$2x^3 - 3x^2 + x - 5$ is continuous for all x.

(D) A rational function is continuous for all x except those values that make a denominator 0.

$\frac{x^2 + 1}{x - 1}$ is continuous for all x except $x = 1$, a value that makes the denominator 0.

(E) For n an odd positive integer greater than 1, $\sqrt[n]{f(x)}$ is continuous wherever $f(x)$ is continuous.

$\sqrt[3]{x^2}$ is continuous for all x.

(F) For n an even positive integer, $\sqrt[n]{f(x)}$ is continuous wherever $f(x)$ is continuous and nonnegative.

$\sqrt[4]{x}$ is continuous on the interval $[0, \infty)$.

Notice that Theorem 1C follows from 1A, 1B, and the general continuity properties stated above. Also, note that Theorem 1D follows from 1C and the general continuity properties, since a rational function is a function that can be expressed as the quotient of two polynomials.

Example 3 **Using Continuity Properties** Using Theorem 1 and the general properties of continuity, determine where each function is continuous.

(A) $f(x) = x^2 - 2x + 1$ (B) $f(x) = \frac{x}{(x + 2)(x - 3)}$

(C) $f(x) = \sqrt[3]{x^2 - 4}$ (D) $f(x) = \sqrt{x - 2}$

SOLUTION
(A) Since f is a polynomial function, f is continuous for all x.
(B) Since f is a rational function, f is continuous for all x except -2 and 3 (values that make the denominator 0).
(C) The polynomial function $x^2 - 4$ is continuous for all x. Since $n = 3$ is odd, f is continuous for all x.
(D) The polynomial function $x - 2$ is continuous for all x and nonnegative for $x \geqslant 2$. Since $n = 2$ is even, f is continuous for $x \geqslant 2$, or on the interval $[2, \infty)$.

Matched Problem 3 Using Theorem 1 and the general properties of continuity, determine where each function is continuous.

(A) $f(x) = x^4 + 2x^2 + 1$ (B) $f(x) = \frac{x^2}{(x + 1)(x - 4)}$

(C) $f(x) = \sqrt{x - 4}$ (D) $f(x) = \sqrt[3]{x^3 + 1}$

INFINITE LIMITS

A function is discontinuous at any point c where $\lim_{x\to c} f(x)$ does not exist. For example, if the one-sided limits are different at $x = c$, then the limit does not exist and the function is discontinuous at $x = c$ (see Fig. 3). Another situation where a limit may fail to exist involves functions whose values become very large as x approaches c. The special symbol ∞ is often used to describe this type of behavior. To illustrate this case, consider the functions

$$f(x) = \frac{1}{x} \quad \text{and} \quad g(x) = \frac{1}{x^2}$$

The graph of f in Figure 6A (page 612) indicates that the values of $f(x)$ are very large negative numbers if x is near 0 on the left and very large positive numbers if x is near 0 on the right. In neither case does the limit exist. However, it is convenient to use the special symbol ∞ to describe the nature of the graph at $x = 0$. Thus, we write

$$\lim_{x\to 0^-} f(x) = -\infty \quad \text{and} \quad \lim_{x\to 0^+} f(x) = \infty$$

Since these two statements represent different types of behavior, we cannot write a single limit statement to describe the nature of the graph at $x = 0$.

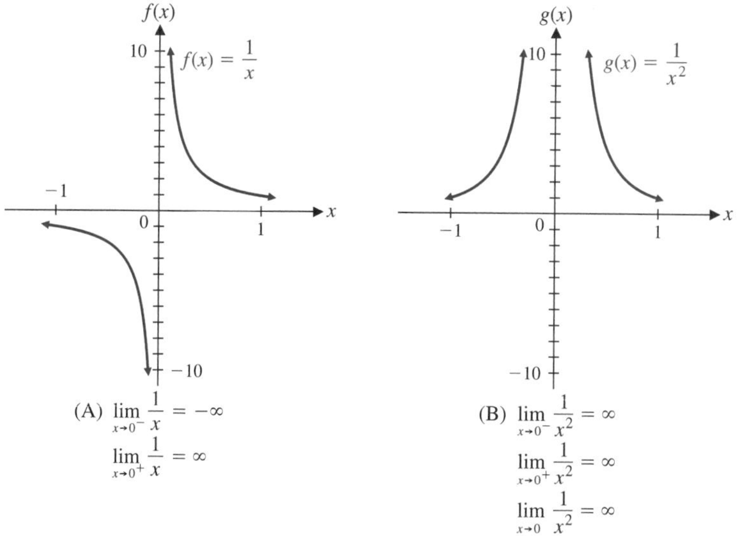

FIGURE 6

The graph of g in Figure 6B shows that the values of $g(x)$ are very large positive numbers when x is near 0 on either side of 0. Once again, the limit of $g(x)$ as x approaches 0 does not exist, but we can describe the behavior of the graph of g near 0 by writing

$$\lim_{x \to 0} g(x) = \infty$$

For both functions f and g, the line $x = 0$ (the vertical axis) is a *vertical asymptote.* These ideas are summarized in the following box.

Vertical Asymptotes

If the limit of a function f fails to exist as x approaches c from the left because the values of $f(x)$ are becoming very large positive numbers (or very large negative numbers), we say that*

$$\lim_{x \to c^-} f(x) = \infty \quad (\text{or } -\infty)$$

If this happens as x approaches c from the right, we say that

$$\lim_{x \to c^+} f(x) = \infty \quad (\text{or } -\infty)$$

If both one-sided limits exhibit the same behavior, we say that

$$\lim_{x \to c} f(x) = \infty \quad (\text{or } -\infty)$$

If any of the above hold, the line $x = c$ is a **vertical asymptote** for the graph of $y = f(x)$.

*The precise definition of this limit statement is as follows: $\lim_{x \to c^-} f(x) = \infty$ if for each $N > 0$, there exists a $d > 0$ such that $f(x) > N$ whenever $c - d < x < c$. Similar statements can be made for limits from the right, unrestricted limits, and limits involving very large negative numbers.

Example 4 **Limits at Points of Discontinuity** For the function

$$f(x) = \frac{1 - x}{x^4 - x^2}$$

use ∞ and $-\infty$, as appropriate, to describe the behavior at each point of discontinuity, and identify all vertical asymptotes.

SOLUTION First, we factor the denominator and identify the points of discontinuity:

$$x^4 - x^2 = x^2(x^2 - 1) = x^2(x - 1)(x + 1)$$

Thus, f is discontinuous at $x = -1, 0$, and 1. We use a numerical approach to investigate the behavior of $f(x)$ near each of these discontinuities.

Behavior at $x = -1$ [values of $f(x)$ rounded to the nearest integer]:

Table 1

x	-1.01	-1.001	$-1.0001 \rightarrow$	-1	$\leftarrow -0.9999$	-0.999	-0.99
$f(x)$	98	998	$9{,}998 \rightarrow \infty$		$-\infty \leftarrow -10{,}002$	$-1{,}002$	-102

Examining the values of $f(x)$ near $x = -1$ (Table 1), we see that the values of $f(x)$ are large positive numbers for x near -1 on the left and large negative numbers for x near -1 on the right. Thus, we must use one-sided limits to describe the behavior at $x = -1$:

$$\lim_{x \to -1^-} \frac{1 - x}{x^4 - x^2} = \infty \quad \text{and} \quad \lim_{x \to -1^+} \frac{1 - x}{x^4 - x^2} = -\infty$$

This shows that the line $x = -1$ is a vertical asymptote for the graph of $y = f(x)$.

Behavior at $x = 0$ [values of $f(x)$ rounded to the nearest integer]:

Table 2

x	-0.01	-0.001	$\rightarrow 0 \leftarrow$	0.001	0.01
$f(x)$	$-10{,}101$	$-1{,}001{,}001 \rightarrow$	$-\infty$	$\leftarrow -999{,}001$	$-9{,}901$

Since the values of $f(x)$ for x near 0, on either side of 0, are very large negative numbers (Table 2), we can use a single limit statement to describe the behavior at $x = 0$:

$$\lim_{x \to 0} \frac{1 - x}{x^4 - x^2} = -\infty$$

Thus, the line $x = 0$ (the y axis) is a vertical asymptote for the graph of $y = f(x)$.

Behavior at $x = 1$ [values of $f(x)$ rounded to three decimal places]:

Table 3

x	0.9	0.99	$0.999 \rightarrow$	1	$\leftarrow 1.001$	1.01	1.1
$f(x)$	-0.650	-0.513	$-0.501 \rightarrow$	-0.5	$\leftarrow -0.499$	-0.488	-0.394

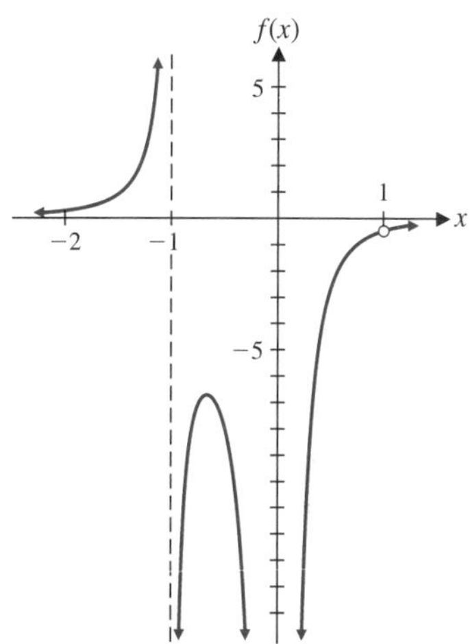

FIGURE 7

$f(x) = \frac{1 - x}{x^4 - x^2}$

The values in Table 3 suggest that $\lim_{x \to 1} f(x)$ exists. We confirm this by using algebraic simplification (notice that this limit is an indeterminate form):

$$\begin{aligned} \lim_{x \to 1} \frac{1 - x}{x^4 - x^2} &= \lim_{x \to 1} \frac{1 - x}{x^2(x + 1)(x - 1)} && \frac{0}{0} \text{ indeterminate form} \\ &= \lim_{x \to 1} \frac{-1}{x^2(x + 1)} && \frac{1 - x}{x - 1} = -1, x \neq 1 \\ &= -0.5 \end{aligned}$$

Since this limit exists, there is no vertical asymptote at $x = 1$.

The graph of $y = f(x)$ (Fig. 7) illustrates the behavior indicated by all these limit statements.

$$\lim_{x \to -1^-} f(x) = \infty \qquad \lim_{x \to 0} f(x) = -\infty$$
$$\lim_{x \to -1^+} f(x) = -\infty \qquad \lim_{x \to 1} f(x) = -0.5$$

We will have much more to say about sketching and analyzing graphs involving asymptotes in Section 9-4.

Matched Problem 4 For the function

$$f(x) = \frac{x - 3}{x^2 - 4x + 3}$$

use ∞ and $-\infty$, as appropriate, to describe the behavior at each point of discontinuity, and identify all vertical asymptotes.

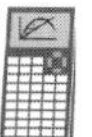

CAUTION

Figure 8A shows the graph of the function f from Example 4 on a graphing utility. It appears that the graphing utility has also drawn the vertical asymptote at $x = -1$, but this is not the case. As we saw in Example 4, points close to -1 on the left have large positive y coordinates, while points close to -1 on the right have large negative y coordinates. For the given x range on this particular graphing utility, there are no points on the screen with x coordinate -1. The graphing utility simply connected the last point to the left of -1 with the first point to the right of -1. Since these points are not visible on the screen, this gives the appearance of a vertical asymptote. Figure 8B shows the graph of the same function with a much larger y range. Now these two points are visible and, in fact, the graph appears to be continuous at $x = -1$, which we know is not true. When you graph functions with vertical asymptotes on a graphing utility, you should proceed as we did in Example 4 to identify the asymptotes first. You cannot depend on the graphing utility to identify asymptotes.

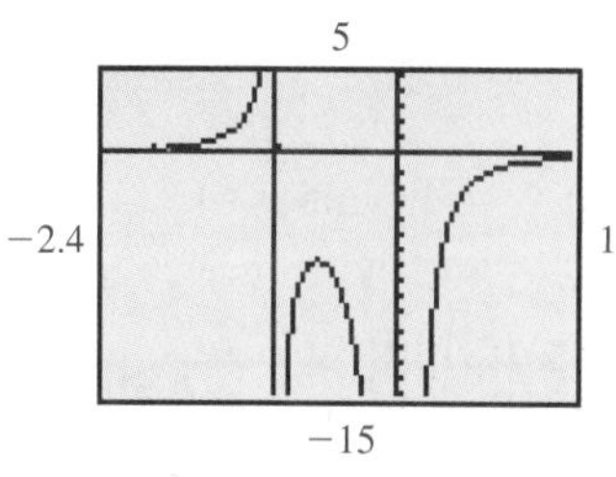

(A)

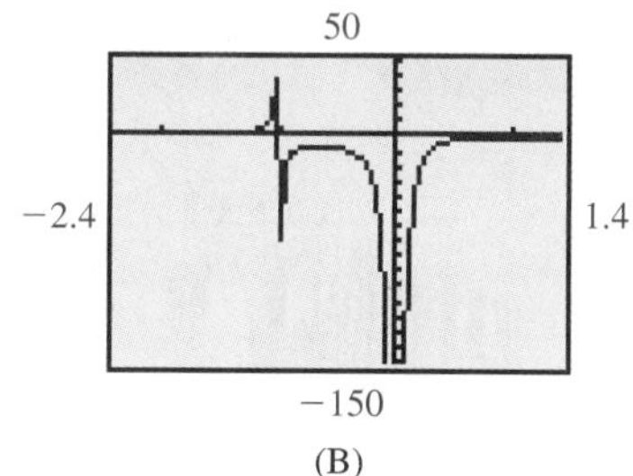

(B)

FIGURE 8

Solving Inequalities Using Continuity Properties

One of the basic tools for analyzing graphs in calculus is a special line graph called a *sign chart.* We will make extensive use of these charts in later sections. In the following discussion, we use continuity properties to develop a simple and efficient procedure for constructing sign charts.

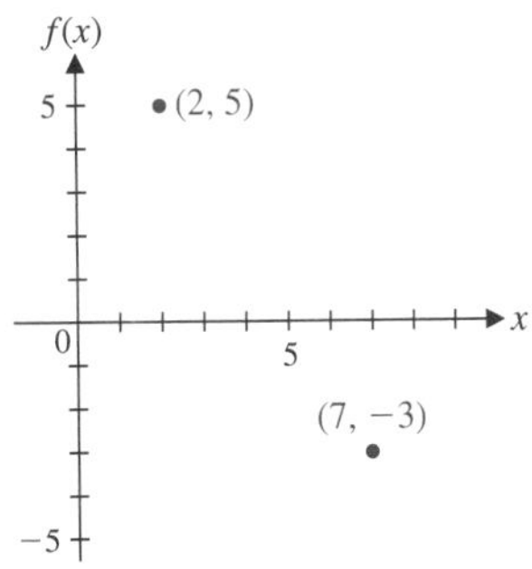

FIGURE 9

Suppose a function f is continuous over the interval $(1, 8)$ and $f(x) \neq 0$ for any x in $(1, 8)$. Also suppose $f(2) = 5$, a positive number. Is it possible for $f(x)$ to be negative for any x in the interval $(1, 8)$? The answer is "no." If $f(7)$ were -3, for example, as shown in Figure 9, how would it be possible to join the points $(2, 5)$ and $(7, -3)$ with the graph of a continuous function without crossing the x axis between 1 and 8 at least once? [Crossing the x axis would violate our assumption that $f(x) \neq 0$ for any x in $(1, 8)$.] Thus, we conclude that $f(x)$ must be positive for all x in $(1, 8)$. If $f(2)$ were negative, then, using the same type of reasoning, $f(x)$ would have to be negative over the whole interval $(1, 8)$.

In general, **if f is continuous and $f(x) \neq 0$ on the interval (a, b), then $f(x)$ cannot change sign on (a, b).** This is the essence of Theorem 2.

Theorem 2 ■■ Sign Properties on an Interval (a, b)

If f is continuous on (a, b) and $f(x) \neq 0$ for all x in (a, b), then either $f(x) > 0$ for all x in (a, b) or $f(x) < 0$ for all x in (a, b). ■■

Theorem 2 provides the basis for an effective method of solving many types of inequalities. Example 5 illustrates the process.

Example 5 **Solving an Inequality** Solve: $\dfrac{x + 1}{x - 2} > 0$

Solution We start by using the left side of the inequality to form the function f:

$$f(x) = \frac{x + 1}{x - 2}$$

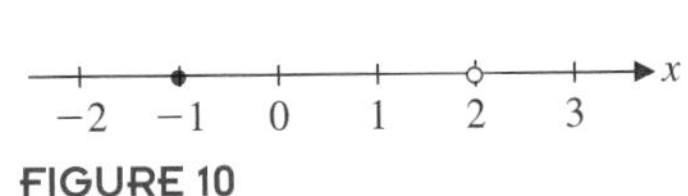

FIGURE 10

The rational function f is discontinuous at $x = 2$, and $f(x) = 0$ for $x = -1$ (a fraction is 0 when the numerator is 0 and the denominator is not 0). We plot $x = 2$ and $x = -1$, which we call *partition numbers,* on a real number line (Fig. 10). (Note that the dot at 2 is open, because the function is not defined at $x = 2$.) The partition numbers 2 and -1 determine three open intervals: $(-\infty, -1)$, $(-1, 2)$, and $(2, \infty)$. The function f is continuous and nonzero on each of these intervals. From Theorem 2 we know that $f(x)$ does not change sign on any of these intervals. Thus, we can find the sign of $f(x)$ on each of these intervals by selecting a **test number** in each interval and evaluating $f(x)$ at that number. Since any number in each subinterval will do, we choose test numbers that are easy to evaluate: -2, 0, and 3. Table 4 shows the results.

Table 4

x	-2	0	3
$f(x)$	$\frac{1}{4}$ +	$-\frac{1}{2}$ −	4 +

Test numbers

The sign of $f(x)$ at each test number is the same as the sign of $f(x)$ over the interval containing that test number. Using this information, we construct a **sign chart** for $f(x)$:

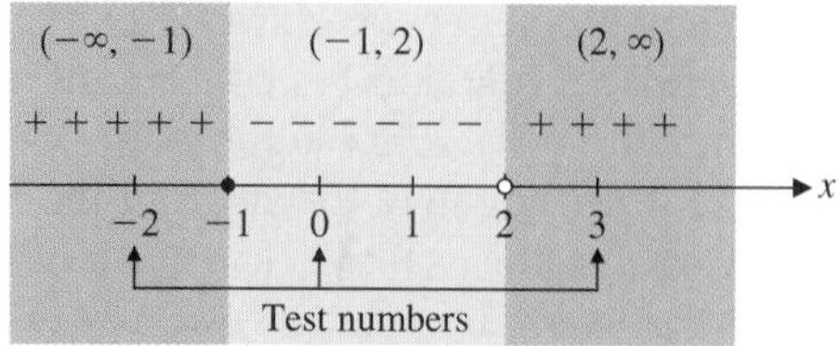

Now using the sign chart, we can easily write the solution for the given nonlinear inequality:

$f(x) > 0$ for $x < -1$ or $x > 2$ *Inequality notation*
$(-\infty, -1) \cup (2, \infty)$ *Interval notation* ■

Most of the inequalities we will encounter will involve strict inequalities ($>$ or $<$). If it is necessary to solve inequalities of the form $\geqslant$ or $\leqslant$, we simply include the end point x of any interval if f is defined at x and $f(x)$ satisfies the given inequality. For example, referring to the sign chart in Example 5, the solution of the inequality

$\frac{x+1}{x-2} \geqslant 0$ is $x \leqslant -1$ or $x > 2$ *Inequality notation*
$(-\infty, -1] \cup (2, \infty)$ *Interval notation*

In general, given a function f, we will call all values x such that f is discontinuous at x or $f(x) = 0$ **partition numbers. Partition numbers determine open intervals where $f(x)$ does not change sign.** By using a test number from each interval, we can construct a sign chart for $f(x)$ on the real number line. It is then an easy matter to determine where $f(x) < 0$ or $f(x) > 0$; that is, to solve the inequality $f(x) < 0$ or $f(x) > 0$.

We summarize the procedure for constructing sign charts in the following box:

Constructing Sign Charts

Given a function f:

Step 1. Find all partition numbers. That is:

(A) Find all numbers where f is discontinuous. (Rational functions are discontinuous for values of x that make a denominator 0.)
(B) Find all numbers where $f(x) = 0$. (For a rational function, this occurs where the numerator is 0 and the denominator is not 0.)

Step 2. Plot the numbers found in step 1 on a real number line, dividing the number line into intervals.

Step 3. Select a test number in each open interval determined in step 2, and evaluate $f(x)$ at each test number to determine whether $f(x)$ is positive (+) or negative (−) in each interval.

(continued)

Constructing Sign Charts (*continued*)

Step 4. Construct a sign chart using the real number line in step 2. This will show the sign of $f(x)$ on each open interval.

[*Note:* From the sign chart, it is easy to find the solution for the inequality $f(x) < 0$ or $f(x) > 0$.]

Matched Problem 5 Solve: $\dfrac{x^2 - 1}{x - 3} < 0$

Explore–Discuss 3

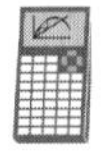

Let $y_1 = (x + 1)/(x - 2)$ and $y_2 = y_1/|y_1|$. Figure 11 shows the graph of y_2 on a graphing utility. Discuss the relationship between this graph and the sign chart constructed in the solution of Example 5.

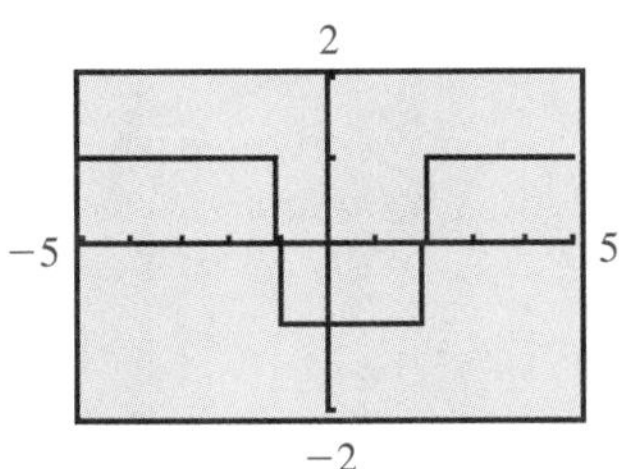

FIGURE 11

Answers to Matched Problems

1. f is not continuous at $x = -3, -1, 2,$ and 4.
   $x = -3$: $\lim_{x\to-3} f(x) = 3$, but $f(-3)$ does not exist
   $x = -1$: $f(-1) = 1$, but $\lim_{x\to-1} f(x)$ does not exist
   $x = 2$: $\lim_{x\to2} f(x) = 5$, but $f(2) = 3$
   $x = 4$: $\lim_{x\to4} f(x)$ does not exist, and $f(4)$ does not exist
2. (A) f is continuous at $x = 1$, since $\lim_{x\to1} f(x) = 2 = f(1)$.
   (B) g is not continuous at $x = 1$, since $g(1)$ is not defined.
   (C) h is not continuous at $x = 2$ for two reasons: $h(2)$ does not exist and $\lim_{x\to2} h(x)$ does not exist.
   h is continuous at $x = 0$, since $\lim_{x\to0} h(x) = -1 = h(0)$.
3. (A) Since f is a polynomial function, f is continuous for all x.
   (B) Since f is a rational function, f is continuous for all x except -1 and 4 (values that make the denominator 0).
   (C) The polynomial function $x - 4$ is continuous for all x and nonnegative for $x \geqslant 4$. Since $n = 2$ is even, f is continuous for $x \geqslant 4$, or on the interval $[4, \infty)$.
   (D) The polynomial function $x^3 + 1$ is continuous for all x. Since $n = 3$ is odd, f is continuous for all x.
4. f has a vertical asymptote at $x = 1$, since $\lim_{x\to1^-} f(x) = -\infty$ and $\lim_{x\to1^+} f(x) = \infty$; f is discontinuous at $x = 3$, since $f(3)$ does not exist, but there is no vertical asymptote at $x = 3$, since $\lim_{x\to3} f(x) = 0.5$.
5. $-\infty < x < -1$ or $1 < x < 3$; $(-\infty, -1) \cup (1, 3)$

Exercise 9-1

A *In Problems 1–6, sketch a possible graph of a function that satisfies the given conditions at $x = 1$, and discuss the continuity of f at $x = 1$.*

1. $f(1) = 2$ and $\lim_{x\to 1} f(x) = 2$
2. $f(1) = -2$ and $\lim_{x\to 1} f(x) = 2$
3. $f(1) = 2$ and $\lim_{x\to 1} f(x) = -2$
4. $f(1) = -2$ and $\lim_{x\to 1} f(x) = -2$
5. $f(1) = -2$, $\lim_{x\to 1^-} f(x) = 2$, and $\lim_{x\to 1^+} f(x) = -2$
6. $f(1) = 2$, $\lim_{x\to 1^-} f(x) = 2$, and $\lim_{x\to 1^+} f(x) = -2$

In Problems 7–10, sketch a possible graph of a function that is continuous for all x except $x = 1$ and satisfies the given conditions at $x = 1$.

7. $\lim_{x\to 1} f(x) = -\infty$
8. $\lim_{x\to 1} f(x) = \infty$
9. $\lim_{x\to 1^-} f(x) = -\infty$ and $\lim_{x\to 1^+} f(x) = \infty$
10. $\lim_{x\to 1^-} f(x) = \infty$ and $\lim_{x\to 1^+} f(x) = -\infty$

Use Theorem 1 to determine where each function in Problems 11–16 is continuous.

11. $f(x) = 2x - 3$
12. $g(x) = 3 - 5x$
13. $h(x) = \dfrac{2}{x - 5}$
14. $k(x) = \dfrac{x}{x + 3}$
15. $g(x) = \dfrac{x - 5}{(x - 3)(x + 2)}$
16. $F(x) = \dfrac{1}{x(x + 7)}$

B *Problems 17–20 refer to the function f shown in the graph. Use the graph to estimate limits as outlined below.*

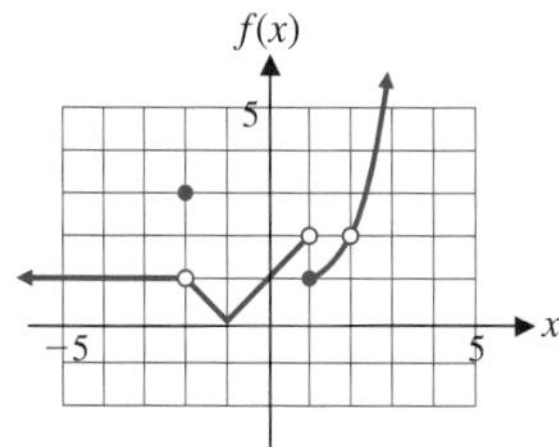

For each value of c:
(A) Find $\lim_{x\to c^-} f(x)$, $\lim_{x\to c^+} f(x)$, $\lim_{x\to c} f(x)$, and $f(c)$.
(B) Is f continuous at $x = c$? Explain.

17. $c = 1$
18. $c = 2$
19. $c = -2$
20. $c = -1$

21. Given the following function f:
$$f(x) = \begin{cases} 2 & \text{if } x \text{ is an integer} \\ 1 & \text{if } x \text{ is not an integer} \end{cases}$$
(A) Graph f. (B) $\lim_{x\to 2} f(x) = ?$ (C) $f(2) = ?$
(D) Is f continuous at $x = 2$?
(E) Where is f discontinuous?

22. Given the following function g:
$$g(x) = \begin{cases} -1 & \text{if } x \text{ is an even integer} \\ 1 & \text{if } x \text{ is not an even integer} \end{cases}$$
(A) Graph g. (B) $\lim_{x\to 1} g(x) = ?$ (C) $g(1) = ?$
(D) Is g continuous at $x = 1$?
(E) Where is g discontinuous?

In Problems 23–32, use $-\infty$ or ∞ where appropriate to describe the behavior at each point of discontinuity, and identify all vertical asymptotes.

23. $f(x) = \dfrac{1}{x + 3}$
24. $g(x) = \dfrac{x}{4 - x}$
25. $h(x) = \dfrac{x^2 + 4}{x^2 - 4}$
26. $k(x) = \dfrac{x^2 - 9}{x^2 + 9}$
27. $F(x) = \dfrac{x^2 - 4}{x^2 + 4}$
28. $G(x) = \dfrac{x^2 + 9}{9 - x^2}$
29. $H(x) = \dfrac{x^2 - 2x - 3}{x^2 - 4x + 3}$
30. $K(x) = \dfrac{x^2 + 2x - 3}{x^2 - 4x + 3}$
31. $T(x) = \dfrac{8x - 16}{x^4 - 8x^3 + 16x^2}$
32. $S(x) = \dfrac{6x + 9}{x^4 + 6x^3 + 9x^2}$

In Problems 33–38, solve each inequality using a sign chart. Express answers in inequality and interval notation.

33. $x^2 - x - 12 < 0$
34. $x^2 - 2x - 8 < 0$
35. $x^2 + 21 > 10x$
36. $x^2 + 7x > -10$
37. $\dfrac{x^2 + 5x}{x - 3} > 0$
38. $\dfrac{x - 4}{x^2 + 2x} < 0$

In Problems 39–44, use a graphing utility to approximate the partition numbers of each function $f(x)$ to two decimal places. Then solve the following inequalities:
(A) $f(x) > 0$ (B) $f(x) < 0$
Express answers in interval notation.

39. $f(x) = x^3 - 3x^2 - 2x + 5$
40. $f(x) = x^3 + 3x^2 - 4x - 8$

41. $f(x) = x^4 - 6x^2 + 3x + 5$

42. $f(x) = x^4 - 4x^2 - 2x + 2$

43. $f(x) = \dfrac{3 + 6x - x^3}{x - 1}$

44. $f(x) = \dfrac{x^3 - 5x + 1}{x + 1}$

Use Theorem 1 to determine where each function in Problems 45–52 is continuous. Express the answer in interval notation.

45. $F(x) = 2x^8 - 3x^4 + 5$ **46.** $h(x) = \dfrac{x^4 - 3x + 5}{x^2 + 2x}$

47. $g(x) = \sqrt{x - 5}$ **48.** $f(x) = \sqrt{3 - x}$

49. $K(x) = \sqrt[3]{x - 5}$ **50.** $H(x) = \sqrt[3]{3 - x}$

51. $f(x) = \dfrac{x^2 - 1}{x^2 - 3x + 2}$ **52.** $k(x) = \dfrac{x^2 - 4}{x^2 + x - 2}$

In Problems 53–58, graph f, locate all points of discontinuity, and discuss the behavior of f at these points.

53. $f(x) = \begin{cases} 1 + x & \text{if } x < 1 \\ 5 - x & \text{if } x \geqslant 1 \end{cases}$

54. $f(x) = \begin{cases} x^2 & \text{if } x \leqslant 1 \\ 2x & \text{if } x > 1 \end{cases}$

55. $f(x) = \begin{cases} 1 + x & \text{if } x \leqslant 2 \\ 5 - x & \text{if } x > 2 \end{cases}$

56. $f(x) = \begin{cases} x^2 & \text{if } x \leqslant 2 \\ 2x & \text{if } x > 2 \end{cases}$

57. $f(x) = \begin{cases} -x & \text{if } x < 0 \\ 1 & \text{if } x = 0 \\ x & \text{if } x > 0 \end{cases}$

58. $f(x) = \begin{cases} 1 & \text{if } x < 0 \\ 0 & \text{if } x = 0 \\ 1 + x & \text{if } x > 0 \end{cases}$

In Problems 59–62, locate all points of discontinuity of f, using a graphing utility as an aid, and discuss the behavior of f at these points. [Hint: Select Xmin and Xmax so that the suspected point of discontinuity is the midpoint of the graphing interval (Xmin, Xmax).]

59. $f(x) = x + \dfrac{|2x - 4|}{x - 2}$ **60.** $f(x) = x + \dfrac{|3x + 9|}{x + 3}$

61. $f(x) = \dfrac{x^2 - 1}{|x| - 1}$ **62.** $f(x) = \dfrac{x^3 - 8}{|x| - 2}$

C

63. Use the graph of the function g to answer the following questions:
(A) Is g continuous on the open interval $(-1, 2)$?
(B) Is g continuous from the right at $x = -1$? That is, does $\lim_{x\to-1^+} g(x) = g(-1)$?
(C) Is g continuous from the left at $x = 2$? That is, does $\lim_{x\to2^-} g(x) = g(2)$?
(D) Is g continuous on the closed interval $[-1, 2]$?

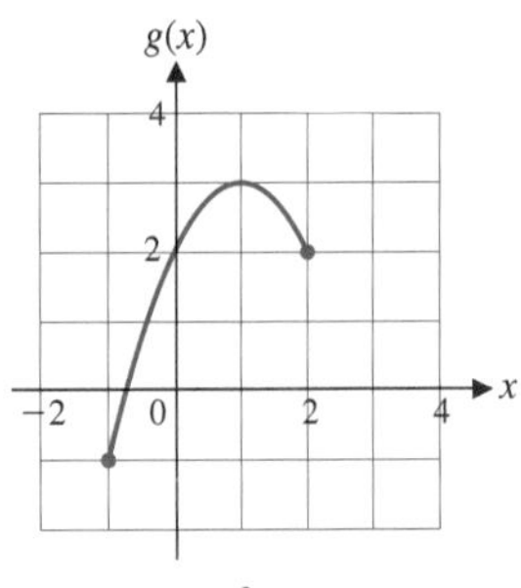

$g(x) = -x^2 + 2x + 2$

Figure for 63

64. Use the graph of the function f to answer the following questions:
(A) Is f continuous on the open interval $(0, 3)$?
(B) Is f continuous from the right at $x = 0$? That is, does $\lim_{x\to0^+} f(x) = f(0)$?
(C) Is f continuous from the left at $x = 3$? That is, does $\lim_{x\to3^-} f(x) = f(3)$?
(D) Is f continuous on the closed interval $[0, 3]$?

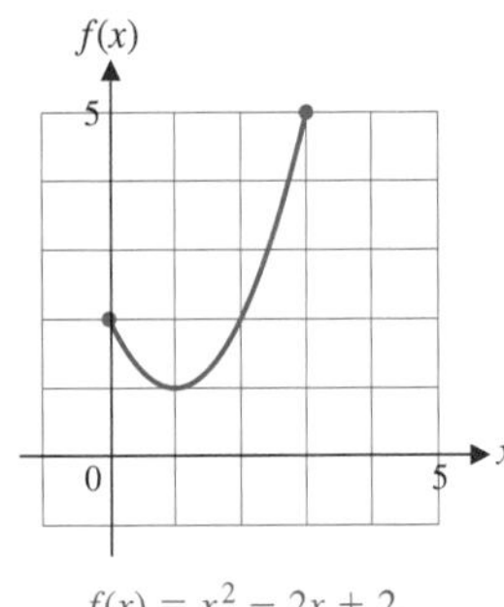

$f(x) = x^2 - 2x + 2$

Figure for 64

*Problems 65 and 66 refer to the **greatest integer function,** which is denoted by* $[\![x]\!]$ *and is defined as follows:*

$$[\![x]\!] = \text{Greatest integer} \leqslant x$$

For example,

$$\begin{aligned} [\![-3.6]\!] &= \text{Greatest integer} \leqslant -3.6 = -4 \\ [\![2]\!] &= \text{Greatest integer} \leqslant 2 = 2 \\ [\![2.5]\!] &= \text{Greatest integer} \leqslant 2.5 = 2 \end{aligned}$$

The graph of $f(x) = [\![x]\!]$ is shown. There, we can see that

$$\begin{aligned} [\![x]\!] &= -2 && \text{for} \quad -2 \leq x < -1 \\ [\![x]\!] &= -1 && \text{for} \quad -1 \leq x < 0 \\ [\![x]\!] &= 0 && \text{for} \quad 0 \leq x < 1 \\ [\![x]\!] &= 1 && \text{for} \quad 1 \leq x < 2 \\ [\![x]\!] &= 2 && \text{for} \quad 2 \leq x < 3 \end{aligned}$$

and so on.

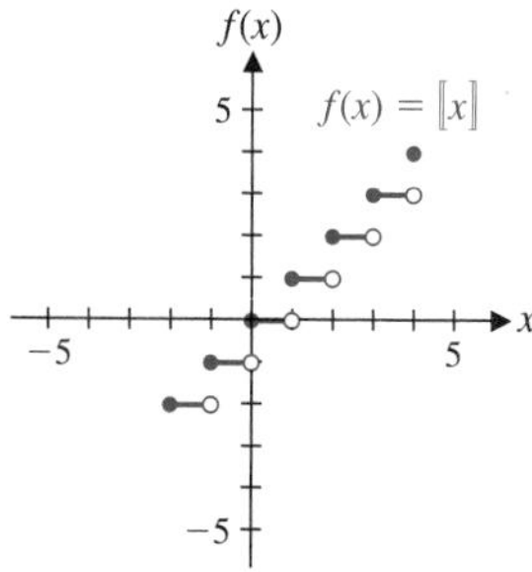

Figure for 65 and 66

65. (A) Is f continuous from the right at $x = 0$?
(B) Is f continuous from the left at $x = 0$?
(C) Is f continuous on the open interval $(0, 1)$?
(D) Is f continuous on the closed interval $[0, 1]$?
(E) Is f continuous on the half-closed interval $[0, 1)$?

66. (A) Is f continuous from the right at $x = 2$?
(B) Is f continuous from the left at $x = 2$?
(C) Is f continuous on the open interval $(1, 2)$?
(D) Is f continuous on the closed interval $[1, 2]$?
(E) Is f continuous on the half-closed interval $[1, 2)$?

In Problems 67–70, sketch a possible graph for a function f that is continuous for all real numbers and satisfies the given conditions. Find the x intercepts for f.

67. $f(x) < 0$ on $(-\infty, -5)$ and $(2, \infty)$; $f(x) > 0$ on $(-5, 2)$

68. $f(x) > 0$ on $(-\infty, -4)$ and $(3, \infty)$; $f(x) < 0$ on $(-4, 3)$

69. $f(x) < 0$ on $(-\infty, -6)$ and $(-1, 4)$; $f(x) > 0$ on $(-6, -1)$ and $(4, \infty)$

70. $f(x) > 0$ on $(-\infty, -3)$ and $(2, 7)$; $f(x) < 0$ on $(-3, 2)$ and $(7, \infty)$

71. The functions $f(x) = 2/(1 - x)$ satisfies $f(0) = 2$ and $f(2) = -2$. Is f equal to 0 anywhere on the interval $(-1, 3)$? Does this contradict Theorem 2? Explain.

72. The function $f(x) = 6/(x - 4)$ satisfies $f(2) = -3$ and $f(7) = 2$. Is f equal to 0 anywhere on the interval $(0, 9)$? Does this contradict Theorem 2? Explain.

73. The function f is continuous and never 0 on the interval $(0, 4)$, and continuous and never 0 on the interval $(4, 8)$. Also, $f(2) = 3$ and $f(6) = -3$. Discuss the validity of the following statement and illustrate your conclusions with graphs: Either $f(4) = 0$ or f is discontinuous at $x = 4$.

74. The function f is continuous and never 0 on the interval $(-3, 1)$, and continuous and never 0 on the interval $(1, 4)$. Also, $f(-2) = -3$ and $f(3) = 4$. Discuss the validity of the following statement and illustrate your conclusions with graphs: Either $f(1) = 0$ or f is discontinuous at $x = 1$.

Applications

Business & Economics

75. *Postal rates.* First-class postage in 1998 was \$0.32 for the first ounce (or any fraction thereof) and \$0.23 for each additional ounce (or fraction thereof) up to 11 ounces. If $P(x)$ is the amount of postage for a letter weighing x ounces, then we can write

$$P(x) = \begin{cases} \$0.32 & \text{if } 0 < x \leq 1 \\ \$0.55 & \text{if } 1 < x \leq 2 \\ \$0.78 & \text{if } 2 < x \leq 3 \\ \text{and so on} \end{cases}$$

(A) Graph P for $0 < x \leq 5$.
(B) Find $\lim_{x \to 4.5} P(x)$ and $P(4.5)$.
(C) Find $\lim_{x \to 4} P(x)$ and $P(4)$.
(D) Is P continuous at $x = 4.5$? At $x = 4$?

76. *Telephone rates.* A person placing a station-to-station call on Saturday from San Francisco to New York is charged \$0.30 for the first minute (or any fraction thereof) and \$0.20 for each additional minute (or fraction thereof). If the length of a call is x minutes, then the long-distance charge $R(x)$ is

$$R(x) = \begin{cases} \$0.30 & \text{if } 0 < x \le 1 \\ \$0.50 & \text{if } 1 < x \le 2 \\ \$0.70 & \text{if } 2 < x \le 3 \\ \text{and so on} \end{cases}$$

(A) Graph R for $0 < x \le 6$.
(B) Find $\lim_{x\to 2.5} R(x)$ and $R(2.5)$.
(C) Find $\lim_{x\to 2} R(x)$ and $R(2)$.
(D) Is R continuous at $x = 2.5$? At $x = 2$?

77. *Postal rates.* Discuss the differences between the function $Q(x) = 0.32 + 0.23[\![x]\!]$ and the function $P(x)$ defined in Problem 75.

78. *Telephone rates.* Discuss the differences between the function $S(x) = 0.30 + 0.20[\![x]\!]$ and the function $R(x)$ defined in Problem 76.

79. *Pricing.* An office products firm sells custom-printed pencils for companies to use for promotional purposes. The minimum order is 150 pencils, and discounts are given for volume purchases, as shown in Table 5. If x is the number of pencils ordered, then the price per pencil is \$0.49 for $150 \le x < 250$, \$0.39 for $250 \le x < 500$, and so on.
(A) Let $y = p(x)$ represent the price per pencil. Graph $y = p(x)$ for $150 \le x \le 1{,}500$.
(B) Identify the discontinuities of p and discuss the behavior at each discontinuity.
(C) Let $y = C(x)$ be the total cost for an order of x pencils. Graph $y = C(x)$ for $150 \le x \le 1{,}500$.
(D) Identify the discontinuities of C and discuss the behavior at each discontinuity.

Table 5

QUANTITY ORDERED	150	250	500	1,000 or more
PRICE PER PENCIL	\$0.49	\$0.39	\$0.29	\$0.24

80. *Pricing.* The office products firm in Problem 79 also sells custom-printed pens. The minimum order is 200 pens, and discounts are given for volume purchases, as shown in Table 6. Let x represent the number of pens ordered.
(A) Let $y = p(x)$ represent the price per pen. Graph $y = p(x)$ for $200 \le x \le 900$.
(B) Identify the discontinuities of p and discuss the behavior at each discontinuity.
(C) Let $y = C(x)$ be the total cost for an order of x pens. Graph $y = C(x)$ for $200 \le x \le 900$.
(D) Identify the discontinuities of C and discuss the behavior at each discontinuity.

Table 6

QUANTITY ORDERED	200	300	500	700 or more
PRICE PER PEN	\$1.19	\$1.14	\$1.04	\$0.93

81. *Income.* A personal computer salesperson receives a base salary of \$1,000 per month and a commission of 5% of all sales over \$10,000 during the month. If the monthly sales are \$20,000 or more, the salesperson is given an additional \$500 bonus. Let $E(s)$ represent the person's earnings during the month as a function of the monthly sales s.
(A) Graph $E(s)$ for $0 \le s \le 30{,}000$.
(B) Find $\lim_{s\to 10{,}000} E(s)$ and $E(10{,}000)$.
(C) Find $\lim_{s\to 20{,}000} E(s)$ and $E(20{,}000)$.
(D) Is E continuous at $s = 10{,}000$? At $s = 20{,}000$?

82. *Equipment rental.* An office equipment rental and leasing company rents typewriters for \$10 per day (and any fraction thereof) or for \$50 per 7 day week. Let $C(x)$ be the cost of renting a typewriter for x days.
(A) Graph $C(x)$ for $0 \le x \le 10$.
(B) Find $\lim_{x\to 4.5} C(x)$ and $C(4.5)$.
(C) Find $\lim_{x\to 8} C(x)$ and $C(8)$.
(D) Is C continuous at $x = 4.5$? At $x = 8$?

Life Sciences

83. *Animal supply.* A medical laboratory raises its own rabbits. The number of rabbits $N(t)$ available at any time t depends on the number of births and deaths. When a birth or death occurs, the function N generally has a discontinuity, as shown in the figure.

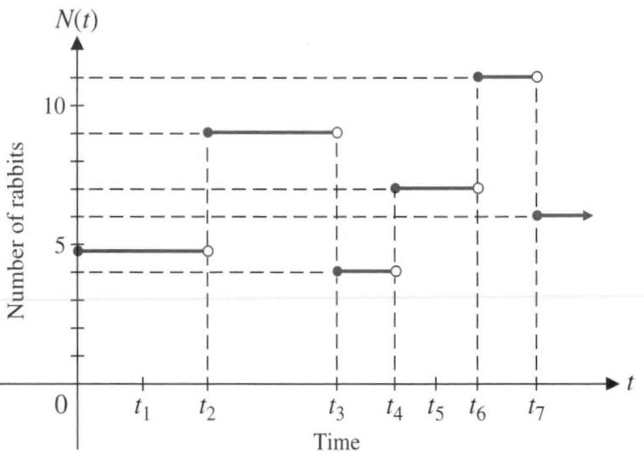

Figure for 83

(A) Where is the function N discontinuous?
(B) $\lim_{t\to t_5} N(t) = ?$; $N(t_5) = ?$
(C) $\lim_{t\to t_3} N(t) = ?$; $N(t_3) = ?$

Social Sciences

84. *Learning.* The graph might represent the history of a particular person learning the material on limits and continuity in this book. At time t_2, the student's mind goes blank during a quiz. At time t_4, the instructor explains a concept particularly well, and suddenly, a big jump in understanding takes place.
(A) Where is the function p discontinuous?
(B) $\lim_{t \to t_1} p(t) = ?; \quad p(t_1) = ?$
(C) $\lim_{t \to t_2} p(t) = ?; \quad p(t_2) = ?$
(D) $\lim_{t \to t_4} p(t) = ?; \quad p(t_4) = ?$

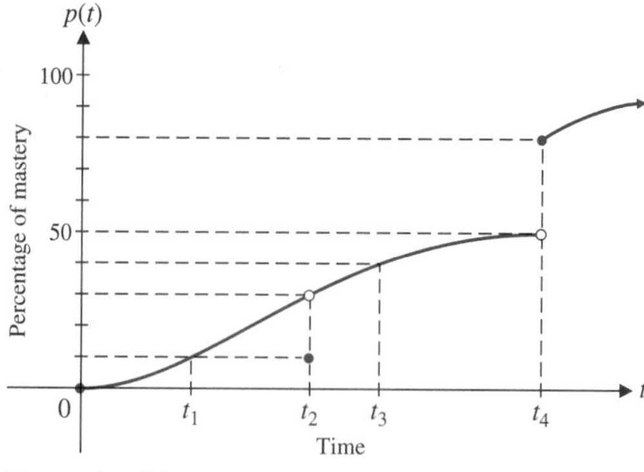

Figure for 84

SECTION 9-2

First Derivative and Graphs

- INCREASING AND DECREASING FUNCTIONS
- LOCAL EXTREMA
- FIRST-DERIVATIVE TEST
- ANALYZING GRAPHS

Since the derivative is associated with the slope of the graph of a function at a point, we might expect that it is also associated with other properties of a graph. As we will see in this and the next section, the derivative can tell us a great deal about the shape of the graph of a function. In addition, this investigation will lead to methods for finding absolute maximum and minimum values for functions that do not require graphing. Manufacturing companies can use these methods to find production levels that will minimize cost or maximize profit. Pharmacologists can use them to find levels of drug dosages that will produce maximum sensitivity to a drug. And so on.

INCREASING AND DECREASING FUNCTIONS

Explore–Discuss 1

Figure 1 shows the graph of $y = f(x)$ and a sign chart for $f'(x)$, where

$$f(x) = x^3 - 3x$$

and

$$f'(x) = 3x^2 - 3 = 3(x + 1)(x - 1)$$

Discuss the relationship between the graph of f and the sign of $f'(x)$ over each interval where $f'(x)$ has a constant sign. Also, describe the behavior of the graph of f at each partition number for f'.

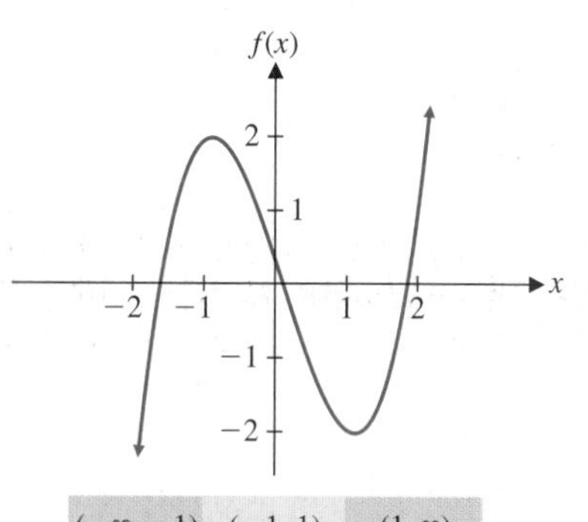

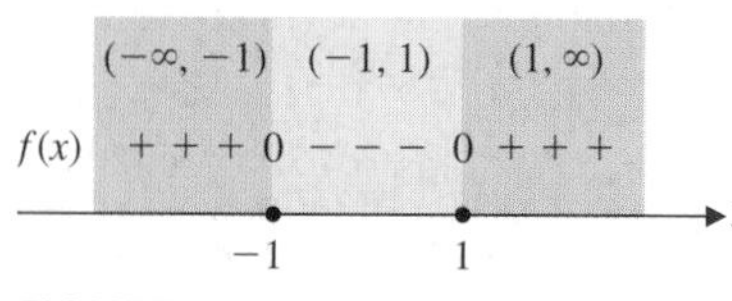

FIGURE 1

Graphs of functions generally have *rising* and *falling* sections as we scan graphs from left to right. Referring to the graph of $f(x) = x^3 - 3x$ in Figure 1, we see that on the interval $(-\infty, -1)$, the graph of f is *rising*, $f(x)$ is *increasing,** and the slope of the graph is positive $[f'(x) > 0]$. On the other hand, on the interval $(-1, 1)$, the graph of f is *falling,* $f(x)$ is *decreasing,* and the slope of the graph is negative $[f'(x) < 0]$. Finally, on the interval $(1, \infty)$, once again the graph of f is rising, $f(x)$ is increasing, and $f'(x) > 0$. At $x = -1$ and $x = 1$, the slope of the graph is 0 $[f'(x) = 0]$ and the tangent lines are horizontal.

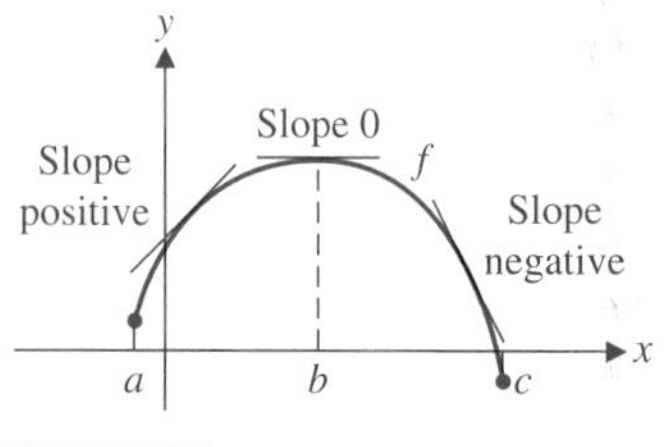

FIGURE 2

In general, if $f'(x) > 0$ (is positive) on the interval (a, b) (Fig. 2), then $f(x)$ increases ($\nearrow$) and the graph of f rises as we move from left to right over the interval; if $f'(x) < 0$ (is negative) on an interval (a, b), then $f(x)$ decreases ($\searrow$) and the graph of f falls as we move from left to right over the interval. We summarize these important results in the box.

Increasing and Decreasing Functions

For the interval (a, b):

$f'(x)$	$f(x)$	GRAPH OF f	EXAMPLES
+	Increases $\nearrow$	Rises $\nearrow$	
−	Decreases $\searrow$	Falls $\searrow$	

Explore–Discuss 2

The graphs of $f(x) = x^2$ and $g(x) = |x|$ are shown in Figure 3. Both functions change from decreasing to increasing at $x = 0$. Discuss the relationship between the graph of each function at $x = 0$ and the derivative of the function at $x = 0$.

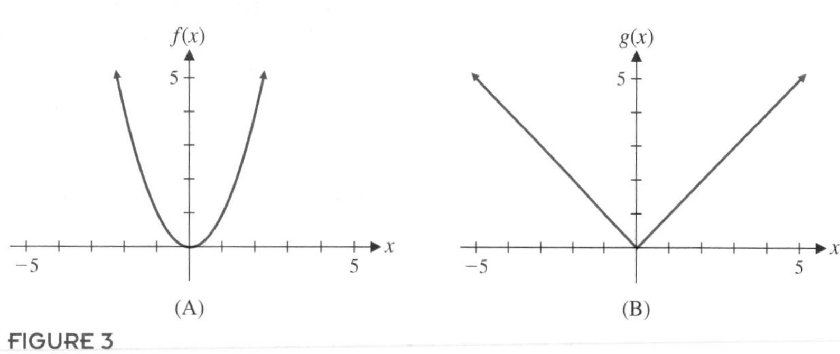

FIGURE 3

*Formally, we say that the function f is **increasing** on an interval (a, b) if $f(x_2) > f(x_1)$ whenever $a < x_1 < x_2 < b$; and f is **decreasing** on (a, b) if $f(x_2) < f(x_1)$ whenever $a < x_1 < x_2 < b$.

Example 1 **Finding Intervals Where a Function Is Increasing or Decreasing**
Given the function $f(x) = 8x - x^2$:

(A) Which values of x correspond to horizontal tangent lines?
(B) For which values of x is $f(x)$ increasing? Decreasing?
(C) Sketch a graph of f. Add any horizontal tangent lines.

Solution (A) $f'(x) = 8 - 2x = 0$

$$x = 4$$

Thus, a horizontal tangent line exists at $x = 4$ only.

(B) We will construct a sign chart for $f'(x)$ to determine which values of x make $f'(x) > 0$ and which values make $f'(x) < 0$. Recall from Section 9-1 that the partition numbers for a function are the points where the function is 0 or discontinuous. Thus, when constructing a sign chart for $f'(x)$, we must locate all points where $f'(x) = 0$ or $f'(x)$ is discontinuous. From part (A) we know that $f'(x) = 8 - 2x = 0$ at $x = 4$. Since $f'(x) = 8 - 2x$ is a polynomial, it is continuous for all x. Thus, 4 is the only partition number. We construct a sign chart for the intervals $(-\infty, 4)$ and $(4, \infty)$, using test numbers 3 and 5:

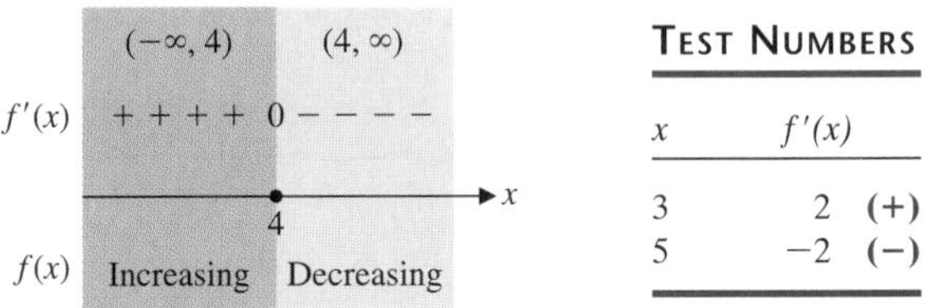

Test Numbers

x	$f'(x)$
3	2 (+)
5	−2 (−)

Thus, $f(x)$ is increasing on $(-\infty, 4)$ and decreasing on $(4, \infty)$.

(C)

x	$f(x)$
0	0
2	12
4	16
6	12
8	0

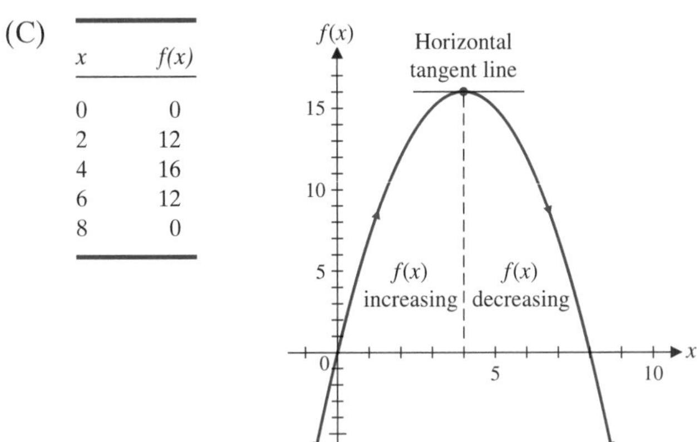

Matched Problem 1 Repeat Example 1 for $f(x) = x^2 - 6x + 10$.

As Example 1 illustrates, construction of a sign chart will play an important role in using the derivative to analyze and sketch the graph of a function f. The partition numbers for f' are central to the construction of these sign charts and also to the analysis of the graph of $y = f(x)$. We already know that if $f'(c) = 0$, then the graph of $y = f(x)$ will have a horizontal tangent line at $x = c$. But the partition numbers for f' also include the numbers c where $f'(c)$ does not exist.* There are two possibilities at this type of number: $f(c)$ does not exist, or $f(c)$ exists, but the slope of the tangent line at $x = c$ is undefined.

Critical Values of f

The values of x in the domain of f where $f'(x) = 0$ or $f'(x)$ does not exist are called the **critical values** of f. The critical values of f are always partition numbers for f', but f' may have partition numbers that are not critical values.

It is important to understand that although f' may not be defined at a critical value c, f must be defined at c.

Critical values of a function f are always in the domain of f.

Example 2 will illustrate the relationship between critical values and partition numbers.

Example 2 **Partition Numbers and Critical Values** For each function, find the partition numbers for f', the critical values for f, and determine the intervals where f is increasing and those where f is decreasing.

(A) $f(x) = 1 + x^3$ (B) $f(x) = (1 - x)^{1/3}$ (C) $f(x) = \dfrac{1}{x - 2}$

SOLUTION (A) $f(x) = 1 + x^3$ $\qquad f'(x) = 3x^2 = 0$

$$x = 0$$

The only partition number for f' is $x = 0$. Since 0 is in the domain of f, $x = 0$ is also the only critical value for f.

Sign chart for $f'(x) = 3x^2$ (partition number is 0):

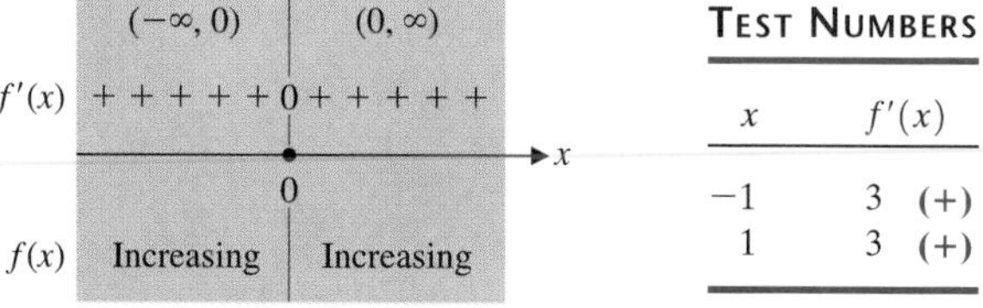

TEST NUMBERS

x	$f'(x)$
−1	3 (+)
1	3 (+)

*We are assuming that $f'(c)$ does not exist at any point of discontinuity of f'. There do exist functions where f' is discontinuous at $x = c$, yet $f'(c)$ exists. However, we will not consider such functions in this text.

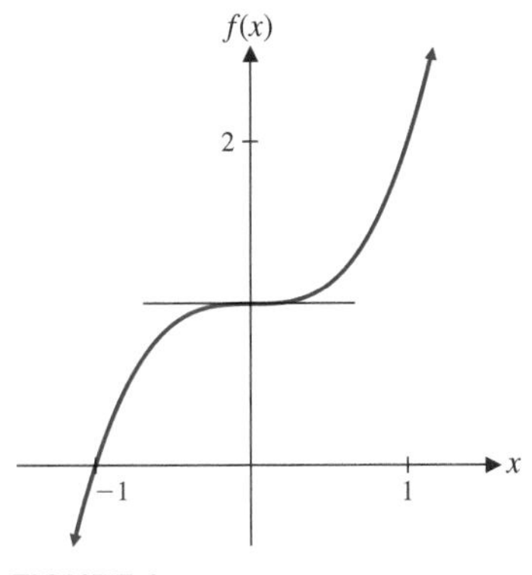

FIGURE 4

The sign chart indicates that $f(x)$ is increasing on $(-\infty, 0)$ and $(0, \infty)$. Since f is continuous at $x = 0$, it follows that $f(x)$ is increasing for all x. The graph of f is shown in Figure 4.

(B) $f(x) = (1 - x)^{1/3} \qquad f'(x) = -\frac{1}{3}(1 - x)^{-2/3} = \frac{-1}{3(1 - x)^{2/3}}$

To find partition numbers for f', we note that f' is continuous for all x except for values of x for which the denominator is 0; that is, $f'(1)$ does not exist and f' is discontinuous at $x = 1$. Since the numerator is the constant -1, $f'(x) \neq 0$ for any value of x. Thus, $x = 1$ is the only partition number for f'. Since 1 is in the domain of f, $x = 1$ is also the only critical value of f. When constructing the sign chart for f' we use the abbreviation ND to note the fact that $f'(x)$ is *not defined* at $x = 1$.

Sign chart for $f'(x) = -1/[3(1 - x)^{2/3}]$ (partition number is 1):

	$(-\infty, 1)$	$(1, \infty)$
$f'(x)$	- - - - ND	- - - -
x	1	
$f(x)$	Decreasing	Decreasing

TEST NUMBERS

x	$f'(x)$
0	$-\frac{1}{3}$ (−)
2	$-\frac{1}{3}$ (−)

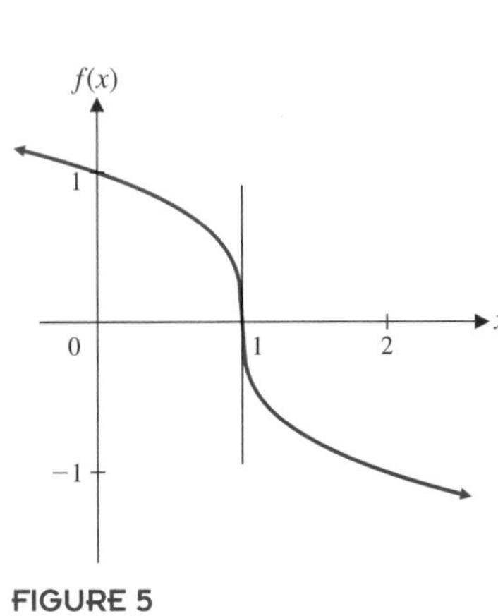

FIGURE 5

The sign chart indicates that f is decreasing on $(-\infty, 1)$ and $(1, \infty)$. Since f is continuous at $x = 1$, it follows that $f(x)$ is decreasing for all x. Thus, **a continuous function can be decreasing (or increasing) on an interval containing values of x where $f'(x)$ does not exist.** The graph of f is shown in Figure 5. Notice that the undefined derivative at $x = 1$ results in a vertical tangent line at $x = 1$. In general, **a vertical tangent will occur at $x = c$ if f is continuous at $x = c$ and $|f'(x)|$ becomes larger and larger as x approaches c.**

(C) $f(x) = \frac{1}{x - 2} \qquad f'(x) = \frac{-1}{(x - 2)^2}$

To find the partition numbers for f', note that $f'(x) \neq 0$ for any x and f' is not defined at $x = 2$. Thus, $x = 2$ is the only partition number for f'. However, $x = 2$ is *not* in the domain of f. Consequently, $x = 2$ is not a critical value of f. This function has no critical values.

Sign chart for $f'(x) = -1/(x - 2)^2$ (partition number is 2):

	$(-\infty, 2)$	$(2, \infty)$
$f'(x)$	- - - - ND	- - - -
x	2	
$f(x)$	Decreasing	Decreasing

TEST NUMBERS

x	$f'(x)$
1	−1 (−)
3	−1 (−)

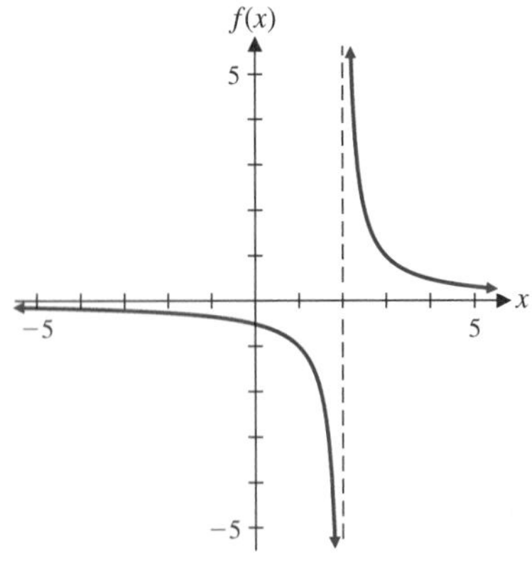

FIGURE 6

Thus, f is decreasing on $(-\infty, 2)$ and $(2, \infty)$. See the graph of f in Figure 6.

Matched Problem 2 ➡ For each function, find the partition numbers for f', the critical values for f, and determine the intervals where f is increasing and those where f is decreasing.

(A) $f(x) = 1 - x^3$ (B) $f(x) = (1 + x)^{1/3}$ (C) $f(x) = \frac{1}{x}$ ■

Explore–Discuss 3

A student examined the sign chart in Example 2C and concluded that $f(x) = 1/(x - 2)$ is decreasing for all x except $x = 2$. But $f(1) = -1 < f(3) = 1$, which seems to indicate that f is increasing. Discuss the difference between the correct answer in Example 2C and the student's answer. Explain why the student's description of where f is decreasing is unacceptable.

CAUTION

Example 2C illustrates two important ideas.

1. Do not assume all partition numbers for the derivative f' are critical values of the function f. A partition number must also be in the domain of f in order to be a critical value.
2. The values where a function is increasing or decreasing must always be expressed in terms of open intervals that are subsets of the domain of the function.

■ LOCAL EXTREMA

When the graph of a continuous function changes from rising to falling, a high point, or *local maximum,* occurs; and when the graph changes from falling to rising, a low point, or *local minimum,* occurs. In Figure 7, high points occur at c_3 and c_6, and low points occur at c_2 and c_4. In general, we call $f(c)$ a **local maximum** if there exists an interval (m, n) containing c such that

$$f(x) \leq f(c) \qquad \text{for all } x \text{ in } (m, n)$$

Note that this inequality need only hold for values of x near c; hence, the use of the term *local.*

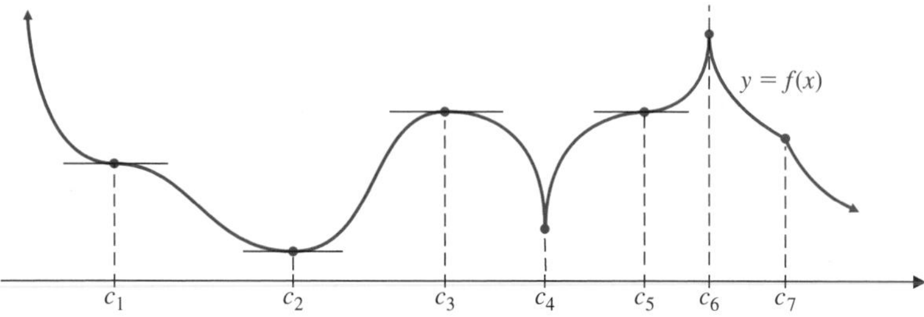

FIGURE 7

The quantity $f(c)$ is called a **local minimum** if there exists an interval (m, n) containing x such that

$$f(x) \geq f(c) \qquad \text{for all } x \text{ in } (m, n)$$

The quantity $f(c)$ is called a **local extremum** if it is either a local maximum or a local minimum. A point on a graph where a local extremum occurs is also called a **turning point.** Thus, in Figure 7, we see that local maxima occur at c_3 and c_6, local minima occur at c_2 and c_4, and all four values produce local extrema. Also note that the local maximum $f(c_3)$ is not the highest point on the graph in Figure 7. Later in this chapter we will consider the problem of finding the highest and lowest points on a graph. For now, we are concerned only with locating local extrema.

How can we locate local maxima and minima if we are given the equation for a function and not its graph? The key is to examine the critical values of the function. The local extrema of the function f in Figure 7 occur either at points where the derivative is 0 (c_2 and c_3) or at points where the derivative does not exist (c_4 and c_6). In other words, local extrema occur only at critical values of f. Theorem 1 shows that this is true in general.

Theorem 1 ■■ EXISTENCE OF LOCAL EXTREMA

If f is continuous on the interval (a, b), c is a number in (a, b), and $f(c)$ is a local extremum, then either $f'(c) = 0$ or $f'(c)$ does not exist (is not defined). ■■

Theorem 1 states that a local extremum can occur only at a critical value, but it does not imply that every critical value produces a local extremum. In Figure 7, c_1 and c_5 are critical values (the slope is 0), but the function does not have a local maximum or local minimum at either of these values.

Our strategy for finding local extrema is now clear. We find all critical values for f and test each one to see if it produces a local maximum, a local minimum, or neither.

■ FIRST-DERIVATIVE TEST

If $f'(x)$ exists on both sides of a critical value c, then the sign of $f'(x)$ can be used to determine whether the point $(c, f(c))$ is a local maximum, a local minimum, or neither. The various possibilities are summarized in the box at the top of the next page and illustrated in Figure 8.

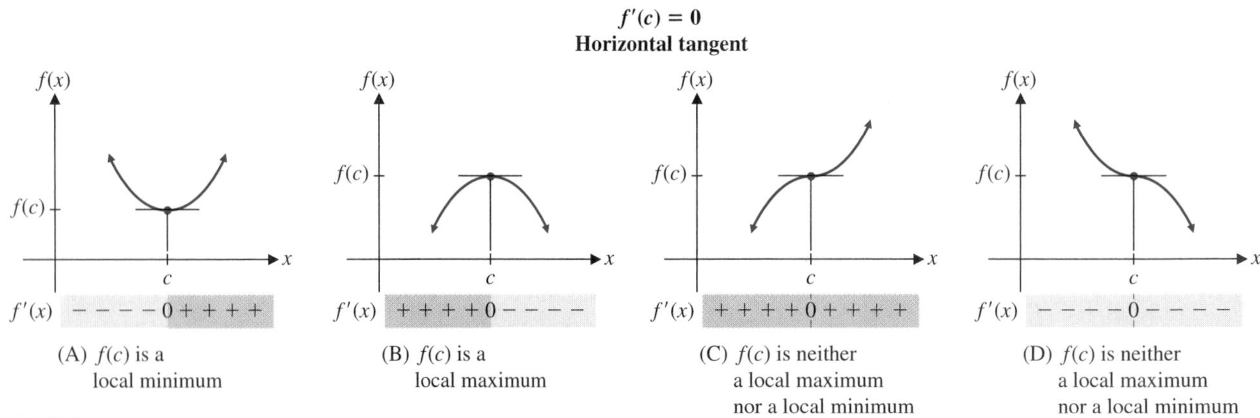

FIGURE 8
Local extrema

First-Derivative Test for Local Extrema

Let c be a critical value of f [$f(c)$ defined and either $f'(c) = 0$ or $f'(c)$ not defined]. Construct a sign chart for $f'(x)$ close to and on either side of c.

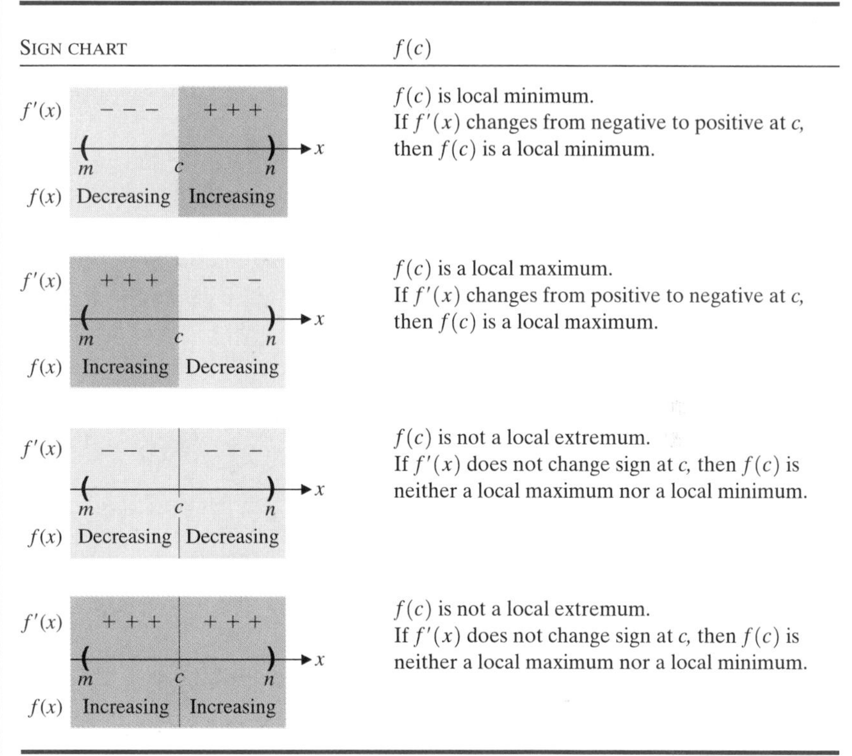

SIGN CHART
f(c)
f′(x) − − − + + +
m c n x
f(x) Decreasing Increasing
f(c) is local minimum.
If f′(x) changes from negative to positive at c, then f(c) is a local minimum.
f′(x) + + + − − −
m c n x
f(x) Increasing Decreasing
f(c) is a local maximum.
If f′(x) changes from positive to negative at c, then f(c) is a local maximum.
f′(x) − − − − − −
m c n x
f(x) Decreasing Decreasing
f(c) is not a local extremum.
If f′(x) does not change sign at c, then f(c) is neither a local maximum nor a local minimum.
f′(x) + + + + + +
m c n x
f(x) Increasing Increasing
f(c) is not a local extremum.
If f′(x) does not change sign at c, then f(c) is neither a local maximum nor a local minimum.

$f'(c)$ is not defined but $f(c)$ is defined

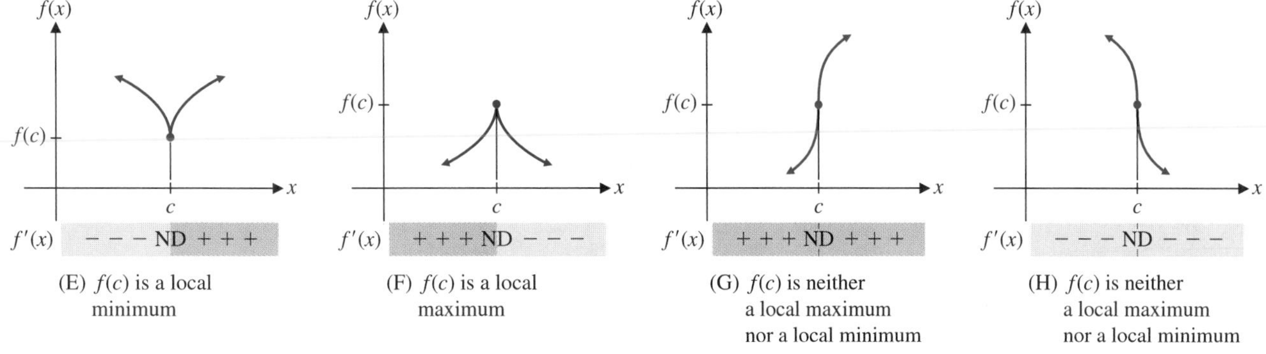

(E) $f(c)$ is a local minimum

(F) $f(c)$ is a local maximum

(G) $f(c)$ is neither a local maximum nor a local minimum

(H) $f(c)$ is neither a local maximum nor a local minimum

Example 3 **Locating Local Extrema** Given $f(x) = x^3 - 6x^2 + 9x + 1$:

(A) Find the critical values of f. (B) Find the local maxima and minima.
(C) Sketch the graph of f.

SOLUTION (A) Find all numbers x in the domain of f where $f'(x) = 0$ or $f'(x)$ does not exist.

$$\begin{aligned} f'(x) = 3x^2 - 12x + 9 &= 0 \\ 3(x^2 - 4x + 3) &= 0 \\ 3(x - 1)(x - 3) &= 0 \\ x = 1 \quad \text{or} \quad x &= 3 \end{aligned}$$

$f'(x)$ exists for all x; the critical values are $x = 1$ and $x = 3$.

(B) The easiest way to apply the first-derivative test for local maxima and minima is to construct a sign chart for $f'(x)$ for all x. Partition numbers for $f'(x)$ are $x = 1$ and $x = 3$ (which also happen to be critical values for f).

Sign chart for $f'(x) = 3(x - 1)(x - 3)$:

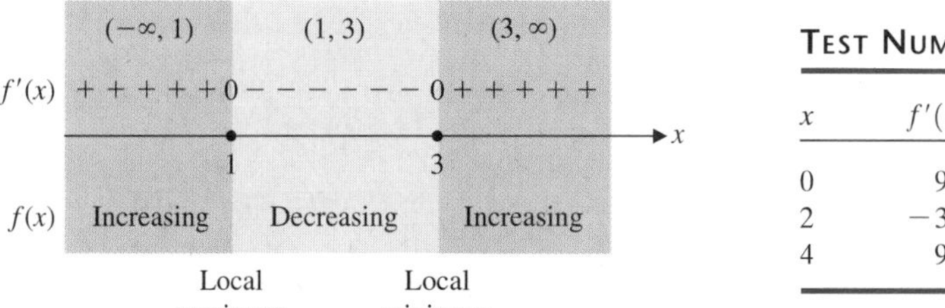

TEST NUMBERS

x	$f'(x)$
0	9 (+)
2	−3 (−)
4	9 (+)

The sign chart indicates that f increases on $(-\infty, 1)$, has a local maximum at $x = 1$, decreases on $(1, 3)$, has a local minimum at $x = 3$, and increases on $(3, \infty)$. These facts are summarized in the following table:

x	$f'(x)$	$f(x)$	GRAPH OF f
$(-\infty, 1)$	+	Increasing	Rising
$x = 1$	0	Local maximum	Horizontal tangent
$(1, 3)$	−	Decreasing	Falling
$x = 3$	0	Local minimum	Horizontal tangent
$(3, \infty)$	+	Increasing	Rising

(C) We sketch a graph of f using the information from part (B) and point-by-point plotting.

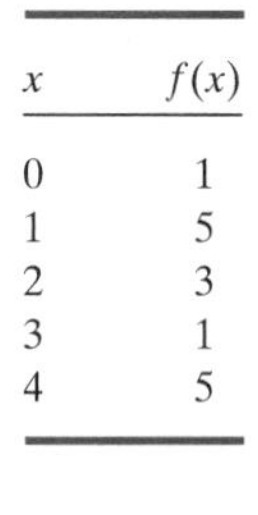

x	$f(x)$
0	1
1	5
2	3
3	1
4	5

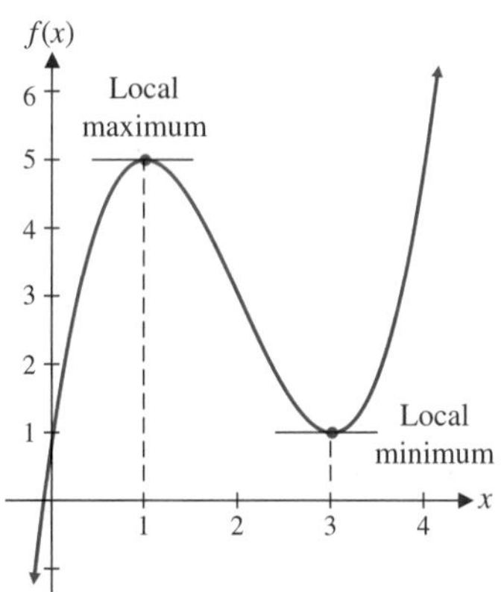

Matched Problem 3 ➡ Given $f(x) = x^3 - 9x^2 + 24x - 10$:

(A) Find the critical values of f.
(B) Find the local maxima and minima.
(C) Sketch a graph of f.

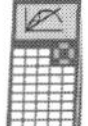

REMARK

Local extrema are easy to recognize on a graphing utility. Figure 9A shows the graph of the function in Example 3. The local maximum and minimum are clearly visible, but how do we determine their location? Two methods for approximating local extrema on a graphing utility are stated below. Consult the manual for your graphing utility for details. You might also want to see if your graphing utility will graph a numerical approximation to the derivative of a function. If it does, then you do not have to find and enter the derivative in the graphing utility.

1. Graph the derivative and use built-in root approximation routines to find the critical values. Examining the graph in Figure 9B, we see that $x = 1$ is the critical value that corresponds to the local maximum in Figure 9A. The critical value for the local minimum is found in the same way.
2. Graph the function and use built-in routines that approximate local maxima and minima. Figure 9C shows that the local minimum occurs at $x = 3$ (after rounding). The local maximum is found in a similar manner.

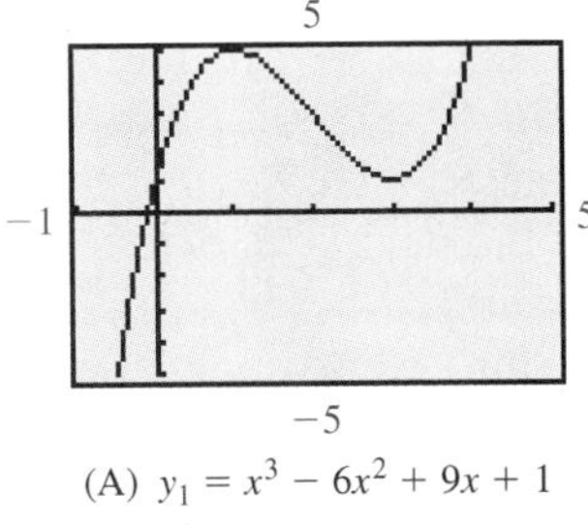

(A) $y_1 = x^3 - 6x^2 + 9x + 1$

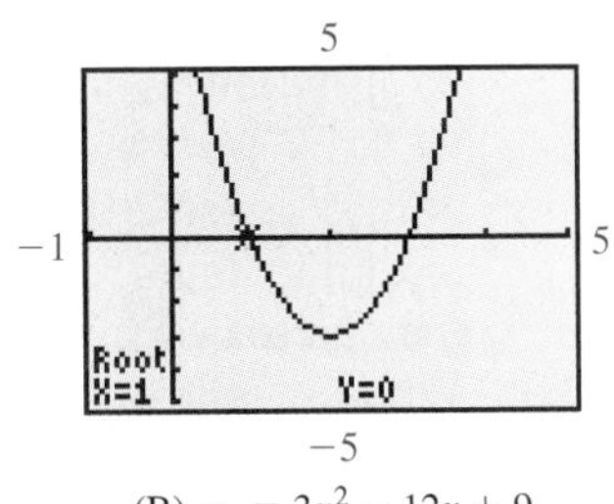

(B) $y_2 = 3x^2 - 12x + 9$

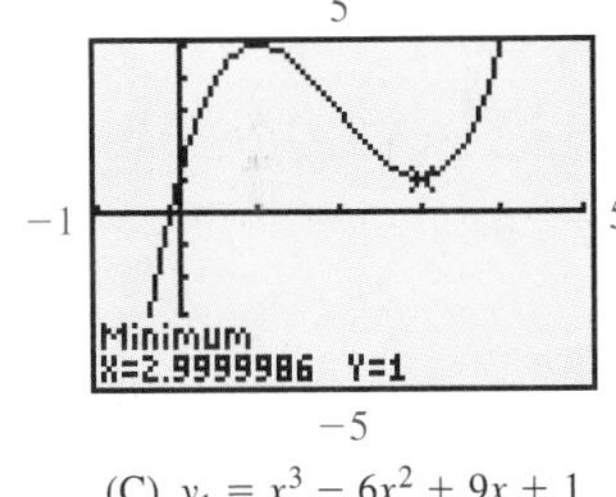

(C) $y_1 = x^3 - 6x^2 + 9x + 1$

FIGURE 9
Approximating local extrema with a graphing utility

How can you tell if you have found all the local extrema for a function? In general, this can be a difficult question to answer. However, in the case of a polynomial function, there is an easily determined upper limit on the number of local extrema. Since the local extrema are the x intercepts of the derivative, this limit is a consequence of the number of x intercepts of a polynomial. The relevant information is summarized in the following theorem, which is stated without proof:

Theorem 2 INTERCEPTS AND LOCAL EXTREMA FOR POLYNOMIAL FUNCTIONS

If $f(x) = a_n x^n + a_{n-1}x^{n-1} + \cdots + a_1 x + a_0$, $a_n \neq 0$, is an nth-degree polynomial, then f has at most n x intercepts and at most $n - 1$ local extrema.

Theorem 2 does not guarantee that every nth-degree polynomial has exactly $n - 1$ local extrema; it says only that there can never be more than $n - 1$ local extrema. For example, the third-degree polynomial in Example 3 has two local extrema, while the third-degree polynomial in Example 2A does not have any.

ANALYZING GRAPHS

In addition to providing information for hand-sketching graphs, the derivative is also an important tool for analyzing graphs and discussing the interplay between a function and its rate of change. The next two examples illustrate this process in the context of some applications to economics.

Example 4 **Agricultural Exports and Imports** Over the past several decades, the United States has exported more agricultural products than it has imported, maintaining a positive balance of trade in this area. However, the trade balance fluctuated considerably during this period. The graph in Figure 10 approximates the rate of change of the trade balance over a 15 year period, where $B(t)$ is the trade balance (in billions of dollars) and t is time (in years).

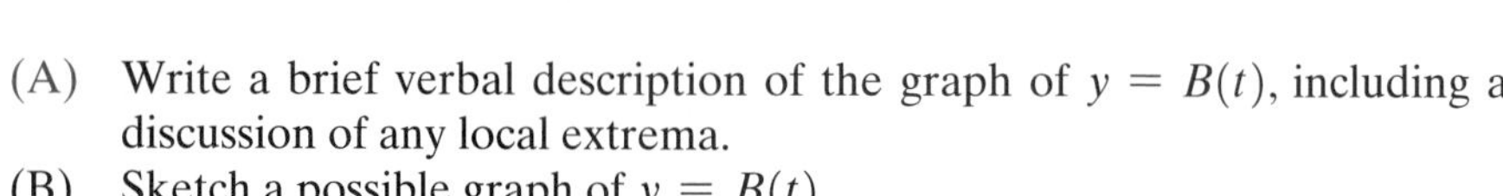

FIGURE 10
Rate of change of the balance of trade

(A) Write a brief verbal description of the graph of $y = B(t)$, including a discussion of any local extrema.

(B) Sketch a possible graph of $y = B(t)$.

SOLUTION (A) The graph of the derivative $y = B'(t)$ contains the same essential information as a sign chart. That is, we see that $B'(t)$ is positive on $(0, 4)$, 0 at $t = 4$, negative on $(4, 12)$, 0 at $t = 12$, and positive on $(12, 15)$. Hence, the trade balance increases for the first 4 years to a local maximum, decreases for the next 8 years to a local minimum, and then increases for the final 3 years.

(B) Without additional information concerning the actual values of $y = B(t)$, we cannot produce an accurate graph. However, we can sketch a possible graph that illustrates the important features, as shown in Figure 11. The absence of a scale on the vertical axis is a consequence of the lack of information about the values of $B(t)$.

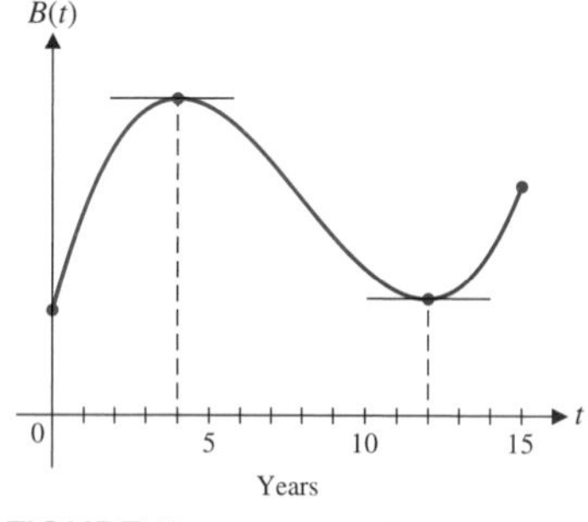

FIGURE 11
Balance of trade

Matched Problem 4 The graph in Figure 12 approximates the rate of change of the U.S. share of the total world production of motor vehicles over a 20 year period, where $S(t)$ is the U.S. share (as a percentage) and t is time (in years).

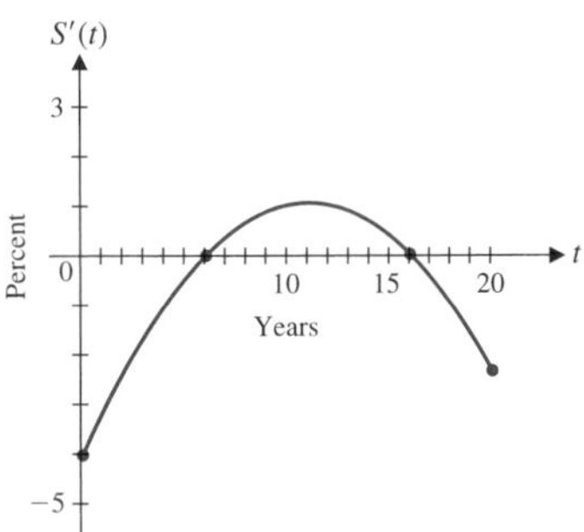

FIGURE 12

(A) Write a brief verbal description of the graph of $y = S(t)$, including a discussion of any local extrema.
(B) Sketch a possible graph of $y = S(t)$.

Example 5 **Revenue Analysis** The graph of the total revenue $R(x)$ (in dollars) from the sale of x bookcases is shown in Figure 13.

(A) Write a brief verbal description of the graph of the marginal revenue function $y = R'(x)$, including a discussion of any x intercepts.
(B) Sketch a possible graph of $y = R'(x)$.

SOLUTION (A) The graph of $y = R(x)$ indicates that $R(x)$ increases on $(0, 550)$, has a local maximum at $x = 550$, and decreases on $(550, 1{,}000)$. Consequently, the marginal revenue function $R'(x)$ must be positive on $(0, 550)$, 0 at $x = 550$, and negative on $(550, 1{,}000)$.
(B) A possible graph of $y = R'(x)$ illustrating the information summarized in part (A) is shown in Figure 14.

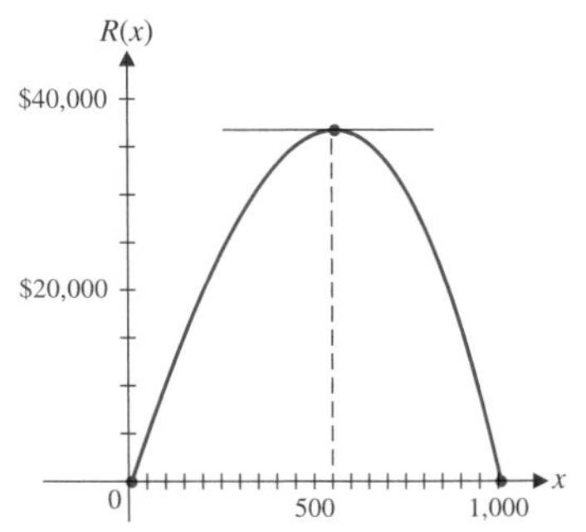

FIGURE 13
Revenue

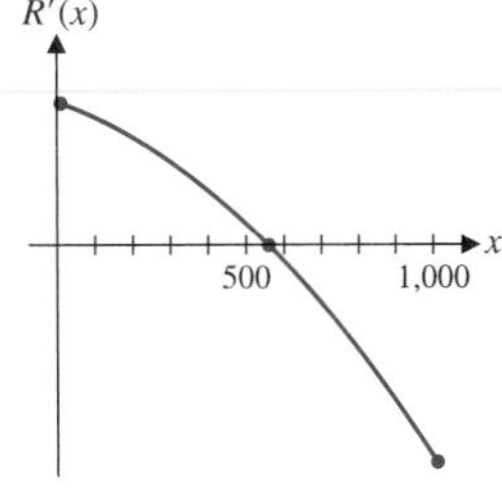

FIGURE 14
Marginal revenue

Matched Problem 5 The graph of the total revenue $R(x)$ (in dollars) from the sale of x desks is shown in Figure 15.

(A) Write a brief verbal description of the graph of the marginal revenue function $y = R'(x)$, including a discussion of any x intercepts.
(B) Sketch a possible graph of $y = R'(x)$.

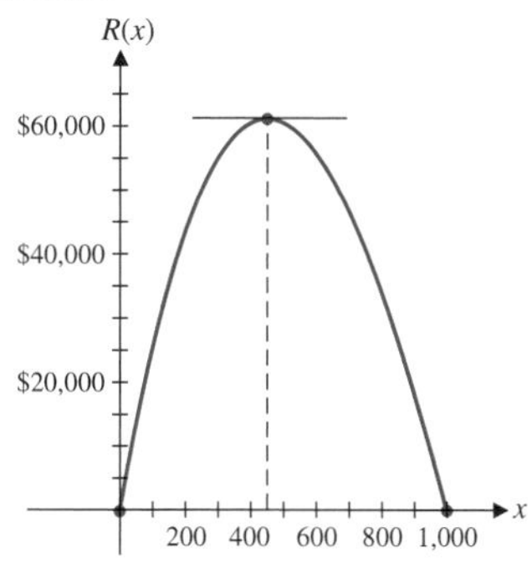

FIGURE 15

Comparing Examples 4 and 5, we see that we were able to obtain more information about the function from the graph of its derivative (Example 4), than we were when the process was reversed (Example 5). In the next section we will introduce some ideas that will enable us to extract additional information about the derivative from the graph of the function.

Answers to Matched Problems

1. (A) Horizontal tangent line at $x = 3$.
   (B) Decreasing on $(-\infty, 3)$; increasing on $(3, \infty)$
   (C)

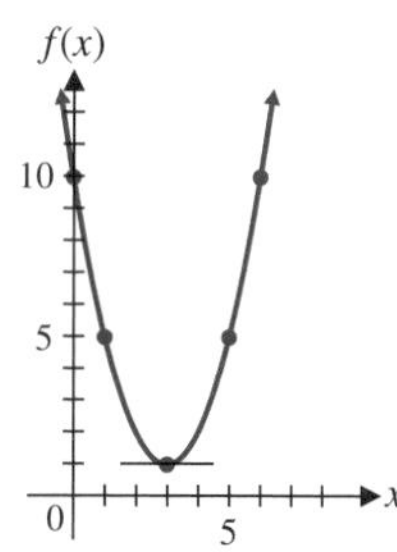

2. (A) Partition number: $x = 0$; critical value: $x = 0$; decreasing for all x
   (B) Partition number: $x = -1$; critical value: $x = -1$; increasing for all x
   (C) Partition number: $x = 0$; no critical values; decreasing on $(-\infty, 0)$ and $(0, \infty)$
3. (A) Critical values: $x = 2, x = 4$
   (B) Local maximum at $x = 2$; local minimum at $x = 4$
   (C)

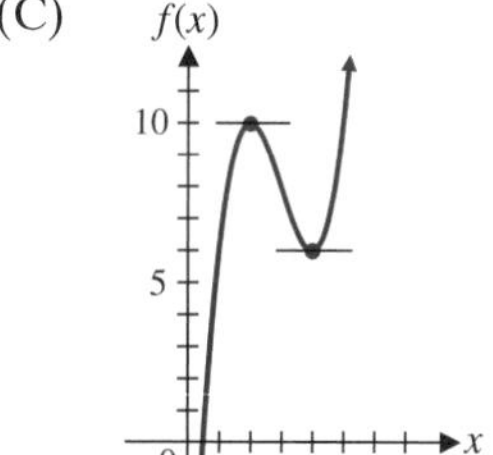

4. (A) The U.S. share of the world market decreases for 6 years to a local minimum, increases for the next 10 years to a local maximum, and then decreases for the final 4 years.
   (B)

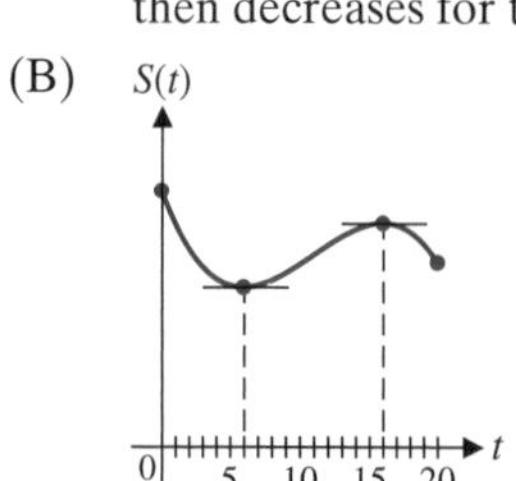

5. (A) The marginal revenue is positive on $(0, 450)$, 0 at $x = 450$, and negative on $(450, 1{,}000)$.
   (B)

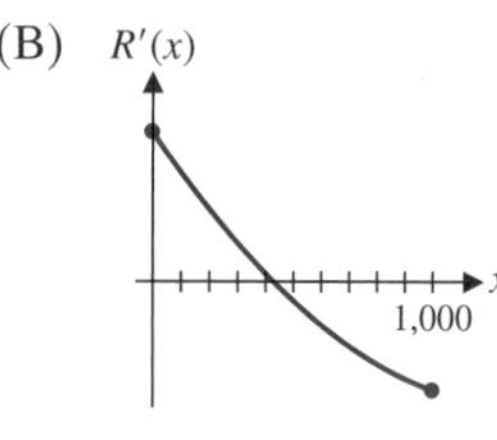

EXERCISE 9-2

A *Problems 1–8 refer to the graph of $y = f(x)$ below.*

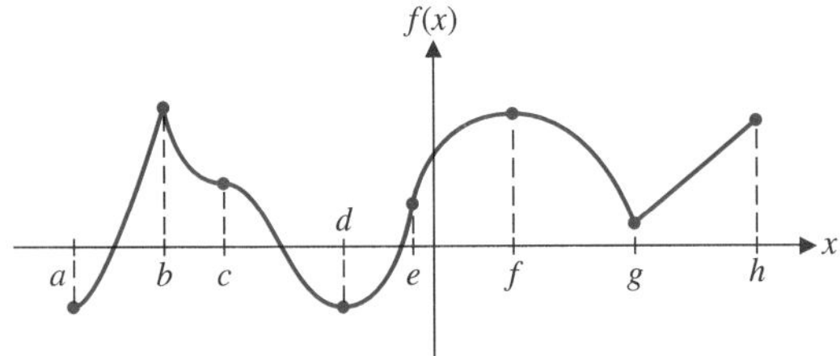

1. Identify the intervals over which $f(x)$ is increasing.
2. Identify the intervals over which $f(x)$ is decreasing.
3. Identify the intervals over which $f'(x) < 0$.
4. Identify the intervals over which $f'(x) > 0$.
5. Identify the x coordinates of the points where $f'(x) = 0$.
6. Identify the x coordinates of the points where $f'(x)$ does not exist.
7. Identify the x coordinates of the points where $f(x)$ has a local maximum.
8. Identify the x coordinates of the points where $f(x)$ has a local minimum.

In Problems 9 and 10, $f(x)$ is continuous on $(-\infty, \infty)$ and has critical values at $x = a, b, c,$ and d. Use the sign chart for $f'(x)$ to determine whether f has a local maximum, a local minimum, or neither at each critical value.

9.

f′(x) + + + 0 − − − ND − − − ND + + + 0 + + +

a b c d x

10.

f′(x) − − − 0 − − − 0 + + + ND + + + ND − − −

a b c d x

In Problems 11–18, $f(x)$ is continuous on $(-\infty, \infty)$. Use the given information to sketch the graph of f.

11.

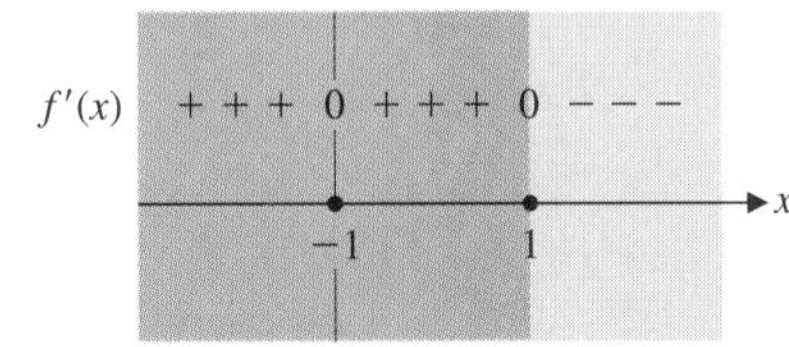

x	−2	−1	0	1	2
$f(x)$	−1	1	2	3	1

12.

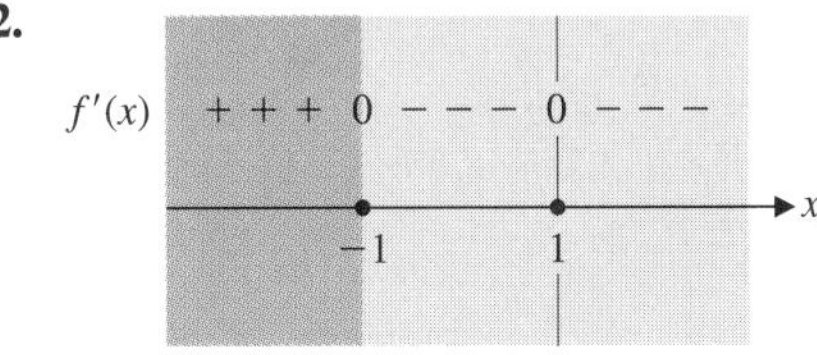

x	−2	−1	0	1	2
$f(x)$	1	3	2	1	−1

13.

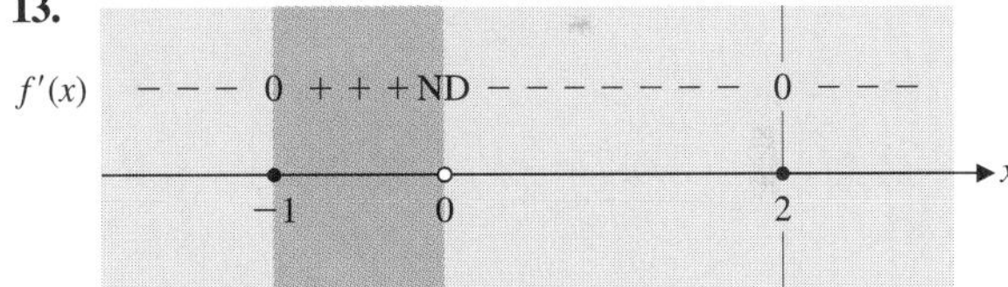

x	−2	−1	0	2	4
$f(x)$	2	1	2	1	0

14.

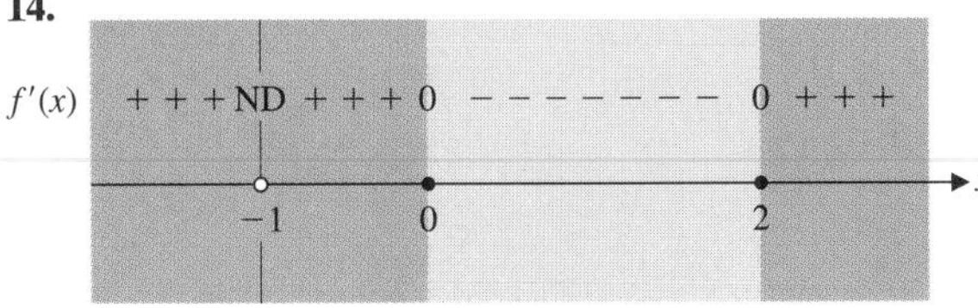

x	−2	−1	0	2	3
$f(x)$	−3	0	2	−1	0

15. $f(-2) = 4, f(0) = 0, f(2) = -4;$
$f'(-2) = 0, f'(0) = 0, f'(2) = 0;$
$f'(x) > 0$ on $(-\infty, -2)$ and $(2, \infty)$;
$f'(x) < 0$ on $(-2, 0)$ and $(0, 2)$

16. $f(-2) = -1, f(0) = 0, f(2) = 1;$
$f'(-2) = 0, f'(2) = 0;$
$f'(x) > 0$ on $(-\infty, -2)$, $(-2, 2)$, and $(2, \infty)$

17. $f(-1) = 2, f(0) = 0, f(1) = -2;$
$f'(-1) = 0, f'(1) = 0$, $f'(0)$ is not defined;
$f'(x) > 0$ on $(-\infty, -1)$ and $(1, \infty)$;
$f'(x) < 0$ on $(-1, 0)$ and $(0, 1)$

18. $f(-1) = 2, f(0) = 0, f(1) = 2;$
$f'(-1) = 0, f'(1) = 0$, $f'(0)$ is not defined;
$f'(x) > 0$ on $(-\infty, -1)$ and $(0, 1)$;
$f'(x) < 0$ on $(-1, 0)$ and $(1, \infty)$

B *Problems 19–24 involve functions f_1–f_6 and their derivatives g_1–g_6. Use the graphs shown in Figures (A) and (B) to match each function f_i with its derivative g_j.*

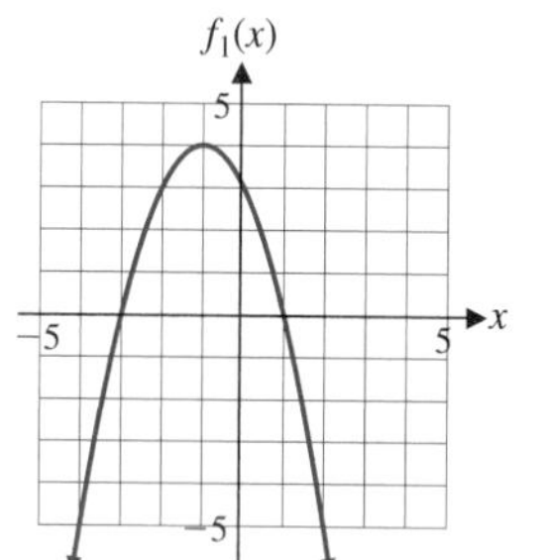

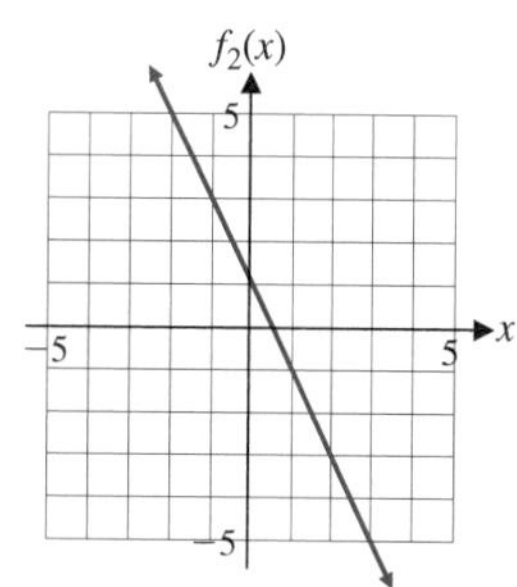

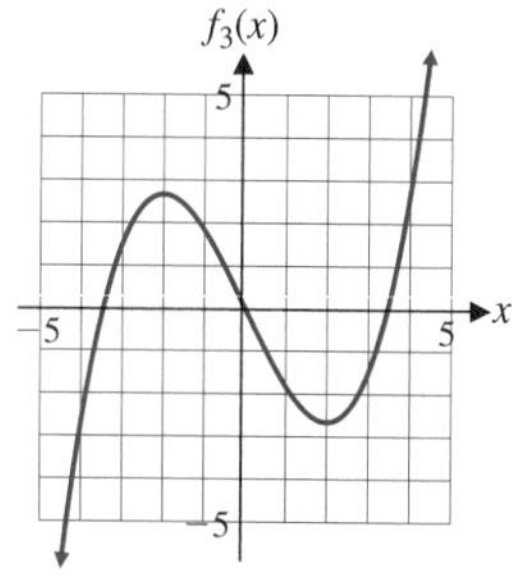

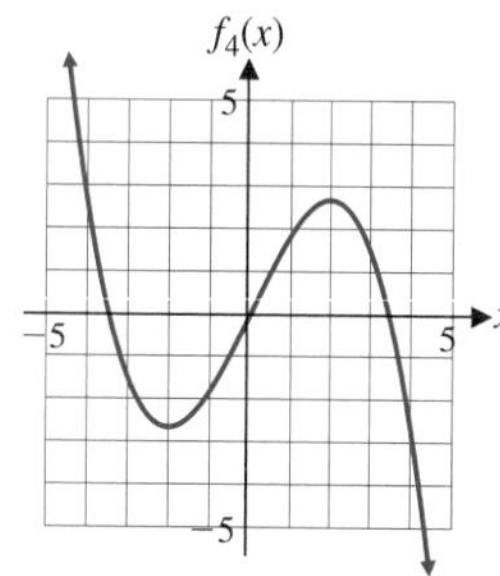

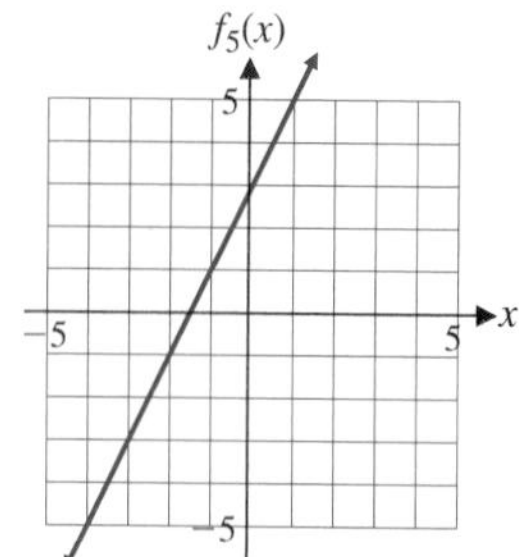

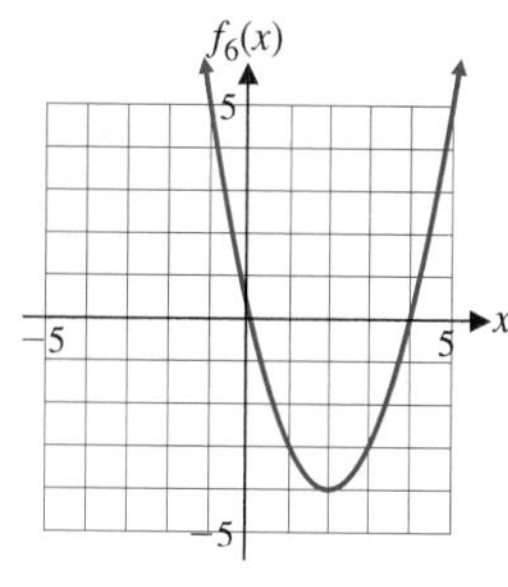

Figure (A) for 19–24

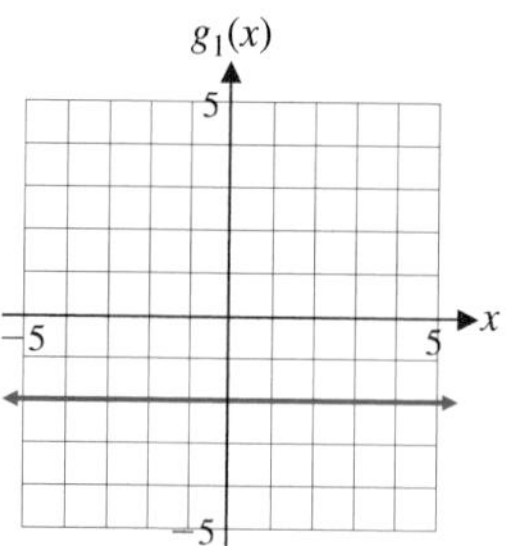

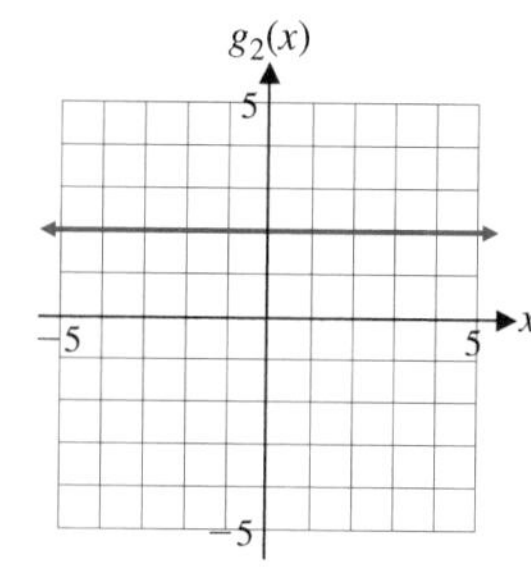

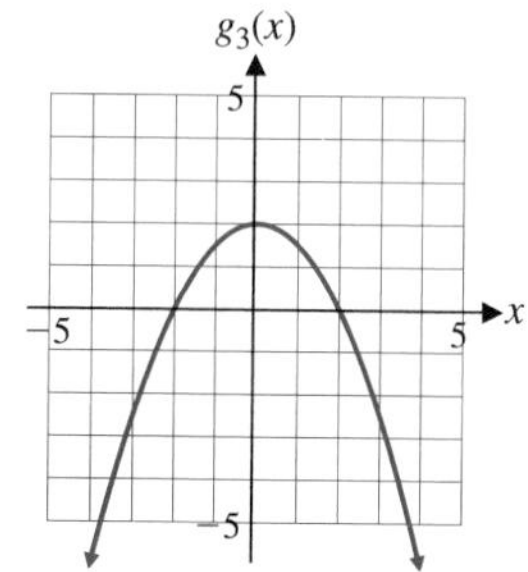

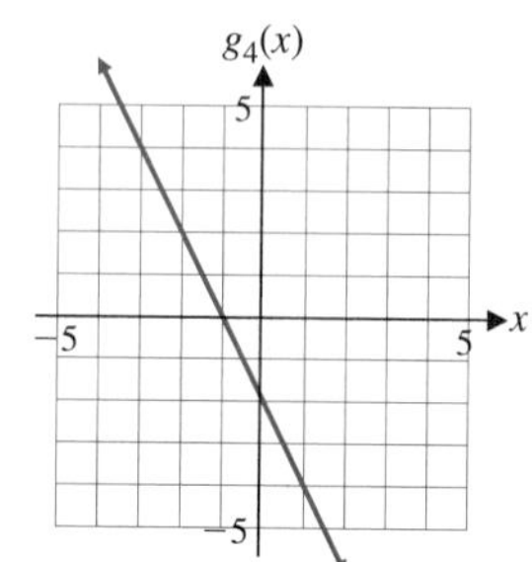

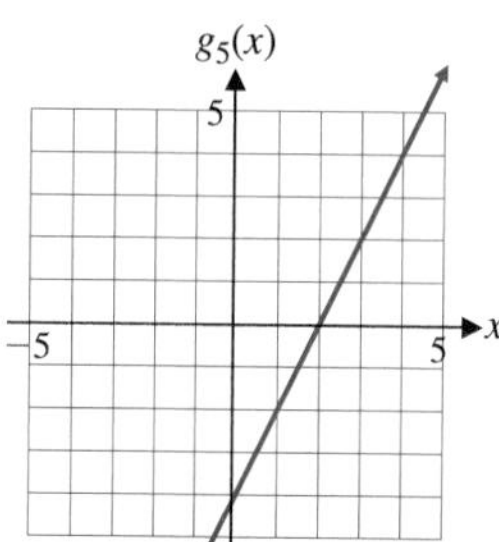

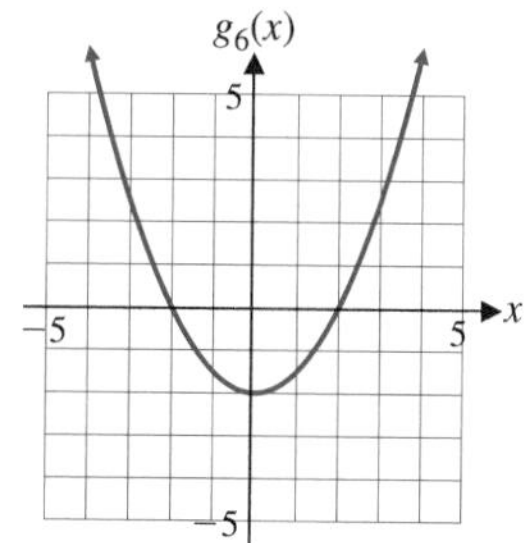

Figure (B) for 19–24

19. f_1 20. f_2 21. f_3
22. f_4 23. f_5 24. f_6

In Problems 25–38, find the intervals where $f(x)$ is increasing, the intervals where $f(x)$ is decreasing, and the local extrema.

25. $f(x) = x^2 - 16x + 12$
26. $f(x) = x^2 + 6x + 7$
27. $f(x) = 4 + 10x - x^2$
28. $f(x) = 5 + 8x - 2x^2$
29. $f(x) = 2x^3 + 4$
30. $f(x) = 2 - 3x^3$
31. $f(x) = 2 - 6x - 2x^3$
32. $f(x) = x^3 + 9x + 7$
33. $f(x) = x^3 - 12x + 8$
34. $f(x) = 3x - x^3$
35. $f(x) = x^3 - 3x^2 - 24x + 7$
36. $f(x) = x^3 + 3x^2 - 9x + 5$
37. $f(x) = 2x^2 - x^4$
38. $f(x) = x^4 - 8x^2 + 3$

In Problems 39–44, find the intervals where $f(x)$ is increasing, the intervals where $f(x)$ is decreasing, and sketch the graph. Add horizontal tangent lines.

39. $f(x) = 4 + 8x - x^2$

40. $f(x) = 2x^2 - 8x + 9$

41. $f(x) = x^3 - 3x + 1$

42. $f(x) = x^3 - 12x + 2$

43. $f(x) = 10 - 12x + 6x^2 - x^3$

44. $f(x) = x^3 + 3x^2 + 3x$

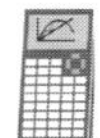

In Problems 45–50, use a graphing utility to approximate the critical values of $f(x)$ to two decimal places. Find the intervals where $f(x)$ is increasing, the intervals where $f(x)$ is decreasing, and the local extrema.

45. $f(x) = x^4 - 2x^2 + 3x$

46. $f(x) = x^4 - x^2 - 4x$

47. $f(x) = x^4 - 3x^3 + 2x$

48. $f(x) = x^4 + 3x^3 - 3x$

49. $f(x) = 0.25x^4 + 2x^3 + 2.5x^2 - 11x$

50. $f(x) = 0.25x^4 - x^3 - 2x^2 + 9x + 9$

In Problems 51–54, use the given graph of $y = f'(x)$ to find the intervals where f is increasing, the intervals where f is decreasing, and the local extrema. Sketch a possible graph for $y = f(x)$.

51.

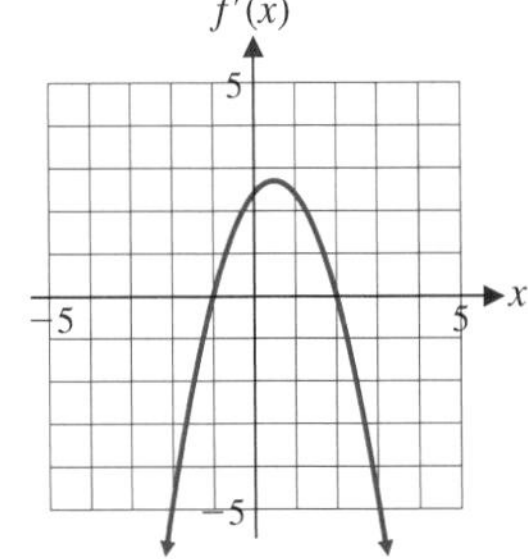

52.

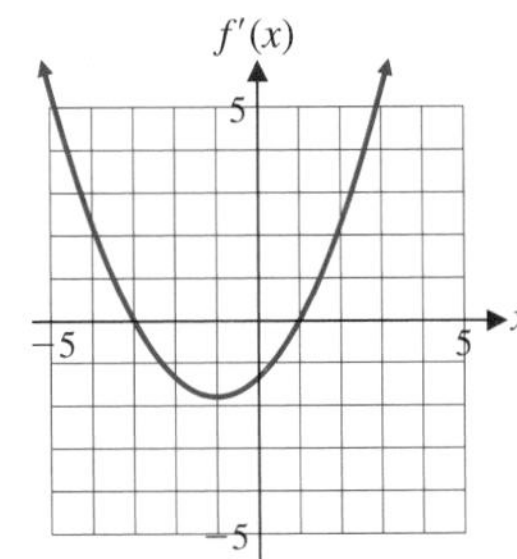

53.

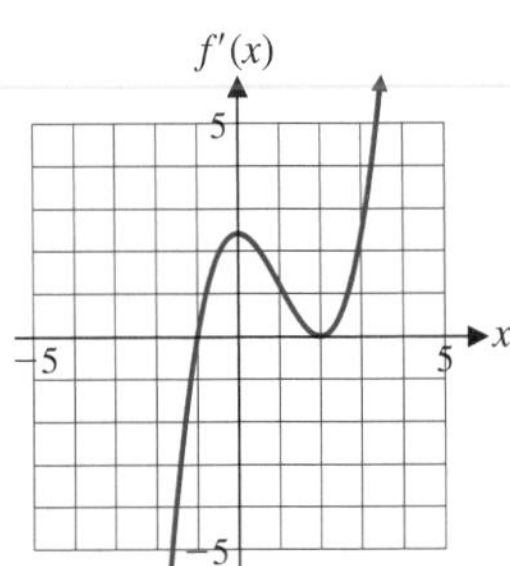

54.

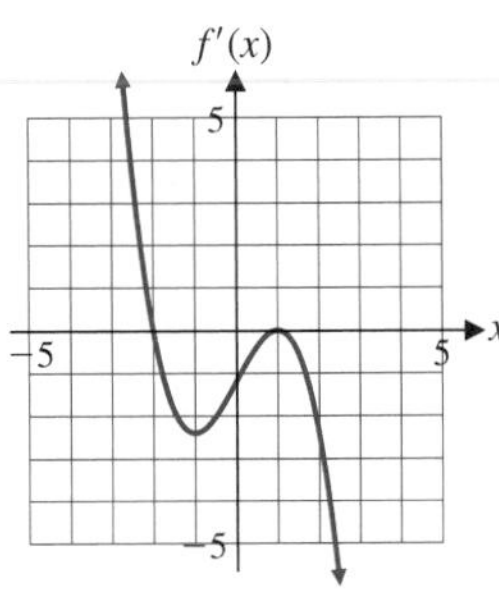

In Problems 55–58, use the given graph of $y = f(x)$ to find the intervals where $f'(x) > 0$, the intervals where $f'(x) < 0$, and the values of x for which $f'(x) = 0$. Sketch a possible graph for $y = f'(x)$.

55.

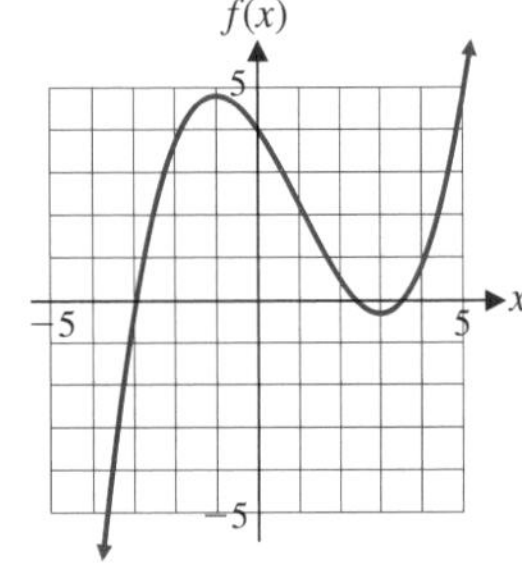

56.

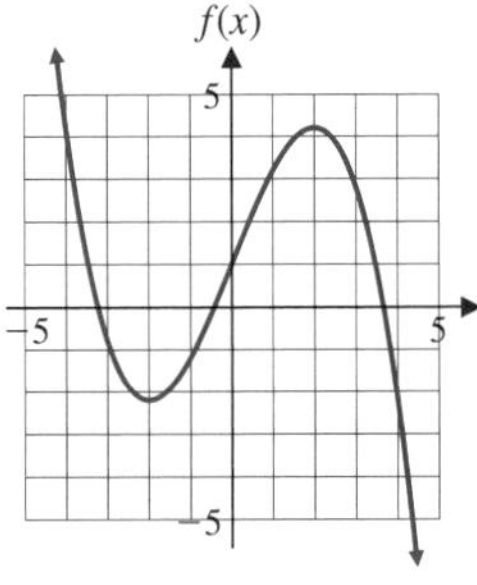

57.

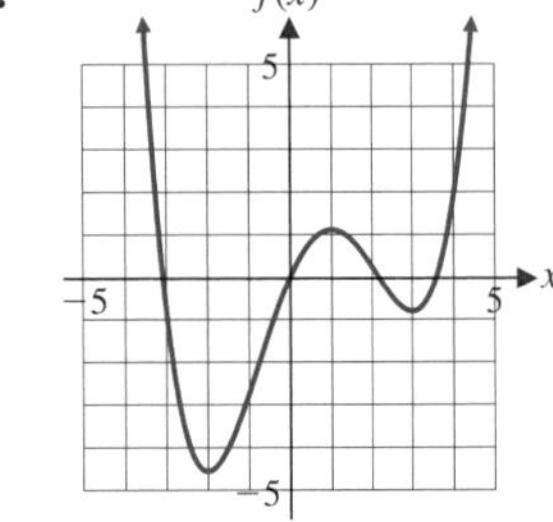

58.

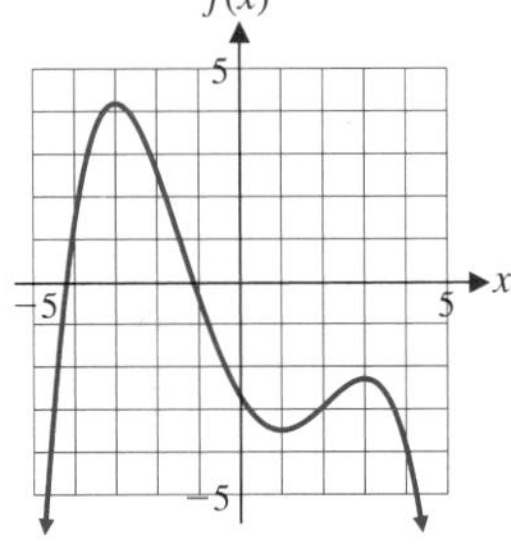

C *In Problems 59–72, find the critical values, the intervals where $f(x)$ is increasing, the intervals where $f(x)$ is decreasing, and the local extrema. Do not graph.*

59. $f(x) = \dfrac{x - 1}{x + 2}$

60. $f(x) = \dfrac{x + 2}{x - 3}$

61. $f(x) = x + \dfrac{4}{x}$

62. $f(x) = \dfrac{9}{x} + x$

63. $f(x) = 1 + \dfrac{1}{x} + \dfrac{1}{x^2}$

64. $f(x) = 3 - \dfrac{4}{x} - \dfrac{2}{x^2}$

65. $f(x) = \dfrac{x^2}{x - 2}$

66. $f(x) = \dfrac{x^2}{x + 1}$

67. $f(x) = x^4(x - 6)^2$

68. $f(x) = x^3(x - 5)^2$

69. $f(x) = 3(x - 2)^{2/3} + 4$

70. $f(x) = 6(4 - x)^{2/3} + 4$

71. $f(x) = 2\sqrt{x} - x, x > 0$

72. $f(x) = x - 4\sqrt{x}, x > 0$

73. Let $f(x) = x^3 + kx$, where k is a constant. Discuss the number of local extrema and the shape of the graph of f if:
(A) $k > 0$ (B) $k < 0$ (C) $k = 0$

74. Let $f(x) = x^4 + kx^2$, where k is a constant. Discuss the number of local extrema and the shape of the graph of f if:
(A) $k > 0$ (B) $k < 0$ (C) $k = 0$

Applications

Business & Economics

75. *Profit analysis.* The graph of the total profit $P(x)$ (in dollars) from the sale of x cordless electric screwdrivers is shown in the figure.

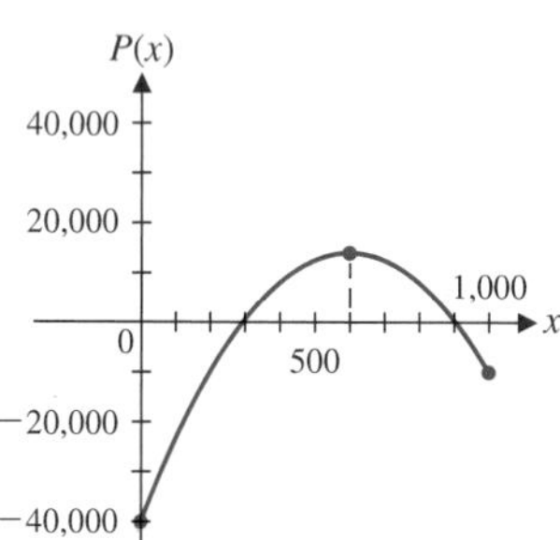

Figure for 75

(A) Write a brief verbal description of the graph of the marginal profit function $y = P'(x)$, including a discussion of any x intercepts.
(B) Sketch a possible graph of $y = P'(x)$.

76. *Revenue analysis.* The graph of the total revenue $R(x)$ (in dollars) from the sale of x cordless electric screwdrivers is shown in the figure.

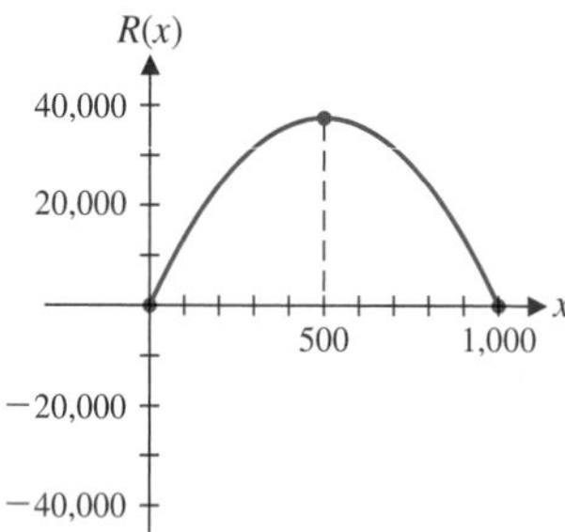

Figure for 76

(A) Write a brief verbal description of the graph of the marginal revenue function $y = R'(x)$, including a discussion of any x intercepts.
(B) Sketch a possible graph of $y = R'(x)$.

77. *Price analysis.* The graph in the figure approximates the rate of change of the price of bacon over a 70 month period, where $B(t)$ is the price of a pound of sliced bacon (in dollars) and t is time (in months).

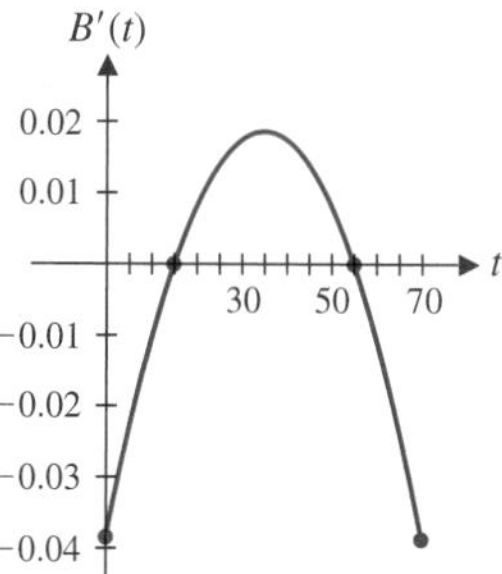

Figure for 77

(A) Write a brief verbal description of the graph of $y = B(t)$, including a discussion of any local extrema.
(B) Sketch a possible graph of $y = B(t)$.

78. *Price analysis.* The graph in the figure approximates the rate of change of the price of eggs over a 70 month period, where $E(t)$ is the price of a dozen eggs (in dollars) and t is time (in months).

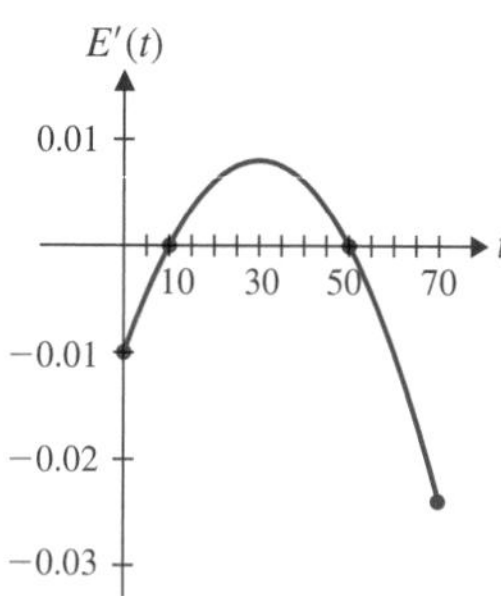

Figure for 78

(A) Write a brief verbal description of the graph of $y = E(t)$, including a discussion of any local extrema.
(B) Sketch a possible graph of $y = E(t)$.

79. *Average cost.* A manufacturer incurs the following costs in producing x toasters in one day for $0 < x < 150$: fixed costs, \$320; unit production cost, \$20 per toaster; equipment maintenance and repairs, $0.05x^2$ dollars. Thus, the cost of manufacturing x toasters in one day is given by

$$C(x) = 0.05x^2 + 20x + 320 \qquad 0 < x < 150$$

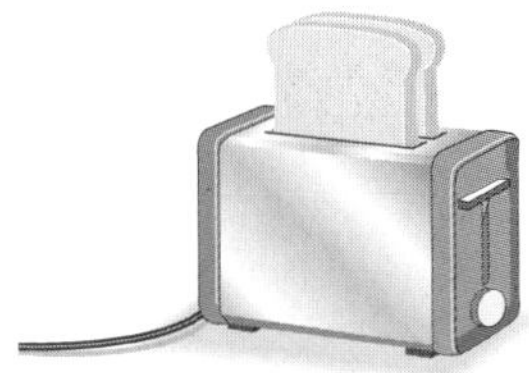

(A) What is the average cost, $\overline{C}(x)$, per toaster if x toasters are produced in one day?

(B) Find the critical values for $\overline{C}(x)$, the intervals where the average cost per toaster is decreasing, the intervals where the average cost per toaster is increasing, and the local extrema. Do not graph.

80. *Average cost.* A manufacturer incurs the following costs in producing x blenders in one day for $0 < x < 200$: fixed costs, \$450; unit production cost, \$30 per blender; equipment maintenance and repairs, $0.08x^2$ dollars.

(A) What is the average cost, $\overline{C}(x)$, per blender if x blenders are produced in one day?

(B) Find the critical values for $\overline{C}(x)$, the intervals where the average cost per blender is decreasing, the intervals where the average cost per blender is increasing, and the local extrema. Do not graph.

81. *Marginal analysis.* Show that profit will be increasing over production intervals (a, b) for which marginal revenue is greater than marginal cost.
[*Hint:* $P(x) = R(x) - C(x)$]

82. *Marginal analysis.* Show that profit will be decreasing over production intervals (a, b) for which marginal revenue is less than marginal cost.

Life Sciences

83. *Medicine.* A drug is injected into the bloodstream of a patient through the right arm. The concentration of the drug in the bloodstream of the left arm t hours after the injection is approximated by

$$C(t) = \frac{0.14t}{t^2 + 1} \qquad 0 < t < 24$$

Find the critical values for $C(t)$, the intervals where the concentration of the drug is increasing, the intervals where the concentration of the drug is decreasing, and the local extrema. Do not graph.

84. *Medicine.* The concentration $C(t)$, in milligrams per cubic centimeter, of a particular drug in a patient's bloodstream is given by

$$C(t) = \frac{0.16t}{t^2 + 4t + 4} \qquad 0 < t < 12$$

where t is the number of hours after the drug is taken orally. Find the critical values for $C(t)$, the intervals where the concentration of the drug is increasing, the intervals where the concentration of the drug is decreasing, and the local extrema. Do not graph.

Social Sciences

85. *Politics.* Public awareness of a Congressional candidate before and after a successful campaign was approximated by

$$P(t) = \frac{8.4t}{t^2 + 49} + 0.1 \qquad 0 < t < 24$$

where t is time (in months) after the campaign started and $P(t)$ is the fraction of people in the Congressional district who could recall the candidate's (and later, Congressman's) name. Find the critical values for $P(t)$, the time intervals where the fraction is increasing, the time intervals where the fraction is decreasing, and the local extrema. Do not graph.

SECTION 9-3 Second Derivative and Graphs

- CONCAVITY
- INFLECTION POINTS
- SECOND-DERIVATIVE TEST
- ANALYZING GRAPHS

In the preceding section, we saw that the derivative can be used to determine when a graph is rising and falling. Now we want to see what the *second derivative* (the derivative of the derivative) can tell us about the shape of a graph.

■ Concavity

Consider the functions

$$f(x) = x^2 \qquad \text{and} \qquad g(x) = \sqrt{x}$$

for x in the interval $(0, \infty)$. Since

$$f'(x) = 2x > 0 \qquad \text{for } 0 < x < \infty$$

and

$$g'(x) = \frac{1}{2\sqrt{x}} > 0 \qquad \text{for } 0 < x < \infty$$

both functions are increasing on $(0, \infty)$.

Explore–Discuss 1

(A) Discuss the difference in the shapes of the graphs of f and g shown in Figure 1.

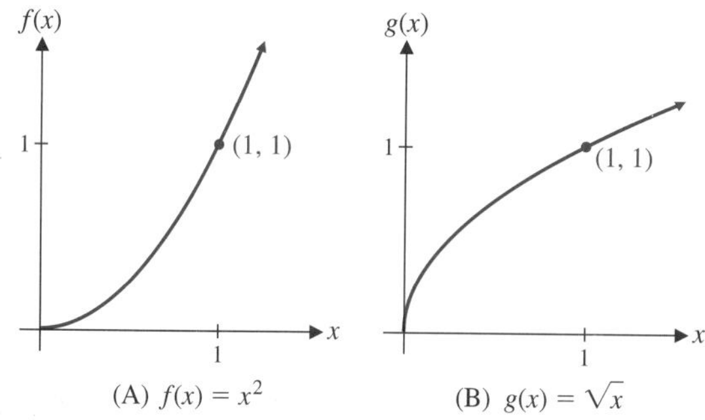

(A) $f(x) = x^2$ (B) $g(x) = \sqrt{x}$

FIGURE 1

(B) Complete the following table and discuss the relationship between the values of the derivatives of f and g and the shapes of their graphs.

x	0.25	0.5	0.75	1
$f'(x)$				
$g'(x)$				

We use the term *concave upward* to describe a graph that opens upward and *concave downward* to describe a graph that opens downward. Thus, the graph of f in Figure 1A is concave upward, and the graph of g in Figure 1B is concave downward. Finding a mathematical formulation of concavity will help us sketch and analyze graphs.

It will be instructive to examine the slopes of f and g at various points on their graphs (see Fig. 2). We can make two observations about each graph. Looking at the graph of f in Figure 2A, we see that $f'(x)$ (the slope of the tangent line) is *increasing* and that the graph lies *above* each tangent line. Looking at Figure 2B, we see that $g'(x)$ is *decreasing* and that the graph lies *below* each tangent line.

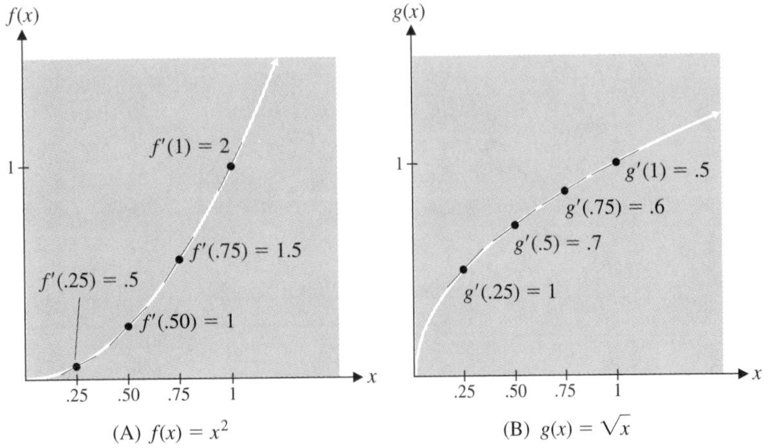

(A) $f(x) = x^2$

(B) $g(x) = \sqrt{x}$

FIGURE 2

With these ideas in mind, we state the general definition of concavity:

The graph of a function f is concave upward on the interval (a, b) if $f'(x)$ is *increasing* on (a, b) and is concave downward on the interval (a, b) if $f'(x)$ is *decreasing* on (a, b).

Geometrically, the graph is concave upward on (a, b) if it lies above its tangent lines in (a, b) and is concave downward on (a, b) if it lies below its tangent lines in (a, b).

How can we determine when $f'(x)$ is increasing or decreasing? In the preceding section, we used the derivative of a function to determine when that function is increasing or decreasing. Thus, to determine when the function $f'(x)$ is increasing or decreasing, we use the derivative of $f'(x)$. The derivative of the derivative of a function is called the *second derivative* of the function. Various notations for the second derivative are given in the following box:

Second Derivative

For $y = f(x)$, the **second derivative** of f, provided it exists, is

$$f''(x) = \frac{d}{dx} f'(x)$$

Other notations for $f''(x)$ are

$$\frac{d^2y}{dx^2} \qquad y''$$

Returning to the functions f and g discussed at the beginning of this section, we have

$$f(x) = x^2 \qquad g(x) = \sqrt{x} = x^{1/2}$$

$$f'(x) = 2x \qquad g'(x) = \frac{1}{2}x^{-1/2} = \frac{1}{2\sqrt{x}}$$

$$f''(x) = \frac{d}{dx}2x = 2 \qquad g''(x) = \frac{d}{dx}\frac{1}{2}x^{-1/2} = -\frac{1}{4}x^{-3/2} = -\frac{1}{4\sqrt{x^3}}$$

For $x > 0$, we see that $f''(x) > 0$; thus, $f'(x)$ is increasing and the graph of f is concave upward (see Fig. 2A). For $x > 0$, we also see that $g''(x) < 0$; thus, $g'(x)$ is decreasing and the graph of g is concave downward (see Fig. 2B). These ideas are summarized in the following box:

Concavity

For the interval (a, b):

$f''(x)$	$f'(x)$	GRAPH OF $y = f(x)$	EXAMPLES
+	Increasing	Concave upward	
−	Decreasing	Concave downward	

Be careful not to confuse concavity with falling and rising. As Figure 3 illustrates, a graph that is concave upward on an interval may be falling, rising, or both falling and rising on that interval. A similar statement holds for a graph that is concave downward.

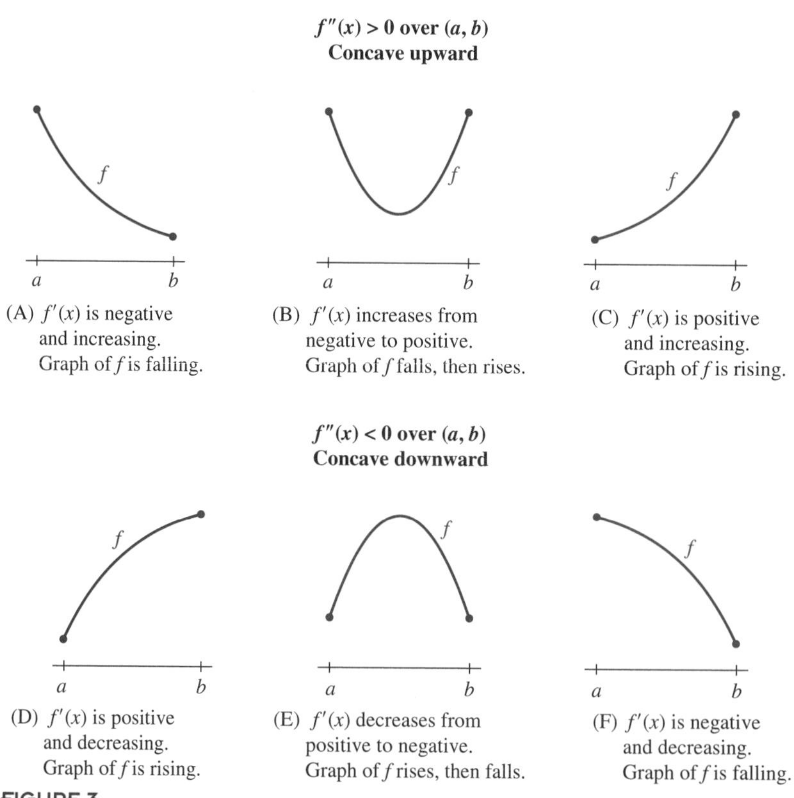

$f''(x) > 0$ over (a, b)
Concave upward
f
a
b
(A) $f'(x)$ is negative and increasing. Graph of f is falling.
(B) $f'(x)$ increases from negative to positive. Graph of f falls, then rises.
(C) $f'(x)$ is positive and increasing. Graph of f is rising.
$f''(x) < 0$ over (a, b)
Concave downward
(D) $f'(x)$ is positive and decreasing. Graph of f is rising.
(E) $f'(x)$ decreases from positive to negative. Graph of f rises, then falls.
(F) $f'(x)$ is negative and decreasing. Graph of f is falling.

FIGURE 3
Concavity

Example 1 **Determining Concavity of a Graph** Let $f(x) = x^3$. Find the intervals where the graph of f is concave upward and the intervals where the graph of f is concave downward. Sketch a graph of f.

SOLUTION To determine concavity, we must determine the sign of $f''(x)$.

$$f(x) = x^3 \qquad f'(x) = 3x^2 \qquad f''(x) = 6x$$

Sign chart for $f''(x) = 6x$ (partition number is 0):

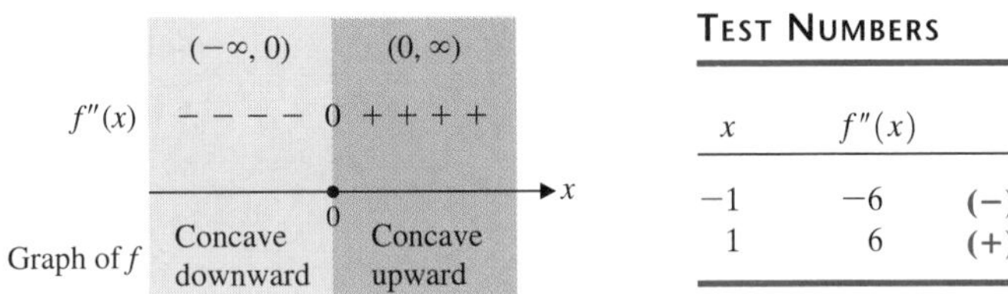

TEST NUMBERS

x	$f''(x)$	
-1	-6	$(-)$
1	6	$(+)$

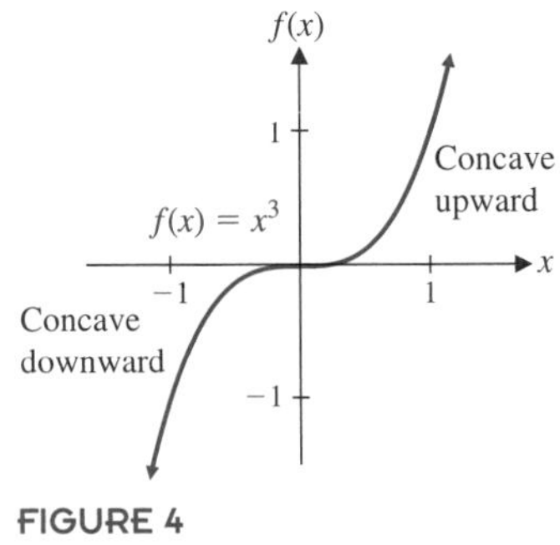

FIGURE 4

Thus, the graph of f is concave downward on $(-\infty, 0)$ and concave upward on $(0, \infty)$. The graph of f (without going through other graphing details) is shown in Figure 4.

Matched Problem 1 Repeat Example 1 for $f(x) = 1 - x^3$.

The graph in Example 1 changes from concave downward to concave upward at the point $(0, 0)$. This point is called an *inflection point.*

INFLECTION POINTS

Explore–Discuss 2

Discuss the relationship between the change in concavity of each of the following functions at $x = 0$ and the second derivative at and near 0.

(A) $f(x) = x^3$ (B) $g(x) = x^{4/3}$ (C) $h(x) = x^4$

In general, an **inflection point** is a point on the graph of the function where the concavity changes (from upward to downward or from downward to upward). In order for the concavity to change at a point, $f''(x)$ must change sign at that point. But in Section 9-1, we saw that the partition numbers* identify the points where a function can change sign. Thus, we have the following theorem:

Theorem 1 INFLECTION POINTS

If $y = f(x)$ is continuous on (a, b) and has an inflection point at $x = c$, then either $f''(c) = 0$ or $f''(c)$ does not exist.

*As we did with the first derivative, we will assume that if f'' is discontinuous at c, then $f''(c)$ does not exist.

Note that inflection points can occur only at partition numbers of f'', but not every partition number of f'' produces an inflection point. Two additional requirements must be satisfied for an inflection point:

A partition number c for f'' produces an inflection point for the graph of f only if:

1. **$f''(x)$ changes sign at c**
2. **c is in the domain of f**

Figure 5 illustrates several typical cases.

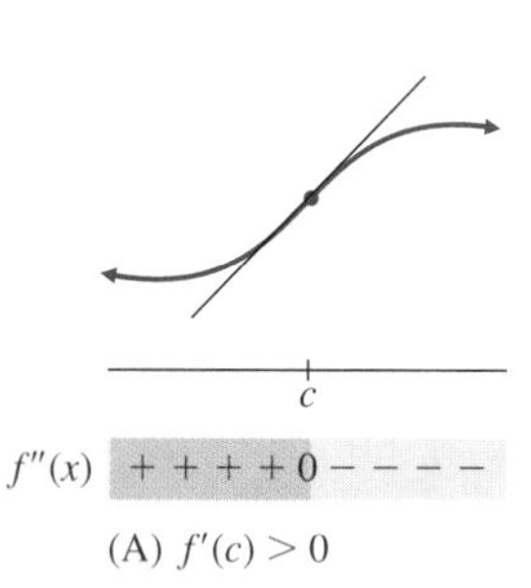

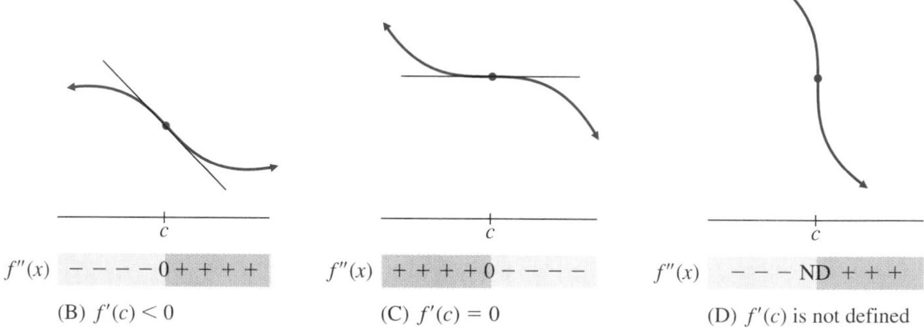

FIGURE 5
Inflection points

If $f'(c)$ exists and $f''(x)$ changes sign at $x = c$, then the tangent line at an inflection point $(c, f(c))$ will always lie below the graph on the side that is concave upward and above the graph on the side that is concave downward (see Figs. 5A, B, and C).

Example 2 **Locating Inflection Points** Find the inflection point(s) of

$$f(x) = x^3 - 6x^2 + 9x + 1$$

Solution Since inflection point(s) occur at values of x where $f''(x)$ changes sign, we construct a sign chart for $f''(x)$.

$$f(x) = x^3 - 6x^2 + 9x + 1$$
$$f'(x) = 3x^2 - 12x + 9$$
$$f''(x) = 6x - 12 = 6(x - 2)$$

Sign chart for $f''(x) = 6(x - 2)$ (partition number is 2):

	$(-\infty, 2)$		$(2, \infty)$
$f''(x)$	− − − −	0	+ + + +
x		2	
Graph of f	Concave downward	Inflection point	Concave upward

Test Numbers

x	$f''(x)$	
1	−6	(−)
3	6	(+)

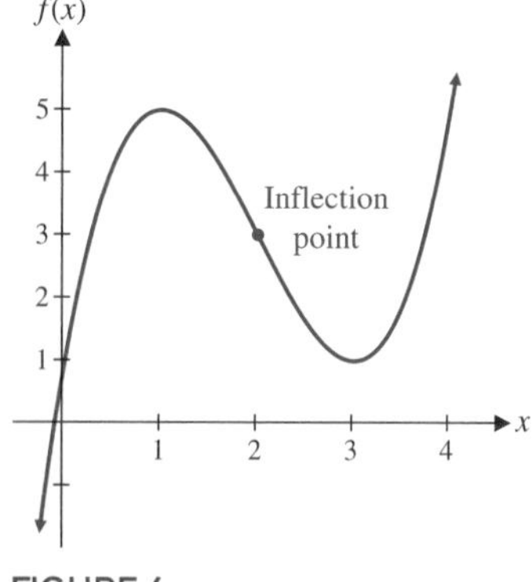

FIGURE 6

From the sign chart, we see that the graph of f has an inflection point at $x = 2$. The graph of f is shown in Figure 6. (See also Example 3 in Section 9-2.)

Matched Problem 2 Find the inflection point(s) of $f(x) = x^3 - 9x^2 + 24x - 10$. (See the answer to Matched Problem 3 in Section 9-2 for the graph of f.)

REMARK

Inflection points can be difficult to recognize on a graphing utility, but they are easily located using root approximation routines. Examining the graph of the function $f(x)$ from Example 2 on a graphing utility (Fig. 7A), it is clear that there must be an inflection point somewhere between the local maximum at $x = 1$ and the local minimum at $x = 3$. Graphing the second derivative and using a built-in root approximation routine (Fig. 7B) shows that this inflection point occurs at $x = 2$. Many graphing utilities also graph a numerical approximation to the second derivative that can be used to find inflection points. Consult the manual for your graphing utility for details.

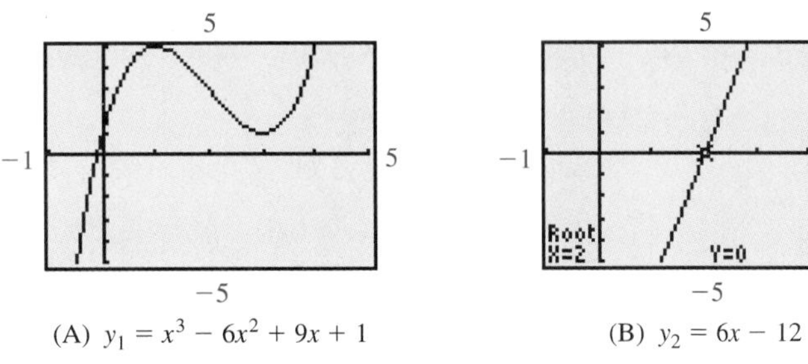

(A) $y_1 = x^3 - 6x^2 + 9x + 1$ (B) $y_2 = 6x - 12$

FIGURE 7

It is important to remember that the partition numbers of f'' are only candidates for inflection points. The function f must be defined at $x = c$, and the second derivative must change sign at $x = c$ in order for the graph to have an inflection point at $x = c$. For example, consider

$$f(x) = x^4 \qquad g(x) = \frac{1}{x}$$

$$f'(x) = 4x^3 \qquad g'(x) = -\frac{1}{x^2}$$

$$f''(x) = 12x^2 \qquad g''(x) = \frac{2}{x^3}$$

In each case, $x = 0$ is a partition number for the second derivative, but neither graph has an inflection point at $x = 0$. Function f does not have an inflection point at $x = 0$ because $f''(x)$ does not change sign at $x = 0$ (see Fig. 8A). Function g does not have an inflection point at $x = 0$ because $g(0)$ is not defined (see Fig. 8B).

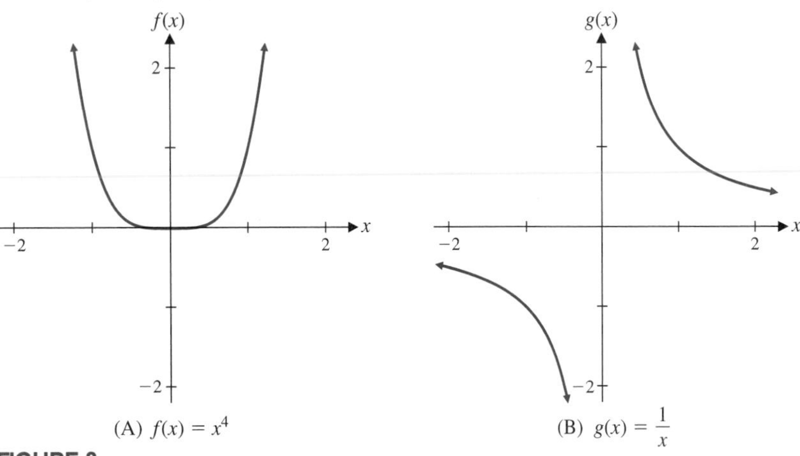

(A) $f(x) = x^4$ (B) $g(x) = \frac{1}{x}$

FIGURE 8

SECOND-DERIVATIVE TEST

Now we want to see how the second derivative can be used to find local extrema. Suppose f is a function satisfying $f'(c) = 0$ and $f''(c) > 0$. First, note that if $f''(c) > 0$, then it follows from the properties of limits* that $f''(x) > 0$ in some interval (m, n) containing c. Thus, the graph of f must be concave upward in this interval. But this implies that $f'(x)$ is increasing in this interval. Since $f'(c) = 0, f'(x)$ must change from negative to positive at $x = c$, and $f(c)$ is a local minimum (see Fig. 9). Reasoning in the same fashion, we conclude that if $f'(c) = 0$ and $f''(c) < 0$, then $f(c)$ is a local maximum. Of course, it is possible that both $f'(c) = 0$ and $f''(c) = 0$. In this case, the second derivative cannot be used to determine the shape of the graph around $x = c$; $f(c)$ may be a local minimum, a local maximum, or neither.

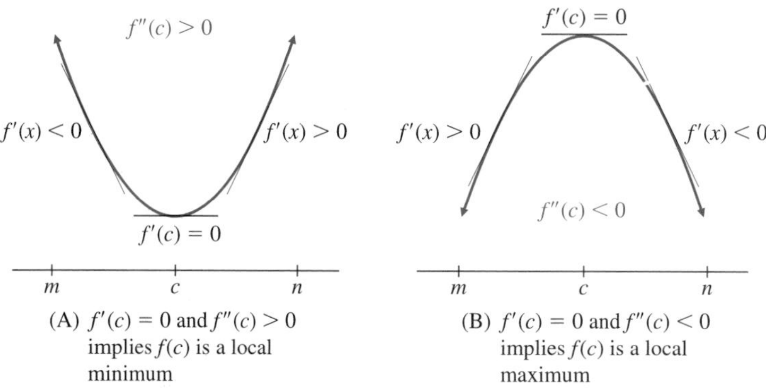

(A) $f'(c) = 0$ and $f''(c) > 0$ implies $f(c)$ is a local minimum

(B) $f'(c) = 0$ and $f''(c) < 0$ implies $f(c)$ is a local maximum

FIGURE 9
The second derivative and local extrema

The sign of the second derivative thus provides a simple test for identifying local maxima and minima. This test is most useful when we do not want to draw the graph of the function. If we are interested in drawing the graph and have already constructed the sign chart for $f'(x)$, then the first-derivative test can be used to identify the local extrema.

Second-Derivative Test for Local Maxima and Minima

Let c be a critical value for $f(x)$.

$f'(c)$	$f''(c)$	GRAPH OF f IS	$f(c)$	EXAMPLE
0	+	Concave upward	Local minimum	∪
0	−	Concave downward	Local maximum	∩
0	0	?	Test fails	

The first-derivative test must be used whenever $f''(c) = 0$ or $f''(c)$ does not exist.

*Actually, we are assuming that $f''(x)$ is continuous in an interval containing c. It is very unlikely that we will encounter a function for which $f''(c)$ exists but $f''(x)$ is not continuous in an interval containing c.

Example 3 **Testing for Local Extrema** Find the local maxima and minima for each function. Use the second-derivative test when it applies.

(A) $f(x) = x^3 - 6x^2 + 9x + 1$ (B) $f(x) = \frac{1}{6}x^6 - 4x^5 + 25x^4$

SOLUTION (A) Take first and second derivatives and find critical values:

$$\begin{aligned} f(x) &= x^3 - 6x^2 + 9x + 1 \\ f'(x) &= 3x^2 - 12x + 9 = 3(x-1)(x-3) \\ f''(x) &= 6x - 12 = 6(x-2) \end{aligned}$$

Critical values are $x = 1$ and $x = 3$.

$f''(1) = -6 < 0$ *f has a local maximum at x = 1.*
$f''(3) = 6 > 0$ *f has a local minimum at x = 3.*

(B)

$$\begin{aligned} f(x) &= \tfrac{1}{6}x^6 - 4x^5 + 25x^4 \\ f'(x) &= x^5 - 20x^4 + 100x^3 = x^3(x-10)^2 \\ f''(x) &= 5x^4 - 80x^3 + 300x^2 \end{aligned}$$

Critical values are $x = 0$ and $x = 10$.

$f''(0) = 0$
$f''(10) = 0$ *The second-derivative test fails at both critical values, so the first-derivative test must be used.*

Sign chart for $f'(x) = x^3(x-10)^2$ (partition numbers are 0 and 10):

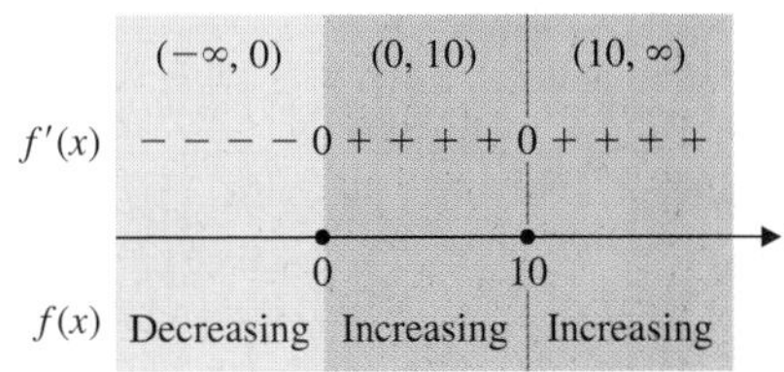

TEST NUMBERS

x	$f'(x)$	
−1	−121	(−)
1	81	(+)
11	1,331	(+)

From the sign chart, we see that $f(x)$ has a local minimum at $x = 0$ and does not have a local extremum at $x = 10$.

Matched Problem 3 Find the local maxima and minima for each function. Use the second-derivative test when it applies.

(A) $f(x) = x^3 - 9x^2 + 24x - 10$ (B) $f(x) = 10x^6 - 24x^5 + 15x^4$

A common error is to assume that $f''(c) = 0$ implies that $f(c)$ is not a local extremum. As Example 3B illustrates, if $f''(c) = 0$, then $f(c)$ may or may not be a local extremum. **The first-derivative test *must* be used whenever $f''(c) = 0$ or $f''(c)$ does not exist.**

ANALYZING GRAPHS

In the next two examples, we will combine increasing/decreasing properties with concavity properties to analyze the graph of a function.

Example 4 **Analyzing a Graph** Figure 10 shows the graph of the derivative of a function f. Use this graph to discuss the graph of f. Include a sketch of a possible graph of f.

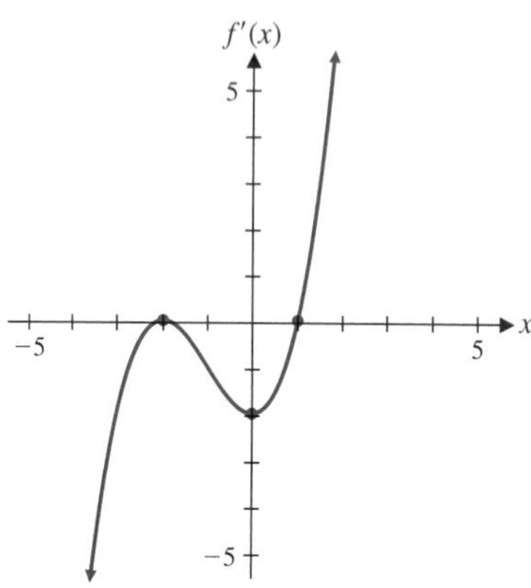

FIGURE 10

Solution The sign of the derivative determines where the original function is increasing and decreasing, and the increasing/decreasing properties of the derivative determine the concavity of the original function. The relevant information obtained from the graph of f' is summarized in Table 1, and a possible graph of f is shown in Figure 11.

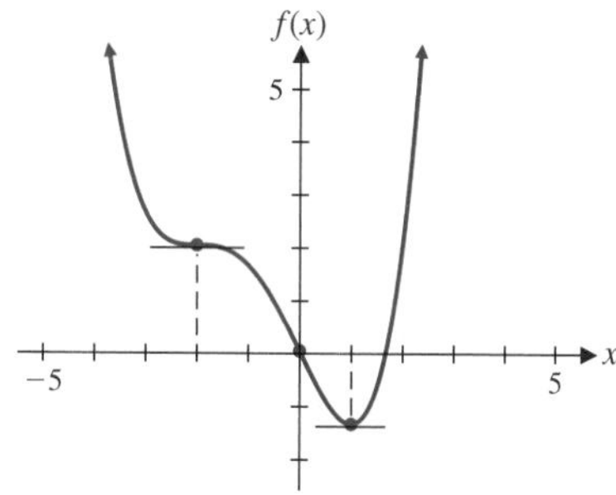

FIGURE 11

Table 1

x	$f'(x)$ (*Fig. 10*)	$f(x)$ (*Fig. 11*)
$-\infty < x < -2$	Negative and increasing	Decreasing and concave upward
$x = -2$	Local maximum	Inflection point
$-2 < x < 0$	Negative and decreasing	Decreasing and concave downward
$x = 0$	Local minimum	Inflection point
$0 < x < 1$	Negative and increasing	Decreasing and concave upward
$x = 1$	x intercept	Local minimum
$1 < x < \infty$	Positive and increasing	Increasing and concave upward

Matched Problem 4 Figure 12 shows the graph of the derivative of a function f. Use this graph to discuss the graph of f. Include a sketch of a possible graph of f.

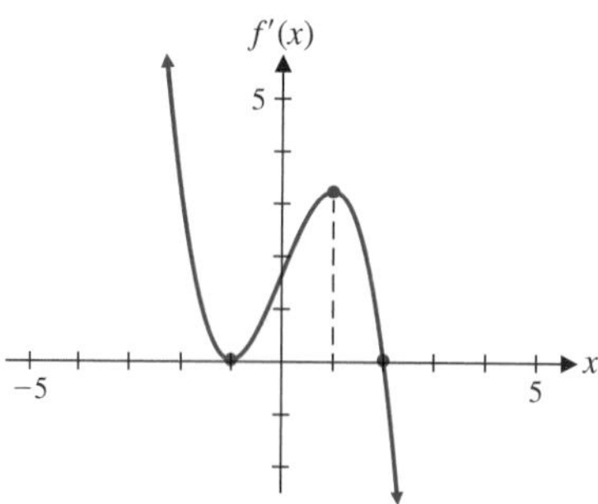

FIGURE 12

Example 5 **Maximum Rate of Change** Using past records, a company estimates that it will sell $N(x)$ items after spending $\$x$ thousand on advertising, as given by

$$N(x) = 2{,}000 - 2x^3 + 60x^2 - 450x \qquad 5 \le x \le 15$$

When is the rate of change of sales with respect to advertising expenditures increasing? Decreasing? What is the maximum rate of change? Graph N and N' on the same coordinate system and interpret.

SOLUTION The rate of change of sales with respect to advertising expenditures is

$$N'(x) = -6x^2 + 120x - 450 = -6(x - 5)(x - 15)$$

To determine when this rate is increasing and decreasing, we find $N''(x)$, the derivative of $N'(x)$:

$$N''(x) = -12x + 120 = 12(10 - x)$$

The information obtained by analyzing the signs of $N'(x)$ and $N''(x)$ is summarized in Table 2 (sign charts are omitted).

Table 2

x	$N''(x)$	$N'(x)$	$N'(x)$	$N(x)$
$5 < x < 10$	+	+	Increasing	Increasing, concave upward
$x = 10$	0	+	Local maximum	Inflection point
$10 < x < 15$	−	+	Decreasing	Increasing, concave downward

Thus, we see that $N'(x)$, the rate of change of sales, is increasing on $(5, 10)$ and decreasing on $(10, 15)$. Both N and N' are graphed in Figure 13. An examination of the graph of $N'(x)$ shows that the maximum rate of change is $N'(10) = 150$. Notice that $N'(x)$ has a local maximum and $N(x)$ has an inflection point at $x = 10$. This value of x is referred to as the **point of diminishing returns,** since the rate of change of sales begins to decrease at this point.

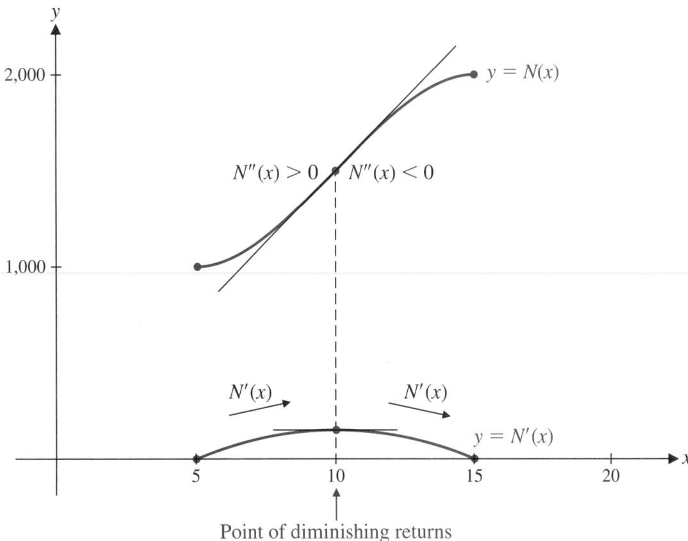

FIGURE 13

Matched Problem 5 Repeat Example 5 for $N(x) = 5{,}000 - x^3 + 60x^2 - 900x$, $10 \leqslant x \leqslant 30$.

Answers to Matched Problems

1. Concave upward on $(-\infty, 0)$
   Concave downward on $(0, \infty)$

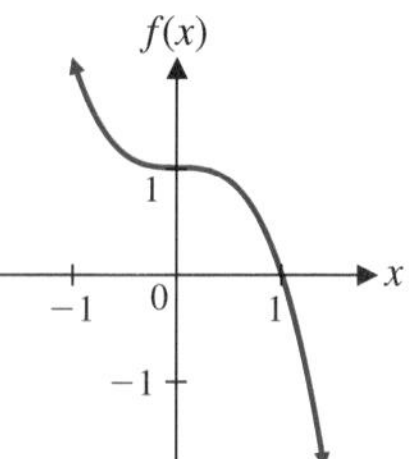

2. Inflection point at $x = 3$
3. (A) $f(2)$ is a local maximum; $f(4)$ is a local minimum
   (B) $f(0)$ is a local minimum; no local extremum at $x = 1$
4.

x	$f'(x)$	$f(x)$
$-\infty < x < -1$	Positive and decreasing	Increasing and concave downward
$x = -1$	Local minimum	Inflection point
$-1 < x < 1$	Positive and increasing	Increasing and concave upward
$x = 1$	Local maximum	Inflection point
$1 < x < 2$	Positive and decreasing	Increasing and concave downward
$x = 2$	x intercept	Local maximum
$2 < x < \infty$	Negative and decreasing	Decreasing and concave downward

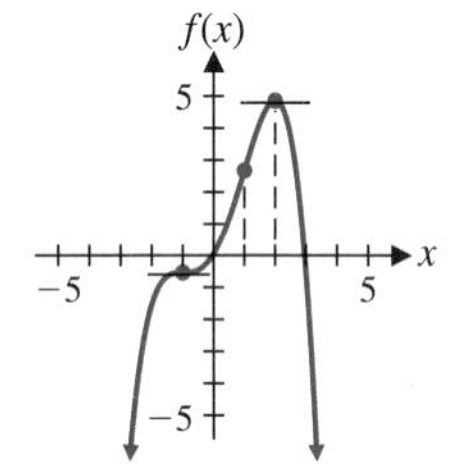

5. $N'(x)$ is increasing on $(10, 20)$, decreasing on $(20, 30)$; maximum rate of change is $N'(20) = 300$; $x = 20$ is point of diminishing returns

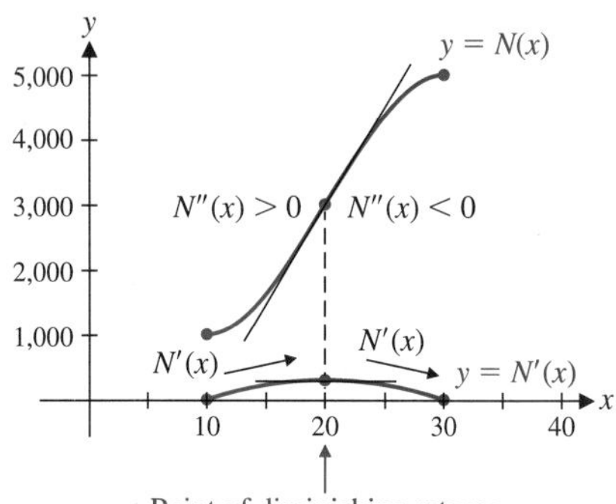

EXERCISE 9-3

A *Problems 1–8 refer to the graph of $y = f(x)$ below.*

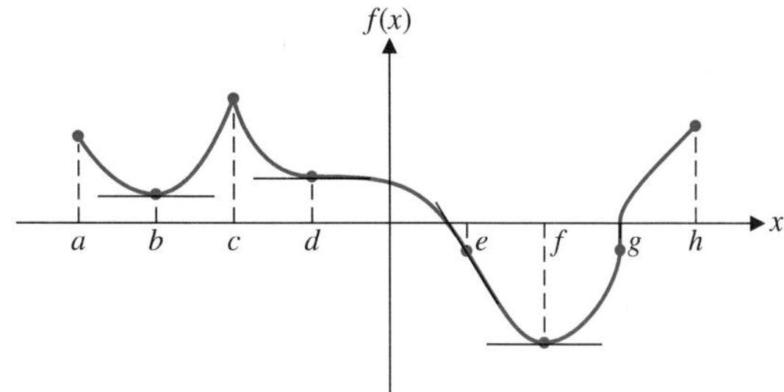

1. Identify intervals over which the graph of f is concave upward.
2. Identify intervals over which the graph of f is concave downward.
3. Identify intervals over which $f''(x) < 0$.
4. Identify intervals over which $f''(x) > 0$.
5. Identify intervals over which $f'(x)$ is increasing.
6. Identify intervals over which $f'(x)$ is decreasing.
7. Identify x coordinates of inflection points.
8. Identify x coordinates of local extrema for $f'(x)$.

In Problems 9–12, match the indicated conditions with one of the graphs (A)–(D) shown in the figure.

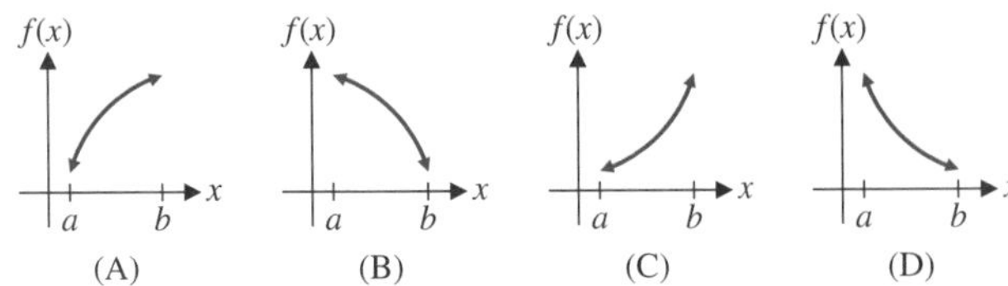

9. $f'(x) > 0$ and $f''(x) > 0$ on (a, b)

10. $f'(x) > 0$ and $f''(x) < 0$ on (a, b)

11. $f'(x) < 0$ and $f''(x) > 0$ on (a, b)

12. $f'(x) < 0$ and $f''(x) > 0$ on (a, b)

In Problems 13–18, describe the graph of f at the given point relative to the existence of a local maximum or minimum with one of the following phrases: "Local maximum," "Local minimum," "Neither," or "Unable to determine from the given information." Assume that $f(x)$ is continuous on $(-\infty, \infty)$.

13. $(2, f(2))$ if $f'(2) = 0$ and $f''(2) > 0$

14. $(4, f(4))$ if $f'(4) = 1$ and $f''(4) < 0$

15. $(-3, f(-3))$ if $f'(-3) = 0$ and $f''(-3) = 0$

16. $(-1, f(-1))$ if $f'(-1) = 0$ and $f''(-1) < 0$

17. $(6, f(6))$ if $f'(6) = 1$ and $f''(6)$ does not exist

18. $(5, f(5))$ if $f'(5) = 0$ and $f''(5)$ does not exist

In Problems 19–26, $f(x)$ is continuous on $(-\infty, \infty)$. Use the given information to sketch the graph of f.

19.

x	−4	−2	−1	0	2	4
$f(x)$	0	3	1.5	0	−1	−3

f′(x) + + + 0 − − − − − − − − 0 − − −
−2 2 x

f″(x) − − − − − − − − 0 + + + 0 − − −
−1 2 x

20.

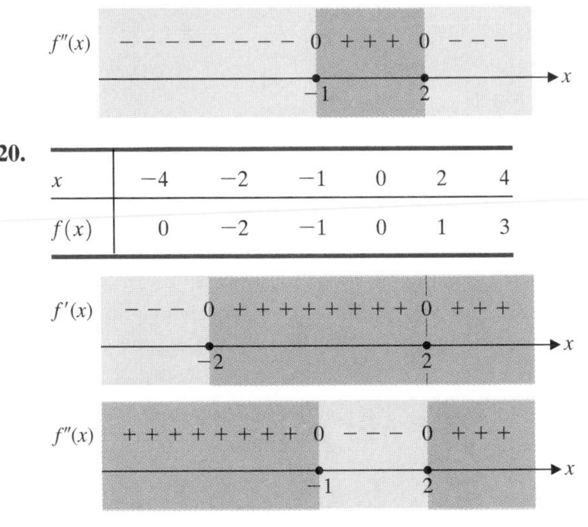

x	−4	−2	−1	0	2	4
$f(x)$	0	−2	−1	0	1	3

21.

x	−3	0	1	2	4	5
$f(x)$	−4	0	2	1	−1	0

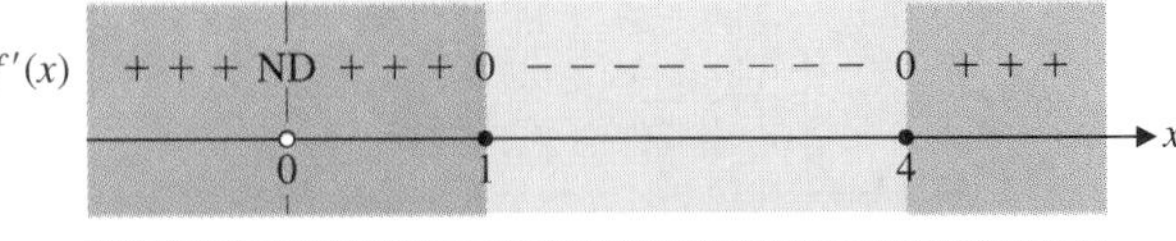

22.

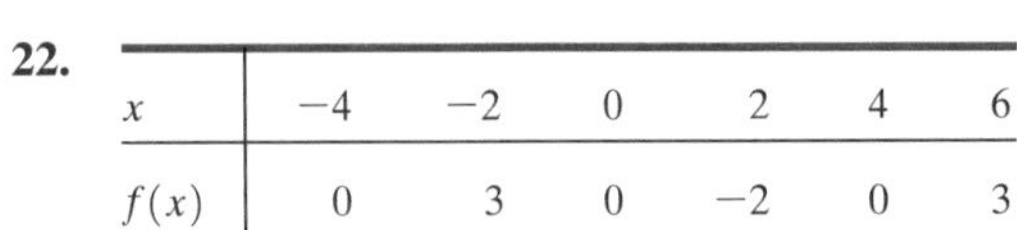

x	−4	−2	0	2	4	6
$f(x)$	0	3	0	−2	0	3

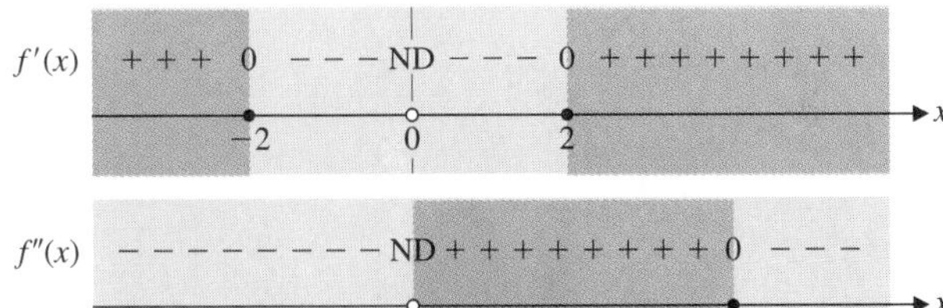

23. $f(0) = 2, f(1) = 0, f(2) = -2$;
$f'(0) = 0, f'(2) = 0$;
$f'(x) > 0$ on $(-\infty, 0)$ and $(2, \infty)$;
$f'(x) < 0$ on $(0, 2)$;
$f''(1) = 0$;
$f''(x) > 0$ on $(1, \infty)$;
$f'(x) < 0$ on $(-\infty, 1)$

24. $f(-2) = -2, f(0) = 1, f(2) = 4$;
$f'(-2) = 0, f'(2) = 0$;
$f'(x) > 0$ on $(-2, 2)$;
$f'(x) < 0$ on $(-\infty, -2)$ and $(2, \infty)$;
$f''(0) = 0$;
$f''(x) > 0$ on $(-\infty, 0)$;
$f'(x) < 0$ on $(0, \infty)$

25. $f(-1) = 0, f(0) = -2, f(1) = 0$;
$f'(0) = 0$, $f'(-1)$ and $f'(1)$ are not defined;
$f'(x) > 0$ on $(0, 1)$ and $(1, \infty)$;
$f'(x) < 0$ on $(-\infty, -1)$ and $(-1, 0)$;
$f''(-1)$ and $f''(1)$ are not defined;
$f''(x) > 0$ on $(-1, 1)$;
$f''(x) < 0$ on $(-\infty, -1)$ and $(1, \infty)$

26. $f(0) = -2, f(1) = 0, f(2) = 4$;
$f'(0) = 0, f'(2) = 0$, $f'(1)$ is not defined;
$f'(x) > 0$ on $(0, 1)$ and $(1, 2)$;
$f'(x) < 0$ on $(-\infty, 0)$ and $(2, \infty)$;
$f''(1)$ is not defined;
$f''(x) > 0$ on $(-\infty, 1)$;
$f''(x) < 0$ on $(1, \infty)$

In Problems 27–34, find the indicated derivative for each function.

27. $f''(x)$ for $f(x) = x^3 - 2x^2 - 1$

28. $g''(x)$ for $g(x) = x^4 - 3x^2 + 5$

29. d^2y/dx^2 for $y = 2x^5 - 3$

30. d^2y/dx^2 for $y = 3x^4 - 7x$

31. $f''(x)$ for $f(x) = 3x^{-1} + 2x^{-2} + 5$

32. $f''(x)$ for $f(x) = x^2 - x^{1/3}$

33. y'' for $y = (x^2 - 1)^3$

34. y'' for $y = (x^2 + 4)^4$

B *In Problems 35–46, find all local maxima and minima using the second-derivative test whenever it applies (do not graph). If the second-derivative test fails, use the first-derivative test.*

35. $f(x) = 2x^2 - 8x + 6$ **36.** $f(x) = 6x - x^2 + 4$

37. $f(x) = 2x^3 - 3x^2 - 12x - 5$

38. $f(x) = 2x^3 + 3x^2 - 12x - 1$

39. $f(x) = 3 - x^3 + 3x^2 - 3x$

40. $f(x) = x^3 + 6x^2 + 12x + 2$

41. $f(x) = x^4 - 8x^2 + 10$

42. $f(x) = x^4 - 18x^2 + 50$

43. $f(x) = x^6 + 3x^4 + 2$ **44.** $f(x) = 4 - x^6 - 6x^4$

45. $f(x) = x + \dfrac{16}{x}$ **46.** $f(x) = x + \dfrac{25}{x}$

In Problems 47–52, find the intervals where the graph of f is concave upward, the intervals where the graph is concave downward, and the inflection points.

47. $f(x) = x^2 - 4x + 5$ **48.** $f(x) = 9 + 3x - 4x^2$

49. $f(x) = x^3 - 18x^2 + 10x - 11$

50. $f(x) = x^3 + 24x^2 + 15x - 12$

51. $f(x) = x^4 - 24x^2 + 10x - 5$

52. $f(x) = -x^4 - 2x^3 + 12x^2 + 15$

In Problems 53–60, find local maxima, local minima, and inflection points. Sketch the graph of each function. Include tangent lines at each local extremum and inflection point.

53. $f(x) = x^3 - 6x^2 + 16$

54. $f(x) = x^3 - 9x^2 + 15x + 10$

55. $f(x) = x^3 + x + 2$

56. $f(x) = 1 - 3x - x^3$

57. $f(x) = (2 - x)^3 + 1$

58. $f(x) = (1 + x)^3 - 1$

59. $f(x) = x^3 - 12x$

60. $f(x) = 27x - x^3$

In Problems 61–64, use the graph of $y = f'(x)$ to discuss the graph of $y = f'(x)$. Organize your conclusions in a table (see Example 4, page 648), and sketch a possible graph of $y = f'(x)$.

61.

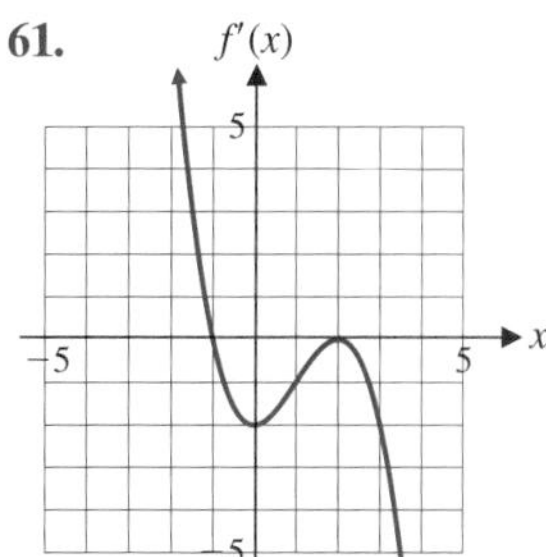

62.

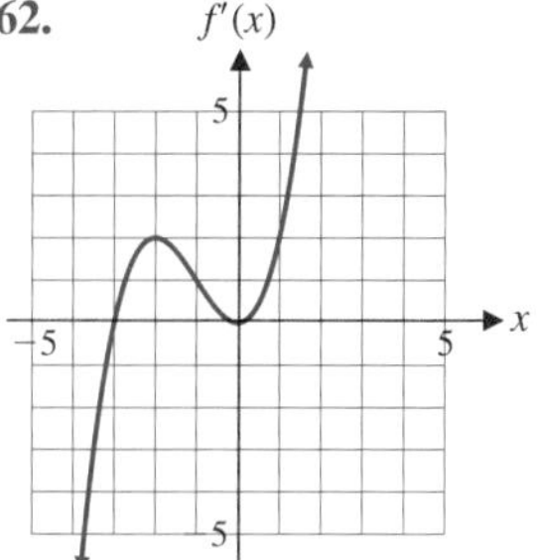

63.

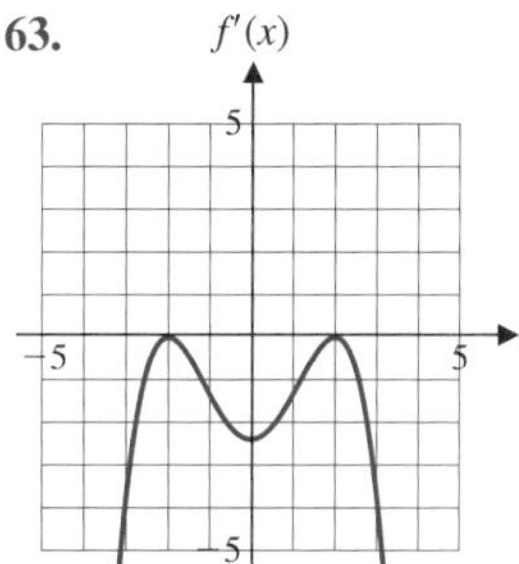

64.

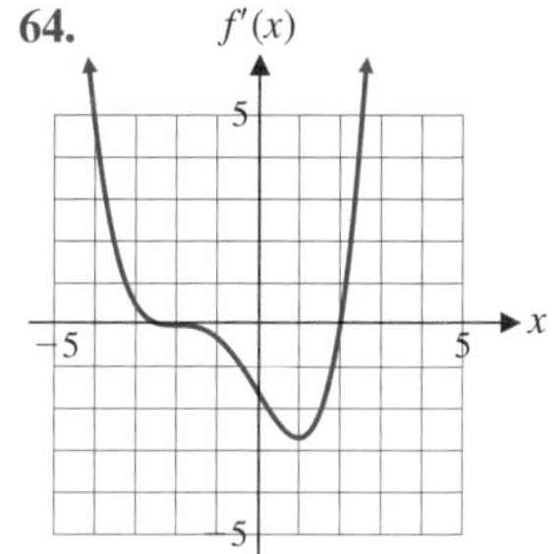

Problems 65–70 require the use of a graphing utility. Approximate the x coordinates of the inflection points of f to two decimal places. Find the intervals where the graph of f is concave upward and the intervals where the graph is concave downward.

65. $f(x) = 0.25x^4 - 2x^3 + 1.5x^2 + 21x + 15$

66. $f(x) = 0.25x^4 - 3.5x^3 + 12x^2 + 21x$

67. $f(x) = x^5 + 2x^4 + 4x^2 - 5$

68. $f(x) = x^5 - 3x^4 + x^3 - x^2 + 10$

69. $f(x) = x^5 - 3x^4 - x^3 + 7x^2 - 2$

70. $f(x) = x^5 - 2x^4 - 3x^3 + 4x^2 + 4x + 5$

In Problems 71–74, assume that f, f′, and f″ are continuous for all real numbers.

71. Explain how you can locate inflection points for the graph of $y = f(x)$ by examining the graph of $y = f'(x)$.

72. Explain how you can determine where $f'(x)$ is increasing or decreasing by examining the graph of $y = f(x)$.

73. Explain how you can locate local maxima and minima for the graph of $y = f'(x)$ by examining the graph of $y = f(x)$.

74. Explain how you can locate local maxima and minima for the graph of $y = f(x)$ by examining the graph of $y = f'(x)$.

C *Find the inflection points in Problems 75–78. Do not graph.*

75. $f(x) = \dfrac{1}{x^2 + 12}$

76. $f(x) = \dfrac{x^2}{x^2 + 12}$

77. $f(x) = \dfrac{x}{x^2 + 12}$

78. $f(x) = \dfrac{x^3}{x^2 + 12}$

Applications

Business & Economics

79. *Inflation.* One commonly used measure of inflation is the annual rate of change of the Consumer Price Index (CPI). A newspaper headline proclaims that the rate of change of inflation for consumer prices is increasing. What does this say about the shape of the graph of the CPI?

80. *Inflation.* Another commonly used measure of inflation is the annual rate of change of the Producers Price Index (PPI). A government report states that the rate of change of inflation for producer prices is decreasing. What does this say about the shape of the graph of the PPI?

81. *Cost analysis.* A company manufactures a variety of lighting fixtures at different locations. The total cost $C(x)$ (in dollars) of producing x desk lamps per week at plant A is shown in the figure. Discuss the shape of the graph of the marginal cost function $C'(x)$ and interpret in terms of the efficiency of the production process at this plant.

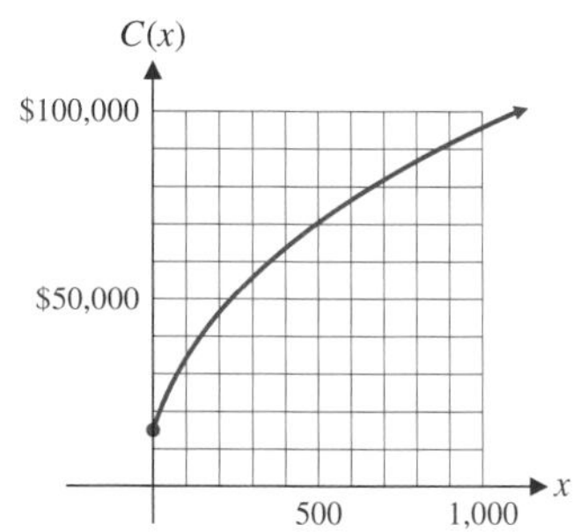

Figure for 81
Production costs at plant A

82. *Cost analysis.* The company in Problem 81 produces the same lamp at another plant. The total cost $C(x)$ (in dollars) of producing x desk lamps per week at plant B is shown in the figure. Discuss the shape of the graph of the marginal cost function $C'(x)$ and interpret in terms of the efficiency of the production process at plant B. Compare the production processes at these two plants.

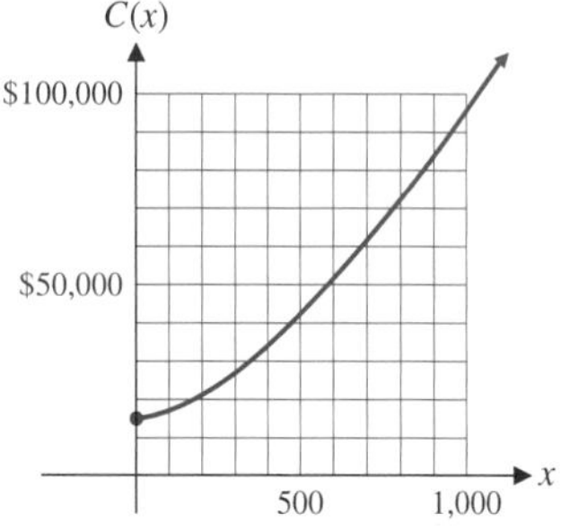

Figure for 82
Production costs at plant B

83. *Revenue.* The marketing research department for a computer company used a large city to test market their new product. They found that the relationship between price p (dollars per unit) and the demand x (units per week) was given approximately by

$$p = 1{,}296 - 0.12x^2 \qquad 0 < x < 80$$

Thus, the weekly revenue can be approximated by

$$R(x) = xp = 1{,}296x - 0.12x^3 \qquad 0 < x < 80$$

(A) Find the local extrema for the revenue function.
(B) Over which intervals is the graph of the revenue function concave upward? Concave downward?

84. *Profit.* Suppose the cost equation for the company in Problem 83 is

$$C(x) = 830 + 396x$$

(A) Find the local extrema for the profit function.
(B) Over which intervals is the graph of the profit function concave upward? Concave downward?

85. *Advertising.* A company estimates that it will sell $N(x)$ units of a product after spending $\$x$ thousand on advertising, as given by

$$N(x) = -3x^3 + 225x^2 - 3{,}600x + 17{,}000 \qquad 10 \le x \le 40$$

(A) When is the rate of change of sales $N'(x)$ increasing? Decreasing?
(B) Find the inflection points for the graph of N.
(C) Graph N and N' on the same coordinate system.
(D) What is the maximum rate of change of sales?

86. *Advertising.* A company estimates that it will sell $N(x)$ units of a product after spending $\$x$ thousand on advertising, as given by

$$N(x) = -2x^3 + 90x^2 - 750x + 2{,}000 \qquad 5 \le x \le 25$$

(A) When is the rate of change of sales $N'(x)$ increasing? Decreasing?
(B) Find the inflection points for the graph of N.
(C) Graph N and N' on the same coordinate system.
(D) What is the maximum rate of change of sales?

87. C *Advertising.* An automobile dealer uses television advertising to promote car sales. Using past records, the dealer arrived at the data in the table, where x is the number of ads placed monthly and y is the number of cars sold that month.

Number of Ads x	Number of Cars y
10	325
12	339
20	417
30	546
35	615
40	682
50	795

(A) Enter the data in a graphing utility and find a cubic regression equation for the number of cars sold monthly as a function of the number of ads.
(B) How many ads should the dealer place each month to maximize the rate of change of sales with respect to the number of ads, and how many cars can the dealer expect to sell with this number of ads? Round answers to the nearest integer.

88. C *Advertising.* A music store uses radio advertising to promote sales of CD's. The store manager used past records to determine the data in the table, where x is the number of ads placed monthly and y is the number of CD's sold that month.

Number of Ads x	Number of CD's y
10	345
14	488
20	746
30	1,228
40	1,671
50	1,955

(A) Enter the data in a graphing utility and find a cubic regression equation for the number of CD's sold monthly as a function of the number of ads.
(B) How many ads should the store manager place each month to maximize the rate of change of sales with respect to the number of ads, and how many CD's can the manager expect to sell with this number of ads? Round answers to the nearest integer.

Life Sciences

89. *Population growth—bacteria.* A drug that stimulates reproduction is introduced into a colony of bacteria. After t minutes, the number of bacteria is given approximately by

$$N(t) = 1{,}000 + 30t^2 - t^3 \qquad 0 \le t \le 20$$

(A) When is the rate of growth $N'(t)$ increasing? Decreasing?
(B) Find the inflection points for the graph of N.
(C) Sketch the graphs of N and N' on the same coordinate system.
(D) What is the maximum rate of growth?

90. *Drug sensitivity.* One hour after x milligrams of a particular drug are given to a person, the change in body temperature $T(x)$, in degrees Fahrenheit, is given by

$$T(x) = x^2\left(1 - \frac{x}{9}\right) \qquad 0 \le x \le 6$$

The rate $T'(x)$ at which $T(x)$ changes with respect to the size of the dosage x is called the *sensitivity* of the body to the dosage.

(A) When is $T'(x)$ increasing? Decreasing?
(B) Where does the graph of T have inflection points?
(C) Sketch the graphs of T and T' on the same coordinate system.
(D) What is the maximum value of $T'(x)$?

Social Sciences

91. *Learning.* The time T (in minutes) it takes a person to learn a list of length n is

$$T(n) = 0.08n^3 - 1.2n^2 + 6n \qquad n \ge 0$$

(A) When is the rate of change of T with respect to the length of the list increasing? Decreasing?
(B) Where does the graph of T have inflection points? Graph T and T' on the same coordinate system.
(C) What is the minimum value of $T'(n)$?

SECTION 9-4

Curve Sketching Techniques: Unified and Extended

- LIMITS AT INFINITY
- VERTICAL ASYMPTOTES
- GRAPHING STRATEGY
- USING THE STRATEGY
- APPLICATION

In this section we will apply, in a systematic way, all the graphing concepts discussed in the preceding three sections. Before we do this, we need to discuss the behavior of a graph as x increases or decreases without bound and also at certain points of discontinuity.

LIMITS AT INFINITY

An important element in the analysis of the graph of a function is the behavior of the function as x increases or decreases without bound. Recall from Section 9-1 that we used the special symbol ∞ to describe limits that increase or decrease without bound. We will write $x \to \infty$ to indicate that x is increasing without bound and $x \to -\infty$ to indicate that x is decreasing without bound.

We begin by considering power functions of the form x^p and $1/x^p$, where p is a positive real number.

Explore–Discuss 1

(A) Complete the following table:

x	100	1,000	10,000	100,000	1,000,000
x^2					
$1/x^2$					

(B) Describe verbally the behavior of x^2 as x increases without bound. Then use limit notation to describe this behavior.

(C) Repeat part (B) for $1/x^2$.

If p is a positive real number, then x^p increases as x increases, and it can be shown that there is no upper bound on the values of x^p. We indicate this by writing

$$x^p \to \infty \quad \text{as} \quad x \to \infty \qquad \text{or} \qquad \lim_{x\to\infty} x^p = \infty$$

Since the reciprocals of very large numbers are very small numbers, it follows that $1/x^p$ approaches 0 as x increases without bound. We indicate this behavior by writing

$$\frac{1}{x^p} \to 0 \quad \text{as} \quad x \to \infty \qquad \text{or} \qquad \lim_{x\to\infty} \frac{1}{x^p} = 0$$

Figure 1 illustrates this behavior for $f(x) = x^2$ and $g(x) = 1/x^2$.

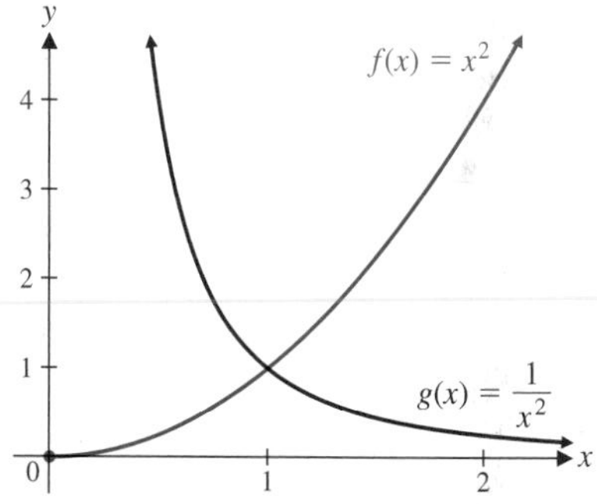

FIGURE 1
$\lim_{x\to\infty} f(x) = \infty$
$\lim_{x\to\infty} g(x) = 0$

Limits of power forms as x decreases without bound behave in a similar manner, with two important differences. First, if x is negative, then x^p is not a real number for all values of p. For example, $x^{1/2} = \sqrt{x}$ is not a real number for negative values of x. Second, if x^p represents a real number for all real x, then it may approach ∞ or $-\infty$, depending on the value of p. For example,

$$\lim_{x\to-\infty} x^2 = \infty \qquad \text{but} \qquad \lim_{x\to-\infty} x^3 = -\infty$$

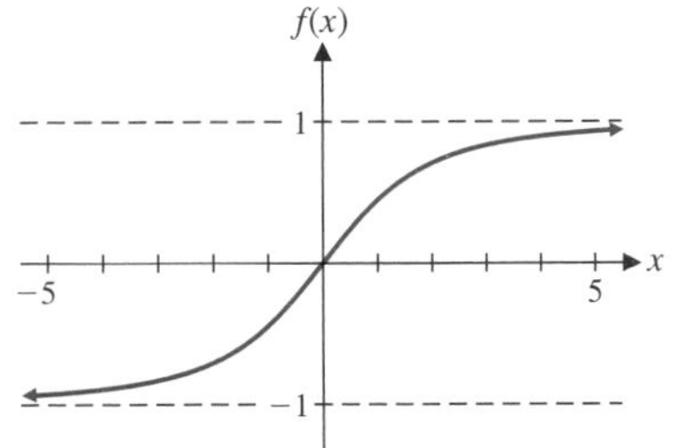

FIGURE 2

$\lim_{x\to-\infty} f(x) = -1$

$\lim_{x\to\infty} f(x) = 1$

For the function g in Figure 1, the line $y = 0$ (the x axis) is called a *horizontal asymptote*. In general, a line $y = b$ is a **horizontal asymptote** for the graph of $y = f(x)$ if $f(x)$ approaches b as x either decreases without bound or increases without bound. Symbolically, $y = b$ is a horizontal asymptote if either

$$\lim_{x\to-\infty} f(x) = b \qquad \text{or} \qquad \lim_{x\to\infty} f(x) = b$$

In the first case, the graph of f will be close to the horizontal line $y = b$ for large (in absolute value) negative x (see Fig. 2). In the second case, the graph will be close to the horizontal line $y = b$ for large positive x.

Theorem 1 summarizes the various possibilities for limits of power functions as x increases or decreases without bound.

Theorem 1 ■■ LIMITS AT INFINITY FOR POWER FUNCTIONS

If p is a positive real number and k is any real constant, then

1. $\lim_{x\to-\infty} \dfrac{k}{x^p} = 0$ **2.** $\lim_{x\to\infty} \dfrac{k}{x^p} = 0$

3. $\lim_{x\to-\infty} kx^p = \pm\infty$ **4.** $\lim_{x\to\infty} kx^p = \pm\infty$

provided that x^p names a real number for negative values of x. The limits in 3 and 4 will be either $-\infty$ or ∞, depending on k and p. ■■

It is important to understand that the symbol ∞ does not represent an actual number that x is approaching, but is used only to indicate that x is increasing with no upper limit on its size. And, as we discussed in Section 9-1, when we say that a limit is equal to ∞, we are using ∞ to describe the behavior of a limit that does not exist. Finally, we note that **limit properties 1–6, listed in Theorem 1 in Section 8-2, are valid if we replace the statement $x \to c$ with $x \to \infty$ or $x \to -\infty$.**

Now we want to use the ideas discussed above to investigate limits at infinity for polynomial functions and rational functions. We begin with an example involving a polynomial function.

Example 1 **Limit of a Polynomial Function at Infinity** Let

$$p(x) = 2x^3 - x^2 - 7x + 3$$

Discuss the limit of $p(x)$ as x approaches ∞ and as x approaches $-\infty$.

SOLUTION Since limits of power functions of the form $1/x^p$ approach 0 as x approaches ∞ or $-\infty$, it is convenient to work with these reciprocal forms whenever possible. If we factor out the term involving the highest power of x, we can write $p(x)$ as

$$p(x) = 2x^3\left(1 - \frac{1}{2x} - \frac{7}{2x^2} + \frac{3}{2x^3}\right)$$

Using Theorem 1 and limit properties from Section 8-2, we have

$$\lim_{x\to\infty}\left(1 - \frac{1}{2x} - \frac{7}{2x^2} + \frac{3}{2x^3}\right) = 1 - 0 - 0 + 0 = 1$$

Thus, for large values of x,

$$\left(1 - \frac{1}{2x} - \frac{7}{2x^2} + \frac{3}{2x^3}\right) \approx 1$$

and

$$p(x) = 2x^3\left(1 - \frac{1}{2x} - \frac{7}{2x^2} + \frac{3}{2x^3}\right) \approx 2x^3$$

Since $2x^3 \to \infty$ as $x \to \infty$, we can conclude that

$$\lim_{x\to\infty}(2x^3 - x^2 - 7x + 3) = \lim_{x\to\infty} 2x^3\left(1 - \frac{1}{2x} - \frac{7}{2x^2} + \frac{3}{2x^3}\right) = \infty$$

Similarly, $2x^3 \to -\infty$ as $x \to -\infty$ implies that

$$\lim_{x\to-\infty}(2x^3 - x^2 - 7x + 3) = \lim_{x\to-\infty} 2x^3\left(1 - \frac{1}{2x} - \frac{7}{2x^2} + \frac{3}{2x^3}\right) = -\infty$$

Thus, we can conclude that the behavior of $p(x)$ for large values is the same as the behavior of the highest-degree term, $2x^3$.

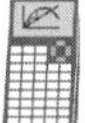

Figure 3 illustrates the behavior described in Example 1 on a graphing utility.

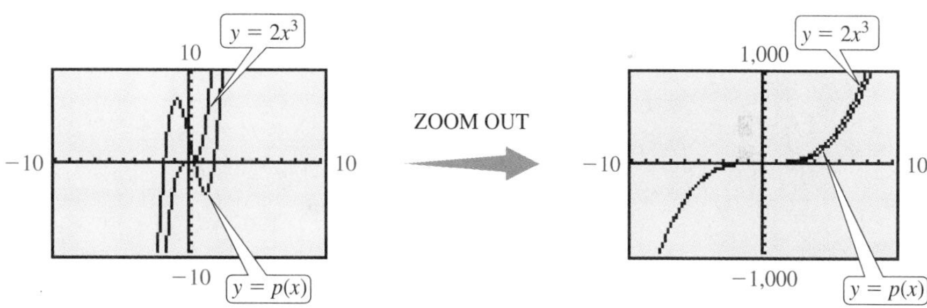

FIGURE 3
$p(x) = 2x^3 - x^2 - 7x + 3$

Matched Problem 1 Let $p(x) = -4x^4 + 2x^3 + 3x$. Discuss the limit of $p(x)$ as x approaches ∞ and as x approaches $-\infty$.

The ideas introduced in Example 1 are generalized to any polynomial in Theorem 2.

Theorem 2 LIMITS AT INFINITY FOR POLYNOMIAL FUNCTIONS

If

$$p(x) = a_nx^n + a_{n-1}x^{n-1} + \cdots + a_1x + a_0 \qquad a_n \neq 0, \quad n \geqslant 1$$

(continued)

then

$$\lim_{x\to\infty} p(x) = \lim_{x\to\infty} a_n x^n = \pm\infty$$

and

$$\lim_{x\to-\infty} p(x) = \lim_{x\to-\infty} a_n x^n = \pm\infty$$

Each limit will be either $-\infty$ or ∞, depending on a_n and n. ■■

A polynomial of degree 0 is a constant function, $p(x) = a_0$, and its limit as x approaches ∞ or $-\infty$ is the number a_0. For any polynomial of degree 1 or greater, Theorem 2 states that the limit as x approaches ∞ or $-\infty$ cannot be equal to a number. Thus, it follows that **polynomials of degree 1 or greater never have horizontal asymptotes.**

Explore–Discuss 2

Write a verbal description of the relationship between a polynomial and its highest-degree term as stated in Theorem 2.

Since a rational function is the ratio of two polynomials, it is not surprising that reciprocals of powers of x can also be used to analyze limits at infinity for rational functions. For example, consider the rational function

$$f(x) = \frac{2x^2 - 5x + 9}{3x^2 + 4x - 8}$$

Factoring the highest-degree term out of the numerator and the denominator, we have

$$f(x) = \frac{2x^2\left(1 - \dfrac{5}{2x} + \dfrac{9}{2x^2}\right)}{3x^2\left(1 + \dfrac{4}{3x} - \dfrac{8}{3x^2}\right)} = \frac{2}{3} \cdot \frac{1 - \dfrac{5}{2x} + \dfrac{9}{2x^2}}{1 + \dfrac{4}{3x} - \dfrac{8}{3x^2}}$$

Using Theorem 1, we have

$$\lim_{x\to\infty} f(x) = \lim_{x\to\infty}\left(\frac{2}{3} \cdot \frac{1 - \dfrac{5}{2x} + \dfrac{7}{2x^2}}{1 - \dfrac{4}{3x} + \dfrac{8}{3x^2}}\right) = \boxed{\frac{2}{3} \cdot \frac{1 - 0 + 0}{1 - 0 + 0}} = \frac{2}{3}$$

and

$$\lim_{x\to-\infty} f(x) = \lim_{x\to-\infty}\left(\frac{2}{3} \cdot \frac{1 - \dfrac{5}{2x} + \dfrac{7}{2x^2}}{1 - \dfrac{4}{3x} + \dfrac{8}{3x^2}}\right) = \boxed{\frac{2}{3} \cdot \frac{1 - 0 + 0}{1 - 0 + 0}} = \frac{2}{3}$$

Thus, for this rational function, the behavior as x approaches infinity is determined by the ratio of the highest-degree term in the numerator to the highest-degree term in the denominator. Theorem 3 generalizes this to any rational function and lists the three possible outcomes. This provides us with a useful tool for analyzing the behavior of rational functions as x approaches infinity.

Theorem 3 ■■ LIMITS AT INFINITY AND HORIZONTAL ASYMPTOTES FOR RATIONAL FUNCTIONS

If

$$f(x) = \frac{a_m x^m + a_{m-1}x^{m-1} + \cdots + a_1 x + a_0}{b_n x^n + b_{n-1}x^{n-1} + \cdots + b_1 x + b_0} \qquad a_m \neq 0, \quad b_n \neq 0$$

then

$$\lim_{x\to\infty} f(x) = \lim_{x\to\infty} \frac{a_m x^m}{b_n x^n} \qquad \text{and} \qquad \lim_{x\to-\infty} f(x) = \lim_{x\to-\infty} \frac{a_m x^m}{b_n x^n}$$

There are three possible cases for these limits:

1. If $m < n$, then $\lim_{x\to\infty} f(x) = \lim_{x\to-\infty} f(x) = 0$, and the line $y = 0$ (the x axis) is a horizontal asymptote for $f(x)$.
2. If $m = n$, then $\lim_{x\to\infty} f(x) = \lim_{x\to-\infty} f(x) = a_m/b_n$, and the line $y = a_m/b_n$ is a horizontal asymptote for $f(x)$.
3. If $m > n$, then each limit will be ∞ or $-\infty$, depending on m, n, a_m, and b_n, and $f(x)$ does not have a horizontal asymptote. ■■

Notice in cases 1 and 2 of Theorem 3 that the limit is the same if x approaches ∞ or $-\infty$. Thus, **a rational function can have at most one horizontal asymptote.**

Explore–Discuss 3

Case 1 in Theorem 3 can be stated verbally as follows:

> If the degree of the numerator of a rational function is less than the degree of the denominator, then the x axis is a horizontal asymptote.

Write similar verbal statements for cases 2 and 3.

Example 2 **Horizontal Asymptotes for Rational Functions** Find all horizontal asymptotes for each function:

(A) $f(x) = \dfrac{5x^3 - 2x^2 + 1}{4x^3 + 2x - 7}$ (B) $f(x) = \dfrac{3x^4 - x^2 + 1}{8x^6 - 10}$

(C) $f(x) = \dfrac{2x^5 - x^3 - 1}{6x^3 + 2x^2 - 7}$

SOLUTION (A) $\dfrac{a_m x^m}{b_n x^n} = \dfrac{5x^3}{4x^3} = \dfrac{5}{4}$

The line $y = \frac{5}{4}$ is a horizontal asymptote for $f(x)$ (Theorem 3, case 2).

(B) $\dfrac{a_m x^m}{b_n x^n} = \dfrac{3x^4}{8x^6} = \dfrac{3}{8x^2}$

The line $y = 0$ (the x axis) is a horizontal asymptote for $f(x)$ (Theorem 3, case 1).

(C) $\dfrac{a_m x^m}{b_n x^n} = \dfrac{2x^5}{6x^3} = \dfrac{x^2}{3}$

Thus, $f(x)$ does not have a horizontal asymptote (Theorem 3, case 3).

Matched Problem 2 Find all horizontal asymptotes for each function:

(A) $f(x) = \dfrac{4x^3 - 5x + 8}{2x^4 - 7}$ (B) $f(x) = \dfrac{5x^6 + 3x}{2x^5 - x - 5}$

(C) $f(x) = \dfrac{2x^3 - x + 7}{4x^3 + 3x^2 - 100}$

Vertical Asymptotes

Theorem 3 added an important tool for locating horizontal asymptotes to our mathematical toolbox. But in Section 9-1 we discussed another type of asymptote—a vertical asymptote. Theorem 4 provides a simple and effective tool for locating vertical asymptotes.

Theorem 4 Locating Vertical Asymptotes

Let $f(x) = n(x)/d(x)$, where both n and d are continuous at $x = c$. If, at $x = c$, the denominator $d(x)$ is 0 and the numerator $n(x)$ is not 0, then the line $x = c$ is a vertical asymptote for the graph of f.

[*Note:* Since a rational function is a ratio of two polynomial functions and polynomial functions are continuous for all real numbers, this theorem includes rational functions as a special case.]

If $f(x) = n(x)/d(x)$ and both $n(c) = 0$ and $d(c) = 0$, then the limit of $f(x)$ as x approaches c involves an indeterminate form and Theorem 4 does not apply:

$$\lim_{x \to c} f(x) = \lim_{x \to c} \frac{n(x)}{d(x)} \quad \frac{0}{0} \text{ indeterminate form}$$

Algebraic simplification is often useful in this situation.

Example 3 **Locating Vertical Asymptotes** Find all vertical asymptotes for

$$f(x) = \frac{x^2 + x - 2}{x^2 - 1}$$

SOLUTION Let $n(x) = x^2 + x - 2$ and $d(x) = x^2 - 1$. Factoring the denominator, we see that

$$d(x) = x^2 - 1 = (x - 1)(x + 1)$$

Since $d(-1) = 0$ and $n(-1) = -2 \neq 0$, Theorem 4 tells us that the line $x = -1$ is a vertical asymptote. On the other hand, $d(1) = 0$ but $n(1) = 0$

also, so Theorem 4 does not apply at $x = 1$. We use algebraic simplification to investigate the behavior of the function at $x = 1$:

$$\begin{aligned}\lim_{x\to 1} f(x) &= \lim_{x\to 1} \frac{x^2 + x - 2}{x^2 - 1}\\ &= \lim_{x\to 1} \frac{(x-1)(x+2)}{(x-1)(x+1)}\\ &= \lim_{x\to 1} \frac{x+2}{x+1}\\ &= \frac{3}{2}\end{aligned}$$

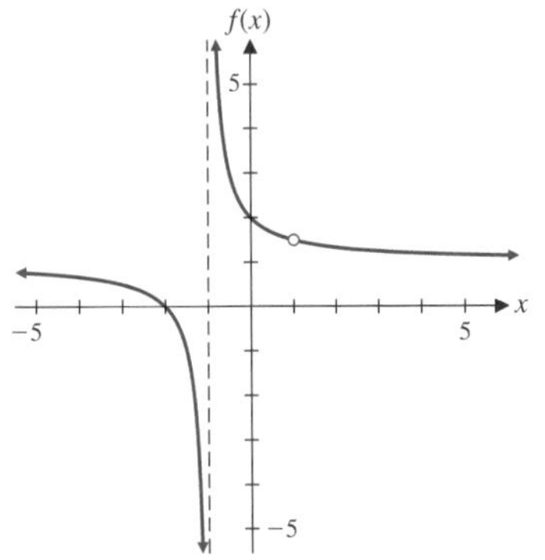

FIGURE 4
$f(x) = \dfrac{x^2 + x - 2}{x^2 - 1}$

Since the limit exists as x approaches 1, f does not have a vertical asymptote at $x = 1$ (see Fig. 4).

Matched Problem 3 Find all vertical asymptotes for: $f(x) = \dfrac{x - 3}{x^2 - 4x + 3}$

Graphing Strategy

We now have powerful tools to determine the shape of a graph of a function—even before we plot any points. We can accurately sketch the graphs of many functions using these tools and point-by-point plotting as necessary (often, very little point-by-point plotting is required). These same tools can be used to analyze a graph produced on a graphing utility or other electronic device. We organize these tools in the graphing strategy summarized in the next box.

A Graphing Strategy for $y = f(x)$

Step 1. *Analyze* $f(x)$:

(A) Find the domain of f. [The domain of f is the set of all real numbers x that produce real values for $f(x)$.]
(B) Find intercepts. [The y intercept is $f(0)$, if it exists; the x intercepts are the solutions to $f(x) = 0$, if they exist.]
(C) Find asymptotes. [Use Theorems 3 and 4, if they apply; otherwise, calculate limits at points of discontinuity and as x increases and decreases without bound.]

Step 2. *Analyze* $f'(x)$: Find any critical values for $f(x)$ and any partition numbers for $f'(x)$. [Remember, every critical value for $f(x)$ is also a partition number for $f'(x)$, but some partition numbers for $f'(x)$ may not be critical values for $f(x)$.] Construct a sign chart for $f'(x)$, determine the intervals where $f(x)$ is increasing and decreasing, and find local maxima and minima.

Step 3. *Analyze* $f''(x)$: Construct a sign chart for $f''(x)$, determine where the graph of f is concave upward and concave downward, and find any inflection points.

Step 4. *Sketch the graph of f:* Draw asymptotes and locate intercepts, local maxima and minima, and inflection points. Sketch in what you know from steps 1–3. In regions of uncertainty, use point-by-point plotting to complete the graph.

■ Using the Strategy

Some examples will illustrate the use of the graphing strategy.

Example 4 **Using the Graphing Strategy** Analyze the function

$$f(x) = x^4 - 2x^3$$

following the graphing strategy. State all the pertinent information, and sketch the graph of f.

SOLUTION **Step 1.** *Analyze* $f(x)$: $f(x) = x^4 - 2x^3$

(A) Domain: All real x

(B) y intercept: $f(0) = 0$

x intercept:

$$\begin{aligned} f(x) &= 0 \\ x^4 - 2x^3 &= 0 \\ x^3(x - 2) &= 0 \\ x &= 0, 2 \end{aligned}$$

(C) Asymptotes: Since f is a polynomial, there are no horizontal or vertical asymptotes.

Step 2. *Analyze* $f'(x)$: $f'(x) = 4x^3 - 6x^2 = 4x^2\left(x - \frac{3}{2}\right)$

Critical values for $f(x)$: 0 and $\frac{3}{2}$

Partition numbers for $f'(x)$: 0 and $\frac{3}{2}$

Sign chart for $f'(x)$:

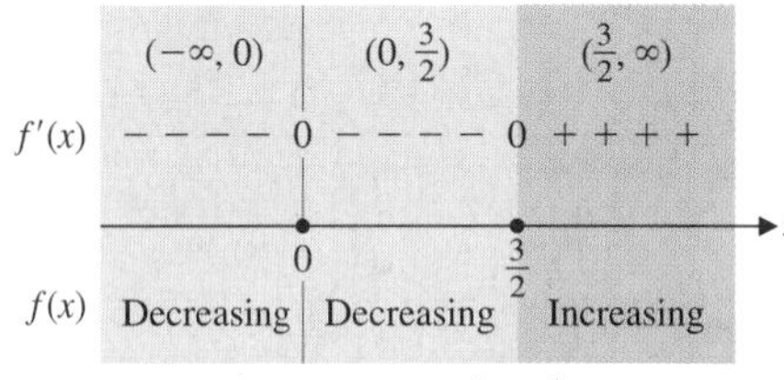

Local minimum

Test Numbers

x	$f'(x)$	
−1	−10	(−)
1	−2	(−)
2	8	(+)

Thus, $f(x)$ is decreasing on $\left(-\infty, \frac{3}{2}\right)$, increasing on $\left(\frac{3}{2}, \infty\right)$, and has a local minimum at $x = \frac{3}{2}$.

Step 3. *Analyze* $f''(x)$: $f''(x) = 12x^2 - 12x = 12x(x - 1)$

Partition numbers for $f''(x)$: 0 and 1

Sign chart for $f''(x)$:

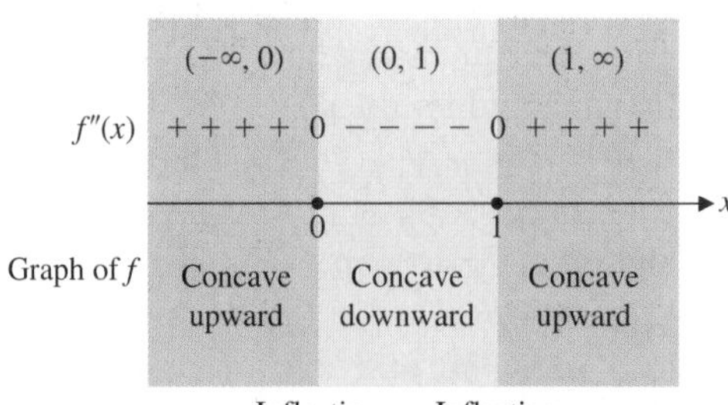

Inflection point Inflection point

Test Numbers

x	$f''(x)$	
−1	24	(+)
$\frac{1}{2}$	−3	(−)
2	24	(+)

Thus, the graph of f is concave upward on $(-\infty, 0)$ and $(1, \infty)$, concave downward on $(0, 1)$, and has inflection points at $x = 0$ and $x = 1$.

Step 4. *Sketch the graph of f:*

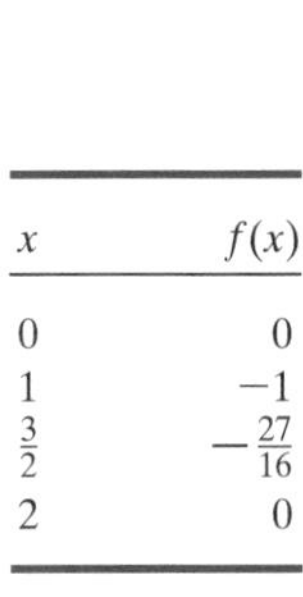

x	$f(x)$
0	0
1	-1
$\frac{3}{2}$	$-\frac{27}{16}$
2	0

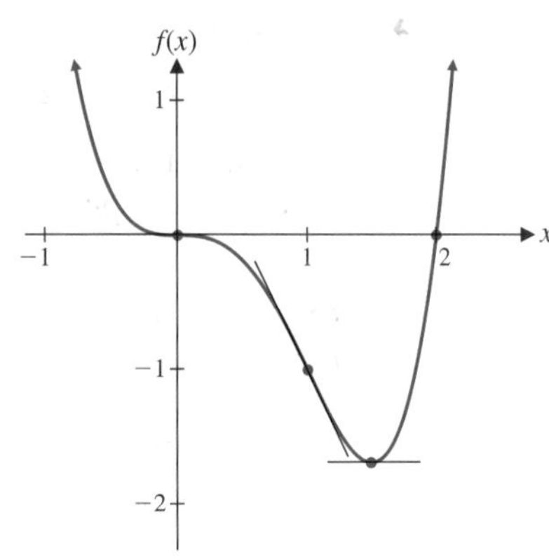

Matched Problem 4 ➡ Analyze the function $f(x) = x^4 + 4x^3$ following the graphing strategy. State all the pertinent information, and sketch the graph of f.

Example 5 ➡ **Using the Graphing Strategy** Analyze the function $f(x) = (x - 1)/(x - 2)$ following the graphing strategy. State all the pertinent information, and sketch the graph of f.

SOLUTION **Step 1.** *Analyze $f(x)$:* $f(x) = \dfrac{x - 1}{x - 2}$

(A) Domain: All real x, except $x = 2$

(B) y intercept: $f(0) = \dfrac{0 - 1}{0 - 2} = \dfrac{1}{2}$

x intercepts: Since a fraction is 0 when its numerator is 0 and the denominator is not 0, the x intercept is $x = 1$.

(C) Horizontal asymptote: $\dfrac{a_m x^m}{b_n x^n} = \dfrac{x}{x} = 1$

Thus, the line $y = 1$ is a horizontal asymptote.

Vertical asymptote: The denominator is 0 for $x = 2$, and the numerator is not 0 for this value. Therefore, the line $x = 2$ is a vertical asymptote.

Step 2. *Analyze $f'(x)$:* $f'(x) = \dfrac{(x - 2)(1) - (x - 1)(1)}{(x - 2)^2} = \dfrac{-1}{(x - 2)^2}$

Critical values for $f(x)$: None

Partition number for $f'(x)$: $x = 2$

Sign chart for $f'(x)$:

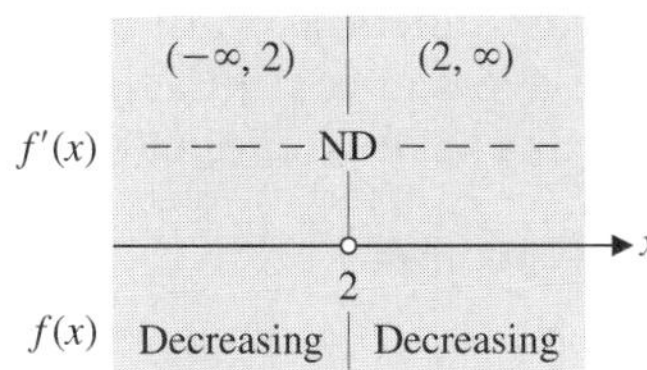

TEST NUMBERS

x	$f'(x)$	
1	-1	$(-)$
3	-1	$(-)$

Thus, $f(x)$ is decreasing on $(-\infty, 2)$ and $(2, \infty)$. There are no local extrema.

Step 3. *Analyze* $f''(x)$: $f''(x) = \dfrac{2}{(x-2)^3}$

Partition number for $f''(x)$: $x = 2$

Sign chart for $f''(x)$:

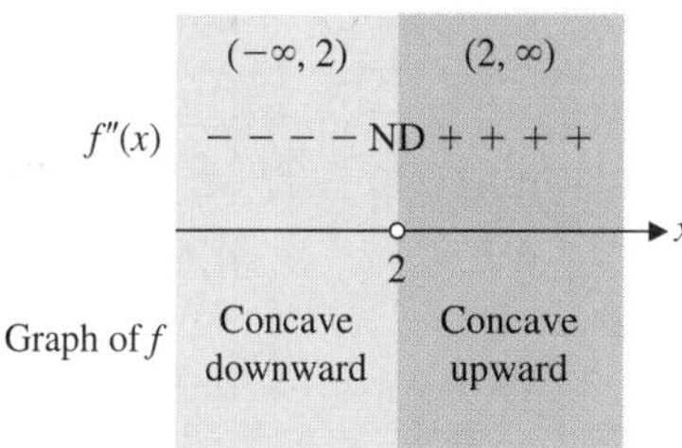

Test Numbers

x	$f''(x)$	
1	-2	$(-)$
3	2	$(+)$

Thus, the graph of f is concave downward on $(-\infty, 2)$ and concave upward on $(2, \infty)$. Since $f(2)$ is not defined, there is no inflection point at $x = 2$, even though $f''(x)$ changes sign at $x = 2$.

Step 4. *Sketch the graph of f:* Insert intercepts and asymptotes, and plot a few additional points (for functions with asymptotes, plotting additional points is often helpful). Then sketch the graph:

x	$f(x)$
-2	$\frac{3}{4}$
0	$\frac{1}{2}$
1	0
$\frac{3}{2}$	-1
$\frac{5}{2}$	3
3	2
4	$\frac{3}{2}$

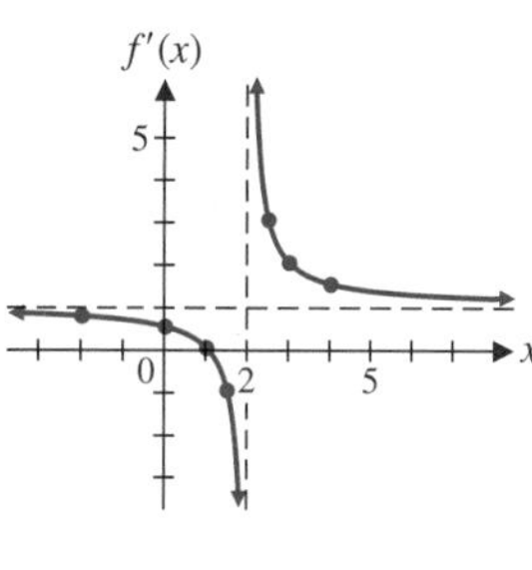

Matched Problem 5 ➡ Analyze the function $f(x) = 2x/(1 - x)$ following the graphing strategy. State all the pertinent information, and sketch the graph of f.

■ Application

Example 6 ➡ **Average Cost** Given the cost function $C(x) = 5{,}000 + 0.5x^2$, where x is the number of items produced, use the graphing strategy to analyze the graph of the average cost function. State all the pertinent information and sketch the graph of the average cost function. Find the marginal cost function and graph it on the same set of coordinate axes.

Solution The average cost function is

$$\overline{C}(x) = \frac{5{,}000 + 0.5x^2}{x} = \frac{5{,}000}{x} + 0.5x$$

Step 1. *Analyze* $\overline{C}(x)$:

(A) Domain: Since negative values of x do not make sense and $\overline{C}(0)$ is not defined, the domain is the set of positive real numbers.

(B) Intercepts: None

(C) Horizontal asymptote: $\dfrac{a_m x^m}{b_n x^n} = \dfrac{0.5x^2}{x} = 0.5x$

Thus, there is no horizontal asymptote.

Vertical asymptote: The line $x = 0$ is a vertical asymptote since the denominator is 0 and the numerator is not 0 for $x = 0$.

Oblique asymptotes: Some graphs have asymptotes that are neither vertical nor horizontal. These are called **oblique asymptotes.** If x is a large positive number, then $5{,}000/x$ is very small and

$$\overline{C}(x) = \frac{5{,}000}{x} + 0.5x \approx 0.5x$$

That is,

$$\lim_{x\to\infty}[\overline{C}(x) - 0.5x] = \lim_{x\to\infty}\frac{5{,}000}{x} = 0$$

This implies that the graph of $y = \overline{C}(x)$ approaches the line $y = 0.5x$ as x approaches ∞. This line is an oblique asymptote for the graph of $y = \overline{C}(x)$.

Step 2. *Analyze* $\overline{C}'(x)$:

$$\begin{aligned}\overline{C}'(x) &= -\frac{5{,}000}{x^2} + 0.5\\ &= \frac{0.5x^2 - 5{,}000}{x^2}\\ &= \frac{0.5(x - 100)(x + 100)}{x^2}\end{aligned}$$

Critical value for $\overline{C}(x)$: 100

Partition number for $\overline{C}'(x)$: 0 and 100

Sign chart for $\overline{C}'(x)$:

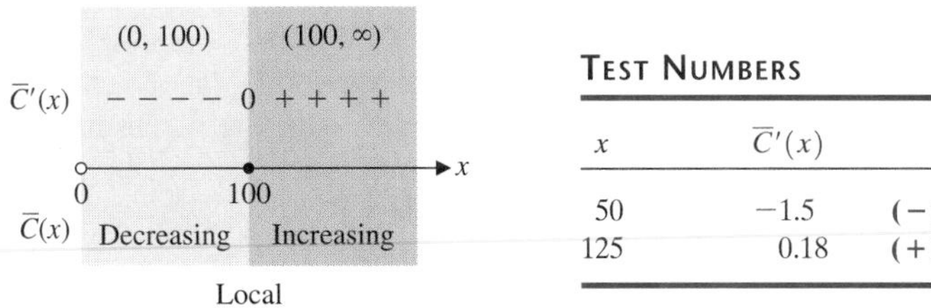

TEST NUMBERS

x	$\overline{C}'(x)$	
50	−1.5	(−)
125	0.18	(+)

Thus, $\overline{C}(x)$ is decreasing on $(0, 100)$, increasing on $(100, \infty)$, and has a local minimum at $x = 100$.

Step 3. *Analyze* $\overline{C}''(x)$: $\overline{C}''(x) = \dfrac{10{,}000}{x^3}$

$\overline{C}''(x)$ is positive for all positive x; therefore, the graph of $y = \overline{C}(x)$ is concave upward on $(0, \infty)$.

Step 4. *Sketch the graph of $\overline{C}$:* The graph of $\overline{C}$ is shown in Figure 5.

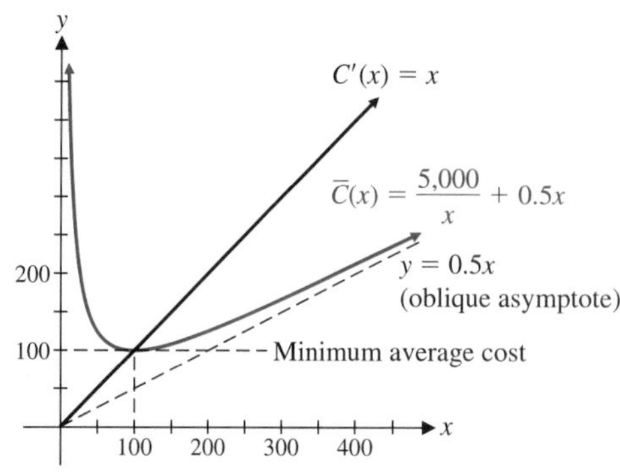

FIGURE 5

The marginal cost function is $C'(x) = x$. The graph of this linear function is also shown in Figure 5.

The graph in Figure 5 illustrates an important principle in economics:

The minimum average cost occurs when the average cost is equal to the marginal cost.

Matched Problem 6 Given the cost function $C(x) = 1{,}600 + 0.25x^2$, where x is the number of items produced:

(A) Use the graphing strategy to analyze the graph of the average cost function. State all the pertinent information and sketch the graph of the average cost function. Find the marginal cost function and graph it on the same set of coordinate axes. Include any oblique asymptotes.

(B) Find the minimum average cost.

Answers to Matched Problems

1. $p(x) = -4x^4\left(1 - \frac{1}{2x} - \frac{3}{4x^3}\right) \approx -4x^4$ for x large (in absolute value); $\lim_{x\to-\infty} p(x) = -\infty$, $\lim_{x\to\infty} p(x) = -\infty$

2. (A) $y = 0$ (B) No horizontal asymptote (C) $y = \frac{1}{2}$

3. $x = 1$

4. Domain: $(-\infty, \infty)$
y intercept: $f(0) = 0$; x intercepts: $-4, 0$
Asymptotes: No horizontal or vertical asymptotes
Decreasing on $(-\infty, -3)$; increasing on $(-3, \infty)$; local minimum at $x = -3$
Concave upward on $(-\infty, -2)$ and $(0, \infty)$; concave downward on $(-2, 0)$
Inflection points at $x = -2$ and $x = 0$

x	$f(x)$
−4	0
−3	−27
−2	−16
0	0

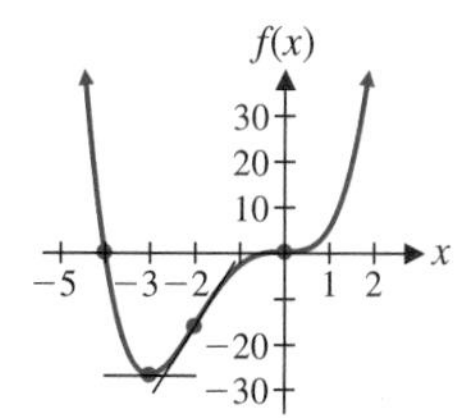

5. Domain: All real x, except $x = 1$
y intercept: $f(0) = 0$; x intercept: 0
Horizontal asymptote: $y = -2$; vertical asymptote: $x = 1$
Increasing on $(-\infty, 1)$ and $(1, \infty)$
Concave upward on $(-\infty, 1)$; concave downward on $(1, \infty)$

x	$f(x)$
-1	-1
0	0
$\frac{1}{2}$	2
$\frac{3}{2}$	-6
2	-4
5	$-\frac{5}{2}$

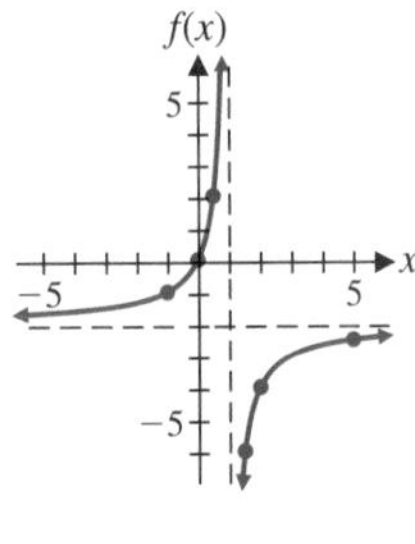

6. (A) Domain: $(0, \infty)$
Intercepts: None
Vertical asymptote: $x = 0$; oblique asymptote: $y = 0.25x$
Decreasing on $(0, 80)$; increasing on $(80, \infty)$; local minimum at $x = 80$
Concave upward on $(0, \infty)$

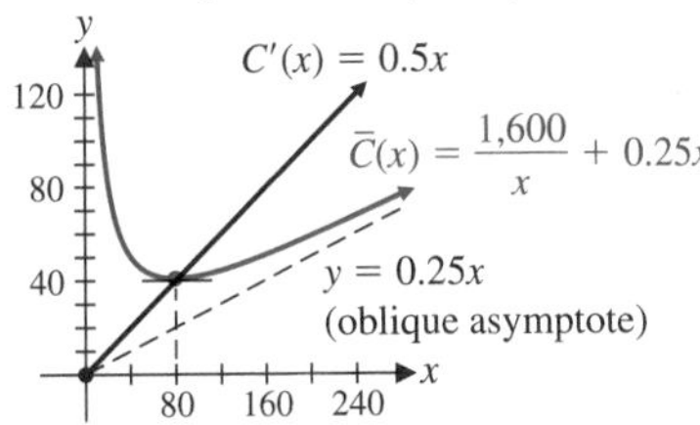

(B) Minimum average cost is 40 at $x = 80$.

Exercise 9-4

A *Problems 1–14 refer to the graph of $y = f(x)$ shown in the figure below.*

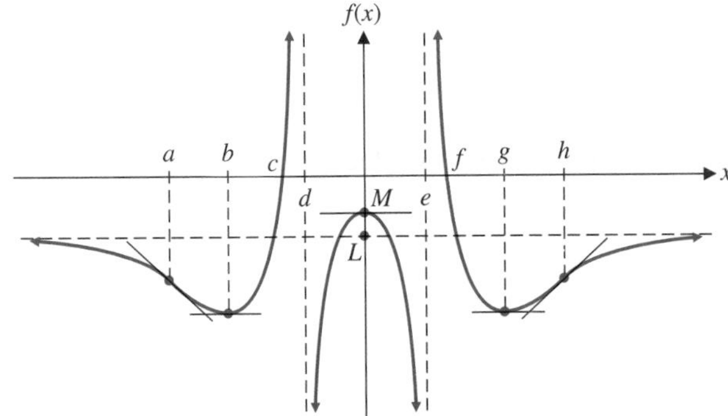

1. Identify the intervals over which $f'(x) < 0$.
2. Identify the intervals over which $f'(x) > 0$.
3. Identify the intervals over which $f(x)$ is increasing.
4. Identify the intervals over which $f(x)$ is decreasing.
5. Identify the x coordinate(s) of the point(s) where $f(x)$ has a local maximum.
6. Identify the x coordinate(s) of the point(s) where $f(x)$ has a local minimum.
7. Identify the intervals over which $f''(x) < 0$.
8. Identify the intervals over which $f''(x) > 0$.
9. Identify the intervals over which the graph of f is concave upward.
10. Identify the intervals over which the graph of f is concave downward.
11. Identify the x coordinate(s) of the inflection point(s).
12. Identify the horizontal asymptote(s).
13. Identify the vertical asymptote(s).
14. Identify the x and y intercepts.

In Problems 15–22, use the given information to sketch the graph of f. Assume that f is continuous on its domain and that all intercepts are included in the table of values.

15. Domain: All real x; $\lim_{x\to\pm\infty} f(x) = 2$

x	-4	-2	0	2	4
$f(x)$	0	-2	0	-2	0

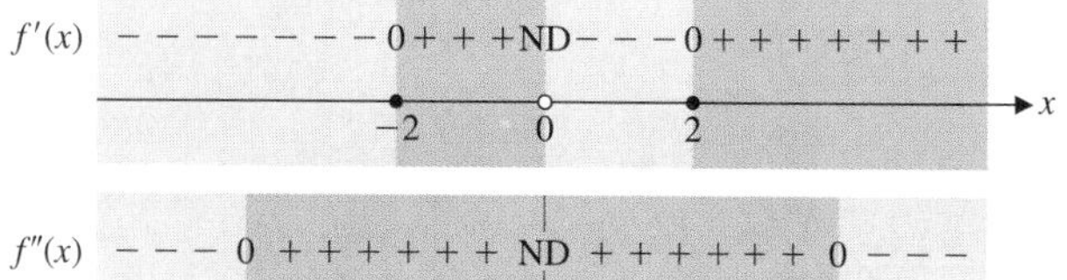

f″(x) – – – 0 + + + + + + ND + + + + + + 0 – – –

x: –4, 0, 4

16. Domain: All real x; $\lim_{x\to-\infty} f(x) = -3$; $\lim_{x\to\infty} f(x) = 3$

x	-2	-1	0	1	2
$f(x)$	0	2	0	-2	0

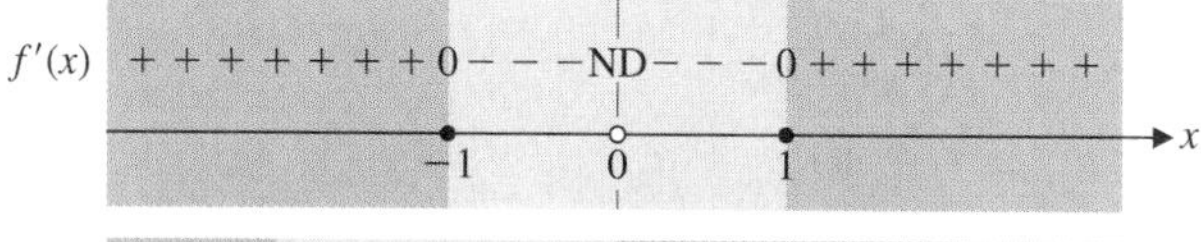

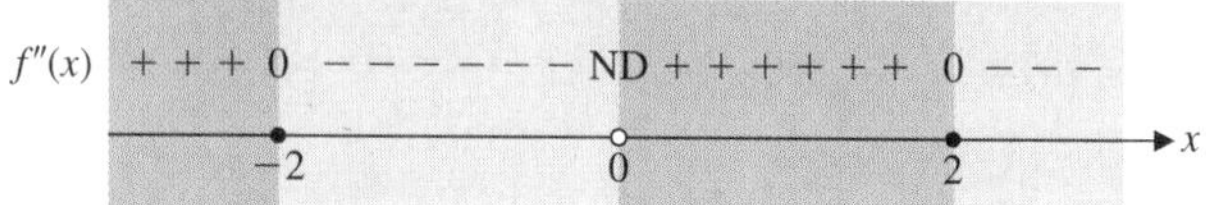

17. Domain: All real x, except $x = -2$; $\lim_{x\to-2^-} f(x) = \infty$; $\lim_{x\to-2^+} f(x) = -\infty$; $\lim_{x\to\infty} f(x) = 1$

x	-4	0	4	6
$f(x)$	0	0	3	2

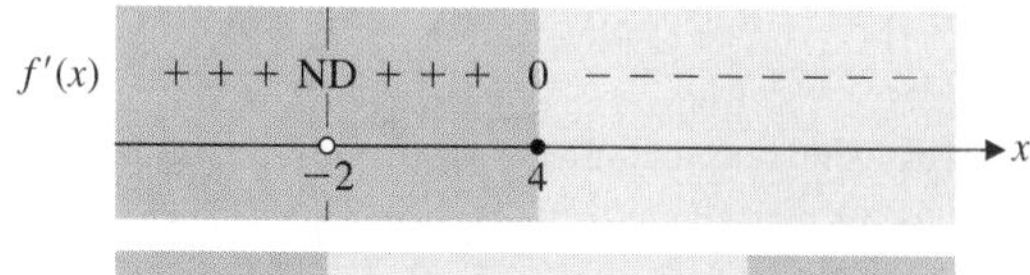

f″(x) + + + ND – – – – – – – – – 0 + + +

x: –2, 6

18. Domain: All real x, except $x = 1$; $\lim_{x\to1^-} f(x) = \infty$; $\lim_{x\to1^+} f(x) = \infty$; $\lim_{x\to\infty} f(x) = -2$

x	-4	-2	0	2
$f(x)$	0	-2	0	0

f′(x) – – – 0 + + + ND – – –

x: –2, 1

f″(x) + + + + + + + + + ND + + +

x: 1

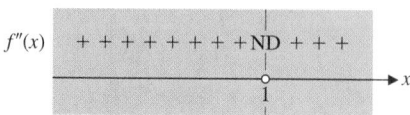

19. Domain: All real x, except $x = -1$;
$f(-3) = 2, f(-2) = 3, f(0) = -1, f(1) = 0$;
$f'(x) > 0$ on $(-\infty, -1)$ and $(-1, \infty)$;
$f''(x) > 0$ on $(-\infty, -1)$; $f''(x) < 0$ on $(-1, \infty)$;
vertical asymptote: $x = -1$; horizontal asymptote: $y = 1$

20. Domain: All real x, except $x = 1$;
$f(0) = -2, f(2) = 0$;
$f'(x) < 0$ on $(-\infty, 1)$ and $(1, \infty)$;
$f''(x) < 0$ on $(-\infty, 1)$;
$f''(x) > 0$ on $(1, \infty)$;
vertical asymptote: $x = 1$; horizontal asymptote: $y = -1$

21. Domain: All real x, except $x = -2$ and $x = 2$;
$f(-3) = -1, f(0) = 0, f(3) = 1$;
$f'(x) < 0$ on $(-\infty, -2)$ and $(2, \infty)$;
$f'(x) > 0$ on $(-2, 2)$;
$f''(x) < 0$ on $(-\infty, -2)$ and $(-2, 0)$;
$f''(x) > 0$ on $(0, 2)$ and $(2, \infty)$;
vertical asymptotes: $x = -2$ and $x = 2$;
horizontal asymptote: $y = 0$

22. Domain: All real x, except $x = -1$ and $x = 1$;
$f(-2) = 1, f(0) = 0, f(2) = 1$;
$f'(x) > 0$ on $(-\infty, -1)$ and $(0, 1)$;
$f'(x) < 0$ on $(-1, 0)$ and $(1, \infty)$;
$f''(x) > 0$ on $(-\infty, -1), (-1, 1)$, and $(1, \infty)$;
vertical asymptotes: $x = -1$ and $x = 1$;
horizontal asymptote: $y = 0$

B *In Problems 23–32 evaluate the indicated limit. Use $-\infty$ or ∞ where appropriate.*

23. $\lim_{x\to\infty}(4x^3 + 2x - 9)$

24. $\lim_{x\to\infty}(-2x^5 + 5x^3 + 3)$

25. $\lim_{x\to-\infty}(-3x^6 + 9x^5 + 4)$

26. $\lim_{x\to-\infty}(7x^4 - 4x^3 + 7x)$

27. $\lim_{x\to\infty} \dfrac{4x + 7}{5x - 9}$

28. $\lim_{x\to\infty} \dfrac{2 - 3x^3}{7 + 4x^3}$

29. $\lim_{x\to\infty} \dfrac{5x^2 + 11}{7x - 2}$

30. $\lim_{x\to\infty} \dfrac{5x + 11}{7x^3 - 2}$

31. $\lim_{x\to-\infty} \dfrac{7x^4 - 14x^2}{6x^5 + 3}$

32. $\lim_{x\to-\infty} \dfrac{4x^7 - 8x}{6x^4 + 9x^2}$

In Problems 33–44, find all horizontal and vertical asymptotes.

33. $f(x) = \dfrac{2x}{x + 2}$ **34.** $f(x) = \dfrac{3x + 2}{x - 4}$

35. $f(x) = \dfrac{x^2 + 1}{x^2 - 1}$ **36.** $f(x) = \dfrac{x^2 - 1}{x^2 + 2}$

37. $f(x) = \dfrac{x^3}{x^2 + 6}$ **38.** $f(x) = \dfrac{x}{x^2 - 4}$

39. $f(x) = \dfrac{x^2}{x - 3}$ **40.** $f(x) = \dfrac{x + 5}{x^2}$

41. $f(x) = \dfrac{2x^2 + 3x - 2}{x^2 - x - 2}$ **42.** $f(x) = \dfrac{x^2 + 7x + 12}{2x^2 + 5x - 12}$

43. $f(x) = \dfrac{2x^2 - 5x + 2}{x^2 - x - 2}$ **44.** $f(x) = \dfrac{x^2 - x - 12}{2x^2 + 5x - 12}$

In Problems 45–54, summarize the pertinent information obtained by applying the graphing strategy and sketch the graph of $y = f(x)$.

45. $f(x) = x^3 - 6x^2$

46. $f(x) = 3x^2 - x^3$

47. $f(x) = (x + 4)(x - 2)^2$

48. $f(x) = (2 - x)(x + 1)^2$

49. $f(x) = 8x^3 - 2x^4$ **50.** $f(x) = x^4 - 4x^3$

51. $f(x) = \dfrac{x + 3}{x - 3}$ **52.** $f(x) = \dfrac{2x - 4}{x + 2}$

53. $f(x) = \dfrac{x}{x - 2}$ **54.** $f(x) = \dfrac{2 + x}{3 - x}$

55. Theorem 2 states that

$$\lim_{x\to\infty}(a_nx^n + a_{n-1}x^{n-1} + \cdots + a_0) = \pm\infty$$

What conditions must n and a_n satisfy for the limit to be ∞? For the limit to be $-\infty$?

56. Theorem 2 also states that

$$\lim_{x\to-\infty}(a_nx^n + a_{n-1}x^{n-1} + \cdots + a_0) = \pm\infty$$

What conditions must n and a_n satisfy for the limit to be ∞? For the limit to be $-\infty$?

57. Let $p(x) = x^3 - 2x^2$.
(A) Find $\lim_{x\to\infty} p'(x)$ and $\lim_{x\to\infty} p''(x)$. Describe the shape of the graph of $y = p(x)$ for large positive values of x.
(B) Find $\lim_{x\to-\infty} p'(x)$ and $\lim_{x\to-\infty} p''(x)$. Describe the shape of the graph of $y = p(x)$ for large (in absolute value) negative values of x.

58. Repeat Problem 57 for $p(x) = x^4 - 2x^3$.

C *In Problems 59 and 60, show that the line $y = x$ is an oblique asymptote for the graph of $y = f(x)$, summarize the pertinent information obtained by applying the graphing strategy, and sketch the graph of $y = f(x)$.*

59. $f(x) = x + \dfrac{1}{x}$ **60.** $f(x) = x - \dfrac{1}{x}$

In Problems 61–72, summarize the pertinent information obtained by applying the graphing strategy and sketch the graph of $y = f(x)$.

61. $f(x) = x^3 - x$

62. $f(x) = x^3 + x$

63. $f(x) = (x^2 + 3)(9 - x^2)$

64. $f(x) = (x^2 + 3)(x^2 - 1)$

65. $f(x) = (x^2 - 4)^2$

66. $f(x) = (x^2 - 1)(x^2 - 5)$

67. $f(x) = 2x^6 - 3x^5$

68. $f(x) = 3x^5 - 5x^4$

69. $f(x) = \dfrac{x}{x^2 - 4}$

70. $f(x) = \dfrac{1}{x^2 - 4}$

71. $f(x) = \dfrac{1}{1 + x^2}$

72. $f(x) = \dfrac{x^2}{1 + x^2}$

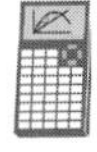

In Problems 73–80, apply steps 1–3 of the graphing strategy to $f(x)$. Use a graphing utility to approximate (to two decimal places) x intercepts, critical values, and x coordinates of inflection points. Summarize all the pertinent information.

73. $f(x) = x^4 - 5x^3 + 3x^2 + 8x - 5$

74. $f(x) = x^4 + 2x^3 - 5x^2 - 4x + 4$

75. $f(x) = x^4 - 21x^3 + 100x^2 + 20x + 100$

76. $f(x) = x^4 - 12x^3 + 28x^2 + 76x - 50$

77. $f(x) = -x^4 - x^3 + 2x^2 - 2x + 3$

78. $f(x) = -x^4 + x^3 + x^2 + 6$

79. $f(x) = 0.1x^5 + 0.3x^4 - 4x^3 - 5x^2 + 40x + 30$

80. $f(x) = x^5 + 4x^4 - 7x^3 - 20x^2 + 20x - 20$

Applications

Business & Economics

81. *Revenue.* The marketing research department for a computer company used a large city to test market their new product. They found that the relationship between price p (dollars per unit) and the demand x (units sold per week) was given approximately by

$$p = 1{,}296 - 0.12x^2 \qquad 0 \le x \le 80$$

Thus, the weekly revenue can be approximated by

$$R(x) = xp = 1{,}296x - 0.12x^3 \qquad 0 \le x \le 80$$

Graph the revenue function R.

82. *Profit.* Suppose the cost function $C(x)$ (in dollars) for the company in Problem 81 is

$$C(x) = 830 + 396x$$

(A) Write an equation for the profit $P(x)$.
(B) Graph the profit function P.

83. *Pollution.* In Silicon Valley (California), a number of computer-related manufacturing firms were found to be contaminating underground water supplies with toxic chemicals stored in leaking underground containers. A water quality control agency ordered the companies to take immediate corrective action and to contribute to a monetary pool for testing and cleanup of the underground contamination. Suppose the required monetary pool (in millions of dollars) for the testing and cleanup is estimated to be given by

$$P(x) = \frac{2x}{1 - x} \qquad 0 \le x < 1$$

where x is the percentage (expressed as a decimal fraction) of the total contaminant removed.

(A) Where is $P(x)$ increasing? Decreasing?
(B) Where is the graph of P concave upward? Downward?
(C) Find any horizontal and vertical asymptotes.
(D) Find the x and y intercepts.
(E) Sketch a graph of P.

84. *Employee training.* A company producing computer components has established that on the average a new employee can assemble $N(t)$ components per day after t days of on-the-job training, as given by

$$N(t) = \frac{100t}{t + 9} \qquad t \ge 0$$

(A) Where is $N(t)$ increasing? Decreasing?
(B) Where is the graph of N concave upward? Downward?
(C) Find any horizontal and vertical asymptotes.
(D) Find the intercepts.
(E) Sketch a graph of N.

85. *Replacement time.* An office copier has an initial price of \$3,200. A maintenance/service contract costs \$300 for the first year and increases \$100 per year thereafter. It can be shown that the total cost of the copier (in dollars) after n years is given by

$$C(n) = 3{,}200 + 250n + 50n^2$$

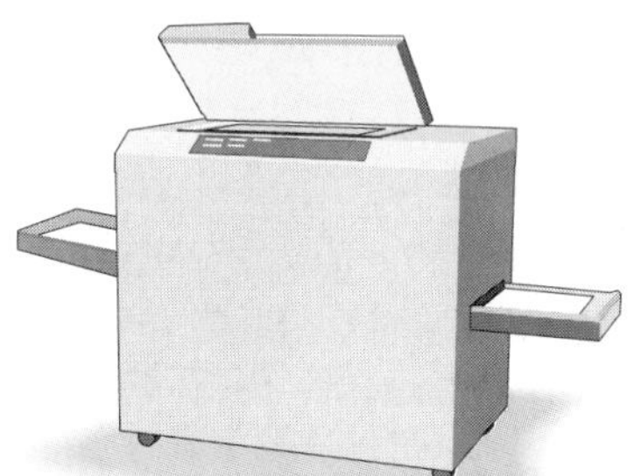

(A) Write an expression for the average cost per year, $\overline{C}(n)$, for n years.
(B) Graph the average cost function found in part (A).
(C) When is the average cost per year minimum? (This is frequently referred to as the **replacement time** for this piece of equipment.)

86. *Construction costs.* The management of a manufacturing plant wishes to add a fenced-in rectangular storage yard of 20,000 square feet, using the plant building as one side of the yard (see the figure). If x is the distance (in feet) from the building to the fence parallel to the building, then show that the length of the fence required for the yard is given by

$$L(x) = 2x + \frac{20{,}000}{x} \qquad x > 0$$

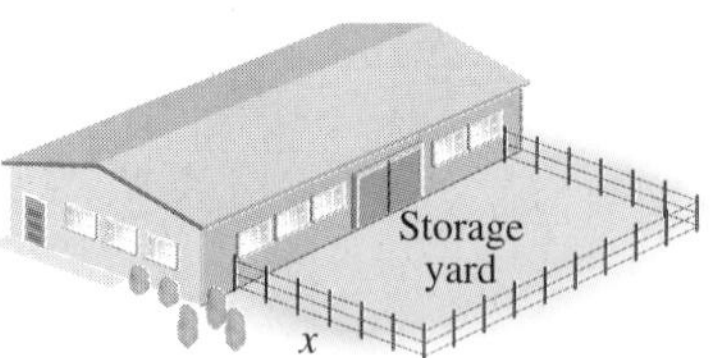

Figure for 86

(A) Graph L.
(B) What are the dimensions of the rectangle requiring the least amount of fencing?

87. *Average and marginal costs.* The total daily cost (in dollars) of producing x park benches is given by

$$C(x) = 1{,}000 + 5x + 0.1x^2$$

(A) Sketch the graphs of the average cost function and the marginal cost function on the same set of coordinate axes. Include any oblique asymptotes.
(B) Find the minimum average cost.

88. *Average and marginal costs.* The total daily cost (in dollars) of producing x picnic tables is given by

$$C(x) = 500 + 2x + 0.2x^2$$

(A) Sketch the graphs of the average cost function and the marginal cost function on the same set of coordinate axes. Include any oblique asymptotes.
(B) Find the minimum average cost.

89. *Minimizing average costs.* The data in the table give the total daily costs y (in dollars) of producing x pepperoni pizzas at various production levels.

NUMBER OF PIZZAS x	TOTAL COSTS (\$) y
50	395
100	475
150	640
200	910
250	1,140
300	1,450

(A) Enter the data in a graphing utility and find a quadratic regression equation for the total costs.
(B) Use the regression equation from part (A) to find the minimum average cost (to the nearest cent) and the corresponding production level (to the nearest integer).

90.  *Minimizing average costs.* The data in the table give the total daily costs y (in dollars) of producing x deluxe pizzas at various production levels.

NUMBER OF PIZZAS x	TOTAL COSTS (\$) y
50	595
100	755
150	1,110
200	1,380
250	1,875
300	2,410

(A) Enter the data in a graphing utility and find a quadratic regression equation for the total costs.
(B) Use the regression equation from part (A) to find the minimum average cost (to the nearest cent) and the corresponding production level (to the nearest integer).

Life Sciences

91. *Medicine.* A drug is injected into the bloodstream of a patient through her right arm. The concentration of the drug in the bloodstream of the left arm t hours after the injection is given by

$$C(t) = \frac{0.14t}{t^2 + 1}$$

Graph C.

92. *Physiology.* In a study on the speed of muscle contraction in frogs under various loads, researchers W. O. Fems and J. Marsh found that the speed of contraction decreases with increasing loads. More precisely, they found that the relationship between speed of contraction S (in centimeters per second) and load w (in grams) is given approximately by

$$S(w) = \frac{26 + 0.06w}{w} \qquad w \geq 5$$

Graph S.

Social Sciences

93. *Psychology—retention.* An experiment on retention is conducted in a psychology class. Each student in the class is given 1 day to memorize the same list of 30 special characters. The lists are turned in at the end of the day, and for each succeeding day for 30 days each student is asked to turn in a list of as many of the symbols as can be recalled. Averages are taken, and it is found that

$$N(t) = \frac{5t + 20}{t} \qquad t \geq 1$$

provides a good approximation of the average number of symbols, $N(t)$ retained after t days. Graph N.

SECTION 9-5 Optimization; Absolute Maxima and Minima

- ABSOLUTE MAXIMA AND MINIMA
- APPLICATIONS

We are now ready to consider one of the most important applications of the derivative, namely, the use of derivatives to find the *absolute maximum* or *minimum* value of a function. As we mentioned earlier, an economist may be interested in the price or production level of a commodity that will bring a maximum profit; a doctor may be interested in the time it takes for a drug to reach its maximum concentration in the bloodstream after an injection; and a city planner might be interested in the location of heavy industry in a city to produce minimum pollution in residential and business areas. Before we launch an attack on problems of this type, which are called *optimization problems,* we have to say a few words about the procedures needed to find absolute maximum and absolute minimum values of functions. We have most of the tools we need from the previous sections.

Absolute Maxima and Minima

First, what do we mean by *absolute maximum* and *absolute minimum?* We say that $f(c)$ is an **absolute maximum** of f if

$$f(c) \geq f(x)$$

for all x in the domain of f. Similarly, $f(c)$ is called an **absolute minimum** of f if

$$f(c) \leq f(x)$$

for all x in the domain of f. Figure 1 illustrates some typical examples.

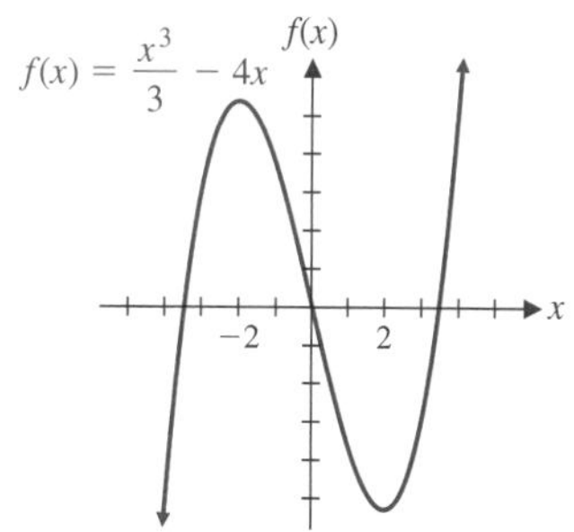

(A) No absolute maximum or minimum
One local maximum at $x = -2$
One local minimum at $x = 2$

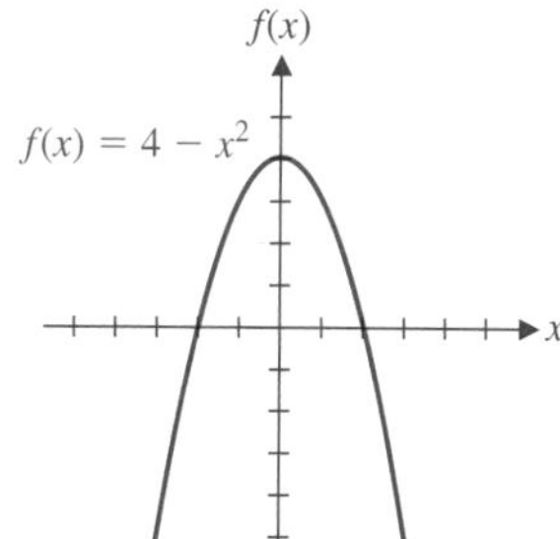

(B) Absolute maximum at $x = 0$
No absolute minimum

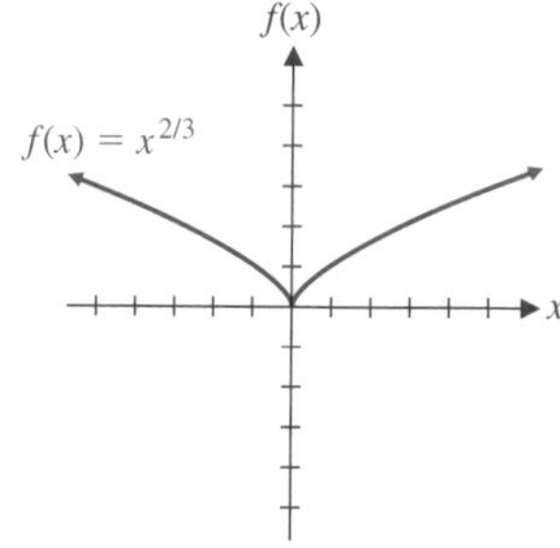

(C) Absolute minimum at $x = 0$
No absolute maximum

FIGURE 1

Explore–Discuss 1

Functions f, g, and h, along with their graphs, are shown in Figure 2.

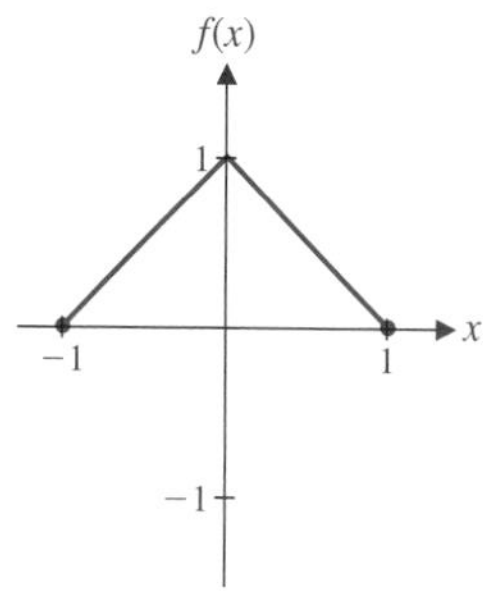

(A) $f(x) = 1 - |x|$

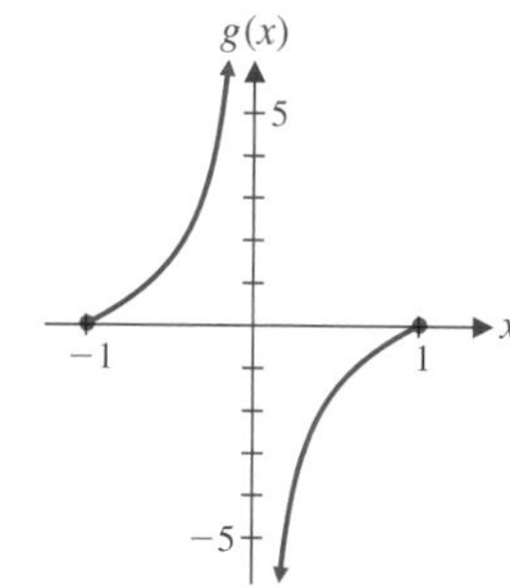

(B) $g(x) = x - \frac{1}{x}$

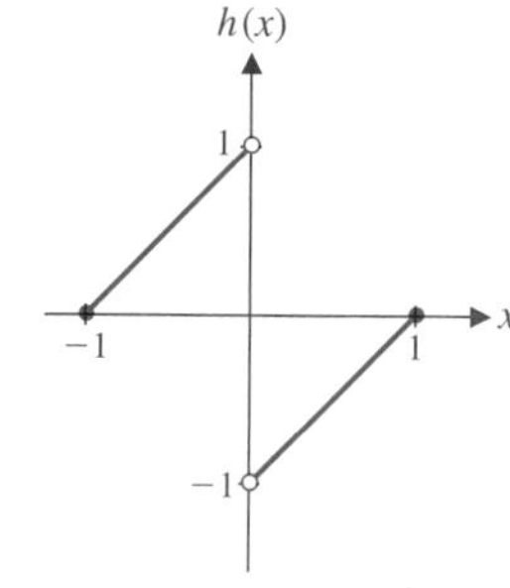

(C) $h(x) = x - \frac{|x|}{x}$

FIGURE 2

(A) Which of these functions are continuous on $[-1, 1]$?
(B) Find the absolute maximum and the absolute minimum of each function on $[-1, 1]$, if they exist, and the corresponding values of x that produce these absolute extrema.

(continued)

Explore–Discuss 1 (continued)

(C) Suppose that a function p is continuous on $[-1, 1]$ and satisfies $p(-1) = 0$ and $p(1) = 0$. Sketch a possible graph for p. Does the function you graphed have an absolute maximum? An absolute minimum? Can you modify your sketch so that p does not have an absolute maximum or an absolute minimum on $[-1, 1]$?

In many practical problems, the domain of a function is restricted because of practical or physical considerations. If the domain is restricted to some closed interval, as is often the case, then Theorem 1 can be proved.

Theorem 1 ■■ EXTREME VALUE THEOREM

A function f that is continuous on a closed interval $[a, b]$ (see Section 9-1) has both an absolute maximum value and an absolute minimum value on that interval. ■■

It is important to understand that the absolute maximum and minimum values depend on both the function f and the interval $[a, b]$. Figure 3 illustrates four cases.

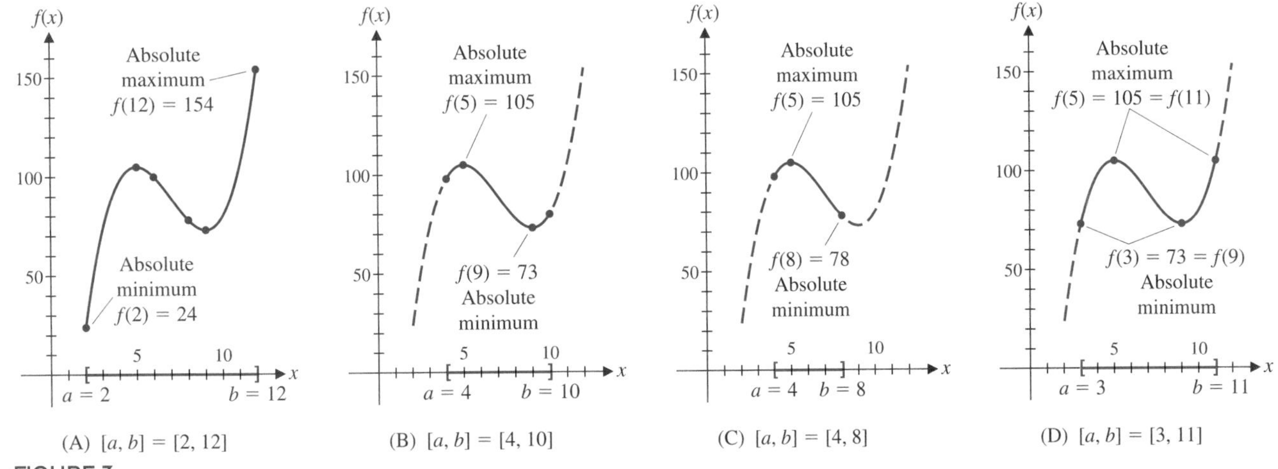

FIGURE 3
Absolute extrema for $f(x) = x^3 - 21x^2 + 135x - 170$ for various closed intervals

In all four cases illustrated in Figure 3, the absolute maximum value and absolute minimum value both occur at a critical value or an end point. Both the absolute maximum value and the absolute minimum value are unique, but each can occur at more than one point in the interval (Fig. 3D). In general:

Absolute extrema (if they exist) must always occur at critical values or at end points.

Thus, to find the absolute maximum or minimum value of a continuous function on a closed interval, we simply identify the end points and the critical values in the interval, evaluate the function at each, and then choose the largest and smallest values out of this group.

Steps in Finding Absolute Maximum and Minimum Values of a Continuous Function f on a Closed Interval $[a, b]$

Step 1. Check to make certain that f is continuous over $[a, b]$.
Step 2. Find the critical values in the interval (a, b).
Step 3. Evaluate f at the end points a and b and at the critical values found in step 2.
Step 4. The absolute maximum $f(x)$ on $[a, b]$ is the largest of the values found in step 3.
Step 5. The absolute minimum $f(x)$ on $[a, b]$ is the smallest of the values found in step 3.

Example 1 ➡ **Finding Absolute Extrema** Find the absolute maximum and absolute minimum values of

$$f(x) = x^3 + 3x^2 - 9x - 7$$

on each of the following intervals:

(A) $[-6, 4]$ (B) $[-4, 2]$ (C) $[-2, 2]$

SOLUTION (A) The function is continuous for all values of x.

$$f'(x) = 3x^2 + 6x - 9 = 3(x - 1)(x + 3)$$

Thus, $x = -3$ and $x = 1$ are critical values in the interval $(-6, 4)$. Evaluate f at the end points and critical values (-6, -3, 1, and 4), and choose the maximum and minimum from these:

$f(-6) = -61$ Absolute minimum
$f(-3) = 20$
$f(1) = -12$
$f(4) = 69$ Absolute maximum

(B) Interval: $[-4, 2]$

x	$f(x)$	
-4	13	
-3	20	Absolute maximum
1	-12	Absolute minimum
2	-5	

(C) Interval: $[-2, 2]$

x	$f(x)$	
-2	15	Absolute maximum
1	-12	Absolute minimum
2	-5	

The critical value $x = -3$ is not included in this table, because it is not in the interval $[-2, 2]$.

Matched Problem 1 ➡ Find the absolute maximum and absolute minimum values of

$$f(x) = x^3 - 12x$$

on each of the following intervals:

(A) $[-5, 5]$ (B) $[-3, 3]$ (C) $[-3, 1]$

Now, suppose we want to find the absolute maximum or minimum value of a function that is continuous on an interval that is not closed. Since Theorem 1 no longer applies, we cannot be certain that the absolute maximum or minimum value exists. Figure 4 illustrates several ways that functions can fail to have absolute extrema.

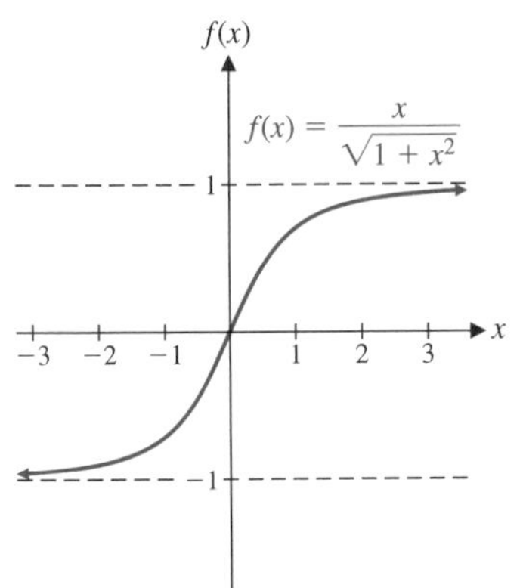

(A) No absolute extrema on $(-\infty, \infty)$:
$-1 < f(x) < 1$ for all x
$[f(x) \neq 1$ or -1 for any $x]$

$f(x) = 8x - 2x^2 - 3$
$1 < x < 2$

(B) No absolute extrema on $(1, 2)$:
$3 < f(x) < 5$ for $x \in (1, 2)$
$[f(x) \neq 3$ or 5 for any $x \in (1, 2)]$

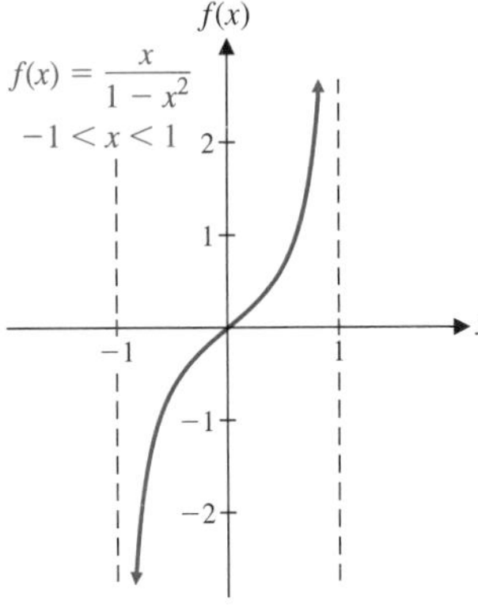

(C) No absolute extrema on $(-1, 1)$:
Graph has vertical asymptotes
at $x = -1$ and $x = 1$

FIGURE 4
Functions with no absolute extrema

In general, the best procedure to follow when the interval is not a closed interval (that is, not of the form $[a, b]$) is to sketch the graph of the function. However, one special case that occurs frequently in applications can be analyzed without drawing a graph. It often happens that f is continuous on an interval I and has only one critical value c in the interval I (here, I can be any type of interval—open, closed, or half-closed). If this is the case and if $f''(c)$ exists, then we have the second-derivative test for absolute extrema given in the box below.

Second-Derivative Test for Absolute Maximum and Minimum When f Is Continuous on an Interval I and c Is the Only Critical Value in I

$f'(c)$	$f''(c)$	$f(c)$	EXAMPLE
0	+	Absolute mininum	x, I
0	−	Absolute maximum	x, I
0	0	Test fails	

Example 2 **Finding an Absolute Extremum on an Open Interval** Find the absolute minimum value of

$$f(x) = x + \frac{4}{x}$$

on the interval $(0, \infty)$.

SOLUTION

$$f'(x) = 1 - \frac{4}{x^2} = \frac{x^2 - 4}{x^2} = \frac{(x-2)(x+2)}{x^2} \qquad f''(x) = \frac{8}{x^3}$$

The only critical value in the interval $(0, \infty)$ is $x = 2$. Since $f''(2) = 1 > 0$, $f(2) = 4$ is the absolute minimum value of f on $(0, \infty)$.

Matched Problem 2 Find the absolute maximum value of

$$f(x) = 12 - x - \frac{9}{x}$$

on the interval $(0, \infty)$.

Applications

Now we want to solve some applied problems that involve absolute extrema. Before beginning, we outline in the next box the steps to follow in solving this type of problem. The first step is the most difficult one. The techniques used to solve optimization problems are best illustrated through a series of examples.

A Strategy for Solving Applied Optimization Problems

Step 1. Introduce variables and a function f, including the domain I of f, and then construct a mathematical model of the form

Maximize (or minimize) $f(x)$ on the interval I

Step 2. Find the absolute maximum (or minimum) value of $f(x)$ on the interval I and the value(s) of x where this occurs.

Step 3. Use the solution to the mathematical model to answer the questions asked in the problem.

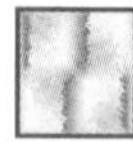

Example 3 **Maximize Revenue and Profit** A company manufactures and sells x transistor radios per week. If the weekly cost and price–demand equations are

$$C(x) = 5{,}000 + 2x$$
$$p = 10 - 0.001x \qquad 0 \leq x \leq 10{,}000$$

find the following for each week:

(A) The maximum revenue

(B) The maximum profit, the production level that will realize the maximum profit, and the price that the company should charge for each radio to realize the maximum profit

SOLUTION (A) The revenue received for selling x radios at \$$p$ per radio is

$$\begin{aligned} R(x) &= xp \\ &= x(10 - 0.001x) \\ &= 10x - 0.001x^2 \end{aligned}$$

Thus, the mathematical model is

$$\begin{aligned} \text{Maximize} \quad R(x) &= 10x - 0.001x^2 \qquad 0 \leq x \leq 10{,}000 \\ R'(x) &= 10 - 0.002x \\ 10 - 0.002x &= 0 \\ x &= 5{,}000 \qquad \text{Only critical value} \end{aligned}$$

Use the second-derivative test for absolute extrema:

$$R''(x) = -0.002 < 0 \qquad \text{for all } x$$

Thus, the maximum revenue is

$$\text{Max } R(x) = R(5{,}000) = \$25{,}000$$

(B) Profit = Revenue − Cost

$$\begin{aligned} P(x) &= R(x) - C(x) \\ &= 10x - 0.001x^2 - 5{,}000 - 2x \\ &= 8x - 0.001x^2 - 5{,}000 \end{aligned}$$

The mathematical model is

$$\begin{aligned} \text{Maximize} \quad P(x) &= 8x - 0.001x^2 - 5{,}000 \qquad 0 \leq x \leq 10{,}000 \\ P'(x) &= 8 - 0.002x \\ 8 - 0.002x &= 0 \\ x &= 4{,}000 \\ P''(x) &= -0.002 < 0 \qquad \text{for all } x \end{aligned}$$

Since $x = 4{,}000$ is the only critical value and $P''(x) < 0$,

$$\text{Max } P(x) = P(4{,}000) = \$11{,}000$$

Using the price–demand equation with $x = 4{,}000$, we find

$$p = 10 - 0.001(4{,}000) = \$6$$

Thus, a maximum profit of \$11,000 per week is realized when 4,000 radios are produced weekly and sold for \$6 each. Notice that this is not the same level of production that produces the maximum revenue. ▪

All the results in Example 3 are illustrated in Figure 5 (page 678). We also note that profit is maximum when

$$P'(x) = R'(x) - C'(x) = 0$$

that is, when the marginal revenue is equal to the marginal cost (the rate of increase in revenue is the same as the rate of increase in cost at the 4,000 output level—notice that the slopes of the two curves are the same at this point).

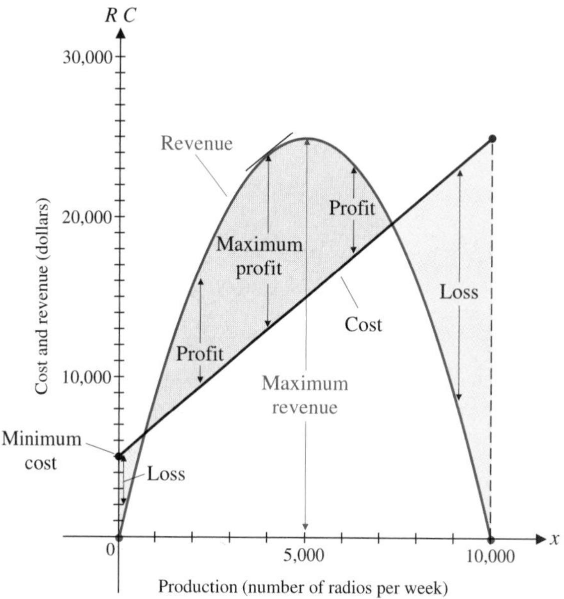

FIGURE 5

Matched Problem 3 Repeat Example 3 for

$$C(x) = 90{,}000 + 30x$$

$$p = 300 - \frac{x}{30} \qquad 0 \le x \le 9{,}000$$

Example 4 **Maximizing Profit** The government has decided to tax the company in Example 3 \$2 for each radio produced. Taking into account this additional cost, how many radios should the company manufacture each week in order to maximize its weekly profit? What is the maximum weekly profit? How much should the company charge for the radios to realize the maximum weekly profit?

SOLUTION The tax of \$2 per unit changes the company's cost equation:

$$\begin{aligned} C(x) &= \text{Original cost} + \text{Tax} \\ &= 5{,}000 + 2x + 2x \\ &= 5{,}000 + 4x \end{aligned}$$

The new profit function is

$$\begin{aligned} P(x) &= R(x) - C(x) \\ &= 10x - 0.001x^2 - 5{,}000 - 4x \\ &= 6x - 0.001x^2 - 5{,}000 \end{aligned}$$

Thus, we must solve the following:

$$\begin{aligned}
\text{Maximize} \quad P(x) &= 6x - 0.001x^2 - 5{,}000 \qquad 0 \le x \le 10{,}000 \\
P'(x) &= 6 - 0.002x \\
6 - 0.002x &= 0 \\
x &= 3{,}000 \\
P''(x) &= -0.002 < 0 \qquad \text{for all } x \\
\text{Max } P(x) &= P(3{,}000) = \$4{,}000
\end{aligned}$$

Using the price–demand equation with $x = 3{,}000$, we find

$$p = 10 - 0.001(3{,}000) = \$7$$

Thus, the company's maximum profit is $4,000 when 3,000 radios are produced and sold weekly at a price of $7.

Even though the tax caused the company's cost to increase by $2 per radio, the price that the company should charge to maximize its profit increases by only $1. The company must absorb the other $1 with a resulting decrease of $7,000 in maximum profit.

Matched Problem 4 ➡ Repeat Example 4 if

$$C(x) = 90{,}000 + 30x$$

$$p = 300 - \frac{x}{30} \qquad 0 \le x \le 9{,}000$$

and the government decides to tax the company $20 for each unit produced. Compare the results with the results in Matched Problem 3B.

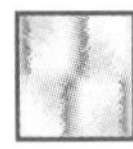

Example 5 ➡ **Maximizing Yield** A walnut grower estimates from past records that if 20 trees are planted per acre, each tree will average 60 pounds of nuts per year. If for each additional tree planted per acre (up to 15) the average yield per tree drops 2 pounds, how many trees should be planted to maximize the yield per acre? What is the maximum yield?

SOLUTION Let x be the number of additional trees planted per acre. Then

$$\begin{aligned}
20 + x &= \text{Total number of trees per acre} \\
60 - 2x &= \text{Yield per tree} \\
\text{Yield per acre} &= (\text{Total number of trees per acre})(\text{Yield per tree}) \\
Y(x) &= (20 + x)(60 - 2x) \\
&= 1{,}200 + 20x - 2x^2 \qquad 0 \le x \le 15
\end{aligned}$$

Thus, we must solve the following:

$$\begin{aligned}
\text{Maximize} \quad Y(x) &= 1{,}200 + 20x - 2x^2 \qquad 0 \le x \le 15 \\
Y'(x) &= 20 - 4x \\
20 - 4x &= 0 \\
x &= 5 \\
Y''(x) &= -4 < 0 \qquad \text{for all } x
\end{aligned}$$

Hence,

$$\text{Max } Y(x) = Y(5) = 1{,}250 \text{ pounds per acre}$$

Thus, a maximum yield of 1,250 pounds of nuts per acre is realized if 25 trees are planted per acre.

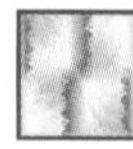

Matched Problem 5 Repeat Example 5 starting with 30 trees per acre and a reduction of 1 pound per tree for each additional tree planted.

Explore–Discuss 2

In Example 5, letting x be the number of *additional* trees planted per acre produced a simple and direct solution to the problem. However, this is not the most obvious choice for a variable. Suppose we proceed as follows:

Let x be the total number of trees planted per acre and let y be the yield per tree. Then the yield per acre is given by xy.

(A) Find y when $x = 20$ and when $x = 21$. Find the equation of the line through these two points.
(B) Use the equation from part (A) to express the yield per acre in terms of either x or y, and use this expression to solve Example 5.
(C) Compare this method of solution to the one used in Example 5 with respect to ease of comprehension and ease of computation.

Example 6 **Maximizing Area** A farmer wants to construct a rectangular pen next to a barn 60 feet long, using all of the barn as part of one side of the pen. Find the dimensions of the pen with the largest area that the farmer can build if:

(A) 160 feet of fencing material is available
(B) 250 feet of fencing material is available

Solution (A) We begin by constructing and labeling the figure shown in the margin. The area of the pen is

$$A = (x + 60)y$$

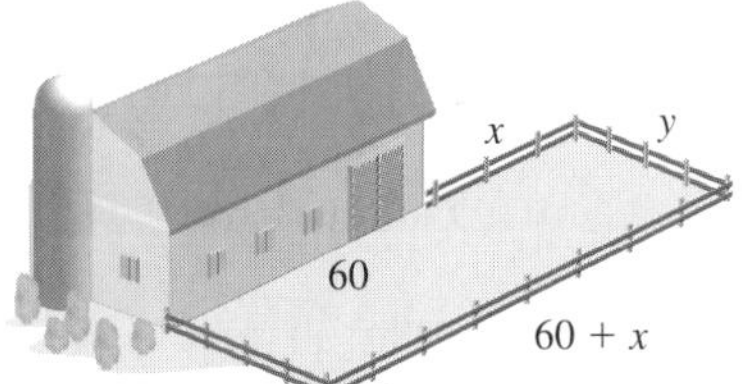

Before we can maximize the area, we must determine a relationship between x and y in order to express A as a function of one variable. In this case, x and y are related to the total amount of available fencing material:

$$\begin{aligned} x + y + 60 + x + y &= 160 \\ 2x + 2y &= 100 \\ y &= 50 - x \end{aligned}$$

Thus,

$$A(x) = (x + 60)(50 - x)$$

Now we need to determine the permissible values of x; that is, the domain of the function A. Since the farmer wants to use all of the barn as part of one side of the pen, x cannot be negative. Since y is the other dimension of the pen, y cannot be negative. Thus,

$$\begin{aligned} y = 50 - x &\geq 0 \\ 50 &\geq x \end{aligned}$$

The domain of A is $[0, 50]$. Thus, we must solve the following:

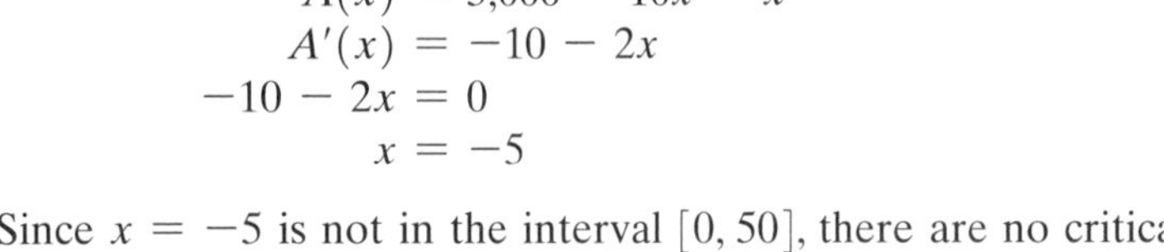

$$\begin{aligned} \text{Maximize} \quad A(x) &= (x + 60)(50 - x) \qquad 0 \leq x \leq 50 \\ A(x) &= 3{,}000 - 10x - x^2 \\ A'(x) &= -10 - 2x \\ -10 - 2x &= 0 \\ x &= -5 \end{aligned}$$

Since $x = -5$ is not in the interval $[0, 50]$, there are no critical values in the interval. $A(x)$ is continuous on $[0, 50]$, so the absolute maximum must occur at one of the end points.

$$\begin{aligned} A(0) &= 3{,}000 \quad \text{Maximum area} \\ A(50) &= 0 \end{aligned}$$

If $x = 0$, then $y = 50$. Thus, the dimensions of the pen with largest area are 60 feet by 50 feet.

(B) If 250 feet of fencing material is available, then

$$\begin{aligned} x + y + 60 + x + y &= 250 \\ 2x + 2y &= 190 \\ y &= 95 - x \end{aligned}$$

The model becomes

$$\begin{aligned} \text{Maximize} \quad A(x) &= (x + 60)(95 - x) \qquad 0 \leq x \leq 95 \\ A(x) &= 5{,}700 + 35x - x^2 \\ A'(x) &= 35 - 2x \\ 35 - 2x &= 0 \\ x &= \tfrac{35}{2} = 17.5 \quad \text{The only critical value} \\ A''(x) &= -2 < 0 \qquad \text{for all } x \\ \text{Max } A(x) &= A(17.5) = 6{,}006.25 \\ y &= 95 - 17.5 = 77.5 \end{aligned}$$

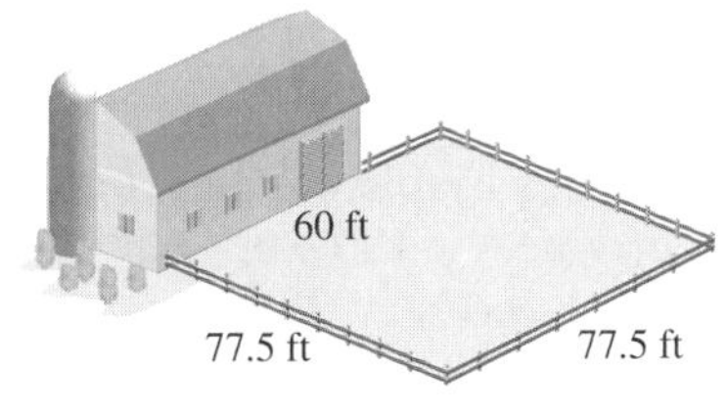

This time, the dimensions of the pen with the largest area are 77.5 feet by 77.5 feet.

Matched Problem 6 Repeat Example 6 if the barn is 80 feet long.

REMARK

A graphing utility is a convenient tool for locating critical values and solutions at end points. The graphs of the area functions from Examples 6A and 6B (see Fig. 6) confirm the results we obtained in our solutions.

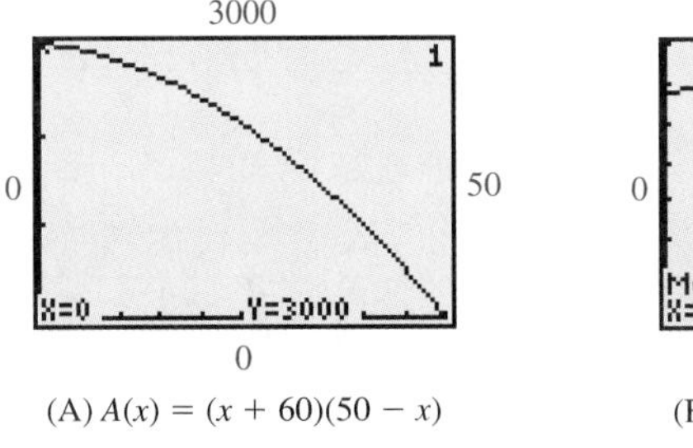

(A) $A(x) = (x + 60)(50 - x)$

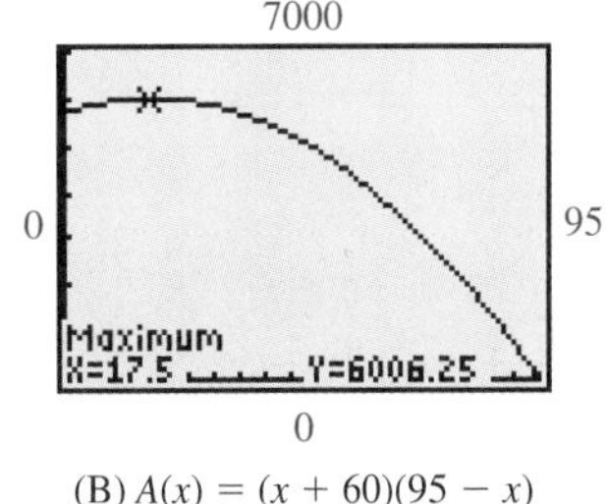

(B) $A(x) = (x + 60)(95 - x)$

FIGURE 6

Example 7 **Inventory Control** A recording company anticipates that there will be a demand for 20,000 copies of a certain compact disk (CD) during the following year. It costs the company \$0.50 to store a CD for 1 year. Each time it must make additional CD's, it costs \$200 to set up the equipment. How many CD's should the company make during each production run in order to minimize its total storage and set-up costs?

SOLUTION This type of problem is called an **inventory control problem.** One of the basic assumptions made in such problems is that the demand is uniform. For example, if there are 250 working days in a year, then the daily demand would be $20{,}000 \div 250 = 80$ CD's. The company could decide to produce all 20,000 CD's at the beginning of the year. This would certainly minimize the set-up costs, but would result in very large storage costs. At the other extreme, it could produce 80 CD's each day. This would minimize the storage costs, but would result in very large set-up costs. Somewhere between these two extremes is the optimal solution that will minimize the total storage and set-up costs. Let

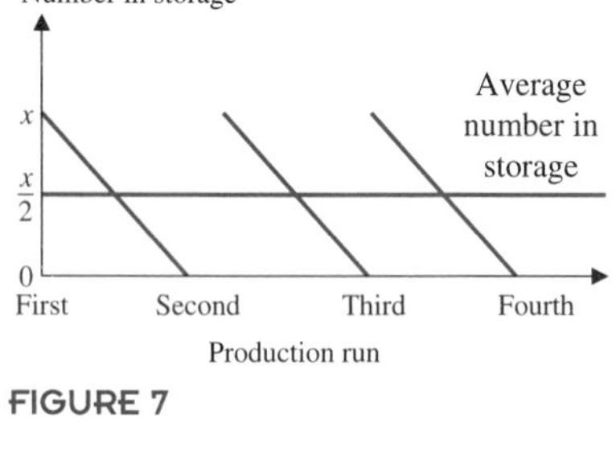

FIGURE 7

$$x = \text{Number of CD's manufactured during each production run}$$
$$y = \text{Number of production runs}$$

It is easy to see that the total set-up cost for the year is $200y$, but what is the total storage cost? If the demand is uniform, then the number of CD's in storage between production runs will decrease from x to 0, and the average number in storage each day is $x/2$. This result is illustrated in Figure 7.

Since it costs \$0.50 to store a CD for 1 year, the total storage cost is $0.5(x/2) = 0.25x$ and the total cost is

$$\text{Total cost} = \text{Set-up cost} + \text{Storage cost}$$
$$C = 200y + 0.25x$$

In order to write the total cost C as a function of one variable, we must find a relationship between x and y. If the company produces x CD's in each of y production runs, then the total number of CD's produced is xy. Thus,

$$xy = 20{,}000$$
$$y = \frac{20{,}000}{x}$$

Certainly, x must be at least 1 and cannot exceed 20,000. Thus, we must solve the following:

$$\text{Minimize} \quad C(x) = 200\left(\frac{20{,}000}{x}\right) + 0.25x \qquad 1 \leqslant x \leqslant 20{,}000$$

$$C(x) = \frac{4{,}000{,}000}{x} + 0.25x$$

$$C'(x) = -\frac{4{,}000{,}000}{x^2} + 0.25$$

$$-\frac{4{,}000{,}000}{x^2} + 0.25 = 0$$

$$x^2 = \frac{4{,}000{,}000}{0.25}$$

$$x^2 = 16{,}000{,}000$$

$$x = 4{,}000 \qquad -4{,}000 \text{ is not a critical value, since } 1 \leqslant x \leqslant 20{,}000$$

$$C''(x) = \frac{8{,}000{,}000}{x^3} > 0 \qquad \text{for } x \in (1, 20{,}000)$$

Thus,

$$\text{Min } C(x) = C(4{,}000) = 2{,}000$$
$$y = \frac{20{,}000}{4{,}000} = 5$$

The company will minimize its total cost by making 4,000 CD's five times during the year.

Matched Problem 7 ➡ Repeat Example 7 if it costs \$250 to set up a production run and \$0.40 to store a CD for 1 year.

Answers to Matched Problems

1. (A) Absolute maximum: $f(5) = 65$; absolute minimum: $f(-5) = -65$
(B) Absolute maximum: $f(-2) = 16$; absolute minimum: $f(2) = -16$
(C) Absolute maximum: $f(-2) = 16$; absolute minimum: $f(1) = -11$

2. $f(3) = 6$

3. (A) Max $R(x) = R(4{,}500) = \$675{,}000$
(B) Max $P(x) = P(4{,}050) = \$456{,}750$; $p = \$165$

4. Max $P(x) = P(3{,}750) = \$378{,}750$; $p = \$175$; price increases \$10, profit decreases \$78,000

5. Max $Y(x) = Y(15) = 2{,}025$ lb/acre

6. (A) 80 ft by 40 ft (B) 82.5 ft by 82.5 ft

7. Make 5,000 CD's four times during the year

EXERCISE 9-5

A *Problems 1–10 refer to the graph of $y = f(x)$ shown below. Find the absolute minimum and the absolute maximum over the indicated interval.*

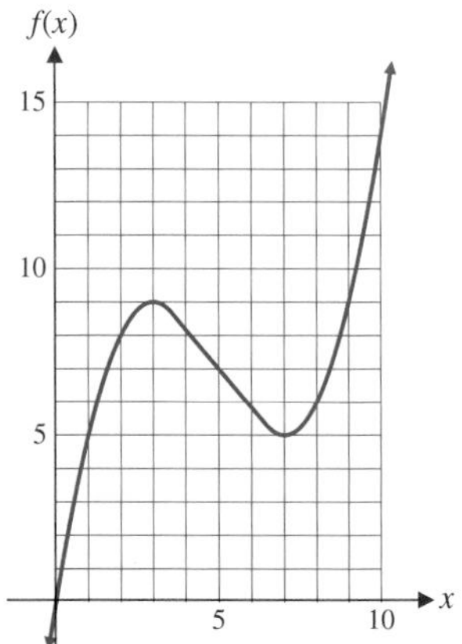

1. $[0, 10]$ **2.** $[2, 8]$ **3.** $[0, 8]$ **4.** $[2, 10]$

5. $[1, 10]$ **6.** $[0, 9]$ **7.** $[1, 9]$ **8.** $[0, 2]$

9. $[2, 5]$ **10.** $[5, 8]$

In Problems 11–16, find the absolute maximum and minimum, if either exists, for each function.

11. $f(x) = x^2 - 4x + 5$ **12.** $f(x) = x^2 + 6x + 7$

13. $f(x) = 10 + 8x - x^2$ **14.** $f(x) = 6 - 8x - x^2$

15. $f(x) = 1 - x^3$ **16.** $f(x) = 1 - x^4$

B *In Problems 17–22, find the indicated extremum of each function, if it exists.*

17. Absolute maximum value of

$$f(x) = 24 - 2x - \frac{8}{x} \qquad x > 0$$

18. Absolute minimum value of

$$f(x) = 3x + \frac{27}{x} \qquad x > 0$$

19. Absolute minimum value of

$$f(x) = 5 + 3x + \frac{12}{x^2} \qquad x > 0$$

20. Absolute maximum value of

$$f(x) = 10 - 2x - \frac{27}{x^2} \qquad x > 0$$

21. Absolute maximum value (rounded to two decimal places) of

C

$$f(x) = 15 + 9x - x^2 - \frac{2}{x} \qquad x > 0$$

22. Absolute minimum value (rounded to two decimal places) of

C

$$f(x) = 5 + x + 2x^2 + \frac{3}{x} \qquad x > 0$$

In Problems 23–26, find the absolute maximum and minimum, if either exists, for each function on the indicated intervals.

23. $f(x) = x^3 - 6x^2 + 9x - 6$
(A) $[-1, 5]$ (B) $[-1, 3]$ (C) $[2, 5]$

24. $f(x) = 2x^3 - 3x^2 - 12x + 24$
(A) $[-3, 4]$ (B) $[-2, 3]$ (C) $[-2, 1]$

25. $f(x) = (x - 1)(x - 5)^3 + 1$
(A) $[0, 3]$ (B) $[1, 7]$ (C) $[3, 6]$

26. $f(x) = x^4 - 8x^2 + 16$
(A) $[-1, 3]$ (B) $[0, 2]$ (C) $[-3, 4]$

C *Preliminary word problems:*

27. How would you divide a 10 inch line so that the product of the two lengths is a maximum?

28. What quantity should be added to 5 and subtracted from 5 in order to produce the maximum product of the results?

29. Find two numbers whose difference is 30 and whose product is a minimum.

30. Find two positive numbers whose sum is 60 and whose product is a maximum.

31. Find the dimensions of a rectangle with perimeter 100 centimeters that has maximum area. Find the maximum area.

32. Find the dimensions of a rectangle of area 225 square centimeters that has the least perimeter. What is the perimeter?

Problems 33–36 refer to a rectangular area enclosed by a fence that costs \$B per foot. Discuss the existence of a solution and the economical implications of each optimization problem.

33. Given a fixed area, minimize the cost of the fencing.

34. Given a fixed area, maximize the cost of the fencing.

35. Given a fixed amount to spend on fencing, maximize the enclosed area.

36. Given a fixed amount to spend on fencing, minimize the enclosed area.

APPLICATIONS

Business & Economics

37. *Maximum revenue and profit.* A company manufactures and sells x television sets per month. The monthly cost and price–demand equations are

$$C(x) = 72{,}000 + 60x$$

$$p = 200 - \frac{x}{30} \qquad 0 \le x \le 6{,}000$$

(A) Find the maximum revenue.
(B) Find the maximum profit, the production level that will realize the maximum profit, and the price the company should charge for each television set.
(C) If the government decides to tax the company \$5 for each set it produces, how many sets should the company manufacture each month in order to maximize its profit? What is the maximum profit? What should the company charge for each set?

38. *Maximum revenue and profit.* Repeat Problem 37 for

$$C(x) = 60{,}000 + 60x$$

$$p = 200 - \frac{x}{50} \qquad 0 \le x \le 10{,}000$$

39. *Maximum profit.* The table contains price–demand and total cost data for the production of radial arm saws, where p is the wholesale price (in dollars) of a saw for an annual demand of x saws and C is the total cost (in dollars) of producing x saws.

C

x	p (\$)	C (\$)
950	240	130,000
1,200	210	150,000
1,800	160	180,000
2,050	120	190,000

(A) Find a quadratic regression equation for the price–demand data using x as the independent variable.
(B) Find a linear regression equation for the cost data using x as the independent variable.
(C) What is the maximum profit? What is the wholesale price per saw that should be charged to realize the maximum profit? Round answers to the nearest dollar.

40. *Maximum profit.* The table contains price–demand and total cost data for the production of airbrushes, where p is the wholesale price (in dollars) of an airbrush for an annual demand of x airbrushes and C is the total cost (in dollars) of producing x airbrushes.

C

x	p ($)	C ($)
2,300	98	145,000
3,300	84	170,000
4,500	67	190,000
5,200	51	210,000

(A) Find a quadratic regression equation for the price–demand data using x as the independent variable.
(B) Find a linear regression equation for the cost data using x as the independent variable.
(C) What is the maximum profit? What is the wholesale price per airbrush that should be charged to realize the maximum profit? Round answers to the nearest dollar.

41. *Car rental.* A car rental agency rents 200 cars per day at a rate of $30 per day. For each $1 increase in rate, 5 fewer cars are rented. At what rate should the cars be rented to produce the maximum income? What is the maximum income?

42. *Rental income.* A 300 room hotel in Las Vegas is filled to capacity every night at $80 a room. For each $1 increase in rent, 3 fewer rooms are rented. If each rented room costs $10 to service per day, how much should the management charge for each room to maximize gross profit? What is the maximum gross profit?

43. *Agriculture.* A commercial cherry grower estimates from past records that if 30 trees are planted per acre, each tree will yield an average of 50 pounds of cherries per season. If for each additional tree planted per acre (up to 20), the average yield per tree is reduced by 1 pound, how many trees should be planted per acre to obtain the maximum yield per acre? What is the maximum yield?

44. *Agriculture.* A commercial pear grower must decide on the optimum time to have fruit picked and sold. If the pears are picked now, they will bring 30¢ per pound, with each tree yielding an average of 60 pounds of salable pears. If the average yield per tree increases 6 pounds per tree per week for the next 4 weeks, but the price drops 2¢ per pound per week, when should the pears be picked to realize the maximum return per tree? What is the maximum return?

45. *Manufacturing.* A candy box is to be made out of a piece of cardboard that measures 8 by 12 inches. Squares of equal size will be cut out of each corner, and then the ends and sides will be folded up to form a rectangular box. What size square should be cut from each corner to obtain a maximum volume?

46. *Packaging.* A parcel delivery service will deliver a package only if the length plus girth (distance around) does not exceed 108 inches.

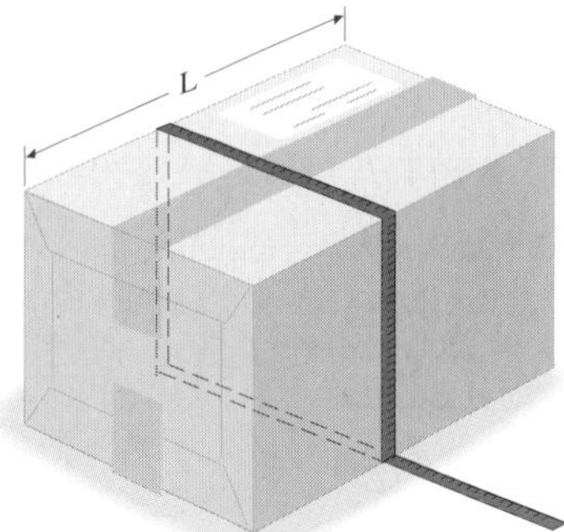

Figure for 46

(A) Find the dimensions of a rectangular box with square ends that satisfies the delivery service's restriction and has maximum volume. What is the maximum volume?
(B) Find the dimensions (radius and height) of a cylindrical container that meets the delivery service's requirement and has maximum volume. What is the maximum volume?

47. *Construction costs.* A fence is to be built to enclose a rectangular area of 800 square feet. The fence along three sides is to be made of material that costs $6 per foot. The material for the fourth side costs $18 per foot. Find the dimensions of the rectangle that will allow the most economical fence to be built.

48. *Construction costs.* The owner of a retail lumber store wants to construct a fence to enclose an outdoor storage area adjacent to the store, using all of the store as part of one side of the area (see the figure). Find the dimensions that will enclose the largest area if:
(A) 240 feet of fencing material is used.
(B) 400 feet of fencing material is used.

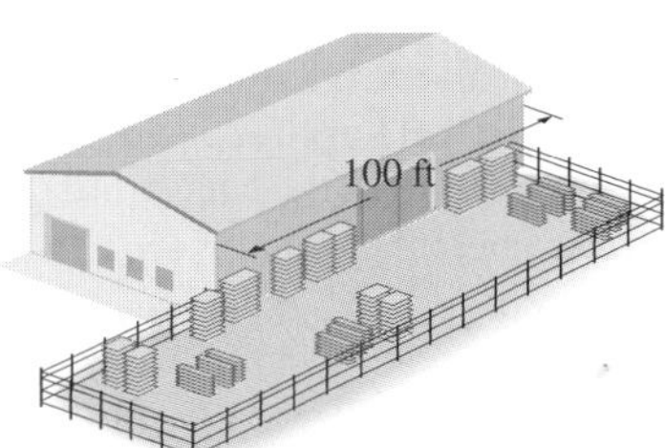

Figure for 48

49. *Inventory control.* A publishing company sells 50,000 copies of a certain book each year. It costs the company \$1 to store a book for 1 year. Each time it must print additional copies, it costs the company \$1,000 to set up the presses. How many books should the company produce during each printing in order to minimize its total storage and set-up costs?

50. *Operational costs.* The cost per hour for fuel to run a train is $v^2/4$ dollars, where v is the speed of the train in miles per hour. (Note that the cost goes up as the square of the speed.) Other costs, including labor, are \$300 per hour. How fast should the train travel on a 360 mile trip to minimize the total cost for the trip?

51. *Construction costs.* A freshwater pipeline is to be run from a source on the edge of a lake to a small resort community on an island 5 miles off-shore, as indicated in the figure.

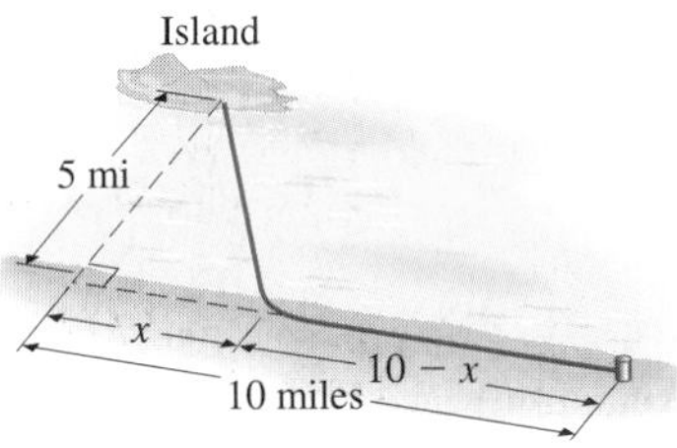

Figure for 51

(A) If it costs 1.4 times as much to lay the pipe in the lake as it does on land, what should x be (in miles) to minimize the total cost of the project?

(B) If it costs only 1.1 times as much to lay the pipe in the lake as it does on land, what should x be to minimize the total cost of the project? [*Note:* Compare with Problem 56.]

52. *Manufacturing costs.* A manufacturer wants to produce cans that will hold 12 ounces (approximately 22 cubic inches) in the form of a right circular cylinder. Find the dimensions (radius of an end and height) of the can that will use the smallest amount of material. Assume the circular ends are cut out of squares, with the corner portions wasted, and the sides are made from rectangles, with no waste.

Life Sciences

53. *Bacteria control.* A recreational swimming lake is treated periodically to control harmful bacteria growth. Suppose t days after a treatment, the concentration of bacteria per cubic centimeter is given by

$$C(t) = 30t^2 - 240t + 500 \qquad 0 \le t \le 8$$

How many days after a treatment will the concentration be minimal? What is the minimum concentration?

54. *Drug concentration.* The concentration $C(t)$, in milligrams per cubic centimeter, of a particular drug in a patient's bloodstream is given by

$$C(t) = \frac{0.16t}{t^2 + 4t + 4}$$

where t is the number of hours after the drug is taken. How many hours after the drug is given will the concentration be maximum? What is the maximum concentration?

55. *Laboratory management.* A laboratory uses 500 white mice each year for experimental purposes. It costs \$4 to feed a mouse for 1 year. Each time mice are ordered from a supplier, there is a service charge of \$10 for processing the order. How many mice should be ordered each time in order to minimize the total cost of feeding the mice and of placing the orders for the mice?

56. *Bird flights.* Some birds tend to avoid flights over large bodies of water during daylight hours. (It is speculated that more energy is required to fly over water than land, because air generally rises over land and falls over water during the day.) Suppose an adult bird with this tendency is taken from its nesting area on the edge of a large lake to an island 5 miles off-shore and is then released (see the figure).

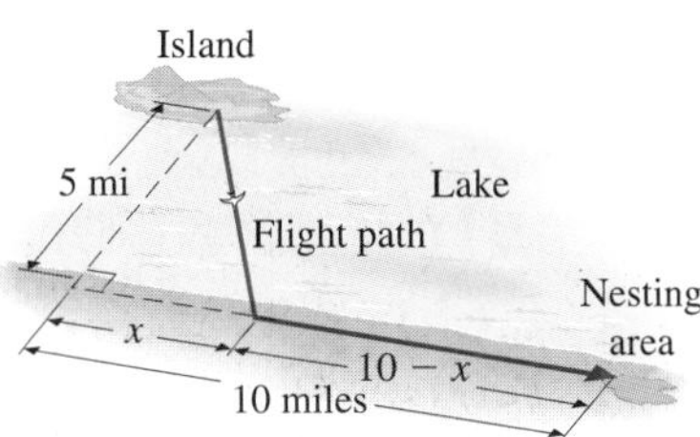

Figure for 56

(A) If it takes 1.4 times as much energy to fly over water as land, how far up-shore (x, in miles) should the bird head in order to minimize the total energy expended in returning to the nesting area?

(B) If it takes only 1.1 times as much energy to fly over water as land, how far up-shore should the bird head in order to minimize the total energy expended in returning to the nesting area? [*Note:* Compare with Problem 51.]

57. *Botany.* If it is known from past experiments that the height (in feet) of a given plant after t months is given approximately by

$$H(t) = 4t^{1/2} - 2t \qquad 0 \le t \le 2$$

how long, on the average, will it take a plant to reach its maximum height? What is the maximum height?

58. *Pollution.* Two heavy industrial areas are located 10 miles apart, as indicated in the figure. If the concentration of particulate matter (in parts per million) decreases as the reciprocal of the square of the distance from the source, and area A_1 emits eight times the particulate matter as A_2, then the concentration of particulate matter at any point between the two areas is given by

$$C(x) = \frac{8k}{x^2} + \frac{k}{(10 - x)^2} \qquad 0.5 \leq x \leq 9.5, \quad k > 0$$

How far from A_1 will the concentration of particulate matter between the two areas be at a minimum?

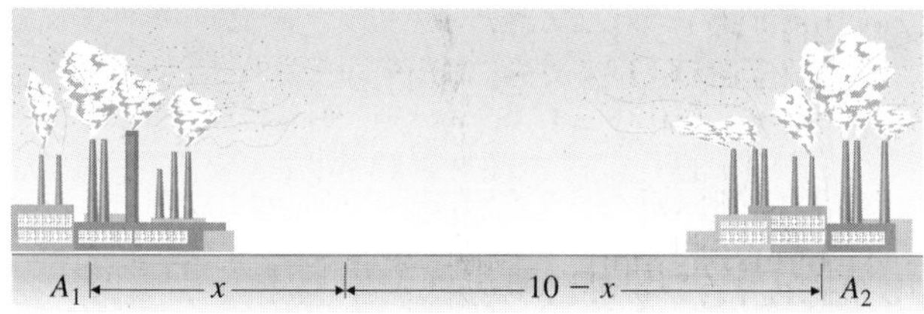

Figure for 58

Social Sciences

59. *Politics.* In a newly incorporated city, it is estimated that the voting population (in thousands) will increase according to

$$N(t) = 30 + 12t^2 - t^3 \qquad 0 \leq t \leq 8$$

where t is time in years. When will the rate of increase be most rapid?

60. *Learning.* A large grocery chain found that, on the average, a checker can recall $P\%$ of a given price list x hours after starting work, as given approximately by

$$P(x) = 96x - 24x^2 \qquad 0 \leq x \leq 3$$

At what time x does the checker recall a maximum percentage? What is the maximum?

Important Terms and Symbols

9-1 *Continuity and Graphs.* Continuous at a point; continuous on an open interval; discontinuous at a point; continuous on the right; continuous on the left; continuous on a closed interval; general continuity properties; continuity properties of specific functions; vertical asymptote; solving inequalities using continuity properties; sign chart; test number; partition number

$\lim_{x \to c} f(x) = f(c)$ if f is continuous at $x = c$;

$\lim_{x \to c} f(x) = \infty$; $\lim_{x \to c} f(x) = -\infty$

9-2 *First Derivative and Graphs.* Increasing and decreasing functions; rising and falling graphs; critical values; local extremum; local maximum; local minimum; turning point; first-derivative test for local extrema

9-3 *Second Derivative and Graphs.* Concave upward; concave downward; second derivative; concavity and the second derivative; inflection point; second-derivative test for local maxima and minima; point of diminishing returns

$f''(x)$; $\frac{d^2y}{dx^2}$; y''

9-4 *Curve Sketching Techniques: Unified and Extended.* Horizontal asymptote; limits at infinity for power functions, polynomials, and rational functions; locating vertical asymptotes; graphing strategy [analyze $f(x)$ to find domain, intercepts, horizontal and vertical asymptotes; analyze $f'(x)$ to find increasing and decreasing regions, local extrema; analyze $f''(x)$ to find concave upward and concave downward regions, inflection points]; oblique asymptote

9-5 *Optimization; Absolute Maxima and Minima.* Absolute maximum; absolute minimum; absolute extrema of a function on a closed interval; second-derivative test for absolute maximum and minimum; optimization problems; inventory control

Review Exercise

Work through all the problems in this chapter review and check your answers in the back of the book. Answers to all review problems are there along with section numbers in italics to indicate where each type of problem is discussed. Where weaknesses show up, review appropriate sections in the text.

A *Problems 1–8 refer to the graph of $y = f(x)$ below. Identify the points or intervals on the x axis that produce the indicated behavior.*

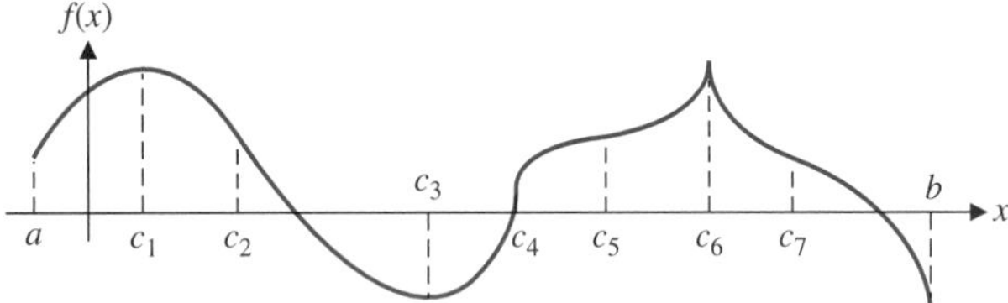

1. $f(x)$ is increasing
2. $f'(x) < 0$
3. Graph of f is concave downward
4. Local minima
5. Absolute maxima
6. $f'(x)$ appears to be 0
7. $f'(x)$ does not exist
8. Inflection points

In Problems 9–11, use the graph of the function f shown in the figure to answer each question.

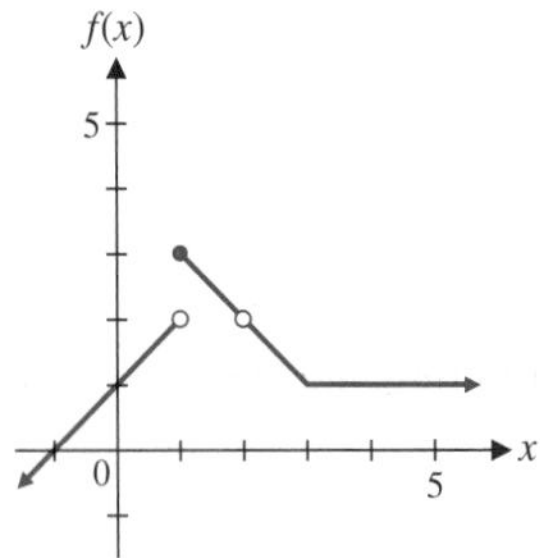

9. (A) $\lim_{x\to 1} f(x) = ?$ (B) $f(1) = ?$
(C) Is f continuous at $x = 1$?

10. (A) $\lim_{x\to 2} f(x) = ?$ (B) $f(2) = ?$
(C) Is f continuous at $x = 2$?

11. (A) $\lim_{x\to 3} f(x) = ?$ (B) $f(3) = ?$
(C) Is f continuous at $x = 3$?

In Problems 12 and 13, use the given information to sketch the graph of f. Assume that f is continuous on its domain and that all intercepts are included in the given information.

12. Domain: All real x

x	-3	-2	-1	0	2	3
$f(x)$	0	3	2	0	-3	0

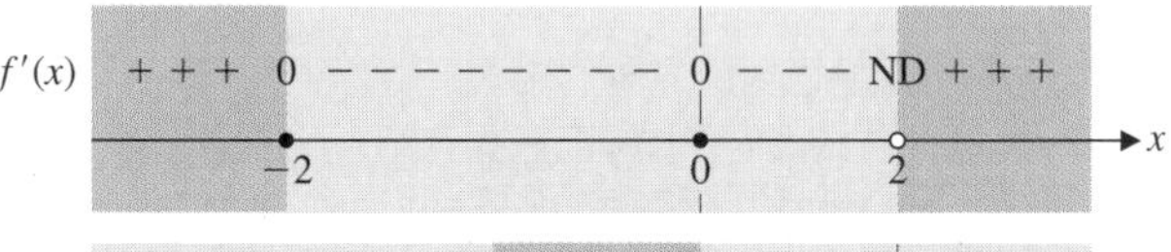

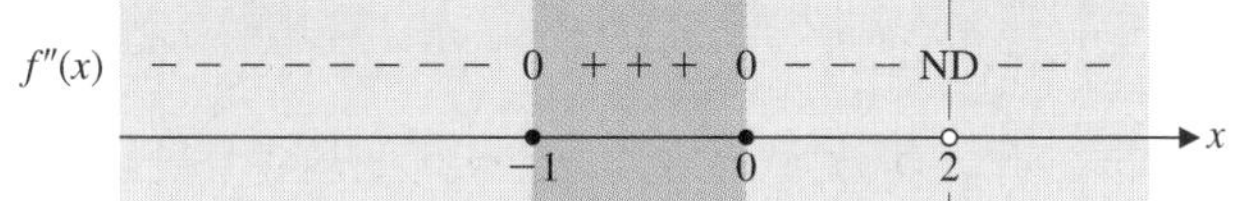

13. Domain: All real x;
$f(-2) = 1, f(0) = 0, f(2) = 1$;
$f'(0) = 0; f'(x) < 0$ on $(-\infty, 0)$;
$f'(x) > 0$ on $(0, \infty)$;
$f''(-2) = 0, f''(2) = 0$;
$f''(x) < 0$ on $(-\infty, -2)$ and $(2, \infty)$;
$f''(x) > 0$ on $(-2, 2)$;
$\lim_{x\to-\infty} f(x) = 2$; $\lim_{x\to\infty} f(x) = 2$

14. Find $f''(x)$ for $f(x) = x^4 + 5x^3$.

15. Find y'' for $y = 3x + \frac{4}{x}$.

B *Problems 16 and 17 refer to the function f in the figure.*

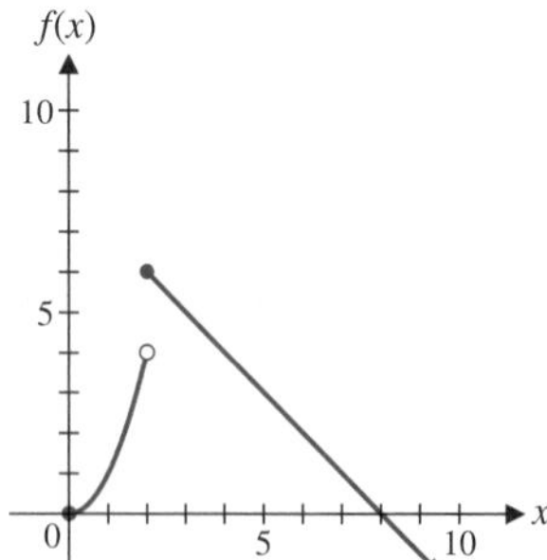

Figure for 16 and 17: $f(x) = \begin{cases} x^2 & \text{if } 0 \leqslant x < 2 \\ 8 - x & \text{if } x \geqslant 2 \end{cases}$

16. (A) $\lim_{x\to 2^-} f(x) = ?$ (B) $\lim_{x\to 2^+} f(x) = ?$
(C) $\lim_{x\to 2} f(x) = ?$ (D) $f(2) = ?$
(E) Is f continuous at $x = 2$?

17. (A) $\lim_{x\to 5^-} f(x) = ?$ (B) $\lim_{x\to 5^+} f(x) = ?$
(C) $\lim_{x\to 5} f(x) = ?$ (D) $f(5) = ?$
(E) Is f continuous at $x = 5$?

In Problems 18–20, solve each inequality. Express the answer in interval notation. Use a graphing utility in Problem 20 to approximate partition numbers to two decimal places.

18. $x^2 - x < 12$

19. $\dfrac{x - 5}{x^2 + 3x} > 0$

20. $x^3 + x^2 - 4x - 2 > 0$
C

Problems 21–24 refer to the function

$$f(x) = x^3 - 18x^2 + 81x$$

21. Using $f(x)$:
(A) Determine the domain of f.
(B) Find any intercepts for the graph of f.
(C) Find any horizontal or vertical asymptotes for the graph of f.

22. Using $f'(x)$:
(A) Find critical values for $f(x)$.
(B) Find partition numbers for $f'(x)$.
(C) Find intervals over which $f(x)$ is increasing; decreasing.
(D) Find any local maxima and minima.

23. Using $f''(x)$:
(A) Find intervals over which the graph of f is concave upward; concave downward.
(B) Find any inflection points.

24. Use the results of Problems 21–23 to graph f.

Problems 25–28 refer to the function

$$y = f(x) = \frac{3x}{x + 2}$$

25. Using $f(x)$:
(A) Determine the domain of f.
(B) Find any intercepts for the graph of f.
(C) Find any horizontal or vertical asymptotes for the graph of f.

26. Using $f'(x)$:
(A) Find critical values for $f(x)$.
(B) Find partition numbers for $f'(x)$.
(C) Find intervals over which $f(x)$ is increasing; decreasing.
(D) Find any local maxima and minima.

27. Using $f''(x)$:
(A) Find intervals over which the graph of f is concave upward; concave downward.
(B) Find any inflection points.

28. Use the results of Problems 25–27 to graph f.

29. Use the graph of $y = f'(x)$ shown below to discuss the graph of $y = f(x)$. Organize your conclusions in a table (see Example 4, page 648). Sketch a possible graph for $y = f(x)$.

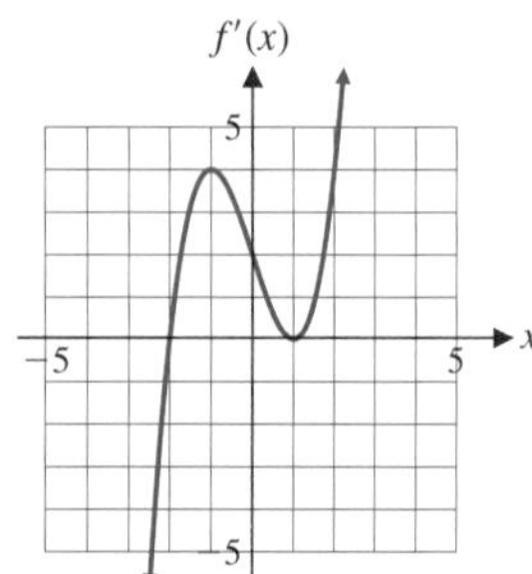

Figure for 29 and 30

30. Refer to the graph of $y = f'(x)$ above. Which of the following could be the graph of $y = f''(x)$?

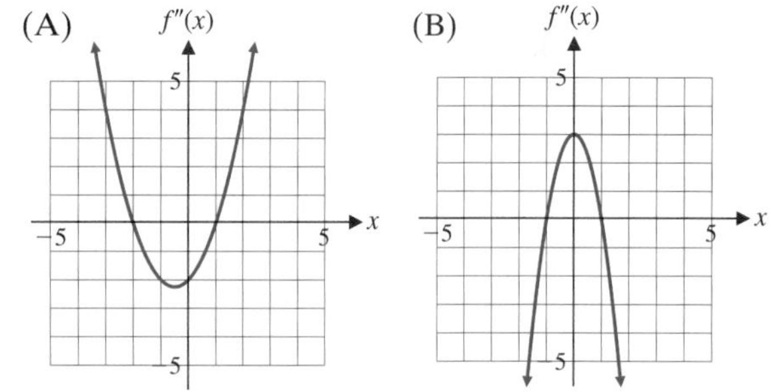

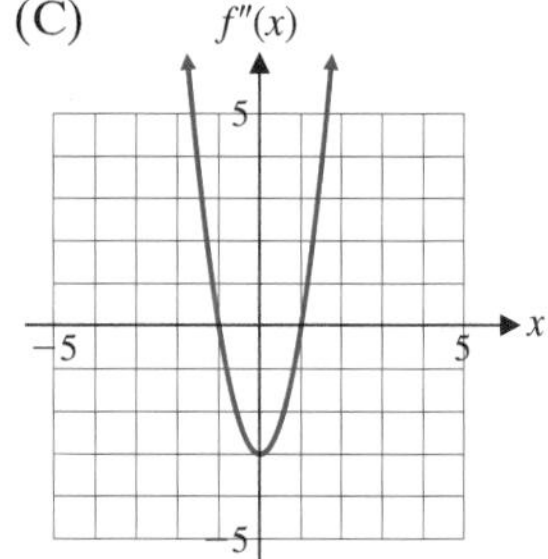

31. Use the second-derivative test to find any local extrema for

$$f(x) = x^3 - 6x^2 - 15x + 12$$

32. Find the absolute maximum and absolute minimum, if either exists, for

$$y = f(x) = x^3 - 12x + 12 \qquad -3 \leq x \leq 5$$

33. Find the absolute minimum, if it exists, for

$$y = f(x) = x^2 + \frac{16}{x^2} \qquad x > 0$$

In Problems 34–38, determine where f is continuous. Express the answer in interval notation.

34. $f(x) = 2x^2 - 3x + 1$

35. $f(x) = \dfrac{1}{x + 5}$

36. $f(x) = \dfrac{x - 3}{x^2 - x - 6}$

37. $f(x) = \sqrt{x - 3}$

38. $f(x) = \sqrt[3]{1 - x^2}$

In Problems 39–48, evaluate the indicated limit. Use $-\infty$ or ∞ where appropriate.

39. $\lim_{x \to 2^-} \dfrac{x}{2 - x}$

40. $\lim_{x \to 2^+} \dfrac{x}{2 - x}$

41. $\lim_{x \to 2} \dfrac{x}{2 - x}$

42. $\lim_{x \to 2} \dfrac{x}{(2 - x)^2}$

43. $\lim_{x \to \infty}(-3x^5)$

44. $\lim_{x \to -\infty}(3x^2 - 4x^3)$

45. $\lim_{x \to \infty}(4x^6 - 2x^5)$

46. $\lim_{x \to \infty} \dfrac{6x^3 + 4x^2 + 5}{3x^3 + 2x + 7}$

47. $\lim_{x \to \infty} \dfrac{6x^4 + 4x^2 + 5}{3x^3 + 2x + 7}$

48. $\lim_{x \to \infty} \dfrac{6x^3 + 4x^2 + 5}{3x^4 + 2x + 7}$

Find horizontal and vertical asymptotes, if they exist, in Problems 49 and 50.

49. $f(x) = \dfrac{x}{x^2 + 9}$

50. $f(x) = \dfrac{x^3}{x^2 - 9}$

51. Let $y = f(x)$ be a polynomial function with local minima at $x = a$ and $x = b$, $a < b$. Must f have at least one local maximum between a and b? Justify your answer.

52. The derivative of $f(x) = x^{-1}$ is $f'(x) = -x^{-2}$. Since $f'(x) < 0$ for $x \neq 0$, is it correct to say that $f(x)$ is decreasing for all x except $x = 0$? Explain.

53. Discuss the difference between a partition number for $f'(x)$ and a critical value for $f(x)$, and illustrate with examples.

C

54. Find the absolute maximum for $f'(x)$ if

$$f(x) = 6x^2 - x^3 + 8$$

Graph f and f' on the same coordinate system for $0 \leq x \leq 4$.

55. Find two positive numbers whose product is 400 and whose sum is a minimum. What is the minimum sum?

56. Let $f(x) = (x - 1)^3(x + 3)$. Apply the graphing strategy discussed in Section 9-4, summarize the pertinent information, and sketch the graph.

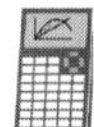

In Problems 57 and 58, apply steps 1–3 of the graphing strategy discussed in Section 9-4 and summarize the pertinent information. Round any approximate values to two decimal places.

57. $f(x) = x^4 + x^3 - 4x^2 - 3x + 4$

58. $f(x) = 0.25x^4 - 5x^3 + 31x^2 - 70x$

Applications

Business & Economics

59. *Pricing.* An office products firm gives volume discounts on purchases of three-ring binders, as shown in the table.

QUANTITY ORDERED	1	12	24 or more
PRICE PER BINDER	$1.99	$1.76	$1.49

(A) Let $y = p(x)$ represent the price per binder if x binders are ordered. Graph $y = p(x)$ for $1 \leq x \leq 30$.

(B) Identify the discontinuities of p and discuss the behavior at each discontinuity.

(C) Let $y = C(x)$ be the total cost for an order of x binders. Graph $y = C(x)$ for $1 \leq x \leq 30$.

(D) Identify the discontinuities of C and discuss the behavior at each discontinuity.

60. *Price analysis.* The graph in the figure approximates the rate of change of the price of tomatoes over a 60 month period, where $p(t)$ is the price of a pound of tomatoes and t is time (in months).

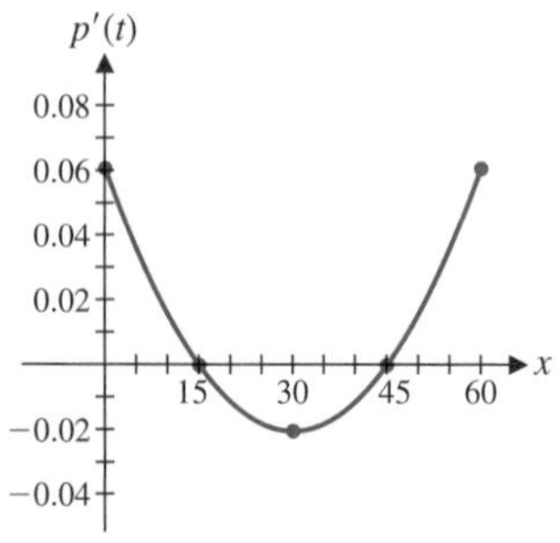

Figure for 60

(A) Write a brief verbal description of the graph of $y = p(t)$, including a discussion of local extrema and inflection points.
(B) Sketch a possible graph of $y = p(t)$.

61. *Maximum revenue and profit.* A company manufactures and sells x electric stoves per month. The monthly cost and price–demand equations are

$$C(x) = 350x + 50{,}000$$
$$p = 500 - 0.025x \qquad 0 \le x \le 20{,}000$$

(A) Find the maximum revenue.
(B) How many stoves should the company manufacture each month to maximize its profit? What is the maximum monthly profit? How much should the company charge for each stove?
(C) If the government decides to tax the company \$20 for each stove it produces, how many stoves should the company manufacture each month to maximize its profit? What is the maximum monthly profit? How much should the company charge for each stove?

62. *Construction.* A fence is to be built to enclose a rectangular area. The fence along three sides is to be made of material that costs \$5 per foot. The material for the fourth side costs \$15 per foot.
(A) If the area is 5,000 square feet, find the dimensions of the rectangle that will allow the most economical fence to be built.
(B) If \$3,000 is available for the fencing, find the dimensions of the rectangle that will enclose the most area.

63. *Rental income.* A 200 room hotel in Fresno is filled to capacity every night at a rate of \$40 per room. For each \$1 increase in the nightly rate, 4 fewer rooms are rented. If each rented room costs \$8 a day to service, how much should the management charge per room in order to maximize gross profit? What is the maximum gross profit?

64. *Inventory control.* A computer store sells 7,200 boxes of floppy disks annually. It costs the store \$0.20 to store a box of disks for 1 year. Each time it reorders disks, the store must pay a \$5.00 service charge for processing the order. How many times during the year should the store order disks in order to minimize the total storage and reorder costs?

65. *Average cost.* The total cost of producing x units per month is given by

$$C(x) = 4{,}000 + 10x + 0.1x^2$$

Find the minimum average cost. Graph the average cost and the marginal cost functions on the same coordinate system. Include any oblique asymptotes.

66. C *Average cost.* The data in the table give the total daily cost y (in dollars) of producing x dozen chocolate chip cookies at various production levels.

DOZENS OF COOKIES x	TOTAL COST y (\$)
50	119
100	187
150	248
200	382
250	505
300	695

(A) Enter the data in a graphing utility and find a quadratic regression equation for the total cost.
(B) Use the regression equation from part (A) to find the minimum average cost (to the nearest cent) and the corresponding production level (to the nearest integer).

67. C *Advertising.* A chain of appliance stores uses television ads to promote the sales of refrigerators. Analyzing past records produced the data in the table, where x is the number of ads placed monthly and y is the number of refrigerators sold that month.

NUMBER OF ADS x	NUMBER OF REFRIGERATORS y
10	271
20	427
25	526
30	629
45	887
48	917

(A) Enter the data in a graphing utility, set the utility to display two decimal places, and find a cubic regression equation for the number of refrigerators sold monthly as a function of the number of ads.
(B) How many ads should be placed each month to maximize the rate of change of sales with respect to the number of ads, and how many refrigerators can be expected to be sold with this number of ads? Round answers to the nearest integer.

Life Sciences

68. *Bacteria control.* If t days after a treatment, the bacteria count per cubic centimeter in a body of water is given by

$$C(t) = 20t^2 - 120t + 800 \qquad 0 \le t \le 9$$

in how many days will the count be a minimum?

Social Sciences

69. *Politics.* In a new suburb, it is estimated that the number of registered voters will grow according to

$$N = 10 + 6t^2 - t^3 \qquad 0 \le t \le 5$$

where t is time in years and N is in thousands. When will the rate of increase be maximum?

Group Activity 1 Maximizing Profit

A company manufactures and sells x air-conditioners per month. The monthly cost and price–demand equations are

$$C(x) = 180x + 20{,}000$$
$$p = 220 - 0.001x \qquad 0 \le x \le 100{,}000$$

(A) How many air-conditioners should the company manufacture each month in order to maximize its monthly profit? What is the maximum monthly profit, and what should the company charge for each air-conditioner to realize the maximum monthly profit?

(B) Repeat part (A) if the government decides to tax the company at the rate of \$18 per air-conditioner produced. How much revenue will the government receive from the tax on these air-conditioners?

(C) Repeat part (A) if the government raises the tax to \$23 per air-conditioner. Discuss the effect of this tax increase on the government's tax revenue.

(D) Repeat part (A) if the government sets the tax rate at \$$t$ per air-conditioner. What value of t will maximize the government's tax revenue? What is the government's maximum tax revenue?

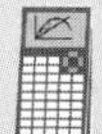

Group Activity 2 Minimizing Construction Costs

Two resort communities are located on separate islands in a lake, as indicated in Figure 1. A communications company wants to run underground cables from a single source on the shore to each of the islands. Since the high cost of laying cable under water is directly proportional to the amount of cable used, the company wants to position the source of the cables so that the amount of cable used is minimized.

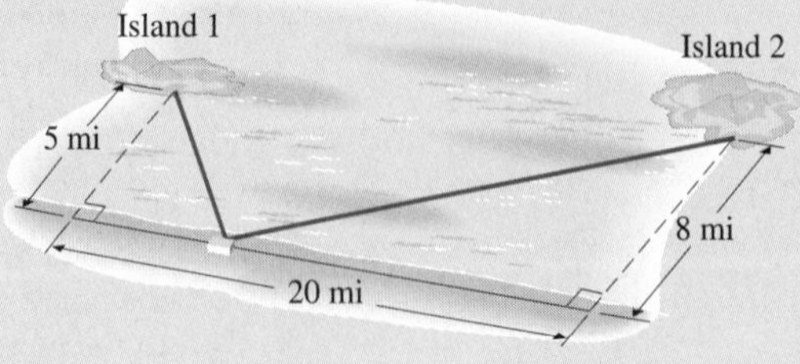

FIGURE 1

(A) Assume that the shoreline is straight and position a coordinate system as indicated in Figure 2. Let f be the function that represents the total length of cable used. Express f in terms of x (Fig. 2), graph f and f' (use the numerical derivative of f to graph f'), and find the value of x (in miles) that will minimize the total cable length.

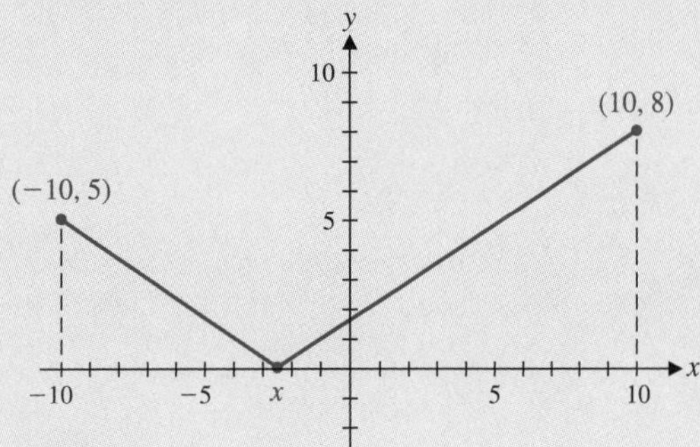

FIGURE 2

(B) Repeat part (A) under the assumption that the shoreline is in the shape of the parabola $y = 0.02x^2 - 2$ (Fig. 3).

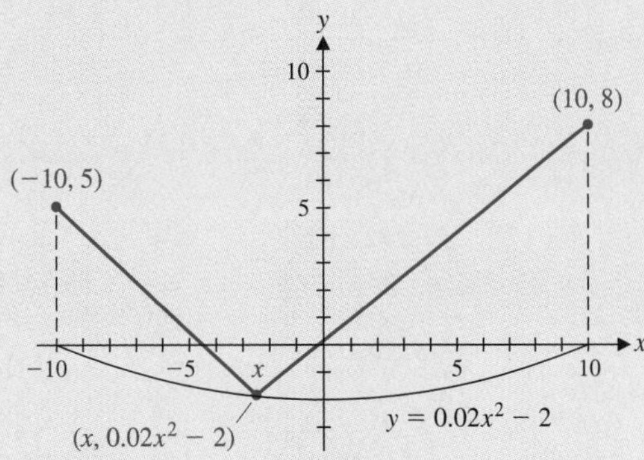

FIGURE 3

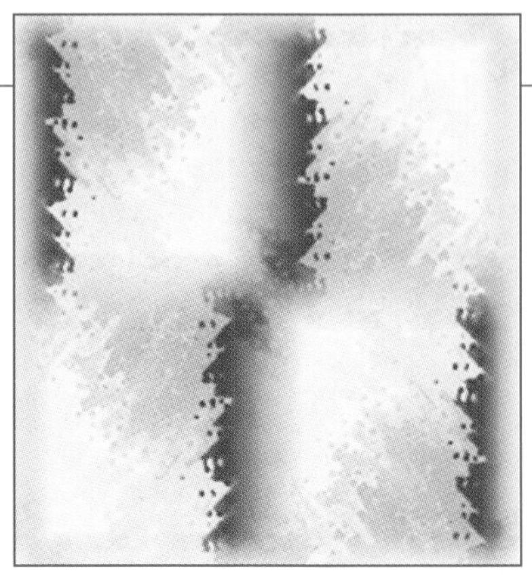

Chapter 10

Additional Derivative Topics

INTRODUCTION

In this chapter we complete our discussion of derivatives by first looking at the differentiation of forms that involve the exponential and logarithmic functions and then considering some additional topics and applications involving all the different types of functions we have encountered thus far. You will probably find it helpful to review some of the important properties of the exponential and logarithmic functions given in Chapter 2 before proceeding further.

Section 10-1 The Constant e and Continuous Compound Interest

- The Constant e
- Continuous Compound Interest

In Chapter 2, both the exponential function with base e and continuous compound interest were introduced informally. Now, with limit concepts at our disposal, we can give precise definitions of e and continuous compound interest.

The Constant e

The special irrational number e is a particularly suitable base for both exponential and logarithmic functions. The reasons for choosing this number as a base will become clear as we develop differentiation formulas for the exponential function e^x and the natural logarithmic function $\ln x$.

In precalculus treatments (Chapter 2), the number e is informally defined as an irrational number that can be approximated by the expression $[1 + (1/n)]^n$ by taking n sufficiently large. Now we will use the limit concept to formally define e as either of the following two limits:

The Number e

$$e = \lim_{n\to\infty}\left(1 + \frac{1}{n}\right)^n \quad \text{or, alternatively,} \quad e = \lim_{s\to 0}(1 + s)^{1/s}$$

$$e = 2.718\ 281\ 828\ 459\ldots$$

We will use both these limit forms. [*Note:* If $s = 1/n$, then as $n \to \infty, s \to 0$.]

The proof that the indicated limits exist and represent an irrational number between 2 and 3 is not easy and is omitted here. Many people reason (incorrectly) that the limits are 1, since "$1 + s$ approaches 1 as $s \to 0$, and 1 to any power is 1." A little experimentation with a calculator can convince you otherwise. Consider the following table of values for s and $f(s) = (1 + s)^{1/s}$ and the graph shown in Figure 1 for s close to 0.

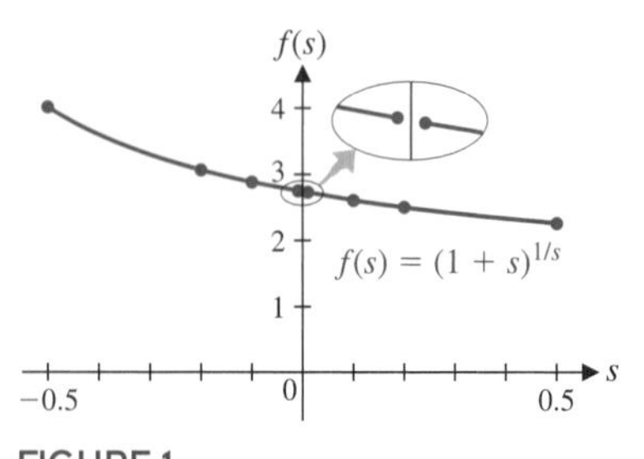

FIGURE 1

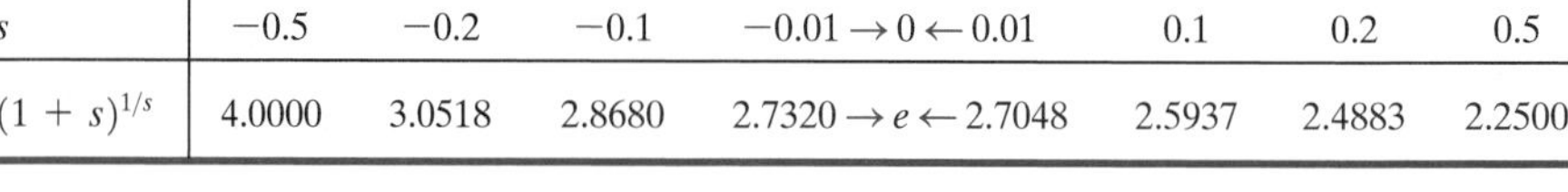

s approaches 0 from the left $\rightarrow$ 0 $\leftarrow$ s approaches 0 from the right

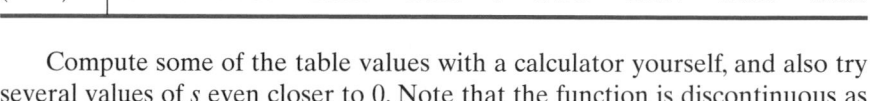

s	-0.5	-0.2	-0.1	$-0.01 \to 0 \leftarrow 0.01$	0.1	0.2	0.5
$(1 + s)^{1/s}$	4.0000	3.0518	2.8680	$2.7320 \to e \leftarrow 2.7048$	2.5937	2.4883	2.2500

Compute some of the table values with a calculator yourself, and also try several values of s even closer to 0. Note that the function is discontinuous as $s = 0$.

Exactly who discovered e is still being debated. It is named after the great mathematician Leonhard Euler (1707–1783), who computed e to twenty-three decimal places using $[1 + (1/n)]^n$.

Continuous Compound Interest

Now we will see how e appears quite naturally in the important application of compound interest. Let us start with simple interest, move on to compound interest, and then on to continuous compound interest.

If a principal P is borrowed at an annual rate of r,* then after t years at simple interest the borrower will owe the lender an amount A given by

$$A = P + Prt = P(1 + rt) \quad \text{Simple interest} \tag{1}$$

On the other hand, if interest is compounded n times a year, then the borrower will owe the lender an amount A given by

$$A = P\left(1 + \frac{r}{n}\right)^{nt} \quad \text{Compound interest} \tag{2}$$

*If r is the interest rate written as a decimal, then $100r\%$ is the rate using %. For example, if $r = 0.12$, then we have $100r\% = 100(0.12)\% = 12\%$. The expressions 0.12 and 12% are therefore equivalent. Unless stated otherwise, all formulas in this book use r in decimal form.

where r/n is the interest rate per compounding period and nt is the number of compounding periods. Suppose P, r, and t in (2) are held fixed and n is increased. Will the amount A increase without bound, or will it tend to approach some limiting value?

Let us perform a calculator experiment before we attack the general limit problem. If $P = \$100$, $r = 0.06$, and $t = 2$ years, then

$$A = 100\left(1 + \frac{0.06}{n}\right)^{2n}$$

We compute A for several values of n in Table 1. The biggest gain appears in the first step; then the gains slow down as n increases. In fact, it appears that A might be tending to approach \$112.75 as n gets larger and larger.

Table 1

COMPOUNDING FREQUENCY	n	$A = 100\left(1 + \frac{0.06}{n}\right)^{2n}$
Annually	1	\$112.3600
Semiannually	2	112.5509
Quarterly	4	112.6493
Weekly	52	112.7419
Daily	365	112.7486
Hourly	8,760	112.7496

Explore–Discuss 1

(A) Suppose \$1,000 is deposited in a savings account that earns 6% simple interest. How much will be in the account after 2 years?

(B) Suppose \$1,000 is deposited in a savings account that earns compound interest at a rate of 6% per year. How much will be in the account after 2 years if interest is compounded annually? Semiannually? Quarterly? Weekly?

(C) How frequently must interest be compounded at the 6% rate in order to have \$1,150 in the account after 2 years?

Now we turn back to the general problem for a moment. Keeping P, r, and t fixed in equation (2), we compute the following limit and observe an interesting and useful result:

$$\lim_{n\to\infty} P\left(1 + \frac{r}{n}\right)^{nt} = P \lim_{n\to\infty}\left(1 + \frac{r}{n}\right)^{(n/r)rt}$$

Insert r/r in the exponent and let $s = r/n$ Note that $n \to \infty$ implies $s \to 0$.

$$= P \lim_{s\to 0}[(1 + s)^{1/s}]^{rt}$$

Use the limit property given in the footnote below.*

$$= P[\lim_{s\to 0}(1 + s)^{1/s}]^{rt}$$

$\lim_{s\to 0}(1 + s)^{1/s} = e$

$$= Pe^{rt}$$

The resulting formula is called the **continuous compound interest formula,** a very important and widely used formula in business and economics.

*The following new limit property is used: If $\lim_{x\to c} f(x)$ exists, then $\lim_{x\to c}[f(x)]^p = [\lim_{x\to c} f(x)]^p$, provided the last expression names a real number.

Continuous Compound Interest

$$A = Pe^{rt}$$

where

P = Principal
r = Annual nominal interest rate compounded continuously
t = Time in years
A = Amount at time t

Example 1 **Computing Continuously Compounded Interest** If \$100 is invested at an annual nominal rate of 6% compounded continuously, what amount will be in the account after 2 years? How much interest will be earned?

SOLUTION

$$\begin{aligned} A &= Pe^{rt} \\ &= 100e^{(0.06)(2)} \quad \text{6\% is equivalent to } r = 0.06. \\ &\approx \$112.7497 \end{aligned}$$

(Compare this result with the values calculated in Table 1.) The interest earned is \$112.7497 − \$100 = \$12.7497.

Matched Problem 1 What amount (to the nearest cent) will an account have after 5 years if \$100 is invested at an annual nominal rate of 8% compounded annually? Semiannually? Continuously?

Example 2 **Graphing the Growth of an Investment** If \$100 is invested at 12% compounded continuously,* graph the amount in the account relative to time for a period of 10 years.

SOLUTION We want to graph

$$A = 100e^{0.12t} \qquad 0 \leq t \leq 10$$

We construct a table of values using a calculator, graph the points from the table, and join the points with a smooth curve.

t	A (\$)
0	100
1	113
2	127
3	143
4	162
5	182
6	205
7	232
8	261
9	294
10	332

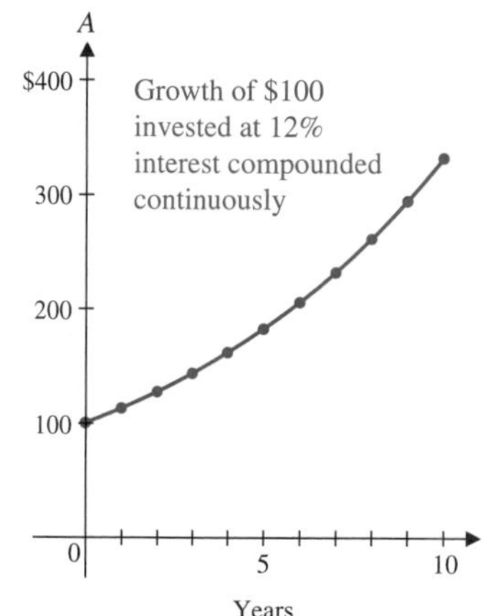

*Following common usage, we will often write "at 12% compounded continuously," understanding that this means "at an annual nominal rate of 12% compounded continuously."

Figure 2 shows the graph from Example 2 on a graphing utility.

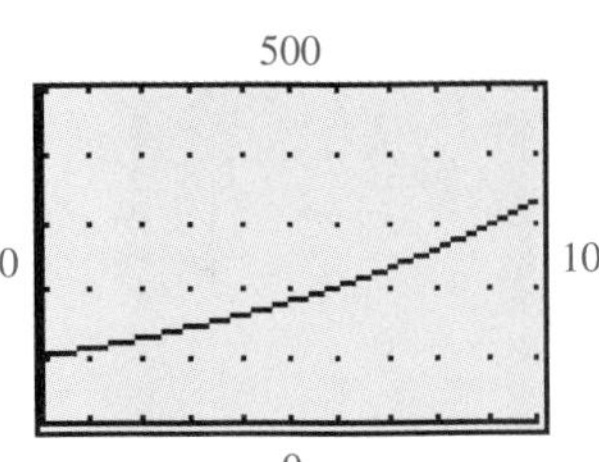

FIGURE 2

Matched Problem 2 If \$5,000 is invested at 20% compounded continuously, graph the amount in the account relative to time for a period of 10 years.

Example 3 **Computing Growth Time** How long will it take an investment of \$5,000 to grow to \$8,000 if it is invested at 12% compounded continuously?

SOLUTION Starting with the continuous compound interest formula $A = Pe^{rt}$, we must solve for t:

$$
\begin{aligned}
A &= Pe^{rt} \\
8{,}000 &= 5{,}000e^{0.12t} && \text{Divide both sides by 5,000 and reverse the equation.} \\
e^{0.12t} &= 1.6 && \text{Take the natural logarithm of both sides—recall that} \\
\ln e^{0.12t} &= \ln 1.6 && \log_b b^x = x. \\
0.12t &= \ln 1.6 \\
t &= \frac{\ln 1.6}{0.12} \\
t &\approx 3.92 \text{ years}
\end{aligned}
$$

Figure 3 shows an alternative method for solving Example 3 on a graphing utility.

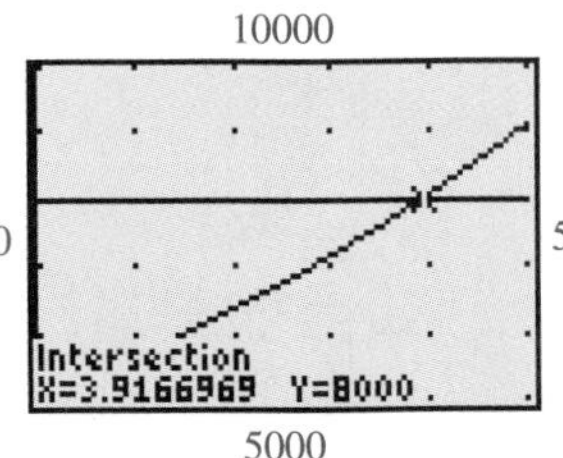

FIGURE 3

$y_1 = 5{,}000e^{0.12x}$

$y_2 = 8{,}000$

Matched Problem 3 How long will it take an investment of \$10,000 to grow to \$15,000 if it is invested at 9% compounded continuously?

Example 4 **Computing Doubling Time** How long will it take money to double if it is invested at 18% compounded continuously?

SOLUTION Starting with the continuous compound interest formula $A = Pe^{rt}$, we must solve for t given $A = 2P$ and $r = 0.18$:

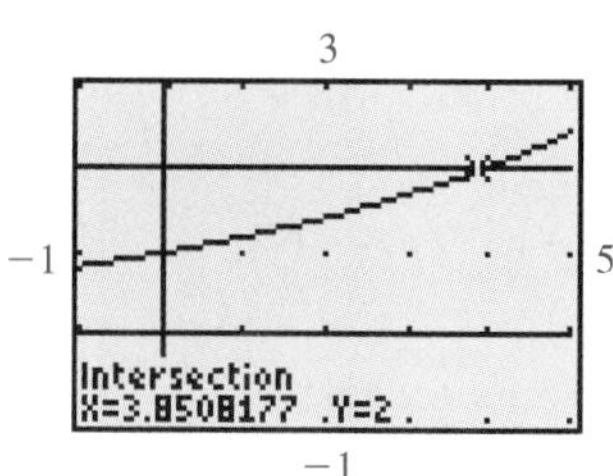

FIGURE 4
$y_1 = e^{0.18x}$
$y_2 = 2$

$$
\begin{aligned}
2P &= Pe^{0.18t} && \text{Divide both sides by } P \text{ and reverse the equation.} \\
e^{0.18t} &= 2 && \text{Take the natural logarithm of both sides.} \\
\ln e^{0.18t} &= \ln 2 \\
0.18t &= \ln 2 \\
t &= \frac{\ln 2}{0.18} \\
t &\approx 3.85 \text{ years}
\end{aligned}
$$

Figure 4 illustrates the solution of $e^{0.18t} = 2$ from Example 4 on a graphing utility.

Matched Problem 4 How long will it take money to triple if it is invested at 12% compounded continuously?

Explore–Discuss 2

You are considering three options for investing $10,000: at 7% compounded annually, at 6% compounded monthly, and at 5% compounded continuously.

(A) Which option would be the best for investing $10,000 for 8 years?
(B) How long would you need to invest your money for the third option to be the best?

Answers to Matched Problems

1. $146.93; $148.02; $149.18

2. $A = 5{,}000e^{0.2t}$

t	A ($)
0	5,000
1	6,107
2	7,459
3	9,111
4	11,128
5	13,591
6	16,601
7	20,276
8	24,765
9	30,248
10	36,945

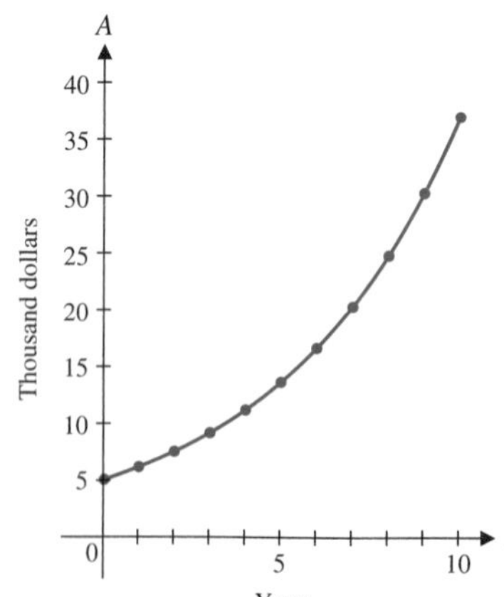

3. 4.51 yr

4. 9.16 yr

EXERCISE 10-1

A *Use a calculator to evaluate A to the nearest cent in Problems 1 and 2.*

1. $A = \$1{,}000e^{0.1t}$ for $t = 2, 5$, and 8

2. $A = \$5{,}000e^{0.08t}$ for $t = 1, 4$, and 10

3. If \$6,000 is invested at 10% compounded continuously, graph the amount in the account relative to time for a period of 8 years.

4. If \$4,000 is invested at 8% compounded continuously, graph the amount in the account relative to time for a period of 6 years.

B *In Problems 5–10, solve for t or r to two decimal places.*

5. $2 = e^{0.06t}$

6. $2 = e^{0.03t}$

7. $3 = e^{0.1t}$

8. $3 = e^{0.25t}$

9. $2 = e^{5r}$

10. $3 = e^{10r}$

C *In Problems 11 and 12, complete each table to five decimal places using a calculator.*

11.

n	$[1 + (1/n)]^n$
10	2.593 74
100	
1,000	
10,000	
100,000	
1,000,000	
10,000,000	
↓	↓
∞	$e = 2.718\ 281\ 828\ 459\ldots$

12.

s	$(1 + s)^{1/s}$
0.01	2.704 81
−0.01	
0.001	
−0.001	
0.000 1	
−0.000 1	
0.000 01	
−0.000 01	
↓	↓
0	$e = 2.718\ 281\ 828\ 459\ldots$

13. Use a calculator and a table of values to investigate

$$\lim_{n\to\infty}(1 + n)^{1/n}$$

Do you think this limit exists? If so, what do you think it is?

14. Use a calculator and a table of values to investigate

$$\lim_{s\to 0^+}\left(1 + \frac{1}{s}\right)^s$$

Do you think this limit exists? If so, what do you think it is?

15. **C** It can be shown that the number e satisfies the inequality

$$\left(1 + \frac{1}{n}\right)^n < e < \left(1 + \frac{1}{n}\right)^{n+1} \qquad n \geq 1$$

Illustrate this graphically by graphing

$$y_1 = (1 + 1/n)^n$$
$$y_2 = 2.718\ 281\ 828 \approx e$$
$$y_3 = (1 + 1/n)^{n+1}$$

in the same viewing window for $1 \leq n \leq 20$.

16. **C** It can be shown that

$$e^s = \lim_{n\to\infty}\left(1 + \frac{s}{n}\right)^n$$

for any real number s. Illustrate this graphically for $s = 2$ by graphing

$$y_1 = (1 + 2/n)^n$$
$$y_2 = 7.389\ 056\ 099 \approx e^2$$

in the same viewing window for $1 \leq n \leq 50$.

APPLICATIONS

Business & Economics

17. *Continuous compound interest.* If \$20,000 is invested at an annual nominal rate of 12% compounded continuously, how much will it be worth in 8.5 years?

18. *Continuous compound interest.* Assume \$1 had been invested at an annual nominal rate of 4% compounded continuously at the time of the birth of Christ. What would be the value of the account in solid gold Earths in the year 2000? Assume that the Earth weighs approximately 2.11×10^{26} ounces and that gold will be worth \$1,000 an ounce in the year 2000. What would be the value of the account in dollars at simple interest?

19. *Present value.* A note will pay \$20,000 at maturity 10 years from now. How much should you be willing to pay for the note now if money is worth 7% compounded continuously?

20. *Present value.* A note will pay \$50,000 at maturity 5 years from now. How much should you be willing to pay for the note now if money is worth 8% compounded continuously?

21. *Continuous compound interest.* An investor bought stock for \$20,000. Four years later, the stock was sold for \$30,000. If interest is compounded continuously, what annual nominal rate of interest did the original \$20,000 investment earn?

22. *Continuous compound interest.* A family paid \$40,000 cash for a house. Ten years later, they sold the house for \$100,000. If interest is compounded continuously, what annual nominal rate of interest did the original \$40,000 investment earn?

23. *Present value.* Solving $A = Pe^{rt}$ for P, we obtain

$$P = Ae^{-rt}$$

which is the present value of the amount A due in t years if money earns interest at an annual nominal rate r compounded continuously.
(A) Graph $P = 10{,}000e^{-0.08t}$, $0 \le t \le 50$.
(B) $\lim_{t\to\infty} 10{,}000e^{-0.08t} = ?$ [Guess, using part (A).]

[*Conclusion:* The longer the duration of time until the amount A is due, the smaller its present value, as we would expect.]

24. *Present value.* Referring to Problem 23, in how many years will the \$10,000 have to be due in order for its present value to be \$5,000?

25. *Doubling time.* How long will it take money to double if it is invested at 25% compounded continuously?

26. *Doubling time.* How long will it take money to double if it is invested at 5% compounded continuously?

27. *Doubling rate.* At what nominal rate compounded continuously must money be invested to double in 5 years?

28. *Doubling rate.* At what nominal rate compounded continuously must money be invested to double in 3 years?

29. C *Growth time.* A man with \$20,000 to invest decides to diversify his investments by placing \$10,000 in an account that earns 7.2% compounded continuously and \$10,000 in an account that earns 8.4% compounded annually. Use graphical approximation methods to determine how long it will take for his total investment in the two accounts to grow to \$35,000.

30. C *Growth time.* A woman invests \$5,000 in an account that earns 8.8% compounded continuously and \$7,000 in an account that earns 9.6% compounded annually. Use graphical approximation methods to determine how long it will take for her total investment in the two accounts to grow to \$20,000.

31. *Doubling times.*
(A) Show that the doubling time t (in years) at an annual rate r compounded continuously is given by

$$t = \frac{\ln 2}{r}$$

(B) Graph the doubling-time equation from part (A) for $0.02 \le r \le 0.30$. Are these restrictions on r reasonable? Explain.
(C) Determine the doubling times (in years, to two decimal places) for $r = 5\%, 10\%, 15\%, 20\%, 25\%$, and 30%.

32. *Doubling rates.*

(A) Show that the rate r that doubles an investment at continuously compounded interest in t years is given by

$$r = \frac{\ln 2}{t}$$

(B) Graph the doubling-rate equation from part (A) for $1 \leq t \leq 20$. Are these restrictions on t reasonable? Explain.

(C) Determine the doubling rates for $t = 2, 4, 6, 8, 10,$ and 12 years.

Life Sciences

33. *Radioactive decay.* A mathematical model for the decay of radioactive substances is given by

$$Q = Q_0e^{rt}$$

where

Q_0 = Amount of the substance at time $t = 0$
r = Continuous compound rate of decay
t = Time in years
Q = Amount of the substance at time t

If the continuous compound rate of decay of radium per year is $r = -0.000\ 433\ 2$, how long will it take a certain amount of radium to decay to half the original amount? (This period of time is the *half-life* of the substance.)

34. *Radioactive decay.* The continuous compound rate of decay of carbon-14 per year is $r = -0.000\ 123\ 8$. How long will it take a certain amount of carbon-14 to decay to half the original amount? (Use the radioactive decay model in Problem 33.)

35. *Radioactive decay.* A cesium isotope has a half-life of 30 years. What is the continuous compound rate of decay? (Use the radioactive decay model in Problem 33.)

36. *Radioactive decay.* A strontium isotope has a half-life of 90 years. What is the continuous compound rate of decay? (Use the radioactive decay model in Problem 33.)

Social Sciences

37. *World population.* A mathematical model for world population growth over short periods of time is given by

$$P = P_0e^{rt}$$

where

P_0 = Population at time $t = 0$
r = Continuous compound rate of growth
t = Time in years
P = Population at time t

How long will it take the world population to double if it continues to grow at its current continuous compound rate of 2% per year?

38. *World population.* Repeat Problem 37 under the assumption that the world population is growing at a continuous compound rate of 1% per year.

39. *Population growth.* Some underdeveloped nations have population doubling times of 20 years. At what continuous compound rate is the population growing? (Use the population growth model in Problem 37.)

40. *Population growth.* Some developed nations have population doubling times of 120 years. At what continuous compound rate is the population growing? (Use the population growth model in Problem 37.)

41. *World population.* If the world population is now 5 billion (5×10^9) people and if it continues to grow at a continuous compound rate of 2% per year, how long will it be before there is only 1 square yard of land per person? (The Earth has approximately 1.68×10^{14} square yards of land.)

SECTION 10-2 Derivatives of Logarithmic and Exponential Functions

- DERIVATIVE FORMULAS FOR $\ln x$ AND e^x
- GRAPHING TECHNIQUES
- APPLICATION

In this section, we discuss derivative formulas for $\ln x$ and e^x. Out of all the possible choices for bases for the logarithmic and exponential functions, $\log_b x$ and b^x, it turns out (as we will see in this and the next section) that the simplest derivative formulas occur when the base b is chosen to be e.

Explore–Discuss 1

(A) Using the graph of $f(x) = 2^x$ in Figure 1, sketch tangent lines for several values of x, estimate their slopes, and sketch a graph of $f'(x)$. Is the graph of the derivative above or below the graph of $f(x) = 2^x$?

(B) Using the graph of $g(x) = 3^x$ in Figure 2, sketch tangent lines for several values of x, estimate their slopes, and sketch a graph of $g'(x)$. Is the graph of the derivative above or below the graph of $g(x) = 3^x$?

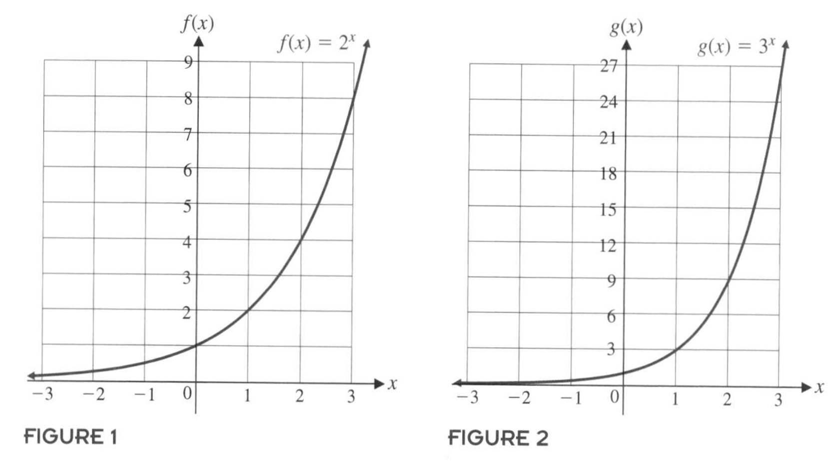

FIGURE 1

FIGURE 2

Explore–Discuss 1 led to the discovery that for $f(x) = 2^x$, the graphs of $f(x)$ and $f'(x)$ look much the same, with $f'(x)$ always below $f(x)$. Similarly, for $g(x) = 3^x$, the graphs of $g(x)$ and $g'(x)$ look much the same, with $g'(x)$ always above $g(x)$. These observations suggest that there may be an exponential function with base between 2 and 3 that is its own derivative. In this section we will show that the exponential function with base $e = 2.718\ldots$ has exactly this property—that is,

$$\frac{d}{dx} e^x = e^x$$

We will also obtain the derivative formula

$$\frac{d}{dx} \ln x = \frac{1}{x}$$

for the natural logarithm function.

■ Derivative Formulas for $\ln x$ and e^x

We are now ready to derive a formula for the derivative of

$$f(x) = \ln x = \log_e x \qquad x > 0$$

using the definition of the derivative

$$f'(x) = \lim_{h\to 0} \frac{f(x+h) - f(x)}{h}$$

and the two-step process discussed in Section 8-3.

Step 1. Simplify the difference quotient first:

$$\begin{aligned}
\frac{f(x+h) - f(x)}{h} &= \frac{\ln(x+h) - \ln x}{h} \\
&= \frac{1}{h}[\ln(x+h) - \ln x] && \text{Use } \ln A - \ln B = \ln \frac{A}{B}. \\
&= \frac{1}{h} \ln \frac{x+h}{x} && \text{Multiply by } 1 = x/x \text{ to change form.} \\
&= \frac{x}{x} \cdot \frac{1}{h} \ln \frac{x+h}{x} \\
&= \frac{1}{x}\left[\frac{x}{h} \ln\left(1 + \frac{h}{x}\right)\right] && \text{Use } p \ln A = \ln A^p. \\
&= \frac{1}{x} \ln\left(1 + \frac{h}{x}\right)^{x/h}
\end{aligned}$$

Step 2. Find the limit: Let $s = h/x$. For x fixed, if $h \to 0$, then $s \to 0$. Thus,

$$\begin{aligned}
\frac{d}{dx} \ln x &= \lim_{h\to 0} \frac{f(x+h) - f(x)}{h} \\
&= \lim_{h\to 0}\left[\frac{1}{x} \ln\left(1 + \frac{h}{x}\right)^{x/h}\right] && \text{Let } s = h/x. \text{ Note that } h \to 0 \text{ implies } s \to 0. \\
&= \frac{1}{x} \lim_{s\to 0}[\ln(1+s)^{1/s}] && \text{Use the new limit property given in the footnote below.*} \\
&= \frac{1}{x} \ln\left[\lim_{s\to 0}(1+s)^{1/s}\right] && \text{Use the definition of } e. \\
&= \frac{1}{x} \ln e && \ln e = \log_e e = 1 \\
&= \frac{1}{x}
\end{aligned}$$

Thus,

$$\boldsymbol{\frac{d}{dx} \ln x = \frac{1}{x}}$$

In the next section, we will show that, in general,

$$\frac{d}{dx} \log_b x = \frac{1}{\ln b}\left(\frac{1}{x}\right)$$

which is a somewhat more complicated result than the above—unless $b = e$.

---

*The following new limit property is used: If $\lim_{x\to c} f(x)$ exists and is positive, then $\lim_{x\to c}[\ln f(x)] = \ln[\lim_{x\to c} f(x)]$.

In the process of finding the derivative of e^x, we will use (without proof) the fact that

$$\lim_{h \to 0}\left(\frac{e^h - 1}{h}\right) = 1$$

Explore–Discuss 2

Compute

$$\frac{e^h - 1}{h}$$

for the following values of h: $-0.1, -0.01, -0.001, -0.0001, 0.0001, 0.001, 0.01, 0.1$. Do your calculations make it reasonable to conclude that

$$\lim_{h \to 0} \frac{e^h - 1}{h} = 1?$$

Discuss.

We now apply the two-step process to the exponential function $f(x) = e^x$.

Step 1. Simplify the difference quotient first:

$$\frac{f(x + h) - f(x)}{h} = \frac{e^{x+h} - e^x}{h} \quad \text{Use } e^{a+b} = e^a e^b.$$

$$= \frac{e^x e^h - e^x}{h} \quad \text{Factor out } e^x.$$

$$= e^x\left(\frac{e^h - 1}{h}\right)$$

Step 2. Compute the limit of the result in step 1:

$$\frac{d}{dx} e^x = \lim_{h \to 0} \frac{f(x + h) - f(x)}{h}$$

$$= \lim_{h \to 0} e^x\left(\frac{e^h - 1}{h}\right)$$

$$= e^x \lim_{h \to 0}\left(\frac{e^h - 1}{h}\right) \quad \text{Use the assumed limit given above.}$$

$$= e^x \cdot 1 = e^x$$

Thus,

$$\boldsymbol{\frac{d}{dx} e^x = e^x}$$

In the next section, we will show that

$$\frac{d}{dx} b^x = b^x \ln b$$

which is, again, a somewhat more complicated result than the above—unless $b = e$.

The two results just obtained explain why e^x is so widely used that it is sometimes referred to as *the* exponential function. These two new and important derivative formulas are restated in the box for reference.

Derivatives of the Natural Logarithmic and Exponential Functions

$$\frac{d}{dx}\ln x = \frac{1}{x} \qquad \frac{d}{dx}e^x = e^x$$

These new derivatives formulas can be combined with the rules of differentiation discussed in Chapter 8 to differentiate a wide variety of functions.

Example 1 **Finding Derivatives** Find $f'(x)$ for:

(A) $f(x) = 2e^x + 3\ln x$ (B) $f(x) = \dfrac{e^x}{x^3}$

(C) $f(x) = (\ln x)^4$ (D) $f(x) = \ln x^4$

SOLUTION (A) $f'(x) = 2\dfrac{d}{dx}e^x + 3\dfrac{d}{dx}\ln x$

$$= 2e^x + 3\left(\frac{1}{x}\right) = 2e^x + \frac{3}{x}$$

(B) $$f'(x) = \frac{x^3\dfrac{d}{dx}e^x - e^x\dfrac{d}{dx}x^3}{(x^3)^2} \quad \text{Quotient rule}$$

$$= \frac{x^3e^x - e^x3x^2}{x^6} = \frac{x^2e^x(x-3)}{x^6} = \frac{e^x(x-3)}{x^4}$$

(C) $$\frac{d}{dx}(\ln x)^4 = 4(\ln x)^3\frac{d}{dx}\ln x \quad \text{Power rule for functions}$$

$$= 4(\ln x)^3\left(\frac{1}{x}\right) = \frac{4(\ln x)^3}{x}$$

(D) $$\frac{d}{dx}\ln x^4 = \frac{d}{dx}(4\ln x) \quad \text{Property of logarithms}$$

$$= 4\left(\frac{1}{x}\right) = \frac{4}{x}$$

Matched Problem 1 Find $f'(x)$ for:

(A) $f(x) = 4\ln x - 5e^x$ (B) $f(x) = x^2e^x$

(C) $f(x) = \ln x^3$ (D) $f(x) = (\ln x)^3$

CAUTION

$$\frac{d}{dx}e^x \neq xe^{x-1} \qquad \frac{d}{dx}e^x = e^x$$

The power rule cannot be used to differentiate the exponential function. The power rule applies to exponential forms x^n where the exponent is a constant and the base is a variable. In the exponential form e^x, the base is a constant and the exponent is a variable.

Graphing Techniques

Using the techniques discussed in Chapter 9, we can use first and second derivatives to gain useful information about the graphs of $y = \ln x$ and $y = e^x$. With the derivative formulas given above, we can construct Table 1.

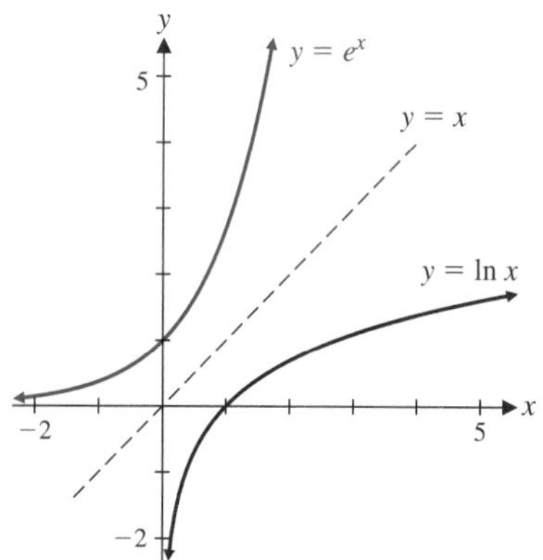

FIGURE 3
e^x is continuous on $(-\infty, \infty)$
$\ln x$ is continuous on $(0, \infty)$

Table 1

$\ln x$		e^x	
$y = \ln x$	$x > 0$	$y = e^x$	$-\infty < x < \infty$
$y' = 1/x > 0$	$x > 0$	$y' = e^x > 0$	$-\infty < x < \infty$
$y'' = -1/x^2 < 0$	$x > 0$	$y'' = e^x > 0$	$-\infty < x < \infty$

From the table, we can see that both functions are increasing throughout their respective domains, the graph of $y = \ln x$ is always concave downward, and the graph of $y = e^x$ is always concave upward. It can be shown that the y axis is a vertical asymptote for the graph of $y = \ln x$ ($\lim_{x\to 0^+} \ln x = -\infty$), and the x axis is a horizontal asymptote for the graph of $y = e^x$ ($\lim_{x\to -\infty} e^x = 0$). Both equations are graphed in Figure 3.

Notice that if we fold the page along the dashed line $y = x$, the two graphs match exactly (see Section 2-3). Also notice that both graphs are unbounded as $x \to \infty$. Comparing each graph with the graph of $y = x$ (the dashed line), we conclude that e^x grows more rapidly than x and $\ln x$ grows more slowly than x. In fact, the following limits can be established:

Growth of Exponential and Logarithmic Functions

$$\lim_{x\to\infty} \frac{x^p}{e^x} = 0, \quad p > 0 \qquad \text{and} \qquad \lim_{x\to\infty} \frac{\ln x}{x^p} = 0, \quad p > 0$$

The limits in the box indicate that e^x grows more rapidly than any positive power of x, and $\ln x$ grows more slowly than any positive power of x.

Now we will apply graphing techniques to a slightly more complicated function.

Example 2 **Graphing Strategy** Analyze the function $f(x) = xe^x$ following the steps in the graphing strategy discussed in Section 9-4. State all the pertinent information and sketch the graph of *f*.

SOLUTION **Step 1.** *Analyze* $f(x)$: $f(x) = xe^x$

(A) Domain: All real numbers
(B) *y* intercept: $f(0) = 0$

x intercept: $xe^x = 0$ for $x = 0$ only, since $e^x > 0$ for all *x* (see Fig. 3).
(C) Vertical asymptotes: None

Horizontal asymptotes: We have not developed limit techniques for functions of this type to determine the behavior of $f(x)$ as $x \to -\infty$ and $x \to \infty$. However, the following tables of values suggest the nature of the graph of *f* as $x \to -\infty$ and $x \to \infty$:

x	1	5	10	$\to\infty$
$f(x)$	2.72	742.07	220,264.66	$\to\infty$

x	-1	-5	-10	$\to -\infty$
$f(x)$	-0.37	-0.03	$-0.000\ 45$	$\to 0$

Step 2. *Analyze* $f'(x)$:

$$\begin{aligned} f'(x) &= x\frac{d}{dx}e^x + e^x\frac{d}{dx}x \\ &= xe^x + e^x = e^x(x + 1) \end{aligned}$$

Critical value for $f(x)$: -1

Partition number for $f'(x)$: -1

Sign chart for $f'(x)$:

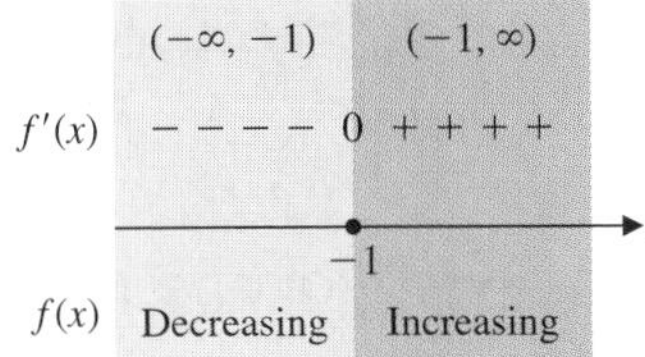

TEST NUMBERS

x	$f'(x)$	
-2	$-e^{-2}$	(−)
0	1	(+)

Thus, $f(x)$ decreases on $(-\infty, -1)$, has a local minimum at $x = -1$, and increases on $(-1, \infty)$. (Since $e^x > 0$ for all *x*, we do not have to evaluate e^{-2} to conclude that $-e^{-2} < 0$ when using the test number -2.)

Step 3. *Analyze* $f''(x)$:

$$\begin{aligned} f''(x) &= e^x\frac{d}{dx}(x + 1) + (x + 1)\frac{d}{dx}e^x \\ &= e^x + (x + 1)e^x = e^x(x + 2) \end{aligned}$$

Sign chart for $f''(x)$ (partition number is -2):

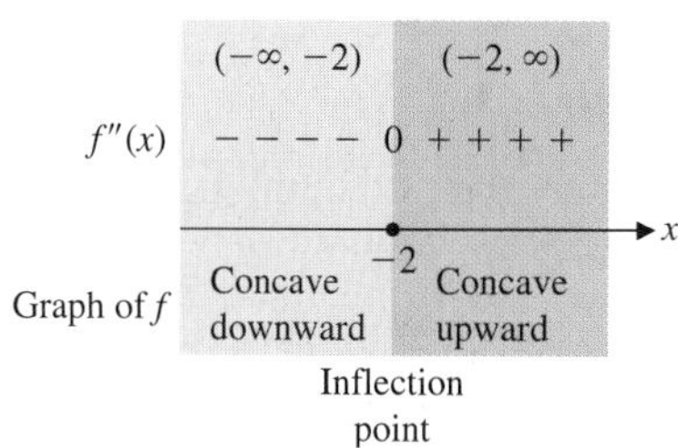

Test Numbers

x	$f''(x)$	
-3	$-e^{-3}$	$(-)$
-1	e^{-1}	$(+)$

Thus, the graph of f is concave downward on $(-\infty, -2)$, has an inflection point at $x = -2$, and is concave upward on $(-2, \infty)$.

Step 4. *Sketch the graph of f using the information from steps 1–3:*

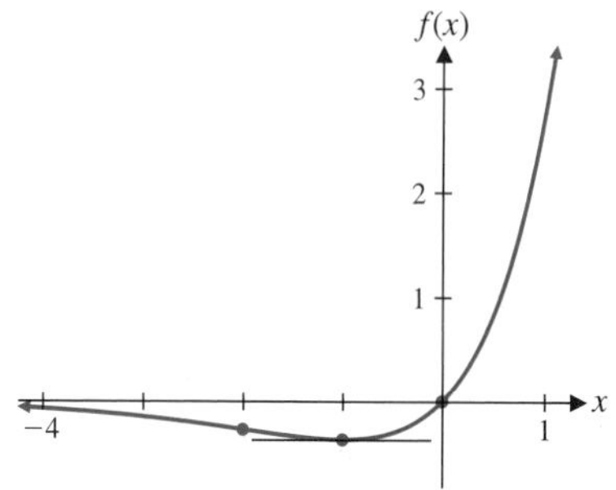

x	$f(x)$
-2	-0.27
-1	-0.37
0	0

Matched Problem 2 Analyze the function $f(x) = x \ln x$. State all the pertinent information and sketch the graph of f.

Application

Example 3

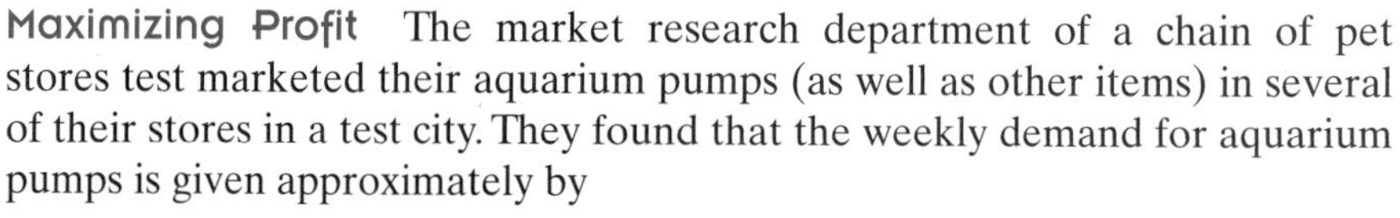

Maximizing Profit The market research department of a chain of pet stores test marketed their aquarium pumps (as well as other items) in several of their stores in a test city. They found that the weekly demand for aquarium pumps is given approximately by

$$p = 12 - 2 \ln x \qquad 0 < x < 90$$

where x is the number of pumps sold each week and \$$p$ is the price of one pump. If each pump costs the chain \$3, how should it be priced in order to maximize the weekly profit?

Solution Although we want to find the price that maximizes the weekly profit, it will be easier to first find the number of pumps that will maximize the weekly profit. The revenue equation is

$$R(x) = xp = 12x - 2x \ln x$$

The cost equation is

$$C(x) = 3x$$

and the profit equation is

$$\begin{aligned} P(x) &= R(x) - C(x) \\ &= 12x - 2x \ln x - 3x \\ &= 9x - 2x \ln x \end{aligned}$$

Thus, we must solve the following:

$$\text{Maximize} \quad P(x) = 9x - 2x \ln x \qquad 0 < x < 90$$

$$\begin{aligned} P'(x) &= 9 - 2x\left(\frac{1}{x}\right) - 2 \ln x \\ &= 7 - 2 \ln x = 0 \\ 2 \ln x &= 7 \\ \ln x &= 3.5 \\ x &= e^{3.5} \\ P''(x) &= -2\left(\frac{1}{x}\right) = -\frac{2}{x} \end{aligned}$$

Since $x = e^{3.5}$ is the only critical value and $P''(e^{3.5}) < 0$, the maximum weekly profit occurs when $x = e^{3.5} \approx 33$ and $p = 12 - 2 \ln e^{3.5} = \$5$.

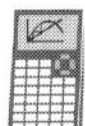

Figure 4 shows an alternative method for finding the maximum value of $P(x)$ in Example 3 on a graphing utility.

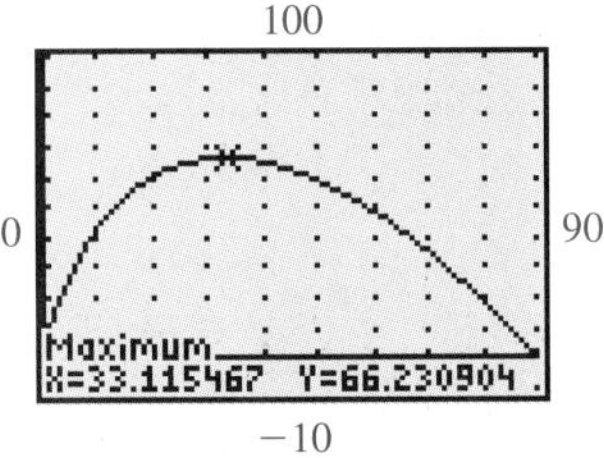

FIGURE 4
$y_1 = 9x - 2x \ln x$

Matched Problem 3 Repeat Example 3 if each pump costs the chain \$3.50.

Answers to Matched Problems

1. (A) $(4/x) - 5e^x$ (B) $xe^x(x + 2)$ (C) $3/x$ (D) $3(\ln x)^2/x$

2. Domain: $(0, \infty)$
y intercept: None [$f(0)$ is not defined]
x intercept: $x = 1$
Increasing on (e^{-1}, ∞)
Decreasing on $(0, e^{-1})$
Local minimum at $x = e^{-1} \approx 0.368$
Concave upward on $(0, \infty)$

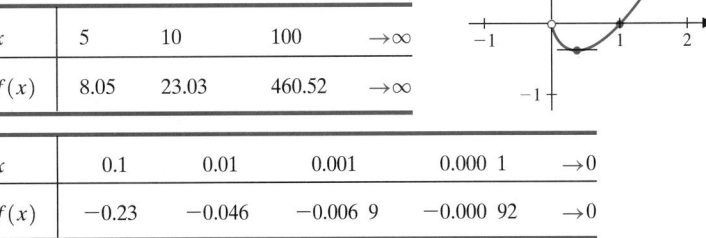

x	5	10	100	$\to\infty$
$f(x)$	8.05	23.03	460.52	$\to\infty$

x	0.1	0.01	0.001	0.000 1	$\to 0$
$f(x)$	−0.23	−0.046	−0.006 9	−0.000 92	$\to 0$

3. Maximum profit occurs for $x = e^{3.25} \approx 26$ and $p = \$5.50$.

Exercise 10-2

For many of the problems in this exercise set, the answers in the back of the book include both an unsimplified form and a simplified form. When checking your work, first check that you applied the rules correctly, and then check that you performed the algebraic simplification correctly.

A *In Problems 1–32, find $f'(x)$ and simplify.*

1. $f(x) = 6e^x - 7\ln x$
2. $f(x) = 4e^x + 5\ln x$
3. $f(x) = 2x^e + 3e^x$
4. $f(x) = 4e^x - ex^e$
5. $f(x) = \ln x^5$
6. $f(x) = (\ln x)^5$
7. $f(x) = (\ln x)^2$
8. $f(x) = \ln x^2$

B

9. $f(x) = x^4 \ln x$
10. $f(x) = x^3 \ln x$
11. $f(x) = x^3 e^x$
12. $f(x) = x^4 e^x$
13. $f(x) = \dfrac{e^x}{x^2 + 9}$
14. $f(x) = \dfrac{e^x}{x^2 + 4}$
15. $f(x) = \dfrac{\ln x}{x^4}$
16. $f(x) = \dfrac{\ln x}{x^3}$
17. $f(x) = (x + 2)^3 \ln x$
18. $f(x) = (x - 1)^2 \ln x$
19. $f(x) = (x + 1)^3 e^x$
20. $f(x) = (x - 2)^3 e^x$
21. $f(x) = \dfrac{x^2 + 1}{e^x}$
22. $f(x) = \dfrac{x + 1}{e^x}$
23. $f(x) = x(\ln x)^3$
24. $f(x) = x(\ln x)^2$
25. $f(x) = (4 - 5e^x)^3$
26. $f(x) = (5 - \ln x)^4$
27. $f(x) = \sqrt{1 + \ln x}$
28. $f(x) = \sqrt{1 + e^x}$
29. $f(x) = xe^x - e^x$
30. $f(x) = x\ln x - x$
31. $f(x) = 2x^2 \ln x - x^2$
32. $f(x) = x^2 e^x - 2xe^x + 2e^x$

In Problems 33–36, find the equation of the line tangent to the graph of $y = f(x)$ at the indicated value of x.

33. $f(x) = e^x;\quad x = 1$
34. $f(x) = e^x;\quad x = 2$
35. $f(x) = \ln x;\quad x = e$
36. $f(x) = \ln x;\quad x = 1$

37. A student claims that the tangent line to the graph of $f(x) = e^x$ at $x = 3$ passes through the point $(2, 0)$ (see the figure). Is she correct? Will the tangent line at $x = 4$ pass through $(3, 0)$? Explain.

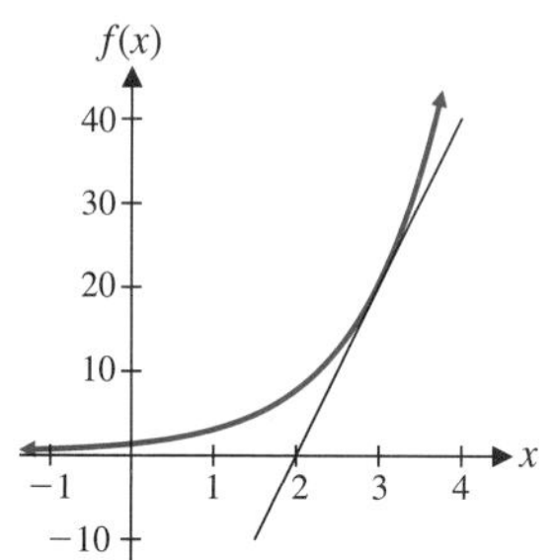

Figure for 37

38. A student claims that the tangent line to the graph of $g(x) = \ln x$ at $x = 3$ passes through the origin (see the figure). Is he correct? Will the tangent line at $x = 4$ pass through the origin? Explain.

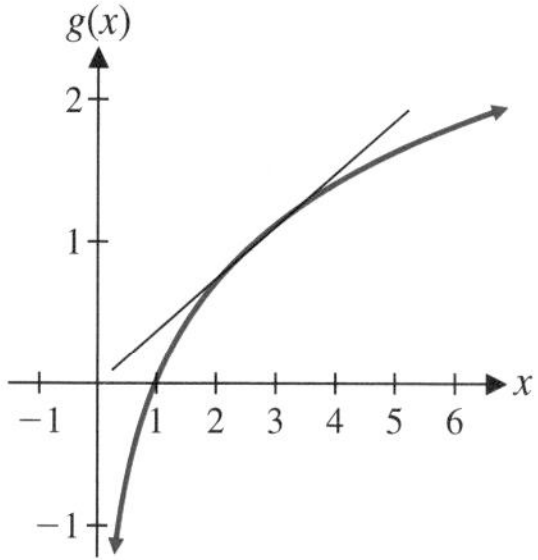

Figure for 38

Problems 39 and 40 make use of the following familiar properties of exponential and logarithmic functions (see Sections 2-2 and 2-3):

$$(a^b)^c = a^{bc} \qquad \ln ab = \ln a + \ln b$$

39. Find the derivatives of $y = e^{2x}$, $y = e^{3x}$, and $y = e^{4x}$. What do you think would be the derivative of $y = e^{kx}$ for any constant k? Justify your answer.
40. Find the derivatives of $y = \ln(2x)$, $y = \ln(3x)$, and $y = \ln(4x)$. What do you think would be the derivative of $y = \ln(kx)$ for any positive constant k? Justify your answer.

C *In Problems 41–46, find the indicated extremum of each function for $x > 0$.*

41. Absolute maximum value of
$$f(x) = 4x - x\ln x$$
42. Absolute minimum value of
$$f(x) = x\ln x - 3x$$

43. Absolute minimum value of
$$f(x) = \frac{e^x}{x}$$

44. Absolute maximum value of
$$f(x) = \frac{x^2}{e^x}$$

45. Absolute maximum value of
$$f(x) = \frac{1 + 2\ln x}{x}$$

46. Absolute minimum value of
$$f(x) = \frac{1 - 5\ln x}{x}$$

In Problems 47–54, apply the graphing strategy discussed in Section 9-4 to f, summarize the pertinent information, and sketch the graph.

47. $f(x) = 1 - e^x$
48. $f(x) = 1 - \ln x$
49. $f(x) = x - \ln x$
50. $f(x) = e^x - x$
51. $f(x) = (3 - x)e^x$
52. $f(x) = (x - 2)e^x$
53. $f(x) = x^2 \ln x$
54. $f(x) = \dfrac{\ln x}{x}$

Problems 55–58 require the use of a graphing utility. Approximate the critical values of $f(x)$ to two decimal places and find the intervals where $f(x)$ is increasing, the intervals where $f(x)$ is decreasing, and the local extrema.

55. $f(x) = e^x - 2x^2$
56. $f(x) = e^x + x^2$
57. $f(x) = 20\ln x - e^x$
58. $f(x) = x^2 - 3x\ln x$

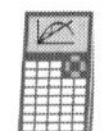

In Problems 59–62, use graphical approximation methods to find the points of intersection of $f(x)$ and $g(x)$ (to two decimal places).

59. $f(x) = e^x$; $g(x) = x^4$
[Note that there are three points of intersection and that e^x is greater than x^4 for large values of x.]

60. $f(x) = e^x$; $g(x) = x^5$
[Note that there are two points of intersection and that e^x is greater than x^5 for large values of x.]

61. $f(x) = \ln x$; $g(x) = x^{1/5}$
[Note that $\ln x$ is less than $x^{1/5}$ for large values of x.]

62. $f(x) = \ln x$; $g(x) = x^{1/4}$
[Note that $\ln x$ is less than $x^{1/4}$ for large values of x.]

APPLICATIONS

Business & Economics

63. *Maximum profit.* A national food service runs food concessions for sporting events throughout the country. Their marketing research department chose a particular football stadium to test market a new jumbo hot dog. It was found that the demand for the new hot dog is given approximately by
$$p = 5 - \ln x \qquad 5 \le x \le 50$$
where x is the number of hot dogs (in thousands) that can be sold during one game at a price of \$$p$. If the concessionaire pays \$1 for each hot dog, how should the hot dogs be priced to maximize the profit per game?

64. *Maximum profit.* On a national tour of a rock band, the demand for T-shirts is given by
$$p = 15 - 4\ln x \qquad 1 \le x \le 40$$
where x is the number of T-shirts (in thousands) that can be sold during a single concert at a price of \$$p$. If the shirts cost the band \$5 each, how should they be priced in order to maximize the profit per concert?

65. C *Maximum profit.* A regional chain of department stores has collected the data in the table, showing weekly sales of a certain brand of jeans. The same jeans have been offered at various prices ranging from the regular price of \$35.99 to the lowest sale price of \$23.99. Each pair of this brand costs the chain \$20.00. Use logarithmic regression ($p = a + b\ln x$) to find the price (to the nearest cent) that will maximize profit.

PAIRS OF JEANS x	PRICE PER PAIR (\$) p
21,543	23.99
14,029	25.99
12,130	27.99
9,169	29.99
6,964	31.99
5,506	33.99
4,187	35.99

66. *Maximum profit.* A mail order company specializes in the sale of athletic shoes. Its market research team has determined the weekly demand shown in the table for a particular model of basketball shoe when priced at various levels. The company is able to purchase this model from the manufacturer for \$62 per pair. Use logarithmic regression ($p = a + b \ln x$) to find the price (to the nearest cent) that will maximize profit.

DEMAND x	PRICE PER PAIR (\$) p
4,312	74
3,064	79
2,499	84
2,047	89
1,823	94
1,781	99

67. *Minimum average cost.* The cost of producing x units of a product is given by

$$C(x) = 600 + 100x - 100 \ln x \qquad x \geq 1$$

Find the minimum average cost.

68. *Minimum average cost.* The cost of producing x units of a product is given by

$$C(x) = 1{,}000 + 200x - 200 \ln x \qquad x \geq 1$$

Find the minimum average cost.

69. *Maximizing revenue.* A cosmetic company is planning the introduction and promotion of a new lipstick line. The marketing research department, after test marketing the new line in a carefully selected large city, found that the demand in that city is given approximately by

$$p = 10e^{-x} \qquad 0 \leq x \leq 2$$

where x thousand lipsticks were sold per week at a price of \$$p$ each.

(A) At what price will the weekly revenue $R(x) = xp$ be maximum? What is the maximum weekly revenue in the test city?

(B) Graph R for $0 \leq x \leq 2$.

70. *Maximizing revenue.* Repeat Problem 69 using the demand equation $p = 12e^{-x}, 0 \leq x \leq 2$.

Life Sciences

71. *Blood pressure.* An experiment was set up to find a relationship between weight and systolic blood pressure in normal children. Using hospital records for 5,000 normal children, it was found that the systolic blood pressure was given approximately by

$$P(x) = 17.5(1 + \ln x) \qquad 10 \leq x \leq 100$$

where $P(x)$ is measured in millimeters of mercury and x is measured in pounds. What is the rate of change of blood pressure with respect to weight at the 40 pound weight level? At the 90 pound weight level?

72. *Blood pressure.* Graph the systolic blood pressure equation in Problem 71.

73. *Drug concentration.* The concentration of a drug in the bloodstream t hours after injection is given approximately by

$$C(t) = 4.35e^{-t} \qquad 0 \leq t \leq 5$$

where $C(t)$ is concentration in milligrams per milliliter.

(A) What is the rate of change of concentration after 1 hour? After 4 hours?

(B) Graph C.

74. *Water pollution.* The use of iodine crystals is a popular way of making small quantities of nonpotable water safe to drink. Crystals placed in a 1 ounce bottle of water will dissolve until the solution is saturated. After saturation, half of this solution is poured into a quart container of nonpotable water, and after about an hour, the water is usually safe to drink. The half-empty 1 ounce bottle is then refilled to be used again in the same way. Suppose the concentration of iodine in the 1 ounce bottle t minutes after the crystals are introduced can be approximated by

$$C(t) = 250(1 - e^{-t}) \qquad t \geq 0$$

where $C(t)$ is the concentration of iodine in micrograms per milliliter.

(A) What is the rate of change of the concentration after 1 minute? After 4 minutes?

(B) Graph C for $0 \leq t \leq 5$.

Social Sciences

75. *Psychology—stimulus/response.* In psychology, the Weber–Fechner law for stimulus response is

$$R = k \ln\left(\frac{S}{S_0}\right)$$

where R is the response, S is the stimulus, and S_0 is the lowest level of stimulus that can be detected. Find dR/dS.

76. *Psychology—learning.* A mathematical model for the average of a group of people learning to type is given by

$$N(t) = 10 + 6 \ln t \qquad t \geq 1$$

where $N(t)$ is the number of words per minute typed after t hours of instruction and practice (2 hours per day, 5 days per week). What is the rate of learning after 10 hours of instruction and practice? After 100 hours?

SECTION 10-3 Chain Rule: General Form

- Composite Functions
- Chain Rule
- Generalized Derivative Rules
- Other Logarithmic and Exponential Functions

Explore–Discuss 1

(A) We have shown that the function $f(x) = e^x$ is its own derivative. Use the graph of $g(x) = e^{x^2}$ in Figure 1 to explain why $g(x)$ is not its own derivative.

(B) Use the graph of $h(x) = 2xe^{x^2}$ in Figure 2 to explain why $h(x)$ could be the derivative of $g(x)$. [We will show in this section that $h(x)$ is indeed the derivative of $g(x)$.]

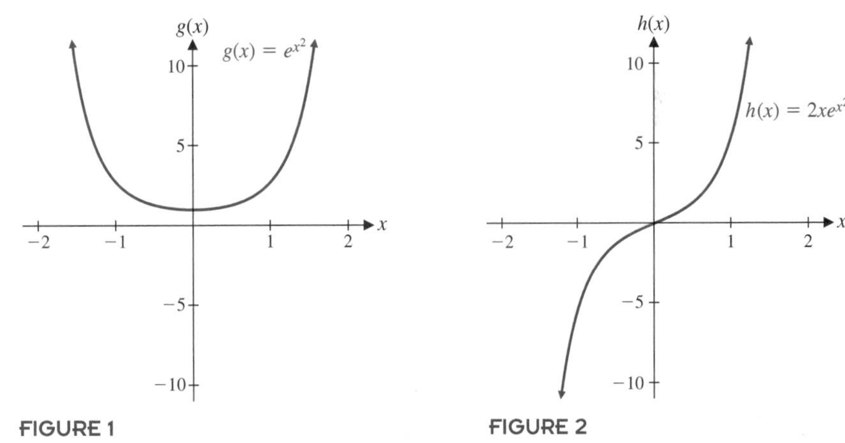

FIGURE 1

FIGURE 2

In Section 8-6, we introduced the power form of the chain rule:

$$\frac{d}{dx}[u(x)]^n = n[u(x)]^{n-1}u'(x)$$

For example,

$$\frac{d}{dx}(x^2 - 3)^5 = 5(x^2 - 3)^4 \frac{d}{dx}(x^2 - 3)$$
$$= 10x(x^2 - 3)^4$$

This general power rule is a special case of one of the most important derivative rules of all—the *chain rule*—which will enable us to determine the derivatives of some fairly complicated functions in terms of derivatives of more elementary functions.

Suppose you were asked to find the derivative of

$$m(x) = \ln(2x + 1) \qquad \text{or} \qquad n(x) = e^{3x^2-1}$$

We have formulas for computing derivatives of $\ln x$ and e^x, and polynomial functions in general, but not in the indicated combinations. The chain rule is used to compute derivatives of functions that are *compositions* of more elementary functions whose derivatives are known. We therefore start this section with a brief review of *composite functions.*

■ Composite Functions

Let us look at function m more closely:

$$m(x) = \ln(2x + 1)$$

The function m is a combination of the natural logarithm function and a linear function. To see this more clearly, let

$$y = f(u) = \ln u \qquad \text{and} \qquad u = g(x) = 2x + 1$$

Then we can express y as a function of x as follows:

$$y = f(u) = f[g(x)] = \ln(2x + 1) = m(x)$$

The function m is said to be the *composite* of the two simpler functions f and g. (Loosely speaking, we can think of m as a function of a function.) In general, we have the following:

Composite Functions

A function m is a **composite** of functions f and g if

$$m(x) = f[g(x)]$$

The domain of m is the set of all numbers x such that x is in the domain of g and $g(x)$ is in the domain of f.

Example 1 ➡ **Composite Functions** Let $f(u) = e^u$, $g(x) = 3x^2 + 1$, and $m(v) = v^{3/2}$. Find:

(A) $f[g(x)]$ (B) $g[f(u)]$ (C) $m[g(x)]$

SOLUTION
(A) $f[g(x)] = e^{g(x)} = e^{3x^2+1}$
(B) $g[f(u)] = 3[f(u)]^2 + 1 = 3(e^u)^2 + 1 = 3e^{2u} + 1$
(C) $m[g(x)] = [g(x)]^{3/2} = (3x^2 + 1)^{3/2}$

Matched Problem 1 ➡ Let $f(u) = \ln u$, $g(x) = 2x^3 + 4$, and $m(v) = v^{-5}$. Find:

(A) $f[g(x)]$ (B) $g[f(u)]$ (C) $m[g(x)]$

Example 2 **Composite Functions** Write each function as a composition of the natural logarithm or exponential function and a polynomial.

(A) $y = \ln(x^3 - 2x^2 + 1)$ (B) $y = e^{x^2+4}$

SOLUTION (A) Let

$$y = f(u) = \ln u$$
$$u = g(x) = x^3 - 2x^2 + 1$$

Check: $y = f[g(x)] = \ln[g(x)] = \ln(x^3 - 2x^2 + 1)$

(B) Let

$$y = f(u) = e^u$$
$$u = g(x) = x^2 + 4$$

Check: $y = f[g(x)] = e^{g(x)} = e^{x^2+4}$

Matched Problem 2 Repeat Example 2 for:

(A) $y = e^{2x^3+7}$ (B) $y = \ln(x^4 + 10)$

CHAIN RULE

The word *chain* in the name *chain rule* comes from the fact that a function formed by composition (such as those in Example 1) involves a chain of functions—that is, a function of a function. The *chain rule* will enable us to compute the derivative of a composite function in terms of the derivatives of the functions making up the composition.

Suppose

$$y = m(x) = f[g(x)]$$

is a composite of f and g, where

$$y = f(u) \qquad \text{and} \qquad u = g(x)$$

We would like to express the derivative dy/dx in terms of the derivatives of f and g. From the definition of a derivative (see Section 8-3), we have

$$\frac{dy}{dx} = \lim_{h\to 0} \frac{m(x+h) - m(x)}{h}$$

Substitute $m(x + h) = f[g(x + h)]$ and $m(x) = f[g(x)]$.

$$= \lim_{h\to 0} \frac{f[g(x+h)] - f[g(x)]}{h}$$

Multiply by $1 = \frac{g(x+h) - g(x)}{g(x+h) - g(x)}$.

$$= \lim_{h\to 0}\left[\frac{f[g(x+h)] - f[g(x)]}{h} \cdot \frac{g(x+h) - g(x)}{g(x+h) - g(x)}\right]$$

$$= \lim_{h\to 0}\left[\frac{f[g(x+h)] - f[g(x)]}{g(x+h) - g(x)} \cdot \frac{g(x+h) - g(x)}{h}\right] \qquad (1)$$

We recognize the second factor in (1) as the difference quotient for $g(x)$. To interpret the first factor as the difference quotient for $f(u)$, we let $k = g(x + h) - g(x)$. Since $u = g(x)$, we can write

$$u + k = g(x) + g(x + h) - g(x) = g(x + h)$$

Substituting in (1), we now have

$$\frac{dy}{dx} = \lim_{h\to 0}\left[\frac{f(u+k)-f(u)}{k} \cdot \frac{g(x+h)-g(x)}{h}\right] \tag{2}$$

If we assume that $k = [g(x+h) - g(x)] \to 0$ as $h \to 0$, then we can find the limit of each difference quotient in (2):

$$\begin{aligned}\frac{dy}{dx} &= \left[\lim_{k\to 0}\frac{f(u+k)-f(u)}{k}\right]\left[\lim_{h\to 0}\frac{g(x+h)-g(x)}{h}\right]\\ &= f'(u)g'(x)\\ &= \frac{dy}{du}\frac{du}{dx}\end{aligned}$$

This result is correct under rather general conditions, and is called the *chain rule,* but our "derivation" is superficial, because it ignores a number of hidden problems. Since a formal proof of the chain rule is beyond the scope of this book, we simply state it as follows:

Chain Rule

If $y = f(u)$ and $u = g(x)$, define the composite function

$$y = m(x) = f[g(x)]$$

Then

$$\frac{dy}{dx} = \frac{dy}{du}\frac{du}{dx} \qquad \text{provided } \frac{dy}{du} \text{ and } \frac{du}{dx} \text{ exist}$$

Or, equivalently,

$$m'(x) = f'[g(x)]g'(x) \qquad \text{provided } f'[g(x)] \text{ and } g'(x) \text{ exist}$$

Example 3 **Using the Chain Rule** Find dy/dx, given:

(A) $y = \ln(x^2 - 4x + 2)$ (B) $y = e^{2x^3+5}$ (C) $y = (3x^2 + 1)^{3/2}$

SOLUTION (A) Let $y = \ln u$ and $u = x^2 - 4x + 2$. Then

$$\begin{aligned}\frac{dy}{dx} &= \frac{dy}{du}\frac{du}{dx} \quad {}^{*}\\ &= \frac{1}{u}(2x-4)\\ &= \frac{1}{x^2-4x+2}(2x-4) \qquad \text{Since } u = x^2 - 4x + 2\\ &= \frac{2x-4}{x^2-4x+2}\end{aligned}$$

*After some experience with the chain rule, the steps in the dashed boxes are usually done mentally.

(B) Let $y = e^u$ and $u = 2x^3 + 5$. Then

$$\frac{dy}{dx} = \frac{dy}{du}\frac{du}{dx}$$

$$= e^u(6x^2)$$

$$= 6x^2e^{2x^3+5} \quad \text{Since } u = 2x^3 + 5$$

(C) We have two methods:

Method 1. *Chain Rule—General Form.* Let $y = u^{3/2}$ and $u = 3x^2 + 1$. Then

$$\frac{dy}{dx} = \frac{dy}{du}\frac{du}{dx}$$

$$= \tfrac{3}{2}u^{1/2}(6x)$$

$$= \tfrac{3}{2}(3x^2 + 1)^{1/2}(6x) \quad \text{Since } u = 3x^2 + 1$$

$$= 9x(3x^2 + 1)^{1/2} \quad \text{or} \quad 9x\sqrt{3x^2 + 1}$$

Method 2. *Chain Rule—Power Form (General Power Rule).*

$$\frac{d}{dx}(3x^2 + 1)^{3/2} = \tfrac{3}{2}(3x^2 + 1)^{1/2}\frac{d}{dx}(3x^2 + 1) \quad \frac{d}{dx}[u(x)]^n = n[u(x)]^{n-1}\frac{d}{dx}u(x)$$

$$= \tfrac{3}{2}(3x^2 + 1)^{1/2}(6x)$$

$$= 9x(3x^2 + 1)^{1/2} \quad \text{or} \quad 9x\sqrt{3x^2 + 1}$$

The general power rule stated in Section 8-6 can be derived using the chain rule as follows: Given $y = [u(x)]^n$, let $y = v^n$ and $v = u(x)$. Then

$$\frac{dy}{dx} = \frac{dy}{dv}\frac{dv}{dx}$$

$$= nv^{n-1}\frac{d}{dx}u(x)$$

$$= n[u(x)]^{n-1}\frac{d}{dx}u(x) \quad \text{Since } v = u(x)$$

Matched Problem 3 Find dy/dx, given:

(A) $y = e^{3x^4+6}$ (B) $y = \ln(x^2 + 9x + 4)$ (C) $y = (2x^3 + 4)^{-5}$ (Use two methods.)

The chain rule can be extended to compositions of three or more functions. For example, if $y = f(w)$, $w = g(u)$, and $u = h(x)$, then

$$\frac{dy}{dx} = \frac{dy}{dw}\frac{dw}{du}\frac{du}{dx}$$

Example 4 **Using the Chain Rule** For $y = h(x) = e^{1+(\ln x)^2}$, find dy/dx.

SOLUTION Note that h is of the form $y = e^w$, where $w = 1 + u^2$ and $u = \ln x$. Thus,

$$\begin{aligned}
\frac{dy}{dx} &= \frac{dy}{dw}\frac{dw}{du}\frac{du}{dx} \\
&= e^w(2u)\left(\frac{1}{x}\right) \\
&= e^{1+u^2}(2u)\left(\frac{1}{x}\right) && \text{Since } w = 1 + u^2 \\
&= e^{1+(\ln x)^2}(2\ln x)\left(\frac{1}{x}\right) && \text{Since } u = \ln x \\
&= \frac{2}{x}(\ln x)e^{1+(\ln x)^2}
\end{aligned}$$

Matched Problem 4 For $y = h(x) = [\ln(1 + e^x)]^3$, find dy/dx.

GENERALIZED DERIVATIVE RULES

In practice, it is not necessary to introduce additional variables when using the chain rule, as we did in Examples 3 and 4. Instead, the chain rule can be used to extend the derivative rules for specific functions to general derivative rules for compositions. This is what we did above when we showed that the general power rule is a consequence of the chain rule. The same technique can be applied to functions of the form $y = e^{f(x)}$ and $y = \ln[f(x)]$ (see Problems 71 and 72 in Exercise 10-3). The results are summarized in the following box:

General Derivative Rules

$$\frac{d}{dx}[f(x)]^n = n[f(x)]^{n-1}f'(x) \tag{3}$$

$$\frac{d}{dx}\ln[f(x)] = \frac{1}{f(x)}f'(x) \tag{4}$$

$$\frac{d}{dx}e^{f(x)} = e^{f(x)}f'(x) \tag{5}$$

For power, natural logarithm, or exponential forms, we can either use the chain rule discussed earlier or these special differentiation formulas, which are based on the chain rule. Use whichever is easier for you. In Example 5, we will use the general derivative rules.

Example 5 **Using General Derivative Rules**

(A) $\frac{d}{dx} e^{2x} = e^{2x} \frac{d}{dx} 2x$ Using (5)

$= e^{2x}(2) = 2e^{2x}$

(B) $\frac{d}{dx} \ln(x^2 + 9) = \frac{1}{x^2 + 9} \frac{d}{dx}(x^2 + 9)$ Using (4)

$= \frac{1}{x^2 + 9} 2x = \frac{2x}{x^2 + 9}$

(C) $\frac{d}{dx}(1 + e^{x^2})^3 = 3(1 + e^{x^2})^2 \frac{d}{dx}(1 + e^{x^2})$ Using (3)

$= 3(1 + e^{x^2})^2 e^{x^2} \frac{d}{dx} x^2$ Using (5)

$= 3(1 + e^{x^2})^2 e^{x^2}(2x)$

$= 6xe^{x^2}(1 + e^{x^2})^2$

Matched Problem 5 Find:

(A) $\frac{d}{dx} \ln(x^3 + 2x)$ (B) $\frac{d}{dx} e^{3x^2+2}$ (C) $\frac{d}{dx}(2 + e^{-x^2})^4$

Other Logarithmic and Exponential Functions

In most applications involving logarithmic or exponential functions, the number e is the preferred base. However, there are situations where it is convenient to use a base other than e. Derivatives of $y = \log_b x$ and $y = b^x$ can be obtained by expressing these functions in terms of the natural logarithmic and exponential functions.

We begin by finding a relationship between $\log_b x$ and $\ln x$ for any base b, $b > 0$ and $b \neq 1$. Some of you may prefer to remember the process, and others the formula.

$y = \log_b x$ Change to exponential form.

$b^y = x$ Take the natural logarithm of both sides.

$\ln b^y = \ln x$ Recall that $\ln b^y = y \ln b$.

$y \ln b = \ln x$ Solve for y.

$y = \frac{1}{\ln b} \ln x$

Thus,

$$\log_b x = \frac{1}{\ln b} \ln x \quad \text{Change-of-base formula*} \tag{6}$$

Differentiating both sides of (6), we have

$$\frac{d}{dx} \log_b x = \frac{1}{\ln b} \frac{d}{dx} \ln x = \frac{1}{\ln b}\left(\frac{1}{x}\right)$$

*Equation (6) is a special case of the **general change-of-base formula** for logarithms (which can be derived in the same way): $\log_b x = (\log_a x)/(\log_a b)$.

Explore–Discuss 2

(A) The graphs of $f(x) = \log_2 x$ and $g(x) = \log_4 x$ are shown in Figure 3. Which graph belongs to which function?

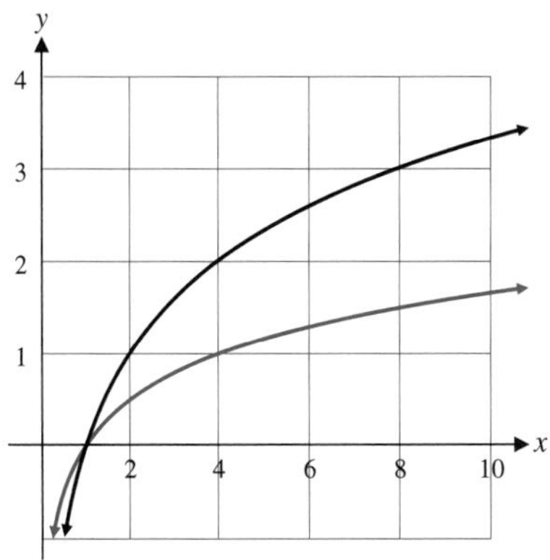

FIGURE 3

(B) Sketch graphs of $f'(x)$ and $g'(x)$.

(C) The function $f(x)$ is related to $g(x)$ in the same way that $f'(x)$ is related to $g'(x)$. What is that relationship?

Example 6 **Differentiating Logarithmic Functions** Find $f'(x)$ for:

(A) $f(x) = \log_2 x$ (B) $f(x) = \log(1 + x^3)$

SOLUTION (A) $f(x) = \log_2 x = \dfrac{1}{\ln 2}\ln x$ *Using (6)*

$$f'(x) = \frac{1}{\ln 2}\left(\frac{1}{x}\right)$$

(B) $f(x) = \log(1 + x^3)$ *Recall that $\log r = \log_{10} r$.*

$$= \frac{1}{\ln 10}\ln(1 + x^3) \quad \text{Using (6)}$$

$$f'(x) = \frac{1}{\ln 10}\left(\frac{1}{1 + x^3}3x^2\right) = \frac{1}{\ln 10}\left(\frac{3x^2}{1 + x^3}\right)$$

Matched Problem 6 Find $f'(x)$ for:

(A) $f(x) = \log x$ (B) $f(x) = \log_3(x + x^2)$

Now we want to find a relationship between b^x and e^x for any base b, $b > 0$ and $b \neq 1$.

$$\begin{aligned} y &= b^x && \text{Take the natural logarithm of both sides.} \\ \ln y &= \ln b^x \\ &= x \ln b && \text{If } \ln A = B, \text{ then } A = e^B. \\ y &= e^{x \ln b} \end{aligned}$$

Thus,

$$b^x = e^{x \ln b} \tag{7}$$

Differentiating both sides of (7), we have

$$\frac{d}{dx} b^x = e^{x \ln b} \ln b = b^x \ln b$$

Example 7 **Differentiating Exponential Functions** Find $f'(x)$ for:

(A) $f(x) = 2^x$ (B) $f(x) = 10^{x^5+x}$

SOLUTION (A) $f(x) = 2^x = e^{x \ln 2}$ Using (7)
$f'(x) = e^{x \ln 2} \ln 2 = 2^x \ln 2$

(B) $f(x) = 10^{x^5+x} = e^{(x^5+x)\ln 10}$ Using (7)
$f'(x) = e^{(x^5+x)\ln 10}(5x^4 + 1)\ln 10$
$= 10^{x^5+x}(5x^4 + 1)\ln 10$

Matched Problem 7 Find $f'(x)$ for:

(A) $f(x) = 5^x$ (B) $f(x) = 4^{x^2+3x}$

Answers to Matched Problems

1. (A) $\ln(2x^3 + 4)$ (B) $2(\ln u)^3 + 4$ (C) $(2x^3 + 4)^{-5}$
2. (A) $y = f(u) = e^u; u = g(x) = 2x^3 + 7$
   (B) $y = f(u) = \ln u; u = g(x) = x^4 + 10$
3. (A) $12x^3e^{3x^4+6}$ (B) $\frac{2x + 9}{x^2 + 9x + 4}$ (C) $-30x^2(2x^3 + 4)^{-6}$
4. $\frac{3e^x[\ln(1 + e^x)]^2}{1 + e^x}$
5. (A) $\frac{3x^2 + 2}{x^3 + 2x}$ (B) $6xe^{3x^2+2}$ (C) $-8xe^{-x^2}(2 + e^{-x^2})^3$
6. (A) $\frac{1}{\ln 10}\left(\frac{1}{x}\right)$ (B) $\frac{1}{\ln 3}\left(\frac{1 + 2x}{x + x^2}\right)$
7. (A) $5^x \ln 5$ (B) $4^{x^2+3x}(2x + 3)\ln 4$

EXERCISE 10-3

A *In Problems 1–6, find $f[g(x)]$.*

1. $f(u) = u^4; g(x) = 7x - 5$
2. $f(u) = u^6; g(x) = 4 - 2x^2$
3. $f(u) = \ln u; g(x) = 2x + 5$
4. $f(u) = \ln u; g(x) = x^2 + 1$
5. $f(u) = e^u; g(x) = x^3$
6. $f(u) = e^u; g(x) = x^2$

Write each composite function in Problems 7–12 in the form $y = f(u)$ and $u = g(x)$.

7. $y = (2x + 5)^3$
8. $y = (3x - 7)^5$
9. $y = \ln(2x^2 + 7)$
10. $y = \ln(x^2 - 2x + 5)$
11. $y = e^{x^2-2}$
12. $y = e^{3x^3+5x}$

In Problems 13–18, express y in terms of x. Use the chain rule to find dy/dx, and then express dy/dx in terms of x.

13. $y = u^2; u = 2 + e^x$
14. $y = u^3; u = 3 - \ln x$
15. $y = e^u; u = 2 - x^4$
16. $y = e^u; u = x^6 + 5x^2$
17. $y = \ln u; u = 4x^5 - 7$
18. $y = \ln u; u = 2 + 3x^4$

In Problems 19–52, find the derivative and simplify. (For some of these problems, the answers in the back of the book include both an unsimplified form and a simplified form. When checking your work, first check that you applied the rules correctly, and then check that you performed the algebraic simplification correctly.)

19. $\frac{d}{dx}\ln(x-3)$ **20.** $\frac{d}{dw}\ln(w+100)$

21. $\frac{d}{dt}\ln(3-2t)$ **22.** $\frac{d}{dy}\ln(4-5y)$

23. $\frac{d}{dx}3e^{2x}$ **24.** $\frac{d}{dy}2e^{3y}$

25. $\frac{d}{dt}2e^{-4t}$ **26.** $\frac{d}{dr}6e^{-3r}$

B

27. $\frac{d}{dx}100e^{-0.03x}$ **28.** $\frac{d}{dt}1{,}000e^{0.06t}$

29. $\frac{d}{dx}\ln(x+1)^4$ **30.** $\frac{d}{dx}\ln(x+1)^{-3}$

31. $\frac{d}{dx}(2e^{2x}-3e^x+5)$ **32.** $\frac{d}{dt}(1+e^{-t}-e^{-2t})$

33. $\frac{d}{dx}e^{3x^2-2x}$ **34.** $\frac{d}{dx}e^{x^3-3x^2+1}$

35. $\frac{d}{dt}\ln(t^2+3t)$ **36.** $\frac{d}{dx}\ln(x^3-3x^2)$

37. $\frac{d}{dx}\ln(x^2+1)^{1/2}$ **38.** $\frac{d}{dx}\ln(x^4+5)^{3/2}$

39. $\frac{d}{dt}[\ln(t^2+1)]^4$ **40.** $\frac{d}{dw}[\ln(w^3-1)]^2$

41. $\frac{d}{dx}(e^{2x}-1)^4$ **42.** $\frac{d}{dx}(e^{x^2}+3)^5$

43. $\frac{d}{dx}\frac{e^{2x}}{x^2+1}$ **44.** $\frac{d}{dx}\frac{e^{x+1}}{x+1}$

45. $\frac{d}{dx}(x^2+1)e^{-x}$ **46.** $\frac{d}{dx}(1-x)e^{2x}$

47. $\frac{d}{dx}(e^{-x}\ln x)$ **48.** $\frac{d}{dx}\frac{\ln x}{e^x+1}$

49. $\frac{d}{dx}\frac{1}{\ln(1+x^2)}$ **50.** $\frac{d}{dx}\frac{1}{\ln(1-x^3)}$

51. $\frac{d}{dx}\sqrt[3]{\ln(1-x^2)}$ **52.** $\frac{d}{dt}\sqrt[5]{\ln(1-t^5)}$

53. Describe the relationship between the graphs of $y=\ln x$ and $y=\log_b x$ for $b>1$.

54. Describe the relationship between the graphs of $y=\ln x$ and $y=\log_b x$ for $0<b<1$.

C

In Problems 55–60, apply the graphing strategy discussed in Section 9-4 to f, state the pertinent information, and sketch the graph.

55. $f(x)=1-e^{-x}$ **56.** $f(x)=2-3e^{-2x}$

57. $f(x)=\ln(1-x)$ **58.** $f(x)=\ln(2x+4)$

59. $f(x)=e^{-0.5x^2}$ **60.** $f(x)=\ln(x^2+4)$

In Problems 61 and 62, express y in terms of x. Use the chain rule to find dy/dx, and express dy/dx in terms of x.

61. $y=1+w^2;\ w=\ln u;\ u=2+e^x$

62. $y=\ln w;\ w=1+e^u;\ u=x^2$

In Problems 63–70, find each derivative.

63. $\frac{d}{dx}\log_2(3x^2-1)$ **64.** $\frac{d}{dx}\log(x^3-1)$

65. $\frac{d}{dx}10^{x^2+x}$ **66.** $\frac{d}{dx}8^{1-2x^2}$

67. $\frac{d}{dx}\log_3(4x^3+5x+7)$ **68.** $\frac{d}{dx}\log_5(5^{x^2-1})$

69. $\frac{d}{dx}2^{x^3-x^2+4x+1}$ **70.** $\frac{d}{dx}10^{\ln x}$

71. Use the chain rule to derive the formula:

$$\frac{d}{dx}\ln[f(x)]=\frac{1}{f(x)}f'(x)$$

72. Use the chain rule to derive the formula:

$$\frac{d}{dx}e^{f(x)}=e^{f(x)}f'(x)$$

73. (A) Use the graphs of $f(x)=\ln(x^2+1)$ and $g(x)=1/(x^2+1)$ (see the figure) to explain why $g(x)$ is not the derivative of $f(x)$.
(B) Compute $f'(x)$ and sketch its graph.

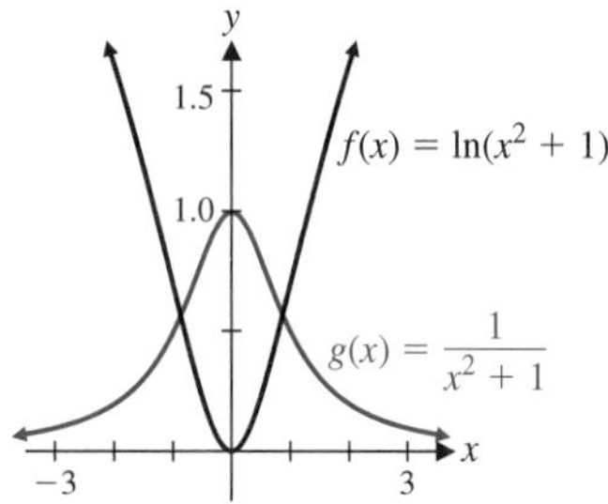

Figure for 73

74. (A) How many times is the chain rule used in computing the derivative of $f(x)=e^{\sqrt{3x}}$?
(B) Construct a function $g(x)$ such that the chain rule is used three times in computing its derivative.

75. Suppose a student reasons that the functions $f(x) = \ln[5(x^2 + 3)^4]$ and $g(x) = 4\ln(x^2 + 3)$ must have the same derivative, since he has entered $f(x)$, $g(x)$, $f'(x)$, and $g'(x)$ into a graphing utility, but only three graphs appear (see the figure). Is his reasoning correct? Are $f'(x)$ and $g'(x)$ the same function? Explain.

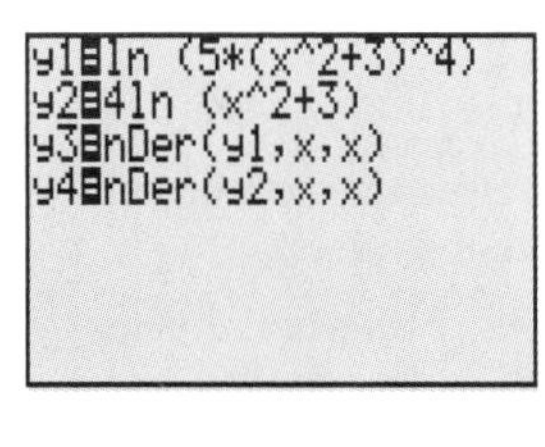

(A)

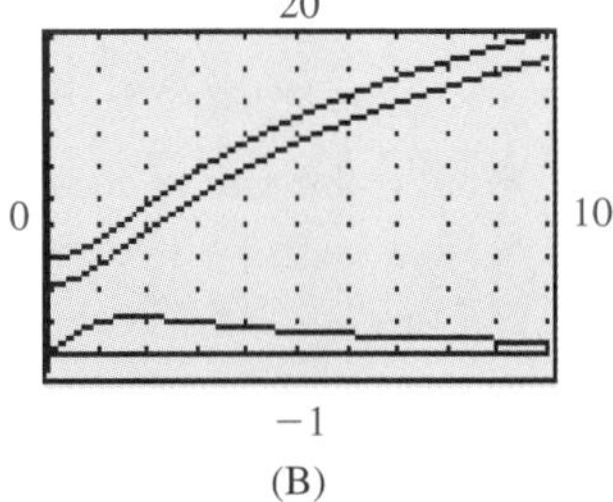

(B)

Figure for 75

76. Suppose a student reasons that the functions $f(x) = (x + 1)\ln(x + 1) - x$ and $g(x) = (x + 1)^{1/3}$ must have the same derivative, since she has entered $f(x)$, $g(x)$, $f'(x)$, and $g'(x)$ into a graphing utility, but only three graphs appear (see the figure). Is her reasoning correct? Are $f'(x)$ and $g'(x)$ the same function? Explain.

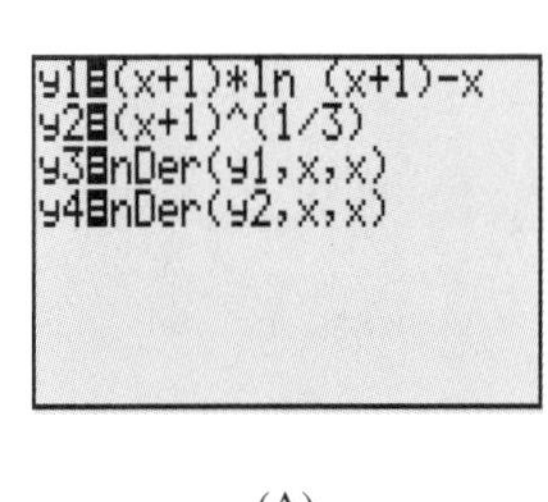

(A)

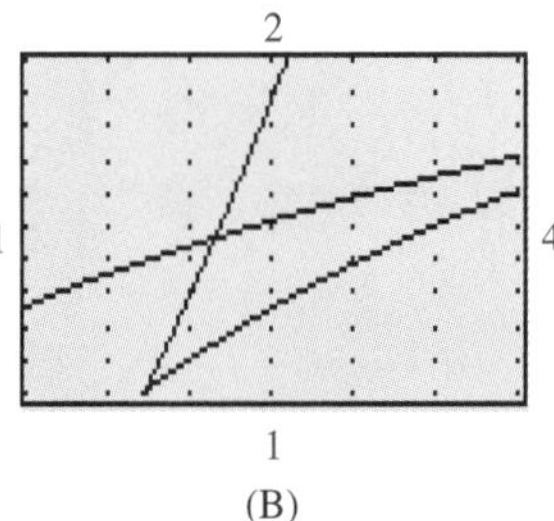

(B)

Figure for 76

Applications

Business & Economics

77. *Maximum revenue.* Suppose the price–demand equation for x units of a commodity is determined from empirical data to be

$$p = 100e^{-0.05x}$$

where x units are sold per day at a price of $\$p$ each. Find the production level and price that maximize revenue. What is the maximum revenue?

78. *Maximum revenue.* Repeat Problem 77 using the price–demand equation

$$p = 10e^{-0.04x}$$

79. *Maximum profit.* Refer to Problem 77. If the daily fixed cost is \$400 and the cost per unit is \$6, use approximation techniques to find the production level and the price that maximize profit. What is the maximum profit? [*Hint:* Graph $y = P(x)$ and $y = P'(x)$ in the same viewing window.]

80. *Maximum profit.* Refer to Problem 78. If the daily fixed cost is \$30 and the cost per unit is \$0.70, use approximation techniques to find the production level and the price that maximize profit. What is the maximum profit? [*Hint:* Graph $y = P(x)$ and $y = P'(x)$ in the same viewing window.]

81. *Maximum profit.* A mail-order company specializing in computer equipment has collected the data in the table, showing the weekly demand x for Data-Link modems at various prices p. The company purchases the modems from the manufacturer for \$100 each. Use exponential regression ($p = ab^x$) to find the price (to the nearest cent) that will maximize the weekly profit.

Demand x	Price per modem (\$) p
412	169.95
488	149.95
575	139.95
722	129.95
786	119.95

82. *Maximum profit.* The mail-order company in Problem 81 also sells 100 megabyte disks. The data in the table show the weekly demand x for these disks at various price levels p. The company purchases the disks from the manufacturer for \$6 each. Use exponential regression ($p = ab^x$) to find the price (to the nearest cent) that will maximize the weekly profit.

Demand x	Price per disk (\$) p
578	16.95
942	14.95
1,218	13.95
1,758	11.95
2,198	10.95

83. *Salvage value.* The salvage value S (in dollars) of a company airplane after t years is estimated to be given by

$$S(t) = 300{,}000e^{-0.1t}$$

What is the rate of depreciation (in dollars per year) after 1 year? 5 years? 10 years?

84. *Resale value.* The resale value R (in dollars) of a company car after t years is estimated to be given by

$$R(t) = 20{,}000e^{-0.15t}$$

What is the rate of depreciation (in dollars per year) after 1 year? 2 years? 3 years?

85. *Promotion and maximum profit.* A recording company has produced a new compact disk featuring a very popular recording group. Before launching a national sales campaign, the marketing research department chose to test market the CD in a bellwether city. Their interest is in determining the length of a sales campaign that will maximize total profits. From empirical data, the research department estimates that the proportion of a target group of 50,000 persons buying the CD after t days of television promotion is given by $1 - e^{-0.03t}$. If \$4 is received for each CD sold, then the total revenue after t days of promotion will be approximated by

$$R(t) = (4)(50{,}000)(1 - e^{-0.03t}) \qquad t \geq 0$$

Television promotion costs are

$$C(t) = 4{,}000 + 3{,}000t \qquad t \geq 0$$

(A) How many days of television promotion should be used to maximize total profit? What is the maximum total profit? What percentage of the target market will have purchased the CD when the maximum profit is reached?

(B) Graph the profit function.

86. *Promotion and maximum profit.* Repeat Problem 85 using the revenue equation

$$R(t) = (3)(60{,}000)(1 - e^{-0.04t})$$

Life Sciences

87. *Blood pressure and age.* A research group using hospital records developed the following approximate mathematical model relating systolic blood pressure and age:

$$P(x) = 40 + 25\ln(x + 1) \qquad 0 \leq x \leq 65$$

where $P(x)$ is pressure measured in millimeters of mercury and x is age in years. What is the rate of change of pressure at the end of 10 years? At the end of 30 years? At the end of 60 years?

88. *Biology.* A yeast culture at room temperature (68°F) is placed in a refrigerator maintaining a constant temperature of 38°F. After t hours, the temperature T of the culture is given approximately by

$$T = 30e^{-0.58t} + 38 \qquad t \geq 0$$

What is the rate of change of temperature of the culture at the end of 1 hour? At the end of 4 hours?

89. *Bacterial growth.* A single cholera bacterium divides every 0.5 hour to produce two complete cholera bacteria. If we start with a colony of 5,000 bacteria, then after t hours there will be

$$A(t) = 5{,}000 \cdot 2^{2t}$$

bacteria. Find $A'(t)$, $A'(1)$, and $A'(5)$, and interpret the results.

90. *Bacterial growth.* Repeat Problem 89 for a starting colony of 1,000 bacteria where a single bacterium divides every 0.25 hour.

Social Sciences

91. *Sociology.* Daniel Lowenthal, a sociologist at Columbia University, made a 5 year study on the sale of popular records relative to their position in the top 20. He found that the average number of sales $N(n)$ of the nth ranking record was given approximately by

$$N(n) = N_1e^{-0.09(n-1)} \qquad 1 \leq n \leq 20$$

where N_1 was the number of sales of the top record on the list at a given time. Graph N for $N_1 = 1{,}000{,}000$ records.

92. *Political science.* Thomas W. Casstevens, a political scientist at Oakland University, has studied legislative turnover. He (with others) found that the number $N(t)$ of continuously serving members of an elected legislative body remaining t years after an election is given approximately by a function of the form

$$N(t) = N_0e^{-ct}$$

In particular, for the 1965 election for the U.S. House of Representatives, it was found that

$$N(t) = 434e^{-0.0866t}$$

What is the rate of change after 2 years? After 10 years?

SECTION 10-4

Implicit Differentiation

- SPECIAL FUNCTION NOTATION
- IMPLICIT DIFFERENTIATION

SPECIAL FUNCTION NOTATION

The equation

$$y = 2 - 3x^2 \tag{1}$$

defines a function f with y as a dependent variable and x as an independent variable. Using function notation, we would write

$$y = f(x) \qquad \text{or} \qquad f(x) = 2 - 3x^2$$

In order to reduce to a minimum the number of symbols involved in a discussion, we will often write equation (1) in the form

$$y = 2 - 3x^2 = y(x)$$

where y is *both* a dependent variable and a function symbol. This is a convenient notation, and no harm is done as long as one is aware of the double role of y. Other examples are

$$x = 2t^2 - 3t + 1 = x(t)$$
$$z = \sqrt{u^2 - 3u} = z(u)$$
$$r = \frac{1}{(s^2 - 3s)^{2/3}} = r(s)$$

This type of notation will simplify much of the discussion and work that follows.

Until now we have considered functions involving only one independent variable. There is no reason to stop there. The concept can be generalized to functions involving two or more independent variables, and this will be done in detail in Chapter 13. For now, we will "borrow" the notation for a function involving two independent variables. For example,

$$F(x, y) = x^2 - 2xy + 3y^2 - 5$$

specifies a function F involving two independent variables.

IMPLICIT DIFFERENTIATION

Consider the equation

$$3x^2 + y - 2 = 0 \tag{2}$$

and the equation obtained by solving (2) for y in terms of x,

$$y = 2 - 3x^2 \tag{3}$$

Both equations define the same function using x as the independent variable and y as the dependent variable. For (3), we can write

$$y = f(x)$$

where

$$f(x) = 2 - 3x^2 \tag{4}$$

and we have an **explicit** (clearly stated) rule that enables us to determine y for each value of x. On the other hand, the y in equation (2) is the same y as in equation (3), and equation (2) **implicitly** gives (implies though does not plainly express) y as a function of x. Thus, we say that equations (3) and (4) define the function f explicitly, and equation (2) defines f implicitly.

The direct use of an equation that defines a function implicitly to find the derivative of the dependent variable with respect to the independent variable is called **implicit differentiation.** Let us differentiate (2) implicitly and (3) directly, and compare results.

Starting with

$$3x^2 + y - 2 = 0$$

we think of y as a function of x—that is, $y = y(x)$—and write

$$3x^2 + y(x) - 2 = 0$$

Then we differentiate both sides with respect to x:

$$\frac{d}{dx}[3x^2 + y(x) - 2)] = \frac{d}{dx}0$$

$$\frac{d}{dx}3x^2 + \frac{d}{dx}y(x) - \frac{d}{dx}2 = 0$$

$$6x + y' - 0 = 0$$

Since y is a function of x, but is not explicitly given we simply write $\frac{d}{dx}y(x) = y'$ to indicate its derivative.

Now, we solve for y':

$$y' = -6x$$

Note that we get the same result if we start with equation (3) and differentiate directly:

$$y = 2 - 3x^2$$
$$y' = -6x$$

Why are we interested in implicit differentiation? In general, why do we not solve for y in terms of x and differentiate directly? The answer is that there are many equations of the form

$$F(x, y) = 0 \tag{5}$$

that are either difficult or impossible to solve for y explicitly in terms of x (try it for $x^2y^5 - 3xy + 5 = 0$ or for $e^y - y = 3x$, for example). But it can be shown that, under fairly general conditions on F, equation (5) will define one or more functions where y is a dependent variable and x is an independent variable. To find y' under these conditions, we differentiate (5) implicitly.

Explore–Discuss 1

(A) How many tangent lines are there to the graph in Figure 1 when $x = 0$? When $x = 1$? When $x = 2$? When $x = 4$? When $x = 6$?

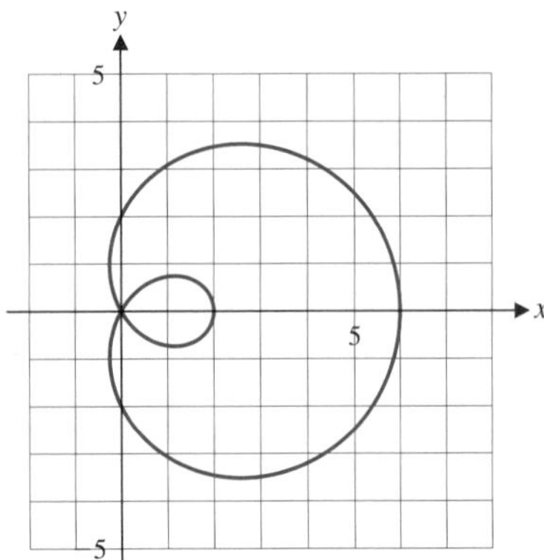

FIGURE 1

(B) Sketch the tangent lines referred to in part (A) and estimate each of their slopes.

(C) Explain why the graph in Figure 1 is not the graph of a function.

Example 1 **Differentiating Implicitly** Given

$$F(x, y) = x^2 + y^2 - 25 = 0 \tag{6}$$

find y' and the slope of the graph at $x = 3$.

Solution We start with the graph of $x^2 + y^2 - 25 = 0$ (a circle, as shown in Fig. 2) so that we can interpret our results geometrically. From the graph, it is clear that equation (6) does not define a function. But with a suitable restriction on the variables, equation (6) can define two or more functions. For example, the upper half and the lower half of the circle each define a function. A point on each half-circle that corresponds to $x = 3$ is found by substituting $x = 3$ into (6) and solving for y:

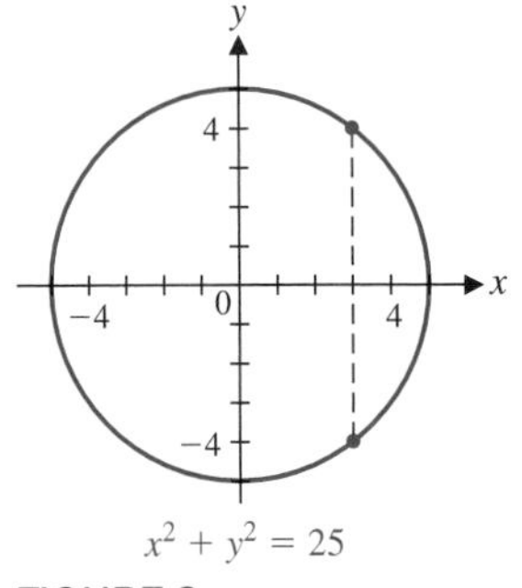

FIGURE 2

$$\begin{aligned} x^2 + y^2 - 25 &= 0 \\ (3)^2 + y^2 &= 25 \\ y^2 &= 16 \\ y &= \pm 4 \end{aligned}$$

Thus, the point $(3, 4)$ is on the upper half-circle, and the point $(3, -4)$ is on the lower half-circle. We will use these results in a moment. We now differentiate (6) implicitly, treating y as a function of x; that is, $y = y(x)$:

$$x^2 + y^2 - 25 = 0$$

$$x^2 + [y(x)]^2 - 25 = 0$$

$$\frac{d}{dx}\{x^2 + [y(x)]^2 - 25\} = \frac{d}{dx}0$$

$$\frac{d}{dx}x^2 + \frac{d}{dx}[y(x)]^2 - \frac{d}{dx}25 = 0 \qquad \text{Use the chain rule.}$$

$$2x + 2[y(x)]^{2-1}y'(x) - 0 = 0$$

$$2x + 2yy' = 0 \qquad \text{Solve for } y' \text{ in terms of } x \text{ and } y.$$

$$y' = -\frac{2x}{2y}$$

$$y' = -\frac{x}{y} \qquad \text{Leave the answer in terms of } x \text{ and } y.$$

We have found y' without first solving $x^2 + y^2 - 25 = 0$ for y in terms of x. And by leaving y' in terms of x and y, we can use $y' = -x/y$ to find y' for *any* point on the graph of $x^2 + y^2 - 25 = 0$ (except where $y = 0$). In particular, for $x = 3$, we found that $(3, 4)$ and $(3, -4)$ are on the graph; thus, the slope of the graph at $(3, 4)$ is

$$y'|_{(3,4)} = -\tfrac{3}{4} \qquad \text{The slope of the graph at } (3, 4)$$

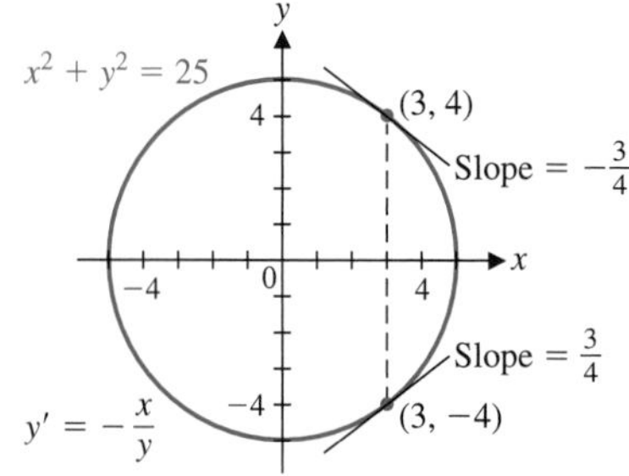

FIGURE 3

and the slope at $(3, -4)$ is

$$y'|_{(3,-4)} = -\tfrac{3}{-4} = \tfrac{3}{4} \qquad \text{The slope of the graph at } (3, -4)$$

The symbol

$$y'|_{(a,b)}$$

is used to indicate that we are evaluating y' at $x = a$ and $y = b$.

The results are interpreted geometrically on the original graph as shown in Figure 3.

In Example 1, the fact that y' is given in terms of both x and y is not a great disadvantage. We have only to make certain that when we want **to evaluate y' for a particular value of x and y, say, (x_0, y_0), the ordered pair must satisfy the original equation.**

Matched Problem 1 Graph $x^2 + y^2 - 169 = 0$, find y' by implicit differentiation, and find the slope of the graph when $x = 5$.

Example 2 **Differentiating Implicitly** Find the equation(s) of the tangent line(s) to the graph of

$$y - xy^2 + x^2 + 1 = 0 \tag{7}$$

at the point(s) where $x = 1$.

Solution We first find y when $x = 1$:

$$\begin{aligned} y - xy^2 + x^2 + 1 &= 0 \\ y - (1)y^2 + (1)^2 + 1 &= 0 \\ y - y^2 + 2 &= 0 \\ y^2 - y - 2 &= 0 \\ (y - 2)(y + 1) &= 0 \\ y &= -1, 2 \end{aligned}$$

Thus, there are two points on the graph of (7) where $x = 1$; namely, $(1, -1)$ and $(1, 2)$. We next find the slope of the graph at these two points by differentiating (7) implicitly:

$$\begin{aligned} y - xy^2 + x^2 + 1 &= 0 \\ \frac{d}{dx}y - \frac{d}{dx}xy^2 + \frac{d}{dx}x^2 + \frac{d}{dx}1 &= \frac{d}{dx}0 \\ y' - (x \cdot 2yy' + y^2) + 2x &= 0 \\ y' - 2xyy' - y^2 + 2x &= 0 \\ y' - 2xyy' &= y^2 - 2x \\ (1 - 2xy)y' &= y^2 - 2x \\ y' &= \frac{y^2 - 2x}{1 - 2xy} \end{aligned}$$

Use the product rule and the chain rule for $\frac{d}{dx}xy^2$.

Solve for y' by getting all terms involving y' on one side.

Now, find the slope at each point:

$$y'\big|_{(1,-1)} = \frac{(-1)^2 - 2(1)}{1 - 2(1)(-1)} = \frac{1 - 2}{1 + 2} = \frac{-1}{3} = -\frac{1}{3}$$

$$y'\big|_{(1,2)} = \frac{(2)^2 - 2(1)}{1 - 2(1)(2)} = \frac{4 - 2}{1 - 4} = \frac{2}{-3} = -\frac{2}{3}$$

Equation of tangent line at $(1, -1)$:

$$\begin{aligned} y - y_1 &= m(x - x_1) \\ y + 1 &= -\tfrac{1}{3}(x - 1) \\ y + 1 &= -\tfrac{1}{3}x + \tfrac{1}{3} \\ y &= -\tfrac{1}{3}x - \tfrac{2}{3} \end{aligned}$$

Equation of tangent line at $(1, 2)$:

$$\begin{aligned} y - y_1 &= m(x - x_1) \\ y - 2 &= -\tfrac{2}{3}(x - 1) \\ y - 2 &= -\tfrac{2}{3}x + \tfrac{2}{3} \\ y &= -\tfrac{2}{3}x + \tfrac{8}{3} \end{aligned}$$

Matched Problem 2 Repeat Example 2 for $x^2 + y^2 - xy - 7 = 0$ at $x = 1$.

Explore–Discuss 2

The slopes of the tangent lines to $y^2 + 3xy + 4x = 9$ when $x = 0$ can be found in either of the following ways: (1) by differentiating the equation implicitly; or (2) by solving for y explicitly in terms of x (using the quadratic formula), and then computing the derivative. Which of the two methods is more efficient? Explain.

Example 3 **Differentiating Implicitly** Find x' for $x = x(t)$ defined implicitly by

$$t \ln x = xe^t - 1$$

and evaluate x' at $(t, x) = (0, 1)$.

SOLUTION It is important to remember that x is the dependent variable and t is the independent variable. Therefore, we differentiate both sides of the equation with respect to t (using product and chain rules where appropriate), and then solve for x':

$$t \ln x = xe^t - 1$$ Differentiate implicitly with respect to t.

$$\frac{d}{dt}(t \ln x) = \frac{d}{dt}(xe^t) - \frac{d}{dt}1$$

$$t\frac{x'}{x} + \ln x = xe^t + x'e^t$$ Clear fractions.

$$x \cdot t\frac{x'}{x} + x \cdot \ln x = x \cdot xe^t + x \cdot e^t x'$$ $x \neq 0$

$$tx' + x \ln x = x^2e^t + xe^tx'$$ Solve for x'.

$$tx' - xe^tx' = x^2e^t - x \ln x$$ Factor out x'.

$$(t - xe^t)x' = x^2e^t - x \ln x$$

$$x' = \frac{x^2e^t - x \ln x}{t - xe^t}$$

Now, we evaluate x' at $(t, x) = (0, 1)$, as requested:

$$x'|_{(0,1)} = \frac{(1)^2e^0 - 1 \ln 1}{0 - 1e^0}$$

$$= \frac{1}{-1} = -1$$

Matched Problem 3 Find x' for $x = x(t)$ defined implicitly by

$$1 + x \ln t = te^x$$

and evaluate x' at $(t, x) = (1, 0)$.

Answers to Matched Problems

1. $y' = -x/y$; when $x = 5$, $y = \pm 12$, thus, $y'|_{(5,12)} = -\frac{5}{12}$ and $y'|_{(5,-12)} = \frac{5}{12}$

2. $y' = \dfrac{y - 2x}{2y - x}$; $y = \frac{4}{5}x - \frac{14}{5}$, $y = \frac{1}{5}x + \frac{14}{5}$

3. $x' = \dfrac{te^x - x}{t \ln t - t^2e^x}$; $x'|_{(1,0)} = -1$

EXERCISE 10-4

A *In Problems 1–4, find y' two ways:*

(A) Differentiate the given equation implicitly and then solve for y'.

(B) Solve the given equation for y and then differentiate directly.

1. $2x + 3y - 7 = 0$

2. $5x - 4y + 11 = 0$

3. $x^2 - 2y - 4 = 0$

4. $x^3 + 5y + 1 = 0$

In Problems 5–22, find y' without solving for y in terms of x (use implicit differentiation). Evaluate y' at the indicated point.

5. $y - 3x^2 + 5 = 0$; $(1, -2)$

6. $3x^4 + y - 2 = 0$; $(1, -1)$

7. $y^2 - 3x^2 + 8 = 0$; $(2, 2)$

8. $3y^2 + 2x^3 - 14 = 0$; $(1, 2)$

9. $y^2 + y - x = 0$; $(2, 1)$

10. $2y^3 + y^2 - x = 0$; $(3, 1)$

B

11. $xy - 6 = 0$; $(2, 3)$
12. $3xy - 2x - 2 = 0$; $(2, 1)$
13. $2xy + y + 2 = 0$; $(-1, 2)$
14. $2y + xy - 1 = 0$; $(-1, 1)$
15. $x^2y - 3x^2 - 4 = 0$; $(2, 4)$
16. $2x^3y - x^3 + 5 = 0$; $(-1, 3)$
17. $e^y = x^2 + y^2$; $(1, 0)$
18. $x^2 - y = 4e^y$; $(2, 0)$
19. $x^3 - y = \ln y$; $(1, 1)$
20. $\ln y = 2y^2 - x$; $(2, 1)$
21. $x \ln y + 2y = 2x^3$; $(1, 1)$
22. $xe^y - y = x^2 - 2$; $(2, 0)$

In Problems 23 and 24, find x' for $x = x(t)$ defined implicitly by the given equation. Evaluate x' at the indicated point.

23. $x^2 - t^2x + t^3 + 11 = 0$; $(-2, 1)$
24. $x^3 - tx^2 - 4 = 0$; $(-3, -2)$

Problems 25 and 26 refer to the equation and graph shown in the figure.

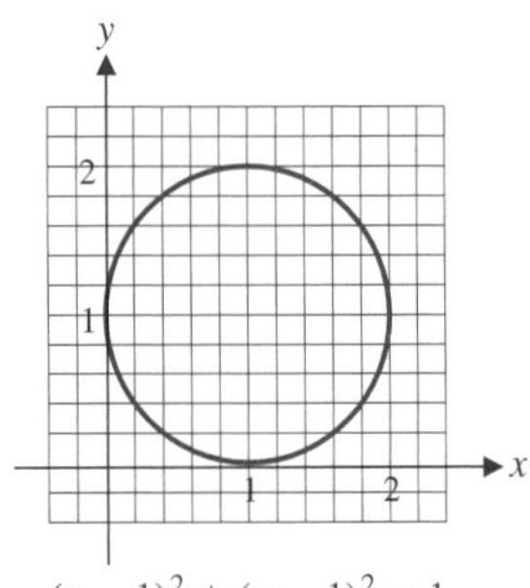

$(x - 1)^2 + (y - 1)^2 = 1$

25. Use implicit differentiation to find the slopes of the tangent lines at the points on the graph where $x = 1.6$. Check your answers by visually estimating the slopes on the graph in the figure.
26. Find the slopes of the tangent lines at the points on the graph where $x = 0.2$. Check your answers by visually estimating the slopes on the graph in the figure.

In Problems 27–30, find the equation(s) of the tangent line(s) to the graphs of the indicated equations at the point(s) with the given value of x.

27. $xy - x - 4 = 0$; $x = 2$
28. $3x + xy + 1 = 0$; $x = -1$
29. $y^2 - xy - 6 = 0$; $x = 1$
30. $xy^2 - y - 2 = 0$; $x = 1$
31. If $xe^y = 1$, find y' in two ways: first by implicit differentiation, then by solving for y explicitly in terms of x. Which method do you prefer? Explain.
32. Explain the difficulty in solving $x^3 + y + xe^y = 1$ for y as an explicit function of x. Find the slope of the tangent line to the graph of the equation at the point $(0, 1)$.

C *In Problems 33–40, find y' and the slope of the tangent line to the graph of each equation at the indicated point.*

33. $(1 + y)^3 + y = x + 7$; $(2, 1)$
34. $(y - 3)^4 - x = y$; $(-3, 4)$
35. $(x - 2y)^3 = 2y^2 - 3$; $(1, 1)$
36. $(2x - y)^4 - y^3 = 8$; $(-1, -2)$
37. $\sqrt{7 + y^2} - x^3 + 4 = 0$; $(2, 3)$
38. $6\sqrt{y^3 + 1} - 2x^{3/2} - 2 = 0$; $(4, 2)$
39. $\ln(xy) = y^2 - 1$; $(1, 1)$
40. $e^{xy} - 2x = y + 1$; $(0, 0)$
41. **C** Find the equation(s) of the tangent line(s) at the point(s) on the graph of the equation

$$y^3 - xy - x^3 = 2$$

where $x = 1$. Round all approximate values to two decimal places.

42. **C** Refer to the equation in Problem 41. Find the equation(s) of the tangent line(s) at the point(s) on the graph where $y = -1$. Round all approximate values to two decimal places.

Applications

Business & Economics

For the demand equations in Problems 43–46, find the rate of change of p with respect to x by differentiating implicitly (x is the number of items that can be sold at a price of \$p.)

43. $x = p^2 - 2p + 1{,}000$
44. $x = p^3 - 3p^2 + 200$
45. $x = \sqrt{10{,}000 - p^2}$
46. $x = \sqrt[3]{1{,}500 - p^3}$

Life Sciences

47. *Biophysics.* In biophysics, the equation

$$(L + m)(V + n) = k$$

is called the *fundamental equation of muscle contraction,* where m, n, and k are constants, and V is the velocity of the shortening of muscle fibers for a muscle subjected to a load of L. Find dL/dV using implicit differentiation.

48. *Biophysics.* In Problem 47, find dV/dL using implicit differentiation.

SECTION 10-5 Related Rates

The workers in a union are concerned that the rate at which wages are increasing is lagging behind the rate of increase in the company's profits. An automobile dealer wants to predict how badly an anticipated increase in interest rates will decrease his rate of sales. An investor is studying the connection between the rate of increase in the Dow Jones Average and the rate of increase in the Gross Domestic Product over the past 50 years.

In each of these situations there are two quantities—wages and profits in the first instance, for example—that are changing with respect to time. We would like to discover the precise relationship between the rates of increase (or decrease) of the two quantities. We will begin our discussion of such *related rates* by considering some familiar situations in which the two quantities are distances, and the two rates are velocities.

Example 1 **Related Rates and Motion** A 26 foot ladder is placed against a wall (Fig. 1). If the top of the ladder is sliding down the wall at 2 feet per second, at what rate is the bottom of the ladder moving away from the wall when the bottom of the ladder is 10 feet away from the wall?

SOLUTION Many people reason that since the ladder is of constant length, the bottom of the ladder will move away from the wall at the same rate that the top of the ladder is moving down the wall. This is not the case, as we will see.

FIGURE 1

At any moment in time, let x be the distance of the bottom of the ladder from the wall, and let y be the distance of the top of the ladder on the wall (see Fig. 1). Both x and y are changing with respect to time and can be thought of as functions of time; that is, $x = x(t)$ and $y = y(t)$. Furthermore, x and y are related by the Pythagorean relationship:

$$x^2 + y^2 = 26^2 \tag{1}$$

Differentiating (1) implicitly with respect to time t, and using the chain rule where appropriate, we obtain

$$2x\frac{dx}{dt} + 2y\frac{dy}{dt} = 0 \tag{2}$$

The rates dx/dt and dy/dt are related by equation (2); hence, this type of problem is referred to as a **related rates problem.**

Now our problem is to find dx/dt when $x = 10$ feet, given that $dy/dt = -2$ (y is decreasing at a constant rate of 2 feet per second). We have all the quantities we need in equation (2) to solve for dx/dt, except y. When $x = 10$, y can be found using (1):

$$10^2 + y^2 = 26^2$$
$$y = \sqrt{26^2 - 10^2} = 24 \text{ feet}$$

Substitute $dy/dt = -2, x = 10$, and $y = 24$ into (2); then solve for dx/dt:

$$2(10)\frac{dx}{dt} + 2(24)(-2) = 0$$
$$\frac{dx}{dt} = \frac{-2(24)(-2)}{2(10)} = 4.8 \text{ feet per second}$$

Thus, the bottom of the ladder is moving away from the wall at a rate of 4.8 feet per second.

Matched Problem 1 Again, a 26 foot ladder is placed against a wall (Fig. 1). If the bottom of the ladder is moving away from the wall at 3 feet per second, at what rate is the top moving down when the top of the ladder is 24 feet up the wall?

Explore–Discuss 1

(A) For which values of x and y in Example 1 is dx/dt equal to 2 (that is, the same rate at which the ladder is sliding down the wall)?
(B) When is dx/dt greater than 2? Less than 2?

Suggestions for Solving Related Rates Problems

Step 1. Sketch a figure if helpful.
Step 2. Identify all relevant variables, including those whose rates are given and those whose rates are to be found.
Step 3. Express all given rates and rates to be found as derivatives.
Step 4. Find an equation connecting the variables in step 2.
Step 5. Implicitly differentiate the equation found in step 4, using the chain rule where appropriate, and substitute in all given values.
Step 6. Solve for the derivative that will give the unknown rate.

Example 2 **Related Rates and Motion** Suppose two motorboats leave from the same point at the same time. If one travels north at 15 miles per hour and the other travels east at 20 miles per hour, how fast will the distance between them be changing after 2 hours?

SOLUTION First, draw a picture, as shown in Figure 2.

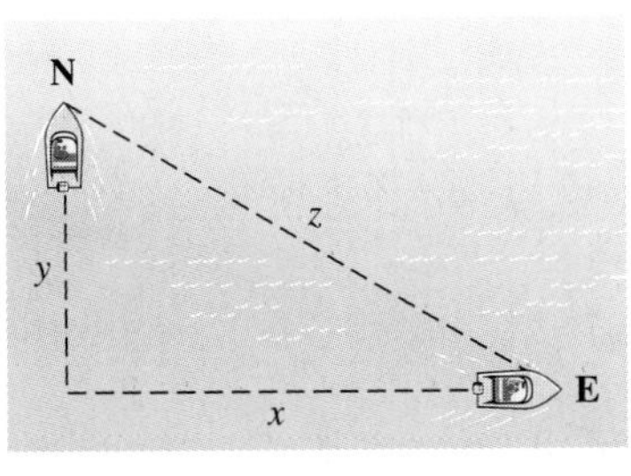

FIGURE 2

All variables, x, y, and z, are changing with time. Hence, they can be thought of as functions of time; $x = x(t)$, $y = y(t)$, and $z = z(t)$, given implicitly. It now makes sense to take derivatives of each variable with respect to time. From the Pythagorean theorem,

$$z^2 = x^2 + y^2 \tag{3}$$

We also know that

$$\frac{dx}{dt} = 20 \text{ miles per hour} \quad \text{and} \quad \frac{dy}{dt} = 15 \text{ miles per hour}$$

We would like to find dz/dt at the end of 2 hours; that is, when $x = 40$ miles and $y = 30$ miles. To do this, we differentiate both sides of (3) with respect to t and solve for dz/dt:

$$2z\frac{dz}{dt} = 2x\frac{dx}{dt} + 2y\frac{dy}{dt} \tag{4}$$

We have everything we need except z. When $x = 40$ and $y = 30$, we find z from (3) to be 50. Substituting the known quantities into (4), we obtain

$$2(50)\frac{dz}{dt} = 2(40)(20) + 2(30)(15)$$
$$\frac{dz}{dt} = 25 \text{ miles per hour}$$

Thus, the boats will be separating at a rate of 25 miles per hour.

Matched Problem 2 Repeat Example 2 for the situation at the end of 3 hours.

Example 3 **Related Rates and Motion** Suppose a point is moving along the graph of $x^2 + y^2 = 25$ (Fig. 3). When the point is at $(-3, 4)$, its x coordinate is increasing at the rate of 0.4 unit per second. How fast is the y coordinate changing at that moment?

SOLUTION Since both x and y are changing with respect to time, we can think of each as a function of time:

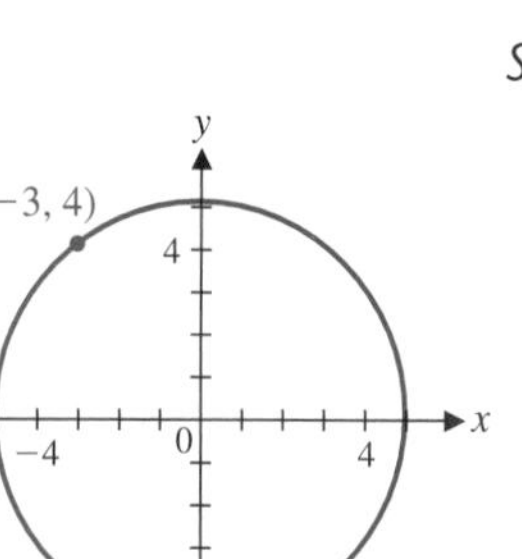

FIGURE 3

$$x = x(t) \qquad \text{and} \qquad y = y(t)$$

but restricted so that

$$x^2 + y^2 = 25 \tag{5}$$

Our problem is now to find dy/dt, given $x = -3$, $y = 4$, and $dx/dt = 0.4$. Implicitly differentiating both sides of (5) with respect to t, we have

$$x^2 + y^2 = 25$$
$$2x\frac{dx}{dt} + 2y\frac{dy}{dt} = 0 \qquad \text{Divide both sides by 2.}$$
$$x\frac{dx}{dt} + y\frac{dy}{dt} = 0 \qquad \text{Substitute } x = -3,\ y = 4, \text{ and } dx/dt = 0.4, \text{ and solve for } dy/dt.$$
$$(-3)(0.4) + 4\frac{dy}{dt} = 0$$
$$\frac{dy}{dt} = 0.3 \text{ unit per second}$$

Matched Problem 3 A point is moving on the graph of $y^3 = x^2$. When the point is at $(-8, 4)$, its y coordinate is decreasing at 2 units per second. How fast is the x coordinate changing at that moment?

Example 4 **Related Rates and Business** Suppose that for a company manufacturing transistor radios, the cost, revenue, and profit equations are given by

$$C = 5{,}000 + 2x \qquad \text{Cost equation}$$
$$R = 10x - 0.001x^2 \qquad \text{Revenue equation}$$
$$P = R - C \qquad \text{Profit equation}$$

where the production output in 1 week is x radios. If production is increasing at the rate of 500 radios per week when production is 2,000 radios, find the rate of increase in:

(A) Cost (B) Revenue (C) Profit

SOLUTION If production x is a function of time (it must be, since it is changing with respect to time), then C, R, and P must also be functions of time. These functions are implicitly (rather than explicitly) given. Letting t represent time in weeks, we differentiate both sides of each of the three equations above with respect to t, and then substitute $x = 2{,}000$ and $dx/dt = 500$ to find the desired rates.

(A)

$$C = 5{,}000 + 2x \qquad \text{Think: } C = C(t) \text{ and } x = x(t).$$
$$\frac{dC}{dt} = \frac{d}{dt}(5{,}000) + \frac{d}{dt}(2x) \qquad \text{Differentiate both sides with respect to } t.$$
$$\frac{dC}{dt} = 0 + 2\frac{dx}{dt} = 2\frac{dx}{dt}$$

Since $dx/dt = 500$ when $x = 2{,}000$,

$$\frac{dC}{dt} = 2(500) = \$1{,}000 \text{ per week}$$

Cost is increasing at a rate of $1,000 per week.

(B)

$$R = 10x - 0.001x^2$$
$$\frac{dR}{dt} = \frac{d}{dt}(10x) - \frac{d}{dt}0.001x^2$$
$$\frac{dR}{dt} = 10\frac{dx}{dt} - 0.002x\frac{dx}{dt}$$
$$\frac{dR}{dt} = (10 - 0.002x)\frac{dx}{dt}$$

Since $dx/dt = 500$ when $x = 2{,}000$,

$$\frac{dR}{dt} = [10 - 0.002(2{,}000)](500) = \$3{,}000 \text{ per week}$$

Revenue is increasing at a rate of $3,000 per week.

(C)

$$P = R - C$$
$$\frac{dP}{dt} = \frac{dR}{dt} - \frac{dC}{dt}$$
$$= \$3{,}000 - \$1{,}000 \qquad \text{Results from parts (A) and (B)}$$
$$= \$2{,}000 \text{ per week}$$

Profit is increasing at a rate of $2,000 per week.

Matched Problem 4 Repeat Example 4 for a production level of 6,000 radios per week.

Explore–Discuss 2

(A) In Example 4 suppose that $x(t) = 500t + 500$. Find the time and production level at which the profit is maximized.
(B) Suppose that $x(t) = t^2 + 492t + 16$. Find the time and production level at which the profit is maximized.
(C) Explain why it is unnecessary to know a formula for $x(t)$ in order to determine the production level at which the profit is maximized.

Answers to Matched Problems

1. $dy/dt = -1.25$ ft/sec 2. $dz/dt = 25$ mi/hr 3. $dx/dt = 6$ units/sec
4. (A) $dC/dt = \$1,000/\text{wk}$ (B) $dR/dt = -\$1,000/\text{wk}$
(C) $dP/dt = -\$2,000/\text{wk}$

Exercise 10-5

A *In Problems 1–6, assume $x = x(t)$ and $y = y(t)$. Find the indicated rate, given the other information.*

1. $y = 2x^2 - 1$; $dy/dt = ?$; $dx/dt = 2$ when $x = 30$
2. $y = 2x^{1/2} + 3$; $dy/dt = ?$; $dx/dt = 8$ when $x = 4$
3. $x^2 + y^2 = 25$; $dy/dt = ?$; $dx/dt = -3$ when $x = 3$ and $y = 4$
4. $y^2 + x = 11$; $dx/dt = ?$; $dy/dt = -2$ when $x = 2$ and $y = 3$
5. $x^2 + xy + 2 = 0$; $dy/dt = ?$; $dx/dt = -1$ when $x = 2$ and $y = -3$
6. $y^2 + xy - 3x = -3$; $dx/dt = ?$; $dy/dt = -2$ when $x = 1$ and $y = 0$

B

7. A point is moving on the graph of $xy = 36$. When the point is at $(4, 9)$, its x coordinate is increasing at 4 units per second. How fast is the y coordinate changing at that moment?
8. A point is moving on the graph of $4x^2 + 9y^2 = 36$. When the point is at $(3, 0)$, its y coordinate is decreasing at 2 units per second. How fast is its x coordinate changing at that moment?
9. A boat is being pulled toward a dock as indicated in the figure. If the rope is being pulled in at 3 feet per second, how fast is the distance between the dock and the boat decreasing when it is 30 feet from the dock?

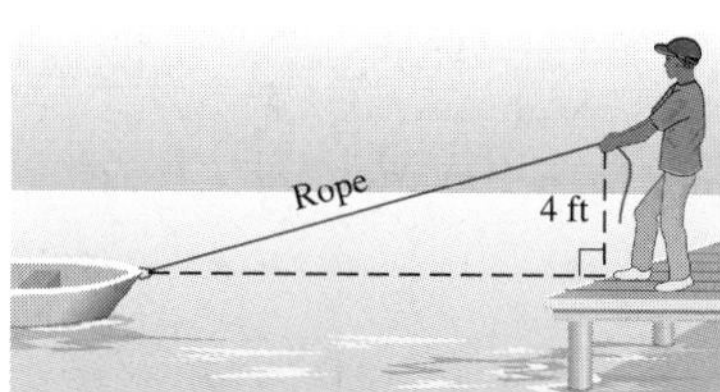

Figure for 9 and 10

10. Refer to Problem 9. Suppose the distance between the boat and the dock is decreasing at 3.05 feet per second. How fast is the rope being pulled in when the boat is 10 feet from the dock?
11. A rock is thrown into a still pond and causes a circular ripple. If the radius of the ripple is increasing at 2 feet per second, how fast is the area changing when the radius is 10 feet? [Use $A = \pi R^2$, $\pi \approx 3.14$.]
12. Refer to Problem 11. How fast is the circumference of a circular ripple changing when the radius is 10 feet? [Use $C = 2\pi R$, $\pi \approx 3.14$.]
13. The radius of a spherical balloon is increasing at the rate of 3 centimeters per minute. How fast is the volume changing when the radius is 10 centimeters? [Use $V = \frac{4}{3}\pi R^3$, $\pi \approx 3.14$.]
14. Refer to Problem 13. How fast is the surface area of the sphere increasing? [Use $S = 4\pi R^2$, $\pi \approx 3.14$.]
15. Boyle's law for enclosed gases states that if the volume is kept constant, then the pressure P and temperature T are related by the equation

$$\frac{P}{T} = k$$

where k is a constant. If the temperature is increasing at 3 kelvins per hour, what is the rate of change of pressure when the temperature is 250 K and the pressure is 500 pounds per square inch?

16. Boyle's law for enclosed gases states that if the temperature is kept constant, then the pressure P and volume V of the gas are related by the equation

$$VP = k$$

where k is a constant. If the volume is decreasing by 5 cubic inches per second, what is the rate of change of the pressure when the volume is 1,000 cubic inches and the pressure is 40 pounds per square inch?

17. A 10 foot ladder is placed against a vertical wall. Suppose the bottom slides away from the wall at a constant rate of 3 feet per second. How fast is the top sliding down the wall (negative rate) when the bottom is 6 feet from the wall? [*Hint:* Use the Pythagorean theorem: $a^2 + b^2 = c^2$, where c is the length of the hypotenuse of a right triangle and a and b are the lengths of the two shorter sides.]

18. A weather balloon is rising vertically at the rate of 5 meters per second. An observer is standing on the ground 300 meters from the point where the balloon was released. At what rate is the distance between the observer and the balloon changing when the balloon is 400 meters high?

C

19. A street light is on top of a 20 foot pole. A person who is 5 feet tall walks away from the pole at the rate of 5 feet per second. At what rate is the tip of the person's shadow moving away from the pole when he is 20 feet from the pole?

20. Refer to Problem 19. At what rate is the person's shadow growing when he is 20 feet from the pole?

21. Helium is pumped into a spherical balloon at a constant rate of 4 cubic feet per second. How fast is the radius increasing after 1 minute? After 2 minutes? Is there any time at which the radius is increasing at a rate of 100 feet per second? Explain.

22. A point is moving along the x axis at a constant rate of 5 units per second. At which point is its distance from (0, 1) increasing at a rate of 2 units per second? At 4 units per second? At 5 units per second? At 10 units per second? Explain.

23. C A point is moving on the graph of $y = e^x + x + 1$ in such a way that its x coordinate is always increasing at a rate of 3 units per second. How fast is the y coordinate changing when the point crosses the x axis?

24. C A point is moving on the graph of $x^3 + y^2 = 1$ in such a way that its y coordinate is always increasing at a rate of 2 units per second. At which point(s) is the x coordinate increasing at a rate of 1 unit per second?

APPLICATIONS

Business & Economics

25. *Cost, revenue, and profit rates.* Suppose that for a company manufacturing calculators, the cost, revenue, and profit equations are given by

$$C = 90{,}000 + 30x \qquad R = 300x - \frac{x^2}{30}$$

$$P = R - C$$

where the production output in 1 week is x calculators. If production is increasing at a rate of 500 calculators per week when production output is 6,000 calculators, find the rate of increase (decrease) in:

(A) Cost (B) Revenue (C) Profit

26. *Cost, revenue, and profit rates.* Repeat Problem 25 for

$$C = 72{,}000 + 60x \qquad R = 200x - \frac{x^2}{30}$$

$$P = R - C$$

where production is increasing at a rate of 500 calculators per week at a production level of 1,500 calculators.

27. *Advertising.* A retail store estimates that weekly sales s and weekly advertising costs x (both in dollars) are related by

$$s = 60{,}000 - 40{,}000e^{-0.0005x}$$

The current weekly advertising costs are \$2,000, and these costs are increasing at the rate of \$300 per week. Find the current rate of change of sales.

28. *Advertising.* Repeat Problem 27 for

$$s = 50{,}000 - 20{,}000e^{-0.0004x}$$

29. *Price–demand.* The price p (in dollars) and demand x for a product are related by

$$2x^2 + 5xp + 50p^2 = 80{,}000$$

(A) If the price is increasing at a rate of \$2 per month when the price is \$30, find the rate of change of the demand.

(B) If the demand is decreasing at a rate of 6 units per month when the demand is 150 units, find the rate of change of the price.

30. *Price–demand.* Repeat Problem 29 for

$$x^2 + 2xp + 25p^2 = 74{,}500$$

Life Sciences

31. *Pollution.* An oil tanker aground on a reef is leaking oil that forms a circular oil slick about 0.1 foot thick (see the figure). To estimate the rate dV/dt (in cubic feet per minute) at which the oil is leaking from the tanker, it was found that the radius of the slick was increasing at 0.32 foot per minute ($dR/dt = 0.32$) when the radius R was 500 feet. Find dV/dt, using $\pi \approx 3.14$.

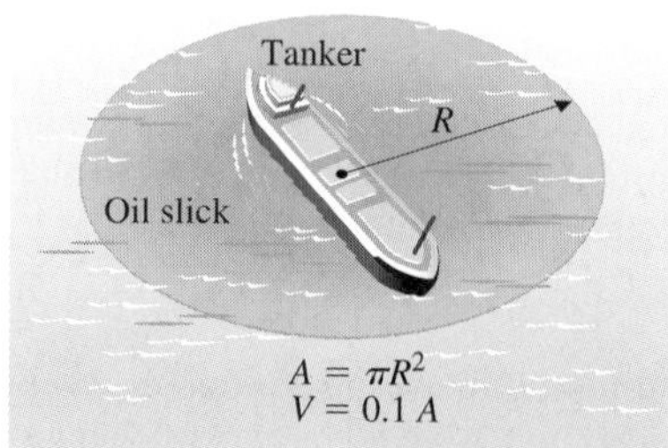

Figure for 31

Social Sciences

32. *Learning.* A person who is new on an assembly line performs an operation in T minutes after x performances of the operation, as given by

$$T = 6\left(1 + \frac{1}{\sqrt{x}}\right)$$

If $dx/dt = 6$ operations per hour, where t is time in hours, find dT/dt after 36 performances of the operation.

Important Terms and Symbols

10-1 *The Constant e and Continuous Compound Interest.* Definition of e; continuous compound interest

$$A = Pe^{rt}$$

10-2 *Derivatives of Logarithmic and Exponential Functions.* Derivative formulas for the natural logarithmic and exponential functions; graph properties of $y = \ln x$ and $y = e^x$

10-3 *Chain Rule: General Form.* Composite functions; chain rule; general derivative formulas; derivative formulas for $y = \log_b x$ and $y = b^x$

10-4 *Implicit Differentiation.* Special function notation; function explicitly defined; function implicitly defined; implicit differentiation

$$y = f(x); \quad y = y(x); \quad F(x, y) = 0; \quad y'|_{(a,\, b)}$$

10-5 *Related Rates.* Related rates

$$x = x(t); \quad y = y(t)$$

Additional Rules of Differentiation

$$\frac{d}{dx}\ln x = \frac{1}{x} \qquad \frac{d}{dx}e^x = e^x \qquad \frac{d}{dx}b^x = \frac{d}{dx}e^{x\ln b} = e^{x\ln b}\ln b = b^x \ln b$$

$$\frac{d}{dx}\ln[f(x)] = \frac{1}{f(x)}f'(x) \qquad \frac{d}{dx}e^{f(x)} = e^{f(x)}f'(x) \qquad \frac{d}{dx}[f(x)]^n = n[f(x)]^{n-1}f'(x)$$

$$\frac{d}{dx}\log_b x = \frac{1}{\ln b}\frac{d}{dx}\ln x = \frac{1}{\ln b}\left(\frac{1}{x}\right) \qquad \frac{dy}{dx} = \frac{dy}{du}\frac{du}{dx}, \quad \frac{dy}{dx} = \frac{dy}{dw}\frac{dw}{du}\frac{du}{dx}, \quad \text{and so on}$$

Review Exercise

Work through all the problems in this chapter review and check your answers in the back of the book. Answers to all review problems are there along with section numbers in italics to indicate where each type of problem is discussed. Where weaknesses show up, review appropriate sections in the text.

A

1. Use a calculator to evaluate $A = 2{,}000e^{0.09t}$ to the nearest cent for $t = 5, 10$, and 20.

Find the indicated derivatives in Problems 2–4.

2. $\dfrac{d}{dx}(2 \ln x + 3e^x)$

3. $\dfrac{d}{dx} e^{2x-3}$

4. y' for $y = \ln(2x + 7)$

5. Let $y = \ln u$ and $u = 3 + e^x$.
   (A) Express y in terms of x.
   (B) Use the chain rule to find dy/dx, and then express dy/dx in terms of x.

6. Find y' for $y = y(x)$ defined implicitly by the equation $2y^2 - 3x^3 - 5 = 0$, and evaluate at $(x, y) = (1, 2)$.

7. For $y = 3x^2 - 5$, where $x = x(t)$ and $y = y(t)$, find dy/dt if $dx/dt = 3$ when $x = 12$.

B

8. Graph $y = 100e^{-0.1x}$.

9. Use a calculator and a table of values to investigate

$$\lim_{n \to \infty}\left(1 + \frac{2}{n}\right)^n$$

   Do you think the limit exists? If so, what do you think it is?

Find the indicated derivatives in Problems 10–15.

10. $\dfrac{d}{dz}[(\ln z)^7 + \ln z^7]$

11. $\dfrac{d}{dx}(x^6 \ln x)$

12. $\dfrac{d}{dx}\dfrac{e^x}{x^6}$

13. y' for $y = \ln(2x^3 - 3x)$

14. $f'(x)$ for $f(x) = e^{x^3 - x^2}$

15. dy/dx for $y = e^{-2x} \ln 5x$

16. Find the equation of the line tangent to the graph of $y = f(x) = 1 + e^{-x}$ at $x = 0$. At $x = -1$.

17. Find y' for $y = y(x)$ defined implicitly by the equation $x^2 - 3xy + 4y^2 = 23$, and find the slope of the graph at $(-1, 2)$.

18. Find x' for $x = x(t)$ defined implicitly by $x^3 - 2t^2x + 8 = 0$, and evaluate at $(t, x) = (-2, 2)$.

19. Find y' for $y = y(x)$ defined implicitly by $x - y^2 = e^y$, and evaluate at $(1, 0)$.

20. Find y' for $y = y(x)$ defined implicitly by $\ln y = x^2 - y^2$, and evaluate at $(1, 1)$.

21. A point is moving on the graph of $y^2 - 4x^2 = 12$ so that its x coordinate is decreasing at 2 units per second when $(x, y) = (1, 4)$. Find the rate of change of the y coordinate.

22. A 17 foot ladder is placed against a wall. If the foot of the ladder is pushed toward the wall at 0.5 foot per second, how fast is the top rising when the foot of the ladder is 8 feet from the wall?

23. Water from a water heater is leaking onto a floor. A circular pool is created whose area is increasing at the rate of 24 square inches per minute. How fast is the radius R of the pool increasing when the radius is 12 inches? $[A = \pi R^2]$

C

In Problems 24–27, find the absolute maximum value of $f(x)$ for $x > 0$.

24. $f(x) = 11x - 2x \ln x$

25. $f(x) = 10xe^{-2x}$

26. $f(x) = 3x - x^2 + e^{-x}$

27. $f(x) = \dfrac{\ln x}{e^x}$

In Problem 28 and 29, apply the graphing strategy discussed in Section 9-4 to f, summarize the pertinent information, and sketch the graph.

28. $f(x) = 5 - 5e^{-x}$

29. $f(x) = x^3 \ln x$

30. Let $y = w^3$, $w = \ln u$, and $u = 4 - e^x$.
   (A) Express y in terms of x.
   (B) Use the chain rule to find dy/dx, and then express dy/dx in terms of x.

Find the indicated derivatives in Problems 31–33.

31. y' for $y = 5^{x^2-1}$

32. $\dfrac{d}{dx}\log_5(x^2 - x)$

33. $\dfrac{d}{dx}\sqrt{\ln(x^2 + x)}$

34. Find y' for $y = y(x)$ defined implicitly by the equation $e^{xy} = x^2 + y + 1$, and evaluate at $(0, 0)$.

35. A rock is thrown into a still pond and causes a circular ripple. Suppose the radius is increasing at a constant rate of 3 feet per second. Show that the area does not increase at a constant rate. When is the rate of increase of the area the smallest? The largest? Explain.

36. A point moves along the graph of $y = x^3$ in such a way that its y coordinate is increasing at a constant rate of 5 units per second. Does the x coordinate ever increase at a faster rate than the y coordinate? Explain.

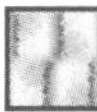

APPLICATIONS

Business & Economics

37. *Doubling time.* How long will it take money to double if it is invested at 5% interest compounded.
(A) Annually? (B) Continuously?

38. *Continuous compound interest.* If $100 is invested at 10% interest compounded continuously, the amount (in dollars) at the end of t years is given by
$$A = 100e^{0.1t}$$
Find $A'(t)$, $A'(1)$, and $A'(10)$.

39. *Marginal analysis.* The price–demand equation for 14 cubic foot refrigerators at an appliance store is
$$p(x) = 1{,}000e^{-0.02x}$$
where x is the monthly demand and p is the price in dollars. Find the marginal revenue equation.

40. *Maximum revenue.* For the price–demand equation in Problem 39, find the production level and price per unit that produces the maximum revenue. What is the maximum revenue?

41. *Maximum revenue.* Graph the revenue function from Problems 39 and 40 for $0 \le x \le 100$.

42. *Maximum profit.* Refer to Problem 39. If the refrigerators cost the store $220 each, find the price (to the nearest cent) that maximizes the profit. What is the maximum profit (to the nearest dollar)?

43. *Maximum profit.* The data in the table show the daily demand x for cream puffs at a state fair at various price levels p. If it costs $1 to make a cream puff, use logarithmic regression ($p = a + b \ln x$) to find the price (to the nearest cent) that maximizes profit.

DEMAND x	PRICE PER CREAM PUFF ($) p
3,125	1.99
3,879	1.89
5,263	1.79
5,792	1.69
6,748	1.59
8,120	1.49

44. *Minimum average cost.* The cost of producing x units of a product is given by
$$C(x) = 200 + 50x - 50 \ln x \qquad x \ge 1$$
Find the minimum average cost.

45. *Demand equation.* Given the demand equation
$$x = \sqrt{5{,}000 - 2p^3}$$
find the rate of change of p with respect to x by implicit differentiation (x is the number of items that can be sold at a price of $\$p$ per item).

46. *Rate of change of revenue.* A company is manufacturing a new video game and can sell all it manufactures. The revenue (in dollars) is given by
$$R = 36x - \frac{x^2}{20}$$
where the production output in 1 day is x games. If production is increasing at 10 games per day when production is 250 games per day, find the rate of increase in revenue.

Life Sciences

47. *Drug concentration.* The concentration of a drug in the bloodstream t hours after injection is given approximately by
$$C(t) = 5e^{-0.3t}$$
where $C(t)$ is concentration in milligrams per milliliter. What is the rate of change of concentration after 1 hour? After 5 hours?

48. *Wound healing.* A circular wound on an arm is healing at the rate of 45 square millimeters per day (the area of the wound is decreasing at this rate). How fast is the radius R of the wound decreasing when $R = 15$ millimeters? $[A = \pi R^2]$

Social Sciences

49. *Psychology—learning.* In a computer assembly plant, a new employee, on the average, is able to assemble
$$N(t) = 10(1 - e^{-0.4t})$$

units after t days of on-the-job-training.

(A) What is the rate of learning after 1 day? After 5 days?

(B) Graph N for $0 \leq t \leq 10$.

50. *Learning.* A new worker on the production line performs an operation in T minutes after x performances of the operation, as given by

$$T = 2\left(1 + \frac{1}{x^{3/2}}\right)$$

If, after performing the operation 9 times, the rate of improvement is $dx/dt = 3$ operations per hour, find the rate of improvement in time dT/dt in performing each operation.

Group Activity 1 Elasticity of Demand

We normally expect that by lowering the price of a commodity we will increase demand, and that by raising the price we will decrease demand. To measure the sensitivity of demand to changes in price, we define the **elasticity of demand,** denoted by $E(p)$, to be the limit of the relative change in demand x divided by the relative change in price p. Therefore, if $x = f(p)$ is the price–demand equation, then

$$\begin{aligned} E(p) &= \lim_{h\to 0} \frac{\dfrac{f(p+h) - f(p)}{f(p)}}{\dfrac{h}{p}} \\ &= \lim_{h\to 0}\left[\frac{p}{f(p)} \cdot \frac{f(p+h) - f(p)}{h}\right] \\ &= \frac{p}{f(p)} f'(p) \end{aligned}$$

Since an increase in price normally produces a decrease in demand, we expect $E(p)$ to be negative. If $E(p) = -1$, a 10% increase in price produces a 10% decrease in demand, and we say that demand has **unit elasticity.** If $E(p) = -4$, a 10% increase in price produces a 40% decrease in demand; we express such sensitivity of demand to price by saying that demand is **elastic** whenever $E(p) < -1$. If $E(p) = -0.25$, a 10% increase in price produces only a 2.5% decrease in demand; we express such insensitivity to price by saying that demand is **inelastic** whenever $E(p) > -1$.

(A) Consider the price–demand equation

$$x = f(p) = 10{,}000 - 500p \qquad 0 \leq p \leq 20$$

Calculate the elasticity of demand at $p = 4, 7, 10, 13$, and 16, and describe each case as elastic, inelastic, or of unit elasticity.

(B) Graph the elasticity of demand $E(p)$ and the revenue $R(p)$ as functions of price, and graph the line $y = -1$. Use the graphs to explain the connection between elasticity and the increasing/decreasing behavior of the revenue function.

(C) Repeat part (B) for the price–demand equation

$$x = f(p) = \sqrt{144 - 2p} \qquad 0 \leq p \leq 72$$

(D) Suppose that $x = g(p)$ is any price–demand equation for which $g(p)$ is differentiable. Let the associated revenue function be $R(p) = xp = g(p)p$. Show that $R'(p) = g(p)[E(p) + 1]$. Explain why revenue increases with price if and only if demand is inelastic.

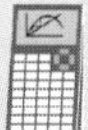

Group Activity 2 Point of Diminishing Returns*

Table 1 shows the total number of copies of a new spreadsheet for personal computers that have been sold x months after the product has been introduced. The **point of diminishing returns** occurs at the value of x where the rate of change of sales assumes its maximum value. In Section 9-3, we saw that a cubic polynomial model can be used to describe this situation.

Table 1

MONTH x	TOTAL SALES (THOUSANDS OF COPIES)
4	71
8	182
12	305
16	405
20	450

(A) Find a cubic regression model $P(x)$ for the data in Table 1, and find the point of diminishing returns (correct to two decimal places).

The cubic polynomial model provides a good description of the data for $4 \leq x \leq 20$, but not for $x > 20$. A more sophisticated model that can be used to predict the behavior for $x > 20$ is provided by a **logistic growth function** of the form

$$L(x) = \frac{c}{1 + ae^{-bx}} \tag{1}$$

(B) Find a logistic regression model $L(x)$ for the data in Table 1, and find the point of diminishing returns (correct to two decimal places) for this model.

(C) Graph $P(x)$ and $L(x)$ for Xmin = 4, Xmax = 20, Ymin = 0, and Ymax = 500. Are there any significant differences in the descriptions of the data provided by these two functions?

(D) Repeat part (C) with Xmax increased to 30.

(E) Find $\lim_{x\to\infty} L(x)$ (correct to the nearest integer). The value of this limit is called the **carrying capacity** of the model. Compare the carrying capacity with the value of $L(x)$ at the point of diminishing returns found in part (B).

(F) Sales data for a new word processor are given in Table 2. Find a logistic regression model for the data, the point of diminishing returns (correct to two decimal places), and the carrying capacity (correct to the nearest integer).

(G) For any function of the form in (1), show that the carrying capacity is c, the point of diminishing returns is $(\ln a)/b$, and the value of the function at the point of diminishing returns is $c/2$.

Table 2

MONTH x	TOTAL SALES (THOUSANDS OF COPIES)
4	151
8	204
12	327
16	480
20	530

*This group activity requires a graphing utility that supports logistic regression.

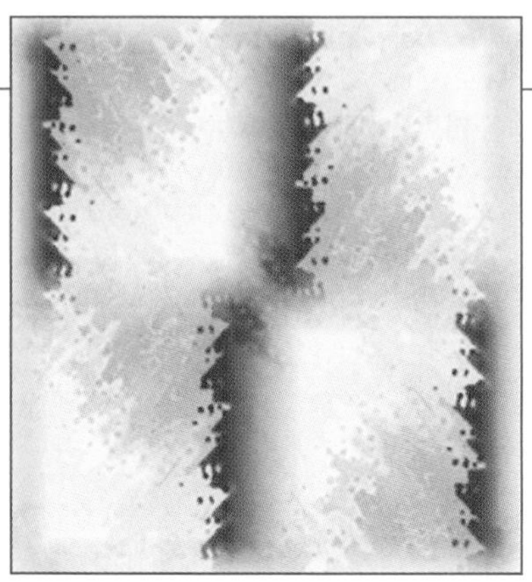

CHAPTER 11

INTEGRATION

INTRODUCTION

The last three chapters dealt with differential calculus. We now begin the development of the second main part of calculus, called *integral calculus.* Two types of integrals will be introduced, the *indefinite integral* and the *definite integral,* each quite different from the other. But through the remarkable *fundamental theorem of calculus,* we will show that not only are the two integral forms intimately related, but both are intimately related to differentiation.

SECTION 11-1 Antiderivatives and Indefinite Integrals

- Antiderivatives
- A Geometric–Numeric Look at Antiderivatives
- Antiderivatives and Indefinite Integrals: Algebraic Forms
- Antiderivatives and Indefinite Integrals: Exponential and Logarithmic Forms
- Applications

Many operations in mathematics have reverses—compare addition and subtraction, multiplication and division, and powers and roots. We now know how to find the derivatives of many functions. The reverse operation, *antidifferentiation* (the reconstruction of a function from its derivative) will receive our attention in this and the next two sections. A function F is an **antiderivative** of a function f if $F'(x) = f(x)$. We will look at the problem geometrically, numerically, and algebraically, and we will use an algebraic approach to develop special antiderivative formulas in much the same way as we developed derivative formulas.

ANTIDERIVATIVES

Explore–Discuss 1

(A) Find three antiderivatives of $2x$.
(B) How many antiderivatives of $2x$ exist, and how are they related to each other?
(C) What notation would you use to represent all antiderivatives of $2x$?

What is an antiderivative of x^2?

$$\frac{x^3}{3} \quad \text{is an antiderivative of} \quad x^2$$

since

$$\frac{d}{dx}\left(\frac{x^3}{3}\right) = x^2$$

Note also that

$$\frac{d}{dx}\left(\frac{x^3}{3} + 2\right) = x^2 \qquad \frac{d}{dx}\left(\frac{x^3}{3} - \pi\right) = x^2 \qquad \frac{d}{dx}\left(\frac{x^3}{3} + \sqrt{5}\right) = x^2$$

Hence,

$$\frac{x^3}{3} + 2 \qquad \frac{x^3}{3} - \pi \qquad \frac{x^3}{3} + \sqrt{5}$$

are also antiderivatives of x^2, since each has x^2 as a derivative. In fact, it appears that

$$\frac{x^3}{3} + C \qquad \text{for any real number } C$$

is an antiderivative of x^2, since

$$\frac{d}{dx}\left(\frac{x^3}{3} + C\right) = x^2$$

Thus, antidifferentiation of a given function does not lead to a unique function, but to a whole family of functions.

Does the expression

$$\frac{x^3}{3} + C \qquad \text{with } C \text{ any real number}$$

include all antiderivatives of x^2? Theorem 1 (which we state without proof) indicates that the answer is yes.

Theorem 1 ON ANTIDERIVATIVES

If the derivatives of two functions are equal on an open interval (a, b), then the functions differ by at most a constant. Symbolically: If F and G are differentiable functions on the interval (a, b) and $F'(x) = G'(x)$ for all x in (a, b), then $F(x) = G(x) + k$ for some constant k.

A GEOMETRIC–NUMERIC LOOK AT ANTIDERIVATIVES

Explore–Discuss 2

Given $f'(x) = 1$:

(A) What does $f'(x) = 1$ tell us about graphs of any antiderivative function f?

(B) Sketch, in the same coordinate system, graphs of three different antiderivatives $y = f(x)$ satisfying $f(0) = -2, f(0) = 0$, and $f(0) = 2$. What are the geometric relationships among the graphs?

Table 1

PROPERTIES OF THE GRAPH OF $y = f(x)$ DETERMINED FROM THE GRAPH OF $y = f'(x)$

$f'(x)$	GRAPH OF $y = f(x)$
Positive	Increasing
Negative	Decreasing
Increasing	Concave upward
Decreasing	Concave downward

We can visualize certain characteristics of an antiderivative by using slopes. If we are given the graph of a derivative function $y = f'(x)$, we can sketch a possible graph of an antiderivative function f showing its characteristic shape. Recall the properties listed in Table 1.

Example 1 **Analyzing the Graph of $y = f(x)$ Given the Graph of $y = f'(x)$**

Given the graph of $y = f'(x)$ in Figure 1, sketch possible graphs of three antiderivative functions f such that $f(0) = -1, f(0) = 0$, and $f(0) = 2$, all using the same set of coordinate axes.

SOLUTION Table 2 summarizes the information obtained from the graph of $y = f'(x)$.

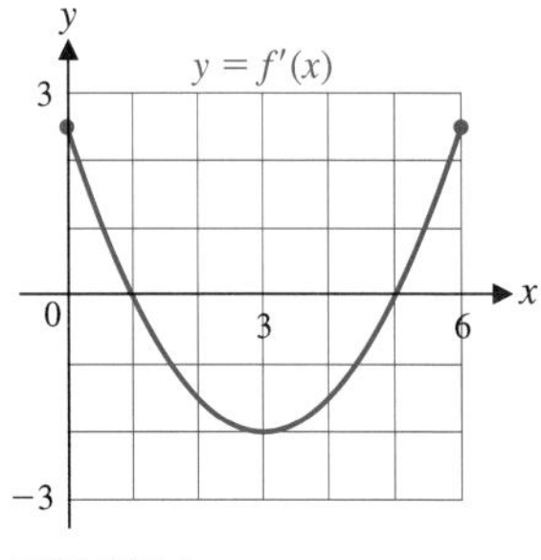

FIGURE 1

Table 2

x	$f'(x)$	GRAPH OF $y = f(x)$
$0 < x < 1$	Positive and decreasing	Increasing and concave downward
$x = 1$	x intercept	Local maximum
$1 < x < 3$	Negative and decreasing	Decreasing and concave downward
$x = 3$	Local minimum	Inflection point
$3 < x < 5$	Negative and increasing	Decreasing and concave upward
$x = 5$	x intercept	Local minimum
$5 < x < 6$	Positive and increasing	Increasing and concave upward

If $f(x)$ is one antiderivative of $f'(x)$, since all antiderivatives are given by $y = f(x) + C$, **the graph of any antiderivative function can be obtained from another by a vertical translation.** We know where each of the three graphs of $y = f(x)$ start and the general shape over the interval $(0, 6)$.

Figure 2 shows possible graphs of three antiderivative functions having the required characteristics.

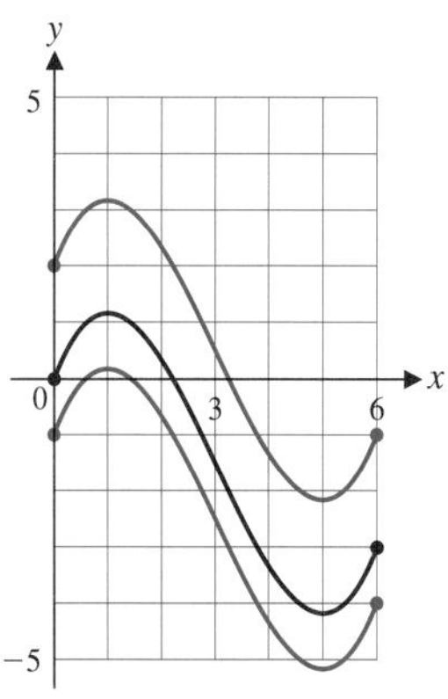

FIGURE 2

Matched Problem 1 Given the graph of $y = f'(x)$ in Figure 3, sketch possible graphs of three antiderivative functions f such that $f(0) = -1, f(0) = 0$, and $f(0) = 1$, all using the same set of coordinate axes.

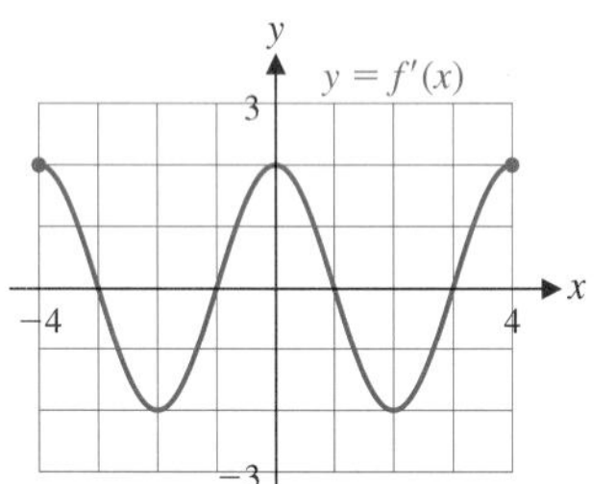

FIGURE 3

ANTIDERIVATIVES AND INDEFINITE INTEGRALS: ALGEBRAIC FORMS

Theorem 1 states that if the derivatives of two functions are equal, then the functions differ by at most a constant. We use the symbol

$$\int f(x)\,dx$$

called the **indefinite integral,** to represent the family of all antiderivatives of $f(x)$, and write

$$\int f(x)\,dx = F(x) + C \qquad \text{if} \qquad F'(x) = f(x)$$

The symbol $\int$ is called an **integral sign,** and the function $f(x)$ is called the **integrand.** The symbol dx indicates that the antidifferentiation is performed with respect to the variable x. (We will have more to say about the symbols $\int$ and dx later in this chapter.) The arbitrary constant C is called the **constant of integration.** Referring to the preceding discussion, we can write

$$\int x^2\,dx = \frac{x^3}{3} + C \qquad \text{since} \qquad \frac{d}{dx}\left(\frac{x^3}{3} + C\right) = x^2$$

Of course, variables other than x can be used in indefinite integrals. For example,

$$\int t^2\,dt = \frac{t^3}{3} + C \qquad \text{since} \qquad \frac{d}{dt}\left(\frac{t^3}{3} + C\right) = t^2$$

or

$$\int u^2\,du = \frac{u^3}{3} + C \qquad \text{since} \qquad \frac{d}{du}\left(\frac{u^3}{3} + C\right) = u^2$$

The fact that indefinite integration and differentiation are reverse operations, except for the addition of the constant of integration, can be expressed symbolically as

$$\frac{d}{dx}\left[\int f(x)\,dx\right] = f(x)$$ The derivative of the indefinite integral of $f(x)$ is $f(x)$.

and

$$\int F'(x)\,dx = F(x) + C$$ The indefinite integral of the derivative of $F(x)$ is $F(x) + C$.

Just as with differentiation, we can develop formulas and special properties that will enable us to find indefinite integrals of many frequently encountered functions. To start, we list some formulas that can be established using the definitions of antiderivative and indefinite integral, and the properties of derivatives considered in Chapter 8.

Indefinite Integral Formulas and Properties

For k and C constants:

1. $\int k\,dx = kx + C$
2. $\displaystyle\int x^n\,dx = \frac{x^{n+1}}{n+1} + C, \quad n \neq -1$
3. $\int kf(x)\,dx = k\int f(x)\,dx$
4. $\int [f(x) \pm g(x)]\,dx = \int f(x)\,dx \pm \int g(x)\,dx$

We will establish formula 2 and property 3 here (the others may be shown to be true in a similar manner). To establish formula 2, we simply differentiate the right side to obtain the integrand on the left side. Thus,

$$\frac{d}{dx}\left(\frac{x^{n+1}}{n+1} + C\right) = \frac{(n+1)x^n}{n+1} + 0 = x^n \qquad n \neq -1$$

(Notice that formula 2 cannot be used when $n = -1$; that is, when the integrand is x^{-1} or $1/x$. The indefinite integral of $x^{-1} = 1/x$ will be considered later in this section.)

To establish property 3, let F be a function such that $F'(x) = f(x)$. Then

$$k\int f(x)\,dx = k\int F'(x)\,dx = k[F(x) + C_1] = kF(x) + kC_1$$

and since $[kF(x)]' = kF'(x) = kf(x)$, we have

$$\int kf(x)\,dx = \int kF'(x)\,dx = kF(x) + C_2$$

But $kF(x) + kC_1$ and $kF(x) + C_2$ describe the same set of functions, since C_1 and C_2 are arbitrary real numbers. Thus, property 3 is established.

CAUTION

It is important to remember that property 3 states that **a constant factor can be moved across an integral sign. A variable factor cannot be moved across an integral sign:**

CONSTANT FACTOR	*VARIABLE FACTOR*
$\int 5x^{1/2}\,dx = 5\int x^{1/2}\,dx$	$\int xx^{1/2}\,dx \neq x\int x^{1/2}\,dx$

Now let us put the formulas and properties to use.

Example 2 **Using Indefinite Integral Properties and Formulas**

(A) $\int 5\,dx = 5x + C$

(B) $\int x^4\,dx = \frac{x^{4+1}}{4+1} + C = \frac{x^5}{5} + C$

(C) $\int 5t^7\,dt = 5\int t^7\,dt = 5\frac{t^8}{8} + C = \frac{5}{8}t^8 + C$

(D)

$$\begin{aligned}\int(4x^3 + 2x - 1)\,dx &= \int 4x^3\,dx + \int 2x\,dx - \int dx \\ &= 4\int x^3\,dx + 2\int x\,dx - \int dx \\ &= \frac{4x^4}{4} + \frac{2x^2}{2} - x + C \\ &= x^4 + x^2 - x + C\end{aligned}$$

Property 4 can be extended to the sum and difference of an arbitrary number of functions.

(E) $\int \frac{3\,dx}{x^2} = \int 3x^{-2}\,dx = \frac{3x^{-2+1}}{-2+1} + C = -3x^{-1} + C$

(F)

$$\begin{aligned}\int 5\sqrt[3]{u^2}\,du &= 5\int u^{2/3}\,du = 5\frac{u^{(2/3)+1}}{\frac{2}{3}+1} + C \\ &= 5\frac{u^{5/3}}{\frac{5}{3}} + C = 3u^{5/3} + C\end{aligned}$$

To check any of the results in Example 2, we differentiate the final result to obtain the integrand in the original indefinite integral. When you evaluate an indefinite integral, do not forget to include the arbitrary constant C.

Matched Problem 2 Find each of the following:

(A) $\int dx$ (B) $\int 3t^4\,dt$ (C) $\int (2x^5 - 3x^2 + 1)\,dx$

(D) $\int 4\sqrt[5]{w^3}\,dw$ (E) $\int \left(2x^{2/3} - \frac{3}{x^4}\right)dx$

Example 3 **Using Indefinite Integral Properties and Formulas**

(A)

$$\begin{aligned}\int \frac{x^3 - 3}{x^2}\,dx &= \int\left(\frac{x^3}{x^2} - \frac{3}{x^2}\right)dx \\ &= \int (x - 3x^{-2})\,dx \\ &= \int x\,dx - 3\int x^{-2}\,dx \\ &= \frac{x^{1+1}}{1+1} - 3\frac{x^{-2+1}}{-2+1} + C \\ &= \tfrac{1}{2}x^2 + 3x^{-1} + C\end{aligned}$$

(B) $$\int\left(\frac{2}{\sqrt[3]{x}} - 6\sqrt{x}\right)dx = \int (2x^{-1/3} - 6x^{1/2})\,dx$$

$$= 2\int x^{-1/3}\,dx - 6\int x^{1/2}\,dx$$

$$= 2\frac{x^{(-1/3)+1}}{-\frac{1}{3}+1} - 6\frac{x^{(1/2)+1}}{\frac{1}{2}+1} + C$$

$$= 2\frac{x^{2/3}}{\frac{2}{3}} - 6\frac{x^{3/2}}{\frac{3}{2}} + C$$

$$= 3x^{2/3} - 4x^{3/2} + C$$

Matched Problem 3 Find each indefinite integral.

(A) $\displaystyle\int \frac{x^4 - 8x^3}{x^2}\,dx$ (B) $\displaystyle\int \left(8\sqrt[3]{x} - \frac{6}{\sqrt{x}}\right)dx$

Antiderivatives and Indefinite Integrals: Exponential and Logarithmic Forms

We now give indefinite integral formulas for e^x and $1/x$. (Recall that the form $x^{-1} = 1/x$ is not covered by formula 2, given earlier.)

Indefinite Integral Formulas

5. $\int e^x\,dx = e^x + C$

6. $\displaystyle\int \frac{1}{x}\,dx = \ln|x| + C, \quad x \neq 0$

Formula 5 follows immediately from the derivative formula for the exponential function discussed in the last chapter. Because of the absolute value, formula 6 does not follow directly from the derivative formula for the natural logarithm function. Let us show that

$$\frac{d}{dx}\ln|x| = \frac{1}{x} \qquad x \neq 0$$

We consider two cases, $x > 0$ and $x < 0$:

Case 1. $x > 0$:

$$\frac{d}{dx}\ln|x| = \frac{d}{dx}\ln x \qquad \text{Since } |x| = x \text{ for } x > 0$$

$$= \frac{1}{x}$$

Case 2. $x < 0$:

$$\frac{d}{dx}\ln|x| = \frac{d}{dx}\ln(-x) \qquad \text{Since } |x| = -x \text{ for } x < 0$$
$$= \frac{1}{-x}\frac{d}{dx}(-x)$$
$$= \frac{-1}{-x} = \frac{1}{x}$$

Thus,

$$\frac{d}{dx}\ln|x| = \frac{1}{x} \qquad x \neq 0$$

and hence,

$$\int \frac{1}{x}\,dx = \ln|x| + C \qquad x \neq 0$$

What about the indefinite integral of ln x? We postpone a discussion of $\int \ln x\,dx$ until Section 12-3, where we will be able to find it using a technique called *integration by parts.*

Example 4 **Exponential and Logarithmic Forms**

$$\int\left(2e^x + \frac{3}{x}\right)dx = 2\int e^x\,dx + 3\int \frac{1}{x}\,dx$$
$$= 2e^x + 3\ln|x| + C$$

Matched Problem 4 Find: $\int\left(\frac{5}{x} - 4e^x\right)dx$

[*Note:* More general forms of exponential and logarithmic functions are considered in Section 11-2.]

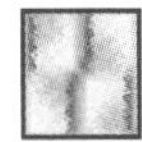

APPLICATIONS

Let us now consider some applications of the indefinite integral to see why we are interested in finding antiderivatives of functions.

Example 5 **Curves** Find the equation of the curve that passes through (2, 5) if its slope is given by $dy/dx = 2x$ at any point x.

SOLUTION We are interested in finding a function $y = f(x)$ such that

$$\frac{dy}{dx} = 2x \qquad (1)$$

and

$$y = 5 \quad \text{when} \quad x = 2 \qquad (2)$$

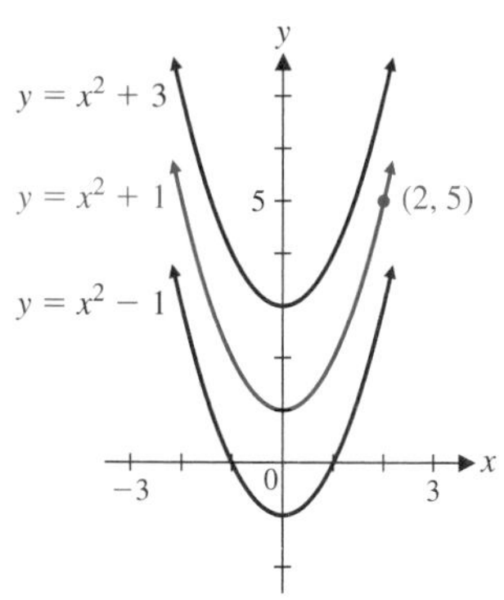

FIGURE 4
$y = x^2 + C$

If $dy/dx = 2x$, then

$$y = \int 2x\, dx$$
$$= x^2 + C \tag{3}$$

Since $y = 5$ when $x = 2$, we determine the *particular value of C* so that

$$5 = 2^2 + C$$

Thus, $C = 1$, and

$$y = x^2 + 1$$

is the *particular antiderivative* out of all those possible from (3) that satisfies both (1) and (2). See Figure 4.

Note how Example 5 differs from Example 1, where we reconstructed a possible graph of a function using the graph of its derivative. In Example 5 we actually found an equation for the antiderivative whose graph passes through (2, 5). Consequently, a precise graph of the antiderivative could be produced.

Explore–Discuss 3

Graph the derivative function $y = f'(x) = 2x$ in Example 5 and use the information from the graph to confirm the shape of the graphs of the antiderivative functions shown in Figure 4.

Matched Problem 5 Find the equation of the curve that passes through (2, 6) if its slope is given by $dy/dx = 3x^2$ at any point x.

In certain situations, it is easier to determine the rate at which something happens than how much of it has happened in a given length of time (for example, population growth rates, business growth rates, rate of healing of a wound, rates of learning or forgetting). If a rate function (derivative) is given and we know the value of the dependent variable for a given value of the independent variable, then—if the rate function is not too complicated—we can often find the original function by integration.

Example 6 **Cost Function** If the marginal cost of producing x units is given by

$$C'(x) = 0.3x^2 + 2x$$

and the fixed cost is \$2,000, find the cost function $C(x)$ and the cost of producing 20 units.

Solution Recall that marginal cost is the derivative of the cost function and that fixed cost is cost at a 0 production level. Thus, the mathematical problem is to find $C(x)$ given

$$C'(x) = 0.3x^2 + 2x \qquad C(0) = 2{,}000$$

We now find the indefinite integral of $0.3x^2 + 2x$ and determine the arbitrary integration constant using $C(0) = 2{,}000$:

$$C'(x) = 0.3x^2 + 2x$$
$$C(x) = \int (0.3x^2 + 2x)\, dx$$
$$= 0.1x^3 + x^2 + K$$

Since C represents the cost, we use K for the constant of integration.

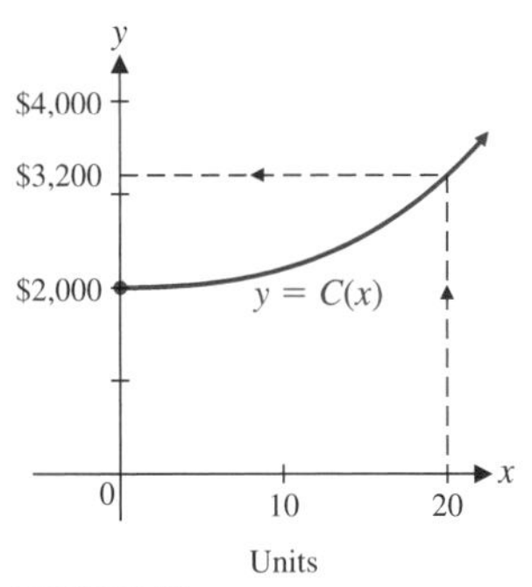

FIGURE 5

But

$$C(0) = (0.1)0^3 + 0^2 + K = 2{,}000$$

Thus, $K = 2{,}000$, and the particular cost function is

$$C(x) = 0.1x^3 + x^2 + 2{,}000$$

We now find $C(20)$, the cost of producing 20 units:

$$\begin{aligned} C(20) &= (0.1)20^3 + 20^2 + 2{,}000 \\ &= \$3{,}200 \end{aligned}$$

See Figure 5 for a geometric representation.

Matched Problem 6 Find the revenue function $R(x)$ when the marginal revenue is

$$R'(x) = 400 - 0.4x$$

and no revenue results at a 0 production level. What is the revenue at a production level of 1,000 units?

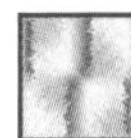

Example 7 **Advertising** An FM radio station is launching an aggressive advertising campaign in order to increase the number of daily listeners. The station currently has 27,000 daily listeners, and management expects the number of daily listeners, $S(t)$, to grow at the rate of

$$S'(t) = 60t^{1/2}$$

listeners per day, where t is the number of days since the campaign began. How long should the campaign last if the station wants the number of daily listeners to grow to 41,000?

Solution We must solve the equation $S(t) = 41{,}000$ for t, given that

$$S'(t) = 60t^{1/2} \quad \text{and} \quad S(0) = 27{,}000$$

First, we use integration to find $S(t)$:

$$\begin{aligned} S(t) &= \int 60t^{1/2}\,dt \\ &= 60\frac{t^{3/2}}{\frac{3}{2}} + C \\ &= 40t^{3/2} + C \end{aligned}$$

Since

$$S(0) = 40(0)^{3/2} + C = 27{,}000$$

we have $C = 27{,}000$, and

$$S(t) = 40t^{3/2} + 27{,}000$$

Now we solve the equation $S(t) = 41{,}000$ for t:

$$\begin{aligned} 40t^{3/2} + 27{,}000 &= 41{,}000 \\ 40t^{3/2} &= 14{,}000 \\ t^{3/2} &= 350 \\ t &= 350^{2/3} \qquad \text{Use a calculator.} \\ &= 49.664\ 419\ldots \end{aligned}$$

Thus, the advertising campaign should last approximately 50 days.

A graphing utility offers an alternative approach to solving $S(t) = 41{,}000$ in Example 7 (see Fig. 6).

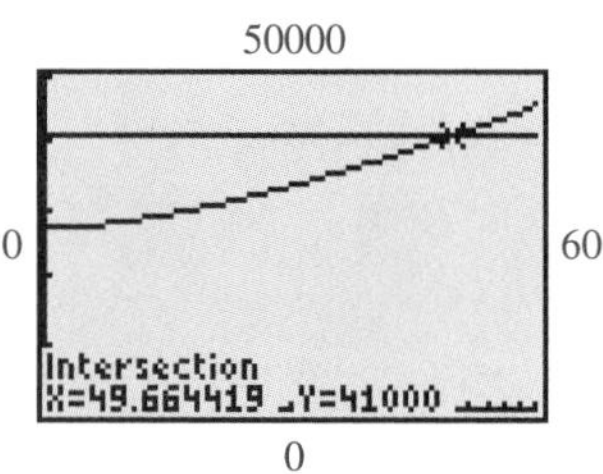

FIGURE 6
$y_1 = 40x^{1.5} + 27{,}000$
$y_2 = 41{,}000$

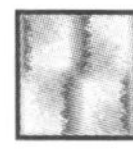

Matched Problem 7 The current monthly circulation of the magazine *Computing News* is 640,000 copies. Due to competition from a new magazine in the same field, the monthly circulation of *Computing News*, $C(t)$, is expected to decrease at the rate of

$$C'(t) = -6{,}000t^{1/3}$$

copies per month, where t is the time in months since the new magazine began publication. How long will it take for the circulation of *Computing News* to decrease to 460,000 copies per month?

CAUTION

1. $\displaystyle\int e^x\,dx \neq \frac{e^{x+1}}{x+1} + C$

   The power rule applies only to power functions of the form x^n where the exponent n is a real constant not equal to -1 and the base x is the variable. The function e^x is an exponential function with variable exponent x and constant base e. The correct form is

   $$\int e^x\,dx = e^x + C$$

2. $\displaystyle\int x(x^2 + 2)\,dx \neq \frac{x^2}{2}\left(\frac{x^3}{3} + 2x\right) + C$

   The integral of a product is not equal to the product of the integrals. The correct form is

   $$\int x(x^2 + 2)\,dx = \int (x^3 + 2x)\,dx = \frac{x^4}{4} + x^2 + C$$

3. Not all elementary functions have elementary antiderivatives. But finding one when it exists can markedly simplify the solution of some problems.

Answers to Matched Problems

1.

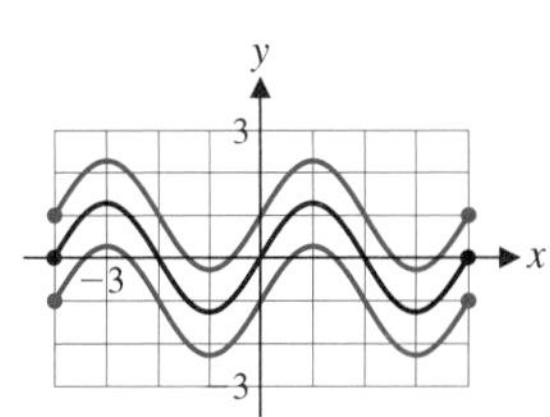

2. (A) $x + C$ (B) $\frac{3}{5}t^5 + C$
(C) $(x^6/3) - x^3 + x + C$
(D) $\frac{5}{2}w^{8/5} + C$
(E) $\frac{6}{5}x^{5/3} + x^{-3} + C$

3. (A) $\frac{1}{3}x^3 - 4x^2 + C$ (B) $6x^{4/3} - 12x^{1/2} + C$ 4. $5\ln|x| - 4e^x + C$
5. $y = x^3 - 2$ 6. $R(x) = 400x - 0.2x^2$; $R(1{,}000) = \$200{,}000$
7. $t = (40)^{3/4} \approx 16$ mo

EXERCISE 11-1

A *In Problems 1–18, find each indefinite integral. (Check by differentiating.)*

1. $\int 7\,dx$
2. $\int \pi\,dx$
3. $\int x^6\,dx$
4. $\int x^3\,dx$
5. $\int x^{-2}\,dx$
6. $\int x^{-5}\,dx$
7. $\int 8t^3\,dt$
8. $\int 10t^4\,dt$
9. $\int (2u + 1)\,du$
10. $\int (1 - 2u)\,du$
11. $\int (3x^2 + 2x - 5)\,dx$
12. $\int (2 + 4x - 6x^2)\,dx$
13. $\int (s^4 - 8s^5)\,ds$
14. $\int (t^5 + 6t^3)\,dt$
15. $\int 3e^t\,dt$
16. $\int 2e^t\,dt$
17. $\int 2z^{-1}\,dz$
18. $\int \frac{3}{s}\,ds$

Each graph in Problems 19–22 is the graph of a derivative function $y = f'(x)$. For each, sketch possible graphs of three antiderivative functions f satisfying $f(0) = -2$, $f(0) = 0$, and $f(0) = 2$, respectively, all on the same coordinate system.

19.

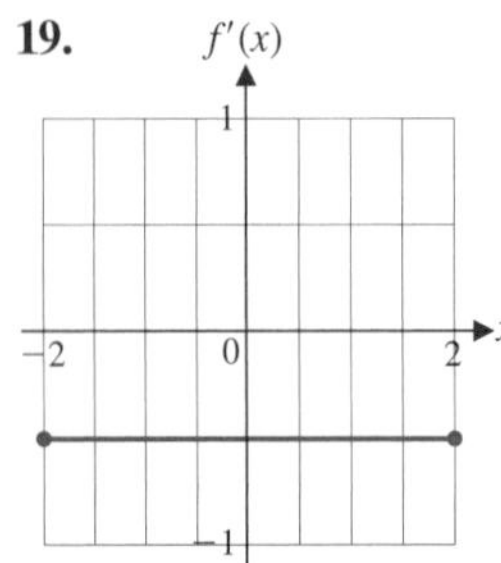

20.

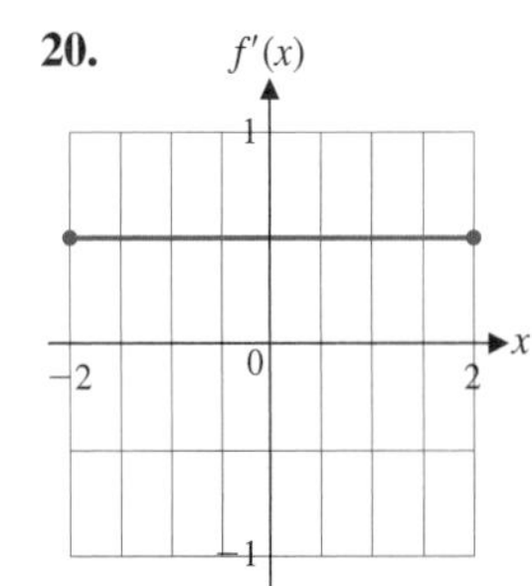

21.

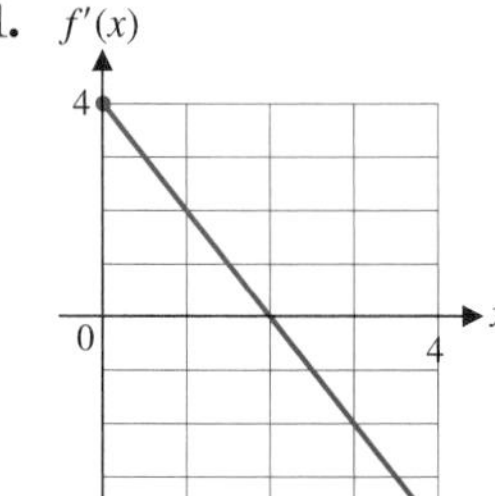

22.

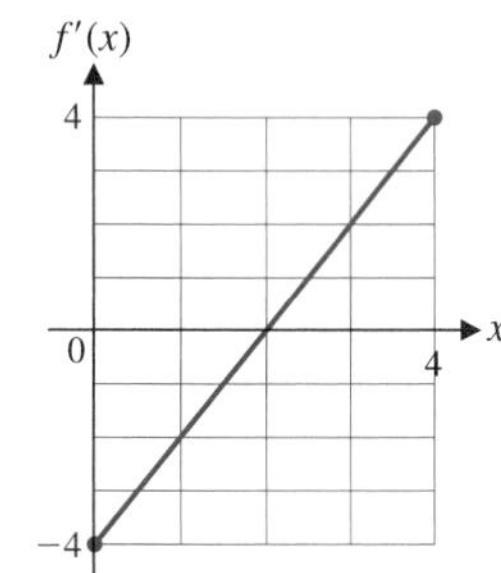

In Problems 23–32, find all the antiderivatives for each derivative.

23. $\dfrac{dy}{dx} = 200x^4$
24. $\dfrac{dx}{dt} = 42t^5$
25. $\dfrac{dP}{dx} = 24 - 6x$
26. $\dfrac{dy}{dx} = 3x^2 - 4x^3$
27. $\dfrac{dy}{du} = 2u^5 - 3u^2 - 1$
28. $\dfrac{dA}{dt} = 3 - 12t^3 - 9t^5$
29. $\dfrac{dy}{dx} = e^x + 3$
30. $\dfrac{dy}{dx} = x - e^x$
31. $\dfrac{dx}{dt} = 5t^{-1} + 1$
32. $\dfrac{du}{dv} = \dfrac{4}{v} + \dfrac{v}{4}$

B *In Problems 33 and 34, discuss the validity of each statement. If the statement is always true, explain why. If not, give a counterexample.*

33. (A) If n is an integer, then $x^{n+1}/(n + 1)$ is an antiderivative of x^n.
(B) The function $f(x) = \pi$ is an antiderivative of the function $g(x) = 0$.

34. (A) $\int \frac{d}{dx}(x^4 + x^2)\,dx = x^4 + x^2 + C$

(B) $\frac{d}{dx}\left(\int x^2\,dx\right) = x^2 + C$

Which sets of graphs in Problems 35–38 could be the graphs of three antiderivative functions from a family of antiderivative functions? Explain.

35.

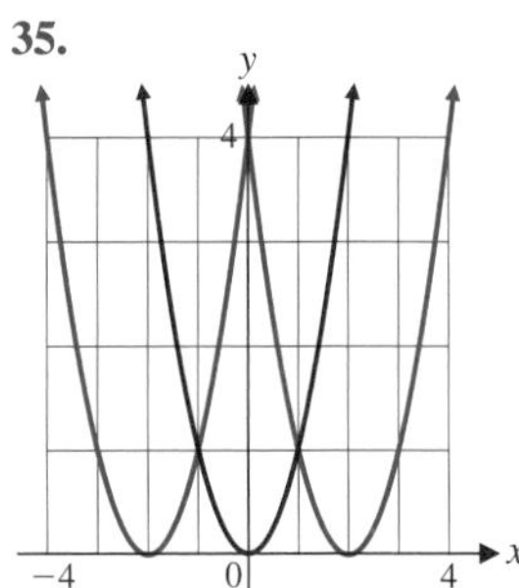

36.

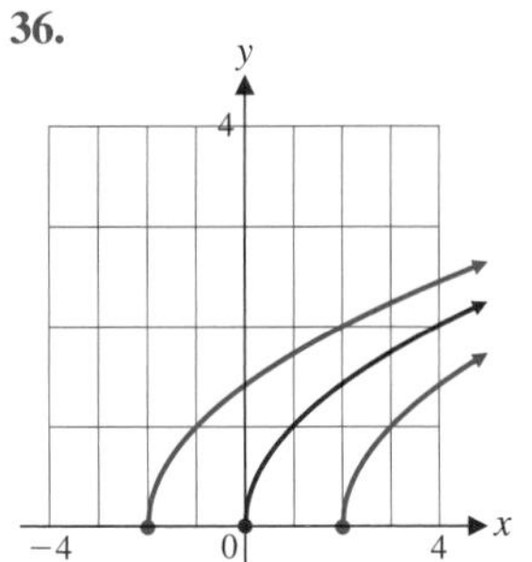

37.

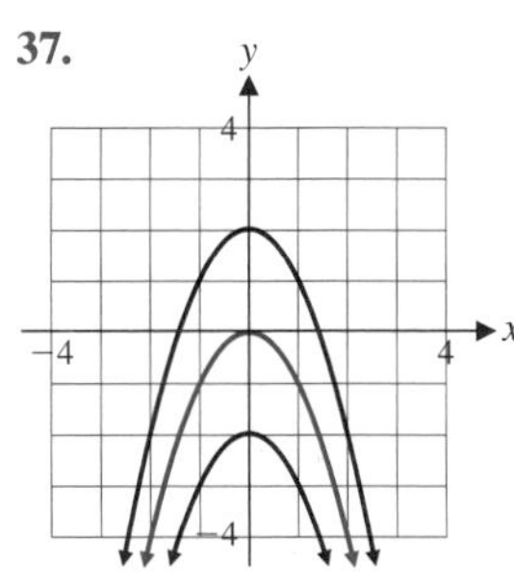

38.

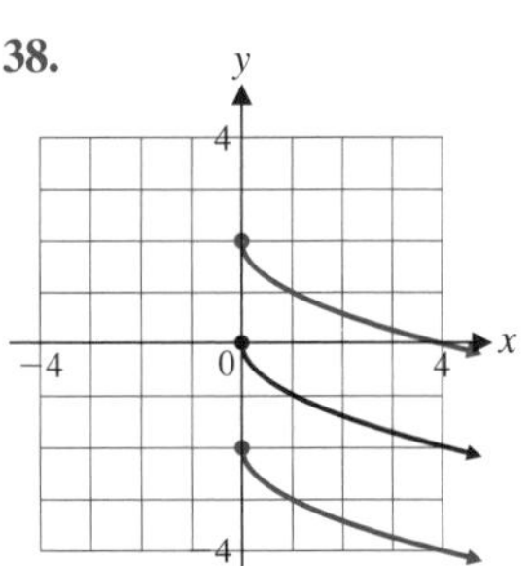

In Problems 39–60, find each indefinite integral. (Check by differentiation.)

39. $\int 6x^{1/2}\,dx$

40. $\int 8t^{1/3}\,dt$

41. $\int 8x^{-3}\,dx$

42. $\int 12u^{-4}\,du$

43. $\int \frac{du}{\sqrt{u}}$

44. $\int \frac{dt}{\sqrt[3]{t}}$

45. $\int \frac{dx}{4x^3}$

46. $\int \frac{6\,dm}{m^2}$

47. $\int \frac{du}{2u^5}$

48. $\int \frac{dy}{3y^4}$

49. $\int \left(3x^2 - \frac{2}{x^2}\right) dx$

50. $\int \left(4x^3 + \frac{2}{x^3}\right) dx$

51. $\int \left(10x^4 - \frac{8}{x^5} - 2\right) dx$

52. $\int \left(\frac{6}{x^4} - \frac{2}{x^3} + 1\right) dx$

53. $\int \left(3\sqrt{x} + \frac{2}{\sqrt{x}}\right) dx$

54. $\int \left(\frac{2}{\sqrt[3]{x}} - \sqrt[3]{x^2}\right) dx$

55. $\int \left(\sqrt[3]{x^2} - \frac{4}{x^3}\right) dx$

56. $\int \left(\frac{12}{x^5} - \frac{1}{\sqrt[3]{x^2}}\right) dx$

57. $\int \frac{e^x - 3x}{4}\,dx$

58. $\int \frac{e^x - 3x^2}{2}\,dx$

59. $\int (2z^{-3} + z^{-2} + z^{-1})\,dz$

60. $\int (3x^{-2} - x^{-1})\,dx$

Each graph in Problems 61–64 is the graph of a derivative function $y = f'(x)$. For each, sketch possible graphs of three antiderivative functions f satisfying $f(0) = -2$, $f(0) = 0$, and $f(0) = 2$, respectively, all on the same coordinate system.

61.

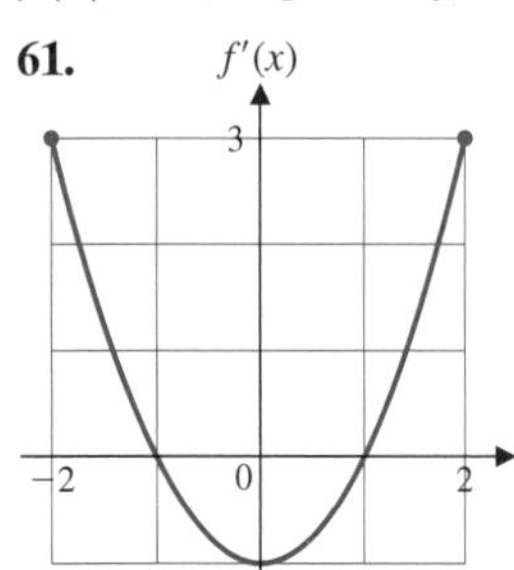

62.

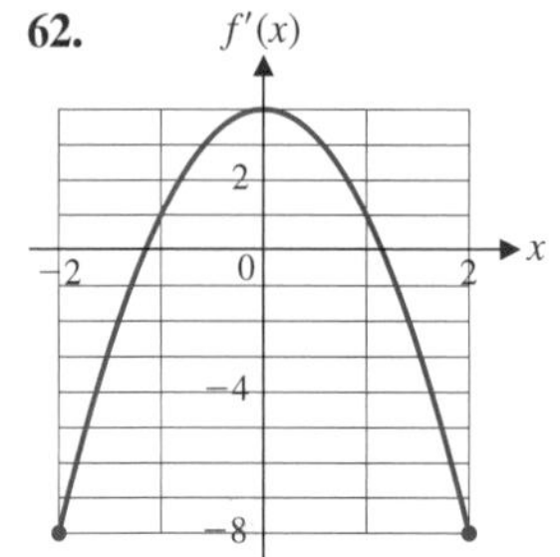

63.

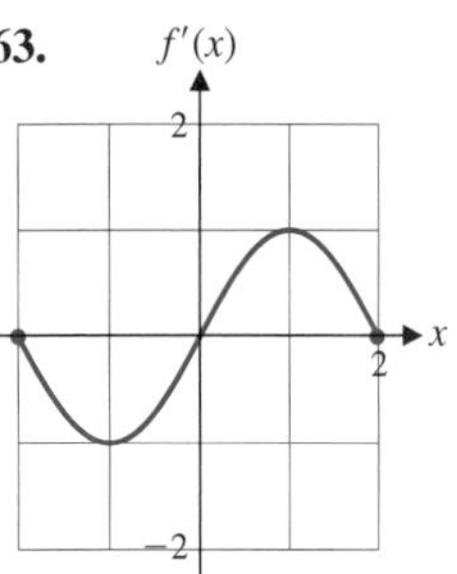

64.

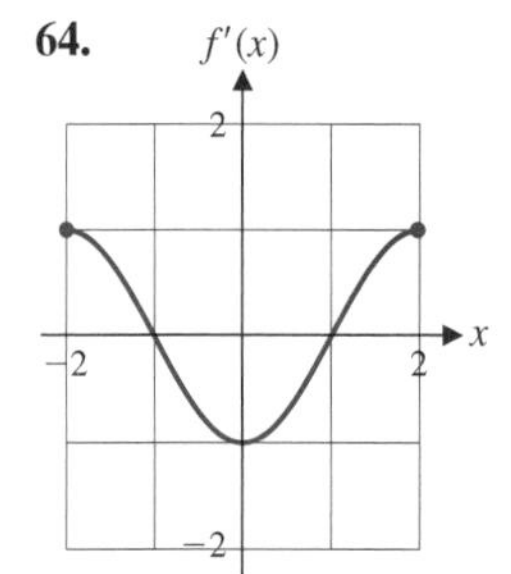

In Problems 65–74, find the particular antiderivative of each derivative that satisfies the given condition.

65. $\frac{dy}{dx} = 2x - 3;\ y(0) = 5$

66. $\frac{dy}{dx} = 5 - 4x;\ y(0) = 20$

67. $C'(x) = 6x^2 - 4x;\ C(0) = 3{,}000$

68. $R'(x) = 600 - 0.6x;\ R(0) = 0$

69. $\frac{dx}{dt} = \frac{20}{\sqrt{t}};\ x(1) = 40$

70. $\frac{dR}{dt} = \frac{100}{t^2};\ R(1) = 400$

71. $\frac{dy}{dx} = 2x^{-2} + 3x^{-1} - 1;\ y(1) = 0$

72. $\frac{dy}{dx} = 3x^{-1} + x^{-2};\ y(1) = 1$

73. $\frac{dx}{dt} = 4e^t - 2;\ x(0) = 1$

74. $\frac{dy}{dt} = 5e^t - 4;\ y(0) = -1$

75. Find the equation of the curve that passes through (2, 3) if its slope is given by

$$\frac{dy}{dx} = 4x - 3$$

for each x.

76. Find the equation of the curve that passes through (1, 3) if its slope is given by

$$\frac{dy}{dx} = 12x^2 - 12x$$

for each x.

C *In Problems 77–82, find each indefinite integral.*

77. $\int \frac{2x^4 - x}{x^3}\,dx$

78. $\int \frac{x^{-1} - x^4}{x^2}\,dx$

79. $\int \frac{x^5 - 2x}{x^4}\,dx$

80. $\int \frac{1 - 3x^4}{x^2}\,dx$

81. $\int \frac{x^2e^x - 2x}{x^2}\,dx$

82. $\int \frac{1 - xe^x}{x}\,dx$

For each derivative in Problems 83–88, find an antiderivative that satisfies the given condition.

83. $\frac{dM}{dt} = \frac{t^2 - 1}{t^2};\ M(4) = 5$

84. $\frac{dR}{dx} = \frac{1 - x^4}{x^3};\ R(1) = 4$

85. $\frac{dy}{dx} = \frac{5x + 2}{\sqrt[3]{x}};\ y(1) = 0$

86. $\frac{dx}{dt} = \frac{\sqrt{t^3} - t}{\sqrt{t^3}};\ x(9) = 4$

87. $p'(x) = -\frac{10}{x^2};\ p(1) = 20$

88. $p'(x) = \frac{10}{x^3};\ p(1) = 15$

In Problems 89–92, find the derivative or indefinite integral as indicated.

89. $\frac{d}{dx}\left(\int x^3\,dx\right)$

90. $\frac{d}{dt}\left(\int \frac{\ln t}{t}\,dt\right)$

91. $\int \frac{d}{dx}(x^4 + 3x^2 + 1)\,dx$

92. $\int \frac{d}{du}(e^{u^2})\,du$

Applications

Business & Economics

93. *Cost function.* The marginal average cost for producing x digital sports watches is given by

$$\overline{C}'(x) = -\frac{1{,}000}{x^2} \qquad \overline{C}(100) = 25$$

where $\overline{C}(x)$ is the average cost in dollars. Find the average cost function and the cost function. What are the fixed costs?

94. *Paper and paperboard production.* In spite of the prediction of a paperless computerized office, paper and paperboard production in the United States has steadily increased. In 1990 the production was 80.3 million short tons, and since 1970 production has been growing at a rate given by

$$f'(t) = 0.048t + 0.95$$

where t is years after 1970 (data from American Paper Institute, Inc.). Noting that $f(20) = 80.3$, find $f(t)$. Also find $f(0)$ and $f(30)$, production levels for the years 1970 and 2000.

95. *Manufacturing costs.* The graph of the marginal cost function from the manufacturing of x thousand watches per month [where cost $C(x)$ is in thousands of dollars per month] is given in the figure.

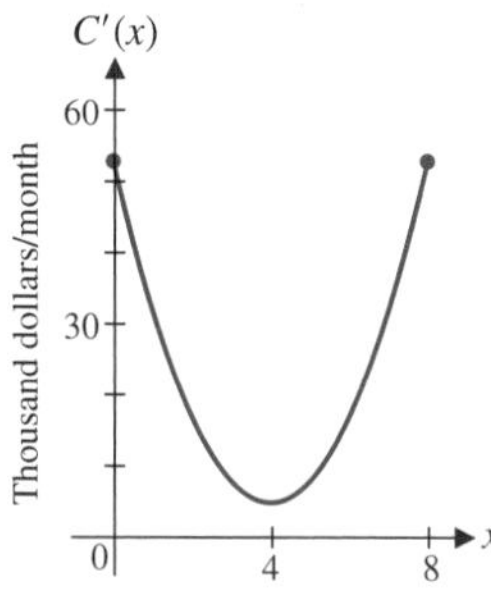

Figure for 95

(A) Using the graph shown, verbally describe the shape of the graph of the cost function $C(x)$ as x increases from 0 to 8,000 watches per month.

(B) Given the equation of the marginal cost function,

$$C'(x) = 3x^2 - 24x + 53$$

find the cost function if monthly fixed costs at 0 output are \$30,000. What is the cost for manufacturing 4,000 watches per month? 8,000 watches per month?

(C) Graph the cost function for $0 \leq x \leq 8$. [Check the shape of the graph relative to the analysis in part (A).]

(D) Why do you think that the graph of the cost function is steeper at both ends than in the middle?

96. *Revenue.* The graph of the marginal revenue function from the sale of x digital sports watches is given in the figure.

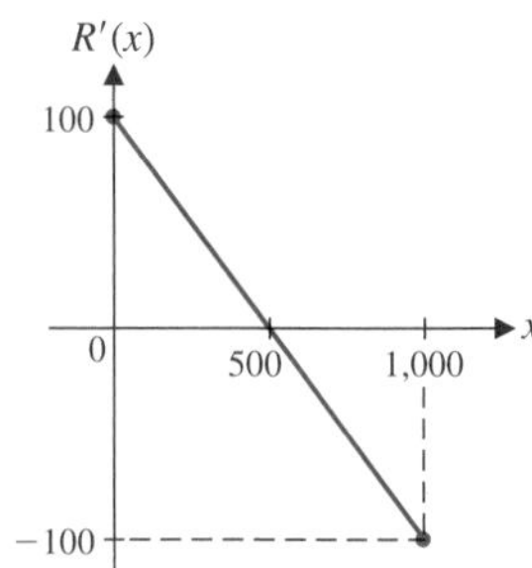

Figure for 96

(A) Using the graph shown, verbally describe the shape of the graph of the revenue function $R(x)$ as x increases from 0 to 1,000.

(B) Find the equation of the marginal revenue function (the linear function shown in the figure).

(C) Find the equation of the revenue function that satisfies $R(0) = 0$. Graph the revenue function over the interval [0, 1,000]. [Check the shape of the graph relative to the analysis in part (A).]

(D) Find the price–demand equation and determine the price when the demand is 700 units.

97. *Sales analysis.* The monthly sales of a particular personal computer are expected to decline at the rate of

$$S'(t) = -25t^{2/3}$$

computers per month, where t is time in months and $S(t)$ is the number of computers sold each month. The company plans to stop manufacturing this computer when the monthly sales reach 800 computers. If the monthly sales now ($t = 0$) are 2,000 computers, find $S(t)$. How long will the company continue to manufacture this computer?

98. *Sales analysis.* The rate of change of the monthly sales of a new home video game cartridge is given by

$$S'(t) = 500t^{1/4} \qquad S(0) = 0$$

where t is the number of months since the game was released and $S(t)$ is the number of cartridges sold each month. Find $S(t)$. When will the monthly sales reach 20,000 cartridges?

99. C *Sales analysis.* Repeat Problem 97 if $S'(t) = -25t^{2/3} - 70$ and all other information remains the same. Use a graphing utility to approximate the solution to the equation $S(t) = 800$ to two decimal places.

100. C *Sales analysis.* Repeat Problem 98 if $S'(t) = 500t^{1/4} + 300$ and all other information remains the same. Use a graphing utility to approximate the solution to the equation $S(t) = 20{,}000$ to two decimal places.

101. *Labor costs and learning.* A defense contractor is starting production on a new missile control system. On the basis of data collected while assembling the first 16 control systems, the production manager obtained the following function describing the rate of labor use:

$$g(x) = 2{,}400x^{-1/2}$$

where $g(x)$ is the number of labor-hours required to assemble the xth unit of the control system. For example, after assembling 16 units, the rate of assembly is 600 labor-hours per unit, and after assembling 25 units, the rate of assembly is 480 labor-hours per unit. The more units assembled, the more efficient the process because of learning. If 19,200 labor-hours are required to assemble the first 16 units, how many labor-hours, $L(x)$, will be required to assemble the first x units? The first 25 units?

102. *Labor costs and learning.* If the rate of labor use in Problem 101 is

$$g(x) = 2{,}000x^{-1/3}$$

and if the first 8 control units require 12,000 labor-hours, how many labor-hours, $L(x)$, will be required for the first x control units? The first 27 control units?

Life Sciences

103. *Weight–height.* For an average person, the rate of change of weight W (in pounds) with respect to height h (in inches) is given approximately by

$$\frac{dW}{dh} = 0.0015h^2$$

Find $W(h)$ if $W(60) = 108$ pounds. Also, find the weight for a person who is 5 feet 10 inches tall.

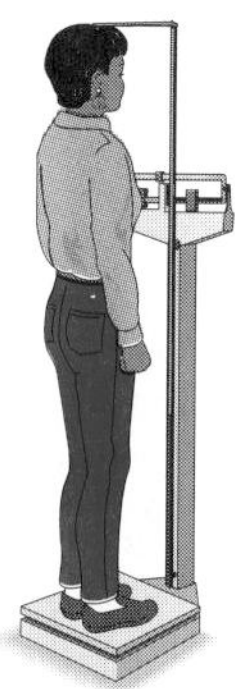

104. *Wound healing.* If the area A of a healing wound changes at a rate given approximately by

$$\frac{dA}{dt} = -4t^{-3} \qquad 1 \leq t \leq 10$$

where t is time in days and $A(1) = 2$ square centimeters, what will the area of the wound be in 10 days?

Social Sciences

105. *Urban growth.* The rate of growth of the population, $N(t)$, of a newly incorporated city t years after incorporation is estimated to be

$$\frac{dN}{dt} = 400 + 600\sqrt{t} \qquad 0 \leq t \leq 9$$

If the population was 5,000 at the time of incorporation, find the population 9 years later.

106. *Learning.* A beginning high school language class was chosen for an experiment in learning. Using a list of 50 words, the experiment involved measuring the rate of vocabulary memorization at different times during a continuous 5 hour study session. It was found that the average rate of learning for the whole class was inversely proportional to the time spent studying and was given approximately by

$$V'(t) = \frac{15}{t} \qquad 1 \leq t \leq 5$$

If the average number of words memorized after 1 hour of study was 15 words, what was the average number of words learned after t hours of study for $1 \leq t \leq 5$? After 4 hours of study? (Round answer to the nearest whole number.)

Section 11-2 Integration by Substitution

- Reversing the Chain Rule
- Integration by Substitution
- Additional Substitution Techniques
- Application

Many of the indefinite integral formulas introduced in the preceding section are based on corresponding derivative formulas studied earlier. We now consider indefinite integral formulas and procedures based on the chain rule for differentiation.

Reversing the Chain Rule

Recall the chain rule:

$$\frac{d}{dx}f[g(x)] = f'[g(x)]g'(x)$$

The expression on the right is formed by taking the derivative of the outside function f and multiplying it by the derivative of the inside function g. If we recognize an integrand as a chain-rule form $f'[g(x)]g'(x)$, then we can easily find an antiderivative and its indefinite integral:

Reversing the Chain Rule

$$\int f'[g(x)]g'(x)\,dx = f[g(x)] + C \qquad (1)$$

Explore–Discuss 1

(A) Which of the following has e^{x^3-1} as an antiderivative?

$x^2e^{x^3-1}$ $\qquad 3x^2e^{x^3-1}$ $\qquad 3xe^{x^3-1}$

(B) Which of the following would have e^{x^3-1} as an antiderivative if it were multiplied by a constant factor? A variable factor?

$3xe^{x^3-1}$ $\qquad x^2e^{x^3-1}$

We are interested in finding the indefinite integral:

$$\int 3x^2e^{x^3-1}\,dx \qquad (2)$$

The integrand appears to be the chain-rule form $e^{g(x)}g'(x)$, which is the derivative of $e^{g(x)}$. Since

$$\frac{d}{dx}e^{x^3-1} = 3x^2e^{x^3-1}$$

it follows that

$$\int 3x^2e^{x^3-1}\,dx = e^{x^3-1} + C \qquad (3)$$

How does the following indefinite integral differ from (2)?

$$\int x^2e^{x^3-1}\,dx \qquad (4)$$

It is missing the constant factor 3. That is, $x^2e^{x^3-1}$ is within a constant factor of being the derivative of e^{x^3-1}. But because a constant factor can be moved across the integral sign, this causes us little trouble in finding the indefinite integral of $x^2e^{x^3-1}$. We introduce the constant factor 3, and at the same time multiply by $\frac{1}{3}$ and move the $\frac{1}{3}$ factor outside the integral sign. This is equivalent to multiplying the integrand in (4) by 1.

$$\int x^2e^{x^3-1}\,dx = \int \frac{3}{3}x^2e^{x^3-1}\,dx$$

$$= \frac{1}{3}\int 3x^2e^{x^3-1}\,dx = \frac{1}{3}e^{x^3-1} + C \qquad (5)$$

The derivative of the right side of (5) is the integrand of the indefinite integral (4). You should check this.

How does the following indefinite integral differ from (2)?

$$\int 3xe^{x^3-1}\,dx \qquad (6)$$

It is missing a variable factor x. This is more serious. As tempting as it might be, we *cannot* adjust (6) by introducing the variable factor x and moving $1/x$ outside the integral sign, as we did with the constant 3 in (5). If we could move $1/x$ across the integral sign, what would stop us from moving the whole integrand across the integral sign? Then, indefinite integration would become a trivial exercise and would not give us the results we want—antiderivatives of the integrand.

SUMMARY OF THE ABOVE INTEGRAL FORMS

$$\int 3x^2 e^{x^3-1}\,dx = e^{x^3-1} + C$$ Integrand is a chain-rule form.

$$\int x^2 e^{x^3-1}\,dx = \frac{1}{3}e^{x^3-1} + C$$ Integrand can be adjusted to a chain-rule form.

$$\int 3x e^{x^3-1}\,dx = ?$$ Integrand cannot be adjusted to be a chain-rule form.

CAUTION

A constant factor can be moved across an integral sign, but a variable factor cannot.

There is nothing wrong with educated guessing when looking for an antiderivative of a given function, and you are encouraged to do so. You have only to check the result by differentiation. And if you are right, you go on your way; if you are wrong, you simply try another approach.

In Section 10-3, we saw that the chain rule extends the derivative formulas for x^n, e^x, and $\ln x$ to derivative formulas for $[f(x)]^n$, $e^{f(x)}$, and $\ln[f(x)]$. The chain rule can also be used to extend the indefinite integral formulas discussed in Section 11-1. Some general formulas are summarized in the following box:

General Indefinite Integral Formulas

1. $\int [f(x)]^n f'(x)\,dx = \frac{[f(x)]^{n+1}}{n+1} + C, \quad n \neq -1$
2. $\int e^{f(x)} f'(x)\,dx = e^{f(x)} + C$
3. $\int \frac{1}{f(x)} f'(x)\,dx = \ln|f(x)| + C$

Each formula can be verified by using the chain rule to show that the derivative of the function on the right is the integrand on the left. For example,

$$\frac{d}{dx}[e^{f(x)} + C] = e^{f(x)} f'(x)$$

verifies formula 2.

Example 1 **Reversing the Chain Rule**

(A) $\int (3x+4)^{10}(3)\,dx = \frac{(3x+4)^{11}}{11} + C$ Formula 1 with $f(x) = 3x + 4$ and $f'(x) = 3$

Check: $\frac{d}{dx}\frac{(3x+4)^{11}}{11} = 11\frac{(3x+4)^{10}}{11}\frac{d}{dx}(3x+4) = (3x+4)^{10}(3)$

(B) $\int e^{x^2}(2x)\,dx = e^{x^2} + C$ Formula 2 with $f(x) = x^2$ and $f'(x) = 2x$

Check: $\frac{d}{dx}e^{x^2} = e^{x^2}\frac{d}{dx}x^2 = e^{x^2}(2x)$

(C) $\int \frac{1}{1 + x^3}3x^2\,dx = \ln|1 + x^3| + C$ Formula 3 with $f(x) = 1 + x^3$ and $f'(x) = 3x^2$

Check: $\frac{d}{dx}\ln|1 + x^3| = \frac{1}{1 + x^3}\frac{d}{dx}(1 + x^3) = \frac{1}{1 + x^3}3x^2$

Matched Problem 1 Find each indefinite integral.

(A) $\int (2x^3 - 3)^{20}(6x^2)\,dx$ (B) $\int e^{5x}(5)\,dx$

(C) $\int \frac{1}{4 + x^2}2x\,dx$

INTEGRATION BY SUBSTITUTION

The key step in using formulas 1, 2, and 3 is recognizing the form of the integrand. Some people find it difficult to identify $f(x)$ and $f'(x)$ in these formulas and prefer to use a *substitution* to simplify the integrand. The *method of substitution,* which we now discuss, becomes increasingly useful as one progresses in studies of integration.

We start by introducing the idea of the *differential.* We represented the derivative by the symbol *dy/dx* taken as a whole. We now define *dy* and *dx* as two separate quantities with the property that their ratio is still equal to $f'(x)$:

Differentials

If $y = f(x)$ defines a differentiable function, then:

1. The **differential *dx*** of the independent variable x is an arbitrary real number.
2. The **differential *dy*** of the dependent variable y is defined as the product of $f'(x)$ and dx—that is, as

$$dy = f'(x)\,dx$$

Differentials involve mathematical subtleties that are treated carefully in advanced mathematics courses. Here, we are interested in them mainly as a bookkeeping device to aid in the process of finding indefinite integrals. We can always check the results by differentiating.

Example 2 **Differentials**

(A) If $y = f(x) = x^2$, then

$$dy \boxed{= f'(x)\,dx} = 2x\,dx$$

(B) If $u = g(x) = e^{3x}$, then

$$du \boxed{= g'(x)\,dx} = 3e^{3x}\,dx$$

(C) If $w = h(t) = \ln(4 + 5t)$, then

$$dw \boxed{= h'(t)\,dt} = \frac{5}{4 + 5t}\,dt$$

Matched Problem 2 (A) Find dy for $y = f(x) = x^3$. (B) Find du for $u = h(x) = \ln(2 + x^2)$.
(C) Find dv for $v = g(t) = e^{-5t}$.

The **method of substitution** is developed through the following examples.

Example 3 **Using Substitution** Find $\int (x^2 + 2x + 5)^5(2x + 2)\,dx$.

SOLUTION If

$$u = x^2 + 2x + 5$$

then the differential of u is

$$du = (2x + 2)\,dx$$

Notice that du is one of the factors in the integrand. Substitute u for $x^2 + 2x + 5$ and du for $(2x + 2)\,dx$ to obtain

$$\begin{aligned}\int (x^2 + 2x + 5)^5(2x + 2)\,dx &= \int u^5\,du \\ &= \frac{u^6}{6} + C \\ &= \frac{1}{6}(x^2 + 2x + 5)^6 + C \quad \text{Since } u = x^2 + 2x + 5\end{aligned}$$

CHECK

$$\begin{aligned}\frac{d}{dx}\frac{1}{6}(x^2 + 2x + 5)^6 &= \frac{1}{6}(6)(x^2 + 2x + 5)^5\frac{d}{dx}(x^2 + 2x + 5) \\ &= (x^2 + 2x + 5)^5(2x + 2)\end{aligned}$$

Matched Problem 3 Find $\int (x^2 - 3x + 7)^4(2x - 3)\,dx$ by substitution.

The substitution method is also called the **change-of-variable method,** since u replaces the variable x in the process. Substituting $u = f(x)$ and $du = f'(x)\,dx$ in formulas 1, 2, and 3 produces the general indefinite integral formulas in the following box:

General Indefinite Integral Formulas

4. $\int u^n\,du = \dfrac{u^{n+1}}{n+1} + C, \quad n \neq -1$
5. $\int e^u\,du = e^u + C$
6. $\int \dfrac{1}{u}\,du = \ln|u| + C$

These formulas are valid if u is an independent variable or if u is a function of another variable and du is its differential with respect to that variable.

The substitution method for evaluating certain indefinite integrals is outlined in the following box:

Integration by Substitution

Step 1. Select a substitution that appears to simplify the integrand. In particular, try to select u so that du is a factor in the integrand.

Step 2. Express the integrand entirely in terms of u and du, completely eliminating the original variable and its differential.

Step 3. Evaluate the new integral, if possible.

Step 4. Express the antiderivative found in step 3 in terms of the original variable.

Example 4 **Using Substitution** Use a substitution to find the following:

(A) $\int (3x+4)^6(3)\,dx$ (B) $\int e^{t^2}(2t)\,dt$

SOLUTION (A) If we let $u = 3x + 4$, then $du = 3\,dx$, and

$$\begin{aligned}\int (3x+4)^6(3)\,dx &= \int u^6\,du && \text{Use formula 4.}\\ &= \frac{u^7}{7} + C\\ &= \frac{(3x+4)^7}{7} + C && \text{Since } u = 3x+4\end{aligned}$$

Check: $\dfrac{d}{dx}\dfrac{(3x+4)^7}{7} = \dfrac{7(3x+4)^6}{7}\dfrac{d}{dx}(3x+4) = (3x+4)^6(3)$

(B) If we let $u = t^2$, then $du = 2t\,dt$, and

$$\begin{aligned}\int e^{t^2}(2t)\,dt &= \int e^u\,du && \text{Use formula 5.}\\ &= e^u + C\\ &= e^{t^2} + C && \text{Since } u = t^2\end{aligned}$$

Check: $\dfrac{d}{dt}e^{t^2} = e^{t^2}\dfrac{d}{dt}t^2 = e^{t^2}(2t)$ ■

Matched Problem 4 Use a substitution to find each indefinite integral.

(A) $\int (2x^3 - 3)^4(6x^2)\,dx$ (B) $\int e^{5w}(5)\,dw$

CAUTION

Integration by substitution is an effective procedure for some indefinite integrals, but not all. Substitution is not helpful for $\int e^{x^2}\,dx$ or $\int (\ln x)\,dx$, for example.

ADDITIONAL SUBSTITUTION TECHNIQUES

In order to use the substitution method, **the integrand must be expressed entirely in terms of u and du.** In some cases, the integrand will have to be modified before making a substitution and using one of the integration formulas. Example 5 illustrates this process.

Example 5 **Substitution Techniques** Integrate:

(A) $\displaystyle\int \frac{1}{4x+7}\,dx$ (B) $\int te^{-t^2}\,dt$ (C) $\int 4x^2\sqrt{x^3+5}\,dx$

SOLUTION (A) If $u = 4x + 7$, then $du = 4\,dx$. We are missing a factor of 4 in the integrand to match formula 6 exactly. Recalling that a constant factor can be moved across an integral sign, we proceed as follows:

$$\begin{aligned}\int \frac{1}{4x+7}\,dx &= \int \frac{1}{4x+7}\frac{\mathbf{4}}{\mathbf{4}}\,dx \\ &= \frac{1}{4}\int \frac{1}{4x+7}4\,dx && \text{Substitute } u = 4x+7 \text{ and } du = 4\,dx. \\ &= \frac{1}{4}\int \frac{1}{u}\,du && \text{Use formula 6.} \\ &= \tfrac{1}{4}\ln|u| + C \\ &= \tfrac{1}{4}\ln|4x+7| + C && \text{Since } u = 4x+7\end{aligned}$$

Check:

$$\frac{d}{dx}\frac{1}{4}\ln|4x+7| = \frac{1}{4}\frac{1}{4x+7}\frac{d}{dx}(4x+7) = \frac{1}{4}\frac{1}{4x+7}4 = \frac{1}{4x+7}$$

(B) If $u = -t^2$, then $du = -2t\,dt$. Proceed as in part (A):

$$\begin{aligned}\int te^{-t^2}\,dt &= \int e^{-t^2}\frac{\mathbf{-2}}{\mathbf{-2}}t\,dt \\ &= -\frac{1}{2}\int e^{-t^2}(-2t)\,dt && \text{Substitute } u = -t^2 \text{ and } du = -2t\,dt. \\ &= -\frac{1}{2}\int e^u\,du && \text{Use formula 5.} \\ &= -\tfrac{1}{2}e^u + C \\ &= -\tfrac{1}{2}e^{-t^2} + C && \text{Since } u = -t^2\end{aligned}$$

Check: $\frac{d}{dt}(-\tfrac{1}{2}e^{-t^2}) = -\tfrac{1}{2}e^{-t^2}\frac{d}{dt}(-t^2) = -\tfrac{1}{2}e^{-t^2}(-2t) = te^{-t^2}$

(C) $\int 4x^2\sqrt{x^3 + 5}\,dx = 4\int \sqrt{x^3 + 5}\,(x^2)\,dx$ — Move the 4 across the integral sign and proceed as before.

$$= 4\int \sqrt{x^3 + 5}\,\frac{3}{3}(x^2)\,dx$$

$$= \frac{4}{3}\int \sqrt{x^3 + 5}(3x^2)\,dx$$ Substitute $u = x^3 + 5$ and $du = 3x^2\,dx$.

$$= \frac{4}{3}\int \sqrt{u}\,du$$

$$= \frac{4}{3}\int u^{1/2}\,du$$ Use formula 4.

$$= \frac{4}{3}\frac{u^{3/2}}{\frac{3}{2}} + C$$

$$= \tfrac{8}{9}u^{3/2} + C$$

$$= \tfrac{8}{9}(x^3 + 5)^{3/2} + C$$ Since $u = x^3 + 5$

Check: $\frac{d}{dx}\left[\tfrac{8}{9}(x^3 + 5)^{3/2}\right] = \tfrac{4}{3}(x^3 + 5)^{1/2}\frac{d}{dx}(x^3 + 5)$

$$= \tfrac{4}{3}(x^3 + 5)^{1/2}(3x^2) = 4x^2\sqrt{x^3 + 5}$$

Matched Problem 5 Integrate:

(A) $\int e^{-3x}\,dx$ (B) $\int \frac{x}{x^2 - 9}\,dx$ (C) $\int 5t^2(t^3 + 4)^{-2}\,dt$

Even if it is not possible to find a substitution that makes an integrand match one of the integration formulas exactly, a substitution may sufficiently simplify the integrand so that other techniques can be used.

Example 6 **Substitution Techniques** Find: $\int \frac{x}{\sqrt{x + 2}}\,dx$

SOLUTION Proceeding as before, if we let $u = x + 2$, then $du = dx$ and

$$\int \frac{x}{\sqrt{x + 2}}\,dx = \int \frac{x}{\sqrt{u}}\,du$$

Notice that this substitution is not yet complete, because we have not expressed the integrand entirely in terms of u and du. As we noted earlier, only a constant factor can be moved across an integral sign, so we cannot move x outside the integral sign (as much as we would like to). Instead, we must return to the original substitution, solve for x in terms of u, and use the resulting equation to complete the substitution:

$$u = x + 2$$ Solve for x in terms of u.

$$u - 2 = x$$ Substitute this expression for x.

Thus,

$$\begin{aligned}
\int \frac{x}{\sqrt{x+2}}\,dx &= \int \frac{u-2}{\sqrt{u}}\,du \qquad \text{Simplify the integrand.}\\
&= \int \frac{u-2}{u^{1/2}}\,du\\
&= \int (u^{1/2} - 2u^{-1/2})\,du\\
&= \int u^{1/2}\,du - 2\int u^{-1/2}\,du\\
&= \frac{u^{3/2}}{\frac{3}{2}} - 2\frac{u^{1/2}}{\frac{1}{2}} + C\\
&= \tfrac{2}{3}(x+2)^{3/2} - 4(x+2)^{1/2} + C \qquad \text{Since } u = x + 2
\end{aligned}$$

CHECK

$$\begin{aligned}
\frac{d}{dx}\left[\tfrac{2}{3}(x+2)^{3/2} - 4(x+2)^{1/2}\right] &= (x+2)^{1/2} - 2(x+2)^{-1/2}\\
&= \frac{x+2}{(x+2)^{1/2}} - \frac{2}{(x+2)^{1/2}}\\
&= \frac{x}{(x+2)^{1/2}}
\end{aligned}$$

Matched Problem 6 Find: $\int x\sqrt{x+1}\,dx$

■ APPLICATION

Example 7

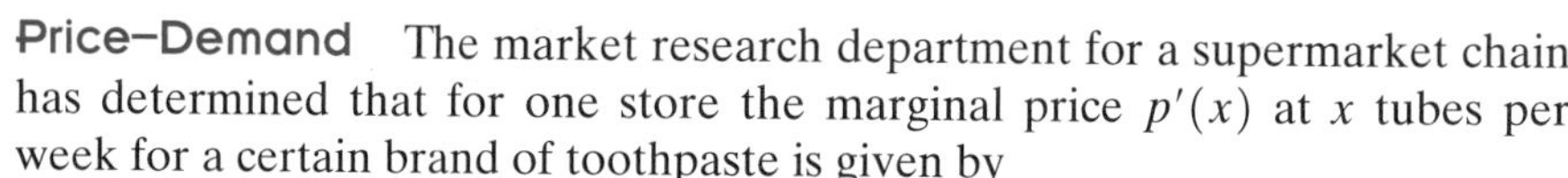

Price–Demand The market research department for a supermarket chain has determined that for one store the marginal price $p'(x)$ at x tubes per week for a certain brand of toothpaste is given by

$$p'(x) = -0.015e^{-0.01x}$$

Find the price–demand equation if the weekly demand is 50 tubes when the price of a tube is \$2.35. Find the weekly demand when the price of a tube is \$1.89.

SOLUTION

$$\begin{aligned}
p(x) &= \int -0.015e^{-0.01x}\,dx\\
&= -0.015\int e^{-0.01x}\,dx\\
&= -0.015\int e^{-0.01x}\frac{\mathbf{-0.01}}{\mathbf{-0.01}}\,dx\\
&= \frac{-0.015}{-0.01}\int e^{-0.01x}(-0.01)\,dx \qquad \text{Substitute } u = -0.01x \text{ and } du = -0.01\,dx.\\
&= 1.5\int e^{u}\,du\\
&= 1.5e^{u} + C\\
&= 1.5e^{-0.01x} + C \qquad \text{Since } u = -0.01x
\end{aligned}$$

We find C by noting that

$$\begin{aligned} p(50) = 1.5e^{-0.01(50)} + C &= \$2.35 \\ C &= \$2.35 - 1.5e^{-0.5} \quad \text{Use a calculator.} \\ C &= \$2.35 - 0.91 \\ C &= \$1.44 \end{aligned}$$

Thus,

$$p(x) = 1.5e^{-0.01x} + 1.44$$

To find the demand when the price is \$1.89, we solve $p(x) = \$1.89$ for x:

$$\begin{aligned} 1.5e^{-0.01x} + 1.44 &= 1.89 \\ 1.5e^{-0.01x} &= 0.45 \\ e^{-0.01x} &= 0.3 \\ -0.01x &= \ln 0.3 \\ x &= -100 \ln 0.3 \approx 120 \text{ tubes} \end{aligned}$$

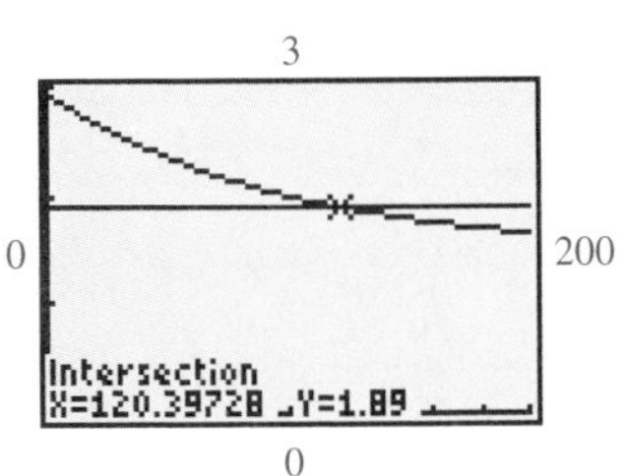

FIGURE 1
$y_1 = 1.5e^{-0.01x} + 1.44$
$y_2 = 1.89$

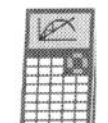

A graphing utility offers an alternative approach to solving $p(x) = 1.89$ in Example 7 (see Fig. 1).

Matched Problem 7 The marginal price $p'(x)$ at a supply level of x tubes per week for a certain brand of toothpaste is given by

$$p'(x) = 0.001e^{0.01x}$$

Find the price–supply equation if the supplier is willing to supply 100 tubes per week at a price of \$1.65 each. How many tubes would the supplier be willing to supply at a price of \$1.98 each?

Explore–Discuss 2

In each of the following examples explain why $\neq$ is used; then work the problem correctly using either an appropriate substitution or another method.

1. $$\begin{aligned} \int (x^2+3)^2\,dx &= \int (x^2+3)^2 \frac{2x}{2x}\,dx \\ &\neq \frac{1}{2x}\int (x^2+3)^2(2x)\,dx \end{aligned}$$

2. $$\begin{aligned} \int \frac{1}{10x+3}\,dx &= \int \frac{1}{u}\,dx \qquad u = 10x+3 \\ &\neq \ln|u| + C \end{aligned}$$

We conclude with two final cautions (the first was stated earlier, but is worth repeating):

CAUTION

1. A variable cannot be moved across an integral sign!
2. An integral must be expressed entirely in terms of u and du before applying integration formulas 4, 5, and 6.

Answers to Matched Problems

1. (A) $\frac{1}{21}(2x^3 - 3)^{21} + C$ (B) $e^{5x} + C$
   (C) $\ln|4 + x^2| + C$ or $\ln(4 + x^2) + C$, since $4 + x^2 > 0$
2. (A) $dy = 3x^2\,dx$ (B) $du = \frac{2x}{2 + x^2}\,dx$ (C) $dv = -5e^{-5t}\,dt$
3. $\frac{1}{5}(x^2 - 3x + 7)^5 + C$ 4. (A) $\frac{1}{5}(2x^3 - 3)^5 + C$ (B) $e^{5w} + C$
5. (A) $-\frac{1}{3}e^{-3x} + C$ (B) $\frac{1}{2}\ln|x^2 - 9| + C$ (C) $-\frac{5}{3}(t^3 + 4)^{-1} + C$
6. $\frac{2}{5}(x + 1)^{5/2} - \frac{2}{3}(x + 1)^{3/2} + C$ 7. $p(x) = 0.1e^{0.01x} + 1.38$; 179 tubes

EXERCISE 11-2

A *In Problems 1–40, find each indefinite integral, and check the result by differentiating.*

1. $\int (x^2 - 4)^5(2x)\,dx$
2. $\int (x^3 + 1)^4(3x^2)\,dx$
3. $\int e^{4x}(4)\,dx$
4. $\int e^{-3x}(-3)\,dx$
5. $\int (x - 2)^8\,dx$
6. $\int (x + 5)^{-4}\,dx$
7. $\int \frac{1}{2t + 3}2\,dt$
8. $\int \frac{1}{5t - 7}5\,dt$

B

9. $\int (3x - 2)^7\,dx$
10. $\int (5x + 3)^9\,dx$
11. $\int (x^2 + 3)^7 x\,dx$
12. $\int (x^3 - 5)^4 x^2\,dx$
13. $\int 10e^{-0.5t}\,dt$
14. $\int 4e^{0.01t}\,dt$
15. $\int \frac{1}{10x + 7}\,dx$
16. $\int \frac{1}{100 - 3x}\,dx$
17. $\int xe^{2x^2}\,dx$
18. $\int x^2e^{4x^3}\,dx$
19. $\int \frac{x^2}{x^3 + 4}\,dx$
20. $\int \frac{x}{x^2 - 2}\,dx$
21. $\int \frac{t}{(3t^2 + 1)^4}\,dt$
22. $\int \frac{t^2}{(t^3 - 2)^5}\,dt$
23. $\int \frac{x^2}{(4 - x^3)^2}\,dx$
24. $\int \frac{x}{(5 - 2x^2)^5}\,dx$
25. $\int x\sqrt{x + 4}\,dx$
26. $\int x\sqrt{x - 9}\,dx$
27. $\int \frac{x}{\sqrt{x - 3}}\,dx$
28. $\int \frac{x}{\sqrt{x + 5}}\,dx$
29. $\int x(x - 4)^9\,dx$
30. $\int x(x + 6)^8\,dx$
31. $\int e^{2x}(1 + e^{2x})^3\,dx$
32. $\int e^{-x}(1 - e^{-x})^4\,dx$
33. $\int \frac{1 + x}{4 + 2x + x^2}\,dx$
34. $\int \frac{x^2 - 1}{x^3 - 3x + 7}\,dx$
35. $\int (2x + 1)e^{x^2+x+1}\,dx$
36. $\int (x^2 + 2x)e^{x^3+3x^2}\,dx$
37. $\int (e^x - 2x)^3(e^x - 2)\,dx$
38. $\int (x^2 - e^x)^4(2x - e^x)\,dx$
39. $\int \frac{x^3 + x}{(x^4 + 2x^2 + 1)^4}\,dx$
40. $\int \frac{x^2 - 1}{(x^3 - 3x + 7)^2}\,dx$

In Problems 41–46, imagine that the indicated "solutions" were given to you by a student whom you are tutoring in this class.
(A) How would you have the student check each solution?
(B) Is the solution right or wrong? If the solution is wrong, explain what is wrong and how it can be corrected.
(C) Show a correct solution for each incorrect solution, and check the result by differentiation.

41. $\int \frac{1}{2x - 3}\,dx = \ln|2x - 3| + C$
42. $\int \frac{x}{x^2 + 5}\,dx = \ln|x^2 + 5| + C$
43. $\int x^3e^{x^4}\,dx = e^{x^4} + C$
44. $\int e^{4x-5}\,dx = e^{4x-5} + C$
45. $\int 2(x^2 - 2)^2\,dx = \frac{(x^2 - 2)^2}{3x} + C$
46. $\int (-10x)(x^2 - 3)^{-4}\,dx = (x^2 - 3)^{-5} + C$

C *In Problems 47–58, find each indefinite integral, and check the result by differentiating.*

47. $\int x\sqrt{3x^2 + 7}\,dx$
48. $\int x^2\sqrt{2x^3 + 1}\,dx$
49. $\int x(x^3 + 2)^2\,dx$
50. $\int x(x^2 + 2)^2\,dx$
51. $\int x^2(x^3 + 2)^2\,dx$
52. $\int (x^2 + 2)^2\,dx$
53. $\int \frac{x^3}{\sqrt{2x^4 + 3}}\,dx$
54. $\int \frac{x^2}{\sqrt{4x^3 - 1}}\,dx$
55. $\int \frac{(\ln x)^3}{x}\,dx$
56. $\int \frac{e^x}{1 + e^x}\,dx$
57. $\int \frac{1}{x^2}e^{-1/x}\,dx$
58. $\int \frac{1}{x\ln x}\,dx$

In Problems 59–64, find the antiderivative of each derivative.

59. $\dfrac{dx}{dt} = 7t^2(t^3 + 5)^6$

60. $\dfrac{dm}{dn} = 10n(n^2 - 8)^7$

61. $\dfrac{dy}{dt} = \dfrac{3t}{\sqrt{t^2 - 4}}$

62. $\dfrac{dy}{dx} = \dfrac{5x^2}{(x^3 - 7)^4}$

63. $\dfrac{dp}{dx} = \dfrac{e^x + e^{-x}}{(e^x - e^{-x})^2}$

64. $\dfrac{dm}{dt} = \dfrac{\ln(t - 5)}{t - 5}$

Use substitution techniques to derive the integration formulas in Problems 65 and 66. Then check your work by differentiation.

65. $\displaystyle\int e^{au}\, du = \frac{1}{a} e^{au} + C, \quad a \neq 0$

66. $\displaystyle\int \frac{1}{au + b}\, du = \frac{1}{a} \ln|au + b| + C, \quad a \neq 0$

Applications

Business & Economics

67. *Price–demand equation.* The marginal price for a weekly demand of x bottles of baby shampoo in a drug store is given by

$$p'(x) = \frac{-6{,}000}{(3x + 50)^2}$$

Find the price–demand equation if the weekly demand is 150 when the price of a bottle of shampoo is \$4. What is the weekly demand when the price is \$2.50?

68. *Price–supply equation.* The marginal price at a supply level of x bottles of baby shampoo per week is given by

$$p'(x) = \frac{300}{(3x + 25)^2}$$

Find the price–supply equation if the distributor of the shampoo is willing to supply 75 bottles a week at a price of \$1.60 per bottle. How many bottles would the supplier be willing to supply at a price of \$1.75 per bottle?

69. *Cost function.* The weekly marginal cost of producing x pairs of tennis shoes is given by

$$C'(x) = 12 + \frac{500}{x + 1}$$

where $C(x)$ is cost in dollars. If the fixed costs are \$2,000 per week, find the cost function. What is the average cost per pair of shoes if 1,000 pairs of shoes are produced each week?

70. *Revenue function.* The weekly marginal revenue from the sale of x pairs of tennis shoes is given by

$$R'(x) = 40 - 0.02x + \frac{200}{x + 1} \qquad R(0) = 0$$

where $R(x)$ is revenue in dollars. Find the revenue function. Find the revenue from the sale of 1,000 pairs of shoes.

71. *Marketing.* An automobile company is ready to introduce a new line of cars with a national sales campaign. After test marketing the line in a carefully selected city, the marketing research department estimates that sales (in millions of dollars) will increase at the monthly rate of

$$S'(t) = 10 - 10e^{-0.1t} \qquad 0 \le t \le 24$$

t months after the national campaign has started.

(A) What will be the total sales, $S(t)$, t months after the beginning of the national campaign if we assume 0 sales at the beginning of the campaign?

(B) What are the estimated total sales for the first 12 months of the campaign?

C (C) When will the estimated total sales reach \$100 million? Use a graphing utility to approximate the answer to two decimal places.

72. *Marketing.* Repeat Problem 71 if the monthly rate of increase in sales is found to be approximated by

$$S'(t) = 20 - 20e^{-0.05t} \qquad 0 \le t \le 24$$

73. *Oil production.* Using data from the first 3 years of production as well as geological studies, the management of an oil company estimates that oil will be pumped from a producing field at a rate given by

$$R(t) = \frac{100}{t + 1} + 5 \qquad 0 \le t \le 20$$

where $R(t)$ is the rate of production (in thousands of barrels per year) t years after pumping begins. How many barrels of oil, $Q(t)$, will the field produce the first t years if $Q(0) = 0$? How many barrels will be produced the first 9 years?

74. C *Oil production.* Assume that the rate in Problem 73 is found to be

$$R(t) = \frac{120t}{t^2 + 1} + 3 \qquad 0 \le t \le 20$$

(A) When is the rate of production greatest?

(B) How many barrels of oil, $Q(t)$, will the field produce the first t years if $Q(0) = 0$? How many barrels will be produced the first 5 years?

(C) How long (to the nearest tenth of a year) will it take to produce a total of a quarter of a million barrels of oil?

Life Sciences

75. *Biology.* A yeast culture is growing at the rate of $W'(t) = 0.2e^{0.1t}$ grams per hour. If the starting culture weighs 2 grams, what will be the weight of the culture, $W(t)$, after t hours? After 8 hours?

76. *Medicine.* The rate of healing for a skin wound (in square centimeters per day) is approximated by $A'(t) = -0.9e^{-0.1t}$. If the initial wound has an area of 9 square centimeters, what will its area, $A(t)$, be after t days? After 5 days?

77. C *Pollution.* A contaminated lake is treated with a bactericide. The rate of increase in harmful bacteria t days after the treatment is given by

$$\frac{dN}{dt} = -\frac{2{,}000t}{1 + t^2} \qquad 0 \leq t \leq 10$$

where $N(t)$ is the number of bacteria per milliliter of water (since dN/dt is negative, the count of harmful bacteria is decreasing).

(A) Find the minimum value of dN/dt.

(B) If the initial count was 5,000 bacteria per milliliter, find $N(t)$ and then find the bacteria count after 10 days.

(C) When (to two decimal places) is the bacteria count 1,000 bacteria per milliliter?

78. *Pollution.* An oil tanker aground on a reef is losing oil and producing an oil slick that is radiating outward at a rate given approximately by

$$\frac{dR}{dt} = \frac{60}{\sqrt{t + 9}} \qquad t \geq 0$$

where R is the radius (in feet) of the circular slick after t minutes. Find the radius of the slick after 16 minutes if the radius is 0 when $t = 0$.

Social Sciences

79. *Learning.* In a particular business college, it was found that an average student enrolled in an advanced typing class progressed at a rate of $N'(t) = 6e^{-0.1t}$ words per minute per week, t weeks after enrolling in a 15 week course. If at the beginning of the course a student could type 40 words per minute, how many words per minute, $N(t)$, would the student be expected to type t weeks into the course? After completing the course?

80. *Learning.* In the same business college, it was also found that an average student enrolled in a beginning shorthand class progressed at a rate of $N'(t) = 12e^{-0.06t}$ words per minute per week, t weeks after enrolling in a 15 week course. If at the beginning of the course a student could take dictation in shorthand at 0 words per minute, how many words per minute, $N(t)$, would the student be expected to handle t weeks into the course? After completing the course?

81. *College enrollment.* The projected rate of increase in enrollment in a new college is estimated by

$$\frac{dE}{dt} = 5{,}000(t + 1)^{-3/2} \qquad t \geq 0$$

where $E(t)$ is the projected enrollment in t years. If enrollment is 2,000 now ($t = 0$), find the projected enrollment 15 years from now.

SECTION 11-3 Differential Equations—Growth and Decay

- DIFFERENTIAL EQUATIONS AND SLOPE FIELDS
- CONTINUOUS COMPOUND INTEREST REVISITED
- EXPONENTIAL GROWTH LAW
- POPULATION GROWTH; RADIOACTIVE DECAY; LEARNING
- A COMPARISON OF EXPONENTIAL GROWTH PHENOMENA

In the previous section, we considered equations of the form

$$\frac{dy}{dx} = 6x^2 - 4x \qquad p'(x) = -400e^{-0.04x}$$

These are examples of *differential equations.* In general, an equation is a **differential equation** if it involves an unknown function and one or more of its derivatives. Other examples of differential equations are

$$\frac{dy}{dx} = ky \qquad y'' - xy' + x^2 = 5 \qquad \frac{dy}{dx} = 2xy$$

The first and third equations are called **first-order** equations because each involves a first derivative, but no higher derivative. The second equation is called a **second-order** equation because it involves a second derivative, and no higher derivatives. Finding solutions to different types of differential equations (functions that satisfy the equation) is the subject matter for whole books and courses on this topic. Here, we will consider only a few very special but very important first-order equations that have immediate and significant applications.

We start by looking at some first-order equations geometrically—in terms of *slope fields.* We then reconsider continuous compound interest as modeled by a first-order differential equation, and from this we will be able to generalize the approach to a wide variety of other types of growth phenomena.

■ Differential Equations and Slope Fields

We introduce the concept of *slope field* through an example. Suppose we are given the first-order differential equation

$$\frac{dy}{dx} = 0.2y \tag{1}$$

A function f is a solution of (1) if $y = f(x)$ satisfies equation (1) for all values of x in the domain of f. Geometrically interpreted, equation (1) gives us the slope of a solution curve that passes through the point (x, y). For example, if $y = f(x)$ is a solution of (1) that passes through the point $(0, 2)$, then the slope of f at $(0, 2)$ is given by

$$\frac{dy}{dx} = 0.2(2) = 0.4$$

We indicate this by drawing a short segment of the tangent line at the point $(0, 2)$, as shown in Figure 1A. This procedure is repeated for points $(-3, 1)$ and $(2, 3)$, as also shown in Figure 1A. Assuming that the graph of f passes through all three points, we sketch an approximate graph of f in Figure 1B.

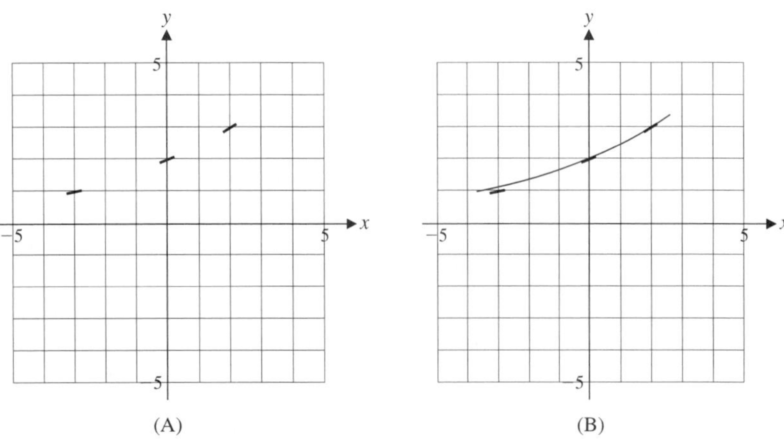

FIGURE 1

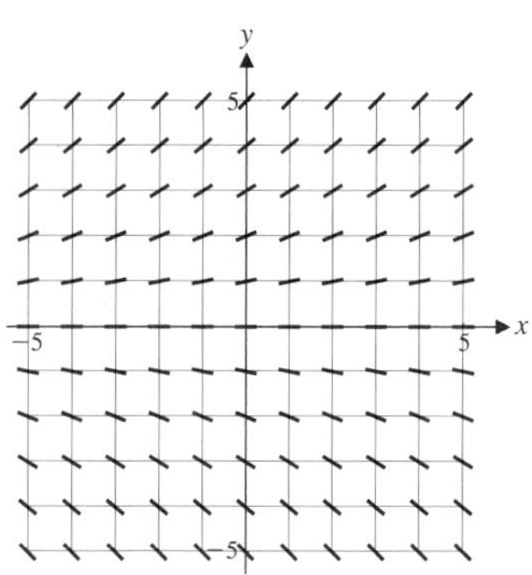

FIGURE 2

If we continue the process of drawing tangent line segments at each point in the grid in Figure 1—a task easily handled by computers but not by hand—we obtain a *slope field.* A slope field for differential equation (1), drawn by a computer, is shown in Figure 2. In general, a **slope field** for a first-order differential equation is obtained by drawing tangent line segments determined by the equation at each point in a grid. In a more advanced treatment of the subject, one can find out a lot about the shape and behavior of solution curves of first-order differential equations by looking at slope fields. Our interests are more modest here.

Explore–Discuss 1

(A) In Figure 1A (or a copy), draw tangent line segments for a solution curve of differential equation (1) that passes through $(-3, -1)$, $(0, -2)$, and $(2, -3)$.

(B) In Figure 1B (or a copy), sketch an approximate graph of the solution curve that passes through these three points. (Repeat the tangent line segments first.)

(C) Of all the elementary functions discussed in the first two chapters, make a conjecture as to what type of function appears to be a solution to differential equation (1).

In Explore–Discuss 1, if you guessed that solutions to (1) are exponential functions, you are to be congratulated. We now show that

$$y = Ce^{0.2x} \tag{2}$$

is a solution to (1) for any real number C. [Later in this section we will show how to find (2) directly from (1).] To do this, we substitute $y = Ce^{0.2x}$ into (1) to see if the left side is equal to the right side for all real x:

$$\frac{dy}{dx} = 0.2y$$

$$\textit{Left side:}\quad \frac{dy}{dx} = \frac{d}{dx}(Ce^{0.2x}) = 0.2Ce^{0.2x}$$

$$\textit{Right side:}\quad 0.2y = 0.2Ce^{0.2x}$$

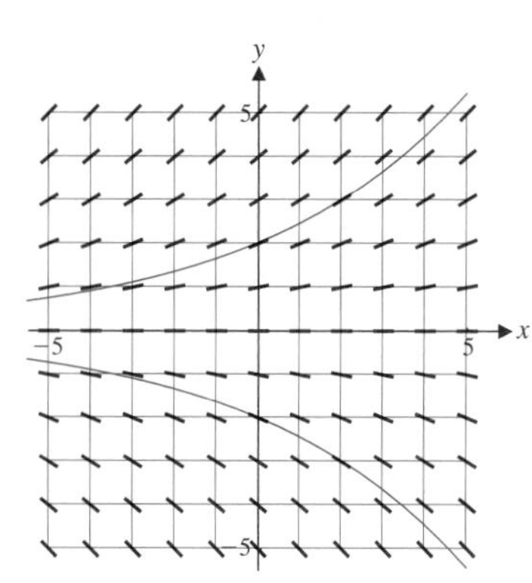

FIGURE 3

Thus, (2) is a solution of (1) for C any real number. Which values of C will produce solution curves that pass through $(0, 2)$ and $(0, -2)$, respectively? Substituting the coordinates of each point into (2) and solving for C (a task left the reader), we obtain

$$y = 2e^{0.2x} \qquad \text{and} \qquad y = -2e^{0.2x} \tag{3}$$

as can be easily checked. The graphs of equations (3) are shown in Figure 3 and confirm the results in Figure 1B.

Explore–Discuss 2

(A) In Figure 3 (or a copy), sketch in an approximate solution curve that passes through $(0, 3)$ and one that passes through $(0, -3)$.

(B) Use a graphing utility to graph $y = Ce^{0.2x}$ for $C = -4, -3, -2, 2, 3, 4$, all in the same viewing window. Notice how these solution curves follow the flow of the tangent line segments in the slope field in Figure 2.

As indicated above, drawing slope fields by hand is not a task for human beings—a 20 by 20 grid would require drawing 400 tangent line segments! Repetitive tasks of this type are what computers are for. A few problems in Exercise 11-3 involve interpreting slope fields, not drawing them.

Continuous Compound Interest Revisited

Let P be the initial amount of money deposited in an account, and let A be the amount in the account at any time t. Instead of assuming that the money in the account earns a particular rate of interest, suppose we say that the rate of growth of the amount of money in the account at any time t is proportional to the amount present at that time. Since dA/dt is the rate of growth of A with respect to t, we have

$$\frac{dA}{dt} = rA \qquad A(0) = P \qquad A, P > 0 \tag{4}$$

where r is an appropriate constant. We would like to find a function $A = A(t)$ that satisfies these conditions. Multiplying both sides of equation (4) by $1/A$, we obtain

$$\frac{1}{A}\frac{dA}{dt} = r$$

Now we integrate each side with respect to t:

$$\int \frac{1}{A}\frac{dA}{dt}\,dt = \int r\,dt \qquad \frac{dA}{dt}\,dt = A'(t)\,dt = dA$$

$$\int \frac{1}{A}\,dA = \int r\,dt$$

$$\ln|A| = rt + C \qquad |A| = A, \text{ since } A > 0$$

$$\ln A = rt + C$$

We convert this last equation into the equivalent exponential form

$$A = e^{rt+C} \qquad \text{Definition of logarithmic function: } y = \ln x \text{ if and only if } x = e^y$$

$$= e^C e^{rt} \qquad \text{Property of exponents: } b^m b^n = b^{m+n}$$

Since $A(0) = P$, we evaluate $A(t) = e^C e^{rt}$ at $t = 0$ and set it equal to P:

$$A(0) = e^C e^0 = e^C = P$$

Hence, $e^C = P$, and we can rewrite $A = e^C e^{rt}$ in the form

$$A = Pe^{rt}$$

This is the same continuous compound interest formula obtained in Section 10-1, where the principal P is invested at an annual nominal rate r compounded continuously for t years.

Exponential Growth Law

In general, if the rate of change with respect to time of a quantity Q is proportional to the amount present and $Q(0) = Q_0$, then proceeding in exactly the same way as above, we obtain the following:

Exponential Growth Law

If $\dfrac{dQ}{dt} = rQ$ and $Q(0) = Q_0$, then $Q = Q_0 e^{rt}$,

where

Q_0 = Amount at $t = 0$
r = Continuous compound growth rate (expressed as a decimal)
t = Time
Q = Quantity at time t

The constant r in the exponential growth law is sometimes called the **growth constant,** or the **growth rate.** The last term can be misleading, since the rate of growth of Q with respect to time is dQ/dt, not r. Notice that if $r < 0$, then $dQ/dt < 0$ and Q is decreasing. This type of growth is called **exponential decay.**

Once we know that the rate of growth of something is proportional to the amount present, then we know it has exponential growth and we can use the results summarized in the box without having to solve the differential equation each time. The exponential growth law applies not only to money invested at interest compounded continuously, but also to many other types of problems—population growth, radioactive decay, natural resource depletion, and so on.

Population Growth; Radioactive Decay; Learning

The world population is growing at an ever-increasing rate, as illustrated in Figure 4. **Population growth** over certain periods of time often can be approximated by the exponential growth law described above.

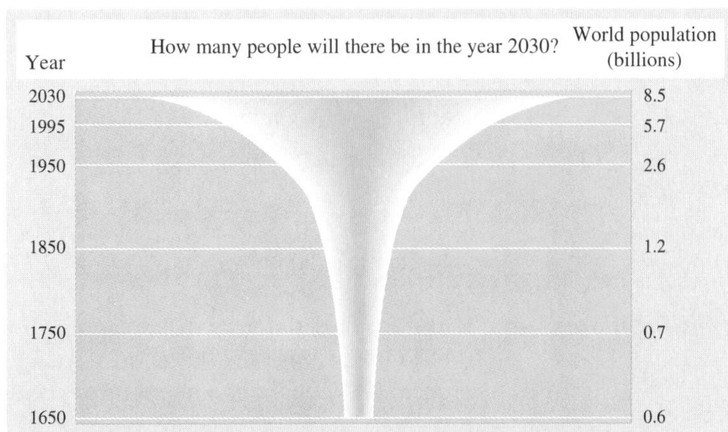

FIGURE 4
The population explosion
Source: World Bank, 1994; *World Almanac,* 1994

Example 1 **Population Growth** India had a population of about 0.9 billion people in 1995 ($t = 0$) and a growth rate of 1.3% per year, which we will assume is compounded continuously. Let P represent the population (in billions) t years after 1995.

(A) Find an equation that represents India's population growth after 1995, assuming the 1.3% growth rate continues.
(B) What is the estimated population (to the nearest tenth of a billion) for India in the year 2030?
(C) Graph the equation found in part (A) from 1995 to 2030.

SOLUTION (A) The exponential growth law applies, and we have

$$\frac{dP}{dt} = 0.013P \qquad P(0) = 0.9$$

Thus,

$$P = 0.9e^{0.013t} \tag{5}$$

(B) Using equation (5), we can estimate the population in India in 2030 ($t = 35$):

$$P = 0.9e^{0.013(35)} = 1.4 \text{ billion people}$$

(C) The graph is shown in Figure 5.

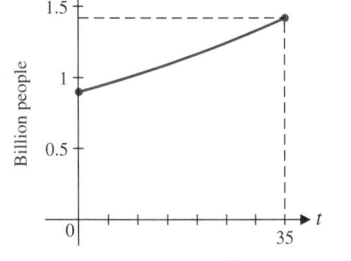

FIGURE 5

Matched Problem 1 Assuming the same continuous compound growth rate as in Example 1, what will India's population be (to the nearest tenth of a billion) in the year 2012?

Example 2 **Population Growth** If the exponential growth law applies to Canada's population growth, at what continuous compound growth rate will the population double over the next 100 years?

SOLUTION The problem is to find r, given $P = 2P_0$ and $t = 100$:

$$\begin{aligned} P &= P_0e^{rt} \\ 2P_0 &= P_0e^{100r} \\ 2 &= e^{100r} \\ 100r &= \ln 2 \\ r &= \frac{\ln 2}{100} \\ &\approx 0.0069 \quad \text{or} \quad 0.69\% \end{aligned}$$

Take the natural logarithm of both sides and reverse the equation.

Matched Problem 2 ➡ If the exponential growth law applies to population growth in Nigeria, find the doubling time (to the nearest year) of the population if it continues to grow at 2.1% per year compounded continuously.

We now turn to another type of exponential growth—**radioactive decay.** In 1946, Willard Libby (who later received a Nobel prize in chemistry) found that as long as a plant or animal is alive, radioactive carbon-14 is maintained at a constant level in its tissues. Once the plant or animal is dead, however, the radioactive carbon-14 diminishes by radioactive decay at a rate proportional to the amount present. Thus,

$$\frac{dQ}{dt} = rQ \qquad Q(0) = Q_0$$

and we have another example of the exponential growth law. The continuous compound rate of decay for radioactive carbon-14 has been found to be 0.000 123 8; thus, $r = -0.000\ 123\ 8$, since decay implies a negative continuous compound growth rate.

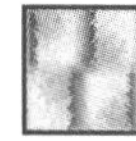

Example 3 ➡ **Archaeology** A piece of human bone was found at an archaeological site in Africa. If 10% of the original amount of radioactive carbon-14 was present, estimate the age of the bone (to the nearest 100 years).

SOLUTION Using the exponential growth law for

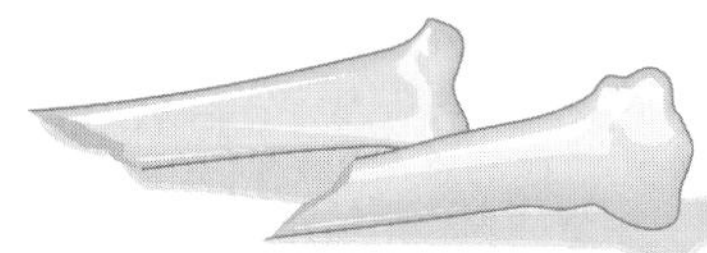

$$\frac{dQ}{dt} = -0.000\ 123\ 8Q \qquad Q(0) = Q_0$$

we find that

$$Q = Q_0e^{-0.0001238t}$$

and our problem is to find t so that $Q = 0.1Q_0$ (since the amount of carbon-14 present now is 10% of the amount present, Q_0, at the death of the person). Thus,

$$\begin{aligned} 0.1Q_0 &= Q_0e^{-0.0001238t} \\ 0.1 &= e^{-0.0001238t} \\ \ln 0.1 &= \ln e^{-0.0001238t} \\ t &= \frac{\ln 0.1}{-0.000\ 123\ 8} \approx 18{,}600 \text{ years} \end{aligned}$$

1

0 30000

Intersection
X=18599.233 Y=.1

−0.5

FIGURE 6
$y_1 = e^{-0.0001238x}$; $y_2 = 0.1$

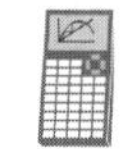

See Figure 6 for a graphical solution to Example 3.

Matched Problem 3 Estimate the age of the bone in Example 3 (to the nearest 100 years) if 50% of the original amount of carbon-14 is present.

In learning certain skills such as typing and swimming, a mathematical model often used is one that assumes there is a maximum skill attainable, say, M, and the rate of improving is proportional to the difference between that achieved, y, and that attainable, M. Mathematically,

$$\frac{dy}{dt} = k(M - y) \qquad y(0) = 0$$

We solve this type of problem using the same technique that was used to obtain the exponential growth law. First, multiply both sides of the first equation by $1/(M - y)$ to obtain

$$\frac{1}{M - y}\frac{dy}{dt} = k$$

and then integrate each side with respect to t:

$$\int \frac{1}{M - y}\frac{dy}{dt}\,dt = \int k\,dt$$

$$-\int \frac{1}{M-y}\left(-\frac{dy}{dt}\right) dt = \int k\,dt \qquad \text{Substitute } u = M - y \text{ and } du = -dy = -\frac{dy}{dt}\,dt.$$

$$-\int \frac{1}{u}\,du = \int k\,dt$$

$$-\ln|u| = kt + C \qquad \text{Substitute } M - y = u.$$

$$-\ln(M - y) = kt + C \qquad \text{Absolute value signs are not required. (Why?)}$$

$$\ln(M - y) = -kt - C$$

Change this last equation to equivalent exponential form:

$$M - y = e^{-kt-C}$$
$$M - y = e^{-C}e^{-kt}$$
$$y = M - e^{-C}e^{-kt}$$

Now, $y(0) = 0$; hence,

$$y(0) = M - e^{-C}e^{0} = 0$$

Solving for e^{-C}, we obtain

$$e^{-C} = M$$

and our final solution is

$$y = M - Me^{-kt} = M(1 - e^{-kt})$$

Example 4 **Learning** For a particular person who is learning to swim, it is found that the distance y (in feet) the person is able to swim in 1 minute after t hours of practice is given approximately by

$$y = 50(1 - e^{-0.04t})$$

What is the rate of improvement (to two decimal places) after 10 hours of practice?

SOLUTION

$$y = 50 - 50e^{-0.04t}$$
$$y'(t) = 2e^{-0.04t}$$
$$y'(10) = 2e^{-0.04(10)} \approx 1.34 \text{ feet per hour of practice}$$

Matched Problem 4 In Example 4, what is the rate of improvement (to two decimal places) after 50 hours of practice?

A Comparison of Exponential Growth Phenomena

The graphs and equations given in Table 1 compare several widely used growth models. These are divided basically into two groups: unlimited growth and limited growth. Following each equation and graph is a short (and necessarily incomplete) list of areas in which the models are used. This only touches on a subject that has been extensively developed and which you are likely to encounter in greater depth in the future.

Table 1

Exponential Growth

Description	Model	Solution	Graph	Uses
Unlimited growth: Rate of growth is proportional to the amount present	$\frac{dy}{dt} = ky$ $k, t > 0$ $y(0) = c$	$y = ce^{kt}$	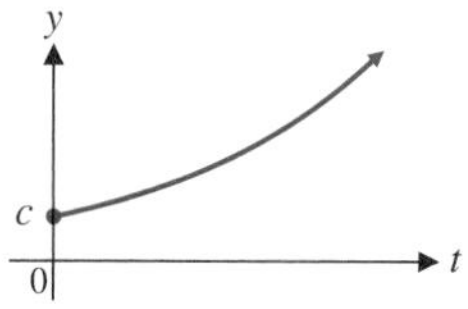	• Short-term population growth (people, bacteria, etc.) • Growth of money at continuous compound interest • Price–supply curves
Exponential decay: Rate of growth is proportional to the amount present	$\frac{dy}{dt} = -ky$ $k, t > 0$ $y(0) = c$	$y = ce^{-kt}$	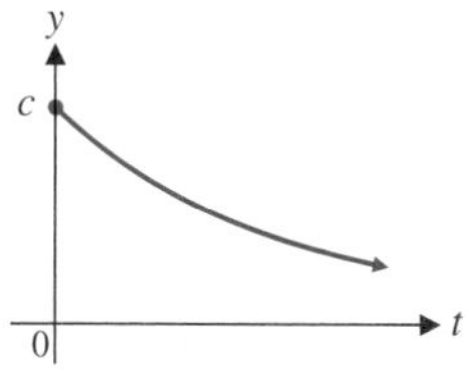	• Depletion of natural resources • Radioactive decay • Light absorption in water • Price–demand curves • Atmospheric pressure (t is altitude)
Limited growth: Rate of growth is proportional to the difference between the amount present and a fixed limit	$\frac{dy}{dt} = k(M - y)$ $k, t > 0$ $y(0) = 0$	$y = M(1 - e^{-kt})$	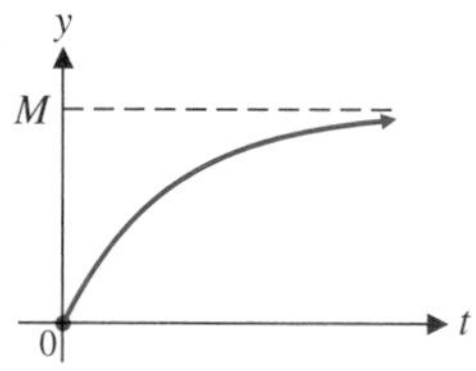	• Sales fads (for example, skateboards) • Depreciation of equipment • Company growth • Learning
Logistic growth: Rate of growth is proportional to the amount present and to the difference between the amount present and a fixed limit	$\frac{dy}{dt} = -ky(M - y)$ $k, t > 0$ $y(0) = \frac{M}{1 + c}$	$y = \frac{M}{1 + ce^{-kMt}}$	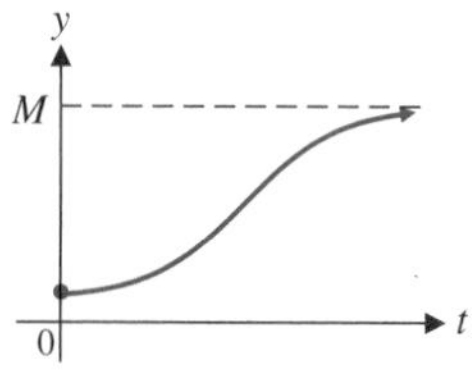	• Long-term population growth • Epidemics • Sales of new products • Rumor spread • Company growth

Answers to Matched Problems **1.** 1.1 billion people **2.** 33 yr **3.** 5,600 yr **4.** 0.27 ft/hr

EXERCISE 11-3

A *In Problems 1–8, find the general or particular solution, as indicated, for each differential equation.*

1. $\frac{dy}{dx} = 4.16x$

2. $\frac{dy}{dx} = x^{-2}$

3. $\frac{dy}{dx} = e^{0.5x}$

4. $\frac{dy}{dx} = \frac{2}{x}$

5. $\frac{dy}{dx} = x^2 - x;\ y(0) = 0$

6. $\frac{dy}{dx} = \sqrt{x};\ y(0) = 0$

7. $\frac{dy}{dx} = -2xe^{-x^2};\ y(0) = 3$

8. $\frac{dy}{dx} = e^{x-3};\ y(3) = -5$

B *Problems 9–14 refer to the following slope fields:*

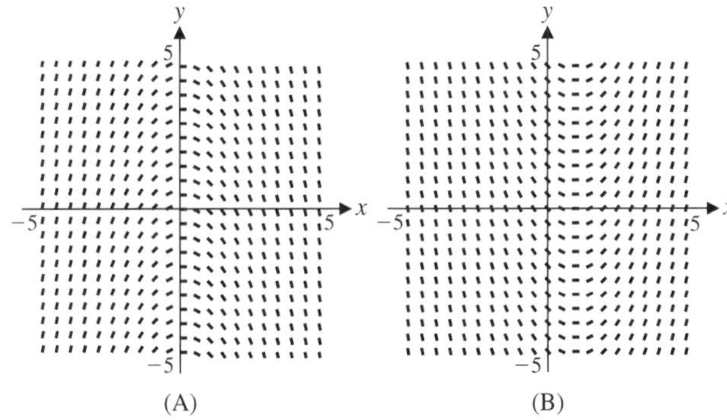

9. Which slope field is associated with the differential equation $dy/dx = x - 1$? Briefly justify your answer.

10. Which slope field is associated with the differential equation $dy/dx = -x$? Briefly justify your answer.

11. Solve the differential equation $dy/dx = x - 1$, and find the particular solution that passes through $(0, -2)$.

12. Solve the differential equation $dy/dx = -x$, and find the particular solution that passes through $(0, 3)$.

13. Graph the particular solution found in Problem 11 in the appropriate figure above (or a copy).

14. Graph the particular solution found in Problem 12 in the appropriate figure above (or a copy).

In Problems 15–20, find the general or particular solution, as indicated, for each differential equation.

15. $\frac{dy}{dx} = -0.8y$

16. $\frac{dy}{dx} = 0.6y$

17. $\frac{dy}{dx} = 0.07y;\ y(0) = 1{,}000$

18. $\frac{dy}{dx} = -0.004y;\ y(0) = 5$

19. $\frac{dx}{dt} = -x$

20. $\frac{dy}{dt} = y$

C *Problems 21–28 refer to the following slope fields:*

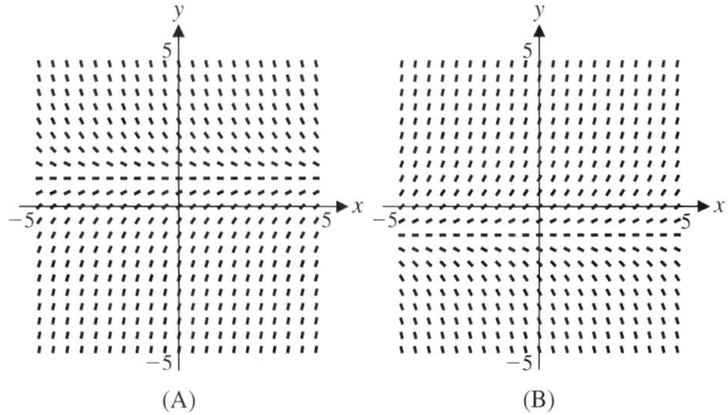

21. Which slope field is associated with the differential equation $dy/dx = 1 - y$? Briefly justify your answer.

22. Which slope field is associated with the differential equation $dy/dx = y + 1$? Briefly justify your answer.

23. Show that $y = 1 - Ce^{-x}$ is a solution of the differential equation $dy/dx = 1 - y$ for any real number C. Find the particular solution that passes through $(0, 0)$.

24. Show that $y = Ce^{x} - 1$ is a solution of the differential equation $dy/dx = y + 1$ for any real number C. Find the particular solution that passes through $(0, 0)$.

25. Graph the particular solution found in Problem 23 in the appropriate figure above (or a copy).

26. Graph the particular solution found in Problem 24 in the appropriate figure above (or a copy).

27. C Use a graphing utility to graph $y = 1 - Ce^{-x}$ for $C = -2, -1, 1,$ and 2 for $-5 \leq x \leq 5, -5 \leq y \leq 5$, all in the same viewing window. Observe how the solution curves go with the flow of the tangent line segments in the corresponding slope field shown above.

28. C Use a graphing utility to graph $y = Ce^{x} - 1$ for $C = -2, -1, 1,$ and 2 for $-5 \leq x \leq 5, -5 \leq y \leq 5$, all in the same viewing window. Observe how the solution curves go with the flow of the tangent line segments in the corresponding slope field shown above.

In Problems 29–36, use a graphing utility to graph the given examples of the various cases in Table 1.

29. Unlimited growth:
$y = 1{,}000e^{0.08t}$
$0 \leq t \leq 15$
$0 \leq y \leq 3{,}500$

30. Unlimited growth:
$y = 5{,}250e^{0.12t}$
$0 \leq t \leq 10$
$0 \leq y \leq 20{,}000$

31. Exponential decay:
$p = 100e^{-0.05x}$
$0 \leq x \leq 30$
$0 \leq p \leq 100$

32. Exponential decay:
$p = 1{,}000e^{-0.08x}$
$0 \leq x \leq 40$
$0 \leq p \leq 1{,}000$

33. Limited growth:
$N = 100(1 - e^{-0.05t})$
$0 \leq t \leq 100$
$0 \leq N \leq 100$

34. Limited growth:
$N = 1{,}000(1 - e^{-0.07t})$
$0 \leq t \leq 70$
$0 \leq N \leq 1{,}000$

35. Logistic growth:
$N = \dfrac{1{,}000}{1 + 999e^{-0.4t}}$
$0 \leq t \leq 40$
$0 \leq N \leq 1{,}000$

36. Logistic growth:
$N = \dfrac{400}{1 + 99e^{-0.4t}}$
$0 \leq t \leq 30$
$0 \leq N \leq 400$

37. Show that the rate of logistic growth, $dy/dt = ky(M - y)$, has its maximum value when $y = M/2$.

38. Find the value of t for which the logistic function

$$y = \frac{M}{1 + ce^{-kMt}}$$

is equal to $M/2$.

Applications

Business & Economics

39. *Continuous compound interest.* Find the amount A in an account after t years if

$$\frac{dA}{dt} = 0.08A \qquad \text{and} \qquad A(0) = 1{,}000$$

40. *Continuous compound interest.* Find the amount A in an account after t years if

$$\frac{dA}{dt} = 0.12A \qquad \text{and} \qquad A(0) = 5{,}250$$

41. *Continuous compound interest.* Find the amount A in an account after t years if

$$\frac{dA}{dt} = rA \qquad A(0) = 8{,}000 \qquad A(2) = 9{,}020$$

42. *Continuous compound interest.* Find the amount A in an account after t years if

$$\frac{dA}{dt} = rA \qquad A(0) = 5{,}000 \qquad A(5) = 7{,}460$$

43. *Price–demand.* The marginal price dp/dx at x units of demand per week is proportional to the price p. There is no weekly demand at a price of \$100 per unit $[p(0) = 100]$, and there is a weekly demand of 5 units at a price of \$77.88 per unit $[p(5) = 77.88]$.
(A) Find the price–demand equation.
(B) At a demand of 10 units per week, what is the price?
(C) Graph the price–demand equation for $0 \leq x \leq 25$.

44. *Price–supply.* The marginal price dp/dx at x units of supply per day is proportional to the price p. There is no supply at a price of \$10 per unit $[p(0) = 10]$, and there is a daily supply of 50 units at a price of \$12.84 per unit $[p(50) = 12.84]$.
(A) Find the price–supply equation.
(B) At a supply of 100 units per day, what is the price?
(C) Graph the price–supply equation for $0 \leq x \leq 250$.

45. *Advertising.* A company is trying to expose a new product to as many people as possible through television advertising. Suppose the rate of exposure to new people is proportional to the number of those who have not seen the product out of L possible viewers. No one is aware of the product at the start of the campaign, and after 10 days 40% of L are aware of the product. Mathematically,

$$\frac{dN}{dt} = k(L - N) \qquad N(0) = 0 \qquad N(10) = 0.4L$$

(A) Solve the differential equation.
(B) What percent of L will have been exposed after 5 days of the campaign?
(C) How many days will it take to expose 80% of L?
(D) Graph the solution found in part (A) for $0 \leq t \leq 90$.

46. *Advertising.* Suppose the differential equation for Problem 45 is

$$\frac{dN}{dt} = k(L - N) \qquad N(0) = 0 \qquad N(10) = 0.1L$$

(A) Interpret $N(10) = 0.1L$ verbally.
(B) Solve the differential equation.
(C) How many days will it take to expose 50% of L?
(D) Graph the solution found in part (B) for $0 \leq t \leq 300$.

Life Sciences

47. *Biology.* For relatively clear bodies of water, light intensity is reduced according to

$$\frac{dI}{dx} = -kI \qquad I(0) = I_0$$

where I is the intensity of light at x feet below the surface. For the Sargasso Sea off the West Indies, $k = 0.009\ 42$. Find I in terms of x, and find the depth at which the light is reduced to half of that at the surface.

48. *Blood pressure.* It can be shown under certain assumptions that blood pressure P in the largest artery in the human body (the aorta) changes between beats with respect to time t according to

$$\frac{dP}{dt} = -aP \qquad P(0) = P_0$$

where a is a constant. Find $P = P(t)$ that satisfies both conditions.

49. *Drug concentration.* A single injection of a drug is administered to a patient. The amount Q in the body then decreases at a rate proportional to the amount present, and for a particular drug the rate is 4% per hour. Thus,

$$\frac{dQ}{dt} = -0.04Q \qquad Q(0) = Q_0$$

where t is time in hours.

(A) If the initial injection is 3 milliliters $[Q(0) = 3]$, find $Q = Q(t)$ satisfying both conditions.

(B) How many milliliters (to two decimal places) are in the body after 10 hours?

(C) How many hours (to two decimal places) will it take for only 1 milliliter of the drug to be left in the body?

(D) Graph the solution found in part (A).

50. *Simple epidemic.* A community of 1,000 individuals is assumed to be homogeneously mixed. One individual who has just returned from another community has influenza. Assume the home community has not had influenza shots and all are susceptible. One mathematical model for an influenza epidemic assumes that influenza tends to spread at a rate in direct proportion to the number who have it, N, and to the number who have not yet contracted it—in this case, $1{,}000 - N$. Mathematically,

$$\frac{dN}{dt} = kN(1{,}000 - N) \qquad N(0) = 1$$

where N is the number of people who have contracted influenza after t days. For $k = 0.0004$, it can be shown that $N(t)$ is given by

$$N(t) = \frac{1{,}000}{1 + 999e^{-0.4t}}$$

(A) How many people have contracted influenza after 10 days? After 20 days?

(B) How many days will it take until half the community has contracted influenza?

(C) Find $\lim_{t \to \infty} N(t)$.

C (D) Graph $N = N(t)$ for $0 \le t \le 30$.

51. *Nuclear accident.* One of the dangerous radioactive isotopes detected after the Chernobyl nuclear accident in 1986 was cesium-137. If 93.3% of the cesium-137 emitted during the accident is still present 3 years later, find the continuous compound rate of decay of this isotope.

52. *Insecticides.* Many countries have banned the use of the insecticide DDT because of its long-term adverse effects. Five years after a particular country stopped using DDT, the amount of DDT in the ecosystem had declined to 75% of the amount present at the time of the ban. Find the continuous compound rate of decay of DDT.

Social Sciences

53. *Archaeology.* A skull from an ancient tomb was discovered and was found to have 5% of the original amount of radioactive carbon-14 present. Estimate the age of the skull. (See Example 3.)

54. *Learning.* For a particular person learning to type, it was found that the number of words per minute, N, the person was able to type after t hours of practice was given approximately by

$$N = 100(1 - e^{-0.02t})$$

See Table 1 (limited growth) for a characteristic graph. What is the rate of improvement after 10 hours of practice? After 40 hours of practice?

55. *Small group analysis.* In a study on small group dynamics, sociologists Stephan and Mischler found that, when the members of a discussion group of 10 were ranked according to the number of times each participated, the number of times $N(k)$ the kth-ranked person participated was given approximately by

$$N(k) = N_1 e^{-0.11(k-1)} \qquad 1 \le k \le 10$$

where N_1 is the number of times the 1st-ranked person participated in the discussion. If, in a particular discussion group of 10 people, $N_1 = 180$, estimate how many times the 6th-ranked person participated. The 10th-ranked person.

56. *Perception.* One of the oldest laws in mathematical psychology is the Weber–Fechner law (discovered in the middle of the nineteenth century). It concerns a person's sensed perception of various strengths of stimulation involving weights, sound, light, shock, taste, and so on.

One form of the law states that the rate of change of sensed sensation S with respect to stimulus R is inversely proportional to the strength of the stimulus R. Thus,

$$\frac{dS}{dR} = \frac{k}{R}$$

where k is a constant. If we let R_0 be the threshold level at which the stimulus R can be detected (the least amount of sound, light, weight, and so on, that can be detected), then it is appropriate to write

$$S(R_0) = 0$$

Find a function S in terms of R that satisfies the above conditions.

57. *Rumor spread.* A group of 400 parents, relatives, and friends are waiting anxiously at Kennedy Airport for a student charter flight to return after a year in Europe. It is stormy and the plane is late. A particular parent thought he had heard that the plane's radio had gone out and related this news to some friends, who in turn passed it on to others, and so on. Sociologists have studied rumor propagation and have found that a rumor tends to spread at a rate in direct proportion to the number who have heard it, x, and to the number who have not, $P - x$, where P is the total population. Mathematically, for our case, $P = 400$ and

$$\frac{dx}{dt} = 0.001x(400 - x) \qquad x(0) = 1$$

where t is time (in minutes). From this, it can be shown that

$$x(t) = \frac{400}{1 + 399e^{-0.4t}}$$

See Table 1 (logistic growth) for a characteristic graph.

(A) How many people have heard the rumor after 5 minutes? 20 minutes?

(B) Find $\lim_{t\to\infty} x(t)$.

C (C) Graph $x = x(t)$ for $0 \leq t \leq 30$.

58. *Rumor spread.* In Problem 57, how long (to the nearest minute) will it take for half of the group of 400 to have heard the rumor?

Section 11-4 A Geometric–Numeric Introduction to the Definite Integral

- Area
- Rate, Area, and Distance
- Rate, Area, and Total Change

The first three sections of this chapter were focused on the *indefinite integral*—the set of all antiderivatives of a given function. In this and the next section we introduce a related form called the *definite integral.* Our approach in this section will be intuitive and informal. In the next section these concepts will be made more precise.

We introduce the *definite integral* by considering the area bounded by the graph of a function f, the x axis, and the vertical lines $x = a$ and $x = b$, as shown in Figure 1. If $f(x)$ is positive, then we refer to this area as the area under the graph of $y = f(x)$ from $x = a$ to $x = b$.

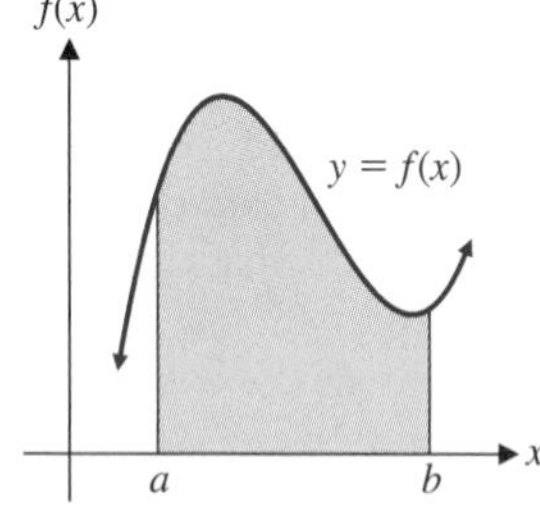

FIGURE 1
Area under the graph of $y = f(x)$

We will denote the shaded area in Figure 1 by the **definite integral symbol:**

$$\int_a^b f(x)\,dx = \begin{pmatrix}\text{Area under the graph}\\ \text{from } x = a \text{ to } x = b\end{pmatrix}$$

This symbol is precisely defined in Section 11-5, where we will also see how the definite integral of a function over an interval $[a, b]$ is related to the indefinite integral (antiderivative) of the function through the fundamental theorem of calculus. To keep things simple at the start, we restrict $f(x)$ to positive values—this restriction will be relaxed shortly. We also assume all functions referred to are continuous over any interval under consideration.

Surprisingly, out of this analysis of area we will be able to find the total change in a function over an interval from its derivative. For example, given a rate function (derivative) or a table of rate values, we will be able to determine the distance that a steel ball dropped from a bridge travels in a given time interval, or the additional costs resulting from an increase in production.

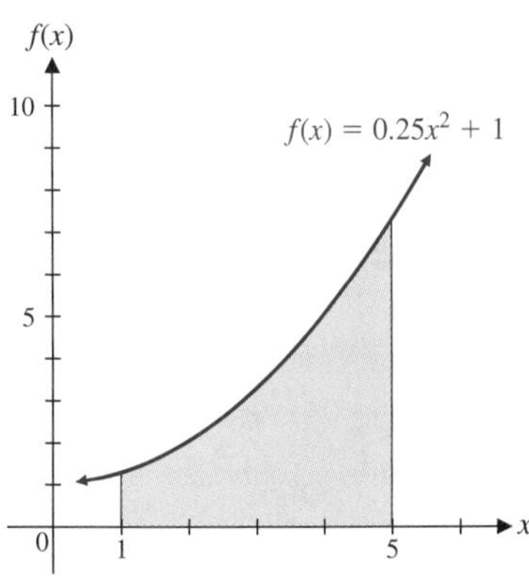

FIGURE 2
What is the shaded area?

■ AREA

How do we find the shaded area in Figure 2? That is, how do we find the area bounded by the graph of $f(x) = 0.25x^2 + 1$, the x axis, and the vertical lines $x = 1$ and $x = 5$? [This cumbersome description is usually shortened to "the area under the graph of $f(x) = 0.25x^2 + 1$ from $x = 1$ to $x = 5$."] Our standard geometric area formulas do not apply directly, but the formula for the area of a rectangle can be used indirectly. To see how, we look at a method of approximating the area under the graph by using rectangles. This method will give us any accuracy desired, which is quite different from finding the area exactly. Our first area approximation is made by dividing the interval [1, 5] on the x axis into four equal parts of length

$$\Delta x = \frac{5 - 1}{4} = 1$$

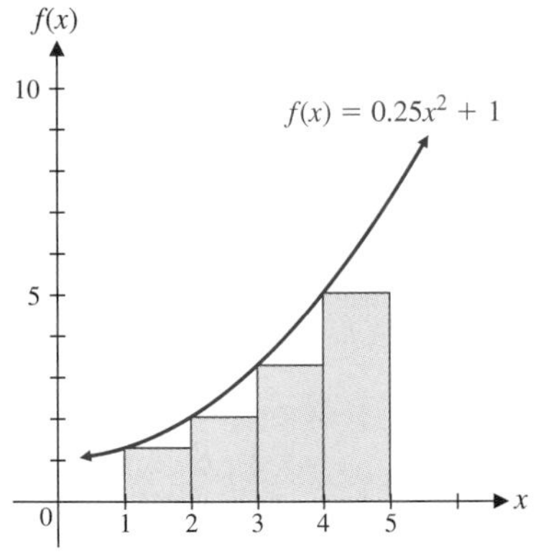

(It is customary to denote the length of the subintervals by Δx, which is read "delta x," since Δ is the Greek letter delta.) We then place a rectangle on each subinterval with a height determined by the function evaluated at the left end point of the subinterval (see Fig. 3).

FIGURE 3
Left rectangles

Summing the areas of the rectangles in Figure 3, we obtain a **left sum** of four rectangles, denoted by L_4, as follows:

$$\begin{aligned} L_4 &= f(1) \cdot 1 + f(2) \cdot 1 + f(3) \cdot 1 + f(4) \cdot 1 \\ &= 1.25 + 2.00 + 3.25 + 5 = 11.5 \end{aligned}$$

From Figure 3, since $f(x)$ is increasing, it is clear that the left sum L_4 underestimates the area, and we can write

$$11.5 = L_4 < \int_1^5 (0.25x^2 + 1)\,dx = \text{Area}$$

Explore–Discuss 1

If $f(x)$ were decreasing over the interval [1, 5], would the left sum L_4 over- or underestimate the actual area under the curve? Explain.

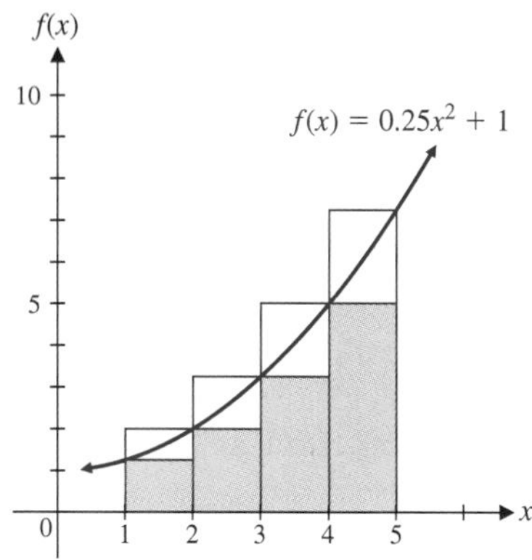

FIGURE 4
Left and right rectangles

Now suppose we use the right end point of each subinterval to obtain the height of the rectangle placed on top of it. Superimposing this result on top of Figure 3, we obtain Figure 4.

Summing the areas of the higher rectangles in Figure 4, we obtain the **right sum** of the four rectangles, denoted by R_4, as follows (compare R_4 below with L_4 above and note that R_4 can be obtained from L_4 by deleting one rectangle area and adding one more):

$$\begin{aligned} R_4 &= f(2)\cdot 1 + f(3)\cdot 1 + f(4)\cdot 1 + f(5)\cdot 1 \\ &= 2.00 + 3.25 + 5.00 + 7.25 = 17.5 \end{aligned}$$

From Figure 4, since $f(x)$ is increasing, it is clear that the right sum R_4 overestimates the area, and we conclude that the actual area is between 11.5 and 17.5. That is,

$$11.5 = L_4 < \int_1^5 (0.25x^2 + 1)\, dx < R_4 = 17.5 \qquad (1)$$

Since the actual area lies between the left estimate and the right estimate, the average should be even closer:

$$\text{Average} = \frac{L_4 + R_4}{2} = \frac{11.5 + 17.5}{2} = 14.5$$

Explore–Discuss 2

If $f(x)$ in Figure 4 were decreasing over the interval $[1, 5]$, would the right sum R_4 over- or underestimate the actual area under the curve? Explain.

The first approximation of the area under the curve in (1) is fairly coarse, but the method outlined can be continued with increasingly accurate results by dividing the interval $[1, 5]$ into more and more equal subintervals. Of course, this is not a job for hand calculation, but a job computers are designed to do.* Figure 5 shows left and right rectangle approximations for 16 equal subdivisions.

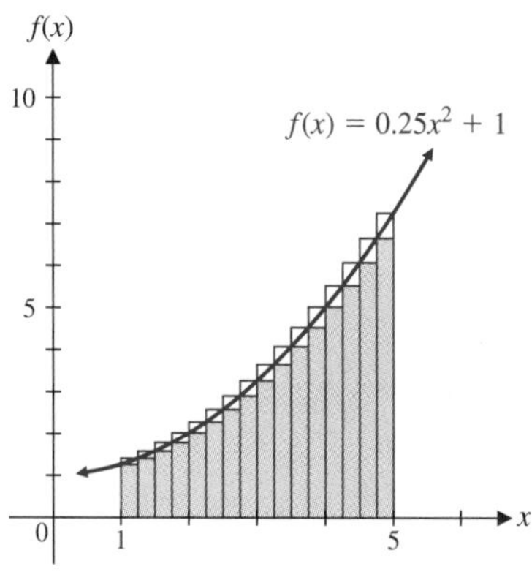

FIGURE 5

For this case,

$$\begin{aligned} \Delta x &= \frac{5 - 1}{16} = 0.25 \\ L_{16} &= f(1)\cdot\Delta x + f(1.25)\cdot\Delta x + \cdots + f(4.75)\cdot\Delta x \\ &= 13.59 \\ R_{16} &= f(1.25)\cdot\Delta x + f(1.50)\cdot\Delta x + \cdots + f(5)\cdot\Delta x \\ &= 15.09 \end{aligned}$$

Thus, we now know that the area under the curve is between 13.59 and 15.09. That is,

*The computer software that accompanies this text will perform these calculations (see the Preface).

$$13.59 = L_{16} < \int_1^5 (0.25x^2 + 1)\,dx < R_{16} = 15.09 \tag{2}$$

$$\text{Average} = \frac{L_{16} + R_{16}}{2} = \frac{13.59 + 15.09}{2} = 14.34$$

For 100 equal subdivisions, computer calculations give us

$$14.214 = L_{100} < \int_1^5 (0.25x^2 + 1)\,dx < R_{100} = 14.454 \tag{3}$$

$$\text{Average} = \frac{L_{100} + R_{100}}{2} = \frac{14.214 + 14.454}{2} = 14.334$$

For an approximation process to be useful, we need an estimate of the maximum possible error produced by the process. We define the **error in an approximation** to be the absolute value of the difference between the approximated value and the actual value.

Explore–Discuss 3

If a and b are the coordinates of two points on a real number line, then the distance between these two points is given by $|b - a|$. The following inequalities involving distance on a real number line are very useful for estimating errors in approximations.

1. If x is between a and b on a number line, then

$$|x - a| \leq |b - a| \quad \text{and} \quad |x - b| \leq |b - a|$$

2. If $A = \dfrac{a + b}{2}$, then

$$|x - A| \leq \tfrac{1}{2}|b - a|$$

Interpret each of these inequalities both verbally and geometrically on a real number line.

A function is **monotone** over an interval $[a, b]$ if it is either increasing over $[a, b]$ or decreasing over $[a, b]$. The following important observation about monotone functions forms the basis for estimating the error when a left sum or a right sum is used to approximate an area.

If f is a monotone function, then the area under the graph of $f(x)$ always lies between the left sum L_n and the right sum R_n for any integer n.

Consider a positive monotone function $y = f(x)$, as shown in Figure 6 (at the top of the next page). If $\int_a^b f(x)\,dx$ is the actual area under the graph of $y = f(x)$ from $x = a$ to $x = b$, then this area lies between L_n and R_n. Thus, approximating this area with either L_n or R_n produces an error that is less than the absolute value of the difference between the left sum and the right sum, $|R_n - L_n|$. This difference is just the sum of the areas of the white rectangles shown in Figure 6. (Think about this.)

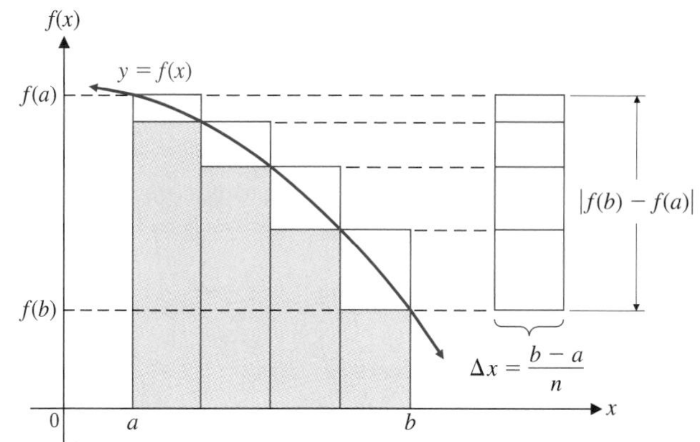

FIGURE 6
$|R_n - L_n| = |f(b) - f(a)|\Delta x$

Thus,

$$\text{Error} \leqslant |R_n - L_n| = |f(b) - f(a)|\Delta x = |f(b) - f(a)|\frac{b-a}{n}$$

If the approximation is the average of left and right sums, then the bound for the error is cut in half. These results are summarized in the following box for convenient reference:

Error Bounds for Left and Right Sums and Their Average (Monotone Functions)

If $f(x)$ is monotonic on the interval $[a, b]$ and

$$I = \int_a^b f(x)\,dx, \quad L_n = \text{Left sum}, \quad R_n = \text{Right sum}, \quad A_n = \frac{L_n + R_n}{2}$$

then the following error bounds hold:

$$|I - L_n| \leqslant |f(b) - f(a)|\frac{b-a}{n}$$

$$|I - R_n| \leqslant |f(b) - f(a)|\frac{b-a}{n}$$

$$|I - A_n| \leqslant |f(b) - f(a)|\frac{b-a}{2n}$$

These formulas not only enable us to compute error bounds for a particular approximation, L_n, R_n, or A_n, but more importantly, they can be used to tell us in advance how large n should be to produce an error less than a specified amount.

Example 1 **Approximating Areas** Given the function $f(x) = 9 - 0.25x^2$, we are interested in approximating the area under $y = f(x)$ from $x = 2$ to $x = 5$.

(A) Graph the function over the interval $[0, 6]$; then draw left and right rectangles for the interval $[2, 5]$ with $n = 6$.

(B) Calculate L_6, R_6, and A_6; then calculate error bounds for each.
(C) How large should n be chosen for L_n, R_n, and A_n for the approximation of $\int_2^5 (9 - 0.25x^2)\,dx$ to be within 0.05 of the true value?

SOLUTION (A) $\Delta x = 0.5$:

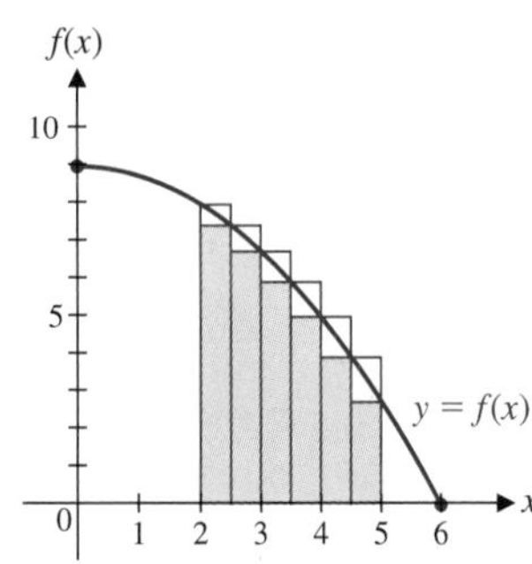

(B) $L_6 = f(2)\cdot\Delta x + f(2.5)\cdot\Delta x + f(3)\cdot\Delta x + f(3.5)\cdot\Delta x + f(4)\cdot\Delta x + f(4.5)\cdot\Delta x = 18.53$

$R_6 = f(2.5)\cdot\Delta x + f(3)\cdot\Delta x + f(3.5)\cdot\Delta x + f(4)\cdot\Delta x + f(4.5)\cdot\Delta x + f(5)\cdot\Delta x = 15.91$

$$A_6 = \frac{L_6 + R_6}{2} = 17.22$$

Error bound for L_6 and R_6:

$$\text{Error} \leqslant |f(5) - f(2)|\frac{5-2}{6} = |2.75 - 8|(0.5) = 2.625$$

Error bound for A_6:

$$\text{Error} \leqslant \frac{2.625}{2} = 1.3125$$

(C) For L_n and R_n, find n such that Error $\leqslant 0.05$:

$$|f(b) - f(a)|\frac{b-a}{n} \leqslant 0.05$$
$$|2.75 - 8|\frac{3}{n} \leqslant 0.05$$
$$15.75 \leqslant 0.05n$$
$$n \geqslant \frac{15.75}{0.05} = 315$$

For A_n, find n such that Error $\leqslant 0.05$:

$$|f(b) - f(a)|\frac{b-a}{2n} \leqslant 0.05$$
$$|2.75 - 8|\frac{3}{2n} \leqslant 0.05$$
$$7.875 \leqslant 0.05n$$
$$n \geqslant \frac{7.875}{0.05} = 157.5$$
$$n \geqslant 158$$

Round up to the next integer.

Matched Problem 1 Given the function $f(x) = 8 - 0.5x^2$, we are interested in approximating the area under $y = f(x)$ from $x = 1$ to $x = 3$.

(A) Graph the function over the interval $[0, 4]$; then draw left and right rectangles for the interval $[1, 3]$ with $n = 4$.
(B) Calculate L_4, R_4, and A_4; then calculate error bounds for each.
(C) How large should n be chosen for L_n, R_n, and A_n for the approximation of $\int_2^5 (8 - 0.5x^2)\,dx$ to be within 0.5 of the true value?

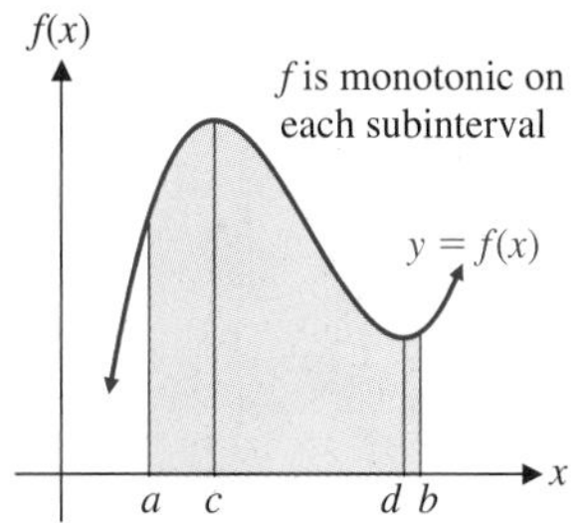

FIGURE 7
A nonmonotonic function

If f is not monotonic, then the area may not lie between the left sum and the right sum, and our error estimates cannot be applied to the entire interval. However, we can usually divide the interval $[a, b]$ into subintervals such that f is monotonic on each, and then proceed as above (see Fig. 7).

The discussion until now assumed that the graph of $y = f(x)$ is above the x axis; that is, $f(x)$ is positive on the interval $[a, b]$. If part (or all) of the graph lies below the x axis, then for some (or all) values of x, $f(x)$ will be negative and $f(x)\,\Delta x$ will be the negative of the area of a rectangle lying below the x axis (Fig. 8). In this case, the left sum and the right sum, L_n and R_n, are the sums of the areas of the rectangles above the x axis and the negatives of the areas of the rectangles below the x axis. It is customary to refer to these as **signed areas,** since the positive quantities represent areas and the negative quantities represent negatives of areas.

This observation leads to the following expanded interpretation of the *definite integral symbol:*

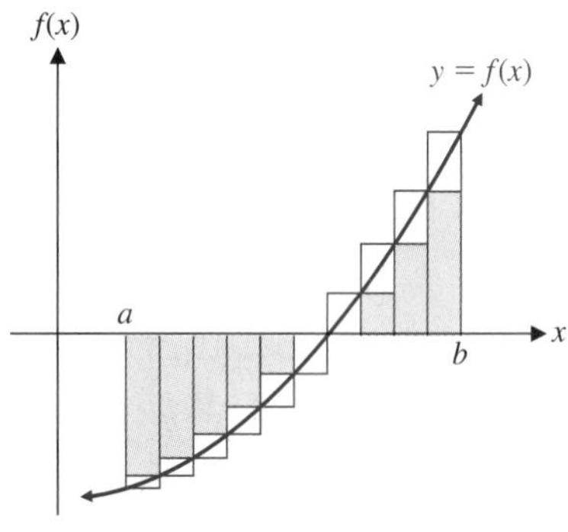

FIGURE 8
Signed areas

Definite Integral Symbol for Functions with Negative Values

If $f(x)$ is positive for some values of x on $[a, b]$ and negative for others, then the **definite integral symbol**

$$\int_a^b f(x)\,dx$$

represents the cumulative sum of the signed areas between the graph of $y = f(x)$ and the x axis where the areas above the x axis are counted positively and the areas below the x axis are counted negatively (see Fig. 9, where A and B are actual areas of the indicated regions).

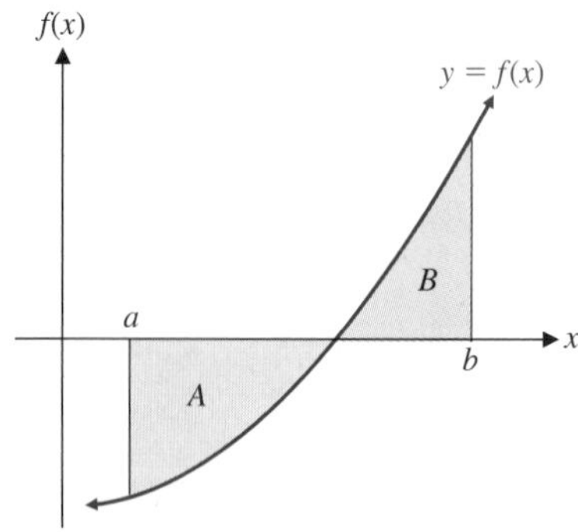

FIGURE 9
$\int_a^b f(x)\,dx = -A + B$

If a function $f(x)$ is monotonic and has both negative and positive values on the interval $[a, b]$, as in Figures 8 and 9, then the process for approximating $\int_a^b f(x)\,dx$ by L_n, R_n, or A_n, including the use of the associated error bound formulas, proceeds unchanged from that used above when $f(x)$ was restricted to positive quantities.

■ Rate, Area, and Distance

We now turn our attention to a seemingly unrelated problem involving distance, rate, and time. In our introduction to the derivative in Chapter 8, we started with the concept

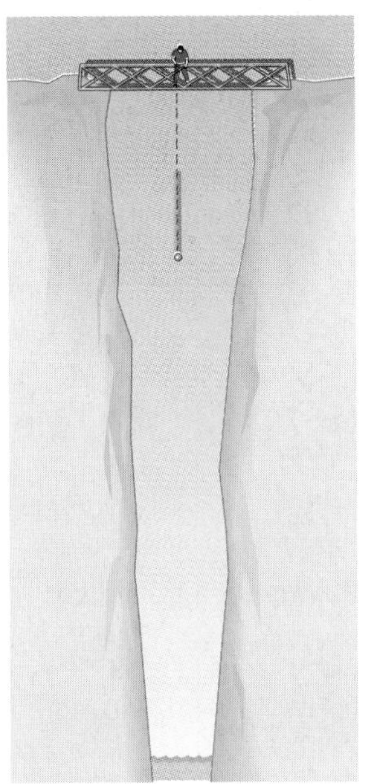

FIGURE 10
Steel ball dropped from a bridge

$$\text{Average rate} = \frac{\text{Total distance}}{\text{Elapsed time}} \qquad (4)$$

and then developed the concept of instantaneous rate as the limit of a difference quotient:

$$\begin{aligned}\frac{ds}{dt} &= \lim_{h\to 0} \frac{s(t+h) - s(t)}{h} \\ &= \lim_{\Delta t\to 0} \frac{s(t+\Delta t) - s(t)}{\Delta t}\end{aligned}$$

The increment notation Δt is sometimes used in place of h in the difference quotient.

where s is the position of a moving object at time t. Now we are going to reverse the process. That is, we will start with a rate function, either in the form of an equation or a table, and work back to distance. Central to the process is equation (4) in the form

$$\text{Distance} = (\text{Rate})(\text{Time}) \qquad (5)$$

A concrete example should make the process clear.

From physics it can be shown (neglecting air resistance) that a small steel ball dropped from a bridge (Fig. 10) will have an instantaneous rate of descent given approximately by

$$r(t) = 32t$$

where $r(t)$ is the rate in feet per second at the end of t seconds. (Note that the rate at which the ball falls increases with time due to gravity, as you would expect.) How far will the ball fall between the first and fourth second? To answer this question, we look at the area under the curve $r(t) = 32t$ from $t = 1$ to $t = 4$. But how do we get distance from area? To make the connection, we divide the interval $[1, 4]$ into three equal parts and estimate the area under the curve using L_3 and R_3, as shown in Figure 11.

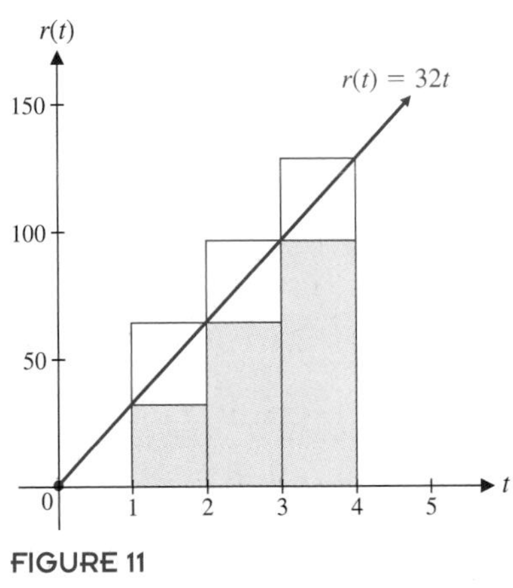

FIGURE 11
$\Delta x = 1$

$$\text{Area of first left rectangle} = r(1)\,\Delta t$$

But

$$\begin{aligned}r(1)\,\Delta t &= (\text{Rate at end of 1st second}) \times (1 \text{ second}) \\ &= \text{Distance traveled from } t = 1 \text{ to } t = 2 \text{ at a constant rate of } r(1)\end{aligned}$$

Since $r(t)$ is increasing over the interval $[1, 2]$, $r(1)$ is the minimum rate on this interval. Thus, $r(1)\,\Delta t$ is an underestimate of the actual distance the steel ball travels from $t = 1$ to $t = 2$.

$$\text{Area of first right rectangle} = r(2)\,\Delta t$$

But

$$\begin{aligned}r(2)\,\Delta t &= (\text{Rate at end of 2nd second}) \times (1 \text{ second}) \\ &= \text{Distance traveled from } t = 1 \text{ to } t = 2 \text{ at a constant rate of } r(2)\end{aligned}$$

Since $r(t)$ is increasing over the interval $[1, 2]$, $r(2)$ is the maximum rate on this interval. Thus, $r(2)\,\Delta t$ is an overestimate of the actual distance the steel ball travels from $t = 1$ to $t = 2$.

The same analysis applies to the second and third rectangles, and we conclude:

$$L_3 \leqslant \begin{pmatrix}\text{Distance traveled} \\ \text{from } t = 1 \text{ to } t = 4\end{pmatrix} \leqslant R_3$$

Proceeding in the same way for arbitrary n, we can write

$$L_n \leqslant \begin{pmatrix}\text{Distance traveled} \\ \text{from } t = 1 \text{ to } t = 4\end{pmatrix} \leqslant R_n$$

This suggests that the actual area under the rate function curve $r(t) = 32t$ from $t = 1$ to $t = 4$ represents the exact distance the steel ball travels from $t = 1$ to $t = 4$; that is,

$$\int_1^4 r(t)\,dt = \begin{pmatrix}\text{Distance traveled} \\ \text{from } t = 1 \text{ to } t = 4\end{pmatrix}$$

In general, it can be shown that:

Rate, Area, and Distance

If $r = r(t)$ is a positive rate function for an object moving on a line, then

$$\int_a^b r(t)\,dt = \begin{pmatrix}\text{Net distance traveled} \\ \text{from } t = a \text{ to } t = b\end{pmatrix}$$

$$= \begin{pmatrix}\text{Total change in position} \\ \text{from } t = a \text{ to } t = b\end{pmatrix}$$

Example 2 **Distance and Area**

(A) Using the rate function $r(t) = 32t$ from the above discussion, estimate the distance the ball falls from $t = 1$ to $t = 4$ using A_6. Calculate an error bound for this estimate.

(B) How large should n be so that the error in using A_n is not greater than 0.5 foot?

SOLUTION (A)

$$\Delta t = \frac{b - a}{n} = \frac{4 - 1}{6} = 0.5$$

$$L_6 = r(1)\,\Delta t + r(1.5)\,\Delta t + r(2)\,\Delta t + r(2.5)\,\Delta t + r(3)\,\Delta t + r(3.5)\,\Delta t$$
$$= 216$$

$$R_6 = r(1.5)\,\Delta t + r(2)\,\Delta t + r(2.5)\,\Delta t + r(3)\,\Delta t + r(3.5)\,\Delta t + r(4)\,\Delta t$$
$$= 264$$

$$A_6 = \frac{216 + 264}{2} = 240 \text{ feet}$$ Approximate distance traveled from $t = 1$ to $t = 4$

$$\text{Error} \leqslant |r(4) - r(1)|\,\frac{4 - 1}{2 \cdot 6} = 24 \text{ feet}$$

Thus,

$$\begin{pmatrix}\text{Distance traveled} \\ \text{from } t = 1 \text{ to } t = 4\end{pmatrix} = \int_1^4 r(t)\,dt = 240 \pm 24 \text{ feet}$$

[*Note:* $\int_1^4 r(t)\,dt = 240 \pm 24$ feet is a common and convenient way of representing $240 - 24 \le \int_1^4 r(t)\,dt \le 240 + 24$, and we will use this form where appropriate.]

(B) Solve the error bound inequality:

$$|r(4) - r(1)|\frac{4-1}{2n} \le 0.5$$
$$\frac{144}{n} \le 0.5$$
$$n \ge 288$$

Matched Problem 2 ➡ (A) Using the rate function $r(t) = 32t$, approximate the distance traveled from $t = 2$ to $t = 5$ using A_6. Calculate an error bound for this estimate.

(B) How large should n be so that the error in using A_n is not greater than 1 foot?

RATE, AREA, AND TOTAL CHANGE

Recall from Chapter 8 that marginal cost (or revenue or profit) is the instantaneous rate of change of cost (or revenue or profit) relative to production at a given production level. We now investigate how total change in cost is related to marginal cost and area through a concrete example.

Table 1
MARGINAL COSTS

x	5	10	15	20	25
$C'(x)$	24	20	17	15	14

A company in Florida manufactures cruising power boats. Marginal costs at various production levels are given in Table 1, where x is the number of boats produced per month and the marginal cost $C'(x)$ is in thousands of dollars.

We are interested in estimating the additional cost (total change in cost) in going from a production level of 5 boats per month to 25 boats per month. We make this estimate by graphing the values in Table 1 along with left and right rectangles, as shown in Figure 12.

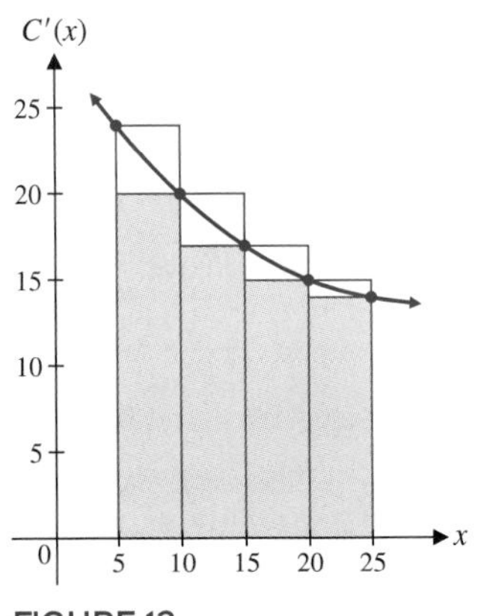

FIGURE 12
$\Delta x = 5$

We assume that a continuous decreasing curve representing the marginal cost function passes through each of the Table 1 points as indicated, but we do not know the equation of this curve. (This situation occurs very frequently in real-world applications.) Nevertheless, we can still estimate the total change in cost by proceeding as we did above in finding distance, given a rate.

$$\text{Area of first left rectangle} = C'(5)\,\Delta x$$

But

$$C'(5)\,\Delta x = \begin{pmatrix}\text{Approximate cost of producing} \\ \text{the next boat at an output} \\ \text{level of 5 boats}\end{pmatrix} \times 5$$
$$= \text{Approximate additional cost of increasing production from 5 to 10 boats per month}$$

Since $C'(x)$ is decreasing over the interval $[5, 10]$, $C'(5)$ is the maximum rate on this interval. Thus, $C'(5)\,\Delta x$ is an overestimate of the additional cost of increasing production from 5 to 10 boats per month.

$$\text{Area of first right rectangle} = C'(10)\,\Delta x$$

But

$$\begin{aligned} C'(10)\,\Delta x &= \begin{pmatrix}\text{Approximate cost of producing}\\ \text{the next boat at an output}\\ \text{level of 10 boats}\end{pmatrix} \times 5 \\ &= \text{Approximate additional cost of increasing} \\ &\quad\ \text{production from 5 to 10 boats per month}\end{aligned}$$

Since $C'(x)$ is decreasing over the interval $[5, 10]$, $C'(10)$ is the minimum rate on this interval. Thus, $C'(10)\,\Delta x$ is an underestimate of the additional cost of increasing production from 5 to 10 boats per month.

The same analysis applies to the second, third, and fourth rectangles, and we conclude

$$R_4 \leqslant \begin{pmatrix}\text{Additional cost of increasing}\\ \text{production from 5 to 25 boats}\\ \text{per month}\end{pmatrix} \leqslant L_4$$

This suggests that the actual area under the graph of the marginal cost function $y = C'(x)$—if we had the whole graph—would represent the additional cost of increasing production from 5 to 25 boats per month, and we would be able to write

$$\int_5^{25} C'(x)\,dx = \begin{pmatrix}\text{Additional cost of increasing}\\ \text{production from 5 to 25 boats}\\ \text{per month}\end{pmatrix}$$

In general, it can be shown that:

Rate, Area, and Total Change

If $y = F'(x)$ is a rate function (derivative), then the cumulated sum of the signed areas between the graph of $y = F'(x)$ and the x axis from $x = a$ to $x = b$ represents the total net change in $F(x)$ from $x = a$ to $x = b$. Symbolically,

$$\int_a^b F'(x)\,dx = \begin{pmatrix}\text{Total net change in } F(x)\\ \text{from } x = a \text{ to } x = b\end{pmatrix}$$

Informally, a **definite integral** produces the total change in a function from its rate of change. We are very close to a landmark result in mathematics, the *fundamental theorem of calculus,* which will be discussed in Section 11-5.

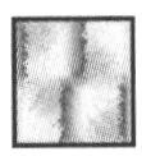

Example 3 **Change in Cost and Area** Refer to Table 1 and Figure 12 (page 793), and use A_4 to estimate the additional cost (total change in cost) in going from a production level of 5 boats per month to 25 boats per month. Calculate an error bound for this estimate.

SOLUTION

$$\begin{aligned} \Delta x &= 5 \\ L_4 &= C'(5)\,\Delta x + C'(10)\,\Delta x + C'(15)\,\Delta x + C'(20)\,\Delta x \\ &= \$380 \text{ thousand} \quad \text{Overestimate} \\ R_4 &= C'(10)\,\Delta x + C'(15)\,\Delta x + C'(20)\,\Delta x + C'(25)\,\Delta x \\ &= \$330 \text{ thousand} \quad \text{Underestimate} \\ A_4 &= \frac{380 + 330}{2} = \$355 \text{ thousand} \end{aligned}$$

Error bound estimate for A_4:

$$\text{Error} \leqslant |C'(5) - C'(25)|\frac{25 - 5}{2 \cdot 4} = \$25 \text{ thousand}$$

Thus, the additional cost (total change in cost) in going from a production level of 5 boats per month to 25 boats per month is

$$\int_5^{25} C'(x)\,dx = \$355{,}000 \pm \$25{,}000$$

Matched Problem 3 Refer to Table 1 and Figure 12 (page 793), and use A_2 to estimate the additional cost (total change in cost) in going from a production level of 10 boats per month to 20 boats per month. Calculate an error bound for this estimate.

Answers to Matched Problems

1. (A) $\Delta x = 0.5$:

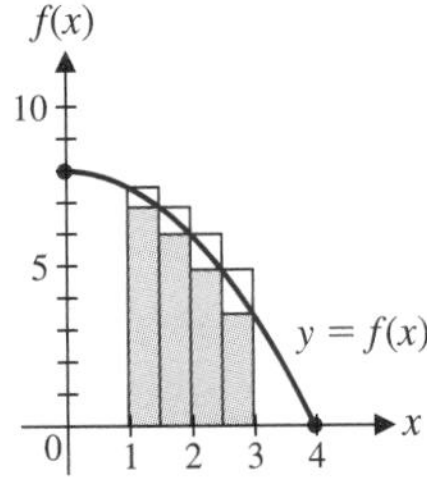

(B) $L_4 = 12.625$, $R_4 = 10.625$, $A_4 = 11.625$;
Error for L_4 and $R_4 = 2$; A_4 error bound $= 1$

(C) $n > 16$ for L_n and R_n; $n > 8$ for A_n

2. (A) $\begin{pmatrix}\text{Distance traveled} \\ \text{from } t = 2 \text{ to } t = 5\end{pmatrix} = 336 \pm 24$ ft

(B) $n \geqslant 144$

3. $\begin{pmatrix}\text{Additional cost of increasing} \\ \text{production from 10 to 20 boats} \\ \text{per month}\end{pmatrix} = \$172{,}500 \pm \$12{,}500$

EXERCISE 11-4

A *Problems 1–10 involve estimating the area under the curves in figures A–D from $x = 1$ to $x = 4$. For each figure, divide the interval $[1, 4]$ into three equal subintervals.*

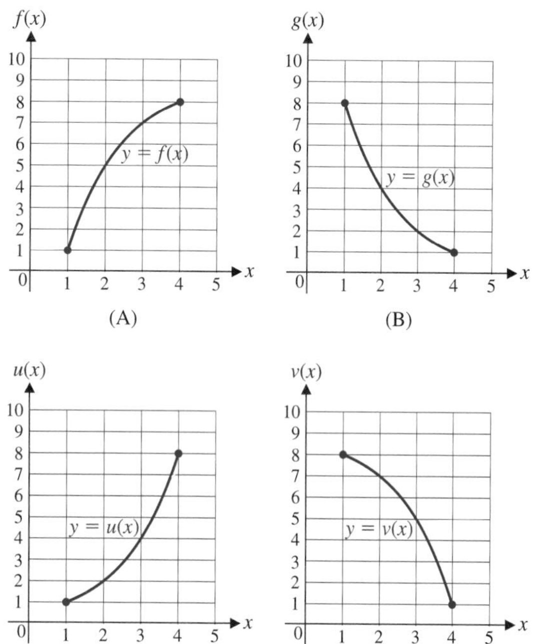

1. Draw in left and right rectangles for Figures A and B.
2. Draw in left and right rectangles for Figures C and D.
3. Using the results of Problem 1, compute L_3, R_3, and A_3 for Figure A and for Figure B.
4. Using the results of Problem 2, compute L_3, R_3, and A_3 for Figure C and for Figure D.
5. Replace the question marks with L_3 and R_3 as appropriate. Explain your choice.

$$? \leq \int_1^4 f(x)\,dx \leq ? \qquad ? \leq \int_1^4 g(x)\,dx \leq ?$$

6. Replace the question marks with L_3 and R_3 as appropriate. Explain your choice.

$$? \leq \int_1^4 u(x)\,dx \leq ? \qquad ? \leq \int_1^4 v(x)\,dx \leq ?$$

7. Compute error bounds for L_3, R_3, and A_3 found in Problem 3 for both figures.
8. Compute error bounds for L_3, R_3, and A_3 found in Problem 4 for both figures.
9. Using results from the preceding problems, interpret the following in geometric language:

$$\int_1^4 f(x)\,dx = 16.5 \pm 3.5$$

10. Using results from the preceding problems, interpret the following in geometric language:

$$\int_1^4 u(x)\,dx = 10.5 \pm 3.5$$

11. An arrow is shot vertically into the air with an initial rate of 128 feet per second. Neglecting air resistance, its rate of ascent $r(t)$ (in feet per second) at the end of t seconds is shown in the figure.

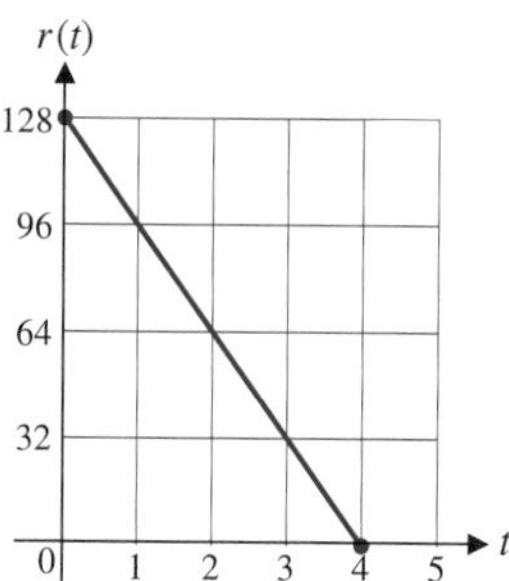

Figure for 11

(A) Estimate the height the arrow reaches using left and right sums and their average over four equal subdivisions. (Read values directly from the graph.) Calculate error bounds for these estimates.
(B) Explain how the areas of the rectangles used in the left or right sum approximations in part (A) are related to distance.
(C) How many equal subdivisions on the interval $[0, 4]$ should be used to have A_n approximate the height the arrow reaches within 1 foot?

12. Referring to Problem 11, use left and right sums and their average using two equal subdivisions to estimate the distance the arrow travels during the first 2 seconds of flight. Estimate error bounds for each estimate.

B *In Problems 13 and 14, discuss the validity of each statement. If the statement is always true, explain why. If not, give a counterexample.*

13. (A) The function $f(x) = x^2 - 2x$ is monotone over $[0, 2]$.
(B) The function $f(x) = x^2 - 2x$ is monotone over $[1, 3]$.

14. (A) If f is a monotone function on $[a, b]$, then the area under the graph of $f(x)$ is greater than the left sum L_n and less than the right sum R_n for any positive integer n.

(B) If the area under the graph of $f(x)$ on $[a, b]$ is equal to both the left sum L_n and the right sum R_n for some positive integer n, then f is a constant function.

Problems 15 and 16 refer to the figure below showing two parcels of land along a river.

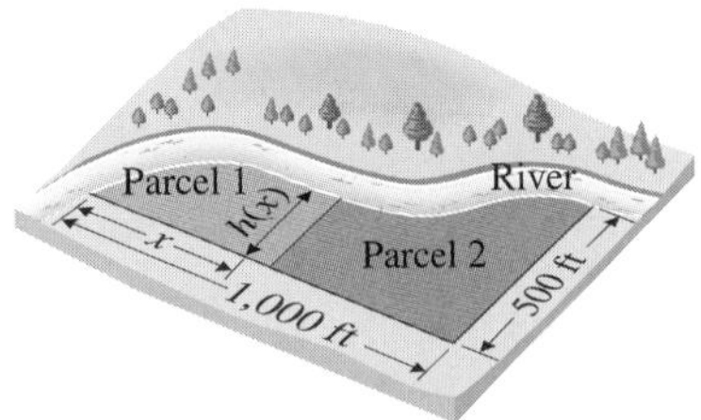

15. You are interested in purchasing both parcels of land shown in the figure and wish to make a quick check on their combined area. There is no equation for the river frontage, so you use the average of the left and right sums of rectangles covering the area. The 1,000 foot baseline is divided into ten equal parts. At the end of each subinterval, a measurement is made from the baseline to the river, and the results are tabulated. Let x be the distance from the left end of the baseline and let $h(x)$ be the distance from the baseline to the river at x. Estimate the combined area of both parcels using A_{10}, and calculate an error bound for this estimate. How many subdivisions of the baseline would be required so that the error in using A_n would not exceed 2,500 square feet?

x	0	100	200	300	400	500
$h(x)$	0	183	235	245	260	286

x	600	700	800	900	1,000
$h(x)$	322	388	453	489	500

16. Estimate the separate area of each parcel in Problem 15 using A_5 for each, and calculate an error bound for each estimate. The baseline length of each parcel is 500 feet.

17. A car is traveling at 75 miles per hour (110 feet per second), and the driver slams on the brakes to stop. During the 7 seconds it takes to stop, the speed of the car at the end of each second is recorded by a special instrument as follows:

t	0	1	2	3	4	5	6	7
$r(t)$	110	85	63	45	29	16	5	0

where $r(t)$ is the speed (in feet per second) at the end of t seconds, $0 \le t \le 7$.

(A) Estimate the distance required to stop using the average of left and right sums over seven equal subdivisions. Calculate an error bound for this estimate.

(B) How many equal subdivisions should be used on the interval $[0, 7]$ to have A_n approximate the stopping distance within 5 feet?

18. In Problem 17, estimate the distance traveled during the first 3 seconds after applying the brakes using the average of left and right sums over three equal subdivisions. Calculate an error bound for this estimate. How many equal subdivisions should be used on the interval $[0, 3]$ to have A_n approximate the distance within 10 feet?

19. The figure shows the graph of a function that is not monotonic over the interval $[0, 5]$. We wish to estimate the area under the curve over this interval.

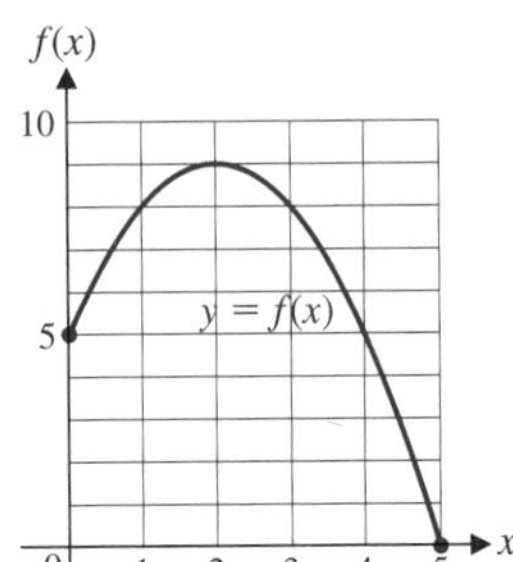

Figure for 19

(A) Find a point P on the x axis that divides $[0, 5]$ into two parts such that $f(x)$ is monotonic over each.

(B) Divide the interval $[0, 5]$ into five equal subintervals, and sketch left and right rectangles over the interval.

(C) Describe the behavior of left and right rectangles to the left and to the right of P relative to under- and overestimating the true area.

(D) Find numbers N_1 and N_2 by combining appropriate left and right rectangle sums so that

$$N_1 \le \int_0^5 f(x)\,dx \le N_2$$

20. Find numbers N_1 and N_2 by combining appropriate left and right rectangle sums from Problem 19 so that

$$N_1 \le \int_1^4 f(x)\,dx \le N_2$$

C *Problems 21 and 22 refer to the figure below.*

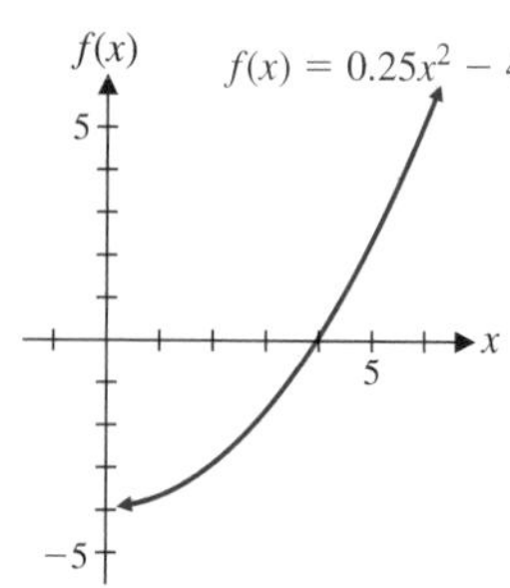

21. Approximate $\int_2^5 (0.25x^2 - 4)\,dx$ using L_6, R_6, and A_6. Compute error bounds for each. (Round all values to two decimal places.) Describe in geometric terms what the definite integral over the interval $[2, 5]$ represents.

22. Approximate $\int_1^6 (0.25x^2 - 4)dx$ using L_5, R_5, and A_5. Compute error bounds for each. (Round all values to two decimal places.) Describe in geometric terms what the definite integral over the interval $[1, 6]$ represents.

Use a graphing utility to determine the intervals on which each function is monotone.

23. $f(x) = e^{-x^2}$

24. $f(x) = \dfrac{3}{1 + 2e^{-x}}$

25. $f(x) = x^4 - 2x^2 + 3$

26. $f(x) = e^{x^2}$

The left sum L_n or the right sum R_n is used to approximate each definite integral in Problems 27–30 to the indicated accuracy. How large must n be chosen in each case? (Each function is monotonic over the indicated interval.)

27. $\int_1^2 x^x\,dx = R_n \pm 0.05$

28. $\int_0^3 2^x\,dx = L_n \pm 0.5$

29. $\int_0^2 e^{-x^2}\,dx = L_n \pm 0.005$

30. $\int_0^1 e^{t^2}\,dt = R_n \pm 0.05$

31. If f is monotonic on the interval $[a, b]$, give an argument why we are justified in writing

$$\int_a^b f(x)\,dx = \lim_{n\to\infty} L_n$$

32. If f is monotonic on the interval $[a, b]$, give an argument why we are justified in writing

$$\int_a^b f(x)\,dx = \lim_{n\to\infty} R_n$$

Applications

Business & Economics

33. *Cost.* A company manufactures mountain bikes. The research department produced the marginal cost graph shown, where $C'(x)$ is in dollars and x is the number of bikes produced per month. Estimate the increase in cost going from a production level of 300 bikes per month to 900 bikes per month. Use the average of the left and right sums over two equal subintervals, and compute an error bound for this estimate.

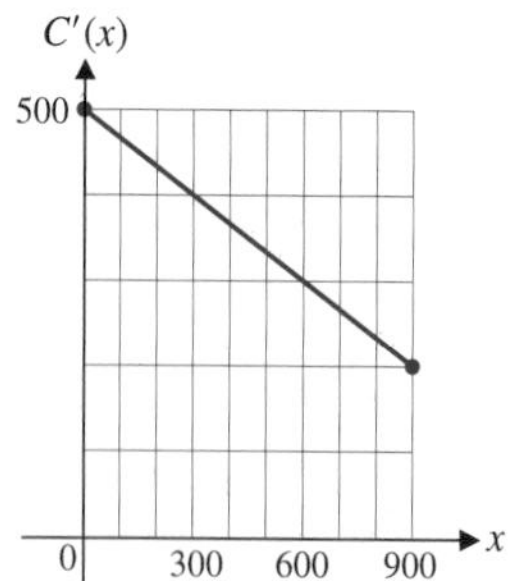

Figure for 33 and 34

34. *Cost.* Referring to Problem 33, estimate the increase in cost going from a production level of 0 bikes per month to 600 bikes per month using left and right sums over two equal subintervals. Replace the question marks with the values of L_2 or R_2 as appropriate:

$$? \leq \int_0^{600} C'(x)\,dx \leq ?$$

35. *Continuous compound interest.* Ten thousand dollars is deposited in an account where the rate of change of the amount in the account is given by

$$A'(t) = 800e^{0.08t}$$

t years after the initial deposit. By how much will the account change from the end of the second year to the end of the sixth year? Estimate the amount by using left and right sums over four equal subintervals. (Calculate all quantities to the nearest dollar.) Replace the question marks with the values of L_4 or R_4 as appropriate:

$$? \leq \int_2^6 800e^{0.08t}\,dt \leq ?$$

36. *Continuous compound interest.* One thousand dollars is deposited in an account where the rate of change of the amount in the account is given by

$$A'(t) = 50e^{0.05t}$$

t years after the initial deposit. By how much will the account change from the end of the first year to the end of the fifth year? Estimate the amount by using the average of the left sum and the right sum over four equal subintervals. Calculate an error bound for this estimate. (Calculate all quantities to the nearest dollar.)

37. *Employee training.* A company producing computer components has established that, on the average, a new employee can assemble $N(t)$ components per day after t days of on-the-job training, as indicated in the following table (a new employee's productivity continuously increases with time on the job):

t	0	20	40	60	80	100	120
$N(t)$	10	51	68	76	81	84	86

Use the average of left and right sums to estimate the total number of units produced by a new employee over the first 60 days. Over the second 60 days. Use three equal subintervals for each. Calculate an error bound for each estimate.

38. *Employee training.* For a new employee in Problem 37, use left and right sums to estimate the number of units produced over the time interval [20, 100]. Replace the question marks with the values of L_4 or R_4 as appropriate:

$$? \le \int_{20}^{100} N(t)\,dt \le ?$$

39. *Revenue.* The research department for a market chain in a city established the following price–demand equation for a premium beer sold by the six-pack:

$$p = 8 - \frac{x}{50} \qquad 0 \le x \le 300 \qquad \text{Price–demand}$$

where x is the number of six-packs sold per day at a price of p dollars each. From the price–demand equation we obtain the revenue and marginal revenue equations and their graphs shown in the figure.

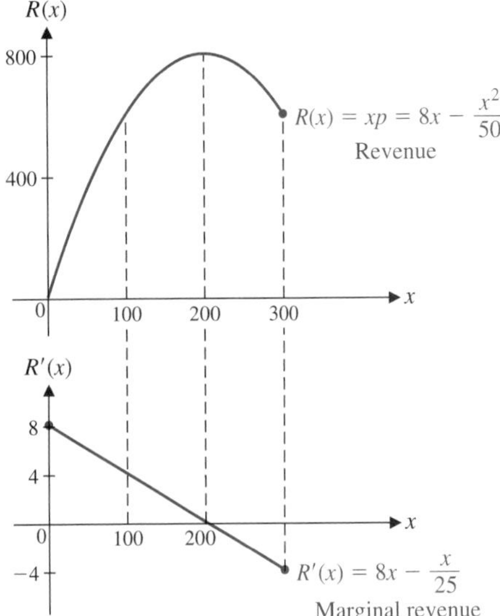

Figure for 39 and 40

(A) Interpret $\int_{100}^{200} R'(x)\,dx$ geometrically and relative to change in revenue.

(B) Approximate $\int_{100}^{200} R'(x)\,dx$ using the average of the left and right sums over four equal subintervals. Calculate an error bound for this estimate.

(C) Evaluate $R(200) - R(100)$. How does this quantity relate to part (A)?

40. *Revenue.* Refer to Problem 39.

(A) Interpret $\int_{100}^{300} R'(x)\,dx$ geometrically and relative to change in revenue. Guess what the value should be from the graph.

(B) Approximate $\int_{100}^{300} R'(x)\,dx$ using the average of left and right sums over four equal subintervals. Calculate an error bound for this estimate.

(C) Evaluate $R(300) - R(100)$. How does this quantity relate to part (A)?

Life Sciences

41. *Medicine.* The rate of healing $A'(t)$ (in square centimeters per day) for a certain type of abrasive skin wound is given approximately by the following table:

t	0	1	2	3	4	5
$A'(t)$	0.90	0.81	0.74	0.67	0.60	0.55

t	6	7	8	9	10
$A'(t)$	0.49	0.45	0.40	0.36	0.33

$A(t)$ is the area of the wound in square centimeters after t days of healing.

(A) Approximate the area of the wound that will be healed in the first 5 days of healing using the left and right sums over five equal subintervals.

(B) Replace the question marks with values of L_5 and R_5 as appropriate:

$$? \le \int_0^5 A'(t)\,dt \le ?$$

42. *Medicine.* Refer to Problem 41. Approximate the area of the wound that will be healed in the second 5 days of healing. Use the average of the left and right sums over five equal subintervals. Calculate an error bound for this estimate.

Social Sciences

43. *Learning.* During a special study on learning, a psychologist found that, on the average, the rate of learning a list of special symbols in a code, $N'(x)$, after x days of practice was given approximately by the following table values:

x	0	2	4	6	8	10	12
$N'(x)$	29	26	23	21	19	17	15

$N(x)$ is the number of special symbols learned after x days of practice. Approximate the number of code symbols learned from the sixth to the twelfth day using the average of the left and right sums over three equal subintervals. Calculate an error bound for this estimate.

44. *Learning.* For the data in Problem 43, approximate the number of code symbols learned during the first 6 days of practice using left and right sums over three equal subintervals. Replace the question marks with values of L_3 and R_3 as appropriate:

$$? \le \int_0^6 N'(x)\,dx \le ?$$

SECTION 11-5 Definite Integral as a Limit of a Sum; Fundamental Theorem of Calculus

- DEFINITE INTEGRAL AS A LIMIT OF A SUM
- FUNDAMENTAL THEOREM OF CALCULUS
- RECOGNIZING A DEFINITE INTEGRAL—AVERAGE VALUE

The previous section presented an intuitive and informal introduction to the definite integral and the fundamental theorem of calculus. There, we used the definite integral symbol $\int_a^b f(x)\,dx$ to represent the cumulative sums of signed areas between the graph of $y = f(x)$ and the x axis from $x = a$ to $x = b$. See Figure 1, where A, B, and C represent actual areas of the respective regions.

f(x)
y = f(x)
A
B
C
a
b
x

FIGURE 1

$\int_a^b f(x)\,dx = A - B + C$

We also found that if $f(x)$ represents a rate, then $\int_a^b f(x)\,dx$ can be interpreted as the total net change in an antiderivative of $f(x)$ from $x = a$ to $x = b$. In particular, if $f(x) = C'(x)$, the marginal cost for producing x items, then $\int_a^b C'(x)\,dx$ represents the total change in cost going from a production level of a units to b units.

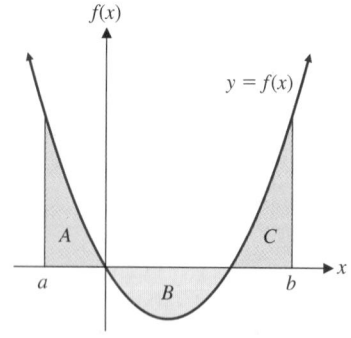

DEFINITE INTEGRAL AS A LIMIT OF A SUM

We now make some of the concepts in the preceding section more precise. We start by giving a precise description of left and right sums: If we partition the interval $[a, b]$ into n equal subintervals of length $\Delta x = (b - a)/n$ with end points $a = x_0, x_1, \ldots, x_n = b$, then, using **summation notation** (see Appendix B-1):

$$\begin{aligned} L_n = \textbf{Left sum} &= \sum_{k=1}^{n} f(x_{k-1})\,\Delta x \\ &= f(x_0)\,\Delta x + f(x_1)\,\Delta x + \cdots + f(x_{n-1})\,\Delta x \\ R_n = \textbf{Right sum} &= \sum_{k=1}^{n} f(x_k)\,\Delta x \\ &= f(x_1)\,\Delta x + f(x_2)\,\Delta x + \cdots + f(x_n)\,\Delta x \end{aligned}$$

Other rectangle sums, such as the *midpoint sum*, can also be used to approximate $\int_a^b f(x)\,dx$. With the midpoint sum, the height of each rectangle is found by evaluating $f(x)$ at the midpoint of each subinterval instead of at an end point. The **midpoint sum** is given as follows:

$$\begin{aligned} M_n = \textbf{Midpoint sum} &= \sum_{k=1}^{n} f\left(\frac{x_{k-1} + x_k}{2}\right)\Delta x \\ &= f\left(\frac{x_0 + x_1}{2}\right)\Delta x + f\left(\frac{x_1 + x_2}{2}\right)\Delta x + \cdots + f\left(\frac{x_{n-1} + x_n}{2}\right)\Delta x \end{aligned}$$

Figure 2 shows the use of the midpoint sum to approximate $\int_1^6 (0.25x^2 - 4)\,dx$ for $n = 5$, 10, and 20 (all done on a computer, of course).

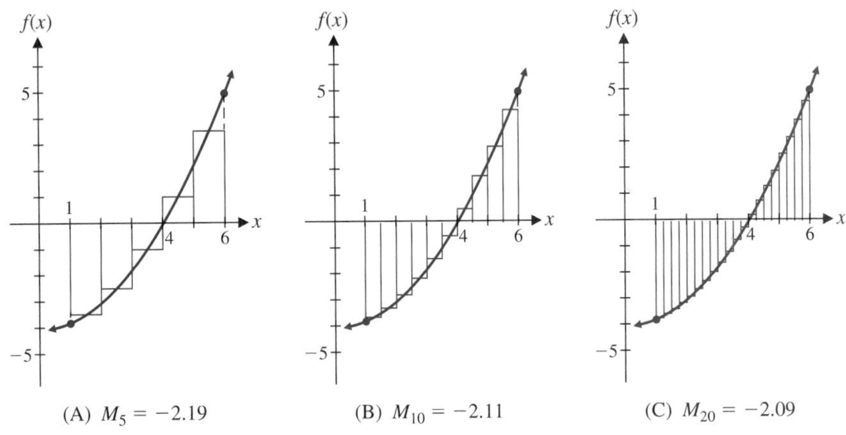

FIGURE 2
Midpoint sums $[f(x) = 0.25x^2 - 4]$

Explore–Discuss 1

In Figure 2, let A represent the exact area below the x axis from $x = 1$ to $x = 4$, and let B represent the exact area above the x axis from $x = 4$ to $x = 6$. Make a reasonable guess for the following limit in terms of areas A and B, and explain the reasoning behind your guess.

$$\lim_{n\to\infty} M_n = ?$$

We are now ready to give a general definition of the *definite integral* in a form that not only covers the discussions above and in the preceding section as special cases, but also opens the concept up to many other interpretations and applications.

Definition of a Definite Integral

Let f be a continuous function defined on the closed interval $[a, b]$, and let

1. $a = x_0 < x_1 < \cdots < x_{n-1} < x_n = b$
2. $\Delta x_k = x_k - x_{k-1} \qquad \text{for } k = 1, 2, \ldots, n$
3. $\Delta x_k \to 0 \quad \text{as} \quad n \to \infty$
4. $x_{k-1} \leqslant c_k \leqslant x_k \qquad \text{for } k = 1, 2, \ldots, n$

Then

$$\int_a^b f(x)\,dx = \lim_{n\to\infty} \sum_{k=1}^{n} f(c_k)\,\Delta x_k$$

$$= \lim_{n\to\infty}[f(c_1)\,\Delta x_1 + f(c_2)\,\Delta x_2 + \cdots + f(c_n)\,\Delta x_n]$$

is called a **definite integral** of f from a to b. The **integrand** is $f(x)$, the **lower limit** is a, and the **upper limit** is b.

In the definition of a definite integral, we divide the closed interval $[a, b]$ into n subintervals of arbitrary length in such a way that the length of each subinterval $\Delta x_k = x_k - x_{k-1}$ tends to 0 as n increases without bound. From each of the n subintervals we then select a point c_k and form the sum

$$\sum_{k=1}^{n} f(c_k)\,\Delta x_k = f(c_1)\,\Delta x_1 + f(c_2)\,\Delta x_2 + \cdots + f(c_n)\,\Delta x_n$$

which is called a **Riemann sum** [named after the celebrated German mathematician Georg Riemann (1826–1866)].

Under the conditions stated in the definition, it can be shown that the limit of the Riemann sum always exists, and it is a real number. The limit is independent of the nature of the subdivisions of $[a, b]$ as long as condition 3 holds, and it is independent of the choice of c_k as long as condition 4 holds.* It is important to remember that irrespective of its original interpretation in a practical problem, **a definite integral can always be interpreted geometrically in terms of signed areas just as a derivative can always be interpreted geometrically in terms of slope.**

The left, right, and midpoint sums discussed above are particularly simple Riemann sums with subintervals all the same length and c_k chosen in a regular manner as an end point of a subinterval or the midpoint of the subinterval. Any one of these special sums (as well as others) can be used to approximate a definite integral. To make these approximations useful, we need formulas for error bounds that include nonmonotonic functions. We state these formulas in the box at the top of the next page without proof.

*In this text we consider only limits of Riemann sums for which the n subintervals of $[a, b]$ have the same length. Condition 3 is then automatically satisfied.

Error Bounds for L_n, R_n, and M_n

Let

$$I = \int_a^b f(x)\,dx$$

L_n = Left sum, R_n = Right sum, M_n = Midpoint sum

Left and Right Sum Error Bound

If $|f'(x)| \leqslant B_1$ for all x on $[a, b]$, then

$$|I - L_n| \leqslant \frac{B_1(b-a)^2}{2n} \qquad |I - R_n| \leqslant \frac{B_1(b-a)^2}{2n}$$

Midpoint Error Bound

If $|f''(x)| \leqslant B_2$ for all x on $[a, b]$, then

$$|I - M_n| \leqslant \frac{B_2(b-a)^3}{24n^2}$$

Example 1 **Using Estimates for Error Bounds** Given

$$I = \int_{-1}^{3} (4x - x^3)\,dx$$

For the integrand $f(x) = 4x - x^3$, use the graphs of $y = f'(x)$ and $y = f''(x)$ in Figure 3 as needed.

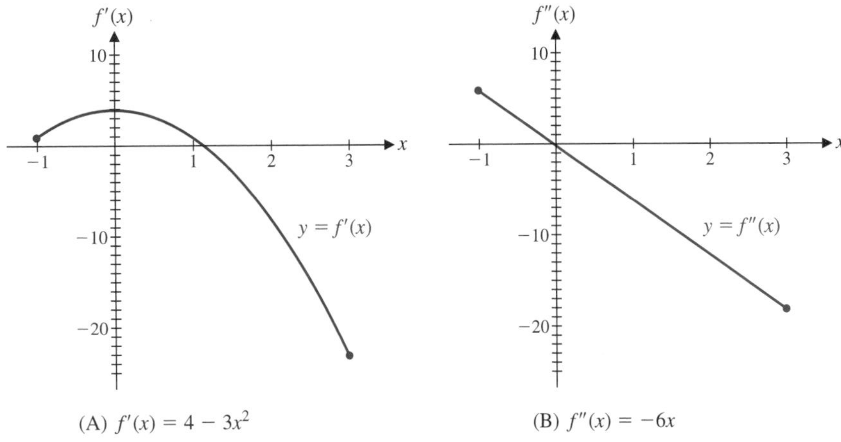

(A) $f'(x) = 4 - 3x^2$ (B) $f''(x) = -6x$

FIGURE 3

(A) How large should n be chosen for either L_n or R_n to approximate I with an error no greater than 0.5?

(B) How large should n be chosen for M_n to approximate I with an error no greater than 0.5?

SOLUTION (A) Choose n such that

$$\frac{B_1(b-a)^2}{2n} \leqslant 0.5$$

where B_1 is any upper bound for $|f'(x)|$ on the interval $[-1, 3]$. [*Caution:* We are looking for an upper bound for $|f'(x)|$, not for $f'(x)$, on the interval $[-1, 3]$.] From the graph of $y = f'(x)$ in Figure 3A, we see that B_1 can be chosen as any number 23 or larger. We choose $B_1 = 23$. [Graphs of $y = f'(x)$ are often used to get an upper bound for $|f'(x)|$ on an interval $[a, b]$ if an upper bound is not easily obtained algebraically or is not obvious.] Thus,

$$\frac{23[3-(-)1]^2}{2n} \leqslant 0.5$$

$$\frac{23(8)}{n} \leqslant 0.5$$

$$n \geqslant \frac{23(8)}{0.5} = 368$$

(B) Choose n such that

$$\frac{B_2(b-a)^3}{24n^2} \leqslant 0.5$$

where B_2 is any upper bound for $|f''(x)|$ on the interval $[-1, 3]$. From the graph of $y = f''(x)$ in Figure 3B, we see that B_2 can be chosen as any number 18 or larger. We choose $B_2 = 18$. Thus,

$$\frac{18[3-(-1)]^3}{24n^2} \leqslant 0.5$$

$$\frac{18(64)}{24(0.5)} \leqslant n^2$$

$$n \geqslant \sqrt{\frac{18(64)}{24(0.5)}} \quad \text{Since } n \text{ is positive}$$

$$= 9.8 \quad \text{or} \quad 10$$

[*Note:* The midpoint sum only required 10 subintervals to get the same accuracy that either the left or right sum achieved using 368 subintervals. In both cases, the accuracy may be considerably better than indicated, but these error bound formulas guarantee the results for the indicated values of n. Another approximation sum that is usually more accurate for a given value of n, *Simpson's rule,* is discussed in Group Activity 1.]

Matched Problem 1 Given $I = \int_{-1}^{2} (x^2 - 4x)\, dx$:

(A) How large should n be chosen for either L_n or R_n to approximate I with an error no greater than 0.1?

(B) How large should n be chosen for M_n to approximate I with an error no greater than 0.1?

Explore–Discuss 2

Describe the difference between the *definite integral* $\int_a^b f(x)\,dx$ and the *indefinite integral* $\int f(x)\,dx$ in terms of what each represents.

In the next box we state several useful properties of the definite integral. Note that properties 3 and 4 parallel properties 3 and 4 given earlier in Section 11-1 for the indefinite integral.

Definite Integral Properties

1. $\int_a^a f(x)\,dx = 0$
2. $\int_a^b f(x)\,dx = -\int_b^a f(x)\,dx$
3. $\int_a^b kf(x)\,dx = k\int_a^b f(x)\,dx$, k a constant
4. $\int_a^b [f(x) \pm g(x)]\,dx = \int_a^b f(x)\,dx \pm \int_a^b g(x)\,dx$
5. $\int_a^b f(x)\,dx = \int_a^c f(x)\,dx + \int_c^b f(x)\,dx$

Most of these properties follow directly from the definition of the definite integral. Example 2 illustrates how they can be used.

Example 2 **Using Properties of the Definite Integral** If

$$\int_0^2 x\,dx = 2 \qquad \int_0^2 x^2\,dx = \frac{8}{3} \qquad \int_2^3 x^2\,dx = \frac{19}{3}$$

then:

(A) $\displaystyle\int_0^2 12x^2\,dx = 12\int_0^2 x^2\,dx = 12\left(\frac{8}{3}\right) = 32$

(B) $\displaystyle\int_0^2 (2x - 6x^2)\,dx = 2\int_0^2 x\,dx - 6\int_0^2 x^2\,dx = 2(2) - 6\left(\frac{8}{3}\right) = -12$

(C) $\displaystyle\int_3^2 x^2\,dx = -\int_2^3 x^2\,dx = -\frac{19}{3}$

(D) $\int_5^5 3x^2\,dx = 0$

(E) $\displaystyle\int_0^3 3x^2\,dx = 3\int_0^2 x^2\,dx + 3\int_2^3 x^2\,dx = 3\left(\frac{8}{3}\right) + 3\left(\frac{19}{3}\right) = 27$

Matched Problem 2 Using the same integral values given in Example 2, find:

(A) $\int_2^3 6x^2\,dx$ (B) $\int_0^2 (9x^2 - 4x)\,dx$ (C) $\int_2^0 3x\,dx$
(D) $\int_{-2}^{-2} 3x\,dx$ (E) $\int_0^3 12x^2\,dx$

FUNDAMENTAL THEOREM OF CALCULUS

We are now ready to present one of the most important results in mathematics, the *fundamental theorem of calculus.* We start by reviewing the marginal cost interpretation of the definite integral discussed earlier.

If $C'(x)$ represents marginal cost at an output of x units, then $\int_a^b C'(x)\,dx$ represents the total change in cost as production changes from

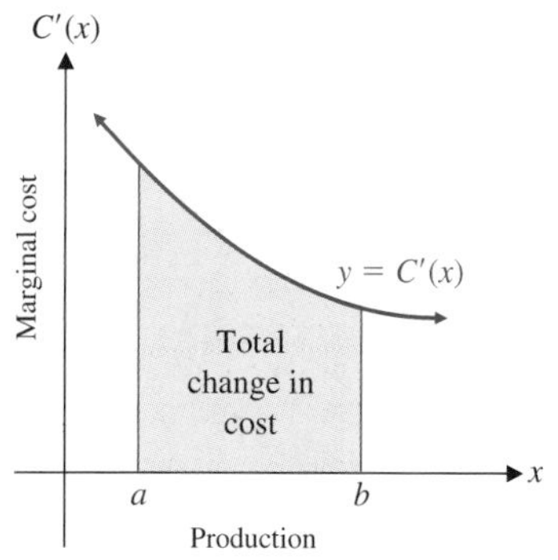

FIGURE 4
$\int_a^b C'(x)\,dx$ = Total change in cost

$x = a$ units to $x = b$ units (see Fig. 4). We can also get the total change in cost by using $C(b) - C(a)$, where $C(x)$ is the cost function, an antiderivative of the marginal cost function $C'(x)$. Thus, it appears that we can write

$$\int_a^b C'(x)\,dx = C(b) - C(a)$$

In general, since $F'(x)$ is an instantaneous rate of change at x, then:

$\int_a^b$ **$F'(x)\,dx$ represents the total change in $F(x)$ from $x = a$ to $x = b$.**

But this is also given by $F(b) - F(a)$. Thus, it appears that

$$\int_a^b F'(x)\,dx = F(b) - F(a)$$

It turns out that this remarkable result is true under rather general conditions, which we now state:

Fundamental Theorem of Calculus

If f is a continuous function on the closed interval $[a, b]$ and F is any antiderivative of f, then

$$\int_a^b f(x)\,dx = F(x)\Big|_a^b = F(b) - F(a) \qquad F'(x) = f(x)$$

If we can find an antiderivative of $f(x)$, then we can evaluate $\int_a^b f(x)\,dx$, the limit of a Riemann sum over the interval $[a, b]$, exactly and quickly by evaluating an antiderivative of $f(x)$ at the end points a and b and taking the difference. Of course, if $f(x)$ is given in table or graph form without an equation, or if an antiderivative of $f(x)$ cannot be found or does not exist (a common occurrence in the real world), then we must resort to the use of an approximation procedure such as those described earlier. The symbol $F(x)|_a^b$ is used to represent the **net change** in $F(x)$ from $x = a$ to $x = b$ and is included as a convenient intermediate step in the evaluation of a definite integral by the fundamental theorem.

Now you know why we studied techniques of indefinite integration before this section—so we would have methods of finding antiderivatives of large classes of elementary functions for use with the fundamental theorem. It is important to remember that:

Any antiderivative of $f(x)$ can be used in the fundamental theorem. One generally chooses the simplest antiderivative by letting $C = 0$, since any other value of C will drop out when computing the difference $F(b) - F(a)$.

A variety of definite integrals are evaluated using the fundamental theorem in the following examples.

Example 3 **Evaluating Definite Integrals** Evaluate: $\displaystyle\int_1^2 \left(2x + 3e^x - \frac{4}{x}\right) dx$

SOLUTION
$$\begin{aligned}\int_1^2 \left(2x + 3e^x - \frac{4}{x}\right) dx &= 2\int_1^2 x\,dx + 3\int_1^2 e^x\,dx - 4\int_1^2 \frac{1}{x}\,dx \\ &= 2\frac{x^2}{2}\bigg|_1^2 + 3e^x\bigg|_1^2 - 4\ln|x|\bigg|_1^2 \\ &= (2^2 - 1^2) + (3e^2 - 3e^1) - (4\ln 2 - 4\ln 1) \\ &= 3 + 3e^2 - 3e - 4\ln 2 \approx 14.24\end{aligned}$$

Matched Problem 3 Evaluate: $\displaystyle\int_1^3 \left(4x - 2e^x + \frac{5}{x}\right) dx$

The evaluation of a definite integral is a two-step process: First, find an antiderivative, and then find the net change in that antiderivative. If *substitution techniques* are required to find the antiderivative, there are two different ways to proceed. The next example illustrates both methods.

Example 4 **Definite Integrals and Substitution Techniques** Evaluate:

$$\int_0^5 \frac{x}{x^2 + 10}\,dx$$

SOLUTION We will solve this problem using substitution in two different ways:

Method 1. Use substitution in an indefinite integral to find an antiderivative as a function of x; then evaluate the definite integral:

$$\begin{aligned}\int \frac{x}{x^2 + 10}\,dx &= \frac{1}{2}\int \frac{1}{x^2 + 10}\,2x\,dx &&\text{Substitute } u = x^2 + 10 \text{ and } du = 2x\,dx. \\ &= \frac{1}{2}\int \frac{1}{u}\,du \\ &= \tfrac{1}{2}\ln|u| + C \\ &= \tfrac{1}{2}\ln(x^2 + 10) + C &&\text{Since } u = x^2 + 10 > 0\end{aligned}$$

We choose $C = 0$ and use the antiderivative $\frac{1}{2}\ln(x^2 + 10)$ to evaluate the definite integral:

$$\begin{aligned}\int_0^5 \frac{x}{x^2 + 10}\,dx &= \frac{1}{2}\ln(x^2 + 10)\bigg|_0^5 \\ &= \tfrac{1}{2}\ln 35 - \tfrac{1}{2}\ln 10 \approx 0.626\end{aligned}$$

Method 2. Substitute directly in the definite integral, changing both the variable of integration and the limits of integration: In the definite integral

$$\int_0^5 \frac{x}{x^2 + 10}\,dx$$

the upper limit is $x = 5$ and the lower limit is $x = 0$. When we make the substitution $u = x^2 + 10$ in this definite integral, we must change the limits of integration to the corresponding values of u:

$$\begin{aligned}x &= 5 \quad \text{implies} \quad u = 5^2 + 10 = 35 &&\text{New upper limit} \\ x &= 0 \quad \text{implies} \quad u = 0^2 + 10 = 10 &&\text{New lower limit}\end{aligned}$$

Thus, we have

$$\begin{aligned}\int_0^5 \frac{x}{x^2+10}\,dx &= \frac{1}{2}\int_0^5 \frac{1}{x^2+10}2x\,dx\\ &= \frac{1}{2}\int_{10}^{35}\frac{1}{u}\,du\\ &= \frac{1}{2}\left(\ln|u|\,\Big|_{10}^{35}\right)\\ &= \tfrac{1}{2}(\ln 35 - \ln 10) \approx 0.626\end{aligned}$$

Matched Problem 4 Use both methods described in Example 4 to evaluate: $\displaystyle\int_0^1 \frac{1}{2x+4}\,dx$

Example 5 **Definite Integrals and Substitution** Use method 2 described in Example 4 to evaluate

$$\int_{-4}^{1}\sqrt{5-t}\,dt$$

SOLUTION If $u = 5 - t$, then $du = -dt$, and

$t = 1$	implies	$u = 5 - 1 = 4$	New upper limit
$t = -4$	implies	$u = 5 - (-4) = 9$	New lower limit

Notice that the lower limit for u is larger than the upper limit. Be careful not to reverse these two values when substituting in the definite integral.

$$\begin{aligned}\int_{-4}^{1}\sqrt{5-t}\,dt &= -\int_{-4}^{1}\sqrt{5-t}\,(-dt)\\ &= -\int_9^4 \sqrt{u}\,du\\ &= -\int_9^4 u^{1/2}\,du\\ &= -\left(\frac{u^{3/2}}{\frac{3}{2}}\,\Bigg|_9^4\right)\\ &= -\left[\tfrac{2}{3}(4)^{3/2} - \tfrac{2}{3}(9)^{3/2}\right]\\ &= -\left[\tfrac{16}{3} - \tfrac{54}{3}\right] = \tfrac{38}{3} \approx 12.667\end{aligned}$$

Matched Problem 5 Use method 2 described in Example 4 to evaluate: $\displaystyle\int_2^5 \frac{1}{\sqrt{6-t}}\,dt$

Explore–Discuss 3

Explain why $\neq$ is used in each and finish each correctly:

1. $\displaystyle\int_0^2 e^x\,dx = e^x\Big|_0^2 \neq e^2$

2. $\displaystyle\int_2^5 \frac{1}{2x+3}\,dx \neq \frac{1}{2}\int_2^5 \frac{1}{u}\,du \qquad u = 2x+3, \quad du = 2\,dx$

Example 6 **Change in Profit** A company manufactures x television sets per month. The monthly marginal profit (in dollars) is given by

$$P'(x) = 165 - 0.1x \qquad 0 \leq x \leq 4{,}000$$

The company is currently manufacturing 1,500 sets per month, but is planning to increase production. Find the total change in the monthly profit if monthly production is increased to 1,600 sets.

SOLUTION

$$\begin{aligned} P(1{,}600) - P(1{,}500) &= \int_{1{,}500}^{1{,}600} (165 - 0.1x)\,dx \\ &= (165x - 0.05x^2)\Big|_{1{,}500}^{1{,}600} \\ &= [165(1{,}600) - 0.05(1{,}600)^2] - [165(1{,}500) - 0.05(1{,}500)^2] \\ &= 136{,}000 - 135{,}000 \\ &= 1{,}000 \end{aligned}$$

Thus, increasing monthly production from 1,500 units to 1,600 units will increase the monthly profit by \$1,000.

Matched Problem 6 Repeat Example 6 if

$$P'(x) = 300 - 0.2x \qquad 0 \leq x \leq 3{,}000$$

and monthly production is increased from 1,400 sets to 1,500 sets.

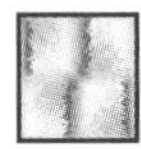

Example 7 **Useful Life** An amusement company maintains records for each video game it installs in an arcade. Suppose that $C(t)$ and $R(t)$ represent the total accumulated costs and revenues (in thousands of dollars), respectively, t years after a particular game has been installed and that

$$C'(t) = 2 \qquad R'(t) = 9e^{-0.5t}$$

The value of t for which $C'(t) = R'(t)$ is called the **useful life** of the game.

(A) Find the useful life of the game to the nearest year.
(B) Find the total profit accumulated during the useful life of the game.

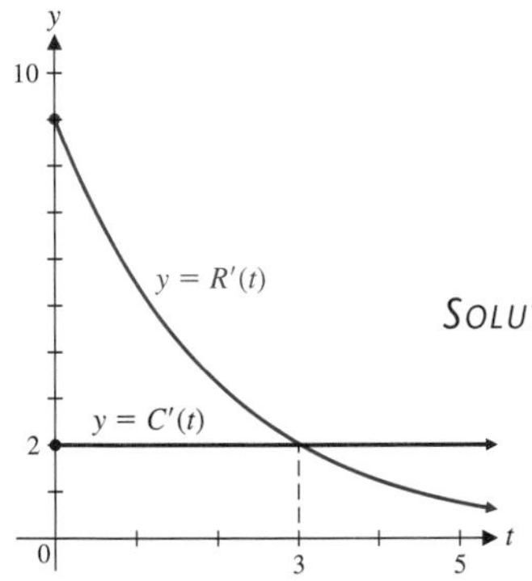

SOLUTION (A)

$$\begin{aligned} R'(t) &= C'(t) \\ 9e^{-0.5t} &= 2 \\ e^{-0.5t} &= \tfrac{2}{9} && \text{Convert to equivalent logarithmic form.} \\ -0.5t &= \ln \tfrac{2}{9} \\ t &= -2 \ln \tfrac{2}{9} \approx 3 \text{ years} \end{aligned}$$

Thus, the game has a useful life of 3 years. This is illustrated graphically in Figure 5.

FIGURE 5
Useful life

(B) The total profit accumulated during the useful life of the game is

$$\begin{aligned}
P(3) - P(0) &= \int_0^3 P'(t)\,dt \\
&= \int_0^3 [R'(t) - C'(t)]\,dt \\
&= \int_0^3 (9e^{-0.5t} - 2)\,dt \\
&= \left(\frac{9}{-0.5}e^{-0.5t} - 2t\right)\Big|_0^3 \qquad \textit{Recall: } \int e^{ax}\,dx = \frac{1}{a}e^{ax} + C \\
&= (-18e^{-0.5t} - 2t)|_0^3 \\
&= (-18e^{-1.5} - 6) - (-18e^0 - 0) \\
&= 12 - 18e^{-1.5} \approx 7.984 \quad \text{or} \quad \$7,984
\end{aligned}$$

Matched Problem 7 Repeat Example 7 if $C'(t) = 1$ and $R'(t) = 7.5e^{-0.5t}$.

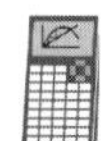

Example 8 **Numerical Integration on a Graphing Utility** Evaluate (to three decimal places): $\int_{-1}^{2} e^{-x^2}\,dx$

FIGURE 6

Solution The integrand e^{-x^2} does not have an elementary antiderivative, so we are unable to use the fundamental theorem to evaluate the definite integral. Instead, we use a numerical integration routine that has been preprogrammed in a graphing utility (consult your user's manual for specific details). Such a routine is an approximation algorithm, more powerful than the left sum and right sum methods discussed in Section 11-4. From Figure 6,

$$\int_{-1}^{2} e^{-x^2}\,dx = 1.629$$

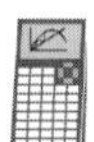

Matched Problem 8 Evaluate (to three decimal places): $\displaystyle\int_{1.5}^{4.3} \frac{x}{\ln x}\,dx$

Recognizing a Definite Integral—Average Value

Recall that the derivative of a function f was defined in Section 8-3 by

$$f'(x) = \lim_{h \to 0} \frac{f(x + h) - f(x)}{h}$$

This form is generally not easy to compute directly, but it is easy to recognize it in certain practical problems (slope, instantaneous velocity, rates of change, and so on). Once we know that we are dealing with a derivative, we then proceed to try to compute the derivative using derivative formulas and rules.

Similarly, evaluating a definite integral using the definition

$$\int_a^b f(x)\,dx = \lim_{n \to \infty}[f(c_1)\,\Delta x_1 + f(c_2)\,\Delta x_2 + \cdots + f(c_n)\,\Delta x_n] \tag{1}$$

is generally not easy; but the form on the right occurs naturally in many practical problems. We can use the fundamental theorem to evaluate the definite integral (once it is recognized) if an antiderivative can be found; otherwise,

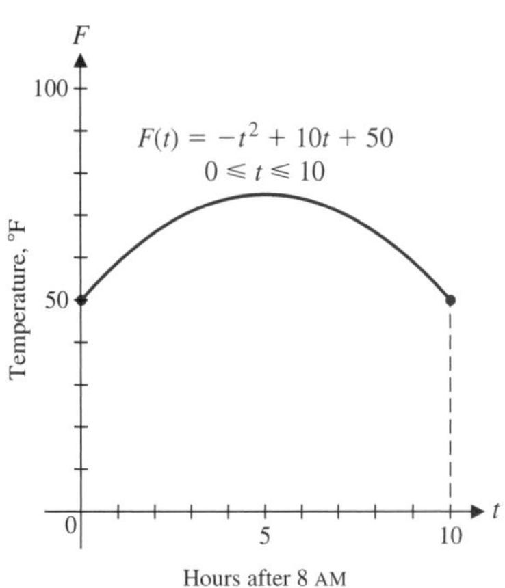

FIGURE 7

we will approximate it using a rectangle sum. We will now illustrate these points by finding the *average value* of a continuous function.

Suppose the temperature F (in degrees Fahrenheit) in the middle of a small shallow lake from 8 AM ($t = 0$) to 6 PM ($t = 10$) during the month of May is given approximately as shown in Figure 7.

How can we compute the average temperature from 8 AM to 6 PM? We know that the average of a finite number of values $a_1, a_2, \ldots, a_n$ is given by

$$\text{Average} = \frac{a_1 + a_2 + \cdots + a_n}{n}$$

But how can we handle a continuous function with infinitely many values? It would seem reasonable to divide the time interval $[0, 10]$ into n equal subintervals, compute the temperature at a point in each subinterval, and then use the average of these values as an approximation of the average value of the continuous function $F = F(t)$ over $[0, 10]$. We would expect the approximations to improve as n increases. In fact, we would be inclined to define the limit of the average of n values as $n \to \infty$ as the *average value of F over* $[0, 10]$, if the limit exists. This is exactly what we will do:

$$\begin{pmatrix}\text{Average temperature}\\ \text{for } n \text{ values}\end{pmatrix} = \frac{1}{n}[F(t_1) + F(t_2) + \cdots + F(t_n)] \tag{2}$$

where t_k is a point in the kth subinterval. We will call the limit of (2) as $n \to \infty$ the *average temperature over the time interval* $[0, 10]$.

Form (2) looks sort of like form (1), but we are missing the Δt_k. We take care of this by multiplying (2) by $(b - a)/(b - a)$, which will change the form of (2) without changing its value:

$$\begin{aligned}\frac{b-a}{b-a}\cdot\frac{1}{n}[F(t_1) + F(t_2) + \cdots + F(t_n)] &= \frac{1}{b-a}\cdot\frac{b-a}{n}[F(t_1) + F(t_2) + \cdots + F(t_n)]\\ &= \frac{1}{b-a}\left[F(t_1)\frac{b-a}{n} + F(t_2)\frac{b-a}{n} + \cdots + F(t_n)\frac{b-a}{n}\right]\\ &= \frac{1}{b-a}[F(t_1)\,\Delta t + F(t_2)\,\Delta t + \cdots + F(t_n)\,\Delta t]\end{aligned}$$

Thus,

$$\begin{aligned}\begin{pmatrix}\text{Average temperature}\\ \text{over } [a, b] = [0, 10]\end{pmatrix} &= \lim_{n\to\infty}\left\{\frac{1}{b-a}[F(t_1)\,\Delta t + F(t_2)\,\Delta t + \cdots + F(t_n)\,\Delta t]\right\}\\ &= \frac{1}{b-a}\left\{\lim_{n\to\infty}[F(t_1)\,\Delta t + F(t_2)\,\Delta t + \cdots + F(t_n)\,\Delta t]\right\}\end{aligned}$$

Now the limit inside the braces is of form (1)—that is, a definite integral. Thus,

$$\begin{aligned}\begin{pmatrix}\text{Average temperature}\\ \text{over } [a, b] = [0, 10]\end{pmatrix} &= \frac{1}{b-a}\int_a^b F(t)\,dt\\ &= \frac{1}{10-0}\int_0^{10}(-t^2 + 10t + 50)\,dt\\ &= \frac{1}{10}\left(-\frac{t^3}{3} + 5t^2 + 50t\right)\Bigg|_0^{10}\\ &= \frac{200}{3} \approx 67°\text{F}\end{aligned}$$

We now evaluate the definite integral using the fundamental theorem.

In general, proceeding as above for an arbitrary continuous function f over an interval $[a, b]$, we obtain the following general formula:

Average Value of a Continuous Function f Over $[a, b]$

$$\frac{1}{b - a} \int_a^b f(x)\, dx$$

Explore–Discuss 4

In Figure 8 the rectangle has the same area as the area under the graph of $y = f(x)$ from $x = a$ to $x = b$. Explain how the average value of $f(x)$ over the interval $[a, b]$ is related to the height of the rectangle.

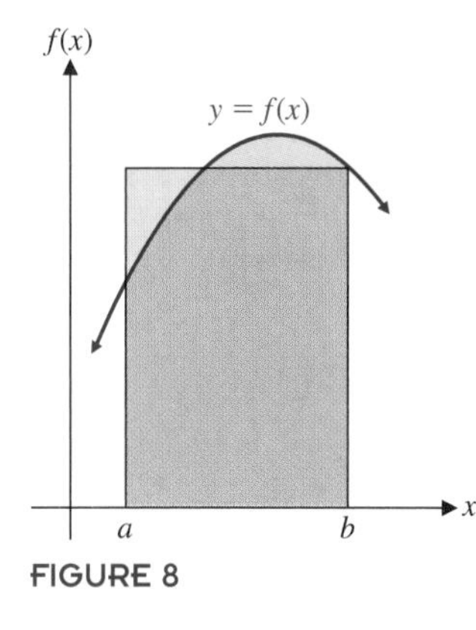

FIGURE 8

Example 9 **Average Value of a Function** Find the average value of $f(x) = x - 3x^2$ over the interval $[-1, 2]$.

SOLUTION

$$\frac{1}{b - a} \int_a^b f(x)\, dx = \frac{1}{2 - (-1)} \int_{-1}^2 (x - 3x^2)\, dx$$
$$= \frac{1}{3}\left(\frac{x^2}{2} - x^3\right)\Bigg|_{-1}^2 = -\frac{5}{2}$$

Matched Problem 9 Find the average value of $g(t) = 6t^2 - 2t$ over the interval $[-2, 3]$.

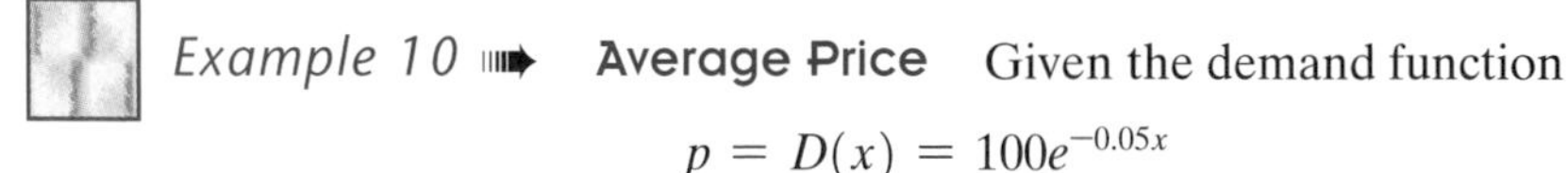

Example 10 **Average Price** Given the demand function

$$p = D(x) = 100e^{-0.05x}$$

find the average price (in dollars) over the demand interval $[40, 60]$.

Solution

$$\text{Average price} = \frac{1}{b-a}\int_a^b D(x)\,dx$$
$$= \frac{1}{60-40}\int_{40}^{60} 100e^{-0.05x}\,dx$$
$$= \frac{100}{20}\int_{40}^{60} e^{-0.05x}\,dx \qquad \text{Use } \int e^{ax}\,dx = \frac{1}{a}e^{ax},\ a \neq 0.$$
$$= -\frac{5}{0.05}e^{-0.05x}\Big|_{40}^{60}$$
$$= 100(e^{-2} - e^{-3}) \approx \$8.55$$

Matched Problem 10 Given the supply equation

$$p = S(x) = 10e^{0.05x}$$

find the average price (in dollars) over the supply interval [20, 30].

Answers to Matched Problems

1. (A) $n \geqslant 270$ (B) $n \geqslant 5$
2. (A) 38 (B) 16 (C) −6 (D) 0 (E) 108
3. $16 + 2e - 2e^3 + 5\ln 3 \approx -13.241$ **4.** $\frac{1}{2}(\ln 6 - \ln 4) \approx 0.203$
5. 2 **6.** \$1,000
7. (A) $-2\ln\frac{2}{15} \approx 4$ yr (B) $11 - 15e^{-2} \approx 8.970$ or \$8,970
8. 8.017 **9.** 13 **10.** \$35.27

Exercise 11-5

A *Problems 1–6 refer to the following figure with the indicated areas:*

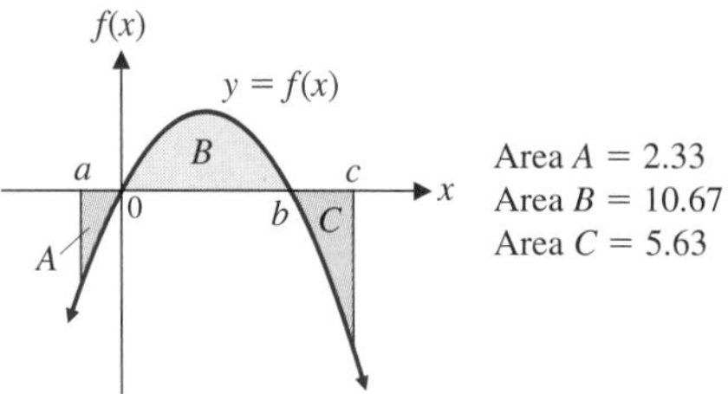

1. $\int_a^0 f(x)\,dx = ?$
2. $\int_b^c f(x)\,dx = ?$
3. $\int_a^b f(x)\,dx = ?$
4. $\int_0^c f(x)\,dx = ?$
5. $\int_0^b \frac{f(x)}{10}\,dx = ?$
6. $\int_0^b 10f(x)\,dx = ?$

Evaluate the integrals in Problems 7–20.

7. $\int_2^3 2x\,dx$
8. $\int_1^2 3x^2\,dx$
9. $\int_3^4 5\,dx$
10. $\int_{12}^{20} dx$
11. $\int_1^3 (2x - 3)\,dx$
12. $\int_1^3 (6x + 5)\,dx$
13. $\int_{-3}^4 (4 - x^2)\,dx$
14. $\int_{-1}^2 (x^2 - 4x)\,dx$
15. $\int_0^1 24x^{11}\,dx$
16. $\int_0^2 30x^5\,dx$
17. $\int_0^1 e^{2x}\,dx$
18. $\int_{-1}^1 e^{5x}\,dx$
19. $\int_1^{3.5} 2x^{-1}\,dx$
20. $\int_1^2 \frac{dx}{x}$

B *Problems 21–24 refer to the figure for Problems 1–6 and the indicated areas.*

21. $\int_b^0 f(x)\,dx = ?$
22. $\int_0^a f(x)\,dx = ?$
23. $\int_c^0 f(x)\,dx = ?$
24. $\int_b^a f(x)\,dx = ?$
25. In Example 2 of Section 11-4 we used the rate function $r(t) = 32t$ and left and right sums to estimate the distance that a steel ball dropped from a bridge falls over the time interval [1, 4]. Find this distance using a definite integral and the fundamental theorem of calculus.
26. Repeat Problem 25 for the time interval [2, 5].

Evaluate the integrals in Problems 27–44.

27. $\int_1^2 (2x^{-2} - 3)\,dx$

28. $\int_1^2 (5 - 16x^{-3})\,dx$

29. $\int_1^4 3\sqrt{x}\,dx$

30. $\int_4^{25} \frac{2}{\sqrt{x}}\,dx$

31. $\int_2^3 12(x^2 - 4)^5 x\,dx$

32. $\int_0^1 32(x^2 + 1)^7 x\,dx$

33. $\int_3^9 \frac{1}{x - 1}\,dx$

34. $\int_2^8 \frac{1}{x + 1}\,dx$

35. $\int_{-5}^{10} e^{-0.05x}\,dx$

36. $\int_{-10}^{25} e^{-0.01x}\,dx$

37. $\int_{-6}^{0} \sqrt{4 - 2x}\,dx$

38. $\int_{-4}^{2} \frac{1}{\sqrt{8 - 2x}}\,dx$

39. $\int_{-1}^{7} \frac{x}{\sqrt{x + 2}}\,dx$

40. $\int_0^3 x\sqrt{x + 1}\,dx$

41. $\int_0^1 (e^{2x} - 2x)^2(e^{2x} - 1)\,dx$

42. $\int_0^1 \frac{2e^{4x} - 3}{e^{2x}}\,dx$

43. $\int_{-2}^{-1} (x^{-1} + 2x)\,dx$

44. $\int_{-3}^{-1} (-3x^{-2} + x^{-1})\,dx$

In Problems 45–52:

(A) Find the average value of each function over the indicated interval.

(B) Use a graphing utility to graph the function and its average value over the indicated interval in the same viewing window.

45. $f(x) = 500 - 50x$; $[0, 10]$

46. $g(x) = 2x + 7$; $[0, 5]$

47. $f(t) = 3t^2 - 2t$; $[-1, 2]$

48. $g(t) = 4t - 3t^2$; $[-2, 2]$

49. $f(x) = \sqrt[3]{x}$; $[1, 8]$

50. $g(x) = \sqrt{x + 1}$; $[3, 8]$

51. $f(x) = 4e^{-0.2x}$; $[0, 10]$

52. $f(x) = 64e^{0.08x}$; $[0, 10]$

Problems 53–58 refer to the figure below.

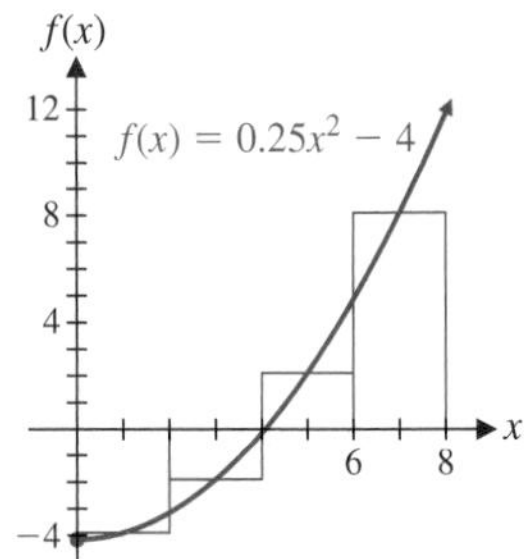

53. Use the midpoint sum M_4 to approximate the integral $I = \int_0^8 (0.25x^2 - 4)\,dx$. Calculate an error bound for this estimate.

54. Use the midpoint sum M_3 to approximate the integral $\int_0^6 (0.25x^2 - 4)\,dx$. Calculate an error bound for this estimate.

55. Evaluate $I = \int_0^8 (0.25x^2 - 4)\,dx$ using the fundamental theorem. Use the midpoint sum M_4 from Problem 53 to calculate the actual error $|I - M_4|$. Does this error lie within the error bound computed in Problem 53? Explain.

56. Evaluate $I = \int_0^6 (0.25x^2 - 4)\,dx$ using the fundamental theorem. Use the midpoint sum M_3 from Problem 54 to calculate the actual error $|I - M_3|$. Does this error lie within the error bound computed in Problem 54? Explain.

57. How large must n be chosen to have a midpoint sum M_n approximate $I = \int_0^8 (0.25x^2 - 4)\,dx$ with an error that does not exceed 0.005?

58. How large must n be chosen to have a midpoint sum M_n approximate $I = \int_0^6 (0.25x^2 - 4)\,dx$ with an error that does not exceed 0.05?

C *Write Problems 59–62 in the form $\int_a^b f(x)\,dx$ and evaluate using the fundamental theorem of calculus.*

59. $\lim_{n\to\infty} \left[(1 - c_1^2)\,\Delta x + (1 - c_2^2)\,\Delta x + \cdots + (1 - c_n^2)\,\Delta x\right]$, where $\Delta x = \frac{5 - 2}{n}$ and $c_k = 2 + k\frac{3}{n}$, $k = 1, 2, \ldots, n$

60. $\lim_{n\to\infty} \left[(c_1^2 - 3)\,\Delta x + (c_2^2 - 3)\,\Delta x + \cdots + (c_n^2 - 3)\,\Delta x\right]$, where $\Delta x = \frac{10 - 0}{n}$ and $c_k = 0 + k\frac{10}{n}$, $k = 1, 2, \ldots, n$

61. $\lim_{n\to\infty} \left[(3c_1^2 - 2c_1 + 3)\,\Delta x + (3c_2^2 - 2c_2 + 3)\,\Delta x + \cdots + (3c_n^2 - 2c_n + 3)\,\Delta x\right]$, where $\Delta x = \frac{12 - 2}{n}$ and $c_k = 2 + k\frac{10}{n}$, $k = 1, 2, \ldots, n$

62. $\lim_{n\to\infty} \left[(4c_1^3 + 3c_1^2 - 5)\,\Delta x + (4c_2^3 + 3c_2^2 - 5)\,\Delta x + \cdots + (4c_n^3 + 3c_n^2 - 5)\,\Delta x\right]$, where $\Delta x = \frac{2 - 1}{n}$ and $c_k = 1 + k\frac{1}{n}$, $k = 1, 2, \ldots, n$

Evaluate the integrals in Problems 63–68.

63. $\int_2^3 x\sqrt{2x^2 - 3}\,dx$

64. $\int_0^1 x\sqrt{3x^2 + 2}\,dx$

65. $\int_0^1 \frac{x - 1}{x^2 - 2x + 3}\,dx$

66. $\int_1^2 \frac{x + 1}{2x^2 + 4x + 4}\,dx$

67. $\int_{-1}^{1} \frac{e^{-x} - e^x}{(e^{-x} + e^x)^2}\,dx$

68. $\int_6^7 \frac{\ln(t - 5)}{t - 5}\,dt$

Problems 69–72 refer to the following: In a more advanced treatment of exponential and logarithmic functions, the natural logarithmic function is defined in terms of a definite integral,

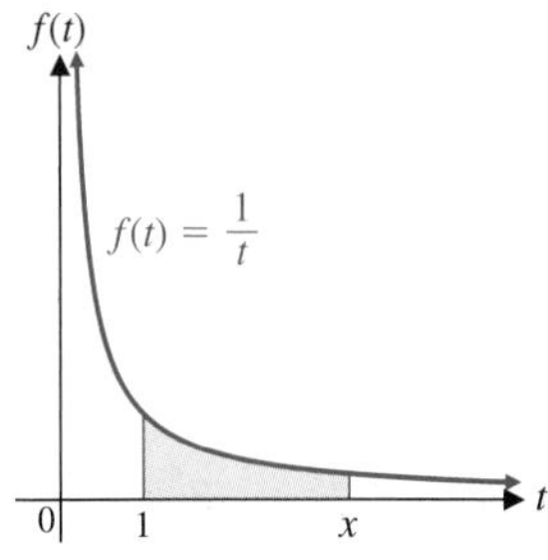

$$\ln x = \int_1^x \frac{1}{t}\,dt \qquad x > 0$$

69. Round all calculations to four decimal places.
(A) Approximate

$$\ln 2 = \int_1^2 \frac{1}{t}\,dt$$

using the midpoint sum with $n = 5$. Calculate an error bound.
(B) Evaluate ln 2 directly on your calculator.
(C) Calculate the actual error from using the midpoint sum. [Use the results from parts (A) and (B) and Error $= |\ln 2 - M_5|$.] Is this within the error bound determined in part (A)?

70. Round all calculations to four decimal places.
(A) Approximate

$$\ln 3 = \int_1^3 \frac{1}{t}\,dt$$

using the midpoint sum with $n = 10$. Calculate an error bound.
(B) Evaluate ln 3 directly on your calculator.
(C) Calculate the actual error from using the midpoint sum. [Use the results from parts (A) and (B) and Error $= |\ln 3 - M_{10}|$.] Is this within the error bound determined in part (A)?

71. In using the midpoint sum in Problem 69, how large should n be chosen so that the error in approximating ln 2 does not exceed 0.0005?

72. In using the midpoint sum in Problem 70, how large should n be chosen so that the error in approximating ln 3 does not exceed 0.0005?

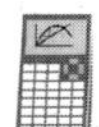

Evaluate each definite integral in Problems 73–76 (to three decimal places) using a numerical integration routine.

73. $\int_{-1.1}^{2.3} e^{x^2}\,dx$

74. $\displaystyle\int_1^{15} \frac{\ln x}{x^3}\,dx$

75. $\displaystyle\int_6^7 \frac{5}{\sqrt{x^3 + 4x + 2}}\,dx$

76. $\displaystyle\int_{1.5}^{8.9} \frac{x \ln x}{1 + e^x}\,dx$

Applications

Business & Economics

77. *Cost.* A company manufactures mountain bikes. The research department produced the following marginal cost function:

$$C'(x) = 500 - \frac{x}{3} \qquad 0 \le x \le 900$$

where $C'(x)$ is in dollars and x is the number of bikes produced per month. Compute the increase in cost going from a production level of 300 bikes per month to 900 bikes per month. Set up a definite integral and evaluate.

78. *Cost.* Referring to Problem 77, compute the increase in cost going from a production level of 0 bikes per month to 600 bikes per month. Set up a definite integral and evaluate.

79. *Salvage value.* A new piece of industrial equipment will depreciate in value rapidly at first, then less rapidly as time goes on. Suppose the rate (in dollars per year) at which the book value of a new milling machine changes is given approximately by

$$V'(t) = f(t) = 500(t - 12) \qquad 0 \le t \le 10$$

where $V(t)$ is the value of the machine after t years. What is the total loss in value of the machine in the first 5 years? In the second 5 years? Set up appropriate integrals and solve.

80. *Maintenance costs.* Maintenance costs for an apartment house generally increase as the building gets older. From past records, a managerial service determines that the rate of increase in maintenance costs (in dollars per year) for a particular apartment complex is given approximately by

$$M'(x) = f(x) = 90x^2 + 5{,}000$$

where x is the age of the apartment complex in years and $M(x)$ is the total (accumulated) cost of maintenance for x years. Write a definite integral that will give the total maintenance costs from the end of the second year to the end of the seventh year after the apartment complex was built, and evaluate it.

81. *Employee training.* A company producing computer components has established that, on the average, a new employee can assemble $N(t)$ components per day after t days of on-the-job training, as indicated in the table (a new employee's productivity usually increases with time on the job up to a leveling off point):

t	0	20	40	60	80	100	120
$N(t)$	10	51	68	76	81	84	85

(A) Find a quadratic regression equation for the data, and graph it and the data set in the same viewing window.
(B) Use the regression equation and a numerical integration routine on a graphing utility to approximate the number of units assembled by a new employee during the first 100 days on the job.

82. *Employee training.* Refer to Problem 81.
(A) Find a cubic regression equation for the data, and graph it and the data set in the same viewing window.
(B) Use the regression equation and a numerical integration routine on a graphing utility to approximate the number of units assembled by a new employee during the second 60 days on the job.

83. *Useful life.* The total accumulated costs $C(t)$ and revenues $R(t)$ (in thousands of dollars), respectively, for a coin-operated photocopying machine satisfy

$$C'(t) = \tfrac{1}{11}t \quad \text{and} \quad R'(t) = 5te^{-t^2}$$

where t is time in years. Find the useful life of the machine to the nearest year. What is the total profit accumulated during the useful life of the machine?

84. *Useful life.* The total accumulated costs $C(t)$ and revenues $R(t)$ (in thousands of dollars), respectively, for a coal mine satisfy

$$C'(t) = 3 \quad \text{and} \quad R'(t) = 15e^{-0.1t}$$

where t is the number of years the mine has been in operation. Find the useful life of the mine to the nearest year. What is the total profit accumulated during the useful life of the mine?

85. *Average cost.* The total cost (in dollars) of manufacturing x auto body frames is $C(x) = 60{,}000 + 300x$.
(A) Find the average cost per unit if 500 frames are produced. [*Hint:* Recall that $\overline{C}(x)$ is the average cost per unit.]
(B) Find the average value of the cost function over the interval $[0, 500]$.
(C) Discuss the difference between parts (A) and (B).

86. *Average cost.* The total cost (in dollars) of printing x dictionaries is $C(x) = 20{,}000 + 10x$.
(A) Find the average cost per unit if 1,000 dictionaries are produced.
(B) Find the average value of the cost function over the interval $[0, 1{,}000]$.
(C) Discuss the difference between parts (A) and (B).

87. *Cost.* The marginal cost at various levels of output per month for a company that manufactures watches is shown in the table. The output x is given in thousands of units per month, and $C(x)$ is given in thousands of dollars per month.

x	0	1	2	3	4	5	6	7	8
$C'(x)$	58	30	18	9	5	7	17	33	51

(A) Find a quadratic regression equation for the data, and graph it and the data set in the same viewing window.
(B) Use the regression equation and a numerical integration routine on a graphing utility to approximate (to the nearest dollar) the increased cost in going from a production level of 2 thousand watches per month to 8 thousand watches per month.

88. *Cost.* Refer to Problem 87.
(A) Find a cubic regression equation for the data, and graph it and the data set in the same viewing window.
(B) Use the regression equation and a numerical integration routine on a graphing utility to approximate (to the nearest dollar) the increased cost in going from a production level of 1 thousand watches per month to 7 thousand watches per month.

89. *Supply function.* Given the supply function

$$P = S(x) = 10(e^{0.02x} - 1)$$

find the average price (in dollars) over the supply interval $[20, 30]$.

90. *Demand function.* Given the demand function

$$p = D(x) = \frac{1{,}000}{x}$$

find the average price (in dollars) over the demand interval $[400, 600]$.

91. *Labor costs and learning.* A defense contractor is starting production on a new missile control system. On the basis of data collected while assembling the first 16 control systems, the production manager obtained the following function for rate of labor use:

$$g(x) = 2{,}400x^{-1/2}$$

where $g(x)$ is the number of labor-hours required to assemble the xth unit of a control system. Approximately how many labor-hours will be required to assemble the 17th through the 25th control units? [*Hint:* Let $a = 16$ and $b = 25$.]

92. *Labor costs and learning.* If the rate of labor use in Problem 91 is

$$g(x) = 2{,}000x^{-1/3}$$

approximately how many labor-hours will be required to assemble the 9th through the 27th control units? [*Hint:* Let $a = 8$ and $b = 27$.]

93. *Inventory.* A store orders 600 units of a product every 3 months. If the product is steadily depleted to 0 by the end of each 3 months, the inventory on hand, I, at any time t during the year is illustrated as shown in the figure.

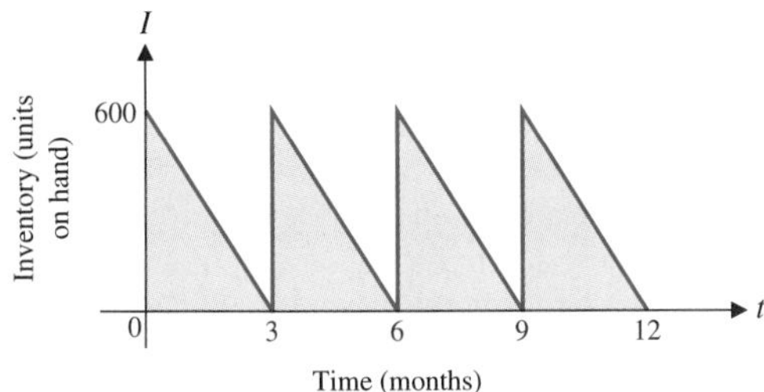

Figure for 93

(A) Write an inventory function (assume it is continuous) for the first 3 months. [The graph is a straight line joining (0, 600) and (3, 0).]

(B) What is the average number of units on hand for a 3 month period?

94. Repeat Problem 93 with an order of 1,200 units every 4 months.

95. *Oil production.* Using data from the first 3 years of production as well as geological studies, the management of an oil company estimates that oil will be pumped from a producing field at a rate given by

$$R(t) = \frac{100}{t + 1} + 5 \qquad 0 \leq t \leq 20$$

where $R(t)$ is the rate of production (in thousands of barrels per year) t years after pumping begins. Approximately how many barrels of oil will the field produce during the first 10 years of production? From the end of the 10th year to the end of the 20th year of production?

96. *Oil production.* In Problem 95, if the rate is found to be

$$R(t) = \frac{120t}{t^2 + 1} + 3 \qquad 0 \leq t \leq 20$$

approximately how many barrels of oil will the field produce during the first 5 years of production? The second 5 years of production?

97. *Profit.* Let $R(t)$ and $C(t)$ represent the total accumulated revenues and costs (in dollars), respectively, for a producing oil well, where t is time in years. The graphs of the derivatives of R and C over a 5 year period are shown in the figure. Use the midpoint sum with $n = 5$ to approximate the total accumulated profits from the well over this 5 year period. Estimate necessary function values from the graphs.

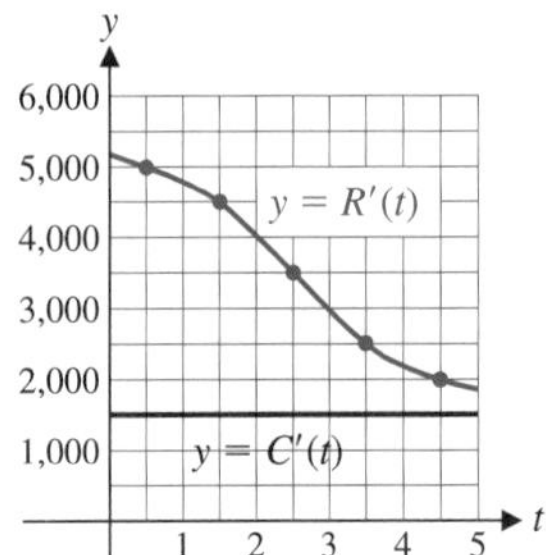

Figure for 97 and 98

98. *Revenue.* Use the figure and the midpoint sum with $n = 5$ to approximate the average annual revenue from the oil well.

99. *Real estate.* A surveyor produced the table below by measuring the vertical distance (in feet) across a piece of real estate at 600 foot intervals, starting at 300 (see the figure). Use these values and the midpoint sum to estimate the area of the property.

x	300	900	1,500	2,100
$f(x)$	900	1,700	1,700	900

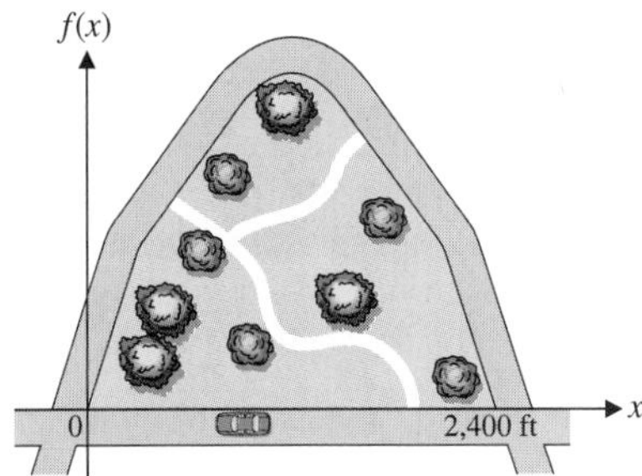

Figure for 99

100. *Real estate.* Repeat Problem 99 for the following table of measurements:

x	200	600	1,000	1,400	1,800	2,200
$f(x)$	600	1,400	1,800	1,800	1,400	600

Life Sciences

101. *Biology.* A yeast culture weighing 2 grams is removed from a refrigerator unit and is expected to grow at the rate of $W'(t) = 0.2e^{0.1t}$ grams per hour at a higher controlled temperature. How much will the weight of the culture increase during the first 8 hours of growth? How much will the weight of the culture increase from the end of the 8th hour to the end of the 16th hour of growth?

102. *Medicine.* The rate of healing for a skin wound (in square centimeters per day) is given approximately by $A'(t) = -0.9e^{-0.1t}$. The initial wound has an area of 9 square centimeters. How much will the area change during the first 5 days? The second 5 days?

103. *Temperature.* If the temperature $C(t)$ in an aquarium is made to change according to

$$C(t) = t^3 - 2t + 10 \qquad 0 \leq t \leq 2$$

(in degrees Celsius) over a 2 hour period, what is the average temperature over this period?

104. *Medicine.* A drug is injected into the bloodstream of a patient through her right arm. The concentration of the drug in the bloodstream of the left arm t hours after the injection is given by

$$C(t) = \frac{0.14t}{t^2 + 1}$$

What is the average concentration of the drug in the bloodstream of the left arm during the first hour after the injection? During the first 2 hours after the injection?

105. *Medicine—respiration.* Physiologists use a machine called a pneumotachograph to produce a graph of the rate of flow $R(t)$ of air into the lungs (inspiration) and out of the lungs (expiration). The figure gives the graph of the inspiration phase of the breathing cycle for an individual at rest. The area under this graph represents the total volume of air inhaled during the inspiration phase. Use the midpoint sum with $n = 3$ to approximate the area under the graph. Estimate the necessary function values from the graph.

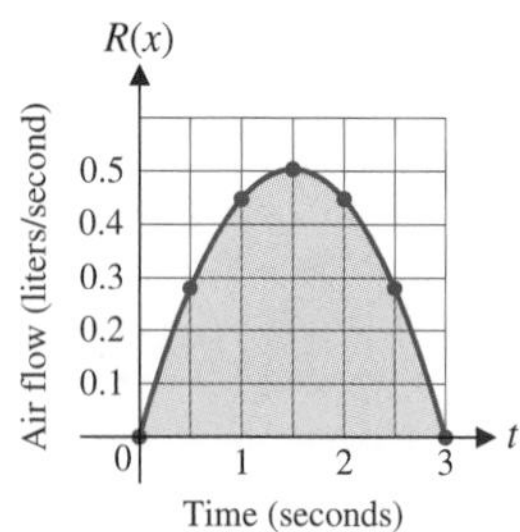

Figure for 105 and 106

106. *Medicine—respiration.* Use the result obtained in Problem 105 to approximate the average volume of air in the lungs during the inspiration phase.

Social Sciences

107. *Politics.* Public awareness of a Congressional candidate before and after a successful campaign was approximated by

$$P(t) = \frac{8.4t}{t^2 + 49} + 0.1 \qquad 0 \leq t \leq 24$$

where t is time in months after the campaign started and $P(t)$ is the fraction of people in the Congressional district who could recall the candidate's name. What is the average fraction of people who could recall the candidate's name during the first 7 months after the campaign began? During the first 2 years after the campaign began?

108. *Population composition.* Because of various factors (such as birth rate expansion, then contraction; family flights from urban areas; and so on), the number of children in a large city was found to increase and then decrease rather drastically. If the number of children over a 6 year period was found to be given approximately by

$$N(t) = -\tfrac{1}{4}t^2 + t + 4 \qquad 0 \leq t \leq 6$$

what was the average number of children in the city over the 6 year time period? [Assume $N = N(t)$ is continuous.]

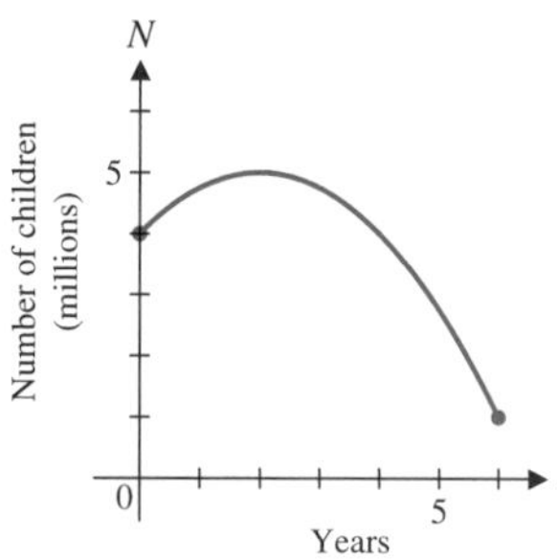

Figure for 108

IMPORTANT TERMS AND SYMBOLS

11-1 *Antiderivatives and Indefinite Integrals.* Antiderivative; indefinite integral; integral sign; integrand; constant of integration

11-2 *Integration by Substitution.* Reversing the chain rule; general indefinite integral formulas; differentials; method of substitution; change of variable

11-3 *Differential Equations—Growth and Decay.* Differential equation; first-order; second-order; slope field; continuous compound interest; exponential growth law; growth constant; growth rate; exponential decay; population growth; radioactive decay; unlimited growth; limited growth; logistic growth

$$\frac{dQ}{dt} = rQ; \quad Q = Q_0 e^{rt}$$

11-4 *A Geometric–Numeric Introduction to the Definite Integral.* Area under a curve; definite integral symbol; left sum; right sum; monotone functions; error bounds for left and right sums and their average; signed areas; rate, area, and distance; rate, area, and total change

$$\int_a^b f(x)\,dx; \quad L_n; \quad R_n; \quad A_n$$

11-5 *Definite Integral as a Limit of a Sum; Fundamental Theorem of Calculus.* Left sum; right sum; midpoint sum; general error bounds for left, right, and midpoint sums; definite integral as a limit of a Riemann sum; integrand; lower limit; upper limit; properties of definite integrals; fundamental theorem of calculus; useful life; recognizing a definite integral; average value of a continuous function

$$L_n; \quad R_n; \quad M_n;$$

$$\int_a^b f(x)\,dx; \quad F(x)\Big|_a^b = F(b) - F(a);$$

$$\sum_{k=1}^{n} f(c_k)\,\Delta x_k; \quad \frac{1}{b-a}\int_a^b f(x)\,dx$$

INTEGRATION FORMULAS AND PROPERTIES

$$\int k\,dx = kx + C$$

$$\int kf(x)\,dx = k\int f(x)\,dx$$

$$\int [f(x) \pm g(x)]\,dx = \int f(x)\,dx \pm \int g(x)\,dx$$

$$\int u^n\,du = \frac{u^{n+1}}{n+1} + C, \quad n \neq -1$$

$$\int e^u\,du = e^u + C$$

$$\int e^{au}\,du = \frac{1}{a}e^{au} + C, \quad a \neq 0$$

$$\int \frac{1}{u}\,du = \ln|u| + C, \quad u \neq 0$$

$$\int_a^a f(x)\,dx = 0$$

$$\int_a^b f(x)\,dx = -\int_b^a f(x)\,dx$$

$$\int_a^b kf(x)\,dx = k\int_a^b f(x)\,dx, \quad k \text{ a constant}$$

$$\int_a^b [f(x) \pm g(x)]\,dx = \int_a^b f(x)\,dx \pm \int_a^b g(x)\,dx$$

$$\int_a^b f(x)\,dx = \int_a^c f(x)\,dx + \int_c^b f(x)\,dx$$

REVIEW EXERCISE

Work through all the problems in this chapter review and check your answers in the back of the book. Answers to all review problems are there along with section numbers in italics to indicate where each type of problem is discussed. Where weaknesses show up, review appropriate sections in the text.

A *Find each integral in Problems 1–6.*

1. $\int (3t^2 - 2t)\,dt$

2. $\int_2^5 (2x - 3)\,dx$

3. $\int (3t^{-2} - 3)\,dt$

4. $\int_1^4 x\,dx$

5. $\int e^{-0.5x}\,dx$

6. $\int_1^5 \frac{2}{u}\,du$

In Problems 7 and 8, find the derivative or indefinite integral as indicated.

7. $\dfrac{d}{dx}\left(\int e^{-x^2}\,dx\right)$ **8.** $\int \dfrac{d}{dx}\left(\sqrt{4+5x}\right)dx$

9. Find a function $y = f(x)$ that satisfies both conditions:

$$\frac{dy}{dx} = 3x^2 - 2 \qquad f(0) = 4$$

10. From the graph of $y = f'(x)$ shown in the figure, verbally describe the shape of the graph of an antiderivative function f. How would the graph of one antiderivative of $f'(x)$ differ from another?

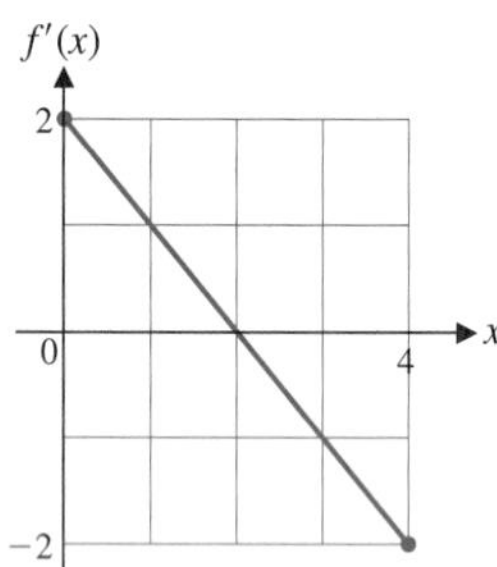

Figure for 10 and 11

11. From the graph of $y = f'(x)$ shown in the figure, sketch possible graphs of three antiderivative functions f such that $f(0) = -2$, $f(0) = 0$, and $f(0) = 2$, all on the same set of coordinate axes.

12. Find all antiderivatives of:

(A) $\dfrac{dy}{dx} = 8x^3 - 4x - 1$ (B) $\dfrac{dx}{dt} = e^t - 4t^{-1}$

13. Approximate $\int_1^5 (x^2 + 1)\,dx$ using a midpoint sum with $n = 2$. Calculate an error bound for this approximation.

14. Evaluate the integral in Problem 13 using the fundamental theorem of calculus, and calculate the actual error $|I - M_2|$ produced in using M_2.

15. Use the table of values below and a midpoint sum with $n = 4$ to approximate $\int_1^{17} f(x)\,dx$.

x	3	7	11	15
$f(x)$	1.2	3.4	2.6	0.5

16. Find the average value of $f(x) = 6x^2 + 2x$ over the interval $[-1, 2]$.

17. Describe a rectangle that would have the same area as the area under the graph of $f(x) = 6x^2 + 2x$ from $x = -1$ to $x = 2$ (see Problem 16).

B *Use the graph and actual areas of the indicated regions in the figure to evaluate the integrals in Problems 18–25:*

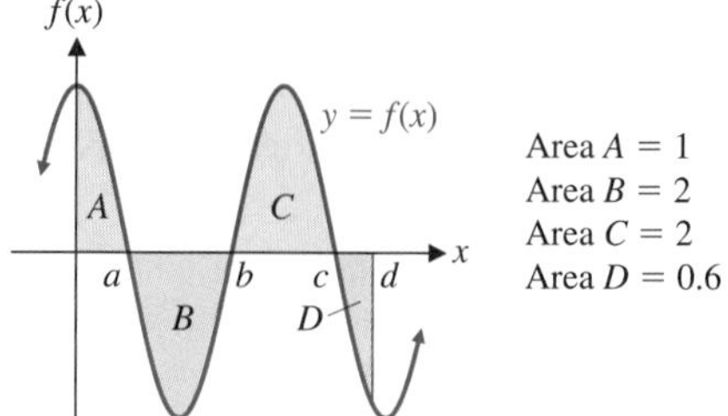

18. $\int_a^b 5f(x)\,dx$ **19.** $\int_b^c \dfrac{f(x)}{5}\,dx$ **20.** $\int_b^d f(x)\,dx$

21. $\int_a^c f(x)\,dx$ **22.** $\int_0^d f(x)\,dx$ **23.** $\int_b^a f(x)\,dx$

24. $\int_c^b f(x)\,dx$ **25.** $\int_d^0 f(x)\,dx$

26. For the graph of $y = f'(x)$ shown, verbally describe the shape of the graph of an antiderivative function f. How would the graph of one antiderivative of $f'(x)$ differ from another?

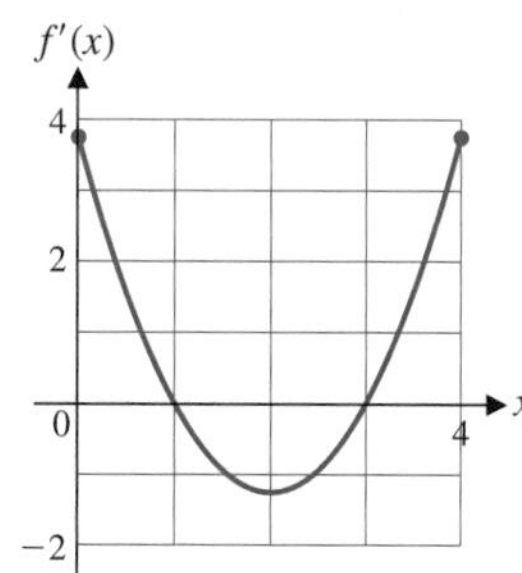

Figure for 26 and 27

27. For the graph of $y = f'(x)$ shown, sketch possible graphs of three antiderivative functions f such that $f(0) = -1$, $f(0) = 0$, and $f(0) = 1$, all on the same set of coordinate axes.

Problems 28–33 refer to the slope field shown in the figure:

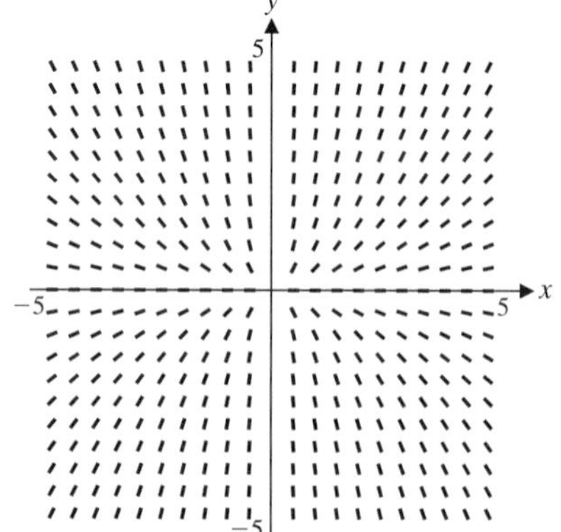

28. (A) For $dy/dx = (2y)/x$, what is the slope of a solution curve at $(2, 1)$? At $(-2, -1)$?

(B) For $dy/dx = (2x)/y$, what is the slope of a solution curve at $(2, 1)$? At $(-2, -1)$?

29. Is the slope field shown in the figure for $dy/dx = (2x)/y$ or for $dy/dx = (2y)/x$? Explain.

30. Show that $y = Cx^2$ is a solution of $dy/dx = (2y)/x$ for any real number C.

31. Referring to Problem 30, find the particular solution of $dy/dx = (2y)/x$ that passes through $(2, 1)$. Through $(-2, -1)$.

32. Graph the two particular solutions found in Problem 31 in the slope field shown above (or a copy).

33. C Use a graphing utility to graph in the same viewing window graphs of $y = Cx^2$ for $C = -2, -1, 1$, and 2 for $-5 \leq x \leq 5$ and $-5 \leq y \leq 5$.

Find each integral in Problems 34–44.

34. $\int \sqrt[3]{6x - 5}\, dx$

35. $\int_0^1 10(2x - 1)^4\, dx$

36. $\int \left(\frac{2}{x^2} - 2xe^{x^2}\right) dx$

37. $\int_0^4 x\sqrt{x^2 + 4}\, dx$

38. $\int (e^{-2x} + x^{-1})\, dx$

39. $\int_0^{10} 10e^{-0.02x}\, dx$

40. $\int_0^3 \frac{x}{1 + x^2}\, dx$

41. $\int_0^3 \frac{x}{(1 + x^2)^2}\, dx$

42. $\int x^3(2x^4 + 5)^5\, dx$

43. $\int \frac{e^{-x}}{e^{-x} + 3}\, dx$

44. $\int \frac{e^x}{(e^x + 2)^2}\, dx$

45. Find a function $y = f(x)$ that satisfies both conditions:

$$\frac{dy}{dx} = 3x^{-1} - x^{-2} \qquad f(1) = 5$$

46. Find the equation of the curve that passes through $(2, 10)$ if its slope is given by

$$\frac{dy}{dx} = 6x + 1$$

for each x.

Problems 47–50 refer to the following: A toy rocket is shot vertically into the air with an initial rate of 160 feet per second. Neglecting air resistance, its rate of ascent, r(t) (in feet per second), at the end of t seconds is shown in the figure.

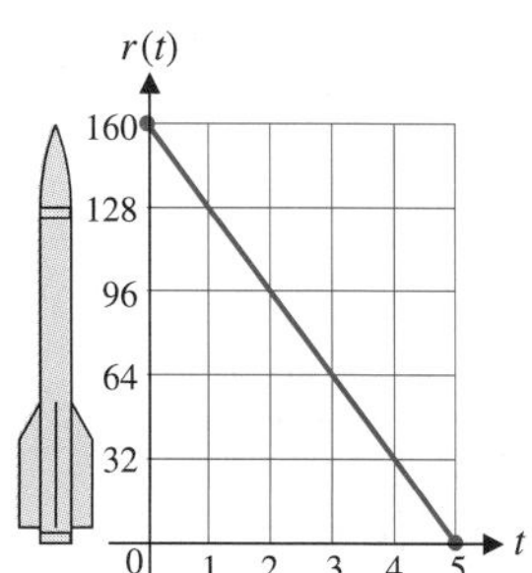

47. Estimate the height the rocket reaches using left and right sums and their average over five equal subdivisions. (Read values directly from the graph.) Calculate error bounds for these estimates.

48. Explain how the areas of the rectangles used in the left or right sum approximations in Problem 47 are related to distance.

49. How many equal subdivisions should be used on the interval $[0, 5]$ to have A_n approximate the height the rocket reaches within 1 foot?

50. Set up a definite integral that represents the height the rocket reaches and evaluate it. [*Hint:* First find the equation of the rate function.]

51. (A) Find the average value of $f(x) = 3\sqrt{x}$ over the interval $[1, 9]$.
(B) Graph $f(x) = 3\sqrt{x}$ and its average over the interval $[1, 9]$ in the same coordinate system.

C *Find each integral in Problems 52–56.*

52. $\int \frac{(\ln x)^2}{x}\, dx$

53. $\int x(x^3 - 1)^2\, dx$

54. $\int \frac{x}{\sqrt{6 - x}}\, dx$

55. $\int_0^7 x\sqrt{16 - x}\, dx$

56. $\int_{-1}^1 x(x + 1)^4\, dx$

57. Find a function $y = f(x)$ that satisfies both conditions:

$$\frac{dy}{dx} = 9x^2e^{x^3} \qquad f(0) = 2$$

58. Solve the differential equation:

$$\frac{dN}{dt} = 0.06N \qquad N(0) = 800 \qquad N > 0$$

Problems 59–63 involve estimating the value of the definite integral

$$I = \int_0^1 e^{-x^2}\, dx$$

(The integrand does not have an elementary antiderivative.)

59. Graph $f(x) = e^{-x^2}$ over the interval $[0, 1]$.

60. Approximate I using a midpoint sum with $n = 5$.

61. C Calculate $f''(x)$ and show that 2 is an upper bound for $|f''(x)|$ by graphing $|f''(x)|$ on the interval $[0, 1]$ using a graphing utility.

62. Use the results from Problem 61 to calculate an error bound for the approximation M_5 in Problem 60.

63. Using the result from Problem 62, determine how large n should be so that M_n approximates I with an error that does not exceed 0.0005.

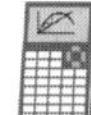

Graph Problems 64–67 on a graphing utility and identify each as unlimited growth, exponential decay, limited growth, or logistic growth:

64. $N = 50(1 - e^{-0.07t}); 0 \le t \le 80, 0 \le N \le 60$

65. $p = 500e^{-0.03x}; 0 \le x \le 100, 0 \le p \le 500$

66. $A = 200e^{0.08t}; 0 \le t \le 20, 0 \le A \le 1{,}000$

67. $N = \dfrac{100}{1 + 9e^{-0.3t}}; 0 \le t \le 25, 0 \le N \le 100$

Evaluate each definite integral in Problems 68–70 (to three decimal places) using a numerical integration routine.

68. $\displaystyle\int_3^8 \frac{\sqrt{7 + 2x}}{4 + x}\, dx$

69. $\int_{0.1}^{1.2} x2^x\, dx$

70. $\int_1^3 x^4 \ln x\, dx$

Applications

Business & Economics

71. *Cost.* A company manufactures downhill skis. The research department produced the marginal cost graph shown in the figure, where $C'(x)$ is in dollars and x is the number of pairs of skis produced per week. Estimate the increase in cost going from a production level of 200 to 600 pairs of skis per week. Use left and right sums over two equal subintervals. Replace the question marks with the values of L_2 and R_2 as appropriate:

$$? \le \int_{200}^{600} C'(x)\, dx \le ?$$

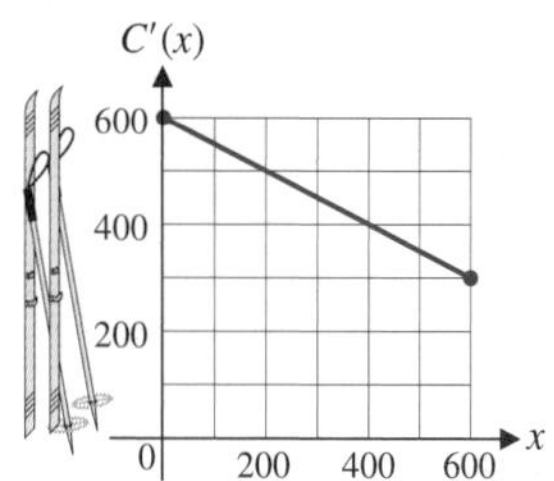

Figure for 71

72. *Cost.* Explain how the increase in production cost is related to the left or right rectangles used in Problem 71.

73. *Cost.* Assuming that the marginal cost function in Problem 71 is linear, find its equation and write a definite integral that represents the increase in costs going from a production level of 200 to 600 pairs of skis per week. Evaluate the definite integral.

74. *Profit and production.* The weekly marginal profit for an output of x units is given approximately by

$$P'(x) = 150 - \frac{x}{10} \qquad 0 \le x \le 40$$

What is the total change in profit for a production change from 10 units per week to 40 units? Set up a definite integral and evaluate it.

75. *Profit function.* If the marginal profit for producing x units per day is given by

$$P'(x) = 100 - 0.02x \qquad P(0) = 0$$

where $P(x)$ is the profit in dollars, find the profit function P and the profit on 10 units of production per day.

76. *Resource depletion.* An oil well starts out producing oil at the rate of 60,000 barrels of oil per year, but the production rate is expected to decrease by 4,000 barrels per year. Thus, if $P(t)$ is the total production (in thousands of barrels) in t years, then

$$P'(t) = f(t) = 60 - 4t \qquad 0 \le t \le 15$$

Write a definite integral that will give the total production after 15 years of operation and evaluate it.

77. *Inventory.* Suppose the inventory of a certain item t months after the first of the year is given approximately by

$$I(t) = 10 + 36t - 3t^2 \qquad 0 \le t \le 12$$

What is the average inventory for the second quarter of the year?

78. *Price–supply.* Given the price–supply function

$$p = S(x) = 8(e^{0.05x} - 1)$$

find the average price (in dollars) over the supply interval $[40, 50]$.

79. *Employee training.* A company producing sound system equipment has found that, on the average, a new employee can assemble $N(t)$ components per day after t days of on-the-job training, as indicated in the following table:

t	0	10	20	30	40	50
$N(t)$	5	10	14	17	19	20

(A) Use the average of left and right sums, A_5, for five equal subdivisions to estimate the total number of units assembled by a new employee the first 50 days of employment. Calculate an error bound for the estimate.

C (B) Find a quadratic regression equation for the data, and graph it and the data set in the same viewing window.

C (C) Use the regression equation and a numerical integration routine on a graphing utility to approximate the number of units assembled by a new employee the first 50 days of employment.

80. *Useful life.* The total accumulated costs $C(t)$ and revenues $R(t)$ (in thousands of dollars), respectively, for a coal mine satisfy

$$C'(t) = 3 \qquad \text{and} \qquad R'(t) = 20e^{-0.1t}$$

where t is the number of years the mine has been in operation. Find the useful life of the mine to the nearest year. What is the total profit accumulated during the useful life of the mine?

81. *Marketing.* The market research department for an automobile company estimates that the sales (in millions of dollars) of a new automobile will increase at the monthly rate of

$$S'(t) = 4e^{-0.08t} \qquad 0 \le t \le 24$$

t months after the introduction of the automobile. What will be the total sales $S(t)$, t months after the automobile is introduced, if we assume that there were 0 sales at the time the automobile entered the marketplace? What are the estimated total sales during the first 12 months after the introduction of the automobile? How long will it take for the total sales to reach $40 million?

Life Sciences

82. *Pollution.* In an industrial area, the concentration $C(t)$ of particulate matter (in parts per million) during a 12 hour period is given in the figure. Use a midpoint sum with $n = 6$ to approximate the average concentration during this 12 hour period. Estimate the necessary function values from the graph.

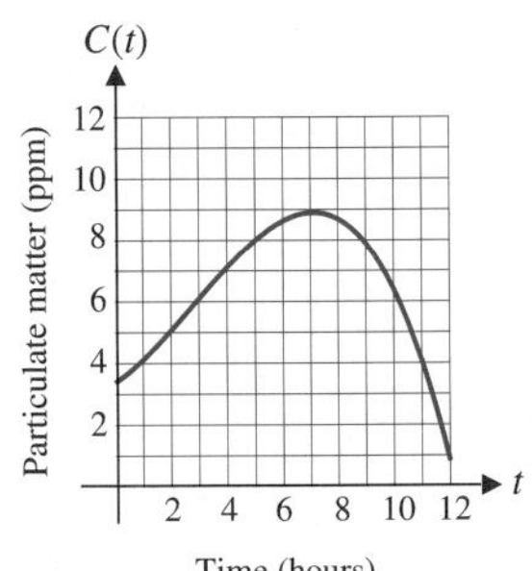

83. *Wound healing.* The area of a small, healing surface wound changes at a rate given approximately by

$$\frac{dA}{dt} = -5t^{-2} \qquad 1 \le t \le 5$$

where t is time in days and $A(1) = 5$ square centimeters. What will the area of the wound be in 5 days?

84. *Pollution.* An environmental protection agency estimates that the rate of seepage of toxic chemicals from a waste dump (in gallons per year) is given by

$$R(t) = \frac{1{,}000}{(1 + t)^2}$$

where t is time in years since the discovery of the seepage. Find the total amount of toxic chemicals that seep from the dump during the first 4 years after the seepage is discovered.

85. *Population.* According to the World Bank, the population in the Americas (north, central, and south) in 1995 was about 770 million, and was growing at the rate of about 1% compounded continuously.

(A) If the population continues to grow at this rate, what is the estimated population in the Americas for the year 2030?

(B) At the indicated growth rate, how long will it take the population in the Americas to double?

Social Sciences

86. *Archaeology.* The continuous compound rate of decay for carbon-14 is $r = -0.000\ 123\ 8$. A piece of animal bone found at an archaeological site contains 4% of the original amount of carbon-14. Estimate the age of the bone.

87. *Learning.* In a particular business college, it was found that an average student enrolled in a typing class progressed at a rate of $N'(t) = 7e^{-0.1t}$ words per minute t weeks after enrolling in a 15 week course. If at the beginning of the course a student could type 25 words per minute, how many words per minute, $N(t)$, would the student be expected to type t weeks into the course? After completing the course?

Group Activity 1 Simpson's Rule

Introduction to Simpson's Rule

The left sum L_n, right sum R_n, average A_n of left and right sums, and midpoint sum M_n can all be used as numerical integration devices to approximate definite integrals. If a function f is increasing, then the right sum overestimates and the left sum underestimates the definite integral (Fig. 1).

$$L_n \leq \int_a^b f(x)\,dx \leq R_n$$

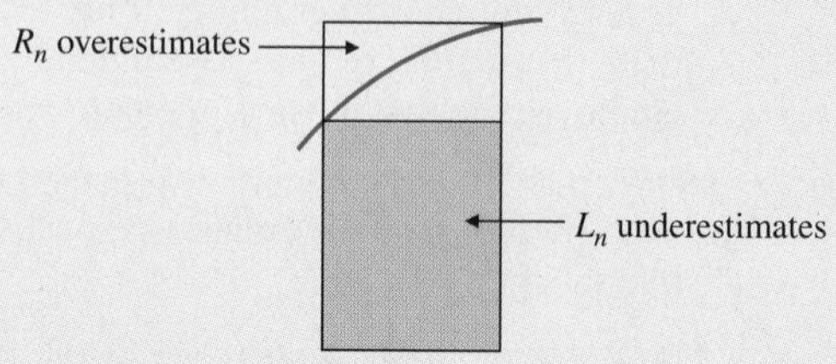

R_n overestimates

L_n underestimates

FIGURE 1

The average A_n of left and right sums is a better estimate for the definite integral of a monotone function than either L_n or R_n. It is often called a *trapezoidal sum,* and denoted by T_n, since the average area of a left rectangle and right rectangle is the area of a trapezoid (the smaller, shaded trapezoid in Fig. 2). Adding the following expressions for L_n and R_n and dividing by 2 gives the expression for $T_n = A_n$ called the **trapezoidal rule.** The midpoint sum M_n, like T_n, is also a better estimate for the definite integral of a monotone function than either the left or right sum.

$$L_n = [f(x_0) + f(x_1) + \cdots + f(x_{n-1})]\,\Delta x \qquad \text{Left sum}$$
$$R_n = [f(x_1) + f(x_2) + \cdots + f(x_n)]\,\Delta x \qquad \text{Right sum}$$
$$T_n = [f(x_0) + 2f(x_1) + \cdots + 2f(x_{n-1}) + f(x_n)]\,\frac{\Delta x}{2} \qquad \text{Trapezoidal rule}$$
$$M_n = \left[f\left(\frac{x_0 + x_1}{2}\right) + f\left(\frac{x_1 + x_2}{2}\right) + \cdots + f\left(\frac{x_{n-1} + x_n}{2}\right)\right]\Delta x \qquad \text{Midpoint sum}$$

A midpoint sum rectangle has the same area as the corresponding tangent line trapezoid (the larger trapezoid in Fig. 2). It appears from Figure 2, and can be proved in general, that the trapezoidal sum error is about double the midpoint sum error when the graph of the function is concave up or concave down.

$$T_n \leq \int_a^b f(x)\,dx \leq M_n$$

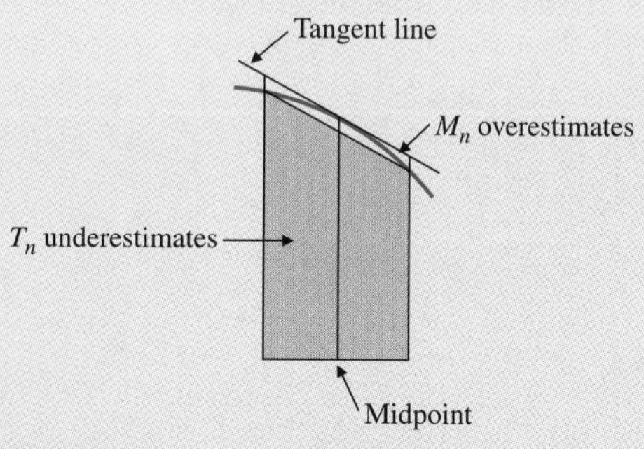

Tangent line

M_n overestimates

T_n underestimates

Midpoint

FIGURE 2

This suggests that a weighted average of the two estimates, with the midpoint sum being counted double the trapezoidal sum, might be an even better estimate than either separately. This weighted average is called **Simpson's rule,** and is given symbolically as follows:

$$S_{2n} = \frac{2M_n + T_n}{3} \quad \text{Simpson's rule} \tag{1}$$

The trapezoidal rule involves integrand values at $n + 1$ points (including a and b), and the midpoint sum involves integrand values at the n midpoints. So Simpson's rule involves integrand values at $2n + 1$ points (including a and b). To apply Simpson's rule, we divide the interval $[a, b]$ into $2n$ equal subdivisions, evaluate the integrand at each subdivision point, and then use n equal subdivisions for each of the sums M_n and T_n. Thus, **Simpson's rule always requires an even number of equal subdivisions of $[a, b]$.**

(A) For $\Delta x = (b - a)/4$, show, starting with equation (1) in the form

$$S_4 = \frac{2M_2 + T_2}{3}$$

that

$$S_4 = [f(x_0) + 4f(x_1) + 2f(x_2) + 4f(x_3) + f(x_4)]\frac{\Delta x}{3}$$

Error Comparisons

One way to compare the effectiveness of a numerical integration technique is to find the error produced for a given value of n and to see what happens to the error as n is increased. We do this for the five sums discussed above, L_n, R_n, M_n, T_n, and S_{2n}, by using these sums to approximate a definite integral of known value and looking at the corresponding errors.

(B) Given the definite integral

$$I = \int_4^{12} \frac{dx}{x} = 1.099 \quad \text{To three decimal places}$$

use the values of the integrand in Table 1 to complete Table 2. Round all answers to three decimal places.

Table 1

INTEGRAND VALUES

x	4	5	6	7	8	9	10	11	12
$1/x$	0.250	0.200	0.167	0.143	0.125	0.111	0.100	0.091	0.083

Table 2

ERROR

	$L_n - I$	$R_n - I$	$M_n - I$	$T_n - I$	$S_{2n} - I$
$n = 2$	0.401				0.002
$n = 4$			−0.009		

Table 2 suggests that the relationships between error and the change in n shown in Table 3 exist. This can be confirmed in general.

Table 3

ERROR AND CHANGE IN n

	FACTOR BY WHICH ERROR IS CHANGED		
CHANGE IN n	L_n, R_n	M_n, T_n	S_{2n}
$2n$	$\frac{1}{2}$	$\frac{1}{4}$	$\frac{1}{16}$
$10n$	$\frac{1}{10}$ Adds one more digit of accuracy	$\frac{1}{10^2}$ Adds two more digits of accuracy	$\frac{1}{10^4}$ Adds four more digits of accuracy

We have already noted that L_n and R_n give exact results if the integrand is a constant function. Also, M_n and T_n give exact results for constant and linear functions. What about Simpson's rule? Simpson's rule gives exact results if the integrand is any polynomial of degree 3 or less. It should now be clear why Simpson's rule is so popular. However, Simpson's rule is generally not used in numerical integration routines. More powerful algorithms exist. For example, the Gauss–Kronrod algorithm is used in a number of popular graphing utilities and gives exact results for polynomials of degree 5 or less.

(C) Use Simpson's rule to approximate the definite integral

$$I = \int_2^{10} \frac{x}{\ln x}\, dx$$

for $2n = 4$ and for $2n = 8$. Compute S_{2n} to three decimal places.

(D) Using more powerful algorithms, it is known that $I = 27.159$ to three decimal places. Compute the error $|I - S_{2n}|$ for the two estimates in part (C).

Application

We now turn to an example where Simpson's rule is used to estimate cost. (Note that it is not necessary to find a regression equation, as we would in order to use a numerical integration routine on a graphing utility.)

A company manufactures and wholesales a popular brand of sunglasses. Their financial research department collected the data in Table 4, showing marginal cost $C'(x)$ (in dollars) at selected production levels of x pairs of sunglasses per hour.

Table 4

MARGINAL COSTS

t	50	100	150	200	250	300	350	400	450
$C'(x)$	21.80	16.95	15.31	14.50	14.00	13.66	13.42	13.25	13.11

(E) If the company is now producing 50 pairs of sunglasses per hour and increases production to 450 pairs per hour, estimate the total increase in costs per hour using L_n and R_n with an appropriate value of n. How are L_n and R_n related to the actual increase in cost?

(F) Use Simpson's rule to estimate the increase in cost in part (E).

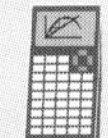

Group Activity 2 Bell-Shaped Curves

One of the most important functions in probability and statistics is the **normal probability density function** and its bell-shaped graph or **normal curve,** as shown in Figure 1.

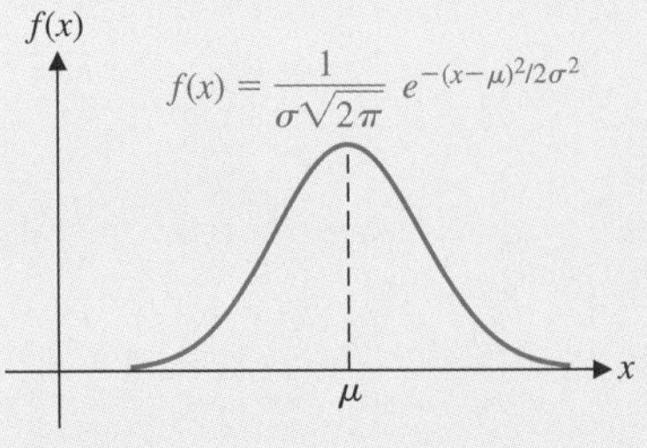

FIGURE 1
A normal curve

It can be shown that the total area under the curve from $-\infty$ to ∞, for μ any real number and σ any positive real number, is always 1. Thus, the area under the curve over an interval $[a, b]$ is the percentage of the total area that is under the curve between a and b. We will interpret the normal probability density function through an example.

A manufacturer of 100 watt light bulbs tests a large sample and finds that the average life of these bulbs is 5 hundred hours ($\mu = 5$) with a *standard deviation* of 1 hundred hours ($\sigma = 1$). (Standard deviation measures the dispersion of the normal probability density function about the mean or average. A small standard deviation is associated with a tall, narrow normal curve, and a large standard deviation is related to a low, flat normal curve. You will see this below.) For $\mu = 5$ and $\sigma = 1$, the probability density function and corresponding normal curve are shown in Figure 2. The area under the curve between 5 hundred hours and 6 hundred hours represents the percentage of light bulbs in the manufacturing process that will have a life between 5 and 6 hundred hours; that is, the probability of a light bulb drawn at random having a life between 5 and 6 hundred hours.

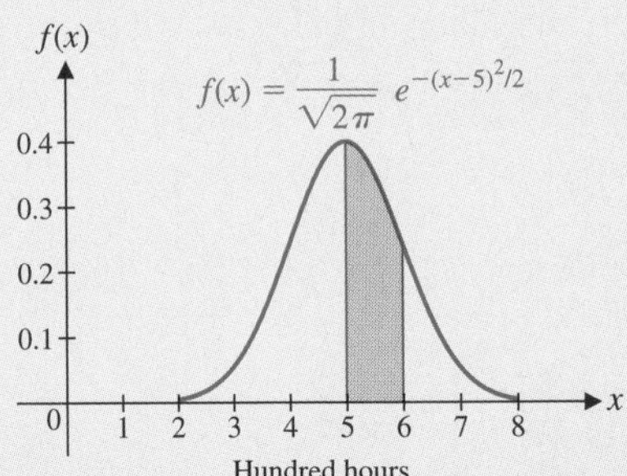

FIGURE 2

(A) Write a definite integral that represents the probability of a light bulb drawn at random having a life between 5 and 6 hundred hours.

(B) Approximate the definite integral in part (A) with a midpoint sum using five equal subintervals. Explain what the result means relative to the original problem.

(C) Compute $f''(x)$. Use a graphing utility to graph $y = |f''(x)|$ over the interval $[5, 6]$, and use the graph to show that 0.4 is an upper bound for $|f''(x)|$ on this interval.

(D) Calculate an error bound for the estimate M_5 in part (B) using the result from part (C).

(E) Using the result from part (C), how large should n be chosen so that the error in using M_n is no greater than 0.000 05?

(F) If the area under the normal curve from 5 to ∞ is 0.5, what is the probability of selecting a light bulb at random that has a life greater than 600 hours? Explain how you arrived at your answer.

(G) What is the probability of selecting a light bulb at random that has a life less than 500 hours? Explain how you arrived at your answer.

(H) Use a numerical integration routine on a graphing utility to find the probability of selecting a light bulb at random that has a life between 450 and 550 hours. That has a life between 350 and 650 hours.

(I) Graph normal probability density functions with $\mu = 8$ and $\sigma = 1, 2,$ and 3, in the same viewing window. What effect does changing σ have on the shape of the curve?

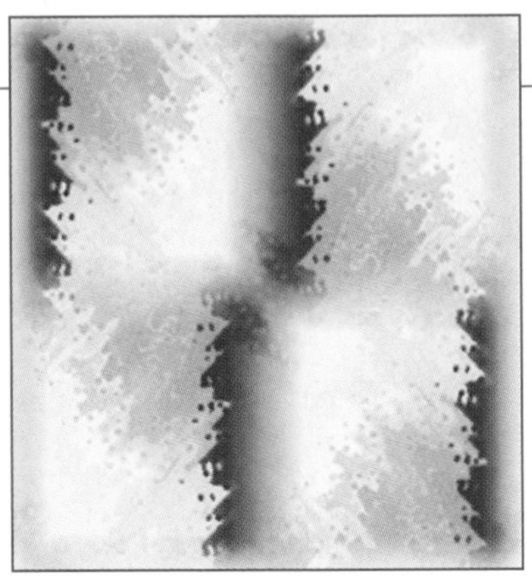

CHAPTER 12

ADDITIONAL INTEGRATION TOPICS

INTRODUCTION

This chapter contains additional topics on integration. Since they are essentially independent of one another, they may be taken up in any order, and certain sections may be omitted if desired.

SECTION 12-1

Area between Curves

- AREA BETWEEN A CURVE AND THE x AXIS
- AREA BETWEEN TWO CURVES
- APPLICATION: INCOME DISTRIBUTION

In the last chapter we found that the definite integral $\int_a^b f(x)\,dx$ represents the sum of the signed areas between the graph of $y = f(x)$ and the x axis from $x = a$ to $x = b$, where the areas above the x axis are counted positively and the areas below the x axis are counted negatively (see Fig. 1). In this section we are interested in using the definite integral to find the actual area between a curve and the x axis or the actual area between two curves. These areas are always nonnegative quantities—**area measure is never negative.**

FIGURE 1

$$\int_a^b f(x)\,dx = -A + B$$

AREA BETWEEN A CURVE AND THE x AXIS

In Figure 1, A represents the area between $y = f(x)$ and the x axis from $x = a$ to $x = c$, and B represents the area between $y = f(x)$ and the x axis from $x = c$ to $x = b$. Both A and B are positive quantities. Since $f(x) \geqslant 0$ on the interval $[c, b]$,

$$\int_c^b f(x)\,dx = B$$

And since $f(x) \leqslant 0$ on the interval $[a, c]$,

$$\int_a^c f(x)\,dx = -A$$

or

$$A = -\int_a^c f(x)\,dx = \int_a^c [-f(x)]\,dx$$

Thus:

The area between the graph of a negative function and the x axis is equal to the definite integral of the negative of the function.

Explore–Discuss 1

Sketch a graph of a function f such that $f(x) \leqslant 0$ over the interval $[1, 5]$. (No equation is necessary.) Sketch a graph of $y = -f(x)$ over the same interval in the same coordinate system. Explain how these figures relate to the above discussion.

All the above interpretation of the definite integral relative to area, as we saw in the last section, is based on the definition of the definite integral as the limit of a Riemann sum:

$$\int_a^b f(x)\,dx = \lim_{n\to\infty} \sum_{k=1}^{n} f(c_k)\,\Delta x_k$$

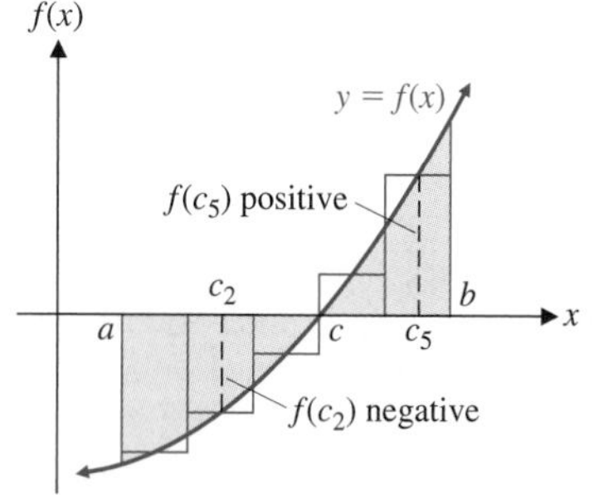

FIGURE 2
Area and Riemann sums

Figure 2 shows a particular Riemann sum, the midpoint sum M_5, over the interval $[a, b]$. The product $f(c_k)\,\Delta x_k$ is negative for any rectangle on $[a, c]$ and positive for any rectangle on $[c, b]$. Thus, $f(c_k)\,\Delta x_k$ represents the negative of the area of a rectangle on $[a, c]$, and $-f(c_k)\,\Delta x_k$ represents the actual area of the rectangle. Consequently, for the interval $[a, c]$, where $f(x) \leqslant 0$,

$$\text{Area} = \lim_{n\to\infty} \sum_{k=1}^{n} [-f(c_k)]\,\Delta x_k = \int_a^c [-f(x)]\,dx$$

But for the interval $[c, b]$, where $f(x) \geq 0$,

$$\text{Area} = \lim_{n\to\infty} \sum_{k=1}^{n} f(c_k)\,\Delta x_k = \int_c^b f(x)\,dx$$

In summary:

Area between a Curve and the x Axis

For a function f that is continuous over $[a, b]$, the area between $y = f(x)$ and the x axis from $x = a$ to $x = b$ can be found using definite integrals as follows:

For $f(x) \geq 0$ over $[a, b]$: $\quad \text{Area} = \int_a^b f(x)\,dx$

For $f(x) \leq 0$ over $[a, b]$: $\quad \text{Area} = \int_a^b [-f(x)]\,dx$

If $f(x)$ is positive for some values of x and negative for others on an interval (as in Fig. 1), the area between the graph of f and the x axis can be obtained by dividing the interval into subintervals over which f is always positive or always negative, finding the area over each subinterval, and then summing these areas.

Example 1 **Area between a Curve and the x Axis** Find the area bounded by $f(x) = 6x - x^2$ and $y = 0$ for $1 \leq x \leq 4$.

SOLUTION We sketch a graph of the region first (Fig. 3). (The solution of every area problem should begin with a sketch.) Since $f(x) \geq 0$ on $[1, 4]$,

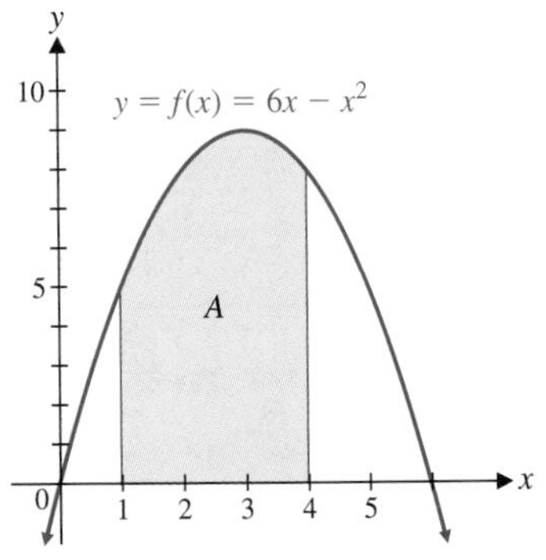

$$A = \int_1^4 (6x - x^2)\,dx = \left(3x^2 - \frac{x^3}{3}\right)\Big|_1^4$$

$$= \left[3(4)^2 - \frac{(4)^3}{3}\right] - \left[3(1)^2 - \frac{(1)^3}{3}\right]$$

$$= 48 - \tfrac{64}{3} - 3 + \tfrac{1}{3}$$

$$= 48 - 21 - 3$$

$$= 24$$

FIGURE 3

Matched Problem 1 Find the area bounded by $f(x) = x^2 + 1$ and $y = 0$ for $-1 \leq x \leq 3$.

Example 2 **Area between a Curve and the x Axis** Find the area between the graph of $f(x) = x^2 - 2x$ and the x axis over the indicated intervals:

(A) $[1, 2]$ (B) $[-1, 1]$

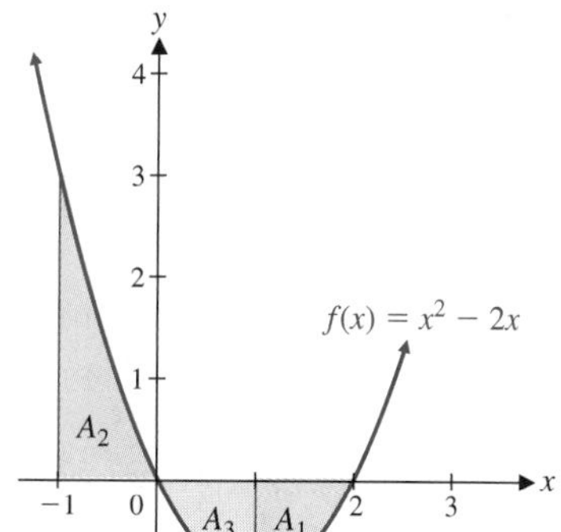

FIGURE 4

Solution We begin by sketching the graph of *f*, as shown in Figure 4.

(A) From the graph, we see that $f(x) \leq 0$ for $1 \leq x \leq 2$, so we integrate $-f(x)$:

$$\begin{aligned} A_1 &= \int_1^2 [-f(x)]\,dx \\ &= \int_1^2 (2x - x^2)\,dx \\ &= \left(x^2 - \frac{x^3}{3}\right)\Bigg|_1^2 \\ &= \left[(2)^2 - \frac{(2)^3}{3}\right] - \left[(1)^2 - \frac{(1)^3}{3}\right] \\ &= 4 - \tfrac{8}{3} - 1 + \tfrac{1}{3} = \tfrac{2}{3} \approx 0.667 \end{aligned}$$

(B) Since the graph shows that $f(x) \geq 0$ on $[-1, 0]$ and $f(x) \leq 0$ on $[0, 1]$, the computation of this area will require two integrals:

$$\begin{aligned} A &= A_2 + A_3 \\ &= \int_{-1}^0 f(x)\,dx + \int_0^1 [-f(x)]\,dx \\ &= \int_{-1}^0 (x^2 - 2x)\,dx + \int_0^1 (2x - x^2)\,dx \\ &= \left(\frac{x^3}{3} - x^2\right)\Bigg|_{-1}^0 + \left(x^2 - \frac{x^3}{3}\right)\Bigg|_0^1 \\ &= \tfrac{4}{3} + \tfrac{2}{3} = 2 \end{aligned}$$

Matched Problem 2 Find the area between the graph of $f(x) = x^2 - 9$ and the x axis over the indicated intervals:

(A) $[0, 2]$ (B) $[2, 4]$

Area between Two Curves

Now we want to consider the area bounded by $y = f(x)$ and $y = g(x)$, where $f(x) \geq g(x) \geq 0$, for $a \leq x \leq b$, as indicated in Figure 5.

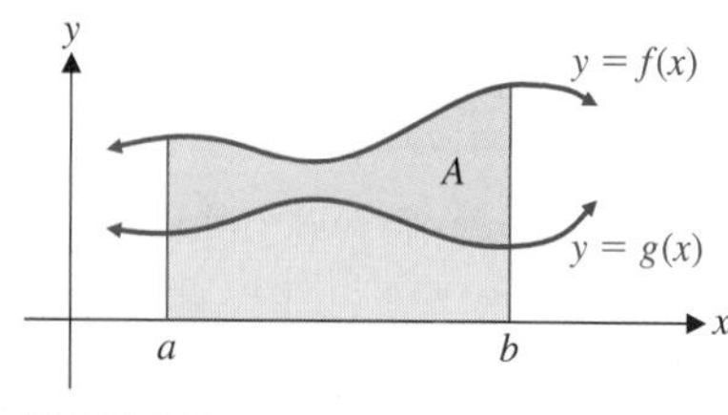

FIGURE 5

$$\begin{pmatrix}\text{Area } A \text{ between} \\ f(x) \text{ and } g(x)\end{pmatrix} = \begin{pmatrix}\text{Area} \\ \text{under } f(x)\end{pmatrix} - \begin{pmatrix}\text{Area} \\ \text{under } g(x)\end{pmatrix}$$

Areas are from $x = a$ to $x = b$ above the x axis.

$$= \int_a^b f(x)\,dx - \int_a^b g(x)\,dx$$

Use definite integral property 4 (Section 11-5).

$$= \int_a^b [f(x) - g(x)]\,dx$$

It can be shown that the preceding result does not require $f(x)$ or $g(x)$ to remain positive over the interval $[a, b]$. A more general result is stated in the box:

Area between Two Curves

If f and g are continuous and $f(x) \geq g(x)$ over the interval $[a, b]$, then the area bounded by $y = f(x)$ and $y = g(x)$ for $a \leq x \leq b$ is given exactly by

$$A = \int_a^b [f(x) - g(x)]\, dx$$

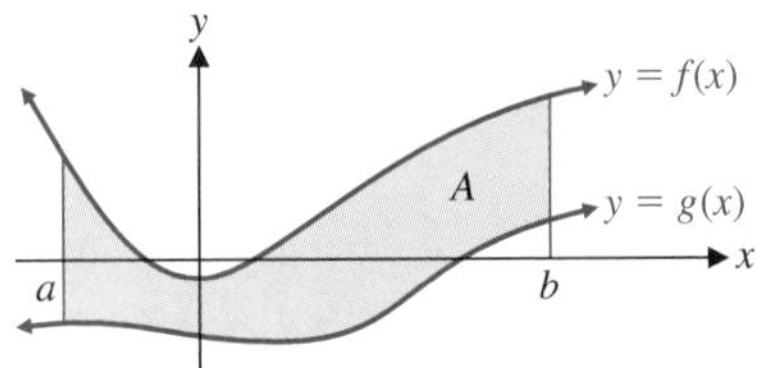

Explore–Discuss 2

A Riemann sum for the integral representing the area between the graphs of $y = f(x)$ and $y = g(x)$ has the form

$$\sum_{k=1}^{n} [f(c_k) - g(c_k)]\, \Delta x_k$$

If $f(x) \geq g(x)$, then each term in this sum represents the area of a rectangle with height $f(c_k) - g(c_k)$ and width Δx. Discuss the relationship between these rectangles and the area between the graphs of $y = f(x)$ and $y = g(x)$.

Example 3 **Area between Two Curves** Find the area bounded by the graphs of $f(x) = \frac{1}{2}x + 3$, $g(x) = -x^2 + 1$, $x = -2$, and $x = 1$.

SOLUTION We first sketch the area (Fig. 6), and then set up and evaluate an appropriate definite integral. We observe from the graph that $f(x) \geq g(x)$ for $-2 \leq x \leq 1$, so

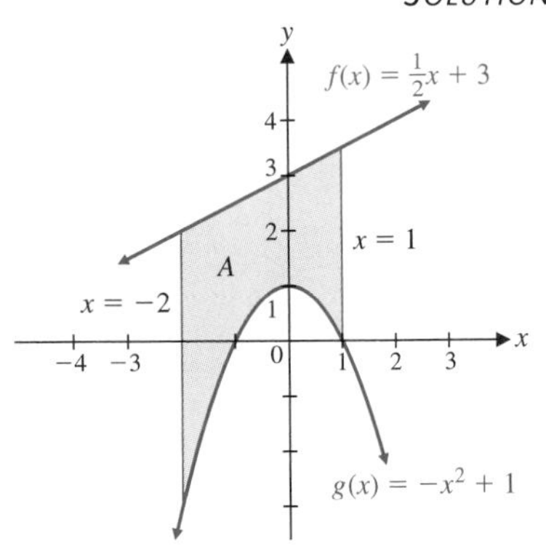

FIGURE 6

$$A = \int_{-2}^{1} [f(x) - g(x)]\, dx = \int_{-2}^{1} \left[\left(\frac{x}{2} + 3 \right) - (-x^2 + 1) \right] dx$$

$$= \int_{-2}^{1} \left(x^2 + \frac{x}{2} + 2 \right) dx$$

$$= \left(\frac{x^3}{3} + \frac{x^2}{4} + 2x \right) \Bigg|_{-2}^{1}$$

$$= \left(\frac{1}{3} + \frac{1}{4} + 2 \right) - \left(\frac{-8}{3} + \frac{4}{4} - 4 \right) = \frac{33}{4} = 8.25$$

Matched Problem 3 Find the area bounded by $f(x) = x^2 - 1$, $g(x) = -\frac{1}{2}x - 3$, $x = -1$, and $x = 2$.

Example 4 **Area between Two Curves** Find the area bounded by $f(x) = 5 - x^2$ and $g(x) = 2 - 2x$.

Solution First, graph f and g on the same coordinate system, as shown in Figure 7. Since the statement of the problem does not include any limits on the values of x, we must determine the appropriate values from the graph. The graph of f is a parabola and the graph of g is a line, as shown. The area bounded by these two graphs extends from the intersection point on the left to the intersection point on the right. To find these intersection points, we solve the equation $f(x) = g(x)$ for x:

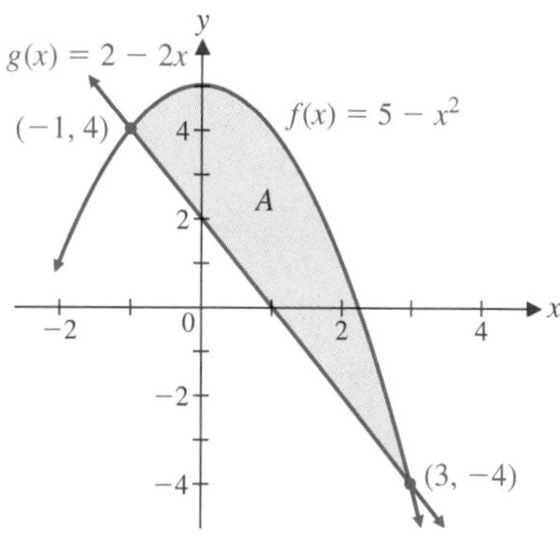

FIGURE 7

$$
\begin{aligned}
f(x) &= g(x)\\
5 - x^2 &= 2 - 2x\\
x^2 - 2x - 3 &= 0\\
x &= -1, 3
\end{aligned}
$$

You should check these values in the original equations. (Note that the area between the graphs for $x < -1$ is unbounded on the left, and the area between the graphs for $x > 3$ is unbounded on the right.) Figure 7 shows that $f(x) \geq g(x)$ over the interval $[-1, 3]$, so we have

$$
\begin{aligned}
A &= \int_{-1}^{3} [f(x) - g(x)]\,dx = \int_{-1}^{3} [5 - x^2 - (2 - 2x)]\,dx\\
&= \int_{-1}^{3} (3 + 2x - x^2)\,dx\\
&= \left(3x + x^2 - \frac{x^3}{3}\right)\Bigg|_{-1}^{3}\\
&= \left[3(3) + (3)^2 - \frac{(3)^3}{3}\right] - \left[3(-1) + (-1)^2 - \frac{(-1)^3}{3}\right] = \frac{32}{3} \approx 10.667
\end{aligned}
$$

Matched Problem 4 Find the area bounded by $f(x) = 6 - x^2$ and $g(x) = x$.

Example 5 **Area between Two Curves** Find the area bounded by $f(x) = x^2 - x$ and $g(x) = 2x$ for $-2 \leq x \leq 3$.

Solution The graphs of f and g are shown in Figure 8. Examining the graph, we see that $f(x) \geq g(x)$ on the interval $[-2, 0]$, but $g(x) \geq f(x)$ on the interval $[0, 3]$. Thus, two integrals are required to compute this area:

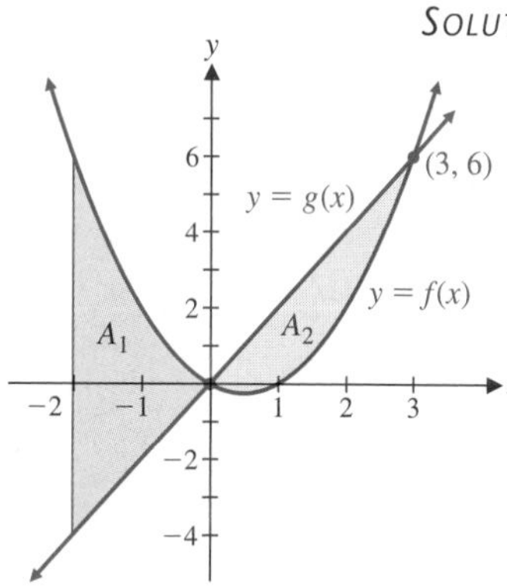

FIGURE 8

$$
\begin{aligned}
A_1 &= \int_{-2}^{0} [f(x) - g(x)]\,dx \quad f(x) \geq g(x) \text{ on } [-2, 0]\\
&= \int_{-2}^{0} [x^2 - x - 2x]\,dx\\
&= \int_{-2}^{0} (x^2 - 3x)\,dx\\
&= \left(\frac{x^3}{3} - \frac{3}{2}x^2\right)\Bigg|_{-2}^{0}\\
&= (0) - \left[\frac{(-2)^3}{3} - \frac{3}{2}(-2)^2\right] = \frac{26}{3} \approx 8.667
\end{aligned}
$$

$$\begin{aligned} A_2 &= \int_0^3 [g(x) - f(x)]\,dx \quad g(x) \geq f(x) \text{ on } [0,3] \\ &= \int_0^3 [2x - (x^2 - x)]\,dx \\ &= \int_0^3 (3x - x^2)\,dx \\ &= \left(\frac{3}{2}x^2 - \frac{x^3}{3}\right)\Bigg|_0^3 \\ &= \left[\frac{3}{2}(3)^2 - \frac{(3)^3}{3}\right] - (0) = \frac{9}{2} = 4.5 \end{aligned}$$

The total area between the two graphs is

$$A = A_1 + A_2 = \tfrac{26}{3} + \tfrac{9}{2} = \tfrac{79}{6} \approx 13.167$$

Matched Problem 5 ➡ Find the area bounded by $f(x) = 2x^2$ and $g(x) = 4 - 2x$ for $-2 \leq x \leq 2$.

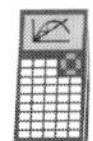

Example 6 ➡ **Computing Areas Using a Numerical Integration Routine** Find the area (to three decimal places) bounded by $f(x) = e^{-x^2}$ and $g(x) = x^2 - 1$.

SOLUTION First, we use a graphing utility to graph the functions f and g and find their intersection points (see Fig. 9A). We see that the graph of f is bell-shaped and the graph of g is a parabola, and note that $f(x) \geq g(x)$ on the interval $[-1.131, 1.131]$. Now we compute the area A by a numerical integration routine (see Fig. 9B):

$$A = \int_{-1.131}^{1.131} [e^{-x^2} - (x^2 - 1)]\,dx = 2.876$$

2

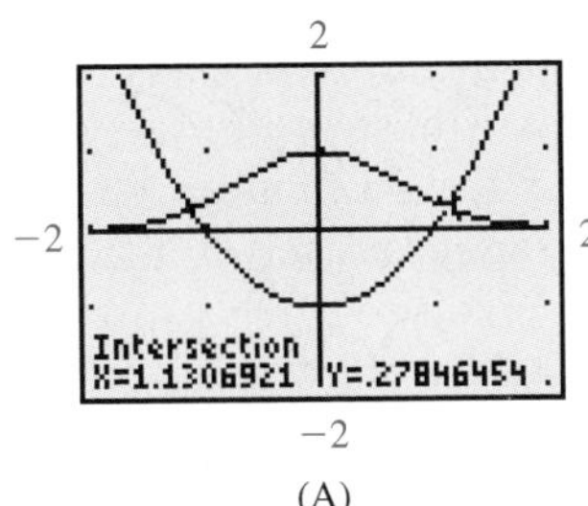

−2 2

−2

(A)

fnInt(e^(-X^2)-(X^2-1),X,-1.131,1.131)
2.875501719

(B)

FIGURE 9

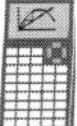

Matched Problem 6 ➡ Find the area (to three decimal places) bounded by the graphs of $f(x) = x^2 \ln x$ and $g(x) = 3x - 3$.

■ APPLICATION: INCOME DISTRIBUTION

The U.S. Bureau of the Census compiles and analyzes a great deal of data having to do with the distribution of income among families in the United States. For 1992 the Bureau reported that the lowest 20% of families received 4% of all family income, and the top 20% received 45%. Table 1 and Figure 10 give a detailed picture of the distribution of family income in 1992.

Table 1

FAMILY INCOME DISTRIBUTION IN THE UNITED STATES IN 1992

INCOME LEVEL	x	y
Under $17,000	0.20	0.04
Under $30,000	0.40	0.15
Under $44,000	0.60	0.31
Under $64,000	0.80	0.55

Source: U.S. Bureau of the Census

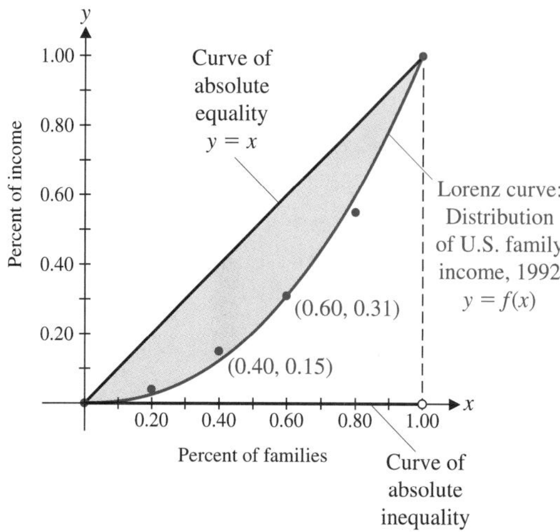

FIGURE 10
Lorenz chart

The graph of $y = f(x)$ through the data points in Figure 10 is called a **Lorenz curve** and is generally found using *regression analysis,* a technique of fitting a particular elementary function to a data set over a given interval. The variable x **represents the cumulative percentage of families at or below a given income level** and y **represents the cumulative percentage of total family income received.** For example, data point $(0.40, 0.15)$ in Table 1 or on the Lorenz curve in Figure 10 indicates that the bottom 40% of families (those with incomes under $30,000) receive 15% of the total income for all families; data point $(0.60, 0.31)$ indicates that the bottom 60% of families receive 31% of the total income for all families; and so on.

Absolute equality of income would occur if the area between the Lorenz curve and $y = x$ were 0. In this case, the Lorenz curve would be $y = x$ and all families would receive equal shares of the total income. That is, 5% of the families would receive 5% of the income, 20% of the families would receive 20% of the income, 65% of the families would receive 65% of the income, and so on. The maximum possible area between a Lorenz curve and $y = x$ is $\frac{1}{2}$, the area of the triangle below $y = x$. In this case, we would have **absolute inequality**—all the income would be in the hands of one family and the rest would have none. In actuality, Lorenz curves lie between these two extremes. But as the shaded area increases, the greater the inequality of income distribution.

The ratio of the area bounded by $y = x$ and the Lorenz curve $y = f(x)$ to the area of the triangle under the line $y = x$ from $x = 0$ to $x = 1$, is called the **index of income concentration.** The area bounded by $y = x$ and $y = f(x)$ is given by $\int_0^1 [x - f(x)]\, dx$, and the area of the triangle below $y = x$ is $\frac{1}{2}$. Thus, we have the following:

Index of Income Concentration

If $y = f(x)$ is the equation of a Lorenz curve, then

$$\textbf{Index of income concentration} = 2\int_0^1 [x - f(x)]\,dx$$

The index of income concentration is always a number between 0 and 1:

A measure of 0 indicates absolute equality—all individuals share equally in the income. A measure of 1 indicates absolute inequality—one individual has all the income and the rest have none.

The closer the index is to 0, the closer the income is to being equally distributed. The closer the index is to 1, the closer the income is to being concentrated in a few hands. The index of income concentration is used to compare income distributions at various points in time, between different groups of people, before and after taxes are paid, between different countries, and so on.

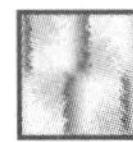

Example 7 **Distribution of Income** The Lorenz curve for the distribution of income in a certain country in 1990 is given by $f(x) = x^{2.6}$. Economists predict that the Lorenz curve for the country in the year 2010 will be given by $g(x) = x^{1.8}$. Find the index of income concentration for each curve, and interpret the results.

Solution The Lorenz curves are shown in Figure 11.

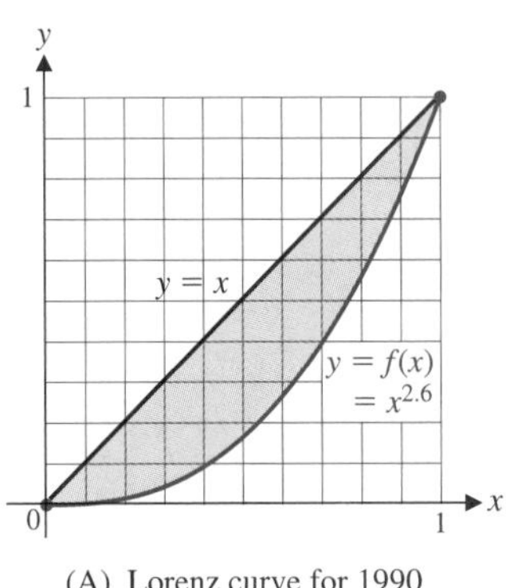

(A) Lorenz curve for 1990

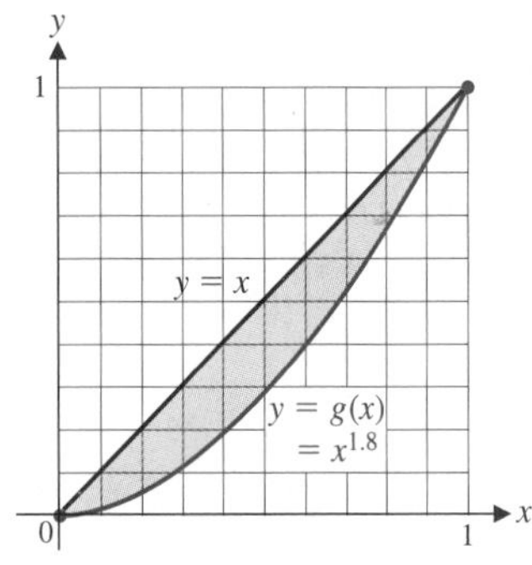

(B) Projected Lorenz curve for 2010

FIGURE 11

The index of income concentration in 1990 is (see Fig. 11A)

$$2\int_0^1 [x - f(x)]\,dx = 2\int_0^1 [x - x^{2.6}]\,dx = 2\left(\frac{1}{2}x^2 - \frac{1}{3.6}x^{3.6}\right)\Big|_0^1$$
$$= 2\left(\frac{1}{2} - \frac{1}{3.6}\right) \approx 0.444$$

The projected index of income concentration in 2010 is (see Fig. 11B)

$$2\int_0^1 [x - g(x)]\,dx = 2\int_0^1 [x - x^{1.8}]\,dx = 2\left(\frac{1}{2}x^2 - \frac{1}{2.8}x^{2.8}\right)\Big|_0^1$$
$$= 2\left(\frac{1}{2} - \frac{1}{2.8}\right) \approx 0.286$$

If this projection is correct, the index of income concentration will decrease, and income will be more equally distributed in the year 2010 than in 1990.

Matched Problem 7 Repeat Example 7 if the projected Lorenz curve in the year 2010 is given by $g(x) = x^{3.8}$.

Answers to Matched Problems

1. $A = \int_{-1}^{3} (x^2 + 1)\,dx = \frac{40}{3} \approx 13.333$
2. (A) $A = \int_{0}^{2} (9 - x^2)\,dx = \frac{46}{3} \approx 15.333$
   (B) $A = \int_{2}^{3} (9 - x^2)\,dx + \int_{3}^{4} (x^2 - 9)\,dx = 6$
3. $A = \int_{-1}^{2} \left[(x^2 - 1) - \left(-\frac{x}{2} - 3\right)\right] dx = \frac{39}{4} = 9.75$
4. $A = \int_{-3}^{2} [(6 - x^2) - x]\,dx = \frac{125}{6} \approx 20.833$
5. $A = \int_{-2}^{1} [(4 - 2x) - 2x^2]\,dx + \int_{1}^{2} [2x^2 - (4 - 2x)]\,dx = \frac{38}{3} \approx 12.667$
6. 0.443
7. Index of income concentration ≈ 0.583; income will be less equally distributed in 2010.

EXERCISE 12-1

A *Problems 1–6 refer to Figures A–D. Set up definite integrals in Problems 1–4 that represent the indicated shaded area.*

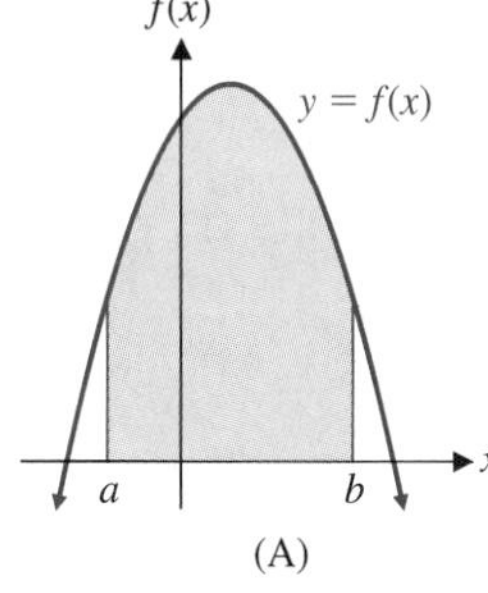

(A)

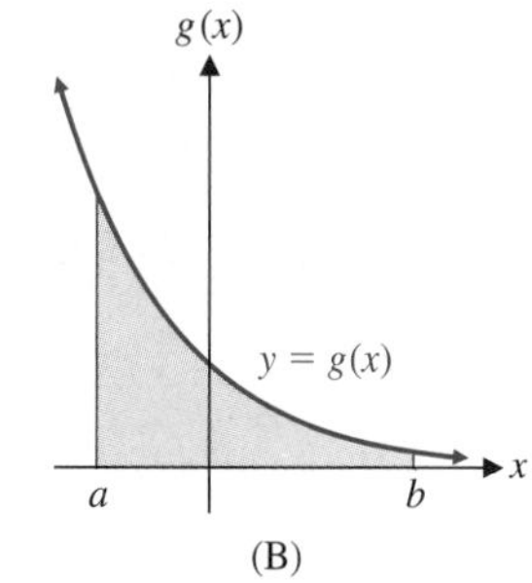

(B)

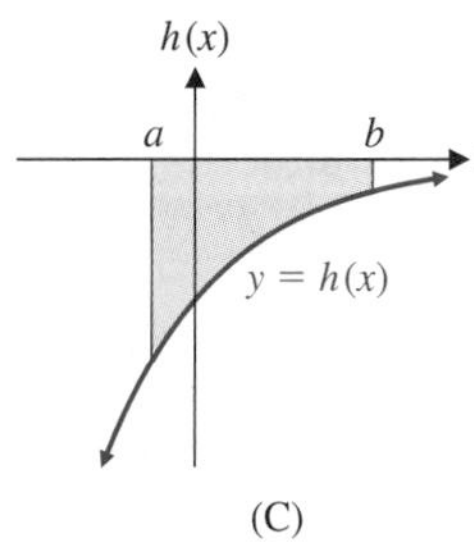

(C)

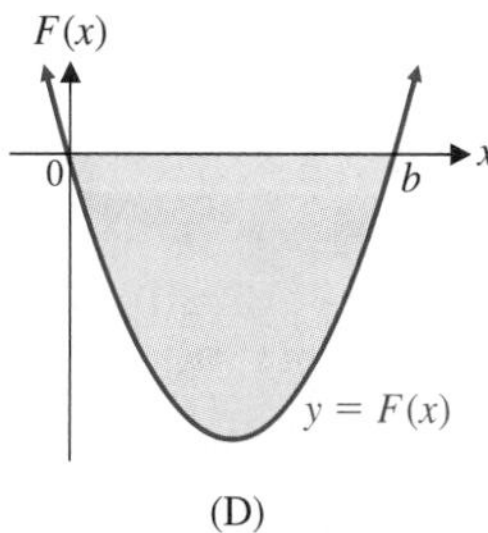

(D)

1. Shaded area in Figure B
2. Shaded area in Figure A
3. Shaded area in Figure C
4. Shaded area in Figure D
5. Explain why $\int_a^b h(x)\,dx$ does not represent the area between the graph of $y = h(x)$ and the x axis from $x = a$ to $x = b$ in Figure C.
6. Explain why $\int_a^b [-h(x)]\,dx$ represents the area between the graph of $y = h(x)$ and the x axis from $x = a$ to $x = b$ in Figure C.

In Problems 7–16, find the area bounded by the graphs of the indicated equations over the given intervals. Compute answers to three decimal places.

7. $y = -2x - 1;\ y = 0,\ 0 \le x \le 4$
8. $y = 2x - 4;\ y = 0,\ -2 \le x \le 1$
9. $y = x^2 + 2;\ y = 0,\ -1 \le x \le 0$
10. $y = 3x^2 + 1;\ y = 0,\ -2 \le x \le 0$
11. $y = x^2 - 4;\ y = 0,\ -1 \le x \le 2$
12. $y = 3x^2 - 12;\ y = 0,\ -2 \le x \le 1$
13. $y = e^x;\ y = 0,\ -1 \le x \le 2$
14. $y = e^{-x};\ y = 0,\ -2 \le x \le 1$
15. $y = -1/t;\ y = 0,\ 0.5 \le t \le 1$
16. $y = -1/t;\ y = 0,\ 0.1 \le t \le 1$

B *Problems 17–26 refer to Figures A and B. Set up definite integrals in Problems 17–24 that represent the indicated shaded areas over the given intervals.*

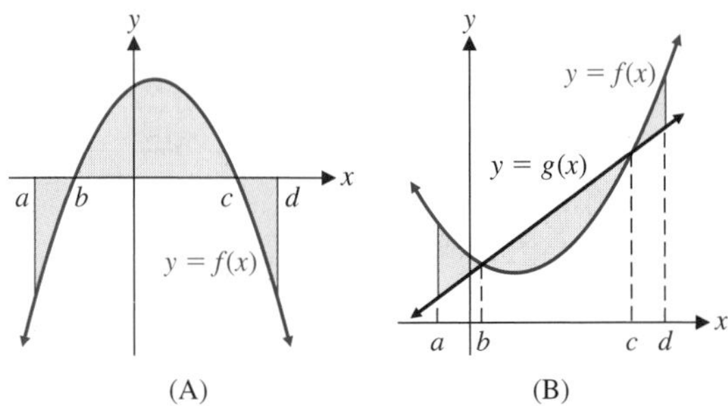

17. Over interval $[a, b]$ in Figure A
18. Over interval $[c, d]$ in Figure A
19. Over interval $[b, d]$ in Figure A
20. Over interval $[a, c]$ in Figure A
21. Over interval $[c, d]$ in Figure B
22. Over interval $[a, b]$ in Figure B
23. Over interval $[a, c]$ in Figure B
24. Over interval $[b, d]$ in Figure B

25. Referring to Figure B, explain how you would use definite integrals and the functions f and g to find the area bounded by the two functions from $x = a$ to $x = d$.
26. Referring to Figure A, explain how you would use definite integrals to find the area between the graph of $y = f(x)$ and the x axis from $x = a$ to $x = d$.

In Problems 27–42, find the area bounded by the graphs of the indicated equations over the given intervals (when stated). Compute answers to three decimal places.

27. $y = -x; y = 0, -2 \leq x \leq 1$
28. $y = -x + 1; y = 0, -1 \leq x \leq 2$
29. $y = x^2 - 4; y = 0, 0 \leq x \leq 3$
30. $y = 4 - x^2; y = 0, 0 \leq x \leq 4$
31. $y = 4 - x^2; y = 0, -3 \leq x \leq 4$
32. $y = x^2 - 4; y = 0, -4 \leq x \leq 3$
33. $y = -2x + 8; y = 12; -1 \leq x \leq 2$
34. $y = 2x + 6; y = 3; -1 \leq x \leq 2$
35. $y = 3x^2; y = 12$
36. $y = x^2; y = 9$
37. $y = 4 - x^2; y = -5$
38. $y = x^2 - 1; y = 3$
39. $y = x^2 + 1; y = 2x - 2; -1 \leq x \leq 2$
40. $y = x^2 - 1; y = x - 2; -2 \leq x \leq 1$
41. $y = e^{0.5x}; y = -\frac{1}{x}; 1 \leq x \leq 2$
42. $y = \frac{1}{x}; y = -e^x; 0.5 \leq x \leq 1$

In Problems 43–46, use a graphing utility to graph the equations and find relevant intersection points. Then find the area bounded by the curves. Compute answers to three decimal places.

43. $y = 3 - 5x - 2x^2; y = 2x^2 + 3x - 2$
44. $y = 3 - 2x^2; y = 2x^2 - 4x$
45. $y = -0.5x + 2.25; y = \frac{1}{x}$
46. $y = x - 4.25; y = -\frac{1}{x}$

C *In Problems 47–54, find the area bounded by the graphs of the indicated equations over the given intervals (when stated). Compute answers to three decimal places.*

47. $y = 10 - 2x; y = 4 + 2x; 0 \leq x \leq 4$
48. $y = 3x; y = x + 5; 0 \leq x \leq 5$
49. $y = x^3; y = 4x$
50. $y = x^3 + 1; y = x + 1$
51. $y = x^3 - 3x^2 - 9x + 12; y = x + 12$
52. $y = x^3 - 6x^2 + 9x; y = x$
53. $y = x^4 - 4x^2 + 1; y = x^2 - 3$
54. $y = x^4 - 6x^2; y = 4x^2 - 9$

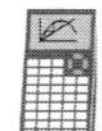

In Problems 55–60, use a graphing utility to graph the equations and find relevant intersection points. Then find the area bounded by the curves. Compute answers to three decimal places.

55. $y = x^3 - x^2 + 2; y = -x^3 + 8x - 2$
56. $y = 2x^3 + 2x^2 - x; y = -2x^3 - 2x^2 + 2x$
57. $y = e^{-x}; y = 3 - 2x$
58. $y = 2 - (x + 1)^2; y = e^{x+1}$
59. $y = e^x; y = 5x - x^3$
60. $y = 2 - e^x; y = x^3 + 3x^2$

In Problems 61–64, use a numerical integration routine on a graphing utility to find the area bounded by the graphs of the indicated equations over the given interval (when stated). Compute answers to three decimal places.

61. $y = e^{-x}; y = \sqrt{\ln x}; 2 \leq x \leq 5$
62. $y = x^2 + 3x + 1; y = e^{e^x}; -3 \leq x \leq 0$
63. $y = e^{x^2}; y = x + 2$
64. $y = \ln(\ln x); y = 0.01x$

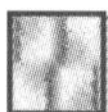

APPLICATIONS

In the following applications it is helpful to sketch graphs to get a clearer understanding of each problem and to interpret results. A graphing utility will prove useful if you have one, but it is not necessary.

Business & Economics

65. *Oil production.* Using data from the first 3 years of production as well as geological studies, the management of an oil company estimates that oil will be pumped from a producing field at a rate given by

$$R(t) = \frac{100}{t + 10} + 10 \qquad 0 \le t \le 15$$

where $R(t)$ is the rate of production (in thousands of barrels per year) t years after pumping begins. Find the area between the graph of R and the t axis over the interval $[5, 10]$ and interpret the results.

66. *Oil production.* In Problem 65, if the rate is found to be

$$R(t) = \frac{100t}{t^2 + 25} + 4 \qquad 0 \le t \le 25$$

find the area between the graph of R and the t axis over the interval $[5, 15]$ and interpret the results.

67. *Useful life.* An amusement company maintains records for each video game it installs in an arcade. Suppose that $C(t)$ and $R(t)$ represent the total accumulated costs and revenues (in thousands of dollars), respectively, t years after a particular game has been installed. If

$$C'(t) = 2 \qquad \text{and} \qquad R'(t) = 9e^{-0.3t}$$

find the area between the graphs of C' and R' over the interval on the t axis from 0 to the useful life of the game and interpret the results.

68. *Useful life.* Repeat Problem 67 if

$$C'(t) = 2t \qquad \text{and} \qquad R'(t) = 5te^{-0.1t^2}$$

69. *Income distribution.* As part of a study of the effects of World War II on the economy of the United States, an economist used data from the U.S. Bureau of the Census to produce the following Lorenz curves for distribution of income in the United States in 1935 and in 1947:

$$f(x) = x^{2.4} \qquad \text{Lorenz curve for 1935}$$
$$g(x) = x^{1.6} \qquad \text{Lorenz curve for 1947}$$

Find the index of income concentration for each Lorenz curve and interpret the results.

70. *Income distribution.* Using data from the U.S. Bureau of the Census, an economist produced the following Lorenz curves for distribution of income in the United States in 1962 and in 1972:

$$f(x) = \tfrac{3}{10}x + \tfrac{7}{10}x^2 \qquad \text{Lorenz curve for 1962}$$
$$g(x) = \tfrac{1}{2}x + \tfrac{1}{2}x^2 \qquad \text{Lorenz curve for 1972}$$

Find the index of income concentration for each Lorenz curve and interpret the results.

71. *Distribution of wealth.* Lorenz curves also can be used to provide a relative measure of the distribution of the total assets of a country. Using data in a report by the U.S. Congressional Joint Economic Committee, an economist produced the following Lorenz curves for the distribution of total assets in the United States in 1963 and in 1983:

$$f(x) = x^{10} \qquad \text{Lorenz curve for 1963}$$
$$g(x) = x^{12} \qquad \text{Lorenz curve for 1983}$$

Find the index of income concentration for each Lorenz curve and interpret the results.

72. *Income distribution.* The government of a small country is planning sweeping changes in the tax structure in order to provide a more equitable distribution of income. The Lorenz curves for the current income distribution and for the projected income distribution after enactment of the tax changes are given below. Find the index of income concentration for each Lorenz curve. Will the proposed changes provide a more equitable income distribution? Explain.

$$f(x) = x^{2.3} \qquad \text{Current Lorenz curve}$$
$$g(x) = 0.4x + 0.6x^2 \qquad \text{Projected Lorenz curve after changes in tax laws}$$

73. C *Distribution of wealth.* The data in the table describe the distribution of wealth in a country:

x	0	0.20	0.40	0.60	0.80	1
y	0	0.12	0.31	0.54	0.78	1

(A) Use quadratic regression to find the equation of a Lorenz curve for the data.

(B) Use the regression equation and a numerical integration routine to approximate the index of income concentration.

74. C *Distribution of wealth.* Refer to Problem 73.

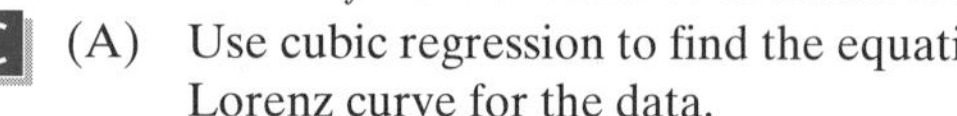

(A) Use cubic regression to find the equation of a Lorenz curve for the data.

(B) Use the cubic regression equation and a numerical integration routine to approximate the index of income concentration.

Life Sciences

75. *Biology.* A yeast culture is growing at a rate of $W'(t) = 0.3e^{0.1t}$ grams per hour. Find the area between the graph of W' and the t axis over the interval $[0, 10]$ and interpret the results.

76. *Natural resource depletion.* The instantaneous rate of change of the demand for lumber in the United States since 1970 ($t = 0$) in billions of cubic feet per year is estimated to be given by

$$Q'(t) = 12 + 0.006t^2 \qquad 0 \leq t \leq 50$$

Find the area between the graph of Q' and the t axis over the interval $[15, 20]$ and interpret the results.

Social Sciences

77. *Learning.* A beginning high school language class was chosen for an experiment on learning. Using a list of 50 words, the experiment involved measuring the rate of vocabulary memorization at different times during a continuous 5 hour study session. It was found that the average rate of learning for the whole class was inversely proportional to the time spent studying and was given approximately by

$$V'(t) = \frac{15}{t} \qquad 1 \leq t \leq 5$$

Find the area between the graph of V' and the t axis over the interval $[2, 4]$ and interpret the results.

78. *Learning.* Repeat Problem 77 if $V'(t) = 13/t^{1/2}$ and the interval is changed to $[1, 4]$.

SECTION 12-2 Applications in Business and Economics

- Probability Density Functions
- Continuous Income Stream
- Future Value of a Continuous Income Stream
- Consumers' and Producers' Surplus

This section contains a number of important applications of the definite integral from business and economics. Included are three independent topics: probability density functions, continuous income streams, and consumers' and producers' surplus. Any of the three may be covered as time and interests dictate, and in any order.

Probability Density Functions

We will now take a brief look at the use of the definite integral to determine probabilities. Our approach will be intuitive and informal. A more formal treatment of the subject requires the use of the special "improper" integral form $\int_{-\infty}^{\infty} f(x)\,dx$, which we have not defined or discussed at this point.

Suppose an experiment is designed in such a way that any real number x on the interval $[c, d]$ is a possible outcome. For example, x may represent an IQ score, the height of a person in inches, or the life of a light bulb in hours. Technically, we refer to x as a *continuous random variable.*

In certain situations it is possible to find a function f with x as an independent variable such that the function f can be used to determine the probability that the outcome x of an experiment will be in the interval $[c, d]$. Such a function, called a **probability density function,** must satisfy the following three conditions (see Fig. 1):

1. $f(x) \geqslant 0$ for all real x
2. The area under the graph of $f(x)$ over the interval $(-\infty, \infty)$ is exactly 1.
3. If $[c, d]$ is a subinterval of $(-\infty, \infty)$, then

$$\text{Probability}(c \leqslant x \leqslant d) = \int_c^d f(x)\,dx$$

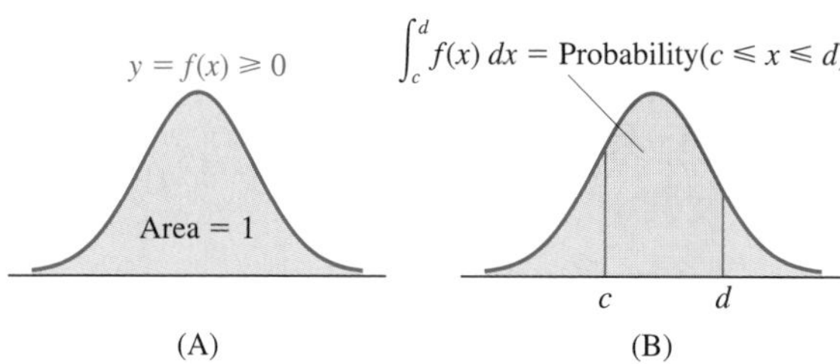

FIGURE 1
Probability density function

Explore–Discuss 1

Let: $f(x) = \begin{cases} \frac{1}{4}e^{-x/4} & \text{if } x \geqslant 0 \\ 0 & \text{otherwise} \end{cases}$

(A) Explain why $f(x) \geqslant 0$ over the interval $(-\infty, \infty)$.
(B) Find $\int_0^{10} f(x)\,dx$, $\int_0^{20} f(x)\,dx$, and $\int_0^{30} f(x)\,dx$.
(C) On the basis of part (B), what would you conjecture the area under the graph of $f(x)$ over the interval $(-\infty, \infty)$ to be equal to?

Example 1 **Duration of Telephone Calls** Suppose the length of telephone calls (in minutes) in a public telephone booth is a continuous random variable with probability density function shown in Figure 2.

$$f(t) = \begin{cases} \frac{1}{4}e^{-t/4} & \text{if } t \geqslant 0 \\ 0 & \text{otherwise} \end{cases}$$

(A) Determine the probability that a call selected at random will last between 2 and 3 minutes.
(B) Find b (to two decimal places) so that the probability of a call selected at random lasting between 2 and b minutes is .5.

FIGURE 2

SOLUTION (A)
$$\begin{aligned} \text{Probability}(2 \leqslant t \leqslant 3) &= \int_2^3 \tfrac{1}{4}e^{-t/4}\,dt \\ &= (-e^{-t/4})\big|_2^3 \\ &= -e^{-3/4} + e^{-1/2} \approx .13 \end{aligned}$$

(B) We want to find b such that Probability$(2 \leq t \leq b) = .5$.

$$\int_2^b \tfrac{1}{4}e^{-t/4}\,dt = .5$$

$$-e^{-b/4} + e^{-1/2} = .5 \qquad \text{Solve for } b.$$

$$e^{-b/4} = e^{-.5} - .5$$

$$-\frac{b}{4} = \ln(e^{-.5} - .5)$$

$$b = 8.96 \text{ minutes}$$

Thus, the probability of a call selected at random lasting from 2 to 8.96 minutes is .5.

Matched Problem 1

(A) In Example 1, find the probability that a call selected at random will last 4 minutes or less.

(B) Find b (to two decimal places) so that the probability of a call selected at random lasting b minutes or less is .9

$f(x) = \frac{1}{\sigma\sqrt{2\pi}} e^{-(x-\mu)^2/2\sigma^2}$

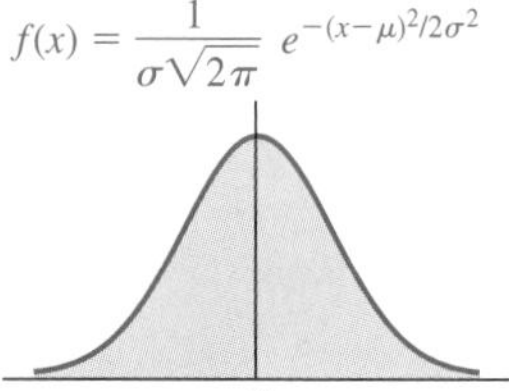

FIGURE 3
Normal curve

Group Activity 2 at the end of the last chapter investigated one of the most important probability density functions, the **normal probability density function** defined below and graphed in Figure 3.

$$f(x) = \frac{1}{\sigma\sqrt{2\pi}} e^{-(x-\mu)^2/2\sigma^2} \qquad \begin{array}{l}\mu \text{ is the mean.}\\ \sigma \text{ is the standard deviation.}\end{array}$$

It can be shown (but not easily) that the area under the normal curve in Figure 3 over the interval $(-\infty, \infty)$ is exactly 1. Since $\int e^{-x^2}\,dx$ is nonintegrable in terms of elementary functions (that is, the antiderivative cannot be expressed as a finite combination of simple functions), probabilities such as

$$\text{Probability}(c \leq x \leq d) = \frac{1}{\sigma\sqrt{2\pi}} \int_c^d e^{-(x-\mu)^2/2\sigma^2}\,dx$$

are generally determined by making an appropriate substitution in the integrand and then using a table of areas under the standard normal curve (that is, the normal curve with $\mu = 0$ and $\sigma = 1$). Such tables are readily available in most mathematical handbooks. A table can be constructed by using a rectangle rule, as discussed in Section 11-5; however, computers that employ refined techniques are generally used for this purpose. Some calculators have the capability of computing normal curve areas directly.

Continuous Income Stream

We start with a simple example having an obvious solution and generalize the concept to examples having less obvious solutions.

Suppose an aunt has established a trust that pays you \$2,000 a year for 10 years. What is the total amount you will receive from the trust by the end of the 10th year? Since there are 10 payments of \$2,000 each, you will receive

$$10 \times \$2{,}000 = \$20{,}000$$

We now look at the same problem from a different point of view, a point of view that will be useful in more complex problems. Let us assume that the income stream is continuous at a rate of \$2,000 per year. In Figure 4 the area under the graph of $f(t) = 2{,}000$ from 0 to t represents the income accumulated t years after the start. For example, for $t = \frac{1}{4}$ year, the income would be $\frac{1}{4}(2{,}000) = \$500$; for $t = \frac{1}{2}$ year, the income would be $\frac{1}{2}(2{,}000) = \$1{,}000$; for $t = 1$ year, the income would be $1(2{,}000) = \$2{,}000$; for $t = 5.3$ years, the income would be $5.3(2{,}000) = \$10{,}600$; and for $t = 10$ years, the income would be $10(2{,}000) = \$20{,}000$. The total income over a 10 year period—that is, the area under the graph of $f(t) = 2{,}000$ from 0 to 10—is also given by the definite integral

$$\int_0^{10} 2{,}000\,dt = 2{,}000t\Big|_0^{10} = 2{,}000(10) - 2{,}000(0) = \$20{,}000$$

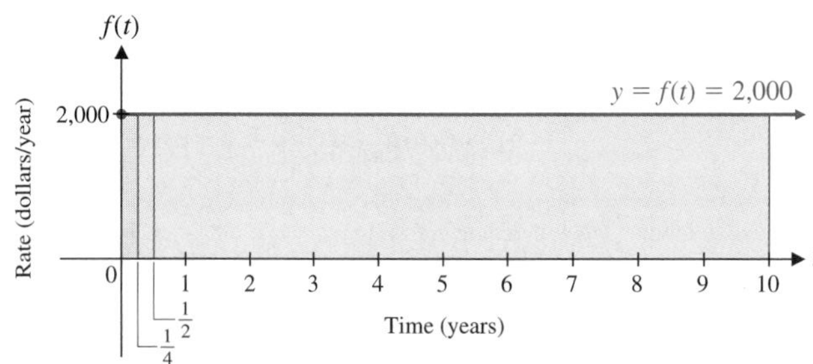

FIGURE 4
Continuous income stream

We now apply the idea of a continuous income stream to a less obvious problem.

Example 2 **Continuous Income Stream** The rate of change of the income produced by a vending machine located at an airport is given by

$$f(t) = 5{,}000e^{0.04t}$$

where t is time in years since the installation of the machine. Find the total income produced by the machine during the first 5 years of operation.

SOLUTION The area under the graph of the rate of change function from 0 to 5 represents the total change in income over the first 5 years (Fig. 5), and hence is given by a definite integral:

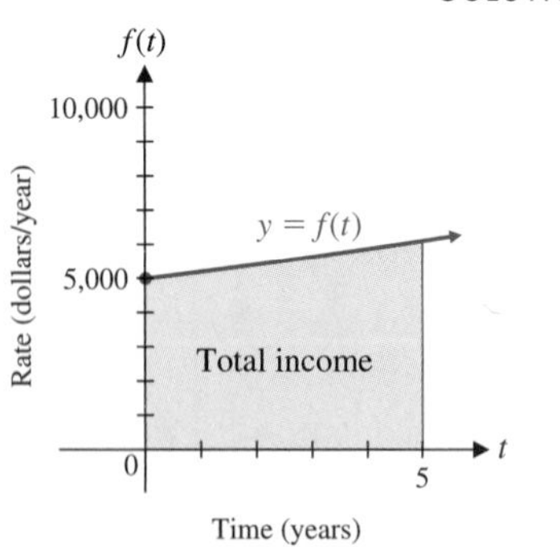

FIGURE 5
Continuous income stream

$$\begin{aligned}\text{Total income} &= \int_0^5 5{,}000e^{0.04t}\,dt\\ &= 125{,}000e^{0.04t}\Big|_0^5\\ &= 125{,}000e^{0.04(5)} - 125{,}000e^{0.04(0)}\\ &= 152{,}675 - 125{,}000\\ &= \$27{,}675 \quad \text{Rounded to the nearest dollar}\end{aligned}$$

Thus, the vending machine produces a total income of \$27,675 during the first 5 years of operation.

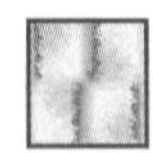

Matched Problem 2 Referring to Example 2, find the total income produced (to the nearest dollar) during the second 5 years of operation.

In reality, income from a vending machine is not usually received as a single payment at the end of each year, even though the rate is given as a yearly rate. Income is usually collected on a daily or weekly basis. In problems of this type it is convenient to assume that income is actually received in a **continuous stream;** that is, we assume that income is a continuous function of time and the rate of change is an instantaneous rate. The rate of change is called the **rate of flow** of the continuous income stream. In general, we have the following:

Total Income for a Continuous Income Stream

If $f(t)$ is the rate of flow of a continuous income stream, then the **total income** produced during the time period from $t = a$ to $t = b$ is

$$\text{Total income} = \int_a^b f(t)\,dt$$

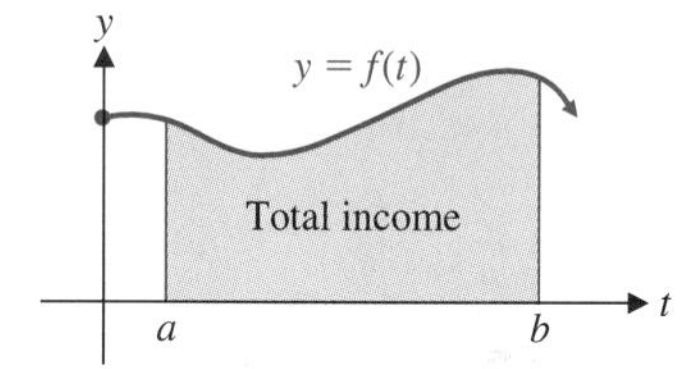

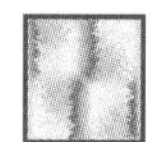

■ Future Value of a Continuous Income Stream

In Section 10-1, we discussed the continuous compound interest formula

$$A = Pe^{rt}$$

where P is the principal (or present value), A is the amount (or future value), r is the annual rate of continuous compounding (expressed as a decimal), and t is time in years. For example, if money is worth 12% compounded continuously, then the future value of a \$10,000 investment in 5 years is (to the nearest dollar)

$$A = 10{,}000e^{0.12(5)} = \$18{,}221$$

Now we want to apply the future value concept to the income produced by a continuous income stream. Suppose $f(t)$ is the rate of flow of a continuous income stream, and the income produced by this continuous income stream is invested as soon as it is received at a rate r, compounded continuously. We already know how to find the total income produced after T years, but how can we find the total of the income produced and the interest earned by this income? Since the income is received in a continuous flow, we cannot just use the formula $A = Pe^{rt}$. This formula is valid only for a single deposit P, not for a continuous flow of income. Instead, we use a Riemann sum approach that will allow us to apply the formula $A = Pe^{rt}$ repeatedly. To begin, we divide the time interval $[0, T]$ into n equal subintervals of length Δt and choose an arbitrary point c_k in each subinterval, as illustrated in Figure 6.

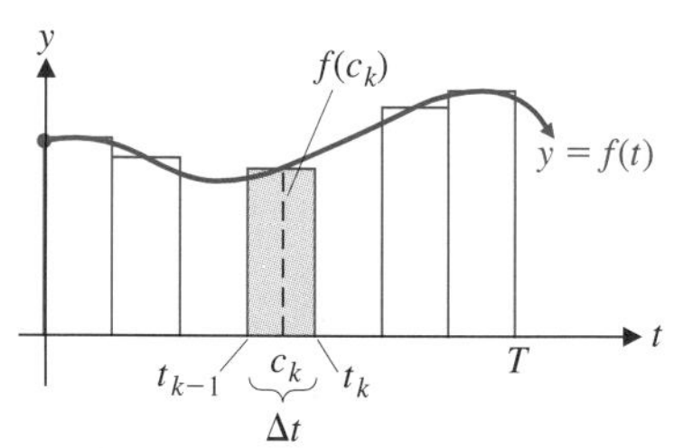

FIGURE 6

The total income produced during the time period from $t = t_{k-1}$ to $t = t_k$ is equal to the area under the graph of $f(t)$ over this subinterval and is approximately equal to $f(c_k)\,\Delta t$, the area of the shaded rectangle in Figure 6. The income received during this time period will earn interest for approximately $T - c_k$ years. Thus, using the future value formula $A = Pe^{rt}$ with

$P = f(c_k)\,\Delta t$ and $t = T - c_k$, the future value of the income produced during the time period from $t = t_{k-1}$ to $t = t_k$ is approximately equal to

$$f(c_k)\,\Delta t e^{(T-c_k)r}$$

The total of these approximate future values over n subintervals is then

$$f(c_1)\,\Delta t e^{(T-c_1)r} + f(c_2)\,\Delta t e^{(T-c_2)r} + \cdots + f(c_n)\,\Delta t e^{(T-c_n)r} = \sum_{k=1}^{n} f(c_k)e^{r(T-c_k)}\,\Delta t$$

This has the form of a Riemann sum, and the limit of this sum is a definite integral. (See the definition of definite integral in Section 11-5.) Thus, the *future value, FV,* of the income produced by the continuous income stream is given by

$$FV = \int_0^T f(t)e^{r(T-t)}\,dt$$

Since r and T are constants, we also can write

$$FV = \int_0^T f(t)e^{rT}e^{-rt}\,dt = e^{rT}\int_0^T f(t)e^{-rt}\,dt \qquad (1)$$

This last form is preferable, since the integral is usually easier to evaluate than the first form.

Future Value of a Continuous Income Stream

If $f(t)$ is the rate of flow of a continuous income stream, $0 \le t \le T$, and if the income is continuously invested at a rate r compounded continuously, then the **future value, *FV*,** at the end of T years is given by

$$FV = \int_0^T f(t)e^{r(T-t)}\,dt = e^{rT}\int_0^T f(t)e^{-rt}\,dt$$

The future value of a continuous income stream is the total value of all money produced by the continuous income stream (income and interest) at the end of T years.

We return to the trust set up for you by your aunt. Suppose the $2,000 per year you receive from the trust is invested as soon as it is received at 8% compounded continuously. We consider the trust income a continuous income stream with a flow rate of $2,000 per year. What is its future value (to the nearest dollar) by the end of the 10th year? Using the definite integral for future value from the box, we have

$$\begin{aligned}
FV &= e^{rT}\int_0^T f(t)e^{-rt}\,dt \\
FV &= e^{0.08(10)}\int_0^{10} 2{,}000e^{-0.08t}\,dt \qquad r = 0.08,\ T = 10,\ f(t) = 2{,}000 \\
&= 2{,}000e^{0.8}\int_0^{10} e^{-0.08t}\,dt \\
&= 2{,}000e^{0.8}\left[\frac{e^{-0.08t}}{-0.08}\right]\Bigg|_0^{10} \\
&= -25{,}000e^{0.8}\left[e^{-0.08(10)} - e^{-0.08(0)}\right] = \$30{,}639
\end{aligned}$$

Thus, at the end of 10 years you will have received \$30,639, including interest. How much is interest? Since you received \$20,000 in income from the trust, the interest is the difference between the future value and income. Thus,

$$\$30{,}639 - \$20{,}000 = \$10{,}639$$

is the interest earned by the income received from the trust over the 10 year period.

Explore–Discuss 2

Suppose that a trust fund is set up to pay you \$2,000 per year for 20 years with the continuous income stream invested at 8% compounded continuously. When will its future value be equal to \$50,000?

We now apply the same analysis to Example 2, the slightly more involved vending machine problem.

Example 3 **Future Value of a Continuous Income Stream** Using the continuous income rate of flow for the vending machine in Example 2,

$$f(t) = 5{,}000e^{0.04t}$$

find the future value of this income stream at 12% compounded continuously for 5 years, and find the total interest earned. Compute answers to the nearest dollar.

SOLUTION Using the formula

$$FV = e^{rT} \int_0^T f(t)e^{-rt}\, dt$$

with $r = 0.12, T = 5$, and $f(t) = 5{,}000e^{0.04t}$, we have

$$\begin{aligned} FV &= e^{0.12(5)} \int_0^5 5{,}000e^{0.04t}e^{-0.12t}\, dt \\ &= 5{,}000e^{0.6} \int_0^5 e^{-0.08t}\, dt \\ &= 5{,}000e^{0.6}\left(\frac{e^{-0.08t}}{-0.08}\right)\Bigg|_0^5 \\ &= 5{,}000e^{0.6}(-12.5e^{-0.4} + 12.5) \\ &= \$37{,}545 \quad \text{Rounded to the nearest dollar} \end{aligned}$$

Thus, the future value of the income stream at 12% compounded continuously at the end of 5 years is \$37,545.

In Example 2, we saw that the total income produced by this vending machine over a 5 year period was \$27,675. The difference between the future value and income is interest. Thus,

$$\$37{,}545 - \$27{,}675 = \$9{,}870$$

is the interest earned by the income produced by the vending machine during the 5 year period.

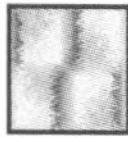

Matched Problem 3 Repeat Example 3 if the interest rate is 9% compounded continuously.

■ Consumers' and Producers' Surplus

Let $p = D(x)$ be the price–demand equation for a product, where x is the number of units of the product that consumers will purchase at a price of \$$p$ per unit. Suppose $\overline{p}$ is the current price and $\overline{x}$ is the number of units that can be sold at that price. The price–demand curve in Figure 7 shows that if the price is higher than $\overline{p}$, then the demand x is less than $\overline{x}$, but some consumers are still willing to pay the higher price. Consumers who are willing to pay more than $\overline{p}$ but are still able to buy the product at $\overline{p}$ have saved money. We want to determine the total amount saved by all the consumers who are willing to pay a price higher than $\overline{p}$ for this product.

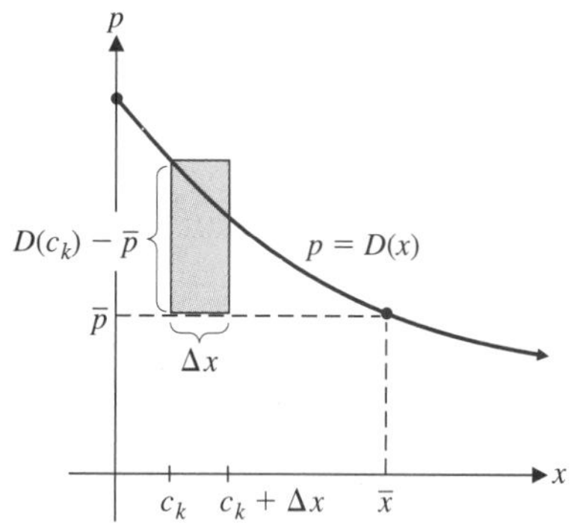

FIGURE 7

To do this, consider the interval $[c_k, c_k + \Delta x]$, where $c_k + \Delta x < \overline{x}$. If the price remained constant over this interval, then the savings on each unit would be the difference between $D(c_k)$, the price consumers are willing to pay, and $\overline{p}$, the price they actually pay. Since Δx represents the number of units purchased by consumers over the interval, the total savings to consumers over this interval is approximately equal to

$$[D(c_k) - \overline{p}]\,\Delta x \quad \text{(Savings per unit)} \times \text{(Number of units)}$$

which is the area of the shaded rectangle shown in Figure 7. If we divide the interval $[0, \overline{x}]$ into n equal subintervals, then the total savings to consumers is approximately equal to

$$[D(c_1) - \overline{p}]\Delta x + [D(c_2) - \overline{p}]\Delta x + \cdots + [D(c_n) - \overline{p}]\Delta x = \sum_{k=1}^{n} [D(c_k) - \overline{p}]\Delta x$$

which we recognize as a Riemann sum for the following integral:

$$\int_0^{\overline{x}} [D(x) - \overline{p}]\,dx$$

Thus, we define the *consumers' surplus* to be this integral.

Consumers' Surplus

If $(\overline{x}, \overline{p})$ is a point on the graph of the price–demand equation $p = D(x)$ for a particular product, then the **consumers' surplus, *CS*,** at a price level of $\overline{p}$ is

$$CS = \int_0^{\overline{x}} [D(x) - \overline{p}]\,dx$$

which is the area between $p = \overline{p}$ and $p = D(x)$ from $x = 0$ to $x = \overline{x}$, as shown in the margin.

The consumers' surplus represents the total savings to consumers who are willing to pay more than $\overline{p}$ for the product but are still able to buy the product for $\overline{p}$.

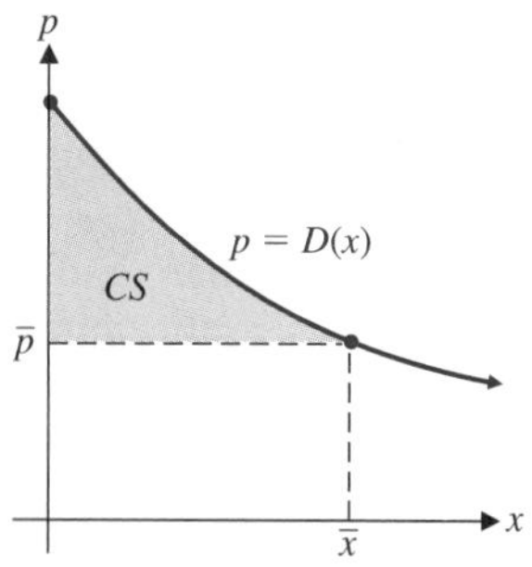

Example 4 **Consumers' Surplus** Find the consumers' surplus at a price level of \$8 for the price–demand equation

$$p = D(x) = 20 - 0.05x$$

SOLUTION **Step 1.** Find $\overline{x}$, the demand when the price is $\overline{p} = 8$:

$$\begin{aligned} \overline{p} &= 20 - 0.05\overline{x} \\ 8 &= 20 - 0.05\overline{x} \\ 0.05\overline{x} &= 12 \\ \overline{x} &= 240 \end{aligned}$$

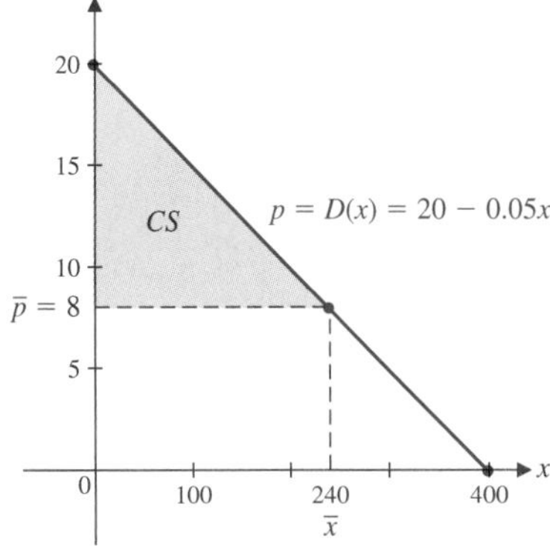

FIGURE 8

Step 2. Sketch a graph, as shown in Figure 8.

Step 3. Find the consumers' surplus (the shaded area in the graph):

$$\begin{aligned} CS &= \int_0^{\overline{x}} [D(x) - \overline{p}]\,dx \\ &= \int_0^{240} (20 - 0.05x - 8)\,dx \\ &= \int_0^{240} (12 - 0.05x)\,dx \\ &= (12x - 0.025x^2)\big|_0^{240} \\ &= 2{,}880 - 1{,}440 = \$1{,}440 \end{aligned}$$

Thus, the total savings to consumers who are willing to pay a higher price for the product is \$1,440.

Matched Problem 4 Repeat Example 4 for a price level of \$4.

If $p = S(x)$ is the price–supply equation for a product, $\overline{p}$ is the current price, and $\overline{x}$ is the current supply, then some suppliers are still willing to supply some units at a lower price than $\overline{p}$. The additional money that these suppliers gain from the higher price is called the *producers' surplus* and can be expressed in terms of a definite integral (proceeding as we did for the consumers' surplus).

Producers' Surplus

If $(\overline{x}, \overline{p})$, is a point on the graph of the price–supply equation $p = S(x)$, then the **producers' surplus, *PS*,** at a price level of $\overline{p}$ is

$$PS = \int_0^{\overline{x}} [\overline{p} - S(x)]\,dx$$

which is the area between $p = \overline{p}$ and $p = S(x)$ from $x = 0$ to $x = \overline{x}$, as shown in the margin.

The producers' surplus represents the total gain to producers who are willing to supply units at a lower price than $\overline{p}$ but are still able to supply units at $\overline{p}$.

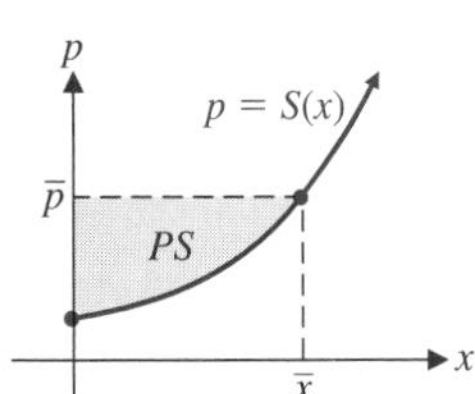

Example 5 **Producers' Surplus** Find the producers' surplus at a price level of \$20 for the price–supply equation

$$p = S(x) = 2 + 0.0002x^2$$

SOLUTION **Step 1.** Find $\overline{x}$, the supply when the price is $\overline{p} = 20$:

$$\begin{aligned} \overline{p} &= 2 + 0.0002\overline{x}^2 \\ 20 &= 2 + 0.0002\overline{x}^2 \\ 0.0002\overline{x}^2 &= 18 \\ \overline{x}^2 &= 90{,}000 \\ \overline{x} &= 300 \end{aligned}$$

There is only one solution since $\overline{x} \geqslant 0$.

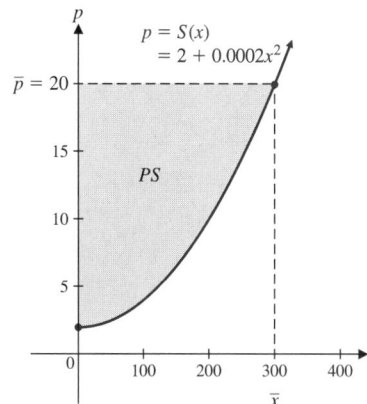

Step 2. Sketch a graph, as shown in Figure 9.

FIGURE 9

Step 3. Find the producers' surplus (the shaded area in the graph):

$$\begin{aligned} PS &= \int_0^{\overline{x}} [\overline{p} - S(x)]\,dx = \int_0^{300} [20 - (2 + 0.0002x^2)]\,dx \\ &= \int_0^{300} (18 - 0.0002x^2)\,dx = \left(18x - 0.0002\frac{x^3}{3}\right)\bigg|_0^{300} \\ &= 5{,}400 - 1{,}800 = \$3{,}600 \end{aligned}$$

Thus, the total gain to producers who are willing to supply units at a lower price is \$3,600.

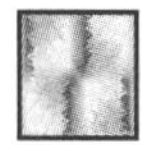

Matched Problem 5 Repeat Example 5 for a price level of \$4.

In a free competitive market, the price of a product is determined by the relationship between supply and demand. If $p = D(x)$ and $p = S(x)$ are the price–demand and price–supply equations, respectively, for a product and if $(\overline{x}, \overline{p})$ is the point of intersection of these equations, then $\overline{p}$ is called the **equilibrium price** and $\overline{x}$ is called the **equilibrium quantity.** If the price stabilizes at the equilibrium price $\overline{p}$, then this is the price level that will determine both the consumers' surplus and the producers' surplus.

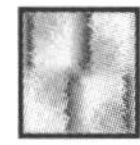

Example 6 **Equilibrium Price and Consumers' and Producers' Surplus** Find the equilibrium price and then find the consumers' surplus and producers' surplus at the equilibrium price level if

$$p = D(x) = 20 - 0.05x \qquad \text{and} \qquad p = S(x) = 2 + 0.0002x^2$$

SOLUTION **Step 1.** Find the equilibrium point. Set $D(x)$ equal to $S(x)$ and solve:

$$\begin{aligned} D(x) &= S(x) \\ 20 - 0.05x &= 2 + 0.0002x^2 \\ 0.0002x^2 + 0.05x - 18 &= 0 \\ x^2 + 250x - 90{,}000 &= 0 \\ x &= 200, -450 \end{aligned}$$

Since x cannot be negative, the only solution is $x = 200$. The equilibrium price can be determined by using $D(x)$ or $S(x)$. We will use both to check our work:

$$\begin{aligned} \overline{p} &= D(200) \\ &= 20 - 0.05(200) = 10 \end{aligned} \qquad \begin{aligned} \overline{p} &= S(200) \\ &= 2 + 0.0002(200)^2 = 10 \end{aligned}$$

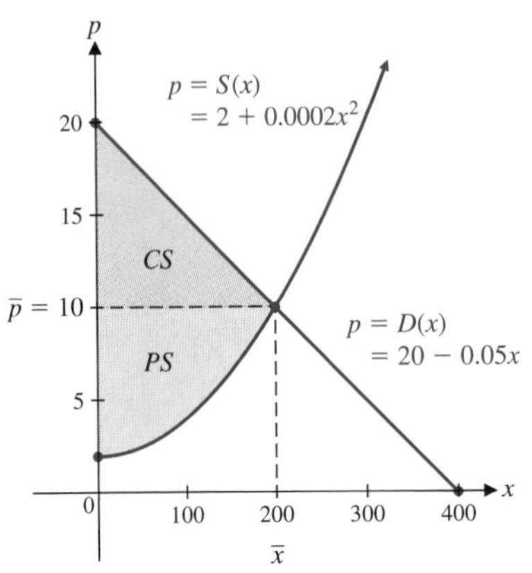

FIGURE 10

Thus, the equilibrium price is $\overline{p} = 10$, and the equilibrium quantity is $\overline{x} = 200$.

Step 2. Sketch a graph, as shown in Figure 10.

Step 3. Find the consumers' surplus:

$$\begin{aligned}
CS &= \int_0^{\overline{x}} [D(x) - \overline{p}]\, dx \\
&= \int_0^{200} (20 - 0.05x - 10)\, dx \\
&= \int_0^{200} (10 - 0.05x)\, dx \\
&= (10x - 0.025x^2)\big|_0^{200} \\
&= 2{,}000 - 1{,}000 = \$1{,}000
\end{aligned}$$

Step 4. Find the producers' surplus:

$$\begin{aligned}
PS &= \int_0^{\overline{x}} [\overline{p} - S(x)]\, dx \\
&= \int_0^{200} [10 - (2 + 0.0002x^2)]\, dx \\
&= \int_0^{200} (8 - 0.0002x^2)\, dx \\
&= \left(8x - 0.0002\frac{x^3}{3}\right)\bigg|_0^{200} \\
&= 1{,}600 - \tfrac{1{,}600}{3} \approx \$1{,}067 \qquad \text{Rounded to the nearest dollar}
\end{aligned}$$

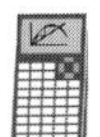

A graphing utility offers an alternative approach to finding the equilibrium point for Example 6 (Fig. 11A). A numerical integration routine can then be used to find the consumers' and producers' surplus (Fig. 11B).

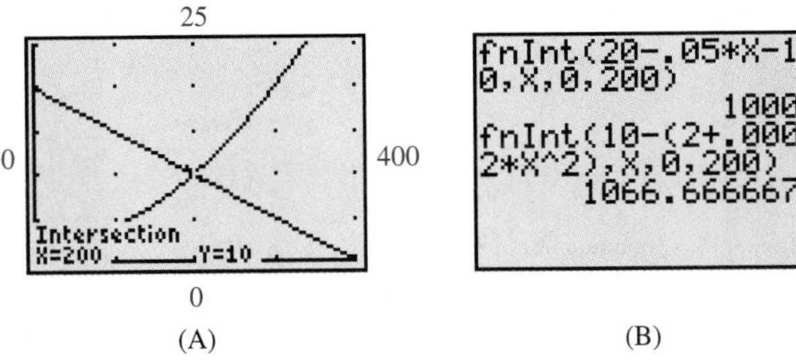

FIGURE 11

Matched Problem 6 ➡ Repeat Example 6 for

$$p = D(x) = 25 - 0.001x^2 \qquad \text{and} \qquad p = S(x) = 5 + 0.1x$$

Answers to Matched Problems

1. (A) .63 (B) 9.21 min **2.** \$33,803
3. FV = \$34,691; interest = \$7,016 **4.** \$2,560 **5.** \$133
6. $\overline{p} = 15$; CS = \$667; PS = \$500

EXERCISE 12-2

A *In Problems 1–4, evaluate each definite integral to two decimal places.*

1. $\int_0^5 e^{-0.08t}\,dt$
2. $\int_0^5 e^{0.08(5-t)}\,dt$
3. $\int_0^{30} e^{0.06t}e^{0.12(30-t)}\,dt$
4. $\int_0^{20} 1{,}000e^{0.03t}e^{0.15(20-t)}\,dt$

B *In Problems 5 and 6, explain which of (A), (B), and (C) are equal before evaluating the expressions. Then evaluate each expression to two decimal places.*

5. (A) $\int_0^8 e^{0.07(8-t)}\,dt$ (B) $\int_0^8 (e^{0.56} - e^{0.07t})\,dt$
   (C) $e^{0.56}\int_0^8 e^{-0.07t}\,dt$
6. (A) $\int_0^{10} 2{,}000e^{0.05t}e^{0.12(10-t)}\,dt$
   (B) $2{,}000e^{1.2}\int_0^{10} e^{-0.07t}\,dt$
   (C) $2{,}000e^{0.05}\int_0^{10} e^{0.12(10-t)}\,dt$

APPLICATIONS

Business & Economics

Unless stated to the contrary, compute all monetary answers to the nearest dollar.

7. The life expectancy (in years) of a certain brand of clock radio is a continuous random variable with probability density function

$$f(x) = \begin{cases} 2/(x+2)^2 & \text{if } x \geq 0 \\ 0 & \text{otherwise} \end{cases}$$

   (A) Find the probability that a randomly selected clock radio lasts at most 6 years.
   (B) Find the probability that a randomly selected clock radio lasts from 6 to 12 years.
   (C) Graph $y = f(x)$ for $[0, 12]$ and show the shaded region for part (A).

8. The shelf-life (in years) of a certain brand of flashlight batteries is a continuous random variable with probability density function

$$f(x) = \begin{cases} 1/(x+1)^2 & \text{if } x \geq 0 \\ 0 & \text{otherwise} \end{cases}$$

   (A) Find the probability that a randomly selected battery has a shelf-life of 3 years or less.
   (B) Find the probability that a randomly selected battery has a shelf-life of from 3 to 9 years.
   (C) Graph $y = f(x)$ for $[0, 10]$ and show the shaded region for part (A).

9. In Problem 7, find d so that the probability of a randomly selected clock radio lasting d years or less is .8.

10. In Problem 8, find d so that the probability of a randomly selected battery lasting d years or less is .5.

11. A manufacturer guarantees a product for 1 year. The time to failure of the product after it is sold is given by the probability density function

$$f(t) = \begin{cases} .01e^{-.01t} & \text{if } t \geq 0 \\ 0 & \text{otherwise} \end{cases}$$

    where t is time in months. What is the probability that a buyer chosen at random will have a product failure:
    (A) During the warranty period?
    (B) During the second year after purchase?

12. In a certain city, the daily use of water (in hundreds of gallons) per household is a continuous random variable with probability density function

$$f(x) = \begin{cases} .15e^{-.15x} & \text{if } x \geq 0 \\ 0 & \text{otherwise} \end{cases}$$

    Find the probability that a household chosen at random will use:
    (A) At most 400 gallons of water per day
    (B) Between 300 and 600 gallons of water per day

13. In Problem 11, what is the probability that the product will last at least 1 year? [*Hint:* Recall that the total area under the probability density function curve is 1.]

14. In Problem 12, what is the probability that a household will use more than 400 gallons of water per day? [See the hint in Problem 13.]

15. Find the total income produced by a continuous income stream in the first 5 years if the rate of flow is $f(t) = 2{,}500$.

16. Find the total income produced by a continuous income stream in the first 10 years if the rate of flow is $f(t) = 3{,}000$.

17. Interpret the results in Problem 15 with both a graph and a verbal description of the graph.

18. Interpret the results in Problem 16 with both a graph and a verbal description of the graph.

19. Find the total income produced by a continuous income stream in the first 3 years if the rate of flow is $f(t) = 400e^{0.05t}$.

20. Find the total income produced by a continuous income stream in the first 2 years if the rate of flow is $f(t) = 600e^{0.06t}$.

21. Interpret the results in Problem 19 with both a graph and a verbal description of the graph.

22. Interpret the results in Problem 20 with both a graph and a verbal description of the graph.

23. Starting at age 25, you deposit \$2,000 a year into an IRA account for retirement. Treat the yearly deposits into the account as a continuous income stream. If money in the account earns 5% compounded continuously, how much will be in the account 40 years later when you retire at age 65? How much of the final amount is interest?

24. Suppose in Problem 23 that you start the IRA deposits at age 30, but the account earns 6% compounded continuously. Treat the yearly deposits into the account as a continuous income stream. How much will be in the account 35 years later when you retire at age 65? How much of the final amount is interest?

25. Find the future value at 10% interest compounded continuously for 4 years for the continuous income stream with rate of flow $f(t) = 1{,}500e^{-0.02t}$.

26. Find the future value at 7% interest compounded continuously for 6 years for the continuous income stream with rate of flow $f(t) = 2{,}000e^{0.06t}$.

27. Compute the interest earned in Problem 25.

28. Compute the interest earned in Problem 26.

29. An investor is presented with a choice of two investments, an established clothing store and a new computer store. Each choice requires the same initial investment and each produces a continuous income stream of 10% compounded continuously. The rate of flow of income from the clothing store is $f(t) = 12{,}000$, and the rate of flow of income from the computer store is expected to be $g(t) = 10{,}000e^{0.05t}$. Compare the future values of these investments to determine which is the better choice over the next 5 years.

30. Refer to Problem 29. Which investment is the better choice over the next 10 years?

31. An investor has \$10,000 to invest in either a bond that matures in 5 years or a business that will produce a continuous stream of income over the next 5 years with rate of flow $f(t) = 2{,}000$. If both the bond and the continuous income stream earn 8% compounded continuously, which is the better investment?

32. Refer to Problem 31. Which is the better investment if the rate of the income from the business is $f(t) = 3{,}000$?

33. A business is planning to purchase a piece of equipment that will produce a continuous stream of income for 8 years with rate of flow $f(t) = 9{,}000$. If the continuous income stream earns 12% compounded continuously, what single deposit into an account earning the same interest rate will produce the same future value as the continuous income stream? (This deposit is called the **present value** of the continuous income stream.)

34. Refer to Problem 33. Find the present value of a continuous income stream at 8% compounded continuously for 12 years if the rate of flow is $f(t) = 1{,}000e^{0.03t}$.

35. Find the future value at a rate r compounded continuously for T years for a continuous income stream with rate of flow $f(t) = k$, where k is a constant.

36. Find the future value at a rate r compounded continuously for T years for a continuous income stream with rate of flow $f(t) = ke^{ct}$, where c and k are constants, $c \neq r$.

37. Find the consumers' surplus at a price level of $\overline{p} = \$150$ for the price–demand equation

$$p = D(x) = 400 - 0.05x$$

38. Find the consumers' surplus at a price level of $\overline{p} = \$120$ for the price–demand equation

$$p = D(x) = 200 - 0.02x$$

39. Interpret the results in Problem 37 with both a graph and a verbal description of the graph.

40. Interpret the results in Problem 38 with both a graph and a verbal description of the graph.

41. Find the producers' surplus at a price level of $\overline{p} = \$67$ for the price–supply equation

$$p = S(x) = 10 + 0.1x + 0.0003x^2$$

42. Find the producers' surplus at a price level of $\overline{p} = \$55$ for the price–supply equation

$$p = S(x) = 15 + 0.1x + 0.003x^2$$

43. Interpret the results in Problem 41 with both a graph and a verbal description of the graph.

44. Interpret the results in Problem 42 with both a graph and a verbal description of the graph.

In Problems 45–52, find the consumers' surplus and the producers' surplus at the equilibrium price level for the given price–demand and price–supply equations. Include a graph that identifies the consumers' surplus and the producers' surplus. Round all values to the nearest integer.

45. $p = D(x) = 50 - 0.1x$; $p = S(x) = 11 + 0.05x$

46. $p = D(x) = 25 - 0.004x^2$; $p = S(x) = 5 + 0.004x^2$

47. $p = D(x) = 80e^{-0.001x}$; $p = S(x) = 30e^{0.001x}$

48. $p = D(x) = 185e^{-0.005x}$; $p = S(x) = 25e^{0.005x}$

49. C $p = D(x) = 80 - 0.04x$; $p = S(x) = 30e^{0.001x}$

50. C $p = D(x) = 190 - 0.2x$; $p = S(x) = 25e^{0.005x}$

51. C $p = D(x) = 80e^{-0.001x}$; $p = S(x) = 15 + 0.0001x^2$

52. C $p = D(x) = 185e^{-0.005x}$; $p = S(x) = 20 + 0.002x^2$

53. C The tables give price–demand and price–supply data for the sale of soybeans at a grain market, where x is the number of bushels of soybeans (in thousands of bushels) and p is the price per bushel (in dollars):

PRICE–DEMAND

x	$p = D(x)$
0	6.70
10	6.59
20	6.52
30	6.47
40	6.45

PRICE–SUPPLY

x	$p = S(x)$
0	6.43
10	6.45
20	6.48
30	6.53
40	6.62

Use quadratic regression to model the price–demand data and linear regression to model the price–supply data.

(A) Find the equilibrium quantity (to three decimal places) and equilibrium price (to the nearest cent).

(B) Use a numerical integration routine to find the consumers' surplus and producers' surplus at the equilibrium price level.

54. C Repeat Problem 53 using quadratic regression to model both sets of data.

SECTION 12-3 Integration by Parts

In Section 11-1, we said we would return later to the indefinite integral

$$\int \ln x\, dx$$

since none of the integration techniques considered up to that time could be used to find an antiderivative for $\ln x$. We will now develop a very useful technique, called *integration by parts*, that will enable us to find not only the above integral, but also many others, including integrals such as

$$\int x \ln x\, dx \qquad \text{and} \qquad \int xe^x\, dx$$

The technique of integration by parts is also used to derive many integration formulas that are tabulated in mathematical handbooks. Some of these handbook formulas are discussed in the next section.

The method of integration by parts is based on the product formula for derivatives. If f and g are differentiable functions, then

$$\frac{d}{dx}[f(x)g(x)] = f(x)g'(x) + g(x)f'(x)$$

which can be written in the equivalent form

$$f(x)g'(x) = \frac{d}{dx}[f(x)g(x)] - g(x)f'(x)$$

Integrating both sides, we obtain

$$\int f(x)g'(x)\, dx = \int \frac{d}{dx}[f(x)g(x)]\, dx - \int g(x)f'(x)\, dx$$

The first integral to the right of the equal sign is $f(x)g(x) + C$. (Why?) We will leave out the constant of integration for now, since we can add it after integrating the second integral to the right of the equal sign. So we have

$$\int f(x)g'(x)\,dx = f(x)g(x) - \int g(x)f'(x)\,dx$$

This equation can be transformed into a more convenient form by letting $u = f(x)$ and $v = g(x)$; then $du = f'(x)\,dx$ and $dv = g'(x)\,dx$. Making these substitutions, we obtain the **integration by parts formula:**

Integration by Parts Formula

$$\int u\,dv = uv - \int v\,du$$

This formula can be very useful when the integral on the left is difficult or impossible to integrate using standard formulas. If u and dv are chosen with care—this is the crucial part of the process—then the integral on the right side may be easier to integrate than the one on the left. The formula provides us with another tool that is helpful in many, but not all cases. We are able to easily check the results by differentiating to get the original integrand, a good habit to get into. Several examples will demonstrate the use of the formula.

Example 1 **Integration by Parts** Find $\int xe^x\,dx$ using integration by parts and check the result.

SOLUTION First, write the integration by parts formula:

$$\int u\,dv = uv - \int v\,du \tag{1}$$

Now try to identify u and dv in $\int xe^x\,dx$ so that the integral $\int v\,du$ on the right side of (1) is easier to integrate than $\int u\,dv = \int xe^x\,dx$ on the left side. There are essentially two reasonable choices in selecting u and dv in $\int xe^x\,dx$:

Choice 1 $\qquad$ Choice 2

$$\int \overbrace{x}^{u}\,\overbrace{e^x\,dx}^{dv} \qquad \int \overbrace{e^x}^{u}\,\overbrace{x\,dx}^{dv}$$

We will pursue choice 1 and leave choice 2 for you to explore (see Explore–Discuss 1 following this example).

From choice 1, $u = x$ and $dv = e^x\,dx$. Looking at formula (1), we need du and v to complete the right side. It is convenient to proceed with the following arrangement: Let

$$u = x \qquad dv = e^x\,dx$$

then

$$du = dx \qquad \int dv = \int e^x\,dx$$
$$v = e^x$$

Any constant may be added to v, but we will always choose 0 for simplicity. The general arbitrary constant of integration will be added at the end of the process.

Substituting these results in formula (1), we obtain

$$\int u\,dv = uv - \int v\,du$$

$$\int xe^x\,dx = xe^x - \int e^x\,dx \quad \text{The right integral is easy to integrate.}$$

$$= xe^x - e^x + C \quad \text{Now add the arbitrary constant } C.$$

CHECK $\frac{d}{dx}(xe^x - e^x + C) = xe^x + e^x - e^x = xe^x$

Explore–Discuss 1 Pursue choice 2 in Example 1 using the integration by parts formula, and explain why this choice does not work out.

Matched Problem 1 Find: $\int xe^{2x}\,dx$

Example 2 **Integration by Parts** Find: $\int x \ln x\,dx$

SOLUTION As before, we have essentially two choices in choosing u and dv:

$$\overset{\text{Choice 1}}{\int \underbrace{x}_{u}\ \underbrace{\ln x\,dx}_{dv}} \qquad \overset{\text{Choice 2}}{\int \underbrace{\ln x}_{u}\ \underbrace{x\,dx}_{dv}}$$

Choice 1 is rejected, since we do not as yet know how to find an antiderivative of $\ln x$. So we move to choice 2 and choose $u = \ln x$ and $dv = x\,dx$; then proceed as in Example 1. Let

$$u = \ln x \qquad dv = x\,dx$$

then

$$du = \frac{1}{x}\,dx \qquad \int dv = \int x\,dx$$

$$v = \frac{x^2}{2}$$

Substitute these results into the integration by parts formula:

$$\int u\,dv = uv - \int v\,du$$

$$\int x \ln x\,dx = (\ln x)\left(\frac{x^2}{2}\right) - \int \left(\frac{x^2}{2}\right)\left(\frac{1}{x}\right)dx$$

$$= \frac{x^2}{2}\ln x - \int \frac{x}{2}\,dx \quad \text{An easy integral to evaluate}$$

$$= \frac{x^2}{2}\ln x - \frac{x^2}{4} + C$$

CHECK $\frac{d}{dx}\left(\frac{x^2}{2}\ln x - \frac{x^2}{4} + C\right) = x\ln x + \left(\frac{x^2}{2}\cdot\frac{1}{x}\right) - \frac{x}{2} = x\ln x$

Matched Problem 2 ➡ Find: $\int x \ln 2x\, dx$

As you should have discovered in Explore–Discuss 1, some choices for u and dv will lead to integrals that are more complicated than the original integral. This does not mean there is an error in the calculations or the integration by parts formula. It simply means that the particular choice of u and dv does not change the problem into one we can solve. When this happens, we must look for a different choice of u and dv. In some problems, it is possible that no choice will work. These observations and some guidelines for selecting u and dv are summarized in the box.

Integration by Parts: Selection of u and dv

For $\int u\, dv = uv - \int v\, du$:

1. The product $u\, dv$ must equal the original integrand.
2. It must be possible to integrate dv (preferably by using standard formulas or simple substitutions).
3. The new integral $\int v\, du$ should not be any more involved than the original integral $\int u\, dv$.
4. For integrals involving $x^p e^{ax}$, try

$$u = x^p \qquad \text{and} \qquad dv = e^{ax}\, dx$$

5. For integrals involving $x^p(\ln x)^q$, try

$$u = (\ln x)^q \qquad \text{and} \qquad dv = x^p\, dx$$

In some cases, repeated use of the integration by parts formula will lead to the evaluation of the original integral. The next example provides an illustration of such a case.

Example 3 ➡ **Repeated Use of Integration by Parts** Find: $\int x^2 e^{-x}\, dx$

SOLUTION Following suggestion 4 in the box, we choose

$$u = x^2 \qquad dv = e^{-x}\, dx$$

Then,

$$du = 2x\, dx \qquad v = -e^{-x}$$

and

$$\int x^2 e^{-x}\, dx = x^2(-e^{-x}) - \int (-e^{-x})2x\, dx$$

$$= -x^2 e^{-x} + 2\int x e^{-x}\, dx \qquad (2)$$

The new integral is not one we can evaluate by standard formulas, but it is simpler than the original integral. Applying the integration by parts formula to it will produce an even simpler integral. For the integral $\int xe^{-x}\, dx$, we choose

$$u = x \qquad dv = e^{-x}\, dx$$

Then,

$$du = dx \qquad v = -e^{-x}$$

and

$$\begin{aligned} \int xe^{-x}\,dx &= x(-e^{-x}) - \int(-e^{-x})\,dx \\ &= -xe^{-x} + \int e^{-x}\,dx \\ &= -xe^{-x} - e^{-x} \quad \text{Choose 0 for the constant.} \end{aligned} \tag{3}$$

Substituting (3) into (2), we have

$$\begin{aligned} \int x^2e^{-x}\,dx &= -x^2e^{-x} + 2(-xe^{-x} - e^{-x}) + C \\ &= -x^2e^{-x} - 2xe^{-x} - 2e^{-x} + C \end{aligned}$$ Add an arbitrary constant here.

CHECK $\frac{d}{dx}(-x^2e^{-x} - 2xe^{-x} - 2e^{-x} + C) = x^2e^{-x} - 2xe^{-x} + 2xe^{-x} - 2e^{-x} + 2e^{-x}$
$= x^2e^{-x}$

Matched Problem 3 ➡ Find: $\int x^2e^{2x}\,dx$

Example 4 ➡ **Using Integration by Parts** Find $\int_1^e \ln x\,dx$ and interpret geometrically.

SOLUTION First, we will find $\int \ln x\,dx$, and then return to the definite integral. Following suggestion 5 in the box (with $p = 0$), we choose

$$u = \ln x \qquad dv = dx$$

Then,

$$du = \frac{1}{x}\,dx \qquad v = x$$

Hence,

$$\begin{aligned} \int \ln x\,dx &= (\ln x)(x) - \int (x)\frac{1}{x}\,dx \\ &= x\ln x - x + C \end{aligned}$$

Note that this is the important result we mentioned at the beginning of this section. Now, we have

$$\begin{aligned} \int_1^e \ln x\,dx &= (x\ln x - x)\Big|_1^e \\ &= (e\ln e - e) - (1\ln 1 - 1) \\ &= (e - e) - (0 - 1) \\ &= 1 \end{aligned}$$

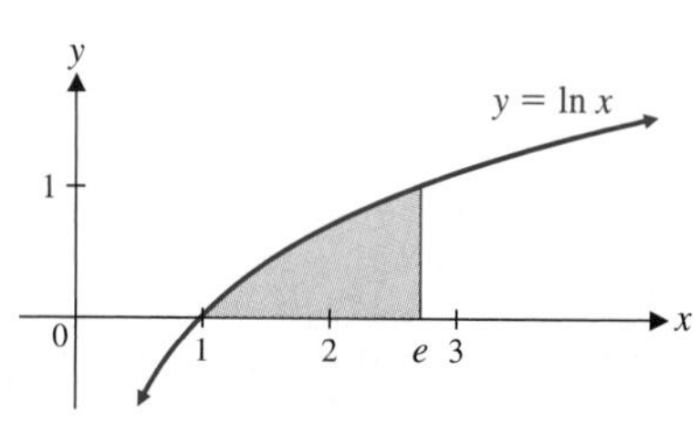

FIGURE 1

The integral represents the area under the curve $y = \ln x$ from $x = 1$ to $x = e$, as shown in Figure 1.

Matched Problem 4 ➡ Find: $\int_1^2 \ln 3x\,dx$

Explore–Discuss 2

Try using the integration by parts formula for $\int e^{x^2}\,dx$ and explain why it does not solve the problem.

Answer to Matched Problems

1. $\frac{x}{2}e^{2x} - \frac{1}{4}e^{2x} + C$
2. $\frac{x^2}{2}\ln 2x - \frac{x^2}{4} + C$
3. $\frac{x^2}{2}e^{2x} - \frac{x}{2}e^{2x} + \frac{1}{4}e^{2x} + C$
4. $2\ln 6 - \ln 3 - 1 \approx 1.4849$

EXERCISE 12-3

A *In Problems 1–4, integrate using the method of integration by parts. Assume $x > 0$ whenever the natural logarithm function is involved.*

1. $\int xe^{3x}\,dx$
2. $\int xe^{4x}\,dx$
3. $\int x^2\ln x\,dx$
4. $\int x^3\ln x\,dx$

B

5. If you want to use integration by parts to find $\int (x+1)^5(x+2)\,dx$, which is the better choice for u: $u = (x+1)^5$ or $u = x + 2$? Explain your choice, and then integrate.
6. If you want to use integration by parts to find $\int (5x-7)(x-1)^4\,dx$, which is the better choice for u: $u = 5x - 7$ or $u = (x-1)^4$? Explain your choice, and then integrate.

Problems 7–20 are mixed—some require integration by parts and others can be solved using techniques we have considered earlier. Integrate as indicated, assuming $x > 0$ whenever the natural logarithm function is involved.

7. $\int xe^{-x}\,dx$
8. $\int (x-1)e^{-x}\,dx$
9. $\int xe^{x^2}\,dx$
10. $\int xe^{-x^2}\,dx$
11. $\int_0^1 (x-3)e^x\,dx$
12. $\int_0^1 (x+1)e^x\,dx$
13. $\int_1^3 \ln 2x\,dx$
14. $\int_1^2 \ln\left(\frac{x}{2}\right)dx$
15. $\int \frac{2x}{x^2+1}\,dx$
16. $\int \frac{x^2}{x^3+5}\,dx$
17. $\int \frac{\ln x}{x}\,dx$
18. $\int \frac{e^x}{e^x+1}\,dx$
19. $\int \sqrt{x}\ln x\,dx$
20. $\int \frac{\ln x}{\sqrt{x}}\,dx$

In Problems 21–24, illustrate each integral graphically and describe what the integral represents in terms of areas.

21. Problem 11
22. Problem 12
23. Problem 13
24. Problem 14

C *Problems 25–42 are mixed—some may require use of the integration by parts formula along with techniques we have considered earlier; others may require repeated use of the integration by parts formula. Assume $g(x) > 0$ whenever $\ln g(x)$ is involved.*

25. $\int x^2e^x\,dx$
26. $\int x^3e^x\,dx$
27. $\int xe^{ax}\,dx,\ a \neq 0$
28. $\int \ln(ax)\,dx,\ a > 0$
29. $\int_1^e \frac{\ln x}{x^2}\,dx$
30. $\int_1^2 x^3e^{x^2}\,dx$
31. $\int_0^2 \ln(x+4)\,dx$
32. $\int_0^2 \ln(4-x)\,dx$
33. $\int xe^{x-2}\,dx$
34. $\int xe^{x+1}\,dx$
35. $\int x\ln(1+x^2)\,dx$
36. $\int x\ln(1+x)\,dx$
37. $\int e^x\ln(1+e^x)\,dx$
38. $\int \frac{\ln(1+\sqrt{x})}{\sqrt{x}}\,dx$
39. $\int (\ln x)^2\,dx$
40. $\int x(\ln x)^2\,dx$
41. $\int (\ln x)^3\,dx$
42. $\int x(\ln x)^3\,dx$

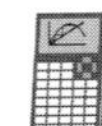

In Problems 43–46, use a graphing utility to graph each equation over the indicated interval, and find the area between the curve and the x axis over that interval. Find answers to two decimal places.

43. $y = x - 2 - \ln x;\ 1 \leqslant x \leqslant 4$
44. $y = 6 - x^2 - \ln x;\ 1 \leqslant x \leqslant 4$
45. $y = 5 - xe^x;\ 0 \leqslant x \leqslant 3$
46. $y = xe^x + x - 6;\ 0 \leqslant x \leqslant 3$

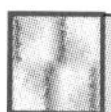

Applications

Business & Economics

47. *Profit.* If the marginal profit (in millions of dollars per year) is given by

$$P'(t) = 2t - te^{-t}$$

find the total profit earned over the first 5 years of operation (to the nearest million dollars) by use of an appropriate definite integral.

48. *Production.* An oil field is estimated to produce oil at a rate of $R(t)$ thousand barrels per month t months from now, as given by

$$R(t) = 10te^{-0.1t}$$

Find the total production in the first year of operation (to the nearest thousand barrels) by use of an appropriate definite integral.

49. *Profit.* Interpret the results in Problem 47 with both a graph and a verbal description of the graph.

50. *Production.* Interpret the results in Problem 48 with both a graph and a verbal description of the graph.

51. *Continuous income stream.* Find the future value at 8% compounded continuously for 5 years of a continuous income stream with a rate of flow of

$$f(t) = 1{,}000 - 200t$$

52. *Continuous income stream.* Find the interest earned at 10% compounded continuously for 4 years for a continuous income stream with a rate of flow of

$$f(t) = 1{,}000 - 250t$$

53. *Income distribution.* Find the index of income concentration for the Lorenz curve with equation

$$y = xe^{x-1}$$

54. *Income distribution.* Find the index of income concentration for the Lorenz curve with equation

$$y = x^2e^{x-1}$$

55. *Income distribution.* Interpret the results in Problem 53 with both a graph and a verbal description of the graph.

56. *Income distribution.* Interpret the results in Problem 54 with both a graph and a verbal description of the graph.

57. C *Sales analysis.* The monthly sales of a particular personal computer are expected to decline at the rate of

$$S'(t) = -4te^{0.1t}$$

computers per month, where t is time in months and $S(t)$ is the number of computers sold each month. The company plans to stop manufacturing this computer when the monthly sales reach 800 computers. If the monthly sales now ($t = 0$) are 2,000 computers, find $S(t)$. How long, to the nearest month, will the company continue to manufacture this computer?

58. C *Sales analysis.* The rate of change of the monthly sales of a new home video game cartridge is given by

$$S'(t) = 350\ln(t + 1) \qquad S(0) = 0$$

where t is the number of months since the game was released and $S(t)$ is the number of cartridges sold each month. Find $S(t)$. When, to the nearest month, will the monthly sales reach 15,000 cartridges?

59. *Consumers' surplus.* Find the consumers' surplus (to the nearest dollar) at a price level of $\overline{p} = \$2.089$ for the price–demand equation

$$p = D(x) = 9 - \ln(x + 4)$$

Use $\overline{x}$ computed to the nearest higher unit.

60. *Producers' surplus.* Find the producers' surplus (to the nearest dollar) at a price level of $\overline{p} = \$26$ for the price–supply equation

$$p = S(x) = 5\ln(x + 1)$$

Use $\overline{x}$ computed to the nearest higher unit.

61. *Consumers' surplus.* Interpret the results in Problem 59 with both a graph and a verbal description of the graph.

62. *Producers' surplus.* Interpret the results in Problem 60 with both a graph and a verbal description of the graph.

Life Sciences

63. *Pollution.* The concentration of particulate matter (in parts per million) t hours after a factory ceases operation for the day is given by

$$C(t) = \frac{20\ln(t + 1)}{(t + 1)^2}$$

Find the average concentration for the time period from $t = 0$ to $t = 5$.

64. *Medicine.* After a person takes a pill, the drug contained in the pill is assimilated into the bloodstream. The rate of assimilation t minutes after taking the pill is

$$R(t) = te^{-0.2t}$$

Find the total amount of the drug that is assimilated into the bloodstream during the first 10 minutes after the pill is taken.

Social Sciences

65.  *Learning.* In a particular business college, it was found that an average student enrolled in an advanced typing class progressed at a rate of

$$N'(t) = (t + 6)e^{-0.25t}$$

words per minute per week, t weeks after enrolling in a 15 week course. If at the beginning of the course a student could type 40 words per minute, how many words per minute, $N(t)$, would the student be expected to type t weeks into the course? How long, to the nearest week, should it take the student to achieve the 70 word per minute level? How many words per minute should the student be able to type by the end of the course?

66.  *Learning.* In the same business college, it was also found that an average student enrolled in a beginning shorthand class progressed at a rate of

$$N'(t) = (t + 10)e^{-0.1t}$$

words per minute per week, t weeks after enrolling in a 15 week course. If at the beginning of the course a student had no knowledge of shorthand (that is, could take dictation in shorthand at 0 words per minute), how many words per minute, $N(t)$, would the student be expected to handle t weeks into the course? How long, to the nearest week, should it take the student to achieve 90 words per minute? How many words per minute should the student be able to handle by the end of the course?

67. *Politics.* The number of voters (in thousands) in a certain city is given by

$$N(t) = 20 + 4t - 5te^{-0.1t}$$

where t is time in years. Find the average number of voters during the time period from $t = 0$ to $t = 5$.

SECTION 12-4 Integration Using Tables

- Using a Table of Integrals
- Substitution and Integral Tables
- Reduction Formulas
- Application

A **table of integrals** is a list of integration formulas used to evaluate integrals. People who frequently evaluate complex integrals may refer to tables that contain hundreds of formulas. Tables of this type are included in mathematical handbooks available in most college book stores. Table II of Appendix C contains a short list of integral formulas illustrating the types found in more extensive tables. Some of these formulas can be derived using the integration techniques discussed earlier, while others require techniques we have not considered. However, it is possible to verify each formula by differentiating the right side.

Using a Table of Integrals

The formulas in Table II (and in larger integral tables) are organized by categories, such as "Integrals Involving $a + bu$," "Integrals Involving $\sqrt{u^2 - a^2}$," and so on. The variable u is the variable of integration. All other symbols represent constants. To use a table to evaluate an integral, you must first find the category that most closely agrees with the form of the integrand and then find a formula in that category that can be made to match the integrand exactly by assigning values to the constants in the formula. The following examples illustrate this process.

Example 1 **Integration Using Tables** Use Table II to find: $\int \frac{x}{(5 + 2x)(4 - 3x)}\,dx$

SOLUTION Since the integrand

$$f(x) = \frac{x}{(5 + 2x)(4 - 3x)}$$

is a rational function involving terms of the form $a + bu$ and $c + du$, we examine formulas 15–20 in Table II to see if any of the integrands in these formulas can be made to match $f(x)$ exactly. Comparing the integrand in formula 16 with $f(x)$, we see that this integrand will match $f(x)$ if we let $a = 5$, $b = 2$, $c = 4$, and $d = -3$. Letting $u = x$ and substituting for a, b, c, and d in formula 16, we have

$$\int \frac{u}{(a + bu)(c + du)}\,du = \frac{1}{ad - bc}\left(\frac{a}{b}\ln|a + bu| - \frac{c}{d}\ln|c + du|\right) \qquad \text{Formula 16}$$

$$\int \frac{x}{(\underset{a}{5} + \underset{b}{2}x)(\underset{c}{4} \underset{d}{-} 3x)}\,dx = \frac{1}{5\cdot(-3) - 2\cdot 4}\left(\frac{5}{2}\ln|5 + 2x| - \frac{4}{-3}\ln|4 - 3x|\right) + C$$

$$a\cdot d - b\cdot c = 5\cdot(-3) - 2\cdot 4 = -23$$

$$= -\tfrac{5}{46}\ln|5 + 2x| - \tfrac{4}{69}\ln|4 - 3x| + C$$

Notice that the constant of integration C is not included in any of the formulas in Table II. However, you must still include C in all antiderivatives.

Matched Problem 1 Use Table II to find: $\int \frac{1}{(5 + 3x)^2(1 + x)}\,dx$

Example 2 **Integration Using Tables** Evaluate: $\int_3^4 \frac{1}{x\sqrt{25 - x^2}}\,dx$

SOLUTION First, we use Table II to find

$$\int \frac{1}{x\sqrt{25 - x^2}}\,dx$$

Since the integrand involves the expression $\sqrt{25 - x^2}$, we examine formulas 29–31 and select formula 29 with $a^2 = 25$ and $a = 5$:

$$\int \frac{1}{u\sqrt{a^2 - u^2}}\,du = -\frac{1}{a}\ln\left|\frac{a + \sqrt{a^2 - u^2}}{u}\right| \qquad \text{Formula 29}$$

$$\int \frac{1}{x\sqrt{25 - x^2}}\,dx = -\frac{1}{5}\ln\left|\frac{5 + \sqrt{25 - x^2}}{x}\right| + C$$

Thus,

$$\int_3^4 \frac{1}{x\sqrt{25 - x^2}}\,dx = \frac{1}{5}\ln\left|\frac{5 + \sqrt{25 - x^2}}{x}\right|\Bigg|_3^4$$

$$= -\frac{1}{5}\ln\left|\frac{5 + 3}{4}\right| + \frac{1}{5}\ln\left|\frac{5 + 4}{3}\right|$$

$$= -\tfrac{1}{5}\ln 2 + \tfrac{1}{5}\ln 3 = \tfrac{1}{5}\ln 1.5 \approx 0.0811$$

Matched Problem 2 ➡ Evaluate: $\int_6^8 \frac{1}{x^2\sqrt{100 - x^2}}\,dx$

Substitution and Integral Tables

As Examples 1 and 2 illustrate, if the integral we want to evaluate can be made to match one in the table exactly, then evaluating the indefinite integral consists of simply substituting the correct values of the constants into the formula. What happens if we cannot match an integral with one of the formulas in the table? In many cases, a substitution will change the given integral into one that corresponds to a table entry. The following examples illustrate several frequently used substitutions.

Example 3 ➡ **Integration Using Substitution and Tables** Find: $\int \frac{x^2}{\sqrt{16x^2 - 25}}\,dx$

SOLUTION In order to relate this integral to one of the formulas involving $\sqrt{u^2 - a^2}$ (formulas 40–45), we observe that if $u = 4x$, then

$$u^2 = 16x^2 \qquad \text{and} \qquad \sqrt{16x^2 - 25} = \sqrt{u^2 - 25}$$

Thus, we will use the substitution $u = 4x$ to change this integral into one that appears in the table:

$$\int \frac{x^2}{\sqrt{16x^2 - 25}}\,dx = \frac{1}{4}\int \frac{\frac{1}{16}u^2}{\sqrt{u^2 - 25}}\,du = \frac{1}{64}\int \frac{u^2}{\sqrt{u^2 - 25}}\,du$$

Substitution: $u = 4x$, $du = 4\,dx$, $x = \frac{1}{4}u$

This last integral can be evaluated by using formula 44 with $a = 5$:

$$\int \frac{u^2}{\sqrt{u^2 - a^2}}\,du = \frac{1}{2}\left(u\sqrt{u^2 - a^2} + a^2 \ln|u + \sqrt{u^2 - a^2}|\right)$$

Formula 44

$$\int \frac{x^2}{\sqrt{16x^2 - 25}}\,dx = \frac{1}{64}\int \frac{u^2}{\sqrt{u^2 - 25}}\,du$$

Use formula 44 with $a = 5$.

$$= \tfrac{1}{128}\left(u\sqrt{u^2 - 25} + 25 \ln|u + \sqrt{u^2 - 25}|\right) + C$$

Substitute $u = 4x$.

$$= \tfrac{1}{128}\left(4x\sqrt{16x^2 - 25} + 25 \ln|4x + \sqrt{16x^2 - 25}|\right) + C$$

Matched Problem 3 ➡ Find: $\int \sqrt{9x^2 - 16}\,dx$

Example 4 ➡ **Integration Using Substitution and Tables** Find: $\int \frac{x}{\sqrt{x^4 + 1}}\,dx$

SOLUTION None of the formulas in the table involve fourth powers; however, if we let $u = x^2$, then

$$\sqrt{x^4 + 1} = \sqrt{u^2 + 1}$$

and this form does appear in formulas 32–39. Thus, we substitute $u = x^2$:

$$\int \frac{1}{\sqrt{x^4 + 1}}\,x\,dx = \frac{1}{2}\int \frac{1}{\sqrt{u^2 + 1}}\,du$$

Substitution: $u = x^2$, $du = 2x\,dx$

We recognize the last integral as formula 36 with $a = 1$:

$$\int \frac{1}{\sqrt{u^2 + a^2}}\, du = \ln|u + \sqrt{u^2 + a^2}| \qquad \text{Formula 36}$$

$$\int \frac{x}{\sqrt{x^4 + 1}}\, dx = \frac{1}{2}\int \frac{1}{\sqrt{u^2 + 1}}\, du \qquad \text{Use formula 36 with } a = 1.$$

$$= \tfrac{1}{2}\ln|u + \sqrt{u^2 + 1}| + C \qquad \text{Substitute } u = x^2.$$

$$= \tfrac{1}{2}\ln|x^2 + \sqrt{x^4 + 1}| + C$$

Matched Problem 4 Find: $\int x\sqrt{x^4 + 1}\, dx$

Reduction Formulas

Example 5 **Using Reduction Formulas** Use Table II to find: $\int x^2 e^{3x}\, dx$

Solution Since the integrand involves the function e^{3x}, we examine formulas 46–48 and conclude that formula 47 can be used for this problem. Letting $u = x$, $n = 2$, and $a = 3$ in formula 47, we have

$$\int u^n e^{au}\, du = \frac{u^n e^{au}}{a} - \frac{n}{a}\int u^{n-1} e^{au}\, du \qquad \text{Formula 47}$$

$$\int x^2 e^{3x}\, dx = \frac{x^2 e^{3x}}{3} - \frac{2}{3}\int x e^{3x}\, dx$$

Notice that the expression on the right still contains an integral, but the exponent of x has been reduced by 1. Formulas of this type are called **reduction formulas** and are designed to be applied repeatedly until an integral that can be evaluated is obtained. Applying formula 47 to $\int xe^{3x}\, dx$ with $n = 1$, we have

$$\int x^2 e^{3x}\, dx = \frac{x^2 e^{3x}}{3} - \frac{2}{3}\left(\frac{xe^{3x}}{3} - \frac{1}{3}\int e^{3x}\, dx\right)$$

$$= \frac{x^2 e^{3x}}{3} - \frac{2xe^{3x}}{9} + \frac{2}{9}\int e^{3x}\, dx$$

This last expression contains an integral that is easy to evaluate:

$$\int e^{3x}\, dx = \tfrac{1}{3}e^{3x}$$

After making a final substitution and adding a constant of integration, we have

$$\int x^2 e^{3x}\, dx = \frac{x^2 e^{3x}}{3} - \frac{2xe^{3x}}{9} + \frac{2}{27}e^{3x} + C$$

Matched Problem 5 Use Table II to find: $\int (\ln x)^2\, dx$

Application

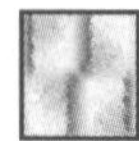

Example 6 **Producers' Surplus** Find the producers' surplus at a price level of \$20 for the price–supply equation

$$p = S(x) = \frac{5x}{500 - x}$$

SOLUTION **Step 1.** Find $\overline{x}$, the supply when the price is $\overline{p} = 20$:

$$\overline{p} = \frac{5\overline{x}}{500 - \overline{x}}$$
$$20 = \frac{5\overline{x}}{500 - \overline{x}}$$
$$10{,}000 - 20\overline{x} = 5\overline{x}$$
$$10{,}000 = 25\overline{x}$$
$$\overline{x} = 400$$

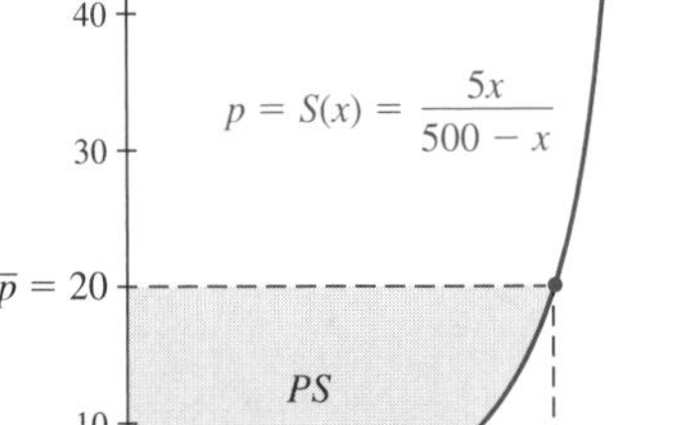

FIGURE 1

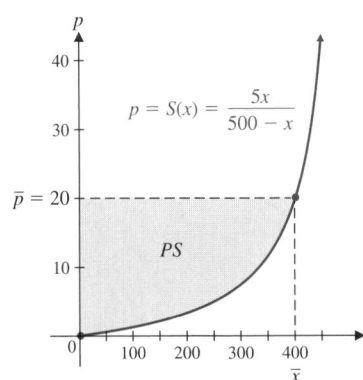

Step 2. Sketch a graph, as shown in Figure 1.

Step 3. Find the producers' surplus (the shaded area in the graph):

$$PS = \int_0^{\overline{x}} [\overline{p} - S(x)]\,dx$$
$$= \int_0^{400} \left(20 - \frac{5x}{500 - x}\right) dx$$
$$= \int_0^{400} \frac{10{,}000 - 25x}{500 - x}\,dx$$

Use formula 20 with $a = 10{,}000$, $b = -25$, $c = 500$, and $d = -1$:

$$\int \frac{a + bu}{c + du}\,du = \frac{bu}{d} + \frac{ad - bc}{d^2} \ln|c + du| \quad \text{Formula 20}$$

$$PS = (25x + 2{,}500 \ln|500 - x|)\Big|_0^{400}$$
$$= 10{,}000 + 2{,}500 \ln|100| - 2{,}500 \ln|500|$$
$$\approx \$5{,}976$$

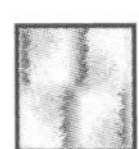

Matched Problem 6 ➡ Find the consumers' surplus at a price level of \$10 for the price–demand equation

$$p = D(x) = \frac{20x - 8{,}000}{x - 500}$$

Explore–Discuss 1

Use algebraic manipulation, including algebraic long division, on the integrand in Example 6 to show that

$$\frac{5\overline{x}}{500 - \overline{x}} = \frac{-5\overline{x}}{\overline{x} - 500} = -5 - \frac{2{,}500}{\overline{x} - 500}$$

Use this result to find the indefinite integral in Example 6 without resorting to table formulas.

Answers to Matched Problems

1. $\frac{1}{2}\left(\frac{1}{5 + 3x}\right) + \frac{1}{4}\ln\left|\frac{1 + x}{5 + 3x}\right| + C$ **2.** $\frac{7}{1{,}200} \approx 0.0058$

3. $\frac{1}{6}\left(3x\sqrt{9x^2 - 16} - 16\ln|3x + \sqrt{9x^2 - 16}|\right) + C$

4. $\frac{1}{4}\left(x^2\sqrt{x^4 + 1} + \ln|x^2 + \sqrt{x^4 + 1}|\right) + C$ **5.** $x(\ln x)^2 - 2x\ln x + 2x + C$

6. $3{,}000 + 2{,}000 \ln 200 - 2{,}000 \ln 500 \approx \$1{,}167$

EXERCISE 12-4

A *Use Table II to find each indefinite integral in Problems 1–14.*

1. $\displaystyle\int \frac{1}{x(1+x)}\,dx$

2. $\displaystyle\int \frac{1}{x^2(1+x)}\,dx$

3. $\displaystyle\int \frac{1}{(3+x)^2(5+2x)}\,dx$

4. $\displaystyle\int \frac{x}{(5+2x)^2(2+x)}\,dx$

5. $\displaystyle\int \frac{x}{\sqrt{16+x}}\,dx$

6. $\displaystyle\int \frac{1}{x\sqrt{16+x}}\,dx$

7. $\displaystyle\int \frac{1}{x\sqrt{1-x^2}}\,dx$

8. $\displaystyle\int \frac{\sqrt{9-x^2}}{x}\,dx$

9. $\displaystyle\int \frac{1}{x\sqrt{x^2+4}}\,dx$

10. $\displaystyle\int \frac{1}{x^2\sqrt{x^2-16}}\,dx$

11. $\int x^2 \ln x\,dx$

12. $\int x^3 \ln x\,dx$

13. $\displaystyle\int \frac{1}{1+e^x}\,dx$

14. $\displaystyle\int \frac{1}{5+2e^{3x}}\,dx$

Evaluate each definite integral in Problems 15–20. Use Table II to find the antiderivative.

15. $\displaystyle\int_1^3 \frac{x^2}{3+x}\,dx$

16. $\displaystyle\int_2^6 \frac{x}{(6+x)^2}\,dx$

17. $\displaystyle\int_0^7 \frac{1}{(3+x)(1+x)}\,dx$

18. $\displaystyle\int_0^7 \frac{x}{(3+x)(1+x)}\,dx$

19. $\displaystyle\int_0^4 \frac{1}{\sqrt{x^2+9}}\,dx$

20. $\displaystyle\int_4^5 \sqrt{x^2-16}\,dx$

B *In Problems 21–32, use substitution techniques and Table II to find each indefinite integral.*

21. $\displaystyle\int \frac{\sqrt{4x^2+1}}{x^2}\,dx$

22. $\int x^2\sqrt{9x^2-1}\,dx$

23. $\displaystyle\int \frac{x}{\sqrt{x^4-16}}\,dx$

24. $\int x\sqrt{x^4-16}\,dx$

25. $\displaystyle\int x^2\sqrt{x^6+4}\,dx$

26. $\displaystyle\int \frac{x^2}{\sqrt{x^6+4}}\,dx$

27. $\displaystyle\int \frac{1}{x^3\sqrt{4-x^4}}\,dx$

28. $\displaystyle\int \frac{\sqrt{x^4+4}}{x}\,dx$

29. $\displaystyle\int \frac{e^x}{(2+e^x)(3+4e^x)}\,dx$

30. $\displaystyle\int \frac{e^x}{(4+e^x)^2(2+e^x)}\,dx$

31. $\displaystyle\int \frac{\ln x}{x\sqrt{4+\ln x}}\,dx$

32. $\displaystyle\int \frac{1}{x\ln x\sqrt{4+\ln x}}\,dx$

C *In Problems 33–38, use Table II to find each indefinite integral.*

33. $\int x^2e^{5x}\,dx$

34. $\int x^2e^{-4x}\,dx$

35. $\int x^3e^{-x}\,dx$

36. $\int x^3e^{2x}\,dx$

37. $\int (\ln x)^3\,dx$

38. $\int (\ln x)^4\,dx$

Problems 39–46 are mixed—some require the use of Table II and others can be solved using techniques we considered earlier.

39. $\int_3^5 x\sqrt{x^2-9}\,dx$

40. $\int_3^5 x^2\sqrt{x^2-9}\,dx$

41. $\displaystyle\int_2^4 \frac{1}{x^2-1}\,dx$

42. $\displaystyle\int_2^4 \frac{x}{(x^2-1)^2}\,dx$

43. $\displaystyle\int \frac{x+1}{x^2+2x}\,dx$

44. $\displaystyle\int \frac{x+1}{x^2+x}\,dx$

45. $\displaystyle\int \frac{x+1}{x^2+3x}\,dx$

46. $\displaystyle\int \frac{x^2+1}{x^2+3x}\,dx$

In Problems 47–50, find the area bounded by the graphs of $y = f(x)$ and $y = g(x)$ to two decimal places. Use a graphing utility to approximate intersection points to two decimal places.

47. $f(x) = \dfrac{10}{\sqrt{x^2+1}};\ g(x) = x^2 + 3x$

48. $f(x) = \sqrt{1+x^2};\ g(x) = 5x - x^2$

49. $f(x) = x\sqrt{4+x};\ g(x) = 1 + x$

50. $f(x) = \dfrac{x}{\sqrt{x+4}};\ g(x) = x - 2$

Applications

Use Table II to evaluate all integrals involved in any solutions of Problems 51–74.

Business & Economics

51. *Consumers' surplus.* Find the consumers' surplus at a price level of $\overline{p} = \$15$ for the price–demand equation

$$p = D(x) = \frac{7{,}500 - 30x}{300 - x}$$

52. *Producers' surplus.* Find the producers' surplus at a price level of $\overline{p} = \$20$ for the price–supply equation

$$p = S(x) = \frac{10x}{300 - x}$$

53. *Consumers' surplus.* For Problem 51, graph the price–demand equation and the price level equation $\overline{p} = 15$ in the same coordinate system. What region represents the consumers' surplus?

54. *Producers' surplus.* For Problem 52, graph the price–supply equation and the price level equation $\overline{p} = 20$ in the same coordinate system. What region represents the producers' surplus?

55. C *Cost.* A company manufactures downhill skis. It has fixed costs of \$25,000 and a marginal cost given by

$$C'(x) = \frac{250 + 10x}{1 + 0.05x}$$

where $C(x)$ is the total cost at an output of x pairs of skis. Find the cost function $C(x)$ and determine the production level (to the nearest unit) that produces a cost of \$150,000. What is the cost (to the nearest dollar) for a production level of 850 pairs of skis?

56. C *Cost.* A company manufactures a portable CD player. It has fixed costs of \$11,000 per week and a marginal cost given by

$$C'(x) = \frac{65 + 20x}{1 + 0.4x}$$

where $C(x)$ is the total cost per week at an output of x players per week. Find the cost function $C(x)$ and determine the production level (to the nearest unit) that produces a cost of \$52,000 per week. What is the cost (to the nearest dollar) for a production level of 700 players per week?

57. *Continuous income stream.* Find the future value at 10% compounded continuously for 10 years for the continuous income stream with rate of flow $f(t) = 50t^2$.

58. *Continuous income stream.* Find the interest earned at 8% compounded continuously for 5 years for the continuous income stream with rate of flow $f(t) = 200t$.

59. *Income distribution.* Find the index of income concentration for the Lorenz curve with equation

$$y = \tfrac{1}{2}x\sqrt{1 + 3x}$$

60. *Income distribution.* Find the index of income concentration for the Lorenz curve with equation

$$y = \tfrac{1}{2}x^2\sqrt{1 + 3x}$$

61. *Income distribution.* For Problem 59, graph $y = x$ and the Lorenz curve over the interval $[0, 1]$. Discuss the effect of the area bounded by $y = x$ and the Lorenz curve getting smaller relative to the equitable distribution of income.

62. *Income distribution.* For Problem 60, graph $y = x$ and the Lorenz curve over the interval $[0, 1]$. Discuss the effect of the area bounded by $y = x$ and the Lorenz curve getting larger relative to the equitable distribution of income.

63. *Marketing.* After test marketing a new high-fiber cereal, the market research department of a major food producer estimates that monthly sales (in millions of dollars) will grow at the monthly rate of

$$S'(t) = \frac{t^2}{(1 + t)^2}$$

t months after the cereal is introduced. If we assume 0 sales at the time the cereal is first introduced, find the total sales, $S(t)$, t months after the cereal is introduced. Find the total sales during the first 2 years this cereal is on the market.

64. *Average price.* At a discount department store, the price–demand equation for premium motor oil is given by

$$p = D(x) = \frac{50}{\sqrt{100 + 6x}}$$

where x is the number of cans of oil that can be sold at a price of \$$p$. Find the average price over the demand interval $[50, 250]$.

65. *Marketing.* In Problem 63, show the sales over the first 2 years geometrically, and verbally describe the geometric representation.

66. *Price–demand.* In Problem 64, graph the price–demand equation and the line representing the average price in the same coordinate system over the interval $[50, 250]$. Describe how the areas under the two curves over the interval $[50, 250]$ are related.

67. *Profit.* The marginal profit for a small car agency that C sells x cars per week is given by

$$P'(x) = x\sqrt{2 + 3x}$$

where $P(x)$ is the profit in dollars. The agency's profit on the sale of only 1 car per week is $-\$2{,}000$. Find the profit function and the number of cars that must be sold (to the nearest unit) to produce a profit of \$13,000 per week. How much weekly profit (to the nearest dollar) will the agency have if 80 cars are sold per week?

68. *Revenue.* The marginal revenue for a company that C manufactures and sells x solar-powered calculators per week is given by

$$R'(x) = \frac{x}{\sqrt{1 + 2x}} \qquad R(0) = 0$$

where $R(x)$ is the revenue in dollars. Find the revenue function and the number of calculators that must be sold (to the nearest unit) to produce \$10,000 in revenue per week. How much weekly revenue (to the nearest dollar) will the company have if 1,000 calculators are sold per week?

Life Sciences

69. *Pollution.* An oil tanker aground on a reef is losing oil and producing an oil slick that is radiating outward at a rate given approximately by

$$\frac{dR}{dt} = \frac{100}{\sqrt{t^2 + 9}} \qquad t \geq 0$$

where R is the radius (in feet) of the circular slick after t minutes. Find the radius of the slick after 4 minutes if the radius is 0 when $t = 0$.

70. *Pollution.* The concentration of particulate matter (in parts per million) during a 24 hour period is given approximately by

$$C(t) = t\sqrt{24 - t} \qquad 0 \leq t \leq 24$$

where t is time in hours. Find the average concentration during the time period from $t = 0$ to $t = 24$.

Social Sciences

71. *Learning.* A person learns N items at a rate given approximately by

$$N'(t) = \frac{60}{\sqrt{t^2 + 25}} \qquad t \geq 0$$

where t is the number of hours of continuous study. Determine the total number of items learned in the first 12 hours of continuous study.

72. *Politics.* The number of voters (in thousands) in a metropolitan area is given approximately by

$$f(t) = \frac{500}{2 + 3e^{-t}} \qquad t \geq 0$$

where t is time in years. Find the average number of voters during the time period from $t = 0$ to $t = 10$.

73. *Learning.* Interpret Problem 71 geometrically. Verbally describe the geometric interpretation.

74. *Politics.* In Problem 72, graph $y = f(t)$ and the line representing the average number of voters over the interval $[0, 10]$ in the same coordinate system. Describe how the areas under the two curves over the interval $[0, 10]$ are related.

IMPORTANT TERMS AND SYMBOLS

12-1 *Area between Curves.* Area between a curve and the x axis; area between two curves; distribution of income; Lorenz curve; index of income concentration; absolute equality; absolute inequality

12-2 *Applications in Business and Economics.* Probability density function; normal probability density function; continuous income stream; rate of flow; total income; future value of a continuous income stream; consumers' surplus; producers' surplus; equilibrium price; equilibrium quantity

12-3 *Integration by Parts.* Integration by parts; selection of u and dv

$$\int u\,dv = uv - \int v\,du$$

12-4 *Integration Using Tables.* Table of integrals; substitution and integral tables; reduction formulas

REVIEW EXERCISE

Work through all the problems in this chapter review and check your answers in the back of the book. Answers to all review problems are there along with section numbers in italics to indicate where each type of problem is discussed. Where weaknesses show up, review appropriate sections in the text.

Compute all numerical answers to three decimal places unless directed otherwise.

A *In Problems 1–3, set up definite integrals that represent the shaded areas in the figure over the indicated intervals.*

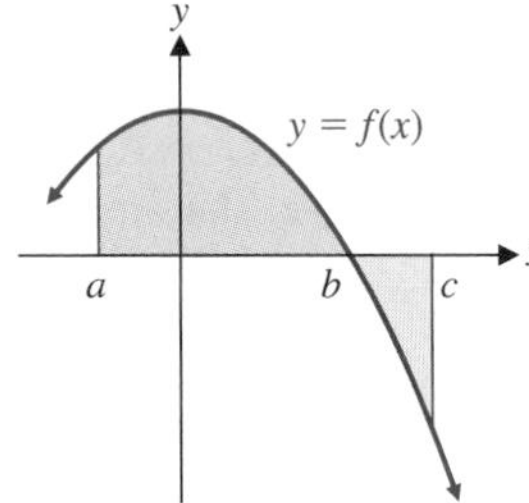

1. Interval $[a, b]$ **2.** Interval $[b, c]$ **3.** Interval $[a, c]$

4. Sketch a graph of the area between the graphs of $y = \ln x$ and $y = 0$ over the interval $[0.5, e]$, and find the area.

In Problems 5–10, evaluate each integral.

5. $\int xe^{4x}\,dx$ **6.** $\int x \ln x\,dx$

7. $\int \frac{\ln x}{x}\,dx$ **8.** $\int \frac{x}{1 + x^2}\,dx$

9. $\int \frac{1}{x(1 + x)^2}\,dx$ **10.** $\int \frac{1}{x^2\sqrt{1 + x}}\,dx$

B *In Problems 11–14, set up definite integrals that represent the shaded areas in the figure over the indicated intervals.*

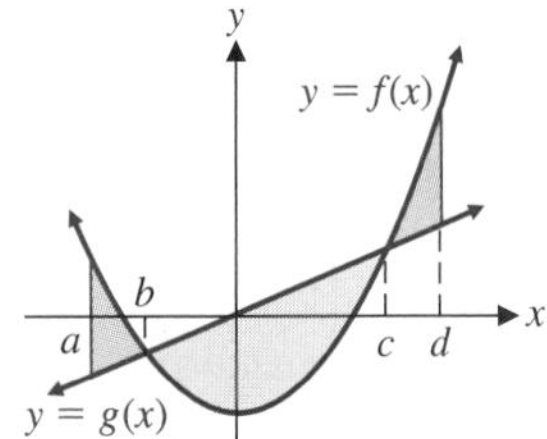

11. Interval $[a, b]$ **12.** Interval $[b, c]$

13. Interval $[b, d]$ **14.** Interval $[a, d]$

15. Sketch a graph of the area bounded by the graphs of $y = x^2 - 6x + 9$ and $y = 9 - x$, and find the area.

In Problems 16–21, evaluate each integral.

16. $\int_0^1 xe^x\,dx$ **17.** $\int_0^3 \frac{x^2}{\sqrt{x^2 + 16}}\,dx$

18. $\int \sqrt{9x^2 - 49}\,dx$ **19.** $\int te^{-0.5t}\,dt$

20. $\int x^2 \ln x\,dx$ **21.** $\int \frac{1}{1 + 2e^x}\,dx$

22. Sketch a graph of the area bounded by the indicated graphs and find the area. In part (B), approximate intersection points and area to two decimal places.

(A) $y = x^3 - 6x^2 + 9x$; $y = x$

C (B) $y = x^3 - 6x^2 + 9x$; $y = x + 1$

C *In Problems 23–30, evaluate each integral.*

23. $\int \frac{(\ln x)^2}{x}\,dx$

24. $\int x(\ln x)^2\,dx$

25. $\int \frac{x}{\sqrt{x^2 - 36}}\,dx$

26. $\int \frac{x}{\sqrt{x^4 - 36}}\,dx$

27. $\int_0^4 x\ln(10 - x)\,dx$

28. $\int (\ln x)^2\,dx$

29. $\int xe^{-2x^2}\,dx$

30. $\int x^2e^{-2x}\,dx$

31. C Use a numerical integration routine on a graphing utility to find the area (to three decimal places) in the first quadrant that is below the graph of

$$y = \frac{6}{2 + 5e^{-x}}$$

and above the graph of $y = 0.2x + 1.6$.

APPLICATIONS

Business & Economics

32. *Product warranty.* A manufacturer warrants a product for parts and labor for 1 year, and for parts only for a second year. The time to a failure of the product after it is sold is given by the probability density function

$$f(t) = \begin{cases} 0.21e^{-0.21t} & \text{if } t \geq 0 \\ 0 & \text{otherwise} \end{cases}$$

What is the probability that a buyer chosen at random will have a product failure:

(A) During the first year of warranty?
(B) During the second year of warranty?

33. *Product warranty.* Graph the probability density function for Problem 32 over the interval $[0, 3]$, interpret part (B) of Problem 32 geometrically, and verbally describe the geometric representation.

34. C *Revenue function.* The weekly marginal revenue from the sale of x hair dryers is given by

$$R'(x) = 65 - 6\ln(x + 1) \qquad R(0) = 0$$

where $R(x)$ is the revenue in dollars. Find the revenue function and the production level (to the nearest unit) for a revenue of \$20,000 per week. What is the weekly revenue (to the nearest dollar) at a production level of 1,000 hair dryers per week?

35. *Continuous income stream.* The rate of flow (in dollars per year) of a continuous income stream for a 5 year period is given by

$$f(t) = 2{,}500e^{0.05t} \qquad 0 \leq t \leq 5$$

(A) Graph $y = f(t)$ over $[0, 5]$ and shade in the area that represents the total income received from the end of the first year to the end of the fourth year.
(B) Find the total income received, to the nearest dollar, from the end of the first year to the end of the fourth year.

36. *Future value of a continuous income stream.* The continuous income stream in Problem 35 is invested as it is received at 15% compounded continuously.

(A) Find the future value (to the nearest dollar) at the end of the 5 year period.
(B) Find the interest earned (to the nearest dollar) during this 5 year period.

37. *Income distribution.* An economist produced the following Lorenz curves for the current income distribution and the projected income distribution 10 years from now in a certain country:

$$f(x) = 0.1x + 0.9x^2 \quad \text{Current Lorenz curve}$$
$$g(x) = x^{1.5} \quad \text{Projected Lorenz curve}$$

(A) Graph $y = x$ and the current Lorenz curve on one set of coordinate axes for $[0, 1]$, and graph $y = x$ and the projected Lorenz curve on another set of coordinate axes over the same interval.
(B) Looking at the areas bounded by the Lorenz curves and $y = x$, can you say that the income will be more or less equitably distributed 10 years from now?
(C) Compute the index of income concentration (to one decimal place) for the current and projected curves. Now what can you say about the distribution of income 10 years from now? More equitable or less?

38. *Consumers' and producers' surplus.* Find the consumers' surplus and the producers' surplus at the equilibrium price level for each pair of price–demand and price–supply equations. Include a graph that identifies the consumers' surplus and the producers' surplus. Round all values to the nearest integer.

(A) $p = D(x) = 70 - 0.2x$;
$p = S(x) = 13 + 0.0012x^2$
C (B) $p = D(x) = 70 - 0.2x$; $p = S(x) = 13e^{0.006x}$

39.  *Producers' surplus.* The table below gives price–supply data for the sale of hogs at a livestock market, where x is the number of pounds (in thousands) and p is the price per pound (in cents).

PRICE–SUPPLY

x	$p = S(x)$
0	43.50
10	46.74
20	50.05
30	54.72
40	59.18

(A) Using quadratic regression to model the data, find the demand at a price of 52.50 cents per pound.

(B) Use a numerical integration routine to find the producers' surplus (to the nearest dollar) at a price level of 52.50 cents per pound.

Life Sciences

40. *Drug assimilation.* The rate at which the body eliminates a drug (in milliliters per hour) is given by

$$R(t) = \frac{60t}{(t + 1)^2(t + 2)}$$

where t is the number of hours since the drug was administered. How much of the drug is eliminated during the first hour after it was administered? During the fourth hour?

41. C With the aid of a graphing utility, illustrate Problem 40 geometrically.

42. *Medicine.* For a particular doctor, the length of time (in hours) spent with a patient per office visit has the probability density function

$$f(t) = \begin{cases} \dfrac{\frac{4}{3}}{(t + 1)^2} & \text{if } 0 \leq t \leq 3 \\ 0 & \text{otherwise} \end{cases}$$

(A) What is the probability that this doctor will spend less than 1 hour with a randomly selected patient?

(B) What is the probability that this doctor will spend more than 1 hour with a randomly selected patient?

43. *Medicine.* Illustrate part (B) in Problem 42 geometrically. Describe the geometric interpretation verbally.

Social Sciences

44. *Politics.* The rate of change of the voting population of a city with respect to time t (in years) is estimated to be

$$N'(t) = \frac{100t}{(1 + t^2)^2}$$

where $N(t)$ is in thousands. If $N(0)$ is the current voting population, how much will this population increase during the next 3 years?

45. *Psychology.* Rats were trained to go through a maze by rewarding them with a food pellet upon successful completion. After the seventh successful run, it was found that the probability density function for length of time (in minutes) until success on the eighth trial is given by

$$f(t) = \begin{cases} .5e^{-.5t} & \text{if } t \geq 0 \\ 0 & \text{otherwise} \end{cases}$$

What is the probability that a rat selected at random after seven successful runs will take 2 or more minutes to complete the eighth run successfully? [Recall that the area under a probability density function curve from $-\infty$ to ∞ is 1.]

Group Activity 1 *Analysis of Income Concentration from Raw Data*

This group activity may be done without the use of a graphing utility, but additional insight into mathematical modeling will be gained if one is available.

We start with raw data on income distribution supplied in table form by the U.S. Bureau of the Census (Table 1, page 872). From the raw data in the table, we wish to compare income distribution among Whites and income distribution among Blacks in the United States in 1992. The approach will be numeric, geometric, and symbolic. We will first organize in table form the data that correspond to data points for a Lorenz curve. We will then find Lorenz curves of the form $f(x) = x^p$ for each set of data points. We will interpret the income distribution geometrically by graphing the Lorenz curves and $y = x$ for both sets of data. Finally, we compute the index of income concentration for Blacks and for Whites, and interpret the results. (See the discussion of Lorenz curves in Section 12-1 for relevant background material.)

Table 1

INCOME DISTRIBUTION BY POPULATION FIFTHS

FAMILIES, 1992 RACE	UPPER LIMIT OF EACH FIFTH* Lowest	Second	Third	Fourth	Top 5%†
TOTAL	\$16,960	\$30,000	\$44,200	\$64,300	\$106,509
WHITE	19,000	32,000	46,250	66,252	109,900
BLACK	7,531	15,609	26,800	44,200	75,619

FAMILIES, 1992 RACE	PERCENTAGE DISTRIBUTION OF TOTAL INCOME Lowest fifth	Second fifth	Third fifth	Fourth fifth	Highest fifth	Top 5%
TOTAL	4.4	10.5	16.5	24.0	44.6	17.6
WHITE	4.9	10.9	16.7	23.7	43.8	17.3
BLACK	3.0	8.2	15.0	25.0	48.8	18.5

*The highest fifth does not have an upper limit.
†Lower limit for top 5%.
Source: U.S. Bureau of the Census, U.S. Department of Commerce

Table 2

BLACK FAMILY INCOME DISTRIBUTION IN 1992

INCOME LEVEL	x	y
Under \$8,000	0.20	0.03
Under		0.11
Under		
Under \$44,000	0.80	

Table 3

WHITE FAMILY INCOME DISTRIBUTION IN 1992

INCOME LEVEL	x	y
Under \$19,000	0.20	0.05
Under		0.16
Under		
Under \$66,000	0.80	

Part 1. *Numeric Analysis.* Complete Tables 2 and 3 using the data in Table 1. Round income levels to the nearest thousand dollars and represent percents in decimal form to two decimal places. Remember that x represents the cumulative percentage of families in a given category and y represents the corresponding cumulative percentage of income received by these families. Verbally describe the meaning of the last two lines in Tables 2 and 3 after they are completed.

Part 2. *Geometric Analysis*

(A) Sketch separate graphs on suitable graph paper for each table by plotting the data points from the x and y columns in Tables 2 and 3. Also, graph the line $y = x$ over the interval $[0, 1]$ in each graph.

(B) Find p (to one decimal place) so that the graph of the function $f(x) = x^p$ goes through the point (0.20, 0.03) in Table 2. Plot this curve on the corresponding graph, $0 \leq x \leq 1$. Also, find q (to one decimal place) so that the function $g(x) = x^q$ goes through the point (0.20, 0.05) in Table 3. Plot this curve on the corresponding graph, $0 \leq x \leq 1$. Repeat this process for each of the remaining points in each table. Compare all the graphs based on Table 2 and select the value of p that best fits the data (by eye). Do the same for all the graphs based on Table 3.

(C) From the final graphs chosen in part (B) can you draw any conclusions about whether income is distributed more equitably among Blacks or Whites? Verbally support your conclusions.

Part 3. *Symbolic Analysis.* Use the values of p and q you determined in part 2B to compute the index of income concentration for Blacks and for Whites, and interpret.

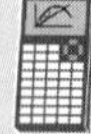

Group Activity 2 *Grain Exchange*

The tables below give price–demand and price–supply data for the sale of wheat at a grain market, where x is the number of bushels of wheat (in thousands) and p is the price per bushel (in cents).

PRICE–DEMAND

x	$p = D(x)$
20	345
25	336
30	323
35	320
40	318
45	307

PRICE–SUPPLY

x	$p = S(x)$
20	311
25	312
30	321
35	323
40	326
45	338

(A) Find quadratic, logarithmic, and exponential regression models for each set of data.

(B) Use the sum of squares of the residuals (see Group Activity 2 in Chapter 2) to compare the models from part (A).

(C) Using the models that best fit the data, approximate the supply and demand supported by a price of \$3.50 per bushel; by a price of \$3.25 per bushel; by a price of \$3.00 per bushel.

(D) Using the models that best fit the data, find the equilibrium quantity and equilibrium price.

(E) Use a numerical integration routine to find the consumers' surplus and the producers' surplus at the equilibrium price level.

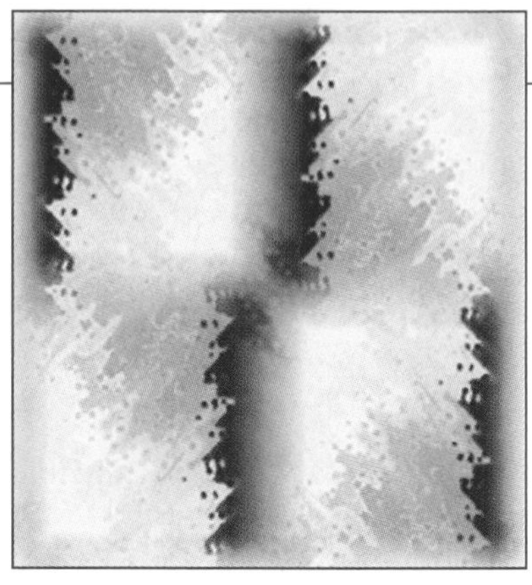

CHAPTER 13

MULTIVARIABLE CALCULUS

SECTION 13-1

Functions of Several Variables

- FUNCTIONS OF TWO OR MORE INDEPENDENT VARIABLES
- EXAMPLES OF FUNCTIONS OF SEVERAL VARIABLES
- THREE-DIMENSIONAL COORDINATE SYSTEMS

FUNCTIONS OF TWO OR MORE INDEPENDENT VARIABLES

In Section 1-1, we introduced the concept of a function with one independent variable. Now we will broaden the concept to include functions with more than one independent variable. We start with an example.

A small manufacturing company produces a standard type of surfboard and no other products. If fixed costs are \$500 per week and variable costs are \$70 per board produced, then the weekly cost function is given by

$$C(x) = 500 + 70x \tag{1}$$

where x is the number of boards produced per week. The cost function is a function of a single independent variable x. For each value of x from the domain of C there exists exactly one value of $C(x)$ in the range of C.

Now, suppose the company decides to add a high-performance competition board to its line. If the fixed costs for the competition board are \$200 per week and the variable costs are \$100 per board, then the cost function (1) must be modified to

$$C(x, y) = 700 + 70x + 100y \tag{2}$$

where $C(x, y)$ is the cost for weekly output of x standard boards and y competition boards. Equation (2) is an example of a function with two independent variables, x and y. Of course, as the company expands its product line even further, its weekly cost function must be modified to include more and more independent variables, one for each new product produced.

In general, an equation of the form

$$z = f(x, y)$$

describes a **function of two independent variables** if for each permissible ordered pair (x, y), there is one and only one value of z determined by $f(x, y)$. The variables x and y are **independent variables,** and the variable z is a **dependent variable.** The set of all ordered pairs of permissible values of x and y is the **domain** of the function, and the set of all corresponding values $f(x, y)$ is the **range** of the function. Unless otherwise stated, we will assume that the domain of a function specified by an equation of the form $z = f(x, y)$ is the set of all ordered pairs of real numbers (x, y) such that $f(x, y)$ is also a real number. It should be noted, however, that certain conditions in practical problems often lead to further restrictions of the domain of a function.

We can similarly define functions of three independent variables, $w = f(x, y, z)$; of four independent variables, $u = f(w, x, y, z)$; and so on. In this chapter, we will primarily concern ourselves with functions of two independent variables.

Example 1 **Evaluating a Function of Two Independent Variables** For the cost function $C(x, y) = 700 + 70x + 100y$ described earlier, find $C(10, 5)$.

Solution

$$\begin{aligned} C(10, 5) &= 700 + 70(10) + 100(5) \\ &= \$1{,}900 \end{aligned}$$

Matched Problem 1 Find $C(20, 10)$ for the cost function in Example 1.

Example 2 **Evaluating a Function of Three Independent Variables** For the function $f(x, y, z) = 2x^2 - 3xy + 3z + 1$, find $f(3, 0, -1)$.

Solution

$$\begin{aligned} f(3, 0, -1) &= 2(3)^2 - 3(3)(0) + 3(-1) + 1 \\ &= 18 - 0 - 3 + 1 = 16 \end{aligned}$$

Matched Problem 2 Find $f(-2, 2, 3)$ for f in Example 2.

Example 3 **Revenue, Cost, and Profit Functions** The surfboard company discussed at the beginning of this section has determined that the demand equations for the two types of boards they produce are given by

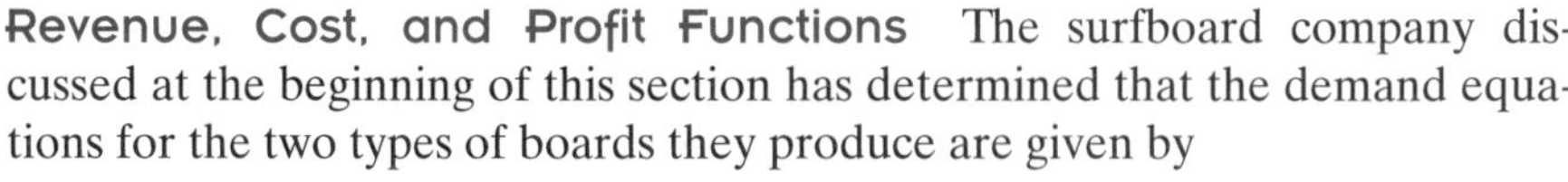

$$\begin{aligned} p &= 210 - 4x + y \\ q &= 300 + x - 12y \end{aligned}$$

where p is the price of the standard board, q is the price of the competition board, x is the weekly demand for standard boards, and y is the weekly demand for competition boards.

(A) Find the weekly revenue function $R(x, y)$, and evaluate $R(20, 10)$.
(B) If the weekly cost function is

$$C(x, y) = 700 + 70x + 100y$$

find the weekly profit function $P(x, y)$, and evaluate $P(20, 10)$.

Solution (A)

$$\text{Revenue} = \begin{pmatrix}\text{Demand for}\\ \text{standard}\\ \text{boards}\end{pmatrix} \times \begin{pmatrix}\text{Price of a}\\ \text{standard}\\ \text{board}\end{pmatrix} + \begin{pmatrix}\text{Demand for}\\ \text{competition}\\ \text{boards}\end{pmatrix} \times \begin{pmatrix}\text{Price of a}\\ \text{competition}\\ \text{board}\end{pmatrix}$$

$$\begin{aligned} R(x, y) &= xp + yq \\ &= x(210 - 4x + y) + y(300 + x - 12y) \\ &= 210x + 300y - 4x^2 + 2xy - 12y^2 \\ R(20, 10) &= 210(20) + 300(10) - 4(20)^2 + 2(20)(10) - 12(10)^2 \\ &= \$4{,}800 \end{aligned}$$

(B)

$$\begin{aligned} \text{Profit} &= \text{Revenue} - \text{Cost} \\ P(x, y) &= R(x, y) - C(x, y) \\ &= 210x + 300y - 4x^2 + 2xy - 12y^2 - 700 - 70x - 100y \\ &= 140x + 200y - 4x^2 + 2xy - 12y^2 - 700 \\ P(20,10) &= 140(20) + 200(10) - 4(20)^2 + 2(20)(10) - 12(10)^2 - 700 \\ &= \$1{,}700 \end{aligned}$$

Matched Problem 3 Repeat Example 3 if the demand and cost equations are given by

$$\begin{aligned} p &= 220 - 6x + y \\ q &= 300 + 3x - 10y \\ C(x, y) &= 40x + 80y + 1{,}000 \end{aligned}$$

Examples of Functions of Several Variables

A number of concepts we have already considered can be thought of in terms of functions of two or more variables. We list a few of these below.

Area of a rectangle $A(x, y) = xy$

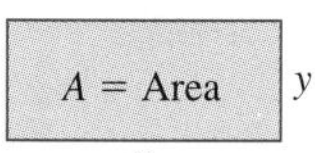

Volume of a box $V(x, y, z) = xyz$

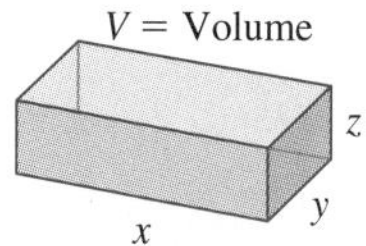

Volume of a right circular cylinder $V(r, h) = \pi r^2 h$

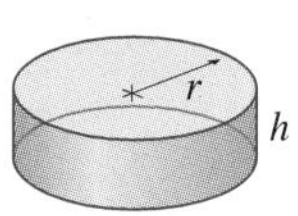

Simple interest $A(P, r, t) = P(1 + rt)$

A = Amount
P = Principle
r = Annual rate
t = Time in years

Compound interest	$A(P, r, t, n) = P\left(1 + \frac{r}{n}\right)^{nt}$	A = Amount P = Principle r = Annual rate t = Time in years n = Compound periods per year
IQ	$Q(M, C) = \frac{M}{C}(100)$	Q = IQ = Intelligence quotient M = MA = Mental age C = CA = Chronological age
Resistance for blood flow in a vessel (Poiseuille's law)	$R(L, r) = k\frac{L}{r^4}$	R = Resistance L = Length of vessel r = Radius of vessel k = Constant

Example 4 **Package Design** A company uses a box with a square base and an open top for one of its products (see the figure). If x is the length (in inches) of each side of the base and y is the height (in inches), find the total amount of material $M(x, y)$ required to construct one of these boxes, and evaluate $M(5, 10)$.

SOLUTION

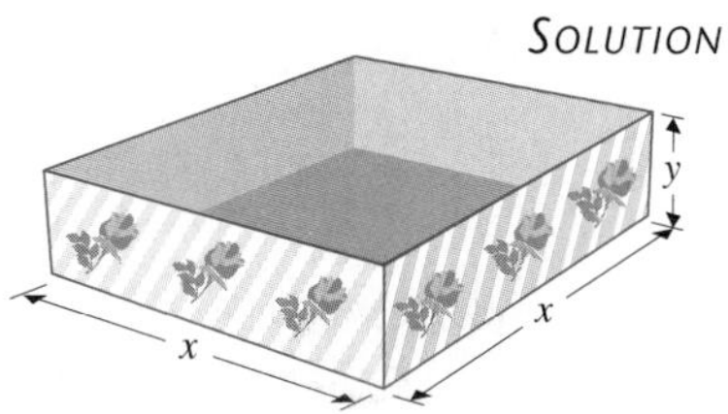

$$\begin{aligned} \text{Area of base} &= x^2 \\ \text{Area of one side} &= xy \\ \text{Total material} &= (\text{Area of base}) + 4(\text{Area of one side}) \\ M(x, y) &= x^2 + 4xy \\ M(5, 10) &= (5)^2 + 4(5)(10) \\ &= 225 \text{ square inches} \end{aligned}$$

Matched Problem 4 For the box in Example 4, find the volume $V(x, y)$, and evaluate $V(5, 10)$.

The next example concerns the **Cobb–Douglas production function,**

$$f(x, y) = kx^m y^n$$

where k, m, and n are positive constants with $m + n = 1$. Economists use this function to describe the number of units $f(x, y)$ produced from the utilization of x units of labor and y units of capital (for equipment such as tools, machinery, buildings, and so on). Cobb–Douglas production functions are also used to describe the productivity of a single industry, of a group of industries producing the same product, or even of an entire country.

Example 5 **Productivity** The productivity of a steel manufacturing company is given approximately by the function

$$f(x, y) = 10x^{0.2}y^{0.8}$$

with the utilization of x units of labor and y units of capital. If the company uses 3,000 units of labor and 1,000 units of capital, how many units of steel will be produced?

Solution The number of units of steel produced is given by

$$f(3{,}000, 1{,}000) = 10(3{,}000)^{0.2}(1{,}000)^{0.8} \quad \text{Use a calculator.}$$
$$\approx 12{,}457 \text{ units}$$

Matched Problem 5 Refer to Example 5. Find the steel production if the company uses 1,000 units of labor and 2,000 units of capital.

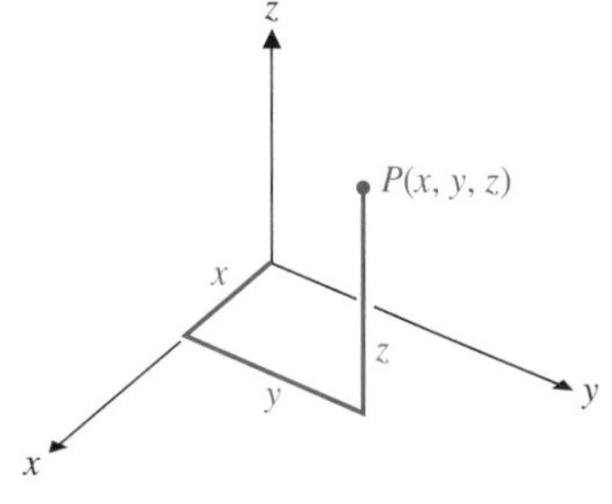

FIGURE 1
Rectangular coordinate system

Three-Dimensional Coordinate Systems

We now take a brief look at some graphs of functions of two independent variables. Since functions of the form $z = f(x, y)$ involve two independent variables, x and y, and one dependent variable, z, we need a *three-dimensional coordinate system* for their graphs. A **three-dimensional coordinate system** is formed by three mutually perpendicular number lines intersecting at their origins (see Fig. 1). In such a system, every ordered **triplet of numbers (x, y, z)** can be associated with a unique point, and conversely.

Example 6 **Three-Dimensional Coordinates** Locate $(-3, 5, 2)$ in a rectangular coordinate system.

Solution

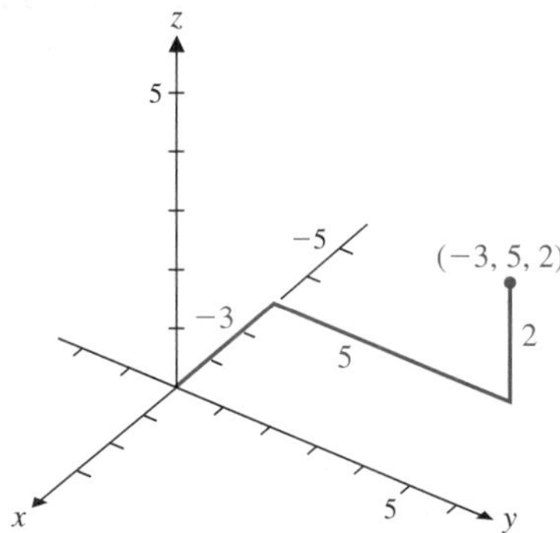

Matched Problem 6 Find the coordinates of the corners A, C, G, and D of the rectangular box shown in the figure.

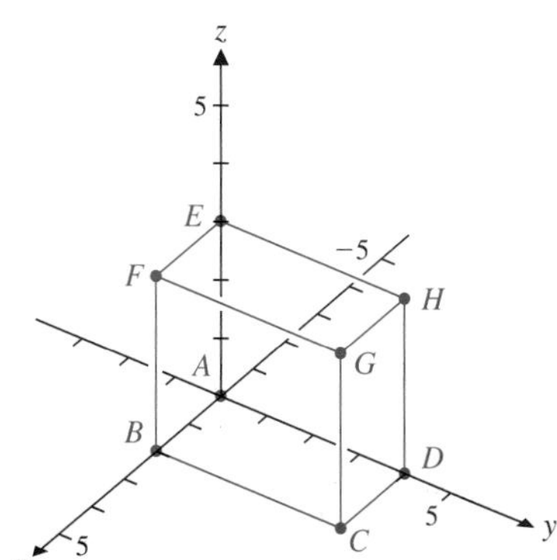

Explore–Discuss 1

Imagine that you are facing the front of a classroom whose rectangular walls meet at right angles. Suppose that the point of intersection of the floor, front wall, and left side wall is the origin of a three-dimensional coordinate system in which every point in the room has nonnegative coordinates. Then the plane $z = 0$ (or, equivalently, the xy plane) can be described as "the floor," and the plane $z = 2$ can be described as "the plane parallel to, but 2 units above, the floor." Give similar descriptions of the following planes:

(A) $x = 0$ (B) $x = 3$ (C) $y = 0$
(D) $y = 4$ (E) $x = -1$

What does the graph of $z = x^2 + y^2$ look like? If we let $x = 0$ and graph $z = 0^2 + y^2 = y^2$ in the yz plane, we obtain a parabola; if we let $y = 0$ and graph $z = x^2 + 0^2 = x^2$ in the xz plane, we obtain another parabola. It can be shown that the graph of $z = x^2 + y^2$ is either one of these parabolas rotated around the z axis (see Fig. 2). This cup-shaped figure is a *surface* and is called a **paraboloid.**

In general, the graph of any function of the form $z = f(x, y)$ is called a **surface.** The graph of such a function is the graph of all ordered triplets of numbers (x, y, z) that satisfy the equation. Graphing functions of two independent variables is often a very difficult task, and the general process will not be dealt with in this book. We present only a few simple graphs to suggest extensions of earlier geometric interpretations of the derivative and local maxima and minima to functions of two variables. Note that $z = f(x, y) = x^2 + y^2$ appears (see Fig. 2) to have a local minimum at $(x, y) = (0, 0)$. Figure 3 shows a local maximum at $(x, y) = (0, 0)$.

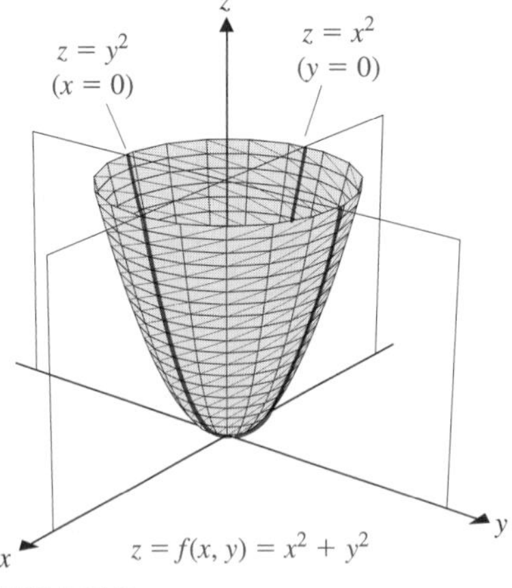

FIGURE 2
Paraboloid

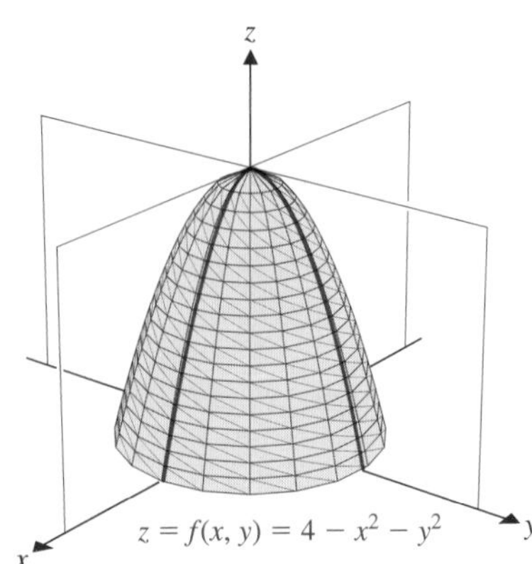

FIGURE 3
Local maximum: $f(0, 0) = 4$

Figure 4 shows a point at $(x, y) = (0, 0)$, called a **saddle point,** which is neither a local minimum nor a local maximum. Note that if $x = 0$, then the saddle point is a local minimum, and if $y = 0$, then the saddle point is a local maximum. More will be said about local maxima and minima in Section 13-3.

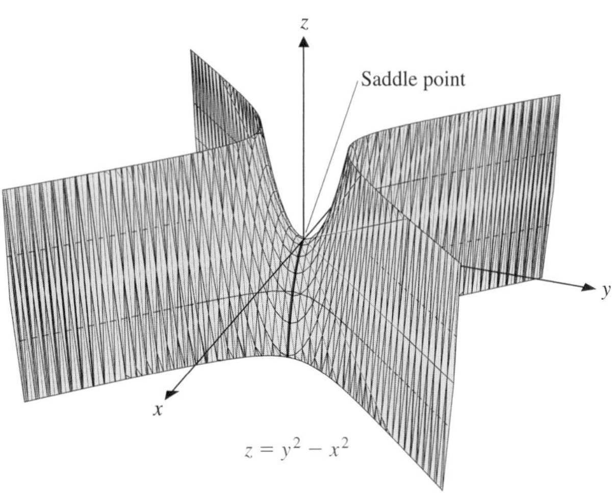

FIGURE 4
Saddle point at $(0, 0, 0)$

Explore–Discuss 2

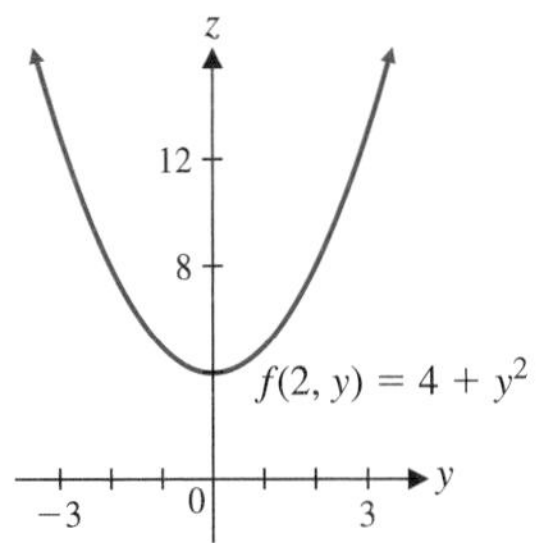

FIGURE 5

(A) Let $f(x, y) = x^2 + y^2$. The cross-section of the surface $z = f(x, y)$ by the plane $x = 2$ is the graph of $z = f(2, y) = 4 + y^2$, which is a parabola (see Fig. 5). Explain why each cross-section of $z = f(x, y)$ by a plane parallel to the yz plane is a parabola that opens upward. Explain why each cross-section of $z = f(x, y)$ by a plane parallel to the xz plane is a parabola that opens upward.

(B) Let $g(x, y) = y^2 - x^2$. Explain why each cross-section of $z = g(x, y)$ by a plane parallel to the yz plane is a parabola that opens upward (see Fig. 4). Explain why each cross-section of $z = f(x, y)$ by a plane parallel to the xz plane is a parabola that opens downward.

Some graphing utilities are designed to draw graphs (like those of Figs. 2, 3, and 4) of functions of two independent variables. Others, like the graphing calculator used for the displays in this text, are designed to draw graphs of functions of just a single independent variable. When using the latter type of calculator, we can graph cross-sections by planes parallel to the xz plane or yz plane to gain insight into the graph of a function of two independent variables.

Example 7 **Graphing Cross-Sections**

(A) Describe the cross-sections of $f(x, y) = 2x^2 + y^2$ in the planes $y = 0$, $y = 1, y = 2, y = 3$, and $y = 4$.

(B) Describe the cross-sections of $f(x, y) = 2x^2 + y^2$ in the planes $x = 0$, $x = 1, x = 2, x = 3$, and $x = 4$.

SOLUTION (A) The cross-section of $f(x, y) = 2x^2 + y^2$ by the plane $y = 0$ is the graph of the function $f(x, 0) = 2x^2$ in this plane. We can examine the shape of this cross-section by graphing $y_1 = 2x^2$ on a graphing utility (Fig. 6). Similarly, the graphs of $y_2 = f(x, 1) = 2x^2 + 1$, $y_3 = f(x, 2) = 2x^2 + 4$, $y_4 = f(x, 3) = 2x^2 + 9$, and $y_5 = f(x, 4) = 2x^2 + 16$ show the shapes of the other four cross-sections (see Fig. 6). Each of these is a parabola that opens upward. Note the correspondence between the graphs in Figure 6 and the actual cross-sections of $f(x, y) = 2x^2 + y^2$ shown in Figure 7.

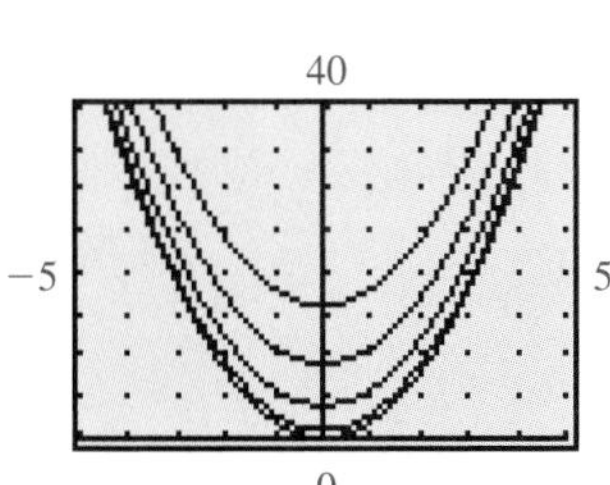

FIGURE 6
$y_1 = 2x^2$
$y_2 = 2x^2 + 1$
$y_3 = 2x^2 + 4$
$y_4 = 2x^2 + 9$
$y_5 = 2x^2 + 16$

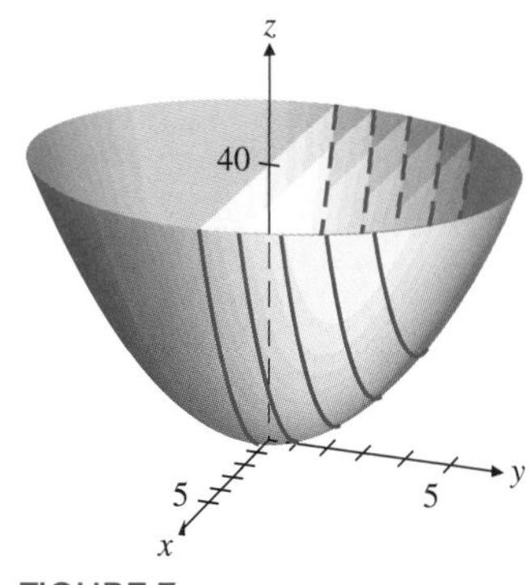

FIGURE 7

40
−5
5
0

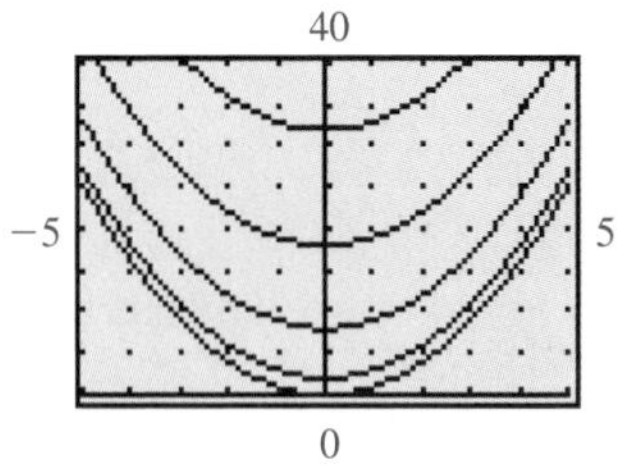

FIGURE 8
$y_1 = x^2$
$y_2 = 2 + x^2$
$y_3 = 8 + x^2$
$y_4 = 18 + x^2$
$y_5 = 32 + x^2$

(B) The five cross-sections are represented by the graphs of the functions $f(0, y) = y^2, f(1, y) = 2 + y^2, f(2, y) = 8 + y^2, f(3, y) = 18 + y^2$, and $f(4, y) = 32 + y^2$. These five functions are graphed in Figure 8. (Note that changing the name of the independent variable from y to x for graphing purposes does not affect the graph displayed.) Each of the five cross-sections is a parabola that opens upward.

Matched Problem 7

(A) Describe the cross-sections of $g(x, y) = y^2 - x^2$ in the planes $y = 0, y = 1, y = 2, y = 3$, and $y = 4$.

(B) Describe the cross-sections of $g(x, y) = y^2 - x^2$ in the planes $x = 0, x = 1, x = 2, x = 3$, and $x = 4$.

Answers to Matched Problems

1. \$3,100 **2.** 30

3. (A) $R(x, y) = 220x + 300y - 6x^2 + 4xy - 10y^2$; $R(20, 10) = \$4{,}800$
(B) $P(x, y) = 180x + 220y - 6x^2 + 4xy - 10y^2 - 1{,}000$; $P(20, 10) = \$2{,}200$

4. $V(x, y) = x^2y$; $V(5, 10) = 250$ in.^{3} **5.** 17,411 units

6. $A(0, 0, 0)$; $C(2, 4, 0)$; $G(2, 4, 3)$; $D(0, 4, 0)$

7. (A) Each cross-section is a parabola that opens downward.
(B) Each cross-section is a parabola that opens upward.

EXERCISE 13-1

A *For the functions*

$$f(x, y) = 10 + 2x - 3y \quad \textit{and} \quad g(x, y) = x^2 - 3y^2$$

find the indicated values in Problems 1–10.

1. $f(0, 0)$ **2.** $f(2, 1)$ **3.** $f(-3, 1)$

4. $f(2, -7)$ **5.** $g(0, 0)$ **6.** $g(0, -1)$

7. $g(2, -1)$ **8.** $g(-1, 2)$ **9.** $g(3, 4) - 6f(3, 4)$

10. $f(-2, 5) + 9g(-2, 5)$

B *Find the indicated value of the given function in Problems 11–24.*

11. $A(2, 3)$ for $A(x, y) = xy$

12. $V(2, 4, 3)$ for $V(x, y, z) = xyz$

13. $Q(12, 8)$ for $Q(M, C) = \dfrac{M}{C}(100)$

14. $T(50, 17)$ for $T(V, x) = \dfrac{33V}{x + 33}$

15. $V(2, 4)$ for $V(r, h) = \pi r^2 h$

16. $S(4, 2)$ for $S(x, y) = 5x^2y^3$

17. $R(1, 2)$ for $R(x, y) = -5x^2 + 6xy - 4y^2 + 200x + 300y$

18. $P(2, 2)$ for $P(x, y) = -x^2 + 2xy - 2y^2 - 4x + 12y + 5$

19. $R(6, 0.5)$ for $R(L, r) = 0.002\dfrac{L}{r^4}$

20. $L(2{,}000, 50)$ for $L(w, v) = (1.25 \times 10^{-5})wv^2$

21. $A(100, 0.06, 3)$ for $A(P, r, t) = P + Prt$

22. $A(10, 0.04, 3, 2)$ for $A(P, r, t, n) = P\left(1 + \dfrac{r}{n}\right)^{tn}$

23. $A(100, 0.08, 10)$ for $A(P, r, t) = Pe^{rt}$

24. $A(1{,}000, 0.06, 8)$ for $A(P, r, t) = Pe^{rt}$

25. Let $F(x, y) = x^2 + e^x y - y^2$. Find all values of x such that $F(x, 2) = 0$. **C**

26. Let $G(a, b, c) = a^3 + b^3 + c^3 - (ab + ac + bc) - 6$. Find all values of b such that $G(2, b, 1) = 0$. **C**

C

27. For the function $f(x, y) = x^2 + 2y^2$, find:

$$\frac{f(x + h, y) - f(x, y)}{h}$$

28. For the function $f(x, y) = x^2 + 2y^2$, find:

$$\frac{f(x, y + k) - f(x, y)}{k}$$

29. For the function $f(x, y) = 2xy^2$, find:

$$\frac{f(x + h, y) - f(x, y)}{h}$$

30. For the function $f(x, y) = 2xy^2$, find:

$$\frac{f(x, y + k) - f(x, y)}{k}$$

31. Find the coordinates of E and F in the figure for Matched Problem 6 (in the text).

32. Find the coordinates of B and H in the figure for Matched Problem 6 (in the text).

In Problems 33–38, use a graphing utility as necessary to explore the graphs of the indicated cross-sections.

33. Let $f(x, y) = x^2$.
(A) Explain why the cross-sections of the surface $z = f(x, y)$ by planes parallel to $y = 0$ are parabolas.
(B) Describe the cross-sections of the surface by the planes $x = 0, x = 1$, and $x = 2$.
(C) Describe the surface $z = f(x, y)$.

34. Let $f(x, y) = \sqrt{4 - y^2}$.
(A) Explain why the cross-sections of the surface $z = f(x, y)$ by planes parallel to $x = 0$ are semicircles of radius 2.
(B) Describe the cross-sections of the surface by the planes $y = 0, y = 2$, and $y = 3$.
(C) Describe the surface $z = f(x, y)$.

35. Let $f(x, y) = \sqrt{36 - x^2 - y^2}$.
(A) Describe the cross-sections of the surface $z = f(x, y)$ by the planes $y = 1, y = 2, y = 3, y = 4$, and $y = 5$.
(B) Describe the cross-sections of the surface by the planes $x = 0, x = 1, x = 2, x = 3, x = 4$, and $x = 5$.
(C) Describe the surface $z = f(x, y)$.

36. Let $f(x, y) = 100 + 10x + 25y - x^2 - 5y^2$.
(A) Describe the cross-sections of the surface $z = f(x, y)$ by the planes $y = 0, y = 1, y = 2$, and $y = 3$.
(B) Describe the cross-sections of the surface by the planes $x = 0, x = 1, x = 2$, and $x = 3$.
(C) Describe the surface $z = f(x, y)$.

37. Let $f(x, y) = e^{-(x^2+y^2)}$.
(A) Explain why $f(a, b) = f(c, d)$ whenever (a, b) and (c, d) are points on the same circle centered at the origin in the xy plane.
(B) Describe the cross-sections of the surface $z = f(x, y)$ by the planes $x = 0, y = 0$, and $x = y$.
(C) Describe the surface $z = f(x, y)$.

38. Let $f(x, y) = 4 - \sqrt{x^2 + y^2}$.
(A) Explain why $f(a, b) = f(c, d)$ whenever (a, b) and (c, d) are points on the same circle with center at the origin in the xy plane.
(B) Describe the cross-sections of the surface $z = f(x, y)$ by the planes $x = 0, y = 0$, and $x = y$.
(C) Describe the surface $z = f(x, y)$.

APPLICATIONS

Business & Economics

39. *Cost function.* A small manufacturing company produces two models of a surfboard: a standard model and a competition model. If the standard model is produced at a variable cost of \$70 each, the competition model at a variable cost of \$100 each, and the total fixed costs per month are \$2,000, then the monthly cost function is given by

$$C(x, y) = 2{,}000 + 70x + 100y$$

where x and y are the numbers of standard and competition models produced per month, respectively. Find $C(20, 10)$, $C(50, 5)$, and $C(30, 30)$.

40. *Advertising and sales.* A company spends \$$x$ thousand per week on newspaper advertising and \$$y$ thousand per week on television advertising. Its weekly sales are found to be given by

$$S(x, y) = 5x^2y^3$$

Find $S(3, 2)$ and $S(2, 3)$.

41. *Revenue function.* A supermarket sells two brands of coffee: brand A at \$$p$ per pound and brand B at \$$q$ per pound. The daily demand equations for brands A and B are, respectively,

$$x = 200 - 5p + 4q$$
$$y = 300 + 2p - 4q$$

(both in pounds). Find the daily revenue function $R(p, q)$. Evaluate $R(2, 3)$ and $R(3, 2)$.

42. *Revenue, cost, and profit functions.* A company manufactures ten-speed and three-speed bicycles. The weekly demand and cost equations are

$$p = 230 - 9x + y$$
$$q = 130 + x - 4y$$
$$C(x, y) = 200 + 80x + 30y$$

where \$$p$ is the price of a ten-speed bicycle, \$$q$ is the price of a three-speed bicycle, x is the weekly demand for ten-speed bicycles, y is the weekly demand for three-speed bicycles, and $C(x, y)$ is the cost function. Find the weekly revenue function $R(x, y)$ and the weekly profit function $P(x, y)$. Evaluate $R(10, 15)$ and $P(10, 15)$.

43. *Productivity.* The Cobb–Douglas production function for a petroleum company is given by

$$f(x, y) = 20x^{0.4}y^{0.6}$$

where x is the utilization of labor and y is the utilization of capital. If the company uses 1,250 units of labor and 1,700 units of capital, how many units of petroleum will be produced?

44. *Productivity.* The petroleum company in Problem 43 is taken over by another company that decides to double both the units of labor and the units of capital utilized in the production of petroleum. Use the Cobb–Douglas production function given in Problem 43 to find the amount of petroleum that will be produced by this increased utilization of labor and capital. What is the effect on productivity of doubling both the units of labor and the units of capital?

45. *Future value.* At the end of each year, \$2,000 is invested into an IRA earning 9% compounded annually.
(A) How much will be in the account at the end of 30 years? Use the annuity formula

$$F(P, i, n) = P\frac{(1 + i)^n - 1}{i}$$

where

P = Periodic payment
i = Rate per period
n = Number of payments (periods)
$F = FV$ = Future value

C (B) Use graphical approximation methods to determine the rate of interest that would produce \$500,000 in the account at the end of 30 years.

46. *Package design.* The packaging department in a company has been asked to design a rectangular box with no top and a partition down the middle (see the figure). Let x, y, and z be the dimensions (in inches).

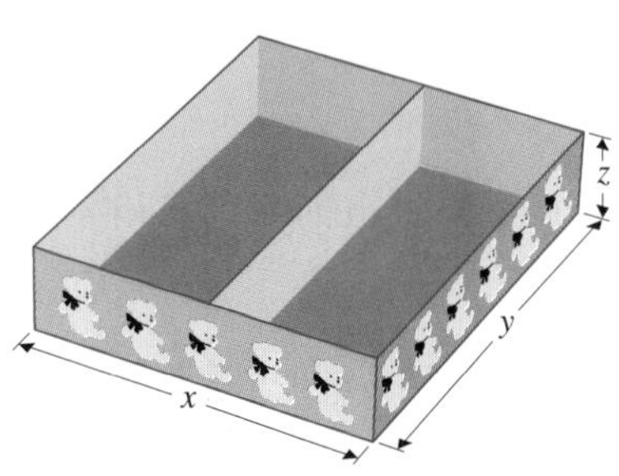

Figure for 46

(A) Find the total amount of material $M(x, y, z)$ used in constructing one of these boxes, and evaluate $M(10, 12, 6)$.

C (B) Suppose the box is to have a square base and a volume of 720 cubic inches. Use graphical approximation methods to determine the dimensions that require the least amount of material.

Life Sciences

47. *Marine biology.* In using scuba diving gear, a marine biologist estimates the time of a dive according to the equation

$$T(V, x) = \frac{33V}{x + 33}$$

where

T = Time of dive in minutes
V = Volume of air, at sea level pressure, compressed into tanks
x = Depth of dive in feet

Find $T(70, 47)$ and $T(60, 27)$.

48. *Blood flow.* Poiseuille's law states that the resistance, R, for blood flowing in a blood vessel varies directly as the length of the vessel, L, and inversely as the fourth power of its radius, r. Stated as an equation,

$$R(L, r) = k\frac{L}{r^4} \qquad k \text{ a constant}$$

Find $R(8, 1)$ and $R(4, 0.2)$.

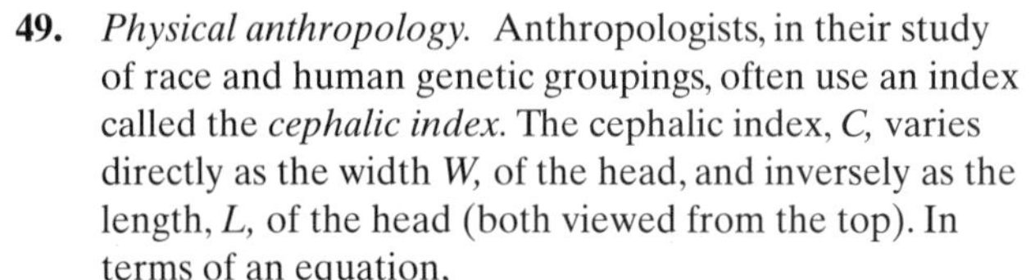

49. *Physical anthropology.* Anthropologists, in their study of race and human genetic groupings, often use an index called the *cephalic index.* The cephalic index, C, varies directly as the width W, of the head, and inversely as the length, L, of the head (both viewed from the top). In terms of an equation,

$$C(W, L) = 100\frac{W}{L}$$

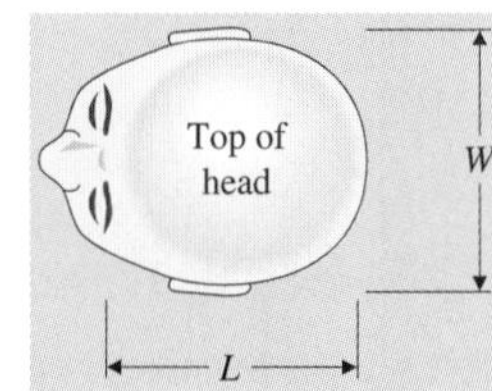

where

W = Width in inches
L = Length in inches

Find $C(6, 8)$ and $C(8.1, 9)$.

Social Sciences

50. *Safety research.* Under ideal conditions, if a person driving a car slams on the brakes and skids to a stop, the length of the skid marks (in feet) is given by the formula

$$L(w, v) = kwv^2$$

where

k = Constant
w = Weight of car in pounds
v = Speed of car in miles per hour

For $k = 0.000\ 013\ 3$, find $L(2{,}000, 40)$ and $L(3{,}000, 60)$.

51. *Psychology.* The intelligence quotient (IQ) is defined to be the ratio of mental age (MA), as determined by certain tests, and chronological age (CA), multiplied by 100. Stated as an equation,

$$Q(M, C) = \frac{M}{C} \cdot 100$$

where

Q = IQ $\qquad M$ = MA $\qquad C$ = CA

Find $Q(12, 10)$ and $Q(10, 12)$.

Section 13-2 Partial Derivatives

- Partial Derivatives
- Second-Order Partial Derivatives

Partial Derivatives

We know how to differentiate many kinds of functions of one independent variable and how to interpret the results. What about functions with two or more independent variables? Let us return to the surfboard example considered at the beginning of the chapter.

For the company producing only the standard board, the cost function was

$$C(x) = 500 + 70x$$

Differentiating with respect to x, we obtain the marginal cost function

$$C'(x) = 70$$

Since the marginal cost is constant, \$70 is the change in cost for a 1 unit increase in production at any output level.

For the company producing two types of boards, a standard model and a competition model, the cost function was

$$C(x, y) = 700 + 70x + 100y$$

Now suppose we differentiate with respect to x, holding y fixed, and denote this by $C_x(x, y)$; or suppose we differentiate with respect to y, holding x fixed, and denote this by $C_y(x, y)$. Differentiating in this way, we obtain

$$C_x(x, y) = 70 \qquad C_y(x, y) = 100$$

Each of these is called a **partial derivative,** and, in this example, each represents marginal cost. The first is the change in cost due to a 1 unit increase in production of the standard board with the production of the competition model held fixed. The second is the change in cost due to a 1 unit increase in production of the competition board with the production of the standard board held fixed.

In general, if $z = f(x, y)$, then the **partial derivative of f with respect to x,** denoted by $\partial z/\partial x$, f_x, or $f_x(x, y)$, is defined by

$$\frac{\partial z}{\partial x} = \lim_{h \to 0} \frac{f(x + h, y) - f(x, y)}{h}$$

provided the limit exists. We recognize this as the ordinary derivative of f with respect to x, holding y constant. Thus, we are able to continue to use all the derivative rules and properties discussed in Chapters 8–10 for partial derivatives.

Similarly, the **partial derivative of f with respect to y,** denoted by $\partial z/\partial y$, f_y, or $f_y(x, y)$, is defined by

$$\frac{\partial z}{\partial y} = \lim_{k \to 0} \frac{f(x, y + k) - f(x, y)}{k}$$

which is the ordinary derivative with respect to y, holding x constant.

Parallel definitions and interpretations hold for functions with three or more independent variables.

Example 1 ➡ **Partial Derivatives** For $z = f(x, y) = 2x^2 - 3x^2y + 5y + 1$, find:

(A) $\partial z/\partial x$ (B) $f_x(2, 3)$

SOLUTION (A) $z = 2x^2 - 3x^2y + 5y + 1$

Differentiating with respect to x, holding y constant (that is, treating y as a constant), we obtain

$$\frac{\partial z}{\partial x} = 4x - 6xy$$

(B) $$f(x, y) = 2x^2 - 3x^2y + 5y + 1$$

First, differentiate with respect to x. From part (A) we have

$$f_x(x, y) = 4x - 6xy$$

Then evaluate at (2, 3):

$$f_x(2, 3) = 4(2) - 6(2)(3) = -28$$

In Example 1B, an alternative approach would be to substitute $y = 3$ into $f(x, y)$ and graph the function $f(x, 3) = -7x^2 + 16$, which represents the cross-section of the surface $z = f(x, y)$ by the plane $y = 3$. Then determine the slope of the tangent line when $x = 2$. Again, we conclude that $f_x(2, 3) = -28$ (see Fig. 1).

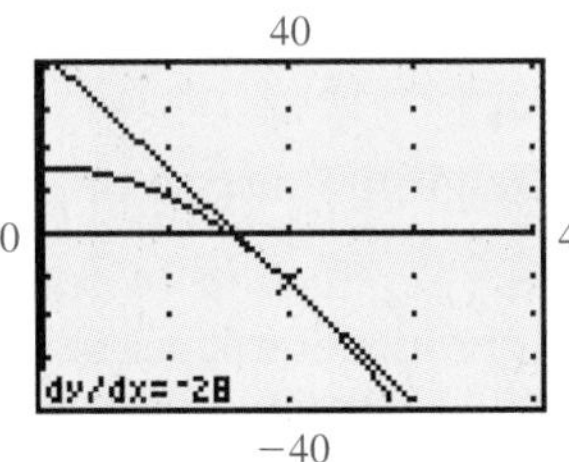

FIGURE 1
$y_1 = -7x^2 + 16$

Matched Problem 1 ➡ For f in Example 1, find:

(A) $\partial z/\partial y$ (B) $f_y(2, 3)$

Example 2 ➡ **Partial Derivatives Using the Chain Rule** For $z = f(x, y) = e^{x^2+y^2}$, find:

(A) $\partial z/\partial x$ (B) $f_y(2, 1)$

SOLUTION (A) Using the chain rule [thinking of $z = e^u$, $u = u(x)$; y is held constant], we obtain

$$\frac{\partial z}{\partial x} = e^{x^2+y^2}\frac{\partial(x^2 + y^2)}{\partial x} = 2xe^{x^2+y^2}$$

(B) $$f_y(x, y) = e^{x^2+y^2}\frac{\partial(x^2 + y^2)}{\partial y} = 2ye^{x^2+y^2}$$

$$f_y(2, 1) = 2(1)e^{(2)^2+(1)^2} = 2e^5$$

Matched Problem 2 ➡ For $z = f(x, y) = (x^2 + 2xy)^5$, find:

(A) $\partial z/\partial y$ (B) $f_x(1, 0)$

Example 3 ➡ **Profit** The profit function for the surfboard company in Example 3 in Section 13-1 was

$$P(x, y) = 140x + 200y - 4x^2 + 2xy - 12y^2 - 700$$

Find $P_x(15, 10)$ and $P_x(30, 10)$, and interpret the results.

SOLUTION

$$P_x(x, y) = 140 - 8x + 2y$$
$$P_x(15, 10) = 140 - 8(15) + 2(10) = 40$$
$$P_x(30, 10) = 140 - 8(30) + 2(10) = -80$$

At a production level of 15 standard and 10 competition boards per week, increasing the production of standard boards by 1 unit and holding the production of competition boards fixed at 10 will increase profit by approximately \$40. At a production level of 30 standard and 10 competition boards per week, increasing the production of standard boards by 1 unit and holding the production of competition boards fixed at 10 will decrease profit by approximately \$80.

Matched Problem 3 For the profit function in Example 3, find $P_y(25, 10)$ and $P_y(25, 15)$, and interpret the results.

Explore–Discuss 1

Let $P(x, y)$ be the profit function of Example 3 and Matched Problem 3.

(A) Assume the production of competition boards remains fixed at 10. Which production level of standard boards will yield a maximum profit? Calculate that maximum profit.

(B) Assume the production of standard boards remains fixed at 25. Which production level of competition boards will yield a maximum profit? Calculate that maximum profit.

Example 4 **Productivity** The productivity of a major computer manufacturer is given approximately by the Cobb–Douglas production function

$$f(x, y) = 15x^{0.4}y^{0.6}$$

with the utilization of x units of labor and y units of capital. The partial derivative $f_x(x, y)$ represents the rate of change of productivity with respect to labor and is called the **marginal productivity of labor.** The partial derivative $f_y(x, y)$ represents the rate of change of productivity with respect to capital and is called the **marginal productivity of capital.** If the company is currently utilizing 4,000 units of labor and 2,500 units of capital, find the marginal productivity of labor and the marginal productivity of capital. For the greatest increase in productivity, should the management of the company encourage increased use of labor or increased use of capital?

SOLUTION

$$f_x(x, y) = 6x^{-0.6}y^{0.6}$$
$$f_x(4{,}000, 2{,}500) = 6(4{,}000)^{-0.6}(2{,}500)^{0.6}$$
$$\approx 4.53 \qquad \text{Marginal productivity of labor}$$

$$f_y(x, y) = 9x^{0.4}y^{-0.4}$$
$$f_y(4{,}000, 2{,}500) = 9(4{,}000)^{0.4}(2{,}500)^{-0.4}$$
$$\approx 10.86 \qquad \text{Marginal productivity of capital}$$

At the current level of utilization of 4,000 units of labor and 2,500 units of capital, each 1 unit increase in labor utilization (keeping capital utilization fixed at 2,500 units) will increase production by approximately 4.53 units, and each 1 unit increase in capital utilization (keeping labor utilization fixed at 4,000 units) will increase production by approximately 10.86 units. Thus, the management of the company should encourage increased use of capital.

Matched Problem 4 The productivity of an airplane manufacturing company is given approximately by the Cobb–Douglas production function

$$f(x, y) = 40x^{0.3}y^{0.7}$$

(A) Find $f_x(x, y)$ and $f_y(x, y)$.
(B) If the company is currently using 1,500 units of labor and 4,500 units of capital, find the marginal productivity of labor and the marginal productivity of capital.
(C) For the greatest increase in productivity, should the management of the company encourage increased use of labor or increased use of capital?

Partial derivatives have simple geometric interpretations, as indicated in Figure 2. If we hold x fixed, say, $x = a$, then $f_y(a, y)$ is the slope of the curve obtained by intersecting the plane $x = a$ with the surface $z = f(x, y)$. A similar interpretation is given to $f_x(x, b)$.

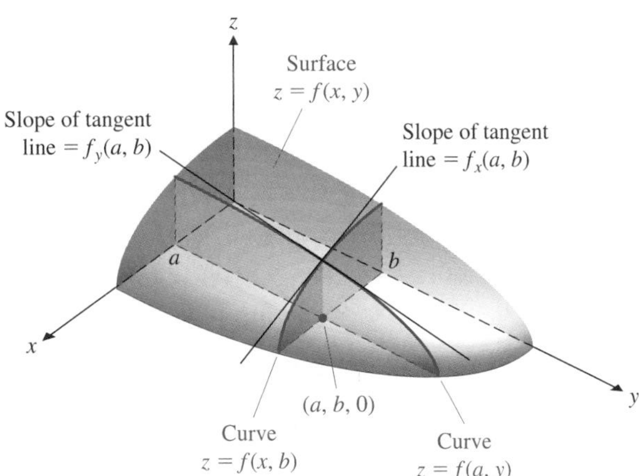

FIGURE 2

Explore–Discuss 2

Let $f(x, y) = x^2y - 2xy^2 + 3$ and consider the surface $z = f(x, y)$.

(A) The cross-section of the surface by the plane $x = 2$ is the graph of $f(2, y) = 4y - 4y^2 + 3$, which is a parabola. Use an ordinary derivative to find the slope of the tangent line to this parabola when $y = \frac{1}{2}$.
(B) Use partial derivatives to confirm your answer to part (A).
(C) Explain why the cross-section of the surface by the plane $y = 1$ is also a parabola. Use an ordinary derivative to find the slope of the tangent line to this parabola when $x = 3$.
(D) Use partial derivatives to confirm your answer to part (C).

Second-Order Partial Derivatives

The function

$$z = f(x, y) = x^4y^7$$

has two **first-order partial derivatives,**

$$\frac{\partial z}{\partial x} = f_x = f_x(x, y) = 4x^3y^7 \quad \text{and} \quad \frac{\partial z}{\partial y} = f_y = f_y(x, y) = 7x^4y^6$$

Each of these partial derivatives, in turn, has two partial derivatives called **second-order partial derivatives** of $z = f(x, y)$. Generalizing the various notations we have for first-order partial derivatives, the four second-order partial derivatives of $z = f(x, y) = x^4y^7$ are written as

Equivalent notations

$$f_{xx} = f_{xx}(x, y) = \frac{\partial^2 z}{\partial x^2} = \frac{\partial}{\partial x}\left(\frac{\partial z}{\partial x}\right) = \frac{\partial}{\partial x}(4x^3y^7) = 12x^2y^7$$

$$f_{xy} = f_{xy}(x, y) = \frac{\partial^2 z}{\partial y\,\partial x} = \frac{\partial}{\partial y}\left(\frac{\partial z}{\partial x}\right) = \frac{\partial}{\partial y}(4x^3y^7) = 28x^3y^6$$

$$f_{yx} = f_{yx}(x, y) = \frac{\partial^2 z}{\partial x\,\partial y} = \frac{\partial}{\partial x}\left(\frac{\partial z}{\partial y}\right) = \frac{\partial}{\partial x}(7x^4y^6) = 28x^3y^6$$

$$f_{yy} = f_{yy}(x, y) = \frac{\partial^2 z}{\partial y^2} = \frac{\partial}{\partial y}\left(\frac{\partial z}{\partial y}\right) = \frac{\partial}{\partial y}(7x^4y^6) = 42x^4y^5$$

In the mixed partial derivative $\partial^2 z/\partial y\,\partial x = f_{xy}$, we started with $z = f(x, y)$ and first differentiated with respect to x (holding y constant). Then we differentiated with respect to y (holding x constant). In the other mixed partial derivative, $\partial^2 z/\partial x\,\partial y = f_{yx}$, the order of differentiation was reversed; however, the final result was the same—that is, $f_{xy} = f_{yx}$. Although it is possible to find functions for which $f_{xy} \neq f_{yx}$, such functions rarely occur in applications involving partial derivatives. Thus, for all the functions in this text, we will assume that $f_{xy} = f_{yx}$.

In general, we have the following definitions:

Second-Order Partial Derivatives

If $z = f(x, y)$, then

$$f_{xx} = f_{xx}(x, y) = \frac{\partial^2 z}{\partial x^2} = \frac{\partial}{\partial x}\left(\frac{\partial z}{\partial x}\right)$$

$$f_{xy} = f_{xy}(x, y) = \frac{\partial^2 z}{\partial y\,\partial x} = \frac{\partial}{\partial y}\left(\frac{\partial z}{\partial x}\right)$$

$$f_{yx} = f_{yx}(x, y) = \frac{\partial^2 z}{\partial x\,\partial y} = \frac{\partial}{\partial x}\left(\frac{\partial z}{\partial y}\right)$$

$$f_{yy} = f_{yy}(x, y) = \frac{\partial^2 z}{\partial y^2} = \frac{\partial}{\partial y}\left(\frac{\partial z}{\partial y}\right)$$

Example 5 **Second-Order Partial Derivatives** For $z = f(x, y) = 3x^2 - 2xy^3 + 1$, find:

(A) $\dfrac{\partial^2 z}{\partial x\,\partial y}, \dfrac{\partial^2 z}{\partial y\,\partial x}$ (B) $\dfrac{\partial^2 z}{\partial x^2}$ (C) $f_{yx}(2, 1)$

SOLUTION (A) First differentiate with respect to y and then with respect to x:

$$\frac{\partial z}{\partial y} = -6xy^2 \qquad \frac{\partial^2 z}{\partial x\,\partial y} = \frac{\partial}{\partial x}\left(\frac{\partial z}{\partial y}\right) = \frac{\partial}{\partial x}(-6xy^2) = -6y^2$$

First differentiate with respect to x and then with respect to y:

$$\frac{\partial z}{\partial x} = 6x - 2y^3 \qquad \frac{\partial^2 z}{\partial y\,\partial x} = \frac{\partial}{\partial y}\left(\frac{\partial z}{\partial x}\right) = \frac{\partial}{\partial y}(6x - 2y^3) = -6y^2$$

(B) Differentiate with respect to x twice:

$$\frac{\partial z}{\partial x} = 6x - 2y^3 \qquad \frac{\partial^2 z}{\partial x^2} = \frac{\partial}{\partial x}\left(\frac{\partial z}{\partial x}\right) = 6$$

(C) First find $f_{yx}(x, y)$; then evaluate at $(2, 1)$. Again, remember that f_{yx} means to differentiate with respect to y first and then with respect to x. Thus,

$$f_y(x, y) = -6xy^2 \qquad f_{yx}(x, y) = -6y^2$$

and

$$f_{yx}(2, 1) = -6(1)^2 = -6$$

Matched Problem 5 For $z = f(x, y) = x^3y - 2y^4 + 3$, find:

(A) $\dfrac{\partial^2 z}{\partial y\,\partial x}$ (B) $\dfrac{\partial^2 z}{\partial y^2}$

(C) $f_{xy}(2, 3)$ (D) $f_{yx}(2, 3)$

Answers to Matched Problems

1. (A) $\partial z/\partial y = -3x^2 + 5$ (B) $f_y(2, 3) = -7$

2. (A) $10x(x^2 + 2xy)^4$ (B) 10

3. $P_y(25, 10) = 10$: at a production level of $x = 25$ and $y = 10$, increasing y by 1 unit and holding x fixed at 25 will increase profit by approx. \$10; $P_y(25, 15) = -110$: at a production level of $x = 25$ and $y = 15$, increasing y by 1 unit and holding x fixed at 25 will decrease profit by approx. \$110

4. (A) $f_x(x, y) = 12x^{-0.7}y^{0.7}$; $f_y(x, y) = 28x^{0.3}y^{-0.3}$
(B) Marginal productivity of labor ≈ 25.89;
Marginal productivity of capital ≈ 20.14
(C) Labor

5. (A) $3x^2$ (B) $-24y^2$
(C) 12 (D) 12

EXERCISE 13-2

A *For $z = f(x, y) = 10 + 3x + 2y$, find each of the following:*

1. $\partial z/\partial x$

2. $\partial z/\partial y$

3. $f_y(1, 2)$

4. $f_x(1, 2)$

For $z = f(x, y) = 3x^2 - 2xy^2 + 1$, find each of the following:

5. $\partial z/\partial y$

6. $\partial z/\partial x$

7. $\dfrac{\partial^2 z}{\partial y^2}$

8. $\dfrac{\partial^2 z}{\partial x^2}$

9. $f_x(2, 3)$

10. $f_y(2, 3)$

For $S(x, y) = 5x^2y^3$, find each of the following:

11. $S_x(x, y)$

12. $S_y(x,y)$

13. $S_y(2, 1)$

14. $S_x(2, 1)$

15. $S_{xy}(x, y)$

16. $S_{yx}(x, y)$

B *For $C(x, y) = x^2 - 2xy + 2y^2 + 6x - 9y + 5$, find each of the following:*

17. $C_x(x, y)$

18. $C_y(x, y)$

19. $C_x(2, 2)$

20. $C_y(2, 2)$

21. $C_{xy}(x, y)$

22. $C_{yx}(x,y)$

23. $C_{xx}(x, y)$

24. $C_{yy}(x, y)$

For $z = f(x, y) = e^{2x+3y}$, find each of the following:

25. $\dfrac{\partial z}{\partial x}$

26. $\dfrac{\partial z}{\partial y}$

27. $\dfrac{\partial^2 z}{\partial x\, \partial y}$

28. $\dfrac{\partial^2 z}{\partial y\, \partial x}$

29. $f_{xy}(1, 0)$

30. $f_{yx}(0, 1)$

31. $f_{xx}(0, 1)$

32. $f_{yy}(1, 0)$

In Problems 33–42, find $f_x(x, y)$ and $f_y(x, y)$ for each function f.

33. $f(x, y) = (x^2 - y^3)^3$

34. $f(x, y) = \sqrt{2x - y^2}$

35. $f(x, y) = (3x^2y - 1)^4$

36. $f(x, y) = (3 + 2xy^2)^3$

37. $f(x, y) = \ln(x^2 + y^2)$

38. $f(x, y) = \ln(2x - 3y)$

39. $f(x, y) = y^2e^{xy^2}$

40. $f(x, y) = x^3e^{x^2y}$

41. $f(x, y) = \dfrac{x^2 - y^2}{x^2 + y^2}$

42. $f(x, y) = \dfrac{2x^2y}{x^2 + y^2}$

43. (A) Let $f(x, y) = y^3 + 4y^2 - 5y + 3$. Show that $\partial f/\partial x = 0$.
(B) Explain why there are an infinite number of functions $g(x, y)$ such that $\partial g/\partial x = 0$.

44. (A) Find an example of a function $f(x, y)$ such that $\partial f/\partial x = 3$ and $\partial f/\partial y = 2$.
(B) How many such functions are there? Explain.

In Problems 45–50, find $f_{xx}(x, y)$, $f_{xy}(x, y)$, $f_{yx}(x, y)$, and $f_{yy}(x, y)$ for each function f.

45. $f(x, y) = x^2y^2 + x^3 + y$

46. $f(x, y) = x^3y^3 + x + y^2$

47. $f(x, y) = \dfrac{x}{y} - \dfrac{y}{x}$

48. $f(x, y) = \dfrac{x^2}{y} - \dfrac{y^2}{x}$

49. $f(x, y) = xe^{xy}$

50. $f(x, y) = x\ln(xy)$

C

51. For

$$P(x, y) = -x^2 + 2xy - 2y^2 - 4x + 12y - 5$$

find all values of x and y such that

$$P_x(x, y) = 0 \quad \text{and} \quad P_y(x, y) = 0$$

simultaneously.

52. For

$$C(x, y) = 2x^2 + 2xy + 3y^2 - 16x - 18y + 54$$

find all values of x and y such that

$$C_x(x, y) = 0 \quad \text{and} \quad C_y(x, y) = 0$$

simultaneously.

53. For C

$$F(x, y) = x^3 - 2x^2y^2 - 2x - 4y + 10$$

find all values of x and y such that

$$F_x(x, y) = 0 \quad \text{and} \quad F_y(x, y) = 0$$

simultaneously.

54. For 

$$G(x, y) = x^2 \ln y - 3x - 2y + 1$$

find all values of x and y such that

$$G_x(x, y) = 0 \quad \text{and} \quad G_y(x, y) = 0$$

simultaneously.

55. Let $f(x, y) = 3x^2 + y^2 - 4x - 6y + 2$.
(A) Find the minimum value of $f(x, y)$ when $y = 1$.
(B) Explain why the answer to part (A) is not the minimum value of the function $f(x, y)$.

56. Let $f(x, y) = 5 - 2x + 4y - 3x^2 - y^2$.
(A) Find the maximum value of $f(x, y)$ when $x = 2$.
(B) Explain why the answer to part (A) is not the maximum value of the function $f(x, y)$.

57. Let $f(x, y) = 4 - x^4y + 3xy^2 + y^5$.
C (A) Use graphical approximation methods to find c (to three decimal places) such that $f(c, 2)$ is the maximum value of $f(x, y)$ when $y = 2$.
(B) Find $f_x(c, 2)$ and $f_y(c, 2)$.

58. Let $f(x, y) = e^x + 2e^y + 3xy^2 + 1$.
C (A) Use graphical approximation methods to find d (to three decimal places) such that $f(1, d)$ is the minimum value of $f(x, y)$ when $x = 1$.
(B) Find $f_x(1, d)$ and $f_y(1, d)$.

In Problems 59 and 60, show that the function f satisfies $f_{xx}(x, y) + f_{yy}(x, y) = 0$.

59. $f(x, y) = \ln(x^2 + y^2)$ 60. $f(x, y) = x^3 - 3xy^2$

61. For $f(x, y) = x^2 + 2y^2$, find:
(A) $\lim_{h \to 0} \frac{f(x + h, y) - f(x, y)}{h}$
(B) $\lim_{k \to 0} \frac{f(x, y + k) - f(x, y)}{k}$

62. For $f(x, y) = 2xy^2$, find:
(A) $\lim_{h \to 0} \frac{f(x + h, y) - f(x, y)}{h}$
(B) $\lim_{k \to 0} \frac{f(x, y + k) - f(x, y)}{k}$

APPLICATIONS

Business & Economics

63. *Profit function.* A firm produces two types of calculators, x of type A and y of type B each week. The weekly revenue and cost functions (in dollars) are

$$R(x, y) = 80x + 90y + 0.04xy - 0.05x^2 - 0.05y^2$$
$$C(x, y) = 8x + 6y + 20{,}000$$

Find $P_x(1{,}200, 1{,}800)$ and $P_y(1{,}200, 1{,}800)$, and interpret the results.

64. *Advertising and sales.* A company spends \$$x$ per week on newspaper advertising and \$$y$ per week on television advertising. Its weekly sales were found to be given by

$$S(x, y) = 10x^{0.4}y^{0.8}$$

Find $S_x(3{,}000, 2{,}000)$ and $S_y(3{,}000, 2{,}000)$, and interpret the results.

65. *Demand equations.* A supermarket sells two brands of coffee, brand A at \$$p$ per pound and brand B at \$$q$ per pound. The daily demand equations for brands A and B are, respectively,

$$x = 200 - 5p + 4q$$
$$y = 300 + 2p - 4q$$

Find $\partial x/\partial p$ and $\partial y/\partial p$, and interpret the results.

66. *Revenue and profit functions.* A company manufactures ten-speed and three-speed bicycles. The weekly demand and cost functions are

$$p = 230 - 9x + y$$
$$q = 130 + x - 4y$$
$$C(x, y) = 200 + 80x + 30y$$

where \$$p$ is the price of a ten-speed bicycle, \$$q$ is the price of a three-speed bicycle, x is the weekly demand for ten-speed bicycles, y is the weekly demand for three-speed bicycles, and $C(x, y)$ is the cost function. Find $R_x(10, 5)$ and $P_x(10, 5)$, and interpret the results.

67. *Productivity.* The productivity of a certain third-world country is given approximately by the function

$$f(x, y) = 10x^{0.75}y^{0.25}$$

with the utilization of x units of labor and y units of capital.
(A) Find $f_x(x, y)$ and $f_y(x, y)$.
(B) If the country is now using 600 units of labor and 100 units of capital, find the marginal productivity of labor and the marginal productivity of capital.
(C) For the greatest increase in the country's productivity, should the government encourage increased use of labor or increased use of capital?

68. *Productivity.* The productivity of an automobile manufacturing company is given approximately by the function

$$f(x, y) = 50\sqrt{xy} = 50x^{0.5}y^{0.5}$$

with the utilization of x units of labor and y units of capital.
(A) Find $f_x(x, y)$ and $f_y(x, y)$.
(B) If the company is now using 250 units of labor and 125 units of capital, find the marginal productivity of labor and the marginal productivity of capital.
(C) For the greatest increase in the company's productivity, should the management encourage increased use of labor or increased use of capital?

Problems 69–72 refer to the following: If a decrease in demand for one product results in an increase in demand for another product, then the two products are said to be **competitive, or substitute, products.** *(Real whipping cream and imitation whipping cream are examples of competitive, or substitute, products.) If a decrease in demand for one product results in a decrease in demand for another product, then the two products are said to be* **complementary products.** *(Fishing boats and outboard motors are examples of complementary products.) Partial derivatives can be used to test whether two products are competitive, complementary, or neither. We start with demand functions for two products where the demand for either depends on the prices for both:*

$$x = f(p, q) \quad \text{Demand function for product A}$$
$$y = g(p, q) \quad \text{Demand function for product B}$$

The variables x and y represent the number of units demanded of products A and B, respectively, at a price p for 1 unit of product A and a price q for 1 unit of product B. Normally, if the price of A increases while the price of B is held constant, then the demand for A will decrease; that is, $f_p(p, q) < 0$. *Then, if A and B are competitive products, the demand for B will increase; that is,* $g_p(p, q) > 0$. *Similarly, if the price of B increases while the price of A is held constant, then the demand for B will decrease; that is,* $g_q(p, q) < 0$. *And if A and B are competitive products, then the demand for A will increase; that is,* $f_q(p, q) > 0$. *Reasoning similarly for complementary products, we arrive at the following test:*

TEST FOR COMPETITIVE AND COMPLEMENTARY PRODUCTS

PARTIAL DERIVATIVES			PRODUCTS A AND B
$f_q(p, q) > 0$	and	$g_p(p, q) > 0$	Competitive (Substitute)
$f_q(p, q) < 0$	and	$g_p(p, q) < 0$	Complementary
$f_q(p, q) \geq 0$	and	$g_p(p, q) \leq 0$	Neither
$f_q(p, q) \leq 0$	and	$g_p(p, q) \geq 0$	Neither

Use this test in Problems 69–72 to determine whether the indicated products are competitive, complementary, or neither.

69. *Product demand.* The weekly demand equations for the sale of butter and margarine in a supermarket are

$$x = f(p, q) = 8{,}000 - 0.09p^2 + 0.08q^2 \quad \text{Butter}$$
$$y = g(p, q) = 15{,}000 + 0.04p^2 - 0.3q^2 \quad \text{Margarine}$$

70. *Product demand.* The daily demand equations for the sale of brand A coffee and brand B coffee in a supermarket are

$$x = f(p, q) = 200 - 5p + 4q \quad \text{Brand A coffee}$$
$$y = g(p, q) = 300 + 2p - 4q \quad \text{Brand B coffee}$$

71. *Product demand.* The monthly demand equations for the sale of skis and ski boots in a sporting goods store are

$$x = f(p, q) = 800 - 0.004p^2 - 0.003q^2 \quad \text{Skis}$$
$$y = g(p, q) = 600 - 0.003p^2 - 0.002q^2 \quad \text{Ski boots}$$

72. *Product demand.* The monthly demand equations for the sale of tennis rackets and tennis balls in a sporting goods store are

$$x = f(p, q) = 500 - 0.5p - q^2 \quad \text{Tennis rackets}$$
$$y = g(p, q) = 10{,}000 - 8p - 100q^2 \quad \text{Tennis balls (cans)}$$

Life Sciences

73. *Medicine.* The following empirical formula relates the surface area A (in square inches) of an average human body to its weight w (in pounds) and its height h (in inches):

$$A = f(w, h) = 15.64w^{0.425}h^{0.725}$$

Knowing the surface area of a human body is useful, for example, in studies pertaining to hypothermia (heat loss due to exposure).

(A) Find $f_w(w, h)$ and $f_h(w, h)$.

(B) For a 65 pound child who is 57 inches tall, find $f_w(65, 57)$ and $f_h(65, 57)$, and interpret the results.

74. *Blood flow.* Poiseuille's law states that the resistance, R, for blood flowing in a blood vessel varies directly as the length of the vessel, L, and inversely as the fourth power of its radius, r. Stated as an equation,

$$R(L, r) = k\frac{L}{r^4} \qquad k \text{ a constant}$$

Find $R_L(4, 0.2)$ and $R_r(4, 0.2)$, and interpret the results.

Social Sciences

75. *Physical anthropology.* Anthropologists, in their study of race and human genetic groupings, often use the cephalic index, C, which varies directly as the width, W, of the head, and inversely as the length, L, of the head (both viewed from the top). In terms of an equation,

$$C(W, L) = 100\frac{W}{L}$$

where

$W =$ Width in inches $\qquad L =$ Length in inches

Find $C_W(6, 8)$ and $C_L(6, 8)$, and interpret the results.

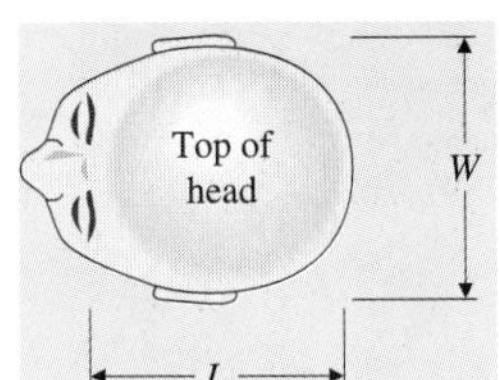

76. *Safety research.* Under ideal conditions, if a person driving a car slams on the brakes and skids to a stop, the length of the skid marks (in feet) is given by the formula

$$L(w, v) = kwv^2$$

where

$k =$ Constant
$w =$ Weight of car in pounds
$v =$ Speed of car in miles per hour

For $k = 0.000\ 013\ 3$, find $L_w(2{,}500, 60)$ and $L_v(2{,}500, 60)$, and interpret the results.

SECTION 13-3 Maxima and Minima

We are now ready to undertake a brief but useful analysis of local maxima and minima for functions of the type $z = f(x, y)$. Basically, we are going to extend the second-derivative test developed for functions of a single independent variable. To start, we assume that all second-order partial derivatives exist for the function f in some circular region in the xy plane. This guarantees that the surface $z = f(x, y)$ has no sharp points, breaks, or ruptures. In other words, we are dealing only with smooth surfaces with no edges (like the edge of a box); or breaks (like an earthquake fault); or sharp points (like the bottom point of a golf tee). See Figure 1.

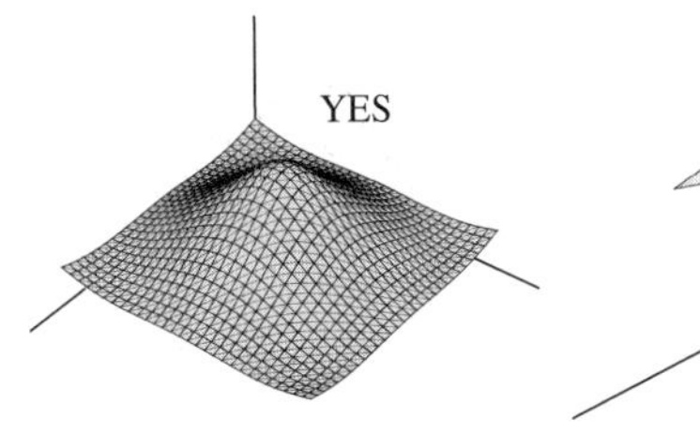

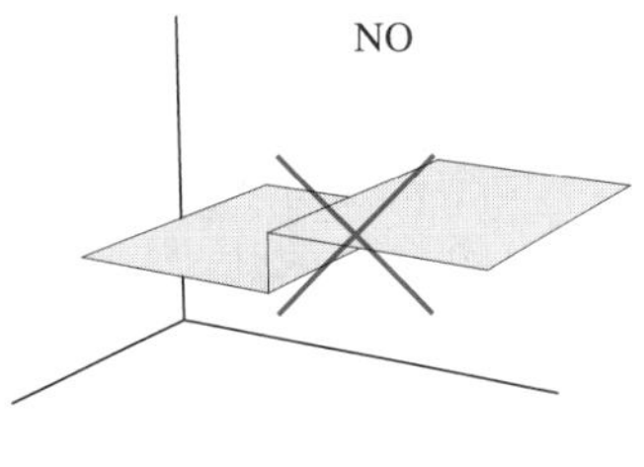

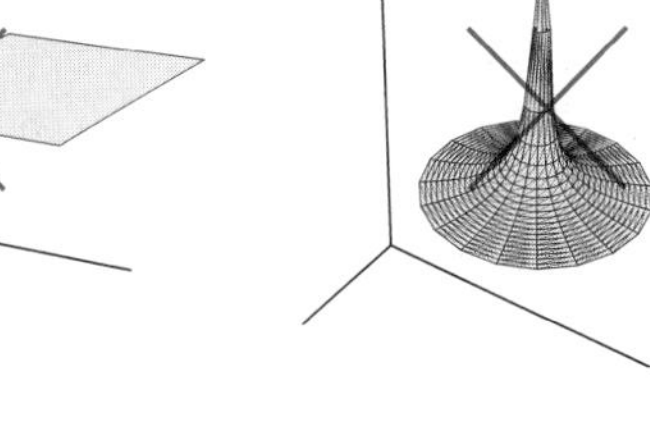

FIGURE 1

In addition, we will not concern ourselves with boundary points or absolute maxima–minima theory. In spite of these restrictions, the procedure we are now going to describe will help us solve a large number of useful problems.

What does it mean for $f(a, b)$ to be a local maximum or a local minimum? We say that **$f(a, b)$ is a local maximum** if there exists a circular region in the domain of f with (a, b) as the center, such that

$$f(a, b) \geq f(x, y)$$

for all (x, y) in the region. Similarly, we say that **$f(a, b)$ is a local minimum** if there exists a circular region in the domain of f with (a, b) as the center, such that

$$f(a, b) \leqslant f(x, y)$$

for all (x, y) in the region. Figure 2A illustrates a local maximum, Figure 2B a local minimum, and Figure 2C a **saddle point,** which is neither a local maximum nor a local minimum.

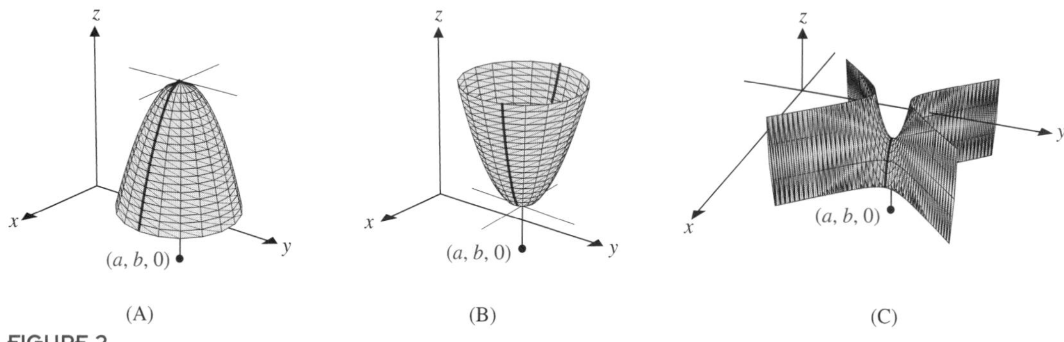

FIGURE 2

What happens to $f_x(a, b)$ and $f_y(a, b)$ if $f(a, b)$ is a local minimum or a local maximum and the partial derivatives of f exist in a circular region containing (a, b)? Figure 2 suggests that $f_x(a, b) = 0$ and $f_y(a, b) = 0$, since the tangent lines to the given curves are horizontal. Theorem 1 indicates that our intuitive reasoning is correct.

Theorem 1 LOCAL EXTREMA AND PARTIAL DERIVATIVES

Let $f(a, b)$ be a local extremum (a local maximum or a local minimum) for the function f. If both f_x and f_y exist at (a, b) then

$$f_x(a, b) = 0 \quad \text{and} \quad f_y(a, b) = 0 \qquad (1)$$

The converse of this theorem is false. That is, if $f_x(a, b) = 0$ and $f_y(a, b) = 0$, then $f(a, b)$ may or may not be a local extremum; for example, the point $(a, b, f(a, b))$ may be a saddle point (see Fig. 2C).

Theorem 1 gives us *necessary* (but not *sufficient*) conditions for $f(a, b)$ to be a local extremum. We thus find all points (a, b) such that $f_x(a, b) = 0$ and $f_y(a, b) = 0$ and test these further to determine whether $f(a, b)$ is a local extremum or a saddle point. Points (a, b) such that conditions (1) hold are called **critical points.**

Explore–Discuss 1

(A) Let $f(x, y) = y^2 + 1$. Explain why $f(x, y)$ has a local minimum at every point on the x axis. Verify that every point on the x axis is a critical point. Explain why the graph of $z = f(x, y)$ could be described as a "trough."

(B) Let $g(x, y) = x^3$. Show that every point on the y axis is a critical point. Explain why no point on the y axis is a local extremum. Explain why the graph of $z = g(x, y)$ could be described as a "slide."

The next theorem, using second-derivative tests, gives us *sufficient* conditions for a critical point to produce a local extremum or a saddle point. (As was the case with Theorem 1, we state this theorem without proof.)

Theorem 2 ■■ SECOND-DERIVATIVE TEST FOR LOCAL EXTREMA

Given:

1. $z = f(x, y)$
2. $f_x(a, b) = 0$ and $f_y(a, b) = 0$ [(a, b) is a critical point]
3. All second-order partial derivatives of f exist in some circular region containing (a, b) as a center.
4. $A = f_{xx}(a, b)$, $B = f_{xy}(a, b)$, $C = f_{yy}(a, b)$

Then:

Case 1. If $AC - B^2 > 0$ and $A < 0$, then $f(a, b)$ is a local maximum.
Case 2. If $AC - B^2 > 0$ and $A > 0$, then $f(a, b)$ is a local minimum.
Case 3. If $AC - B^2 < 0$, then f has a saddle point at (a, b).
Case 4. If $AC - B^2 = 0$, the test fails. ■■

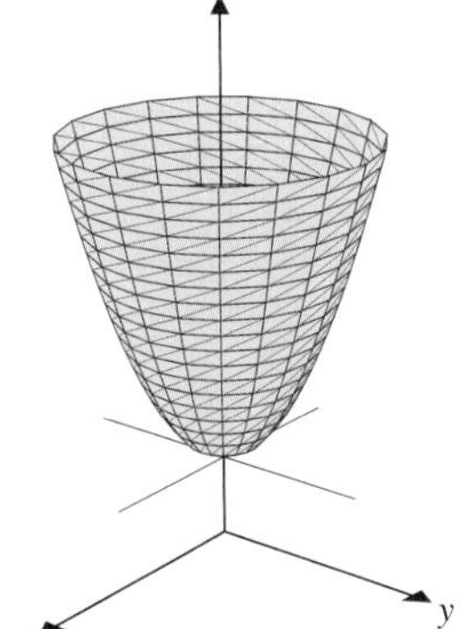

FIGURE 3

To illustrate the use of Theorem 2, we will first find the local extremum for a very simple function whose solution is almost obvious: $z = f(x, y) = x^2 + y^2 + 2$. From the function f itself and its graph (Fig. 3), it is clear that a local minimum is found at $(0, 0)$. Let us see how Theorem 2 confirms this observation.

Step 1. Find critical points: Find (x, y) such that $f_x(x, y) = 0$ and $f_y(x, y) = 0$ simultaneously:

$$f_x(x, y) = 2x = 0 \qquad f_y(x, y) = 2y = 0$$
$$x = 0 \qquad\qquad\qquad y = 0$$

The only critical point is $(a, b) = (0, 0)$.
Step 2. Compute $A = f_{xx}(0, 0)$, $B = f_{xy}(0, 0)$, and $C = f_{yy}(0, 0)$:

$$f_{xx}(x, y) = 2 \quad \text{thus} \quad A = f_{xx}(0, 0) = 2$$
$$f_{xy}(x, y) = 0 \quad \text{thus} \quad B = f_{xy}(0, 0) = 0$$
$$f_{yy}(x, y) = 2 \quad \text{thus} \quad C = f_{yy}(0, 0) = 2$$

Step 3. Evaluate $AC - B^2$ and try to classify the critical point $(0, 0)$ using Theorem 2:

$$AC - B^2 = (2)(2) - (0)^2 = 4 > 0 \quad \text{and} \quad A = 2 > 0$$

Therefore, case 2 in Theorem 2 holds. That is, $f(0, 0) = 2$ is a local minimum.

We will now use Theorem 2 in the following examples to analyze extrema without the aid of graphs.

Example 1 ➡ **Finding Local Extrema** Use Theorem 2 to find local extrema for

$$f(x, y) = -x^2 - y^2 + 6x + 8y - 21$$

SOLUTION **Step 1.** Find critical points: Find (x, y) such that $f_x(x, y) = 0$ and $f_y(x, y) = 0$ simultaneously:

$$f_x(x, y) = -2x + 6 = 0 \qquad f_y(x, y) = -2y + 8 = 0$$
$$x = 3 \qquad y = 4$$

The only critical point is $(a, b) = (3, 4)$.

Step 2. Compute $A = f_{xx}(3, 4)$, $B = f_{xy}(3, 4)$, and $C = f_{yy}(3, 4)$:

$$f_{xx}(x, y) = -2 \quad \text{thus} \quad A = f_{xx}(3, 4) = -2$$
$$f_{xy}(x, y) = 0 \quad \text{thus} \quad B = f_{xy}(3, 4) = 0$$
$$f_{yy}(x, y) = -2 \quad \text{thus} \quad C = f_{yy}(3, 4) = -2$$

Step 3. Evaluate $AC - B^2$ and try to classify the critical point $(3, 4)$ using Theorem 2:

$$AC - B^2 = (-2)(-2) - (0)^2 = 4 > 0 \quad \text{and} \quad A = -2 < 0$$

Therefore, case 1 in Theorem 2 holds. That is, $f(3, 4) = 4$ is a local maximum.

Matched Problem 1 Use Theorem 2 to find local extrema for:

$$f(x, y) = x^2 + y^2 - 10x - 2y + 36$$

Example 2 **Finding Local Extrema—Multiple Critical Points** Use Theorem 2 to find local extrema for:

$$f(x, y) = x^3 + y^3 - 6xy$$

SOLUTION **Step 1.** Find critical points for $f(x, y) = x^3 + y^3 - 6xy$:

$$f_x(x, y) = 3x^2 - 6y = 0 \qquad \text{Solve for } y.$$
$$6y = 3x^2$$
$$y = \tfrac{1}{2}x^2 \qquad (2)$$
$$f_y(x, y) = 3y^2 - 6x = 0$$
$$3y^2 = 6x \qquad \text{Use (2) to eliminate } y.$$
$$3(\tfrac{1}{2}x^2)^2 = 6x$$
$$\tfrac{3}{4}x^4 = 6x \qquad \text{Solve for } x.$$
$$3x^4 - 24x = 0$$
$$3x(x^3 - 8) = 0$$
$$x = 0 \quad \text{or} \quad x = 2$$
$$y = 0 \qquad y = \tfrac{1}{2}(2)^2 = 2$$

The critical points are $(0, 0)$ and $(2, 2)$.

Since there are two critical points, steps 2 and 3 must be performed twice.

TEST $(0, 0)$ **Step 2.** Compute $A = f_{xx}(0, 0)$, $B = f_{xy}(0, 0)$, and $C = f_{yy}(0, 0)$:

$$f_{xx}(x, y) = 6x \quad \text{thus} \quad A = f_{xx}(0, 0) = 0$$
$$f_{xy}(x, y) = -6 \quad \text{thus} \quad B = f_{xy}(0, 0) = -6$$
$$f_{yy}(x, y) = 6y \quad \text{thus} \quad C = f_{yy}(0, 0) = 0$$

Step 3. Evaluate $AC - B^2$ and try to classify the critical point $(0, 0)$ using Theorem 2:

$$AC - B^2 = (0)(0) - (-6)^2 = -36 < 0$$

Therefore, case 3 in Theorem 2 applies. That is, f has a saddle point at $(0, 0)$.

Now we will consider the second critical point, (2, 2).

TEST (2, 2) **Step 2.** Compute $A = f_{xx}(2, 2)$, $B = f_{xy}(2, 2)$, and $C = f_{yy}(2, 2)$:

$$\begin{aligned} f_{xx}(x, y) &= 6x && \text{thus} && A = f_{xx}(2, 2) = 12 \\ f_{xy}(x, y) &= -6 && \text{thus} && B = f_{xy}(2, 2) = -6 \\ f_{yy}(x, y) &= 6y && \text{thus} && C = f_{yy}(2, 2) = 12 \end{aligned}$$

Step 3. Evaluate $AC - B^2$ and try to classify the critical point (2, 2) using Theorem 2:

$$AC - B^2 = (12)(12) - (-6)^2 = 108 > 0 \quad \text{and} \quad A = 12 > 0$$

Thus, case 2 in Theorem 2 applies, and $f(2, 2) = -8$ is a local minimum.

Our conclusions in Example 2 may be confirmed geometrically by graphing cross-sections of the function f. The cross-sections of f by the planes $y = 0$, $x = 0$, $y = x$, and $y = -x$ [each of these planes contains $(0, 0)$] are represented by the graphs of the functions $f(x, 0) = x^3$, $f(0, y) = y^3$, $f(x, x) = 2x^3 - 6x^2$, and $f(x, -x) = 6x^2$, respectively, as shown in Figure 4A (note that the first two functions have the same graph). The cross-sections of f by the planes $y = 2$, $x = 2$, $y = x$, and $y = 4 - x$ [each of these planes contains $(2, 2)$] are represented by the graphs of the functions $f(x, 2) = x^3 - 12x + 8$, $f(2, y) = y^3 - 12y + 8$, $f(x, x) = 2x^3 - 6x^2$, and $f(x, 4 - x) = x^3 + (4 - x)^3 + 6x^2 - 24x$, respectively, as shown in Figure 4B (the first two functions have the same graph). Figure 4B illustrates the fact that since f has a local minimum at (2, 2), each of the cross-sections of f through (2, 2) has a local minimum of -8 at (2, 2). Figure 4A, on the other hand, indicates that some cross-sections of f through (0, 0) have a local minimum, some a local maximum, and some neither one, at (0, 0).

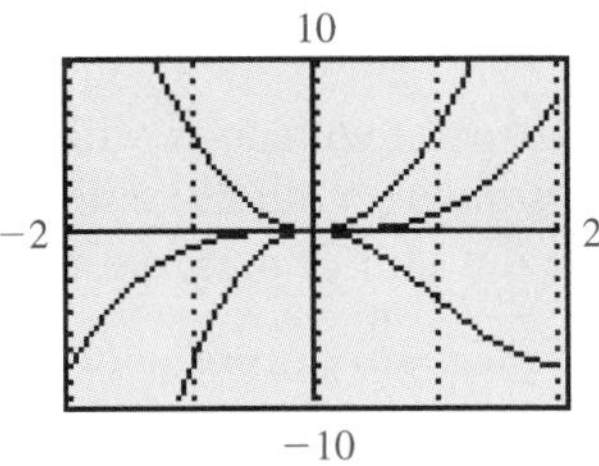

(A) $y_1 = x^3$
$y_2 = 2x^3 - 6x^2$
$y_3 = 6x^2$

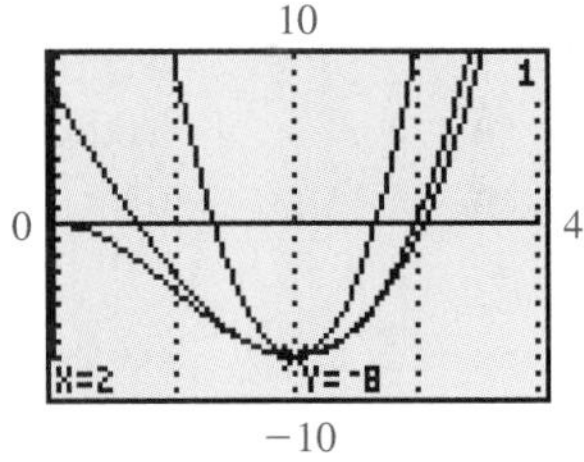

(B) $y_1 = x^3 - 12x + 8$
$y_2 = 2x^3 - 6x^2$
$y_3 = x^3 + (4 - x)^3 + 6x^2 - 24x$

FIGURE 4

Matched Problem 2 Use Theorem 2 to find local extrema for: $f(x, y) = x^3 + y^2 - 6xy$

Explore–Discuss 2

Let $f(x, y) = x^4 + y^2 + 3$, $g(x, y) = 10 - x^2 - y^4$, and $h(x, y) = x^3 + y^2$.

(A) Show that each function has only (0, 0) as a critical point.
(B) Explain why $f(x, y)$ has a minimum at (0, 0), $g(x, y)$ has a maximum at (0, 0), and $h(x, y)$ has neither a minimum nor a maximum at (0, 0).
(C) Are the results of part (B) consequences of Theorem 2? Explain.

 *Example 3* ➠ **Profit** Suppose the surfboard company discussed earlier has developed the yearly profit equation

$$P(x, y) = -22x^2 + 22xy - 11y^2 + 110x - 44y - 23$$

where x is the number (in thousands) of standard surfboards produced per year, y is the number (in thousands) of competition surfboards produced per year, and P is profit (in thousands of dollars). How many of each type of board should be produced per year to realize a maximum profit? What is the maximum profit?

SOLUTION **Step 1.** Find critical points:

$$P_x(x, y) = -44x + 22y + 110 = 0$$
$$P_y(x, y) = 22x - 22y - 44 = 0$$

Solving this system, we obtain (3, 1) as the only critical point.

Step 2. Compute $A = P_{xx}(3, 1)$, $B = P_{xy}(3, 1)$, and $C = P_{yy}(3, 1)$:

$$P_{xx}(x, y) = -44 \quad \text{thus} \quad A = P_{xx}(3, 1) = -44$$
$$P_{xy}(x, y) = 22 \quad \text{thus} \quad B = P_{xy}(3, 1) = 22$$
$$P_{yy}(x, y) = -22 \quad \text{thus} \quad C = P_{yy}(3, 1) = -22$$

Step 3. Evaluate $AC - B^2$ and try to classify the critical point (3, 1) using Theorem 2:

$$AC - B^2 = (-44)(-22) - 22^2 = 484 > 0 \quad \text{and} \quad A = -44 < 0$$

Therefore, case 1 in Theorem 2 applies. That is, $P(3, 1) = 120$ is a local maximum. A maximum profit of \$120,000 is obtained by producing and selling 3,000 standard boards and 1,000 competition boards per year.

 *Matched Problem 3* ➠ Repeat Example 3 with:

$$P(x, y) = -66x^2 + 132xy - 99y^2 + 132x - 66y - 19$$

 *Example 4* ➠ **Package Design** The packaging department in a company has been asked to design a rectangular box with no top and a partition down the middle. The box must have a volume of 48 cubic inches. Find the dimensions that will minimize the amount of material used to construct the box.

SOLUTION Refer to Figure 5. The amount of material used in constructing this box is

$$M = \underset{\text{Base}}{xy} + \underset{\text{Front, back}}{2xz} + \underset{\text{Sides, partition}}{3yz} \tag{3}$$

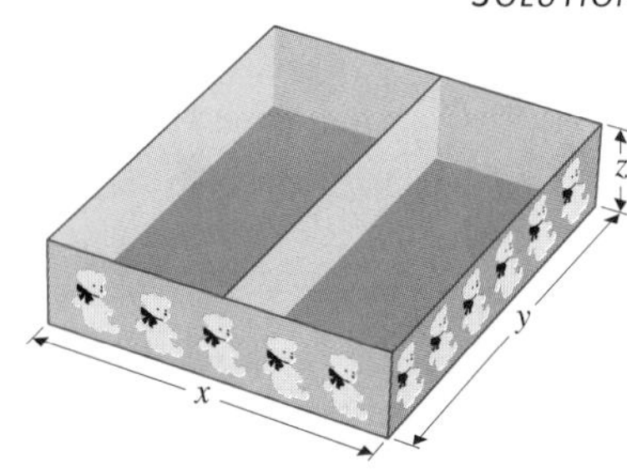

FIGURE 5

The volume of the box is

$$V = xyz = 48 \tag{4}$$

Since Theorem 2 applies only to functions with two independent variables, we must use (4) to eliminate one of the variables in (3):

$$M = xy + 2xz + 3yz \qquad \text{Substitute } z = 48/xy.$$
$$= xy + 2x\left(\frac{48}{xy}\right) + 3y\left(\frac{48}{xy}\right)$$
$$= xy + \frac{96}{y} + \frac{144}{x}$$

Thus, we must find the minimum value of

$$M(x, y) = xy + \frac{96}{y} + \frac{144}{x} \qquad x > 0 \quad \text{and} \quad y > 0$$

Step 1. Find critical points:

$$M_x(x, y) = y - \frac{144}{x^2} = 0$$

$$y = \frac{144}{x^2} \qquad (5)$$

$$M_y(x, y) = x - \frac{96}{y^2} = 0$$

$$x = \frac{96}{y^2}$$ *Solve for y^2.*

$$y^2 = \frac{96}{x}$$ *Use (5) to eliminate y and solve for x.*

$$\left(\frac{144}{x^2}\right)^2 = \frac{96}{x}$$

$$\frac{20{,}736}{x^4} = \frac{96}{x}$$ *Multiply both sides by $x^4/96$ (recall, $x > 0$).*

$$x^3 = \frac{20{,}736}{96} = 216$$

$$x = 6$$ *Use (5) to find y.*

$$y = \frac{144}{36} = 4$$

Thus, (6, 4) is the only critical point.

Step 2. Compute $A = M_{xx}(6, 4)$, $B = M_{xy}(6, 4)$, and $C = M_{yy}(6, 4)$:

$$M_{xx}(x, y) = \frac{288}{x^3} \quad \text{thus} \quad A = M_{xx}(6, 4) = \tfrac{288}{216} = \tfrac{4}{3}$$

$$M_{xy}(x, y) = 1 \quad \text{thus} \quad B = M_{xy}(6, 4) = 1$$

$$M_{yy}(x, y) = \frac{192}{y^3} \quad \text{thus} \quad C = M_{yy}(6, 4) = \tfrac{192}{64} = 3$$

Step 3. Evaluate $AC - B^2$ and try to classify the critical point (6, 4) using Theorem 2:

$$AC - B^2 = \left(\tfrac{4}{3}\right)(3) - (1)^2 = 3 > 0 \quad \text{and} \quad A = \tfrac{4}{3} > 0$$

Therefore, case 2 in Theorem 2 applies; $M(x, y)$ has a local minimum at (6, 4). If $x = 6$ and $y = 4$, then

$$z = \frac{48}{xy} = \frac{48}{(6)(4)} = 2$$

Thus, the dimensions that will require the minimum amount of material are 6 inches by 4 inches by 2 inches.

Matched Problem 4 ➡ If the box in Example 4 must have a volume of 384 cubic inches, find the dimensions that will require the least amount of material.

Answers to Matched Problems

1. $f(5, 1) = 10$ is a local minimum
2. f has a saddle point at (0, 0); $f(6, 18) = -108$ is a local minimum
3. Local maximum for $x = 2$ and $y = 1$; $P(2, 1) = 80$; a maximum profit of $80,000 is obtained by producing and selling 2,000 standard boards and 1,000 competition boards
4. 12 in. by 8 in. by 4 in.

EXERCISE 13-3

A *In Problems 1–4, find $f_x(x, y)$ and $f_y(x, y)$ and explain, using Theorem 1, why $f(x, y)$ has no local extrema.*

1. $f(x, y) = 4x + 5y - 6$
2. $f(x, y) = 10 - 2x - 3y + x^2$
3. $f(x, y) = 3.7 - 1.2x + 6.8y + 0.2y^3 + x^4$
4. $f(x, y) = x^3 - y^2 + 7x + 3y + 1$

Find local extrema in Problems 5–24 using Theorem 2.

5. $f(x, y) = 6 - x^2 - 4x - y^2$
6. $f(x, y) = 3 - x^2 - y^2 + 6y$
7. $f(x, y) = x^2 + y^2 + 2x - 6y + 14$
8. $f(x, y) = x^2 + y^2 - 4x + 6y + 23$

B

9. $f(x, y) = xy + 2x - 3y - 2$
10. $f(x, y) = x^2 - y^2 + 2x + 6y - 4$
11. $f(x, y) = -3x^2 + 2xy - 2y^2 + 14x + 2y + 10$
12. $f(x, y) = -x^2 + xy - 2y^2 + x + 10y - 5$
13. $f(x, y) = 2x^2 - 2xy + 3y^2 - 4x - 8y + 20$
14. $f(x, y) = 2x^2 - xy + y^2 - x - 5y + 8$

C

15. $f(x, y) = e^{xy}$
16. $f(x, y) = x^2y - xy^2$
17. $f(x, y) = x^3 + y^3 - 3xy$
18. $f(x, y) = 2y^3 - 6xy - x^2$
19. $f(x, y) = 2x^4 + y^2 - 12xy$
20. $f(x, y) = 16xy - x^4 - 2y^2$
21. $f(x, y) = x^3 - 3xy^2 + 6y^2$
22. $f(x, y) = 2x^2 - 2x^2y + 6y^3$
23. $f(x, y) = y^3 + 2x^2y^2 - 3x - 2y + 8$ C
24. $f(x, y) = x \ln y + x^2 - 4x - 5y + 3$ C
25. Explain why $f(x, y) = x^2$ has an infinite number of local extrema.
26. (A) Find the local extrema of the functions $f(x, y) = x + y$, $g(x, y) = x^2 + y^2$, and $h(x, y) = x^3 + y^3$.
    (B) Discuss the local extrema of the function $k(x) = x^n + y^n$, where n is a positive integer.
27. C (A) Show that $(0, 0)$ is a critical point for the function $f(x, y) = x^4e^y + x^2y^4 + 1$, but that the second-derivative test for local extrema fails.
    (B) Use cross-sections, as in Example 2, to decide whether f has a local maximum, a local minimum, or a saddle point at $(0, 0)$.
28. C (A) Show that $(0, 0)$ is a critical point for the function $g(x, y) = e^{xy^2} + x^2y^3 + 2$, but that the second-derivative test for local extrema fails.
    (B) Use cross-sections, as in Example 2, to decide whether g has a local maximum, a local minimum, or a saddle point at $(0, 0)$.

APPLICATIONS

Business & Economics

29. *Product mix for maximum profit.* A firm produces two types of calculators, x thousand of type A and y thousand of type B per year. If the revenue and cost equations for the year are (in millions of dollars)

$$R(x, y) = 2x + 3y$$
$$C(x, y) = x^2 - 2xy + 2y^2 + 6x - 9y + 5$$

determine how many of each type of calculator should be produced per year to maximize profit. What is the maximum profit?

30. *Automation–labor mix for minimum cost.* The annual labor and automated equipment cost (in millions of dollars) for a company's production of television sets is given by

$$C(x, y) = 2x^2 + 2xy + 3y^2 - 16x - 18y + 54$$

where x is the amount spent per year on labor and y is the amount spent per year on automated equipment (both in millions of dollars). Determine how much should be spent on each per year to minimize this cost. What is the minimum cost?

31. *Maximizing profit.* A department store sells two brands of inexpensive calculators. The store pays \$6 for each brand A calculator and \$8 for each brand B calculator. The research

department has estimated the following weekly demand equations for these two competitive products:

$$x = 116 - 30p + 20q \quad \textit{Demand equation for brand A}$$
$$y = 144 + 16p - 24q \quad \textit{Demand equation for brand B}$$

where p is the selling price for brand A and q is the selling price for brand B.

(A) Determine the demands x and y when $p = \$10$ and $q = \$12$; when $p = \$11$ and $q = \$11$.

(B) How should the store price each calculator to maximize weekly profits? What is the maximum weekly profit? [*Hint:* $C = 6x + 8y$, $R = px + qy$, and $P = R - C$.]

32. *Maximizing profit.* A store sells two brands of color print film. The store pays \$2 for each roll of brand A film and \$3 for each roll of brand B film. A consulting firm has estimated the following daily demand equations for these two competitive products:

$$x = 75 - 40p + 25q \quad \textit{Demand equation for brand A}$$
$$y = 80 + 20p - 30q \quad \textit{Demand equation for brand B}$$

where p is the selling price for brand A and q is the selling price for brand B.

(A) Determine the demands x and y when $p = \$4$ and $q = \$5$; when $p = \$4$ and $q = \$4$.

(B) How should the store price each brand of film to maximize daily profits? What is the maximum daily profit? [*Hint:* $C = 2x + 3y$, $R = px + qy$, and $P = R - C$.]

33. *Minimizing cost.* A satellite television reception station is to be located at $P(x, y)$ so that the sum of the squares of the distances from P to the three towns A, B, and C is minimum (see the figure). Find the coordinates of P. This location will minimize the cost of providing satellite cable television for all three towns.

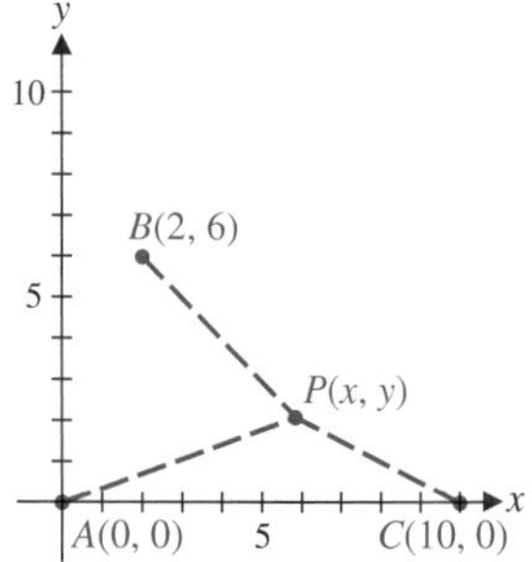

Figure for 33

34. *Minimizing cost.* Repeat Problem 33, replacing the coordinates of B with $B(6, 9)$ and the coordinates of C with $C(9, 0)$.

35. *Minimum material.* A rectangular box with no top and two parallel partitions (see the figure) is to be made to hold a volume of 64 cubic inches. Find the dimensions that will require the least amount of material.

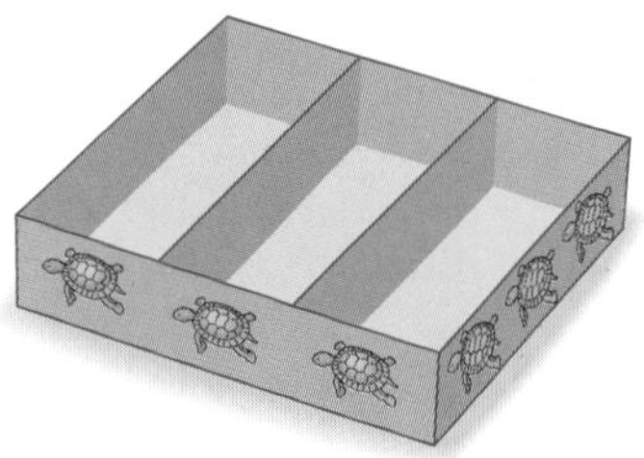

Figure for 35

36. *Minimum material.* A rectangular box with no top and two intersecting partitions (see the figure) is to be made to hold a volume of 72 cubic inches. What should its dimensions be in order to use the least amount of material in its construction?

Figure for 36

37. *Maximum volume.* A mailing service states that a rectangular package shall have the sum of the length and girth not to exceed 120 inches (see the figure). What are the dimensions of the largest (in volume) mailing carton that can be constructed meeting these restrictions?

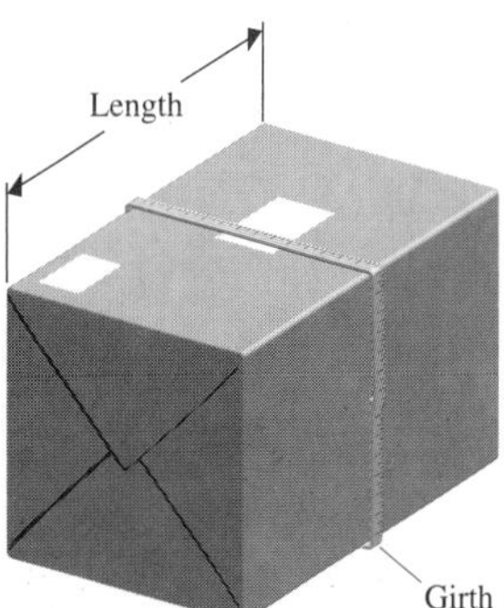

Figure for 37

38. *Maximum shipping volume.* A shipping box is to be reinforced with steel bands in all three directions, as indicated in the figure. A total of 150 inches of steel tape are to be used, with 6 inches of waste because of a 2 inch overlap in each direction. Find the dimensions of the box with maximum volume that can be taped as indicated.

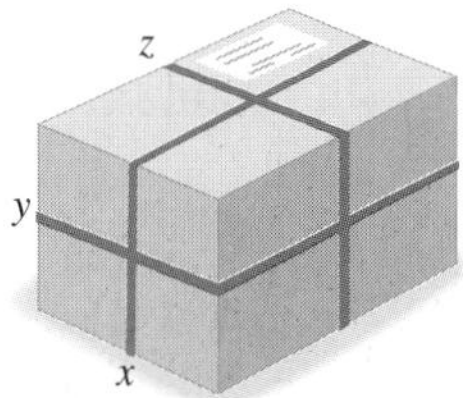

Figure for 38

SECTION 13-4 Maxima and Minima Using Lagrange Multipliers

- FUNCTIONS OF TWO INDEPENDENT VARIABLES
- FUNCTIONS OF THREE INDEPENDENT VARIABLES

FUNCTIONS OF TWO INDEPENDENT VARIABLES

We will now consider a particularly powerful method of solving a certain class of maxima–minima problems. The method is due to Joseph Louis Lagrange (1736–1813), an eminent eighteenth century French mathematician, and it is called the **method of Lagrange multipliers.** We introduce the method through an example; then we will formalize the discussion in the form of a theorem.

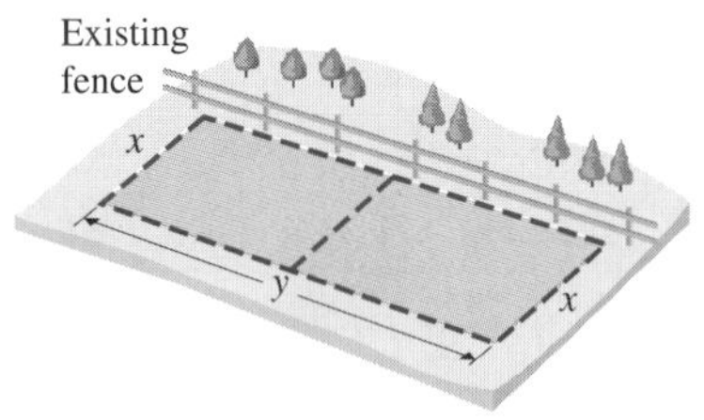

FIGURE 1

A rancher wants to construct two feeding pens of the same size along an existing fence (see Fig. 1). If the rancher has 720 feet of fencing materials available, how long should x and y be in order to obtain the maximum total area? What is the maximum area?

The total area is given by

$$f(x, y) = xy$$

which can be made as large as we like, providing there are no restrictions on x and y. But there are restrictions on x and y, since we have only 720 feet of fencing. That is, x and y must be chosen so that

$$3x + y = 720$$

This restriction on x and y, called a **constraint,** leads to the following maxima–minima problem:

$$\text{Maximize} \quad f(x, y) = xy \qquad (1)$$
$$\text{Subject to} \quad 3x + y = 720 \quad \text{or} \quad 3x + y - 720 = 0 \qquad (2)$$

This problem is a special case of a general class of problems of the form

$$\text{Maximize (or Minimize)} \quad z = f(x, y) \qquad (3)$$
$$\text{Subject to} \quad g(x, y) = 0 \qquad (4)$$

Of course, we could try to solve (4) for y in terms of x, or for x in terms of y, then substitute the result into (3), and use methods developed in Section 9-5 for functions of a single variable. But what if (4) were more complicated than (2), and solving for one variable in terms of the other was either very difficult or impossible? In the method of Lagrange multipliers, we will work with $g(x, y)$ directly and avoid having to solve (4) for one variable in terms of the other. In addition, the method generalizes to functions of arbitrarily many variables subject to one or more constraints.

Now, to the method. We form a new function F, using functions f and g in (3) and (4), as follows:

$$F(x, y, \lambda) = f(x, y) + \lambda g(x, y) \qquad (5)$$

where λ (the Greek letter lambda) is called a **Lagrange multiplier.** Theorem 1 gives the basis for the method.

Theorem 1 ■■ THE METHOD OF LAGRANGE MULTIPLIERS FOR FUNCTIONS OF TWO VARIABLES

Any local maxima or minima of the function $z = f(x, y)$ subject to the constraint $g(x, y) = 0$ will be among those points (x_0, y_0) for which (x_0, y_0, λ_0) is a solution to the system

$$F_x(x, y, \lambda) = 0$$
$$F_y(x, y, \lambda) = 0$$
$$F_\lambda(x, y, \lambda) = 0$$

where $F(x, y, \lambda) = f(x, y) + \lambda g(x, y)$, provided all the partial derivatives exist. ■■

We now solve the fence problem using the method of Lagrange multipliers.

Step 1. Formulate the problem in the form of equations (3) and (4):

$$\text{Maximize} \quad f(x, y) = xy$$
$$\text{Subject to} \quad g(x, y) = 3x + y - 720 = 0$$

Step 2. Form the function F, introducing the Lagrange multiplier λ:

$$F(x, y, \lambda) = f(x, y) + \lambda g(x, y)$$
$$= xy + \lambda(3x + y - 720)$$

Step 3. Solve the system $F_x = 0$, $F_y = 0$, $F_\lambda = 0$. (The solutions are called **critical points** for F.)

$$F_x = y + 3\lambda = 0$$
$$F_y = x + \lambda = 0$$
$$F_\lambda = 3x + y - 720 = 0$$

From the first two equations, we see that

$$y = -3\lambda$$
$$x = -\lambda$$

Substitute these values for x and y into the third equation and solve for λ:

$$-3\lambda - 3\lambda = 720$$
$$-6\lambda = 720$$
$$\lambda = -120$$

Thus,

$$y = -3(-120) = 360 \text{ feet}$$
$$x = -(-120) = 120 \text{ feet}$$

and $(x_0, y_0, \lambda_0) = (120, 360, -120)$ is the only critical point for F.

Step 4. According to Theorem 1, if the function $f(x, y)$, subject to the constraint $g(x, y) = 0$, has a local maximum or minimum, it must occur at $x = 120$, $y = 360$. Although it is possible to develop a test similar to Theorem 2 in Section 13-3 to determine the nature of this local extremum, we will not do so. [Note that Theorem 2 cannot be applied to $f(x, y)$ at

(120, 360), since this point is not a critical point of the unconstrained function $f(x, y)$.] We will simply assume that the maximum value of $f(x, y)$ must occur for $x = 120$, $y = 360$. Thus,

$$\text{Max } f(x, y) = f(120, 360) = (120)(360) = 43{,}200 \text{ square feet}$$

The key steps in applying the method of Lagrange multipliers are listed in the following box:

Method of Lagrange Multipliers—Key Steps

Step 1. Formulate the problem in the form

$$\begin{aligned} &\text{Maximize (or Minimize)} \quad z = f(x, y) \\ &\text{Subject to} \quad g(x, y) = 0 \end{aligned}$$

Step 2. Form the function F:

$$F(x, y, \lambda) = f(x, y) + \lambda g(x, y)$$

Step 3. Find the critical points for F; that is, solve the system

$$\begin{aligned} F_x(x, y, \lambda) &= 0 \\ F_y(x, y, \lambda) &= 0 \\ F_\lambda(x, y, \lambda) &= 0 \end{aligned}$$

Step 4. If (x_0, y_0, λ_0) is the only critical point of F, then we assume that (x_0, y_0) will always produce the solution to the problems we consider. If F has more than one critical point, then we evaluate $z = f(x, y)$ at (x_0, y_0) for each critical point (x_0, y_0, λ_0) of F. For the problems we consider, we assume that the largest of these values is the maximum value of $f(x, y)$, subject to the constraint $g(x, y) = 0$, and the smallest is the minimum value of $f(x, y)$, subject to the constraint $g(x, y) = 0$.

Example 1 ➠ **Minimization Subject to a Constraint** Minimize $f(x, y) = x^2 + y^2$ subject to $x + y = 10$.

SOLUTION **Step 1.** Minimize $f(x, y) = x^2 + y^2$
Subject to $g(x, y) = x + y - 10 = 0$

Step 2. $F(x, y, \lambda) = x^2 + y^2 + \lambda(x + y - 10)$

Step 3.

$$\begin{aligned} F_x &= 2x + \lambda = 0 \\ F_y &= 2y + \lambda = 0 \\ F_\lambda &= x + y - 10 = 0 \end{aligned}$$

From the first two equations, $x = -\lambda/2$ and $y = -\lambda/2$. Substituting these into the third equation, we obtain

$$\begin{aligned} -\frac{\lambda}{2} - \frac{\lambda}{2} &= 10 \\ -\lambda &= 10 \\ \lambda &= -10 \end{aligned}$$

The only critical point is $(x_0, y_0, \lambda_0) = (5, 5, -10)$.

Step 4. Since $(5, 5, -10)$ is the only critical point for F, we conclude that (see step 4 in the box)

$$\text{Min } f(x, y) = f(5, 5) = (5)^2 + (5)^2 = 50$$

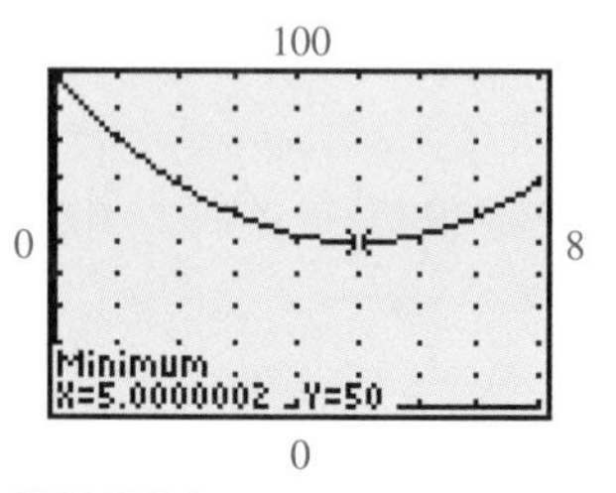

FIGURE 2
$h(x) = x^2 + (10 - x)^2$

Since $g(x, y)$ in Example 1 has a relatively simple form, an alternative to the method of Lagrange multipliers is to solve $g(x, y) = 0$ for y, and then substitute into $f(x, y)$ to obtain the function $h(x) = f(x, 10 - x) = x^2 + (10 - x)^2$ in the single variable x. Then minimize h (see Fig. 2). Again, we conclude that Min $f(x, y) = f(5, 5) = 50$. This technique depends on being able to solve the constraint for one of the two variables, and thus is not always available as an alternative to the method of Lagrange multipliers.

Matched Problem 1 Maximize $f(x, y) = 25 - x^2 - y^2$ subject to $x + y = 4$.

Figures 3 and 4 illustrate the results obtained in Example 1 and Matched Problem 1, respectively.

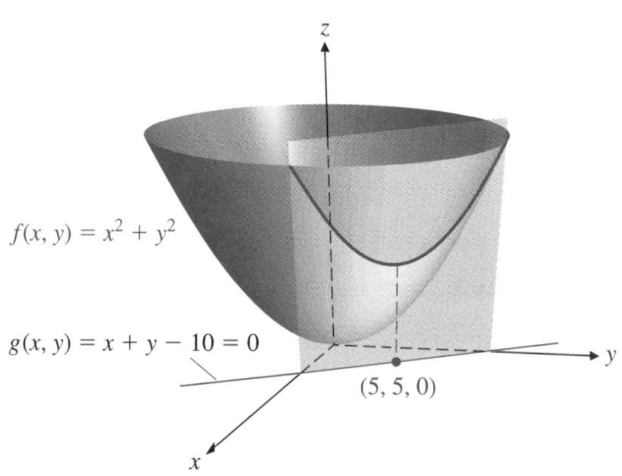

FIGURE 3

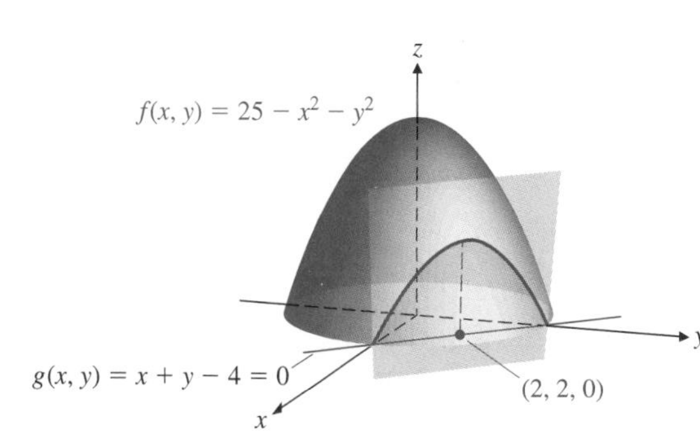

FIGURE 4

Explore–Discuss 1

Consider the problem of minimizing $f(x, y) = 3x^2 + 5y^2$ subject to the constraint $g(x, y) = 2x + 3y - 6 = 0$.

(A) Compute the value of $f(x, y)$ when x and y are integers, $0 \leq x \leq 3, 0 \leq y \leq 2$. Record your answers next to the points (x, y) in Figure 5.
(B) Graph the constraint $g(x, y) = 0$.
(C) Estimate the minimum value of f on the basis of your graph and the computations from part (A).
(D) Use the method of Lagrange multipliers to solve the minimization problem.

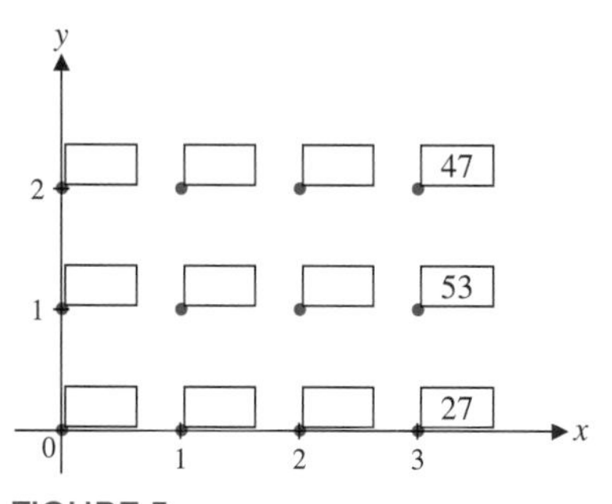

FIGURE 5

Example 2 **Productivity** The Cobb–Douglas production function for a new product is given by

$$N(x, y) = 16x^{0.25}y^{0.75}$$

where x is the number of units of labor and y is the number of units of capital required to produce $N(x, y)$ units of the product. Each unit of labor costs \$50 and each unit of capital costs \$100. If \$500,000 has been budgeted for the production of this product, how should this amount be allocated between labor and capital in order to maximize production? What is the maximum number of units that can be produced?

SOLUTION The total cost of using x units of labor and y units of capital is $50x + 100y$. Thus, the constraint imposed by the \$500,000 budget is

$$50x + 100y = 500{,}000$$

Step 1. Maximize $N(x, y) = 16x^{0.25}y^{0.75}$
Subject to $g(x, y) = 50x + 100y - 500{,}000 = 0$

Step 2. $F(x, y, \lambda) = 16x^{0.25}y^{0.75} + \lambda(50x + 100y - 500{,}000)$

Step 3.
$$F_x = 4x^{-0.75}y^{0.75} + 50\lambda = 0$$
$$F_y = 12x^{0.25}y^{-0.25} + 100\lambda = 0$$
$$F_\lambda = 50x + 100y - 500{,}000 = 0$$

From the first two equations,

$$\lambda = -\tfrac{2}{25}x^{-0.75}y^{0.75} \quad \text{and} \quad \lambda = -\tfrac{3}{25}x^{0.25}y^{-0.25}$$

Thus,

$$-\tfrac{2}{25}x^{-0.75}y^{0.75} = -\tfrac{3}{25}x^{0.25}y^{-0.25}$$ Multiply both sides by $x^{0.75}y^{0.25}$.

$$-\tfrac{2}{25}y = -\tfrac{3}{25}x$$ (We can assume $x \neq 0$ and $y \neq 0$.)

$$y = \tfrac{3}{2}x$$

Now, substitute for y in the third equation and solve for x:

$$50x + 100\left(\tfrac{3}{2}x\right) - 500{,}000 = 0$$
$$200x = 500{,}000$$
$$x = 2{,}500$$

Thus,

$$y = \tfrac{3}{2}(2{,}500) = 3{,}750$$

and

$$\lambda = -\tfrac{2}{25}(2{,}500)^{-0.75}(3{,}750)^{0.75} \approx -0.1084$$

The only critical point of F is $(2{,}500, 3{,}750, -0.1084)$.

Step 4. Since F has only one critical point, we conclude that maximum productivity occurs when 2,500 units of labor and 3,750 units of capital are used (see step 4 in the method of Lagrange multipliers). Thus,

$$\begin{aligned}\text{Max } N(x, y) &= N(2{,}500, 3{,}750)\\ &= 16(2{,}500)^{0.25}(3{,}750)^{0.75}\\ &\approx 54{,}216 \text{ units}\end{aligned}$$

The negative of the value of the Lagrange multiplier found in step 3 is called the **marginal productivity of money** and gives the approximate increase in production for each additional dollar spent on production. In Example 2, increasing the production budget from \$500,000 to \$600,000 would result in an approximate increase in production of

$$0.1084(100{,}000) = 10{,}840 \text{ units}$$

Note that simplifying the constraint equation

$$50x + 100y - 500{,}000 = 0$$

to

$$x + 2y - 10{,}000 = 0$$

before forming the function $F(x, y, \lambda)$ would make it difficult to interpret $-\lambda$ correctly. Thus, **in marginal productivity problems, the constraint equation should not be simplified.**

Matched Problem 2 ➡ The Cobb–Douglas production function for a new product is given by

$$N(x, y) = 20x^{0.5}y^{0.5}$$

where x is the number of units of labor and y is the number of units of capital required to produce $N(x, y)$ units of the product. Each unit of labor costs \$40 and each unit of capital costs \$120.

(A) If \$300,000 has been budgeted for the production of this product, how should this amount be allocated in order to maximize production? What is the maximum production?

(B) Find the marginal productivity of money in this case, and estimate the increase in production if an additional \$40,000 is budgeted for production.

Explore–Discuss 2

Consider the problem of maximizing $f(x, y) = 4 - x^2 - y^2$ subject to the constraint $g(x, y) = y - x^2 + 1 = 0$.

(A) Explain why $f(x, y) = 3$ whenever (x, y) is a point on the circle of radius 1 centered at the origin. What is the value of $f(x, y)$ when (x, y) is a point on the circle of radius 2 centered at the origin? On the circle of radius 3 centered at the origin? (See Fig. 6.)

(B) Explain why some points on the parabola $y - x^2 + 1 = 0$ lie inside the circle $x^2 + y^2 = 1$.

(C) In light of part (B), would you guess that the maximum value of $f(x, y)$ subject to the constraint is greater than 3? Explain.

(D) Use Lagrange multipliers to solve the maximization problem.

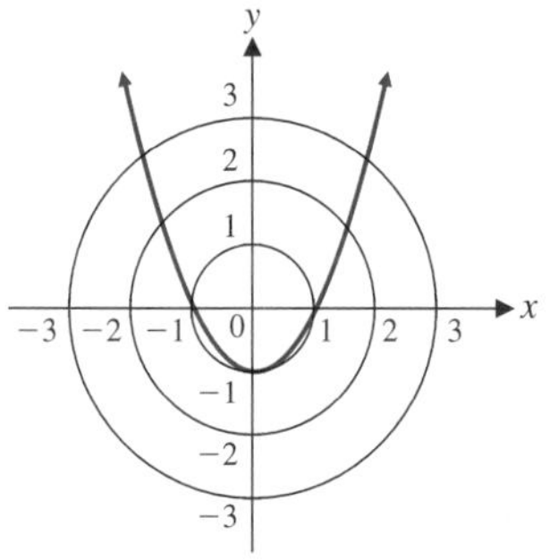

FIGURE 6

■ Functions of Three Independent Variables

We have indicated that the method of Lagrange multipliers can be extended to functions with arbitrarily many independent variables with one or more constraints. We now state a theorem for functions with three independent variables and one constraint, and consider an example that will demonstrate the advantage of the method of Lagrange multipliers over the method used in Section 13-3.

Theorem 2 ■■ THE METHOD OF LAGRANGE MULTIPLIERS FOR FUNCTIONS OF THREE VARIABLES

Any local maxima or minima of the function $w = f(x, y, z)$ subject to the constraint $g(x, y, z) = 0$ will be among the set of points (x_0, y_0, z_0) for which $(x_0, y_0, z_0, \lambda_0)$ is a solution to the system

$$\begin{aligned} F_x(x, y, z, \lambda) &= 0 \\ F_y(x, y, z, \lambda) &= 0 \\ F_z(x, y, z, \lambda) &= 0 \\ F_\lambda(x, y, z, \lambda) &= 0 \end{aligned}$$

where $F(x, y, z, \lambda) = f(x, y, z) + \lambda g(x, y, z)$, provided all the partial derivatives exist. ■■

Example 3 ➡ **Package Design** A rectangular box with an open top and one partition is to be constructed from 162 square inches of cardboard (Fig. 7). Find the dimensions that will result in a box with the largest possible volume.

SOLUTION We must maximize

$$V(x, y, z) = xyz$$

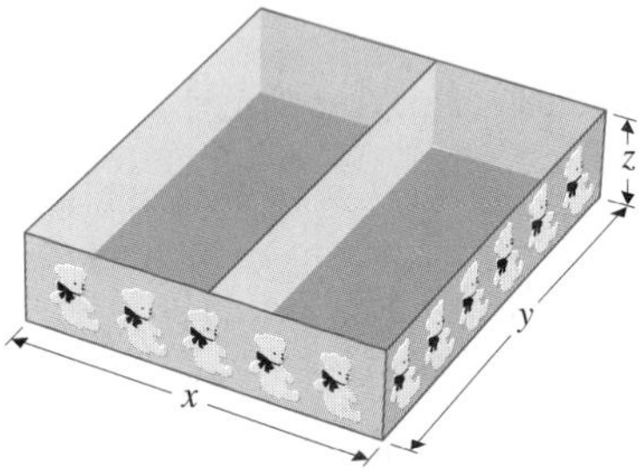

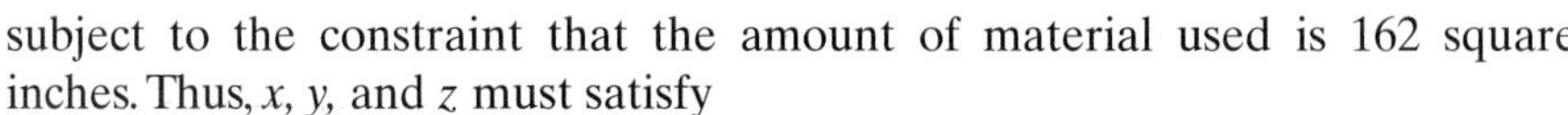

subject to the constraint that the amount of material used is 162 square inches. Thus, x, y, and z must satisfy

$$xy + 2xz + 3yz = 162$$

FIGURE 7

Step 1. Maximize $V(x, y, z) = xyz$
Subject to $g(x, y, z) = xy + 2xz + 3yz - 162 = 0$

Step 2. $F(x, y, z, \lambda) = xyz + \lambda(xy + 2xz + 3yz - 162)$

Step 3.
$$\begin{aligned} F_x &= yz + \lambda(y + 2z) = 0 \\ F_y &= xz + \lambda(x + 3z) = 0 \\ F_z &= xy + \lambda(2x + 3y) = 0 \\ F_\lambda &= xy + 2xz + 3yz - 162 = 0 \end{aligned}$$

From the first two equations, we can write

$$\lambda = \frac{-yz}{y + 2z} \qquad \lambda = \frac{-xz}{x + 3z}$$

Eliminating λ, we have

$$\begin{aligned} \frac{-yz}{y + 2z} &= \frac{-xz}{x + 3z} \\ -xyz - 3yz^2 &= -xyz - 2xz^2 \\ 3yz^2 &= 2xz^2 && \text{We can assume } z \neq 0. \\ 3y &= 2x \\ x &= \tfrac{3}{2}y \end{aligned}$$

From the second and third equations,

$$\lambda = \frac{-xz}{x + 3z} \qquad \lambda = \frac{-xy}{2x + 3y}$$

Eliminating λ, we have

$$\begin{aligned}
\frac{-xz}{x + 3z} &= \frac{-xy}{2x + 3y} \\
-2x^2z - 3xyz &= -x^2y - 3xyz \\
2x^2z &= x^2y \qquad \text{We can assume } x \neq 0. \\
2z &= y \\
z &= \tfrac{1}{2}y
\end{aligned}$$

Substituting $x = \frac{3}{2}y$ and $z = \frac{1}{2}y$ in the fourth equation, we have

$$\begin{aligned}
\left(\tfrac{3}{2}y\right)y + 2\left(\tfrac{3}{2}y\right)\left(\tfrac{1}{2}y\right) + 3y\left(\tfrac{1}{2}y\right) - 162 &= 0 \\
\tfrac{3}{2}y^2 + \tfrac{3}{2}y^2 + \tfrac{3}{2}y^2 &= 162 \\
y^2 &= 36 \qquad \text{We can assume } y > 0. \\
y &= 6 \\
x &= \tfrac{3}{2}(6) = 9 \qquad \text{Using } x = \tfrac{3}{2}y \\
z &= \tfrac{1}{2}(6) = 3 \qquad \text{Using } z = \tfrac{1}{2}y
\end{aligned}$$

and, finally,

$$\lambda = \frac{-(6)(3)}{6 + 2(3)} = -\frac{3}{2} \qquad \text{Using } \lambda = \frac{-yz}{y + 2z}$$

Thus, the only critical point of F with x, y, and z all positive is $\left(9, 6, 3, -\frac{3}{2}\right)$.

Step 4. The box with maximum volume has dimensions 9 inches by 6 inches by 3 inches.

Matched Problem 3 A box of the same type as described in Example 3 is to be constructed from 288 square inches of cardboard. Find the dimensions that will result in a box with the largest possible volume.

Suppose we had decided to solve Example 3 by the method used in Section 13-3. First, we would have to solve the material constraint for one of the variables, say, z:

$$z = \frac{162 - xy}{2x + 3y}$$

Then we would eliminate z in the volume function and maximize

$$V(x, y) = xy\frac{162 - xy}{2x + 3y}$$

Using the method of Lagrange multipliers allows us to avoid the formidable task of finding the partial derivatives of V.

Answers to Matched Problems

1. Max $f(x, y) = f(2, 2) = 17$ (see Fig. 4)
2. (A) 3,750 units of labor and 1,250 units of capital; Max $N(x, y) = N(3{,}750, 1{,}250) \approx 43{,}301$ units
   (B) Marginal productivity of money ≈ 0.1443; increase in production $\approx 5{,}774$ units
3. 12 in. by 8 in. by 4 in.

EXERCISE 13-4

A *Use the method of Lagrange multipliers in Problems 1–4.*

1. Maximize $f(x, y) = 2xy$
Subject to $x + y = 6$

2. Minimize $f(x, y) = 6xy$
Subject to $y - x = 6$

3. Minimize $f(x, y) = x^2 + y^2$
Subject to $3x + 4y = 25$

4. Maximize $f(x, y) = 25 - x^2 - y^2$
Subject to $2x + y = 10$

B *In Problems 5 and 6, use Theorem 1 to explain why no maxima or minima exist.*

5. Minimize $f(x, y) = 4y - 3x$
Subject to $2x + 5y = 3$

6. Maximize $f(x, y) = 6x + 5y + 24$
Subject to $3x + 2y = 4$

Use the method of Lagrange multipliers in Problems 7–16.

7. Find the maximum and minimum of $f(x, y) = 2xy$ subject to $x^2 + y^2 = 18$.

8. Find the maximum and minimum of $f(x, y) = x^2 - y^2$ subject to $x^2 + y^2 = 25$.

9. Maximize the product of two numbers if their sum must be 10.

10. Minimize the product of two numbers if their difference must be 10.

C

11. Minimize $f(x, y, z) = x^2 + y^2 + z^2$
Subject to $2x - y + 3z = -28$

12. Maximize $f(x, y, z) = xyz$
Subject to $2x + y + 2z = 120$

13. Maximize and minimize $f(x, y, z) = x + y + z$
Subject to $x^2 + y^2 + z^2 = 12$

14. Maximize and minimize $f(x, y, z) = 2x + 4y + 4z$
Subject to $x^2 + y^2 + z^2 = 9$

15. Maximize $f(x, y) = y + xy^2$
C Subject to $x + y^2 = 1$

16. Maximize and minimize $f(x, y) = x + e^y$
C Subject to $x^2 + y^2 = 1$

In Problems 17 and 18, use Theorem 1 to explain why no maxima or minima exist.

17. Maximize $f(x, y) = e^x + 3e^y$
Subject to $x - 2y = 6$

18. Minimize $f(x, y) = x^3 + 2y^3$
Subject to $6x - 2y = 1$

19. Consider the problem of maximizing $f(x, y)$ subject to $g(x, y) = 0$, where $g(x, y) = y - 5$. Explain how the maximization problem can be solved without using the method of Lagrange multipliers.

20. Consider the problem of minimizing $f(x, y)$ subject to $g(x, y) = 0$, where $g(x, y) = 4x - y + 3$. Explain how the minimization problem can be solved without using the method of Lagrange multipliers.

21. Consider the problem of maximizing $f(x, y) = e^{-(x^2+y^2)}$
C subject to the constraint $g(x, y) = x^2 + y - 1 = 0$.
(A) Solve the constraint equation for y, and then substitute into $f(x, y)$ to obtain a function $h(x)$ of the single variable x. Graph h and solve the original maximization problem by maximizing h (round answers to three decimal places).
(B) Confirm your answer using the method of Lagrange multipliers.

22. Consider the problem of minimizing
C

$$f(x, y) = x^2 + 2y^2$$

subject to the constraint $g(x, y) = ye^{x^2} - 1 = 0$.
(A) Solve the constraint equation for y, and then substitute into $f(x, y)$ to obtain a function $h(x)$ of the single variable x. Graph h and solve the original minimization problem by minimizing h (round answers to three decimal places).
(B) Confirm your answer using the method of Lagrange multipliers.

APPLICATIONS

Business & Economics

23. *Budgeting for least cost.* A manufacturing company produces two models of a television set, x units of model A and y units of model B per week, at a cost (in dollars) of

$$C(x, y) = 6x^2 + 12y^2$$

If it is necessary (because of shipping considerations) that

$$x + y = 90$$

how many of each type of set should be manufactured per week to minimize cost? What is the minimum cost?

24. *Budgeting for maximum production.* A manufacturing firm has budgeted \$60,000 per month for labor and materials. If \$$x$ thousand is spent on labor and \$$y$ thousand is spent on materials, and if the monthly output (in units) is given by

$$N(x, y) = 4xy - 8x$$

how should the \$60,000 be allocated to labor and materials in order to maximize N? What is the maximum N?

25. *Productivity.* A consulting firm for a manufacturing company arrived at the following Cobb–Douglas production function for a particular product:

$$N(x, y) = 50x^{0.8}y^{0.2}$$

where x is the number of units of labor and y is the number of units of capital required to produce $N(x, y)$ units of the product. Each unit of labor costs \$40 and each unit of capital costs \$80.

(A) If \$400,000 is budgeted for production of the product, determine how this amount should be allocated to maximize production, and find the maximum production.

(B) Find the marginal productivity of money in this case, and estimate the increase in production if an additional \$50,000 is budgeted for the production of this product.

26. *Productivity.* The research department for a manufacturing company arrived at the following Cobb–Douglas production function for a particular product:

$$N(x, y) = 10x^{0.6}y^{0.4}$$

where x is the number of units of labor and y is the number of units of capital required to produce $N(x, y)$ units of the product. Each unit of labor costs \$30 and each unit of capital costs \$60.

(A) If \$300,000 is budgeted for production of the product, determine how this amount should be allocated to maximize production, and find the maximum production.

(B) Find the marginal productivity of money in this case, and estimate the increase in production if an additional \$80,000 is budgeted for the production of this product.

27. *Maximum volume.* A rectangular box with no top and two intersecting partitions is to be constructed from 192 square inches of cardboard (see the figure). Find the dimensions that will maximize the volume.

Figure for 27

28. *Maximum volume.* A mailing service states that a rectangular package shall have the sum of the length and girth not to exceed 120 inches (see the figure). What are the dimensions of the largest (in volume) mailing carton that can be constructed meeting these restrictions?

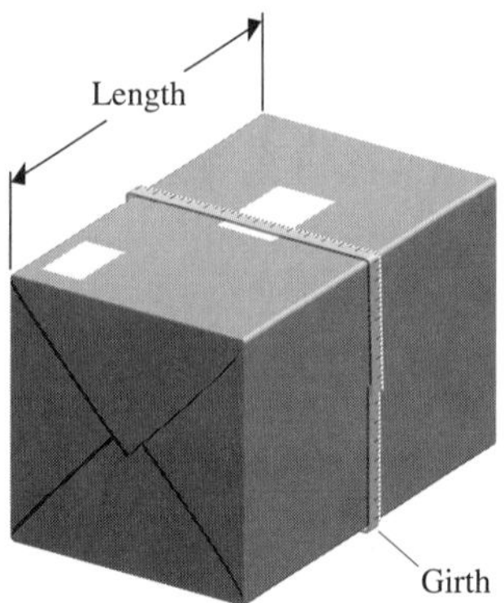

Figure for 28

Life Sciences

29. *Agriculture.* Three pens of the same size are to be built along an existing fence (see the figure). If 400 feet of fencing are available, what length should x and y be to produce the maximum total area? What is the maximum area?

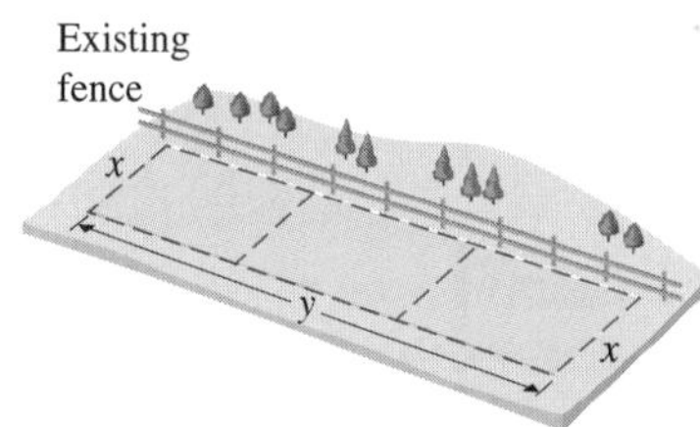

Figure for 29

30. *Diet and minimum cost.* A group of guinea pigs is to receive 25,600 calories per week. Two available foods produce $200xy$ calories for a mixture of x kilograms of type M food and y kilograms of type N food. If type M costs \$1 per kilogram and type N costs \$2 per kilogram, how much of each type of food should be used to minimize weekly food costs? What is the minimum cost? [*Note:* $x \geq 0, y \geq 0$]

SECTION 13-5 Method of Least Squares

- LEAST SQUARES APPROXIMATION
- APPLICATIONS

LEAST SQUARES APPROXIMATION

Table 1

NUMBER OF UNITS x (*hundreds*)	COST y (*thousand dollars*)
2	4
5	6
6	7
9	8

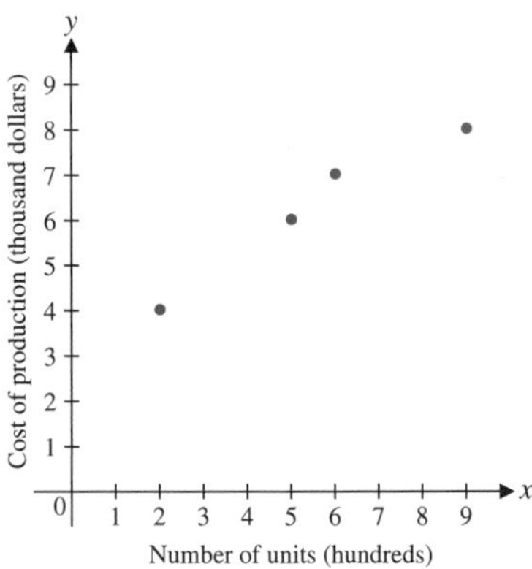

FIGURE 1

Regression analysis is the process of fitting an elementary function to a set of data points using the **method of least squares.** The mechanics of using regression techniques were introduced in Chapter 2. Now, using the optimization techniques of Section 13-3, we are able to develop and explain the mathematical foundation of the method of least squares. We begin with **linear regression,** the process of finding the equation of the line that is the "best" approximation to a set of data points.

A manufacturer wants to approximate the cost function for a product. The value of the cost function has been determined for certain levels of production, as listed in Table 1. Although these points do not all lie on a line (see Fig. 1), they are very close to being linear. The manufacturer would like to approximate the cost function by a linear function; that is, determine values a and b so that the line

$$y = ax + b$$

is, in some sense, the "best" approximation to the cost function.

What do we mean by "best"? Since the line $y = ax + b$ will not go through all four points, it is reasonable to examine the differences between the y coordinates of the points listed in the table and the y coordinates of the corresponding points on the line. Each of these differences is called the **residual** at that point (see Fig. 2). For example, at $x = 2$, the point from Table 1 is $(2, 4)$ and the point on the line is $(2, 2a + b)$, so the residual is

$$4 - (2a + b) = 4 - 2a - b$$

All the residuals are listed in Table 2.

Table 2

x	y	$ax + b$	RESIDUAL
2	4	$2a + b$	$4 - 2a - b$
5	6	$5a + b$	$6 - 5a - b$
6	7	$6a + b$	$7 - 6a - b$
9	8	$9a + b$	$8 - 9a - b$

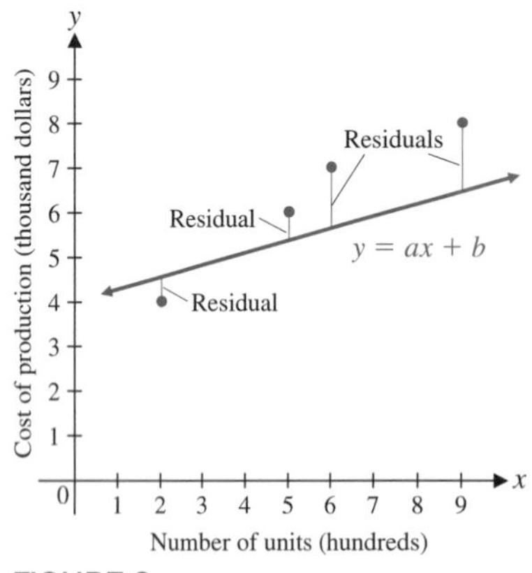

FIGURE 2

Our criterion for the "best" approximation is the following: Determine the values of a and b that *minimize the sum of the squares* of the residuals.

The resulting line is called the **least squares line,** or the **regression line.** To this end, we minimize

$$F(a, b) = (4 - 2a - b)^2 + (6 - 5a - b)^2 + (7 - 6a - b)^2 + (8 - 9a - b)^2$$

Step 1. Find critical points:

$$\begin{aligned} F_a(a, b) &= 2(4 - 2a - b)(-2) + 2(6 - 5a - b)(-5) \\ &\quad + 2(7 - 6a - b)(-6) + 2(8 - 9a - b)(-9) \\ &= -304 + 292a + 44b = 0 \\ F_b(a, b) &= 2(4 - 2a - b)(-1) + 2(6 - 5a - b)(-1) \\ &\quad + 2(7 - 6a - b)(-1) + 2(8 - 9a - b)(-1) \\ &= -50 + 44a + 8b = 0 \end{aligned}$$

After dividing each equation by 2, we solve the system

$$\begin{aligned} 146a + 22b &= 152 \\ 22a + 4b &= 25 \end{aligned}$$

obtaining $(a, b) = (0.58, 3.06)$ as the only critical point.

Step 2. Compute $A = F_{aa}(a, b)$, $B = F_{ab}(a, b)$, and $C = F_{bb}(a, b)$:

$$\begin{aligned} F_{aa}(a, b) &= 292 \quad \text{thus} \quad A = F_{aa}(0.58, 3.06) = 292 \\ F_{ab}(a, b) &= 44 \quad \text{thus} \quad B = F_{ab}(0.58, 3.06) = 44 \\ F_{bb}(a, b) &= 8 \quad \text{thus} \quad C = F_{bb}(0.58, 3.06) = 8 \end{aligned}$$

Step 3. Evaluate $AC - B^2$ and try to classify the critical point (a, b) using Theorem 2 in Section 13-3:

$$AC - B^2 = (292)(8) - (44)^2 = 400 > 0 \quad \text{and} \quad A = 292 > 0$$

Therefore, case 2 in Theorem 2 applies, and $F(a, b)$ has a local minimum at the critical point $(0.58, 3.06)$.

Thus, the least squares line for the given data is

$$y = 0.58x + 3.06 \quad \text{Least squares line}$$

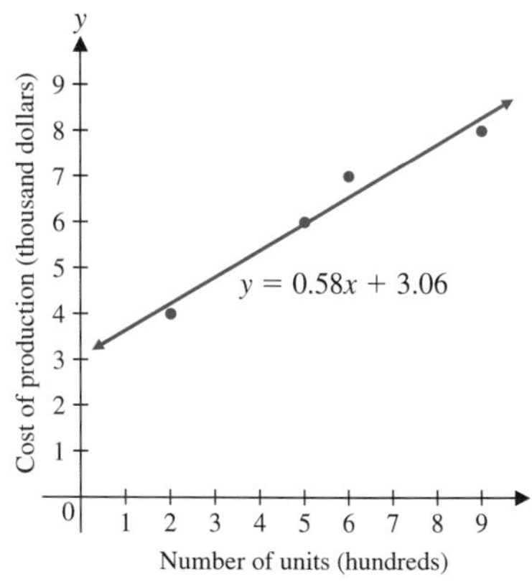

FIGURE 3

The sum of the squares of the residuals is minimized for this choice of a and b (see Fig. 3).

This linear function can now be used by the manufacturer to estimate any of the quantities normally associated with the cost function—such as costs, marginal costs, average costs, and so on. For example, the cost of producing 2,000 units is approximately

$$y = (0.58)(20) + 3.06 = 14.66 \quad \text{or} \quad \$14{,}660$$

The marginal cost function is

$$\frac{dy}{dx} = 0.58$$

The average cost function is

$$\bar{y} = \frac{0.58x + 3.06}{x}$$

In general, if we are given a set of n points $(x_1, y_1), (x_2, y_2), \ldots, (x_n, y_n)$, we want to determine the line $y = ax + b$ for which the sum of the squares of the residuals is minimized. Using summation notation, the sum of the squares of the residuals is given by

$$F(a, b) = \sum_{k=1}^{n} (y_k - ax_k - b)^2$$

Note that in this expression the variables are a and b, and the x_k and y_k are all known values. To minimize $F(a, b)$, we thus compute the partial derivatives with respect to a and b and set them equal to 0:

$$F_b(a, b) = \sum_{k=1}^{n} 2(y_k - ax_k - b)(-x_k) = 0$$
$$F_b(a, b) = \sum_{k=1}^{n} 2(y_k - ax_k - b)(-1) = 0$$

Dividing each equation by 2 and simplifying, we see that the coefficients a and b of the least squares line $y = ax + b$ must satisfy the following system of *normal equations:*

$$\left(\sum_{k=1}^{n} x_k^2\right)a + \left(\sum_{k=1}^{n} x_k\right)b = \sum_{k=1}^{n} x_k y_k$$
$$\left(\sum_{k=1}^{n} x_k\right)a + nb = \sum_{k=1}^{n} y_k$$

Solving this system for a and b produces the formulas given in the box.

Least Squares Approximation

For a set of n points $(x_1, y_1), (x_2, y_2), \ldots, (x_n, y_n)$, the coefficients of the least squares line $y = ax + b$ are the solutions of the system of **normal equations**

$$\left(\sum_{k=1}^{n} x_k^2\right)a + \left(\sum_{k=1}^{n} x_k\right)b = \sum_{k=1}^{n} x_k y_k \tag{1}$$
$$\left(\sum_{k=1}^{n} x_k\right)a + nb = \sum_{k=1}^{n} y_k$$

and are given by the formulas

$$a = \frac{n\left(\sum_{k=1}^{n} x_k y_k\right) - \left(\sum_{k=1}^{n} x_k\right)\left(\sum_{k=1}^{n} y_k\right)}{n\left(\sum_{k=1}^{n} x_k^2\right) - \left(\sum_{k=1}^{n} x_k\right)^2} \tag{2}$$

$$b = \frac{\sum_{k=1}^{n} y_k - a\left(\sum_{k=1}^{n} x_k\right)}{n} \tag{3}$$

Table 3

	x_k	y_k	$x_k y_k$	x_k^2
	2	4	8	4
	5	6	30	25
	6	7	42	36
	9	8	72	81
Totals	22	25	152	146

Now we return to the data in Table 1 and tabulate the sums required for the normal equations and their solution in Table 3.

The normal equations (1) are then

$$146a + 22b = 152$$
$$22a + 4b = 25$$

The solution to the normal equations given by equations (2) and (3) is

$$a = \frac{4(152) - (22)(25)}{4(146) - (22)^2} = 0.58$$
$$b = \frac{25 - 0.58(22)}{4} = 3.06$$

Compare these results with step 1 on page 915. Note that Table 3 provides a convenient format for the computation of step 1 above.

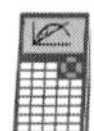

Many graphing utilities have a linear regression feature that solves the system of normal equations obtained by setting the partial derivatives of the sum of squares of the residuals equal to 0. Therefore, in practice, we simply enter the given data points and use the linear regression feature to determine the line $y = ax + b$ that best fits the data (see Fig. 4). There is no need to compute partial derivatives, or even to tabulate sums (as in Table 3).

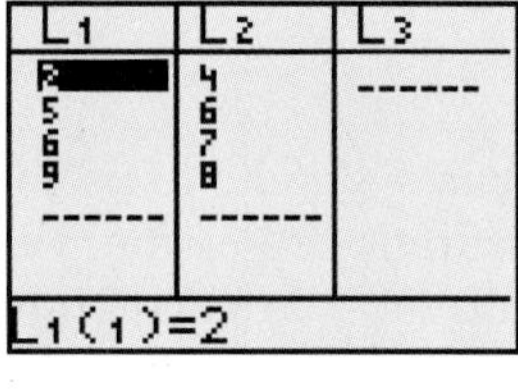

(A)

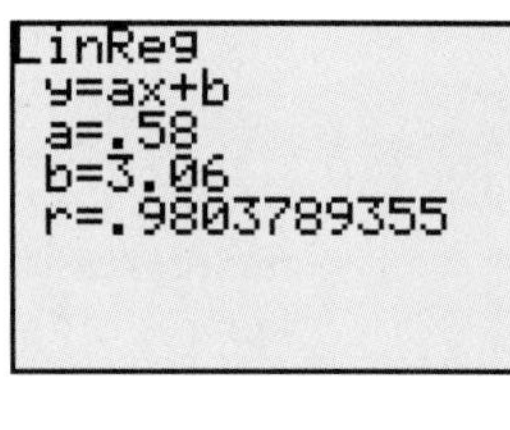

(B)

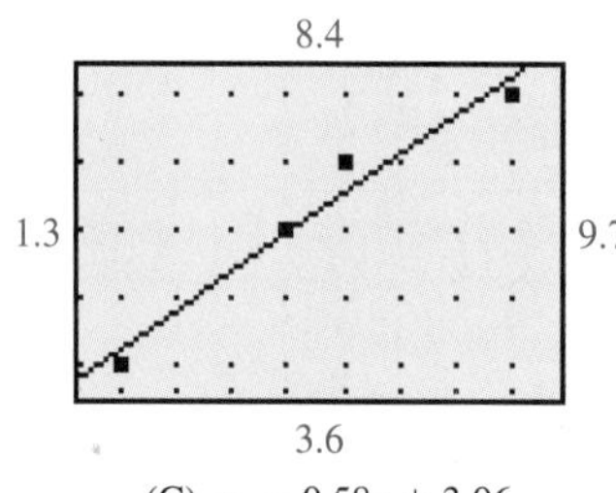

(C) $y_1 = 0.58x + 3.06$

FIGURE 4

Explore–Discuss 1

(A) Plot the four points $(0, 0)$, $(0, 1)$, $(10, 0)$, and $(10, 1)$. Which line would you guess "best" fits these four points? Use formulas (2) and (3) to test your conjecture.

(B) Plot the four points $(0, 0)$, $(0, 10)$, $(1, 0)$, and $(1, 10)$. Which line would you guess "best" fits these four points? Use formulas (2) and (3) to test your conjecture.

(C) If either of your conjectures was wrong, explain how your reasoning was mistaken.

The method of least squares can also be applied to find the quadratic equation $y = ax^2 + bx + c$ that best fits a set of data points. In this case, the sum of the squares of the residuals is a function of three variables:

$$F(a, b, c) = \sum_{k=1}^{n} (y_k - ax_k^2 - bx_k - c)^2$$

There are now three partial derivatives to compute and set equal to 0:

$$F_a(a, b, c) = \sum_{k=1}^{n} 2(y_k - ax_k^2 - bx_k - c)(-x_k^2) = 0$$
$$F_b(a, b, c) = \sum_{k=1}^{n} 2(y_k - ax_k^2 - bx_k - c)(-x_k) = 0$$
$$F_c(a, b, c) = \sum_{k=1}^{n} 2(y_k - ax_k^2 - bx_k - c)(-1) = 0$$

The resulting set of three linear equations in the three variables *a, b,* and *c* is called the *set of normal equations for quadratic regression.*

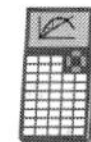

A quadratic regression feature on a calculator is designed to solve such normal equations after the given set of points has been entered. Figure 5 illustrates the computation for the data of Table 1.

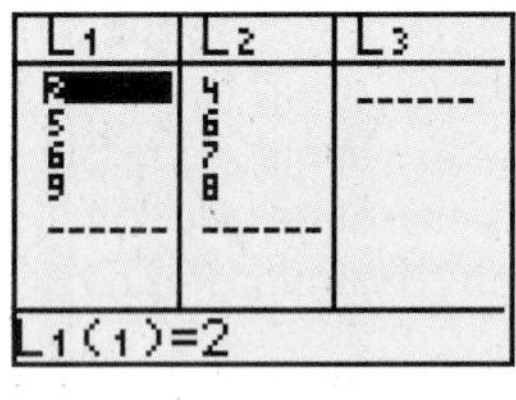

(A)

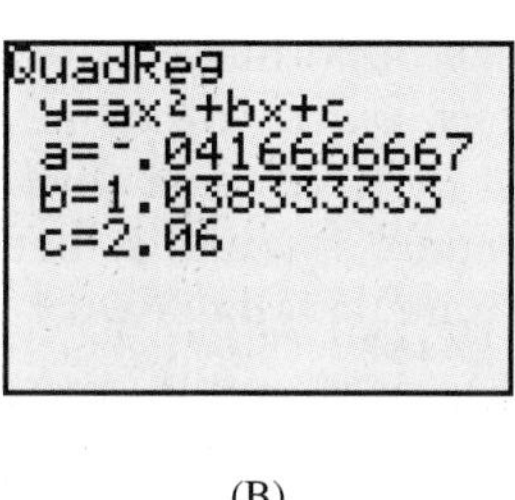

(B)

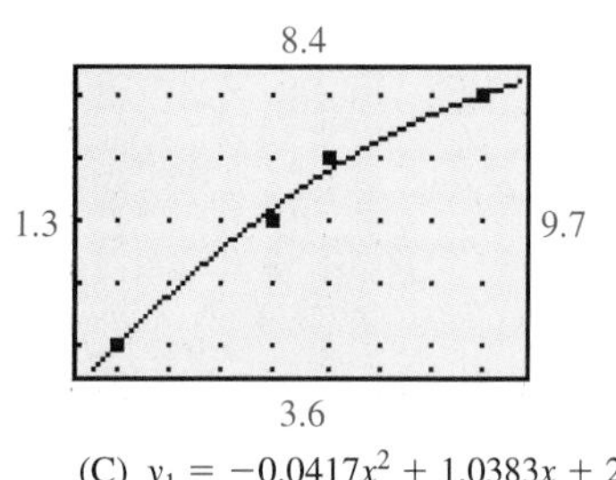

(C) $y_1 = -0.0417x^2 + 1.0383x + 2.06$

FIGURE 5

Explore–Discuss 2

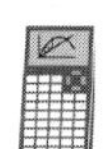

(A) Use the graphs in Figures 4 and 5 to predict which technique, linear regression or quadratic regression, yields the smaller sum of squares of the residuals for the data of Table 1. Explain.

(B) Confirm your prediction by computing the sum of squares of the residuals in each case.

The method of least squares can also be applied to other regression equations—for example, cubic, quartic, logarithmic, exponential, and power regression models. Details are explored in some of the exercises at the end of this section.

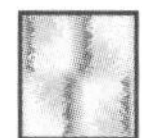

Applications

Example 1 **Educational Testing** Table 4 lists the midterm and final examination scores for 10 students in a calculus course.

Table 4

Midterm	Final	Midterm	Final
49	61	78	77
53	47	83	81
67	72	85	79
71	76	91	93
74	68	99	99

(A) Use formulas (1), (2), and (3) to find the normal equations and the least squares line for the data given in Table 4.

C (B) Use the linear regression feature on a graphing utility to find and graph the least squares line.

(C) Use the least squares line to predict the final examination score for a student who scored 95 on the midterm examination.

SOLUTION (A) Table 5 shows a convenient way to compute all the sums in the formulas for *a* and *b*.

Table 5

	x_k	y_k	$x_k y_k$	x_k^2
	49	61	2,989	2,401
	53	47	2,491	2,809
	67	72	4,824	4,489
	71	76	5,396	5,041
	74	68	5,032	5,476
	78	77	6,006	6,084
	83	81	6,723	6,889
	85	79	6,715	7,225
	91	93	8,463	8,281
	99	99	9,801	9,801
Totals	750	753	58,440	58,496

From the last line in Table 5, we have

$$\sum_{k=1}^{10} x_k = 750 \qquad \sum_{k=1}^{10} y_k = 753 \qquad \sum_{k=1}^{10} x_k y_k = 58{,}440 \qquad \sum_{k=1}^{10} x_k^2 = 58{,}496$$

and the normal equations are

$$\begin{aligned} 58{,}496a + 750b &= 58{,}440 \\ 750a + 10b &= 753 \end{aligned}$$

Using formulas (2) and (3),

$$a = \frac{10(58{,}440) - (750)(753)}{10(58{,}496) - (750)^2} = \frac{19{,}650}{22{,}460} \approx 0.875$$

$$b = \frac{753 - 0.875(750)}{10} = 9.675$$

The least squares line is given (approximately) by

$$y = 0.875x + 9.675$$

(B) We enter the data and use the linear regression feature, as shown in Figure 6.

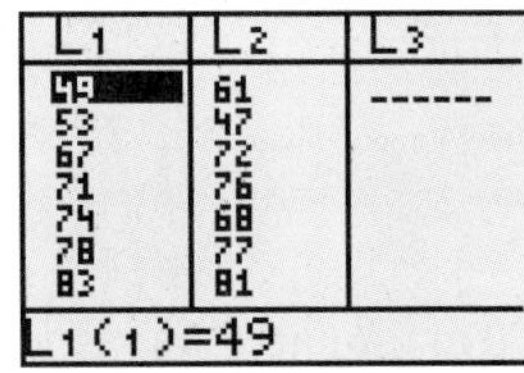

(A)

(B)

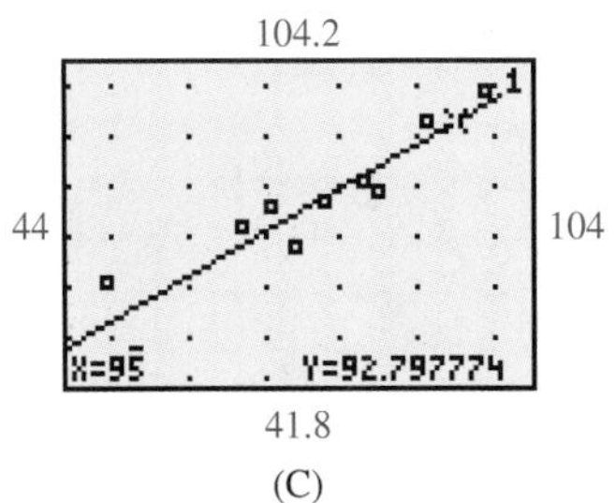

(C)

FIGURE 6

(C) If $x = 95$, then $y = 0.875(95) + 9.683 \approx 92.8$ is the predicted score on the final exam. This is also indicated in Figure 6C. If we assume the exam score must be an integer, we would predict a score of 93.

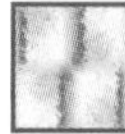

Matched Problem 1 Repeat Example 1 for the scores listed in Table 6.

Table 6

MIDTERM	FINAL	MIDTERM	FINAL
54	50	84	80
60	66	88	95
75	80	89	85
76	68	97	94
78	71	99	86

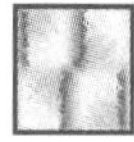

Example 2 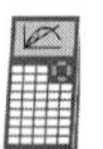 **Energy Consumption** The use of fuel oil for home heating in the United States has declined steadily for several decades. Table 7 lists the percentage of occupied housing units in the United States that were heated by fuel oil for various years between 1960 and 1991. Use the data in the table and linear regression to predict the percentage of occupied housing units in the United States that will be heated by fuel oil in the year 2000.

Table 7

OCCUPIED HOUSING UNITS HEATED BY FUEL OIL

YEAR	PERCENT	YEAR	PERCENT
1960	32.4	1985	14.1
1970	26.0	1987	14.0
1975	22.5	1989	13.3
1980	18.1	1991	12.3
1983	14.9		

SOLUTION We enter the data, with $x = 0$ representing 1960, $x = 10$ representing 1970, etc., and use the linear regression feature, as shown in Figure 7.

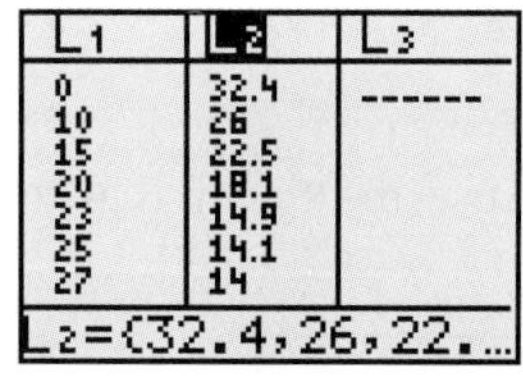

(A)

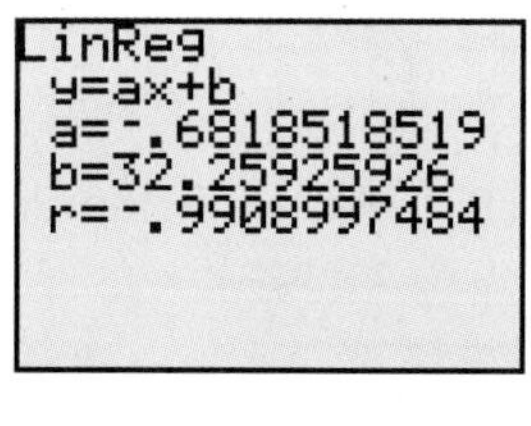

(B)

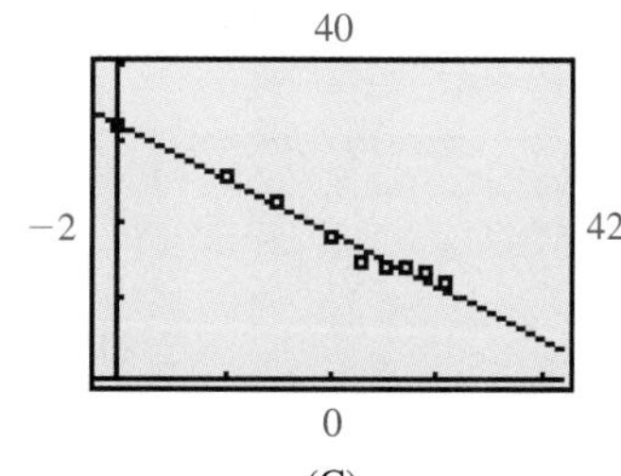

(C)

FIGURE 7

Figure 7 indicates that the least squares line is $y = -0.68x + 32.26$. The estimated percentage of occupied housing units heated by fuel oil in the year 2000 (corresponding to $x = 40$) is thus $-0.68(40) + 32.26 = 5.06\%$.

Matched Problem 2 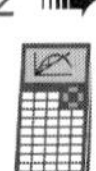 In 1950, coal was still a major source of fuel for home energy consumption, and the percentage of occupied housing units heated by fuel oil was only 22.1. Add the data for 1950 to the data for Example 2, and compute the new least squares line and the new estimate for the percentage of occupied housing units heated by fuel oil in the year 2000. Discuss the discrepancy between the two estimates.

Explore–Discuss 3

Table 8

t (hours)	A (milligrams)
0	50.0
5	46.4
10	43.1
15	39.9
20	37.1

The data in Table 8 give the amount A (in milligrams) of the radioactive isotope gallium-67, used in the diagnosis of malignant tumors, at time t (in hours).

(A) Since the data describe radioactive decay, we would expect the relationship between A and t to be exponential—that is, $A = ae^{bt}$ for some constants a and b. Show that if $A = ae^{bt}$, then the relationship between $\ln A$ and t is linear.

(B) Compute $\ln A$ for each data point and find the line that "best" fits the data $(t, \ln A)$.

(C) Determine the "best" values of a and b.

(D) Confirm your results by using the exponential regression feature on a graphing utility.

Answers to Matched Problems

1. (A) $y = 0.85x + 9.47$

(B)

L1	L2	L3
54	50	------
60	66	
75	80	
76	68	
78	71	
84	80	
88	95	

L1(1)=54

```
LinReg
y=ax+b
a=.8504140787
b=9.466873706
r=.8959772625
```

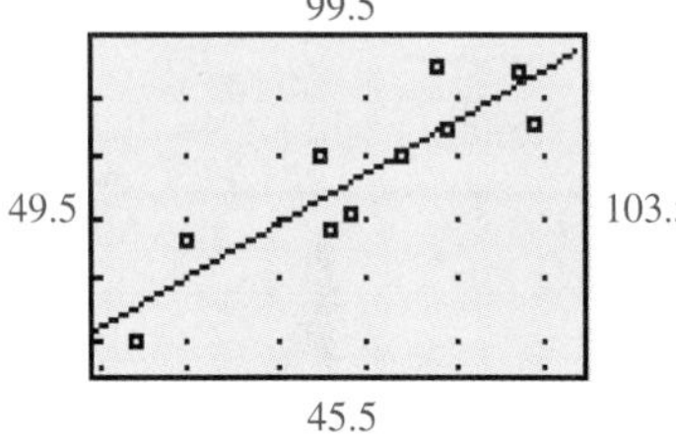

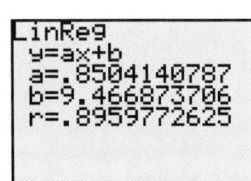

(C) 90.3

2. $y = -0.40x + 25.75$; 9.75%

Exercise 13-5

A *In Problems 1–6, find the least squares line. Graph the data and the least squares line.*

1.

x	y
1	1
2	3
3	4
4	3

2.

x	y
1	−2
2	−1
3	3
4	5

3.

x	y
1	8
2	5
3	4
4	0

4.

x	y
1	20
2	14
3	11
4	3

5.

x	y
1	3
2	4
3	5
4	6

6.

x	y
1	2
2	3
3	3
4	2

B *In Problems 7–14, find the least squares line and use it to estimate y for the indicated value of x. Round answers to two decimal places.*

7.

x	y
1	3
2	1
2	2
3	0

Estimate y when $x = 2.5$.

8.

x	y
1	0
3	1
3	6
3	4

Estimate y when $x = 3$.

9.

x	y
0	10
5	22
10	31
15	46
20	51

Estimate y when $x = 25$.

10.

x	y
−5	60
0	50
5	30
10	20
15	15

Estimate y when $x = 20$.

11.

x	y
−1	14
1	12
3	8
5	6
7	5

Estimate y when $x = 2$.

12.

x	y
2	−4
6	0
10	8
14	12
18	14

Estimate y when $x = 15$.

13.

x	y	x	y
0.5	25	9.5	12
2	22	11	11
3.5	21	12.5	8
5	21	14	5
6.5	18	15.5	1

Estimate y when $x = 8$.

14.

x	y	x	y
0	−15	12	11
2	−9	14	13
4	−7	16	19
6	−7	18	25
8	−1	20	33

Estimate y when $x = 10$.

C

15. To find the coefficients of the parabola

$$y = ax^2 + bx + c$$

that is the "best" fit for the points (1, 2), (2, 1), (3, 1), and (4, 3), minimize the sum of the squares of the residuals

$$\begin{aligned} F(a, b, c) = {} & (a + b + c - 2)^2 \\ & + (4a + 2b + c - 1)^2 \\ & + (9a + 3b + c - 1)^2 \\ & + (16a + 4b + c - 3)^2 \end{aligned}$$

by solving the system of normal equations

$$F_a(a, b, c) = 0 \qquad F_b(a, b, c) = 0 \qquad F_c(a, b, c) = 0$$

for a, b, and c. Graph the points and the parabola.

16. Repeat Problem 15 for the points (−1, −2), (0, 1), (1, 2), and (2, 0).

Problems 17 and 18 refer to the system of normal equations and the formulas for a and b given in the text.

17. Verify formulas (2) and (3) by solving the system of normal equations (1) for a and b.

18. If

$$\bar{x} = \frac{1}{n}\sum_{k=1}^{n} x_k \quad \text{and} \quad \bar{y} = \frac{1}{n}\sum_{k=1}^{n} y_k$$

are the averages of the x and y coordinates, respectively, show that the point $(\bar{x}, \bar{y})$ satisfies the equation of the least squares line $y = ax + b$.

19. (A) Suppose that $n = 5$ and that the x coordinates of the data points $(x_1, y_1), (x_2, y_2), \ldots, (x_n, y_n)$ are −2, −1, 0, 1, 2. Show that system (1) implies that

$$a = \frac{\sum x_k y_k}{\sum x_k^2}$$

and that b is equal to the average of the values of y_k.

(B) Show that the conclusion of part (A) holds whenever the average of the x coordinates of the data points is 0.

20. (A) Give an example of a set of six data points such that half of the points lie above the least squares line and half lie below.

(B) Give an example of a set of six data points such that just one of the points lies above the least squares line and five lie below.

21. C (A) Find the linear and quadratic functions that best fit the data points (0, 1.3), (1, 0.6), (2, 1.5), (3, 3.6), and (4, 7.4). (Round coefficients to two decimal places.)

(B) Which of the two functions best fits the data? Explain.

22. C (A) Find the linear, quadratic, and logarithmic functions that best fit the data points (1, 3.2), (2, 4.2), (3, 4.7), (4, 5.0), and (5, 5.3). (Round coefficients to two decimal places.)

(B) Which of the three functions best fits the data? Explain.

23. Describe the normal equations for cubic regression. How many equations are there? What are the variables? What techniques could be used to solve them?

24. Describe the normal equations for quartic regression. How many equations are there? What are the variables? What techniques could be used to solve them?

APPLICATIONS

Business & Economics

25. *Production.* Data for passenger car production in Mexico for the years 1985–1991 are given in the table.

PASSENGER CAR PRODUCTION IN MEXICO

YEAR	THOUSANDS PER MONTH	YEAR	THOUSANDS PER MONTH
1985	23.8	1989	37.9
1986	16.5	1990	51.2
1987	19.0	1991	61.1
1988	29.0		

(A) Find the least squares line for the data using $x = 0$ for 1985.

(B) Use the least squares line to estimate the monthly production (in thousands) in the year 1998.

26. *Purchasing.* Data for the purchase of U.S. auto parts by Japanese car makers (for their cars made in the United States) are given in the table.

PURCHASE OF U.S. AUTO PARTS BY JAPANESE CAR MAKERS

YEAR	BILLION DOLLARS	YEAR	BILLION DOLLARS
1986	2.1	1991	8.5
1987	2.5	1992	11.2
1988	3.9	1993	12.9
1989	5.6	1994	17.0
1990	7.1		

(A) Find the least squares line for the data using $x = 0$ for 1986.

(B) Use the least squares line to estimate the purchase of auto parts by Japanese car makers (for their cars made in the United States) in the year 2000.

27. *Maximizing profit.* The market research department for a drugstore chain chose two summer resort areas to test market a new sun screen lotion packaged in 4 ounce plastic bottles. After a summer of varying the selling price and recording the monthly demand, the research department arrived at the demand table given below, where y is the number of bottles purchased per month (in thousands) at x dollars per bottle.

x	y
5.0	2.0
5.5	1.8
6.0	1.4
6.5	1.2
7.0	1.1

(A) Find a demand equation using the method of least squares.

(B) If each bottle of sun screen costs the drugstore chain \$4, how should it be priced to achieve a maximum monthly profit? [*Hint:* Use the result of part (A), with $C = 4y$, $R = xy$, and $P = R - C$.]

28. *Maximizing profit.* A market research consultant for a supermarket chain chose a large city to test market a new brand of mixed nuts packaged in 8 ounce cans. After a year of varying the selling price and recording the monthly demand, the consultant arrived at the demand table given below, where y is the number of cans purchased per month (in thousands) at x dollars per can.

x	y
4.0	4.2
4.5	3.5
5.0	2.7
5.5	1.5
6.0	0.7

(A) Find a demand equation using the method of least squares.

(B) If each can of nuts costs the supermarket chain \$3, how should it be priced to achieve a maximum monthly profit?

Life Sciences

29. *Medicine.* If a person dives into cold water, a neural reflex response automatically shuts off blood circulation to the skin and muscles and reduces the pulse rate. A medical research team conducted an experiment using a group of ten 2-year-olds. A child's face was placed momentarily in cold water, and the corresponding reduction in pulse rate was recorded. The data for the average reduction in heart rate for each temperature are summarized in the table.

Water temperature (°F)	Pulse rate reduction
50	15
55	13
60	10
65	6
70	2

(A) If T is water temperature (in degrees Fahrenheit) and P is pulse rate reduction (in beats per minute), use the method of least squares to find a linear equation relating T and P.

(B) Use the equation found in part (A) to find P when $T = 57$.

30. *Biology.* In biology there is an approximate rule, called the *bioclimatic rule for temperate climates,* that has been known for a couple of hundred years. This rule states that in spring and early summer, periodic phenomena such as blossoming of flowers, appearance of insects, and ripening of fruit usually come about 4 days later for each 500 feet of altitude. Stated as a formula,

$$d = 8h \qquad 0 \leqslant h \leqslant 4$$

where d is the change in days and h is the altitude (in thousands of feet). To test this rule, an experiment was set up to record the difference in blossoming time of the same type of apple tree at different altitudes. A summary of the results is given in the table.

h	d
0	0
1	7
2	18
3	28
4	33

(A) Use the method of least squares to find a linear equation relating h and d. Does the bioclimatic rule, $d = 8h$, appear to be approximately correct?

(B) How much longer will it take this type of apple tree to blossom at 3.5 thousand feet than at sea level? [Use the linear equation found in part (A).]

Social Sciences

31. *Political science.* Association of economic class and party affiliation did not start with Roosevelt's New Deal; it goes back to the time of Andrew Jackson (1767–1845). Paul Lazarsfeld of Columbia University published an article in the November 1950 issue of *Scientific American* in which he discusses statistical investigations of the relationships between economic class and party affiliation. The data in the table are taken from this article.

Political Affiliations, 1836

Ward	Average assessed value per person (in \$100)	Democratic votes (%)
12	1.7	51
3	2.1	49
1	2.3	53
5	2.4	36
2	3.6	65
11	3.7	35
10	4.7	29
4	6.2	40
6	7.1	34
9	7.4	29
8	8.7	20
7	11.9	23

(A) If A represents the average assessed value per person in a given ward in 1836 and D represents the percentage of people in that ward voting Democratic in 1836, use the method of least squares to find a linear equation relating A and D.

(B) If the average assessed value per person in a ward had been \$300, what is the predicted percentage of people in that ward that would have voted Democratic?

32. *Education.* The table lists the high school grade-point averages (GPA's) of 10 students, along with their grade-point averages after one semester of college.

High school GPA	College GPA	High school GPA	College GPA
2.0	1.5	3.0	2.3
2.2	1.5	3.1	2.5
2.4	1.6	3.3	2.9
2.7	1.8	3.4	3.2
2.9	2.1	3.7	3.5

(A) Find the least squares line for the data.

(B) Estimate the college GPA for a student with a high school GPA of 3.5.

(C) Estimate the high school GPA necessary for a college GPA of 2.7.

33.  *Olympic Games.* The table gives the winning heights in the pole vault in the Olympic Games from 1896 to 1992.

OLYMPIC POLE VAULT WINNING HEIGHT

YEAR	HEIGHT (FT)	YEAR	HEIGHT (FT)
1896	10.81	1952	14.93
1900	10.82	1956	14.96
1904	11.50	1960	15.43
1906	11.60	1964	16.73
1908	12.17	1968	17.71
1912	12.96	1972	18.04
1920	13.46	1976	18.04
1924	12.96	1980	18.96
1928	13.78	1984	18.85
1932	14.16	1988	18.35
1936	14.27	1992	19.02
1948	14.10		

(A) Use a graphing utility to find the least squares line for the data using $x = 0$ for 1896.
(B) Estimate the winning height in the pole vault in the Olympic Games of 2008.

34.  *Tuition.* The table gives resident and nonresident tuition for a random sample of state universities.

TUITION AT STATE UNIVERSITIES (1994)

RESIDENT	NONRESIDENT	RESIDENT	NONRESIDENT
$2,290	$ 7,480	$2,540	$11,332
4,365	14,069	2,904	10,603
3,876	9,648	4,618	9,664
1,900	5,600	2,835	6,735
2,376	5,810	3,328	7,580
2,187	5,367	2,088	7,052
2,799	8,292	1,798	5,970
1,750	4,941	1,679	6,438

(A) Use a graphing utility to find the least squares line for the data.
(B) Estimate the nonresident tuition at a state university where the resident tuition is $4,000.

SECTION 13-6

Double Integrals Over Rectangular Regions

- INTRODUCTION
- DEFINITION OF THE DOUBLE INTEGRAL
- AVERAGE VALUE OVER RECTANGULAR REGIONS
- VOLUME AND DOUBLE INTEGRALS

INTRODUCTION

We have generalized the concept of differentiation to functions with two or more independent variables. How can we do the same with integration, and how can we interpret the results? Let us first look at the operation of antidifferentiation. We can antidifferentiate a function of two or more variables with respect to one of the variables by treating all the other variables as though they were constants. Thus, this operation is the reverse operation of partial differentiation, just as ordinary antidifferentiation is the reverse operation of ordinary differentiation. We write $\int f(x, y)\, dx$ to indicate that we are to antidifferentiate $f(x, y)$ with respect to x, holding y fixed; we write $\int f(x, y)\, dy$ to indicate that we are to antidifferentiate $f(x, y)$ with respect to y, holding x fixed.

Example 1 **Partial Antidifferentiation** Evaluate:

(A) $\int (6xy^2 + 3x^2)\, dy$ (B) $\int (6xy^2 + 3x^2)\, dx$

SOLUTION (A) Treating x as a constant and using the properties of antidifferentiation from Section 11-1, we have

$$\begin{aligned}\int (6xy^2 + 3x^2)\,dy &= \int 6xy^2\,dy + \int 3x^2\,dy \\ &= 6x \int y^2\,dy + 3x^2 \int dy \\ &= 6x\left(\frac{y^3}{3}\right) + 3x^2(y) + C(x) \\ &= 2xy^3 + 3x^2y + C(x)\end{aligned}$$

The dy tells us we are looking for the antiderivative of $6xy^2 + 3x^2$ with respect to y only, holding x constant.

Notice that the constant of integration actually can be *any function of x alone,* since, for any such function

$$\frac{\partial}{\partial y} C(x) = 0$$

Check: We can verify that our answer is correct by using partial differentiation:

$$\begin{aligned}\frac{\partial}{\partial y}[2xy^3 + 3x^2y + C(x)] &= 6xy^2 + 3x^2 + 0 \\ &= 6xy^2 + 3x^2\end{aligned}$$

(B) Now we treat y as a constant:

$$\begin{aligned}\int (6xy^2 + 3x^2)\,dx &= \int 6xy^2\,dx + \int 3x^2\,dx \\ &= 6y^2 \int x\,dx + 3 \int x^2\,dx \\ &= 6y^2\left(\frac{x^2}{2}\right) + 3\left(\frac{x^3}{3}\right) + E(y) \\ &= 3x^2y^2 + x^3 + E(y)\end{aligned}$$

This time, the antiderivative contains an arbitrary function $E(y)$ of y alone.

Check:

$$\begin{aligned}\frac{\partial}{\partial x}[3x^2y^2 + x^3 + E(y)] &= 6xy^2 + 3x^2 + 0 \\ &= 6xy^2 + 3x^2\end{aligned}$$

Matched Problem 1 Evaluate:

(A) $\int (4xy + 12x^2y^3)\,dy$ (B) $\int (4xy + 12x^2y^3)\,dx$

Now that we have extended the concept of antidifferentiation to functions with two variables, we also can evaluate definite integrals of the form

$$\int_a^b f(x, y)\,dx \quad \text{or} \quad \int_c^d f(x, y)\,dy$$

Example 2 **Evaluating a Partial Antiderivative** Evaluate, substituting the limits of integration in y if dy is used and in x if dx is used:

(A) $\int_0^2 (6xy^2 + 3x^2)\,dy$ (B) $\int_0^1 (6xy^2 + 3x^2)\,dx$

SOLUTION (A) From Example 1A, we know that

$$\int (6xy^2 + 3x^2)\,dy = 2xy^3 + 3x^2y + C(x)$$

According to properties of the definite integral for a function of one variable, we can use any antiderivative to evaluate the definite integral. Thus, choosing $C(x) = 0$, we have

$$\begin{aligned}\int_0^2 (6xy^2 + 3x^2)\,dy &= (2xy^3 + 3x^2y)\Big|_{y=0}^{y=2}\\ &= [2x(2)^3 + 3x^2(2)] - [2x(0)^3 + 3x^2(0)]\\ &= 16x + 6x^2\end{aligned}$$

(B) From Example 1B, we know that

$$\int (6xy^2 + 3x^2)\,dx = 3x^2y^2 + x^3 + E(y)$$

Thus, choosing $E(y) = 0$, we have

$$\begin{aligned}\int_0^1 (6xy^2 + 3x^2)\,dx &= (3x^2y^2 + x^3)\Big|_{x=0}^{x=1}\\ &= [3y^2(1)^2 + (1)^3] - [3y^2(0)^2 + (0)^3]\\ &= 3y^2 + 1\end{aligned}$$

Matched Problem 2 Evaluate:

(A) $\int_0^1 (4xy + 12x^2y^3)\,dy$ (B) $\int_0^3 (4xy + 12x^2y^3)\,dx$

Notice that integrating and evaluating a definite integral with integrand $f(x, y)$ with respect to y produces a function of x alone (or a constant). Likewise, integrating and evaluating a definite integral with integrand $f(x, y)$ with respect to x produces a function of y alone (or a constant). Each of these results, involving at most one variable, can now be used as an integrand in a second definite integral.

Example 3 **Evaluating Iterated Integrals** Evaluate:

(A) $\int_0^1 \left[\int_0^2 (6xy^2 + 3x^2)\,dy\right] dx$ (B) $\int_0^2 \left[\int_0^1 (6xy^2 + 3x^2)\,dx\right] dy$

SOLUTION (A) Example 2A showed that

$$\int_0^2 (6xy^2 + 3x^2)\,dy = 16x + 6x^2$$

Thus,

$$\begin{aligned}\int_0^1 \left[\int_0^2 (6xy^2 + 3x^2)\,dy\right] dx &= \int_0^1 (16x + 6x^2)\,dx\\ &= (8x^2 + 2x^3)\Big|_{x=0}^{x=1}\\ &= [8(1)^2 + 2(1)^3] - [8(0)^2 + 2(0)^3] = 10\end{aligned}$$

(B) Example 2B showed that

$$\int_0^1 (6xy^2 + 3x^2)\,dx = 3y^2 + 1$$

Thus,

$$\int_0^2\left[\int_0^1 (6xy^2 + 3x^2)\,dx\right]dy = \int_0^2 (3y^2 + 1)\,dy$$
$$= (y^3 + y)\Big|_{y=0}^{y=2}$$
$$= [(2)^3 + 2] - [(0)^3 + 0] = 10$$

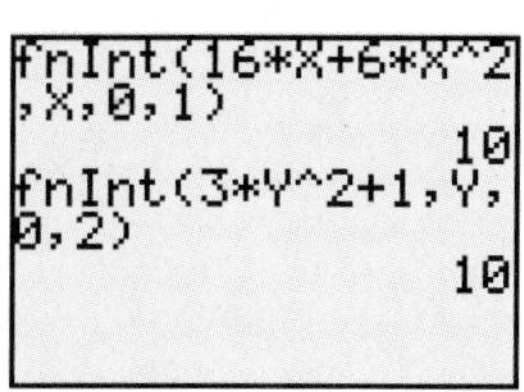

FIGURE 1

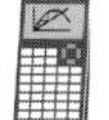

A numerical integration routine can be used as an alternative to the fundamental theorem of calculus to evaluate the last integrals in Examples 3A and 3B, $\int_0^1 (16x + 6x^2)\,dx$ and $\int_0^2 (3y^2 + 1)\,dy$, since the integrand in each case is a function of a single variable (see Fig. 1).

Matched Problem 3 Evaluate:

(A) $\int_0^3\left[\int_0^1 (4xy + 12x^2y^3)\,dy\right]dx$

(B) $\int_0^1\left[\int_0^3 (4xy + 12x^2y^3)\,dx\right]dy$

Definition of the Double Integral

Notice that the answers in Examples 3A and 3B are identical. This is not an accident. In fact, it is this property that enables us to define the *double integral,* as follows:

Double Integral

The **double integral** of a function $f(x, y)$ over a rectangle

$$R = \{(x, y) \mid a \le x \le b,\quad c \le y \le d\}$$

is

$$\iint_R f(x, y)\,dA = \int_a^b\left[\int_c^d f(x, y)\,dy\right]dx$$
$$= \int_c^d\left[\int_a^b f(x, y)\,dx\right]dy$$

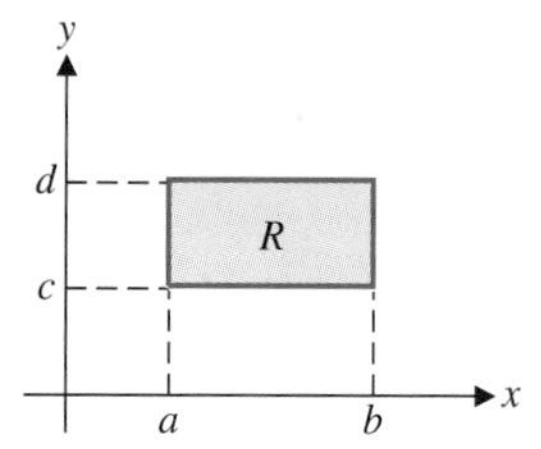

In the double integral $\iint_R f(x, y)\,dA$, $f(x, y)$ is called the **integrand** and R is called the **region of integration.** The expression dA indicates that this is an integral over a two-dimensional region. The integrals

$$\int_a^b\left[\int_c^d f(x, y)\,dy\right]dx \quad \text{and} \quad \int_c^d\left[\int_a^b f(x, y)\,dx\right]dy$$

are referred to as **iterated integrals** (the brackets are often omitted), and the order in which dx and dy are written indicates the order of integration. This is not the most general definition of the double integral over a rectangular region; however, it is equivalent to the general definition for all the functions we will consider.

Example 4 **Evaluating a Double Integral** Evaluate:

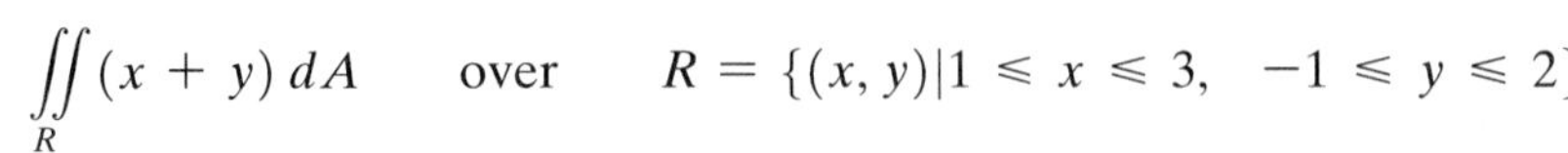

$$\iint_R (x + y)\, dA \qquad \text{over} \qquad R = \{(x, y) \mid 1 \le x \le 3, \quad -1 \le y \le 2\}$$

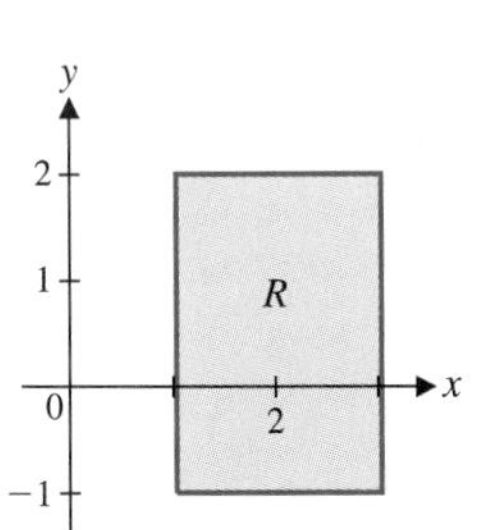

FIGURE 2

SOLUTION Region R is illustrated in Figure 2. We can choose either order of iteration. As a check, we will evaluate the integral both ways:

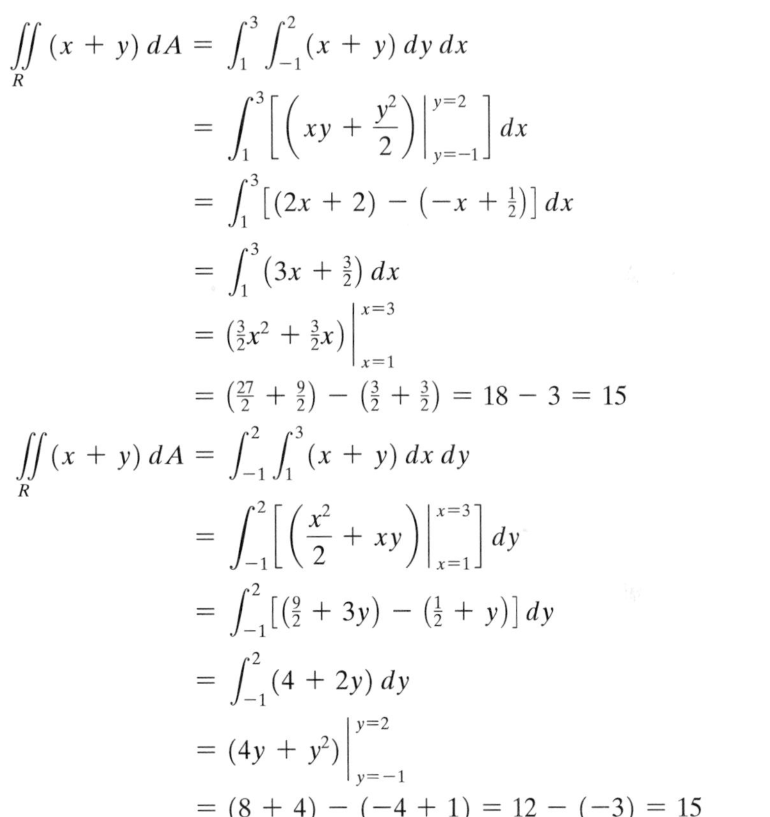

$$\begin{aligned}
\iint_R (x + y)\, dA &= \int_1^3 \int_{-1}^2 (x + y)\, dy\, dx \\
&= \int_1^3 \left[\left(xy + \frac{y^2}{2} \right) \Big|_{y=-1}^{y=2} \right] dx \\
&= \int_1^3 [(2x + 2) - (-x + \tfrac{1}{2})]\, dx \\
&= \int_1^3 (3x + \tfrac{3}{2})\, dx \\
&= (\tfrac{3}{2}x^2 + \tfrac{3}{2}x) \Big|_{x=1}^{x=3} \\
&= (\tfrac{27}{2} + \tfrac{9}{2}) - (\tfrac{3}{2} + \tfrac{3}{2}) = 18 - 3 = 15
\end{aligned}$$

$$\begin{aligned}
\iint_R (x + y)\, dA &= \int_{-1}^2 \int_1^3 (x + y)\, dx\, dy \\
&= \int_{-1}^2 \left[\left(\frac{x^2}{2} + xy \right) \Big|_{x=1}^{x=3} \right] dy \\
&= \int_{-1}^2 [(\tfrac{9}{2} + 3y) - (\tfrac{1}{2} + y)]\, dy \\
&= \int_{-1}^2 (4 + 2y)\, dy \\
&= (4y + y^2) \Big|_{y=-1}^{y=2} \\
&= (8 + 4) - (-4 + 1) = 12 - (-3) = 15
\end{aligned}$$

Matched Problem 4 Evaluate both ways:

$$\iint_R (2x - y)\, dA \qquad \text{over} \qquad R = \{(x, y) \mid -1 \le x \le 5, \quad 2 \le y \le 4\}$$

Example 5 **The Double Integral of an Exponential Function** Evaluate:

$$\iint_R 2xe^{x^2+y}\, dA \qquad \text{over} \qquad R = \{(x, y) \mid 0 \le x \le 1, \quad -1 \le y \le 1\}$$

Solution Region R is illustrated in Figure 3.

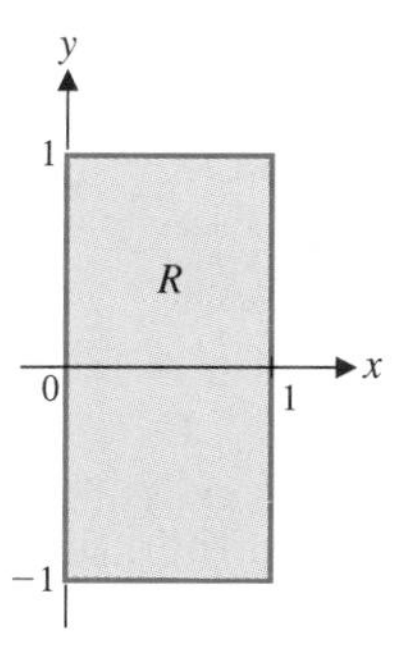

FIGURE 3

$$\iint_R 2xe^{x^2+y}\,dA = \int_{-1}^{1}\int_0^1 2xe^{x^2+y}\,dx\,dy$$
$$= \int_{-1}^{1}\left[\left(e^{x^2+y}\right)\Big|_{x=0}^{x=1}\right]dy$$
$$= \int_{-1}^{1}(e^{1+y} - e^y)\,dy$$
$$= (e^{1+y} - e^y)\Big|_{y=-1}^{y=1}$$
$$= (e^2 - e) - (e^0 - e^{-1})$$
$$= e^2 - e - 1 + e^{-1}$$

Matched Problem 5 Evaluate: $\displaystyle\iint_R \frac{x}{y^2}e^{x/y}\,dA$ over $R = \{(x, y)|0 \le x \le 1,\ \ 1 \le y \le 2\}$

Average Value Over Rectangular Regions

In Section 11-5, the average value of a function $f(x)$ over an interval $[a, b]$ was defined as

$$\frac{1}{b - a}\int_a^b f(x)\,dx$$

This definition is easily extended to functions of two variables over rectangular regions, as shown in the box. Notice that the denominator in the expression given in the box, $(b - a)(d - c)$, is simply the area of the rectangle R.

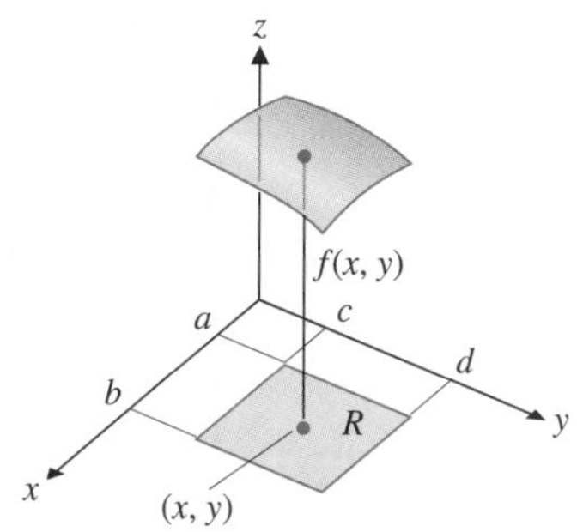

Average Value Over Rectangular Regions

The **average value** of the function $f(x, y)$ over the rectangle

$$R = \{(x, y)|a \le x \le b,\ \ c \le y \le d\}$$

is

$$\frac{1}{(b - a)(d - c)}\iint_R f(x, y)\,dA$$

Example 6 **Average Value** Find the average value of $f(x, y) = 4 - \frac{1}{2}x - \frac{1}{2}y$ over the rectangle $R = \{(x, y)|0 \le x \le 2,\ \ 0 \le y \le 2\}$.

SOLUTION Region R is illustrated in Figure 4.

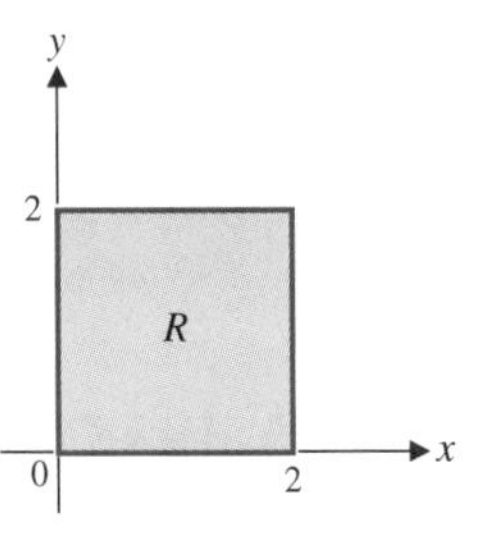

FIGURE 4

$$\frac{1}{(b-a)(d-c)}\iint_R f(x, y)\, dA = \frac{1}{(2-0)(2-0)}\iint_R \left(4 - \frac{1}{2}x - \frac{1}{2}y\right) dA$$

$$= \frac{1}{4}\int_0^2 \int_0^2 \left(4 - \tfrac{1}{2}x - \tfrac{1}{2}y\right) dy\, dx$$

$$= \frac{1}{4}\int_0^2 \left[\left(4y - \tfrac{1}{2}xy - \tfrac{1}{4}y^2\right)\Big|_{y=0}^{y=2}\right] dx$$

$$= \frac{1}{4}\int_0^2 (7 - x)\, dx$$

$$= \tfrac{1}{4}\left(7x - \tfrac{1}{2}x^2\right)\Big|_{x=0}^{x=2}$$

$$= \tfrac{1}{4}(12) = 3$$

Figure 5 illustrates the surface $z = f(x, y)$, and our calculations show that 3 is the average of the z values over the region R.

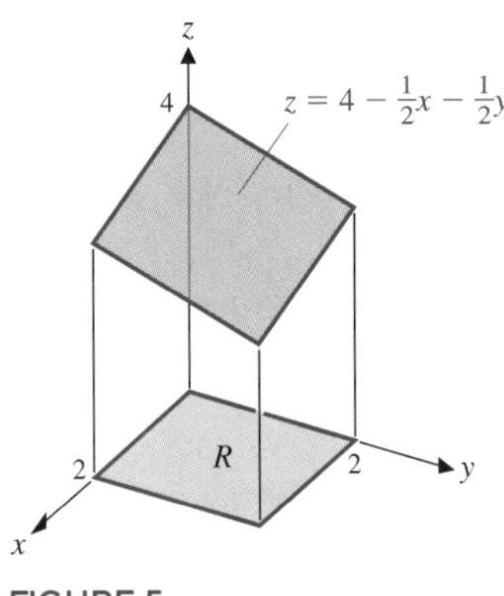

FIGURE 5

Matched Problem 6 Find the average value of $f(x, y) = x + 2y$ over the rectangle

$$R = \{(x, y) | 0 \leq x \leq 2, \quad 0 \leq y \leq 1\}$$

Explore–Discuss 1

(A) Which of the two functions, $f(x, y) = 4 - x^2 - y^2$ or $g(x, y) = 4 - x - y$, would you guess has the greater average value over the rectangle $R = \{(x, y) | 0 \leq x \leq 1, \quad 0 \leq y \leq 1\}$? Explain.

(B) Use double integrals to check the correctness of your guess in part (A).

VOLUME AND DOUBLE INTEGRALS

One application of the definite integral of a function with one variable is the calculation of areas, so it is not surprising that the definite integral of a function of two variables can be used to calculate volumes of solids.

Volume Under a Surface

If $f(x, y) \geqslant 0$ over a rectangle $R = \{(x, y)|a \leqslant x \leqslant b, \quad c \leqslant y \leqslant d\}$, then the volume of the solid formed by graphing f over the rectangle R is given by

$$V = \iint_R f(x, y)\, dA$$

A proof of the statement in the box is left to a more advanced text.

Example 7 **Volume** Find the volume of the solid under the graph of $f(x, y) = 1 + x^2 + y^2$ over the rectangle $R = \{(x, y)|0 \leqslant x \leqslant 1, \quad 0 \leqslant y \leqslant 1\}$.

SOLUTION Figure 6 shows the region R, and Figure 7 illustrates the volume under consideration.

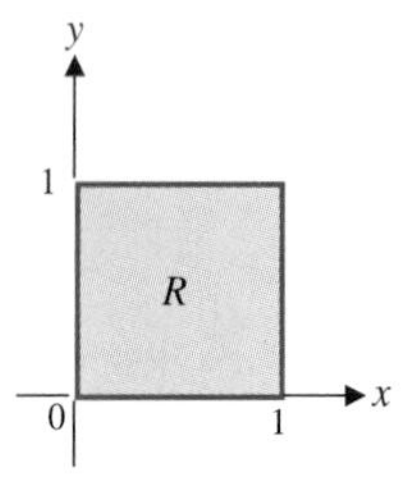

FIGURE 6

$$\begin{aligned} V &= \iint_R (1 + x^2 + y^2)\, dA \\ &= \int_0^1 \int_0^1 (1 + x^2 + y^2)\, dx\, dy \\ &= \int_0^1 \left[\left(x + \tfrac{1}{3}x^3 + xy^2 \right) \Big|_{x=0}^{x=1} \right] dy \\ &= \int_0^1 \left(\tfrac{4}{3} + y^2 \right) dy \\ &= \left(\tfrac{4}{3}y + \tfrac{1}{3}y^3 \right) \Big|_{y=0}^{y=1} = \tfrac{5}{3} \text{ cubic units} \end{aligned}$$

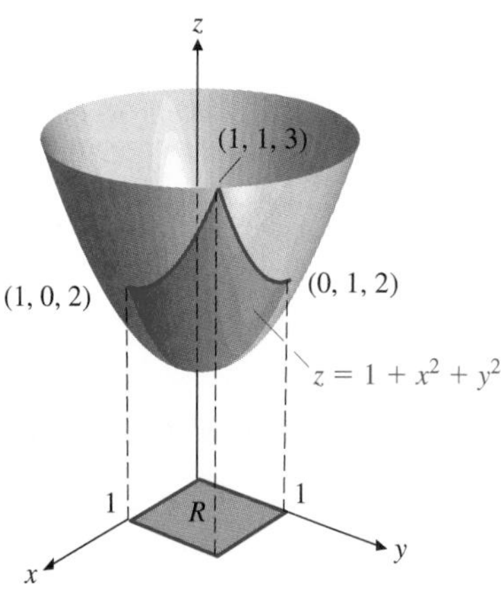

FIGURE 7

Matched Problem 7 ➡ Find the volume of the solid under the graph of $f(x, y) = 1 + x + y$ over the rectangle $R = \{(x, y)|0 \leqslant x \leqslant 1, \quad 0 \leqslant y \leqslant 2\}$.

Explore–Discuss 2

Consider the solid under the graph of $f(x, y) = 4 - y^2$ and above the rectangle $R = \{(x, y)|0 \leqslant x \leqslant 3, \quad 0 \leqslant y \leqslant 2\}$.

(A) Explain why each cross-section of the solid by a plane parallel to the yz plane has the same area, and compute that area.
(B) Compute the areas of the cross-sections of the solid by the planes $y = 0$, $y = \frac{1}{2}$, and $y = 1$.
(C) Compute the volume of the solid in two different ways.

Answers to Matched Problems

1. (A) $2xy^2 + 3x^2y^4 + C(x)$ (B) $2x^2y + 4x^3y^3 + E(y)$
2. (A) $2x + 3x^2$ (B) $18y + 108y^3$ **3.** (A) 36 (B) 36
4. 12 **5.** $e - 2e^{1/2} + 1$ **6.** 2 **7.** 5 cubic units

EXERCISE 13-6

A *In Problems 1–8, find each antiderivative. Then use the antiderivative to evaluate the definite integral.*

1. (A) $\int 12x^2y^3\,dy$ (B) $\int_0^1 12x^2y^3\,dy$

2. (A) $\int 12x^2y^3\,dx$ (B) $\int_{-1}^2 12x^2y^3\,dx$

3. (A) $\int (4x + 6y + 5)\,dx$
(B) $\int_{-2}^3 (4x + 6y + 5)\,dx$

4. (A) $\int (4x + 6y + 5)\,dy$
(B) $\int_1^4 (4x + 6y + 5)\,dy$

5. (A) $\displaystyle\int \frac{x}{\sqrt{y + x^2}}\,dx$ (B) $\displaystyle\int_0^2 \frac{x}{\sqrt{y + x^2}}\,dx$

6. (A) $\displaystyle\int \frac{x}{\sqrt{y + x^2}}\,dy$ (B) $\displaystyle\int_1^5 \frac{x}{\sqrt{y + x^2}}\,dy$

7. (A) $\displaystyle\int \frac{\ln x}{xy}\,dy$ (B) $\displaystyle\int_1^{e^2} \frac{\ln x}{xy}\,dy$

8. (A) $\displaystyle\int \frac{\ln x}{xy}\,dx$ (B) $\displaystyle\int_1^{e} \frac{\ln x}{xy}\,dx$

B *In Problems 9–16, evaluate each iterated integral. (See the indicated problem for the evaluation of the inner integral.)*

9. $\int_{-1}^2 \int_0^1 12x^2y^3\,dy\,dx$
(See Problem 1.)

10. $\int_0^1 \int_{-1}^2 12x^2y^3\,dx\,dy$
(See Problem 2.)

11. $\int_1^4 \int_{-2}^3 (4x + 6y + 5)\,dx\,dy$
(See Problem 3.)

12. $\int_{-2}^3 \int_1^4 (4x + 6y + 5)\,dy\,dx$
(See Problem 4.)

13. $\displaystyle\int_1^5 \int_0^2 \frac{x}{\sqrt{y + x^2}}\,dx\,dy$
(See Problem 5.)

14. $\displaystyle\int_0^2 \int_1^5 \frac{x}{\sqrt{y + x^2}}\,dy\,dx$
(See Problem 6.)

15. $\displaystyle\int_1^e \int_1^{e^2} \frac{\ln x}{xy}\,dy\,dx$
(See Problem 7.)

16. $\displaystyle\int_1^{e^2} \int_1^e \frac{\ln x}{xy}\,dx\,dy$
(See Problem 8.)

Use both orders of iteration to evaluate each double integral in Problems 17–20.

17. $\displaystyle\iint_R xy\,dA$; $R = \{(x, y)|0 \leqslant x \leqslant 2, \quad 0 \leqslant y \leqslant 4\}$

18. $\displaystyle\iint_R \sqrt{xy}\,dA$; $R = \{(x, y)|1 \leqslant x \leqslant 4, \quad 1 \leqslant y \leqslant 9\}$

19. $\displaystyle\iint_R (x + y)^5\,dA$;
$R = \{(x, y)|-1 \leqslant x \leqslant 1, \quad 1 \leqslant y \leqslant 2\}$

20. $\displaystyle\iint_R xe^y\,dA$; $R = \{(x, y)|-2 \leqslant x \leqslant 3, \quad 0 \leqslant y \leqslant 2\}$

In Problems 21–24, find the average value of each function over the given rectangle.

21. $f(x, y) = (x + y)^2$;
$R = \{(x, y) | 1 \le x \le 5, \quad -1 \le y \le 1\}$

22. $f(x, y) = x^2 + y^2$;
$R = \{(x, y) | -1 \le x \le 2, \quad 1 \le y \le 4\}$

23. $f(x, y) = x/y$; $R = \{(x, y) | 1 \le x \le 4, \quad 2 \le y \le 7\}$

24. $f(x, y) = x^2y^3$;
$R = \{(x, y) | -1 \le x \le 1, \quad 0 \le y \le 2\}$

In Problems 25–28, find the volume of the solid under the graph of each function over the given rectangle.

25. $f(x, y) = 2 - x^2 - y^2$;
$R = \{(x, y) | 0 \le x \le 1, \quad 0 \le y \le 1\}$

26. $f(x, y) = 5 - x$;
$R = \{(x, y) | 0 \le x \le 5, \quad 0 \le y \le 5\}$

27. $f(x, y) = 4 - y^2$;
$R = \{(x, y) | 0 \le x \le 2, \quad 0 \le y \le 2\}$

28. $f(x, y) = e^{-x-y}$;
$R = \{(x, y) | 0 \le x \le 1, \quad 0 \le y \le 1\}$

C *Evaluate each double integral in Problems 29–32. Select the order of integration carefully—each problem is easy to do one way and difficult the other.*

29. $\iint_R xe^{xy}\, dA$; $R = \{(x, y) | 0 \le x \le 1, \quad 1 \le y \le 2\}$

30. $\iint_R xye^{x^2y}\, dA$; $R = \{(x, y) | 0 \le x \le 1, \quad 1 \le y \le 2\}$

31. $\iint_R \dfrac{2y + 3xy^2}{1 + x^2}\, dA$;
$R = \{(x, y) | 0 \le x \le 1, \quad -1 \le y \le 1\}$

32. $\iint_R \dfrac{2x + 2y}{1 + 4y + y^2}\, dA$;
$R = \{(x, y) | 1 \le x \le 3, \quad 0 \le y \le 1\}$

33. Show that $\int_0^2 \int_0^2 (1 - y)\, dx\, dy = 0$. Does the double integral represent the volume of a solid? Explain.

34. (A) Find the average values of the functions $f(x, y) = x + y$, $g(x, y) = x^2 + y^2$, and $h(x, y) = x^3 + y^3$ over the rectangle

$$R = \{(x, y) | 0 \le x \le 1, \quad 0 \le y \le 1\}$$

(B) Does the average value of $k(x, y) = x^n + y^n$ over the rectangle

$$R_1 = \{(x, y) | 0 \le x \le 1, \quad 0 \le y \le 1\}$$

increase or decrease as n increases? Explain.

(C) Does the average value of $k(x, y) = x^n + y^n$ over the rectangle

$$R_2 = \{(x, y) | 0 \le x \le 2, \quad 0 \le y \le 2\}$$

increase or decrease as n increases? Explain.

35. Let $f(x, y) = x^3 + y^2 - e^{-x} - 1$.
(A) Find the average value of $f(x, y)$ over the rectangle $R = \{(x, y) | -2 \le x \le 2, \quad -2 \le y \le 2\}$.
C (B) Graph the set of all points (x, y) in R for which $f(x, y) = 0$.
(C) For which points (x, y) in R is $f(x, y)$ greater than 0? Less than 0? Explain.

36. C Find the dimensions of the square S centered at the origin for which the average value of $f(x, y) = x^2e^y$ over S is equal to 100.

Applications

Business & Economics

37. *Multiplier principle.* Suppose Congress enacts a one-time-only 10% tax rebate that is expected to infuse $\$y$ billion, $5 \le y \le 7$, into the economy. If every individual and corporation is expected to spend a proportion x, $0.6 \le x \le 0.8$, of each dollar received, then by the **multiplier principle** in economics, the total amount of spending S (in billions of dollars) generated by this tax rebate is given by

$$S(x, y) = \frac{y}{1 - x}$$

What is the average total amount of spending for the indicated ranges of the values of x and y? Set up a double integral and evaluate.

38. *Multiplier principle.* Repeat Problem 37 if $6 \le y \le 10$ and $0.7 \le x \le 0.9$.

39. *Cobb–Douglas production function.* If an industry invests x thousand labor-hours, $10 \leq x \leq 20$, and $\$y$ million, $1 \leq y \leq 2$, in the production of N thousand units of a certain item, then N is given by

$$N(x, y) = x^{0.75}y^{0.25}$$

What is the average number of units produced for the indicated ranges of x and y? Set up a double integral and evaluate.

40. *Cobb–Douglas production function.* Repeat Problem 39 for

$$N(x, y) = x^{0.5}y^{0.5}$$

where $10 \leq x \leq 30$ and $1 \leq y \leq 3$.

Life Sciences

41. *Population distribution.* In order to study the population distribution of a certain species of insects, a biologist has constructed an artificial habitat in the shape of a rectangle 16 feet long and 12 feet wide. The only food available to the insects in this habitat is located at its center. The biologist has determined that the concentration C of insects per square foot at a point d units from the food supply (see the figure) is given approximately by

$$C = 10 - \tfrac{1}{10}d^2$$

What is the average concentration of insects throughout the habitat? Express C as a function of x and y, set up a double integral, and evaluate.

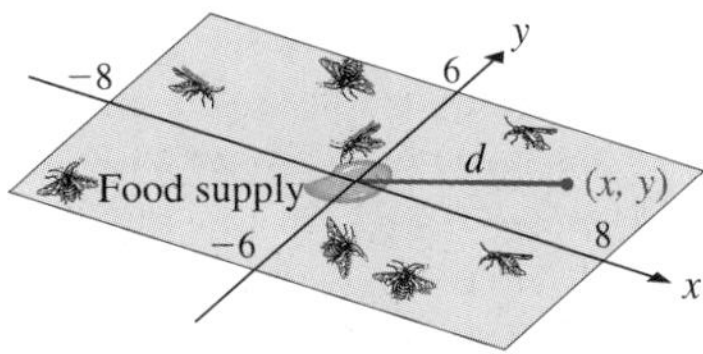

Figure for 41

42. *Population distribution.* Repeat Problem 41 for a square habitat that measures 12 feet on each side, where the insect concentration is given by

$$C = 8 - \tfrac{1}{10}d^2$$

43. *Pollution.* A heavy industrial plant located in the center of a small town emits particulate matter into the atmosphere. Suppose the concentration of particulate matter (in parts per million) at a point d miles from the plant (see the figure) is given by

$$C = 100 - 15d^2$$

If the boundaries of the town form a rectangle 4 miles long and 2 miles wide, what is the average concentration of particulate matter throughout the city? Express C as a function of x and y, set up a double integral, and evaluate.

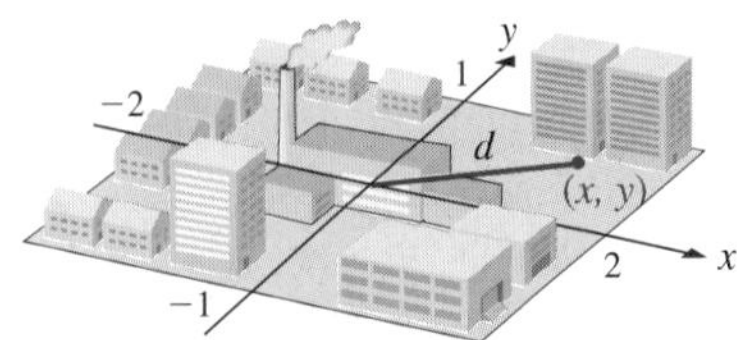

Figure for 43

44. *Pollution.* Repeat Problem 43 if the boundaries of the town form a rectangle 8 miles long and 4 miles wide and the concentration of particulate matter is given by

$$C = 100 - 3d^2$$

Social Sciences

45. *Safety research.* Under ideal conditions, if a person driving a car slams on the brakes and skids to a stop, the length of the skid marks (in feet) is given by the formula

$$L = 0.000\ 013\ 3xy^2$$

where x is the weight of the car (in pounds) and y is the speed of the car (in miles per hour). What is the average length of the skid marks for cars weighing between 2,000 and 3,000 pounds and traveling at speeds between 50 and 60 miles per hour? Set up a double integral and evaluate.

46. *Safety research.* Repeat Problem 45 for cars weighing between 2,000 and 2,500 pounds and traveling at speeds between 40 and 50 miles per hour.

47. *Psychology.* The intelligence quotient Q for an individual with mental age x and chronological age y is given by

$$Q(x, y) = 100\frac{x}{y}$$

In a group of sixth graders, the mental age varies between 8 and 16 years and the chronological age varies between 10 and 12 years. What is the average intelligence quotient for this group? Set up a double integral and evaluate.

48. *Psychology.* Repeat Problem 47 for a group with mental ages between 6 and 14 years and chronological ages between 8 and 10 years.

Important Terms and Symbols

13-1 *Functions of Several Variables.* Functions of two independent variables; functions of several independent variables; Cobb–Douglas production function; three-dimensional coordinate system; triplet of numbers (x, y, z); surface; paraboloid; saddle point

$$z = f(x, y); \quad w = f(x, y, z)$$

13-2 *Partial Derivatives.* Partial derivative of f with respect to x; partial derivative of f with respect to y; marginal productivity of labor; marginal productivity of capital; second-order partial derivatives

$$\frac{\partial z}{\partial x}; \quad \frac{\partial z}{\partial y}; \quad f_x(x, y); \quad f_y(x, y);$$

$$\frac{\partial^2 z}{\partial x^2} = f_{xx}(x, y); \quad \frac{\partial^2 z}{\partial x\,\partial y} = f_{yx}(x, y);$$

$$\frac{\partial^2 z}{\partial y\,\partial x} = f_{xy}(x, y); \quad \frac{\partial^2 z}{\partial y^2} = f_{yy}(x, y)$$

13-3 *Maxima and Minima.* Local maximum; local minimum; saddle point; critical point; second-derivative test

13-4 *Maxima and Minima Using Lagrange Multipliers.* Constraint; Lagrange multiplier; critical point; method of Lagrange multipliers for functions of two variables; marginal productivity of money; method of Lagrange multipliers for functions of three variables

13-5 *Method of Least Squares.* Least squares approximation; regression analysis; method of least squares; linear regression; residual; least squares line; regression line; normal equations

13-6 *Double Integrals Over Rectangular Regions.* Double integral; integrand; region of integration; iterated integral; average value over rectangular region; volume under a surface

$$\iint_R f(x, y)\,dA = \int_a^b \left[\int_c^d f(x, y)\,dy\right] dx$$

$$= \int_c^d \left[\int_a^b f(x, y)\,dx\right] dy;$$

$$\frac{1}{(b - a)(d - c)} \iint_R f(x, y)\,dA;$$

$$V = \iint_R f(x, y)\,dA$$

Review Exercise

Work through all the problems in this chapter review and check your answers in the back of the book. Answers to all review problems are there along with section numbers in italics to indicate where each type of problem is discussed. Where weaknesses show up, review appropriate sections in the text.

A

1. For $f(x, y) = 2{,}000 + 40x + 70y$, find $f(5, 10)$, $f_x(x, y)$, and $f_y(x, y)$.
2. For $z = x^3y^2$, find $\partial^2 z/\partial x^2$ and $\partial^2 z/\partial x\,\partial y$.
3. Evaluate: $\int (6xy^2 + 4y)\,dy$
4. Evaluate: $\int (6xy^2 + 4y)\,dx$
5. Evaluate: $\int_0^1 \int_0^1 4xy\,dy\,dx$
6. For $f(x, y) = 6 + 5x - 2y + 3x^2 + x^3$, find $f_x(x, y)$ and $f_y(x, y)$, and explain why $f(x, y)$ has no local extrema.

B

7. For $f(x, y) = 3x^2 - 2xy + y^2 - 2x + 3y - 7$, find $f(2, 3)$, $f_y(x, y)$, and $f_y(2, 3)$.
8. For $f(x, y) = -4x^2 + 4xy - 3y^2 + 4x + 10y + 81$, find $[f_{xx}(2, 3)][f_{yy}(2, 3)] - [f_{xy}(2, 3)]^2$
9. If $f(x, y) = x + 3y$ and $g(x, y) = x^2 + y^2 - 10$, find the critical points of $F(x, y, \lambda) = f(x, y) + \lambda g(x, y)$.
10. Use the least squares line for the data in the table to estimate y when $x = 10$.

x	y
2	12
4	10
6	7
8	3

11. For $R = \{(x, y)|-1 \le x \le 1, \quad 1 \le y \le 2\}$, evaluate the following in two ways:

$$\iint_R (4x + 6y)\, dA$$

C

12. For $f(x, y) = e^{x^2+2y}$, find f_x, f_y, and f_{xy}.

13. For $f(x, y) = (x^2 + y^2)^5$, find f_x and f_{xy}.

14. Find all critical points and test for extrema for

$$f(x, y) = x^3 - 12x + y^2 - 6y$$

15. Use Lagrange multipliers to maximize $f(x, y) = xy$ subject to $2x + 3y = 24$.

16. Use Lagrange multipliers to minimize $f(x, y, z) = x^2 + y^2 + z^2$ subject to $2x + y + 2z = 9$.

17. Find the least squares line for the data in the table.

x	y	x	y
10	50	60	80
20	45	70	85
30	50	80	90
40	55	90	90
50	65	100	110

18. Find the average value of $f(x, y) = x^{2/3}y^{1/3}$ over the rectangle

$$R = \{(x, y)|-8 \le x \le 8, \quad 0 \le y \le 27\}$$

19. Find the volume of the solid under the graph of $z = 3x^2 + 3y^2$ over the rectangle

$$R = \{(x, y)|0 \le x \le 1, \quad -1 \le y \le 1\}$$

20. Without doing any computation, predict the average value of $f(x, y) = x + y$ over the rectangle $R = \{(x, y)|-10 \le x \le 10, \quad -10 \le y \le 10\}$. Then check the correctness of your prediction by evaluating a double integral.

21. C (A) Find the dimensions of the square S centered at the origin such that the average value of

$$f(x, y) = \frac{e^x}{y + 10}$$

over S is equal to 5.

(B) Is there a square centered at the origin over which

$$f(x, y) = \frac{e^x}{y + 10}$$

has average value 0.05? Explain.

22. Explain why the function $f(x, y) = 4x^3 - 5y^3$, subject to the constraint $3x + 2y = 7$, has no maxima or minima.

APPLICATIONS

Business & Economics

23. *Maximizing profit.* A company produces x units of product A and y units of product B (both in hundreds per month). The monthly profit equation (in thousands of dollars) is found to be

$$P(x, y) = -4x^2 + 4xy - 3y^2 + 4x + 10y + 81$$

(A) Find $P_x(1, 3)$ and interpret the results.
(B) How many of each product should be produced each month to maximize profit? What is the maximum profit?

24. *Minimizing material.* A rectangular box with no top and six compartments (see the figure) is to have a volume of 96 cubic inches. Find the dimensions that will require the least amount of material.

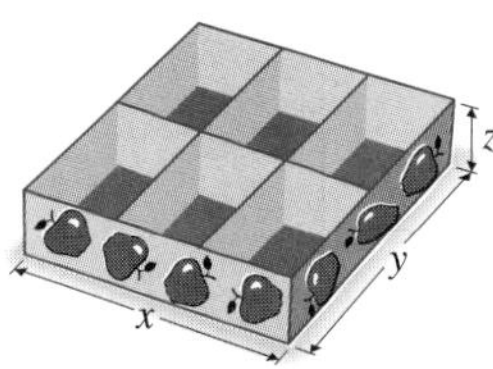

Figure for 24

25. *Profit.* A company's annual profits (in millions of dollars) over a 5 year period are given in the table. Use the least squares line to estimate the profit for the sixth year.

YEAR	PROFIT
1	2
2	2.5
3	3.1
4	4.2
5	4.3

26. *Productivity.* The Cobb–Douglas production function for a product is

$$N(x, y) = 10x^{0.8}y^{0.2}$$

where x is the number of units of labor and y is the number of units of capital required to produce N units of the product.

(A) Find the marginal productivity of labor and the marginal productivity of capital at $x = 40$ and $y = 50$. For the greatest increase in productivity, should management encourage increased use of labor or increased use of capital?

(B) If each unit of labor costs \$100, each unit of capital costs \$50, and \$10,000 is budgeted for production of this product, use the method of Lagrange multipliers to determine the allocations of labor and capital that will maximize the number of units produced and find the maximum production. Find the marginal productivity of money and approximate the increase in production that would result from an increase of \$2,000 in the amount budgeted for production.

(C) If $50 \leq x \leq 100$ and $20 \leq y \leq 40$, find the average number of units produced. Set up a definite integral and evaluate.

Life Sciences

27. *Marine biology.* The function used for timing dives with scuba gear is

$$T(V, x) = \frac{33V}{x + 33}$$

where T is the time of the dive in minutes, V is the volume of air (in cubic feet, at sea level pressure) compressed into tanks, and x is the depth of the dive in feet. Find $T_x(70, 17)$ and interpret the results.

28. *Pollution.* A heavy industrial plant located in the center of a small town emits particulate matter into the atmosphere. Suppose the concentration of particulate matter (in parts per million) at a point d miles from the plant is given by

$$C = 100 - 24d^2$$

If the boundaries of the town form a square 4 miles long and 4 miles wide, what is the average concentration of particulate matter throughout the town? Express C as a function of x and y, set up a double integral, and evaluate.

Social Sciences

29. *Sociology.* Joseph Cavanaugh, a sociologist, found that the number of long-distance telephone calls, n, between two cities in a given period of time varied (approximately) jointly as the populations P_1 and P_2 of the two cities, and varied inversely as the distance, d, between the two cities. In terms of an equation for a time period of 1 week,

$$n(P_1, P_2, d) = 0.001\frac{P_1P_2}{d}$$

Find $n(100{,}000, 50{,}000, 100)$.

30. *Education.* At the beginning of the semester, students in a foreign language course are given a proficiency exam. The same exam is given at the end of the semester. The results for 5 students are given in the table. Use the least squares line to estimate the score on the second exam for a student who scored 40 on the first exam.

FIRST EXAM	SECOND EXAM
30	60
50	75
60	80
70	85
90	90

31. *Population density.* The table gives the population per square mile in the United States for the years 1900–1990.

U.S. POPULATION DENSITY

YEAR	POPULATION (PER SQUARE MILE)	YEAR	POPULATION (PER SQUARE MILE)
1900	25.6	1950	50.7
1910	31.0	1960	50.6
1920	35.6	1970	57.4
1930	41.2	1980	64.0
1940	44.2	1990	70.3

(A) Find the least squares line for the data.

(B) Use linear regression to estimate the population density in the United States in the year 2000.

C (C) Use quadratic regression and exponential regression to make the estimate of part (B).

32. *Life expectancy.* The table gives life expectancies for males and females in a sample of Central and South American countries.

Life Expectancies for Central and South American Countries

Males	Females	Males	Females
62.30	67.50	70.15	74.10
68.05	75.05	62.93	66.58
72.40	77.04	68.43	74.88
63.39	67.59	66.68	72.80
55.11	59.43		

(A) Find the least squares line for the data.

(B) Use linear regression to estimate the life expectancy of a female in a Central or South American country in which the life expectancy for males is 60 years.

C (C) Use quadratic and logarithmic regression to make the estimate of part (B).

33. C *Women in the work force.* It is reasonable to conjecture from the data given in the table that many Japanese women tend to leave the work force to marry and have children, but then reenter the work force when the children are grown.

Women in the Work Force in Japan (1993)

Age	Percentage of Women Employed	Age	Percentage of Women Employed
22	75	47	72
27	64	52	66
32	52	57	57
37	61	62	40
42	70		

(A) Explain why you might expect cubic regression to provide a better fit to the data than linear or quadratic regression.

(B) Investigate your expectation by plotting the data points and graphing the curve of best fit using linear, quadratic, and cubic regression.

Group Activity 1 City Planning

A city planning commission is seeking to identify prime locations for a new zoo and a new hospital. The city's economy is heavily dependent on two industrial plants located relatively near the city center. Both emit particulate matter into the atmosphere, and the resulting air pollution is of concern to the commission and will influence its decisions. The consensus of the commission is that the new zoo should be built in the least polluted area within the city limits, and the new hospital should be built in the least polluted location within 2 miles of the city center.

The boundaries of the city form a rectangle 10 miles from east to west and 6 miles from north to south. When a coordinate system is chosen with the origin at the center of the rectangle (the city center), industrial plant 1 has coordinates $(-1, 1)$ and industrial plant 2 has coordinates $(1, 0)$. At a point (x, y), the concentration of particulate matter (in parts per million) due to emissions from plant 1 is given by $C_1 = 200 - 3(d_1)^2$, where d_1 is the distance from (x, y) to plant 1. Similarly, the concentration due to emissions from plant 2 is given by $C_2 = 200 - 3(d_2)^2$, where d_2 is the distance from (x, y) to plant 2.

(A) Find the point within the city limits that has the greatest concentration of particulate matter.

(B) Find the points on the city boundaries that have the greatest and least concentrations of particulate matter.

(C) Find the average concentration of particulate matter throughout the city.

(D) Find the points on the circle of radius 2 miles, centered at the origin, that have the greatest and least concentrations of particulate matter.

(E) Determine the optimal locations for the city's new zoo and new hospital.

Group Activity 2 Numerical Integration of Multivariable Functions

A definite integral $\int_a^b f(x)\,dx$ is a limit of Riemann sums of the form

$$\sum_{k=1}^{n} f(c_k)\,\Delta x_k$$

(see Section 11-5). Analogously, a double integral $\iint_R f(x, y)\,dx\,dy$ over a rectangle $R = \{(x, y) | a \leq x \leq b, \quad c \leq x \leq d\}$ is a limit of sums of the form

$$\sum_{j=1}^{m} \sum_{k=1}^{n} f(c_j, d_k)\,\Delta x_j\,\Delta y_k$$

where $a = x_0 < x_1 < x_2 < \cdots < x_m = b, c = y_0 < y_1 < y_2 < \cdots < y_n = d$, $x_{j-1} \leq c_j \leq x_j$ for $j = 1, 2, \ldots, m$, and $y_{k-1} \leq d_k \leq y_k$ for $k = 1, 2, \ldots, n$.

It follows that numerical integration methods such as Simpson's rule (see Group Activity 1 in Chapter 11) can also be used to approximate double integrals. We illustrate with an example.

Example 1 **Using Simpson's Rule to Approximate Double Integrals** Use Simpson's rule to approximate

$$\int_0^4 \int_0^6 (x^2y + 2y^2 + 3x + 1)\,dy\,dx$$

by partitioning [0, 4] into four equal subintervals and [0, 6] into six equal subintervals.

SOLUTION We evaluate the function $f(x, y) = x^2y + 2y^2 + 3x + 1$ at each of the 35 intersection points of the grid determined by the two partitions (see Table 1).

Table 1

	$y = 0$	1	2	3	4	5	6	R_j
$x = 0$	1	3	9	19	33	51	73	150
1	4	7	14	25	40	59	82	186
2	7	13	23	37	55	77	103	258
3	10	21	36	55	78	105	136	366
4	13	31	53	79	109	143	181	510

Then we apply Simpson's rule for six equal subdivisions of an interval to each row of the table to compute the entries in the column labeled R_j. For example, letting $g(y) = f(x_0, y)$, we have

$$\begin{aligned} R_0 &= [g(y_0) + 4g(y_1) + 2g(y_2) + 4g(y_3) + 2g(y_4) + 4g(y_5) + g(y_6)]\frac{\Delta y}{3} \\ &= [(1) + 4(3) + 2(9) + 4(19) + 2(33) + 4(51) + 1(73)]\tfrac{1}{3} \\ &= 150 \end{aligned}$$

Finally, we apply Simpson's rule for four equal subdivisions of an interval to the values R_0, R_1, R_2, R_3, R_4:

$$\begin{aligned} S &= [R_0 + 4R_1 + 2R_2 + 4R_3 + R_4]\frac{\Delta x}{3} \\ &= [150 + 4(186) + 2(258) + 4(366) + 510]\tfrac{1}{3} \\ &= 1{,}128 \end{aligned}$$

Therefore, the value of the double integral $\int_0^4 \int_0^6 (x^2y + 2y^2 + 3x + 1)\,dy\,dx$ is approximately 1,128.

In Example 1, the double integral can be evaluated as an iterated integral. The exact value of the double integral is also 1,128. This is not surprising since Simpson's rule gives exact results for polynomials of degree 3 or less. But the technique illustrated above can also be applied when it is impossible to find antiderivatives, or when we have only a table of values (and not a formula) for $f(x, y)$.

(A) Evaluate the double integral of Example 1 as an iterated integral.

(B) Show that the sum S in Example 1 may also be found by applying Simpson's rule to each column of Table 1, obtaining $C_0, C_1, \ldots, C_6$ and then applying Simpson's rule to $C_0, C_1, \ldots, C_6$.

(C) Approximate

$$\int_0^3 \int_0^3 e^{-(x^2+y^2)}\,dx\,dy$$

by partitioning each of the intervals $[0, 3]$ into six equal subintervals, evaluating the function $f(x, y) = e^{-(x^2+y^2)}$ (to four decimal places) at each of the 49 intersection points of the grid determined by the two partitions, and applying Simpson's rule.

(D) Repeat part (C), but use a numerical integration routine on a graphing utility, rather than Simpson's rule, to approximate the R_j.

(E) It can be shown that the volume under the graph of $f(x, y) = e^{-(x^2+y^2)}$ and above the entire first quadrant is $\pi/4$. Compare your answers to parts (C) and (D) with this value and explain any discrepancy.

APPENDIX A

Basic Algebra Review

INTRODUCTION

Appendix A reviews some important basic algebra concepts usually studied in earlier courses. The material may be studied systematically before commencing with the rest of the book or reviewed as needed. The Self-Test on Basic Algebra that precedes Section A-1 may be taken to locate areas of weakness. All the answers to the self-test are in the answer section in the back of the book and are keyed to the sections in Appendix A where the related topics are discussed.

SELF-TEST ON BASIC ALGEBRA

Work through all the problems in this self-test and check your answers in the back of the book. All answers are there and are keyed to relevant sections in Appendix A. Where weaknesses show up, review appropriate sections in the appendix.

1. Indicate true (T) or false (F):
(A) $7 \notin \{4, 6, 8\}$ (B) $\{8\} \subset \{4, 6, 8\}$
(C) $\varnothing \in \{4, 6, 8\}$ (D) $\varnothing \subset \{4, 6, 8\}$

2. Replace each question mark with an appropriate expression that will illustrate the use of the indicated real number property:
(A) Commutative $(\cdot)$: $x(y + z) = ?$
(B) Associative $(+)$: $2 + (x + y) = ?$
(C) Distributive: $(2 + 3)x = ?$

Problems 3–7 refer to the following polynomials:
(A) $3x - 4$
(B) $x + 2$
(C) $3x^2 + x - 8$
(D) $x^3 + 8$

3. Add all four.

4. Subtract the sum of (A) and (C) from the sum of (B) and (D).

5. Multiply (C) and (D).

6. What is the degree of (D)?

7. What is the coefficient of the second term in (C)?

In Problems 8–13, perform the indicated operations and simplify.

8. $5x^2 - 3x[4 - 3(x - 2)]$

9. $(2x + y)(3x - 4y)$

10. $(2a - 3b)^2$

11. $(2x - y)(2x + y) - (2x - y)^2$

12. $(m^2 + 2mn - n^2)(m^2 - 2mn - n^2)$

13. $(x - 2y)^3$

14. Write in scientific notation:
(A) 4,065,000,000,000 (B) 0.0073

15. Write in standard decimal form:
(A) 2.55×10^8 (B) 4.06×10^{-4}

16. If $U = \{2, 4, 5, 6, 8\}$, $M = \{2, 4, 5\}$, and $N = \{5, 6\}$, find:
(A) $M \cup N$ (B) $M \cap N$
(C) $(M \cup N)'$ (D) $M \cap N'$

17. Given the Venn diagram shown in the figure, how many elements are in each of the following sets:
(A) $A \cup B$ (B) $A \cap B$
(C) $(A \cup B)'$ (D) $A \cap B'$

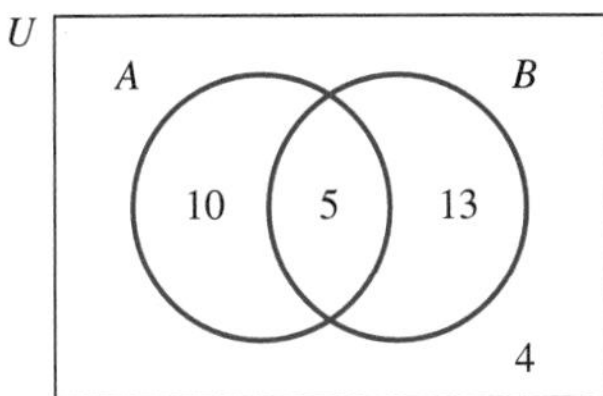

18. Indicate true (T) or false (F):
(A) A natural number is a rational number.
(B) A number with a repeating decimal expansion is an irrational number.

19. Give an example of an integer that is not a natural number.

Simplify Problems 20–28 and write answers using positive exponents only. All variables represent positive real numbers.

20. $6(xy^3)^5$

21. $\dfrac{9u^8v^6}{3u^4v^8}$

22. $(2 \times 10^5)(3 \times 10^{-3})$

23. $(x^{-3}y^2)^{-2}$

24. $u^{5/3}u^{2/3}$

25. $(9a^4b^{-2})^{1/2}$

26. $\dfrac{5^0}{3^2} + \dfrac{3^{-2}}{2^{-2}}$

27. $(x^{1/2} + y^{1/2})^2$

28. $(3x^{1/2} - y^{1/2})(2x^{1/2} + 3y^{1/2})$

Write Problems 29–34 in completely factored form relative to the integers. If a polynomial cannot be factored further relative to the integers, say so.

29. $12x^2 + 5x - 3$

30. $8x^2 - 18xy + 9y^2$

31. $t^2 - 4t - 6$

32. $6n^3 - 9n^2 - 15n$

33. $(4x - y)^2 - 9x^2$

34. $2x^2 + 4xy - 5y^2$

In Problems 35–40, perform the indicated operations and reduce to lowest terms. Represent all compound fractions as simple fractions reduced to lowest terms.

35. $\dfrac{2}{5b} - \dfrac{4}{3a^3} - \dfrac{1}{6a^2b^2}$

36. $\dfrac{3x}{3x^2 - 12x} + \dfrac{1}{6x}$

37. $\dfrac{x}{x^2 - 16} - \dfrac{x + 4}{x^2 - 4x}$

38. $\dfrac{y - 2}{y^2 - 4y + 4} \div \dfrac{y^2 + 2y}{y^2 + 4y + 4}$

39. $\dfrac{\dfrac{1}{7 + h} - \dfrac{1}{7}}{h}$

40. $\dfrac{x^{-1} + y^{-1}}{x^{-2} - y^{-2}}$

41. Each statement illustrates the use of one of the following real number properties or definitions. Indicate which one.

Commutative (+, •)	Associative (+, •)	Distributive
Identity (+, •)	Inverse (+, •)	Subtraction
Division	Negatives	Zero

(A) $(-7) - (-5) = (-7) + [-(-5)]$
(B) $5u + (3v + 2) = (3v + 2) + 5u$
(C) $(5m - 2)(2m + 3) = (5m - 2)2m + (5m - 2)3$
(D) $9 \cdot (4y) = (9 \cdot 4)y$
(E) $\dfrac{u}{-(v - w)} = -\dfrac{u}{v - w}$
(F) $(x - y) + 0 = (x - y)$

42. In a freshman class of 100 students, 70 are taking English, 45 are taking math, and 25 are taking both English and math.
(A) How many students are taking either English or math?
(B) How many students are taking English and not math?

43. Change to rational exponent form:
$6\sqrt[5]{x^2} - 7\sqrt[4]{(x - 1)^3}$

44. Change to radical form: $2x^{1/2} - 3x^{2/3}$

45. Write in the form $ax^p + bx^q$, where a and b are real numbers and p and q are rational numbers:

$$\frac{4\sqrt{x} - 3}{2\sqrt{x}}$$

In Problems 46 and 47, rationalize the denominator.

46. $\dfrac{3x}{\sqrt{3x}}$

47. $\dfrac{x-5}{\sqrt{x}-\sqrt{5}}$

In Problems 48 and 49, rationalize the numerator.

48. $\dfrac{\sqrt{x}-5}{x-5}$

49. $\dfrac{\sqrt{u+h}-\sqrt{u}}{h}$

Solve Problems 50–54 for x.

50. $\dfrac{x}{12}-\dfrac{x-3}{3}=\dfrac{1}{2}$

51. $x^2 = 5x$

52. $3x^2 - 21 = 0$

53. $x^2 - x - 20 = 0$

54. $2x = 3 + \dfrac{1}{x}$

In Problems 55–57, solve and graph on a real number line.

55. $2(x+4) > 5x - 4$

56. $1 - \dfrac{x-3}{3} \leqslant \dfrac{1}{2}$

57. $-2 \leqslant \dfrac{x}{2} - 3 < 3$

In Problems 58 and 59, solve for y in terms of x.

58. $2x - 3y = 6$

59. $xy - y = 3$

60. If $A \cap B = A$, then is it always true that $A \subset B$? Explain.

APPLICATIONS

61. *Economics.* If the Gross Domestic Product (GDP) was \$5,951,000,000,000 for the United States in 1992 and the population was 255,100,000, determine the GDP per person using scientific notation. Express the answer in scientific notation and in standard decimal form to the nearest dollar.

62. *Investment.* An investor has \$60,000 to invest. If part is invested at 8% and the rest at 14%, how much should be invested at each rate to yield 12% on the total amount?

63. *Break-even analysis.* A producer of educational videos is producing an instructional video. The producer estimates it will cost \$72,000 to shoot the video and \$12 per unit to copy and distribute the tape. If the wholesale price of the tape is \$30, how many tapes must be sold for the producer to break even?

SECTION A-1 Sets

- SET PROPERTIES AND SET NOTATION
- SET OPERATIONS
- APPLICATION

In this section we will review a few key ideas from set theory. Set concepts and notation not only help us talk about certain mathematical ideas with greater clarity and precision, but are indispensable to a clear understanding of probability.

SET PROPERTIES AND SET NOTATION

We can think of a **set** as any collection of objects specified in such a way that we can tell whether any given object is or is not in the collection. Capital letters, such as *A*, *B*, and *C*, are often used to designate particular sets. Each object in a set is called a **member,** or **element,** of the set. Symbolically:

$a \in A$	means	"a is an element of set A"
$a \notin A$	means	"a is not an element of set A"

A set without any elements is called the **empty,** or **null, set.** For example, the set of all people over 20 feet tall is an empty set. Symbolically:

$\varnothing$	represents	"the empty, or null, set"

A set is usually described either by listing all its elements between braces { } or by enclosing a rule within braces that determines the elements of the set. Thus, if $P(x)$ is a statement about x, then

$S = \{x \mid P(x)\}$	means	"S is the set of all x such that $P(x)$ is true"

Recall that the vertical bar within the braces is read "such that." The following example illustrates the rule and listing methods of representing sets.

Example 1 **Representing Sets**

$$\begin{aligned} &\text{Rule} & &\text{Listing} \\ \{x \mid x \text{ is a weekend day}\} &= \{\text{Saturday, Sunday}\} \\ \{x \mid x^2 = 4\} &= \{-2, 2\} \\ \{x \mid x \text{ is an odd positive counting number}\} &= \{1, 3, 5, \ldots\} \end{aligned}$$

The three dots (...) in the last set given in Example 1 indicate that the pattern established by the first three entries continues indefinitely. The first two sets in Example 1 are **finite sets** (we intuitively know that the elements can be counted, and there is an end); the last set is an **infinite set** (we intuitively know that there is no end in counting the elements). When listing the elements in a set, we do not list an element more than once, and the order in which the elements are listed does not matter.

Matched Problem 1 Let G be the set of all numbers such that $x^2 = 9$.

(A) Denote G by the rule method. (B) Denote G by the listing method.
(C) Indicate whether the following are true or false: $3 \in G$; $9 \notin G$.

If each element of a set A is also an element of set B, we say that A is a **subset** of B. For example, the set of all women students in a class is a subset of the whole class. Note that the definition allows a set to be a subset of itself. If set A and set B have exactly the same elements, then the two sets are said to be **equal.** Symbolically:

$A \subset B$	means	"A is a subset of B"
$A = B$	means	"A and B have exactly the same elements"
$A \not\subset B$	means	"A is not a subset of B"
$A \neq B$	means	"A and B do not have exactly the same elements"

It can be proved that:

$\varnothing$ is a subset of every set.

Example 2 **Set Notation** If $A = \{-3, -1, 1, 3\}$, $B = \{3, -3, 1, -1\}$, and $C = \{-3, -2, -1, 0, 1, 2, 3\}$, then each of the following statements is true:

$$\begin{array}{lll} A = B & A \subset C & A \subset B \\ C \neq A & C \not\subset A & B \subset A \\ \varnothing \subset A & \varnothing \subset C & \varnothing \notin A \end{array}$$

Matched Problem 2 Given $A = \{0, 2, 4, 6\}$, $B = \{0, 1, 2, 3, 4, 5, 6\}$, and $C = \{2, 6, 0, 4\}$, indicate whether the following relationships are true (T) or false (F):

(A) $A \subset B$ (B) $A \subset C$ (C) $A = C$
(D) $C \subset B$ (E) $B \not\subset A$ (F) $\varnothing \subset B$

Example 3 **Subsets** List all the subsets of the set $\{a, b, c\}$.

SOLUTION $\{a, b, c\}$, $\{a, b\}$, $\{a, c\}$, $\{b, c\}$, $\{a\}$, $\{b\}$, $\{c\}$, $\varnothing$

Matched Problem 3 List all the subsets of the set $\{1, 2\}$.

Explore–Discuss 1

Which of the following statements are true?

(A) $\varnothing \subset \varnothing$ (B) $\varnothing \in \varnothing$ (C) $\varnothing = \{0\}$ (D) $\varnothing \subset \{0\}$

Set Operations

The **union** of sets A and B, denoted by $A \cup B$, is the set of all elements formed by combining all the elements of A and all the elements of B into one set. Symbolically:

Union

$$A \cup B = \{x | x \in A \textbf{ or } x \in B\}$$

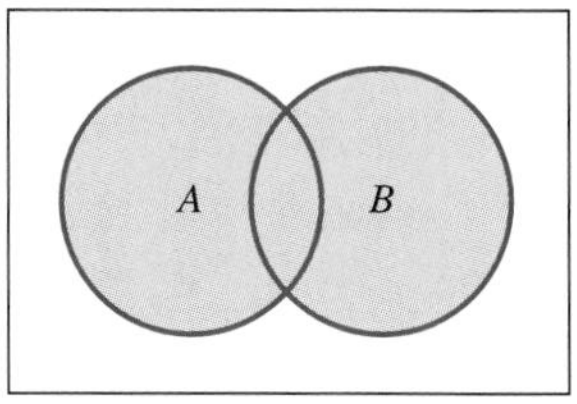

FIGURE 1
$A \cup B$ is the shaded region.

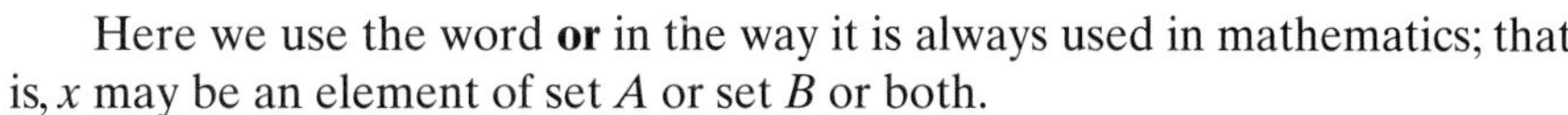

Here we use the word **or** in the way it is always used in mathematics; that is, x may be an element of set A or set B or both.

Venn diagrams are useful in visualizing set relationships. The union of two sets can be illustrated as shown in Figure 1. Note that

$$A \subset A \cup B \qquad \text{and} \qquad B \subset A \cup B$$

The **intersection** of sets A and B, denoted by $A \cap B$, is the set of elements in set A that are also in set B. Symbolically:

Intersection

$$A \cap B = \{x | x \in A \textbf{ and } x \in B\}$$

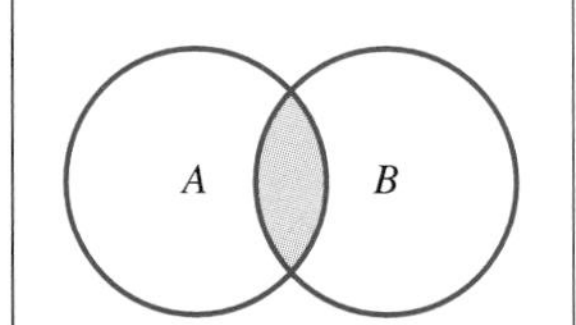

FIGURE 2
$A \cap B$ is the shaded region.

This relationship is easily visualized in the Venn diagram shown in Figure 2. Note that

$$A \cap B \subset A \qquad \text{and} \qquad A \cap B \subset B$$

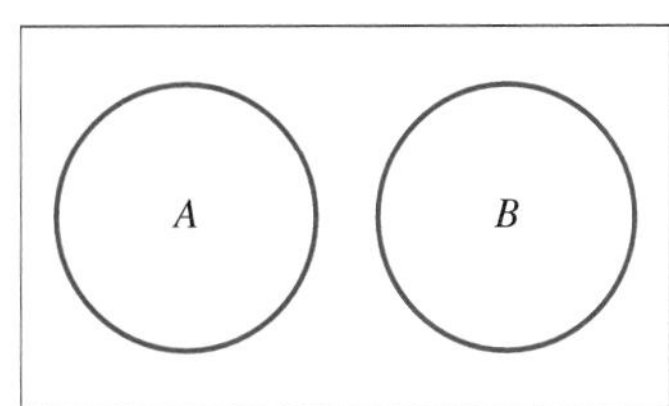

FIGURE 3
$A \cap B = \varnothing$; A and B are disjoint.

If $A \cap B = \varnothing$, then the sets A and B are said to be **disjoint;** this is illustrated in Figure 3.

The set of all elements under consideration is called the **universal set** $\boldsymbol{U}$. Once the universal set is determined for a particular case, all other sets under discussion must be subsets of U.

We now define one more operation on sets, called the *complement.* The **complement** of A (relative to U), denoted by A', is the set of elements in U that are not in A (see Fig. 4). Symbolically:

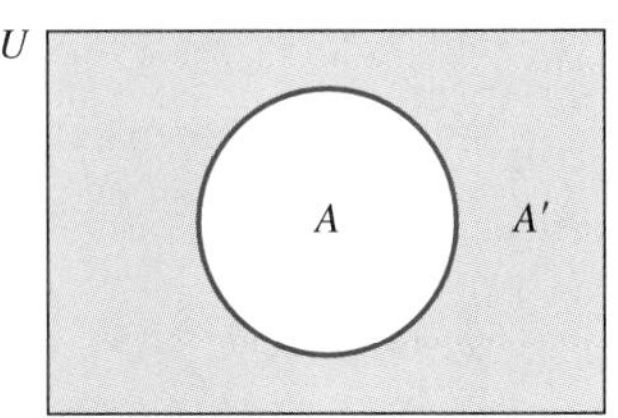

FIGURE 4
The complement of A is A'.

Complement

$$A' = \{x \in U \mid x \notin A\}$$

Example 4 **Union, Intersection, and Complement** If $A = \{3, 6, 9\}$, $B = \{3, 4, 5, 6, 7\}$, $C = \{4, 5, 7\}$, and $U = \{1, 2, 3, 4, 5, 6, 7, 8, 9\}$, then

$$A \cup B = \{3, 4, 5, 6, 7, 9\}$$
$$A \cap B = \{3, 6\}$$
$$A \cap C = \varnothing \qquad \text{A and C are disjoint}$$
$$B' = \{1, 2, 8, 9\}$$

Matched Problem 4 If $R = \{1, 2, 3, 4\}$, $S = \{1, 3, 5, 7\}$, $T = \{2, 4\}$, and $U = \{1, 2, 3, 4, 5, 6, 7, 8, 9\}$, find:

(A) $R \cup S$ (B) $R \cap S$ (C) $S \cap T$ (D) S'

■ APPLICATION

Example 5 **Marketing Survey** In a survey of 100 randomly chosen students, a marketing questionnaire included the following three questions:

1. Do you own a TV?
2. Do you own a car?
3. Do you own a TV and a car?

Seventy-five answered yes to 1, 45 answered yes to 2, and 35 answered yes to 3.

(A) How many students owned either a car or a TV?
(B) How many students did not own either a car or a TV?

SOLUTION Venn diagrams are very useful for this type of problem. If we let

U = Set of students in sample (100)
T = Set of students who own TV's (75)
C = Set of students who own cars (45)
$T \cap C$ = Set of students who own cars and TV's (35)

then:

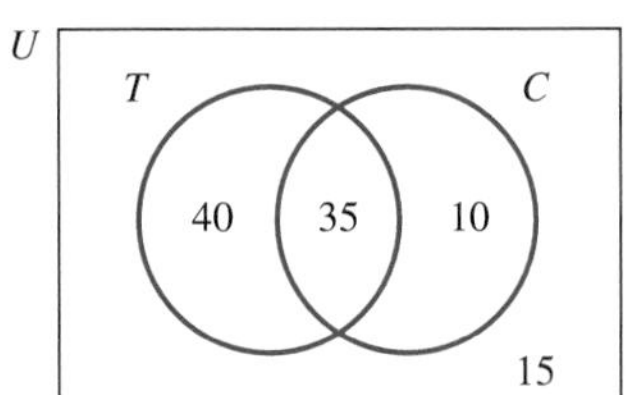

Place the number in the intersection first; then work outward:
40 = 75 − 35
10 = 45 − 35
15 = 100 − (40 + 35 + 10)

(A) The number of students who own either a car or a TV is the number of students in the set $T \cup C$. You might be tempted to say that this is just the number of students in T plus the number of students in C, $75 + 45 = 120$, but this sum is larger than the sample we started with! What is wrong? We have actually counted the number in the intersection (35) twice. The correct answer, as seen in the Venn diagram, is

$$40 + 35 + 10 = 85$$

(B) The number of students who do not own either a car or a TV is the number of students in the set $(T \cup C)'$; that is, 15.

Matched Problem 5 Referring to Example 5:

(A) How many students owned a car but not a TV?
(B) How many students did not own both a car and a TV?

Note in Example 5 and Matched Problem 5 that the word **and** is associated with intersection and the word **or** is associated with union.

Explore–Discuss 2

In Example 5, find the number of students in the set $(T \cup C) \cap C'$. Describe this set verbally and with a Venn diagram.

Answers to Matched Problems

1. (A) $\{x \mid x^2 = 9\}$ (B) $\{-3, 3\}$ (C) True; True
2. All are true **3.** $\{1, 2\}, \{1\}, \{2\}, \varnothing$
4. (A) $\{1, 2, 3, 4, 5, 7\}$ (B) $\{1, 3\}$ (C) $\varnothing$ (D) $\{2, 4, 6, 8, 9\}$
5. (A) 10, the number in $T' \cap C$ (B) 65, the number in $(T \cap C)'$

Exercise A-1

A *Indicate true (T) or false (F) in Problems 1–8.*

1. $4 \in \{2, 3, 4\}$
2. $6 \notin \{2, 3, 4\}$
3. $\{2, 3\} \subset \{2, 3, 4\}$
4. $\{3, 2, 4\} = \{2, 3, 4\}$
5. $\{3, 2, 4\} \subset \{2, 3, 4\}$
6. $\{3, 2, 4\} \in \{2, 3, 4\}$
7. $\varnothing \subset \{2, 3, 4\}$
8. $\varnothing = \{0\}$

In Problems 9–20, write the resulting set using the listing method.

9. $\{1, 3, 5\} \cup \{2, 3, 4\}$
10. $\{3, 4, 6, 7\} \cup \{3, 4, 5\}$
11. $\{1, 3, 4\} \cap \{2, 3, 4\}$
12. $\{3, 4, 6, 7\} \cap \{3, 4, 5\}$
13. $\{1, 5, 9\} \cap \{3, 4, 6, 8\}$
14. $\{6, 8, 9, 11\} \cap \{3, 4, 5, 7\}$

B

15. $\{x \mid x - 2 = 0\}$

16. $\{x \mid x + 7 = 0\}$

17. $\{x \mid x^2 = 49\}$

18. $\{x \mid x^2 = 100\}$

19. $\{x \mid x$ is an odd number between 1 and 9, inclusive$\}$

20. $\{x \mid x$ is a month starting with $M\}$

21. For $U = \{1, 2, 3, 4, 5\}$ and $A = \{2, 3, 4\}$, find A'.

22. For $U = \{7, 8, 9, 10, 11\}$ and $A = \{7, 11\}$, find A'.

Problems 23–34 refer to the Venn diagram below. How many elements are in each of the indicated sets?

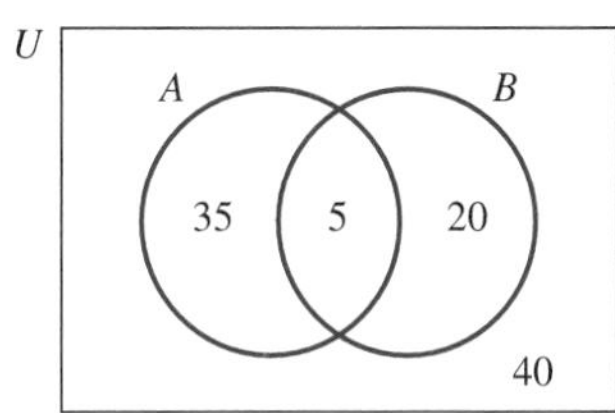

23. A

24. U

25. A'

26. B'

27. $A \cup B$

28. $A \cap B$

29. $A' \cap B$

30. $A \cap B'$

31. $(A \cap B)'$

32. $(A \cup B)'$

33. $A' \cap B'$

34. U'

35. If $R = \{1, 2, 3, 4\}$ and $T = \{2, 4, 6\}$, find:
(A) $\{x \mid x \in R$ **or** $x \in T\}$ (B) $R \cup T$

36. If $R = \{1, 3, 4\}$ and $T = \{2, 4, 6\}$, find:
(A) $\{x \mid x \in R$ **and** $x \in T\}$ (B) $R \cap T$

37. For $P = \{1, 2, 3, 4\}$, $Q = \{2, 4, 6\}$, and $R = \{3, 4, 5, 6\}$, find $P \cup (Q \cap R)$.

38. For P, Q, and R in Problem 37, find $P \cap (Q \cup R)$.

C

Venn diagrams may be of help in Problems 39–44.

39. If $A \cup B = B$, can we always conclude that $A \subset B$?

40. If $A \cap B = B$, can we always conclude that $B \subset A$?

41. If A and B are arbitrary sets, can we always conclude that $A \cap B \subset B$?

42. If $A \cap B = \varnothing$, can we always conclude that $B = \varnothing$?

43. If $A \subset B$ and $x \in A$, can we always conclude that $x \in B$?

44. If $A \subset B$ and $x \in B$, can we always conclude that $x \in A$?

45. How many subsets does each of the following sets have? Also, try to discover a formula in terms of n for a set with n elements.
(A) $\{a\}$ (B) $\{a, b\}$
(C) $\{a, b, c\}$ (D) $\{a, b, c, d\}$

46. How do the sets $\varnothing$, $\{\varnothing\}$, and $\{0\}$ differ from each other?

Applications

Business & Economics

Marketing survey. Problems 47–58 refer to the following survey: A marketing survey of 1,000 car commuters found that 600 answered yes to listening to the news, 500 answered yes to listening to music, and 300 answered yes to listening to both. Let

N = *Set of commuters in the sample who listen to news*
M = *Set of commuters in the sample who listen to music*

Following the procedures in Example 5, find the number of commuters in each set described below.

47. $N \cup M$

48. $N \cap M$

49. $(N \cup M)'$

50. $(N \cap M)'$

51. $N' \cap M$

52. $N \cap M'$

53. Set of commuters who listen to either news or music

54. Set of commuters who listen to both news and music

55. Set of commuters who do not listen to either news or music

56. Set of commuters who do not listen to both news and music

57. Set of commuters who listen to music but not news

58. Set of commuters who listen to news but not music

59. *Committee selection.* The management of a company, a president and three vice-presidents, denoted by the set $\{P, V_1, V_2, V_3\}$, wish to select a committee of 2 people from among themselves. How many ways can this committee be formed? That is, how many 2-person subsets can be formed from a set of 4 people?

60. *Voting coalition.* The management of the company in Problem 59 decides for or against certain measures as follows: The president has 2 votes and each vice-president has 1 vote. Three favorable votes are needed to pass a measure. List all minimal winning coalitions; that is, list all subsets of $\{P, V_1, V_2, V_3\}$ that represent exactly 3 votes.

Life Sciences

Blood types. *When receiving a blood transfusion, a recipient must have all the antigens of the donor. A person may have one or more of the three antigens A, B, and Rh, or none at all. Eight blood types are possible, as indicated in the following Venn diagram, where U is the set of all people under consideration:*

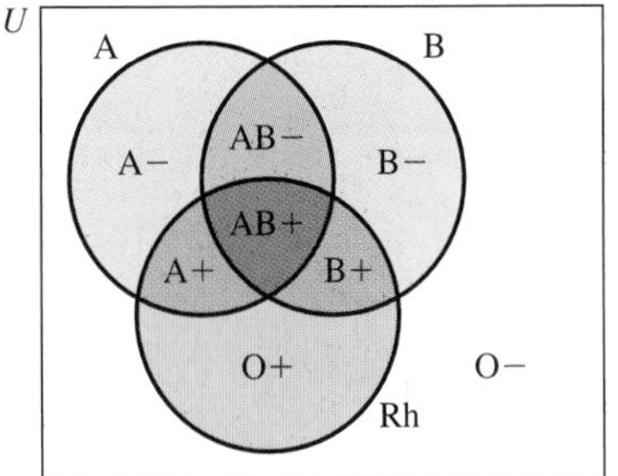

Blood Types

A−
A+
B−
B+
AB−
AB+
O−
O+

An A− person has A antigens but no B or Rh; an O+ person has Rh but neither A nor B; an AB− person has A and B antigens but no Rh; and so on.

Using the Venn diagram, indicate which of the eight blood types are included in the sets in Problems 61–68.

61. $A \cap Rh$ **62.** $A \cap B$ **63.** $A \cup Rh$

64. $A \cup B$ **65.** $(A \cup B)'$ **66.** $(A \cup B \cup Rh)'$

67. $A' \cap B$ **68.** $Rh' \cap A$

Social Sciences

Group structures. *R. D. Luce and A. D. Perry, in a study on group structure (Psychometrika, 1949, 14:95–116), used the idea of sets to formally define the notion of a clique within a group. Let G be the set of all persons in the group and let $C \subset G$. Then C is a clique provided that:*

1. *C contains at least 3 elements.*
2. *For every $a, b \in C$, $a\ \mathbf{R}\ b$ and $b\ \mathbf{R}\ a$.*
3. *For every $a \notin C$, there is at least one $b \in C$ such that $a \not\mathbf{R}\ b$ or $b \not\mathbf{R}\ a$ or both.*

[Note: *Interpret "$a\ \mathbf{R}\ b$" to mean "a relates to b," "a likes b," "a is as wealthy as b," and so on. Of course, "$a \not\mathbf{R}\ b$" means "a does not relate to b," and so on.]*

69. Translate statement 2 into ordinary English.

70. Translate statement 3 into ordinary English.

Section A-2 Algebra and Real Numbers

- The Set of Real Numbers
- The Real Number Line
- Basic Real Number Properties
- Further Properties
- Fraction Properties

The rules for manipulating and reasoning with symbols in algebra depend, in large measure, on properties of the real numbers. In this section we will look at some of the important properties of this number system. To make our discussions here and elsewhere in the text clearer and more precise, we will occasionally make use of simple *set* concepts and notation. Refer to Section A-1 if you are not yet familiar with the basic ideas concerning sets.

The Set of Real Numbers

What number system have you been using most of your life? The *real number system.* Informally, a **real number** is any number that has a decimal representation. Table 1 describes the set of real numbers and some of its important subsets. Figure 1 illustrates how these sets of numbers are related.

Table 1
The Set of Real Numbers

Symbol	Name	Description	Examples
N	Natural numbers	Counting numbers (also called positive integers)	$1, 2, 3, \ldots$
Z	Integers	Natural numbers, their negatives, and 0	$\ldots, -2, -1, 0, 1, 2, \ldots$
Q	Rational numbers	Numbers that can be represented as a/b, where a and b are integers and $b \neq 0$; decimal representations are repeating or terminating	$-4, 0, 1, 25, \frac{-3}{5}, \frac{2}{3}, 3.67, -0.33\overline{3}, 5.272\ \overline{727}$*
I	Irrational numbers	Numbers that can be represented as nonrepeating and nonterminating decimal numbers	$\sqrt{2}, \pi, \sqrt[3]{7}, 1.414\ 213\ldots, 2.718\ 281\ 82\ldots$
R	Real numbers	Rational and irrational numbers	

*The overbar indicates that the number (or block of numbers) repeats indefinitely. The space after every third digit is used to help keep track of the number of decimal places.

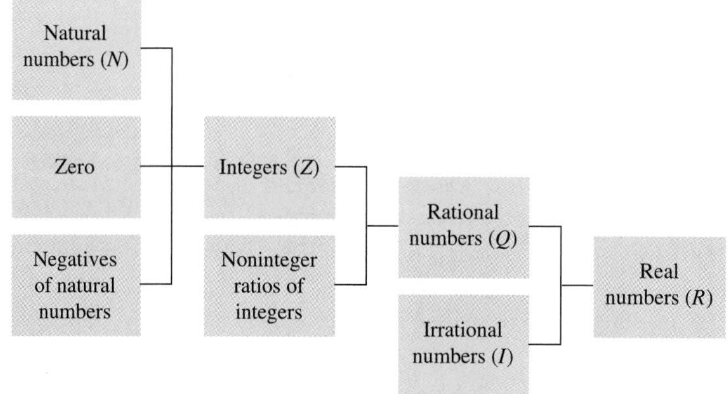

FIGURE 1
Real numbers and important subsets

The set of integers contains all the natural numbers and something else—their negatives and 0. The set of rational numbers contains all the integers and something else—noninteger ratios of integers. And the set of real numbers contains all the rational numbers and something else—irrational numbers.

The Real Number Line

A one-to-one correspondence exists between the set of real numbers and the set of points on a line. That is, each real number corresponds to exactly one point, and each point corresponds to exactly one real number. A line with a real number associated with each point, and vice versa, as shown in Figure 2, is called a **real number line,** or simply a **real line.** Each number associated with a point is called the **coordinate** of the point.

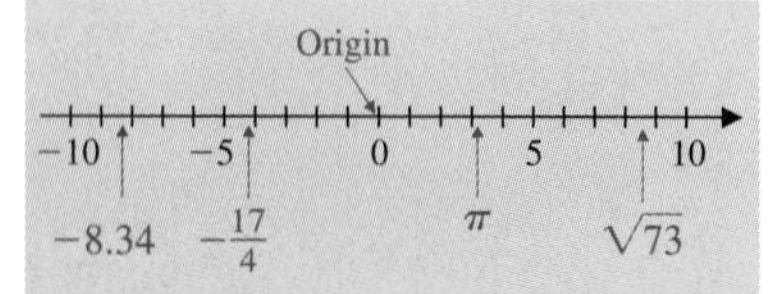

FIGURE 2
The real number line

The point with coordinate 0 is called the **origin.** The arrow on the right end of the line indicates a positive direction. The coordinates of all points to the right of the origin are called **positive real numbers,** and those to the left of the origin are called **negative real numbers.** The real number 0 is neither positive nor negative.

■ BASIC REAL NUMBER PROPERTIES

We now take a look at some of the basic properties of the real number system that enable us to convert algebraic expressions into *equivalent forms.* These assumed properties become operational rules in the algebra of real numbers.

Basic Properties of the Set of Real Numbers

Let a, b, and c be arbitrary elements in the set of real numbers R.

Addition Properties

Associative: $(a + b) + c = a + (b + c)$

Commutative: $a + b = b + a$

Identity: 0 is the additive identity; that is, $0 + a = a + 0 = a$ for all a in R, and 0 is the only element in R with this property.

Inverse: For each a in R, $-a$ is its unique additive inverse; that is, $a + (-a) = (-a) + a = 0$, and $-a$ is the only element in R relative to a with this property.

Multiplication Properties

Associative: $(ab)c = a(bc)$

Commutative: $ab = ba$

Identity: 1 is the multiplicative identity; that is, $(1)a = a(1) = a$ for all a in R, and 1 is the only element in R with this property.

Inverse: For each a in R, $a \neq 0$, $1/a$ is its unique multiplicative inverse; that is, $a(1/a) = (1/a)a = 1$, and $1/a$ is the only element in R relative to a with this property.

Distributive Properties

$$a(b + c) = ab + ac \qquad (a + b)c = ac + bc$$

Do not be intimidated by the names of these properties. Most of the ideas presented here are quite simple. In fact, you have been using many of these properties in arithmetic for a long time.

You are already familiar with the **commutative properties** for addition and multiplication. They indicate that the order in which the addition or multiplication of two numbers is performed does not matter. For example,

$$7 + 2 = 2 + 7 \quad \text{and} \quad 3 \cdot 5 = 5 \cdot 3$$

Is there a commutative property relative to subtraction or division? That is, does $a - b = b - a$ or does $a \div b = b \div a$ for all real numbers a and b (division by 0 excluded)? The answer is no, since, for example,

$$8 - 6 \neq 6 - 8 \quad \text{and} \quad 10 \div 5 \neq 5 \div 10$$

When computing

$$3 + 2 + 6 \quad \text{or} \quad 3 \cdot 2 \cdot 6$$

why do we not need parentheses to indicate which two numbers are to be added or multiplied first? The answer is to be found in the **associative properties.** These properties allow us to write

$$(3 + 2) + 6 = 3 + (2 + 6) \quad \text{and} \quad (3 \cdot 2) \cdot 6 = 3 \cdot (2 \cdot 6)$$

so it does not matter how we group numbers relative to either operation. Is there an associative property for subtraction or division? The answer is no, since, for example,

$$(12 - 6) - 2 \neq 12 - (6 - 2) \quad \text{and} \quad (12 \div 6) \div 2 \neq 12 \div (6 \div 2)$$

Evaluate each side of each equation to see why.

Relative to addition, commutativity and associativity permit us to change the order of addition at will and insert or remove parentheses as we please. The same is true for multiplication, but not for subtraction and division.

What number added to a given number will give that number back again? What number times a given number will give that number back again? The answers are 0 and 1, respectively. Because of this, 0 and 1 are called the **identity elements** for the real numbers. Hence, for any real numbers a and b,

$$0 + 5 = 5 \quad \text{and} \quad (a + b) + 0 = a + b$$
$$1 \cdot 4 = 4 \quad \text{and} \quad (a + b) \cdot 1 = a + b$$

We now consider **inverses.** For each real number a, there is a unique real number $-a$ such that $a + (-a) = 0$. The number $-a$ is called the **additive inverse** of a, or the **negative** of a. For example, the additive inverse (or negative) of 7 is -7, since $7 + (-7) = 0$. The additive inverse (or negative) of -7 is $-(-7) = 7$, since $-7 + [-(-7)] = 0$. It is important to remember that:

$-a$ is not necessarily a negative number; it is positive if a is negative and negative if a is positive.

For each nonzero real number a, there is a unique real number $1/a$ such that $a(1/a) = 1$. The number $1/a$ is called the **multiplicative inverse** of a, or the **reciprocal** of a. For example, the multiplicative inverse (or reciprocal) of 4 is $\frac{1}{4}$, since $4(\frac{1}{4}) = 1$. (Also note that 4 is the multiplicative inverse of $\frac{1}{4}$.) The number 0 has no multiplicative inverse.

We now turn to the **distributive properties,** which involve both multiplication and addition. Consider the following two computations:

$$5(3 + 4) = 5 \cdot 7 = 35 \qquad 5 \cdot 3 + 5 \cdot 4 = 15 + 20 = 35$$

Thus,

$$5(3 + 4) = 5 \cdot 3 + 5 \cdot 4$$

and we say that multiplication by 5 *distributes* over the sum $(3 + 4)$. In general, **multiplication distributes over addition** in the real number system. Two more illustrations are

$$9(m + n) = 9m + 9n \qquad (7 + 2)u = 7u + 2u$$

Example 1 **Real Number Properties** State the real number property that justifies the indicated statement.

	STATEMENT	PROPERTY ILLUSTRATED
(A)	$x(y + z) = (y + z)x$	Commutative ($\cdot$)
(B)	$5(2y) = (5 \cdot 2)y$	Associative ($\cdot$)
(C)	$2 + (y + 7) = 2 + (7 + y)$	Commutative ($+$)
(D)	$4z + 6z = (4 + 6)z$	Distributive
(E)	If $m + n = 0$, then $n = -m$.	Inverse ($+$)

Matched Problem 1 State the real number property that justifies the indicated statement.

(A) $8 + (3 + y) = (8 + 3) + y$
(B) $(x + y) + z = z + (x + y)$
(C) $(a + b)(x + y) = a(x + y) + b(x + y)$
(D) $5xy + 0 = 5xy$
(E) If $xy = 1$, $x \neq 0$, then $y = 1/x$.

FURTHER PROPERTIES

Subtraction and *division* can be defined in terms of addition and multiplication, respectively:

Subtraction and Division

For all real numbers a and b:

Subtraction: $a - b = a + (-b)$ $\qquad 7 - (-5) = 7 + [-(-5)] = 7 + 5 = 12$

Division: $a \div b = a\left(\frac{1}{b}\right)$, $b \neq 0$ $\qquad 9 \div 4 = 9\left(\frac{1}{4}\right) = \frac{9}{4}$

Thus, to subtract b from a, add the negative (the additive inverse) of b to a. To divide a by b, multiply a by the reciprocal (the multiplicative inverse) of b. Note that division by 0 is not defined, since 0 does not have a reciprocal. Thus:

0 can never be used as a divisor!

The following properties of negatives can be proved using the preceding assumed properties and definitions.

Properties of Negatives

For all real numbers a and b:

1. $-(-a) = a$
2. $(-a)b = -(ab) = a(-b) = -ab$
3. $(-a)(-b) = ab$
4. $(-1)a = -a$
5. $\frac{-a}{b} = -\frac{a}{b} = \frac{a}{-b}$, $b \neq 0$
6. $\frac{-a}{-b} = -\frac{-a}{b} = -\frac{a}{-b} = \frac{a}{b}$, $b \neq 0$

We now state two important properties involving 0:

Zero Properties

For all real numbers a and b:

1. $a \cdot 0 = 0$ $0 \cdot 0 = 0$ $(-35)(0) = 0$
2. $ab = 0$ if and only if $a = 0$ or $b = 0$
   If $(3x + 2)(x - 7) = 0$, then either $3x + 2 = 0$ or $x - 7 = 0$.

Explore–Discuss 1

In general, a set of numbers is closed under an operation if performing the operation on numbers in the set always produces another number in the set. For example, the real numbers R are closed under addition, multiplication, subtraction, and division, excluding division by 0. Replace each **?** in the following tables with T (true) or F (false), and illustrate each false statement with an example. (See Table 1 for the definitions of the sets N, Z, Q, I, and R.)

	Closed under addition	Closed under multiplication
N	?	?
Z	?	?
Q	?	?
I	?	?
R	T	T

	Closed under subtraction	Closed under division*
N	?	?
Z	?	?
Q	?	?
I	?	?
R	T	T

*Excluding division by 0.

Fraction Properties

Recall that the quotient $a \div b\ (b \neq 0)$ written in the form a/b is called a **fraction.** The quantity a is called the **numerator,** and the quantity b is called the **denominator.**

Fraction Properties

For all real numbers a, b, c, d, and k (division by 0 excluded):

1. $\frac{a}{b} = \frac{c}{d}$ if and only if $ad = bc$ $\frac{4}{6} = \frac{6}{9}$ since $4 \cdot 9 = 6 \cdot 6$

2. $\frac{ka}{kb} = \frac{a}{b}$ $\frac{7 \cdot 3}{7 \cdot 5} = \frac{3}{5}$

3. $\frac{a}{b} \cdot \frac{c}{d} = \frac{ac}{bd}$ $\frac{3}{5} \cdot \frac{7}{8} = \frac{3 \cdot 7}{5 \cdot 8}$

4. $\frac{a}{b} \div \frac{c}{d} = \frac{a}{b} \cdot \frac{d}{c}$ $\frac{2}{3} \div \frac{5}{7} = \frac{2}{3} \cdot \frac{7}{5}$

5. $\frac{a}{b} + \frac{c}{b} = \frac{a + c}{b}$ $\frac{3}{6} + \frac{5}{6} = \frac{3 + 5}{6}$

6. $\frac{a}{b} - \frac{c}{b} = \frac{a - c}{b}$ $\frac{7}{8} - \frac{3}{8} = \frac{7 - 3}{8}$

7. $\frac{a}{b} + \frac{c}{d} = \frac{ad + bc}{bd}$ $\frac{2}{3} + \frac{3}{5} = \frac{2 \cdot 5 + 3 \cdot 3}{3 \cdot 5}$

Answers to Matched Problem **1.** (A) Associative (+) (B) Commutative (+) (C) Distributive (D) Identity (+) (E) Inverse (•)

EXERCISE A-2

All variables represent real numbers.

A *In Problems 1–6, replace each question mark with an appropriate expression that will illustrate the use of the indicated real number property.*

1. Commutative property (•): $uv = ?$

2. Commutative property (+): $x + 7 = ?$

3. Associative property (+): $3 + (7 + y) = ?$

4. Associative property (•): $x(yz) = ?$

5. Identity property (•): $1(u + v) = ?$

6. Identity property (+): $0 + 9m = ?$

In Problems 7–26, indicate true (T) or false (F).

7. $5(8m) = (5 \cdot 8)m$

8. $a + cb = a + bc$

9. $5x + 7x = (5 + 7)x$

10. $uv(w + x) = uvw + uvx$

11. $7 - 11 = 7 + (-11)$

12. $8 \div (-5) = 8(\frac{1}{-5})$

13. $(x + 3) + 2x = 2x + (x + 3)$

14. $(4x + 3) + (x + 2) = 4x + [3 + (x + 2)]$

15. $\dfrac{2x}{-(x + 3)} = -\dfrac{2x}{x + 3}$

16. $-\dfrac{2x}{-(x - 3)} = \dfrac{2x}{x - 3}$

17. $(-3)(\frac{1}{-3}) = 1$

18. $(-0.5) + (0.5) = 0$

19. $-x^2y^2 = (-1)x^2y^2$

20. $[-(x + 2)](-x) = (x + 2)x$

21. $\dfrac{a}{b} + \dfrac{c}{d} = \dfrac{a + c}{b + d}$

22. $\dfrac{k}{k + b} = \dfrac{1}{1 + b}$

23. $(x + 8)(x + 6) = (x + 8)x + (x + 8)6$

24. $u(u - 2v) + v(u - 2v) = (u + v)(u - 2v)$

25. If $(x - 2)(2x + 3) = 0$, then either $x - 2 = 0$ or $2x + 3 = 0$.

26. If either $x - 2 = 0$ or $2x + 3 = 0$, then $(x - 2)(2x + 3) = 0$.

B

27. If $uv = 1$, does either u or v have to be 1? Explain.

28. If $uv = 0$, does either u or v have to be 0? Explain.

29. Indicate whether the following are true (T) or false (F):
(A) All integers are natural numbers.
(B) All rational numbers are real numbers.
(C) All natural numbers are rational numbers.

30. Indicate whether the following are true (T) or false (F):
(A) All natural numbers are integers.
(B) All real numbers are irrational.
(C) All rational numbers are real numbers.

31. Give an example of a real number that is not a rational number.

32. Give an example of a rational number that is not an integer.

33. Given the sets of numbers N (natural numbers), Z (integers), Q (rational numbers), and R (real numbers), indicate to which set(s) each of the following numbers belongs:
(A) 8 (B) $\sqrt{2}$ (C) -1.414 (D) $\frac{-5}{2}$

34. Given the sets of numbers N, Z, Q, and R (see Problem 33), indicate to which set(s) each of the following numbers belongs:
(A) -3 (B) 3.14 (C) π (D) $\frac{2}{3}$

35. Indicate true (T) or false (F), and for each false statement find real number replacements for a, b, and c that will provide a counterexample. For all real numbers a, b, and c:
(A) $(a + b) + c = a + (b + c)$
(B) $(a - b) - c = a - (b - c)$
(C) $a(bc) = (ab)c$
(D) $(a \div b) \div c = a \div (b \div c)$

36. Indicate true (T) or false (F), and for each false statement find real number replacements for a and b that will provide a counterexample. For all real numbers a and b:
(A) $a + b = b + a$
(B) $a - b = b - a$
(C) $ab = ba$
(D) $a \div b = b \div a$

C

37. If $c = 0.151\ 515\ldots$, then $100c = 15.151\ 5\ldots$ and

$$\begin{aligned} 100c - c &= 15.151\ 5\ldots - 0.151\ 515\ldots \\ 99c &= 15 \\ c &= \tfrac{15}{99} = \tfrac{5}{33} \end{aligned}$$

Proceeding similarly, convert the repeating decimal 0.090 909... into a fraction. (All repeating decimals are rational numbers, and all rational numbers have repeating decimal representations.)

38. Repeat Problem 37 for 0.181 818....

Use a calculator to express each number in Problems 39 and 40 as a decimal to the capacity of your calculator. Observe the repeating decimal representation of the rational numbers and the nonrepeating decimal representation of the irrational numbers.

39. (A) $\frac{13}{6}$ (B) $\sqrt{21}$ (C) $\frac{7}{16}$ (D) $\frac{29}{111}$

40. (A) $\frac{8}{9}$ (B) $\frac{3}{11}$ (C) $\sqrt{5}$ (D) $\frac{11}{8}$

Section A-3 Operations on Polynomials

- Natural Number Exponents
- Polynomials
- Combining Like Terms
- Addition and Subtraction
- Multiplication
- Combined Operations

This section covers basic operations on *polynomials*, a mathematical form that is encountered frequently. Our discussion starts with a brief review of natural number exponents. Integer and rational exponents and their properties will be discussed in detail in subsequent sections. (Natural numbers, integers, and rational numbers are important parts of the real number system; see Table 1 and Figure 1 in Appendix A-2.)

Natural Number Exponents

We define a **natural number exponent** as follows:

Natural Number Exponent

For n a natural number and b any real number,

$b^n = b \cdot b \cdot \cdots \cdot b$ n factors of b

$3^5 = 3 \cdot 3 \cdot 3 \cdot 3 \cdot 3$ 5 factors of 3

where n is called the **exponent** and b is called the **base.**

Along with this definition, we state the **first property of exponents:**

Theorem 1 First Property of Exponents

For any natural numbers m and n, and any real number b:

$$b^m b^n = b^{m+n} \qquad (2t^4)(5t^3) = 2 \cdot 5t^{4+3} = 10t^7$$

Polynomials

Algebraic expressions are formed by using constants and variables and the algebraic operations of addition, subtraction, multiplication, division, raising to powers, and taking roots. Special types of algebraic expressions are called *polynomials.* A **polynomial in one variable** x is constructed by adding or subtracting constants and terms of the form ax^n, where a is a real number and n is a natural number. A **polynomial in two variables** x and y is constructed by adding and subtracting constants and terms of the form $ax^m y^n$, where a is a real number and m and n are natural numbers. Polynomials in three and more variables are defined in a similar manner.

POLYNOMIALS		NOT POLYNOMIALS	
8	0	$\frac{1}{x}$	$\frac{x - y}{x^2 + y^2}$
$3x^3 - 6x + 7$	$6x + 3$		
$2x^2 - 7xy - 8y^2$	$9y^3 + 4y^2 - y + 4$	$\sqrt{x^3 - 2x}$	$2x^{-2} - 3x^{-1}$
$2x - 3y + 2$	$u^5 - 3u^3v^2 + 2uv^4 - v^4$		

Polynomial forms are encountered frequently in mathematics. For the efficient study of polynomials it is useful to classify them according to their *degree.* If a term in a polynomial has only one variable as a factor, then the **degree of the term** is the power of the variable. If two or more variables are present in a term as factors, then the **degree of the term** is the sum of the powers of the variables. The **degree of a polynomial** is the degree of the nonzero term with the highest degree in the polynomial. Any nonzero constant is defined to be a **polynomial of degree 0.** The number 0 is also a polynomial but is not assigned a degree.

Example 1 **Degree**

(A) The degree of the first term in $5x^3 + \sqrt{3}x - \frac{1}{2}$ is 3, the degree of the second term is 1, the degree of the third term is 0, and the degree of the whole polynomial is 3 (the same as the degree of the term with the highest degree).

(B) The degree of the first term in $8u^3v^2 - \sqrt{7}uv^2$ is 5, the degree of the second term is 3, and the degree of the whole polynomial is 5.

Matched Problem 1

(A) Given the polynomial $6x^5 + 7x^3 - 2$, what is the degree of the first term? The second term? The third term? The whole polynomial?

(B) Given the polynomial $2u^4v^2 - 5uv^3$, what is the degree of the first term? The second term? The whole polynomial?

In addition to classifying polynomials by degree, we also call a single-term polynomial a **monomial,** a two-term polynomial a **binomial,** and a three-term polynomial a **trinomial.**

COMBINING LIKE TERMS

The concept of *coefficient* plays a central role in the process of combining *like terms.* A constant in a term of a polynomial, including the sign that precedes it, is called the **numerical coefficient,** or simply, the **coefficient,** of the term. If a constant does not appear, or only a + sign appears, the coefficient is understood to be 1. If only a − sign appears, the coefficient is understood to be −1. Thus, given the polynomial

$$5x^4 - x^3 - 3x^2 + x - 7 = 5x^4 + (-1)x^3 + (-3)x^2 + 1x + (-7)$$

the coefficient of the first term is 5, the coefficient of the second term is −1, the coefficient of the third term is −3, the coefficient of the fourth term is 1, and the coefficient of the fifth term is −7.

The following distributive properties are fundamental to the process of combining *like terms.*

Distributive Properties of Real Numbers

1. $a(b + c) = (b + c)a = ab + ac$
2. $a(b - c) = (b - c)a = ab - ac$
3. $a(b + c + \cdots + f) = ab + ac + \cdots + af$

Two terms in a polynomial are called **like terms** if they have exactly the same variable factors to the same powers. The numerical coefficients may or may not be the same. Since constant terms involve no variables, all constant terms are like terms. If a polynomial contains two or more like terms, these terms can be combined into a single term by making use of distributive properties. The following example illustrates the reasoning behind the process:

$$\begin{aligned} 3x^2y - 5xy^2 + x^2y - 2x^2y &= 3x^2y + x^2y - 2x^2y - 5xy^2 \\ &= (3x^2y + 1x^2y - 2x^2y) - 5xy^2 \\ &= (3 + 1 - 2)x^2y - 5xy^2 \\ &= 2x^2y - 5xy^2 \end{aligned}$$

Note the use of distributive properties.

It should be clear that free use is made of the real number properties discussed in Appendix A-2. The steps shown in the dashed box are usually done mentally, and the process is quickly mechanized as follows:

Like terms in a polynomial are combined by adding their numerical coefficients.

How can we simplify expressions such as $4(x - 2y) - 3(2x - 7y)$? We clear the expression of parentheses using distributive properties, and combine like terms:

$$\begin{aligned} 4(x - 2y) - 3(2x - 7y) &= 4x - 8y - 6x + 21y \\ &= -2x + 13y \end{aligned}$$

Example 2 **Removing Parentheses** Remove parentheses and simplify:

(A) $$\begin{aligned} 2(3x^2 - 2x + 5) + (x^2 + 3x - 7) &= 2(3x^2 - 2x + 5) + 1(x^2 + 3x - 7) \\ &= 6x^2 - 4x + 10 + x^2 + 3x - 7 \\ &= 7x^2 - x + 3 \end{aligned}$$

(B) $$\begin{aligned} &(x^3 - 2x - 6) - (2x^3 - x^2 + 2x - 3) \\ &= 1(x^3 - 2x - 6) + (-1)(2x^3 - x^2 + 2x - 3) \\ &= x^3 - 2x - 6 - 2x^3 + x^2 - 2x + 3 \\ &= -x^3 + x^2 - 4x - 3 \end{aligned}$$

Be careful with the sign here.

(C) $$\begin{aligned} [3x^2 - (2x + 1)] - (x^2 - 1) &= [3x^2 - 2x - 1] - (x^2 - 1) \\ &= 3x^2 - 2x - 1 - x^2 + 1 \\ &= 2x^2 - 2x \end{aligned}$$

Remove inner parentheses first.

Matched Problem 2 ➡ Remove parentheses and simplify:

(A) $3(u^2 - 2v^2) + (u^2 + 5v^2)$
(B) $(m^3 - 3m^2 + m - 1) - (2m^3 - m + 3)$
(C) $(x^3 - 2) - [2x^3 - (3x + 4)]$

Addition and Subtraction

Addition and subtraction of polynomials can be thought of in terms of removing parentheses and combining like terms, as illustrated in Example 2. Horizontal and vertical arrangements are illustrated in the next two examples. You should be able to work either way, letting the situation dictate your choice.

Example 3 ➡ **Adding Polynomials** Add horizontally and vertically:

$$x^4 - 3x^3 + x^2, \quad -x^3 - 2x^2 + 3x, \quad \text{and} \quad 3x^2 - 4x - 5$$

SOLUTION Add horizontally:

$$\begin{aligned}&(x^4 - 3x^3 + x^2) + (-x^3 - 2x^2 + 3x) + (3x^2 - 4x - 5)\\&\quad = x^4 - 3x^3 + x^2 - x^3 - 2x^2 + 3x + 3x^2 - 4x - 5\\&\quad = x^4 - 4x^3 + 2x^2 - x - 5\end{aligned}$$

Or vertically, by lining up like terms and adding their coefficients:

$$\begin{array}{r}x^4 - 3x^3 + x^2 \qquad\qquad\\ -x^3 - 2x^2 + 3x \qquad\\ 3x^2 - 4x - 5\\ \hline x^4 - 4x^3 + 2x^2 - x - 5\end{array}$$

Matched Problem 3 ➡ Add horizontally and vertically:

$$3x^4 - 2x^3 - 4x^2, \quad x^3 - 2x^2 - 5x, \quad \text{and} \quad x^2 + 7x - 2$$

Example 4 ➡ **Subtracting Polynomials** Subtract $4x^2 - 3x + 5$ from $x^2 - 8$, both horizontally and vertically.

SOLUTION

$$\begin{aligned}&(x^2 - 8) - (4x^2 - 3x + 5)\\&\quad = x^2 - 8 - 4x^2 + 3x - 5\\&\quad = -3x^2 + 3x - 13\end{aligned}$$

or

$$\begin{array}{r}x^2 \qquad\quad - 8\\ -4x^2 + 3x - 5\\ \hline -3x^2 + 3x - 13\end{array}$$

← Change signs and add.

Matched Problem 4 ➡ Subtract $2x^2 - 5x + 4$ from $5x^2 - 6$, both horizontally and vertically.

Multiplication

Multiplication of algebraic expressions involves the extensive use of distributive properties for real numbers, as well as other real number properties.

Example 5 ➡ **Multiplying Polynomials** Multiply: $(2x - 3)(3x^2 - 2x + 3)$

SOLUTION

$$\begin{aligned}(2x - 3)(3x^2 - 2x + 3) &= 2x(3x^2 - 2x + 3) - 3(3x^2 - 2x + 3)\\&= 6x^3 - 4x^2 + 6x - 9x^2 + 6x - 9\\&= 6x^3 - 13x^2 + 12x - 9\end{aligned}$$

Or, using a vertical arrangement,

$$
\begin{array}{r}
3x^2 - 2x + 3 \\
2x - 3 \\
\hline
6x^3 - 4x^2 + 6x \\
- 9x^2 + 6x - 9 \\
\hline
6x^3 - 13x^2 + 12x - 9
\end{array}
$$

Matched Problem 5 ➡ Multiply: $(2x - 3)(2x^2 + 3x - 2)$

Thus, to multiply two polynomials, multiply each term of one by each term of the other, and combine like terms.

Products of binomial factors occur frequently, so it is useful to develop procedures that will enable us to write down their products by inspection. To find the product $(2x - 1)(3x + 2)$, we proceed as follows:

$$(2x - 1)(3x + 2) = 6x^2 + 4x - 3x - 2$$
$$= 6x^2 + x - 2$$

The inner and outer products are like terms, so combine into a single term.

To speed the process, we do the step in the dashed box mentally.

Products of certain binomial factors occur so frequently that it is useful to learn formulas for their products. The following formulas are easily verified by multiplying the factors on the left:

Special Products

1. $(a - b)(a + b) = a^2 - b^2$
2. $(a + b)^2 = a^2 + 2ab + b^2$
3. $(a - b)^2 = a^2 - 2ab + b^2$

Explore–Discuss 1

(A) Explain the relationship between special product formula 1 and the areas of the rectangles in the figure.

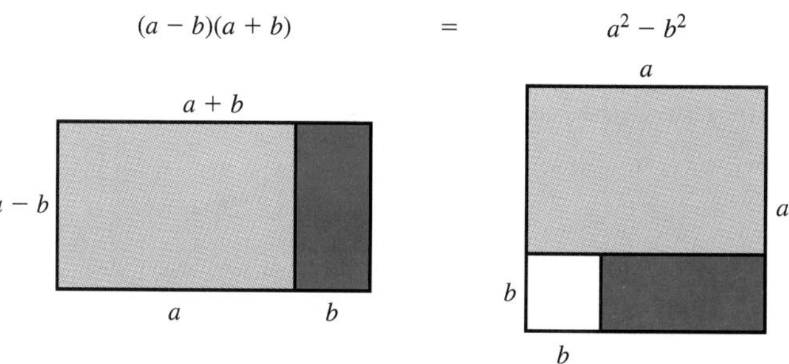

(B) Construct similar figures to provide geometric interpretations for special product formulas 2 and 3.

Example 6 **Special Products** Multiply mentally where possible:

(A) $(2x - 3y)(5x + 2y)$ (B) $(3a - 2b)(3a + 2b)$
(C) $(5x - 3)^2$ (D) $(m + 2n)^3$

SOLUTION

(A) $(2x - 3y)(5x + 2y) = 10x^2 + 4xy - 15xy - 6y^2$
$= 10x^2 - 11xy - 6y^2$

(B) $(3a - 2b)(3a + 2b) = (3a)^2 - (2b)^2$
$= 9a^2 - 4b^2$

(C) $(5x - 3)^2 = (5x)^2 - 2(5x)(3) + 3^2$
$= 25x^2 - 30x + 9$

(D)
$$\begin{aligned}(m + 2n)^3 &= (m + 2n)^2(m + 2n)\\ &= (m^2 + 4mn + 4n^2)(m + 2n)\\ &= m^2(m + 2n) + 4mn(m + 2n) + 4n^2(m + 2n)\\ &= m^3 + 2m^2n + 4m^2n + 8mn^2 + 4mn^2 + 8n^3\\ &= m^3 + 6m^2n + 12mn^2 + 8n^3\end{aligned}$$

Matched Problem 6 Multiply mentally where possible:

(A) $(4u - 3v)(2u + v)$ (B) $(2xy + 3)(2xy - 3)$
(C) $(m + 4n)(m - 4n)$ (D) $(2u - 3v)^2$
(E) $(2x - y)^3$

Combined Operations

We complete this section by considering several examples that use all the operations just discussed. Note that in simplifying, we usually remove grouping symbols starting from the inside. That is, we remove parentheses ( ) first, then brackets [ ], and finally braces { }, if present. Also:

> **Multiplication and division precede addition and subtraction, and taking powers precedes multiplication and division.**

Example 7 **Combined Operations** Perform the indicated operations and simplify:

(A)
$$\begin{aligned}3x - \{5 - 3[x - x(3 - x)]\} &= 3x - \{5 - 3[x - 3x + x^2]\}\\ &= 3x - \{5 - 3x + 9x - 3x^2\}\\ &= 3x - 5 + 3x - 9x + 3x^2\\ &= 3x^2 - 3x - 5\end{aligned}$$

(B)
$$\begin{aligned}(x - 2y)(2x + 3y) - (2x + y)^2 &= 2x^2 - xy - 6y^2 - (4x^2 + 4xy + y^2)\\ &= 2x^2 - xy - 6y^2 - 4x^2 - 4xy - y^2\\ &= -2x^2 - 5xy - 7y^2\end{aligned}$$

Matched Problem 7 Perform the indicated operations and simplify:

(A) $2t - \{7 - 2[t - t(4 + t)]\}$
(B) $(u - 3v)^2 - (2u - v)(2u + v)$

Answers to Matched Problems

1. (A) 5, 3, 0, 5 (B) 6, 4, 6
2. (A) $4u^2 - v^2$ (B) $-m^3 - 3m^2 + 2m - 4$ (C) $-x^3 + 3x + 2$
3. $3x^4 - x^3 - 5x^2 + 2x - 2$ 4. $3x^2 + 5x - 10$ 5. $4x^3 - 13x + 6$
6. (A) $8u^2 - 2uv - 3v^2$ (B) $4x^2y^2 - 9$ (C) $m^2 - 16n^2$ (D) $4u^2 - 12uv - 9v^2$ (E) $8x^3 - 12x^2y + 6xy^2 - y^3$
7. (A) $-2t^2 - 4t - 7$ (B) $-3u^2 - 6uv + 10v^2$

EXERCISE A-3

A *Problems 1–8 refer to the following polynomials:*
(A) $2x - 3$ *(B)* $2x^2 - x + 2$ *(C)* $x^3 + 2x^2 - x + 3$

1. What is the degree of (C)?
2. What is the degree of (A)?
3. Add (B) and (C).
4. Add (A) and (B).
5. Subtract (B) from (C).
6. Subtract (A) from (B).
7. Multiply (B) and (C).
8. Multiply (A) and (C).

In Problems 9–30, perform the indicated operations and simplify.

9. $2(u - 1) - (3u + 2) - 2(2u - 3)$
10. $2(x - 1) + 3(2x - 3) - (4x - 5)$
11. $4a - 2a[5 - 3(a + 2)]$
12. $2y - 3y[4 - 2(y - 1)]$
13. $(a + b)(a - b)$
14. $(m - n)(m + n)$
15. $(3x - 5)(2x + 1)$
16. $(4t - 3)(t - 2)$
17. $(2x - 3y)(x + 2y)$
18. $(3x + 2y)(x - 3y)$
19. $(3y + 2)(3y - 2)$
20. $(2m - 7)(2m + 7)$
21. $(3m + 7n)(2m - 5n)$
22. $(6x - 4y)(5x + 3y)$
23. $(4m + 3n)(4m - 3n)$
24. $(3x - 2y)(3x + 2y)$
25. $(3u + 4v)^2$
26. $(4x - y)^2$
27. $(a - b)(a^2 + ab + b^2)$
28. $(a + b)(a^2 - ab + b^2)$
29. $(4x + 3y)^2$
30. $(3x + 2)^2$

B *In Problems 31–44, perform the indicated operations and simplify.*

31. $m - \{m - [m - (m - 1)]\}$
32. $2x - 3\{x + 2[x - (x + 5)] + 1\}$
33. $(x^2 - 2xy + y^2)(x^2 + 2xy + y^2)$
34. $(2x^2 + x - 2)(x^2 - 3x + 5)$
35. $(3a - b)(3a + b) - (2a - 3b)^2$
36. $(2x - 1)^2 - (3x + 2)(3x - 2)$
37. $(m - 2)^2 - (m - 2)(m + 2)$
38. $(x - 3)(x + 3) - (x - 3)^2$
39. $(x - 2y)(2x + y) - (x + 2y)(2x - y)$
40. $(3m + n)(m - 3n) - (m + 3n)(3m - n)$
41. $(u + v)^3$
42. $(x - y)^3$
43. $(x - 2y)^3$
44. $(2m - n)^3$
45. Subtract the sum of the last two polynomials from the sum of the first two: $2x^2 - 4xy + y^2$, $3xy - y^2$, $x^2 - 2xy - y^2$, $-x^2 + 3xy - 2y^2$
46. Subtract the sum of the first two polynomials from the sum of the last two: $3m^2 - 2m + 5$, $4m^2 - m$, $3m^2 - 3m - 2$, $m^3 + m^2 + 2$

C *In Problems 47–50, perform the indicated operations and simplify.*

47. $(2x - 1)^3 - 2(2x - 1)^2 + 3(2x - 1) + 7$
48. $2(x - 2)^3 - (x - 2)^2 - 3(x - 2) - 4$
49. $2\{(x - 3)(x^2 - 2x + 1) - x[3 - x(x - 2)]\}$
50. $-3x\{x[x - x(2 - x)] - (x + 2)(x^2 - 3)\}$
51. If you are given two polynomials, one of degree m and the other of degree n, where m is greater than n, what is the degree of their product?
52. What is the degree of the sum of the two polynomials in Problem 51?
53. How does the answer to Problem 51 change if the two polynomials can have the same degree?
54. How does the answer to Problem 52 change if the two polynomials can have the same degree?
55. Show by example that, in general, $(a + b)^2 \neq a^2 + b^2$. Discuss possible conditions on a and b that would make this a valid equation.
56. Show by example that, in general, $(a - b)^2 \neq a^2 - b^2$. Discuss possible conditions on a and b that would make this a valid equation.

APPLICATIONS

Business & Economics

57. *Investment.* You have \$10,000 to invest, part at 9% and the rest at 12%. If x is the amount invested at 9%, write an algebraic expression that represents the total annual income from both investments. Simplify the expression.

58. *Investment.* A person has \$100,000 to invest. If $\$x$ are invested in a money market account yielding 7% and twice that amount in certificates of deposit yielding 9%, and if the rest is invested in high-grade bonds yielding 11%, write an algebraic expression that represents the total annual income from all three investments. Simplify the expression.

59. *Gross receipts.* Four thousand tickets are to be sold for a musical show. If x tickets are to be sold for \$10 each and three times that number for \$30 each, and if the rest are sold for \$50 each, write an algebraic expression that represents the gross receipts from ticket sales, assuming all tickets are sold. Simplify the expression.

60. *Gross receipts.* Six thousand tickets are to be sold for a concert, some for \$9 each and the rest for \$15 each. If x is the number of \$9 tickets sold, write an algebraic expression that represents the gross receipts from ticket sales, assuming all tickets are sold. Simplify the expression.

Life Sciences

61. *Nutrition.* Food mix A contains 2% fat, and food mix B contains 6% fat. A 10 kilogram diet mix of foods A and B is formed. If x kilograms of food A are used, write an algebraic expression that represents the total number of kilograms of fat in the final food mix. Simplify the expression.

62. *Nutrition.* Each ounce of food M contains 8 units of calcium, and each ounce of food N contains 5 units of calcium. A 160 ounce diet mix is formed using foods M and N. If x is the number of ounces of food M used, write an algebraic expression that represents the total number of units of calcium in the diet mix. Simplify the expression.

SECTION A-4 Factoring Polynomials

- COMMON FACTORS
- FACTORING BY GROUPING
- FACTORING SECOND-DEGREE POLYNOMIALS
- SPECIAL FACTORING FORMULAS
- COMBINED FACTORING TECHNIQUES

A polynomial is written in factored form if it is written as the product of two or more polynomials. The following polynomials are written in factored form:

$$4x^2y - 6xy^2 = 2xy(2x - 3y) \qquad 2x^3 - 8x = 2x(x - 2)(x + 2)$$
$$x^2 - x - 6 = (x - 3)(x + 2) \qquad 5m^2 + 20 = 5(m^2 + 4)$$

Unless stated to the contrary, we will limit our discussion of factoring of polynomials to polynomials with integer coefficients.

A polynomial with integer coefficients is said to be **factored completely** if each factor cannot be expressed as the product of two or more polynomials with integer coefficients, other than itself or 1. All the polynomials above, as we will see by the conclusion of this section, are factored completely.

Writing polynomials in completely factored form is often a difficult task. But accomplishing it can lead to the simplification of certain algebraic expressions and to the solution of certain types of equations and inequalities. The distributive properties for real numbers are central to the factoring process.

■ Common Factors

Generally, a first step in any factoring procedure is to factor out all factors common to all terms.

Example 1 ➡ **Common Factors** Factor out all factors common to all terms:

(A) $3x^3y - 6x^2y^2 - 3xy^3$ (B) $3y(2y + 5) + 2(2y + 5)$

Solution (A) $3x^3y - 6x^2y^2 - 3xy^3 = (\mathbf{3xy})x^2 - (\mathbf{3xy})2xy - (\mathbf{3xy})y^2$

$= 3xy(x^2 - 2xy - y^2)$

(B) $3y(2y + 5) + 2(2y + 5) = 3y\mathbf{(2y + 5)} + 2\mathbf{(2y + 5)}$

$= (3y + 2)(2y + 5)$

Matched Problem 1 ➡ Factor out all factors common to all terms:

(A) $2x^3y - 8x^2y^2 - 6xy^3$ (B) $2x(3x - 2) - 7(3x - 2)$

■ Factoring by Grouping

Occasionally, polynomials can be factored by grouping terms in such a way that we obtain results that look like Example 1B. We can then complete the factoring following the steps used in that example. This process will prove useful in the next subsection, where an efficient method is developed for factoring a second-degree polynomial as the product of two first-degree polynomials, if such factors exist.

Example 2 ➡ **Factoring by Grouping** Factor by grouping:

(A) $3x^2 - 3x - x + 1$ (B) $4x^2 - 2xy - 6xy + 3y^2$
(C) $y^2 + xz + xy + yz$

Solution (A) $3x^2 - 3x - x + 1$

$= (3x^2 - 3x) - (x - 1)$ Group the first two and the last two terms.

$= 3x\mathbf{(x - 1)} - \mathbf{(x - 1)}$ Factor out any common factors from each group. The common factor $(x - 1)$ can be taken out, and the factoring is complete.

$= \mathbf{(x - 1)}(3x - 1)$

(B) $4x^2 - 2xy - 6xy + 3y^2 = (4x^2 - 2xy) - (6xy - 3y^2)$

$= 2x\mathbf{(2x - y)} - 3y\mathbf{(2x - y)}$

$= \mathbf{(2x - y)}(2x - 3y)$

(C) If we group the first two terms and the last two terms of $y^2 + xz + xy + yz$, as in parts (A) and (B), no common factor can be taken out of each group to complete the factoring. However, if the two middle terms are reversed, we can proceed as before:

$$\begin{aligned} y^2 + xz + xy + yz &= y^2 + xy + xz + yz \\ &= (y^2 + xy) + (xz + yz) \\ &= y(y + x) + z(x + y) \\ &= y\mathbf{(x + y)} + z\mathbf{(x + y)} \\ &= \mathbf{(x + y)}(y + z) \end{aligned}$$

Matched Problem 2 Factor by grouping:

(A) $6x^2 + 2x + 9x + 3$ (B) $2u^2 + 6uv - 3uv - 9v^2$
(C) $ac + bd + bc + ad$

FACTORING SECOND-DEGREE POLYNOMIALS

We now turn our attention to factoring second-degree polynomials of the form

$$2x^2 - 5x - 3 \quad \text{and} \quad 2x^2 + 3xy - 2y^2$$

into the product of two first-degree polynomials with integer coefficients. Since many second-degree polynomials with integer coefficients cannot be factored in this way, it would be useful to know ahead of time that the factors we are seeking actually exist. The factoring approach we use, involving the *ac test,* determines at the beginning whether first-degree factors with integer coefficients do exist. Then, if they exist, the test provides a simple method for finding them.

ac Test for Factorability

If in polynomials of the form

$$ax^2 + bx + c \quad \text{or} \quad ax^2 + bxy + cy^2 \tag{1}$$

the product ac has two integer factors p and q whose sum is the coefficient b of the middle term; that is, if integers p and q exist so that

$$pq = ac \quad \text{and} \quad p + q = b \tag{2}$$

then the polynomials have first-degree factors with integer coefficients. If no integers p and q exist that satisfy (2), then the polynomials in (1) will not have first-degree factors with integer coefficients.

If integers p and q exist that satisfy (2) in the *ac* test, then the factoring always can be completed as follows: Using $b = p + q$, split the middle terms in (1) to obtain

$$ax^2 + bx + c = ax^2 + px + qx + c$$
$$ax^2 + bxy + cy^2 = ax^2 + pxy + qxy + cy^2$$

Complete the factoring by grouping the first two terms and the last two terms as in Example 2. This process always works, and it does not matter if the two middle terms on the right are interchanged.

Several examples should make the process clear. After a little practice, you will perform many of the steps mentally and will find the process fast and efficient.

Example 3 **Factoring Second-Degree Polynomials** Factor, if possible, using integer coefficients:

(A) $4x^2 - 4x - 3$ (B) $2x^2 - 3x - 4$ (C) $6x^2 - 25xy + 4y^2$

SOLUTION (A) $4x^2 - 4x - 3$

Step 1. Use the ac test to test for factorability. Comparing $4x^2 - 4x - 3$ with $ax^2 + bx + c$, we see that $a = 4, b = -4$, and $c = -3$. Multiply a and c to obtain

$$ac = (4)(-3) = -12$$

pq
$(1)(-12)$
$(-1)(12)$
$(2)(-6)$
$(-2)(6)$
$(3)(-4)$
$(-3)(4)$

All factor pairs of $-12 = ac$

List all pairs of integers whose product is -12, as shown in the margin. These are called **factor pairs** of -12. Then try to find a factor pair that sums to $b = -4$, the coefficient of the middle term in $4x^2 - 4x - 3$. (In practice, this part of step 1 is often done mentally and can be done rather quickly.) Notice that the factor pair 2 and -6 sums to -4. Thus, by the ac test, $4x^2 - 4x - 3$ has first-degree factors with integer coefficients.

Step 2. Split the middle term, using $b = p + q$, and complete the factoring by grouping. Using $-4 = 2 + (-6)$, we split the middle term in $4x^2 - 4x - 3$ and complete the factoring by grouping:

$$\begin{aligned} 4x^2 - 4x - 3 &= 4x^2 + 2x - 6x - 3 \\ &= (4x^2 + 2x) - (6x + 3) \\ &= 2x(\mathbf{2x + 1}) - 3(\mathbf{2x + 1}) \\ &= (\mathbf{2x + 1})(2x - 3) \end{aligned}$$

The result can be checked by multiplying the two factors to obtain the original polynomial.

(B) $2x^2 - 3x - 4$

Step 1. Use the ac test to test for factorability:

$$ac = (2)(-4) = -8$$

pq
$(-1)(8)$
$(1)(-8)$
$(-2)(4)$
$(2)(-4)$

All factor pairs of $-8 = ac$

Does -8 have a factor pair whose sum is -3? None of the factor pairs listed in the margin sums to $-3 = b$, the coefficient of the middle term in $2x^2 - 3x - 4$. According to the ac test, we can conclude that $2x^2 - 3x - 4$ does not have first-degree factors with integer coefficients, and we say that the polynomial is **not factorable.**

(C) $6x^2 - 25xy + 4y^2$

Step 1. Use the ac test to test for factorability:

$$ac = (6)(4) = 24$$

Mentally checking through the factor pairs of 24, keeping in mind that their sum must be $-25 = b$, we see that if $p = -1$ and $q = -24$, then

$$pq = (-1)(-24) = 24 = ac$$

and

$$p + q = (-1) + (-24) = -25 = b$$

Thus, the polynomial is factorable.

Step 2. Split the middle term, using $b = p + q$, and complete the factoring by grouping. Using $-25 = (-1) + (-24)$, we split the middle term in $6x^2 - 25xy + 4y^2$ and complete the factoring by grouping:

$$\begin{aligned} 6x^2 - 25xy + 4y^2 &= 6x^2 - xy - 24xy + 4y^2 \\ &= (6x^2 - xy) - (24xy - 4y^2) \\ &= x(6x - y) - 4y(6x - y) \\ &= (6x - y)(x - 4y) \end{aligned}$$

The check is left to the reader.

Matched Problem 3 Factor, if possible, using integer coefficients:

(A) $2x^2 + 11x - 6$ (B) $4x^2 + 11x - 6$ (C) $6x^2 + 5xy - 4y^2$

Special Factoring Formulas

The factoring formulas listed below will enable us to factor certain polynomial forms that occur frequently. These formulas can be established by multiplying the factors on the right.

Special Factoring Formulas

Perfect square:	1. $u^2 + 2uv + v^2 = (u + v)^2$
Perfect square:	2. $u^2 - 2uv + v^2 = (u - v)^2$
Difference of squares:	3. $u^2 - v^2 = (u - v)(u + v)$
Difference of cubes:	4. $u^3 - v^3 = (u - v)(u^2 + uv + v^2)$
Sum of cubes:	5. $u^3 + v^3 = (u + v)(u^2 - uv + v^2)$

CAUTION

Notice that $u^2 + v^2$ is not included in the list of special factoring formulas. In general,

$$u^2 + v^2 \neq (au + bv)(cu + dv)$$

for any choice of real number coefficients a, b, c, and d.

Example 4 **Factoring** Factor completely:

(A) $4m^2 - 12mn + 9n^2$ (B) $x^2 - 16y^2$ (C) $z^3 - 1$
(D) $m^3 + n^3$ (E) $a^2 - 4(b + 2)^2$

SOLUTION (A) $4m^2 - 12mn + 9n^2 = (2m - 3n)^2$

(B) $x^2 - 16y^2 = x^2 - (4y)^2 = (x - 4y)(x + 4y)$

(C) $z^3 - 1 = (z - 1)(z^2 + z + 1)$

(D) $m^3 + n^3 = (m + n)(m^2 - mn + n^2)$

(E) $a^2 - 4(b + 2)^2 = [a - 2(b + 2)][a + 2(b + 2)]$

Matched Problem 4 Factor completely:

(A) $x^2 + 6xy + 9y^2$ (B) $9x^2 - 4y^2$ (C) $8m^3 - 1$
(D) $x^3 + y^3z^3$ (E) $9(m - 3)^2 - 4n^2$

Explore–Discuss 1

(A) Verify the following factoring formulas for $u^4 - v^4$:

$$\begin{aligned} u^4 - v^4 &= (u - v)(u + v)(u^2 + v^2) \\ &= (u - v)(u^3 + u^2v + uv^2 + v^3) \end{aligned}$$

(B) Discuss the pattern in the following formulas:

$$\begin{aligned} u^2 - v^2 &= (u - v)(u + v) \\ u^3 - v^3 &= (u - v)(u^2 + uv + v^2) \\ u^4 - v^4 &= (u - v)(u^3 + u^2v + uv^2 + v^3) \end{aligned}$$

(C) Use the pattern you discovered in part (B) to write similar formulas for $u^5 - v^5$ and $u^6 - v^6$. Verify your formulas by multiplication.

Combined Factoring Techniques

We complete this section by considering several factoring problems that involve combinations of the preceding techniques. Generally speaking: **when factoring a polynomial, we first take out all factors common to all terms, if they are present.** Then we continue, using techniques discussed above, until the polynomial is in a completely factored form.

Example 5 **Combined Factoring Techniques** Factor completely:

(A) $3x^3 - 48x$ (B) $3u^4 - 3u^3v - 9u^2v^2$ (C) $3m^4 - 24mn^3$
(D) $3x^4 - 5x^2 + 2$

SOLUTION
(A) $3x^3 - 48x = 3x(x^2 - 16) = 3x(x - 4)(x + 4)$
(B) $3u^4 - 3u^3v - 9u^2v^2 = 3u^2(u^2 - uv - 3v^2)$
(C) $3m^4 - 24mn^3 = 3m(m^3 - 8n^3) = 3m(m - 2n)(m^2 + 2mn + 4n^2)$
(D) $3x^4 - 5x^2 + 2 = (3x^2 - 2)(x^2 - 1) = (3x^2 - 2)(x - 1)(x + 1)$

Matched Problem 5 Factor completely:

(A) $18x^3 - 8x$ (B) $4m^3n - 2m^2n^2 + 2mn^3$ (C) $2t^4 - 16t$
(D) $2y^4 - 5y^2 - 12$

Answers to Matched Problems

1. (A) $2xy(x^2 - 4xy - 3y^2)$ (B) $(2x - 7)(3x - 2)$
2. (A) $(3x + 1)(2x + 3)$ (B) $(u + 3v)(2u - 3v)$ (C) $(a + b)(c + d)$
3. (A) $(2x - 1)(x + 6)$ (B) Not factorable (C) $(3x + 4y)(2x - y)$
4. (A) $(x + 3y)^2$ (B) $(3x - 2y)(3x + 2y)$
(C) $(2m - 1)(4m^2 + 2m + 1)$ (D) $(x + yz)(x^2 - xyz + y^2z^2)$
(E) $[3(m - 3) - 2n][3(m - 3) + 2n]$
5. (A) $2x(3x - 2)(3x + 2)$ (B) $2mn(2m^2 - mn + n^2)$
(C) $2t(t - 2)(t^2 + 2t + 4)$ (D) $(2y^2 + 3)(y - 2)(y + 2)$

EXERCISE A-4

A *In Problems 1–8, factor out all factors common to all terms.*

1. $6m^4 - 9m^3 - 3m^2$
2. $6x^4 - 8x^3 - 2x^2$
3. $8u^3v - 6u^2v^2 + 4uv^3$
4. $10x^3y + 20x^2y^2 - 15xy^3$
5. $7m(2m - 3) + 5(2m - 3)$
6. $5x(x + 1) - 3(x + 1)$
7. $a(3c + d) - 4b(3c + d)$
8. $2w(y - 2z) - x(y - 2z)$

In Problems 9–18, factor by grouping.

9. $2x^2 - x + 4x - 2$
10. $x^2 - 3x + 2x - 6$
11. $3y^2 - 3y + 2y - 2$
12. $2x^2 - x + 6x - 3$
13. $2x^2 + 8x - x - 4$
14. $6x^2 + 9x - 2x - 3$
15. $wy - wz + xy - xz$
16. $ac + ad + bc + bd$
17. $am - bn - bm + an$
18. $ab + 6 + 2a + 3b$

B *In Problems 19–56, factor completely. If a polynomial cannot be factored, say so.*

19. $3y^2 - y - 2$
20. $2x^2 + 5x - 3$
21. $u^2 - 2uv - 15v^2$
22. $x^2 - 4xy - 12y^2$
23. $m^2 - 6m - 3$
24. $x^2 + x - 4$
25. $w^2x^2 - y^2$
26. $25m^2 - 16n^2$
27. $9m^2 - 6mn + n^2$
28. $x^2 + 10xy + 25y^2$
29. $y^2 + 16$
30. $u^2 + 81$
31. $4z^2 - 28z + 48$
32. $6x^2 + 48x + 72$
33. $2x^4 - 24x^3 + 40x^2$
34. $2y^3 - 22y^2 + 48y$
35. $4xy^2 - 12xy + 9x$
36. $16x^2y - 8xy + y$
37. $6m^2 - mn - 12n^2$
38. $6s^2 + 7st - 3t^2$
39. $4u^3v - uv^3$
40. $x^3y - 9xy^3$
41. $2x^3 - 2x^2 + 8x$
42. $3m^3 - 6m^2 + 15m$
43. $r^3 - t^3$
44. $m^3 + n^3$
45. $a^3 + 1$
46. $c^3 - 1$

C

47. $(x + 2)^2 - 9y^2$
48. $(a - b)^2 - 4(c - d)^2$
49. $5u^2 + 4uv - 2v^2$
50. $3x^2 - 2xy - 4y^2$
51. $6(x - y)^2 + 23(x - y) - 4$
52. $4(A + B)^2 - 5(A + B) - 6$
53. $y^4 - 3y^2 - 4$
54. $m^4 - n^4$
55. $27a^2 + a^5b^3$
56. $s^4t^4 - 8st$

SECTION A-5 Operations on Rational Expressions

- REDUCING TO LOWEST TERMS
- MULTIPLICATION AND DIVISION
- ADDITION AND SUBTRACTION
- COMPOUND FRACTIONS

We now turn our attention to fractional forms. A quotient of two algebraic expressions (division by 0 excluded) is called a **fractional expression.** If both the numerator and the denominator are polynomials, the fractional expression is called a **rational expression.** Some examples of rational expressions are

$$\frac{1}{x^3 + 2x} \qquad \frac{5}{x} \qquad \frac{x + 7}{3x^2 - 5x + 1} \qquad \frac{x^2 - 2x + 4}{1}$$

In this section we will discuss basic operations on rational expressions, including multiplication, division, addition, and subtraction.

Since variables represent real numbers in the rational expressions we will consider, the properties of real number fractions summarized in Appendix A-2 will play a central role in much of the work that we will do.

Even though not always explicitly stated, we always assume that variables are restricted so that division by 0 is excluded.

For example, given the rational expression

$$\frac{2x + 5}{x(x + 2)(x - 3)}$$

the variable x is understood to be restricted from being 0, -2, or 3, since these values would cause the denominator to be 0.

Reducing to Lowest Terms

Central to the process of reducing rational expressions to *lowest terms* is the *fundamental property of fractions*, which we restate here for convenient reference:

Fundamental Property of Fractions

If a, b, and k are real numbers with $b, k \neq 0$, then

$$\frac{ka}{kb} = \frac{a}{b} \qquad \frac{5 \cdot 2}{5 \cdot 7} = \frac{2}{7} \qquad \frac{x(x + 4)}{2(x + 4)} = \frac{x}{2}, \quad x \neq -4$$

Using this property from left to right to eliminate all common factors from the numerator and the denominator of a given fraction is referred to as **reducing a fraction to lowest terms.** We are actually dividing the numerator and denominator by the same nonzero common factor.

Using the property from right to left—that is, multiplying the numerator and denominator by the same nonzero factor—is referred to as **raising a fraction to higher terms.** We will use the property in both directions in the material that follows.

Example 1 **Reducing to Lowest Terms** Reduce each rational expression to lowest terms.

(A)
$$\frac{6x^2 + x - 1}{2x^2 - x - 1} = \frac{(2x + 1)(3x - 1)}{(2x + 1)(x - 1)} \qquad \text{Factor numerator and denominator completely.}$$
$$= \frac{3x - 1}{x - 1} \qquad \text{Divide numerator and denominator by the common factor } (2x + 1).$$

(B)
$$\frac{x^4 - 8x}{3x^3 - 2x^2 - 8x} = \frac{x(x - 2)(x^2 + 2x + 4)}{x(x - 2)(3x + 4)}$$
$$= \frac{x^2 + 2x + 4}{3x + 4}$$

Matched Problem 1 Reduce each rational expression to lowest terms.

(A) $\dfrac{x^2 - 6x + 9}{x^2 - 9}$ (B) $\dfrac{x^3 - 1}{x^2 - 1}$

Multiplication and Division

Since we are restricting variable replacements to real numbers, multiplication and division of rational expressions follow the rules for multiplying and dividing real number fractions summarized in Appendix A-2.

Multiplication and Division

If a, b, c, and d are real numbers, then:

1. $\dfrac{a}{b} \cdot \dfrac{c}{d} = \dfrac{ac}{bd}, \quad b, d \neq 0$ $\qquad \dfrac{3}{5} \cdot \dfrac{x}{x+5} = \dfrac{3x}{5(x+5)}$
2. $\dfrac{a}{b} \div \dfrac{c}{d} = \dfrac{a}{b} \cdot \dfrac{d}{c}, \quad b, c, d \neq 0$ $\qquad \dfrac{3}{5} \div \dfrac{x}{x+5} = \dfrac{3}{5} \cdot \dfrac{x+5}{x}$

Explore–Discuss 1

Write a verbal description of the process of multiplying two rational expressions. Do the same for the quotient of two rational expressions.

Example 2 **Multiplication and Division** Perform the indicated operations and reduce to lowest terms.

(A) $\dfrac{10x^3y}{3xy + 9y} \cdot \dfrac{x^2 - 9}{4x^2 - 12x}$

Factor numerators and denominators. Then divide any numerator and any denominator with a like common factor.

$$= \dfrac{\overset{5x^2}{\cancel{10x^3y}}}{\underset{3 \cdot 1}{\cancel{3y(x+3)}}} \cdot \dfrac{\overset{1 \cdot 1}{\cancel{(x-3)(x+3)}}}{\underset{2 \cdot 1}{\cancel{4x(x-3)}}}$$

$$= \dfrac{5x^2}{6}$$

(B) $\dfrac{4 - 2x}{4} \div (x - 2) = \dfrac{\overset{1}{\cancel{2}}(2-x)}{\underset{2}{\cancel{4}}} \cdot \dfrac{1}{x-2}$ $\qquad x - 2 = \dfrac{x-2}{1}$

$$= \dfrac{2 - x}{2(x-2)} = \dfrac{\overset{-1}{\cancel{-(x-2)}}}{2\underset{1}{\cancel{(x-2)}}}$$

$b - a = -(a - b)$, a useful change in some problems

$$= -\dfrac{1}{2}$$

Matched Problem 2 Perform the indicated operations and reduce to lowest terms.

(A) $\dfrac{12x^2y^3}{2xy^2 + 6xy} \cdot \dfrac{y^2 + 6y + 9}{3y^3 + 9y^2}$ (B) $(4 - x) \div \dfrac{x^2 - 16}{5}$

ADDITION AND SUBTRACTION

Again, because we are restricting variable replacements to real numbers, addition and subtraction of rational expressions follow the rules for adding and subtracting real number fractions.

Addition and Subtraction

For a, b, and c real numbers:

1. $\dfrac{a}{b} + \dfrac{c}{b} = \dfrac{a + c}{b}, \quad b \neq 0 \qquad \dfrac{x}{x + 5} + \dfrac{8}{x + 5} = \dfrac{x + 8}{x + 5}$
2. $\dfrac{a}{b} - \dfrac{c}{b} = \dfrac{a - c}{b}, \quad b \neq 0 \qquad \dfrac{x}{3x^2y^2} - \dfrac{x + 7}{3x^2y^2} = \dfrac{x - (x + 7)}{3x^2y^2}$

Thus, we add rational expressions with the same denominators by adding or subtracting their numerators and placing the result over the common denominator. If the denominators are not the same, we raise the fractions to higher terms, using the fundamental property of fractions to obtain common denominators, and then proceed as described.

Even though any common denominator will do, our work will be simplified if the *least common denominator (LCD)* is used. Often, the LCD is obvious, but if it is not, the steps in the next box describe how to find it.

The Least Common Denominator (LCD)

The LCD of two or more rational expressions is found as follows:

1. Factor each denominator completely, including integer factors.
2. Identify each different factor from all the denominators.
3. Form a product using each different factor to the highest power that occurs in any one denominator. This product is the LCD.

Example 3 **Addition and Subtraction** Combine into a single fraction and reduce to lowest terms.

(A) $\dfrac{3}{10} + \dfrac{5}{6} - \dfrac{11}{45}$ (B) $\dfrac{4}{9x} - \dfrac{5x}{6y^2} + 1$ (C) $\dfrac{1}{x - 1} - \dfrac{1}{x} - \dfrac{2}{x^2 - 1}$

SOLUTION (A) To find the LCD, factor each denominator completely:

$$\left.\begin{aligned} 10 &= 2 \cdot 5 \\ 6 &= 2 \cdot 3 \\ 45 &= 3^2 \cdot 5 \end{aligned}\right\} \quad \text{LCD} = 2 \cdot 3^2 \cdot 5 = 90$$

Now use the fundamental property of fractions to make each denominator 90:

$$\frac{3}{10} + \frac{5}{6} - \frac{11}{45} = \frac{\mathbf{9} \cdot 3}{\mathbf{9} \cdot 10} + \frac{\mathbf{15} \cdot 5}{\mathbf{15} \cdot 6} - \frac{\mathbf{2} \cdot 11}{\mathbf{2} \cdot 45}$$

$$= \frac{27}{90} + \frac{75}{90} - \frac{22}{90}$$

$$= \frac{27 + 75 - 22}{90} = \frac{80}{90} = \frac{8}{9}$$

(B) $\left.\begin{array}{l} 9x = 3^2x \\ 6y^2 = 2 \cdot 3y^2 \end{array}\right\}$ LCD $= 2 \cdot 3^2xy^2 = 18xy^2$

$$\frac{4}{9x} - \frac{5x}{6y^2} + 1 = \frac{\mathbf{2y^2} \cdot 4}{\mathbf{2y^2} \cdot 9x} - \frac{\mathbf{3x} \cdot 5x}{\mathbf{3x} \cdot 6y^2} + \frac{\mathbf{18xy^2}}{\mathbf{18xy^2}}$$

$$= \frac{8y^2 - 15x^2 + 18xy^2}{18xy^2}$$

(C) $\dfrac{1}{x - 1} - \dfrac{1}{x} - \dfrac{2}{x^2 - 1}$

$$= \frac{1}{x - 1} - \frac{1}{x} - \frac{2}{(x - 1)(x + 1)} \qquad \text{LCD} = x(x - 1)(x + 1)$$

$$= \frac{x(x + 1) - (x - 1)(x + 1) - 2x}{x(x - 1)(x + 1)}$$

$$= \frac{x^2 + x - x^2 + 1 - 2x}{x(x - 1)(x + 1)}$$

$$= \frac{1 - x}{x(x - 1)(x + 1)}$$

$$= \frac{\overset{-1}{-\cancel{(x - 1)}}}{x\underset{1}{\cancel{(x - 1)}}(x + 1)} = \frac{-1}{x(x + 1)}$$

Matched Problem 3 Combine into a single fraction and reduce to lowest terms.

(A) $\dfrac{5}{28} - \dfrac{1}{10} + \dfrac{6}{35}$ (B) $\dfrac{1}{4x^2} - \dfrac{2x + 1}{3x^3} + \dfrac{3}{12x}$

(C) $\dfrac{2}{x^2 - 4x + 4} + \dfrac{1}{x} - \dfrac{1}{x - 2}$

Explore–Discuss 2

What is the value of $\dfrac{\frac{16}{4}}{2}$?

What is the result of entering $16 \div 4 \div 2$ on a calculator?

What is the difference between $16 \div (4 \div 2)$ and $(16 \div 4) \div 2$?

How could you use fraction bars to distinguish between these two cases when writing $\dfrac{16}{\frac{4}{2}}$?

Compound Fractions

A fractional expression with fractions in its numerator, denominator, or both is called a **compound fraction.** It is often necessary to represent a compound fraction as a **simple fraction**—that is (in all cases we will consider), as the quotient of two polynomials. The process does not involve any new concepts. It is a matter of applying old concepts and processes in the correct sequence.

Example 4 **Simplifying Compound Fractions** Express as a simple fraction reduced to lowest terms:

(A) $\dfrac{\dfrac{1}{5+h} - \dfrac{1}{5}}{h}$ (B) $\dfrac{\dfrac{y}{x^2} - \dfrac{x}{y^2}}{\dfrac{y}{x} - \dfrac{x}{y}}$

Solution We will simplify the expressions in parts (A) and (B) using two different methods—each is suited to the particular type of problem.

(A) We simplify this expression by combining the numerator into a single fraction and using division of rational forms.

$$\frac{\dfrac{1}{5+h} - \dfrac{1}{5}}{h} = \left[\frac{1}{5+h} - \frac{1}{5}\right] \div \frac{h}{1}$$

$$= \frac{5 - 5 - h}{5(5+h)} \cdot \frac{1}{h}$$

$$= \frac{-h}{5(5+h)h} = \frac{-1}{5(5+h)}$$

(B) The method used here makes effective use of the fundamental property of fractions in the form

$$\frac{a}{b} = \frac{ka}{kb} \qquad b, k \neq 0$$

Multiply the numerator and denominator by the LCD of all fractions in the numerator and denominator—in this case, x^2y^2:

$$\frac{x^2y^2\left(\dfrac{y}{x^2} - \dfrac{x}{y^2}\right)}{x^2y^2\left(\dfrac{y}{x} - \dfrac{x}{y}\right)} = \frac{x^2y^2\dfrac{y}{x^2} - x^2y^2\dfrac{x}{y^2}}{x^2y^2\dfrac{y}{x} - x^2y^2\dfrac{x}{y}} = \frac{y^3 - x^3}{xy^3 - x^3y}$$

$$= \frac{\overset{1}{\cancel{(y-x)}}(y^2 + xy + x^2)}{xy\underset{1}{\cancel{(y-x)}}(y+x)}$$

$$= \frac{y^2 + xy + x^2}{xy(y+x)} \quad \text{or} \quad \frac{x^2 + xy + y^2}{xy(x+y)}$$

Matched Problem 4 Express as a simple fraction reduced to lowest terms:

(A) $\dfrac{\dfrac{1}{2+h}-\dfrac{1}{2}}{h}$ (B) $\dfrac{\dfrac{a}{b}-\dfrac{b}{a}}{\dfrac{a}{b}+2+\dfrac{b}{a}}$

Answers to Matched Problems

1. (A) $\dfrac{x-3}{x+3}$ (B) $\dfrac{x^2+x+1}{x+1}$ **2.** (A) $2x$ (B) $\dfrac{-5}{x+4}$

3. (A) $\dfrac{1}{4}$ (B) $\dfrac{3x^2-5x-4}{12x^3}$ (C) $\dfrac{4}{x(x-2)^2}$

4. (A) $\dfrac{-1}{2(2+h)}$ (B) $\dfrac{a-b}{a+b}$

EXERCISE A-5

A *In Problems 1–18, perform the indicated operations and reduce answers to lowest terms.*

1. $\dfrac{d^5}{3a}\div\left(\dfrac{d^2}{6a^2}\cdot\dfrac{a}{4d^3}\right)$

2. $\left(\dfrac{d^5}{3a}\div\dfrac{d^2}{6a^2}\right)\cdot\dfrac{a}{4d^3}$

3. $\dfrac{x^2}{12}+\dfrac{x}{18}-\dfrac{1}{30}$

4. $\dfrac{2y}{18}-\dfrac{-1}{28}-\dfrac{y}{42}$

5. $\dfrac{4m-3}{18m^3}+\dfrac{3}{4m}-\dfrac{2m-1}{6m^2}$

6. $\dfrac{3x+8}{4x^2}-\dfrac{2x-1}{x^3}-\dfrac{5}{8x}$

7. $\dfrac{x^2-9}{x^2-3x}\div(x^2-x-12)$

8. $\dfrac{2x^2+7x+3}{4x^2-1}\div(x+3)$

9. $\dfrac{2}{x}-\dfrac{1}{x-3}$

10. $\dfrac{3}{m}-\dfrac{2}{m+4}$

11. $\dfrac{3}{x^2-1}-\dfrac{2}{x^2-2x+1}$

12. $\dfrac{1}{a^2-b^2}+\dfrac{1}{a^2+2ab+b^2}$

13. $\dfrac{x+1}{x-1}-1$

14. $m-3-\dfrac{m-1}{m-2}$

15. $\dfrac{3}{a-1}-\dfrac{2}{1-a}$

16. $\dfrac{5}{x-3}-\dfrac{2}{3-x}$

17. $\dfrac{2x}{x^2-16}-\dfrac{x-4}{x^2+4x}$

18. $\dfrac{m+2}{m^2-2m}-\dfrac{m}{m^2-4}$

B *In Problems 19–30, perform the indicated operations and reduce answers to lowest terms. Represent any compound fractions as simple fractions reduced to lowest terms.*

19. $\dfrac{x^2}{x^2+2x+1}+\dfrac{x-1}{3x+3}-\dfrac{1}{6}$

20. $\dfrac{y}{y^2-y-2}-\dfrac{1}{y^2+5y-14}-\dfrac{2}{y^2+8y+7}$

21. $\dfrac{2-x}{2x+x^2}\cdot\dfrac{x^2+4x+4}{x^2-4}$

22. $\dfrac{9-m^2}{m^2+5m+6}\cdot\dfrac{m+2}{m-3}$

23. $\dfrac{c+2}{5c-5}-\dfrac{c-2}{3c-3}+\dfrac{c}{1-c}$

24. $\dfrac{x+7}{ax-bx}+\dfrac{y+9}{by-ay}$

25. $\dfrac{1+\dfrac{3}{x}}{x-\dfrac{9}{x}}$

26. $\dfrac{1-\dfrac{y^2}{x^2}}{1-\dfrac{y}{x}}$

27. $\dfrac{\dfrac{1}{2(x+h)}-\dfrac{1}{2x}}{h}$

28. $\dfrac{\dfrac{1}{x+h}-\dfrac{1}{x}}{h}$

29. $\dfrac{\dfrac{x}{y}-2+\dfrac{y}{x}}{\dfrac{x}{y}-\dfrac{y}{x}}$

30. $\dfrac{1+\dfrac{2}{x}-\dfrac{15}{x^2}}{1+\dfrac{4}{x}-\dfrac{5}{x^2}}$

In Problems 31–38, imagine that the indicated "solutions" were given to you by a student whom you were tutoring in this class.
(A) Is the solution correct? If the solution is incorrect, explain what is wrong and how it can be corrected.
(B) Show a correct solution for each incorrect solution.

31. $\dfrac{x^2 + 4x + 3}{x + 3} = \dfrac{x^2 + 4x}{x} = x + 4$

32. $\dfrac{x^2 - 3x - 4}{x - 4} = \dfrac{x^2 - 3x}{x} = x - 3$

33. $\dfrac{(x + h)^2 - x^2}{h} = (x + 1)^2 - x^2 = 2x + 1$

34. $\dfrac{(x + h)^3 - x^3}{h} = (x + 1)^3 - x^3 = 3x^2 + 3x + 1$

35. $\dfrac{x^2 - 3x}{x^2 - 2x - 3} + x - 3 = \dfrac{x^2 - 3x + x - 3}{x^2 - 2x - 3} = 1$

36. $\dfrac{2}{x - 1} - \dfrac{x + 3}{x^2 - 1} = \dfrac{2x + 2 - x - 3}{x^2 - 1} = \dfrac{1}{x + 1}$

37. $\dfrac{2x^2}{x^2 - 4} - \dfrac{x}{x - 2} = \dfrac{2x^2 - x^2 - 2x}{x^2 - 4} = \dfrac{x}{x + 2}$

38. $x + \dfrac{x - 2}{x^2 - 3x + 2} = \dfrac{x + x - 2}{x^2 - 3x + 2} = \dfrac{2}{x - 2}$

C *Represent the compound fractions in Problems 39–42 as simple fractions reduced to lowest terms.*

39. $\dfrac{\dfrac{1}{3(x + h)^2} - \dfrac{1}{3x^2}}{h}$

40. $\dfrac{\dfrac{1}{(x + h)^2} - \dfrac{1}{x^2}}{h}$

41. $1 - \dfrac{1}{1 - \dfrac{1}{1 - \dfrac{1}{x}}}$

42. $2 - \dfrac{1}{1 - \dfrac{2}{a + 2}}$

Section A-6 Integer Exponents and Scientific Notation

- Integer Exponents
- Scientific Notation

We now review basic operations on integer exponents and scientific notation and its use.

Integer Exponents

Definitions for **integer exponents** are listed below.

Definition of a^n

For n an integer and a a real number:

1. For n a positive integer,

   $a^n = a \cdot a \cdot \cdots \cdot a$ $\quad$ n factors of a $\quad$ $5^4 = 5 \cdot 5 \cdot 5 \cdot 5$

2. For $n = 0$,

   $a^0 = 1 \quad a \neq 0 \quad 12^0 = 1$
   0^0 is not defined.

3. For n a negative integer,

   $a^n = \dfrac{1}{a^{-n}} \quad a \neq 0 \quad a^{-3} = \dfrac{1}{a^{-(-3)}} = \dfrac{1}{a^3}$

   [If n is negative, then $(-n)$ is positive.]
   Note: It can be shown that for *all* integers n,

   $a^{-n} = \dfrac{1}{a^n}$ and $a^n = \dfrac{1}{a^{-n}} \quad a \neq 0 \quad a^5 = \dfrac{1}{a^{-5}}, \; a^{-5} = \dfrac{1}{a^5}$

The following integer exponent properties are very useful in manipulating integer exponent forms.

Theorem 1

EXPONENT PROPERTIES

For n and m integers and a and b real numbers:

1. $a^m a^n = a^{m+n}$ $\qquad a^8 a^{-3} = a^{8+(-3)} = a^5$
2. $(a^n)^m = a^{mn}$ $\qquad (a^{-2})^3 = a^{3(-2)} = a^{-6}$
3. $(ab)^m = a^m b^m$ $\qquad (ab)^{-2} = a^{-2}b^{-2}$
4. $\left(\dfrac{a}{b}\right)^m = \dfrac{a^m}{b^m} \quad b \neq 0$ $\qquad \left(\dfrac{a}{b}\right)^5 = \dfrac{a^5}{b^5}$
5. $\dfrac{a^m}{a^n} = a^{m-n} = \dfrac{1}{a^{n-m}} \quad a \neq 0$ $\qquad \dfrac{a^{-3}}{a^7} = \dfrac{1}{a^{7-(-3)}} = \dfrac{1}{a^{10}}$

Explore–Discuss 1

Property 1 in Theorem 1 can be expressed verbally as follows:

To find the product of two exponential forms with the same base, add the exponents and use the same base.

Express the other properties in Theorem 1 verbally. Decide which you find easier to remember—a formula or a verbal description.

Exponent forms are frequently encountered in algebraic applications. You should sharpen your skills in using these forms by reviewing the above basic definitions and properties and the examples that follow.

Example 1 **Simplifying Exponent Forms** Simplify, and express the answers using positive exponents only.

(A) $(2x^3)(3x^5) = 2 \cdot 3x^{3+5} = 6x^8$

(B) $x^5x^{-9} = x^{-4} = \dfrac{1}{x^4}$

(C) $\dfrac{x^5}{x^7} = x^{5-7} = x^{-2} = \dfrac{1}{x^2}$

or $\dfrac{x^5}{x^7} = \dfrac{1}{x^{7-5}} = \dfrac{1}{x^2}$

(D) $\dfrac{x^{-3}}{y^{-4}} = \dfrac{y^4}{x^3}$

(E) $(u^{-3}v^2)^{-2} = (u^{-3})^{-2}(v^2)^{-2} = u^6v^{-4} = \dfrac{u^6}{v^4}$

(F) $\left(\dfrac{y^{-5}}{y^{-2}}\right)^{-2} = \dfrac{(y^{-5})^{-2}}{(y^{-2})^{-2}} = \dfrac{y^{10}}{y^4} = y^6$

(G) $\dfrac{4m^{-3}n^{-5}}{6m^{-4}n^3} = \dfrac{2m^{-3-(-4)}}{3n^{3-(-5)}} = \dfrac{2m}{3n^8}$

Matched Problem 1 Simplify, and express the answers using positive exponents only.

(A) $(3y^4)(2y^3)$ (B) m^2m^{-6} (C) $(u^3v^{-2})^{-2}$

(D) $\left(\dfrac{y^{-6}}{y^{-2}}\right)^{-1}$ (E) $\dfrac{8x^{-2}y^{-4}}{6x^{-5}y^2}$

Example 2 **Converting to a Simple Fraction** Write as a simple fraction with positive exponents:

$$\frac{1 - x}{x^{-1} - 1}$$

SOLUTION First note that

$$\frac{1 - x}{x^{-1} - 1} \neq \frac{x(1 - x)}{-1} \qquad \text{A common error}$$

The original expression is a complex fraction, and we proceed to simplify it as follows:

$$\frac{1 - x}{x^{-1} - 1} = \frac{1 - x}{\frac{1}{x} - 1} \qquad \text{Multiply numerator and denominator by } x \text{ to clear internal fractions.}$$

$$= \frac{x(1 - x)}{x\left(\frac{1}{x} - 1\right)}$$

$$= \frac{x(1 - x)}{1 - x} = x$$

Matched Problem 2 Write as a simple fraction with positive exponents: $\frac{1 + x^{-1}}{1 - x^{-2}}$

Scientific Notation

In the real world, one often encounters very large numbers. For example:

The public debt in the United States in 1992, to the nearest billion dollars, was

$4,065,000,000,000

The world population in the year 2000, to the nearest million, is projected to be

6,166,000,000

Very small numbers are also encountered:

The sound intensity of a normal conversation is

0.000 000 000 316 watt per square centimeter*

It is generally troublesome to write and work with numbers of this type in standard decimal form. The first and last example cannot even be entered into many calculators as they are written. But with exponents

*We write 0.000 000 000 316 in place of 0.000000000316, because it is then easier to keep track of the number of decimal places. We follow this convention when there are more than five decimal places to the right of the decimal.

defined for all integers, we can now express any finite decimal form as the product of a number between 1 and 10 and an integer power of 10; that is, in the form

$$a \times 10^n \qquad 1 \leq a < 10, \quad a \text{ in decimal form}, \quad n \text{ an integer}$$

A number expressed in this form is said to be in **scientific notation.** The following are some examples of numbers in standard decimal notation and in scientific notation:

DECIMAL AND SCIENTIFIC NOTATION

$7 = 7 \times 10^0$	$0.5 = 5 \times 10^{-1}$
$67 = 6.7 \times 10$	$0.45 = 4.5 \times 10^{-1}$
$580 = 5.8 \times 10^2$	$0.0032 = 3.2 \times 10^{-3}$
$43{,}000 = 4.3 \times 10^4$	$0.000\ 045 = 4.5 \times 10^{-5}$
$73{,}400{,}000 = 7.34 \times 10^7$	$0.000\ 000\ 391 = 3.91 \times 10^{-7}$

Note that the power of 10 used corresponds to the number of places we move the decimal to form a number between 1 and 10. The power is positive if the decimal is moved to the left and negative if it is moved to the right. Positive exponents are associated with numbers greater than or equal to 10; negative exponents are associated with positive numbers less than 1; and a zero exponent is associated with a number that is 1 or greater, but less than 10.

Example 3 ➡ **Scientific Notation**

(A) Write each number in scientific notation:

7,320,000 and 0.000 000 54

(B) Write each number in standard decimal form:

4.32×10^6 and 4.32×10^{-5}

SOLUTION (A)

$7{,}320{,}000 = 7.320\ 000. \times 10^6 = 7.32 \times 10^6$

6 places left
Positive exponent

$0.000\ 000\ 54 = 0.000\ 000\ 5.4 \times 10^{-7} = 5.4 \times 10^{-7}$

7 places right
Negative exponent

(B) $4.32 \times 10^6 = 4{,}320{,}000$ (6 places right; Positive exponent 6)

$4.32 \times 10^{-5} = \dfrac{4.32}{10^5} = 0.000\ 043\ 2$ (5 places left; Negative exponent −5)

Matched Problem 3 ➡ (A) Write each number in scientific notation: 47,100; 2,443,000,000; 1.45
(B) Write each number in standard decimal form: 3.07×10^8; 5.98×10^{-6}

Explore–Discuss 2

Scientific and graphing calculators can calculate in either standard decimal mode or scientific notation mode. If the result of a computation in decimal mode is either too large or too small to be displayed, then most calculators will automatically display the answer in scientific notation. Read the manual for your calculator and experiment with some operations on very large and very small numbers in both decimal mode and scientific notation mode. For example, show that:

$$\frac{216{,}700{,}000{,}000}{0.000\ 000\ 000\ 000\ 078\ 8} = 2.75 \times 10^{24}$$

Answers to Matched Problems

1. (A) $6y^7$ (B) $\frac{1}{m^4}$ (C) $\frac{v^4}{u^6}$ (D) y^4 (E) $\frac{4x^3}{3y^6}$ **2.** $\frac{x}{x-1}$

3. (A) 4.7×10^4; 2.443×10^9; 1.45×10^0 (B) 307,000,000; 0.000 005 98

Exercise A-6

A *In Problems 1–14, simplify and express answers using positive exponents only. Variables are restricted to avoid division by 0.*

1. $2x^{-9}$ **2.** $3y^{-5}$ **3.** $\frac{3}{2w^{-7}}$

4. $\frac{5}{4x^{-9}}$ **5.** $2x^{-8}x^5$ **6.** $3c^{-9}c^4$

7. $\frac{w^{-8}}{w^{-3}}$ **8.** $\frac{m^{-11}}{m^{-5}}$ **9.** $5v^8v^{-8}$

10. $7d^{-4}d^4$ **11.** $(a^{-3})^2$ **12.** $(b^4)^{-3}$

13. $(x^6y^{-3})^{-2}$ **14.** $(a^{-3}b^4)^{-3}$

Write each number in Problems 15–20 in scientific notation.

15. 82,300,000,000 **16.** 5,380,000

17. 0.783 **18.** 0.019

19. 0.000 034 **20.** 0.000 000 007 832

Write each number in Problems 21–28 in standard decimal notation.

21. 4×10^4 **22.** 9×10^6

23. 7×10^{-3} **24.** 2×10^{-5}

25. 6.171×10^7 **26.** 3.044×10^3

27. 8.08×10^{-4} **28.** 1.13×10^{-2}

B *In Problems 29–38, simplify and express answers using positive exponents only.*

29. $(22 + 31)^0$ **30.** $(2x^3y^4)^0$

31. $\frac{10^{-3} \cdot 10^4}{10^{-11} \cdot 10^{-2}}$ **32.** $\frac{10^{-17} \cdot 10^{-5}}{10^{-3} \cdot 10^{-14}}$

33. $(5x^2y^{-3})^{-2}$ **34.** $(2m^{-3}n^2)^{-3}$

35. $\frac{8 \times 10^{-3}}{2 \times 10^{-5}}$ **36.** $\frac{18 \times 10^{12}}{6 \times 10^{-4}}$

37. $\frac{8x^{-3}y^{-1}}{6x^2y^{-4}}$ **38.** $\frac{9m^{-4}n^3}{12m^{-1}n^{-1}}$

In Problems 39–42, write each expression in the form $ax^p + bx^q$ or $ax^p + bx^q + cx^r$, where a, b, and c are real numbers and p, q, and r are integers. For example,

$$\frac{2x^4 - 3x^2 + 1}{2x^3} = \frac{2x^4}{2x^3} - \frac{3x^2}{2x^3} + \frac{1}{2x^3} = x - \frac{3}{2}x^{-1} + \frac{1}{2}x^{-3}$$

39. $\frac{7x^5 - x^2}{4x^5}$ **40.** $\frac{5x^3 - 2}{3x^2}$

41. $\frac{3x^4 - 4x^2 - 1}{4x^3}$ **42.** $\frac{2x^3 - 3x^2 + x}{2x^2}$

Write each expression in Problems 43–46 with positive exponents only, and as a single fraction reduced to lowest terms.

43. $\dfrac{3x^2(x-1)^2 - 2x^3(x-1)}{(x-1)^4}$

44. $\dfrac{5x^4(x+3)^2 - 2x^5(x+3)}{(x+3)^4}$

45. $2x^{-2}(x-1) - 2x^{-3}(x-1)^2$

46. $2x(x+3)^{-1} - x^2(x+3)^{-2}$

In Problems 47–50, convert each number to scientific notation and simplify. Express the answer in both scientific notation and in standard decimal form.

47. $\dfrac{9{,}600{,}000{,}000}{(1{,}600{,}000)(0.000\ 000\ 25)}$

48. $\dfrac{(60{,}000)(0.000\ 003)}{(0.0004)(1{,}500{,}000)}$

49. $\dfrac{(1{,}250{,}000)(0.000\ 38)}{0.0152}$

50. $\dfrac{(0.000\ 000\ 82)(230{,}000)}{(625{,}000)(0.0082)}$

51. What is the result of entering 2^{3^2} on a calculator?

52. Refer to Problem 51. What is the difference between $2^{(3^2)}$ and $(2^3)^2$? Which agrees with the value of 2^{3^2} obtained with a calculator?

53. If $n = 0$, then property 1 in Theorem 1 implies that $a^m a^0 = a^{m+0} = a^m$. Explain how this helps motivate the definition of a^0.

54. If $m = -n$, then property 1 in Theorem 1 implies that $a^{-n}a^n = a^0 = 1$. Explain how this helps motivate the definition of a^{-n}.

C *Write the fractions in Problems 55–58 as simple fractions reduced to lowest terms.*

55. $\dfrac{u+v}{u^{-1}+v^{-1}}$

56. $\dfrac{x^{-1}-y^{-1}}{x-y}$

57. $\dfrac{b^{-2}-c^{-2}}{b^{-3}-c^{-3}}$

58. $\dfrac{xy^{-2}-yx^{-2}}{y^{-1}-x^{-1}}$

APPLICATIONS

Business & Economics

Problems 59 and 60 refer to Table 1.

Table 1

ASSETS OF THE FIVE LARGEST U.S. COMMERCIAL BANKS, 1992

BANK	ASSETS ($)
Citicorp, New York	213,701,000,000
Bank of America, San Francisco	180,646,000,000
Chemical Banking, New York	139,655,000,000
J. P. Morgan, New York	102,941,000,000
Chase Manhattan, New York	95,862,000,000

59. *Financial assets*
(A) Write Citicorp's assets in scientific notation.
(B) After converting to scientific notation, determine the ratio of the assets of Citicorp to the assets of Chase Manhattan. Write the answer in standard decimal form to four decimal places.
(C) Repeat part (B) with the banks reversed.

60. *Financial assets*
(A) Write Bank of America's assets in scientific notation.
(B) After converting to scientific notation, determine the ratio of the assets of Bank of America to the assets of J. P. Morgan. Write the answer in standard decimal form to four decimal places.
(C) Repeat part (B) with the banks reversed.

Problems 61 and 62 refer to Table 2.

Table 2

U.S. PUBLIC DEBT, INTEREST ON DEBT, AND POPULATION

YEAR	PUBLIC DEBT ($)	INTEREST ON DEBT ($)	POPULATION
1982	1,142,000,000,000	117,400,000,000	231,500,000
1992	4,065,000,000,000	292,300,000,000	255,100,000

61. *Public debt.* Carry out the following computations using scientific notation, and write final answers in standard decimal form.
(A) What was the per capita debt in 1992 (to the nearest dollar)?
(B) What was the per capita interest paid on the debt in 1992 (to the nearest dollar)?
(C) What was the percentage interest paid on the debt in 1992 (to two decimal places)?

62. *Public debt.* Carry out the following computations using scientific notation, and write final answers in standard decimal form.
(A) What was the per capita debt in 1982 (to the nearest dollar)?
(B) What was the per capita interest paid on the debt in 1982 (to the nearest dollar)?
(C) What was the percentage interest paid on the debt in 1982 (to two decimal places)?

Life Sciences

Air pollution. *Air quality standards establish maximum amounts of pollutants considered acceptable in the air. The amounts are frequently given in parts per million (ppm). A standard of 30 ppm also can be expressed as follows:*

$$30 \text{ ppm} = \frac{30}{1{,}000{,}000} = \frac{3 \times 10}{10^6}$$
$$= 3 \times 10^{-5} = 0.000\ 03 = 0.003\%$$

In Problems 63 and 64, express the given standard:
(A) In scientific notation
(B) In standard decimal notation
(C) As a percent

63. 9 ppm, the standard for carbon monoxide, when averaged over a period of 8 hours

64. 0.03 ppm, the standard for sulfur oxides, when averaged over a year

Social Sciences

65. *Crime.* In 1992, the United States had a violent crime rate of 757.5 per 100,000 people and a population of 255.1 million people. How many violent crimes occurred that year? Compute the answer using scientific notation and convert the answer to standard decimal form (to the nearest thousand).

66. *Population density.* The United States had a 1992 population of 255.1 million people and a land area of 3,539,000 square miles. What was the population density? Compute the answer using scientific notation and convert the answer to standard decimal form (to one decimal place).

SECTION A-7 Rational Exponents and Radicals

- *n*TH ROOTS OF REAL NUMBERS
- RATIONAL EXPONENTS AND RADICALS
- PROPERTIES OF RADICALS

Square roots may now be generalized to *nth roots,* and the meaning of exponent may be generalized to include all rational numbers.

*n*TH ROOTS OF REAL NUMBERS

Consider a square of side r with area 36 square inches. We can write

$$r^2 = 36$$

and conclude that side r is a number whose square is 36. We say that r is a **square root** of b if $r^2 = b$. Similarly, we say that r is a **cube root** of b if $r^3 = b$. And, in general:

> **nth Root**
>
> For any natural number n,
>
> r is an ***n*th root** of b if $r^n = b$

Thus, 4 is a square root of 16, since $4^2 = 16$, and -2 is a cube root of -8, since $(-2)^3 = -8$. Since $(-4)^2 = 16$, we see that -4 is also a square root of 16. It can be shown that any positive number has two real square roots, two real 4th roots, and, in general, two real nth roots if n is even. Negative numbers have no real square roots, no real 4th roots, and, in general, no real nth roots if n is even. The reason is that no real number raised to an even power can be negative. For odd roots the situation is simpler. Every real number has exactly one real cube root, one real 5th root, and, in general, one real nth root if n is odd.

Additional roots can be considered in the *complex number system*. But in this book we restrict our interest to *real roots of real numbers*, and "root" will always be interpreted to mean "real root."

Rational Exponents and Radicals

We now turn to the question of what symbols to use to represent nth roots. For n a natural number greater than 1, we use

$$b^{1/n} \qquad \text{or} \qquad \sqrt[n]{b}$$

to represent a **real nth root of b.** The exponent form is motivated by the fact that $(b^{1/n})^n = b$ if exponent laws are to continue to hold for rational exponents. The other form is called an **nth root radical.** In the expression below, the symbol $\sqrt{}$ is called a **radical,** n is the **index** of the radical, and a is the **radicand:**

Index → $\sqrt[n]{a}$ ← Radical
↑ Radicand

When the index is 2, it is usually omitted. That is, when dealing with square roots, we simply use $\sqrt{b}$ rather than $\sqrt[2]{b}$. If there are two real nth roots, both $b^{1/n}$ and $\sqrt[n]{b}$ denote the positive root, called the **principal nth root.**

Example 1 **Finding nth Roots** Evaluate each of the following:

(A) $4^{1/2}$ and $\sqrt{4}$ (B) $-4^{1/2}$ and $-\sqrt{4}$ (C) $(-4)^{1/2}$ and $\sqrt{-4}$
(D) $8^{1/3}$ and $\sqrt[3]{8}$ (E) $(-8)^{1/3}$ and $\sqrt[3]{-8}$ (F) $-8^{1/3}$ and $-\sqrt[3]{8}$

Solution (A) $4^{1/2} = \sqrt{4} = 2 \quad (\sqrt{4} \neq \pm 2)$ (B) $-4^{1/2} = -\sqrt{4} = -2$
(C) $(-4)^{1/2}$ and $\sqrt{-4}$ are not real numbers
(D) $8^{1/3} = \sqrt[3]{8} = 2$ (E) $(-8)^{1/3} = \sqrt[3]{-8} = -2$
(F) $-8^{1/3} = -\sqrt[3]{8} = -2$

Matched Problem 1 Evaluate each of the following:

(A) $16^{1/2}$ (B) $-\sqrt{16}$ (C) $\sqrt[3]{-27}$ (D) $(-9)^{1/2}$ (E) $(\sqrt[4]{81})^3$

Common Error

The symbol $\sqrt{4}$ represents the single number 2, not ± 2. Do not confuse $\sqrt{4}$ with the solutions of the equation $x^2 = 4$, which are usually written in the form $x = \pm\sqrt{4} = \pm 2$.

We now define b^r for any rational number $r = m/n$.

Rational Exponents

If m and n are natural numbers without common prime factors, b is a real number, and b is nonnegative when n is even, then

$$b^{m/n} = \begin{cases} (b^{1/n})^m = (\sqrt[n]{b})^m \\ (b^m)^{1/n} = \sqrt[n]{b^m} \end{cases} \qquad \begin{matrix} 8^{2/3} = (8^{1/3})^2 = (\sqrt[3]{8})^2 = 2^2 = 4 \\ 8^{2/3} = (8^2)^{1/3} = \sqrt[3]{8^2} = \sqrt[3]{64} = 4 \end{matrix}$$

and

$$b^{-m/n} = \frac{1}{b^{m/n}} \quad b \neq 0 \qquad 8^{-2/3} = \frac{1}{8^{2/3}} = \frac{1}{4}$$

Note that the two definitions of $b^{m/n}$ are equivalent under the indicated restrictions on m, n, and b.

All the properties listed for integer exponents in Theorem 1 in Section A-6 also hold for rational exponents, provided b is nonnegative when n is even. Unless stated to the contrary, all variables in the rest of the discussion represent positive real numbers.

Example 2 **From Rational Exponent Form to Radical Form and Vice Versa** Change rational exponent form to radical form.

(A) $x^{1/7} = \sqrt[7]{x}$

(B) $(3u^2v^3)^{3/5} = \sqrt[5]{(3u^2v^3)^3}$ or $(\sqrt[5]{3u^2v^3})^3$ The first is usually preferred.

(C) $y^{-2/3} = \dfrac{1}{y^{2/3}} = \dfrac{1}{\sqrt[3]{y^2}}$ or $\sqrt[3]{y^{-2}}$ or $\sqrt[3]{\dfrac{1}{y^2}}$

Change radical form to rational exponent form.

(D) $\sqrt[5]{6} = 6^{1/5}$

(E) $-\sqrt[3]{x^2} = -x^{2/3}$

(F) $\sqrt{x^2 + y^2} = (x^2 + y^2)^{1/2}$ Note that $(x^2 + y^2)^{1/2} \neq x + y$. Why?

Matched Problem 2 Convert to radical form.

(A) $u^{1/5}$ (B) $(6x^2y^5)^{2/9}$ (C) $(3xy)^{-3/5}$

Convert to rational exponent form.

(D) $\sqrt[4]{9u}$ (E) $-\sqrt[7]{(2x)^4}$ (F) $\sqrt[3]{x^3 + y^3}$

Example 3 **Working with Rational Exponents** Simplify each and express answers using positive exponents only. If rational exponents appear in final answers, convert to radical form.

(A) $(3x^{1/3})(2x^{1/2}) = 6x^{1/3+1/2} = 6x^{5/6} = 6\sqrt[6]{x^5}$

(B) $(-8)^{5/3} = [(-8)^{1/3}]^5 = (-2)^5 = -32$

(C) $(2x^{1/3}y^{-2/3})^3 = 8xy^{-2} = \dfrac{8x}{y^2}$

(D) $\left(\dfrac{4x^{1/3}}{x^{1/2}}\right)^{1/2} = \dfrac{4^{1/2}x^{1/6}}{x^{1/4}} = \dfrac{2}{x^{1/4-1/6}} = \dfrac{2}{x^{1/12}} = \dfrac{2}{\sqrt[12]{x}}$

Matched Problem 3 Simplify each and express answers using positive exponents only. If rational exponents appear in final answers, convert to radical form.

(A) $9^{3/2}$ (B) $(-27)^{4/3}$ (C) $(5y^{1/4})(2y^{1/3})$ (D) $(2x^{-3/4}y^{1/4})^4$

(E) $\left(\dfrac{8x^{1/2}}{x^{2/3}}\right)^{1/3}$

Explore–Discuss 1

In each of the following, evaluate both radical forms:

$$16^{3/2} = \sqrt{16^3} = (\sqrt{16})^3$$
$$27^{2/3} = \sqrt[3]{27^2} = (\sqrt[3]{27})^2$$

Which radical conversion form is easier to use if you are performing the calculations by hand?

Example 4 **Working with Rational Exponents** Multiply, and express answers using positive exponents only.

(A) $3y^{2/3}(2y^{1/3} - y^2)$ (B) $(2u^{1/2} + v^{1/2})(u^{1/2} - 3v^{1/2})$

SOLUTION (A) $3y^{2/3}(2y^{1/3} - y^2) = 6y^{2/3+1/3} - 3y^{2/3+2}$
$= 6y - 3y^{8/3}$

(B) $(2u^{1/2} + v^{1/2})(u^{1/2} - 3v^{1/2}) = 2u - 5u^{1/2}v^{1/2} - 3v$

Matched Problem 4 Multiply, and express answers using positive exponents only.

(A) $2c^{1/4}(5c^3 - c^{3/4})$ (B) $(7x^{1/2} - y^{1/2})(2x^{1/2} + 3y^{1/2})$

Example 5 **Working with Rational Exponents** Write the following expression in the form $ax^p + bx^q$, where a and b are real numbers and p and q are rational numbers:

$$\frac{2\sqrt{x} - 3\sqrt[3]{x^2}}{2\sqrt[3]{x}}$$

SOLUTION

$$\frac{2\sqrt{x} - 3\sqrt[3]{x^2}}{2\sqrt[3]{x}} = \frac{2x^{1/2} - 3x^{2/3}}{2x^{1/3}}$$ *Change to rational exponent form.*

$$= \frac{2x^{1/2}}{2x^{1/3}} - \frac{3x^{2/3}}{2x^{1/3}}$$ *Separate into two fractions.*

$$= x^{1/6} - 1.5x^{1/3}$$

Matched Problem 5 Write the following expression in the form $ax^p + bx^q$, where a and b are real numbers and p and q are rational numbers:

$$\frac{5\sqrt[3]{x} - 4\sqrt{x}}{2\sqrt{x^3}}$$

Properties of Radicals

Changing or simplifying radical expressions is aided by several properties of radicals that follow directly from the properties of exponents considered earlier.

Properties of Radicals

If c, n, and m are natural numbers greater than or equal to 2, and if x and y are positive real numbers, then:

1. $\sqrt[n]{x^n} = x$ $\qquad \sqrt[3]{x^3} = x$
2. $\sqrt[n]{xy} = \sqrt[n]{x}\sqrt[n]{y}$ $\qquad \sqrt[5]{xy} = \sqrt[5]{x}\sqrt[5]{y}$
3. $\sqrt[n]{\dfrac{x}{y}} = \dfrac{\sqrt[n]{x}}{\sqrt[n]{y}}$ $\qquad \sqrt[4]{\dfrac{x}{y}} = \dfrac{\sqrt[4]{x}}{\sqrt[4]{y}}$

Example 6 **Applying Properties of Radicals** Simplify using properties of radicals:

(A) $\sqrt[4]{(3x^4y^3)^4}$ (B) $\sqrt[4]{8}\sqrt[4]{2}$ (C) $\sqrt[3]{\dfrac{xy}{27}}$

SOLUTION

(A) $\sqrt[4]{(3x^4y^3)^4} = 3x^4y^3$ — Property 1

(B) $\sqrt[4]{8}\sqrt[4]{2} = \sqrt[4]{16} = \sqrt[4]{2^4} = 2$ — Properties 2 and 1

(C) $\sqrt[3]{\dfrac{xy}{27}} = \dfrac{\sqrt[3]{xy}}{\sqrt[3]{27}} = \dfrac{\sqrt[3]{xy}}{3}$ or $\dfrac{1}{3}\sqrt[3]{xy}$ — Properties 3 and 1

Matched Problem 6 Simplify using properties of radicals:

(A) $\sqrt[7]{(x^3 + y^3)^7}$ (B) $\sqrt[3]{8y^3}$ (C) $\dfrac{\sqrt[3]{16x^4y}}{\sqrt[3]{2xy}}$

Explore–Discuss 2

Multiply:

$$(\sqrt{a} - \sqrt{b})(\sqrt{a} + \sqrt{b})$$

How can this product be used to simplify radical forms?

A question arises regarding the best form in which a radical expression should be left. There are many answers, depending on what use we wish to make of the expression. In deriving certain formulas, it is sometimes useful to clear either a denominator or a numerator of radicals. The process is referred to as **rationalizing** the denominator or numerator. Examples 7 and 8 illustrate the rationalizing process.

Example 7 **Rationalizing Denominators** Rationalize each denominator:

(A) $\dfrac{6x}{\sqrt{2x}}$ (B) $\dfrac{6}{\sqrt{7} - \sqrt{5}}$ (C) $\dfrac{x - 4}{\sqrt{x} + 2}$

SOLUTION

(A) $\dfrac{6x}{\sqrt{2x}} = \dfrac{6x}{\sqrt{2x}} \cdot \dfrac{\sqrt{2x}}{\sqrt{2x}} = \dfrac{6x\sqrt{2x}}{2x} = 3\sqrt{2x}$

(B) $$\frac{6}{\sqrt{7}-\sqrt{5}} = \frac{6}{\sqrt{7}-\sqrt{5}}\cdot\frac{\sqrt{7}+\sqrt{5}}{\sqrt{7}+\sqrt{5}} = \frac{6(\sqrt{7}+\sqrt{5})}{2} = 3(\sqrt{7}+\sqrt{5})$$

(C) $$\frac{x-4}{\sqrt{x}+2} = \frac{x-4}{\sqrt{x}+2}\cdot\frac{\sqrt{x}-2}{\sqrt{x}-2} = \frac{(x-4)(\sqrt{x}-2)}{x-4} = \sqrt{x}-2$$

Matched Problem 7 ➡ Rationalize each denominator:

(A) $\frac{12ab^2}{\sqrt{3ab}}$ (B) $\frac{9}{\sqrt{6}+\sqrt{3}}$ (C) $\frac{x^2-y^2}{\sqrt{x}-\sqrt{y}}$

Example 8 ➡ **Rationalizing Numerators** Rationalize each numerator:

(A) $\frac{\sqrt{2}}{2\sqrt{3}}$ (B) $\frac{3+\sqrt{m}}{9-m}$ (C) $\frac{\sqrt{2+h}-\sqrt{2}}{h}$

SOLUTION (A) $$\frac{\sqrt{2}}{2\sqrt{3}} = \frac{\sqrt{2}}{2\sqrt{3}}\cdot\frac{\sqrt{2}}{\sqrt{2}} = \frac{2}{2\sqrt{6}} = \frac{1}{\sqrt{6}}$$

(B) $$\frac{3+\sqrt{m}}{9-m} = \frac{3+\sqrt{m}}{9-m}\cdot\frac{3-\sqrt{m}}{3-\sqrt{m}} = \frac{9-m}{(9-m)(3-\sqrt{m})} = \frac{1}{3-\sqrt{m}}$$

(C) $$\frac{\sqrt{2+h}-\sqrt{2}}{h} = \frac{\sqrt{2+h}-\sqrt{2}}{h}\cdot\frac{\sqrt{2+h}+\sqrt{2}}{\sqrt{2+h}+\sqrt{2}} = \frac{h}{h(\sqrt{2+h}+\sqrt{2})} = \frac{1}{\sqrt{2+h}+\sqrt{2}}$$

Matched Problem 8 ➡ Rationalize each numerator:

(A) $\frac{\sqrt{3}}{3\sqrt{2}}$ (B) $\frac{2-\sqrt{n}}{4-n}$ (C) $\frac{\sqrt{3+h}-\sqrt{3}}{h}$

Answers to Matched Problems

1. (A) 4 (B) -4 (C) -3 (D) Not a real number (E) 27
2. (A) $\sqrt[5]{u}$ (B) $\sqrt[9]{(6x^2y^5)^2}$ or $(\sqrt[9]{6x^2y^5})^2$ (C) $1/\sqrt[5]{(3xy)^3}$
(D) $(9u)^{1/4}$ (E) $-(2x)^{4/7}$ (F) $(x^3+y^3)^{1/3}$ (not $x+y$)
3. (A) 27 (B) 81 (C) $10y^{7/12} = 10\sqrt[12]{y^7}$ (D) $16y/x^3$
(E) $2/x^{1/18} = 2/\sqrt[18]{x}$
4. (A) $10c^{13/4} - 2c$ (B) $14x + 19x^{1/2}y^{1/2} - 3y$ **5.** $2.5x^{-7/6} - 2x^{-1}$
6. (A) $x^3 + y^3$ (B) $2y$ (C) $2x$
7. (A) $4b\sqrt{3ab}$ (B) $3(\sqrt{6}-\sqrt{3})$ (C) $(x+y)(\sqrt{x}+\sqrt{y})$
8. (A) $\frac{1}{\sqrt{6}}$ (B) $\frac{1}{2+\sqrt{n}}$ (C) $\frac{1}{\sqrt{3+h}+\sqrt{3}}$

EXERCISE A-7

A *Change each expression in Problems 1–6 to radical form. Do not simplify.*

1. $6x^{3/5}$ **2.** $7y^{2/5}$ **3.** $(4xy^3)^{2/5}$

4. $(7x^2y)^{5/7}$ **5.** $(x^2 + y^2)^{1/2}$ **6.** $x^{1/2} + y^{1/2}$

Change each expression in Problems 7–12 to rational exponent form. Do not simplify.

7. $5\sqrt[4]{x^3}$ **8.** $7m\sqrt[5]{n^2}$ **9.** $\sqrt[5]{(2x^2y)^3}$

10. $\sqrt[9]{(3m^4n)^2}$ **11.** $\sqrt[3]{x} + \sqrt[3]{y}$ **12.** $\sqrt[3]{x^2 + y^3}$

In Problems 13–24, find rational number representations for each, if they exist.

13. $25^{1/2}$ **14.** $64^{1/3}$ **15.** $16^{3/2}$

16. $16^{3/4}$ **17.** $-36^{1/2}$ **18.** $-32^{3/5}$

19. $(-36)^{1/2}$ **20.** $(-32)^{3/5}$ **21.** $(\frac{4}{25})^{3/2}$

22. $(\frac{8}{27})^{2/3}$ **23.** $9^{-3/2}$ **24.** $8^{-2/3}$

In Problems 25–34, simplify each expression and write answers using positive exponents only. All variables represent positive real numbers.

25. $x^{4/5}x^{-2/5}$ **26.** $y^{-3/7}y^{4/7}$ **27.** $\dfrac{m^{2/3}}{m^{-1/3}}$

28. $\dfrac{x^{1/4}}{x^{3/4}}$ **29.** $(8x^3y^{-6})^{1/3}$ **30.** $(4u^{-2}v^4)^{1/2}$

31. $\left(\dfrac{4x^{-2}}{y^4}\right)^{-1/2}$ **32.** $\left(\dfrac{w^4}{9x^{-2}}\right)^{-1/2}$ **33.** $\dfrac{8x^{-1/3}}{12x^{1/4}}$

34. $\dfrac{6a^{3/4}}{15a^{-1/3}}$

Simplify each expression in Problems 35–40 using properties of radicals. All variables represent positive real numbers.

35. $\sqrt[5]{(2x + 3)^5}$ **36.** $\sqrt[3]{(7 + 2y)^3}$ **37.** $\sqrt{18x^3}\sqrt{2x^3}$

38. $\sqrt{2a^3}\sqrt{32a^5}$ **39.** $\dfrac{\sqrt{6x}\sqrt{10}}{\sqrt{15x}}$ **40.** $\dfrac{\sqrt{8}\sqrt{12y}}{\sqrt{6y}}$

B *In Problems 41–48, multiply, and express answers using positive exponents only.*

41. $3x^{3/4}(4x^{1/4} - 2x^8)$

42. $2m^{1/3}(3m^{2/3} - m^6)$

43. $(3u^{1/2} - v^{1/2})(u^{1/2} - 4v^{1/2})$

44. $(a^{1/2} + 2b^{1/2})(a^{1/2} - 3b^{1/2})$

45. $(5m^{1/2} + n^{1/2})(5m^{1/2} - n^{1/2})$

46. $(2x^{1/2} - 3y^{1/2})(2x^{1/2} + 3y^{1/2})$

47. $(3x^{1/2} - y^{1/2})^2$

48. $(x^{1/2} + 2y^{1/2})^2$

Write each expression in Problems 49–54 in the form $ax^p + bx^q$, where a and b are real numbers and p and q are rational numbers.

49. $\dfrac{\sqrt[3]{x^2} + 2}{2\sqrt[3]{x}}$ **50.** $\dfrac{12\sqrt{x} - 3}{4\sqrt{x}}$ **51.** $\dfrac{2\sqrt[4]{x^3} + \sqrt[3]{x}}{3x}$

52. $\dfrac{3\sqrt[3]{x^2} + \sqrt{x}}{5x}$ **53.** $\dfrac{2\sqrt[3]{x} - \sqrt{x}}{4\sqrt{x}}$ **54.** $\dfrac{x^2 - 4\sqrt{x}}{2\sqrt[3]{x}}$

Rationalize the denominators in Problems 55–60.

55. $\dfrac{12mn^2}{\sqrt{3mn}}$ **56.** $\dfrac{14x^2}{\sqrt{7x}}$ **57.** $\dfrac{2}{\sqrt{x} - 2}$

58. $\dfrac{3}{\sqrt{x} + 4}$ **59.** $\dfrac{7(x - y)^2}{\sqrt{x} - \sqrt{y}}$ **60.** $\dfrac{3a - 3b}{\sqrt{a} + \sqrt{b}}$

Rationalize the numerators in Problems 61–66.

61. $\dfrac{\sqrt{5xy}}{5x^2y^2}$ **62.** $\dfrac{\sqrt{3mn}}{3mn}$

63. $\dfrac{\sqrt{x + h} - \sqrt{x}}{h}$ **64.** $\dfrac{\sqrt{2(a + h)} - \sqrt{2a}}{h}$

65. $\dfrac{\sqrt{t} - \sqrt{x}}{t - x}$ **66.** $\dfrac{\sqrt{x} - \sqrt{y}}{\sqrt{x} + \sqrt{y}}$

Problems 67–70 illustrate common errors involving rational exponents. In each case, find numerical examples that show that the left side is not always equal to the right side.

67. $(x + y)^{1/2} \neq x^{1/2} + y^{1/2}$

68. $(x^3 + y^3)^{1/3} \neq x + y$

69. $(x + y)^{1/3} \neq \dfrac{1}{(x + y)^3}$

70. $(x + y)^{-1/2} \neq \dfrac{1}{(x + y)^2}$

C *In Problems 71–76, simplify by writing each expression as a simple or single fraction reduced to lowest terms and without negative exponents.*

71. $-\frac{1}{2}(x - 2)(x + 3)^{-3/2} + (x + 3)^{-1/2}$

72. $2(x - 2)^{-1/2} - \frac{1}{2}(2x + 3)(x - 2)^{-3/2}$

73. $\dfrac{(x - 1)^{1/2} - x(\frac{1}{2})(x - 1)^{-1/2}}{x - 1}$

74. $\dfrac{(2x-1)^{1/2}-(x+2)(\frac{1}{2})(2x-1)^{-1/2}(2)}{2x-1}$

75. $\dfrac{(x+2)^{2/3}-x(\frac{2}{3})(x+2)^{-1/3}}{(x+2)^{4/3}}$

76. $\dfrac{2(3x-1)^{1/3}-(2x+1)(\frac{1}{3})(3x-1)^{-2/3}(3)}{(3x-1)^{2/3}}$

In Problems 77–82, evaluate using a calculator. (Refer to the instruction book for your calculator to see how exponential forms are evaluated.)

77. $22^{3/2}$ **78.** $15^{5/4}$ **79.** $827^{-3/8}$

80. $103^{-3/4}$ **81.** $37.09^{7/3}$ **82.** $2.876^{8/5}$

In Problems 83 and 84, evaluate each expression on a calculator and determine which pairs have the same value. Verify these results algebraically.

83. (A) $\sqrt{3}+\sqrt{5}$ (B) $\sqrt{2+\sqrt{3}}+\sqrt{2-\sqrt{3}}$
(C) $1+\sqrt{3}$ (D) $\sqrt[3]{10+6\sqrt{3}}$
(E) $\sqrt{8+\sqrt{60}}$ (F) $\sqrt{6}$

84. (A) $2\sqrt[3]{2+\sqrt{5}}$ (B) $\sqrt{8}$
(C) $\sqrt{3}+\sqrt{7}$ (D) $\sqrt{3+\sqrt{8}}+\sqrt{3-\sqrt{8}}$
(E) $\sqrt{10+\sqrt{84}}$ (F) $1+\sqrt{5}$

SECTION A-8 Linear Equations and Inequalities in One Variable

- LINEAR EQUATIONS
- LINEAR INEQUALITIES
- APPLICATIONS

The equation

$$3-2(x+3)=\frac{x}{3}-5$$

and the inequality

$$\frac{x}{2}+2(3x-1)\geq 5$$

are both first-degree in one variable. In general, a **first-degree, or linear, equation** in one variable is any equation that can be written in the form

Standard form: $ax+b=0 \qquad a\neq 0$ (1)

If the equality symbol, =, in (1) is replaced by $<, >, \leq$, or $\geq$, then the resulting expression is called a **first-degree, or linear, inequality.**

A **solution** of an equation (or inequality) involving a single variable is a number that, when substituted for the variable, makes the equation (or inequality) true. The set of all solutions is called the **solution set.** When we say that we **solve an equation** (or inequality), we mean that we find its solution set.

Knowing what is meant by the solution set is one thing; finding it is another. We start by recalling the idea of equivalent equations and equivalent inequalities. If we perform an operation on an equation (or inequality) that produces another equation (or inequality) with the same solution set, then the two equations (or inequalities) are said to be **equivalent.** The basic idea in solving equations and inequalities is to perform operations on these forms that produce simpler equivalent forms, and to continue the process until we obtain an equation or inequality with an obvious solution.

Linear Equations

Linear equations are generally solved using the following equality properties:

Equality Properties

An equivalent equation will result if:

1. The same quantity is added to or subtracted from each side of a given equation.
2. Each side of a given equation is multiplied by or divided by the same nonzero quantity.

Several examples should remind you of the process of solving equations.

Example 1 **Solving a Linear Equation** Solve and check:

$$8x - 3(x - 4) = 3(x - 4) + 6$$

Solution

$$\begin{aligned} 8x - 3(x - 4) &= 3(x - 4) + 6 \\ 8x - 3x + 12 &= 3x - 12 + 6 \\ 5x + 12 &= 3x - 6 \\ 2x &= -18 \\ x &= -9 \end{aligned}$$

Check

$$\begin{aligned} 8x - 3(x - 4) &= 3(x - 4) + 6 \\ 8(-9) - 3[(-9) - 4] &\stackrel{?}{=} 3[(-9) - 4] + 6 \\ -72 - 3(-13) &\stackrel{?}{=} 3(-13) + 6 \\ -33 &\stackrel{\checkmark}{=} -33 \end{aligned}$$

Matched Problem 1 Solve and check: $3x - 2(2x - 5) = 2(x + 3) - 8$

Explore–Discuss 1

According to equality property 2, multiplying both sides of an equation by a nonzero number always produces an equivalent equation. What is the smallest positive number that you could use to multiply both sides of the following equation in order to produce an equivalent equation without fractions?

$$\frac{x + 1}{3} - \frac{x}{4} = \frac{1}{2}$$

Example 2 **Solving a Linear Equation** Solve and check: $\frac{x + 2}{2} - \frac{x}{3} = 5$

Solution What operations can we perform on

$$\frac{x + 2}{2} - \frac{x}{3} = 5$$

to eliminate the denominators? If we can find a number that is exactly divisible by each denominator, then we can use the multiplication property of equality to clear the denominators. The LCD (least common denominator) of the fractions, 6, is exactly what we are looking for! Actually, any common denominator will do, but the LCD results in a simpler equivalent equation. Thus, we multiply both sides of the equation by 6:

$$6\left(\frac{x+2}{2} - \frac{x}{3}\right) = 6 \cdot 5$$

$$\overset{3}{\cancel{6}} \cdot \frac{(x+2)}{\underset{1}{\cancel{2}}} - \overset{2}{\cancel{6}} \cdot \frac{x}{\underset{1}{\cancel{3}}} = 30$$

$$\begin{aligned} 3(x+2) - 2x &= 30 \\ 3x + 6 - 2x &= 30 \\ x &= 24 \end{aligned}$$

CHECK

$$\begin{aligned} \frac{x+2}{2} - \frac{x}{3} &= 5 \\ \frac{\mathbf{24}+2}{2} - \frac{\mathbf{24}}{3} &\overset{?}{=} 5 \\ 13 - 8 &\overset{?}{=} 5 \\ 5 &\overset{\checkmark}{=} 5 \end{aligned}$$

Matched Problem 2 Solve and check: $\frac{x+1}{3} - \frac{x}{4} = \frac{1}{2}$

In many applications of algebra, formulas or equations must be changed to alternative equivalent forms. The following examples are typical.

Example 3 **Solving a Formula for a Particular Variable** Solve the amount formula for simple interest, $A = P + Prt$, for:

(A) r in terms of the other variables
(B) P in terms of the other variables

SOLUTION (A)

$$\begin{aligned} A &= P + Prt && \text{Reverse equation.} \\ P + Prt &= A && \text{Now isolate } r \text{ on the left side.} \\ Prt &= A - P && \text{Divide both members by } Pt. \\ r &= \frac{A - P}{Pt} \end{aligned}$$

(B)

$$\begin{aligned} A &= P + Prt && \text{Reverse equation.} \\ P + Prt &= A && \text{Factor out } P \text{ (note the use of the distributive property).} \\ P(1 + rt) &= A && \text{Divide by } (1 + rt). \\ P &= \frac{A}{1 + rt} \end{aligned}$$

Matched Problem 3 Solve $M = Nt + Nr$ for:
(A) t (B) N

LINEAR INEQUALITIES

Before we start solving linear inequalities, let us recall what we mean by < (less than) and > (greater than). If a and b are real numbers, then we write

$a < b$ a is less than b

if there exists a positive number p such that $a + p = b$. Certainly, we would expect that if a positive number was added to any real number, the sum would be larger than the original. That is essentially what the definition states. If $a < b$, we may also write

$$b > a \quad b \text{ is greater than } a$$

Example 4 **Inequalities**

(A) $3 < 5$ Since $3 + 2 = 5$
(B) $-6 < -2$ Since $-6 + 4 = -2$
(C) $0 > -10$ Since $-10 < 0$

Matched Problem 4 Replace each question mark with either $<$ or $>$.

(A) 2 ? 8 (B) −20 ? 0 (C) −3 ? −30

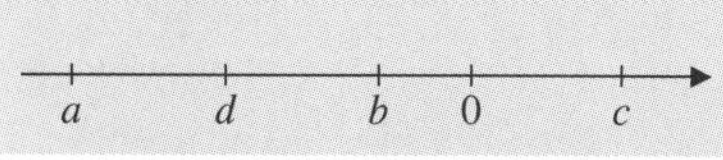

FIGURE 1
$a < b, c > d$

The inequality symbols have a very clear geometric interpretation on the real number line. If $a < b$, then a is to the left of b on the number line; if $c > d$, then c is to the right of d (Fig. 1). Check this geometric property with the inequalities in Example 4.

Explore–Discuss 2

Replace ? with $<$ or $>$ in each of the following:

(A) $-1 \ ? \ 3$ and $2(-1) \ ? \ 2(3)$
(B) $-1 \ ? \ 3$ and $-2(-1) \ ? \ -2(3)$
(C) $12 \ ? \ -8$ and $\frac{12}{4} \ ? \ \frac{-8}{4}$
(D) $12 \ ? \ -8$ and $\frac{12}{-4} \ ? \ \frac{-8}{-4}$

Based on these examples, describe verbally the effect of multiplying both sides of an inequality by a number.

Now let us turn to the problem of solving linear inequalities in one variable. Recall that a **solution** of an inequality involving one variable is a number that, when substituted for the variable, makes the inequality true. The set of all solutions is called the **solution set.** When we say that we **solve an inequality,** we mean that we find its solution set. The procedures used to solve linear inequalities in one variable are almost the same as those used to solve linear equations in one variable but with one important exception, as noted in property 3 below.

Inequality Properties

An equivalent inequality will result and the **sense will remain the same** if each side of the original inequality:

1. Has the same real number added to or subtracted from it.
2. Is multiplied or divided by the same positive number.

(continued)

Inequality Properties *(continued)*

An equivalent inequality will result and the **sense will reverse** if each side of the original inequality:

3. Is multiplied or divided by the same negative number.

Note: Multiplication by 0 and division by 0 are not permitted.

Thus, we can perform essentially the same operations on inequalities that we perform on equations, with the exception that **the sense of the inequality reverses if we multiply or divide both sides by a negative number.** Otherwise, the sense of the inequality does not change. For example, if we start with the true statement

$$-3 > -7$$

and multiply both sides by 2, we obtain

$$-6 > -14$$

and the sense of the inequality stays the same. But if we multiply both sides of $-3 > -7$ by -2, then the left side becomes 6 and the right side becomes 14, so we must write

$$6 < 14$$

to have a true statement. Thus, the sense of the inequality reverses.

If $a < b$, the double inequality $a < x < b$ means that $\boldsymbol{x > a}$ **and** $\boldsymbol{x < b}$; that is, x is between a and b. Other variations, as well as a useful interval notation, are given in Table 1. Note that an end point on a line graph has a square bracket through it if it is included in the inequality and a parenthesis through it if it is not.

Table 1

INTERVAL NOTATION	INEQUALITY NOTATION	LINE GRAPH
$[a, b]$	$a \leq x \leq b$	[ ] at a, b, x
$[a, b)$	$a \leq x < b$	[ ) at a, b, x
$(a, b]$	$a < x \leq b$	( ] at a, b, x
(a, b)	$a < x < b$	( ) at a, b, x
$(-\infty, a]$	$x \leq a$	] at a, x
$(-\infty, a)$	$x < a$	) at a, x
$[b, \infty)$*	$x \geq b$	[at b, x
(b, ∞)	$x > b$	(at b, x

*The symbol ∞ (read "infinity") is not a number. When we write $[b, \infty)$, we are simply referring to the interval starting at b and continuing indefinitely to the right. We would never write $[b, \infty]$.

Example 5 **Interval and Inequality Notation, and Line Graphs**

(A) Write $[-2, 3)$ as a double inequality and graph.
(B) Write $x \geqslant -5$ in interval notation and graph.

SOLUTION (A) $[-2, 3)$ is equivalent to $-2 \leqslant x < 3$.

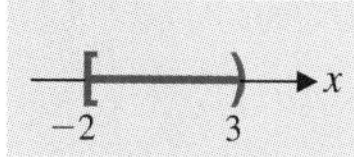

(B) $x \geqslant -5$ is equivalent to $[-5, \infty)$.

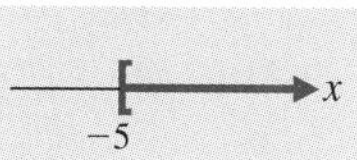

Matched Problem 5 (A) Write $(-7, 4]$ as a double inequality and graph.
(B) Write $x < 3$ in interval notation and graph.

Explore–Discuss 3

The solution to Example 5B shows the graph of the inequality $x \geqslant -5$. What is the graph of $x < -5$? What is the corresponding interval? Describe the relationship between these sets.

Example 6 **Solving a Linear Inequality** Solve and graph:

$$2(2x + 3) < 6(x - 2) + 10$$

SOLUTION

$$\begin{aligned} 2(2x + 3) &< 6(x - 2) + 10 \\ 4x + 6 &< 6x - 12 + 10 \\ 4x + 6 &< 6x - 2 \\ -2x + 6 &< -2 \\ -2x &< -8 \\ x &> 4 \quad \text{or} \quad (4, \infty) \end{aligned}$$

Notice that the sense of the inequality reverses when we divide both sides by −2.

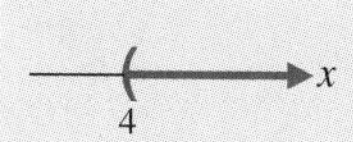

Notice that in the graph of $x > 4$, we use a parenthesis through 4, since the point 4 is not included in the graph.

Matched Problem 6 Solve and graph: $3(x - 1) \leqslant 5(x + 2) - 5$

Example 7 **Solving a Double Inequality** Solve and graph: $-3 < 2x + 3 \leqslant 9$

SOLUTION We are looking for all numbers x such that $2x + 3$ is between −3 and 9, including 9 but not −3. We proceed as above except that we try to isolate x in the middle:

$$-3 < 2x + 3 \leqslant 9$$

$$-3 - \mathbf{3} < 2x + 3 - \mathbf{3} \leqslant 9 - \mathbf{3}$$

$$-6 < 2x \leqslant 6$$

$$\frac{-6}{2} < \frac{2x}{2} \leqslant \frac{6}{2}$$

$$-3 < x \leqslant 3 \quad \text{or} \quad (-3, 3]$$

Matched Problem 7 Solve and graph: $-8 \leqslant 3x - 5 < 7$

Note that a linear equation usually has exactly one solution, while a linear inequality usually has infinitely many solutions.

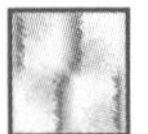

Applications

To realize the full potential of algebra, we must be able to translate real-world problems into mathematical forms. In short, we must be able to do word problems.

The first example below involves the important concept of **break-even analysis,** which is encountered in several places in this text. Any manufacturing company has **costs, *C*,** and **revenues, *R*.** The company will have a **loss** if $R < C$, will **break even** if $R = C$, and will have a **profit** if $R > C$. Costs involve **fixed costs,** such as plant overhead, product design, set-up, and promotion; and **variable costs,** which are dependent upon the number of items produced at a certain cost per item.

Example 8 **Break-Even Analysis** A recording company produces compact disks (CD's). One-time fixed costs for a particular CD are \$24,000, which includes costs such as recording, album design, and promotion. Variable costs amount to \$6.20 per CD and include the manufacturing, distribution, and royalty costs for each disk actually manufactured and sold to a retailer. The CD is sold to retail outlets at \$8.70 each. How many CD's must be manufactured and sold for the company to break even?

Solution Let x = Number of CD's manufactured and sold
C = Cost of producing x CD's
R = Revenue (return) on sales of x CD's

The company breaks even if $R = C$, with

$$\begin{aligned} C &= \text{Fixed costs} + \text{Variable costs} \\ &= \$24{,}000 + \$6.20x \\ R &= \$8.70x \end{aligned}$$

Find x such that $R = C$; that is, such that

$$\begin{aligned} 8.7x &= 24{,}000 + 6.2x \\ 2.5x &= 24{,}000 \\ x &= 9{,}600 \text{ CD's} \end{aligned}$$

Check For $x = 9{,}600$,

$$\begin{aligned} C &= 24{,}000 + 6.2x & R &= 8.7x \\ &= 24{,}000 + 6.2(9{,}600) & &= 8.7(9{,}600) \\ &= \$83{,}520 & &= \$83{,}520 \end{aligned}$$

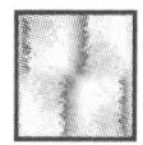

Matched Problem 8 ➡ What is the break-even point in Example 8 if fixed costs are \$18,000, variable costs are \$5.20 per CD, and the CD's are sold to retailers for \$7.60 each?

Algebra has many different types of applications—so many, in fact, that no single approach applies to all. However, the following suggestions may help you get started:

Suggestions for Solving Word Problems

1. Read the problem very carefully.
2. Write down important facts and relationships.
3. Identify unknown quantities in terms of a single letter, if possible.
4. Write an equation (or inequality) relating the unknown quantities and the facts in the problem.
5. Solve the equation (or inequality).
6. Write all solutions requested in the original problem.
7. Check the solution(s) in the original problem.

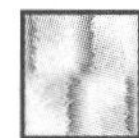

Example 9 ➡ **Consumer Price Index** The Consumer Price Index (CPI) is a measure of the average change in prices over time from a designated reference period, which equals 100. The index is based on prices of basic consumer goods and services, and is published at regular intervals by the Bureau of Labor Statistics. Table 2 lists the CPI for several years from 1960 to 1992. What net annual salary in 1992 would have the same purchasing power as a net annual salary of \$13,000 in 1960? Compute the answer to the nearest dollar.

Table 2
CPI (1982–1984 = 100)

YEAR	INDEX
1960	29.6
1970	38.8
1980	82.4
1990	130.7
1992	140.3

SOLUTION To have the same purchasing power, the ratio of a salary in 1992 to a salary in 1960 would have to be the same as the ratio of the CPI in 1992 to the CPI in 1960. Thus, if x is the net annual salary in 1992, we solve the equation

$$\frac{x}{13{,}000} = \frac{140.3}{29.6}$$

$$x = 13{,}000 \cdot \frac{140.3}{29.6}$$

$$= \$61{,}618 \text{ per year}$$

Matched Problem 9 ➡ What net annual salary in 1970 would have had the same purchasing power as a net annual salary of \$75,000 in 1992? Compute the answer to the nearest dollar.

Answers to Matched Problems

1. $x = 4$ **2.** $x = 2$ **3.** (A) $t = \frac{M - Nr}{N}$ (B) $N = \frac{M}{t + r}$

4. (A) $<$ (B) $<$ (C) $>$

5. (A) $-7 < x \leq 4$; (number line: −7 to 4, x) (B) $(-\infty, 3)$ (number line: 3, x)

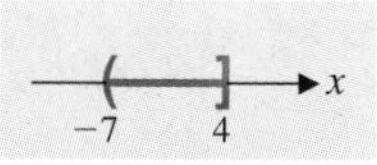

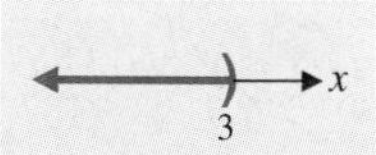

6. $x \geq -4$ or $[-4, \infty)$

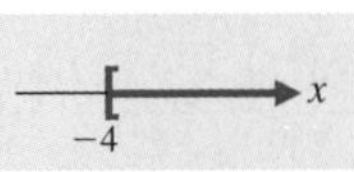

7. $-1 \leq x < 4$ or $[-1, 4)$

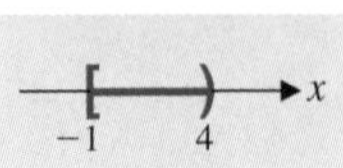

8. 7,500 CD's **9.** $20,741

EXERCISE A-8

A *Solve Problems 1–6.*

1. $2m + 9 = 5m - 6$

2. $3y - 4 = 6y - 19$

3. $x + 5 < -4$

4. $x - 3 > -2$

5. $-3x \geq -12$

6. $-4x \leq 8$

Solve Problems 7–10 and graph.

7. $-4x - 7 > 5$

8. $-2x + 8 < 4$

9. $2 \leq x + 3 \leq 5$

10. $-3 < y - 5 < 8$

Solve Problems 11–26.

11. $\frac{y}{7} - 1 = \frac{1}{7}$

12. $\frac{m}{5} - 2 = \frac{3}{5}$

13. $\frac{x}{3} > -2$

14. $\frac{y}{-2} \leq -1$

15. $\frac{y}{3} = 4 - \frac{y}{6}$

16. $\frac{x}{4} = 9 - \frac{x}{2}$

B

17. $10x + 25(x - 3) = 275$

18. $-3(4 - x) = 5 - (x + 1)$

19. $3 - y \leq 4(y - 3)$

20. $x - 2 \geq 2(x - 5)$

21. $\frac{x}{5} - \frac{x}{6} = \frac{6}{5}$

22. $\frac{y}{4} - \frac{y}{3} = \frac{1}{2}$

23. $\frac{m}{5} - 3 < \frac{3}{5} - m$

24. $u - \frac{2}{3} > \frac{u}{3} + 2$

25. $0.1(x - 7) + 0.05x = 0.8$

26. $0.4(u + 5) - 0.3u = 17$

Solve Problems 27–30 and graph.

27. $2 \leq 3x - 7 < 14$

28. $-4 \leq 5x + 6 < 21$

29. $-4 \leq \frac{9}{5}C + 32 \leq 68$

30. $-1 \leq \frac{2}{3}t + 5 \leq 11$

C *Solve Problems 31–38 for the indicated variable.*

31. $3x - 4y = 12$; for y

32. $y = -\frac{2}{3}x + 8$; for x

33. $Ax + By = C$; for y $(B \neq 0)$

34. $y = mx + b$; for m

35. $F = \frac{9}{5}C + 32$; for C

36. $C = \frac{5}{9}(F - 32)$; for F

37. $A = Bm - Bn$; for B

38. $U = 3C - 2CD$; for C

Solve Problems 39 and 40 and graph.

39. $-3 \leq 4 - 7x < 18$

40. $-1 < 9 - 2u \leq 5$

41. What can be said about the signs of the numbers a and b in each case?

(A) $ab > 0$ (B) $ab < 0$

(C) $\frac{a}{b} > 0$ (D) $\frac{a}{b} < 0$

42. What can be said about the signs of the numbers a, b, and c in each case?

(A) $abc > 0$ (B) $\frac{ab}{c} < 0$

(C) $\frac{a}{bc} > 0$ (D) $\frac{a^2}{bc} < 0$

43. Replace each question mark with $<$ or $>$, as appropriate:

(A) If $a - b = 2$, then a ? b.

(B) If $c - d = -1$, then c ? d.

44. For what c and d is $c + d < c - d$?

45. If both a and b are positive numbers and b/a is greater than 1, then is $a - b$ positive or negative?

46. If both a and b are negative numbers and b/a is greater than 1, then is $a - b$ positive or negative?

APPLICATIONS

Business & Economics

47. *Puzzle.* A jazz concert brought in $165,000 on the sale of 8,000 tickets. If the tickets sold for $15 and $25 each, how many of each type of ticket were sold?

48. *Puzzle.* An all-day parking meter takes only dimes and quarters. If it contains 100 coins with a total value of $14.50, how many of each type of coin are in the meter?

49. *Investing.* You have $12,000 to invest. If part is invested at 10% and the rest at 15%, how much should be invested at each rate to yield 12% on the total amount?

50. *Investing.* An investor has $20,000 to invest. If part is invested at 8% and the rest at 12%, how much should be invested at each rate to yield 11% on the total amount?

51. *Inflation.* If the price change of cars parallels the change in the CPI (see Table 2 in Example 9), what would a car sell for (to the nearest dollar) in 1992 if a comparable model sold for $5,000 in 1970?

52. *Inflation.* If the price change in houses parallels the CPI (see Table 2 in Example 9), what would a house valued at $200,000 in 1992 be valued at (to the nearest dollar) in 1960?

53. *Break-even analysis.* A publisher for a promising new novel figures fixed costs (overhead, advances, promotion, copy editing, typesetting, and so on) at $55,000, and variable costs (printing, paper, binding, shipping) at $1.60 for each book produced. If the book is sold to distributors for $11 each, what is the break-even point for the publisher?

54. *Break-even analysis.* The publisher of a new book called *Muscle-Powered Sports* figures fixed costs at $92,000 and variable costs at $2.10 for each book produced. If the book is sold to distributors for $15 each, how many must be sold for the publisher to break even?

55. *Break-even analysis.* The publisher in Problem 53 finds that rising prices for paper increase the variable costs to $2.10 per book.

(A) Discuss possible strategies the company might use to deal with this increase in costs.

(B) If the company continues to sell the books for $11, how many books must they sell now to make a profit?

(C) If the company wants to start making a profit at the same production level as before the cost increase, how much should they sell the book for now?

56. *Break-even analysis.* The publisher in Problem 54 finds that rising prices for paper increase the variable costs to $2.70 per book.

(A) Discuss possible strategies the company might use to deal with this increase in costs.

(B) If the company continues to sell the books for $15, how many books must they sell now to make a profit?

(C) If the company wants to start making a profit at the same production level as before the cost increase, how much should they sell the book for now?

Life Sciences

57. *Wildlife management.* A naturalist for a fish and game department estimated the total number of rainbow trout in a certain lake using the popular capture–mark–recapture technique. He netted, marked, and released 200 rainbow trout. A week later, allowing for thorough mixing, he again netted 200 trout and found 8 marked ones among them. Assuming that the proportion of marked fish in the second sample was the same as the proportion of all marked fish in the total population, estimate the number of rainbow trout in the lake.

58. *Ecology.* If the temperature for a 24 hour period at an Antarctic station ranged between −49°F and 14°F (that is, $-49 \leq F \leq 14$),what was the range in degrees Celsius? [*Note:* $F = \frac{9}{5}C + 32$.]

Social Sciences

59. *Psychology.* The IQ (intelligence quotient) is found by dividing the mental age (MA), as indicated on standard tests, by the chronological age (CA) and multiplying by 100. For example, if a child has a mental age of 12 and a chronological age of 8, the calculated IQ is 150. If a 9-year-old girl has an IQ of 140, compute her mental age.

60. *Anthropology.* In their study of genetic groupings, anthropologists use a ratio called the *cephalic index.*

This is the ratio of the width of the head to its length (looking down from above) expressed as a percentage. Symbolically,

$$C = \frac{100W}{L}$$

where C is the cephalic index, W is the width, and L is the length. If an Indian tribe in Baja California (Mexico) had an average cephalic index of 66 and the average width of their heads was 6.6 inches, what was the average length of their heads?

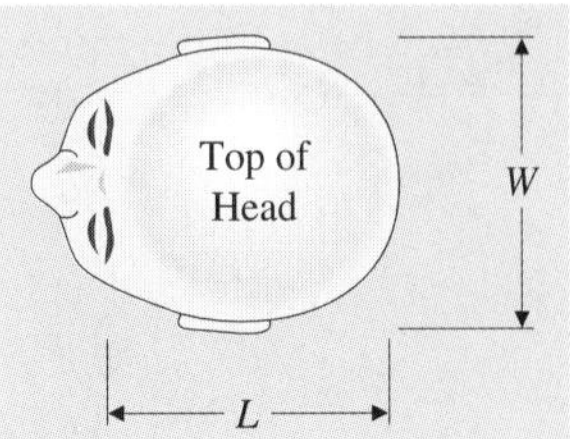

Figure for 60

SECTION A-9 Quadratic Equations

- SOLUTION BY SQUARE ROOT
- SOLUTION BY FACTORING
- QUADRATIC FORMULA
- THE QUADRATIC FORMULA AND FACTORING
- APPLICATION: SUPPLY AND DEMAND

A **quadratic equation** in one variable is any equation that can be written in the form

$$\boldsymbol{ax^2 + bx + c = 0 \qquad a \neq 0}$$

where x is a variable and a, b, and c are constants. We will refer to this form as the **standard form.** The equations

$$5x^2 - 3x + 7 = 0 \qquad \text{and} \qquad 18 = 32t^2 - 12t$$

are both quadratic equations, since they are either in the standard form or can be transformed into this form.

We will restrict our review to finding real solutions to quadratic equations.

SOLUTION BY SQUARE ROOT

The easiest type of quadratic equation to solve is the special form where the first-degree term is missing:

$$\boldsymbol{ax^2 + c = 0 \qquad a \neq 0}$$

The method of solution of this special form makes direct use of the square root property:

Square Root Property

If $a^2 = b$, then $a = \pm\sqrt{b}$.

Explore–Discuss 1

Determine whether each of the following pairs of equations are equivalent. Explain.

(A) $x^2 = 4$ and $x = 2$
(B) $x^2 = 4$ and $x = -2$
(C) $x = \sqrt{4}$ and $x = 2$
(D) $x = \sqrt{4}$ and $x = -2$
(E) $x = -\sqrt{4}$ and $x = -2$

Example 1 **Square Root Method** Use the square root property to solve each equation.

(A) $x^2 - 7 = 0$ (B) $2x^2 - 10 = 0$ (C) $3x^2 + 27 = 0$
(D) $(x - 8)^2 = 9$

SOLUTION

(A)
$$x^2 - 7 = 0$$
$$x^2 = 7 \quad \text{What real number squared is 7?}$$
$$x = \pm\sqrt{7} \quad \text{Short for } \sqrt{7} \text{ and } -\sqrt{7}$$

(B)
$$2x^2 - 10 = 0$$
$$2x^2 = 10$$
$$x^2 = 5 \quad \text{What real number squared is 5?}$$
$$x = \pm\sqrt{5}$$

(C)
$$3x^2 + 27 = 0$$
$$3x^2 = -27$$
$$x^2 = -9 \quad \text{What real number squared is } -9\text{?}$$

No real solution, since no real number squared is negative.

(D)
$$(x - 8)^2 = 9$$
$$x - 8 = \pm\sqrt{9}$$
$$x - 8 = \pm 3$$
$$x = 8 \pm 3 = 5 \quad \text{or} \quad 11$$

Matched Problem 1 Use the square root property to solve each equation.

(A) $x^2 - 6 = 0$ (B) $3x^2 - 12 = 0$ (C) $x^2 + 4 = 0$
(D) $(x + 5)^2 = 1$

Solution by Factoring

If the left side of a quadratic equation when written in standard form can be factored, then the equation can be solved very quickly. The method of solution by factoring rests on the following important property of real numbers (see Appendix A-2):

If a and b are real numbers, then $ab = 0$ if and only if $a = 0$ or $b = 0$ (or both).

Example 2 **Factoring Method** Solve by factoring using integer coefficients, if possible.

(A) $3x^2 - 6x - 24 = 0$ (B) $3y^2 = 2y$ (C) $x^2 - 2x - 1 = 0$

Solution (A)

$$\begin{aligned} 3x^2 - 6x - 24 &= 0 && \text{Divide both sides by 3, since 3 is a factor of each coefficient.} \\ x^2 - 2x - 8 &= 0 && \text{Factor the left side, if possible.} \\ (x - 4)(x + 2) &= 0 \\ x - 4 = 0 \quad \text{or} \quad x + 2 &= 0 \\ x = 4 \quad \text{or} \quad x &= -2 \end{aligned}$$

(B)

$$\begin{aligned} 3y^2 &= 2y \\ 3y^2 - 2y &= 0 && \text{We lose the solution } y = 0 \text{ if both sides are divided by } y \\ y(3y - 2) &= 0 && (3y^2 = 2y \text{ and } 3y = 2 \text{ are not equivalent).} \\ y = 0 \quad \text{or} \quad 3y - 2 &= 0 \\ 3y &= 2 \\ y &= \tfrac{2}{3} \end{aligned}$$

(C) $x^2 - 2x - 1 = 0$

This equation cannot be factored using integer coefficients. We will solve this type of equation by another method, considered below.

Matched Problem 2 Solve by factoring using integer coefficients, if possible.

(A) $2x^2 + 4x - 30 = 0$ (B) $2x^2 = 3x$ (C) $2x^2 - 8x + 3 = 0$

Note that an equation such as $x^2 = 25$ can be solved by either the square root or the factoring method, and the results are the same (as they should be). Solve this equation both ways and compare.

Also, note that the factoring method can be extended to higher-degree polynomial equations. Consider the following:

$$\begin{aligned} x^3 - x &= 0 \\ x(x^2 - 1) &= 0 \\ x(x - 1)(x + 1) &= 0 \\ x = 0 \quad \text{or} \quad x - 1 &= 0 \quad \text{or} \quad x + 1 = 0 \\ \text{Solution:} \quad x &= 0, 1, -1 \end{aligned}$$

Check these solutions in the original equation.

The factoring and square root methods are fast and easy to use when they apply. However, there are quadratic equations that look simple but cannot be solved by either method. For example, as was noted in Example 2C, the polynomial in

$$x^2 - 2x - 1 = 0$$

cannot be factored using integer coefficients. This brings us to the well-known and widely used *quadratic formula.*

Quadratic Formula

There is a method called *completing the square* that will work for all quadratic equations. After briefly reviewing this method, we will then use it to develop the famous quadratic formula—a formula that will enable us to solve any quadratic equation quite mechanically.

Explore–Discuss 2

Replace ? in each of the following with a number that makes the equation valid.

(A) $(x + 1)^2 = x^2 + 2x + ?$ (B) $(x + 2)^2 = x^2 + 4x + ?$
(C) $(x + 3)^2 = x^2 + 6x + ?$ (D) $(x + 4)^2 = x^2 + 8x + ?$

Replace ? in each of the following with a number that makes the trinomial a perfect square.

(E) $x^2 + 10x + ?$ (F) $x^2 + 12x + ?$ (G) $x^2 + bx + ?$

The method of **completing the square** is based on the process of transforming a quadratic equation in standard form,

$$ax^2 + bx + c = 0$$

into the form

$$(x + A)^2 = B$$

where A and B are constants. Then, this last equation can be solved easily (if it has a real solution) by the square root method discussed above.

Consider the equation from Example 2C:

$$x^2 - 2x - 1 = 0 \qquad (1)$$

Since the left side does not factor using integer coefficients, we add 1 to each side to remove the constant term from the left side:

$$x^2 - 2x = 1 \qquad (2)$$

Now we try to find a number that we can add to each side to make the left side a square of a first-degree polynomial. Note the following square of a binomial:

$$(x + m)^2 = x^2 + 2mx + m^2$$

We see that the third term on the right is the square of one-half the coefficient of x in the second term on the right. To complete the square in equation (2), we add the square of one-half the coefficient of x, $\left(-\frac{2}{2}\right)^2 = 1$, to each side. (This rule works only when the coefficient of x^2 is 1, that is, $a = 1$.) Thus,

$$x^2 - 2x + \mathbf{1} = 1 + \mathbf{1}$$

The left side is the square of $x - 1$, and we write

$$(x - 1)^2 = 2$$

What number squared is 2?

$$x - 1 = \pm\sqrt{2}$$
$$x = 1 \pm \sqrt{2}$$

And equation (1) is solved!

Let us try the method on the general quadratic equation

$$ax^2 + bx + c = 0 \qquad a \neq 0 \qquad (3)$$

and solve it once and for all for x in terms of the coefficients a, b, and c. We start by multiplying both sides of (3) by $1/a$ to obtain

$$x^2 + \frac{b}{a}x + \frac{c}{a} = 0$$

Add $-c/a$ to both sides:

$$x^2 + \frac{b}{a}x = -\frac{c}{a}$$

Now we complete the square on the left side by adding the square of one-half the coefficient of x, that is, $(b/2a)^2 = b^2/4a^2$, to each side:

$$x^2 + \frac{b}{a}x + \frac{b^2}{4a^2} = \frac{b^2}{4a^2} - \frac{c}{a}$$

Writing the left side as a square and combining the right side into a single fraction, we obtain

$$\left(x + \frac{b}{2a}\right)^2 = \frac{b^2 - 4ac}{4a^2}$$

Now we solve by the square root method:

$$x + \frac{b}{2a} = \pm\sqrt{\frac{b^2 - 4ac}{4a^2}}$$

$$x = -\frac{b}{2a} \pm \frac{\sqrt{b^2 - 4ac}}{2a}$$

Since $\pm\sqrt{4a^2} = \pm 2a$ for any real number a

When this is written as a single fraction, it becomes the **quadratic formula:**

Quadratic Formula

If $ax^2 + bx + c = 0, a \neq 0$, then

$$x = \frac{-b \pm \sqrt{b^2 - 4ac}}{2a}$$

This formula is generally used to solve quadratic equations when the square root or factoring methods do not work. The quantity $b^2 - 4ac$ under the radical is called the **discriminant,** and it gives us the useful information about solutions listed in Table 1.

Table 1

$b^2 - 4ac$	$ax^2 + bx + c = 0$
Positive	Two real solutions
Zero	One real solution
Negative	No real solutions

Example 3 **Quadratic Formula Method** Solve $x^2 - 2x - 1 = 0$ using the quadratic formula.

SOLUTION

$$x^2 - 2x - 1 = 0$$

$$x = \frac{-b \pm \sqrt{b^2 - 4ac}}{2a} \qquad a = 1, b = -2, c = -1$$

$$= \frac{-(-2) \pm \sqrt{(-2)^2 - 4(1)(-1)}}{2(1)}$$

$$= \frac{2 \pm \sqrt{8}}{2} = \frac{2 \pm 2\sqrt{2}}{2} = 1 \pm \sqrt{2} \approx -0.414 \quad \text{or} \quad 2.414$$

CHECK

$$x^2 - 2x - 1 = 0$$

When $x = 1 + \sqrt{2}$,

$$(1 + \sqrt{2})^2 - 2(1 + \sqrt{2}) - 1 = 1 + 2\sqrt{2} + 2 - 2 - 2\sqrt{2} - 1 = 0$$

When $x = 1 - \sqrt{2}$,

$$(1 - \sqrt{2})^2 - 2(1 - \sqrt{2}) - 1 = 1 - 2\sqrt{2} + 2 - 2 + 2\sqrt{2} - 1 = 0$$

Matched Problem 3 Solve $2x^2 - 4x - 3 = 0$ using the quadratic formula.

If we try to solve $x^2 - 6x + 11 = 0$ using the quadratic formula, we obtain

$$x = \frac{6 \pm \sqrt{-8}}{2}$$

which is not a real number. (Why?)

THE QUADRATIC FORMULA AND FACTORING

As in Section A-4, we restrict our interest in factoring to polynomials with integer coefficients. If a polynomial cannot be factored as a product of lower-degree polynomials with integer coefficients, we say that the polynomial is **not factorable in the integers.**

Suppose you were asked to factor

$$x^2 - 19x - 372 \qquad (4)$$

The larger the coefficients, the more difficult the process of applying the *ac* test discussed in Section A-4. The quadratic formula provides a simple and efficient method of factoring a second-degree polynomial with integer coefficients as the product of two first-degree polynomials with integer coefficients, if the factors exist. We illustrate the method using equation (4), and generalize the process from this experience.

We start by solving the corresponding quadratic equation using the quadratic formula:

$$x^2 - 19x - 372 = 0$$

$$x = \frac{-(-19) \pm \sqrt{(-19)^2 - 4(1)(-372)}}{2}$$

$$= \frac{19 \pm \sqrt{1{,}849}}{2}$$

$$= \frac{19 \pm 43}{2} = -12 \quad \text{or} \quad 31$$

Now, we write

$$x^2 - 19x - 372 = [x - (-12)](x - 31) = (x + 12)(x - 31)$$

Multiplying the two factors on the right produces the second-degree polynomial on the left.

What is behind this procedure? The following two theorems justify and generalize the process:

Theorem 1 ■■ FACTORABILITY THEOREM

A second-degree polynomial, $ax^2 + bx + c$, with integer coefficients can be expressed as the product of two first-degree polynomials with integer coefficients if and only if $\sqrt{b^2 - 4ac}$ is an integer. ■■

Theorem 2 ■■ FACTOR THEOREM

If r_1 and r_2 are solutions to $ax^2 + bx + c = 0$, then

$$ax^2 + bx + c = a(x - r_1)(x - r_2)$$ ■■

Example 4 **Factoring with the Aid of the Discriminant** Factor, if possible, using integer coefficients:

(A) $4x^2 - 65x + 264$ (B) $2x^2 - 33x - 306$

SOLUTION (A) $4x^2 - 65x + 264$

Step 1. Test for factorability:

$$\sqrt{b^2 - 4ac} = \sqrt{(-65)^2 - 4(4)(264)} = 1$$

Since the result is an integer, the polynomial has first-degree factors with integer coefficients.

Step 2. Factor, using the factor theorem. Find the solutions to the corresponding quadratic equation using the quadratic formula:

$$4x^2 - 65x + 264 = 0$$

From step 1

$$x = \frac{-(-65) \pm 1}{2 \cdot 4} = \frac{33}{4} \quad \text{or} \quad 8$$

Thus,

$$4x^2 - 65x + 264 = 4\left(x - \frac{33}{4}\right)(x - 8)$$
$$= (4x - 33)(x - 8)$$

(B) $2x^2 - 33x - 306$

Step 1. Test for factorability:

$$\sqrt{b^2 - 4ac} = \sqrt{(-33)^2 - 4(2)(-306)} = \sqrt{3{,}537}$$

Since $\sqrt{3{,}537}$ is not an integer, the polynomial is not factorable in the integers.

Matched Problem 4 Factor, if possible, using integer coefficients:

(A) $3x^2 - 28x - 464$ (B) $9x^2 + 320x - 144$

Application: Supply and Demand

Supply and demand analysis is a very important part of business and economics. In general, producers are willing to supply more of an item as the price of an item increases, and less of an item as the price decreases. Similarly, buyers are willing to buy less of an item as the price increases, and more of an item as the price decreases. Thus, we have a dynamic situation where the price, supply, and demand fluctuate until a price is reached at which the supply is equal to the demand. In economic theory, this point is called the **equilibrium point**—if the price increases from this point, the supply will increase and the demand will decrease; if the price decreases from this point, the supply will decrease and the demand will increase.

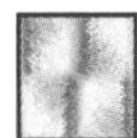

Example 5 **Supply and Demand** At a large beach resort in the summer, the weekly supply and demand equations for folding beach chairs are

$$p = \frac{x}{140} + \frac{3}{4} \quad \text{Supply equation}$$
$$p = \frac{5{,}670}{x} \quad \text{Demand equation}$$

The supply equation indicates that the supplier is willing to sell x units at a price of p dollars per unit. The demand equation indicates that consumers are willing to buy x units at a price of p dollars per unit. How many units are required for supply to equal demand? At what price will supply equal demand?

Solution Set the right side of the supply equation equal to the right side of the demand equation and solve for x:

$$\frac{x}{140} + \frac{3}{4} = \frac{5{,}670}{x} \quad \text{Multiply by } 140x \text{, the LCD.}$$
$$x^2 + 105x = 793{,}800 \quad \text{Write in standard form.}$$
$$x^2 + 105x - 793{,}800 = 0 \quad \text{Use the quadratic formula.}$$
$$x = \frac{-105 \pm \sqrt{105^2 - 4(1)(-793{,}800)}}{2}$$
$$x = 840 \text{ units}$$

The negative root is discarded, since a negative number of units cannot be produced or sold. Substitute $x = 840$ back into either the supply equation or the demand equation to find the equilibrium price (we use the demand equation):

$$p = \frac{5{,}670}{x} = \frac{5{,}670}{840} = \$6.75$$

Thus, at a price of \$6.75 the supplier is willing to supply 840 chairs and consumers are willing to buy 840 chairs during a week.

Matched Problem 5 ➡ Repeat Example 5 if near the end of summer the supply and demand equations are

$$p = \frac{x}{80} - \frac{1}{20} \quad \text{Supply equation}$$

$$p = \frac{1{,}264}{x} \quad \text{Demand equation}$$

Answers to Matched Problems

1. (A) $\pm\sqrt{6}$ (B) ± 2 (C) No real solution (D) $-6, -4$

2. (A) $-5, 3$ (B) $0, \frac{3}{2}$ (C) Cannot be factored using integer coefficients

3. $(2 \pm \sqrt{10})/2$

4. (A) Cannot be factored using integer coefficients (B) $(9x - 4)(x + 36)$

5. 320 chairs at \$3.95 each

EXERCISE A-9

Find only real solutions in the problems below. If there are no real solutions, say so.

A *Solve Problems 1–4 by the square root method.*

1. $2x^2 - 22 = 0$ **2.** $3m^2 - 21 = 0$

3. $(x - 1)^2 = 4$ **4.** $(x + 2)^2 = 9$

Solve Problems 5–8 by factoring.

5. $2u^2 - 8u - 24 = 0$ **6.** $3x^2 - 18x + 15 = 0$

7. $x^2 = 2x$ **8.** $n^2 = 3n$

Solve Problems 9–12 by using the quadratic formula.

9. $x^2 - 6x - 3 = 0$ **10.** $m^2 + 8m + 3 = 0$

11. $3u^2 + 12u + 6 = 0$ **12.** $2x^2 - 20x - 6 = 0$

B *Solve Problems 13–30 by using any method.*

13. $2x^2 = 4x$ **14.** $2x^2 = -3x$

15. $4u^2 - 9 = 0$ **16.** $9y^2 - 25 = 0$

17. $8x^2 + 20x = 12$ **18.** $9x^2 - 6 = 15x$

19. $x^2 = 1 - x$ **20.** $m^2 = 1 - 3m$

21. $2x^2 = 6x - 3$ **22.** $2x^2 = 4x - 1$

23. $y^2 - 4y = -8$ **24.** $x^2 - 2x = -3$

25. $(x + 4)^2 = 11$ **26.** $(y - 5)^2 = 7$

27. $\frac{3}{p} = p$ **28.** $x - \frac{7}{x} = 0$

29. $2 - \frac{2}{m^2} = \frac{3}{m}$ **30.** $2 + \frac{5}{u} = \frac{3}{u^2}$

In Problems 31–38, factor, if possible, as the product of two first-degree polynomials with integer coefficients. Use the quadratic formula and the factor theorem.

31. $x^2 + 40x - 84$ **32.** $x^2 - 28x - 128$

33. $x^2 - 32x + 144$ **34.** $x^2 + 52x + 208$

35. $2x^2 + 15x - 108$ **36.** $3x^2 - 32x - 140$

37. $4x^2 + 241x - 434$ **38.** $6x^2 - 427x - 360$

C

39. Solve $A = P(1 + r)^2$ for r in terms of A and P; that is, isolate r on the left side of the equation (with coefficient 1) and end up with an algebraic expression on the right side involving A and P but not r. Write the answer using positive square roots only.

40. Solve $x^2 + mx + n = 0$ for x in terms of m and n.

41. Consider the quadratic equation

$$x^2 + 4x + c = 0$$

where c is a real number. Discuss the relationship between the values of c and the three types of roots listed in Table 1.

42. Consider the quadratic equation

$$x^2 - 2x + c = 0$$

where c is a real number. Discuss the relationship between the values of c and the three types of roots listed in Table 1.

APPLICATIONS

Business & Economics

43. *Supply and demand.* A company wholesales a certain brand of shampoo in a particular city. Their marketing research department established the following weekly supply and demand equations:

$$p = \frac{x}{450} + \frac{1}{2} \quad \text{Supply equation}$$

$$p = \frac{6{,}300}{x} \quad \text{Demand equation}$$

How many units are required for supply to equal demand? At what price per bottle will supply equal demand?

44. *Supply and demand.* An importer sells a certain brand of automatic camera to outlets in a large metropolitan area. During the summer, the weekly supply and demand equations are

$$p = \frac{x}{6} + 9 \quad \text{Supply equation}$$

$$p = \frac{24{,}840}{x} \quad \text{Demand equation}$$

How many units are required for supply to equal demand? At what price will supply equal demand?

45. *Interest rate.* If P dollars is invested at $100r$ percent compounded annually, at the end of 2 years it will grow to $A = P(1 + r)^2$. At what interest rate will \$100 grow to \$144 in 2 years? [*Note:* If $A = 144$ and $P = 100$, find r.]

46. *Interest rate.* Using the formula in Problem 45, determine the interest rate that will make \$1,000 grow to \$1,210 in 2 years.

Life Sciences

47. *Ecology.* An important element in the erosive force of moving water is its velocity. To measure the velocity v (in feet per second) of a stream, we position a hollow L-shaped tube with one end under the water pointing upstream and the other end pointing straight up a couple of feet out of the water. The water will then be pushed up the tube a certain distance h (in feet) above the surface of the stream. Physicists have shown that $v^2 = 64h$. Approximately how fast is a stream flowing if $h = 1$ foot? If $h = 0.5$ foot?

Social Sciences

48. *Safety research.* It is of considerable importance to know the least number of feet d in which a car can be stopped, including reaction time of the driver, at various speeds v (in miles per hour). Safety research has produced the formula $d = 0.044v^2 + 1.1v$. If it took a car 550 feet to stop, estimate the car's speed at the moment the stopping process was started.

APPENDIX B

Special Topics

SECTION B-1

Sequences, Series, and Summation Notation

- SEQUENCES
- SERIES AND SUMMATION NOTATION

If someone asked you to list all natural numbers that are perfect squares, you might begin by writing

$$1, \ 4, \ 9, \ 16, \ 25, \ 36$$

But you would soon realize that it is impossible to actually list all the perfect squares, since there are an infinite number of them. However, you could represent this collection of numbers in several different ways. One common method is to write

$$1, 4, 9, \ldots, n^2, \ldots \qquad n \in N$$

where N is the set of natural numbers. A list of numbers such as this is generally called a *sequence*. Sequences and related topics form the subject matter of this section.

SEQUENCES

Consider the function f given by

$$f(n) = 2n + 1 \tag{1}$$

where the domain of f is the set of natural numbers N. Note that

$$f(1) = 3, \quad f(2) = 5, \quad f(3) = 7, \quad \ldots$$

The function f is an example of a sequence. In general, a **sequence** is a function with domain a set of successive integers. Instead of the standard function notation used in equation (1), sequences are usually defined in terms of a special notation.

The range value $f(n)$ is usually symbolized more compactly with a symbol such as a_n. Thus, in place of equation (1), we write

$$a_n = 2n + 1$$

and the domain is understood to be the set of natural numbers unless something is said to the contrary or the context indicates otherwise. The elements in the range are called **terms of the sequence;** a_1 is the first term, a_2 is the second term, and a_n is the ***n*th term,** or **general term.**

$$\begin{aligned} a_1 &= 2(1) + 1 = 3 && \text{First term} \\ a_2 &= 2(2) + 1 = 5 && \text{Second term} \\ a_3 &= 2(3) + 1 = 7 && \text{Third term} \\ &\vdots \\ a_n &= 2n + 1 && \text{General term} \end{aligned}$$

The ordered list of elements

$$3,\ 5,\ 7, \ldots, 2n + 1, \ldots$$

obtained by writing the terms of the sequence in their natural order with respect to the domain values is often informally referred to as a sequence. A sequence also may be represented in the abbreviated form $\{a_n\}$, where a symbol for the nth term is written within braces. For example, we could refer to the sequence $3, 5, 7, \ldots, 2n + 1, \ldots$ as the sequence $\{2n + 1\}$.

If the domain of a sequence is a finite set of successive integers, then the sequence is called a **finite sequence.** If the domain is an infinite set of successive integers, then the sequence is called an **infinite sequence.** The sequence $\{2n + 1\}$ discussed above is an infinite sequence.

Example 1 **Writing the Terms of a Sequence** Write the first four terms of each sequence:

(A) $a_n = 3n - 2$ (B) $\left\{\frac{(-1)^n}{n}\right\}$

SOLUTION (A) $1, 4, 7, 10$ (B) $-1, \frac{1}{2}, \frac{-1}{3}, \frac{1}{4}$

Matched Problem 1 Write the first four terms of each sequence:

(A) $a_n = -n + 3$ (B) $\left\{\frac{(-1)^n}{2^n}\right\}$

Explore–Discuss 1

(A) A multiple-choice test question asked for the next term in the sequence

$$2, 4, 8, \ldots$$

and gave the following choices:

(1) 16 (2) 14 (3) $\frac{25}{2}$

Which is the correct answer?

(B) Compare the first four terms of the following sequences:

(1) $a_n = 2^n$ (2) $b_n = n^2 - n + 2$ (3) $c_n = 5n + \frac{6}{n} - 9$

Now, which of the choices in part (A) appears to be correct?

Now that we have seen how to use the general term to find the first few terms in a sequence, we consider the reverse problem. That is, can a sequence be defined just by listing the first three or four terms of the sequence? And can we then use these initial terms to find a formula for the nth term? In general, without other information, the answer to the first question is no. As Explore–Discuss 1 illustrates, many different sequences may start off with the same terms. Simply listing the first three terms (or any other finite number of terms) does not specify a particular sequence. In fact, it can be shown that given any list of m numbers, there are an infinite number of sequences whose first m terms agree with these given numbers.

What about the second question? That is, given a few terms, can we find the general formula for at least one sequence whose first few terms agree with the given terms? The answer to this question is a qualified yes. If we can observe a simple pattern in the given terms, then we usually can construct a general term that will produce that pattern. The next example illustrates this approach.

Example 2 **Finding the General Term of a Sequence** Find the general term of a sequence whose first four terms are:

(A) $3, 4, 5, 6, \ldots$ (B) $5, -25, 125, -625, \ldots$

SOLUTION (A) Since these terms are consecutive integers, one solution is $a_n = n$, $n \geqslant 3$. If we want the domain of the sequence to be all natural numbers, then another solution is $b_n = n + 2$.

(B) Each of these terms can be written as the product of a power of 5 and a power of -1:

$$\begin{aligned} 5 &= (-1)^0 5^1 = a_1 \\ -25 &= (-1)^1 5^2 = a_2 \\ 125 &= (-1)^2 5^3 = a_3 \\ -625 &= (-1)^3 5^4 = a_4 \end{aligned}$$

If we choose the domain to be all natural numbers, then a solution is

$$a_n = (-1)^{n-1} 5^n$$

Matched Problem 2 Find the general term of a sequence whose first four terms are:

(A) $3, 6, 9, 12, \ldots$ (B) $1, -2, 4, -8, \ldots$

In general, there is usually more than one way of representing the nth term of a given sequence (see the solution of Example 2A). However, unless something is stated to the contrary, we assume the domain of the sequence is the set of natural numbers N.

Series and Summation Notation

If $a_1, a_2, a_3, \ldots, a_n, \ldots$ is a sequence, then the expression

$$a_1 + a_2 + a_3 + \cdots + a_n + \cdots$$

is called a **series.** If the sequence is finite, the corresponding series is a **finite series.** If the sequence is infinite, the corresponding series is an

infinite series. We will consider only finite series in this section. For example,

$$1, 3, 5, 7, 9 \qquad \text{Finite sequence}$$
$$1 + 3 + 5 + 7 + 9 \qquad \text{Finite series}$$

Notice that we can easily evaluate this series by adding the five terms:

$$1 + 3 + 5 + 7 + 9 = 25$$

Series are often represented in a compact form called **summation notation.** Consider the following examples:

$$\sum_{k=3}^{6} k^2 = 3^2 + 4^2 + 5^2 + 6^2$$
$$= 9 + 16 + 25 + 36 = 86$$
$$\sum_{k=0}^{2} (4k + 1) = (4 \cdot 0 + 1) + (4 \cdot 1 + 1) + (4 \cdot 2 + 1)$$
$$= 1 + 5 + 9 = 15$$

In each case, the terms of the series on the right are obtained from the expression on the left by successively replacing the **summing index *k*** with integers, starting with the number indicated below the **summation sign** Σ and ending with the number that appears above Σ. The summing index may be represented by letters other than k and may start at any integer and end at any integer greater than or equal to the starting integer. Thus, if we are given the finite sequence

$$\frac{1}{2}, \frac{1}{4}, \frac{1}{8}, \ldots, \frac{1}{2^n}$$

the corresponding series is

$$\frac{1}{2} + \frac{1}{4} + \frac{1}{8} + \cdots + \frac{1}{2^n} = \sum_{j=1}^{n} \frac{1}{2^j}$$

where we have used j for the summing index.

Example 3 **Summation Notation** Write

$$\sum_{k=1}^{5} \frac{k}{k^2 + 1}$$

without summation notation. Do not evaluate the sum.

SOLUTION

$$\sum_{k=1}^{5} \frac{k}{k^2 + 1} = \frac{1}{1^2 + 1} + \frac{2}{2^2 + 1} + \frac{3}{3^2 + 1} + \frac{4}{4^2 + 1} + \frac{5}{5^2 + 1}$$
$$= \frac{1}{2} + \frac{2}{5} + \frac{3}{10} + \frac{4}{17} + \frac{5}{26}$$

Matched Problem 3 Write

$$\sum_{k=1}^{5} \frac{k + 1}{k}$$

without summation notation. Do not evaluate the sum.

Explore–Discuss 2

(A) Find the smallest value of n for which the value of the series

$$\sum_{k=1}^{n} \frac{k}{k^2 + 1}$$

is greater than 3.

(B) Find the smallest value of n for which the value of the series

$$\sum_{j=1}^{n} \frac{1}{2^j}$$

is greater than 0.99. Greater than 0.999.

If the terms of a series are altenately positive and negative, we call the series an **alternating series.** The next example deals with the representation of such a series.

Example 4 **Summation Notation** Write the alternating series

$$\frac{1}{2} - \frac{1}{4} + \frac{1}{6} - \frac{1}{8} + \frac{1}{10} - \frac{1}{12}$$

using summation notation with:

(A) The summing index k starting at 1
(B) The summing index j starting at 0

Solution (A) $(-1)^{k+1}$ provides the alternation of sign, and $1/(2k)$ provides the other part of each term. Thus, we can write

$$\frac{1}{2} - \frac{1}{4} + \frac{1}{6} - \frac{1}{8} + \frac{1}{10} - \frac{1}{12} = \sum_{k=1}^{6} \frac{(-1)^{k+1}}{2k}$$

(B) $(-1)^j$ provides the alternation of sign, and $1/[2(j + 1)]$ provides the other part of each term. Thus, we can write

$$\frac{1}{2} - \frac{1}{4} + \frac{1}{6} - \frac{1}{8} + \frac{1}{10} - \frac{1}{12} = \sum_{j=0}^{5} \frac{(-1)^j}{2(j + 1)}$$

Matched Problem 4 Write the alternating series

$$1 - \frac{1}{3} + \frac{1}{9} - \frac{1}{27} + \frac{1}{81}$$

using summation notation with:

(A) The summing index k starting at 1
(B) The summing index j starting at 0

Summation notation provides a compact notation for the sum of any list of numbers, even if the numbers are not generated by a formula. For example, suppose the results of an examination taken by a class of 10 students are given in the following list:

87, 77, 95, 83, 86, 73, 95, 68, 75, 86

If we let $a_1, a_2, a_3, \ldots, a_{10}$ represent these 10 scores, then the average test score is given by

$$\frac{1}{10}\sum_{k=1}^{10} a_k = \frac{1}{10}(87 + 77 + 95 + 83 + 86 + 73 + 95 + 68 + 75 + 86)$$
$$= \frac{1}{10}(825) = 82.5$$

More generally, in statistics, the **arithmetic mean** $\bar{a}$ of a list of n numbers $a_1, a_2, \ldots, a_n$ is defined as

$$\bar{a} = \frac{1}{n}\sum_{k=1}^{n} a_k$$

Example 5 **Arithmetic Mean** Find the arithmetic mean of 3, 5, 4, 7, 4, 2, 3, and 6.

SOLUTION $\bar{a} = \frac{1}{8}\sum_{k=1}^{8} a_k = \frac{1}{8}(3 + 5 + 4 + 7 + 4 + 2 + 3 + 6) = \frac{1}{8}(34) = 4.25$

Matched Problem 5 Find the arithmetic mean of 9, 3, 8, 4, 3, and 6.

Answers to Matched Problems

1. (A) 2, 1, 0, −1 (B) $\frac{-1}{2}, \frac{1}{4}, \frac{-1}{8}, \frac{1}{16}$

2. (A) $a_n = 3n$ (B) $a_n = (-2)^{n-1}$ **3.** $2 + \frac{3}{2} + \frac{4}{3} + \frac{5}{4} + \frac{6}{5}$

4. (A) $\sum_{k=1}^{5} \frac{(-1)^{k-1}}{3^{k-1}}$ (B) $\sum_{j=0}^{4} \frac{(-1)^j}{3^j}$ **5.** 5.5

EXERCISE B-1

A *Write the first four terms for each sequence in Problems 1–6.*

1. $a_n = 2n + 3$ **2.** $a_n = 4n - 3$

3. $a_n = \frac{n+2}{n+1}$ **4.** $a_n = \frac{2n+1}{2n}$

5. $a_n = (-3)^{n+1}$ **6.** $a_n = (-\frac{1}{4})^{n-1}$

7. Write the 10th term of the sequence in Problem 1.

8. Write the 15th term of the sequence in Problem 2.

9. Write the 99th term of the sequence in Problem 3.

10. Write the 200th term of the sequence in Problem 4.

In Problems 11–16, write each series in expanded form without summation notation, and evaluate.

11. $\sum_{k=1}^{6} k$ **12.** $\sum_{k=1}^{5} k^2$ **13.** $\sum_{k=4}^{7} (2k - 3)$

14. $\sum_{k=0}^{4} (-2)^k$ **15.** $\sum_{k=0}^{3} \frac{1}{10^k}$ **16.** $\sum_{k=1}^{4} \frac{1}{2^k}$

Find the arithmetic mean of each list of numbers in Problems 17–20.

17. 5, 4, 2, 1, and 6 **18.** 7, 9, 9, 2, and 4

19. 96, 65, 82, 74, 91, 88, 87, 91, 77, and 74

20. 100, 62, 95, 91, 82, 87, 70, 75, 87, and 82

B *Write the first five terms of each sequence in Problems 21–26.*

21. $a_n = \frac{(-1)^{n+1}}{2^n}$ **22.** $a_n = (-1)^n(n - 1)^2$

23. $a_n = n[1 + (-1)^n]$ **24.** $a_n = \frac{1 - (-1)^n}{n}$

25. $a_n = \left(-\frac{3}{2}\right)^{n-1}$ **26.** $a_n = \left(-\frac{1}{2}\right)^{n+1}$

In Problems 27–42, find the general term of a sequence whose first four terms agree with the given terms.

27. −2, −1, 0, 1, . . . **28.** 4, 5, 6, 7, . . .

29. 4, 8, 12, 16, . . . **30.** −3, −6, −9, −12, . . .

31. $\frac{1}{2}, \frac{3}{4}, \frac{5}{6}, \frac{7}{8}, \ldots$ **32.** $\frac{1}{2}, \frac{2}{3}, \frac{3}{4}, \frac{4}{5}, \ldots$

33. $1, -2, 3, -4, \ldots$ **34.** $-2, 4, -8, 16, \ldots$

35. $1, -3, 5, -7, \ldots$ **36.** $3, -6, 9, -12, \ldots$

37. $1, \frac{2}{5}, \frac{4}{25}, \frac{8}{125}, \ldots$ **38.** $\frac{4}{3}, \frac{16}{9}, \frac{64}{27}, \frac{256}{81}, \ldots$

39. $x, x^2, x^3, x^4, \ldots$ **40.** $1, 2x, 3x^2, 4x^3, \ldots$

41. $x, -x^3, x^5, -x^7, \ldots$ **42.** $x, \frac{x^2}{2}, \frac{x^3}{3}, \frac{x^4}{4}, \ldots$

Write each series in Problems 43–50 in expanded form without summation notation. Do not evaluate.

43. $\sum_{k=1}^{5} (-1)^{k+1}(2k - 1)^2$ **44.** $\sum_{k=1}^{4} \frac{(-2)^{k+1}}{2k + 1}$

45. $\sum_{k=2}^{5} \frac{2^k}{2k + 3}$ **46.** $\sum_{k=3}^{7} \frac{(-1)^k}{k^2 - k}$

47. $\sum_{k=1}^{5} x^{k-1}$ **48.** $\sum_{k=1}^{3} \frac{1}{k} x^{k+1}$

49. $\sum_{k=0}^{4} \frac{(-1)^k x^{2k+1}}{2k + 1}$ **50.** $\sum_{k=0}^{4} \frac{(-1)^k x^{2k}}{2k + 2}$

Write each series in Problems 51–54 using summation notation with:
(A) The summing index k starting at $k = 1$
(B) The summing index j starting at $j = 0$

51. $2 + 3 + 4 + 5 + 6$ **52.** $1^2 + 2^2 + 3^2 + 4^2$

53. $1 - \frac{1}{2} + \frac{1}{3} - \frac{1}{4}$ **54.** $1 - \frac{1}{3} + \frac{1}{5} - \frac{1}{7} + \frac{1}{9}$

Write each series in Problems 55–58 using summation notation with the summing index k starting at $k = 1$.

55. $2 + \frac{3}{2} + \frac{4}{3} + \cdots + \frac{n + 1}{n}$

56. $1 + \frac{1}{2^2} + \frac{1}{3^2} + \cdots + \frac{1}{n^2}$

57. $\frac{1}{2} - \frac{1}{4} + \frac{1}{8} - \cdots + \frac{(-1)^{n+1}}{2^n}$

58. $1 - 4 + 9 - \cdots + (-1)^{n+1}n^2$

C *Some sequences are defined by a* ***recursion formula****—that is, a formula that defines each term of the sequence in terms of one or more of the preceding terms. For example, if $\{a_n\}$ is defined by*

$$a_1 = 1 \quad \text{and} \quad a_n = 2a_{n-1} + 1 \quad \text{for } n \geq 2$$

then

$$a_2 = 2a_1 + 1 = 2 \cdot 1 + 1 = 3$$
$$a_3 = 2a_2 + 1 = 2 \cdot 3 + 1 = 7$$
$$a_4 = 2a_3 + 1 = 2 \cdot 7 + 1 = 15$$

and so on. In Problems 59–62, write the first five terms of each sequence.

59. $a_1 = 2$ and $a_n = 3a_{n-1} + 2$ for $n \geq 2$

60. $a_1 = 3$ and $a_n = 2a_{n-1} - 2$ for $n \geq 2$

61. $a_1 = 1$ and $a_n = 2a_{n-1}$ for $n \geq 2$

62. $a_1 = 1$ and $a_n = -\frac{1}{3}a_{n-1}$ for $n \geq 2$

If A is a positive real number, then the terms of the sequence defined by

$$a_1 = \frac{A}{2} \quad \text{and} \quad a_n = \frac{1}{2}\left(a_{n-1} + \frac{A}{a_{n-1}}\right) \quad \text{for } n \geq 2$$

can be used to approximate $\sqrt{A}$ to any decimal place accuracy desired. In Problems 63 and 64, compute the first four terms of this sequence for the indicated value of A, and compare the fourth term with the value of $\sqrt{A}$ obtained from a calculator.

63. $A = 2$ **64.** $A = 6$

Section B-2 Arithmetic and Geometric Sequences

- Arithmetic and Geometric Sequences
- *n*th-Term Formulas
- Sum Formulas for Finite Arithmetic Series
- Sum Formulas for Finite Geometric Series
- Sum Formula for Infinite Geometric Series
- Applications

For most sequences it is difficult to sum an arbitrary number of terms of the sequence without adding term by term. But particular types of sequences—*arithmetic sequences* and *geometric sequences*—have certain properties that lead to convenient and useful formulas for the sums of the corresponding *arithmetic series* and *geometric series*.

■ Arithmetic and Geometric Sequences

The sequence $5, 7, 9, 11, 13, \ldots, 5 + 2(n - 1), \ldots$, where each term after the first is obtained by adding 2 to the preceding term, is an example of an arithmetic sequence. The sequence $5, 10, 20, 40, 80, \ldots, 5(2)^{n-1}, \ldots$, where each term after the first is obtained by multiplying the preceding term by 2, is an example of a geometric sequence.

Arithmetic Sequence

A sequence of numbers

$$a_1, a_2, a_3, \ldots, a_n, \ldots$$

is called an **arithmetic sequence** if there is a constant d, called the **common difference,** such that

$$a_n - a_{n-1} = d$$

That is,

$$a_n = a_{n-1} + d \qquad \text{for every } n > 1$$

Geometric Sequence

A sequence of numbers

$$a_1, a_2, a_3, \ldots, a_n, \ldots$$

is called a **geometric sequence** if there exists a nonzero constant r, called a **common ratio,** such that

$$\frac{a_n}{a_{n-1}} = r$$

That is,

$$a_n = ra_{n-1} \qquad \text{for every } n \geqslant 1$$

Explore–Discuss 1

(A) Describe verbally all arithmetic sequences with common difference 2.
(B) Describe verbally all geometric sequences with common ratio 2.

Example 1 **Recognizing Arithmetic and Geometric Sequences** Which of the following can be the first four terms of an arithmetic sequence? Of a geometric sequence?

(A) $1, 2, 3, 5, \ldots$ (B) $-1, 3, -9, 27, \ldots$
(C) $3, 3, 3, 3, \ldots$ (D) $10, 8.5, 7, 5.5, \ldots$

SOLUTION

(A) Since $2 - 1 \neq 5 - 3$, there is no common difference, so the sequence is not an arithmetic sequence. Since $2/1 \neq 3/2$, there is no common ratio, so the sequence is not geometric either.

(B) The sequence is geometric with common ratio -3. It is not arithmetic.

(C) The sequence is arithmetic with common difference 0, and is also geometric with common ratio 1.

(D) The sequence is arithmetic with common difference -1.5. It is not geometric.

Matched Problem 1 Which of the following can be the first four terms of an arithmetic sequence? Of a geometric sequence?

(A) $8, 2, 0.5, 0.125, \ldots$ (B) $-7, -2, 3, 8, \ldots$

(C) $1, 5, 25, 100, \ldots$

*n*TH-TERM FORMULAS

If $\{a_n\}$ is an arithmetic sequence with common difference d, then

$$a_2 = a_1 + d$$
$$a_3 = a_2 + d = a_1 + 2d$$
$$a_4 = a_3 + d = a_1 + 3d$$

This suggests that:

nth Term of an Arithmetic Sequence

$$a_n = a_1 + (n - 1)d \qquad \text{for all } n > 1 \tag{1}$$

Similarly, if $\{a_n\}$ is a geometric sequence with common ratio r, then

$$a_2 = a_1 r$$
$$a_3 = a_2 r = a_1 r^2$$
$$a_4 = a_3 r = a_1 r^3$$

This suggests that:

nth Term of a Geometric Sequence

$$a_n = a_1 r^{n-1} \qquad \text{for all } n > 1 \tag{2}$$

Example 2 **Finding Terms in Arithmetic and Geometric Sequences**

(A) If the 1st and 10th terms of an arithmetic sequence are 3 and 30, respectively, find the 40th term of the sequence.

(B) If the 1st and 10th terms of a geometric sequence are 3 and 30, find the 40th term to three decimal places.

SOLUTION (A) First use formula (1) with $a_1 = 3$ and $a_{10} = 30$ to find d:

$$\begin{aligned} a_n &= a_1 + (n-1)d \\ a_{10} &= a_1 + (10-1)d \\ 30 &= 3 + 9d \\ d &= 3 \end{aligned}$$

Now find a_{40}:

$$a_{40} = 3 + 39 \cdot 3 = 120$$

(B) First use formula (2) with $a_1 = 3$ and $a_{10} = 30$ to find r:

$$\begin{aligned} a_n &= a_1 r^{n-1} \\ a_{10} &= a_1 r^{10-1} \\ 30 &= 3r^9 \\ r^9 &= 10 \\ r &= 10^{1/9} \end{aligned}$$

Now find a_{40}:

$$a_{40} = 3(10^{1/9})^{39} = 3(10^{39/9}) = 64{,}633.041$$

Matched Problem 2 (A) If the 1st and 15th terms of an arithmetic sequence are -5 and 23, respectively, find the 73rd term of the sequence.

(B) Find the 8th term of the geometric sequence

$$\frac{1}{64}, \frac{-1}{32}, \frac{1}{16}, \ldots$$

■ SUM FORMULAS FOR FINITE ARITHMETIC SERIES

If $a_1, a_2, a_3, \ldots, a_n$ is a finite arithmetic sequence, then the corresponding series $a_1 + a_2 + a_3 + \cdots + a_n$ is called a *finite arithmetic series.* We will derive two simple and very useful formulas for the sum of a finite arithmetic series. Let d be the common difference of the arithmetic sequence $a_1, a_2, a_3, \ldots, a_n$ and let S_n denote the sum of the series $a_1 + a_2 + a_3 + \cdots + a_n$. Then

$$S_n = a_1 + (a_1 + d) + \cdots + [a_1 + (n-2)d] + [a_1 + (n-1)d]$$

Reversing the order of the sum, we obtain

$$S_n = [a_1 + (n-1)d] + [a_1 + (n-2)d] + \cdots + (a_1 + d) + a_1$$

Something interesting happens if we combine these last two equations by addition (adding corresponding terms on the right sides):

$$2S_n = [2a_1 + (n-1)d] + [2a_1 + (n-1)d] + \cdots + [2a_1 + (n-1)d] + [2a_1 + (n-1)d]$$

All the terms on the right side are the same, and there are n of them. Thus,

$$2S_n = n[2a_1 + (n-1)d]$$

and we have the following general formula:

Sum of a Finite Arithmetic Series—First Form

$$S_n = \frac{n}{2}[2a_1 + (n-1)d] \qquad (3)$$

Replacing

$$[a_1 + (n - 1)d] \quad \text{in} \quad \frac{n}{2}[a_1 + a_1 + (n - 1)d]$$

by a_n from equation (1), we obtain a second useful formula for the sum:

Sum of a Finite Arithmetic Series—Second Form

$$S_n = \frac{n}{2}(a_1 + a_n) \qquad (4)$$

Example 3 **Finding a Sum** Find the sum of the first 30 terms in the arithmetic sequence:

3, 8, 13, 18, . . .

SOLUTION Use formula (3) with $n = 30$, $a_1 = 3$, and $d = 5$:

$$S_{30} = \frac{30}{2}[2 \cdot 3 + (30 - 1)5] = 2{,}265$$

Matched Problem 3 Find the sum of the first 40 terms in the arithmetic sequence:

15, 13, 11, 9, . . .

Example 4 **Finding a Sum** Find the sum of all the even numbers between 31 and 87.

SOLUTION First, find n using equation (1):

$$\begin{aligned} a_n &= a_1 + (n - 1)d \\ 86 &= 32 + (n - 1)2 \\ n &= 28 \end{aligned}$$

Now find S_{28} using formula (4):

$$S_n = \frac{n}{2}(a_1 + a_n)$$

$$S_{28} = \frac{28}{2}(32 + 86) = 1{,}652$$

Matched Problem 4 Find the sum of all the odd numbers between 24 and 208.

Sum Formulas for Finite Geometric Series

If $a_1, a_2, a_3, \ldots, a_n$ is a finite geometric sequence, then the corresponding series $a_1 + a_2 + a_3 + \cdots + a_n$ is called a *finite geometric series.* As with arithmetic series, we can derive two simple and very useful formulas for the sum of a finite geometric series. Let r be the common ratio of the geometric sequence $a_1, a_2, a_3, \ldots, a_n$ and let S_n denote the sum of the series $a_1 + a_2 + a_3 + \cdots + a_n$. Then

$$S_n = a_1 + a_1r + a_1r^2 + \cdots + a_1r^{n-2} + a_1r^{n-1}$$

If we multiply both sides by r, we obtain

$$rS_n = a_1r + a_1r^2 + a_1r^3 + \cdots + a_1r^{n-1} + a_1r^n$$

Now combine these last two equations by subtraction to obtain

$$rS_n - S_n = (a_1r + a_1r^2 + a_1r^3 + \cdots + a_1r^{n-1} + a_1r^n) - (a_1 + a_1r + a_1r^2 + \cdots + a_1r^{n-2} + a_1r^{n-1})$$
$$(r - 1)S_n = a_1r^n - a_1$$

Notice how many terms drop out on the right side. Solving for S_n, we have:

Sum of a Finite Geometric Series—First Form

$$S_n = \frac{a_1(r^n - 1)}{r - 1} \qquad r \neq 1 \qquad (5)$$

Since $a_n = a_1r^{n-1}$, or $ra_n = a_1r^n$, formula (5) also can be written in the form:

Sum of a Finite Geometric Series—Second Form

$$S_n = \frac{ra_n - a_1}{r - 1} \qquad r \neq 1 \qquad (6)$$

Example 5 **Finding a Sum** Find the sum of the first ten terms of the geometric sequence:

$$1, 1.05, 1.05^2, \ldots$$

SOLUTION Use formula (5) with $a_1 = 1$, $r = 1.05$, and $n = 10$:

$$S_n = \frac{a_1(r^n - 1)}{r - 1}$$
$$S_{10} = \frac{1(1.05^{10} - 1)}{1.05 - 1}$$
$$\approx \frac{0.6289}{0.05} \approx 12.58$$

Matched Problem 5 Find the sum of the first eight terms of the geometric sequence:

$$100, 100(1.08), 100(1.08)^2, \ldots$$

Sum Formula for Infinite Geometric Series

Explore–Discuss 2

(A) For any n, the sum of a finite geometric series with $a_1 = 5$ and $r = \frac{1}{2}$ is given by [see formula (5)]

$$S_n = \frac{5[(\frac{1}{2})^n - 1]}{\frac{1}{2} - 1} = 10 - 10\left(\frac{1}{2}\right)^n$$

Discuss the behavior of S_n as n increases.

(B) Repeat part (A) if $a_1 = 5$ and $r = 2$.

Given a geometric series, what happens to the sum S_n of the first n terms as n increases without stopping? To answer this question, let us write formula (5) in the form

$$S_n = \frac{a_1 r^n}{r - 1} - \frac{a_1}{r - 1}$$

It is possible to show that if $-1 < r < 1$, then r^n will approach 0 as n increases. Thus, the first term above will approach 0 and S_n can be made as close as we please to the second term, $-a_1/(r - 1)$ [which can be written as $a_1/(1 - r)$], by taking n sufficiently large. Thus, if the common ratio r is between -1 and 1, we define the sum of an infinite geometric series to be:

Sum of an Infinite Geometric Series

$$S_\infty = \frac{a_1}{1 - r} \qquad -1 < r < 1 \tag{7}$$

If $r \leqslant -1$ or $r \geqslant 1$, then an infinite geometric series has no sum.

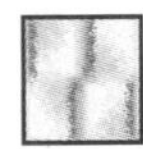

■ Applications

Example 6 ➡ **Loan Repayment** A person borrows \$3,600 and agrees to repay the loan in monthly installments over a period of 3 years. The agreement is to pay 1% of the unpaid balance each month for using the money and \$100 each month to reduce the loan. What is the total cost of the loan over the 3 years?

Solution Let us look at the problem relative to a time line:

	\$3,600	\$3,500	\$3,400	. . .	\$200	\$100		Unpaid balance
0	1	2	3	. . .	34	35	36	Months
	0.01(3,600) = 36	0.01(3,500) = 35	0.01(3,400) = 34	. . .	0.01(300) = 3	0.01(200) = 2	0.01(100) = 1	1% of unpaid balance

The total cost of the loan is

$$1 + 2 + \cdots + 34 + 35 + 36$$

The terms form a finite arithmetic series with $n = 36$, $a_1 = 1$, and $a_{36} = 36$, so we can use formula (4):

$$S_n = \frac{n}{2}(a_1 + a_n)$$

$$S_{36} = \frac{36}{2}(1 + 36) = \$666$$

And we conclude that the total cost of the loan over the period of 3 years is \$666. ▪

Matched Problem 6 Repeat Example 6 with a loan of \$6,000 over a period of 5 years.

Example 7 **Economy Stimulation** The government has decided on a tax rebate program to stimulate the economy. Suppose you receive \$600 and you spend 80% of this, and each of the people who receive what you spend also spend 80% of what they receive, and this process continues without end. According to the **multiplier principle** in economics, the effect of your \$600 tax rebate on the economy is multiplied many times. What is the total amount spent if the process continues as indicated?

SOLUTION We need to find the sum of an infinite geometric series with the first amount spent being $a_1 = (0.8)(\$600) = \$480$ and $r = 0.8$. Using formula (7), we obtain

$$S_\infty = \frac{a_1}{1 - r}$$
$$= \frac{\$480}{1 - 0.8} = \$2{,}400$$

Thus, assuming the process continues as indicated, we would expect the \$600 tax rebate to result in about \$2,400 of spending.

Matched Problem 7 Repeat Example 7 with a tax rebate of \$1,000.

Answers to Matched Problems

1. (A) The sequence is geometric with $r = \frac{1}{4}$. It is not arithmetic.
(B) The sequence is arithmetic with $d = 5$. It is not geometric.
(C) The sequence is neither arithmetic nor geometric.
2. (A) 139 (B) −2 **3.** −960 **4.** 10,672
5. 1,063.66 **6.** \$1,830 **7.** \$4,000

EXERCISE B-2

A *In Problems 1 and 2, determine whether the indicated sequence can be the first three terms of an arithmetic or geometric sequence, and, if so, find the common difference or common ratio and the next two terms of the sequence.*

1. (A) $-11, -16, -21, \ldots$ (B) $2, -4, 8, \ldots$
(C) $1, 4, 9, \ldots$ (D) $\frac{1}{2}, \frac{1}{6}, \frac{1}{18}, \ldots$

2. (A) $5, 20, 100, \ldots$ (B) $-5, -5, -5, \ldots$
(C) $7, 6.5, 6, \ldots$ (D) $512, 256, 128, \ldots$

B *Let $a_1, a_2, a_3, \ldots, a_n, \ldots$ be an arithmetic sequence. In Problems 3–8, find the indicated quantities.*

3. $a_1 = 7; d = 4; a_2 = ?; a_3 = ?$

4. $a_1 = -2; d = -3; a_2 = ?; a_3 = ?$

5. $a_1 = 2; d = 4; a_{21} = ?; S_{31} = ?$

6. $a_1 = 8; d = -10; a_{15} = ?; S_{23} = ?$

7. $a_1 = 18; a_{20} = 75; S_{20} = ?$

8. $a_1 = 203; a_{30} = 261; S_{30} = ?$

Let $a_1, a_2, a_3, \ldots, a_n, \ldots$ be a geometric sequence. In Problems 9–18, find the indicated quantities.

9. $a_1 = 3; r = -2; a_2 = ?; a_3 = ?; a_4 = ?$

10. $a_1 = 32; r = -\frac{1}{2}; a_2 = ?; a_3 = ?; a_4 = ?$

11. $a_1 = 1; a_7 = 729; r = -3; S_7 = ?$

12. $a_1 = 3; a_7 = 2{,}187; r = 3; S_7 = ?$

13. $a_1 = 100; r = 1.08; a_{10} = ?$

14. $a_1 = 240; r = 1.06; a_{12} = ?$

15. $a_1 = 100; a_9 = 200; r = ?$

16. $a_1 = 100; a_{10} = 300; r = ?$

17. $a_1 = 500; r = 0.6; S_{10} = ?; S_\infty = ?$

18. $a_1 = 8{,}000; r = 0.4; S_{10} = ?; S_\infty = ?$

19. $S_{41} = \sum_{k=1}^{41} (3k + 3) = ?$ **20.** $S_{50} = \sum_{k=1}^{50} (2k - 3) = ?$

21. $S_8 = \sum_{k=1}^{8} (-2)^{k-1} = ?$ **22.** $S_8 = \sum_{k=1}^{8} 2^k = ?$

23. Find the sum of all the odd integers between 12 and 68.

24. Find the sum of all the even integers between 23 and 97.

25. Find the sum of each infinite geometric sequence (if it exists).
(A) $2, 4, 8, \ldots$ (B) $2, -\frac{1}{2}, \frac{1}{8}, \ldots$

26. Repeat Problem 25 for:
(A) $16, 4, 1, \ldots$ (B) $1, -3, 9, \ldots$

C

27. Find $f(1) + f(2) + f(3) + \cdots + f(50)$ if $f(x) = 2x - 3$.

28. Find $g(1) + g(2) + g(3) + \cdots + g(100)$ if $g(t) = 18 - 3t$.

29. Find $f(1) + f(2) + \cdots + f(10)$ if $f(x) = (\frac{1}{2})^x$.

30. Find $g(1) + g(2) + \cdots + g(10)$ if $g(x) = 2^x$.

31. Show that the sum of the first n odd positive integers is n^2, using appropriate formulas from this section.

32. Show that the sum of the first n even positive integers is $n + n^2$, using formulas in this section.

Applications

Business & Economics

33. *Loan repayment.* If you borrow $4,800 and repay the loan by paying $200 per month to reduce the loan and 1% of the unpaid balance each month for the use of the money, what is the total cost of the loan over 24 months?

34. *Loan repayment.* If you borrow $5,400 and repay the loan by paying $300 per month to reduce the loan and 1.5% of the unpaid balance each month for the use of the money, what is the total cost of the loan over 18 months?

35. *Economy stimulation.* The government, through a subsidy program, distributes $5,000,000. If we assume each individual or agency spends 70% of what is received, and 70% of this is spent, and so on, how much total increase in spending results from this government action? (Let $a_1 = \$3{,}500{,}000$.)

36. *Economy stimulation.* Due to reduced taxes, an individual has an extra $1,200 in spendable income. If we assume that the individual spends 65% of this on consumer goods, and the producers of these goods in turn spend 65% on consumer goods, and that this process continues indefinitely, what is the total amount spent (to the nearest dollar) on consumer goods?

37. *Compound interest.* If $1,000 is invested at 5% compounded annually, the amount A present after n years forms a geometric sequence with common ratio $1 + 0.05 = 1.05$. Use a geometric sequence formula to find the amount A in the account (to the nearest cent) after 10 years. After 20 years. [*Hint:* Use a time line.]

38. *Compound interest.* If $\$P$ is invested at $100r\%$ compounded annually, the amount A present after n years forms a geometric sequence with common ratio $1 + r$. Write a formula for the amount present after n years. [*Hint:* Use a time line.]

Section B-3 The Binomial Theorem

- **Factorial**
- **Binomial Theorem—Development**

The binomial form

$$(a + b)^n$$

where n is a natural number, appears more frequently than you might expect. The coefficients in the expansion play an important role in probability studies. The *binomial formula,* which we will derive informally, enables us to expand $(a + b)^n$ directly for n any natural number. Since the formula involves *factorials,* we digress for a moment here to introduce this important concept.

FACTORIAL

For n a natural number, ***n* factorial,** denoted by $\boldsymbol{n!}$, is the product of the first n natural numbers. **Zero factorial** is defined to be 1. That is:

$$n! = n \cdot (n-1) \cdot \cdots \cdot 2 \cdot 1$$
$$1! = 1$$
$$0! = 1$$

It is also useful to note that:

$$n! = n \cdot (n-1)! \qquad n \geq 1$$

Example 1 **Factorial Forms** Evaluate each:

(A) $5! = 5 \cdot 4 \cdot 3 \cdot 2 \cdot 1 = 120$ (B) $\dfrac{8!}{7!} = \dfrac{8 \cdot 7!}{7!} = 8$

(C) $\dfrac{10!}{7!} = \dfrac{10 \cdot 9 \cdot 8 \cdot \cancel{7!}}{\cancel{7!}} = 720$

Matched Problem 1 Evaluate each:

(A) $4!$ (B) $\dfrac{7!}{6!}$ (C) $\dfrac{8!}{5!}$

The following important formula involving factorials has applications in many areas of mathematics and statistics. We will use this formula to provide a more concise form for the expressions encountered later in this discussion.

For n and r integers satisfying $0 \leq r \leq n$,

$$C_{n,r} = \frac{n!}{r!(n-r)!}$$

Example 2 **Evaluating $C_{n,r}$**

(A) $C_{9,2} = \dfrac{9!}{2!(9-2)!} = \dfrac{9!}{2!7!} = \dfrac{9 \cdot 8 \cdot \cancel{7!}}{2 \cdot \cancel{7!}} = 36$

(B) $C_{5,5} = \dfrac{5!}{5!(5-5)!} = \dfrac{5!}{5!0!} = \dfrac{5!}{5!} = 1$

Matched Problem 2 Find:

(A) $C_{5,2}$ (B) $C_{6,0}$

BINOMIAL THEOREM—DEVELOPMENT

Let us expand $(a + b)^n$ for several values of n to see if we can observe a pattern that leads to a general formula for the expansion for any natural number n:

$$\begin{aligned}
(a + b)^1 &= a + b \\
(a + b)^2 &= a^2 + 2ab + b^2 \\
(a + b)^3 &= a^3 + 3a^2b + 3ab^2 + b^3 \\
(a + b)^4 &= a^4 + 4a^3b + 6a^2b^2 + 4ab^3 + b^4 \\
(a + b)^5 &= a^5 + 5a^4b + 10a^3b^2 + 10a^2b^3 + 5ab^4 + b^5
\end{aligned}$$

OBSERVATIONS

1. The expansion of $(a + b)^n$ has $(n + 1)$ terms.
2. The power of a decreases by 1 for each term as we move from left to right.
3. The power of b increases by 1 for each term as we move from left to right.
4. In each term, the sum of the powers of a and b always equals n.
5. Starting with a given term, we can get the coefficient of the next term by multiplying the coefficient of the given term by the exponent of a and dividing by the number that represents the position of the term in the series of terms. For example, in the expansion of $(a + b)^4$ above, the coefficient of the third term is found from the second term by multiplying 4 and 3, and then dividing by 2 [that is, the coefficient of the third term $= (4 \cdot 3)/2 = 6$].

We now postulate these same properties for the general case:

$$\begin{aligned}
(a + b)^n &= a^n + \frac{n}{1}a^{n-1}b + \frac{n(n-1)}{1 \cdot 2}a^{n-2}b^2 + \frac{n(n-1)(n-2)}{1 \cdot 2 \cdot 3}a^{n-3}b^3 + \cdots + b^n \\
&= \frac{n!}{0!(n-0)!}a^n + \frac{n!}{1!(n-1)!}a^{n-1}b + \frac{n!}{2!(n-2)!}a^{n-2}b^2 + \frac{n!}{3!(n-3)!}a^{n-3}b^3 + \cdots + \frac{n!}{n!(n-n)!}b^n \\
&= C_{n,0}a^n + C_{n,1}a^{n-1}b + C_{n,2}a^{n-2}b^2 + C_{n,3}a^{n-3}b^3 + \cdots + C_{n,n}b^n
\end{aligned}$$

And we are led to the formula in the binomial theorem (a formal proof requires mathematical induction, which is beyond the scope of this book):

> **Binomial Theorem**
>
> For all natural numbers n,
>
> $$(a + b)^n = C_{n,0}a^n + C_{n,1}a^{n-1}b + C_{n,2}a^{n-2}b^2 + C_{n,3}a^{n-3}b^3 + \cdots + C_{n,n}b^n$$

Example 3 **Using the Binomial Theorem** Use the binomial formula to expand $(u + v)^6$.

SOLUTION

$$\begin{aligned}
(u + v)^6 &= C_{6,0}u^6 + C_{6,1}u^5v + C_{6,2}u^4v^2 + C_{6,3}u^3v^3 + C_{6,4}u^2v^4 + C_{6,5}uv^5 + C_{6,6}v^6 \\
&= u^6 + 6u^5v + 15u^4v^2 + 20u^3v^3 + 15u^2v^4 + 6uv^5 + v^6
\end{aligned}$$

Matched Problem 3 ➡ Use the binomial formula to expand $(x + 2)^5$.

Example 4 ➡ **Using the Binomial Theorem** Use the binomial formula to find the sixth term in the expansion of $(x - 1)^{18}$.

SOLUTION

$$\text{Sixth term} = C_{18,5}x^{13}(-1)^5 = \frac{18!}{5!(18-5)!}x^{13}(-1) = -8{,}568x^{13}$$

Matched Problem 4 ➡ Use the binomial formula to find the fourth term in the expansion of $(x - 2)^{20}$.

Explore–Discuss 1

(A) Use the formula for $C_{n,r}$ to find

$$C_{6,0} + C_{6,1} + C_{6,2} + C_{6,3} + C_{6,4} + C_{6,5} + C_{6,6}$$

(B) Write the binomial theorem for $n = 6$, $a = 1$, and $b = 1$, and compare with the results of part (A).

(C) For any natural number n, find

$$C_{n,0} + C_{n,1} + C_{n,2} + \cdots + C_{n,n}$$

Answers to Matched Problems

1. (A) 24 (B) 7 (C) 336 **2.** (A) 10 (B) 1
3. $x^5 + 10x^4 + 40x^3 + 80x^2 + 80x + 32$
4. $-9{,}120x^{17}$

EXERCISE B-3

A *In Problems 1–20, evaluate each expression.*

1. $6!$ **2.** $7!$ **3.** $\frac{10!}{9!}$

4. $\frac{20!}{19!}$ **5.** $\frac{12!}{9!}$ **6.** $\frac{10!}{6!}$

7. $\frac{5!}{2!3!}$ **8.** $\frac{7!}{3!4!}$ **9.** $\frac{6!}{5!(6-5)!}$

10. $\frac{7!}{4!(7-4)!}$ **11.** $\frac{20!}{3!17!}$ **12.** $\frac{52!}{50!2!}$

B

13. $C_{5,3}$ **14.** $C_{7,3}$ **15.** $C_{6,5}$ **16.** $C_{7,4}$

17. $C_{5,0}$ **18.** $C_{5,5}$ **19.** $C_{18,15}$ **20.** $C_{18,3}$

Expand each expression in Problems 21–26 using the binomial formula.

21. $(a + b)^4$ **22.** $(m + n)^5$ **23.** $(x - 1)^6$

24. $(u - 2)^5$ **25.** $(2a - b)^5$ **26.** $(x - 2y)^5$

Find the indicated term in each expansion in Problems 27–32.

27. $(x - 1)^{18}$; 5th term **28.** $(x - 3)^{20}$; 3rd term

29. $(p + q)^{15}$; 7th term **30.** $(p + q)^{15}$; 13th term

31. $(2x + y)^{12}$; 11th term **32.** $(2x + y)^{12}$; 3rd term

C

33. Show that $C_{n,0} = C_{n,n}$.

34. Show that $C_{n,r} = C_{n,n-r}$.

35. The triangle below is called **Pascal's triangle.** Can you guess what the next two rows at the bottom are? Compare these numbers with the coefficients of binomial expansions.

```
        1
      1   1
    1   2   1
  1   3   3   1
1   4   6   4   1
```

APPENDIX C

Tables

Table I

Basic Geometric Formulas

1. Similar Triangles

(A) Two triangles are similar if two angles of one triangle have the same measure as two angles of the other.

(B) If two triangles are similar, their corresponding sides are proportional:

$$\frac{a}{a'} = \frac{b}{b'} = \frac{c}{c'}$$

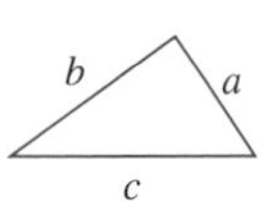

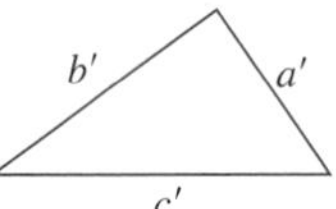

2. Pythagorean Theorem

$$c^2 = a^2 + b^2$$

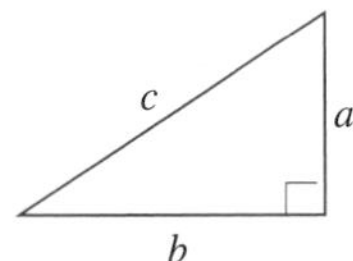

3. Rectangle

$A = ab$ Area

$P = 2a + 2b$ Perimeter

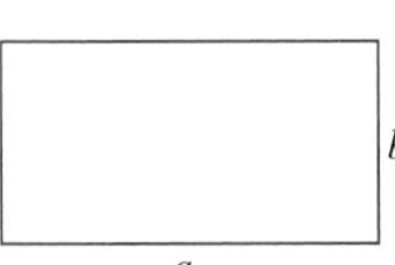

4. Parallelogram

$h =$ Height

$A = ah = ab\sin\theta$ Area

$P = 2a + 2b$ Perimeter

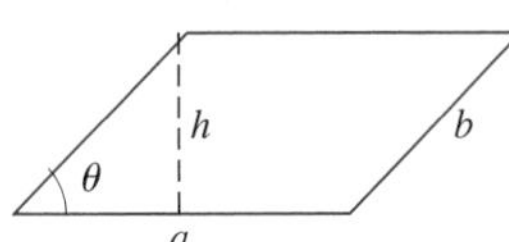

5. Triangle

$h =$ Height

$A = \frac{1}{2}hc$ Area

$P = a + b + c$ Perimeter

$s = \frac{1}{2}(a + b + c)$ Semiperimeter

$A = \sqrt{s(s - a)(s - b)(s - c)}$ Area—Heron's formula

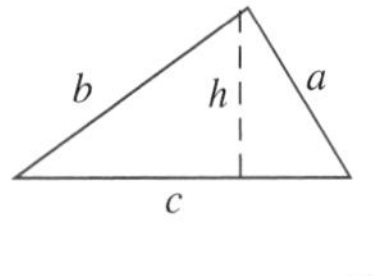

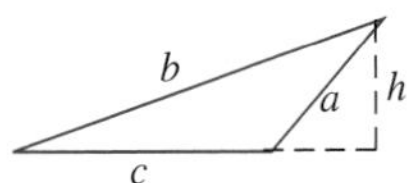

6. Trapezoid

Base a is parallel to base b.

$h =$ Height

$A = \frac{1}{2}(a + b)h$ Area

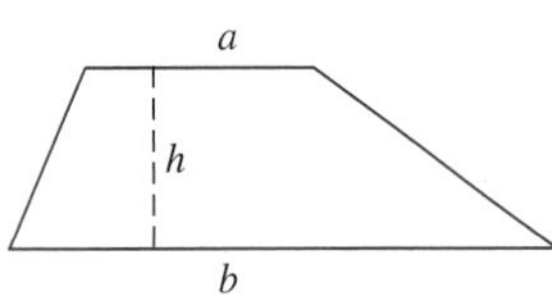

Table I *(Continued)*

7. Circle

$R =$ Radius
$D =$ Diameter
$D = 2R$
$A = \pi R^2 = \frac{1}{4}\pi D^2$ Area
$C = 2\pi R = \pi D$ Circumference
$\dfrac{C}{D} = \pi$ For all circles
$\pi \approx 3.141\ 59$

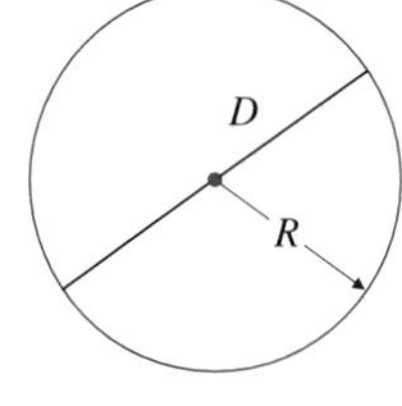

8. Rectangular Solid

$V = abc$ Volume
$T = 2ab + 2ac + 2bc$ Total surface area

9. Right Circular Cylinder

$R =$ Radius of base
$h =$ Height
$V = \pi R^2 h$ Volume
$S = 2\pi Rh$ Lateral surface area
$T = 2\pi R(R + h)$ Total surface area

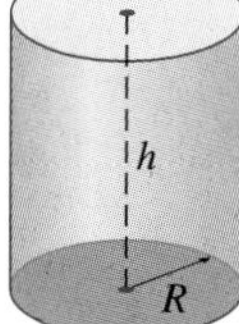

10. Right Circular Cone

$R =$ Radius of base
$h =$ Height
$s =$ Slant height
$V = \frac{1}{3}\pi R^2 h$ Volume
$S = \pi Rs = \pi R\sqrt{R^2 + h^2}$ Lateral surface area
$T = \pi R(R + s) = \pi R(R + \sqrt{R^2 + h^2})$ Total surface area

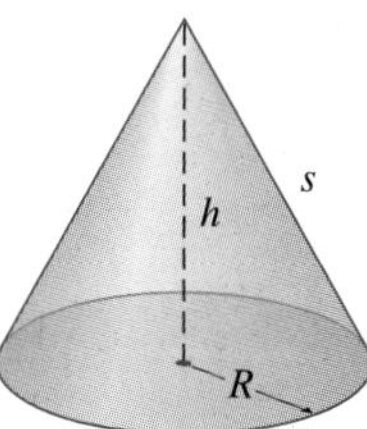

11. Sphere

$R =$ Radius
$D =$ Diameter
$D = 2R$
$V = \frac{4}{3}\pi R^3 = \frac{1}{6}\pi D^3$ Volume
$S = 4\pi R^2 = \pi D^2$ Surface area

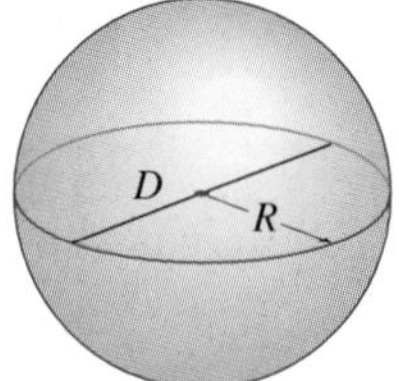

Table II

INTEGRATION FORMULAS

[*Note:* The constant of integration is omitted for each integral, but must be included in any particular application of a formula. The variable u is the variable of integration; all other symbols represent constants.]

Integrals Involving u^n

1. $\displaystyle\int u^n\,du = \frac{u^{n+1}}{n+1}, \quad n \neq -1$

2. $\displaystyle\int u^{-1}\,du = \int \frac{1}{u}\,du = \ln|u|$

Integrals Involving $a + bu$, $a \neq 0$ and $b \neq 0$

3. $\displaystyle\int \frac{1}{a+bu}\,du = \frac{1}{b}\ln|a+bu|$

4. $\displaystyle\int \frac{u}{a+bu}\,du = \frac{u}{b} - \frac{a}{b^2}\ln|a+bu|$

5. $\displaystyle\int \frac{u^2}{a+bu}\,du = \frac{(a+bu)^2}{2b^3} - \frac{2a(a+bu)}{b^3} + \frac{a^2}{b^3}\ln|a+bu|$

6. $\displaystyle\int \frac{u}{(a+bu)^2}\,du = \frac{1}{b^2}\left(\ln|a+bu| + \frac{a}{a+bu}\right)$

7. $\displaystyle\int \frac{u^2}{(a+bu)^2}\,du = \frac{(a+bu)}{b^3} - \frac{a^2}{b^3(a+bu)} - \frac{2a}{b^3}\ln|a+bu|$

8. $\displaystyle\int u(a+bu)^n\,du = \frac{(a+bu)^{n+2}}{(n+2)b^2} - \frac{a(a+bu)^{n+1}}{(n+1)b^2}, \quad n \neq -1, -2$

9. $\displaystyle\int \frac{1}{u(a+bu)}\,du = \frac{1}{a}\ln\left|\frac{u}{a+bu}\right|$

10. $\displaystyle\int \frac{1}{u^2(a+bu)}\,du = -\frac{1}{au} + \frac{b}{a^2}\ln\left|\frac{a+bu}{u}\right|$

11. $\displaystyle\int \frac{1}{u(a+bu)^2}\,du = \frac{1}{a(a+bu)} + \frac{1}{a^2}\ln\left|\frac{u}{a+bu}\right|$

12. $\displaystyle\int \frac{1}{u^2(a+bu)^2}\,du = -\frac{a+2bu}{a^2u(a+bu)} + \frac{2b}{a^3}\ln\left|\frac{a+bu}{u}\right|$

Integrals Involving $a^2 - u^2$, $a > 0$

13. $\displaystyle\int \frac{1}{u^2-a^2}\,du = \frac{1}{2a}\ln\left|\frac{u-a}{u+a}\right|$

14. $\displaystyle\int \frac{1}{a^2-u^2}\,du = \frac{1}{2a}\ln\left|\frac{u+a}{u-a}\right|$

Integrals Involving $(a + bu)$ and $(c + du)$, $b \neq 0$, $d \neq 0$, and $ad - bc \neq 0$

15. $\displaystyle\int \frac{1}{(a+bu)(c+du)}\,du = \frac{1}{ad-bc}\ln\left|\frac{c+du}{a+bu}\right|$

16. $\displaystyle\int \frac{u}{(a+bu)(c+du)}\,du = \frac{1}{ad-bc}\left(\frac{a}{b}\ln|a+bu| - \frac{c}{d}\ln|c+du|\right)$

17. $\displaystyle\int \frac{u^2}{(a+bu)(c+du)}\,du = \frac{1}{bd}u - \frac{1}{ad-bc}\left(\frac{a^2}{b^2}\ln|a+bu| - \frac{c^2}{d^2}\ln|c+du|\right)$

Table II *(Continued)*

18. $\displaystyle\int \frac{1}{(a+bu)^2(c+du)}\,du = \frac{1}{ad-bc}\frac{1}{a+bu} + \frac{d}{(ad-bc)^2}\ln\left|\frac{c+du}{a+bu}\right|$

19. $\displaystyle\int \frac{u}{(a+bu)^2(c+du)}\,du = -\frac{a}{b(ad-bc)}\frac{1}{a+bu} - \frac{c}{(ad-bc)^2}\ln\left|\frac{c+du}{a+bu}\right|$

20. $\displaystyle\int \frac{a+bu}{c+du}\,du = \frac{bu}{d} + \frac{ad-bc}{d^2}\ln|c+du|$

Integrals Involving $\sqrt{a+bu}$, $a \neq 0$ and $b \neq 0$

21. $\displaystyle\int \sqrt{a+bu}\,du = \frac{2\sqrt{(a+bu)^3}}{3b}$

22. $\displaystyle\int u\sqrt{a+bu}\,du = \frac{2(3bu-2a)}{15b^2}\sqrt{(a+bu)^3}$

23. $\displaystyle\int u^2\sqrt{a+bu}\,du = \frac{2(15b^2u^2-12abu+8a^2)}{105b^3}\sqrt{(a+bu)^3}$

24. $\displaystyle\int \frac{1}{\sqrt{a+bu}}\,du = \frac{2\sqrt{a+bu}}{b}$

25. $\displaystyle\int \frac{u}{\sqrt{a+bu}}\,du = \frac{2(bu-2a)}{3b^2}\sqrt{a+bu}$

26. $\displaystyle\int \frac{u^2}{\sqrt{a+bu}}\,du = \frac{2(3b^2u^2-4abu+8a^2)}{15b^3}\sqrt{a+bu}$

27. $\displaystyle\int \frac{1}{u\sqrt{a+bu}}\,du = \frac{1}{\sqrt{a}}\ln\left|\frac{\sqrt{a+bu}-\sqrt{a}}{\sqrt{a+bu}+\sqrt{a}}\right|, \quad a > 0$

28. $\displaystyle\int \frac{1}{u^2\sqrt{a+bu}}\,du = -\frac{\sqrt{a+bu}}{au} - \frac{b}{2a\sqrt{a}}\ln\left|\frac{\sqrt{a+bu}-\sqrt{a}}{\sqrt{a+bu}+\sqrt{a}}\right|, \quad a > 0$

Integrals Involving $\sqrt{a^2-u^2}$, $a > 0$

29. $\displaystyle\int \frac{1}{u\sqrt{a^2-u^2}}\,du = -\frac{1}{a}\ln\left|\frac{a+\sqrt{a^2-u^2}}{u}\right|$

30. $\displaystyle\int \frac{1}{u^2\sqrt{a^2-u^2}}\,du = -\frac{\sqrt{a^2-u^2}}{a^2u}$

31. $\displaystyle\int \frac{\sqrt{a^2-u^2}}{u}\,du = \sqrt{a^2-u^2} - a\ln\left|\frac{a+\sqrt{a^2-u^2}}{u}\right|$

Integrals Involving $\sqrt{u^2+a^2}$, $a > 0$

32. $\displaystyle\int \sqrt{u^2+a^2}\,du = \frac{1}{2}\left(u\sqrt{u^2+a^2} + a^2\ln|u+\sqrt{u^2+a^2}|\right)$

33. $\displaystyle\int u^2\sqrt{u^2+a^2}\,du = \frac{1}{8}\left[u(2u^2+a^2)\sqrt{u^2+a^2} - a^4\ln|u+\sqrt{u^2+a^2}|\right]$

34. $\displaystyle\int \frac{\sqrt{u^2+a^2}}{u}\,du = \sqrt{u^2+a^2} - a\ln\left|\frac{a+\sqrt{u^2+a^2}}{u}\right|$

35. $\displaystyle\int \frac{\sqrt{u^2+a^2}}{u^2}\,du = -\frac{\sqrt{u^2+a^2}}{u} + \ln|u+\sqrt{u^2+a^2}|$

Table II *(Continued)*

36. $\displaystyle\int \frac{1}{\sqrt{u^2 + a^2}}\,du = \ln|u + \sqrt{u^2 + a^2}|$

37. $\displaystyle\int \frac{1}{u\sqrt{u^2 + a^2}}\,du = \frac{1}{a}\ln\left|\frac{u}{a + \sqrt{u^2 + a^2}}\right|$

38. $\displaystyle\int \frac{u^2}{\sqrt{u^2 + a^2}}\,du = \frac{1}{2}(u\sqrt{u^2 + a^2} - a^2\ln|u + \sqrt{u^2 + a^2}|)$

39. $\displaystyle\int \frac{1}{u^2\sqrt{u^2 + a^2}}\,du = -\frac{\sqrt{u^2 + a^2}}{a^2u}$

Integrals Involving $\sqrt{u^2 - a^2}$, $a > 0$

40. $\displaystyle\int \sqrt{u^2 - a^2}\,du = \frac{1}{2}(u\sqrt{u^2 - a^2} - a^2\ln|u + \sqrt{u^2 - a^2}|)$

41. $\displaystyle\int u^2\sqrt{u^2 - a^2}\,du = \frac{1}{8}[u(2u^2 - a^2)\sqrt{u^2 - a^2} - a^4\ln|u + \sqrt{u^2 - a^2}|]$

42. $\displaystyle\int \frac{\sqrt{u^2 - a^2}}{u^2}\,du = -\frac{\sqrt{u^2 - a^2}}{u} + \ln|u + \sqrt{u^2 - a^2}|$

43. $\displaystyle\int \frac{1}{\sqrt{u^2 - a^2}}\,du = \ln|u + \sqrt{u^2 - a^2}|$

44. $\displaystyle\int \frac{u^2}{\sqrt{u^2 - a^2}}\,du = \frac{1}{2}(u\sqrt{u^2 - a^2} + a^2\ln|u + \sqrt{u^2 - a^2}|)$

45. $\displaystyle\int \frac{1}{u^2\sqrt{u^2 - a^2}}\,du = \frac{\sqrt{u^2 - a^2}}{a^2u}$

Integrals Involving e^{au}, $a \neq 0$

46. $\displaystyle\int e^{au}\,du = \frac{e^{au}}{a}$

47. $\displaystyle\int u^n e^{au}\,du = \frac{u^n e^{au}}{a} - \frac{n}{a}\int u^{n-1}e^{au}\,du$

48. $\displaystyle\int \frac{1}{c + de^{au}}\,du = \frac{u}{c} - \frac{1}{ac}\ln|c + de^{au}|, \quad c \neq 0$

Integrals Involving $\ln u$

49. $\displaystyle\int \ln u\,du = u\ln u - u$

50. $\displaystyle\int \frac{\ln u}{u}\,du = \frac{1}{2}(\ln u)^2$

51. $\displaystyle\int u^n\ln u\,du = \frac{u^{n+1}}{n+1}\ln u - \frac{u^{n+1}}{(n+1)^2}, \quad n \neq -1$

52. $\displaystyle\int (\ln u)^n\,du = u(\ln u)^n - n\int (\ln u)^{n-1}\,du$

Table II *(Continued)*

Integrals Involving Trigonometric Functions of au, $a \neq 0$

53. $\displaystyle\int \sin au\, du = -\frac{1}{a}\cos au$

54. $\displaystyle\int \cos au\, du = \frac{1}{a}\sin au$

55. $\displaystyle\int \tan au\, du = -\frac{1}{a}\ln|\cos au|$

56. $\displaystyle\int \cot au\, du = \frac{1}{a}\ln|\sin au|$

57. $\displaystyle\int \sec au\, du = \frac{1}{a}\ln|\sec au + \tan au|$

58. $\displaystyle\int \csc au\, du = \frac{1}{a}\ln|\csc au - \cot au|$

59. $\displaystyle\int (\sin au)^2\, du = \frac{u}{2} - \frac{1}{4a}\sin 2au$

60. $\displaystyle\int (\cos au)^2\, du = \frac{u}{2} + \frac{1}{4a}\sin 2au$

61. $\displaystyle\int (\sin au)^n\, du = -\frac{1}{an}(\sin au)^{n-1}\cos au + \frac{n-1}{n}\int (\sin au)^{n-2}\, du, \quad n \neq 0$

62. $\displaystyle\int (\cos au)^n\, du = \frac{1}{an}\sin au(\cos au)^{n-1} + \frac{n-1}{n}\int (\cos au)^{n-2}\, du, \quad n \neq 0$

ANSWERS

Chapter 1

Exercise 1-1

1. Function **3.** Not a function **5.** Function **7.** Function **9.** Not a function **11.** Function **13.** 4
15. -5 **17.** -6 **19.** -2 **21.** -12 **23.** -1 **25.** -6 **27.** 12 **29.** $\frac{3}{4}$ **31.** $y = 0$ **33.** $y = 4$
35. $x = -5, 0, 4$ **37.** $x = -6$ **39.** All real numbers **41.** All real numbers except -4
43. All real numbers except -4 and 1 **45.** $x \leq 7$ **47.** $x < 7$
49. $f(2) = 0$, and 0 is a number; therefore, $f(2)$ exists. On the other hand, $f(3)$ is not defined, since the denominator would be 0; therefore, we say that $f(3)$ does not exist.
51. $g(x) = 2x^3 - 5$ **53.** $G(x) = 2\sqrt{x} - x^2$
55. Function f multiplies the domain element by 2 and subtracts 3 from the result.
57. Function F multiplies the cube of the domain element by 3 and subtracts twice the square root of the domain element from the result.
59. A function with domain R **61.** A function with domain R **63.** Not a function; for example, when $x = 1$, $y = \pm 3$
65. A function with domain all real numbers except $x = 4$ **67.** Not a function; for example, when $x = 4$, $y = \pm 3$
69. 4 **71.** $h - 1$ **73.** 4 **75.** $8a + 4h - 7$ **77.** $3a^2 + 3ah + h^2$ **79.** $\dfrac{1}{\sqrt{a+h} + \sqrt{a}}$
81. $P(w) = 2w + \dfrac{50}{w}, w > 0$ **83.** $A(l) = l(50 - l), 0 \leq l \leq 50$

85.

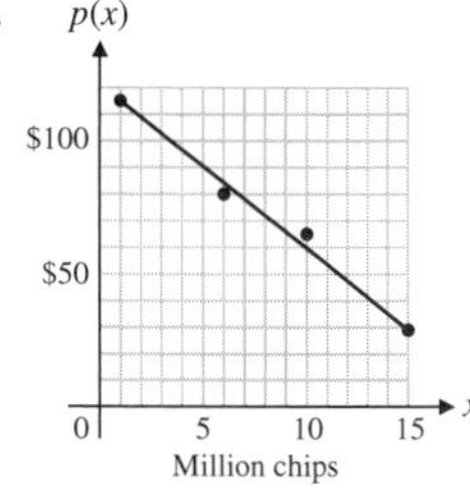

$p(8) = 71$ dollars per chip;
$p(11) = 53$ dollars per chip

87. (A) $R(x) = xp(x) = x(119 - 6x)$ million dollars;
Domain: $1 \leq x \leq 15$

(B)

x (MILLION)	R(x) (MILLION $)
1	113
3	303
6	498
9	585
12	564
15	435

(C)

89. (A) $P(x) = R(x) - C(x) = x(119 - 6x) - (234 + 23x)$ million dollars;
Domain: $1 \leq x \leq 15$

(B)

x (MILLION)	P(x) (MILLION $)
1	−144
3	0
6	126
9	144
12	54
15	−144

(C)

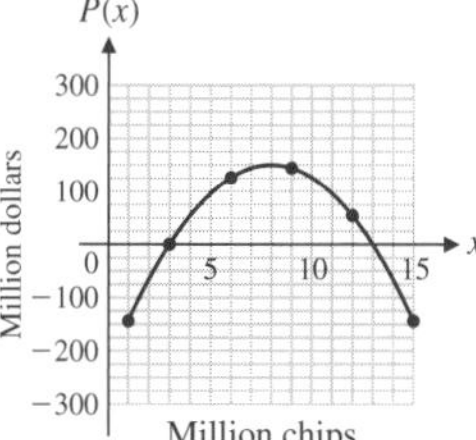

91. (A) $V(x) = x(8 - 2x)(12 - 2x)$
(B) $0 \leq x \leq 4$

(C)

x	V(x)
1	60
2	64
3	36

(D)

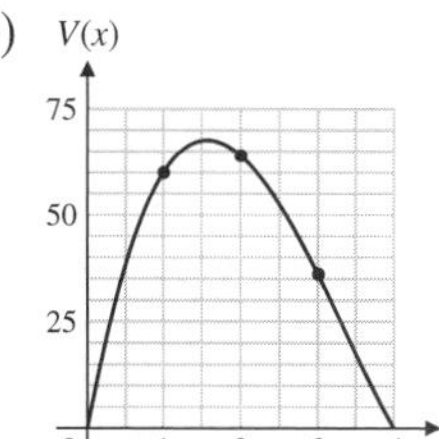

93. (A) The graph indicates that there is a value of x near 2 that will produce a volume of 65. The table shows $x = 1.9$ to one decimal place:

x	1.8	1.9	2
$V(x)$	66.5	65.4	64

(B) $x = 1.93$ to two decimal places

X	Y1	
1.9	65.436	
1.91	65.307	
1.92	65.176	
1.93	65.04	
1.94	64.902	
1.95	64.76	
1.96	64.614	

X=1.93

95. $v = \dfrac{75 - w}{15 + w}$; 1.9032 cm/sec

Exercise 1-2

1. Domain: All real numbers; Range: All real numbers
3. Domain: All real numbers; Range: $(-\infty, 0]$
5. Domain: $[0, \infty)$; Range: $(-\infty, 0]$
7. Domain: All real numbers; Range: All real numbers

9.

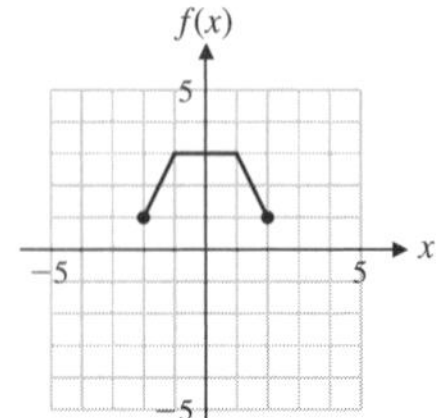

11.

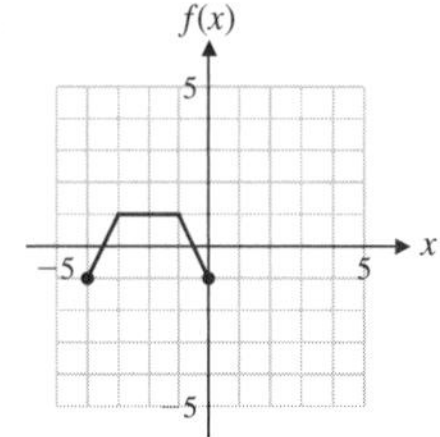

13.

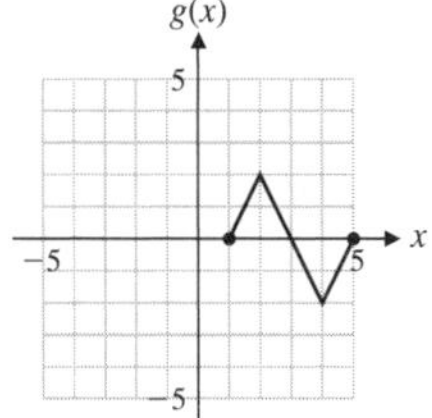

15.

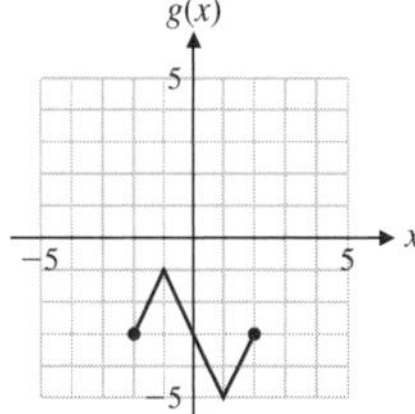

17.

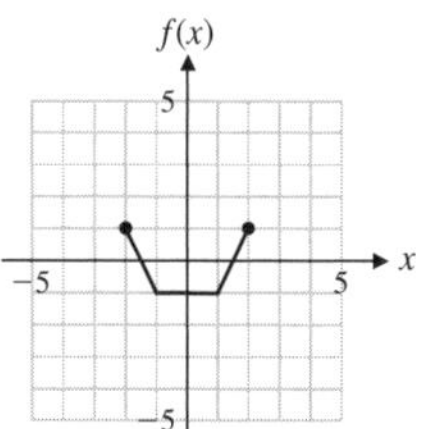

19.

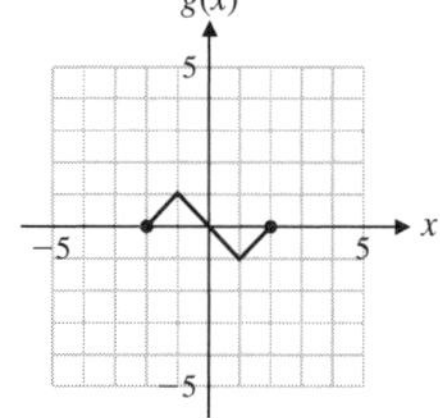

21. The graph of $g(x) = -|x + 3|$ is the graph of $y = |x|$ reflected in the x axis and shifted 3 units to the left.

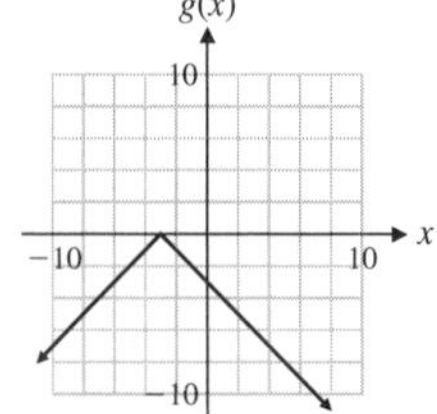

23. The graph of $f(x) = (x - 4)^2 - 3$ is the graph of $y = x^2$ shifted 4 units to the right and 3 units down.

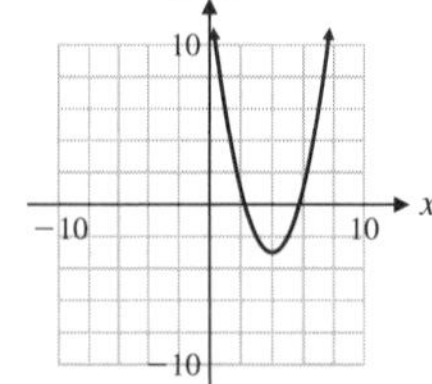

25. The graph of $f(x) = 7 - \sqrt{x}$ is the graph of $y = \sqrt{x}$ reflected in the x axis and shifted 7 units up.

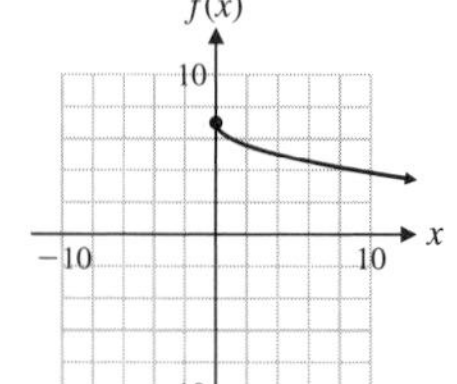

27. The graph of $h(x) = -3|x|$ is the graph of $y = |x|$ reflected in the x axis and vertically expanded by a factor of 3.

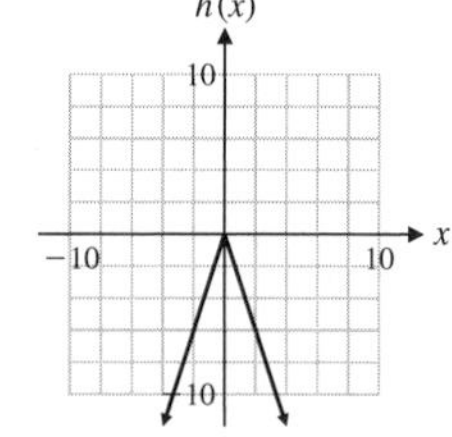

29. The graph of the basic function $y = x^2$ is shifted 2 units to the left and 3 units down. Equation: $y = (x + 2)^2 - 3$.
31. The graph of the basic function $y = x^2$ is reflected in the x axis and shifted 3 units to the right and 2 units up. Equation: $y = 2 - (x - 3)^2$.
33. The graph of the basic function $y = \sqrt{x}$ is reflected in the x axis and shifted 4 units up. Equation: $y = 4 - \sqrt{x}$.
35. The graph of the basic function $y = x^3$ is shifted 2 units to the left and 1 unit down. Equation: $y = (x + 2)^3 - 1$.
37. $g(x) = \sqrt{x - 2} - 3$ **39.** $g(x) = -|x + 3|$ **41.** $g(x) = -(x - 2)^3 - 1$

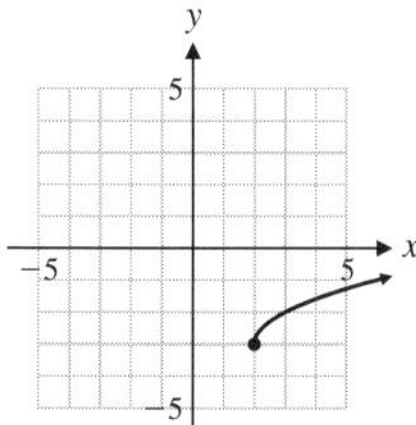
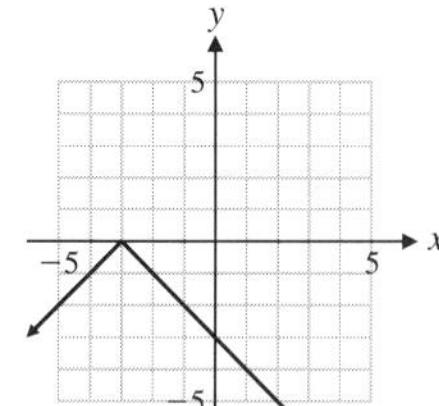
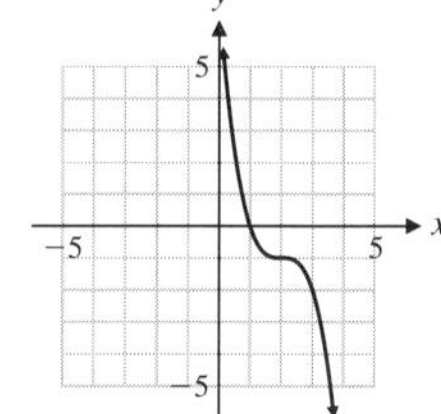

43. Discontinuous at $x = 0$ **45.** No points of discontinuity **47.** Discontinuous at $x = -1$ and $x = 1$

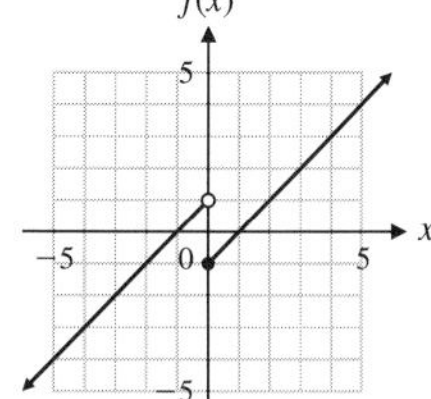
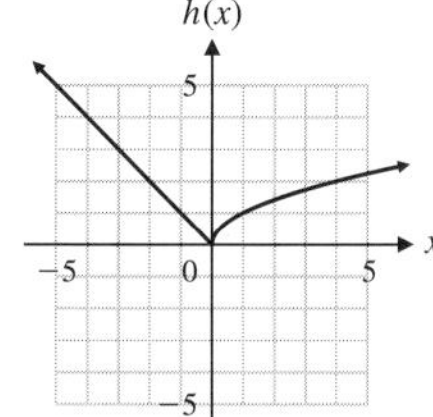
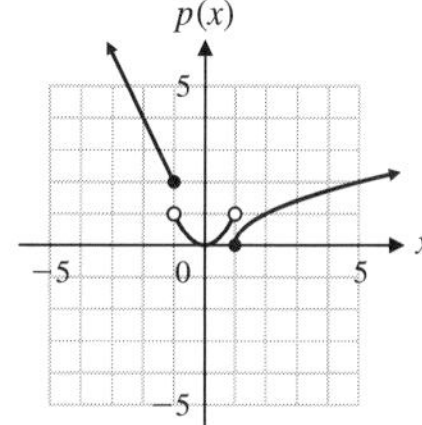

49. The graph of the basic function $y = |x|$ is reflected in the x axis and vertically contracted by a factor of 0.5. Equation: $y = -0.5|x|$.
51. The graph of the basic function $y = x^2$ is reflected in the x axis and vertically expanded by a factor of 2. Equation: $y = -2x^2$.
53. The graph of the basic function $y = \sqrt[3]{x}$ is reflected in the x axis and vertically expanded by a factor of 3. Equation: $y = -3\sqrt[3]{x}$.
55. Reversing the order does not change the result.
57. Reversing the order can change the result.
59. Reversing the order can change the result.
61. (A) The graph of the basic function $y = \sqrt{x}$ is reflected in the x axis, vertically expanded by a factor of 4, and shifted up 115 units.

(B)

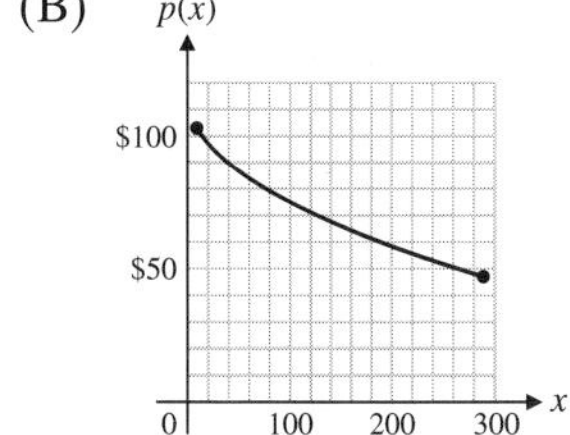

63. (A) The graph of the basic function $y = x^3$ is vertically contracted by a factor of 0.000 48 and shifted right 500 units and up 60,000 units.

(B)

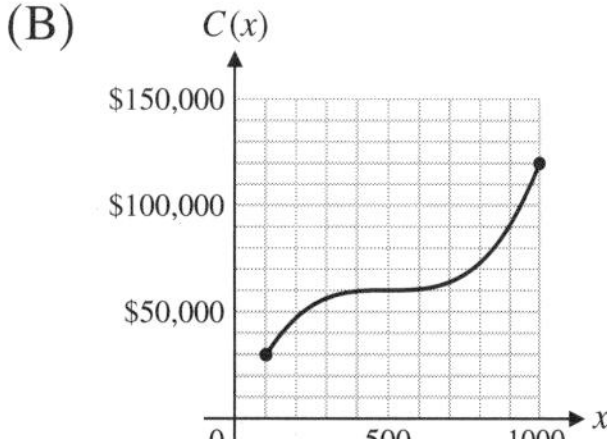

65. (A) $C(x) = \begin{cases} 10.06x & \text{if } 1 \leq x < 4 \\ 8.52x & \text{if } 4 \leq x < 10 \\ 7.31x & \text{if } 10 \leq x \end{cases}$

(B)

(C) Since $C(3) = \$30.18$ is less than $C(4) = \$34.08$, the customer would have to decide if a fourth box for an additional \$3.90 was desirable. However, since $C(9) = \$76.68$ is greater than $C(10) = \$73.10$, it is always to the customer's advantage to purchase 10 boxes instead of 9 boxes.

67. (A) $C(x) = \begin{cases} 0.5x & \text{if } 0 \leq x \leq 50 \\ 0.35x + 7.5 & \text{if } 50 < x \leq 100 \\ 0.2x + 22.5 & \text{if } 100 < x \end{cases}$

(B)

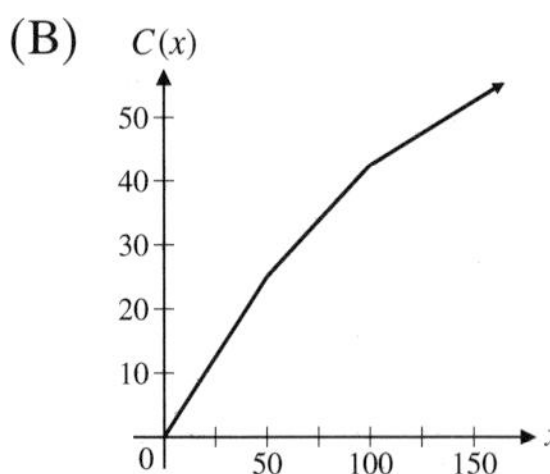

(C) In this case, the discount mileage rates do not apply to previous mileage, so there is no break in the graph at the points where the mileage rate changes.

69. (A) The graph of the basic function $y = x$ is vertically expanded by a factor of 5.5 and shifted down 220 units.

(B)

71. (A) The graph of the basic function $y = \sqrt{x}$ is vertically expanded by a factor of 7.08.

(B)

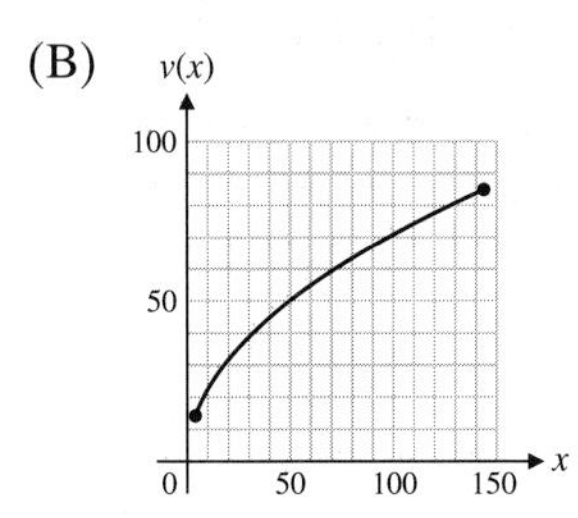

Exercise 1-3

1. (D) **3.** (C); slope is 0

5.

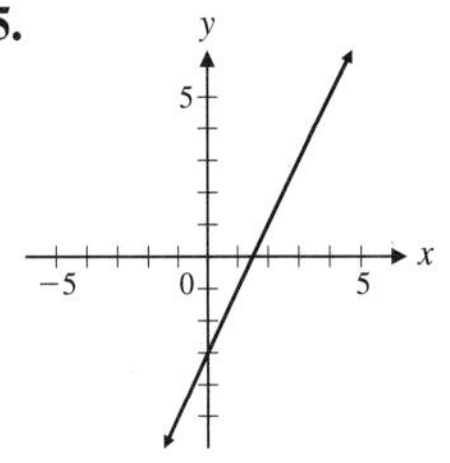

7.

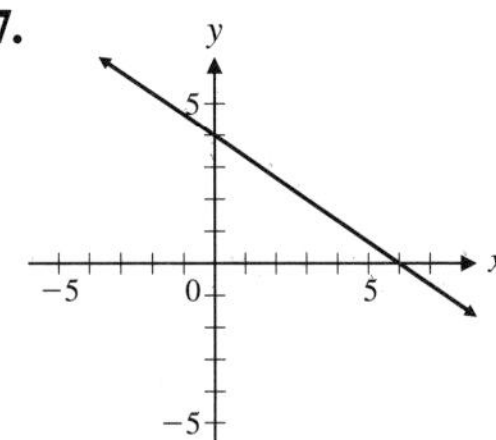

9. Slope = 2; y intercept = −3
11. Slope = $-\frac{2}{3}$; y intercept = 2
13. $y = -2x + 4$
15. $y = -\frac{3}{5}x + 3$

17.

19.

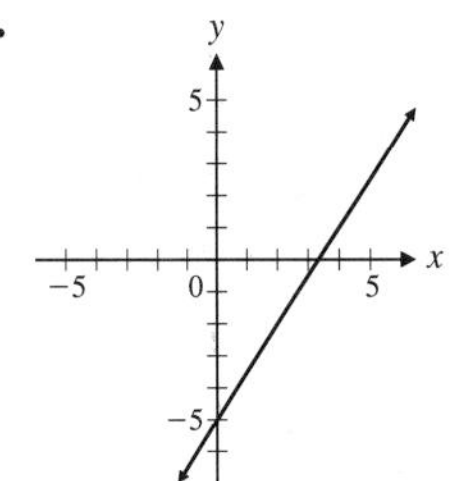

21.

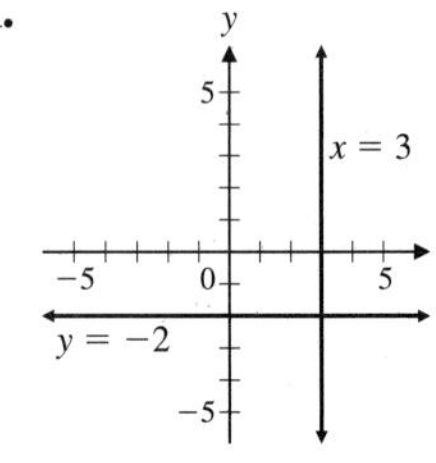

23. $y = -3x + 5;\ m = -3$
25. $y = -\frac{2}{3}x + 4;\ m = -\frac{2}{3}$

27. (A)

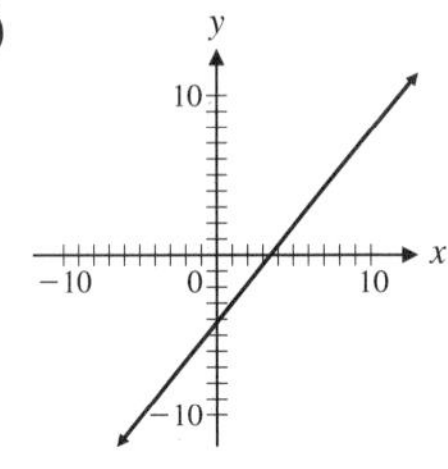

(B) x intercept: 3.5; y intercept: -4.2

(C)

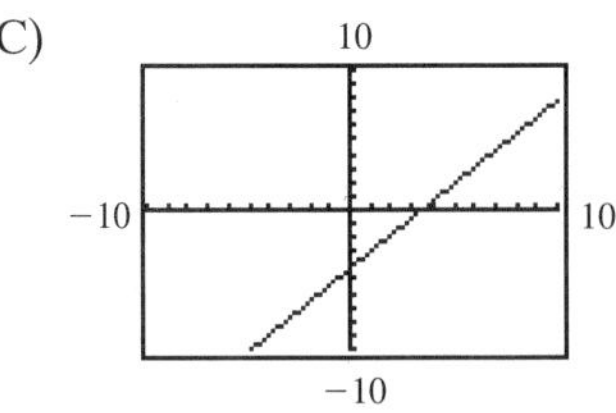

(D) x intercept: 3.5; y intercept: -4.2

(E) $x > 3.5$ or $(3.5, \infty)$

29. $x = 3$; $y = -5$ **31.** $x = -1$; $y = -3$
33. $y + 1 = -3(x - 4)$; $y = -3x + 11$
35. $y + 5 = \frac{2}{3}(x + 6)$; $y = \frac{2}{3}x - 1$ **37.** $y = 0x - 5$, or $y = -5$ **39.** $\frac{1}{3}$
41. $-\frac{1}{5}$ **43.** Not defined **45.** 0 **47.** $(y - 3) = \frac{1}{3}(x - 1)$; $x - 3y = -8$
49. $(y + 2) = -\frac{1}{5}(x + 5)$; $x + 5y = -15$ **51.** $x = 2$ **53.** $y = 3$
55. A linear function **57.** Not a function **59.** A constant function

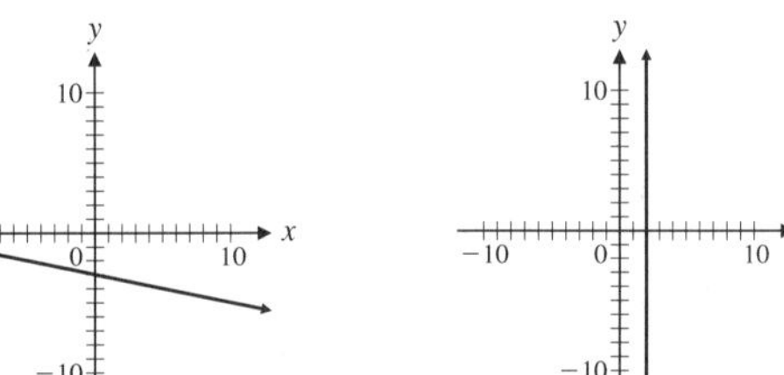

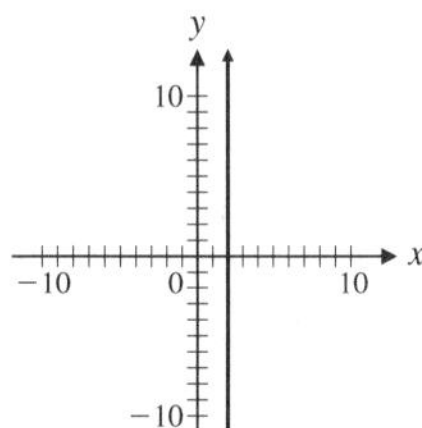

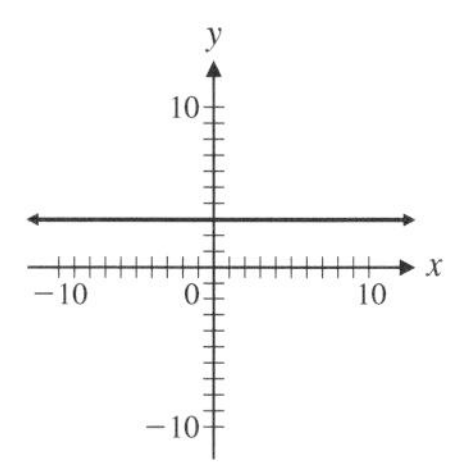

61. The graphs have the same y intercept, $(0, 2)$.

63. (A)

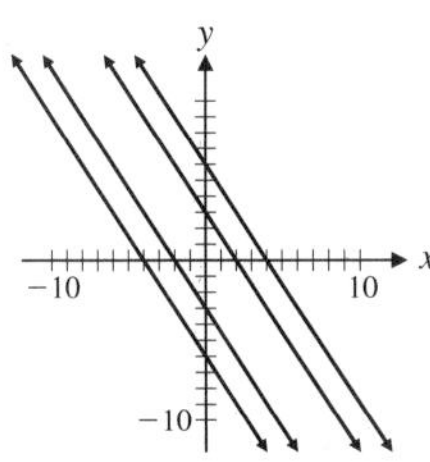

(B) Varying C produces a family of parallel lines. This is verified by observing that varying C does not change the slope of the lines but changes the intercepts.

65. The graph of g is the same as the graph of f for x satisfying $mx + b \geqslant 0$ and the reflection of the graph of f in the x axis for x satisfying $mx + b < 0$. The function g is never linear.

67. (A) \$130; \$220

(B)

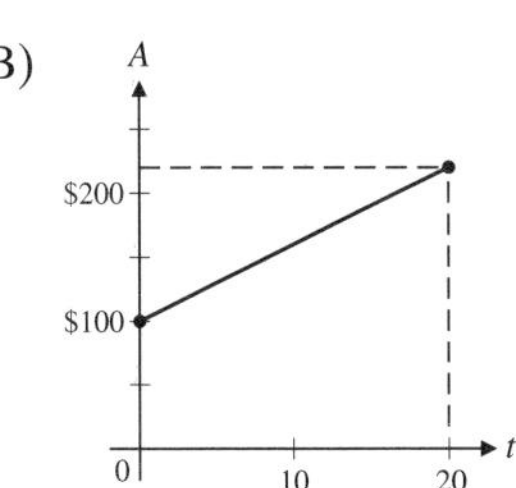

(C) The slope is 6. The amount in the account is growing at the rate of \$6 per year.

69. (A) $C(x) = 180x + 200$

(B) \$2,360

(C)

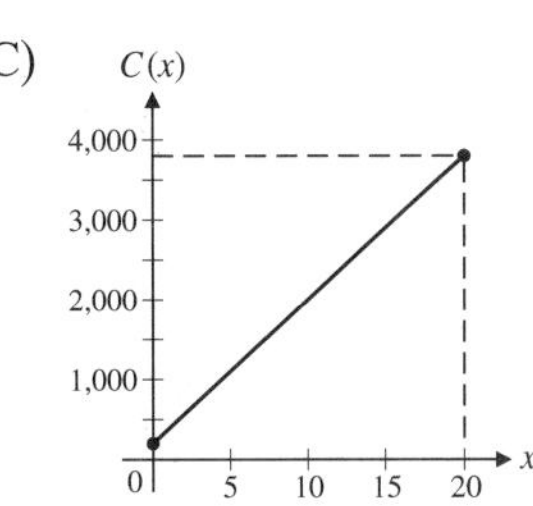

71. (A)

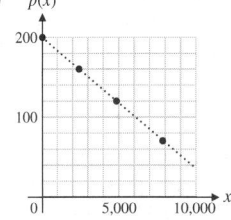

(B) $p(x) = -\frac{1}{60}x + 200$

(C) $p(3{,}000) = \$150$

(D) The slope is $-\frac{1}{60} \approx -0.02$. The price decreases \$0.02 for each unit increase in demand.

73. (A)

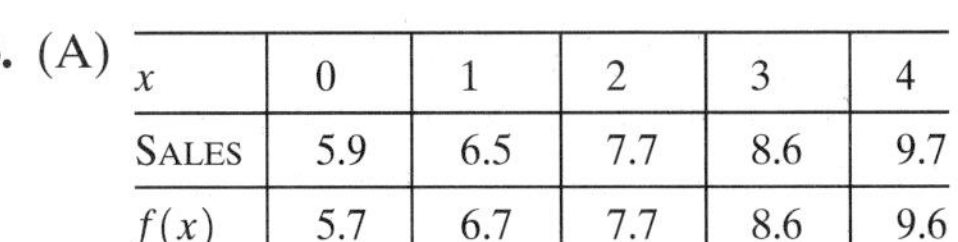

x	0	1	2	3	4
SALES	5.9	6.5	7.7	8.6	9.7
$f(x)$	5.7	6.7	7.7	8.6	9.6

(B)

(C) \$10.6 billion; \$17.4 billion

(D) The sales are \$5.9 billion in 1988 and increase at approximately \$0.97 billion per year for the next 4 years.

75. $0.2x + 0.1y = 20$

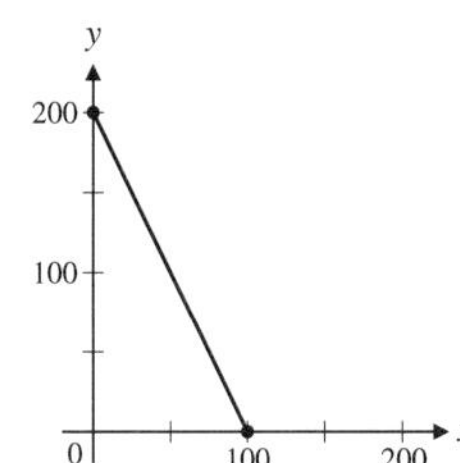

77. (A) 64 g; 35 g

(B)

(C) $-\frac{1}{5}$

Exercise 1-4

1. (A), (C), (E), (F) **3.** (A) m (B) g (C) f (D) n

5. (A) x intercepts: 1, 3; y intercept: -3
(B) Vertex: (2, 1)
(C) Maximum: 1
(D) Range: $y \leqslant 1$ or $(-\infty, 1]$
(E) Increasing interval: $x \leqslant 2$ or $(-\infty, 2]$
(F) Decreasing interval: $x \geqslant 2$ or $[2, \infty)$

7. (A) x intercepts: $-3, -1$; y intercept: 3
(B) Vertex: $(-2, -1)$
(C) Minimum: -1
(D) Range: $y \geqslant -1$ or $[-1, \infty)$
(E) Increasing interval: $x \geqslant -2$ or $[-2, \infty)$
(F) Decreasing interval: $x \leqslant -2$ or $(-\infty, -2]$

9. (A) x intercepts: 1, 3; y intercept: -3
(B) Vertex: (2, 1)
(C) Maximum: 1
(D) Range: $y \leqslant 1$ or $(-\infty, 1]$

11. (A) x intercepts: $-3, -1$; y intercept: 3 (B) Vertex: $(-2, -1)$ (C) Minimum: -1 (D) Range: $y \geqslant -1$ or $[-1, \infty)$

13. $y = -[x - (-2)]^2 + 5$ or $y = -(x + 2)^2 + 5$ **15.** $y = (x - 1)^2 - 3$

17. $f(x) = (x - 4)^2 - 3$ (A) x intercepts: 2.3, 5.7; y intercept: 13 (B) Vertex: $(4, -3)$ (C) Minimum: -3 (D) Range: $y \geqslant -3$ or $[-3, \infty)$

19. $M(x) = -(x + 3)^2 + 10$ (A) x intercepts: $-6.2, 0.2$; y intercept: 1 (B) Vertex: $(-3, 10)$ (C) Maximum: 10 (D) Range: $y \leqslant 10$ or $(-\infty, 10]$

21. $G(x) = 0.5(x - 4)^2 + 2$ (A) x intercepts: none; y intercept: 10 (B) Vertex: (4, 2) (C) Minimum: 2 (D) Range: $y \geqslant 2$ or $[2, \infty)$

23. (A) $-4.87, 8.21$ (B) $-3.44, 6.78$ (C) No solution **25.** The vertex of the parabola is on the x axis.

27. $g(x) = 0.25(x - 3)^2 - 9.25$ (A) x intercepts: $-3.08, 9.08$; y intercept: -7 (B) Vertex: $(3, -9.25)$ (C) Minimum: -9.25 (D) Range: $y \geqslant -9.25$ or $[-9.25, \infty)$

29. $f(x) = -0.12(x - 4)^2 + 3.12$ (A) x intercepts: $-1.1, 9.1$; y intercept: 1.2 (B) Vertex: (4, 3.12) (C) Maximum: 3.12 (D) Range: $y \leqslant 3.12$ or $(-\infty, 3.12]$

31. $x = -5.37, 0.37$ **33.** $-1.37 < x < 2.16$ **35.** $x \leqslant -0.74$ or $x \geqslant 4.19$

37. Axis: $x = 2$; Vertex: (2, 4); Range: $y \geqslant 4$ or $[4, \infty)$; no x intercepts

39. (A)

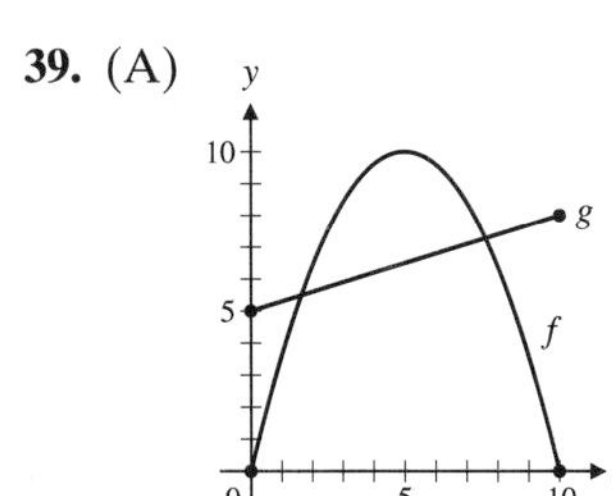

(B) 1.64, 7.61
(C) $1.64 < x < 7.61$
(D) $0 \leqslant x < 1.64$ or $7.61 < x \leqslant 10$

41. (A)

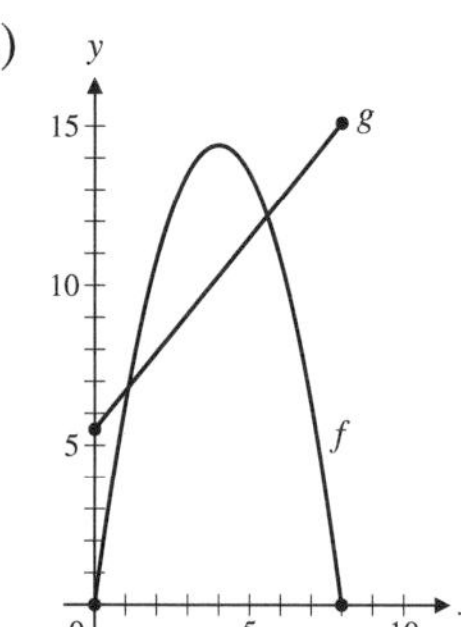

(B) 1.10, 5.57
(C) $1.10 < x < 5.57$
(D) $0 \leqslant x < 1.10$ or $5.57 < x \leqslant 8$

43. $f(x) = x^2 + 1$ and $g(x) = -(x - 4)^2 - 1$ are two examples. The graphs do not cross the x axis.

39. (A) x intercept: $\pm\sqrt{3}$; y intercept: $-\frac{2}{3}$ (B) Vertical asymptotes: $x = -3$, $x = 3$; Horizontal asymptote: $y = -2$

(C)

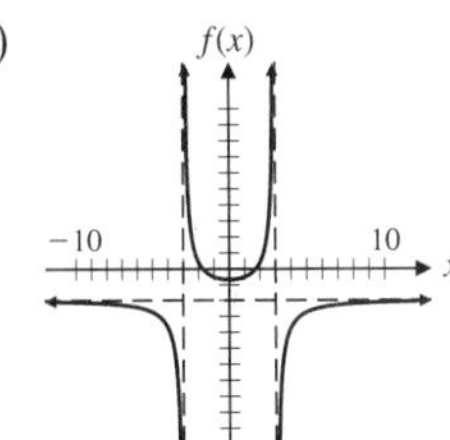

(D)

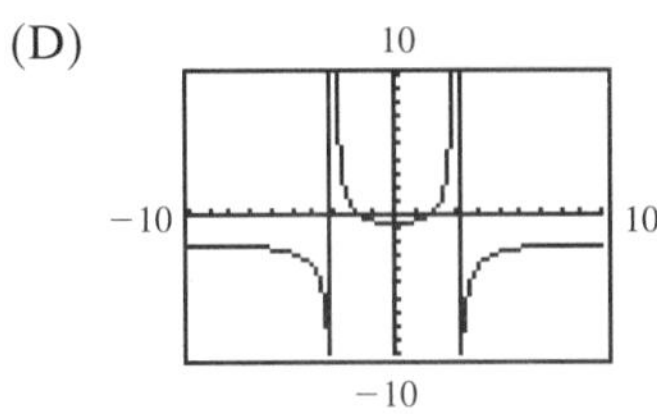

41. (A) x intercept: 0; y intercept: 0 (B) Vertical asymptotes: $x = -3$, $x = 2$; Horizontal asymptote: $y = 0$

(C)

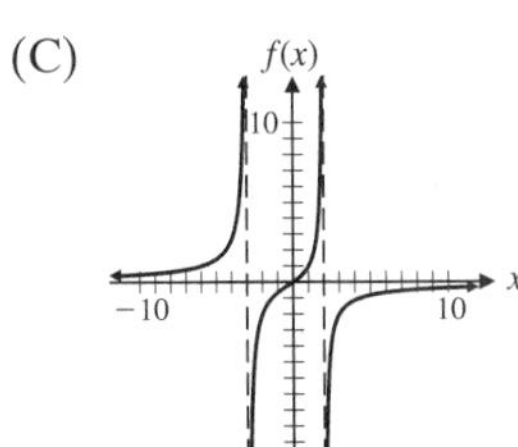

(D)

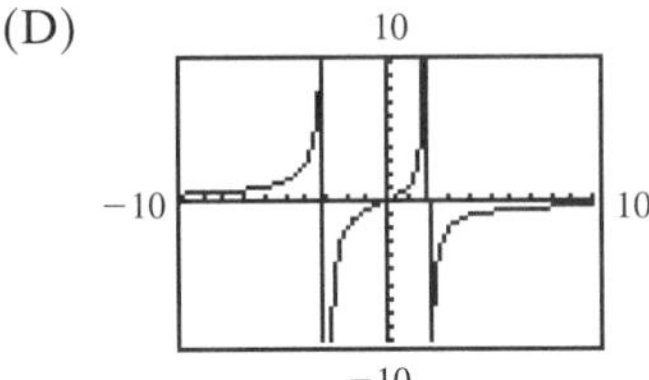

43. $f(x) = x^2 - x - 2$

45. $f(x) = 4x - x^3$

47. (A) $C(x) = 180x + 200$

(B) $\overline{C}(x) = \dfrac{180x + 200}{x}$

(C)

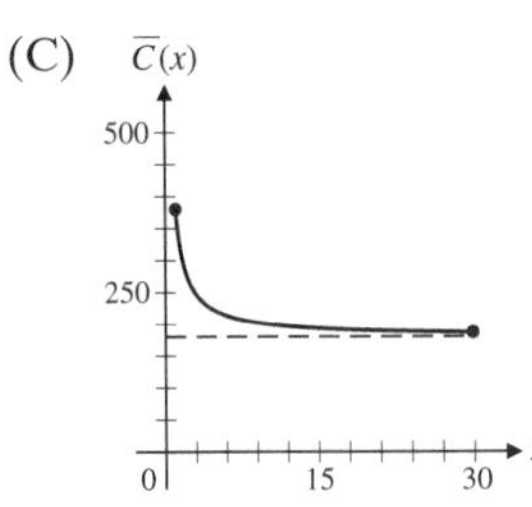

(D) $180 per board

49. 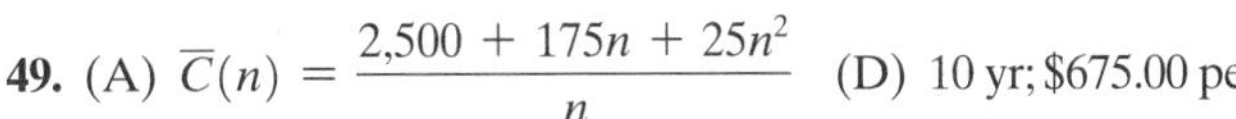

(A) $\overline{C}(n) = \dfrac{2{,}500 + 175n + 25n^2}{n}$ (D) 10 yr; $675.00 per yr

(B)

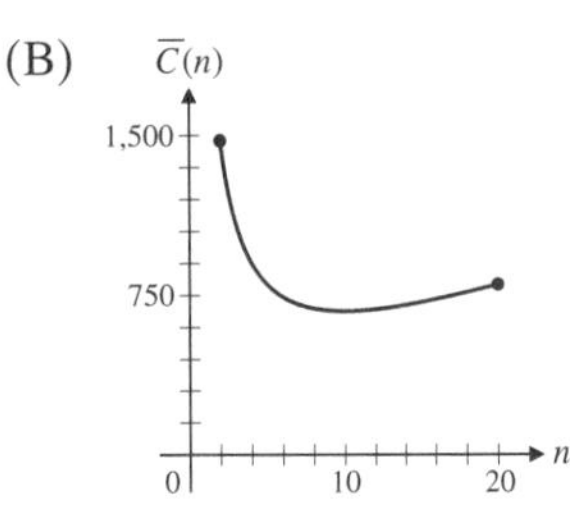

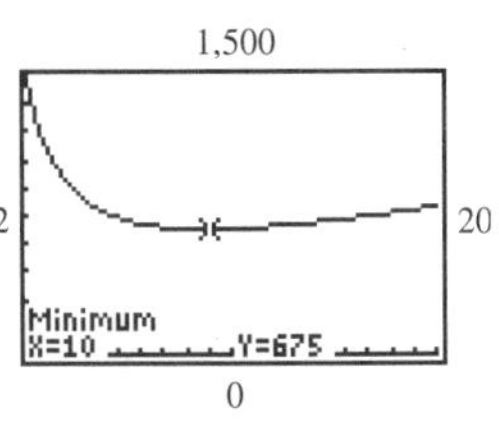

(C) 10 yr; $675.00 per yr

51. 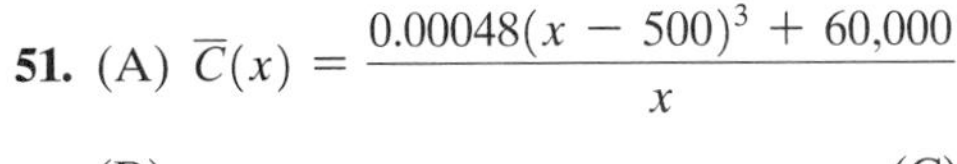

(A) $\overline{C}(x) = \dfrac{0.00048(x - 500)^3 + 60{,}000}{x}$

(B)

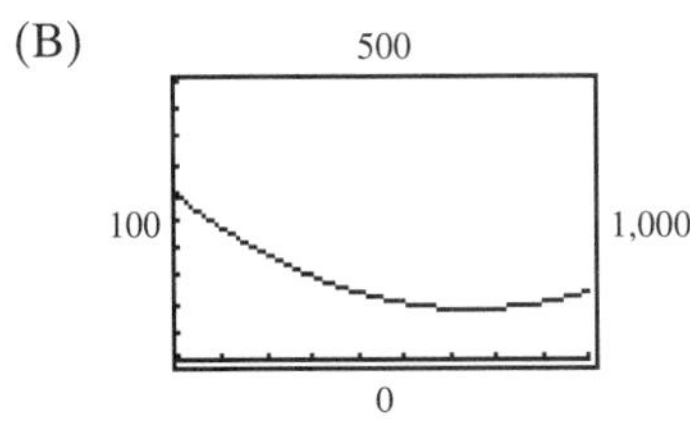

(C) 750 cases per month; $90 per case

53. (A)

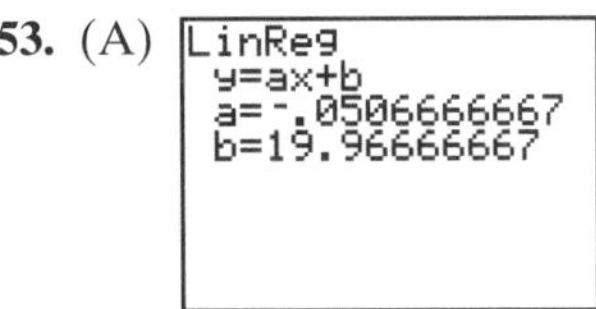

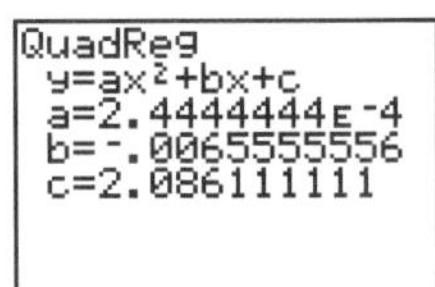

(B) $\overline{x} = 195$; $\overline{p} = \$10.09$

55. (A)

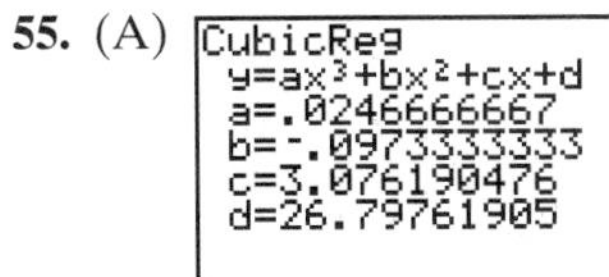

(B) $1,072.8 billion

57. (A) 0.06 cm/sec

(B)

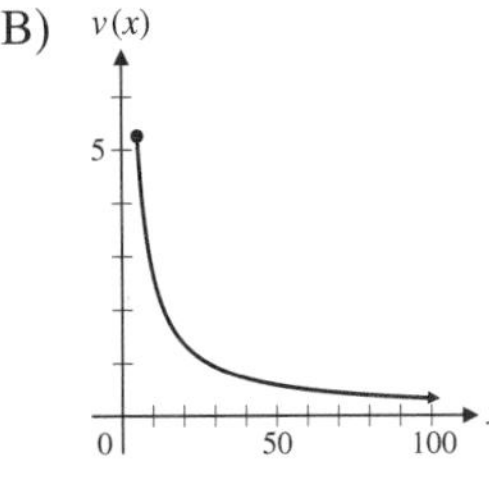

59. (A)

CubicReg
y=ax³+bx²+cx+d
a=-3.86532E-4
b=.0244083694
c=-.3914694565
d=10.87777778

(B) 7.5 marriages per 1,000 population

Exercise 2-2

1.
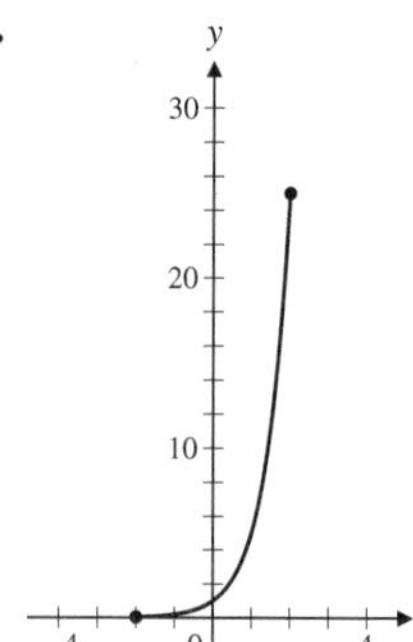

3.
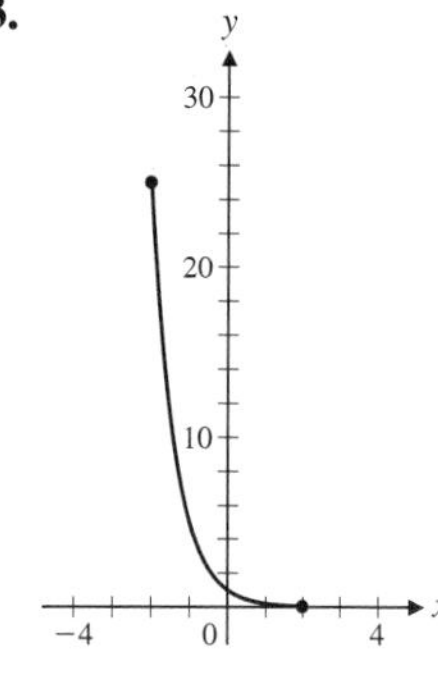

5.
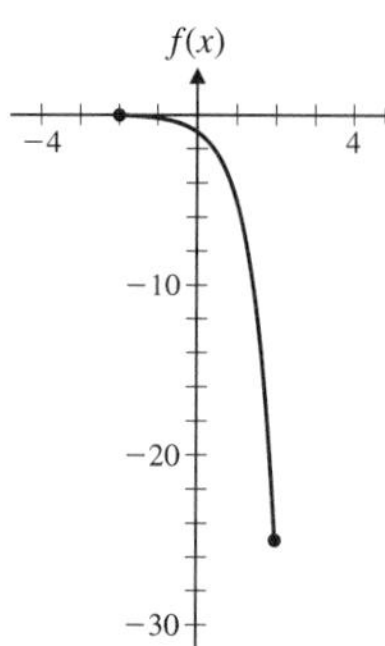

7.
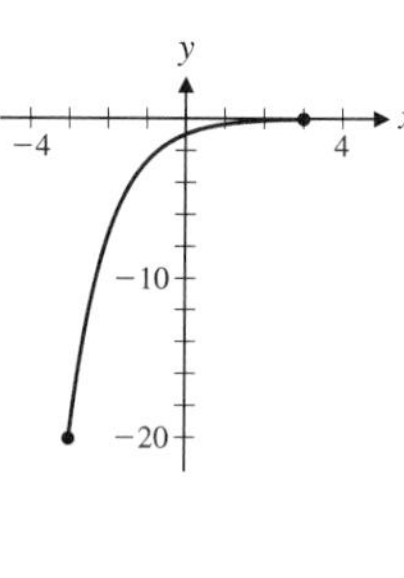

9.
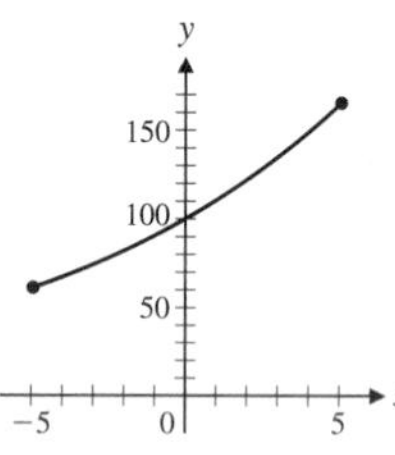

11.
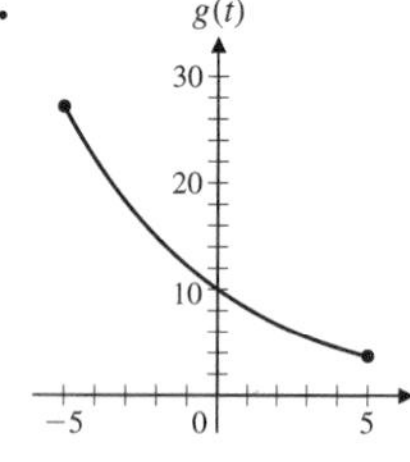

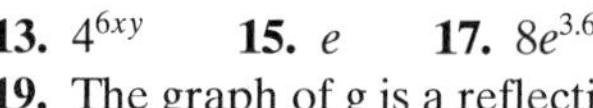
13. 4^{6xy} **15.** e **17.** $8e^{3.6t}$

19. The graph of g is a reflection of the graph f in the x axis.
21. The graph of g is the graph of f shifted 1 unit to the left.
23. The graph of g is the graph of f shifted 1 unit up.

25. The graph of g is the graph of f vertically expanded by a factor of 2 and shifted to the left 2 units.

27.
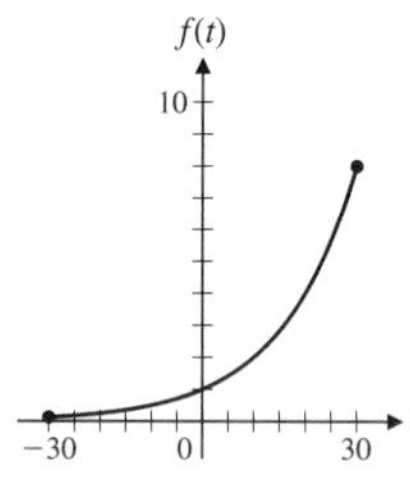

29.
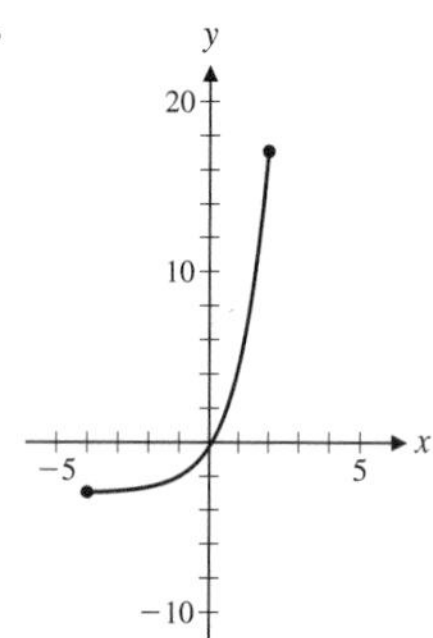

31.
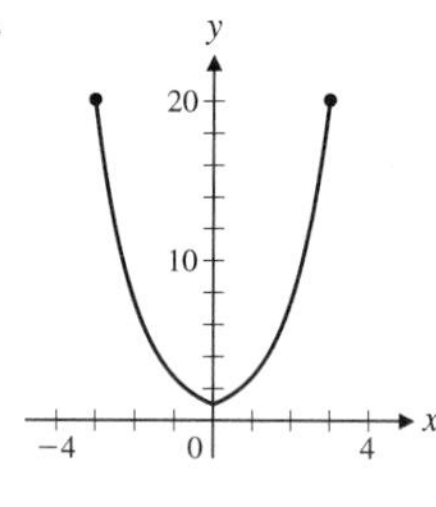

33.
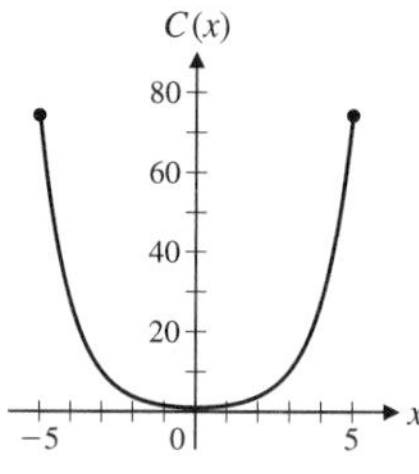

35.
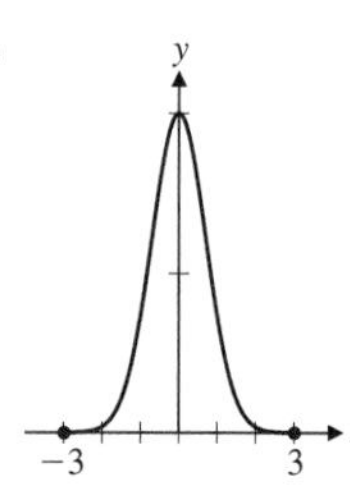

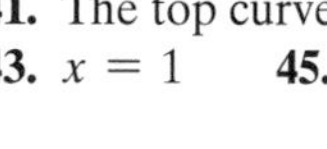
37. $a = 1, -1$ **39.** The top curve belongs to f; the bottom curve belongs to g.
41. The top curve belongs to g; the bottom curve belongs to f.
43. $x = 1$ **45.** $x = -1, 6$ **47.** $x = 3$ **49.** $x = 3$ **51.** $x = -3, 0$

53.
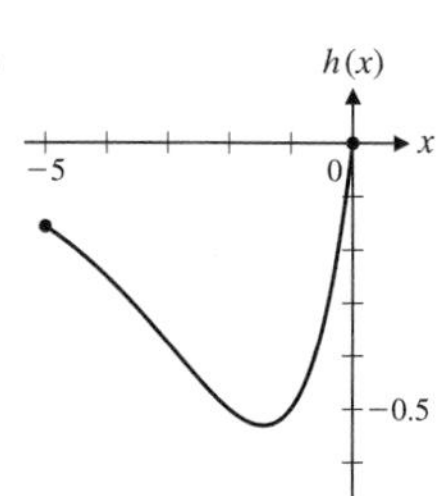

55.
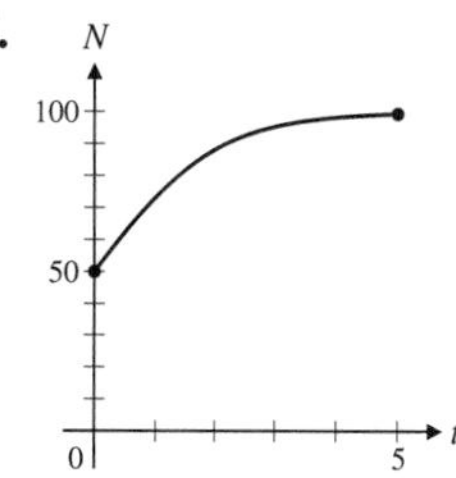

57. $x = 1.40$ **59.** $x = -0.73$
61. (A) $2,633.56 (B) $7,079.54 **63.** (A) $11,871.65 (B) $20,427.93 **65.** $9,217
67. (A) $10,850.88 (B) $10,838.29 **69.** $28,847.49

71. N approaches 2 as t increases without bound.

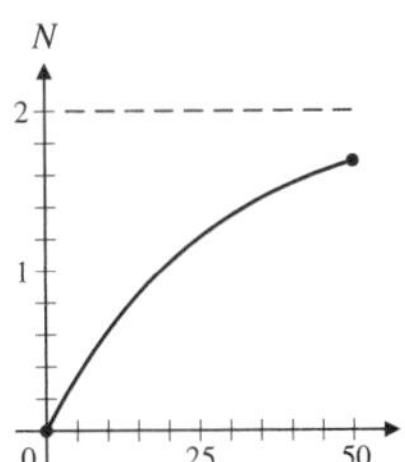

73. (A) 1995: \$673,000; 2010: \$5,162,000

```
ExpReg
 y=a*b^x
 a=87.70214922
 b=1.145496294
```

75. (A) 10%
(B) 1%

77. (A) $N = 40{,}000e^{0.21t}$
(B) 215,000; 613,000
(C)

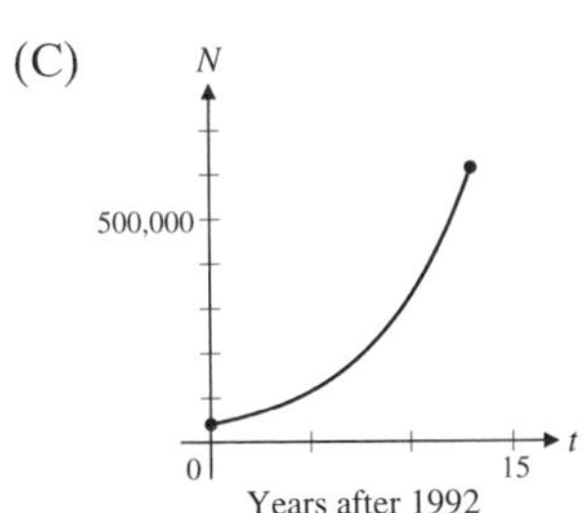

79. (A) $P = 5.7e^{0.0114t}$
(B) 6.8 billion; 8.5 billion
(C)

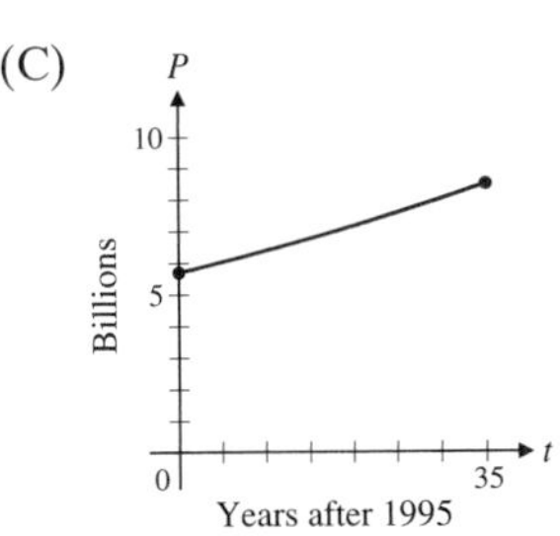

81. (A) 1995: 7,244,000; 1996: 13,507,000

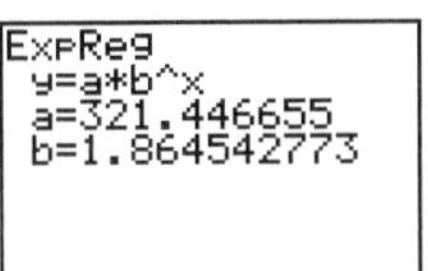

Exercise 2-3

1. $27 = 3^3$ **3.** $10^0 = 1$ **5.** $8 = 4^{3/2}$ **7.** $\log_7 49 = 2$ **9.** $\log_4 8 = \frac{3}{2}$ **11.** $\log_b A = u$ **13.** 0 **15.** 1
17. 1 **19.** 3 **21.** -3 **23.** 3 **25.** $\log_b P - \log_b Q$ **27.** $5 \log_b L$ **29.** $\log_b p - \log_b q - \log_b r - \log_b s$
31. $x = 9$ **33.** $y = 2$ **35.** $b = 10$ **37.** $x = 2$ **39.** $y = -2$ **41.** $b = 100$ **43.** $5 \log_b x - 3 \log_b y$
45. $\frac{1}{3} \log_b N$ **47.** $2 \log_b x + \frac{1}{3} \log_b y$ **49.** $\log_b 50 - 0.2t \log_b 2$ **51.** $\log_b P + t \log_b(1 + r)$ **53.** $\log_e 100 - 0.01t$
55. $x = 2$ **57.** $x = 8$ **59.** $x = 7$ **61.** No solution

63.

65. The graph of $y = \log_2(x - 2)$ is the graph of $y = \log_2 x$ shifted to the right 2 units.
67. Domain: $(-1, \infty)$; Range: All real numbers
69. (A) 3.547 43 (B) -2.160 32 (C) 5.626 29 (D) -3.197 04
71. (A) 13.4431 (B) 0.0089 (C) 16.0595 (D) 0.1514
73. 1.0792 **75.** 1.4595 **77.** 30.6589 **79.** 18.3559

81. Increasing: $(0, \infty)$

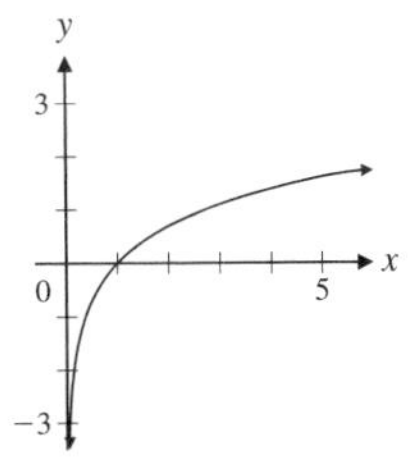

83. Decreasing: $(0, 1]$
Increasing: $[1, \infty)$

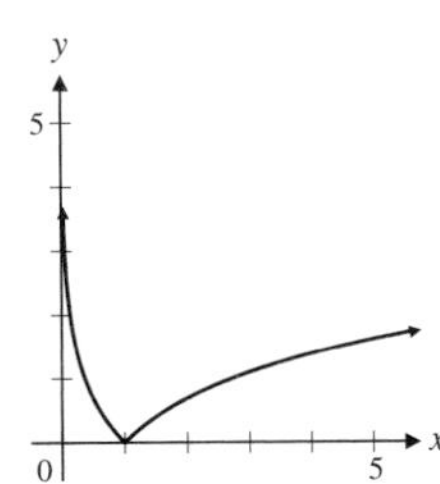

85. Increasing: $(-2, \infty)$

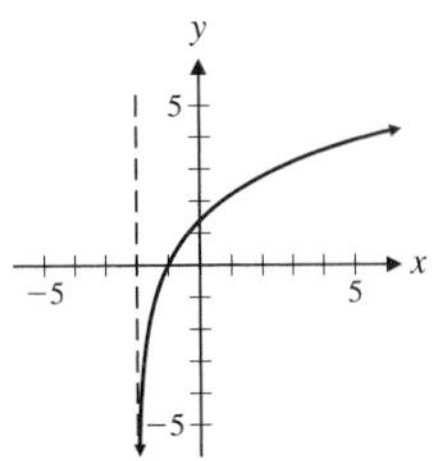

87. Increasing: $(0, \infty)$

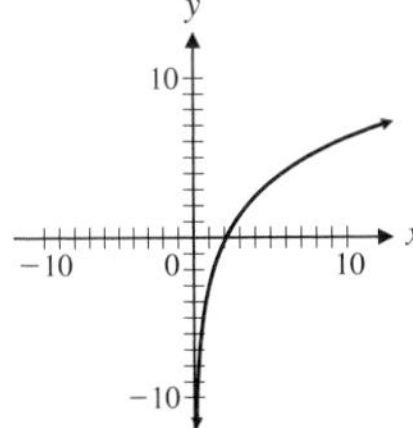

89. The calculator interprets log 13/7 as $\dfrac{\log 13}{7}$, not as $\log \frac{13}{7}$. **91.** Because $b^0 = 1$ for any permissible base b ($b > 0, b \neq 1$).

93. $y = c \cdot 10^{0.8x}$ **95.** $x > \sqrt{x} > \ln x$ for $1 < x \leq 16$ **97.** 12 yr **99.** 9.87 yr; 9.80 yr
101. 8.664% compounded continuously

103. (A) 5,373 (B) 7,220

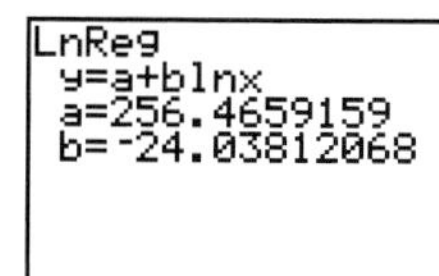

LnReg
y=a+blnx
a=-127.8085281
b=20.01315349

107. (A) 1996: 123.1 bushels/acre; 2010: 141.5 bushels/acre

LnReg
y=a+blnx
a=-492.3028128
b=134.8088483

109. 901 yr

Review Exercise

1. $v = \ln u$ *(2-3)* **2.** $y = \log x$ *(2-3)* **3.** $M = e^N$ *(2-3)* **4.** $u = 10^v$ *(2-3)* **5.** 5^{2x} *(2-2)* **6.** e^{2u^2} *(2-2)*
7. $x = 9$ *(2-3)* **8.** $x = 6$ *(2-3)* **9.** $x = 4$ *(2-3)* **10.** $x = 2.157$ *(2-3)* **11.** $x = 13.128$ *(2-3)*
12. $x = 1{,}273.503$ *(2-3)* **13.** $x = 0.318$ *(2-3)* **14.** (A) 3 (B) 2 (C) 3 (D) 1 (E) 1 (F) 1 *(2-1)*
15. (A) 4 (B) 3 (C) 4 (D) 0 (E) 1 (F) 1 *(2-1)* **16.** (A) 2 (B) 3 (C) Positive *(2-1)*
17. (A) 3 (B) 4 (C) Negative *(2-1)*
18. (A) x intercept: -4; y intercept: -2 (B) All real numbers, except $x = 2$ (C) Vertical asymptote: $x = 2$; Horizontal asymptote: $y = 1$

(D) (E) *(2-1)*

19. (A) x intercept: $\frac{4}{3}$; y intercept: -2 (B) All real numbers, except $x = -2$ (C) Vertical asymptote: $x = -2$; Horizontal asymptote: $y = 3$

(D) (E) *(2-1)*

20. $x = 8$ *(2-3)* **21.** $x = 3$ *(2-3)* **22.** $x = 3$ *(2-2)* **23.** $x = -1, 3$ *(2-2)* **24.** $x = 0, \frac{3}{2}$ *(2-2)* **25.** $x = -2$ *(2-3)*
26. $x = \frac{1}{2}$ *(2-3)* **27.** $x = 27$ *(2-3)* **28.** $x = 13.3113$ *(2-3)* **29.** $x = 158.7552$ *(2-3)* **30.** $x = 0.0097$ *(2-3)*
31. $x = 1.4359$ *(2-3)* **32.** $x = 1.4650$ *(2-3)* **33.** $x = 92.1034$ *(2-3)* **34.** $x = 9.0065$ *(2-3)* **35.** $x = 2.1081$ *(2-3)*
36. $x = 2.8074$ *(2-3)* **37.** $x = -1.0387$ *(2-3)* **38.** They look very much alike. *(2-1)*

39. (A) (B) *(2-1)*

40. $x = -1.14, 6.78$ *(2-1)* **41.** $(-1.54, -0.79)$; $(0.69, 0.99)$ *(2-2, 2-3)* **42.** $1 + 2e^x - e^{-x}$ *(2-2)* **43.** $2e^{-2x} - 2$ *(2-2)*

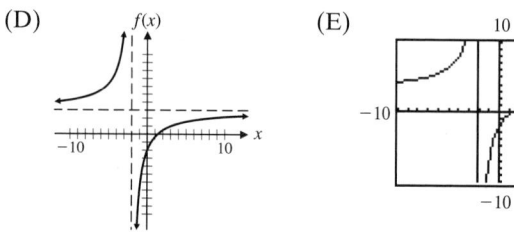

44. Increasing: $[-2, 4]$ *(2-2)* **45.** Decreasing $[0, \infty)$ *(2-2)* **46.** Increasing: $(-1, 10]$ *(2-3)*

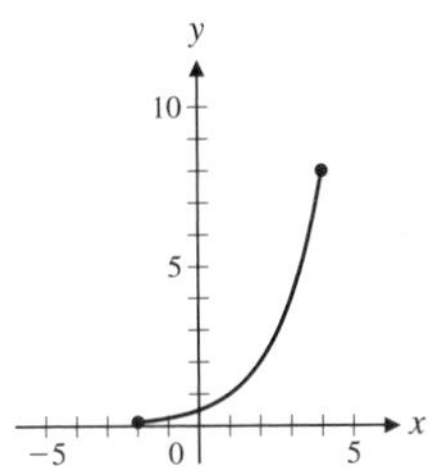

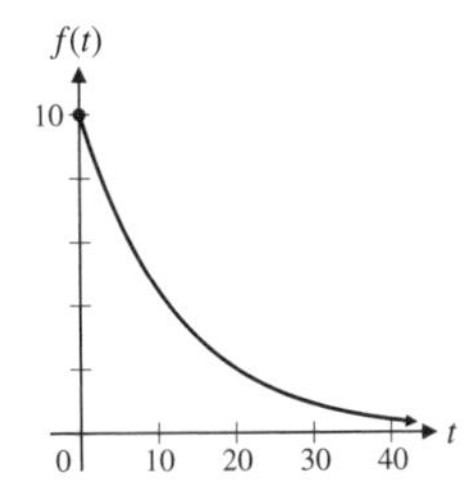

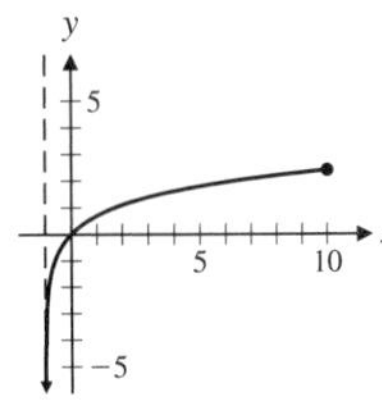

47. $\log 10^{\pi} = \pi$ and $10^{\log\sqrt{2}} = \sqrt{2}$; $\ln e^{\pi} = \pi$ and $e^{\ln\sqrt{2}} = \sqrt{2}$ *(2-3)* **48.** $x = 2$ *(2-3)* **49.** $x = 2$ *(2-3)*
50. $x = 1$ *(2-3)* **51.** $x = 300$ *(2-3)* **52.** $y = ce^{-5t}$ *(2-3)*
53. If $\log_1 x = y$, then $1^y = x$; that is, $1 = x$ for all positive real numbers x, which is not possible. *(2-3)*
54. \$10,263.65 *(2-2)* **55.** \$10,272.17 *(2-2)* **56.** 8 yr *(2-2, 2-3)* **57.** 6.93 yr *(2-2, 2-3)*

58. (A) $C(x) = 40x + 300$; $\overline{C}(x) = \dfrac{40x + 300}{x}$

(B)

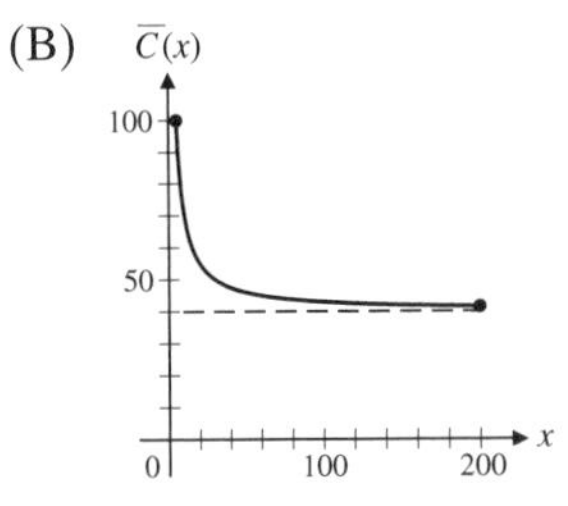

(C) $y = 40$ (D) \$40 per pair *(2-1)*

59. (A) $\overline{C}(x) = \dfrac{20x^3 - 360x^2 + 2{,}300x - 1{,}000}{x}$

(B)

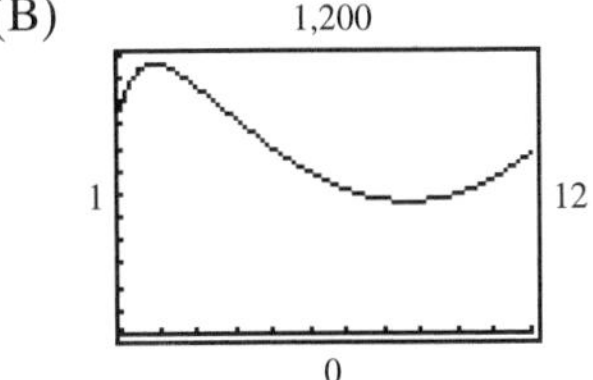

(C) 8.667 thousand cases (8,667)
(D) \$567 per case *(2-1)*

60. (A) 2,833 sets (B) 4,836

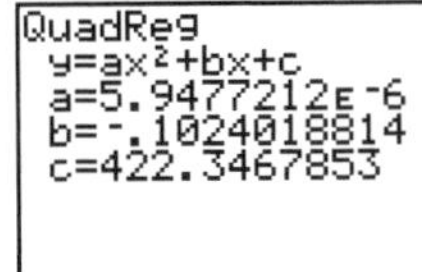
QuadReg
y=ax²+bx+c
a=5.9477212E-6
b=-.1024018814
c=422.3467853

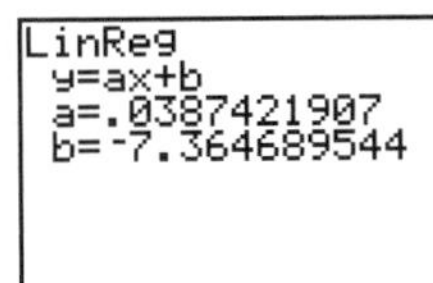
LinReg
y=ax+b
a=.0387421907
b=-7.364689544

(D) Equilibrium price: \$131.59; Equilibrium quantity: 3,587 cookware sets *(2-1)*
61. (A) $N = 2^{2t}$ or $N = 4^t$ (B) 15 days *(2-2, 2-3)* **62.** $k = 0.009\ 42$; 489 ft *(2-2, 2-3)*
63. (A) 1996: 1,724 million bushels; 2010: 2,426 million bushels. *(2-3)*

LnReg
y=a+blnx
a=-21796.9294
b=5153.244133

64. 23.4 yr *(2-2, 2-3)* **65.** 23.1 yr *(2-2, 2-3)* **66.** (A) 1996: \$207 billion; 2010: \$886 billion (B) Midway through 2004 *(2-2)*

ExpReg
y=a*b^x
a=39.23474084
b=1.109502846

CHAPTER 3

Exercise 3-1

1. 0.095; $\frac{1}{6}$ yr **3.** 18%; $\frac{5}{12}$ yr **5.** $I = \$20$ **7.** $r = 0.08$ or 8% **9.** $A = \$112$ **11.** $P = \$888.89$
13. $r = I/Pt$ **15.** $P = A/(1 + rt)$
17. The graphs are linear, all with y intercept \$1,000; their slopes are 40, 80, and 120, respectively. **19.** \$140 **21.** \$9.23
23. \$7,685.00 **25.** 10.125% **27.** 18% **29.** \$1,680; 36% **31.** 9.126% **33.** \$9,693.91 **35.** 13.538%
37. 21.335% **39.** 11.161%

Exercise 3-2

1. $A = \$112.68$ **3.** $A = \$3,433.50$ **5.** $P = \$2,419.99$ **7.** $P = \$7,351.04$ **9.** 0.75% per month
11. 1.75% per quarter **13.** 9.6% compounded monthly **15.** 9% compounded semiannually
17. (A) \$126.25; \$26.25 (B) \$126.90; \$26.90 (C) \$127.05; \$27.05 **19.** (A) \$7,147.51 (B) \$10,217.39
21. All three graphs are increasing, curve upward, and have the same y intercept; the greater the interest rate, the greater the increase. The amounts at the end of 8 years are \$1,376.40, \$1,892.46, and \$2,599.27, respectively.

23.

PERIOD	INTEREST	AMOUNT
0		\$1,000.00
1	\$97.50	\$1,097.50
2	\$107.01	\$1,204.51
3	\$117.44	\$1,321.95
4	\$128.89	\$1,450.84
5	\$141.46	\$1,592.29
6	\$155.25	\$1,747.54

25. (A) \$6,755.64 (B) \$4,563.87
27. (A) 10.38% (B) 12.68%
29. $5\frac{1}{2}$ yr
31. $n \approx 12$
33. (A) $7\frac{1}{4}$ yr (B) 6 yr
35. \$22,702.60
37. \$196,993.25
39. \$14.26/ft^2/mo
41. 18 yr

43. 9% compounded monthly, since its effective rate is 9.38%; the effective rate of 9.3% compounded annually is 9.3%
45. (A) In 1998, 222 years after the signing, it would be worth \$76,139.26.
(B) If interest were compounded monthly, daily, or continuously, it would be worth \$77,409.05, \$78,033.73, or \$78,055.09, respectively.
(C)

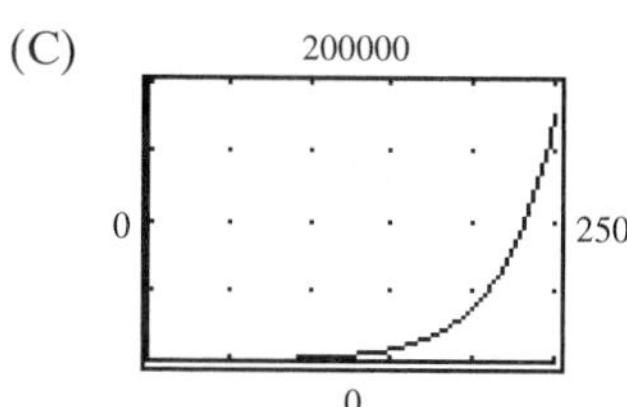

47. 2 yr, 10 mo **49.** \$328,791.70 **51.** 4.952 yr; 4.959 yr

53.

YEARS	EXACT RATE	RULE OF 72
6	12.2	12.0
7	10.4	10.3
8	9.1	9.0
9	8.0	8.0
10	7.2	7.2
11	6.5	6.5
12	5.9	6.0

55. 23 quarters

57. To maximize earnings, choose 10% simple interest for investments lasting less than 11 years and 7% compound interest for investments lasting 11 years or more.
59. 7.14% **61.** $5,935.34 **63.** 9.08% **65.** (A) 8.60% (B) 8.60% (C) 8.57% **67.** 14.48% **69.** 7.95%

Exercise 3-3

1. $FV = \$13{,}435.19$ **3.** $FV = \$60{,}401.98$ **5.** $PMT = \$123.47$ **7.** $PMT = \$310.62$ **9.** $n = 17$ **11.** $i = 0.09$
13. Value: $30,200.99; Interest: $10,200.99 **15.** $20,931.01 **17.** $331.46 **19.** $625.28

21.

PERIOD	AMOUNT	INTEREST	BALANCE
1	$1,000.00	$0.00	$1,000.00
2	$1,000.00	$83.20	$2,083.20
3	$1,000.00	$173.32	$3,256.52
4	$1,000.00	$270.94	$4,527.46
5	$1,000.00	$376.69	$5,904.15

23. First year: $50.76; second year: $168.09; third year: $296.42
25. (A) $413,092 (B) $393,965
27. $177.46; $1,481.92
29. 3 yr, 5 mo
31. 11.58%
33. 7.13%
35. After 11 quarterly payments

Exercise 3-4

1. $PV = \$3{,}458.41$ **3.** $PV = \$4{,}606.09$ **5.** $PMT = \$199.29$ **7.** $PMT = \$586.01$ **9.** $n = 29$ **11.** $i = 0.029$
13. $109,421.92 **15.** $14,064.67; $16,800.00 **17.** (A) $36.59/mo; $58.62 interest (B) $38.28/mo; $89.04 interest
19. $273.69/mo; $7,705.68 interest

21.

PAYMENT NUMBER	PAYMENT	INTEREST	UNPAID BALANCE REDUCTION	UNPAID BALANCE
0				$5,000.00
1	$ 758.05	$ 225.00	$ 533.05	4,466.95
2	758.05	201.01	557.04	3,909.91
3	758.05	175.95	582.10	3,327.81
4	758.05	149.75	608.30	2,719.51
5	758.05	122.38	635.67	2,083.84
6	758.05	93.77	664.28	1,419.56
7	758.05	63.88	694.17	725.39
8	758.03	32.64	725.39	0.00
Totals	$6,064.38	$1,064.38	$5,000.00	

23. First year interest = $625.07; Second year interest = $400.91; Third year interest = $148.46
25. $85,846.38; $128,153.62 **27.** $143.85/mo; $904.80
29. Monthly payment = $841.39
(A) $70,952.33 (B) $55,909.02 (C) $36,813.32
31. (A) Monthly payment = $395.04;
Total interest = $64,809.60
(B) 114 mo, or 9.5 yr; Interest saved = $38,375.04
33. (A) 157 (B) 243 (C) The withdrawals continue forever.
35. (A) Monthly withdrawals: $1,229.66;
Total interest: $185,338.80
(B) Monthly deposits: $162.65
37. $29,799
39. All three graphs are decreasing, curve downward, and have the same x intercept; the unpaid balances are always in the ratio 2:3:4. The monthly payments are $402.31, $603.47, and $804.62, with total interest amounting to $94,831.60, $142,249.20, and $189,663.20, respectively.
41. 14.45% **43.** 10.21%

Review Exercise

1. $A = \$104.50$ *(3-1)* **2.** $P = \$800$ *(3-1)* **3.** $t = 0.75$ yr, or 9 mo *(3-1)* **4.** $r = 6\%$ *(3-1)*
5. $A = \$1{,}393.68$ *(3-2)* **6.** $P = \$3{,}193.50$ *(3-2)* **7.** $FV = \$69{,}770.03$ *(3-3)* **8.** $PMT = \$115.00$ *(3-3)*
9. $PV = \$33{,}944.27$ *(3-4)* **10.** $PMT = \$166.07$ *(3-4)* **11.** $n \approx 16$ *(3-2)* **12.** $n \approx 41$ *(3-3)*
13. $3,350.00; $350.00 *(3-1)* **14.** $27,551.32 *(3-2)* **15.** $9,422.24 *(3-2)*

16. (A)

PERIOD	INTEREST	AMOUNT
0		\$400.00
1	\$21.60	\$421.60
2	\$22.77	\$444.37
3	\$24.00	\$468.36
4	\$25.29	\$493.65

(3-2)

(B)

PERIOD	INTEREST	PAYMENT	BALANCE
1		\$100.00	\$100.00
2	\$5.40	\$100.00	\$205.40
3	\$11.09	\$100.00	\$316.49
4	\$17.09	\$100.00	\$433.58

(3-3)

17. To maximize earnings, choose 13% simple interest for investments lasting less than 9 years and 9% compound interest for investments lasting 9 years or more. *(3-2)*
18. \$164,402 *(3-2)*
19. 9% compounded quarterly, since its effective rate is 9.31%, while the effective rate of 9.25% compounded annually is 9.25% *(3-2)*
20. \$27,971.23; \$8,771.23 *(3-3)* **21.** \$11.64 *(3-1)* **22.** \$10,210.25 *(3-2)* **23.** \$6,268.21 *(3-2)* **24.** 15% *(3-1)*
25. \$10,318.91; \$2,281.09 *(3-4)* **26.** 2 yr, 3 mo *(3-2)* **27.** 5 yr, 10 mo; 3 yr, 11 mo *(3-2)*
28. (A) \$571,499 (B) \$1,973,277 *(3-3)* **29.** 10.45% *(3-2)* **30.** 20% *(3-1)* **31.** \$10,988.22 *(3-4)*
32. (A) \$115,573.86 (B) \$359.64 (C) \$171,228.80 *(3-4)* **33.** \$99.85/mo; \$396.36 *(3-4)* **34.** \$526.28/mo *(3-3)*
35. 6.93 yr; 7.27 yr *(3-2)* **36.** \$175.28; \$2,516.80 *(3-4)* **37.** 5 yr, 10 mo *(3-3)* **38.** 18 yr *(3-4)* **39.** 28.8% *(3-1)*

40.

PAYMENT NUMBER	PAYMENT	INTEREST	UNPAID BALANCE REDUCTION	UNPAID BALANCE
0				\$1,000.00
1	\$ 265.82	\$25.00	\$ 240.82	759.18
2	265.82	18.98	246.84	512.34
3	265.82	12.81	253.01	259.33
4	265.81	6.48	259.33	0.00
Totals	\$1,063.27	\$63.27	\$1,000.00	

(3-4)

41. 1 yr, 1 mo *(3-2)*
42. \$55,347.48; \$185,830.24 *(3-3)*
43. West Lake S & L: 9.800%; Security S & L: 9.794% *(3-2)*
44. 8.24% *(3-1)*
45. 43 *(3-3)*
46. (A) \$1,435.63 (B) \$74,397.48 (C) \$11,625.04 *(3-4)*
47. The certificate would be worth \$81,041.86 when the 360th payment is made. By reducing the principal the loan would be paid off in 238 months. If the monthly payment were then invested at 8.75% compounded monthly, it would be worth \$105,167.89 at the time of the 360th payment. *(3-2, 3-3, 3-4)*
48. The lower rate would save \$15,485.03 in interest payments. *(3-4)* **49.** \$3,176.14 *(3-2)* **50.** 8.37% *(3-2)*
51. \$4,744.73 *(3-1)* **52.** \$6,697.11 *(3-4)* **53.** 9.38% *(3-2)* **54.** (A) \$398,807 (B) \$374,204 *(3-3)*
55. (A) \$746.79/mo; \$896.76/mo (B) \$73,558.78; \$41,482.19 *(3-4)* **56.** \$19,239 *(3-4)* **57.** 33.52% *(3-4)*
58. (A) 10.74% (B) 15 yr; 40 yr *(3-3)*

CHAPTER 4

Exercise 4-1

1. Graph (B); no solution **3.** Graph (A); $x = -3, y = 1$ **5.** $x = 2, y = 4$ **7.** No solution (parallel lines)
9. $x = 4, y = 5$ **11.** $x = 1, y = 4$ **13.** $u = 2, v = -3$ **15.** $m = 8, n = 6$ **17.** $x = 1, y = -5$
19. No solution (inconsistent) **21.** $x = -\frac{4}{3}, y = 1$ **23.** Infinitely many solutions (dependent) **25.** $x = \frac{4}{7}, x = \frac{3}{7}$
27. $x = 1.1, y = 0.3$ **29.** $x = -\frac{5}{4}, y = \frac{5}{3}$ **31.** (2.41, 1.12) **33.** (−3.31, −2.24)

35.

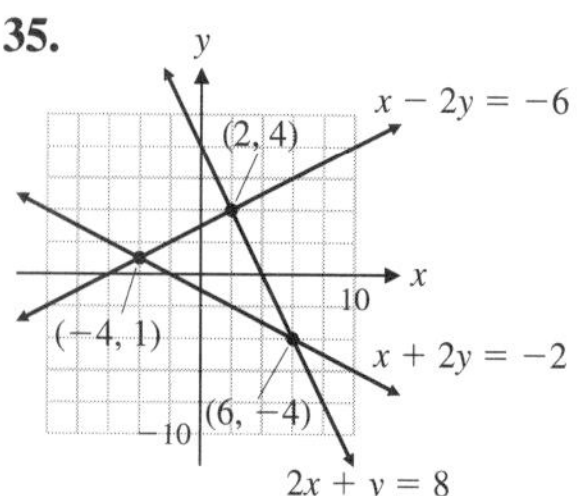

37.

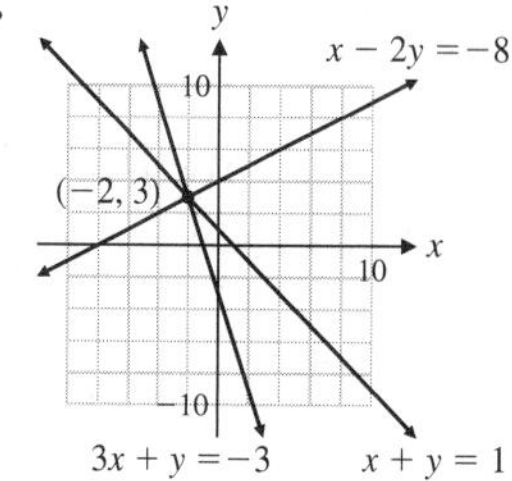

39.

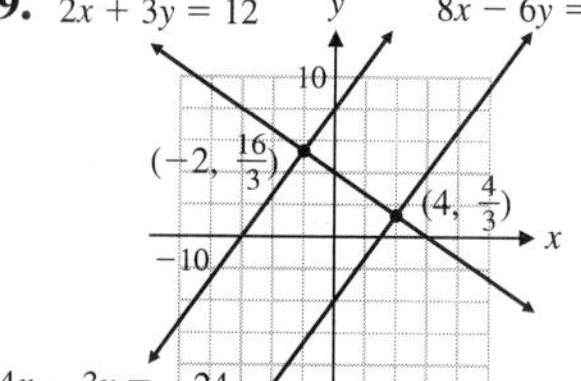

41. (A) (20, −24) (B) (−4, 6) (C) No solution

43. (A) Supply: 143 T-shirts; Demand: 647 T-shirts
(B) Supply: 857 T-shirts; Demand: 353 T-shirts
(C) Equilibrium price = \$6.50; Equilibrium quantity = 500 T-shirts
(D)

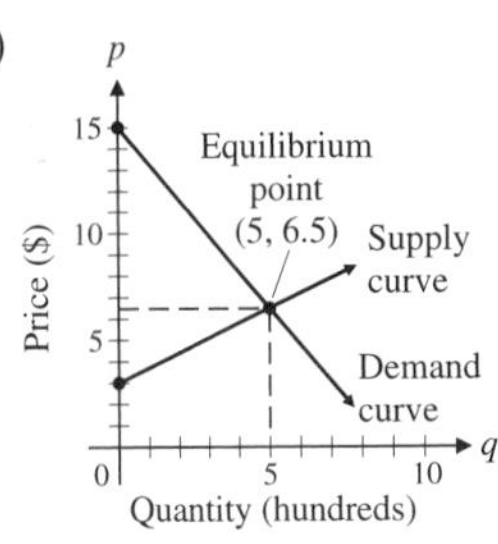

45. (A) $p = 0.001q + 0.15$
(B) $p = -0.002q + 1.74$
(C) Equilibrium price = \$0.68;
Equilibrium quantity = 530
(D)

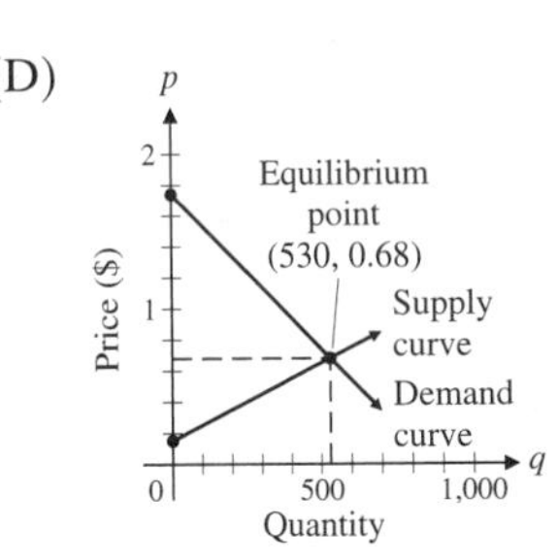

47. (A) For $x = 120$ units,
Cost = \$216,000 = Revenue
(B)

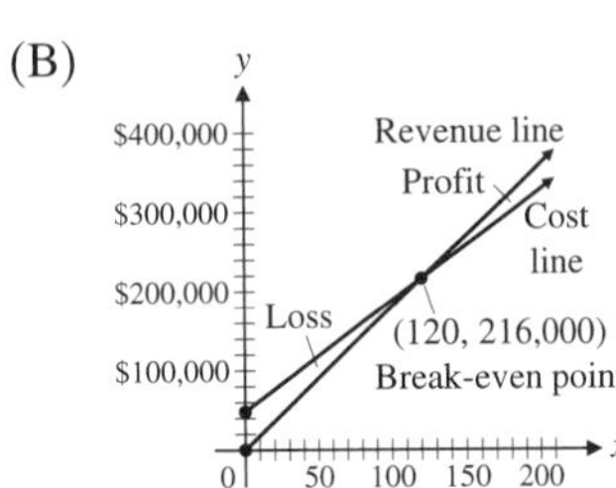

49. (A) For $x = 1{,}920$ tapes,
Cost = \$38,304 = Revenue
(B)

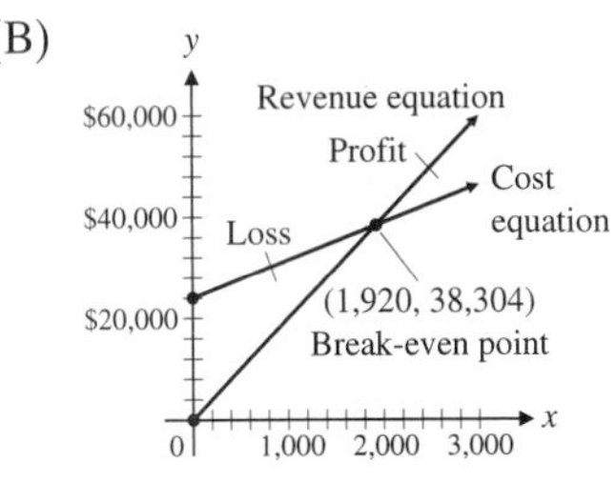

51. Base price = \$17.95; Surcharge = \$2.45/lb
53. 5,720 lb robust blend; 6,160 lb mild blend
55. Mix *A:* 80 g; Mix *B:* 60 g

57. (A)

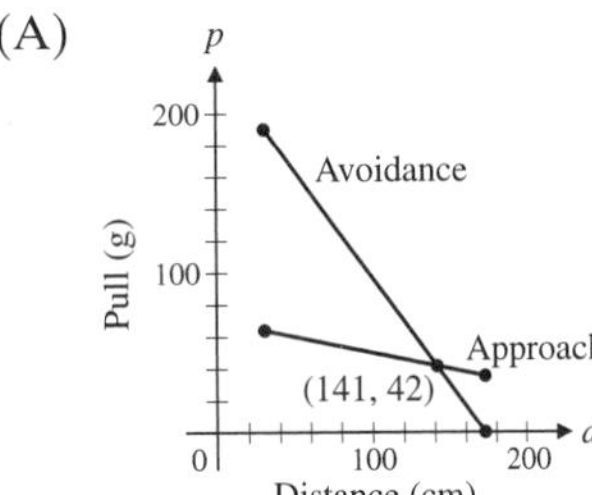

(B) $d = 141$ cm (approx.)
(C) Vacillate

Exercise 4-2

1. A is 2×3; C is 1×3 **3.** C **5.** B **7.** $a_{12} = -4$; $a_{23} = -5$ **9.** $-1, 8, 0$
11. (A) 2×4 (B) 1 (C) $e_{23} = 7$; $f_{12} = -6$

13. $\left[\begin{array}{cc|c} 4 & -6 & -8 \\ 1 & -3 & 2 \end{array}\right]$ **15.** $\left[\begin{array}{cc|c} -4 & 12 & -8 \\ 4 & -6 & -8 \end{array}\right]$ **17.** $\left[\begin{array}{cc|c} 1 & -3 & 2 \\ 8 & -12 & -16 \end{array}\right]$

19. $\left[\begin{array}{cc|c} 1 & -3 & 2 \\ 0 & 6 & -16 \end{array}\right]$ **21.** $\left[\begin{array}{cc|c} 1 & -3 & 2 \\ 2 & 0 & -12 \end{array}\right]$ **23.** $\left[\begin{array}{cc|c} 1 & -3 & 2 \\ 3 & -3 & -10 \end{array}\right]$

25. $\frac{1}{3}R_2 \to R_2$ **27.** $6R_1 + R_2 \to R_2$ **29.** $\frac{1}{3}R_2 + R_1 \to R_1$
31. $x_1 = 3, x_2 = 2$; each pair of lines has the same intersection point.

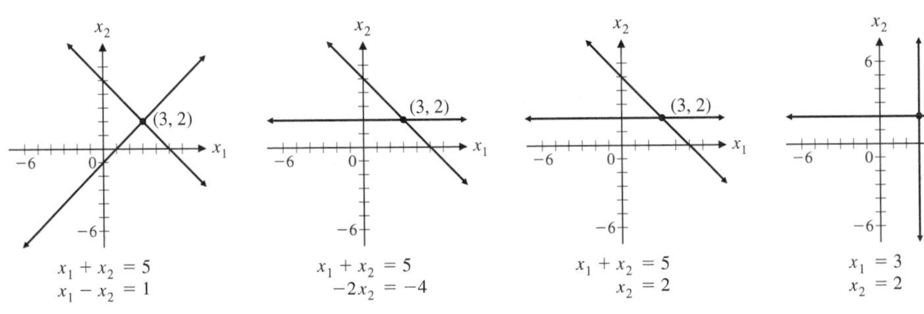

33. $x_1 = 3, x_2 = 1$
35. $x_1 = 2, x_2 = 1$
37. $x_1 = 2, x_2 = 4$
39. No solution
41. $x_1 = 1, x_2 = 4$

43. Infinitely many solutions: $x_2 = s$, $x_1 = 2s - 3$ for any real number s
45. Infinitely many solutions; $x_2 = s$, $x_1 = \frac{1}{2}s + \frac{1}{2}$ for any real number s
47. $x_1 = -1, x_2 = 3$ **49.** No solution **51.** Infinitely many solutions: $x_2 = t$, $x_1 = \frac{3}{2}t + 2$ for any real number t
53. $x_1 = 2, x_2 = -1$ **55.** $x_1 = 2, x_2 = -1$ **57.** $x_1 = 1.1, x_2 = 0.3$ **59.** $x_1 = -23.125, x_2 = 7.8125$
61. $x_1 = 3.225, x_2 = -6.9375$

Exercise 4-3

1. Reduced form **3.** Not reduced form; $R_2 \leftrightarrow R_3$ **5.** Not reduced form; $\frac{1}{3}R_2 \rightarrow R_2$
7. Reduced form **9.** Not reduced form; $2R_2 + R_1 \rightarrow R_1$ **11.** $x_1 = -2, x_2 = 3, x_3 = 0$
13. $x_1 = 2t + 3, x_2 = -t - 5, x_3 = t$ for t any real number **15.** No solution
17. $x_1 = 2s + 3t - 5, x_2 = s, x_3 = -3t + 2, x_4 = t$ for s and t any real numbers **19.** $\left[\begin{array}{cc|c} 1 & 0 & -7 \\ 0 & 1 & 3 \end{array}\right]$

21. $\left[\begin{array}{ccc|c} 1 & 0 & 0 & -5 \\ 0 & 1 & 0 & 4 \\ 0 & 0 & 1 & -2 \end{array}\right]$ **23.** $\left[\begin{array}{ccc|c} 1 & 0 & 2 & -\frac{5}{3} \\ 0 & 1 & -2 & \frac{1}{3} \\ 0 & 0 & 0 & 0 \end{array}\right]$ **25.** $x_1 = -2, x_2 = 3, x_3 = 1$ **27.** $x_1 = 0, x_2 = -2, x_3 = 2$

29. $x_1 = 2t + 3, x_2 = t - 2, x_3 = t$ for t any real number **31.** $x_1 = 1, x_2 = 2$ **33.** No solution
35. $x_1 = t - 1, x_2 = 2t + 2, x_3 = t$ for t any real number **37.** $x_1 = -2s + t + 1, x_2 = s, x_3 = t$ for s and t any real numbers
39. No solution **41.** $x_1 = 2.5t - 4, x_2 = t, x_3 = -5$ for t any real number **43.** $x_1 = 1, x_2 = -2, x_3 = 1$
45. (A) Dependent system with two parameters and an infinite number of solutions
(B) Dependent system with one parameter and an infinite number of solutions
(C) Independent system with a unique solution (D) Impossible
47. The system has no solution for $k = -3$ and a unique solution for all other values of k.
49. The system has an infinite number of solutions for $k = 3$ and a unique solution for all other values of k.
51. $x_1 = 2s - 3t + 3, x_2 = s + 2t + 2, x_3 = s, x_4 = t$ for s and t any real numbers
53. $x_1 = -0.5, x_2 = 0.2, x_3 = 0.3, x_4 = -0.4$
55. $x_1 = 2s - 1.5t + 1, x_2 = s, x_3 = -t + 1.5, x_4 = 0.5t - 0.5, x_5 = t$ for s and t any real numbers

57. (A) x_1 = Number of one-person boats
x_2 = Number of two-person boats
x_3 = Number of four-person boats
$$\begin{aligned} 0.5x_1 + x_2 + 1.5x_3 &= 380 \\ 0.6x_1 + 0.9x_2 + 1.2x_3 &= 330 \\ 0.2x_1 + 0.3x_2 + 0.5x_3 &= 120 \end{aligned}$$
20 one-person boats, 220 two-person boats, and 100 four-person boats

(B)
$$\begin{aligned} 0.5x_1 + x_2 + 1.5x_3 &= 380 \\ 0.6x_1 + 0.9x_2 + 1.2x_3 &= 330 \end{aligned}$$
$t - 80$ one-person boats, $420 - 2t$ two-person boats, and t four-person boats, where t is an integer satisfying $80 \leqslant t \leqslant 210$

(C)
$$\begin{aligned} 0.5x_1 + x_2 &= 380 \\ 0.6x_1 + 0.9x_2 &= 330 \\ 0.2x_1 + 0.3x_2 &= 120 \end{aligned}$$
There is no production schedule that will use all the labor-hours in all departments.

59. x_1 = Number of 6,000 gal tank cars
x_2 = Number of 8,000 gal tank cars
x_3 = Number of 18,000 gal tank cars
$$\begin{aligned} x_1 + x_2 + x_3 &= 24 \\ 6{,}000x_1 + 8{,}000x_2 + 18{,}000x_3 &= 250{,}000 \end{aligned}$$
$5t - 29$ 6,000 gal tank cars, $53 - 6t$ 8,000 gal tank cars, and t 18,000 gal tank cars, where $t = 6, 7$, or 8

61. x_1 = Federal income tax
x_2 = State income tax
x_3 = Local income tax
$$\begin{aligned} x_1 + 0.5x_2 + 0.5x_3 &= 3{,}825{,}000 \\ 0.2x_1 + x_2 + 0.2x_3 &= 1{,}530{,}000 \\ 0.1x_1 + 0.1x_2 + x_3 &= 765{,}000 \end{aligned}$$
Tax liability is 57.65%.

63. x_1 = Taxable income of company A
x_2 = Taxable income of company B
x_3 = Taxable income of company C
$$\begin{aligned} x_1 - 0.08x_2 - 0.03x_3 - 0.07x_4 &= 2.272 \\ -0.12x_1 + x_2 - 0.11x_3 - 0.13x_4 &= 2.106 \\ -0.11x_1 - 0.09x_2 + x_3 - 0.08x_4 &= 2.736 \\ -0.06x_1 - 0.02x_2 - 0.14x_3 + x_4 &= 3.168 \end{aligned}$$
Taxable incomes are \$2,927,000 for company A, \$3,372,000 for company B, \$3,675,000 for company C, and \$3,926,000 for company D.

65. (A) x_1 = Number of ounces of food A
x_2 = Number of ounces of food B
x_3 = Number of ounces of food C
$$\begin{aligned} 30x_1 + 10x_2 + 20x_3 &= 340 \\ 10x_1 + 10x_2 + 20x_3 &= 180 \\ 10x_1 + 30x_2 + 20x_3 &= 220 \end{aligned}$$
8 oz of food A, 2 oz of food B, and 4 oz of food C

(B)
$$\begin{aligned} 30x_1 + 10x_2 &= 340 \\ 10x_1 + 10x_2 &= 180 \\ 10x_1 + 30x_2 &= 220 \end{aligned}$$
There is no combination that will meet all the requirements.

(C)
$$\begin{aligned} 30x_1 + 10x_2 + 20x_3 &= 340 \\ 10x_1 + 10x_2 + 20x_3 &= 180 \end{aligned}$$
8 oz of food A, $10 - 2t$ oz of food B, and t oz of food C, where $0 \leqslant t \leqslant 5$

67. x_1 = Number of barrels of mix A
x_2 = Number of barrels of mix B
x_3 = Number of barrels of mix C
x_4 = Number of barrels of mix D

$$30x_1 + 30x_2 + 30x_3 + 60x_4 = 900$$
$$50x_1 + 75x_2 + 25x_3 + 25x_4 = 750$$
$$30x_1 + 20x_2 + 20x_3 + 50x_4 = 700$$

$10 - t$ barrels of mix A, $t - 5$ barrels of mix B, $25 - 2t$ barrels of mix C, and t barrels of mix D, where t is an integer satisfying $5 \leqslant t \leqslant 10$

69. x_1 = Number of hours for company A
x_2 = Number of hours for company B
$30x_1 + 20x_1 = 600$
$10x_1 + 20x_1 = 400$
Company A: 10 hr; company B: 15 hr

71. (A) 6th St. and Washington Ave.: $x_1 + x_2 = 1{,}200$; 6th St. and Lincoln Ave.: $x_2 + x_3 = 1{,}000$; 5th St. and Lincoln Ave.: $x_3 + x_4 = 1{,}300$
(B) $x_1 = 1{,}500 - t$, $x_2 = t - 300$, $x_3 = 1{,}300 - t$, and $x_4 = t$, where $300 \leqslant t \leqslant 1{,}300$
(C) 1,300; 300 (D) Washington Ave.: 500; 6th St.: 700; Lincoln Ave.: 300

Exercise 4-4

1. $\begin{bmatrix} -1 & 0 \\ 5 & -3 \end{bmatrix}$ **3.** Not defined **5.** $\begin{bmatrix} 5 & -7 \\ -5 & 2 \\ 0 & 4 \end{bmatrix}$ **7.** $\begin{bmatrix} 5 & -10 & 0 & 20 \\ -15 & 10 & -5 & 30 \end{bmatrix}$ **9.** $\begin{bmatrix} 5 \\ -3 \end{bmatrix}$

11. $\begin{bmatrix} 2 & 4 \\ 1 & -5 \end{bmatrix}$ **13.** $\begin{bmatrix} 1 & -5 \\ -2 & -4 \end{bmatrix}$ **15.** $[-7]$ **17.** $\begin{bmatrix} -15 & 6 \\ -20 & 8 \end{bmatrix}$ **19.** $[11]$ **21.** $\begin{bmatrix} 3 & -2 & -4 \\ 6 & -4 & -8 \\ -9 & 6 & 12 \end{bmatrix}$

23. $\begin{bmatrix} -12 & 12 & 18 \\ 20 & -18 & -6 \end{bmatrix}$ **25.** Not defined **27.** $\begin{bmatrix} 11 & 2 \\ 4 & 27 \end{bmatrix}$ **29.** $\begin{bmatrix} 6 & 4 \\ 0 & -3 \end{bmatrix}$ **31.** $\begin{bmatrix} -1.3 & -0.7 \\ -0.2 & -0.5 \\ 0.1 & 1.1 \end{bmatrix}$

33. $\begin{bmatrix} -66 & 69 & 39 \\ 92 & -18 & -36 \end{bmatrix}$ **35.** Not defined **37.** $\begin{bmatrix} -18 & 48 \\ 54 & -34 \end{bmatrix}$ **39.** $\begin{bmatrix} -26 & -15 & -25 \\ -4 & -18 & 4 \\ 2 & 43 & -19 \end{bmatrix}$

41. B^n approaches $\begin{bmatrix} 0.25 & 0.75 \\ 0.25 & 0.75 \end{bmatrix}$; AB^n approaches $[0.25 \quad 0.75]$ **43.** $a = -1, b = 1, c = 3, d = -5$ **45.** $x = 2, y = -3$

47. $x = 3, y = 2$ **49.** $a = 3, b = 4, c = 1, d = 2$ **51.** All are true.

53.

	Guitar	Banjo
Materials	\$33	\$26
Labor	\$57	\$77

55. Markup

	Basic car	Air	AM/FM radio	Cruise control
Model A	\$1,180	\$82	\$54	\$30
Model B	\$1,075	\$98	\$81	\$30
Model C	\$1,715	\$106	\$108	\$36

57. (A) \$11.80 (B) \$30.30 (C) MN gives the labor costs per boat at each plant.

(D) MN =

	Plant I	Plant II
One-person boat	\$11.80	\$13.80
Two-person boat	\$18.50	\$21.60
Four-person boat	\$26.00	\$30.30

59. (A) $A^2 = \begin{bmatrix} 0 & 0 & 2 & 0 & 0 \\ 1 & 0 & 0 & 0 & 1 \\ 0 & 1 & 0 & 2 & 0 \\ 1 & 0 & 0 & 0 & 1 \\ 0 & 0 & 1 & 0 & 0 \end{bmatrix}$

There is one way to travel from Baltimore to Atlanta with one intermediate connection; there are two ways to travel from Atlanta to Chicago with one intermediate connection. In general, the elements in A^2 indicate the number of different ways to travel from the ith city to the jth city with one intermediate connection.

(B) $A^3 = \begin{bmatrix} 2 & 0 & 0 & 0 & 2 \\ 0 & 1 & 0 & 2 & 0 \\ 0 & 0 & 3 & 0 & 0 \\ 0 & 1 & 0 & 2 & 0 \\ 1 & 0 & 0 & 0 & 1 \end{bmatrix}$

There is one way to travel from Denver to Baltimore with two intermediate connections; there are two ways to travel from Atlanta to El Paso with two intermediate connections. In general, the elements in A^3 indicate the number of different ways to travel from the ith city to the jth city with two intermediate connections.

(C) $A + A^2 + A^3 + A^4 = \begin{bmatrix} 2 & 3 & 2 & 5 & 2 \\ 1 & 1 & 4 & 2 & 1 \\ 4 & 1 & 3 & 2 & 4 \\ 1 & 1 & 4 & 2 & 1 \\ 1 & 1 & 1 & 3 & 1 \end{bmatrix}$

It is possible to travel from any origin to any destination with at most three intermediate connections.

61. (A) 70 g (B) 30 g
(C) MN gives the amount (in grams) of protein, carbohydrate, and fat in 20 oz of each mix.

(D) $MN = \begin{bmatrix} 70 & 60 & 50 \\ 380 & 360 & 340 \\ 50 & 40 & 30 \end{bmatrix}$ (columns: Mix X, Mix Y, Mix Z; rows: Protein, Carbohydrate, Fat)

63. (A) \$2,650 (B) \$4,400
(C) NM gives the total cost per town.

(D) $NM = \begin{bmatrix} \$2{,}650 \\ \$4{,}400 \end{bmatrix}$ (Cost per town; rows: Berkeley, Oakland)

(E) $[1 \;\; 1]N = [3{,}000 \;\; 1{,}300 \;\; 13{,}000]$ (Telephone call, House call, Letter)

(F) $N\begin{bmatrix} 1 \\ 1 \\ 1 \end{bmatrix} = \begin{bmatrix} 6{,}500 \\ 10{,}800 \end{bmatrix}$ (Total contacts; rows: Berkeley, Oakland)

65. (A) $\begin{bmatrix} 0 & 0 & 1 & 1 & 1 & 0 \\ 1 & 0 & 0 & 1 & 1 & 0 \\ 0 & 1 & 0 & 1 & 0 & 0 \\ 0 & 0 & 0 & 0 & 0 & 1 \\ 0 & 0 & 1 & 1 & 0 & 1 \\ 1 & 1 & 1 & 0 & 0 & 0 \end{bmatrix}$ (B) $\begin{bmatrix} 0 & 1 & 2 & 3 & 1 & 2 \\ 1 & 0 & 2 & 3 & 2 & 2 \\ 1 & 1 & 0 & 2 & 1 & 1 \\ 1 & 1 & 1 & 0 & 0 & 1 \\ 1 & 2 & 2 & 2 & 0 & 2 \\ 2 & 2 & 2 & 3 & 2 & 0 \end{bmatrix}$ (C) $BC = \begin{bmatrix} 9 \\ 10 \\ 6 \\ 4 \\ 9 \\ 11 \end{bmatrix}$ where $C = \begin{bmatrix} 1 \\ 1 \\ 1 \\ 1 \\ 1 \\ 1 \end{bmatrix}$

(D) Frank, Bart, Aaron and Elvis (tie), Charles, Dan

Exercise 4-5

1. $\begin{bmatrix} 2 & -3 \\ 4 & 5 \end{bmatrix}$ **3.** $\begin{bmatrix} 2 & -3 \\ 4 & 5 \end{bmatrix}$ **5.** $\begin{bmatrix} -2 & 1 & 3 \\ 2 & 4 & -2 \\ 5 & 1 & 0 \end{bmatrix}$ **7.** $\begin{bmatrix} -2 & 1 & 3 \\ 2 & 4 & -2 \\ 5 & 1 & 0 \end{bmatrix}$

9. Yes **11.** No **13.** Yes **15.** No **17.** Yes

19. $\begin{bmatrix} -1 & 0 \\ -3 & 1 \end{bmatrix}$ **21.** $\begin{bmatrix} 3 & -2 \\ -1 & 1 \end{bmatrix}$ **23.** $\begin{bmatrix} 7 & -3 \\ -2 & 1 \end{bmatrix}$ **25.** $\begin{bmatrix} -5 & -12 & 3 \\ -2 & -4 & 1 \\ 2 & 5 & -1 \end{bmatrix}$

27. $\begin{bmatrix} 6 & -2 & -1 \\ -5 & 2 & 1 \\ -3 & 1 & 1 \end{bmatrix}$ **29.** $\begin{bmatrix} -2 & -3 \\ 3 & 4 \end{bmatrix}$ **31.** Does not exist **33.** $\begin{bmatrix} 1.5 & -0.5 \\ -2 & 1 \end{bmatrix}$

35. $\begin{bmatrix} 1 & 2 & 2 \\ -2 & -3 & -4 \\ -1 & -2 & -1 \end{bmatrix}$ **37.** Does not exist **39.** $\begin{bmatrix} -3 & -2 & 1.5 \\ 4 & 3 & -2 \\ 3 & 2 & -1.25 \end{bmatrix}$ **41.** $\begin{bmatrix} -1.75 & -0.375 & 0.5 \\ -5.5 & -1.25 & 1 \\ 0.5 & 0.25 & 0 \end{bmatrix}$

45. M^{-1} exists if and only if all the elements on the main diagonal are nonzero.

47. $A^{-1} = A; A^2 = I$ **49.** $A^{-1} = A; A^2 = I$

51. $\begin{bmatrix} 0.5 & -0.3 & 0.85 & -0.25 \\ 0 & 0.1 & 0.05 & -0.25 \\ -1 & 0.9 & -1.55 & 0.75 \\ -0.5 & 0.4 & -1.3 & 0.5 \end{bmatrix}$ **53.** $\begin{bmatrix} 1.75 & 5.25 & 8.75 & -1 & -18.75 \\ 1.25 & 3.75 & 6.25 & -1 & -13.25 \\ -4.75 & -13.25 & -22.75 & 3 & 48.75 \\ -1.375 & -4.625 & -7.875 & 1 & 16.375 \\ 3.25 & 8.75 & 15.25 & -2 & -32.25 \end{bmatrix}$

55. 36 44 59 86 61 82 68 95 25 37 49 64 63 81 47 66 43 62

57. GONE WITH THE WIND

59. 27 19 82 9 89 50 114 101 70 123 50 100 75 72 89 63 96 99 69 99 52 105 106 56 97 59 127 110 81 132

61. THE GREATEST SHOW ON EARTH

Exercise 4-6

1. $\begin{aligned} 3x_1 + x_2 &= 5 \\ 2x_1 - x_2 &= -4 \end{aligned}$ **3.** $\begin{aligned} -3x_1 + x_2 \quad &= 3 \\ 2x_1 \quad + x_3 &= -4 \\ -x_1 + 3x_2 - 2x_3 &= 2 \end{aligned}$ **5.** $\begin{bmatrix} 3 & -4 \\ 2 & 1 \end{bmatrix}\begin{bmatrix} x_1 \\ x_2 \end{bmatrix} = \begin{bmatrix} 1 \\ 5 \end{bmatrix}$ **7.** $\begin{bmatrix} 1 & -3 & 2 \\ -2 & 3 & 0 \\ 1 & 1 & 4 \end{bmatrix}\begin{bmatrix} x_1 \\ x_2 \\ x_3 \end{bmatrix} = \begin{bmatrix} -3 \\ 1 \\ -2 \end{bmatrix}$

9. $x_1 = -8, x_2 = 2$ **11.** $x_1 = 0, x_2 = 4$ **13.** $x_1 = 3, x_2 = -2$ **15.** $x_1 = 11, x_2 = 4$

17. (A) $x_1 = -3, x_2 = 2$ (B) $x_1 = -1, x_2 = 2$ (C) $x_1 = -8, x_2 = 3$ **19.** (A) $x_1 = 17, x_2 = -5$ (B) $x_1 = 7, x_2 = -2$ (C) $x_1 = 24, x_2 = -7$ **21.** (A) $x_1 = 1, x_2 = 0, x_3 = 0$ (B) $x_1 = -7, x_2 = -2, x_3 = 3$ (C) $x_1 = 17, x_2 = 5, x_3 = -7$

23. (A) $x_1 = 8, x_2 = -6, x_3 = -2$ (B) $x_1 = -6, x_2 = 6, x_3 = 2$ (C) $x_1 = 20, x_2 = -16, x_3 = -10$ **25.** $x_1 = 2t + 2.5, x_2 = t$, t any real number **27.** No solution **29.** $x_1 = 13t + 3, x_2 = 8t + 1, x_3 = t$, t any real number

31. $X = (A - B)^{-1}C$ **33.** $X = (A + I)^{-1}C$ **35.** $X = (A + B)^{-1}(C + D)$

37. (A) $x_1 = 1, x_2 = 0$ (B) $x_1 = -2{,}000, x_2 = 1{,}000$ (C) $x_1 = 2{,}001, x_2 = -1{,}000$

39. $x_1 = 18.2, x_2 = 27.9, x_3 = -15.2$ **41.** $x_1 = 24, x_2 = 5, x_3 = -2, x_4 = 15$

43. (A) x_1 = Number of \$4 tickets sold
x_2 = Number of \$8 tickets sold
$x_1 + x_2 = 10{,}000$
$4x_1 + 8x_2 = k_1$ Return
Concert 1: 6,000 \$4 tickets and 4,000 \$8 tickets
Concert 2: 5,000 \$4 tickets and 5,000 \$8 tickets
Concert 3: 3,000 \$4 tickets and 7,000 \$8 tickets
(B) No
(C) \$40,000 + $4t$, where t is an integer satisfying $0 \leq t \leq 10{,}000$

45. x_1 = Number of hours plant A operates
x_2 = Number of hours plant B operates
$10x_1 + 8x_2 = k_1$ Number of car frames produced
$5x_1 + 8x_2 = k_2$ Number of truck frames produced
Order 1: 280 hr at plant A and 25 hr at plant B
Order 2: 160 hr at plant A and 150 hr at plant B
Order 3: 80 hr at plant A and 225 hr at plant B

47. x_1 = President's bonus
x_2 = Executive vice-president's bonus
x_3 = Associative vice-president's bonus
x_4 = Assistant vice-president's bonus

$$\begin{aligned} x_1 + 0.03x_2 + 0.03x_3 + 0.03x_4 &= 60{,}000 \\ 0.025x_1 + x_2 + 0.025x_3 + 0.025x_4 &= 50{,}000 \\ 0.02x_1 + 0.02x_2 + x_3 + 0.02x_4 &= 40{,}000 \\ 0.015x_1 + 0.015x_2 + 0.015x_3 + x_4 &= 30{,}000 \end{aligned}$$

President: \$56,600; Executive vice-president: \$47,000; Associate vice-president: \$37,400; Assistant vice-president: \$27,900

49. (A) x_1 = Number of ounces of mix A
x_2 = Number of ounces of mix B
$0.2x_1 + 0.1x_2 = k_1$ Protein
$0.02x_1 + 0.06x_2 = k_2$ Fat
Diet 1: 60 oz mix A and 80 oz mix B
Diet 2: 20 oz mix A and 60 oz mix B
Diet 3: 0 oz mix A and 100 oz mix B
(B) No

Exercise 4-7

1. 40¢ from A; 20¢ from E **3.** $\begin{bmatrix} 0.6 & -0.2 \\ -0.2 & 0.9 \end{bmatrix}; \begin{bmatrix} 1.8 & 0.4 \\ 0.4 & 1.2 \end{bmatrix}$ **5.** $X = \begin{bmatrix} x_1 \\ x_2 \end{bmatrix} = \begin{bmatrix} 16.4 \\ 9.2 \end{bmatrix}$

7. 20¢ from A; 10¢ from B; 10¢ from E **11.** Agriculture: \$18 billion; Building: \$15.6 billion; Energy: \$22.4 billion

13. $\begin{bmatrix} 1.4 & 0.4 \\ 0.6 & 1.6 \end{bmatrix}; \begin{bmatrix} 24 \\ 46 \end{bmatrix}$ **15.** $\begin{bmatrix} 1.58 & 0.24 & 0.58 \\ 0.4 & 1.2 & 0.4 \\ 0.22 & 0.16 & 1.22 \end{bmatrix}; \begin{bmatrix} 38.6 \\ 18 \\ 17.4 \end{bmatrix}$

17. (A) Agriculture: \$80 million; Manufacturing: \$64 million
(B) The final demand for agriculture increases to \$54 million and the final demand for manufacturing decreases to \$38 million.

19. The total output of the energy sector should be 75% of the total output of the mining sector.

21. Each element should be between 0 and 1, inclusive. **23.** Coal: \$28 billion; Steel: \$26 billion

25. Agriculture: \$148 million; Tourism: \$146 million

27. Agriculture: \$40.1 billion; Manufacturing: \$29.4 billion; Energy: \$34.4 billion

29. Year 1: Agriculture: \$65 billion; Energy: \$83 billion; Labor: \$71 billion; Manufacturing: \$88 billion.
Year 2: Agriculture: \$81 billion; Energy: \$97 billion; Labor: \$83 billion; Manufacturing: \$99 billion.
Year 3: Agriculture: \$117 billion; Energy: \$124 billion; Labor: \$106 billion; Manufacturing: \$120 billion.

Review Exercise

1. $x = 4, y = 4$ *(4-1)* **2.** $x = 4, y = 4$ *(4-1)*
3. (A) Not in reduced form; $R_1 \leftrightarrow R_2$ (B) Not in reduced form; $\frac{1}{3}R_2 \to R_2$
(C) Reduced form (D) Not in reduced form; $(-1)R_2 + R_1 \to R_1$ *(4-3)*
4. (A) $2 \times 5, 3 \times 2$ (B) $a_{24} = 3, a_{15} = 2, b_{31} = -1, b_{22} = 4$ (C) AB is not defined; BA is defined *(4-2, 4-4)*
5. $x_1 = 8, x_2 = 2$ *(4-6)* **6.** $\begin{bmatrix} 3 & 3 \\ 4 & 2 \end{bmatrix}$ *(4-4)* **7.** Not defined *(4-4)* **8.** $\begin{bmatrix} -3 & 0 \\ 1 & -1 \end{bmatrix}$ *(4-4)* **9.** $\begin{bmatrix} 4 & 3 \\ 7 & 4 \end{bmatrix}$ *(4-4)*
10. Not defined *(4-4)* **11.** $\begin{bmatrix} 5 \\ 5 \end{bmatrix}$ *(4-4)* **12.** $\begin{bmatrix} 2 & 3 \\ 4 & 6 \end{bmatrix}$ *(4-4)* **13.** $[8]$ *(4-4)* **14.** Not defined *(4-4)*
15. $\begin{bmatrix} -2 & 3 \\ 3 & -4 \end{bmatrix}$ *(4-5)* **16.** $x_1 = 9, x_2 = -11$ *(4-1)* **17.** $x_1 = 9, x_2 = -11$ *(4-2)*
18. $x_1 = 9, x_2 = -11; x_1 = 16, x_2 = -19; x_1 = -2, x_2 = 4$ *(4-6)* **19.** Not defined *(4-4)* **20.** $\begin{bmatrix} 10 & -8 \\ 4 & 6 \end{bmatrix}$ *(4-4)*
21. $\begin{bmatrix} -2 & 8 \\ 8 & 6 \end{bmatrix}$ *(4-4)* **22.** $\begin{bmatrix} -2 & -1 & -3 \\ 4 & 2 & 6 \\ 6 & 3 & 9 \end{bmatrix}$ *(4-4)* **23.** $[9]$ *(4-4)* **24.** $\begin{bmatrix} 10 & -5 & 1 \\ -1 & -4 & -5 \\ 1 & -7 & -2 \end{bmatrix}$ *(4-4)*
25. $\begin{bmatrix} -\frac{5}{2} & 2 & -\frac{1}{2} \\ 1 & -1 & 1 \\ \frac{1}{2} & 0 & -\frac{1}{2} \end{bmatrix}$ *(4-5)* **26.** (A) $x_1 = 2, x_2 = 1, x_3 = -1$
(B) $x_1 = -5t - 12, x_2 = 3t + 7, x_3 = t$ for t any real number *(4-3)*
27. $x_1 = 2, x_2 = 1, x_3 = -1; x_1 = 1, x_2 = -2, x_3 = 1; x_1 = -1, x_2 = 2, x_3 = -2$ *(4-6)*
28. The system has an infinite number of solutions for $k = 3$ and a unique solution for any other value of k. *(4-3)*
29. $(I - M)^{-1} = \begin{bmatrix} 1.4 & 0.3 \\ 0.8 & 1.6 \end{bmatrix}; X = \begin{bmatrix} 48 \\ 56 \end{bmatrix}$ *(4-7)* **30.** $x = 3.46, y = 1.69$ *(4-1)* **31.** $\begin{bmatrix} -0.9 & -0.1 & 5 \\ 0.8 & 0.2 & -4 \\ 0.1 & -0.1 & 0 \end{bmatrix}$ *(4-5)*
32. $x_1 = 1{,}400, x_2 = 3{,}200, x_3 = 2{,}400$ *(4-6)* **33.** $x_1 = 1{,}400, x_2 = 3{,}200, x_3 = 2{,}400$ *(4-3)*
34. $(I - M)^{-1} = \begin{bmatrix} 1.3 & 0.4 & 0.7 \\ 0.2 & 1.6 & 0.3 \\ 0.1 & 0.8 & 1.4 \end{bmatrix}; X = \begin{bmatrix} 81 \\ 49 \\ 62 \end{bmatrix}$ *(4-7)*
35. (A) Unique solution
(B) Either no solution or an infinite number of solutions *(4-6)*
36. (A) Unique solution
(B) No solution
(C) Infinite number of solutions *(4-3)*
37. (B) is the only correct solution. *(4-6)*
38. (A) $C = 243{,}000 + 22.45x; R = 59.95x$
(B) $x = 6{,}480$ machines; $R = C = \$388{,}476$
(C) Profit occurs if $x > 6{,}480$;
loss occurs if $x < 6{,}480$.

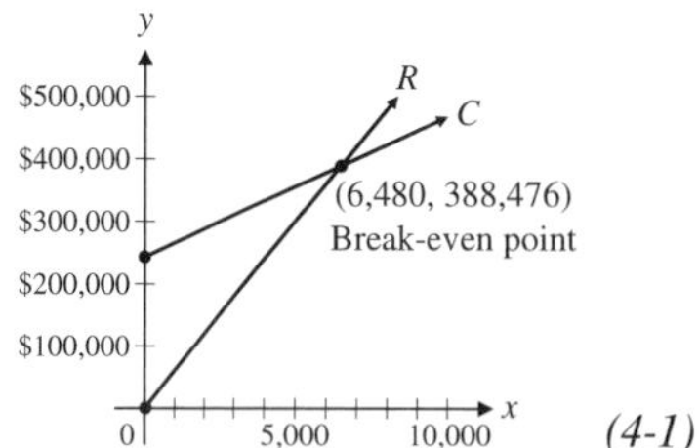

(4-1)

39. x_1 = Number of tons of ore A
x_2 = Number of tons of ore B
$0.01x_1 + 0.02x_2 = 4.5$
$0.02x_1 + 0.05x_2 = 10$
$x_1 = 250$ tons of ore A,
$x_2 = 100$ tons of ore B *(4-3)*

40. (A) $\overset{X}{\begin{bmatrix} x_1 \\ x_2 \end{bmatrix}} = \overset{A^{-1}}{\begin{bmatrix} 500 & -200 \\ -200 & 100 \end{bmatrix}} \begin{bmatrix} 4.5 \\ 10 \end{bmatrix} = \begin{bmatrix} 250 \\ 100 \end{bmatrix}$
$x_1 = 250$ tons of ore A, $x_2 = 100$ tons of ore B
(B) $\overset{X}{\begin{bmatrix} x_1 \\ x_2 \end{bmatrix}} = \overset{A^{-1}}{\begin{bmatrix} 500 & -200 \\ -200 & 100 \end{bmatrix}} \begin{bmatrix} 2.3 \\ 5 \end{bmatrix} = \begin{bmatrix} 150 \\ 40 \end{bmatrix}$ *(4-6)*
$x_1 = 150$ tons of ore A, $x_2 = 40$ tons of ore B

41. x_1 = Number of model A trucks purchased
x_2 = Number of model B trucks purchased
x_3 = Number of model C trucks purchased

$$\begin{aligned} x_1 + \quad x_2 + \quad x_3 &= 12 \\ 18{,}000x_1 + 22{,}000x_2 + 30{,}000x_3 &= 300{,}000 \end{aligned}$$

$2t - 9$ model A trucks, $21 - 3t$ model B trucks, and t model C trucks, where $t = 5, 6,$ or 7 *(4-3)*

42. (A) Elements in MN give the cost of materials for each alloy from each supplier.

(B) $MN = \begin{bmatrix} \$7{,}620 & \$7{,}530 \\ \$13{,}880 & \$13{,}930 \end{bmatrix}$ (columns: Supplier A, Supplier B; rows: Alloy 1, Alloy 2)

(C) $[1 \quad 1]MN = [\$21{,}500 \quad \$21{,}460]$ (Supplier A, Supplier B) Total material costs *(4-4)*

43. (A) \$6.35
(B) Elements in MN give the total labor costs for each calculator at each plant.
(C) $MN = \begin{bmatrix} \$3.65 & \$3.00 \\ \$6.35 & \$5.20 \end{bmatrix}$ (columns: CA, TX; rows: Model A, Model B) *(4-4)*

44. x_1 = Amount invested at 5%
x_2 = Amount invested at 10%

$$\begin{aligned} x_1 + \quad x_2 &= 5{,}000 \\ 0.05x_1 + 0.1x_2 &= 400 \end{aligned}$$

\$2,000 at 5%, \$3,000 at 10% *(4-3)*

45. \$2,000 at 5% and \$3,000 at 10% *(4-6)*

46. No to both. The annual yield must be between \$250 and \$500 inclusive. *(4-6)*

47. x_1 = Number of \$8 tickets
x_2 = Number of \$12 tickets
x_3 = Number of \$20 tickets

$$\begin{aligned} x_1 + \quad x_2 + \quad x_3 &= 25{,}000 \\ 8x_1 + 12x_2 + 20x_3 &= k_1 \quad \text{Return requested} \\ x_1 \qquad\quad - \quad x_3 &= 0 \end{aligned}$$

Concert 1: 5,000 \$8 tickets, 15,000 \$12 tickets, and 5,000 \$20 tickets
Concert 2: 7,500 \$8 tickets, 10,000 \$12 tickets, and 7,500 \$20 tickets
Concert 3: 10,000 \$8 tickets, 5,000 \$12 tickets, and 10,000 \$20 tickets *(4-6)*

48.
$$\begin{aligned} x_1 + \quad x_2 + \quad x_3 &= 25{,}000 \\ 8x_1 + 12x_2 + 20x_3 &= k_1 \quad \text{Return requested} \end{aligned}$$

Concert 1: $2t - 5{,}000$ \$8 tickets, $30{,}000 - 3t$ \$12 tickets, and t \$20 tickets, where t is an integer satisfying $2{,}500 \le t \le 10{,}000$
Concert 2: $2t - 7{,}500$ \$8 tickets, $32{,}500 - 3t$ \$12 tickets, and t \$20 tickets, where t is an integer satisfying $3{,}750 \le t \le 10{,}833$
Concert 3: $2t - 10{,}000$ \$8 tickets, $35{,}000 - 3t$ \$12 tickets, and t \$20 tickets, where t is an integer satisfying $5{,}000 \le t \le 11{,}666$ *(4-3)*

49. (A) Agriculture: \$80 billion; Fabrication: \$60 billion (B) Agriculture: \$135 billion; Fabrication: \$145 billion *(4-7)*

50. GRAPHING UTILITY *(4-5)*

51. (A)
1st & Elm: $x_1 + x_4 = 1{,}300$
2nd & Elm: $x_1 - x_2 = 400$
2nd & Oak: $x_2 + x_3 = 700$
1st & Oak: $x_3 - x_4 = -200$

(B) $x_1 = 1{,}300 - t$, $x_2 = 900 - t$, $x_3 = t - 200$, $x_4 = t$, where $200 \le t \le 900$
(C) 900; 200
(D) Elm St.: 800; 2nd St.: 400; Oak St.: 300 *(4-3)*

52. (A) $M = \begin{bmatrix} 0 & 1 & 0 & 0 \\ 0 & 0 & 1 & 1 \\ 1 & 0 & 0 & 1 \\ 1 & 0 & 0 & 0 \end{bmatrix}$ (rows and columns: A, B, C, D)

(B) $M + M^2 = \begin{bmatrix} 0 & 1 & 1 & 1 \\ 2 & 0 & 1 & 2 \\ 2 & 1 & 0 & 1 \\ 1 & 1 & 0 & 0 \end{bmatrix}$ (rows and columns: A, B, C, D)

Rank: B, C, A, D *(4-4)*

Chapter 5

Exercise 5-1

1.

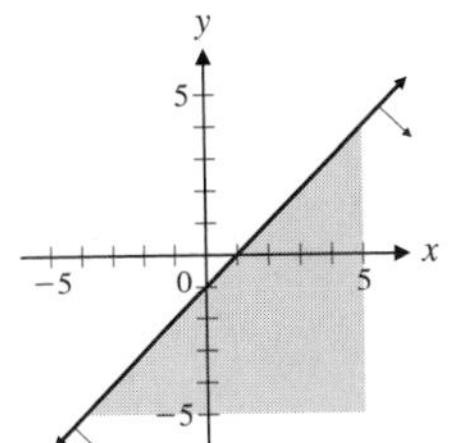

3.

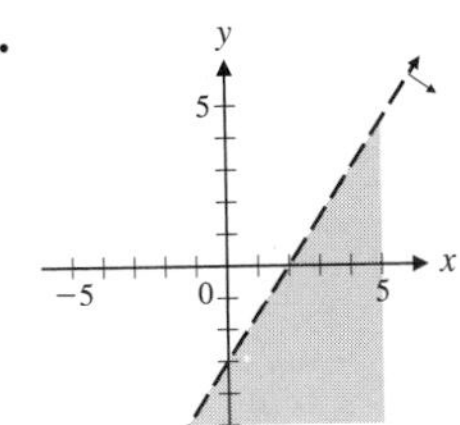

5.

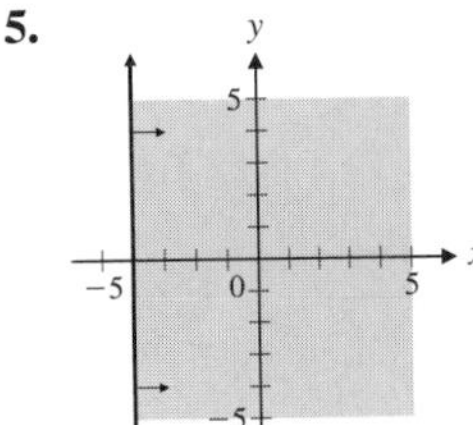

7.

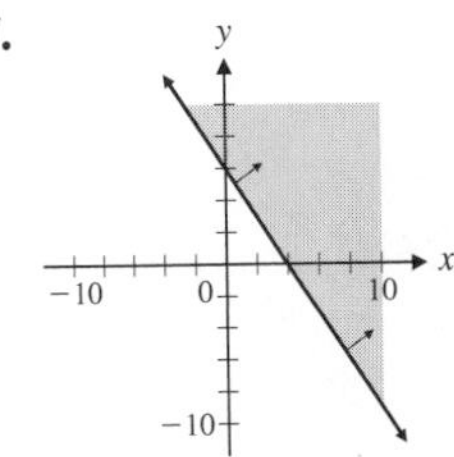

9.

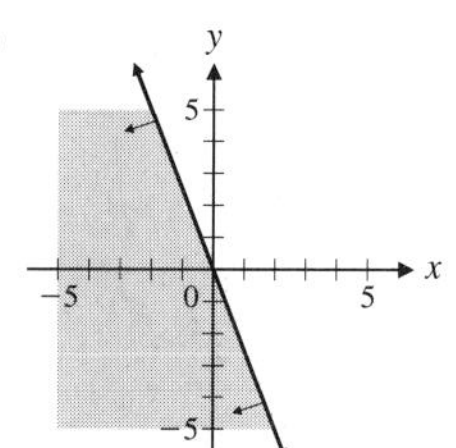

11. IV **13.** I **15.**

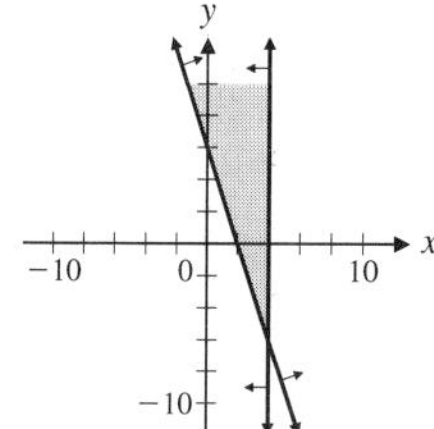

17.

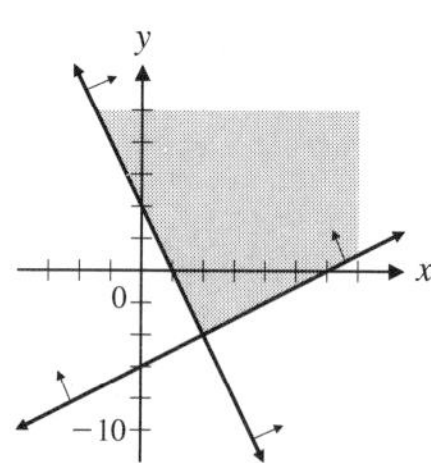

19. (A) Solution region is the double-shaded region.

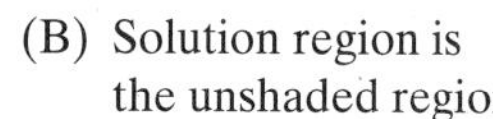
(B) Solution region is the unshaded region.

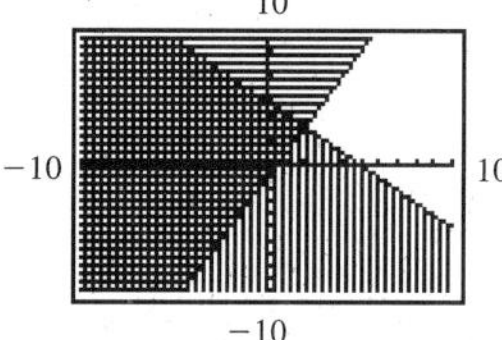

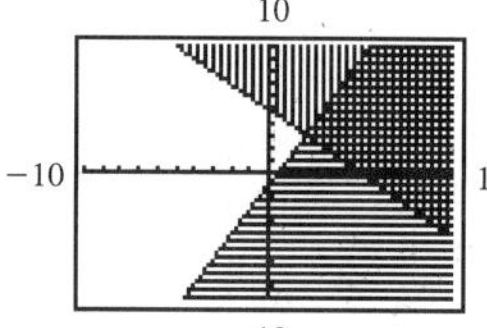

21. (A) Solution region is the double-shaded region. (B) Solution region is the unshaded region.

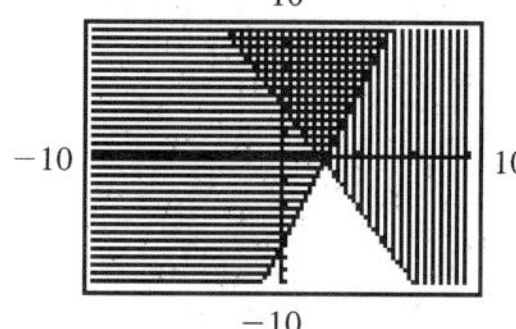

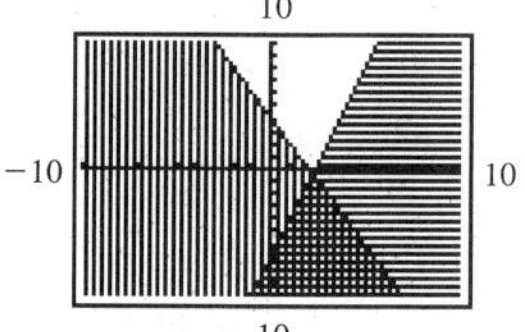

23. IV; (8, 0), (18, 0), (6, 4) **25.** I; (0, 16), (6, 4), (18, 0) **27.** Bounded

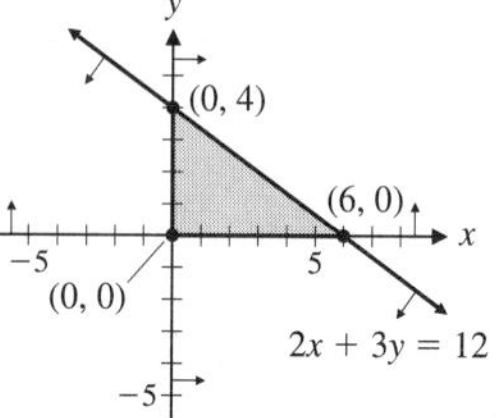

29. Bounded

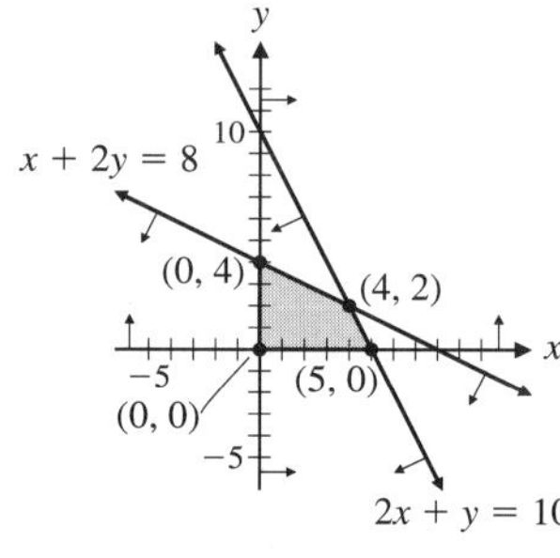

31. Unbounded

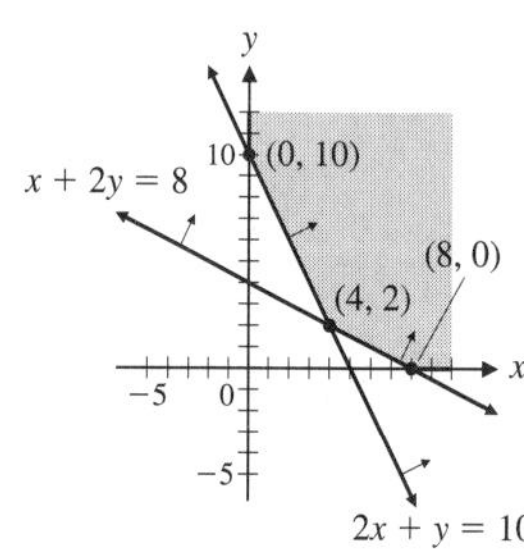

33. Bounded

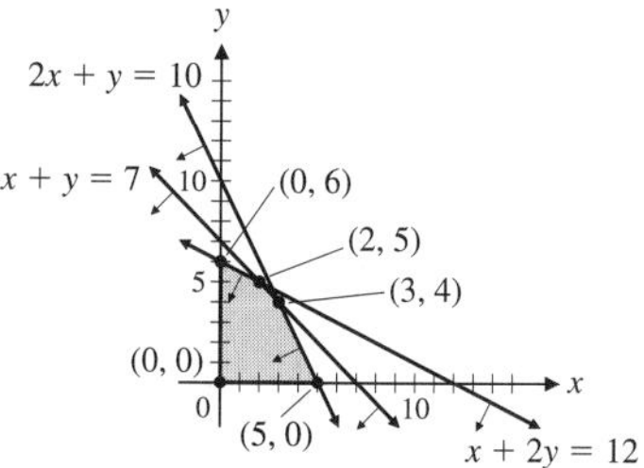

35. Unbounded

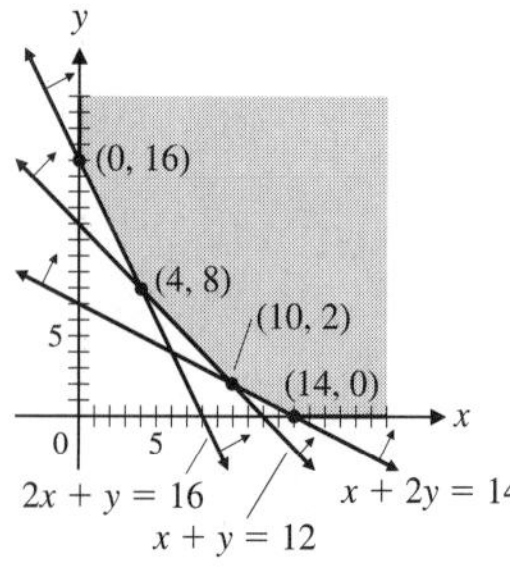

37. Bounded

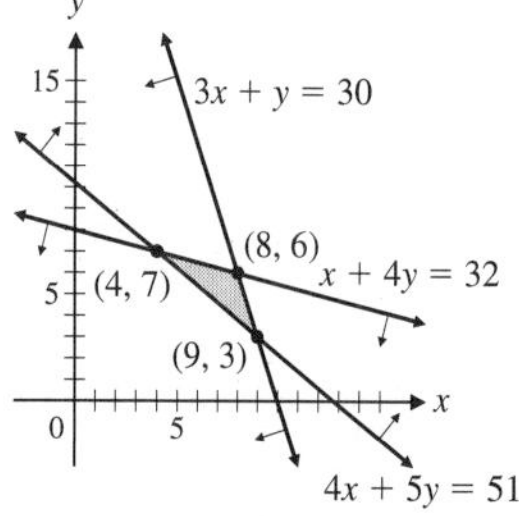

39. Empty

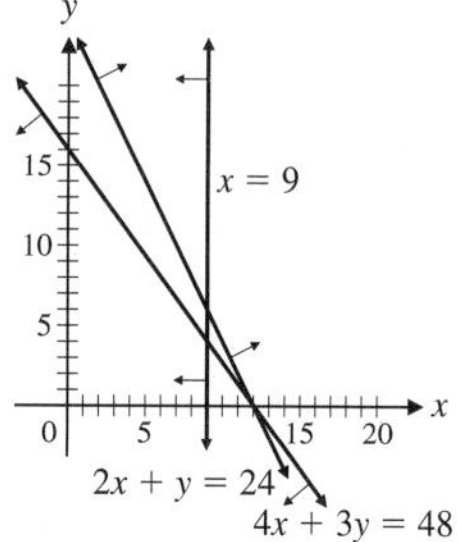

41. Unbounded

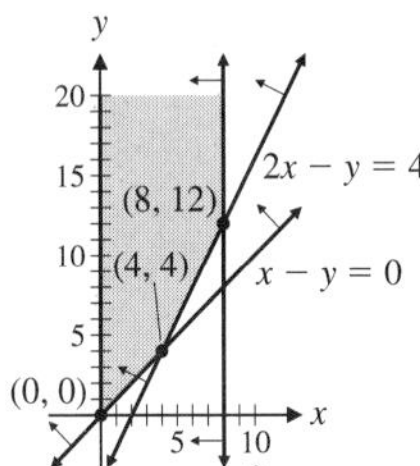

43. Bounded

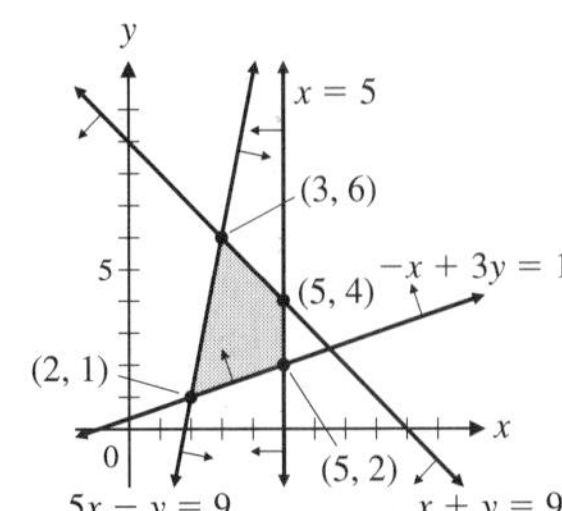

45. Bounded

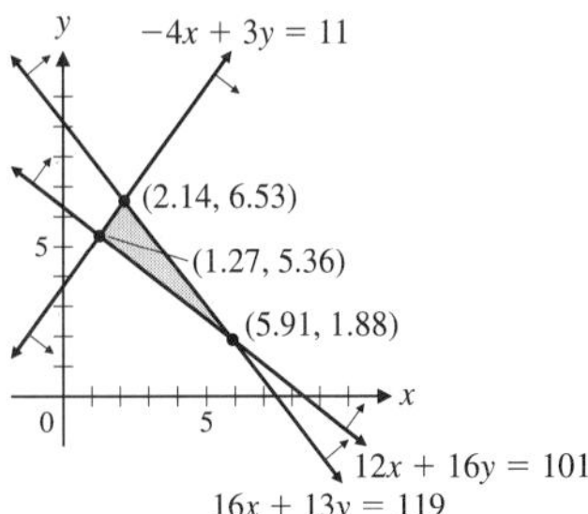

47. (A) $3x + 4y = 36$ and $3x + 2y = 30$ intersect at $(8, 3)$; $3x + 4y = 36$ and $x = 0$ intersect at $(0, 9)$; $3x + 4y = 36$ and $y = 0$ intersect at $(12, 0)$; $3x + 2y = 30$ and $x = 0$ intersect at $(0, 15)$; $3x + 2y = 30$ and $y = 0$ intersect at $(10, 0)$; $x = 0$ and $y = 0$ intersect at $(0, 0)$
(B) $(8, 3), (0, 9), (10, 0), (0, 0)$

49. $6x + 4y \leq 108$
$x + y \leq 24$
$x \geq 0$
$y \geq 0$

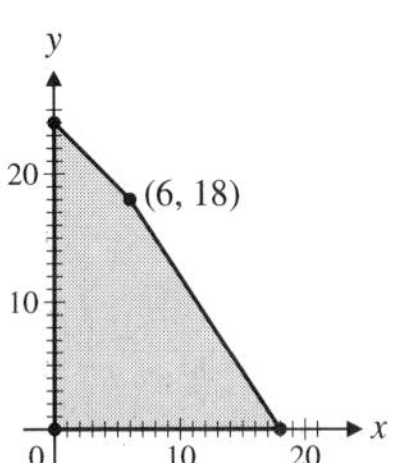

51. (A) All production schedules in the feasible region that are on the graph of $50x + 60y = 1{,}100$ will result in a profit of $1,100.
(B) There are many possible choices. For example, producing 5 trick skis and 15 slalom skis will produce a profit of $1,150. All the production schedules in the feasible region that are on the graph of $50x + 60y = 1{,}150$ will result in a profit of $1,150.

53. $20x + 10y \geq 460$
$30x + 30y \geq 960$
$5x + 10y \geq 220$
$x \geq 0$
$y \geq 0$

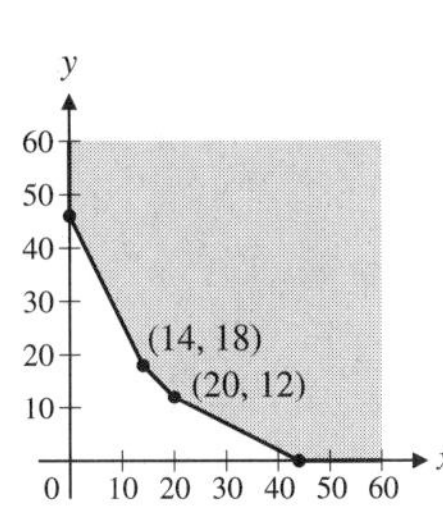

55. $10x + 20y \leq 800$
$20x + 10y \leq 640$
$x \geq 0$
$y \geq 0$

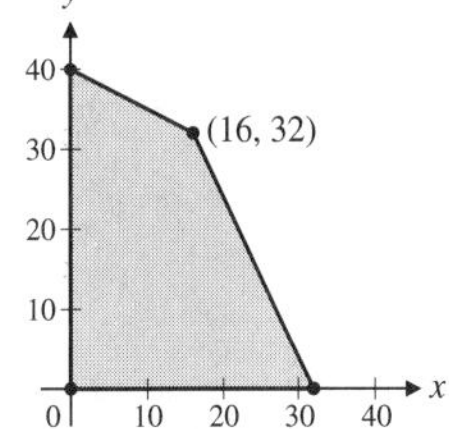

Exercise 5-2

1. 16 **3.** 84 **5.** 32 **7.** 36 **9.** Max $P = 30$ at $x_1 = 4$ and $x_2 = 2$
11. Min $z = 14$ at $x_1 = 4$ and $x_2 = 2$; no max
13. Max $P = 260$ at $x_1 = 2$ and $x_2 = 5$ **15.** Min $z = 140$ at $x_1 = 14$ and $x_2 = 0$; no max
17. Min $P = 20$ at $x_1 = 0$ and $x_2 = 2$; Max $P = 150$ at $x_1 = 5$ and $x_2 = 0$
19. Feasible region empty; no optimal solutions
21. Min $P = 140$ at $x_1 = 3$ and $x_2 = 8$; Max $P = 260$ at $x_1 = 8$ and $x_2 = 10$, at $x_1 = 12$ and $x_2 = 2$, or at any point on the line segment from $(8, 10)$ to $(12, 2)$
23. Max $P = 26{,}000$ at $x_1 = 400$ and $x_2 = 600$ **25.** Max $P = 5{,}507$ at $x_1 = 6.62$ and $x_2 = 4.25$
27. Max $z = 2$ at $x_1 = 4$ and $x_2 = 2$; Min z does not exist
29. (A) $2a < b$ (B) $\frac{1}{3}a < b < 2a$ (C) $b < \frac{1}{3}a$ (D) $b = 2a$ (E) $b = \frac{1}{3}a$
31. (A) Max profit = $780 when 6 trick skis and 18 slalom skis are produced.
(B) Max profit decreases to $720 when 18 trick skis and no slalom skis are produced.
(C) Max profit increases to $1,080 when no trick skis and 24 slalom skis are produced.

33. (A) Plant *A:* 5 days; Plant *B:* 4 days; min cost \$8,600
(B) Plant *A:* 10 days; Plant *B:* 0 days; min cost \$6,000
(C) Plant *A:* 0 days; Plant *B:* 10 days; min cost \$8,000

35. 7 buses, 15 vans; min cost \$9,900

37. \$40,000 in CD's and \$20,000 in mutual fund; max return is \$3,800

39. (A) Max $P =$ \$450 when 750 gal are produced using the old process exclusively.
(B) Max $P =$ \$380 when 400 gal are produced using the old process and 700 gal are produced using the new process.
(C) Max $P =$ \$288 when 1,440 gal are produced using the new process exclusively.

41. (A) 150 bags brand *A*, 100 bags brand *B;* max nitrogen 1,500 lb
(B) 0 bags brand *A*, 250 bags brand *B;* min nitrogen 750 lb

43. 20 yd^3 *A*, 12 yd^3 *B;* \$1,020 **45.** 48; 16 mice, 32 rats

Exercise 5-3

1. (A) 2 (B) 2 basic and 3 nonbasic variables (C) 2 linear equations with 2 variables

3. (A) 5 (B) 4 (C) 5 basic and 4 nonbasic variables (D) 5 linear equations with 5 variables

5.

	Nonbasic	Basic	Feasible?
(A)	x_1, x_2	s_1, s_2	Yes
(B)	x_1, s_1	x_2, s_2	Yes
(C)	x_1, s_2	x_2, s_1	No
(D)	x_2, s_1	x_1, s_2	No
(E)	x_2, s_2	x_1, s_1	Yes
(F)	s_1, s_2	x_1, x_2	Yes

7.

	x_1	x_2	s_1	s_2	Feasible?
(A)	0	0	50	40	Yes
(B)	0	50	0	−60	No
(C)	0	20	30	0	Yes
(D)	25	0	0	15	Yes
(E)	40	0	−30	0	No
(F)	20	10	0	0	Yes

9.

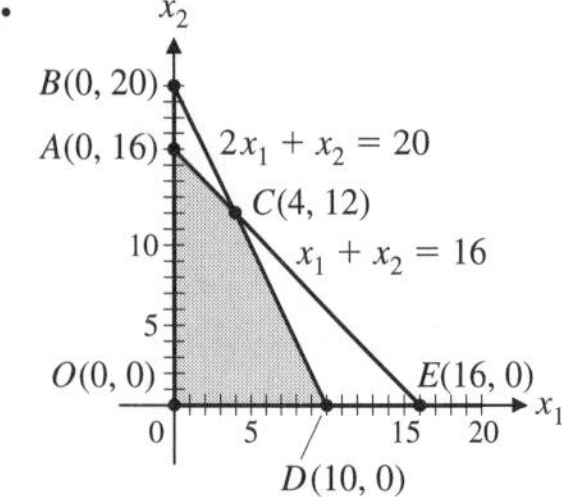

$$\begin{aligned} x_1 + x_2 + s_1 \quad &= 16 \\ 2x_1 + x_2 \quad + s_2 &= 20 \end{aligned}$$

x_1	x_2	s_1	s_2	Intersection Point	Feasible?
0	0	16	20	*O*	Yes
0	16	0	4	*A*	Yes
0	20	−4	0	*B*	No
16	0	0	−12	*E*	No
10	0	6	0	*D*	Yes
4	12	0	0	*C*	Yes

11.

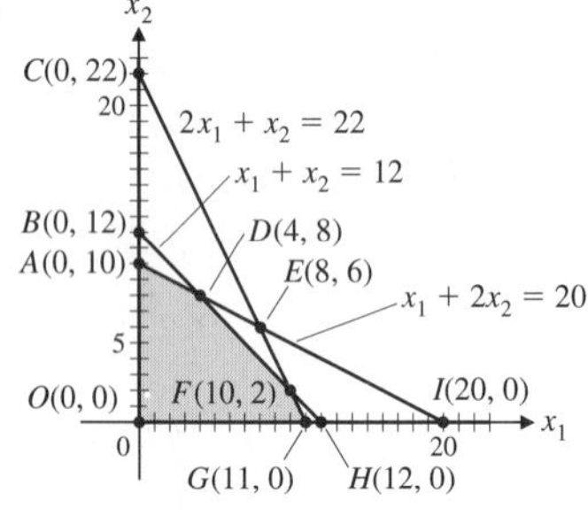

$$\begin{aligned} 2x_1 + x_2 + s_1 \quad\quad &= 22 \\ x_1 + x_2 \quad + s_2 \quad &= 12 \\ x_1 + 2x_2 \quad\quad + s_3 &= 20 \end{aligned}$$

x_1	x_2	s_1	s_2	s_3	Intersection Point	Feasible?
0	0	22	12	20	*O*	Yes
0	22	0	−10	−24	*C*	No
0	12	10	0	−4	*B*	No
0	10	12	2	0	*A*	Yes
11	0	0	1	9	*G*	Yes
12	0	−2	0	8	*H*	No
20	0	−18	−8	0	*I*	No
10	2	0	0	6	*F*	Yes
8	6	0	−2	0	*E*	No
4	8	6	0	0	*D*	Yes

Exercise 5-4

1. (A) Basic: x_2, s_1, P; nonbasic: x_1, s_2 (B) $x_1 = 0, x_2 = 12, s_1 = 15, s_2 = 0, P = 20$ (C) Additional pivot required

3. (A) Basic: x_2, x_3, s_3, P; nonbasic: x_1, s_1, s_2 (B) $x_1 = 0, x_2 = 15, x_3 = 5, s_1 = 0, s_2 = 0, s_3 = 12, P = 45$ (C) No optimal solution

5.

$$\begin{array}{r|ccccc|c} & \overset{\text{Enter}}{\downarrow} & & & & & \\ & x_1 & x_2 & s_1 & s_2 & P & \\ \hline \text{Exit}\rightarrow s_1 & \textcircled{1} & 4 & 1 & 0 & 0 & 4 \\ s_2 & 3 & 5 & 0 & 1 & 0 & 24 \\ \hdashline P & -8 & -5 & 0 & 0 & 1 & 0 \end{array}$$

$$\sim \begin{array}{r|ccccc|c} x_1 & 1 & 4 & 1 & 0 & 0 & 4 \\ s_2 & 0 & -7 & -3 & 1 & 0 & 12 \\ \hdashline P & 0 & 27 & 8 & 0 & 1 & 32 \end{array}$$

7.

$$\begin{array}{r|cccccc|c} & \overset{\text{Enter}}{\downarrow} & & & & & & \\ & x_1 & x_2 & s_1 & s_2 & s_3 & P & \\ \hline \text{Exit}\rightarrow x_2 & \textcircled{2} & 1 & 1 & 0 & 0 & 0 & 4 \\ s_2 & 3 & 0 & 1 & 1 & 0 & 0 & 8 \\ s_3 & 0 & 0 & 2 & 0 & 1 & 0 & 2 \\ \hdashline P & -4 & 0 & -3 & 0 & 0 & 1 & 5 \end{array}$$

$$\sim \begin{array}{r|cccccc|c} x_1 & 1 & \frac{1}{2} & \frac{1}{2} & 0 & 0 & 0 & 2 \\ s_2 & 0 & -\frac{3}{2} & -\frac{1}{2} & 1 & 0 & 0 & 2 \\ s_3 & 0 & 0 & 2 & 0 & 1 & 0 & 2 \\ \hdashline P & 0 & 2 & -1 & 0 & 0 & 1 & 13 \end{array}$$

9. (A)

$$\begin{aligned} 2x_1 + x_2 + s_1 \qquad\quad &= 10 \\ x_1 + 3x_2 \qquad + s_2 \quad &= 10 \\ -15x_1 - 10x_2 \qquad\quad + P &= 0 \end{aligned}$$

(B)

$$\begin{array}{r|ccccc|c} & \overset{\text{Enter}}{\downarrow} & & & & & \\ & x_1 & x_2 & s_1 & s_2 & P & \\ \hline \text{Exit}\rightarrow s_1 & \textcircled{2} & 1 & 1 & 0 & 0 & 10 \\ s_2 & 1 & 3 & 0 & 1 & 0 & 10 \\ \hdashline P & -15 & -10 & 0 & 0 & 1 & 0 \end{array}$$

(C) Max $P = 80$ at $x_1 = 4$ and $x_2 = 2$

11. (A)

$$\begin{aligned} 2x_1 + x_2 + s_1 \qquad\quad &= 10 \\ x_1 + 3x_2 \qquad + s_2 \quad &= 10 \\ -30x_1 - x_2 \qquad\quad + P &= 0 \end{aligned}$$

(B)

$$\begin{array}{r|ccccc|c} & \overset{\text{Enter}}{\downarrow} & & & & & \\ & x_1 & x_2 & s_1 & s_2 & P & \\ \hline \text{Exit}\rightarrow s_1 & \textcircled{2} & 1 & 1 & 0 & 0 & 10 \\ s_2 & 1 & 3 & 0 & 1 & 0 & 10 \\ \hdashline P & -30 & -1 & 0 & 0 & 1 & 0 \end{array}$$

(C) Max $P = 150$ at $x_1 = 5$ and $x_2 = 0$

13. Max $P = 260$ at $x_1 = 2$ and $x_2 = 5$

15. No optimal solution exists.

17. Max $P = 7$ at $x_1 = 3$ and $x_2 = 5$

19. Max $P = 58$ at $x_1 = 12$, $x_2 = 0$, and $x_3 = 2$

21. Max $P = 17$ at $x_1 = 4$, $x_2 = 3$, and $x_3 = 0$

23. Max $P = 22$ at $x_1 = 1$, $x_2 = 6$, and $x_3 = 0$

25. Max $P = 26{,}000$ at $x_1 = 400$ and $x_2 = 600$

27. Max $P = 450$ at $x_1 = 0$, $x_2 = 180$, and $x_3 = 30$

29. Max $P = 88$ at $x_1 = 24$ and $x_2 = 8$

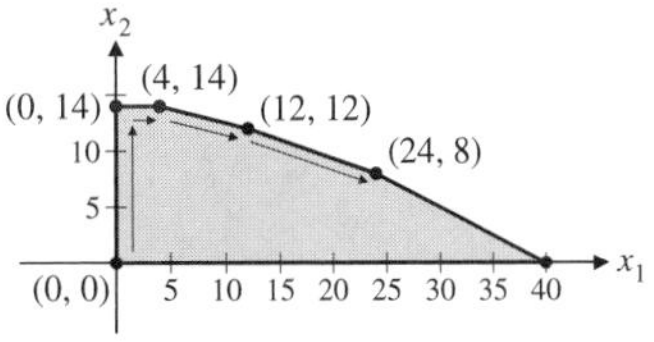

x_2
(0, 14)
(4, 14)
(12, 12)
(24, 8)
(0, 0)
5 10 15 20 25 30 35 40
x_1

31. Choosing either column produces the same optimal solution: Max $P = 13$ at $x_1 = 3$ and $x_2 = 10$.

33. Choosing Column 1: Max $P = 60$ at $x_1 = 12$, $x_2 = 8$, and $x_3 = 0$. Choosing Column 2: Max $P = 60$ at $x_1 = 0$, $x_2 = 20$, and $x_3 = 0$.

35. Let: x_1 = Number of A components
x_2 = Number of B components
x_3 = Number of C components

$$\begin{aligned} \text{Maximize} \quad & P = 7x_1 + 8x_2 + 10x_3 \\ \text{Subject to} \quad & 2x_1 + 3x_2 + 2x_3 \le 1{,}000 \\ & x_1 + x_2 + 2x_3 \le 800 \\ & x_1, x_2, x_3 \ge 0 \end{aligned}$$

200 A components, 0 B components, and 300 C components; Max profit is \$4,400

37. Let: x_1 = Amount invested in government bonds
x_2 = Amount invested in mutual funds
x_3 = Amount invested in money market funds

$$\begin{aligned} \text{Maximize} \quad & P = 0.08x_1 + 0.13x_2 + 0.15x_3 \\ \text{Subject to} \quad & x_1 + x_2 + x_3 \le 100{,}000 \\ & -x_1 + x_2 + x_3 \le 0 \\ & x_1, x_2, x_3 \ge 0 \end{aligned}$$

\$50,000 in government bonds, \$0 in mutual funds, and \$50,000 in money market funds; Max return is \$11,500

39. Let: x_1 = Number of ads placed in daytime shows
x_2 = Number of ads placed in prime-time shows
x_3 = Number of ads placed in late-night shows

Maximize $P = 14{,}000x_1 + 24{,}000x_2 + 18{,}000x_3$
Subject to
$$x_1 + x_2 + x_3 \le 15$$
$$1{,}000x_1 + 2{,}000x_2 + 1{,}500x_3 \le 20{,}000$$
$$x_1, x_2, x_3 \ge 0$$

10 daytime ads, 5 prime-time ads, and 0 late-night ads; Max number of potential customers is 260,000

41. (A) Let: x_1 = Number of colonial houses
x_2 = Number of split-level houses
x_3 = Number of ranch houses

Maximize $P = 20{,}000x_1 + 18{,}000x_2 + 24{,}000x_3$
Subject to
$$\tfrac{1}{2}x_1 + \tfrac{1}{2}x_2 + x_3 \le 30$$
$$60{,}000x_1 + 60{,}000x_2 + 80{,}000x_3 \le 3{,}200{,}000$$
$$4{,}000x_1 + 3{,}000x_2 + 4{,}000x_3 \le 180{,}000$$
$$x_1, x_2, x_3 \ge 0$$

20 colonial, 20 split-level, and 10 ranch houses; Max profit is \$1,000,000

(B) 0 colonial, 40 split-level, and 10 ranch houses; Max profit is \$960,000; 20,000 labor-hours are not used

(C) 45 colonial, 0 split-level, and 0 ranch houses; Max profit is \$1,125,000; 7.5 acres of land and \$500,000 of capital are not used

43. (A) Let: x_1 = Number of boxes of assortment I
x_2 = Number of boxes of assortment II
x_3 = Number of boxes of assortment III

Maximize $P = 4x_1 + 3x_2 + 5x_3$
Subject to
$$4x_1 + 12x_2 + 8x_3 \le 4{,}800$$
$$4x_1 + 4x_2 + 8x_3 \le 4{,}000$$
$$12x_1 + 4x_2 + 8x_3 \le 5{,}600$$
$$x_1, x_2, x_3 \ge 0$$

200 boxes of assortment I, 100 boxes of assortment II, and 350 boxes of assortment III; Max profit is \$2,850

(B) 100 boxes of assortment I, 0 boxes of assortment II, and 550 boxes of assortment III; Max profit is \$3,150; 200 fruit-filled candies are not used

(C) 0 boxes of assortment I, 50 boxes of assortment II, and 675 boxes of assortment III; Max profit is \$3,525; 400 fruit-filled candies are not used

45. Let: x_1 = Number of grams of food *A*
x_2 = Number of grams of food *B*
x_3 = Number of grams of food *C*

Maximize $P = 3x_1 + 4x_2 + 5x_3$
Subject to
$$x_1 + 3x_2 + 2x_3 \le 30$$
$$2x_1 + x_2 + 2x_3 \le 24$$
$$x_1, x_2, x_3 \ge 0$$

0 g food *A*, 3 g food *B*, and 10.5 g food *C*; Max protein is 64.5 units

47. Let: x_1 = Number of undergraduate students
x_2 = Number of graduate students
x_3 = Number of faculty members

Maximize $P = 18x_1 + 25x_2 + 30x_3$
Subject to
$$x_1 + x_2 + x_3 \le 20$$
$$100x_1 + 150x_2 + 200x_3 \le 3{,}200$$
$$x_1, x_2, x_3 \ge 0$$

0 undergraduate students, 16 graduate students, and 4 faculty members; Max number of interviews is 520

Exercise 5-5

1. $\begin{bmatrix} -5 \\ 0 \\ 3 \\ -1 \\ 8 \end{bmatrix}$ **3.** $[1 \quad -2 \quad 0 \quad 4]$ **5.** $\begin{bmatrix} 2 & 5 \\ 1 & 2 \\ -6 & 0 \\ 0 & 1 \\ -1 & 3 \end{bmatrix}$ **7.** $\begin{bmatrix} 1 & 0 & 8 & 4 \\ 2 & 2 & 0 & -1 \\ -1 & -7 & 1 & 3 \end{bmatrix}$

9. (A) Maximize $P = 4y_1 + 5y_2$
Subject to
$$y_1 + 2y_2 \le 8$$
$$3y_1 + y_2 \le 9$$
$$y_1, y_2 \ge 0$$

(B)
$$y_1 + 2y_2 + x_1 = 8$$
$$3y_1 + y_2 + x_2 = 9$$
$$-4y_1 - 5y_2 + P = 0$$

(C)
$$\begin{array}{ccccc|c} y_1 & y_2 & x_1 & x_2 & P & \\ 1 & 2 & 1 & 0 & 0 & 8 \\ 3 & 1 & 0 & 1 & 0 & 9 \\ \hline -4 & -5 & 0 & 0 & 1 & 0 \end{array}$$

11. (A) Max $P = 121$ at $y_1 = 3$ and $y_2 = 5$ (B) Min $C = 121$ at $x_1 = 1$ and $x_2 = 2$

13. (A) Maximize $P = 13y_1 + 12y_2$
Subject to
$$4y_1 + 3y_2 \le 9$$
$$y_1 + y_2 \le 2$$
$$y_1, y_2 \ge 0$$

(B) Min $C = 26$ at $x_1 = 0$ and $x_2 = 13$

15. (A) Maximize $P = 15y_1 + 8y_2$
Subject to
$$2y_1 + y_2 \le 7$$
$$3y_1 + 2y_2 \le 12$$
$$y_1, y_2 \ge 0$$

(B) Min $C = 54$ at $x_1 = 6$ and $x_2 = 1$

17. (A) Maximize $P = 8y_1 + 4y_2$
Subject to
$$\begin{aligned} 2y_1 - 2y_2 &\leqslant 11 \\ y_1 + 3y_2 &\leqslant 4 \\ y_1, y_2 &\geqslant 0 \end{aligned}$$
(B) Min $C = 32$ at $x_1 = 0$ and $x_2 = 8$

19. (A) Maximize $P = 6y_1 + 4y_2$
Subject to
$$\begin{aligned} -3y_1 + y_2 &\leqslant 7 \\ y_1 - 2y_2 &\leqslant 9 \\ y_1, y_2 &\geqslant 0 \end{aligned}$$
(B) No optimal solution exists.

21. Min $C = 24$ at $x_1 = 8$ and $x_2 = 0$ **23.** Min $C = 20$ at $x_1 = 0$ and $x_2 = 4$

25. Min $C = 140$ at $x_1 = 14$ and $x_2 = 0$ **27.** Min $C = 44$ at $x_1 = 6$ and $x_2 = 2$

29. Min $C = 43$ at $x_1 = 0$, $x_2 = 1$, and $x_3 = 3$ **31.** No optimal solution exists.

33. 2 variables and 4 problem constraints **35.** 2 constraints and any number of variables

37. No; the dual problem is not a standard maximization problem. **39.** Yes; multiply both sides of the inequality by -1.

41. Min $C = 44$ at $x_1 = 0$, $x_2 = 3$, and $x_3 = 5$ **43.** Min $C = 166$ at $x_1 = 0$, $x_2 = 12$, $x_3 = 20$, and $x_4 = 3$

45. (A) Let: x_1 = Numbers of hours the Cedarburg plant is operated
x_2 = Number of hours the Grafton plant is operated
x_3 = Number of hours the West Bend plant is operated

Minimize $C = 70x_1 + 75x_2 + 90x_3$
Subject to
$$\begin{aligned} 20x_1 + 10x_2 + 20x_3 &\geqslant 300 \\ 10x_1 + 20x_2 + 20x_3 &\geqslant 200 \\ x_1, x_2, x_3 &\geqslant 0 \end{aligned}$$
Cedarburg plant 10 hr per day, West Bend plant 5 hr per day, Grafton plant not used; Min cost is \$1,150
(B) West Bend plant 15 hr per day, Cedarburg and Grafton plants not used; Min cost is \$1,350
(C) Grafton plant 10 hr per day, West Bend plant 10 hr per day, Cedarburg plant not used; Min cost is \$1,650

47. (A) Let: x_1 = Number of single-sided drives ordered from Associated Electronics
x_2 = Number of double-sided drives ordered from Associated Electronics
x_3 = Number of single-sided drives ordered from Digital Drives
x_4 = Number of double-sided drives ordered from Digital Drives

Minimize $C = 250x_1 + 350x_2 + 290x_3 + 320x_4$
Subject to
$$\begin{aligned} x_1 + x_2 \qquad\qquad &\leqslant 1{,}000 \\ x_3 + x_4 &\leqslant 2{,}000 \\ x_1 \qquad + x_3 \qquad &\geqslant 1{,}200 \\ x_2 + \qquad x_4 &\geqslant 1{,}600 \\ x_1, x_2, x_3, x_4 &\geqslant 0 \end{aligned}$$
1,000 single-sided drives from Associated Electronics, 200 single-sided and 1,600 double-sided drives from Digital Drives; Min cost is \$820,000

49. Let: x_1 = Number of ounces of food L
x_2 = Number of ounces of food M
x_3 = Number of ounces of food N

Minimize $C = 20x_1 + 24x_2 + 18x_3$
Subject to
$$\begin{aligned} 20x_1 + 10x_2 + 10x_3 &\geqslant 300 \\ 10x_1 + 10x_2 + 10x_3 &\geqslant 200 \\ 10x_1 + 15x_2 + 10x_3 &\geqslant 240 \\ x_1, x_2, x_3 &\geqslant 0 \end{aligned}$$
10 oz *L*, 8 oz *M*, 2 oz *N*; Min cholesterol intake is 428 units

51. Let: x_1 = Number of students bused from North Division to Central
x_2 = Number of students bused from North Division to Washington
x_3 = Number of students bused from South Division to Central
x_4 = Number of students bused from South Division to Washington

Minimize $C = 5x_1 + 2x_2 + 3x_3 + 4x_4$
Subject to
$$\begin{aligned} x_1 + x_2 \qquad\qquad &\geqslant 300 \\ x_3 + x_4 &\geqslant 500 \\ x_1 \qquad + x_3 \qquad &\leqslant 400 \\ x_2 \qquad + x_4 &\leqslant 500 \\ x_1, x_2, x_3, x_4 &\geqslant 0 \end{aligned}$$
300 students bused from North Division to Washington, 400 from South Division to Central, and 100 from South Division to Washington; Min cost is \$2,200

Exercise 5-6

1. (A) Maximize $P = 5x_1 + 2x_2 - Ma_1$
Subject to
$$\begin{aligned} x_1 + 2x_2 + s_1 \qquad\qquad &= 12 \\ x_1 + x_2 \qquad - s_2 + a_1 &= 4 \\ x_1, x_2, s_1, s_2, a_1 &\geqslant 0 \end{aligned}$$
(B)
$$\begin{array}{c} \\ \\ \\ \end{array}\begin{matrix} x_1 & x_2 & s_1 & s_2 & a_1 & P \end{matrix}$$
$$\left[\begin{array}{cccccc|c} 1 & 2 & 1 & 0 & 0 & 0 & 12 \\ 1 & 1 & 0 & -1 & 1 & 0 & 4 \\ \hline -M-5 & -M-2 & 0 & M & 0 & 1 & -4M \end{array}\right]$$
(C) $x_1 = 12, x_2 = 0, s_1 = 0, s_2 = 8, a_1 = 0, P = 60$ (D) Max $P = 60$ at $x_1 = 12$ and $x_2 = 0$

3. (A) Maximize $P = 3x_1 + 5x_2 - Ma_1$
Subject to
$$\begin{aligned} 2x_1 + x_2 + s_1 \quad &= 8 \\ x_1 + x_2 \quad + a_1 &= 6 \\ x_1, x_2, s_1, a_1 &\geq 0 \end{aligned}$$

(B)
$$\begin{bmatrix} x_1 & x_2 & s_1 & a_1 & P & \\ 2 & 1 & 1 & 0 & 0 & 8 \\ 1 & 1 & 0 & 1 & 0 & 6 \\ \hline -M-3 & -M-5 & 0 & 0 & 1 & -6M \end{bmatrix}$$

(C) $x_1 = 0, x_2 = 6, s_1 = 2, a_1 = 0, P = 30$ (D) Max $P = 30$ at $x_1 = 0$ and $x_2 = 6$

5. (A) Maximize $P = 4x_1 + 3x_2 - Ma_1$
Subject to
$$\begin{aligned} -x_1 + 2x_2 + s_1 \quad &= 2 \\ x_1 + x_2 \quad - s_2 + a_1 &= 4 \\ x_1, x_2, s_1, s_2, a_1 &\geq 0 \end{aligned}$$

(B)
$$\begin{bmatrix} x_1 & x_2 & s_1 & s_2 & a_1 & P & \\ -1 & 2 & 1 & 0 & 0 & 0 & 2 \\ 1 & 1 & 0 & -1 & 1 & 0 & 4 \\ \hline -M-4 & -M-3 & 0 & M & 0 & 1 & -4M \end{bmatrix}$$

(C) No optimal solution exists. (D) No optimal solution exists.

7. (A) Maximize $P = 5x_1 + 10x_2 - Ma_1$
Subject to
$$\begin{aligned} x_1 + x_2 + s_1 \quad &= 3 \\ 2x_1 + 3x_2 \quad - s_2 + a_1 &= 12 \\ x_1, x_2, s_1, s_2, a_1 &\geq 0 \end{aligned}$$

(B)
$$\begin{bmatrix} x_1 & x_2 & s_1 & s_2 & a_1 & P & \\ 1 & 1 & 1 & 0 & 0 & 0 & 3 \\ 2 & 3 & 0 & -1 & 1 & 0 & 12 \\ \hline -2M-5 & -3M-10 & 0 & M & 0 & 1 & -12M \end{bmatrix}$$

(C) $x_1 = 0, x_2 = 3, s_1 = 0, s_2 = 0, a_1 = 3, P = -3M + 30$ (D) No optimal solution exists.

9. Min $P = 1$ at $x_1 = 3$ and $x_2 = 5$; Max $P = 16$ at $x_1 = 8$ and $x_2 = 0$ **11.** Max $P = 44$ at $x_1 = 2$ and $x_2 = 8$

13. No optimal solution exists. **15.** Min $C = -9$ at $x_1 = 0, x_2 = \frac{7}{4}$, and $x_3 = \frac{3}{4}$

17. Max $P = 32$ at $x_1 = 0, x_2 = 4$, and $x_3 = 2$ **19.** Max $P = 65$ at $x_1 = \frac{35}{2}, x_2 = 0$, and $x_3 = \frac{15}{2}$

21. Max $P = 120$ at $x_1 = 20, x_2 = 0$, and $x_3 = 20$

23. Problem 5—Unbounded feasible region: Problem 7—Empty feasible region:

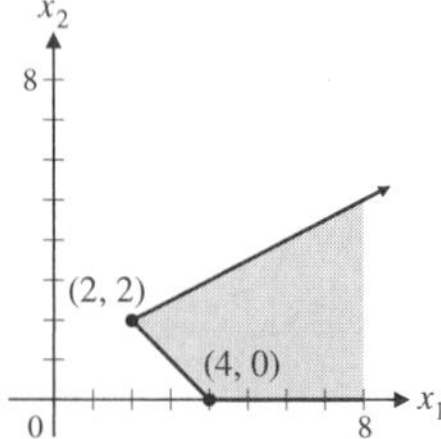

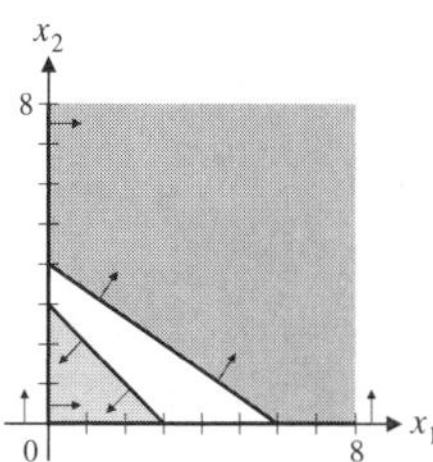

25. Min $C = -30$ at $x_1 = 0, x_2 = \frac{3}{4}$, and $x_3 = 0$ **27.** Max $P = 17$ at $x_1 = \frac{49}{5}, x_2 = 0$, and $x_3 = \frac{22}{5}$

29. Min $C = \frac{135}{2}$ at $x_1 = \frac{15}{4}, x_2 = \frac{3}{4}$, and $x_3 = 0$ **31.** Max $P = 372$ at $x_1 = 28, x_2 = 4$, and $x_3 = 0$

33. Let: x_1 = Number of 16k modules manufactured daily
x_2 = Number of 64k modules manufactured daily

Maximize $P = 18x_1 + 30x_2$
Subject to
$$\begin{aligned} 10x_1 + 15x_2 &\leq 2{,}200 \\ 2x_1 + 4x_2 &\leq 500 \\ x_1 \quad &\geq 50 \\ x_1, x_2 &\geq 0 \end{aligned}$$

130 of the 16k modules and 60 of the 64k modules; Max profit is $4,140

35. Let: x_1 = Number of ads placed in the *Sentinel*
x_2 = Number of ads placed in the *Journal*
x_3 = Number of ads placed in the *Tribune*

Minimize $C = 200x_1 + 200x_2 + 100x_3$
Subject to
$$\begin{aligned} x_1 + x_2 + x_3 &\leq 10 \\ 2{,}000x_1 + 500x_2 + 1{,}500x_3 &\geq 16{,}000 \\ x_1, x_2, x_3 &\geq 0 \end{aligned}$$

2 ads in the *Sentinel,* 0 ads in the *Journal,* 8 ads in the *Tribune;* Min cost is $1,200

37. Let: x_1 = Number of bottles of brand A
x_2 = Number of bottles of brand B
x_3 = Number of bottles of brand C

Minimize $C = 0.6x_1 + 0.4x_2 + 0.9x_3$
Subject to
$$\begin{aligned} 10x_1 + 10x_2 + 20x_3 &\geq 100 \\ 2x_1 + 3x_2 + 4x_3 &\leq 24 \\ x_1, x_2, x_3 &\geq 0 \end{aligned}$$

0 bottles of A, 4 bottles of B, 3 bottles of C; Min cost is $4.30

39. Let: x_1 = Number of cubic yards of mix A
x_2 = Number of cubic yards of mix B
x_3 = Number of cubic yards of mix C

$$\begin{aligned} &\text{Maximize} && P = 12x_1 + 16x_2 + 8x_3 \\ &\text{Subject to} && 12x_1 + 8x_2 + 16x_3 \leq 700 \\ &&& 16x_1 + 8x_2 + 16x_3 \geq 800 \\ &&& x_1, x_2, x_3 \geq 0 \end{aligned}$$

25 yd^3 A, 50 yd^3 B, 0 yd^3 C: Max is 1,100 lb

41. Let: x_1 = Number of car frames produced at the Milwaukee plant
x_2 = Number of truck frames produced at the Milwaukee plant
x_3 = Number of car frames produced at the Racine plant
x_4 = Number of truck frames produced at the Racine plant

$$\begin{aligned} &\text{Maximize} && P = 50x_1 + 70x_2 + 50x_3 + 70x_4 \\ &\text{Subject to} && x_1 + x_3 \leq 250 \\ &&& x_2 + x_4 \leq 350 \\ &&& x_1 + x_2 \leq 300 \\ &&& x_3 + x_4 \leq 200 \\ &&& 150x_1 + 200x_2 \leq 50{,}000 \\ &&& 135x_3 + 180x_4 \leq 35{,}000 \\ &&& x_1, x_2, x_3, x_4 \geq 0 \end{aligned}$$

43. Let: x_1 = Number of barrels of A used in regular gasoline
x_2 = Number of barrels of A used in premium gasoline
x_3 = Numbers of barrels of B used in regular gasoline
x_4 = Number of barrels of B used in premium gasoline
x_5 = Numbers of barrels of C used in regular gasoline
x_6 = Number of barrels of C used in premium gasoline

$$\begin{aligned} &\text{Maximize} && P = 10x_1 + 18x_2 + 8x_3 + 16x_4 + 4x_5 + 12x_6 \\ &\text{Subject to} && x_1 + x_2 \leq 40{,}000 \\ &&& x_3 + x_4 \leq 25{,}000 \\ &&& x_5 + x_6 \leq 15{,}000 \\ &&& x_1 + x_3 + x_5 \geq 30{,}000 \\ &&& x_2 + x_4 + x_6 \geq 25{,}000 \\ &&& -5x_1 + 5x_3 + 15x_5 \geq 0 \\ &&& -15x_2 - 5x_4 + 5x_6 \geq 0 \\ &&& x_1, x_2, x_3, x_4, x_5, x_6 \geq 0 \end{aligned}$$

45. Let: x_1 = Percentage invested in high-tech funds
x_2 = Percentage invested in global funds
x_3 = Percentage invested in corporate bonds
x_4 = Percentage invested in municipal bonds
x_5 = Percentage invested in CD's

$$\begin{aligned} &\text{Maximize} && P = 0.11x_1 + 0.1x_2 + 0.09x_3 + 0.08x_4 + 0.05x_5 \\ &\text{Subject to} && x_1 + x_2 + x_3 + x_4 + x_5 = 1 \\ &&& 2.7x_1 + 1.8x_2 + 1.2x_3 + 0.5x_4 \leq 1.8 \\ &&& x_5 \geq 0.2 \\ &&& x_1, x_2, x_3, x_4, x_5 \geq 0 \end{aligned}$$

47. Let: x_1 = Number of ounces of food L
x_2 = Number of ounces of food M
x_3 = Number of ounces of food N

$$\begin{aligned} &\text{Minimize} && C = 0.4x_1 + 0.6x_2 + 0.8x_3 \\ &\text{Subject to} && 30x_1 + 10x_2 + 30x_3 \geq 400 \\ &&& 10x_1 + 10x_2 + 10x_3 \geq 200 \\ &&& 10x_1 + 30x_2 + 20x_3 \geq 300 \\ &&& 8x_1 + 4x_2 + 6x_3 \leq 150 \\ &&& 60x_1 + 40x_2 + 50x_3 \leq 900 \\ &&& x_1, x_2, x_3 \geq 0 \end{aligned}$$

49. Let: x_1 = Number of students from town A enrolled in school I
x_2 = Number of students from town A enrolled in school II
x_3 = Number of students from town B enrolled in school I
x_4 = Number of students from town B enrolled in school II
x_5 = Number of students from town C enrolled in school I
x_6 = Number of students from town C enrolled in school II

$$\begin{aligned} &\text{Minimize} && C = 4x_1 + 8x_2 + 6x_3 + 4x_4 + 3x_5 + 9x_6 \\ &\text{Subject to} && x_1 + x_2 = 500 \\ &&& x_3 + x_4 = 1{,}200 \\ &&& x_5 + x_6 = 1{,}800 \\ &&& x_1 + x_3 + x_5 \leq 2{,}000 \\ &&& x_2 + x_4 + x_6 \leq 2{,}000 \\ &&& x_1 + x_3 + x_5 \geq 1{,}400 \\ &&& x_2 + x_4 + x_6 \geq 1{,}400 \\ &&& x_1 \leq 300 \\ &&& x_2 \leq 300 \\ &&& x_3 \leq 720 \\ &&& x_4 \leq 720 \\ &&& x_5 \leq 1{,}080 \\ &&& x_6 \leq 1{,}080 \\ &&& x_1, x_2, x_3, x_4, x_5, x_6 \geq 0 \end{aligned}$$

Review Exercise

1. Bounded *(5-1)*

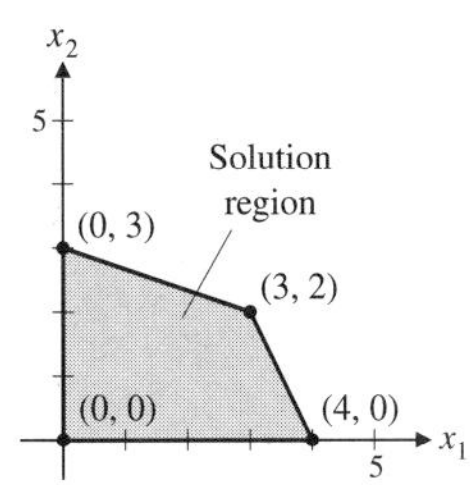

2. Unbounded *(5-1)*

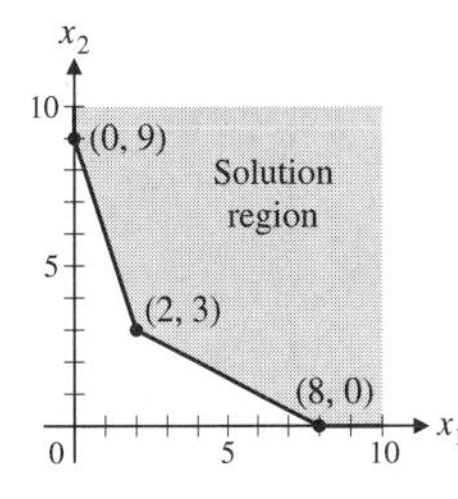

3. Max $P = 24$ at $x_1 = 4$ and $x_2 = 0$ *(5-2)*

4. $2x_1 + x_2 + s_1 = 8$
$x_1 + 2x_2 + s_2 = 10$ *(5-3)*

5. 2 basic and 2 nonbasic variables *(5-3)*

6.

x_1	x_2	s_1	s_2	FEASIBLE?
0	0	8	10	Yes
0	8	0	−6	No
0	5	3	0	Yes
4	0	0	6	Yes
10	0	−12	0	No
2	4	0	0	Yes

(5-3)

7. Enter ↓ (x_1 column); Exit → (s_1 row)

$$\begin{array}{c} \\ s_1 \\ s_2 \\ P \end{array}\left[\begin{array}{ccccc|c} x_1 & x_2 & s_1 & s_2 & P & \\ \textcircled{2} & 1 & 1 & 0 & 0 & 8 \\ 1 & 2 & 0 & 1 & 0 & 10 \\ \hline -6 & -2 & 0 & 0 & 1 & 0 \end{array}\right]$$ *(5-4)*

8. Max $P = 24$ at $x_1 = 4$ and $x_2 = 0$ *(5-4)*

9. Basic variables: x_2, s_2, s_3, P; nonbasic variables: x_1, x_3, s_1

Enter ↓ (x_1 column); Exit → (s_3 row)

$$\begin{array}{c} \\ x_2 \\ s_2 \\ s_3 \\ P \end{array}\left[\begin{array}{ccccccc|c} x_1 & x_2 & x_3 & s_1 & s_2 & s_3 & P & \\ 2 & 1 & 3 & -1 & 0 & 0 & 0 & 20 \\ 3 & 0 & 4 & 1 & 1 & 0 & 0 & 30 \\ \textcircled{2} & 0 & 5 & 2 & 0 & 1 & 0 & 10 \\ \hline -8 & 0 & -5 & 3 & 0 & 0 & 1 & 50 \end{array}\right] \sim \begin{array}{c} \\ x_2 \\ s_2 \\ x_1 \\ P \end{array}\left[\begin{array}{ccccccc|c} x_1 & x_2 & x_3 & s_1 & s_2 & s_3 & P & \\ 0 & 1 & -2 & -3 & 0 & -1 & 0 & 10 \\ 0 & 0 & -\frac{7}{2} & -2 & 1 & -\frac{3}{2} & 0 & 15 \\ 1 & 0 & \frac{5}{2} & 1 & 0 & \frac{1}{2} & 0 & 5 \\ \hline 0 & 0 & 15 & 11 & 0 & 4 & 1 & 90 \end{array}\right]$$ *(5-4)*

10. (A) $x_1 = 0, x_2 = 2, s_1 = 0, s_2 = 5, P = 12$; additional pivoting required
(B) $x_1 = 0, x_2 = 0, s_1 = 0, s_2 = 7, P = 22$; no optimal solution exists
(C) $x_1 = 6, x_2 = 0, s_1 = 15, s_2 = 0, P = 10$; optimal solution *(5-4)*

11. Min $C = 40$ at $x_1 = 0$ and $x_2 = 20$ *(5-2)*

12. Maximize $P = 15y_1 + 20y_2$
Subject to $y_1 + 2y_2 \le 5$
$3y_1 + y_2 \le 2$
$y_1, y_2 \ge 0$ *(5-5)*

13.
$$\begin{aligned} y_1 + 2y_2 + x_1 \qquad\qquad &= 5 \\ 3y_1 + y_2 \qquad + x_2 \qquad &= 2 \\ -15y_1 - 20y_2 \qquad\qquad + P &= 0 \end{aligned}$$ *(5-5)*

14.
$$\left[\begin{array}{ccccc|c} y_1 & y_2 & x_1 & x_2 & P & \\ 1 & 2 & 1 & 0 & 0 & 5 \\ 3 & 1 & 0 & 1 & 0 & 2 \\ \hline -15 & -20 & 0 & 0 & 1 & 0 \end{array}\right]$$ *(5-5)*

15. Max $P = 40$ at $y_1 = 0$ and $y_2 = 2$ *(5-4)*

16. Min $C = 40$ at $x_1 = 0$ and $x_2 = 20$ *(5-5)*

17. Max $P = 26$ at $x_1 = 2$ and $x_2 = 5$ *(5-2)*

18. Max $P = 26$ at $x_1 = 2$ and $x_2 = 5$ *(5-4)*

19. Min $C = 51$ at $x_1 = 9$ and $x_2 = 3$ *(5-2)*

20. Maximize $P = 10y_1 + 15y_2 + 3y_3$
Subject to $y_1 + y_2 \le 3$
$y_1 + 2y_2 + y_3 \le 8$
$y_1, y_2, y_3 \ge 0$ *(5-5)*

21. Min $C = 51$ at $x_1 = 9$ and $x_2 = 3$ *(5-5)*

22. No optimal solution exists. *(5-4)*

23. Max $P = 23$ at $x_1 = 4, x_2 = 1$, and $x_3 = 0$ *(5-4)*

24. (A) Modified problem:
Maximize $P = x_1 + 3x_2 - Ma_1$
Subject to $x_1 + x_2 - s_1 + a_1 = 6$
$x_1 + 2x_2 + s_2 = 8$
$x_1, x_2, s_1, s_2, a_1 \ge 0$

(B) Preliminary simplex tableau:

$$\left[\begin{array}{cccccc|c} x_1 & x_2 & s_1 & a_1 & s_2 & P & \\ 1 & 1 & -1 & 1 & 0 & 0 & 6 \\ 1 & 2 & 0 & 0 & 1 & 0 & 8 \\ \hline -1 & -3 & 0 & M & 0 & 1 & 0 \end{array}\right]$$

Initial simplex tableau:

$$\left[\begin{array}{cccccc|c} x_1 & x_2 & s_1 & a_1 & s_2 & P & \\ 1 & 1 & -1 & 1 & 0 & 0 & 6 \\ 1 & 2 & 0 & 0 & 1 & 0 & 8 \\ \hline -M-1 & -M-3 & M & 0 & 0 & 1 & -6M \end{array}\right]$$

24. (C) $x_1 = 4, x_2 = 2, s_1 = 0, a_1 = 0, s_2 = 0, P = 10$
(D) Since $a_1 = 0$, the optimal solution to the original problem is Max $P = 10$ at $x_1 = 4$ and $x_2 = 2$ *(5-6)*

25. (A) Modified problem:

$$\begin{aligned} &\text{Maximize} && P = x_1 + x_2 - Ma_1 \\ &\text{Subject to} && x_1 + x_2 - s_1 + a_1 = 5 \\ &&& x_1 + 2x_2 + s_2 = 4 \\ &&& x_1, x_2, s_1, s_2, a_1 \geq 0 \end{aligned}$$

(B) Preliminary simplex tableau:

$$\begin{array}{cccccc|c} x_1 & x_2 & s_1 & a_1 & s_2 & P & \\ \hline 1 & 1 & -1 & 1 & 0 & 0 & 5 \\ 1 & 2 & 0 & 0 & 1 & 0 & 4 \\ \hline -1 & -1 & 0 & M & 0 & 1 & 0 \end{array}$$

Initial simplex tableau:

$$\begin{array}{cccccc|c} x_1 & x_2 & s_1 & a_1 & s_2 & P & \\ \hline 1 & 1 & -1 & 1 & 0 & 0 & 5 \\ 1 & 2 & 0 & 0 & 1 & 0 & 4 \\ \hline -M-1 & -M-1 & M & 0 & 0 & 1 & -5M \end{array}$$

(C) $x_1 = 4, x_2 = 0, s_1 = 0, s_2 = 0, a_1 = 1, P = -M + 4$
(D) Since $a_1 \neq 0$, the original problem has no optimal solution. *(5-6)*

26.
$$\begin{aligned} &\text{Maximize} && P = 2x_1 + 3x_2 + x_3 - Ma_1 - Ma_2 \\ &\text{Subject to} && x_1 - 3x_2 + x_3 + s_1 = 7 \\ &&& x_1 + x_2 - 2x_3 - s_2 + a_1 = 2 \\ &&& 3x_1 + 2x_2 - x_3 + a_2 = 4 \\ &&& x_1, x_2, x_3, s_1, s_2, a_1, a_2 \geq 0 \end{aligned}$$
(5-6)

27. The geometric method solves maximization and minimization problems involving two decision variables. If the feasible region is bounded, there are no restrictions on the coefficients or constants. If the feasible region is unbounded and the coefficients of the objective function are positive, then a minimization problem has a solution, but a maximization does not. *(5-2)*

28. The basic simplex method with slack variables solves standard maximization problems involving $\leq$ constraints with nonnegative constants on the right side. *(5-4)*

29. The dual method solves minimization problems with positive coefficients in the objective function. *(5-5)*

30. The big M method solves any linear programming problem. *(5-6)*

31. Max $P = 36$ at $x_1 = 6, x_2 = 8$ *(5-2)*

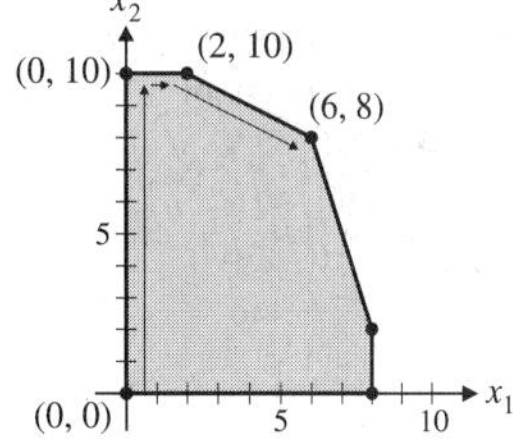

32. Min $C = 15$ at $x_1 = 3$ and $x_2 = 3$ *(5-5)*

33. Min $C = 15$ at $x_1 = 3$ and $x_2 = 3$ *(5-6)*

34. Min $C = 9{,}960$ at $x_1 = 0, x_2 = 240, x_3 = 400$, and $x_4 = 60$ *(5-5)*

35. (A) Let: x_1 = Number of regular sails
x_2 = Number of competition sails

$$\begin{aligned} &\text{Maximize} && P = 100x_1 + 200x_2 \\ &\text{Subject to} && 2x_1 + 3x_2 \leq 150 \\ &&& 4x_1 + 10x_2 \leq 380 \\ &&& x_1, x_2 \geq 0 \end{aligned}$$

Max $P = \$8{,}500$ when 45 regular and 20 competition sails are produced.
(B) Max profit increases to \$9,880 when 38 competition and no regular sails are produced.
(C) Max profit decreases to \$7,500 when no competition and 75 regular sails are produced. *(5-2)*

36. (A) Let: x_1 = Amount invested in oil stock
x_2 = Amount invested in steel stock
x_3 = Amount invested in government bonds

$$\begin{aligned} &\text{Maximize} && P = 0.12x_1 + 0.09x_2 + 0.05x_3 \\ &\text{Subject to} && x_1 + x_2 + x_3 \leq 150{,}000 \\ &&& x_1 \leq 50{,}000 \\ &&& x_1 + x_2 - x_3 \leq 25{,}000 \\ &&& x_1, x_2, x_3 \geq 0 \end{aligned}$$

Max return is \$12,500 when \$50,000 is invested in oil stock, \$37,500 is invested in steel stock, and \$62,500 in government bonds.
(B) Max return is \$13,625 when \$87,500 is invested in steel stock and \$62,500 in government bonds. *(5-4)*

37. Let: x_1 = Number of motors shipped from factory A to plant X
x_2 = Number of motors shipped from factory A to plant Y
x_3 = Number of motors shipped from factory B to plant X
x_4 = Number of motors shipped from factory B to plant Y

$$\begin{aligned} \text{Minimize } & C = 5x_1 + 8x_2 + 9x_3 + 7x_4 \\ \text{Subject to } & x_1 + x_2 \leq 1{,}500 \\ & x_3 + x_4 \leq 1{,}000 \\ & x_1 + x_3 \geq 900 \\ & x_2 + x_4 \geq 1{,}200 \\ & x_1, x_2, x_3, x_4 \geq 0 \end{aligned}$$

Min C = \$13,100 when 900 motors are shipped from factory A to plant X, 200 motors are shipped from factory A to plant Y, and 1,000 motors are shipped from factory B to plant Y *(5-5)*

38. Let: x_1 = Number of pounds of long-grain rice used in brand A
x_2 = Number of pounds of long-grain rice used in brand B
x_3 = Number of pounds of wild rice used in brand A
x_4 = Number of pounds of wild rice used in brand B

$$\begin{aligned} \text{Maximize } & P = 0.8x_1 + 0.5x_2 - 1.9x_3 - 2.2x_4 \\ \text{Subject to } & 0.1x_1 - 0.9x_3 \leq 0 \\ & 0.05x_2 - 0.95x_4 \leq 0 \\ & x_1 + x_2 \leq 8{,}000 \\ & x_3 + x_4 \leq 500 \\ & x_1, x_2, x_3, x_4 \geq 0 \end{aligned}$$

Max profit is \$3,350 when 1,350 lb long-grain rice and 150 lb wild rice are used to produce 1,500 lb brand A, and 6,650 lb long-grain rice and 350 lb wild rice are used to produce 7,000 lb brand B. *(5-4)*

39. (A) Let: x_1 = Number of grams of mix A
x_2 = Number of grams of mix B

$$\begin{aligned} \text{Minimize } & C = 0.04x_1 + 0.09x_2 \\ \text{Subject to } & 2x_1 + 5x_2 \geq 850 \\ & 2x_1 + 4x_2 \geq 800 \\ & 4x_1 + 5x_2 \geq 1{,}150 \\ & x_1, x_2 \geq 0 \end{aligned}$$

Min C = \$16.50 when 300 g mix A and 50 g mix B are used.
(B) The minimum cost decreases to \$13.00 when 100 g mix A and 150 g mix B are used.
(C) The minimum cost increases to \$17.00 when 425 g mix A and no mix B are used. *(5-2)*

Chapter 6

Exercise 6-1

1. 115 **3.** 300 **5.** 165 **7.** 210 **9.** 95 **11.** 260

13. (A) 4 ways: **15.** (A) 12 combined outcomes:

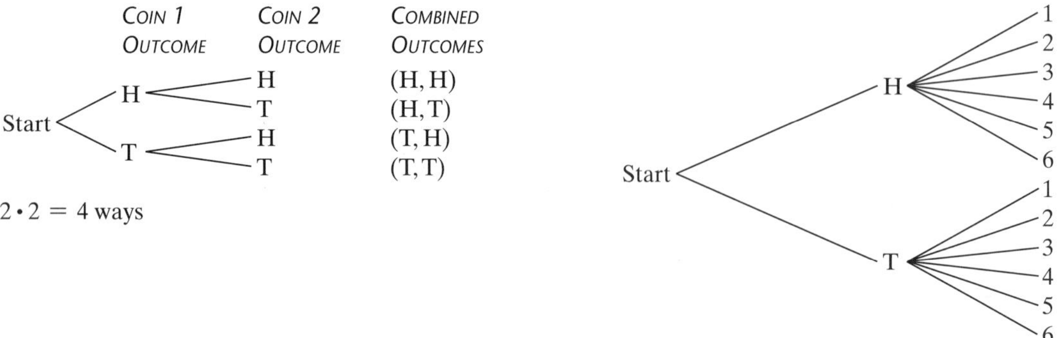

(B) $2 \cdot 2 = 4$ ways

(B) $2 \cdot 6 = 12$ combined outcomes

17. $n(A \cap B') = 60, n(A \cap B) = 20, n(A' \cap B) = 30, n(A' \cap B') = 90$

19. $n(A \cap B') = 5, n(A \cap B) = 20, n(A' \cap B) = 35, n(A' \cap B') = 40$

21.

	A	A'	*Totals*
B	30	60	90
B'	40	70	110
Totals	70	130	200

23.

	A	A'	*Totals*
B	20	35	55
B'	25	20	45
Totals	45	55	100

25. (A) True (B) False
27. $5 \cdot 3 \cdot 4 \cdot 2 = 120$
29. $6 \cdot 5 \cdot 4 \cdot 3 = 360$; $6 \cdot 6 \cdot 6 \cdot 6 = 1{,}296$; $6 \cdot 5 \cdot 5 \cdot 5 = 750$

31. $10 \cdot 9 \cdot 8 \cdot 7 \cdot 6 = 30{,}240$; $10 \cdot 10 \cdot 10 \cdot 10 \cdot 10 = 100{,}000$; $10 \cdot 9 \cdot 9 \cdot 9 \cdot 9 = 65{,}610$
33. $26 \cdot 26 \cdot 26 \cdot 10 \cdot 10 \cdot 10 = 17{,}576{,}000$; $26 \cdot 25 \cdot 24 \cdot 10 \cdot 9 \cdot 8 = 11{,}232{,}000$
35. No, the same 8 combined choices are available either way. **37.** 14 **39.** 14
41. (A) 6 combined outcomes:

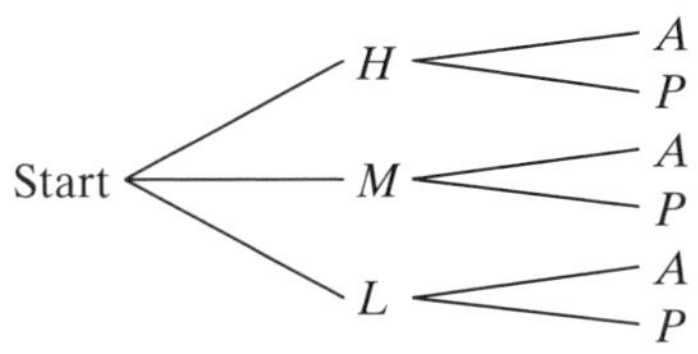

(B) $3 \cdot 2 = 6$

43. 12 **45.** (A) 1,010 (B) 190 (C) 270 **47.** 1,570
49. (A) 102 (B) 689 (C) 1,470 (D) 1,372

51. (A) 12 classifications:

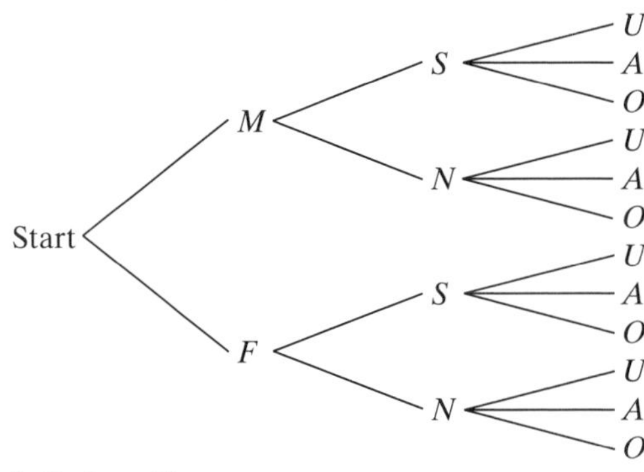

(B) $2 \cdot 2 \cdot 3 = 12$

53. 2,905

Exercise 6-2

1. 24 **3.** 9 **5.** 990 **7.** 10 **9.** 35 **11.** 1 **13.** 60 **15.** 6,497,400 **17.** 10 **19.** 270,725
21. (A) Permutation (B) Combination **23.** $P_{10,3} = 10 \cdot 9 \cdot 8 = 720$ **25.** $C_{7,3} = 35$; $P_{7,3} = 210$
27. The factorial function $x!$ grows much faster than the exponential function 3^x, which in turn grows much faster than the cubic function x^3.
29. $C_{13,5} = 1{,}287$ **31.** $C_{13,5}C_{13,2} = 100{,}386$ **33.** $C_{8,3}C_{10,4}C_{7,2} = 246{,}960$
35. The numbers are the same read up or down, since $C_{n,r} = C_{n,n-r}$.
37. (A) $C_{8,2} = 28$ (B) $C_{8,3} = 56$ (C) $C_{8,4} = 70$ **39.** $P_{5,2} = 20$; $P_{5,3} = 60$; $P_{5,4} = 120$; $P_{5,5} = 120$
41. (A) $P_{8,5} = 6{,}720$ (B) $C_{8,5} = 56$ (C) $2 \cdot C_{6,4} = 30$
43. For many calculators $k = 69$, but your calculator may be different. **45.** $C_{24,12} = 2{,}704{,}156$
47. (A) $C_{24,3} = 2{,}024$ (B) $C_{19,3} = 969$
49. (A) $C_{30,10} = 30{,}045{,}015$ (B) $C_{8,2}C_{12,5}C_{10,3} = 2{,}661{,}120$
51. (A) $C_{6,3}C_{5,2} = 200$ (B) $C_{6,4}C_{5,1} = 75$ (C) $C_{6,5} = 6$ (D) $C_{11,5} = 462$ (E) $C_{6,4}C_{5,1} + C_{6,5} = 81$
53. 336; 512 **55.** $P_{4,2} = 12$

Exercise 6-3

1. Occurrence of E is certain. **3.** $\frac{1}{2}$
5. (A) Reject; no probability can be negative (B) Reject; $P(J) + P(G) + P(P) + P(S) \neq 1$ (C) Acceptable
7. $P(J) + P(P) = .56$ **9.** $\frac{1}{8}$ **11.** $1/P_{10,3} \approx .0014$ **13.** $C_{26,5}/C_{52,5} \approx .025$ **15.** $C_{12,5}/C_{52,5} \approx .000\ 305$
17. $S = \{$All days in a year, 365, excluding leap year$\}$; $\frac{1}{365}$, assuming each day is as likely as any other day for a person to be born

19. $1/P_{5,5} = 1/5! = .008\ 33$ **21.** $\frac{1}{36}$ **23.** $\frac{5}{36}$ **25.** $\frac{1}{6}$ **27.** $\frac{7}{9}$ **29.** 0 **31.** $\frac{1}{3}$ **33.** $\frac{2}{9}$ **35.** $\frac{2}{3}$
37. $\frac{1}{4}$ **39.** $\frac{1}{4}$ **41.** $\frac{3}{4}$
43. (A) Yes (B) Yes, because we would expect, on average, 20 heads in 40 flips; $P(\text{H}) = \frac{37}{40} = .925$; $P(\text{T}) = \frac{3}{40} = .075$
45. $\frac{1}{9}$ **47.** $\frac{1}{3}$ **49.** $\frac{1}{9}$ **51.** $\frac{4}{9}$ **53.** $C_{16,5}/C_{52,5} \approx .001\ 68$ **55.** $48/C_{52,5} \approx .000\ 018\ 5$ **57.** $4/C_{52,5} \approx .000\ 001\ 5$
59. $C_{4,2}C_{4,3}/C_{52,5} \approx .000\ 009$ **61.** (A) $\frac{7}{50} = .14$ (B) $\frac{1}{6} \approx .167$ (C) Answer depends on results of simulation.
63. (A) Represent the outcomes H and T by 1 and 2, respectively, and select 500 random integers from the integers 1 and 2. (B) Answer depends on results of simulation. (C) Each is $\frac{1}{2} = .5$
65. (A) $1/P_{12,4} \approx .000\ 084$ (B) $1/12^4 \approx .000\ 048$
67. (A) $C_{6,3}C_{5,2}/C_{11,5} \approx .433$ (B) $C_{6,4}C_{5,1}/C_{11,5} \approx .162$ (C) $C_{6,5}C_{11,5} \approx .013$ (D) $(C_{6,4}C_{5,1} + C_{6,5})/C_{11,5} \approx .175$
69. (A) $1/P_{8,3} \approx .0030$ (B) $1/8^3 \approx .0020$ **71.** (A) $P_{6,2}/P_{11,2} \approx .273$ (B) $(C_{5,3} + C_{6,1}C_{5,2})/C_{11,3} \approx .424$

Exercise 6-4

1. .997 **3.** (1); $\frac{1}{2}$ **5.** (2); $\frac{7}{10}$ **7.** .4 **9.** .25 **11.** .05 **13.** .2 **15.** .6 **17.** .65 **19.** $\frac{1}{4}$ **21.** $\frac{11}{36}$
23. (A) $\frac{3}{5}; \frac{5}{3}$ (B) $\frac{1}{3}; \frac{3}{1}$ (C) $\frac{2}{3}; \frac{3}{2}$ (D) $\frac{11}{9}; \frac{9}{11}$ **25.** (A) $\frac{3}{11}$ (B) $\frac{11}{18}$ (C) $\frac{4}{5}$ or .8 (D) .49 **27.** (A) False (B) True
29. 1:1 **31.** 7:1 **33.** 2:1 **35.** 1:2 **37.** (A) $\frac{1}{8}$ (B) \$8 **39.** (A) .31; $\frac{31}{69}$ (B) .6; $\frac{3}{2}$ **41.** $\frac{11}{26}; \frac{11}{15}$
43. $\frac{7}{13}; \frac{7}{6}$ **45.** .78 **47.** $\frac{250}{1,000} = .25$
49. Either events A, B, and C are mutually exclusive, or events A and B are not mutually exclusive and the other pairs of events are mutually exclusive.
51. There are fewer calculator steps, and, in addition, 365! produces an overflow error on many calculators, while $P_{365,n}$ does not produce an overflow error for many values of n.
53. $P(E) = 1 - \dfrac{12!}{(12-n)!12^n}$
57. (A) $\frac{10}{50} + \frac{10}{50} = \frac{20}{50} = .4$ (B) $\frac{6}{36} + \frac{5}{36} = \frac{11}{36} \approx .306$ (C) Answer depends on results of simulation.
59. (A) $P(C \cup S) = P(C) + P(S) - P(C \cap S) = .45 + .75 - .35 = .85$ (B) $P(C' \cap S') = .15$
61. (A) $P(M_1 \cup A) = P(M_1) + P(A) - P(M_1 \cap A) = .2 + .3 - .05 = .45$
(B) $P[(M_2 \cap A') \cup (M_3 \cap A')] = P(M_2 \cap A') + P(M_3 \cap A') = .2 + .35 = .55$
63. $P(K' \cap D') = .9$ **65.** .83 **67.** $P(A \cap S) = \frac{50}{1,000} = .05$
69. (A) $P(U \cup N) = .22; \frac{11}{39}$ (B) $P[(D \cap A) \cup (R \cap A)] = .3; \frac{7}{3}$ **71.** $1 - C_{15,3}/C_{20,3} \approx .6$

Exercise 6-5

1. .50 **3.** .20 **5.** .10 **7.** .06 **9.** .50 **11.** .30 **13.** Independent **15.** Dependent
17. (A) $\frac{1}{2}$ (B) $2(\frac{1}{2})^8 \approx .007\ 81$ **19.** (A) $\frac{1}{4}$ (B) Dependent **21.** (A) .18 (B) .26
23. (A) Independent and not mutually exclusive (B) Dependent and mutually exclusive. **25.** $(\frac{1}{2})(\frac{1}{2}) = \frac{1}{4}; \frac{1}{2} + \frac{1}{2} - \frac{1}{4} = \frac{3}{4}$
27. (A) $(\frac{1}{4})(\frac{13}{51}) \approx .0637$ (B) $(\frac{1}{4})(\frac{1}{4}) = .0625$ **29.** (A) $\frac{3}{13}$ (B) Independent **31.** (A) Dependent (B) Independent

33.

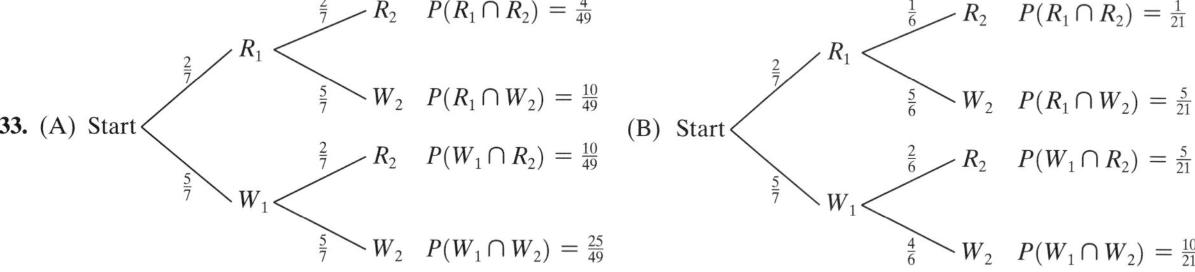

35. (A) $\frac{24}{49}$ (B) $\frac{11}{21}$ **37.** (A) False (B) False **39.** $\frac{5}{18}$ **41.** (A) .167 (B) .25 (C) .25
45. $P(A|A) = P(A \cap A)/P(A) = P(A)/P(A) = 1$ **47.** $P(A)P(B) \neq 0 = P(A \cap B)$

49. (A)

	H	*S*	*B*	*Totals*
Y	.400	.180	.020	.600
N	.150	.120	.130	.400
Totals	.550	.300	.150	1.000

(B) $P(Y|H) = \dfrac{.400}{.550} \approx .727$ (C) $P(Y|B) = \dfrac{.020}{.150} \approx .133$
(D) $P(S) = .300$; $P(S|Y) = .300$ (E) $P(H) = .550$; $P(H|Y) \approx .667$
(F) $P(B \cap N) = .130$ (G) Yes (H) No (I) No

51. (A) .167 (B) .25 (C) .25

53. (A)

	C	C'	*Totals*
R	.06	.44	.50
R'	.02	.48	.50
Totals	.08	.92	1.00

(B) Dependent
(C) $P(C|R) = .12$ and $P(C) = .08$; since $P(C|R) > P(C)$, the red dye should be banned.
(D) $P(C|R) = .04$ and $P(C) = .08$; since $P(C|R) < P(C)$, the red dye should not be banned, since it appears to prevent cancer.

55. (A)

	A	B	C	*Totals*
F	.130	.286	.104	.520
F'	.120	.264	.096	.480
Totals	.250	.550	.200	1.000

(B) $P(A|F) = \frac{.130}{.520} = .250$; $P(A|F') = \frac{.120}{.480} = .250$
(C) $P(C|F) = \frac{.104}{.520} = .200$; $P(C|F') = \frac{.096}{.480} = .200$
(D) $P(A) = .250$ (E) $P(B) = .550$; $P(B|F') = .550$
(F) $P(F \cap C) = .104$
(G) No; A, B, and C are independent of F and F'.

Exercise 6-6

1. $(.6)(.8) = .48$ **3.** $(.6)(.8) + (.4)(.3) = .60$ **5.** .80 **7.** .417 **9.** .375 **11.** .222 **13.** .50 **15.** .278

17. .125 **19.** .50 **21.** .375 **23.**

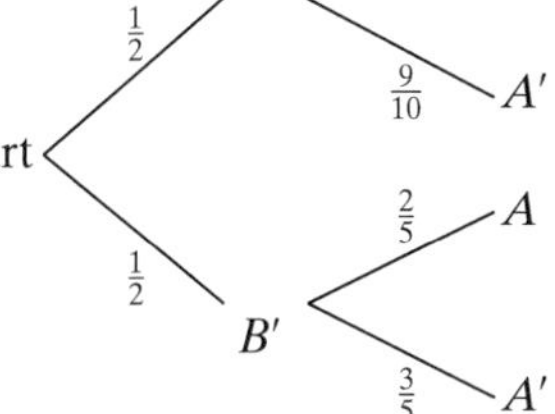

25. .25 **27.** .333 **29.** .50 **31.** .745

33. M and U are independent, which follows from any one of the following:
$P(M|U) = c = P(M)$; $P(U|M) = a = P(U)$; $P(M \cap U) = ca = P(M)P(U)$.

35. (A) True (B) True **37.** .235 **39.** $\frac{P(U_1 \cap R)}{P(R)} + \frac{P(U_1' \cap R)}{P(R)} = \frac{P(U_1 \cap R) + P(U_1' \cap R)}{P(R)} = \frac{P(R)}{P(R)} = 1$

41. .913; .226 **43.** .091; .545; .364 **45.** .667; .000 412 **47.** .231; .036 **49.** .941; .0588

Exercise 6-7

1. $E(X) = -.1$

3. Probability distribution:

x_i	0	1	2
p_i	$\frac{1}{4}$	$\frac{1}{2}$	$\frac{1}{4}$

$E(X) = 1$

5. Payoff table:

x_i	\$1	−\$1
p_i	$\frac{1}{2}$	$\frac{1}{2}$

$E(X) = 0$; game is fair

7. Payoff table:

x_i	−\$3	−\$2	−\$1	\$0	\$1	\$2
p_i	$\frac{1}{6}$	$\frac{1}{6}$	$\frac{1}{6}$	$\frac{1}{6}$	$\frac{1}{6}$	$\frac{1}{6}$

$E(X) = -50¢$; game is not fair

9. −\$0.50 **11.** −\$0.036; \$0.036
13. \$40. Let x = amount you should lose if a 6 turns up. Set up a payoff table; then set the expected value of the game equal to zero and solve for x.
15. Win \$1 **17.** A_2; \$210 **19.** Payoff table:

x_i	\$35	−\$1
p_i	$\frac{1}{38}$	$\frac{37}{38}$

$E(X) = -5.26¢$

21. .002 **23.** Payoff table:

x_i	\$499	\$99	\$19	\$4	−\$1
p_i	.0002	.0006	.001	.004	.9942

$E(X) = -80¢$

25. (A)

x_i	0	1	2
p_i	$\frac{7}{15}$	$\frac{7}{15}$	$\frac{1}{15}$

(B) .60

27. (A)

x_i	−$5	$195	$395	$595
p_i	.985	.0149	.000 059 9	.000 000 06

(B) $E(X) \approx -\$2$

29. (A) −$92 (B) The value per game is $\frac{-\$92}{200} = -\$0.46$, compared with an expected value of −$0.0526.
(C) The simulated gain or loss depends on the results of the simulation; the expected loss is $26.32.

31. Payoff table:

x_i	$4,850	−$150
p_i	.01	.99

$E(X) = -\$100$

33. Site *A*, with $E(X) = \$3.6$ million
35. 1.54
37. For A_1, $E(X) = \$4$, and for A_2, $E(X) = \$4.80$; A_2 is better

Review Exercise

1. (A) 12 combined outcomes:

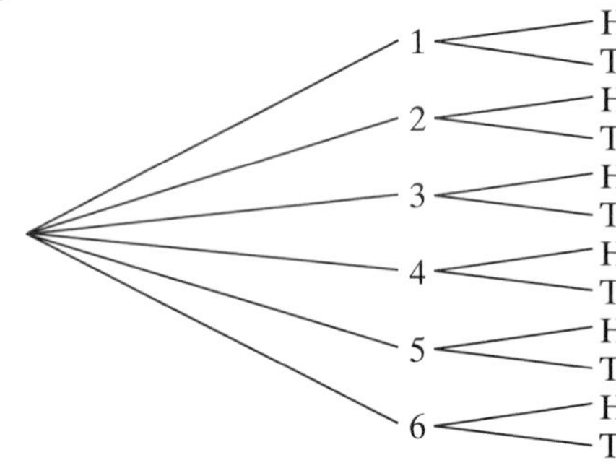

(B) 6 • 2 = 12 *(6-1)*

2. (A) 65 (B) 75 (C) 35 (D) 105 (E) 150 (F) 85 (G) 115 (H) 45 *(6-1)*
3. 15; 30 *(6-2)* **4.** 6 • 5 • 4 • 3 • 2 • 1 = 720 *(6-1)*
5. $P_{6,6} = 6! = 720$ *(6-2)*
6. $C_{13,5}/C_{52,5} \approx .0005$ *(6-3)*
7. $1/P_{15,2} \approx .0048$ *(6-3)*
8. $1/P_{10,3} \approx .0014$; $1/C_{10,3} \approx .0083$ *(6-3)*
9. .05 *(6-3)*
10. Payoff table:

x_i	−$2	−$1	$0	$1	$2
p_i	$\frac{1}{5}$	$\frac{1}{5}$	$\frac{1}{5}$	$\frac{1}{5}$	$\frac{1}{5}$

$E(X) = 0$; game is fair *(6-7)*

11. (A) .7 (B) .6 *(6-4)* **12.** $P(R \cup G) = .8$; odds for $R \cup G$ are 8 to 2 *(6-4)* **13.** $\frac{5}{11} \approx .455$ *(6-4)*
14. .27 *(6-5)* **15.** .20 *(6-5)* **16.** .02 *(6-5)* **17.** .03 *(6-5)* **18.** .15 *(6-5)* **19.** .1304 *(6-5)* **20.** .1*(6-5)*
21. No, since $P(T|Z) \neq P(T)$ *(6-5)* **22.** Yes, since $P(S \cap X) = P(S)P(X)$ *(6-5)* **23.** .4 *(6-5)* **24.** .2 *(6-5)*
25. .3 *(6-5)* **26.** .08 *(6-5)* **27.** .18 *(6-5)* **28.** .26 *(6-5)* **29.** .31 *(6-6)* **30.** .43 *(6-6)*
31. (A) $\frac{10}{32}$ (B) $\frac{1}{4}$
(C) As the sample in part (A) is increased in size, approximate empirical probabilities should approach the theoretical probabilities. *(6-3)*
32. (A) True (B) False *(6-4)* **33.** (A) False (B) False *(6-5)*
34. Payoff table:

x_i	$5	−$4	$2
p_i	.25	.5	.25

$E(X) = -25¢$; game is not fair *(6-7)*

35. (A) $\frac{1}{3}$ (B) $\frac{2}{9}$ *(6-5)*
36. (A) $\frac{2}{13}$; 2 to 11 (B) $\frac{4}{13}$; 4 to 9 (C) $\frac{12}{13}$; 12 to 1 *(6-4)*
37. (A) 1 to 8 (B) $8 *(6-4)*
38. (A) $P(2 \text{ heads}) = .21$, $P(1 \text{ head}) = .48$, $P(0 \text{ heads}) = .31$
(B) $P(2 \text{ heads}) = .25$, $P(1 \text{ head}) = .50$, $P(0 \text{ heads}) = .25$
(C) 2 heads, 250; 1 head, 500; 0 heads, 250 *(6-3, 6-7)*
39. 5 children, 15 grandchildren, and 30 great grandchildren, for a total of 50 descendants *(6-1, 6-2)*
40. $\frac{1}{2}$; since the coin has no memory, the 10th toss is independent of the preceding 9 tosses. *(6-5)*

41. (A)

x_i	2	3	4	5	6	7	8	9	10	11	12
p_i	$\frac{1}{36}$	$\frac{2}{36}$	$\frac{3}{36}$	$\frac{4}{36}$	$\frac{5}{36}$	$\frac{6}{36}$	$\frac{5}{36}$	$\frac{4}{36}$	$\frac{3}{36}$	$\frac{2}{36}$	$\frac{1}{36}$

(B) $E(X) = 7$ *(6-7)*

42. $A = \{(1, 3), (2, 2), (3, 1), (2, 6), (3, 5), (4, 4), (5, 3), (6, 2), (6, 6)\}$;
$B = \{(1, 5), (2, 4), (3, 3), (4, 2), (5, 1), (6, 6)\}$; $P(A) = \frac{1}{4}$; $P(B) = \frac{1}{6}$; $P(A \cap B) = \frac{1}{36}$; $P(A \cup B) = \frac{7}{18}$ *(6-4)*
43. (1) Probability of an event cannot be negative; (2) sum of probabilities of simple events must be 1; (3) probability of an event cannot be greater than 1. *(6-3)*

44.

	A	A'	*Totals*
B	15	30	45
B'	35	20	55
Totals	50	50	100

(6-4)

45. (A) .6 (B) $\frac{5}{6}$ *(6-5)* **46.** (A) $\frac{1}{13}$ (B) Independent *(6-5)*
47. 336; 512; 392 *(6-1)* **48.** (A) $P_{6,3} = 120$ (B) $C_{5,2} = 10$ *(6-2)*
49. $C_{25,12} = C_{25,13} = 5{,}200{,}300$ *(6-2)* **50.** (A) $\frac{6}{25}$ (B) $\frac{3}{10}$ *(6-5)*
51. Part (B) *(6-5)* **52.** (A) 1.2 (B) 1.2 *(6-7)*
53. (A) $\frac{3}{5}$ (B) $\frac{1}{3}$ (C) $\frac{7}{15}$ (D) $\frac{9}{14}$ (E) $\frac{5}{8}$ (F) $\frac{3}{10}$ *(6-5, 6-6)* **54.** No *(6-5)*
55. (A) $C_{13,5}/C_{52,5}$ (B) $C_{13,3} \cdot C_{13,2}/C_{52,5}$ *(6-3)* **56.** $C_{8,2}/C_{10,4} = \frac{2}{15}$ *(6-3)* **57.** $N_1 \cdot N_2 \cdot N_3$ *(6-1)*
58. Events S and H are mutually exclusive. Hence, $P(S \cap H) = 0$, while $P(S) \neq 0$ and $P(H) \neq 0$. Therefore, $P(S \cap H) \neq P(S)P(H)$, which implies S and H are dependent. *(6-5)*
59. (A) $\frac{9}{50} = .18$ (B) $\frac{9}{36} = .25$
(C) The empirical probability depends on the results of the simulation; the theoretical probability is $\frac{5}{36} \approx .139$. *(6-3)*
60. The empirical probability depends on the results of the simulation; the theoretical probability is $\frac{2}{52} \approx .038$. *(6-3)*
61. (A) .350 (B) $\frac{3}{8} = .375$ (C) 375 *(6-3)* **62.** $-.0172$; .0172; no *(6-7)*
63. (A) $P_{10,3} = 720$ (B) $P_{6,3}/P_{10,3} = \frac{1}{6}$ (C) $C_{10,3} = 120$ (D) $(C_{6,3} + C_{6,2} \cdot C_{4,1})/C_{10,3} = \frac{2}{3}$ *(6-1, 6-2, 6-3)*
64. 33 *(6-1)* **65.** $1 - C_{7,3}/C_{10,3} = \frac{17}{24}$ *(6-3)* **66.** $\frac{12}{51} \approx .235$ *(6-5)* **67.** $\frac{12}{51} \approx .235$ *(6-5)*
68. (A)

x_i	2	3	4	5	6
p_i	$\frac{9}{36}$	$\frac{12}{36}$	$\frac{10}{36}$	$\frac{4}{36}$	$\frac{1}{36}$

(B) $E(X) = \frac{10}{3}$ *(6-7)*
69. $E(X) \approx -\$0.167$; no; $\$(10/3) \approx \$3.33$ *(6-7)* **70.** $2^5 = 32$; 6 *(6-1)*
71. (A) $\frac{1}{4}$; 1 to 3 (B) \$3 *(6-4, 6-6)* **72.** $1 - 10!/(5!10^5) \approx .70$ *(6-4)*
73. $P_{5,5} = 120$ *(6-1)* **74.** (A) 610 (B) 390 (C) 270 *(6-1)*
75. (A) .8 (B) .2 (C) .5 *(6-1, 6-4)*
76. $P(A \cap P) = P(A)P(P|A) = .34$ *(6-5)*
77. (A) $P(A) = .290$; $P(B) = .290$; $P(A \cap B) = .100$; $P(A|B) = .345$; $P(B|A) = .345$
(B) No, since $P(A \cap B) \neq P(A)P(B)$.
(C) $P(C) = .880$; $P(D) = .120$; $P(C \cap D) = 0$; $P(C|D) = 0$; $P(D|C) = 0$
(D) Yes, since $C \cap D = \varnothing$; dependent, since $P(C \cap D) = 0$ and $P(C)P(D) \neq 0$ *(6-5)*
78. Plan *A*: $E(X) = \$7.6$ million; plan *B*: $E(X) = \$7.8$ million; plan *B* *(6-7)*
79. Payoff table:

x_i	\$270	$-\$30$
p_i	.08	.92

$E(X) = -\$6$ *(6-7)*

80. $1 - (C_{10,4}/C_{12,4}) \approx .576$ *(6-2, 6-4)*
81. (A)

x_i	0	1	2
p_i	$\frac{12}{22}$	$\frac{9}{22}$	$\frac{1}{22}$

(B) $E(X) = \frac{1}{2}$ *(6-7)*

82. .955 *(6-6)* **83.** $\frac{6}{7} \approx .857$ *(6-6)*

Chapter 7

Exercise 7-1

1. $S_1 = \overset{A \quad B}{[.8 \quad .2]}$; the probability of being in state A after one trial is .8, and the probability of being in state B after one trial is .2.

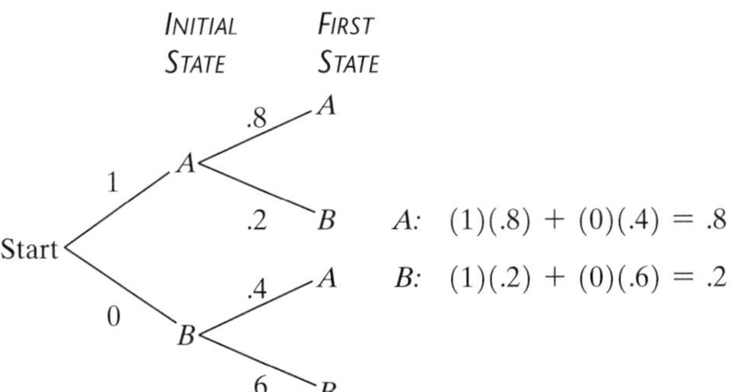

3. $S_1 = \overset{A \quad B}{[.6 \quad .4]}$; the probability of being in state A after one trial is .6, and the probability of being in state B after one trial is .4.

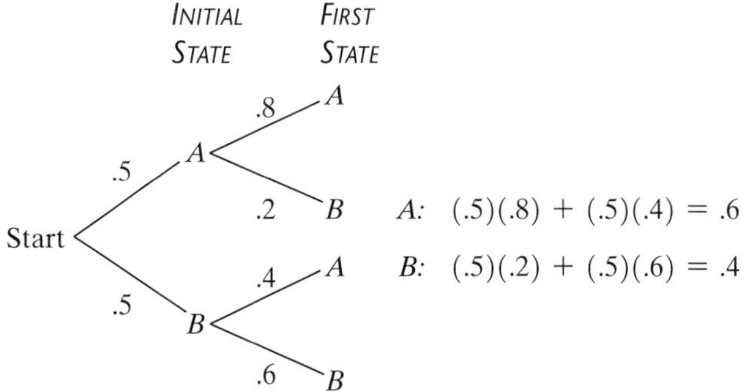

5. $S_2 = \overset{A\quad\ \ B}{[.72\quad .28]}$; the probability of being in state A after two trials is .72, and the probability of being in state B after two trials is .28.

7. $S_2 = \overset{A\quad\ \ B}{[.64\quad .36]}$; the probability of being in state A after two trials is .64, and the probability of being in state B after two trials is .36.

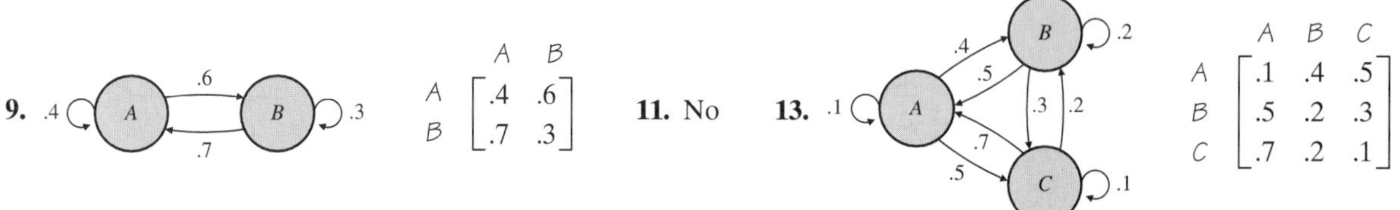

9. $\begin{array}{c|cc} & A & B \\ \hline A & .4 & .6 \\ B & .7 & .3 \end{array}$ **11.** No **13.** $\begin{array}{c|ccc} & A & B & C \\ \hline A & .1 & .4 & .5 \\ B & .5 & .2 & .3 \\ C & .7 & .2 & .1 \end{array}$

15. $a = .5, b = .6, c = .7$ **17.** $a = .7, b = 1, c = .2$ **19.** No **21.** $\begin{array}{c|cc} & A & B \\ \hline A & .3 & .7 \\ B & .9 & .1 \end{array}$

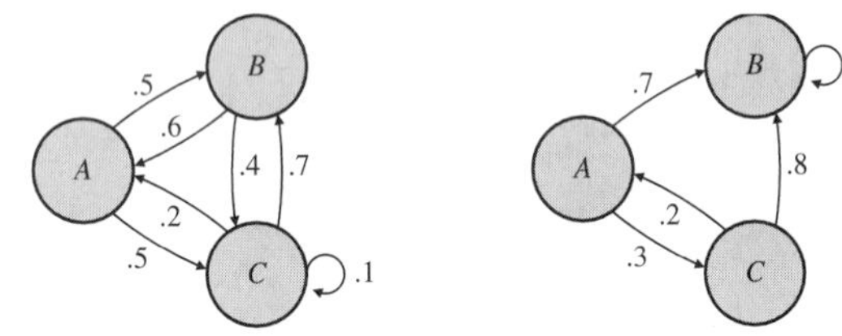

23. $\begin{array}{c|ccc} & A & B & C \\ \hline A & .6 & .1 & .3 \\ B & .2 & .3 & .5 \\ C & 0 & 0 & 1 \end{array}$ **25.** .35 **27.** .212

29. $S_2 = \overset{A\quad\ \ B\quad\ \ C}{[.43\quad .35\quad .22]}$; the probabilities of going from state A to states A, B, and C in two trials

31. $S_3 = \overset{A\quad\ \ \ B\quad\ \ \ C}{[.212\quad .298\quad .49]}$; the probabilities of going from state C to states A, B, and C in three trials

33. $n = 9$ **35.** $P^4 = \begin{array}{c|cc} & A & B \\ \hline A & .4375 & .5625 \\ B & .375 & .625 \end{array}$; $S_4 = \overset{A\quad\ \ \ B}{[.425\quad .575]}$ **37.** $P^4 = \begin{array}{c|ccc} & A & B & C \\ \hline A & .36 & .16 & .48 \\ B & .6 & 0 & .4 \\ C & .4 & .24 & .36 \end{array}$; $S_4 = \overset{A\quad\ \ \ B\quad\ \ \ C}{[.452\quad .152\quad .396]}$

41. (A) $\begin{array}{c|cccc} & A & B & C & D \\ \hline A & .0154 & .3534 & .0153 & .6159 \\ B & 0 & 1 & 0 & 0 \\ C & .0102 & .2962 & .0103 & .6833 \\ D & 0 & 0 & 0 & 1 \end{array}$ (B) .6159 (C) .2962 (D) 0

45. (A) $[.25\quad .75]$ (B) $[.25\quad .75]$ (C) $[.25\quad .75]$ (D) $[.25\quad .75]$
(E) All the state matrices appear to approach the same matrix, $S = [.25\quad .75]$, regardless of the values in the initial-state matrix.

47. $Q = \begin{bmatrix} .25 & .75 \\ .25 & .75 \end{bmatrix}$; the rows of Q are the same as the matrix S from Problem 45.

49. (A) R = Rain, R' = No rain

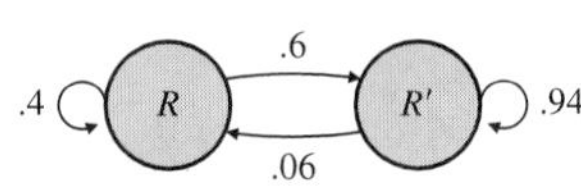

(B) $\begin{matrix} & R & R' \\ R & .4 & .6 \\ R' & .06 & .94 \end{matrix}$ (C) Saturday: .196; Sunday: .12664

51. (A)

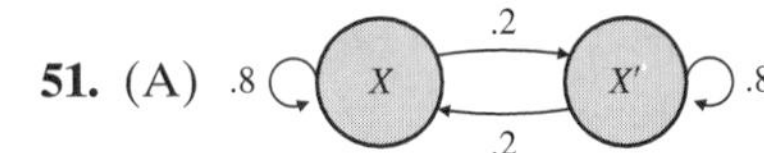

(B) $\begin{matrix} & X & X' \\ X & .8 & .2 \\ X' & .2 & .8 \end{matrix}$

(C) 32%; 39.2%

53. (A) N = National Property, U = United Family, O = Other companies

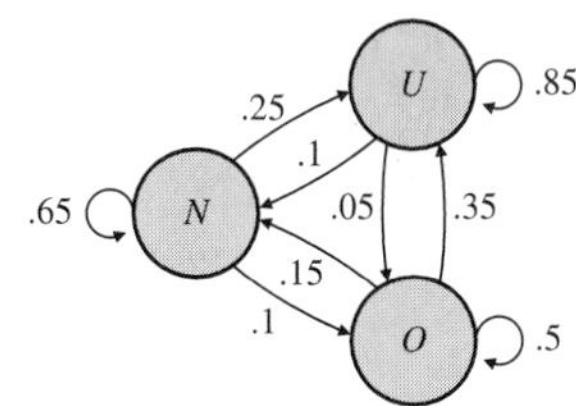

(B) $\begin{matrix} & N & U & O \\ N & .65 & .25 & .1 \\ U & .1 & .85 & .05 \\ O & .15 & .35 & .5 \end{matrix}$

(C) 38.5%; 32% (D) 45%; 53.65%

55. (A) B = Beginning agent, I = Intermediate agent, T = Terminated agent, Q = Qualified agent

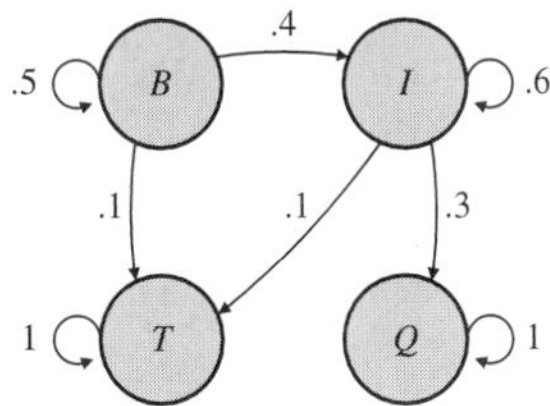

(B) $\begin{matrix} & B & I & T & Q \\ B & .5 & .4 & .1 & 0 \\ I & 0 & .6 & .1 & .3 \\ T & 0 & 0 & 1 & 0 \\ Q & 0 & 0 & 0 & 1 \end{matrix}$ (C) .12; .3612

57. (A) $\begin{matrix} & HMO & PPO & FFS \\ HMO & .8 & .15 & .05 \\ PPO & .2 & .7 & .1 \\ FFS & .25 & .3 & .45 \end{matrix}$

(B) HMO: 34.75%; PPO: 37%; FFS: 28.25%
(C) HMO: 42.2625%; PPO: 39.5875%; FFS: 18.15%

59. (A) $\begin{matrix} & H & R \\ H & .88 & .12 \\ R & .05 & .95 \end{matrix}$ H = Homeowner, R = Renter

(B) H: 42.433% (C) H: 40.219%

Exercise 7-2

1. Regular **3.** Not regular **5.** Regular **7.** Not regular **9.** Regular **11.** Not regular **13.** Regular

15. $S = [.4 \quad .6]$; $\overline{P} = \begin{bmatrix} .4 & .6 \\ .4 & .6 \end{bmatrix}$ **17.** $S = [.375 \quad .625]$; $\overline{P} = \begin{bmatrix} .375 & .625 \\ .375 & .625 \end{bmatrix}$

19. $S = [.3 \quad .5 \quad .2]$; $\overline{P} = \begin{bmatrix} .3 & .5 & .2 \\ .3 & .5 & .2 \\ .3 & .5 & .2 \end{bmatrix}$ **21.** $S = [.6 \quad .24 \quad .16]$; $\overline{P} = \begin{bmatrix} .6 & .24 & .16 \\ .6 & .24 & .16 \\ .6 & .24 & .16 \end{bmatrix}$

23. (A) True (B) False **25.** $S = [.3553 \quad .6447]$ **27.** $S = [.3636 \quad .4091 \quad .2273]$

29. (A) .4 Red —.6→ Blue .8; Blue —.2→ Red

(B) $\begin{matrix} & Red & Blue \\ Red & .4 & .6 \\ Blue & .2 & .8 \end{matrix}$

(C) $[.25 \quad .75]$; in the long run, the red urn will be selected 25% of the time and the blue urn 75% of the time.

31. (A) The state matrices alternate between $[.2 \quad .8]$ and $[.8 \quad .2]$; hence, they do not approach any one matrix.
(B) The state matrices are all equal to S_0; hence, S_0 is a stationary matrix.
(C) The powers of P alternate between P and I, the 2×2 identity; hence, they do not approach a limiting matrix.
(D) Parts (B) and (C) of Theorem 1 are not valid for this matrix. Since P is not regular, this is not a contradiction.

33. (A) Since P is not regular, it may have more than one stationary matrix.
(B) $[.5 \quad 0 \quad .5]$ is another stationary matrix.
(C) P has an infinite number of stationary matrices.

35. $\overline{P} = \begin{bmatrix} 1 & 0 & 0 \\ .25 & 0 & .75 \\ 0 & 0 & 1 \end{bmatrix}$; each row of $\overline{P}$ is a stationary matrix for P.

37. (A) .39; .3; .284; .277
(B) Each entry of the second column of P^{k+1} is the product of a row of P and the second column of P^k. Each entry of the latter is $\leq M_k$, so the product is $\leq M_k$.

39. 72.5%

41. (A) $S_1 = [.516 \quad .484]$; $S_2 = [.577 \quad .423]$

(B)

YEAR	DATA (%)	MODEL (%)
1970	43.3	43.3
1980	51.5	51.6
1990	57.5	57.7

(C) 74.1%

43. GTT: 25%; NCJ: 25%; Dash: 50%

45. Poor: 20%; satisfactory: 40%; preferred: 40%

47. 51%

49. Stationary matrix $= [.25 \quad .50 \quad .25]$

51. (A) $[.25 \quad .75]$ (B) 42.5%; 51.25%
(C) 60% rapid transit; 40% automobile

53. (A) $S_1 = [.334 \quad .666]$; $S_2 = [.343 \quad .657]$

(B)

YEAR	DATA (%)	MODEL (%)
1970	30.9	30.9
1980	33.3	33.4
1990	34.4	34.3

(C) 35%

Exercise 7-3

1. B, C **3.** No absorbing states **5.** A, D **7.** B is an absorbing state; absorbing chain
9. C is an absorbing state; not an absorbing chain

11. $\begin{array}{c|ccc} & B & A & C \\ \hline B & 1 & 0 & 0 \\ A & .5 & .2 & .3 \\ C & .1 & .5 & .4 \end{array}$ **13.** $\begin{array}{c|cccc} & B & D & A & C \\ \hline B & 1 & 0 & 0 & 0 \\ D & 0 & 1 & 0 & 0 \\ A & .4 & .1 & .3 & .2 \\ C & .4 & .3 & 0 & .3 \end{array}$ **15.** $\begin{array}{c|ccc} & B & C & A \\ \hline B & 1 & 0 & 0 \\ C & 0 & 1 & 0 \\ A & .3 & .5 & .2 \end{array}$ **17.** $\begin{array}{c|cccc} & B & D & A & C \\ \hline B & 1 & 0 & 0 & 0 \\ D & 0 & 1 & 0 & 0 \\ A & .2 & .4 & .1 & .3 \\ C & .2 & .1 & .5 & .2 \end{array}$

19. $\overline{P} = \begin{array}{c|ccc} & A & B & C \\ \hline A & 1 & 0 & 0 \\ B & 0 & 1 & 0 \\ C & .2 & .8 & 0 \end{array}$; $P(C \text{ to } A) = .2$; $P(C \text{ to } B) = .8$. It will take an average of 2 trials to go from C to either A or B.

21. $\overline{P} = \begin{array}{c|ccc} & A & B & C \\ \hline A & 1 & 0 & 0 \\ B & 1 & 0 & 0 \\ C & 1 & 0 & 0 \end{array}$; $P(B \text{ to } A) = 1$; $P(C \text{ to } A) = 1$. It will take an average of 4 trials to go from B to A and an average of 3 trials to go from C to A.

23. $\overline{P} = \begin{array}{c} \\ A \\ B \\ C \\ D \end{array}\begin{array}{c} \begin{array}{cccc} A & B & C & D \end{array} \\ \begin{bmatrix} 1 & 0 & 0 & 0 \\ 0 & 1 & 0 & 0 \\ .36 & .64 & 0 & 0 \\ .44 & .56 & 0 & 0 \end{bmatrix} \end{array}$; $P(C \text{ to } A) = .36$; $P(C \text{ to } B) = .64$; $P(D \text{ to } A) = .44$; $P(D \text{ to } B) = .56$. It will take an average of 3.2 trials to go from C to either A or B and an average of 2.8 trials to go from D to either A or B.

25. (A) $[.2 \quad .8 \quad 0]$ (B) $[.26 \quad .74 \quad 0]$ **27.** (A) $[1 \quad 0 \quad 0]$ (B) $[1 \quad 0 \quad 0]$

29. (A) $[.44 \quad .56 \quad 0 \quad 0]$ (B) $[.36 \quad .64 \quad 0 \quad 0]$ (C) $[.408 \quad .592 \quad 0 \quad 0]$ (D) $[.384 \quad .616 \quad 0 \quad 0]$

31. (A) True (B) False

33. $\begin{array}{c} \\ A \\ B \\ C \\ D \end{array}\begin{array}{c} \begin{array}{cccc} A & B & C & D \end{array} \\ \begin{bmatrix} 1 & 0 & 0 & 0 \\ 0 & 1 & 0 & 0 \\ .6375 & .3625 & 0 & 0 \\ .7375 & .2625 & 0 & 0 \end{bmatrix} \end{array}$ **35.** $\begin{array}{c} \\ A \\ B \\ C \\ D \\ E \end{array}\begin{array}{c} \begin{array}{ccccc} A & B & C & D & E \end{array} \\ \begin{bmatrix} 1 & 0 & 0 & 0 & 0 \\ 0 & 1 & 0 & 0 & 0 \\ .0875 & .9125 & 0 & 0 & 0 \\ .1875 & .8125 & 0 & 0 & 0 \\ .4375 & .5625 & 0 & 0 & 0 \end{bmatrix} \end{array}$ **37.** $\begin{array}{c} \\ A \\ B \\ C \\ D \end{array}\begin{array}{c} \begin{array}{cccc} A & B & C & D \end{array} \\ \begin{bmatrix} 0 & .52 & 0 & .48 \\ 0 & 1 & 0 & 0 \\ 0 & .36 & 0 & .64 \\ 0 & 0 & 0 & 1 \end{bmatrix} \end{array}$

43. (A) .370; .297; .227; .132; .045 (B) For large k, all entries of Q^k are close to 0.

45. (A) 75% (B) 12.5% (C) 7.5 months **47.** (A) Company A: 30%, company B: 15%, company C: 55% (B) 5 yr

49. (A) 91.52% (B) 4.96% (C) 6.32 days **51.** (A) .375 (B) 1.75 exits

Review Exercise

1. $S_1 = \begin{array}{c} \begin{array}{cc} A & B \end{array} \\ [.32 \quad .68] \end{array}$; $S_2 = \begin{array}{c} \begin{array}{cc} A & B \end{array} \\ [.328 \quad .672] \end{array}$. The probability of being in state A after one trial is .32 and after two trials is .328; the probability of being in state B after one trial is .68 and after two trials is .672. *(7-1)*

2. State A is absorbing; chain is absorbing. *(7-2, 7-3)* **3.** No absorbing states; chain is regular. *(7-2, 7-3)*

4. No absorbing states; chain is neither. *(7-2, 7-3)* **5.** States B and C are absorbing; chain is absorbing. *(7-2, 7-3)*

6. States A and B are absorbing; chain is neither. *(7-2, 7-3)*

7. $\begin{array}{c} \\ A \\ B \\ C \end{array}\begin{array}{c} \begin{array}{ccc} A & B & C \end{array} \\ \begin{bmatrix} 0 & 1 & 0 \\ .1 & 0 & .9 \\ 0 & 1 & 0 \end{bmatrix} \end{array}$; no absorbing states; chain is neither. *(7-1, 7-2, 7-3)*

8. $\begin{array}{c} \\ A \\ B \\ C \end{array}\begin{array}{c} \begin{array}{ccc} A & B & C \end{array} \\ \begin{bmatrix} 0 & 1 & 0 \\ .1 & .2 & .7 \\ 0 & 0 & 1 \end{bmatrix} \end{array}$; C is absorbing; chain is absorbing. *(7-1, 7-2, 7-3)*

9. $\begin{array}{c} \\ A \\ B \\ C \end{array}\begin{array}{c} \begin{array}{ccc} A & B & C \end{array} \\ \begin{bmatrix} 0 & 0 & 1 \\ .1 & .2 & .7 \\ 0 & 1 & 0 \end{bmatrix} \end{array}$; no absorbing states; chain is regular. *(7-1, 7-2, 7-3)*

10. $\begin{array}{c} \\ A \\ B \\ C \\ D \end{array}\begin{array}{c} \begin{array}{cccc} A & B & C & D \end{array} \\ \begin{bmatrix} .3 & .2 & 0 & .5 \\ 0 & 1 & 0 & 0 \\ 0 & 0 & .2 & .8 \\ 0 & 0 & .3 & .7 \end{bmatrix} \end{array}$; B is absorbing; chain is neither. *(7-1, 7-2, 7-3)*

11. 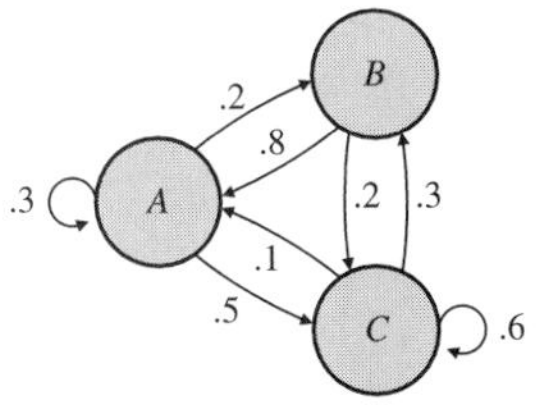 $\begin{array}{c} \\ A \\ B \\ C \end{array}\begin{array}{c} \begin{array}{ccc} A & B & C \end{array} \\ \begin{bmatrix} .3 & .2 & .5 \\ .8 & 0 & .2 \\ .1 & .3 & .6 \end{bmatrix} \end{array}$ *(7-1)* **12.** (A) .3 (B) .675 *(7-1)*

13. $S = \begin{array}{c} \begin{array}{cc} A & B \end{array} \\ [.25 \quad .75] \end{array}$; $\overline{P} = \begin{array}{c} \\ A \\ B \end{array}\begin{array}{c} \begin{array}{cc} A & B \end{array} \\ \begin{bmatrix} .25 & .75 \\ .25 & .75 \end{bmatrix} \end{array}$ *(7-2)* **14.** $S = \begin{array}{c} \begin{array}{ccc} A & B & C \end{array} \\ [.4 \quad .48 \quad .12] \end{array}$; $\overline{P} = \begin{array}{c} \\ A \\ B \\ C \end{array}\begin{array}{c} \begin{array}{ccc} A & B & C \end{array} \\ \begin{bmatrix} .4 & .48 & .12 \\ .4 & .48 & .12 \\ .4 & .48 & .12 \end{bmatrix} \end{array}$ *(7-2)*

15. $\begin{array}{c} \\ A \\ B \\ C \end{array}\begin{array}{c} \begin{array}{ccc} A & B & C \end{array} \\ \begin{bmatrix} 1 & 0 & 0 \\ 0 & 1 & 0 \\ .75 & .25 & 0 \end{bmatrix} \end{array}$; $P(C \text{ to } A) = .75$; $P(C \text{ to } B) = .25$. It takes an average of 2.5 trials to go from C to an absorbing state. *(7-3)*

16. $\begin{array}{c} \\ A \\ B \\ C \\ D \end{array}\begin{array}{c} \begin{array}{cccc} A & B & C & D \end{array} \\ \begin{bmatrix} 1 & 0 & 0 & 0 \\ 0 & 1 & 0 & 0 \\ .2 & .8 & 0 & 0 \\ .3 & .7 & 0 & 0 \end{bmatrix} \end{array}$; $P(C \text{ to } A) = .2$; $P(C \text{ to } B) = .8$; $P(D \text{ to } A) = .3$; $P(D \text{ to } B) = .7$. It takes an average of 2 trials to go from C to an absorbing state and an average of 3 trials to go from D to an absorbing state. *(7-3)*

21. $\begin{array}{c} \\ B \\ D \\ A \\ C \end{array}\begin{array}{c} \begin{array}{cccc} B & D & A & C \end{array} \\ \begin{bmatrix} 1 & 0 & 0 & 0 \\ 0 & 1 & 0 & 0 \\ .1 & .1 & .6 & .2 \\ .2 & .2 & .3 & .3 \end{bmatrix} \end{array}$ *(7-3)*

22. (A) $\begin{array}{c} \begin{array}{ccc} A & B & C \end{array} \\ \begin{bmatrix} .1 & .4 & .5 \end{bmatrix} \end{array}$ (B) $\begin{array}{c} \begin{array}{ccc} A & B & C \end{array} \\ \begin{bmatrix} .1 & .4 & .5 \end{bmatrix} \end{array}$ *(7-3)*

23. (A) $\begin{array}{c} \begin{array}{ccc} A & B & C \end{array} \\ \begin{bmatrix} .25 & .75 & 0 \end{bmatrix} \end{array}$ (B) $\begin{array}{c} \begin{array}{ccc} A & B & C \end{array} \\ \begin{bmatrix} .55 & .45 & 0 \end{bmatrix} \end{array}$ *(7-3)*

24. No. Each row of P would contain a 0 and a 1, but none of the four matrices with this property is regular. *(7-2)*

25. Yes; for example, $P = \begin{bmatrix} 0 & 0 & 1 \\ 0 & 0 & 1 \\ .2 & .3 & .5 \end{bmatrix}$ is regular. *(7-2)*

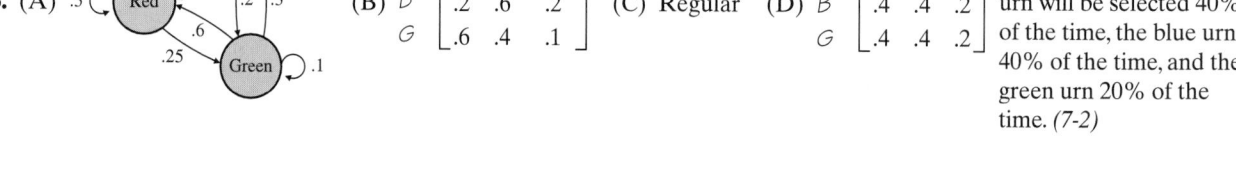

26. (A)

(B) $\begin{array}{c} \\ R \\ B \\ G \end{array}\begin{array}{c} \begin{array}{ccc} R & B & G \end{array} \\ \begin{bmatrix} .5 & .25 & .25 \\ .2 & .6 & .2 \\ .6 & .4 & .1 \end{bmatrix} \end{array}$ (C) Regular (D) $\begin{array}{c} \\ R \\ B \\ G \end{array}\begin{array}{c} \begin{array}{ccc} R & B & G \end{array} \\ \begin{bmatrix} .4 & .4 & .2 \\ .4 & .4 & .2 \\ .4 & .4 & .2 \end{bmatrix} \end{array}$ In the long run, the red urn will be selected 40% of the time, the blue urn 40% of the time, and the green urn 20% of the time. *(7-2)*

27. (A)

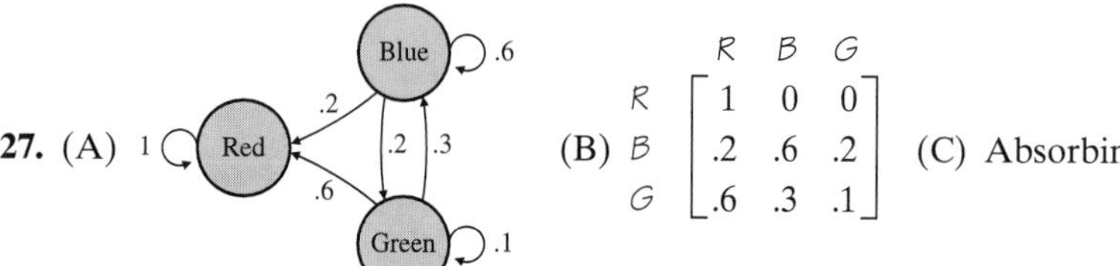

(B) $\begin{array}{c} \\ R \\ B \\ G \end{array}\begin{array}{c} \begin{array}{ccc} R & B & G \end{array} \\ \begin{bmatrix} 1 & 0 & 0 \\ .2 & .6 & .2 \\ .6 & .3 & .1 \end{bmatrix} \end{array}$ (C) Absorbing (D) $\begin{array}{c} \\ R \\ B \\ G \end{array}\begin{array}{c} \begin{array}{ccc} R & B & G \end{array} \\ \begin{bmatrix} 1 & 0 & 0 \\ 1 & 0 & 0 \\ 1 & 0 & 0 \end{bmatrix} \end{array}$ Once the red urn is selected, the blue and green urns will never be selected again. It will take an average of 3.67 trials to reach the red urn from the blue urn and an average of 2.33 trials to reach the red urn from the green urn. *(7-3)*

29. No such chain exists. *(7-2, 7-3)* **30.** No such chain exists. *(7-2, 7-3)* **31.** No such chain exists. *(7-2)*

32. $S = \begin{bmatrix} 1 & 0 & 0 \end{bmatrix}$ and $S' = \begin{bmatrix} 0 & 1 & 0 \end{bmatrix}$ are both stationary matrices for $P = \begin{array}{c} \\ A \\ B \\ C \end{array}\begin{array}{c} \begin{array}{ccc} A & B & C \end{array} \\ \begin{bmatrix} 1 & 0 & 0 \\ 0 & 1 & 0 \\ .6 & .3 & .1 \end{bmatrix} \end{array}$ *(7-3)*

33. $P = \begin{array}{c} \\ A \\ B \end{array}\begin{array}{c} \begin{array}{cc} A & B \end{array} \\ \begin{bmatrix} 0 & 1 \\ 1 & 0 \end{bmatrix} \end{array}$ *(7-2, 7-3)* **34.** No such chain exists. *(7-2)* **35.** No such chain exists. *(7-3)*

36. No limiting matrix *(7-2, 7-3)* **37.** $P = \begin{matrix} & A & B & C & D \\ A & .392 & .163 & .134 & .311 \\ B & .392 & .163 & .134 & .311 \\ C & .392 & .163 & .134 & .311 \\ D & .392 & .163 & .134 & .311 \end{matrix}$ *(7-2)*

38. (A) 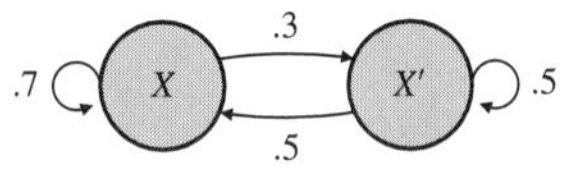

(B) $\begin{matrix} & X & X' \\ X & .7 & .3 \\ X' & .5 & .5 \end{matrix}$ (C) $\begin{matrix} X & X' \\ [.2 & .8] \end{matrix}$

(D) $\begin{matrix} X & X' \\ [.54 & .46] \end{matrix}$; 54% of the consumers will purchase brand X on the next purchase. (E) $\begin{matrix} X & X' \\ [.625 & .375] \end{matrix}$ (F) 62.5% *(7-2)*

39. (A) Brand *A:* 24%, brand *B:* 32%, brand *C:* 44% (B) 4 yr *(7-3)*

40. (A) $S_1 = \begin{matrix} V & V' \\ [.585 & .415] \end{matrix}$; $S_2 = \begin{matrix} V & V' \\ [.728 & .272] \end{matrix}$ (B)

YEAR	DATA (%)	MODEL (%)
1984	10.6	10.6
1988	58.0	58.5
1992	72.5	72.8

(C) 79% *(7-2)*

41. (A) 63.75% (B) 15% (C) 8.75 yr *(7-3)* **42.** $\overline{P} = \begin{matrix} & \text{Red} & \text{Pink} & \text{White} \\ \text{Red} & 1 & 0 & 0 \\ \text{Pink} & 1 & 0 & 0 \\ \text{White} & 1 & 0 & 0 \end{matrix}$ *(7-3)*

43. (A) $S_1 = \begin{matrix} F & F' \\ [.022 & .978] \end{matrix}$; $S_2 = \begin{matrix} F & F' \\ [.0175 & .9825] \end{matrix}$ (B)

YEAR	DATA (%)	MODEL (%)
1980	2.8	2.8
1985	2.3	2.2
1990	1.9	1.75

(C) 0.4% *(7-2)*

Chapter 8

Exercise 8-1

1. 45 **3.** 15 **5.** 15 **9.**

h	−0.1	−0.01	−0.001 → 0 ← 0.001	0.01	0.1
$\frac{f(1+h) - f(1)}{h}$	5.7	5.97	5.997 → 6 ← 6.003	6.03	6.3

11. 6 **13.** 6 ft/sec **15.** 6 **19.** 100 ft/sec **21.** 80 ft/sec **23.** (A) 4 (B) 3 (C) 2 (D)

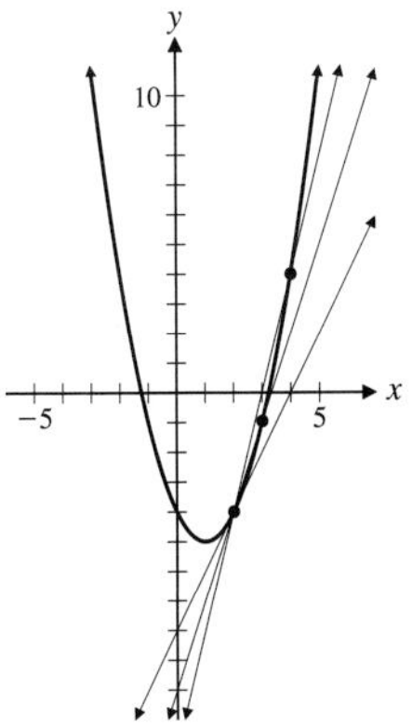

25. Slope at $x = -1$ is 1; slope at $x = 3$ is -2.

27. Slope at $x = -3$ is -5; slope at $x = -1$ is 0; slope at $x = 1$ is -1; slope at $x = 3$ is 4.

29.

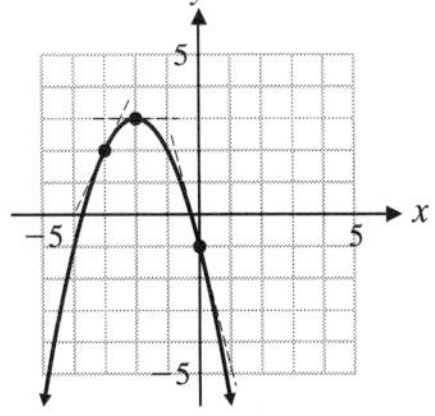

31.

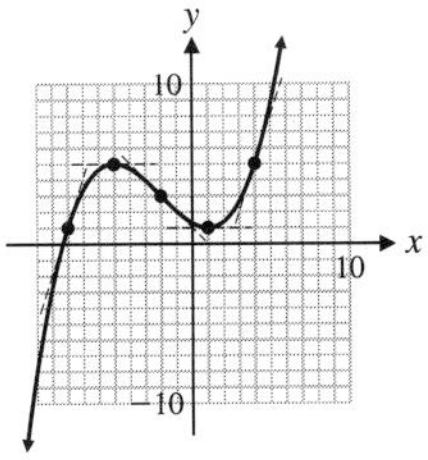

33. 8 **35.** $\frac{1}{4}$

37.

x	-3	-2	-1	0	1	2	3
Slope	-3	-2	-1	0	1	2	3

Slope function: $y = x$

39. The slope of the line is m. The slope of the graph at any point on the graph is also m.

41. $\dfrac{f(1+h) - f(1)}{h} = 6 + 3h \to 6$ as $h \to 0$

43.

h	-0.1	-0.01	$-0.001 \to$	0	$\leftarrow 0.001$	0.01	0.1
$\dfrac{f(0+h) - f(0)}{h}$	-1	-1	-1	$\to -1 \neq 1 \leftarrow$	1	1	1

The slope of the graph is not defined at (0, 0).

45. (A) -0.12 hr/yr (B) \$0.37/yr

47. (A)

L1	L2	L3
0	37.1	------
5	36.1	
10	35.3	
15	34.9	
20	34.5	
------	------	

L1={0,5,10,15,2...

QuadReg
y=ax²+bx+c
a=.0045714286
b=-.2194285714
c=37.08857143

(B) -0.13 hr/yr

49. (A) At a production level of 1,000 car seats, the revenue is \$35,000 and is increasing at the rate of \$10 per seat.
(B) At a production level of 1,300 car seats, the revenue is \$35,750 and is decreasing at the rate of \$5 per seat.

51. In 1990, the annual production was 11,360,000 metric tons and was decreasing at the rate of 130,000 metric tons per year.

53. (A) \$67.5 billion/yr (B) \$1.7 billion/yr

55. (A)

L1	L2	L3
0	476.9	------
1	526.2	
2	583.6	
3	652.4	
4	728.6	
------	------	

L1={0,1,2,3,4}

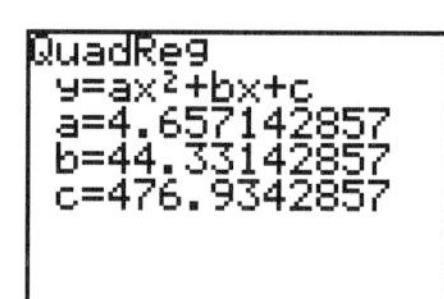
QuadReg
y=ax²+bx+c
a=4.657142857
b=44.33142857
c=476.9342857

(B) \$62.96 billion/yr

57. In 1990, the number of male infant deaths per 100,000 births was 9.8 and was decreasing at the rate of 0.42 death per 100,000 births per year.

Exercise 8-2

1.

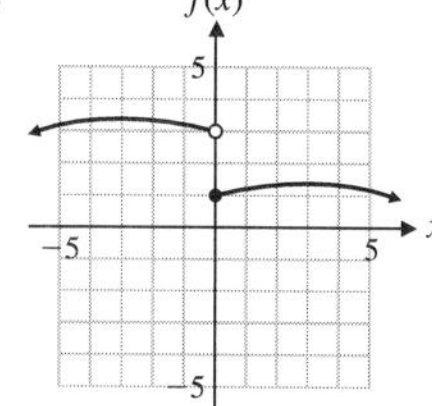

3.

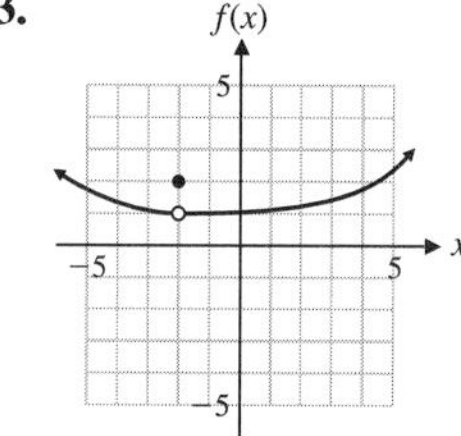

5. (A) 2 (B) 2 (C) 2 (D) 2
7. (A) 1 (B) 2 (C) Does not exist (D) 2 (E) No
9. (A) 1 (B) 2 (C) Does not exist (D) Does not exist (E) No
11. (A) 1 (B) 1 (C) 1 (D) 3 (E) Yes, define $g(3) = 1$.
13. -4 **15.** 36 **17.** $\frac{5}{9}$ **19.** $\sqrt{5}$ **21.** 1.4 **23.** 47
25. $\frac{5}{3}$ **27.** 243 **29.** 3 **31.** Does not exist **33.** -6
35. 1 **37.** -1 **39.** Does not exist **41.** 1 **43.** 0.5
45. -1 **47.** -5 **49.** $\frac{2}{3}$ **51.** 0 **53.** 3 **55.** 4
57. $1/(2\sqrt{2})$ **59.** Does not exist
65. 80 ft/sec **67.** $-1, 0, 2$
69. (A) The limit does not exist. The values of $1/x$ are large negative numbers when x is close to 0 on the left.
(B) The limit does not exist. The values of $1/x$ are large positive numbers when x is close to 0 on the right.

61. Slope = 4

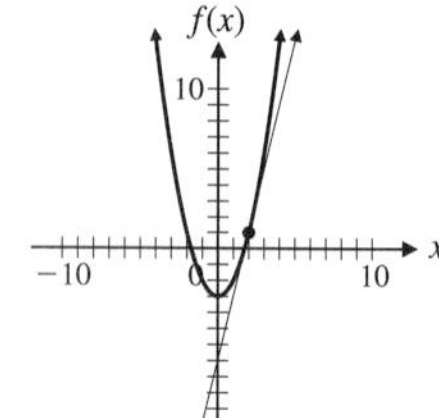

63. Slope = $\frac{1}{4}$

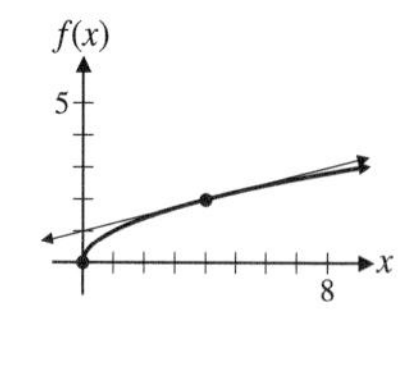

71. (A) $\lim_{x\to 1^-} f(x) = 2$
$\lim_{x\to 1^+} f(x) = 3$

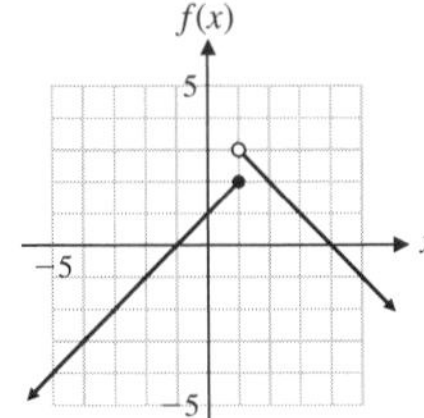

(B) $\lim_{x\to 1^-} f(x) = 3$
$\lim_{x\to 1^+} f(x) = 2$

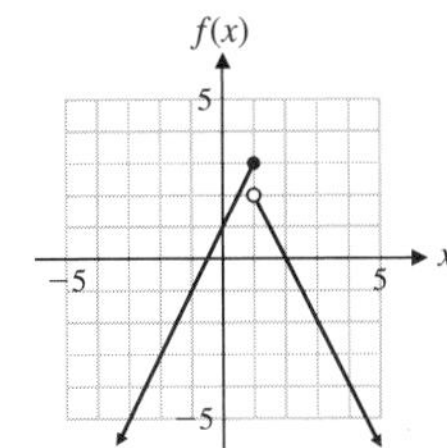

(C) $\lim_{x\to 1^-} f(x) = 2.5$
$\lim_{x\to 1^+} f(x) = 2.5$

(D) The graph in (A) is broken when it jumps from $(1, 2)$ up to $(1, 3)$. The graph in (B) is also broken when it jumps down from $(1, 3)$ to $(1, 2)$. The graph in (C) is one continuous piece, with no breaks or jumps.

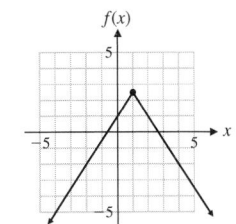

73. $2a$ **75.** $1/(2\sqrt{a})$
77. (A) $-1, 1, 3$ (B) $2a - 3$
(C) The slopes are the same as in part (A). In part (A), each value of a required a new limit operation. In part (B), a single limit operation produces the slope for all values of a.
79. 10 **81.** 0.693 **83.** 2.718 **85.** Typical values of n are 95 on a TI-81, 94 on a TI-82 and TI-83, and 126 on a TI-85.
87. $\lim_{x\to -2^-} f(x) = \lim_{x\to -2^+} f(x) = 2$
$\lim_{x\to 2^-} f(x) = \lim_{x\to 2^+} f(x) = 2$

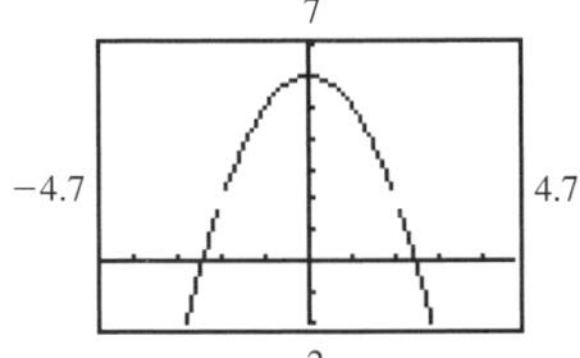

89. $\lim_{x\to -3^-} f(x) = -3, \lim_{x\to -3^+} f(x) = 3$
$\lim_{x\to 3^-} f(x) = -3, \lim_{x\to 3^+} f(x) = 3$

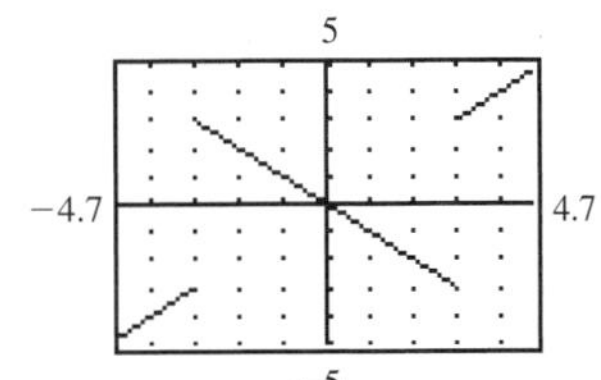

91. (A) At a production level of 900 units, the revenue is $99,000 and is increasing at the rate of $20 per jigsaw.
(B) At a production level of 1,200 units, the revenue is $96,000 and is decreasing at the rate of $40 per jigsaw.
93. In 1990, the revolving-credit debt was $233.1 billion and was increasing at the rate of $23.8 billion per year.
95. The alveolar pressure at 9,000 feet is 70.3% of that at sea level and is decreasing at the rate of 2.94% per thousand feet.
97. (A) The school-aged population in 1970 was 50 million and was increasing at the rate of 0.3 million per year.
(B) The school-aged population in 1990 was 44 million and was decreasing at the rate of 0.9 million per year.

Exercise 8-3

1. (A) -3; slope of the secant line through $(1, f(1))$ and $(2, f(2))$
(B) $-2 - h$; slope of the secant line through $(1, f(1))$ and $(1 + h, f(1 + h))$
(C) -2; slope of the tangent line at $(1, f(1))$
3. $4x - 3$ **5.** $3x^2 - 2x$ **7.** $f'(x) = 0; f'(1) = 0, f'(2) = 0, f'(3) = 0$
9. $f'(x) = 2; f'(1) = 2, f'(2) = 2, f'(3) = 2$ **11.** $f'(x) = -2x; f'(1) = -2, f'(2) = -4, f'(3) = -6$
13. $-2, 1, 3$ **15.** $f'(-3) = -1, f'(3) = 2$
17. (A) 5 (B) $3 + h$ (C) 3 (D) $y = 3x - 1$ **19.** (A) 5 m/sec (B) $3 + h$ m/sec (C) 3 m/sec
21. $f'(x) = 6 - 2x; f'(1) = 4, f'(2) = 2, f'(3) = 0$
23. $f'(x) = 1/(2\sqrt{x}); f'(1) = \frac{1}{2}, f'(2) = 1/(2\sqrt{2}), f'(3) = 1/(2\sqrt{3})$
25. $f'(x) = 1/x^2; f'(1) = 1, f'(2) = \frac{1}{4}, f'(3) = \frac{1}{9}$ **27.** Yes **29.** No **31.** No **33.** Yes
35. (A) $f'(x) = 2x - 4$ (B) $-4, 0, 4$ **37.** $v = f'(x) = 8x - 2$; 6 ft/sec, 22 ft/sec, 38 ft/sec

(C)

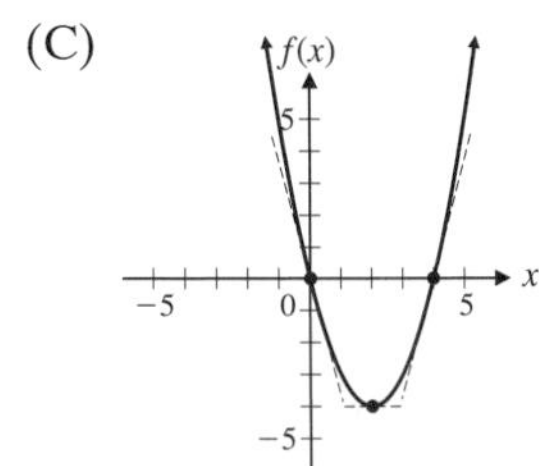

39.

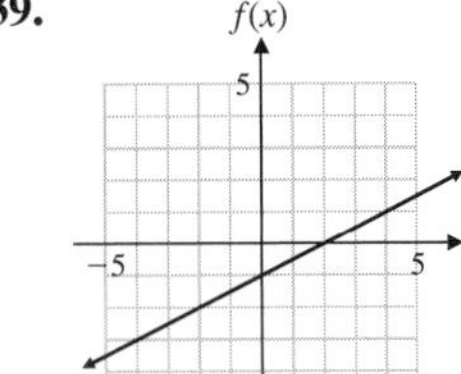

41.

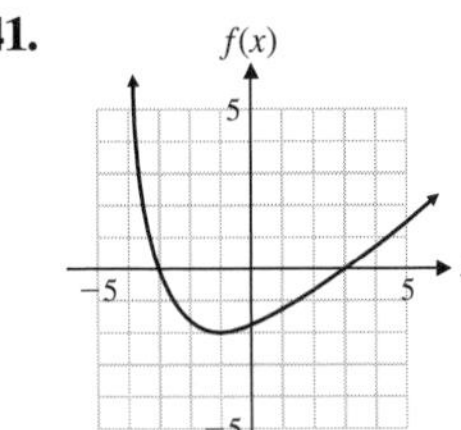

43. 0.69 **45.** 0.71

47. (A) The graphs of g and h are vertical translations of the graph of f. All three functions should have the same derivative.
(B) $2x$

49. (A) The slope of the graph of f is 0 at any point on the graph.

51. f is nondifferentiable at $x = 1$ **53.** f is differentiable for all real numbers **55.** No **57.** No

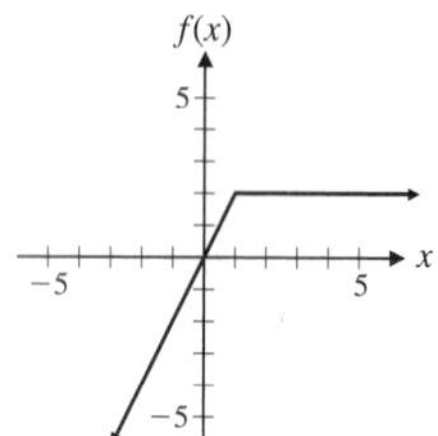

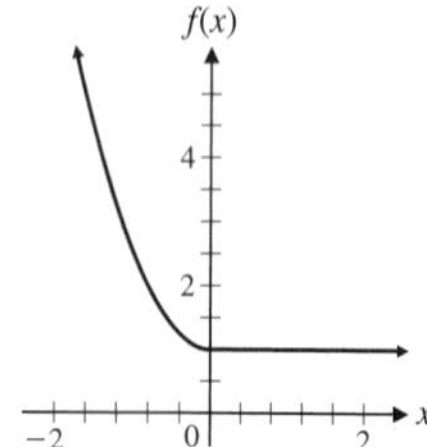

59. (A) $2 - 2x$ (B) 2 (C) -2

61. (A) $S'(t) = 1/\sqrt{t + 10}$
(B) $S(15) = 10$; $S'(15) = 0.2$. After 15 months, the total sales are \$10 million and are increasing at the rate of \$0.2 million, or \$200,000, per month.
(C) The estimated total sales are \$10.2 million after 16 months and \$10.4 million after 17 months.

63. (A) $A(5) = 133.82$; $A'(5) = 7.80$
(B) After 5 years, the original \$100 investment has grown to \$133.82 and is continuing to grow at the rate of \$7.80 per year.

65. (A) In March, the price of the stock was \$80 and was increasing at the rate of \$5 per month.
(B) The stock reached its highest price in April when the rate of change was 0.

67. (A)

```
QuadReg
 y=ax²+bx+c
 a=-.0583731517
 b=-1.509819783
 c=50.23597577
```

(B) In 1995, 14.5% of the households had black-and-white television sets, and this number was decreasing at the rate of 3.3% per year.

69. (A) $P'(t) = 12 - 2t$
(B) $P(3) = 107$; $P'(3) = 6$. After 3 hours, the ozone level is 107 ppb and is increasing at the rate of 6 ppb per hour.

71. (A) $P(30) = 47.1$; $P'(30) = 0.9$
(B) In 2010, there will be 47.1 million people aged 65 or older, and the number in this group will be growing at the rate of 900,000 per year.
(C) The estimated population is 48 million in 2011 and 48.9 million in 2012.

Exercise 8-4

1. 0 **3.** $12x^{11}$ **5.** 1 **7.** $-7x^{-8} = -7/x^8$ **9.** $\frac{5}{2}x^{3/2}$ **11.** $-5x^{-6} = -5/x^6$ **13.** $8x^3$ **15.** $1.2x^5$ **17.** $x^4/3$
19. 12 **21.** 2 **23.** 9 **25.** $-10x^{-6} = -10/x^6$ **27.** $-16x^{-5} = -16/x^5$ **29.** $x^{-3} = 1/x^3$
31. $-x^{-2/3} = -1/x^{2/3}$ **33.** $4.8x - 3.5$ **35.** $15x^4 - 6x^2$ **37.** $-12x^{-5} - 4x^{-3} = \dfrac{-12}{x^5} - \dfrac{4}{x^3}$
39. $-\dfrac{1}{2}x^{-2} + 2x^{-4} = \dfrac{-1}{2x^2} + \dfrac{2}{x^4}$ **41.** $2x^{-1/3} - \dfrac{5}{3}x^{-2/3} = \dfrac{2}{x^{1/3}} - \dfrac{5}{3x^{2/3}}$ **43.** $-\dfrac{9}{5}x^{-8/5} + 3x^{-3/2} = \dfrac{-9}{5x^{8/5}} + \dfrac{3}{x^{3/2}}$
45. $-\dfrac{1}{3}x^{-4/3} = \dfrac{-1}{3x^{4/3}}$ **47.** $-0.6x^{-3/2} + 6.4x^{-3} + 1 = -\dfrac{0.6}{x^{3/2}} + \dfrac{6.4}{x^3} + 1$
49. (A) $f'(x) = 6 - 2x$ (B) $f'(2) = 2$; $f'(4) = -2$ (C) $y = 2x + 4$; $y = -2x + 16$ (D) $x = 3$
51. (A) $f'(x) = 12x^3 - 12x$ (B) $f'(2) = 72$; $f'(4) = 720$ (C) $y = 72x - 127$; $y = 720x - 2{,}215$ (D) $x = -1, 0, 1$
53. (A) $v = f'(x) = 176 - 32x$ (B) $f'(0) = 176$ ft/sec; $f'(3) = 80$ ft/sec (C) 5.5 sec
55. (A) $v = f'(x) = 3x^2 - 18x + 15$ (B) $f'(0) = 15$ ft/sec; $f'(3) = -12$ ft/sec (C) $x = 1$ sec, $x = 5$ sec
57. $f'(x) = 2x - 3 - 2x^{-1/2} = 2x - 3 - \dfrac{2}{x^{1/2}}$; $x = 2.18$ **59.** $f'(x) = 4\sqrt[3]{x} - 3x - 3$; $x = -2.90$
61. $f'(x) = 0.2x^3 + 0.3x^2 - 3x - 1.6$; $x = -4.46, -0.52, 3.48$ **63.** $f'(x) = 0.8x^3 - 9.36x^2 + 32.5x - 28.25$; $x = 1.30$

65. $x = -b/(2a)$ **67.** (A) $x^3 + x$ (B) x^3 (C) $x^3 - x$ **69.** $-20x^{-2} = -20/x^2$
71. $2x - 3 - 10x^{-3} = 2x - 3 - (10/x^3)$
75. The domain of $f'(x)$ is all real numbers except $x = 0$. At $x = 0$, the graph of $y = f(x)$ is smooth, but it has a vertical tangent.
77. (A) $C'(x) = 60 - (x/2)$
(B) $C'(60) = \$30$/racket. At a production level of 60 rackets, the rate of change of total cost relative to production is \$30 per racket; thus, the cost of producing 1 more racket at this level of production is approx. \$30.
(C) \$29.75; the marginal cost of \$30 per racket found in part (B) is a close approximation to this value.
(D) $C'(80) = \$20$ per racket. At a production level of 80 rackets, the rate of change of total cost relative to production is \$20 per racket; thus, the cost of producing 1 more racket at this level of production is approx. \$20.

79. (A)

```
CubicReg
y=ax³+bx²+cx+d
a=7.445341406
b=-218.3357137
c=1657.552951
d=2545.706745
```

(B) $L(16) = 3{,}700$; $L'(16) = 400$. In 1996, 3,700 limousines were produced and the annual production was increasing at the rate of 400 limousines per year.

81. The approximate cost of producing the 101st oven is greater than that of the 401st oven. Since these marginal costs are decreasing, the manufacturing process is becoming more efficient.
83. (A) $N'(x) = 3{,}780/x^2$
(B) $N'(10) = 37.8$. At the \$10,000 level of advertising, sales are increasing at the rate of 37.8 boats per \$1,000 spent on advertising.
$N'(20) = 9.45$. At the \$20,000 level of advertising, sales are increasing at the rate of 9.45 boats per \$1,000 spent on advertising.
85. (A) -1.37 beats/min (B) -0.58 beat/min **87.** (A) 25 items/hr (B) 8.33 items/hr

Exercise 8-5

1. $2x^3(2x) + (x^2 - 2)(6x^2) = 10x^4 - 12x^2$ **3.** $(x - 3)(2) + (2x - 1)(1) = 4x - 7$
5. $\frac{(x-3)(1) - x(1)}{(x-3)^2} = \frac{-3}{(x-3)^2}$ **7.** $\frac{(x-2)(2) - (2x+3)(1)}{(x-2)^2} = \frac{-7}{(x-2)^2}$
9. $(x^2 + 1)(2) + (2x - 3)(2x) = 6x^2 - 6x + 2$ **11.** $(0.4x + 2)(0.5) + (0.5x - 5)(0.4) = 0.4x - 1$
13. $\frac{(2x-3)(2x) - (x^2+1)(2)}{(2x-3)^2} = \frac{2x^2 - 6x - 2}{(2x-3)^2}$ **15.** $(x^2 + 2)2x + (x^2 - 3)2x = 4x^3 - 2x$
17. $\frac{(x^2-3)2x - (x^2+2)2x}{(x^2-3)^2} = \frac{-10x}{(x^2-3)^2}$ **19.** 8 **21.** 1 **23.** $\frac{1}{8}$
25. $(2x + 1)(2x - 3) + (x^2 - 3x)(2) = 6x^2 - 10x - 3$
27. $(2.5x - x^2)(4) + (4x + 1.4)(2.5 - 2x) = -12x^2 + 17.2x + 3.5$
29. $\frac{(x^2+2x)(5) - (5x-3)(2x+2)}{(x^2+2x)^2} = \frac{-5x^2 + 6x + 6}{(x^2+2x)^2}$
31. $\frac{(x^2-1)(2x-3) - (x^2-3x+1)(2x)}{(x^2-1)^2} = \frac{3x^2 - 4x + 3}{(x^2-1)^2}$
33. $f'(x) = (1 + 3x)(-2) + (5 - 2x)(3)$; $y = -11x + 29$ **35.** $f'(x) = \frac{(3x-4)(1) - (x-8)(3)}{(3x-4)^2}$; $y = 5x - 13$
37. $f'(x) = (2x - 15)(2x) + (x^2 + 18)(2) = 6(x - 2)(x - 3)$; $x = 2$, $x = 3$
39. $f'(x) = \frac{(x^2+1)(1) - x(2x)}{(x^2+1)^2} = \frac{1 - x^2}{(x^2+1)^2}$; $x = -1$, $x = 1$ **41.** $7x^6 - 3x^2$ **43.** $-27x^{-4} = -\frac{27}{x^4}$
45. $(2x^4 - 3x^3 + x)(2x - 1) + (x^2 - x + 5)(8x^3 - 9x^2 + 1) = 12x^5 - 25x^4 + 52x^3 - 42x^2 - 2x + 5$
47. $\frac{(4x^2+5x-1)(6x-2) - (3x^2-2x+3)(8x+5)}{(4x^2+5x-1)^2} = \frac{23x^2 - 30x - 13}{(4x^2+5x-1)^2}$
49. $9x^{1/3}(3x^2) + (x^3 + 5)(3x^{-2/3}) = \frac{30x^3 + 15}{x^{2/3}}$ **51.** $\frac{(x^2-3)(2x^{-2/3}) - 6x^{1/3}(2x)}{(x^2-3)^2} = \frac{-10x^2 - 6}{(x^2-3)^2x^{2/3}}$

53. $x^{-2/3}(3x^2 - 4x) + (x^3 - 2x^2)(-\frac{2}{3}x^{-5/3}) = -\frac{8}{3}x^{1/3} + \frac{7}{3}x^{4/3}$

55. $\dfrac{(x^2 + 1)[(2x^2 - 1)(2x) + (x^2 + 3)(4x)] - (2x^2 - 1)(x^2 + 3)(2x)}{(x^2 + 1)^2} = \dfrac{4x^5 + 8x^3 + 16x}{(x^2 + 1)^2}$

59. $f'(x) = n[u(x)]^{n-1}u'(x)$ **61.** $x = -1.49, x = 0.48, x = 1.77$ **63.** $x = -2.51, x = -1.48, x = 1.12, x = 2.87$

65. (A) $S'(t) = \dfrac{(t^2 + 50)(180t) - 90t^2(2t)}{(t^2 + 50)^2} = \dfrac{9{,}000t}{(t^2 + 50)^2}$
(B) $S(10) = 60$; $S'(10) = 4$. After 10 months, the total sales are 60,000 CD's, and sales are increasing at the rate of 4,000 CD's per month.
(C) Approx. 64,000 CD's

67. (A) $\dfrac{dx}{dp} = \dfrac{(0.1p + 1)(0) - 4{,}000(0.1)}{(0.1p + 1)^2} = \dfrac{-400}{(0.1p + 1)^2}$
(B) $x = 800$; $dx/dp = -16$. At a price level of \$40, the demand is 800 CD players per week, and demand is decreasing at the rate of 16 players per dollar.
(C) Approx. 784 CD players

69. (A) $C'(t) = \dfrac{(t^2 + 1)(0.14) - 0.14t(2t)}{(t^2 + 1)^2} = \dfrac{0.14 - 0.14t^2}{(t^2 + 1)^2}$
(B) $C'(0.5) = 0.0672$. After 0.5 hr, concentration is increasing at the rate of 0.0672 mg/cc/hr.
$C'(3) = -0.0112$. After 3 hr, concentration is decreasing at the rate of 0.0112 mg/cc/hr.

71. (A) $N'(x) = \dfrac{(x + 32)(100) - (100x + 200)}{(x + 32)^2} = \dfrac{3{,}000}{(x + 32)^2}$ (B) $N'(4) = 2.31$; $N'(68) = 0.30$

Exercise 8-6

1. 3 **3.** $(-4x)$ **5.** $(2 + 6x)$ **7.** $6(2x + 5)^2$ **9.** $-8(5 - 2x)^3$ **11.** $5(4 + 0.2x)^4(0.2) = (4 + 0.2x)^4$

13. $30x(3x^2 + 5)^4$ **15.** $8(x^3 - 2x^2 + 2)^7(3x^2 - 4x)$ **17.** $(2x - 5)^{-1/2} = \dfrac{1}{(2x - 5)^{1/2}}$

19. $-8x^3(x^4 + 1)^{-3} = \dfrac{-8x^3}{(x^4 + 1)^3}$ **21.** $f'(x) = 6(2x - 1)^2$; $y = 6x - 5$; $x = \frac{1}{2}$

23. $f'(x) = 2(4x - 3)^{-1/2} = \dfrac{2}{(4x - 3)^{1/2}}$; $y = \dfrac{2}{3}x + 1$; none **25.** $24x(x^2 - 2)^3$

27. $-6(x^2 + 3x)^{-4}(2x + 3) = \dfrac{-6(2x + 3)}{(x^2 + 3x)^4}$ **29.** $x(x^2 + 8)^{-1/2} = \dfrac{x}{(x^2 + 8)^{1/2}}$ **31.** $(3x + 4)^{-2/3} = \dfrac{1}{(3x + 4)^{2/3}}$

33. $\dfrac{1}{4}(0.8x + 3.6)^{-3/4}(0.8) = \dfrac{0.2}{(0.8x + 3.6)^{3/4}}$ **35.** $(x^2 - 4x + 2)^{-1/2}(x - 2) = \dfrac{x - 2}{(x^2 - 4x + 2)^{1/2}}$

37. $-2(2x + 4)^{-2} = \dfrac{-2}{(2x + 4)^2}$ **39.** $-15x^2(x^3 + 4)^{-6} = \dfrac{-15x^2}{(x^3 + 4)^6}$ **41.** $(-8x + 4)(4x^2 - 4x + 1)^{-2} = \dfrac{-4}{(2x - 1)^3}$

43. $-2(2x - 3)(x^2 - 3x)^{-3/2} = \dfrac{-2(2x - 3)}{(x^2 - 3x)^{3/2}}$

45. $f'(x) = (4 - x)^3 - 3x(4 - x)^2 = 4(4 - x)^2(x - 1)$; $y = -16x + 48$

47. $f'(x) = \dfrac{(2x - 5)^3 - 6x(2x - 5)^2}{(2x - 5)^6} = \dfrac{-4x - 5}{(2x - 4)^4}$; $y = -17x + 54$

49. $f'(x) = (2x + 2)^{1/2} + x(2x + 2)^{-1/2} = \dfrac{3x + 2}{(2x + 2)^{1/2}}$; $y = \frac{5}{2}x - \frac{1}{2}$

51. $f'(x) = 2x(x - 5)^3 + 3x^2(x - 5)^2 = 5x(x - 5)^2(x - 2)$; $x = 0, 2, 5$

53. $f'(x) = \dfrac{(2x + 5)^2 - 4x(2x + 5)}{(2x + 5)^4} = \dfrac{5 - 2x}{(2x + 5)^3}$; $x = \frac{5}{2}$

55. $f'(x) = (x^2 - 8x + 20)^{-1/2}(x - 4) = \dfrac{x - 4}{(x^2 - 8x + 20)^{1/2}}$; $x = 4$ **57.** $x = -2.90, 0.16, 1.41$

59. $x = -2.71, -0.19, 1.00, 1.90$ **61.** $x = -0.64, 0.83, 2.81$

63. $18x^2(x^2 + 1)^2 + 3(x^2 + 1)^3 = 3(x^2 + 1)^2(7x^2 + 1)$

65. $\dfrac{24x^5(x^3-7)^3-(x^3-7)^4 6x^2}{4x^6}=\dfrac{3(x^3-7)^3(3x^3+7)}{2x^4}$

67. $(2x-3)^2[12x(2x^2+1)^2]+(2x^2+1)^3[4(2x-3)]=4(2x^2+1)^2(2x-3)(8x^2-9x+1)$

69. $4x^3(x^2-1)^{-1/2}+8x(x^2-1)^{1/2}=\dfrac{12x^3-8x}{(x^2-1)^{1/2}}$

71. $\dfrac{(x-3)^{1/2}(2)-x(x-3)^{-1/2}}{x-3}=\dfrac{x-6}{(x-3)^{3/2}}$

73. $\frac{1}{2}[(2x-1)^3(x^2+3)^4]^{-1/2}[8x(2x-1)^3(x^2+3)^3+6(x^2+3)^4(2x-1)^2]=(2x-1)^{1/2}(x^2+3)(11x^2-4x+9)$

75. (A) $C'(x)=(2x+16)^{-1/2}=\dfrac{1}{(2x+16)^{1/2}}$
(B) $C'(24)=\frac{1}{8}$, or \$12.50. At a production level of 24 calculators, total cost is increasing at the rate of \$12.50 per calculator; also, the cost of producing the 25th calculator is approx. \$12.50.
$C'(42)=\frac{1}{10}$, or \$10.00. At a production level of 42 calculators, total cost is increasing at the rate of \$10.00 per calculator; also, the cost of producing the 43rd calculator is approx. \$10.00.

77. (A) $\dfrac{dx}{dp}=40(p+25)^{-1/2}=\dfrac{40}{(p+25)^{1/2}}$
(B) $x=400$ and $dx/dp=4$. At a price of \$75, the supply is 400 speakers per week, and supply is increasing at the rate of 4 speakers per dollar.

79. $4{,}000(1+\frac{1}{12}r)^{47}$ **81.** $\dfrac{(4\times 10^6)x}{(x^2-1)^{5/3}}$

83. (A) $f'(n)=n(n-2)^{-1/2}+2(n-2)^{1/2}$
(B) $f'(11)=\frac{29}{3}=9.67$. When the list contains 11 items, the learning time is increasing at the rate of 9.67 min per item.
$f'(27)=\frac{77}{5}=15.4$. When the list contains 27 items, the learning time is increasing at the rate of 15.4 min per item.

Exercise 8-7

1. (A) \$29.50 (B) \$30

3. (A) \$420
(B) $\overline{C}'(500)=-0.24$. At a production level of 500 frames, average cost is decreasing at the rate of 24¢ per frame.
(C) Approx. \$419.76

5. (A) $R'(1{,}600)=20$. At a production level of 1,600 radios, revenue is increasing at the rate of \$20 per radio.
(B) $R'(2{,}500)=-25$. At a production level of 2,500 radios, revenue is decreasing at the rate of \$25 per radio.

7. (A) \$4.50 (B) \$5

9. (A) $P'(450)=0.5$. At a production level of 450 cassettes, profit is increasing at the rate of 50¢ per cassette.
(B) $P'(750)=-2.5$. At a production level of 750 cassettes, profit is decreasing at the rate of \$2.50 per cassette.

11. (A) \$13.50
(B) $\overline{P}'(50)=\$0.27$. At a production level of 50 mowers, the average profit per mower is increasing at the rate of \$0.27 per mower.
(C) Approx. \$13.77

13. (A) $C'(x)=60$ (B) $R(x)=200x-(x^2/30)$ (C) $R'(x)=200-(x/15)$
(D) $R'(1{,}500)=100$. At a production level of 1,500 saws, revenue is increasing at the rate of \$100 per saw.
$R'(4{,}500)=-100$. At a production level of 4,500 saws, revenue is decreasing at the rate of \$100 per saw.
(E) Break-even points; (600, 108,000) and (3,600, 288,000)

(F) $P(x)=-(x^2/30)+140x-72{,}000$
(G) $P'(x)=-(x/15)+140$
(H) $P'(1{,}500)=40$. At a production level of 1,500 saws, profit is increasing at the rate of \$40 per saw.
$P'(3{,}000)=-60$. At a production level of 3,000 saws, profit is decreasing at the rate of \$60 per saw.

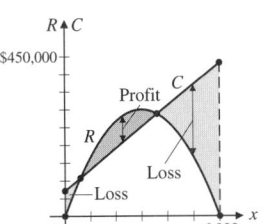

15. (A) $p = 20 - 0.02x$ (B) $R(x) = 20x - 0.02x^2$ (C) $C(x) = 4x + 1{,}400$
(D) Break-even points: (100, 1,800) and (700, 4,200) (E) $P(x) = 16x - 0.02x^2 - 1{,}400$

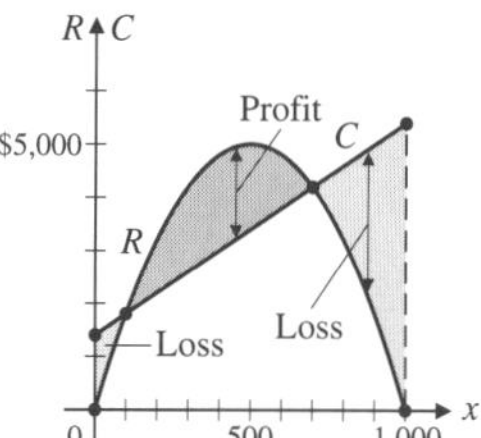

(F) $P'(250) = 6$. At a production level of 250 toasters, profit is increasing at the rate of \$6 per toaster.
$P'(475) = -3$. At a production level of 475 toasters, profit is decreasing at the rate of \$3 per toaster.

17. (A) $x = 500$ (B) $P(x) = 176x - 0.2x^2 - 21{,}900$
(C) $x = 440$
(D) Break-even points: (150, 25,500) and (730, 39,420); x intercepts for $P(x)$: $x = 150$ and $x = 730$

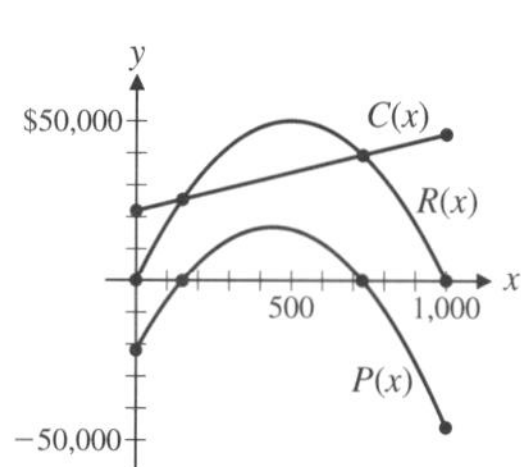

19. (A) $R(x) = 20x - x^{3/2}$
(B) Break-even points: (44, 588), (258, 1,016)

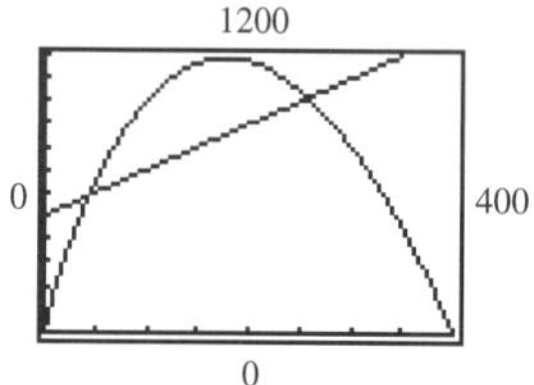

21. (A)

```
QuadReg
 y=ax²+bx+c
 a=1.4101002E-5
 b=-.2732556676
 c=1320.924694
```

(B) Fixed costs ≈ \$721,680
Variable costs ≈ \$121

```
LinReg
 y=ax+b
 a=120.7047281
 b=721680.1282
 r=.9934384133
```

(C) (713, 807,703), (5,423, 1,376,227)
(D) $\$254 \leqslant p \leqslant \$1{,}133$

Review Exercise

1. (A) 16 (B) 8 (C) 8 (D) 4 (E) 4 (F) 4 *(8-1, 8-3, 8-4)* **2.** $f'(-1) \approx -2; f'(1) \approx 1$ *(8-1, 8-3)*
3. (A) 22 (B) 8 (C) 2 *(8-2)* **4.** (A) 1 (B) 1 (C) 1 (D) 1 *(8-2)*
5. (A) 2 (B) 3 (C) Does not exist (D) 3 *(8-2)* **6.** (A) 4 (B) 4 (C) 4 (D) Does not exist *(8-2)*
7. $3x^2 + 2x$ *(8-3)* **8.** (A) −11 (B) −14 (C) $\frac{5}{2}$ (D) −8 *(8-4, 8-5, 8-6)* **9.** $(6x + 4)$ *(8-6)* **10.** $12x^3 - 4x$ *(8-4)*
11. $x^{-1/2} - 3 = \dfrac{1}{x^{1/2}} - 3$ *(8-4)* **12.** 0 *(8-4)* **13.** $-x^{-3} + x = \dfrac{-1}{x^3} + x$ *(8-4)* **14.** $-2x^{-5} + x^3 = \dfrac{-2}{x^5} + x^3$ *(8-4)*
15. $(2x - 1)(3) + (3x + 2)(2) = 12x + 1$ *(8-5)* **16.** $(x^2 - 1)(3x^2) + (x^3 - 3)(2x) = 5x^4 - 3x^2 - 6x$ *(8-5)*
17. $(0.2x - 1.5)(0.5) + (0.5x + 0.4)(0.2) = 0.2x - 0.67$ *(8-5)* **18.** $\dfrac{(x^2 + 2)2 - 2x(2x)}{(x^2 + 2)^2} = \dfrac{4 - 2x^2}{(x^2 + 2)^2}$ *(8-5)*
19. $(-1)(3x + 2)^{-2}(3) = \dfrac{-3}{(3x + 2)^2}$ *(8-6)* **20.** $3(2x - 3)^2(2) = 6(2x - 3)^2$ *(8-6)*
21. $-2(x^2 + 2)^{-3}(2x) = \dfrac{-4x}{(x^2 + 2)^3}$ *(8-6)* **22.** (A) 3 (B) $2 + 0.5h$ (C) 2 *(8-1, 8-3)*

23. *(8-1)*

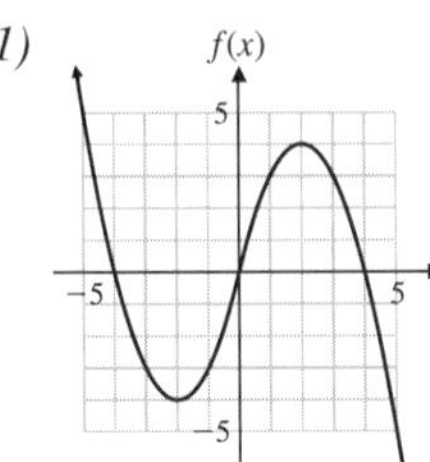

24. $12x^3 + 6x^{-4} = 12x^3 + \frac{6}{x^4}$ *(8-4)*
25. $(2x^2 - 3x + 2)(2x + 2) + (x^2 + 2x - 1)(4x - 3) = 8x^3 + 3x^2 - 12x + 7$ *(8-5)*
26. $\frac{(x-1)^2(2) - (2x-3)(2)(x-1)}{(x-1)^4} = \frac{4-2x}{(x-1)^3}$ *(8-5)*
27. $x^{-1/2} - 2x^{-3/2} = \frac{1}{x^{1/2}} - \frac{2}{x^{3/2}}$ *(8-4)*

28. $0.6x^{-2/3} - 0.3x^{-4/3} = \frac{0.6}{x^{2/3}} - \frac{0.3}{x^{4/3}}$ *(8-4)*
29. $(x^2 - 1)[2(2x + 1)(2)] + (2x + 1)^2(2x) = 2(2x + 1)(4x^2 + x - 2)$ *(8-5, 8-6)*
30. $\frac{1}{3}(x^3 - 5)^{-2/3}(3x^2) = \frac{x^2}{(x^3 - 5)^{2/3}}$ *(8-6)* **31.** $-8x^{-3} = \frac{-8}{x^3}$ *(8-4)*
32. $\frac{(2x-3)(4)(x^2+2)^3(2x) - (x^2+2)^4(2)}{(2x-3)^2} = \frac{2(x^2+2)^3(7x^2 - 12x - 2)}{(2x-3)^2}$ *(8-5, 8-6)*
33. (A) $m = f'(1) = 2$ (B) $y = 2x + 3$ *(8-3, 8-4)* **34.** (A) $m = f'(1) = 16$ (B) $y = 16x - 12$ *(8-3, 8-5)*
35. $x = 5$ *(8-4)* **36.** $x = -5, x = 3$ *(8-5)* **37.** $x = -2, x = 2$ *(8-5)* **38.** $x = 0, x = 3, x = \frac{15}{2}$ *(8-5)*
39. $x = -1.37, 0.60, 1.52$ *(8-4)* **40.** $x = -2.96, -2.20, 1.85, 3.31$ *(8-5)* **41.** $x = -1.89, -0.36, 0.74$ *(8-5, 8-6)*
42. (A) $v = f'(x) = 32x - 4$ (B) $f'(3) = 92$ ft/sec *(8-4)* **43.** (A) $v = f'(x) = 96 - 32x$ (B) $x = 3$ sec *(8-4)*
44.

h	−0.1	−0.01	−0.001 → 0	← 0.001	0.01	0.1
$\frac{f(h) - f(0)}{h}$	1.49	1.60	1.61 → 1.61	← 1.61	1.62	1.75

1.61 *(8-1, 8-3)*

45. −1.39 *(8-1, 8-3)*

46. (A) The graph of g is the graph of f shifted 4 units to the right, and the graph of h is the graph of f shifted 3 units to the left:

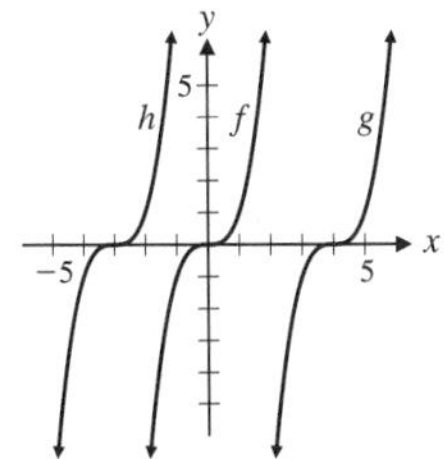

(B) The graph of g' is the graph of f' shifted 4 units to the right, and the graph of h' is the graph of f' shifted 3 units to the left:

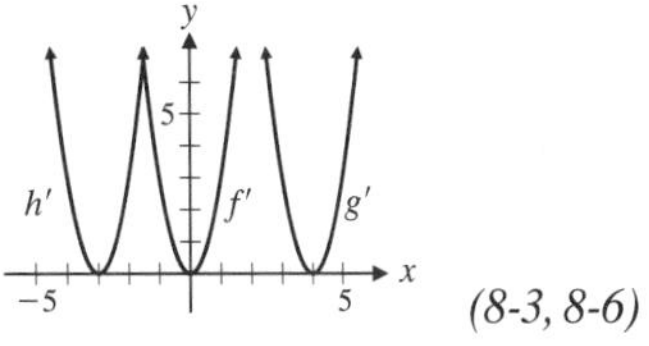

(8-3, 8-6)

47. (A) The graph of g is a horizontal translation of the graph of f, and the graph of g' is a horizontal translation of the graph of f'.
(B) The graph of g is a vertical translation of the graph of f, and the graph of g' is the same as the graph of f'. *(8-3)*
48. $\frac{3}{8}$ *(8-2)* **49.** 16 *(8-2)* **50.** −1 *(8-2)* **51.** $\frac{1}{6}$ *(8-2)* **52.** −1 *(8-2)* **53.** 1 *(8-2)* **54.** Does not exist *(8-2)*
55. 4 *(8-2)* **56.** 4 *(8-2)* **57.** $\frac{-1}{(x+2)^2}$ *(8-2)*
58. (A) $\lim_{x\to -2^-} f(x) = -6$; $\lim_{x\to -2^+} f(x) = 6$; $\lim_{x\to -2} f(x)$ does not exist (B) $\lim_{x\to 0} f(x) = 4$
(C) $\lim_{x\to 2^-} f(x) = 2$; $\lim_{x\to 2^+} f(x) = -2$; $\lim_{x\to 2} f(x)$ does not exist *(8-2)*
59. $2x - 1$ *(8-3)* **60.** $1/(2\sqrt{x})$ *(8-3)* **61.** No *(8-3)* **62.** No *(8-3)* **63.** No *(8-3)* **64.** Yes *(8-3)*
65. $(x - 4)^4(3)(x + 3)^2 + (x + 3)^3(4)(x - 4)^3 = 7x(x - 4)^3(x + 3)^2$ *(8-5, 8-6)*
66. $\frac{(2x+1)^4(5x^4) - x^5(4)(2x+1)^3(2)}{(2x+1)^8} = \frac{x^4(2x+5)}{(2x+1)^5}$ *(8-5, 8-6)*

67. $\dfrac{x(\frac{1}{2})(x^2-1)^{-1/2}(2x)-(x^2-1)^{1/2}}{x^2}=\dfrac{1}{x^2(x^2-1)^{1/2}}$ *(8-5, 8-6)*

68. $\dfrac{(x^2+4)^{1/2}-x(\frac{1}{2})(x^2+4)^{-1/2}(2x)}{x^2+4}=\dfrac{4}{(x^2+4)^{3/2}}$ *(8-5, 8-6)*

69. The domain of $f'(x)$ is all real numbers except $x=0$. At $x=0$, the graph of $y=f(x)$ is smooth, but it has a vertical tangent. *(8-3)*

70. (A) $\lim_{x\to 1^-} f(x)=1$; $\lim_{x\to 1^+} f(x)=-1$ (B) $\lim_{x\to 1^-} f(x)=-1$; $\lim_{x\to 1^+} f(x)=1$ (C) $m=1$

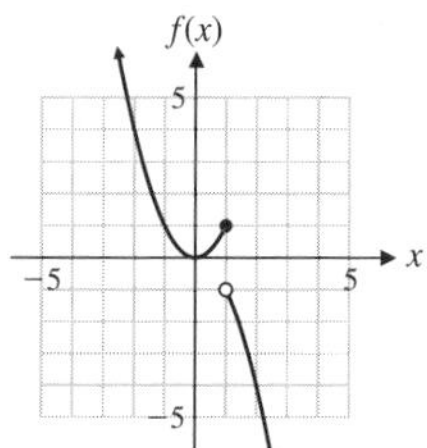

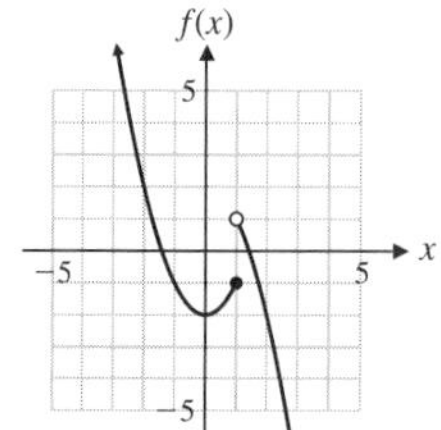

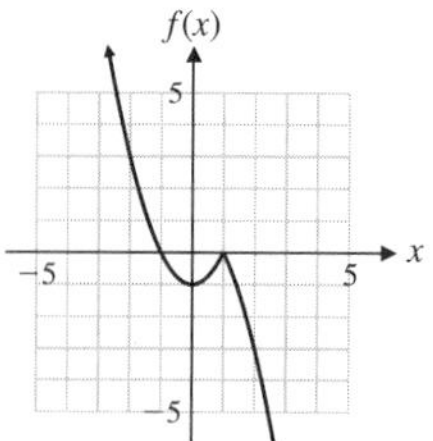

(D) The graphs in (A) and (B) have jumps at $x=1$; the graph in (C) does not. *(8-2)*

71. (A) 1 (B) −1 (C) Does not exist (D) No *(8-3)* **72.** (A) \$179.90 (B) \$180 *(8-7)*

73. (A) $C(100)=9{,}500$; $C'(100)=50$. At a production level of 100 bicycles, the total cost is \$9,500, and cost is increasing at the rate of \$50 per bicycle.

(B) $\overline{C}(100)=95$; $\overline{C}'(100)=-0.45$. At a production level of 100 bicycles, the average cost is \$95, and average cost is decreasing at a rate of \$0.45 per bicycle. *(8-7)*

74. The approximate cost of producing the 201st printer is greater than that of the 601st printer. Since these marginal costs are decreasing, the manufacturing process is becoming more efficient. *(8-7)*

75. (A) $C'(x)=2$; $\overline{C}(x)=2+\dfrac{9{,}000}{x}$; $\overline{C}'(x)=\dfrac{-9{,}000}{x^2}$

(B) $R(x)=xp=25x-0.01x^2$; $R'(x)=25-0.02x$; $\overline{R}(x)=25-0.01x$; $\overline{R}'(x)=-0.01$

(C) $P(x)=R(x)-C(x)=23x-0.01x^2-9{,}000$; $P'(x)=23-0.02x$; $\overline{P}(x)=23-0.01x-\dfrac{9{,}000}{x}$;

$\overline{P}'(x)=-0.01+\dfrac{9{,}000}{x^2}$

(D) (500, 10,000) and (1,800, 12,600)

(E) $P'(1{,}000)=3$. Profit is increasing at the rate of \$3 per umbrella.
$P'(1{,}150)=0$. Profit is flat.
$P'(1{,}400)=-5$. Profit is decreasing at the rate of \$5 per umbrella.

(F) 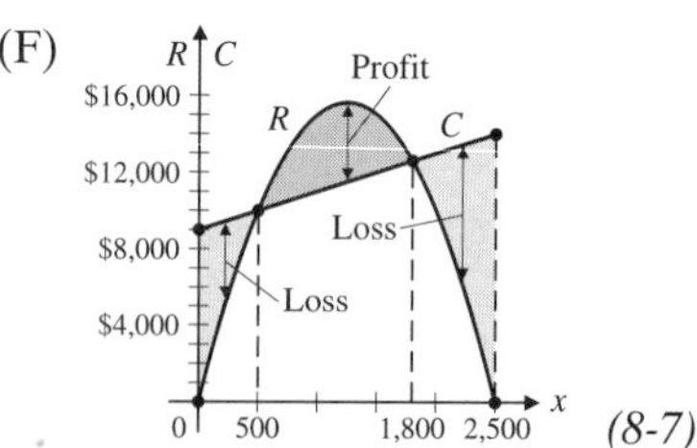

(8-7)

76. (A) 2 components/day (B) 3.2 components/day *(8-5)*

77. $N(5)=15$; $N'(5)=3.833$. After 5 months, the total sales are 15,000 pools, and sales are increasing at the rate of 3,833 pools per month. *(8-6)*

78. (A) $A(10)=9{,}836$; $A'(10)\approx 665$

(B) After 10 years, the amount in the account is \$9,836, and it is growing at the rate of \$665 per year. *(8-3)*

79. (A)

```
CubicReg
 y=ax³+bx²+cx+d
 a=.0015333333
 b=-.0921904762
 c=1.590238095
 d=12.06190476
```

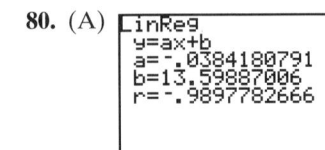

(B) $N(35) = 20.5$; $N'(35) = 0.8$. In 1995 consumption was 20.5 trillion ft^3 of natural gas, and this number was increasing at the rate of 0.8 trillion ft^3/yr. *(8-3)*

80. (A)

```
LinReg
 y=ax+b
 a=-.0384180791
 b=13.59887006
 r=-.9897782666
```

(B) Fixed costs: \$484.21; variable costs per kringle: \$2.11

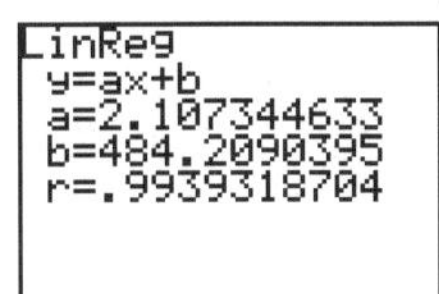

(C) (51, 591.15), (248, 1,007.62)
(D) $\$4.07 < p < \$11.64$ *(8-7)*

81. $C'(9) = -1$ ppm/m; $C'(99) = -0.001$ ppm/m *(8-6)*
82. $F(3) = 100$; $F'(3) = -0.25$. After 3 hr, the body temperature is 100°F, and the temperature is decreasing at the rate of 0.25°F/hr. *(8-6)*
83. (A) 10 items/hr (B) 5 items/hr *(8-4)*
84. (A) $M(50) \approx 66.9$; $M'(50) \approx 0.7$
(B) In 2010, there will be 66.9 million married couples, and this number will be growing at the rate of 0.7 million = 700,000 couples per year.
(C) 2011: 67.6 million couples; 2012: 68.3 million couples *(8-3)*

Chapter 9

Exercise 9-1

1. f is continuous at $x = 1$, since $\lim_{x\to 1} f(x) = f(1)$

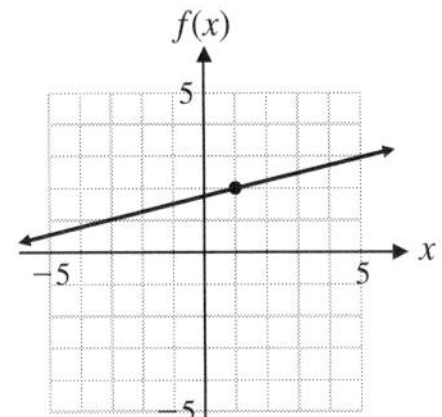

3. f is discontinuous at $x = 1$, since $\lim_{x\to 1} f(x) \neq f(1)$

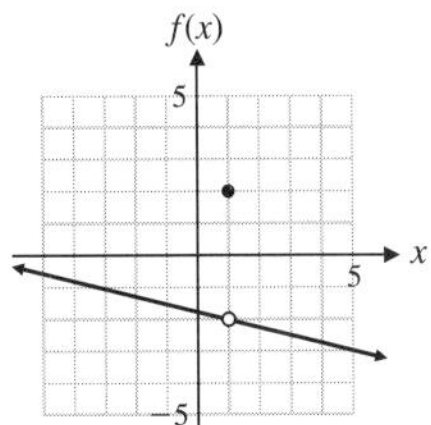

5. f is discontinuous at $x = 1$, since $\lim_{x\to 1} f(x)$ does not exist

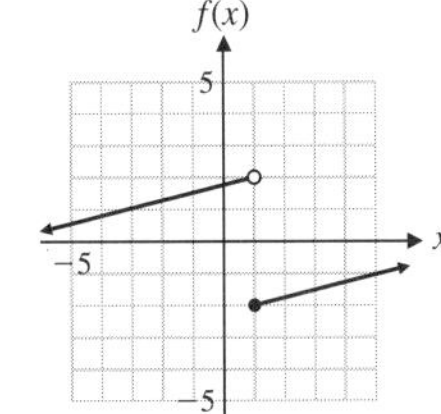

7.

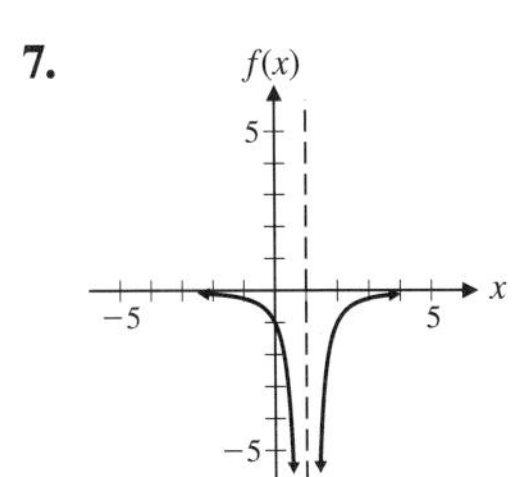

9.

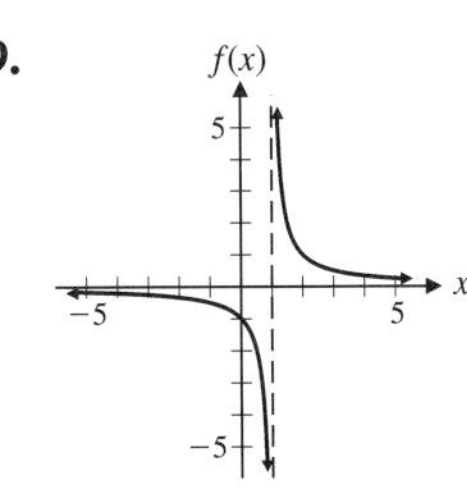

11. All x **13.** All x, except $x = 5$
15. All x, except $x = -2$ and $x = 3$
17. (A) $\lim_{x\to 1^-} f(x) = 2$; $\lim_{x\to 1^+} f(x) = 1$; $\lim_{x\to 1} f(x)$ does not exist; $f(1) = 1$
(B) No, because $\lim_{x\to 1} f(x)$ does not exist.
19. (A) $\lim_{x\to -2^-} f(x) = 1$; $\lim_{x\to -2^+} f(x) = 1$; $\lim_{x\to -2} f(x) = 1$; $f(-2) = 3$
(B) No, because $\lim_{x\to -2} f(x) \neq f(-2)$.

21. (A)
(B) 1
(C) 2
(D) No
(E) All integers

23. f is discontinuous at $x = -3$; $\lim_{x\to -3^-} f(x) = -\infty$ and $\lim_{x\to -3^+} f(x) = \infty$; the line $x = -3$ is a vertical asymptote
25. h is discontinuous at $x = -2$; $\lim_{x\to -2^-} h(x) = \infty$ and $\lim_{x\to -2^+} h(x) = -\infty$; the line $x = -2$ is a vertical asymptote; h is discontinuous at $x = 2$; $\lim_{x\to 2^-} h(x) = -\infty$ and $\lim_{x\to 2^+} h(x) = \infty$; the line $x = 2$ is a vertical asymptote
27. F is continuous for all real numbers x and has no vertical asymptotes.

29. H is discontinuous at $x = 1$; $\lim_{x\to1^-} H(x) = -\infty$ and $\lim_{x\to1^+} H(x) = \infty$; the line $x = 1$ is a vertical asymptote; H is discontinuous at $x = 3$; $\lim_{x\to3} H(x) = 2$, but $H(3)$ does not exist; there is no vertical asymptote at $x = 3$
31. T is discontinuous at $x = 0$; $\lim_{x\to0^-} T(x) = -\infty$ and $\lim_{x\to0^+} T(x) = -\infty$; the line $x = 0$ (the y axis) is a vertical asymptote; T is discontinuous at $x = 4$; $\lim_{x\to4^-} T(x) = \infty$ and $\lim_{x\to4^+} T(x) = \infty$; the line $x = 4$ is a vertical asymptote
33. $-3 < x < 4$; $(-3, 4)$ **35.** $x < 3$ or $x > 7$; $(-\infty, 3) \cup (7, \infty)$ **37.** $-5 < x < 0$ or $x > 3$; $(-5, 0) \cup (3, \infty)$
39. (A) $(-1.33, 1.20) \cup (3.13, \infty)$ (B) $(-\infty, -1.33) \cup (1.20, 3.13)$
41. (A) $(-\infty, -2.53) \cup (-0.72, \infty)$ (B) $(-2.53, -0.72)$
43. (A) $(-2.15, -0.52) \cup (1, 2.67)$ (B) $(-\infty, -2.15) \cup (-0.52, 1) \cup (2.76, \infty)$ **45.** $(-\infty, \infty)$ **47.** $[5, \infty)$
49. $(-\infty, \infty)$ **51.** $(-\infty, 1), (1, 2), (2, \infty)$
53. Since $\lim_{x\to1^-} f(x) = 2$ and $\lim_{x\to1^+} f(x) = 4$, $\lim_{x\to1} f(x)$ does not exist and f is not continuous at $x = 1$.

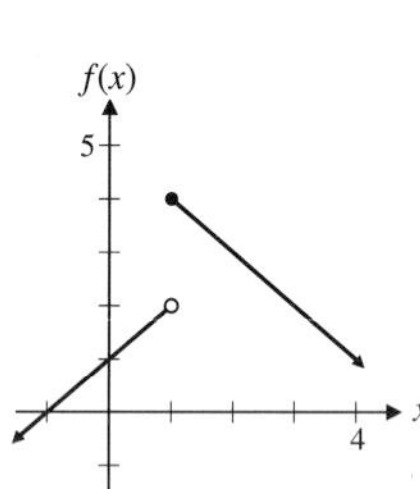

55. This function is continuous for all x.

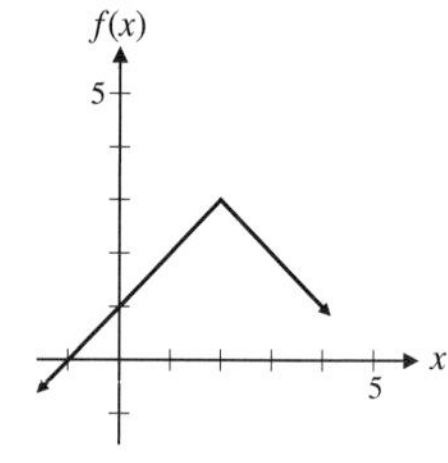

57. Since $\lim_{x\to0} f(x) = 0$ and $f(0) = 1$, $\lim_{x\to0} f(x) \neq f(0)$ and f is not continuous at $x = 0$.

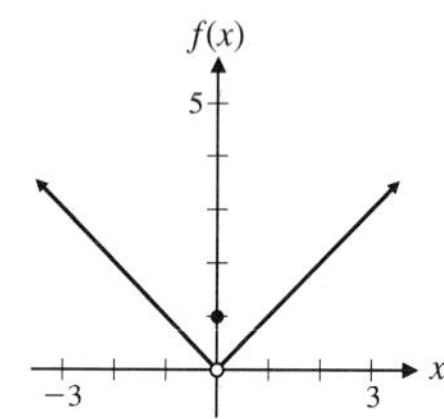

59. Since $\lim_{x\to2^-} f(x) = 0$ and $\lim_{x\to2^+} f(x) = 4$, $\lim_{x\to2} f(x)$ does not exist. Furthermore, $f(2)$ is not defined. Thus, f is not continuous at $x = 2$.
61. Since $f(-1)$ and $f(1)$ are not defined, f is not continuous at $x = -1$ and $x = 1$, even though $\lim_{x\to-1} f(x) = 2$ and $\lim_{x\to1} f(x) = 2$.
63. (A) Yes (B) Yes (C) Yes (D) Yes **65.** (A) Yes (B) No (C) Yes (D) No (E) Yes
67. x intercepts: $x = -5, 2$

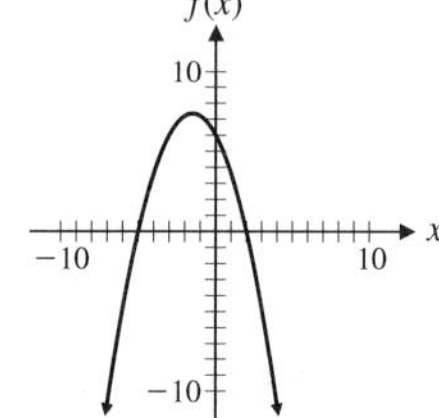

69. x intercepts: $x = -6, -1, 4$

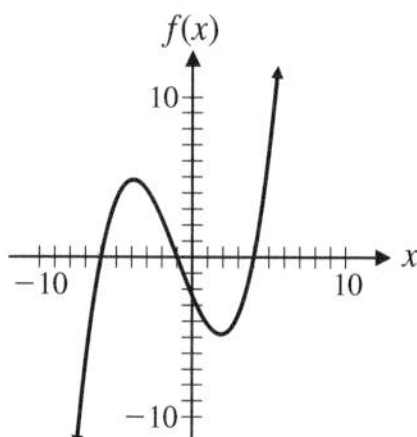

71. No, but this does not contradict Theorem 2, since f is discontinuous at $x = 1$.

73. The following sketches illustrate that either condition is possible. Theorem 2 implies that one of these two conditions must occur.

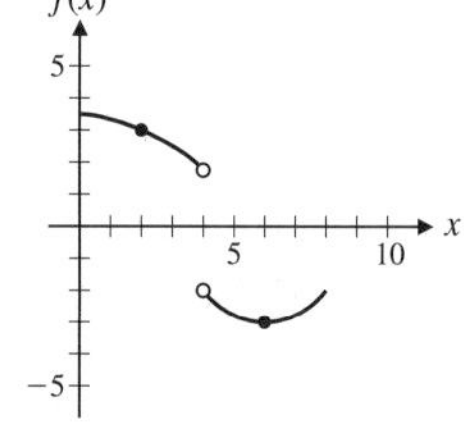

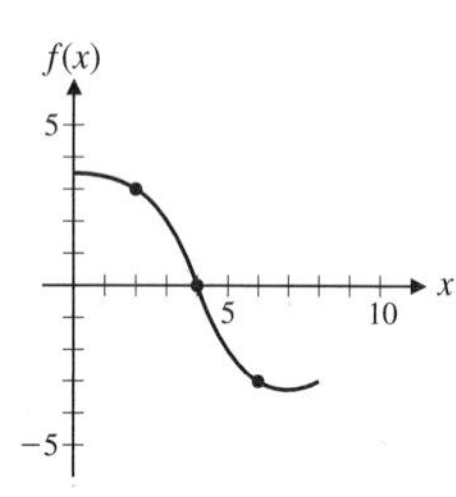

75. (A)

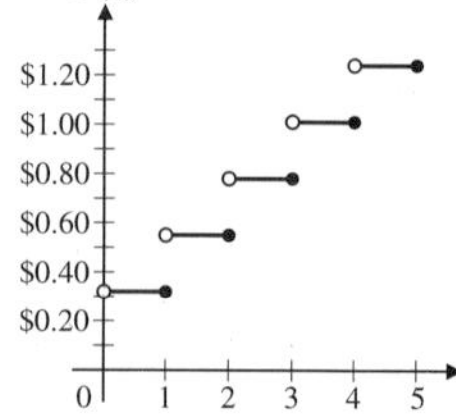

(B) $\lim_{x\to4.5} P(x) = \$1.24$; $P(4.5) = \$1.24$
(C) $\lim_{x\to4} P(x)$ does not exist; $P(4) = \$1.01$
(D) Continuous at $x = 4.5$; not continuous at $x = 4$

77. If x is a positive integer, then $Q(x) = P(x) + 0.23$; $Q(x) = P(x)$ for all other values of x in the domain of P.

79. (A)

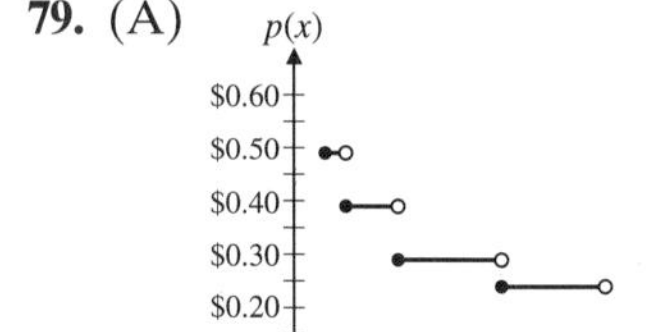

(C)

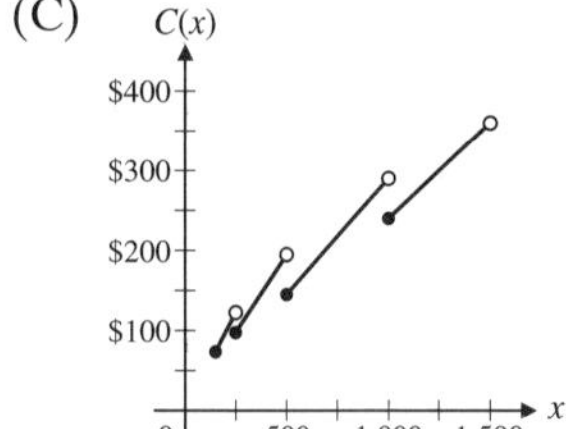

(B) p is discontinuous at $x = 250, 500$, and 1,000. In each case, the limit from the left is greater than the limit from the right, reflecting the corresponding drop in price at these order quantities.

(D) C is discontinuous at $x = 250, 500$, and 1,000. In each case, the limit from the left is greater than the limit from the right, reflecting savings to the customer due to the corresponding drop in price at these order quantities.

81. (A)

(B) $\lim_{s\to 10{,}000} E(s) = \$1{,}000$; $E(10{,}000) = \$1{,}000$
(C) $\lim_{s\to 20{,}000} E(s)$ does not exist; $E(20{,}000) = \$2{,}000$
(D) Yes; no

83. (A) t_2, t_3, t_4, t_6, t_7
(B) $\lim_{t\to t_5} N(t) = 7$; $N(t_5) = 7$
(C) $\lim_{t\to t_3} N(t)$ does not exist; $N(t_3) = 4$

Exercise 9-2

1. (a, b); (d, f); (g, h) **3.** (b, c); (c, d); (f, g) **5.** c, d, f **7.** b, f
9. Local maximum at $x = a$; local minimum at $x = c$; no local extrema at $x = b$ and $x = d$

11.

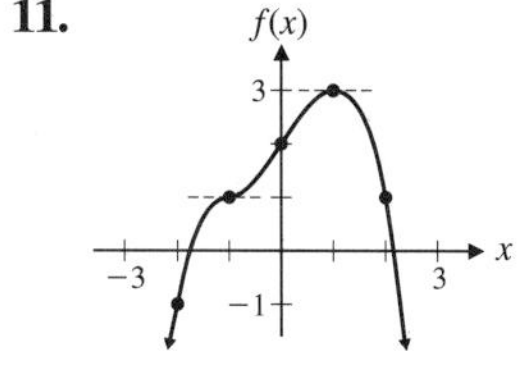

13.

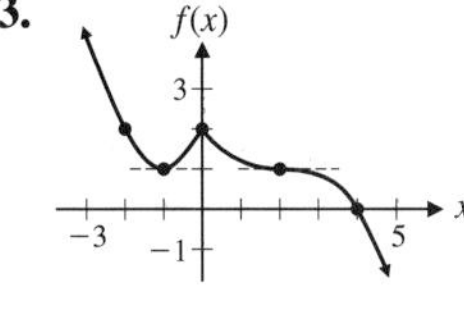

15.

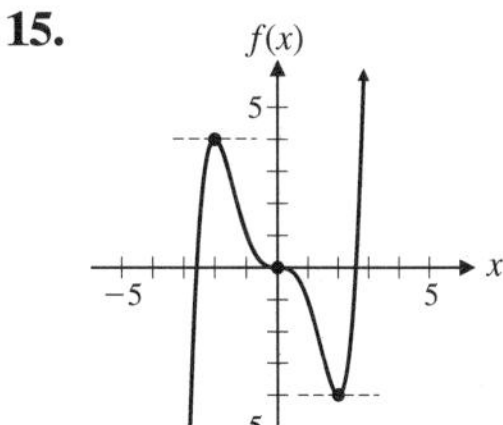

17.

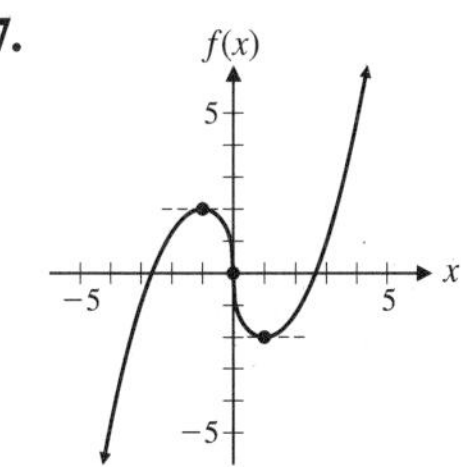

19. g_4 **21.** g_6 **23.** g_2 **25.** Decreasing on $(-\infty, 8)$; increasing on $(8, \infty)$; local minimum at $x = 8$
27. Increasing on $(-\infty, 5)$; decreasing on $(5, \infty)$; local maximum at $x = 5$ **29.** Increasing for all x; no local extrema
31. Decreasing for all x; no local extrema
33. Increasing on $(-\infty, -2)$ and $(2, \infty)$; decreasing on $(-2, 2)$; local maximum at $x = -2$; local minimum at $x = 2$
35. Increasing on $(-\infty, -2)$ and $(4, \infty)$; decreasing on $(-2, 4)$; local maximum at $x = -2$; local minimum at $x = 4$
37. Increasing on $(-\infty, -1)$ and $(0, 1)$; decreasing on $(-1, 0)$ and $(1, \infty)$; local maxima at $x = -1$ and $x = 1$; local minimum at $x = 0$

39. Increasing on $(-\infty, 4)$
Decreasing on $(4, \infty)$
Horizontal tangent at $x = 4$

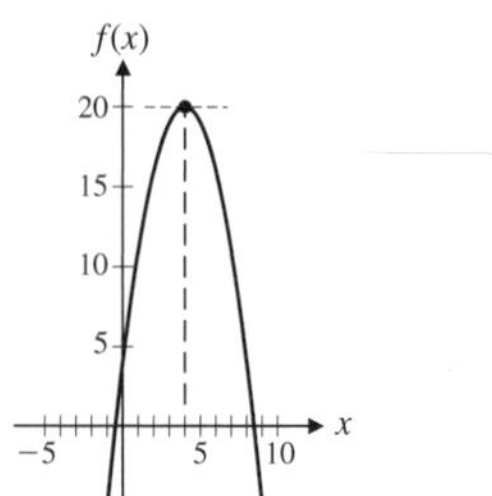

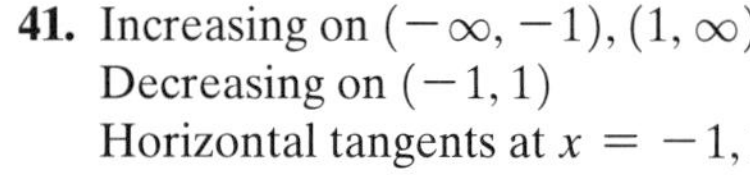

41. Increasing on $(-\infty, -1), (1, \infty)$
Decreasing on $(-1, 1)$
Horizontal tangents at $x = -1, 1$

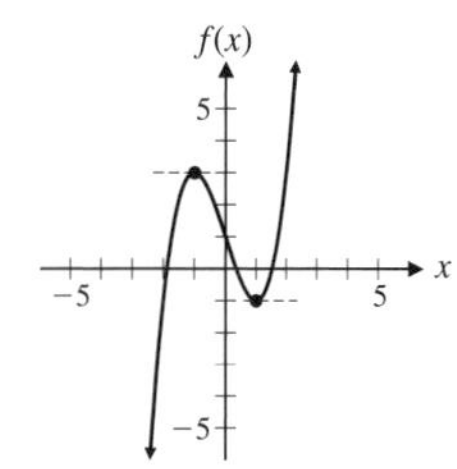

43. Decreasing for all x
Horizontal tangent at $x = 2$

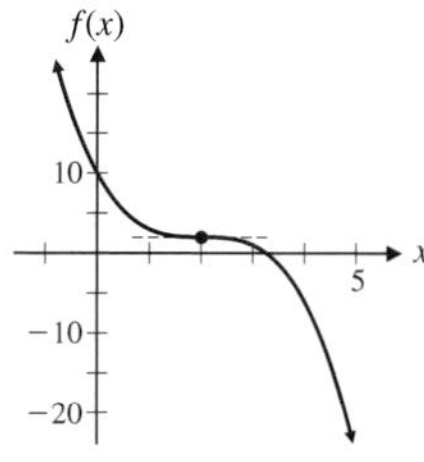

45. Critical value: $x = -1.26$; increasing on $(-1.26, \infty)$; decreasing on $(-\infty, -1.26)$; local minimum at $x = -1.26$

47. Critical values: $x = -0.43$, $x = 0.54$, $x = 2.14$; increasing on $(-0.43, 0.54)$ and $(2.14, \infty)$; decreasing on $(-\infty, -0.43)$ and $(0.54, 2.14)$; local maximum at $x = 0.54$; local minima at $x = -0.43$ and $x = 2.14$

49. Critical values: $x = -4.17$, $x = -2.78$, $x = 0.95$; increasing on $(-4.17, -2.78)$ and $(0.95, \infty)$; decreasing on $(-\infty, -4.17)$ and $(-2.78, 0.95)$; local maximum at $x = -2.78$; local minima at $x = -4.17$ and $x = 0.95$

51. Increasing on $(-1, 2)$; decreasing on $(-\infty, -1)$ and $(2, \infty)$; local minimum at $x = -1$; local maximum at $x = 2$

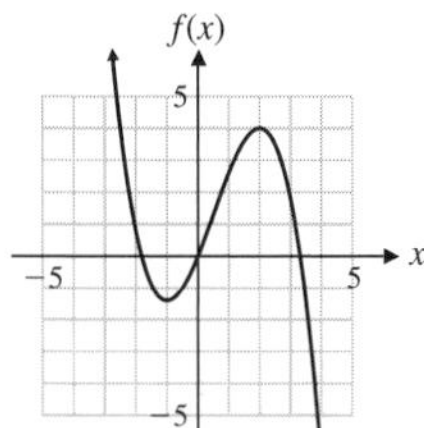

53. Increasing on $(-1, 2)$ and $(2, \infty)$; decreasing on $(-\infty, -1)$; local minimum at $x = -1$

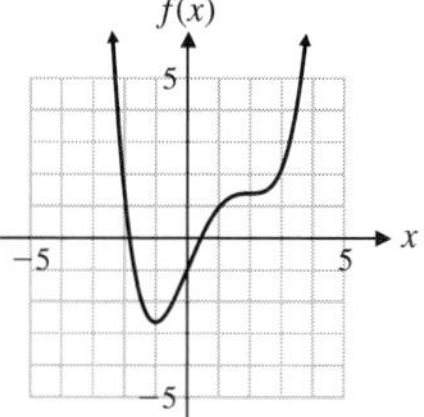

55. $f'(x) > 0$ on $(-\infty, -1)$ and $(3, \infty)$; $f'(x) < 0$ on $(-1, 3)$; $f'(x) = 0$ at $x = -1$ and $x = 3$

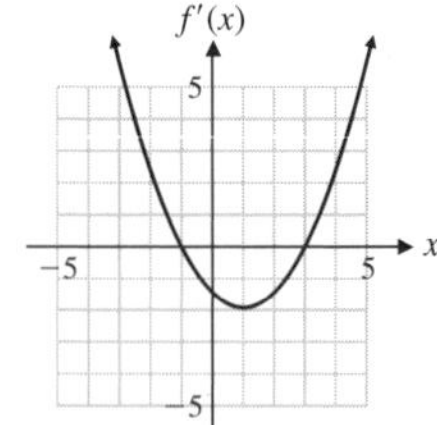

57. $f'(x) > 0$ on $(-2, 1)$ and $(3, \infty)$; $f'(x) < 0$ on $(-\infty, -2)$ and $(1, 3)$; $f'(x) = 0$ at $x = -2$, $x = 1$, and $x = 3$

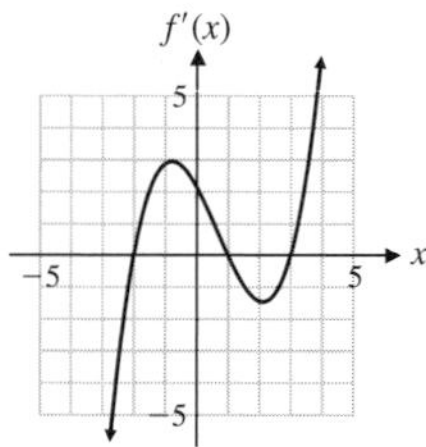

59. No critical values; increasing on $(-\infty, -2)$ and $(-2, \infty)$; no local extrema

61. Critical values: $x = -2$, $x = 2$; increasing on $(-\infty, -2)$ and $(2, \infty)$; decreasing on $(-2, 0)$ and $(0, 2)$; local maximum at $x = -2$; local minimum at $x = 2$

63. Critical value: $x = -2$; increasing on $(-2, 0)$; decreasing on $(-\infty, -2)$ and $(0, \infty)$; local minimum at $x = -2$

65. Critical values: $x = 0$, $x = 4$; increasing on $(-\infty, 0)$ and $(4, \infty)$; decreasing on $(0, 2)$ and $(2, 4)$; local maximum at $x = 0$; local minimum at $x = 4$

67. Critical values: $x = 0$, $x = 4$, $x = 6$; increasing on $(0, 4)$ and $(6, \infty)$; decreasing on $(-\infty, 0)$ and $(4, 6)$; local maximum at $x = 4$; local minima at $x = 0$ and $x = 6$
69. Critical value: $x = 2$; increasing on $(2, \infty)$; decreasing on $(-\infty, 2)$; local minimum at $x = 2$
71. Critical value: $x = 1$; increasing on $(0, 1)$; decreasing on $(1, \infty)$; local maximum at $x = 1$
73. (A) There are no critical values and no local extrema. The function is increasing for all x.
(B) There are two critical values, $x = \pm\sqrt{-k/3}$. The function increases on $(-\infty, -\sqrt{-k/3})$ to a local maximum at $x = -\sqrt{-k/3}$, decreases on $(-\sqrt{-k/3}, \sqrt{-k/3})$ to a local minimum at $x = \sqrt{-k/3}$, and increases on $(\sqrt{-k/3}, \infty)$.
(C) The only critical value is $x = 0$. There are no local extrema. The function is increasing for all x.
75. (A) The marginal profit is positive on $(0, 600)$, 0 at $x = 600$, and negative on $(600, 1{,}000)$.
(B)

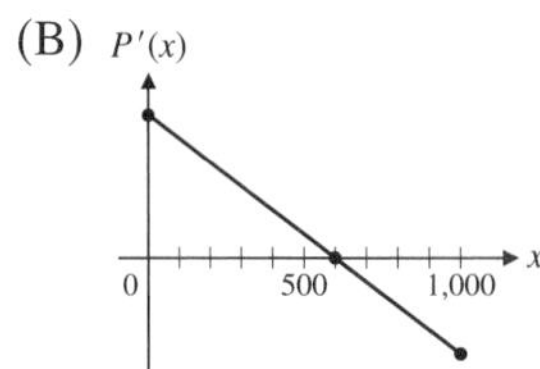

77. (A) The price decreases for the first 15 months to a local minimum, increases for the next 40 months to a local maximum, and then decreases for the remaining 15 months.
(B)

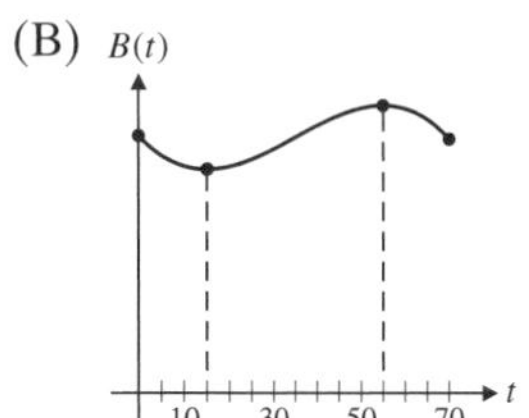

79. (A) $\overline{C}(x) = 0.05x + 20 + \dfrac{320}{x}$
(B) Critical value: $x = 80$; decreasing for $0 < x < 80$; increasing for $80 < x < 150$; local minimum at $x = 80$
81. $P(x)$ is increasing over (a, b) if $P'(x) = R'(x) - C'(x) > 0$ over (a, b); that is, if $R'(x) > C'(x)$ over (a, b).
83. Critical value: $t = 1$; increasing for $0 < t < 1$; decreasing for $1 < t < 24$; local maximum at $t = 1$
85. Critical value: $t = 7$; increasing for $0 < t < 7$; decreasing for $7 < t < 24$; local maximum at $t = 7$

Exercise 9-3

1. $(a, c), (c, d), (e, g)$ **3.** $(d, e), (g, h)$ **5.** $(a, c), (c, d), (e, g)$ **7.** d, e, g **9.** (C) **11.** (D)
13. Local minimum **15.** Unable to determine **17.** Neither

19.

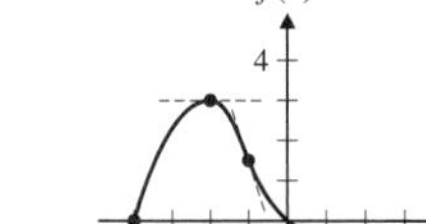
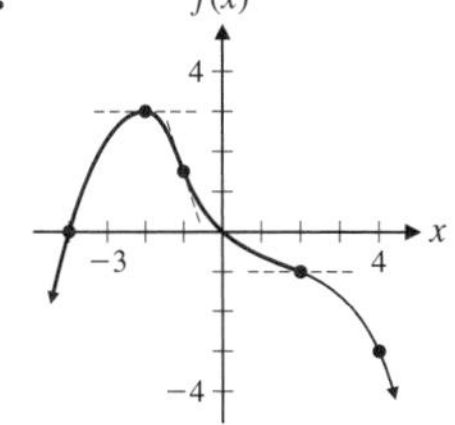

21.

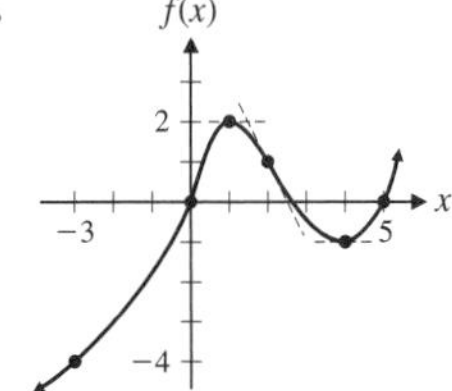

23.

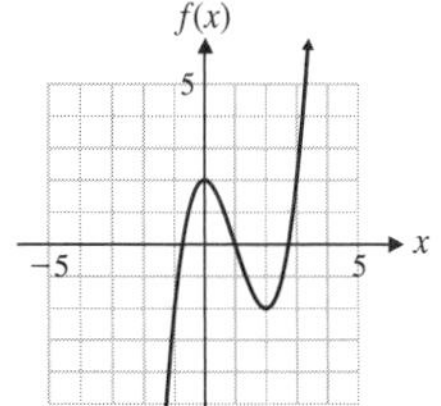

25.

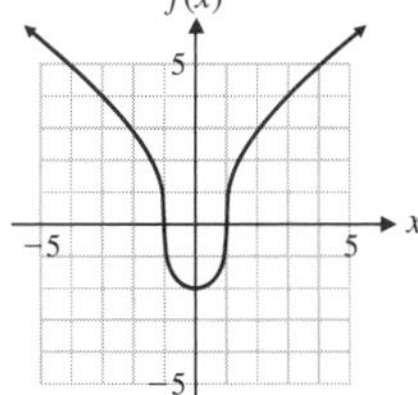

27. $6x - 4$ **29.** $40x^3$ **31.** $6x^{-3} + 12x^{-4}$ **33.** $24x^2(x^2 - 1) + 6(x^2 - 1)^2 = 6(x^2 - 1)(5x^2 - 1)$
35. $f(2) = -2$ is a local minimum **37.** $f(-1) = 2$ is a local maximum; $f(2) = -25$ is a local minimum
39. No local extrema **41.** $f(-2) = -6$ is a local minimum; $f(0) = 10$ is a local maximum; $f(2) = -6$ is a local minimum
43. $f(0) = 2$ is a local minimum **45.** $f(-4) = -8$ is a local maximum; $f(4) = 8$ is a local minimum.
47. Concave upward for all x; no inflection points
49. Concave upward on $(6, \infty)$; concave downward on $(-\infty, 6)$; inflection point at $x = 6$
51. Concave upward on $(-\infty, -2)$ and $(2, \infty)$; concave downward on $(-2, 2)$; inflection points at $x = -2$ and $x = 2$

53. Local maximum at $x = 0$
Local minimum at $x = 4$
Inflection point at $x = 2$

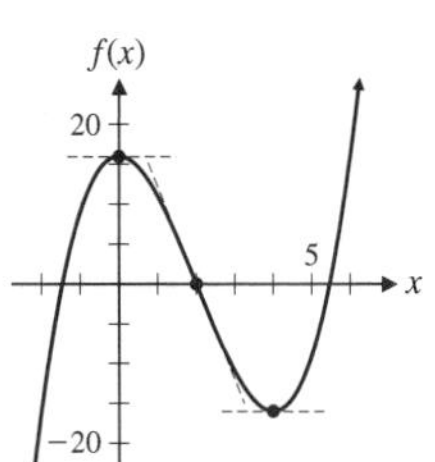

55. Inflection point at $x = 0$

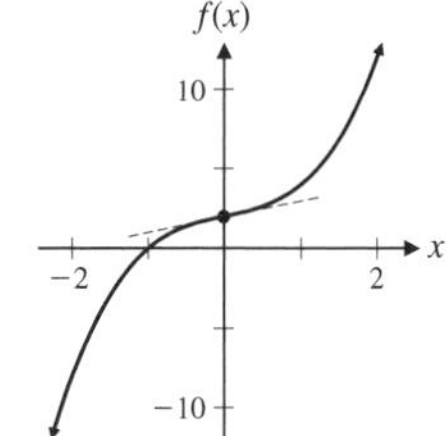

57. Infection point at $x = 2$

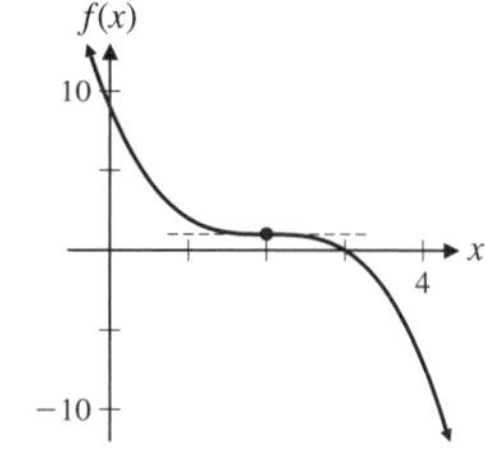

59. Local maximum at $x = -2$
Local minimum at $x = 2$
Inflection point at $x = 0$

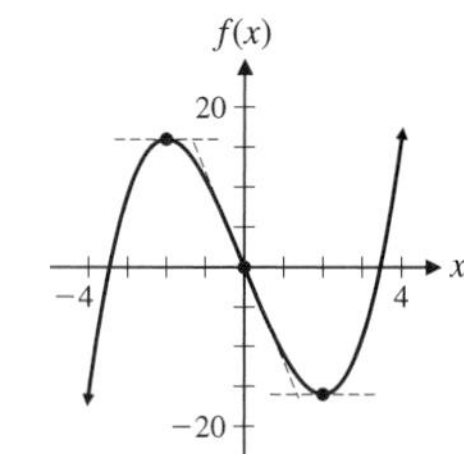

61.

x	$f'(x)$	$f(x)$
$-\infty < x < -1$	Positive and decreasing	Increasing and concave downward
$x = -1$	x intercept	Local maximum
$-1 < x < 0$	Negative and decreasing	Decreasing and concave downward
$x = 0$	Local minimum	Inflection point
$0 < x < 2$	Negative and increasing	Decreasing and concave upward
$x = 2$	Local maximum	Inflection point
$2 < x < \infty$	Negative and decreasing	Decreasing and concave downward

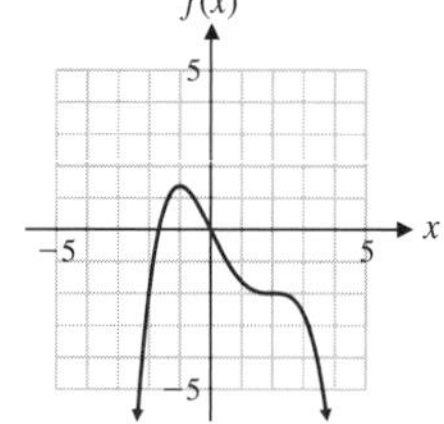

63.

x	$f'(x)$	$f(x)$
$-\infty < x < -2$	Negative and increasing	Decreasing and concave upward
$x = -2$	Local maximum	Inflection point
$-2 < x < 0$	Negative and decreasing	Decreasing and concave downward
$x = 0$	Local minimum	Inflection point
$0 < x < 2$	Negative and increasing	Decreasing and concave upward
$x = 2$	Local maximum	Inflection point
$2 < x < \infty$	Negative and decreasing	Decreasing and concave downward

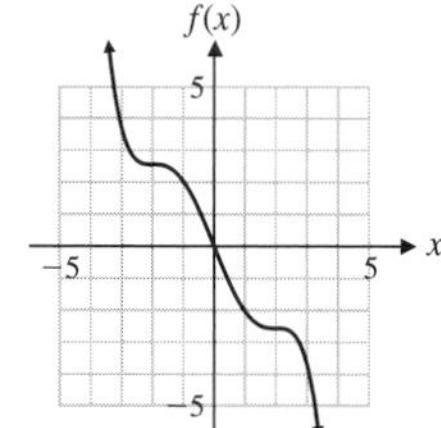

65. Inflection points at $x = 0.27$ and $x = 3.73$; concave upward on $(-\infty, 0.27)$ and $(3.73, \infty)$; concave downward on $(0.27, 3.73)$

67. Inflection point at $x = -1.40$; concave upward on $(-1.40, \infty)$; concave downward on $(-\infty, -1.40)$

69. Inflection points at $x = -0.61$, $x = 0.66$, and $x = 1.74$; concave upward on $(-0.61, 0.66)$ and $(1.74, \infty)$; concave downward on $(-\infty, -0.61)$ and $(0.66, 1.74)$

71. If $f'(x)$ has a local extremum at $x = c$, then $f'(x)$ must change from increasing to decreasing or from decreasing to increasing at $x = c$. Thus, the graph of $y = f(x)$ must change concavity at $x = c$, and there must be an inflection point at $x = c$.

73. If there is an inflection point on the graph of $y = f(x)$ at $x = c$, then $f(x)$ must change concavity at $x = c$. Consequently, $f'(x)$ must change from increasing to decreasing or from decreasing to increasing at $x = c$, and $x = c$ is a local extremum for $f'(x)$.

75. Inflection points at $x = -2$ and $x = 2$ **77.** Inflection points at $x = -6$, $x = 0$, and $x = 6$

79. The graph of the CPI is concave upward.

81. The graph of $y = C'(x)$ is positive and decreasing. Since since marginal costs are decreasing, the production process is becoming more efficient as production increases.

83. (A) Local maximum at $x = 60$ (B) Concave downward on the whole interval $(0, 80)$

85. (A) Increasing on $(10, 25)$; decreasing on $(25, 40)$
(B) Inflection point at $x = 25$

(C)

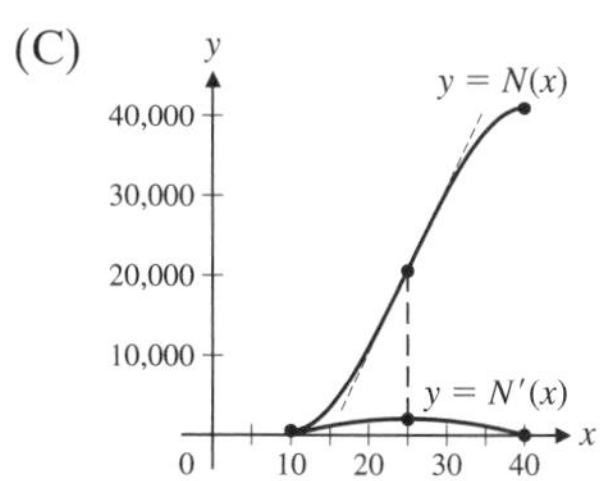

(D) Max $N'(x) = N'(25) = 2{,}025$

87. (A)

```
CubicReg
 y=ax³+bx²+cx+d
 a=-.005
 b=.485
 c=-1.85
 d=300
```

(B) 32 ads to sell 574 cars per month

89. (A) Increasing on $(0, 10)$; decreasing on $(10, 20)$
(B) Inflection point at $t = 10$

(C)

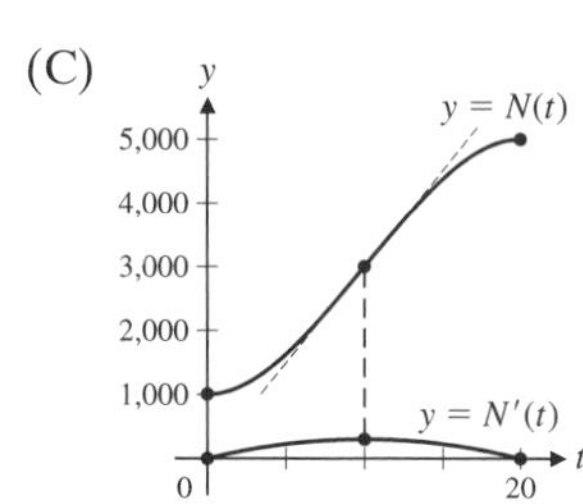

(D) $N'(10) = 300$

91. (A) Increasing on $(5, \infty)$; decreasing on $(0, 5)$
(B) Inflection point at $n = 5$

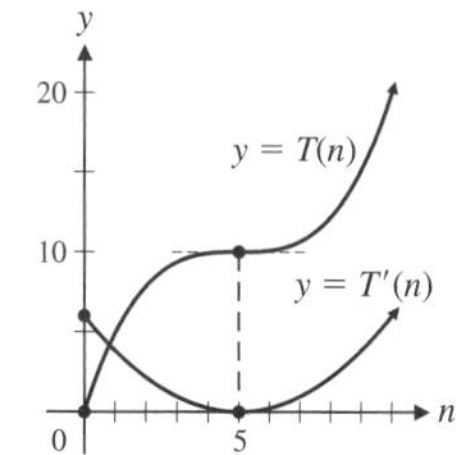

(C) $T'(5) = 0$

Exercise 9-4

1. $(-\infty, b), (0, e), (e, g)$ **3.** $(b, d), (d, 0), (g, \infty)$ **5.** $x = 0$ **7.** $(-\infty, a), (d, e), (h, \infty)$ **9.** $(a, d), (e, h)$
11. $x = a, x = h$ **13.** $x = d, x = e$

15.

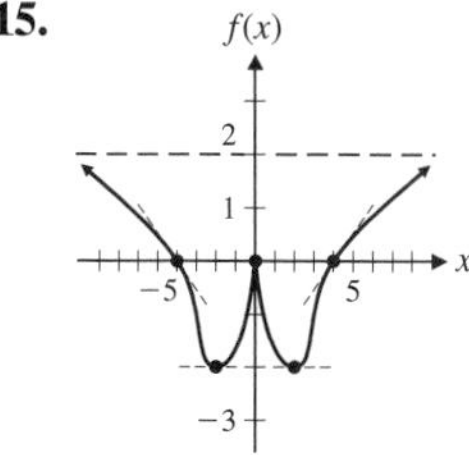

17.

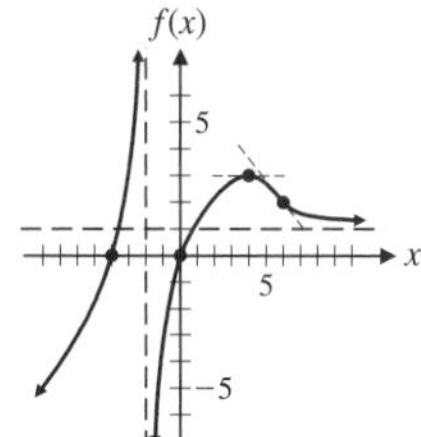

19.

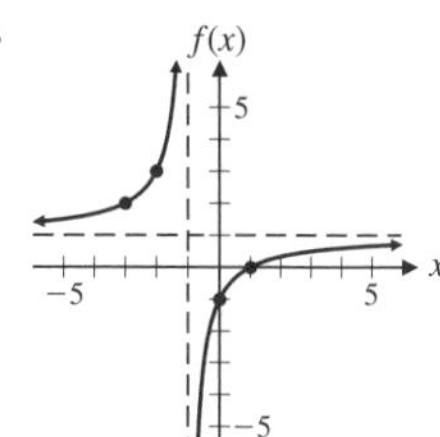

21.

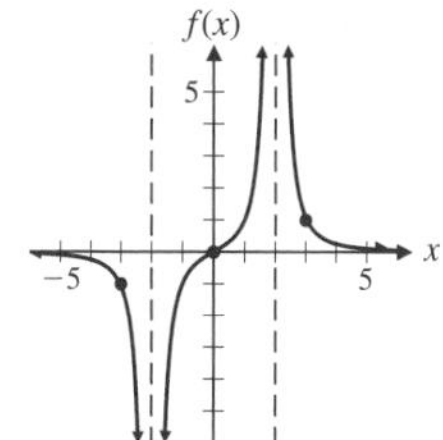

23. ∞ **25.** $-\infty$ **27.** $\frac{4}{5}$ **29.** ∞ **31.** 0
33. Horizontal asymptote: $y = 2$; vertical asymptote: $x = -2$
35. Horizontal asymptote: $y = 1$; vertical asymptotes: $x = -1$ and $x = 1$
37. No horizontal or vertical asymptotes
39. No horizontal asymptote; vertical asymptote: $x = 3$
41. Horizontal asymptote: $y = 2$; vertical asymptotes: $x = -1$ and $x = 2$
43. Horizontal asymptote: $y = 2$; vertical asymptote: $x = -1$

45. Domain: All real numbers
y intercept: 0; x intercepts: 0, 6
Increasing on $(-\infty, 0)$ and $(4, \infty)$
Decreasing on $(0, 4)$
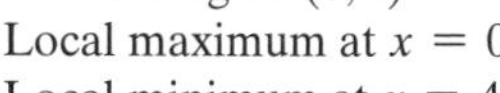
Local maximum at $x = 0$
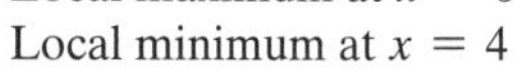
Local minimum at $x = 4$
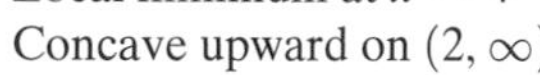
Concave upward on $(2, \infty)$
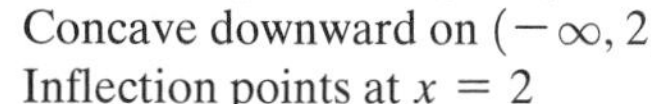
Concave downward on $(-\infty, 2)$
Inflection points at $x = 2$

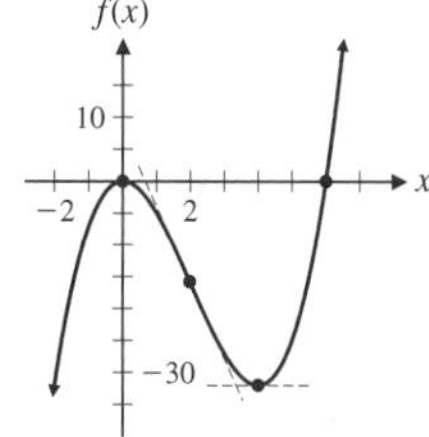

47. Domain: All real numbers
y intercept: 16; x intercepts: $-4, 2$
Increasing on $(-\infty, -2)$ and $(2, \infty)$
Decreasing on $(-2, 2)$
Local maximum at $x = -2$
Local minimum at $x = 2$
Concave upward on $(0, \infty)$
Concave downward on $(-\infty, 0)$
Inflection point at $x = 0$

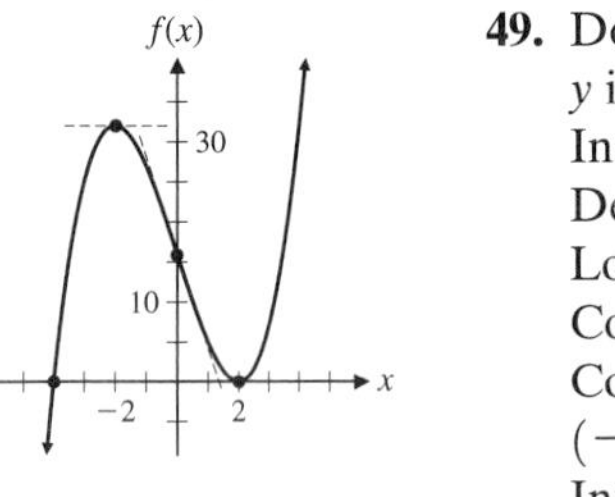

49. Domain: All real numbers
y intercept: 0; x intercepts: 0, 4
Increasing on $(-\infty, 3)$
Decreasing on $(3, \infty)$
Local maximum at $x = 3$
Concave upward on $(0, 2)$
Concave downward on $(-\infty, 0)$ and $(2, \infty)$
Inflection points at $x = 0$ and $x = 2$

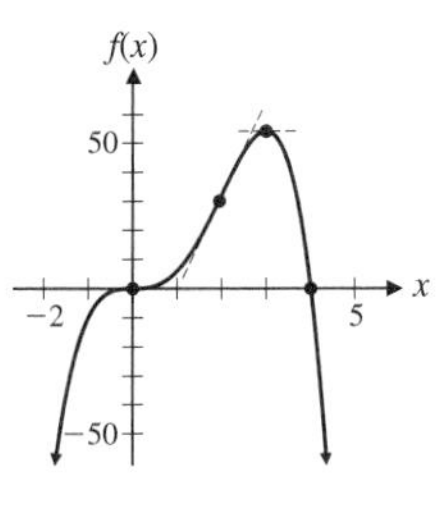

51. Domain: All real numbers, except 3
y intercept: -1; x intercept: -3
Horizontal asymptote: $y = 1$
Vertical asymptote: $x = 3$
Decreasing on $(-\infty, 3)$ and $(3, \infty)$
Concave upward on $(3, \infty)$
Concave downward on $(-\infty, 3)$

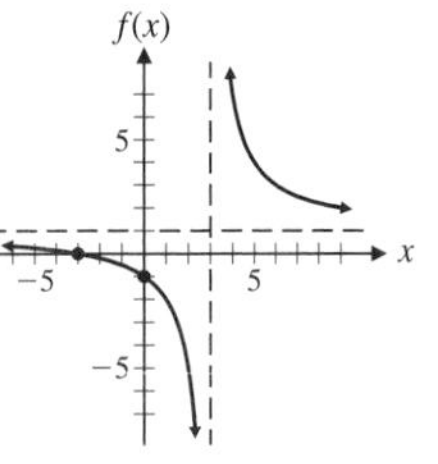

53. Domain: All real numbers, except 2
y intercept: 0; x intercept: 0
Horizontal asymptote: $y = 1$
Vertical asymptote: $x = 2$
Decreasing on $(-\infty, 2)$ and $(2, \infty)$
Concave downward on $(-\infty, 2)$
Concave upward on $(2, \infty)$

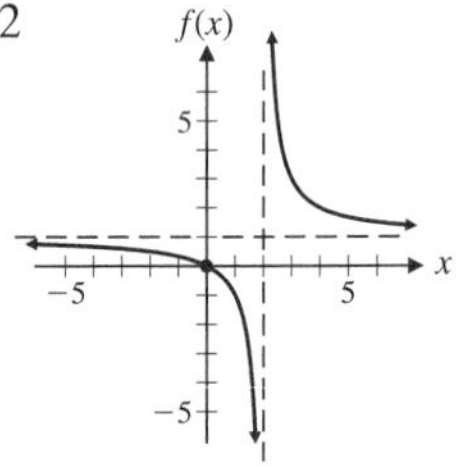

55. For any $n \geqslant 1$, the limit will be ∞ if $a > 0$ and $-\infty$ if $a < 0$.

57. (A) $\lim_{x\to\infty} p'(x) = \infty$; $\lim_{x\to\infty} p''(x) = \infty$. The graph of $y = p(x)$ is increasing and concave upward for large positive values of x.
(B) $\lim_{x\to-\infty} p'(x) = \infty$; $\lim_{x\to-\infty} p''(x) = -\infty$. The graph of $y = p(x)$ is increasing and concave downward for large negative values of x.

59. Domain: All real numbers, except 0
Vertical asymptote: $x = 0$
Increasing on $(-\infty, -1)$ and $(1, \infty)$
Decreasing on $(-1, 0)$ and $(0, 1)$
Local maximum at $x = -1$
Local minimum at $x = 1$
Concave upward on $(0, \infty)$
Concave downward on $(-\infty, 0)$

61. Domain: All real numbers
y intercept: 0;
x intercepts: $-1, 0, 1$
Increasing on $(-\infty, -\sqrt{3}/3)$ and $(\sqrt{3}/3, \infty)$
Decreasing on $(-\sqrt{3}/3, \sqrt{3}/3)$
Local maximum at $x = -\sqrt{3}/3$
Local minimum at $x = \sqrt{3}/3$
Concave downward on $(-\infty, 0)$
Concave upward on $(0, \infty)$
Inflection point at $x = 0$

63. Domain: All real numbers
y intercept: 27; x intercepts: $-3, 3$
Increasing on $(-\infty, -\sqrt{3})$ and $(0, \sqrt{3})$
Decreasing on $(-\sqrt{3}, 0)$ and $(\sqrt{3}, \infty)$
Local maxima at $x = -\sqrt{3}$ and $x = \sqrt{3}$
Local minimum at $x = 0$
Concave upward on $(-1, 1)$
Concave downward on $(-\infty, -1)$ and $(1, \infty)$
Inflection points at $x = -1$ and $x = 1$

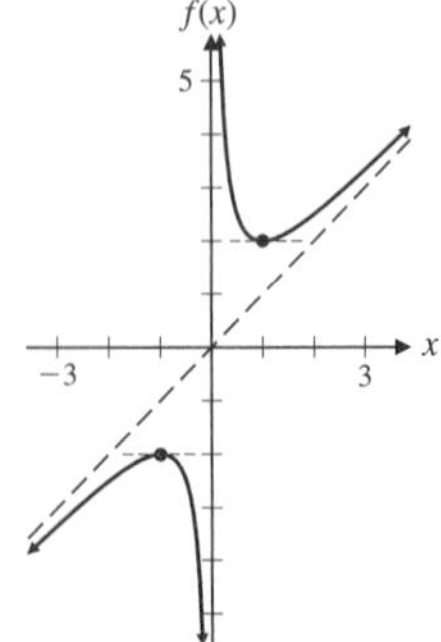

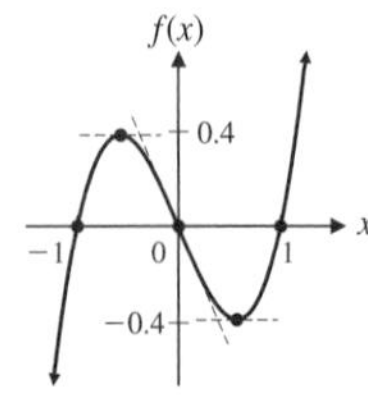

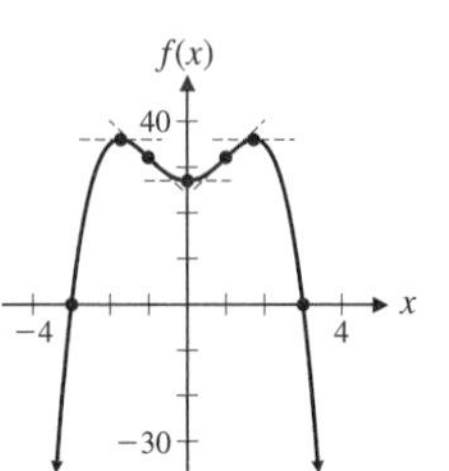

65. Domain: All real numbers
y intercept: 16; x intercepts: $-2, 2$
Decreasing on $(-\infty, -2)$ and $(0, 2)$
Increasing on $(-2, 0)$ and $(2, \infty)$
Local minima at $x = -2$ and $x = 2$
Local maximum at $x = 0$
Concave upward on $(-\infty, -2\sqrt{3}/3)$ and $(2\sqrt{3}/3, \infty)$
Concave downward on $(-2\sqrt{3}/3, 2\sqrt{3}/3)$
Inflection points at $x = -2\sqrt{3}/3$ and $x = 2\sqrt{3}/3$

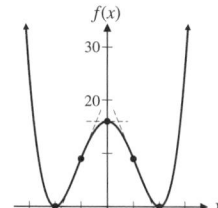

67. Domain: All real numbers
y intercept: 0; x intercepts: 0, 1.5
Decreasing on $(-\infty, 0)$ and $(0, 1.25)$
Increasing on $(1.25, \infty)$
Local minimum at $x = 1.25$
Concave upward on $(-\infty, 0)$ and $(1, \infty)$
Concave downward on $(0, 1)$
Inflection points at $x = 0$ and $x = 1$

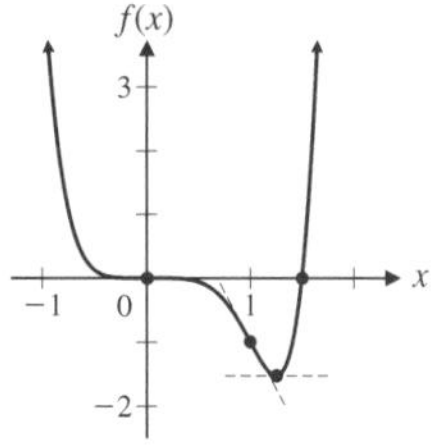

69. Domain: All real numbers, except ± 2
y intercept: 0; x intercept: 0
Horizontal asymptote: $y = 0$
Vertical asymptotes: $x = -2, x = 2$
Decreasing on $(-\infty, -2)$, $(-2, 2)$, and $(2, \infty)$
Concave upward on $(-2, 0)$ and $(2, \infty)$
Concave downward on $(-\infty, -2)$ and $(0, 2)$
Inflection point at $x = 0$

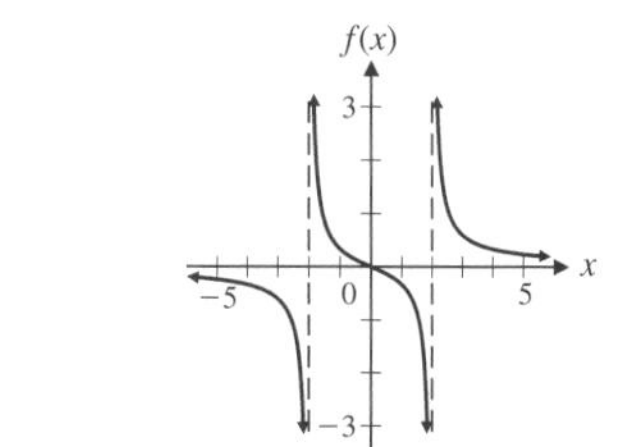

71. Domain: All real numbers
y intercept: 1
Horizontal asymptote: $y = 0$
Increasing on $(-\infty, 0)$
Decreasing on $(0, \infty)$
Local maximum at $x = 0$
Concave upward on $(-\infty, -\sqrt{3}/3)$ and $(\sqrt{3}/3, \infty)$
Concave downward on $(-\sqrt{3}/3, \sqrt{3}/3)$
Inflection points at $x = -\sqrt{3}/3$ and $x = \sqrt{3}/3$

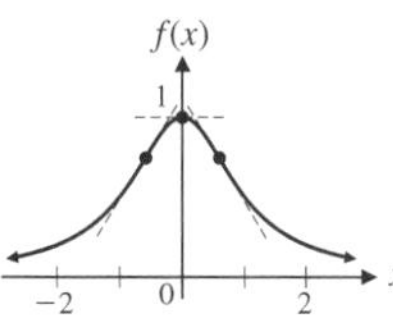

73. Domain: All real numbers
x intercepts: $-1.18, 0.61, 1.87, 3.71$;
y intercept: -5
Decreasing on $(-\infty, -0.53)$ and $(1.24, 3.04)$
Increasing on $(-0.53, 1.24)$ and $(3.04, \infty)$
Local minima at $x = -0.53$ and $x = 3.04$
Local maximum at $x = 1.24$
Concave upward on $(-\infty, 0.22)$ and $(2.28, \infty)$
Concave downward on $(0.22, 2.28)$
Inflection points at $x = 0.22$ and $x = 2.28$

75. Domain: All real numbers
y intercept: 100; x intercepts: 8.01, 13.36
Increasing on $(-0.10, 4.57)$ and $(11.28, \infty)$
Decreasing on $(-\infty, -0.10)$ and $(4.57, 11.28)$
Local maximum at $x = 4.57$
Local minima at $x = -0.10$ and $x = 11.28$
Concave upward on $(-\infty, 1.95)$ and $(8.55, \infty)$
Concave downward on $(1.95, 8.55)$
Inflection points at $x = 1.95$ and $x = 8.55$

77. Domain: All real numbers
x intercepts: $-2.40, 1.16$; y intercept: 3
Increasing on $(-\infty, -1.58)$
Decreasing on $(-1.58, \infty)$
Local maximum at $x = -1.58$
Concave downward on $(-\infty, -0.88)$ and $(0.38, \infty)$
Concave upward on $(-0.88, 0.38)$
Inflection points at $x = -0.88$ and $x = 0.38$

79. Domain: All real numbers
x intercepts: $-6.68, -3.64, -0.72$; y intercept: 30
Decreasing on $(-5.59, -2.27)$ and $(1.65, 3.82)$
Increasing on $(-\infty, -5.59)$, $(-2.27, 1.65)$, and $(3.82, \infty)$
Local minima at $x = -2.27$ and $x = 3.82$
Local maxima at $x = -5.59$ and $x = 1.65$
Concave upward on $(-4.31, -0.40)$ and $(2.91, \infty)$
Concave downward on $(-\infty, -4.31)$ and $(-0.40, 2.91)$
Inflection points at $x = -4.31$, $x = -0.40$, and $x = 2.91$

81.

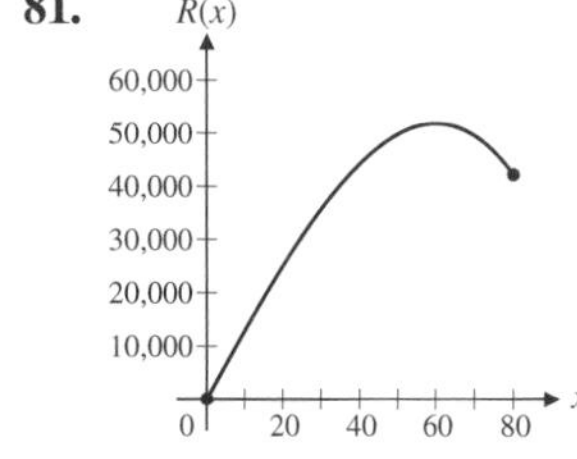

83. (A) Increasing on $(0, 1)$
(B) Concave upward on $(0, 1)$
(C) $x = 1$ is a vertical asymptote
(D) The origin is both an x and a y intercept
(E)

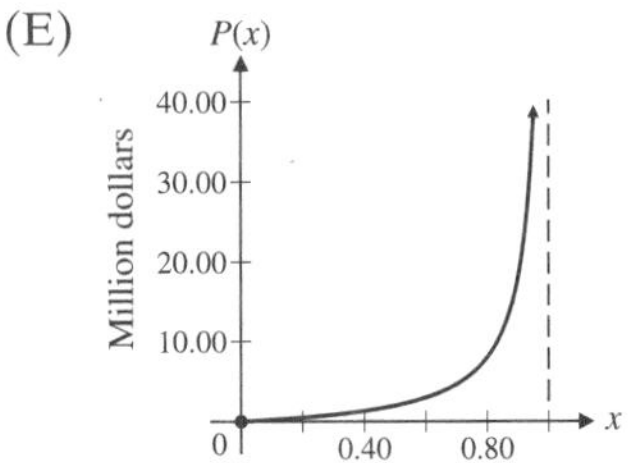

85. (A) 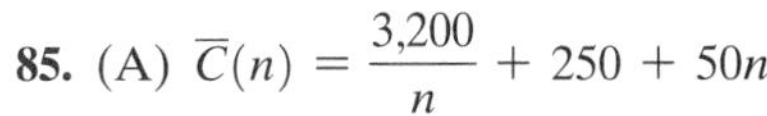 $\overline{C}(n) = \dfrac{3{,}200}{n} + 250 + 50n$

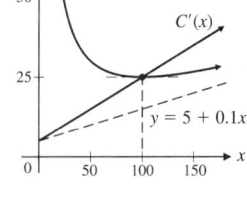

(B)

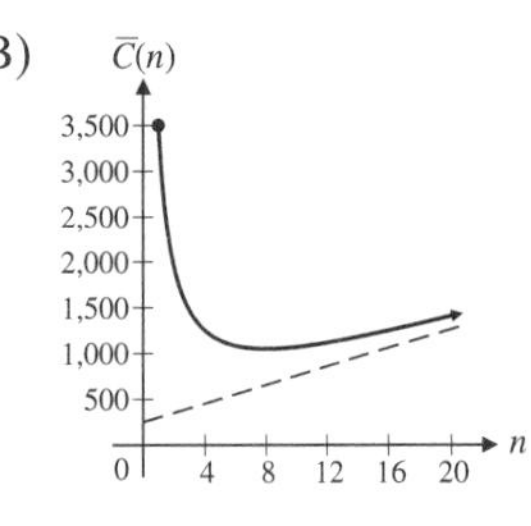

(C) 8 yr

87. (A)

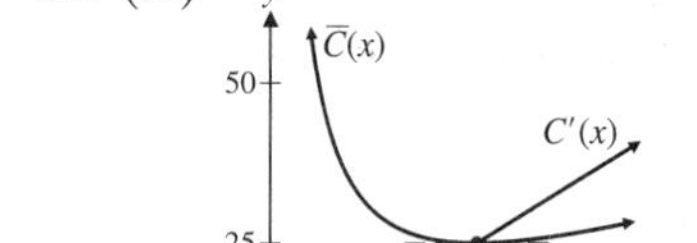

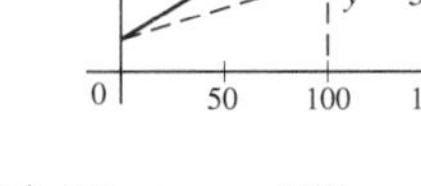

(B) 25 at $x = 100$

89. (A)

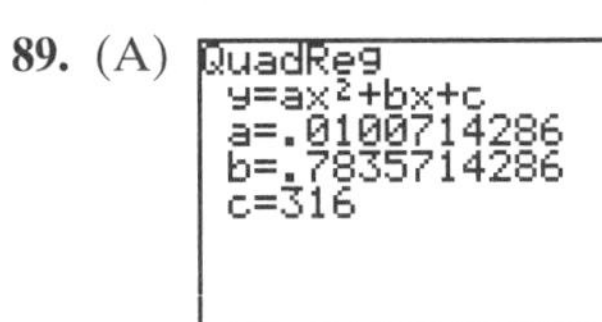

(B) Minimum average cost is \$4.35 when 177 pizzas are produced daily.

91.

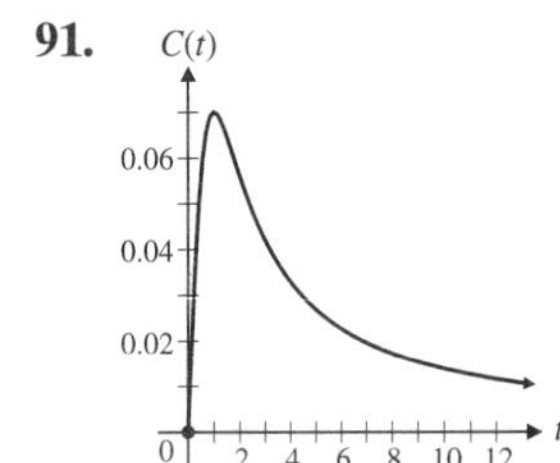

93.

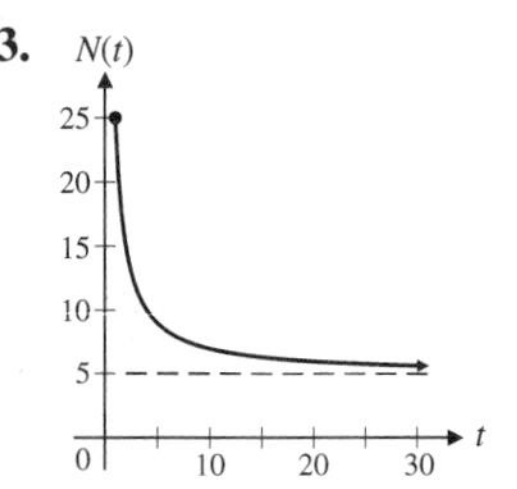

Exercise 9-5

1. Min $f(x) = f(0) = 0$; Max $f(x) = f(10) = 14$ **3.** Min $f(x) = f(0) = 0$; Max $f(x) = f(3) = 9$
5. Min $f(x) = f(1) = f(7) = 5$; Max $f(x) = f(10) = 14$
7. Min $f(x) = f(1) = f(7) = 5$; Max $f(x) = f(3) = f(9) = 9$
9. Min $f(x) = f(5) = 7$; Max $f(x) = f(3) = 9$ **11.** Min $f(x) = f(2) = 1$; no maximum
13. Max $f(x) = f(4) = 26$; no minimum **15.** No absolute extrema exist. **17.** Max $f(x) = f(2) = 16$
19. Min $f(x) = f(2) = 14$ **21.** Max $f(x) = f(4.55) = 34.81$
23. (A) Max $f(x) = f(5) = 14$; Min $f(x) = f(-1) = -22$ (B) Max $f(x) = f(1) = -2$; Min $f(x) = f(-1) = -22$
(C) Max $f(x) = f(5) = 14$; Min $f(x) = f(3) = -6$
25. (A) Max $f(x) = f(0) = 126$; Min $f(x) = f(2) = -26$ (B) Max $f(x) = f(7) = 49$; Min $f(x) = f(2) = -26$
(C) Max $f(x) = f(6) = 6$; Min $f(x) = f(3) = -15$
27. Exactly in half **29.** 15 and -15 **31.** A square of side 25 cm; maximum area $= 625\text{ cm}^2$
33. If x and y are the dimensions of the rectangle and A is the fixed area, the model is: Minimize $C = 2Bx + 2AB/x, x > 0$. This mathematical problem always has a solution. This agrees with our economic intuition that there should be a cheapest way to build the fence.
35. If x and y are the dimensions of the rectangle and C is the fixed amount to be spent, the model is: Maximize $A = x(C - 2Bx)/(2B), 0 \le x \le C/(2B)$. This mathematical problem always has a solution. This agrees with our economic intuition that there should be a largest area that can be enclosed with a fixed amount of fencing.
37. (A) Max $R(x) = R(3{,}000) = \$300{,}000$
(B) Maximum profit is \$75,000 when 2,100 sets are manufactured and sold for \$130 each.
(C) Maximum profit is \$64,687.50 when 2,025 sets are manufactured and sold for \$132.50 each.

39. (A)

```
QuadReg
 y=ax²+bx+c
 a=-2.352941E-5
 b=-.0325964781
 c=288.9535407
```

(B)

```
LinReg
 y=ax+b
 a=53.50318471
 b=82245.22293
```

(C) The maximum profit is \$118,996 when the price per saw is \$195.

41. \$35; \$6,125 **43.** 40 trees; 1,600 lb **45.** $(10 - 2\sqrt{7})/3 = 1.57$ in. squares
47. 20 ft by 40 ft (with the expensive side being one of the short sides) **49.** 10,000 books in 5 printings
51. (A) $x = 5.1$ mi (B) $x = 10$ mi **53.** 4 days; 20 bacteria/cm^3 **55.** 50 mice per order **57.** 1 month; 2 ft
59. 4 yr from now

Review Exercise

1. $(a, c_1), (c_3, c_6)$ *(9-2, 9-3)* **2.** $(c_1, c_3), (c_6, b)$ *(9-2, 9-3)* **3.** $(a, c_2), (c_4, c_5), (c_7, b)$ *(9-2, 9-3)* **4.** c_3 *(9-2)* **5.** c_6 *(9-5)*
6. c_1, c_3, c_5 *(9-2)* **7.** c_6 *(9-2)* **8.** c_2, c_4, c_5, c_7 *(9-3)* **9.** (A) Does not exist (B) 3 (C) No *(9-1)*
10. (A) 2 (B) Not defined (C) No *(9-1)* **11.** (A) 1 (B) 1 (C) Yes *(9-1)*

12. *(9-4)* **13.** *(9-4)*

14. $f''(x) = 12x^2 + 30x$ *(9-3)* **15.** $y'' = 8/x^3$ *(9-3)* **16.** (A) 4 (B) 6 (C) Does not exist (D) 6 (E) No *(9-1)*
17. (A) 3 (B) 3 (C) 3 (D) 3 (E) Yes *(9-1)* **18.** $(-3, 4)$ *(9-1)* **19.** $(-3, 0) \cup (5, \infty)$ *(9-1)*
20. $(-2.34, -0.47) \cup (1.81, \infty)$ *(9-1)*
21. (A) All real numbers (B) y intercept: 0; x intercepts: 0, 9
(C) No horizontal or vertical asymptotes *(9-4)*
22. (A) 3, 9 (B) 3, 9
(C) Increasing on $(-\infty, 3)$ and $(9, \infty)$; decreasing on $(3, 9)$
(D) Local maximum at $x = 3$; local minimum at $x = 9$ *(9-4)*
23. (A) Concave downward on $(-\infty, 6)$; concave upward on $(6, \infty)$
(B) Inflection point at $x = 6$ *(9-4)*

24. *(9-4)*

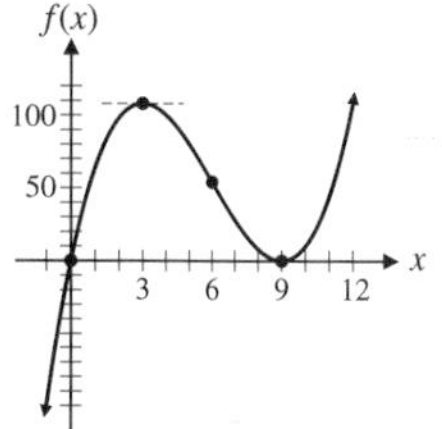

25. (A) All real numbers, except -2 (B) y intercept: 0; x intercept: 0
(C) Horizontal asymptote: $y = 3$; vertical asymptote: $x = -2$ *(9-4)*
26. (A) None (B) -2 (C) Increasing on $(-\infty, -2)$ and $(-2, \infty)$ (D) None *(9-4)*
27. (A) Concave upward on $(-\infty, -2)$; concave downward on $(-2, \infty)$
(B) No inflection points *(9-4)*

28. *(9-4)*

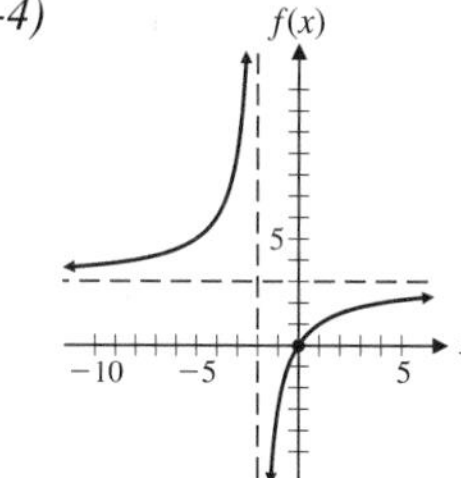

29.

x	$f'(x)$	$f(x)$
$-\infty < x < -2$	Negative and increasing	Decreasing and concave upward
$x = -2$	x intercept	Local minimum
$-2 < x < -1$	Positive and increasing	Increasing and concave upward
$x = -1$	Local maximum	Inflection point
$-1 < x < 1$	Positive and decreasing	Increasing and concave downward
$x = 1$	Local minimum	Inflection point
$1 < x < \infty$	Positive and increasing	Increasing and concave upward

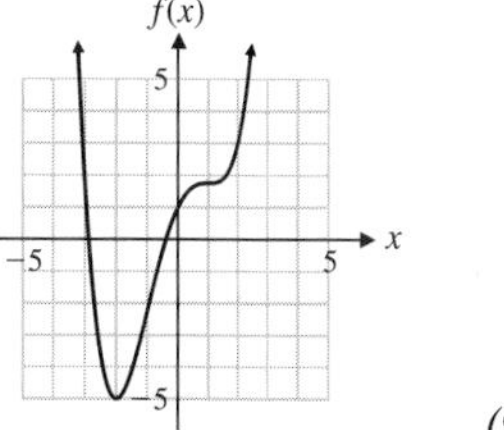

(9-3)

30. (C) *(9-3)* **31.** Local maximum at $x = -1$; local minimum at $x = 5$ *(9-3)*
32. Min $f(x) = f(2) = -4$; Max $f(x) = f(5) = 77$ *(9-5)* **33.** Min $f(x) = f(2) = 8$ *(9-5)*
34. $(-\infty, \infty)$ *(9-1)* **35.** $(-\infty, -5), (-5, \infty)$ *(9-1)* **36.** $(-\infty, -2), (-2, 3), (3, \infty)$ *(9-1)* **37.** $[3, \infty)$ *(9-1)*

38. $(-\infty, \infty)$ *(9-1)* **39.** ∞ *(9-1)* **40.** $-\infty$ *(9-1)* **41.** Does not exist *(9-1)* **42.** ∞ *(9-1)* **43.** $-\infty$ *(9-4)*
44. ∞ *(9-4)* **45.** ∞ *(9-4)* **46.** 2 *(9-4)* **47.** ∞ *(9-4)* **48.** 0 *(9-4)*
49. Horizontal asymptote; $y = 0$; no vertical asymptotes *(9-4)*
50. No horizontal asymptotes; vertical asymptotes: $x = -3$ and $x = 3$ *(9-4)*
51. Yes. Since f is continuous on $[a, b]$, f has an absolute maximum on $[a, b]$. But each end point is a local minimum; hence, the absolute maximum must occur between a and b. *(9-5)*
52. No, increasing/decreasing properties apply to intervals in the domain of f. It is correct to say that $f(x)$ is decreasing on $(-\infty, 0)$ and $(0, \infty)$. *(9-2)*
53. A critical value for $f(x)$ is a partition number for $f'(x)$ that is also in the domain of f. For example, if $f(x) = x^{-1}$, then 0 is a partition number for $f'(x) = -x^{-2}$, but 0 is not a critical value for $f(x)$ since 0 is not in the domain of f. *(9-2)*
54. Max $f'(x) = f'(2) = 12$ *(9-5)*
55. Each number is 20; minimum sum is 40 *(9-5)*
56. Domain: all real numbers
y intercept: -3; x intercepts: $-3, 1$
No vertical or horizontal asymptotes
Increasing on $(-2, \infty)$
Decreasing on $(-\infty, -2)$
Local minimum at $x = -2$
Concave upward on $(-\infty, -1)$ and $(1, \infty)$
Concave downward on $(-1, 1)$
Inflection points at $x = -1$ and $x = 1$ *(9-4)*

57. Domain: All real numbers
x intercepts: 0.79, 1.64; y intercept: 4
Increasing on $(-1.68, -0.35)$ and $(1.28, \infty)$
Decreasing on $(-\infty, -1.68)$ and $(-0.35, 1.28)$
Local minima at $x = -1.68$ and $x = 1.28$
Local maximum at $x = -0.35$
Concave downward on $(-1.10, 0.60)$
Concave upward on $(-\infty, -1.10)$ and $(0.60, \infty)$
Inflection points at $x = -1.10$ and $x = 0.60$ *(9-4)*

58. Domain: All real numbers
x intercepts; 0, 11.10; y intercept: 0
Increasing on $(1.87, 4.19)$ and $(8.94, \infty)$
Decreasing on $(-\infty, 1.87)$ and $(4.19, 8.94)$
Local maximum at $x = 4.19$
Local minima at $x = 1.87$ and $x = 8.94$
Concave upward on $(-\infty, 2.92)$ and $(7.08, \infty)$
Concave downward on $(2.92, 7.08)$
Inflection points at $x = 2.92$ and $x = 7.08$ *(9-4)*

59. (A)

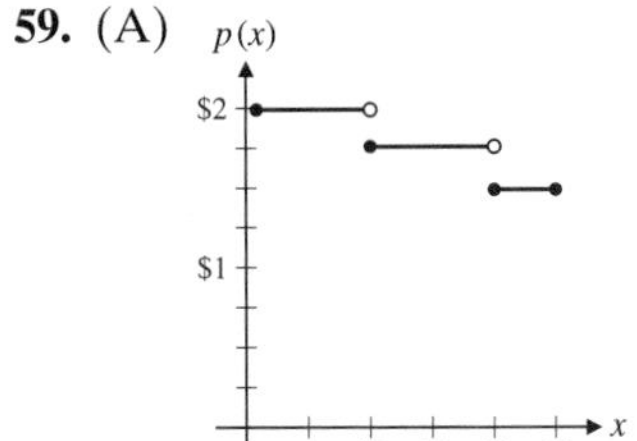

(B) p is discontinuous at $x = 12$ and $x = 24$. In each case, the limit from the left is greater than the limit from the right, reflecting the corresponding drop in price at these order quantities.

(C)

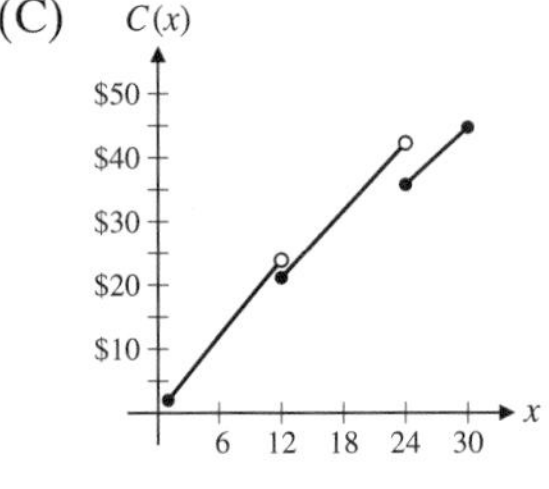

(D) C is discontinuous at $x = 12$ and $x = 24$. In each case, the limit from the left is greater than the limit from the right, reflecting savings to the customer due to the corresponding drop in price at these order quantities. *(9-1)*

60. (A) For the first 15 months, the graph of the price is increasing and concave downward, with a local maximum at $t = 15$. For the next 15 months, the graph of the price is decreasing and concave downward, with an inflection point at $t = 30$. For the next 15 months, the graph of the price is decreasing and concave upward, with a local minimum at $t = 45$. For the remaining 15 months, the graph of the price is increasing and concave upward.

(B)

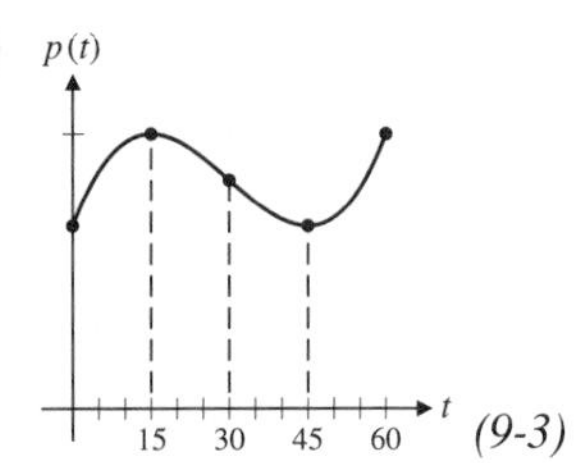

(9-3)

61. (A) Max $R(x) = R(10{,}000) = \$2{,}500{,}000$
(B) Maximum profit is \$175,000 when 3,000 stoves are manufactured and sold for \$425 each.
(C) Maximum profit is \$119,000 when 2,600 stoves are manufactured and sold for \$435 each. *(9-5)*
62. (A) The expensive side is 50 ft; the other side is 100 ft. (B) The expensive side is 75 ft; the other side is 150 ft. *(9-5)*
63. \$49; \$6,724 *(9-5)* **64.** 12 orders/yr *(9-5)*

65. Min $\overline{C}(x) = \overline{C}(200) = 50$

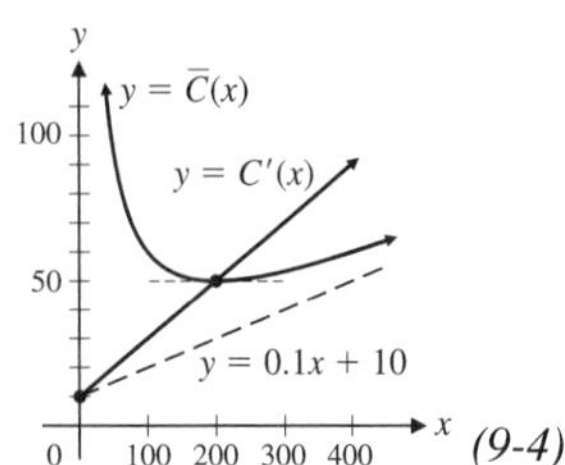

(9-4)

66. (A)

```
QuadReg
 y=ax²+bx+c
 a=.0061285714
 b=.1224285714
 c=102.2
```

(B) Min $\overline{C}(x) = \overline{C}(129) = \$1.71$ *(9-4)*

67. (A)

```
CubicReg
 y=ax³+bx²+cx+d
 a=-.01
 b=.83
 c=-2.3
 d=221
```

(B) 28 ads to sell 588 refrigerators per month *(9-4)*

68. 3 days *(9-2)*

69. 2 yr from now *(9-2)*

CHAPTER 10

Exercise 10-1

1. \$1,221.40; \$1,648.72; \$2,225.54

3.

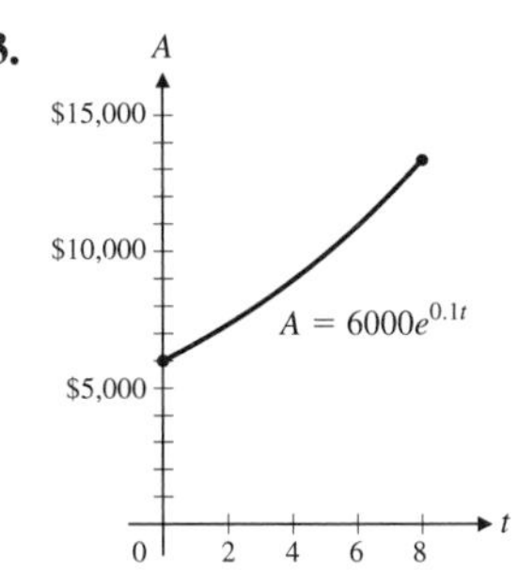

5. 11.55

7. 10.99

9. 0.14

11.

n	$[1 + (1/n)]^n$
10	2.593 74
100	2.704 81
1,000	2.716 92
10,000	2.718 15
100,000	2.718 27
1,000,000	2.718 28
10,000,000	2.718 28
↓	↓
∞	$e = 2.718\ 281\ 828\ 459\ldots$

13. $\lim_{n\to\infty}(1 + n)^{1/n} = 1$

15.

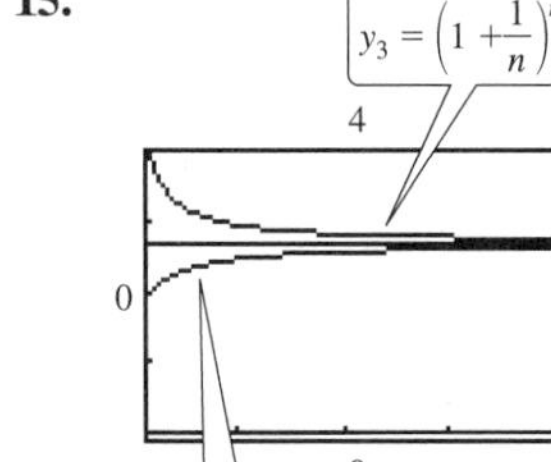

17. \$55,463.90

19. \$9,931.71

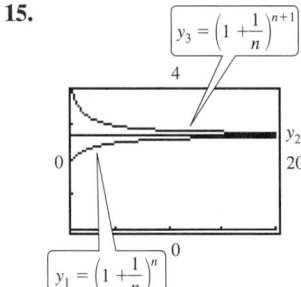

21. $r = \frac{1}{4}\ln 1.5 \approx 0.1014$ or 10.14%

23. (A)

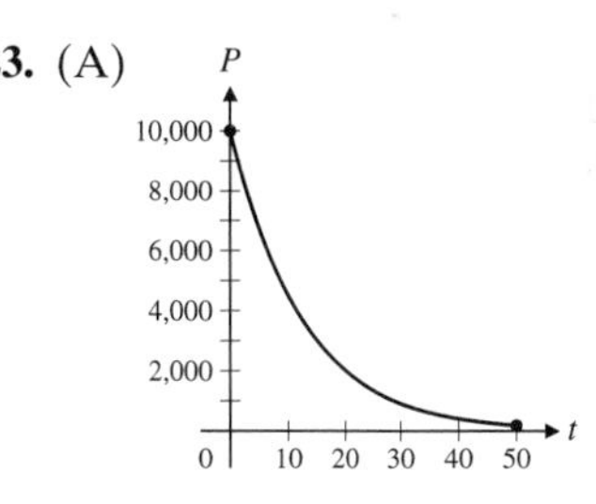

(B) $\lim_{t\to\infty} 10{,}000e^{-0.08t} = 0$

25. 2.77 yr

27. 13.86%

29. 7.3 yr

31. (A)
$$A = Pe^{rt}$$
$$2P = Pe^{rt}$$
$$2 = e^{rt}$$
$$rt = \ln 2$$
$$t = \frac{\ln 2}{r}$$

(B)

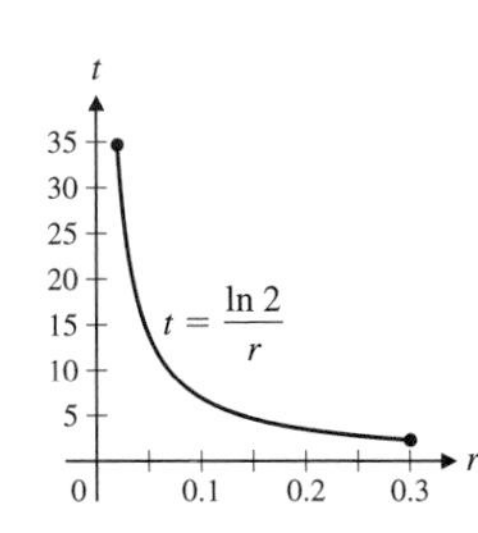

Although r could be any positive number, the restrictions on r are reasonable in the sense that most investments would be expected to earn a return of between 2% and 30%.

(C) The doubling times (in years) are 13.86, 6.93, 4.62, 3.47, 2.77, and 2.31, respectively.

33. $t = -(\ln 0.5)/0.000\ 433\ 2 \approx 1{,}600$ yr **35.** $r = (\ln 0.5)/30 \approx -0.0231$ **37.** 34.66 yr **39.** 3.47%

41. Approx. 521 yr

Exercise 10-2

1. $6e^x - \frac{7}{x}$ **3.** $2ex^{e-1} + 3e^x$ **5.** $\frac{5}{x}$ **7.** $\frac{2 \ln x}{x}$ **9.** $x^3 + 4x^3 \ln x = x^3(1 + 4 \ln x)$

11. $x^3e^x + 3x^2e^x = x^2e^x(x + 3)$ **13.** $\frac{(x^2 + 9)e^x - 2xe^x}{(x^2 + 9)^2} = \frac{e^x(x^2 - 2x + 9)}{(x^2 + 9)^2}$ **15.** $\frac{x^3 - 4x^3 \ln x}{x^8} = \frac{1 - 4 \ln x}{x^5}$

17. $3(x + 2)^2 \ln x + \frac{(x + 2)^3}{x} = (x + 2)^2\left(3 \ln x + \frac{x + 2}{x}\right)$ **19.** $(x + 1)^3e^x + 3(x + 1)^2e^x = (x + 1)^2e^x(x + 4)$

21. $\frac{2xe^x - (x^2 + 1)e^x}{(e^x)^2} = \frac{2x - x^2 - 1}{e^x}$ **23.** $(\ln x)^3 + 3(\ln x)^2 = (\ln x)^2(\ln x + 3)$

25. $3(4 - 5e^x)^2(-5e^x) = -15e^x(4 - 5e^x)^2$ **27.** $\frac{1}{2}(1 + \ln x)^{-1/2}\left(\frac{1}{x}\right) = \frac{1}{2x(1 + \ln x)^{1/2}}$ **29.** $xe^x + e^x - e^x = xe^x$

31. $2x^2\left(\frac{1}{x}\right) + 4x \ln x - 2x = 4x \ln x$ **33.** $y = ex$ **35.** $y = \frac{1}{e}x$

37. Yes, she is correct. In fact, for any real number c, the tangent line to $y = e^x$ at the point (c, e^c) has equation $y - e^c = e^c(x - c)$, and thus the tangent line passes through the point $(c - 1, 0)$.

39. $\frac{d}{dx} e^{kx} = ke^{kx}$ for any constant k

41. Max $f(x) = f(e^3) = e^3 \approx 20.086$ **43.** Min $f(x) = f(1) = e \approx 2.718$ **45.** Max $f(x) = f(e^{1/2}) = 2e^{-1/2} \approx 1.213$

47. Domain: All real numbers
y intercept: 0; x intercept: 0
Horizontal asymptote: $y = 1$
Decreasing on $(-\infty, \infty)$
Concave downward on $(-\infty, \infty)$

49. Domain: $(0, \infty)$
Vertical asymptote: $x = 0$
Increasing on $(1, \infty)$
Decreasing on $(0, 1)$
Local minimum at $x = 1$
Concave upward on $(0, \infty)$

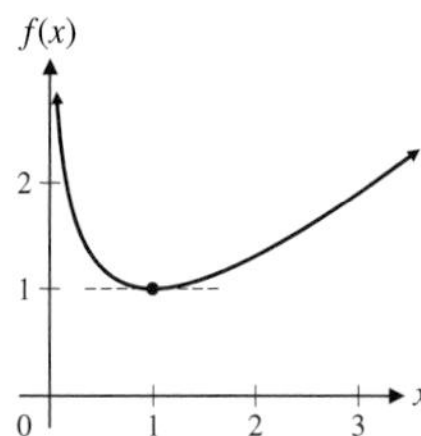

51. Domain: All real numbers
y intercept: 3; x intercept: 3
Horizontal asymptote: $y = 0$
Increasing on $(-\infty, 2)$
Decreasing on $(2, \infty)$
Local maximum at $x = 2$
Concave upward on $(-\infty, 1)$
Concave downward on $(1, \infty)$
Inflection point at $x = 1$

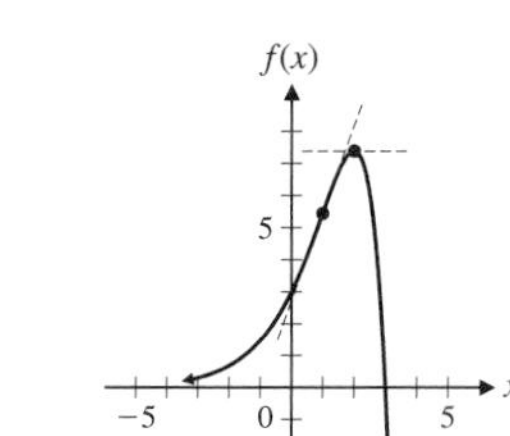

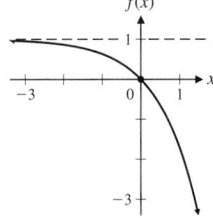

53. Domain: $(0, \infty)$
x intercept: 1
Increasing on $(e^{-1/2}, \infty)$
Decreasing on $(0, e^{-1/2})$
Local minimum at $x = e^{-1/2}$
Concave upward on $(e^{-3/2}, \infty)$
Concave downward on $(0, e^{-3/2})$
Inflection point at $x = e^{-3/2}$

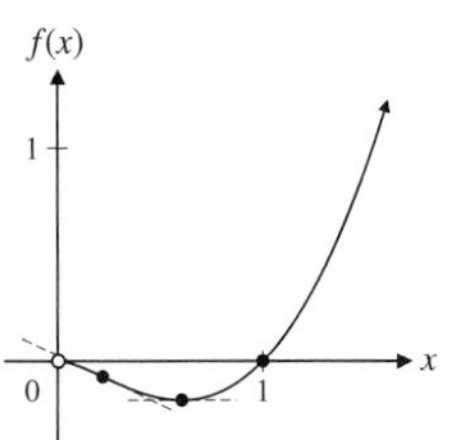

55. Critical values: $x = 0.36$, $x = 2.15$
Increasing on $(-\infty, 0.36)$ and $(2.15, \infty)$
Decreasing on $(0.36, 2.15)$
Local maximum at $x = 0.36$
Local minimum at $x = 2.15$

57. Critical value: $x = 2.21$
Increasing on $(0, 2.21)$
Decreasing on $(2.21, \infty)$
Local maximum at $x = 2.21$

59. $(-0.82, 0.44), (1.43, 4.18), (8.61, 5503.66)$ **61.** $(3.65, 1.30), (332{,}105.11, 12.71)$ **63.** $p = \$2$
65. $p = \$27.57$ **67.** Min $\overline{C}(x) = \overline{C}(e^7) \approx \$99.91$
69. (A) At \$3.68 each, the maximum revenue will be \$3,680/wk (in the test city).

(B)

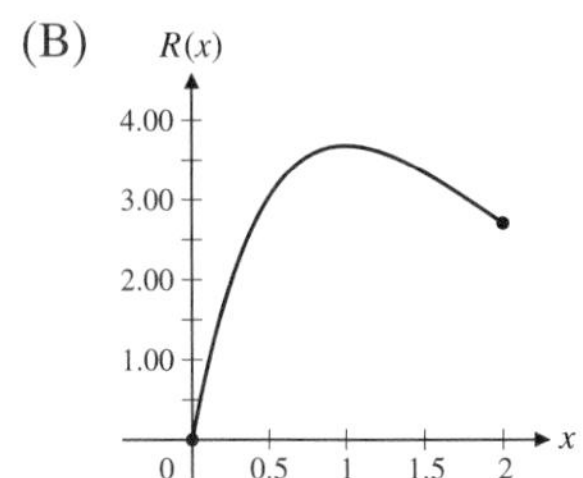

71. At the 40 lb weight level, blood pressure would increase at the rate of 0.44 mm of mercury/lb of weight gain. At the 90 lb weight level, blood pressure would increase at the rate of 0.19 mm of mercury/lb of weight gain.
73. (A) After 1 hr, the concentration is decreasing at the rate of 1.60 mg/ml/hr; after 4 hr, the concentration is decreasing at the rate of 0.08 mg/ml/hr.

(B)

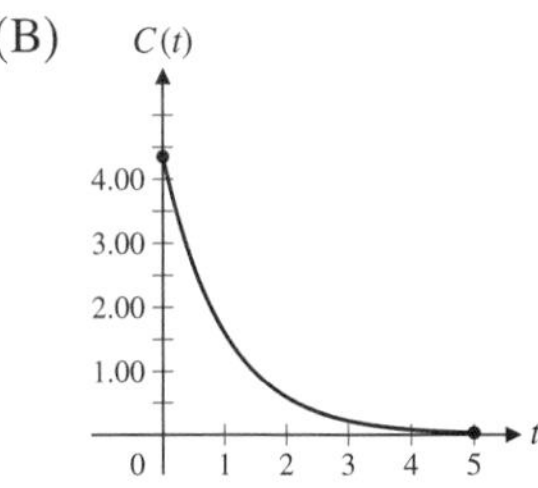

75. $dR/dS = k/S$

Exercise 10-3

1. $(7x - 5)^4$ **3.** $\ln(2x + 5)$ **5.** e^{x^3} **7.** $y = u^3; u = 2x + 5$ **9.** $y = \ln u; u = 2x^2 + 7$
11. $y = e^u; u = x^2 - 2$ **13.** $y = (2 + e^x)^2; dy/dx = 2e^x(2 + e^x)$
15. $y = e^{2-x^4}; dy/dx = -4x^3e^{2-x^4}$ **17.** $y = \ln(4x^5 - 7); \dfrac{dy}{dx} = \dfrac{20x^4}{4x^5 - 7}$ **19.** $\dfrac{1}{x - 3}$ **21.** $\dfrac{-2}{3 - 2t}$ **23.** $6e^{2x}$
25. $-8e^{-4t}$ **27.** $-3e^{-0.03x}$ **29.** $\dfrac{4}{x + 1}$ **31.** $4e^{2x} - 3e^x$ **33.** $(6x - 2)e^{3x^2-2x}$ **35.** $\dfrac{2t + 3}{t^2 + 3t}$ **37.** $\dfrac{x}{x^2 + 1}$
39. $\dfrac{4[\ln(t^2 + 1)]^3(2t)}{t^2 + 1} = \dfrac{8t[\ln(t^2 + 1)]^3}{t^2 + 1}$ **41.** $4(e^{2x} - 1)^3(2e^{2x}) = 8e^{2x}(e^{2x} - 1)^3$
43. $\dfrac{(x^2 + 1)(2e^{2x}) - e^{2x}(2x)}{(x^2 + 1)^2} = \dfrac{2e^{2x}(x^2 - x + 1)}{(x^2 + 1)^2}$
45. $(x^2 + 1)(-e^{-x}) + e^{-x}(2x) = e^{-x}(2x - x^2 - 1)$
47. $\dfrac{e^{-x}}{x} - e^{-x}\ln x = \dfrac{e^{-x}(1 - x\ln x)}{x}$
49. $\dfrac{-2x}{(1 + x^2)[\ln(1 + x^2)]^2}$ **51.** $\dfrac{-2x}{3(1 - x^2)[\ln(1 - x^2)]^{2/3}}$
53. The graph of $y = \log_b x$ is a vertical expansion of the graph of $y = \ln x$ if $1 < b < e$, and a vertical contraction if $b > e$.

55. Domain: $(-\infty, \infty)$
y intercept: 0; x intercept: 0
Horizontal asymptote: $y = 1$
Increasing on $(-\infty, \infty)$
Concave downward on $(-\infty, \infty)$

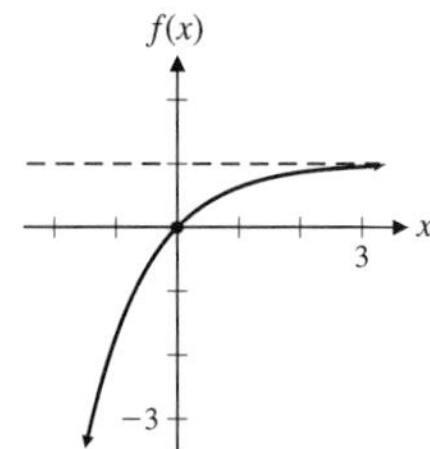

57. Domain: $(-\infty, 1)$
y intercept: 0; x intercept: 0
Vertical asymptote: $x = 1$
Decreasing on $(-\infty, 1)$
Concave downward on $(-\infty, 1)$

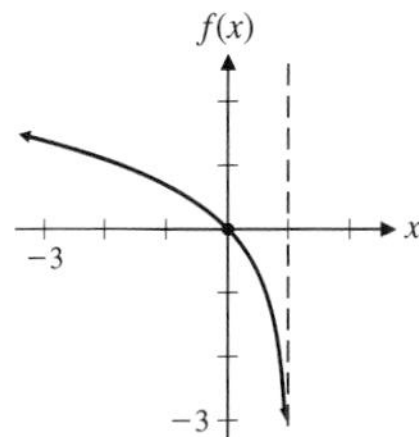

59. Domain: $(-\infty, \infty)$
y intercept: 1
Horizontal asymptote: $y = 0$
Increasing on $(-\infty, 0)$
Decreasing on $(0, \infty)$
Local maximum at $x = 0$
Concave upward on $(-\infty, -1)$ and $(1, \infty)$
Concave downward on $(-1, 1)$
Inflection points at $x = -1$ and $x = 1$

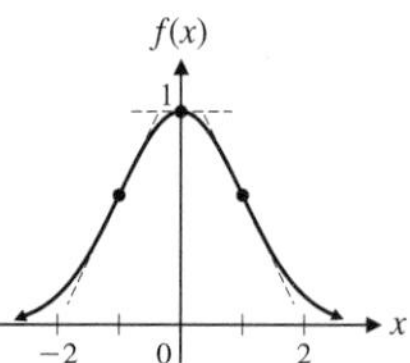

61. $y = 1 + [\ln(2 + e^x)]^2; \dfrac{dy}{dx} = \dfrac{2e^x \ln(2 + e^x)}{2 + e^x}$

63. $\dfrac{1}{\ln 2}\left(\dfrac{6x}{3x^2 - 1}\right)$

65. $(2x + 1)(10^{x^2+x})(\ln 10)$

67. $\dfrac{12x^2 + 5}{(4x^3 + 5x + 7)\ln 3}$

69. $2^{x^3-x^2+4x+1}(3x^2 - 2x + 4)\ln 2$

73. (A) $g(x)$ is not negative when $f(x)$ is decreasing
(B) $f'(x) = \dfrac{2x}{x^2 + 1}$

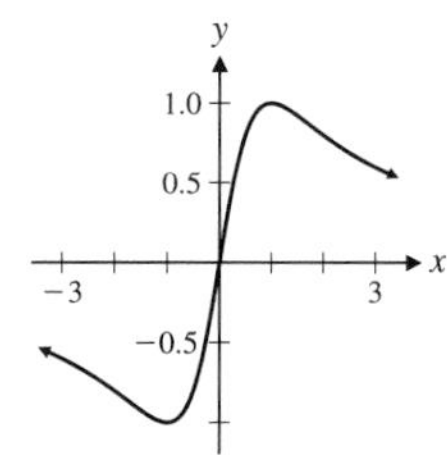

75. $f'(x) = g'(x) = \dfrac{8x}{x^2 + 3}$

77. A maximum revenue of \$735.80 is realized at a production level of 20 units at \$36.79 each.

79. A maximum profit of \$224.61 is realized at a production level of 17 units at \$42.74 each.

81. \$159.68

83. $-$\$27,145/yr; $-$\$18,196/yr; $-$\$11,036/yr

85. (A) 23 days; \$26,685; about 50%
(B)

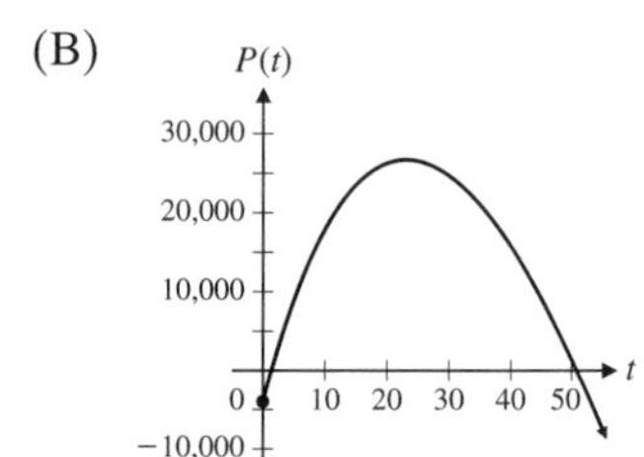

87. 2.27 mm of mercury/yr; 0.81 mm of mercury/yr; 0.41 mm of mercury/yr

89. $A'(t) = 2(\ln 2)5{,}000e^{2t \ln 2} = 10{,}000(\ln 2)2^{2t}$; $A'(1) = 27{,}726$ bacteria/hr (rate of change at the end of the first hour); $A'(5) = 7{,}097{,}827$ bacteria/hr (rate of change at the end of the fifth hour)

91.

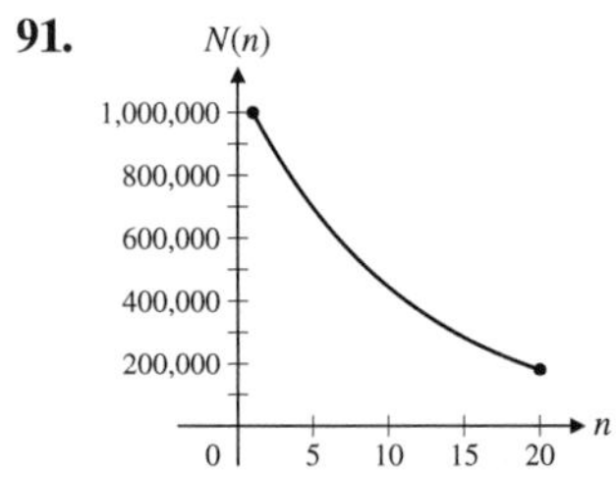

Exercise 10-4

1. $y' = -\frac{2}{3}$ **3.** $y' = x$ **5.** $y' = 6x; 6$ **7.** $y' = \frac{3x}{y}; 3$ **9.** $y' = \frac{1}{2y + 1}; \frac{1}{3}$ **11.** $y' = -\frac{y}{x}; -\frac{3}{2}$

13. $y' = -\frac{2y}{2x + 1}; 4$ **15.** $y' = \frac{6 - 2y}{x}; -1$ **17.** $y' = \frac{2x}{e^y - 2y}; 2$ **19.** $y' = \frac{3x^2y}{y + 1}; \frac{3}{2}$

21. $y' = \frac{6x^2y - y \ln y}{x + 2y}; 2$ **23.** $x' = \frac{2tx - 3t^2}{2x - t^2}; 8$ **25.** $y'\Big|_{(1.6,\,1.8)} = -\frac{3}{4}; y'\Big|_{(1.6,\,0.2)} = \frac{3}{4}$ **27.** $y = -x + 5$

29. $y = \frac{2}{3}x - \frac{12}{5}; y = \frac{3}{5}x + \frac{12}{5}$ **31.** $y' = -\frac{1}{x}$ **33.** $y' = \frac{1}{3(1 + y)^2 + 1}; \frac{1}{13}$ **35.** $y' = \frac{3(x - 2y)^2}{6(x - 2y)^2 + 4y}; \frac{3}{10}$

37. $y' = \frac{3x^2(7 + y^2)^{1/2}}{y}; 16$ **39.** $y' = \frac{y}{2xy^2 - x}; 1$ **41.** $y = 0.63x + 1.04$ **43.** $p' = \frac{1}{2p - 2}$

45. $p' = -\frac{\sqrt{10{,}000 - p^2}}{p}$ **47.** $\frac{dL}{dV} = \frac{-(L + m)}{V + n}$

Exercise 10-5

1. 240 **3.** $\frac{9}{4}$ **5.** $\frac{1}{2}$ **7.** Decreasing at 9 units/sec **9.** Approx. -3.03 ft/sec **11.** $dA/dt \approx 126$ ft²/sec
13. 3,768 cm³/min **15.** 6 lb/in.²/hr **17.** $-\frac{9}{4}$ ft/sec **19.** $\frac{20}{3}$ ft/sec
21. 0.0214 ft/sec; 0.0135 ft/sec; yes, at $t = 0.000\ 19$ sec **23.** 3.835 units/sec
25. (A) $dC/dt = \$15{,}000$/wk (B) $dR/dt = -\$50{,}000$/wk (C) $dP/dt = -\$65{,}000$/wk **27.** $ds/dt = \$2{,}207$/wk
29. (A) $dx/dt = -12.73$ units/month (B) $dp/dt = \$1.53$/month **31.** Approx. 100 ft³/min

Review Exercise

1. \$3,136.62; \$4,919.21; \$12,099.29 *(10-1)* **2.** $\frac{2}{x} + 3e^x$ *(10-2)* **3.** $2e^{2x-3}$ *(10-2)* **4.** $\frac{2}{2x + 7}$ *(10-2)*

5. (A) $y = \ln(3 + e^x)$ (B) $\frac{dy}{dx} = \frac{e^x}{3 + e^x}$ *(10-3)* **6.** $y' = \frac{9x^2}{4y}; \frac{9}{8}$ *(10-4)* **7.** $dy/dt = 216$ *(10-4)*

8. Domain: All real numbers
y intercept: 100
Horizontal asymptote: $y = 0$
Decreasing on $(-\infty, \infty)$
Concave upward on $(-\infty, \infty)$ *(10-2)*

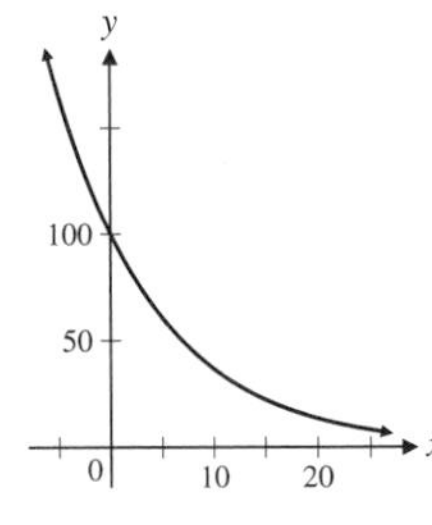

9. $\lim_{n\to\infty}\left(1 + \frac{2}{n}\right)^n = e^2 \approx 7.389\ 06$ *(10-1)*

10. $\frac{7[(\ln z)^6 + 1]}{z}$ *(10-3)* **11.** $x^5(1 + 6 \ln x)$ *(10-2)*

12. $\frac{e^x(x - 6)}{x^7}$ *(10-2)* **13.** $\frac{6x^2 - 3}{2x^3 - 3x}$ *(10-3)*

14. $(3x^2 - 2x)e^{x^3-x^2}$ *(10-3)*

15. $\frac{1 - 2x \ln 5x}{xe^{2x}}$ *(10-3)*

16. $y = -x + 2; y = -ex + 1$ *(10-2)*

17. $y' = \frac{3y - 2x}{8y - 3x}; \frac{8}{19}$ *(10-4)* **18.** $x' = \frac{4tx}{3x^2 - 2t^2}; -4$ *(10-4)* **19.** $y' = \frac{1}{e^y + 2y}; 1$ *(10-4)*

20. $y' = \frac{2xy}{1 + 2y^2}; \frac{2}{3}$ *(10-4)* **21.** $dy/dt = -2$ units/sec *(10-5)* **22.** 0.27 ft/sec *(10-5)*
23. $dR/dt = 1/\pi \approx 0.318$ in./min *(10-5)*
24. Max $f(x) = f(e^{4.5}) = 2e^{4.5} \approx 180.03$ *(10-2)* **25.** Max $f(x) = f(0.5) = 5e^{-1} \approx 1.84$ *(10-3)*
26. Max $f(x) = f(1.373) = 2.487$ *(10-2)* **27.** Max $f(x) = f(1.763) = 0.097$ *(10-2)*

28. Domain: All real numbers
y intercept: 0; x intercept: 0
Horizontal asymptote: $y = 5$
Increasing on $(-\infty, \infty)$
Concave downward on
$(-\infty, \infty)$

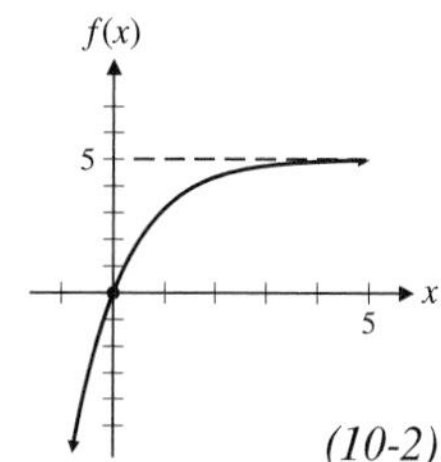

(10-2)

29. Domain: $(0, \infty)$
x intercept: 1
Increasing on $(e^{-1/3}, \infty)$
Decreasing on $(0, e^{-1/3})$
Local minimum at $x = e^{-1/3}$
Concave upward on $(e^{-5/6}, \infty)$
Concave downward on $(0, e^{-5/6})$
Inflection point at $x = e^{-5/6}$

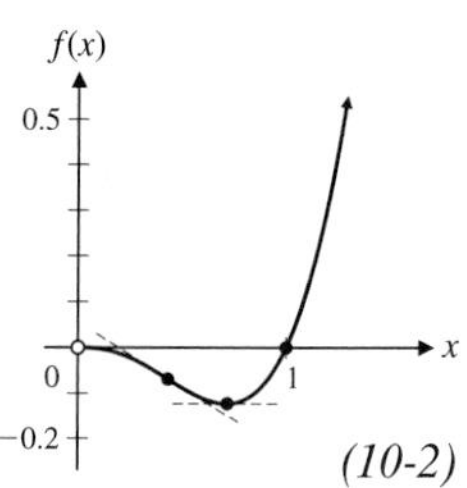

(10-2)

30. (A) $y = [\ln(4 - e^x)]^3$ (B) $\frac{dy}{dx} = \frac{-3e^x[\ln(4 - e^x)]^2}{4 - e^x}$ *(10-3)* **31.** $2x(5^{x^2-1})(\ln 5)$ *(10-3)* **32.** $\left(\frac{1}{\ln 5}\right)\frac{2x - 1}{x^2 - x}$ *(10-3)*
33. $\frac{2x + 1}{2(x^2 + x)\sqrt{\ln(x^2 + x)}}$ *(10-3)* **34.** $y' = \frac{2x - e^{xy}y}{xe^{xy} - 1}$; 0 *(10-4)*
35. The rate of increase of area is proportional to the radius R, so it is smallest when $R = 0$, and has no largest value. *(10-5)*
36. Yes, for $-\sqrt{3}/3 < x < \sqrt{3}/3$ *(10-5)* **37.** (A) 15 yr (B) 13.9 yr *(10-1)*
38. $A'(t) = 10e^{0.1t}$; $A'(1) = \$11.05$/yr; $A'(10) = \$27.18$/yr *(10-1)* **39.** $R'(x) = (1{,}000 - 20x)e^{-0.02x}$ *(10-3)*
40. A maximum revenue of \$18,394 is realized at a production level of 50 units at \$367.88 each. *(10-3)*

41. *(10-3)*

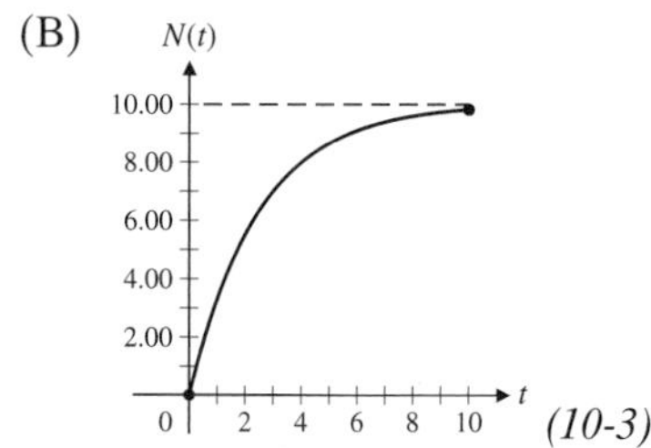

42. \$549.15; \$9,684 *(10-3)* **43.** \$1.52 *(10-2)*
44. Min $\overline{C}(x) = \overline{C}(e^5) \approx \$49.66$ *(10-2)* **45.** $p' = \frac{-(5{,}000 - 2p^3)^{1/2}}{3p^2}$ *(10-4)*
46. $dR/dt = \$110$/day *(10-5)* **47.** -1.111 mg/ml/hr; -0.335 mg/ml/hr *(10-3)*
48. $dR/dt = -3/(2\pi)$; approx. 0.477 mm/day *(10-5)*

49. (A) Increasing at the rate of 2.68 units/day at the end of 1 day of training; increasing at the rate of 0.54 unit/day after 5 days of training

(B)

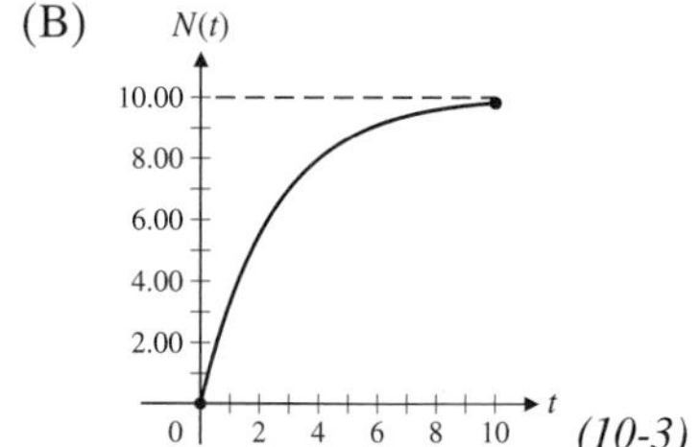

(10-3)

50. $dT/dt = -1/27 \approx -0.037$ min/operation hour *(10-5)*

Chapter 11

Exercise 11-1

1. $7x + C$
3. $(x^7/7) + C$
5. $-x^{-1} + C$
7. $2t^4 + C$ **9.** $u^2 + u + C$
11. $x^3 + x^2 - 5x + C$
13. $(s^5/5) - \frac{4}{3}s^6 + C$
15. $3e^t + C$
17. $2 \ln|z| + C$

19.

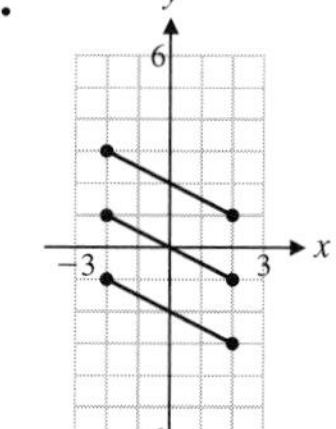

21.

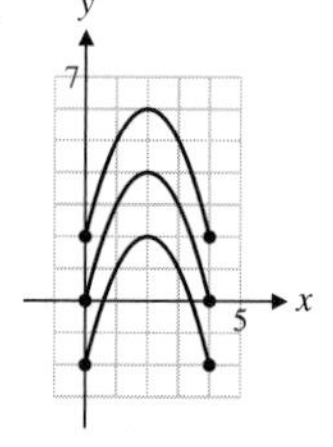

23. $y = 40x^5 + C$ **25.** $P = 24x - 3x^2 + C$ **27.** $y = \frac{1}{3}u^6 - u^3 - u + C$
29. $y = e^x + 3x + C$ **31.** $x = 5\ln|t| + t + C$ **33.** (A) False (B) True
35. No, since one graph cannot be obtained from another by a vertical translation.
37. Yes, since one graph can be obtained from another by a vertical translation.
39. $4x^{3/2} + C$ **41.** $-4x^{-2} + C$ **43.** $2\sqrt{u} + C$ **45.** $-(x^{-2}/8) + C$ **47.** $-(u^{-4}/8) + C$
49. $x^3 + 2x^{-1} + C$ **51.** $2x^5 + 2x^{-4} - 2x + C$ **53.** $2x^{3/2} + 4x^{1/2} + C$ **55.** $\frac{3}{5}x^{5/3} + 2x^{-2} + C$
57. $(e^x/4) - (3x^2/8) + C$ **59.** $-z^{-2} - z^{-1} + \ln|z| + C$

61.

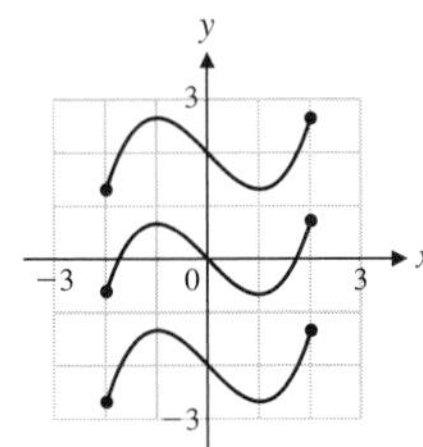

63.

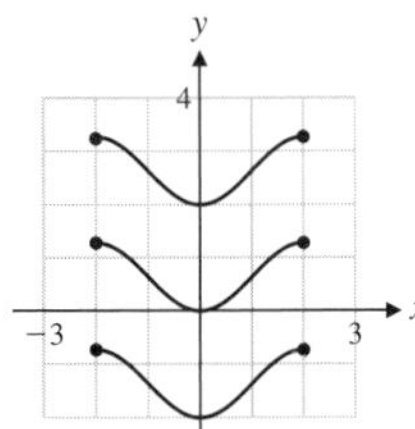

65. $y = x^2 - 3x + 5$ **67.** $C(x) = 2x^3 - 2x^2 + 3{,}000$ **69.** $x = 40\sqrt{t}$ **71.** $y = -2x^{-1} + 3\ln|x| - x + 3$
73. $x = 4e^t - 2t - 3$ **75.** $y = 2x^2 - 3x + 1$ **77.** $x^2 + x^{-1} + C$ **79.** $\frac{1}{2}x^2 + x^{-2} + C$ **81.** $e^x - 2\ln|x| + C$
83. $M = t + t^{-1} + \frac{3}{4}$ **85.** $y = 3x^{5/3} + 3x^{2/3} - 6$ **87.** $p(x) = 10x^{-1} + 10$ **89.** x^3 **91.** $x^4 + 3x^2 + C$
93. $\overline{C}(x) = 15 + \frac{1{,}000}{x}$; $C(x) = 15x + 1{,}000$; $C(0) = \$1{,}000$
95. (A) The cost function increases from 0 to 8, is concave downward from 0 to 4, and is concave upward from 4 to 8. There is an inflection point at $x = 4$.
(B) $C(x) = x^3 - 12x^2 + 53x + 30$; $C(4) = \$114{,}000$; $C(8) = \$198{,}000$
(C)

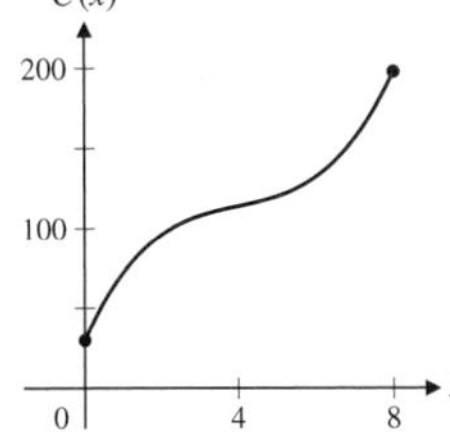

(D) Manufacturing plants are often inefficient at low and high levels of production.

97. $S(t) = 2{,}000 - 15t^{5/3}$; $80^{3/5} \approx 14$ mo **99.** $S(t) = 2{,}000 - 15t^{5/3} - 70t$; $t \approx 8.92$ mo
101. $L(x) = 4{,}800x^{1/2}$; $L(25) = 24{,}000$ labor-hours **103.** $W(h) = 0.0005h^3$; $W(70) = 171.5$ lb **105.** 19,400

Exercise 11-2

1. $\frac{1}{6}(x^2 - 4)^6 + C$ **3.** $e^{4x} + C$ **5.** $\frac{(x-2)^9}{9} + C$ **7.** $\ln|2t + 3| + C$ **9.** $\frac{1}{24}(3x - 2)^8 + C$
11. $\frac{1}{16}(x^2 + 3)^8 + C$ **13.** $-20e^{-0.5t} + C$ **15.** $\frac{1}{10}\ln|10x + 7| + C$ **17.** $\frac{1}{4}e^{2x^2} + C$ **19.** $\frac{1}{3}\ln|x^3 + 4| + C$
21. $-\frac{1}{18}(3t^2 + 1)^{-3} + C$ **23.** $\frac{1}{3}(4 - x^3)^{-1} + C$ **25.** $\frac{2}{5}(x + 4)^{5/2} - \frac{8}{3}(x + 4)^{3/2} + C$
27. $\frac{2}{3}(x - 3)^{3/2} + 6(x - 3)^{1/2} + C$ **29.** $\frac{1}{11}(x - 4)^{11} + \frac{2}{5}(x - 4)^{10} + C$ **31.** $\frac{1}{8}(1 + e^{2x})^4 + C$
33. $\frac{1}{2}\ln|4 + 2x + x^2| + C$ **35.** $e^{x^2+x+1} + C$ **37.** $\frac{1}{4}(e^x - 2x)^4 + C$ **39.** $-\frac{1}{12}(x^4 + 2x^2 + 1)^{-3} + C$
41. (A) Differentiate the right side to get the integrand on the left side.
(B) Wrong, since $\frac{d}{dx}[\ln|2x - 3| + C] = \frac{2}{2x - 3} \neq \frac{1}{2x - 3}$. If $u = 2x - 3$, then $du = 2\,dx$. The integrand was not adjusted for the missing constant factor 2.
(C) $\int \frac{1}{2x - 3}\,dx = \frac{1}{2}\int \frac{2}{2x - 3}\,dx = \frac{1}{2}\ln|2x - 3| + C$ *Check:* $\frac{d}{dx}\left[\frac{1}{2}\ln|2x - 3| + C\right] = \frac{1}{2x - 3}$

43. (A) Differentiate the right side to get the integrand on the left side.

(B) Wrong, since $\frac{d}{dx}[e^{x^4} + C] = 4x^3e^{x^4} \neq x^3e^{x^4}$. If $u = x^4$, then $du = 4x^3\,dx$. The integrand was not adjusted for the missing constant factor 4.

(C) $\int x^3e^{x^4}\,dx = \frac{1}{4}\int 4x^3e^{x^4}\,dx = \frac{1}{4}e^{x^4} + C$ *Check:* $\frac{d}{dx}\left[\frac{1}{4}e^{x^4} + C\right] = x^3e^{x^4}$

45. (A) Differentiate the right side to get the integrand on the left side.

(B) Wrong, since $\frac{d}{dx}\left[\frac{(x^2-2)^2}{3x} + C\right] = \frac{3x^4 - 4x^2 - 4}{3x^2} \neq 2(x^2-2)^2$. If $u = x^2 - 2$, then $du = 2x\,du$. It appears that the student moved a variable factor across the integral sign as follows (which is *not* valid):
$\int 2(x^2-2)^2\,dx = \frac{1}{x}\int 2x(x^2-2)^2\,dx$.

(C) $\int 2(x^2-2)^2\,dx = \int (2x^4 - 8x^2 + 8)\,dx = \frac{2}{5}x^5 - \frac{8}{3}x^3 + 8x + C$

Check: $\frac{d}{dx}\left[\frac{2}{5}x^5 - \frac{8}{3}x^3 + 8x + C\right] = 2x^4 - 8x^2 + 8 = 2(x^2-2)^2$

47. $\frac{1}{9}(3x^2+7)^{3/2} + C$ **49.** $\frac{1}{8}x^8 + \frac{4}{5}x^5 + 2x^2 + C$ **51.** $\frac{1}{9}(x^3+2)^3 + C$ **53.** $\frac{1}{4}(2x^4+3)^{1/2} + C$ **55.** $\frac{1}{4}(\ln x)^4 + C$
57. $e^{-1/x} + C$ **59.** $x = \frac{1}{3}(t^3+5)^7 + C$ **61.** $y = 3(t^2-4)^{1/2} + C$ **63.** $p = -(e^x - e^{-x})^{-1} + C$
67. $p(x) = 2{,}000/(3x+50)$; 250 bottles **69.** $C(x) = 12x + 500\ln(x+1) + 2{,}000$; $\overline{C}(1{,}000) = \$17.45$
71. (A) $S(t) = 10t + 100e^{-0.1t} - 100$, $0 \le t \le 24$ (B) $S(12) \approx \$50$ million (C) 18.41 mo
73. $Q(t) = 100\ln(t+1) + 5t$, $0 \le t \le 20$; $Q(9) \approx 275$ thousand barrels **75.** $W(t) = 2e^{0.1t}$; $W(8) \approx 4.45$ g
77. (A) $-1{,}000$ bacteria/ml/day (B) $N(t) = 5{,}000 - 1{,}000\ln(1+t^2)$; 385 bacteria/ml (C) 7.32 days
79. $N(t) = 100 - 60e^{-0.1t}$, $0 \le t \le 15$; $N(15) \approx 87$ words/min
81. $E(t) = 12{,}000 - 10{,}000(t+1)^{-1/2}$; $E(15) = 9{,}500$ students

Exercise 11-3

1. $y = 2.08x^2 + C$ **3.** $y = 2e^{0.5x} + C$ **5.** $y = \frac{x^3}{3} - \frac{x^2}{2}$ **7.** $y = e^{-x^2} + 3$

9. Figure B. When $x = 1$, the slope $dy/dx = 1 - 1 = 0$ for any y. When $x = 0$, the slope $dy/dx = 0 - 1 = -1$ for any y. Both are consistent with the slope field shown in Figure B.

11. $y = \frac{x^2}{2} - x + C$; $y = \frac{x^2}{2} - x - 2$

13.

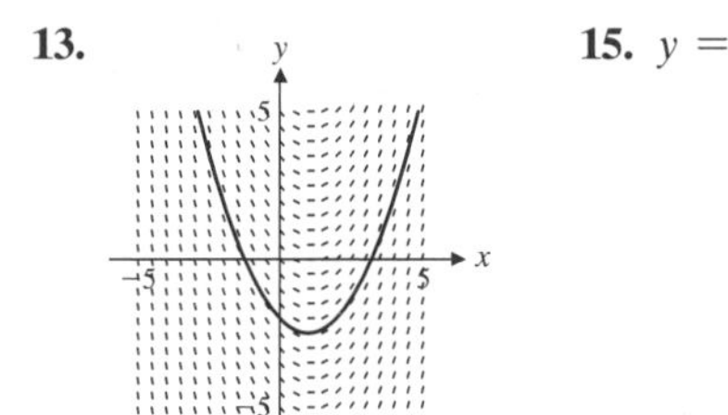

15. $y = Ce^{-0.8x}$ **17.** $y = 1{,}000e^{0.07x}$ **19.** $x = Ce^{-t}$

21. Figure A. When $y = 1$, the slope $dy/dx = 1 - 1 = 0$ for any x. When $y = 2$, the slope $dy/dx = 1 - 2 = -1$ for any x. Both are consistent with the slope field shown in Figure A.

23. $y = 1 - e^{-x}$ **25.**

27.

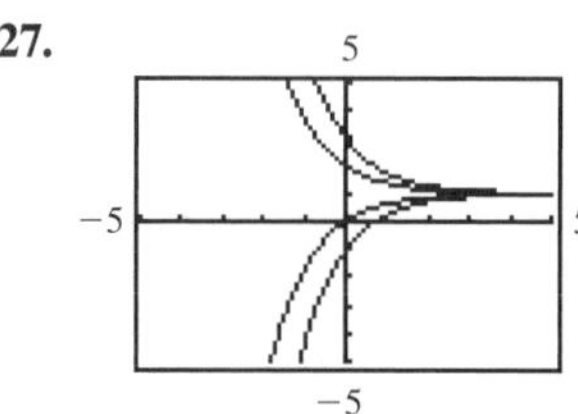

29.

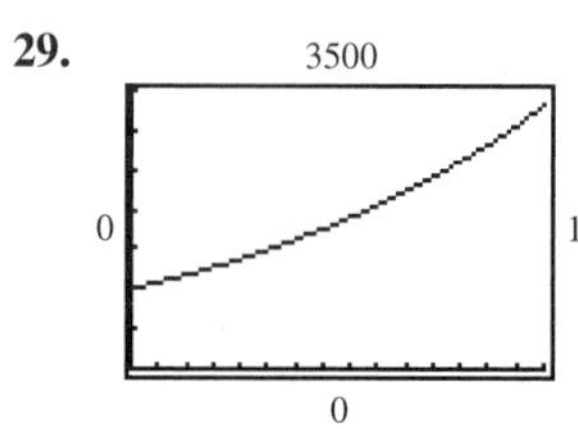

31.

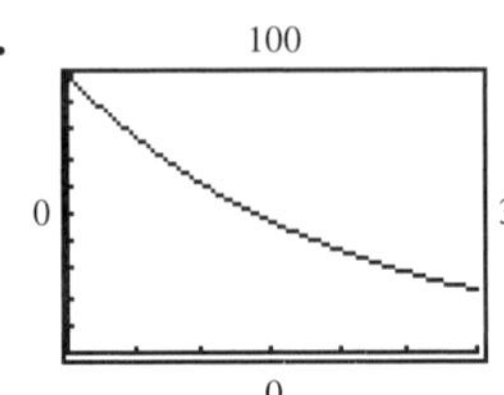

33.

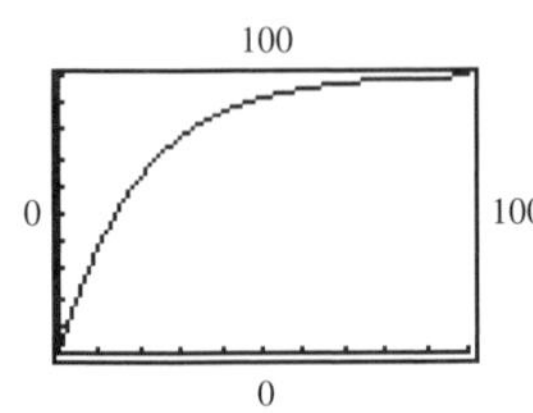

35.

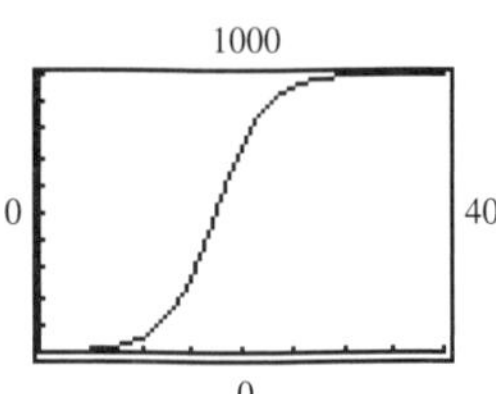

37. Apply the second-derivative test to $f(y) = ky(M - y)$. **39.** $A = 1{,}000e^{0.08t}$ **41.** $A = 8{,}000e^{0.06t}$

43. (A) $p(x) = 100e^{-0.05x}$
(B) \$60.65 per unit
(C)

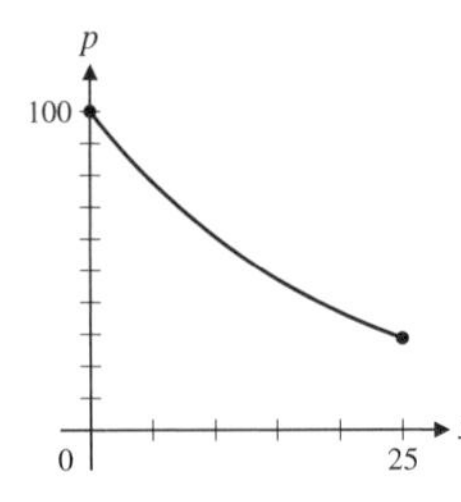

45. (A) $N = L(1 - e^{-0.051t})$
(B) 22.5%
(C) 32 days
(D)

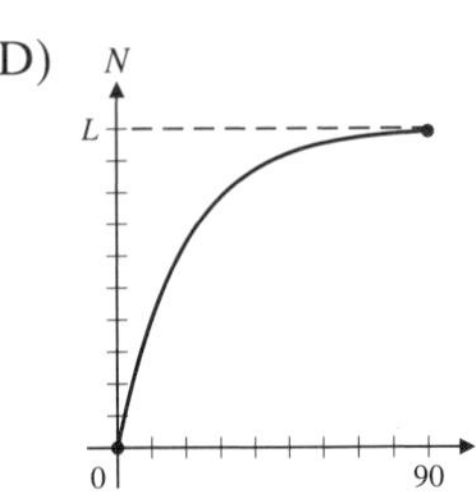

47. $I = I_0e^{-0.00942x}$; $x \approx 74$ ft

49. (A) $Q = 3e^{-0.04t}$
(B) $Q(10) = 2.01$ ml
(C) 27.47 hr
(D)

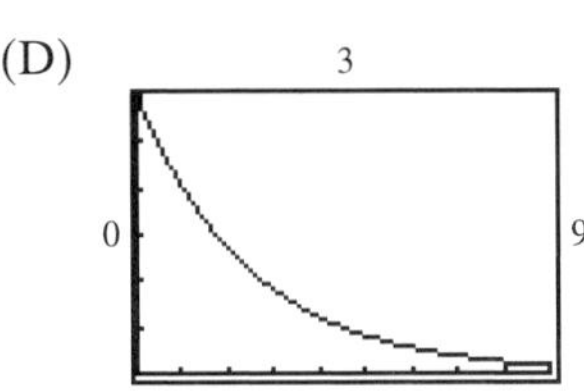

51. $-0.023\ 117$
53. Approx. 24,200 yr
55. 104 times; 67 times

57. (A) 7 people; 353 people
(B) 400
(C)

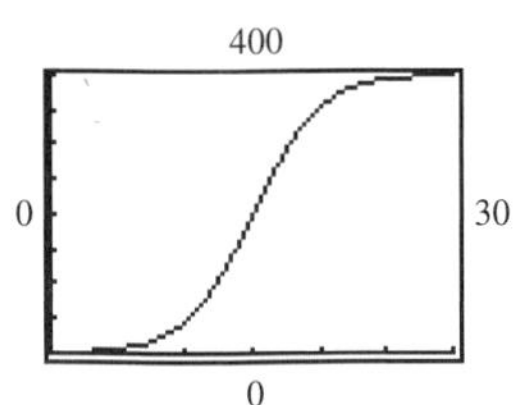

Exercise 11-4

1.

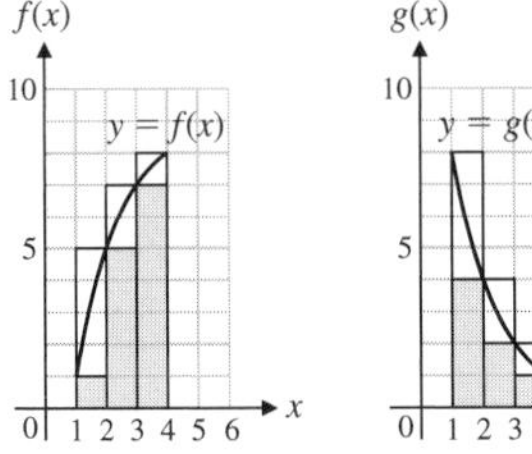

3. Figure A: $L_3 = 13$; $R_3 = 20$, $A_3 = 16.5$; Figure B: $L_3 = 14$, $R_3 = 7$, $A_3 = 10.5$

5. $L_3 \leq \int_1^4 f(x)\,dx \leq R_3$; $R_3 \leq \int_1^4 g(x)\,dx \leq L_3$; since $f(x)$ is increasing, L_3 underestimates the area and R_3 overestimates the area; since $g(x)$ is decreasing, the reverse is true.

7. Both figures: Error bound for L_3 and R_3 is 7; error bound for A_3 is 3.5.

9. The exact area under the graph of $y = f(x)$ is within 3.5 units (either way) of the average of the left sum and right sum estimates, $A_3 = 16.5$.

11. (A) $L_4 = 320$, $R_4 = 192$, $A_4 = 256$; error bound for L_4 and R_4 is 128; error bound for A_4 is 64
(B) The height of each rectangle represents an instantaneous rate, and the base of each rectangle represents a time interval; rate times time is distance.
(C) $n > 256$

13. (A) False (B) True **15.** $A_{10} = 311{,}100$ ft^2; error bound is 25,000 ft^2; $n \geq 100$

17. (A) $A_7 = 298$ ft; error bound $= 55$ ft (B) $n > 77$

19. (A) $P = 2$

(B)

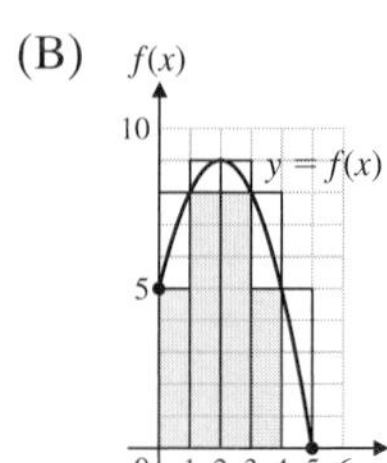

(C) To the left of $P = 2$, the left rectangles underestimate the true area and the right rectangles overestimate the true area. To the right of $P = 2$, the left rectangles overestimate the true area and the right rectangles underestimate the true area.

(D) $26 \leq \int_0^5 f(x)\,dx \leq 39$

21. $L_6 = -3.53$, $R_6 = -0.91$, $A_6 = -2.22$; error bound for L_6 and R_6 is 2.63; error bound for A_6 is 1.31. Geometrically, the definite integral over the interval $[2, 5]$ is the sum of the areas between the curve and the x axis from $x = 2$ to $x = 5$, with the areas below the x axis counted negatively and those above the x axis counted positively.
23. $(-\infty, 0], [0, \infty)$ **25.** $(-\infty, -1], [-1, 0], [0, 1], [1, \infty)$ **27.** $n \geq 60$ **29.** $n \geq 394$

31. If we let I represent the definite integral on the left side, then we know that $|I - L_n| \leq |f(b) - f(a)|\frac{b - a}{n}$. Since the limit of the right side is 0 as $n \to \infty$, the result follows.
33. $A_2 = \$180,000$; error bound $= \$30,000$ **35.** $L_4 = \$4,251$, $R_4 = \$4,605$; $\$4,251 \leq \int_2^6 800e^{0.08t}\,dt \leq \$4,605$
37. First 60 days: $A_3 = 3,240$ units; error bound $= 660$ units; Second 60 days: $A_3 = 4,920$ units; error bound $= 100$ units
39. (A) $\int_{100}^{200} R'(x)\,dx$ represents the area under the marginal revenue curve from $x = 100$ to $x = 200$; it also represents the total change in revenue going from sales of 100 six-packs per day to sales of 200 six-packs per day.
(B) $A_4 = \$200$; error bound $= \$50$
(C) \$200; both $R(200) - R(100)$ and $\int_{100}^{200} R'(x)\,dx$ represent total change in revenue going from sales of 100 six-packs per day to sales of 200 six-packs per day. This suggests that $\int_{100}^{200} R'(x)\,dx = R(200) - R(100)$.
41. (A) $L_5 = 3.72\text{ cm}^2$, $R_5 = 3.37\text{ cm}^2$ (B) $3.37 \leq \int_0^5 A'(t)\,dt \leq 3.72$
43. $A_3 = 108$ code symbols; error bound $= 6$ code symbols

Exercise 11-5

1. -2.33 **3.** 8.34 **5.** 1.067 **7.** 5 **9.** 5 **11.** 2 **13.** $-\frac{7}{3} \approx -2.333$ **15.** 2
17. $\frac{1}{2}(e^2 - 1) \approx 3.195$ **19.** $2 \ln 3.5 \approx 2.506$ **21.** -10.67 **23.** -5.04 **25.** 240 ft **27.** -2
29. 14 **31.** $5^6 = 15,625$ **33.** $\ln 4 \approx 1.386$ **35.** $20(e^{0.25} - e^{-0.5}) \approx 13.550$ **37.** $\frac{56}{3} \approx 18.667$
39. $\frac{28}{3} \approx 9.333$ **41.** $\frac{1}{6}[(e^2 - 2)^3 - 1] \approx 25.918$ **43.** $-3 - \ln 2 \approx -3.693$

45. (A) Average $f(x) = 250$

(B)

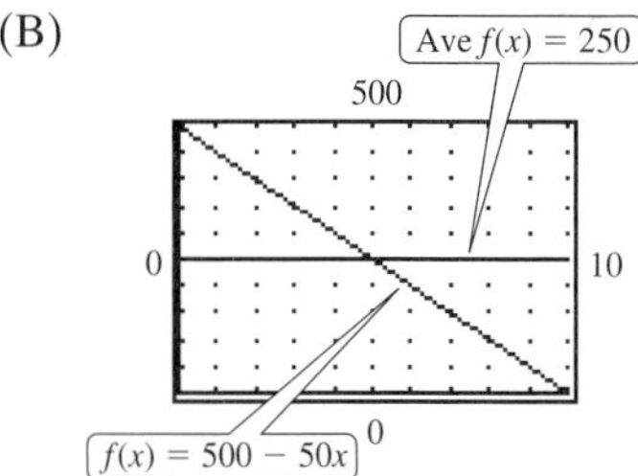

47. (A) Average $f(t) = 2$

(B)

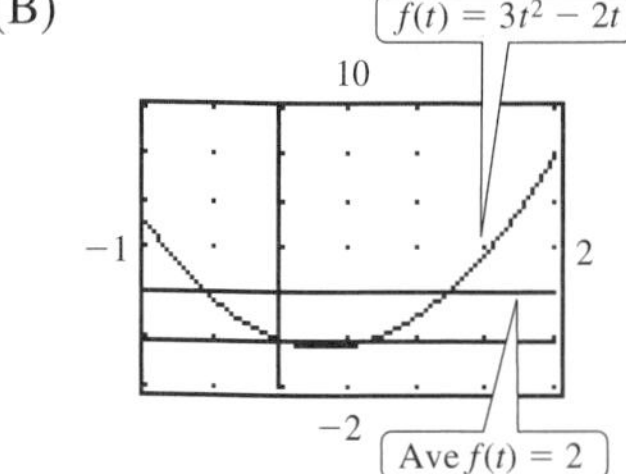

49. (A) Average $f(x) = \frac{45}{28} \approx 1.61$

(B)

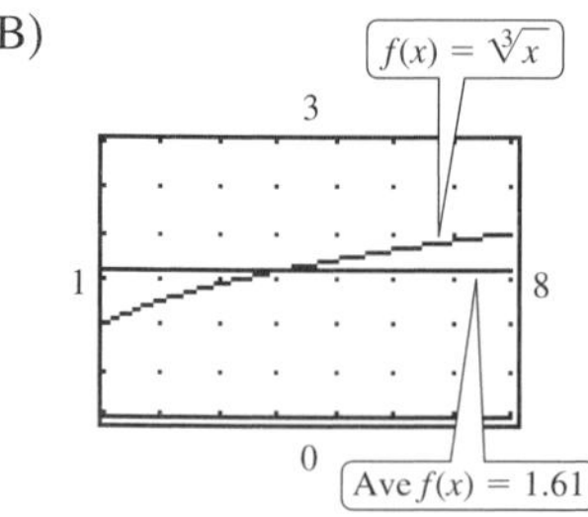

51. (A) Average $f(x) = 2(1 - e^{-2}) \approx 1.73$ **53.** $I = 10 \pm 0.67$ **55.** 10.67; $|I - M_4| = 0.67$; yes **57.** $n \geq 47$

(B)

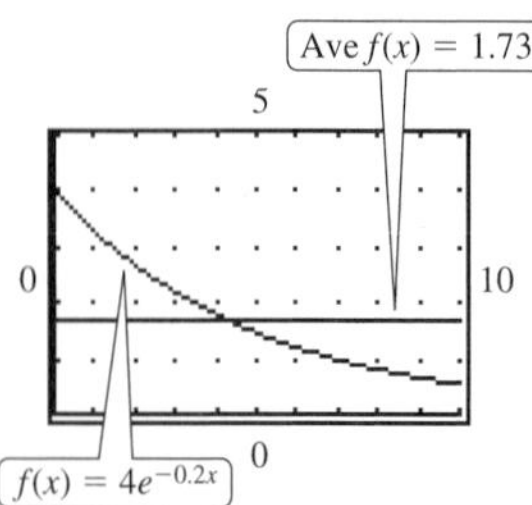

59. $\int_2^5 (1 - x^2)\,dx = -36$ **61.** $\int_2^{12} (3x^2 - 2x + 3)\,dx = 1,610$
63. $\frac{1}{6}(15^{3/2} - 5^{3/2}) \approx 7.819$ **65.** $\frac{1}{2}(\ln 2 - \ln 3) \approx -0.203$ **67.** 0
69. (A) $\ln 2 = 0.6919 \pm 0.0033$ (B) $\ln 2 = 0.6931$ (C) 0.0012; yes
71. $n \geq 13$ **73.** 51.163 **75.** 0.288 **77.** $\displaystyle\int_{300}^{900} \left(500 - \frac{x}{3}\right) dx = \$180,000$
79. $\int_0^5 500(t - 12)\,dt = -\$23,750$; $\int_5^{10} 500(t - 12)\,dt = -\$11,250$

81. (A)

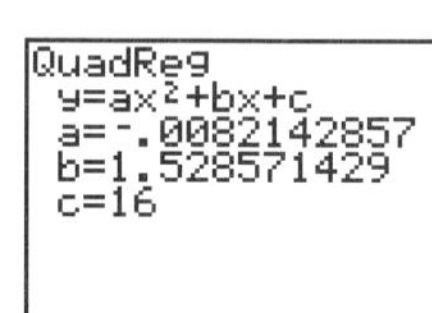

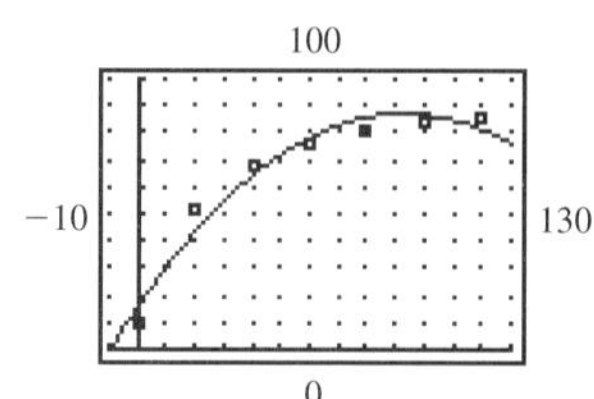

(B) 6,505 **83.** Useful life $= \sqrt{\ln 55} \approx 2$ yr; Total profit $= \frac{51}{22} - \frac{5}{2}e^{-4} \approx 2.272$ or \$2,272 **85.** (A) \$420 (B) \$135,000

87. (A)

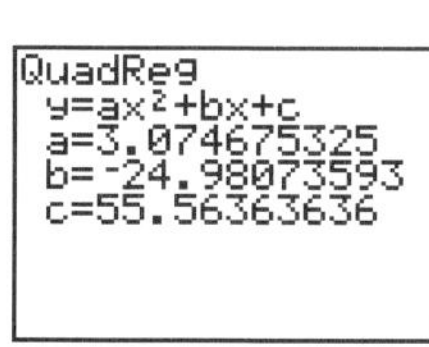

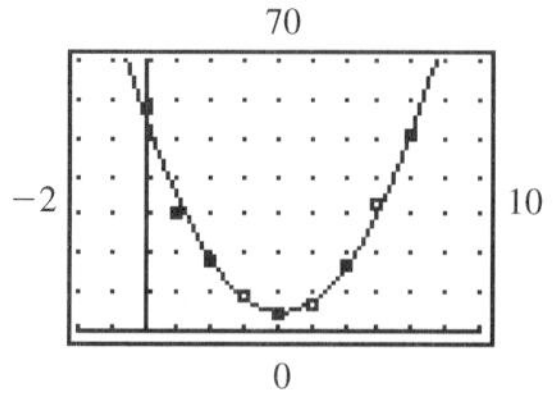

(B) \$100,505 **89.** $50e^{0.6} - 50e^{0.4} - 10 \approx \$6.51$ **91.** 4,800 labor-hours **93.** (A) $I = -200t + 600$ (B) $\frac{1}{3}\int_0^3 (-200t + 600)\, dt = 300$

95. $100 \ln 11 + 50 \approx 290$ thousand barrels; $100 \ln 21 - 100 \ln 11 + 50 \approx 115$ thousand barrels **97.** \$10,000 **99.** 3,120,000 ft^2 **101.** $2e^{0.8} - 2 \approx 2.45$ g; $2e^{1.6} - 2e^{0.8} \approx 5.45$ g **103.** 10°C **105.** 1.1 liters **107.** $0.6 \ln 2 + 0.1 \approx 0.516$; $(4.2 \ln 625 + 2.4 - 4.2 \ln 49)/24 \approx 0.546$

Review Exercise

1. $t^3 - t^2 + C$ *(11-1)* **2.** 12 *(11-5)* **3.** $-3t^{-1} - 3t + C$ *(11-1)* **4.** $\frac{15}{2} = 7.5$ *(11-5)* **5.** $-2e^{-0.5x} + C$ *(11-2)* **6.** $2 \ln 5 \approx 3.219$ *(11-5)* **7.** e^{-x^2} *(11-1)* **8.** $\sqrt{4 + 5x} + C$ *(11-1)* **9.** $y = f(x) = x^3 - 2x + 4$ *(11-3)* **10.** Increasing on [0, 2]; decreasing on [2, 4]; concave downward on [0, 4]; local maximum at $x = 2$. Antiderivative graphs differ by a vertical translation. *(11-1)*

11. *(11-1)*

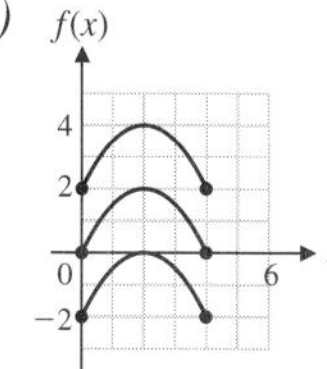

12. (A) $2x^4 - 2x^2 - x + C$ (B) $e^t - 4 \ln|t| + C$ *(11-1)* **13.** $\int_1^5 (x^2 + 1)\, dx = 44 \pm 1.333$ *(11-5)* **14.** 45.333; $|I - M_2| = 1.333$ *(11-5)* **15.** 30.8 *(11-5)* **16.** 7 *(11-5)* **17.** Width $= 2 - (-1) = 3$; Height $=$ Average $f(x) = 7$ *(11-5)* **18.** −10 *(11-4, 11-5)* **19.** 0.4 *(11-4, 11-5)* **20.** 1.4 *(11-4, 11-5)* **21.** 0 *(11-4, 11-5)* **22.** 0.4 *(11-4, 11-5)* **23.** 2 *(11-4, 11-5)* **24.** −2 *(11-4, 11-5)* **25.** −0.4 *(11-4, 11-5)*

26. Increasing on [0, 1] and [3, 4]; decreasing on [1, 3]; concave downward on [0, 2]; concave upward on [2, 4]; local maximum at $x = 1$; local minimum at $x = 3$; inflection point at $x = 2$; graphs of antiderivatives differ by a vertical translation. *(11-1)*

27. *(11-1)*

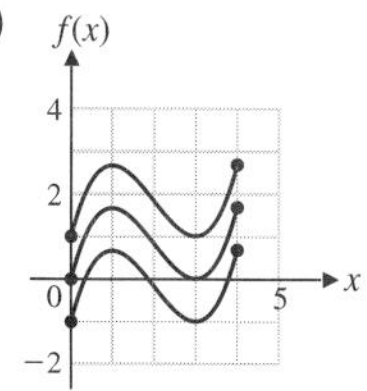

28. (A) 1; 1 (B) 4; 4 *(11-3)*

29. $dy/dx = (2y)/x$; the slopes computed in Problem 28A are compatible with the slope field shown. *(11-3)* **31.** $y = \frac{1}{4}x^2$; $y = -\frac{1}{4}x^2$ *(11-3)*

32. *(11-3)*

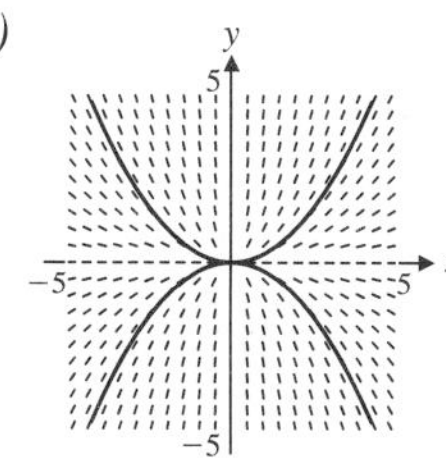

33. *(11-3)*

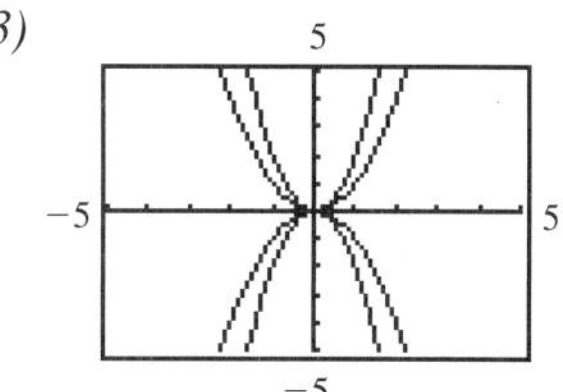

34. $\frac{1}{8}(6x - 5)^{4/3} + C$ *(11-1, 11-2)* **35.** 2 *(11-5)* **36.** $-2x^{-1} - e^{x^2} + C$ *(11-2)* **37.** $(20^{3/2} - 8)/3 \approx 27.148$ *(11-5)* **38.** $-\frac{1}{2}e^{-2x} + \ln|x| + C$ *(11-2)* **39.** $-500(e^{-0.2} - 1) \approx 90.635$ *(11-5)* **40.** $\frac{1}{2} \ln 10 \approx 1.151$ *(11-5)* **41.** 0.45 *(11-5)* **42.** $\frac{1}{48}(2x^4 + 5)^6 + C$ *(11-2)*

43. $-\ln(e^{-x} + 3) + C$ *(11-2)* **44.** $-(e^x + 2)^{-1} + C$ *(11-2)* **45.** $y = f(x) = 3\ln|x| + x^{-1} + 4$ *(11-2, 11-3)*
46. $y = 3x^2 + x - 4$ *(11-3)*
47. $L_5 = 480$ ft; $R_5 = 320$ ft; $A_5 = 400$ ft; error bound for L_5 and $R_5 = 160$; error bound for $A_5 = 80$ *(11-4)*
48. The height of each rectangle represents an instantaneous rate, and the base of each rectangle represents a time interval; rate times time is distance. *(11-4)*
49. $n > 80$ *(11-4)* **50.** Height $= \int_0^5 (160 - 32t)\,dt = 400$ ft *(11-4)*

51. (A) Average $f(x) = 6.5$ (B)

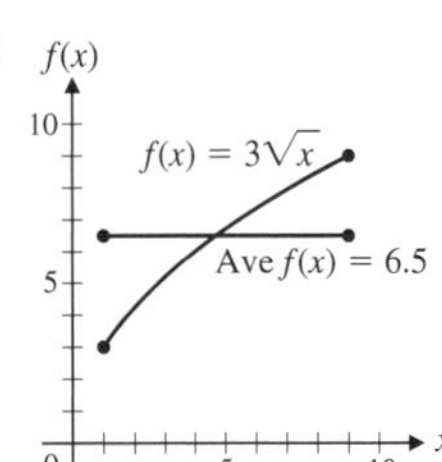

(11-5)

52. $\frac{1}{3}(\ln x)^3 + C$ *(11-2)*
53. $\frac{1}{8}x^8 - \frac{2}{5}x^5 + \frac{1}{2}x^2 + C$ *(11-2)*
54. $\frac{2}{3}(6 - x)^{3/2} - 12(6 - x)^{1/2} + C$ *(11-2)*
55. $\frac{1{,}234}{15} \approx 82.267$ *(11-5)*
56. $\frac{64}{15} \approx 4.267$ *(11-5)*
57. $y = 3e^{x^3} - 1$ *(11-3)*
58. $N = 800e^{0.06t}$ *(11-3)*

59. *(11-5)*

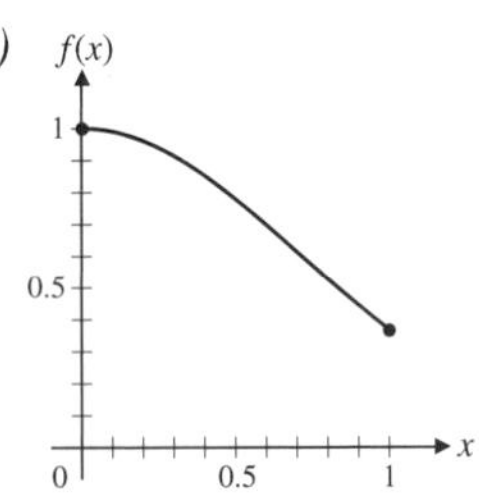

60. $M_5 = 0.748\ 05$ *(11-5)* **61.** $f''(x) = (4x^2 - 2)e^{-x^2}$

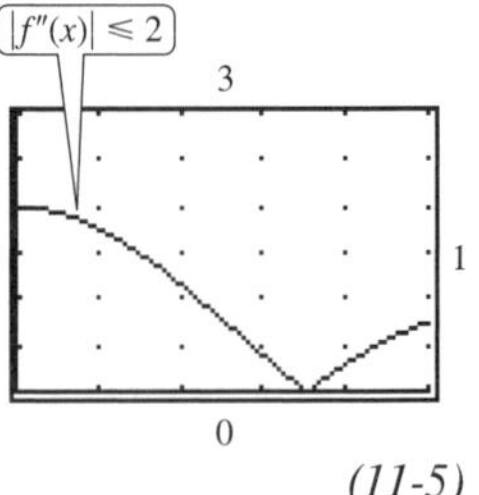

(11-5)

62. $I = \int_0^1 e^{-x^2}\,dx = M_5 \pm 0.003\ 34 = 0.748\ 05 \pm 0.003\ 34$ *(11-5)* **63.** $n \geq 13$ *(11-5)* **64.** Limited growth

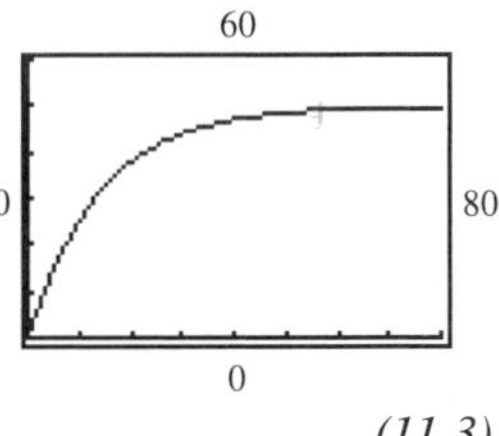

(11-3)

65. Exponential decay

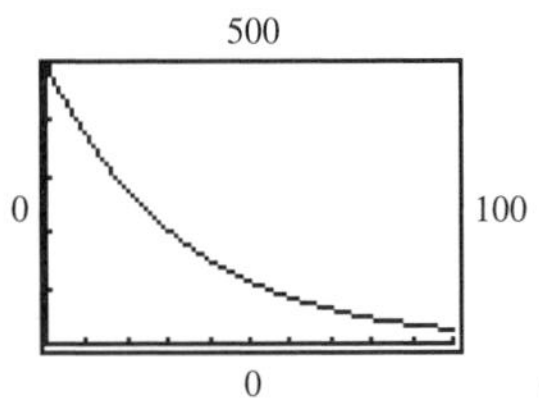

(11-3)

66. Unlimited growth

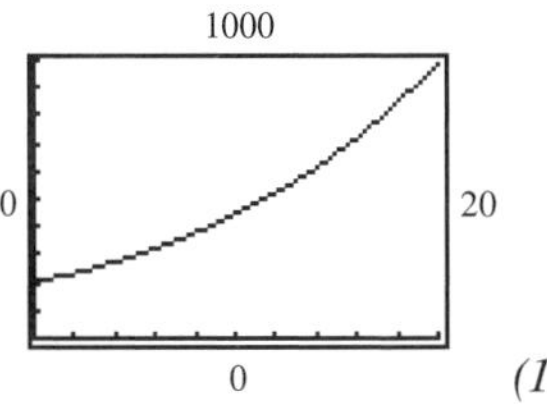

(11-3)

67. Logistic growth

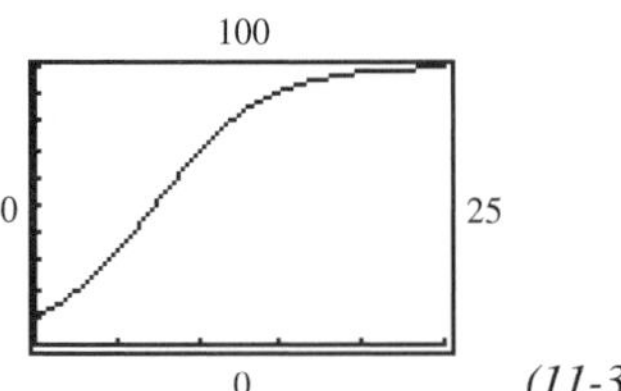

(11-3)

68. 2.251 *(11-5)* **69.** 1.272 *(11-5)* **70.** 43.713 *(11-5)*
71. $L_2 = \$180{,}000$; $R_2 = \$140{,}000$; $\$140{,}000 \leq \int_{200}^{600} C'(x)\,dx \leq \$180{,}000$ *(11-4)*
72. The height of the rectangle, $C'(x)$, represents the marginal cost at a production level of x units—that is, the approximate cost per unit at that production level. The width of the rectangle represents the number of units involved in the increase in production over $x = 200$. Thus, the cost per unit times the number of units equals the increase in production costs. The approximation improves as n increases. *(11-4)*
73. $\int_{200}^{600} \left(600 - \frac{x}{2}\right) dx = \$160{,}000$ *(11-5)* **74.** $\int_{10}^{40} \left(150 - \frac{x}{10}\right) dx = \$4{,}425$ *(11-5)*
75. $P(x) = 100x - 0.01x^2$; $P(10) = \$999$ *(11-3)* **76.** $\int_0^{15} (60 - 4t)\,dt = 450$ thousand barrels *(11-5)*
77. 109 items *(11-5)* **78.** $16e^{2.5} - 16e^2 - 8 \approx \$68.70$ *(11-5)*

79. (A) $A_5 = 725$; error bound $= 75$ (B)

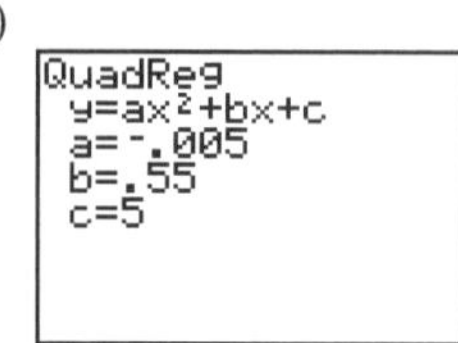

25
−10 60
0

(C) 729 *(11-4, 11-5)*

80. Useful life $= 10 \ln \frac{20}{3} \approx 19$ yr; total profit $= 143 - 200e^{-1.9} \approx 113.086$ or \$113,086 *(11-5)*
81. $S(t) = 50 - 50e^{-0.08t}$; $50 - 50e^{-0.96} \approx$ \$31 million; $-(\ln 0.2)/0.8 \approx 20$ mo *(11-3)* **82.** 6.5 ppm *(11-5)*
83. 1 cm^2 *(11-3)* **84.** 800 gal *(11-5)* **85.** (A) 1,093 million (B) About 70 years *(11-3)*
86. $\dfrac{-\ln 0.04}{0.000\ 123\ 8} \approx 26{,}000$ yr *(11-3)* **87.** $N(t) = 95 - 70e^{-0.1t}$; $N(15) \approx 79$ words/min *(11-3)*

Chapter 12

Exercise 12-1

1. $\int_a^b g(x)\,dx$ **3.** $\int_a^b [-h(x)]\,dx$
5. Since the shaded region in Figure C is below the x axis, $h(x) \leq 0$; thus, $\int_a^b h(x)\,dx$ represents the negative of the area of the region.
7. 20 **9.** $\frac{7}{3} \approx 2.333$ **11.** 9 **13.** 7.021 **15.** 0.693 **17.** $\int_a^b [-f(x)]\,dx$ **19.** $\int_b^c f(x)\,dx + \int_c^d [-f(x)]\,dx$
21. $\int_c^d [f(x) - g(x)]\,dx$ **23.** $\int_a^b [f(x) - g(x)]\,dx + \int_b^c [g(x) - f(x)]\,dx$
25. Find the intersection points by solving $f(x) = g(x)$ on the interval $[a, d]$ to determine b and c. Then observe that $f(x) \geq g(x)$ over $[a, b]$, $g(x) \geq f(x)$ over $[b, c]$, and $f(x) \geq g(x)$ over $[c, d]$. Thus,

$$\text{Area} = \int_a^b [f(x) - g(x)]\,dx + \int_b^c [g(x) - f(x)]\,dx + \int_c^d [f(x) - g(x)]\,dx.$$

27. 2.5 **29.** 7.667 **31.** 23.667 **33.** 15 **35.** 32 **37.** 36 **39.** 9 **41.** 2.832 **43.** 18 **45.** 1.858
47. 17 **49.** 8 **51.** 101.75 **53.** 8 **55.** 17.979 **57.** 5.113 **59.** 8.290 **61.** 3.166 **63.** 1.385
65. Total production from the end of the 5th year to the end of the 10th year is $50 + 100 \ln 20 - 100 \ln 15 \approx 79$ thousand barrels.
67. Total profit over the 5 yr useful life of the game is $20 - 30e^{-1.5} \approx 13.306$ or \$13,306.
69. 1935: 0.412; 1947: 0.231; income was more equally distributed in 1947.
71. 1963: 0.818; 1983: 0.846; total assets were less equally distributed in 1983.
73. (A) $f(x) = 0.3125x^2 + 0.7175x - 0.015$ (B) 0.104 **75.** Total weight gain during the first 10 hr is $3e - 3 \approx 5.15$ g.
77. Average number of words learned during the second 2 hr is $15 \ln 4 - 15 \ln 2 \approx 10$.

Exercise 12-2

1. 4.12 **3.** 509.14 **5.** (A) 10.72 (B) 3.28 (C) 10.72 **7.** (A) .75 (B) .11 (C)

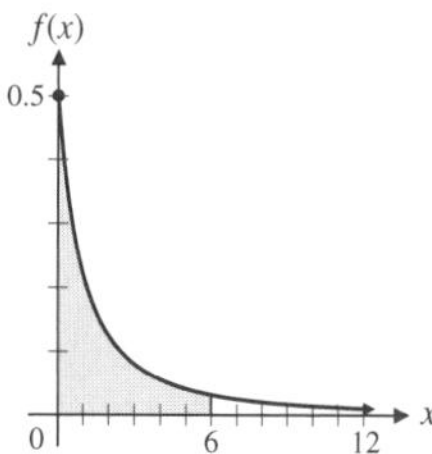

9. 8 yr **11.** (A) .11 (B) .10 **13.** $P(t \geq 12) = 1 - P(0 \leq t \leq 12) = .89$

15. \$12,500 **17.**

f(t)
y = f(t)
2,500
Total income
0 5 t

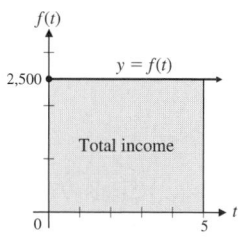

If $f(t)$ is the rate of flow of a continuous income stream, then the total income produced from 0 to 5 yr is the area under the graph of $y = f(t)$ from $t = 0$ to $t = 5$.

19. $8{,}000(e^{0.15} - 1) \approx \$1{,}295$

21. 

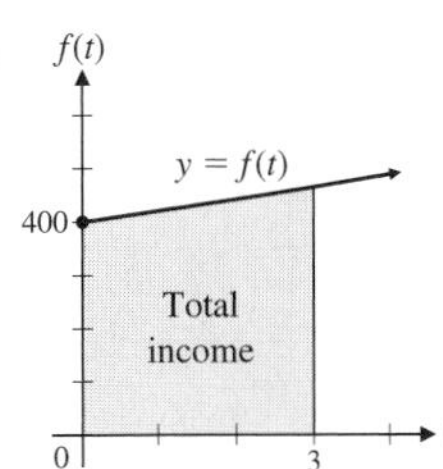

If $f(t)$ is the rate of flow of a continuous income stream, then the total income produced from 0 to 3 yr is the area under the graph of $y = f(t)$ from $t = 0$ to $t = 3$.

23. \$255,562; \$175,562 **25.** $12{,}500(e^{0.4} - e^{-0.08}) \approx \$7{,}109$ **27.** \$1,343

29. Clothing store: $FV = 120{,}000(e^{0.5} - 1) \approx \$77{,}847$; Computer store: $FV = 200{,}000(e^{0.5} - e^{0.25}) \approx \$72{,}939$; the clothing store is the better investment.

31. Bond: $FV = 10{,}000e^{0.4} \approx \$14{,}918$; Business: $FV = 25{,}000(e^{0.4} - 1) \approx \$12{,}296$; the bond is the better investment.

33. \$46,283 **35.** $\frac{k}{r}(e^{rT} - 1)$

37. \$625,000 **39.** 

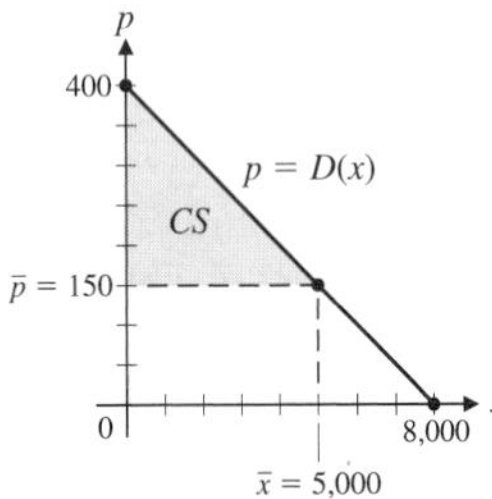

The shaded area is the consumers' surplus and represents the total savings to consumers who are willing to pay more than \$150 for a product but are still able to buy the product for \$150.

41. \$9,900 **43.**

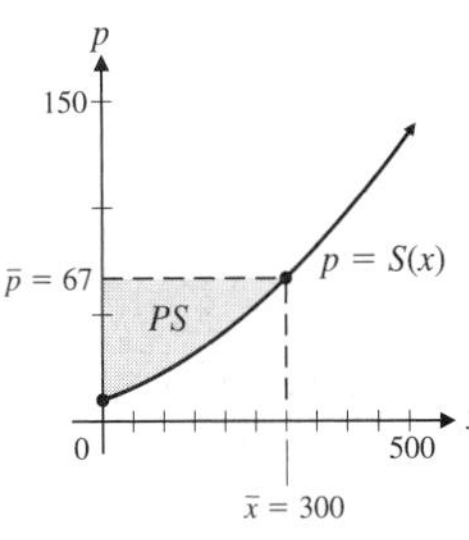

The area of the region PS is the producers' surplus and represents the total gain to producers who are willing to supply units at a lower price than \$67 but are still able to supply the product at \$67.

45. $CS = \$3{,}380$; $PS = \$1{,}690$ **47.** $CS = \$6{,}980$; $PS = \$5{,}041$ **49.** $CS = \$7{,}810$; $PS = \$8{,}336$

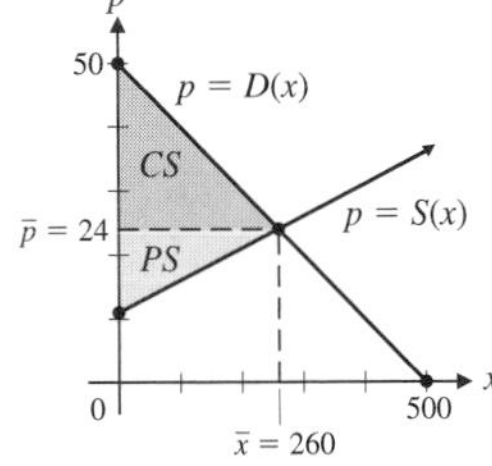

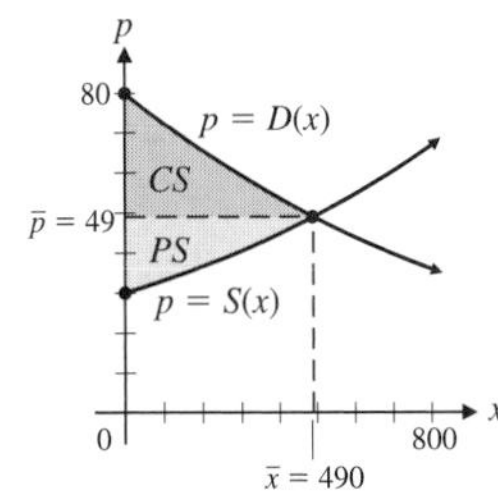

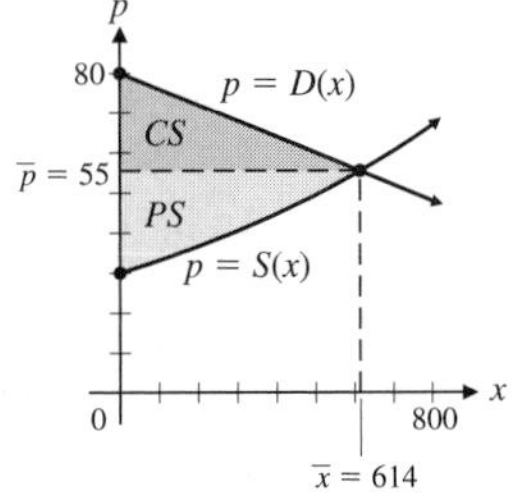

51. $CS = \$8{,}544$; $PS = \$11{,}507$ **53.** (A) $\bar{x} = 21.457$; $\bar{p} = \$6.51$ (B) $CS = 1.774$ or \$1,774; $PS = 1.087$ or \$1,087

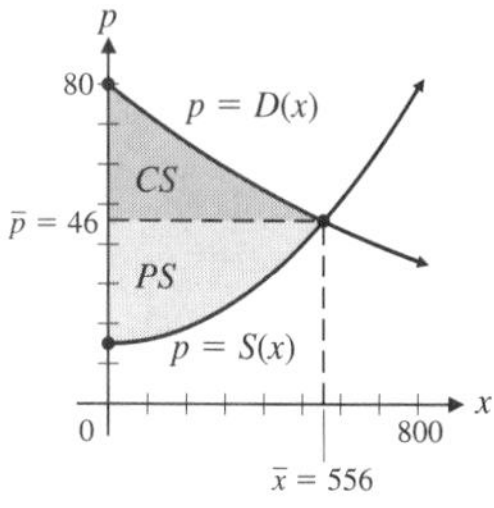

Exercise 12-3

1. $\frac{1}{3}xe^{3x} - \frac{1}{9}e^{3x} + C$ **3.** $\frac{x^3}{3}\ln x - \frac{x^3}{9} + C$ **5.** $u = x + 2$; $\frac{(x+2)(x+1)^6}{6} - \frac{(x+1)^7}{42} + C$ **7.** $-xe^{-x} - e^{-x} + C$
9. $\frac{1}{2}e^{x^2} + C$ **11.** $(xe^x - 4e^x)|_0^1 = -3e + 4 \approx -4.1548$ **13.** $(x\ln 2x - x)|_1^3 = (3\ln 6 - 3) - (\ln 2 - 1) \approx 2.6821$
15. $\ln(x^2 + 1) + C$ **17.** $(\ln x)^2/2 + C$ **19.** $\frac{2}{3}x^{3/2}\ln x - \frac{4}{9}x^{3/2} + C$

21.

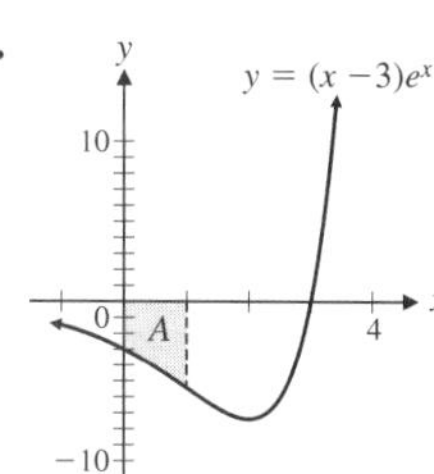

The integral represents the negative of the area between the graph of $y = (x - 3)e^x$ and the x axis from $x = 0$ to $x = 1$.

23.

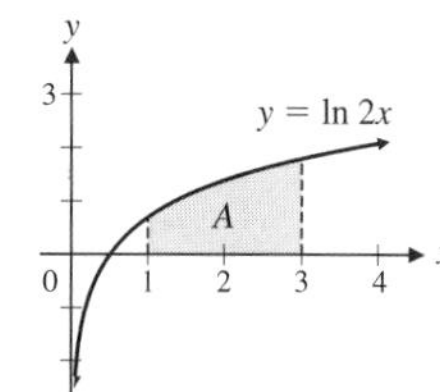

The integral represents the area between the graph of $y = \ln 2x$ and the x axis from $x = 1$ to $x = 3$.

25. $(x^2 - 2x + 2)e^x + C$ **27.** $\frac{xe^{ax}}{a} - \frac{e^{ax}}{a^2} + C$ **29.** $\left(-\frac{\ln x}{x} - \frac{1}{x}\right)\Big|_1^e = -\frac{2}{e} + 1 \approx 0.2642$
31. $6\ln 6 - 4\ln 4 - 2 \approx 3.205$ **33.** $xe^{x-2} - e^{x-2} + C$ **35.** $\frac{1}{2}(1 + x^2)\ln(1 + x^2) - \frac{1}{2}(1 + x^2) + C$
37. $(1 + e^x)\ln(1 + e^x) - (1 + e^x) + C$ **39.** $x(\ln x)^2 - 2x\ln x + 2x + C$

41. $x(\ln x)^3 - 3x(\ln x)^2 + 6x\ln x - 6x + C$ **43.** 1.56 **45.** 34.98 **47.** $\int_0^5 (2t - te^{-t})\,dt = \$24$ million

49.

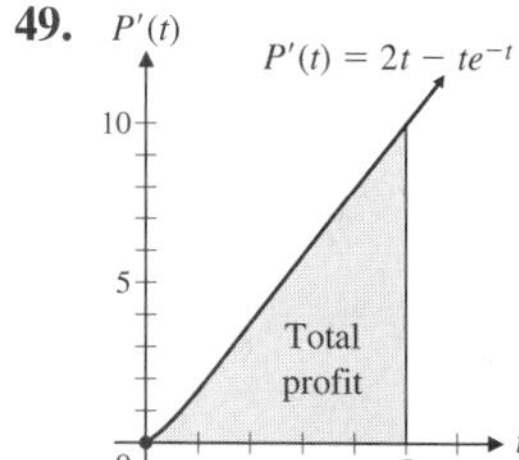

The total profit for the first 5 yr (in millions of dollars) is the same as the area under the marginal profit function, $P'(t) = 2t - te^{-t}$, from $t = 0$ to $t = 5$.

51. \$3,278 **53.** 0.264

55.

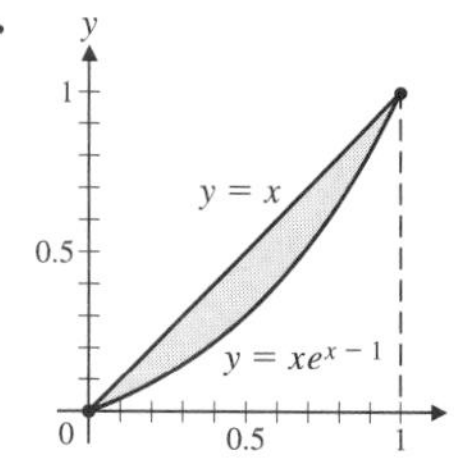

The area bounded by $y = x$ and the Lorenz curve $y = xe^{x-1}$, divided by the area under the graph of $y = x$ from $x = 0$ to $x = 1$ is the index of income concentration. The closer this index is to 0, the more equally distributed income is; the closer this index is to 1, the more concentrated income is in a few hands.

57. $S(t) = 1{,}600 + 400e^{0.1t} - 40te^{0.1t}$; 15 mo **59.** \$977

61.

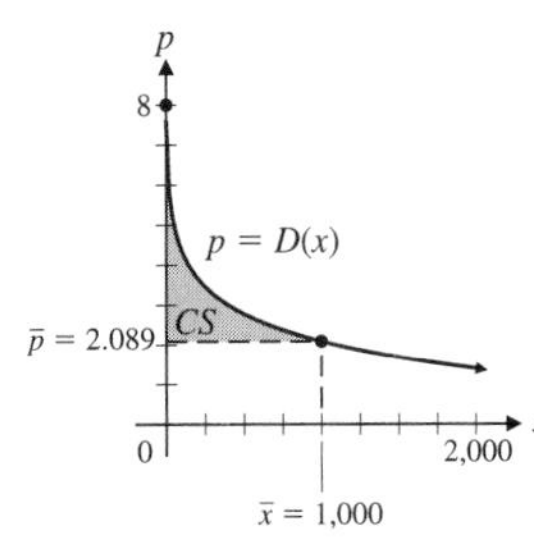

The area bounded by the price–demand equation, $p = 9 - \ln(x + 4)$, and the price equation, $y = \bar{p} = 2.089$, from $x = 0$ to $x = \bar{x} = 1{,}000$, represents the consumers' surplus. This is the amount consumers who are willing to pay more than \$2,089 save.

63. 2.1388 ppm **65.** $N(t) = -4te^{-0.25t} - 40e^{-0.25t} + 80$; 8 wk; 78 words/min **67.** 20,980

Exercise 12-4

1. $\ln\left|\frac{x}{1+x}\right| + C$ **3.** $\frac{1}{3+x} + 2\ln\left|\frac{5+2x}{3+x}\right| + C$ **5.** $\frac{2(x-32)}{3}\sqrt{16+x} + C$ **7.** $-\ln\left|\frac{1+\sqrt{1-x^2}}{x}\right| + C$
9. $\frac{1}{2}\ln\left|\frac{2}{2+\sqrt{x^2+4}}\right| + C$ **11.** $\frac{1}{3}x^3\ln x - \frac{1}{9}x^3 + C$ **13.** $x - \ln|1 + e^x| + C$ **15.** $9\ln\frac{3}{2} - 2 \approx 1.6492$
17. $\frac{1}{2}\ln\frac{12}{5} \approx 0.4377$ **19.** $\ln 3 \approx 1.0986$ **21.** $-\frac{\sqrt{4x^2+1}}{x} + 2\ln|2x + \sqrt{4x^2+1}| + C$
23. $\frac{1}{2}\ln|x^2 + \sqrt{x^4 - 16}| + C$ **25.** $\frac{1}{6}\left(x^3\sqrt{x^6+4} + 4\ln|x^3 + \sqrt{x^6+4}|\right) + C$ **27.** $-\frac{\sqrt{4-x^4}}{8x^2} + C$
29. $\frac{1}{5}\ln\left|\frac{3+4e^x}{2+e^x}\right| + C$ **31.** $\frac{2}{3}(\ln x - 8)\sqrt{4 + \ln x} + C$ **33.** $\frac{1}{5}x^2e^{5x} - \frac{2}{25}xe^{5x} + \frac{2}{125}e^{5x} + C$
35. $-x^3e^{-x} - 3x^2e^{-x} - 6xe^{-x} - 6e^{-x} + C$ **37.** $x(\ln x)^3 - 3x(\ln x)^2 + 6x\ln x - 6x + C$
39. $\frac{64}{3}$ **41.** $\frac{1}{2}\ln\frac{9}{5} \approx 0.2939$ **43.** $\frac{1}{2}\ln|x^2 + 2x| + C$
45. $\frac{2}{3}\ln|3 + x| + \frac{1}{3}\ln|x| + C$ **47.** 31.38 **49.** 5.48 **51.** $3{,}000 + 1{,}500\ln\frac{1}{3} \approx \$1{,}352$ **53.**

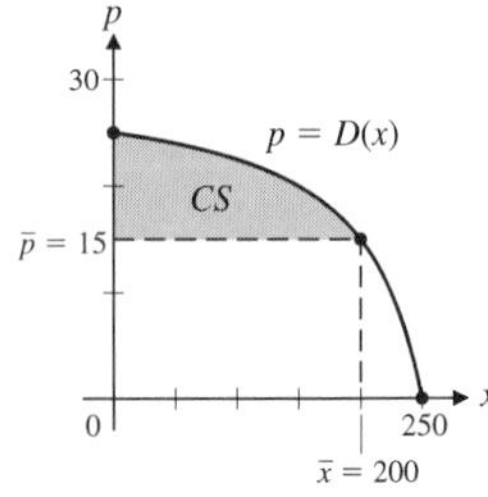

55. $C(x) = 200x + 1{,}000\ln(1 + 0.05x) + 25{,}000$; 608; \$198,773 **57.** $100{,}000e - 250{,}000 \approx \$21{,}828$ **59.** 0.1407

61.

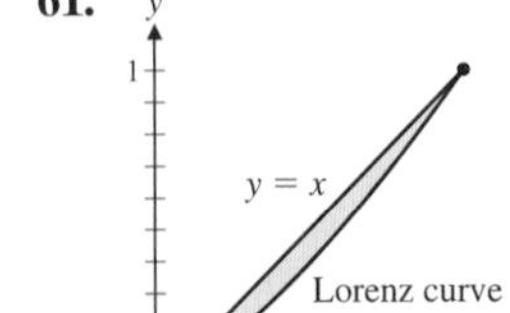

As the area bounded by the two curves gets smaller, the Lorenz curve approaches $y = x$ and the distribution of income approaches perfect equality—all individuals share equally in the income available.

63. $S(t) = 1 + t - \frac{1}{1+t} - 2\ln|1 + t|$; $24.96 - 2\ln 25 \approx \$18.5$ million

65.

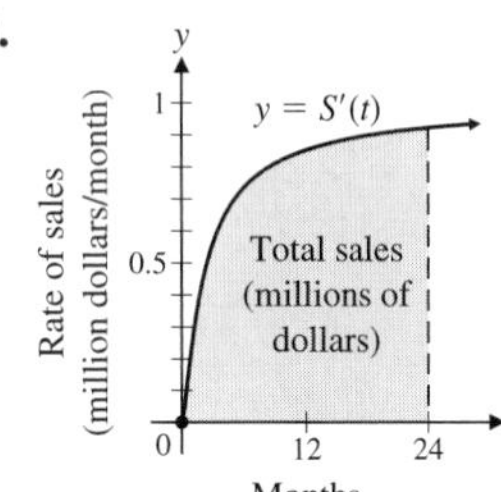

The total sales (in millions of dollars) over the first 2 yr (24 mo) is the area under the graph of $y = S'(t)$ from $t = 0$ to $t = 24$.

67. $P(x) = \frac{2(9x - 4)}{135}(2 + 3x)^{3/2} - 2{,}000.83$; 54; \$37,932 **69.** $100 \ln 3 \approx 110$ ft **71.** $60 \ln 5 \approx 97$ items

73.

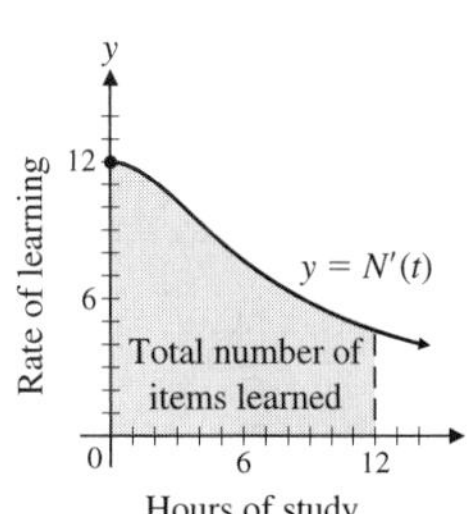

The area under the graph of $y = N'(t)$ from $t = 0$ to $t = 12$ represents the total number of items learned in that time interval.

Review Exercise

1. $\int_a^b f(x)\, dx$ *(12-1)* **2.** $\int_b^c [-f(x)]\, dx$ *(12-1)* **3.** $\int_a^b f(x)\, dx + \int_b^c [-f(x)]\, dx$ *(12-1)* **4.** Area $= 1.153$

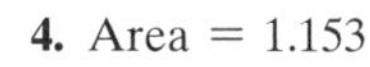

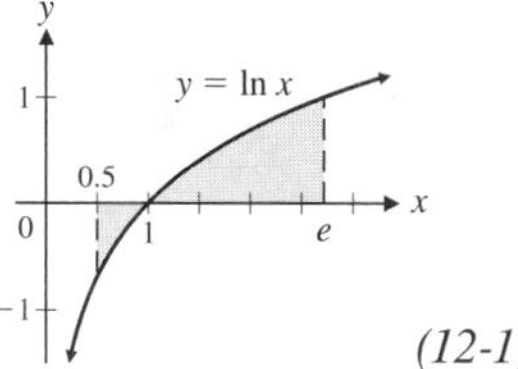

(12-1)

5. $\frac{1}{4}xe^{4x} - \frac{1}{16}e^{4x} + C$ *(12-3, 12-4)* **6.** $\frac{1}{2}x^2 \ln x - \frac{1}{4}x^2 + C$ *(12-3, 12-4)* **7.** $\frac{(\ln x)^2}{2} + C$ *(11-2)* **8.** $\frac{\ln(1 + x^2)}{2} + C$ *(11-2)*

9. $\frac{1}{1 + x} + \ln\left|\frac{x}{1 + x}\right| + C$ *(12-4)* **10.** $-\frac{\sqrt{1 + x}}{x} - \frac{1}{2}\ln\left|\frac{\sqrt{1 + x} - 1}{\sqrt{1 + x} + 1}\right| + C$ *(12-4)* **11.** $\int_a^b [f(x) - g(x)]\, dx$ *(12-1)*

12. $\int_b^c [g(x) - f(x)]\, dx$ *(12-1)* **13.** $\int_b^c [g(x) - f(x)]\, dx + \int_c^d [f(x) - g(x)]\, dx$ *(12-1)*

14. $\int_a^b [f(x) - g(x)]\, dx + \int_b^c [g(x) - f(x)]\, dx + \int_c^d [f(x) - g(x)]\, dx$ *(12-1)*

15. Area $= 20.833$

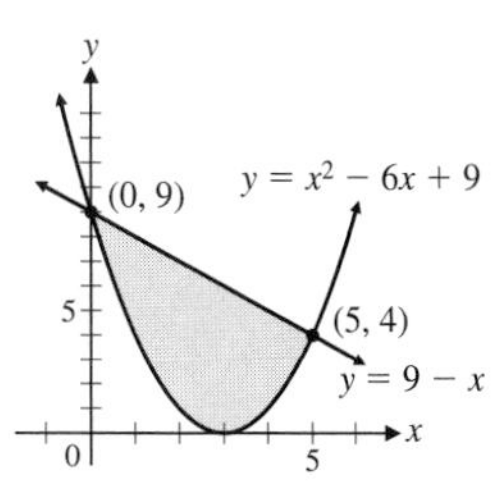

(12-1)

16. 1 *(12-3, 12-4)*

17. $\frac{15}{2} - 8 \ln 8 + 8 \ln 4 \approx 1.955$ *(12-4)*

18. $\frac{1}{6}(3x\sqrt{9x^2 - 49} - 49 \ln|3x + \sqrt{9x^2 - 49}|) + C$ *(12-4)*

19. $-2te^{-0.5t} - 4e^{-0.5t} + C$ *(12-3, 12-4)*

20. $\frac{1}{3}x^3 \ln x - \frac{1}{9}x^3 + C$ *(12-3, 12-4)*

21. $x - \ln|1 + 2e^x| + C$ *(12-4)*

22. (A) Area = 8

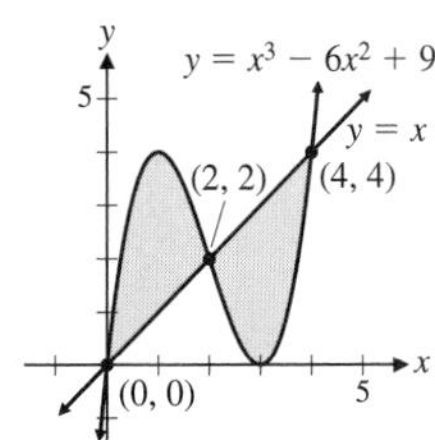

(B) Area = 8.38

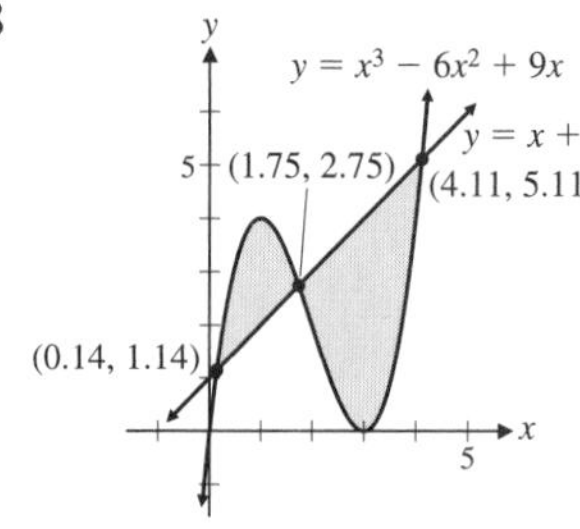

(12-1)

23. $\frac{1}{3}(\ln x)^3 + C$ *(11-2)* **24.** $\frac{1}{2}x^2(\ln x)^2 - \frac{1}{2}x^2 \ln x + \frac{1}{4}x^2 + C$ *(12-3, 12-4)* **25.** $\sqrt{x^2 - 36} + C$ *(11-2)*
26. $\frac{1}{2}\ln|x^2 + \sqrt{x^4 - 36}| + C$ *(12-4)* **27.** $50 \ln 10 - 42 \ln 6 - 24 \approx 15.875$ *(12-3, 12-4)*
28. $x(\ln x)^2 - 2x \ln x + 2x + C$ *(12-3, 12-4)* **29.** $-\frac{1}{4}e^{-2x^2} + C$ *(11-2)* **30.** $-\frac{1}{2}x^2e^{-2x} - \frac{1}{2}xe^{-2x} - \frac{1}{4}e^{-2x} + C$ *(12-3, 12-4)*
31. 1.703 *(12-1)* **32.** (A) .189 (B) .154 *(12-2)*

33.

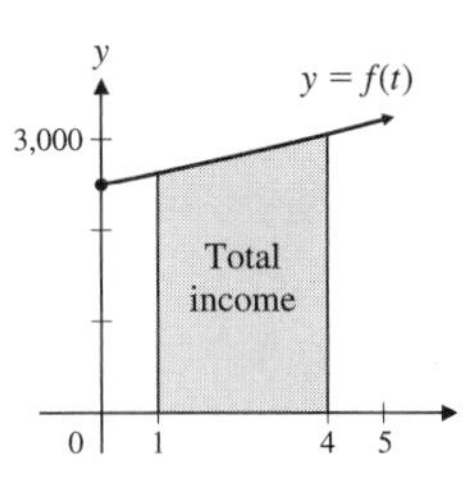

The probability that the product will fail during the second year of warranty is the area under the probability density function $y = f(t)$ from $t = 1$ to $t = 2$. *(12-2)*

34. $R(x) = 65x - 6[(x + 1)\ln(x + 1) - x]$; 618/wk; $29,506 *(12-3)*

35. (A)

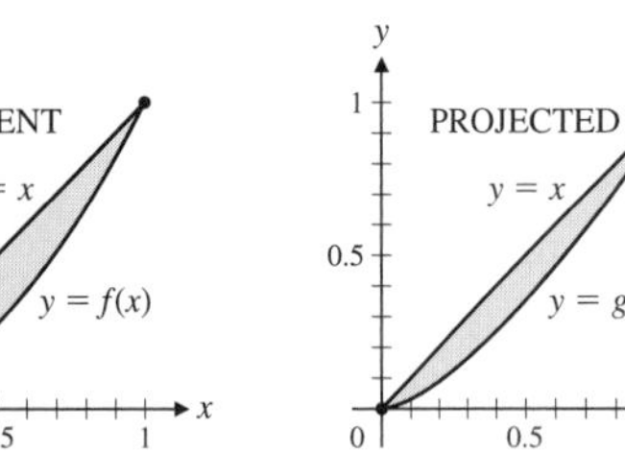

(B) $8,507 *(12-2)* **36.** (A) $20,824 (B) $6,623 *(12-2)*

37. (A)

(B) More equitably distributed, since the area bounded by the two curves will have decreased.
(C) Current = 0.3; projected = 0.2; income will be more equitably distributed 10 years from now. *(12-1)*

38. (A) CS = $2,250; PS = $2,700 (B) CS = $2,890; PS = $2,278

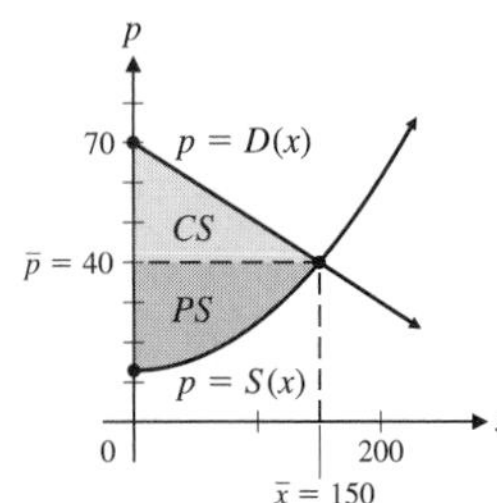

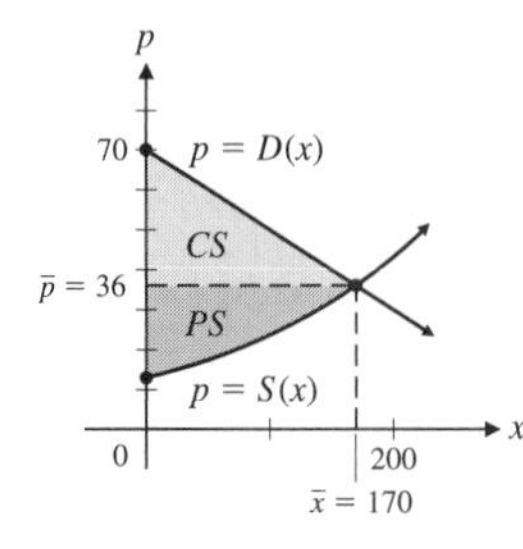

(12-2)

39. (A) 25.403 or 25,403 lb
(B) PS = 121.6 or $1,216 *(12-2)*
40. 4.522 ml; 1.899 ml *(11-5, 12-4)*

41.

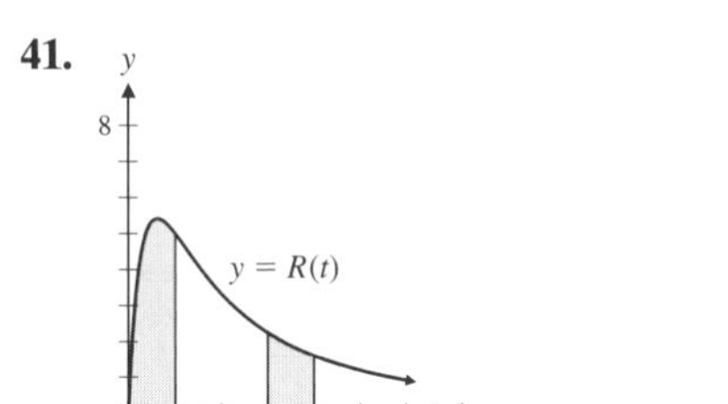

(11-5, 12-1)

42. .667; .333 *(12-2)*

43.

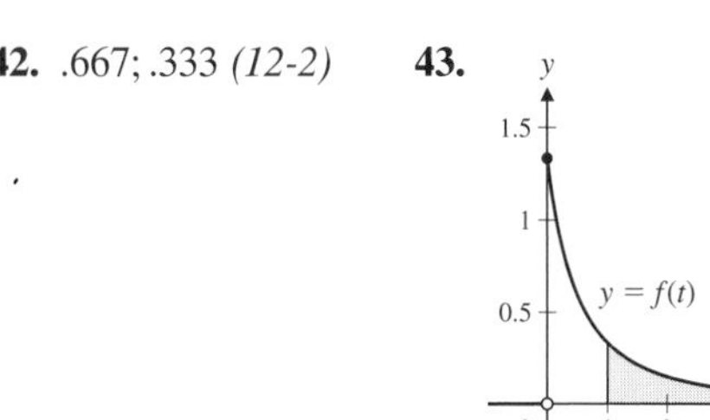

The probability that the doctor will spend more than an hour with a randomly selected patient is the area under the probability density function $y = f(t)$ from $t = 1$ to $t = 3$. *(12-2)*

44. 45 thousand *(11-5, 12-1)* **45.** .368 *(12-2)*

CHAPTER 13

Exercise 13-1

1. 10 **3.** 1 **5.** 0 **7.** 1 **9.** −63 **11.** 6 **13.** 150 **15.** 16π **17.** 791 **19.** 0.192 **21.** 118
23. $100e^{0.8} = 222.55$ **25.** −1.926, 0.599 **27.** $2x + h$ **29.** $2y^2$ **31.** $E(0, 0, 3)$; $F(2, 0, 3)$
33. (A) In the plane $y = c$, c any constant, $z = x^2$.
(B) The y axis; the horizontal line parallel to the y axis and passing through the point (1, 0, 1); the horizontal line parallel to the y axis and passing through the point (2, 0, 4)
(C) A parabolic "trough" lying on top of the y axis
35. (A) Upper semicircles whose centers lie on the y axis
(B) Upper semicircles whose centers lie on the x axis
(C) The upper hemisphere of radius 6 with center at the origin
37. (A) $a^2 + b^2$ and $c^2 + d^2$ both equal the square of the radius of the circle.
(B) Bell-shaped curves with maximum values of 1 at the origin
(C) A bell, with maximum value 1 at the origin, extending infinitely far in all directions.
39. \$4,400; \$6,000; \$7,100 **41.** $R(p, q) = -5p^2 + 6pq - 4q^2 + 200p + 300q$; $R(2, 3) = \$1{,}280$; $R(3, 2) = \$1{,}175$
43. 30,065 units **45.** (A) \$272,615.08 (B) 12.2% **47.** $T(70, 47) \approx 29$ min; $T(60, 27) = 33$ min
49. $C(6, 8) = 75$; $C(8.1, 9) = 90$ **51.** $Q(12, 10) = 120$; $Q(10, 12) \approx 83$

Exercise 13-2

1. 3 **3.** 2 **5.** $-4xy$ **7.** $-4x$ **9.** −6 **11.** $10xy^3$ **13.** 60 **15.** $30xy^2$ **17.** $2x - 2y + 6$ **19.** 6
21. −2 **23.** 2 **25.** $2e^{2x+3y}$ **27.** $6e^{2x+3y}$ **29.** $6e^2$ **31.** $4e^3$
33. $f_x(x, y) = 6x(x^2 - y^3)^2$; $f_y(x, y) = -9y^2(x^2 - y^3)^2$ **35.** $f_x(x, y) = 24xy(3x^2y - 1)^3$; $f_y(x, y) = 12x^2(3x^2y - 1)^3$
37. $f_x(x, y) = 2x/(x^2 + y^2)$; $f_y(x, y) = 2y/(x^2 + y^2)$ **39.** $f_x(x, y) = y^4e^{xy^2}$; $f_y(x, y) = 2xy^3e^{xy^2} + 2ye^{xy^2}$
41. $f_x(x, y) = 4xy^2/(x^2 + y^2)^2$; $f_y(x, y) = -4x^2y/(x^2 + y^2)^2$
45. $f_{xx}(x, y) = 2y^2 + 6x$; $f_{xy}(x, y) = 4xy = f_{yx}(x, y)$; $f_{yy}(x, y) = 2x^2$
47. $f_{xx}(x, y) = -2y/x^3$; $f_{xy}(x, y) = (-1/y^2) + (1/x^2) = f_{yx}(x, y)$; $f_{yy}(x, y) = 2x/y^3$
49. $f_{xx}(x, y) = (2y + xy^2)e^{xy}$; $f_{xy}(x, y) = (2x + x^2y)e^{xy} = f_{yx}(x, y)$; $f_{yy}(x, y) = x^3e^{xy}$ **51.** $x = 2$ and $y = 4$
53. $x = 1.200$ and $y = -0.695$ **55.** (A) $-\frac{13}{3}$ (B) The function $f(0, y)$, for example, has values less than $-\frac{13}{3}$.
57. (A) $c = 1.145$ (B) $f_x(c, 2) = 0$; $f_y(c, 2) = 92.021$
59. $f_{xx}(x, y) + f_{yy}(x, y) = (2y^2 - 2x^2)/(x^2 + y^2)^2 + (2x^2 - 2y^2)/(x^2 + y^2)^2 = 0$ **61.** (A) $2x$ (B) $4y$
63. $P_x(1{,}200, 1{,}800) = 24$; profit will increase approx. \$24 per unit increase in production of type A calculators at the (1,200, 1,800) output level; $P_y(1{,}200, 1{,}800) = -48$; profit will decrease approx. \$48 per unit increase in production of type B calculators at the (1,200, 1,800) output level
65. $\partial x/\partial p = -5$: a \$1 increase in the price of brand A will decrease the demand for brand A by 5 lb at any price level (p, q); $\partial y/\partial p = 2$: a \$1 increase in the price of brand A will increase the demand for brand B by 2 lb at any price level (p, q)
67. (A) $f_x(x, y) = 7.5x^{-0.25}y^{0.25}$; $f_y(x, y) = 2.5x^{0.75}y^{-0.75}$
(B) Marginal productivity of labor $= f_x(600, 100) \approx 4.79$; Marginal productivity of capital $= f_y(600, 100) \approx 9.58$
(C) Capital
69. Competitive **71.** Complementary

73. (A) $f_w(w, h) = 6.65w^{-0.575}h^{0.725}$; $f_h(w, h) = 11.34w^{0.425}h^{-0.275}$
(B) $f_w(65, 57) = 11.31$: for a 65 lb child 57 in. tall, the rate of change in surface area is 11.31 in.^{2} for each pound gained in weight (height is held fixed); $f_h(65, 57) = 21.99$: for a child 57 in. tall, the rate of change in surface area is 21.99 in.^{2} for each inch gained in height (weight is held fixed)

75. $C_W(6, 8) = 12.5$: index increases approx. 12.5 units for 1 in. increase in width of head (length held fixed) when $W = 6$ and $L = 8$; $C_L(6, 8) = -9.38$: index decreases approx. 9.38 units for 1 in. increase in length (width held fixed) when $W = 6$ and $L = 8$

Exercise 13-3

1. $f_x(x, y) = 4$; $f_y(x, y) = 5$; the functions $f_x(x, y)$ and $f_y(x, y)$ never have the value 0.
3. $f_x(x, y) = -1.2 + 4x^3$; $f_y(x, y) = 6.8 + 0.6y^2$; the function $f_y(x, y)$ never has the value 0.
5. $f(-2, 0) = 10$ is a local maximum **7.** $f(-1, 3) = 4$ is a local minimum **9.** f has a saddle point at $(3, -2)$
11. $f(3, 2) = 33$ is a local maximum **13.** $f(2, 2) = 8$ is a local minimum **15.** f has a saddle point at $(0, 0)$
17. f has a saddle point at $(0, 0)$; $f(1, 1) = -1$ is a local minimum
19. f has a saddle point at $(0, 0)$; $f(3, 18) = -162$ and $f(-3, -18) = -162$ are local minima
21. The test fails at $(0, 0)$; f has saddle points at $(2, 2)$ and $(2, -2)$ **23.** f has a saddle point at $(0.614, -1.105)$
25. $f(x, y)$ is nonnegative and equals 0 when $x = 0$, so f has a local minimum at each point of the y axis.
27. (B) Local minimum **29.** 2,000 type A and 4,000 type B; Max $P = P(2, 4) = \$15$ million
31. (A) When $p = \$10$ and $q = \$12$, $x = 56$ and $y = 16$; when $p = \$11$ and $q = \$11$, $x = 6$ and $y = 56$.
(B) A maximum weekly profit of \$288 is realized for $p = \$10$ and $q = \$12$.
33. $P(x, y) = P(4, 2)$ **35.** 8 in. by 4 in. by 2 in. **37.** 20 in. by 20 in. by 40 in.

Exercise 13-4

1. Max $f(x, y) = f(3, 3) = 18$ **3.** Min $f(x, y) = f(3, 4) = 25$
5. $F_x = -3 + 2\lambda = 0$ and $F_y = 4 + 5\lambda = 0$ have no simultaneous solution.
7. Max $f(x, y) = f(3, 3) = f(-3, -3) = 18$; Min $f(x, y) = f(3, -3) = f(-3, 3) = -18$
9. Maximum product is 25 when each number is 5 **11.** Min $f(x, y, z) = f(-4, 2, -6) = 56$
13. Max $f(x, y, z) = f(2, 2, 2) = 6$; Min $f(x, y, z) = f(-2, -2, -2) = -6$ **15.** Max $f(x, y) = f(0.217, 0.885) = 1.055$
17. $F_x = e^x + \lambda = 0$ and $F_y = 3e^y - 2\lambda = 0$ have no simultaneous solution.
19. Maximize $f(x, 5)$, a function of just one independent variable.
21. (A) Max $f(x, y) = f(0.707, 0.5) = f(-0.707, 0.5) = 0.47$
23. 60 of model A and 30 of model B will yield a minimum cost of \$32,400 per week
25. (A) 8,000 units of labor and 1,000 units of capital; Max $N(x, y) = N(8{,}000, 1{,}000) \approx 263{,}902$ units
(B) Marginal productivity of money ≈ 0.6598; increase in production $\approx 32{,}990$ units
27. 8 in. by 8 in. by $\frac{8}{3}$ in. **29.** $x = 50$ ft and $y = 200$ ft; maximum area is 10,000 ft^2

Exercise 13-5

1.
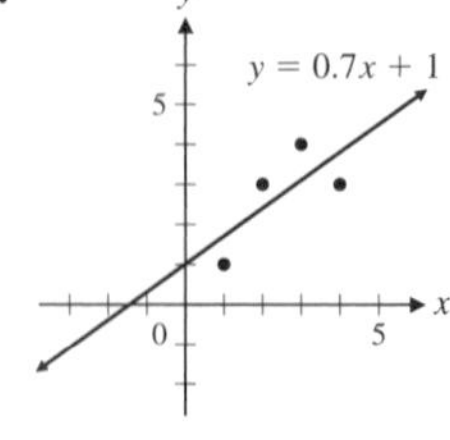

3.
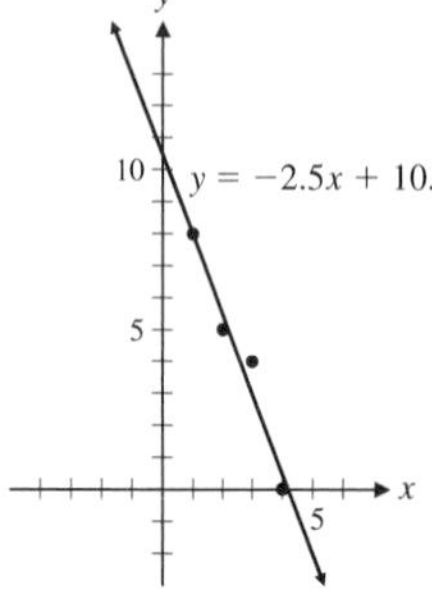

5.
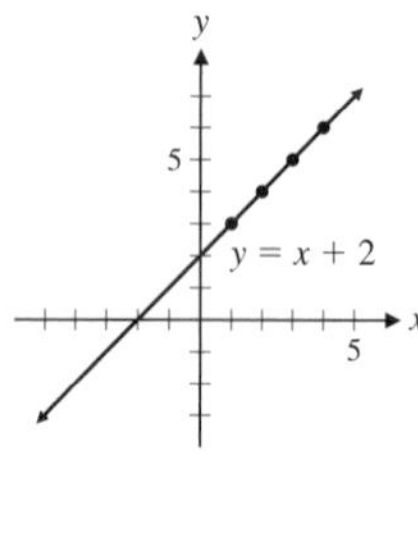

7. $y = -1.5x + 4.5$; $y = 0.75$ when $x = 2.5$ **9.** $y = 2.12x + 10.8$; $y = 63.8$ when $x = 25$
11. $y = -1.2x + 12.6$; $y = 10.2$ when $x = 2$ **13.** $y = -1.53x + 26.67$; $y = 14.4$ when $x = 8$

15.

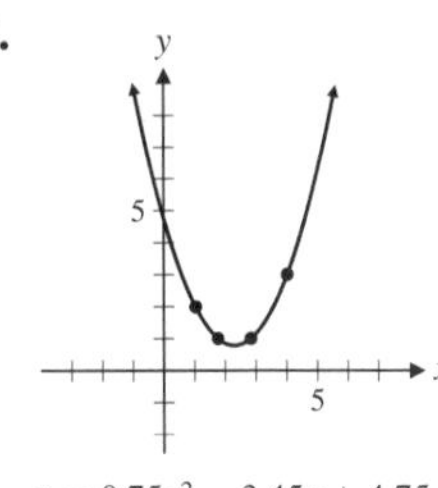

21. (A) $y = 1.52x - 0.16$; $y = 0.73x^2 - 1.39x + 1.30$ (B) The quadratic function.
23. The normal equations form a system of 4 linear equations in the 4 variables *a, b, c,* and *d,* which can be solved using Gauss–Jordan elimination.
25. (A) $y = 7.15x + 12.62$ (B) 105.57 thousand per month
27. (A) $y = -0.48x + 4.38$ (B) \$6.56/bottle
29. (A) $P = -0.66T + 48.8$ (B) 11.18 beats/min
31. (A) $D = -3.1A + 54.6$ (B) 45%
33. (A) $y = 0.08653x + 10.81$ (B) 20.50 ft

Exercise 13-6

1. (A) $3x^2y^4 + C(x)$ (B) $3x^2$ **3.** (A) $2x^2 + 6xy + 5x + E(y)$ (B) $35 + 30y$
5. (A) $\sqrt{y + x^2} + E(y)$ (B) $\sqrt{y + 4} - \sqrt{y}$ **7.** (A) $\dfrac{\ln x \ln y}{x} + C(x)$ (B) $\dfrac{2 \ln x}{x}$ **9.** 9 **11.** 330
13. $(56 - 20\sqrt{5})/3$ **15.** 1 **17.** 16 **19.** 49 **21.** $\frac{1}{8}\int_1^5 \int_{-1}^1 (x + y)^2\,dy\,dx = \frac{32}{3}$
23. $\frac{1}{15}\int_1^4 \int_2^7 (x/y)\,dy\,dx = \frac{1}{2}\ln\frac{7}{2} \approx 0.6264$ **25.** $\frac{4}{3}$ cubic units **27.** $\frac{32}{3}$ cubic units
29. $\int_0^1 \int_1^2 xe^{xy}\,dy\,dx = \frac{1}{2} + \frac{1}{2}e^2 - e$ **31.** $\int_0^1 \int_{-1}^1 \dfrac{2y + 3xy^2}{1 + x^2}\,dy\,dx = \ln 2$

35. (A) $\frac{1}{3} + \frac{1}{4}e^{-2} - \frac{1}{4}e^2$ (B)

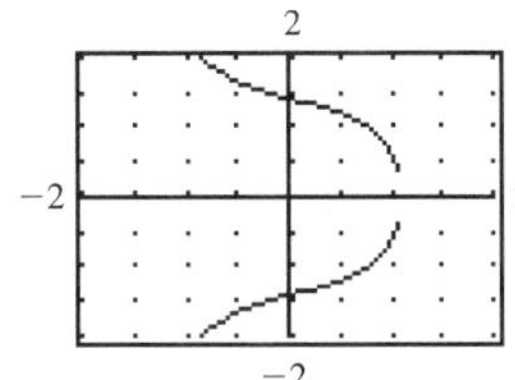

(C) Points to the right of the graph in part (B) are greater than 0; points to the left of the graph are less than 0.

37. $\dfrac{1}{0.4}\displaystyle\int_{0.6}^{0.8}\int_5^7 \frac{y}{1 - x}\,dy\,dx = 30 \ln 2 \approx \$20.8$ billion
39. $\frac{1}{10}\int_{10}^{20}\int_1^2 x^{0.75}y^{0.25}\,dy\,dx = \frac{8}{175}(2^{1.25} - 1)(20^{1.75} - 10^{1.75}) \approx 8.375$ or 8,375 units
41. $\frac{1}{192}\int_{-8}^{8}\int_{-6}^{6}\left[10 - \frac{1}{10}(x^2 + y^2)\right]dy\,dx = \frac{20}{3}$ insects/ft^2 **43.** $\frac{1}{8}\int_{-2}^{2}\int_{-1}^{1}[100 - 15(x^2 + y^2)]\,dy\,dx = 75$ ppm
45. $\frac{1}{10{,}000}\int_{2{,}000}^{3{,}000}\int_{50}^{60} 0.000\ 013\ 3xy^2\,dy\,dx \approx 100.86$ ft **47.** $\frac{1}{16}\int_8^{16}\int_{10}^{12} 100\dfrac{x}{y}\,dy\,dx = 600 \ln 1.2 \approx 109.4$

Review Exercise

1. $f(5, 10) = 2{,}900$; $f_x(x, y) = 40$; $f_y(x, y) = 70$ *(13-1, 13-2)* **2.** $\partial^2 z/\partial x^2 = 6xy^2$; $\partial^2 z/\partial x\,\partial y = 6x^2y$ *(13-2)*
3. $2xy^3 + 2y^2 + C(x)$ *(13-6)* **4.** $3x^2y^2 + 4xy + E(y)$ *(13-6)* **5.** 1 *(13-6)*
6. $f_x(x, y) = 5 + 6x + 3x^2$; $f_y(x, y) = -2$; the function $f_y(x, y)$ never has the value 0. *(13-3)*
7. $f(2, 3) = 7$; $f_y(x, y) = -2x + 2y + 3$; $f_y(2, 3) = 5$ *(13-1, 13-2)*
8. $(-8)(-6) - (4)^2 = 32$ *(13-2)* **9.** $(1, 3, -\frac{1}{2}), (-1, -3, \frac{1}{2})$ *(13-4)*
10. $y = -1.5x + 15.5$; $y = 0.5$ when $x = 10$ *(13-5)* **11.** 18 *(13-6)*
12. $f_x(x, y) = 2xe^{x^2+2y}$; $f_y(x, y) = 2e^{x^2+2y}$; $f_{xy}(x, y) = 4xe^{x^2+2y}$ *(13-2)*
13. $f_x(x, y) = 10x(x^2 + y^2)^4$; $f_{xy}(x, y) = 80xy(x^2 + y^2)^3$ *(13-2)*
14. $f(2, 3) = -25$ is a local minimum; f has a saddle point at $(-2, 3)$ *(13-3)* **15.** Max $f(x, y) = f(6, 4) = 24$ *(13-4)*
16. Min $f(x, y, z) = f(2, 1, 2) = 9$ *(13-4)* **17.** $y = \frac{116}{165}x + \frac{100}{3}$ *(13-5)* **18.** $\frac{27}{5}$ *(13-6)*
19. 4 cubic units *(13-6)* **20.** 0 *(13-6)* **21.** (A) 12.56 (B) No *(13-6)*
22. $F_x = 12x^2 + 3\lambda = 0$, $F_y = -15y^2 + 2\lambda = 0$, and $F_\lambda = 3x + 2y - 7 = 0$ have no simultaneous solution. *(13-4)*

23. (A) $P_x(1, 3) = 8$; profit will increase \$8,000 for a 100 unit increase in product A if production of product B is held fixed at an output level of $(1, 3)$.
(B) For 200 units of A and 300 units of B, $P(2, 3) =$ \$100 thousand is a local maximum. *(13-2, 13-3)*
24. 8 in. by 6 in. by 2 in. *(13-3)* **25.** $y = 0.63x + 1.33$; profit in sixth year is \$5.11 million *(13-4)*
26. (A) Marginal productivity of labor ≈ 8.37; marginal productivity of capital ≈ 1.67; management should encourage increased use of labor.
(B) 80 units of labor and 40 units of capital; Max $N(x, y) = N(80, 40) \approx 696$ units; marginal productivity of money ≈ 0.0696; increase in production ≈ 139 units
(C) $\frac{1}{1{,}000}\int_{50}^{100}\int_{20}^{40} 10x^{0.8}y^{0.2}\,dy\,dx = \frac{(40^{1.2} - 20^{1.2})(100^{1.8} - 50^{1.8})}{216} = 621$ items *(13-4)*
27. $T_x(70, 17) = -0.924$ min/ft increase in depth when $V = 70$ ft^3 and $x = 17$ ft *(13-2)*
28. $\frac{1}{16}\int_{-2}^{2}\int_{-2}^{2}[100 - 24(x^2 + y^2)]\,dy\,dx = 36$ ppm *(13-6)* **29.** 50,000 *(13-1)*
30. $y = \frac{1}{2}x + 48$; $y = 68$ when $x = 40$ *(13-5)*
31. (A) $y = 0.4709x + 25.87$ (B) 72.96 people/mi^2 (C) 74.41 people/mi^2; 80.13 people/mi^2 *(13-5)*
32. (A) $y = 1.069x + 0.522$ (B) 64.68 yr (C) 64.78 yr; 64.80 yr *(13-5)*

Appendix A

Self-Test on Basic Algebra

1. (A) T (B) T (C) F (D) T *(A-1)* **2.** (A) $(y + z)x$ (B) $(2 + x) + y$ (C) $2x + 3x$ *(A-2)*
3. $x^3 + 3x^2 + 5x - 2$ *(A-3)* **4.** $x^3 - 3x^2 - 3x + 22$ *(A-3)* **5.** $3x^5 + x^4 - 8x^3 + 24x^2 + 8x - 64$ *(A-3)*
6. 3 *(A-3)* **7.** 1 *(A-3)* **8.** $14x^2 - 30x$ *(A-3)* **9.** $6x^2 - 5xy - 4y^2$ *(A-3)* **10.** $4a^2 - 12ab + 9b^2$ *(A-3)*
11. $4xy - 2y^2$ *(A-3)* **12.** $m^4 - 6m^2n^2 + n^2$ *(A-3)* **13.** $x^3 - 6x^2y + 12xy^2 - 8y^3$ *(A-3)*
14. (A) 4.065×10^{12} (B) 7.3×10^{-3} *(A-6)* **15.** (A) 255,000,000 (B) 0.000 406 *(A-6)*
16. (A) $\{2, 4, 5, 6\}$ (B) $\{5\}$ (C) $\{8\}$ (D) $\{2, 4\}$ *(A-1)* **17.** (A) 28 (B) 5 (C) 4 (D) 10 *(A-1)*
18. (A) T (B) F *(A-2)* **19.** 0 and -3 are two examples of infinitely many. *(A-2)* **20.** $6x^5y^{15}$ *(A-6)*
21. $3u^4/v^2$ *(A-6)* **22.** 6×10^2 *(A-6)* **23.** x^6/y^4 *(A-6)* **24.** $u^{7/3}$ *(A-7)* **25.** $3a^2/b$ *(A-7)* **26.** $\frac{5}{9}$ *(A-6)*
27. $x + 2x^{1/2}y^{1/2} + y$ *(A-7)* **28.** $6x + 7x^{1/2}y^{1/2} - 3y$ *(A-7)* **29.** $(3x - 1)(4x + 3)$ *(A-4)*
30. $(4x - 3y)(2x - 3y)$ *(A-4)* **31.** Not factorable relative to the integers *(A-4)*
32. $3n(2n - 5)(n + 1)$ *(A-4)* **33.** $(x - y)(7x - y)$ *(A-4)*
34. Not factorable relative to the integers *(A-4)* **35.** $\frac{12a^3b - 40b^3 - 5a}{30a^3b^2}$ *(A-5)* **36.** $\frac{7x - 4}{6x(x - 4)}$ *(A-5)*
37. $\frac{-8(x + 2)}{x(x - 4)(x + 4)}$ *(A-5)* **38.** $\frac{y + 2}{y(y - 2)}$ *(A-5)* **39.** $\frac{-1}{7(7 + h)}$ *(A-5)* **40.** $\frac{xy}{y - x}$ *(A-7)*
41. (A) Subtraction (B) Commutative (+) (C) Distributive (D) Associative (•) (E) Negatives (F) Identity (+) *(A-2)*
42. (A) 90 (B) 45 *(A-1)* **43.** $6x^{2/5} - 7(x - 1)^{3/4}$ *(A-7)* **44.** $2\sqrt{x} - 3\sqrt[3]{x^2}$ *(A-7)* **45.** $2 - \frac{3}{2}x^{-1/2}$ *(A-7)*
46. $\sqrt{3x}$ *(A-7)* **47.** $\sqrt{x} + \sqrt{5}$ *(A-7)* **48.** $\frac{1}{\sqrt{x} - 5}$ *(A-7)* **49.** $\frac{1}{\sqrt{u + h} + \sqrt{u}}$ *(A-7)* **50.** $x = 2$ *(A-8)*
51. $x = 0, 5$ *(A-9)* **52.** $x = \pm\sqrt{7}$ *(A-9)* **53.** $x = -4, 5$ *(A-9)* **54.** $x = (3 \pm \sqrt{17})/4$ *(A-9)*

55. $x < 4$ or $(-\infty, 4)$ 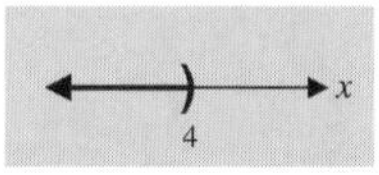

(A-8) **56.** $x \geq \frac{9}{2}$ or $[\frac{9}{2}, \infty)$ 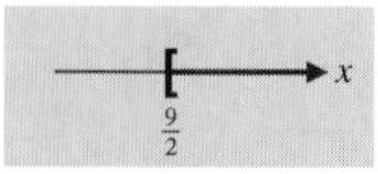

(A-8)

57. $2 \leq x < 12$ or $[2, 12)$ 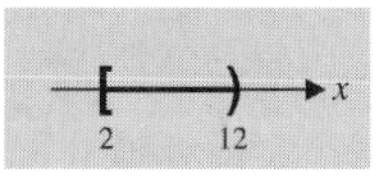

(A-8) **58.** $y = \frac{2}{3}x - 2$ *(A-8)* **59.** $y = 3/(x - 1)$ *(A-8)* **60.** Yes *(A-1)*

61. $2.3328 \times 10^4 =$ \$23,328 per person *(A-6)* **62.** \$20,000 at 8%; \$40,000 at 14% *(A-8)* **63.** 4,000 tapes *(A-8)*

Exercise A-1

1. T **3.** T **5.** T **7.** T **9.** $\{1, 2, 3, 4, 5\}$ **11.** $\{3, 4\}$ **13.** $\varnothing$ **15.** $\{2\}$ **17.** $\{-7, 7\}$
19. $\{1, 3, 5, 7, 9\}$ **21.** $A' = \{1, 5\}$ **23.** 40 **25.** 60 **27.** 60 **29.** 20 **31.** 95 **33.** 40
35. (A) $\{1, 2, 3, 4, 6\}$ (B) $\{1, 2, 3, 4, 6\}$ **37.** $\{1, 2, 3, 4, 6\}$ **39.** Yes **41.** Yes **43.** Yes
45. (A) 2 (B) 4 (C) 8 (D) 16; 2^n **47.** 800 **49.** 200 **51.** 200 **53.** 800 **55.** 200 **57.** 200
59. 6 **61.** A+, AB+ **63.** A−, A+, B+, AB− AB+, O+ **65.** O+, O− **67.** B−, B+
69. Everybody in the clique relates to each other.

Exercise A-2

1. vu **3.** $(3 + 7) + y$ **5.** $u + v$ **7.** T **9.** T **11.** T **13.** T **15.** T **17.** T **19.** T **21.** F
23. T **25.** T **27.** No **29.** (A) F (B) T (C) T **31.** $\sqrt{2}$ and π are two examples of infinitely many.
33. (A) N, Z, Q, R (B) R (C) Q, R (D) Q, R
35. (A) T (B) F, since, for example, $(8 - 4) - 2 \neq 8 - (4 - 2)$. (C) T
(D) F, since, for example, $(8 \div 4) \div 2 \neq 8 \div (4 \div 2)$.
37. $\frac{1}{11}$ **39.** (A) 2.166 666 666... (B) 4.582 575 69... (C) 0.437 500 000... (D) 0.261 261 261...

Exercise A-3

1. 3 **3.** $x^3 + 4x^2 - 2x + 5$ **5.** $x^3 + 1$ **7.** $2x^5 + 3x^4 - 2x^3 + 11x^2 - 5x + 6$ **9.** $-5u + 2$
11. $6a^2 + 6a$ **13.** $a^2 - b^2$ **15.** $6x^2 - 7x - 5$ **17.** $2x^2 + xy - 6y^2$ **19.** $9y^2 - 4$ **21.** $6m^2 - mn - 35n^2$
23. $16m^2 - 9n^2$ **25.** $9u^2 + 24uv + 16v^2$ **27.** $a^3 - b^3$ **29.** $16x^2 + 24xy + 9y^2$ **31.** 1 **33.** $x^4 - 2x^2y^2 + y^4$
35. $5a^2 + 12ab - 10b^2$ **37.** $-4m + 8$ **39.** $-6xy$ **41.** $u^3 + 3u^2v + 3uv^2 + v^3$
43. $x^3 - 6x^2y + 12xy^2 - 8y^3$ **45.** $2x^2 - 2xy + 3y^2$ **47.** $8x^3 - 20x^2 + 1$ **49.** $4x^3 - 14x^2 + 8x - 6$
51. $m + n$ **53.** No change **55.** $(1 + 1)^2 \neq 1^2 + 1^2$; either a or b must be 0
57. $0.09x + 0.12(10{,}000 - x) = 1{,}200 - 0.03x$ **59.** $10x + 30(3x) + 50(4{,}000 - x - 3x) = 200{,}000 - 100x$
61. $0.02x + 0.06(10 - x) = 0.6 - 0.04x$

Exercise A-4

1. $3m^2(2m^2 - 3m - 1)$ **3.** $2uv(4u^2 - 3uv + 2v^2)$ **5.** $(7m + 5)(2m - 3)$ **7.** $(a - 4b)(3c + d)$
9. $(2x - 1)(x + 2)$ **11.** $(y - 1)(3y + 2)$ **13.** $(x + 4)(2x - 1)$ **15.** $(w + x)(y - z)$ **17.** $(a - b)(m + n)$
19. $(3y + 2)(y - 1)$ **21.** $(u - 5v)(u + 3v)$ **23.** Not factorable **25.** $(wx - y)(wx + y)$ **27.** $(3m - n)^2$
29. Not factorable **31.** $4(z - 3)(z - 4)$ **33.** $2x^2(x - 2)(x - 10)$ **35.** $x(2y - 3)^2$ **37.** $(2m - 3n)(3m + 4n)$
39. $uv(2u - v)(2u + v)$ **41.** $2x(x^2 - x + 4)$ **43.** $(r - t)(r^2 + rt + t^2)$ **45.** $(a + 1)(a^2 - a + 1)$
47. $[(x + 2) - 3y][(x + 2) + 3y]$ **49.** Not factorable **51.** $(6x - 6y - 1)(x - y + 4)$
53. $(y - 2)(y + 2)(y^2 + 1)$ **55.** $a^2(3 + ab)(9 - 3ab + a^2b^2)$

Exercise A-5

1. $8d^6$ **3.** $\frac{15x^2 + 10x - 6}{180}$ **5.** $\frac{15m^2 + 14m - 6}{36m^3}$ **7.** $\frac{1}{x(x - 4)}$ **9.** $\frac{x - 6}{x(x - 3)}$ **11.** $\frac{x - 5}{(x - 1)^2(x + 1)}$
13. $\frac{2}{x - 1}$ **15.** $\frac{5}{a - 1}$ **17.** $\frac{x^2 + 8x - 16}{x(x - 4)(x + 4)}$ **19.** $\frac{7x^2 - 2x - 3}{6(x + 1)^2}$ **21.** $-\frac{1}{x}$ **23.** $\frac{-17c + 16}{15(c - 1)}$ **25.** $\frac{1}{x - 3}$
27. $\frac{-1}{2x(x + h)}$ **29.** $\frac{x - y}{x + y}$ **31.** (A) Incorrect (B) $x + 1$ **33.** (A) Incorrect (B) $2x + h$
35. (A) Incorrect (B) $\frac{x^2 - x - 3}{x + 1}$ **37.** (A) Correct **39.** $\frac{-2x - h}{3(x + h)^2x^2}$ **41.** x

Exercise A-6

1. $2/x^9$ **3.** $3w^7/2$ **5.** $2/x^3$ **7.** $1/w^5$ **9.** 5 **11.** $1/a^6$ **13.** y^6/x^{12} **15.** 8.23×10^{10} **17.** 7.83×10^{-1}
19. 3.4×10^{-5} **21.** 40,000 **23.** 0.007 **25.** 61,710,000 **27.** 0.000 808 **29.** 1 **31.** 10^{14} **33.** $y^6/25x^4$
35. 4×10^2 **37.** $4y^3/3x^5$ **39.** $\frac{7}{4} - \frac{1}{4}x^{-3}$ **41.** $\frac{3}{4}x - \frac{1}{4}x^{-1} - \frac{1}{4}x^{-3}$ **43.** $\frac{x^2(x - 3)}{(x - 1)^3}$ **45.** $\frac{2(x - 1)}{x^3}$

47. 2.4×10^{10}; 24,000,000,000 **49.** 3.125×10^4; 31,250 **51.** 64 **55.** uv **57.** $\dfrac{bc(c + b)}{c^2 + bc + b^2}$
59. (A) 2.13701×10^{11} (B) 2.2293 (C) 0.4486 **61.** (A) \$15,935 (B) \$1,146 (C) 7.19%
63. (A) 9×10^{-6} (B) 0.000 009 (C) 0.0009% **65.** 1,932,000

Exercise A-7

1. $6\sqrt[5]{x^3}$ **3.** $\sqrt[5]{(4xy^3)^2}$ **5.** $\sqrt{x^2 + y^2}$ (not $x + y$) **7.** $5x^{3/4}$ **9.** $(2x^2y)^{3/5}$ **11.** $x^{1/3} + y^{1/3}$ **13.** 5 **15.** 64
17. -6 **19.** Not a rational number (not even a real number) **21.** $\frac{8}{125}$ **23.** $\frac{1}{27}$ **25.** $x^{2/5}$ **27.** m **29.** $2x/y^2$
31. $xy^2/2$ **33.** $2/3x^{7/12}$ **35.** $2x + 3$ **37.** $6x^3$ **39.** 2 **41.** $12x - 6x^{35/4}$ **43.** $3u - 13u^{1/2}v^{1/2} + 4v$
45. $25m - n$ **47.** $9x - 6x^{1/2}y^{1/2} + y$ **49.** $\frac{1}{2}x^{1/3} + x^{-1/3}$ **51.** $\frac{2}{3}x^{-1/4} + x^{-2/3}$ **53.** $\frac{1}{2}x^{-1/6} - \frac{1}{4}$ **55.** $4n\sqrt{3mn}$
57. $\dfrac{2\sqrt{x} - 2}{x - 2}$ **59.** $7(x - y)(\sqrt{x} + \sqrt{y})$ **61.** $\dfrac{1}{xy\sqrt{5xy}}$ **63.** $\dfrac{1}{\sqrt{x + h} + \sqrt{x}}$ **65.** $\dfrac{1}{\sqrt{t} + \sqrt{x}}$
67. $x = y = 1$ is one of many choices. **69.** $x = y = 1$ is one of many choices. **71.** $\dfrac{x + 8}{2(x + 3)^{3/2}}$ **73.** $\dfrac{x - 2}{2(x - 1)^{3/2}}$
75. $\dfrac{x + 6}{3(x + 2)^{5/3}}$ **77.** 103.2 **79.** 0.0805 **81.** 4,588 **83.** (A) and (E); (B) and (F); (C) and (D)

Exercise A-8

1. $m = 5$ **3.** $x < -9$ **5.** $x \leq 4$ **7.** $x < -3$ or $(-\infty, -3)$

9. $-1 \leq x \leq 2$ or $[-1, 2]$ **11.** $y = 8$ **13.** $x > -6$ **15.** $y = 8$ **17.** $x = 10$

19. $y \geq 3$ **21.** $x = 36$ **23.** $m < 3$ **25.** $x = 10$ **27.** $3 \leq x < 7$ or $[3, 7)$

29. $-20 \leq C \leq 20$ or $[-20, 20]$ **31.** $y = \frac{3}{4}x - 3$

33. $y = -(A/B)x + (C/B) = (-Ax + C)/B$ **35.** $C = \frac{5}{9}(F - 32)$ **37.** $B = A/(m - n)$

39. $-2 < x \leq 1$ or $(-2, 1]$

41. (A) and (C): $a > 0$ and $b > 0$, or $a < 0$ and $b < 0$ (B) and (D): $a > 0$ and $b < 0$, or $a < 0$ and $b > 0$
43. (A) $>$ (B) $<$ **45.** Negative **47.** 5,000 \$15 tickets; 3,000 \$25 tickets **49.** \$7,200 at 10%; \$4,800 at 15%
51. \$18,080 **53.** 5,851 books **55.** (B) 6,180 books (C) At least \$11.50 **57.** 5,000 **59.** 12.6 yr

Exercise A-9

1. $\pm\sqrt{11}$ **3.** $-1, 3$ **5.** $-2, 6$ **7.** $0, 2$ **9.** $3 \pm 2\sqrt{2}$ **11.** $-2 \pm \sqrt{2}$ **13.** $0, 2$ **15.** $\pm\frac{3}{2}$
17. $\frac{1}{2}, -3$ **19.** $(-1 \pm \sqrt{5})/2$ **21.** $(3 \pm \sqrt{3})/2$ **23.** No real solution **25.** $-4 \pm \sqrt{11}$ **27.** $\pm\sqrt{3}$
29. $-\frac{1}{2}, 2$ **31.** $(x - 2)(x + 42)$ **33.** Not factorable in the integers **35.** $(2x - 9)(x + 12)$
37. $(4x - 7)(x + 62)$ **39.** $r = \sqrt{A/P} - 1$
41. If $c < 4$, there are two distinct real roots; if $c = 4$, there is one real double root; and if $c > 4$, there are no real roots.
43. 1,575 bottles at \$4 each **45.** 0.2, or 20% **47.** 8 ft/sec; $4\sqrt{2}$ or 5.66 ft/sec

Appendix B

Exercise B-1

1. 5, 7, 9, 11 **3.** $\frac{3}{2}, \frac{4}{3}, \frac{5}{4}, \frac{6}{5}$ **5.** 9, −27, 81, −243 **7.** 23 **9.** $\frac{101}{100}$ **11.** $1 + 2 + 3 + 4 + 5 + 6 = 21$
13. $5 + 7 + 9 + 11 = 32$ **15.** $1 + \frac{1}{10} + \frac{1}{100} + \frac{1}{1{,}000} = \frac{1{,}111}{1{,}000}$ **17.** 3.6 **19.** 82.5 **21.** $\frac{1}{2}, -\frac{1}{4}, \frac{1}{8}, -\frac{1}{16}, \frac{1}{32}$
23. 0, 4, 0, 8, 0 **25.** $1, -\frac{3}{2}, \frac{9}{4}, -\frac{27}{8}, \frac{81}{16}$ **27.** $a_n = n - 3$ **29.** $a_n = 4n$ **31.** $a_n = (2n - 1)/2n$
33. $a_n = (-1)^{n+1}n$ **35.** $a_n = (-1)^{n+1}(2n - 1)$ **37.** $a_n = \left(\frac{2}{5}\right)^{n-1}$ **39.** $a_n = x^n$ **41.** $a_n = (-1)^{n+1}x^{2n-1}$
43. $1 - 9 + 25 - 49 + 81$ **45.** $\frac{4}{7} + \frac{8}{9} + \frac{16}{11} + \frac{32}{13}$ **47.** $1 + x + x^2 + x^3 + x^4$ **49.** $x - \dfrac{x^3}{3} + \dfrac{x^5}{5} - \dfrac{x^7}{7} + \dfrac{x^9}{9}$
51. (A) $\sum_{k=1}^{5} (k + 1)$ (B) $\sum_{j=0}^{4} (j + 2)$ **53.** (A) $\sum_{k=1}^{4} \dfrac{(-1)^{k+1}}{k}$ (B) $\sum_{j=0}^{3} \dfrac{(-1)^j}{j + 1}$ **55.** $\sum_{k=1}^{n} \dfrac{k + 1}{k}$ **57.** $\sum_{k=1}^{n} \dfrac{(-1)^{k+1}}{2^k}$
59. 2, 8, 26, 80, 242 **61.** 1, 2, 4, 8, 16 **63.** $1, \frac{3}{2}, \frac{17}{12}, \frac{577}{408}$; $a_4 = \frac{577}{408} \approx 1.414\ 216$, $\sqrt{2} \approx 1.414\ 214$

Exercise B-2

1. (A) Arithmetic, with $d = -5$; −26, −31 (B) Geometric, with $r = 2$; −16, 32 (C) Neither
(D) Geometric, with $r = \frac{1}{3}$; $\frac{1}{54}, \frac{1}{162}$
3. $a_2 = 11, a_3 = 15$ **5.** $a_{21} = 82, S_{31} = 1{,}922$ **7.** $S_{20} = 930$ **9.** $a_2 = -6, a_3 = 12, a_4 = -24$ **11.** $S_7 = 547$
13. $a_{10} = 199.90$ **15.** $r = 1.09$ **17.** $S_{10} = 1{,}242, S_\infty = 1{,}250$ **19.** 2,706 **21.** −85 **23.** 1,120
25. (A) Does not exist (B) $S_\infty = \frac{8}{5} = 1.6$ **27.** 2,400 **29.** 0.999
31. Use $a_1 = 1$ and $d = 2$ in $S_n = (n/2)[2a_1 + (n - 1)d]$. **33.** $48 + $46 + ⋯ + $4 + $2 = $600
35. About $11,670,000 **37.** $1,628.89; $2,653.30

Exercise B-3

1. 720 **3.** 10 **5.** 1,320 **7.** 10 **9.** 6 **11.** 1,140 **13.** 10 **15.** 6 **17.** 1 **19.** 816
21. $C_{4,0}a^4 + C_{4,1}a^3b + C_{4,2}a^2b^2 + C_{4,3}ab^3 + C_{4,4}b^4 = a^4 + 4a^3b + 6a^2b^2 + 4ab^3 + b^4$
23. $x^6 - 6x^5 + 15x^4 - 20x^3 + 15x^2 - 6x + 1$ **25.** $32a^5 - 80a^4b + 80a^3b^2 - 40a^2b^3 + 10ab^4 - b^5$ **27.** $3{,}060x^{14}$
29. $5{,}005p^9q^6$ **31.** $264x^2y^{10}$ **33.** $C_{n,0} = \dfrac{n!}{0!n!} = 1$; $C_{n,n} = \dfrac{n!}{n!0!} = 1$ **35.** 1 5 10 10 5 1; 1 6 15 20 15 6 1

INDEX

Applications Index

Business & Economics

Life Sciences